Where cutting edge science meets state of the art learning.

The Fifteenth Edition of the world-renowned *Brock Biology of Microorganisms* introduces today's students to cutting edge microbiology research and ensures core concept mastery, enhanced by MasteringMicrobiology®.

NEW & REVISED! Microbiology Now chapter opening features highlight cutting edge, engaging research that is important to how we understand microbiology today. Paired assessments in MasteringMicrobiology engage students in the course material and foster deep concept mastery.

Microbial Growth and Its Control

5

microbiology now

Picking Apart a Microbial Consortium

In nature, certain metabolic processes are carried out by microbes that team up to get the job done, a cozy arrangement called a consortium. Such is the case with the oxidation of methane (CH_4) linked to the reduction of sulfate (SO_4^{2-}) in anoxic marine sediments. The overall reaction ($CH_4 + SO_4^{2-} \rightarrow HCO_3^- + HS^- + H_2O$) is exergonic and the small amount of energy released is shared between two distinct microbes. The methane oxidizer in the consortium is a species of *Archaea* nicknamed ANME (for anaerobic *methanotroph*, blue in photo), and its sulfate-reducing partner is a species of *Bacteria* (brown in photo). The consortium is thought to play a key role in the carbon cycle as a major methane sink, and thus a detailed picture of how it works is important to our understanding of the global carbon economy, climate change, and marine biogeochemistry.

Researchers have tried for years to separate the consortium into its components but always found that methane oxidation required both organisms. However, some researchers hypothesized that it might be possible to replace the sulfate reducer with an artificial electron acceptor and that this might unlock the consortium and allow the methanotroph to grow in pure culture. Using an electron acceptor called AQDS, the scientists discovered that they could turn off sulfate reduction in the consortium while maintaining CH_4 oxidation. During this process, the methanotroph used electrons from CH_4 to reduce AQDS rather than passing them on to its sulfate-reducing partner. Several other electron acceptors known to support anaerobic respiration also sustained methane oxidation, giving hope that ANME may eventually be obtained in pure culture.

The ability to grow a microbe in pure culture is the "gold standard" for the study of its physiology, biochemistry, regulation, and several other aspects of its biology. In the case of the ANME–sulfate reducer consortium, several physiologies were active at once, and resolving these many reactions proved to be a major scientific challenge. However, if further work shows that ANME can be removed from the consortium and grown in pure culture, detailed aspects of its biology can be studied that were not possible when the organism was tightly coupled to its partner in the consortium (photo).

Source: Scheller, S., H. Yu, G.L. Chadwick, S.E. McGlynn, and V.J. Orphan. 2016. Artificial electron acceptors decouple archaeal methane oxidation from sulfate reduction. *Science* 351: 703–706.

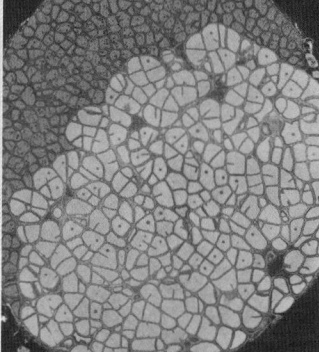

Microbial Infection and Pathogenesis

25

microbiology now

The Microbial Community That Thrives on Your Teeth

Few people have such superb oral hygiene that they lack dental plaque, the microbial biofilm that forms on and between teeth and along or below the gumline. If not removed regularly, dental plaque invariably leads to dental caries (cavities), the condition in which portions of tooth enamel and dentin break down from the onslaught of bacterial activities. Dental plaque and dental caries develop from the natural tendency of oral bacteria such as *Streptococcus mutans* and its close relative *S. sobrinus* to attach firmly to the teeth and gums and ferment sucrose (table sugar) to lactic acid, which attacks the teeth and slowly rots them away.

Until recently, dental plaque was thought to consist largely of the aforementioned streptococci. Both species could easily be isolated from dental plaque and both light and electron microscopy typically showed large numbers of cocci in chains, a hallmark of the genus *Streptococcus*. But a recent molecular ecology study of the microbial diversity of dental plaque revealed that this material is composed of more than just streptococci and develops in a precisely structured way.

The photo here is a light micrograph of a section through human dental plaque stained by fluorescence in situ hybridization (FISH). Different oligonucleotides, each specific for a different major phylum of *Bacteria* and containing a distinct fluorescent dye, were allowed to hybridize to the ribosomal RNA in cells in the plaque and then observed by fluorescence microscopy. Surprisingly, instead of seeing primarily streptococci, the researchers saw a diverse and highly organized microbial community. The micrograph shows streptococci (stained green) located primarily at the periphery of the plaque beyond several other bacteria that combine to form a scaffold emerging from the tooth surface. These include *Corynebacterium* (purple), *Capnocytophaga* (red), *Fusobacterium* (yellow), *Leptotrichia* (blue-green), and *Haemophilus* (orange), among others. A major conclusion that emerged from this study was that the scaffolding microbes likely function to position the streptococci out into the oral cavity where sucrose should be more available.

New views of old problems often reveal surprising results. In the case of dental plaque, FISH technology has revealed a whole new microbial world in a habitat previously thought to be dominated by only two species of well-characterized bacteria.

Source: Mark Welch, J.L., et al. 2016. Biogeography of a human oral microbiome at the micron scale. *Proc. Natl. Acad. Sci. (USA) 113*: doi: 10.1073/pnas.1522149113.

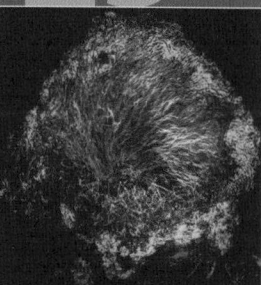

757

Microbiology today and tomorrow.

Genomics, and the various "omics" it has spawned, support content throughout the Fifteenth Edition ensuring that today's students understand the transformation that biology, and specifically microbiology, has undergone – and preparing them for the fast paced nature of the science.

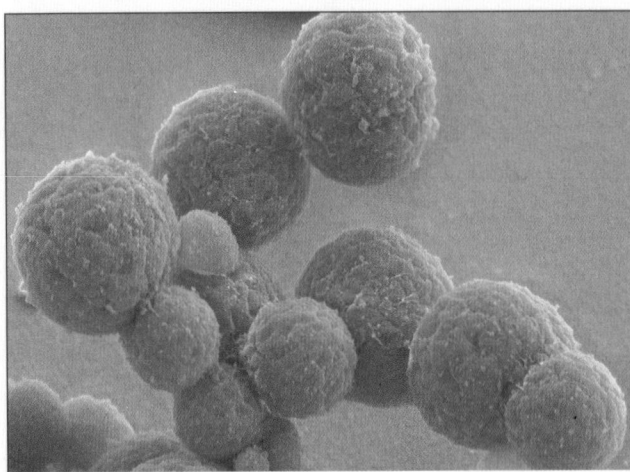

microbiology now

Creation of a New Life Form: Design of a Minimal Cell

A cell's genome is its blueprint for life. However, what is the bare minimum number of genes needed to sustain a free-living cell? This is a question that microbiologists at the J. Craig Venter Institute (JCVI) have attempted to answer ever since they sequenced the genomes of several *Mycoplasma* species in the 1990s. Because *Mycoplasma* species are parasitic bacteria, their genomes are already reduced in size and hence provide an excellent foundation for creating a "minimal cell." However, little did the scientists at JCVI suspect that it would take 20 years to satisfy their scientific curiosity!

Instead of beginning by genetically manipulating a *Mycoplasma* species, microbiologists at JCVI wanted to have more control. To begin unraveling the genetic requirements for life, they first generated a synthetic self-replicating *Mycoplasma* (described in this chapter). The genome of this pioneering synthetic life form was synthesized from scratch based on its known genome sequence. The synthetic cell did not possess a "designer genome," or even a minimal one; it simply contained its own genome but one completely constructed in the laboratory. This breakthrough in synthetic biology provided the technology needed for microbiologists to create designer genomes.

Using comparative genomics and prior knowledge about specific gene sequences, microbiologists at JCVI continued their work by designing and synthesizing several minimal genomes that they hypothesized would sustain life. To their dismay, none of these resulted in a viable cell. So instead, they generated modules of DNA corresponding to a *Mycoplasma* genome and sewed different combinations together to form synthetic genomes. Once viable cells were obtained from transplanting these genomes, nonessential genes from the smallest genome were identified by transposon mutagenesis. After removing these unnecessary genes, a synthetic minimal cell coined JCVI-syn3.0 was created (see photo). This autonomous life form possesses a 531-kilobase genome encoding 473 genes; JCVI-syn3.0 thus contains a genome smaller than any other free-living cell.

While this work showcases the amazing advancements in synthetic biology and the potential for creating designer cells with novel functions, a surprising mystery surrounds this minimal cell: The roles for almost a third of JCVI-syn3.0's genes remain unknown, highlighting how much we still need to learn about the genetic foundation of a living cell.

Source: Hutchison, C.A. 3rd, et al. 2016. Design and synthesis of a minimal bacterial genome. *Science* 351(6280): aad6253. Photo provided by Clyde Hutchison and J. Craig Venter, JCVI and Thomas Deerinck and Mark Ellisman, NCMIR.

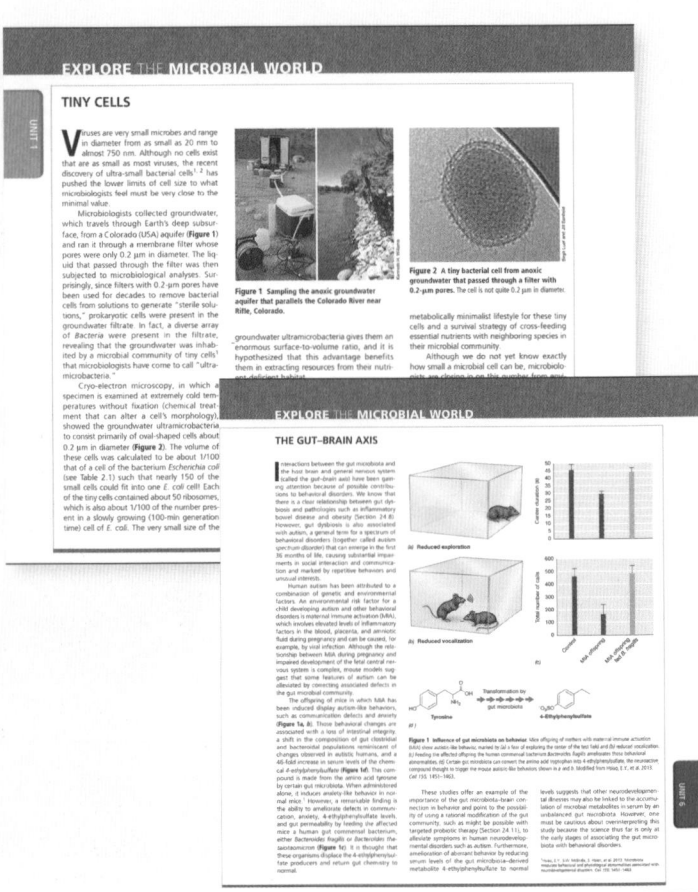

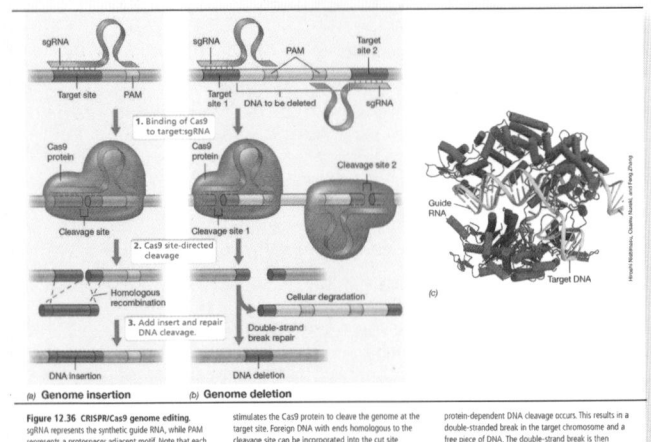

Figure 12.36 CRISPR/Cas9 genome editing. sgRNA represents the synthetic guide RNA; PAM represents a protospacer adjacent motif. Note that each genome target site must possess a PAM sequence for DNA cleavage to occur. (a) Insertion of foreign DNA into a targeted site of the genome. The binding of the sgRNA to bind to a single target site on the genome through complementarity. This binding of the sgRNA to the DNA stimulates the Cas9 protein to cleave the genome at the target site. Foreign DNA with ends homologous to the cleavage site can be incorporated into the cut site through homologous recombination. This results in a genomic insertion. (b) Deletion of a genomic region. Two separate target sites flanking the DNA to be deleted are selected. After the design, addition, and binding of sgRNAs corresponding to these regions, Cas9 protein-dependent DNA cleavage occurs. This results in a double-stranded break in the target chromosome and a free piece of DNA. The double-strand break is then ligated by the cell's DNA double-strand break repair pathway, while the free piece of genomic DNA is degraded. This results in a genomic deletion. (c) Crystal structure of the *Streptococcus pyogenes* Cas9 protein. The target DNA is shown in green and the sgRNA in orange.

Authoritative. Accurate. Accessible.

The Fifteenth Edition continues its legacy of authoritative, accessible writing; beautiful and clear art; and student-focused pedagogy, engaging learners in the science.

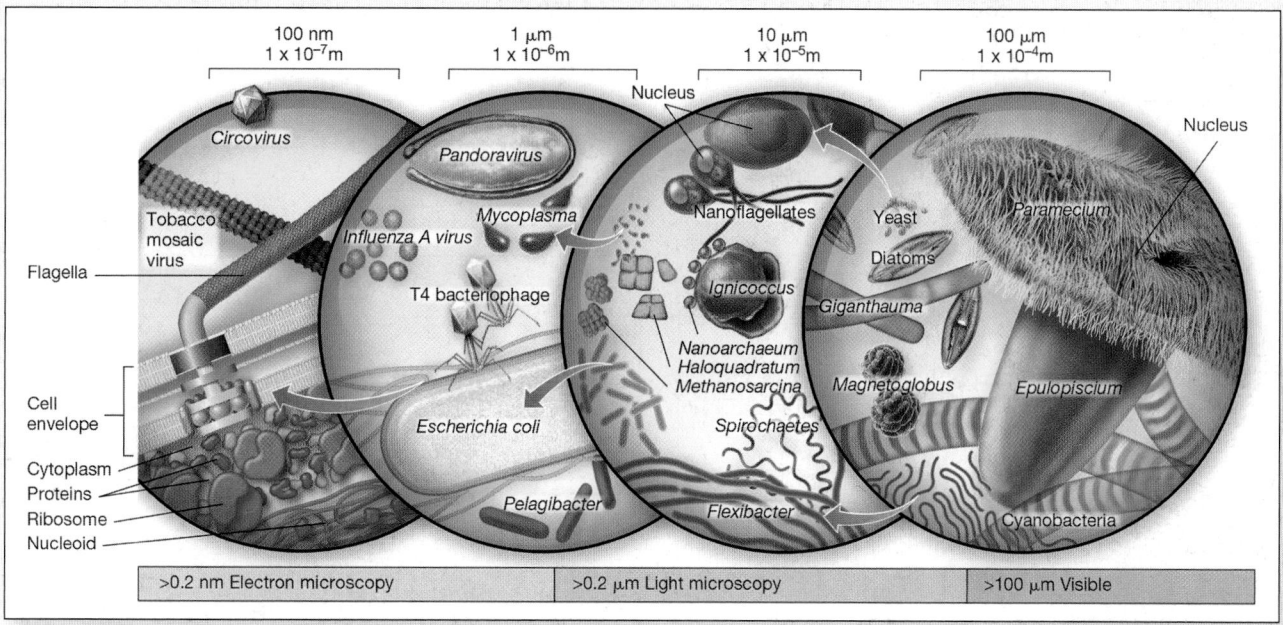

Student focused pedagogy informs the organization and design of each chapter feature

TABLE 9.6 Some omics terminology

DNA	**Genome** the total complement of genetic information of a cell or a virus
	Metagenome the total genetic complement of all the cells present in a particular environment
	Epigenome the total number of possible epigenetic changes
	Methylome the total number of methylated sites on the DNA (whether epigenetic or not)
	Mobilome the total number of mobile genetic elements in a cell
RNA	**Transcriptome** the total RNA produced in an organism under a specific set of conditions
Protein	**Proteome** the total set of proteins encoded by a genome; sometimes also used in place of *translatome*
	Translatome the total set of proteins present under specified conditions
	Interactome the total set of interactions between proteins (or other macromolecules)
	Secretome the total set of proteins secreted by a cell
Metabolites	**Metabolome** the total complement of small molecules and metabolic intermediates
	Glycome the total complement of sugars and other carbohydrates
Organisms	**Microbiome** the total complement of microorganisms in an environment (including those associated with a higher organism)
	Virome the total complement of viruses in an environment
	Mycobiome the total complement of fungi in a natural environment

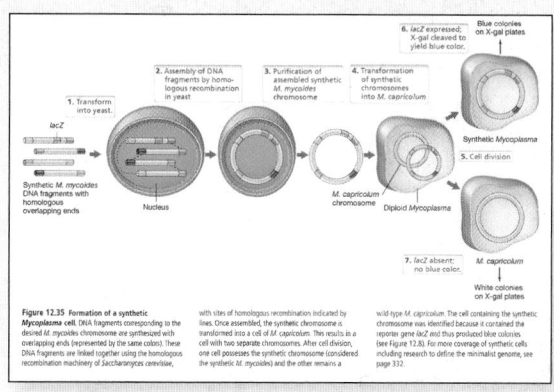

Figure 12.35 Formation of a synthetic *Mycoplasma* cell. DNA fragments corresponding to the desired *M. mycoides* chromosome are synthesized with overlapping ends (represented by the same color). These DNA fragments are linked together using the homologous recombination machinery of *Saccharomyces cerevisiae*, with sites of homologous recombination indicated by lines. Once assembled, the synthetic chromosome is transformed into a cell of *M. capricolum*. This results in a cell with two separate chromosomes. After cell division, one cell possesses the synthetic chromosome (considered the synthetic *M. mycoides*) and the other remains a wild-type *M. capricolum*. The cell containing the synthetic chromosome was identified because it contained the reporter gene *lacZ* and thus produced blue colonies (see Figure 12.8). For more coverage of synthetic cells including research to define the minimalist genome, see page 332.

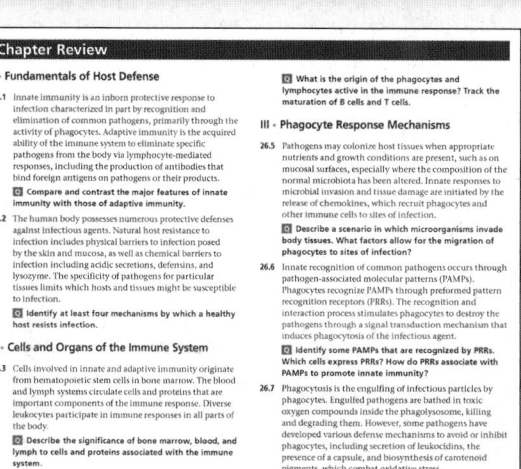

Chapter Review

I • Fundamentals of Host Defense

26.1 Innate immunity is an inborn protective response to infection characterized in part by recognition and elimination of common pathogens, primarily through the activity of phagocytes. Adaptive immunity is the acquired ability of the immune system to eliminate specific pathogens from the body via lymphocyte-mediated responses, including the production of antibodies that bind foreign antigens on pathogens or their products.

 Compare and contrast the major features of innate immunity with those of adaptive immunity.

26.2 The human body possesses numerous protective defenses against infectious agents. Natural host resistance to infection includes physical barriers to infection posed by the skin and mucosa, as well as chemical barriers to infection including acidic secretions, defensins, and lysozyme. The specificity of pathogens for particular tissues limits which hosts and tissues might be susceptible to infection.

 Identify at least four mechanisms by which a healthy host resists infection.

II • Cells and Organs of the Immune System

26.3 Cells involved in innate and adaptive immunity originate from hematopoietic stem cells in bone marrow. The blood and lymph systems circulate cells and proteins that are important components of the immune response. Diverse leukocytes participate in immune responses in all parts of the body.

 Describe the significance of bone marrow, blood, and lymph to cells and proteins associated with the immune system.

 What is the origin of the phagocytes and lymphocytes active in the immune response? Track the maturation of B cells and T cells.

III • Phagocyte Response Mechanisms

26.5 Pathogens may colonize host tissues when appropriate nutrients and growth conditions are present, such as on mucosal surfaces, especially where the composition of the normal microbiota has been altered. Innate responses to microbial invasion and tissue damage are initiated by the release of chemokines, which recruit phagocytes and other immune cells to sites of infection.

 Describe a scenario in which microorganisms invade body tissues. What factors allow for the migration of phagocytes to sites of infection?

26.6 Innate recognition of common pathogens occurs through pathogen-associated molecular patterns (PAMPs). Phagocytes recognize PAMPs through preformed pattern recognition receptors (PRRs). The recognition and interaction process stimulates phagocytes to destroy the pathogens through a signal transduction mechanism that induces phagocytosis of the infectious agent.

 Identify some PAMPs that are recognized by PRRs. Which cells express PRRs? How do PRRs associate with PAMPs to promote innate immunity?

26.7 Phagocytosis is the engulfing of infectious particles by phagocytes. Engulfed pathogens are bathed in toxic oxygen compounds inside the phagolysosome, killing and degrading them. However, some pathogens have developed various defense mechanisms to avoid or inhibit phagocytes, including secretion of leukocidins, the presence of a capsule, and biosynthesis of carotenoid pigments, which combat oxidative stress.

Continuous Learning
Before, During, and After Class

MasteringMicrobiology improves results by engaging students before, during, and after class.

BEFORE CLASS

Reading Questions, art-based activities and MCAT Prep, along with Quantitative Questions, prepare students for in-depth class discussion.

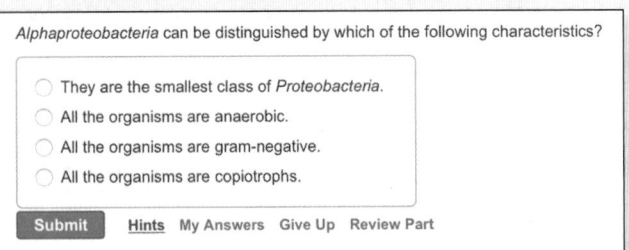

Alphaproteobacteria can be distinguished by which of the following characteristics?

- They are the smallest class of *Proteobacteria*.
- All the organisms are anaerobic.
- All the organisms are gram-negative.
- All the organisms are copiotrophs.

Submit   **Hints**   My Answers   Give Up   Review Part

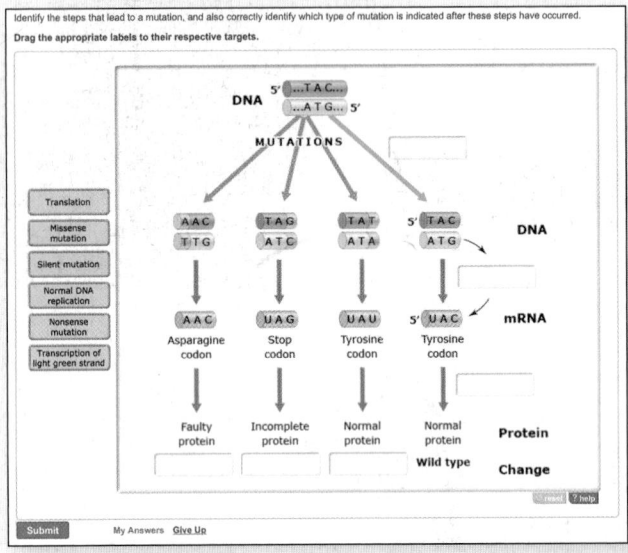

Identify the steps that lead to a mutation, and also correctly identify which type of mutation is indicated after these steps have occurred.

Drag the appropriate labels to their respective targets.

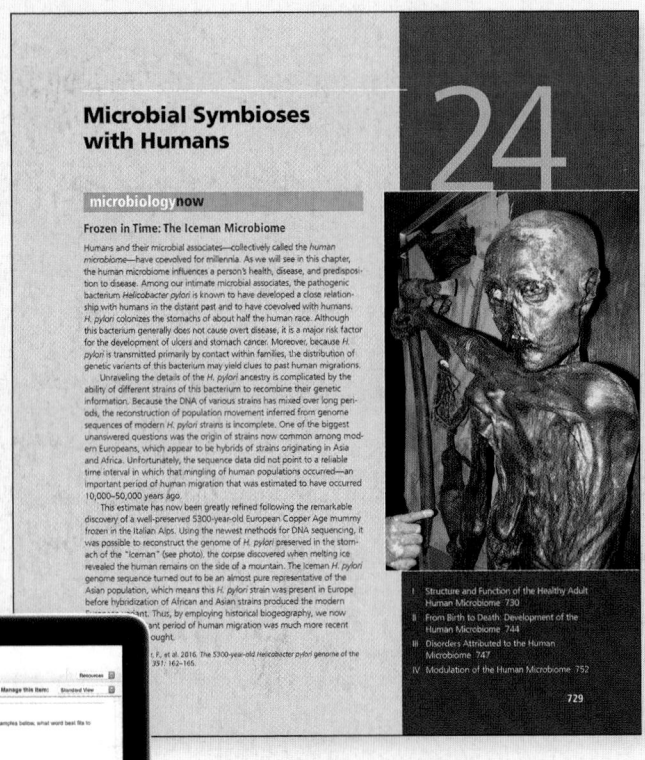

Microbial Symbioses with Humans

24

microbiologynow

Frozen in Time: The Iceman Microbiome

Instructors can further encourage students to apply microbiology principles to today's research by assigning MicrobiologyNow Coaching Activities.

with MasteringMicrobiology®

DURING CLASS

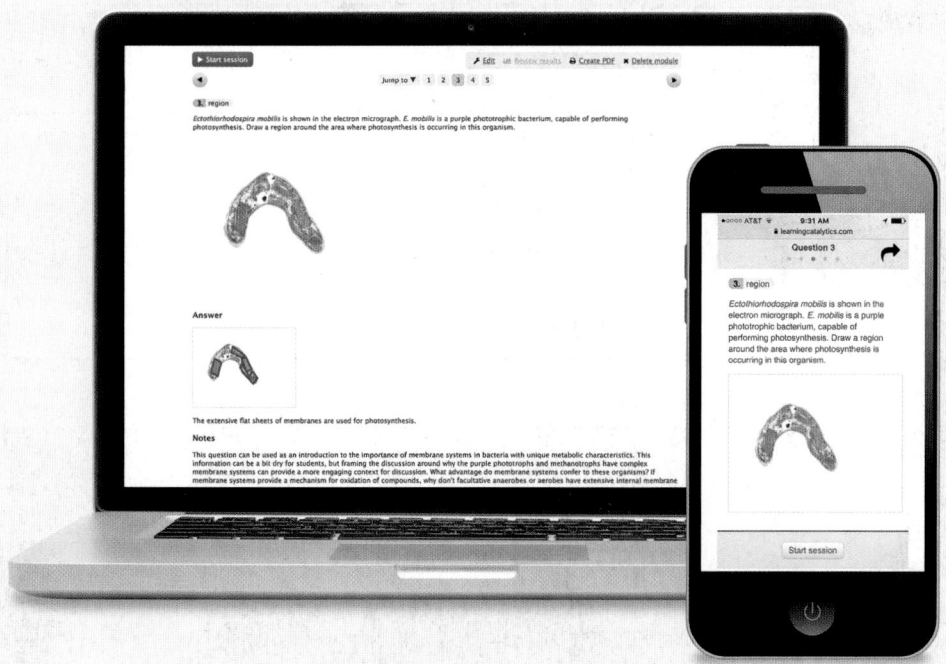

A wide variety of interactive coaching activities as well as high-level assessments can be assigned after class to continue student learning and concept mastery.

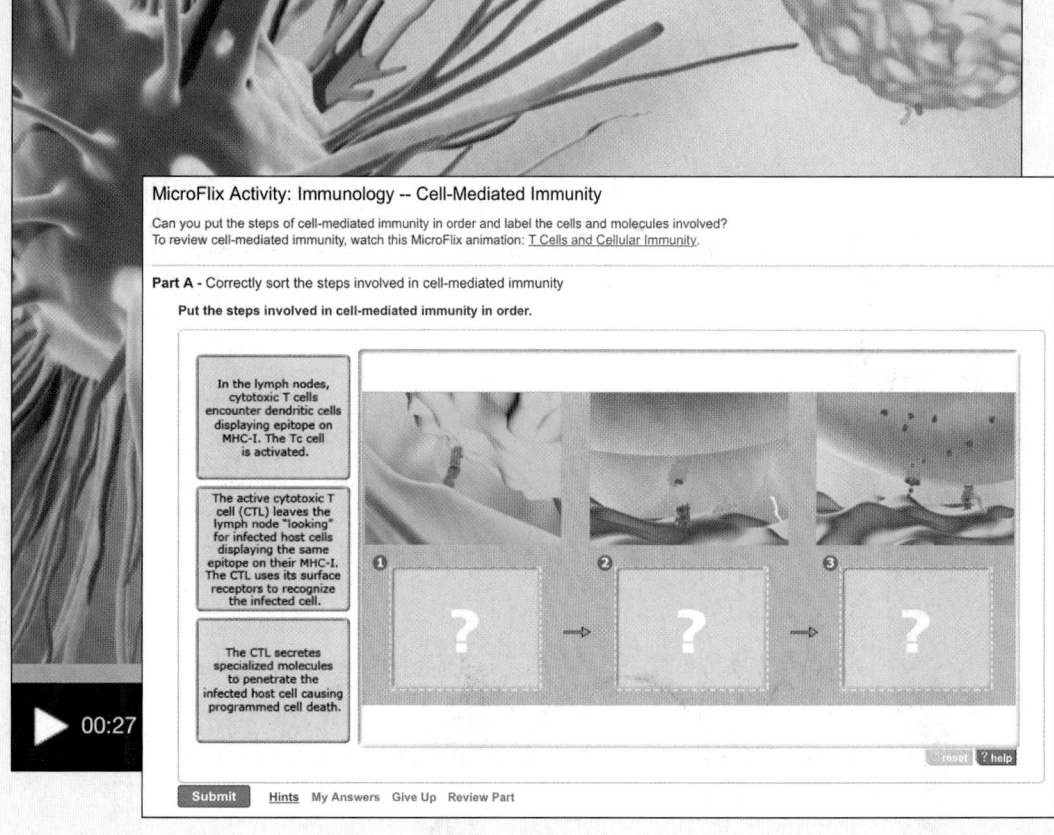

Visualize Microbiology

AFTER CLASS

NEW! Interactive Microbiology is a dynamic suite of interactive tutorials and animations that teach key concepts in microbiology, including Operons; Biofilms and Quorum Sensing; Aerobic Respiration in Bacteria; Complement; and more. Students actively engage with each topic via a case study and learn by manipulating variables, predicting outcomes, and answering formative and summative assessment questions.

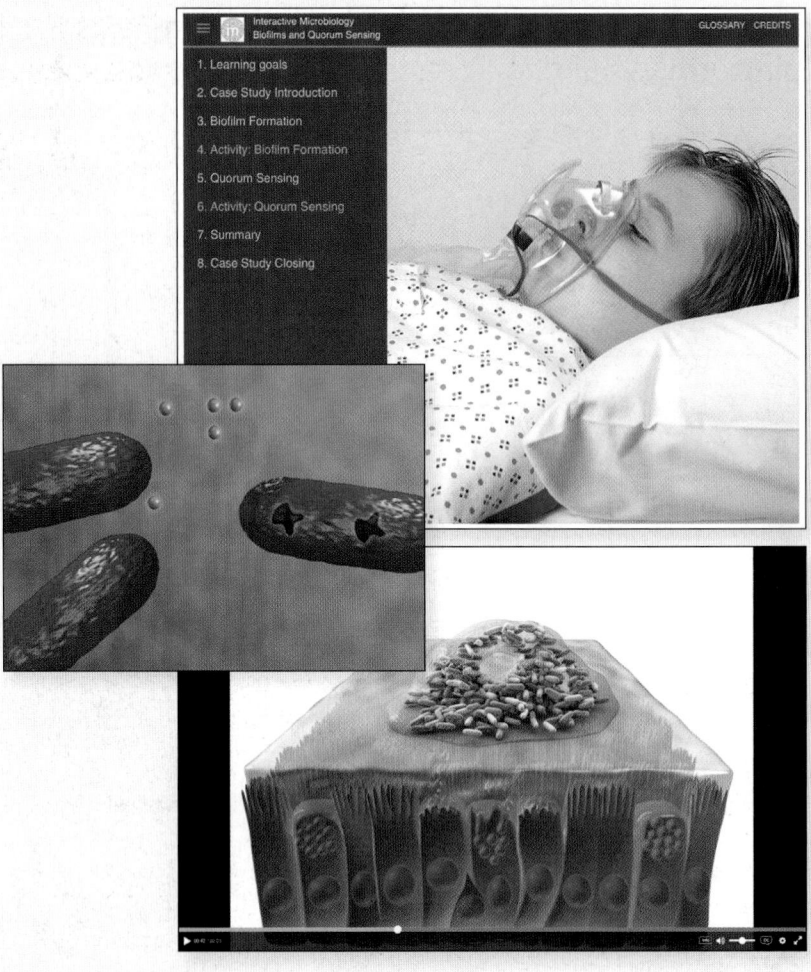

MicroLab Tutors, Lab Technique Videos and Lab Practical Assessments ensure students connect lecture concepts with lab techniques and protocol.

Access the complete textbook online and offline with eText 2.0

NEW! The **Fifteenth Edition** is available in Pearson's fully-accessible eText 2.0 platform, providing a seamless text and media experience, regardless of device.*

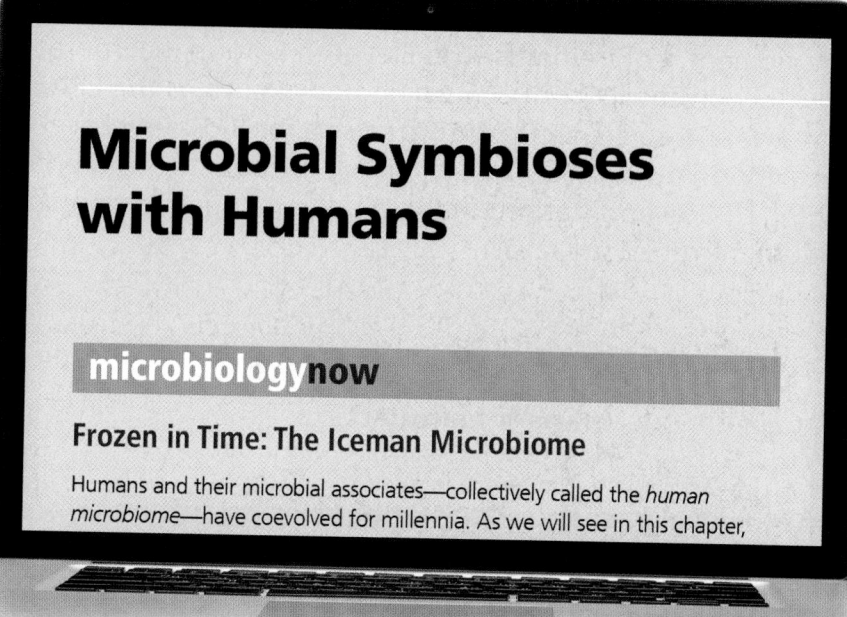

Microbial Symbioses with Humans

microbiologynow

Frozen in Time: The Iceman Microbiome

Humans and their microbial associates—collectively called the *human microbiome*—have coevolved for millennia. As we will see in this chapter,

NEW! eText 2.0's mobile app offers offline access and can be downloaded for most iOS and Android phones and tablets from the iTunes or Google Play stores.

Powerful interactive and customization functions include instructor and student note-taking, highlighting, bookmarking, search, and links to glossary terms.

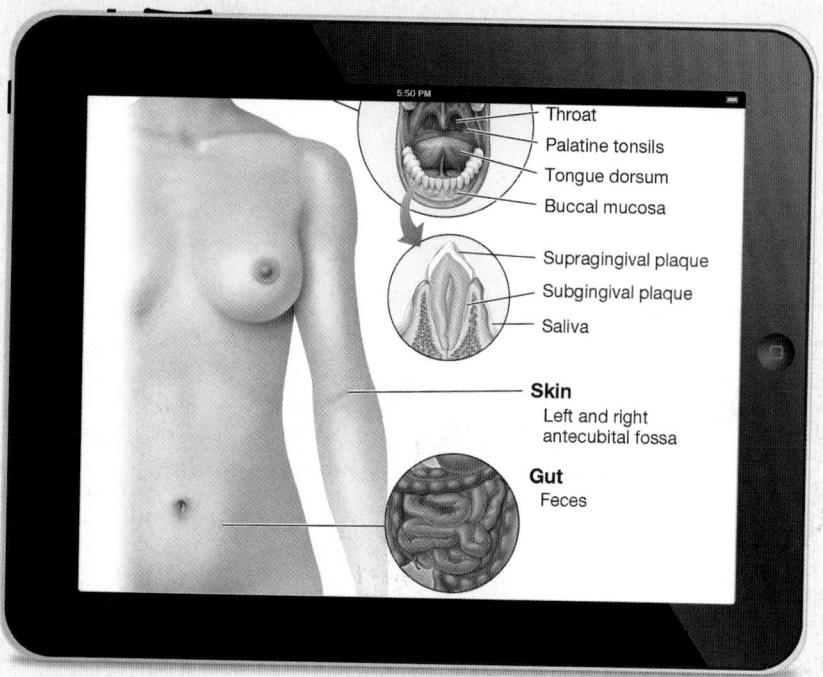

*The eText 2.0 edition will be available for Fall 2017 classes.

Additional Support for Students and Instructors

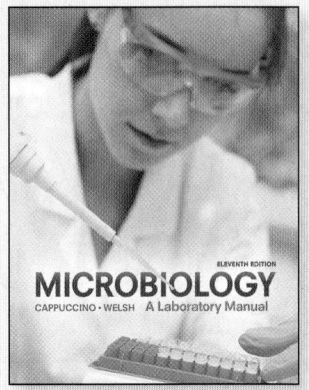

NEW! *Microbiology: A Laboratory Manual,* **Eleventh Edition**
James G. Cappuccino and Chad T. Welsh

Flexible and comprehensive, *Microbiology: A Laboratory Manual,* is known for its thorough coverage, straightforward procedures and minimal equipment requirements. The Eleventh Edition incorporates **UPDATED** safety protocols from governing bodies such as the EPA, ASM and AOAC and offers alternate organisms for Biosafety Level 1 and 2 labs. **NEW** labs on Food Safety, ample introductory material and engaging clinical applications make this lab manual appropriate for all modern microbiology labs!

Laboratory Experiments in Microbiology,
Eleventh Edition
Ted R. Johnson and Christine L. Case

Techniques in Microbiology: A Student Handbook
John M. Lammert

Consult your Pearson Representative for package ISBNs and ordering information.

The Instructor Resources Area in MasteringMicrobiology includes the following downloadable tools:

The Instructor Resource DVD offering a wealth of media resources including:

- all of the figures, photos, and tables from the text in JPEG and PowerPoint® formats, in labelled and unlabeled versions, and with customizable labels and leader lines

- Step-edit Powerpoint slides that present multi-step process figures step-by-step

- Clicker Questions and Quiz Show Game questions that encourage class interaction

- MicroFlix™ animations, Interactive Microbiology, and Microbiology Animations

- MicroLab Tutors and Lab Technique Videos to help prepare students for lab and make the connection between lecture and lab

- Customizable PowerPoint® lecture outlines save valuable class prep time

A comprehensive Instructor's Manual including chapter summaries to aid in class preparation as well as the answers to the end-of-chapter review and application questions.

Test Bank provides over 3,000 customizable questions available in Microsoft® Word and TestGen® formats.

BROCK BIOLOGY OF
MICROORGANISMS

FIFTEENTH EDITION

MICHAEL T. MADIGAN
Southern Illinois University Carbondale

KELLY S. BENDER
Southern Illinois University Carbondale

DANIEL H. BUCKLEY
Cornell University

W. MATTHEW SATTLEY
Indiana Wesleyan University

DAVID A. STAHL
University of Washington Seattle

 Pearson

330 Hudson Street, NY NY 10030

Editor-in Chief: Serina Beauparlant
Senior Courseware Portfolio Manager: Kelsey Churchman
Managing Producer: Nancy Tabor
Content and Design Manager: Michele Mangelli, Mangelli Productions, LLC
Courseware Director, Content Development: Barbara Yien
Art development editor: Kim Brucker
Editorial Coordinator: Kate Abderholden
Rich Media Content Producer: Lucinda Bingham
Copyeditor: Anita Hueftle
Proofreader: Martha Ghent
Compositor: iEnergizer Aptara®, Ltd.

Art Coordinator: Jean Lake
Production Coordinator: Karen Gulliver
Indexer: Steele/Katigbak
Interior Designer: Hespenheide Design
Cover Designer: Hespenheide Design
Illustrators: Imagineering STA Media Services, Inc.
Rights & Permissions Manager: Ben Ferrini
Photo Researcher: Kristin Piljay
Manufacturing Buyer: Stacey Weinberger
Product Marketing Manager: Christa Pelaez
Field Marketing Manager: Kelly Galli

Cover Photo Credits: Kristen Earle, Gabriel Billings, KC Huang, and Justin Sonnenburg, Stanford University School of Medicine (fluoresence micrograph); Clyde Hutchison and J. Craig Venter, JCVI, and Thomas Deerinck and Mark Ellisman, National Center for Microscopy and Imaging Research, University of California, San Diego (scanning electron micrograph)

Library of Congress Cataloging-in-Publication Data

Names: Madigan, Michael T., 1949- | Bender, Kelly S., 1977- | Buckley, Daniel
    H., 1972 | Sattley, W. Matthew, 1975- | Stahl, David A., 1949-
Title: Brock biology of microorganisms.
Other titles: Biology of microorganisms
Description: Fifteenth edition/Michael T. Madigan, Southern Illinois
    University Carbondale, Kelly S. Bender, Southern Illinois University
    Carbondale, Daniel H. Buckley, Cornell University, W. Matthew Sattley,
    Indiana Wesleyan University, David A. Stahl, University of Washington
    Seattle. | Boston : Pearson, [2018] | Includes index.
Identifiers: LCCN 2016039954 | ISBN 9780134261928 | ISBN 0134261925
Subjects: LCSH: Microbiology.
Classification: LCC QR41.2 .B77 2018 | DDC 579--dc23 LC record available at
https://lccn.loc.gov/2016039954

ISBN 10: 0-13-426192-5; ISBN 13: 978-0-13-426192-8 (Student edition)
ISBN 10: 0-13-460236-6; ISBN 13: 978-0-13-460236-3 (Instructor's Review Copy)
2 17
www.pearsonhighered.com

About the Authors

Michael T. Madigan received his B.S. in Biology and Education from Wisconsin State University–Stevens Point (1971) and his M.S. (1974) and Ph.D. (1976) in Bacteriology from the University of Wisconsin–Madison in the laboratory of Thomas Brock. Following a postdoc at Indiana University with Howard Gest, Mike moved to Southern Illinois University Carbondale, where he taught courses in introductory microbiology and bacterial diversity as a professor of microbiology for 33 years. In 1988 Mike was selected as the Outstanding Teacher in the College of Science and in 1993, the Outstanding Researcher. In 2001 he received the SIUC Outstanding Scholar Award. In 2003 he received the Carski Award for Distinguished Undergraduate Teaching from the American Society for Microbiology, and he is an elected Fellow of the American Academy of Microbiology. Mike's research is focused on bacteria that inhabit extreme environments, and for the past 20 years his emphasis has been Antarctic microbiology. Mike has co-edited a major treatise on phototrophic bacteria and served for 10 years as chief editor of the journal *Archives of Microbiology*. He currently serves on the editorial board of the journals *Environmental Microbiology* and *Antonie van Leeuwenhoek*. Mike's other interests include forestry, swimming, reading, and caring for his dogs and horses. He lives on a quiet lake with his wife, Nancy, three dogs (Kato, Nut, and Merry), and three horses (Eddie, Gwen, and Georgie).

Kelly S. Bender received her B.S. in Biology from Southeast Missouri State University (1999) and her Ph.D. (2003) in Molecular Biology, Microbiology, and Biochemistry from Southern Illinois University Carbondale. Her dissertation research focused on the genetics of perchlorate-reducing bacteria. During her postdoctoral fellowship, Kelly worked on the genetic regulation of sulfate-reducing bacteria in the laboratory of Judy Wall at the University of Missouri–Columbia. She also completed a transatlantic biotechnology fellowship at Uppsala University in Sweden researching regulatory small RNAs in bacteria. In 2006, Kelly returned to her alma mater, Southern Illinois University Carbondale, as an Assistant Professor in the Department of Microbiology and in 2012 was tenured and promoted to Associate Professor. Her lab studies a range of topics including regulation in sulfate-reducing bacteria and the microbial community dynamics of sites impacted by acid mine drainage. Kelly teaches courses in introductory microbiology and microbial diversity, has served on numerous federal grant review panels, and is an active member of the American Society for Microbiology (ASM). Her other interests include spending time with her daughter, Violet, and husband, Dick.

Daniel H. Buckley is a Professor at Cornell University in the School of Integrative Plant Science. He earned his B.S. in Microbiology (1994) at the University of Rochester and his Ph.D. in Microbiology (2000) at Michigan State University. His graduate research focused on the ecology of soil microbial communities and was conducted in the laboratory of Thomas M. Schmidt in affiliation with the Center for Microbial Ecology. Dan's postdoctoral research examined linkages between microbial diversity and biogeochemistry in marine microbial mats and stromatolites and was conducted in the laboratory of Pieter T. Visscher at the University of Connecticut. Dan joined the Cornell faculty in 2003. His research program investigates the ecology and evolution of microbial communities in soils with a focus on the causes and consequences of microbial diversity. He has taught both introductory and advanced courses in microbiology, microbial diversity, and microbial genomics. He received a National Science Foundation Faculty Early Career Development (CAREER) award in 2005 for excellence in integrating research and education. He has served as Director of the Graduate Field of Soil and Crop Sciences at Cornell and Co-Director of the Microbial Diversity summer course of the Marine Biological Laboratory in Woods Hole, Massachusetts. He currently serves on the editorial boards of *Applied and Environmental Microbiology* and *Environmental Microbiology*. Dan lives in Ithaca, New York, with his wife, Merry, and sons, Finn and Colin.

W. Matthew Sattley received his B.A. in Biology in 1998 from Blackburn College (Illinois) and his Ph.D. (2006) in Molecular Biology, Microbiology, and Biochemistry from Southern Illinois University Carbondale. His graduate studies focused on the microbiology of sulfur cycling and other biogeochemical processes in permanently ice-covered lakes of Antarctica. In his postdoctoral research at Washington University in Saint Louis, he studied the physiology and genomics of anoxygenic phototrophic bacteria in Robert Blankenship's laboratory. Matt then accepted a faculty appointment to the Department of Biology at MidAmerica Nazarene University (Kansas), where he supervised undergraduate research and taught courses in microbiology, environmental science, and cell biology. In 2010, Matt transitioned to the Division of Natural Sciences at Indiana Wesleyan University, where he is a Professor of Biology and Director of the Hodson Summer Research Institute, a faculty-led summer research program for undergraduate students in the Natural Sciences. His research group investigates the ecology, diversity, and genomics of bacteria that inhabit extreme environments. Matt is a member of the American Society for Microbiology (including its Indiana Branch) and the Indiana Academy of Science, and he currently serves as an expert reviewer for the undergraduate microbiology research journal *Fine Focus*. Matt lives in Marion, Indiana, with his wife, Ann, and sons, Josiah and Samuel. Outside of teaching and research, Matt enjoys playing drums, reading, motorcycling, and talking baseball and cars with his boys.

David A. Stahl received his B.S. degree in Microbiology from the University of Washington, Seattle, and completed graduate studies in microbial phylogeny and evolution with Carl Woese in the Department of Microbiology at the University of Illinois at Urbana–Champaign. Subsequent work as a postdoctoral fellow with Norman Pace, then at the National Jewish Hospital in Colorado, involved early applications of 16S rRNA-based sequence analysis to the study of natural microbial communities. In 1984 Dave joined the faculty at the University of Illinois with appointments in Veterinary Medicine, Microbiology, and Civil Engineering. In 1994 he moved to the Department of Civil Engineering at Northwestern University, and in 2000 returned to the University of Washington as professor in the Departments of Civil and Environmental Engineering and Microbiology. Dave is known for his work in microbial evolution, ecology, and systematics, and received the 1999 Bergey Award and the 2006 ASM Procter & Gamble Award in Applied and Environmental Microbiology. Dave is an elected fellow of the American Academy of Microbiology and a member of the National Academy of Engineering. His main research interests surround the biogeochemistry of nitrogen and sulfur and the microbial communities that sustain the associated nutrient cycles. His laboratory was first to culture ammonia-oxidizing *Archaea,* a group believed to be the key mediators of this process in the nitrogen cycle. Dave has taught several courses in environmental microbiology, was one of the founding editors of the journal *Environmental Microbiology,* and has served on many advisory committees. Outside the lab, Dave enjoys hiking, bicycling, spending time with family, reading a good science fiction book, and—with his wife, Lin—renovating an old farmhouse on Bainbridge Island.

Dedications

Michael T. Madigan
dedicates this edition to students who have drawn inspiration from his textbook to make some aspect of microbiology their life's work.

Kelly S. Bender
dedicates this book to the memory of her grandmother, Alberta, whose biggest regret in life was not being able to attend school past the fifth grade.

Daniel H. Buckley
dedicates this book to the memory of his mother, Judy, who taught me to see joy and wonder, even in the smallest of things.

W. Matthew Sattley
dedicates this book to his amazing wife, Ann, for her endless support and understanding.

David A. Stahl
dedicates this book to his wife, Lin. My love, and one that helps me keep the important things in perspective.

Preface

Welcome to an exciting new edition of *Brock Biology of Microorganisms* (*BBOM*). This Fifteenth Edition is the strongest yet and presents microbiology in the context of the excitement this science generates today. For three generations, students and instructors have relied on the accuracy, authority, consistency, and up-to-date presentation of *BBOM* to learn or teach the principles of modern microbiology. Both students and instructors will benefit from the Fifteenth Edition in at least four major ways: (1) from the use of cutting-edge research to illustrate basic concepts; (2) from the seamless integration of molecular and ecological microbiology with evolution, diversity, the immune system, and infectious diseases; (3) from the visually stunning art program and spectacular photos; and (4) from the wide assortment of teaching and learning tools that accompany the book itself.

Veteran authors Madigan, Bender, Buckley, and Stahl welcome new coauthor Matt Sattley to the Fifteenth Edition. Matt, a professor at Indiana Wesleyan University, teaches both general microbiology and health professions microbiology and did a great job of reorganizing and refreshing our coverage of immunology and related areas. With an extremely strong author team that employs experts in each of our major areas of emphasis, we sincerely feel that *BBOM* 15e is the best learning resource available in microbiology today.

What's New in the 15th Edition?

The Fifteenth Edition guides students through the six major themes of microbiology as outlined by the American Society for Microbiology Conference on Undergraduate Education (ASMCUE): Evolution, Cell Structure and Function, Metabolic Pathways, Information Flow and Genetics, Microbial Systems, and the Impact of Microorganisms. With enhanced and revised artwork complemented with over 90 new color photos, *BBOM* 15e presents microbiology as the visual science it is. Thirty-three new MicrobiologyNow chapter-opening vignettes were composed for this edition, each designed to introduce a chapter's theme through a recent discovery published in the microbiology literature. Several new Explore the Microbial World features were also developed for this edition, each designed to give students a feel for exciting special topics in microbiology and to fuel their scientific curiosity.

Genomics, and all of the various "omics" it has spawned, support content in every chapter of *BBOM* 15e, reflecting how the omics revolution has transformed all of biology, especially microbiology. Mastering the principles of the dynamic field of microbiology today requires an understanding of the supportive molecular biology. Hence, we have constructed *BBOM* 15e in a way that provides both the foundation for the science and the science itself. The result is a robust and modern treatment of microbiology that now includes exciting new chapters devoted to microbial systems

biology, synthetic biology, the human microbiome, and the molecular biology of microbial growth.

To strengthen the learning experience, each section summary in the chapter review is followed immediately by a review question to better link concept review with concept mastery. *BBOM* 15e is supported by MasteringMicrobiology™, Pearson's online homework, tutorial, and assessment system that assists students in pacing their learning and keeps instructors current on class performance. MasteringMicrobiology includes chapter-specific reading quizzes, MicrobiologyNow, Clinical Case and MicroCareer coaching activities, animation quizzes, MCAT Prep questions, and many additional study and assessment tools, including tutorials and assessments for the microbiology lab. Collectively, the content and presentation of *BBOM* 15e, coupled with the powerful learning tools of MasteringMicrobiology, create an unparalleled educational experience in microbiology.

Revision Highlights

Chapter 1

- The book begins with a revised and reorganized kickoff chapter that weaves introductory concepts in microbiology within an historical narrative. Foundational aspects of microbiology are now presented in the context of the major discoveries that have expanded our knowledge of the microbial world.
- Some highlights: introducing the principles of microscopy in a historical context; a new section on molecular biology and the importance of microbes in understanding the unity of life; the contributions of Carl Woese and the use of rRNA sequences to develop the universal tree of life; an introduction to the viral world; spectacular new summary art that explores the diversity of microbial life across a wide range of spatial scales.

Chapter 2

- Microbial cell structure and function are key pillars of microbiology, and this newly reworked and streamlined chapter offers a thorough introduction to comparative cell structure and provides the instructor with all of the tools necessary for effective classroom presentations. Coverage of nutrient transport systems has been moved to Chapter 3 to better present this topic in its proper context.
- Some highlights: a new Explore the Microbial World entitled "Tiny Cells"; unique attachment structures of *Archaea*; new coverage of archaella.

Chapter 3

- The essential features of microbial metabolism necessary for understanding how microbes transform energy are laid out in a logical sequence and at just the right level for introductory students. With the material on membrane transport now located here, the uptake of nutrients is highlighted as the initial step of any metabolic process.

- Some highlights: new coverage of the macromolecular composition of a cell; a more complete picture of energy transformation and the importance of free energy change; coverage of the citric acid cycle prior to (rather than following) discussion of the proton motive force.

Chapter 4

- Chapter 4 has been reorganized to provide the streamlined view of molecular biology necessary for both supporting and understanding virtually all aspects of microbiology today.
- Some highlights: new coverage of coupled transcription and translation in *Bacteria* and *Archaea*; new material on the assembly of cofactor-containing enzymes; stronger coverage of types I–VI secretion systems in gram-negative bacteria; updated art throughout.

Chapter 5

- Unit 2 is all about growth and begins with the Chapter 5 presentation of the essential principles of microbial growth and cultivation. Coverage of microbial growth control balances this chapter with a practical view of how microbial growth can be suppressed for both health and aesthetic reasons.
- Some highlights: new material on budding cell division and on biofilms; reworked chemostat coverage better explains continuous culture and its connection to basic growth principles; new coverage on how the environment affects growth previews the extensive coverage of microbial ecology and environmental microbiology later in the book.

Chapter 6

- This chapter on microbial regulation includes broad coverage of the classic forms of regulation but has been streamlined by moving the regulation of cell differentiation and biofilm formation to Chapter 7; this allowed for enhanced coverage of hot new areas in metabolic regulation such as regulation by anti-sigma factors and transcriptional regulation in *Archaea*.
- Some highlights: new coverage of the global phosphate regulon; new coverage of dual-acting transcriptional regulators in *Archaea* and how the stringent response affects the ecology of bacteria as diverse as *Escherichia coli*, *Caulobacter crescentus*, and *Mycobacterium tuberculosis*; updated art throughout.

Chapter 7

- A new chapter focused on the molecular biology of microbial growth showcases the orchestrated events leading to cell division and surveys the molecular processes targeted by antibiotics. Coverage of peptidoglycan synthesis, developmental stages in various *Bacteria,* and biofilm formation—previously scattered through the book—has been consolidated here to unite their common underlying themes.

- Some highlights: An introduction to the powerful tool of super-resolution microscopy includes several spectacular examples of how this breakthrough in resolution has remolded our view of molecular events in microbial growth; expanded coverage of biofilm formation; new coverage of bacterial persistence, a growing problem in medical microbiology; updated art throughout.

Chapter 8

- The introductory virology chapter is now included in the microbial growth unit and provides an introduction to the structure, replication, and lifestyles of viruses without overshadowing these important principles with the extensive diversity of the viral world, now covered in Chapter 10.
- Some highlights: discussion of the parallels between bacterial growth and viral replication; expanded coverage of how host cell growth is impacted by viral infection; high-resolution viral images; updated art throughout.

Chapter 9

- This revolutionary chapter on microbial systems biology kicks off our unit on genomics and genetics by underscoring the importance of microbial genome sequences and the field of functional "omics" to modern microbiology today. The chapter also includes examples of how systems biology can be used to model an organism's response to its environment.
- Some highlights: how functional and metabolic predictions are gleaned from genomic analyses; expanded coverage of RNA-Seq and metabolomic analyses; coverage of all of the common "omics" and how they relate to one another; new coverage of the systems biology of the important pathogen *Mycobacterium tuberculosis* and other systems biology studies related to human health; metagenomics and metabolomics of human skin; updated and spectacular new art and photos throughout.

Chapter 10

- Chapter 10, entitled "Viral Genomics, Diversity, and Ecology," now includes coverage of viral ecology and diversity that was previously in Chapter 8. The many diverse genomes and replication schemes of viruses form the foundation for coverage of the diversity and ecological activities of viruses.
- Some highlights: the viral "immune system" of *Bacteria* and *Archaea*—CRISPR; large viruses and viral evolution; the human virome; beneficial prions; viral host preferences; updated and new art throughout.

Chapter 11

- Chapter 11, "Genetics of *Bacteria* and *Archaea*," has been streamlined to focus on the essential concepts of mutation and gene transfer in prokaryotic cells. New high-resolution images have been included to illustrate gene transfer processes.
- Some highlights: new coverage on the utility of transposon mutagenesis; a spectacular photo series illustrating the concept of competence; new coverage on defective bacteriophages as "gene transfer agents"; updated art throughout.

Chapter 12

- This highly reorganized chapter entitled "Biotechnology and Synthetic Biology" covers the essential tools of biotechnology and discusses commercial products produced by genetically engineered microbes. New coverage presents the remarkable advances in synthetic biology and CRISPR genome editing.
- Some highlights: engineering microbes to produce biofuels; expanded coverage of synthetic pathways and synthetic cells; new coverage of the biocontainment of genetically modified organisms; updated art throughout.

Chapter 13

- Chapter 13 sets the stage for our unit on evolution and diversity by revealing how nucleic acid sequences have revealed the true diversity of the microbial world. The chapter has also been revised and reorganized to increase the emphasis on the origin and diversification of life and microbial systematics.
- Some highlights: revised text places phylogeny into firm context with microbial systematics; how the tree of life and molecular sequences form the foundation of our understanding of the origin and diversification of the three domains; revised coverage of phylogenetic tree construction and what such trees can tell us about microbial evolution.

Chapter 14

- Our discussion of microbial metabolism has been revised and reorganized to highlight the modularity of microbial metabolism and to include coverage of newly discovered microbial metabolisms.
- Some highlights: a new section on assimilatory processes of autotrophy and nitrogen fixation; grouping respiratory processes by electron donor, electron acceptor, or one-carbon metabolisms; new art depicting electron flow in oxygenic photosynthesis, sulfur chemolithotrophy, and acetogenesis; discussion of the role of flavin-based electron bifurcation in energy conservation; coverage of the exciting discoveries of intra-aerobic methanotrophy and interspecies electron transfer in anaerobic methane oxidation.

Chapters 15 and 16

- These chapters, covering functional and phylogenetic diversity of *Bacteria*, respectively, have been updated and streamlined in spots to provide the highly organized view of bacterial diversity that offers instructors the freedom to present this subject in the way that best suits their course needs.
- Some highlights: functional diversity organized by metabolism, unique morphologies, and other special properties shows how functional diversity is often unlinked to phylogenetic diversity; phylogenetic diversity organized around the major phyla of *Bacteria* shows how phylogenetic diversity is often unlinked to metabolic properties.

Chapter 17

- Chapter 17, entitled "Diversity of *Archaea*," has been updated to include new coverage of recent discoveries in archaeal diversity including the fact that *Archaea* are widespread in nature and not just restricted to extreme environments.
- Some highlights: updated coverage of methanogenic *Archaea* to include the extensive diversity characteristic of this group; new coverage of the evolutionary origins and distribution of methanogens within the archaeal domain; the latest story on *Archaea* and the upper temperature limit for life.

Chapter 18

- Coverage of the microbial eukaryotes has been revised to include significant new advances in our understanding of the phylogeny of *Eukarya*.
- Some highlights: a new phylogenetic tree of *Eukarya*; updated terminology throughout; the "SAR" lineages; the new understanding of fungal diversity that incorporates the *Microsporidia* as a deeply divergent fungal group.

Chapter 19

- This chapter begins a new unit on ecology and environmental microbiology. The modern tools of the microbial ecologist are described with examples of how each has helped sculpt the science.
- Some highlights: complete coverage of the omics revolution and how it is being exploited to solve complex problems in microbial ecology; Raman microspectroscopy and its use for nondestructive molecular and isotopic analyses of single cells; high-throughput cultivation methods and how they can be used to bring novel microbes into laboratory culture.

Chapter 20

- The properties and microbial diversity of major microbial ecosystems including soils and both freshwater and marine systems are compared and contrasted in an exciting new way.
- Some highlights: new environmental census data for deep marine sediments reveal the novel *Archaea* and *Bacteria* living thousands of meters below the seafloor; expanded coverage of the links between terrestrial and marine microorganisms and climate change.

Chapter 21

- Extensive coverage of the major nutrient cycles in nature and the microbes that catalyze them presented in a fashion that allows the cycles to be taught as individual entities or as interrelated metabolic loops.
- Some highlights: new coverage of how humans are affecting the nitrogen and carbon cycles; microbial respiration of solid metal oxides in the iron and manganese cycles including the concept of "microbial wires" that can carry electrons over great distances; how microbes contribute to mercury contamination of aquatic life.

Chapter 22

- A newly revised chapter on the "built environment" shows how humans create new microbial habitats through construction of buildings, supporting infrastructure, and habitat modification.

- Some highlights: coverage of the effects microbes have on wastewater treatment, mining and acid mine drainage, the corrosion of metals, and the degradation of stone and concrete; the pathogens of most concern in drinking water and how we eliminate them; the major microbes that inhabit our household and work environments.

Chapter 23

- A chapter devoted to nonhuman microbial symbioses describes the major microbial partners that live in symbiotic or other types of close associations with plants and animals.
- Some highlights: using our knowledge of plant and animal symbioses to develop microbially centered insect pest controls; revealing the common symbiotic mechanism used by certain bacteria and fungi to provide plants with key nutrients.

Chapter 24

- A new chapter devoted exclusively to the human microbiome kicks off our unit on microbe–human interactions and the immune system by introducing the dramatic advances in our understanding of the microbes that inhabit the human body and their relationship to health and disease.
- Some highlights: extensive coverage of "who lives where (and why)" in and on the human body; how the new understanding of our intimate microbial partners was used to develop novel microbial-based disease therapies; mapping the biogeography of our skin microbiota using new molecular techniques; how gut microbes likely influence both our health and behavior; a new Explore the Microbial World entitled "The Gut–Brain Axis."

Chapter 25

- This heavily reworked and more visually appealing chapter is devoted exclusively to microbial infection and pathogenesis. Major topics in the first part include microbial adherence, colonization, invasion and pathogenicity, and virulence and attenuation. The second part is focused on the destructive enzymes and toxins produced by pathogenic bacteria. Microbial and host factors are compared as to how each can tip the balance toward health or disease.
- Some highlights: eight new color photos bring host–microbe relationships into better focus; new coverage of dental caries is supported by a spectacular fluorescent micrograph that reveals the previously hidden diversity of this disease; increased coverage of microbial infection and the compromised host.

Chapter 26

- Coverage of the immune response has been completely reorganized to provide a fresh take on immune mechanisms. Concepts of innate and adaptive immunity are now organized into separate chapters (26 and 27, respectively) that provide a more teachable format and enhance the student experience. The new organization provides a natural progression to the updated topics in clinical microbiology and immunology presented in Chapter 28.

- Some highlights: extensively revised and reorganized text and vibrant new artwork clearly illustrate the roles of inflammation, fever, and interferons in the innate immune response; stronger, clearer coverage of the complement system, including extensive new artwork, helps clarify its important role in innate immunity.

Chapter 27

- Fundamental concepts of the adaptive immune response are now reorganized into a dedicated chapter and presented in a thoroughly revised and more streamlined format.
- Some highlights: beautifully enhanced art and new photos more clearly orient students to key concepts including clonal selection and deletion of B cells and T cells, antibody structure, and antigen binding and presentation.

Chapter 28

- Clear and concise new text now includes automated culture systems, antibody precipitation, and monoclonal antibody production, as well as a reorganized treatment of antimicrobial drugs. Both reimagined and totally new art supported by 20 new color photos brightly illustrate complex topics and enhance the visual experience.
- Some highlights: how a clinical microbiology laboratory actually functions; an exciting new Explore the Microbial World feature on MRSA describes how emerging resistance to antibiotics in *Staphylococcus aureus* has led to high global incidence of what is now a virtually untreatable bacterial pathogen.

Chapter 29

- A significantly reworked and streamlined discussion of epidemiology kicks off our unit on infectious diseases with a visual presentation of the everyday language of epidemiology and then closely integrates this terminology throughout the chapter. Fewer lengthy tables are presented and visual appeal is greater, while the essential concepts of disease spread and control remain the major themes of the chapter.
- Some highlights: updated and new coverage of emerging infectious diseases and current pandemics, including HIV/AIDS, cholera, and influenza; the key role of the epidemiologist in tracking disease outbreaks and maintaining public health.

Chapter 30

- This is the first of four chapters on microbial diseases grouped by their modes of transmission; this approach emphasizes the common ecology of these diseases despite differences in etiology. Classical as well as emerging and reemerging bacterial and viral diseases transmitted person to person are the focus of this highly visual chapter.
- Some highlights: several new photos add to the already extensive visual showcase of infectious diseases; new coverage of Ebola describes why this pathogen is so dangerous and the extraordinary precautions healthcare workers must take to prevent infection; new coverage of hepatitis, a widespread disease with serious implications.

Chapter 31

- Vectorborne microbial diseases are becoming more and more common worldwide and are covered in detail in this visually appealing chapter. From diseases with high mortality, such as rabies and hantavirus syndromes, to those with high incidence and low mortality but significant side effects, such as Lyme and West Nile diseases, all of the major vectorborne infectious diseases found today are consolidated in one place.

- Some highlights: new coverage of Zika and Chikungunya diseases and their relationship to dengue and yellow fevers; updated coverage of Lyme, West Nile, and *Coxiella* (Q fever) infections supported by new color photos.

Chapter 32

- Food- and waterborne illnesses are still common, even in developed countries. This chapter consolidates these topics to emphasize their "common source" modes of transmission while differentiating the major pathogens seen in each vehicle.

- Some highlights: a clearer distinction between food infections and food poisonings; new coverage of the potentially fatal food-borne infection caused by the intracellular pathogenic bacterium *Listeria*.

Chapter 33

- Major infectious diseases caused by eukaryotic microbes—fungi, parasites, and pathogenic helminths—are organized into one highly visual chapter. With climate change affecting infectious disease ecology, many of these diseases previously found only in tropical or subtropical countries are now creeping northward.

- Some highlights: new emphasis on the different modes of transmission (food, water, vector) of major eukaryotic pathogens; new coverage of river blindness and trichinosis as common filariases.

Acknowledgments

A textbook is a complex undertaking and can only emerge from the combined contributions of a large book team. In addition to the authors, the team is composed of folks both inside and outside of Pearson. Senior Courseware Portfolio Manager Kelsey Churchman (Pearson) paved the way for the Fifteenth Edition of *Brock Biology of Microorganisms* and provided the resources necessary for the authors to produce a spectacular revision in a timely fashion. The authors thank Kelsey for her dedication to making *BBOM* a first-class textbook and in orchestrating the Fifteenth Edition from start to finish.

Michele Mangelli (Mangelli Productions) oversaw the book's writing and review process as well as the production process. Michele had several roles in the final product. She oversaw both manuscript preparation and review, and assembled and managed the production team. Michele's strong efforts in these regards kept the entire book team on mission, on budget, and on schedule—and were accomplished in her typically helpful, friendly, and accommodating manner. Both cover and interior designs were created by Gary Hespenheide (Hespenheide Design). The artistic magic of Gary is clearly visible in the outstanding text and cover designs of *BBOM* 15e; his talents have made this book easy to use and navigate and an exciting and fun read. The art team at Imagineering Art (Toronto) did an outstanding job in helping the authors link art with text and provided many helpful suggestions and options for art presentation, consistency, and style. Thanks go out to Michele, Gary, and Imagineering for their tireless work on *BBOM* 15e.

Many other people were part of the book production, editorial, or marketing team including Karen Gulliver, Jean Lake, Kim Brucker, Kristin Piljay, Betsy Dietrich, Martha Ghent, Susan Wang, Ann Paterson, Christa Pelaez, and Kelly Galli. Karen was our excellent and highly efficient production editor; she kept manuscript and pages moving smoothly through the wheels of production and graciously tolerated the authors' many requests and time constraints. Jean was our art coordinator, tracking and routing art and handling interactions between the art studio and the authors to ensure quality control and a timely schedule. Kim did a superb job in art development, for select figures, by helping the authors visualize art layout possibilities and transforming some mighty rough "roughs" into spectacular art. Betsy and Martha worked with Jean and Karen to ensure an art program and text free of both bloopers and subtle errors. Kristin was our photo researcher who dug out some of the hard-to-find specialty photos that grace *BBOM* 15e. Susan and Ann were dedicated accuracy reviewers at the final manuscript stage and made numerous very helpful comments. Lauren headed up the marketing team for *BBOM* 15e, and along with Kelsey, composed the striking walk-through highlights that precede this preface. The authors thank Karen, Jean, Kim, Kristin, Betsy, Martha, Susan, Ann, Christa, and Kelly for their strong efforts to put the book in front of you that you see today.

Special thanks go to Anita Hueftle, our spectacular copyeditor and a key part of the book team. Anita is not only a master wordsmith; her amazing gift of being able to keep track of *where* everything was said in this book and *how* everything was said in this book was extremely impressive to witness during the *BBOM* 15e maturation process. Simply put, Anita significantly improved the accuracy, clarity, readability, and consistency of *BBOM* 15e. Thank you kindly, Anita; you are the best copy editor an author team could hope for.

We are also grateful to the top-notch educators who constructed the MasteringMicrobiology program that accompanies this text; these include: Ann Paterson, Narveen Jandu, Jennifer Hatchel, Emily Booms, Barbara May, Ronald Porter, Eileen Gregory, Erin McClelland, Candice Damiani, Susan Gibson, Ines Rauschenbach, Lee Kurtz, Vicky McKinley, Clifton Franklund, Benjamin Rohe, Ben Rowley, and Helen Walter.

And last but not least, no textbook in microbiology could be published without reviewing of the manuscript and the gift of new photos from experts in the field. We are therefore extremely grateful for the assistance of the many individuals who kindly provided manuscript reviews, unpublished results, and new photos. Complete photo credits in this book are found either alongside a photo or in the photo credits listed before the index. Reviewers and photo suppliers included:

Jônatas Abrahão, *Universidade Federal de Minas Gerais (Brazil)*
Sue Katz Amburn *(Rogers State University)*
James Archer, *Centers for Disease Control and Prevention*
Mark Asnicar, *Indiana Wesleyan University*
Hubert Bahl, *Universität Rostock (Germany)*
Jenn Baker, *Indiana Wesleyan University*
Jill Banfield, *University of California, Berkeley*
Jeremy Barr, *San Diego State University*
J. Thomas Beatty, *University of British Columbia (Canada)*
R. Howard Berg, *Danforth Plant Science Center, St. Louis*
James Berger, *Johns Hopkins School of Medicine*
Robert Blankenship, *Washington University in St. Louis*
Melanie Blokesch, *Swiss Federal Institute of Technology Lausanne (Switzerland)*
Antje Boetius, *Max Planck Institute for Marine Microbiology (Germany)*
F. C. Boogerd, *Vrije Universiteit (The Netherlands)*
Emily Booms *(Northeastern Illinois University)*
Timothy Booth, *Public Health Agency of Canada*
Gary Borisy, *The Forsyth Institute*
Amina Bouslimani, *University of California, San Diego*
Laurie Bradley *(Hudson Valley CC)*
Samir Brahmachari, *Institute of Genomics and Integrative Biology (India)*
Yves Brun, *Indiana University*
Linda Bruslind *(Oregon State University)*
Heiki Bücking, *South Dakota State University*
Gustavo Caetano-Anollés, *University of Illinois*
Elisabeth Carniel, *Institut Pasteur (France)*
Luis R. Comolli, *Lawrence Berkeley National Laboratory*
Wei Dai, *Baylor College of Medicine*

Holger Daims, *University of Vienna (Austria)*

Christina Davis, *Davis Photography, Logan, Ohio*

Thomas Deerinck, *National Center for Microscopy and Imaging Research, University of California, San Diego*

Cees Dekker, *Delft University of Technology (The Netherlands)*

Pieter Dorrestein, *University of California, San Diego*

Paul Dunlap, *University of Michigan*

Michelle Dunstone, *Monash University (Australia)*

Harald Engelhardt, *Max Planck Institute of Biochemistry (Germany)*

Thijs Ettema, *Uppsala University (Sweden)*

Babu Fathepure *(Oklahoma State University)*

Jingyi Fei, *University of Illinois*

Derek J. Fisher, *Southern Illinois University*

Patrick Forterre, *Institut Pasteur (France)*

Patricia Foster, *Indiana University*

Melitta Franceschini, *South Tyrol Museum of Archaeology (Italy)*

James Frederickson, *Pacific Northwest National Laboratory*

Jed Fuhrman, *University of Southern California*

Eric Gillock *(Fort Hays State University)*

Heidi Goodrich-Blair, *University of Wisconsin*

James Golden, *University of California, San Diego*

Cynthia Goldsmith, *Centers for Disease Control, Atlanta*

Eric Grafman, *Centers for Disease Control Public Health Image Library*

Peter Graumann, *Universität Marburg (Germany)*

Claudia Gravekamp, *Albert Einstein College of Medicine*

A.D. Grossman, *Massachusetts Institute of Technology*

Ricardo Guerrero, *University of Barcelona (Spain)*

Maria J. Harrison, *Cornell University*

Stephen Harrison, *Harvard Medical School*

Ryan Hartmaier, *University of Pittsburgh*

Zhili He, *University of Oklahoma*

Monique Heijmans, *Wageningen University (The Netherlands)*

Bart Hoogenboom, *London Centre for Nanotechnology (England)*

Matthias Horn, *University of Vienna (Austria)*

M.D. Shakhawat Hossain, *University of Missouri*

Ji-Fan Hu, *Stanford University*

Jenni Hultman, *University of Helsinki (Finland)*

Rustem Ismagilov, *California Institute of Technology*

Christian Jogler, *Leibniz-Institut DSMZ (Germany)*

Robert Kelly, *North Carolina State University*

Takehiko Kenzaka, *Osaka Ohtani University (Japan)*

Jan-Ulrich Kreft, *University of Birmingham (England)*

Misha Kudryashev, *University of Basel (Switzerland)*

Alberto Lerner, *CHROMagar (France)*

Jennifer Li-Pook-Than, *Stanford University*

Jun Liu, *University of Texas Health Science Center*

Martin Loose, *Institute of Science and Technology (Austria)*

Brigit Luef, *Norwegian University of Science and Technology (Norway)*

Marina Lusic, *University Hospital Heidelberg (Germany)*

Liang Ma, *California Institute of Technology*

Terry Machen, *University of California, Berkeley*

Sergei Markov *(Austin Peay State University)*

Stephen Mayfield, *University of California, San Diego*

John McCutcheon, *University of Montana*

Michael Minnick *(University of Montana)*

William E. Moerner, *Stanford University*

Robert Moir, *Massachusetts General Hospital and Harvard Medical School*

Christine Moissl-Eichinger, *Medical University Graz (Austria)*

Nancy Moran, *University of Texas*

Katsuhiko Murakami, *The Pennsylvania State University*

Dieter Oesterhelt, *Max Planck Institute of Biochemistry (Germany)*

George O'Toole, *Dartmouth University*

Joshua Quick, *University of Birmingham (England)*

Nicolás Pinel, *Universidad de Antioquia (Colombia)*

Rodrigo Reyes-Lamothe, *McGill University (Canada)*

Charisse Sallade, *Indiana Wesleyan University*

Bernhard Schink, *University of Konstanz (Germany)*

Christa Schleper, *University of Vienna (Austria)*

Matthew Schrenk *(Michigan State University)*

Hubert Schriebl, *Schriebl Photography, Londonderry, Vermont*

Howard Shuman, *University of Chicago*

Gary Siuzdak, *Scripps Center for Metabolomics*

Justin L. Sonnenburg, *Stanford University School of Medicine*

Rochelle Soo, *University of Queensland (Australia)*

John Stark *(Utah State University)*

Andrzej Stasiak, *University of Lausanne (Switzerland)*

S. Patricia Stock, *University of Arizona*

María Suárez Diez, *Wageningen University (The Netherlands)*

Lei Sun, *Purdue University*

Andreas Teske, *University of North Carolina*

Tammy Tobin *(Susquehanna University)*

Stephan Uphoff, *Oxford University (England)*

Joyce Van Eck, *Cornell University*

Gunter Wegener, *Max Planck Institute of Marine Microbiology (Germany)*

Jessica Mark Welch, *Marine Biological Laboratory, Woods Hole*

Mari Winkler, *University of Washington*

Cynthia Whitchurch, *University of Technology Sydney (Australia)*

Conrad Woldringh, *University of Amsterdam (The Netherlands)*

Steven Yannone, *Cinder Biological*

Shige Yoshimura, *Kyoto University (Japan)*

Feng Zhang, *Massachusetts Institute of Technology*

Joeseph Zhou, *University of Oklahoma*

Steve Zinder, *Cornell University*

As hard as a publishing team may try, no textbook can ever be completely error-free. Although we are confident the reader will be hard pressed to find errors in *BBOM* 15e, any errors that do exist, either of commission or omission, are the responsibility of the authors. In past editions, users have been kind enough to contact us when they spot an error so we can fix it in a subsequent printing. Users should feel free to continue to contact the authors directly about any errors, concerns, questions, or suggestions they have about the book. We are always happy to hear from our readers; through the years, your comments have helped make the book stronger.

Michael T. Madigan (madigan@siu.edu)
Kelly S. Bender (bender@siu.edu)
Daniel H. Buckley (dbuckley@cornell.edu)
W. Matthew Sattley (matthew.sattley@indwes.edu)
David A. Stahl (dastahl@uw.edu)

Brief Contents

Contents

4 Molecular Information Flow and Protein Processing 102

UNIT 2 Microbial Growth and Regulation

5 Microbial Growth and Its Control 137

6 Microbial Regulatory Systems 173

11 Genetics of *Bacteria* and *Archaea* 306

microbiologynow **Killing and Stealing: DNA Uptake by the Predator *Vibrio cholerae*** 306

12 Biotechnology and Synthetic Biology 332

microbiologynow **Creation of a New Life Form: Design of a Minimal Cell** 332

UNIT 4 Microbial Evolution and Diversity

13 Microbial Evolution and Systematics 363

microbiologynow *Lokiarchaeota* and the Origin of *Eukarya* 363

UNIT 6 Microbe–Human Interactions and the Immune System

ASM Recommended Curriculum Guidelines for Undergraduate Microbiology

The American Society for Microbiology (ASM) endorses a concept-based curriculum for undergraduate microbiology, emphasizing skills and concepts that have lasting importance beyond the classroom and laboratory. The ASM (in its *Curriculum Guidelines for Understanding Microbiology Education*) recommends deep understanding of 27 key concepts, 4 scientific thinking competencies, and 7 key skills. These guidelines follow scientific literacy reports and recommendations from the American Association for the Advancement of Science and the Howard Hughes Medical Institute by encouraging an active learning, student-based course. Consider these guiding statements as you progress through this book and master principles, problem solving, and laboratory skills in microbiology.

ASM Guideline Concepts and Statements

Evolution: Chapters 1, 9–14, 21, 28, 29

- Cells, organelles (e.g., mitochondria and chloroplasts), and all major metabolic pathways evolved from early prokaryotic cells.
- Mutations and horizontal gene transfer, with the immense variety of microenvironments, have selected for a huge diversity of microorganisms.
- Human impact on the environment influences the evolution of microorganisms (e.g., emerging diseases and the selection of antibiotic resistance).
- Traditional concept of species is not readily applicable to microbes due to asexual reproduction and the frequent occurrence of horizontal gene transfer.
- Evolutionary relatedness of organisms is best reflected in phylogenetic trees.

Cell Structure and Function: Chapters 1, 2, 8, 10, 14

- Structure and function of microorganisms have been revealed by the use of microscopy (including bright-field, phase contrast, fluorescence, and electron).
- Bacteria have unique cell structures that can be targets for antibiotics, immunity, and phage infection.
- *Bacteria* and *Archaea* have specialized structures (e.g., flagella, endospores, and pili) that often confer critical capabilities.
- While microscopic eukaryotes (for example, fungi, protozoa, and algae) carry out some of the same processes as bacteria, many of the cellular properties are fundamentally different.
- Replication cycles of viruses (lytic and lysogenic) differ among viruses and are determined by their unique structures.

Metabolic Pathways: Chapters 1, 3, 5, 7, 12, 14

- *Bacteria and Archaea* exhibit extensive, and often unique, metabolic diversity (e.g., nitrogen fixation, methane production, anoxygenic photosynthesis).
- Interactions of microorganisms among themselves and with their environment are determined by their metabolic abilities (e.g., quorum sensing, oxygen consumption, nitrogen transformations).
- Survival and growth of any microorganism in a given environment depends on its metabolic characteristics.
- Growth of microorganisms can be controlled by physical, chemical, mechanical, or biological means.

Information Flow and Genetics: Chapters 1, 4, 6–9, 11

- Genetic variations can impact microbial functions (e.g., in biofilm formation, pathogenicity, and drug resistance).
- Although the central dogma is universal in all cells, the processes of replication, transcription, and translation differ in *Bacteria*, *Archaea*, and eukaryotes.
- Regulation of gene expression is influenced by external and internal molecular cues and/or signals.
- Synthesis of viral genetic material and proteins is dependent on host cells.
- Cell genomes can be manipulated to alter cell function.

Microbial Systems: Chapters 1, 9, 14–18, 23–33

- Microorganisms are ubiquitous and live in diverse and dynamic ecosystems.
- Many bacteria in nature live in biofilm communities.
- Microorganisms and their environment interact with and modify each other.
- Microorganisms, cellular and viral, can interact with both human and nonhuman hosts in beneficial, neutral, or detrimental ways.

Impact of Microorganisms: Chapters 1, 5, 7, 12, 19–22

- Microbes are essential for life as we know it and the processes that support life (e.g., in biogeochemical cycles and plant and/or animal microbiota).
- Microorganisms provide essential models that give us fundamental knowledge about life processes.
- Humans utilize and harness microorganisms and their products.
- Because the true diversity of microbial life is largely unknown, its effects and potential benefits have not been fully explored.

The Microbial World

microbiology**now**

Microorganisms, Our Constant Companions

Microorganisms are everywhere, and though small, their activities have tremendous impacts on everything in our biosphere. As you learn more about microbiology you will realize that many of our day-to-day interactions with the world are influenced, for better or worse, by microbial life. Indeed, hundreds of trillions of bacteria are working within your body now to digest your last meal. The total complement of microbial cells in and on your body—your microbiome—contains thousands of species each adapted to grow best in a particular part of your body. For example, your gut microbiome encodes enzymes that help digest your food and synthesize vitamins critical to your health. The composition of your microbiome changes in response to your diet, your genes, your health, and the medicines you take. Our microbiome is absolutely essential to our health and well-being, yet we are only beginning to understand the diverse ways in which we depend upon our gut microorganisms.

Our knowledge of the microbial world is highly dependent on technological developments. Recent advances in microscopic techniques have made it possible to visualize microbes in intimate association with the lining of the gut (see photo). This image, generated by laser scanning confocal microscopy, reveals a cross section from the colon of a mouse and shows the dense and complex microbial community residing within the gut. Mice, which can be raised in a germ-free condition and then inoculated with a human gut microbiome, are used as a model system to explore microbiome function. Fluorescent stains identify the mucus layer (green) and host cell nuclei (blue) of the gut epithelium. *Bacteria* of the phylum *Firmicutes* stain yellow and those of the family *Bacteroidaceae* stain pink; all other bacteria stain red. Changes in diet alter the thickness of the mucus layer and the potential for microbial interactions with the epithelium. Such interactions can change gut function and possibly lead to inflammation.

In the chapters that follow, the exciting science of microbiology will unfold, and you will see that there is still much to learn about the inner workings of the microbial world.

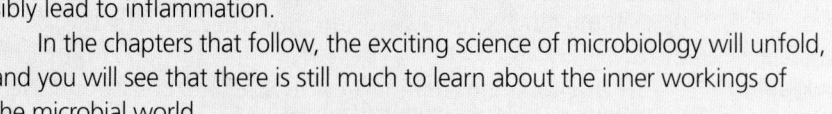

Source: Earle, K.A., et al. 2015. Quantitative imaging of gut microbiota spatial organization. *Cell Host & Microbe 18:* 478–488.

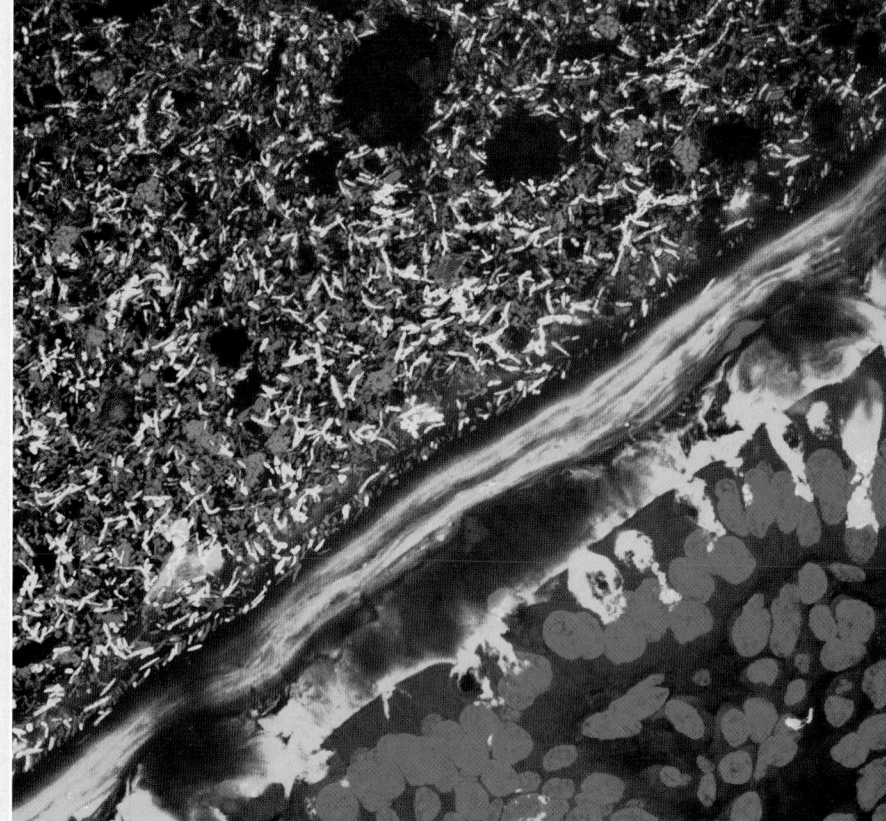

I • Exploring the Microbial World

1.1 Microorganisms, Tiny Titans of the Earth

Microorganisms (also called *microbes*) are life forms too small to be seen by the unaided human eye. These microscopic organisms are diverse in form and function and they inhabit every environment on Earth that supports life. Many microbes are undifferentiated single-celled organisms, but some can form complex structures, and some are even multicellular. Microorganisms typically live in complex **microbial communities (Figure 1.1)**, and their activities are regulated by interactions with each other, with

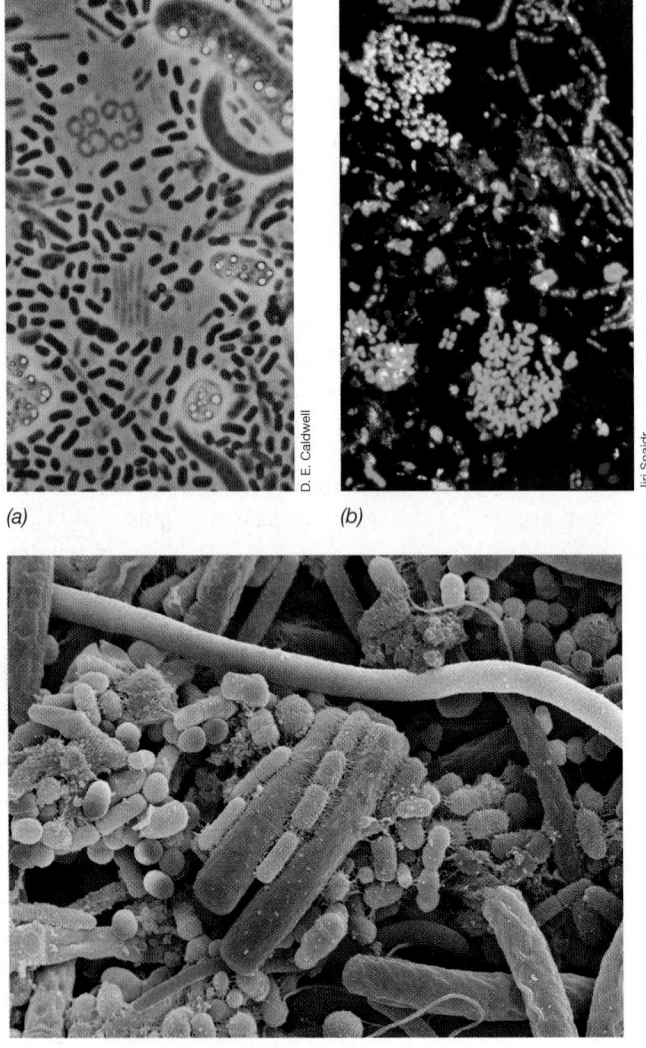

(a)

(b)

D. E. Caldwell

Jiri Snaidr

(c)

Figure 1.1 Microbial communities. *(a)* A bacterial community that developed in the depths of a small Michigan lake, showing cells of various phototrophic bacteria. The bacteria were visualized using phase-contrast microscopy. *(b)* A bacterial community in a sewage sludge sample. The sample was stained with a series of dyes, each of which stained a specific bacterial group. From *Journal of Bacteriology 178:* 3496–3500, Fig. 2b. © 1996 American Society for Microbiology. *(c)* Scanning electron micrograph of a microbial community scraped from a human tongue.

their environments, and with other organisms. The science of microbiology is all about microorganisms, who they are, how they work, and what they do.

Microorganisms were teeming on the land and in the seas for billions of years before the appearance of plants and animals, and their diversity is staggering. Microorganisms represent a major fraction of Earth's biomass, and their activities are essential to sustaining life. Indeed, the very oxygen (O_2) we breathe is the result of microbial activities. Plants and animals are immersed in a world of microbes, and their evolution and survival are heavily influenced by microbial activities, by microbial symbioses, and by pathogens—those microbes that cause disease. Microorganisms are woven into the fabric of human life as well, from infectious diseases, to the food we eat, the water we drink, the fertility of our soils, the health of our animals, and even the fuel we put in automobiles. Microbiology is the study of the dominant form of life on Earth, and the effect that microbes have on our planet and all of the living things that call it home.

Microbiologists have many tools for studying microorganisms. Microbiology was born of the microscope, and microscopy is foundational to microbiology. Microbiologists have developed an array of methods for visualizing microorganisms, and these microscopic techniques are essential to microbiology. The cultivation of microorganisms is also foundational to microbiology. A microbial **culture** is a collection of cells that have been grown in or on a nutrient medium. A **medium** (plural, media) is a liquid or solid nutrient mixture that contains all of the nutrients required for a microorganism to grow. In microbiology, we use the word **growth** to refer to the increase in cell number as a result of cell division. A single microbial cell placed on a solid nutrient medium can grow and divide into millions of cells that form a visible **colony (Figure 1.2)**. The formation of visible colonies makes it easier to see and grow microorganisms. Comprehension of the microbial basis of disease and microbial biochemical diversity has relied on the ability to grow microorganisms in the laboratory.

The ability to grow microorganisms rapidly under controlled conditions makes them highly useful for experiments that probe the fundamental processes of life. Most discoveries relating to the molecular and biochemical basis of life have been made using microorganisms. The study of molecules and their interactions is essential to defining the workings of microbial cells, and the tools of molecular biology and biochemistry are foundational to microbiology. Molecular biology has also provided a variety of tools to study microorganisms without need for their cultivation in the laboratory. These molecular tools have greatly expanded our knowledge of microbial ecology and diversity. Finally, the tools of genomics and molecular genetics are also cornerstones of modern microbiology and allow microbiologists to study the genetic basis of life, how genes evolve, and how they regulate the activities of cells.

This chapter begins our journey into the microbial world. Here we will begin to discover what microorganisms are, what they do, and how they can be studied. We will also place microbiology in historical context, as a process of scientific discovery.

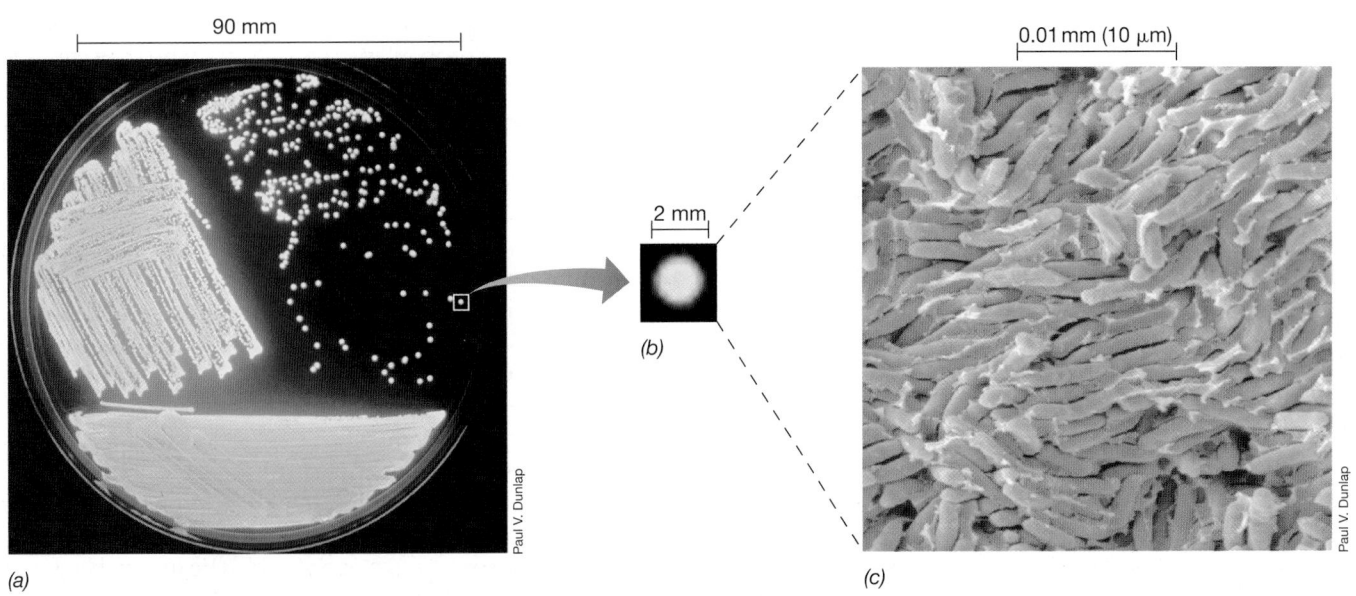

90 mm

0.01 mm (10 μm)

2 mm

(a)

(b)

(c)

Paul V. Dunlap

Figure 1.2 Microbial cells. *(a)* Bioluminescent (light-emitting) colonies of the bacterium *Photobacterium* grown in laboratory culture on a Petri plate. *(b)* A single colony can contain more than 10 million (10^7) individual cells. *(c)* Scanning electron micrograph of cells of *Photobacterium*.

MINIQUIZ

- In what ways are microorganisms important to humans?
- Why are microbial cells useful for understanding the basis of life?
- What is a microbial colony and how is one formed?

1.2 Structure and Activities of Microbial Cells

Microbial cells are living compartments that interact with their environment and with other cells in dynamic ways. In Chapter 2 we will examine the structure of cells in detail and relate specific structures to specific functions. Here we present a snapshot of microbial structure and activities. We purposely exclude viruses in most of this discussion because although they resemble cells in many ways, viruses are not cells but instead a special category of microorganism. We consider the structure, diversity, and activities of viruses in Section 1.14 and in Chapters 8 and 10.

Elements of Microbial Structure

All cells have much in common and contain many of the same components (**Figure 1.3**). All cells have a permeability barrier called the **cytoplasmic membrane** that separates the inside of the cell, the **cytoplasm**, from the outside. The cytoplasm is an aqueous mixture of **macromolecules** (for example proteins, lipids, nucleic acids, and polysaccharides), small organic molecules (mostly the precursors of macromolecules), various inorganic ions, and ribosomes. **Ribosomes** are the structures responsible for protein synthesis and are found in all cells. Some cells have a **cell wall** that lends structural strength to a cell. The cell wall is a relatively permeable structure located outside the cell membrane and is a much stronger layer than the membrane itself. Plant cells

and most microorganisms have cell walls, whereas animal cells typically do not.

Examination of cell structure reveals there are two major structural classes of cells, called **prokaryotic** cells and **eukaryotic** cells (Figure 1.3). Eukaryotic cells are found in the phylogenetic domain *Eukarya*. This group includes plants and animals as well as diverse microbial eukaryotes such as algae, protozoa, and fungi. Eukaryotic cells contain an assortment of membrane-enclosed cytoplasmic structures called **organelles** (Figure 1.3*b*). These include, most prominently, the DNA-containing nucleus but also mitochondria and chloroplasts, organelles that specialize in supplying the cell with energy, and various other organelles.

Prokaryotic cells are found in the domains *Bacteria* and *Archaea*. Prokaryotic cells have few internal structures, they lack a nucleus, and they typically lack organelles (Figure 1.3*a*). The prokaryotic cell structure evolved prior to the evolution of the eukaryotic cell (Section 1.3). While *Archaea* and *Bacteria* both contain exclusively prokaryotic cells, these groups have diverged greatly and we will see later that the *Archaea* actually share many molecular and genetic characteristics with cells of *Eukarya*.

Genes, Genomes, Nucleus, and Nucleoid

In addition to a cytoplasmic membrane and ribosomes, all cells also possess a DNA **genome**. The genome is the complement of all genes in a cell. A gene is a segment of DNA that encodes a protein or an RNA molecule. The genome is the living blueprint of an organism; the characteristics, activities, and very survival of a cell are governed by its genome.

The genomes of prokaryotic cells and eukaryotic cells are organized differently. In eukaryotes, DNA is present as several linear molecules within the membrane-enclosed **nucleus**. By contrast, the genomes of *Bacteria* and *Archaea* are typically closed circular chromosomes (though a few prokaryotes have linear

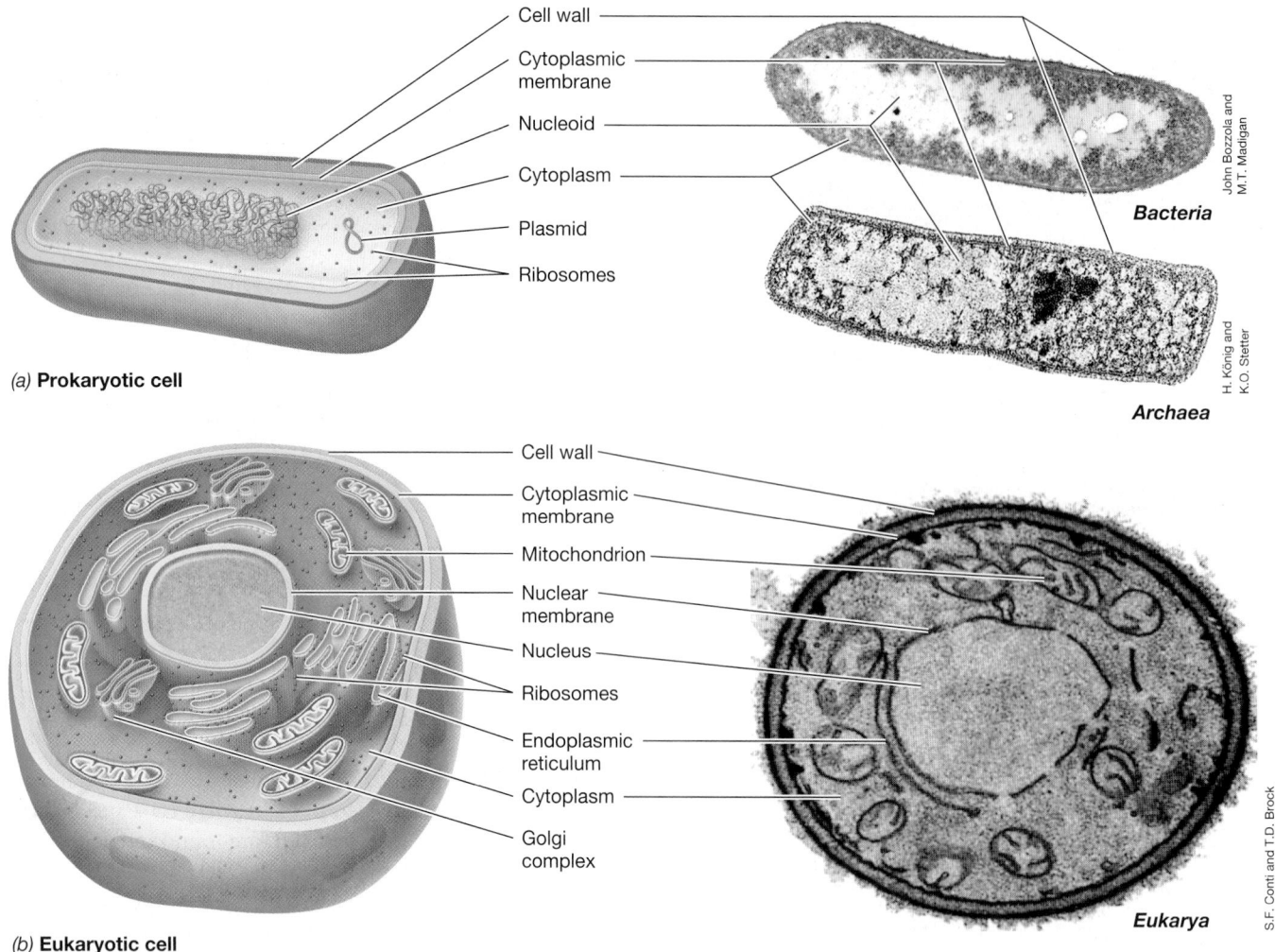

Cell wall

Cytoplasmic membrane

Nucleoid

Cytoplasm

Plasmid

Ribosomes

(a) **Prokaryotic cell**

John Bozzola and M.T. Madigan

Bacteria

H. König and K.O. Stetter

Archaea

Cell wall

Cytoplasmic membrane

Mitochondrion

Nuclear membrane

Nucleus

Ribosomes

Endoplasmic reticulum

Cytoplasm

Golgi complex

S.F. Conti and T.D. Brock

Eukarya

(b) **Eukaryotic cell**

Figure 1.3 Microbial cell structure. *(a)* (Left) Diagram of a prokaryotic cell. (Right) Electron micrograph of *Heliobacterium modesticaldum* (*Bacteria*, cell is about 1 μm in diameter) and *Thermoproteus neutrophilus* (*Archaea*, cell is about 0.5 μm in diameter). *(b)* (Left) Diagram of a eukaryotic cell. (Right) Electron micrograph of a cell of *Saccharomyces cerevisiae* (*Eukarya*, cell is about 8 μm in diameter).

chromosomes). The chromosome aggregates within the prokaryotic cell to form the **nucleoid**, a mass visible in the electron microscope (Figure 1.3*a*). Most prokaryotic cells have only a single chromosome, but many also contain one or more small circles of DNA distinct from that of the chromosome, called **plasmids**. Plasmids typically contain genes that are not essential and confer some special property on the cell (such as a unique metabolism, or antibiotic resistance). The genomes of *Bacteria* and *Archaea* are typically small and compact and most contain between 500 and 10,000 genes encoded by 0.5 to 10 million base pairs. Eukaryotic cells typically have much larger and much less compact genomes than prokaryotic cells. A human cell, for example, contains approximately 3 billion base pairs, which encode about 20,000–25,000 genes.

Activities of Microbial Cells

What activities do microbial cells carry out? We will see that in nature, microbial cells typically live in groups called microbial communities (Figure 1.1). **Figure 1.4** considers some of the ongoing

cellular activities within the microbial community. All cells show some form of **metabolism** by taking up nutrients from the environment and transforming them into new cell materials and waste products. During these transformations, energy is conserved to support synthesis of new structures. Production of these new structures culminates in the division of the cell to form two cells. Microbial growth results from successive rounds of cell division.

During metabolism and growth, genes are decoded to form proteins that regulate cellular processes. **Enzymes**, those proteins that have catalytic activity, are required to carry out reactions that supply the energy and precursors necessary for the biosynthesis of all cell components. Enzymes and other proteins are synthesized during *gene expression* in the sequential processes of transcription and translation (Figure 1.4). **Transcription** is the process by which the information on DNA is copied into an RNA molecule, and **translation** is the process whereby the information on an RNA molecule is used by a ribosome to synthesize a protein (Chapter 4). Gene expression and enzyme activity in a microbial cell are coordinated and highly regulated to ensure that the cell remains

Properties of *all* cells:

Metabolism

Cells take up nutrients, transform them, and expel wastes.
1. **Genetic** (replication, transcription, translation)
2. **Catalytic** (energy, biosyntheses)

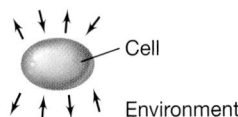

Growth

Nutrients from the environment are converted into new cell materials to form new cells.

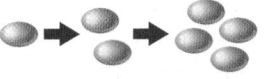

Evolution

Cells evolve to display new properties. Phylogenetic trees capture evolutionary relationships.

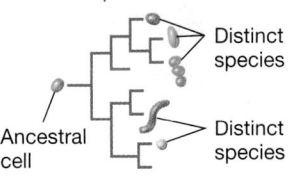

Properties of *some* cells

Differentiation

Some cells can form new cell structures such as a spore.

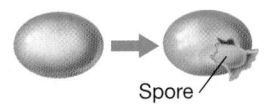

Communication

Cells interact with each other by chemical messengers.

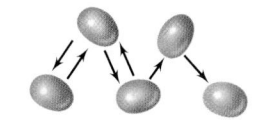

Genetic exchange

Cells can exchange genes by several mechanisms.

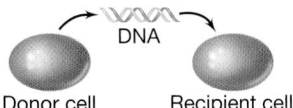

Motility

Some cells are capable of self-propulsion.

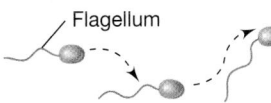

Figure 1.4 The properties of microbial cells. Major activities ongoing in cells in the microbial community are depicted.

optimally tuned to its surroundings. Ultimately, microbial growth requires replication of the genome through the process of **DNA replication**, followed by cell division. All cells carry out the processes of transcription, translation, and DNA replication.

Microorganisms have the ability to sense and respond to changes in their local environment. Many microbial cells are capable of **motility**, typically by self-propulsion (Figure 1.4). Motility allows cells to relocate in response to environmental conditions. Some microbial cells undergo **differentiation**, which may result in the formation of modified cells specialized for growth, dispersal, or survival. Cells respond to chemical signals in their environment, including those produced by other cells of either the same or different species, and these signals often trigger new cellular activities. Microbial cells thus exhibit **intercellular communication**; they are "aware" of their neighbors and can respond accordingly. Many prokaryotic cells can also exchange genes with neighboring cells, either of the same species or of a different species, in the process of **horizontal gene transfer**.

Evolution (Figure 1.4) results when genes in a population of cells change in sequence and frequency over time, leading to descent with modification. The evolution of microorganisms can

be very rapid relative to the evolution of plants and animals. For example, the indiscriminate use of antibiotics in human and veterinary medicine has selected for the proliferation of antibiotic resistance in pathogenic bacteria. The rapid pace of microbial evolution can be attributed in part to the ability of microorganisms to grow very quickly and to acquire new genes though the process of horizontal gene transfer.

Not all of the processes depicted in Figure 1.4 occur in all cells. Metabolism, growth, and evolution, however, are universal and will be major areas of emphasis throughout this book.

MINIQUIZ

- What structures are universal to all type of cells?
- What processes are universal to all types of cells?
- What structures can be used to distinguish between prokaryotic cells and eukaryotic cells?

1.3 Microorganisms and the Biosphere

Microbes are the oldest form of life on Earth, and they have evolved to perform critical functions that sustain the biosphere. In this section we will learn how microbes have changed our planet and how they continue to do so.

A Brief History of Life on Earth

Earth is about 4.6 billion years old, and microbial cells first appeared between 3.8 and 4.3 billion years ago (**Figure 1.5**). During the first 2 billion years of Earth's existence, its atmosphere was anoxic (O_2 was absent), and only nitrogen (N_2), carbon dioxide (CO_2), and a few other gases were present. Only microorganisms capable of anaerobic metabolisms (that is, metabolisms that do not require O_2) could survive under these conditions.

The evolution of phototrophic microorganisms—organisms that harvest energy from sunlight—occurred within 1 billion years of the formation of Earth (Figure 1.5a). The first phototrophs were anoxygenic (non-oxygen-producing), such as the purple sulfur bacteria and green sulfur bacteria we know today (**Figure 1.6**). *Cyanobacteria* (oxygenic phototrophs) (Figure 1.6f) evolved nearly a billion years later (Figure 1.5a) and began the slow process of oxygenating Earth's atmosphere. These early phototrophs lived in structures called *microbial mats*, which are still found on Earth today (Figure 1.6a–c). After the oxygenation of Earth's atmosphere, multicellular life forms eventually evolved, culminating in the plants and animals we know today. But plants and animals have only existed for about half a billion years. The timeline of life on Earth (Figure 1.5a) shows that *80% of life's history was exclusively microbial*, and thus in many ways, Earth can be considered a microbial planet.

As evolutionary events unfolded, three major lineages of microbial cells—the *Bacteria*, the *Archaea*, and the *Eukarya* (Figure 1.5b)—were distinguished. These three major cell lineages are called **domains**, and all known cellular organisms belong to one of these three domains. All cellular organisms also share certain characteristics and genes. For example, approximately 60 genes are universally present in cells of all domains. Examination of these genes reveals that all three domains have descended from a

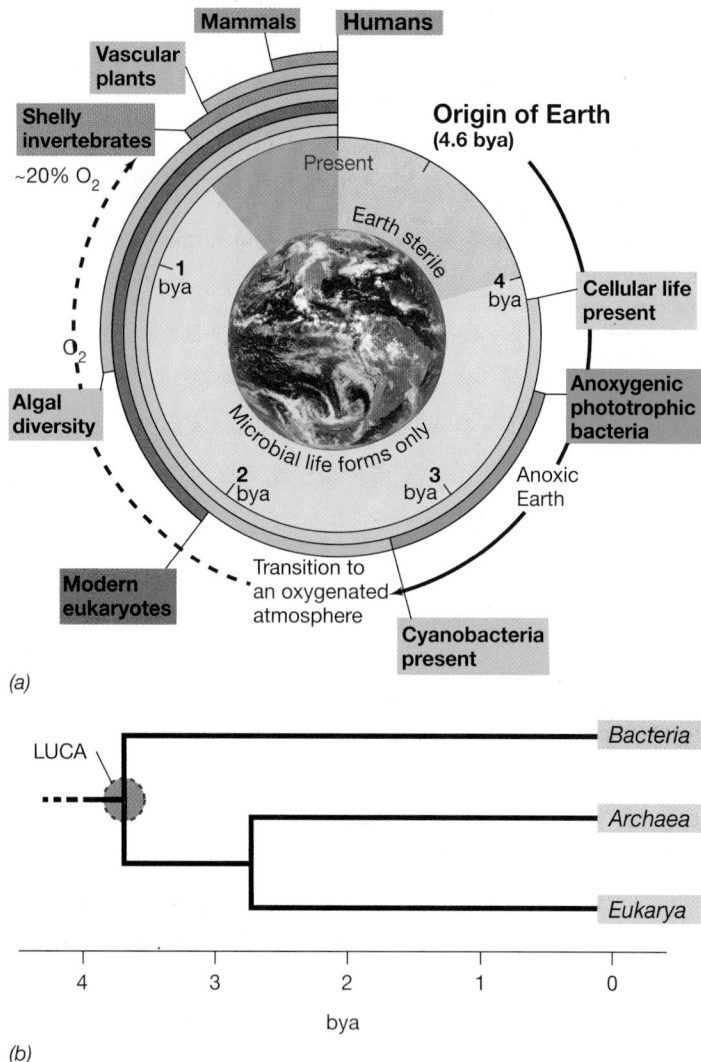

(a)

(b)

Figure 1.5   A summary of life on Earth through time and origin of the cellular domains. *(a)* At its origin, Earth was sterile. Cellular life was present on Earth by 3.8 billion years ago (bya). Cyanobacteria began the slow oxygenation of Earth about 3 bya, but current levels of O_2 in the atmosphere were not achieved until 500–800 million years ago. *(b)* The three domains of cellular organisms are *Bacteria*, *Archaea*, and *Eukarya*. *Archaea* and *Eukarya* diverged long before eukaryotic cells appear in the fossil record. LUCA, last universal common ancestor.

common ancestor, the *last universal common ancestor* (LUCA, Figure 1.5*b*). Over enormous periods of time, microorganisms derived from these three domains have evolved to fill every suitable environment on Earth.

Microbial Abundance and Activity in the Biosphere

Microorganisms are present everywhere on Earth that will support life. They constitute a major fraction of global biomass and are key reservoirs of nutrients essential for life. There are an estimated 2×10^{30} microbial cells on Earth. To put this number in context, the universe in all its vast extent is estimated to contain merely 7×10^{22} stars. The total amount of carbon present in all microbial cells is a significant fraction of Earth's biomass (**Figure 1.7**). Moreover, the total amount of nitrogen and phosphorus (essential nutrients for life) within microbial cells is nearly four

times that in all plant and animal cells combined. Microbes also represent a major fraction of the total DNA in the biosphere (about 31%), and their genetic diversity far exceeds that of plants and animals (see Figure 1.36).

Microbes are even abundant in habitats that are much too harsh for other forms of life, such as volcanic hot springs, glaciers and ice-covered regions, high-salt environments, extremely acidic or alkaline habitats, and deep in the sea or deep in the earth at extremely high pressure. Such microorganisms are called **extremophiles** and their properties define the physiochemical limits to life as we know it (**Table 1.1**). We will revisit many of these organisms in later chapters and discover the special structural and biochemical properties that allow them to thrive under extreme conditions.

All ecosystems are influenced greatly by microbial activities. The metabolic activities of microorganisms can change the habitats in which they live, both chemically and physically, and these changes can affect other organisms. For example, excess nutrients added to a habitat can cause aerobic (O_2-consuming) microorganisms to grow rapidly and consume O_2, rendering the habitat anoxic (O_2-free). Many human activities release nutrients into the coastal oceans, thereby stimulating excessive microbial growth, which can cause enormous anoxic zones in these waters. These "dead zones" cause massive mortality of fish and shellfish in coastal oceans worldwide, because most aquatic animals require O_2 and die if it is not available. Only by understanding microorganisms and microbiology can we predict and minimize the effects of human activity on the biosphere that sustains us.

Though diverse habitats are influenced strongly by microorganisms, their contributions are often overlooked because of their small sizes. Within the human body, for example, there are between one and ten microbial cells (mainly of *Bacteria*) for every human cell and more than 200 microbial genes for every human gene. These microbes provide nutritional and other benefits that are essential to human health. In later chapters we will return to a consideration of the ways in which microorganisms affect animals, plants, and the entire global ecosystem. This is the science of **microbial ecology**, perhaps the most exciting subdiscipline of microbiology today. We will see that microbes are important to myriad issues of global importance to humans including climate change, agricultural productivity, and even energy policy.

— **MINIQUIZ** —

- How old is Earth and when did cells first appear on Earth?
- Name the three domains of life.
- Why were cyanobacteria so important in the evolution of life on Earth?

1.4 The Impact of Microorganisms on Human Society

Microbiologists have made great strides in discovering how microorganisms function, and application of this knowledge has greatly advanced human health and welfare. Besides understanding

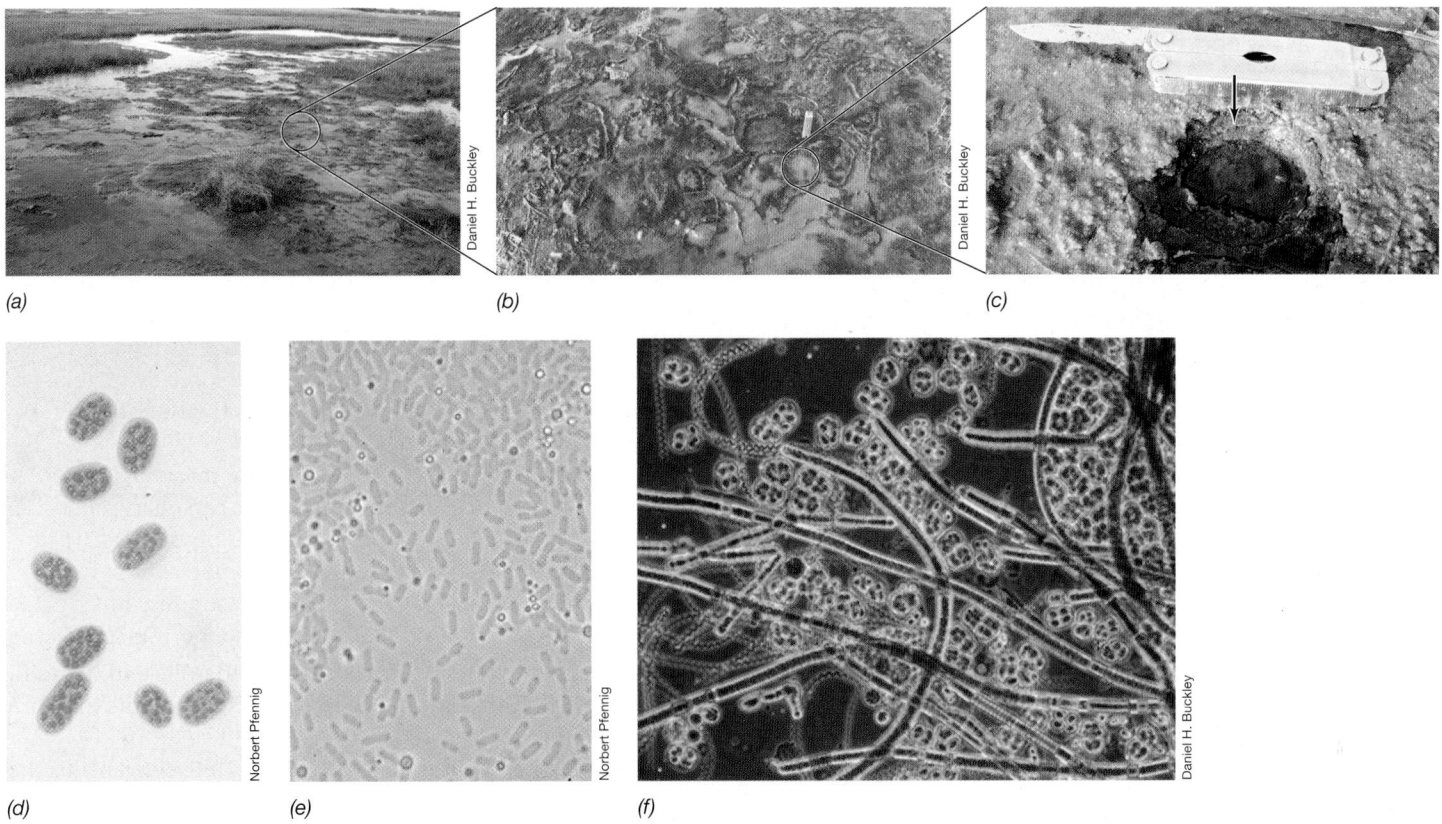

(a)   (b)   (c)

(d)   (e)   (f)

Figure 1.6 Phototrophic microorganisms.
The earliest phototrophs lived in microbial mats.
(a) Photosynthetic microbial mats in the Great Sippewissett Marsh, a salt marsh in Massachusetts, USA. *(b)* Mats develop a cohesive structure that forms at the sediment surface. *(c)* A slice through the mat shows colored layers that form due to the presence of photopigments. Cyanobacteria form the green layer nearest the surface, purple sulfur bacteria form the purple and yellow layers below, and green sulfur bacteria form the bottommost green layer. The scale on the knife is in cm. *(d)* Purple sulfur bacteria, *(e)* green sulfur bacteria, and *(f)* cyanobacteria imaged by bright-field and phase-contrast microscopy. Purple and green sulfur bacteria are anoxygenic phototrophs that appeared on Earth long before oxygenic phototrophs evolved (see Figure 1.5*a*).

microorganisms as agents of disease, microbiology has made great advances in understanding the important roles microorganisms play in food and agriculture, and microbiologists have been able to exploit microbial activities to produce valuable human products, generate energy, and clean up the environment.

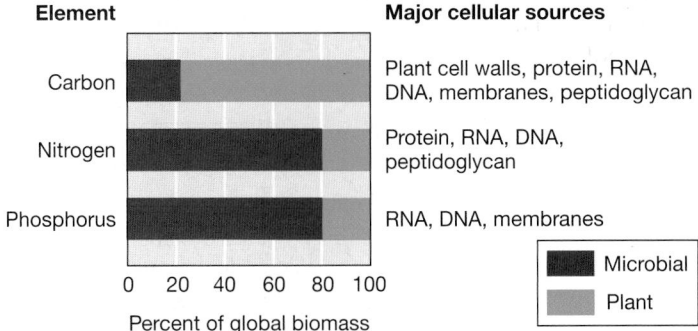

Figure 1.7 Contribution of microbial cells to global biomass. Microorganisms comprise a significant fraction of the carbon (C) and a majority of the nitrogen (N) and phosphorus (P) in the biomass of all organisms on Earth. C, N, and P are the macronutrients required in the greatest quantity by living organisms. Animal biomass is a minor fraction (<0.1%) of total global biomass and is not shown.

Microorganisms as Agents of Disease

The statistics summarized in **Figure 1.8** show how microbiologists and clinical medicine have combined to conquer infectious diseases in the past 100 years. At the beginning of the twentieth century, the major causes of human death were infectious diseases caused by bacterial and viral **pathogens**. In those days children and the aged in particular succumbed in large numbers to microbial diseases. Today, however, infectious diseases are much less deadly, at least in developed countries. Control of infectious disease has come from a combination of advances including our increased understanding of disease processes, improved sanitary and public health practices, active vaccine campaigns, and the widespread use of antimicrobial agents, such as antibiotics. As we will see later in this chapter, the development of microbiology as a science can be traced to pioneering studies of infectious disease.

While pathogens and infectious disease remain a major threat to humanity, and combating these harmful organisms remains a major focus of microbiology, most microorganisms are not harmful to humans. In fact, most microorganisms are beneficial, and in many cases are even essential to human welfare and the functioning of the planet. We turn our attention to these microorganisms and microbial activities now.

TABLE 1.1 Classes and examples of extremophiles[a]

Extreme	Descriptive term	Genus/species	Domain	Habitat	Minimum	Optimum	Maximum
Temperature							
High	Hyperthermophile	*Methanopyrus kandleri*	*Archaea*	Undersea hydrothermal vents	90°C	106°C	122°C[b]
Low	Psychrophile	*Psychromonas ingrahamii*	*Bacteria*	Sea ice	−12°C[c]	5°C	10°C
pH							
Low	Acidophile	*Picrophilus oshimae*	*Archaea*	Acidic hot springs	−0.06	0.7[d]	4
High	Alkaliphile	*Natronobacterium gregoryi*	*Archaea*	Soda lakes	8.5	10[e]	12
Pressure	Barophile (piezophile)	*Moritella yayanosii*	*Bacteria*	Deep ocean sediments	500 atm	700 atm[f]	>1000 atm
Salt (NaCl)	Halophile	*Halobacterium salinarum*	*Archaea*	Salterns	15%	25%	32% (saturation)

[a]The organisms listed are the current "record holders" for growth in laboratory culture at the extreme condition listed.
[b]Anaerobe showing growth at 122°C only under several atmospheres of pressure.
[c]The permafrost bacterium *Planococcus halocryophilus* can grow at −15°C and metabolize at −25°C. However, the organism grows optimally at 25°C and grows up to 37°C and thus is not a true psychrophile.
[d]*P. oshimae* is also a thermophile, growing optimally at 60°C.
[e]*N. gregoryi* is also an extreme halophile, growing optimally at 20% NaCl.
[f]*M. yayanosii* is also a psychrophile, growing optimally near 4°C.

Microorganisms, Agriculture, and Human Nutrition

Agriculture benefits from the cycling of key plant nutrients by microorganisms. For example, legumes are a diverse family of plants that include major crop species such as beans, peas, and lentils, among others. Legumes live in close association with bacteria that form structures called *nodules* on their roots. In the nodules, these bacteria convert atmospheric nitrogen (N_2) into ammonia (NH_3) through the process of *nitrogen fixation*. NH_3 is the major nutrient found in fertilizer and is used as a nitrogen source for plant growth (**Figure 1.9**). In this way bacteria allow legumes to make their own fertilizer, thereby reducing the need for farmers to apply fertilizers produced industrially. Bacteria regulate nutrient cycles, such as the nitrogen cycle and the sulfur cycle (Figure 1.9), transforming and recycling nutrients that form the basis of soil fertility.

Also of major agricultural importance are microorganisms that inhabit the *rumen* of ruminant animals, such as cattle and sheep. The rumen is a microbial ecosystem in which microbial communities digest and ferment the polysaccharide cellulose (Figure 1.9*d*), the major component of plant cell walls. Without these symbiotic microorganisms, ruminants could not thrive on cellulose-rich (but otherwise nutrient-poor) food such as grass and hay. Many domesticated and wild herbivorous mammals—including deer, bison, camels, giraffes, and goats—are also ruminants.

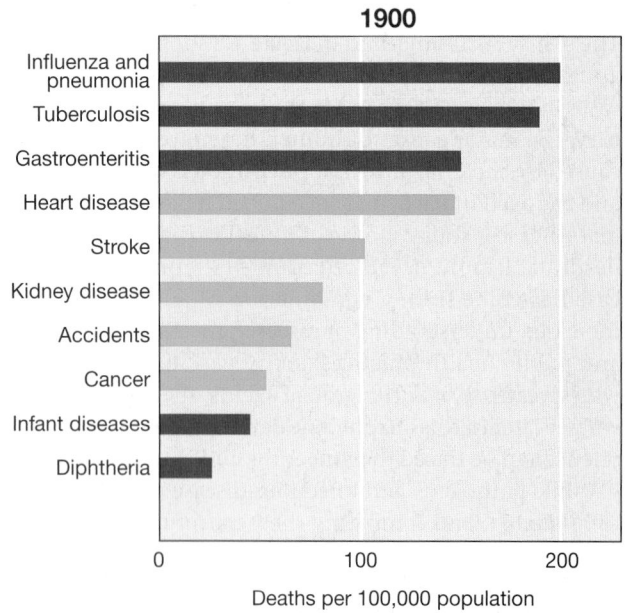

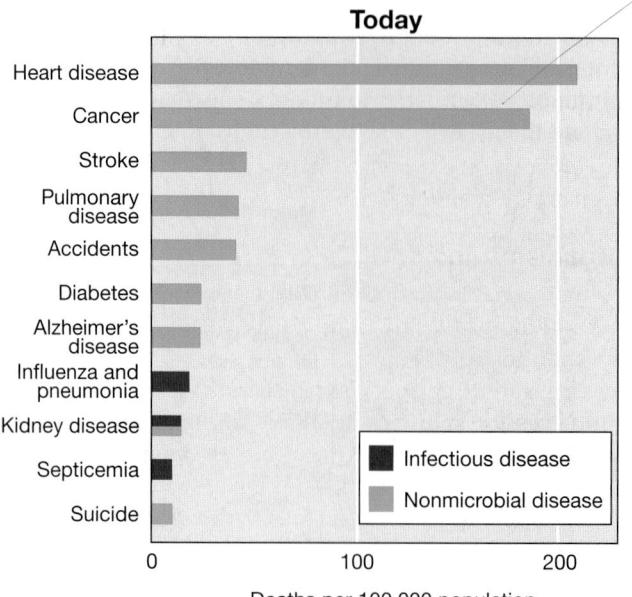

Figure 1.8 Death rates for the leading causes of death in the United States: 1900 and today. Infectious diseases were the leading causes of death in 1900, whereas today they account for relatively few deaths. Kidney diseases can be caused by microbial infections or systemic sources (diabetes, cancers, toxicities, metabolic diseases, etc.). Data are from the United States National Center for Health Statistics and the Centers for Disease Control and Prevention.

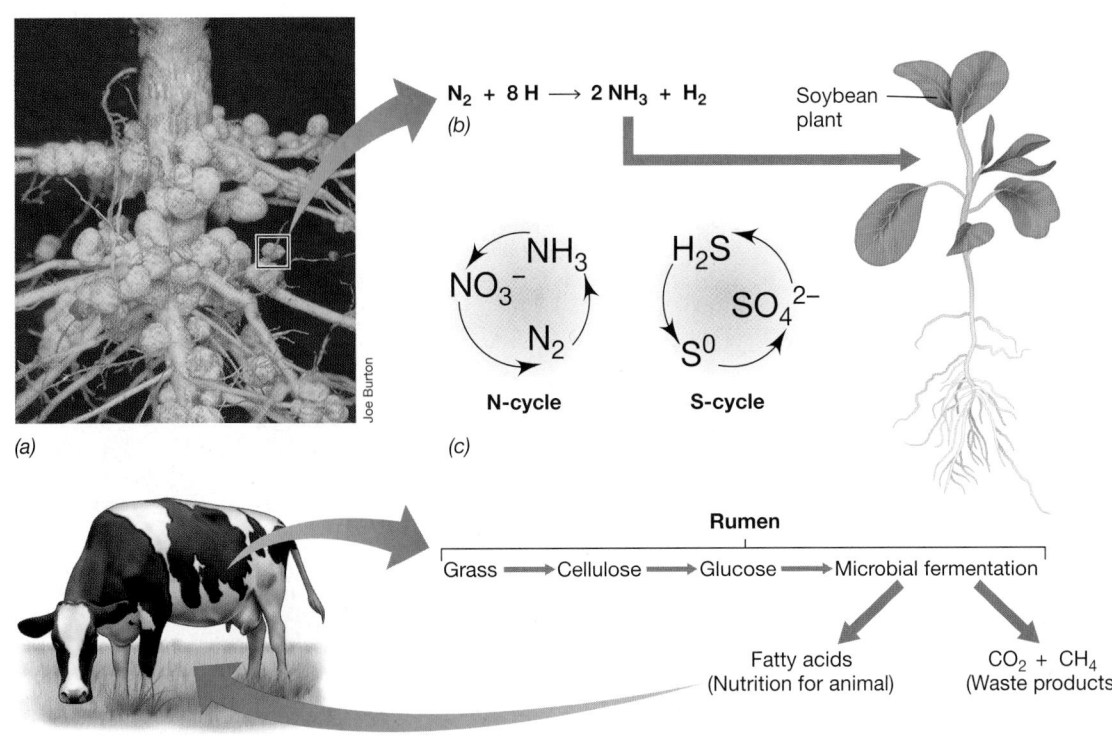

Joe Burton

$$N_2 + 8\,H \longrightarrow 2\,NH_3 + H_2$$
(b)

Soybean plant

N-cycle   S-cycle

(c)

Rumen

Grass → Cellulose → Glucose → Microbial fermentation

Fatty acids
(Nutrition for animal)

$CO_2 + CH_4$
(Waste products)

(d)

Figure 1.9 Microorganisms in modern agriculture. *(a, b)* Root nodules on this soybean plant contain bacteria that fix molecular nitrogen (N_2) for use by the plant. *(c)* The nitrogen and sulfur cycles, key nutrient cycles in nature. *(d)* Ruminant animals. Microorganisms in the rumen of the cow convert cellulose from grass into fatty acids that can be used by the animal. The other products are not so desirable, as CO_2 and CH_4 are the major gases that cause global warming.

The human gastrointestinal (GI) tract lacks a rumen, but complex carbohydrates (which can represent 10–30% of food energy) are digested by the **gut microbiome**. The colon, or large intestine (**Figure 1.10**), follows the stomach and small intestine in the digestive tract, and it contains about 10^{11} microbial cells per gram of colonic contents. Microbial cell numbers are low in the very acidic (pH 2) stomach (about 10^4 per gram) but increase to about 10^8 per gram near the end of the small intestine (pH 5) and then reach maximal numbers in the colon (pH 7) (Figure 1.10). The colon contains diverse microbial species that assist in the digestion of complex carbohydrates, and that synthesize vitamins and other nutrients essential to host nutrition. The gut microbiome develops from birth, but it can change over time with the human host. The composition of the gut microbiome has major effects on GI function and human health.

Microorganisms and Food

Microbes are intimately associated with the foods we eat. Microbial growth in food can cause food spoilage and foodborne disease. The manner in which we harvest and store food (e.g., canning, refrigeration, drying, salting, etc.), the ways in which we cook it, and even the spices we use, have all been fundamentally influenced by microbes in order to minimize microbial growth and eliminate harmful organisms. Microbial food safety and prevention of food spoilage is a major focus of the food industry and a major cause of economic loss every year.

While some microbes can cause foodborne disease and food spoilage, not all microorganisms in foods are harmful. Indeed, beneficial microbes have been used for thousands of years to improve food safety and to preserve foods (**Figure 1.11**). For example, cheeses, yogurt, and buttermilk are all produced by the microbial fermentation of dairy products to produce acids that improve shelf life and prevent the growth of foodborne pathogens. Such microbial fermentations are used to produce a variety of foods including sauerkraut, kimchi, pickles, and certain sausages. Even the production of chocolate and coffee rely on microbial fermentation. Moreover, baked goods and alcoholic beverages rely on the fermentative activities of yeast, which generate carbon dioxide (CO_2) to raise the dough and alcohol as a key component (Figure 1.11), respectively. Fermentation products affect the flavor and taste of foods, and can prevent spoilage as well as the growth of deleterious organisms.

Microorganisms and Industry

Microorganisms play important roles in all manner of human industry. Microbes can grow in almost any habitat containing liquid water, including structures made by humans. For example, microbes often grow on submerged surfaces, forming *biofilms*. Biofilms that grow in pipes and drains can cause fouling and blockages in factory settings and pipelines, in sewers, and even in water distribution systems. In addition, biofilms that grow on ships' hulls can cause marked reductions in speed and efficiency. Biofilms can even grow in tanks that store oil and fuel, leading to spoilage of these products. We will learn that biofilms are also of great importance in medicine, as biofilms that form on implanted medical devices (↩ Figure 5.4*a*) can cause infections that are extremely difficult to treat.

Microorganisms can be harnessed to produce commercially valuable products. In *industrial microbiology*, naturally occurring microorganisms are grown on a massive scale to make large amounts of products at relatively low cost, such as antibiotics, enzymes, and certain chemicals. By contrast, *biotechnology* employs genetically engineered microorganisms to synthesize products of high value, such as insulin or other human proteins, usually on a small scale.

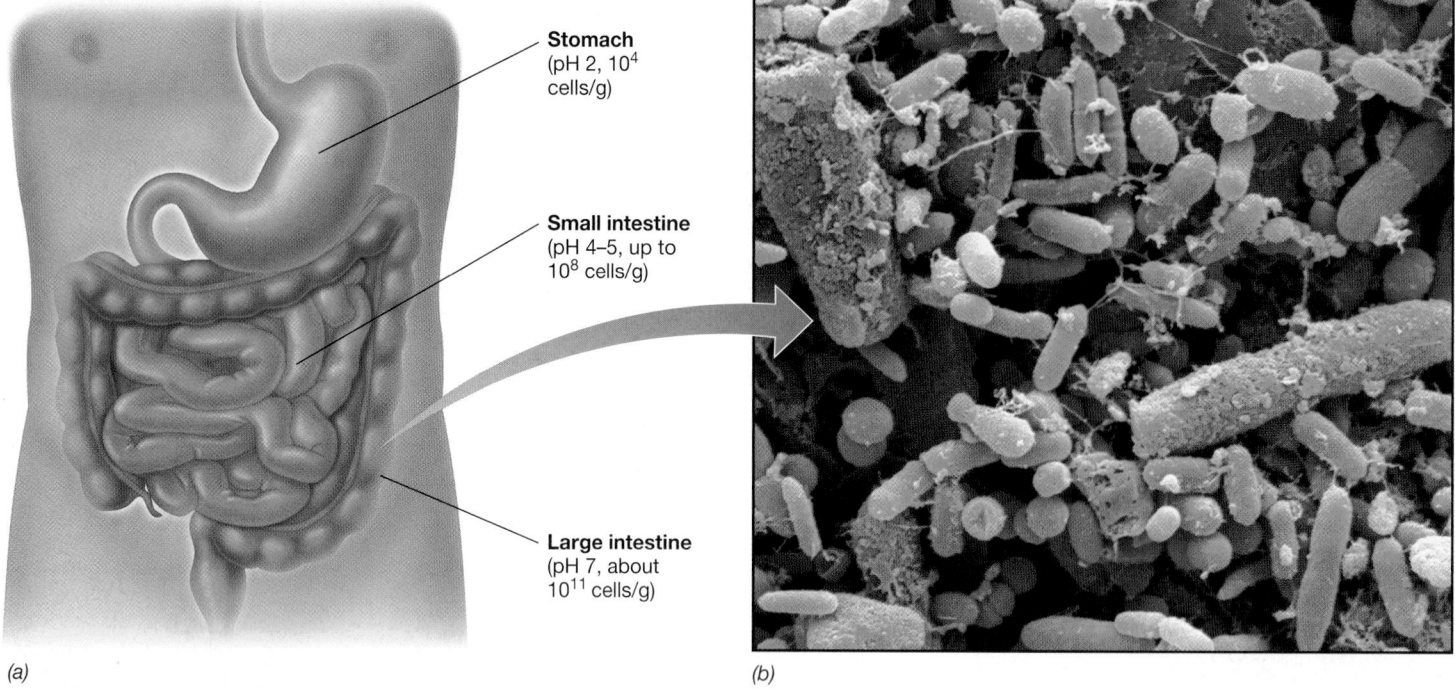

(a)

(b)

Stomach
(pH 2, 10^4
cells/g)

Small intestine
(pH 4–5, up to
10^8 cells/g)

Large intestine
(pH 7, about
10^{11} cells/g)

Figure 1.10 The human gastrointestinal tract. *(a)* Diagram of the human GI tract showing the major organs. *(b)* Scanning electron micrograph of microbial cells in the human colon (large intestine). Cell numbers in the colon can reach as high as 10^{11} per gram. As well as high *numbers* of cells, the microbial *diversity* in the colon is also quite high.

Microorganisms can also be used to produce *biofuels*. For example, natural gas (methane, CH_4) is a product of the anaerobic metabolism of a group of *Archaea* called *methanogens*. Ethyl alcohol (ethanol) is a major fuel supplement, which is produced by the microbial fermentation of glucose obtained from carbon-rich feedstocks such as sugarcane, corn, or rapidly growing grasses (**Figure 1.12**). Microorganisms can even convert waste materials, such as domestic refuse, animal wastes, and cellulose, into ethanol and methane.

Microorganisms are also used to clean up wastes. Wastewater treatment is essential to sanitation and human health. Wastewater treatment relies on microbes to treat water contaminated with human waste so that it can be reused or returned safely to the environment. Waterborne disease such as cholera and typhoid can proliferate in the absence of proper wastewater treatment. Microbes can also be used to clean up industrial pollution in a process called *bioremediation*. In bioremediation, microorganisms are used to transform spilled oil, solvents, pesticides, heavy metals, and other environmentally toxic pollutants. Bioremediation accelerates the cleanup process either by adding special microorganisms to a polluted environment or by adding nutrients that stimulate indigenous microorganisms to degrade the pollutants. In either case the goal is to accelerate disappearance of the pollutant.

As these examples show, the influence of microorganisms on humans is great and their activities are essential for the functioning of the planet. Or, as the eminent French chemist and early microbiologist Louis Pasteur so aptly put it: "The role of the infinitely small in nature is infinitely large." Microscopes provide an essential portal though which microbiologists such as Pasteur gazed into the world of microbes. We therefore continue our introduction to the microbial world with an overview of microscopy.

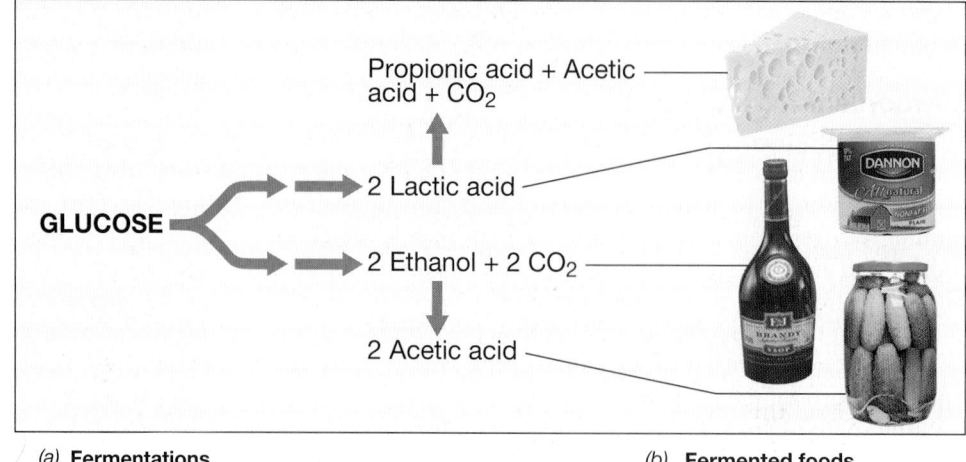

Propionic acid + Acetic acid + CO_2

2 Lactic acid

GLUCOSE

2 Ethanol + 2 CO_2

2 Acetic acid

(a) **Fermentations**

(b) **Fermented foods**

Figure 1.11 Fermented foods. *(a)* Major fermentations in various fermented foods. It is the fermentation product (ethanol, or lactic, propionic, or acetic acids) that both preserves the food and renders in it a characteristic flavor. *(b)* Photo of several fermented foods showing the characteristic fermentation product in each.

(a)   *(b)*

Figure 1.12 Ethanol as a biofuel. *(a)* Major crop plants used as feedstocks for biofuel ethanol production.
Top: switchgrass, a source of cellulose. Bottom: corn, a source of cornstarch. Both cellulose and starch are composed
of glucose, which is fermented to ethanol by yeast. *(b)* An ethanol plant in the United States. Ethanol produced by
fermentation is distilled and then stored in the tanks.

MINIQUIZ

- How do microbes contribute to the nutrition of animals such as humans and cows?
- Describe several ways in which microorganisms are important in the food and agricultural industries.
- What is wastewater treatment and why is it important?

II • Microscopy and the Origins of Microbiology

Historically, the science of microbiology has taken its greatest leaps forward as new tools are developed and old tools improve. The microscope is the microbiologist's oldest and most fundamental tool for studying microorganisms. Indeed, microbiology did not exist before the invention of the microscope. Many forms of microscopy are available and some are extremely powerful. Throughout this text you will see images of microorganisms that were taken through the microscope using a variety of different techniques. So let's take a moment to explore how microscopy can be used to visualize microbial cells, starting at the very beginning with the invention of the microscope.

1.5 Light Microscopy and the Discovery of Microorganisms

Although the existence of creatures too small to be seen with the naked eye had been suspected for centuries, their discovery had to await invention of the microscope. The English mathematician and natural historian Robert Hooke (1635–1703) was an excellent microscopist. In his famous book *Micrographia* (1665), the first book devoted to microscopic observations, Hooke illustrated many microscopic images including the fruiting structures of molds (**Figure 1.13**). This was the first known description of microorganisms.

The first person to see bacteria, the smallest microbial cells, was the Dutch draper and amateur microscopist Antoni van Leeuwenhoek (1632–1723). Van Leeuwenhoek constructed extremely simple microscopes containing a single lens to examine various natural substances for microorganisms (**Figure 1.14**). These microscopes were crude by today's standards, but by careful manipulation and focusing, van Leeuwenhoek was able to see bacteria. He discovered bacteria in 1676 while studying pepper–water infusions, and reported his observations in a series of letters to the prestigious Royal Society of London, which published them in English translation in 1684. Drawings of some of van Leeuwenhoek's "wee animalcules," as he referred to them, are shown in Figure 1.14*b*, and a photo taken through a van Leeuwenhoek microscope is shown in Figure 1.14*c*.

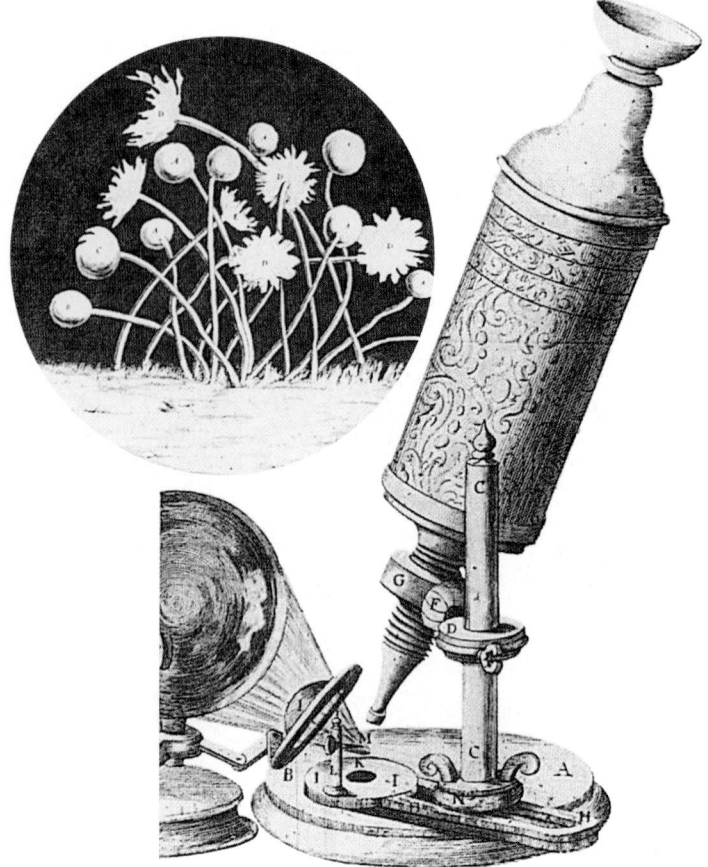

Figure 1.13 Robert Hooke and early microscopy. A drawing of the microscope used by Robert Hooke in 1664. The lens was fitted at the end of an adjustable bellows (G) and light focused on the specimen by a separate lens (1). Inset: Hooke's drawing of a bluish mold he found degrading a leather surface; the round structures contain spores of the mold.

Van Leeuwenhoek's microscope was a *light* microscope, and his design used a simple lens that could magnify an image at least 266 times. In a light microscope the sample is illuminated with visible light. **Magnification** describes the capacity of a microscope to enlarge an image. All microscopes employ lenses that provide magnification. Magnification, however, is not the limiting factor in our ability to see small objects. It is resolution that governs our ability to see the very small. **Resolution** is the ability to distinguish two adjacent objects as distinct and separate. The limit of resolution for a light microscope is about 0.2 μm (μm is the abbreviation for micrometer, 10^{-6} m). What this means is that two objects that are closer together than 0.2 μm cannot be resolved as distinct and separate.

Microscopy has improved considerably since the days of van Leeuwenhoek. Several types of light microscopy are now available, including *bright-field, phase-contrast, differential interference contrast, dark-field,* and *fluorescence*. With the modern compound light microscope, light from a light source is focused on the specimen by the condenser (**Figure 1.15**), and this light passes through the sample and is collected by the lenses. The modern compound light microscope contains two types of lenses, *objective* and *ocular,* that function in combination to magnify the image. Microscopes used in microbiology have ocular lenses that magnify 10–30× and objective lenses that magnify 10–100× (Figure 1.15*b*). The total magnification of a compound light microscope is the *product* of the magnification of its objective and ocular lenses (Figure 1.15*b*). Magnification of 1000× is required to resolve objects 0.2 μm in diameter, which is the limit of resolution for most light microscopes (increasing magnification beyond 1000× provides little improvement in the resolution of a light microscope).

The limit of resolution for a light microscope is a function of the wavelength of light used and the light-gathering ability of the objective lens, a property known as its *numerical aperture*. There is a correlation between the magnification of a lens and its numerical aperture; lenses with higher magnification typically have higher numerical apertures. The diameter of the smallest object resolvable by any lens is equal to 0.5λ/numerical aperture, where λ is the wavelength of light used. With objectives that have a very high numerical aperture (such as the 100× objective), an optical grade oil is placed between the microscope slide and the objective. Lenses on which oil is used are called *oil-immersion* lenses. Immersion oil increases the light-gathering ability of a lens, that is, it increases the amount of light that is collected and viewed by the lens.

In light microscopy, specimens are visualized because of differences in **contrast** that exist between them and their surroundings. In bright-field microscopy, contrast results when cells absorb or scatter light

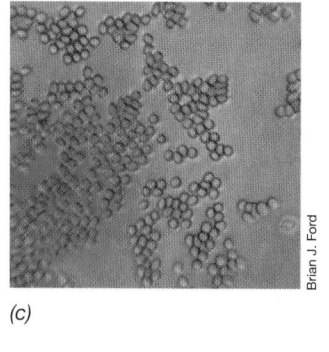

Figure 1.14 The van Leeuwenhoek microscope. *(a)* A replica of Antoni van Leeuwenhoek's microscope. *(b)* Van Leeuwenhoek's drawings of bacteria, published in 1684. Even from these simple drawings we can recognize several shapes of common bacteria: A, C, F, and G, rods; E, cocci; H, packets of cocci. *(c)* Photomicrograph of a human blood smear taken through a van Leeuwenhoek microscope. Red blood cells are clearly apparent.

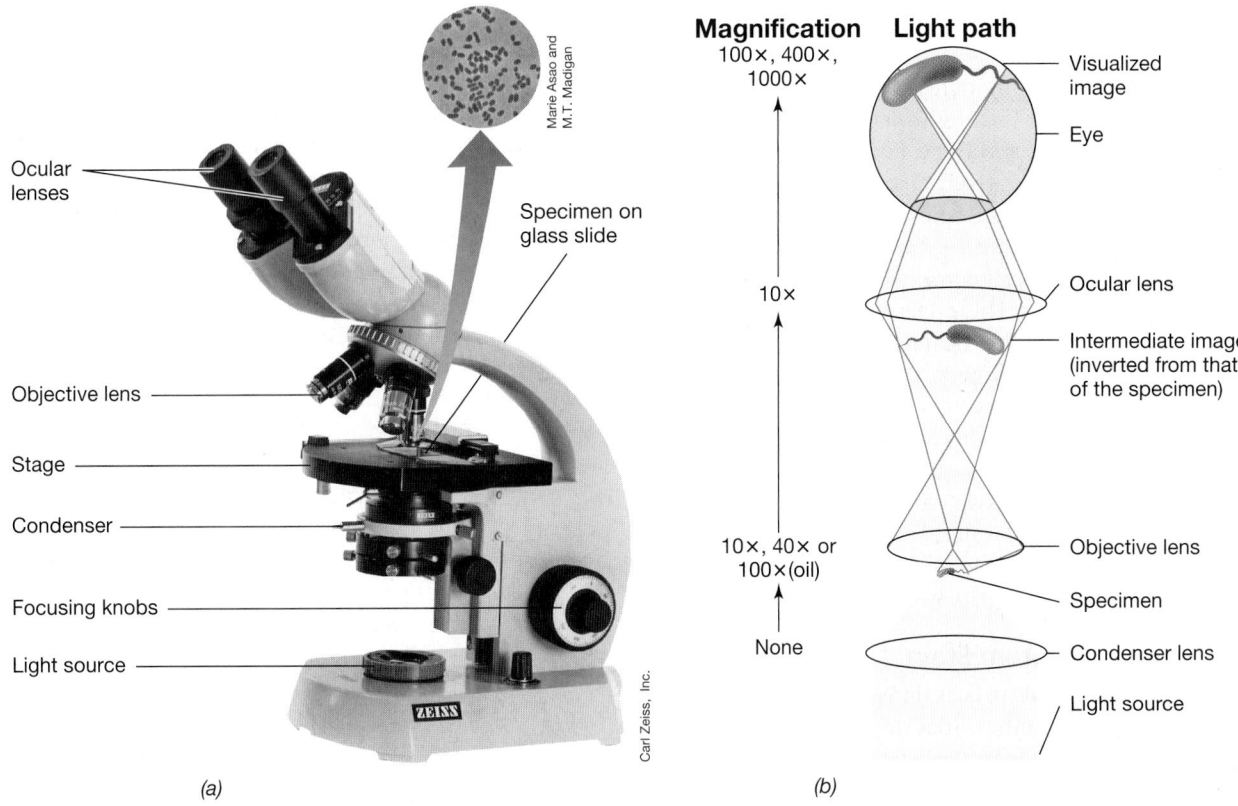

(a)

(b)

Figure 1.15 Microscopy. *(a)* A compound light microscope (inset photomicrograph of unstained cells taken through a phase-contrast light microscope). *(b)* Path of light through a compound light microscope. Figure 1.19 compares cells visualized by bright field with those visualized by phase contrast.

differently from their surroundings. Bacterial cells typically lack contrast, that is, their optical properties are similar to the surrounding medium, and hence they are difficult to see well with the bright-field microscope. Pigmented microorganisms are an exception because the color of the organism adds contrast, thus improving visualization by bright-field optics (**Figure 1.16**). For cells lacking pigments there are several ways to boost contrast, and we consider these methods in the next section.

MINIQUIZ

- Define the terms magnification and resolution.
- What is the limit of resolution for a bright-field microscope? What defines this limit?

1.6 Improving Contrast in Light Microscopy

Contrast is necessary in light microscopy to distinguish microorganisms from their surroundings. Cells can be stained to improve contrast, and staining is commonly used to visualize bacteria with bright-field microscopy. In addition to staining, other methods of light microscopy have been developed to improve contrast, such as phase contrast, differential interference contrast, dark field, and fluorescence.

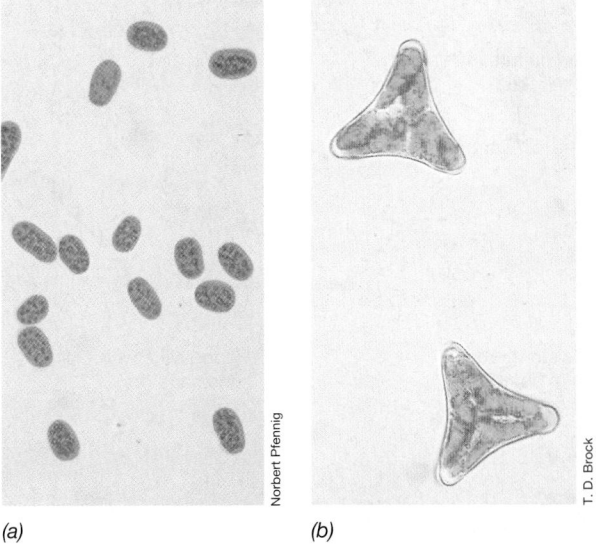

(a)   (b)

Figure 1.16 Bright-field photomicrographs of pigmented microorganisms. *(a)* Purple phototrophic bacteria (*Bacteria*). The bacterial cells are about 5 μm wide. *(b)* A green alga (eukaryote). The green structures are chloroplasts. The algal cells are about 15 μm wide. Purple bacteria are anoxygenic phototrophs, whereas algae are oxygenic phototrophs. Both groups contain photosynthetic pigments but only oxygenic phototrophs produce O_2 (Section 1.3 and Figure 1.5*a*).

Staining: Increasing Contrast for Bright-Field Microscopy

Dyes can be used to stain cells and increase their contrast so that they can be more easily seen in the bright-field microscope. Each class of dye has an affinity for specific cellular materials. Many dyes used in microbiology are positively charged, and for this reason they are called *basic dyes*. Examples of basic dyes include methylene blue, crystal violet, and safranin. Basic dyes bind strongly to negatively charged cell components, such as nucleic acids and acidic polysaccharides. These dyes also stain the surfaces of cells, because cell surfaces tend to be negatively charged. These properties make basic dyes useful general-purpose stains that non-specifically stain most bacterial cells.

To perform a simple stain, one begins with dried preparations of cells (**Figure 1.17**). A clean glass slide containing a dried suspension of cells is flooded for a minute or two with a dilute solution of a basic dye, rinsed several times in water, and blotted dry. Because their cells are so small, it is common to observe dried, stained preparations of bacterial cells with a high-power (oil-immersion) lens.

Differential Stains: The Gram Stain

Stains that render different kinds of cells different colors are called *differential* stains. An important differential-staining procedure used in microbiology is the **Gram stain** (**Figure 1.18**). On the basis of their reaction in the Gram stain, bacteria can be divided into two major groups: **gram-positive** and **gram-negative**. After Gram staining, gram-positive bacteria appear purple-violet and gram-negative bacteria appear pink (Figure 1.18*b*). The color difference in the Gram stain arises because of differences in the cell wall structure of gram-positive and gram-negative cells (↩ Section 2.4). Staining with a basic dye such as crystal violet renders cells purple in color. Cells are then treated with ethanol, which decolorizes gram-negative cells but not gram-positive cells. Finally, cells are counterstained with a different-colored stain, typically the red stain safranin. As a result, gram-positive and gram-negative cells can be distinguished microscopically by their different colors (Figure 1.18*b*).

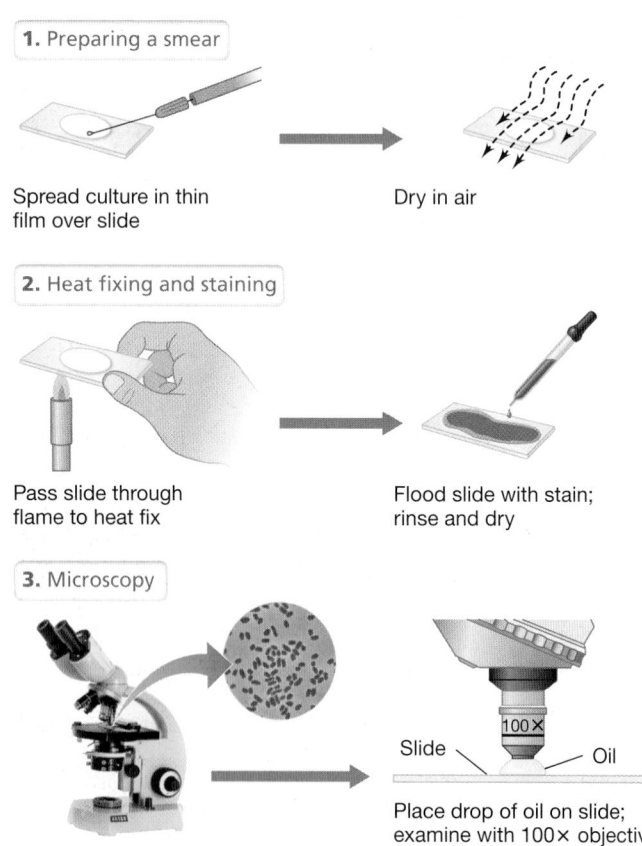

Figure 1.17 Staining cells for microscopic observation. Stains improve the contrast between cells and their background. Step 3 center: Same cells as shown in Figure 1.15 inset but stained with a basic dye.

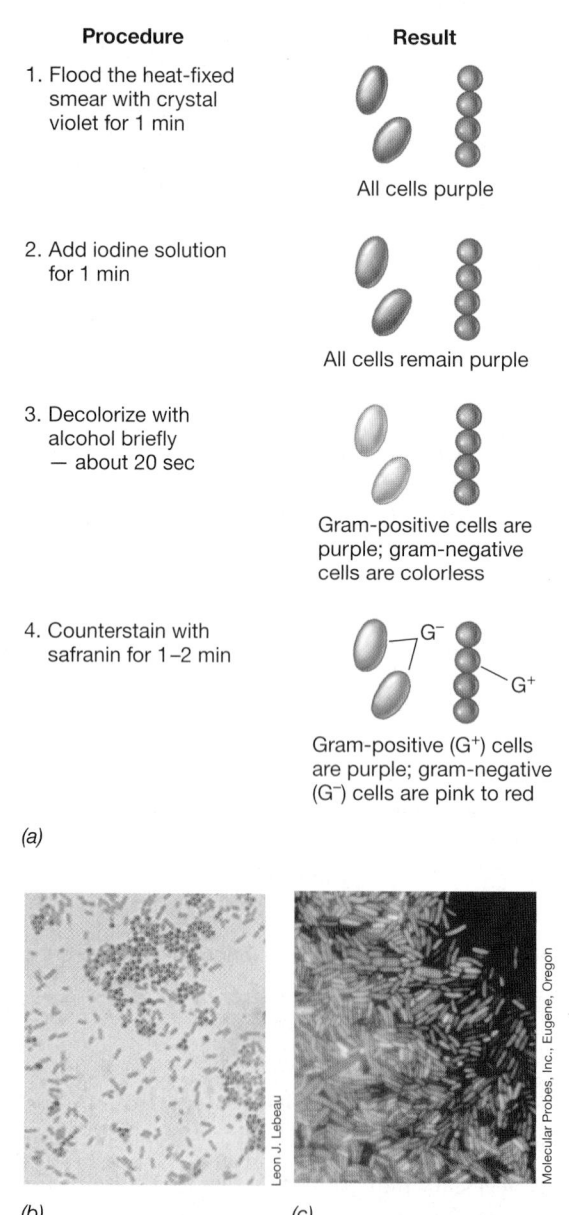

Procedure	Result
1. Flood the heat-fixed smear with crystal violet for 1 min	All cells purple
2. Add iodine solution for 1 min	All cells remain purple
3. Decolorize with alcohol briefly — about 20 sec	Gram-positive cells are purple; gram-negative cells are colorless
4. Counterstain with safranin for 1–2 min	Gram-positive (G⁺) cells are purple; gram-negative (G⁻) cells are pink to red

(a)

(b)   *(c)*

Leon J. Lebeau

Molecular Probes, Inc., Eugene, Oregon

Figure 1.18 The Gram stain. *(a)* Steps in the procedure. *(b)* Microscopic observation of gram-positive (purple) and gram-negative (pink) bacteria. The organisms are *Staphylococcus aureus* and *Escherichia coli*, respectively. *(c)* Cells of *Pseudomonas aeruginosa* (gram-negative, green) and *Bacillus cereus* (gram-positive, orange) stained with a one-step fluorescent staining method. This method allows for differentiating gram-positive from gram-negative cells in a single staining step.

The Gram stain is the most common staining procedure used in microbiology, and it is often performed to begin the characterization of a new bacterium. If a fluorescence microscope is available, the Gram stain can be reduced to a one-step procedure; gram-positive and gram-negative cells fluoresce different colors when treated with a special chemical (Figure 1.18c).

Phase-Contrast and Dark-Field Microscopy

Although staining is widely used in light microscopy, staining often kills cells and can distort their features. Two forms of light microscopy improve image contrast of unstained (and thus live) cells. These are phase-contrast microscopy and dark-field microscopy (**Figure 1.19**). The phase-contrast microscope in particular is widely used in teaching and research for the observation of living preparations.

Phase-contrast microscopy is based on the principle that cells differ in refractive index (that is, the ability of a material to alter the speed of light) from their surroundings. Light passing through a cell thus differs in phase from light passing through the surrounding liquid. This subtle difference is amplified by a device in the objective lens of the phase-contrast microscope called the *phase ring*, resulting in a dark image on a light background (Figure 1.19b; see also inset to Figure 1.15a). The ring consists of a phase plate that amplifies the variation in phase to produce the higher-contrast image.

In the dark-field microscope, light does not pass through the specimen. Instead, light is directed from the sides of the specimen and only light that is scattered when it hits the specimen can reach the lens. Thus, the specimen appears light on a dark background (Figure 1.19c). Dark-field microscopy often has better resolution than light microscopy, and some objects can be resolved by dark-field that cannot be resolved by bright-field or even by phase-contrast microscopes. Dark-field microscopy is a particularly good way to observe microbial motility, as bundles of flagella (the structures responsible for swimming motility) are often resolvable with this technique.

Fluorescence Microscopy

The fluorescence microscope visualizes specimens that fluoresce. In fluorescence microscopy, cells are made to fluoresce (to emit light) by illuminating them from above with light of a single color. Filters are used so that only fluorescent light is seen, and thus cells appear to glow in a black background (**Figure 1.20**).

Cells fluoresce either because they contain naturally fluorescent substances such as chlorophyll (autofluorescence, Figure 1.20b, d) or because they have been stained with a fluorescent dye (Figure 1.20e). DAPI (4′,6-*di*amidino-2-*p*henyl*i*ndole) is a widely used fluorescent dye that stains cells bright blue because it complexes with the cell's DNA (Figure 1.20e). DAPI can be used to visualize cells in their natural habitats, such as soil, water, food, or a clinical specimen. Fluorescence microscopy using DAPI is widely used in clinical diagnostic microbiology and also in microbial ecology for enumerating bacteria in a natural environment or in a cell suspension (Figure 1.20e).

MINIQUIZ

- What color will a gram-negative cell be after Gram staining by the conventional method?
- What major advantage does phase-contrast microscopy have over staining?
- How can cells be made to fluoresce?

1.7 Imaging Cells in Three Dimensions

Thus far we have only considered forms of microscopy in which the rendered images are two-dimensional. Two methods of light microscopy can render a more three-dimensional image, and in this section we explore these forms of microscopy.

Differential Interference Contrast Microscopy

Differential interference contrast (DIC) microscopy is a form of light microscopy that employs a polarizer in the condenser to produce polarized light (light in a single plane). The polarized light then passes through a prism that generates two distinct beams. These beams pass through the specimen and enter the objective lens, where they are recombined into one. Because the two beams pass through substances that differ in refractive index, the combined beams are not totally in phase but instead interfere with each other. This optical effect provides a three-dimensional perspective, which enhances subtle differences in cell structure.

Using DIC microscopy, cellular structures such as the nucleus of eukaryotic cells (**Figure 1.21**), or endospores, vacuoles, and inclusions of bacterial cells, appear more three-dimensional than in

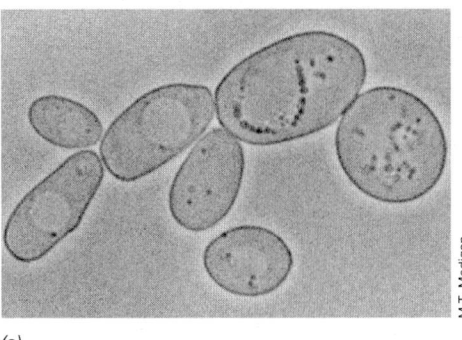
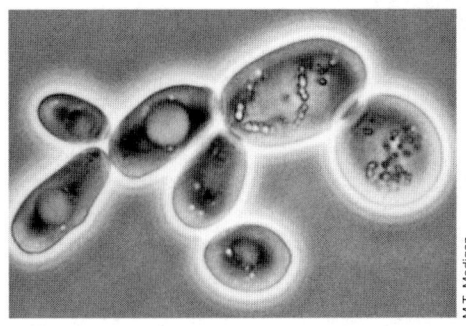
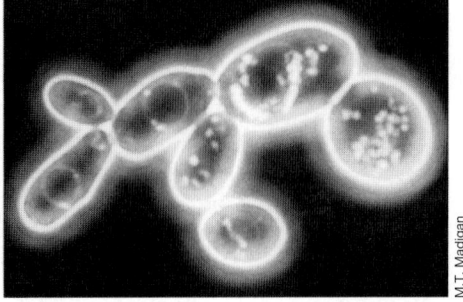

(a)   (b)   (c)

Figure 1.19 Cells visualized by different types of light microscopy. The same field of cells of the yeast *Saccharomyces cerevisiae* visualized by (a) bright-field microscopy, (b) phase-contrast microscopy, and (c) dark-field microscopy. Cells average 8–10 μm wide.

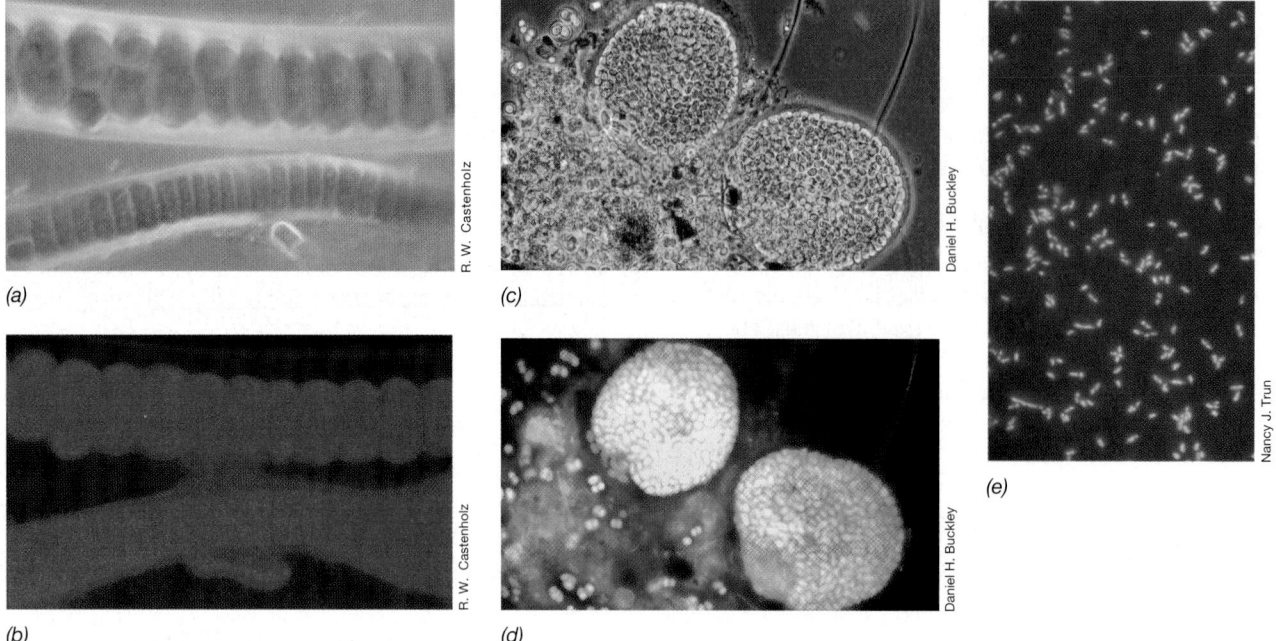

Figure 1.20 Fluorescence microscopy. *(a, b, c, d)* Cyanobacteria. The same cells are observed by phase-contrast microscopy *(a, c)* and by fluorescence microscopy *(b, d)*. The cells fluoresce because they contain chlorophyll *a* and other pigments. The image in *b* was generated using a filter specific for the fluorescence of chlorophyll *a*, while the image in *d* was generated using a permissive filter that shows fluorescence from a range of pigments that occur naturally in cyanobacteria. *(e)* Fluorescence photomicrograph of cells of *Escherichia coli* made fluorescent by staining with the fluorescent dye DAPI, which binds to DNA.

other forms of light microscopy. DIC microscopy is typically used on unstained cells as it can reveal internal cell structures that are nearly invisible by bright-field microscopy without the need for staining (compare Figure 1.19*a* with Figure 1.21).

Confocal Scanning Laser Microscopy

A confocal scanning laser microscope (CSLM) is a computer-controlled microscope that couples a laser to a fluorescence microscope. The laser generates a high-contrast, three-dimensional image and allows the viewer to access several planes of focus in the specimen (**Figure 1.22**). The laser beam is precisely adjusted such that only a particular layer within a specimen is in perfect focus at one time. Cells struck by the laser fluoresce to generate the image as in fluorescence microscopy (Section 1.6).

Cells in CSLM preparations can also be stained with fluorescent dyes to make them more distinct (Figure 1.22*a*). The laser then scans up and down through the layers of the sample, generating an image for each layer. A computer assembles the pictures to compose the many layers into a single high-resolution, three-dimensional image. Thus, for a relatively thick specimen such as a bacterial biofilm (Figure 1.22*a*), not only can cells on the surface of the biofilm be observed, as with conventional light microscopy, but cells in the various layers are also observed by adjusting the plane of focus of the laser beam. CSLM is particularly useful when thick specimens need to be examined.

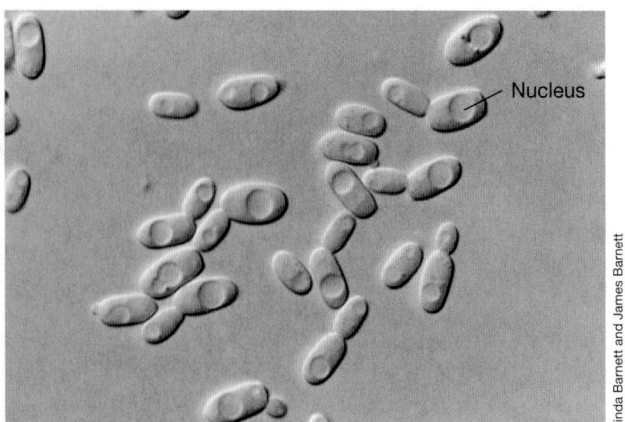

Figure 1.21 Differential interference contrast microscopy. The yeast cells are about 8 μm wide. Note the clearly visible nucleus and compare to the bright-field image of yeast cells in Figure 1.19*a*.

Nucleus

--- MINIQUIZ ---

- What structure in eukaryotic cells is more easily seen in DIC than in bright-field microscopy? (*Hint:* Compare Figures 1.19*a* and 1.21).

- Why is CSLM able to view different layers in a thick preparation while bright-field microscopy cannot?

1.8 Probing Cell Structure: Electron Microscopy

Electron microscopes use electrons instead of visible light (photons) to image cells and cell structures. In the electron microscope, electromagnets function as lenses, and the whole system operates in a vacuum (**Figure 1.23**). Electron microscopes are fitted with

(a)

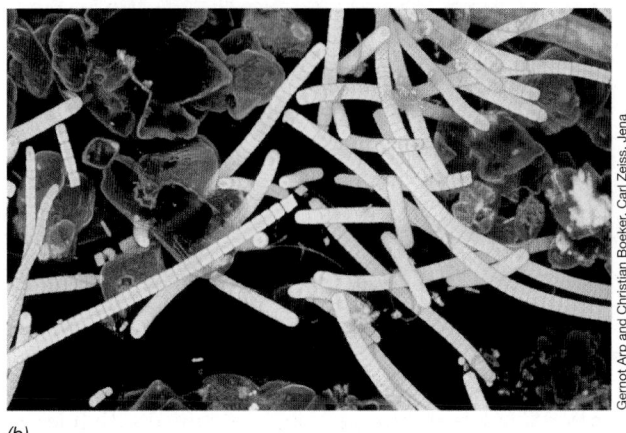

(b)

Subramanian Karthikeyan

Gernot Arp and Christian Boeker, Carl Zeiss, Jena

Figure 1.22 Confocal scanning laser microscopy. *(a)* Confocal image of a microbial biofilm community. The green, rod-shaped cells are *Pseudomonas aeruginosa* experimentally introduced into the biofilm. Cells of different colors are present at different depths in the biofilm. *(b)* Confocal image of a filamentous cyanobacterium growing in a soda lake. Cells are about 5 μm wide.

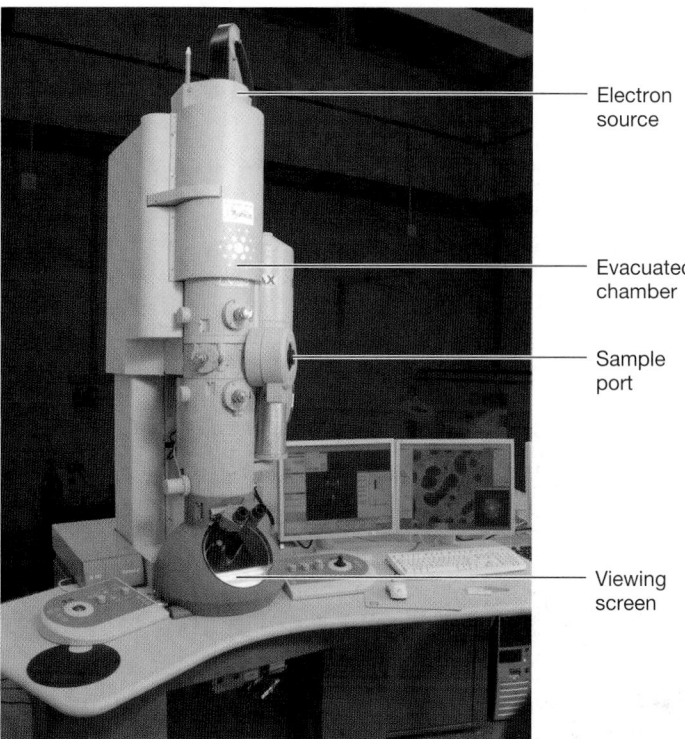

Electron source

Evacuated chamber

Sample port

Viewing screen

Figure 1.23 The electron microscope. This instrument encompasses both transmission and scanning electron microscope functions.

cameras to allow a photograph, called an *electron micrograph*, to be taken. Two types of electron microscopy are in routine use in microbiology: transmission and scanning.

Transmission Electron Microscopy

The *transmission electron microscope* (TEM) is used to examine cells and cell structure at very high magnification and resolution. The resolving power of a TEM is much greater than that of the light microscope, even allowing one to view structures at the molecular level (**Figure 1.24**). This is because the wavelength of electrons is much shorter than the wavelength of visible light, and, as we have learned, wavelength affects resolution (Section 1.5). For example, whereas the resolving power of a light microscope is about 0.2 *micrometer*, the resolving power of a TEM is about 0.2 *nanometer*, a thousandfold improvement. With such powerful resolution, objects as small as individual protein and nucleic acid molecules can be visualized by transmission electron microscopy (Figure 1.24*b*).

Unlike photons, electrons are very poor at penetrating; even a single cell is too thick to penetrate with an electron beam. Consequently, to view the internal structure of a cell, *thin sections* of the cell are needed, and the sections must be stabilized and stained with various chemicals to make them visible. A single bacterial cell, for instance, is cut into extremely thin (20–60 nm) slices, which are then examined individually by TEM (Figure 1.24*a*). To obtain sufficient contrast, the sections are treated with stains such as osmic acid, or permanganate, uranium, lanthanum, or lead salts. Because these substances are composed of atoms of high atomic weight, they scatter electrons well and thus improve contrast. If only the *external* features of an organism are to be observed, thin sections are unnecessary. Intact cells or cell components can be observed directly in the TEM by a technique called *negative staining* (Figure 1.24*b*).

Scanning Electron Microscopy

For optimal three-dimensional imaging of cells, a *scanning electron microscope* (SEM) is used. In scanning electron microscopy, the specimen is coated with a thin film of a heavy metal, typically gold. An electron beam then scans back and forth across the specimen. Electrons scattered from the metal coating are collected and projected on a monitor to produce an image (Figure 1.24*c*). In the SEM, even fairly large specimens can be observed, and the depth of field (the portion of the image that remains in sharp focus) is extremely good. A wide range of magnifications can be obtained with the SEM, from as low as 15× up to about 100,000×, but only the *surface* of an object is typically visualized.

Electron micrographs taken by either TEM or SEM are black-and-white images. Although the original image contains the maximum amount of scientific information that is available, color is often added to scanning electron micrographs by manipulating

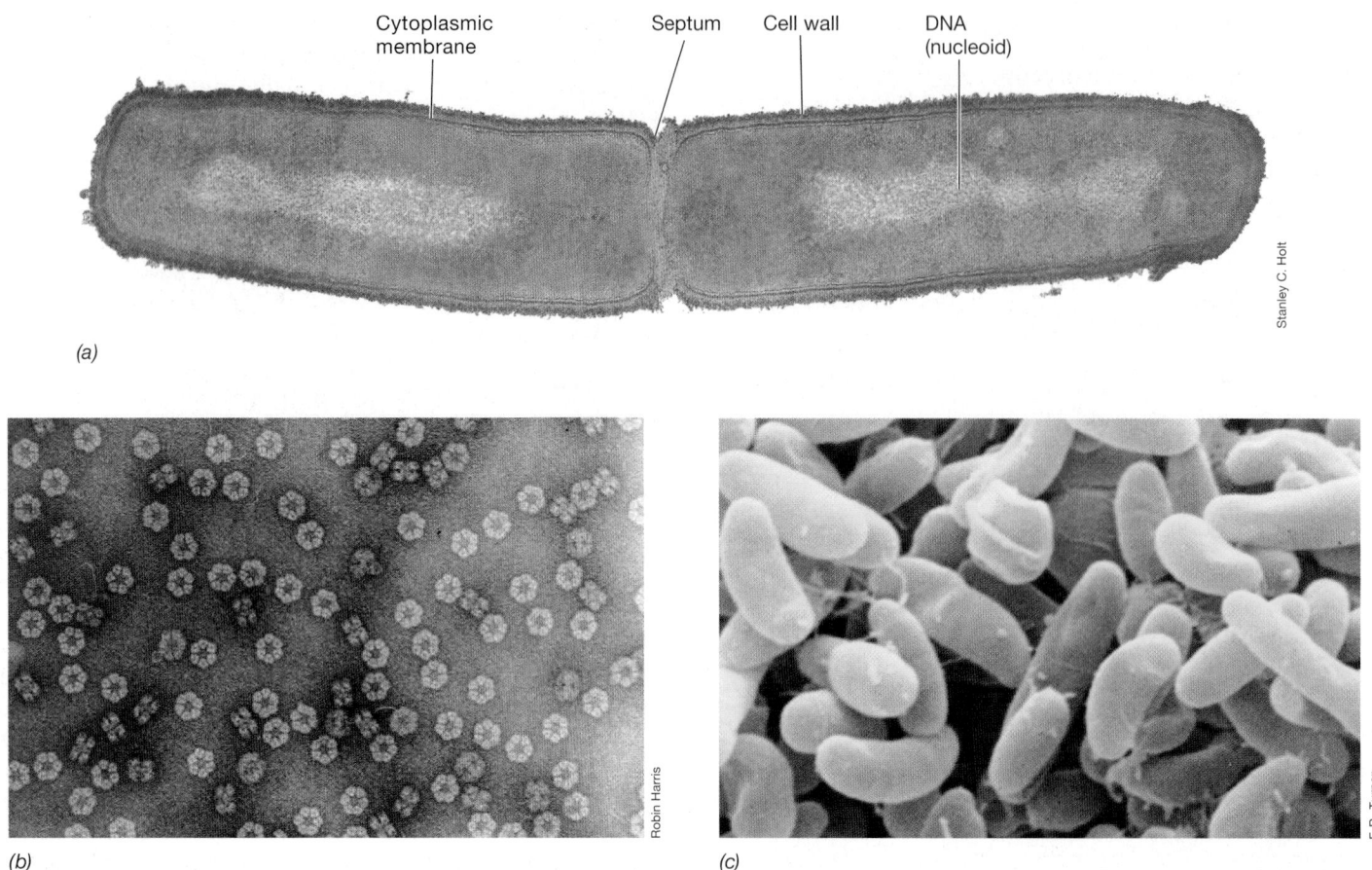

Cytoplasmic membrane Septum Cell wall DNA (nucleoid)

(a)

(b)

(c)

Figure 1.24 Electron micrographs. *(a)* Micrograph of a thin section of a dividing bacterial cell, taken by transmission electron microscopy (TEM). The cell is about 0.8 μm wide. *(b)* TEM of negatively stained molecules of hemoglobin. Each hexagonal-shaped molecule is about 25 nanometers (nm) in diameter and consists of two doughnut-shaped rings, a total of 15 nm wide. *(c)* Scanning electron micrograph (SEM) of bacterial cells. A single cell is about 0.75 μm wide.

them in a computer. However, such false color does not improve resolution of a micrograph. In this book, false color will be used sparingly in electron micrographs so as to present the micrographs in their original scientific context.

— **MINIQUIZ** —

- What is an electron micrograph? Why do electron micrographs have greater resolution than light micrographs?

- What type of electron microscope would be used to view a cluster of cells? What type would be used to observe internal cell structure?

III • Microbial Cultivation Expands the Horizon of Microbiology

Following the discovery of microorganisms driven by microscopic methods, major discoveries in microbiology were fueled by advances in microbial cultivation. Important advances included the development of **aseptic technique**, which is a collection of practices that allow for the preparation and maintenance of **sterile** (that is, without the presence of living organisms) nutrient media and solutions (⟲ Figure 5.12). Aseptic technique is essential for the isolation and maintenance of pure cultures of bacteria. **Pure cultures** are those that contain cells from only a single type of microorganism and are of great value for the study of microorganisms. Finally, **enrichment culture techniques**, which allow for the isolation from nature of microbes having particular metabolic characteristics, facilitate the discovery of diverse microorganisms.

Advances in microbial cultivation are directly responsible for success in fighting infectious disease, the discovery of microbial diversity, and the use of microbes as model systems to discover the fundamental properties of all living cells. Important advances in microbial cultivation occurred in the nineteenth century as microbiologists sought to answer two major questions of that time: (1) Does spontaneous generation occur? (2) What is the nature of infectious disease? Answers to these seminal questions emerged from the work of two giants in the field of microbiology: the French chemist Louis Pasteur and the German physician Robert Koch. We begin with the work of Pasteur.

1.9 Pasteur and Spontaneous Generation

Pasteur was a chemist by training and was one of the first to recognize that many of what were thought to be strictly chemical reactions were actually catalyzed by microorganisms. Pasteur studied the chemistry of crystal formation and he used microscopes to examine crystal structure. His training in chemistry and microscopy prepared him to make a series of foundational discoveries to further the science of microbiology.

The Microbial Basis of Fermentation

Early in his career, Pasteur studied crystals formed during the production of alcohol. Through careful microscopic observation of tartaric acid crystals formed in wine, he observed two types of crystals that had mirror-image structures (**Figure 1.25**). He separated these by hand and observed that each type of crystal bent a beam of polarized light in a different direction. In this way he discovered that

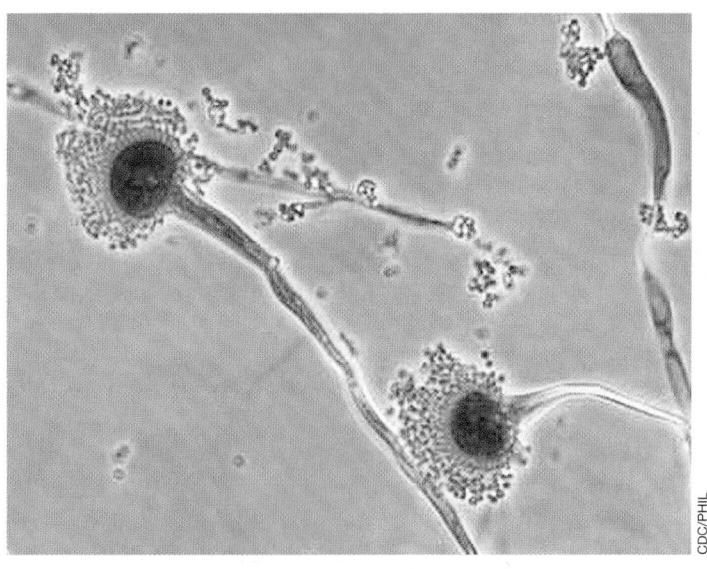

(a)

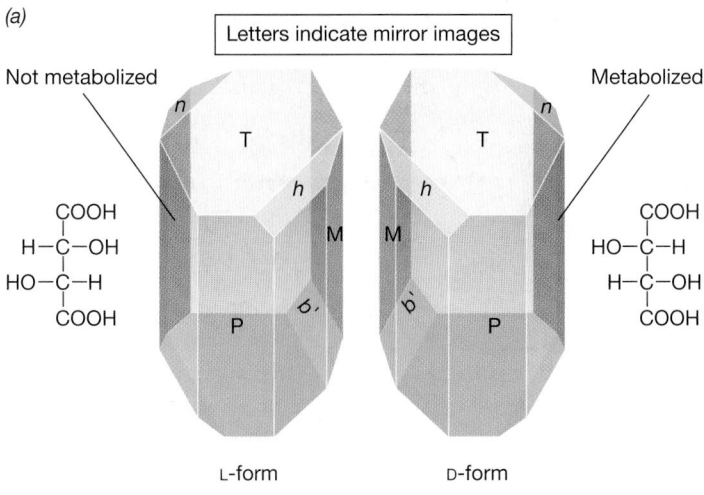

Letters indicate mirror images

Not metabolized

Metabolized

L-form   D-form

(b)

Figure 1.25 Louis Pasteur and optical isomers. (a) Light micrograph of cells of the mold *Aspergillus*. (b) Pasteur's drawings of crystals of tartaric acid. Left-handed L-form crystals bend light to the left, and right-handed D-form crystals bend light to the right. Note that the two crystals are mirror images of one another, a hallmark of optical isomers. Pasteur found that only D-tartrate was metabolized by *Aspergillus*.

chemically identical substances can have *optical isomers*, which have different molecular structures that can influence their properties. Pasteur went on to discover that microorganisms could discriminate between optical isomers. For example, cultures of the mold *Aspergillus* (Figure 1.25a) metabolized exclusively D-tartrate but not its optical isomer, L-tartrate (Figure 1.25b). The fact that a living organism could discriminate between optical isomers led Pasteur to suspect that many reactions that were thought to be purely chemical in nature were actually catalyzed by specific microorganisms.

While a professor of chemistry, Pasteur encountered a local businessman who produced alcohol industrially from beet juice. The businessman was losing money because many of his vats produced, instead of alcohol, a product that smelled like sour milk, which Pasteur determined to be lactic acid. In the mid-nineteenth century the production of alcohol was thought to be solely a chemical process. Pasteur studied the broth with his microscope, but instead of crystals he observed cells. Pasteur observed that the vats that produced alcohol were full of yeast, but the sour vats were full of rod-shaped bacteria. He hypothesized that these were living organisms whose growth produced either alcohol or lactic acid.

Pasteur needed to grow these organisms to prove his hypothesis. He prepared an extract of yeast cells, deducing that this would contain all of the nutrients that yeast need to grow. He then used a porcelain filter to remove all cells from this yeast extract nutrient medium, rendering it sterile. If he introduced living yeast back into this sterile yeast extract medium he could observe their growth and show the production of alcohol, but if he instead introduced the small rods he then observed lactic acid formation. Heating of these cultures eliminated growth *and* the production of either alcohol or lactic acid. In this way he proved that fermentation is carried out by microorganisms and that different microorganisms perform different fermentation reactions.

During his work on fermentation, Pasteur observed that other organisms would often grow in his yeast extract medium. He deduced that these organisms were being introduced from the air. Pasteur's work on fermentation had prepared him to conduct a series of classic experiments on spontaneous generation, experiments that are forever linked to his name and which helped establish microbiology as a modern science.

Spontaneous Generation

The concept of **spontaneous generation** existed for thousands of years and its basic tenet can be easily grasped. If food or some other perishable material is allowed to stand for some time, it putrefies. When examined microscopically, the putrefied material is teeming with microorganisms. From where do these organisms arise? Prior to Pasteur it was common belief that life arose spontaneously from nonliving materials, that is, by *spontaneous generation*.

Pasteur became a powerful opponent of spontaneous generation. He predicted that microorganisms in putrefying materials were descendants of cells that entered from the air or cells that had been initially present on the decaying materials. Pasteur reasoned that if food were treated in such a way as to destroy all living organisms present—that is, if it were rendered sterile—and if it were kept sterile, it would not putrefy.

Pasteur used heat to kill contaminating microorganisms, and he found that extensive heating of a nutrient solution followed by sealing kept it from putrefying. Proponents of spontaneous generation criticized these experiments by declaring that "fresh air" was necessary for the phenomenon to occur. In 1864 Pasteur countered this objection simply and brilliantly by constructing a swan-necked flask, now called a *Pasteur flask* (**Figure 1.26**). In such a flask, nutrient solutions could be heated to boiling and sterilized. After the flask cooled, air could reenter, but the bend in the neck prevented particulate matter (including microorganisms) from entering the nutrient solution and initiating putrefaction. Nutrient solutions in such flasks remained sterile indefinitely. Microbial growth was observed only after particulate matter from the neck of the flask was allowed to enter the liquid in the flask (Figure 1.26c). This experiment settled the spontaneous generation controversy forever.

Pasteur's work on spontaneous generation demonstrated the importance of sterilization and led to the development of effective sterilization procedures that were eventually standardized and applied widely in microbiology, medicine, and industry. For example, the British physician Joseph Lister (1827–1912) deduced from Pasteur's discoveries that surgical infections were caused by microorganisms. He implemented a range of techniques designed to kill microorganisms and to prevent microbial infection of surgical patients. Lister is credited with the introduction of aseptic techniques for surgeries (1867), and his methods were adopted worldwide; these greatly improved the survival rate of surgical patients. The food industry also benefited from the work of Pasteur, as his principles were quickly adapted for the preservation of milk and many other foods by heat treatment, which we now call *pasteurization*.

Other Accomplishments of Pasteur

Pasteur went on to many other triumphs in microbiology and medicine. Some highlights include his development of vaccines

for the diseases anthrax, fowl cholera, and rabies. Pasteur's work on rabies was his most famous success, culminating in July 1885 with the first administration of a rabies vaccine to a human, a young French boy named Joseph Meister who had been bitten by a rabid dog. In those days, a bite from a rabid animal was invariably fatal. News spread quickly of the success of Meister's vaccination, and of one administered shortly thereafter to a young shepherd boy, Jean-Baptiste Jupille (**Figure 1.27a**). Within a year several thousand people bitten by rabid animals had traveled to Paris to be treated with Pasteur's rabies vaccine.

Pasteur's fame was legendary and led the French government to establish the Pasteur Institute in Paris in 1888 (Figure 1.27b). Originally established as a clinical center for the treatment of rabies and other contagious diseases, the Pasteur Institute today is a major biomedical research center focused on antiserum and vaccine research and production. The medical and veterinary breakthroughs of Pasteur not only were highly significant in their own right but helped solidify the concept of the germ theory of disease, whose principles were being developed at about the same time by a second giant of this era, Robert Koch.

MINIQUIZ

- Define the term sterile. What two methods did Pasteur used to make solutions sterile?
- How did Pasteur's experiments using swan-necked flasks defeat the theory of spontaneous generation?
- Besides ending the controversy over spontaneous generation, what other accomplishments do we credit to Pasteur?

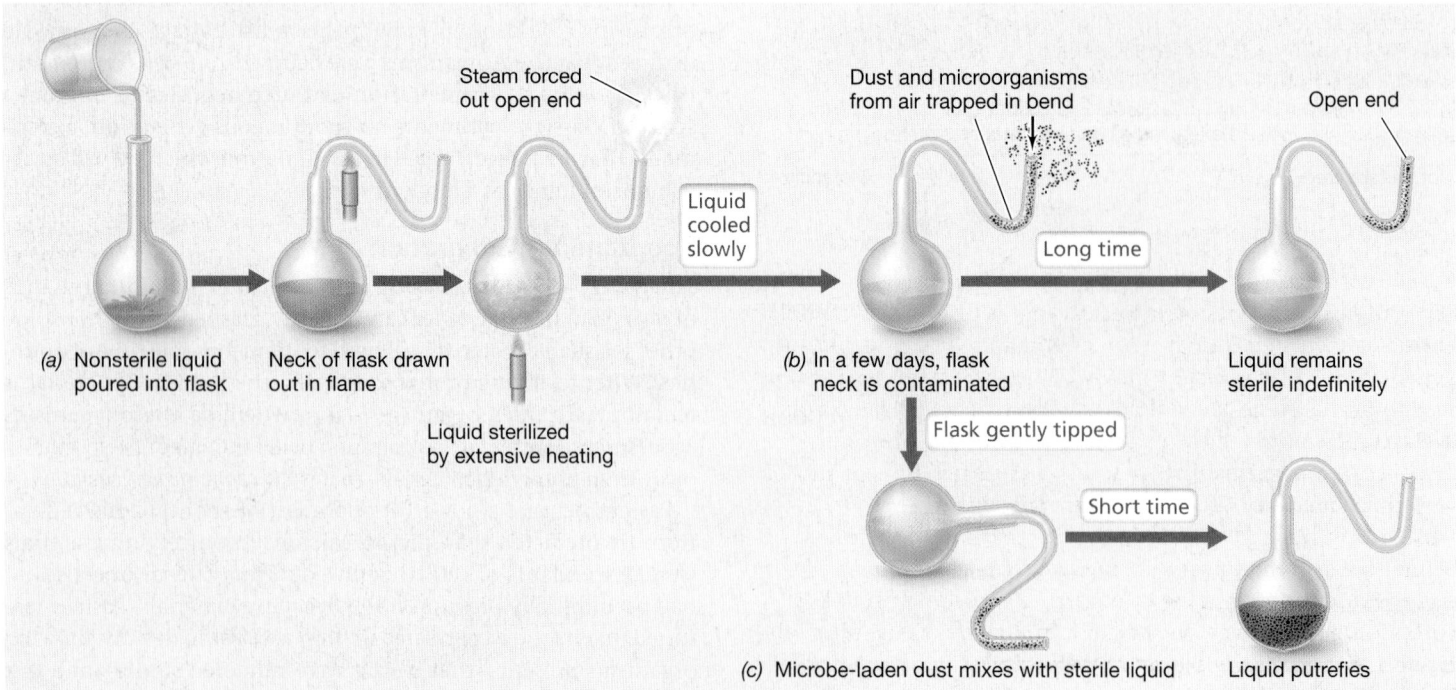

Figure 1.26 The defeat of spontaneous generation: Pasteur's swan-necked flask experiment. In (c) the liquid putrefies because microorganisms enter with the dust. The bend in the flask allowed air to enter (a key objection to Pasteur's sealed flasks) but prevented microorganisms from entering.

(a)

(b)

Figure 1.27 Louis Pasteur and some symbols of his contributions to microbiology. *(a)* A French 5-franc note honoring Pasteur. The shepherd boy Jean-Baptiste Jupille is shown killing a rabid dog that had attacked children. Pasteur's rabies vaccine saved Jupille's life. In France, the franc preceded the euro as a currency. *(b)* Part of the Pasteur Institute, Paris, France. Today this structure, built for Pasteur by the French government, houses a museum that displays some of the original swan-necked flasks used in his experiments and a chapel containing Pasteur's crypt.

1.10 Koch, Infectious Diseases, and Pure Cultures

Proof that some microorganisms can cause disease provided the greatest impetus for the development of microbiology as an independent biological science. As early as the sixteenth century it was suspected that some agent of disease could be transmitted from a diseased person to a healthy person. After microorganisms were discovered, a number of individuals proposed that they caused infectious diseases, but skepticism prevailed and definitive proof was lacking. As early as 1847, the Hungarian physician Ignaz Semmelweis promoted sanitary methods including hand washing as a method for preventing infections. His methods are credited with saving many lives, but he could not prove why these methods worked and his advice was met with scorn by most of the medical community. The work of Pasteur and Lister provided strong evidence that microbes were the cause of infectious disease, but it was not until the work of the German physician Robert Koch

(1843–1910) (**Figure 1.28**) that the germ theory of infectious disease had direct experimental support.

The Germ Theory of Disease and Koch's Postulates

In his early work Koch studied anthrax, a disease of cattle and occasionally of humans. Anthrax is caused by an endospore-forming bacterium called *Bacillus anthracis.* By careful microscopy and staining, Koch established that the bacteria were always present in the blood of an animal that was succumbing to the disease. However, Koch reasoned that the mere *association* of the bacterium with the disease was not actual proof of *cause and effect*, and he seized the opportunity to study cause and effect experimentally using anthrax and laboratory animals. The results of this study formed the standard by which infectious diseases have been studied ever since.

Koch used mice as experimental animals. Using appropriate controls, Koch demonstrated that when a small drop of blood from a mouse with anthrax was injected into a healthy mouse, the latter quickly developed anthrax. He took blood from this second animal, injected it into another, and again observed the characteristic disease symptoms. However, Koch carried this experiment a critically important step further. He discovered that the anthrax bacteria could be grown in a nutrient medium *outside the host* and

Figure 1.28 Robert Koch. The German physician and microbiologist is credited with founding medical microbiology and formulating his famous postulates.

that even after many transfers in laboratory culture, the bacteria still caused the disease when inoculated into a healthy animal.

On the basis of these experiments and others on the causative agent of tuberculosis, Koch formulated a set of rigorous criteria, now known as **Koch's postulates**, for definitively linking cause and effect in an infectious disease. Koch's postulates, summarized in **Figure 1.29**, stressed the importance of *laboratory culture* of the putative infectious agent followed by introduction of the suspected agent into virgin animals and recovery of the pathogen from diseased or dead animals. With these postulates as a guide, Koch, his students, and those that followed them discovered the causative agents of most of the important infectious diseases of humans and domestic animals. These discoveries also led to the development of successful treatments for the prevention and cure of many of these diseases, greatly improving the scientific basis of clinical medicine and human health and welfare (Figure 1.8).

Koch, Pure Cultures, and Microbial Taxonomy

The second of Koch's postulates states that the suspected pathogen must be isolated and grown away from other microorganisms in laboratory culture (Figure 1.29); in microbiology we say that such a culture is *pure*. To accomplish this important goal, Koch and his associates developed several simple but ingenious methods of obtaining and growing bacteria in pure culture, and many of these methods are still used today.

Koch started by using natural surfaces such as a potato slice to obtain pure cultures, but he quickly developed more reliable and reproducible growth media employing liquid nutrient solutions solidified with gelatin, and later with agar, an algal polysaccharide with excellent properties for this purpose. Along with his associate Walther Hesse, Koch observed that when a solid surface was incubated in air, masses of microbial cells called colonies developed, each having a characteristic shape and color (**Figure 1.30**).

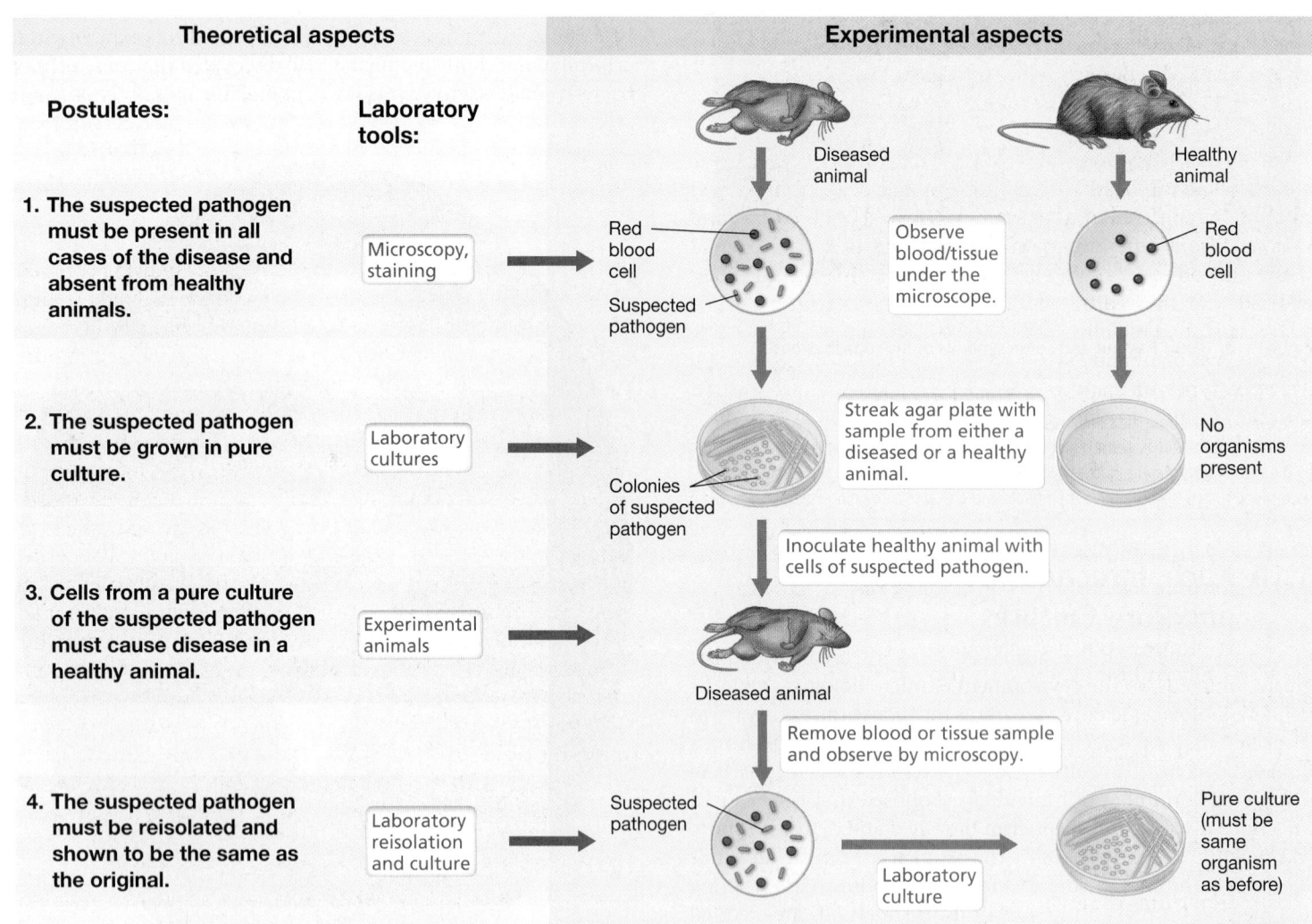

KOCH'S POSTULATES

Figure 1.29 Koch's postulates for proving cause and effect in infectious diseases. Note that following isolation of a pure culture of the suspected pathogen, the cultured organism must both initiate the disease and be recovered from the diseased animal. Establishing the correct conditions for growing the pathogen is essential; otherwise it will be missed.

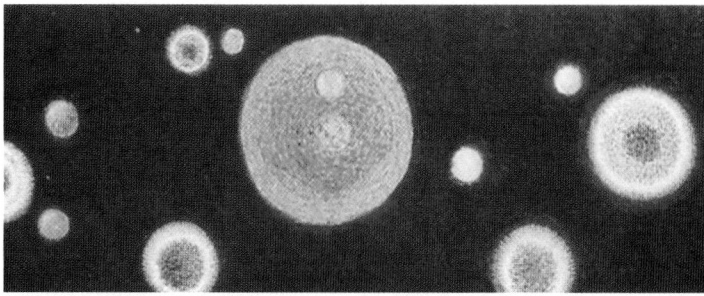

Figure 1.30 A hand-colored photograph taken by Walther Hesse of colonies formed on agar. The colonies include those of molds and bacteria obtained during Hesse's studies of the microbial content of air in Berlin, Germany, in 1882. From Hesse, W. 1884. "Ueber quantitative Bestimmung der in der Luft enthaltenen Mikroorganismen." *Mittheilungen aus dem Kaiserlichen Gesundheitsamte.* 2: 182–207.

He inferred that each colony had arisen from a single bacterial cell that had grown to yield the mass of cells (see also Figure 1.2). Koch reasoned that each colony harbored a pure culture (a population of identical cells), and he quickly realized that solid media provided an easy way to obtain pure cultures. Richard Petri, another associate of Koch, developed the transparent double-sided "Petri dish" in 1887, and this quickly became the standard tool for obtaining pure cultures.

Koch was keenly aware of the implications his pure culture methods had for classifying microorganisms. He observed that colonies that differed in color and size (Figure 1.30) bred true and that cells from different colonies typically differed in size and shape and often in their nutrient requirements as well. Koch realized that these differences were analogous to the criteria taxonomists had established for the classification of larger organisms, such as plant and animal species, and he suggested that the different types of bacteria should be considered as "species, varieties, forms, or other suitable designation." Such insightful thinking was important for the rapid acceptance of microbiology as a new biological science, rooted as biology was in classification during Koch's era.

Koch and Tuberculosis

Koch's crowning scientific accomplishment was his discovery of the causative agent of tuberculosis. At the time Koch began this work (1881), one-seventh of all reported human deaths were caused by tuberculosis (Figure 1.8). There was a strong suspicion that tuberculosis was a contagious disease, but the suspected agent had never been seen, either in diseased tissues or in culture. Following his successful studies of anthrax, Koch set out to demonstrate the cause of tuberculosis, and to this end he brought together all of the methods he had so carefully developed in his previous studies with anthrax: microscopy, staining, pure culture isolation, and an animal model system (Figure 1.29).

The bacterium that causes tuberculosis, *Mycobacterium tuberculosis*, is very difficult to stain because *M. tuberculosis* cells contain large amounts of a waxlike lipid in their cell walls. Nevertheless, Koch devised a staining procedure for *M. tuberculosis* cells in lung tissue samples. Using this method, he observed the blue, rod-shaped cells of *M. tuberculosis* in tubercular tissues but not in healthy tissues (**Figure 1.31**). Obtaining cultures of *M. tuberculosis*

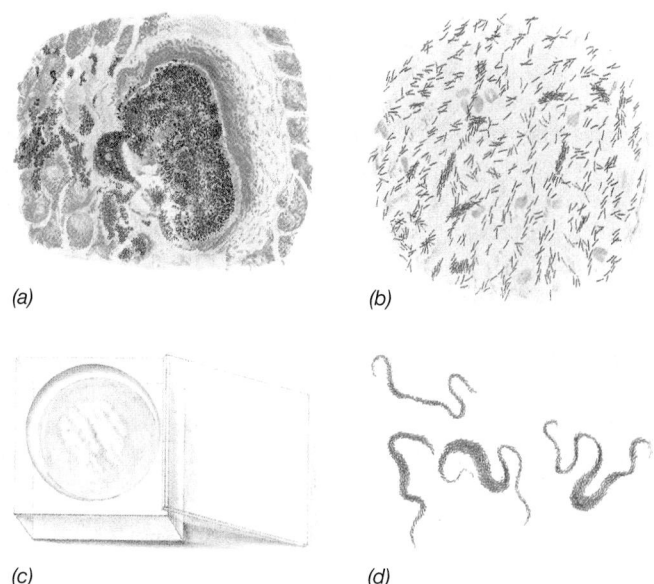

Figure 1.31 Robert Koch's drawings of *Mycobacterium tuberculosis*. (a) Section through infected lung tissue showing cells of *M. tuberculosis* (blue). (b) *M. tuberculosis* cells in a sputum sample from a tubercular patient. (c) Growth of *M. tuberculosis* on a glass plate of coagulated blood serum stored inside a glass box to prevent contamination. (d) *M. tuberculosis* cells taken from the plate in c and observed microscopically; cells appear as long, cordlike forms. Original drawings from Koch, R. 1884. "Die Aetiologie der Tuberkulose." *Mittheilungen aus dem Kaiserlichen Gesundheitsamte 2*: 1–88.

was not easy, but eventually Koch succeeded in growing colonies of this organism on a solidified medium containing blood serum. Under the best of conditions, *M. tuberculosis* grows slowly in culture, but Koch's persistence and patience eventually led to pure cultures of this organism from human and animal sources.

From this point Koch used his postulates (Figure 1.29) to obtain definitive proof that the organism he had isolated was the cause of the disease tuberculosis. Guinea pigs can be readily infected with *M. tuberculosis* and eventually succumb to systemic tuberculosis. Koch showed that tuberculous guinea pigs contained masses of *M. tuberculosis* cells in their lungs and that pure cultures obtained from such animals transmitted the disease to healthy animals. In this way, Koch successfully satisfied all four of his postulates, and the cause of tuberculosis was understood. Koch announced his discovery of the cause of tuberculosis in 1882, and for this accomplishment he was awarded the 1905 Nobel Prize for Physiology or Medicine. Koch had many other triumphs in the growing field of infectious diseases, including the discovery of the causative agent of cholera (the bacterium *Vibrio cholerae*) and the development of methods to diagnose infection with *M. tuberculosis* (the tuberculin skin test).

─── **MINIQUIZ** ───

- How do Koch's postulates ensure that cause and effect of a given disease are clearly differentiated?
- What advantages do solid media offer for the isolation of microorganisms?
- What is a pure culture?

1.11 Discovery of Microbial Diversity

As microbiology entered the twentieth century, its initial focus on basic principles, methods, and medical aspects broadened to include studies of the microbial diversity of soil and water and the metabolic processes that microorganisms carried out in these habitats. Major contributors of this era included the Dutchman Martinus Beijerinck and the Russian Sergei Winogradsky.

Martinus Beijerinck and the Enrichment Culture Technique

Martinus Beijerinck (1851–1931) was a professor at the Delft Polytechnic School in Holland and was originally trained in botany, so he began his career in microbiology studying plants. Beijerinck's greatest contribution to the field of microbiology was his clear formulation of the *enrichment culture technique*.

The media of Pasteur and Koch were rich in nutrients and while they supported the growth of many organisms, they did not select for specific types of organisms. In enrichment culture, microorganisms are isolated by using highly selective media and incubation conditions that favor a particular metabolic group of organisms. For example, Beijerinck devised a precise chemically defined medium to isolate rhizobia and prove that they are responsible for the formation of the root nodules of legumes (Figure 1.9).

Using the enrichment culture technique, Beijerinck isolated the first pure cultures of many soil and aquatic microorganisms, including sulfate-reducing and sulfur-oxidizing bacteria, lactic acid bacteria, green algae, various anaerobic bacteria, and many others. In addition, in his classic studies of "mosaic disease" of tobacco, Beijerinck used selective filters to show that the infectious agent in this disease (a virus) was smaller than a bacterium and that it somehow became incorporated into cells of the living host plant. In this insightful work, Beijerinck described not only the first virus but also the basic principles of virology, which we expand upon in Chapters 8 and 10.

Sergei Winogradsky, Chemolithotrophy, and Nitrogen Fixation

Like Beijerinck, Sergei Winogradsky (1856–1953) was interested in the bacterial diversity of soils and waters and was highly successful in isolating several notable bacteria from natural samples. Winogradsky was particularly interested in bacteria that cycle nitrogen and sulfur compounds, such as the nitrifying bacteria and the sulfur bacteria (**Figure 1.32**). He studied *Beggiatoa*, which are large bacteria commonly observed in marine sediments. He observed that *Beggiatoa* would not grow on the rich nutrient media used by Koch. He designed specific enrichment media to imitate the environment in which *Beggiatoa* lived. He showed that these bacteria catalyze specific chemical transformations in nature and proposed the important concept of **chemolithotrophy**, the oxidation of *inorganic* compounds to yield energy. Winogradsky further showed that these organisms, which he called *lithotrophs* (meaning, literally, "stone eaters"),

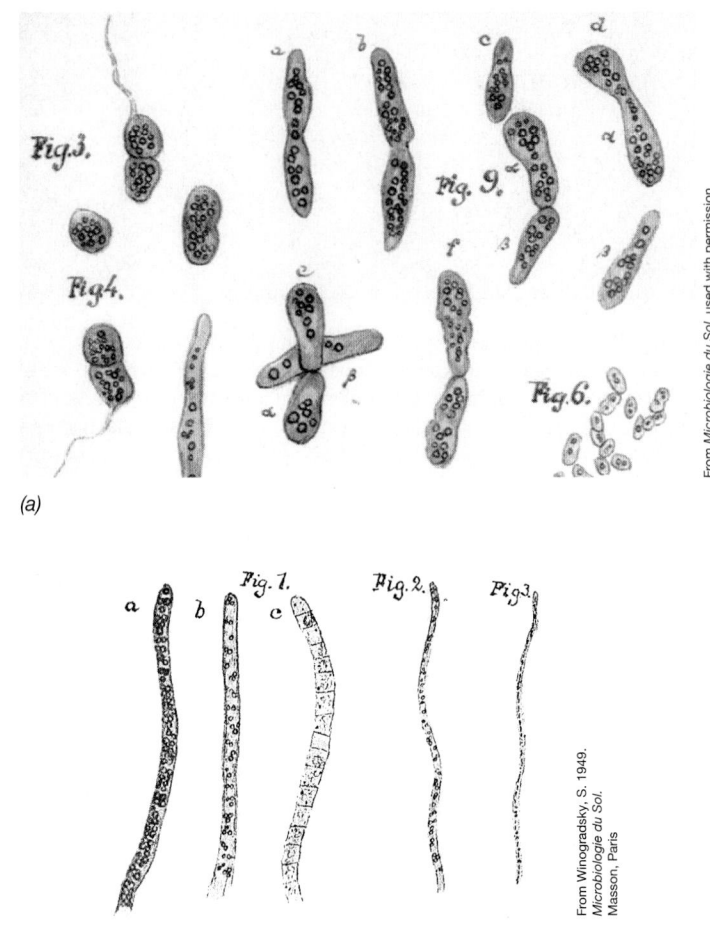

(a)

(b)

Figure 1.32 Sulfur bacteria. The original drawings were made by Sergei Winogradsky in the late 1880s and then copied and hand-colored by his wife Hèléne. *(a)* Purple sulfur phototrophic bacteria. Figures 3 and 4 show cells of *Chromatium okenii* (compare with photomicrographs of *C. okenii* in Figures 1.1a and 1.6d, and 1.16a). *(b) Beggiatoa,* a sulfur chemolithotroph (compare with Figure 15.27a).

are widespread in nature. Winogradsky thus revealed that, like photosynthetic organisms, chemolithotrophic bacteria obtain their carbon from CO_2.

Winogradsky isolated diverse metabolic types of bacteria. Using an enrichment medium that lacked nitrogen, he isolated the anaerobic nitrogen-fixing bacterium *Clostridium pasteurianum*, becoming the first to demonstrate the process of nitrogen fixation. Beijerinck would use a similar technique shortly thereafter to isolate the first aerobic nitrogen-fixing bacterium, *Azotobacter* (**Figure 1.33**). Winogradsky also isolated the first nitrifying bacteria by using an enrichment medium that contained ammonium salts and CO_2.

— **MINIQUIZ** —

- What is meant by the term "enrichment culture"?
- What is meant by the term "chemolithotrophy"? In what way are chemolithotrophs like plants?

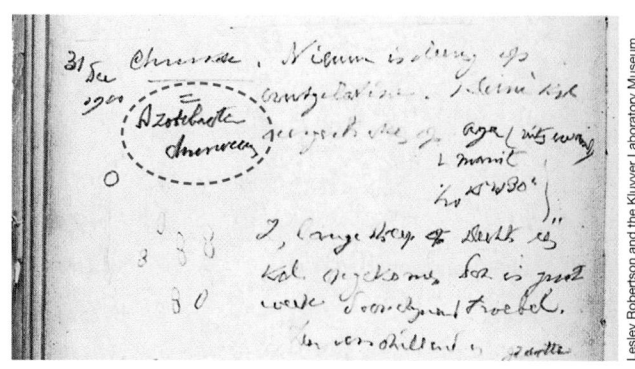

(a)

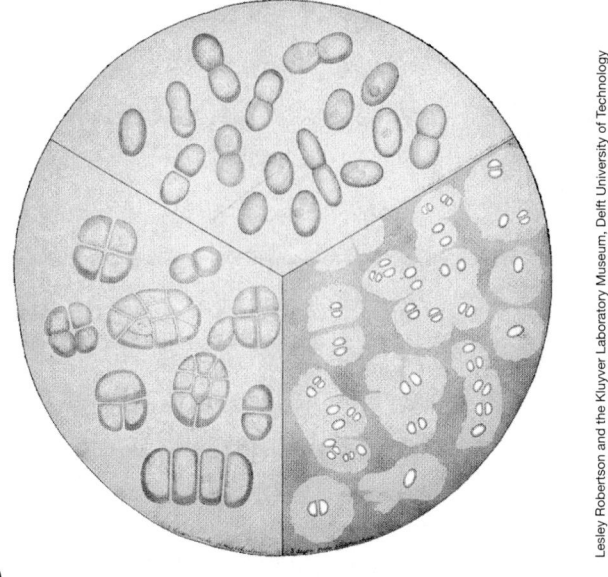

(b)

Figure 1.33 Martinus Beijerinck and *Azotobacter*. *(a)* A page from the laboratory notebook of M. Beijerinck dated 31 December 1900 describing the aerobic nitrogen-fixing bacterium *Azotobacter chroococcum* (name circled in red). Compare Beijerinck's drawings of pairs of *A. chroococcum* cells with the photomicrograph of cells of *Azotobacter* in Figure 15.32a. *(b)* A painting by M. Beijerinck's sister, Henriëtte Beijerinck, showing cells of *A. chroococcum*. Beijerinck used such paintings to illustrate his lectures.

IV • Molecular Biology and the Unity and Diversity of Life

The development of aseptic technique and methods for the enrichment, isolation, and propagation of bacteria at the end of the nineteenth century gave rise to explosive growth in the pace of microbiological discovery. Moreover, microbiologists realized that the ability to grow bacteria rapidly and in controlled laboratory conditions made them excellent model systems in which to explore the fundamental nature of life.

1.12 Molecular Basis of Life

Experiments with bacterial cultures in the twentieth century were critical in describing the foundations of molecular biology, molecular genetics, and biochemistry. Microbiologists came to realize that while microorganisms were incredibly diverse, all cells operated on similar principles.

Unity in Biochemistry

Albert Jan Kluyver (1888–1956) was Beijerinck's successor at what was then called the Delft Institute of Technology. Kluyver recognized that though microbial diversity was tremendous, microorganisms used many of the same biochemical pathways and their metabolic processes faced similar thermodynamic constraints. Kluyver promoted the study of comparative biochemistry to identify the unifying features of all cells. He famously proclaimed, "From elephant to butyric acid bacterium—it is all the same!" This was later reformulated by Jacques Monod (1910–1976) into the expression, "What is true for *E. coli* is also true for the elephant," a statement that proclaimed the importance of working with bacteria to understand the fundamental principles that govern all living things.

The use of microbes as metabolic model systems led to the discovery that certain macromolecules and biochemical reactions are universal, and that to understand their function in one cell is to understand their function in all cells. These discoveries were of central importance to understanding microbial evolution and none were as important as the discovery of DNA as the molecular basis of heredity, a discovery that is less than 80 years old.

Cracking the Code of Life

In the early twentieth century, it was clear that some molecule carried the hereditary information from parent to offspring, but the molecular basis of heredity remained a mystery. Most biologists thought that proteins carried this hereditary information. DNA had been discovered but it was thought to be merely a structural molecule, and too simple in its composition to encode cellular functions. The hunt for the molecular basis of heredity began in earnest with an experiment by Frederick Griffith (1879–1941).

Griffith worked with a virulent strain of *Streptococcus pneumoniae*, a cause of bacterial pneumonia in both humans and mice. This strain, strain S, produced a polysaccharide coat (that is, a capsule, ⮡ Section 2.7) that caused cells to form smooth colonies and conferred the ability to kill infected mice (**Figure 1.34a**). A related strain, strain R, lacked this polysaccharide and produced "rough" colonies that did not cause disease. However, Griffith observed that strain R could be *transformed* to type S, forming smooth colonies and causing disease, when it was mixed with the dead remains of cells of strain S (Figure 1.34a). He reasoned that some molecule that contained genetic information must have been transferred from strain R to strain S in this process, and this experiment showed that genetic transfer could be studied in bacteria.

Later, the Avery–MacLeod–McCarty experiment (1944), named for three scientists at the Rockefeller University, would show that this "transforming principle" is DNA. They treated the dead remains of cells of strain S with chemicals and enzymes that destroyed protein and left behind only DNA. They then repeated Griffith's experiment with the pure DNA of strain S and showed that this DNA was sufficient to cause transformation, causing strain R cells to become S-type cells and virulent (Figure 1.34b). They also demonstrated that transformation failed if the DNA

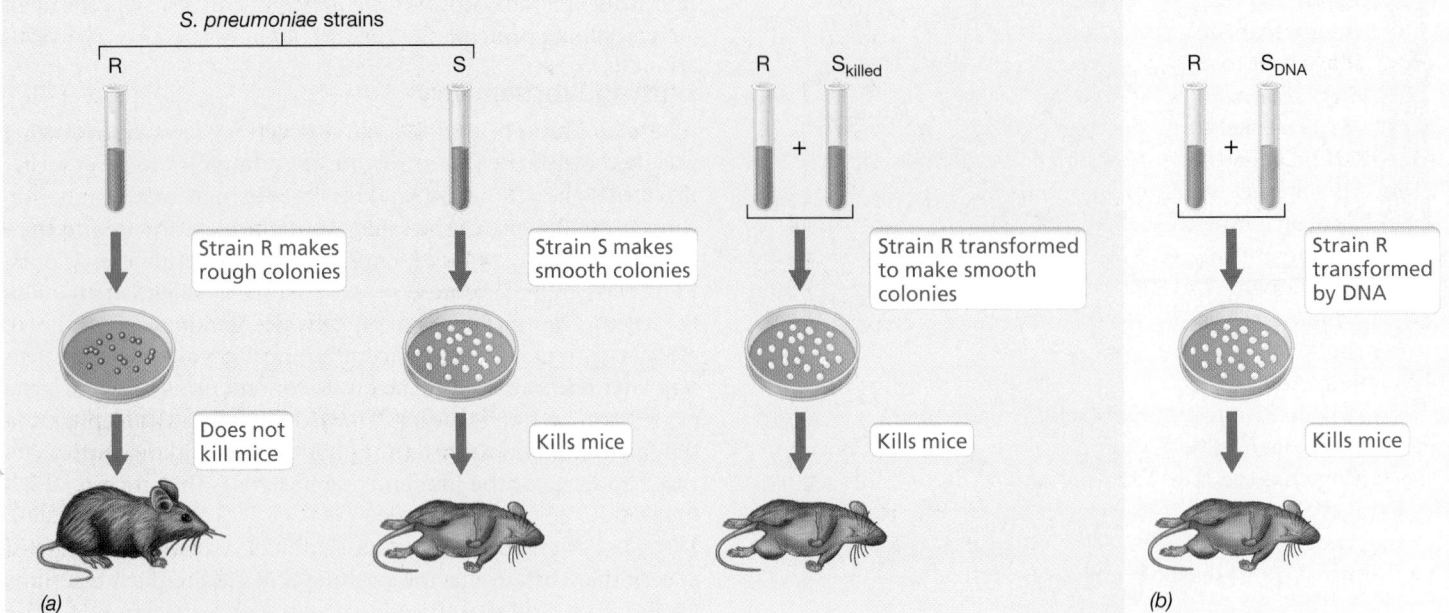

Figure 1.34 Early evidence that DNA is the molecular basis of heredity. (a) Griffith's experiment showed that bacteria can transfer genetic information. *Streptococcus pneumoniae* strain R makes rough colonies and does not kill mice, but strain S makes smooth colonies and does kill mice. Heat-killed cells of strain S do not cause disease, but if these killed cells are mixed with cells of strain R, then strain R is "transformed" to the S type and begins to make smooth colonies and kill mice. (b) The Avery–MacLeod–McCarty experiment showed that DNA contains genetic information. DNA isolated from strain S can transform strain R to cause disease, though the DNA itself does not cause disease. Degraded DNA lacks the ability to transform strain R.

from strain S was degraded. These experiments proved that DNA is the genetic material of cells.

The discovery that DNA is the basis of heredity was followed by intense effort to understand how this molecule stores genetic information. The structure of DNA was ultimately solved by James D. Watson (1928–) and Francis Crick (1916–2004) using X-ray diffraction images of DNA taken by their colleague Rosalind Franklin (1920–1958). They revealed that DNA is composed of a double helix that contains four nitrogenous bases: guanine, cytosine, adenine, and thymine (⊂⊃ Section 4.1). Later research would reveal how the genetic code is read from DNA and translated into a protein alphabet, and these principles are covered in Chapter 4. Once again, however, this research to crack the code of life was enabled by a microbial model system, in this case, the bacterium *Escherichia coli* (commonly called *E. coli*).

Not long after the discovery that genetic information is encoded in the sequence of biological molecules, Emile Zuckerkandl (1922–2013) and Linus Pauling (1901–1994) proposed that molecular sequences could be used to reconstruct evolutionary relationships. They recognized that evolution, as described by Darwin, required variation in offspring and that these variations must be caused by changes in molecular sequences. They predicted that these sequence differences occur randomly in a clocklike fashion over time. This led to the conclusion that the evolutionary history of organisms is inscribed in the sequence of molecules such as DNA. Carl Woese seized upon these insights to pursue the ambitious goal of reconstructing the evolutionary history of all cells.

MINIQUIZ

- Describe the experiments that proved DNA was the transforming principle described by Griffith.
- Why are microbial cells useful tools for basic science?

1.13 Woese and the Tree of Life

Evolutionary relationships between microorganisms remained a mystery until it was discovered that certain molecular sequences maintain a record of evolutionary history. Here we will examine how the sequence of **ribosomal RNA (rRNA)** genes, present in all cells, revolutionized the understanding of microbial evolution and made it possible to construct the first universal tree of life.

Molecular Sequence Data Has Revolutionized Microbial Phylogeny

For over a hundred years, following the 1859 publication of Charles Darwin's *On the Origin of Species*, evolutionary history was studied primarily with the tools of paleontology (through examining fossils) and comparative biology (through comparing the traits of living organisms). These approaches led to progress in understanding the evolution of plants and animals, but they were powerless to explain the evolution of microorganisms. The vast majority of microorganisms do not leave behind fossils, and their morphological and physiological traits provide few clues about their evolutionary history. Moreover, microorganisms do not share any morphological traits with plants and animals; thus it was impossible to create a robust evolutionary framework that included microorganisms.

The first attempt to depict the common evolutionary history of all living cells was published by Ernst Haeckel in 1866 (**Figure 1.35a**). Haeckel correctly suggested that single-cell organisms, which he called *Monera*, were ancestral to other forms of life, but his scheme, which included plants, animals, and protists, did not attempt to resolve evolutionary relationships among microorganisms. The situation was little changed as late as 1967 when Robert Whittaker proposed a five-kingdom classification scheme (Figure 1.35*b*). Whittaker's scheme distinguished the fungi as a distinct lineage, but it was still largely impossible to resolve evolutionary relationships among most microorganisms. Hence, microbial phylogeny had made little progress since Haeckel's day.

Everything changed after the structure of DNA was discovered and it was recognized that evolutionary history is recorded in DNA sequence. Carl Woese (1928–2012), a professor at the University of Illinois (USA), realized in the 1970s that the sequence of ribosomal RNA (rRNA) molecules and the genes that encode them could be used to infer evolutionary relationships between organisms. Ribosomal RNAs are components of ribosomes, the structures that synthesize new proteins in the process of translation (Section 1.2). Woese recognized that genes encoding rRNAs are excellent candidates for phylogenetic analysis because they are (1) universally distributed, (2) functionally constant, (3) highly conserved (that is, slowly changing), and (4) of adequate length to provide a deep view of evolutionary relationships.

Woese compared the sequences of rRNA molecules from many microorganisms. Among the microbes he examined were methanogens. To his astonishment, he found that the rRNA sequences from methanogens were distinct from those of both *Bacteria* and *Eukarya*, the only two domains recognized at that time. He named this new group of prokaryotic cells the *Archaea* (originally *Archaebacteria*) and recognized them as the third domain of life alongside the *Bacteria* and the *Eukarya* (**Figure 1.36b**). More importantly, Woese demonstrated that the analysis of rRNA gene sequences could be used to reveal evolutionary relationships

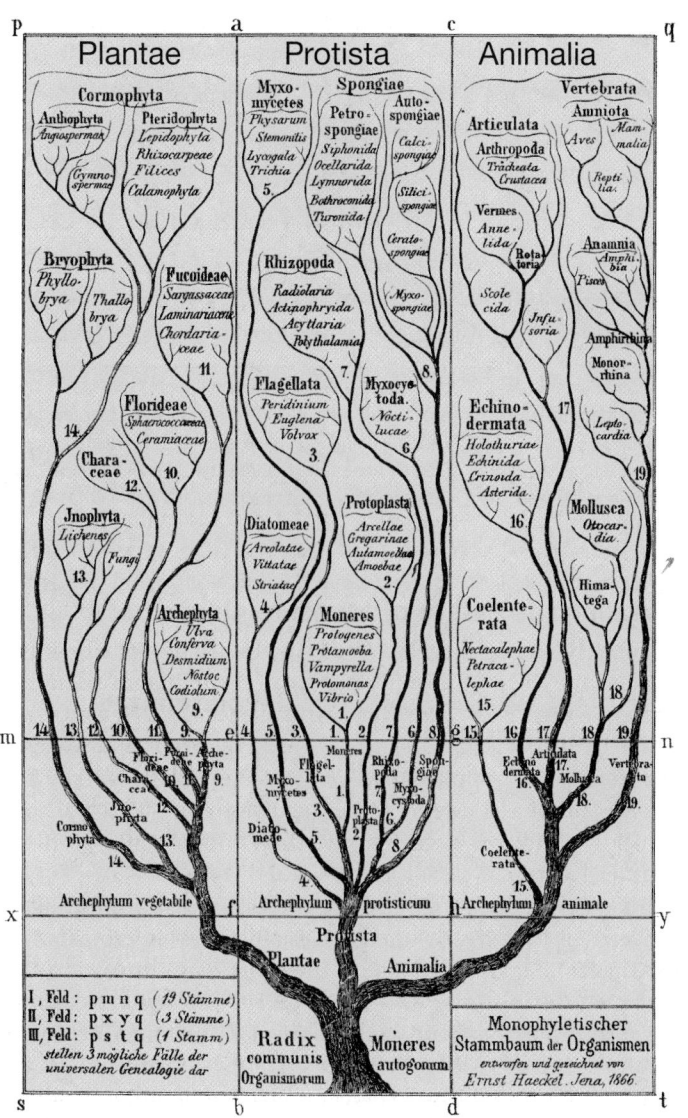

(a) **The Haeckel Tree**

(b) **The Whittaker Tree**

From *Science* 163:150-160, 1967
Reprinted with permission from AAAS

Figure 1.35 Early efforts to depict the universal tree of life. *(a)* Tree of life published in 1866 by Ernst Haeckel in *Generelle Morphologie der Organismen. (b)* Tree of life published by Robert H. Whittaker in 1969. The terms "Monera" and "Moneres" are antiquated terms used to refer to prokaryotic cells. Compare these conceptual trees with the tree generated from rRNA gene sequences in Figure 1.36b.

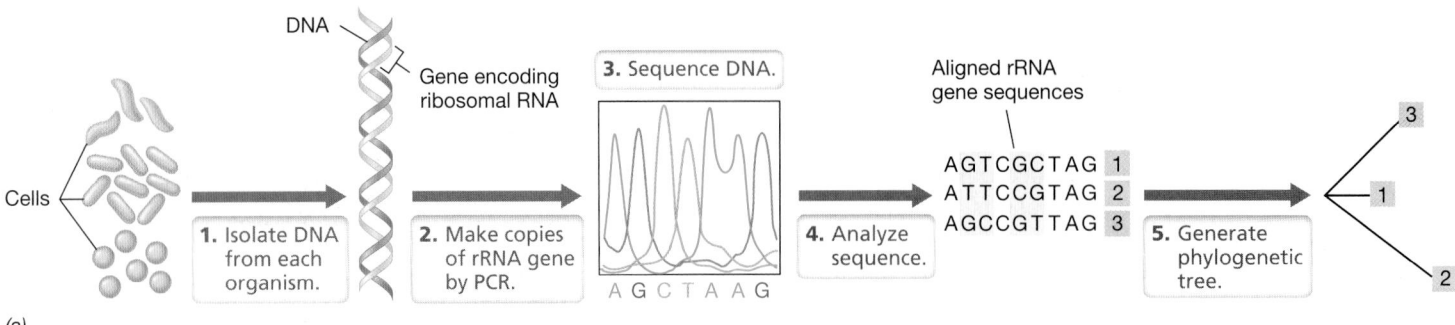

(a)

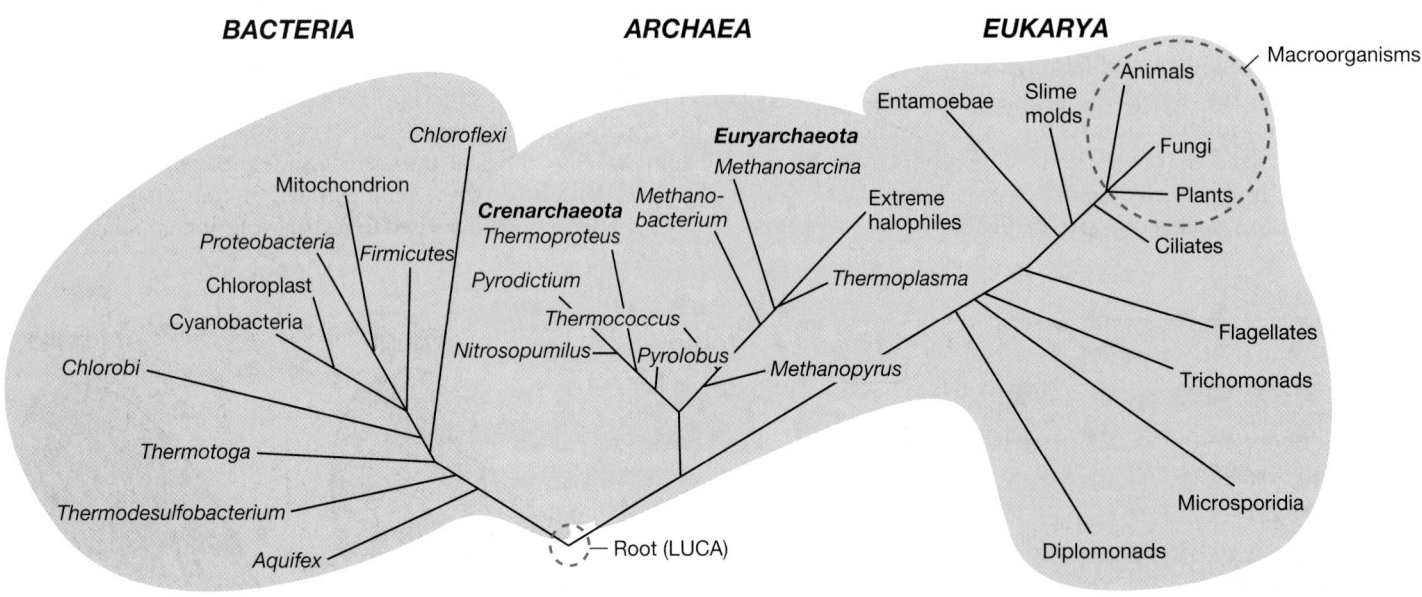

(b)

Figure 1.36 Evolutionary relationships and the phylogenetic tree of life. *(a)* The technology behind ribosomal RNA gene phylogenies. 1. DNA is extracted from cells. 2. Copies of the gene encoding rRNA are made by the polymerase chain reaction (PCR; 🔗 Section 12.1). 3, 4. The gene is sequenced and the sequence aligned with sequences from other organisms. A computer algorithm makes pairwise comparisons at each base and generates a phylogenetic tree, 5, that depicts evolutionary relationships. In the example shown, the sequence differences are highlighted in yellow and are as follows: organism 1 versus organism 2, three differences; 1 versus 3, two differences; 2 versus 3, four differences. Thus organisms 1 and 3 are closer relatives than are 2 and 3 or 1 and 2. *(b)* The phylogenetic tree of life. The tree shows the three domains of organisms and a few representative groups in each domain.

between *all cells*, providing the first effective tool for the evolutionary classification of microorganisms.

The Tree of Life Based on rRNA Genes

The universal tree of life based on rRNA gene sequences (Figure 1.36*b*) is a genealogy of all life on Earth. It is a true **phylogenetic tree**, a diagram that depicts the evolutionary history—the **phylogeny**—of all cells and clearly reveals the three domains. The root of the universal tree represents a point in time when all extant life on Earth shared a common ancestor, the last universal common ancestor, LUCA (Figures 1.5*b* and 1.36*b*). From the last universal common ancestor of all cells, evolution proceeded along two paths to form the domains *Bacteria* and *Archaea*. At some later time, the domain *Archaea* diverged to distinguish the *Eukarya* from the *Archaea* (Figures 1.5*b* and 1.36*b*). The three domains of cellular life are evolutionarily distinct and yet they share features indicative of their common descent from a universal cellular ancestor.

Revealing the Extent of Microbial Diversity

The tools Woese developed to build the tree of life were first used to determine the evolutionary history of microorganisms in pure culture (Figure 1.36*a*). However, Norman Pace (1942–), a professor at the University of Colorado (USA), realized that Woese's approach could be applied to rRNA molecules isolated *directly from the environment* as a way to probe the diversity of microbial communities without first cultivating their component organisms (Chapter 19).

The cultivation-independent methods of rRNA analysis pioneered by Pace greatly improved our picture of microbial diversity (**Figure 1.37**) and have led to the staggering conclusion that most microorganisms on Earth have yet to be brought into laboratory culture! Furthermore, because the ability of microbiologists to culture the microbial diversity that abounds in nature has lagged behind the ability to detect this diversity, microbiology is now in a position to flesh out the true diversity of microbial life.

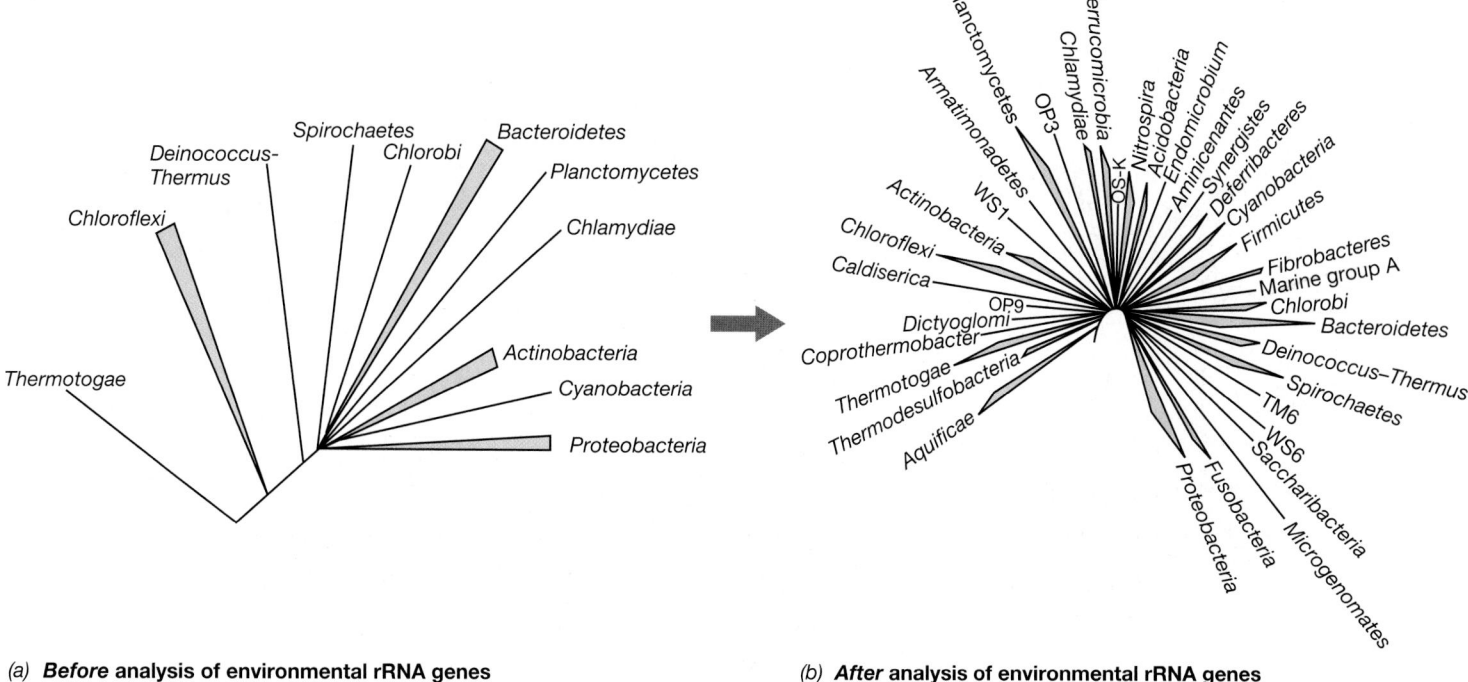

(a) **Before** analysis of environmental rRNA genes

(b) **After** analysis of environmental rRNA genes

Figure 1.37 Analysis of environmental rRNA genes leads to discovery of new phyla of *Bacteria*.
(a) In 1987 Carl Woese described 11 phyla of *Bacteria* from analysis of rRNA genes from cultured species. *(b)* By 1998, analyses of rRNA genes from environmental samples, as described by Norman Pace, had revealed evidence for 36 bacterial phyla. Today there is evidence for more than 80 bacterial phyla.

With an evolutionary framework of the microbial world to guide future research, advances in microbial diversity, both in obtaining cultures and in devising even more powerful methods of assessing diversity, are happening quickly. Besides unveiling the previously hidden concept of three evolutionary domains of life, the contributions of Carl Woese and his associates have given microbiologists the tools they need to understand the scope of microbial diversity at a level similar to biologists' understanding of the diversity of plants and animals.

MINIQUIZ

- What kinds of evidence support the three-domain concept of life?
- What is a phylogenetic tree?
- List three reasons why rRNA genes are suitable for phylogenetic analyses.

1.14 An Introduction to Microbial Life

All cells are unified by the facts that their genetic blueprints are encoded in DNA (Section 1.12) and that evolution is the process by which their blueprints change over time (Section 1.13). We now move on from these fundamental unifying principles to the microbes themselves and take a peek at the diversity of microbial life that evolution has generated.

Microorganisms vary dramatically in size, shape, and structure. And, while much of our focus in this chapter has been on cellular forms of life, not all microbes form cells. In this section we will learn about *Bacteria*, *Archaea*, *Eukarya*, and viruses—the four groups into which all known microorganisms can be classified.

Bacteria

Bacteria have a prokaryotic cell structure (Figure 1.3*a*). Bacteria are often thought of as undifferentiated single cells with a length that ranges from 1 to 10 μm. While bacteria that fit this description are common, the *Bacteria* are actually tremendously diverse in appearance and function. The smallest bacteria are no more than 0.15–0.2 μm in diameter and the largest can be as much as 700 μm long (**Figure 1.38**)! Some bacteria can differentiate to form multiple cell types and others are even multicellular (for example, *Magnetoglobus*, Figure 1.38).

Among the *Bacteria*, 30 major phylogenetic lineages (called phyla) have been described, and some key ones are shown in Figures 1.36 and 1.37. Some of these phyla contain thousands of described species while others contain only a few. More than 90% of bacteria in cultivation belong to one of only four phyla: *Actinobacteria*, *Firmicutes*, *Proteobacteria*, and *Bacteroidetes*. The analysis of rRNA gene sequences and even entire genome sequences from environmental samples reveals that at least 80 bacterial phyla likely exist.

Although species in some bacterial phyla are characterized by unique phenotypic traits, most bacterial phyla contain a wide diversity of species and show tremendous physiological diversity. The *Proteobacteria* illustrate this concept well as they include organisms with a diverse array of physiological traits including

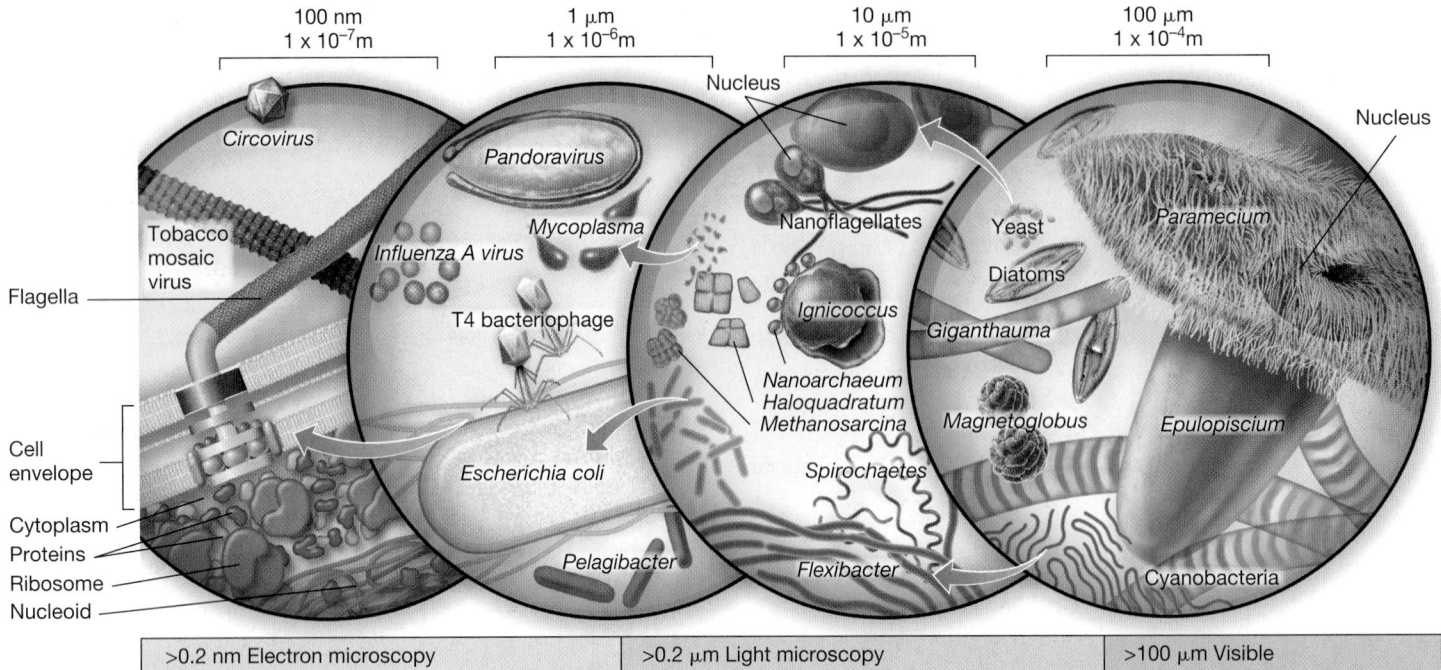

Figure 1.38 Microorganisms vary greatly in size and shape. The smallest known microbe is the circovirus (20 nm) and the largest shown here is the bacterium *Epulopiscium* (700 μm), which represents a 35,000-fold difference in length! Certain protozoa can be even larger than *Epulopiscium* (>2 mm long) and are visible to the unaided eye. Included in the figure are *Eukarya*: *Paramecium* (300 μm × 85 μm), diatoms (*Navicula*, 50 μm × 12 μm), yeast (*Saccharomyces*, 5 μm), and nanoflagellates (*Cafeteria*, 2 μm); *Bacteria*: *Epulopiscium* (700 μm × 80 μm), cyanobacteria (*Oscillatoria*, 10-μm-diameter multicellular filaments), *Magnetoglobus* (multicellular aggregate, 20 μm diameter), *Spirochaetes* (2–10 μm × 0.25 μm), *Flexibacter* (5–100 μm × 0.5 μm filaments), *Escherichia coli* (2 μm × 0.5 μm), *Pelagibacter* (0.4 μm × 0.15 μm), and *Mycoplasma* (0.2 μm); *Archaea*: *Giganthauma* (10-μm-diameter multicellular filament), *Ignicoccus* (6 μm), *Nanoarchaeum* (0.4 μm), *Haloquadratum* (2 μm), *Methanosarcina* (2 μm per cell in packet); and viruses: *Pandoravirus* (1 μm × 0.4 μm), T4 bacteriophage (200 nm × 90 nm), *Influenza A virus* (100 nm), *Tobacco mosaic virus* (300 nm × 20 nm), *Circovirus* (20 nm).

respiration (both with and without oxygen), fermentations of various types, diverse forms of phototrophy, and chemolithotrophic metabolisms using H_2, sulfur or nitrogen compounds, or even metals (as described in Chapters 14 and 15). Species of *Proteobacteria* also possess a wide range of ecological strategies and can be found in all but the hottest and most salty environments on Earth. It is important to remember that while most phyla of plants and animals originated within the last 600 million years (Figure 1.5a), bacterial phyla are billions of years old and this time has allowed for extensive experimentation and diversification. The diversity of *Bacteria* is discussed in detail in Chapters 15 and 16.

Archaea

Like *Bacteria*, *Archaea* also have a prokaryotic cell structure (Figure 1.3a). While *Archaea* are quite diverse in their physiology, cultured isolates have less morphological diversity than *Bacteria*, and most described *Archaea* exist as undifferentiated cells that are 1 to 10 μm in length. The domain *Archaea* consists of five well-described phyla: *Euryarchaeota, Crenarchaeota, Thaumarchaeota, Nanoarchaeota,* and *Korarchaeota*. As for the *Bacteria*, many lineages of *Archaea* are known only from rRNA genes or genome sequences recovered from the environment. Analysis of these environmental DNA sequences indicate more than 12 archaeal phyla likely exist.

Archaea have historically been associated with extreme environments; the first isolates came from hot, salty, or acidic sites. But not all *Archaea* are extremophiles. *Archaea* are indeed common in

the most extreme environments that support life, such as those associated with volcanic systems, and species of *Archaea* hold many of the records that define the chemical and physical limits of life (Table 1.1). But in addition to these, *Archaea* are found widely in nature. For example, methanogens are common in wetlands and in the guts of animals (including humans). Methanogenic *Archaea* produce methane and have a major impact on the greenhouse gas composition of our atmosphere. In addition, species of *Thaumarchaeota* inhabit soils and oceans worldwide and are important contributors to the global nitrogen cycle (Chapter 17).

Archaea are also notable in that this domain lacks any known pathogens or parasites of plants or animals. Most described species of *Archaea* fall within the phyla *Crenarchaeota* and *Euryarchaeota* (Figure 1.36b) while only a handful of species have been described for the *Nanoarchaeota, Korarchaeota,* and *Thaumarchaeota*. We discuss *Archaea* in detail in Chapter 17.

Eukarya

Plants, animals, and fungi are the most well-characterized groups of *Eukarya*. These groups are relatively young in relation to *Bacteria* and *Archaea*, originating during a burst of evolutionary radiation called the Cambrian explosion, which began about 600 million years ago. The first eukaryotes, however, were unicellular microbes. Microbial eukaryotes, which include diverse algae and protozoa, may have first appeared as early as 2 billion years ago, well before the origin of plants, animals, and fungi (Figure 1.5). The major lineages of *Eukarya* are traditionally called kingdoms instead of phyla. There are at least six kingdoms of *Eukarya*, and this diverse domain contains microorganisms as well as the plants and animals.

Microbial eukaryotes vary dramatically in size, shape, and physiology (Figure 1.38). Among the smallest are the nanoflagellates, which are microbial predators that can be as small as 2 μm long. In addition, *Ostreococcus*, a genus of green algae that contains species that are 0.8 μm in diameter, is smaller than many bacteria. The largest single-celled organisms are eukaryotes, but they are hardly microbial. Xenophyophores are amoeba-like, single-celled organisms that live exclusively in the deep ocean. Exploration of the Mariana Trench has revealed xenophyophores up to 10 cm in length. In addition, plasmodial slime molds consisting of a single cytoplasmic compartment can be up to 30 cm in diameter. Microbial eukaryotes include diverse phototrophic organisms, microbial predators, symbionts and parasites, along with a range of other physiological types. In Chapter 18 we consider microbial eukaryotes in detail.

Viruses

Viruses are not found on the tree of life. Indeed, it can be argued that they are not truly alive. Viruses are obligate parasites that can only replicate within the cytoplasm of a host cell. Viruses are not cells, and they lack the cytoplasmic membrane, cytoplasm, and ribosomes found in all forms of cellular life. Viruses cannot conserve energy and they do not carry out metabolic processes; instead, they take over the metabolic systems of infected cells and turn them into vessels for producing more viruses. Unlike cells,

which all have genomes composed of double-stranded DNA, viruses have genomes composed of DNA *or* RNA that can be either double- or single-stranded. Viral genomes are often quite small, with the smallest having only three genes. The small size of most viral genomes means that no genes are conserved among all viruses, or between all viruses and all cells; hence it may be impossible to ever place viruses into the tree of life or build a universal viral phylogenetic tree that includes all viruses.

Viruses are as diverse as the cells they infect, and viruses are known to infect cells from all three domains of life. Viruses are often classified on the basis of structure, genome composition, and host specificity. Viruses that infect bacteria are called *bacteriophages* (or *phages*, for short). Bacteriophages have been used as model systems to explore many aspects of viral biology. While most viruses are considerably smaller than bacterial cells (Figure 1.38), there are also unusually large viruses such as the *Pandoraviruses*, which can be more than 1 micrometer long and have a genome of as many as 2500 genes, larger than that of many bacteria! We will learn more about viruses in Chapters 8 and 10.

— MINIQUIZ —

- How are viruses different from *Bacteria*, *Archaea*, and *Eukarya*?
- What four bacterial phyla contain the most well-characterized species?
- What phylum of *Archaea* is common worldwide in soils and in the oceans?

MasteringMicrobiology®

Visualize, explore, and **think critically** with Interactive Microbiology, MicroLab Tutors, MicroCareers case studies, and more. MasteringMicrobiology offers practice quizzes, helpful animations, and other study tools for lecture and lab to help you master microbiology.

Chapter Review

I • Exploring the Microbial World

1.1   Microorganisms are single-celled microscopic organisms that are essential for the well-being and functioning of other life forms and the planet. The tools of microscopy, microbial cultivation, molecular biology, and genomics are cornerstones of modern microbiology.

> **Q What are bacterial colonies and how are they formed?**

1.2   Prokaryotic and eukaryotic cells differ in cellular architecture, and an organism's characteristics are defined by its complement of genes—its genome. All cells have a cytoplasmic membrane, a cytoplasm, ribosomes, and a double-stranded DNA genome. All cells carry out activities including metabolism, growth, and evolution.

> **Q What cellular structures distinguish prokaryotic and eukaryotic cells? What are some differences between a**

cell wall and a cell membrane? In what types of organisms would you expect to find these structures?

1.3   Diverse microbial populations were widespread on Earth for billions of years before plants and animals appeared. Microbes are abundant in the biosphere and their activities greatly affect the chemical and physical properties of their habitats. *Bacteria, Archaea*, and *Eukarya* are the major phylogenetic lineages (domains) of cells.

> **Q How has Earth changed over its history? How have microorganisms contributed to these changes?**

1.4   Microorganisms can be both beneficial and harmful to humans, although many more microorganisms are beneficial (or even essential) than are harmful. Agriculture, food, energy, and the environment are all affected in major ways by microorganisms.

> **Q How would you convince a friend that microorganisms are much more than just agents of disease? What are some of the benefits that microbes provide?**

II • Microscopy and the Origins of Microbiology

1.5 Microscopes are essential for studying microorganisms. Bright-field microscopy, the most common form of microscopy, employs a microscope with a series of lenses to magnify and resolve the image. The limit of resolution for a light microscope is about 0.2 μm.

> **Q** What is the difference between magnification and resolution? Can either increase without the other?

1.6 An inherent limitation of bright-field microscopy is the lack of contrast between cells and their surroundings. This problem can be overcome by the use of stains or by alternative forms of light microscopy, such as phase contrast or dark field.

> **Q** What is the function of staining in light microscopy? What is the advantage of phase-contrast microscopy over bright-field microscopy?

1.7 Differential interference contrast (DIC) microscopy and confocal scanning laser microscopy allow enhanced three-dimensional imaging or imaging through thick specimens.

> **Q** How is confocal scanning laser microscopy different from fluorescence microscopy? In what ways are they similar? How does differential interference contrast microscopy differ from bright-field microscopy?

1.8 Electron microscopes have far greater resolving power than do light microscopes, the limits of resolution being about 0.2 nm. The two major forms of electron microscopy are transmission, used primarily to observe internal cell structure, and scanning, used to examine the surface of specimens.

> **Q** What are the major differences between electron microscopes and light microscopes? What type of electron microscope would be used to view the three-dimensional features of a cell?

III • Microbial Cultivation Expands the Horizon of Microbiology

1.9 Louis Pasteur devised ingenious experiments proving that living organisms cannot arise spontaneously from nonliving matter. Pasteur introduced many concepts and techniques central to the science of microbiology, including sterilization, and developed a number of key vaccines for humans and other animals.

> **Q** Explain the principle behind the Pasteur flask in studies on spontaneous generation. Why were the results of this experiment inconsistent with the theory of spontaneous generation?

1.10 Robert Koch developed a set of criteria called Koch's postulates for linking cause and effect in infectious diseases. Koch also developed the first reliable and reproducible means for obtaining and maintaining microorganisms in pure culture.

> **Q** What are Koch's postulates and how did they influence the development of microbiology? Why are Koch's postulates still relevant today?

1.11 Martinus Beijerinck and Sergei Winogradsky explored soil and water for microorganisms that carry out important natural processes, such as nutrient cycling and the biodegradation of particular substances. Out of their work came the enrichment culture technique and the concepts of chemolithotrophy and nitrogen fixation.

> **Q** What were the major microbiological interests of Martinus Beijerinck and Sergei Winogradsky? It can be said that both men discovered nitrogen fixation. Explain.

IV • Molecular Biology and the Unity and Diversity of Life

1.12 All cells share certain characteristics, and microorganisms are used as model systems to explore the fundamental processes that define life. The discoveries of DNA as the molecular basis of heredity, and of its structure and function, paved the way for progress in molecular genetics, microbial phylogeny, and genomics.

> **Q** Describe the experiments that proved DNA to be the molecule at the basis of heredity.

1.13 Carl Woese discovered that ribosomal RNA (rRNA) sequences can be used to determine the evolutionary history of microorganisms, and in so doing, he diagrammed the tree of life and discovered the domain *Archaea*. Analysis of rRNA sequences from the environment reveals that microbial diversity is exceptional and that the majority of microorganisms have not yet been cultivated.

> **Q** What insights led to the reconstruction of the tree of life? Which domain, *Archaea* or *Eukarya*, is more closely related to *Bacteria*? What evidence is there to justify your answer?

1.14 The greatest diversity of microorganisms is found in the *Bacteria*, while many extremophiles are found within the *Archaea*. Microbial eukaryotes can vary tremendously in size, with some species being smaller than bacteria. Viruses are acellular and because of this cannot be placed on the tree of life.

> **Q** What features (or lack of features) can be used to distinguish between viruses, *Bacteria*, *Archaea*, and *Eukarya*?

Application Questions

1. Pasteur's experiments on spontaneous generation contributed to the methodology of microbiology, understanding of the origin of life, and techniques for the preservation of food. Explain briefly how Pasteur's experiments affected each of these topics.

2. Describe the lines of proof Robert Koch used to definitively associate the bacterium *Mycobacterium tuberculosis* with the disease tuberculosis. How would his proof have been flawed if any of the tools he developed for studying bacterial diseases had not been available for his study of tuberculosis?

3. Imagine that all microorganisms suddenly disappeared from Earth. From what you have learned in this chapter, why do you think that animals would eventually disappear from Earth? Why would plants disappear? By contrast, if all higher organisms suddenly disappeared, what in Figure 1.5*a* tells you that a similar fate would not befall microorganisms?

Chapter Glossary

Aseptic technique the manipulation of sterile instruments or culture media in such a way as to maintain sterility

Cell wall a rigid layer present outside the cytoplasmic membrane; it confers structural strength on the cell

Chemolithotrophy a form of metabolism in which energy is generated from the oxidation of inorganic compounds

Colony a macroscopically visible population of cells growing on solid medium, arising from a single cell

Contrast the ability to resolve a cell or structure from its surroundings

Culture a collection of microbial cells grown using a nutrient medium

Cytoplasm the fluid portion of a cell, enclosed by the cytoplasmic membrane

Cytoplasmic membrane a semipermeable barrier that separates the cell interior (cytoplasm) from the environment

Differentiation modification of cellular components to form a new structure, such as a spore

Domain one of the three main evolutionary lineages of cells: the *Bacteria*, the *Archaea*, and the *Eukarya*

DNA replication the process by which information from DNA is copied into a new strand of DNA

Enrichment culture technique a method for isolating specific microorganisms from nature using specific culture media and incubation conditions

Enzyme a protein (or in some cases an RNA) catalyst that functions to speed up chemical reactions

Eukaryotic having a membrane-enclosed nucleus and various other membrane-enclosed organelles; cells of *Eukarya*

Evolution a change over time in gene sequence and frequency within a population of organisms, resulting in descent with modification

Extremophiles microorganisms that inhabit environments characterized by extremes of temperature, pH, pressure, or salinity

Genome an organism's full complement of genes

Gram-negative a bacterial cell with a cell wall containing small amounts of peptidoglycan and an outer membrane

Gram-positive a bacterial cell whose cell wall consists chiefly of peptidoglycan; it lacks the outer membrane of gram-negative cells

Gram stain a differential staining procedure that stains cells either purple (gram-positive cells) or pink (gram-negative cells)

Growth in microbiology, an increase in cell number with time

Gut microbiome the microbial communities present in the animal gastrointestinal tract

Horizontal gene transfer the transfer of genes between cells through a process uncoupled from reproduction

Intercellular communication interactions between cells using chemical signals

Koch's postulates a set of criteria for proving that a given microorganism causes a given disease

Macromolecules a polymer of monomeric units, for example proteins, nucleic acids, polysaccharides, and lipids

Magnification the optical enlargement of an image

Medium (plural, media) in microbiology, the liquid or solid nutrient mixture(s) used to grow microorganisms

Metabolism all biochemical reactions in a cell

Microbial community two or more populations of cells that coexist and interact in a habitat

Microbial ecology the study of microorganisms in their natural environments

Microorganism an organism that is too small to be seen by the unaided human eye

Motility the movement of cells by some form of self-propulsion

Nucleoid the aggregated mass of DNA that makes up the chromosome(s) of prokaryotic cells

Nucleus a membrane-enclosed structure in eukaryotic cells that contains the cell's DNA genome

Organelle a bilayer-membrane-enclosed structure such as the mitochondrion, found in eukaryotic cells

Pathogen a disease-causing microorganism

Phylogenetic tree a diagram that depicts the evolutionary history of organisms

Phylogeny the evolutionary history of organisms

Plasmid an extrachromosomal genetic element that is not essential for growth

Prokaryotic lacking a membrane-enclosed nucleus and other organelles; cells of *Bacteria* or *Archaea*

Pure culture a culture containing a single kind of microorganism

Resolution the ability to distinguish two objects as distinct and separate when viewed under the microscope

Ribosomes a structure composed of RNAs and proteins upon which new proteins are made

Ribosomal RNA (rRNA) the types of RNA found in the ribosome

Spontaneous generation the hypothesis that living organisms can originate from nonliving matter

Sterile free of all living organisms (cells) and viruses

Transcription the synthesis of an RNA molecule complementary to one of the two strands of a double-stranded DNA molecule

Translation the synthesis of protein by a ribosome using the genetic information in a messenger RNA as a template

Microbial Cell Structure and Function

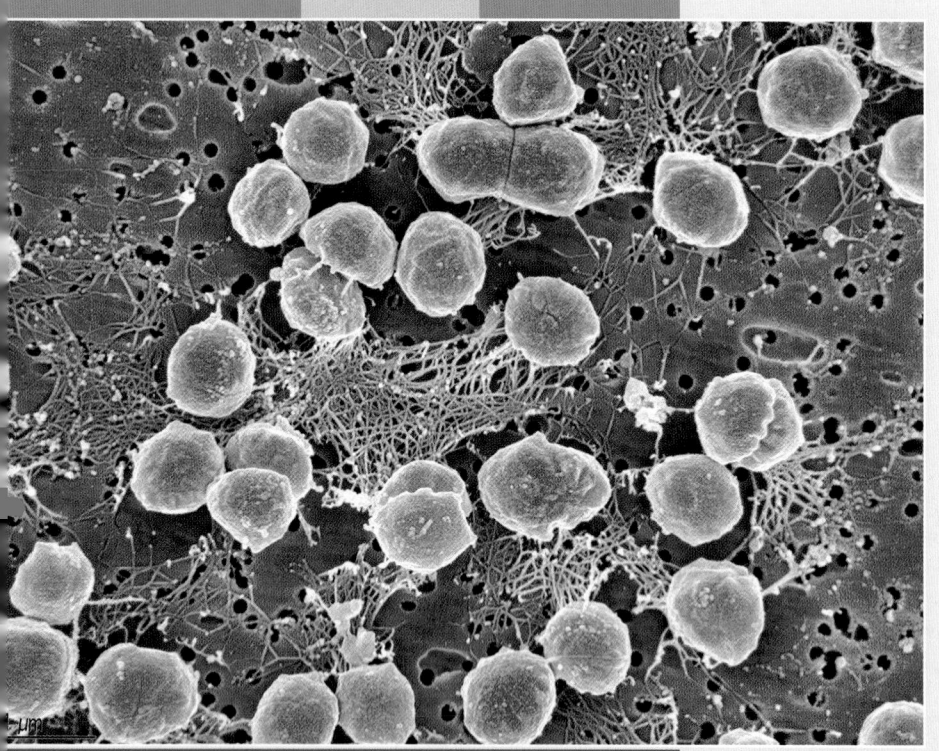

microbiology**now**

The Archaellum: Motility for the *Archaea*

Motility is important for microbes because it allows cells to explore new habitats and exploit their resources. Motility has been studied for over 50 years in the bacterium *Escherichia coli*, and it is with *E. coli* that scientists first discovered that the bacterial flagellum rotates like a propeller and is powered not by ATP but by the proton motive force. Subsequent studies of other motile *Bacteria* indicated that the structure and function of the flagellum was highly conserved.

When *Archaea* were first discovered, it was clear that some species, such as the archaeon *Methanocaldococcus* (see photo), were also motile and that their flagella also functioned by rotating. So it was only natural to assume that archaeal flagella were structurally related to bacterial flagella. But as scientists began to pick apart the archaeal flagellum, they were in for a big surprise.

Studies showed that archaeal flagella—now called *archaella* to distinguish them from flagella—were thinner than their bacterial counterparts and were composed not of one major protein but of several proteins. For example, whereas the flagellar filament (the actual rotating structure) is composed of a single type of protein called *flagellin*, the archaellar filament is composed of at least three proteins, none of which are structurally related to flagellin. Motor proteins in the flagellum and archaellum are also distinct, as is the overall structure of the motor.

Despite these clear differences, genomic studies of motile *Archaea* surprisingly revealed that the archaellum did have a structural counterpart in *Bacteria*—the type IV pilus. In *Bacteria*, type IV pili do not rotate but facilitate "twitching motility" and are often important in attaching pathogenic (disease-causing) bacteria to their host tissues. However, unlike the flagellum, the activities of type IV pili in *Bacteria* are powered by ATP, and ATP also drives rotation of the archaellum.

The archaellum is thus a "rotating type IV pilus" and is a good example of how evolution can modify a single structure to drive different functions. Such discoveries also demonstrate how even firmly entrenched paradigms (such as the structure and function of the bacterial flagellum) can be turned upside down when scientists begin to probe phylogenetically distinct species.

 Source: Albers, S-V., and K.F. Jarrell. 2015. The archaellum: How *Archaea* swim. *Front. Microbiol. 6:* doi: 10.3389/fmicb.2015.00023.

I • Cells of *Bacteria* and *Archaea*

In the opening chapter, we painted a picture of the microbial world using a broad brush. There we considered in a very general way several key aspects of microbiology essential to a modern understanding of the science. In Chapter 2, we move on to begin a more detailed examination of microbial life, with a focus on cell structure and function.

Microscopic examination of microorganisms immediately reveals their shape and size. A variety of cell shapes pervade the microbial world, and although microscopic by their very nature, microbial cells—both prokaryotic and eukaryotic—come in a variety of sizes. Cell shape can be useful for distinguishing different microbial cells and often has ecological significance. Moreover, the very small size of most microbial cells has a profound effect on their ecology and dictates many aspects of their biology. We begin by considering cell shape and then consider cell size.

2.1 Cell Morphology

In microbiology, the term **morphology** means cell shape. Several morphologies are found among *Bacteria* and *Archaea*, and the most common ones are described by terms that are part of the essential lexicon of the microbiologist.

Major Morphologies of Prokaryotic Cells

Common morphologies of prokaryotic cells are shown in **Figure 2.1**. A cell that is spherical or ovoid in morphology is called a *coccus* (plural, cocci). A cylindrically shaped cell is called a *rod* or a *bacillus*. Some cells form curved or loose spiral shapes and are called *spirilla*. The cells of some *Bacteria* and *Archaea* remain together in groups or clusters after cell division, and the arrangements are often characteristic. For instance, some cocci form long chains (for example, the bacterium *Streptococcus*), others occur in three-dimensional cubes (*Sarcina*), and still others in grapelike clusters (*Staphylococcus*).

Some morphological groups are immediately recognizable by the unusual shapes of their individual cells. Examples include the spirochetes, which are tightly coiled *Bacteria*; bacteria that form extensions of their cells as long tubes or stalks (appendaged forms); and filamentous bacteria, which form long, thin cells or chains of cells (Figure 2.1).

The cell morphologies described here are representative but certainly not exhaustive; many variations of these morphologies are known. For example, there can be fat rods, thin rods, short rods, and long rods, a rod simply being a cell—roughly in the shape of a cylinder—that is longer in one dimension than in the other. As we will see, there are even square bacteria and star-shaped bacteria! Cell morphologies thus form a continuum, with some shapes, such as rods and cocci, being very common, whereas others, such as spiral, budding, and filamentous shapes, are less common.

Morphology and Biology

Although cell morphology is easily determined, it is a poor predictor of other properties of a cell. For example, under the microscope many rod-shaped *Archaea* are indistinguishable from rod-shaped *Bacteria*, yet we know they are of different phylogenetic domains (↩ Section 1.13). With rare exceptions, it is impossible to predict the physiology, ecology, phylogeny, pathogenic (disease-causing) potential, or virtually any other major property of a prokaryotic cell by simply knowing its morphology. Nevertheless, cell morphology is an important characteristic that is always noted when describing a particular species of *Bacteria* or *Archaea*.

Why are the cells of a given species the shape they are? Although we know quite a bit about *how* cell shape is controlled, we know relatively little about *why* a particular cell displays the morphology it does. The morphology of a given microbe is undoubtedly the result of the selective forces that have shaped its evolution to

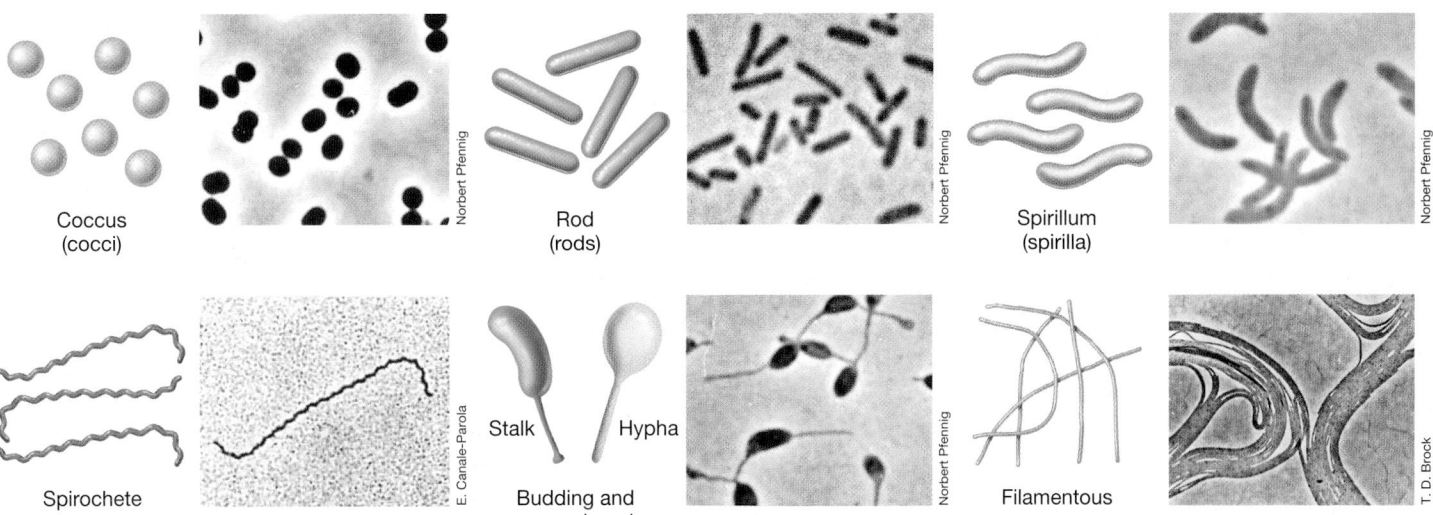

Figure 2.1 Cell morphologies. Beside each drawing is a phase-contrast photomicrograph of cells showing that morphology. Coccus (cell diameter in photomicrograph, 1.5 μm); rod (1 μm); spirillum (1 μm); spirochete (0.25 μm); budding (1.2 μm); filamentous (0.8 μm). All photomicrographs are of species of *Bacteria*. Not all of these morphologies are known among the *Archaea*, but cocci, rods, and spirilla are common.

maximize fitness for competitive success in its habitat. Some examples of these might include evolving an optimal cell shape to maximize nutrient uptake for survival in nutrient-limiting environments (small cells and others with high surface-to-volume ratios, such as appendaged cells), evolving a morphology to exploit swimming motility in viscous environments (helical- or spiral-shaped cells), or evolving a morphology that facilitates gliding motility along a surface (filamentous bacteria) (Figure 2.1).

MINIQUIZ

- How do cocci and rods differ in morphology?
- Using a microscope, could you differentiate a coccus from a spirillum? A pathogen from a nonpathogen?

2.2 The Small World

Cells of *Bacteria* and *Archaea* vary in size from as small as about 0.2 micrometer (μm) in diameter to those more than 700 μm in diameter (Table 2.1). The vast majority of rod-shaped species that have been cultured are between 0.5 and 4 μm wide and less than 15 μm long. A few very large *Bacteria* are known, such as *Epulopiscium fishelsoni*, whose cells exceed 600 μm (0.6 millimeter) in length (**Figure 2.2a**; ⇌ Figure 1.38). This bacterium, phylogenetically related to the endospore-forming bacterium *Clostridium* and found in the gut of the surgeonfish, contains multiple copies of its genome. The many copies are apparently necessary because the volume of an *Epulopiscium* cell is so large (Table 2.1) that a single copy of its genome would be insufficient to support its transcriptional and translational demands.

Cells of the largest known bacterium, the sulfur-oxidizing chemolithotroph *Thiomargarita* (Figure 2.2b), are even larger than those of *Epulopiscium*, about 750 μm in diameter; such cells are just visible to the naked eye. Why these cells are so large is not well understood, although for sulfur bacteria a large cell size may have evolved for storing inclusions of sulfur (used as an energy source). No species of *Archaea* are known that rival *Epulopiscium* or *Thiomargarita* in cell size, but that may simply be because they remain undiscovered.

It is hypothesized that the upper size limit for prokaryotic cells results from the decreasing ability of larger and larger cells to transport nutrients (their surface-to-volume ratio is very small; see the next subsection). Since the metabolic rate of a cell varies inversely with the square of its size, for very large cells, nutrient uptake would eventually limit metabolism to the point that the cell would no longer be competitive with smaller cells.

Very large cells are uncommon in the prokaryotic world. In contrast to *Thiomargarita* or *Epulopiscium* (Figure 2.2), the dimensions of an average rod-shaped bacterium, such as *Escherichia coli*, for example, are about 1–2 μm; these dimensions are typical of cells in the prokaryotic world. By contrast, eukaryotic cells can be as small as 2 to more than 600 μm in diameter, although very small microbial eukaryotes (cells less than about 6 μm in diameter) are uncommon. We explore the world of microbial eukaryotes in Chapter 18.

Surface-to-Volume Ratios, Growth Rates, and Evolution

For a cell, there are advantages to being small. Small cells have more surface area relative to cell volume than do large cells and thus have a higher *surface-to-volume ratio*. To understand this, consider a coccus-shaped cell. The volume of a coccus is a function of the cube of its radius ($V = \frac{4}{3}\pi r^3$), whereas its surface area is a function of the square of the radius ($S = 4\pi r^2$). Therefore, the S/V ratio of a coccus is $3/r$ (**Figure 2.3**). As cell size *increases*, its S/V ratio *decreases*. To illustrate this, consider the S/V ratio for some of the cells of different sizes listed in Table 2.1: *Pelagibacter ubique*, 22; *E. coli*, 4.5; and *E. fishelsoni* (Figure 2.2a), 0.05. These are all rods

---

TABLE 2.1 Cell size and volume of some cells of *Bacteria*, from the largest to the smallest

Organism	Characteristics	Morphology	Size[a] (μm)	Cell volume (μm)^3	Volumes compared to *E. coli*
Thiomargarita namibiensis	Sulfur chemolithotroph	Cocci in chains	750	200,000,000	100,000,000×
Epulopiscium fishelsoni[a]	Chemoorganotroph	Rods with tapered ends	80 × 600	3,000,000	1,500,000×
Beggiatoa species[a]	Sulfur chemolithotroph	Filaments	50 × 160	1,000,000	500,000×
Achromatium oxaliferum	Sulfur chemolithotroph	Cocci	35 × 95	80,000	40,000×
Lyngbya majuscula	Cyanobacterium	Filaments	8 × 80	40,000	20,000×
Thiovulum majus	Sulfur chemolithotroph	Cocci	18	3,000	1,500×
Staphylothermus marinus[a]	Hyperthermophile	Cocci in irregular clusters	15	1,800	900×
Magnetobacterium bavaricum	Magnetotactic bacterium	Rods	2 × 10	30	15×
Escherichia coli	Chemoorganotroph	Rods	1 × 2	2	1×
Pelagibacter ubique[a]	Marine chemoorganotroph	Rods	0.2 × 0.5	0.014	0.007×
Ultra-small bacteria	Uncultured, from groundwater	Variable	<0.2	0.009	0.0045×
Mycoplasma pneumoniae	Pathogenic bacterium	Pleomorphic[b]	0.2	0.005	0.0025×

[a]Where only one number is given, this is the diameter of spherical cells. The values given are for the largest cell size observed in each species. For example, for *T. namibiensis*, an average cell is only about 200 μm in diameter. But on occasion, giant cells of 750 μm are observed. Likewise, an average cell of *S. marinus* is about 1 μm in diameter. The species of *Beggiatoa* here is unclear and *E. fishelsoni*, *M. bavaricum*, and *P. ubique* are not formally recognized names in taxonomy.
[b]*Mycoplasma* is a bacterium that lacks a cell wall and can thus take on many shapes (*pleomorphic* means "many shapes").
Source: Data obtained from Schulz, H.N., and B.B. Jørgensen. 2001. *Annu. Rev. Microbiol.* 55: 105–137, and Luef, B., et al. 2015. *Nature Communications.* doi:10.1038/ncomms7372.

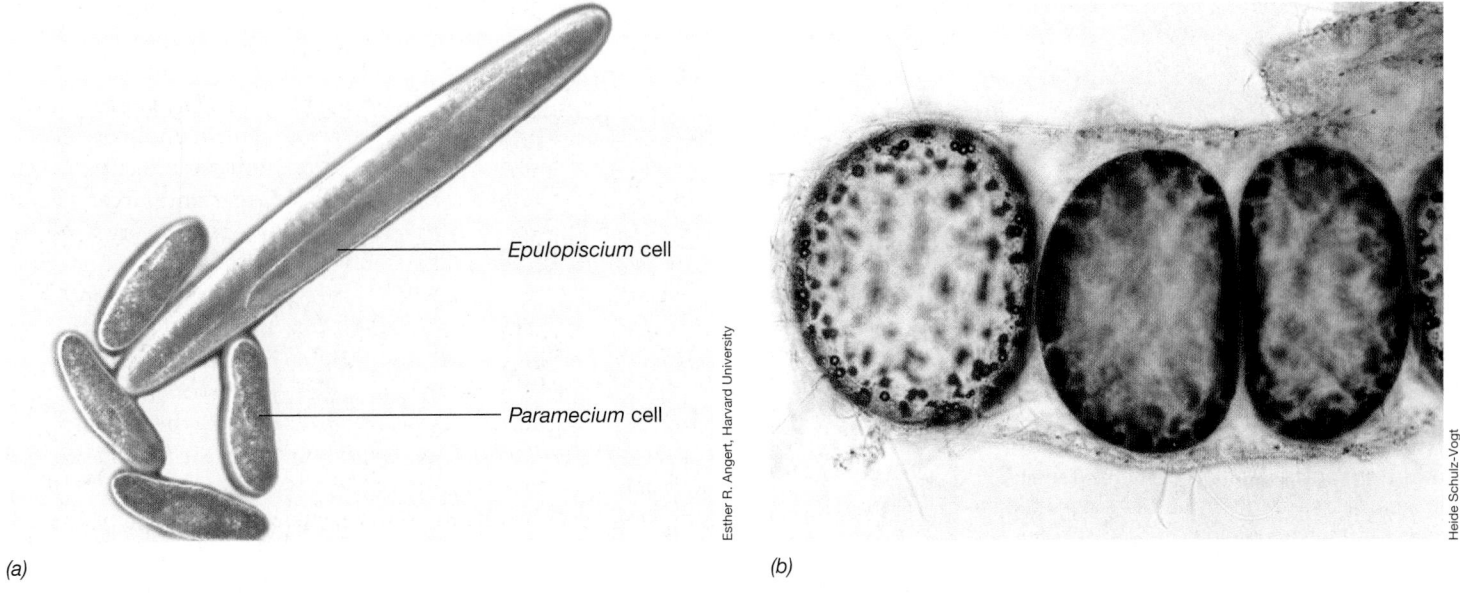

Epulopiscium cell

Paramecium cell

(a)

(b)

Esther R. Angert, Harvard University

Heide Schulz-Vogt

Figure 2.2 Two very large *Bacteria*. *(a) Epulopiscium fishelsoni*. The rod-shaped cell is about 600 μm (0.6 mm) long and 75 μm wide and is shown with four cells of the protist *Paramecium* (a microbial eukaryote), each of which is about 150 μm long. *(b) Thiomargarita namibiensis*, a large sulfur chemolithotroph and currently the largest known of all prokaryotic cells. Cell widths vary from 400 to 750 μm.

rather than cocci, but if it is assumed that a rod-shaped cell is a perfect cylinder, the same S/V principles that hold for cocci also hold for rods; that is, for rods of a given length, cells with a smaller radius have a greater S/V than do cells with a larger radius.

The S/V of a cell controls many of its properties, including its growth rate and evolution. Because how fast a cell can grow depends in part on the rate at which it can exchange nutrients and waste products with its environment, the higher S/V ratio of small cells supports a faster rate of nutrient and waste exchange per unit of cell volume compared with large cells. As a result, free-living smaller cells tend to grow faster than free-living larger cells, and for a given amount of resources (nutrients available to support

growth), a larger population of small cells than of large cells can be supported. This in turn can affect a cell's evolution.

Each time a cell divides, its chromosome replicates. As DNA is replicated, occasional errors, called *mutations*, occur. Because mutation rates are roughly the same in all cells, large or small, the more chromosome replications that occur, the greater the total number of mutations in the cell population. Mutations are the "raw material" of evolution; the larger the pool of mutations, the greater the evolutionary possibilities. Thus, because prokaryotic cells are quite small and are also genetically haploid (they typically have only one copy of each gene, allowing mutations to be expressed immediately), prokaryotic cells can grow faster and evolve more rapidly than can larger cells.

Because of their typically larger size, not only is the S/V ratio of microbial eukaryotes smaller, the diploid character of the eukaryotic cell (a cell has two copies of each gene) allows for a mutation in one gene to be masked by a second, unmutated gene copy. These fundamental differences in size and genetics between prokaryotic and eukaryotic cells help explain why species of *Bacteria* and *Archaea* adapt rapidly to changing environmental conditions and more easily exploit new habitats than do eukaryotic cells. We illustrate this concept in later chapters when we consider the enormous diversity of cells and metabolisms of *Bacteria* and *Archaea* (Chapters 14–17), the rapidity of prokaryotic evolution (Chapter 13), and the ecological ramifications of microbial activities in nature (Chapters 19–23).

Lower Limits to Cell Size

From the foregoing, one would predict that smaller and smaller microbes would have greater and greater selective advantages in nature and that as a consequence, only extremely tiny bacterial cells would exist. However, this is not the case, as there are lower limits to cell size and good reasons why there should be.

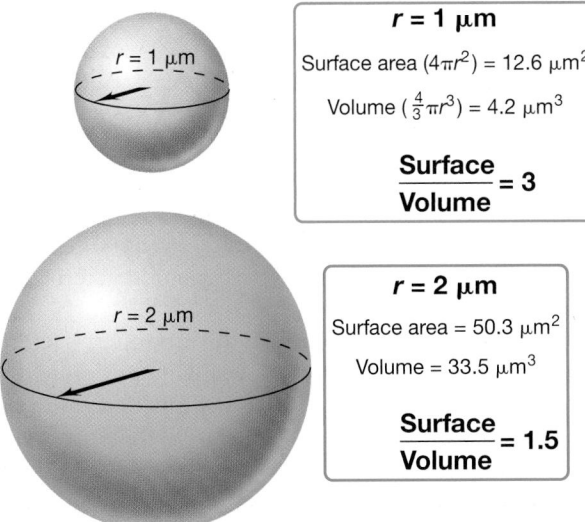

$r = 1$ μm

$r = 1$ μm

Surface area $(4\pi r^2) = 12.6$ μm^2

Volume $(\frac{4}{3}\pi r^3) = 4.2$ μm^3

$$\frac{\text{Surface}}{\text{Volume}} = 3$$

$r = 2$ μm

$r = 2$ μm

Surface area $= 50.3$ μm^2

Volume $= 33.5$ μm^3

$$\frac{\text{Surface}}{\text{Volume}} = 1.5$$

Figure 2.3 Surface area and volume relationships in cells. As a cell increases in size, its S/V ratio decreases.

TINY CELLS

Viruses are very small microbes and range in diameter from as small as 20 nm to almost 750 nm. Although no cells exist that are as small as most viruses, the recent discovery of ultra-small bacterial cells[1, 2] has pushed the lower limits of cell size to what microbiologists feel must be very close to the minimal value.

Microbiologists collected groundwater, which travels through Earth's deep subsurface, from a Colorado (USA) aquifer (**Figure 1**) and ran it through a membrane filter whose pores were only 0.2 μm in diameter. The liquid that passed through the filter was then subjected to microbiological analyses. Surprisingly, since filters with 0.2-μm pores have been used for decades to remove bacterial cells from solutions to generate "sterile solutions," prokaryotic cells were present in the groundwater filtrate. In fact, a diverse array of *Bacteria* were present in the filtrate, revealing that the groundwater was inhabited by a microbial community of tiny cells[1] that microbiologists have come to call "ultramicrobacteria."

Cryo-electron microscopy, in which a specimen is examined at extremely cold temperatures without fixation (chemical treatment that can alter a cell's morphology), showed the groundwater ultramicrobacteria to consist primarily of oval-shaped cells about 0.2 μm in diameter (**Figure 2**). The volume of these cells was calculated to be about 1/100 that of a cell of the bacterium *Escherichia coli* (see Table 2.1) such that nearly 150 of the small cells could fit into one *E. coli* cell! Each of the tiny cells contained about 50 ribosomes, which is also about 1/100 of the number present in a slowly growing (100-min generation time) cell of *E. coli*. The very small size of the

Figure 1 Sampling the anoxic groundwater aquifer that parallels the Colorado River near Rifle, Colorado.

Kenneth H. Williams

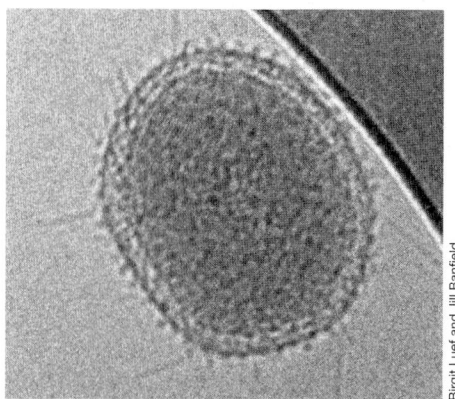

Birgit Luef and Jill Banfield

Figure 2 A tiny bacterial cell from anoxic groundwater that passed through a filter with 0.2-μm pores. The cell is not quite 0.2 μm in diameter.

groundwater ultramicrobacteria gives them an enormous surface-to-volume ratio, and it is hypothesized that this advantage benefits them in extracting resources from their nutrient-deficient habitat.

Despite the fact that the tiny groundwater bacteria have yet to be cultured in the laboratory, much is already known about them because their small genomes—less than 1 Mb in size—were obtained and analyzed.[2] From a phylogenetic perspective, the different species detected were distantly related to major phyla of *Bacteria* known from environmental analyses of diverse environments but which have thus far defied laboratory culture. Further analyses showed that genes encoding the enzymes for several core metabolic pathways widely distributed among microorganisms were absent from the genomes of the groundwater ultramicrobacteria. This suggests a

metabolically minimalist lifestyle for these tiny cells and a survival strategy of cross-feeding essential nutrients with neighboring species in their microbial community.

Although we do not yet know exactly how small a microbial cell can be, microbiologists are closing in on this number from environmental analyses such as the Colorado groundwater study. From the same samples that yielded ultra-small *Bacteria* in this study, ultra-small *Archaea* were also detected and found to contain small and highly reduced genomes.[2] Apparently, a large diversity of very small prokaryotic cells occurs in nature, and from the continued study of these tiny cells, more precise values for both the lower limits to cell size and the minimal genomic requirements for life should emerge.

[1] Luef, B., et al. 2015. *Nature Communications.* doi:10.1038/ncomms7372.
[2] Castelle, C.J., et al. 2015. *Current Biology. 25:* 1–12.

If one calculates the volume needed to house the essential components of a free-living cell—proteins, nucleic acids, ribosomes, and so on—cells 0.15–0.2 μm in diameter are probably the lower limit, or very close to it. Many small prokaryotic cells are known and several have been grown in the laboratory. Open ocean water, for example, contains 10^5–10^6 prokaryotic cells per milliliter, and these tend to be very small cells, 0.2–0.4 μm in diameter. In addition, populations of bacterial and archaeal cells have been discovered in Earth's deep subsurface that are about 0.2 μm in diameter (see Explore the Microbial World, "Tiny Cells"). Collectively, these cells have been referred to as "ultramicrobacteria" to indicate that

they are considerably smaller than typical bacterial cells. We will see later that the cells of many pathogenic bacteria are also very small.

Interestingly, when the genomes of ultramicrobacteria are unraveled, they are typically found to be highly streamlined and to be missing many genes whose products or functions must be supplied to them by other microbial cells or by the host organisms (plants and animals, for example) that harbor them. In these tiny prokaryotic cells, evolutionary success is apparently linked to a minimalist biochemistry and obligate association with one or more other organisms. In some cases, these associations are so specific and so tightly linked that the tiny cell (and in some cases, the second organism) cannot survive without its respective partner. We consider the genomes—both large and small—of *Bacteria* and *Archaea* in many chapters of this book and focus exclusively on genomes and related aspects of the systems biology of microbial cells in Chapter 9.

II • The Cell Membrane and Wall

We consider here the structure and function of two of a cell's most essential structures, the *cytoplasmic membrane* and the *cell wall*. The cytoplasmic membrane plays many roles, chief among them as the "gatekeeper" for the entrance and exit of dissolved substances. The cell wall, by contrast, confers structural strength on the cell in order to keep it from bursting due to osmotic pressure.

2.3 The Cytoplasmic Membrane

The **cytoplasmic membrane** surrounds the *cytoplasm*—the mixture of macromolecules and small molecules inside the cell—and separates it from the environment. The cytoplasmic membrane is physically rather weak but it is an ideal structure for its major cellular function: selective permeability. In order for a cell to grow, nutrients must be transported inwards and waste products outwards. Both of these events occur across the cytoplasmic membrane. A variety of proteins located in the cytoplasmic membrane facilitate these reactions, and many other membrane proteins play important roles in energy metabolism.

The Bacterial Cytoplasmic Membrane

The cytoplasmic membrane of all cells is a phospholipid bilayer containing embedded proteins. Phospholipids are composed of both hydrophobic (water-repelling) and hydrophilic (water-attracting) components (**Figure 2.4**). In *Bacteria* and *Eukarya*, the hydrophobic component consists of fatty acids and the hydrophilic component of a glycerol molecule containing phosphate and one of several other functional groups (such as sugars, ethanolamine, or choline) bonded to the phosphate. The fatty acids point inward toward each other to form a hydrophobic region, while the hydrophilic portion remains exposed to either the environment or the cytoplasm (Figure 2.4b). That is, the *outer* surface of the cytoplasmic membrane faces the environment while the *inner* surface faces the cytoplasm and interacts with the cytoplasmic milieu. This type of membrane structure is called a *lipid bilayer*, or a *unit membrane* because each phospholipid "leaf" forms half of the unit (see Figure 2.5).

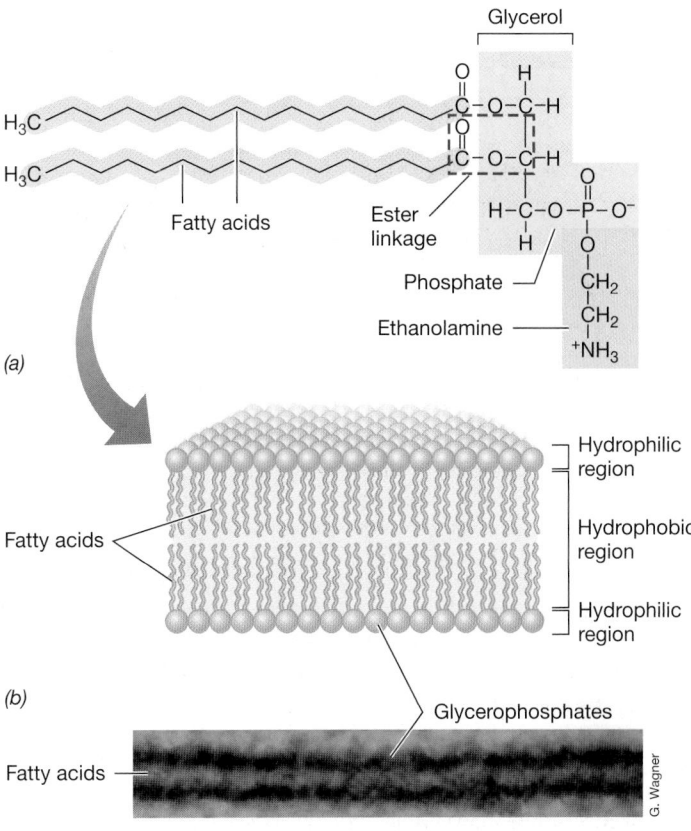

(a)

(b)

(c)

Figure 2.4 Phospholipid bilayer membrane. *(a)* Structure of the phospholipid phosphatidylethanolamine. The side chains are fatty acids and the ester linkage (characteristic of the lipids of *Bacteria* and *Eukarya* but not *Archaea*) is boxed with a red dashed line. *(b)* General architecture of a bilayer membrane; the blue spheres depict glycerol with phosphate and/or other hydrophilic groups. *(c)* Transmission electron micrograph of a membrane. The light inner area is the hydrophobic region of the model membrane shown in *b*.

The cytoplasmic membrane is only 8–10 nanometers wide but can be resolved easily by transmission electron microscopy (Figure 2.4c). In addition, although physically weak, the cytoplasmic membranes of some *Bacteria* are strengthened by sterol-like molecules called *hopanoids*. Sterols are rigid and planar molecules that strengthen the membranes of eukaryotic cells (Section 2.14), many of which lack a cell wall.

A variety of proteins are attached to or integrated into the cytoplasmic membrane; membrane proteins typically have hydrophobic domains that span the membrane and hydrophilic domains that contact the environment or the cytoplasm (**Figure 2.5**). Proteins significantly embedded in the membrane are called *integral* membrane proteins. By contrast, *peripheral* membrane proteins are more loosely attached. Some peripheral membrane proteins are lipoproteins, proteins that contain a hydrophobic lipid tail that anchors the protein into the membrane (Figure 2.5). Peripheral membrane proteins typically interact with integral membrane proteins in important cellular processes such as energy metabolism and transport.

Archaeal Membranes

The cytoplasmic membrane of *Archaea* is structurally similar to those of *Bacteria* and *Eukarya*, but the chemistry is somewhat

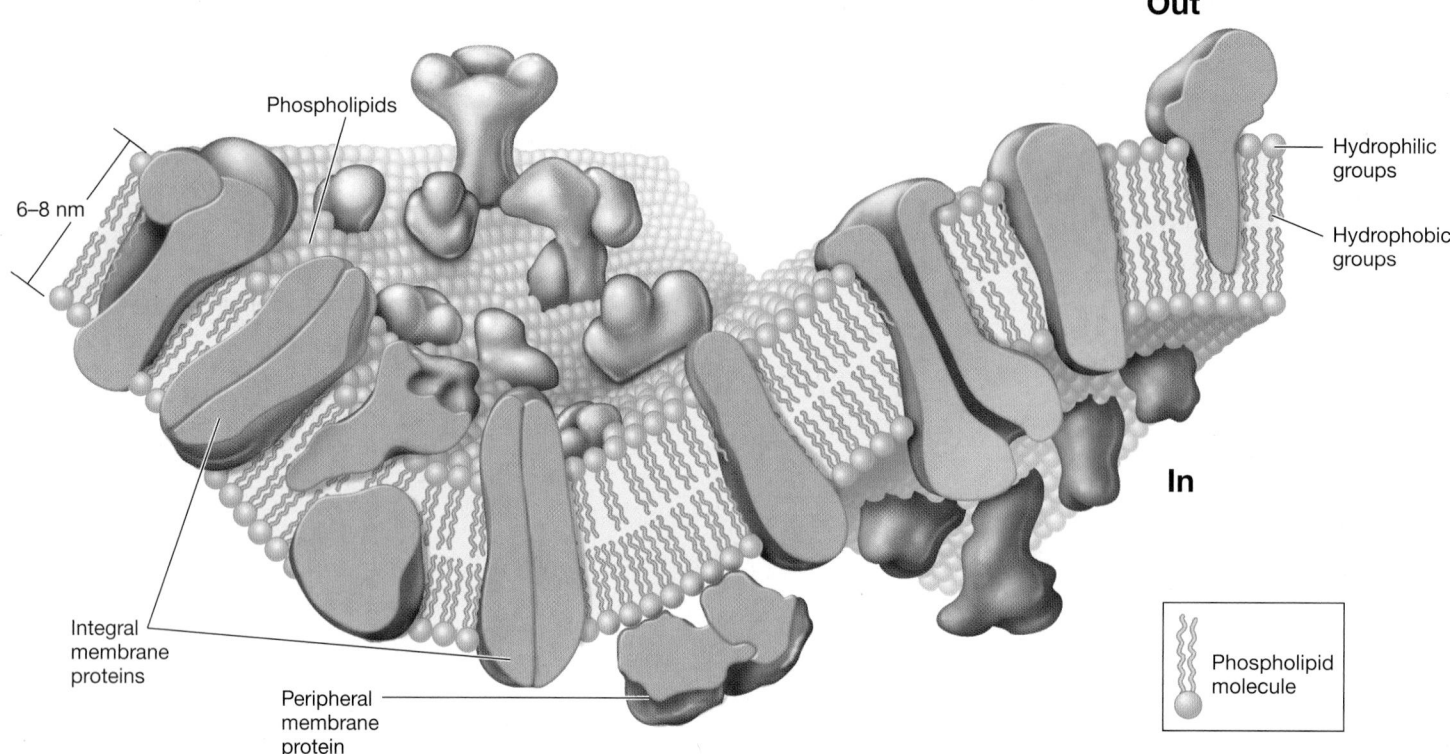

Figure 2.5 Structure of the cytoplasmic membrane. The inner surface (**In**) faces the cytoplasm and the outer surface (**Out**) faces the environment. Phospholipids compose the matrix of the cytoplasmic membrane with proteins embedded (integral) or surface associated (peripheral). The general design of the cytoplasmic membrane is similar in both prokaryotic and eukaryotic cells, although there are chemical differences between different species.

different. In contrast to the lipids of *Bacteria* and *Eukarya* in which *ester* linkages bond *fatty acids* to glycerol (Figure 2.4), the lipids of *Archaea* contain *ether* bonds between glycerol and a hydrophobic side chain that is not a fatty acid (**Figure 2.6**). The hydrophobic region of archaeal membranes is formed from repeating units of the five-carbon hydrocarbon *isoprene*, rather than from fatty acids (compare Figures 2.4 and 2.6).

The cytoplasmic membrane of *Archaea* is constructed from either phosphoglycerol diethers, which have C_{20} side chains (called a *phytanyl* group), or diphosphoglycerol tetraethers (C_{40} side chains, called a *biphytanyl* group) (Figure 2.6). In the tetra-ether lipid structure, the ends of the inwardly pointing phytanyl groups are covalently linked at their termini to form a *lipid mono-layer* (Figure 2.6e) instead of a lipid bilayer (Figure 2.6d) membrane.

Some archaeal lipids contain rings within the hydrocarbon side chains. For example, *crenarchaeol*, a common membrane lipid in cells of *Crenarchaeota* (a major phylum of *Archaea*) contains four C_5 rings and one C_6 ring (Figure 2.6c). These rings affect the chemical properties of the lipids and thus influence membrane function. As in other organisms, the polar head groups in archaeal lipids can be sugars, ethanolamine, or a variety of other molecules. Hopanoids, present in the cytoplasmic membranes of many *Bacteria*, have not been found in the cytoplasmic membranes of *Archaea*.

Despite differences in chemistry between the cytoplasmic membranes of *Archaea* and organisms in the other phylogenetic domains, the fundamental construction of the archaeal cytoplas-mic membrane—inner and outer hydrophilic surfaces and a

hydrophobic interior—is the same as that of membranes in all cells. Obviously, evolution has selected this fundamental design as the best solution to the major functions of the cytoplasmic membrane, an issue we turn to now.

Cytoplasmic Membrane Function

The cytoplasmic membrane has at least three major functions (**Figure 2.7**). First, it is the cell's permeability barrier, preventing the passive leakage of solutes into or out of the cell. Second, the cyto-plasmic membrane anchors several proteins that catalyze a suite of key cell functions. And third, the cytoplasmic membrane of *Bacteria* and *Archaea* plays a major role in energy conservation and consumption.

The cytoplasmic membrane is a barrier to the diffusion of most substances, especially polar or charged molecules. Because the cytoplasmic membrane is so impermeable, most substances that enter or leave the cell must be carried in or out by *transport proteins*. These are not simply ferrying proteins but instead function to *accu-mulate* solutes against the concentration gradient, a process that diffusion alone cannot do (**Figure 2.8**). Transport, which requires energy, ensures that the cytoplasm has sufficient concentrations of the nutrients it needs to perform biochemical reactions efficiently.

Transport proteins typically display high sensitivity and high specificity. If the concentration of a solute is high enough to sat-urate the transporter, which often occurs at the very low concen-trations of nutrients found in nature, the rate of uptake can be near maximal (Figure 2.8). Some nutrients are transported by a

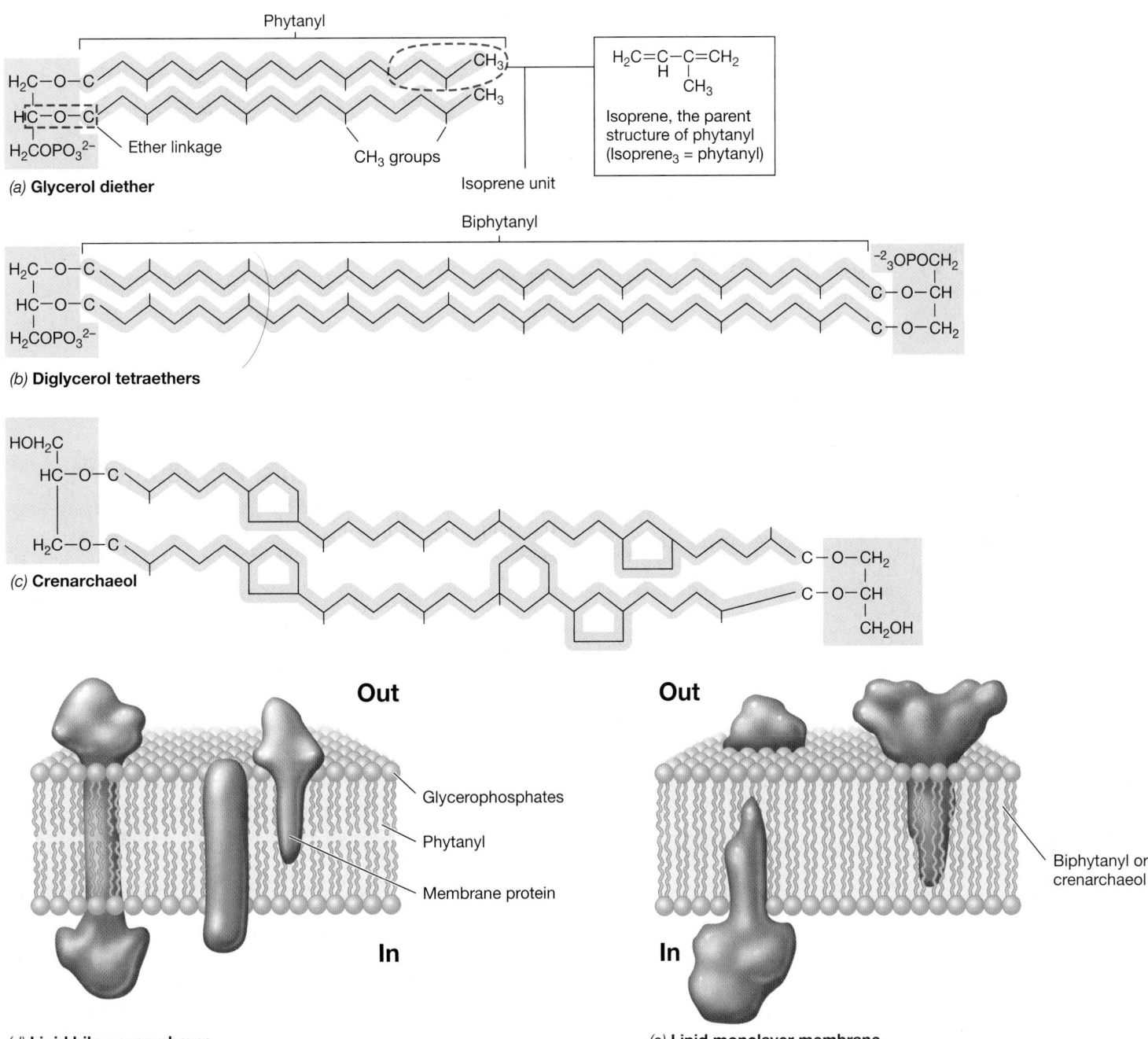

Figure 2.6 Major lipids of *Archaea* and the architecture of archaeal membranes. *(a, b)* Note that the hydrocarbon of the lipid is bonded to the glycerol by an ether linkage (in dashed red box in *a*) in both cases. The hydrocarbon is phytanyl (C_{20}) in *a* and biphytanyl (C_{40}) in *b*; both are multiples of the parent structure, isoprene (in dashed red oval; detailed structure shown in black box). *(c)* A major lipid of *Thaumarchaeota* is crenarchaeol, a lipid containing 5- and 6-carbon rings. *(d, e)* The membrane structure in *Archaea* may form a lipid bilayer or a lipid monolayer (or a mix of both).

low-affinity transporter when present at high external concentration and by a separate, typically higher-affinity, transporter for those present at low concentration (Figure 2.8). Moreover, many transport proteins transport only a single kind of molecule while others carry a related class of molecules, such as different sugars or different amino acids. This economizing reduces the need for separate transport proteins for each different sugar or

amino acid. We revisit the important issue of nutrient transport in Section 3.2, where we focus on transport mechanisms.

In addition to its permeability and transport functions, the cytoplasmic membrane of *Bacteria* and *Archaea* is a major site of both energy conservation and consumption. We discuss in Chapter 3 how the cytoplasmic membrane can be energized when protons (H^+) are separated from hydroxyl ions (OH^-) across the membrane

Functions of the cytoplasmic membrane

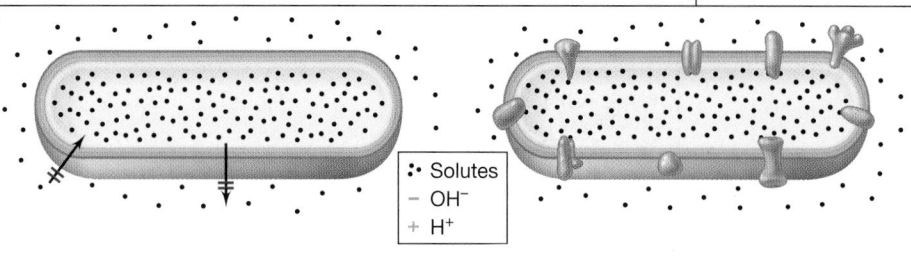
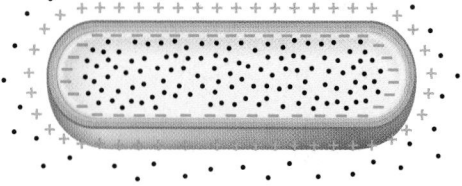

(a) **Permeability barrier:**
Prevents leakage and functions as a gateway for transport of nutrients into, and wastes out of, the cell

(b) **Protein anchor:**
Site of proteins that participate in transport, bioenergetics, and chemotaxis

(c) **Energy conservation:**
Site of generation and dissipation of the proton motive force

Figure 2.7 The major functions of the cytoplasmic membrane. Although physically weak, the cytoplasmic membrane controls at least three critically important cellular functions: preventing leakage, anchoring proteins, and conserving energy.

surface (Figure 2.7c). This charge separation creates an energized state of the membrane called the *proton motive force*, analogous to the potential energy present in a charged battery. Dissipation of the proton motive force can be coupled to several energy-requiring reactions, such as transport, cell locomotion, and the biosynthesis of ATP. In eukaryotic microbial cells, although transport across the cytoplasmic membrane is just as necessary as it is in prokaryotic cells, energy conservation takes place in the membrane systems of the cell's key organelles, the mitochondrion (respiration) and chloroplast (photosynthesis).

MINIQUIZ

- Draw the basic structure of a lipid bilayer and label the hydrophilic and hydrophobic regions. Why is the cytoplasmic membrane a good permeability barrier?
- How are the membrane lipids of *Bacteria* and *Archaea* similar, and how do they differ?
- Describe the major functions of the cytoplasmic membrane.

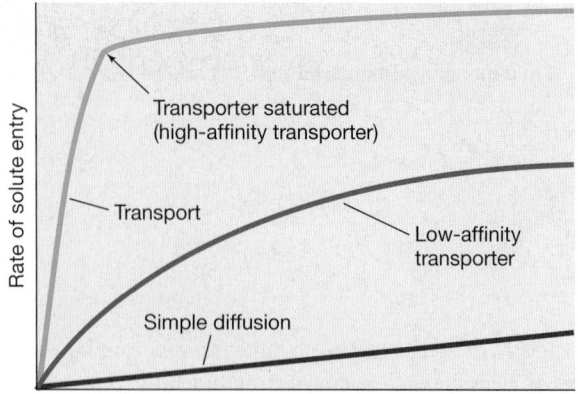

Figure 2.8 The importance of transport in membrane function. In transport, the uptake rate shows saturation at relatively low external concentrations. Both high-affinity and low-affinity transport systems are depicted.

2.4 Bacterial Cell Walls: Peptidoglycan

The cytoplasm of prokaryotic cells maintains a high concentration of dissolved solutes that creates significant osmotic pressure—about 2 atm (203 kPa); this is about the same as the pressure in an automobile tire. To withstand these pressures and prevent bursting—a process called *cell lysis*—most cells of *Bacteria* and *Archaea* have a layer outside the cytoplasmic membrane called the *cell wall*. Besides protecting against osmotic lysis, cell walls also confer shape and rigidity on the cell.

Knowledge of cell wall structure and function is important not only for understanding the biology of microbial cells, but also because certain antibiotics, for example, the penicillins and cephalosporins, target bacterial cell wall synthesis, leaving the cell susceptible to osmotic lysis. Since human cells lack cell walls and are therefore not a target of such antibiotics, these drugs are of obvious benefit for treating bacterial infections.

Cells of *Bacteria* can be divided into two major groups, *gram-positive* and *gram-negative*. The distinction between gram-positive and gram-negative bacteria is based on the Gram stain reaction (⇔ Section 1.6), and differences in cell wall structure play a major role in the reaction. The surface of gram-positive and gram-negative cells as viewed in the electron microscope differs markedly, as shown in **Figure 2.9**. The gram-negative cell wall, or *cell envelope* as it is also called, consists of at least two layers, whereas the gram-positive cell wall is typically thicker and consists primarily of a single type of molecule.

Our focus in this section is on a key molecule found in the cell walls of both gram-positive and gram-negative *Bacteria*. In Section 2.5, we describe some additional wall components present in gram-negative *Bacteria*, and in Section 2.6, we describe the cell walls of *Archaea*.

Structure of Peptidoglycan

The walls of cells of *Bacteria* contain a rigid polysaccharide called **peptidoglycan** that confers structural strength on the cell. Peptidoglycan is found in all *Bacteria* that contain a cell wall, but it is not present in the cell walls of *Archaea* or *Eukarya*. Peptidoglycan is composed of alternating repeats of two modified glucose residues

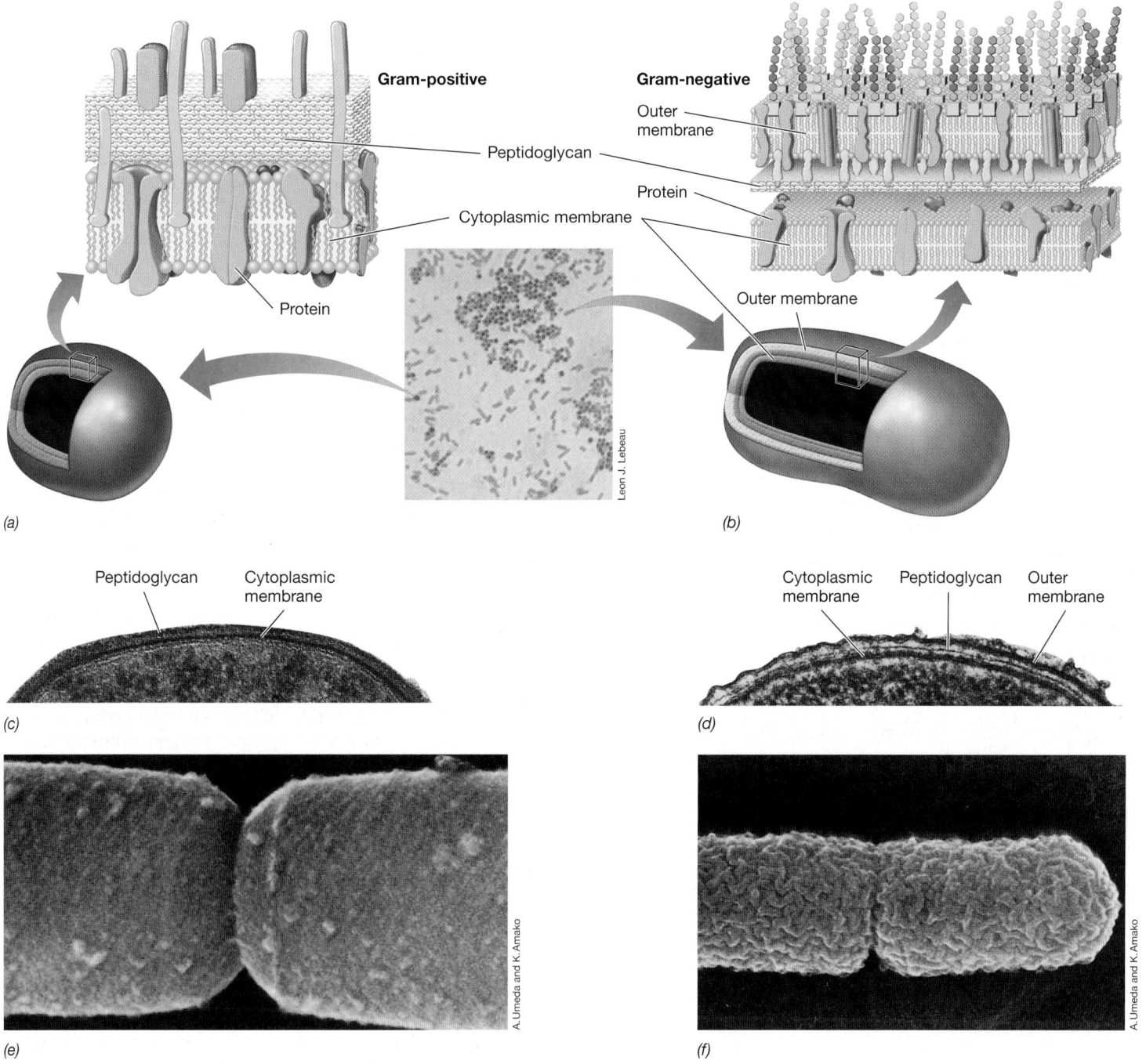

Figure 2.9 Cell walls of *Bacteria*. *(a, b)* Schematic diagrams of gram-positive and gram-negative cell walls; the Gram stain procedure was described in Section 1.6 and shown in Figure 1.18. The photo of Gram-stained bacteria in the center shows cells of *Staphylococcus aureus* (purple, gram-positive) and *Escherichia coli* (pink, gram-negative). *(c, d)* Transmission electron micrographs showing the cell wall of a gram-positive bacterium and a gram-negative bacterium, respectively. *(e, f)* Scanning electron micrographs of gram-positive and gram-negative bacteria, respectively. Note differences in surface texture. Each cell is about 1 μm wide.

called *N-acetylglucosamine* and *N-acetylmuramic acid* along with the amino acids L-alanine, D-alanine, D-glutamic acid, and either L-lysine or diaminopimelic acid (DAP). These constituents are connected in an ordered way to form the *glycan tetrapeptide* (**Figure 2.10**), and long chains of this basic unit form peptidoglycan.

Strands of peptidoglycan are biosynthesized adjacent to one another to form a sheet surrounding the cell, and the individual strands are connected by peptide cross-links; this forms a polymer that is strong in both X and Y directions (**Figure 2.11**). In gram-negative bacteria, the cross-link forms from the amino group of DAP of one glycan strand to the carboxyl group of the terminal D-alanine on the adjacent glycan strand (Figure 2.11*a*). In gram-positive bacteria, the cross-link often contains a short peptide "interbridge," the kinds and numbers of amino acids in the

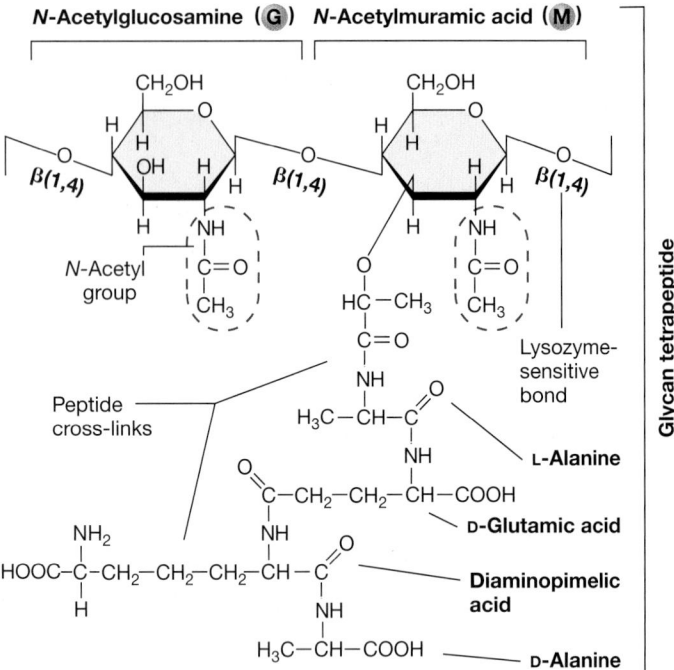

Figure 2.10 Structure of the repeating unit in peptidoglycan, the glycan tetrapeptide. The structure given is that of *Escherichia coli* and most other gram-negative *Bacteria*. In some *Bacteria*, other amino acids are present as cross-linkers.

interbridge varying between species. In the gram-positive bacterium *Staphylococcus aureus*, for example, a bacterium whose cell wall chemistry is well understood, the interbridge consists of five glycines (Figure 2.11*b*). The overall structure of peptidoglycan is shown in Figure 2.11*c*.

Peptidoglycan can be destroyed by *lysozyme*, an enzyme that cleaves the glycosidic bond between *N*-acetylglucosamine and *N*-acetylmuramic acid (Figure 2.11). This weakens the peptidoglycan and can cause cell lysis. Lysozyme is present in human secretions including tears, saliva, and other bodily fluids, and functions as a major line of defense against bacterial infection. When we consider peptidoglycan biosynthesis in Chapter 7, we will see that the antibiotic penicillin also destroys peptidoglycan, but in a different way than lysozyme does. Whereas lysozyme destroys preexisting peptidoglycan, penicillin blocks a key step in its biosynthesis; this also weakens the molecule and leads to osmotic lysis.

An unusual feature of peptidoglycan is the presence of two amino acids of the D *stereoisomer*, D-alanine and D-glutamic acid. Teichoic acids, to be described shortly, also contain D-amino acids. By contrast, proteins are always constructed of L-amino acids. All in all, more than 100 chemically distinct peptidoglycans have been described that vary in their peptide cross-links and/or interbridge structures. By contrast, the glycan portion of all peptidoglycans appears to be universal; only alternating repeats of *N*-acetylglucosamine and *N*-acetylmuramic acid in β-1,4 linkage are known (Figures 2.10 and 2.11).

Overview of the Gram-Positive Cell Wall

As much as 90% of the cell wall of a gram-positive bacterium can consist of peptidoglycan. Although some bacteria have only a

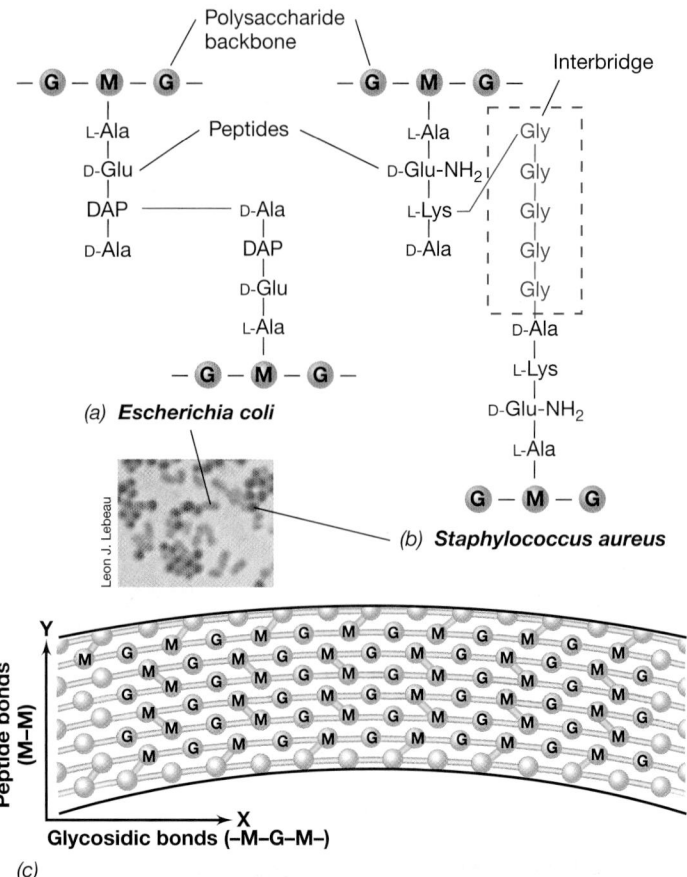

Figure 2.11 Peptidoglycan in *Escherichia coli* and *Staphylococcus aureus*. (a) No interbridge is present in *E. coli* peptidoglycan nor in that of other gram-negative *Bacteria*. (b) The glycine interbridge in the gram-positive bacterium *S. aureus*. (c) Overall structure of peptidoglycan. G, *N*-acetylglucosamine; M, *N*-acetylmuramic acid. Note how glycosidic bonds confer strength on peptidoglycan in the X direction whereas peptide bonds confer strength in the Y direction.

single layer of peptidoglycan, many gram-positive bacteria form several layers of peptidoglycan stacked one upon another (Figure 2.11*c*). It is thought that peptidoglycan is synthesized by the cell in the form of "cables" about 50 nm wide, with each cable consisting of several glycan strands (**Figure 2.12*a***). As peptidoglycan is synthesized, the cables themselves become cross-linked to form an even stronger cell wall structure.

In addition to peptidoglycan, many gram-positive bacteria produce acidic molecules called **teichoic acids** embedded in their cell wall. Teichoic acids are composed of glycerol phosphate or ribitol phosphate with attached molecules of glucose or D-alanine (or both). Individual alcohol molecules are then connected through their phosphate groups to form long strands, and these are then covalently linked to peptidoglycan (Figure 2.12*b*). Teichoic acids also function to bind divalent metal ions, such as Ca^{2+} and Mg^{2+}, prior to their transport into the cell. Some teichoic acids are covalently bonded to membrane lipids rather than to peptidoglycan, and these are called *lipoteichoic* acids. Figure 2.12*c* summarizes the structure of the cell wall of gram-positive *Bacteria* and shows how teichoic acids and lipoteichoic acids are arranged in the overall wall structure.

mycoplasmas contain sterols in their cytoplasmic membranes; these molecules function to add strength and rigidity to the membrane as they do in the cytoplasmic membranes of eukaryotic cells (Section 2.14). *Thermoplasma* membranes contain molecules called *lipoglycans* that serve a similar strengthening function.

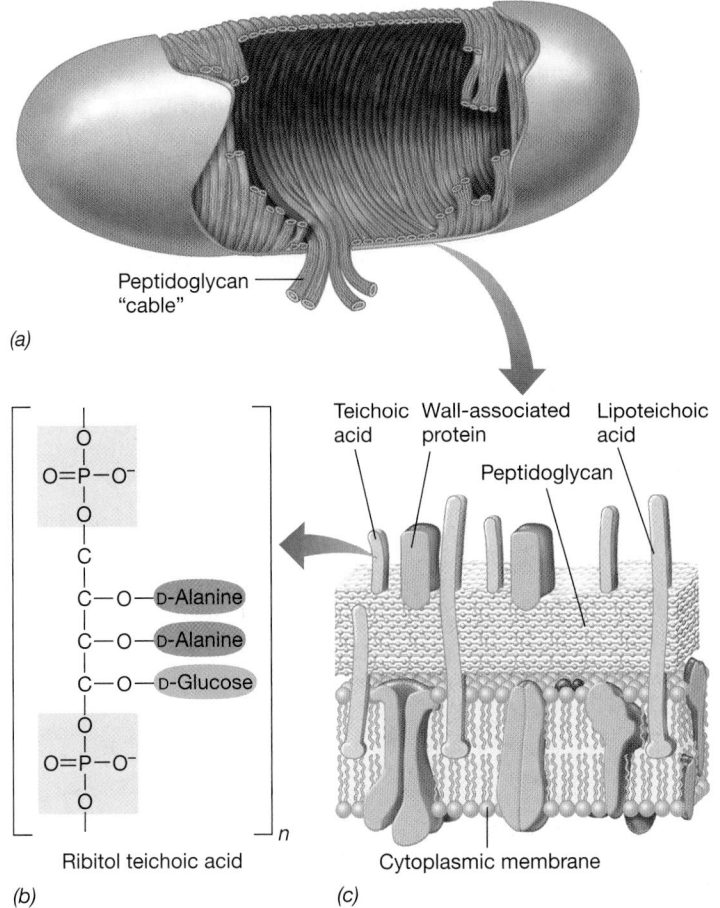

(a)

(b)   (c)

Figure 2.12 Structure of the gram-positive bacterial cell wall. *(a)* Schematic of a gram-positive rod showing the internal architecture of the peptidoglycan "cables." *(b)* Structure of a ribitol teichoic acid. The teichoic acid is a polymer of the repeating ribitol unit shown here. *(c)* Summary diagram of the gram-positive bacterial cell wall.

A very few *Bacteria* and *Archaea* lack cell walls altogether. These include in particular the mycoplasmas, pathogenic *Bacteria* related to gram-positive bacteria that cause a variety of infectious diseases of humans and other animals, and *Thermoplasma* and some of its relatives (*Archaea*). Lacking a cell wall, these cells would be expected to contain unusually tough cytoplasmic membranes, and chemical analyses show that they do. For example, most

---

MINIQUIZ

- Why do bacterial cells need cell walls? Do all bacteria have cell walls?
- Why is peptidoglycan such a strong molecule?
- What do the enzyme lysozyme and the antibiotic penicillin have in common?

2.5 LPS: The Outer Membrane

In gram-negative bacteria, only a small amount of the total cell wall consists of peptidoglycan, as most of the wall is composed of the **outer membrane**. This layer is effectively a second lipid bilayer, but it is not constructed solely of phospholipid and protein, as is the cytoplasmic membrane (Figures 2.4 and 2.5). Instead, the outer membrane also contains polysaccharide, and the lipid and polysaccharide are linked to form a complex. Hence, the outer membrane is often called the **lipopolysaccharide** layer, or simply **LPS** for short.

The outer membrane confers only modest structural strength on the gram-negative cell (peptidoglycan remains the major strengthening agent), but it acts as an effective barrier against many substances such as lipophilic antibiotics and other harmful agents that might otherwise penetrate the cytoplasmic membrane. Indeed, many antibiotics that are clinically useful against gram-positive bacterial pathogens show little to no activity against gram-negative pathogens because of their outer membrane.

Structure and Activity of LPS

The structure of LPS from several bacteria is known, and there are many variations. As seen in **Figure 2.13**, the polysaccharide portion of LPS consists of two components, the *core polysaccharide* and the *O-specific polysaccharide*. In *Salmonella* species, where LPS has been well studied, the core polysaccharide consists of ketodeoxyoctonate (KDO), various seven-carbon sugars (heptoses),

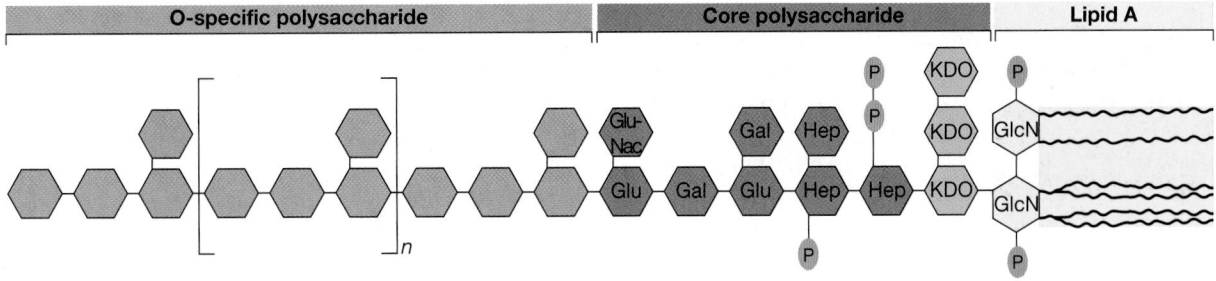

Figure 2.13 Structure of bacterial lipopolysaccharide. The chemistry of lipid A and the polysaccharide components varies among species of gram-negative *Bacteria*, but the major components (lipid A–KDO–core–O-specific) are typically the same. The O-specific polysac-

charide is highly variable among species. KDO, ketodeoxyoctonate; Hep, heptose; Glu, glucose; Gal, galactose; GluNac, *N*-acetylglucosamine; GlcN, glucosamine; P, phosphate. Glucosamine and the lipid A fatty acids are linked through the amine groups of GlcN.

The lipid A portion of LPS can be toxic to animals and comprises the endotoxin complex (⟲ Section 25.7 and Figure 25.18). Compare this figure with Figure 2.14 and follow the LPS components by their color-coding.

glucose, galactose, and *N*-acetylglucosamine. Connected to the core is the O-specific polysaccharide, which typically contains galactose, glucose, rhamnose, and mannose, as well as one or more dideoxyhexoses, such as abequose, colitose, paratose, or tyvelose. These sugars are connected in four- or five-membered sequences, which often are branched. When the sequences repeat, the long O-specific polysaccharide is formed.

The relationship of the LPS layer to the overall gram-negative cell wall is shown in **Figure 2.14**. The lipid portion of the LPS, called *lipid A*, is not a typical glycerol lipid (see Figure 2.4*a*); instead the fatty acids are bonded through the amine groups from a disaccharide composed of glucosamine phosphate. The disaccharide is attached to the core polysaccharide through KDO (Figure 2.13). Fatty acids commonly found in lipid A include caproic (C_6), lauric

(C_{12}), myristic (C_{14}), palmitic (C_{16}), and stearic (C_{18}) acids. LPS replaces much of the phospholipid in the outer half of the outer membrane, and although the outer membrane is technically a lipid bilayer, its many unique components distinguish it from the cytoplasmic membrane. The outer membrane is anchored to the peptidoglycan layer by the *Braun lipoprotein*, a molecule that spans the gap between the LPS layer and the peptidoglycan layer (in the periplasm, discussed in the next subsection) (Figure 2.14*a*).

An important biological activity of LPS is its toxicity to animals. Common gram-negative pathogens for humans include species of *Salmonella*, *Shigella*, and *Escherichia*, among many others, and some of the gastrointestinal symptoms these pathogens elicit are due to their toxic outer membrane components. Toxicity is specifically linked to the LPS layer, in particular, to lipid A. The term

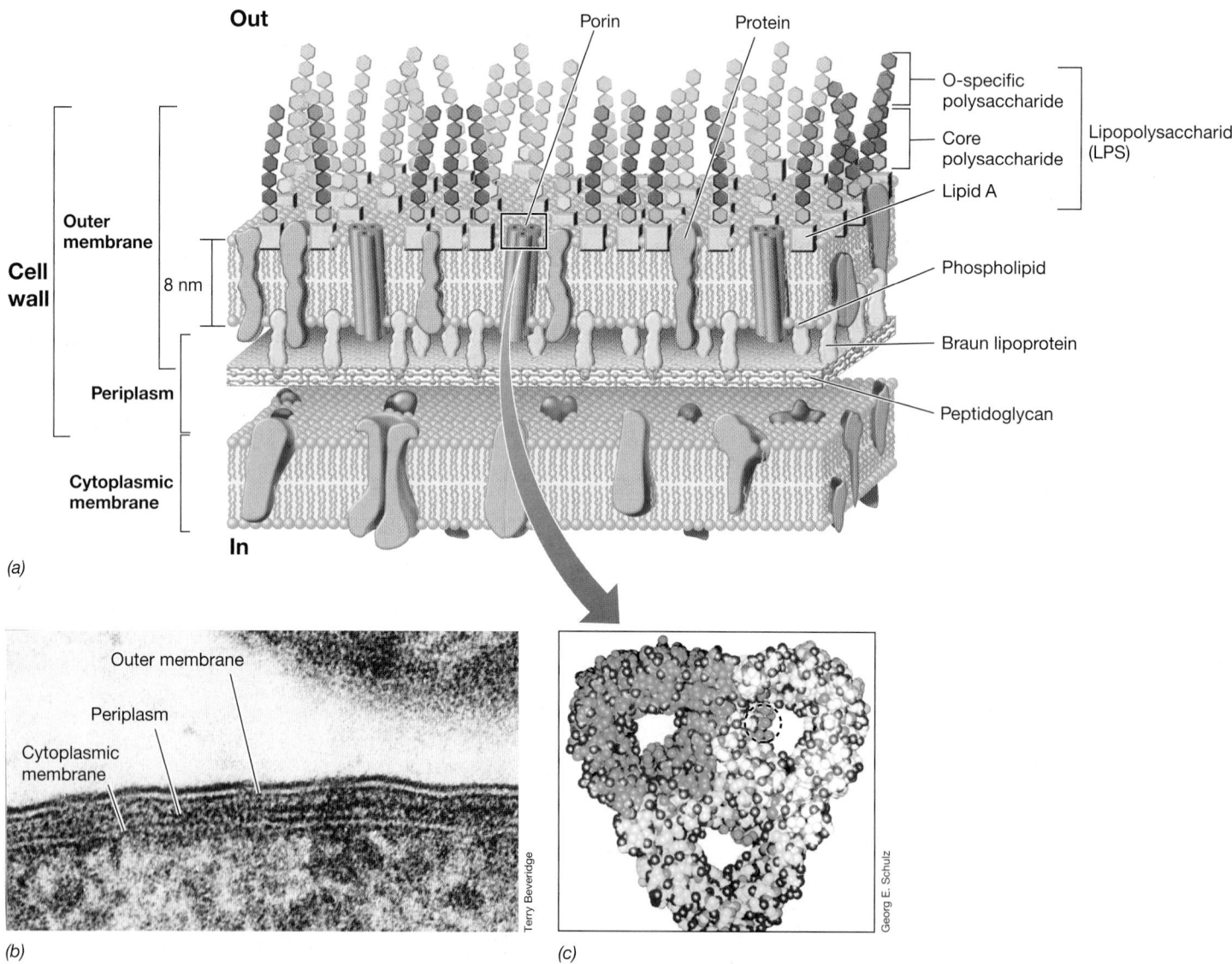

Figure 2.14 The gram-negative bacterial cell wall. *(a)* Arrangement of lipopolysaccharide, lipid A, phospholipid, porins, and Braun lipoprotein in the outer membrane. See Figure 2.13 for details of the structure of LPS. *(b)* Transmission electron micrograph of a cell of *Escherichia coli* showing the cytoplasmic membrane and wall. *(c)* Molecular model of porin proteins. Note the four pores present, one within each of the proteins forming a porin molecule and a smaller central pore (circled) between the porin proteins. The view is perpendicular to the plane of the membrane.

endotoxin refers to this toxic component of LPS. Some endotoxins cause violent symptoms in humans, including gas, diarrhea, and vomiting, and the endotoxins produced by *Salmonella* and enteropathogenic strains of *E. coli* transmitted in contaminated foods are classic examples of this. We discuss major gram-negative enteric pathogens in Chapter 32 and endotoxin in Section 25.7.

The Periplasm and Porins

The outer membrane is impermeable to proteins and other very large molecules. In fact, a major function of the outer membrane is to prevent cellular proteins whose activities must occur outside the cytoplasm from diffusing away from the cell. These extracellular proteins reside in the **periplasm**. This space, located between the outer surface of the cytoplasmic membrane and the inner surface of the outer membrane, spans about 15 nm (Figure 2.14*a, b*).

The periplasm may contain several different classes of proteins. These include hydrolytic enzymes, which function in the initial degradation of polymeric substances; binding proteins, which begin the process of transporting substrates (⇌ Section 3.2); chemoreceptors, which are proteins that govern the chemotaxis response (Section 2.13); and proteins that construct extracellular structures (such as peptidoglycan and the outer membrane) from precursor molecules secreted through the cytoplasmic membrane. Most periplasmic proteins reach the periplasm by way of a protein-exporting system present in the cytoplasmic membrane (⇌ Sections 4.12 and 4.13).

The outer membrane is relatively permeable to small molecules because of proteins called *porins* that function as channels for the entrance and exit of solutes (Figure 2.14*a, c*). Several porins are known, including both specific and nonspecific classes. Nonspecific porins form water-filled channels through which virtually any very small hydrophilic substance can pass. By contrast, specific porins contain a binding site for one or a group of structurally related substances. Porins are transmembrane proteins composed of three identical polypeptides. Besides the channel present in each subunit of a porin, the porin subunits aggregate in such a way that a hole about 1 nm in diameter is formed through which very small molecules can travel (Figure 2.14*c*).

The signature molecule of *Bacteria*—peptidoglycan—is absent from the *Archaea*. Nevertheless, cells of *Archaea* face the same osmotic stresses as do cells of *Bacteria* and thus need to counter these stresses with a cell wall. We consider the cell walls of *Archaea* now and see that the cell wall chemistries of these fascinating microbes occasionally hint at those of *Bacteria* but often confer their structural strength in chemically unique ways.

MINIQUIZ

- Describe the major chemical components in the outer membrane of gram-negative bacteria.
- What is the function of porins and where are they located in a gram-negative cell wall?
- What component of the gram-negative cell has endotoxin properties?

2.6 Archaeal Cell Walls

A variety of cell wall structures are found in *Archaea*, including walls containing polysaccharides, proteins, or glycoproteins or some mixture of these macromolecules.

Pseudomurein and Other Polysaccharide Cell Walls

The cell walls of certain methane-producing *Archaea* (methanogens) contain a molecule that is remarkably similar to peptidoglycan, a polysaccharide called *pseudomurein* (the term "murein" is from the Latin word for wall and was an old term for peptidoglycan) (**Figure 2.15**). The backbone of pseudomurein is formed from alternating repeats of *N*-acetylglucosamine (also present in peptidoglycan) and *N*-acetyltalosaminuronic acid; the latter replaces the *N*-acetylmuramic acid of peptidoglycan (Section 2.4). Pseudomurein also differs from peptidoglycan in that the glycosidic bonds between the sugar derivatives are β-1,3 instead of β-1,4, and the amino acids are all of the L stereoisomer (Figure 2.15).

Because in many respects they are so similar, it is likely that peptidoglycan and pseudomurein are variants of a cell wall polysaccharide originally present in the common ancestor of *Bacteria* and *Archaea*. However, although they are structurally very similar, they differ sufficiently that pseudomurein is immune from destruction by both lysozyme and penicillin, molecules that destroy peptidoglycan (Section 2.4).

Cell walls of some other *Archaea* lack pseudomurein and instead contain other polysaccharides. For example, *Methanosarcina* species have thick polysaccharide walls composed of polymers of glucose, glucuronic acid, galactosamine uronic acid, and acetate. Extremely halophilic (salt-loving) *Archaea* such as *Halococcus*, which are related to *Methanosarcina*, have similar cell walls that contain large amounts of sulfate. The negative charges on the sulfate ion (SO_4^{2-}) bind the abundant Na^+ present in the habitats of *Halococcus*

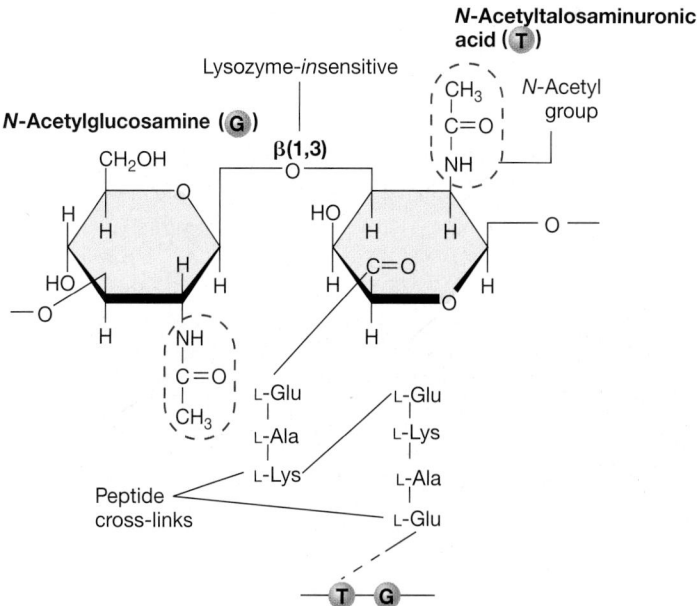

Figure 2.15 Pseudomurein. Structure of pseudomurein, the cell wall polymer of *Methanobacterium* species. Note the similarities and differences between pseudomurein and peptidoglycan (Figures 2.10 and 2.11).

(salt evaporation ponds and saline seas and lakes), and this sulfate–sodium complex helps stabilize the *Halococcus* cell wall.

S-Layers

The most common type of cell wall in *Archaea* is the paracrystalline surface layer, or **S-layer** as it is called. S-layers consist of interlocking molecules of protein or glycoprotein (**Figure 2.16**). The paracrystalline structure of S-layers can form various symmetries, including hexagonal, tetragonal, or trimeric, depending upon the number and structure of the subunits of which it is composed. S-layers have been found in representatives of all major lineages of *Archaea* and also in some species of *Bacteria* (Figure 2.16).

The cell walls of some *Archaea*, for example the methanogen *Methanocaldococcus jannaschii*, consist only of an S-layer. Thus, S-layers are sufficiently strong to withstand osmotic pressures without any other wall components. However, in many organisms S-layers are present in addition to other cell wall components, usually polysaccharides. When an S-layer accompanies other wall components, the S-layer is always the *outermost* wall layer; that is, the layer that is in direct contact with the environment.

Besides serving as protection from osmotic lysis, S-layers undoubtedly have other functions. For example, as the interface between the cell and its environment, it is likely that the S-layer functions as a selective sieve, allowing the passage of low-molecular-weight solutes while excluding large molecules or structures (such as viruses or lytic enzymes). The S-layer may also function to retain proteins near the cell surface that must function outside the cytoplasmic membrane, much as the outer membrane (Section 2.5) retains periplasmic proteins and prevents their drifting away in gram-negative *Bacteria*.

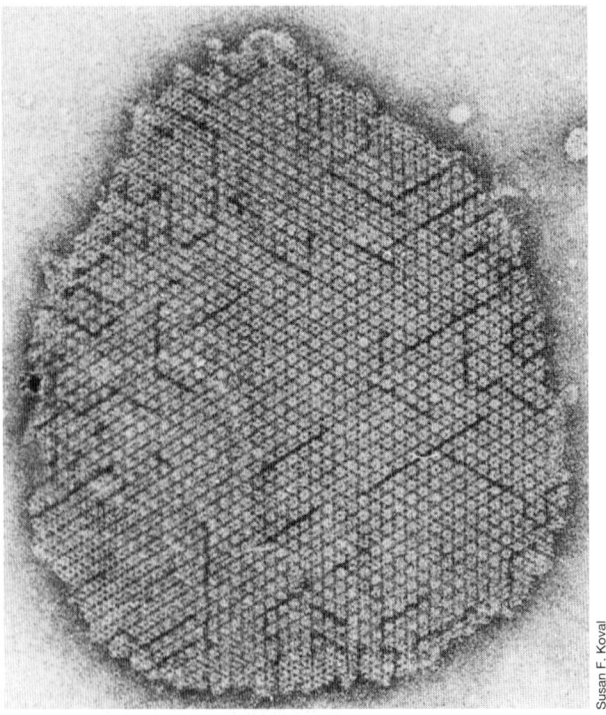

Susan F. Koval

Figure 2.16 The S-layer. Transmission electron micrograph of an S-layer fragment showing its paracrystalline nature. Shown is the S-layer from *Aquaspirillum* (*Bacteria*); this S-layer shows hexagonal symmetry common in S-layers of *Archaea*.

> **MINIQUIZ**
> - How does pseudomurein resemble peptidoglycan? How do the two molecules differ?
> - What is the structure of an S-layer, and what are its functions?

III • Cell Surface Structures and Inclusions

In addition to the cytoplasmic membrane and cell wall, cells of *Bacteria* and *Archaea* may have other layers or structures in contact with the environment and often contain one or more types of cellular inclusions. We examine some of these here.

2.7 Cell Surface Structures

Many *Bacteria* and *Archaea* secrete sticky or slimy materials on their cell surface that consist of either polysaccharide or protein. However, these are not considered part of the cell wall because they do not confer significant structural strength on the cell. The terms "capsule" and "slime layer" are used to describe these layers.

Capsules and Slime Layers

The terms capsule and slime layer are often used interchangeably, but the two terms do not refer to the same thing. If the layer is organized in a tight matrix that excludes small particles and is tightly attached, it is called a **capsule**. Capsules are readily visible by light microscopy if cells are treated with India ink, which stains the background but not the capsule, and can also be seen in the electron microscope (**Figure 2.17b–d**). By contrast, if the layer is more easily deformed and loosely attached, it will not exclude particles and is more difficult to see microscopically. This form is called a *slime layer* and is easily recognized in colonies of slime-forming species such as the lactic acid bacterium *Leuconostoc* (Figure 2.17a).

Outer surface layers have several functions. Surface polysaccharides assist in the attachment of microorganisms to solid surfaces. As we will see later, pathogenic microorganisms that enter the body by specific routes usually do so by first binding specifically to surface components of host tissues; this binding is often facilitated by bacterial cell surface polysaccharides. When the opportunity arises, many bacteria will bind to solid surfaces, often forming a thick layer of cells called a *biofilm*. Extracellular polysaccharides play a key role in the development and maintenance of biofilms as well.

Besides attachment, outer surface layers have other functions. These include acting as virulence factors (molecules that contribute to the pathogenicity of a bacterial pathogen) and preventing dehydration. For example, the causative agent of the diseases anthrax and bacterial pneumonia—*Bacillus anthracis* and *Streptococcus pneumoniae*, respectively—each contain a thick capsule of either protein (*B. anthracis*) or polysaccharide (*S. pneumoniae*). Encapsulated cells of these bacteria avoid destruction by the host's immune system because the immune cells that would otherwise recognize these pathogens as foreign and destroy them are blocked from doing so by the bacterial capsule. In addition to this role in

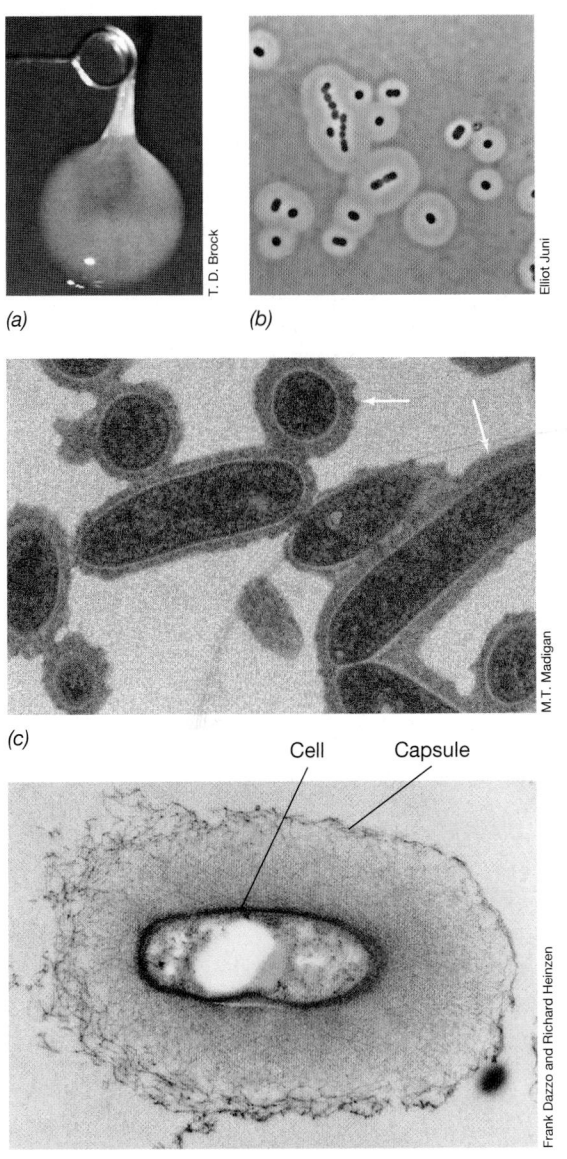

(a)   T. D. Brock

(b)   Elliot Juni

(c)   M.T. Madigan

Cell   Capsule

(d)   Frank Dazzo and Richard Heinzen

Figure 2.17 Bacterial capsules and slime formation. *(a)* A viscid colony of the bacterium *Leuconostoc mesenteroides* (lifted up by an inoculating loop) contains a thick dextran (glucose polymer) slime layer formed by the cells. *(b)* Capsules of *Acinetobacter* species observed by phase-contrast microscopy after negative staining with India ink. India ink does not penetrate the capsule and so the capsule appears as a light area surrounding the cell, which appears black. *(c)* Transmission electron micrograph of a thin section of cells of *Rhodobacter capsulatus* with capsules (arrows) clearly evident; cells are about 0.9 μm wide. *(d)* Transmission electron micrograph of *Rhizobium trifolii* stained with ruthenium red to reveal the capsule. The cell is about 0.7 μm wide.

disease, outer surface layers of virtually any type bind water and because of this likely protect the cell from desiccation in periods of dryness.

Fimbriae, Pili, and Hami

Fimbriae and pili are thin (2–10 nm in diameter) filamentous structures made of protein that extend from the surface of a cell and can have many functions. *Fimbriae* (**Figure 2.18**) enable cells to stick to surfaces, including animal tissues in the case of patho-

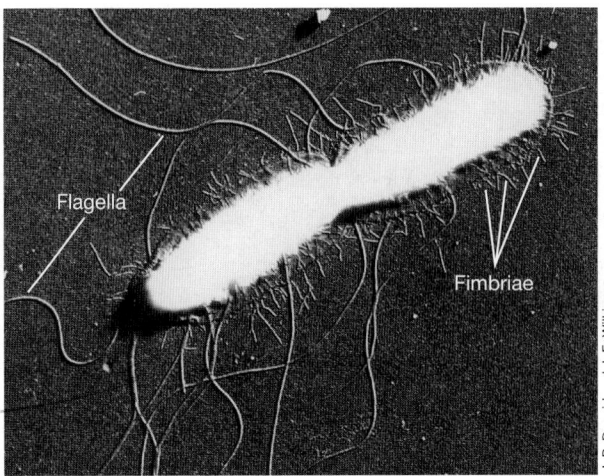

Flagella

Fimbriae

J. P. Duguid and J. F. Wilkinson

Figure 2.18 Fimbriae. Electron micrograph of a dividing cell of *Salmonella enterica (typhi)*, showing flagella and fimbriae. A single cell is about 0.9 μm wide.

genic bacteria, or to form pellicles (thin sheets of cells on a liquid surface) or biofilms on solid surfaces. **Pili** are similar to fimbriae, but are typically longer and only one or a few pili are present on the surface of a cell. All gram-negative bacteria produce pili of one sort or another, and many gram-positive bacteria also contain these structures. Because pili can be receptors for certain types of viruses, they can be easily seen under the electron microscope when they become coated with virus particles (**Figure 2.19**).

Many classes of pili are known, distinguished by their structure and function. Two very important functions of pili include facilitating genetic exchange between cells in a process called *conjugation* (conjugative or sex pili) and enabling the adhesion of pathogens to specific host tissues that they subsequently invade (type IV and other pili). *Type IV pili* not only facilitate specific adhesion but also support an unusual form of cell movement called *twitching motility* in certain bacterial species. On rod-shaped cells that move by twitching, type IV pili are present only at the poles. Twitching motility is a type of gliding motility, movement along a solid surface (Section 2.12). In twitching motility, extension of pili followed by their retraction drags the cell along a solid surface, and ATP supplies the energy necessary for this movement. The motility of certain species of *Pseudomonas* and *Moraxella* are the best-known examples of twitching motility.

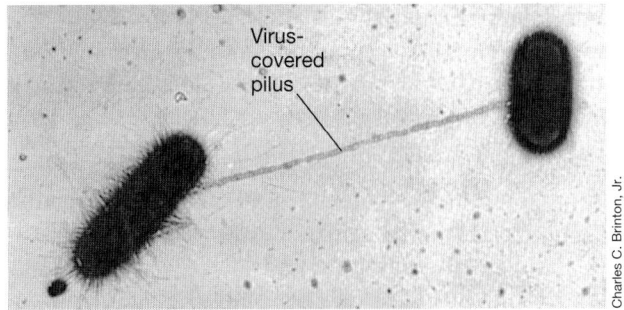

Virus-covered pilus

Charles C. Brinton, Jr.

Figure 2.19 Pili. The pilus on an *Escherichia coli* cell that is undergoing conjugation (a form of genetic transfer, Chapter 11) with a second cell is better resolved because viruses have adhered to it. The cells are about 0.8 μm wide.

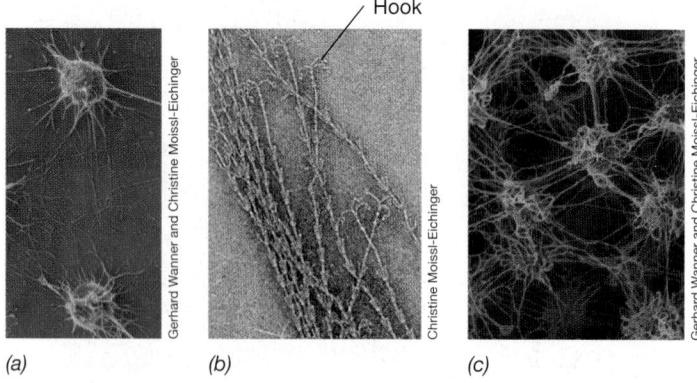

Hook

(a)   (b)   (c)

Gerhard Wanner and Christine Moissl-Eichinger

Christine Moissl-Eichinger

Gerhard Wanner and Christine Moissl-Eichinger

Figure 2.20 Unique attachment structures in the SM1 group of *Archaea*: Hami. *(a)* Cells of SM1 *Archaea* showing the pili-like surface structures called hami. *(b)* Transmission electron micrograph of isolated hami. A hamus "grappling hook" (labeled "Hook" in the micrograph) is about 60 nm in diameter. *(c)* A biofilm of SM1 cells showing the network of hami connecting individual cells.

Type IV pili have also been implicated as key colonization factors for certain human pathogens, including the gram-negative pathogens *Vibrio cholerae* (cholera) and *Neisseria gonorrhoeae* (gonorrhea) and the gram-positive pathogen *Streptococcus pyogenes* (strep throat and scarlet fever). The twitching motility of these organisms assists them in locating specific sites for attachment to initiate the disease process. Type IV pili also mediate genetic transfer by the process of transformation in some bacteria, which, along with conjugation and transduction, are the three known means of horizontal gene transfer in *Bacteria* and *Archaea* (Chapter 11). Type IV pili are also widespread in the *Archaea*, functioning in surface adhesion and cell aggregation events that lead to biofilm formation.

An unusual group of *Archaea*, the SM1 group, forms a unique attachment structure called a *hamus* (plural, hami) that resembles a tiny grappling hook (**Figure 2.20a, b**). The SM1 group inhabits anoxic groundwater in Earth's deep subsurface, and hami function to affix cells to a surface to form a networked biofilm (Figure 2.20c). Hami structurally resemble type IV pili except for their barbed terminus, which functions to attach cells both to surfaces and to each other (Figure 2.20c). The biofilms formed by SM1 *Archaea* are likely an ecological strategy that allows these microbes to more efficiently trap the scarce nutrients present in their deep subsurface habitat. Although cells of the SM1 group are not as small as the groundwater ultramicrobacterial cells described earlier (see Explore the Microbial World, "Tiny Cells"), they are less than 1 μm in diameter and live in a similar nutrient-limiting habitat. Thus, their hami probably play an important role in preventing cells from being washed away in groundwater flowage.

MINIQUIZ

- Could a bacterial cell dispense with a cell wall if it had a capsule? Why or why not?
- How do fimbriae differ from pili, both structurally and functionally?
- How can type IV pili facilitate pathogenesis? What are hami?

2.8 Cell Inclusions

Prokaryotic cells often contain inclusions of one sort or another. Inclusions function as energy reserves and/or carbon reservoirs or have special functions. Inclusions can often be seen in cells with the light microscope and are enclosed by a thin membrane that partitions off the inclusion in the cytoplasm. Storing carbon or other substances in an insoluble form is advantageous because it reduces the osmotic stress that the cell would encounter should the same amount be dissolved in the cytoplasm.

Carbon Storage Polymers

One of the most common inclusion bodies in prokaryotic organisms is **poly-β-hydroxybutyric acid (PHB)**, a lipid that is formed from β-hydroxybutyric acid units. The monomers of PHB polymerize by ester linkage and then the polymer aggregates into granules; the granules can be seen by either light or electron microscopy (**Figure 2.21**).

The monomer in the polymer is usually hydroxybutyrate (C_4) but can vary in length from as short as C_3 to as long as C_{18}. Thus,

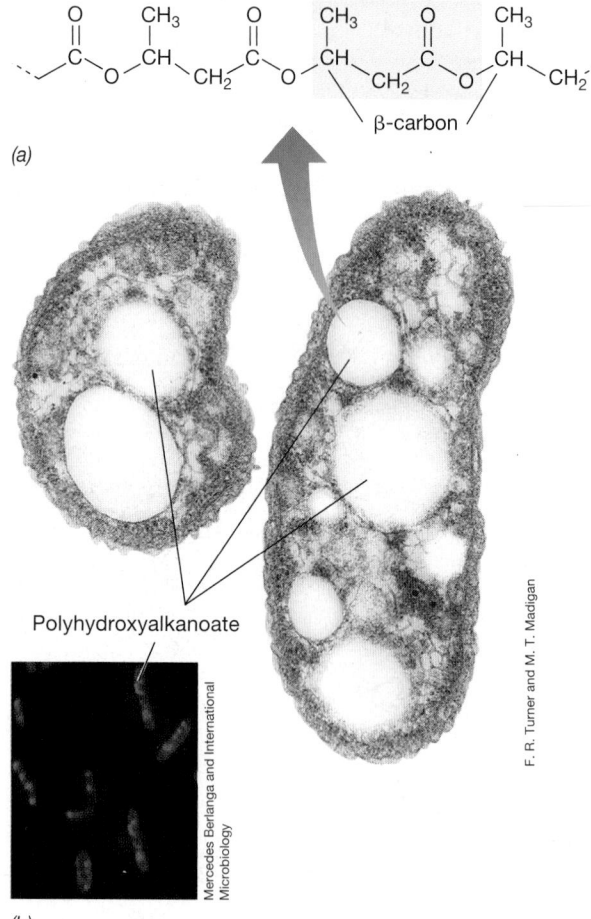

(a)

β-carbon

Polyhydroxyalkanoate

Mercedes Berlanga and International Microbiology

F. R. Turner and M. T. Madigan

(b)

Figure 2.21 Poly-β-hydroxyalkanoates (PHAs). *(a)* Chemistry of poly-β-hydroxybutyrate, a common PHA. A monomeric unit is shown in color. Other PHAs are made by substituting longer-chain hydrocarbons for the –CH_3 group on the β-carbon. *(b)* Electron micrograph of a thin section of cells of a bacterium containing granules of PHB. Color photo: Nile red–stained cells of a PHA-containing bacterium.

the more generic term poly-β-hydroxyalkanoate (PHA) is often used to describe this class of carbon- and energy-storage polymers. PHAs are synthesized by cells when there is an excess of carbon and are broken down as carbon or energy sources when conditions warrant. Many *Bacteria* produce PHAs, as do several extremely halophilic species of *Archaea*.

Another storage inclusion is *glycogen*, which is a polymer of glucose; like PHA, glycogen is a reservoir of both carbon and energy and is produced when carbon is in excess. Glycogen resembles starch, the major storage reserve of plants, but differs slightly from starch in the manner in which the glucose units are linked together.

Polyphosphate, Sulfur, and Carbonate Minerals

Many prokaryotic and eukaryotic microbes accumulate inorganic phosphate (PO_4^{3-}) in the form of *polyphosphate* granules (**Figure 2.22a**). These granules are formed when phosphate is in excess and can be drawn upon as a source of phosphate for nucleic acid and phospholipid biosynthesis when phosphate is limiting. In addition, in some organisms, polyphosphate can be broken down to synthesize the energy-rich compound ATP from ADP.

Many gram-negative *Bacteria* and *Archaea* oxidize reduced sulfur compounds, such as hydrogen sulfide (H_2S); these organisms are the "sulfur bacteria," discovered by the great Russian microbiologist Sergei Winogradsky (⮌ Section 1.11). The oxidation of sulfide generates electrons for use in energy metabolism (chemolithotrophy) or CO_2 fixation (autotrophy). In either case, *elemental sulfur* (S^0) from the oxidation of sulfide may accumulate in the cell in microscopically visible granules (Figure 2.22b). This sulfur remains as long as the source of reduced sulfur from which it was derived is still present. However, as the reduced sulfur source becomes limiting, the S^0 in the granules is oxidized to sulfate (SO_4^{2-}), and the granules slowly disappear. Interestingly, although sulfur globules appear to reside in the cytoplasm (Figure 2.22b), they are actually

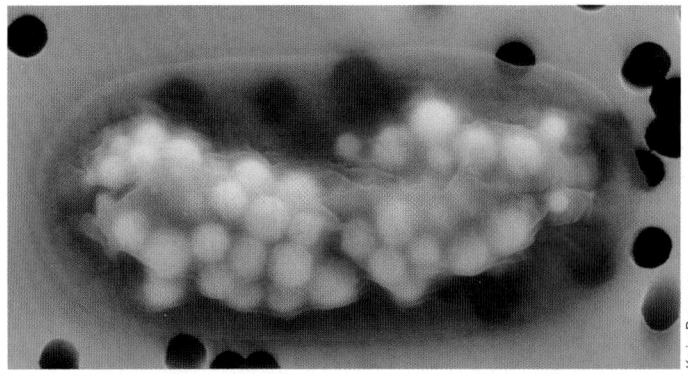

Figure 2.23 Biomineralization by a cyanobacterium. Electron micrograph of a cell of the cyanobacterium *Gloeomargarita* containing granules of the mineral benstonite [$(Ba,Sr)_6(Ca,Mn)_6Mg(CO_3)_{13}$]. A cell is about 2 μm wide.

present in the periplasm (Section 2.5). In these cells the periplasm expands outward to accommodate the growing globules as H_2S is oxidized to S^0 and then contracts inward as S^0 is oxidized to SO_4^{2-} (⮌ Section 14.9).

Filamentous cyanobacteria have long been known to form carbonate minerals on the external surface of their cells. However, some cyanobacteria also form carbonate minerals *inside* the cell, as cell inclusions. For example, the unicellular cyanobacterium *Gloeomargarita* forms intracellular granules of benstonite, a carbonate mineral that contains barium, strontium, and magnesium (**Figure 2.23**). The microbiological process of forming minerals is called *biomineralization*. It is unclear why benstonite is formed by *Gloeomargarita*, although it might function as ballast to maintain cells of this cyanobacterium in their habitat, deep in an alkaline lake. Alternatively (or in addition), the mineral could be a way to sequester carbonate (a source of CO_2) to support autotrophic growth. The biomineralization of several different minerals is catalyzed by various bacteria, but only in the case of *Gloeomargarita* and magnetosomes (to be discussed next) do we see the process yield actual intracellular inclusions.

Magnetic Storage Inclusions: Magnetosomes

Some bacteria can orient themselves within a magnetic field because they contain **magnetosomes**. These structures are biomineralized particles of the magnetic iron oxides magnetite [$Fe(II)Fe(III)_2O_4$] or greigite [$Fe(II)Fe(III)_2S_4$] (**Figure 2.24**). Magnetosomes impart a magnetic dipole on a cell, allowing it to orient itself in a magnetic field. This allows the cell to undergo *magnetotaxis*, the process of migrating along Earth's magnetic field lines. Magnetosomes have been found in several aquatic organisms that grow best at low O_2 concentrations or are anaerobic. It has thus been hypothesized that one function of magnetosomes may be to guide these aquatic cells downward (the direction of Earth's magnetic field) toward the sediments where O_2 is low or absent.

Magnetosome synthesis begins with insertion of magnetosome-specific proteins into the cytoplasmic membrane followed by invagination of the membrane to form a vesicle. The vesicle is then filled with iron—primarily iron in the Fe(II) oxidation state—and biomineralization proceeds through the

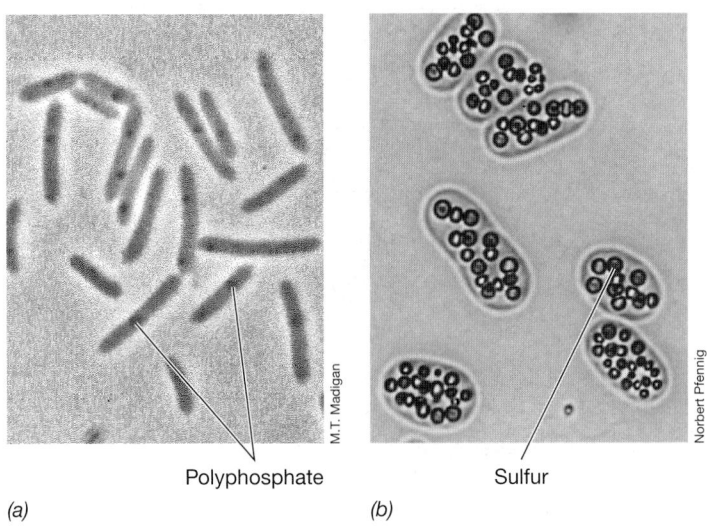

Polyphosphate   Sulfur

(a)   *(b)*

Figure 2.22 Polyphosphate and sulfur storage products. *(a)* Phase-contrast photomicrograph of cells of *Heliobacterium modesticaldum* showing polyphosphate as dark granules; a cell is about 1 μm wide. *(b)* Bright-field photomicrograph of cells of the purple sulfur bacterium *Isochromatium buderi*. The periplasmic inclusions are sulfur globules formed from the oxidation of hydrogen sulfide (H_2S). A cell is about 4 μm wide.

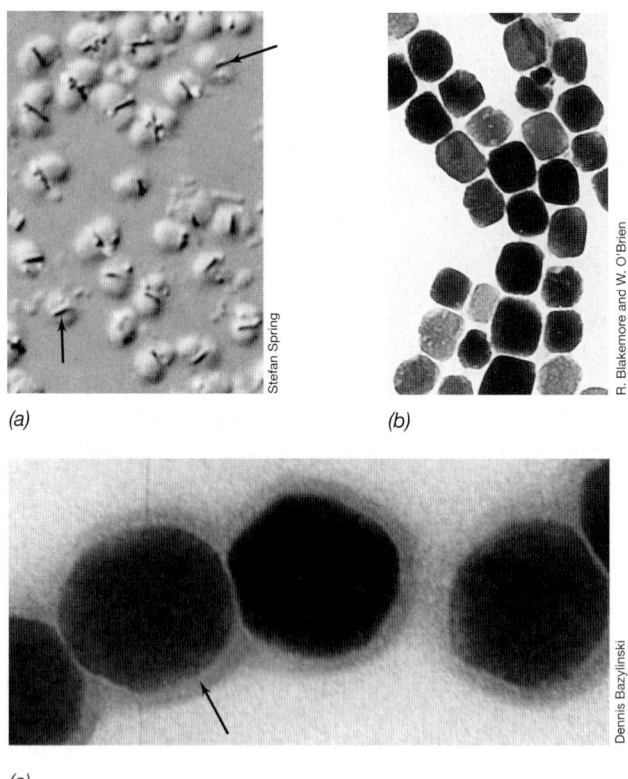

(a)

(b)

(c)

Figure 2.24 Magnetotactic bacteria and magnetosomes. *(a)* Differential interference contrast micrograph of coccoid magnetotactic bacteria; note chains of magnetosomes (arrows). A cell is 2.2 μm wide. *(b)* Magnetosomes isolated from the magnetotactic bacterium *Magnetospirillum magnetotacticum*; each particle is about 50 nm wide. *(c)* Transmission electron micrograph of magnetosomes from an unnamed magnetic coccus. The arrow points to the membrane that surrounds each magnetosome. A single magnetosome is about 90 nm wide.

activities of the magnetosome proteins, which includes an iron oxidase that generates the Fe(III) needed to form the magnetic minerals. The morphology of magnetosomes varies and appears to be species-specific; several morphologies are possible, but square, rectangular, or spike-shaped magnetosomes are most common.

MINIQUIZ

- Under what nutritional conditions would you expect PHAs or glycogen to be produced?

- Why would it be impossible for gram-positive bacteria to store sulfur as gram-negative sulfur-oxidizing chemolithotrophs can?

- How are magnetosomes and the *Gloeomargarita* inclusions similar and how do they differ?

2.9 Gas Vesicles

Some *Bacteria* and *Archaea* are *planktonic*, meaning that they inhabit the water column of lakes and the oceans. Most planktonic organisms move up and down with changes in currents, but some can float because they contain **gas vesicles**, structures that confer buoyancy and allow the cells to position themselves in regions of the water column that best suit their metabolisms.

The most dramatic examples of gas-vesiculate microbes are cyanobacteria that form massive accumulations called *blooms* in lakes or other bodies of water (**Figure 2.25a**). Cyanobacteria are oxygenic phototrophic bacteria (⮂ Section 14.4). Gas-vesiculate cells rise to the surface of the lake and are blown by winds into dense masses. Several other primarily aquatic *Bacteria* and *Archaea* have gas vesicles, but the structures are not found in microbial eukaryotes.

Gas Vesicle Structure

Gas vesicles are conical-shaped structures made of protein; they are hollow yet rigid and of variable length and diameter (Figure 2.25b, c and see Figure 2.26a). Gas vesicles in different species vary in length from about 300 to more than 1000 nm and in width from 45 to 120 nm, but the vesicles of a given species are of constant size. Gas vesicles may number from a few to hundreds per cell and are impermeable to water and solutes but permeable to gases. The presence of gas vesicles in cells can be detected

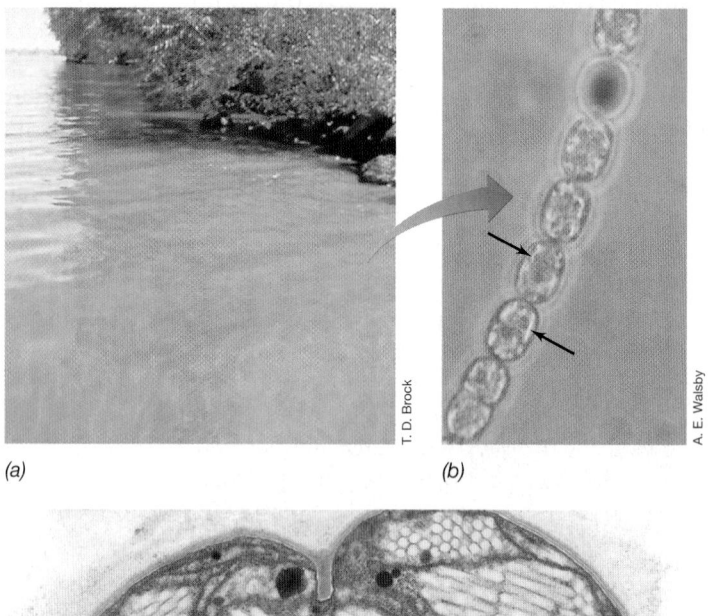

(a)

(b)

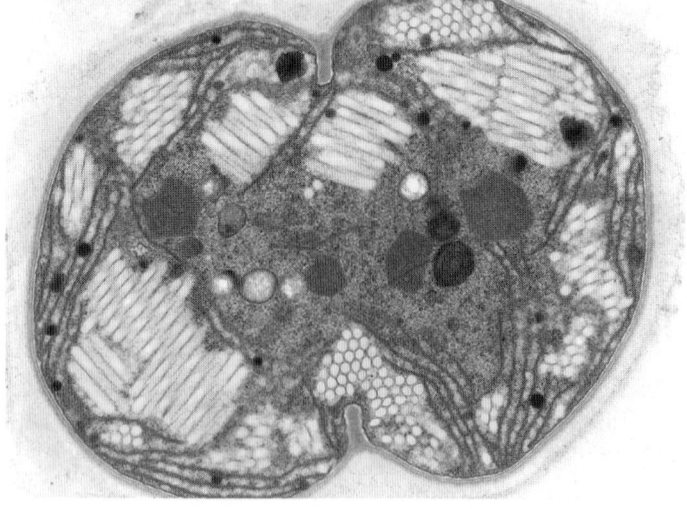

(c)

Figure 2.25 Buoyant cyanobacteria and their gas vesicles. *(a)* Flotation of a bloom of gas-vesiculate cyanobacteria in a freshwater lake. *(b)* Phase-contrast photomicrograph of *Anabaena*. Clusters of gas vesicles form phase-bright gas vacuoles (arrows). *(c)* Transmission electron micrograph of *Microcystis*. Gas vesicles are arranged in bundles, here seen in both longitudinal and cross-section. Both cells are about 5 μm wide.

either by light microscopy, where clusters of vesicles, called *gas vacuoles*, appear as irregular bright inclusions (Figure 2.25*b*), or by transmission electron microscopy of cell thin sections (Figure 2.25*c*).

Gas vesicles are composed of two distinct proteins (**Figure 2.26*b***). The major gas vesicle protein, called *GvpA*, forms the watertight vesicle shell and is a small, hydrophobic, and very rigid protein; multiple copies of GvpA align to form the parallel "ribs" of the vesicle. The rigidity is essential for the structure to resist the pressures exerted on it from outside. A minor protein, called *GvpC*, functions to strengthen the shell of the gas vesicle by cross-linking and binding the ribs at an angle to group several GvpA molecules together (Figure 2.26*b*).

The composition and pressure of the gas inside a gas vesicle is that in which the organism is suspended. Because an inflated gas vesicle has a density only one-tenth that of the cell proper, inflated gas vesicles combine to decrease a cell's density and thereby increase its buoyancy. If and when vesicles are collapsed, buoyancy is lost. Phototrophic bacteria (Chapter 14) in particular can benefit from gas vesicles because they allow cells to adjust their vertical position in a water column to sink or rise to regions where conditions (for example, light intensity) are optimal for photosynthesis.

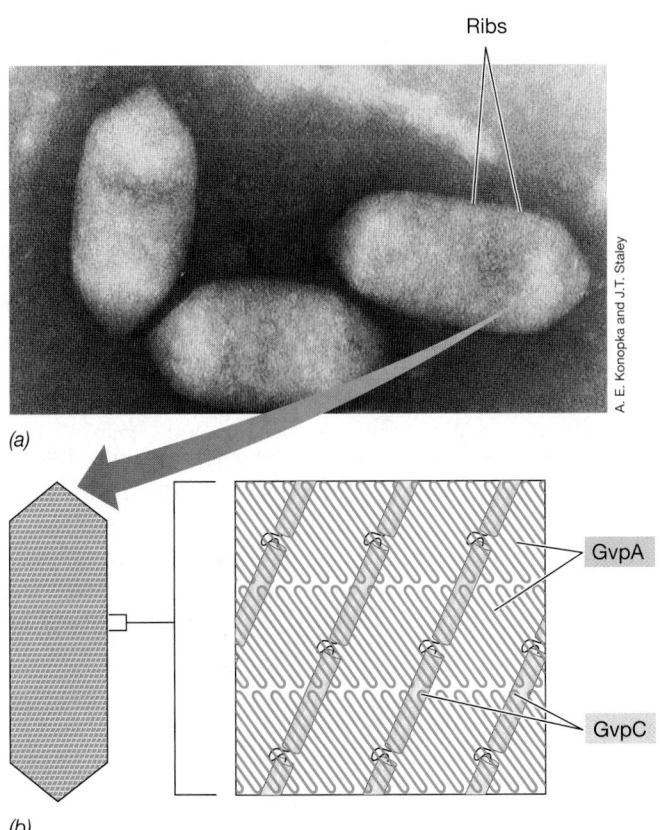

(a)

(b)

A. E. Konopka and J.T. Staley

Figure 2.26 Gas vesicle architecture. *(a)* Transmission electron micrograph of gas vesicles purified from the bacterium *Ancylobacter aquaticus* and examined in negatively stained preparations. A single vesicle is about 100 nm in diameter. *(b)* Model of how gas vesicle proteins GvpA and GvpC interact to form a watertight but gas-permeable structure. GvpA, a rigid β-sheet, makes up the rib, and GvpC, an α-helix structure, is the cross-link. (See Figure 4.30*b, c* for the structures of a β-sheet and an α-helix.)

── MINIQUIZ ──

- What gas is present in a gas vesicle? Why might a photosynthetic cell benefit from controlling its buoyancy?
- How are the two proteins that make up the gas vesicle, GvpA and GvpC, arranged to form such a water-impermeable structure?

2.10 Endospores

Certain species of *Bacteria* produce structures called **endospores** (**Figure 2.27**) during a process called *endosporulation* (or just *sporulation* for short). Endospores (the prefix *endo-* means "within") are highly differentiated cells that are extremely resistant to heat, harsh chemicals, and radiation. Endospores function as survival structures and enable the organism to endure unfavorable growth conditions, including but not limited to extremes of temperature, drying, or nutrient depletion. Endospores can thus be thought of as the dormant stage of a bacterial life cycle: vegetative cell → endospore → vegetative cell. Endospores are easily dispersed by wind, water, or through the animal gut, and hence endospore-forming bacteria are widely distributed in nature.

The endospore-forming bacteria *Bacillus* and *Clostridium* are common in soil and the best-studied representatives. Some endospore-forming bacteria are serious pathogens of humans and other animals, the endospore stage being an effective way of surviving outside the host or until conditions within the host can support disease. Botulism, tetanus, and several foodborne bacterial infections are caused by species of endospore-forming bacteria.

Endospore Formation and Germination

During endospore formation, a vegetative cell is converted into a nongrowing, heat-resistant, and light-refractive structure (**Figure 2.28**). Cells do not sporulate when they are actively growing but only when growth ceases owing to the exhaustion of an essential nutrient. Thus, cells of *Bacillus*, a typical endospore-forming bacterium, cease vegetative growth and begin sporulation when, for example, a key nutrient such as carbon or nitrogen becomes limiting (⇨ Section 7.6).

An endospore can remain dormant for years but can convert back to a vegetative cell rapidly. This process occurs in three

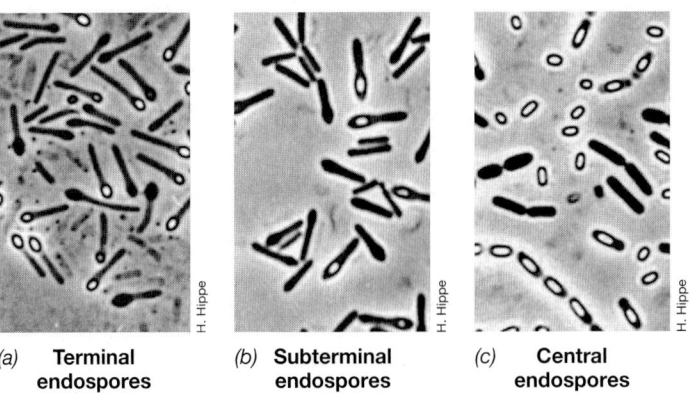

(a) **Terminal endospores**

(b) **Subterminal endospores**

(c) **Central endospores**

H. Hippe

Figure 2.27 The bacterial endospore. Phase-contrast photomicrographs showing different intracellular locations of endospores in different species of bacteria. Endospores appear bright by phase-contrast microscopy.

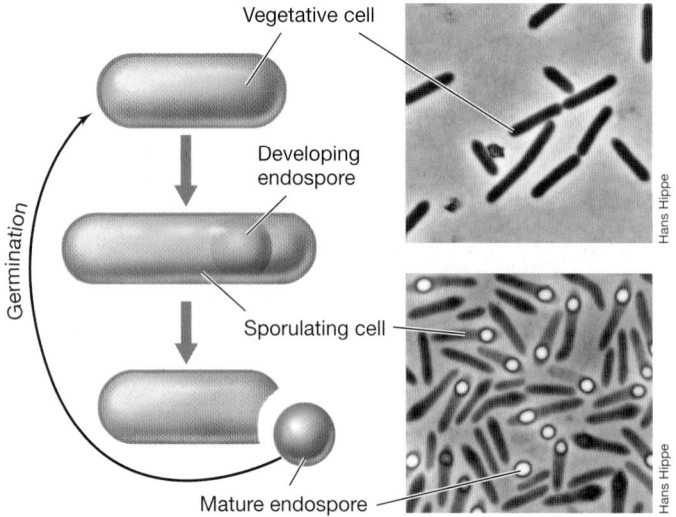

Figure 2.28 The life cycle of an endospore-forming bacterium. The phase-contrast photomicrographs are of cells of *Clostridium pascui*. A cell is about 0.8 μm wide.

steps: *activation, germination,* and *outgrowth* (**Figure 2.29**). Activation occurs when endospores are heated for several minutes at an elevated but sublethal temperature. Activated endospores are then conditioned to germinate when supplied with certain nutrients, such as certain amino acids. Germination, typically a rapid process (occurring in a matter of minutes), is signaled by the loss of refractility of the endospore (Figure 2.29*b*) and loss of resistance to heat and chemicals. The final stage, outgrowth (Figure 2.29*c, d*), involves visible swelling due to water uptake and synthesis of RNA, proteins, and DNA. The vegetative cell emerges from the broken endospore and begins to grow, remaining in vegetative growth until environmental signals once again trigger sporulation.

Endospore Structure and Features

Endospores are visible by light microscopy as strongly refractile structures (Figures 2.27 and 2.29*a*). Endospores are impermeable to most dyes, so occasionally they are seen as unstained regions within cells that have been stained with basic dyes such as methylene blue. To stain endospores, special stains and procedures must be used. In the classical endospore-staining protocol, the stain malachite green is used and is infused into the spore with steam.

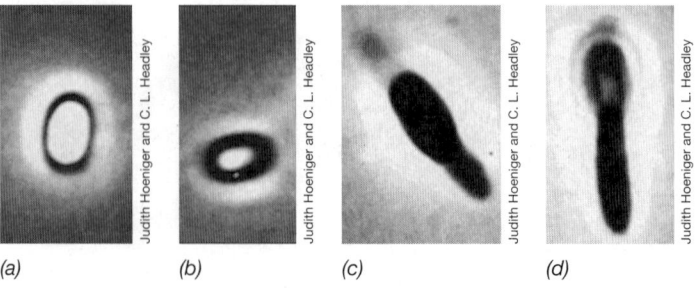

Figure 2.29 Endospore germination in *Bacillus*. Conversion of an endospore into a vegetative cell. The series of phase-contrast photomicrographs shows the sequence of events starting from *(a)* a highly refractile free endospore. *(b)* Activation: Refractility is diminishing. *(c, d)* Outgrowth: The new vegetative cell is emerging.

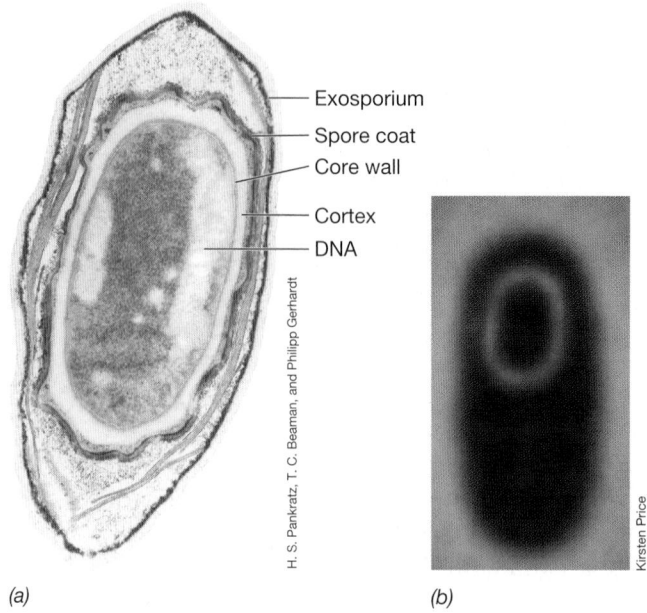

Figure 2.30 Structure of the bacterial endospore. *(a)* Transmission electron micrograph of a thin section through an endospore of *Bacillus megaterium*. *(b)* Fluorescent photomicrograph of a cell of *Bacillus subtilis* undergoing sporulation. The green color is a dye that specifically stains a sporulation protein in the spore coat.

The structure of the endospore as seen with the electron microscope differs distinctly from that of the vegetative cell (**Figure 2.30**). The endospore contains many layers absent from the vegetative cell. The outermost layer is the *exosporium*, a thin protein covering. Moving inward there are several *spore coats*, composed of layers of spore-specific proteins (Figure 2.30*b*). Below the spore coat is the *cortex*, which consists of loosely cross-linked peptidoglycan, and inside the cortex is the *core*, which contains the core wall, cytoplasmic membrane, cytoplasm, nucleoid, ribosomes, and other cellular essentials. Thus, the endospore differs structurally from the vegetative cell primarily in the kinds of structures found outside the core wall.

One substance found in endospores but not in vegetative cells is **dipicolinic acid** (**Figure 2.31***a*), which accumulates in the core. Endospores also contain large amounts of calcium (Ca^{2+}), most of which is complexed with dipicolinic acid (Figure 2.31*b*). The calcium–dipicolinic acid (DPA) complex forms about 10% of the dry weight of the endospore and functions to bind free water within the endospore, helping to dehydrate the developing

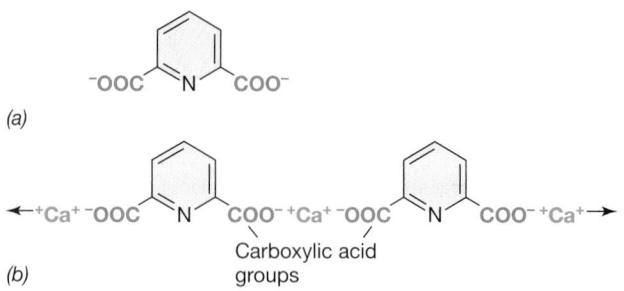

Figure 2.31 Dipicolinic acid (DPA). *(a)* Structure of DPA. *(b)* How Ca^{2+} cross-links DPA molecules to form a complex.

endospore. In addition, the DPA complex inserts between bases in DNA, which helps stabilize DNA against heat denaturation.

The core of the endospore differs significantly from the cytoplasm of the vegetative cell that produced it. The endospore core contains less than one quarter of the water found in the vegetative cell, and thus the consistency of the core cytoplasm is that of a gel. Dehydration of the core greatly increases the heat resistance of macromolecules within the spore. Some bacterial endospores survive heating to temperatures as high as 150°C, although 121°C, the standard for microbiological sterilization (121°C is autoclave temperature, ⮌ Section 5.15), kills the endospores of most species. Dehydration has also been shown to provide endospores with resistance to toxic chemicals, such as hydrogen peroxide (H_2O_2), and causes enzymes in the core to become inactive. In addition to the low water content of the endospore, the pH of the core is about one unit lower than that of the vegetative cell cytoplasm.

The endospore core contains high levels of *small acid-soluble spore proteins* (SASPs). These proteins are only made during the sporulation process and have at least two functions. SASPs bind tightly to DNA in the core and protect it from potential damage from ultraviolet radiation, desiccation, and dry heat. Ultraviolet resistance is conferred when SASPs alter the molecular structure of DNA from the normal "B" form to the (more compact) "A" form. A-form DNA better resists pyrimidine dimer formation by UV radiation, which can cause mutations (⮌ Section 11.4), and resists the denaturing effects of dry heat. In addition, SASPs function as a carbon and energy source for the outgrowth of a new vegetative cell from the endospore during germination.

TABLE 2.2 Differences between endospores and vegetative cells

Characteristic	Vegetative cell	Endospore
Microscopic appearance	Nonrefractile	Refractile
Calcium content	Low	High
Dipicolinic acid	Absent	Present
Enzymatic activity	High	Low
Respiration rate	High	Low or absent
Macromolecular synthesis	Present	Absent
Heat resistance	Low	High
Radiation resistance	Low	High
Resistance to chemicals	Low	High
Lysozyme	Sensitive	Resistant
Water content	High, 80–90%	Low, 10–25% in core
Small acid-soluble spore proteins	Absent	Present

The Sporulation Cycle

Sporulation is a form of cellular differentiation (⮌ Figure 1.4), and many genetically directed changes in the cell occur during the conversion from vegetative growth to sporulation (**Table 2.2**). The structural changes in sporulating cells of *Bacillus* are shown in **Figure 2.32**. Sporulation can be divided into several stages. In

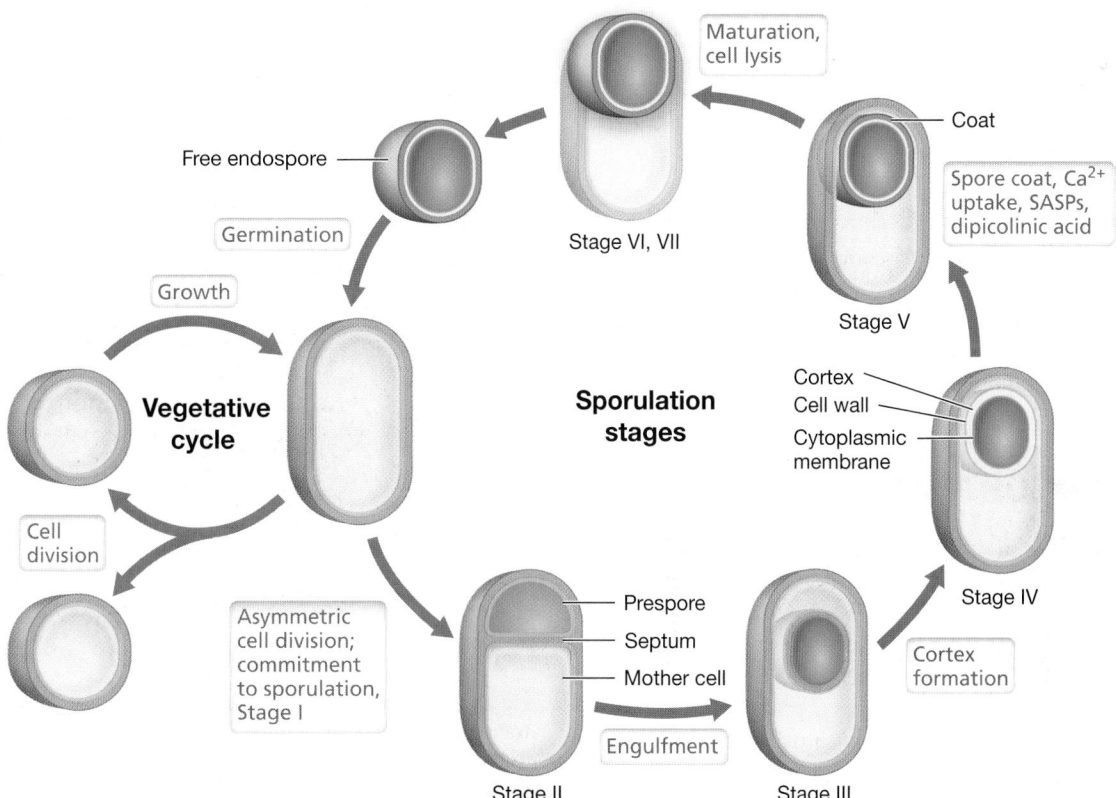

Figure 2.32 Stages in endospore formation. The stages are defined from genetic and microscopic analyses of sporulation in *Bacillus subtilis*, the model organism for studies of sporulation. SASPs, small acid-soluble proteins.

Bacillus subtilis, which has been studied in detail, the conversion of a vegetative cell into an endospore takes about 8 hours and begins with asymmetric cell division (Figure 2.32). Note how key events such as asymmetric cell division, cortex formation, and SASP production take place in a defined sequence and at specific times in the sporulation cycle (Figure 2.32). Genetic studies of mutants of *Bacillus subtilis*, each blocked at one of the stages of endosporulation, indicate that more than 200 spore-specific genes exist.

Endosporulation requires differential protein synthesis. This occurs by the sequential activation of several families of endospore-specific genes and the turning off of many vegetative cell functions. The proteins encoded by sporulation-specific genes catalyze the series of events leading from the moist, metabolizing, vegetative cell to the relatively dry, metabolically inert, but extremely resistant endospore (Table 2.2). In Section 7.6 we examine some of the molecular events that take place during the endosporulation process.

Diversity and Phylogenetic Aspects of Endospore Formation

Nearly 20 genera of *Bacteria* form endospores, although the process has only been studied in detail in a few species. Nevertheless, most of the major events described here, such as the formation of DPA complexes and the production of endospore-specific SASPs, seem universal. From a phylogenetic perspective, the capacity to produce endospores is limited to a particular lineage of the gram-positive bacteria. Despite this, the physiologies of endospore-forming bacteria are highly diverse and include anaerobes, aerobes, phototrophs, and chemolithotrophs. In light of this physiological diversity, the actual triggers for endospore formation may vary with different species and could include signals other than simple nutrient starvation, the major trigger for endospore formation in *Bacillus*. No *Archaea* have been shown to form endospores, suggesting that the capacity to produce endospores evolved sometime after *Bacteria* and *Archaea* diverged about 3.5 billion years ago.

MINIQUIZ

- What is dipicolinic acid and the DPA complex, and where is it found?
- What are SASPs and what is their function?
- What is formed when an endospore germinates?

IV • Cell Locomotion

We finish our survey of prokaryotic structure and function by examining cell locomotion. Many microbial cells can move under their own power. Motility allows cells to reach different parts of their environment, and in nature, a new location may offer additional resources for a cell and spell the difference between life and death.

We examine here the two major types of prokaryotic cell movement, *swimming* and *gliding*. We then consider how motile cells are able to move in a directed fashion toward or away from particular stimuli (phenomena called *taxes*) and present examples of these simple behavioral responses.

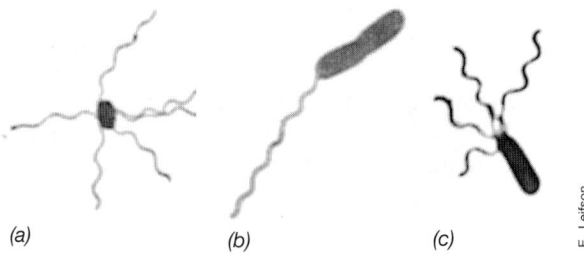

Figure 2.33 Bacterial flagella. Classic light photomicrographs taken by Einar Leifson of bacteria containing different arrangements of flagella. Cells are stained with the Leifson flagella stain. *(a)* Peritrichous. *(b)* Polar. *(c)* Lophotrichous.

2.11 Flagella, Archaella, and Swimming Motility

Many *Bacteria* are motile by swimming due to a structure called the **flagellum** (plural, flagella) (**Figure 2.33**); an analogous structure called the **archaellum** is present in many *Archaea*. Flagella and archaella are tiny rotating machines that function to push or pull the cell through a liquid.

Flagella and Flagellation

Bacterial flagella are long, thin appendages (15–20 nm wide, depending on the species) free at one end and anchored into the

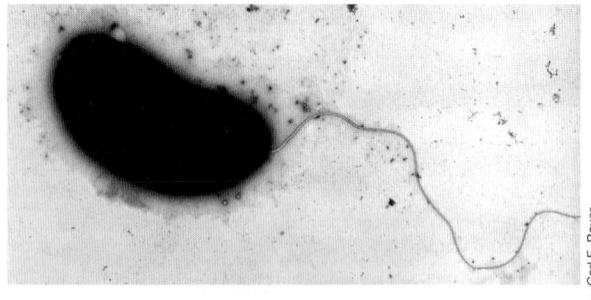

(a)

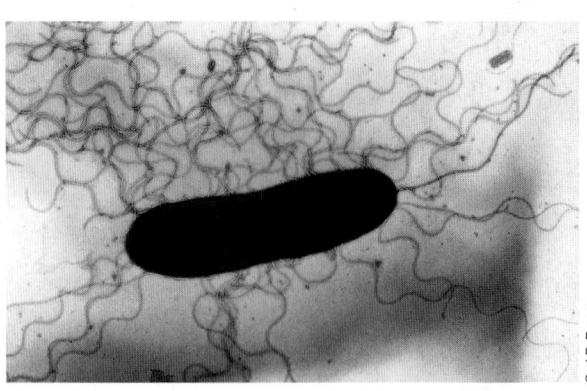

(b)

Figure 2.34 Bacterial flagella as observed by negative staining in the transmission electron microscope. *(a)* A single polar flagellum. *(b)* Peritrichous flagella. Both micrographs are of cells of the phototrophic bacterium *Rhodospirillum centenum*, which are about 1.5 μm wide. Cells of *R. centenum* are normally polarly flagellated but under certain growth conditions form peritrichous flagella. See Figure 2.44b for a photo of colonies of *R. centenum* cells that move toward an increasing gradient of light (phototaxis).

cell at the other end. Flagella can be stained and observed by light microscopy (Figure 2.33) or electron microscopy (**Figure 2.34**).

Flagella can be anchored to a cell in different locations. In **polar flagellation**, the flagella are attached at one or both ends of a cell (Figure 2.33*b*). Occasionally a group of flagella (called a *tuft*) may arise at one end of the cell, a type of polar flagellation called *lophotrichous* (Figure 2.33*c*). Tufts of flagella can sometimes be seen in large unstained cells by dark-field or phase-contrast microscopy (**Figure 2.35**). When a tuft of flagella emerges from both poles of the cell, flagellation is called *amphitrichous*. In **peritrichous flagellation** (Figures 2.33*a* and 2.34*b*), flagella are inserted around the cell surface.

Flagella do not rotate at a constant speed but increase or decrease their rotational speed in relation to the strength of the proton motive force. Flagella can rotate at up to 1000 revolutions per second to support a swimming speed of up to 60 cell-lengths/sec. The fastest known land animal, the cheetah, can move at about 25 body-lengths/sec. Thus, a bacterium swimming at 60 cell-lengths/sec is actually moving over twice as fast—relative to its size—as the fastest animal!

The swimming motions of polarly and lophotrichously flagellated organisms differ from those of peritrichously flagellated organisms, and these can be distinguished microscopically (**Figure 2.36**). Peritrichously flagellated organisms typically move slowly in a straight line. By contrast, polarly flagellated organisms move more rapidly, often spinning around and seemingly dashing from place to place. The different behavior of flagella on polar and peritrichous organisms, including differences in reversibility of the flagellum, is illustrated in Figure 2.36.

Flagella Structure and Activity

Flagella are not straight structures but are helical. The main part of the flagellum, called the *filament*, is composed of many copies of a

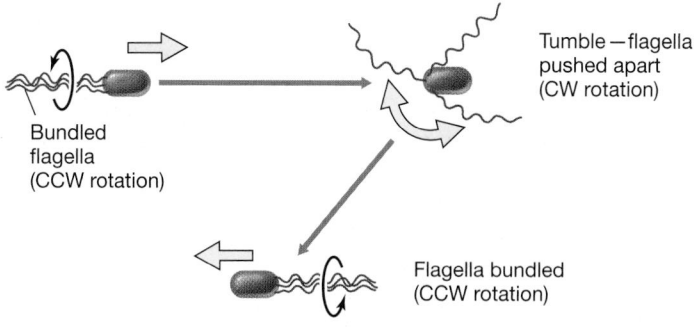

(a) **Peritrichous**

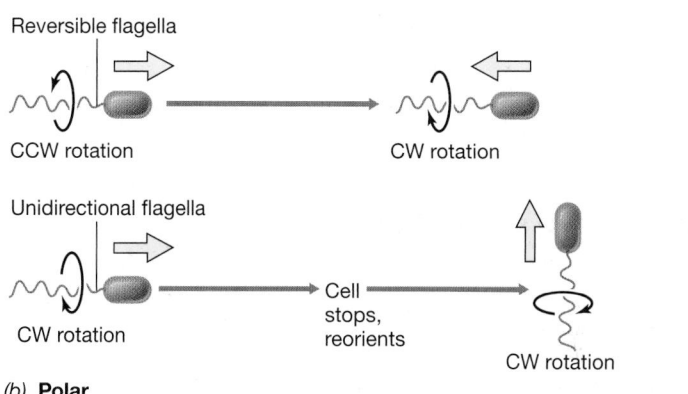

(b) **Polar**

Figure 2.36 Movement in peritrichously and polarly flagellated prokaryotic cells. *(a)* Peritrichous: Forward motion is imparted by all flagella forming into a bundle and rotating counterclockwise (CCW). Clockwise (CW) rotation causes the bundle to break apart and the cell to tumble. A return to counterclockwise rotation leads the cell off in a new direction. *(b)* Polar: Cells change direction by reversing flagellar rotation (thus pulling instead of pushing the cell) or, with unidirectional flagella, by stopping periodically to reorient and then moving forward by clockwise rotation of its flagella. The yellow arrows show the direction the cell is traveling.

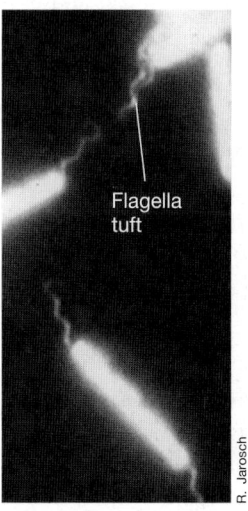

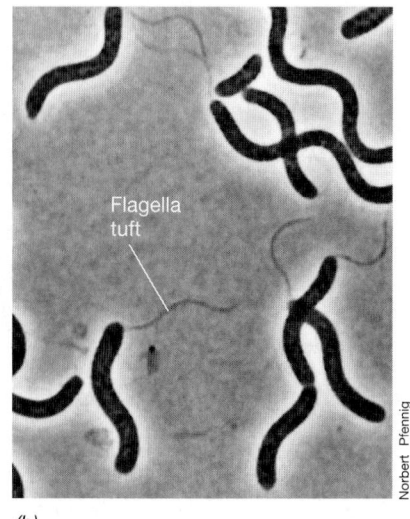

(a)   (b)

Figure 2.35 Bacterial flagella observed in living cells. *(a)* Dark-field photomicrograph of a group of large rod-shaped bacteria with flagellar tufts at each pole (a condition called *amphitrichous flagellation*). A single cell is about 2 μm wide. *(b)* Phase-contrast photomicrograph of cells of the large phototrophic purple bacterium *Rhodospirillum photometricum* with a tuft of lophotrichous flagella that emanate from one of the poles. A cell measures about 4 × 25 μm.

protein called *flagellin*. The amino acid sequence of flagellin is highly conserved in *Bacteria*, suggesting that flagellar motility evolved early and has deep roots within this domain. In addition to the filament, a flagellum consists of several other components. A wider region at the base of the filament called the *hook* consists of a single type of protein and connects the filament to the flagellum motor in the base (**Figure 2.37**).

The flagellum motor is a reversible rotating machine composed of several proteins and is anchored in the cytoplasmic membrane and cell wall. The motor consists of a central rod that passes through a series of rings. In gram-negative bacteria, an outer ring, called the *L ring*, is anchored in the outer membrane (Section 2.5). A second ring, called the *P ring*, is anchored in the peptidoglycan layer. A third set of rings, called the *MS and C rings*, are located within the cytoplasmic membrane and the cytoplasm, respectively (Figure 2.37*a*). In gram-positive bacteria, which lack an outer membrane, only the inner pair of rings is present. Surrounding the inner ring and anchored in the cytoplasmic membrane and peptidoglycan are a series of proteins called *Mot proteins*. Another set of proteins, called *Fli proteins* (Figure 2.37*a*), function as the motor switch, reversing the direction of rotation of the flagella in response to intracellular signals.

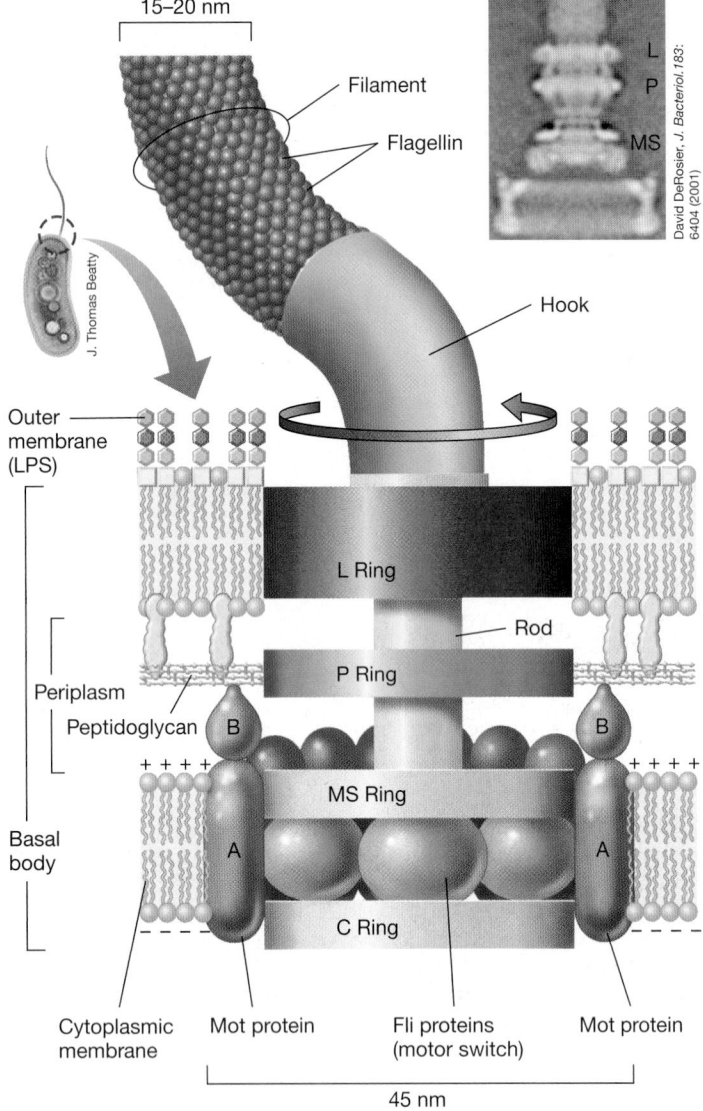

David DeRosier, *J. Bacteriol.183*: 6404 (2001)

J. Thomas Beatty

(a)

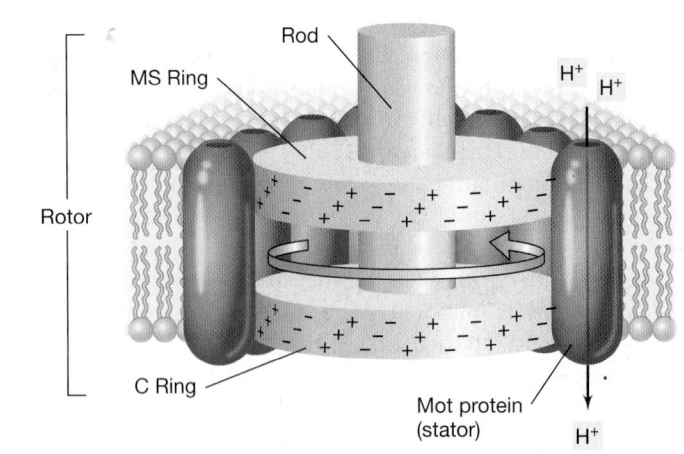

(b)

The flagellum motor contains two main components: the *rotor* and the *stator*. The rotor consists of the central rod and the L, P, C, and MS rings. Collectively, these structures make up the flagellar **basal body** (Figure 2.37). The stator consists of the Mot proteins that surround the rotor and function to generate torque. Rotation of the flagellum occurs at the expense of the proton motive force (Section 2.3), and it is thought that rotation is imparted to the flagellum by a type of "proton turbine" process. In this model, proton translocation through the Mot complex drives rotation of the flagellum, with about 1200 protons being translocated per each rotation of the flagellum (Figure 2.37*b*). Protons flowing through the Mot proteins exert electrostatic forces on helically arranged charges on the rotor proteins. Alternating attractions between positive and negative charges on the rotor as protons flow though the Mot proteins then cause the entire basal body to rotate. Rotational speed of the flagellum is set by the proton flow rate through the Mot proteins, which is a function of the intensity of the proton motive force.

Flagellar Synthesis

Several genes encode the motility apparatus of *Bacteria*. In *Escherichia* and *Salmonella* species, in which motility studies have been extensive, over 50 genes are linked to motility in one way or another. These genes encode the structural proteins of the flagellum and motor apparatus, of course, but also encode proteins that export the structural proteins through the cytoplasmic membrane to the outside of the cell and proteins that regulate the synthesis of new flagella.

A flagellar filament grows not from its base, as does an animal hair, but from its tip. The MS ring is synthesized first and inserted into the cytoplasmic membrane. Then other anchoring proteins are synthesized along with the hook before the filament forms (**Figure 2.38**). Flagellin molecules synthesized in the cytoplasm pass up through a 3-nm channel inside the filament and add on at the terminus to form the mature flagellum. A protein "cap" is present at the end of the growing flagellum. Cap proteins assist flagellin molecules that have diffused through the filament channel to assemble in the proper fashion at the flagellum terminus (Figure 2.38). Approximately 20,000 flagellin protein molecules are needed to make one filament. The flagellum grows more or less continuously until it reaches its final length. Broken flagella still rotate and can be repaired with new flagellin units passed through the filament channel to replace the lost ones.

Figure 2.37 Structure and function of the flagellum in gram-negative *Bacteria.* *(a)* Structure. The L ring is embedded in the LPS and the P ring in peptidoglycan. The MS ring is embedded in the cytoplasmic membrane and the C ring in the cytoplasm. A narrow channel exists in the rod and filament through which flagellin molecules diffuse to reach the site of flagellar synthesis. The Mot proteins function as the flagellar motor, whereas the Fli proteins function as the motor switch. The flagellar motor rotates the filament to propel the cell through the medium. Inset photos: Top left, a cell of the purple sulfur bacterium *Chromatium* containing a tuft of polar flagella; Top right, transmission electron micrograph of a flagellar basal body from *Salmonella enterica* with the various rings labeled. *(b)* Function. A "proton turbine" model explains rotation of the flagellum. Protons, flowing through the Mot proteins, exert forces on charges present on the C and MS rings, thereby spinning the rotor.

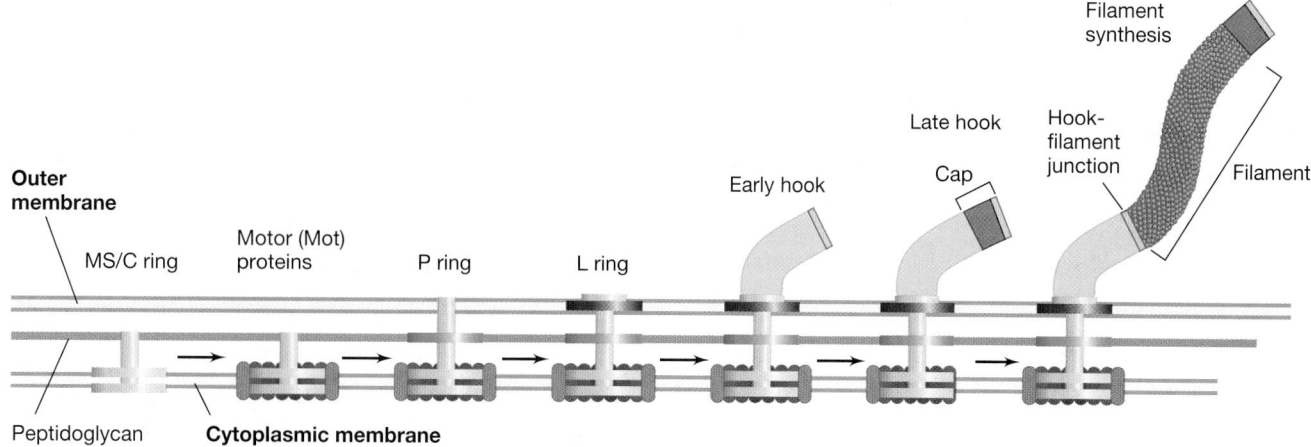

Figure 2.38 Flagella biosynthesis. Synthesis begins with assembly of MS and C rings in the cytoplasmic membrane, followed by the other rings, the hook, and the cap. Flagellin protein flows through the hook to form the filament and is guided into position by cap proteins.

Archaella

As in *Bacteria*, swimming motility is widespread among species of *Archaea* due to rotation of their flagella analog, the archaellum (see also page 34). These structures are roughly half the diameter of flagella, measuring about 10–13 nm in width (**Figure 2.39a**), and impart movement to the cell by rotating, as do flagella. However, unlike *Bacteria*, in which a single type of protein makes up the filament, several different filament proteins are known in *Archaea*, and the genes that encode them bear little sequence homology to genes that encode bacterial flagellin. Depending on the archaeal species, 7–12 genes encode the major proteins that make up the archaellum. Archaella have been particularly well studied in the salt-loving archaeon *Halobacterium*, the heat- and acid-loving archaeon *Sulfolobus*, and the methane-producing archaeon *Methanocaldococcus*.

Studies of swimming cells of *Halobacterium* show that they swim at speeds only about one-tenth that of cells of *Escherichia coli*. This could be due to the smaller diameter of the archaellum compared to the flagellum, as this would be expected to reduce the torque of the structure significantly. However, this hypothesis has been questioned since the discovery that some *Archaea* swim incredibly fast. For example, cells of *Methanocaldococcus* (Figure 2.39c) swim nearly 50 times faster than cells of *Halobacterium* and 10 times faster than cells of *Escherichia coli* (*Bacteria*). In fact, *Methanocaldococcus* swims at nearly 500 cell lengths per second, which makes it the fastest organism on Earth! Thus, the net torque or rotational speeds of archaella from different species of *Archaea* can obviously vary significantly.

The overall structure of the archaellum bears a strong resemblance to that of type IV pili (Figure 2.39b), and it is clear that the archaellum is structurally related to these appendages (Section 2.7).

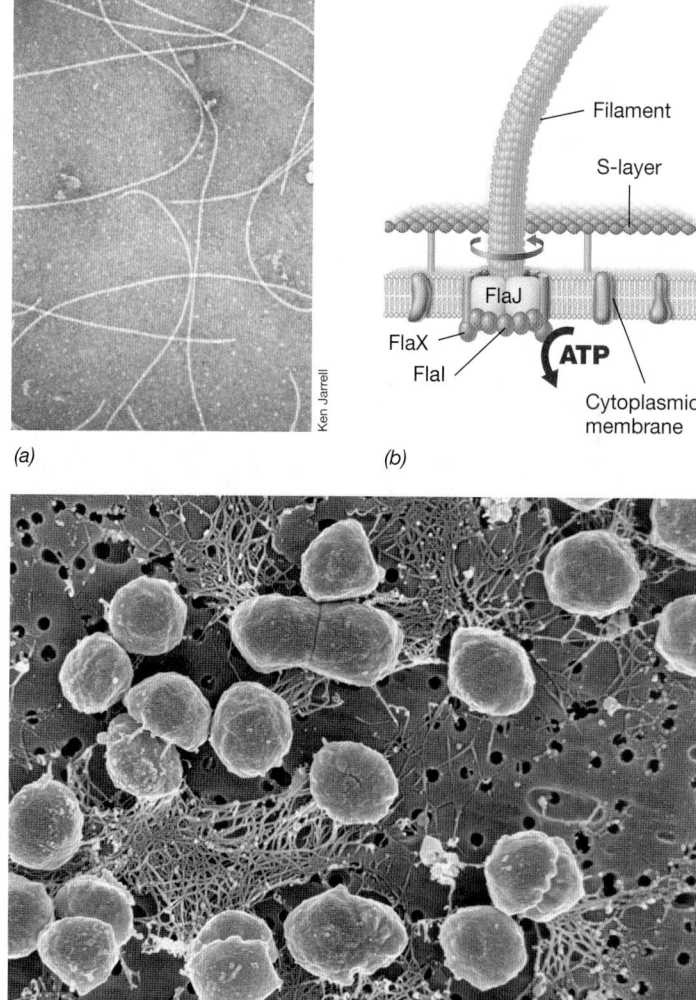

(a)   (b)   (c)

Figure 2.39 Archaella. *(a)* Transmission electron micrograph of archaella isolated from the methanogen *Methanococcus maripaludis*. A single archaellum is about 12 nm wide. *(b)* Depiction of an archaellum embedded in the archaeal cell wall and cytoplasmic membrane. ATP (rather than the proton motive force, see Figure 2.37b) drives archaella rotation. *(c)* Scanning electron micrograph of cells of *Methanocaldococcus jannaschii* containing abundant archaella.

In fact, the archaellum can be considered a rotating type IV pilus capable of both clockwise and counterclockwise rotation. Moreover, in contrast to the flagellum, whose energy requirement is met by dissipation of the proton motive force (Figure 2.37b), rotation of the archaellum is driven by the hydrolysis of ATP. Thus, although flagella and archaella are functionally similar—rotating filaments that drive cell propulsion—their flagellar motors are powered in fundamentally different ways. This suggests that swimming motility evolved separately in *Bacteria* and *Archaea* as these domains diverged some 3.5 billion years ago.

MINIQUIZ

- Cells of *Salmonella* are peritrichously flagellated, those of *Pseudomonas* polarly flagellated, and those of *Spirillum* lophotrichously flagellated. Using a sketch, show how each organism would appear in a flagella stain.
- Compare flagella and archaella in terms of their structure, function, and energy source.

2.12 Gliding Motility

Some bacteria are motile but lack flagella. Most of these nonswimming yet still motile cells move by *gliding*. Unlike flagellar motility, in which cells stop and then start off in a different direction, gliding motility is a slower and smoother form of movement and typically occurs along the long axis of the cell.

Diversity of Gliding Motility

Gliding motility is widely distributed among *Bacteria* but has been well studied in only a few groups. The gliding movement itself—up to 10 μm/sec in some gliding bacteria—is considerably slower than propulsion by flagella but still offers the cell a means of moving about its habitat.

Gliding bacteria are typically filamentous or rod-shaped in morphology, and the gliding process requires that the cells be in contact with a solid surface (**Figure 2.40**). The morphology of colonies of a typical gliding bacterium is distinctive because cells glide out and move away from the center of the colony (Figure 2.40c). Perhaps the best-known gliding bacteria are the filamentous cyanobacteria (Figure 2.40a, b), certain gram-negative bacteria such as *Myxococcus* and other myxobacteria, and species of *Cytophaga* and *Flavobacterium* (Figure 2.40c, d). No gliding *Archaea* are known.

Mechanisms of Gliding Motility

More than one mechanism drives gliding motility. Cyanobacteria glide by secreting a polysaccharide slime from pores onto the outer surface of the cell. The slime contacts both the cell surface and the solid surface against which the cell moves. As the excreted slime adheres to the surface, the cell slides along. The nonphototrophic gliding bacterium *Cytophaga* also glides at the expense of slime excretion, rotating along its long axis as it does.

Cells capable of "twitching motility" also display a form of gliding motility using a mechanism by which repeated extension and retraction of type IV pili (Section 2.7) drag the cell along a surface. The gliding myxobacterium *Myxococcus xanthus* has two forms of gliding motility. One form is driven by type IV pili, whereas the

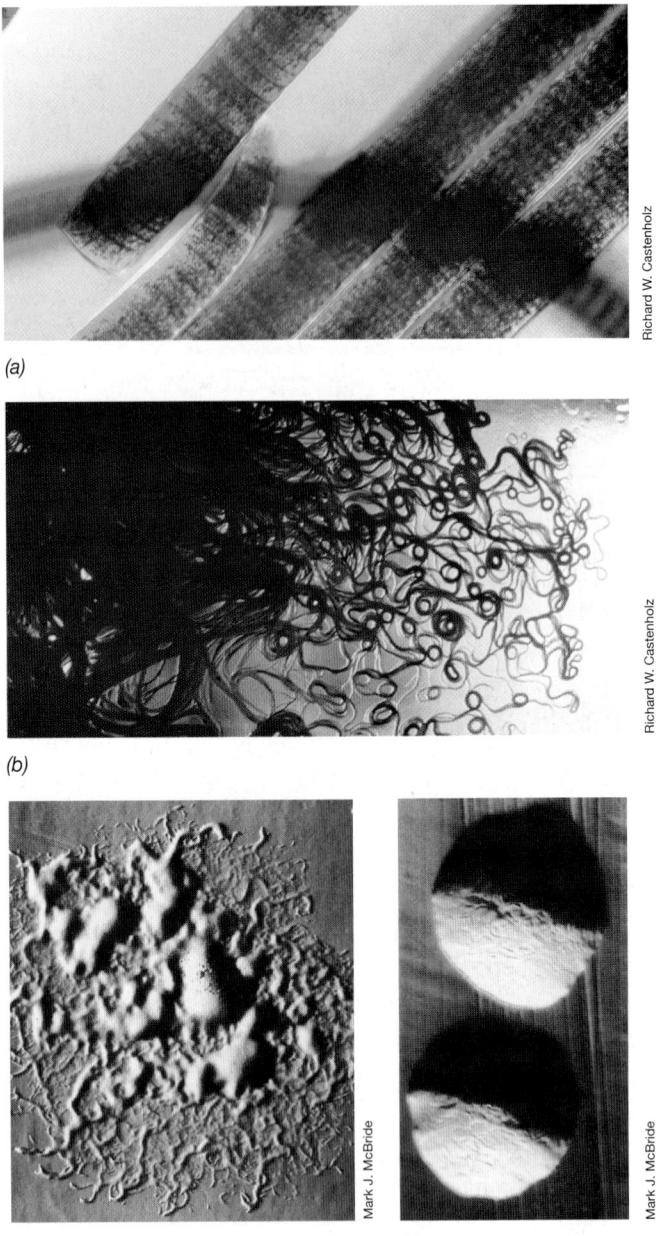

(a)

(b)

(c)   (d)

Richard W. Castenholz

Richard W. Castenholz

Mark J. McBride

Mark J. McBride

Figure 2.40 Gliding bacteria. *(a, b)* The large filamentous cyanobacterium *Oscillatoria* has cells about 35 μm wide. *(b) Oscillatoria* filaments gliding on an agar surface. *(c)* Masses of the bacterium *Flavobacterium johnsoniae* gliding away from the center of the colony (the colony is about 2.7 mm wide). *(d)* Nongliding mutant strain of *F. johnsoniae* showing typical colony morphology of nongliding bacteria (the colonies are 0.7–1 mm in diameter). See also Figure 2.41.

other is distinct from either the type IV pili or the slime extrusion methods. In this form of *M. xanthus* motility, a protein adhesion complex is formed at one pole of the rod-shaped cell and remains at a fixed position on the surface as the cell glides forward. This means that the adhesion complex moves in the direction opposite that of the cell, presumably fueled by some sort of cytoplasmic motility engine.

Neither slime extrusion nor twitching is the mechanism of gliding in other gliding bacteria. In the genus *Flavobacterium* (Figure 2.40c and **Figure 2.41**), for example, no slime is excreted and

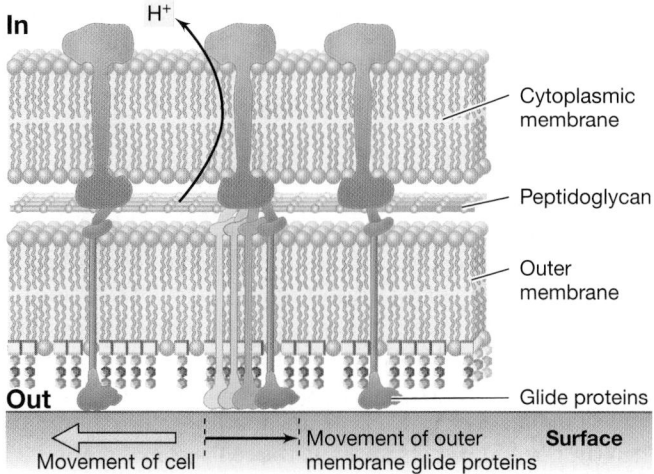

Figure 2.41 Gliding motility in *Flavobacterium johnsoniae*. Tracks (yellow) exist in the peptidoglycan that connect cytoplasmic proteins to outer membrane glide proteins and propel the glide proteins along the solid surface. Note that the glide proteins and the cell proper move in opposite directions.

interact with other cells. Myxobacteria, such as *Myxococcus xanthus*, have a very social and cooperative lifestyle, and gliding motility may play an important role in the intimate cell-to-cell interactions necessary to complete their life cycle (⟳ Section 15.17).

— MINIQUIZ —
- How does gliding motility differ from swimming motility in both mechanism and requirements?
- Contrast the mechanism of gliding motility in a filamentous cyanobacterium and in *Flavobacterium*.

2.13 Chemotaxis and Other Taxes

Cells of *Bacteria* and *Archaea* often encounter gradients of physical or chemical agents in nature and have evolved means to respond to these gradients by moving either toward or away from the agent. Such a directed movement is called a *taxis* (plural, taxes). **Chemotaxis**, a response to chemicals, and **phototaxis**, a response to light, are two well-studied taxes. The ability of a cell to move toward or away from various stimuli has ecological significance in that the directed movement may enhance a cell's access to resources or allow it to avoid harmful substances that could damage or kill it.

Chemotaxis has been well studied in swimming bacteria, and much is known at the genetic level about how the chemical state of a cell's environment is communicated to the flagellum. Our discussion here will thus deal solely with swimming bacteria. However, some gliding bacteria are also chemotactic, and there are phototactic movements in filamentous cyanobacteria (Figure 2.40*a*, *b*). In addition, many swimming species of *Archaea* are also chemotactic, and several of the proteins that control chemotaxis in *Bacteria* have homologs in these *Archaea*. Here we discuss microbial taxes in a general way. In Section 6.7 we examine the molecular mechanism of chemotaxis and its genetics and regulation in *Escherichia coli* as a general model for the control of taxes in both *Bacteria* and *Archaea*.

cells lack type IV pili. Instead of using one of these gliding mechanisms, the movement of proteins on the *Flavobacterium* cell surface supports gliding motility in this organism. Specific motility proteins anchored in the cytoplasmic and outer membranes are thought to propel cells of *Flavobacterium* forward by a ratcheting mechanism (Figure 2.41). Movement of gliding-specific proteins in the cytoplasmic membrane is driven by energy from the proton motive force, and this motion is then transmitted to complementary glide proteins in the outer membrane. Movement of the outer membrane proteins against the solid surface then pulls the cell forward (Figure 2.41).

Like other forms of motility, gliding motility has ecological relevance. Gliding allows a cell to exploit new resources and to

Chemotaxis in Peritrichously Flagellated *Bacteria*

Much research on chemotaxis has been done with the peritrichously flagellated bacterium *E. coli*. To understand how chemotaxis affects the behavior of *E. coli*, consider the situation in which a cell encounters a gradient of some chemical in its environment (**Figure 2.42**). In the absence of the gradient, cells move in a random fashion that includes *runs*, in which the cell is swimming forward in a smooth fashion, and *tumbles*, when the cell stops and jiggles about. During forward movement in a run,

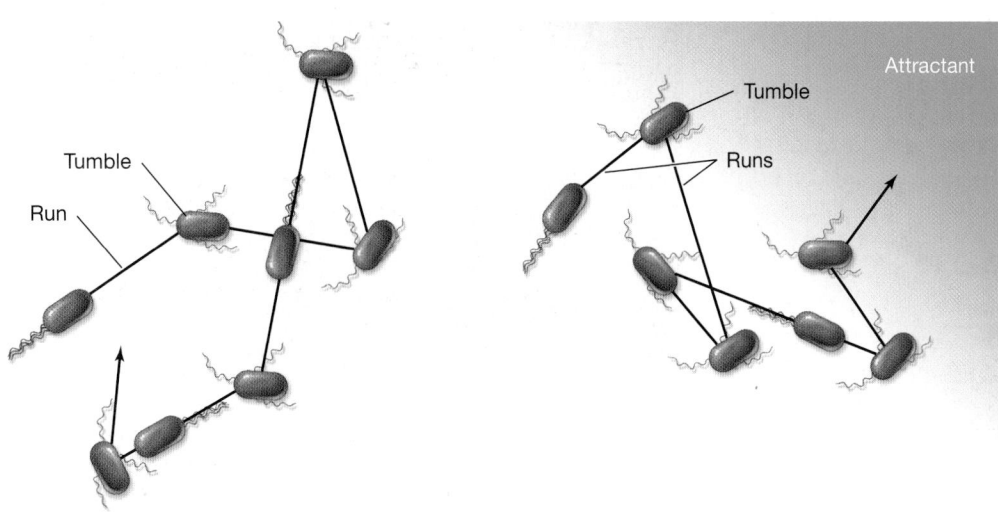

(a) **No attractant present: Random movement**   (b) **Attractant present: Directed movement**

Figure 2.42 Chemotaxis in a peritrichously flagellated bacterium. *(a)* In the absence of a chemical attractant, the cell swims randomly in runs, changing direction during tumbles. *(b)* In the presence of an attractant, runs become biased, and the cell moves up the gradient of the attractant. The attractant gradient is depicted in green, with the highest concentration where the color is most intense.

the flagellar motor rotates counterclockwise. When flagella rotate clockwise, the bundle of flagella pushes apart, forward motion ceases, and the cells tumble (Figure 2.42).

Following a tumble, the direction of the next run is random. Thus, by means of runs and tumbles, the cell moves about its environment in a random fashion. However, if a gradient of a chemical attractant is present, these random movements become biased. If the organism senses that it is moving toward higher concentrations of the attractant, runs become longer and tumbles are less frequent. The result of this behavioral response is that the organism moves up the concentration gradient of the attractant (Figure 2.42b). If the organism senses a repellent, the same mechanism applies, although in this case it is the *decrease* in concentration of the repellent (rather than the *increase* in concentration of an attractant) that promotes runs.

How are chemical gradients sensed? Prokaryotic cells are too small to sense a gradient of a chemical along the length of a single cell. Instead, while moving, cells "monitor" their environment by sampling chemicals periodically and comparing the concentration of a particular chemical with that sensed a few moments before. Bacterial cells thus respond to *temporal* rather than *spatial* differences in the concentration of a chemical as they swim. Sensory information is fed through an elaborate cascade of proteins that eventually affect the direction of rotation of the flagellar motor. The attractants and repellents are sensed by a series of membrane proteins called *chemoreceptors*. These sensory proteins bind the chemicals and begin the process of sensory transduction to the flagellum (⮑ Section 6.7). Chemotaxis can thus be considered a type of *sensory response system*, analogous to sensory responses in the nervous system of animals.

Chemotaxis in Polarly Flagellated *Bacteria*

Chemotaxis in polarly flagellated cells is similar but not identical to that in peritrichously flagellated cells such as *E. coli*. Many polarly flagellated bacteria, such as *Pseudomonas* species, can fully reverse the direction of rotation of their flagella. In so doing, they do not tumble but immediately reverse their direction of movement (Figure 2.36b). However, in the phototrophic bacterium *Rhodobacter*, cells of which have only a single flagellum that can rotate in just one direction, rotation of the flagellum stops periodically. When the flagellum stops rotating, the cell becomes reoriented by Brownian motion (Figure 2.36b). Then as the flagellum begins to rotate again, the cell moves off in a new direction.

Despite this seemingly random activity, cells of *Rhodobacter* are strongly chemotactic to various organic compounds and also to oxygen and light. *Rhodobacter* cannot reverse its flagellar motor and tumble as *E. coli* can, but cells do maintain runs as long as they sense an increasing concentration of attractant. If the cells sense a decreasing concentration of attractant, movement ceases. By such starting and stopping, a cell eventually finds the path of increasing attractant and maintains a run until either its chemoreceptors are saturated or it senses a decrease in the level of attractant.

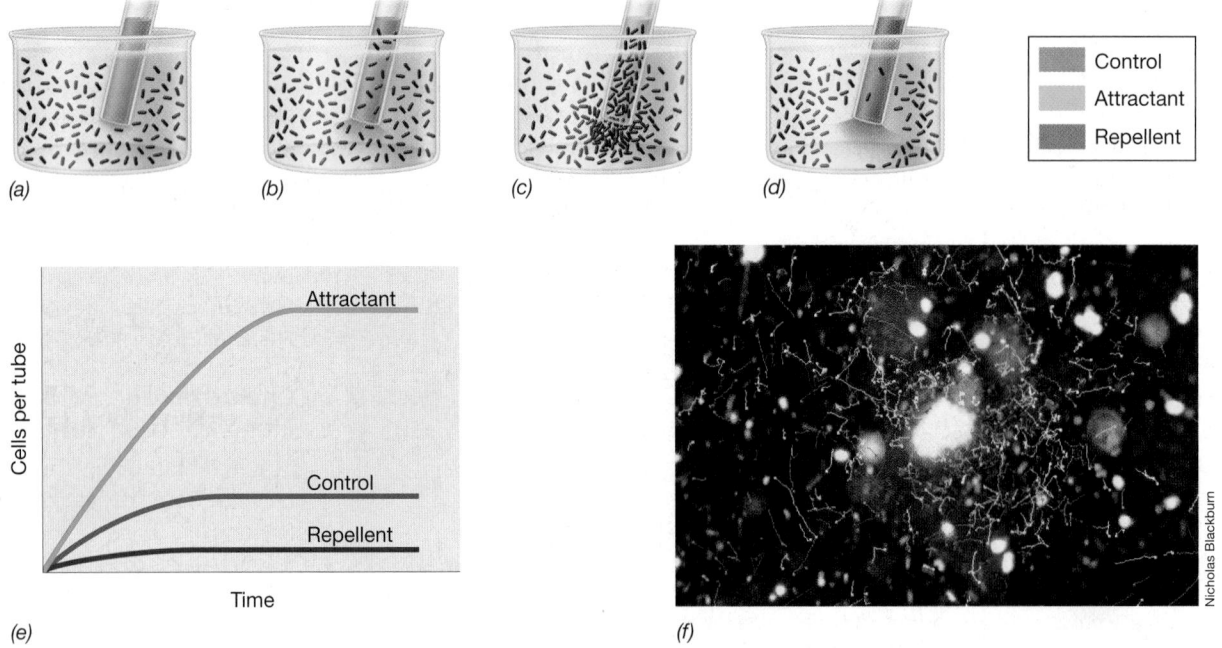

Figure 2.43 Measuring chemotaxis using a capillary tube assay. *(a)* Insertion of the capillary into a bacterial suspension. As the capillary is inserted, a gradient of the chemical begins to form. *(b)* Control capillary contains a salt solution that is neither an attractant nor a repellent. Cell concentration inside the capillary becomes the same as that outside. *(c)* Accumulation of bacteria in a capillary containing an attractant. *(d)* Repulsion of bacteria by a repellent. *(e)* Time course showing cell numbers in capillaries containing various chemicals. *(f)* Tracks of motile bacteria in seawater swarming around an algal cell (large white spot, center) photographed with a tracking video camera system attached to a microscope. The bacterial cells are showing positive aerotaxis by moving toward the oxygen-producing algal cell. The alga is about 60 μm in diameter. The proteins that participate in chemotaxis and the mechanisms by which chemotaxis is regulated, are discussed in detail in Section 6.7.

Measuring Chemotaxis

Bacterial chemotaxis can be demonstrated and quantified by immersing a small glass capillary tube containing an attractant into a suspension of motile bacteria that does not contain the attractant. From the tip of the capillary, a gradient forms into the surrounding medium, with the concentration of chemical gradually decreasing with distance from the tip (**Figure 2.43**). When an attractant is present, chemotactic bacteria will move toward it, forming a swarm around the open tip (Figure 2.43*c*) with many of the bacteria swimming into the capillary itself. Of course, because of random movements some chemotactic bacteria will swim into the capillary even if it contains a solution of the same composition as the medium (control solution, Figure 2.43*b*). However, when an attractant is present, the number of cells within the capillary will be many times higher than external cell numbers. If the capillary is removed after a time period and the cells within it are counted and compared with that of the control, attractants can easily be identified (Figure 2.43*e*).

If the inserted capillary contains a repellent, just the opposite occurs; the cells sense an increasing gradient of repellent and the appropriate chemoreceptors affect flagellar rotation to gradually move the cells away from the repellent. In this case, the number of bacteria within the capillary will be fewer than in the control (Figure 2.43*d*). Using this simple capillary method, it is possible to screen chemicals to see if they are attractants or repellents for a given bacterium.

Chemotaxis can also be observed microscopically. Using a video camera that captures the position of bacterial cells with time and shows the motility tracks of each cell, it is possible to see the chemotactic movements of cells (Figure 2.43*f*). This method has been used to study chemotaxis of mixtures of microbes in natural environments. In nature it is thought that the major chemotactic agents for bacteria are nutrients excreted from larger microbial cells or from live or dead macroorganisms. Algae, for example, produce both organic compounds and oxygen (O_2, from photosynthesis) that can trigger chemotactic movements of bacteria toward the algal cell (Figure 2.43*f*).

Phototaxis and Other Taxes

Many phototrophic microorganisms can move toward light, a process called *phototaxis*. Phototaxis allows a phototrophic organism to position itself most efficiently to receive light for photosynthesis. This can be shown if a light spectrum is spread across a microscope slide on which there are motile phototrophic purple bacteria. On such a slide the bacteria accumulate at wavelengths at which their photosynthetic pigments absorb (**Figure 2.44***a*). These pigments include, in particular, bacteriochlorophylls and carotenoids (Chapter 14).

Two different light-mediated taxes are observed in phototrophic bacteria. One, called *scotophobotaxis*, can be observed only microscopically and occurs when a phototrophic bacterium happens to swim outside the illuminated field of view of the microscope into darkness. Entering darkness negatively affects photosynthesis and thus the energy state of the cell and signals the cell to tumble, reverse direction, and once again swim in a run, thus reentering the light. Scotophobotaxis is presumably a mechanism by which phototrophic purple bacteria avoid entering darkened habitats when they are moving about in illuminated ones, and this likely improves their competitive success.

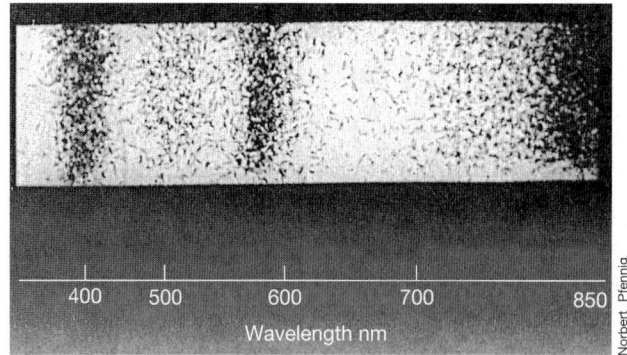

(a)

(b)

Figure 2.44 Phototaxis of phototrophic bacteria. *(a)* Scotophobic accumulation of the phototrophic purple bacterium *Thiospirillum jenense* at wavelengths of light at which its pigments absorb. A light spectrum was displayed on a microscope slide containing a dense suspension of the bacteria; after a period of time, the bacteria had accumulated selectively and the photomicrograph was taken. The wavelengths at which the bacteria accumulated are those at which the photosynthetic pigment bacteriochlorophyll *a* absorbs (compare with ↭ Figure 14.2*b*). *(b)* Phototaxis of an entire colony of the purple phototrophic bacterium *Rhodospirillum centenum*. These strongly phototactic cells move in unison toward the light source at the top. See Figure 2.34 for electron micrographs of flagellated *R. centenum* cells.

Phototaxis differs from scotophobotaxis in that cells move up a light gradient from lower to higher intensities. Phototaxis is analogous to chemotaxis except that the attractant is light instead of a chemical. In some phototactic organisms, such as the highly motile phototrophic purple bacterium *Rhodospirillum centenum* (Figure 2.34), *entire colonies* of cells show phototaxis and move in unison toward the light (Figure 2.44*b*).

Several components of the regulatory system that govern chemotaxis also control phototaxis. A *photoreceptor*, a protein that functions similarly to a chemoreceptor but senses a gradient of light instead of chemicals, is the initial sensor in the phototaxis response. The photoreceptor then interacts with the same cytoplasmic proteins that control flagellar rotation in chemotaxis, maintaining the cell in a run if it is swimming toward an increasing intensity of light. Section 6.7 describes the activities of these proteins in more detail.

Other bacterial taxes, such as movement toward or away from oxygen (*aerotaxis*, see Figure 2.43*f*) or toward or away from conditions of high ionic strength (*osmotaxis*), are known among various swimming bacteria. In some gliding cyanobacteria, *hydrotaxis* (movement toward water), has also been observed. Hydrotaxis allows gliding cyanobacteria that inhabit dry environments, such as desert soils, to glide toward a gradient of increasing hydration.

MINIQUIZ

- Define the word chemotaxis. How does chemotaxis differ from aerotaxis?
- What causes a run versus a tumble?
- How can chemotaxis be measured quantitatively?
- How does scotophobotaxis differ from phototaxis?

V • Eukaryotic Microbial Cells

Compared with prokaryotic cells, microbial eukaryotes typically have structurally more complex and much larger cells. We complete our study of microbial cell structure and function with a consideration of structure/function issues in microbial eukaryotes, common models for the study of eukaryotic biology. Microbial eukaryotes include the fungi, the algae, and the protozoa and other protists. We cover the diversity of microbial eukaryotes in Chapter 18.

2.14 The Nucleus and Cell Division

Eukaryotic cells vary in the complement of organelles they contain, but a unit membrane–enclosed nucleus is universal and a hallmark of the eukaryotic cell. Mitochondria are nearly universal among eukaryotic cells, while pigmented chloroplasts are found only in phototrophic cells. Other structures include the Golgi complex, lysosomes, endoplasmic reticula, and microtubules and microfilaments (**Figure 2.45**). Some microbial eukaryotes have flagella or cilia—structures that confer motility—and a cell wall is present in many, such as the fungi and algae.

Eukaryotic cell membranes contain *sterols*. These molecules, absent from all but a few prokaryotic cells, lend structural strength to the eukaryotic cell, something especially important to those eukaryotes that lack a cell wall, such as the protozoa or animal cells.

The Nucleus

The **nucleus** contains the chromosomes of the eukaryotic cell. DNA within the nucleus is wound around basic (positively charged) proteins called **histones**, which tightly pack the negatively

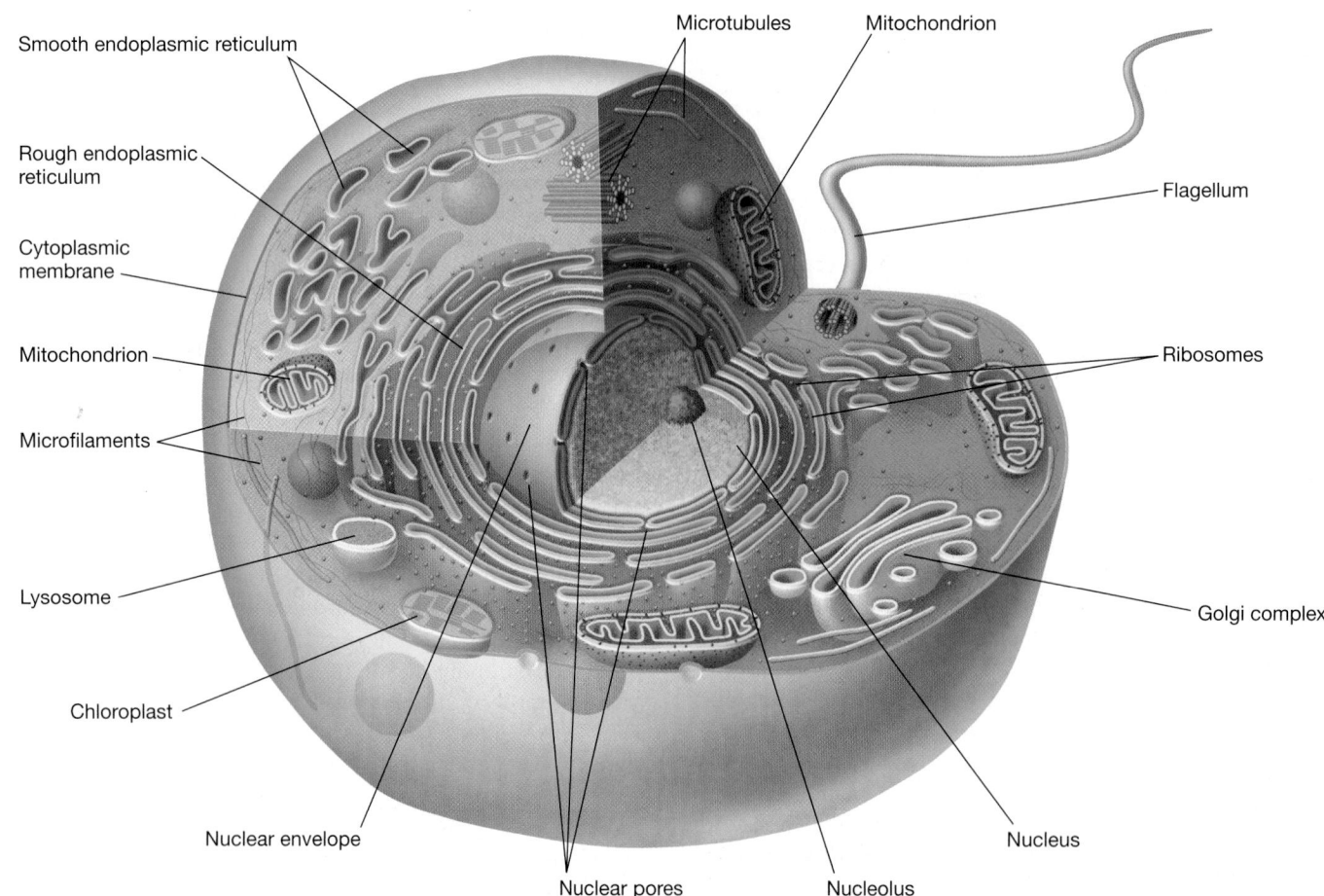

Figure 2.45 Cutaway schematic of a microbial eukaryote. Although all eukaryotic cells contain a nucleus, not all organelles and other structures shown are present in all microbial eukaryotes. Not shown is the cell wall, found in fungi, algae, plants, and a few protists.

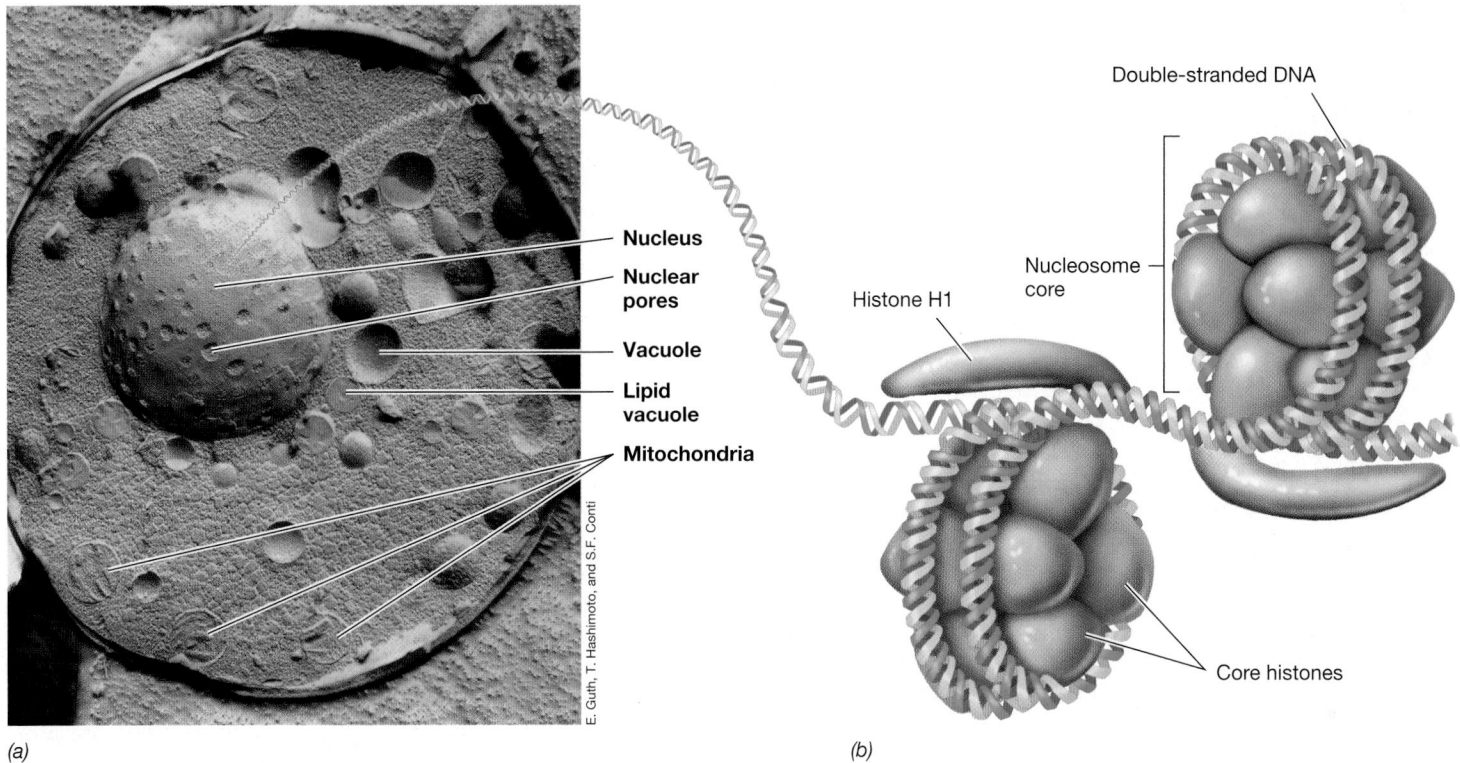

E. Guth, T. Hashimoto, and S.F. Conti

(a)

(b)

Figure 2.46 The nucleus and DNA packaging in eukaryotes. (a) Electron micrograph of a yeast cell prepared in such a way as to reveal a surface view of the nucleus. The cell is about 8 μm wide. (b) Packaging of DNA around histone proteins to form a nucleosome. Nucleosomes are arranged along the DNA strand like beads on a string and aggregate to form chromosomes during the process of mitosis (see Figure 2.47).

charged DNA to form nucleosomes (**Figure 2.46**) and from them, chromosomes. The nucleus is enclosed by a pair of membranes, each with its own function, separated by a space. The innermost membrane is a simple sac while the outermost membrane is in many places continuous with the endoplasmic reticulum. The inner and outer nuclear membranes specialize in interactions with the nucleoplasm and the cytoplasm, respectively. The nuclear membranes contain pores (Figures 2.45 and 2.46*a*), formed from holes where the inner and outer membranes are joined. The pores allow transport proteins to import and export other proteins and nucleic acids into and out of the nucleus, a process called *nuclear transport*.

Within the nucleus is found the *nucleolus* (Figure 2.45), the site of ribosomal RNA (rRNA) synthesis. The nucleolus is rich in RNA, and ribosomal proteins synthesized in the cytoplasm are transported into the nucleolus and combine with rRNA to form the small and large subunits of eukaryotic ribosomes. These are then exported to the cytoplasm, where they associate to form the intact ribosome and function in protein synthesis.

Cell Division

Eukaryotic cells divide by a process in which the chromosomes are replicated, the nucleus is disassembled, the chromosomes are segregated into two sets, and a nucleus is reassembled in each daughter cell (**Figure 2.47**). Whereas prokaryotic cells are genetically haploid, many microbial eukaryotes can exist in either of two genetic states: haploid or diploid. *Diploid* cells have two copies of each chromosome whereas *haploid* cells have only one. For example, the brewer's yeast *Saccharomyces cerevisiae* can exist in the haploid state (16 chromosomes) as well as in the diploid state (32 chromosomes). However, regardless of its genetic state, during cell division the chromosome number is first doubled and later halved to give each daughter cell its correct complement of chromosomes. This is the process of **mitosis**, unique to eukaryotic cells. During mitosis, the chromosomes condense, divide, and are separated into two sets, one for each daughter cell (Figure 2.47).

In contrast to mitosis, **meiosis** converts a diploid cell into several haploid cells. Meiosis consists of two successive cell divisions. In the first meiotic division, homologous chromosomes segregate into separate cells, changing the genetic state from diploid to haploid. The second meiotic division is essentially the same as mitosis, as the two haploid cells divide to form a total of four haploid cells called *gametes*. In higher organisms these are the eggs and sperm; in eukaryotic microorganisms, they are spores or related reproductive structures.

MINIQUIZ
- How is DNA arranged in the chromosomes of eukaryotes?
- What are histones and what do they do?
- What are the major differences between mitosis and meiosis?

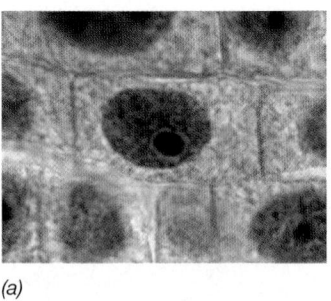
(a)

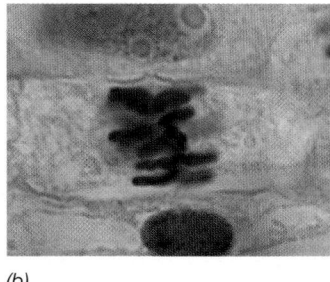
(b)

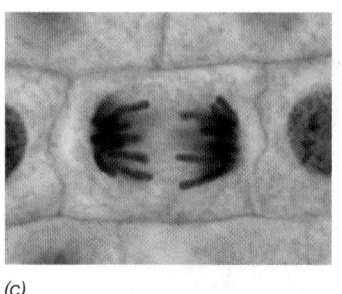
(c)

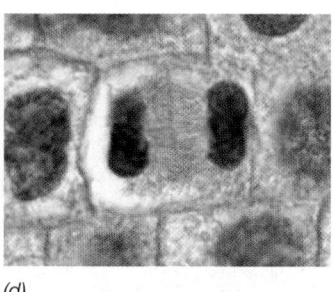
(d)

Figure 2.47 Light micrograph of eukaryotic cells undergoing mitosis. *(a)* Interphase, distinct chromosomes are not apparent. *(b)* Metaphase. Homologous chromosomes are lining up along the cell center. *(c)* Anaphase. Homologous chromosomes are pulling apart. *(d)* Telophase. Chromosomes have separated into the newly forming daughter cells.

2.15 Mitochondria, Hydrogenosomes, and Chloroplasts

Organelles that specialize in energy metabolism in eukaryotes include the mitochondrion or hydrogenosome, and in phototrophic eukaryotes, the chloroplast. These organelles have evolutionary roots within the *Bacteria* and provide ATP to the eukaryotic cell from either the oxidation of organic compounds or from light.

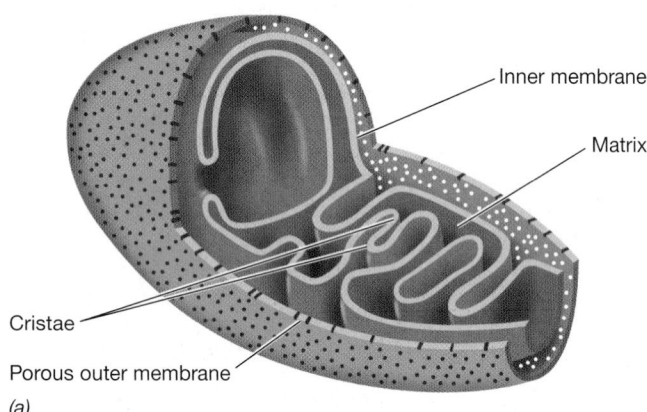

Inner membrane

Matrix

Cristae

Porous outer membrane

(a)

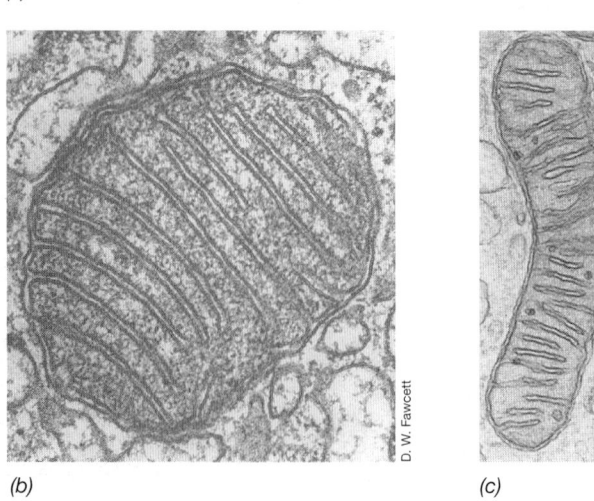

D. W. Fawcett

(b)   *(c)*

Figure 2.48 Structure of the mitochondrion. *(a)* Diagram showing the overall structure of the mitochondrion; note the inner and outer membranes. *(b, c)* Transmission electron micrographs of mitochondria from rat tissue showing the variability in morphology; note the cristae.

Mitochondria

In aerobic eukaryotic cells, respiration occurs in the mitochondrion. **Mitochondria** are of bacterial dimensions and can take on many shapes (**Figure 2.48**). The number of mitochondria per cell depends somewhat on the cell type and size. A yeast cell may have only a few mitochondria per cell, whereas an animal cell may have over a thousand. The mitochondrion is enclosed by a double membrane system. Like the nuclear membrane, the outermost mitochondrial membrane is relatively permeable and contains pores that allow the passage of small molecules. The innermost membrane is less permeable and its structure more closely resembles that of the cytoplasmic membrane of *Bacteria*.

Mitochondria also contain folded internal membranes called **cristae**. These membranes, formed by invagination of the inner membrane, contain the enzymes needed for respiration and ATP production. Cristae also contain transport proteins that regulate the passage of key molecules such as ATP into and out of the *matrix*, the innermost compartment of the mitochondrion (Figure 2.48*a*). The matrix contains enzymes for the oxidation of organic compounds, in particular, enzymes of the citric acid cycle, the major pathway for the combustion of organic compounds to CO_2 (∞ Section 3.9).

Hydrogenosomes

Some eukaryotic microorganisms are killed by O_2 and, like many *Bacteria* and *Archaea*, live an anaerobic lifestyle. Such cells lack mitochondria and some of them contain structures called **hydrogenosomes** (**Figure 2.49**). Although similar in size to mitochondria, hydrogenosomes lack citric acid cycle enzymes and also lack cristae. Microbial eukaryotes that contain hydrogenosomes carry out a strictly fermentative metabolism. Examples include the human parasite *Trichomonas* (∞ Sections 18.3 and 33.4) and various protists that inhabit the rumen of ruminant animals (∞ Section 23.13) or anoxic muds and lake sediments.

The major biochemical reaction in the hydrogenosome is the oxidation of pyruvate to H_2, CO_2, and acetate (Figure 2.49*b*). Some anaerobic eukaryotes have H_2-consuming, methane-producing *Archaea* in their cytoplasm. These *methanogens* consume the H_2 and CO_2 produced by the hydrogenosome and combine them to form methane (CH_4). Because hydrogenosomes are anoxic and cannot respire, they cannot oxidize the acetate produced from

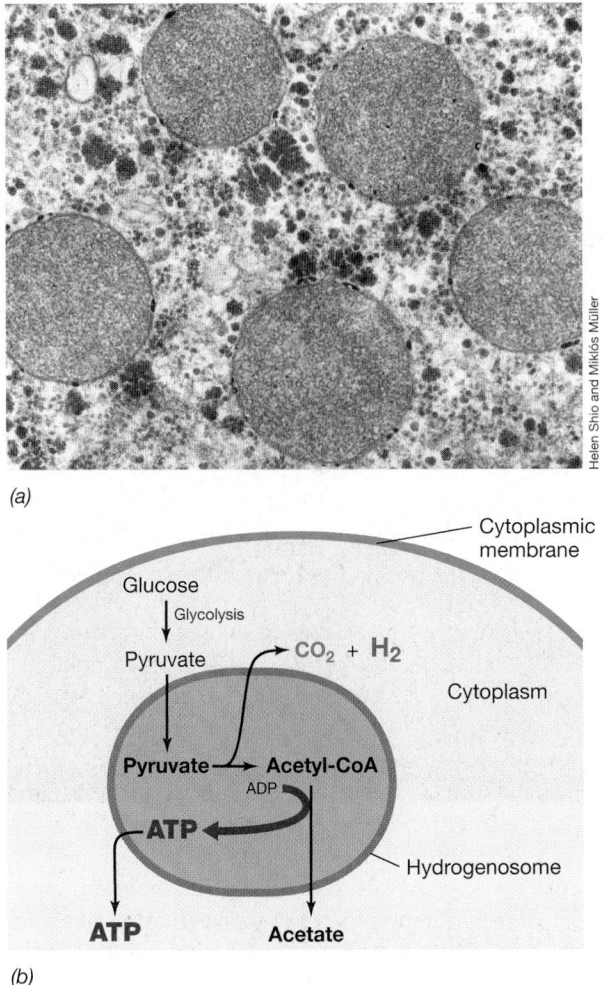

Helen Shio and Miklos Müller

(a)

(b)

Figure 2.49 The hydrogenosome. *(a)* Electron micrograph of a thin section through a cell of the anaerobic protist *Trichomonas vaginalis* showing five hydrogenosomes in cross section. Compare their internal structure with that of mitochondria in Figure 2.48. *(b)* Biochemistry of the hydrogenosome. Pyruvate is taken up by the hydrogenosome, and H_2, CO_2, acetate, and ATP are produced.

pyruvate oxidation as mitochondria do. Acetate is therefore excreted from the hydrogenosome into the cytoplasm of the host cell (Figure 2.49*b*).

Chloroplasts

Chloroplasts are the chlorophyll-containing organelles of phototrophic microbial eukaryotes such as the algae and function to carry out photosynthesis. Chloroplasts are relatively large and readily visible with the light microscope (**Figure 2.50**), and their number per cell varies among species.

Like mitochondria, chloroplasts have a permeable outer membrane and a much less-permeable inner membrane. The innermost membrane surrounds the **stroma**, analogous to the matrix of the mitochondrion (Figure 2.50*c*). The stroma contains the enzyme *ribulose bisphosphate carboxylase* (RubisCO), the key enzyme of the *Calvin cycle*, the series of biosynthetic reactions by which phototrophs convert CO_2 to organic compounds (⮂ Section 14.5). The permeability of the outermost chloroplast membrane allows glucose and ATP produced during

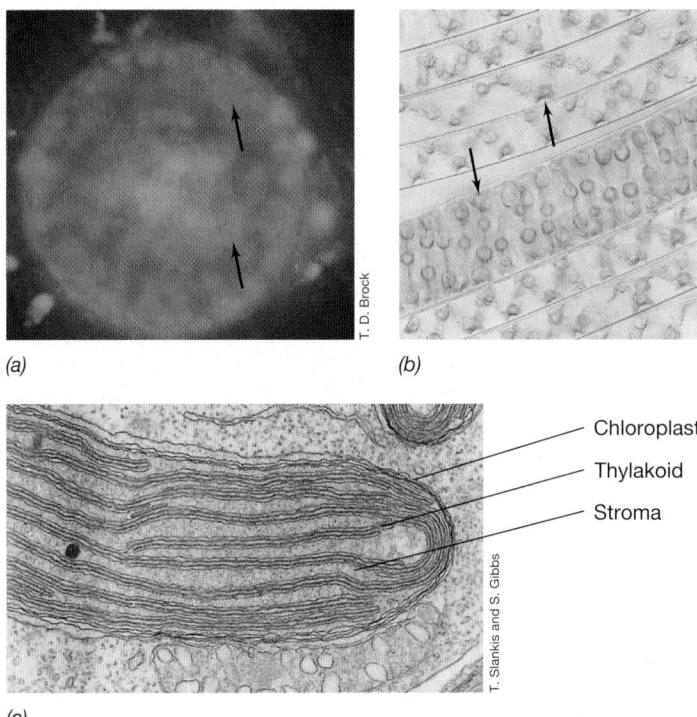

(a)   (b)

T. D. Brock

Chloroplast
Thylakoid
Stroma

T. Slankis and S. Gibbs

(c)

Figure 2.50 Chloroplasts of a diatom and a green alga cell. *(a)* Fluorescence photomicrograph of a diatom shows chlorophyll fluorescence; arrows, chloroplasts. The cell is about 40 μm wide. *(b)* Phase-contrast photomicrograph of the filamentous green alga *Spirogyra* showing the characteristic spiral-shaped chloroplasts (arrows) of this phototroph. A cell is about 20 μm wide. *(c)* Transmission electron micrograph showing a chloroplast of a diatom; note the thylakoids.

photosynthesis to diffuse into the cell cytoplasm where they are used in biosynthesis.

Chlorophyll and all other components needed for ATP synthesis in chloroplasts are located in a series of flattened membrane discs called **thylakoids** (Figure 2.50*c*). Like the cytoplasmic membrane, the thylakoid membrane is highly impermeable and its major function is to form a proton motive force (Figure 2.7*c*) that results in ATP synthesis.

Organelles and Endosymbiosis

On the basis of their relative autonomy, size, and morphological resemblance to bacteria, it was hypothesized over 100 years ago that mitochondria and chloroplasts were descendants of respiratory and phototrophic bacterial cells, respectively. By associating with nonphototrophic eukaryal hosts, the latter gained a new form of energy metabolism while the symbiotic bacterial cells received a stable and supportive growth environment inside the host. Over time, these originally free-living symbionts became an intimate part of the eukaryotic cell. This idea of symbiotic bacteria as the ancestors of the mitochondrion, hydrogenosome, and chloroplast is called the **endosymbiotic hypothesis** (⮂ Sections 13.4 and 18.1) and is now well accepted in biology.

Several lines of evidence support the endosymbiotic hypothesis. These include in particular the fact that mitochondria, hydrogenosomes, and chloroplasts contain their own genomes and ribosomes. The genomes are arranged in a circular fashion as for bacterial chromosomes (⟲ Section 9.3), and the sequence of genes that encode ribosomal RNA (⟲ Figure 1.36) in organelles clearly points to their bacterial origin. Thus, the eukaryotic cell is a genetic chimera containing genes from two domains of life: the host cell (*Eukarya*) and the endosymbiont (*Bacteria*).

MINIQUIZ

- What key reactions occur in the mitochondrion and in the chloroplast, and what key product is made in each?
- Compare and contrast pyruvate metabolism in the mitochondrion and the hydrogenosome.
- What is the endosymbiotic hypothesis and what evidence is there to support it?

2.16 Other Eukaryotic Cell Structures

Besides the nucleus and the mitochondrion (or hydrogenosome), and chloroplasts in phototrophic cells, other cytoplasmic structures are present in microbial eukaryotes. These include the endoplasmic reticulum, the Golgi complex, lysosomes, a variety of tubular structures, and structures responsible for motility. However, unlike mitochondria and chloroplasts, these structures lack DNA and are not of endosymbiotic origin. Cell walls are also present in certain microbial eukaryotes and function as they do in prokaryotic cells to provide shape and protect the cell from osmotic lysis. The exact structure of the cell wall varies with the organism, but various polysaccharides and proteins are commonly observed.

Endoplasmic Reticulum, the Golgi Complex, and Lysosomes

The endoplasmic reticulum (ER) is a network of membranes continuous with the nuclear membrane. Two types of endoplasmic reticulum exist: *rough* ER, which contains attached ribosomes,

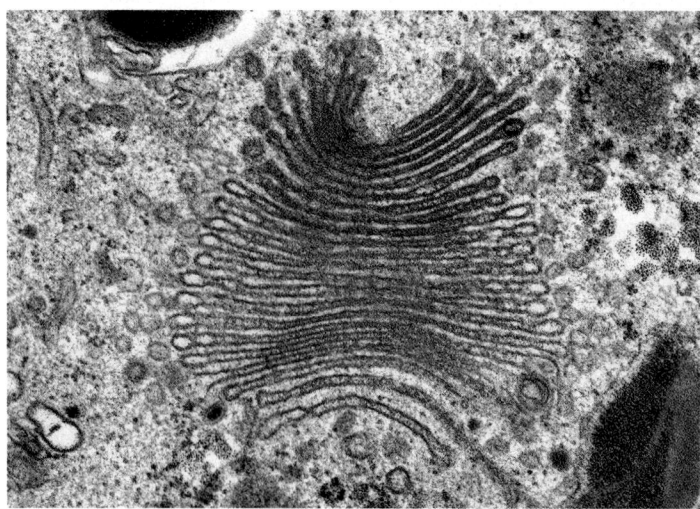

Figure 2.51 The Golgi complex. Transmission electron micrograph of a portion of a eukaryotic cell showing the Golgi complex (colored in gold). Note the multiple folded membranes of the Golgi complex (membrane stacks are 0.5–1.0 μm in diameter).

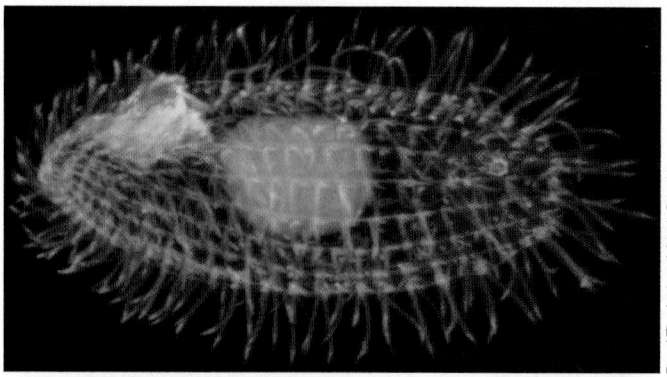

(a)

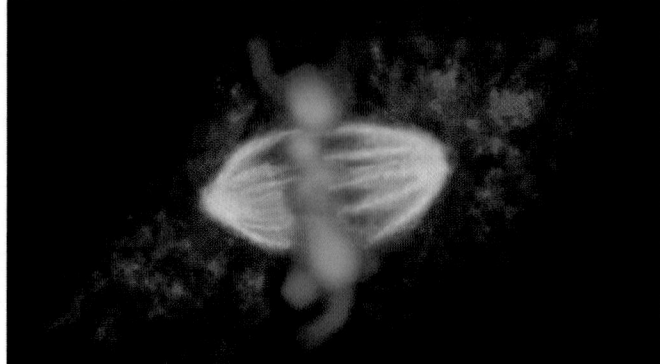

(b)

Microfilaments

(c)

Figure 2.52 Tubulin and microfilaments. *(a)* Fluorescence photomicrograph of a cell of *Tetrahymena* with red- and green-labeled antitubulin antibodies (the two combine to give yellow) and with DAPI, which stains DNA (blue, nucleus). A cell is about 10 μm wide. *(b)* An animal cell showing the role of tubulin (green) in separating chromosomes (blue) during metaphase of mitosis (cytoplasmic proteins stain red). *(c)* Electron microscopic image of the cellular slime mold *Dictyostelium discoideum* showing the network of actin microfilaments that along with microtubules functions as the cell cytoskeleton. Microfilaments are about 7 nm in diameter. *D. discoideum* has been used for decades as an experimental model system for eukaryotic cellular development and cell-to-cell cooperation (⟲ Figures 18.17 and 18.18).

and *smooth* ER, which does not (Figure 2.45). Smooth ER participates in the synthesis of lipids and in some aspects of carbohydrate metabolism. Rough ER, through the activity of its ribosomes, is a major producer of glycoproteins and also produces new membrane material that is transported throughout the cell to enlarge the various membrane systems before cell division.

The Golgi complex is a stack of membrane-bound sacs called *cisternae* (**Figure 2.51**) that arise from preexisting Golgi bodies and function in concert with the ER. In the Golgi complex, products of the ER are chemically modified and sorted into those destined for secretion versus those that will function in other membranous structures in the cell. Many of the modifications made in the Golgi complex are glycosylations (addition of sugar residues) that convert the proteins into glycoproteins that can then be targeted to specific locations in the cell.

Lysosomes (Figure 2.45) are membrane-enclosed compartments that contain digestive enzymes that hydrolyze proteins, fats, and polysaccharides. The lysosome fuses with food that enters the cell in vacuoles and then releases digestive enzymes that break down the foods for biosynthesis and energy generation. Lysosomes also function in degrading damaged cellular components and recycling these materials for new biosyntheses. The lysosome thus allows the cell's lytic activities to be partitioned away from the cytoplasm proper. Following the degradation of macromolecules in the lysosome, the resulting nutrients pass from the lysosome into the cytoplasm for use by cytoplasmic enzymes.

Microtubules, Microfilaments, and Intermediate Filaments

Just as buildings are supported by structural reinforcement, the large size of eukaryotic cells and their ability to move requires structural reinforcement. This internal support network consists of *microtubules, microfilaments*, and *intermediate filaments*; together, these structures form the cell **cytoskeleton** (Figure 2.45).

Microtubules are hollow tubes about 25 nm in diameter and are composed of the proteins α-*tubulin* and β-*tubulin*. Microtubules (**Figure 2.52a**) have many functions including maintaining cell shape and cell motility by cilia and flagella, moving chromosomes during mitosis (Figures 2.47 and 2.52b), and in movement of organelles within the cell. **Microfilaments** (Figure 2.52c) are smaller than microtubules, about 7 nm in diameter, and are polymers of two intertwined strands of the protein *actin*. Microfilaments function in maintaining or changing cell shape, in cell motility by cells that move by amoeboid movement, and during cell division. **Intermediate filaments** are fibrous keratin proteins that form into fibers 8–12 nm in diameter and function in maintaining cell shape and positioning organelles in the cell.

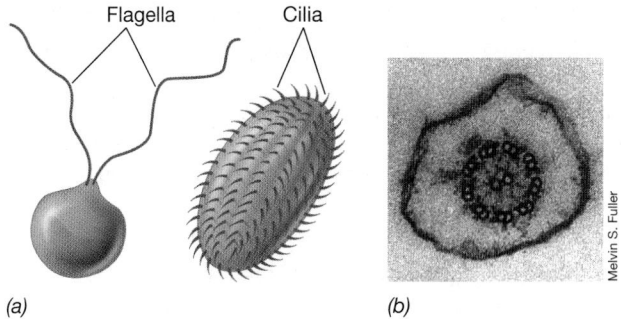

Figure 2.53 Motility organelles in eukaryotic cells: Flagella and cilia. *(a)* Flagella can be present as single or multiple filaments. Cilia are structurally very similar to flagella but much shorter. Eukaryotic flagella move in a whiplike motion. *(b)* Cross section through a flagellum of the fungus *Blastocladiella* showing the outer sheath, the outer nine pairs of microtubules, and the central pair of microtubules.

Flagella and Cilia

Flagella and cilia are present on the surface of many eukaryotic microbes and function as organelles of motility, allowing cells to move by swimming. Motility has survival value, as the ability to move allows motile organisms to move about their habitat and exploit new resources. *Cilia* are essentially short flagella that beat in synchrony to propel the cell—usually quite rapidly—through the medium. *Flagella*, by contrast, are long appendages present singly or in groups that propel the cell along—typically more slowly than by cilia—through a whiplike motion (**Figure 2.53a**). The flagella of eukaryotic cells are structurally quite distinct from bacterial flagella and do not rotate as do the flagella and archaella of *Bacteria* and *Archaea*, respectively (Section 2.11).

In cross section, cilia and flagella appear similar. Each contains a bundle of nine pairs of microtubules surrounding a central pair of microtubules (Figure 2.53b). A protein called *dynein* is attached to the microtubules and uses ATP to drive motility. Movement of flagella and cilia is similar. In both cases, movement is the result of the coordinated sliding of microtubules against one another in a direction toward or away from the base of the cell. This movement confers the whiplike motion on the flagellum or cilium that ultimately results in cell propulsion.

— **MINIQUIZ** —

• Why are the activities in the lysosome best partitioned away from the cytoplasm proper?

• How is the cell's cytoskeleton held together?

• From a functional standpoint, how does the flagellum of eukaryotic cells differ from that of prokaryotic cells?

Chapter Review

I • Cells of *Bacteria* and *Archaea*

2.1 Prokaryotic cells can have many different shapes; rods, cocci, and spirilla are common cell morphologies. Morphology is a poor predictor of other cell properties and is a genetically directed characteristic that has evolved to best serve the ecology of the cell.

Q **What are the major morphologies of prokaryotic cells? Draw cells for each morphology you list.**

2.2 Cells of *Bacteria* and *Archaea* are typically smaller than those of eukaryotes, although a few very large bacteria are known. The typical small size of prokaryotic cells affects their physiology, growth rate, ecology, and evolution. The lower limit for the diameter of a coccus-shaped cell is 0.15–0.2 μm.

Q **How large can a bacterium be? How small? Why is it that we likely know the lower limit more accurately than the upper limit? What are the dimensions of the rod-shaped bacterium *Escherichia coli*?**

II • The Cell Membrane and Wall

2.3 The cytoplasmic membrane is a highly selective permeability barrier constructed of lipids and proteins that form a bilayer, hydrophobic inside and hydrophilic outside. In contrast to *Bacteria* and *Eukarya*, where fatty acids are ester-linked to glycerol, *Archaea* contain ether-linked lipids and some form monolayer instead of bilayer membranes. The major functions of the cytoplasmic membrane are permeability, transport, and energy conservation, and to accumulate nutrients against the concentration gradient, transport systems are necessary.

Q **Describe in a single sentence the structure of a unit membrane. Describe the major structural differences between membranes of *Bacteria* and *Archaea*. How do solutes enter a cell through the membrane?**

2.4 Peptidoglycan is a polysaccharide found only in *Bacteria* that consists of an alternating repeat of *N*-acetylglucosamine and *N*-acetylmuramic acid, the latter cross-linked by tetrapeptides in adjacent strands. The enzyme lysozyme and the antibiotic penicillin both target peptidoglycan, leading to cell lysis.

Q **Why is the rigid layer of the bacterial cell wall called peptidoglycan? What are the structural reasons for the rigidity that is conferred on the cell wall by the peptidoglycan structure?**

2.5 Gram-negative *Bacteria* have an outer membrane consisting of LPS, protein, and lipoprotein. Porins allow for permeability across the outer membrane. The gap between the outer and cytoplasmic membranes is called the periplasm and contains proteins involved in transport, sensing chemicals, and other important cell functions.

Q **List several functions of the outer membrane in gram-negative *Bacteria*. What is the chemical composition of lipid A?**

2.6 Cell walls of *Archaea* are of several types, including pseudomurein, various polysaccharides, and S-layers composed of protein or glycoprotein. As for *Bacteria*, the walls of *Archaea* protect the cell from osmotic lysis.

Q **What cell wall polysaccharide common in *Bacteria* is absent from *Archaea*? What is unusual about S-layers compared to other cell walls? What types of cell walls are found in *Archaea*?**

III • Cell Surface Structures and Inclusions

2.7 Many prokaryotic cells contain capsules, slime layers, pili, or fimbriae. These structures have several functions, including attachment, genetic exchange, and twitching motility. Hami, present on the surface of certain *Archaea*, function as miniature grappling hooks to attach cells to a surface or to one another.

Q **What function(s) do polysaccharide layers outside the cell wall have in prokaryotic cells?**

2.8 Prokaryotic cells can contain inclusions of sulfur, polyphosphate, carbon polymers, or various minerals formed by biomineralization. These substances function as nutrient storage materials or, in the case of magnetosomes, in magnetotaxis.

Q **What types of cytoplasmic inclusions are formed by bacteria? How does an inclusion of poly-β-hydroxybutyric acid differ from a magnetosome in composition and metabolic role?**

2.9 Gas vesicles are gas-filled structures that confer buoyancy on cells of certain species of *Bacteria* and *Archaea*. Gas vesicles are composed of two different proteins arranged to form a gas-permeable but watertight structure.

Q **What is the function of gas vesicles? How are these structures made such that they can remain gas tight?**

2.10 The endospore is a highly resistant and differentiated structure produced by certain gram-positive *Bacteria*. Endospores are highly dehydrated and contain calcium dipicolinate and small acid-soluble spore proteins, both of which are absent from vegetative cells. Endospores can remain dormant indefinitely but can germinate quickly when conditions warrant.

Q **In a few sentences, indicate how the bacterial endospore differs from the vegetative cell in structure, chemical composition, and ability to resist extreme environmental conditions.**

IV • Cell Locomotion

2.11 Swimming motility in prokaryotic cells is due to flagella (*Bacteria*) or archaella (*Archaea*). Both structures are composed of several proteins, are anchored in the cell wall and cytoplasmic membrane, and function by rotation. Flagella and archaella differ in structure and how energy is coupled to rotation.

> **Q** Describe the structure and function of a bacterial flagellum. What is the energy source for the flagellum? How do bacterial flagella differ from archaella in size, composition, and power source?

2.12 Bacteria that move by gliding do not employ rotating flagella but instead creep along a solid surface by employing one of several different mechanisms including polysaccharide excretion, twitching, or rotating glide proteins.

> **Q** Contrast the mechanism for motility in *Flavobacterium* with that in *Escherichia coli*.

2.13 Swimming bacteria respond to chemical and physical gradients in their environment by controlling the lengths of runs and frequency of tumbles. Tumbles are controlled by the direction of rotation of the flagellum, which in turn is controlled by a network of sensory and response proteins.

> **Q** In a few sentences, explain how a swimming bacterium is able to sense the direction of an attractant and move toward it. In the experiment described in Figure 2.43, what is the control and why is it essential?

V • Eukaryotic Microbial Cells

2.14 Microbial eukaryotes contain various organelles including the nucleus, mitochondria (or hydrogenosomes), and chloroplasts. The nucleus contains the cell's DNA wrapped around histone proteins. Microbial eukaryotes divide following the process of mitosis and may undergo meiosis as well if a haploid/diploid life cycle occurs.

> **Q** List at least three features of eukaryotic cells that clearly differentiate them from prokaryotic cells. What are histones and what do they do?

2.15 The mitochondrion and hydrogenosome are energy-generating organelles; mitochondria respire, whereas hydrogenosomes ferment. Chloroplasts generate ATP by photosynthesis and also fix CO_2 into cell material. All of these organelles were once free-living *Bacteria* that later established residence inside cells of *Eukarya* (endosymbiosis).

> **Q** How are the mitochondrion and the hydrogenosome similar structurally? How do they differ? How do they differ metabolically? What major physiological processes occur in the chloroplast? What evidence supports the idea that the major organelles of eukaryotes were once *Bacteria*?

2.16 Endoplasmic reticula are membranous structures in eukaryotes that either contain attached ribosomes (rough ER) or not (smooth ER). Flagella and cilia are means of motility and move by a whiplike mechanism instead of by rotation. Lysosomes specialize in degrading large molecules. Microtubules, microfilaments, and intermediate filaments function as internal cell scaffolds to form the cell cytoskeleton.

> **Q** Describe the major functions of the endoplasmic reticulum, Golgi complex, and lysosomes. What makes up the eukaryotic cell cytoskeleton?

Application Questions

1. Calculate the surface-to-volume ratio of a spherical cell 15 μm in diameter and of a cell 2 μm in diameter. What are the consequences of these differences in surface-to-volume ratio for cell function?

2. Assume you are given two cultures, one of a species of gram-negative *Bacteria* and one of a species of *Archaea*. Discuss at least four different ways you could tell which culture was which.

3. Calculate the amount of time it would take a cell of *Escherichia coli* (1×2 μm) swimming at maximum speed (60 cell lengths per second) to travel all the way up a 3-cm-long capillary tube containing a chemical attractant.

Chapter Glossary

Archaellum a long, thin cellular appendage present in many *Archaea* that rotates and is responsible for swimming motility

Basal body the "motor" portion of the bacterial flagellum, embedded in the cytoplasmic membrane and cell wall

Capsule a polysaccharide or protein outermost layer, usually rather slimy, present on some bacteria

Chemotaxis directed movement of an organism toward (positive chemotaxis) or away from (negative chemotaxis) a chemical gradient

Chloroplast the photosynthetic organelle of phototrophic eukaryotes

Cristae the internal membranes of a mitochondrion

Cytoplasmic membrane the permeability barrier of the cell, separating the cytoplasm from the environment

Cytoskeleton the cellular scaffolding typical of eukaryotic cells in which microtubules, microfilaments, and intermediate filaments define the cell's shape

Dipicolinic acid a substance unique to endospores that confers heat resistance on these structures

Endospore a highly heat-resistant, thick-walled, differentiated structure produced by certain gram-positive *Bacteria*

Endosymbiotic hypothesis the idea that mitochondria and chloroplasts originated from *Bacteria*

Flagellum a long, thin cellular appendage that rotates (in *Bacteria*) or has a whiplike motion (in *Eukarya*) and is responsible for swimming motility

Gas vesicles gas-filled cytoplasmic structures bounded by protein and conferring buoyancy on cells

Histones highly basic proteins that compact and wind DNA in the nucleus of eukaryotic cells

Hydrogenosome an organelle of endosymbiotic origin present in certain microbial eukaryotes that oxidizes pyruvate to H_2, CO_2, and acetate, and couples this to ATP synthesis

Intermediate filament a filamentous polymer of fibrous keratin proteins, supercoiled into thicker fibers, that functions in maintaining cell shape and the positioning of certain organelles in the eukaryotic cell

Lipopolysaccharide (LPS) a combination of lipid with polysaccharide and protein that forms the major portion of the outer membrane in gram-negative *Bacteria*

Lysosome an organelle containing digestive enzymes for hydrolysis of proteins, fats, and polysaccharides

Magnetosome a particle of magnetite (Fe_3O_4) enclosed by a nonunit membrane in the cytoplasm of magnetotactic *Bacteria*

Meiosis the nuclear division that halves the diploid number of chromosomes to the haploid

Microfilament a filamentous polymer of the protein actin that helps maintain the shape of a eukaryotic cell

Microtubule a filamentous polymer of the proteins α-tubulin and β-tubulin that functions in eukaryotic cell shape and motility

Mitochondrion the respiratory organelle of eukaryotic organisms

Mitosis nuclear division in eukaryotic cells in which chromosomes are replicated and partitioned into two daughter cells during cell division

Morphology the *shape* of a cell—rod, coccus, spirillum, and so on

Nucleus the organelle that contains the eukaryotic cell's chromosomes

Outer membrane a phospholipid- and polysaccharide-containing unit membrane that lies external to the peptidoglycan layer in cells of gram-negative *Bacteria*

Peptidoglycan a polysaccharide composed of alternating repeats of *N*-acetylglucosamine and *N*-acetylmuramic acid arranged in adjacent layers and cross-linked by short peptides

Periplasm a gel-like region between the outer surface of the cytoplasmic membrane and the inner surface of the lipopolysaccharide layer of gram-negative *Bacteria*

Peritrichous flagellation having flagella located in many places around the surface of the cell

Phototaxis movement of an organism toward light

Pili thin, filamentous structures that extend from the surface of a cell and, depending on type, facilitate cell attachment, genetic exchange, or twitching motility

Polar flagellation having flagella emanating from one or both poles of the cell

Poly-β-hydroxybutyric acid (PHB) a common storage material of prokaryotic cells consisting of a polymer of β-hydroxybutyrate or another β-alkanoic acid or mixtures of β-alkanoic acids

S-layer an outermost cell surface layer composed of protein or glycoprotein present on some *Bacteria* and *Archaea*

Stroma the lumen of the chloroplast, surrounded by the inner membrane

Teichoic acid a phosphorylated polyalcohol found in the cell wall of some gram-positive *Bacteria*

Thylakoid a membrane layer containing the photosynthetic pigments in chloroplasts

Microbial Metabolism

3

Sugars and Sweets: *Archaea* Do It Their Way

Ever since the 1977 proposal by Carl Woese and George Fox to group life into three domains—*Bacteria, Archaea,* and *Eukarya*—new discoveries with the *Archaea* keep emphasizing their unique biology. Their novel lipids, lack of peptidoglycan, and many eukaryotic-like traits are just some of the fundamental properties that set *Archaea* apart from the other domains of life. However, recent studies have shown that even some "housekeeping biochemistry" in *Archaea* differs from that in the other domains, and a good example is the metabolism of sugars.

Studies of sugar metabolism in *Archaea* have focused on species of the phylum *Crenarchaeota*, most of which grow at extremely high temperatures, and the phylum *Euryarchaeota*, many of which, like the cells of *Halococcus* (see photo), grow at extremely high salt concentrations. Several species in these two phyla can use glucose and some other sugars as carbon and energy sources. But whether sugars are used or not, all *Archaea* must be able to metabolize sugars because they form the backbone of cell structures such as cell walls and macromolecules such as nucleic acids.

Biochemical studies have shown that some of the most fundamental pathways of carbohydrate metabolism—such as glycolysis—are absent in *Archaea* and that instead, modified variants of this mainstream pathway are present. Another important sugar-metabolizing pathway, the pentose phosphate pathway, is absent in some species of *Archaea* but is present in others in a truncated form distinct from the classical pathway of *Bacteria* and *Eukarya*. Several new enzymes have evolved to run these unique archaeal pathways, and their catalytic activities are often regulated in unusual ways.

The discovery that *Archaea* have evolved distinct pathways for such fundamental metabolic transformations is yet another indication that this domain, first established on the basis of gene sequences, is indeed a phylogenetically distinct entity. Now the question arises as to *why* these biochemical pathways in *Archaea* differ so much from those in cells of the two other domains. Stay tuned, as the most exciting part of the story likely lies ahead.

Source: Bräsen, C., D. Esser, B. Rauch, and B. Slebers. 2014. Carbohydrate metabolism in *Archaea*: Current insights into unusual enzymes and pathways and their regulation. *Microbiol. Mol. Biol. Rev. 78*: 89–175.

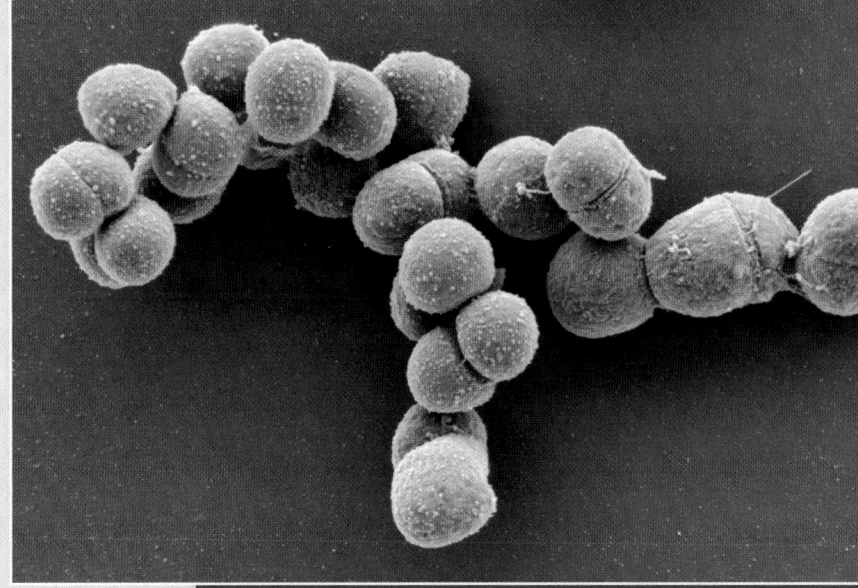

I • Microbial Nutrients and Nutrient Uptake

Metabolism is the series of biochemical reactions by which the cell breaks down or biosynthesizes various metabolites. In order to grow, cells must incorporate nutrients from their environment, transform them into precursor molecules, and then use them to construct a new cell. In this chapter we examine some of these processes, with a focus on three areas: (1) defining the basic nutrients of life, (2) exploring seminal metabolic pathways and alternative metabolic lifestyles, (3) and biosynthesizing the building blocks of macromolecules. We will use some of the principles developed here in Chapter 4, where we explore how informational macromolecules—the nucleic acids and proteins—are biosynthesized, and in Chapter 14, where the enormous metabolic diversity of the microbial world unfolds.

3.1 Feeding the Microbe: Cell Nutrition

Because the metabolic capacities of microbes differ, their nutrient requirements also differ. However, all microbes require a core set of nutrients. Some nutrients, called *macronutrients*, are required in large amounts, while others, called *micronutrients*, are required in minute amounts. We begin by dissecting the cell to reveal its chemical composition and then consider the nutrients that all cells require.

Chemical Makeup of a Cell

Just a handful of the chemical elements predominate in living systems: hydrogen (H), oxygen (O), carbon (C), nitrogen (N), phosphorus (P), sulfur (S), and selenium (Se). C is needed in the largest amount (50% of a cell's dry weight), O and H are next (combined, 25% of dry weight), and N follows (13%). Although required, P, S, K, Mg, and Se combine for less than 5% of a cell's dry weight. In addition to these, at least 50 other elements either are required by one or more microorganisms or, if not required, are still metabolized in some way (**Figure 3.1**).

About 75% of the wet weight of a microbial cell (a single cell of *Escherichia coli* weighs just 10^{-12} g) is water, and the remainder is primarily macromolecules—proteins, nucleic acids, lipids, and polysaccharides (Figure 3.1*b*). The building blocks of these macromolecules are the amino acids, nucleotides, fatty acids, and sugars, respectively. Proteins dominate the macromolecular composition of a cell, and the diversity of proteins exceeds that of all other macromolecules

combined (Figure 3.1*c*). Interestingly, as important as DNA is to a cell (the cell genome), it contributes a very small percentage of a cell's dry weight; RNA is far more abundant (Figure 3.1*c*).

Carbon, Nitrogen, and Other Macronutrients

All cells require carbon and nitrogen in large amounts, and most prokaryotic cells require organic compounds as their source of carbon. Cells obtain organic carbon from the breakdown of polymeric substances or from the direct uptake of their monomeric constituents: the amino acids, fatty acids, organic acids, sugars, nitrogen bases, and aromatic and other organic compounds. Some microbes are autotrophs and can synthesize their own organic compounds from carbon dioxide (CO_2). The bulk of nitrogen available in nature is as ammonia (NH_3), nitrate (NO_3^-), or nitrogen gas (N_2). Virtually all microorganisms can use NH_3 as their nitrogen source and many can also use NO_3^-; some microbes can use organic nitrogen sources, such as amino acids; and a few can use N_2 (the nitrogen-fixing bacteria).

In addition to C and N (and O and H from H_2O), many other macronutrients are needed by cells but typically in smaller amounts (Figure 3.1). Phosphorus is required for nucleic acids and phospholipids and is usually incorporated as phosphate (PO_4^{2-}). Sulfur is present

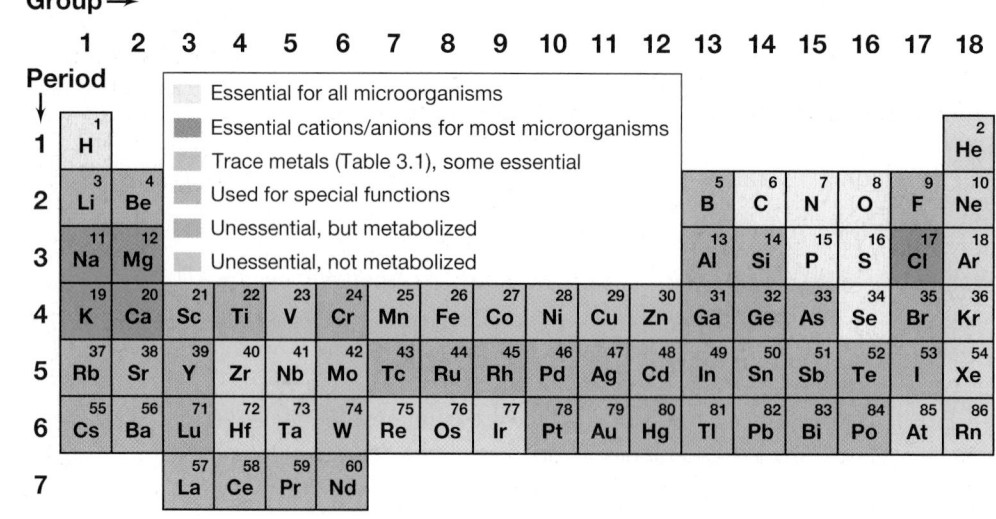

(a)

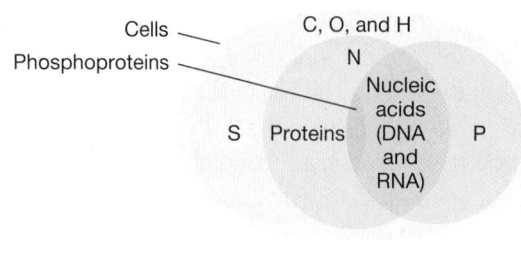

Elemental composition of informational macromolecules

Cells
Phosphoproteins
C, O, and H
N
Nucleic acids (DNA and RNA)
S   Proteins   P

(b)

Macromolecular composition of a cell

Macromolecule	Percent of dry weight
Protein	55
Lipid	9.1
Polysaccharide	5.0
Lipopolysaccharide	3.4
DNA	3.1
RNA	20.5

(c)

Figure 3.1 Elemental and macromolecular composition of a bacterial cell. (a) A microbial periodic table of the elements. With the exception of those shown in row 7, other elements in row 7 and elements in rows beyond row 7 are not known to be metabolized. (b) Distribution of the major elements in informational macromolecules. (c) Relative abundance of macromolecules in a bacterial cell. Data from Escherichia coli *and* Salmonella typhimurium: Cellular and Molecular Biology. ASM, Washington, DC (1996).

in the amino acids cysteine and methionine and also in several vitamins, including thiamine, biotin, and lipoic acid, and is commonly incorporated as sulfate (SO_4^{2-}), sulfide (H_2S), or organic S compounds. Potassium (K) is required for the activity of several enzymes, whereas magnesium (Mg) stabilizes ribosomes, membranes, and nucleic acids and is also required for the activity of many enzymes. Calcium (Ca) and sodium (Na) are essential nutrients for only a few organisms, such as the NaCl requirement of most marine microorganisms.

Micronutrients: Trace Metals and Growth Factors

Microorganisms require several *metals* in very small amounts relative to macronutrients (Figure 3.1); chief among these is iron (Fe), which plays a major role in cellular respiration. Besides iron, many other metals may be required or otherwise metabolized by microorganisms (Figure 3.1*a*). Collectively these metals are called *trace metals* and typically function in the cell as cofactors of certain enzymes (Section 3.5). **Table 3.1** lists the major trace metals and other trace elements of life and examples of enzymes or other molecules in which each is found.

Growth factors differ from trace metals in that they are *organic* (rather than metallic) micronutrients (Table 3.1). Common growth factors include the vitamins, but amino acids, purines, pyrimidines, and several other organic molecules may be growth factors for one or another microorganism. Vitamins are the most frequently required growth factors and a few common ones are shown in Table 3.1. Most vitamins function as enzyme cofactors. Vitamin requirements vary among microorganisms, ranging from none to several. Lactic acid bacteria of the genera *Streptococcus*,

Lactobacillus, and *Leuconostoc* inhabit nutrient-rich habitats such as dairy products and the animal gut (⇔ Sections 16.6 and 24.2); these bacteria are well known for their vitamin requirements, which are even more extensive than those of humans!

If a cell is to grow and divide, it must take up its macronutrients and micronutrients from the environment. But this process is not as easy as it may seem, primarily because of the impermeability of the cytoplasmic membrane and the fact that the concentration of a given nutrient in the cytoplasm must often be much higher than its concentration in the environment. We examine this situation now and see how cells have overcome these fundamental problems.

— MINIQUIZ —

- Which four chemical elements make up the bulk of a cell's dry weight?
- Which two classes of macromolecules contain most of a cell's nitrogen?
- Differentiate between trace metals and growth factors. How are these used by the cell?

3.2 Transporting Nutrients into the Cell

In Chapter 2 we learned how the structure of the cytoplasmic membrane is an effective barrier to leakage; solutes leak neither into nor out of a living cell. However, to fuel metabolism and support growth, cells need to import nutrients and export waste products

---

TABLE 3.1 Micronutrients needed by microorganisms[a]

I. Trace Elements		II. Growth Factors	
Element	**Function**	**Growth factor**	**Function**
Boron (B)	Autoinducer for quorum sensing in bacteria; also found in some polyketide antibiotics	PABA (*p*-aminobenzoic acid)	Precursor of folic acid
Cobalt (Co)	Vitamin B_{12}; transcarboxylase (only in propionic acid bacteria)	Folic acid	One-carbon metabolism; methyl transfers
		Biotin	Fatty acid biosynthesis; some CO_2 fixation reactions
Copper (Cu)	In respiration, cytochrome c oxidase; in photosynthesis, plastocyanin, some superoxide dismutases	B_{12} (Cobalamin)	One-carbon metabolism; synthesis of deoxyribose
Iron (Fe)[b]	Cytochromes; catalases; peroxidases; iron–sulfur proteins; oxygenases; all nitrogenases	B_1 (Thiamine)	Decarboxylation reactions
Manganese (Mn)	Activator of many enzymes; component of certain superoxide dismutases and of the water-splitting enzyme in oxygenic phototrophs (photosystem II)	B_6 (Pyridoxal)	Amino acid/keto acid transformations
		Nicotinic acid (Niacin)	Precursor of NAD^+
		Riboflavin	Precursor of FMN, FAD
Molybdenum (Mo)	Certain flavin-containing enzymes; some nitrogenases, nitrate reductases, sulfite oxidases, DMSO–TMAO reductases; some formate dehydrogenases	Pantothenic acid	Precursor of coenzyme A
		Lipoic acid	Decarboxylation of pyruvate and α-ketoglutarate
Nickel (Ni)	Most hydrogenases; coenzyme F_{430} of methanogens; carbon monoxide dehydrogenase; urease	Vitamin K	Electron transport
Selenium (Se)	Formate dehydrogenase; some hydrogenases; the amino acid selenocysteine	Coenzymes M and B	Methanogenesis[c]
		F_{420} and F_{430}	Methanogenesis[c]
Tungsten (W)	Some formate dehydrogenases; oxotransferases of hyperthermophiles		
Vanadium (V)	Vanadium nitrogenase; bromoperoxidase		
Zinc (Zn)	Carbonic anhydrase; nucleic acid polymerases; many DNA-binding proteins		

[a]Not all trace elements or growth factors are needed by all organisms, and many growth factors are biosynthesized and not required from the environment.
[b]Iron is typically needed in larger amounts than the other trace metals shown.
[c]The production of methane (CH_4) by methanogens (*Archaea*).

on a more or less continuous basis. To do this, several transport systems reside in the cytoplasmic membrane. We consider the most common of these systems here, with a focus on the well-studied transporters widespread in *Bacteria* and *Archaea*.

Active Transport and Transporters

Active transport is the process by which cells accumulate solutes against the concentration gradient, and three basic mechanisms of active transport are found in prokaryotic cells. **Simple transport** consists only of a transmembrane transport protein, **group translocation** employs a series of proteins in the transport event, and **ABC transport systems** consist of three components: a substrate-binding protein, a transmembrane transporter, and an ATP-hydrolyzing protein. Each of these transport systems is energy-driven, be it from the proton motive force, ATP, or some other energy-rich compound (**Figure 3.2**).

The transmembrane component of virtually all transport systems is composed of a polypeptide containing 12 domains that weave back and forth through the membrane to form a channel (see Figure 3.3), and it is through this channel that the solute is actually transported into the cell. Transport is linked to a conformational change in this transmembrane protein that occurs when it binds its specific solute. Like a gate swinging open, this conformational change sweeps the solute into the cell.

Simple Transporters and Group Translocation

Simple transport reactions are driven by the energy inherent in the proton motive force (↩ Sections 2.3 and 3.11). The major transport events catalyzed are either *symport* reactions (where a solute and a proton are cotransported in one direction) or *antiport* reactions (where a solute and a proton are transported in opposite directions) (**Figure 3.3**). A classic example of a simple transporter is the uptake of the disaccharide sugar lactose by way of the *lac permease*, a well-studied symporter in *Escherichia coli*. As each lactose molecule enters the cell, the potential energy in the proton motive force is diminished slightly by the cotransport of a proton (Figure 3.3). The net result is the energy-driven accumulation of lactose in the cytoplasm against the concentration gradient. Many other solutes enter by the activity of their own simple symporters, including phosphate, sulfate, and several different organic compounds.

Group translocation differs from simple transport in two important ways: (1) the transported substance is *chemically modified* during the transport process, and (2) an energy-rich organic compound (rather than the proton motive force) drives the transport event. The best-studied group translocation systems transport the sugars glucose, mannose, and fructose in *E. coli*. During uptake, these compounds are phosphorylated by the *phosphotransferase system*. The phosphotransferase system consists of a family of five proteins that work in concert to transport any given sugar. Before the sugar is transported, the proteins in the phosphotransferase system are themselves alternately phosphorylated and dephosphorylated in a cascading fashion until Enzyme II_c phosphorylates the sugar as it enters the cytoplasm (**Figure 3.4**). A protein called *HPr*, the enzyme that phosphorylates HPr (Enzyme I), and Enzyme II_a are all cytoplasmic proteins. By contrast, Enzyme II_b is a peripheral membrane protein and Enzyme II_c is the transmembrane component.

In the phosphotransferase system, HPr and Enzyme I are nonspecific components and participate in the uptake of several different sugars. By contrast, distinct Enzyme II proteins exist, one set for each different sugar transported. Energy to drive the phosphotransferase system comes from phosphoenolpyruvate, an energy-rich intermediate in glycolysis (Sections 3.7 and 3.8).

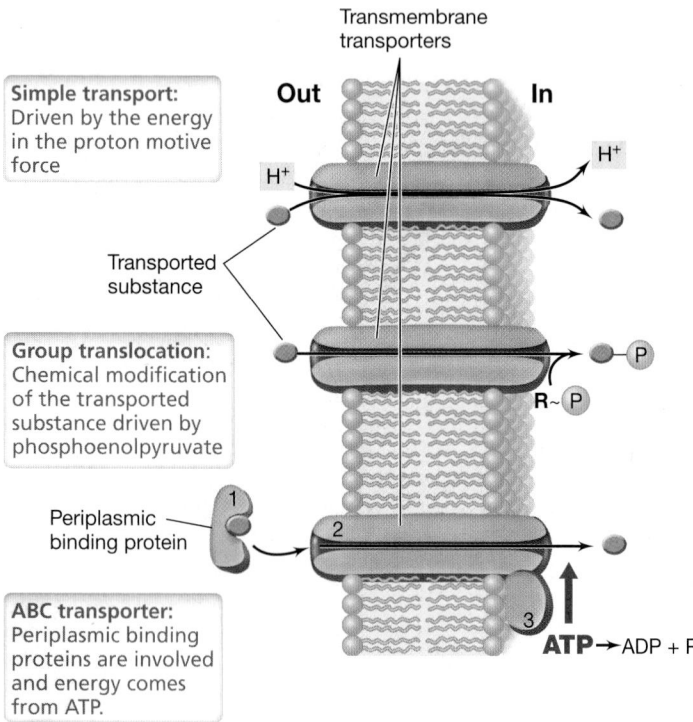

Figure 3.2 The three classes of transport systems. Note how simple transporters and the ABC system transport substances without chemically modifying them, whereas group translocation results in chemical modification (in this case phosphorylation) of the transported substance. The three proteins of the ABC system are labeled 1, 2, and 3.

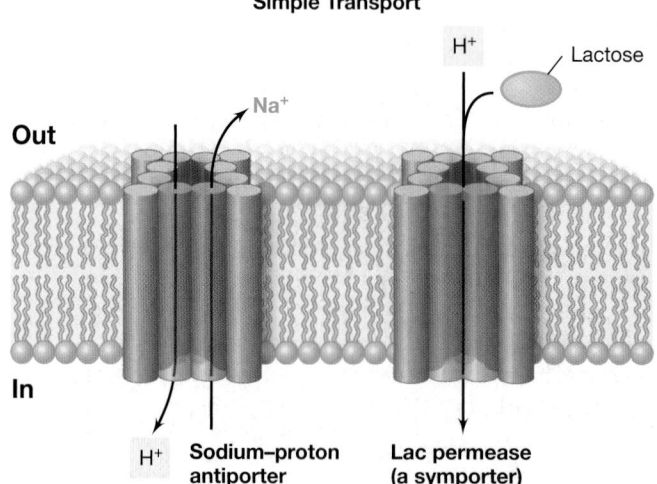

Figure 3.3 Structure of membrane-spanning transporters and symport and antiport events. Transmembrane transporters are composed of a polypeptide that forms 12 α-helices (each shown as a cylinder) that aggregate to form a channel through the membrane. Note how transport is linked to dissipation of the proton motive force.

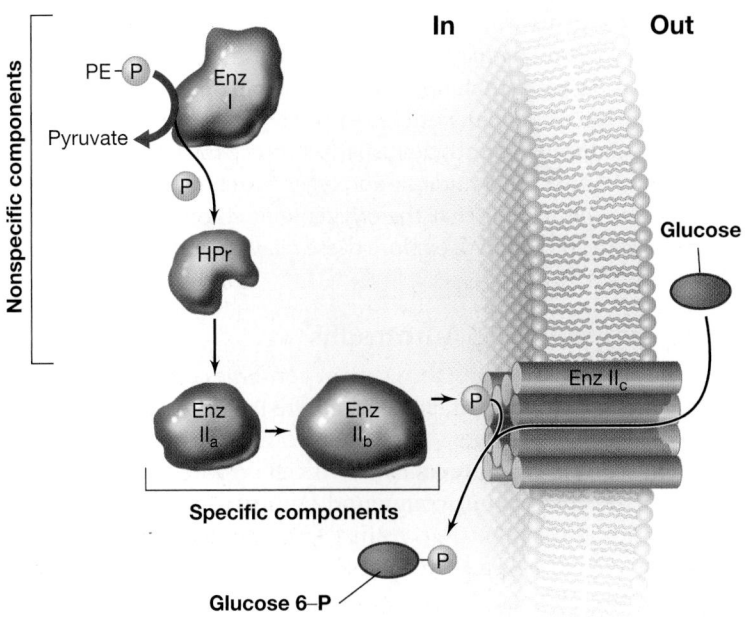

Figure 3.4 Mechanism of the phosphotransferase system of *Escherichia coli.* For glucose transport, the system consists of five proteins: Enzyme (Enz) I, Enzymes II$_a$, II$_b$, and II$_c$, and HPr. A phosphate cascade occurs from phosphoenolpyruvate (PE-P) to Enz II$_c$ and the latter actually transports and phosphorylates the sugar. Proteins HPr and Enz I are nonspecific and participate in the transport of any sugar. By contrast, the three components of Enz II are specific for a particular sugar.

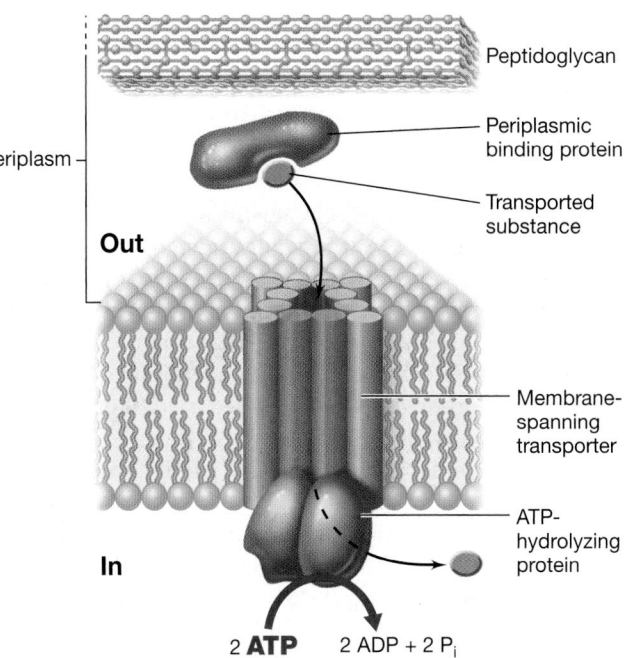

Figure 3.5 Mechanism of an ABC transporter. The periplasmic binding protein has high affinity for substrate, the membrane-spanning proteins form the transport channel, and the cytoplasmic ATP-hydrolyzing proteins supply the energy for the transport event.

Periplasmic Binding Proteins and the ABC System

We learned in Chapter 2 that gram-negative bacteria contain a region called the *periplasm* that lies between the cytoplasmic membrane and the *outer membrane*, part of the gram-negative cell wall (Section 2.5). The periplasm is home to proteins that carry out different functions including transport; the latter is catalyzed by the activity of *periplasmic binding proteins*. Transport systems that employ a periplasmic binding protein along with transmembrane and ATP-hydrolyzing components are called ABC transport systems, the "ABC" standing for *ATP-binding cassette*, a structural feature of proteins that bind ATP (**Figure 3.5**). More than 200 different ABC transport systems have been identified in various *Bacteria*, and these catalyze the uptake of a wide variety of organic and inorganic compounds.

A characteristic property of periplasmic binding proteins is their extremely high substrate affinity. These proteins can bind their specific substrate even when it is present at extremely low concentration; for example, less than 1 micromolar (10^{-6} M). Once its specific substrate is bound, the periplasmic binding protein interacts with its respective transmembrane component to transport the substrate into the cell driven by the energy in ATP (Figures 3.2 and 3.5).

Although gram-positive bacteria and most *Archaea* lack a periplasm, they also have ABC transport systems. In gram-positive bacteria, substrate-binding proteins (the functional equivalent of periplasmic binding proteins) are attached to the external surface of the cytoplasmic membrane. Once they bind their substrate, these proteins interact with their transmembrane component to catalyze the ATP-driven uptake of the substrate. ABC systems are also present in a variety of *Archaea* and are primarily employed for the transport of sugars.

─────────────────── **MINIQUIZ** ───────────────────

• Compare and contrast simple transporters, the phosphotransferase system, and ABC transporters in terms of (1) energy source, (2) chemical alterations of the solute during transport, and (3) number of proteins required.

• Which major characteristic of ABC transport systems makes them ideal for organisms living in nutrient-poor environments?

II • Energetics, Enzymes, and Redox

If a microorganism has all the nutrients it needs and has transported them into its cell, it must next conserve some of the energy released in energy-yielding reactions in order to grow. Here we discuss the different options for energy conservation and use some basic laws of chemistry and physics to guide our understanding of bioenergetics.

3.3 Energy Classes of Microorganisms

Energy-yielding reactions are that part of metabolism called **catabolism**. Here we discuss the various catabolic energy classes of microorganisms, pointing out their similarities and differences.

The terms used to describe the energy classes of microorganisms are important ones and will appear many times in this book.

Chemoorganotrophs, Chemolithotrophs, and Phototrophs

Organisms that conserve energy from chemicals are called *chemotrophs*, and those that use *organic* chemicals are called **chemoorganotrophs** (**Figure 3.6**). Most microorganisms in laboratory culture are chemoorganotrophs. A wide variety of organic compounds can be catabolized during chemoorganotrophic metabolism, and some of the energy released during their oxidation is conserved in the energy-rich bonds of adenosine triphosphate—ATP, the cell's energy currency—or related energy-rich compounds (see Section 3.7 and Figure 3.13).

Many *Bacteria* and *Archaea* can tap the energy available from the oxidation of *inorganic* compounds, and this form of catabolism is called *chemolithotrophy*. Organisms that carry out chemolithotrophic reactions are called **chemolithotrophs** (Figure 3.6). Several inorganic compounds can be oxidized, for example, gaseous hydrogen (H_2), hydrogen sulfide (H_2S), ammonia (NH_4^+), and ferrous iron (Fe^{2+}). Related groups of chemolithotrophs typically specialize in the oxidation of a group of related inorganic compounds, and thus we have the "sulfur" bacteria, the "iron" bacteria, the "nitrifying" bacteria, and so on.

Phototrophs contain chlorophylls and other pigments that convert light energy into ATP and thus, unlike chemotrophs, do not require chemicals as a source of energy. Two forms of phototrophy are known in *Bacteria*. In one form, called *oxygenic photosynthesis*, oxygen (O_2) is produced. Oxygenic photosynthesis is characteristic of cyanobacteria, a major lineage of *Bacteria*, and is also carried out

by algae (eukaryotic microbes). The other form of phototrophy, *anoxygenic photosynthesis*, occurs in at least six phylogenetic lineages of *Bacteria*, including the purple and green bacteria, the heliobacteria, and many others. In anoxygenic photosynthesis, O_2 is not produced. Nevertheless, many strong parallels exist between the mechanisms that underlie anoxygenic and oxygenic photosynthesis, and it is clear that the oxygenic process evolved from the anoxygenic process. We explore these relationships in Chapters 13 and 14.

Heterotrophs and Autotrophs

Regardless of how a microorganism conserves energy, be it from chemicals or from light, all cells require large amounts of carbon in one form or another to make new cell materials (Section 3.1). If an organism is a **heterotroph**, its cell carbon is obtained from one or another organic compound. An **autotroph**, by contrast, uses carbon dioxide (CO_2) as its carbon source. Chemoorganotrophs (Figure 3.6) are by definition also heterotrophs. By contrast, most chemolithotrophs and phototrophs (Figure 3.6) are autotrophs. Autotrophs are also called *primary producers* because they synthesize new organic matter from inorganic carbon (CO_2). Virtually all organic matter on Earth has been synthesized by primary producers, in particular, the phototrophs. The **Calvin cycle** is the major biochemical pathway by which phototrophic organisms incorporate CO_2 into cell material, although we will see in Chapter 14 that many other pathways exist, especially among prokaryotic autotrophs.

MINIQUIZ

- How does a chemoorganotroph differ from a chemolithotroph? A chemotroph from a phototroph?
- How does an autotroph differ from a heterotroph?

3.4 Principles of Bioenergetics

We have just reviewed the options that microbes have in terms of energy conservation: chemoorganotrophy, chemolithotrophy, and phototrophy. But exactly how is the energy available from these processes conserved by the cell? We consider these issues here using chemoorganotrophic metabolism as our model system.

Energy is defined as the ability to do work, and in microbiology, energy transformations are measured in kilojoules (kJ), a unit of heat energy. All chemical reactions in a cell are accompanied by *changes* in energy, energy being either *required* or *released* as a reaction proceeds. To identify which reactions release energy and which require energy, we first need to understand some basic bioenergetic principles.

Basic Bioenergetics

In microbiology we are interested in **free energy** (abbreviated G), which is *the energy available to do work*. Free energy released during a reaction can be conserved by cells in the form of ATP and a handful of other energy-rich substances. The *change* in free energy during a reaction is expressed as $\Delta G^{0\prime}$, where the symbol Δ is read as "change in." The "0" and "prime" in $\Delta G^{0\prime}$ indicate that the

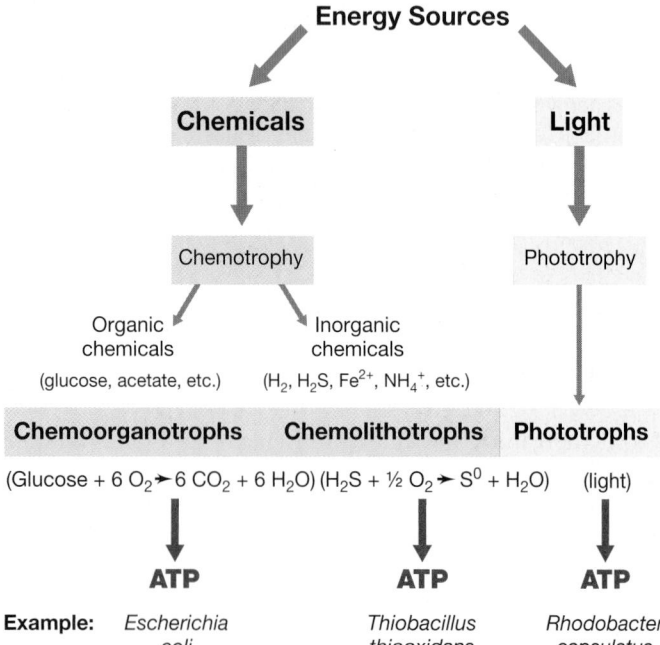

Figure 3.6 Metabolic options for conserving energy by microorganisms. Most organisms employ only one option, but some have two and a few rare species can tap into all three forms of energy conservation.

free-energy value is for *standard conditions*: pH 7 (approximate cytoplasmic conditions), 25°C, 1 atmosphere of pressure, and all reactants and products at molar concentrations.

Consider the reaction

$$A + B \rightarrow C + D$$

If the $\Delta G^{0\prime}$ for this reaction is *negative* in arithmetic sign, then the reaction will proceed with the *release* of free energy; such reactions are said to be **exergonic**. However, if $\Delta G^{0\prime}$ is *positive*, the reaction *requires* energy in order to proceed and such reactions are **endergonic**. Thus, exergonic reactions *release* free energy whereas endergonic reactions *require* free energy. With these essentials in hand, how do we calculate $\Delta G^{0\prime}$?

Free Energy of Formation and Calculating Free-Energy Changes ($\Delta G^{0\prime}$)

To calculate the free-energy yield of a reaction, one first needs to know the free energy inherent in the reactants and products of the reaction. This is the *free energy of formation* (G_f^0), the energy released or required during the formation of a given molecule from the elements. Table 3.2 lists the G_f^0 for a few common substances. By convention, the free energy of formation of the *elements* in their elemental and electrically neutral form (for instance, C, H_2, N_2) is zero. However, the free energies of formation of *compounds* are not zero. If the formation of a compound from its elements proceeds exergonically (free energy released), then the G_f^0 of the compound is negative. If the reaction is endergonic (free energy required), then the G_f^0 of the compound is positive.

For most compounds G_f^0 is negative. This reflects the fact that compounds tend to form spontaneously (that is, with a free-energy release) from their elements. However, the positive G_f^0 for

nitrous oxide (N_2O) (Table 3.2) indicates that this compound does not form spontaneously. Instead, over time it decomposes spontaneously to yield N_2 and O_2. The compounds listed in Table 3.2 are only a small subset of free energy of formation values available from physical chemistry reference sources.

Using free energies of formation, it is possible to calculate $\Delta G^{0\prime}$ of a reaction. For the reaction $A + B \rightarrow C + D$, $\Delta G^{0\prime}$ is calculated by subtracting the sum of the free energies of formation of the reactants ($A + B$) from that of the products ($C + D$). Thus

$$\Delta G^{0\prime} = G_f^0[C + D] - G_f^0[A + B]$$

The value obtained for $\Delta G^{0\prime}$ tells us whether the reaction is exergonic (and can be a potential energy source for the cell) or endergonic (and requires an energy input to proceed). The phrase "products minus reactants" is a simple way to recall how to calculate changes in free energy during chemical reactions.

Before free-energy calculations can be made, it is first necessary to balance the reaction. That is, (1) the total number of each kind of atom and ionic charges must be identical on both sides of the reaction, and (2) the oxidation–reduction state must balance such that all of the electrons removed from one substance are transferred to another substance. Once a reaction is balanced, its $\Delta G^{0\prime}$ can be calculated, and from this, the potential of the reaction as a means of energy conservation for a cell can be assessed.

$\Delta G^{0\prime}$ versus ΔG

Although calculations of $\Delta G^{0\prime}$ are often very accurate estimates of actual free-energy changes, in some cases they are not. We will see when we pick up the bioenergetics theme again in Chapter 14 that the actual concentrations of products and reactants in a microbe's natural habitat, which are rarely at the levels used in calculations of $\Delta G^{0\prime}$, can sometimes change the results of bioenergetic calculations

---

TABLE 3.2 Free energy of formation (G_f^0, kJ/mol) for some common substances[a]

Sugars	Organic and fatty acids	Amino acids and alcohols	Gases and inorganic compounds
Fructose (−951.4)	Acetate (−369.4)	Alanine (−371.5)	O_2, N_2, H_2, S^0, Fe^0 (0)
Glucose (−917.2)	Benzoate (−245.6)	Aspartate (−700.4)	CH_4 (−50.8)
Lactose (−1515.2)	Butyrate (−352.6)	*n*-Butanol (−171.8)	CO_2 (−394.4); CO (−137.4)
Ribose (−369.4)	Caproate (−335.9)	Ethanol (−181.7)	H_2O (−237.2); H^+ (−39.8); OH^- (−198.7)
Sucrose (−757.3)	Citrate (−1168.3)	Glutamate (−699.6)	N_2O (+104.2); NO (+86.6)
	Formate (−351.1)	Glutamine (−529.7)	NO_2^- (−37.2); NO_3^- (−111.3)
	Fumarate (−604.2)	Glycerol (−488.5)	NH_3 (−26.57); NH_4^+ (−79.4)
	Glyoxylate (−468.6)	Mannitol (−942.6)	H_2S (−27.87); HS^- (+12.1)
	Ketoglutarate (−797.5)	Methanol (−175.4)	SO_4^{2-} (−744.6); $S_2O_3^{2-}$ (−513.4)
	Lactate (−517.8)	*n*-Propanol (−175.8)	Fe^{2+} (−78.8); Fe^{3+} (−4.6); FeS (−100.4)
	Malate (−845.1)		
	Propionate (−361.1)		
	Pyruvate (−474.6)		
	Succinate (−690.2)		
	Valerate (−344.3)		

[a]Values for free energy of formation taken from Speight, J. 2005. *Lange's Handbook of Chemistry*, 16th edition, and Thauer, R.K., K. Jungermann, and H. Decker. 1977. Energy conservation in anaerobic chemotrophic bacteria. *Bacteriol. Rev. 41*: 100–180.

in significant ways. In this regard, what is most relevant to a bioenergetic calculation is not $\Delta G^{0\prime}$, but ΔG, the free-energy change that occurs *under the actual conditions* in which the organism is growing. The equation for ΔG takes into account the actual concentrations of reactants and products in the organism's habitat and is expressed as

$$\Delta G = \Delta G^{0\prime} + RT \ln K_{eq}$$

where R and T are physical constants and K_{eq} is the equilibrium constant for the reaction. For the reaction aA + bB $\rightarrow$ cC + dD, $K_{eq} = [C]^c[D]^d/[A]^a[B]^b$, where A and B are reactants and C and D are products; a, b, c, and d are the number of molecules of each; and brackets indicate concentrations.

The reason that ΔG rather than $\Delta G^{0\prime}$, may be a more accurate estimate of bioenergetic processes is that the products of a reaction carried out by one microbe are typically consumed by the activities of other microorganisms. In most cases, this does not significantly affect $\Delta G^{0\prime}$. However, in some cases, the consumption of products is so aggressive that it can drive K_{eq} values to less than 1; the logarithm of such a number will be negative in arithmetic sign. Then, when ΔG is calculated, a reaction that might have been endergonic under standard conditions can become exergonic under the actual conditions present in the microbial habitat. We will see examples of this in Chapter 14, especially in reactions in which H_2—a product of syntrophic metabolisms—is consumed by other microbes to levels that approach zero.

So, although $\Delta G^{0\prime}$ and ΔG are not always identical, at this point in our understanding of microbial bioenergetics the expression $\Delta G^{0\prime}$ is all we need to deal with in order to appreciate energy flow in microbial systems. The main point to keep in mind is that only reactions that are exergonic yield energy that can be conserved by the cell, and this will be our focus in the next few sections.

MINIQUIZ

- What is free energy?
- Does glucose formation from the elements release or require energy?
- Using Table 3.2, calculate $\Delta G^{0\prime}$ for the reaction $CH_4 + \frac{1}{2} O_2 \rightarrow CH_3OH$.

3.5 Catalysis and Enzymes

Free-energy calculations reveal only whether energy is released or required in a given reaction; they say nothing about the *rate* of the reaction. If the rate of a reaction is very slow, it may be of no value to a cell. For example, consider the formation of water from O_2 and H_2. The energetics of this reaction are quite favorable: $H_2 + \frac{1}{2} O_2 \rightarrow H_2O$, $\Delta G^{0\prime} = -237$ kJ. However, if O_2 and H_2 were mixed in a sealed bottle, no measurable amount of water would form, even after years. This is because the bonding of O_2 and H_2 to form H_2O requires that these two gases become reactive. This requires that their bonds be broken and requires a small amount of energy. This energy is called **activation energy**.

Activation energy can be viewed as the minimum energy required for a chemical reaction to begin. For an exergonic reaction, the

situation is as shown in **Figure 3.7**. Although the activation energy barrier is virtually insurmountable in the absence of a *catalyst*—a substance that facilitates a reaction but is not consumed by it—in the presence of a proper catalyst, this barrier is reduced, allowing the reaction to proceed.

Enzymes

Catalysts function by lowering the activation energy of a reaction (Figure 3.7), thereby increasing the reaction rate. Catalysts have no effect on the energetics or the equilibrium of a reaction but only affect the *rate* at which a reaction proceeds. Most cellular reactions will not proceed at significant rates without catalysis.

The major catalysts in cells are **enzymes**, which are proteins (or in a few cases, RNAs) that are highly specific for the reactions they catalyze. This specificity is a function of the precise three-dimensional structure of the enzyme. In an enzyme-catalyzed reaction, the enzyme combines with the reactant, called a *substrate*, forming an *enzyme–substrate complex*. Then, as the reaction proceeds, the *product* is released and the enzyme is returned to its original state, ready to catalyze a new round of the reaction (**Figure 3.8**). The enzyme is generally much larger than the substrate(s), and the portion of the enzyme to which substrate binds is the enzyme's *active site*; the entire enzymatic reaction, from substrate binding to product release, may take only a few milliseconds.

Many enzymes contain small nonprotein molecules that participate in catalysis but are not themselves substrates. These small molecules can be divided into two classes based on the way they associate with the enzyme: *prosthetic groups* and *coenzymes*. Prosthetic groups bind tightly to their enzymes, usually covalently and permanently. The heme group present in cytochromes such as cytochrome *c* (Section 3.10) is an example of a prosthetic group. By contrast, **coenzymes**, with a few exceptions, are loosely and often transiently bound to enzymes; thus, a single coenzyme molecule may associate with a number of different enzymes. Most coenzymes are derivatives of vitamins (Table 3.1).

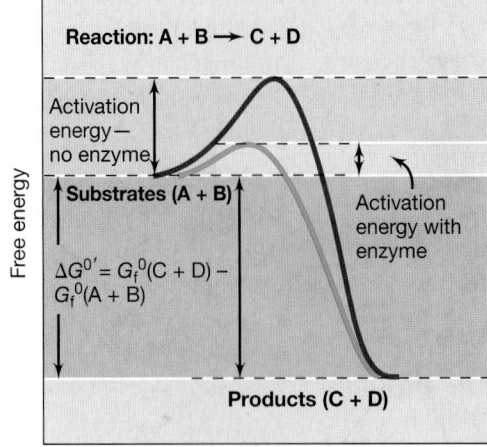

Figure 3.7 Activation energy and catalysis. Even chemical reactions that release energy may not proceed spontaneously if not activated. Once the reactants are activated, the reaction proceeds spontaneously. Catalysts such as enzymes lower the required activation energy.

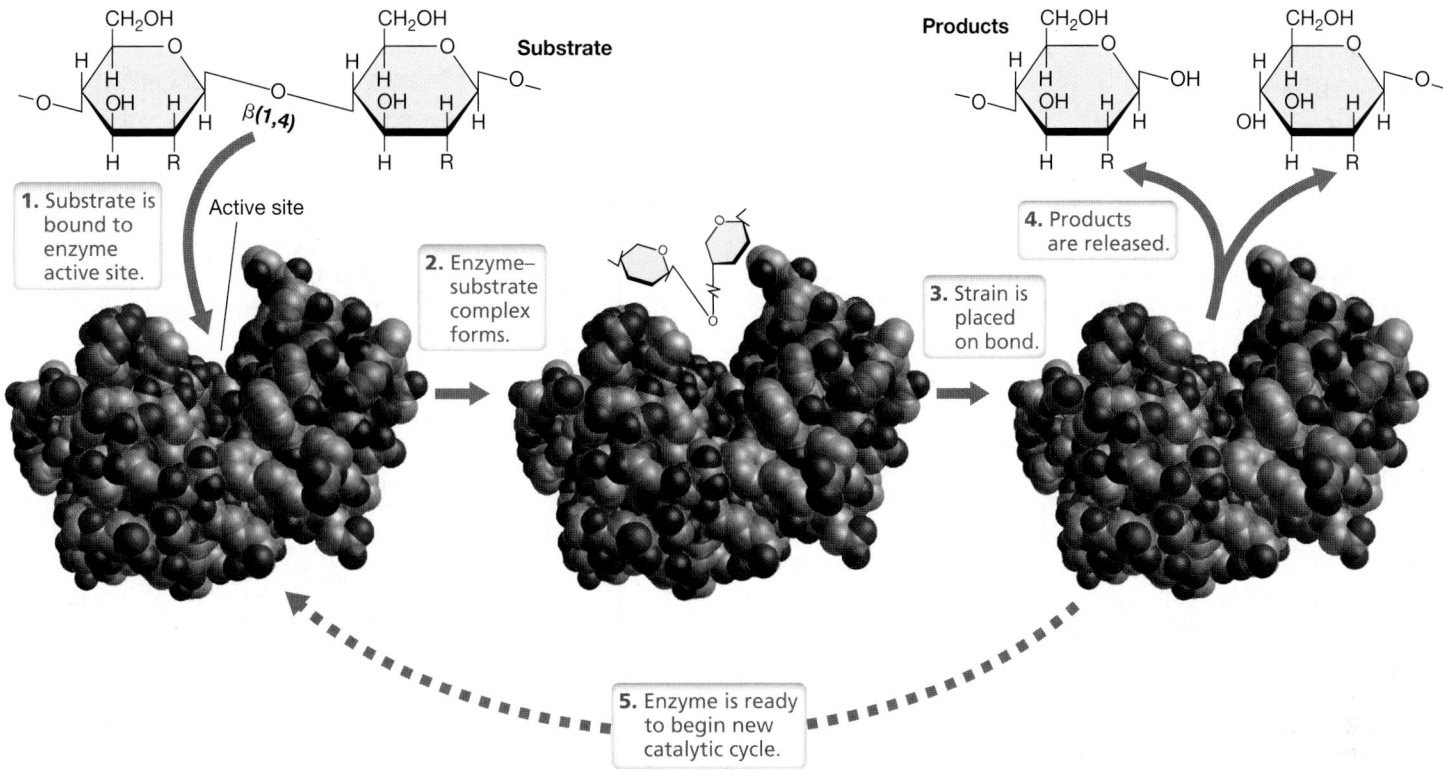

Figure 3.8 The catalytic cycle of an enzyme. The enzyme depicted here, lysozyme, catalyzes the cleavage of the β-1,4-glycosidic bond in the polysaccharide backbone of peptidoglycan. Following substrate binding in the enzyme's active site, strain is placed on the bond, and this favors breakage. Space-filling model of lysozyme courtesy of Richard Feldmann.

Enzyme Catalysis

To catalyze a reaction, an enzyme must bind its substrate and position it properly in its active site. The enzyme–substrate complex (Figure 3.8) aligns reactive groups in the substrate(s) and places strain on specific bonds. This reduces the activation energy required to make the reaction proceed (Figure 3.7). This is shown in Figure 3.8 for the enzyme lysozyme, an enzyme whose substrate is the polysaccharide backbone of peptidoglycan, the bacterial cell wall polymer (⟳ Section 2.4).

The reaction depicted in Figure 3.7 is exergonic. By contrast, some enzymes catalyze endergonic reactions where they convert energy-poor substrates into energy-rich products. In these cases not only must an activation energy barrier (Figure 3.7) be overcome, but sufficient free energy must also be put into the reaction in order to raise the energy level of the substrates to that of the products. This is done by *coupling* the energy-requiring reaction to an energy-yielding one, such as the hydrolysis of ATP or dissipation of the proton motive force (Section 3.11), so that the overall reaction proceeds with a free-energy change that is either negative in arithmetic sign or zero.

Theoretically, all enzymes are reversible in their activity. However, enzymes that catalyze highly exergonic or highly endergonic reactions typically function in only one direction. If a particularly exergonic or endergonic reaction needs to be reversed, a different enzyme usually catalyzes the reverse reaction.

— MINIQUIZ —

- What is the function of a catalyst? What are enzymes made of?
- Where on an enzyme does the substrate bind?
- What is activation energy?

3.6 Electron Donors and Acceptors

Cells conserve energy released from exergonic reactions by coupling the reaction to the biosynthesis of energy-rich compounds, such as ATP. Reactions that release sufficient energy to form ATP require oxidation–reduction biochemistry. An *oxidation* is the removal of an electron (or electrons) from a substance, and a *reduction* is the addition of an electron (or electrons) to a substance. In microbiology, the term *redox* is used as shorthand for oxidation–reduction.

Redox Reactions

Redox reactions occur in pairs. For example, H_2 can release electrons and protons and become oxidized (**Figure 3.9**). However, electrons cannot exist alone in solution; they must be part of atoms or molecules. Thus the oxidation of H_2 is only a *half reaction*, a term that implies the need for a second half reaction to complete the overall reaction. The second half reaction is a reduction, in which another substance is reduced to form the coupled redox reaction (Figure 3.9).

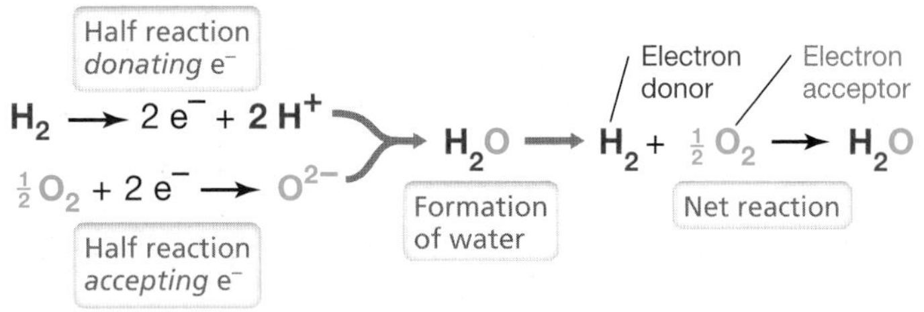

$$H_2 \longrightarrow 2\,e^- + 2\,H^+$$
$$\tfrac{1}{2}\,O_2 + 2\,e^- \longrightarrow O^{2-}$$

Half reaction *donating* e^-

Half reaction *accepting* e^-

$H_2O \longrightarrow H_2 + \tfrac{1}{2}\,O_2 \longrightarrow H_2O$

Electron donor / Electron acceptor

Formation of water

Net reaction

Figure 3.9 Example of an oxidation–reduction reaction. Oxidations are the removal of electrons from a substance, while reductions are the addition of electrons to a substance.

In redox reactions of this type, we refer to the substance *oxidized* (in this case, H_2) as the **electron donor**, and the substance *reduced* (in this case, O_2) as the **electron acceptor** (Figure 3.9). Many different electron donors exist in nature, including a wide variety of organic and inorganic compounds. Many electron acceptors other than O_2 exist as well, including many nitrogen and sulfur compounds, such as NO_3^- and SO_4^{2-}, and many organic compounds (cp Figure 14.25). As we continue our journey through this book we will see that the concept of electron donors and electron acceptors is very important for understanding microbial physiology, diversity, and ecology, and underlies all aspects of cellular energy metabolism.

Reduction Potentials and Redox Couples

Many substances can be *either* an electron donor *or* an electron acceptor, depending on the substance they couple with in a redox reaction. The constituents on each side of the arrow in half reactions are called a *redox couple*, such as $2\,H^+/H_2$, or $\tfrac{1}{2}\,O_2/H_2O$ in Figure 3.9. By convention, in writing a redox couple, the *oxidized* form of the couple is always placed on the left (before the forward slash) followed by the *reduced* form after the forward slash.

Substances differ in their tendency to donate or accept electrons. This tendency is expressed as their **reduction potential** (E_0', standard conditions), a value measured in volts (V) compared with that of a reference substance, H_2 (see Figure 3.10). Moreover, in biology, reduction potentials are listed for half reactions *written as reductions* at pH 7. In the example of Figure 3.9, the E_0' of the $2\,H^+/H_2$ couple is −0.42 V and that of the $\tfrac{1}{2}\,O_2/H_2O$ couple is +0.82 V. We will learn shortly that these values mean that O_2 is an excellent *electron acceptor* and H_2 is an excellent *electron donor*.

When two redox couples react, the *reduced* substance of the couple whose E_0' is *more negative* donates electrons to the *oxidized* substance of the couple whose E_0' is *more positive*. Thus, in the couple $2\,H^+/H_2$, H_2 has a greater tendency to donate electrons than does $2\,H^+$ to accept them, and in the couple $\tfrac{1}{2}\,O_2/H_2O$, H_2O has a poor tendency to donate electrons while O_2 has a strong tendency to accept them. It thus follows that in a redox reaction of H_2 and O_2, H_2 *will be the electron donor* and become oxidized, and O_2 *will be the electron acceptor* and become reduced (Figure 3.9).

Although all half reactions are written as reductions, in an actual reaction between two redox couples, the half reaction with the more negative E_0' proceeds as an oxidation and is therefore written in the opposite direction. For example, in the reaction between H_2 and O_2, H_2 is oxidized and is written in the

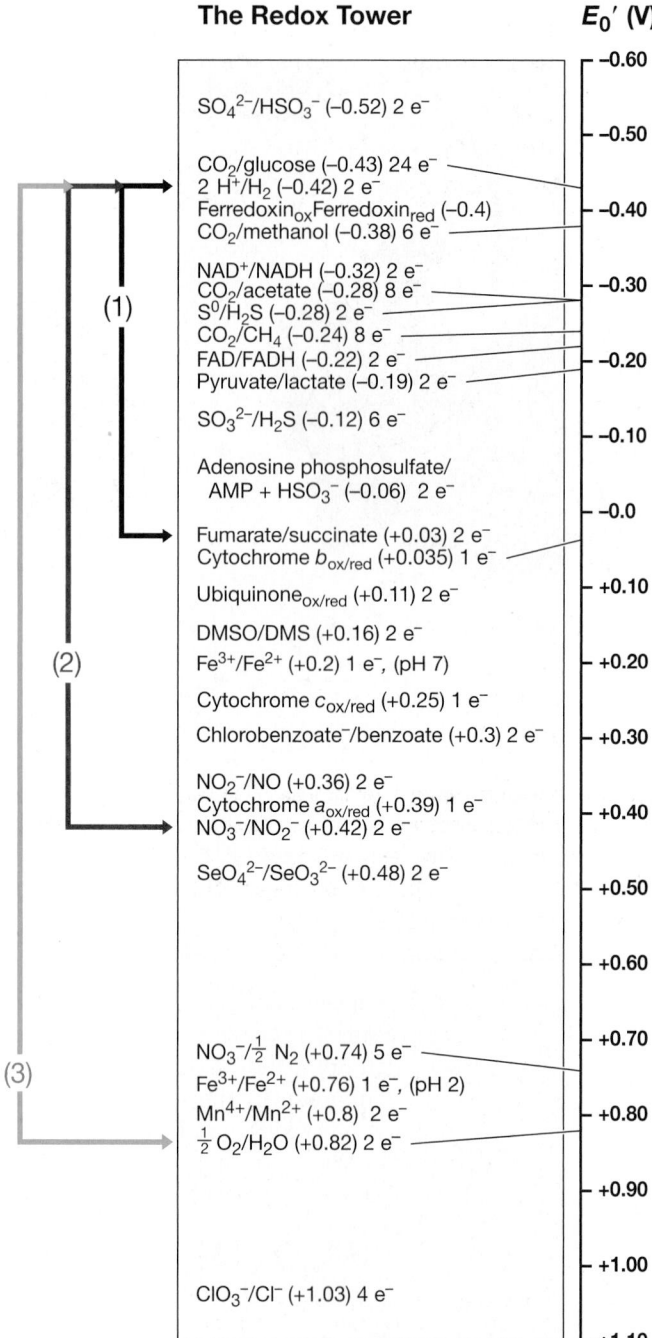

The Redox Tower E_0' **(V)**

SO_4^{2-}/HSO_3^- (−0.52) 2 e^-

$CO_2/glucose$ (−0.43) 24 e^-
$2\,H^+/H_2$ (−0.42) 2 e^-
$Ferredoxin_{ox}Ferredoxin_{red}$ (−0.4)
$CO_2/methanol$ (−0.38) 6 e^-

$NAD^+/NADH$ (−0.32) 2 e^-
$CO_2/acetate$ (−0.28) 8 e^-
S^0/H_2S (−0.28) 2 e^-
CO_2/CH_4 (−0.24) 8 e^-
$FAD/FADH$ (−0.22) 2 e^-
$Pyruvate/lactate$ (−0.19) 2 e^-

SO_3^{2-}/H_2S (−0.12) 6 e^-

Adenosine phosphosulfate/ AMP + HSO_3^- (−0.06) 2 e^-

$Fumarate/succinate$ (+0.03) 2 e^-
$Cytochrome\ b_{ox/red}$ (+0.035) 1 e^-
$Ubiquinone_{ox/red}$ (+0.11) 2 e^-

$DMSO/DMS$ (+0.16) 2 e^-
Fe^{3+}/Fe^{2+} (+0.2) 1 e^-, (pH 7)

$Cytochrome\ c_{ox/red}$ (+0.25) 1 e^-

$Chlorobenzoate^-/benzoate$ (+0.3) 2 e^-

NO_2^-/NO (+0.36) 2 e^-
$Cytochrome\ a_{ox/red}$ (+0.39) 1 e^-
NO_3^-/NO_2^- (+0.42) 2 e^-

SeO_4^{2-}/SeO_3^{2-} (+0.48) 2 e^-

$NO_3^-/\tfrac{1}{2}\,N_2$ (+0.74) 5 e^-
Fe^{3+}/Fe^{2+} (+0.76) 1 e^-, (pH 2)
Mn^{4+}/Mn^{2+} (+0.8) 2 e^-
$\tfrac{1}{2}\,O_2/H_2O$ (+0.82) 2 e^-

ClO_3^-/Cl^- (+1.03) 4 e^-

(1)
(2)
(3)

−0.60, −0.50, −0.40, −0.30, −0.20, −0.10, −0.0, +0.10, +0.20, +0.30, +0.40, +0.50, +0.60, +0.70, +0.80, +0.90, +1.00, +1.10

(1) H_2 + fumarate $\longrightarrow$ succinate $\Delta G^{0'} = -86$ kJ

(2) H_2 + NO_3^- $\longrightarrow$ NO_2^- + H_2O $\Delta G^{0'} = -163$ kJ

(3) H_2 + $\tfrac{1}{2}\,O_2$ $\longrightarrow$ H_2O $\Delta G^{0'} = -237$ kJ

Figure 3.10 The redox tower. Redox couples are arranged from the strongest donors at the top to the strongest acceptors at the bottom. The larger the difference in reduction potential between electron donor and electron acceptor, the more free energy is released. Note the differences in energy yield when H_2 reacts with three different electron acceptors: fumarate (1), nitrate (2), and oxygen (3). The redox tower is placed in the context of anaerobic respiration in Figure 14.25.

TABLE 3.3 Example of free-energy-change calculations using G_f values or electrochemical potentials

For the reaction in which acetate is oxidized completely to CO_2:[a]

$$CH_3COO^- + H^+ + 2\,O_2 \rightarrow 2\,CO_2 + 2\,H_2O$$

1. Calculation from G_f values:
$$\Delta G^{0\prime} = [G_f(\text{products}) - G_f(\text{reactants})]$$
$$= [G_f(2\,CO_2 + 2\,H_2O) - G_f(CH_3COO^- + H^+ + 2\,O_2)]$$
$$= -852\ \text{kJ/reaction}$$

2. Calculation from the Nernst equation:[b]
$$\Delta G^{0\prime} = -nF\Delta E_0{}'$$
$$= [-8(96.5)(1.1)]$$
$$= -849\ \text{kJ/reaction}$$

[a]The reaction is balanced and is an 8-electron oxidation ($n = 8$ in equation 2). G_f^0 values were taken from Table 3.2.
[b]F is the Faraday constant (96.5 kJ/V) and $\Delta E_0{}'$ is calculated from the $E_0{}'$ values in Figure 3.10.

reverse direction from its formal half reaction (Figure 3.9 and see Figure 3.10).

The Redox Tower and Its Relationship to $\Delta G^{0\prime}$

A convenient way of viewing electron transfer reactions is to imagine a vertical tower that represents the entire range of reduction potentials possible for redox couples in nature, from those with the most negative $E_0{}'$ on the top to those with the most positive $E_0{}'$ at the bottom; this is a *redox tower* (**Figure 3.10**). Now imagine electrons from an electron donor near the top of the tower falling and being "caught" by electron acceptors at different levels. The difference in reduction potential between the donor and acceptor redox couples is expressed as $\Delta E_0{}'$.

The further an electron drops before it is caught by an acceptor, the greater is the $\Delta E_0{}'$ between the two redox couples and the greater is the amount of energy released in the net reaction. That is, $\Delta E_0{}'$ is proportional to $\Delta G^{0\prime}$ (Figure 3.10). This relationship is expressed more precisely in the Nernst equation (an expression that equates free energy with electrochemistry), $\Delta G^{0\prime} = -nF\Delta E_0{}'$, where n is the number of electrons transferred and F is the Faraday constant (96.5 kJ/V). We thus see that the $\Delta G^{0\prime}$ of a given redox reaction can be calculated in two ways: from knowledge of the equilibrium constant of the reaction (Section 3.4) or by knowing the $\Delta E_0{}'$ of the two half reactions that make up the full redox reaction. This is illustrated for an example reaction, the oxidation of acetate to CO_2, in **Table 3.3**.

Oxygen (O_2), at the bottom of the redox tower, is the strongest electron acceptor of significance in nature. In the middle of the redox tower, redox couples can be either electron donors or acceptors depending on whom they react with. For instance, the $2\,H^+/H_2$ couple (-0.42 V) can react with the fumarate/succinate couple ($+0.03$ V), the NO_3^-/NO_2^- couple ($+0.42$ V), or the $\frac{1}{2}\,O_2/H_2O$ ($+0.82$ V) couple, with increasing amounts of free energy being released, respectively, because $\Delta G^{0\prime} \propto \Delta E_0{}'$ ($\propto$ means "is proportional to") (Figure 3.10).

Electron Carriers and NAD$^+$/NADH Cycling

Redox reactions are typically facilitated by coenzymes that associate with the redox enzymes that catalyze the reaction. A very common redox coenzyme is nicotinamide adenine dinucleotide, or

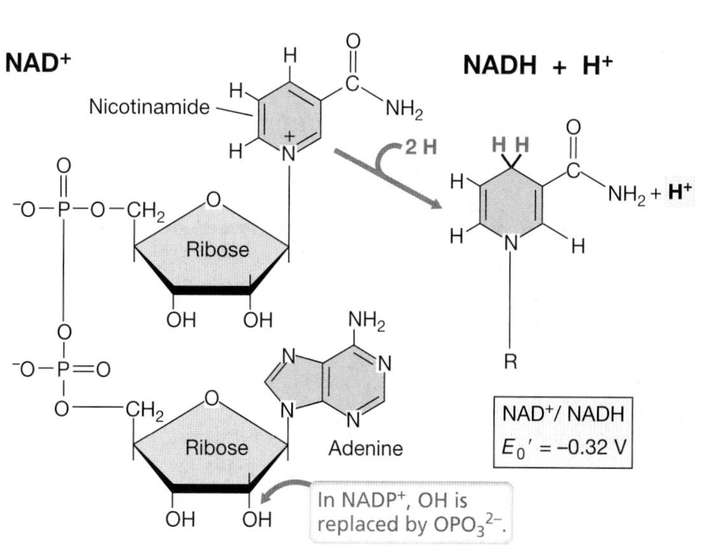

Figure 3.11 The redox coenzyme nicotinamide adenine dinucleotide (NAD$^+$) and NADP$^+$. NAD$^+$ undergoes oxidation–reduction as shown and is freely diffusible. "R" is the adenine dinucleotide portion of NAD$^+$.

NAD$^+$ for short, whose reduced form is NADH (**Figure 3.11**). NAD$^+$/NADH is an electron plus proton carrier, transporting 2 e$^-$ and 2 H$^+$ simultaneously. The reduction potential of the NAD$^+$/NADH couple is -0.32 V, which places it fairly high on the electron tower; that is, NADH is a good electron donor while NAD$^+$ is a rather weak electron acceptor (Figure 3.10).

Coenzymes such as NAD$^+$/NADH increase the diversity of redox reactions that are possible in a cell by allowing many different electron donors and acceptors to interact, with the coenzyme being the redox intermediary that services the different enzymes involved (**Figure 3.12**). For example, electrons removed from an electron donor by an enzyme that oxidizes that donor are used to reduce NAD$^+$ to NADH. The NADH then diffuses away from the enzyme and attaches to a different enzyme that oxidizes NADH back to NAD$^+$ when it reduces an electron acceptor (Figure 3.12).

Electron shuttling mediated by NAD$^+$/NADH is common in microbial catabolism. However, in addition to NAD$^+$/NADH, many other redox coenzymes may participate as electron shuttles. For example, the redox coenzyme nicotinamide adenine dinucleotide phosphate (NADP$^+$) is made from NAD$^+$ by adding a phosphate molecule. NADP$^+$ (and its reduced form NADPH) participates in anabolic redox reactions (biosynthesis of cellular precursors, Sections 3.14 and 3.15), whereas NAD$^+$/NADH participates in catabolic redox reactions, many of which we will explore in Sections 3.8–3.12.

--- MINIQUIZ ---

- In the reaction $H_2 + \frac{1}{2}\,O_2 \rightarrow H_2O$, what is the electron donor and what is the electron acceptor?

- Why is nitrate (NO_3^-) a better electron acceptor than fumarate?

- Is NADH a better electron donor than H_2? Is NAD$^+$ a better acceptor than 2 H$^+$? How does Figure 3.10 tell you this?

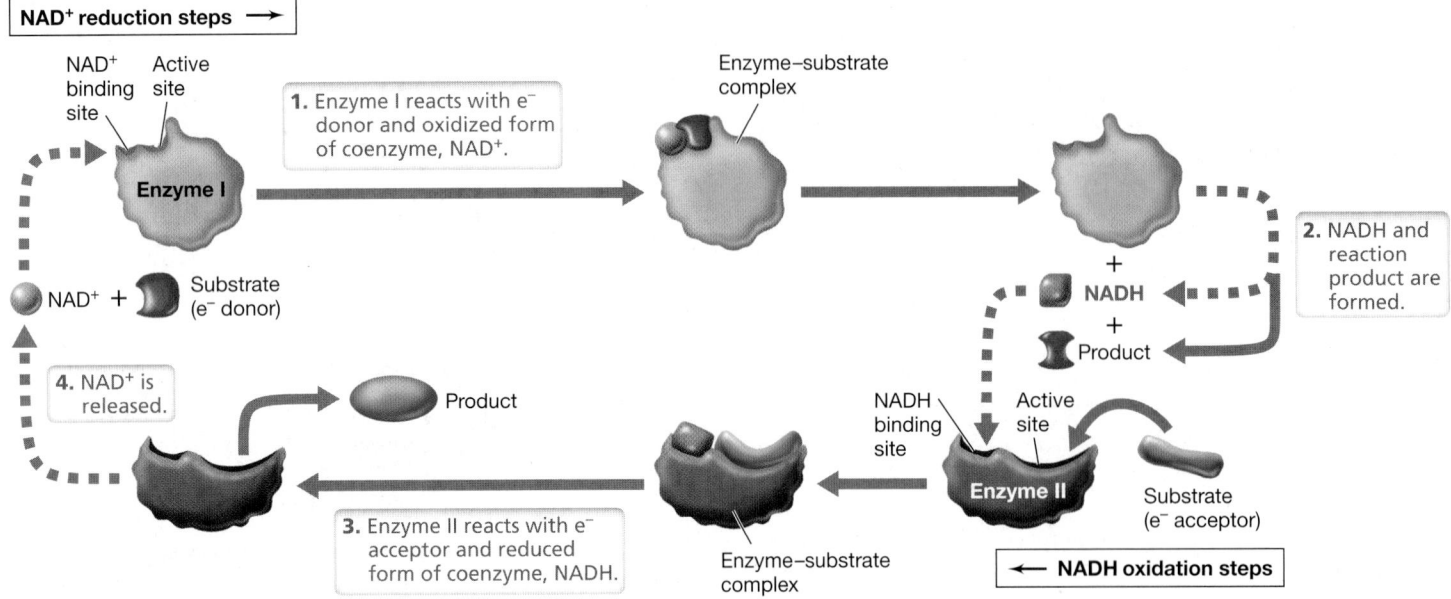

Figure 3.12 NAD⁺/NADH cycling. A schematic example of redox reactions in which two different enzymes are linked by their requirement for either NAD⁺ or NADH.

3.7 Energy-Rich Compounds

The energy released from redox reactions fuels energy-requiring cell functions. But the free energy released in the coupled exergonic redox reaction must first be trapped by the cell and conserved. Energy conservation in cells is accomplished through the formation of a set of compounds containing *energy-rich* phosphate or sulfur bonds. The biosynthesis of these compounds functions as the free-energy trap, and their hydrolysis releases this energy to drive endergonic reactions.

Phosphate can be bonded to organic compounds by either *ester* or *anhydride* bonds, as illustrated in **Figure 3.13**. However, not all phosphate bonds are energy-rich. As seen in this figure, the $\Delta G^{0\prime}$ of hydrolysis of the phosphate *ester* bond in glucose 6-phosphate is −13.8 kJ/mol. By contrast, the $\Delta G^{0\prime}$ of hydrolysis of the phosphate *anhydride* bond in phosphoenolpyruvate is −51.6 kJ/mol, almost four times that of glucose 6-phosphate. Although the phosphate in either compound could be hydrolyzed in energy metabolism, cells typically use compounds whose $\Delta G^{0\prime}$ of phosphate hydrolysis exceeds −30 kJ/mol as energy "currencies" in the cell. Thus, phosphoenolpyruvate is energy-rich, whereas glucose 6-phosphate is not (Figure 3.13).

Adenosine Triphosphate

The most important energy-rich phosphate compound in cells is **adenosine triphosphate (ATP)**. ATP consists of the ribonucleoside adenosine to which three phosphate molecules are bonded in series. From the structure of ATP (Figure 3.13), it can be seen that only two of the phosphate bonds (ATP → ADP + P_i and ADP → AMP + P_i) are phosphoanhydrides and thus have free energies of hydrolysis greater than −30 kJ/mol. By contrast, AMP is not energy-rich because its free energy of hydrolysis is only about half that of ADP or ATP (Figure 3.13).

Although the energy released in ATP hydrolysis is −32 kJ/mol, a caveat must be mentioned here to more precisely state the energy requirements for the synthesis of ATP. In an actively growing *Escherichia coli* cell, the ratio of ATP to ADP is maintained at about 7:1, and this increases the actual free-energy requirements for ATP synthesis. Thus, in an actively growing cell, the actual energy expenditure (that is, the ΔG, Section 3.4) for the synthesis of ATP is more like −55 to −60 kJ/mol. Nevertheless, for the purposes of learning and applying the basic principles of bioenergetics, we will assume that reactions conform to "standard conditions" ($\Delta G^{0\prime}$), and thus that the energy required for the biosynthesis of ATP (from ADP + P_i) *or* hydrolysis of ATP (to ADP + P_i) is 32 kJ/mol.

Coenzyme A

Cells can also use the free energy available in the hydrolysis of a handful of energy-rich compounds other than phosphorylated compounds. These include, in particular, derivatives of *coenzyme A* (for example, acetyl-CoA in Figure 3.13). Coenzyme A derivatives contain energy-rich *thioester* bonds, and hydrolysis of these bonds yields sufficient free energy to be coupled to the synthesis of an energy-rich phosphate bond. For example, in the coupled reaction

$$\text{Acetyl-S-CoA} + H_2O + \text{ADP} + P_i \rightarrow \text{acetate}^- + \text{HS-CoA} + \text{ATP} + H^+$$

the energy released in the hydrolysis of coenzyme A is conserved in the synthesis of ATP. Coenzyme A derivatives (acetyl-CoA is one of many) are especially important in the energetics of anaerobic microorganisms, in particular those whose energy metabolism depends on fermentation (see Table 3.4). We will return to the importance of coenzyme A derivatives in microbial bioenergetics many times in Chapter 14.

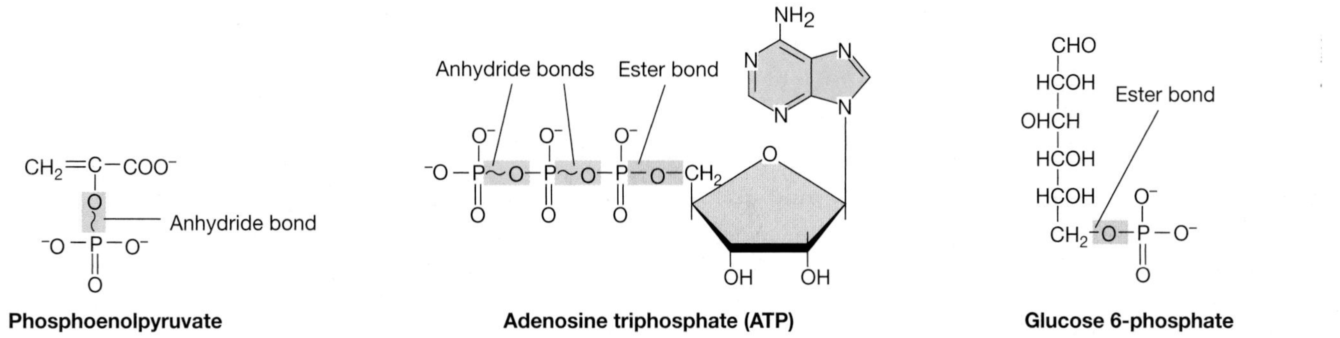

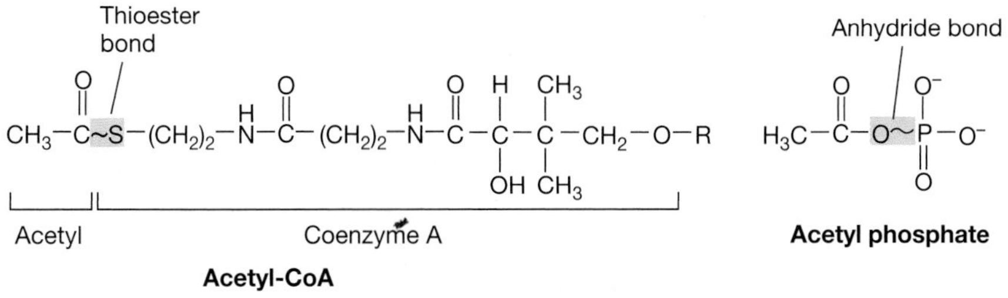

Figure 3.13 Energy-rich bonds in compounds that conserve energy in microbial metabolism. Notice, by referring to the table, the range in free energy of hydrolysis of the phosphate or sulfur bonds highlighted in the compounds. The "R" group of acetyl-CoA is a 3'-phospho ADP group.

Compound	$G^{0'}$ kJ/mol
$\Delta G^{0'} > 30kJ$	
Phosphoenolpyruvate	−51.6
1,3-Bisphosphoglycerate	−52.0
Acetyl phosphate	−44.8
ATP	−31.8
ADP	−31.8
Acetyl-CoA	−35.7
$\Delta G^{0'} < 30kJ$	
AMP	−14.2
Glucose 6-phosphate	−13.8

Energy Storage

ATP is a dynamic molecule in the cell; it is continuously broken down to drive anabolic reactions and resynthesized at the expense of catabolic redox reactions. For long-term energy storage, microorganisms typically produce insoluble polymers that can be catabolized later for the production of ATP.

Examples of energy storage polymers include glycogen (polyglucose), poly-β-hydroxybutyrate and other polyhydroxyalkanoates, and elemental sulfur, stored from the oxidation of H_2S by sulfur chemolithotrophs. These polymers are deposited within the cell as granules that can be seen with the light or electron microscope (⟷ Section 2.8). In eukaryotic microorganisms, starch (also a polymer of glucose) and simple fats are the major reserve materials. In the absence of an available energy source, a cell can break down these polymers to make new cell material or to supply the very low amount of energy, called *maintenance energy*, needed to maintain cell integrity when a cell is in a nongrowing state. This energy requirement, essentially the power necessary to maintain a bacterium in a living state, is thought to be about 1–100 zeptowatts (one zeptowatt equals 10^{-21} W) per cell, an indication of how little energy is actually necessary to keep a bacterial cell alive.

MINIQUIZ

- How much free energy is released when ATP is converted to ADP + P_i or when AMP is converted to adenosine and P_i?

- What is coenzyme A and why is it important?

- During periods of nutrient abundance, how can cells prepare for periods of nutrient starvation?

III • Catabolism: Fermentation and Respiration

We now explore some major catabolic pathways that result in energy conservation in chemoorganotrophs: fermentation and respiration. **Fermentation** is a form of anaerobic catabolism in which organic compounds both donate electrons and accept electrons, and redox balance is achieved without the need for external electron acceptors. By contrast, **respiration** is a form of aerobic or anaerobic catabolism in which an organic or inorganic electron donor is oxidized with O_2 (in aerobic respiration) or some other compound (in anaerobic respiration) functioning as electron acceptors. We will revisit the concepts of fermentation and respiration in detail in Chapter 14 and so focus here only on the basics.

3.8 Glycolysis and Fermentation

A nearly universal pathway for the catabolism of glucose is the *Embden–Meyerhof–Parnas pathway*, better known as **glycolysis**, a series of reactions in which glucose is oxidized to pyruvate. If glucose is ultimately respired, it is first catabolized through glycolysis before pyruvate is further oxidized to CO_2 in the citric acid cycle (Section 3.9). By contrast, if the glucose is fermented, pyruvate is not oxidized completely to CO_2 but instead is used as an electron acceptor to achieve redox balance in glycolysis.

In the series of reactions that are glycolysis, two are redox reactions. During these reactions, free energy is released and is conserved by the simultaneous production of energy-rich compounds (Figure 3.13). ATP is made from these energy-rich compounds by **substrate-level phosphorylation**, a process whereby the energy-rich phosphate bond on the organic compound is transferred directly to ADP to form ATP.

The Three Stages of Glycolysis

Glycolysis can be divided into three stages, each consisting of one or more enzymatic reactions. Stage I consists of "preparatory" reactions; these are not redox reactions and do not release energy but instead form a key intermediate of the pathway. In Stage II, redox reactions occur, energy is conserved, and two molecules of pyruvate are formed. In Stage III, redox balance is achieved and fermentation products are formed (**Figure 3.14**).

To begin glycolysis, glucose is phosphorylated to form glucose 6-phosphate. The latter is then isomerized to fructose 6-phosphate, and a second phosphorylation leads to the production of fructose 1,6-bisphosphate. These steps consume, rather than produce, ATP. The enzyme aldolase then splits fructose 1,6-bisphosphate into two 3-carbon molecules, *glyceraldehyde 3-phosphate* and its isomer, *dihydroxyacetone phosphate*, which is converted into glyceraldehyde 3-phosphate. To this point, all of the reactions, including the consumption of ATP, have proceeded without any redox changes (Figure 3.14).

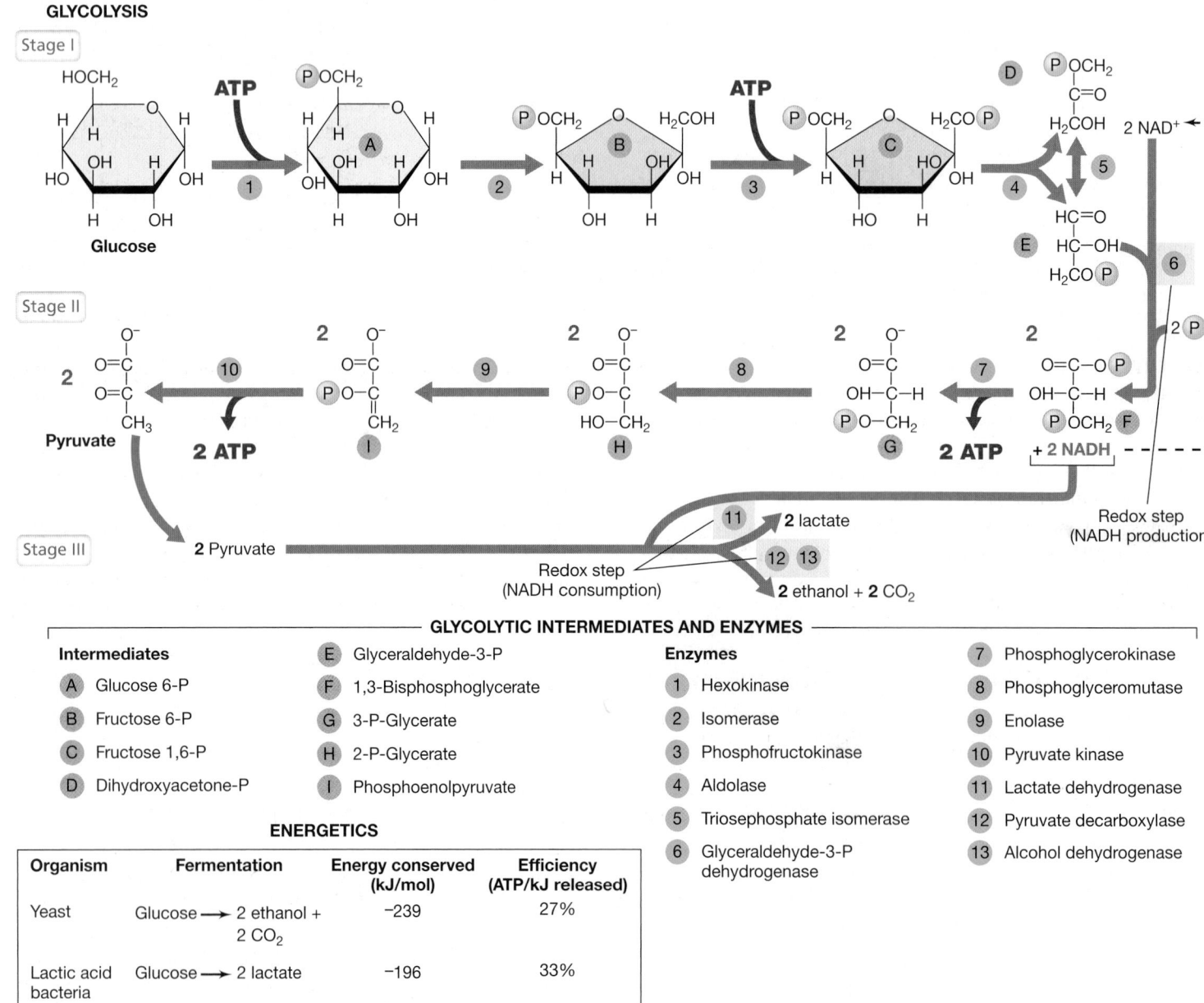

Figure 3.14 Embden–Meyerhof–Parnas pathway (glycolysis). (Top) The sequence of reactions in the catabolism of glucose to pyruvate and then on to fermentation products. Pyruvate is the end product of glycolysis, and fermentation products are made from it. (Bottom) Intermediates, enzymes, and contrasting fermentation balances of yeast and lactic acid bacteria.

The first redox reaction of glycolysis occurs when glyceraldehyde 3-phosphate is oxidized to 1,3-bisphosphoglyceric acid. In this reaction (which occurs twice, once for each of the two glyceraldehyde 3-phosphates), the enzyme glyceraldehyde-3-phosphate dehydrogenase reduces NAD^+ to NADH. Simultaneously, each glyceraldehyde 3-phosphate molecule is phosphorylated by the addition of a molecule of inorganic phosphate. This reaction, in which *inorganic* phosphate is converted to *organic* form, sets the stage for energy conservation, since 1,3-bisphosphoglyceric acid is an energy-rich compound (Figure 3.13). ATP is then synthesized by substrate-level phosphorylation when: (1) each molecule of 1,3-bisphosphoglyceric acid is converted to 3-phosphoglyceric acid, and (2) each molecule of phosphoenolpyruvate is converted to pyruvate (Figure 3.14). During the first two stages of glycolysis, *two* ATP molecules are consumed and *four* ATP molecules are synthesized. Thus, *the net energy yield in glycolysis is two molecules of ATP per molecule of glucose fermented.*

The final stage of glycolysis achieves redox balance. Recall that during the formation of the two 1,3-bisphosphoglyceric acid molecules, two NAD^+ were reduced to NADH (Figure 3.14). This NADH needs to be oxidized (returned to NAD^+) for the next round of glycolysis, and this occurs when pyruvate is reduced by an NADH-containing enzyme to fermentation products; during this reaction, NAD^+ is regenerated (Figure 3.14). In fermentation by yeast, pyruvate is reduced to ethanol (ethyl alcohol) and carbon dioxide (CO_2). By contrast, lactic acid bacteria reduce pyruvate to lactate. Pyruvate can be reduced to many other fermentation products (see next subsection), but the rationale for product formation is the same in each case: NADH must be reoxidized to NAD^+ to keep glycolysis in redox balance (Figure 3.14).

Fermentative Diversity

Not all compounds are inherently fermentable, but sugars—such as glucose and other hexoses as well as most disaccharides and other relatively small sugars—are preeminently fermentable. Since glucose is needed for glycolysis, sugars other than glucose must first be converted to glucose by isomerase enzymes.

Polysaccharides such as cellulose and starch are also fermentable by bacteria that produce enzymes that attack these large molecules and produce sugars from them; if the latter are not glucose, they must first be converted to glucose before they enter glycolysis.

Different types of fermentations are classified by either the substrate fermented or the products formed, and **Table 3.4** summarizes some major fermentations of glucose on the basis of the products formed. All of the organisms listed in Table 3.4 (except for the bacterium *Zymomonas*) use the glycolytic pathway to catabolize glucose, the major difference in the fermentations being in the products formed from pyruvate to achieve redox balance.

In addition to the two (net) ATP produced in glycolysis (Figure 3.14), some of the fermentations listed in Table 3.4 allow for additional ATP synthesis by substrate-level phosphorylation. This is possible if the fermentation product is a fatty acid because the fatty acid is formed from its coenzyme-A precursor. Recall that CoA derivatives of fatty acids, such as acetyl-CoA, are energy-rich (Section 3.7 and Figure 3.13). Thus, when *Clostridium butyricum* forms butyric acid, the final step is

$$Butyryl\text{-}CoA + ADP + P_i \rightarrow butyric\ acid + ATP + CoA$$

The formation of a coenzyme-A derivative during a fermentation increases the yield of ATP, although the yield increase falls far shy of what is possible by glucose respiration (Section 3.9).

Many organic compounds other than sugars can be fermented and may or may not require glycolytic reactions. For instance, some endospore-forming species of *Clostridium* ferment amino acids while others ferment purines and pyrimidines, the products of nucleic acid degradation. Some fermentative anaerobes even ferment aromatic compounds. In most of these cases, the formation of CoA fatty acid derivatives in the fermentative pathway is key to energy conservation.

Some fermentations are catalyzed by only a small group (or even a single species) of anaerobic bacteria. For example, the coupled fermentation of ethanol and acetate is apparently carried out in nature only by the bacterium *Clostridium kluyveri* (Table 3.4).

TABLE 3.4 Common fermentations and some of the organisms carrying them out

Type	Reaction (substrate → products)	Organisms
Alcoholic	Hexose[a] → 2 ethanol + 2 CO_2	Yeast, *Zymomonas*
Homolactic	Hexose → 2 lactate⁻ + 2 H^+	*Streptococcus*, some *Lactobacillus*
Heterolactic	Hexose → lactate⁻ + ethanol + CO_2 + H^+	*Leuconostoc*, some *Lactobacillus*
Propionic acid	3 Lactate⁻ → 2 propionate⁻ + acetate⁻ + CO_2 + H_2O	*Propionibacterium*, *Clostridium propionicum*
Mixed acid[b,c]	Hexose → ethanol + 2,3-butanediol + succinate²⁻ + lactate⁻ + acetate⁻ + formate⁻ + H_2 + CO_2	Enteric bacteria including *Escherichia*, *Salmonella*, *Shigella*, *Klebsiella*, *Enterobacter*
Butyric acid[c]	Hexose → butyrate⁻ + 2 H_2 + 2 CO_2 + H^+	*Clostridium butyricum*
Butanol[c]	2 Hexose → butanol + acetone + 5 CO_2 + 4 H_2	*Clostridium acetobutylicum*
Caproate/Butyrate	6 Ethanol + 3 acetate⁻ → 3 butyrate⁻ + caproate⁻ + 2 H_2 + 4 H_2O + H^+	*Clostridium kluyveri*
Acetogenic	Fructose → 3 acetate⁻ + 3 H^+	*Clostridium aceticum*

[a]Glucose is the starting substrate for glycolysis. However, many other C_6 sugars (hexoses) can be fermented following their conversion to glucose. Except for *Zymomonas*, all organisms catabolize glucose by the glycolytic pathway.
[b]Not all organisms produce all products. In particular, butanediol production is limited to only certain enteric bacteria. The reaction is not balanced.
[c]Other products include some acetate and a small amount of ethanol (butanol fermentation only).

Organisms such as *C. kluyveri* are metabolic specialists, having evolved the capacity to ferment organic compounds not readily catabolized anaerobically by other bacteria. Nevertheless, these fermenters are ecologically important because of their role in the degradation of dead plants, animals, and microorganisms in anoxic environments in nature. We explore these unusual fermentations in more detail in Chapter 14.

Human Benefits of Fermentation and the Fermentation–Respiration Switch

As we have seen, during glycolysis, glucose is consumed, ATP is made, and fermentation products are generated (Figure 3.14). For the organism the crucial product is ATP, fermentation products being merely waste products that must be discarded. However, fermentation products are not waste products to humans. Instead, they are the foundation of the baking and fermented beverage industries (**Figure 3.15**) and are key ingredients in many other fermented foods, such as the lactic and other acids in fermented dairy products (yogurt, sour cream, buttermilk, and the like), cheeses, pickles, and certain sausages and fish products. In the baking and alcohol industries, the metabolic capacities of the key catalyst, the baker's and brewer's yeast *Saccharomyces cerevisiae*, take center stage. However, *S. cerevisiae* can carry out two modes of glucose catabolism, *fermentation*, as we have discussed, and *respiration*, which we consider in the next section.

Cells that can both ferment and respire metabolize in whichever way most benefits them energetically. In this regard, the energy available from a molecule of glucose is much greater if it is respired to CO_2 than if it is fermented. This is because in contrast to CO_2, organic fermentation products such as ethanol or lactic acid discarded by the fermenter still contain a significant amount

Figure 3.15 Common food and beverage products resulting from the alcoholic fermentation of *Saccharomyces cerevisiae*.

of free energy. Thus, when O_2 is available, yeast cells respire glucose rather than ferment it. But the key here is O_2 availability; only when conditions are anoxic do yeasts carry out fermentation. This fact has practical significance. Since the brewer and baker need the *products* of yeast fermentation (Figure 3.15) rather than the cells themselves, care must be taken to ensure that the yeast is forced into a fermentative lifestyle. If air is available, respiration will occur, and we see why this is the case over the course of the next three sections.

— MINIQUIZ —

- Which reactions in glycolysis are redox steps?
- What is the role of NAD^+/NADH in glycolysis?
- Why are fermentation products made during glycolysis? What major fermentation product is present in fermented milk products?

3.9 Respiration: Citric Acid and Glyoxylate Cycles

As an alternative to fermentation, glucose can be respired. During respiration, glucose is first catabolized through glycolysis. But instead of reducing pyruvate to fermentation products and discarding them (Figure 3.14), in respiration, *pyruvate is fully oxidized to CO_2* through activities of the citric acid and glyoxylate cycles, major pathways for respiring organic compounds.

Respiration of Glucose

The pathway by which pyruvate is oxidized to CO_2 is called the **citric acid cycle** (CAC) (**Figure 3.16**). In the CAC, pyruvate is first decarboxylated, leading to the production of CO_2, NADH, and the energy-rich substance *acetyl-CoA* (Figure 3.13). The acetyl group of acetyl-CoA then combines with the four-carbon CAC intermediate oxaloacetate, forming the six-carbon compound citric acid, for which the CAC is named. A sequence of reactions follows, and two additional CO_2 molecules, three more NADH, and one $FADH_2$ are formed per pyruvate oxidized. Ultimately, oxaloacetate is regenerated as the next acetyl acceptor, thus completing the cycle (Figure 3.16).

For each 2 pyruvate oxidized through the citric acid cycle (2 pyruvate are produced from one glucose, Figure 3.14), 6 molecules of CO_2, 8 NADH, and 2 $FADH_2$ are produced (Figure 3.16). $FADH_2$ is a redox prosthetic group that can exist in either an oxidized (FAD) or a reduced ($FADH_2$) form (Section 3.10). And, just as for NADH (Figure 3.12), $FADH_2$ must be reoxidized for the CAC to continue. Both NADH and $FADH_2$ are oxidized in redox reactions that occur in the electron transport chain, a series of reactions that both consumes electrons (through the reduction of O_2) and produces ATP (through the proton motive force) (Sections 3.10 and 3.11).

The combined activities of the CAC and electron transport chain result in the complete oxidation of glucose to CO_2 along with a much greater yield of energy than is possible in fermentation. Whereas only *2 ATP* are produced per glucose fermented in alcoholic or lactic acid fermentations (Figure 3.14), a total of *38 ATP* can be made by aerobically respiring the same glucose molecule to

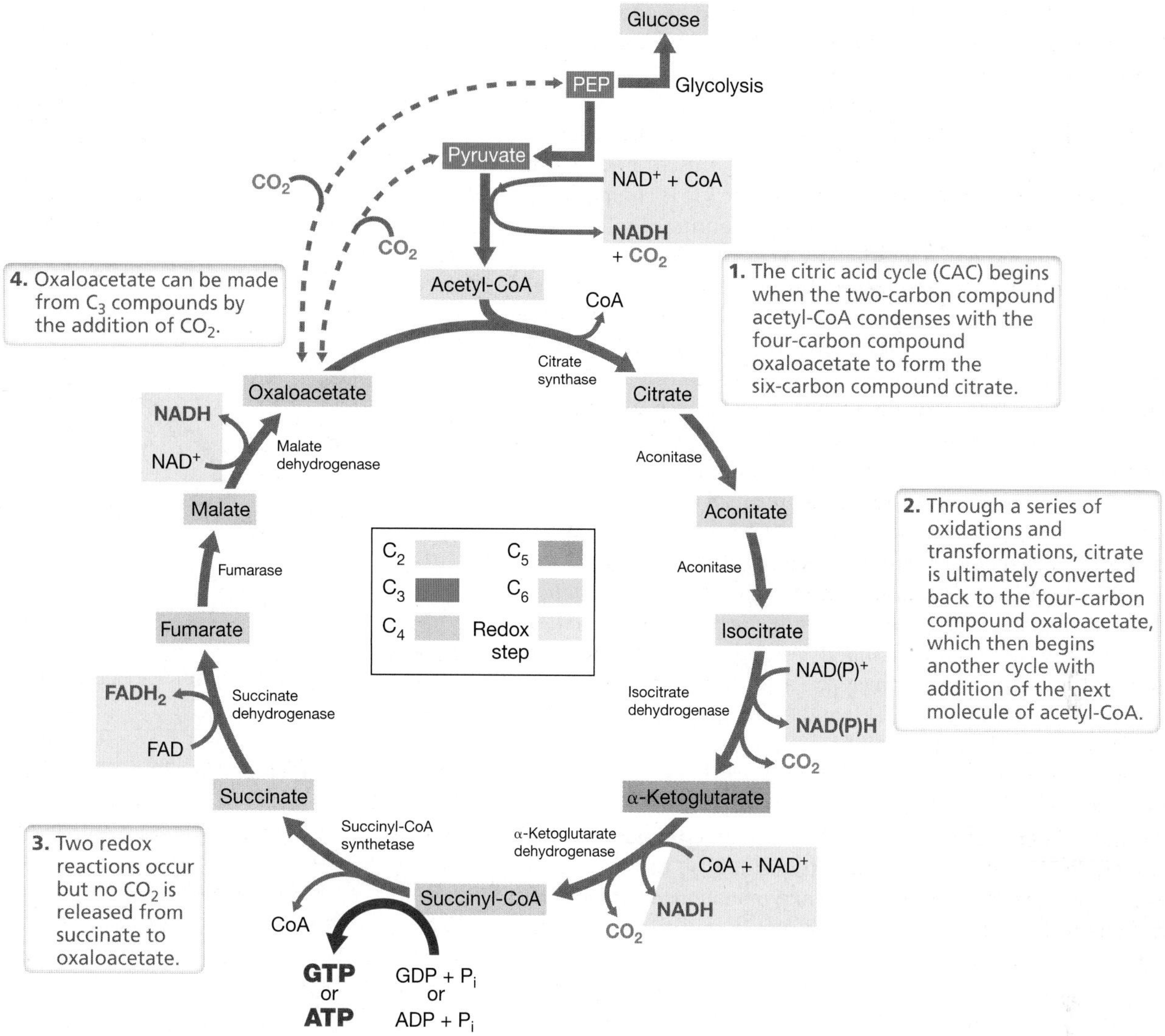

Glucose

PEP — Glycolysis

Pyruvate ← NAD⁺ + CoA

CO_2

NADH + CO_2

4. Oxaloacetate can be made from C_3 compounds by the addition of CO_2.

CO_2

Acetyl-CoA

CoA

1. The citric acid cycle (CAC) begins when the two-carbon compound acetyl-CoA condenses with the four-carbon compound oxaloacetate to form the six-carbon compound citrate.

Citrate synthase

Oxaloacetate

Citrate

NADH

NAD^+

Malate dehydrogenase

Aconitase

Malate

Aconitate

Fumarase

Aconitase

Fumarate

Isocitrate

C_2		C_5	
C_3		C_6	
C_4		Redox step	

2. Through a series of oxidations and transformations, citrate is ultimately converted back to the four-carbon compound oxaloacetate, which then begins another cycle with addition of the next molecule of acetyl-CoA.

FADH₂

FAD

Succinate dehydrogenase

$NAD(P)^+$

NAD(P)H

Isocitrate dehydrogenase

CO_2

Succinate

α-Ketoglutarate

3. Two redox reactions occur but no CO_2 is released from succinate to oxaloacetate.

Succinyl-CoA synthetase

α-Ketoglutarate dehydrogenase

CoA + NAD^+

CoA

Succinyl-CoA

NADH

CO_2

GTP or **ATP**

GDP + P_i or ADP + P_i

ENERGETICS BALANCE SHEET FOR AEROBIC RESPIRATION

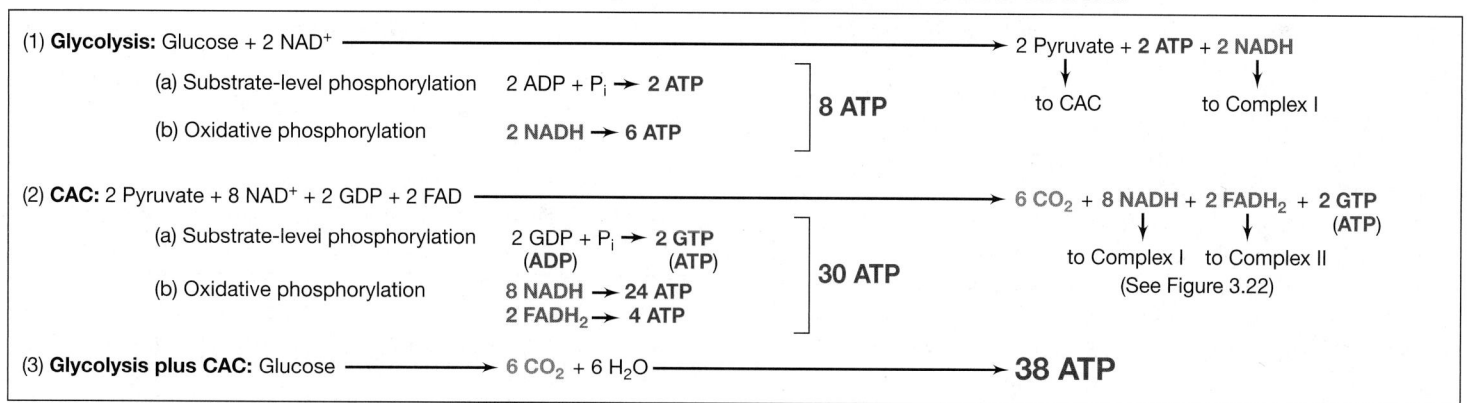

(1) **Glycolysis:** Glucose + 2 NAD^+ ⟶ 2 Pyruvate + **2 ATP** + **2 NADH**

    (a) Substrate-level phosphorylation   2 ADP + P_i ⟶ **2 ATP** ⎤

    (b) Oxidative phosphorylation   **2 NADH ⟶ 6 ATP** ⎦ **8 ATP**

to CAC   to Complex I

(2) **CAC:** 2 Pyruvate + 8 NAD^+ + 2 GDP + 2 FAD ⟶ **6 CO_2 + 8 NADH + 2 FADH₂ + 2 GTP (ATP)**

    (a) Substrate-level phosphorylation   2 GDP + P_i ⟶ **2 GTP** ⎤

              (ADP)   **(ATP)** ⎥

    (b) Oxidative phosphorylation   **8 NADH ⟶ 24 ATP** ⎥ **30 ATP**

                      2 FADH₂ ⟶ 4 ATP ⎦

to Complex I   to Complex II (See Figure 3.22)

(3) **Glycolysis plus CAC:** Glucose ⟶ **6 CO_2 + 6 H_2O** ⟶ **38 ATP**

Figure 3.16 The citric acid cycle. (Top) The citric acid cycle begins when the two-carbon compound acetyl-CoA condenses with the four-carbon compound oxaloacetate to form the six-carbon compound citrate. Through a series of oxidations and transformations, citrate is converted to two CO_2 and the acetyl acceptor molecule, oxaloacetate. (Bottom) The overall balance sheet of fuel (NADH and FADH₂) for the electron transport chain and CO_2 generated in the citric acid cycle. NADH and FADH₂ feed into electron transport chain Complexes I and II, respectively (Figure 3.22). The reactions between PEP and oxaloacetate and between pyruvate and oxaloacetate are typically reversible. These reactions connect the citric acid cycle with glycolysis and are particularly important when citric acid cycle intermediates are drawn off to be used as carbon sources.

CO_2. This is because in aerobic respiration, NADH oxidation yields 3 ATP and $FADH_2$ oxidation yields 2 ATP (Figure 3.16, bottom, and see NADH and $FADH_2$ oxidations in Figure 3.22).

Biosynthesis and the Citric Acid Cycle

Besides its role in combusting pyruvate to CO_2, the citric acid cycle plays another important role in the cell. The cycle is composed of several key organic compounds, small amounts of which are drawn off during growth to produce new cell material. Particularly important in this regard are α-ketoglutarate and oxaloacetate, which are precursors of several amino acids (Section 3.14), and succinyl-CoA, needed to form cytochromes, chlorophyll, and related molecules. Any shortage of oxaloacetate is corrected by the addition of CO_2 (carboxylation) to pyruvate or phosphoenolpyruvate (Figure 3.16).

Oxaloacetate is also an important intermediate because it can be converted to phosphoenolpyruvate (a precursor of glucose) if necessary (Section 3.13). In addition, acetate is important because it provides the raw material for fatty acid biosynthesis (Section 3.15). The CAC thus plays two major roles in the cell: *glucose respiration coupled to energy conservation* and *the biosynthesis of key metabolites*. The same can be said about the glycolytic pathway, as certain intermediates from this pathway are also drawn off for biosynthetic needs as well (Section 3.13) and then replenished from glucose in the next round of glycolysis.

The Glyoxylate Cycle

Citrate, malate, fumarate, and succinate are common natural products, and as for glucose, organisms that use these C_4 or C_6 compounds as electron donors in energy metabolism employ the citric acid cycle for their catabolism. By contrast, two-carbon compounds such as acetate cannot be oxidized by the citric acid cycle alone. This is because the citric acid cycle can continue only if oxaloacetate is regenerated at each turn of the cycle. If oxaloacetate is drawn off to biosynthesize glucose and amino acid precursors, as it must be when cells are growing on nonglucose substrates, the cycle would starve for what it needs to continue functioning (Figure 3.16). Thus, when acetate is used as an electron donor, a variation on the citric acid cycle called the **glyoxylate cycle** (so named because the C_2 compound *glyoxylate* is a key intermediate) becomes active and functions to replenish oxaloacetate used for biosyntheses (**Figure 3.17**).

The glyoxylate cycle is composed of several citric acid cycle enzymatic reactions plus two additional enzymes: *isocitrate lyase*, which cleaves isocitrate into succinate and glyoxylate, and *malate synthase*, which converts glyoxylate and acetyl-CoA to malate (Figure 3.17). The succinate formed can be used for biosynthesis while the glyoxylate combines with acetyl-CoA (C_2) to yield malate (C_4). From malate, the acceptor molecule oxaloacetate is produced and can enter a new round of acetyl-CoA oxidation in the citric acid cycle (Figure 3.17).

Three-carbon compounds such as pyruvate or compounds that are converted to pyruvate (for example, lactate or carbohydrates) also cannot be catabolized through the citric acid cycle alone, but here the glyoxylate cycle is unnecessary. This is because any shortage of C_4 CAC intermediates is compensated by synthesizing

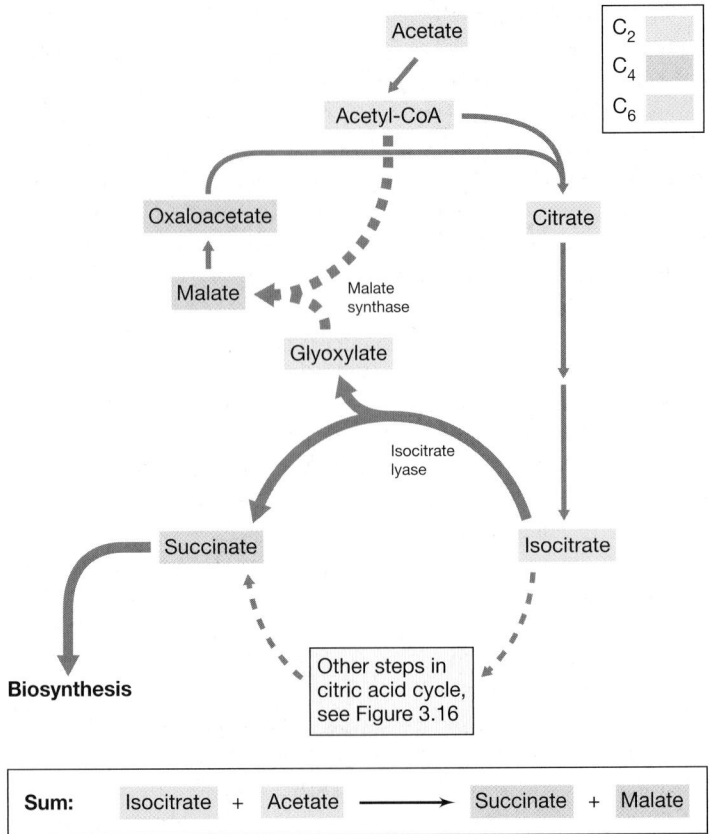

Figure 3.17 The glyoxylate cycle. These reactions occur in conjunction with the citric acid cycle when cells grow on two-carbon electron donors, such as acetate. The glyoxylate cycle regenerates oxaloacetate (from malate) to maintain an acceptor for the citric acid cycle.

oxaloacetate from pyruvate or phosphoenolpyruvate (Figure 3.16). This occurs by carboxylation reactions catalyzed by the enzymes *pyruvate carboxylase* or *phosphoenolpyruvate carboxylase*, respectively.

--- **MINIQUIZ** ---

- How many molecules of CO_2, NADH, and $FADH_2$ are released per pyruvate oxidized in the citric acid cycle?
- What two major roles do the citric acid cycle and glycolysis have in common?
- Why is the glyoxylate cycle necessary for growth on acetate but not on succinate?

3.10 Respiration: Electron Carriers

We have just seen how the citric acid cycle generates CO_2 and the reduced forms of two redox coenzymes, NADH and $FADH_2$ (Figure 3.16). The oxidation of these coenzymes back to NAD^+ and FAD, respectively, is linked to energy conservation through the electron transport chain. In the next section we consider electron transport itself. In this section, we lay the groundwork by considering redox aspects of electron transport and the molecules that participate.

NADH Dehydrogenases and Flavoproteins

Electron transport reactions occur in membranes; in prokaryotic cells this means the cytoplasmic membrane or any internal membranes derived from the cytoplasmic membrane. Several types of oxidation–reduction enzymes participate in electron transport. These include *NADH dehydrogenases, flavoproteins, iron–sulfur proteins*, and *cytochromes*. Also participating are small nonprotein electron carriers called *quinones*. The carriers are arranged in the membrane in order of *increasingly more positive* reduction potential, with NADH dehydrogenase first and the cytochromes last (see Figure 3.22).

NADH dehydrogenases contain an active site that binds NADH. The $2 e^- + 2 H^+$ from NADH are transferred by the dehydrogenase to a flavoprotein, the next carrier in the chain. This generates NAD^+ that is then released from the dehydrogenase and free to react with an enzyme that requires it as coenzyme (Figure 3.12). Flavoproteins contain a derivative of the vitamin riboflavin (**Figure 3.18**). The flavin portion, which is bound to its protein as a prosthetic group (Section 3.5), is reduced as it accepts $2 e^- + 2 H^+$ and oxidized when $2 e^-$ are passed on to the next carrier in the chain (note that flavoproteins *accept* $2 e^- + 2 H^+$ but *donate* only electrons; the fate of the two protons will be considered later). Two types of flavins are commonly found in cells, flavin mononucleotide (FMN, Figure 3.18) and flavin adenine dinucleotide (FAD). The vitamin riboflavin is a source of flavin and is a required growth factor for some organisms (Table 3.1).

Cytochromes, Other Iron Proteins, and Quinones

The cytochromes are proteins that contain heme prosthetic groups (**Figure 3.19**). Cytochromes undergo oxidation and reduction through loss or gain of an electron by the iron atom that exists as either Fe^{2+} or Fe^{3+}. Several classes of cytochromes are known, differing widely in their reduction potentials (Figure 3.10). Different classes of cytochromes are designated by letters, such as cytochrome *a*, cytochrome *b*, cytochrome *c*, and so on, depending upon the type of heme they contain. Occasionally, cytochromes form into complexes with other cytochromes or with iron–sulfur proteins. An important

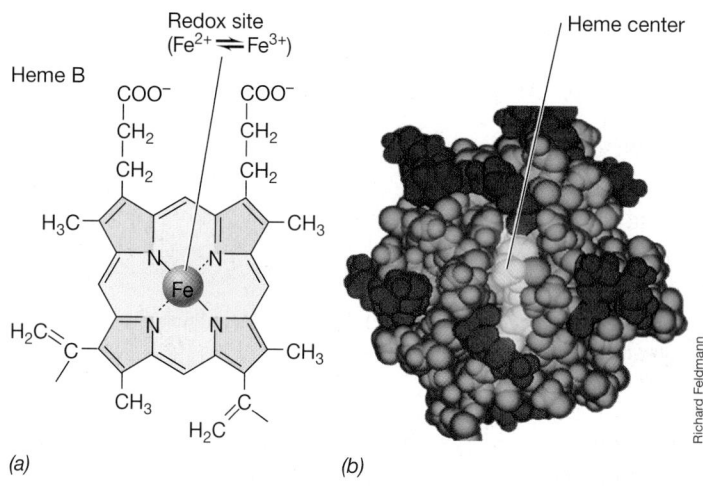

(a) *(b)*

Figure 3.19 Cytochrome and its structure. *(a)* Structure of heme, the iron-containing portion of cytochromes. Cytochromes carry electrons only, and the redox site is the iron atom, which can alternate between the Fe^{2+} and Fe^{3+} oxidation states. *(b)* Space-filling model of cytochrome *c*; heme (light blue) is covalently linked via disulfide bridges to cysteine residues in the protein (dark blue). Cytochromes are tetrapyrroles, composed of four pyrrole rings.

example is the cytochrome bc_1 complex, which contains two different *b*-type cytochromes and one *c*-type cytochrome. The cytochrome bc_1 complex plays an important role in energy metabolism, as we will see in the next section.

In addition to the cytochromes, in which iron is bound to heme, one or more proteins with nonheme iron are also components of electron transport chains. These proteins contain prosthetic groups made up of clusters of iron and sulfur atoms, with Fe_2S_2 and Fe_4S_4 clusters being the most common (**Figure 3.20**). For example, bacterial *ferredoxin*, a nonheme iron–sulfur protein of low reduction potential (about $-0.4V$) (Figure 3.10), contains an Fe_4S_4 cluster. The reduction potentials of iron–sulfur proteins vary from -0.2 to about $-0.45V$, depending on the iron–sulfur cluster present and how the cluster is embedded in the protein. Thus, different iron–sulfur proteins can function at different locations in the electron transport chain. Like cytochromes, iron–sulfur proteins carry electrons only.

Quinones (**Figure 3.21**) are small hydrophobic redox molecules that lack a protein component. Because they are small and hydrophobic, quinones can move about within the membrane. Like the flavins (Figure 3.18), quinones accept $2 e^- + 2 H^+$ but transfer only $2 e^-$ to the next carrier in the chain. Quinones typically function to link iron–sulfur proteins (Figure 3.20) and the initial cytochrome (Figure 3.19) in the electron transport chain. Several types of quinones are known, but ubiquinone (also called coenzyme Q) and menaquinone are the most common quinones and are widely distributed in species of *Bacteria* and *Archaea*.

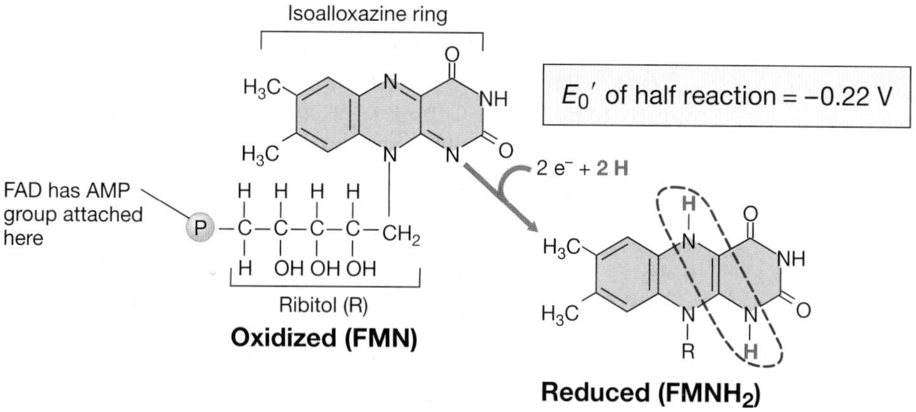

Figure 3.18 Flavin mononucleotide (FMN), a hydrogen atom carrier. The site of oxidation–reduction (dashed red circle) is the same in FMN and the related coenzyme flavin adenine dinucleotide (FAD, not shown). FAD contains an adenosine monophosphate (AMP) group bonded through the phosphate group on FMN.

Isoalloxazine ring

E_0' of half reaction = -0.22 V

$2 e^- + 2 H$

FAD has AMP group attached here

Ribitol (R)

Oxidized (FMN)

Reduced (FMNH$_2$)

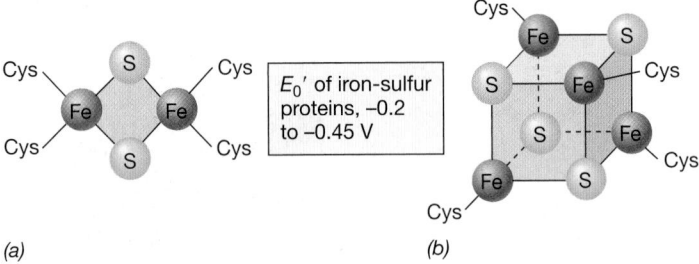

(a)   *(b)*

E_0' of iron-sulfur proteins, −0.2 to −0.45 V

Figure 3.20 Arrangement of the iron–sulfur centers of nonheme iron–sulfur proteins. *(a)* Fe_2S_2 center. *(b)* Fe_4S_4 center. The cysteines (Cys) link the protein to its Fe/S cluster.

MINIQUIZ

- In what major way do quinones differ from other electron carriers in the membrane?
- Which electron carriers described in this section accept $2\,e^- + 2\,H^+$? Which accept electrons only?

3.11 Electron Transport and the Proton Motive Force

Energy conservation in respiration is linked to an energized state of the membrane (⊘ Figure 2.7), and this energized state is established by reactions of the electron transport chain.

Electron Transport

Understanding how electron transport is linked to ATP synthesis requires an appreciation for how the electron transport chain is organized in the cytoplasmic membrane. The electron transport carriers we just discussed (Figures 3.11 and 3.18–3.21) are oriented in the membrane in such a way that when redox reactions occur, protons are separated from electrons across the membrane. Two electrons plus two protons enter the electron transport chain when NADH is oxidized to NAD^+ (through activity of the enzyme NADH dehydrogenase) to begin the process of electron transport. Components of the electron transport chain are arranged in the

Oxidized

E_0' of CoQ (ox/red) ~ 0 V

2 H

Reduced

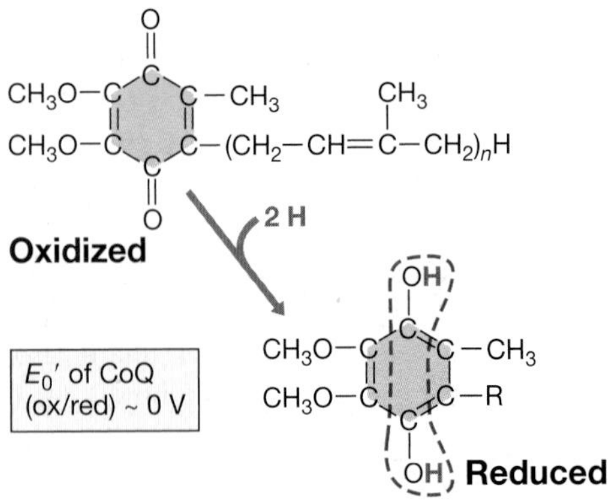

Figure 3.21 Structure of oxidized and reduced forms of ubiquinone (coenzyme Q, or CoQ). The five-carbon unit in the side chain (an isoprenoid) occurs in multiples, typically 6–10. Oxidized ubiquinone requires $2e^-$ and $2H^+$ to become fully reduced (dashed red circles).

membrane in order of their *increasingly positive* reduction potential (Figure 3.10), with the final carrier in the chain donating the electrons plus protons to a terminal electron acceptor such as O_2.

During electron transport, H^+ ions are extruded to the *outer surface* of the membrane. These protons originate from two sources: (1) NADH and (2) the dissociation of H_2O into H^+ and OH^- in the cytoplasm. The extrusion of H^+ to the environment results in the accumulation of OH^- on the inside of the cytoplasmic membrane. However, despite their small size, neither H^+ nor OH^- can diffuse through the membrane because they are charged and highly polar.

As a result of the separation of H^+ and OH^-, the inner and outer surfaces of the membrane differ in charge, pH, and electrochemical potential. This latter is called the **proton motive force (pmf)** and energizes the membrane, much like a battery (⊘ Figure 2.7). Some of the potential energy in the pmf is then conserved when its charged state is dissipated to drive the biosynthesis of ATP. In addition, the energy of the pmf can also be tapped to support other forms of work in the cell, such as nutrient transport, flagellar rotation, and other energy-requiring reactions.

Figure 3.22 depicts the electron transport chain of the bacterium *Paracoccus*, one of many different electron transport schemes known. Three features are characteristic of this and indeed all electron transport chains: (1) the carriers are arranged in order of increasingly more positive E_0', (2) there is some alternation of electron-only and electron-plus-proton carriers in the chain, and (3) the net result is reduction of a terminal electron acceptor (such as O_2) and generation of a proton motive force.

Generation of the Proton Motive Force: Complexes I and II

We now examine the electron transport process in more detail. The proton motive force forms from the activities of flavins, quinones, the cytochrome bc_1 complex, and the terminal protein, cytochrome oxidase. Following the oxidation of NADH + H^+ to form $FMNH_2$, 4 H^+ are released to the outer surface of the membrane when $FMNH_2$ donates 2 e^- to nonheme iron (Fe/S) proteins that form *Complex I* (Figure 3.22). The term *complex* refers to the fact that several proteins are present that function as a unit (in *Escherichia coli*, Complex I contains 14 distinct proteins). Complex I is also called *NADH: quinone oxidoreductase* because the overall reaction is one in which NADH is oxidized and quinone is reduced. Two protons are taken up from the cytoplasm by ubiquinone when it is reduced by the Fe/S protein in Complex I (Figure 3.22).

Complex II bypasses Complex I and feeds electrons from $FADH_2$ directly to quinones. Complex II is also called the *succinate dehydrogenase complex* because succinate (a product of the citric acid cycle, Section 3.9) as well as fatty acids donate electrons (through $FADH_2$) when they are oxidized. However, because Complex II bypasses Complex I (where electrons enter from NADH at a more negative reduction potential), four fewer protons are pumped per 2 e^- that enter from $FADH_2$ at Complex II than those that enter at Complex I (Figure 3.22); this reduces the ATP yield by one per two electrons consumed.

Complexes III and IV: bc_1 and *a*-Type Cytochromes

Reduced ubiquinone (ubiquinol, QH_2) passes electrons one at a time to the cytochrome bc_1 complex (*Complex III*, Figure 3.22).

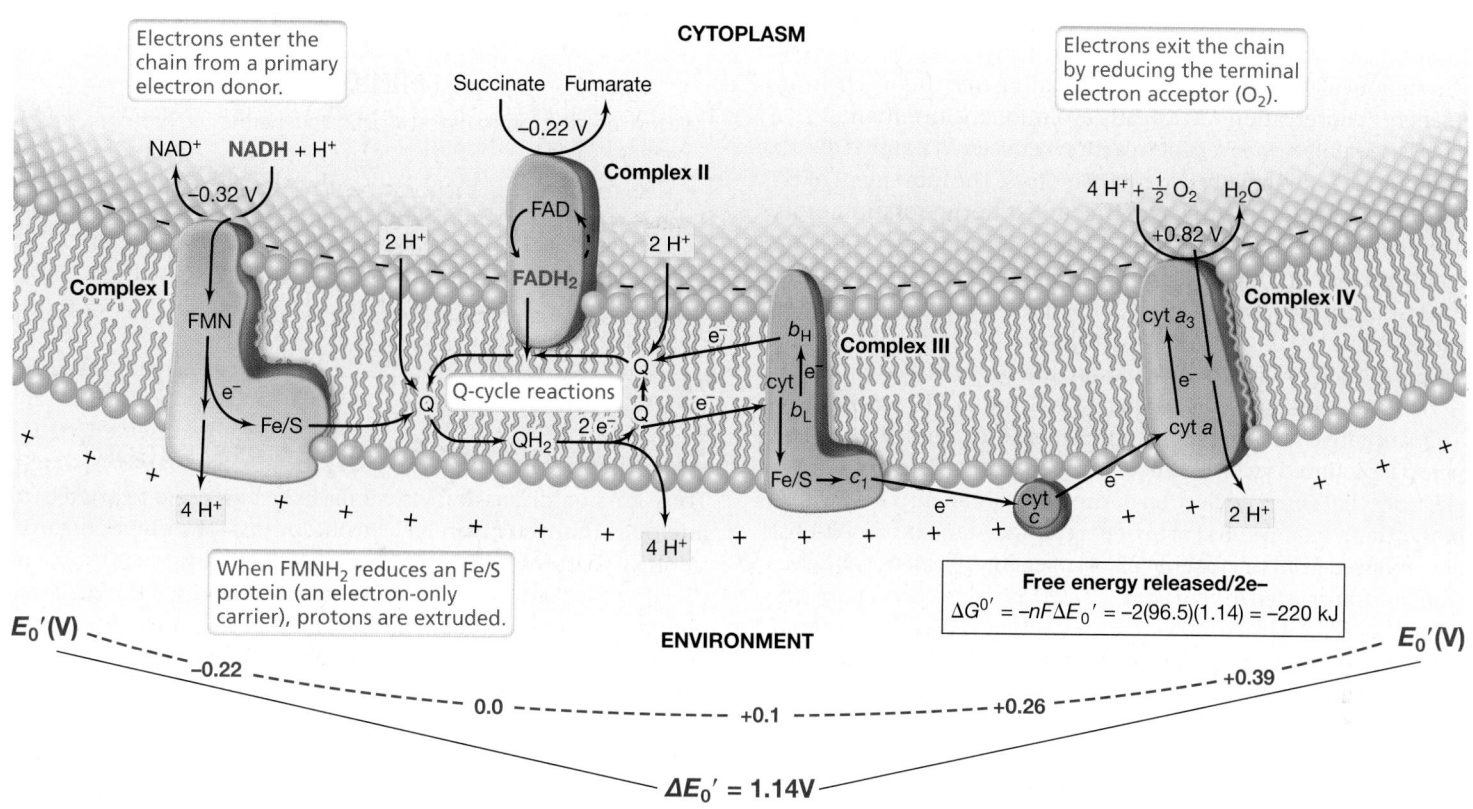

Figure 3.22 **Generation of the proton motive force during aerobic respiration.** The orientation of electron carriers in the cytoplasmic membrane of *Paracoccus denitrificans*. The $^+$ and $^-$ charges at the inner and outer membrane surfaces represent H^+ and OH^-, respectively. Abbreviations: FMN, flavin mononucleotide; FAD, flavin adenine dinucleotide; Q, quinone; Fe/S, iron– sulfur protein; cyt *a, b, c,* cytochromes (b_L and b_H, low- and high-potential *b*-type cytochromes, respectively). At the quinone site, electrons are recycled from Q to bc_1 from reactions of the "Q cycle." Electrons from QH_2 can be split in the bc_1 complex between the Fe/S protein and the *b*-type cytochromes. Electrons that travel through the cytochromes reduce Q (in two, one-electron steps) back to QH_2, thus increasing the number of protons pumped at the Q-bc_1 site. Electrons that travel to the Fe/S protein proceed to reduce cytochrome c_1, and from there cytochrome *c*. Complex II, the succinate dehydrogenase complex, bypasses Complex I and feeds electrons directly into the quinone pool at a more positive E_0' than NADH (see the redox tower in Figure 3.10).

Complex III consists of several proteins that contain two *b*-type hemes of different E_0' (b_L and b_H), one *c*-type heme (c_1), and one iron–sulfur cluster. The bc_1 complex is present in the electron transport chain of virtually every organism that can respire and also plays a role in photosynthetic electron flow in phototrophic organisms (⊅ Sections 14.3 and 14.4).

The major function of the cytochrome bc_1 complex is to transfer e^- from quinones to cytochrome *c*, which is located in the periplasm. Cytochrome *c* functions as a periplasmic shuttle to transfer e^- to the high-redox-potential cytochromes *a* and a_3 (*Complex IV*, Figure 3.22). Complex IV functions as the terminal oxidase and reduces O_2 to H_2O in the final step of the electron transport chain. Complex IV also pumps protons to the outer surface of the membrane, thereby increasing the strength of the proton motive force (Figure 3.22).

Besides transferring e^- to cytochrome *c*, the cytochrome bc_1 complex also interacts with quinones in such a way that on average, two additional H^+ are pumped at the Q-bc_1 site. This happens in a series of electron exchanges between cytochrome bc_1 and Q, called the *Q cycle*. Because quinone and the *b*-type cytochrome in the bc_1 complex have roughly the same E_0' (near 0 V, Figure 3.10), quinone molecules can alternately become oxidized and reduced

using electrons returned to the quinone pool from the bc_1 complex. This mechanism allows on average a total of 4 H^+ (instead of 2 H^+) to be pumped to the outer surface of the membrane at the Q-bc_1 site for every two electrons that enter the chain in Complex I (Figure 3.22). This strengthens the proton motive force, and as we will see now, it is the proton motive force that fuels ATP synthesis.

ATP Synthase

How does the proton motive force generated from electron transport actually drive ATP synthesis? Interestingly, a strong parallel exists between the mechanism of ATP synthesis and the mechanism behind the motor that drives rotation of the bacterial flagellum (⊅ Section 2.11). In analogy to how dissipation of the pmf applies torque that rotates the bacterial flagellum, the pmf also creates torque in the large membrane protein complex that synthesizes ATP. This complex is called **ATP synthase (ATPase)**. The activity of ATPase is driven by the pmf, and the formation of ATP from respiratory electron flow is called **oxidative phosphorylation** (contrast this with substrate-level phosphorylation in fermentation, Section 3.8).

ATPases consist of two components, a multiprotein complex called F_1 that sticks into the cytoplasm and actually catalyzes ATP synthesis, and a membrane-integrated multiprotein complex called F_o that carries out proton-translocation across the membrane (**Figure 3.23**). The structure of ATPase proteins is highly conserved throughout all the domains of life, indicating that this mechanism of energy conservation was an early evolutionary invention.

ATPase catalyzes a reversible reaction between ATP and ADP + P_i, and F_1 and F_o are actually two rotary motors. The movement of H^+ through F_o into the cytoplasm is coupled to the rotation of its c proteins. This generates a torque that is transmitted to F_1 via the coupled rotation of the $\gamma\varepsilon$ subunits (Figure 3.23). The rotation causes conformational changes in the β subunits of F_1 that allows them to bind ADP + P_i. ATP is synthesized when the β subunits return to their original conformation; the free energy captured in the rotated β subunits is released and coupled to ATP synthesis.

Quantitative measures of the number of H^+ consumed by ATPase per ATP produced yield a number between 3 and 4. Hence, per two electrons that enter the electron transport chain, about 3 ATP (whose biosynthesis requires 96 kJ of free energy under standard conditions) are produced from the roughly 220 kJ of energy released as the electrons are transported to O_2 (Figure 3.22). This yields a respiratory efficiency of about 44%, which compares with an efficiency of about 27% for yeast fermenting glucose to ethanol plus CO_2 (Figure 3.14).

ATPases are reversible motors. The hydrolysis of ATP supplies the torque necessary for $\gamma\varepsilon$ to rotate in the opposite direction from that of ATP synthesis and pump H^+ from the cytoplasm through F_o to the environment (Figure 3.23). The net result in this case is *generation of* instead of *dissipation of* the proton motive force. Reversibility of the ATPase explains why strictly fermentative bacteria that lack electron transport chains and are unable to carry out oxidative phosphorylation still contain ATPases. Many important reactions in the cell, such as flagellar rotation and some forms of transport, are coupled to energy from the pmf rather than directly from ATP. Thus, the ATPase of organisms incapable of respiration functions

unidirectionally in the cell to generate a pmf from ATP formed from substrate-level phosphorylation (Section 3.8) in fermentation.

3.12 Options for Energy Conservation

Thus far our discussion of catabolism has been restricted to microbes that use organic electron donors—the chemoorganotrophs. We now briefly consider catabolic diversity and some of the alternatives to fermentation and aerobic respiration. These include *anaerobic respiration, chemolithotrophy,* and *phototrophy* (**Figure 3.24**). We return to this theme of catabolic diversity and explore its many facets in Chapter 14.

Anaerobic Respiration

Under anoxic conditions, electron acceptors other than O_2 support respiration in a wide variety of *Bacteria* and *Archaea* in the process called **anaerobic respiration**. Some of the electron acceptors used in anaerobic respiration include nitrate (NO_3^-, reduced to nitrite, NO_2^-, by *Escherichia coli* or to N_2 by most *Pseudomonas* species), ferric iron (Fe^{3+}, reduced to Fe^{2+} by *Geobacter* and many other species), sulfate (SO_4^{2-}, reduced to hydrogen sulfide, H_2S, by *Desulfovibrio* and other sulfate-reducing species), carbon dioxide (CO_2, reduced to

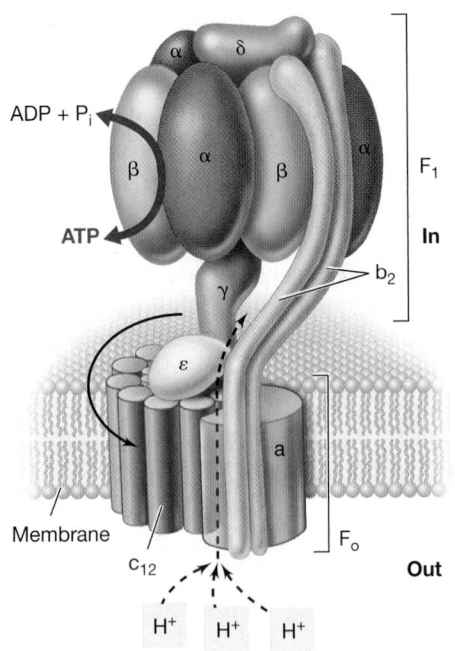

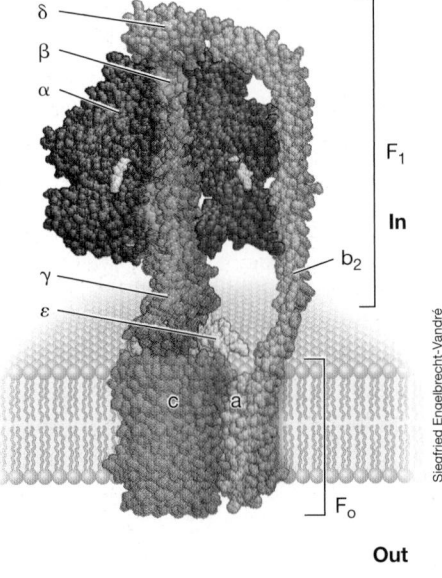

Figure 3.23 Structure and function of the reversible ATP synthase (ATPase) in *Escherichia coli*. *(a)* Schematic. F_1 consists of five different polypeptides forming an $\alpha_3\beta_3\gamma\varepsilon\delta$ complex, the stator. F_1 is the catalytic complex responsible for the interconversion of ADP + P_i and ATP. F_o, the rotor, is integrated in the membrane and consists of three polypeptides in an ab_2c_{12} complex. As protons enter, the dissipation of the proton motive force drives ATP synthesis (3 H^+/ATP). *(b)* Space-filling model. The color-coding corresponds to the art in part *a*. Since proton translocation from outside the cell to inside the cell leads to ATP synthesis by ATPase, it follows that proton translocation from inside to outside in the electron transport chain (Figure 3.22) represents work done on the system and a source of potential energy. All cells, including cells of the *Archaea*, contain ATPases. Also, some bacterial and archaeal ATPases are linked to a sodium (Na^+) rather than a proton (H^+) gradient.

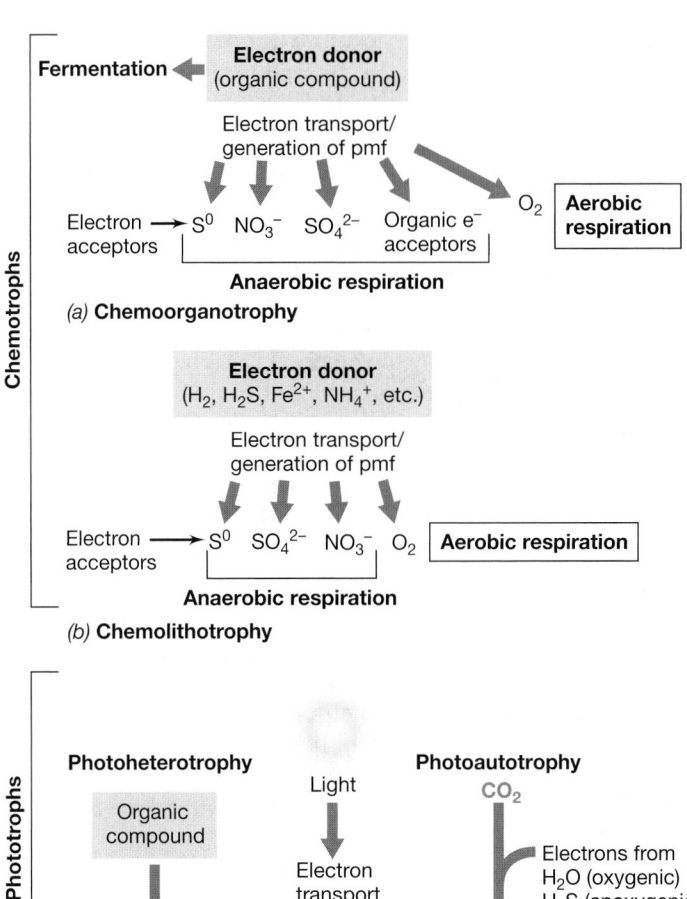

Figure 3.24 Catabolic diversity. (a) Chemoorganotrophs. (b) Chemolithotrophs. (c) Phototrophs. Note the importance of the proton motive force driven by electron transport in energy conservation in both forms of respiration and in photosynthesis.

methane, CH_4, by methanogens or to acetate by acetogens), and even certain organic compounds, such as the citric acid cycle intermediate fumarate (which is reduced to succinate).

Because none of these alternative electron acceptor couples has an E_0' as positive as that of the O_2/H_2O couple (Figure 3.10), less energy is conserved when they are reduced compared with the reduction of O_2 to H_2O (recall that $\Delta G^{0'}$ is proportional to $\Delta E^{0'}$ as was shown in Figure 3.10 and Table 3.3 and discussed in Section 3.6). O_2 is often limiting or even totally absent in many microbial habitats, and so anaerobic respirations can be very important means of energy conservation. As in aerobic respiration (Figure 3.22), anaerobic respirations require electron transport, generate a proton motive force, and employ ATPase to make ATP (Chapter 14).

Chemolithotrophy and Phototrophy

Organisms able to use *inorganic* chemicals as electron donors are called *chemolithotrophs*. Examples of common inorganic electron donors include H_2S, H_2, Fe^{2+}, and NH_4^+. Many of these compounds are the waste products of chemoorganotrophic organisms, and for

this reason, it is common for chemoorganotrophs and chemolithotrophs to coexist in nature.

Chemolithotrophic metabolisms are typically aerobic and begin with the oxidation of the inorganic electron donor and the electrons entering an electron transport chain. Electron flow generates a proton motive force, as we have already seen for the oxidation of organic electron donors by chemoorganotrophs (Figure 3.22). So, respiration by these two groups of organisms is simply a variation on a common theme of oxidative phosphorylation. However, chemoorganotrophs and chemolithotrophs differ significantly in their source of cell carbon. Chemoorganotrophs are heterotrophs and thus use organic compounds (glucose, acetate, and the like) as carbon sources. By contrast, chemolithotrophs use carbon dioxide (CO_2) as a carbon source and are therefore autotrophs (Section 3.3). We discuss autotrophic biosynthetic pathways in Chapter 14.

In the process of photosynthesis, carried out by *phototrophs*, light energy is used instead of a chemical to drive electron flow and generate a proton motive force. During these events, ATPase generates ATP by **photophosphorylation**, the light-driven analog of oxidative phosphorylation (Section 3.11). Most phototrophs assimilate CO_2 as their carbon source and are therefore *photoautotrophs*. However, some phototrophs use organic compounds as carbon sources with light as the energy source; these are the *photoheterotrophs* (Figure 3.24). Despite major differences regarding O_2, oxygenic and anoxygenic phototrophs (Section 3.3) show great parallels in the mechanism by which light drives ATP synthesis, a result of the fact that oxygenic photosynthesis evolved from the simpler anoxygenic system some 3 billion years ago (Chapter 13).

The PMF and Catabolic Diversity

With the exception of fermentation, in which substrate-level phosphorylation occurs (Section 3.8), all other mechanisms of microbial energy conservation are linked to the proton motive force (or as we will see in Chapter 14, a few organisms employ a gradient of sodium ions, Na^+, instead of protons). Whether electrons come from the oxidation of organic or inorganic chemicals or are mediated by light-driven processes, in both respiration and photosynthesis, energy conservation is the result of electron transport reactions and the formation of a pmf. The pmf is then tapped by ATPase to form ATP (Figure 3.23).

Said another way, respiration and anaerobic respiration can be viewed as variations on a theme of *different electron acceptors*, whereas chemoorganotrophy and chemolithotrophy are variations on a theme of *different electron donors*. Phototrophy, as we will see in Chapter 14, is a special case in terms of electron input and output, but the process still draws strong parallels with respiration. Electron transport and the pmf link all of these energy-yielding mechanisms, bringing these seemingly quite different forms of energy conservation into a common focus.

— **MINIQUIZ** —

• In terms of their electron donors, how do chemoorganotrophs differ from chemolithotrophs?

• What is the carbon source for autotrophic organisms?

• Why can it be said that the proton motive force is a unifying theme in most of bacterial metabolism?

IV • Biosyntheses

We conclude Chapter 3 with an overview of how the building blocks of the four classes of cellular macromolecules—sugars (polysaccharides), amino acids (proteins), nucleotides (nucleic acids), and fatty acids (lipids)—are biosynthesized and also look at how polysaccharides and lipids are biosynthesized in a general way. The biosynthesis of *informational* macromolecules—proteins and nucleic acids—is the theme of Chapter 4. Collectively, these biosyntheses are that aspect of metabolism we call **anabolism**.

3.13 Sugars and Polysaccharides

Polysaccharides are key components of microbial cell walls, and cells often store carbon and energy reserves in the form of the polysaccharides glycogen or starch (Chapter 2). How are such large molecules made?

Polysaccharide Biosyntheses and Gluconeogenesis

Polysaccharides are synthesized from *activated* forms of glucose, either uridine diphosphoglucose (*UDPG*; **Figure 3.25a**) or adenosine diphosphoglucose (*ADPG*). UDPG is the precursor of several glucose derivatives used in the biosynthesis of important structural polysaccharides, such as N-acetylglucosamine and N-acetylmuramic acid in peptidoglycan or the lipopolysaccharide component of the gram-negative outer membrane (⊆⊋ Sections 2.4 and 2.5). Polysaccharides are biosynthesized by adding activated glucose to a preexisting polymer fragment. For example, glycogen is synthesized as ADPG + glycogen → ADP glycogen-glucose.

When a cell is growing on a hexose such as glucose, obtaining glucose for polysaccharide synthesis is obviously not a problem. But when the cell is growing on other carbon compounds, glucose must be biosynthesized. This process, called *gluconeogenesis*, uses phosphoenolpyruvate, one of the intermediates of glycolysis, as a starting material and travels backwards through the glycolytic pathway (Figure 3.14) to form glucose. Phosphoenolpyruvate can be synthesized from oxaloacetate, a citric acid cycle intermediate (Figure 3.16). An overview of gluconeogenesis is shown in Figure 3.25b.

Pentose Metabolism and the Pentose Phosphate Pathway

Pentoses (5-carbon sugars) are formed by the removal of one carbon atom from a hexose, typically as CO_2. The pentoses needed for nucleic acid synthesis, ribose (in RNA) and deoxyribose (in DNA), are formed as shown in Figure 3.25c. The enzyme ribonucleotide reductase converts ribose into deoxyribose by reduction of the hydroxyl (–OH) group on the 2′ carbon of the 5-carbon pentose ring. This reaction occurs after, not before, synthesis of nucleotides. Thus, *ribo*nucleotides are biosynthesized, and some of them are later reduced to *deoxy*ribonucleotides for use as precursors of DNA.

The major pathway for pentose production is the **pentose phosphate pathway** (**Figure 3.26**). In this pathway, glucose, a hexose, is oxidized to CO_2, NADPH, and the key intermediate, *ribulose 5-phosphate*; from the latter, several different pentose derivatives can be formed. When pentoses are used as electron donors for energy conservation, they feed directly into the pentose phosphate pathway, typically becoming phosphorylated to form ribose phosphate or a related compound before being further catabolized (Figure 3.26b).

Besides its importance in pentose metabolism, the pentose phosphate pathway is also responsible for producing many other important sugars in the cell, including those containing 4 to 7 carbons. These sugars can eventually be converted to hexoses for either catabolic or biosynthetic purposes (Figures 3.25 and 3.26). A final important role of the pentose phosphate pathway is that it generates NADPH, a coenzyme used in many biosyntheses and in particular as a reductant for the production of deoxyribonucleotides

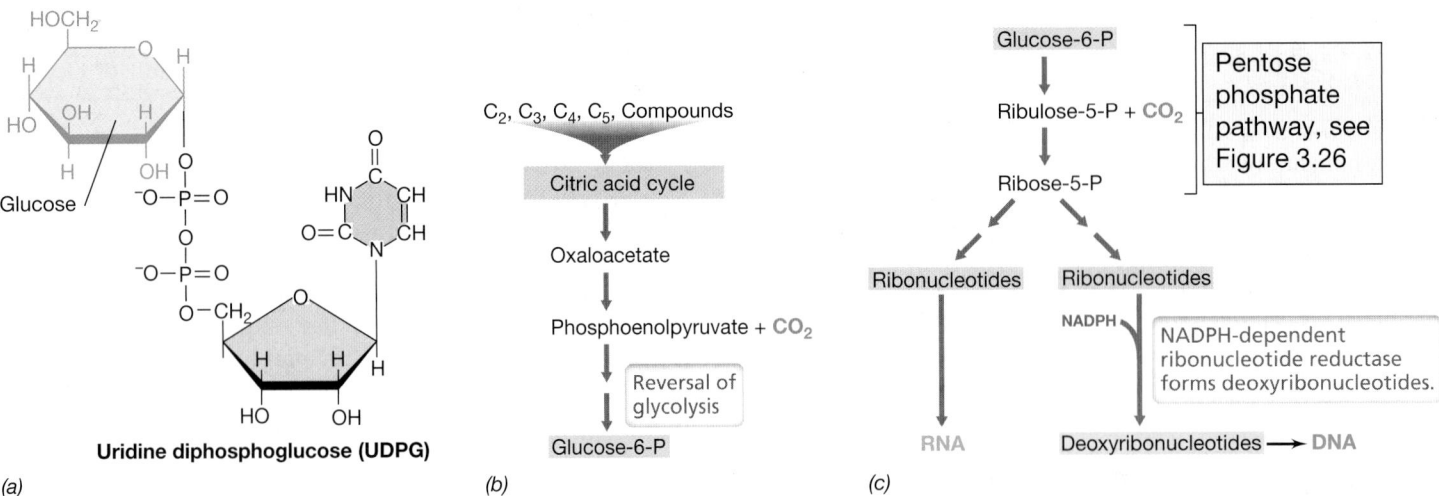

Figure 3.25 Sugar metabolism. *(a)* Polysaccharides are synthesized from activated forms of hexoses such as UDPG. *(b)* Gluconeogenesis. When glucose is needed, it can be biosynthesized from other carbon compounds, generally by the reversal of steps in glycolysis. *(c)* Pentoses for nucleic acid synthesis are formed by decarboxylation of hexoses such as glucose 6-phosphate. Note how the precursors of DNA are produced from the precursors of RNA by the enzyme ribonucleotide reductase.

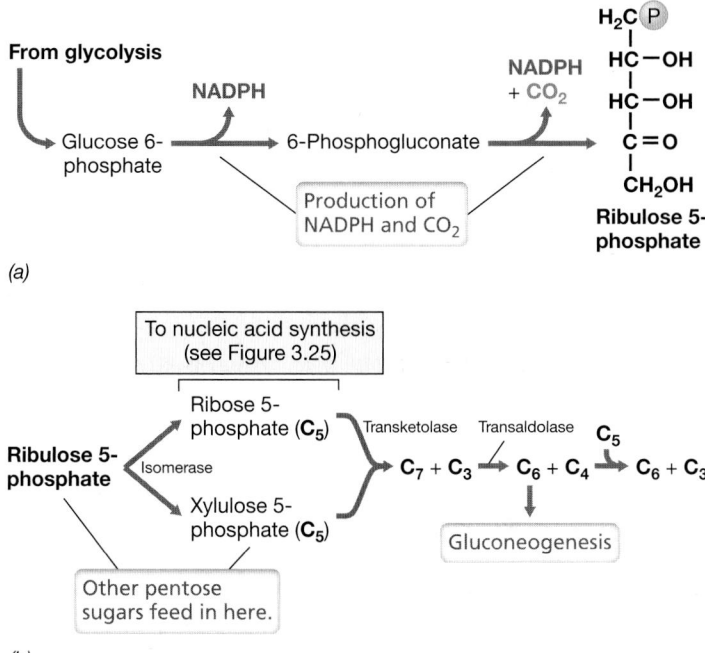

(a)

(b)

Figure 3.26 Pentose phosphate pathway. This pathway generates pentoses from other sugars for biosynthesis and also functions to catabolize pentose sugars. *(a)* Production of the key intermediate, ribulose 5-phosphate. *(b)* Other reactions in the pentose phosphate pathway.

(Figure 3.25*c*) and in the biosynthesis of fatty acids (see Figure 3.30). Although most cells have an exchange mechanism for converting NADH into NADPH, the pentose phosphate pathway is the major means by which NADPH is synthesized directly.

MINIQUIZ

- What form of activated glucose is used in the biosynthesis of glycogen by bacteria?

- What is gluconeogenesis?

- What functions does the pentose phosphate pathway play in the cell?

3.14 Amino Acids and Nucleotides

The monomers in proteins and nucleic acids are the amino acids and nucleotides, respectively. Their biosyntheses are typically multistep biochemical pathways that we need not consider in detail here. Instead, we identify the carbon skeletons needed for the biosynthesis of amino acids and nucleotides, look at their origins in pathways we have already considered, and summarize the mechanisms by which they are made.

Monomers of Proteins: Amino Acids

Organisms that cannot obtain some or all of their amino acids pre-formed from the environment must synthesize them from glucose or other carbon sources. Amino acids are grouped into structurally related *families* that share several biosynthetic steps. The carbon skeletons for amino acids come almost exclusively from intermediates of glycolysis (Figure 3.14) or the citric acid cycle (Figure 3.16) (**Figure 3.27**).

The amino group ($-NH_2$) of amino acids is typically derived from some inorganic nitrogen source, such as ammonia (NH_3). Ammonia is most often incorporated during biosynthesis of the amino acids glutamate or glutamine by the enzymes *glutamate dehydrogenase* and *glutamine synthetase*, respectively (**Figure 3.28**). When NH_3 is present at high levels, glutamate dehydrogenase or other amino acid dehydrogenases are used. However, when NH_3 is present at low levels, glutamine synthetase, with its energy-consuming reaction mechanism (Figure 3.28*b*) and correspondingly high affinity for its substrates, is employed. The enzymes glutamate dehydrogenase and glutamine synthetase are present in most *Bacteria* and *Archaea*.

Once ammonia is incorporated into glutamate or glutamine, the amino group of these amino acids can be transferred to form other nitrogenous compounds. For example, glutamate can donate its amino group to oxaloacetate in a transaminase reaction, producing α-ketoglutarate and aspartate (Figure 3.28*c*). Alternatively, glutamine can react with α-ketoglutarate to form two molecules of glutamate in an aminotransferase reaction (Figure 3.28*d*). The end result of these types of reactions is the shuttling of ammonia into various carbon skeletons from which further biosynthetic reactions occur to form all 22 of the

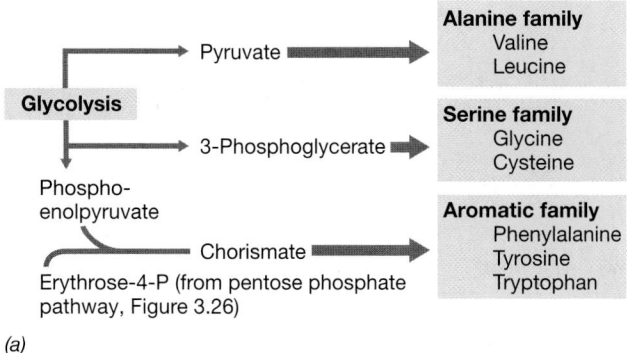

(a)

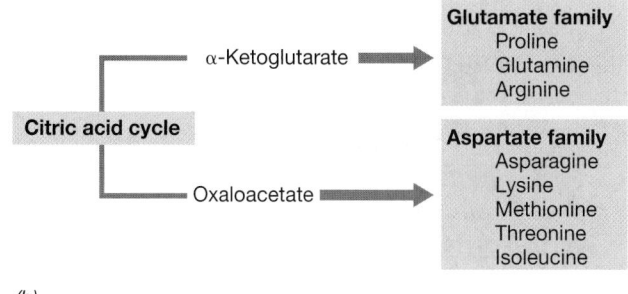

(b)

Figure 3.27 Amino acid families. Glycolysis *(a)* and the citric acid cycle *(b)* provide the carbon skeletons for most amino acids. Synthesis of the various amino acids in a family may require many steps starting with the parent amino acid (shown in bold as the name of the family).

(a) α-Ketoglutarate + NH_3 $\xrightarrow[\text{Glutamate dehydrogenase}]{\text{NADH}}$ Glutamate + NAD^+

(b) Glutamate + NH_3 $\xrightarrow[\text{Glutamine synthetase}]{\text{ATP}}$ Glutamine + ADP + P_i

(c) Glutamate + Oxaloacetate $\xrightarrow{\text{Transaminase}}$ α-Ketoglutarate + Aspartate

(d) Glutamine + α-Ketoglutarate $\xrightarrow[\text{Glutamate synthase}]{\text{NADH}}$ 2 Glutamate + NAD^+

Figure 3.28 Ammonia incorporation in bacteria. Ammonia (NH_3) and the amino groups of all amino acids are shown in green. Two major pathways for NH_3 assimilation in bacteria are those catalyzed by the enzymes (a) glutamate dehydrogenase and (b) glutamine synthetase. (c) Transaminase reactions transfer an amino group from an amino acid to an organic acid. (d) The enzyme glutamate synthase forms two glutamates from one glutamine and one α-ketoglutarate.

amino acids needed to make proteins ($\rightleftarrows$ Figure 4.28) and other nitrogen-containing biomolecules.

Monomers of Nucleic Acids: Nucleotides

The biochemistry behind purine and pyrimidine biosynthesis is quite complex and so only an outline of their biosyntheses is necessary here. Purines are constructed literally atom by atom from several carbon and nitrogen sources, including even CO_2 (**Figure 3.29**). The purine nucleotide skeleton, inosinic acid (Figure 3.29b), is the precursor of the purine nucleotides *adenine* and *guanine*. Once these are synthesized (in their triphosphate forms) and attached to ribose, they are ready to be incorporated into DNA (following ribonucleotide reductase activity, Figure 3.25c) or RNA.

Like the purine ring, the pyrimidine ring is also constructed from several sources (Figure 3.29c). From the pyrimidine nucleotide skeleton uridylate (Figure 3.29d), all of the pyrimidines—*thymine, cytosine,* and *uracil*—are derived. Structures of all of the purines and pyrimidines are shown in Chapter 4 ($\rightleftarrows$ Figure 4.1c).

MINIQUIZ

- What is an amino acid family?
- List the steps required for the cell to incorporate NH_3 into amino acids.
- Which nitrogen bases are purines and which are pyrimidines?

3.15 Fatty Acids and Lipids

Lipids are major components of the cytoplasmic membrane and of the outer membrane of gram-negative bacteria; lipids can also be carbon and energy reserves. Fatty acids are the backbone of microbial lipids. However, recall that fatty acids, per se, are found only in *Bacteria* and *Eukarya*. *Archaea* do not contain fatty acids in

their lipids but instead have hydrophobic isoprenoid side chains that play a similar structural role. All of these concepts were presented in Chapter 2. Our focus here is on the biosynthesis of fatty acids in *Bacteria*.

Fatty Acid Biosynthesis

Fatty acids are biosynthesized two carbon atoms at a time by the activity of a protein called *acyl carrier protein* (ACP). ACP holds the growing fatty acid as it is being constructed and releases it once it has reached its final length (**Figure 3.30**). Although fatty acids are constructed *two* carbons at a time, each C_2 unit originates from the *three*-carbon compound *malonate*, which is attached to the ACP to form malonyl-ACP. As each malonyl residue is donated, one molecule of CO_2 is released (Figure 3.30).

The fatty acid composition of cellular lipids varies from species to species and can also vary in a given organism with differences in growth temperature. Growth at low temperatures promotes the biosynthesis of shorter-chain and unsaturated fatty acids, whereas growth at higher temperatures promotes the biosynthesis of longer-chain and more saturated fatty acids. The most common fatty acids in lipids of *Bacteria* are those with chain lengths of C_{12}–C_{20}.

In addition to saturated, even-carbon-number fatty acids, fatty acids can also be unsaturated, branched, or have an odd number of carbon atoms. Unsaturated fatty acids contain one or more double bonds in the long hydrophobic portion of the molecule. The number and position of these double bonds is often

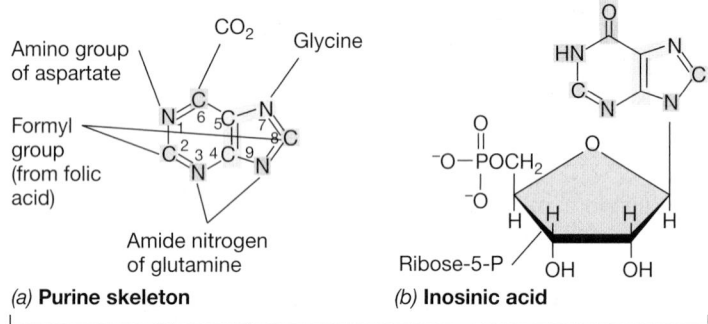

(a) **Purine skeleton**   (b) **Inosinic acid**

Purine biosynthesis

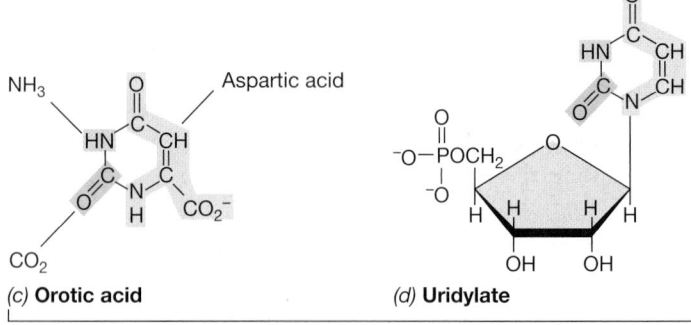

(c) **Orotic acid**   (d) **Uridylate**

Pyrimidine biosynthesis

Figure 3.29 Biosynthesis of purines and pyrimidines. (a) Components of the purine skeleton, labeled with their sources. (b) Inosinic acid, the precursor of all purine nucleotides. (c) Components of the pyrimidine skeleton, orotic acid, labeled with their sources. (d) Uridylate, the precursor of all pyrimidine nucleotides. Uridylate is formed from orotate following a decarboxylation and the addition of ribose 5-phosphate.

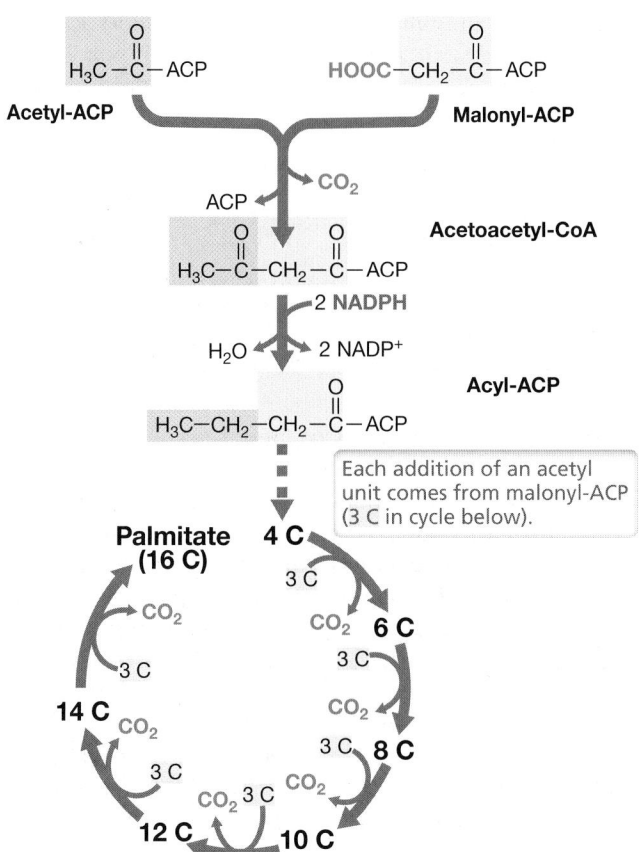

Figure 3.30 The biosynthesis of the C$_{16}$ fatty acid palmitate. The condensation of acetyl-ACP and malonyl-ACP forms acetoacetyl-CoA. Each successive addition of an acetyl unit comes from malonyl-ACP.

species-specific or group-specific, and the double bonds are formed by desaturation of a saturated fatty acid. Branched-chain fatty acids are biosynthesized using a branched-chain initiating molecule, and odd-carbon-number fatty acids (for example, C$_{13}$, C$_{15}$, C$_{17}$) are biosynthesized using an initiating molecule that contains a propionyl (C$_3$) group instead of acetyl.

Lipid Biosynthesis

In the assembly of lipids in cells of *Bacteria* and *Eukarya*, fatty acids are first added to a molecule of glycerol. For simple triglycerides (fats), all three glycerol carbons are esterified with fatty acids. To form complex lipids, one of the carbon atoms in glycerol is embellished with a molecule of phosphate, ethanolamine, carbohydrate, or some other polar substance (⮌ Figure 2.4a). In *Archaea*, although membrane lipids are constructed from isoprene to form the phytanyl (C$_{15}$) or biphytanyl (C$_{30}$) side chains, the glycerol backbone of archaeal membrane lipids contains a polar group (sugar, phosphate, sulfate, or polar organic compound) as for the lipids of *Bacteria* and *Eukarya*. Polar groups are important in lipids for forming the canonical membrane architecture: a hydrophobic interior with hydrophilic inner and outer surfaces (⮌ Section 2.3 and Figures 2.4 and 2.5).

MINIQUIZ

- Explain how fatty acids are constructed two carbon atoms at a time while the immediate donor of these carbons is a three-carbon compound.

- What differences exist in lipids from the three domains of life?

MasteringMicrobiology®

Visualize, explore, and **think critically** with Interactive Microbiology, MicroLab Tutors, MicroCareers case studies, and more. MasteringMicrobiology offers practice quizzes, helpful animations, and other study tools for lecture and lab to help you master microbiology.

Chapter Review

I • Microbial Nutrients and Nutrient Uptake

3.1 Cells are primarily composed of the elements H, O, C, N, P, and S. Nutrients required by a cell in large amounts are called macronutrients while those required in very small amounts, such as trace elements or growth factors, are micronutrients. Proteins are the most abundant class of macromolecules in the cell.

Q Why are carbon and nitrogen macronutrients but cobalt is a micronutrient?

3.2 The active transport of nutrients into the cell is an energy-requiring process driven by ATP (or some other energy-rich compound) or by the proton motive force. At least three classes of transport systems are known: simple, group translocation, and ABC systems. Each functions to accumulate solutes against the concentration gradient.

Q Cells of *Escherichia coli* transport lactose via lac permease, glucose via the phosphotransferase system, and maltose via an ABC-type transporter. For each of these sugars describe: (1) the components of the transport system and (2) the source of energy that drives the transport event.

II • Energetics, Enzymes, and Redox

3.3 All microorganisms conserve energy from either the oxidation of chemicals or from light. Chemoorganotrophs use organic chemicals as their electron donors, while chemolithotrophs use inorganic chemicals. Phototrophic organisms convert light energy into chemical energy (ATP) and include both oxygenic and anoxygenic species.

Q To which energy and carbon utilization classes does *Escherichia coli* belong when it is growing on glucose as

both energy and carbon source? To which classes does *Thiobacillus thioparus* belong when growing on sulfur as energy source and CO_2 as carbon source?

3.4 Chemical reactions in the cell are accompanied by changes in energy, expressed in kilojoules. Reactions either release or consume free energy. $\Delta G^{0\prime}$ is a measure of the energy released or consumed in a reaction under standard conditions and reveals which reactions can be used by an organism to conserve energy.

> **Q** Calculate $\Delta G^{0\prime}$ for the following reaction: glucose + $6\ O_2 \rightarrow 6\ CO_2 + 6\ H_2O$. Is this reaction exergonic or endergonic? Distinguish between $\Delta G^{0\prime}$, ΔG, and G_f^0.

3.5 Enzymes are protein catalysts that increase the rate of biochemical reactions by activating the substrates that bind to their active site. Enzymes are highly specific in the reactions they catalyze, and this specificity resides in the three-dimensional structures of the polypeptide(s) that make up the protein(s).

> **Q** What are enzymes, of what are they composed, and why are they necessary for the cell?

3.6 Oxidation–reduction reactions require electron donors and electron acceptors. The tendency of a compound to accept or release electrons is expressed by its reduction potential ($E_0\prime$). Redox reactions in a cell often employ redox coenzymes such as $NAD^+/NADH$ as electron shuttles.

> **Q** The following is a series of coupled electron donors and electron acceptors (written as donor:acceptor). Using the data in Figure 3.10, order this series from most energy yielding to least energy yielding: $H_2:Fe^{3+}$, $NO:Mn^{4+}$, $H_2S:O_2$, methanol: NO_3^- (producing NO_2^-), $H_2:O_2$, $Fe^{2+}:O_2$, $NO_2^-:Fe^{3+}$ (producing NO_3^-), and $H_2S:NO_3^-$.

3.7 The energy released in redox reactions is conserved in compounds that contain energy-rich phosphate or sulfur bonds. The most common of these compounds is ATP, the prime energy carrier in the cell. Longer-term storage of energy is linked to the formation of polymers, which can be consumed to yield ATP.

> **Q** Why is acetyl phosphate considered an energy-rich compound while glucose 6-phosphate is not?

III • Catabolism: Fermentation and Respiration

3.8 The glycolytic pathway is used to break down glucose to pyruvate and is a widespread mechanism for energy conservation by fermentative anaerobes that employ substrate-level phosphorylation. The pathway releases a small amount of ATP (2–3/glucose) and large amounts of fermentation products. Besides glucose, the fermentation of other sugars, amino acids, nucleotides and polymeric compounds is possible.

> **Q** How is ATP made in fermentation and in respiration? Where in glycolysis is NADH produced, where is it

consumed, and why are fermentation products such as ethanol and lactic acid made?

3.9 Respiration offers an energy yield much greater than that of fermentation. The citric acid cycle generates CO_2 and electrons for the electron transport chain and is also a source of biosynthetic intermediates. The glyoxylate cycle is necessary for the catabolism of two-carbon electron donors, such as acetate.

> **Q** How much more ATP is possible from aerobically respiring glucose instead of fermenting it to lactate? Why is this so? Does the citric acid cycle only have a catabolic function?

3.10 Electron transport chains are composed of membrane-associated redox proteins that are arranged in order of their increasing $E_0\prime$ values. The electron transport chain functions in a concerted fashion to carry electrons from the primary electron donor to the terminal electron acceptor, which is O_2 in aerobic respiration.

> **Q** List some of the key electron carriers found in electron transport chains.

3.11 During electron transport, protons are extruded to the outside of the membrane to form the proton motive force. Key electron carriers include flavins, quinones, the cytochrome bc_1 complex, and other cytochromes. The cell uses the proton motive force to make ATP through the activity of ATPase.

> **Q** What is the proton motive force and how is it generated? How does the proton motive force result in ATP biosynthesis?

3.12 When conditions are anoxic, several electron acceptors can substitute for O_2 in anaerobic respiration. Chemolithotrophs use inorganic compounds as electron donors, whereas phototrophs use light energy. The proton motive force underlies energy conservation in all forms of respiration and photosynthesis.

> **Q** What is the major difference between aerobic respiration and anaerobic respiration? Which metabolic option yields more energy, and why? Give an example of a chemolithotrophic electron donor.

IV • Biosyntheses

3.13 Polysaccharides are important structural components of cells and are biosynthesized from activated forms of their monomers. Gluconeogenesis is the production of glucose from nonsugar precursors.

> **Q** What is the importance of the enzyme ribonucleotide reductase in the metabolism of sugars? What is the difference between "free" and "activated" glucose?

3.14 Amino acids are formed from carbon skeletons to which ammonia is added from glutamate, glutamine, or a few other amino acids. Nucleotides are biosynthesized using carbon skeletons from several different sources.

Q **Name two common enzymes that function to incorporate NH_3 into the cell. How do their reaction mechanisms differ?**

3.15 Fatty acids are synthesized from the three-carbon precursor malonyl-ACP, and fully formed fatty acids are attached to

glycerol to form lipids. Only the lipids of *Bacteria* and *Eukarya* contain fatty acids.

Q **Describe the process by which a fatty acid such as palmitate (a C_{16} straight-chain saturated fatty acid) is synthesized in a cell.**

Application Questions

1. Using the data of Figure 3.10, predict the sequence of electron carriers in the membrane of an organism growing aerobically that has the following electron carriers: ubiquinone, cytochrome aa_3, cytochrome b, NADH, cytochrome c, FAD.

2. Explain the following observation in light of the redox tower: Cells of *Escherichia coli* fermenting glucose grow faster when NO_3^- is supplied to the culture (NO_2^- is produced) and then grow even faster (and stop producing NO_2^-) when the culture is highly aerated.

Chapter Glossary

ABC (ATP-binding cassette) transport system a membrane transport system consisting of three proteins, one of which hydrolyzes ATP; the system transports specific nutrients into the cell

Activation energy the energy required to bring the substrate of an enzyme to the reactive state

Adenosine triphosphate (ATP) a nucleotide that is the primary form in which chemical energy is conserved and utilized in cells

Anabolic reactions (Anabolism) the sum total of all biosynthetic reactions in the cell

Anaerobic respiration a form of respiration in which oxygen is absent and alternative electron acceptors are reduced

ATPase (ATP synthase) a multiprotein enzyme complex embedded in the cytoplasmic membrane that catalyzes the synthesis of ATP coupled to dissipation of the proton motive force

Autotroph an organism capable of biosynthesizing all cell material from CO_2 as the sole carbon source

Calvin cycle the series of biosynthetic reactions by which most phototrophs and many chemolithotrophs convert CO_2 into organic compounds

Catabolic reactions (catabolism) biochemical reactions leading to energy conservation (usually as ATP) by the cell

Chemolithotroph an organism that can grow with inorganic compounds as electron donors in energy metabolism

Chemoorganotroph an organism that obtains its energy from the oxidation of organic compounds

Citric acid cycle a cyclical series of reactions resulting in the conversion of acetate to two molecules of CO_2

Coenzyme a small and loosely bound nonprotein molecule that participates in a reaction as part of an enzyme

Electron acceptor a substance that can accept electrons from an electron donor, becoming reduced in the process

Electron donor a substance that can donate electrons to an electron acceptor, becoming oxidized in the process

Endergonic a reaction that requires free energy

Enzyme a protein that can speed up (catalyze) a specific chemical reaction (a few RNAs are also enzymes)

Exergonic a reaction that releases free energy

Fermentation anaerobic catabolism in which an organic compound is both an electron donor and an electron acceptor and ATP is produced by substrate-level phosphorylation

Free energy (G) energy available to do work; $G^{0'}$ is free energy under standard conditions

Glycolysis a biochemical pathway in which glucose is oxidized to pyruvate, which is either used in respiration or fermented; yields ATP and, in fermentation, various fermentation products (also called the Embden–Meyerhof–Parnas pathway)

Glyoxylate cycle a modification of the citric acid cycle in which isocitrate is cleaved to form succinate and glyoxylate during growth on two-carbon electron donors such as acetate

Group translocation an energy-dependent transport system in which the substance transported is chemically modified during the process of being transported by a series of proteins

Heterotroph an organism that uses organic compounds as a carbon source

Oxidative phosphorylation the production of ATP from a proton motive force formed by electron transport of electrons from organic or inorganic electron donors

Pentose phosphate pathway a series of reactions in which pentoses are catabolized to generate precursors for nucleotide biosynthesis or to synthesize glucose

Photophosphorylation the production of ATP from a proton motive force formed from light-driven electron transport

Phototrophs organisms that use light as their source of energy

Proton motive force (pmf) a source of energy resulting from the separation of protons from hydroxyl ions across the cytoplasmic membrane, generating a membrane electrochemical potential

Reduction potential (E_0') the inherent tendency, measured in volts under standard conditions, of a compound to donate or to accept electrons

Respiration the process in which a compound is oxidized with O_2 (or an O_2 substitute) as the terminal electron acceptor, usually accompanied by ATP production by oxidative phosphorylation

Simple transport system a transporter that consists of only a membrane-spanning protein and is typically driven by energy from the proton motive force

Substrate-level phosphorylation the production of ATP by the direct transfer of an energy-rich phosphate molecule from a phosphorylated organic compound to ADP

4 Molecular Information Flow and Protein Processing

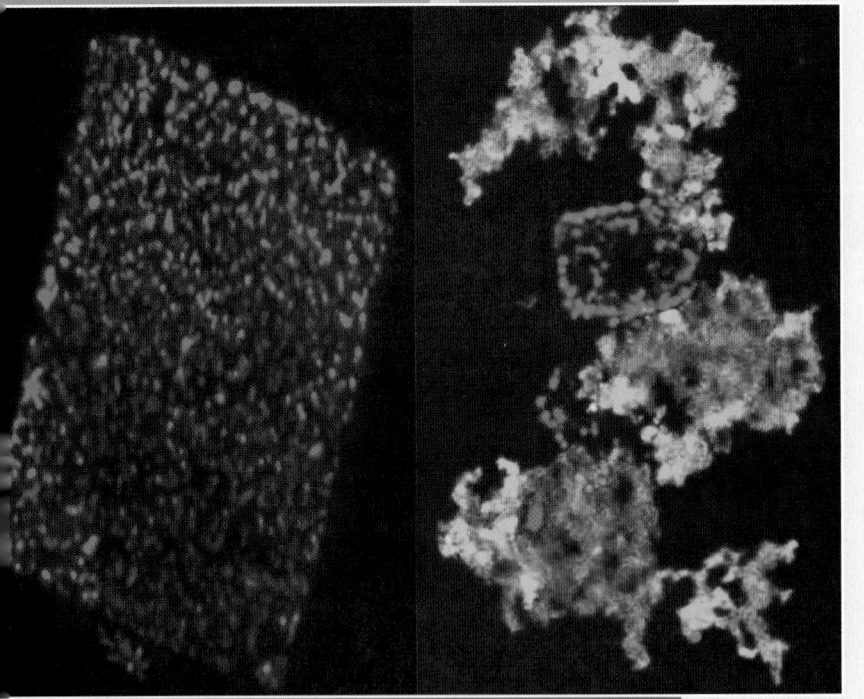

microbiology now

Synthesis of Jumbo Proteins: Secretion of Halomucin

The genetic blueprint of individual cells is responsible for the distinctive attributes and survival mechanisms observed in all life forms. Because the flow of essential biological information—from fairly inert DNA to the synthesis of proteins—requires considerable resources, the discovery of extremely long polypeptides in *Archaea* and *Bacteria* is surprising. While the average protein size in all three domains of life is <500 amino acids, the square-celled archaeon *Haloquadratum walsbyi* is predicted to encode a 9159-amino-acid protein called *halomucin*. As its name suggests, halomucin shares a structural similarity with animal mucin, which protects the eyes and bronchial epithelium from dehydration and chemical stress. Halomucin is also predicted to contain a signal sequence that suggests it is secreted outside of the cell. As you will see in this chapter, considerable energy is required not only to synthesize a protein, but also to get it to its final destination. Does *H. walsbyi* really produce this jumbo protein, and if so, why would it expend so many resources to do so?

H. walsbyi grows under extremely salty conditions and possesses a unique square but thin morphology ($5 \times 5 \times 0.1$ µm). This shape gives cells of *H. walsbyi* the greatest ratio of surface area to volume of all known microbes. Despite the gigantic nature of halomucin, is it possible that the predicted protein is synthesized and then, using its signal sequence, secreted outside of the cell to protect the cell from desiccation?

Scientists have recently answered this question by using fluorescent labeling techniques. The left photo illustrates the unusual cell morphology of *H. walsbyi* by using a red fluorescent stain targeting the abundant carbon storage polymers (poly-β-hydroxyalkanoates) produced by this organism. The right photo shows the same red cell with green-labeled secreted halomucin surrounding the cell. While translation and secretion of this jumbo protein are undoubtedly energetically draining on the cell, the production of halomucin is likely critical for retaining sufficient moisture near the cell surface to allow *H. walsbyi* to survive at the limits of water activity.

Source: Zenke, R., S. von Gronau, H. Bolhuis, M. Gruska, F. Pfeiffer, and D. Oesterhelt. 2015. Fluorescence microscopy visualization of halomucin, a secreted 927 kDa protein surrounding *Haloquadratum walsbyi* cells. *Front. Microbiol. 6*: 249.

Cells can be viewed as both chemical machines and coding devices. As chemical machines, cells transform nutrients into new cell material (Chapter 3). As coding devices, they store, process, and use genetic information. This chapter highlights the coding aspects of a microbe's activities, events that are foundational to the very being of an organism and the entire science of microbiology. Our focus will be on processes as they occur in *Bacteria*, particularly in *Escherichia coli*—the model organism for molecular biology—but we will also consider these processes in *Archaea*, and briefly in eukaryotic cells.

I • Molecular Biology and Genetic Elements

We begin by considering some basic principles of molecular biology—a refresher course, if you will. In Part I we examine the structure of DNA and the basic processes in genetic information flow and then move on to consider the various types of genetic elements. Part I lays the groundwork for a detailed consideration of DNA replication in prokaryotic cells (Part II) and the mechanisms behind RNA and protein synthesis (Parts III and IV).

4.1 DNA and Genetic Information Flow

An overview of the major macromolecules and processes of molecular biology is shown in **Figure 4.1a**. The functional unit of genetic information is the **gene**, and genes make up parts of chromosomes

or other large molecules known collectively as **genetic elements**; the total complement of genetic elements is the **genome**. Genetic information is embedded in the sequence of nucleotides in the nucleic acids **DNA** and **RNA**. DNA carries the cell's genetic blueprint while RNA, produced in transcription, carries a copy of this blueprint. One form of RNA called messenger RNA is converted by translation into defined amino acid sequences in proteins. Collectively, nucleic acids and proteins are called **informational macromolecules** (Figure 4.1a).

The monomers of nucleic acids are called **nucleotides** and so DNA and RNA are **polynucleotides**. A nucleotide has three components: a pentose sugar (either ribose in RNA or deoxyribose in DNA), a nitrogenous base, and a molecule of phosphate, PO_4^{3-} (Figure 4.1b). A **nucleoside** has a pentose sugar and a nitrogenous base but does not include a phosphate group. The nitrogenous bases in nucleic acids are either **pyrimidines** or **purines**. The purines guanine and adenine and the pyrimidine cytosine are present in both DNA and RNA, whereas the pyrimidines thymine and uracil are only present (with minor exceptions) in DNA and RNA, respectively (Figure 4.1c).

Properties of the Double Helix

The nucleic acid backbone is a polymer of alternating sugar and phosphate molecules, and nucleotides are linked by phosphate between the 3′-carbon of one sugar and the 5′-carbon of the next sugar, the **phosphodiester bond** (Figure 4.1b, **Figure 4.2**). The *sequence* of nucleotides in a DNA or RNA molecule is its

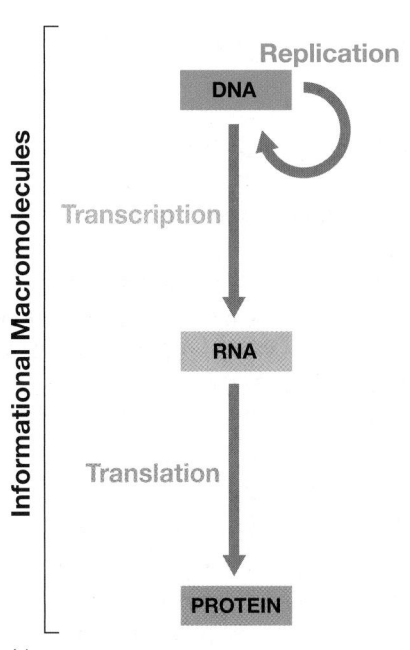

(a)

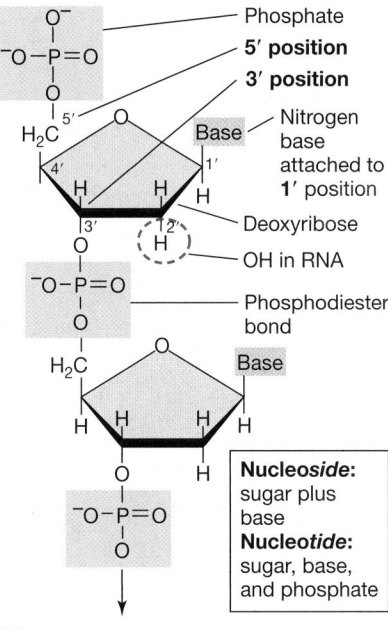

(b)

(c)

Figure 4.1 Genetic information flow and the components of the nucleic acids. *(a)* An overview of the types of informational macromolecules. *(b)* Part of a DNA chain. The numbers on the sugar of the nucleotide contain a prime (′) to differentiate them from the numbering on the rings of the nitrogen bases. In DNA, a hydrogen is present on the 2′-carbon of the pentose sugar. In RNA, an OH group occupies this position. The nucleotides are linked by a phosphodiester bond. *(c)* The nitrogen bases of DNA and RNA and the specific pairing between cytosine (C) and guanine (G) and between thymine (T) and adenine (A) via hydrogen bonds. Uracil (U) instead of thymine is present in RNA. Note the numbering system of the rings in that a pyrimidine base bonds through N-1 to the sugar–phosphate backbone and that a purine base bonds through N-9. Atoms that are found in the major groove of the double helix (see Figure 4.3) and that interact with proteins are highlighted in pink.

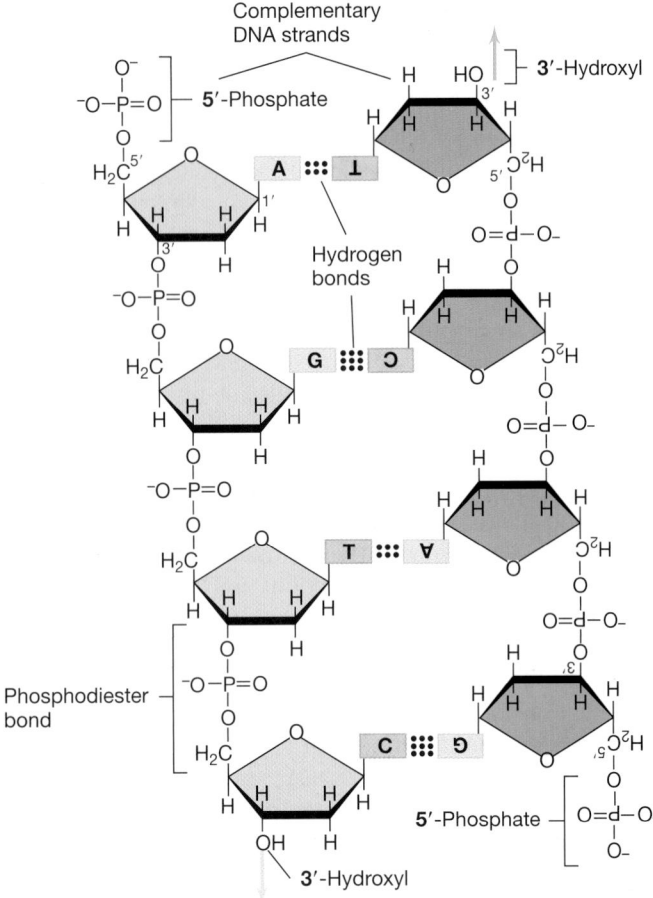

Figure 4.2 DNA structure. Complementary and antiparallel nature of DNA. Note that one chain ends in a 5′-phosphate group, whereas the other ends in a 3′-hydroxyl. The purple bases represent the pyrimidines cytosine (C) and thymine (T), and the yellow bases represent the purines adenine (A) and guanine (G).

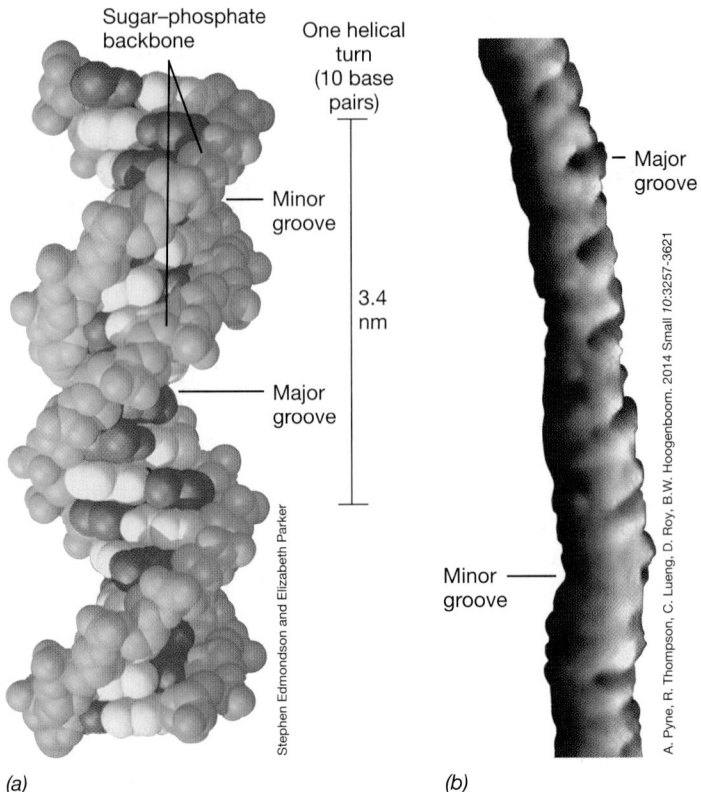

Figure 4.3 Arrangement of the DNA double helix. (a) A computer model of a short segment of DNA showing one of the sugar–phosphate backbones in blue and the other in green. The pyrimidine bases are shown in purple and the purines in yellow. Note the locations of the major and minor grooves. One helical turn contains 10 base pairs. (b) Atomic force microscopy showing the biomolecular structure of a small piece of DNA. Note the locations of the major and minor grooves.

primary structure and encodes the genetic information. In cells, DNA is *double-stranded*, the strands being held together by hydrogen bonds between the bases in the two strands (Figure 4.1c). Specific base pairing, A with T and G with C, ensures that the two strands of DNA are **complementary** in base sequence, and this complementarity is essential for the faithful replication of the molecule. The two strands of the DNA molecule are also arranged in an **antiparallel** fashion; one strand runs 5′ to 3′ (top to bottom), whereas its complement runs 5′ to 3′ (bottom to top) (Figure 4.2).

The complementary and antiparallel strands of DNA are wrapped around each other to form a double helix (**Figure 4.3**). The helix naturally forms two distinct grooves, the *major groove* and the *minor groove*. Most proteins that interact specifically with DNA bind in the major groove, where space is abundant. Because the double helix is a regular structure, some atoms of each base are always exposed in the major groove (and some in the minor groove). These key regions of the DNA that are important in interactions with proteins can be seen in Figure 4.3, and the atoms of the major groove that interact with proteins are indicated in Figure 4.1c.

Size, Shape, and Supercoiling of DNA

The size of a DNA molecule is expressed as its total number of nucleotide base pairs. Thus, a double-stranded DNA molecule consisting of 1000 bases is one kilobase pair (kbp) of DNA. The bacterium *Escherichia coli* has about 4640 kbp (4.64 megabase pairs, Mbp) of DNA in its genome. If this molecule were extended linearly it would be several hundred times longer than the cell itself. To accommodate their genome, cells of *Bacteria* and *Archaea* must compact the DNA, and this is done by the process of *supercoiling* (**Figure 4.4**).

Supercoils are inserted or removed in DNA by enzymes called *topoisomerases*. The activity of supercoiling puts the DNA molecule under torsion (**Figure 4.5**), and DNA can be supercoiled in either a positive or a negative manner. Negative supercoiling results when the DNA is twisted about its axis in the opposite sense from the right-handed double helix and is the form found in most cells. In the *E. coli* chromosome, more than 100 supercoiled domains exist, each stabilized by specific proteins bound to the DNA. Inserting supercoils into DNA requires energy from ATP, whereas releasing supercoils does not. In *Bacteria* and most *Archaea*, the topoisomerase **DNA gyrase** inserts negative supercoils into DNA by making double-strand breaks (Figure 4.5b). We will see in Chapters 5 and 17 that some *Archaea* live at very high

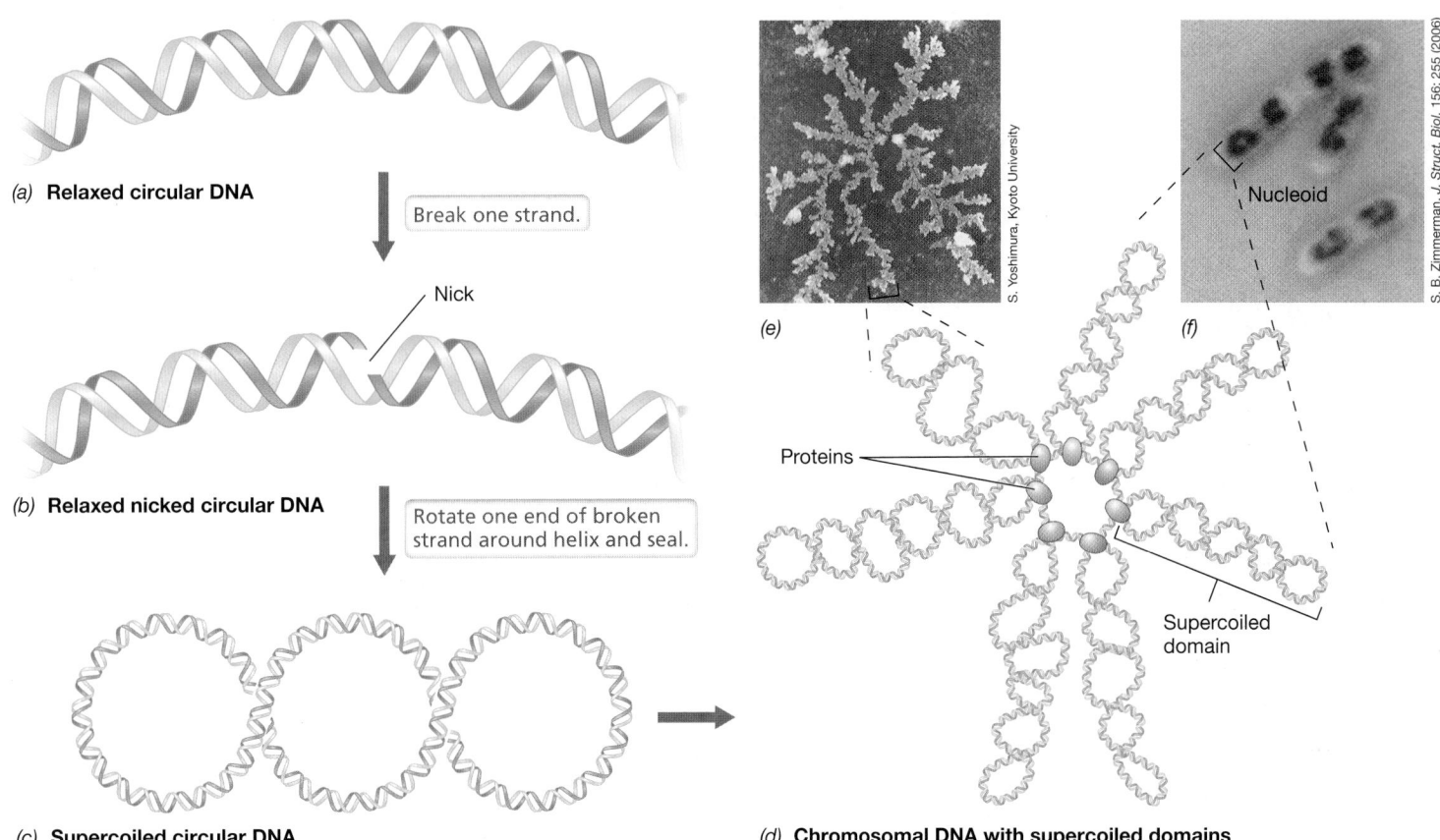

(a) **Relaxed circular DNA**

Break one strand.

Nick

(b) **Relaxed nicked circular DNA**

Rotate one end of broken strand around helix and seal.

(c) **Supercoiled circular DNA**

(d) **Chromosomal DNA with supercoiled domains**

Nucleoid

Proteins

Supercoiled domain

(e)

(f)

S. Yoshimura, Kyoto University

S. B. Zimmerman, *J. Struct. Biol.* 156: 255 (2006)

Figure 4.4 Supercoiled DNA. *(a–c)* Relaxed, nicked, and supercoiled circular DNA. A nick is a break in a phosphodiester bond of one strand. *(d)* In fact, the double-stranded DNA in the bacterial chromosome is arranged not in one supercoil but in several supercoiled domains, as shown here. *(e)* Atomic force microscopy of the *Escherichia coli* nucleoid. *(f)* Simultaneous phase-contrast and fluorescence image of *E. coli* illustrating the location of the nucleoid within growing cells. Cells were treated with a fluorescent dye specific for DNA and the color was inverted to show the nucleoids as black.

temperatures—above the boiling point in some cases. These species have chromosomes that are *positively* (instead of negatively) supercoiled, and this genomic feature helps to maintain DNA structure (that is, it prevents the two strands from melting apart) at such high temperatures (⊂⊃ Section 17.12). Supercoiling is not a feature of eukaryotes since their genomic DNA is linear rather than circular. However, eukaryotic DNA must still be compacted, and this occurs when the DNA is highly wound around histone proteins.

Genes and the Steps in Biological Information Flow

Genetic information flow is a fundamental process in all cells and is the *central dogma of molecular biology* (Figure 4.1*a* and **Figure 4.6**). When genes are *expressed*, the genetic information encoded in DNA is transferred to ribonucleic acid (RNA). While several classes of RNA exist in cells, three main classes of RNA participate in protein synthesis. **Messenger RNAs (mRNAs)** are single-stranded molecules that carry the genetic information from DNA to the ribosome. **Transfer RNAs (tRNAs)** help convert the genetic information in the nucleotide sequences of RNA into a defined sequence of amino acids in proteins. **Ribosomal RNAs (rRNAs)** are important catalytic and structural components of the

ribosome. The molecular processes of genetic information flow can be divided into three stages (Figure 4.6):

1. **Replication.** During replication, the DNA double helix is duplicated. Replication is catalyzed by the enzyme *DNA polymerase.*

2. **Transcription.** The transfer of genetic information from DNA to RNA is called transcription. Transcription is catalyzed by the enzyme *RNA polymerase.*

3. **Translation.** The formation of a polypeptide using the genetic information transferred to mRNA by DNA is a process that occurs on the ribosome.

Many different RNA molecules can be transcribed from a relatively short region of the long DNA molecule. In eukaryotes, each gene is transcribed to yield a single mRNA, whereas a single mRNA molecule may encode several different proteins in *Bacteria* and *Archaea*. However, a linear correspondence exists between the base sequence of a gene and the amino acid sequence of a polypeptide, and as we will see, each group of *three bases* on an mRNA molecule encodes a single amino acid (Section 4.9).

Eukaryotes differ from *Bacteria* and *Archaea* in that the first two steps of the central dogma, replication and transcription (Figures 4.1 and 4.6), occur in the nucleus. Because ribosomes are not present in

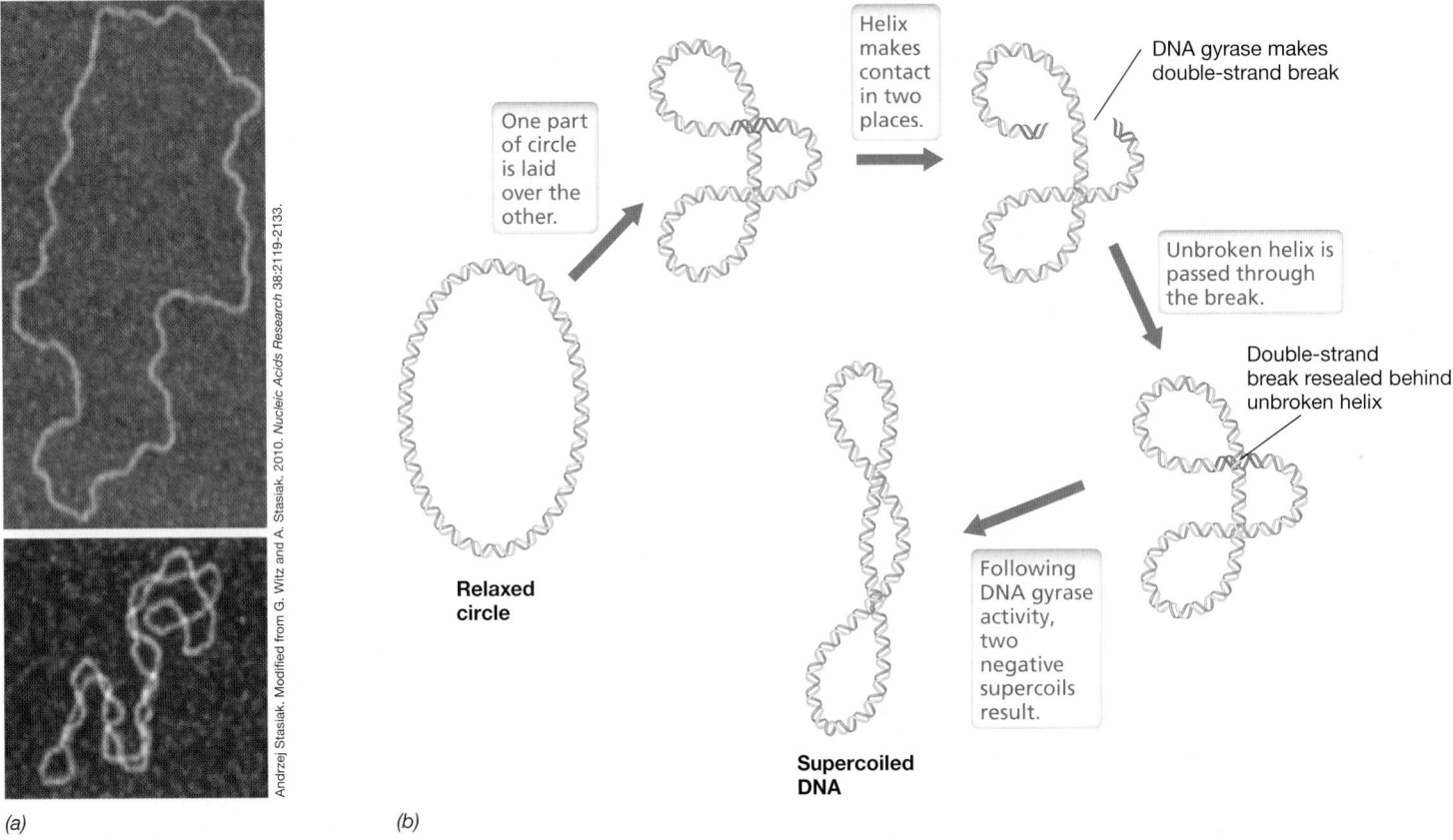

Andrzej Stasiak. Modified from G. Witz and A. Stasiak. 2010. *Nucleic Acids Research* 38:2119-2133.

(a)

(b)

Figure 4.5 DNA gyrase. *(a)* Atomic force microscopy visualization of torsionally relaxed (top) and negatively supercoiled (bottom) plasmid DNA. *(b)* Schematic showing the introduction of negative supercoiling into circular DNA by the activity of DNA gyrase (topoisomerase II), which makes double-strand breaks.

the nucleus, mRNAs as well as other RNAs must be transported outside of the nucleus for translation. By contrast, in prokaryotic cells, mRNAs do not have to be exported from an organelle to be translated. Because of this fundamental difference, transcription and translation in *Bacteria* and *Archaea* can occur simultaneously in a process known as *coupled transcription and translation* (**Figure 4.7**). During this process, a ribosome initiates translation of an mRNA before RNA polymerase has finished synthesizing it. This allows rapidly growing cells to produce proteins at a maximal rate and also allows the cell to rapidly adapt to changes in growth conditions by quickly expressing the new protein sets required.

While the central dogma of molecular biology (Figures 4.1*a* and 4.6) is invariant in *cells*, we will see later that some *viruses* (which are not cells, ↩ Section 1.14) violate the dogma in many interesting ways (Chapters 8 and 10). But for now we move on to consider the different genetic elements present in prokaryotic cells.

MINIQUIZ

- What is a genome and what is it composed of? What is the central dogma of molecular biology?
- Define the terms complementary and antiparallel as they pertain to DNA.
- Why is supercoiling essential to a bacterial cell? What enzyme facilitates this process?

4.2 Genetic Elements: Chromosomes and Plasmids

Structures containing genetic material (DNA in cells but RNA in some viruses) are called *genetic elements*, and the main genetic element in prokaryotic cells is the **chromosome**. However, other genetic elements play important roles in microbes and these include *virus genomes, plasmids, organellar genomes*, and *transposable elements* (**Table 4.1**). Most *Bacteria* and *Archaea* contain a single circular chromosome containing all (or most) of the organism's genes. Although a single chromosome is the rule in prokaryotic cells, there are exceptions, as a few contain two or even three chromosomes. Eukaryotic genomes, by contrast, are composed of two or more chromosomes containing linear DNA. The genomes of viruses consist of *either* DNA or RNA and can be single- or double-stranded and either linear or circular.

Plasmids are circular or linear double-stranded DNA molecules that replicate separately from the chromosome and are typically much smaller than chromosomes. **Transposable elements** are sequences of DNA that are inserted into other DNA molecules but can move from one site on the DNA molecule to another, either within the same molecule or on a different DNA molecule. Chromosomes, plasmids, virus genomes, and any other type of DNA molecule may host a transposable element. Transposable elements are found in both prokaryotic and eukaryotic cells and play important roles in genetic variation (↩ Section 11.11).

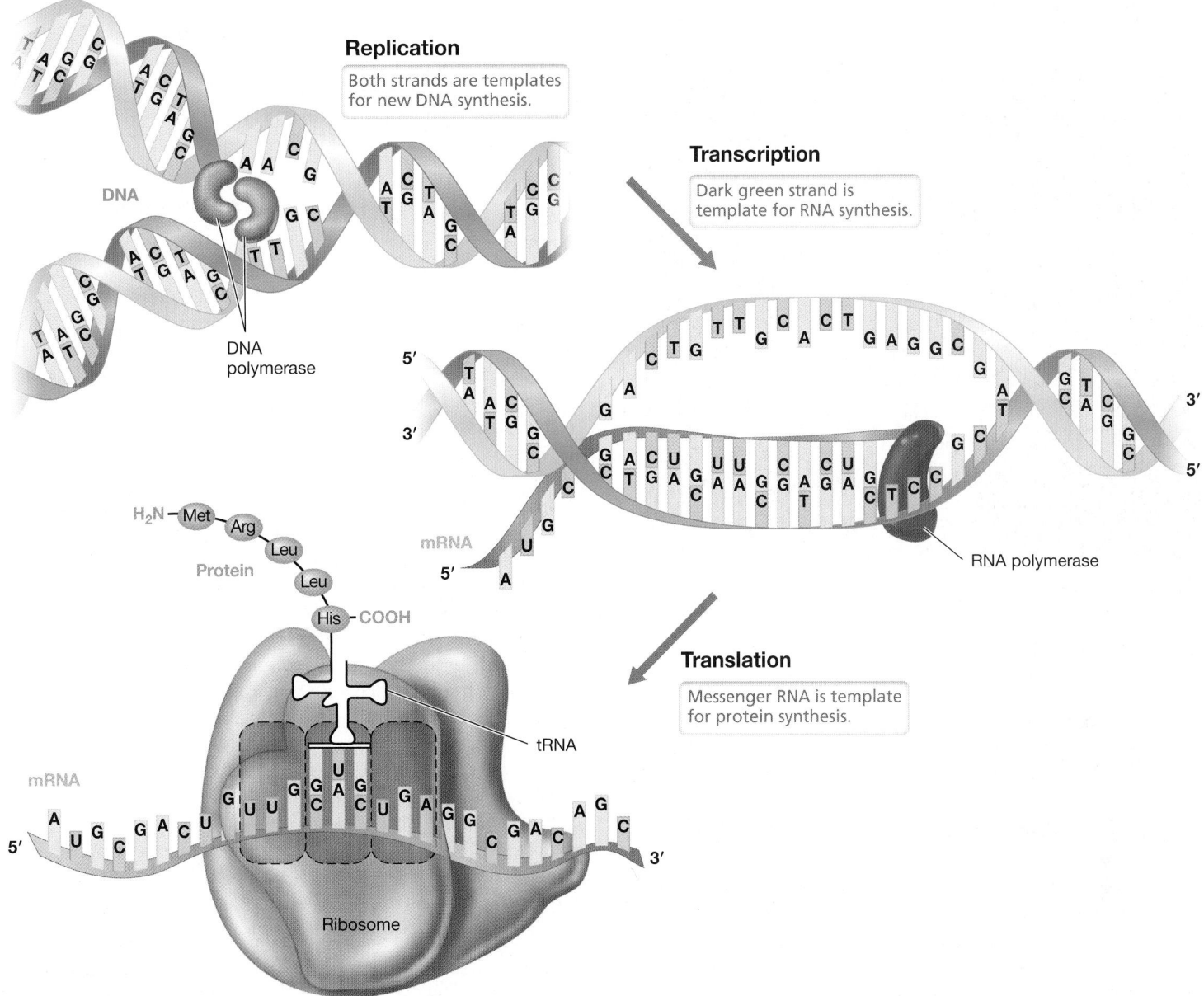

Figure 4.6 Synthesis of the three types of informational macromolecules. Note that for any particular gene only one of the two strands of the DNA double helix is transcribed.

Chromosomal Gene Arrangements and the Operon

Thousands of genomes from species of *Bacteria* and *Archaea* have been completely sequenced, thus revealing the number and location (the genetic map) of the genes they possess. The genetic map of the 4,639,675-bp chromosome of a widely studied strain of *Escherichia coli* is presented in **Figure 4.8**, with only a few of the organism's several thousand genes depicted. Analysis of the *E. coli* genome has revealed 4288 possible protein-encoding genes that account for 88% of the *E. coli* genome. Approximately 1% of the genome encodes tRNAs and rRNAs, and the remaining genes consist of regulatory sequences that may or may not be transcribed (but are not translated) or have other functions. The compact genomes of *Bacteria* and *Archaea* stand in contrast to the genomes of eukaryotes, which typically contain much more DNA than is

needed to encode all the proteins required for cell function. This "extra" DNA in eukaryotes is present as intervening DNA between coding sequences (the intervening sequences are removed before translation) or as repetitive sequences, some of which are repeated hundreds or thousands of times (Chapter 9).

Genetic mapping of the genes encoding the enzymes that function in steps of the same biochemical pathway in *E. coli* has shown that these genes are sometimes clustered. On the genetic map in Figure 4.8, a few such clusters are shown (for example, the *gal, trp*, and *his* clusters); each of these groups is called an **operon**. An operon is transcribed to form a single mRNA that encodes several different proteins and is regulated as a unit. In contrast to these, the genes for many other biochemical pathways in *E. coli* are not clustered. For example, genes for maltose degradation (*mal* genes,

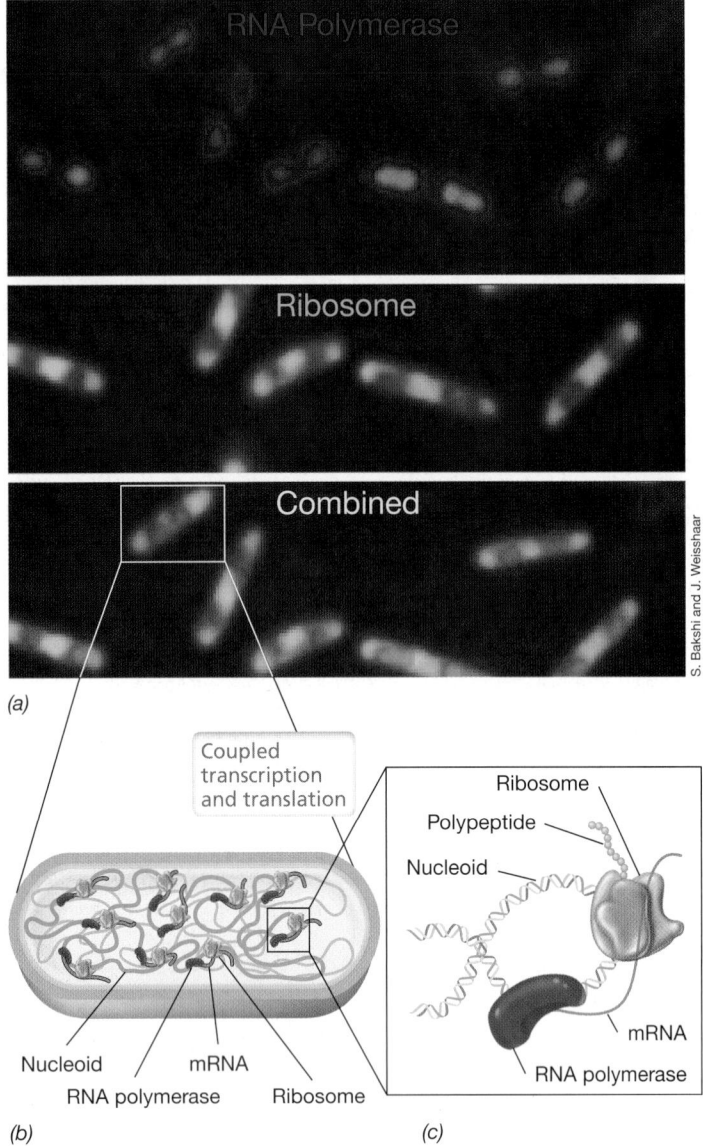

(a)

S. Bakshi and J. Weisshaar

(b)   (c)

Figure 4.7 Coupled transcription and translation in prokaryotic cells.
(a) Fluorescence microscopy and protein tagging of actively growing *Escherichia coli* cells illustrating the position of RNA polymerases and ribosomes performing transcription and translation, respectively. The combined photo (bottom image) shows that transcription and translation are occurring concurrently in the cell. *(b)* Location of the nucleoid, RNA polymerases, mRNA, and ribosomes in the cell during coupled transcription and translation. *(c)* Blowup illustration of a ribosome actively translating an mRNA as it is being synthesized by RNA polymerase.

Figure 4.8) are scattered throughout the chromosome. In fact, analysis of the *E. coli* chromosome has shown that over 70% of the predicted or known transcriptional units are of only a single gene while only 6% of operons have four or more genes. Thus operons, as efficient an arrangement of genes as they may be, appear to be the exception rather than the rule.

Plasmids

Many *Bacteria* and *Archaea* contain plasmids in addition to their chromosome(s). Most plasmids are nonessential since with rare exception they do not contain genes required for growth under all conditions. Thousands of different plasmids are known, and over 300 different plasmids have been isolated from strains of *E. coli* alone. Virtually all plasmids consist of double-stranded DNA and exist in the cytoplasm as free DNA. Most plasmids are circular, but many are linear and vary in size from approximately 1 kbp to more than 1 Mbp.

Typical plasmids are less than 5% of the size of the chromosome (**Figure 4.9**), and some bacteria contain several different plasmids. Moreover, different plasmids may be present in different *copy number*. For example, some plasmids may be present in only one or a few copies per cell, whereas others may be present in over 100 copies. Enzymes that replicate chromosomal DNA also replicate plasmids. Some of the genes encoded on a plasmid function to direct the initiation of plasmid replication and to partition replicated plasmids between daughter cells.

Although by definition plasmids do not encode functions essential to the host, plasmids may carry genes that profoundly influence host cell physiology; for example, plasmid genes may encode enzymes for some special metabolism that ensures survival under certain conditions. Among the most widespread and well-studied groups of plasmids are the resistance plasmids, called *R plasmids*, which confer resistance to antibiotics or other growth inhibitors. The resistance genes encode proteins that either inactivate the antibiotic or protect the cell in some other way, and several antibiotic resistance genes can be encoded on a single R plasmid. Plasmid R100 (**Figure 4.10**), for example, encodes resistance to sulfonamides, streptomycin, spectinomycin, fusidic acid, chloramphenicol, and tetracycline, as well as the toxic heavy metal mercury. Pathogenic bacteria resistant to antibiotics are of considerable medical significance, and their increasing incidence is correlated with the increasing use of antibiotics for treating infectious diseases in humans and animals (Chapter 28).

Pathogenic bacteria express a variety of plasmid-encoded *virulence factors* that assist them in establishing infections. For example,

TABLE 4.1 **Kinds of genetic elements**			
Organism	*Element*	*Type of nucleic acid*	*Description*
Virus	Virus genome	Single- or double-stranded DNA or RNA	Relatively short, circular or linear
Bacteria/Archaea	Chromosome	Double-stranded DNA	Extremely long, usually circular
Eukaryote	Chromosome	Double-stranded DNA	Extremely long, linear
Mitochondrion or chloroplast	Organellar genome	Double-stranded DNA	Medium length, usually circular
All organisms	Plasmid[a]	Double-stranded DNA	Relatively short circular or linear, extrachromosomal
All organisms	Transposable element	Double-stranded DNA	Always found inserted into another DNA molecule

[a]Plasmids are uncommon in eukaryotes.

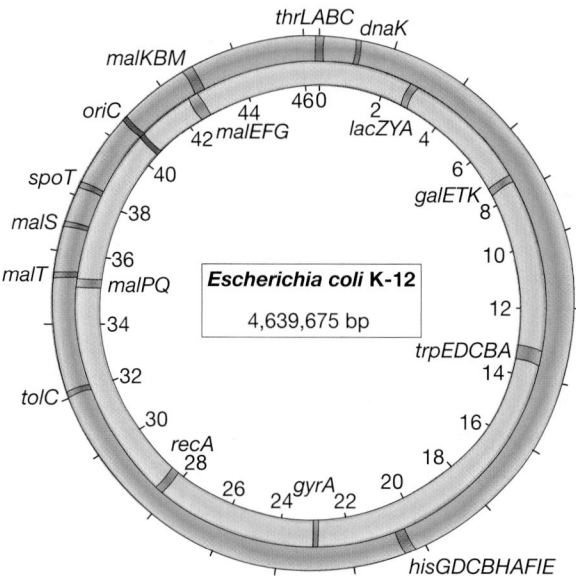

Figure 4.8 The chromosome of *Escherichia coli* strain K-12. Map distances are given in 100 kilobases of DNA. The chromosome contains 4,639,675 base pairs and 4288 open reading frames (genes). Depending on the DNA strand, the locations of a few genes and operons are indicated. Replication (Figure 4.6 and see Figures 4.16 and 4.17) proceeds in both directions from the origin of DNA replication, *oriC*, indicated in red.

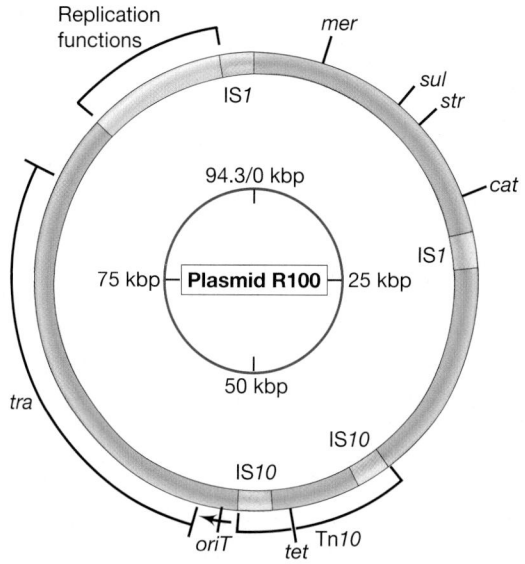

Figure 4.10 Genetic map of the resistance plasmid R100. The inner circle shows the size in kilobase pairs. The outer circle shows the location of major antibiotic resistance genes and other key functions: *mer*, mercuric ion resistance; *sul*, sulfonamide resistance; *str*, streptomycin resistance; *cat*, chloramphenicol resistance; *tet*, tetracycline resistance; *oriT*, origin of conjugative transfer; *tra*, transfer functions. The locations of insertion sequences (IS) and the transposon *Tn10* are also shown. Genes for plasmid replication are found in the region from 88 to 92 kbp.

the ability of a pathogen to attach to and colonize specific host tissues and to produce toxins, enzymes, and other invasive molecules that damage the host are sometimes plasmid encoded. Some bacteria also produce proteins called **bacteriocins** that inhibit or kill

closely related species of bacteria (or even different strains of the same species of bacteria), and the genes encoding these bacteriocins and other proteins that protect the producing organism are typically found on plasmids.

In a few cases plasmids encode properties that are fundamental to the ecology of the bacterium. For example, the ability of the soil bacterium *Rhizobium* to form nitrogen-fixing nodules on the roots of plants (ᴄ⤻ Section 23.3) requires certain functions encoded by plasmids. Other plasmids confer special metabolic properties. For example, the ability to degrade hydrocarbons or toxic pollutants, such as polychlorinated biphenyls (PCBs) and herbicides or other pesticides, is often plasmid encoded. In addition, plasmids play a crucial role in the horizontal gene transfer process called conjugation that we consider in detail later (Chapter 11).

— **MINIQUIZ** —

- Approximately how large is the *Escherichia coli* genome in base pairs? How many genes does it contain?
- What are viruses and plasmids?
- What properties does an R plasmid confer on its host cell?

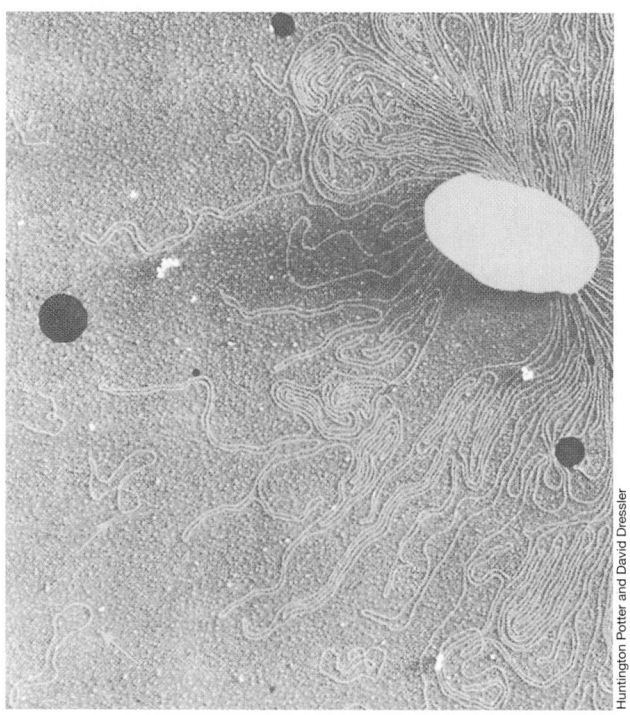

Figure 4.9 The bacterial chromosome and bacterial plasmids, as seen in the electron microscope. The plasmids (arrows) are the circular structures and are much smaller than the main chromosomal DNA. The cell (large, tan structure) was broken gently so the DNA would remain intact.

Huntington Potter and David Dressler

II • Copying the Genetic Blueprint: DNA Replication

D NA replication is necessary for cells to divide, whether to reproduce new organisms, as in unicellular microorganisms, or to produce new cells as part of a multicellular organism. To successfully transmit genetic information from a mother cell to an identical daughter cell, DNA replication must be extremely accurate. We review the basic principles of DNA replication here as a prelude to focusing on the process as it occurs in prokaryotic cells.

4.3 Templates, Enzymes, and the Replication Fork

DNA exists in cells as a double helix of two complementary strands (Figures 4.2 and 4.6), and if the helix is opened up, a new strand can be synthesized as the complement of each parental strand. As shown in **Figure 4.11a**, replication is thus a **semiconservative** process, meaning that the two resulting double helices consist of one new strand and one parental strand. The DNA strand that is used to make a complementary daughter strand is called the *template strand*, and in DNA replication, each parental strand is a template for one newly synthesized strand (Figure 4.11a). The precursor of each new nucleotide in the DNA strand is a deoxynucleoside 5'-triphosphate. During insertion of this molecule, the two terminal phosphates are removed and the remaining phosphate is bonded to a deoxyribose of the growing chain (Figure 4.11b). This addition of the incoming nucleotide requires the presence of a free hydroxyl group, which is available only at the 3' end of the molecule. This leads to the important principle that DNA replication always proceeds *from the 5' end to the 3' end*, the 5'-phosphate of the incoming nucleotide being attached to the 3'-hydroxyl of the previously added nucleotide (Figure 4.11b).

Replication Enzymes

Enzymes that catalyze the polymerization of deoxynucleotides are called **DNA polymerases** (abbreviated DNA Pol), and there are five different enzymes in *Escherichia coli*, DNA Pol I–V. DNA Pol III is the primary enzyme for replicating chromosomal DNA, although DNA Pol I also plays a lesser role. The other DNA polymerases function to repair damaged DNA (↩ Section 11.4). DNA Pol enzymes are just some of the many enzymes that are required for DNA replication (Table 4.2).

All DNA polymerases synthesize DNA in the 5'→3' direction, but none of them can initiate a new chain de novo; they can only add a nucleotide onto a preexisting 3'-OH group. Thus, in order to start a new DNA chain, a **primer**, a nucleic acid molecule to which DNA polymerase can attach the first nucleotide, is required, and this primer is a short stretch of *RNA* rather than DNA (**Figure 4.12**). When the DNA helix is first opened, the enzyme **primase** makes this RNA primer, synthesizing a short stretch (11–12 nucleotides) of RNA complementary in base pairing to the template strand DNA. At the growing end of this RNA primer is a 3'-OH group to which DNA polymerase adds the first deoxyribonucleotide. Continued extension of the molecule thus occurs as DNA rather than RNA (Figure 4.12), and the primer is eventually removed and replaced with DNA (described shortly).

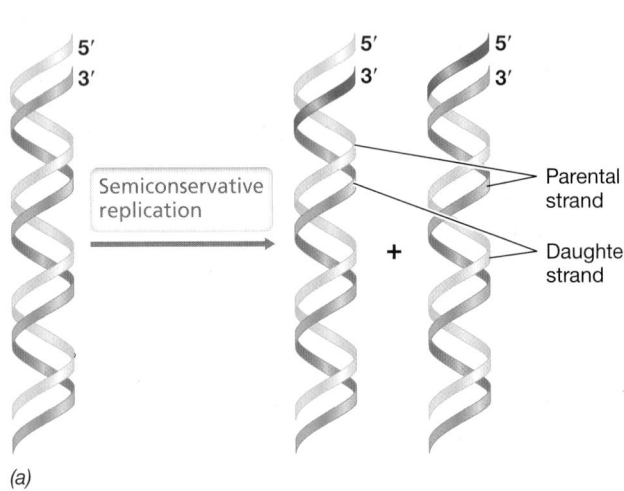

(a)

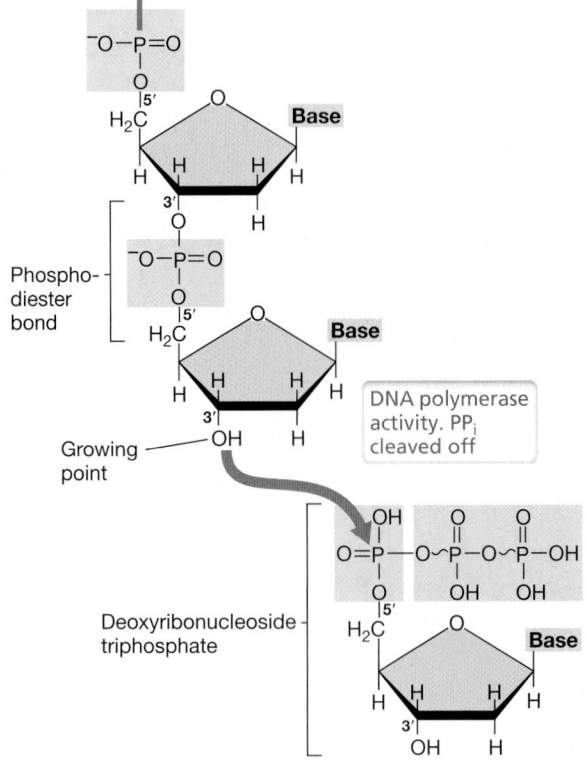

(b)

Figure 4.11 Overview of DNA replication. *(a)* DNA replication is a semiconservative process in all cells. Note that the new double helices each contain one new daughter strand (shown topped in red) and one parental strand. *(b)* Extension of a DNA chain by adding a deoxyribonucleoside triphosphate at the 3' end. Growth proceeds from the 5'-phosphate to the 3'-hydroxyl end. DNA polymerase catalyzes the reaction. The four precursors are deoxythymidine triphosphate (dTTP), deoxyadenosine triphosphate (dATP), deoxyguanosine triphosphate (dGTP), and deoxycytidine triphosphate (dCTP). Upon nucleotide insertion, the two terminal phosphates of the triphosphate are split off as pyrophosphate (PP$_i$). Thus, two energy-rich phosphate bonds are consumed when adding each nucleotide.

TABLE 4.2 Major enzymes that participate in DNA replication in *Bacteria*

Enzyme	Encoding genes	Function
DNA gyrase	*gyrAB*	Replaces supercoils ahead of replisome
Origin-binding protein	*dnaA*	Binds origin of replication to open double helix
Helicase loader	*dnaC*	Loads helicase at origin
Helicase	*dnaB*	Unwinds double helix at replication fork
Single-strand binding protein	*ssb*	Prevents single strands from annealing
Primase	*dnaG*	Primes new strands of DNA
DNA polymerase III		Main polymerizing enzyme
Sliding clamp	*dnaN*	Holds Pol III on DNA
Clamp loader	*holA–E*	Loads Pol III onto sliding clamp
Dimerization subunit (Tau)	*dnaX*	Holds together the two core enzymes for the leading and lagging strands
Polymerase subunit	*dnaE*	Strand elongation
Proofreading subunit	*dnaQ*	Proofreading
DNA polymerase I	*polA*	Excises RNA primer and fills in gaps
DNA ligase	*ligA, ligB*	Seals nicks in DNA
Tus protein	*tus*	Binds terminus and blocks progress of the replication fork
Topoisomerase IV	*parCE*	Unlinking of interlocked circles

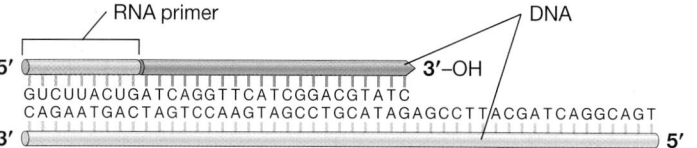

Figure 4.12 The RNA primer. Structure of the RNA–DNA hybrid formed during initiation of DNA synthesis. Orange depicts the RNA primer.

region is immediately covered with copies of single-strand binding protein to stabilize the single-stranded DNA and prevent the double helix from re-forming. DNA synthesis begins at a single site on the chromosome, the origin of replication (*oriC*), where the protein DnaA (Table 4.2) binds and opens up the double helix. Next to assemble is helicase (DnaB), which is helped onto the DNA by a loader protein (DnaC) (Figure 4.13*b*). Two helicases are loaded, one onto each strand, facing in opposite directions. Finally, two primase and two DNA polymerase enzymes are loaded onto the DNA behind the helicases and initiation of DNA replication begins. As replication proceeds, the replication fork appears to move along the DNA (Figure 4.13*a*).

Leading and Lagging Strands and the Replication Process

Figure 4.14 depicts DNA replication at the replication fork. Recall that replication always proceeds from 5′ to 3′ (5′→3′), always adding a new nucleotide to the 3′-OH of the growing chain. On the strand growing from the 5′-PO_4^{2-} to the 3′-OH, called the **leading strand**, DNA synthesis occurs *continuously* because there is always a free 3′-OH at the replication fork to which a new nucleotide can be added; the leading strand must therefore be primed only once. By contrast, on the opposite strand, called the **lagging strand**, DNA synthesis occurs *discontinuously* because there is no 3′-OH at the replication fork to which a new nucleotide can attach; on this strand, primase must synthesize multiple RNA primers in order to provide free 3′-OH groups for DNA Pol III (Figure 4.14). As a result, the lagging strand forms from several short DNA fragments that are combined later to yield a continuous strand of DNA.

Initiation of DNA Synthesis

Before replication can begin, the double helix must be unwound to expose the template strands, the so-called **replication fork**. The enzyme **DNA helicase** unwinds the double helix (using energy from ATP) and exposes a short single-stranded region (**Figure 4.13**). Helicase moves along the DNA and separates the strands just in advance of the replication fork. The single-stranded

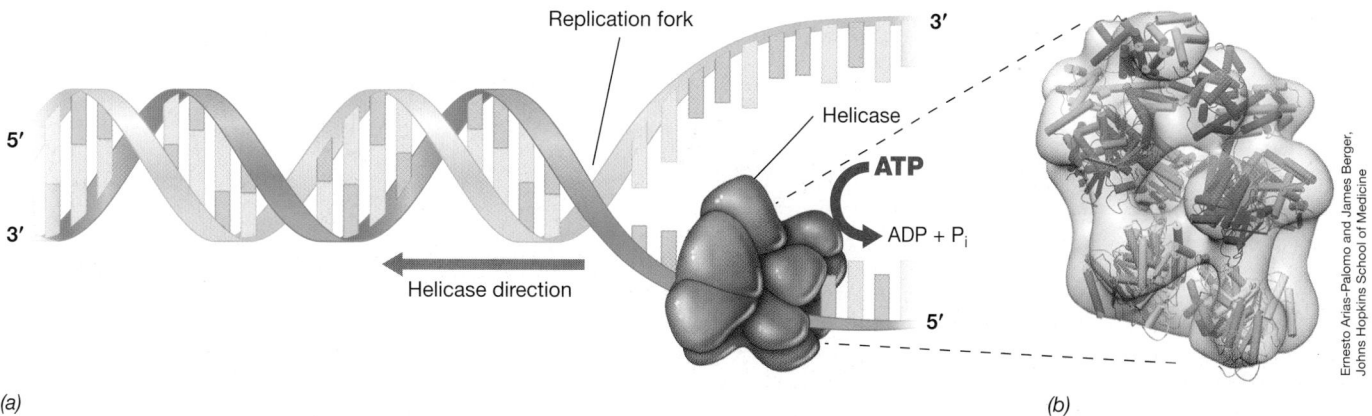

Figure 4.13 DNA helicase unwinding a double helix. *(a)* In this figure, the helicase is denaturing or pulling the two antiparallel strands of DNA apart beginning from the right and moving to the left. *(b)* A three-dimensional model of the helicase (DnaB) along with its loader protein ((DnaC, yellow and orange) based on cryo-electron microscopy.

Ernesto Arias-Palomo and James Berger, Johns Hopkins School of Medicine

Figure 4.14 Events at the DNA replication fork on the nucleoid. Note the polarity and antiparallel nature of the DNA strands. Helicase unwinds the DNA while primase adds the RNA primer. For the steps in introducing and removing supercoils from DNA, see Figure 4.5. Further events in DNA synthesis including sealing replicated fragments are shown in Figure 4.15.

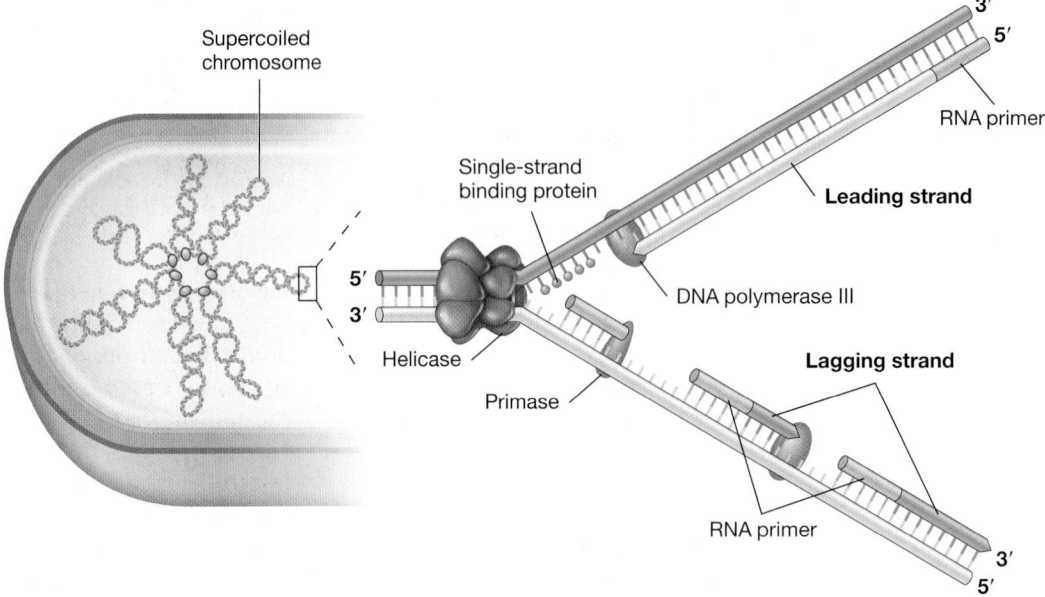

After synthesizing the RNA primer, primase is replaced by DNA Pol III. This enzyme complex (Table 4.2) is held on the DNA by a "sliding clamp," which encircles and slides along the single template strands of DNA. Consequently, the replication fork contains two polymerase core enzymes and two sliding clamps, one set for each strand. After assembly on the lagging strand, the elongation activity of DNA Pol III adds deoxyribonucleotides sequentially until it reaches previously synthesized DNA (**Figure 4.15**); at this point, activity of DNA Pol III stops.

To complete DNA synthesis, DNA Pol I catalyzes two different reactions. Besides synthesizing DNA, Pol I has a $5' \rightarrow 3'$

exonuclease activity that removes the RNA primer (Figure 4.15). When the primer has been excised and replaced with DNA, Pol I is released. The *very last* phosphodiester bond in replicating DNA is made by **DNA ligase**. This enzyme seals nicks in DNAs that have an adjacent $5'$-PO_4^{2-} and $3'$-OH (something that DNA Pol I and Pol III are unable to do), and along with DNA Pol I, it also participates in DNA repair. DNA ligase is also important for sealing genetically manipulated DNA during molecular cloning (⇔ Section 12.2).

We now put DNA synthesis in the context of *Archaea* and *Bacteria* to see how replication events occur around the covalently closed and circular chromosomes typical of these organisms.

MINIQUIZ

- What is the difference between a template strand and a daughter strand of DNA?
- To which end ($5'$ or $3'$) of a newly synthesized strand of DNA does DNA polymerase add a nucleotide?
- In DNA replication, what is the primer composed of and why are there leading and lagging strands?
- What are the functions of DNA Pol I and III and DNA helicase and DNA ligase?

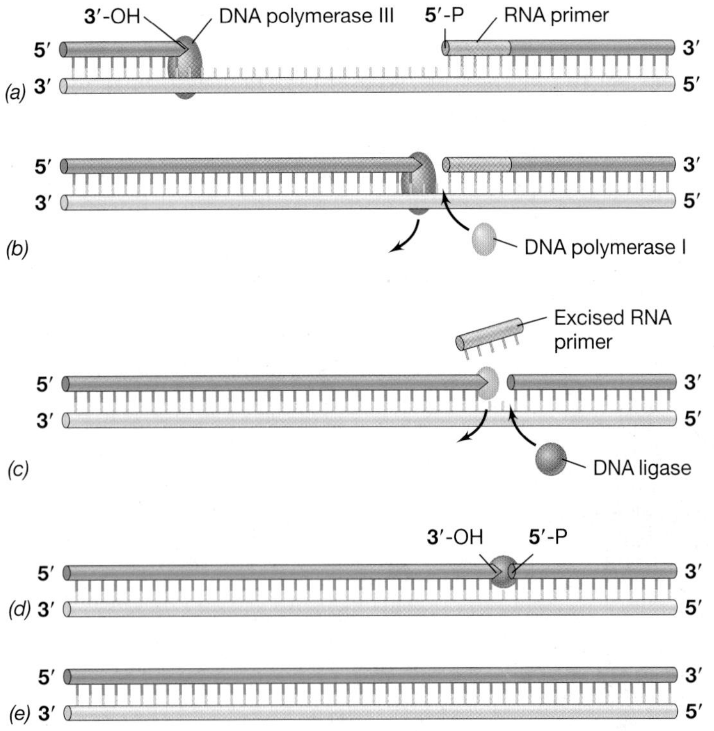

Figure 4.15 Sealing two fragments on the lagging strand. Unlike the leading strand, where synthesis occurs in a continuous fashion, on the lagging strand, DNA fragments need to be sealed to form the intact DNA strand. (a) DNA polymerase III is synthesizing DNA in the $5' \rightarrow 3'$ direction toward the RNA primer of a previously synthesized fragment on the lagging strand. (b) On reaching the fragment, DNA polymerase III leaves and is replaced by DNA polymerase I. (c) DNA polymerase I continues synthesizing DNA while removing the RNA primer from the previous fragment, and DNA ligase replaces DNA polymerase I after the primer has been removed. (d) DNA ligase seals the two fragments together. (e) The final product, complementary and antiparallel double-stranded DNA.

4.4 Bidirectional Replication, the Replisome, and Proofreading

The circular nature of the bacterial and archaeal chromosome accelerates the genomic replication process. In *Escherichia coli*, and probably in all cells with circular chromosomes, DNA replication occurs *bidirectionally* from the origin of replication. There are thus *two* replication forks on each chromosome, each moving in opposite directions. In circular DNA, bidirectional replication leads to the formation in the replicating molecules of characteristic shapes (so-called "theta structures") as synthesis proceeds in both a leading and a lagging fashion on each template strand (**Figure 4.16**). In an actively growing cell of *E. coli*, DNA Pol III adds nucleotides at the rate of about 1000 per second; hence, replication of the entire chromosome takes about 40 min.

The Replisome

Figure 4.14 shows the enzymes that participate in replication, and from such a schematic it may appear that the enzymes are working independently. However, this is not the case. Instead, replication proteins aggregate to form a large replication complex called the **replisome** (**Figure 4.17**). The lagging strand of DNA actually loops out to allow the replisome to move smoothly along both strands, the complex literally pulling the DNA template through it as replication proceeds. In addition to the replisome, helicase and primase form their own subcomplex within the replisome called the *primosome*. This close association facilitates the sequential activities of these two enzymes during the replication process (Figure 4.17). Table 4.2 summarizes the functions of proteins essential for DNA replication in *Bacteria*.

Eventually the work of the replisome is finished, and this is signaled when the replication forks collide at a site located on the opposite side of the chromosome from the origin called the *terminus of replication*. In the terminus region are several DNA sequences called *Ter* sites that are recognized by a protein called Tus, whose function is to block progress of the replication forks. When replication of the circular chromosome is complete, the two circular molecules are linked together, much like the links of a chain. After replication, the DNA is partitioned so that each daughter cell receives a copy of the chromosome; DNA partitioning is facilitated by FtsZ, a protein that orchestrates several key events in the cell division process (Chapter 7).

Fidelity of DNA Replication: Proofreading

DNA replication occurs with a remarkably low error rate. Nevertheless, when errors do occur, a mechanism exists to detect and correct them. Errors in DNA replication introduce *mutations*, changes in DNA sequence. Mutation rates in cells are extremely low, between 10^{-8} and 10^{-11} errors per base pair inserted. This accuracy can be achieved because DNA polymerases effectively get two chances to incorporate the correct base at any given site. The first chance comes when DNA Pol III inserts bases according to the base-pairing rules (Figure 4.1c). The second chance comes when a process called *proofreading* takes place (**Figure 4.18**).

During replication, if an incorrect base has been inserted, a mismatch in base pairing (Figure 4.1c) occurs. Both DNA Pol I and Pol III possess a $3' \rightarrow 5'$ exonuclease activity that can remove such mismatched nucleotides. The polymerase detects the error because incorrect base pairing causes a slight distortion in the topology of the double helix. After the removal of a mismatched nucleotide, the polymerase then gets a second chance to insert the correct nucleotide (Figure 4.18). With the extremely low error rate of DNA polymerases, the chance of inserting the wrong base at the same

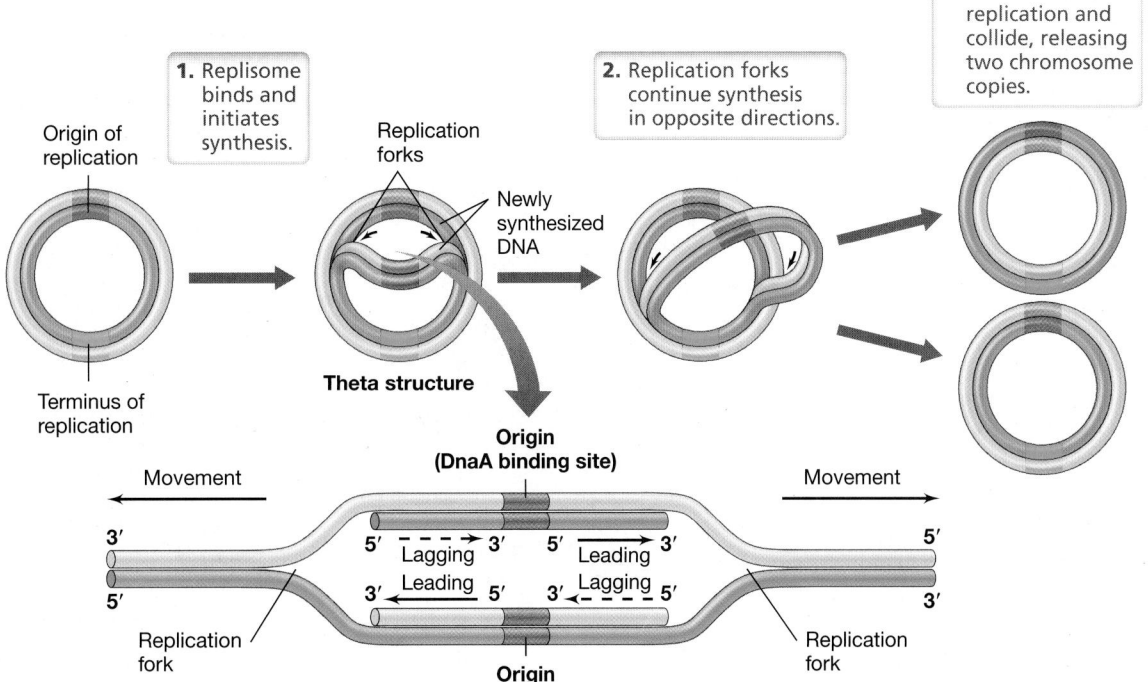

Figure 4.16 Replication of circular DNA: the theta structure. In circular DNA, bidirectional replication from an origin forms an intermediate structure resembling the Greek letter theta (θ). Blowup shows dual replication forks in the circular chromosome. In *Escherichia coli*, the origin of replication is recognized by the DnaA protein and the terminus of replication is recognized by the Tus protein. Note that DNA synthesis is occurring in both a leading and a lagging manner on each of the new daughter strands until the replication forks hit the terminus. Compare this figure with the illustration of the replisome in Figure 4.17.

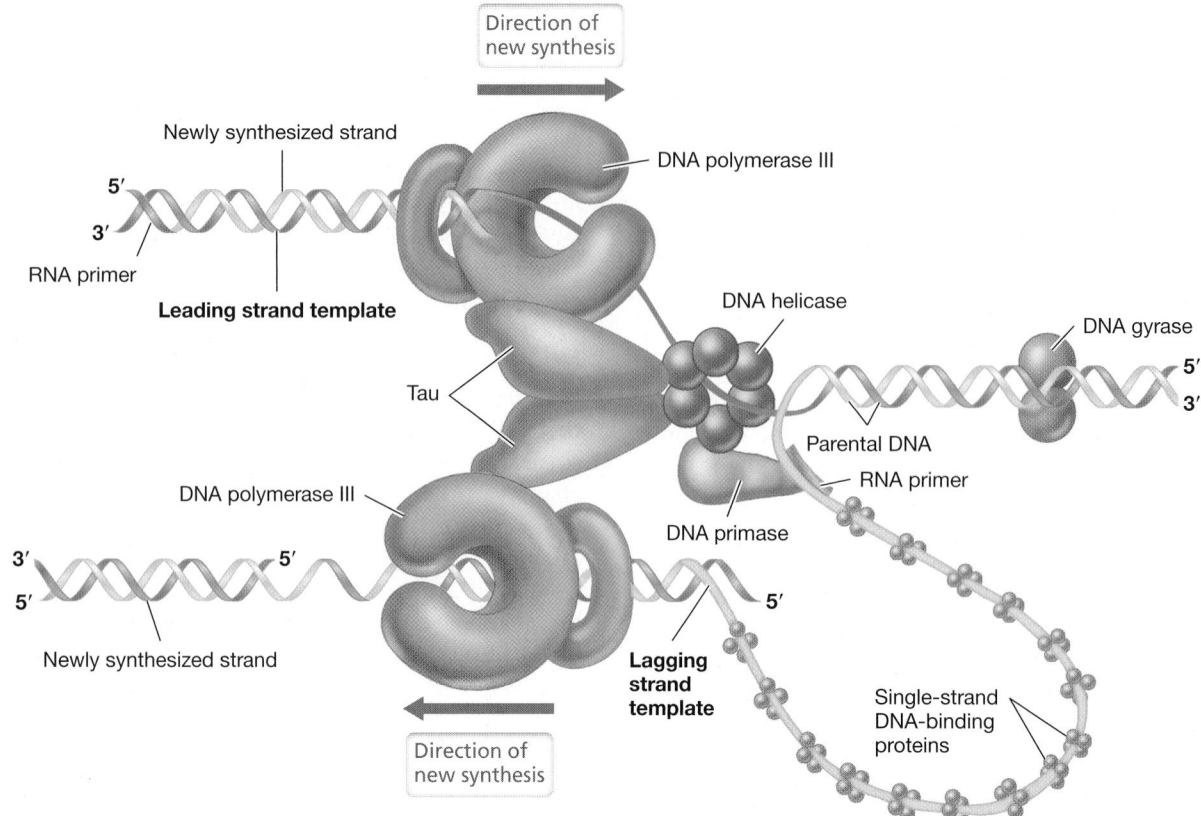

Figure 4.17 The replisome. The replisome consists of two copies of DNA polymerase III and DNA gyrase, plus helicase and primase (together forming the primosome), and many copies of single-strand DNA-binding protein. The Tau subunits hold the two DNA polymerase assemblies and helicase together. Just upstream of the rest of the replisome, DNA gyrase removes supercoils in the DNA to be replicated. Note that the two polymerases are replicating the two individual strands of DNA in opposite directions. Consequently, the lagging-strand template loops around so that the whole replisome moves in the same direction along the chromosome.

site twice is vanishingly small. Exonuclease proofreading occurs in *Bacteria* and *Archaea*, eukaryotes, and viral DNA replication systems.

We now move on from replicating genes to consider *gene expression* as a prelude to examining synthesis of the proteins encoded by the transcribed genes.

MINIQUIZ

- What is the replisome and what are its components?
- How are the activities of the replisome stopped?
- How are errors in DNA replication kept extremely low?

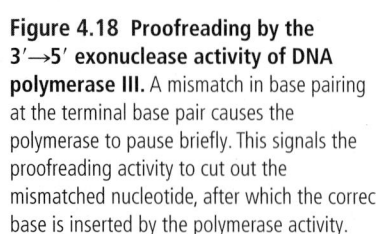

Figure 4.18 Proofreading by the 3′→5′ exonuclease activity of DNA polymerase III. A mismatch in base pairing at the terminal base pair causes the polymerase to pause briefly. This signals the proofreading activity to cut out the mismatched nucleotide, after which the correct base is inserted by the polymerase activity.

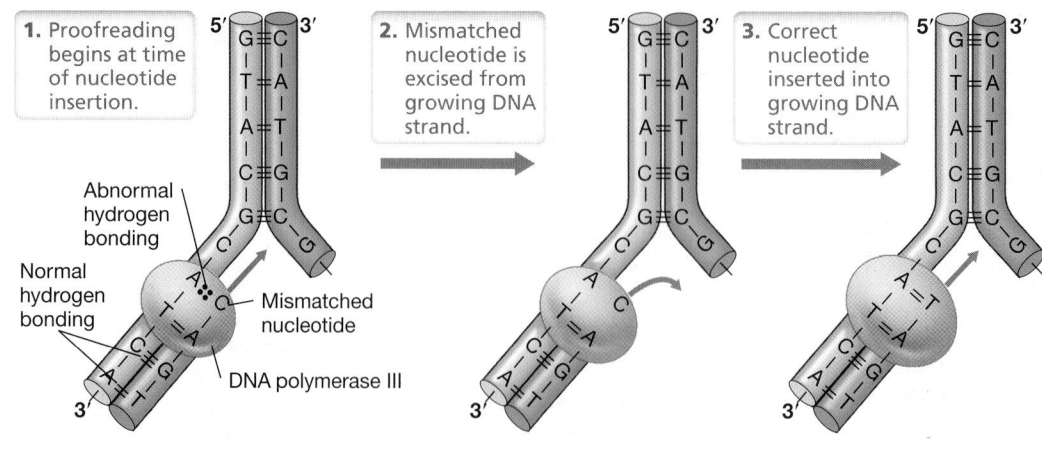

III • RNA Synthesis: Transcription

Transcription—RNA synthesis off of a DNA template—yields three main forms of RNA: *messenger* (mRNA), *transfer* (tRNA), and *ribosomal* (rRNA) (Section 4.1). Several other minor classes of RNA exist but most of these function in regulation (Chapter 6). RNA is both a genetic and a functional molecule. At the genetic level, mRNA encodes genetic information from the genome and carries it to the ribosome. In contrast, rRNAs play both a structural and a functional role in ribosomes, while tRNAs function as the carriers of amino acids to the ribosome for protein synthesis.

There are two key differences in the chemistry of RNA and DNA: (1) RNA contains ribose instead of deoxyribose; and (2) RNA contains uracil instead of thymine. The change from deoxyribose to ribose dramatically affects the chemistry of a nucleic acid, and enzymes that act on DNA typically have no effect on RNA, and vice versa. However, the change from thymine to uracil does not affect base pairing, as these two bases pair with adenine equally well.

With the exception of a few viruses that contain double-stranded RNA genomes (Chapter 10), RNA is a single-stranded molecule. However, the primary structure (sequence of nucleotides) of some RNAs allows them to fold and exploit complementary base pairing. The term **secondary structure** refers to this folding, and the functional role an RNA plays in the cell may depend critically on its secondary structure. For example, messenger RNAs, which are typically unfolded, exist in *Bacteria* (and *Archaea*) for only a few minutes before enzymes called *ribonucleases* degrade them. By contrast, rRNAs and tRNAs (referred to as *stable RNAs*) are long-lived because their secondary structures prevent ribonuclease attack. The rapid turnover of mRNAs in *Bacteria* and *Archaea* allows them to quickly adapt to changing environmental conditions and halt the translation of mRNAs whose products are no longer needed.

We begin with transcription in *Bacteria* and contrast this in the following section with transcriptional events in *Archaea* and *Eukarya*.

4.5 Transcription in *Bacteria*

Transcription is catalyzed by the enzyme **RNA polymerase**. Like DNA polymerase, RNA polymerase forms phosphodiester bonds but between the *ribo*nucleotides rather than *deoxyribo*nucleotides (Figure 4.1*b*). Polymerization is driven by energy released from the hydrolysis of two energy-rich phosphate bonds of the incoming ribonucleoside triphosphates. The mechanism of RNA synthesis is thus quite similar to that of DNA synthesis (Figure 4.11*b*): During elongation of an RNA chain, ribonucleoside triphosphates are added to the 3′-OH of the ribose of the preceding nucleotide. Thus chain growth is 5′→3′ just as in DNA synthesis, and the newly synthesized strand of RNA runs antiparallel to the DNA template strand it was transcribed from. A summary of bacterial transcription is illustrated in **Figure 4.19**.

RNA polymerase uses DNA as a template, but for any given gene, only one of the two strands is transcribed. Unlike DNA polymerase, RNA polymerase can initiate new RNA on its own; no priming is necessary as it is for DNA synthesis (Figure 4.12). Transcription

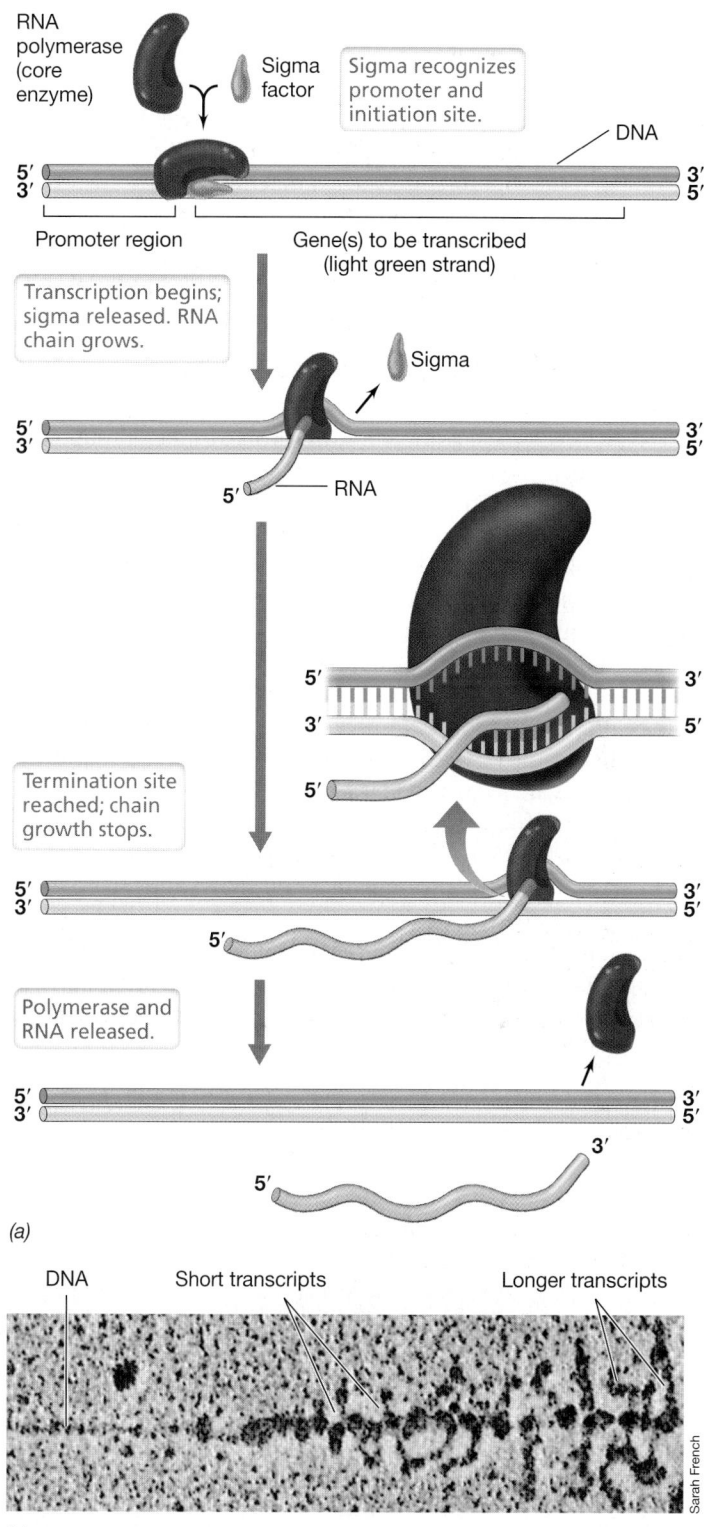

(a)

(b)

Figure 4.19 Transcription. *(a)* Steps in RNA synthesis. The initiation site (promoter) and termination site are specific nucleotide sequences on the DNA. RNA polymerase moves down the DNA chain, temporarily opening the double helix and transcribing one of the DNA strands. *(b)* Electron micrograph illustrates transcription along a gene on the *Escherichia coli* chromosome. Transcription is proceeding from left to right, with the shorter transcripts on the left becoming longer as transcription proceeds.

continues until specific sequences called *transcription terminators* are reached, but unlike DNA replication, which copies the entire genome, transcription occurs on much smaller units of DNA, often as little as a single gene. This system allows the cell to transcribe different genes at different frequencies, depending on the needs of the cell for different proteins. Said a different way, gene expression is a *highly regulated* process. Transcriptional regulation can occur in prokaryotic cells in many different ways, but the different mechanisms have a common outcome: Cell resources are conserved and cell fitness enhanced (Chapter 6).

RNA Polymerases and the Promoter Sequence

The RNA polymerase from *Bacteria*, which has the simplest structure and about which most is known, consists of five different subunits, designated β, β′, α, ω (omega), and σ (sigma), with α present in two copies (**Figure 4.20**). The subunits form an enzyme complex called the *RNA polymerase holoenzyme*. Sigma is not as tightly bound as the other subunits and easily dissociates to yield the *RNA polymerase core enzyme*, $\alpha_2\beta\beta'\omega$. The core enzyme alone synthesizes RNA, and sigma functions only to recognize the appropriate site on the DNA for transcription to begin (sigma dissociates from the holoenzyme once a short sequence of RNA has been formed) (Figure 4.19).

To begin transcription, RNA polymerase must first recognize initiation sites on the DNA; these sites are called **promoters**. In *Bacteria*, promoters are recognized by sigma, and once RNA polymerase has bound to a promoter, transcription can proceed (Figure 4.19). In this process, the DNA helix at the promoter site is opened up by RNA polymerase, and as the polymerase moves, it unwinds the DNA in short segments to expose template DNA. Because some genes reside on one strand of DNA while other genes reside on the other strand of DNA, promoters are present on both strands; as a result, transcription occurs in *opposite directions* on the two different strands of DNA.

Sigma Factors, Consensus Sequences, and Transcriptional Termination

Promoters are specific DNA sequences; **Figure 4.21** shows the sequence of several promoters from *Escherichia coli*. All of these sequences are recognized by the same *E. coli* sigma factor called σ^{70} (the superscript 70 indicates the size of this protein, 70 kilodaltons). Although these sequences are not identical, sigma recognizes two highly conserved regions within the promoter. These conserved sequences are upstream of (prior to) the transcription start site. One is 10 bases upstream, the −10 region, or *Pribnow box*. Although promoter sequences differ slightly, comparison of many −10 regions gives a consensus sequence of TATAAT. The second conserved region is about 35 bases upstream of the start site and its consensus sequence is TTGACA (Figure 4.21). In *E. coli*, promoters that conform most closely to the consensus sequence are more effective in binding RNA polymerase. Such promoters are called *strong promoters* and are very useful in genetic engineering (Chapter 12).

While most genes in *E. coli* require σ^{70} for transcription, several alternative sigma factors exist that recognize different consensus sequences (Table 4.3). Each alternative sigma factor is specific for a group of genes required under special circumstances, and thus the presence or absence of a specific sigma factor is a mechanism for regulating gene expression. That is, by changing the rate of either synthesis or degradation of a particular sigma factor, the cell can control the transcription of entire gene families.

Units of Transcription and Polycistronic mRNA

Genetic information is organized into *transcriptional units*, segments of DNA that are transcribed into a single RNA molecule bounded by their initiation and termination sites. Some transcriptional units contain RNA transcribed from a single gene, whereas

RNA Polymerases

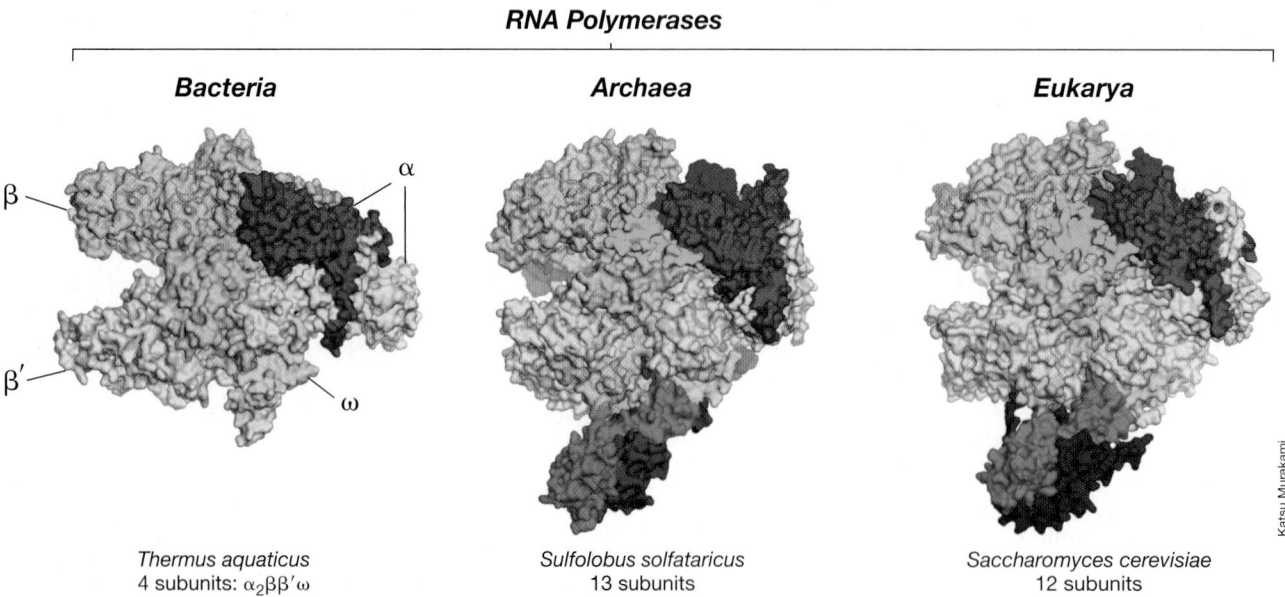

Bacteria	**Archaea**	**Eukarya**

Thermus aquaticus
4 subunits: $\alpha_2\beta\beta'\omega$

Sulfolobus solfataricus
13 subunits

Saccharomyces cerevisiae
12 subunits

Katsu Murakami

Figure 4.20 RNA polymerase from the three domains. Surface representation of multi-subunit cellular RNA polymerase structures from *Bacteria* (left, *Thermus aquaticus* core enzyme), *Archaea* (center, *Sulfolobus solfataricus*), and *Eukarya* (right, *Saccharomyces cerevisiae* RNA Pol II). Orthologous subunits are depicted by the same color. A unique subunit in the *S. solfataricus* RNA polymerase is not shown in this view.

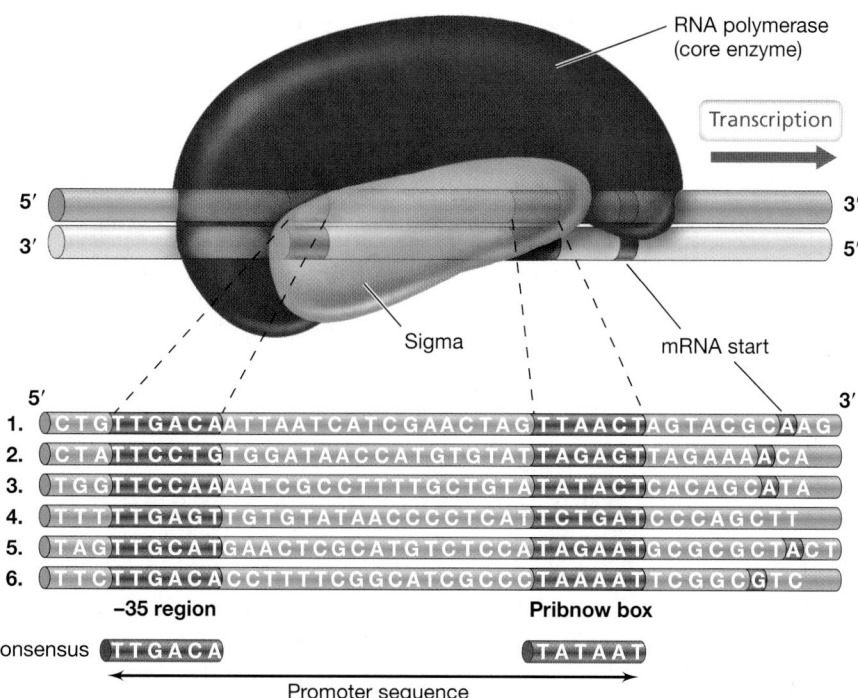

Figure 4.21 The interaction of RNA polymerase with a bacterial promoter. Shown below the RNA polymerase and DNA are six different promoter sequences identified in *Escherichia coli*. The contacts of the RNA polymerase with the −35 region and the Pribnow box (−10 sequence) are shown. Transcription begins at a unique base just downstream from the Pribnow box. Below the actual sequences at the −35 and Pribnow box regions are consensus sequences derived from comparing many promoters. Note that although sigma recognizes the promoter sequences on the 5′→3′ (dark green) strand of DNA, the RNA polymerase core enzyme will actually transcribe the light green strand (that runs 3′→5′) because core enzyme synthesizes only in a 5′→3′ direction.

others are formed from two or more genes (*cotranscribed genes*). Most genes encode proteins, but others encode nontranslated RNAs, such as ribosomal or transfer RNAs. For example, prokaryotic cells produce three size classes of rRNA: 16S, 23S, and 5S (the S refers to *Svedberg units*, a measure of particle size), and their genes are cotranscribed to form a single transcriptional unit that also

includes a tRNA (**Figure 4.22**). This transcriptional unit is subsequently "processed" by proteins that cut them to form the individual rRNAs or tRNAs.

As we have previously considered, genes that encode several enzymes of a particular metabolic pathway in prokaryotic cells, for example the biosynthesis of a particular amino acid, are often clustered together in an *operon* (Section 4.2). Assembling genes for the same biochemical pathway or genes needed under the same conditions into an operon allows their expression to be coordinated. During transcription, RNA polymerase proceeds through the operon and transcribes the entire set of genes into a single mRNA called a *polycistronic mRNA* (**Figure 4.23**). Polycistronic mRNAs contain multiple *open reading frames*, portions of the mRNA that actually encode amino acids (Section 4.9). When this mRNA is translated, several polypeptides are synthesized sequentially by the same ribosome.

Termination of Transcription

In a growing bacterial cell, only those genes that need to be expressed are usually transcribed; therefore, it is critical that transcription end at the correct position. **Termination** of transcription is governed by specific base sequences on the DNA. In *Bacteria* a common termination signal is a GC-rich sequence containing an *inverted repeat* with a central nonrepeating segment. When such a DNA sequence is transcribed, the RNA forms a stem–loop structure by intra-strand base

TABLE 4.3 **Sigma factors in *Escherichia coli***		
Name[a]	Upstream recognition sequence[b]	Function
σ^{70} RpoD	TTGACA	For most genes, major sigma factor for normal growth
σ^{54} RpoN	TTGGCACA	Nitrogen assimilation
σ^{38} RpoS	CCGGCG	Stationary phase, plus oxidative and osmotic stress
σ^{32} RpoH	TNTCNCCTTGAA	Heat shock response
σ^{28} FliA	TAAA	For genes involved in flagella synthesis
σ^{24} RpoE	GAACTT	Response to misfolded proteins in periplasm
σ^{19} FecI	AAGGAAAAT	For certain genes in iron transport

[a]Superscript number indicates size of protein in kilodaltons. Many factors also have other names, for example, σ^{70} is also called σ^{D}.

[b]N = any nucleotide.

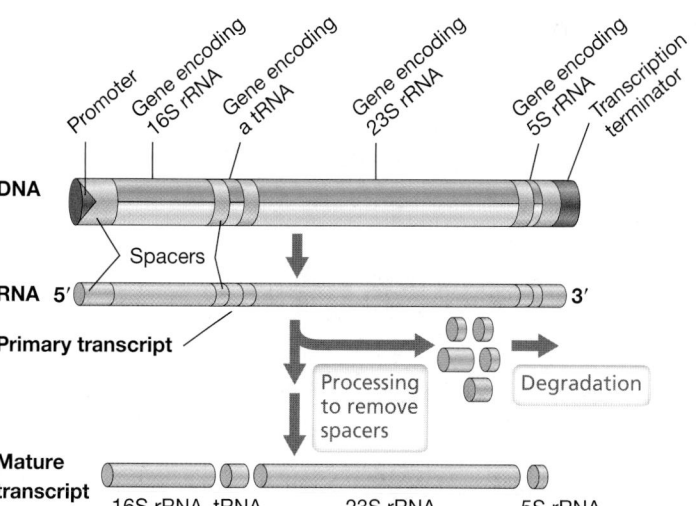

Figure 4.22 A ribosomal rRNA transcription unit from *Bacteria* and its subsequent processing. In *Bacteria*, all rRNA transcription units have the genes in the order 16S rRNA, 23S rRNA, and 5S rRNA (shown approximately to scale). Note that in this particular transcription unit the spacer between the 16S and 23S rRNA genes contains a tRNA gene. In other transcription units this region may contain more than one tRNA gene. Often one or more tRNA genes also follow the 5S rRNA gene and are cotranscribed. *Escherichia coli* contains seven rRNA transcription units.

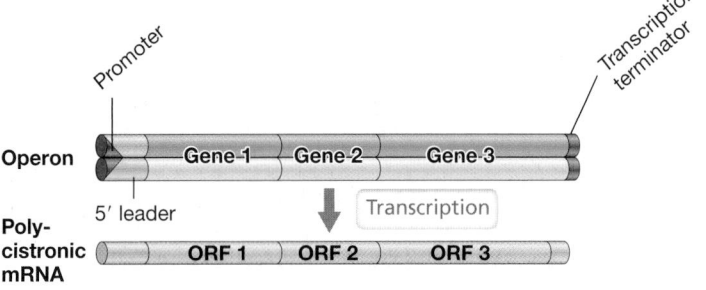

Figure 4.23 Operon and polycistronic mRNA structure. Note that a single promoter controls the three genes within the operon and that the polycistronic mRNA molecule contains an open-reading frame (ORF) corresponding to each gene.

pairing (**Figure 4.24**). Stem–loops followed by a run of adenines in the DNA template (which yield a run of uridines in the mRNA) are strong transcription terminators because a stretch of U:A base pairs are formed that hold the RNA and DNA together. However, this structure is very weak since U:A base pairs have only two hydrogen bonds rather than the three that form in T:A pairs (Figure 4.1*c*). Thus, RNA polymerase pauses at the stem–loop, and the DNA and RNA dissociate at the run of uridines, terminating transcription.

A second mechanism for stopping transcription is catalyzed by the terminator protein Rho. Rho does not bind to RNA polymerase or to the DNA, but binds tightly to RNA and moves down the chain toward the RNA polymerase–DNA complex. Once RNA polymerase has paused at a Rho-dependent termination site (a specific sequence on the DNA template), Rho causes both the RNA and RNA polymerase to be released from the DNA, thus terminating transcription.

Now that we have grasped the essentials of transcription in *Bacteria*, we turn our attention to this crucial cell process in *Archaea*

and *Eukarya*, where the phylogenetic connection between these two domains (⮡ Section 1.13) will be apparent in their mechanisms of transcription.

MINIQUIZ

- What enzyme catalyzes transcription? What is a promoter and what protein recognizes promoters in *Bacteria*?
- What is the role of messenger RNA (mRNA)? What are the other two classes of RNA?
- How does polycistronic mRNA allow for gene families to be controlled as a group?
- What type of structures lead to transcription termination?

4.6 Transcription in *Archaea* and *Eukarya*

Here we discuss key elements of transcription in *Archaea* and *Eukarya* that differ from those of *Bacteria*. Although in both *Archaea* and *Eukarya* the overall flow of genetic information is the same as in *Bacteria*, some details differ, and in eukaryotic cells the presence of the nucleus complicates the routing of genetic information. Many of the details of transcription (and translation) in *Archaea* resemble those in *Eukarya* more closely than *Bacteria*. However, *Archaea* also share some transcriptional similarities with *Bacteria*, such as operons. We begin our discussion at center stage with a consideration of RNA polymerase.

Archaeal and Eukaryotic RNA Polymerases, Promoters, and Terminators

Archaeal and eukaryotic RNA polymerases are similar and more complex than those of *Bacteria*. *Archaea* contain only a single RNA polymerase while eukaryotes have three. The archaeal polymerase most closely resembles eukaryotic RNA polymerase II and is

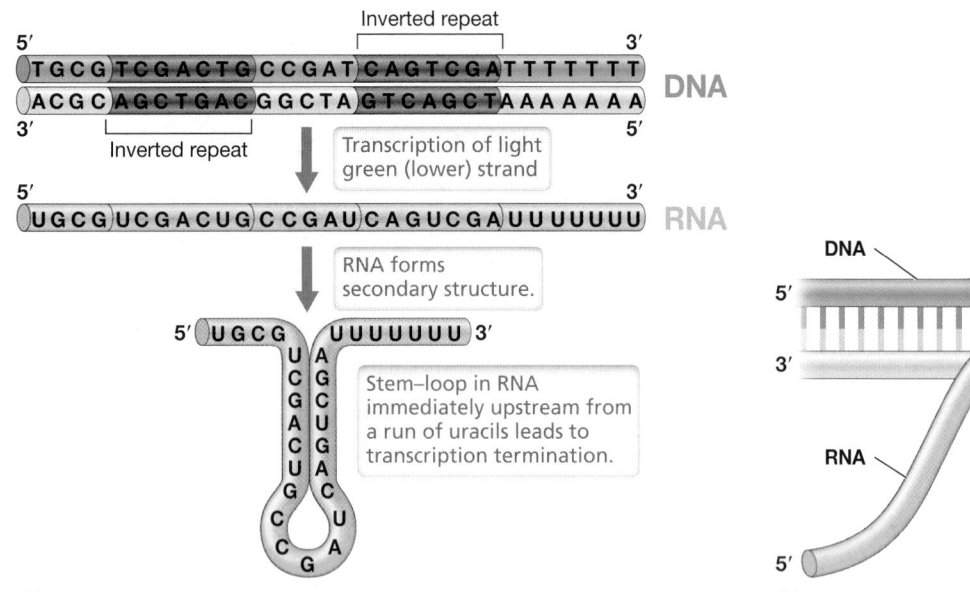

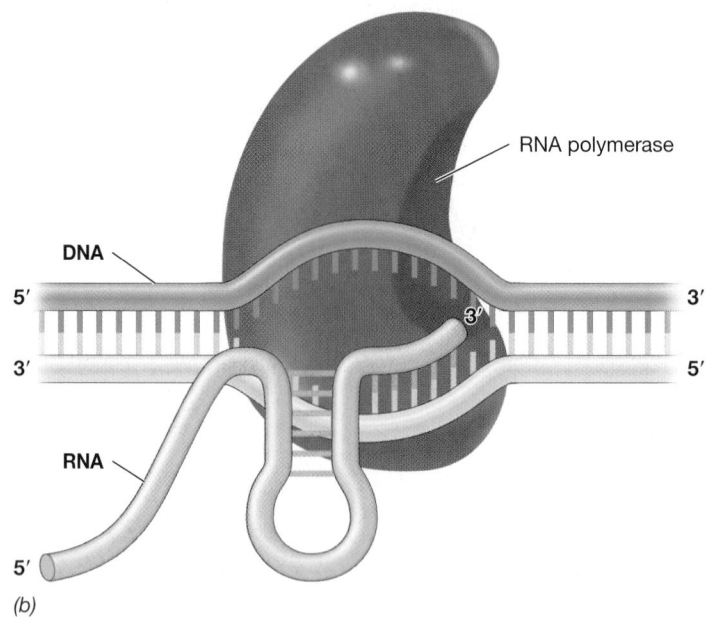

Figure 4.24 Inverted repeats and transcription termination. *(a)* Inverted repeats in transcribed DNA form a stem–loop structure in the RNA that terminates transcription when followed by a run of uracils. *(b)* Schematic indicating the formation of the terminator stem–loop in the RNA within the RNA polymerase.

composed of 11–13 subunits, depending on the species (eukaryotic RNA polymerase II has 12 or more subunits). These contrast with the comparatively simple four-subunit RNA polymerase core enzyme of *Bacteria* (Figure 4.20).

We learned the importance of the promoter and its recognition sequences to the overall process of transcription in Section 4.5. The most important recognition sequence in archaeal and eukaryotic promoters is the 6- to 8-base-pair "TATA" box, located 18–27 nucleotides upstream of the transcriptional start site (**Figure 4.25**). The TATA box is recognized by the *TATA-binding protein* (TBP), one of the many *transcription factors* present in *Archaea* and eukaryotes. Upstream of the TATA box is the *B recognition element* (BRE) sequence that is recognized by archaeal transcription factor B (TFB). In addition, a specific initiator element sequence is located at the start of transcription. Once TBP has bound to the TATA box and TFB has bound to the BRE, then archaeal RNA polymerase can bind and initiate transcription. This process is similar in eukaryotes except that several additional transcription factors are required.

Less is known about transcription termination in *Archaea* than in *Bacteria*, although some archaeal genes have inverted repeats followed by an AT-rich sequence similar to those present in many bacterial transcription terminators (Section 4.5). One other type of suspected transcription terminator lacks inverted repeats but contains repeated runs of thymines. In eukaryotes, the termination process differs depending on the RNA polymerase and often requires a specific termination factor protein.

No Rho-like proteins (Section 4.5) have been found in either *Archaea* or *Eukarya*.

RNA Processing in Eukaryotes and Intervening Sequences in *Archaea*

In contrast to *Bacteria*, *Eukarya* contain many genes that are split into two or more coding regions separated by noncoding regions. The coding sequences are called **exons**, while **introns** are the intervening noncoding regions. The transcripts from these genes thus require alterations—known as **RNA processing**—to form *mature RNAs* suitable for translation. The term **primary transcript** refers to the RNA molecule that is originally transcribed before the introns are removed to form the mature mRNA containing only exons. The process by which introns are removed and exons are joined is called *splicing* (**Figure 4.26**).

RNA splicing occurs in the nucleus by the activity of a macromolecular complex containing both RNA and protein called the **spliceosome**. The proteins of the spliceosome excise the intron(s) from the primary transcript and link the flanking exons to form a contiguous protein-encoding mature mRNA (Figure 4.26). While intervening sequences in genes encoding proteins are extremely rare in *Archaea*, several archaeal tRNA- and rRNA-encoding genes contain introns that must be removed after transcription to generate the mature tRNA or rRNA. In analogy to the introns of eukaryotes, these intervening sequences are called "archaeal introns"; however, their processing is catalyzed by a special ribonuclease rather than by the spliceosome complex.

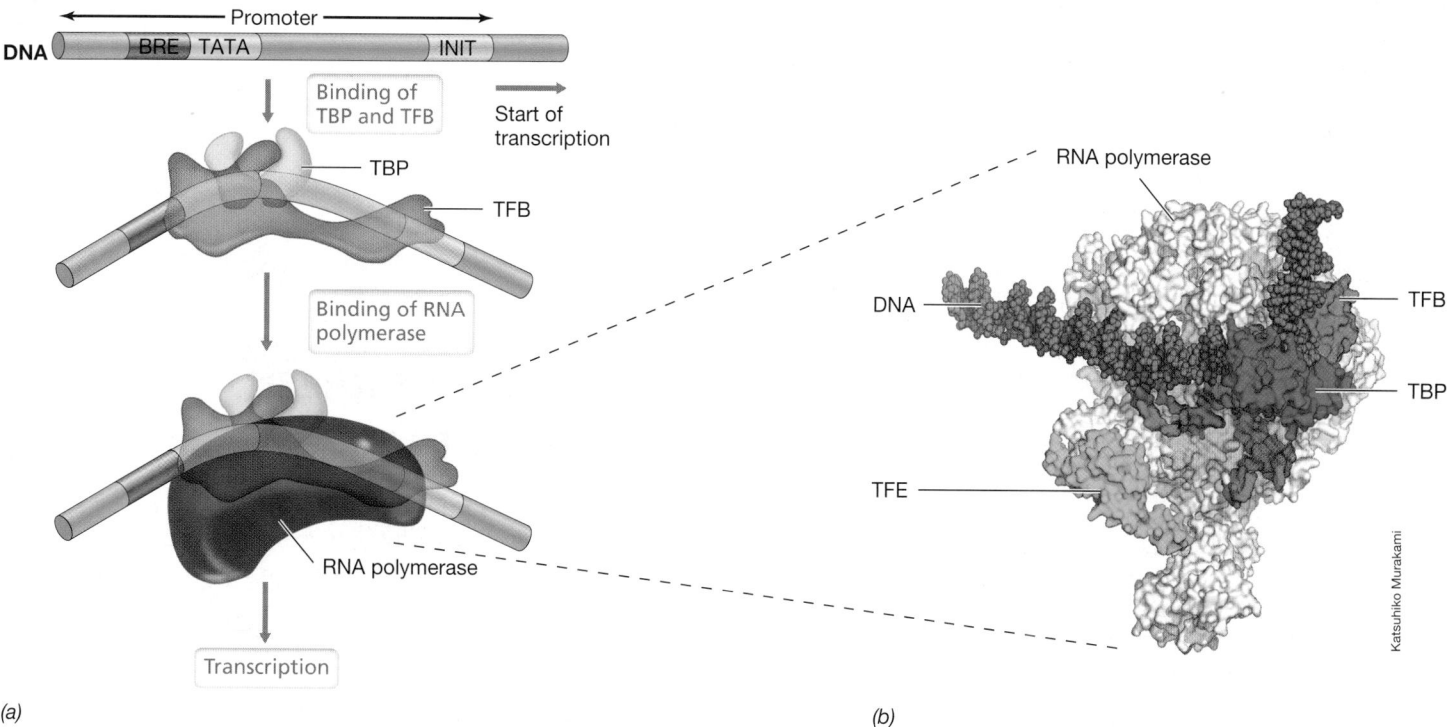

(a)   (b)

Figure 4.25 Promoter architecture and transcription in *Archaea*. *(a)* Three promoter elements are critical for promoter recognition in *Archaea*: the initiator element (INIT), the TATA box, and the B recognition element (BRE). The TATA-binding protein (TBP) binds the TATA box; transcription factor B (TFB) binds to both BRE and INIT. Once both TBP and TFB are in place, RNA polymerase binds. *(b)* Surface representation of the archaeal pre-initiation complex (with TBP and TFB) and including transcription factor E (TFE). TFE is an optional transcription factor frequently associated with the archaeal pre-initiation complex.

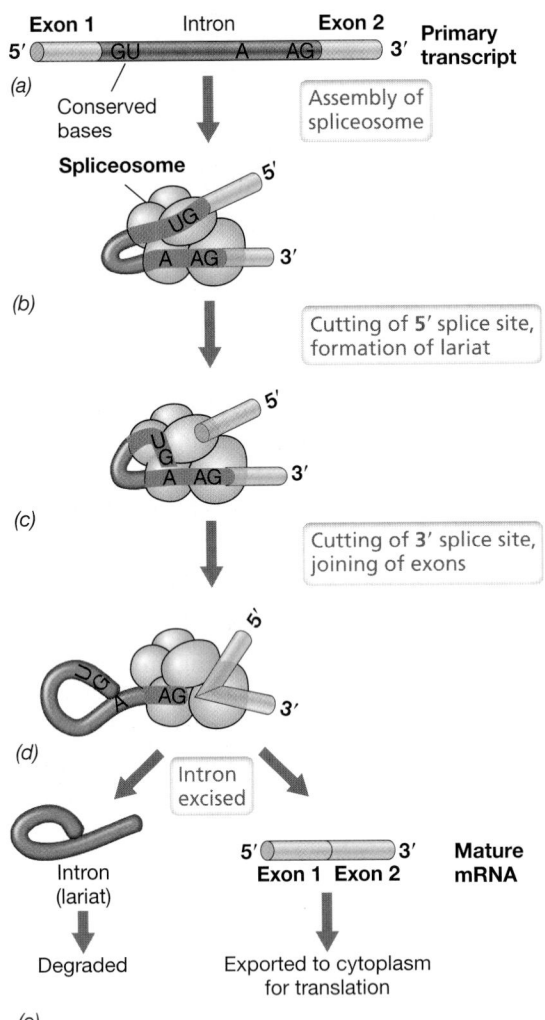

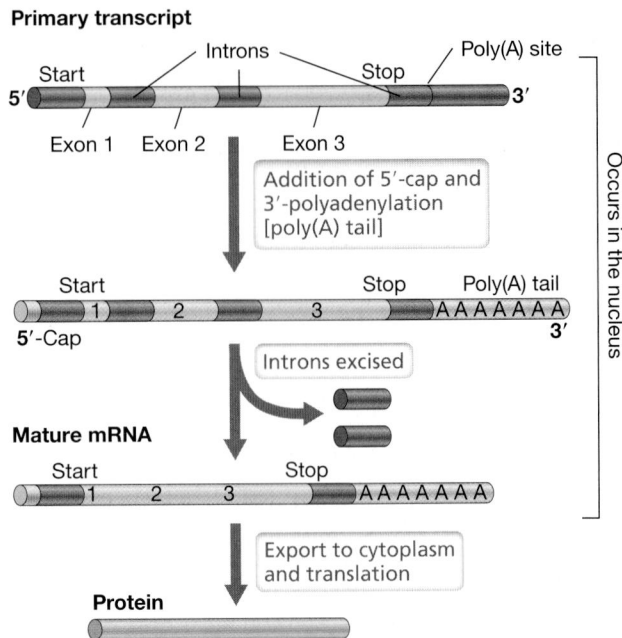

Figure 4.27 Processing of the primary transcript into mature mRNA in eukaryotes. The processing steps include adding a cap at the 5′ end, removing the introns, and clipping the 3′ end of the transcript while adding a poly(A) tail. All these steps are carried out in the nucleus. The location of the start and stop codons to be used during translation are indicated.

We now move on to the culmination of genetic information flow: protein synthesis. Here we will see many events common to all cells with a few exceptions that once again link *Archaea* with *Eukarya*.

Figure 4.26 Activity of the spliceosome. Removal of an intron from the primary transcript of a protein-coding gene in a eukaryote. *(a)* A primary transcript containing a single intron. The sequence GU is conserved at the 5′ splice site, and AG is conserved at the 3′ splice site. There is also an interior A that serves as a branch point. *(b)* Several small ribonucleoprotein particles (shown in light tan) assemble on the RNA to form a spliceosome. Each of these particles contains distinct small RNA molecules that take part in the splicing mechanism. *(c)* The 5′ splice site has been cut with the simultaneous formation of a branch point. *(d)* The 3′ splice site has been cut and the two exons have been joined. Note that overall, two phosphodiester bonds were broken, but two others were formed. *(e)* The final products are the joined exons (the mRNA) and the released intron.

Two other steps in the processing of mRNA in *Eukarya* are unique to this domain, and both steps take place in the nucleus prior to splicing (**Figure 4.27**). The first step, called *capping*, occurs before transcription is complete. Capping is the addition of a methylated guanine nucleotide at the 5′-phosphate end of the mRNA (Figure 4.27). The cap is added in reverse orientation relative to the rest of the mRNA molecule and is needed to initiate translation. The second step, *polyadenylation*, consists of trimming the 3′ end of the primary transcript and adding 100–200 adenine residues, called the *poly(A) tail* (Figure 4.27). The poly(A) tail stabilizes mRNA against nuclease attack and must be removed before the mRNA can be degraded.

MINIQUIZ

- What three major components make up an archaeal promoter?
- What specific eukaryotic enzyme does the archaeal RNA polymerase resemble?
- What steps take place in the processing of eukaryotic RNA?

IV • Protein Synthesis: Translation

Once transcription has occurred, the mRNAs are translated into protein. Translation requires many proteins and RNAs (in addition to mRNA) and a key cellular structure, the ribosome. How these interact to produce a cell's array of proteins is what we consider now, and we begin with a refresher section on the basic properties of proteins.

4.7 Amino Acids, Polypeptides, and Proteins

Proteins play major roles in cell function. Two major classes of proteins are *catalytic proteins* and *structural proteins*. *Enzymes* are the catalysts for chemical reactions that occur in cells (⟐ Section 3.5).

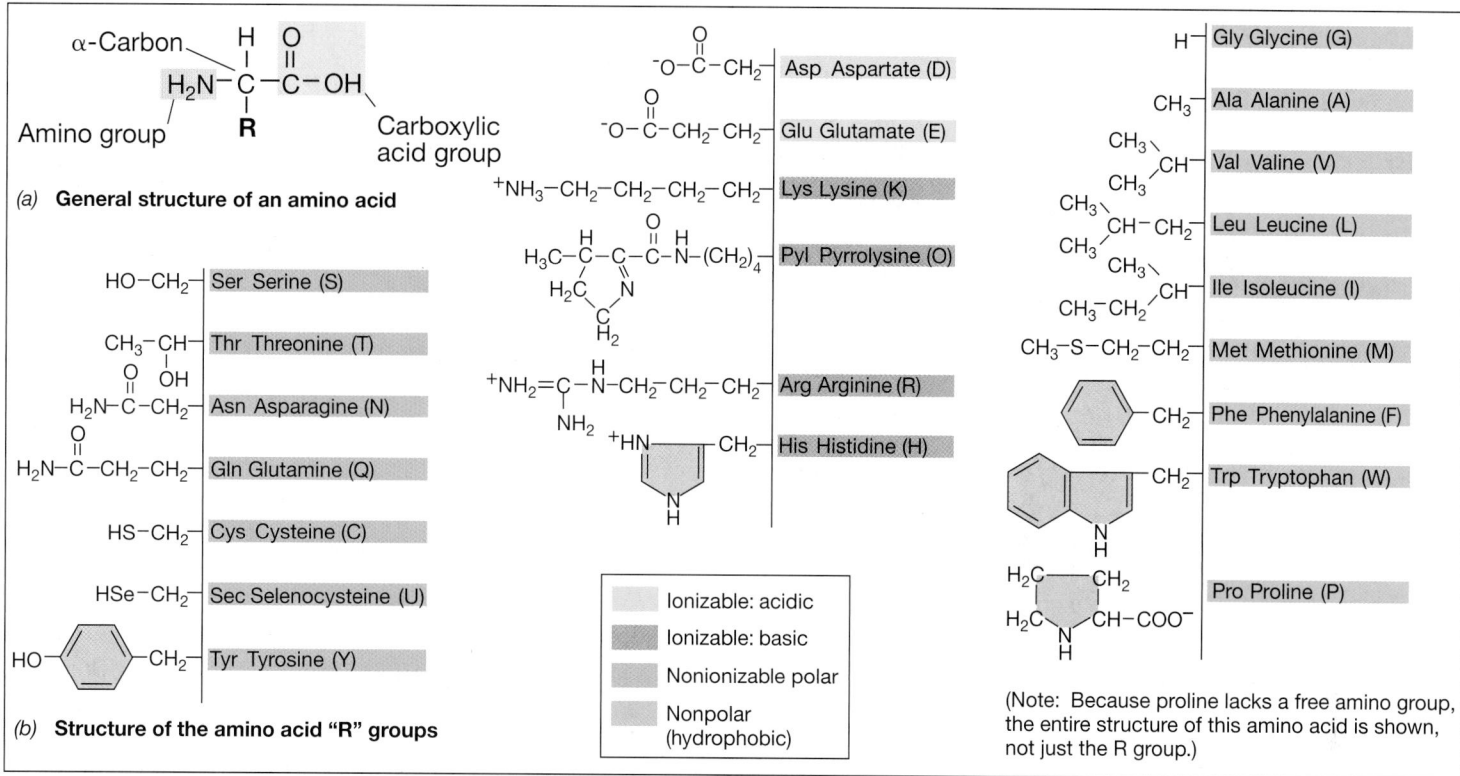

Figure 4.28 Structure of the 22 genetically encoded amino acids. *(a)* General structure. *(b)* R group structure. The three-letter codes for the amino acids are to the left of the names, and the one-letter codes are in parentheses to the right of the names. Pyrrolysine has thus far been found only in certain methanogenic *Archaea* (🔗 Section 17.2).

Structural proteins are parts of the major structures of the cell: membranes, walls, ribosomes, and so on. Regulatory proteins control most cell processes by a variety of mechanisms, including binding to DNA and affecting transcription (Chapter 6). However, all proteins show certain basic features in common.

Composition

Proteins are polymers of **amino acids**, organic compounds that contain both an amino group ($-NH_2$) and a carboxylic acid group ($-COOH$) attached to the α-carbon (**Figure 4.28a**). Bonds between the carboxyl carbon of one amino acid and the amino nitrogen of a second (formed through the elimination of water) are known as **peptide bonds** (**Figure 4.29**). Two amino acids bonded by peptide linkage constitute a *dipeptide*; three amino acids, a *tripeptide*; and so on. When many amino acids are linked, they form a **polypeptide**. A protein consists of one or more polypeptides. The number of amino acids differs greatly from one protein to another, from as few as 15 to as many as 10,000 (see page 102).

Each amino acid has a unique side chain (abbreviated R), which governs the chemical properties of the amino acid. Side chains vary considerably, from as simple as a hydrogen atom in the amino acid glycine to aromatic rings in phenylalanine, tyrosine, and tryptophan (Figure 4.28b). Amino acids with chemically related side chains often show similar chemical properties and are thus grouped into related amino acid "families" (Figure 4.28b). For example, the side chain may contain a carboxylic acid group, as in aspartic acid or glutamic acid, rendering the amino acid *acidic*.

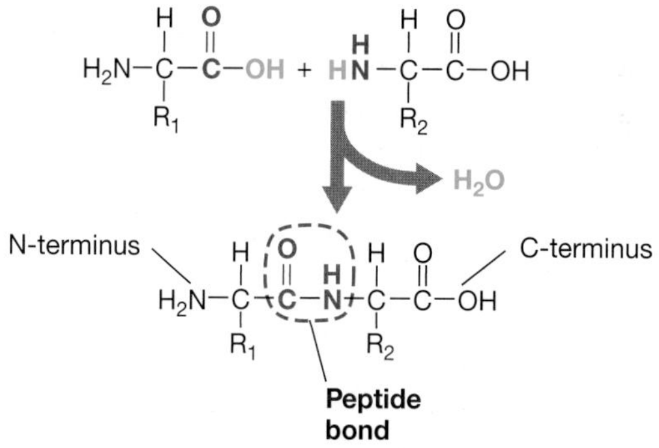

Figure 4.29 Peptide bond formation. R_1 and R_2 refer to the side chains of the amino acids. Note that following peptide bond formation, a free OH group is present at the C-terminus for formation of the next peptide bond.

Others contain additional amino groups, making them positively charged and *basic*. Several amino acids contain hydrophobic side chains and are grouped together as *nonpolar* amino acids. Cysteine contains a sulfhydryl group ($-SH$). Using their sulfhydryl groups, two cysteines can form a disulfide linkage ($R-S-S-R$) that connects two polypeptide chains.

Protein Diversity and Structures

The diversity of chemically distinct amino acids makes possible an enormous number of structurally unique proteins that can have

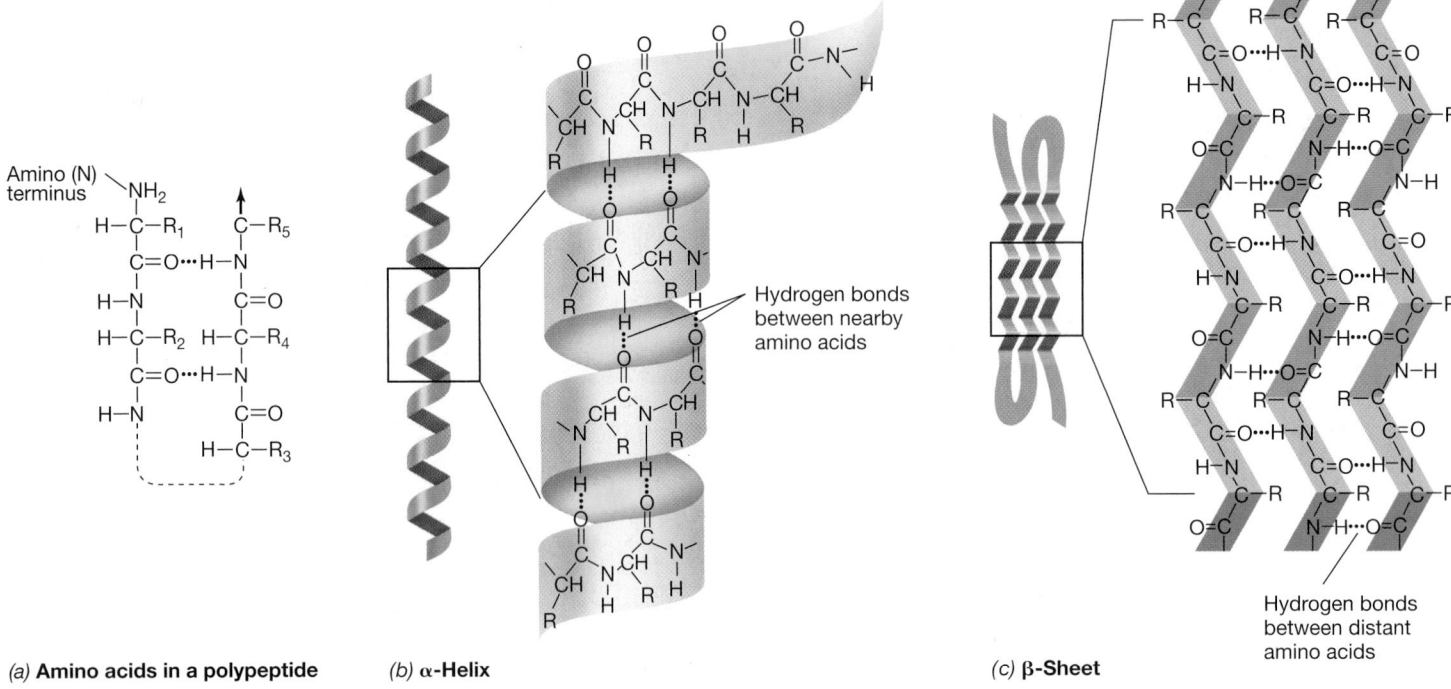

(a) **Amino acids in a polypeptide**   (b) **α-Helix**   (c) **β-Sheet**

Figure 4.30 Secondary structure of polypeptides. *(a)* Hydrogen bonding in protein secondary structure. R represents the side chain of the amino acid. *(b)* α-Helix secondary structure. *(c)* β-Sheet secondary structure. Note that the hydrogen bonding is between atoms in the peptide bonds and does not involve the R groups.

widely different biochemical properties. If one assumes that an average polypeptide contains 300 amino acids, then 22^{300} different polypeptide sequences are theoretically possible. No cell has anywhere near this many different proteins. A cell of *Escherichia coli* contains around 2000 different kinds of proteins, with the exact number being highly dependent on the resources (nutrients) and growth conditions employed.

The linear sequence of amino acids in a polypeptide is its *primary structure*. This ultimately determines the folding pattern of the polypeptide, which in turn determines its biological activity. Even as little as a single amino acid change in the primary structure of a protein can affect its activity. Once formed, a polypeptide proceeds to fold to form a more stable structure. Hydrogen bonding between the oxygen and nitrogen atoms of two peptide bonds generates the *secondary structure*, either as an *α-helix* (imagine a polypeptide wound around a cylinder) or as a *β-sheet* (a repeated "back and forth" type of folding) (**Figure 4.30**). A single polypeptide can contain regions, called *domains*, of α-helix and regions of β-sheet secondary structure. The type of folding and its location in the molecule are determined by the primary structure and the available opportunities for hydrogen bonding.

Interactions between the R groups of the amino acids in a polypeptide generate higher-order structures. A protein's **tertiary structure** depends largely on hydrophobic interactions, with lesser contributions from hydrogen bonds, ionic bonds, and disulfide bonds, and generates the overall three-dimensional shape of the polypeptide (**Figure 4.31**). Many proteins consist of two or more polypeptides and thus show **quaternary structure**. In such proteins, the quaternary structure describes the number and secondary structure of polypeptides (referred to as *subunits*) in the

molecule. Quaternary structures may be stabilized by various interactions and also by disulfide bonds; if cysteines located in two different polypeptides are joined, the disulfide bond links the two subunits.

When proteins are exposed to extremes of heat or pH or to certain chemicals that affect their folding, they may undergo **denaturation**. This results in the loss of a protein's secondary, tertiary, and quaternary structure along with its biological properties. However, because peptide bonds are usually not broken, the denatured polypeptide retains its primary structure. Depending on the severity of the denaturing conditions, the polypeptide may

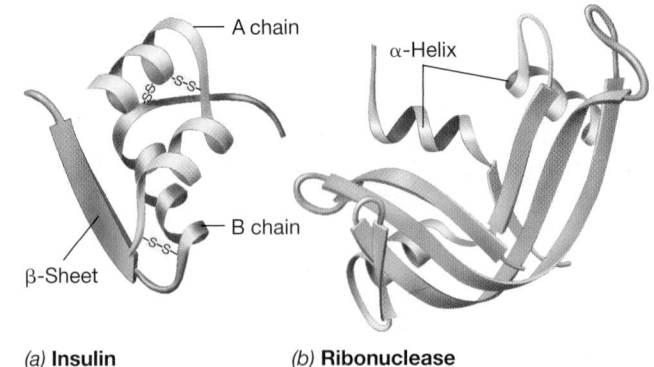

(a) **Insulin**   (b) **Ribonuclease**

Figure 4.31 Tertiary structure of polypeptides. *(a)* Insulin, a protein containing two polypeptide chains; note how the B chain contains both α-helix and β-sheet secondary structure and how disulfide linkages (S–S) help in dictating folding patterns (tertiary structure). *(b)* Ribonuclease, a large protein with several regions of α-helix and β-sheet secondary structure.

properly refold after the denaturant is removed. However, if refolding is not correct, the protein is effectively "dead" and is degraded by proteases to release its amino acids for new protein synthesis.

MINIQUIZ

- Draw the structure of a dipeptide containing the amino acids alanine and tyrosine and outline the peptide bond.
- Differentiate between the different classes of protein structure.
- What is denaturation and why is the process harmful to a cell?

4.8 Transfer RNA

With a primer on proteins behind us, we now consider protein synthesis. But to do so, we must first understand the role of transfer RNA (tRNA). Transfer RNAs function to carry amino acids to the translation machinery. To ensure that they carry the correct amino acid, each tRNA contains a specific three-nucleotide sequence called the **anticodon**, the group of three bases that recognizes a codon (a three-base code for an amino acid) on the mRNA (Section 4.10). The correct amino acid (called the *cognate* amino acid) is linked to a specific tRNA by an enzyme called an **aminoacyl-tRNA synthetase**. For each amino acid, a separate aminoacyl-tRNA synthetase exists that specifically binds both the cognate amino acid and the tRNA that contains the corresponding anticodon, thus ensuring that each tRNA receives its correct amino acid.

General Structure of Transfer RNA

There are about 60 different tRNAs in prokaryotic cells and 100–110 in human cells. Transfer RNAs are short (73–93 nucleotides),

single-stranded molecules that contain extensive secondary structure. Certain base sequences and secondary structures are invariant for tRNAs, whereas other parts are variable. Transfer RNAs also contain some purine and pyrimidine bases that are modified from the bases found in other classes of RNA, and these modifications occur after transcription. These unusual bases include pseudouridine (ψ), inosine, dihydrouridine (D), ribothymidine, methyl guanosine, dimethyl guanosine, and methyl inosine. The mature and active tRNA also contains extensive double-stranded regions formed by internal base pairing when the single-stranded molecule folds back on itself (**Figure 4.32**).

A tRNA is often depicted in the shape of a cloverleaf (Figure 4.32*a*). Some regions of tRNA secondary structure are named after the modified bases found there (for example, the TψC and D loops) or after their functions (for example, the anticodon loop and acceptor stem). The three-dimensional model of a tRNA shown in Figure 4.32*b* is a more realistic view of the molecule and shows how bases that appear widely separated in the cloverleaf model may actually be much closer together when viewed in 3D. This close proximity allows some of the bases in one loop to pair with bases in another loop (Figure 4.32*b*).

At the 3′ end (the acceptor end) of all tRNAs are three unpaired nucleotides. The sequence of these three nucleotides is always *c*ytosine-*c*ytosine-*a*denine (CCA), and they are absolutely essential for function. However, in most organisms the 3′ CCA is not encoded in the tRNA gene on the chromosome, but instead, each nucleotide is added sequentially by a protein called *CCA-adding enzyme*, using CTP and ATP as substrates. The cognate amino acid is then covalently attached to the terminal adenosine of the CCA end of its corresponding tRNA. From this location, the amino acid

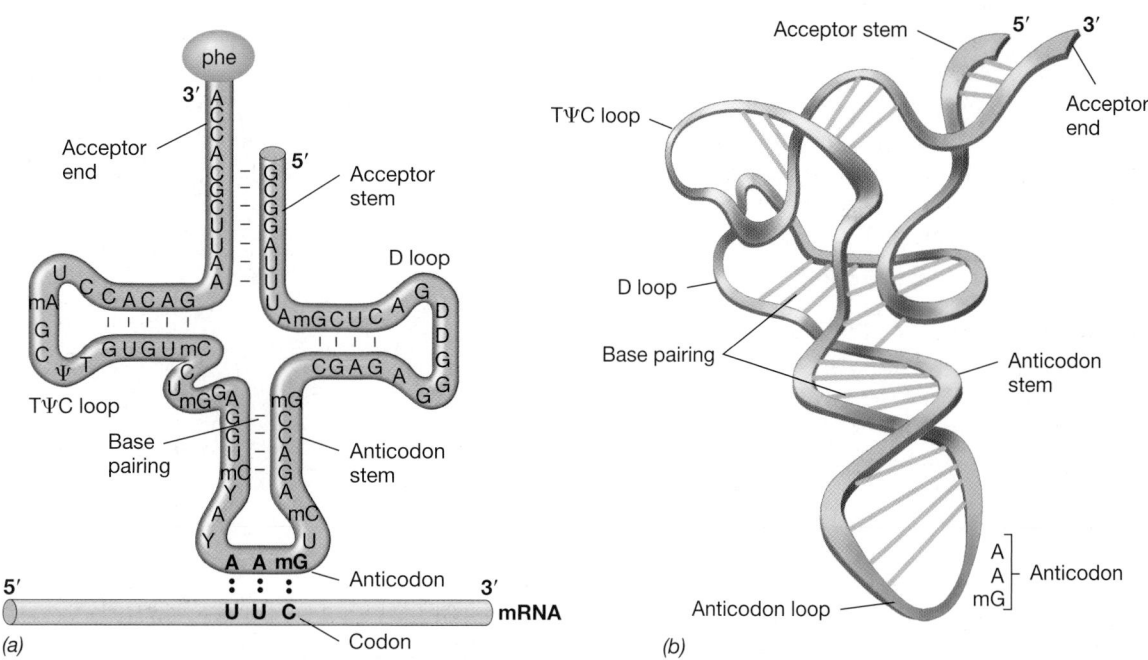

Figure 4.32 Structure of a transfer RNA. *(a)* The conventional cloverleaf structural drawing of yeast phenylalanine tRNA. The amino acid is attached to the ribose of the terminal A at the acceptor end. A, adenine; C, cytosine; U, uracil; G, guanine; T, thymine; ψ, pseudouracil; D, dihydrouracil; m, methyl; Y, a modified purine. *(b)* In fact, the tRNA molecule folds so that the D loop and TψC loops are close together and associate by hydrophobic interactions.

is incorporated into the growing polypeptide chain on the ribosome by a mechanism to be described in Section 4.9.

Recognition, Activation, and Charging of tRNAs

Recognition of the correct tRNA by an aminoacyl-tRNA synthetase is obviously crucial to the fidelity of translation and requires that specific contacts be made between regions of the tRNA and the synthetase (**Figure 4.33**). As might be expected because of its unique sequence, the anticodon of the tRNA is important in recognition by the synthetase. However, other contact sites between the tRNA and the synthetase are also important, including parts of the acceptor stem and D loop of the tRNA (Figure 4.32*a*).

The specific reaction between amino acid and tRNA catalyzed by the aminoacyl-tRNA synthetase begins with *activation* of the amino acid by reaction with ATP:

$$\text{Amino acid} + \text{ATP} \leftrightarrow \text{aminoacyl—AMP} + \text{P—P}$$

The aminoacyl-AMP intermediate formed remains bound to the tRNA synthetase until collision with the appropriate tRNA molecule. Then, as shown in Figure 4.33*a*, the activated amino acid is attached to the CCA stem of its tRNA to form a *charged tRNA*:

$$\text{Aminoacyl—AMP} + \text{tRNA} \leftrightarrow \text{Aaminoacyl—tRNA} + \text{AMP}$$

The pyrophosphate (PP$_i$) formed in the first reaction is split into two molecules of inorganic phosphate. Because ATP is used and AMP is formed in these reactions, a total of *two* energy-rich phosphate bonds are expended to charge a tRNA with its cognate amino acid. After activation and charging, the aminoacyl-tRNA leaves the synthetase. In the next step, it will be bound by a ribosome where actual polypeptide synthesis occurs.

— MINIQUIZ —

- What is the function of the anticodon of a tRNA?
- What is the function of the acceptor stem of a tRNA?
- What steps are required to form a charged tRNA?

4.9 Translation and the Genetic Code

The heart of genetic information transfer is the correspondence between the nucleic acid template and the amino acid sequence of a polypeptide. This correspondence is rooted in the **genetic code**.

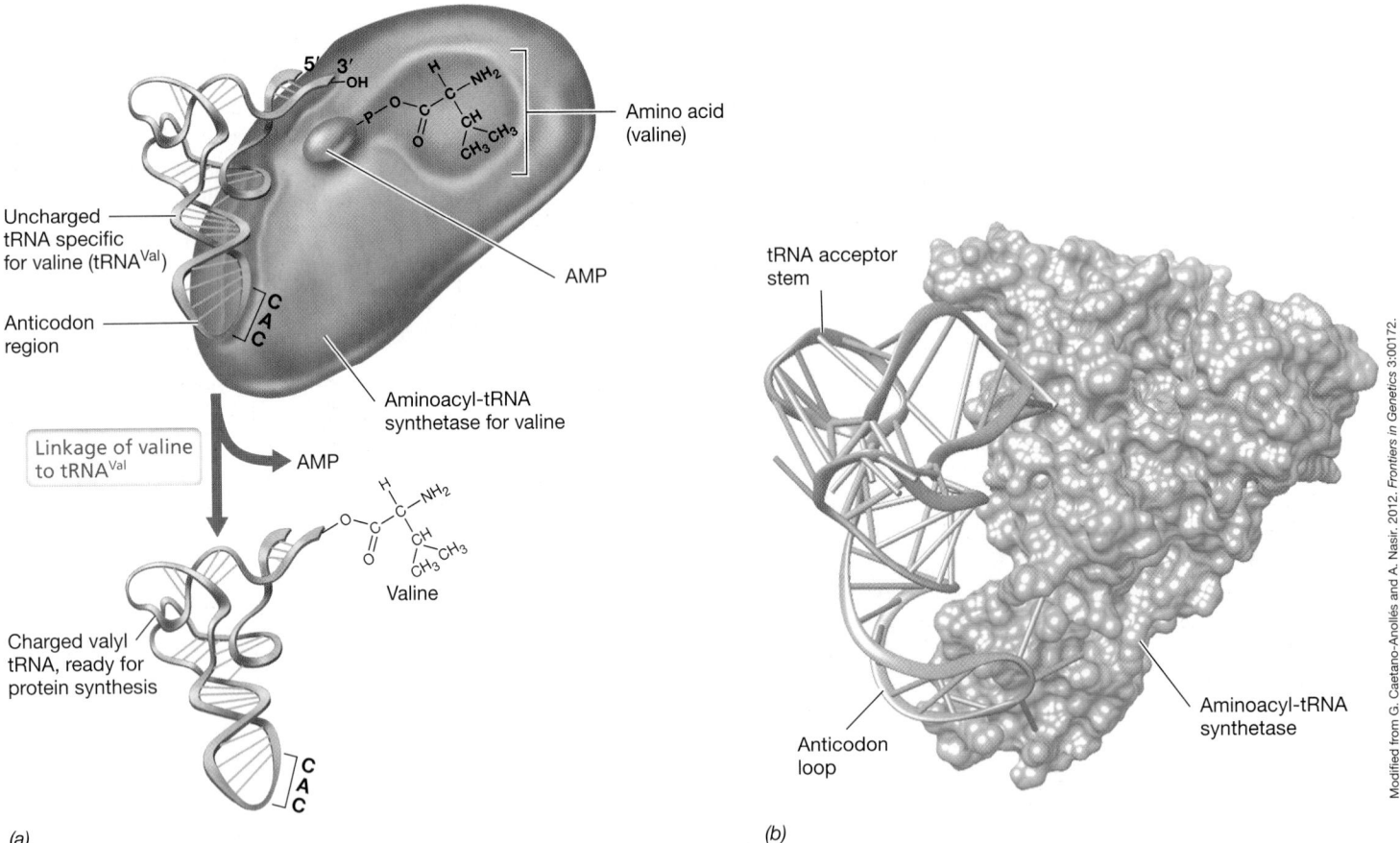

(a) *(b)*

Modified from G. Caetano-Anollés and A. Nasir. 2012. *Frontiers in Genetics* 3:00172.

Figure 4.33 Aminoacyl-tRNA synthetase. *(a)* Mode of activity of an aminoacyl-tRNA synthetase. Recognition of the correct tRNA by a particular synthetase involves contacts between specific nucleic acid sequences in the D loop and acceptor stem of the tRNA and specific amino acids of the synthetase. In this diagram, valyl-tRNA synthetase (specific for the amino acid valine) is shown catalyzing the final step of the reaction, where the valine in valyl-AMP is transferred to tRNA. *(b)* A computer model showing the interaction of the prolyl-tRNA synthetase from *Thermus thermophilus* with its tRNA.

TABLE 4.4 The genetic code as expressed by triplet base sequences of mRNA

Codon	Amino acid	Codon	Amino acid	Codon	Amino acid	Codon	Amino acid
UUU	Phenylalanine	UCU	Serine	UAU	Tyrosine	UGU	Cysteine
UUC	Phenylalanine	UCC	Serine	UAC	Tyrosine	UGC	Cysteine
UUA	Leucine	UCA	Serine	UAA	None (stop signal)	UGA	None (stop signal)
UUG	Leucine	UCG	Serine	UAG	None (stop signal)	UGG	Tryptophan
CUU	Leucine	CCU	Proline	CAU	Histidine	CGU	Arginine
CUC	Leucine	CCC	Proline	CAC	Histidine	CGC	Arginine
CUA	Leucine	CCA	Proline	CAA	Glutamine	CGA	Arginine
CUG	Leucine	CCG	Proline	CAG	Glutamine	CGG	Arginine
AUU	Isoleucine	ACU	Threonine	AAU	Asparagine	AGU	Serine
AUC	Isoleucine	ACC	Threonine	AAC	Asparagine	AGC	Serine
AUA	Isoleucine	ACA	Threonine	AAA	Lysine	AGA	Arginine
AUG (start)[a]	Methionine	ACG	Threonine	AAG	Lysine	AGG	Arginine
GUU	Valine	GCU	Alanine	GAU	Aspartic acid	GGU	Glycine
GUC	Valine	GCC	Alanine	GAC	Aspartic acid	GGC	Glycine
GUA	Valine	GCA	Alanine	GAA	Glutamic acid	GGA	Glycine
GUG	Valine	GCG	Alanine	GAG	Glutamic acid	GGG	Glycine

[a]AUG encodes N-formylmethionine at the beginning of polypeptide chains of *Bacteria*.

An mRNA triplet of three bases, called a **codon**, encodes each specific amino acid (the codons themselves are encoded by the organism's genome). The 64 possible codons (four bases taken three at a time = 4³) are shown in **Table 4.4**. Note that in addition to the codons for amino acids, there are also codons for starting and stopping translation. Here we focus on translation in *Bacteria*.

Properties of the Genetic Code

There are 22 naturally occurring amino acids and because there are 64 codons, several amino acids can be encoded by more than one codon. A code such as this that lacks one-to-one correspondence between "word" (that is, the amino acid) and code (codon) is called a *degenerate code*. A codon is recognized by specific base pairing with a complementary sequence on the *anticodon*, located on a tRNA (Section 4.8). If this base pairing were always the standard pairing of A with U and G with C, then at least one specific tRNA would be needed to recognize each codon. In some cases, this is true. For instance, there are six different tRNAs in *Escherichia coli* for the amino acid leucine, one for each codon (Table 4.4). By contrast, some tRNAs can recognize more than one codon. For example, although there are two lysine codons in *E. coli*, there is only one lysyl tRNA, whose anticodon can base-pair with *either* AAA or AAG. In these cases, the anticodon forms standard base pairs at only the first two positions of the codon and tolerates irregular base pairing at the third position. This phenomenon is called **wobble** and is illustrated in **Figure 4.34**.

If an amino acid is encoded by multiple codons, the codons are typically closely related in base sequence, usually differing at only their third position (Table 4.4) to allow for wobble (Figure 4.34). Interestingly, not all multiple codons for a given amino acid are

used at the same frequency, leading to a **codon bias** that varies from organism to organism. Codon bias is correlated with a corresponding bias in the concentration of different tRNA molecules. That is, a tRNA whose anticodon corresponds to a rarely used codon is typically produced at low levels.

Start and Stop Codons and Reading Frames

Messenger RNA is translated beginning with its **start codon** (AUG, Table 4.4), which encodes a chemically modified methionine in *Bacteria* called *N-formylmethionine* (although AUG at the *beginning* of a coding region encodes *N*-formylmethionine, AUG *within* the coding region encodes methionine). By contrast, *Archaea* and *Eukarya* insert an unmodified methionine as the first amino acid in a polypeptide.

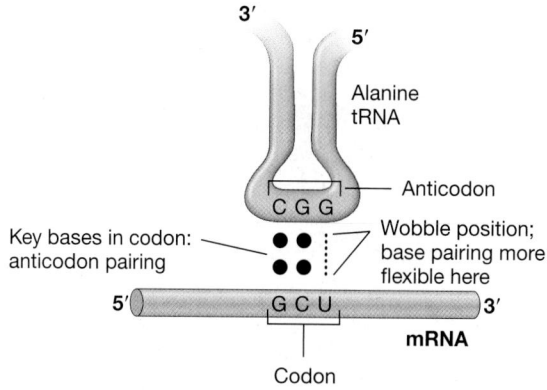

Figure 4.34 The wobble concept. Base pairing is more flexible for the third base of the codon than for the first two. Only a portion of the tRNA is shown here.

With a triplet code it is critical for translation to begin at the correct nucleotide. If it does not, the whole reading frame of the mRNA will be shifted and thus an entirely different (and likely inactive) protein will be made. Alternatively, if the shift introduces a stop codon (Table 4.4) into the reading frame, the polypeptide will terminate prematurely. The reading frame that when translated yields the polypeptide encoded by the gene is called the *0* (zero) *frame*; the other possible reading frames (−1 and +1) do not encode the same amino acid sequence (**Figure 4.35**). Reading frame fidelity is governed by interactions between mRNA and rRNA within the ribosome. In *Bacteria*, ribosomal RNA recognizes a specific AUG on the mRNA as a start codon with the aid of an upstream sequence in the mRNA called the *ribosome-binding site* (RBS) (also called the *Shine–Dalgarno sequence* after its discoverers). This upstream alignment requirement explains why some mRNAs from *Bacteria* can use other start codons, such as GUG. However, because of the RBS, even these unusual start codons direct the incorporation of *N*-formylmethionine as the initiator amino acid (see Figure 4.36).

The codons UAA, UAG, and UGA (Table 4.4) are **stop codons**, and they signal the termination of translation of a protein-coding sequence on the mRNA. Stop codons are also called **nonsense codons**, because they interrupt the "sense" of the growing polypeptide when they terminate translation. In a few rare cases in *Bacteria* and *Archaea*, nonsense codons encode the unusual amino acids selenocysteine and pyrrolysine (Figure 4.28). When this occurs, specific tRNAs are employed whose anticodons read these stop codons. What controls this unusual occurrence is a recognition sequence just downstream of the now-coding stop codon that signals the incorporation of tRNAs containing selenocysteine or pyrrolysine rather than stopping translation. A few other microbes use conventional stop codons to encode amino acids, but in these cases the organisms have simply dispensed with using these particular stop codons as translational stop sites.

If an mRNA can be translated, it is because it contains an **open reading frame** (ORF): a start codon followed by a number of codons and then a stop codon in the same reading frame as the start codon. Using computational methods, a DNA base sequence can be scanned to search for open reading frames. In addition to looking for start and stop codons, computer analyses may include a search for promoters and ribosome-binding sequences as well to confirm the ORF as protein encoding. The search for ORFs is central to the field of genomics (Chapter 9), for if an unknown piece of DNA has been sequenced, the presence of an ORF implies that it can encode protein.

— MINIQUIZ —

- What are start codons and stop codons? Why is it important for the ribosome to read "in frame"?
- What is codon bias?
- If you were given a nucleotide sequence, how would you find ORFs?

4.10 The Mechanism of Protein Synthesis

Protein synthesis is a dynamic process and can be broken down into three major steps: *initiation*, *elongation*, and *termination*. In addition to mRNA, tRNA, and ribosomes, translation requires a number of initiation, elongation, and termination proteins and the energy-rich compound guanosine triphosphate (GTP) to provide the energy for the process.

Ribosomes and the Initiation of Translation

Ribosomes are the sites of protein synthesis. A cell may have many thousands of ribosomes, the number increasing at higher growth rates. Each ribosome consists of two subunits. *Bacteria* and *Archaea* have 30S and 50S ribosomal subunits that yield intact 70S ribosomes. Each ribosomal subunit contains specific ribosomal RNAs and ribosomal proteins. The 30S subunit contains 16S rRNA and 21 proteins, and the 50S subunit contains 5S and 23S rRNA and 31 proteins. Thus, in *Escherichia coli*, there are 52 distinct ribosomal proteins, most present at one copy per ribosome. The ribosome is a highly dynamic structure, and its subunits alternately associate and dissociate during the translational process and also interact with many other proteins. In addition, several cytoplasmic proteins called *translation factors* are essential for translation and interact with the ribosome at various stages of the translational process.

In *Bacteria*, initiation of protein synthesis begins with a free 30S ribosomal subunit (**Figure 4.36**). From this, an *initiation complex* forms consisting of the 30S subunit, mRNA, formylmethionine tRNA (the initiator tRNA in *Bacteria*; after polypeptide completion, the formyl group is removed), and the initiation factors IF1, IF2, and IF3. GTP is also required for this step. Next, a 50S ribosomal subunit is added to the initiation complex to form the active 70S ribosome. Just preceding the start codon on the mRNA is a sequence of three to nine nucleotides that compose the ribosome-binding site (RBS in Figure 4.36). This site is toward the 5′ end of the mRNA and is complementary to base sequences in the 3′ end of the 16S rRNA, which is part of the ribosome. Base pairing between these two RNAs holds the ribosome–mRNA

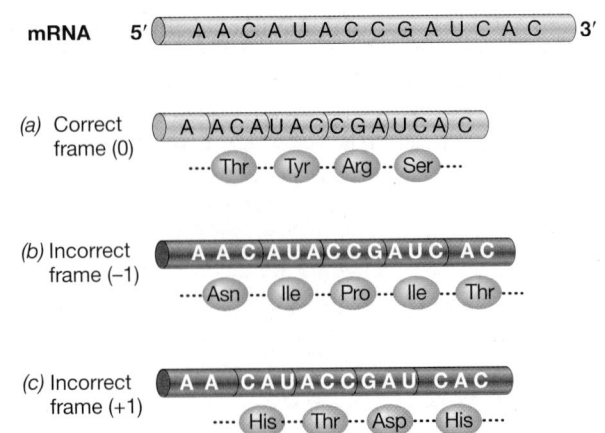

Figure 4.35 Possible reading frames in an mRNA. An interior sequence of an mRNA is shown. *(a)* The amino acids that would be encoded if the ribosome is in the correct reading frame (designated the "0" frame). *(b)* The amino acids that would be encoded by this region of the mRNA if the ribosome were in the −1 reading frame. *(c)* The amino acids that would be encoded if the ribosome were in the +1 reading frame.

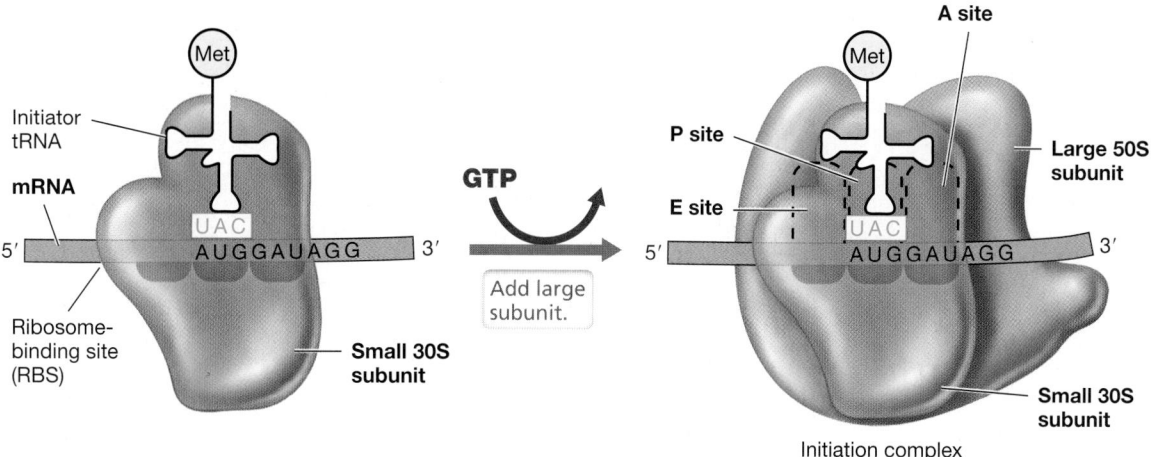

Figure 4.36 The ribosome and initiation of protein synthesis. The mRNA and initiator tRNA, carrying *N*-formylmethionine ("Met"), bind first to the small subunit of the ribosome. Initiation factors (not shown) use energy from GTP to promote the addition of the large ribosomal subunit. The initiator tRNA starts out in the P site (labeled in the structure on the right-hand side of the arrow).

complex securely together in the correct reading frame. Polycistronic mRNA (Section 4.5) contains several RBS sequences, one upstream of each coding sequence. This allows bacterial ribosomes to translate several genes on the same mRNA because the ribosome can locate each initiation site within a message by binding to its RBS.

Elongation, Translocation, and Termination

During translation, the mRNA threads through the ribosome bound to the 30S subunit. The ribosome contains other sites where the tRNAs interact. Two of these sites are located primarily on the 50S subunit and are termed the *A* (*acceptor*) *site* and the *P* (*peptide*) *site* (**Figure 4.37**). The A site is where the incoming charged tRNA first attaches, and the loading of a tRNA into the A site is assisted by the elongation factor EF-Tu. The P site is where the growing polypeptide chain is attached to the prior tRNA. During peptide bond formation, the growing polypeptide chain moves to the tRNA at the A site as a new peptide bond is formed. In addition to EF-Tu, elongation factor EF-Ts, as well as more GTP, is required (Figure 4.37).

Following elongation, the tRNA holding the polypeptide is translocated from the A site to the P site, thus opening the A site for a new charged tRNA; this requires elongation factor EF-G and one molecule of GTP for each translocation event (Figure 4.37). In each translocation, the ribosome advances three nucleotides (one codon) along the mRNA, exposing a new codon at the A site. Translocation pushes the now amino acid–free tRNA to a third site, called the E (exit) site, and it is from here that the tRNA is released from the ribosome. The precision of the translocation step is critical to the accuracy of protein synthesis. That is, the ribosome must move *exactly* one codon at each step or the fidelity of translation will be compromised.

Several ribosomes can translate a single mRNA molecule simultaneously, forming a complex called a *polysome* (**Figure 4.38**). Polysomes increase both the speed and efficiency of translation because each ribosome in the polysome makes a complete polypeptide.

Note in Figure 4.38 how ribosomes in the polysome that are closest to the 5′ end (the beginning) of the mRNA molecule have short polypeptides attached to them because only a few codons have been read, while ribosomes closest to the 3′ end of the mRNA have nearly finished polypeptides.

Translation terminates when the ribosome reaches a stop codon (Table 4.4) because no tRNA binds to a stop codon. Instead, proteins called *release factors* (RFs) recognize the stop codon and cleave the attached polypeptide from the final tRNA, releasing the finished product. Following this, the ribosomal subunits dissociate, and the 30S and 50S subunits are then free to form new initiation complexes (Figure 4.36) and repeat the process.

Role of Ribosomal RNA in Protein Synthesis

Ribosomal RNA plays major roles in all stages of translation, from initiation to termination. By contrast, the primary role of the proteins in the ribosome is to form a scaffold to position key sequences in the ribosomal RNAs. In *Bacteria*, 16S rRNA facilitates initiation by base pairing with the ribosome-binding site on the mRNA, and, along with ribosomal proteins, holds the mRNA in position on either side of the A and P sites. Ribosomal RNA also plays a role in ribosome subunit association, as well as in positioning tRNAs on the ribosome (Figures 4.36 and 4.37). Although charged tRNAs recognize the correct codon by codon–anticodon base pairing (Figure 4.34), they are also bound to the ribosome by interactions between the anticodon stem–loop of the tRNA and specific sequences in the 16S rRNA. Moreover, the acceptor end of the tRNA (Figures 4.36 and 4.37) base-pairs with sequences in 23S rRNA.

In addition to roles in mRNA alignment and translocation along the transcript, rRNA also catalyzes the actual formation of peptide bonds. The peptidyl transferase reaction occurs on the 50S subunit of the ribosome and is catalyzed solely by 23S rRNA. This rRNA also plays a role in translocation and interacts with the elongation factors. Thus, in addition to its role as the structural backbone of the ribosome, ribosomal RNA plays a major catalytic role in the translation process.

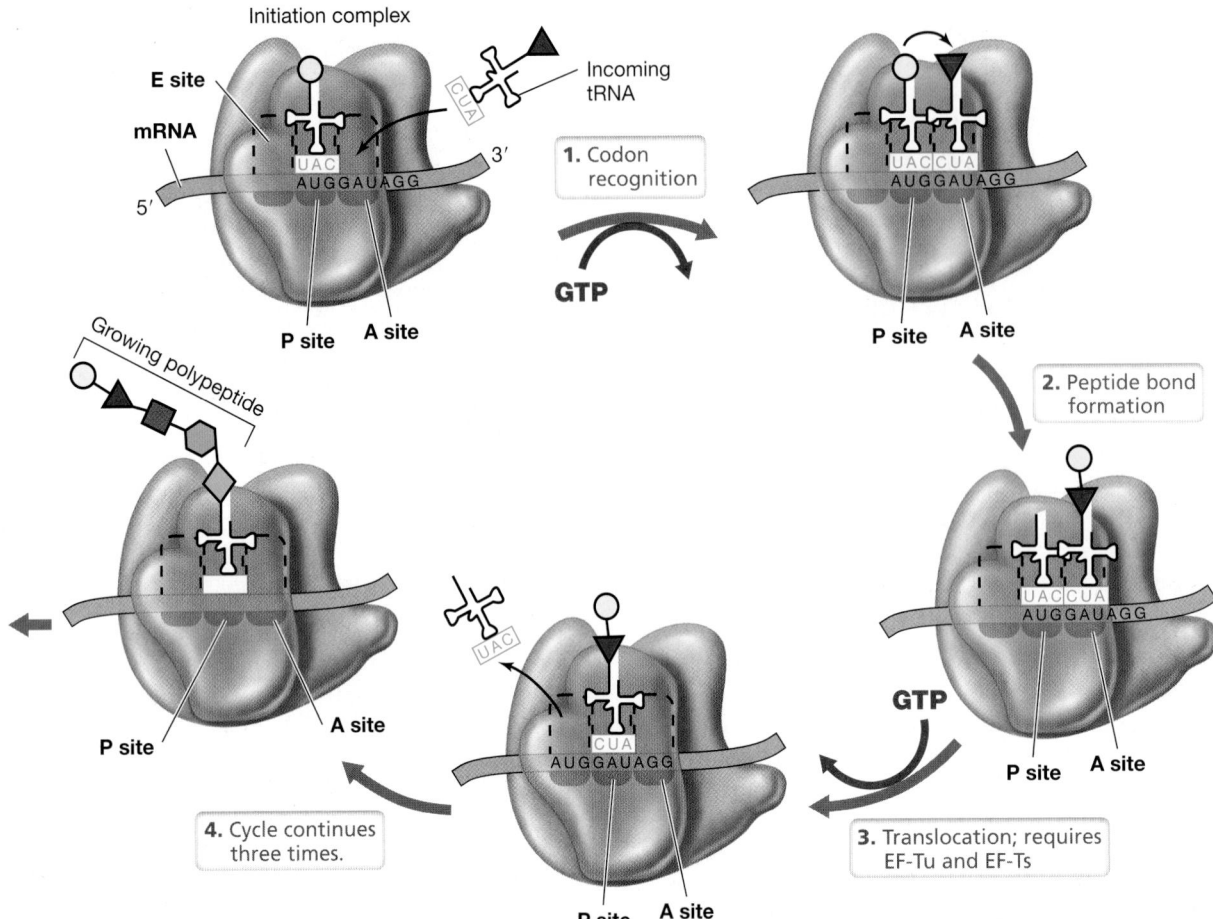

Figure 4.37 Elongation cycle of translation. 1. Elongation factors (not shown) use GTP to install the incoming tRNA into the A site. 2. Peptide bond formation is then catalyzed by the 23S rRNA. 3. Translocation of the ribosome along the mRNA from one codon to the next requires hydrolysis of another GTP and results in movement of the tRNA with the growing peptide to the P site. The outgoing tRNA is released from the E site. 4. The next charged tRNA binds to the A site and the cycle repeats. The genetic code, expressed as codons of mRNA, is shown in Table 4.4.

Freeing Trapped Ribosomes

A defective mRNA that lacks a stop codon cannot be properly translated. Such a defect may arise, for example, from a mutation that removed the stop codon, from defective synthesis of the mRNA, or when partial degradation of an mRNA occurs before it is translated. If a ribosome reaches the end of an mRNA molecule and there is no stop codon, release factor cannot bind and the ribosome cannot be released from the mRNA; the ribosome is effectively "trapped."

To deal with this problem, bacterial cells produce a small RNA molecule called *tmRNA* that frees stalled ribosomes (**Figure 4.39**). The "tm" in its name refers to the fact that tmRNA mimics both tRNA, in that it carries an amino acid (alanine), and mRNA, in that it contains a short stretch of RNA that can be translated. When tmRNA collides with a stalled ribosome, it binds alongside the defective mRNA. Protein synthesis can

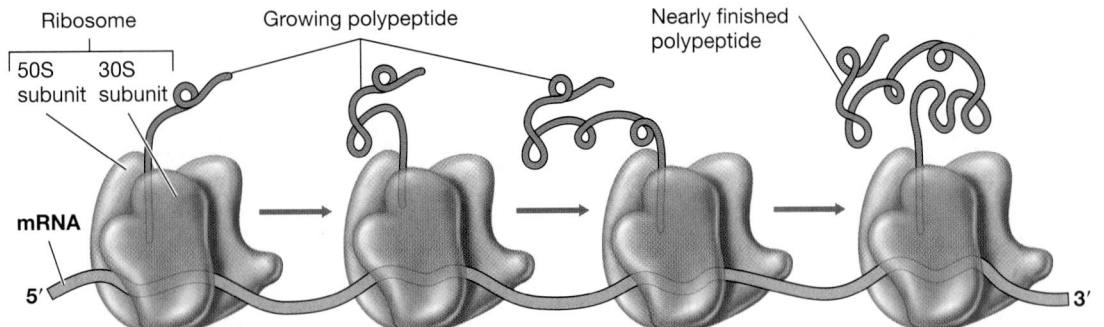

Figure 4.38 Polysomes. Translation by several ribosomes on a single messenger RNA forms the polysome. Note how the ribosomes nearest the 5′ end of the message are at an earlier stage in the translation process than ribosomes nearer the 3′ end, and thus only a relatively short portion of the final polypeptide has been made.

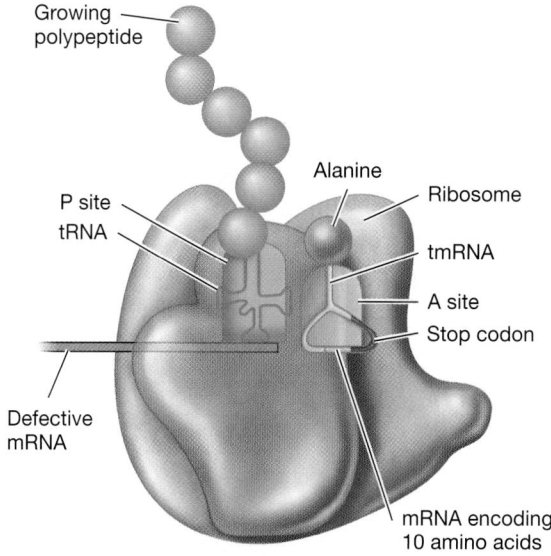

Growing polypeptide

P site tRNA

Alanine

Ribosome

tmRNA

A site

Stop codon

Defective mRNA

mRNA encoding 10 amino acids

Figure 4.39 Freeing of a stalled ribosome by tmRNA. A defective mRNA lacking a stop codon stalls a ribosome that has a partly synthesized polypeptide attached to a tRNA (blue) in the P site. Binding of tmRNA (yellow) in the A site allows translation to continue up to the stop codon provided by the tmRNA.

then proceed, first by adding the alanine on the tmRNA and then by translating the short tmRNA message. The tmRNA contains a stop codon that allows release factor to bind and disassemble the ribosome. The protein made as a result of this rescue operation is defective and is subsequently degraded. This is accomplished by a short sequence of amino acids encoded by tmRNA and added to the end of the defective protein; the sequence is a signal for a specific protease to degrade the protein. Thus, through the activity of tmRNA, stalled ribosomes are not inactivated but instead are freed up to participate in protein synthesis once again.

MINIQUIZ

- What are the components of a ribosome? What functional roles does rRNA play in protein synthesis?
- How is a completed polypeptide chain released from the ribosome?
- How does tmRNA free stalled ribosomes?

V • Protein Processing, Secretion, and Targeting

While translation is the last step in genetic information flow (Figures 4.1*a* and 4.6), some proteins require subsequent *processing* before they become functional. This processing may include assistance in folding or, for some enzymes, the incorporation of *cofactors* or other nonprotein groups. To perform their assigned activities, some proteins must also be targeted to membranes, the periplasm, or even through the membranes of other cells. These processing and targeting activities often require accessory proteins to assist with folding and transport as well as intrinsic signals within the translated protein itself. In the final part of this chapter,

we consider the process of assisted protein folding and mechanisms of protein secretion and targeting.

4.11 Assisted Protein Folding and Chaperones

In Section 4.7 we saw how proteins show several levels of structure including primary, secondary, tertiary, and, in multi-subunit proteins, quaternary structure. Many proteins fold spontaneously into their active form, even while they are being synthesized (Figure 4.38). However, some do not, and these require assistance to achieve an active functional state.

Major Chaperones of *Bacteria*

Bacteria produce a series of proteins called **chaperones** that catalyze a variety of macromolecular folding events. These events include folding proteins that do not fold spontaneously, refolding partially denatured proteins, assembling multiprotein complexes, preventing the improper aggregation of proteins, untangling RNAs, and incorporating cofactors into enzymes. Chaperones are found in all domains of life, and their sequences are highly conserved among all organisms.

Four key chaperones in *Escherichia coli* are the proteins DnaK, DnaJ, GroEL, and GroES. DnaK and DnaJ are ATP-dependent enzymes that bind to newly formed polypeptides and prevent them from folding too quickly, which would increase the risk of improper folding (**Figure 4.40a**). If the DnaKJ complex is unable to fold the protein properly, it may transfer the partially folded protein to the two multi-subunit proteins GroEL and GroES. The protein first enters GroEL, a large, barrel-shaped protein that uses the energy of ATP hydrolysis to fold the protein properly. GroES assists in this (Figure 4.40a). It is estimated that about 100 of the several thousand proteins in a cell of *E. coli* need help in folding from the GroEL–GroES complex, and of these, approximately a dozen are essential for survival of the bacterium.

Other Functions of Chaperones

Chaperones can also refold proteins that have partially denatured in the cell. A protein may denature for many reasons, but often it is because the organism has temporarily experienced high temperatures. Chaperones are thus one type of *heat shock protein*, and their synthesis is greatly accelerated when a cell is stressed by excessive heat (⇨ Section 6.10). The heat shock response is an attempt by the cell to refold its partially denatured proteins for reuse before proteases recognize them as improperly folded and destroy them. In contrast to heat shock proteins, *cold shock proteins* are produced when the cell experiences a sudden shift to very cold temperatures. Cold (but nonfreezing) conditions do not affect most proteins but can affect RNAs. CspA, a major cold shock protein of *E. coli*, is an RNA chaperone instead of a protein chaperone and prevents the spontaneous formation of secondary structures in mRNA that could compromise their ability to be translated. Many other cold shock proteins are known in *E. coli* including ones that refold cold-sensitive proteins rather than acting on mRNAs.

Some chaperones help assemble cofactor-containing enzymes such as those that catalyze redox or electron transport reactions (⇨ Section 3.10). For example, several multi-subunit molybdoenyzmes such as the *E. coli* membrane-bound nitrate reductase require that a molybdenum cofactor (Moco) be inserted during folding, and cofactor insertion is assisted by enzyme-specific chaperone proteins (Figure 4.40b). Nitrate reductase is a key enzyme in anaerobic respiration (⇄ Sections 3.12 and 14.13) and is composed of three subunits: NarGHI. For the enzyme to be functional, the cytoplasmic chaperone protein NarJ must first insert the Moco group into NarG. Once this cofactor has been inserted, the NarG and NarH subunits associate with the membrane-bound NarI subunit to form the active nitrate reductase enzyme complex (Figure 4.40b).

MINIQUIZ

- What are molecular chaperones and why are they necessary?
- What macromolecules are protected by heat shock proteins?
- How do chaperones assist the *Escherichia coli* cell to make nitrate reductase?

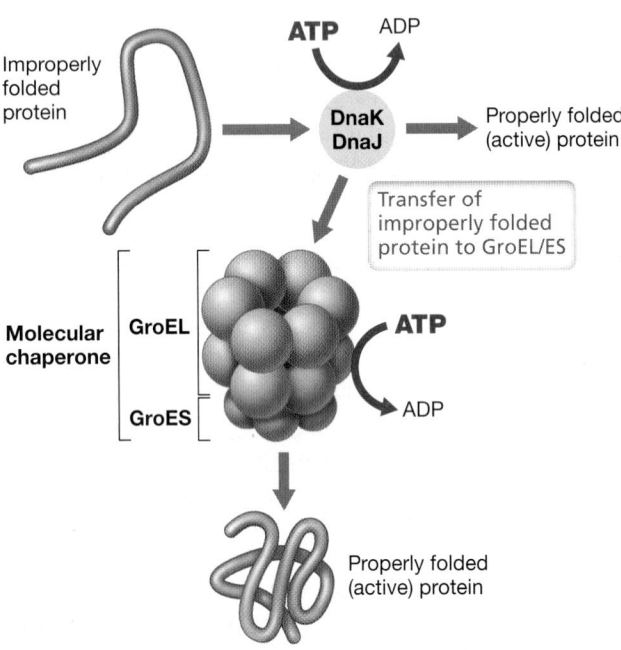

(a)

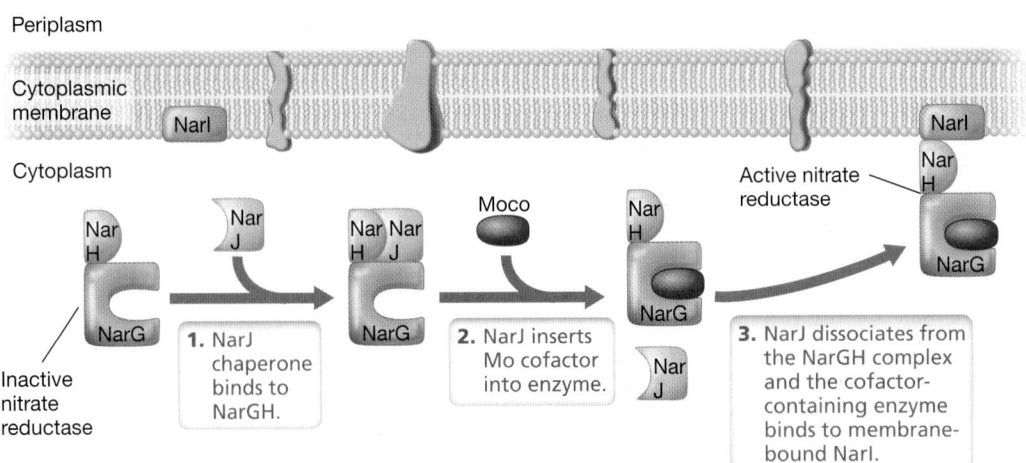

(b)

4.12 Protein Secretion: The Sec and Tat Systems

While many proteins exist in the cell's cytoplasm, some must be transported outside the cytoplasmic membrane into the periplasm (of gram-negative cells) or inserted into the cytoplasmic or outer membranes to facilitate nutrient transport or bioenergetic events. Other proteins such as toxins and extracellular enzymes (exoenzymes) must be secreted from the cell entirely to be active in the environment or to invade another cell. These secreted proteins must get from their site of synthesis on ribosomes into or through the cytoplasmic membrane in order to be functional, and the cell has evolved several systems to get this job done.

The Signal Sequence

Proteins called *translocases* transport specific proteins into or through the membranes of *Bacteria* and *Archaea*. For example, the Sec translocase system both exports unfolded proteins and inserts integral membrane proteins into the cytoplasmic membrane. By contrast, the Tat translocase system transports previously folded proteins through the cytoplasmic membrane. Both the general secretion (Sec) and Tat systems are universal in *Bacteria* and *Archaea*.

Most proteins that must be transported into or through bacterial or archaeal cytoplasmic membranes are synthesized with an amino acid sequence of 15–20 residues—called the **signal sequence**—at the beginning (N-terminus, Figure 4.29) of the protein molecule. Signal sequences are variable, but they typically contain a few positively charged amino acids at the beginning, a central region of hydrophobic residues, and then a more polar region at their end. The signal sequence is so named because it "signals" to a translocase system that this particular protein is to be exported and also helps prevent the protein from completely folding, a process that could interfere with its secretion. Because the signal sequence is the first part of the protein to be synthesized, the early steps in export may actually begin before the protein is completely synthesized (**Figure 4.41**).

Figure 4.40 The activity of chaperone proteins. *(a)* An improperly folded protein can be refolded by either the DnaKJ complex or by the GroEL–GroES complex. In both cases, energy for refolding comes from ATP. *(b)* Formation of an active nitrate reductase enzyme by chaperone-facilitated incorporation of a cofactor. The chaperone protein NarJ carries the molybdenum cofactor (Moco) to the inactive cytoplasmic NarGH complex. After incorporation of Moco, NarJ dissociates and the complex binds to NarI in the membrane becoming an active enzyme.

Sec and Tat Translocases

In the Sec system, unfolded proteins to be exported from the cytoplasm are recognized by either the *SecA protein* or the *signal recognition particle* (*SRP*) (Figure 4.41). Typically, SecA binds proteins that are to be exported into the periplasm, whereas the SRP binds proteins that are to be inserted into the membrane (but not released on the other side). SRPs are found in all cells, but in *Bacteria*, they consist of a single protein and a small, noncoding RNA molecule (4.5S RNA). Both SecA and the SRP deliver proteins to the membrane secretion complex, and after crossing the membrane (Sec-mediated) or inserting into the membrane (SRP-mediated), which is fueled by ATP hydrolysis (Sec) or the proton motive force (Tat), the signal sequence is removed by a protease and the proteins fold to their active form.

Some proteins must be folded before they are translocated because they contain cofactors that must be inserted as the protein folds; for example, iron–sulfur proteins, cytochromes, and other respiratory enzymes (⟳ Section 3.10). Such proteins are processed by the Tat translocase system (for example, nitrate reductase shown in Figure 4.40*b*). Tat stands for "*t*win-*a*rginine *t*ranslocase" because the transported proteins contain a short signal sequence that has a pair of arginine residues. This signal sequence is recognized by TatBC, which carries the folded protein to TatA—the membrane transporter. Following transport, the signal sequence is removed by a protease.

MINIQUIZ

- What is the signal sequence and what does it signal?
- What is the signal recognition particle composed of?
- How do the translocases Sec and Tat differ in the molecules they secrete?

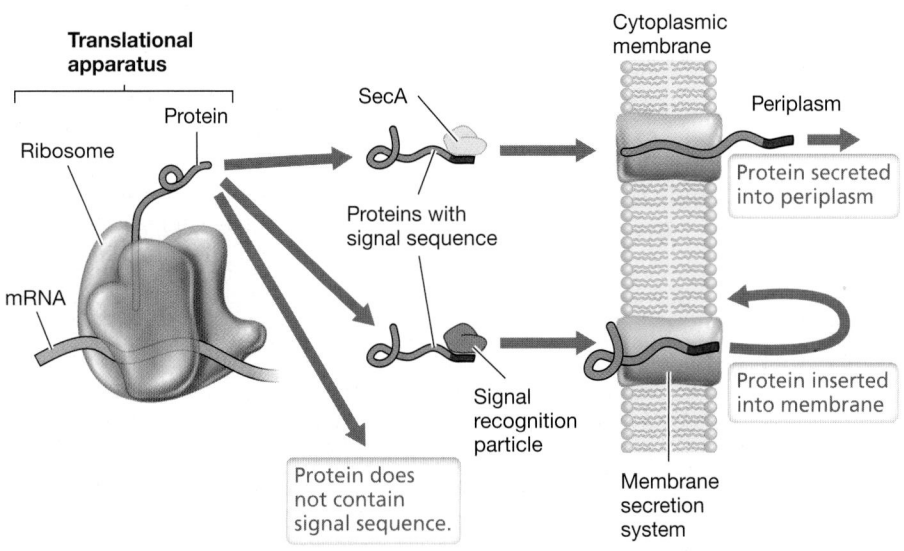

Figure 4.41 Export of proteins via the SecA secretory system. The signal sequence is recognized either by SecA or by the signal recognition particle, which carries the protein to the membrane secretion system. The signal recognition particle binds proteins that are inserted into the membrane, whereas SecA binds proteins that are secreted across the cytoplasmic membrane.

4.13 Protein Secretion: Gram-Negative Systems

In order to insert proteins or other small molecules known as *effectors* into the outer membrane of gram-negative *Bacteria*, or to secrete them outside of the cell, special secretion systems known as types I through VI are used. Gram-positive bacteria and *Archaea* possess secretion systems similar to some of these, but in these organisms, the machinery only has to transport the proteins or effector molecules across the cytoplasmic membrane.

Types I–VI secretion systems facilitate several cellular activities including symbiotic interactions, biofilm formation, extracellular enzyme secretion, transfer of DNA, release of antibiotics, and delivery of proteins into host cells. Thus, molecules secreted by these systems allow bacteria to interact with the environment and other organisms. Each of these systems specifically recognizes its substrate based on amino acid residues and is composed of a large complex of proteins that forms a translocase channel spanning one or more membranes through which the secreted molecule travels (**Figure 4.42**).

Types II and V Secretion Systems

Secretion systems can be grouped into *one-step* and *two-step* classes. Type II and type V systems are two-step translocases because they depend on either the Sec or Tat system (Section 4.12) to transport the secreted protein or a portion of the translocase channel from the cytoplasm through the inner membrane (Figure 4.42*a*). Because types II and V are dependent on the Sec or Tat systems, they generally do not require ATP or the proton motive force to secrete proteins across the outer membrane of the cell.

Type II systems are found in a wide variety of gram-negative pathogenic and nonpathogenic bacteria, and they transport proteins from the periplasm to the extracellular environment. The translocase machinery of type II systems includes a secretion pore in the outer membrane that is anchored to the cytoplasmic membrane by proteins that span the periplasm (Figure 4.42*a*). While the type II system does possess cytoplasmic membrane and periplasmic components, proteins to be secreted do not translocate through these and are instead delivered to the secretion pore in the outer membrane by either the Sec or Tat system. A key to the specificity of the type II system is that the secretion pore only opens to proteins specific to the type II system. Examples of proteins secreted by a type II system include the toxin produced by *Vibrio cholerae* (the causative agent of cholera) and a glucanase exoenzyme produced by *Klebsiella pneumoniae* to degrade large extracellular starch molecules.

Type V systems are the structurally simplest of the secretion systems (Figure 4.42*a*) and are also called *autotransporters* in that the protein to be secreted is fused to a transmembrane protein

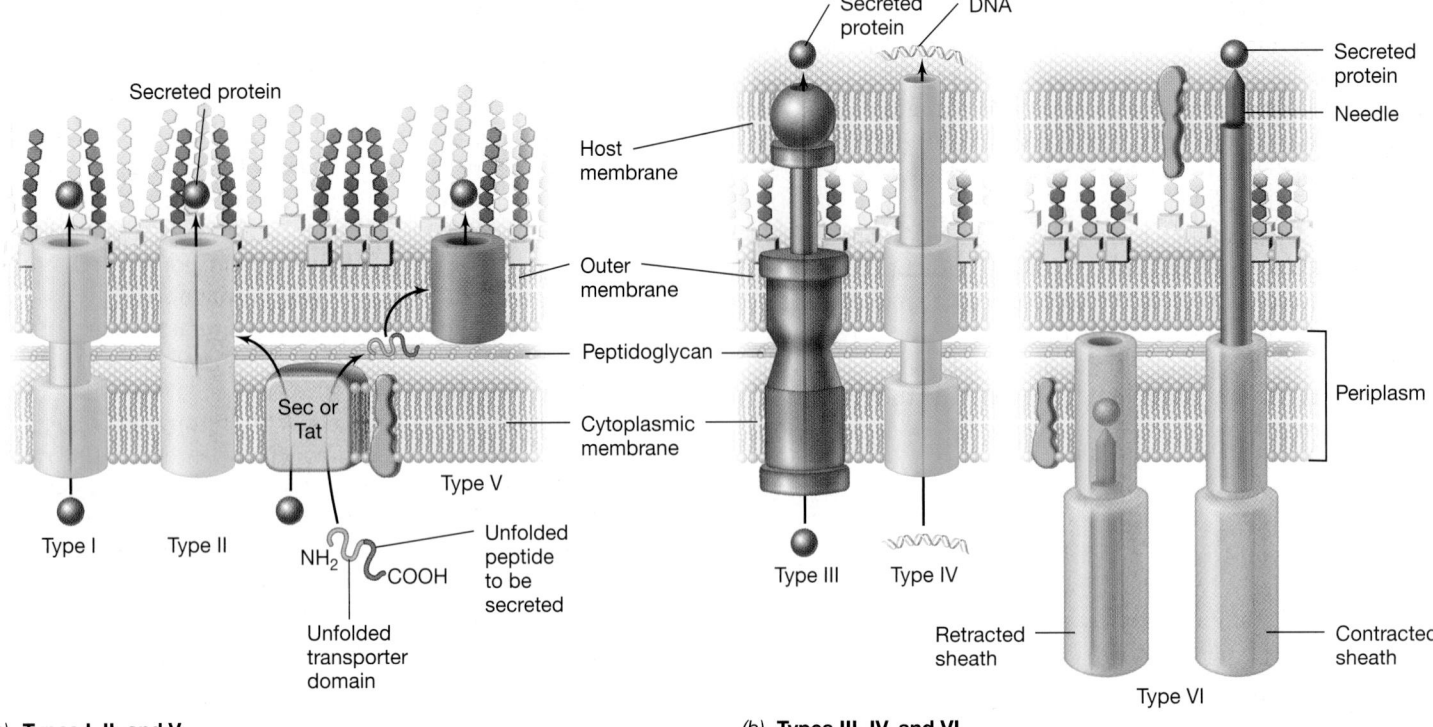

(a) Types I, II, and V

(b) Types III, IV, and VI

Figure 4.42 Secretion systems in gram-negative bacteria. *(a)* Types I, II, and V secrete proteins outside of the bacterial cell. Type I systems secrete proteins in a single step. Types II and V first require the Sec or Tat system to transport the protein to be secreted across the inner membrane. Note that during type V secretion, the Sec system first transports the unfolded transporter domain linked to the unfolded secretion protein through the inner membrane. The translocase then folds in the outer membrane and delivers the folded secretion protein through the outer membrane. *(b)* Types III, IV, and VI secrete molecules outside of the bacterial cell and into a host cell. Type III systems have been termed the injectisome, while type IV systems are similar to a pilus and also secrete DNA into a host cell. Type VI systems contain a sheath or needle in the cytoplasm that contracts to deliver a protein into a host cell.

domain essential to the protein's transport across the outer membrane. After this unfolded multidomain protein is itself transported through the cytoplasmic membrane by the Sec system, a transmembrane domain called the transporter forms a secretion pore in the outer membrane that allows the remainder of the protein (the passenger domain) to pass through and ultimately be secreted outside of the cell (Figure 4.42*a*). Thus proteins to be moved outside of the cell by a type V system are initially transported in the unfolded state, and both the transporter and passenger domains require chaperone proteins (Section 4.11) for proper folding. The folding of the passenger domain (instead of ATP hydrolysis) drives type V secretion. Examples of proteins secreted by the type V system are adhesion proteins used by *E. coli* and *Haemophilus influenzae* (a causative agent of pneumonia) to attach to host cells.

Types I, III, IV, and VI Secretion Systems

The second class of translocases moves proteins through the outer membrane in a single step. Types I, III, IV, and VI are one-step systems because they form channels through both the cytoplasmic and outer membranes and do not require Sec or Tat. *Type I systems* are characterized by three protein components: (1) a cytoplasmic membrane transporter coupled to (2) an outer membrane pore by (3) a membrane fusion protein. The cytoplasmic membrane transporter binds specifically to the protein to be secreted and requires ATP to initiate transport to the outside of the cell (Figure 4.42*a*). While the cytoplasmic membrane transporter is specific to its substrate, a wide range of polypeptide sizes can be secreted. Type I systems secrete small molecules such as a toxic protein (bacteriocin) produced by *E. coli* to kill competing bacterial cells and can also secrete large proteins such as those necessary for biofilm formation on plant surfaces by the soil bacterium *Pseudomonas fluorescens*.

Type III systems are commonly used by pathogenic bacteria not only to secrete toxic proteins outside of the cell but to inject these molecules directly into eukaryotic host cells (Figure 4.42*b*). The entire type III structure is highly complex and composed of over 100 proteins that facilitate substrate recognition, coordinated assembly of the machinery for translocations, and the transport process itself. ATP hydrolysis provides the energy for the secretion and injection. The entire structure, 30–70 nm in length, has been termed the "injectisome" for its similarity to a syringe in both structure and function (**Figure 4.43***a*). Proteins injected into host cells by the type III system often aid in pathogen infection and host invasion. Type III–injected proteins include certain effector molecules of *Chlamydia* (the cause of trachoma and a sexually transmitted disease) and *Salmonella* (a foodborne and waterborne pathogen). However, type III secretion systems are not limited to

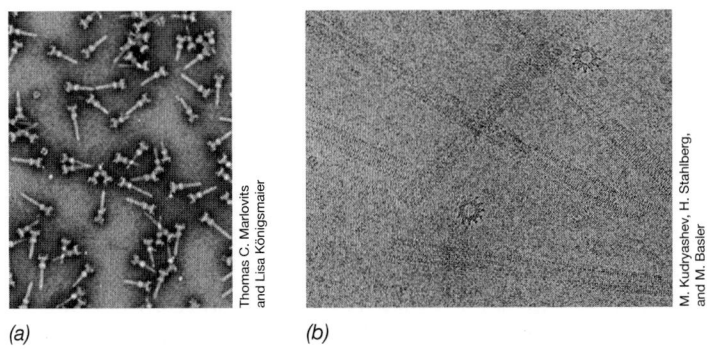

(a)   *(b)*

Thomas C. Marlovits and Lisa Königsmaier

M. Kudryashev, H. Stahlberg, and M. Basler

Figure 4.43 Electron micrographs of secretion machinery. *(a)* Purified type III injectisomes from *Salmonella enterica* (*typhimurium*). *(b)* Purified type VI contractile sheaths from *Vibrio cholerae*.

pathogens. Nitrogen-fixing bacteria deliver molecules critical for establishing a symbiotic relationship with plant roots (root nodules) through type III secretion machinery (⇔ Section 23.3).

Type IV systems (Figure 4.42*b*) are the most ubiquitous and are present in many *Bacteria* and *Archaea*. This system is also used to deliver secreted proteins or other molecules *into* other cells. While this system can be used to transport toxins into host cells, its primary role is to transfer DNA to other cells though the process of conjugation (⇔ Section 11.8), one mechanism that prokaryotic cells have to exchange genes. The type IV translocase is similar to a pilus that extends through the outer membrane into another cell during conjugation. Once the tip of the pilus makes contact with a receptor on the host cell, the DNA is recognized by an inner membrane protein of the donor and then transferred using ATP-facilitated transport (⇔ Section 11.8). Not only can DNA be transferred from one bacterial cell to another using a type IV system, DNA from the plant pathogenic bacterium *Agrobacterium* can be transferred to host plant cells through the system encoded on the Ti plasmid (crown gall disease, ⇔ Section 23.5).

Type VI systems are widely distributed in gram-negative bacteria, and like type IV systems they are capable of delivering a diversity of proteins directly into the cytoplasm of other cells using a one-step, ATP-requiring process similar to type III and IV systems. However, unlike the injectisome or pilus-like structure of the type III and IV translocase systems, the type VI translocase is

cytoplasmic and forms a needle-like protein with a pore-forming protein that contracts all the way through the donor cell's two membranes and directly into a host cell once a substrate molecule is recognized (Figures 4.42*b* and 4.43*b*). The overall delivery process is similar to the mechanism that the tailed bacteriophage T4 uses to deliver DNA into an *E. coli* cell through tail contraction (⇔ Section 8.5). Type VI systems are used by bacteria as weapons to compete with other bacterial cells or to attack eukaryotic cells. In *Pseudomonas aeruginosa*, a type VI system is used to inject a toxin into competing bacterial cells. Similarly, *Vibrio cholerae* uses its type VI system (Figure 4.43*b*) to inject enzymes into competing bacteria; the enzymes degrade the cell wall and membrane. *V. cholerae* also uses the system to deliver toxins to human intestinal cells during a cholera infection.

It should be obvious by now that protein secretion is a crucial aspect of the biology of *Bacteria* and *Archaea* and that these cells have evolved diverse structures and unique mechanisms to deal with protein secretion. However, these elaborate systems are of interest to more than just basic science. Since many of these systems are essential virulence factors for the pathogenic bacteria that produce them, they are currently targets for vaccine development. The rationale here is quite simple: If the activity of these systems could be blocked by the immune system, the pathogen would be unable to establish an infection and cause disease.

We now move on from our focus on the central dogma of biology, protein processing, and protein secretion to tackle the concepts of Chapter 5, where we will examine another process of great importance to a cell's well-being: growth. Without increasing cell numbers, a microbial population will eventually wither away. And as we have just witnessed with the secretion process, aspects of both biochemical unity and diversity underlie the process of microbial growth.

— **MINIQUIZ** —

• Compared with gram-positive bacteria, why is it important for gram-negative bacteria to have additional secretion pathways?

• How do the types I–VI secretion processes differ from Sec and Tat secretion?

• Why is the injectisome so named?

MasteringMicrobiology®   **Visualize, explore, and think critically** with Interactive Microbiology, MicroLab Tutors, MicroCareers case studies, and more. MasteringMicrobiology offers practice quizzes, helpful animations, and other study tools for lecture and lab to help you master microbiology.

Chapter Review

I • Molecular Biology and Genetic Elements

4.1   The informational content of a nucleic acid is determined by the sequence of nitrogen bases along the polynucleotide chain. Both RNA and DNA are informational macromolecules, as are the proteins they encode. DNA forms a double-stranded helix whose strands are complementary and antiparallel. DNA is supercoiled before it is packaged into cells. The three key processes of macromolecular synthesis are: (1) DNA replication; (2) transcription (RNA synthesis); and (3) translation (protein synthesis).

Q Describe the central dogma of molecular biology. With regards to DNA, what is supercoiling and what is meant by the terms antiparallel and complementary?

4.2 In addition to the chromosome, other genetic elements can exist in cells. Plasmids are DNA molecules that exist separately from the chromosome and may confer a selective growth advantage under certain conditions. Viruses contain an RNA or DNA genome, and transposable elements exist as a part of other genetic elements.

Q How are chromosomes and plasmids similar, and how do they differ? What are R plasmids and why are they of medical concern?

II • Copying the Genetic Blueprint: DNA Replication

4.3 Both strands of the DNA helix are templates for the synthesis of new strands (semiconservative replication). The new strands are elongated by addition of deoxyribonucleotides to the 3′ end. DNA polymerases require a primer made of RNA by the enzyme primase, and synthesis begins at an origin of replication. The double helix is unwound by helicase and is stabilized by single-strand binding proteins. Extension of the DNA occurs continuously on the leading strand but discontinuously on the lagging strand, resulting in fragments that must be joined together.

Q What is meant by the term semiconservative replication? What are the functions of DNA Pol I and III, helicase, and primase? How does a leading strand differ from a lagging strand?

4.4 Starting from a single origin on a circular chromosome, two replication forks synthesize DNA simultaneously in both directions until the forks meet at the terminus region. The proteins at the replication fork form a large complex known as the replisome. Most errors in base pairing that occur during replication are corrected by the proofreading functions of DNA polymerases.

Q What is the replisome and what does it contain? Why can replication occur faster on circular DNA than on linear DNA? What is proofreading and why is it important?

III • RNA Synthesis: Transcription

4.5 In *Bacteria*, promoters are recognized by the sigma subunit of RNA polymerase. Alternative sigma factors allow joint regulation of large families of genes in response to growth conditions. Transcription by RNA polymerase continues until specific sites called transcription terminators are reached. These terminators function at the level of RNA. In *Bacteria* and *Archaea* a single mRNA molecule may encode more than one polypeptide. A cluster of genes that are transcribed together from a single promoter is called an operon.

Q How does RNA polymerase know where to begin transcription? How does it know where to end?

4.6 The transcription apparatus and the promoter architecture of *Archaea* and *Eukarya* have many features in common, although the components are usually relatively more simple in *Archaea*. In contrast, the processing of eukaryotic primary transcripts is unique and has three distinct steps: splicing, capping, and adding a poly(A) tail.

Q How does the archaeal RNA polymerase differ from that in *Bacteria*? How does the initiation of transcription in the two domains differ? Why do eukaryotic mRNAs have to be "processed" whereas most prokaryotic RNAs do not?

IV • Protein Synthesis: Translation

4.7 Polypeptide chains contain many amino acids linked via peptide bonds. Twenty-two different amino acids are genetically encoded. The primary structure of a protein is its amino acid sequence, but the higher-order structure (folding) of the polypeptide determines its cellular function.

Q Describe the two types of secondary structure a polypeptide can attain. Which proteins can achieve quaternary structure? Which protein structure(s) are altered by denaturation?

4.8 One or more tRNAs exist for each amino acid incorporated into polypeptides by the ribosome. Enzymes called aminoacyl-tRNA synthetases attach amino acids to their cognate tRNAs.

Q Why are transfer RNAs important in translation? Do genes for tRNAs have promoters, and are tRNAs translated? What are aminoacyl-tRNA synthetases, what are their substrates, and what do they do?

4.9 The genetic code is expressed as RNA, and a single amino acid may be encoded by several different but related codons. In addition to the stop (nonsense) codons, there is also a specific start codon that signals the initiation of translation.

Q Why is the genetic code a degenerate code? What is wobble and how does it accommodate fidelity in the genetic code?

4.10 Translation occurs on the ribosome and requires mRNA and aminoacyl-tRNAs. The ribosome has three sites: acceptor, peptide, and exit (A, P, and E). During each step of translation, the ribosome advances one codon along the mRNA, and the tRNA in the acceptor site moves to the peptide site. Protein synthesis terminates when a stop codon, which does not have a corresponding tRNA, is reached.

Q Where on the ribosome do tRNAs bind, and what is the energy source that supports translocation?

V • Protein Processing, Secretion, and Targeting

4.11 Polypeptides do not remain linear in structure following translation but require proper folding and additional processing for functional activity. Proteins called chaperones help fold some proteins that are unable to fold spontaneously and also assist in the incorporation of cofactors and refolding partially denatured proteins.

Q **What proteins are involved in refolding misfolded proteins? What other functions do they have?**

4.12 Many proteins need to be transported into or through the cytoplasmic membrane. These proteins contain a signal sequence that is recognized by the cellular translocase systems of Sec and Tat.

Q **What are the functions of the Sec and Tat systems and why are they necessary? How do these systems know which proteins to act upon?**

4.13 A large diversity of secretion systems are employed by gram-negative bacteria to secrete proteins into or through their outer membrane and even into other cells, and types I–VI secretion systems function in this regard. Types II and V are two-step translocases and rely on the activity of Sec or Tat, whereas types I, III, IV, and VI are one-step translocases.

Q **What are the major differences between one-step and two-step translocases? Which are used to secrete proteins into other cells?**

Application Questions

1. The genome of the bacterium *Neisseria gonorrhoeae* consists of one double-stranded DNA molecule that contains 2220 kilobase pairs. If 85% of this DNA molecule is made up of the open reading frames of genes encoding proteins, and the average protein is 300 amino acids long, how many protein-encoding genes does *Neisseria* have? What kind of genetic information is present in the other 15% of the DNA?

2. Compare and contrast the activity of DNA and RNA polymerases. What is the function of each? What are the substrates of each? What is the main difference in the behavior of the two polymerases?

3. What would be the result (in terms of protein synthesis) if RNA polymerase initiated transcription one base upstream of its normal starting point? Why? By inspecting Table 4.4, discuss how the genetic code has evolved to help minimize the impact of mutations.

Chapter Glossary

Amino acid one of the 22 different monomers that make up proteins; chemically, a two-carbon carboxylic acid containing an amino group and a characteristic substituent on the alpha carbon

Aminoacyl-tRNA synthetase an enzyme that catalyzes attachment of an amino acid to its cognate tRNA

Anticodon a sequence of three bases in a tRNA molecule that base-pairs with a codon during protein synthesis

Antiparallel in reference to double-stranded DNA, the two strands run in opposite directions (one runs 5′→3′ and the complementary strand 3′→5′)

Bacteriocin a toxic protein secreted by bacteria that inhibits or kills other, related bacteria

Chaperone a protein that helps other proteins fold or refold from a partly denatured state

Chromosome a genetic element, usually circular in prokaryotes, carrying genes essential to cellular function

Codon a sequence of three bases in mRNA that encodes an amino acid

Codon bias nonrandom usage of multiple codons encoding the same amino acid

Complementary nucleic acid sequences that can base-pair with each other

Denaturation loss of the correct folding of a protein, leading (usually) to protein aggregation and loss of biological activity

DNA (deoxyribonucleic acid) a polymer of deoxyribonucleotides linked by phosphodiester bonds that carries genetic information

DNA gyrase an enzyme found in most prokaryotes that introduces negative supercoils in DNA

DNA helicase an enzyme that uses ATP to unwind the double helix of DNA

DNA ligase an enzyme that seals nicks in the backbone of DNA

DNA polymerase an enzyme that synthesizes a new strand of DNA in the 5′→3′ direction using an antiparallel DNA strand as a template

Exons the coding DNA sequences in a split gene (contrast with intron)

Gene a segment of DNA specifying a protein (via mRNA), a tRNA, an rRNA, or any other noncoding RNA

Genetic code the correspondence between nucleic acid sequence and amino acid sequence of proteins

Genetic element a structure that carries genetic information, such as a chromosome, a plasmid, or a virus genome

Genome the total complement of genes contained in a cell or virus

Informational macromolecule any large polymeric molecule that carries genetic information, including DNA, RNA, and protein

Introns the intervening noncoding DNA sequences in a split gene (contrast with exons)

Lagging strand the new strand of DNA that is synthesized in short pieces and then joined together later

Leading strand the new strand of DNA that is synthesized continuously during DNA replication

Messenger RNA (mRNA) an RNA molecule that contains the genetic information to encode one or more polypeptides

Nonsense codon another name for a stop codon

Nucleoside a nitrogenous base (adenine, guanine, cytosine, thymine, or uracil) plus a sugar (either ribose or deoxyribose) but lacking phosphate

Nucleotide a monomer of a nucleic acid containing a nitrogenous base (adenine, guanine, cytosine, thymine, or uracil), one or more molecules of phosphate, and a sugar, either ribose (in RNA) or deoxyribose (in DNA)

Open reading frame (ORF) a sequence of DNA or RNA that could be translated to give a polypeptide

Operon a cluster of genes that are cotranscribed as a single messenger RNA

Peptide bond a type of covalent bond linking amino acids in a polypeptide

Phosphodiester bond a type of covalent bond linking nucleotides together in a polynucleotide

Plasmid an extrachromosomal genetic element that is usually not essential to the cell

Polynucleotide a polymer of nucleotides bonded to one another by covalent bonds called phosphodiester bonds

Polypeptide a polymer of amino acids bonded to one another by peptide bonds

Primary structure the precise sequence of monomers in a macromolecule such as a polypeptide or a nucleic acid

Primary transcript an unprocessed RNA molecule that is the direct product of transcription

Primase the enzyme that synthesizes the RNA primer used in DNA replication

Primer an oligonucleotide to which DNA polymerase attaches the first deoxyribonucleotide during DNA synthesis

Promoter a site on DNA to which RNA polymerase binds to commence transcription

Protein a polypeptide or group of polypeptides forming a molecule of specific biological function

Purine one of the nitrogenous bases of nucleic acids that contain two fused rings; adenine and guanine

Pyrimidine one of the nitrogenous bases of nucleic acids that contain a single ring; cytosine, thymine, and uracil

Quaternary structure in proteins, the number and types of individual polypeptides in the final protein molecule

Replication synthesis of DNA using DNA as a template

Replication fork the site on the chromosome where DNA replication occurs and where the enzymes replicating the DNA are bound to untwisted, single-stranded DNA

Replisome a DNA replication complex that consists of two copies of DNA polymerase III, DNA gyrase, helicase, primase, and copies of single-strand binding protein

Ribosomal RNA (rRNA) types of RNA found in the ribosome; some participate actively in protein synthesis

Ribosome a cytoplasmic particle composed of ribosomal RNA and protein, whose function is to synthesize proteins

RNA (ribonucleic acid) a polymer of ribonucleotides linked by phosphodiester bonds that plays many roles in cells, in particular, during protein synthesis

RNA polymerase an enzyme that synthesizes RNA in the $5' \rightarrow 3'$ direction using a complementary and antiparallel DNA strand as a template

RNA processing the conversion of a primary transcript RNA to its mature form

Secondary structure the initial pattern of folding of a polypeptide or a polynucleotide, usually dictated by opportunities for hydrogen bonding

Semiconservative replication DNA synthesis yielding two new double helices, each consisting of one parental and one progeny strand

Signal sequence a special N-terminal sequence of approximately 20 amino acids that signals that a protein should be incorporated into or exported across the cytoplasmic membrane

Spliceosome a complex of ribonucleoproteins that catalyze the removal of introns from primary RNA transcripts

Start codon a special codon, usually AUG, that signals the start of a protein

Stop codon a codon that signals the end of a protein

Termination stopping the elongation of an RNA molecule at a specific site

Tertiary structure the final folded structure of a polypeptide that has previously attained secondary structure

Transcription the synthesis of RNA using a DNA template

Transfer RNA (tRNA) a small RNA molecule used in translation that possesses an anticodon at one end and has the corresponding amino acid attached to the other end

Translation the synthesis of protein using the genetic information in RNA as a template

Transposable element a piece of DNA able to move (transpose) from one site to another on host DNA molecules

Wobble a less rigid form of base pairing allowed only in codon–anticodon pairing

Microbial Growth and Its Control

microbiology**now**

Picking Apart a Microbial Consortium

In nature, certain metabolic processes are carried out by microbes that team up to get the job done, a cozy arrangement called a consortium. Such is the case with the oxidation of methane (CH_4) linked to the reduction of sulfate (SO_4^{2-}) in anoxic marine sediments. The overall reaction ($CH_4 + SO_4^{2-} \rightarrow HCO_3^- + HS^- + H_2O$) is exergonic and the small amount of energy released is shared between two distinct microbes. The methane oxidizer in the consortium is a species of *Archaea* nicknamed ANME (for *an*aerobic *me*thanotroph, blue in photo), and its sulfate-reducing partner is a species of *Bacteria* (brown in photo). The consortium is thought to play a key role in the carbon cycle as a major methane sink, and thus a detailed picture of how it works is important to our understanding of the global carbon economy, climate change, and marine biogeochemistry.

Researchers have tried for years to separate the consortium into its components but always found that methane oxidation required both organisms. However, some researchers hypothesized that it might be possible to replace the sulfate reducer with an artificial electron acceptor and that this might unlock the consortium and allow the methanotroph to grow in pure culture. Using an electron acceptor called AQDS, the scientists discovered that they could turn off sulfate reduction in the consortium while maintaining CH_4 oxidation. During this process, the methanotroph used electrons from CH_4 to reduce AQDS rather than passing them on to its sulfate-reducing partner. Several other electron acceptors known to support anaerobic respiration also sustained methane oxidation, giving hope that ANME may eventually be obtained in pure culture.

The ability to grow a microbe in pure culture is the "gold standard" for the study of its physiology, biochemistry, regulation, and several other aspects of its biology. In the case of the ANME–sulfate reducer consortium, several physiologies were active at once, and resolving these many reactions proved to be a major scientific challenge. However, if further work shows that ANME can be removed from the consortium and grown in pure culture, detailed aspects of its biology can be studied that were not possible when the organism was tightly coupled to its partner in the consortium (photo).

Source: Scheller, S., H. Yu, G.L. Chadwick, S.E. McGlynn, and V.J. Orphan. 2016. Artificial electron acceptors decouple archaeal methane oxidation from sulfate reduction. *Science* 351: 703–706.

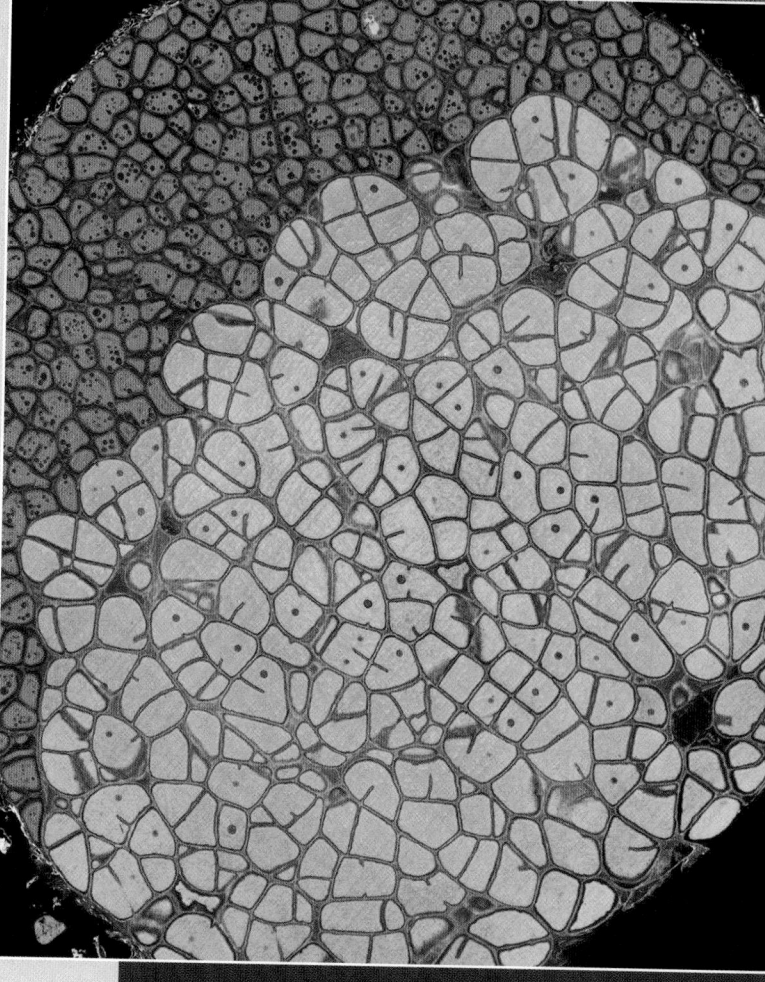

I • Cell Division and Population Growth

In previous chapters we discussed cell structure and function (Chapter 2) and the principles of microbial nutrition and metabolism (Chapter 3). In Chapter 4 we learned the genetic processes that encode the structures and metabolic activities of cells. Here we begin a new unit whose focus is microbial growth and its regulation.

In Chapter 5 we lay the groundwork for the entire unit by presenting the basic principles of exponential growth, how the environment affects growth, and some principles of microbial growth control. Then, after we consider the important topic of microbial regulation in Chapter 6, we will revisit microbial growth in Chapter 7 and reexamine the process from a molecular and regulatory perspective. We finish the unit by delving into the world of viruses and their replication. Although not cells, viruses are critically important microbes whose replication shows parallels with the growth of microbial cells.

5.1 Binary Fission, Budding, and Biofilms

Growth is the result of cell division and is the ultimate process in the life of a microbial cell. In microbiology, growth is defined as *an increase in the number of cells*. Microbial cells have a finite life span, and a species is maintained only as a result of continued growth of its population. As macromolecules accumulate in the cytoplasm of a cell, they assemble into major cell structures, such as the cell wall, cytoplasmic membrane, flagella, ribosomes, enzyme complexes, and so on, eventually leading to the events of cell division itself. In a growing culture of a rod-shaped bacterium such as *Escherichia coli*, cells elongate to approximately twice their original length and then form a partition that constricts the cell into two daughter cells (**Figure 5.1**). This process is called **binary fission** ("binary" to indicate that two cells have arisen from one).

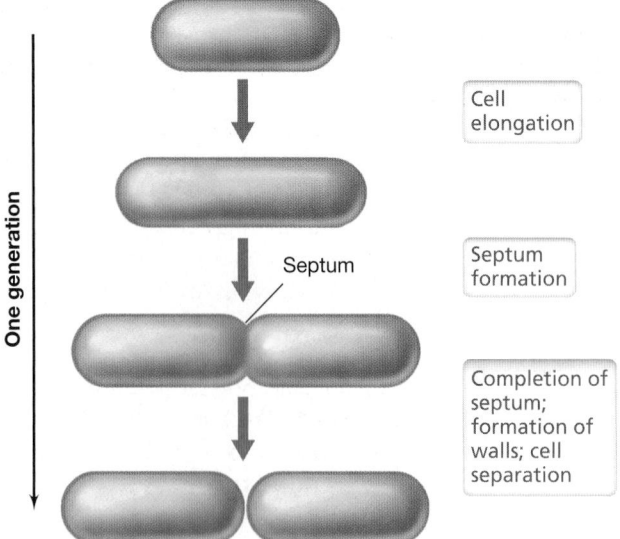

Figure 5.1 Binary fission in a rod-shaped bacterium. Cell numbers (and all components of the cells) double every generation.

Cell elongation

Septum formation

Completion of septum; formation of walls; cell separation

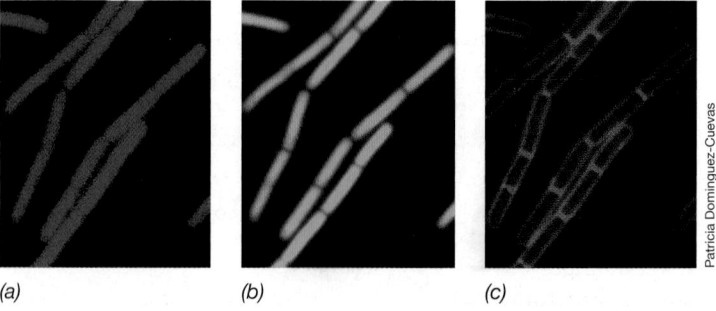

Figure 5.2 Septa. The septum that separates dividing cells of the bacterium *Bacillus subtilis* is clearly visible in this series of fluorescent micrographs. *(a)* DAPI stains the entire cell. *(b)* The green fluorescent protein lights up entire cells. *(c)* A dye that stains only the cytoplasmic membrane shows that septa contain membrane (and cell wall) material.

The partition that forms between dividing cells is called a *septum* and results from the inward growth of the cytoplasmic membrane and cell wall from opposing directions; septum formation continues until the two daughter cells are pinched off. There are some variations in this general pattern of binary fission. In some bacteria, such as *Bacillus subtilis*, a septum forms without cell wall constriction (**Figure 5.2**), while in the budding bacterium *Caulobacter* (see Figure 5.3) constriction occurs but no septum is formed.

Cell Generations and Generation Time

When one cell eventually separates to form two cells, we say that one *generation* has occurred, and the time required for this process is called the **generation time** (Figure 5.1 and see Figure 5.6). During one generation, all cellular constituents increase proportionally. Each daughter cell receives a copy of the chromosome(s) and sufficient copies of ribosomes and all other macromolecular complexes, monomers, and inorganic ions to begin life as an independent entity. Partitioning of the replicated DNA molecule between the two daughter cells depends on the DNA remaining attached to the cytoplasmic membrane during division, with constriction leading to separation of the chromosomes, one to each daughter cell (⟳ Figure 7.1).

The generation time of a given bacterial species is variable and depends on nutritional and genetic factors, and on temperature. Under the best nutritional conditions, the generation time of a laboratory culture of *E. coli* is about 20 min. A few bacteria can grow even faster than this, but many bacteria grow much slower, with generation times of hours or days being more common. In nature, microbial cells probably grow much slower than their maximum rates observed in the laboratory. This is because the conditions and resources necessary for optimal growth in the laboratory are often not present in a natural habitat, and unlike growth in laboratory pure cultures, microbes in nature coexist with other microbes in microbial communities (⟳ Figure 1.1) and must compete with their neighbors for resources and space.

Budding Cell Division

Although cell division in most bacteria occurs by binary fission, in a few bacteria other forms of growth and cell division occur. Budding bacteria are the primary examples here, and these are cells

UNIT 2

that divide as a result of unequal cell growth. In contrast to binary fission that yields two equivalent cells (Figure 5.1), **budding division** forms a totally new daughter cell, with the mother cell retaining its original identity (**Figure 5.3** and ⟳ Section 7.4).

A fundamental difference between budding bacteria and bacteria that divide by binary fission is the formation of new cell wall material from a single point (polar growth) rather than throughout the whole cell (intercalary growth) as in binary fission. An important consequence of polar growth is that large cytoplasmic structures, such as internal membrane complexes, are not partitioned during the cell division process and must be formed de novo in the developing bud. However, this has an advantage in that more complex internal structures can be formed in budding cells than in cells that divide by binary fission, since the latter cells would have to partition these structures between the two daughter cells. Not coincidentally, many budding bacteria, particularly phototrophic and chemolithotrophic species, contain extensive internal membrane systems that house specific enzymes required to perform their particular metabolic specialties.

Some budding bacteria form cytoplasmic extensions such as stalks or hyphae, and classic examples are the genera *Caulobacter* and *Hyphomicrobium* (Figure 5.3). These organisms form cellular extrusions from which new cells bud off. Other budding bacteria such as the aquatic bacterium *Ancalomicrobium* produce multiple appendages that resemble arms extending away from the cell (⟳ Figure 15.57b). The appendages increase the surface-to-volume ratio of the cell (⟳ Section 2.2), which increases its ability to extract nutrients from oligotrophic (very dilute) habitats. Many budding bacteria also have distinctive life cycles, and we consider these and the group as a whole in Section 15.20.

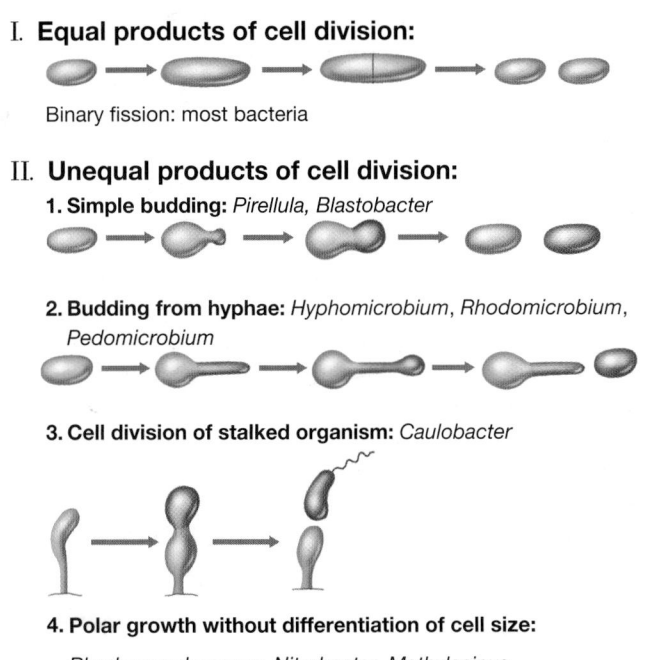

Figure 5.3 Cell division in different morphological forms of bacteria. The contrast is shown between cell division in conventional bacteria (cells that divide by binary fission) and in various budding and stalked bacteria.

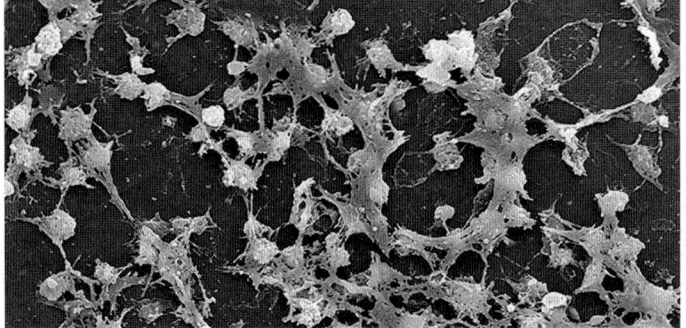

(a)

(b)

Figure 5.4 Biofilms. *(a)* Scanning electron micrograph of a biofilm of cells of *Staphylococcus aureus* that formed on an indwelling catheter. *(b)* A microbial mat of the purple phototrophic bacterium *Thermochromatium tepidum* that developed in a small sulfidic hot spring in Yellowstone National Park.

Biofilms

Whether dividing by binary fission (Figure 5.1) or some form of budding (Figure 5.3), microbial cells can grow either in suspension or attached to surfaces. The suspended lifestyle, called *planktonic growth*, is the way many bacteria live in nature, for example, organisms that inhabit the water column of a lake. However, many other microorganisms show *sessile growth*, meaning that they grow attached to a surface. These attached cells can then develop into **biofilms** (Figure 5.4).

A biofilm is an attached polysaccharide matrix containing embedded bacterial cells (Figure 5.4a). Biofilms form in stages, beginning with the attachment of planktonic cells. This is followed by the production of a sticky matrix and further growth and development to form the tenacious and nearly impenetrable mature biofilm. Some biofilms form multilayered sheets with different organisms present in the individual layers. These biofilms are called *microbial mats*; mats composed of various phototrophic and chemotrophic bacteria are common in the outflows of hot springs (Figure 5.4b) and in marine intertidal regions. Biofilms are a common growth form for bacteria in nature because the intensely interwoven nature of the structure prevents harmful chemicals (for example, antibiotics or other toxic substances) from penetrating. Biofilms are also a barrier to bacterial grazing by protists and prevent cells from being washed away into a potentially less favorable habitat.

Bacterial biofilms affect many aspects of our lives, including human health. For example, biofilms have been implicated in difficult-to-treat infections of implanted medical devices, such as artificial heart valves and joints, and indwelling devices, such as catheters (Figure 5.4a). Moreover, symptoms of the disease cystic fibrosis are caused by a tenacious bacterial biofilm that fills the lungs and prevents gas exchange. Biofilms also cause fouling and plugging of water distribution systems and can form in fuel storage tanks, where they contaminate the fuel by producing souring agents such as H_2S.

We examine biofilms in more detail elsewhere in this book, including Sections 7.9, 20.4, and 20.5.

MINIQUIZ

- Define the term generation. What is meant by the term generation time?
- How do binary fission and budding cell division differ?
- How does the biofilm growth mode differ from that of planktonic cells? Which growth mode better protects the bacterial cells from harm?

5.2 Quantitative Aspects of Microbial Growth

During cell division, one cell becomes two. During the time that it takes for this to occur (the generation time), both total cell *number* and *mass* double (Figure 5.1). As we will see, cell numbers in a bacterial culture can quickly become very large, and so it is necessary to deal with the topic of microbial growth using quantitative methods.

Plotting Growth Data

A growth experiment beginning with a single cell having a generation time of 30 min is presented in **Figure 5.5**. This repetitive pattern, where the number of cells doubles in a constant time interval, is called **exponential growth**. When the cell number from such an experiment is graphed on arithmetic (linear) coordinates as a function of time, one obtains a curve with a continuously increasing slope (Figure 5.5b). By contrast, when the cell number is plotted on a logarithmic ($\log_{10}$) scale as a function of time (a *semilogarithmic* graph), as shown in Figure 5.5b, the points fall on a straight line. This straight-line function reflects the fact that the cells are growing exponentially and that the population is doubling in a constant time interval.

Semilogarithmic graphs are also convenient for estimating the generation time of a culture from actual growth data, since generation times may be inferred directly from the graph as shown in **Figure 5.6**. For example, when two points on the curve that represent one cell doubling on the Y axis are selected and vertical lines drawn from them to intersect the X axis, the time interval measured on the X axis is the generation time (Figure 5.6b).

The Mathematics of Bacterial Growth

The increase in cell number in an exponentially growing bacterial culture can be expressed mathematically as a geometric

Time (h)	Total number of cells	Time (h)	Total number of cells
0	1	4	256 (2^8)
0.5	2	4.5	512 (2^9)
1	4	5	1,024 (2^{10})
1.5	8	5.5	2,048 (2^{11})
2	16	6	4,096 (2^{12})
2.5	32	.	.
3	64	.	.
3.5	128	10	1,048,576 (2^{20})

(a)

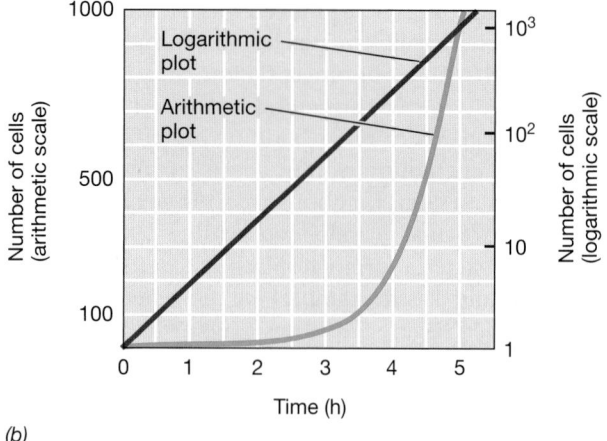

(b)

Figure 5.5 The rate of growth of a microbial culture. (a) Data for a population that doubles every 30 min. (b) Data plotted on arithmetic (left ordinate) and logarithmic (right ordinate) scales.

progression of the number 2. As one cell divides to become two cells, we express this as $2^0 \rightarrow 2^1$. As two cells become four, we express this as $2^1 \rightarrow 2^2$ and so on (Figure 5.5a). A fixed relationship exists between the initial number of cells in a culture and the number present after a period of exponential growth, and this relationship is expressed as

$$N = N_0 2^n$$

where N is the final cell number, N_0 is the initial cell number, and n is the number of generations (a single generation is shown in Figure 5.1) during the period of exponential growth. The generation time (g) of the exponentially growing population is t/n, where t is the duration of exponential growth expressed in days, hours, or minutes. From knowledge of the initial and final cell numbers in an exponentially growing cell population, it is possible to calculate n, and from n and knowledge of t, the generation time, g.

The equation $N = N_0 2^n$ can be expressed in terms of n by taking the logarithms of both sides and doing some simple algebra to yield the expression $n = [3.3(\log N - \log N_0)]$. Using this expression, it is possible to calculate generation times in terms of measurable quantities, N and N_0 (Sections 5.6–5.8 describe methods for quantifying cell numbers). As an example, consider actual growth data from the graph in Figure 5.6b, in which $N = 10^8$, $N_0 = 5 \times 10^7$, and $t = 2$:

$$n = 3.3[\log(10^8) - \log(5 \times 10^7)]$$
$$= 3.3 (8 - 7.69) = 3.3(0.301) = 1$$

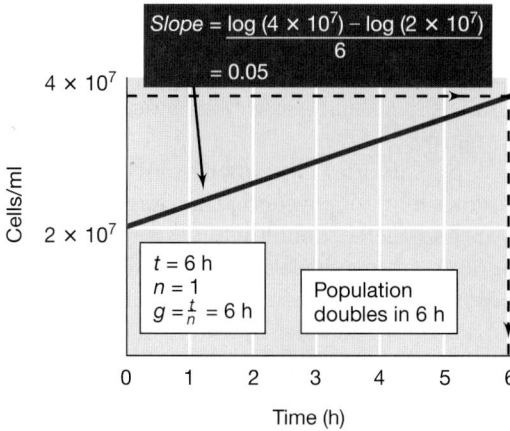

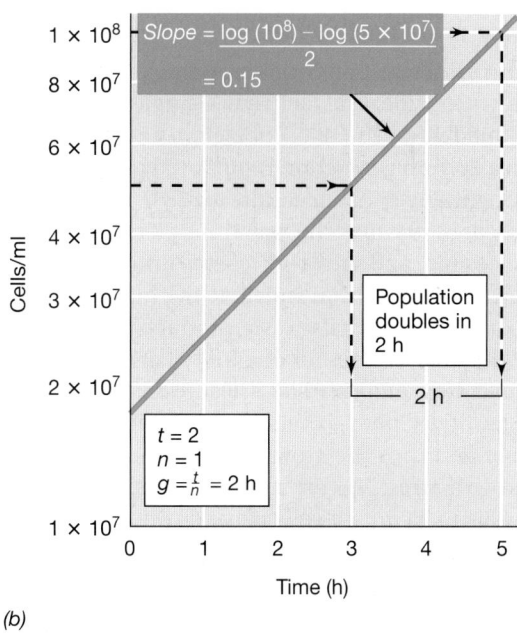

Figure 5.6 Calculating microbial growth parameters. Method of estimating the generation times (*g*) of exponentially growing populations with *g* of (*a*) 6 h and (*b*) 2 h from data plotted on semilogarithmic graphs. The slope of each line is equal to 0.301/*g*, and *n* is the number of generations in the time *t*. All numbers are expressed in scientific notation; that is, 10,000,000 is 1×10^7, 60,000,000 is 6×10^7, and so on.

Thus, in this example, $g = t/n = 2/1 = 2$ h. If exponential growth continued for another 2 h, the cell number would be 2×10^8. Two hours later the cell number would be 4×10^8, and so on. Besides determining the generation time of an exponentially growing culture by inspection of graphical data (Figure 5.6*b*), *g* can be calculated directly from the slope of the straight-line function obtained in a semilogarithmic plot of exponential growth. The slope of this line is equal to $\log(\Delta N)/\Delta t$ and this equals 0.301 *n*/*t* (or 0.301/*g*). In the example of Figure 5.6*b*, the slope would thus be 0.301/2, or 0.15. Since *g* is equal to 0.301/slope, we arrive at the same value of 2 for *g*.

The Instantaneous Growth Rate Constant

Other expressions are often useful in describing exponential growth, and chief among these is the *instantaneous growth rate constant*, abbreviated *k*. The instantaneous growth rate constant expresses the rate at which the population is growing *at any instant* (by contrast, *g* is the mean time required for the cell population to double); *k* is expressed in units of reciprocal hours (h^{-1}).

The instantaneous rate of growth is a function of the number of cells at a given time (*N*) multiplied by *k*, or $dN/dt = kN$. By integration using natural logarithms ($\log_e$), this equation can be reexpressed as $N = N_0 e^{kt}$. Taking the $\log_{10}$ of both sides of this equation converts natural logs to $\log_{10}$ (so that *N* can be plotted against *t* on semilog paper, Figure 5.5*b* and Figure 5.6) and yields the expression $\log N = kt/2.303 + \log N_0$. The slope of this function (*k*/2.303) is also equal to 0.301/*g* (Figure 5.6), and thus *k* can be expressed in terms of *g* by the expression $k = 0.693/g$.

Armed with knowledge of *n* and *t*, one can calculate *g* and *k* for different microorganisms growing under various conditions. This is often useful for optimizing culture conditions for a newly isolated organism and also for testing the positive or negative effect of some treatment on a bacterial culture. For example, comparison with an unamended control allows factors that stimulate or inhibit growth to be identified by measuring their effect on the various growth parameters presented here.

Consequences of Exponential Growth

During exponential growth, the increase in cell number is initially rather slow but increases at an ever-faster rate. In the later stages of exponential growth, this results in an explosive increase in cell numbers. For example, in the experiment shown in Figure 5.5, the rate of cell production in the first 30 min of growth is 1 cell per 30 min. However, between 4 and 4.5 h of growth, the rate of cell production is 256 cells per 30 min, and between 5.5 and 6 h of growth it is 2048 cells per 30 min. Because of this, cell numbers in laboratory cultures of bacteria can quickly become very large, with final population sizes of $>10^9$ cells/ml not uncommon.

Besides being a theoretical consideration, exponential growth can have implications in everyday life. Consider something we have all experienced, the spoilage of milk. The lactic acid bacteria responsible for the soured flavor of spoiled milk contaminate the milk during its collection and exist in fresh, pasteurized milk in low numbers; these organisms grow slowly at refrigerator temperature (4°C) but much faster at room temperature. If a bottle of fresh milk is left to stand at room temperature overnight, some lactic acid is made, but not enough to affect milk quality. However, if week-old milk, which now contains a week's worth of slow bacterial growth (and thus much higher cell numbers), is left standing under the same conditions, a huge amount of lactic acid is made, and spoilage results.

--- **MINIQUIZ** ---

- What is a *semilogarithmic* plot and what information can we derive from it?

- For an exponentially growing culture that increases from 5×10^6 cells/ml to 5×10^8 cells/ml in 8 h, calculate *g*, *n*, and *k* for this culture.

- For testing a bacterium's response to a toxic substance, why would *g* be useful information?

5.3 The Microbial Growth Cycle

The data presented in Figures 5.5 and 5.6 are a reflection of exponentially growing cells. But exponential growth is only part of the *microbial growth cycle*. For several reasons, an organism growing in an enclosed vessel, such as a tube or a flask (a **batch culture**), cannot grow exponentially indefinitely. Instead, a typical *growth curve* for the population is obtained, as illustrated in **Figure 5.7**. The growth curve describes the entire growth cycle and is made up of four phases: lag, exponential, stationary, and death.

Lag and Exponential Phases

When a microbial culture is inoculated into fresh growth media (see Section 5.5), growth begins only after a period of time called the *lag phase*. This interval may be brief or extended depending on the history of the cells used as inocula and the composition of the growth medium and growth conditions (see Sections 5.9–5.14). If an exponentially growing culture is transferred into the same medium under the same conditions of growth, there will be essentially no lag and exponential growth will begin immediately. However, if the inoculum is taken from an old culture, there is usually a lag because the cells are depleted of various essential constituents and time is required for their biosynthesis.

A lag is also observed when a microbial culture is transferred from a nutrient-rich culture medium to one that is nutrient-poor. In order to grow, cells must have a complete complement of enzymes for synthesis of the essential metabolites not present in that medium. Hence, when transferred to a nutrient-poor medium, time is needed for the biosynthesis of new enzymes and for these to produce a sufficient pool of required metabolites before growth can actually begin. These events occur during the lag period.

When a growing cell population doubles at regular intervals (Section 5.2) the cells are said to be in the *exponential phase* of growth. Exponential phase cells are typically in their healthiest state and are thus most desirable for studies of their enzymes or other cell components. Rates of exponential growth vary greatly. Exponential growth rates are influenced by the growth conditions an organism is experiencing as well as genetic characteristics of the organism itself.

In general, prokaryotic cells grow faster than eukaryotes, and small eukaryotes tend to grow faster than large ones. But when all organisms are considered, doubling times for exponential growth vary enormously, from as little as a few minutes to days or weeks. Organisms living under very stressful conditions, such as those in Earth's nutrient-poor deep subsurface, may divide every few months (or even years) (⇔ Sections 20.7 and 20.13). Much plays into an organism's exponential growth rate and it is hard to predict how fast an organism can grow until it is brought into laboratory culture and its ideal growth conditions have been identified.

Stationary and Death Phases

In a batch culture, exponential growth cannot be maintained indefinitely. Consider the fact that a single cell of a bacterium weighing one-trillionth (10^{-12}) of a gram and growing exponentially with a 20-min generation time would produce, if allowed to grow exponentially in batch culture for 48 h, a population of cells that weighed 4000 times the weight of planet Earth! Obviously this is impossible, and growth becomes limited in such cultures because either an essential nutrient in the culture medium is depleted or the organism's waste products accumulate. When exponential growth ceases for one (or both) of these reasons, the population enters *stationary phase* (Figure 5.7).

In the stationary phase, there is no net increase or decrease in cell number and thus the growth rate of the population is zero. Despite growth arrest, energy metabolism and biosynthetic processes in stationary phase cells may continue, but typically at a greatly reduced rate. Some cells may even divide during stationary

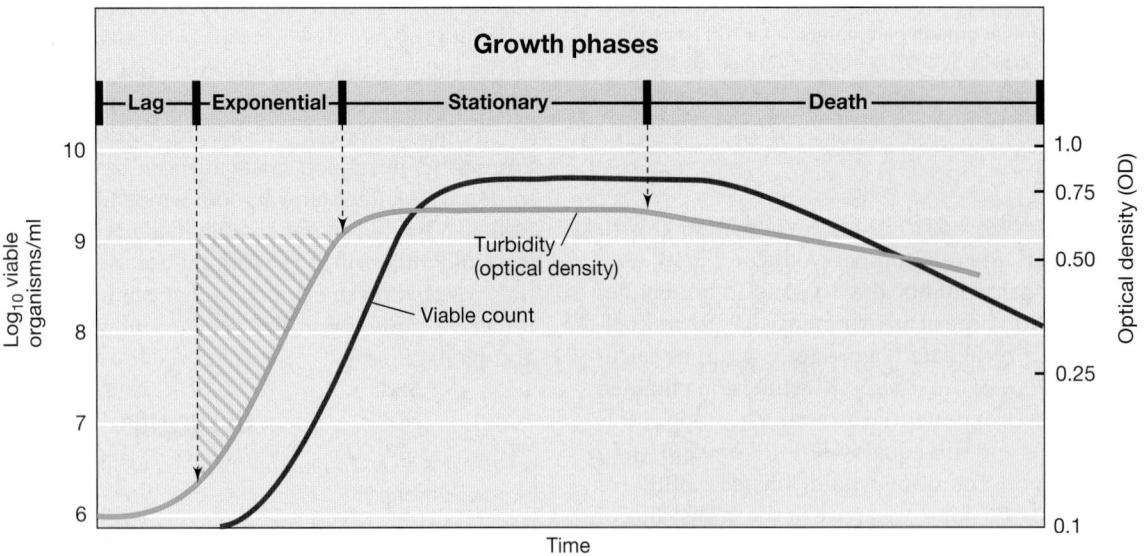

Figure 5.7 Typical growth curve for a bacterial population. A viable count measures the cells in the culture that are capable of reproducing. Optical density (turbidity), a quantitative measure of light scattering by a liquid culture (see Figure 5.16), increases with the increase in cell number.

phase, but no net increase in cell number occurs. This is because some cells in the population grow while others die, the two processes balancing each other out (cryptic growth). Eventually, the population will enter the *death phase* of the growth cycle, which, like the exponential phase, occurs as an exponential function (Figure 5.7). Typically, however, the rate of cell death is much slower than the rate of exponential growth and viable cells may remain in a culture for months or even years.

The phases of bacterial growth shown in Figure 5.7 are reflections of the events in a *population* of cells, not in individual cells. Thus, the terms lag phase, exponential phase, and so on have no meaning with respect to individual cells but only to cell populations. Growth of an individual cell (Section 5.1) is a prerequisite for population growth. But it is population growth that is most relevant to the ecology of microorganisms, because measurable microbial activities require microbial populations, not just an individual microbial cell.

MINIQUIZ

- In which phase of the growth curve do cells divide in a constant time period?
- Under what conditions would a lag phase not occur?
- Why do cells enter stationary phase?

5.4 Continuous Culture

Up to this point our consideration of microbial population growth has been confined to *batch cultures*. The environment in a batch culture is constantly changing because of nutrient consumption and waste production. These limitations can be circumvented in a *continuous culture device*. Unlike a batch culture, which is a *closed* system, a continuous culture is an *open* system. In the continuous culture growth vessel, a known volume of sterile medium is added at a constant rate while an equal volume of spent culture medium (which also contains cells) is removed at the same rate. Once in equilibrium, the culture volume, cell number, and nutrient/waste product status remain constant, and the culture attains *steady state*.

The Chemostat and the Concept of Steady State

The most common type of continuous culture is the **chemostat**, a device wherein both specific growth rate (how fast the cells grow) and cell density (how many cells per ml are obtained) can be controlled independently (**Figure 5.8**). In the chemostat, two factors govern the specific growth rate and cell density, respectively: (1) the *dilution rate* (D) which is expressed as F/V, where F is the flow rate (the rate at which fresh medium is pumped in and spent medium is removed), and V is the culture volume; and (2) the *concentration of a limiting nutrient*, such as a carbon or nitrogen source, present in the sterile medium entering the chemostat vessel.

When a chemostat filled with sterile medium is inoculated, the cells begin growing and increase in number more rapidly than they are removed in the overflow. As cell numbers increase, the level of the limiting nutrient in the culture decreases. This reduction in the growth-limiting nutrient serves as a feedback loop to reduce the specific growth rate, leading to a decrease in cell density as cells are removed in the overflow. However, once the limiting nutrient has decreased to a value just sufficient to support a specific growth rate that compensates for losses of cells through outflow, the chemostat reaches steady state, a condition where cell density and substrate concentration do not change over time.

In steady state, the specific growth rate of the culture is equal to D; that is, the rate of increase in cell numbers due to growth is equal to the rate of decrease in cell numbers due to dilution (outflow). The chemostat steady state is thus a dynamic condition in which cells are continuously growing and continuously being removed. Indeed, this dependency of specific growth rate on substrate concentration (**Figure 5.9**) drives the feedback loop that allows the chemostat to be self-regulating and the experimenter to choose the growth rate of the culture by simply changing the speed of the pump. In a batch culture, the nutrient concentration also affects both growth rate and cell yield, but when the nutrient level exceeds that which supports the maximal growth rate, only cell yield is increased by additional substrate (Figure 5.9).

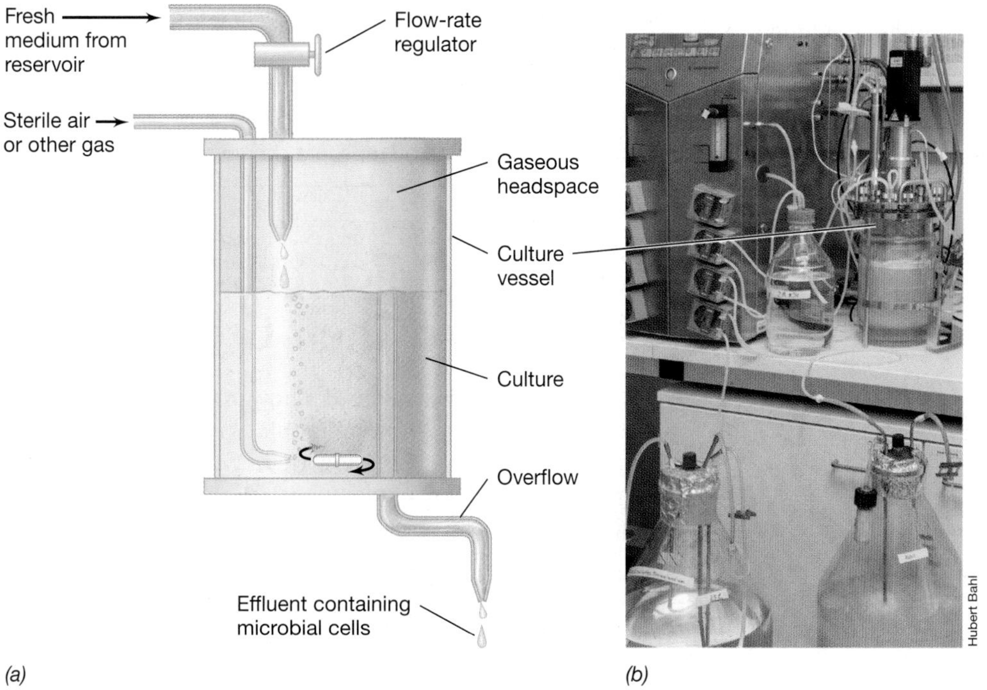

(a)

(b)

Figure 5.8 Continuous culture device (chemostat). The population density is controlled by the concentration of a limiting nutrient in the reservoir, and the growth rate is controlled by the dilution rate; both parameters are set by the experimenter. *(a)* Chemostat components. *(b)* Photo of a chemostat setup.

Fresh medium from reservoir
Flow-rate regulator
Sterile air or other gas
Gaseous headspace
Culture vessel
Culture
Overflow
Effluent containing microbial cells

Hubert Bahl

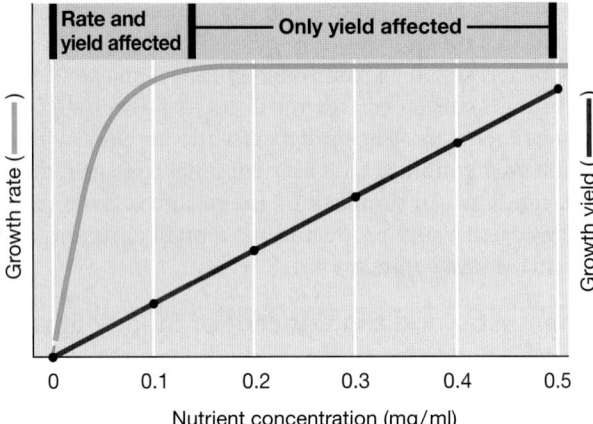

Figure 5.9 The effect of nutrients on growth. Relationship between nutrient concentration, growth rate (green curve), and growth yield (red curve). Only at low nutrient concentrations are both growth rate and growth yield affected.

The effects of varying D in a chemostat culture are shown schematically in **Figure 5.10**. As seen, there are rather wide limits over which D controls growth rate, although at both very low and very high D, the desired steady state with actively growing cells breaks down. In steady state, if the concentration of the limiting nutrient in the inflowing medium is increased at a constant D, cell density will increase but the growth rate will remain the same. Thus, by varying D or the concentration of the growth-limiting nutrient, one can establish dilute (for example, 10^5 cells/ml), moderate (for example, 10^7 cells/ml), or dense (for example, 10^9 cells/ml) cell populations growing at any specific growth rate.

Experimental Uses of the Chemostat

A practical advantage of the chemostat is that a cell population can be maintained in the exponential growth phase for long

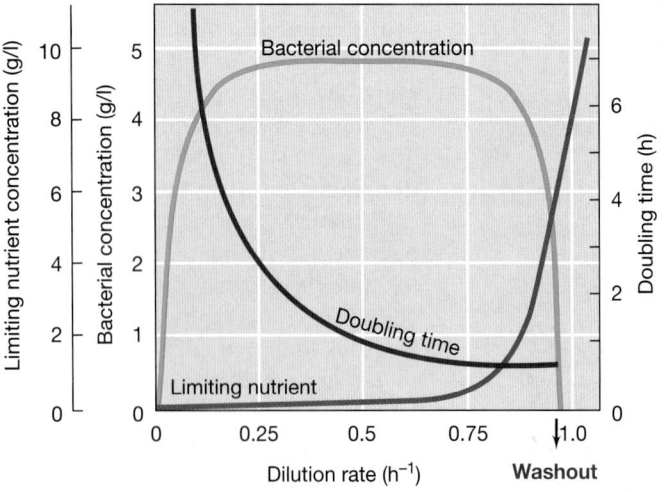

Figure 5.10 Steady-state relationships in the chemostat. The dilution rate (D) is determined from the flow rate and the volume of the culture vessel. Note that at high D, growth cannot balance dilution, and the population washes out. Note also that although both the population density remains constant and the concentration of the growth-limiting nutrient remains near zero during steady state, the specific growth rate (as reflected in the doubling time) can vary over a wide range.

periods—weeks or even months. Exponential phase cells are usually most desirable for physiological experiments. Such cells are available at any time in the chemostat, and the vessel can be repeatedly sampled. Chemostat cultures have been used to study the growth and physiology of cells at submaximal growth rates, and from such studies, several tenets of microbial physiology have emerged; these include the fact that the ribosome content of cells increases in proportion to their specific growth rate and that nutrient concentration controls both specific growth rate and cell yield (Figure 5.9).

The chemostat has also been used in studies of microbial ecology and evolution. For example, because the chemostat can mimic the low substrate concentrations that are often found in nature, it is possible to ask which organisms in mixed cultures of known composition compete best at various specific growth rates or when particular nutrients are limiting. This can be done by monitoring changes in the diversity of the microbial community as a function of variations in D or the limiting nutrient. One can study the evolution of a pure culture in the chemostat by subjecting the culture to a growth or nutrient challenge and asking whether these conditions more rapidly select for particular spontaneous mutants displaying new physiological properties than in batch cultures where all nutrients are in excess.

Chemostats have also been used for the direct enrichment and isolation of bacteria from nature. From a natural sample, one can select a stable population under the chosen conditions of nutrient concentration and D and then slowly increase D until a single organism remains. In this way, microbiologists studying the growth rates of various soil bacteria isolated a bacterium with a 6-min doubling time—the fastest-growing bacterium known!

MINIQUIZ

- How do microorganisms in a chemostat differ from microorganisms in a batch culture?
- What happens in a chemostat if the dilution rate exceeds the maximal specific growth rate of the organism?
- Do pure cultures have to be used in a chemostat?

II • Culturing Microbes and Measuring Their Growth

█ n the next few sections we consider how microbes are grown in laboratory culture and how microbial growth is measured. Culturing microbes and assessing their growth are common events in the daily routine of many microbiologists and microbiology laboratories.

5.5 Growth Media and Laboratory Culture

Laboratory cultures of microorganisms are grown in **culture media**, nutrient solutions tailored to the particular organism to be grown. Because laboratory culture is required for the detailed study of any microorganism, careful attention must be paid to the selection and preparation of media for laboratory culture to be successful. Culture media must be sterilized before use, and sterilization is typically achieved by heating the medium under pressure in an *autoclave*. We discuss the operation and principles of the autoclave in Section 5.15, along with other methods for sterilizing culture media and laboratory devices.

Classes of Culture Media

Two broad classes of culture media are used in microbiology: *defined media* and *complex media*. **Defined media** are prepared by adding precise amounts of pure inorganic or organic chemicals to distilled water. Therefore, the *exact composition* of a defined medium (in both a qualitative and quantitative sense) is known. Of major importance in any culture medium is the carbon source because all cells need large amounts of carbon to make new cell material (↩ Section 3.1). The particular carbon source and its concentration depend on the organism to be cultured. Table 5.1 lists recipes for four different culture media. Some defined media, such as the one listed for *Escherichia coli*, are considered "simple" because they contain only a single carbon source. In such a medium, *E. coli* must biosynthesize all organic molecules from glucose.

For culturing many microorganisms, knowledge of the exact composition of a medium is not essential. In these instances complex media may suffice and may even be advantageous. **Complex media** are made from digests of microbial, animal, or plant products, such as milk protein (casein), beef (beef extract), soybeans

(tryptic soy broth), yeast cells (yeast extract), or any of a number of other highly nutritious substances (Table 5.1). Such digests are commercially available in dehydrated form and need only be hydrated with distilled water to form a culture medium. However, the disadvantage of a complex medium is that its exact nutritional composition is unknown. An *enriched medium*, used for the culture of nutritionally demanding (fastidious) microbes, many of which are pathogens, is a complex medium to which additional highly nutritious substances (such as serum or blood) are added.

Culture media are sometimes prepared that are selective or differential (or both), especially media used in diagnostic microbiology. A *selective medium* contains compounds that inhibit the growth of some microorganisms but not others. For example, selective media are available for the isolation of certain pathogens, such as *Salmonella* or those strains of *E. coli* that cause foodborne illnesses. A *differential medium* is one to which an indicator (typically a dye) is added, which reveals by a color change whether a particular metabolic reaction has occurred during growth. Differential media are useful for distinguishing bacteria and are widely used in clinical diagnostics and microbial taxonomy. Differential and selective media are further discussed in Chapter 28.

Nutritional Requirements and Biosynthetic Capacity

Of the four culture medium recipes in Table 5.1, three are defined and one is complex. The complex medium is easiest to prepare and

TABLE 5.1 Examples of culture media for microorganisms with simple and demanding nutritional requirements[a]

Defined culture medium for *Escherichia coli*	Defined culture medium for *Leuconostoc mesenteroides*	Complex culture medium for either *E. coli* or *L. mesenteroides*	Defined culture medium for *Thiobacillus thioparus*
K_2HPO_4 7 g	K_2HPO_4 0.6 g	Glucose 15 g	KH_2PO_4 0.5 g
KH_2PO_4 2 g	KH_2PO_4 0.6 g	Yeast extract 5 g	NH_4Cl 0.5 g
$(NH_4)_2SO_4$ 1 g	NH_4Cl 3 g	Peptone 5 g	$MgSO_4$ 0.1 g
$MgSO_4$ 0.1 g	$MgSO_4$ 0.1 g	KH_2PO_4 2 g	$CaCl_2$ 0.05 g
$CaCl_2$ 0.02 g	Glucose 25 g	Distilled water 1000 ml	KCl 0.5 g
Glucose 4–10 g	Sodium acetate 25 g	pH 7	$Na_2S_2O_3$ 2 g
Trace elements (Fe, Co, Mn, Zn, Cu, Ni, Mo) 2–10 µg each	Amino acids (alanine, arginine, asparagine, aspartate, cysteine, glutamate, glutamine, glycine, histidine, isoleucine, leucine, lysine, methionine, phenylalanine, proline, serine, threonine, tryptophan, tyrosine, valine) 100–200 µg of each		Trace elements (as in first column) 2–10 µg each
Distilled water 1000 ml			Distilled water 1000 ml
pH 7	Purines and pyrimidines (adenine, guanine, uracil, xanthine) 10 mg of each		pH 7
	Vitamins (biotin, folate, nicotinic acid, pyridoxal, pyridoxamine, pyridoxine, riboflavin, thiamine, pantothenate, *p*-aminobenzoic acid) 0.01–1 mg of each		Carbon source: CO_2 from air
	Trace elements (as in first column) 2–10 µg each		
	Distilled water 1000 ml		
	pH 7		

(a)

(b)

[a]The photos are tubes of *(a)* the defined medium described, and *(b)* the complex medium described. Note how the complex medium is colored from the various organic extracts and digests that it contains. Photo credits: Cheryl L. Broadie and John Vercillo, Southern Illinois University at Carbondale.

supports growth of both *Escherichia coli* and *Leuconostoc mesenteroides*, the examples used in Table 5.1. By contrast, the simple defined medium supports growth of *E. coli* but not of *L. mesenteroides*. Growth of the latter in a defined medium requires the addition of several nutrients not needed by *E. coli*. The nutritional needs of *L. mesenteroides* can be satisfied by preparing either a highly supplemented defined medium, a rather laborious undertaking because of all the individual nutrients that need to be added (Table 5.1), or by preparing a complex medium, a much less demanding operation.

The fourth medium listed in Table 5.1 supports growth of the bacterium *Thiobacillus thioparus* but would not support the growth of any of the other organisms. This is because *T. thioparus* is both a chemolithotroph and an autotroph (⊘ Section 3.3) and thus has no organic carbon requirements. *T. thioparus* derives all of its carbon from CO_2 and conserves energy from the oxidation of the sulfur compound thiosulfate ($Na_2S_2O_3$). Thus, *T. thioparus* has the greatest biosynthetic capacity of all the organisms listed in the table, surpassing even *E. coli* in this regard.

The take-home lesson from Table 5.1 is that different microorganisms may have vastly different nutritional requirements. For successful cultivation, it is necessary to understand an organism's physiology and nutritional requirements and then supply it with the nutrients it needs in both the proper form and amount.

Laboratory Culture

Once a sterile culture medium has been prepared, microbes can be inoculated and the culture can be incubated under conditions that will support growth. In a laboratory situation, inoculation will typically be with a pure culture and into a culture medium that is either liquid (Table 5.1) or solid (**Figure 5.11**). Liquid culture media are solidified with *agar*, an algal polysaccharide. Solid media immobilize cells, allowing them to grow and form visible, isolated masses called *colonies* (Figure 5.11). Microbial colonies are of various shapes and sizes depending on the organism, the culture conditions, the nutrient supply, and other physiological parameters. Some microorganisms produce pigments that cause the colony to be colored (Figure 5.11). Colonies permit the microbiologist to visualize the composition and presumptive purity of a culture. Plates inoculated from a mixed culture (such as a natural sample, Figure 5.11*e*) or from a contaminated pure culture will usually contain more than one colony type.

Once a sterile culture medium has been prepared, it is ready for inoculation. This requires **aseptic technique** (**Figure 5.12**), a series of steps to prevent contamination during manipulations of cultures and sterile culture media, both liquid and solid. With liquid medium, the goal is to transfer a culture while protecting the tube or bottle rim from air currents or contact with nonsterile surfaces (Figure 5.12*a*). With agar Petri plates, the plan is basically the same but with a greater emphasis on keeping the surface of

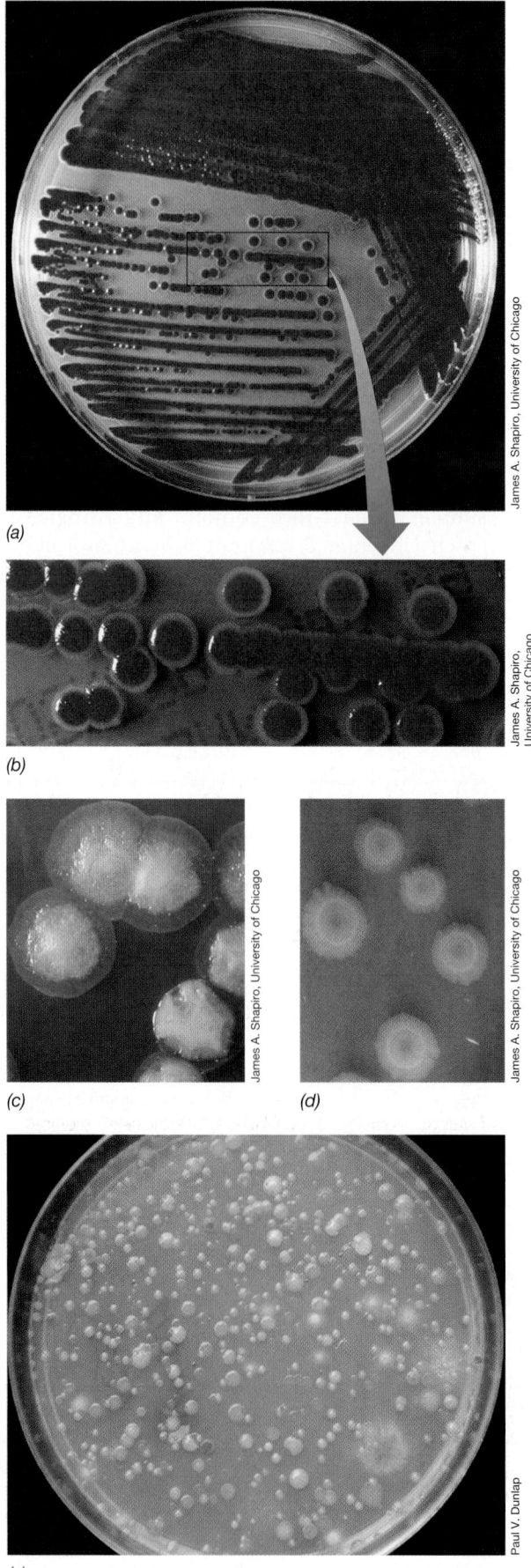

Figure 5.11 Bacterial colonies. Colonies are visible masses of cells formed from the division of one or a few cells and can contain over a billion (10^9) individual cells. *(a) Serratia marcescens*, grown on MacConkey agar. *(b)* Close-up of colonies outlined in part *a*. *(c) Pseudomonas aeruginosa*, grown on trypticase soy agar. *(d) Shigella flexneri*, grown on MacConkey agar. *(e)* An agar plate containing many different bacterial colonies that developed from plating a dilution of seawater.

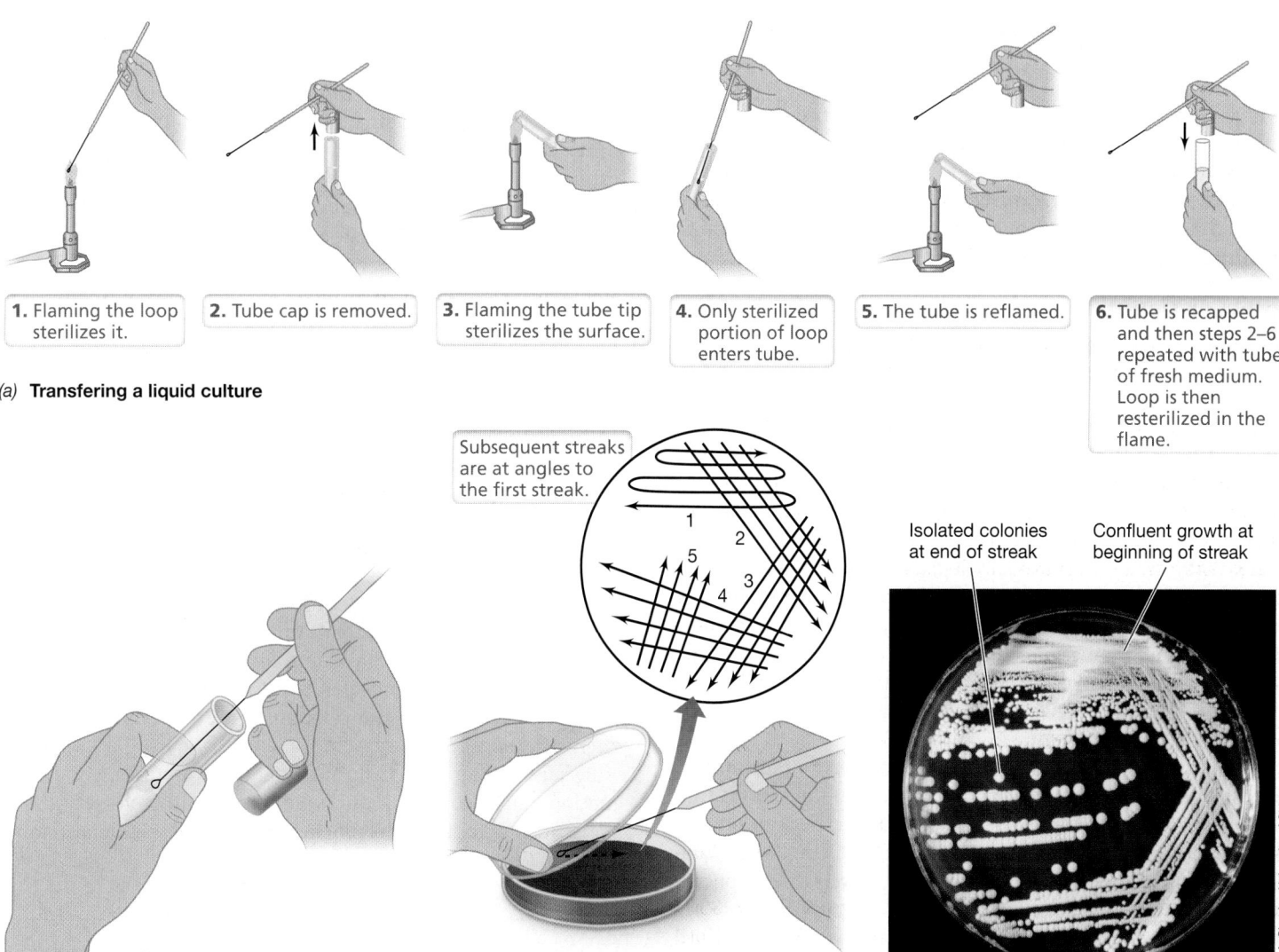

1. Flaming the loop sterilizes it.

2. Tube cap is removed.

3. Flaming the tube tip sterilizes the surface.

4. Only sterilized portion of loop enters tube.

5. The tube is reflamed.

6. Tube is recapped and then steps 2–6 repeated with tube of fresh medium. Loop is then resterilized in the flame.

(a) **Transfering a liquid culture**

Subsequent streaks are at angles to the first streak.

Isolated colonies at end of streak

Confluent growth at beginning of streak

James A. Shapiro, University of Chicago

1. Loop is sterilized and a loopful of inoculum is removed from tube.

2. Initial streak is worked in well in one corner of the agar plate.

3. Appearance of well-streaked plate after incubation shows colonies of the bacterium *Micrococcus luteus* on a blood agar plate.

(b) **Streaking a Petri plate**

Figure 5.12 Aseptic transfer. *(a)* Liquid media: After the tube is recapped at the end, the loop is resterilized. *(b)* Solid media: making a streak plate to obtain pure cultures. The plate cover is opened just enough to permit streaking manipulations. In streaking a plate, the microbial cells are separated by the streaking process to yield widely separated single cells that then grow and divide to form colonies.

the agar protected from aerosols or particulate matter that could drop in (Figure 5.12*b*).

A mastery of aseptic technique is required for maintaining pure cultures, as airborne contaminants are virtually everywhere, even in what may appear to be a very clean microbiology laboratory. Picking an isolated colony and restreaking it is the main method for obtaining pure cultures from mixed liquid or plate (Figure 5.11*e*) cultures and is a common procedure in the microbiology laboratory. Other techniques for obtaining pure cultures have been developed that are especially suited for particular groups of bacteria with unusual growth requirements and these will be discussed in Sections 19.2 and 19.3.

MINIQUIZ

- Why would a complex culture medium for *Leuconostoc mesenteroides* be easier to prepare than a chemically defined medium?

- In which medium shown in Table 5.1, defined or complex, do you think *Escherichia coli* would grow the fastest? Why? *E. coli* will not grow in the medium described for *Thiobacillus thioparus*. Why?

- What is meant by the word sterile? Why is aseptic technique necessary for successful maintenance of pure cultures in the laboratory?

- How many cells could be present in a single bacterial colony?

UNIT 2

5.6 Microscopic Counts of Microbial Cell Numbers

Assessing cell numbers gives quantitative information on the state of a microbial culture or community. Several methods for enumerating a microbial population have been developed, each having their own strengths and caveats. We begin with the classic "total count" carried out by microscopic examination of a culture or natural sample.

Total Cell Count

Total counts of microbial numbers in a culture or natural sample can be done by simply observing and enumerating the cells present by a *microscopic cell count*. Microscopic counts can be performed either on samples dried on slides or on liquid samples. Dried samples can be stained to increase contrast between cells and their background (⇌ Sections 1.6 and 19.3). With liquid samples, counting chambers consisting of a grid with squares of known area etched on the surface of a glass slide are used (**Figure 5.13**). When the coverslip is placed on the chamber, each square on the grid has a precise volume. The number of cells per unit area of grid can be counted under the microscope, giving a measure of the number of cells per small chamber volume. The number of cells per milliliter of suspension is calculated by employing a conversion factor based on the volume of the chamber sample (Figure 5.13).

Microscopic counting is a quick and easy way of estimating microbial cell numbers. However, it has several limitations that restrict its usefulness. For example, without special staining techniques (⇌ Section 19.3), dead cells cannot be distinguished from live cells, and precision is difficult to achieve, even when replicate counts are made. Moreover, small cells are often difficult to see under the microscope, which can lead to erroneous counts, and cell suspensions of low density (less than about 10^6 cells/milliliter) will have few if any cells in a microscope field unless the sample is first concentrated and resuspended in a small volume. Finally, motile cells must be killed (usually with formaldehyde) or otherwise immobilized before counting, and debris in the sample may easily be mistaken for microbial cells.

Microscopic Cell Counts in Microbial Ecology

Despite its many potential caveats, microbial ecologists often use microscopic cell counts on natural samples. But they do so using stains to visualize the cells, often stains that yield phylogenetic or other key information about the cells, such as their metabolic properties.

There are many stains that can be used in a general way. For example, the stain DAPI (⇌ Section 1.6 and Figure 1.20e) stains all cells because it reacts with DNA. Other general fluorescent stains can differentiate live from dead cells by detecting whether the cytoplasmic membrane is intact or not. By contrast, fluorescent stains that are highly specific for certain organisms or groups of related organisms can be prepared by attaching fluorescent dyes to specific nucleic acid probes. For example, *phylogenetic stains* that stain only species of *Bacteria* or only species of *Archaea* can be used in combination with nonspecific stains to determine the proportion of each domain present in a sample (⇌ Section 19.4). Other fluorescent probes have been developed that target genes encoding enzymes that catalyze specific metabolic processes; if a cell is stained by one of these probes, a metabolism can be inferred that may reveal the cell's ecological role in the microbial community. In all of these cases, if cells in the sample are present in only low numbers, for example in a sample of ocean water, this limitation can be overcome by first concentrating the cells on a filter and then counting them after staining.

Because they are easy to do and often yield useful baseline information, microscopic cell counts are common in ecological studies of natural microbial environments. We pursue this theme in more detail in Chapter 19.

MINIQUIZ

- What are some of the problems that can arise when unstained preparations are enumerated in microscopic counts?
- Using microscopic techniques, how could you tell whether *Archaea* were present in an alpine lake where total cell numbers were only 10^5/ml?

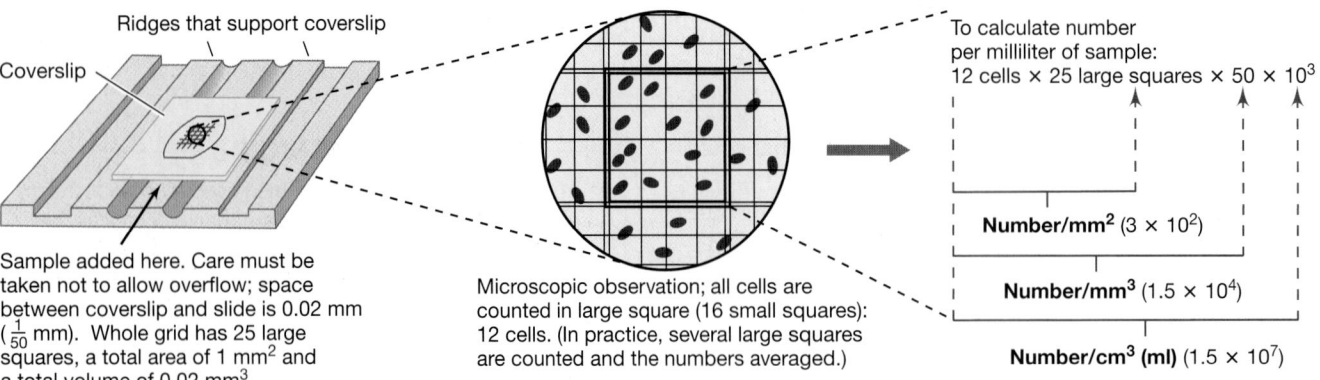

Figure 5.13 Direct microscopic counting procedure using the Petroff–Hausser counting chamber.
A phase-contrast microscope is typically used to count the cells to avoid the necessity for staining.

Ridges that support coverslip

Coverslip

Sample added here. Care must be taken not to allow overflow; space between coverslip and slide is 0.02 mm ($\frac{1}{50}$ mm). Whole grid has 25 large squares, a total area of 1 mm^2 and a total volume of 0.02 mm^3.

Microscopic observation; all cells are counted in large square (16 small squares): 12 cells. (In practice, several large squares are counted and the numbers averaged.)

To calculate number per milliliter of sample:
12 cells × 25 large squares × 50 × 10^3

Number/mm^2 (3 × 10^2)

Number/mm^3 (1.5 × 10^4)

Number/cm^3 (ml) (1.5 × 10^7)

5.7 Viable Counting of Microbial Cell Numbers

A **viable** cell is one that is able to divide and form offspring, and in most cell-counting situations, these are the cells we are most interested in. For these purposes, one would use a **viable count**, also called a **plate count** because agar plates are required. The assumption made in a viable count is that each viable cell will grow and divide to yield one colony, and hence, colony numbers are a reflection of cell numbers (Section 5.5 and Figure 5.11).

Methods for Viable Counts

There are at least two ways of performing a plate count: the *spread-plate* method and the *pour-plate* method (**Figure 5.14**). In the spread-plate method, a volume (usually 0.1 ml or less) of an appropriately diluted culture is spread over the surface of an agar plate using a sterile glass spreader. In the pour-plate method, a known volume (usually 0.1–1.0 ml) of culture is pipetted into a sterile Petri plate. Molten agar medium, tempered to just above gelling temperature (~50°C), is then added and gently mixed before allowing the agar to solidify.

With both the spread-plate and pour-plate methods, it is important that the number of colonies developing on or in the medium not be too many or too few. On crowded plates some cells may not form colonies, and some colonies may fuse, leading to erroneous measurements. If the number of colonies is too small, the statistical significance of the calculated count will be low. The usual practice, which is most valid statistically, is to count colonies only on plates that contain between 30 and 300 colonies.

To obtain a countable colony number, the sample must almost always be diluted. Because one may not know the approximate viable count ahead of time, it is usually necessary to make more than one dilution. Several 10-fold dilutions of the sample are commonly used (**Figure 5.15**). To make a 10-fold (10^{-1}) dilution, one can mix 0.5 ml of sample with 4.5 ml of diluent, or 1.0 ml of sample with 9.0 ml of diluent. If a 100-fold (10^{-2}) dilution is needed, 0.05 ml can be mixed with 4.95 ml of diluent, or 0.1 ml with 9.9 ml of diluent. Alternatively, a 10^{-2} dilution can be achieved by making two successive 10-fold dilutions. With dense cultures, such *serial* dilutions are needed to reach a suitable dilution for plating to yield countable colonies. Thus, if a 10^{-6} ($1/10^6$) dilution is needed, it can be achieved by making three successive 10^{-2} ($1/10^2$) dilutions or six successive 10^{-1} dilutions (Figure 5.15).

Sources of Error in Plate Counting

The number of colonies obtained in a viable count depends not only on the inoculum size and the viability of the culture, but also on the culture medium and the incubation conditions. The colony number can also change with the length of incubation. For example, if a mixed culture is counted, the cells deposited on the plate will not all form colonies at the same rate; if a short incubation time is used, fewer than the maximum number of colonies will be obtained. Furthermore, the size of colonies may vary. If some tiny colonies develop, they may be missed during the counting. With pure cultures, colony development is a more synchronous process and uniform colony morphology is the norm.

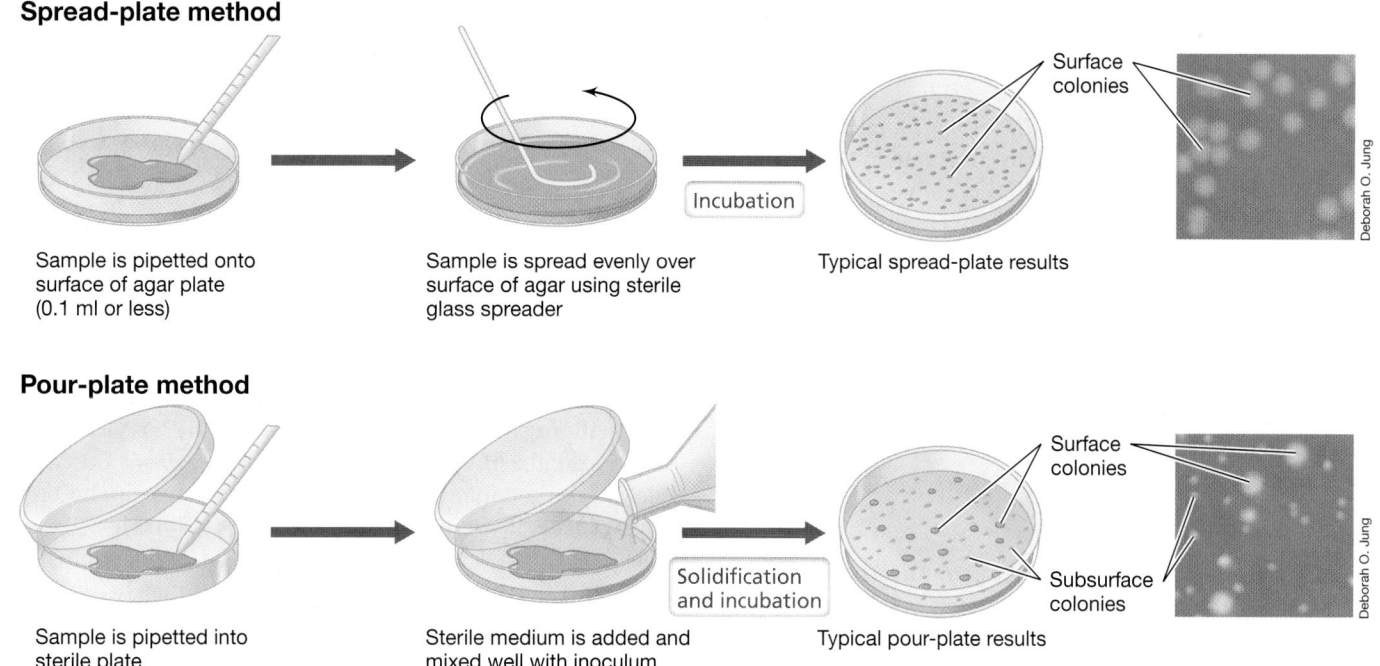

Spread-plate method

Sample is pipetted onto surface of agar plate (0.1 ml or less) → Sample is spread evenly over surface of agar using sterile glass spreader → Incubation → Typical spread-plate results → Surface colonies

Pour-plate method

Sample is pipetted into sterile plate → Sterile medium is added and mixed well with inoculum → Solidification and incubation → Typical pour-plate results → Surface colonies / Subsurface colonies

Figure 5.14 Two methods for the viable count. Only surface colonies form in the spread plate technique. By contrast, in the pour-plate method, colonies form within the agar as well as on the agar surface. On the far right are photos of colonies of *Escherichia coli* formed from cells plated by the spread-plate method (top) or the pour-plate method (bottom).

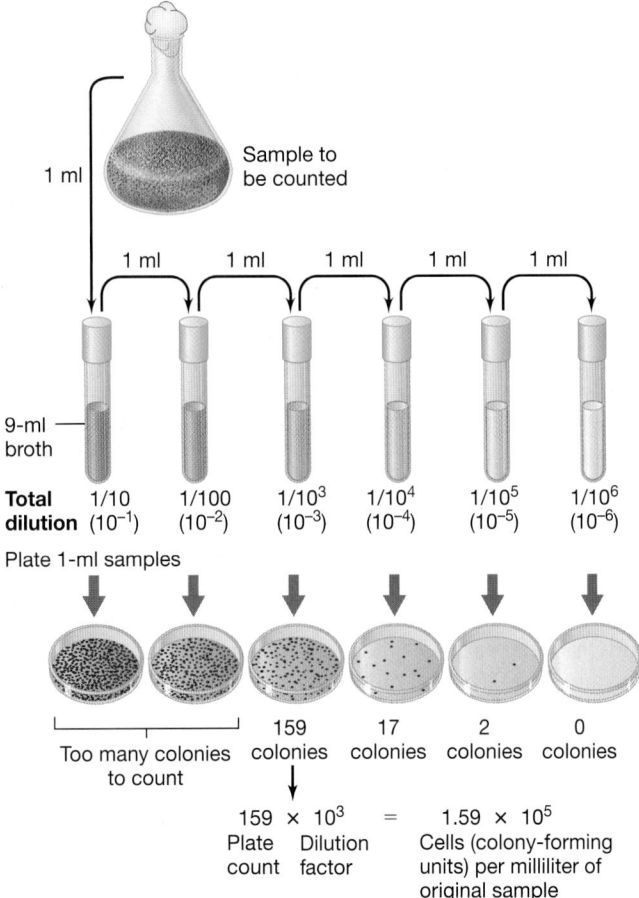

Figure 5.15 Procedure for viable counting using serial dilutions of the sample and the pour-plate method. The sterile liquid used for making dilutions can simply be water, but a solution of mineral salts or actual growth medium may yield a higher recovery. The dilution factor is the reciprocal of the dilution.

Viable counts can be subject to error for several reasons. These include plating inconsistencies, such as inaccurate pipetting of a liquid sample, a nonuniform sample (for example, a sample that has formed cell clumps of differing cell number), insufficient mixing, and heat intolerance if the pour plate method is used. Hence, if accurate counts are to be obtained, great care and consistency must be taken in sample preparation and plating, and replicate plates of key dilutions must be prepared. Because sources of error in plate counting are often considerable, data are typically expressed as the number of *colony-forming units* obtained rather than the actual number of viable cells, to account for clumps containing more than one viable cell.

Applications of the Plate Count

Despite the caveats associated with viable counting, the procedure is quick and easy to do and so is widely used in many subdisciplines of microbiology. For example, in food, dairy, medical, and aquatic microbiology, viable counts are employed routinely. The method has the virtue of high sensitivity, because as few as one viable cell per sample plated can be detected. This feature allows for the sensitive detection of microbial contamination of foods or other materials.

The use of highly selective culture media and growth conditions allows the plate count to be used to target particular species in a sample containing many organisms. For example, a complex medium containing 10% NaCl is very useful in isolating species of potentially pathogenic *Staphylococcus* from skin, because the salt inhibits growth of most other skin bacteria. In practical applications such as in the food industry, viable counting performed using both a complex medium and a selective medium (Section 5.5) on the same sample allows for simultaneous quantitative and qualitative assessments of food quality and safety. That is, with a single food sample, the complex medium yields a total cell count—a relative indicator of freshness and shelf life—while the selective medium indicates the presence or absence of a particular pathogen that may be transmitted in this particular food.

Targeted counting is also common in wastewater and other water analyses. For instance, enteric bacteria such as *Escherichia coli* originate from feces and are easy to recover from natural samples using selective media; if enteric bacteria are detected in a water sample from a swimming site, for example, their presence is a signal that the water contains fecal matter and is therefore unsafe for human contact.

The Great Plate Count Anomaly

Direct microscopic counts of natural samples typically reveal far more microbes than are recoverable on plates of any single culture medium. Thus, although a very sensitive technique, plate counts can be highly unreliable when used to assess total cell numbers of natural samples, such as soil and water. In microbiology, this is referred to as "the great plate count anomaly."

Why do plate counts reveal lower numbers of cells than direct microscopic counts? One obvious factor is that microscopic methods count dead cells, whereas by definition, viable methods do not. More important, however, is the fact that different organisms, even those present in a very small natural sample, may have vastly different requirements for nutrients and growth conditions in laboratory culture (Sections 3.1 and 5.5 and Table 5.1). Thus, one medium and set of growth conditions can only be expected to support the growth of one subset of the total microbial community. If this subset makes up, for example, 10^6 cells/g of a total viable community of 10^9 cells/g, the plate count will reveal only 0.1% of the viable cell population, a vast underestimation of the actual number of organisms present in the sample.

Plate count results thus carry a large caveat. Plate counts targeted to specific organisms using highly selective media (Section 5.5), as in, for example, the microbial analysis of sewage or food, can often yield quite reliable data, since the physiology of the targeted organisms is known and so the recovery of viable cells is near 100%. By contrast, "total" cell counts of the same samples using a single medium and set of growth conditions may be, and usually are, underestimates of actual cell numbers by one to several orders of magnitude. These days, however, counting cells using a cell proxy has become common, as a wide variety of molecular methods have been developed to both detect and quantify specific organisms in natural samples. We explore these methods and what they can tell us in Chapter 19.

MINIQUIZ

- Why is a viable count more sensitive than a microscopic count? What major assumption is made in relating plate count results to cell number?
- Describe how you would dilute a bacterial culture by 10^{-7}.
- Explain the "great plate count anomaly."

5.8 Turbidimetric Measures of Microbial Cell Numbers

During exponential growth, all cellular components increase in proportion to the increase in cell numbers. One such cellular component is cell mass itself. Cells scatter light, and a rapid and widely used technique in microbiology for estimating cell mass is *turbidity*. A cell suspension looks cloudy (turbid) to the eye because cells scatter light that passes through the suspension. The more cells that are present, the more light is scattered, and hence the more turbid the suspension. Because cell mass is proportional to cell number, turbidity measurements can quickly estimate cell numbers in laboratory cultures.

Optical Density and Its Relationship to Cell Numbers

Turbidity is measured with a spectrophotometer, an instrument that passes light through a cell suspension and measures the unscattered light that emerges (**Figure 5.16**). A spectrophotometer employs a prism or diffraction grating to generate incident light of a specific wavelength (Figure 5.16*a*). Commonly used wavelengths

UNIT 2

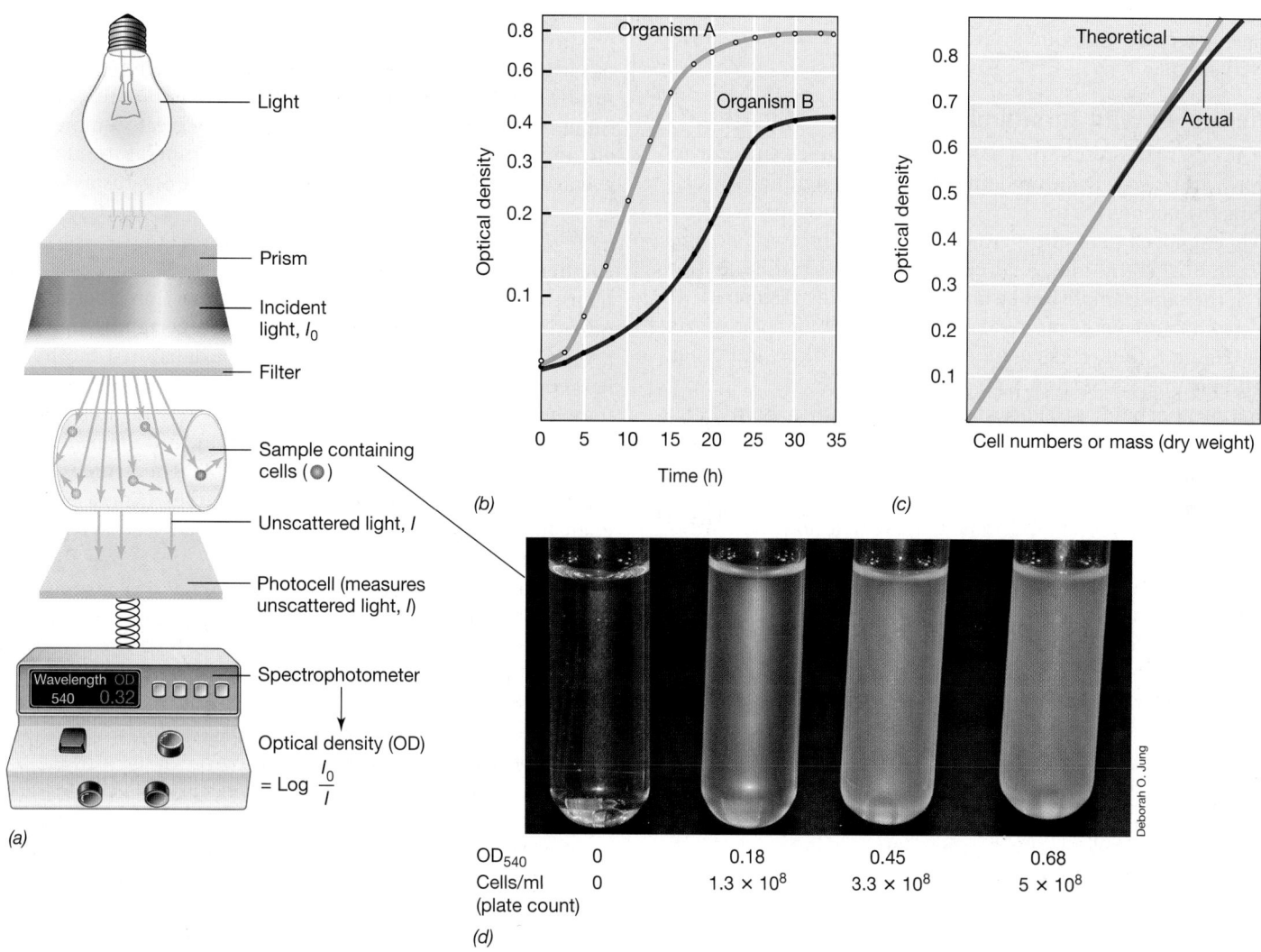

Figure 5.16 Turbidity measurements of microbial growth. (*a*) Measurements of turbidity are made in a spectrophotometer. The photocell measures incident light unscattered by cells in suspension and gives readings in optical density units. (*b*) Typical growth curve data for two organisms growing at different growth rates. For practice, calculate the generation time of the two cultures using the formula $n = 3.3$ ($\log N - \log N_0$) where N and N_0 are two different OD readings with a time interval t between the two. Which organism is growing faster, A or B? (*c*) Relationship between cell number or dry weight and turbidity readings. Note that the one-to-one correspondence between these relationships breaks down at high turbidities. (*d*) Liquid cultures of *Escherichia coli*. The increasing (left to right) optical density (OD$_{540}$) of each culture is shown below the tube, as is the actual cell number measured in a viable count.

for microbial turbidity measurements include 480 nm (blue), 540 nm (green), and 660 nm (red). Sensitivity is best at shorter wavelengths, but measurements of dense cell suspensions are more accurate at longer wavelengths.

The unit of turbidity is *optical density* (OD) at the wavelength specified, for example, OD_{540} for measurements at 540 nm (Figure 5.16). For unicellular organisms, optical density is proportional, within certain limits, to cell number. Turbidity readings can therefore be used as a substitute for total or viable counting methods. However, before this can be done, a standard curve must be prepared that relates cell number (microscopic or viable count) to turbidity. As can be seen in such a plot, proportionality only holds within limits (Figure 5.16c). At high cell densities, light scattered away from the spectrophotometer's photocell by one cell can be scattered back toward the photocell by another, and as a result, the one-to-one correspondence between cell number and turbidity deviates from linearity. Nevertheless, when a standard has been prepared, turbidimetric estimates of bacterial abundance are very useful.

Other Issues with Turbidimetric Growth Estimates

On the one hand, turbidity measurements are quick and easy to perform and can be made without destroying or significantly disturbing the sample. For these reasons, turbidity measurements are widely employed to monitor growth of pure cultures of *Bacteria*, *Archaea*, and many microbial eukaryotes. With turbidimetric assays, the same sample can be checked repeatedly (Figure 5.16d) and the measurements plotted on a semilogarithmic plot versus time (Figure 5.16b) to calculate the generation time and other growth parameters (Section 5.2).

On the other hand, turbidity measurements can occasionally be problematic. Although many microorganisms grow evenly distributed in suspensions in liquid medium, many do not. Some bacteria routinely form small to large clumps, and in such instances, OD measurements may be quite inaccurate as a measure of total microbial mass. In addition, many bacteria form biofilms on the sides of tubes or other growth vessels (Section 5.1). Hence for OD measurements to be an accurate reflection of cell mass (and thus cell numbers) in a liquid culture, clumping and biofilms have to be minimized. This can be accomplished by stirring, shaking, or in some way keeping the cells well mixed during the growth process to prevent the formation of cell aggregates and the sticking of swimming cells to surfaces. Some bacteria are just naturally planktonic and stay well suspended in liquid medium for long periods. But if a solid surface is available, most bacteria will eventually develop a static biofilm, and accurately quantifying cell numbers by turbidity in such a case can be difficult or even impossible.

MINIQUIZ

- List two advantages of using turbidity as a measure of cell growth.
- Describe how you could use a turbidity measurement to tell how many colonies you would expect from plating a culture of a given OD.

III • Environmental Effects on Growth: Temperature

Even when a microbe is provided with an optimal array of its required nutrients, growth is not a sure thing unless the chemical and physical state of its environment is also suitable. Four environmental factors control microbial growth in a major way: temperature, pH, water availability, and oxygen. If any one of these factors is beyond the limits that an organism can tolerate, growth will not occur, even in an ideal culture medium. In Parts III and IV of this chapter we examine these important environmental factors, beginning with temperature, the key factor affecting the growth and survival of microorganisms.

5.9 Temperature Classes of Microorganisms

At either too cold or too hot a temperature, microorganisms will not be able to grow and may even die. The minimum and maximum temperatures supporting growth vary greatly among different organisms and usually reflect the temperature range and average temperature of the environments the organisms inhabit.

Cardinal Temperatures

Temperature affects microorganisms in two opposing ways. As temperatures rise, the rate of enzymatic reactions increases and growth becomes faster. However, above a certain temperature, proteins or other cell components may be denatured or otherwise irreversibly damaged. For every microorganism there is a *minimum* temperature below which growth is not possible, an *optimum* temperature at which growth is most rapid, and a *maximum* temperature above which growth is not possible. These three temperatures, called the **cardinal temperatures** (**Figure 5.17**), are characteristic for any given microorganism and can differ dramat-

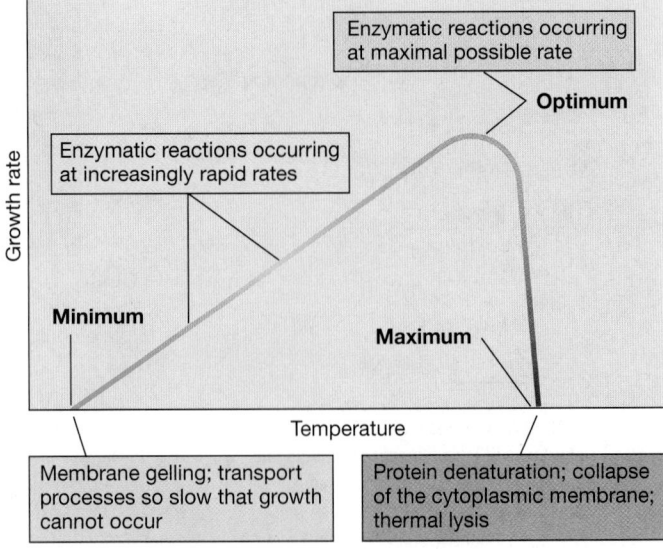

Figure 5.17 The cardinal temperatures: minimum, optimum, and maximum. The actual values may vary greatly for different organisms (see Figure 5.18).

ically between species. For example, some organisms have growth temperature optima near 0°C, whereas the optima for others can be higher than 100°C. The temperature range throughout which microbial growth is possible is even wider than this, from as low as −15°C to at least 122°C. However, no single organism can grow over this whole temperature range, as the range for any given organism is typically less than 40°C.

The maximum growth temperature of an organism reflects the temperature above which denaturation of one or more essential cell components, such as a key enzyme, occurs. The factors controlling an organism's minimum growth temperature are not as clear. However, the cytoplasmic membrane must remain in a semifluid state for nutrient transport and bioenergetic functions to take place. That is, if an organism's cytoplasmic membrane stiffens to the point that it no longer functions properly in transport or can no longer develop or consume a proton motive force, the organism cannot grow. In contrast to the minimum and maximum, the growth temperature *optimum* reflects a state in which all or most cellular components are functioning at their maximum rate and typically lies closer to the maximum than to the minimum (see Figure 5.18).

Temperature Classes of Organisms

Although there is a continuum of organisms, from those with very low temperature optima to those with high temperature optima, it is possible to distinguish four broad classes of microorganisms in relation to their growth temperature optima: **psychrophiles**, with low temperature optima; **mesophiles**, with midrange temperature optima; **thermophiles**, with high temperature optima; and **hyperthermophiles**, with very high temperature optima (**Figure 5.18**).

Mesophiles are widespread in nature and are the most commonly studied microorganisms. Mesophiles are found in the intestines of endothermic (warm-blooded) animals and in terrestrial and aquatic environments in temperate and tropical latitudes. *Escherichia coli* is a typical mesophile, and its cardinal

temperatures have been precisely defined. The optimum temperature for most strains of *E. coli* is near 39°C, the maximum is 48°C, and the minimum is 8°C. Thus, the temperature *range* for *E. coli* is about 40 degrees (Figure 5.18).

Psychrophiles and thermophiles are found in unusually cold and unusually hot environments, respectively. Hyperthermophiles are found in extremely hot habitats such as hot springs, where temperatures can be as hot as 100°C, and deep-sea hydrothermal vents, where temperatures can exceed 100°C. We now consider these fascinating microbes and examine some of the physiological problems they face and some of the biochemical solutions they have evolved to thrive under these extreme conditions.

--- **MINIQUIZ** ---

- How does a hyperthermophile differ from a psychrophile?
- What are the cardinal temperatures for *Escherichia coli*? To what temperature class does it belong?
- *E. coli* can grow at a higher temperature in a complex medium than in a defined medium. Why?

5.10 Microbial Life in the Cold

Because humans live and work in places where temperatures are moderate, it is natural to consider very hot and very cold environments as "extreme." However, many microbial habitats are indeed very hot or very cold, and organisms that inhabit these environments—called *extremophiles* (⇆ Section 1.3 and Table 1.1)—actually thrive in these punishing environments. We consider the biology of these fascinating organisms here and in the next section.

Cold Environments

Much of Earth's surface is cold. The oceans, which make up over half of Earth's surface, have an average temperature of 5°C, and

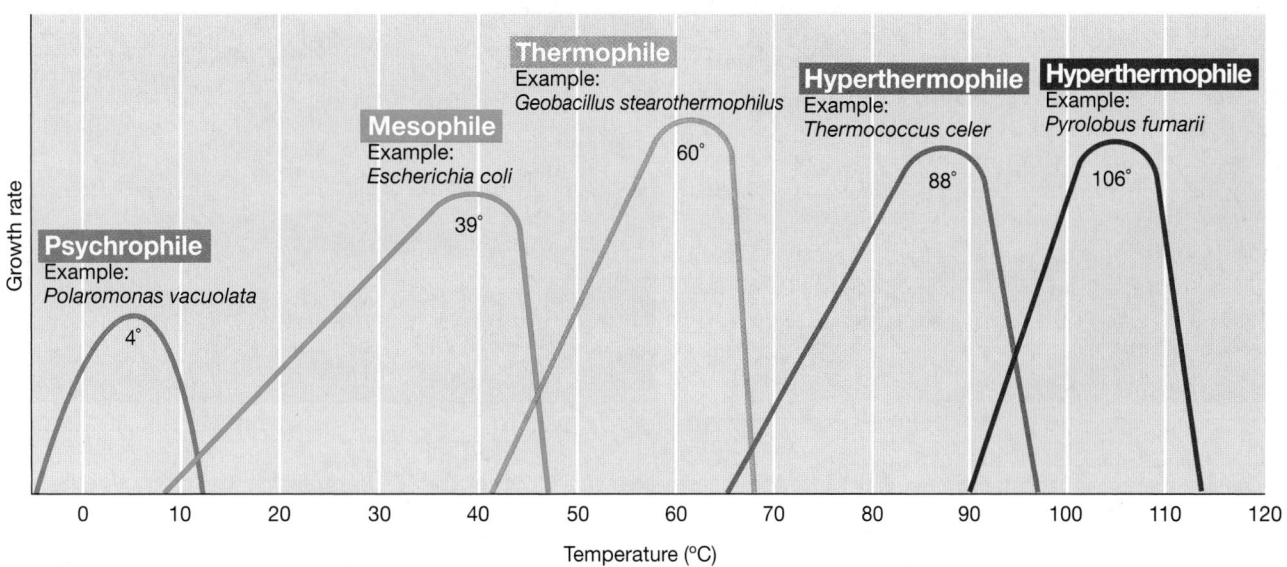

Figure 5.18 Temperature and growth response in different temperature classes of microorganisms. The temperature optimum of each example organism is shown on the graph.

the depths of the open oceans have constant temperatures of 1–3°C. Vast areas of the Arctic and Antarctic are permanently frozen or are unfrozen for only a few weeks in summer (**Figure 5.19**). These cold environments support diverse microbial life, as do glaciers (Figure 5.19*e*), where the networks of liquid water channels that run through and under the glacier are teeming with microorganisms. Even in solidly frozen materials there remain small pockets of liquid water where solutes have concentrated and microorganisms can metabolize and grow, albeit very slowly.

In considering cold environments as microbial habitats, it is important to distinguish between environments that are *constantly* cold and those that are only *seasonally* cold. The latter, characteristic of temperate climates, may have summer temperatures as high as 40°C. A temperate lake, for example, may have ice cover in the winter, but the water may remain at 0°C for only a relatively brief time. By contrast, Antarctic lakes contain a permanent ice cover several meters thick (Figure 5.19*d*), and the water column below the ice in these lakes remains at 0°C or colder year round. Marine sediments and glaciers are also constantly cold, as are subglacial lakes—lakes deep beneath the glacier surface—and all of these are teeming with microbial life. It is thus not surprising that the best examples of microbes well adapted to cold temperatures have emerged from these environments.

Psychrophilic and Psychrotolerant Microorganisms

A psychrophile is a microbe with an optimal growth temperature of 15°C or lower, a maximum growth temperature below 20°C, and a minimum growth temperature of 0°C or lower. By contrast, microbes that grow at 0°C but have optima of 20–40°C are called **psychrotolerant**. Psychrophiles are found in environments that

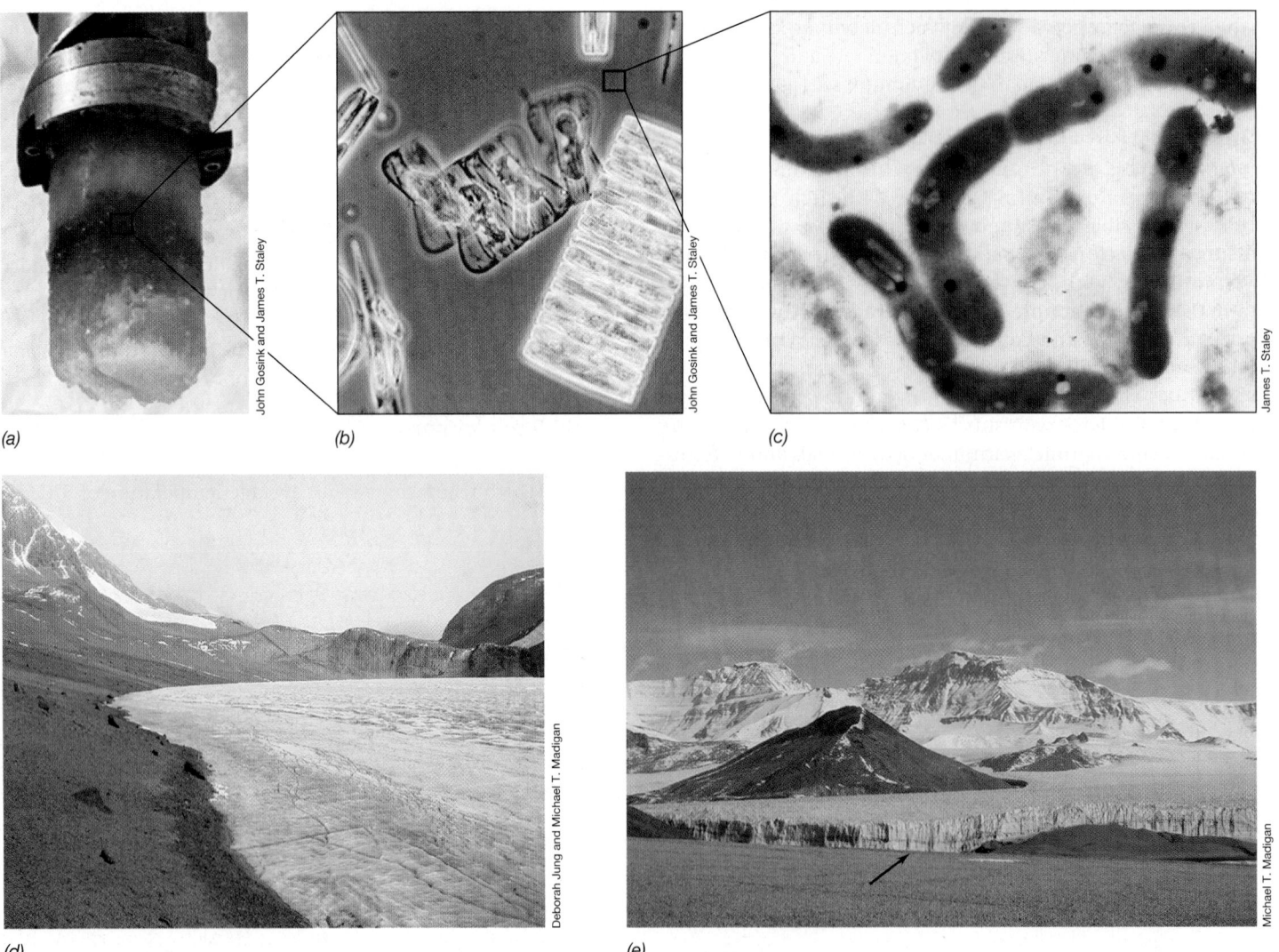

(a) (b) (c)

(d) (e)

Figure 5.19 Antarctic microbial habitats and microorganisms. *(a)* A core of frozen seawater from McMurdo Sound, Antarctica. The core is about 8 cm wide. Note the dense coloration due to pigmented microorganisms. *(b)* Phase-contrast micrograph of phototrophic microorganisms from the core shown in part *a*. Most organisms are either diatoms or green algae (both eukaryotic phototrophs). *(c)* Transmission electron micrograph of *Polaromonas*, a gas vesiculate bacterium that lives in sea ice and grows optimally at 4°C. Cells are about 0.8 μm in diameter. *(d)* Lake Bonney, McMurdo Dry Valleys, Antarctica. Although the lake is permanently ice-covered, the water column under the ice contains a diverse array of *Bacteria* and *Archaea* and microbial eukaryotes. *(e)* Garwood Glacier, McMurdo Dry Valleys, Antarctica. The edge of the glacier (arrow) is about 20 m high. Glaciers and subglacial lakes are teeming with microbial life.

are constantly cold and may even be killed by warming to moderate temperatures. For this reason, the laboratory study of psychrophiles requires that great care be taken to ensure that they never warm up during sampling, transport to the laboratory, isolation, or other manipulations.

Psychrophilic algae and bacteria often grow in dense masses within and under sea ice (frozen seawater that forms seasonally) in polar regions (Figure 5.19a–c). They can also be found on the surfaces of permanent snowfields and glaciers where they impart a distinctive coloration to the surface (**Figure 5.20**). The snow alga *Chlamydomonas nivalis* is an example of this, the carotenoid pigment astaxanthin in its spores being responsible for the brilliant red color of the snow surface (Figure 5.20 inset). This alga grows within the snow as a green-pigmented vegetative cell and then sporulates. As the snow dissipates by melting, erosion, and ablation (evaporation and sublimation), the spores become concentrated on the surface. Related species of snow algae contain different carotenoid pigments, and thus fields of snow algae can also be green, orange, brown, or purple.

Several psychrophilic *Bacteria* and a few psychrophilic *Archaea* have been isolated, and some of these show very low growth temperature optima. The permafrost bacterium *Planococcus halocryophilus* grows slowly at −15°C, the lowest growth temperature documented for any bacterium. However, theoretical considerations of bacterial metabolism suggest that the lower temperature limit for bacterial metabolism is considerably colder than this. For example, microbial respiration (as measured by CO_2 production) has been measured in tundra soils at nearly −40°C. At a temperature of −20°C, pockets of liquid water can exist in "frozen" materials, and studies have shown that enzymes from cold-active bacteria still function under such conditions. Growth at such temperatures, if possible, would be extremely slow. However, if an organism

can grow, even if only at a very low rate, it can remain competitive and maintain a population in its habitat.

Psychrotolerant microorganisms are more widely distributed in nature than are psychrophiles and can be isolated from soils and water in temperate climates as well as from meat, dairy products, cider, vegetables, and fruit stored at standard refrigeration temperatures (4°C). Although psychrotolerant microorganisms grow at 0°C, most do not grow well, and one must often wait weeks before visible growth is seen in laboratory cultures. By contrast, the same organism cultured at 30°C may grow at rates similar to that of many mesophiles. Various *Bacteria*, *Archaea*, and microbial eukaryotes are psychrotolerant.

Molecular Adaptations to Life in the Cold

Psychrophiles produce enzymes that function—often optimally—in the cold and that may be denatured or otherwise inactivated at even very moderate temperatures. The molecular basis for this is not entirely understood, but clearly it is linked to protein structure. Several cold-active enzymes whose structure is known show a greater content of α-helix and lesser content of β-sheet secondary structure (Section 4.7) than do enzymes that show little or no activity in the cold. Because β-sheet secondary structures tend to be more rigid than α-helices, the greater α-helix content of cold-active enzymes allows these proteins greater flexibility for catalyzing their reactions at cold temperatures. Cold-active enzymes also tend to have greater polar and lesser hydrophobic amino acid content (Figure 4.28 for structures of amino acids) and lower numbers of weak bonds, such as hydrogen and ionic bonds, compared with the corresponding enzyme from mesophiles. Collectively, these molecular features are likely to keep cold-active enzymes flexible and functional under cold conditions.

Another characteristic feature of psychrophiles is that their cytoplasmic membranes remain functional at low temperatures. Cytoplasmic membranes from psychrophiles tend to have a higher content of unsaturated and shorter-chain fatty acids, and this helps the membrane remain in a semifluid state at low temperatures to carry out important transport and bioenergetic functions. Some psychrophilic bacteria even contain *polyunsaturated* fatty acids; unlike monounsaturated or fully saturated fatty acids that tend to stiffen at low temperatures, polyunsaturated fatty acids remain flexible even at very cold temperatures.

Other molecular adaptations to cold temperatures include "cold shock" proteins and cryoprotectants, and these are not limited to psychrophiles. Cold shock proteins are a type of molecular chaperone (Section 4.11) and have several functions that include maintaining cold-sensitive proteins in an active form or binding specific mRNAs and facilitating their translation under cold conditions. Cryoprotectants include dedicated antifreeze proteins or specific solutes—such as glycerol or certain sugars—that are produced in large amounts at cold temperatures; these agents help prevent the formation of ice crystals that can puncture the cytoplasmic membrane. Highly psychrophilic bacteria often produce abundant levels of exopolysaccharide cell surface slime, and these slime layers confer cryoprotection as well.

Although freezing temperatures may prevent microbial growth, they do not necessarily cause death. Indeed, just the opposite may occur, and this has been exploited for the preservation of bacterial

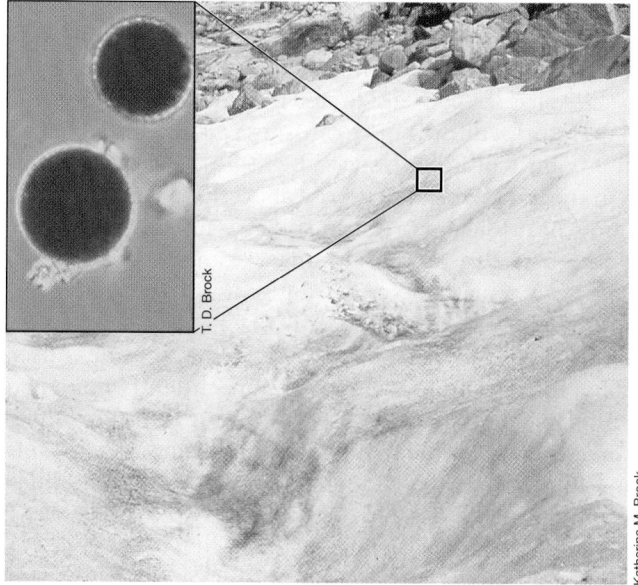

Figure 5.20 Snow algae. Snow bank in the Sierra Nevada, California, with red coloration caused by the presence of snow algae. Inset: photomicrograph of red-pigmented spores of the snow alga *Chlamydomonas nivalis*. Spores are about 18 μm in diameter. The spores germinate to yield motile green algal cells.

cells in microbial culture collections. Cells suspended in growth medium containing 10% dimethyl sulfoxide (DMSO) or glycerol as a cryoprotectant and frozen at −80°C (ultracold freezer) or −196°C (liquid nitrogen) remain viable in the frozen state for years.

MINIQUIZ

- How do psychrotolerant organisms differ from psychrophilic organisms?
- What molecular adaptations to cold temperatures are seen in the cytoplasmic membrane of psychrophiles? Why are they necessary?

5.11 Microbial Life at High Temperatures

Microbial life flourishes in high-temperature environments, from sun-heated soils and pools of water to boiling hot springs, and the organisms that live in these environments are typically highly adapted to their environmental temperature. We examine these organisms now and pick up on them again in several later chapters.

Thermal Environments

Organisms whose growth temperature optimum exceeds 45°C are called *thermophiles* and those whose optimum exceeds 80°C are called *hyperthermophiles* (Figure 5.18). The surface of soils subject to full sunlight can be heated to above 50°C at midday, and some surface soils may warm to as high as 70°C. Fermenting materials such as compost piles and silage can also reach temperatures of 70°C. Thermophiles abound in such environments. The most extreme high-temperature environments in nature, however, are hot springs, and these are home to a huge diversity of thermophiles and hyperthermophiles.

Many terrestrial hot springs have temperatures at or near boiling, while those at the bottom of the ocean, called *hydrothermal vents*, can have temperatures of 350°C or greater. Hot springs are found throughout the world, but they are especially abundant in the western United States, New Zealand, Iceland, Japan, Italy, Indonesia, Central America, and central Africa. The largest concentration of hot springs in the world is in Yellowstone National Park, Wyoming (USA). Although some hot springs vary widely in temperature, many are nearly constant, varying less than a degree or two over many years. In addition, different springs have different chemical compositions and pH values. In habitats hotter than 65°C, only prokaryotic cells can thrive (Table 5.2), but the diversity of *Bacteria* and *Archaea* in such environments is often extensive.

Hyperthermophiles and Thermophiles

A variety of hyperthermophiles inhabit boiling hot springs (**Figure 5.21**), including both chemoorganotrophic and chemolithotrophic species. Growth rates of hyperthermophiles can be studied in the field by immersing a microscope slide into a spring and then examining it microscopically over time. The slide is an excellent surface for microbial attachment and subsequent growth, and so small microbial colonies form (Figure 5.21*b*) and growth rates can be calculated from cell number data. Ecological studies such as

TABLE 5.2 Presently known upper temperature limits for growth of living organisms

Group	Upper temperature limits (°C)
Macroorganisms	
Animals	
Fish and other aquatic vertebrates	38
Insects	45–50
Ostracods (crustaceans)	49–50
Plants	
Vascular plants	45 (60 for one species)
Mosses	50
Microorganisms	
Eukaryotic microorganisms	
Protozoa	56
Algae	55–60
Fungi	60–62
Bacteria and Archaea	
Bacteria	
Cyanobacteria	73
Anoxygenic phototrophs	70–73
Chemoorganotrophs/chemolithotrophs	95
Archaea	
Chemoorganotrophs/chemolithotrophs	122

this have shown that growth rates in boiling springs are often quite high, with generation times (*g*) as short as 1 h not uncommon.

Cultures of diverse hyperthermophiles have been obtained, and a variety of morphological and physiological types of both *Bacteria* and *Archaea* are known. Some hyperthermophilic *Archaea* have growth-temperature optima above 100°C, while no species of *Bacteria* have yet been discovered that grow above 95°C. Growing laboratory cultures of organisms with optima above the boiling point requires pressurized vessels that permit temperatures in the growth medium to rise above 100°C without boiling. The most heat-tolerant organisms known inhabit hydrothermal vents, with the most thermophilic example thus far being *Methanopyrus*, a methane-producing genus of *Archaea* capable of growth at up to 122°C (🔗 Section 17.2).

In contrast to hyperthermophiles, thermophiles (optima 45–80°C) inhabit moderately hot or intermittently hot environments. As boiling water leaves a hot spring, it gradually cools, setting up a thermal gradient. Along this gradient, microorganisms become established, with different species growing in the different temperature ranges (**Figure 5.22a**). By studying the species distribution along such natural thermal gradients, it has been possible to determine the upper temperature limits for various classes of microbes (Table 5.2). Thermophilic *Bacteria* and *Archaea* have also been found in artificial thermal environments, such as hot water heaters. Hot water discharges from power plants and other artificial thermal sources also provide sites where thermophiles can flourish.

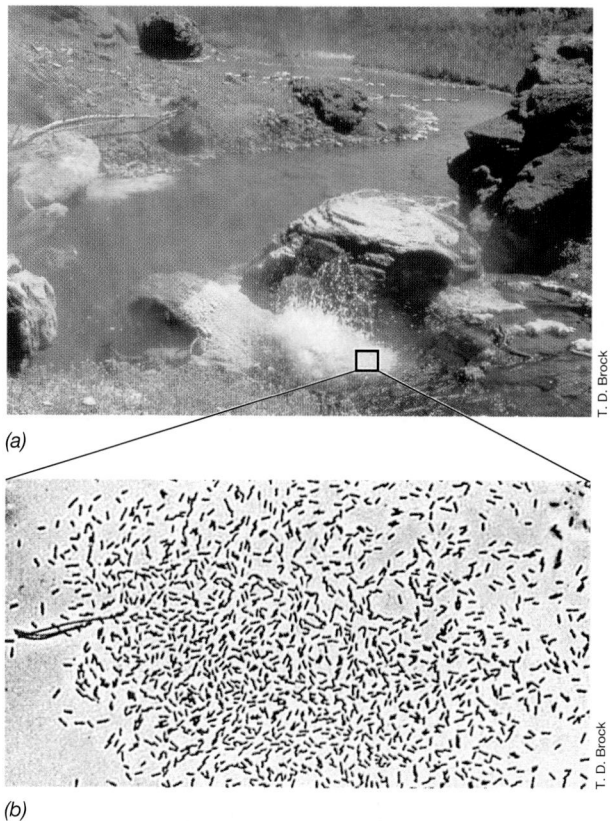

(a)

(b)

Figure 5.21 Growth of hyperthermophiles in boiling water. *(a)* Boulder Spring, a small boiling spring in Yellowstone National Park. This spring is superheated, having a temperature 1–2°C above the boiling point. The mineral deposits around the spring consist mainly of silica and sulfur. *(b)* Photomicrograph of a microcolony of *Archaea* that developed on a microscope slide immersed in such a boiling spring.

Protein and Membrane Stability at High Temperatures

How do thermophiles and hyperthermophiles survive high temperatures? First, their enzymes and other proteins are much more heat-stable than are those of mesophiles and actually function *optimally* at high temperatures. The heat stability of an enzyme from a hyperthermophile is often due to subtle changes in amino acid sequence from the corresponding enzyme from a mesophile, and these changes affect protein structure and function to resist heat denaturation. Heat-stable proteins also typically show increased ionic bonding between basic and acidic amino acids and have highly hydrophobic interiors, factors that also prevent unfolding. Finally, solutes such as di-inositol phosphate, diglycerol phosphate, and mannosylglycerate are produced at high levels in certain hyperthermophiles, and these are thought to help stabilize their proteins against thermal denaturation.

Enzymes from thermophiles and hyperthermophiles have significant commercial uses. Heat-stable enzymes catalyze biochemical reactions at high temperatures and are in general more stable than enzymes from mesophiles, thus prolonging the shelf life of commercial enzyme preparations (Figure 5.22*b*). A classic example of this is the DNA polymerase isolated from *Thermus aquaticus*—*Taq* polymerase—used to automate the repetitive steps in the polymerase chain reaction (PCR), a technique for amplifying

(a)

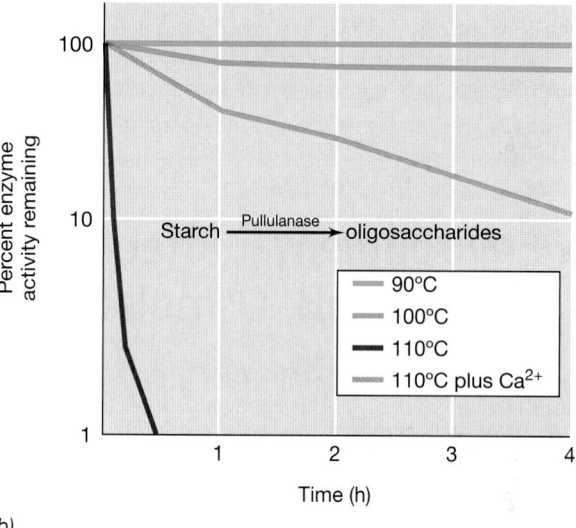

Starch $\xrightarrow{\text{Pullulanase}}$ oligosaccharides

——	90°C
——	100°C
——	110°C
——	110°C plus Ca^{2+}

(b)

Figure 5.22 Hot spring microbes and their heat-stable enzymes. *(a)* Characteristic V-shaped pattern (shown by the dashed white lines) formed by cyanobacteria at the upper temperature for phototrophic life, 70–73°C, in the thermal gradient formed from a boiling hot spring in Yellowstone National Park (USA). The pattern develops because the water cools more rapidly at the edges than in the center of the channel. *(b)* In the spring source, hyperthermophiles thrive, and some have been used as sources of heat-stable enzymes, such as pullulanase from *Pyrococcus* (*Archaea*). Ca^{2+} stabilizes the enzyme above the boiling point.

DNA and a major tool of modern biology (⊘ Section 12.1). Several other heat-stable enzymes are commercially available for specific industrial applications (Figure 5.22*b*).

Besides enzymes and other macromolecules in the cell, the cytoplasmic membranes of thermophiles and hyperthermophiles must be heat-stable. Heat naturally works to peel apart the lipid bilayer that makes up the cytoplasmic membrane (⊘ Section 2.3). In thermophiles and most hyperthermophilic *Bacteria*, the cytoplasmic membrane has a higher content of long-chain and saturated fatty acids and a lower content of unsaturated fatty acids than are found in the cytoplasmic membranes of mesophiles. Saturated fatty acids form a stronger hydrophobic environment than do unsaturated fatty acids, and longer-chain fatty acids have a higher melting point than shorter-chain fatty acids; collectively, these properties increase membrane stability.

Hyperthermophiles, most of which are *Archaea*, do not contain fatty acids in their membranes but instead have C_{40} hydrocarbons composed of repeating units of isoprene bonded by ether linkage to glycerol phosphate (⮌ Figure 2.6). In addition, however, the architecture of the cytoplasmic membranes of many hyperthermophiles takes a unique twist: The membrane forms a lipid *monolayer* rather than a lipid *bilayer* (⮌ Figure 2.6e). The monolayer structure covalently links both halves of the membrane and prevents it from melting at the high growth temperatures of hyperthermophiles. We consider other aspects of heat stability in hyperthermophiles, including that of DNA stability, in Chapter 17.

— MINIQUIZ —

- Which phylogenetic domain includes species with optima of >100°C? What special techniques are required to culture them?
- How does the membrane structure of hyperthermophilic *Archaea* differ from that of *Escherichia coli* and why is this structure helpful for growth at high temperature?
- What is *Taq* polymerase and why is it important?

IV • Environmental Effects on Growth: pH, Osmolarity, and Oxygen

As we have seen, temperature has a major effect on the growth of microorganisms. But many other environmental factors can affect microbial growth as well, including pH, osmolarity, and oxygen.

5.12 Effects of pH on Microbial Growth

Acidity or alkalinity of a solution is expressed by its **pH** on a logarithmic scale in which neutrality is pH 7 (**Figure 5.23**). pH values less than 7 are *acidic* and those greater than 7 are *alkaline*. In analogy to a temperature range (Figure 5.17), every microorganism has a pH range, typically about 2–3 pH units, within which growth is possible. Also, each organism shows a well-defined pH optimum, where growth occurs best. Most natural environments have a pH between 3 and 9, and organisms with pH growth optima in this range are most common. Terms used to describe organisms that grow best in particular pH ranges are shown in **Table 5.3**.

Acidophiles

Organisms that grow optimally at a pH value in the range termed *circumneutral* (pH 5.5 to 7.9) are called **neutrophiles**. For example, the bacterium *Escherichia coli* is a neutrophile (Table 5.3). By contrast, organisms that grow best below pH 5.5 are called **acidophiles**. There are different classes of acidophiles, some growing best at moderately acidic pH and others at very low pH. Many fungi and bacteria grow best at pH 5 or even below, while a more restricted number grow best below pH 3. An even more restricted group grow

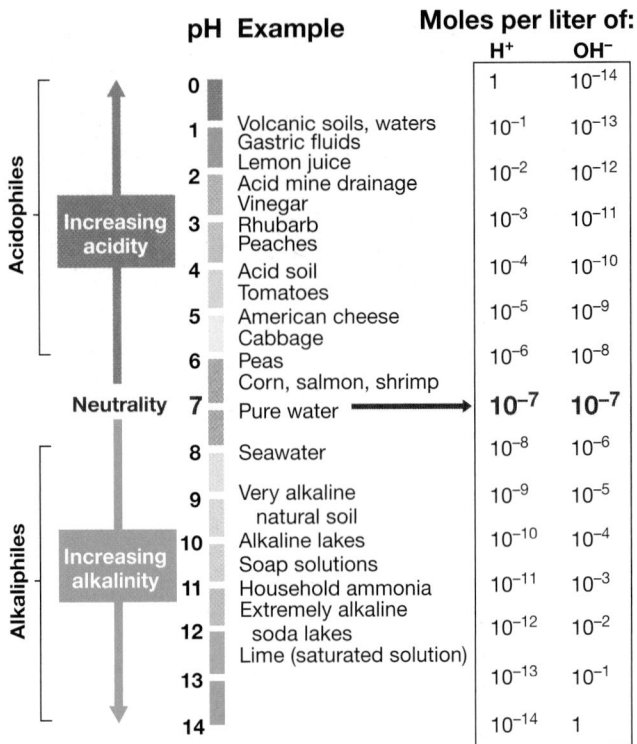

Figure 5.23 The pH scale. Although some microorganisms can live at very low or very high pH, the cell's internal pH remains near neutrality.

best below pH 2 and those with pH optima below 1 are extremely rare. Most acidophiles cannot grow at pH 7 and many cannot grow at pH values more than two units above their optimum.

A critical factor governing acidophily is the stability of the cytoplasmic membrane. When the pH is raised to neutrality, the cytoplasmic membranes of strongly acidophilic bacteria are destroyed and the cells lyse. This indicates that these organisms are not just acid-*tolerant* but that high concentrations of protons are actually *required* for cytoplasmic membrane stability. For example, the most acidophilic microbe known is *Picrophilus oshimae*, a species of *Archaea* that grows optimally at pH 0.7 and 60°C. Above pH 4, cells of *P. oshimae* spontaneously lyse. As one would predict, *P. oshimae* inhabits extremely acidic thermal soils associated with volcanic activity.

TABLE 5.3 Relationships of microorganisms to pH

Physiological class (optima range)	Approximate pH optimum for growth	Example organism[a]
Neutrophile (pH > 5.5 and < 8)	7	*Escherichia coli*
Acidophile (pH < 5.5)	5	*Rhodopila globiformis*
	3	*Acidithiobacillus ferrooxidans*
	1	*Picrophilus oshimae*
Alkaliphile (pH ≥ 8)	8	*Chloroflexus aurantiacus*
	9	*Bacillus firmus*
	10	*Natronobacterium gregoryi*

[a]*Picrophilus* and *Natronobacterium* are *Archaea*; all others are *Bacteria*.

Alkaliphiles

A few extremophiles have very high pH optima for growth, sometimes as high as pH 10, and some of these can still grow, albeit poorly, at even higher pH. Microorganisms showing pH optima of 8 or higher are called **alkaliphiles**. Alkaliphilic microorganisms are typically found in highly alkaline habitats, such as soda lakes and high-carbonate soils. The most well-studied alkaliphilic bacteria are certain *Bacillus* species, such as *Bacillus firmus*. This organism is alkaliphilic but has an unusually broad range for growth, from pH 7.5 to 11. Some extremely alkaliphilic microbes are also halophilic (salt-loving), and most of these are *Archaea* (⌘ Section 17.1). Some phototrophic purple bacteria (⌘ Sections 15.4 and 15.5) are also strongly alkaliphilic. Certain alkaliphiles have commercial uses because they excrete hydrolytic enzymes such as proteases and lipases that maintain their activities at alkaline pH. These enzymes are added to laundry detergents to remove protein and fat stains, respectively, from clothing.

Managing membrane bioenergetics is an obvious problem for alkaliphiles. *B. firmus* uses sodium (Na^+) rather than H^+ to drive transport reactions and rotate its flagellum; that is, it forms a *sodium* motive force instead of a *proton* motive force. Remarkably, however, *B. firmus* uses a proton motive force to drive ATP synthesis even though the external membrane surface is highly alkaline. Exactly how this happens is unclear, although it is thought that hydrogen ions are in some way kept very near the outer surface of the cytoplasmic membrane such that they cannot spontaneously combine with the abundant hydroxyl ions to form water.

Cytoplasmic pH and Buffers

The optimal pH for growth of an organism refers to the *extracellular* environment only; the *intracellular* pH must be maintained at a value consistent with the stability of macromolecules, a range of about 4 pH units from pH 5 to 9. Thus, despite conditions in their habitats, extreme acidophiles and alkaliphiles maintain cytoplasmic pH values nearer to neutrality.

To prevent major shifts in pH during microbial growth in batch cultures, *buffers* are commonly added to culture media along with the nutrients required for growth. However, any given buffer works over only a relatively narrow pH range. For neutrophilic species, potassium phosphate (KH_2PO_4) or sodium bicarbonate ($NaHCO_3$) is often employed. Various organic buffers are available for the growth of acidophiles and alkaliphiles and are widely used for assaying enzymes extracted from cells. The buffer keeps the enzyme solution at optimal pH during the assay, thus ensuring that the enzyme remains catalytically active and unaffected by any protons or hydroxyl ions generated in the enzymatic reaction.

MINIQUIZ

- How does the concentration of H^+ change when a culture medium at pH 5 is adjusted to pH 9?
- What terms are used to describe organisms whose growth pH optimum is very high? Very low?
- In terms of pH, what class of organism is the bacterium *Escherichia coli*?

5.13 Osmolarity and Microbial Growth

Water is the solvent of life, and water availability is an important factor affecting the growth of microorganisms. Water availability not only depends on how moist or dry an environment is but is also a function of the concentration of solutes (salts, sugars, or other substances) dissolved in the water that is present. Solutes bind water, making it less available to organisms. Hence, for organisms to thrive in high-solute environments, physiological adjustments are necessary. Water availability is expressed in terms of **water activity** (a_w), the ratio of the vapor pressure of air in equilibrium with a substance or solution to the vapor pressure of pure water. Values of a_w vary between 0 (no free water) and 1 (pure water); some a_w values are listed in **Table 5.4**.

Water diffuses from regions of high water concentration (low solute concentration) to regions of lower water concentration (higher solute concentration) in the process of osmosis. The cytoplasm of a cell typically has a higher solute concentration than the environment, so the tendency for water is to diffuse into the cell. Under such conditions, the cell is said to be in *positive water balance*, which is the normal state of the cell. However, when a cell is placed in an environment where the solute concentration exceeds that of the cytoplasm, water will flow out of the cell. If a cell has no strategy to counteract this, it will become dehydrated and unable to grow.

Halophiles and Related Organisms

In nature, osmotic effects are of interest mainly in habitats with high concentrations of salts. Seawater contains about 3% sodium chloride (NaCl) plus small amounts of many other minerals and elements. Microorganisms that inhabit marine environments almost always show an NaCl requirement and grow optimally at the a_w of seawater, 0.98 (**Figure 5.24**). Such organisms are called **halophiles**. The requirement for NaCl by halophiles is absolute and cannot be replaced by other salts, such as potassium chloride (KCl), calcium chloride ($CaCl_2$), or magnesium chloride ($MgCl_2$).

Although halophiles require at least some NaCl for growth, the NaCl optimum varies with the organism and is habitat dependent. For example, marine microorganisms typically grow best

TABLE 5.4 Water activity of several substances

Water activity (a_w)	Material	Example organisms[a]
1.000	Pure water	*Caulobacter, Spirillum*
0.995	Human blood	*Streptococcus, Escherichia*
0.980	Seawater	*Pseudomonas, Vibrio*
0.950	Bread	Most gram-positive rods
0.900	Maple syrup, ham	Gram-positive cocci such as *Staphylococcus*
0.850	Salami	*Saccharomyces rouxii* (yeast)
0.800	Fruit cake, jams	*Zygosaccharomyces bailii* (yeast), *Penicillium* (fungus)
0.750	Salt lakes, salted fish	*Halobacterium, Halococcus*
0.700	Cereals, candy, dried fruit	*Xeromyces bisporus* and other xerophilic fungi

[a]Selected examples of *Bacteria* and *Archaea* or fungi capable of growth in culture media adjusted to the stated water activity.

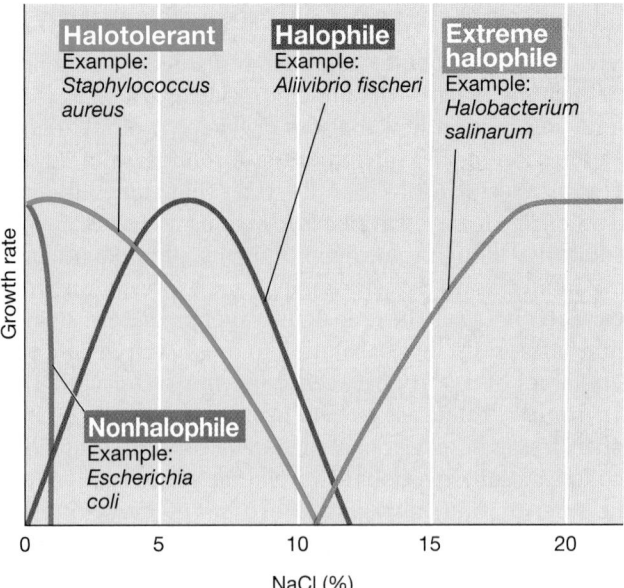

Figure 5.24 Effect of NaCl concentration on growth of microorganisms of different salt tolerances or requirements. The optimum NaCl concentration for marine microorganisms such as *Aliivibrio fischeri* is about 3%; for extreme halophiles, it is between 15 and 30%, depending on the organism.

with 1–4% NaCl, organisms from hypersaline environments (environments that are more salty than seawater) grow best at 3–12% NaCl, and organisms from extremely hypersaline environments require even higher levels of NaCl. Organisms isolated from brackish waters (a mixture of freshwater and seawater) may or may not be halophilic.

In contrast to halophiles, **halotolerant** organisms can tolerate some level of dissolved solutes but grow best in the absence of the added solute (Figure 5.24). Halophiles capable of growth in very salty environments are called **extreme halophiles** (Figure 5.24). These organisms require very high levels of NaCl, typically 15–30%, for optimum growth and are often unable to grow at all at NaCl concentrations below this. Organisms able to live in environments high in sugar are called **osmophiles**, and those able to grow in very dry environments (made dry by lack of water rather than by dissolved solutes) are called **xerophiles**. Examples of these various classes of organisms are given in **Table 5.5**.

From growth data obtained from extremely halophilic representatives of all three domains of life, there appears to be a common lower water activity limit for living organisms, and this limit is 0.61. This lower limit is likely set by the physiochemical constraints on obtaining water in osmotic environments of a_w less than 0.6 that cannot be overcome through biochemical adaptations by the cell. *Matric water activity*, a measure of water bound to a surface, is measured in the same way as osmotic water activity but can drop to significantly lower than 0.6 and still contain viable microbial communities. For example, hyper-arid hot desert soils can have matric a_w values as low as 0.1 during daylight hours. But these environments absorb moisture at night and during rain events, and these increase the water activity to above 0.6, making conditions suitable for microbial metabolism and growth.

Compatible Solutes

When an organism is transferred from a medium of high a_w to one of low a_w, it maintains positive water balance by increasing its internal solute concentration. This is possible either by pumping solutes into the cell from the environment or by synthesizing a cytoplasmic solute (Table 5.5). In either case, the solute must not inhibit biochemical processes in the cell and is thus called a **compatible solute**.

Compatible solutes are highly water-soluble organic molecules and include sugars, alcohols, and amino acid derivatives (Table 5.5). Glycine betaine, an analog of the amino acid glycine, is widely distributed among halophilic bacteria. Other common compatible solutes include sugars such as sucrose and trehalose, dimethylsulfoniopropionate (produced by marine algae), and glycerol, a common solute in xerophilic fungi, organisms that grow at the lowest water activities known (Table 5.5). In contrast to these organic solutes, KCl is the compatible solute of extremely halophilic *Archaea*, such as *Halobacterium* (⮑ Section 17.1), and of a few extremely halophilic *Bacteria*.

The concentration of compatible solute in a cell is a function of the levels of solute in its environment, and adjustments are made in response to the challenge from external solutes. However, in any given organism, the maximal level of compatible solute tolerated is a genetically encoded characteristic. As a result, different organisms have evolved to thrive in habitats of different salinities (Tables 5.4 and 5.5). In fact, organisms designated as *nonhalotolerant, halotolerant, halophilic,* or *extremely halophilic* (Figure 5.24) are to some extent a reflection of their genetic capacity to produce or accumulate compatible solutes.

MINIQUIZ

- What is the a_w of pure water? What is the lower limit of a_w for life?
- What are compatible solutes, and when and why are they needed by the cell? What is the compatible solute of *Halobacterium*?

5.14 Oxygen and Microbial Growth

Oxygen (O_2) is an essential nutrient for many microbes; they are unable to metabolize or grow without it. Other microbes, by contrast, cannot grow in the presence of O_2 and may even be killed by it. We therefore see, just as we did for other environmental factors considered in this chapter, *classes* of microorganisms based on their needs or tolerance of O_2.

Oxygen Classes of Microorganisms

Microorganisms can be grouped according to their relationship with O_2 as outlined in **Table 5.6**. **Aerobes** can grow at full oxygen tensions (air is 21% O_2) and respire O_2 in their metabolism. **Microaerophiles**, by contrast, are aerobes that can use O_2 only when it is present at levels reduced from that in air (microoxic conditions). This is because of the limited capacity of these organisms to respire or because they contain some O_2-sensitive molecule such as an O_2-labile enzyme. Many aerobes are **facultative**, meaning that

TABLE 5.5 Compatible solutes of microorganisms

Organism group and example	Major cytoplasmic compatible solute(s)	Minimum a_w for growth[c]
Most nonphototrophic *Bacteria* (*Escherichia*) and freshwater cyanobacteria (*Anabaena*)	Amino acids (mainly glutamate or proline[a])/ sucrose, trehalose[b]	0.98
Marine cyanobacteria (*Synechococcus*)	α-Glucosylglycerol[b]	0.92
Marine algae (*Phaeocystis*)	Mannitol[b], various glycosides, dimethylsulfoniopropionate	0.92
Halotolerant *Bacteria* (*Staphylococcus*)	Amino acids	0.90
Salt lake cyanobacteria (*Aphanothece*)	Glycine betaine	0.75
Halophilic phototrophic purple *Bacteria* (*Halorhodospira*)	Glycine betaine, ectoine, trehalose[b]	0.75
Extremely halophilic *Archaea* (*Halobacterium*) and some *Bacteria* (*Salinibacter*)	KCl	0.75
Halophilic green algae (*Dunaliella*)	Glycerol	0.75
Haloalkaliphilic *Archaea* (*Natrinema*)	KCl	0.68
Xerophilic and osmophilic yeasts (*Zygosaccharomyces*)	Glycerol	0.62[d]
Xerophilic filamentous fungi (*Xeromyces*)	Glycerol	0.605[d]

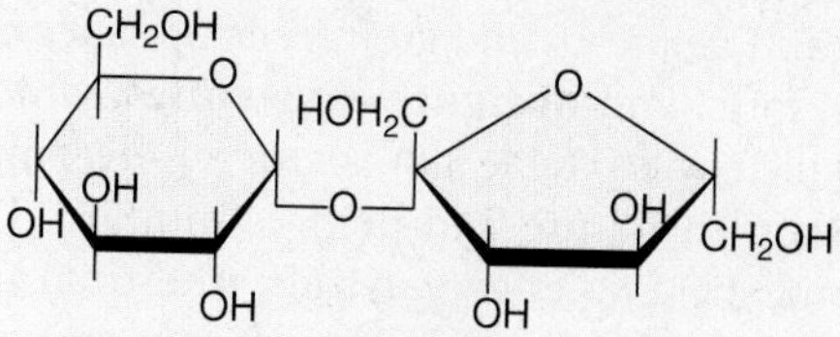

Sucrose

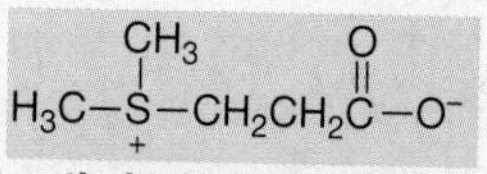

Dimethylsulfoniopropionate

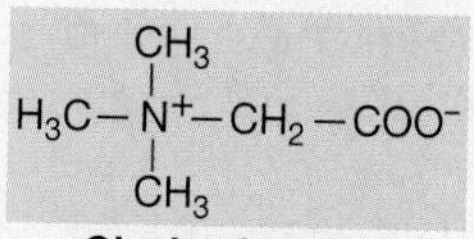

$$H_3C-S^+-CH_2CH_2C(O)-O^-$$
$$\quad\quad CH_3$$

Glycine betaine

$$H_3C-N^+-CH_2-COO^-$$

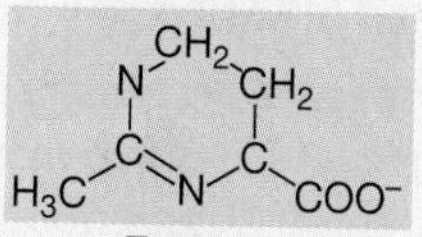

Ectoine

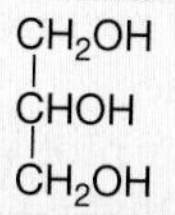

Glycerol

[a]See Figure 4.28 for the structures of amino acids.
[b]Structures not shown. Like sucrose, trehalose is a C_{12} disaccharide; glucosylglycerol is a C_9 alcohol; mannitol is a C_6 alcohol.
[c]To achieve an osmotic a_w lower than about 0.77, solutes other than just NaCl are necessary; for example, other salts ($MgCl_2$, $MgSO_4$, or $CaCl_2$) or non-salts, such as glycerol or sucrose, can extend the lower a_w for growth downward somewhat by additional solutes. For most organisms listed (other than for the xerophiles), the lower a_w for growth can be extended downward somewhat by additional solutes.
[d]Growth of *Zygosaccharomyces* tested in high-sucrose medium. Germination of *Xeromyces* spores tested using matric water potential.

TABLE 5.6 Oxygen relationships of microorganisms

Group	Relationship to O_2	Type of metabolism	Example[a]	Habitat[b]
Aerobes				
Obligate	Required	Aerobic respiration	*Micrococcus luteus* (B)	Skin, dust
Facultative	Not required, but growth better with O_2	Aerobic respiration, anaerobic respiration, fermentation	*Escherichia coli* (B)	Mammalian large intestine
Microaerophilic	Required but at levels lower than atmospheric	Aerobic respiration	*Spirillum volutans* (B)	Lake water
Anaerobes				
Aerotolerant	Not required, and growth no better when O_2 present	Fermentation	*Streptococcus pyogenes* (B)	Upper respiratory tract
Obligate	Harmful or lethal	Fermentation or anaerobic respiration	*Methanobacterium formicicum* (A)	Sewage sludge, anoxic lake sediments

[a]Letters in parentheses indicate phylogenetic status (B, *Bacteria*; A, *Archaea*). Representatives of either domain of prokaryotic cells are known in each category. Most eukaryotes are obligate aerobes, but facultative aerobes (for example, yeast) and obligate anaerobes (for example, certain protozoa and fungi) are known.
[b]Listed are typical habitats of the example organism; many others could be listed.

under the appropriate nutrient and culture conditions they can grow in the absence of O_2.

Some organisms cannot respire oxygen and are called **anaerobes**. There are two kinds of anaerobes: **aerotolerant anaerobes**, which can tolerate O_2 and grow in its presence even though they cannot respire, and **obligate anaerobes**, which are inhibited or even killed by O_2 (Table 5.6). *Anoxic* (O_2-free) microbial habitats are common in nature and include muds and other sediments, bogs, marshes, water-logged soils, intestinal tracts of animals, sewage sludge, the deep subsurface of Earth, and many other environments. Because there are many habitats for anaerobes, they are very common in nature and highly diverse. As far as is known, obligate anaerobiosis is characteristic of only three groups of microorganisms: a wide variety of *Bacteria* and *Archaea*, a few fungi, and a few protozoa.

Some of the best-known prokaryotic anaerobes are *Clostridium*, a genus of gram-positive endospore-forming *Bacteria*, and the methanogens, a group of methane-producing *Archaea*. Among obligate anaerobes, the sensitivity to O_2 varies greatly. Many clostridia, for example, although requiring anoxic conditions for growth, can tolerate traces of O_2 or even full exposure to air. Others, such as the methanogens, are killed rapidly by O_2 exposure.

Culture Techniques for Aerobes and Anaerobes

For the growth of aerobes, it is necessary to provide extensive aeration. This is because the O_2 that is consumed by the organisms during growth is not replaced fast enough by diffusion from the air. Therefore, forced aeration of liquid cultures is needed and can be achieved by either vigorously shaking the flask or tube on a shaker or by bubbling sterilized air into the medium through a fine glass tube or porous glass disc.

For the culture of anaerobes, the problem is not to *provide* O_2 but to *exclude* it. Bottles or tubes filled completely to the top with culture medium and fitted with leakproof closures provide suitably

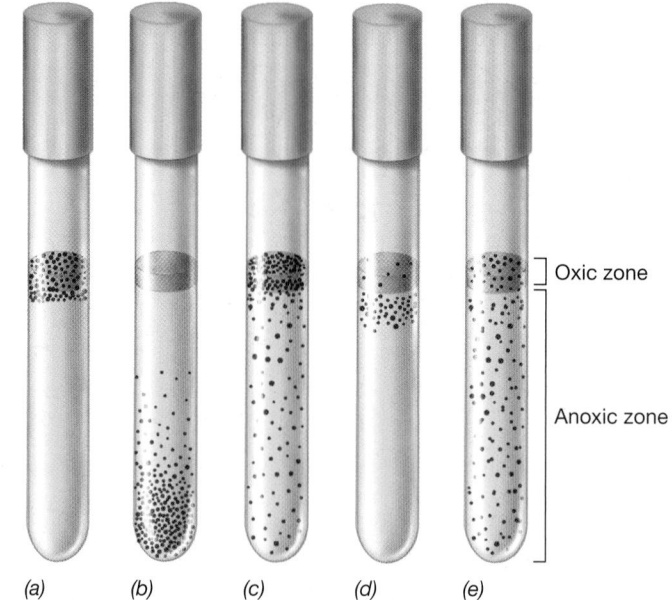

Figure 5.25 Growth versus O_2 concentration. From left to right, aerobic, anaerobic, facultative, microaerophilic, and aerotolerant anaerobe growth, as revealed by the position of microbial colonies (depicted here as black dots) within tubes of thioglycolate broth culture medium. A small amount of agar has been added to keep the liquid from becoming disturbed. The redox dye resazurin, which is pink when oxidized and colorless when reduced, has been added as a redox indicator. *(a)* O_2 penetrates only a short distance into the tube, so obligate aerobes grow only close to the surface. *(b)* Anaerobes, being sensitive to O_2, grow only away from the surface. *(c)* Facultative aerobes are able to grow in either the presence or the absence of O_2 and thus grow throughout the tube. However, growth is better near the surface because these organisms can respire. *(d)* Microaerophiles grow away from the most oxic zone. *(e)* Aerotolerant anaerobes grow throughout the tube. Growth is not better near the surface because these organisms can only ferment. In nature, many different habitats exist for each of these oxygen classes. In addition, a single habitat, such as a soil particle, may support growth of both aerobes and anaerobes (⌘ Section 20.1 and Figure 20.3).

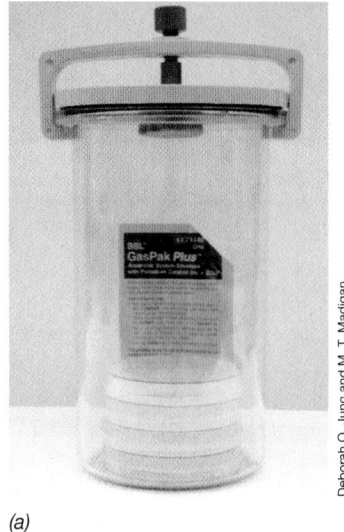

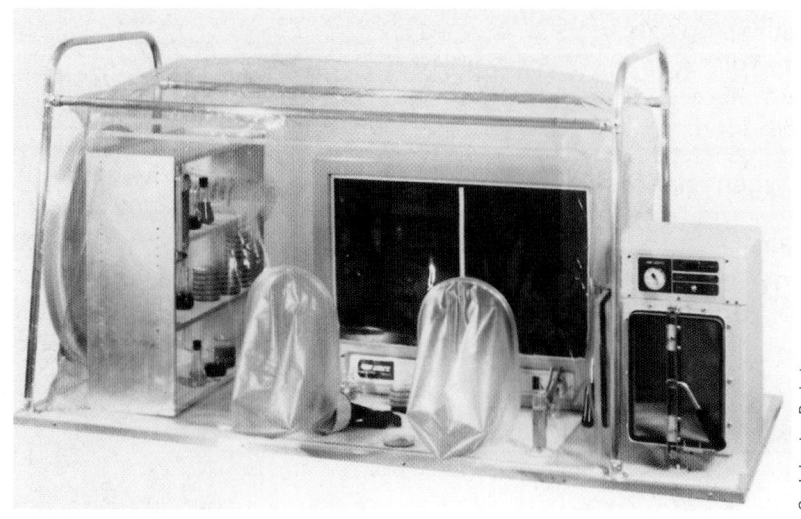

Figure 5.26 Incubation under anoxic conditions. *(a)* Anoxic jar. A chemical reaction in the envelope in the jar generates $H_2 + CO_2$. The H_2 reacts with O_2 in the jar on the surface of a palladium catalyst to yield H_2O; the final atmosphere contains N_2, H_2, and CO_2. *(b)* Anoxic glove bag for manipulating and incubating cultures under anoxic conditions. The airlock on the right, which can be evacuated and filled with O_2-free gas, serves as a port for adding and removing materials to and from the glove bag.

anoxic conditions for organisms that are not overly sensitive to small amounts of O_2. A chemical called a *reducing agent* may be added to such vessels to remove traces of O_2 by reducing it to water (H_2O). An example is thioglycolate, which is present in thioglycolate broth, a medium commonly used to test an organism's requirements for O_2 (**Figure 5.25**).

Thioglycolate broth is a complex medium containing a small amount of agar, making the medium viscous but still fluid. After thioglycolate reacts with O_2 throughout the tube, O_2 can penetrate only near the top of the tube where the medium contacts air. Obligate aerobes grow only at the top of such tubes. Facultative organisms grow throughout the tube but grow best near the top. Microaerophiles grow near the top but not right at the top. Anaerobes grow only near the bottom of the tube, where O_2 cannot penetrate. The redox indicator dye *resazurin* is present in thioglycolate broth to signal oxic regions; the dye is pink when oxidized and colorless when reduced and so gives a visual assessment of the degree of penetration of O_2 into the medium (Figure 5.25).

To remove all traces of O_2 for the culture of strict anaerobes, one can incubate tubes or plates in a glass jar flushed with an O_2-free gas or fitted with an O_2 consumption system (**Figure 5.26a**). For manipulating cultures in an anoxic atmosphere, special enclosures called *anoxic glove bags* permit work with open cultures in completely anoxic atmospheres (Figure 5.26b).

Why Is Oxygen Toxic?

Why are anaerobic microorganisms inhibited in their growth or even killed by oxygen? Molecular oxygen (O_2), per se, is not toxic, but O_2 can be converted to toxic oxygen by-products, and it is these that can harm or kill cells not able to deal with them. These include *superoxide anion* (O_2^-), *hydrogen peroxide* (H_2O_2), and *hydroxyl radical* ($OH\cdot$). All of these are by-products of the reduction of O_2 to H_2O in respiration (**Figure 5.27**). Flavoproteins, quinones, and iron–sulfur proteins, electron carriers found in virtually all cells (🔗 Section 3.10), also catalyze some of these reductions. Thus, regardless of whether it can respire O_2, an organism exposed to O_2 will experience toxic forms of oxygen, and if not destroyed, these molecules can wreak havoc in cells. For example, superoxide anion and $OH\cdot$ are strong oxidizing agents that can oxidize macromolecules and any other organic compounds in the cell. Peroxides such as H_2O_2 can also damage cell components but are not as toxic as O_2^- or $OH\cdot$.

Superoxide Dismutase and Other Enzymes That Destroy Toxic Oxygen

A major requirement for inhabiting an oxic world is to keep toxic oxygen molecules under control. Microbes accomplish this in

---

$$H_2O_2 + H_2O_2 \rightarrow 2\,H_2O + O_2$$

(a) **Catalase**

$$H_2O_2 + \text{NADH} + H^+ \rightarrow 2\,H_2O + \text{NAD}^+$$

(b) **Peroxidase**

$$O_2^- + O_2^- + 2\,H^+ \rightarrow H_2O_2 + O_2$$

(c) **Superoxide dismutase**

$$4\,O_2^- + 4\,H^+ \rightarrow 2\,H_2O + 3\,O_2$$

(d) **Superoxide dismutase/catalase in combination**

$$O_2^- + 2\,H^+ + \text{rubredoxin}_{\text{reduced}} \rightarrow H_2O_2 + \text{rubredoxin}_{\text{oxidized}}$$

(e) **Superoxide reductase**

Figure 5.28 Enzymes that destroy toxic oxygen species. *(a)* Catalases and *(b)* peroxidases are porphyrin-containing proteins, although some flavoproteins may consume toxic oxygen species as well. *(c)* Superoxide dismutases are metal-containing proteins, the metals being copper and zinc, manganese, or iron. *(d)* Combined reaction of superoxide dismutase and catalase. *(e)* Superoxide reductase catalyzes the one-electron reduction of O_2^- to H_2O_2.

much the same way as plants and animals do. Superoxide anion and H_2O_2 are the most abundant toxic oxygen species, and all cells have enzymes that destroy these compounds (**Figure 5.28**). The enzymes catalase and peroxidase attack H_2O_2, forming O_2 and H_2O, respectively (Figure 5.28 and **Figure 5.29**). Superoxide anion is destroyed by the enzyme *superoxide dismutase*, an enzyme that generates H_2O_2 and O_2 from two molecules of O_2^- (Figure 5.28c). Superoxide dismutase and catalase (or peroxidase) thus work in series to convert O_2^- to harmless products (Figure 5.28d).

Aerobes and facultative aerobes typically contain both superoxide dismutase and catalase. Superoxide dismutase is an essential enzyme for aerobes. Some aerotolerant anaerobes lack superoxide dismutase and use protein-free manganese complexes instead to carry out the dismutation of O_2^- to H_2O_2 and O_2. Such a system is not as efficient as superoxide dismutase, but it is sufficient to protect the cells from O_2^- damage. In some strictly anaerobic *Archaea* and *Bacteria*, superoxide dismutase is absent and instead the enzyme *superoxide reductase* functions to remove O_2^-. Unlike superoxide dismutase, superoxide reductase reduces O_2^- to H_2O_2 without the production of O_2 (Figure 5.28e), thus avoiding exposure of the organism to O_2.

Reactants		Products
$O_2 + e^- \rightarrow$	O_2^-	(superoxide)
$O_2^- + e^- + 2\,H^+ \rightarrow$	H_2O_2	(hydrogen peroxide)
$H_2O_2 + e^- + H^+ \rightarrow$	$H_2O + OH\cdot$	(hydroxyl radical)
$OH\cdot + e^- + H^+ \rightarrow$	H_2O	(water)

Outcome:
$$O_2 + 4\,e^- + 4\,H^+ \rightarrow 2\,H_2O$$

Figure 5.27 Four-electron reduction of O_2 to H_2O by stepwise addition of electrons. All the intermediates formed are reactive and toxic to cells; water is not.

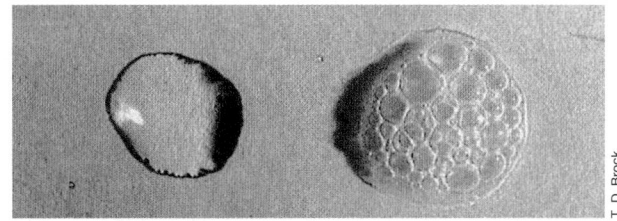

Figure 5.29 Method for testing a microbial culture for the presence of catalase. A heavy loopful of cells from an agar culture was mixed on a slide (right) with a drop of 30% hydrogen peroxide. The immediate appearance of bubbles is indicative of the presence of catalase. The bubbles are O_2 produced by the reaction $H_2O_2 + H_2O_2 \rightarrow 2\,H_2O + O_2$.

MINIQUIZ

- How does an obligate aerobe differ from a facultative aerobe?
- How does a reducing agent work? Give an example of a reducing agent.
- How does superoxide dismutase or superoxide reductase protect a cell?

V • Controlling Microbial Growth

Thus far in this chapter we have discussed microbial growth from the perspective of conditions that *promote* their growth. We close by considering the opposite side of the coin, microbial growth *control*.

Many aspects of microbial growth control have significant practical applications. For example, we wash fresh produce to remove attached microorganisms and we inhibit microbial growth on body surfaces by washing. However, neither of these processes kills or removes all microorganisms. Only **sterilization**—the killing or removal of all microorganisms (including viruses)—ensures that this is the case. In many circumstances, sterility is not required. In others, however, sterilization is absolutely essential.

5.15 General Principles and Growth Control by Heat

The effects of microorganisms can often be controlled by simply limiting or inhibiting growth. Methods for inhibiting microbial growth include *decontamination*, the treatment of an object or surface to make it safe to handle, and **disinfection**, a process that directly targets pathogens although it may not eliminate all microorganisms. Decontamination can be as simple as wiping off food utensils to remove food fragments (and their attached organisms) before using them, while disinfection requires agents called *disinfectants* that actually kill microorganisms or severely inhibit their growth. Physical methods of microbial growth control are used extensively in industry, medicine, and the home, and we consider three classes of physical controls in this section and the next: heat, radiation, and filtration. Of these three, heat is the most widely used method of physically treating an object or substance to render it sterile.

Heat Sterilization

The effectiveness of heat as a sterilant is quantified by the time required for a 10-fold reduction in the viability of a microbial population at a given temperature. This is called the *decimal reduction time* (D). The relationship between D and temperature is exponential, as the logarithm of D plotted against temperature yields a straight line (**Figure 5.30**). Moreover, heat killing proceeds more rapidly as the temperature rises. The type of heat is also important: Moist heat has better penetrating power than dry heat and, at a given temperature, inhibits growth or kills cells more quickly than does dry heat.

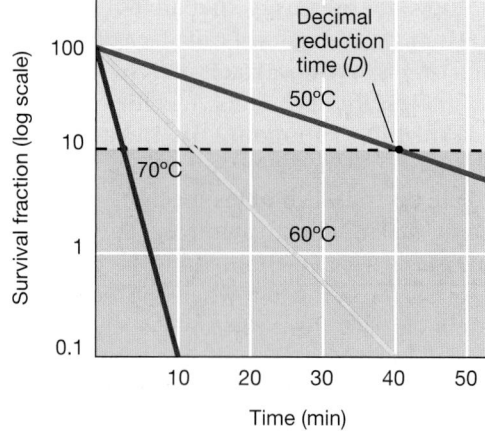

(a)

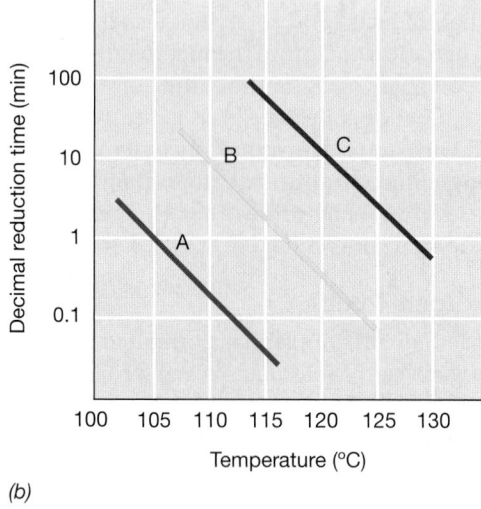

(b)

Figure 5.30 The effect of temperature on the heat killing of microorganisms. (a) The decimal reduction time (D) is the time at which only 10% of the original population of a given organism (in this case, a mesophile) remains viable at a given temperature. For 70°C, D = 3 min; for 60°C, D = 12 min; for 50°C, D = 42 min. (b) D values for model organisms of different temperature classes: A, mesophile; B, thermophile; C, hyperthermophile.

Another way to characterize the heat sensitivity of an organism is to measure its *thermal death time*, the time it takes to kill all cells at a given temperature. To determine the thermal death time, samples of a cell suspension are heated for different times, mixed with culture medium, and incubated. If all the cells have been killed, no growth is observed in the incubated samples. However, unlike a decimal reduction time measurement that is independent of the original cell number, the thermal death time is greatly affected by population size; a longer time is required to kill all cells in a large population than in a small one.

The presence of endospore-forming bacteria in a heat-treated sample can influence both the decimal reduction and thermal death times. Recall that the mature endospore is very dehydrated and contains calcium dipicolinate and small acid-soluble spore proteins (SASPs) that help confer heat stability on the structure (∂ Section 2.10). The medium in which heating takes place also

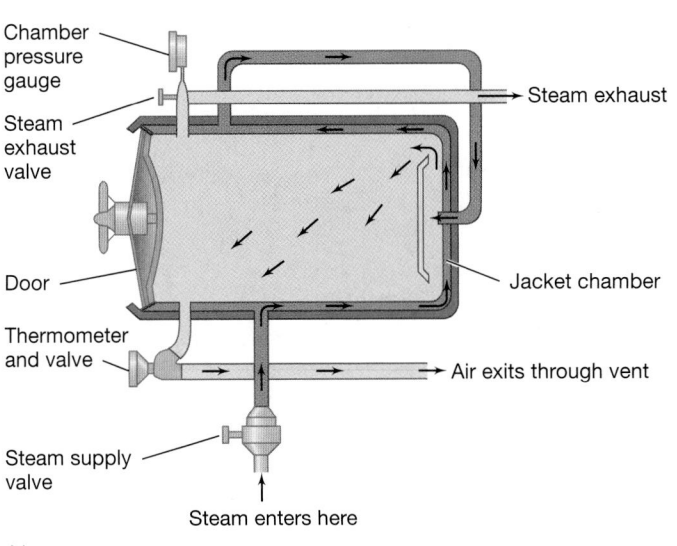

(c)

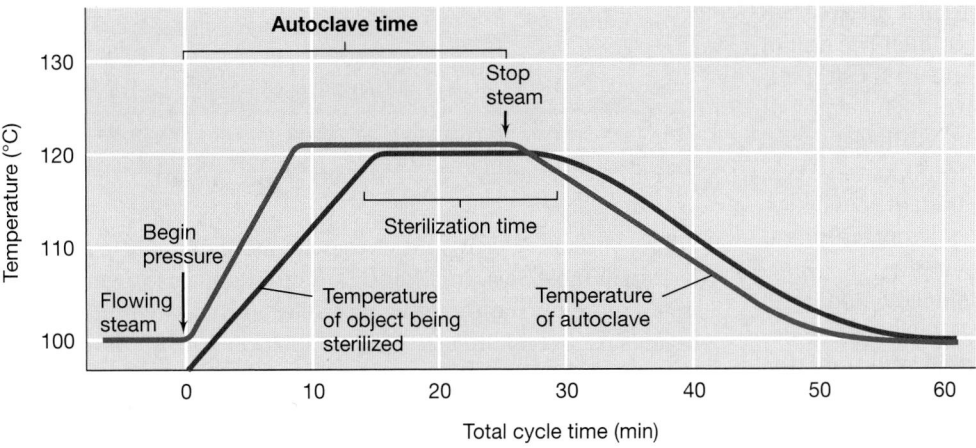

(b)

Figure 5.31 The autoclave and moist heat sterilization. *(a)* The flow of steam through an autoclave. *(b)* A typical autoclave cycle. The temporal heating profile of a fairly bulky object is shown. The temperature of the object rises and falls more slowly than the temperature of the autoclave. The temperature of the object must reach the target temperature and be held for 10–15 min to ensure sterility, regardless of the temperature and time recorded in the autoclave. *(c)* A modern research autoclave. Note the pressure-lock door and the automatic cycle controls on the right panel. The steam inlet and exhaust fittings are on the right side of the autoclave.

boiling point of water at 1 atm. The autoclave places steam under a pressure of 1.1 kg/cm^2 (15 lb/in^2), which yields a temperature of 121°C. At 121°C, the time to achieve sterilization of small amounts of endospore-containing material is about 15 min (Figure 5.31*b*). If the object to be autoclaved is bulky or large volumes of liquids are to be sterilized, heat transfer to the interior is retarded, and thus the total heating time must be extended. Note that it is not the *pressure* inside the autoclave that kills the microorganisms but the high *temperatures* that are achieved when steam is placed under pressure.

Pasteurization, a process named for Louis Pasteur (⇔ Section 1.9), is not the same as sterilization. Pasteurization uses heat to significantly reduce rather than totally eliminate the microorganisms found in liquids, such as milk. At the temperatures and times standardized for the pasteurization of food products, all known pathogenic bacteria are killed. In addition, however, by decreasing the overall microbial load, pasteurization increases the shelf life of perishable liquids.

To pasteurize milk, the liquid is passed through a tubular heat exchanger. Careful control of flow rate and the size and temperature of the heat source raises the temperature of the milk to 71°C for 15 seconds (or even higher temperatures for shorter time periods; see Figure 5.30), after which it is rapidly cooled. This process is called *flash pasteurization*. *Ultrahigh-temperature (UHT) pasteurization* of milk requires heat treatment at 135°C for 1–2 sec and actually sterilizes the milk such that it can be stored at room temperature for long periods without spoilage.

influences the rate of killing, and this is especially relevant in food canning procedures. Microbial death is more rapid at acidic pH, and acid foods such as tomatoes, fruits, and pickles are easier to sterilize than neutral-pH foods such as corn and beans. Moreover, high concentrations of sugars, proteins, and fats decrease heat penetration and usually increase the resistance of organisms to heat, whereas high salt concentrations may either increase or decrease heat resistance, depending on the organism.

The Autoclave and Pasteurization

The **autoclave** is a sealed heating device that uses steam under pressure to kill microorganisms (**Figure 5.31**). Killing of heat-resistant endospores requires heating at temperatures above the

— **MINIQUIZ** —

- Why is heat an effective sterilizing agent?
- What steps are necessary to ensure the sterility of material contaminated with bacterial endospores?
- Distinguish between the sterilization of microbiological media and the pasteurization of dairy products.

5.16 Other Physical Control Methods: Radiation and Filtration

In addition to heat, radiation, in particular ultraviolet (UV) light, X-rays, and gamma rays, are also effective microbial killing agents. However, each type of energy has a different mode of action and killing efficacy and thus their applications vary widely.

Ultraviolet and Ionizing Radiation

Ultraviolet radiation between 220 and 300 nm is absorbed by DNA and can cause mutations or have other serious effects on DNA that lead to death of the exposed organism (⇝ Section 11.4). UV radiation is useful for disinfecting surfaces and air, and is widely used to decontaminate and disinfect the work surface of laboratory laminar flow hoods equipped with a "germicidal" UV light (**Figure 5.32**) and air circulating in hospital and food preparation rooms. However, UV radiation has very poor penetrating power, limiting its use to the disinfection of exposed surfaces or air rather than bulk objects such as canned foods or surgical clothing.

Ionizing radiation is electromagnetic radiation of sufficient energy to produce ions and other reactive molecular species from molecules with which the radiation particles collide. The unit of ionizing radiation is the *roentgen*, and the standard for sterilization is the *absorbed radiation dose*, measured in *rads* (100 erg/g) or *grays* (1 Gy = 100 rad). Ionizing radiation is typically generated from X-ray sources or the radioactive nuclides ^{60}Co and ^{137}Cs. These nuclides produce X-rays or gamma rays, both of which have sufficient energy and penetrating power to kill microorganisms in bulk items such as food products and medical supplies.

Table 5.7 shows the dose necessary for a 10-fold reduction (*D*10) in number of selected microorganisms. The *D*10 value is analogous to the decimal reduction time for heat sterilization, and the killing curve of ionizing radiation yields a similar plot (**Figure 5.33**; compare with Figure 5.30). As is also true of heat treatments, killing endospores with ionizing radiation is more difficult than

TABLE 5.7 Radiation sensitivity of some representative microorganisms		
Type of microorganism	**Characteristics**	**D10ᵃ (Gy)**
Bacteria		
Clostridium botulinum	Gram-positive anaerobe; forms endospores	3,300
Deinococcus radiodurans	Gram-negative, radiation-resistant coccus	2,200
Lactobacillus brevis	Gram-positive, rod-shaped	1,200
Bacillus subtilis	Gram-positive aerobe; forms endospores	600
Escherichia coli	Gram-negative, rod-shaped	300
Salmonella enterica (typhimurium)	Gram-negative, rod-shaped	200
Fungi		
Aspergillus niger	Common mold	500
Saccharomyces cerevisiae	Baker's and brewer's yeast	500
Viruses		
Foot-and-mouth	Pathogen of cloven-hoofed animals	13,000
Coxsackie	Human pathogen	4,500

ᵃD10 is the amount of radiation necessary to reduce the initial population or activity level 10-fold (1 logarithm, see Figure 5.33). Gy, grays. 1 Gy is equivalent to 100 rads. The lethal dose for humans is 10 Gy.

killing vegetative cells, and viruses are more difficult to kill than bacteria (Table 5.7). In addition, microorganisms in general are much more resistant to ionizing radiation than are multicellular organisms. For example, the lethal radiation dose for humans can be as low as 10 Gy if delivered over a short time period.

In the United States, radiation is used for the sterilization of such diverse items as surgical supplies, plastic labware, drugs, and even tissue grafts. Certain foods and food products such as fresh produce, poultry, meat products, and spices are also routinely irradiated to ensure that they are sterile or at least free of pathogens and insects.

Figure 5.32 A laminar flow hood. An ultraviolet light source prevents contamination of the hood when it is not in use. When the hood is in use, air is drawn into the cabinet through a HEPA filter. The filtered air inside the cabinet is exhausted out of the cabinet, preventing contamination of the inside of the hood. The cabinet provides a contaminant-free workspace for microbial and tissue culture manipulations.

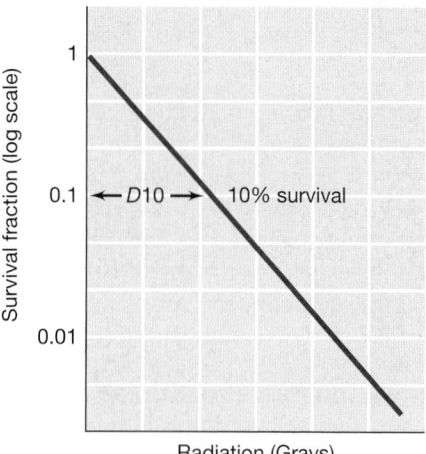

Figure 5.33 Relationship between the survival fraction and the radiation dose of a microorganism. The *D*10, which is the decimal reduction dose, can be interpolated from the data as shown.

Filter Sterilization

Heat is an effective way to decontaminate most liquids, but heat-sensitive liquids not subject to ionizing radiation are typically sterilized by filtration. For sterilization, a filter with pores of average size 0.2 μm is a minimum requirement; however, even such tiny holes will not trap most viruses. Commonly used filter pore sizes for the filter sterilization of small volumes, such as laboratory solutions, are 0.45 μm and 0.2 μm.

Several types of filters are in routine use in microbiology, including depth filters, membrane filters, and nucleopore filters. A depth filter is a fibrous sheet made from an array of overlapping paper or

Figure 5.35 Membrane filters. Disposable, presterilized, and assembled membrane filter units. Left: a filter system designed for small volumes forced through the filter by a syringe. Right: a system employing a peristaltic pump for filtering larger volumes.

glass fibers that traps particles in the network of fibers (**Figure 5.34a**). Depth filters are important in biosafety applications such as in a biological safety cabinet where air, both into and out of the cabinet, flows through a depth filter called a *HEPA filter*, or *h*igh-*e*fficiency *p*articulate *a*ir filter (Figure 5.34a). HEPA filters typically remove 0.3-μm or larger particles from an airstream with an efficiency of greater than 99.9%. This does not ensure sterilization, however.

Membrane filters are the most common filters used for liquid sterilization in the microbiology laboratory (Figure 5.34b and **Figure 5.35**). Membrane filters are composed of high-tensile-strength polymers manufactured in such a way as to contain a large number of tiny pores. Filtration is accomplished by using a syringe or a pump to force the liquid through the filtration apparatus into a sterile collection vessel (Figure 5.35). Another type of membrane filter is the nucleopore filter (Figure 5.34c). Nucleopore filters are made from a thin polycarbonate film that is treated with radiation and then etched with a chemical, yielding very uniform holes (Figure 5.34c). Nucleopore filters are commonly used to isolate specimens for scanning electron microscopy. Microorganisms removed by filtration from a liquid or a natural sample, such as lake water, can then be observed directly on the filter (Figure 5.34d).

(a)

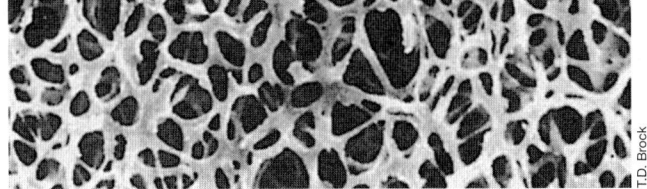

(b)

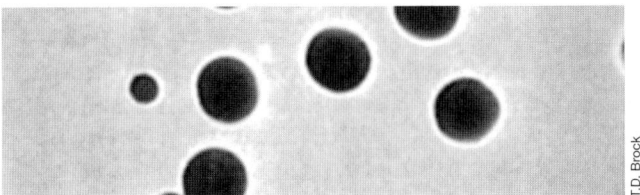

(c)

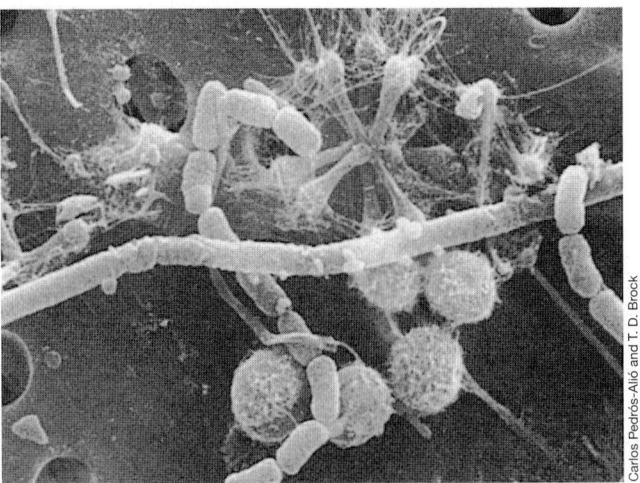

(d)

Figure 5.34 Microbiological filters. Scanning electron micrograph showing the structure of *(a)* a depth filter, *(b)* a conventional membrane filter, and *(c)* a nucleopore filter. *(d)* Scanning electron micrograph of various aquatic microbes trapped on a nucleopore 0.2-μm-pore-size membrane filter.

— MINIQUIZ —

- Define *D*10 and explain why the killing dose for radiation (Table 5.7) is not the same for all bacteria.

- Why is ionizing radiation more effective than UV radiation for sterilization of food products?

- Distinguish between the major types of sterilization filters used in the microbiology laboratory.

5.17 Chemical Control of Microbial Growth

Chemicals are routinely used to control microbial growth, and an **antimicrobial agent** is a natural or synthetic chemical that kills or inhibits the growth of microorganisms. Agents that actually kill are called *-cidal* agents, with a prefix indicating the type of microorganism killed. Thus, **bactericidal**, *fungicidal*, and *viricidal* agents kill bacteria, fungi, and viruses, respectively. Agents that do not kill but only inhibit growth are called *-static* agents, and include **bacteriostatic**, *fungistatic*, and *viristatic* compounds. We focus

UNIT 2

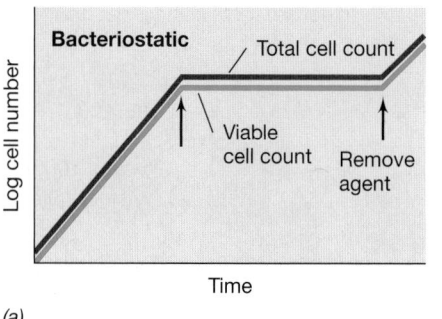
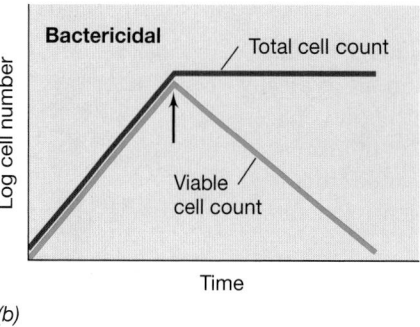
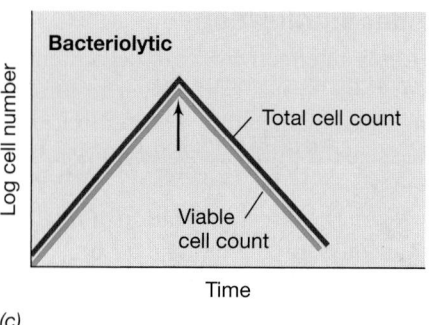

Figure 5.36 Different types of antimicrobial agents. *(a)* Bacteriostatic agents inhibit but do not kill. *(b)* Bactericidal agents kill. *(c)* Bacteriolytic agents lyse cells. At the time indicated by the arrow, a growth-inhibitory concentration of each antimicrobial agent was added to an exponentially growing culture. The turbidity and viable counts shown are characteristic of each type of agent.

here on chemicals commonly used as disinfectants and reserve discussion of the activities of a very important class of chemicals—the antibiotics—for later (⊘ Sections 7.10, 24.10, and 28.10–28.12).

Effect of Antimicrobial Agents on Growth

Antibacterial agents are classified as -static, -cidal, or -lytic (cell lysing) by observing their effects on cultures using viable and turbidimetric growth assays (**Figure 5.36**). Bacteriostatic agents are typically inhibitors of some important biochemical process, such as protein synthesis, and bind relatively weakly; if the agent is removed, the cells can resume growing. Some antibiotics work in this way. By contrast, bactericidal agents, for example formaldehyde, bind tightly to their cellular targets and by definition kill the cell. However, the dead cells are not lysed, and total cell numbers, reflected in the turbidity of the culture, remain constant (Figure 5.36*b*). Bacteriolytic agents kill cells by lysing them, and this affects both viable and total cell numbers (Figure 5.36*c*). An example of a bacteriolytic agent would be a detergent that ruptures the cytoplasmic membrane.

Assaying Antimicrobial Activity

Antimicrobial activity can be measured by determining the smallest amount of the agent needed to inhibit the growth of a test organism, a value called the **minimum inhibitory concentration (MIC)**. One way to determine the MIC of a given agent is to inoculate a series of tubes of liquid growth medium (**Figure 5.37**) containing a test organism and dilutions of the agent. Following incubation, the tubes are scored for growth (turbidimetrically), and the MIC is revealed as the lowest concentration of antimicrobial agent that completely inhibits growth.

Antimicrobial activity can also be assessed using solid media (**Figure 5.38**). Known amounts of an antimicrobial agent are

Maximum growth

Minimum inhibitory concentration

No growth

T. D. Brock

Figure 5.37 Antimicrobial agent susceptibility assay using dilution methods. The assay defines the minimum inhibitory concentration (MIC). A series of increasing concentrations of antimicrobial agent is prepared in the culture medium. Each tube is inoculated with a specific concentration of a test organism, followed by a defined incubation period. Growth, measured as turbidity, occurs in those tubes with antimicrobial agent concentrations below the MIC.

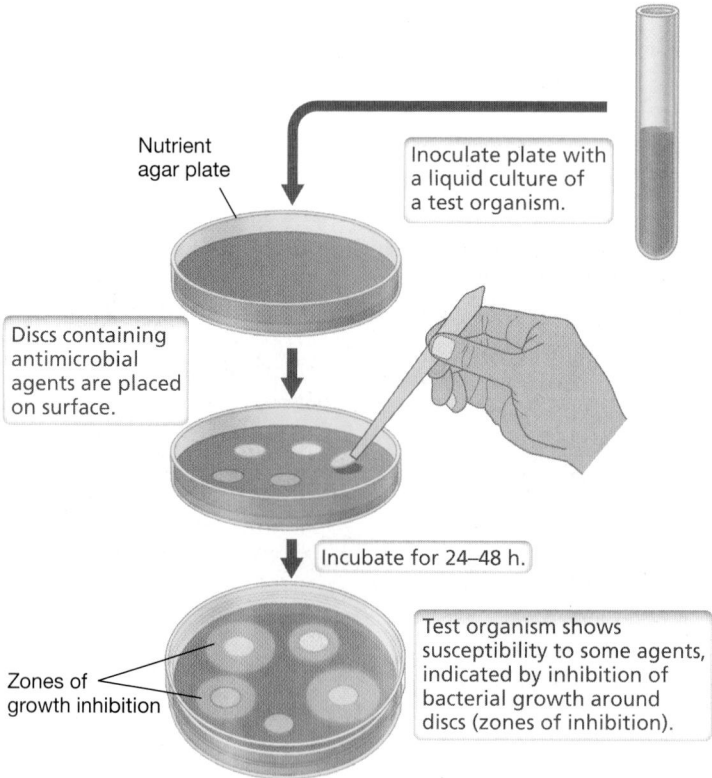

Nutrient agar plate

Inoculate plate with a liquid culture of a test organism.

Discs containing antimicrobial agents are placed on surface.

Incubate for 24–48 h.

Zones of growth inhibition

Test organism shows susceptibility to some agents, indicated by inhibition of bacterial growth around discs (zones of inhibition).

Figure 5.38 Antimicrobial agent susceptibility assay using diffusion methods. The antimicrobial agent diffuses from paper discs into the surrounding agar, inhibiting growth of susceptible microorganisms.

added to filter-paper discs and the discs are arranged on the surface of a uniformly inoculated agar plate. During incubation, the agent diffuses from the disc into the agar, establishing a gradient; the farther a chemical diffuses away from a disc, the lower its concentration. Following an incubation period, a *zone of growth inhibition* forms around discs that released effective chemicals; the zone is a function of several factors, including the amount of antimicrobial agent added to the disc, its solubility and diffusion coefficient, and, most importantly, its overall effectiveness. The disc diffusion assay is routinely used to test clinically isolated pathogenic bacteria for their antibiotic susceptibility (🔗 Section 28.4).

Chemical Antimicrobial Agents

Several antimicrobial agents are used to prevent the growth of human pathogens on inanimate surfaces and on external body surfaces. These include sterilants, disinfectants, sanitizers, and antiseptics (**Table 5.8**).

Sterilants destroy all microorganisms, including endospores. Chemical sterilants are used for decontamination or sterilization in situations where it is impractical or impossible to use heat or radiation. Hospitals, clinics, and laboratories, for example, must routinely decontaminate and sterilize heat-sensitive materials such as thermometers, lensed instruments, polyethylene tubing, catheters, and reusable medical and dental equipment. This process of *cold sterilization*, as it is called, employs gases such as ethylene oxide or aldehydes such as formaldehyde or glutaraldehyde to sterilize the devices.

Disinfectants are chemicals that kill microorganisms but not necessarily endospores and are primarily used on surfaces. For example, phenol and cationic detergents are used to disinfect floors, tables, bench tops, walls, and so on (Table 5.8) and are important agents of infection control in hospitals and other medical settings. **Sanitizers**, by contrast, are less harsh than disinfectants and reduce microbial numbers but do not sterilize. Sanitizers are widely used in the food industry to treat surfaces such as mixing and cooking

TABLE 5.8 Antiseptics, sterilants, disinfectants, and sanitizers[a]

Agent	Mode of action	Use
Antiseptics (germicides)		
Alcohol (60–85% ethanol or isopropanol in water)	Lipid solvent and protein denaturant	Topical antiseptic
Phenol-containing compounds (hexachlorophene, triclosan, chloroxylenol, chlorhexidine)	Disrupts cytoplasmic membrane	Soaps, lotions, cosmetics, deodorants, topical disinfectants; paper, leather, and textile industries
Cationic detergents, especially quaternary ammonium compounds (benzalkonium chloride)	Disrupts cytoplasmic membrane	Soaps, lotions, topical disinfectants; metal and petroleum industries
Hydrogen peroxide (3% solution)	Oxidizing agent	Topical antiseptic
Iodophors (Betadine®)	Iodinates proteins, rendering them nonfunctional; oxidizing agent	Topical antiseptic
Octenidine	Cationic surfactant, disrupts cytoplasmic membrane	Topical antiseptic
Sterilants, disinfectants, and sanitizers		
Alcohol (60–85% ethanol or isopropanol in water)	Lipid solvent and protein denaturant	General-purpose disinfectant for virtually any surface
Cationic detergents (quaternary ammonium compounds, Lysol® and many related disinfectants)	Interact with phospholipids	Disinfectant/sanitizer for medical instruments, food and dairy equipment
Chlorine gas	Oxidizing agent	Disinfectant for drinking water and electrical/nuclear cooling towers
Chlorine compounds (chloramines, sodium hypochlorite, sodium chlorite, chlorine dioxide)	Oxidizing agent	Disinfectant/sanitizer for medical instruments, food/dairy equipment, and in water purification
Copper sulfate	Protein precipitant	Algicide in swimming pools
Ethylene oxide (gas)	Alkylating agent	Sterilant for temperature-sensitive materials such as plastics
Formaldehyde	Alkylating agent	Diluted (3% solution) as surface disinfectant/sterilant; concentrated (37% solution) as sterilant
Glutaraldehyde	Alkylating agent	Disinfectant or sterilant as 2% solution
Hydrogen peroxide	Oxidizing agent	Vapor used as sterilant
Iodophors (Wescodyne®)	Iodinates proteins; oxidizing agent	General disinfectant
OPA (ortho-phthalaldehyde)	Alkylating agent	Powerful disinfectant used for sterilizing medical instruments
Ozone	Strong oxidizing agent	Disinfectant for drinking water
Peroxyacetic acid	Strong oxidizing agent	Disinfectant/sterilant
Phenolic compounds	Protein denaturant	General-purpose disinfectant
Pine oils (Pine-Sol®) (contains phenolics and detergents)	Protein denaturant	General-purpose disinfectant for household surfaces

[a]Alcohols, hydrogen peroxide, and iodophors can be antiseptics, disinfectants, sanitizers, or sterilants depending on concentration, length of exposure, and form of delivery.

equipment, dishes, and utensils, and are also used for dry hand washing when water is unavailable. **Antiseptics**, often called **germicides**, are chemicals that kill or inhibit the growth of microorganisms but are sufficiently nontoxic to animals to be applied to living tissues. Most germicides are used for hand washing or for treating surface wounds (Table 5.8). Certain antiseptics are also effective disinfectants. Ethanol, for example, can be both an antiseptic and a disinfectant, depending on the concentration and exposure time employed.

Several factors affect the efficacy of any chemical antimicrobial agent. For example, many antimicrobial agents are bound and inactivated by organic matter; thus, disinfecting a kitchen countertop littered with spilled foods is more difficult than disinfecting a clean countertop. Furthermore, bacteria often form biofilms (Section 5.1), covering surfaces of tissue or soiled medical devices with microbial cells embedded in polysaccharides. Biofilms may slow or even completely prevent penetration of antimicrobial agents, reducing or negating their effectiveness. Thus, the ultimate efficacy of any antimicrobial agent must be determined empirically and under the actual conditions of use. Only by actually testing the chemical and assaying for microbial growth both before and after treatment can one be confident that the agent is working as it should.

— **MINIQUIZ** —

- Distinguish between the antimicrobial effects of -static, -cidal, and -lytic agents.
- Describe how the minimum inhibitory concentration of an antibacterial agent is determined.
- Distinguish between a sterilant, a disinfectant, and an antiseptic. What is cold sterilization?

MasteringMicrobiology® **Visualize, explore, and think critically** with Interactive Microbiology, MicroLab Tutors, MicroCareers case studies, and more. MasteringMicrobiology offers practice quizzes, helpful animations, and other study tools for lecture and lab to help you master microbiology.

Chapter Review

I • Cell Division and Population Growth

5.1 Microbial growth is defined as an increase in cell numbers and is the final result of the doubling of all cell components prior to actual division that yields two daughter cells. Most microorganisms grow by binary fission but some grow by budding. Biofilms are an alternative growth mode to a suspended (planktonic) lifestyle.

Q **How does the process of binary fission differ from that of budding division?**

5.2 Microbial cells undergo exponential growth, and a semilogarithmic plot of cell numbers with time can reveal the doubling time of the population. Various growth expressions can be calculated from cell number data obtained from an exponentially growing culture. Key expressions here are n, the number of generations; t, time; g, generation time, and k, the instantaneous growth rate constant.

Q **How is the generation time (g) of an exponentially growing culture determined?**

5.3 Microorganisms show a characteristic growth pattern when inoculated into a fresh culture medium. There is usually a lag phase and then growth commences in an exponential fashion. As essential nutrients are depleted and/or toxic products build up, growth ceases and the population enters the stationary phase. Further incubation can lead to cell death.

Q **Describe the growth cycle of a population of bacterial cells from the time this population is first inoculated into fresh medium.**

5.4 The chemostat is an open system used to maintain cell populations in exponential growth for extended periods. In a chemostat, the rate at which a culture is diluted with fresh growth medium controls the doubling time of the population, while the cell density (cells/ml) is controlled by the concentration of a growth-limiting nutrient dissolved in the fresh medium.

Q **How does a chemostat regulate growth rate and cell density independently?**

II • Culturing Microbes and Measuring Their Growth

5.5 Culture media supply the nutritional needs of microorganisms and are either defined or complex. Other media, such as selective, differential, and enriched media, are used for specific purposes. Many microorganisms can be grown in liquid or solid culture media, and pure cultures can be maintained if aseptic technique is practiced.

Q **Why would the following medium not be considered a chemically defined medium: glucose, 5 grams (g); NH_4Cl, 1 g; KH_2PO_4, 1 g; $MgSO_4$, 0.3 g; yeast extract, 5 g; distilled water, 1 liter? What is aseptic technique and why is it necessary?**

5.6 Total cell counts can be done under the microscope using counting chambers and are useful for assessing the total cell numbers in a microbial habitat or laboratory culture. Certain stains can be used to target specific cell populations in a sample.

Q Are total cell counts useful if one does not know the viability of the culture or sample?

5.7 Viable cell counts measure only the living population present in the sample with the assumption that each colony originates from the growth and division of a single cell. Depending on the growth medium and conditions employed, and the nature of the sample, viable counts can be fairly accurate assessments or highly unreliable.

Q How does a viable count differ from a total count?

5.8 Turbidity measurements are a rapid and useful method of measuring microbial growth and are based on the fact that cells in suspension scatter light. However, in order to relate a turbidity value to a cell number, a standard curve plotting these two parameters against one another must first be established.

Q How can turbidity be used as a measure of cell numbers?

III • Environmental Effects on Growth: Temperature

5.9 Temperature is a major environmental factor controlling microbial growth. An organism's cardinal temperatures describe the minimum, optimum, and maximum temperatures at which it grows. Microorganisms can be grouped by their cardinal temperature from cold-loving to heat-loving as psychrophiles, mesophiles, thermophiles, and hyperthermophiles.

Q Examine the graph in Figure 5.17. Why is the optimum temperature for an organism usually closer to its maximum than its minimum?

5.10 Organisms with temperature optima below 15°C are called psychrophiles, and the most extreme representatives inhabit constantly cold environments. Psychrophiles synthesize macromolecules that remain flexible and functional at cold temperatures but that can be unusually sensitive to warm temperatures.

Q Why are psychrophiles more likely to be found in constantly cold rather than intermittently cold environments?

5.11 Organisms with growth temperature optima between 45 and 80°C are called thermophiles while those with optima greater than 80°C are hyperthermophiles. These organisms inhabit hot environments that can have temperatures even above 100°C. Thermophiles and hyperthermophiles produce heat-stable macromolecules.

Q How do cells of hyperthermophiles prevent heat destruction?

IV • Environmental Effects on Growth: pH, Osmolarity, and Oxygen

5.12 The acidity or alkalinity of an environment can greatly affect microbial growth. Some organisms grow best at low or high pH (acidophiles and alkaliphiles, respectively), but most organisms grow best between pH 5.5 and 8. The internal pH of a cell must stay relatively close to neutral to prevent the destruction of macromolecules.

Q Concerning the pH of the environment and of the cell cytoplasm, in what ways are acidophiles and alkaliphiles different? In what ways are they similar?

5.13 The water activity of an aqueous environment is a function of its solute concentration. To survive in high-solute environments, organisms produce or accumulate compatible solutes to maintain positive water balance. Some microorganisms grow best at reduced water potential and some even require high levels of salts for growth.

Q How does a halophile maintain positive water balance while growing in a solution high in NaCl?

5.14 Aerobes require O_2 while anaerobes do not and may even be killed by O_2. Facultative organisms can live with or without O_2. Special techniques are needed to grow aerobic and anaerobic microorganisms. Several toxic forms of oxygen can form in the cell, but enzymes are present that neutralize most of them. Superoxide is a major toxic form of oxygen.

Q Contrast an aerotolerant and an obligate anaerobe in terms of sensitivity to O_2 and ability to grow in the presence of O_2. Compare and contrast the enzymes catalase, superoxide dismutase, and superoxide reductase in regard to their substrates and products.

V • Controlling Microbial Growth

5.15 Sterilization is the killing of all microbes including viruses, and heat is the most widely used method of sterilization. An autoclave employs moist heat under pressure to achieve temperatures above the boiling point of water. Pasteurization does not sterilize liquids, but it reduces the microbial load and kills pathogens.

Q Contrast the terms thermal death time and decimal reduction time. How would the presence of bacterial endospores affect either value? What time and temperature is necessary to ensure sterility in the autoclave?

5.16 Radiation can effectively inhibit or kill microorganisms. Ultraviolet radiation is used for decontaminating surfaces and air, whereas ionizing radiation is used for sterilization where penetration is required. Filters remove microorganisms from air or liquids. Membrane and nucleopore filters are used for sterilization of heat-sensitive liquids and to examine the contents of filtration by microscopy.

Q Describe the effects of a lethal dose of ionizing radiation at the molecular level. What type of filter would be used to filter sterilize a heat-sensitive liquid?

5.17 Chemicals that kill organisms are called -cidal agents while those that arrest growth but do not kill are called -static agents. Antimicrobial agents are tested for efficacy by determining their ability to inhibit growth. Sterilants, disinfectants, and sanitizers are used to decontaminate nonliving material, while antiseptics (germicides) are used to reduce the microbial load on living tissues.

Q Describe the procedure for obtaining the minimum inhibitory concentration (MIC) for a chemical that is bactericidal for *Escherichia coli*. Contrast the action of disinfectants and antiseptics.

Application Questions

1. A medium was inoculated with 5×10^6 cells/ml of *Escherichia coli* cells. Following a 1-h lag, the population grew exponentially for 5 h, after which the population was 5.4×10^9 cells/ml. Calculate g and k for this growth experiment.

2. *Escherichia coli* but not *Pyrolobus fumarii* will grow at 40°C, while *P. fumarii* but not *E. coli* will grow at 110°C. What is happening (or not happening) to prevent growth of each organism at the nonpermissive temperature?

3. In which direction (into or out of the cell) will water flow in cells of *Escherichia coli* (an organism found in your large intestine) suddenly suspended in a solution of 20% NaCl? What if the cells were suspended in distilled water? If growth nutrients were added to each cell suspension, which (if either) would support growth, and why?

Chapter Glossary

Acidophile an organism that grows best at low pH, typically below pH 5.5

Aerobe an organism that can use O_2 in respiration; some require O_2

Aerotolerant anaerobe a microorganism unable to respire O_2 but whose growth is unaffected by it

Alkaliphile an organism that has a growth pH optimum of 8 or higher

Anaerobe an organism that cannot use O_2 in respiration and whose growth is typically inhibited by O_2

Antimicrobial agent a chemical compound that kills or inhibits the growth of microorganisms

Antiseptic (germicide) a chemical agent that kills or inhibits growth of microorganisms and is sufficiently nontoxic to be applied to living tissues

Aseptic technique a series of steps taken to prevent contamination of laboratory cultures and media

Autoclave a sealed heating device that destroys microorganisms with temperature and steam under pressure

Bactericidal agent an agent that kills bacteria

Bacteriostatic agent an agent that inhibits bacterial growth

Batch culture a closed-system microbial culture of fixed volume

Binary fission cell division following enlargement of a cell to twice its minimum size

Biofilm an attached polysaccharide matrix containing bacterial cells

Budding division a cell division process whereby new cell material is produced from a single point instead of along the entire cell

Cardinal temperatures the minimum, maximum, and optimum growth temperatures for a given organism

Chemostat a device that allows for the continuous culture of microorganisms with independent control of both growth rate and cell number

Compatible solute a molecule that is accumulated in the cytoplasm of a cell for adjustment of water activity but that does not inhibit biochemical processes

Complex medium a culture medium of highly nutritious substances for which the exact composition is unknown

Culture medium a nutrient solution required for growing a laboratory culture of a specific microorganism

Defined medium a culture medium for which the exact composition is known

Disinfectant an antimicrobial agent used only on inanimate objects

Disinfection rendering a surface or object free of all pathogenic microorganisms

Exponential growth growth of a microbial population in which cell numbers double within a specific time interval

Extreme halophile a microorganism that requires very large amounts of NaCl, usually greater than 10% and in some cases near saturation, for growth

Facultative with respect to O_2, an organism that can grow in either its presence or absence

Generation time the time required for a population of microbial cells to double

Germicide (antiseptic) a chemical agent that kills or inhibits growth of microorganisms and is sufficiently nontoxic to be applied to living tissues

Growth in microbiology, an increase in cell number

Halophile a microorganism that requires NaCl for growth

Halotolerant a microorganism that does not require NaCl for growth but can grow in the presence of NaCl, in some cases, substantial levels of NaCl

Hyperthermophile a species of *Bacteria* or *Archaea* whose growth temperature optimum is 80°C or greater

Mesophile an organism that grows best at temperatures between 20 and 40°C

Microaerophile an aerobic organism that can grow only when O_2 tensions are reduced from that present in air

Minimum inhibitory concentration (MIC) the minimum concentration of a substance necessary to prevent microbial growth

Neutrophile an organism that grows best at neutral pH, between pH 5.5 and 8

Obligate anaerobe an organism that cannot grow in the presence of O_2

Osmophile an organism that grows best in the presence of high levels of solute, typically a sugar

Pasteurization the heat treatment of milk or another liquid to reduce its total number of microorganisms

pH the negative logarithm of the hydrogen ion (H^+) concentration of a solution

Plate count a method for counting viable cells; the number of colonies on a plate is used as a measure of cell numbers

Psychrophile an organism with a growth temperature optimum of 15°C or lower and a maximum growth temperature below 20°C

Psychrotolerant capable of growing at low temperatures but having an optimum above 20°C

Sanitizer an agent that reduces microorganisms to a safe level, but may not eliminate them

Sterilant a chemical agent that destroys all forms of microbial life

Sterilization the killing or removal of all living organisms and viruses

Thermophile an organism whose growth temperature optimum lies between 45 and 80°C

Viable capable of reproducing

Viable count a measurement of the concentration of live cells in a population

Water activity the ratio of the vapor pressure of air in equilibrium with a solution to the vapor pressure of pure water

Xerophile an organism that is able to live, or that lives best, in very dry environments

Microbial Regulatory Systems

microbiology**now**

A Microbial Hunter: *Pseudomonas aeruginosa* Senses and Scavenges Nutrients from Damaged Tissues

Despite their single-cell nature, many microbes can sense and respond to their surroundings. Motile bacteria are able to move toward nutrients and away from poisons through a process called chemotaxis. Using special chemoreceptors that extend like antennae to the exterior of the cell, bacteria can sense and respond to a wide range of signals including attractants (such as sugars and amino acids) and repellents (such as certain metals and toxins). Binding of sensory molecules to chemoreceptors sends a signal to the cell's flagellar motor to steer the cell toward or away from the molecule.

This internal sense of direction not only enhances a microbe's ability to hunt down food and avoid toxins, but also plays a critical role in the infection process for some pathogens. For example, the opportunistic human pathogen *Pseudomonas aeruginosa* can respond by chemotaxis to over 25 different molecules and is a major factor in the symptoms observed in the disease cystic fibrosis. Cystic fibrosis patients are unable to clear mucus from their lungs, and this mucus provides nutrients to bacteria. Microbiologists have shown that *P. aeruginosa* cells use chemotaxis to scavenge nutrients from damaged lung tissues by sensing amino acids released from injured cells.

The images here show the remarkable time course of live human epithelial cells (blue) and the successful hunt for nutrients by *P. aeruginosa* cells (green) following host cell damage. Before damage (top left photo) and two minutes after scratching the epithelial monolayer (top right photo), very few bacterial cells are present. However, within 5 minutes of wound induction (bottom left photo), *P. aeruginosa* cells migrate to the damaged epithelial cells to devour nutrients released from the cells. Over time, some of the epithelial cells die (red) and the bacterial hunters move on to scavenge more nutrients from living cells (bottom right photo).

It is truly remarkable that all of these events can occur within half an hour, considering that the directed movement of bacterial cells is an orchestrated process that requires a complex regulatory cascade. But this is the pace at which bacterial regulatory events can occur, as we will see many times in this chapter.

Source: Schwarzer, C., H. Fischer, and T. E. Machen. 2016. Chemotaxis and binding of *Pseudomonas aeruginosa* to scratch-wounded human cystic fibrosis airway epithelial cells. *PLoS ONE 11(3):* e0150109.

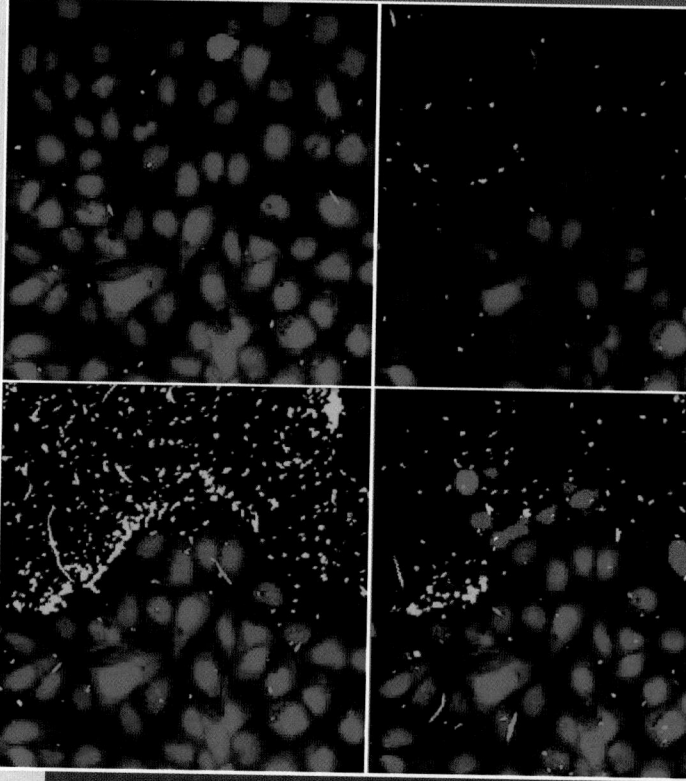

To efficiently orchestrate the numerous reactions that occur in a cell and to maximally use available resources, cells must *regulate* the types, amounts, and activities of proteins and other macromolecules they make. Because some proteins and RNA molecules are needed in the cell at about the same level under all growth conditions, the expression of these molecules is said to be *constitutive*. However, more often a particular protein or RNA is needed under some conditions but not others. For instance, enzymes required for using the sugar lactose are useful only if lactose is available to the cell. Ultimately, regulatory systems work to improve the fitness of the organism and its ability to produce the maximum number of offspring that available resources will allow. In this way, regulation and microbial growth go hand in hand.

Overall, cells use two major approaches to regulate protein function. One controls the *amount* of an enzyme or other protein and the second controls the *activity* of a preformed enzyme or other protein. **Figure 6.1** gives an overview of the regulatory mechanisms a cell has at its disposal. The amount of protein synthesized can be regulated at either the level of transcription, by varying the amount of mRNA made, or at the level of translation, by translating or not translating the mRNA. Collectively, these processes are called **gene expression**. After the protein has been synthesized,

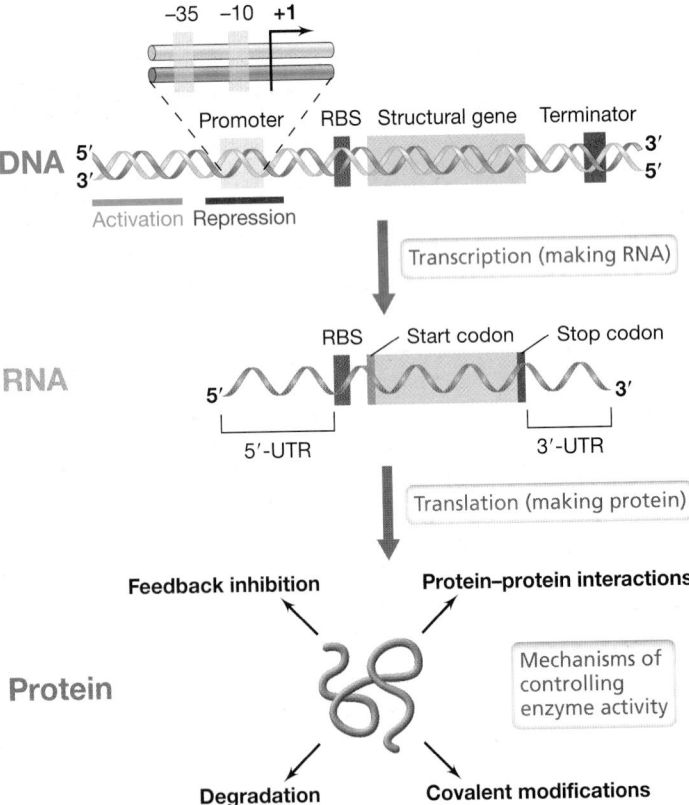

Figure 6.1 Gene expression and regulation of protein activity. For DNA, the promoter and terminator as well as regions related to transcriptional activation and repression are indicated. The 5′ untranslated region (5′-UTR) is a short region between the start of transcription and the start of translation, including the ribosome-binding site (RBS). The 3′ untranslated region (3′-UTR) is a short region between the stop codon and the transcription terminator. These are the regions where translational regulation often occurs. Mechanisms for regulating protein activity after translation are shown in the bottom part of the figure.

post-translational regulatory processes can further regulate the activity of some proteins. In this chapter we discuss how the cell efficiently controls its metabolism by regulating both enzyme synthesis and enzyme activity.

I • DNA-Binding Proteins and Transcriptional Regulation

Although several microbial regulatory mechanisms are possible, our discussion begins with control at the level of *transcription* because this is the major means of regulation in *Bacteria* and *Archaea*.

6.1 DNA-Binding Proteins

For a gene to be transcribed, RNA polymerase must recognize a specific promoter on the DNA (↩ Section 4.5), and small molecules often take part in regulating this process. However, they rarely do so directly. Instead, they typically influence the binding of certain proteins, called *regulatory proteins*, to specific sites on the DNA. This event regulates gene expression by turning transcription either on or off.

Interaction of Proteins with Nucleic Acids

Interactions between proteins and nucleic acids are central to replication, transcription, and translation, and also to the regulation of these processes. Protein–nucleic acid interactions may be specific or nonspecific, depending on whether the protein attaches anywhere along the nucleic acid or binds to a specific site. Most DNA-binding proteins interact with DNA in a sequence-specific manner. Specificity is provided by interactions between specific amino acid side chains of the proteins and specific chemical groups on the nitrogenous bases and the sugar–phosphate backbone of the DNA. Because of its size, the *major groove* of DNA (↩Figure 4.3) is the main site of protein binding, and Figure 4.1*c* identified atoms of the bases in the major groove that are known to interact with proteins. To achieve high specificity, the binding protein must interact simultaneously with several nucleotides.

We have already described a structure in DNA called an *inverted repeat* (↩ Figure 4.24). Such inverted repeats are frequently the locations at which regulatory proteins bind specifically to DNA (**Figure 6.2**). Note that this interaction does not require the formation of stem–loop structures in the DNA. DNA-binding proteins are often homodimeric, meaning they are composed of two identical polypeptide subunits, each subdivided into **domains**—regions of the protein with a specific structure and function. Each subunit has a domain that interacts specifically with a region of DNA in the major groove. When protein dimers interact with inverted repeats on DNA, each subunit binds to one of the inverted repeats. The dimer as a whole thus binds to both DNA strands (Figure 6.2).

Structure of DNA-Binding Proteins

DNA-binding proteins in both *Bacteria* and *Archaea* as well as eukaryotes possess several classes of protein domains that are

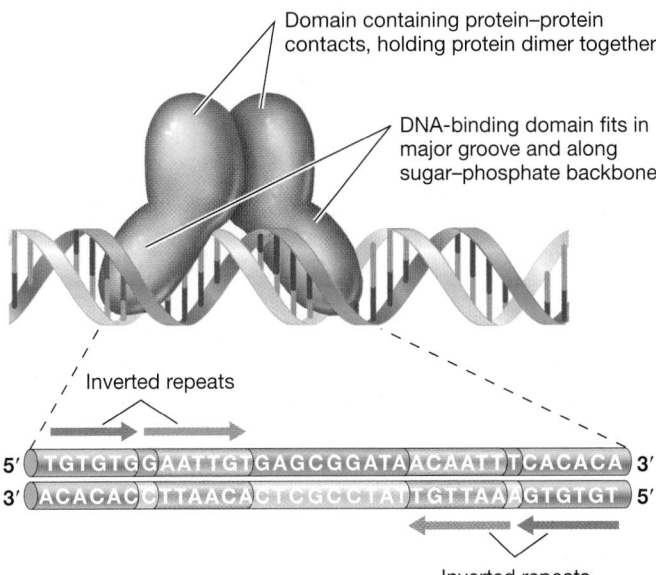

Inverted repeats

```
5'  TGTGTGGAATTGTGAGCGGATAACAATTTCACACA  3'
3'  ACACACCTTAACACTCGCCTATTGTTAAAGTGTGT  5'
```

Inverted repeats

Figure 6.2 DNA-binding proteins. Many DNA-binding proteins are dimers that combine specifically with two sites on the DNA. The specific DNA sequences that interact with the protein are inverted repeats. The nucleotide sequence of the operator gene of the lactose operon (Section 6.2) is shown, and the inverted repeats, which are sites at which the *lac* repressor makes contact with the DNA, are shown in purple and blue boxes.

critical for proper binding to DNA. One of the most common is the *helix-turn-helix* structure (**Figure 6.3a**). This consists of two segments of polypeptide chain that have an α-helix secondary structure connected by a short sequence forming the "turn." The first helix is the *recognition helix*, which interacts specifically with DNA. The second helix, the *stabilizing helix*, stabilizes the first helix by interacting with it by way of hydrophobic interactions. The turn linking the two helices consists of three amino acid residues, the first of which is typically a glycine. Sequences are recognized by noncovalent interactions, including hydrogen bonds and van der Waals contacts, between the recognition helix of the protein and specific chemical groups in the sequence of base pairs on the DNA.

Many different DNA-binding proteins from *Bacteria* contain the helix-turn-helix structure. These include many repressor proteins, such as the *lac* and *trp* repressors of *Escherichia coli* (Section 6.2 and see Figure 6.3 inset), and some proteins of bacterial viruses, such as the bacteriophage lambda repressor (Figure 6.3b). Indeed, over 250 different proteins with this motif bind to DNA to regulate transcription in *E. coli*. Two other types of protein domains are commonly found in DNA-binding proteins. One of these, the *zinc finger*, is frequently found in regulatory proteins in eukaryotes and, as its name implies, binds a zinc ion. The other protein domain commonly found in DNA-binding proteins is the

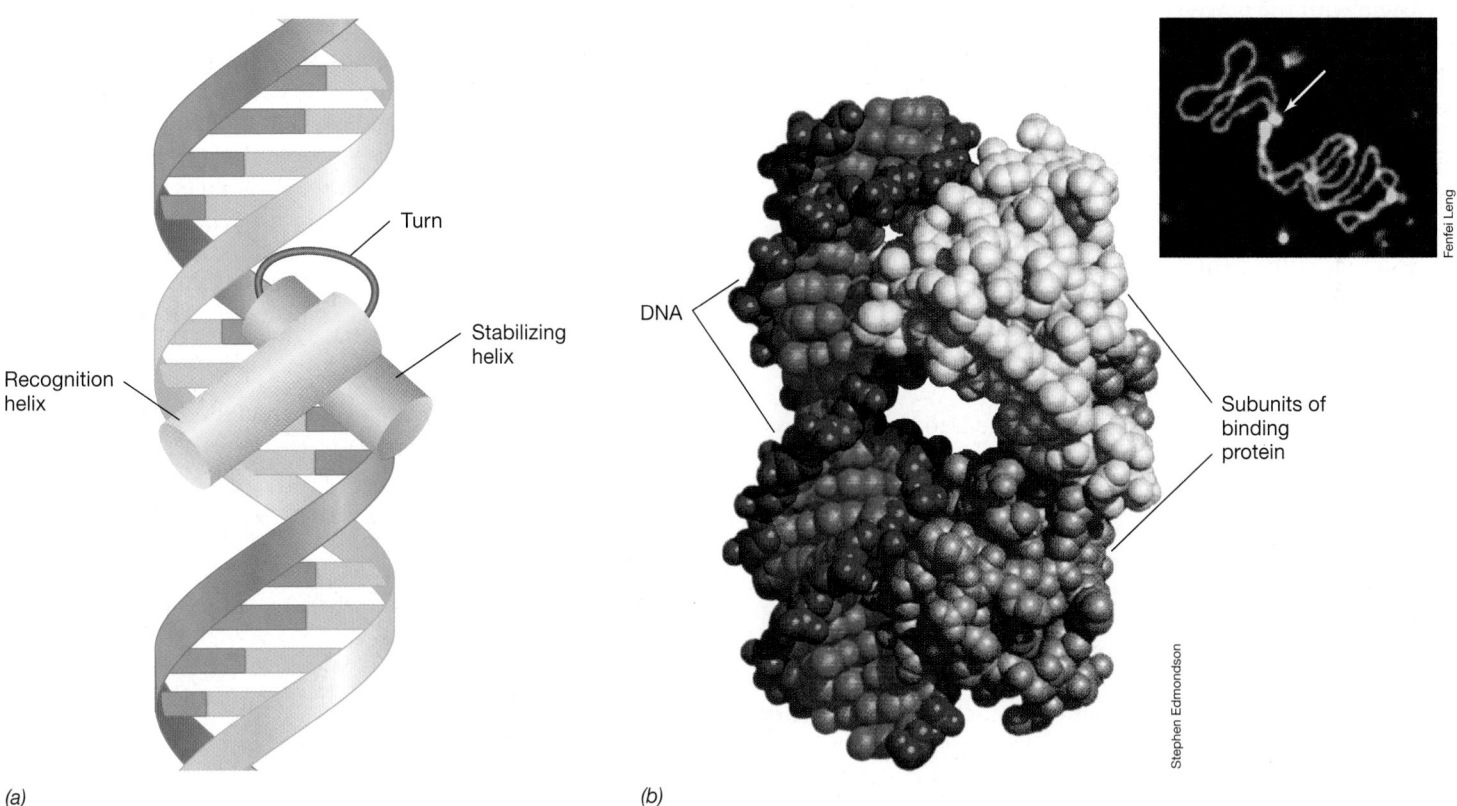

(a) *(b)*

Figure 6.3 The helix-turn-helix structure of some DNA-binding proteins. *(a)* A simple model of the helix-turn-helix structure within a single protein subunit. *(b)* A computer model of both subunits of the bacteriophage lambda repressor bound to its operator. The DNA is red and blue. One subunit of the dimeric repressor is shown in brown and the other in yellow. Each subunit contains a helix-turn-helix structure. The coordinates used to generate this image were downloaded from the Protein Data Bank (http://www.rcsb.org). Inset: Atomic force microscopy showing copies of the LacI repressor protein (arrow) bound to multiple operator sites on a DNA molecule.

leucine zipper, which contains regularly spaced leucine residues that function to hold two recognition helices in the correct orientation to bind DNA.

Once a protein binds at a specific site on the DNA, various outcomes are possible. Some DNA-binding proteins are enzymes that catalyze a specific reaction on the DNA, such as transcription. In other cases, however, the binding event either blocks transcription (*negative regulation*, Section 6.2) or activates it (*positive regulation*, Section 6.3).

MINIQUIZ

- What is a protein domain?
- Why are most DNA-binding proteins specific to certain chemical groups within the DNA?

6.2 Negative Control: Repression and Induction

Transcription is the first step in biological information flow; because of this, it is simple and efficient to control gene expression at this point. If one gene is transcribed more frequently than another, there will be more of its mRNA available for translation and therefore a greater amount of its protein product in the cell. We begin with the processes of repression and induction, simple forms of regulation that govern gene expression at the level of transcription. Here we deal with **negative control** of transcription, control that *prevents* transcription.

Enzyme Repression and Induction

Often the enzymes that catalyze the synthesis of a specific product are not made if the product is already present in the medium in sufficient amounts. For example, in *Escherichia coli* and many other *Bacteria*, the enzymes needed to synthesize the amino acid arginine are made only when arginine is absent from the culture medium; an excess of arginine decreases the synthesis of these enzymes. This is called enzyme **repression**.

As can be seen in **Figure 6.4**, if arginine is added to a culture growing exponentially in a medium devoid of arginine, growth continues at the previous rate, but production of the enzymes for arginine synthesis stops. Note that this is a *specific* effect, as the synthesis of all other enzymes in the cell continues at the previous rate. This is because the enzymes affected by a particular repression event make up only a tiny fraction of the entire complement of proteins in the cell. Enzyme repression is widespread in bacteria as a means of controlling the synthesis of enzymes required for the production of amino acids and the nucleotide precursors purines and pyrimidines. In most cases, the final product of a particular biosynthetic pathway represses the enzymes of the pathway. This ensures that the organism does not waste energy and nutrients synthesizing unneeded enzymes.

Enzyme **induction** is conceptually the opposite of enzyme repression. In enzyme induction, an enzyme is made only when its substrate is *present*. Enzyme repression typically affects biosynthetic (anabolic) enzymes. In contrast, enzyme induction usually affects degradative (catabolic) enzymes. To illustrate induction, consider

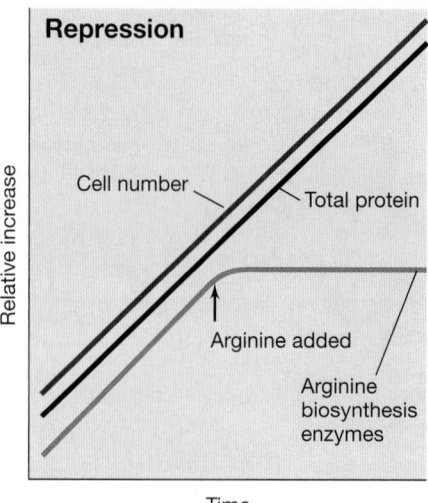

Figure 6.4 Enzyme repression. In a growing bacterial culture, the addition of arginine to the medium specifically represses production of enzymes needed to make arginine. Net protein synthesis is unaffected.

the utilization of the sugar lactose as a carbon and energy source by *E. coli*, the enzymes for which are encoded by the *lac* operon (⇄ Section 4.2). **Figure 6.5** shows the induction of β-galactosidase, the enzyme that cleaves lactose into glucose and galactose. This enzyme is required for *E. coli* to grow on lactose. If lactose is absent, the enzyme is not made, but synthesis begins almost immediately after lactose is added. The three genes in the *lac* operon encode three proteins, including β-galactosidase, that are induced simultaneously upon adding lactose. This type of control mechanism ensures that specific enzymes are synthesized only when needed.

Inducers and Corepressors

A substance that induces enzyme synthesis is called an *inducer* and a substance that represses enzyme synthesis is called a *corepressor*.

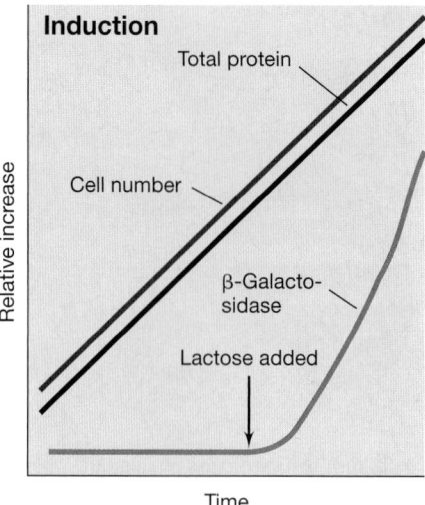

Figure 6.5 Enzyme induction. In a growing bacterial culture, the addition of lactose to the medium specifically induces synthesis of the enzyme β-galactosidase. Net protein synthesis is unaffected.

These substances, which are typically small molecules, are collectively called *effectors*. Interestingly, not all inducers and corepressors are actual substrates or end products of the enzymes involved. For example, structural analogs may induce or repress even though they are not substrates of the enzyme. Isopropylthiogalactoside (IPTG), for instance, is an inducer of β-galactosidase even though IPTG cannot be hydrolyzed by this enzyme. In nature, however, inducers and corepressors are probably normal cell metabolites. Detailed studies of lactose utilization in *E. coli* have shown that the actual inducer of β-galactosidase is not lactose but its isomer allolactose, which is made from lactose.

Mechanism of Repression and Induction

How can inducers and corepressors affect transcription in such a specific manner? They do this indirectly by binding to specific DNA-binding proteins, which, in turn, affect transcription. For an example of a repressible enzyme, we consider the arginine operon (Figure 6.4). **Figure 6.6a** shows transcription of the arginine genes, which proceeds when the cell needs arginine. However, when arginine is plentiful, it becomes a corepressor. As Figure 6.6*b* shows, arginine binds to a specific **repressor protein**, the *arginine repressor*, present in the cell. The repressor protein is **allosteric**; that is, its conformation is altered when the effector molecule binds to it (Section 6.14).

By binding its effector, the repressor protein is *activated* and can then bind to a specific region of the DNA near the promoter of the gene called the *operator*. This region gave its name to the **operon**, a cluster of consecutive genes whose expression is under the control of a single operator (⊖ Section 4.2). All of the genes in an operon are transcribed as a single unit yielding a single mRNA (⊖ Section 4.2). The operator is located downstream of the promoter where synthesis of mRNA is initiated (Figure 6.6). If the repressor binds to the operator, transcription is physically blocked because RNA polymerase can neither bind nor proceed. Hence, the polypeptides encoded by the genes in the operon cannot be synthesized. If the mRNA is polycistronic (⊖ Section 4.5), all the polypeptides encoded by this mRNA will be repressed.

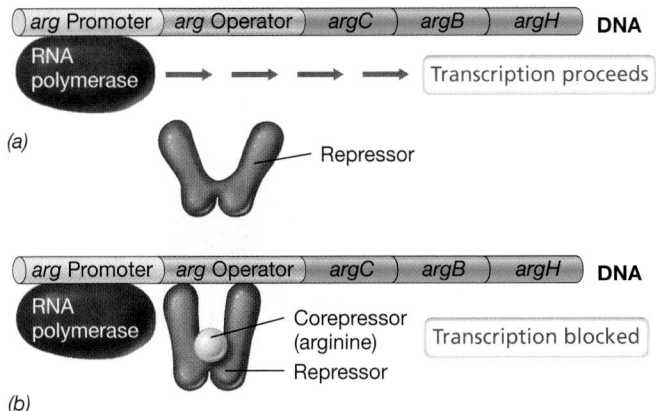

Figure 6.6 Enzyme repression in the arginine operon. *(a)* The operon is transcribed because the repressor is unable to bind to the operator. *(b)* After a corepressor (small molecule) binds to the repressor, the repressor binds to the operator and blocks transcription; mRNA and the proteins it encodes are not made. For the *argCBH* operon, the amino acid arginine is the corepressor that binds to the arginine repressor.

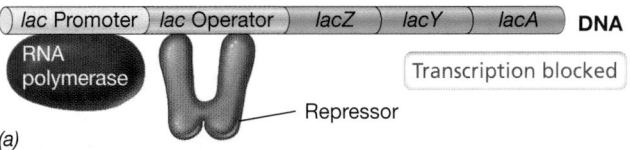

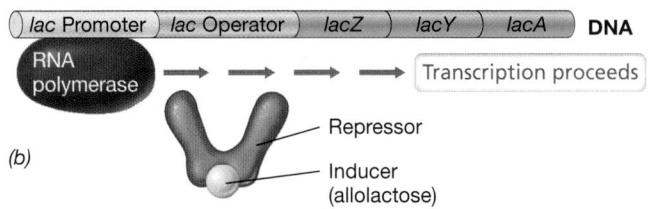

Figure 6.7 Enzyme induction in the lactose operon. *(a)* A repressor protein bound to the operator blocks the binding of RNA polymerase. *(b)* An inducer molecule binds to the repressor and inactivates it so that it no longer can bind to the operator. RNA polymerase then transcribes the DNA and makes an mRNA for that operon. For the *lac* operon, the sugar allolactose is the inducer that binds to the lactose repressor.

Enzyme induction may also be controlled by a repressor. In this case, the repressor protein is *active* in the absence of the inducer, completely blocking transcription. When the inducer is added, it combines with the repressor protein and inactivates it; inhibition is overcome and transcription can proceed (**Figure 6.7**).

All regulatory systems employing repressors have the same underlying mechanism: prevention of mRNA synthesis by the activity of specific repressor proteins that are themselves under the control of specific small effector molecules. And, as previously noted, because the repressor's role is to stop transcription, regulation by repressors is called *negative control*. One point to note is that genes are not turned on and off completely like light switches. DNA-binding proteins vary in concentration and affinity and thus control is quantitative. Even when a gene is "fully repressed" there is often a very low level of basal transcription.

— MINIQUIZ —

- Why is "negative control" so named?
- How does a repressor inhibit the synthesis of a specific mRNA?

6.3 Positive Control: Activation

Negative control relies on a protein (the repressor) to repress mRNA synthesis. By contrast, in **positive control** of transcription, the regulatory protein is an *activator* that activates the binding of RNA polymerase to DNA. An example of positive regulation is the catabolism of the disaccharide sugar maltose in *Escherichia coli*.

Maltose Catabolism in *Escherichia coli*

The enzymes for maltose catabolism in *E. coli* are synthesized only after the addition of maltose to the medium. The expression of these enzymes thus follows the pattern shown for β-galactosidase in Figure 6.5 except that maltose rather than lactose is required to induce gene expression. However, the synthesis of maltose-degrading enzymes is not under negative control, as in the *lac* operon, but under positive control; transcription requires the binding of an **activator protein** to the DNA.

The maltose activator protein cannot bind to the DNA unless it first binds maltose, the inducer. When the maltose activator protein binds to DNA, it allows RNA polymerase to begin transcription (**Figure 6.8**). Like repressor proteins, activator proteins bind specifically to certain chemical groups within the DNA. However, the region on the DNA that is the binding site of the activator is not called an operator (Figures 6.6 and 6.7) but instead an *activator-binding site* (Figure 6.8). Nevertheless, the genes controlled by this activator-binding site are still called an operon.

Binding of Activator Proteins

The promoters of positively controlled operons have nucleotide sequences that bind RNA polymerase weakly and are poor matches to the consensus sequence (⇨ Section 4.5). Thus, even with the correct sigma (σ) factor, the RNA polymerase has difficulty binding to these promoters. The role of the activator protein is to help the RNA polymerase recognize the promoter and begin transcription. For example, the activator protein may modify the structure of the DNA by bending it (**Figure 6.9**), allowing the RNA polymerase to make necessary contacts with nucleotides in the promoter region to begin transcription. Alternatively, the activator protein may interact directly with the RNA polymerase. This can happen either when the activator-binding site is close to the promoter (**Figure 6.10a**) or when it is several hundred base pairs away from the promoter, a situation in which DNA looping is required to make the necessary contacts between protein and nucleic acid (Figure 6.10b).

Many genes in *E. coli* have promoters under positive control and many have promoters under negative control. In addition, many

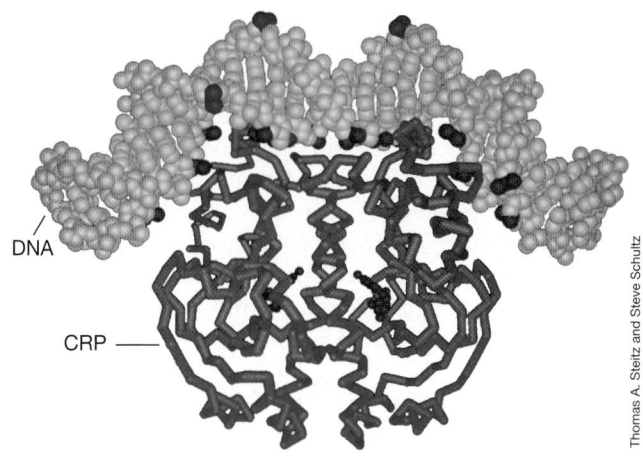

Figure 6.9 Computer model of a positive regulatory protein interacting with DNA. This model shows the cyclic AMP receptor protein (CRP), a regulatory protein that controls several operons. The α-carbon backbone of this protein is shown in blue and purple. The protein is binding to a DNA double helix (green and light blue). Note that binding of the CRP protein to DNA has bent the DNA.

operons have promoters with multiple types of control and some have more than one promoter, each with its own control system! Thus, the simple picture outlined above does not hold for all operons. Multiple control features are common in the operons of virtually all *Bacteria* and *Archaea*, and thus their overall regulation may require a network of interactions.

Operons versus Regulons

In *E. coli*, the genes required for maltose utilization are spread out over the chromosome in several operons, each of which has an activator-binding site to which a copy of the maltose activator protein can bind (**Figure 6.11**). Therefore, the maltose activator

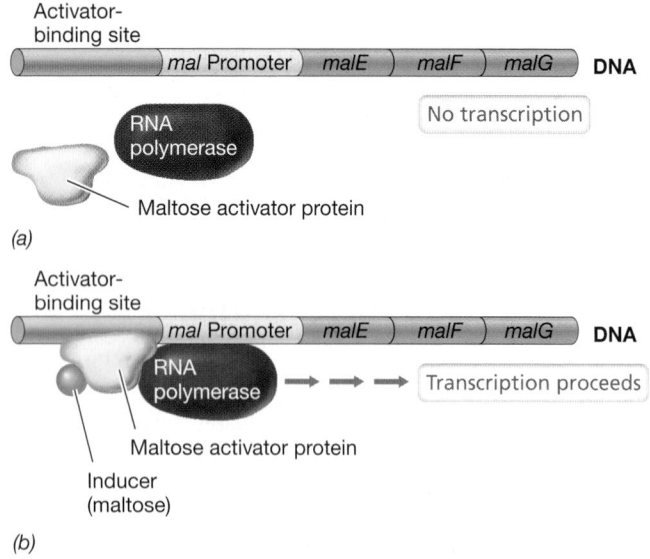

Figure 6.8 Positive control of enzyme induction in the maltose operon. *(a)* In the absence of an inducer, neither the activator protein nor the RNA polymerase can bind to the DNA. *(b)* An inducer molecule (for the *malEFG* operon it is the sugar maltose) binds to the activator protein (MalT), which in turn binds to the activator-binding site. This recruits RNA polymerase to bind to the promoter and begin transcription.

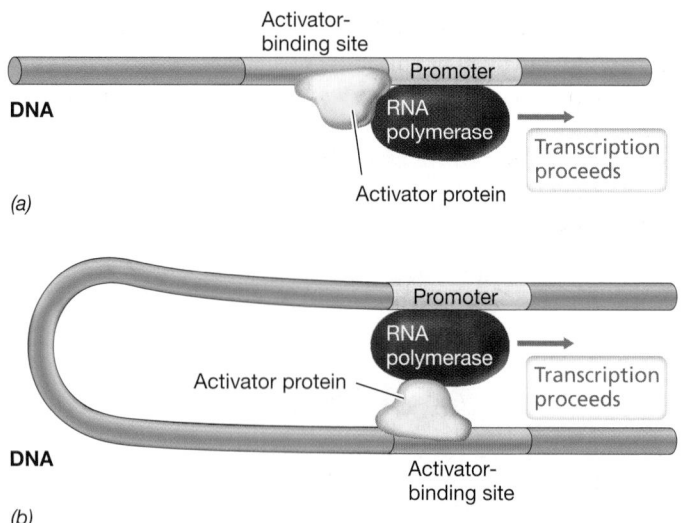

Figure 6.10 Activator protein interactions with RNA polymerase. *(a)* The activator-binding site is near the promoter. *(b)* The activator-binding site is several hundred base pairs from the promoter. In this case, the DNA must be looped to allow the activator and the RNA polymerase to contact.

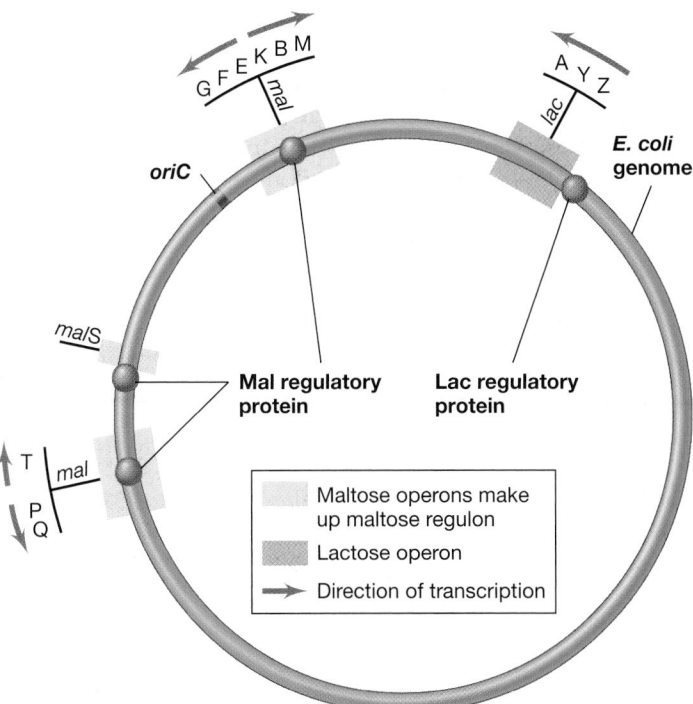

Figure 6.11 Maltose regulon of *Escherichia coli*. The genes and operons required for maltose utilization (*mal*) are dispersed throughout the *E. coli* genome and regulated by the same maltose regulatory protein. Note that the Lac repressor protein binds only to the *lac* operon, which is only located at one position on the chromosome, while the Mal repressor protein binds to multiple operons (the *mal* regulon).

protein actually controls the transcription of more than one operon. When more than one operon is under the control of a single regulatory protein, these operons are collectively called a **regulon**. The enzymes for maltose utilization are encoded by the maltose regulon.

Regulons are known for operons under negative control as well. For example, the arginine biosynthetic enzymes (Section 6.2) are encoded by the arginine regulon, whose operons are all under the control of the arginine repressor protein (only one of the arginine operons was shown in Figure 6.6). In regulon control, a specific DNA-binding protein binds only at those operons it controls regardless of whether it is functioning as an activator or repressor; other operons are not affected.

MINIQUIZ
- Compare and contrast the activities of an activator protein and a repressor protein.
- Distinguish between an operon and a regulon.

6.4 Global Control and the *lac* Operon

An organism often needs to regulate many unrelated genes simultaneously in response to a change in its environment. Regulatory mechanisms that respond to environmental signals by regulating the transcription of many different genes are called *global control systems*. Both the lactose operon and the maltose regulon respond to global controls in addition to their own controls discussed in Sections 6.2 and 6.3. We begin our consideration of global regulation by revisiting the *lac* operon and seeing how cells respond when given more than one sugar.

Catabolite Repression

We have not yet considered the possibility that bacteria might be confronted with several different utilizable carbon sources. For example, *Escherichia coli* can use many different sugars. When given several sugars, including glucose, do cells of *E. coli* use them simultaneously or one at a time? The answer is that *glucose is always used first*. It would be wasteful to induce enzymes for using other sugars when glucose is available, because *E. coli* grows faster on glucose than on other carbon sources. **Catabolite repression** is a mechanism of global control that controls the use of carbon sources if more than one is present.

When cells of *E. coli* are grown in a medium that contains glucose, the synthesis of enzymes needed for the breakdown of other carbon sources (such as lactose or maltose) is repressed, even if those other carbon sources are present. Thus, the presence of a favored carbon source represses the induction of pathways that catabolize other carbon sources. Catabolite repression is sometimes called the "glucose effect" because glucose was the first substance shown to cause this response. But catabolite repression is not always linked to glucose; the key point is that the favored substrate is a better carbon and energy source than other available carbon sources. Thus, catabolite repression ensures that the organism uses the *best* carbon and energy source first.

Why is catabolite repression called *global* control? In *E. coli* and other organisms for which glucose is the best energy source, catabolite repression prevents expression of most other catabolic operons as long as glucose is present. Dozens of catabolic operons are affected, including those for lactose, maltose, a host of other sugars, and most other commonly used carbon and energy sources for *E. coli*. In addition, genes for the synthesis of flagella are controlled by catabolite repression because if bacteria have a good carbon source available, there is no need to swim around in search of nutrients.

One consequence of catabolite repression is that it may lead to two exponential growth phases, a phenomenon called *diauxic growth*. If two usable energy sources are available, the cells first consume the better energy source. Growth stops when the better source is depleted, but then following a lag period, it resumes on the other energy source. Diauxic growth is illustrated in **Figure 6.12** for a culture of *E. coli* grown on a mixture of glucose and lactose. The cells grow more rapidly on glucose than on lactose. Although glucose and lactose are both excellent energy sources for *E. coli*, glucose is superior, and growth is faster.

The proteins of the *lac* operon, including the enzyme β-galactosidase, are required for using lactose and are induced in its presence (Figures 6.5 and 6.7). But the synthesis of these proteins is also subject to catabolite repression. As long as glucose is present, the *lac* operon is not expressed and lactose is not used. However, when glucose is depleted, catabolite repression is abolished and the *lac* operon is expressed; shortly thereafter, cells begin to grow on lactose.

UNIT 2

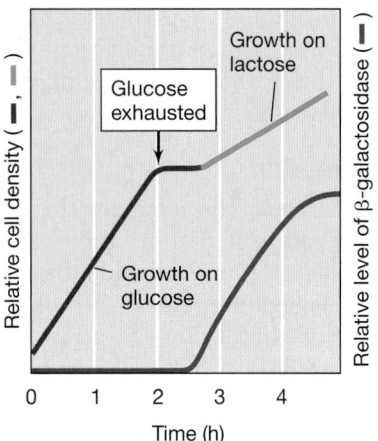

Figure 6.12 Diauxic growth of *Escherichia coli* on a mixture of glucose and lactose. The presence of glucose represses the synthesis of β-galactosidase, the enzyme that cleaves lactose into glucose and galactose. After glucose is depleted, there is a lag during which β-galactosidase is synthesized. Growth then resumes on lactose but at a slower rate, as indicated by the green line.

Cyclic AMP and Cyclic AMP Receptor Protein

Despite its name, catabolite repression relies on an activator protein and is actually a form of positive control (Section 6.3). The activator protein is called the *cyclic AMP receptor protein* (*CRP*). A gene that encodes a catabolite-repressible enzyme is expressed only if CRP binds to DNA in the promoter region. This allows RNA polymerase to bind to the promoter. CRP is an allosteric protein and binds to DNA only if it has first bound a small molecule called *cyclic adenosine monophosphate* (*cyclic AMP* or *cAMP*) (**Figure 6.13**). Like many DNA-binding proteins (Section 6.1), CRP binds to DNA as a dimer.

Cyclic AMP is a key molecule in many metabolic control systems, both in prokaryotic cells and eukaryotes. Because it is derived

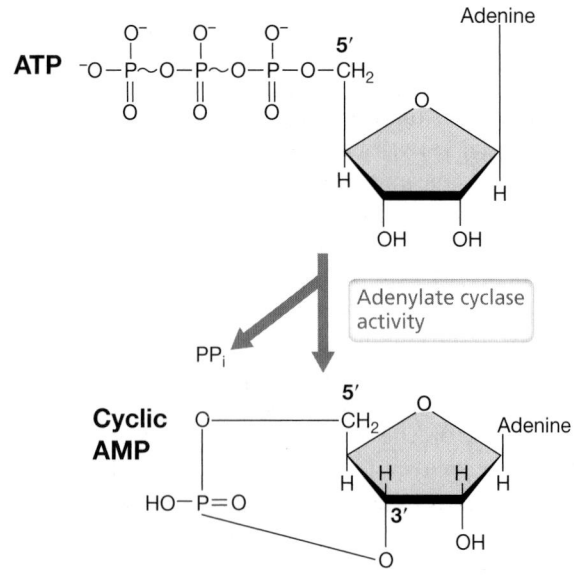

Figure 6.13 Cyclic AMP. Cyclic adenosine monophosphate (cyclic AMP) is made from ATP by the enzyme adenylate cyclase.

from a nucleic acid precursor, it is a **regulatory nucleotide**. Other regulatory nucleotides include cyclic guanosine monophosphate (cyclic GMP; important mostly in eukaryotes), cyclic di-GMP (important in biofilm formation; Section 7.9), and guanosine tetraphosphate (ppGpp, important in the stringent response, Section 6.9). Cyclic AMP is synthesized from ATP by an enzyme called *adenylate cyclase* (Figure 6.13). However, glucose inhibits the synthesis of cyclic AMP and also stimulates cyclic AMP transport out of the cell. When glucose enters the cell, the cyclic AMP level is lowered, CRP cannot bind DNA, and RNA polymerase fails to bind to the promoters of operons subject to catabolite repression. Thus, catabolite repression is an indirect result of the presence of a better energy source (glucose); the direct cause of catabolite repression is a low level of cyclic AMP.

Let us return to the *lac* operon and include catabolite repression. The entire regulatory region of the *lac* operon is diagrammed in **Figure 6.14**. For *lac* genes to be transcribed, two requirements must be met: (1) The level of cyclic AMP must be high enough for the CRP protein to bind to the CRP-binding site (positive control), and (2) lactose or another suitable inducer must be present so that

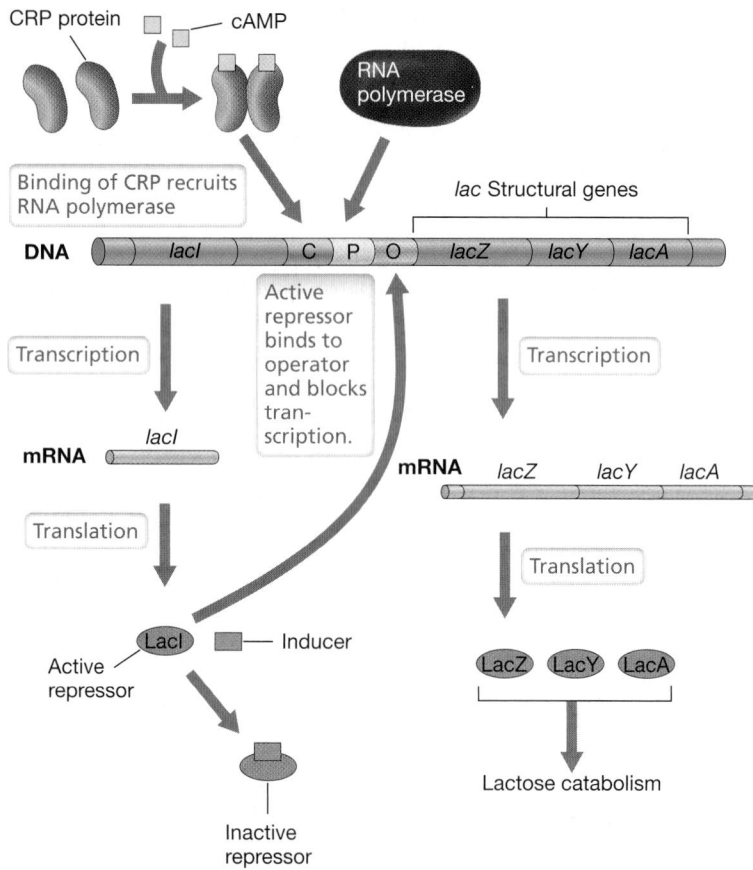

Figure 6.14 Overall regulation of the *lac* system. The *lac* operon consists of *lacZ*, encoding β-galactosidase, *lacY*, encoding lactose permease, and *lacA*, encoding lactose acetylase. The LacI repressor protein is encoded by a separate gene, *lacI*. LacI binds to the operator (O) unless the inducer is present. CRP binds to the C site when activated by cyclic AMP and recruits RNA polymerase to bind to the promoter (P). For the *lac* operon to be transcribed by RNA polymerase, the LacI repressor must be absent (that is, inducer must be present) and cyclic AMP levels must be high (due to the absence of glucose), allowing CRP to bind.

the lactose repressor (LacI protein) does not block transcription by binding to the operator (negative control). If these two conditions are met, the cell is signaled that glucose is absent and lactose is present; then and only then does transcription of the *lac* operon begin.

MINIQUIZ

- Explain how catabolite repression depends on an activator protein.
- What role does cyclic AMP play in glucose regulation?
- Explain how the *lac* operon is both positively and negatively controlled.

6.5 Transcription Controls in *Archaea*

There are two alternative approaches to regulating the activity of RNA polymerase. One strategy, common in *Bacteria*, is to use DNA-binding proteins that either block RNA polymerase activity (repressor proteins) or stimulate RNA polymerase activity (activator proteins). The alternative, common in eukaryotes, is to coordinate numerous DNA-binding proteins known as *transcription factors* to interact with RNA polymerase. Given the greater overall similarity between the mechanism of transcription in *Archaea* and *Eukarya* (↩ Section 4.6), it is perhaps surprising that the regulation of transcription in *Archaea* more closely resembles that of *Bacteria*.

Few repressor or activator proteins from *Archaea* have yet been characterized in detail, but it is clear that *Archaea* have both types of regulatory proteins. Archaeal repressor proteins either block the binding of RNA polymerase itself or block the binding of TBP (TATA-binding protein) and TFB (transcription factor B), proteins that are required for RNA polymerase to bind to the promoter in *Archaea* (↩ Section 4.6). At least some archaeal activator proteins function in just the opposite way, by recruiting TBP to the promoter, thereby facilitating transcription.

Control of Nitrogen Assimilation in *Archaea*

A good example of an archaeal repressor is the NrpR protein from the methanogen *Methanococcus maripaludis* (↩ Section 17.2). NrpR represses genes encoding nitrogen assimilation functions (**Figure 6.15**), such as those for nitrogen fixation and synthesis of the amino acid glutamine. When organic nitrogen is plentiful in the *M. maripaludis* cell, NrpR represses nitrogen assimilation genes. However, if the level of nitrogen becomes limiting, α-ketoglutarate accumulates to high levels. This occurs because α-ketoglutarate, a citric acid cycle intermediate, is also a major acceptor of ammonia during nitrogen assimilation (↩ Section 3.14).

When levels of α-ketoglutarate rise, this signals that ammonia is limiting and that additional pathways need to be activated for obtaining ammonia, such as nitrogen fixation or the high-affinity nitrogen assimilation enzyme glutamine synthetase. Elevated levels of α-ketoglutarate function as an inducer by binding to the NrpR protein. In this state, NrpR loses its affinity for the promoter regions of its target genes and no longer blocks transcription from promoters. In this respect, the NrpR protein resembles the LacI

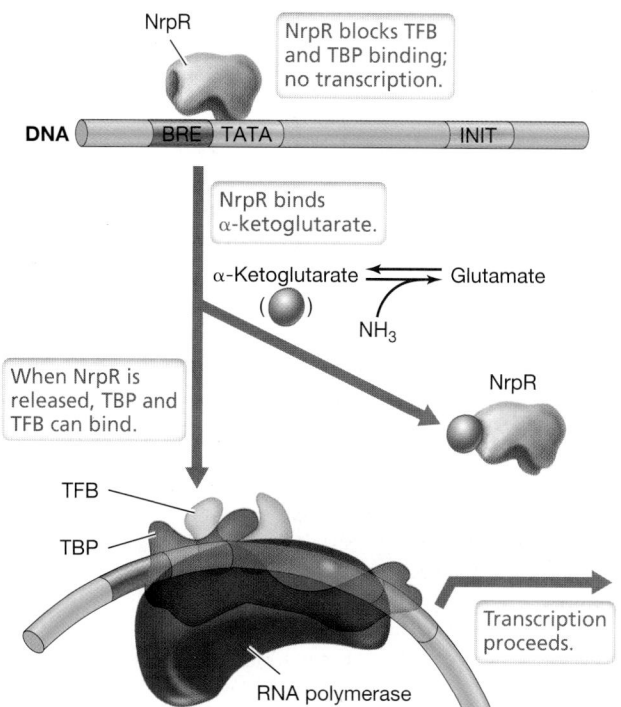

Figure 6.15 Repression of genes for nitrogen metabolism in *Archaea*. The NrpR protein of *Methanococcus maripaludis* acts as a repressor. It blocks the binding of the TFB and TBP proteins, which are required for promoter recognition, to the BRE (B recognition element) site and TATA box, respectively. If there is a shortage of ammonia, α-ketoglutarate is not converted to glutamate. The α-ketoglutarate accumulates and binds to NrpR, releasing it from the DNA. Now TBP and TFB can bind. This in turn allows RNA polymerase to bind and transcribe the operon.

repressor and similar-functioning proteins of *Bacteria* (Section 6.2 and Figure 6.7).

Dual-Acting Transcriptional Regulators in *Archaea*

Some archaeal regulators, such as the TrmB family, can possess *dual functionality* and act as both a repressor and an activator. The TrmB family of transcriptional regulators is widespread in *Archaea*, with more than 250 proteins identified. Members of the TrmB family primarily regulate sugar metabolism and can function as a repressor, an activator, or both. The activity of the allosteric regulator depends on its DNA binding site. In *Pyrococcus furiosus*, a hyperthermophile that does not possess a glucose transporter but can perform glycolysis, TrmBL1 simultaneously functions as a repressor of genes for other sugar transport systems present in this organism and as an activator for gluconeogenesis (glucose synthesis) genes.

As a repressor, TrmBL1 binds to a specific DNA sequence *downstream* of the BRE (B recognition element)/TATA binding sites of maltodextrin and maltose/trehalose uptake genes (**Figure 6.16a**). This binding blocks the recruitment of RNA polymerase to these sites and thus prevents gene expression. If cellular conditions shift and the inducer molecules maltose, maltotriose, or fructose are present, they bind to TrmBL1, resulting in an allosteric change and an inactive repressor that can no longer bind DNA. Without TrmBL1 bound to the DNA, RNA polymerase is able to bind to the

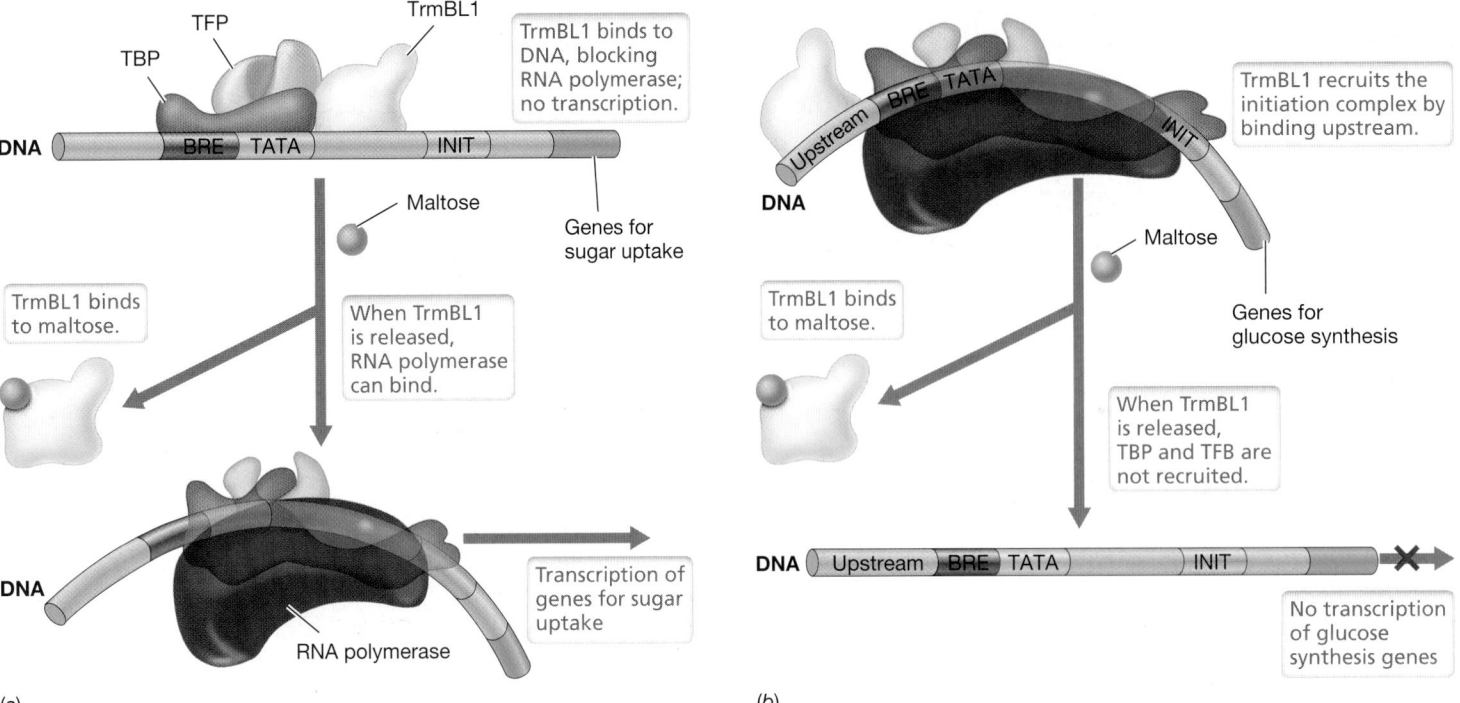

(a)

(b)

Figure 6.16 Dual functionality of the *Pyrococcus furiosus* TrmBL1 regulator. *Pyrococcus furiosus* is a hyperthermophilic species of the *Euryarchaeota* with a growth temperature optimum of 100°C (⟳ Section 17.4). *(a)* The TrmBL1 protein acts as a repressor of genes for sugar uptake. It binds to DNA downstream of the TATA box, blocking the binding of RNA polymerase. Binding of maltose (or another inducer) to TrmBL1 results in release of the regulator from the DNA. Now RNA polymerase can bind and transcribe sugar uptake genes. *(b)* The TrmBL1 protein acts as an activator of genes for glucose synthesis. It binds to a sequence upstream of the promoter region and recruits the transcription initiation complex. This in turn results in transcription of glucose synthesis genes. The presence of maltose results in release of TrmBL1 from the DNA. Without TrmBL1 bound to the DNA, TBP and TFB are not recruited and glucose synthesis genes are not transcribed.

promoter region and derepression of sugar transporter genes occurs (Figure 6.16*a*).

In contrast, TrmBL1 can function as an activator protein by binding to a separate and distinct DNA sequence associated with genes for glucose biosynthesis. This regulatory region differs from the one that TrmBL1 binds to as a repressor as it is located *upstream* of the BRE/TATA binding sites (Figure 6.16*b*). Binding of TrmBL1 to this DNA sequence helps recruit the archaeal transcription initiation complex (TBP, TFB, and RNA polymerase, ⟳ Section 4.6), thus activating gene expression. However, in the same manner as its role as a repressor, TrmBL1 as an allosteric activator will not bind to DNA if the effector molecules maltose, maltotriose, or fructose are present. Without TrmBL1 binding upstream of the BRE/TATA sequences and activating transcription, RNA polymerase and the other needed initiation factors cannot be recruited for transcription of genes necessary for glucose synthesis. The rationale of this dual-functioning control surrounds energy conservation. That is, from an energetic standpoint, this form of regulation prevents expression of the gluconeogenesis pathway when other sugars that can feed into the glycolytic pathway are present in the cell.

The SurR protein of *P. furiosus* is another example of a regulatory protein that functions as both an activator and a repressor, depending on the location of its binding site within the promoter region. SurR controls a catabolic shift of *P. furiosus* from fermentation (which produces H_2) to sulfur (S^0) reduction, a form of anaerobic respiration that produces H_2S (⟳ Sections 3.12 and 14.14). When S^0 is absent, SurR binds to regulatory regions activating the transcription of genes necessary for hydrogenase production so that *P. furiosus* can grow by fermentation. At the same time, SurR functions as a repressor to prevent transcription of genes encoding proteins that participate in sulfur metabolism. However, when S^0 is present, SurR is no longer able to bind to DNA. This inability to bind to DNA is not the result of effector binding to SurR, but is due to the oxidation of cysteine residues within the DNA-binding motif of this regulatory protein that releases it from the DNA. The release of SurR from regulatory regions both promotes expression of genes that participate in S^0 metabolism and represses the expression of hydrogenase genes that encode a hydrogenase enzyme necessary for *P. furiosis* to grow at the expense of fermentation.

— **MINIQUIZ** —

- What is the major difference between transcriptional regulation in *Archaea* and eukaryotes?

- How do transcriptional activators in *Archaea* often differ in mechanism from those in *Bacteria*?

- Explain how the *Pyrococcus furiosus* TrmBL1 transcription regulator is able to act as both an activator and a repressor.

II • Sensing and Signal Transduction

Prokaryotic cells regulate cell metabolism in response to many different environmental fluctuations, including changes in temperature, pH, oxygen or nutrient availability, and even to changes in the number of other cells present. To do this, mechanisms exist by which cells receive signals from the environment and transmit them to the specific target to be regulated. Some of these signals are small molecules that enter the cell and function as effectors. However, in many cases the external signal is not transmitted directly to the regulatory protein but instead is detected by a surface sensing system that transmits the signal to the rest of the regulatory machinery, a process called **signal transduction**.

6.6 Two-Component Regulatory Systems

Because most signal transduction systems contain two parts, they are called **two-component regulatory systems**. Characteristically, such systems consist of a specific **sensor kinase protein** usually located in the cytoplasmic membrane, and a **response regulator protein**, present in the cytoplasm.

A kinase is an enzyme that phosphorylates compounds, typically using phosphate from ATP. Sensor kinases detect a signal from the environment and phosphorylate themselves (a process called *autophosphorylation*) at a specific histidine residue on the protein (**Figure 6.17**). Sensor kinases thus belong to the class of enzymes called *histidine kinases*. The phosphate is then transferred from the sensor to another protein inside the cell, the response regulator. The latter is typically a DNA-binding protein that regulates transcription in either a positive or a negative fashion. In the example shown in Figure 6.17, regulation is negative; the phosphorylated response regulator functions as a repressor that binds DNA, thereby blocking transcription. Once dephosphorylated, the response regulator is released and transcription is permitted.

For a balanced regulatory system to work properly, it must have a *feedback loop*, that is, a way to complete the regulatory circuit and terminate the response. This resets the system for another cycle. This feedback loop employs a *phosphatase*, an enzyme that removes the phosphate from the response regulator at a constant rate. The response regulator itself often catalyzes this reaction, although in some cases separate proteins are needed (Figure 6.17). Phosphatase activity is typically slower than phosphorylation. However, if phosphorylation ceases due to reduced sensor kinase activity, phosphatase activity eventually returns the response regulator to the fully nonphosphorylated state, and the system is reset.

Examples of Two-Component Regulatory Systems

Two-component systems regulate a large number of genes in many different bacteria. Interestingly, two-component systems are either extremely rare or absent in *Archaea* and in *Bacteria* that live as parasites of higher organisms. A few key examples of two-component systems include those that respond to phosphate limitation, nitrogen limitation, and osmotic pressure.

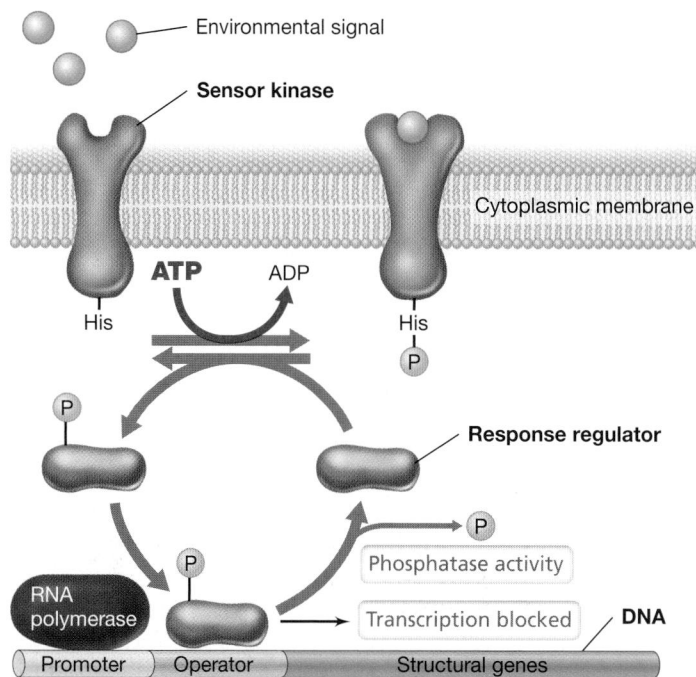

Figure 6.17 The control of gene expression by a two-component regulatory system. One component is a sensor kinase in the cytoplasmic membrane that phosphorylates itself in response to an environmental signal. The phosphoryl group is then transferred to the second component, a response regulator. The phosphorylated form of the response regulator then binds to DNA. In the system shown here, the phosphorylated response regulator is a repressor protein. The phosphatase activity of the response regulator slowly releases the phosphate from the response regulator and resets the system.

In *Escherichia coli* almost 50 different two-component systems are present, and several are listed in **Table 6.1**. In one example, the osmolarity of the environment controls the relative levels of the proteins OmpC and OmpF in the *E. coli* outer membrane. OmpC and OmpF are *porins*, proteins that allow metabolites to cross the outer membrane of gram-negative bacteria (⊜ Section 2.5). If osmotic pressure is *low*, the synthesis of OmpF, a porin with a larger pore, increases; if osmotic pressure is *high*, OmpC, a porin with a smaller pore, is made in larger amounts.

EnvZ, a sensor histidine kinase in the cytoplasmic membrane, detects changes in osmotic pressure. When a shift occurs, EnvZ is autophosphorylated and transfers its phosphate group to OmpR, the response regulator of this system (**Figure 6.18**). Under conditions of *low* osmotic pressure, phosphorylated OmpR (OmpR-P) *activates* transcription of the *ompF* gene. Conversely, when osmotic pressure is *high*, OmpR-P *represses* transcription of *ompF* gene and activates transcription of *ompC* instead (Figure 6.18). The expression of *ompF* is also regulated by an additional control mechanism: regulatory RNA, and we discuss this in Section 6.11.

Two-Component Systems with Multiple Regulators

Some signal transduction systems have more than one regulatory element and their activities can quickly become quite complex. For instance, in the Ntr regulatory system, which regulates nitrogen assimilation in many *Bacteria*, the response regulator is an activator called *nitrogen regulator I* (*NRI*). NRI activates transcription from

TABLE 6.1 Examples of two-component systems that regulate transcription in *Escherichia coli*

System	Environmental signal	Sensor kinase	Response regulator	Primary activity of response regulator[a]
Arc system	Oxygen	ArcB	ArcA	Repressor/activator
Nitrate and nitrite respiration (Nar)	Nitrate and nitrite	NarX	NarL	Activator/repressor
		NarQ	NarP	Activator/repressor
Nitrogen utilization (Ntr)	Shortage of organic nitrogen	NRII (= GlnL)	NRI (= GlnG)	Activator of promoters requiring RpoN/σ^{54}
Pho regulon	Inorganic phosphate	PhoR	PhoB	Activator/repressor
Porin regulation	Osmotic pressure	EnvZ	OmpR	Activator/repressor

[a]Note that many response regulator proteins act as both activators and repressors depending on the genes being regulated. Although ArcA can function as either an activator or a repressor, it functions as a repressor on most operons that it regulates.

promoters recognized by RNA polymerase using the alternative sigma factor σ^{54} (RpoN) (⮂ Section 4.5). The sensor kinase in the Ntr system is a protein called *nitrogen regulator II* (*NRII*), which functions as both a kinase and a phosphatase. The activity of NRII is in turn regulated by another protein called *PII*, whose own activity is regulated by the addition or removal of uridine monophosphate

(UMP) groups (Section 6.15). Under nitrogen starvation conditions, UMP is added to PII, and the resulting PII–UMP complex promotes the kinase activity of NRII and results in the phosphorylation of NRI. Conversely, removal of UMP from PII promotes the phosphatase activity of NRII.

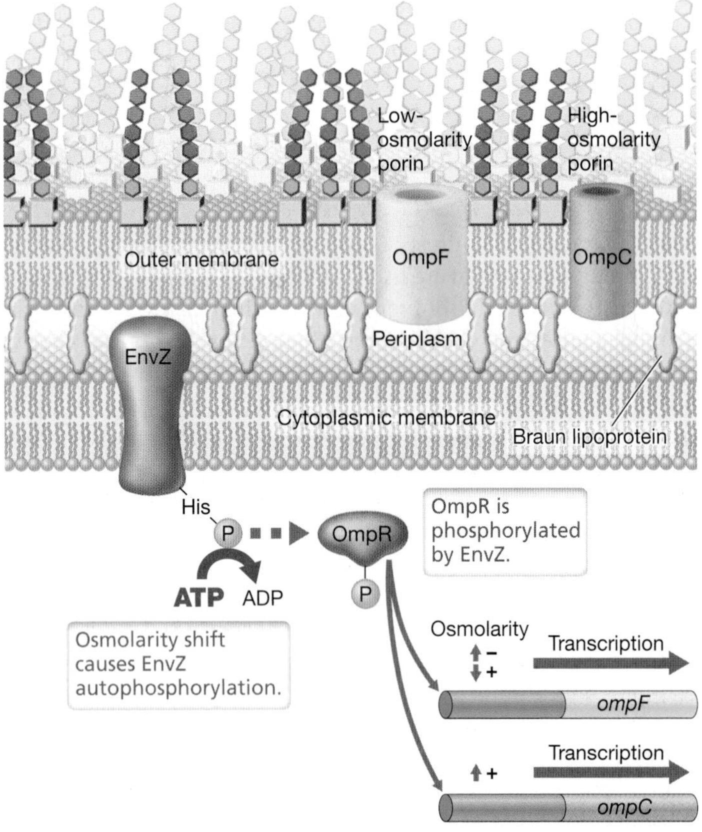

The *Nar regulatory system* (Table 6.1) is another example of a two-component regulatory system with multiple regulators; this system controls a set of genes that allow the use of nitrate (NO_3^-) or nitrite (NO_2^-) (or both) as alternative electron acceptors during anaerobic respiration (⮂ Sections 3.12 and 14.13). The Nar system contains two different sensor kinases and two different response regulators. In addition, all of the genes regulated by this system are themselves controlled by a protein called FNR (*fumarate nitrite regulator*); FNR is a global regulator of genes encoding anaerobic respiration functions (see Table 6.2). This type of regulation in which a hierarchy of systems are deployed in a cascading fashion is common for systems of central importance to cellular metabolism.

Two-component systems closely related to those in *Bacteria* are also present in microbial eukaryotes, such as the yeast *Saccharomyces cerevisiae*, and even in plants. However, most eukaryotic signal transduction pathways rely on phosphorylation of serine, threonine, and tyrosine residues of proteins that are unrelated to those of the bacterial two-component systems that phosphorylate histidine residues (Figures 6.17 and 6.18).

Figure 6.18 Regulation of outer membrane proteins in *Escherichia coli*. The cytoplasmic membrane histidine kinase EnvZ autophosphorylates itself under osmotic pressure changes and then activates the transcriptional regulator OmpR by phosphorylation. OmpR-P binds upstream of the *ompF* gene and activates transcription under low osmotic pressure, but conversely it represses transcription of *ompF* under high osmotic pressure. OmpR-P only activates transcription of the *ompC* gene under conditions of high osmolarity.

─── MINIQUIZ ───

- What are kinases and what is their role in two-component regulatory systems?
- What are phosphatases and what is their role in two-component regulatory systems?

6.7 Regulation of Chemotaxis

We have previously seen that some motile cells of *Bacteria* and *Archaea* respond to challenges such as nutrient limitation and toxin accumulation by moving toward attractants and away from repellents, a behavior called *chemotaxis* (⮂ Section 2.13). We noted there that prokaryotic cells are too small to sense spatial gradients of a chemical, but they can respond to temporal gradients. That is, they can sense the *change* in concentration of a chemical over time rather than the absolute concentration of the chemical stimulus.

Chemotaxis has been well studied in *Bacteria*, which use a modified two-component system to sense temporal changes in attractants or repellents and process this information to regulate the direction of flagellar rotation. This differs from what was considered in the previous section in that the two-component system is used to regulate the activity of *preexisting proteins* (the flagellum protein complex) rather than to control the transcription of genes encoding flagellar proteins.

Response to Signal

The mechanism of chemotaxis depends upon a signal cascade of multiple proteins. Several sensory proteins reside in the cytoplasmic membrane and sense attractants or repellents. These sensor proteins are not themselves sensor kinases but interact with cytoplasmic sensor kinases. These sensory proteins allow the cell to monitor the concentration of various substances over time, and are called *methyl-accepting chemotaxis proteins (MCPs). Escherichia coli* produces five different transmembrane MCPs, each specific for certain compounds. For example, the Tar MCP of *E. coli* senses the attractants aspartate and maltose and the repellents cobalt and nickel. MCPs bind attractants or repellents directly or in some cases indirectly through interactions with periplasmic binding proteins. Binding of an attractant or repellent triggers interactions with cytoplasmic proteins that eventually affect flagellar rotation.

In *E. coli*, thousands of MCPs are often clustered, forming chemoreceptors. These chemoreceptors make contact with the cytoplasmic proteins CheA and CheW (**Figure 6.19**). CheA is the sensor kinase (Section 6.6) for chemotaxis. When an MCP binds a chemical, it changes conformation and, with help from CheW, leads to the autophosphorylation of CheA to form CheA-P. An increase in attractant concentration *decreases* the rate of autophosphorylation, whereas a decrease in attractant or an increase in repellent *increases* the rate of autophosphorylation. CheA-P then passes the phosphate to CheY (forming CheY-P); this is the response regulator (Section 6.6) that controls flagellar rotation. CheA-P can also transfer the phosphate to CheB, which plays a role in adaptation to be described later.

Controlling Flagellar Rotation

CheY is a central protein in the regulatory system because it governs the direction of rotation of the flagellum. Recall that if rotation of the flagellum is counterclockwise, the cell will continue to move in a run (swim smoothly), whereas if the flagellum rotates clockwise, the cell will tumble (move randomly) (⟳ Section 2.13 and Figure 2.36). Once CheY is phosphorylated, it interacts with the flagellar motor to induce clockwise flagellar rotation, which causes tumbling (Figure 6.19). When unphosphorylated, CheY cannot bind to the flagellar motor and the flagellum rotates counterclockwise; this causes the cell to run. Another protein, CheZ, dephosphorylates CheY, returning it to the form that allows runs instead of tumbles. Either an increase in repellents or a decrease in attractants leads to an increase in the level of CheY-P and thus tumbling. By contrast, if the cell is swimming toward attractants, the lower level of CheY-P suppresses tumbles and promotes runs.

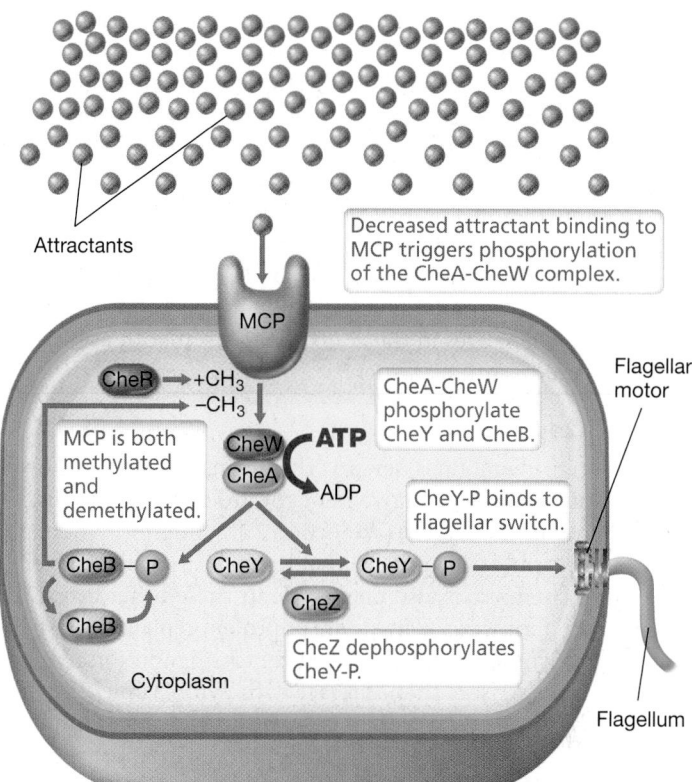

Figure 6.19 Interactions of MCPs, Che proteins, and the flagellar motor in bacterial chemotaxis. The methyl-accepting chemotaxis protein (MCP) forms a complex with the sensor kinase CheA and the coupling protein CheW. This combination triggers autophosphorylation, which can then phosphorylate the response regulators CheB and CheY. Phosphorylated CheY (CheY-P) binds to the flagellar motor switch. CheZ dephosphorylates CheY-P. CheR continually adds methyl groups to the MCP. CheB-P (but not CheB) removes them. The degree of methylation of the MCPs controls their ability to respond to attractants and repellents and leads to adaptation.

Adaptation

Once an organism has successfully responded to a stimulus, it must stop responding and reset the sensory system to await further signals. This is known as *adaptation*. During adaptation of the chemotaxis system, a feedback loop resets the system. This relies on the response regulator CheB, mentioned earlier.

As their name implies, MCPs can be methylated. When MCPs are fully methylated they no longer respond to attractants but are more sensitive to repellents. Conversely, when MCPs are unmethylated they respond strongly to attractants but are insensitive to repellents. Varying the methylation level thus allows adaptation to sensory signals. This is accomplished by methylation and demethylation of the MCPs by CheR and phosphorylated CheB (CheB-P), respectively (Figure 6.19).

If the level of an attractant remains high, the rate of CheA autophosphorylation is low. This leads to unphosphorylated CheY and CheB. Consequently, the cell swims smoothly. Methylation of the MCPs increases during this period because CheB-P is not present to demethylate them. However, MCPs no longer respond to the attractant when they become fully methylated. Therefore, if the level of attractant remains high but constant, the cell begins to tumble. Eventually, CheB becomes phosphorylated and CheB-P demethylates the MCPs. This resets the receptors and they can once again respond to further increases or decreases in level of attractants. Put another way, the cell stops swimming if the attractant concentration is constant. It only continues to swim if even higher levels of attractant are encountered. The course of events is just the opposite for repellents. Fully methylated MCPs respond best to an increasing gradient of repellents and send a signal for cell tumbling to begin. The cell then moves off in a random direction while MCPs are slowly demethylated. With this mechanism for adaptation, chemotaxis successfully achieves the ability to monitor small changes in the concentrations of both attractants and repellents over time.

Other Taxes

In addition to chemotaxis, several other forms of taxis are known, for example, *phototaxis* (movement toward light) and *aerotaxis* (movement toward oxygen) (⟿ Section 2.13). Interestingly, many of the cytoplasmic Che proteins that function to control flagellar activity in chemotaxis also play a role in these other taxes. For example, in phototaxis, a light sensor protein replaces the MCPs of chemotaxis, and in aerotaxis, a redox protein monitors levels of oxygen. These sensors then interact with cytoplasmic Che proteins to begin the cascade of events that direct runs or tumbles. Thus several different kinds of environmental signals converge on the same flagellar control system, and this allows the cell to economize on its regulatory systems.

MINIQUIZ

- What are the primary response regulator and the primary sensor kinase for regulating chemotaxis?
- Why is adaptation during chemotaxis important?
- How does the response of the chemotaxis system to an attractant differ from its response to a repellent?

6.8 Quorum Sensing

Many *Bacteria* respond to the presence in their surroundings of other cells of their own species, and in some species, regulatory pathways are controlled by the density of cells of their own kind. This is a phenomenon called **quorum sensing** (the word "quorum" in this sense means "sufficient numbers"). Quorum-sensing systems have also been detected in a few *Archaea*.

Mechanism of Quorum Sensing

Quorum sensing is a regulatory mechanism that assesses population density. Many bacteria use this approach to ensure that sufficient cell numbers are present before initiating activities that require a certain cell density to work effectively. For example, a pathogenic (disease-causing) bacterium that secretes a toxin will have no effect as a single cell; production of toxin by one cell alone would merely waste resources. However, if a sufficiently large population of cells is present, the coordinated expression of the toxin may successfully cause disease and release resources from the host that can be used to support growth of the pathogen.

Quorum sensing is widespread among gram-negative *Bacteria* but is also found in many gram-positive species as well. Each

species that employs quorum sensing synthesizes a specific signal molecule called an **autoinducer**. This molecule usually diffuses freely across the cell envelope in either direction. Because of this, the autoinducer reaches high concentrations inside the cell only if many cells are nearby, each making the same autoinducer. In the cytoplasm, the autoinducer binds to a specific transcriptional activator protein or a sensor kinase of a two-component system, ultimately triggering transcription of specific genes (**Figure 6.20a**).

While several different classes of autoinducers exist, the first to be identified were the *acyl homoserine lactones* (AHLs) (Figure 6.20b). Several different AHLs, with acyl groups (functional groups containing both carbonyl and alkyl groups) of different lengths, are found in different species of gram-negative bacteria. In addition, many gram-negative bacteria make autoinducer 2 (AI-2; a cyclic furan derivative). This is apparently used as a common autoinducer between many species of bacteria. Gram-positive bacteria generally use certain short peptides as autoinducers.

The phenomenon of quorum sensing was discovered as the mechanism by which light emission in bioluminescent bacteria is regulated (⟿ Section 15.18). Several bacterial species can emit light, including the marine bacterium *Aliivibrio fischeri*. **Figure 6.21** shows bioluminescent colonies of *A. fischeri*. The light is generated by an enzyme called *luciferase*. The *lux* operons encode the proteins needed for bioluminescence. They are under control of the activator protein LuxR and are induced when the concentration of the specific *A. fischeri* AHL, *N*-3-oxohexanoyl homoserine lactone, becomes high enough. This AHL is synthesized by the enzyme encoded by the *luxI* gene.

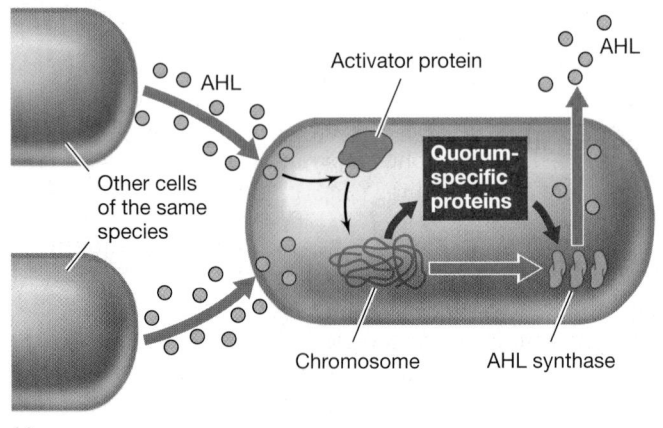

(a)

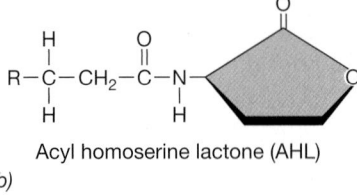

Acyl homoserine lactone (AHL)

(b)

Figure 6.20 Quorum sensing. *(a)* A cell capable of quorum sensing expresses acyl homoserine lactone (AHL) synthase at basal levels. This enzyme makes the cell's specific AHL. When cells of the same species reach a certain density, the concentration of AHL rises sufficiently to bind to the activator protein, which activates transcription of quorum-specific genes. *(b)* General structure of an AHL. Different AHLs are variants of this parent structure. R = alkyl group (C_1–C_{17}); the carbon next to the R group is often modified to a keto group (C = O).

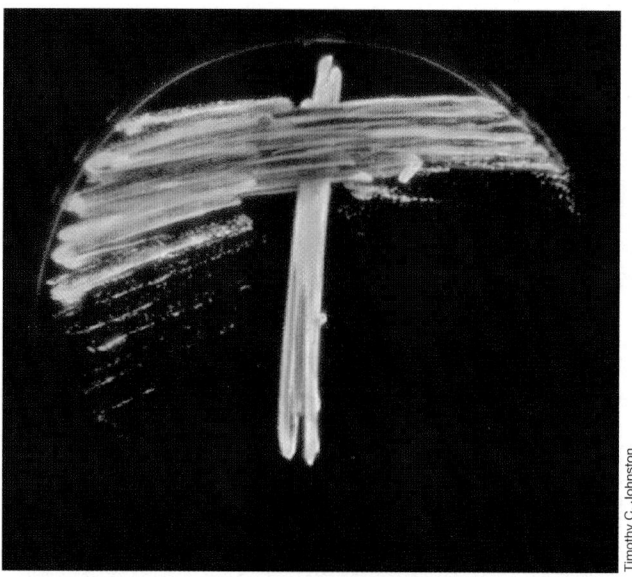

Figure 6.21 Bioluminescent bacteria producing the enzyme luciferase. Cells of the bacterium *Aliivibrio fischeri* were streaked on nutrient agar in a Petri dish and allowed to grow overnight. The photograph was taken in a darkened room using only the light generated by the bacteria.

Quorum sensing also occurs in microbial eukaryotes. For example, in the yeast *Saccharomyces cerevisiae*, specific aromatic alcohols are produced as autoinducers and control the transition between growth of *S. cerevisiae* as single cells and as elongated filaments. Similar transitions are seen in other fungi, some of which cause disease in humans. An example is *Candida*, whose quorum sensing is mediated by the long-chain alcohol farnesol. As the concentration of farnesol increases in this dimorphic fungus, the transition from budding yeast to elongated hyphae is prevented.

Virulence Factors

Various genes are controlled by quorum sensing, including some in pathogenic bacteria. For example, Shiga toxin–producing *Escherichia coli*, such as the notorious foodborne pathogen *E. coli* O157:H7 (🔗 Section 32.7), produces an AHL called AI-3 that induces virulence genes. As the *E. coli* population increases in the intestine, bacterial cells produce AI-3 while host intestinal cells produce the stress hormones epinephrine and norepinephrine. All three of these signal molecules bind to two separate sensor kinases in the *E. coli* cytoplasmic membrane, resulting in the phosphorylation and activation of two transcriptional activator proteins (**Figure 6.22a**). These proteins activate transcription of genes encoding motility functions and secretion of the toxin as well as genes encoding proteins that form lesions on the host intestinal mucosa. This is a rare example of a system that senses both bacterial and eukaryotic chemical signals to regulate gene expression.

The pathogenesis of the gram-positive bacterium *Staphylococcus aureus* (🔗 Section 30.9) requires, among many other things, the production and secretion of small extracellular peptides that damage host cells or that interfere with the host's immune system. The genes encoding these virulence factors are under the control of a quorum-sensing system that uses a small peptide called the

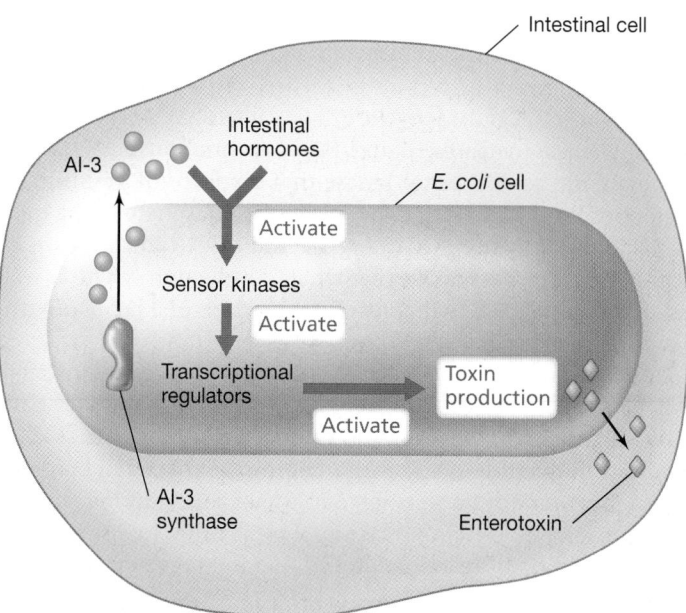

(a) **Virulence factor production in Shiga toxin-producing *Escherichia coli***

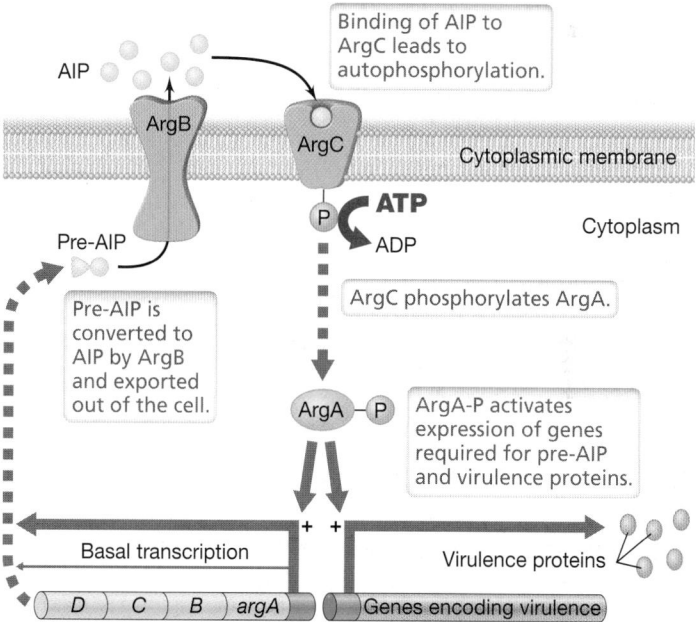

(b) **Virulence factor production in *Staphylococcus aureus***

Figure 6.22 Quorum sensing regulation of virulence factors. The bacteria *Escherichia coli* and *Staphylococcus aureus* can be harmless saprophytes or potent pathogens, depending on the strain. As pathogens, the production of major virulence factors is controlled by quorum sensing. *(a)* As the bacterial population increases, AI-3 produced by *Escherichia coli* and epinephrine and norepinephrine produced by the intestinal cell accumulate and bind to sensor kinases, initiating a cascade of events necessary for production of virulence factors (for example, enterotoxin). *(b)* Basal level transcription of the *argABCD* operon in *Staphylococcus* leads to production of ArgD, the pre-autoinducing peptide (AIP). ArgB trims ArgD into the functional AIP and exports it out of the cell. As the cell population increases, the AIP concentration increases and binds to ArgC, leading to autophosphorylation of ArgC. ArgC-P then activates the transcriptional activator ArgA by transfer of a phosphate group. ArgA-P increases transcription of the *argABCD* operon as well as activating the transcription of an RNA that leads to the production of virulence proteins.

autoinducing peptide (AIP), encoded by the *argD* gene, as the autoinducer. After synthesis of ArgD (pre-AIP), the membrane-bound ArgB protein trims the peptide into its active AIP form and secretes the small peptide outside of the cell (Figure 6.22b). As the cell density of *S. aureus* increases, so does the concentration of AIP. ArgC is a membrane-bound sensor kinase that binds to AIP, resulting in autophosphorylation. ArgC-P transfers its phosphate to the transcriptional activator ArgA. ArgA-P increases transcription of *argABCD* genes that encode the signal transduction system as well as an RNA molecule that controls production of a range of virulence proteins.

Some eukaryotes produce molecules that specifically interfere with bacterial quorum sensing. Most of these found so far have been furanone derivatives containing a halogen atom. These components mimic the AHLs or AI-2 and disrupt bacterial behavior that relies on quorum sensing. Because of this, quorum-sensing disruptors have been proposed as potential drugs to disperse bacterial biofilms (⊂⊃ Sections 5.1 and 7.9) and prevent the expression of virulence genes.

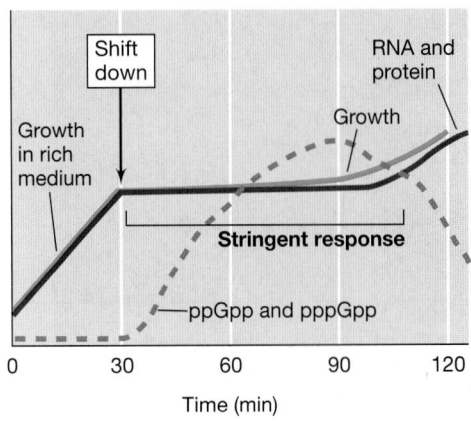

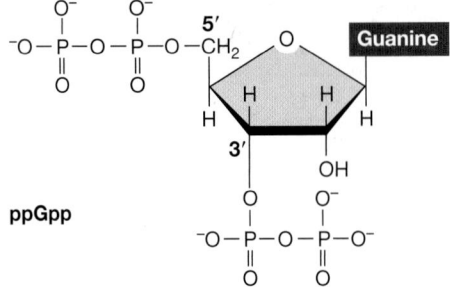

MINIQUIZ

- What advantage do quorum-sensing systems confer on bacterial cells?
- What properties are required for a molecule to function as an autoinducer?
- How do the autoinducers used in quorum sensing by gram-negative bacteria differ from those used by gram-positive bacteria?

6.9 Stringent Response

While originally studied as a response to amino acid starvation, the **stringent response** is now recognized as a widely distributed regulatory mechanism used by bacteria to survive nutrient deprivation, environmental stresses, and antibiotic exposure. Triggering of the stringent response ultimately leads to a shutdown of macromolecule synthesis and the activation of *stress survival pathways* to improve the cell's ability to compete in nature. Because most of the work to elucidate the pathway has been done in *Escherichia coli*, we will focus on this organism. Later we will extend what we know to how the stringent response is triggered in other bacteria in response to environmental stresses.

Mechanism of the Stringent Response

Nutrient levels for microbes in nature can change significantly and rapidly. Such changing conditions can easily be simulated in the laboratory, and much work has been done on the regulation of gene expression following a "shift down" or "shift up" in nutrient status. These include, in particular, the regulatory events triggered by starvation for amino acids or energy.

As a result of a shift down from amino acid excess to limitation, as occurs when a culture is transferred from a rich complex medium to a defined medium with a single carbon source (⊂⊃ Table 5.1), the synthesis of rRNA and tRNA ceases almost immediately, and no new ribosomes are produced (**Figure 6.23a**). Protein and DNA

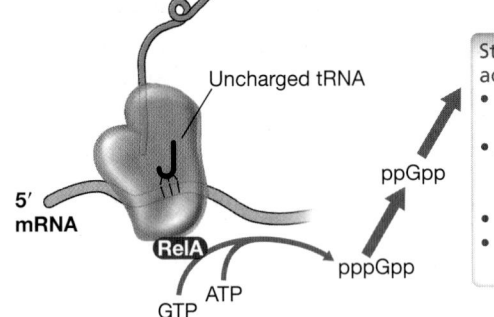

Figure 6.23 The stringent response in *Escherichia coli*. *(a)* Upon nutrient downshift, rRNA, tRNA, and protein syntheses temporarily cease. Sometime later, growth resumes at a decreased rate. *(b)* Structure of guanosine tetraphosphate (ppGpp), a trigger of the stringent response. *(c)* Normal translation, which requires charged tRNAs. *(d)* Synthesis of ppGpp. When cells are starved for amino acids, an uncharged tRNA can bind to the ribosome, which stops ribosome activity. This event triggers the RelA protein to synthesize a mixture of pppGpp and ppGpp.

synthesis are also curtailed, but the biosynthesis of new amino acids is activated. Following such a shift, new proteins must be made to synthesize the amino acids no longer available in the environment, and these proteins are made by existing ribosomes. After a while, rRNA synthesis (and hence, the production of new ribosomes) begins again but at a new rate commensurate with the cell's reduced growth rate (Figure 6.23*a*). This course of events is called the *stringent response* (or stringent control) and, like catabolite repression (Section 6.4), it is another example of global control.

The stringent response is triggered by a mixture of two regulatory nucleotides, *guanosine tetraphosphate* (ppGpp) and *guanosine pentaphosphate* (pppGpp); this mixture is often written as (p)ppGpp (Figure 6.23*b*). In *E. coli*, these nucleotides, which are also called *alarmones*, rapidly accumulate during stress or a shift down from amino acid excess to amino acid starvation. Alarmones are synthesized by a specific protein, called RelA, using ATP as a phosphate donor (Figure 6.23*c, d*). RelA adds two phosphate groups from ATP to GTP or GDP, thus producing pppGpp or ppGpp, respectively. RelA is associated with the 50S subunit of the ribosome and is activated by a signal from the ribosome during amino acid limitation. When the growth of the cell is limited by a shortage of amino acids, the pool of *uncharged* tRNAs increases relative to *charged* tRNAs. Eventually, an uncharged tRNA is inserted into the ribosome instead of a charged tRNA during protein synthesis. When this happens, the ribosome stalls, and this leads to (p)ppGpp synthesis by RelA (Figure 6.23*d*). The protein Gpp converts pppGpp to ppGpp so that ppGpp is the major overall product.

The alarmones ppGpp and pppGpp have global control effects. They strongly inhibit rRNA and tRNA synthesis by binding to RNA polymerase and preventing initiation of transcription of genes for these RNAs in gram-negative bacteria. In gram-positive bacteria, the same alarmones have been shown to interfere with initiating ribonucleotides for transcription. In both gram-negative and gram-positive bacteria, the alarmones activate both the stress response pathways and the biosynthetic operons for certain amino acids. In *E. coli*, the stringent response also inhibits the initiation of new rounds of DNA synthesis and cell division and slows down the synthesis of cell envelope components, such as membrane lipids.

In addition to RelA, a second protein called SpoT helps trigger the stringent response. SpoT can either make (p)ppGpp or degrade it. Under most conditions, SpoT is responsible for degrading (p)ppGpp; however, SpoT synthesizes (p)ppGpp in response to certain stresses or when nutrient deprivation is detected. Thus the stringent response results not only from the absence of precursors for protein synthesis, but also from the lack of energy for biosynthesis.

The Stringent Response and Microbial Ecology

Because the stringent response is a global mechanism that balances the metabolic state of the cell, the environment or habitat of the bacterium ultimately determines the response cascade. When cells of *E. coli* are voided in the feces and face a switch from the nutrient-rich intestine to an open water system, the reduction in nutrients triggers ppGpp synthesis to initiate the stringent response (**Figure 6.24*a***).

In contrast to *E. coli*, the aquatic bacterium *Caulobacter* naturally inhabits oligotrophic (nutrient-poor) freshwaters, and in this

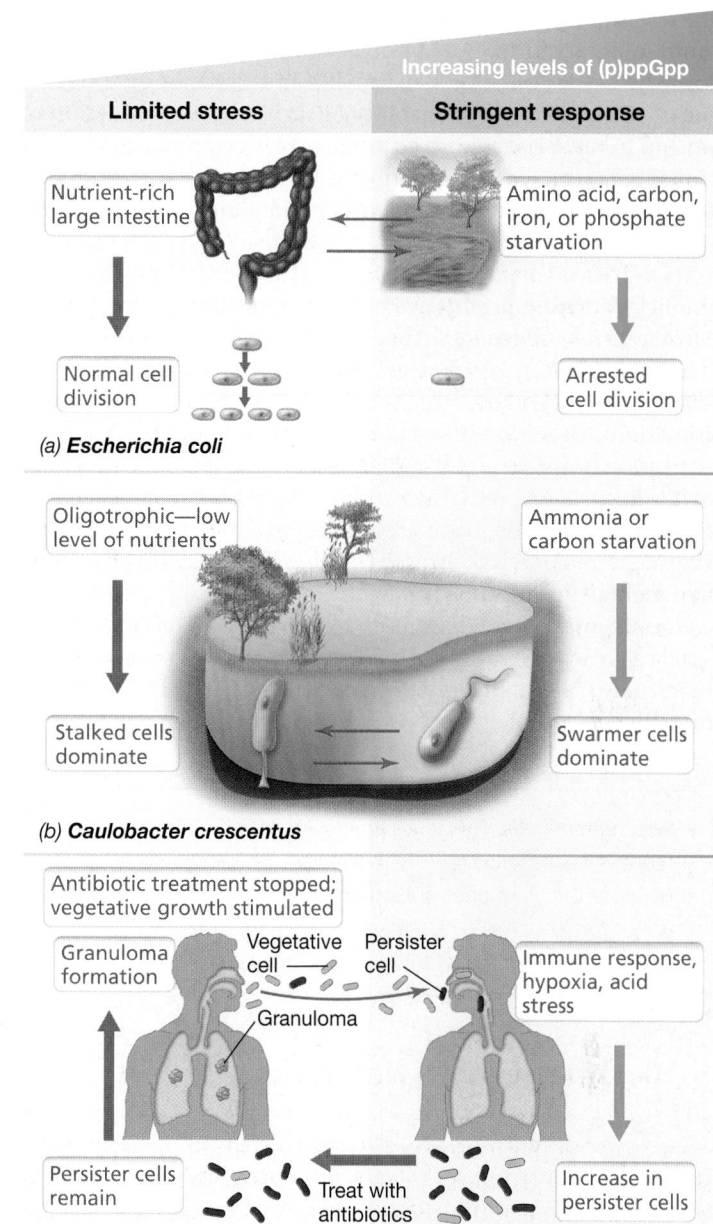

Figure 6.24 Environmental control of stringent response signaling in three separate bacteria, *Escherichia coli* and *Caulobacter crecentus* (both gram-negative), and *Mycobacterium tuberculosis* (gram-positive). *(a) Escherichia coli.* When *E. coli* cells move from the nutrient-rich intestine to an open water source, they experience nutrient limitation. As ppGpp production increases, cell division is inhibited. *(b) Caulobacter crescentus. C. crescentus* naturally inhabits freshwater systems that are limited in nutrients. If cells encounter severe nutrient limitation, the stringent response is induced and ppGpp production increases. This increase leads to the cell morphology changing from stalked cells to swarmer cells that can swim to find more nutrients. *(c) Mycobacterium tuberculosis.* When *M. tuberculosis* cells enter a host, they are exposed to the immune response. This leads to production of ppGpp, which converts some cells to relatively dormant persister cells. These persister cells are resistant to antibiotics and can dominate more sensitive vegetative cells, forming granulomas. Once persister cells are aerosolized and relieved of stress, they revert back to infective vegetative cells. Unlike many other bacterial pathogens that gain antibiotic resistance from mobile genetic elements, horizontal gene transfer is rare in *M. tuberculosis*. Antibiotic resistance in this pathogen is primarily due to chromosomal mutations affecting the target of the antibiotic.

bacterium, the stringent response is triggered by carbon and ammonia starvation, not amino acid limitation. *Caulobacter* can form two types of cells, stalked cells and swarmer cells (Figure 6.24*b*). To increase its probability of survival, the production of ppGpp in *Caulobacter* leads to an increase in swarmer cell formation. Unlike stalked cells, swarmer cells are motile and can therefore swim and perhaps reach a niche containing more nutrients (⇨ Section 7.7) (Figure 6.24*b*).

In a disease example, the stringent response has also been implicated in the persistence of the aerobic bacterial pathogen *Mycobacterium tuberculosis* (tuberculosis, ⇨ Section 30.4) in dormant granulomas. Within the lung of the human host, the local environment is both hypoxic and limited for phosphate, and these conditions trigger the stringent response in cells of *M. tuberculosis*. Activation of the stringent response also leads to a subpopulation of *M. tuberculosis* cells that convert to dormant *persister cells* (⇨ Section 7.11); these cells are resistant to antibiotics and can survive in granulomas that form in the lung (Figure 6.24*c*). Once drug treatment of the infection has greatly reduced cell numbers and the environmental stresses have been relieved, the *M. tuberculosis* persister cells can then revert back to infective cells and trigger chronic tuberculosis, a not uncommon occurrence and one of the reasons drug therapy for tuberculosis is carried out for such a long period.

MINIQUIZ

- Which *Escherichia coli* genes are activated and which are repressed during the stringent response, and why?
- How are the alarmones ppGpp and pppGpp synthesized?
- What are some other conditions that trigger the stringent response in other bacteria?

6.10 Other Global Networks

Catabolite repression (Section 6.4) and quorum sensing (Section 6.8) are both examples of *global regulatory control*. There are several other global control systems in *Escherichia coli* (and probably in all *Bacteria* and *Archaea*), and some key ones are listed in **Table 6.2**. Global control systems regulate many genes comprising more than one regulon (Section 6.3). Global control networks may include activators, repressors, signal molecules, two-component regulatory systems, regulatory RNA (Section 6.11), and alternative sigma (σ) factors (⇨ Section 4.5).

Three well-studied examples of global responses in *Bacteria* include the *phosphate (Pho) regulon*, the **heat shock response**, and the *RpoS regulon*. The Pho regulon illustrates how cells respond to environmental phosphate concentrations and link this response to other metabolic and synthesis pathways, while the heat shock response is widespread in all three domains of life and in *Bacteria* is largely controlled by alternative σ factors. For general bacterial responses to stress, the alternative σ factor RpoS is the *master controller*, as it regulates over 400 different genes.

The Phosphate (Pho) Regulon

Phosphorus is not only essential for DNA, RNA, and membrane synthesis, but it is also critical for energy generation and cell signaling (⇨ Section 3.1). In nature, phosphorus is generally in the form of PO_4^{3-} (inorganic phosphate, abbreviated P_i) and is often the limiting nutrient in many environments. Thus regulatory mechanisms such as the Pho regulon have evolved in *Bacteria* to deal with P_i limitation. This regulon consists of a two-component regulatory system (Section 6.6) that regulates genes encoding extracellular enzymes for obtaining P_i from organic phosphates, P_i transporters, and enzymes for P_i storage using positive control mechanisms (Section 6.3).

In *Streptomyces*, the two-component Pho regulatory system consists of a membrane-bound histidine kinase sensor protein PhoR and a cytoplasmic transcriptional regulator PhoP (PhoB in *E. coli*). While the mechanism for actually *sensing* extracellular P_i levels is unknown, low environmental P_i levels are known to trigger PhoR kinase activity, yielding a phosphorylated PhoP (PhoP-P) (**Figure 6.25**). Once phosphorylated, PhoP-P then binds to conserved promoter regions called Pho boxes scattered across the genome. This binding signals the "general housekeeping" sigma subunit of RNA polymerase to bind, and transcription of genes needed for P_i uptake is *activated*. When P_i is in excess, PhoP is dephosphorylated, which leads to the removal of PhoP from the

---

TABLE 6.2 Examples of global control systems known in *Escherichia coli*[a]

System	Signal	Primary activity of regulatory protein	Number of genes regulated
Aerobic respiration	Presence of O_2	Repressor (ArcA)	> 50
Anaerobic respiration	Lack of O_2	Activator (FNR)	> 70
Catabolite repression	Cyclic AMP level	Activator (CRP)	> 300
Heat shock	Temperature	Alternative sigma factors (RpoH and RpoE)	36
Nitrogen utilization	NH_3 limitation	Activator (NRI)/alternative sigma factor (RpoN)	> 12
Oxidative stress	Oxidizing agents	Activator (OxyR)	> 30
SOS response	Damaged DNA	Repressor (LexA)	> 20
General stress response	Stress conditions	Alternative sigma factor (RpoS)	> 400

[a]For many of the global control systems, regulation is complex. A single regulatory protein can play more than one role. For instance, the regulatory protein for aerobic respiration is a repressor for many promoters but an activator for others, whereas the regulatory protein for anaerobic respiration is an activator protein for many promoters but a repressor for others. Regulation can also be indirect or require more than one regulatory protein. Many genes are regulated by more than one global system.

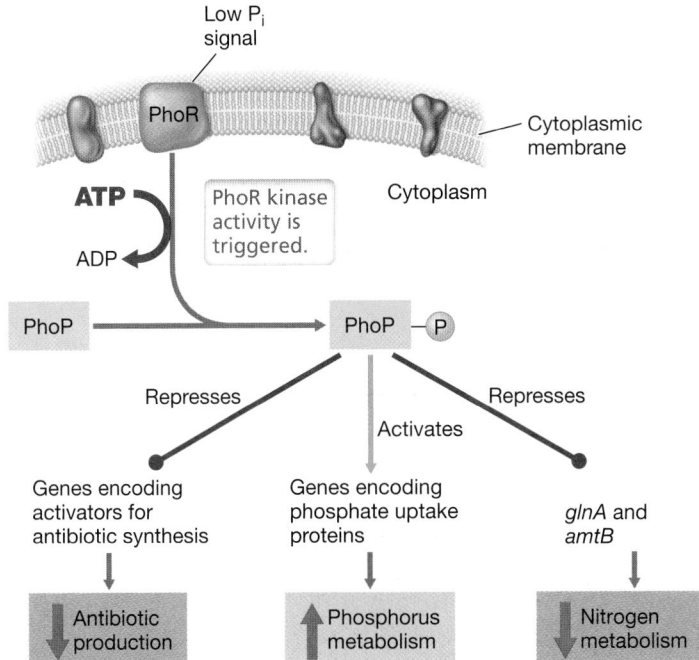

Figure 6.25 The phosphate (Pho) regulon of *Streptomyces*. The genus *Streptomyces* contains species of filamentous gram-positive *Bacteria*, many of which produce antibiotics (ᗺ Section 16.12). If inorganic phosphate (P_i) concentrations are low, a signal is sent to the kinase PhoR to phosphorylate PhoP. PhoP-P recognizes and binds to promoter regions, acting as both an activator and a repressor. This regulation results in increasing phosphorus metabolism by activating phosphate uptake genes and decreasing antibiotic production and nitrogen metabolism by repressing genes regulating antibiotic production and glutamine synthetase (*glnA*) and ammonia transport (*amtB*), respectively.

Pho boxes and decreased transcription of genes encoding phosphate metabolism proteins.

While the Pho regulon is the mechanism many *Bacteria* use to respond to P_i starvation, why is this regulon considered an example of global control? Besides its role as an activator of P_i uptake and storage genes in *Streptomyces* and many other bacteria, PhoP-P also binds some promoters weakly without recruiting a sigma subunit. This binding blocks RNA polymerase and thus leads to repression of the linked gene. Somewhat surprisingly, PhoP-P appears to actually repress more genes than it activates. Genes regulated by P_i limitation and PhoP-P that are not associated with P_i uptake in *Streptomyces* include those whose products are necessary for nitrogen metabolism (glutamine synthetase, *glnA*; ammonium transport, *amtB*) and antibiotic synthesis (Figure 6.25).

The Pho regulon can also control some aspects of pathogenesis. Pathogenic bacteria are exposed to both P_i-limiting and P_i-rich environments within a host depending on the site of infection. While the mechanisms remain unclear, the Pho regulon has been implicated in regulating biofilm formation (ᗺ Sections 5.1 and 7.9), antimicrobial resistance (ᗺ Section 28.12), toxin production (ᗺ Section 25.5), and resistance to acidity in pathogens such as *Vibrio cholerae, Pseudomonas* spp., and pathogenic *E. coli*. Thus the Pho regulon is a barometer of P_i starvation that leads to the regulation of a range of unrelated genes controlling a myriad of cellular functions.

Heat Shock Proteins

Most proteins are relatively stable, even to small increases in temperature. However, some proteins are less stable at elevated temperatures and tend to unfold (denature, ᗺ Section 4.7). Improperly folded proteins are recognized by protease enzymes and are degraded. Consequently, heat stress triggers the synthesis of a set of proteins—the **heat shock proteins**—that help counteract the damage and assist the cell in recovering from stress. Heat shock proteins are induced not only by heat but also by several other stress factors, including the exposure to high levels of certain chemicals, such as ethanol, or exposure to high doses of ultraviolet (UV) radiation.

In most *Bacteria* and *Archaea*, there are three major classes of heat shock protein: Hsp70, Hsp60, and Hsp10. We have encountered these proteins before, although not by these names (ᗺ Section 4.11 and Figure 4.40a). The Hsp70 protein of *E. coli* is DnaK, which prevents aggregation of newly synthesized proteins and stabilizes unfolded proteins. Major representatives of the Hsp60 and Hsp10 families in *E. coli* are the proteins GroEL and GroES, respectively. These are *molecular chaperones* that catalyze the correct refolding of misfolded proteins. Another class of heat shock proteins includes various proteases that degrade denatured or irreversibly aggregated proteins.

Heat Shock Response

In many bacteria, such as *E. coli*, the heat shock response is controlled by the alternative σ factor (ᗺ Section 4.5) RpoH (σ^{32}) (**Figure 6.26**). RpoH controls expression of heat shock proteins and is normally degraded within a minute or two of its synthesis. However, when cells suffer a heat shock, degradation of RpoH is inhibited and its level therefore increases. As a consequence, transcription of those operons whose promoters are recognized by RpoH increases.

The rate of RpoH degradation depends on the level of free DnaK, which inactivates RpoH. In unstressed cells, the level of free DnaK is relatively high and the level of intact RpoH is correspondingly low. However, if heat begins to unfold proteins, DnaK binds preferentially to unfolded proteins and so is no longer free to degrade RpoH. Thus, the more denatured proteins there are, the lower the level of free DnaK and the higher the level of RpoH; the result is heat shock gene expression—the heat shock response. When the environmental stressor has passed, for example upon a temperature downshift, RpoH is once again inactivated by DnaK and the synthesis of heat shock proteins is reduced.

Because heat shock proteins perform vital functions in the cell, there is always a low level of these proteins present, even under optimal conditions. However, the rapid synthesis of heat shock proteins in stressed cells emphasizes their importance for surviving exposure to excessive heat, chemicals, or physical agents. Such stresses can generate large amounts of inactive proteins that need to be refolded (and in the process, reactivated) or degraded to release free amino acids for the synthesis of new proteins.

There is also a heat shock response in *Archaea*, even in species that grow best at very high temperatures. An analog of the bacterial Hsp70 is found in many *Archaea* and is structurally quite similar to that found in gram-positive species of *Bacteria*. Hsp70 is also

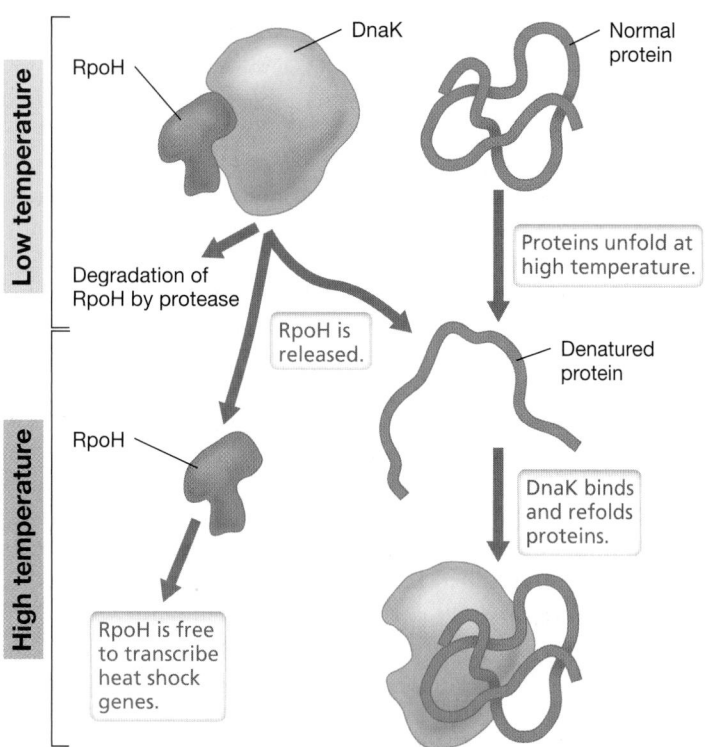

Figure 6.26 Control of heat shock in *Escherichia coli*. The RpoH alternative sigma factor is broken down rapidly by proteases at normal temperatures. This is stimulated by binding of the DnaK chaperone to RpoH. At high temperatures, some proteins are denatured, and DnaK recognizes, binds, and refolds these unfolded polypeptide chains. This removes RpoH from DnaK, which slows the degradation rate. When the level of RpoH rises, the heat shock genes are transcribed.

present in eukaryotes. In addition, other types of heat shock proteins are present in *Archaea* that are unrelated to stress proteins of *Bacteria*.

The General Stress Response: The RpoS Regulon

In nature microorganisms must survive under nutrient-limited conditions and exposure to environmental stressors such as extreme pH and oxidative stress. How do cells cope and adapt to such conditions? While some gram-positive cells undergo sporulation to withstand harsh conditions (⊜ Section 2.10), many gram-negative *Bacteria* possess a general stress response in addition to the stringent response (Section 6.9) to activate cell survival pathways. These pathways are controlled by the alternative sigma factor RpoS (σ^S or σ^{38}, ⊜ Table 4.3). Because RpoS is highly expressed during the transition from exponential to stationary phase, it is also known as the *stationary phase sigma factor*. The RpoS regulon comprises over 400 genes including those associated with nutrient limitation, resistance to DNA damage, biofilm formation, and responses to osmotic, oxidative, and acid stresses. Thus, RpoS not only senses environmental changes but also relays signals to other regulators.

Examples of *E. coli* genes that are recognized by RpoS include *dinB*, which encodes DNA polymerase IV of the SOS DNA repair system (⊜ Section 11.4 and Figure 11.9) and catalase genes necessary for combating reactive oxygen species (⊜ Section 5.14). Because of the expansive nature of the RpoS regulon, this alternative sigma factor is

itself controlled by numerous transcriptional, translational, and post-translational regulatory mechanisms. For example, transcription of the *rpoS* gene increases in response to the presence of the alarmone ppGpp, thus directly linking the RpoS regulon with the stringent response (Section 6.9). Translation of the *rpoS* mRNA is also positively regulated by the presence of small RNAs expressed during stress conditions (Section 6.11). Conversely, the RpoS protein is susceptible to degradation during nonstressful conditions. Overall, the general stress response highlights the global control nature of RpoS and the coordinated processes that must occur for cell survival and adaptation.

MINIQUIZ

- What is the heat shock response?
- Why do cells have more than one type of σ factor?
- How do levels of RpoH control the heat shock response?

III • RNA-Based Regulation

Thus far we have focused on regulatory mechanisms in which proteins sense signals or bind to DNA. In some cases a single protein does both; in other cases, separate proteins carry out these two activities. Nonetheless, all of these mechanisms rely on regulatory *proteins*. However, in some cases *RNA* can regulate gene expression, both at the level of transcription and at the level of translation.

RNA molecules that are not translated to give proteins are collectively known as **noncoding RNA (ncRNA)**. This category includes the rRNA and tRNA molecules that participate in protein synthesis and the RNA present in the signal recognition particle that catalyzes some types of protein secretion (⊜ Sections 4.8, 4.10, and 4.12). Noncoding RNA also includes small RNA molecules necessary for RNA processing, especially the splicing of mRNA in eukaryotes. *Small RNAs* (sRNAs) that range from 40–400 nucleotides long and regulate gene expression are widely distributed in both prokaryotic cells and eukaryotes. In *Escherichia coli*, for example, a number of sRNA molecules regulate various aspects of cell physiology in response to environmental or cellular signals by binding to other RNAs or in some cases to other small molecules; the end result is control of gene expression.

6.11 Regulatory RNAs

Small RNAs (sRNAs) exert their effects by base-pairing directly to other RNA molecules, usually mRNAs, which have regions of complementary sequence. This binding immediately modulates the rate of target mRNA translation because a ribosome cannot translate double-stranded RNA. Thus, sRNAs provide an additional mechanism to regulate the synthesis of a protein once its corresponding mRNA has already been transcribed.

Mechanisms of sRNA Activity

Small RNAs alter the translation of their mRNA target by four different mechanisms (**Figure 6.27**). Some sRNAs base-pair to their target mRNA, changing its secondary structure to either block a

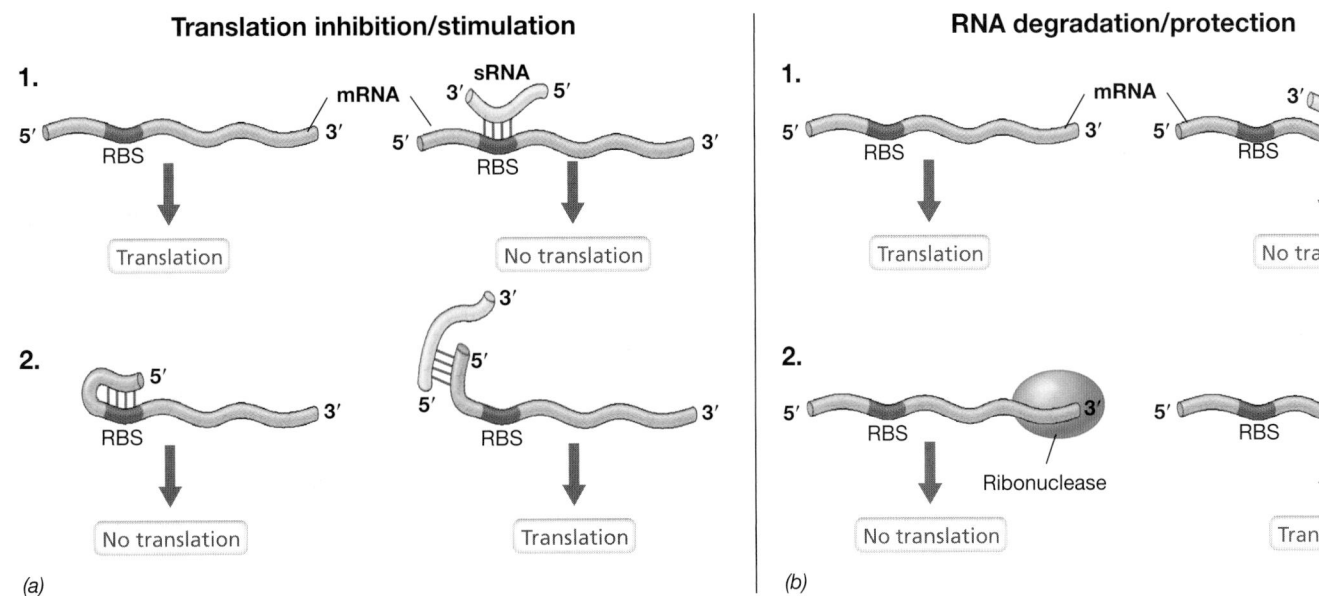

Translation inhibition/stimulation

RNA degradation/protection

(a)   (b)

Figure 6.27 Small RNA mechanisms for modulating the translation of mRNA. *(a)* Binding of a ribosome to mRNA requires that the ribosome-binding site (RBS) of the mRNA be single-stranded. Binding of an sRNA to the RBS (shown in 1) can prevent translation, while the binding of an sRNA to an mRNA whose RBS has secondary structure (shown in 2) can stimulate translation. *(b)* Ribonuclease degrades RNA. Ribonuclease binding to partially double-stranded RNA results in RNA degradation (shown in 1), while sRNA binding at the ribonuclease binding site (shown in 2) can protect the mRNA from degradation. Compare the mechanisms for controlling translation by small RNAs shown in this figure with the mechanism of controlling translation by a riboswitch shown in Figure 6.30.

previously accessible ribosome-binding site (RBS) (⬗ Section 4.9) or to open up a previously blocked RBS, allowing access for the ribosome. These two events decrease or increase expression of the protein encoded by the target mRNA, respectively. The other two mechanisms of sRNA interaction affect mRNA stability; binding of the sRNA to its target can either increase or decrease degradation of the transcript by bacterial ribonucleases, thus modulating protein expression. Increased degradation of an mRNA prevents the synthesis of new protein molecules encoded by that mRNA. Alternatively, increasing the stability of mRNA will lead to higher corresponding protein levels in the cell (Figure 6.27).

Types of Small RNA

Small RNAs (sRNAs) are important modulators of various cellular processes including oxidative, iron, and glucose–phosphate stresses as well as quorum sensing, biofilm formation, and stationary phase events. Transcription of sRNAs is often enhanced under conditions in which their target genes need to be turned off. For example, the RyhB RNA of *Escherichia coli* is transcribed when iron is limiting for growth. RyhB sRNA binds to several distinct target mRNAs that encode proteins needed for iron metabolism or that use iron as cofactors. Binding of RyhB sRNA blocks the RBS of the mRNA and thus inhibits translation (Figure 6.27a). The base-paired RyhB/mRNA molecules are then degraded by ribonucleases, in particular, ribonuclease E, before translation of the mRNA can occur. This forms part of the mechanism by which *E. coli* and related bacteria respond to a shortage of iron. Other responses to iron limitation in *E. coli* include transcriptional controls by repressor and activator proteins (Sections 6.2 and 6.3) that decrease or increase, respectively, the capacity of cells to take up iron or to tap into intracellular iron reserves.

Similarly, SgrS is an sRNA in *E. coli* that is expressed during glucose–phosphate stress to avoid accumulation of glucose 6-phosphate. High levels of glucose 6-phosphate within the cell can block the glycolytic pathway (⬗ Section 3.8) and ultimately lead to cell death if not overcome. When SgrS is expressed, the sRNA molecule binds to the *ptsG* mRNA encoding a glucose transporter. This binding results in a double-stranded RNA complex that triggers degradation of *ptsG* mRNA by ribonuclease E (**Figure 6.28**). This degradation decreases the amount of *ptsG* mRNA that is

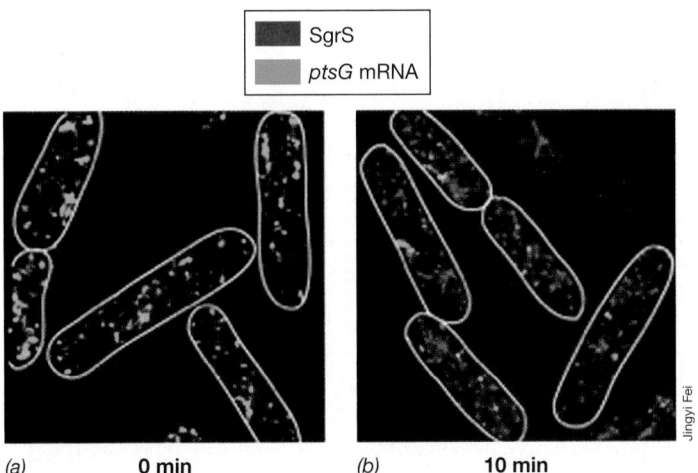

SgrS
ptsG mRNA

(a)   0 min   (b)   10 min

Figure 6.28 The interaction of SgrS sRNA and *ptsG* mRNA during glucose–phosphate stress. Three-dimensional super-resolution image of fluorescently labeled SgrS sRNA (red) and *ptsG* mRNA (green) in cells of *Escherichia coli. (a)* Before glucose–phosphate stress at time 0, *ptsG* mRNA is more abundant than SgrS sRNA. *(b)* After 10 min of glucose–phosphate stress, SgrS levels are higher due to degradation of *ptsG* mRNA.

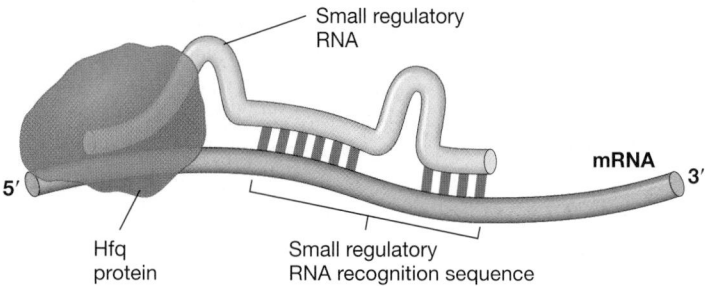

Figure 6.29 The RNA chaperone Hfq holds RNAs together. Binding of sRNA to mRNA often requires the Hfq protein. Small RNA molecules usually have several stem–loop structures. One consequence is that the complementary base sequence that recognizes the mRNA is noncontiguous.

translated, and thus lower levels of the glucose transporter protein are made. This in turn results in decreased levels of glucose 6-phosphate, which allows the glycolytic pathway to remain active.

Because many sRNAs, including RyhB and SgrS, are encoded in intergenic regions and can be spatially separated from their mRNA target, they are called *trans-sRNAs*. As such, these sRNAs usually have limited complementarity to their target molecule and may only base-pair with a 5- to 11-nucleotide stretch. The binding of trans-sRNAs to their targets often depends on a small protein called Hfq (**Figure 6.29**) that binds to both RNA molecules to facilitate their interaction. Hfq forms hexameric rings with RNA-binding sites on both surfaces. Hfq and functionally similar proteins are called *RNA chaperones*, as they help small RNA molecules, including many sRNAs, maintain their correct structure (Figure 6.29).

Small RNAs do not always work by affecting mRNA. For example, replication of the ColE1 plasmid in *Escherichia coli* is regulated by an sRNA that primes DNA synthesis on the plasmid and its anti-sense partner that blocks initiation of DNA synthesis. The level of the antisense RNA determines how often replication is initiated. Some sRNAs also bind to proteins and modulate their activity.

MINIQUIZ

- How do sRNAs alter the translation of target mRNAs?
- Why do trans-sRNAs often require a chaperone protein?
- How does SgrS help *Escherichia coli* prevent a potential metabolic disaster?

6.12 Riboswitches

RNA can carry out many roles once thought to be limited to proteins. In particular, RNA can specifically recognize and bind other molecules, including low-molecular-weight metabolites. It is important to emphasize that such binding does not require complementary base pairing (as does binding of the sRNAs described in the previous section) but results from the folding of the RNA into a specific three-dimensional structure that recognizes the target molecule, much as a protein enzyme recognizes its substrate in its active site. Catalytically active RNAs are called *ribozymes*. Other RNA molecules resemble repressors and activators in binding small metabolites and regulating gene expression; these are the **riboswitches**.

Mechanisms of Riboswitches

Earlier in this chapter we discussed the regulation of gene expression by negative control of transcription (Section 6.2). In this process, a specific metabolite interacts with a specific repressor protein to prevent transcription of genes encoding enzymes for the biosynthetic pathway of the metabolite. In contrast, in a riboswitch there is no regulatory protein. Instead, the metabolite binds directly to the mRNA. Because of this, riboswitches typically exert their control *after* the mRNA has already been synthesized; that is, most riboswitches control *translation* of the mRNA, rather than its *transcription*.

Riboswitch mRNAs contain regions upstream of their coding sequences that can fold into specific three-dimensional structures that bind small molecules. These recognition domains are the "switch" in the riboswitch and exist as two alternative secondary structures, one with the small molecule bound and the other without (**Figure 6.30**). Alternation between the two forms of the riboswitch thus depends on the presence or absence of the small molecule, which in turn controls expression of the mRNA. Riboswitches are known that control the synthesis of enzymes in biosynthetic pathways for various vitamins, a few amino acids, some nitrogen bases, and a precursor in peptidoglycan synthesis (**Table 6.3**).

The metabolite that is bound by the riboswitch is typically the product of a biosynthetic pathway whose constituent enzymes are encoded by the mRNAs that carry the corresponding riboswitches. For example, the thiamine riboswitch that binds the vitamin thiamine pyrophosphate lies upstream of the coding sequences for enzymes that participate in the thiamine biosynthetic pathway. When the pool of thiamine pyrophosphate is sufficient in the cell, this metabolite binds to its specific riboswitch mRNA. The new secondary structure of the riboswitch blocks the ribosome-binding

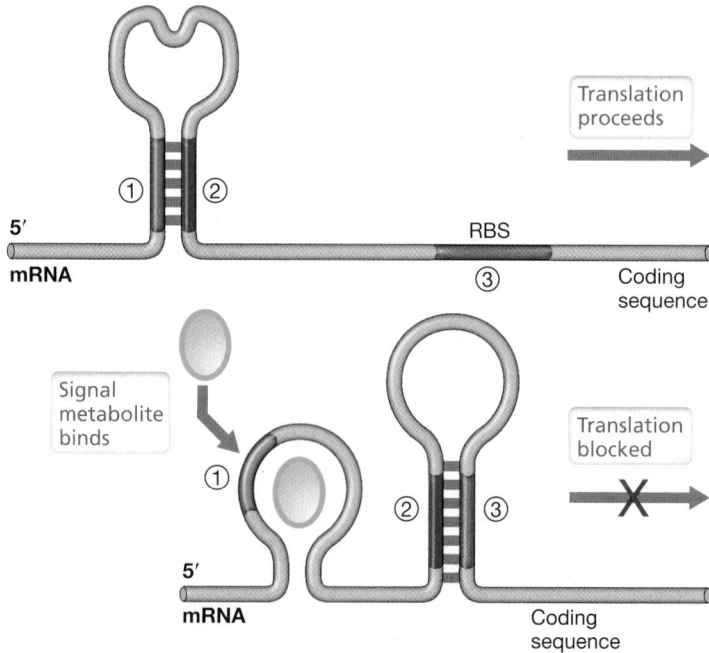

Figure 6.30 Regulation by a riboswitch. Binding of a specific metabolite alters the secondary structure of the riboswitch domain, which is located in the 5′ untranslated region of the mRNA, preventing translation. Numbers indicate regions within the riboswitch that can base-pair together. RBS, ribosome-binding site.

TABLE 6.3 Riboswitches in biosynthetic pathways of *Escherichia coli*

Type	Example of biosynthetic pathway
Vitamins	Cobalamin (B$_{12}$), tetrahydrofolate (folic acid), thiamine
Amino acids	Glutamine, glycine, lysine, methionine
Nitrogen bases of nucleic acids	Adenine, guanine (purine bases)
Others	Flavin mononucleotide (FMN), S-adenosylmethionine (SAM), glucosamine 6-phosphate (peptidoglycan precursor), cyclic di-GMP (biofilm signaling molecule)

site on the mRNA and prevents the mRNA from binding to the ribosome; this prevents translation (Figure 6.30). If the concentration of thiamine pyrophosphate drops sufficiently low, the molecule can dissociate from its riboswitch mRNA. This unfolds the mRNA and exposes the ribosome-binding site, allowing the mRNA to bind to the ribosome and be translated.

Despite being part of the mRNA, some riboswitches nevertheless do control transcription. The mechanism is similar to that seen in attenuation (Section 6.13) where a conformational change in the riboswitch causes premature termination of the synthesis of the mRNA that carries it.

Riboswitches and Evolution

How widespread are riboswitches and how did they evolve? Thus far riboswitches have been found only in some bacteria and a few plants and fungi. However, genomic analyses of *Archaea* indicate that at least some putative riboswitches are likely in this domain, too. Some scientists believe that riboswitches are remnants of the RNA world, a period eons ago before cells, DNA, and protein were present when it is hypothesized that catalytic RNAs were the only self-replicating "life forms." In such an environment, riboswitches may have been a primitive mechanism of metabolic control—a simple means by which RNA life forms could have controlled the synthesis of other RNAs. As proteins evolved, riboswitches might have been the first control mechanisms for their synthesis as well. If this is true, the riboswitches that remain today may be the last vestiges of this simple form of control because, as we have seen in this chapter, metabolic regulation is almost exclusively carried out by way of regulatory *proteins*.

— **MINIQUIZ** —
- What happens when a riboswitch binds the small metabolite that regulates it?
- What are the major differences between a repressor protein and a riboswitch in the control of gene expression?

6.13 Attenuation

Attenuation is a form of transcriptional control in *Bacteria* (and likely in *Archaea* as well) that functions by prematurely terminating mRNA synthesis. That is, in attenuation, control is exerted *after* the initiation of transcription but *before* its completion. Consequently, the number of completed transcripts from an operon is reduced, even though the number of initiated transcripts is not.

The basic principle of attenuation is that the first part of the mRNA to be made, called the *leader*, can fold into two alternative secondary structures. In this respect, the mechanism of attenuation resembles that of riboswitches (Figure 6.30). In attenuation, one mRNA secondary structure allows continued synthesis of the mRNA, whereas the other secondary structure causes premature termination. Folding of the mRNA depends either on events at the ribosome or on the activity of regulatory proteins, depending on the organism. The best examples of attenuation are the regulation of genes controlling the biosynthesis of certain amino acids in gram-negative *Bacteria*. The first to be described was in the tryptophan operon in *Escherichia coli*, and we focus on it here. Because the processes of transcription and translation are spatially separated in eukaryotes (⇄ Section 4.6), attenuation control is absent from *Eukarya*.

Attenuation in the Tryptophan Operon

The tryptophan operon contains structural genes for five proteins of the tryptophan biosynthetic pathway plus the usual promoter and regulatory sequences at the beginning of the operon (**Figure 6.31**). Like many operons, the tryptophan operon has more than one type of regulation. Transcription of the entire tryptophan operon is under negative control (Section 6.2). However, in addition to the promoter and operator regions needed for negative control, there is a sequence in the operon called the *leader sequence* that encodes a short polypeptide, the *leader peptide*. The leader

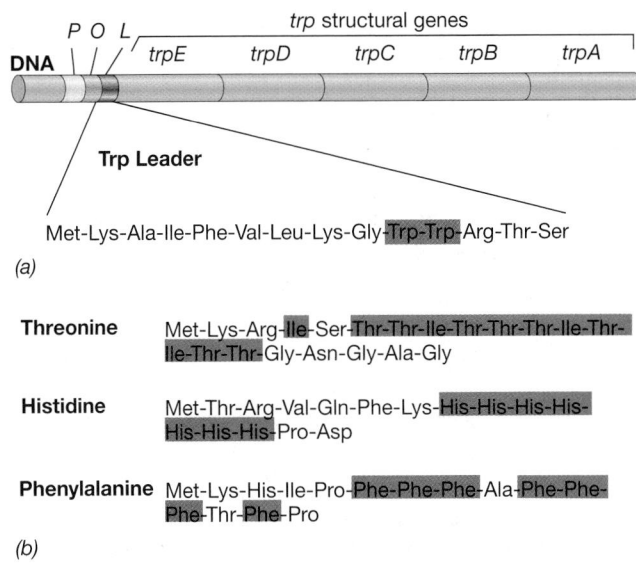

Figure 6.31 Attenuation and leader peptides in *Escherichia coli*. Structure of the tryptophan (*trp*) operon and of the tryptophan leader peptide and other leader peptides in *E. coli*. (a) Arrangement of the *trp* operon. Note that the leader (*L*) encodes a short peptide containing two tryptophan residues near its terminus (there is a stop codon following the Ser codon). The promoter is labeled *P*, and the operator is labeled *O*. The genes labeled *trpE* through *trpA* encode the enzymes needed for tryptophan synthesis. (b) Amino acid sequences of leader peptides of some other amino acid biosynthesis operons. Because isoleucine is made from threonine, it is an important constituent of the threonine leader peptide.

sequence contains tandem tryptophan codons near its terminus and functions as an attenuator (Figure 6.31).

The basis of control of the tryptophan attenuator is as follows. If tryptophan is plentiful in the cell, there will be plenty of charged tryptophan tRNAs and the leader peptide will be synthesized. Synthesis of the leader peptide results in termination of transcription of the remainder of the *trp* operon, which includes the structural genes for the biosynthetic enzymes. On the other hand, if tryptophan is scarce, the tryptophan-rich leader peptide will not be synthesized. If synthesis of the leader peptide is halted by a lack of tryptophan, the rest of the operon is transcribed.

Mechanism of Attenuation

How does translation of the leader peptide regulate transcription of the tryptophan genes downstream? Consider that in prokaryotic cells transcription and translation are simultaneous processes; as mRNA is released from the DNA, the ribosome binds to it and translation begins (⇨ Section 4.1). That is, while *transcription* of downstream DNA sequences is still proceeding, *translation* of transcribed sequences is already underway (**Figure 6.32**).

Transcription is attenuated because a portion of the newly formed mRNA folds into a unique stem–loop that inhibits RNA polymerase activity. The stem–loop structure forms in the mRNA because two stretches of nucleotides near each other are complementary and can thus base-pair. If tryptophan is plentiful, the ribosome translates the leader sequence until it comes to the leader stop codon. The remainder of the leader sequence then forms a stem–loop, a transcription pause site, which is followed by a uracil-rich sequence that actually triggers termination (Figure 6.32a).

If the concentration of tryptophan in the cell is limiting, transcription of genes encoding tryptophan biosynthetic enzymes is needed. Thus, during transcription of the leader, the ribosome pauses at a tryptophan codon because of a shortage of charged tryptophan tRNAs. The presence of the stalled ribosome at this position allows a stem–loop to form (sites 2 and 3 in Figure 6.32b) that differs from the terminator stem–loop. This alternative stem–loop is not a transcription termination signal but instead prevents the terminator stem–loop (sites 3 and 4 in Figure 6.32a) from forming. This allows RNA polymerase to move past the termination site and begin transcription of tryptophan structural genes. Thus, in attenuation control, the rate of transcription is ultimately influenced by the rate of translation.

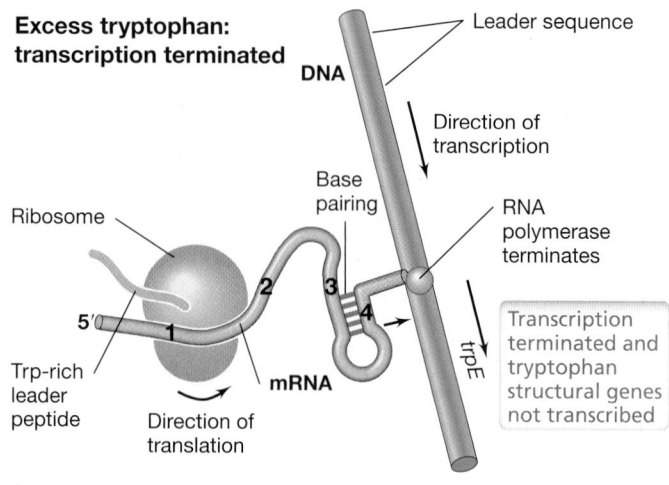

Excess tryptophan: transcription terminated

(a)

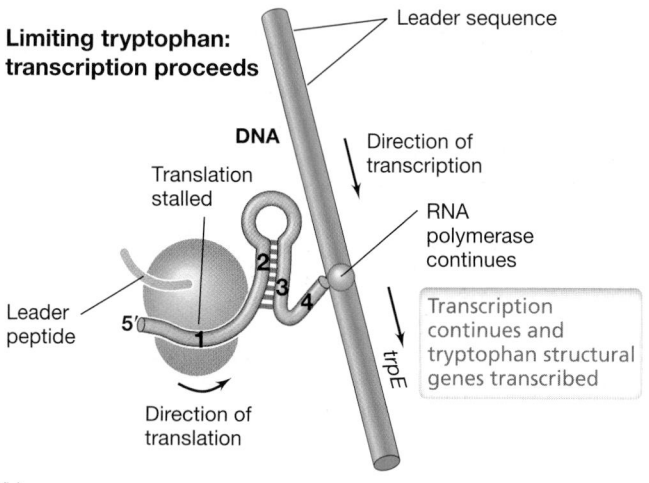

Limiting tryptophan: transcription proceeds

(b)

Figure 6.32 Mechanism of attenuation. Control of transcription of tryptophan (*trp*) operon structural genes by attenuation in *Escherichia coli*. The leader peptide is encoded by regions 1 and 2 of the mRNA. Two regions of the growing mRNA chain are able to form double-stranded loops, shown as 3:4 and 2:3. *(a)* When there is excess tryptophan, the ribosome translates the complete leader peptide, and so region 2 cannot pair with region 3. Regions 3 and 4 then pair to form a loop that terminates transcription. *(b)* If translation is stalled because of tryptophan starvation, a loop forms by pairing of region 2 with region 3, loop 3:4 does not form, and transcription proceeds past the leader sequence.

— MINIQUIZ —

- Why does attenuation control not occur in eukaryotes?
- Explain how the formation of one stem–loop in the RNA can block the formation of another.

IV • Regulation of Enzymes and Other Proteins

We have just explored some of the key mechanisms for regulating the *amount* (or even the complete presence or absence) of an enzyme or other protein within a cell. Here we focus on the mechanisms that cells employ to control the *activity* of enzymes already present in the cell through processes such as feedback inhibition and post-translational regulation.

6.14 Feedback Inhibition

A major means of controlling enzymatic activity is by **feedback inhibition**. This mechanism temporarily shuts off the reactions in an entire biosynthetic pathway. The reactions are shut off because an excess of the end product of the pathway inhibits activity of an early (typically the *first*) enzyme of the pathway. Inhibiting an early step effectively shuts down the entire pathway because no intermediates are generated for subsequent enzymes in the pathway (**Figure 6.33a**). Feedback inhibition is reversible,

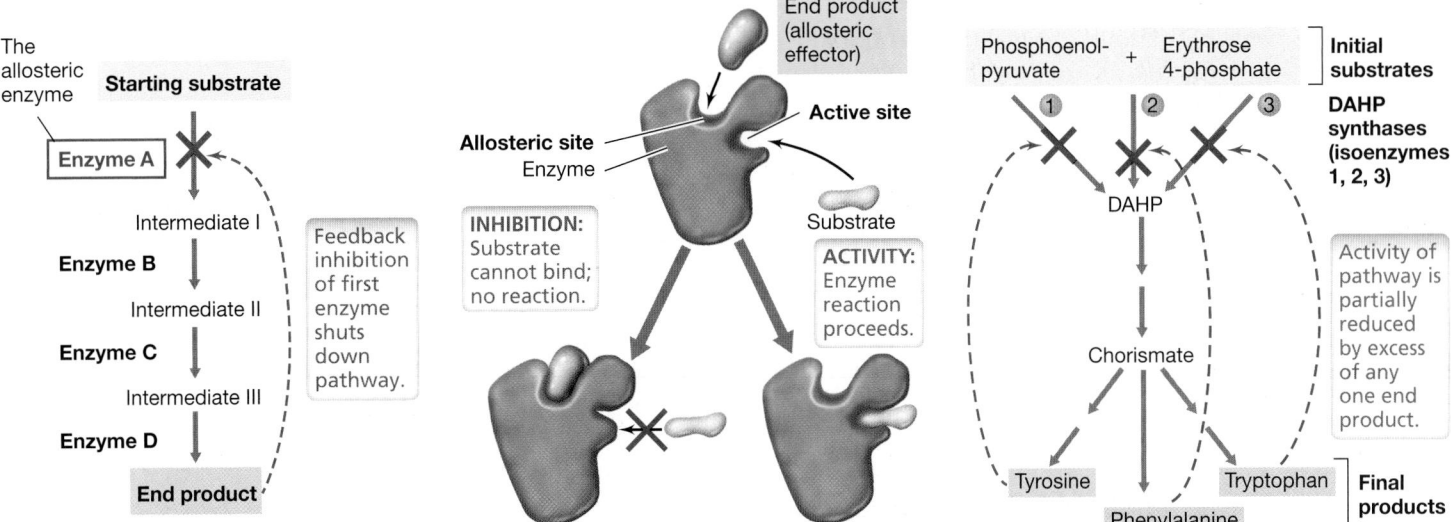

(a) **Feedback inhibition**

(b) **Allosteric inhibition**

(c) **Isoenzyme inhibition**

Figure 6.33 Inhibition of enzyme activity. *(a)* In feedback inhibition, the activity of the first enzyme of the pathway is inhibited by the end product, thus shutting off the production of the three intermediates and the end product. *(b)* The mechanism of allosteric inhibition by the end product of a pathway. When the end product binds at the allosteric site, the conformation of the enzyme is so altered that the substrate can no longer bind to the active site. However, inhibition is reversible, and end product limitation will once again activate the enzyme. *(c)* Inhibition by isoenzymes. In *Escherichia coli*, the pathway leading to the synthesis of the aromatic amino acids contains three isoenzymes of DAHP synthase. Each of these enzymes is feedback-inhibited by one of the aromatic amino acids. However, note how an excess of all three amino acids is required to completely shut off the synthesis of DAHP. In addition to feedback inhibition at the DAHP site, each amino acid feedback inhibits its further metabolism at the chorismate step.

however, because once levels of the end product become limiting, the pathway again becomes functional.

How can the end product of a pathway inhibit the activity of an enzyme whose substrate is quite unrelated to it? This occurs because the inhibited enzyme has two binding sites, the *active site* (where substrate binds, ↩ Section 3.5), and the *allosteric site*, where the end product of the pathway binds. When the end product is in excess, it binds at the allosteric site, changing the conformation of the enzyme such that the substrate can no longer bind at the active site (Figure 6.33*b*). When the concentration of the end product in the cell begins to fall, however, the end product no longer binds to the allosteric site, so the enzyme returns to its catalytic form and once again becomes active.

Isoenzymes

Some biosynthetic pathways controlled by feedback inhibition employ *isoenzymes* ("iso" means "same"). Isoenzymes are different proteins that catalyze the same reaction but are subject to different regulatory controls. Examples are enzymes required for the synthesis of the aromatic amino acids tyrosine, tryptophan, and phenylalanine in *Escherichia coli* (Figure 6.33*c*).

The enzyme 3-deoxy-D-arabino-heptulosonate 7-phosphate (DAHP) synthase plays a central role in aromatic amino acid biosynthesis, whose pathway contains a branch point late in the pathway (Figure 6.33*c*). In *E. coli*, three DAHP synthase isoenzymes catalyze the first reaction in this pathway, each regulated independently by a different one of the end product amino acids. However, unlike the example of feedback inhibition where an end product completely inhibits enzyme activity (Figure 6.33*a, b*), enzyme activity in the DAHP-controlled pathway is diminished incrementally; enzyme activity falls to zero only when *all three* end products are present in excess. This prevents the excess of one or two of the amino acids from shutting down the entire pathway and thereby starving the cell for the third amino acid (Figure 6.33*c*).

MINIQUIZ

- What is feedback inhibition?
- What is the difference between an allosteric site and an active site?
- What is an isoenzyme?

6.15 Post-Translational Regulation

We have already discussed *phosphorylation* and *methylation*, two very common mechanisms for regulating a protein post-translationally, when we considered two-component regulatory systems and chemotaxis, respectively (Sections 6.6 and 6.7). Biosynthetic enzymes can also be regulated by the attachment of other small molecules, such as the nucleotides adenosine monophosphate (AMP), adenosine diphosphate (ADP), and uridine monophosphate (UMP). Throughout this chapter we have also encountered avenues in which the cell regulates the activity of proteins by protein–protein interactions. In contrast to these, some enzymes are regulated by *covalent modification*, typically the attachment or removal of a small molecule to or from the enzyme that subsequently affects its activity, and we consider this regulatory mechanism here.

Regulation of PII Signal Transduction Proteins

PII proteins are a widespread family of signal-transducing proteins (Section 6.6) found in *Bacteria, Archaea,* and the plastids of plants, and they play an important role in nitrogen metabolism by regulating a diverse range of transcription factors, enzymes, and membrane transport proteins. The activity of PII proteins on their regulatory cascade depends on whether they have been covalently modified or not. These modifications range from *uridylylation* (addition of a UMP group), *adenylylation* (addition of AMP), or even *phosphorylation* (in some cyanobacteria). Because these covalent modifications affect protein activity and occur *after* the PII proteins have been fully synthesized, the regulation is considered post-translational. Here we consider PII and its uridylylation by GlnD, a bifunctional enzyme that possesses both uridyltransferase activity and uridyl-removal activity in response to the presence of glutamine (**Figure 6.34**).

One of the main ways the cell determines if ammonia (NH_3) assimilation is necessary is through glutamine sensing by the GlnD protein. If the cellular glutamine pool is low, it signals that NH_3 assimilation is necessary. GlnD transduces this signal through the binding of glutamine itself. When GlnD is not bound to glutamine, the enzyme possesses uridyltransferase activity and adds a UMP group to PII, yielding PII-UMP through post-translational modification.

One protein target of PII-UMP within the nitrogen metabolism pathway is glutamine synthetase adenyltransferase (Figure 6.34*a*). Once uridylylated, PII-UMP stimulates glutamine synthetase adenyltransferase to remove adenyl groups from glutamine synthetase (GS). GS is a key enzyme in ammonia (NH_3) assimilation (⟳ Section 3.14), which must be tightly regulated to conserve energy if cellular levels of nitrogen are high, as sensed by the glutamine pool. As GS becomes less adenylylated, its activity *increases* and NH_3 assimilation increases (Figure 6.34*c*). Once glutamine levels are sufficient, the cell no longer needs to assimilate NH_3. This results in GlnD binding to available glutamine, which stimulates the protein to remove uridyl groups from PII-UMP (Figure 6.34*b*). PII in its unmodified form stimulates glutamine synthetase adenyltransferase to adenylate GS, forming GS-AMP. Fully adenylylated GS is *less active* and thus less NH_3 assimilation occurs.

Why does all this elaborate regulation surround the enzyme GS? The activity of GS requires ATP, and nitrogen assimilation is a major biosynthetic process in the cell. However, when NH_3 is present at high levels in the cell, it can be assimilated into amino acids by enzymes that do not consume ATP; under these conditions, GS remains inactive. When NH_3 levels are low, however, GS becomes catalytically active. By having GS active only when NH_3 is limiting, the cell conserves ATP that would be unnecessarily consumed if GS were active when NH_3 is present in excess.

Inactivation of Sigma Factors

In Section 6.10 we described how the σ factor RpoH is inactivated by DnaK under normal temperature conditions in the heat shock response (Figure 6.26). Proteins known as *anti-sigma* factors can also bind to sigma factors, inactivating them in a form of post-translational regulation.

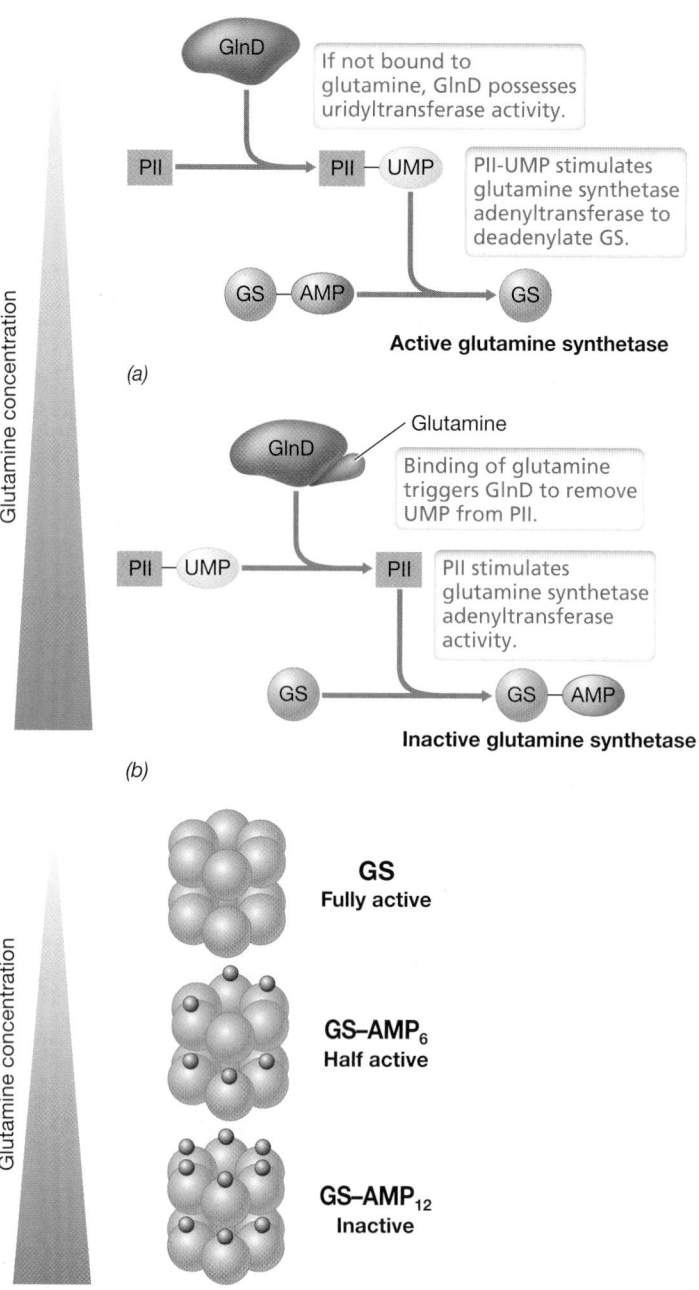

Figure 6.34 Post-translational regulation of PII and glutamine synthetase. *(a)* Uridylylation of PII and activation of glutamine synthetase (GS). In response to low cellular pools of glutamine, the uridyltransferase activity of GlnD is stimulated. This results in PII becoming PII-UMP, which triggers glutamine synthetase adenyltransferase to activate GS through AMP removal. *(b)* Removal of UMP from PII and inactivation of GS. In response to high cellular pools of glutamine, the UMP removal activity of GlnD is activated. This results in PII-UMP becoming PII, which triggers glutamine synthetase adenyltransferase to inactivate GS through adenylylation. *(c)* Control of GS activity. Adenylylated GS subunits are catalytically inactive, so the overall GS activity increases progressively as glutamine levels decrease and more subunits are de-adenylylated.

RpoE (σ^{24} or σ^E, ⟳ Table 4.3) is a sigma factor conserved in many *Bacteria* that responds to extracytoplasmic stress by recognizing promoters for genes that encode products necessary for proper folding, expression, and turnover of proteins of the outer membrane (⟳ Section 2.5). One of these promoters is for *dnaK*.

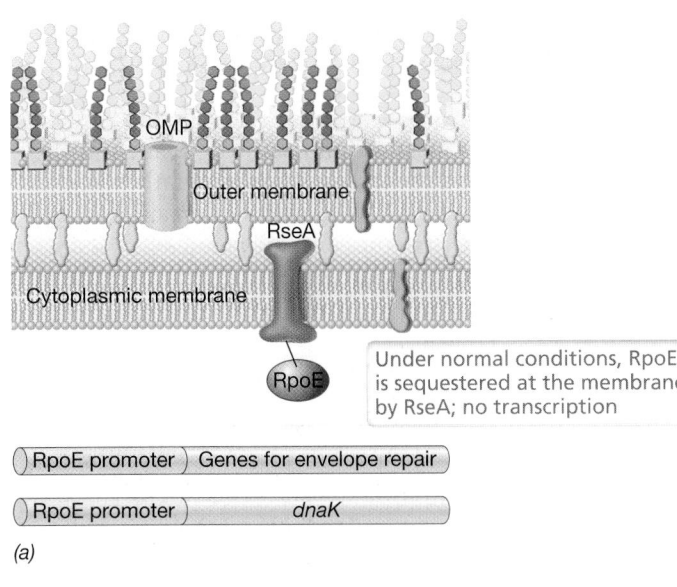

Under normal conditions, RpoE is sequestered at the membrane by RseA; no transcription

| RpoE promoter | Genes for envelope repair |
| RpoE promoter | *dnaK* |

(a)

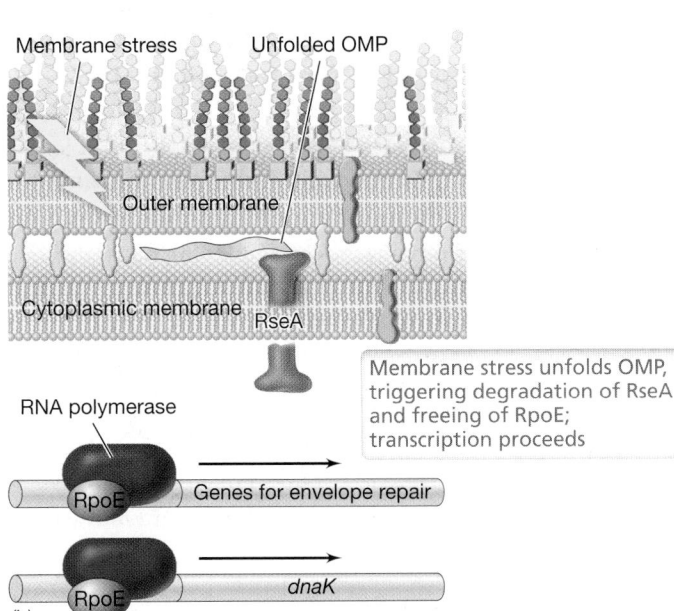

Membrane stress unfolds OMP, triggering degradation of RseA and freeing of RpoE; transcription proceeds

Genes for envelope repair

dnaK

(b)

Figure 6.35 Anti-sigma–sigma factor interactions. *(a)* Sequestration of the RpoE sigma factor by the anti-sigma factor RseA. When the outer membrane is not stressed, cytoplasmic membrane–bound RseA binds to RpoE. This binding prevents the sigma factor from binding to promoters of genes for envelope repair and *dnaK*. *(b)* Inactivation of RseA and release of the RpoE sigma factor. When the outer membrane is exposed to stress, outer membrane proteins (OMPs) are denatured. These unfolded OMPs trigger proteolytic degradation of RseA. Without RseA, the RpoE sigma factor can bind to its promoter regions and transcribe genes for envelope stress and *dnaK* to help repair the membrane and refold proteins.

Under normal nonstressful conditions, the membrane-bound anti-sigma factor RseA sequesters RpoE from binding to the promoters it targets. However, if a cell encounters a condition stressful to the membrane such as heat or osmotic stress, outer membrane proteins are denatured, and this *activates* a protease specific for RseA. Once RseA has been degraded, RpoE is free to activate transcription of genes necessary for envelope repair such as refolding outer membrane proteins and synthesis of phospholipids and lipopolysaccharides (**Figure 6.35**). Release of RpoE also includes increasing the level of DnaK produced so that the chaperone (⇄ Section 4.11) can assist with folding and turnover of outer membrane proteins.

Once the membrane has been repaired, RseA is no longer targeted for degradation and reassociates with the cytoplasmic membrane. In this form, RseA can resume its anti-sigma factor activity by binding and sequestering RpoE from promoter regions. This leads to a *decreased* expression of both genes encoding products needed to respond to membrane stress and *dnaK*. Another well-studied example of regulation by anti-sigma–sigma interactions occurs during the formation of endospores in *Bacillus* when the anti-sigma factor SpoIIAB binds to σ^F, thereby preventing its association with RNA polymerase; we pick up on this control process in Section 7.6 and Figure 7.15.

Regardless of the mechanism, in the final analysis it should be clear that regulating the synthesis and activities of a cell's RNAs and proteins is (1) very important to its biology, (2) possible in many different ways, and (3) a major genetic investment. But the costs to the cell are worth it and are grounded in billions of years of evolutionary refinement. At every turn in a highly competitive world, a microbe's growth and very survival may well depend on its ability to conserve its resources and maximize its progeny.

— **MINIQUIZ** —

- What types of covalent modifications commonly alter the activity of proteins?
- How does GlnD signal to the cell that NH_3 assimilation is necessary?
- Explain the role of an anti-sigma factor.

Chapter Review

I • DNA-Binding Proteins and Transcriptional Regulation

6.1 Proteins can bind to DNA when specific domains of the proteins bind to specific regions of the DNA molecule. In most cases the interactions are sequence-specific. Proteins that bind to DNA are often regulatory proteins that affect gene expression.

Q Describe why a protein that binds to a specific sequence of double-stranded DNA is unlikely to bind to the same sequence if the DNA is single-stranded.

6.2 The amount of a specific enzyme in the cell can be controlled by regulatory proteins that bind to DNA and increase (induce) or decrease (repress) the amount of mRNA that encodes the enzyme. In negative control of transcription, the regulatory protein is called a repressor and it functions by inhibiting mRNA synthesis.

Q Induction is considered the opposite of enzyme repression. Why?

6.3 Positive regulators of transcription are called activator proteins. They bind to activator-binding sites on the DNA and stimulate transcription. Inducers modify the activity of activating proteins. In positive control of enzyme induction, the inducer promotes the binding of the activator protein and thus stimulates transcription.

Q What is the difference between an operon and a regulon?

6.4 Global control systems regulate the expression of many genes simultaneously. Catabolite repression is a global control system that helps cells make the most efficient use of available carbon sources. The *lac* operon is under the control of catabolite repression as well as its own specific negative regulatory system.

Q Describe the mechanism by which cAMP receptor protein (CRP), the regulatory protein for catabolite repression, functions. Use the lactose operon as an example.

6.5 *Archaea* resemble *Bacteria* in using DNA-binding activator and repressor proteins to regulate gene expression at the level of transcription.

Q What are the two mechanisms used by archaeal repressor proteins to repress transcription?

II • Sensing and Signal Transduction

6.6 Signal transduction systems transmit environmental signals to the cell. In *Bacteria* and *Archaea*, signal transduction is typically carried out by a two-component regulatory system that includes a membrane-integrated sensor kinase and a cytoplasmic response regulator. The activity of the response regulator depends on its state of phosphorylation.

Q What are the two components that give their name to the signal transduction system in prokaryotic cells? What is the function of each of the components?

6.7 Chemotactic behavior allows cells to respond to attractants and repellents in their environment. The regulation of chemotaxis affects the activity of proteins rather than their synthesis. Adaptation by methylation allows the system to reset itself to the continued presence of a signal.

Q Adaptation allows the mechanism controlling flagellar rotation to be reset. How is this achieved?

6.8 Quorum sensing allows cells to monitor their environment for cells of their own kind. Quorum sensing depends on the sharing of specific small molecules known as autoinducers. Once a sufficient concentration of the autoinducer is present, specific gene expression is triggered.

Q How can quorum sensing be considered a regulatory mechanism for conserving cell resources?

6.9 The stringent response is employed to survive nutrient limitation and stresses by decreasing expression of genes for macromolecule biosynthesis and activating stress survival pathways.

Q Explain the sequence of molecular events that leads to the synthesis of (p)ppGpp in *Escherichia coli* during the stringent response.

6.10 Cells can control sets of genes by employing transcriptional regulators and alternative sigma factors. These recognize only certain promoters and thus allow transcription of a select category of genes that is most appropriate for the environmental conditions. Cells respond to both excessive temperature and nutritional limitations by expressing sets of genes whose products help the cell overcome stress.

Q Describe the proteins produced when cells of *Escherichia coli* experience a heat shock. Of what value are they to the cell?

III • RNA-Based Regulation

6.11 Cells can control genes in several ways by employing regulatory RNA molecules. One way is to take advantage of base pairing and use sRNA to promote or prevent translation of mRNAs.

Q What are the mechanisms by which regulation by sRNA occurs?

6.12 Riboswitches are sequences at the 5′ ends of certain mRNAs that recognize small molecules and respond by changing their three-dimensional structure to affect translation or transcriptional termination of the mRNA. Riboswitches are mostly employed to control biosynthetic pathways for vitamins, amino acids, purines, and a few other metabolites.

Q What is the mechanism by which a riboswitch regulates translation?

6.13 Attenuation is a mechanism whereby transcription is controlled after initiation of mRNA synthesis. Attenuation mechanisms depend upon alternative stem–loop structures in the mRNA that result in either read-through or stalling of the ribosome.

Q Describe how transcriptional attenuation works. What is actually being "attenuated"?

IV • Regulation of Enzymes and Other Proteins

6.14 In feedback inhibition, an excess of the final product of a biosynthetic pathway inhibits an allosteric enzyme at the beginning of the pathway. Enzyme activity can also be modulated by isoenzymes.

Q Describe how feedback inhibition is reversible.

6.15 Protein activity can be regulated after translation. Reversible covalent modification or interactions with other proteins can modulate protein activity.

Q Which nucleotides are commonly used to covalently modify protein activity?

Application Questions

1. What would happen to regulation from a promoter under negative control if the region where the regulatory protein binds was deleted? What if the promoter was under positive control? Promoters from *Escherichia coli* under positive control are not close matches to the promoter consensus sequence for *E. coli*. Why?

2. Most of the regulatory systems described in this chapter employ regulatory proteins. However, regulatory RNA is also important. Describe how one could achieve negative control of the *lac* operon using either of two different types of regulatory RNA.

3. Many amino acid biosynthetic operons under attenuation control are also under negative control. Considering that the environment of a bacterium can be highly dynamic, what advantage could be conferred by having attenuation as a second layer of control?

4. How would you design a regulatory system to make *Escherichia coli* use succinic acid in preference to glucose? How could you modify it so that *E. coli* prefers to use succinic acid in the light but glucose in the dark?

Chapter Glossary

Activator protein a regulatory protein that binds to specific sites on DNA and stimulates transcription; involved in positive control

Allosteric protein a protein containing an active site for binding substrate and an allosteric site for binding an effector molecule such as the end product of a biochemical pathway

Attenuation a mechanism for controlling gene expression that terminates transcription after initiation but before a full-length messenger RNA is produced

Autoinducer a small signal molecule that takes part in quorum sensing

Catabolite repression the suppression of _alternative catabolic pathways by a preferred source of carbon and energy

Cyclic AMP a regulatory nucleotide that participates in catabolite repression

Domains regions of a protein with specific structure and function

Feedback inhibition a process in which an excess of the end product of a multi-step pathway inhibits activity of the first enzyme in the pathway

Gene expression transcription of a gene followed by translation of the resulting mRNA into protein

Heat shock proteins proteins induced by high temperature (or certain other stresses) that protect against high temperature, especially by refolding partially denatured proteins or by degrading them

Heat shock response response to high temperature that includes the synthesis of heat shock proteins together with other changes in gene expression

Induction production of an enzyme in response to a signal (often the presence of the substrate for the enzyme)

Negative control a mechanism for regulating gene expression in which a repressor protein prevents transcription of genes

Noncoding RNA (ncRNA) RNA that is not translated into protein; examples include ribosomal RNA, transfer RNA, and small regulatory RNAs

Operon two or more genes transcribed into a single RNA and under the control of a single regulatory site

Positive control a mechanism for regulating gene expression in which an activator protein functions to promote transcription of genes

Quorum sensing a regulatory system that monitors the population level and controls gene expression based on cell density

Regulatory nucleotide a nucleotide that functions as a signal rather than being incorporated into RNA or DNA

Regulon a series of operons controlled as a unit

Repression prevention of the synthesis of an enzyme in response to a signal

Repressor protein a regulatory protein that binds to specific sites on DNA and blocks transcription; involved in negative control

Response regulator protein one of the members of a two-component regulatory system; a protein that is phosphorylated by a sensor kinase and then acts as a regulator, often by binding to DNA

Riboswitch an RNA domain, usually in a messenger RNA molecule, that can bind a specific small molecule, altering its own secondary structure; this, in turn, controls translation of the mRNA

Sensor kinase protein one of the members of a two-component regulatory system; a protein that phosphorylates itself in response to an external signal and then transfers the phosphoryl group to a response regulator protein

Signal transduction see *two-component regulatory system*

Stringent response a regulatory mechanism that detects nutrient or environmental stresses and antibiotics

Two-component regulatory system a regulatory system consisting of two proteins: a sensor kinase and a response regulator

7

Molecular Biology of Microbial Growth

microbiology**now**

Explosive Cell Death Promotes Biofilm Formation

To withstand environmental challenges and take best advantage of nutrient cycling, microbes often form slimy, three-dimensional structural communities called biofilms. While the major requirement for a biofilm lifestyle is a surface for attachment, specific phenotypic changes must occur that lead to attached growth instead of a planktonic (free-floating) existence. These changes, as well as the release of sticky substances called extracellular polymeric substances (EPS), require cell-to-cell communication.

EPS is composed of polysaccharides, extracellular DNA, and cytoplasmic proteins and is essential for protecting and holding attached cells together. Often the switch to a biofilm lifestyle can lead to beneficial outcomes for humans, such as in enhanced wastewater treatment. However, biofilm development by pathogens can lead to chronic disease and enhanced antibiotic resistance. This is especially true of the best-studied biofilm-forming bacterium—the opportunistic pathogen *Pseudomonas aeruginosa*.

Because of the robust ability of *P. aeruginosa* to form biofilms resistant to antibiotic treatment, microbiologists have focused on the biofilm developmental cycle. In doing so, they have discovered a previously unknown phenomenon— explosive death in a subpopulation of cells promotes biofilm formation. This explosive death is caused by a lysis enzyme used by an intrinsic virus to disrupt the bacterial cell wall. The DNA of this virus resides as a prophage in the genome of *P. aeruginosa* cells. As the left photo shows, production of the lysis enzyme causes the bacterial cells (purple) to break open, releasing their DNA (yellow). This DNA is critical to forming the EPS that holds the biofilm together. Cell bursting not only leads to the release of DNA (right photo, red), but also to shattered cell membrane fragments that curl together to form vesicles surrounding proteins and DNA that contribute to EPS formation (right photo, small blue circles with red centers).

While exposure to a DNA-damaging agent or an antibiotic normally leads to "activation" of the virus, random expression of the viral lysis gene occurs in a discrete subpopulation of cells. This annihilation of a few cells ultimately benefits the rest of the population. These findings also provide a potential answer to the following conundrum: Why does antibiotic treatment often enhance (rather than diminish) biofilm formation?

Source: Turnbull, L., et al. 2016. Explosive cell lysis as a mechanism for the biogenesis of bacterial membrane vesicles and biofilms. *Nat. Commun. 7:* 1120.

In previous chapters we discussed how prokaryotic cells replicate their chromosome(s) (Chapter 4), grow and divide (Chapter 5), and regulate gene expression (Chapter 6). Here we focus on the key molecular processes that occur in the microbial growth cycle. We also consider cellular differentiation in some model bacteria, revisit the process of biofilm growth, and close by exploring how some common antibiotics affect microbial growth and what weapons bacteria deploy to counter such attacks.

I • Bacterial Cell Division

Most cells divide by binary fission (⮂ Section 5.1 and Figure 5.1), and this process occurs in a defined series of steps such that each daughter cell obtains a copy of the genome. Because the process of cell division in prokaryotic cells requires both temporal and spatial control elements, it is controlled by a cell cycle (**Figure 7.1**).

Temporally, two copies of the cell's genome must exist prior to cell division. Spatially, the two copies must be equally segregated and the septum must form at the correct location in the cell for a successful cell cycle to occur (Figure 7.1). During the division cycle, the cell must also produce new peptidoglycan and cytoskeleton elements to prevent bursting from osmotic forces. To successfully orchestrate all of these events, various regulatory cascades are put into play. In this first part of the chapter we focus on the molecular mechanisms employed by two well-studied gram-negative bacteria, *Escherichia coli* and *Caulobacter crescentus*, and introduce advanced microscopic techniques that have revealed the molecular pathways associated with cell division and cell morphology.

7.1 Visualizing Molecular Growth

While the standard light microscope nicely reveals cell size and shape, more powerful techniques are required to visualize macromolecules such as chromosomes, proteins, and membranes. To observe the tiniest details within a cell, **super-resolution microscopy**, a powerful new form of light microscopy that employs a suite of fluorescent molecules, was developed. Super-resolution techniques are capable of resolving structures as small as 10–50 nm in living cells, allowing for dynamic cellular behaviors to be observed in real time.

Fluorescent Tagging

To visualize the localization of specific proteins and to monitor gene expression, **reporter genes** have long been used. Reporter genes encode proteins that are easy to detect or assay and are fused to genes of interest. For visualizing molecular events, reporter genes that encode fluorescent products such as the **green fluorescent protein (GFP)** are routinely used (**Figure 7.2**). Through the use of fluorescence microscopy (⮂ Section 1.6) and multiple fluorescent reporter proteins, the activity and localization of different macromolecules can be determined simultaneously. Figure 7.2a demonstrates use of GFP and another fluorescent protein to determine the cellular locations of σ^F and σ^E, two spore-specific sigma factors (⮂ Section 4.5) that play key roles in endospore development in *Bacillus* (Section 7.6).

Fluorescent tagging can also be used to resolve different nucleic acids in a cell, such as a chromosome and a plasmid

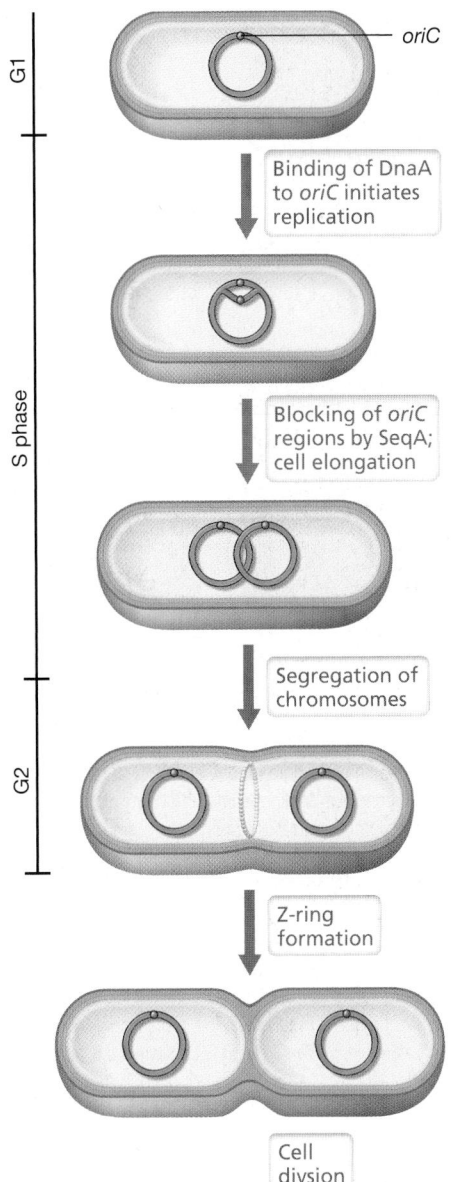

Figure 7.1 Overview of the bacterial cell cycle. Phases analogous to the growth 1 (G1), synthesis (S), and growth 2 (G2) phases of the eukaryotic cell cycle are represented.

(see Figure 7.7). In fact, even different loci within an individual nucleic acid molecule can be observed in a living cell. While DNA or RNA loci are not linked directly to a reporter gene like GFP, nucleic acid–binding proteins specific for target loci can be fused to a reporter gene. If necessary, the corresponding DNA or RNA binding sites can be introduced into the nucleic acid region of interest using recombinant techniques (Chapter 12). Figure 7.2b illustrates how separate "arms" of the *Escherichia coli* chromosome can be visualized in a living cell by introducing specific DNA sequences into the chromosome and tagging the corresponding DNA-binding proteins with different fluorescent reporter genes.

Advances in protein labeling and microscopy techniques have resulted in new fluorescent variants that fold quickly and allow for visualization of both spatial and temporal protein dynamics in

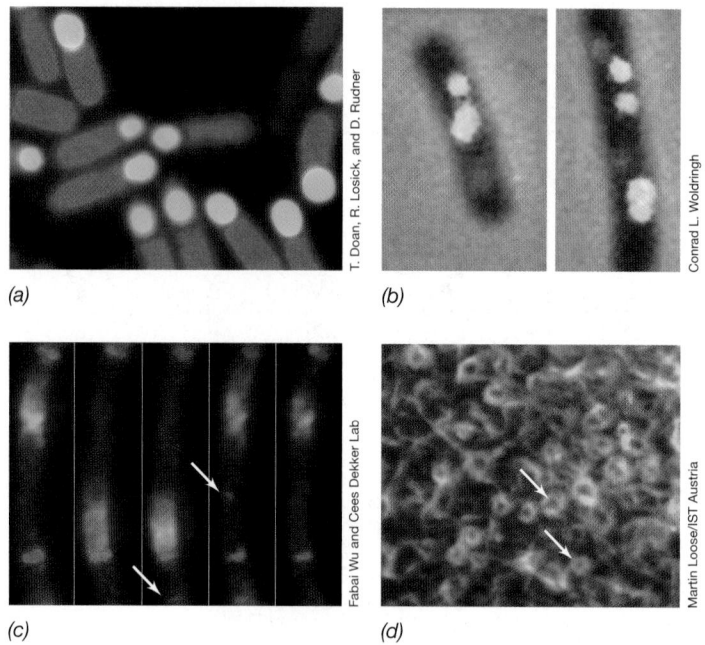

(a)

(b)

(c)

(d)

Figure 7.2　Fluorescence micrographs illustrating molecular growth characteristics. *(a)* During endospore formation in *Bacillus*, alternative sigma factors are localized to specific regions of the cell (Section 7.6). σ^F linked to GFP indicates activity of the protein in the developing endospore (at one end of each cell). σ^E linked to a reporter protein that fluoresces red indicates expression and activity of the protein throughout the mother cell prior to endospore formation. Regions correspond to the model depicted in Figure 7.15*b*. *(b)* Location and orientation of the *Escherichia coli* chromosome during cell cycle progression (Section 7.2). Regions are visible due to tagged DNA-binding proteins targeting the *oriC* (green), left arm (red), and right arm (cyan). Magnification bar represents 2 μm. Modified from Woldringh, C.L., Hansen, F.G., Vischer, N.O.E., and Atlung, T. 2015. *Front. Microbiol. 6:* 448. *(c)* Time-lapse series showing FtsZ (red) and MinD (green) in an elongated *E. coli* cell. Cell division was inhibited by antibiotic treatment and arrows indicate locations where unstable FtsZ proteins polymerize and depolymerize. Modified from Wu, F., Van Rijn, E., Van Schie, B.G.C., Keymer, J.E. and Dekker, C. 2015. *Front. Microbiol. 6:* 607. *(d)* Co-assembly of the cell-division proteins FtsZ (yellow) and FtsA (brown) on a membrane (Section 7.3). Swirls of FtsZ are visible (arrows).

live cells. Figure 7.2*c* illustrates the spatiotemporal dynamics of MinD and FtsZ (proteins of the divisome, see Section 7.3) in a growing cell over a time-lapse series. To visualize membrane-associated proteins, lipid bilayers can be created and applied to microscope slides before introducing fluorescently tagged proteins. This technique has been used to visualize interactions between the cell-division proteins FtsZ and FtsA and the rotating rings of FtsZ (Figure 7.2*d* and Section 7.3).

Super-Resolution Techniques

Although fluorescence microscopy enables observation of tagged internal cellular structures, resolution is insufficient to observe the molecular events occurring during the cell cycle. By contrast, super-resolution techniques can reveal and quantify elements as small as *single molecules* in living cells. Super-resolution techniques employ photoactivatable probes that switch from bright to dark emission states depending on the wavelength of light that strikes them; this allows for only one fluorescent molecule (or target) to be excited at a time. By recording the position of individual molecules over time, a super-resolution image is generated (**Figure 7.3***a*).

Super-resolution techniques can reveal the details of cellular superstructures, for example by precisely mapping the diffusion and movement of individual proteins. Figure 7.3*a* illustrates the use of *photoactivated localization microscopy*, a form of super-resolution microscopy, to map the movement of individual MukB molecules in a living *E. coli* cell. MukBEF is a protein complex that binds to the nucleoid and localizes to distinct regions in the cell to assist with unknotting daughter chromosomes after replication (Section 7.2). Not only can super-resolution microscopy be used to observe structural changes to cellular complexes over time, the micrographs can also be modified to construct

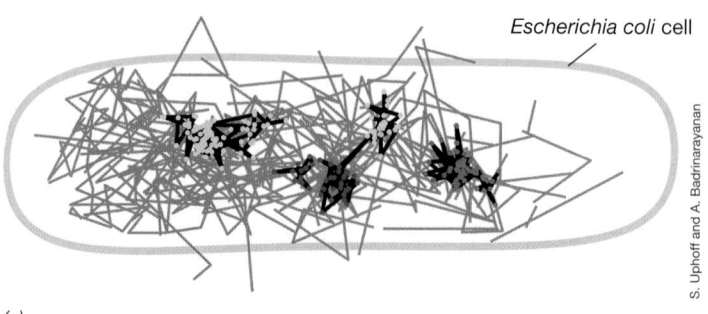

Escherichia coli cell

(a)

(b)

Figure 7.3　Super-resolution imaging of molecular growth characteristics. *(a)* Photoactivated localization micrograph (PALM) mapping the movement of the nucleoid-binding protein MukB in an individual *Escherichia coli* cell. The four different colors represent clusters of nucleoid-bound MukB molecules, while tracks corresponding to their movement are represented by black lines. Gray lines represent tracks of free MukB (Section 7.2). *(b)* Three-dimensional quantitative multicolor subdiffraction imaging of *Caulobacter* cellular structures. PopZ is tagged with a red fluorescent reporter, while crescentin is tagged with a green fluorescent reporter (Section 7.4).

three-dimensional (3D) images. Figure 7.3*b* illustrates the use of a super-resolution method called 3D subdiffraction imaging to observe the arrangement of different protein complexes in the budding bacterium *Caulobacter crescentus*.

By providing spatiotemporal information about individual proteins, super-resolution techniques have been instrumental in proving that like their eukaryotic counterparts, prokaryotic cells contain a highly dynamic subcellular organization and one whose activities during the cellular growth process can be tracked at the molecular level. We now focus on some of these key molecular events.

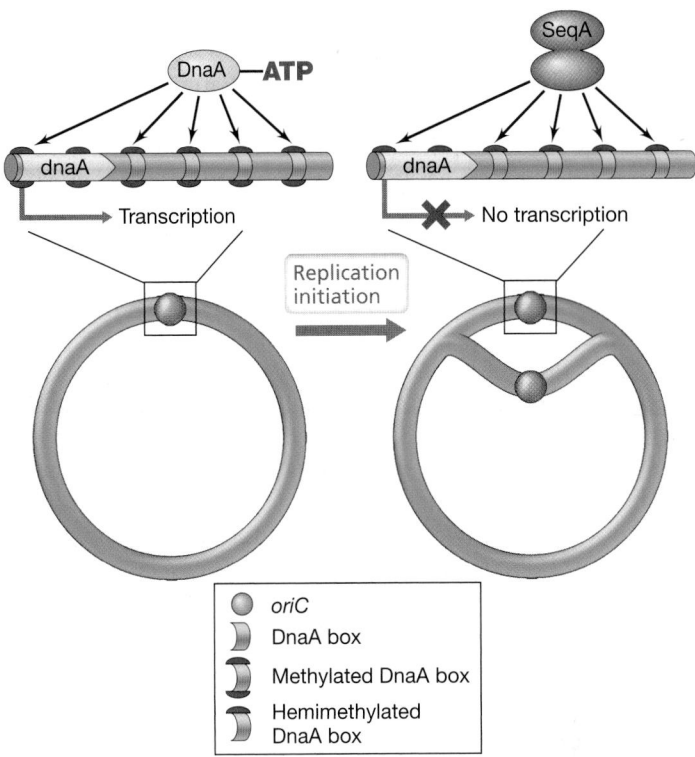

— **MINIQUIZ** —

- What is the utility of a reporter gene?
- What are the advantages of using super-resolution microscopy versus standard fluorescence microscopy?

7.2 Chromosome Replication and Segregation

A prerequisite for bacterial growth is replication of the genome. Chromosome replication must be tightly regulated to coincide with cell division. After genome replication and prior to cell division, molecular mechanisms must also ensure that the daughter chromosomes are segregated (Figure 7.1).

Regulation of Chromosome Replication Initiation

How does the cell ensure that the genome is completely replicated before cell division while also preventing multiple rounds of replication? Several different proteins play a role in initiating and inhibiting chromosome replication in *Escherichia coli*. Here we focus on a key protein in this regard, DnaA.

As discussed in Section 4.3, binding of DnaA to specific DNA sequences within the *oriC* region of the chromosome leads to unwinding of the DNA and loading of the replisome for chromosome replication in *Escherichia coli* (Figure 7.1). DnaA is most active when it is linked to a molecule of ATP, forming DnaA–ATP. To tightly control chromosome replication, multiple regulatory mechanisms come into play to inactivate DnaA–ATP once replication has initiated. These mechanisms include competition for *oriC* binding, repression of *dnaA* expression, titration of DnaA–ATP away from *oriC*, and inactivation of DnaA–ATP.

Before DNA replication initiates, both strands of the chromosome contain methyl groups on the adenine residue of –GATC– sequences within the chromosome. However, immediately after replication has initiated, only the parental strand remains methylated. This results in *hemimethylated DNA* and facilitates a competition for origin binding between DnaA–ATP and a protein called SeqA (**Figure 7.4**). Because hemimethylated *oriC* sequences are strongly bound by SeqA, DnaA–ATP is blocked from binding and reinitiating chromosome replication (Figures 7.1 and 7.4). Approximately 10 min after replication has initiated, GATC sequences within the newly synthesized daughter strand are methylated by a DNA adenine methylase.

Figure 7.4 Binding of the *Escherichia coli oriC* by DnaA and SeqA proteins. DnaA–ATP only binds to fully methylated DnaA boxes, while SeqA binds to hemimethylated DNA. The binding of SeqA upstream of the *dnaA* gene leads to repression of transcription.

As a result both of the chromosomal proximity of the *dnaA* gene to *oriC* and of the promoter region (⊋ Section 4.5) of *dnaA* possessing a GATC sequence, binding of SeqA also plays a role in repressing the expression of *dnaA*. Once replication has initiated, the promoter region for *dnaA* is quickly hemimethylated, and the binding of SeqA prevents RNA polymerase from binding and transcribing *dnaA* (Figure 7.4). Subsequently, expression of *dnaA* is also *autoregulated* by its corresponding protein binding to DnaA boxes within its promoter region.

The final mechanism of preventing DnaA–ATP from binding *oriC* is to decrease the ratio of DnaA–ATP to DnaA–ADP. How does the cell increase the level of DnaA–ADP when ATP dominates over ADP in a growing cell? This is controlled by the ATPase HdaA, which associates with DNA in the proximity of the replisome and specifically targets DnaA–ATP. This enzyme promotes the hydrolysis of ATP associated with DnaA in a replication-dependent manner and thus functions as a final method of regulating the initiation of chromosome replication. As a result of these combined mechanisms, the cellular level of DnaA–ATP oscillates during the cell cycle, reaching the maximum active amount when initiation of chromosome replication is needed and waning thereafter.

Genome Replication in Fast-Growing Cells

As we learned in Chapter 4, the circular nature of the chromosome of *E. coli* (and that of most other *Bacteria* and *Archaea*) creates an opportunity for speeding up DNA replication. This is because replication of circular genomes is *bidirectional* from the origin of replication. During bidirectional replication, synthesis occurs in both leading and lagging directions on each template strand, and this allows DNA to replicate as rapidly as possible (⇔ Figure 4.16). Studies of chromosome replication in *E. coli* have shown that about 40 min is required for genome replication and that this value is independent of the generation time, which in *E. coli* can be as little as 20 min. If an *E. coli* cell is growing at twice the rate that its chromosome is replicating, how does the cell resolve this conundrum?

In cells growing at doubling times shorter than 40 min, *multiple DNA replication forks* are present in each cell. That is, a new round of DNA replication begins before the last round has been completed (**Figure 7.5**), and therefore, some genes are present in more than one copy. This can occur after the DNA in the *oriC* region of the newly synthesized DNA has been methylated, which releases SeqA from the DNA and allows DnaA–ATP to be recruited to reinitiate another round of replication (Figure 7.4). This ensures that at generation times shorter than the time required for replication (a process that occurs at a constant and maximal speed), each daughter cell receives a complete genome at the time of septum formation.

Chromosome Segregation

During cell division, segregation of the chromosomes to the cell poles is needed not only to ensure that each daughter cell gets a copy of the genome, but also to allow for septum formation (Figure 7.1). If the two copies of the nucleoid remained in the center of the cell, nucleoid occlusion (a process that prevents the cell from dividing across the nucleoids) would impede proper cell division. In eukaryotes, mitotic spindles (⇔ Section 2.14) separate replicated chromosomes. In many bacteria including the budding bacterium *Caulobacter*, a similar mechanism known as the *Par (partitioning) system* is used to distribute chromosomes and plasmids equally to progeny cells during growth (**Figure 7.6**). This system is composed of the ParA ATPase, the ParB chromosome-binding protein, and the PopZ complex, as well as a centromere-like *parS* sequence located near *oriC*.

Unlike eukaryotic mitotic spindles, the Par system does not segregate fully replicated chromosomes. Instead, PopZ proteins localize to the old pole of the cell and facilitate anchoring of the chromosome to this location by interacting with ParB bound to the *parS* sequence (Figure 7.6). Once chromosome replication

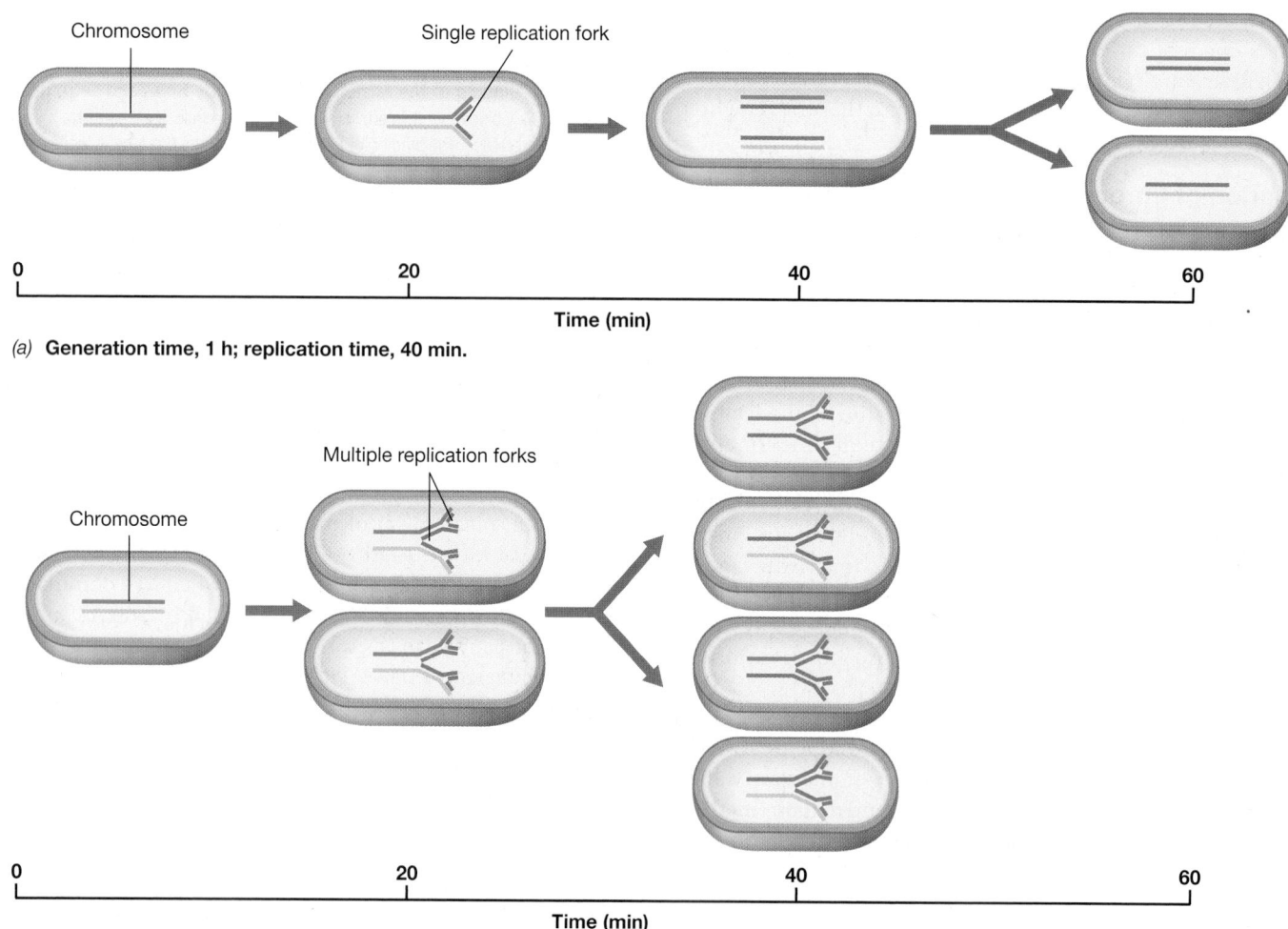

(a) **Generation time, 1 h; replication time, 40 min.**

(b) **Generation time, 20 min; replication time, 40 min.**

Figure 7.5 Genome replication in cells of *Escherichia coli* growing at 60 min or 20 min generation times. In cells doubling every 20 min, multiple replication forks are needed to ensure that each daughter cell gets a complete copy of the genome, which takes 40 min to replicate. In cells doubling every 60 min, multiple replication forks are unnecessary.

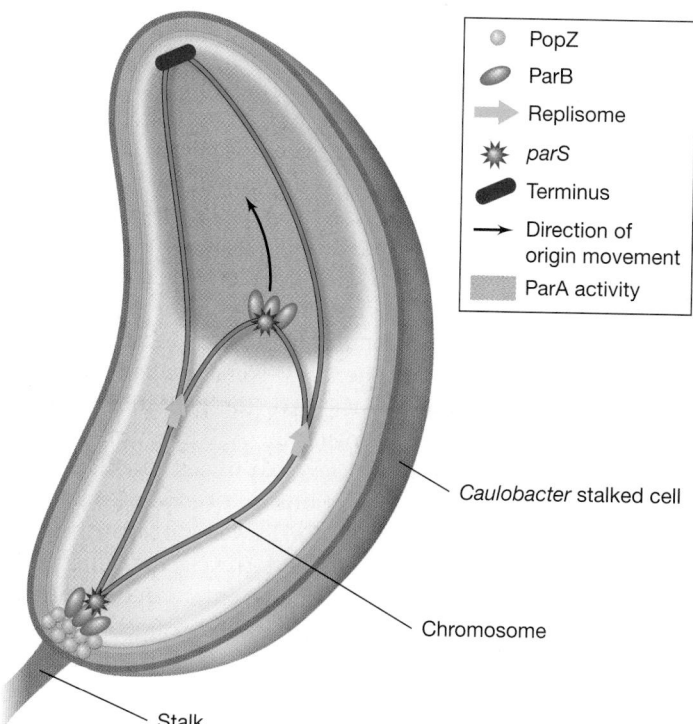

Figure 7.6 Chromosome partitioning in *Caulobacter*. After replication initiates, PopZ localizes to the old pole of the cell. Next, ParB binds to the *parS* sequence on the parental chromosome and the complex associates with PopZ at the pole. As replication continues, the *parS* sequence on the daughter chromosome is bound by ParB and moved to the new cell pole by the activity of diffuse ParA molecules.

initiates, the newly synthesized *parS* sequence binds to another molecule of ParB, which is then pulled to the new cell pole by the ATPase activity of ParA. Not only does PopZ help anchor *parS* of the parent chromosome to the old pole of the cell, it also helps recruit ParA to transfer the newly synthesized *parS* sequence bound to ParB to the new cell pole (Figure 7.6).

While *E. coli* lacks Par, the daughter chromosomes must still be segregated prior to cell division. After replication the resulting circular chromosomes remain interlinked or tangled, much like the links in a chain. This linkage is broken by the *structural maintenance of chromosome* complex, which is composed of a topoisomerase (IV) (⟳ Section 4.1) and MukBEF proteins. Super-resolution images show that the MukBEF proteins move to discrete locations within the nucleoid (Figure 7.3*a*) and recruit a topoisomerase to separate replicated sister chromosomes (a process called *decatenation*) prior to segregation. While the actual segregation process is still not completely understood, the chromosome "arms" appear to remain distinctly separated from one another during replication and the chromosomal loci are pushed to the cell poles in their order of replication (Figure 7.2*b*). This separation of daughter chromosomes in *E. coli* appears to be independent of specific proteins and proceeds instead by the physical action of replication and DNA accumulation.

How are extrachromosomal elements segregated between daughter cells? Although plasmids are not considered essential for cell survival under all conditions, they are replicated using the same cellular machinery as the chromosome (**Figure 7.7**). Various

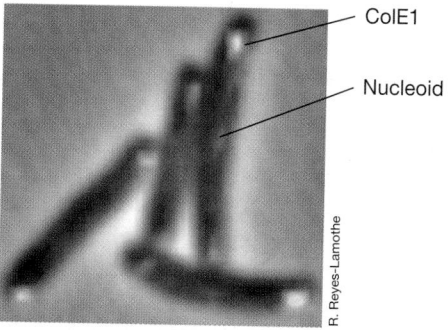

Figure 7.7 Cellular location of ColE-1 plasmid in *Escherichia coli* during replication. The plasmid (yellow) localizes to the cell poles, while the nucleoid remains in the center of the cell during DNA replication. The separate DNA molecules are visible due to fluorescently labeled DNA-binding proteins.

control mechanisms exist to ensure that a relatively constant copy number is disseminated to progeny cells. Replication of large ColE1-type plasmids occurs at the poles instead of in the center of the cell where the nucleoid exists (Figure 7.7). This location helps ensure that efficient transfer to daughter cells occurs during cell division and that stable inheritance is achieved over generations. Other mechanisms for segregating plasmids include partitioning systems similar to the Par system in *Caulobacter* (Figure 7.6).

MINIQUIZ

- How does SeqA prevent DnaA–ATP from binding to the *oriC* regions immediately after chromosome replication initiates?
- Explain how the minimum generation time for the bacterium *Escherichia coli* can be less than the time needed to replicate its chromosome.

7.3 Cell Division and Fts Proteins

Cell division is both spatially and temporally controlled to ensure that each daughter cell has a copy of the genome before the septum seals off the two cells (⟳ Figure 5.1). Here we discuss regulation of septum formation and a series of key proteins that identify the site of cell division and control the overall process.

The Divisome

Several essential proteins play roles in cell division in *Bacteria*. Collectively, these proteins are called *Fts proteins* and a key one, **FtsZ**, plays a crucial role in binary fission. FtsZ is related to tubulin, the important cell-division protein in eukaryotes (⟳ Section 2.16), and is also found in virtually all *Archaea*. Other Fts proteins are found only in *Bacteria* and not in *Archaea*, so our discussion here will be restricted to the *Bacteria*. The gram-negative *Escherichia coli* and the gram-positive *Bacillus subtilis* have been the model bacterial species for the study of cell-division events.

Fts proteins interact in the cell to form a division apparatus called the *divisome*. In rod-shaped cells, formation of the divisome begins with the attachment of molecules of FtsZ in a ring precisely around the center of the cell; this ring becomes the cell-division plane. In a cell of *E. coli*, about 10,000 FtsZ molecules polymerize to form the ring; once formed, the FtsZ ring attracts other divisome proteins, including *FtsA* and *ZipA* (**Figure 7.8**). ZipA is an anchor that connects the FtsZ ring to the cytoplasmic membrane and stabilizes it. FtsA, a protein related to actin, an important cytoskeletal protein in eukaryotes (🔗 Section 2.16), both recruits FtsZ and other divisome proteins and helps connect the ring to the cytoplasmic membrane (Figure 7.2*d*). The divisome forms about three-quarters of the way into the cell-division cycle. However, long before the divisome forms, the cell is already elongating and DNA replication has begun (see Figure 7.9).

The divisome also contains Fts proteins needed for peptidoglycan synthesis, such as FtsI (Figure 7.8). FtsI is one of several *penicillin-binding proteins* present in the cell. Penicillin-binding proteins are so named because their activities are inhibited by the antibiotic penicillin (Section 7.5). The divisome orchestrates synthesis of new cytoplasmic membrane and cell wall material, called the *division septum*, at the center of a rod-shaped cell until the cell reaches twice its original length. Following elongation, the cell divides, yielding two daughter cells (Figure 7.1).

Min Proteins and Cell Division

DNA replicates before the FtsZ ring forms (**Figure 7.9**) because the ring forms in the space between the duplicated nucleoids; before the nucleoids segregate, they effectively block formation of the FtsZ ring in a process known as nucleoid occlusion (Section 7.2). The proteins MinC, MinD, and MinE interact to help guide FtsZ to the cell midpoint. MinD forms a spiral structure on the inner surface of the cytoplasmic membrane and helps to localize MinC to the cytoplasmic membrane. The MinD spiral oscillates back and forth along the long axis of the growing cell and functions to *inhibit* cell division by preventing the FtsZ ring from forming (Figure 7.9). Simultaneously, however, MinE also oscillates from pole to pole, and as it does, it functions to sweep MinC and MinD aside. Hence, because MinC and MinD dwell longer at the poles than anywhere else during their oscillation cycle, the center of the cell will have, on average, the lowest concentration of these proteins. As a result, the cell center becomes the most permissive site for FtsZ binding and so the FtsZ ring forms there. In this unusual series of events, the Min proteins ensure that the divisome forms only at the *cell center* and not at the cell poles (Figure 7.9).

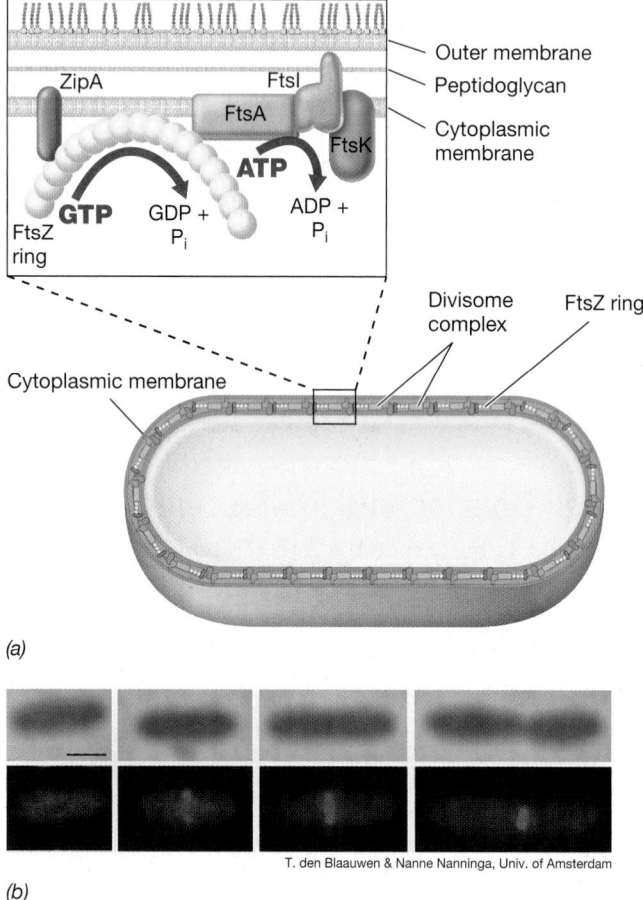

(a)

(b)

T. den Blaauwen & Nanne Nanninga, Univ. of Amsterdam

Figure 7.8 The FtsZ ring and cell division. *(a)* Cutaway view of a rod-shaped cell showing the ring of FtsZ molecules around the division plane. Blowup shows the arrangement of individual divisome proteins. ZipA is an FtsZ anchor, FtsI is a peptidoglycan biosynthesis protein, FtsK assists in chromosome separation, and FtsA is an ATPase. *(b)* Appearance and breakdown of the FtsZ ring during the cell cycle of *Escherichia coli*. Microscopy: upper row, phase-contrast; bottom row, cells stained with a specific reagent against FtsZ. Cell division events: first column, FtsZ ring not yet formed; second column, FtsZ ring appears as nucleoids start to segregate; third column, full FtsZ ring forms as cell elongates; fourth column, breakdown of the FtsZ ring and cell division. Marker bar in upper left photo, 1 μm.

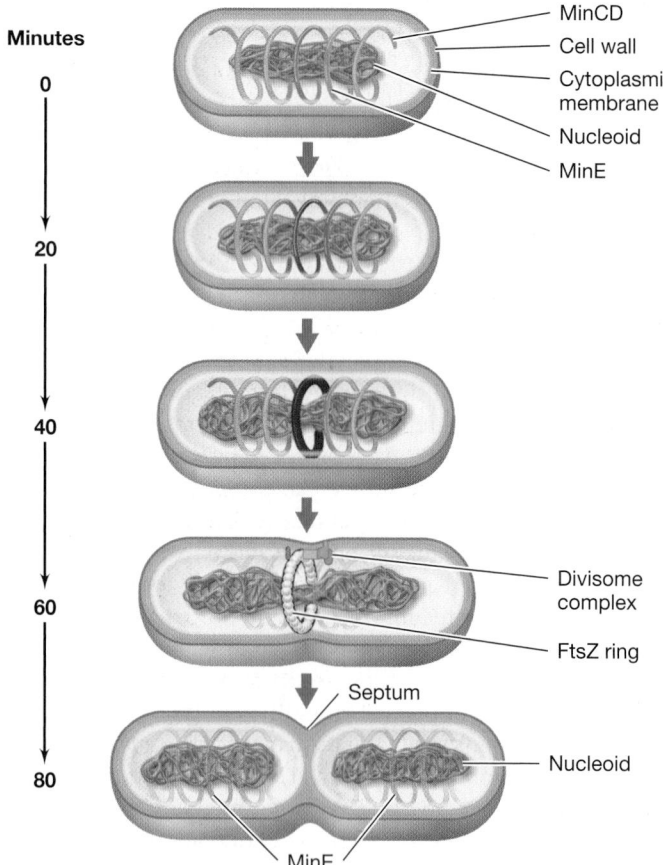

Figure 7.9 DNA replication and cell-division events. The protein MinE directs formation of the FtsZ ring and divisome complex at the cell-division plane. Shown is a schematic for cells of *Escherichia coli* growing with a doubling time of 80 min. MinC and MinD are most abundant at the cell poles during FtsZ ring formation.

As cell elongation continues and septum formation begins, the two copies of the chromosome are pulled apart, each to its own daughter cell (Figure 7.9). The Fts protein *FtsK* and several other proteins assist in this process. As the cell constricts, the FtsZ ring begins to depolymerize, triggering the inward growth of wall materials to form the septum and seal off one daughter cell from the other. The enzymatic activity of FtsZ also hydrolyzes guanosine triphosphate (GTP, an energy-rich compound) to yield the energy necessary to fuel the polymerization and depolymerization of the FtsZ ring (Figures 7.8 and 7.9).

There are significant practical reasons for understanding the details of bacterial cell division because such knowledge could lead to the development of new drugs that target specific steps in the growth of pathogenic bacteria. Like penicillin (a drug that targets bacterial cell wall synthesis), drugs that interfere with the function of specific Fts or other bacterial cell-division proteins could have broad applications in clinical medicine.

MINIQUIZ

- What is the divisome?
- How does FtsZ find the cell midpoint of a rod-shaped cell?

7.4 MreB and Cell Morphology

Just as specific proteins direct cell *division* in bacteria, other specific proteins form the cell *cytoskeleton*. The cytoskeleton is the scaffolding that directs cell *shape*, yielding the spheres, spirals, crescents, and other morphologies typical of bacterial cells (↩ Section 2.1). Interestingly, these shape-determining proteins show significant homology to key cytoskeletal proteins in eukaryotic cells (↩ Section 2.16). Like eukaryotes, bacteria also possess a dynamic and multifaceted cell cytoskeleton.

Cell Shape and MreB

The major shape-determining factor in *Bacteria* is a protein called *MreB*. MreB forms a simple cytoskeleton in *Bacteria* and in a few species of *Archaea*. MreB forms patchlike filaments around the inside of the cell, just below the cytoplasmic membrane (**Figure 7.10**). The MreB cytoskeleton presumably defines cell shape by recruiting other proteins that function in cell wall growth to group into a specific pattern. Inactivation of the gene encoding MreB in rod-shaped bacteria causes the cells to become coccus-shaped. Moreover, most naturally occurring coccoid bacteria lack the *mreB* gene and thus do not make MreB. This suggests that the "default" morphology for a bacterium is a sphere. Variations in the arrangement of MreB filaments in cells of nonspherical bacteria are likely responsible for the different common morphologies of prokaryotic cells.

How does MreB define a cell's shape? The filament structures formed by MreB (Figure 7.10*a*) are not static, but instead passively move from one side to the other within the cytoplasm of a growing cell. MreB filaments localize the synthesis of peptidoglycan (Section 7.5) at points where the rod-shaped filaments contact the cytoplasmic membrane (Figure 7.10*a*). This allows new cell wall to form at several points along the cell rather than from a single

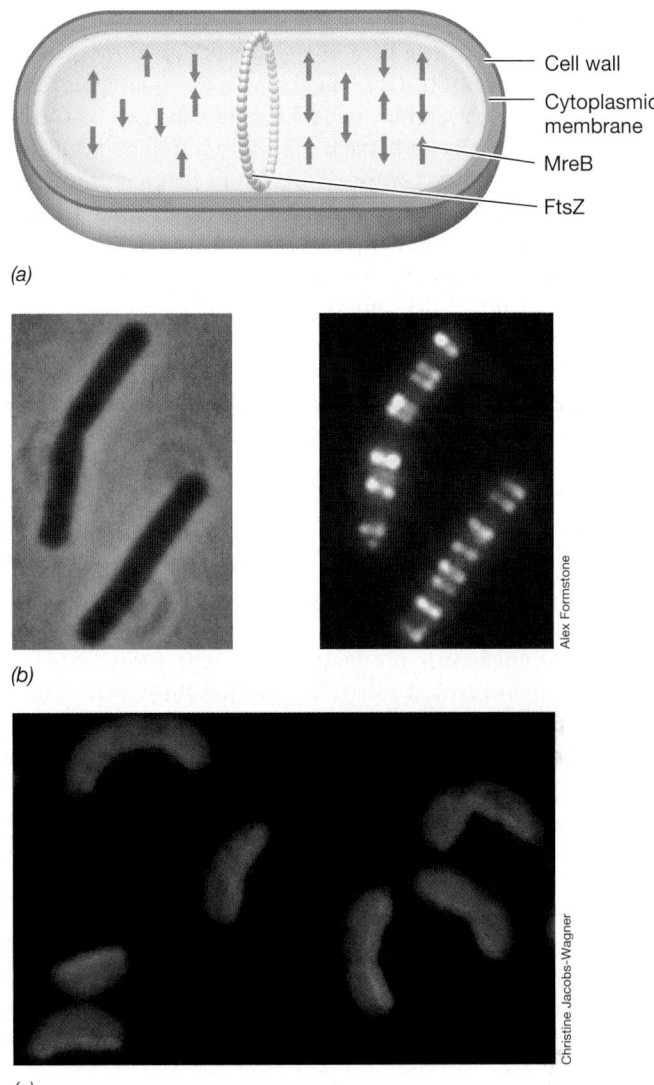

(a)

(b)

(c)

Alex Formstone

Christine Jacobs-Wagner

Figure 7.10 MreB and crescentin as determinants of cell morphology. *(a)* The cytoskeletal protein MreB is an actin analog that moves in tracks perpendicular to the long axis of a rod-shaped cell, making contact with the cytoplasmic membrane in several locations (indicated by arrows). *(b)* Photomicrographs of the same cells of *Bacillus subtilis*. Left, phase-contrast; right, fluorescence. The cells contain a substance that makes the MreB protein fluoresce, shown here as bright white. *(c)* Cells of *Caulobacter crescentus*, a naturally curved (vibrio-shaped) cell. Cells are stained to show the shape-determining protein crescentin (red), which lies along the concave surface of the cell, and with DAPI, which stains DNA and thus the entire cell (blue).

location at the FtsZ site outward, as in spherical bacteria (see Figure 7.12). By moving progressively in tracks perpendicular to the cell cylinder and initiating cell wall synthesis at multiple sites along the cell wall, MreB helps direct new wall synthesis in such a way that a rod-shaped cell elongates only along its long axis with cell wall synthesis dispersed at intervals (Figure 7.10).

Crescentin

In the vibrio-shaped bacterium *Caulobacter*, a shape-determining protein called *crescentin* is present in addition to MreB. Copies of crescentin protein organize into filaments about 10 nm wide that localize onto the concave face of the curved cell (Figure 7.3*b*).

UNIT 2

The arrangement and localization of crescentin filaments are thought to impart the characteristic curved morphology to the *Caulobacter* cell (Figure 7.10c). *Caulobacter* is an aquatic bacterium that undergoes a life cycle in which swimming cells, called *swarmers*, eventually form a stalk and attach to surfaces. Attached cells then undergo cell division to form new swarmer cells that are released to colonize new habitats (see Section 7.7 and Figure 7.16). The steps in this life cycle are highly orchestrated at the genetic level, and *Caulobacter* has been used as a model system for the study of gene expression in cellular differentiation. Although crescentin seems to be unique to *Caulobacter*, proteins similar to crescentin have been found in other curved cells, such as the bacterium that causes peptic ulcers, *Helicobacter pylori*. This suggests that these proteins may be necessary for the formation of curved cells.

Evolution of Cell Division and Cell Shape

How do the determinants of cell shape and cell division in *Bacteria* compare with those in eukaryotes? Despite not being helical in structure, MreB functions similarly to the eukaryotic protein actin. Interestingly, FtsZ is both structurally and functionally related to the eukaryotic protein tubulin. Actin forms structures called *microfilaments* that function as scaffolding in the eukaryotic cell cytoskeleton and in cell division, whereas tubulin forms *microtubules* that are important in mitosis and other processes. In addition, the shape-determining protein crescentin in *Caulobacter* is related to the keratin proteins that make up *intermediate filaments* in eukaryotic cells. Intermediate filaments form part of the eukaryotic cytoskeleton, and genes encoding similar proteins have been found in some other *Bacteria*. It thus appears that several proteins that control cell division and the cell cytoskeleton in eukaryotic cells (Section 2.16) have evolutionary roots in the *Bacteria*. However, with the exception of FtsZ, genes encoding homologs of these proteins appear to be absent from most *Archaea*.

While our focus has been on the filament proteins MreB and crescentin in bacteria that grow by synthesizing new peptidoglycan at the center of the cell (or dispersively), a diversity of morphologies exists in the bacterial world. This is especially evident in species of *Alphaproteobacteria* (**Figure 7.11**), gramnegative bacteria that show not only significant morphological diversity but also significant metabolic diversity. Hence, *Alphaproteobacteria* fill a host of diverse ecological niches in nature, from soil and water to symbiotic plant and animal associations.

Besides growing as rod-shaped cells and producing new cell material as we have seen in Figures 7.8–7.10, the *Alphaproteobacteria* group also contains species that can grow by synthesizing peptidoglycan at the poles (polar elongation), while others grow in a process known as budding. Budding results in diverse cell morphologies (Figure 7.11) and is a consequence of synthesizing peptidoglycan at various regions within the cell. However, the precise positioning of the peptidoglycan-synthesizing machinery in budding cells has yet to be discovered. Figure 7.11 also illustrates a principle we will see repeated many times later in this book: The phylogenetic position of a bacterium cannot be predicted from its morphology, and vice versa.

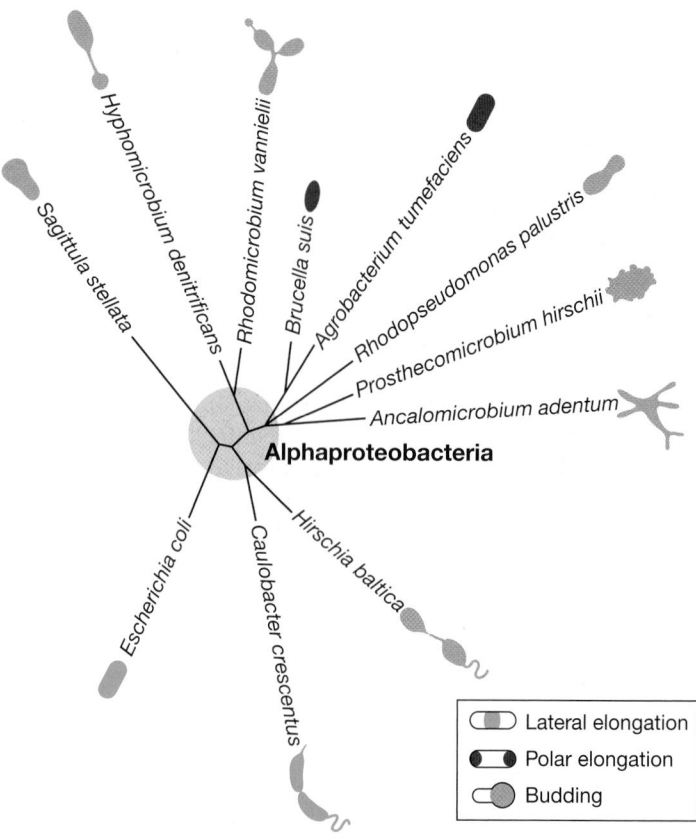

Figure 7.11 Phylogeny and morphology of select diverse bacteria. The tree was constructed from comparative sequences of *rpoC*, a gene encoding one of the subunits of RNA polymerase. Colored shading represents areas of new peptidoglycan synthesis as indicated. Note the extensive morphological diversity within the *Alphaproteobacteria*. Modified from Randich, A.M., and Brun, Y.V. 2015. *Front. Microbiol. 6:* 580.

MINIQUIZ

- How does MreB control the shape of a rod-shaped bacterium?
- What protein is thought to control the shape of cells of *Caulobacter*?
- What relationships exist between cytoskeletal proteins in *Bacteria* and those in eukaryotes?

7.5 Peptidoglycan Biosynthesis

In cells of *Bacteria* that contain peptidoglycan—and virtually all species do—preexisting peptidoglycan has to be temporarily severed to allow newly synthesized peptidoglycan to be inserted during the growth process. In cocci, new cell wall material grows out in opposite directions from the FtsZ ring (**Figure 7.12**). By contrast, as we have just seen in rod-shaped cells, new cell wall grows at several locations along the length of the cell (Figure 7.10a), and it is localized in those cells that divide by budding. However, regardless of morphology, a growing bacterial cell must both synthesize new peptidoglycan and export it outside the cytoplasmic membrane. We consider this problem here.

Insertion of New Peptidoglycan

Peptidoglycan can be thought of as a stress-bearing fabric, much like a thin sheet of rubber. Synthesis of new peptidoglycan during

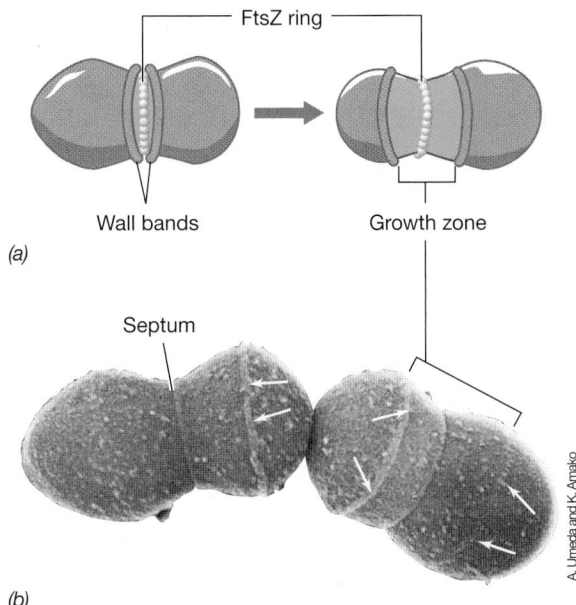

(a)

Septum

(b)

A. Umeda and K. Amako

Figure 7.12 Cell wall synthesis in gram-positive *Bacteria*. *(a)* Localization of cell wall synthesis during cell division. In cocci, cell wall synthesis (shown in green) is localized at only one point (compare with Figure 7.10a). *(b)* Scanning electron micrograph of cells of *Streptococcus hemolyticus* showing wall bands (arrows). A single cell is about 1 μm in diameter.

growth requires the controlled cutting of preexisting peptidoglycan along with the simultaneous insertion of peptidoglycan precursors. A lipid carrier molecule called *bactoprenol* plays a major role in the latter process. Bactoprenol is a hydrophobic C_{55} alcohol that is bonded to an *N*-acetylglucosamine/*N*-acetylmuramic acid/pentapeptide peptidoglycan precursor (**Figure 7.13**). Bactoprenol

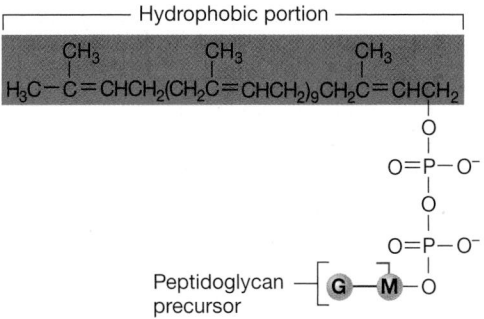

Figure 7.13 Bactoprenol (undecaprenol diphosphate). This highly hydrophobic molecule carries cell wall peptidoglycan precursors through the cytoplasmic membrane.

transports peptidoglycan precursors across the cytoplasmic membrane by rendering them sufficiently hydrophobic to pass through the membrane.

Once outside the cell, the bactoprenol complex interacts with enzymes called *transglycosylases* that insert peptidoglycan precursors into a growing point in the cell wall and catalyze glycosidic bond formation (**Figure 7.14**). Prior to this, small gaps in the existing peptidoglycan are made by enzymes called *autolysins*, enzymes that function to hydrolyze the bonds that connect *N*-acetylglucosamine and *N*-acetylmuramic acid in the peptidoglycan backbone. New cell wall material is then added across the gaps (Figure 7.14a). The junction between new and old peptidoglycan forms a ridge on the cell surface of gram-positive bacteria that can be observed as a *wall band* (Figure 7.12b). Peptidoglycan synthesis must be a highly coordinated process and peptidoglycan precursors must be readily available during autolysin activity. This is because tetrapeptide units must be spliced into existing peptidoglycan immediately

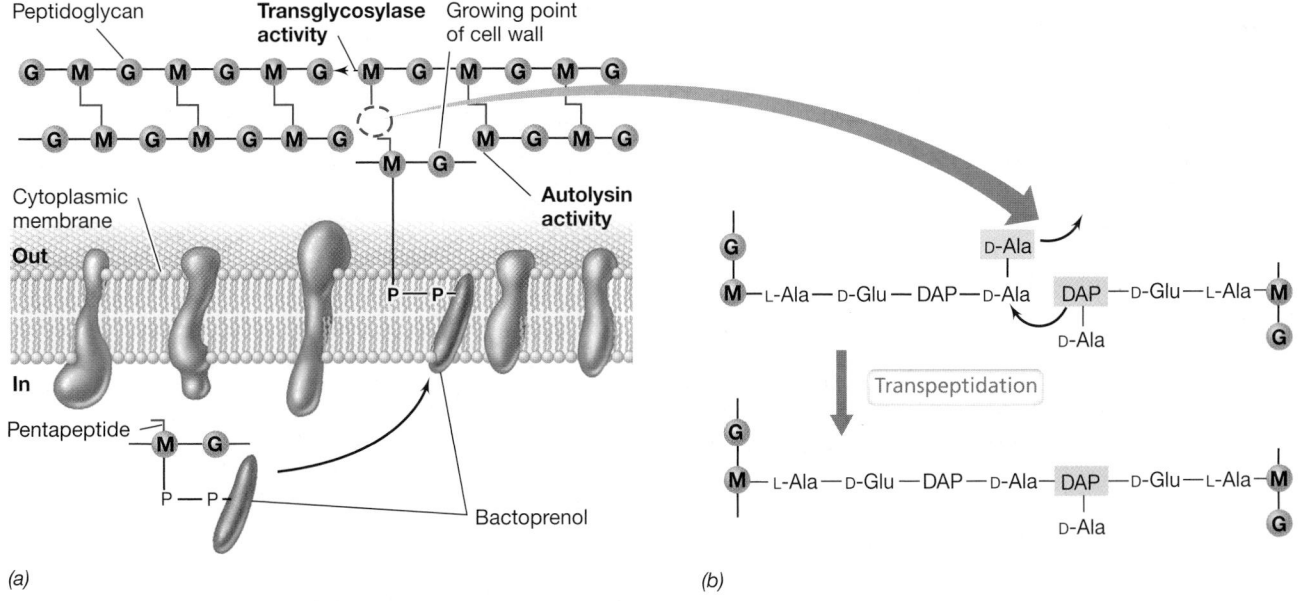

(a)   (b)

Figure 7.14 Peptidoglycan synthesis. *(a)* Transport of peptidoglycan precursors across the cytoplasmic membrane to the growing point of the cell wall. Autolysin breaks glycolytic bonds in preexisting peptidoglycan, while transglycosylase synthesizes them, linking old peptidoglycan with new. *(b)* The transpeptidation reaction that leads to the final cross-linking of two peptidoglycan chains. Penicillin inhibits this reaction.

after autolysin activity in order to prevent a breach in the peptidoglycan fabric at the splice point; a breach could cause spontaneous cell lysis, called *autolysis*.

Transpeptidation

The final step in peptidoglycan synthesis is transpeptidation. Transpeptidation forms the peptide cross-links between muramic acid residues in adjacent glycan chains (⇨ Section 2.4 and Figures 2.10 and 2.11). In gram-negative bacteria such as *Escherichia coli*, cross-links form between diaminopimelic acid (DAP) on one peptide and D-alanine on the adjacent peptide. Although there are two D-alanine residues at the end of the peptidoglycan precursor, only one remains in the final molecule, as the other is removed during transpeptidation (Figure 7.14*b*). This reaction is exergonic (energy-releasing, ⇨ Section 3.4) and supplies the energy necessary to drive transpeptidation forward. In *E. coli*, the protein FtsI (Figure 7.8*a*) functions as a transpeptidase.

Transpeptidation is medically noteworthy because it is the reaction inhibited by the antibiotic penicillin. Several penicillin-binding proteins have been identified in bacteria, including FtsI (Figure 7.8*a*). When penicillin is bound to penicillin-binding proteins, the proteins are inactivated. If transpeptidation is blocked in an otherwise growing cell, the continued activity of autolysins (Figure 7.14) so weakens the peptidoglycan that the cell eventually bursts. The absence of peptidoglycan in eukaryotes (such as humans) is the basis of the clinical efficacy of penicillin—the antibiotic destroys only growing bacteria and thus targets pathogenic bacteria that often grow rapidly in an active infection.

— MINIQUIZ —

- What is the function of bactoprenol?
- What is transpeptidation and why is it important to both the cell and to clinical medicine?

II • Regulation of Development in Model *Bacteria*

N ow that we have an understanding of the basic molecular processes that occur during bacterial growth, we move on to explore how certain bacteria differentiate to form specialized cells and how bacteria can transition from growing in suspensions (planktonic growth) to form multicellular biofilms.

Development and differentiation are largely characteristics one associates with multicellular organisms. Because most *Bacteria* and *Archaea* grow as single cells, few show differentiation. But a few important examples are known and are classical cases of differential gene expression yielding two genetically identical descendants whose functions differ. Here we discuss three well-studied examples of development and differentiation: the formation of endospores in the gram-positive soil bacterium *Bacillus*; the formation of two cell types—motile and stationary—in the gram-negative aquatic bacterium *Caulobacter*; and the formation of heterocysts in the nitrogen-fixing cyanobacterium *Anabaena*. We conclude by considering the formation of biofilms in the gram-negative and pathogenic bacteria *Pseudomonas aeruginosa* and *Vibrio cholerae*.

7.6 Regulation of Endospore Formation

Many microorganisms respond to adverse conditions by converting growing (vegetative) cells into spores (⇨ Section 2.10). Once favorable conditions return, the spore germinates and the organism returns to the vegetative state. Among *Bacteria*, the genus *Bacillus* is well known for the formation of *endospores*, that is, spores formed inside a mother cell. Prior to endospore formation, the cell divides asymmetrically. The smaller cell develops into the endospore, which is surrounded by the larger mother cell. Once development is complete, the mother cell bursts, releasing the endospore.

Endospore Formation: Sporulation Factors

Endospore formation in *Bacillus subtilis* is triggered by unfavorable conditions, such as starvation, desiccation, or growth-inhibitory temperatures. A cell of *B. subtilis* monitors its environment through a group of five sensor kinases. These enzymes function via a phosphotransfer relay system whose mechanism resembles that of a two-component regulatory system (⇨ Section 6.6) but is considerably more complex (**Figure 7.15**). The net result of multiple adverse conditions is the successive phosphorylation of several proteins called *sporulation factors*, culminating with sporulation factor Spo0A. When Spo0A is highly phosphorylated, sporulation proceeds. Spo0A controls the expression of several sporulation-specific genes. The product of one of these, SpoIIE, removes the phosphate from SpoIIAA, and this triggers the latter to remove the anti-sigma factor SpoIIAB; this liberates the sigma factor σ^F, a key step in the sporulation process.

Endospore Formation: Alternative Sigma Factors

Once a cell of *B. subtilis* commits to sporulation, endospore development is controlled by four different σ factors, two of which, σ^F and σ^G, activate genes needed inside the developing endospore (called the *forespore*) and two of which, σ^E and σ^K, activate genes needed in the mother cell surrounding the forespore (Figure 7.15*b*). The sporulation signal, transmitted via Spo0A (see earlier), activates σ^F in the forespore (σ^F is already present in the forespore but is inactive because it is bound by an anti-sigma factor, Figure 7.15*a*). Once free, σ^F is active and can bind to RNA polymerase and promote transcription of genes whose products are needed for the next stage of sporulation (Figures 7.2*a* and 7.15*b*). These include the gene encoding the sigma factor σ^G and the genes for proteins that cross into the mother cell and activate σ^E.

Active σ^E is required for transcription of yet more genes inside the mother cell, including the gene for σ^K. The sigma factors σ^G (in the forespore) and σ^K (in the mother cell) are required for transcription of genes needed even later in the sporulation process (Figure 7.15). Eventually, the many spore coats and other unique structures typical of the endospore (⇨ Section 2.10 and Table 2.2) are formed, and the mature endospore is released.

Nutrients for Endospore Formation

Nutrient limitation is the major trigger of sporulation in *Bacillus* (⇨ Section 2.10). But if this is the case, how do cells obtain

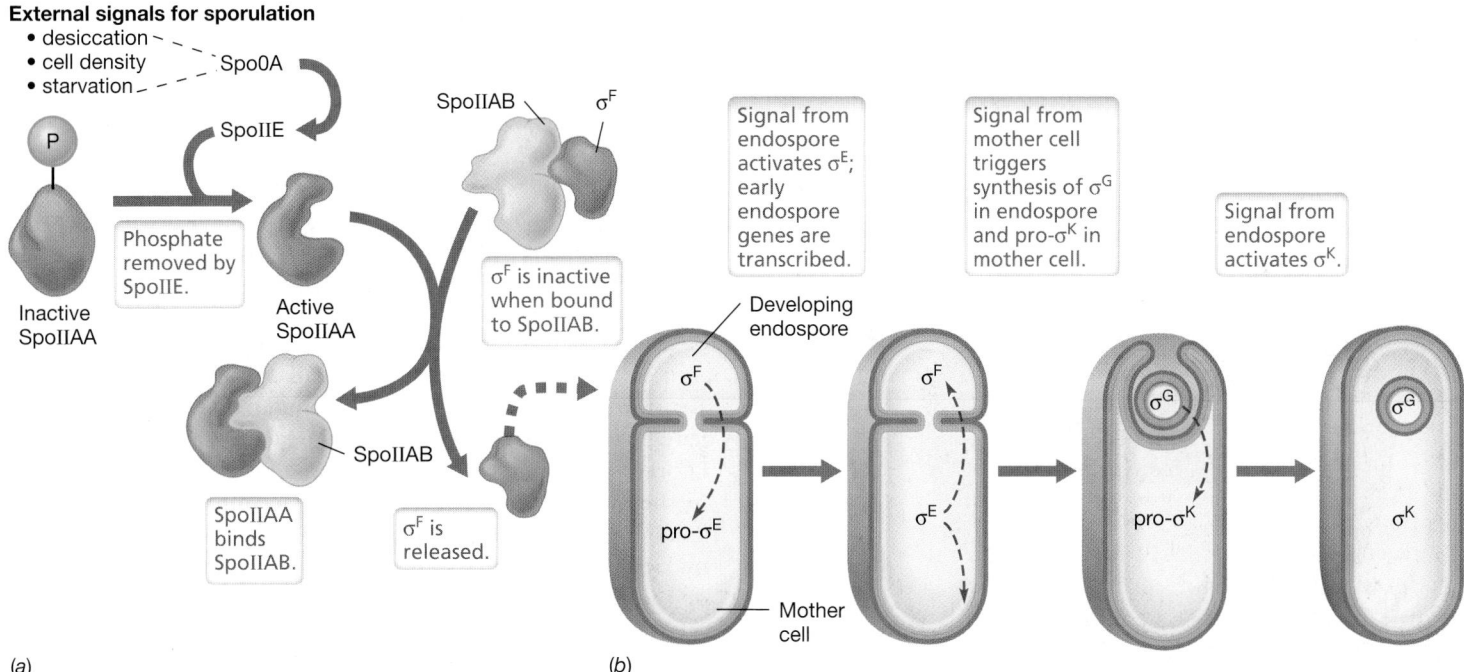

Figure 7.15 Control of endospore formation in *Bacillus*. After an external signal is received, a cascade of sigma (σ) factors controls differentiation. *(a)* Active SpoIIAA binds the anti-σ factor SpoIIAB, thus liberating the first σ factor, σ^F. *(b)* σ^F initiates a cascade of sigma factors, some of which already exist and need to be activated, others of which are not yet present and whose genes must be expressed. These σ factors then promote transcription of genes needed for endospore development.

sufficient nutrients to complete the formation of endospores? One fascinating aspect of the regulation of endospore formation is another regulatory event in which sporulating cells cannibalize cells of their own species. Those *Bacillus* cells in which Spo0A has already become activated secrete a toxic protein that lyses nearby *Bacillus* cells whose Spo0A protein has not yet become activated. This lytic protein is produced along with a second protein that functions to delay sporulation of neighboring cells. Cells committed to sporulation also make an antitoxin protein to protect themselves against the effects of their own toxic protein. When lysed, their sacrificed sister cells are used as a source of nutrients for developing endospores. Shortages of certain key nutrients, in particular phosphate, increase transcription of the gene that encodes the toxic protein.

In sporulation we thus see not only a strategy by which cellular differentiation allows the species to form cells that can withstand adverse conditions, but a strategy in which survival of a few (as opposed to all) cells of the species in a population is a priority and is facilitated by the sacrifice of other cells of the same species.

MINIQUIZ

- How are different sets of genes expressed in the developing endospore and the mother cell?
- What is an anti-sigma factor and how can its effect be overcome?

7.7 *Caulobacter* Differentiation

The gram-negative bacterium *Caulobacter* provides a second example of a cell that divides into two genetically identical daughter cells that are both structurally and functionally distinct and express different sets of genes.

Caulobacter is a genus of *Alphaproteobacteria* (Figure 7.11) that undergoes a simple life cycle and is a common bacterium in oligotrophic (nutrient-poor) aquatic environments (⇄ Section 15.20). In the *Caulobacter* life cycle, free-swimming (swarmer) cells alternate with cells that lack flagella and instead are attached to surfaces by a stalk with a holdfast at its end. The role of the swarmer cells is strictly dispersal, as swarmers cannot divide to form new swarmer cells nor can they replicate their DNA. Conversely, the role of the stalked cell is strictly reproduction. In order to divide, swarmer cells must first differentiate into stalked cells, and to swim, stalked cells must first produce swarmers; this is the nature of the *Caulobacter* life cycle (**Figure 7.16**) (⇄ Section 6.9 and Figure 6.24*b*).

Regulatory Features

The *Caulobacter* cell cycle is controlled by three major regulatory proteins whose concentrations oscillate in succession. Two of these are the transcriptional regulators GcrA and CtrA. The third is DnaA, a protein that functions both in its normal role in initiating DNA replication (Section 7.2) and also as a transcriptional regulator. Each of these regulators is active at a specific stage in the cell cycle, and each controls many other genes that are needed at that particular stage in the cycle.

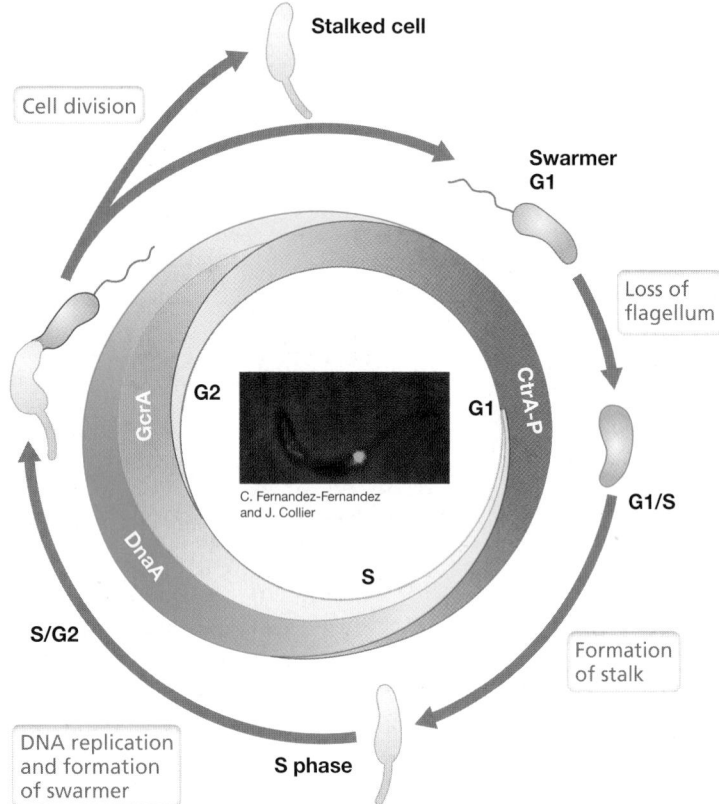

Figure 7.16 Cell cycle regulation in *Caulobacter*. Three global regulators, CtrA, DnaA, and GcrA, oscillate in levels through the cycle as shown. G1 and G2 are the two growth phases and S is the DNA synthesis phase. In G1 swarmer cells, CtrA represses initiation of DNA replication and expression of GcrA. At the G1/S transition, CtrA is degraded and DnaA levels rise. DnaA binds to the origin of replication and initiates replication (see inset photo). GcrA also rises and activates genes for cell division and DNA synthesis. At the S/G2 transition, CtrA levels begin to rise again and shut down GcrA expression. GcrA levels slowly decline in the stalked cell but are rapidly degraded in the swarmer. CtrA is degraded in the stalked cell. Inset: Fused to the green fluorescent protein as a reporter (Section 7.1), a subunit of DNA polymerase is localized in the end of the stalked *Caulobacter* cell where DNA replication occurs. Each cell of the dividing *Caulobacter* pair is about 2 μm long.

CtrA is activated by phosphorylation in emerging swarmer cells in response to external signals. Once phosphorylated, CtrA-P activates genes that encode synthesis of the flagellum and other functions specific to swarmer cells. Conversely, CtrA-P represses the synthesis of GcrA and also inhibits the initiation of DNA replication in swarmer cells by binding to and blocking the origin of replication (Figure 7.16).

As the cell cycle proceeds, CtrA is degraded by a specific protease, and as a consequence, levels of DnaA begin to rise. The absence of CtrA-P allows access to the chromosomal origin of replication, and, as in all *Bacteria*, DnaA binds to the origin and triggers the initiation of DNA replication (Section 7.2). In addition, *Caulobacter* DnaA activates several other genes needed for chromosomal replication. The level of DnaA then falls due to protease degradation, and the level of GcrA rises. The GcrA regulator promotes the elongation phase of chromosome replication, cell division, and the growth of the stalk on the immobile daughter cell. Eventually, GcrA levels fall and CtrA reappears (in the daughter cell destined to swim away) (Figure 7.16) and the cell cycle is repeated.

Caulobacter and the Eukaryotic Cell Cycle

Both external stimuli and internal factors such as nutrient and metabolite levels coordinate the events of the *Caulobacter* cell cycle (👁 Section 6.9). Since its genome has been sequenced and good genetic transfer systems are available, the *Caulobacter* cell cycle has been used as a model for studying cell developmental processes in other organisms as well. This focus is primarily due to the strict cell cycle followed by *Caulobacter*, which resembles the cell cycle of eukaryotic cells in many respects. In fact, the resemblance is so striking that terminology used to describe the eukaryotic cell cycle has been adapted to the *Caulobacter* system.

In eukaryotic cells, phase G1 of cell division is where growth and normal metabolic events occur, and in phase G2 the cell prepares for subsequent mitotic events, which occur in the M phase. Between G1 and G2 is the S phase, where DNA replication occurs. In the *Caulobacter* life cycle there is no mitosis, of course, but analogs of the G1, G2, and S phases are apparent (Figure 7.16), making this bacterium an excellent model for studying cell-division events in higher organisms.

> **MINIQUIZ**
> • Why are the levels of DnaA protein controlled during the *Caulobacter* cell cycle?
> • When do the regulators CtrA and GcrA carry out their main roles during the *Caulobacter* life cycle?

7.8 Heterocyst Formation in *Anabaena*

Cyanobacteria are oxygenic phototrophs that yield oxygen from their photosynthetic metabolism (👁 Sections 14.4 and 15.3). They are also able to perform nitrogen fixation—the reduction of N_2 to NH_3 as a nitrogen source (👁 Section 14.6). Nitrogen fixation is a highly energy-demanding process catalyzed by *nitrogenase*, an enzyme extremely sensitive to oxygen. How then is it possible for both nitrogen fixation and oxygenic photosynthesis to occur simultaneously in the same bacterium? To solve this problem, some filamentous cyanobacteria such as *Anabaena* and *Nostoc* undergo a developmental process to form specialized cells called **heterocysts** that are dedicated to nitrogen fixation.

Heterocyst Formation

Because heterocysts lack photosystem II—the pigment–protein complex that produces O_2 during oxygenic photosynthesis—they are anoxic cells. This anaerobic lifestyle provides a hospitable environment for nitrogenase and thus nitrogen fixation. Heterocysts arise from the differentiation of phototrophic vegetative cells that produce O_2, and typically develop in a regular pattern along a filament (**Figure 7.17a**). As we will see, this *patterning* separates two incompatible metabolic processes while still allowing for necessary nutrient exchanges and growth.

Heterocyst formation requires several morphological and metabolic changes that are regulated by a network of systems that sense both external conditions and intracellular signaling molecules. These changes include the formation of a thickened cell wall to prevent O_2 diffusion into the cell, inactivation of photosystem II,

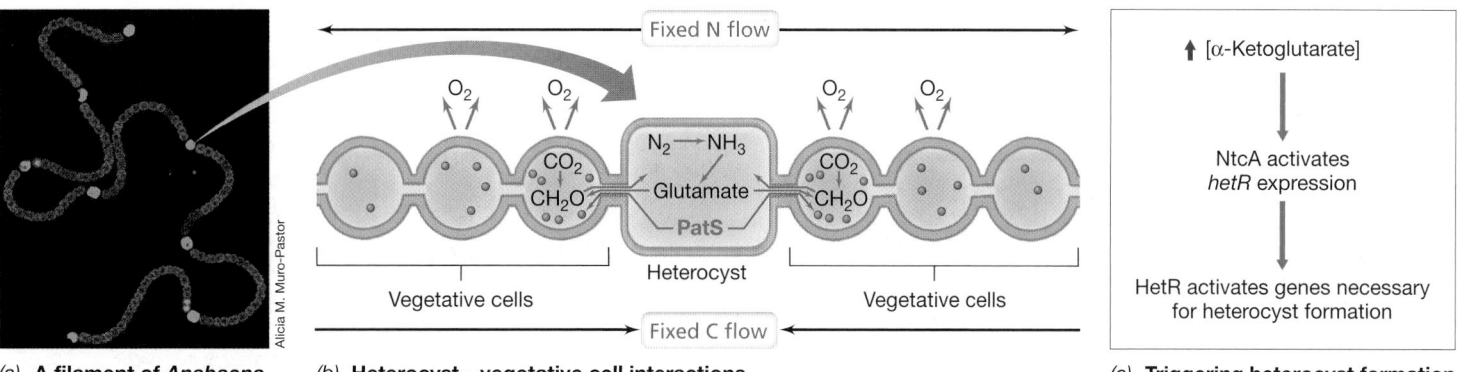

(a) **A filament of *Anabaena***   (b) **Heterocyst—vegetative cell interactions**   (c) **Triggering heterocyst formation**

Figure 7.17 Regulation of heterocyst formation.
(a) Fluorescence microscopy showing *Anabaena* filaments expressing the green fluorescent protein linked to heterocyst-specific genes; vegetative cells are red from chlorophyll *a* fluorescence. *(b)* Molecule dispersion in heterocysts. Fixed carbon from photosynthesis in the vegetative cells is transferred to the heterocyst, while fixed nitrogen produced in the heterocyst is shared with the vegetative cells. The protein PatS, which is synthesized by heterocysts, is also dispersed to neighboring vegetative cells where it inhibits expression of genes necessary for heterocyst formation. *(c)* Cascade of events in the activation of genes necessary for heterocyst formation. The cascade is initiated by an increase in α-ketoglutarate concentration.

expression of nitrogenase, and "patterning" of heterocyst differentiation along the filament (Figure 7.17*a*). Because nutrients can be exchanged between heterocysts and adjacent vegetative cells (Figure 7.17*b*), other regulatory steps are initiated to prevent nearby vegetative cells from undergoing the developmental conversion to heterocysts.

Regulation of Heterocyst Formation

The cascade of events leading to heterocyst formation is triggered by a limitation of fixed nitrogen (nitrate, ammonia, etc.); the limitation is sensed in the vegetative cell as an elevation in levels of α-ketoglutarate, the acceptor molecule for formation of the amino acid glutamate ($\Leftrightarrow$ Section 3.14). When the cell is starved for fixed nitrogen, α-ketoglutarate accumulates and activates the transcriptional global regulator NtcA. NtcA then activates transcription of the gene *hetR*, which encodes HetR, the major transcriptional regulator controlling heterocyst formation. HetR activates a cascade of genes necessary for differentiation of the heterocyst, expression of cytochrome *c* oxidases to remove traces of O_2, as well as expression of the *nif* operon for the synthesis and regulation of nitrogenase (Figure 7.17*c*).

Only specific cells along the filament form heterocysts, and the consistent pattern observed (Figure 7.17*a*) is under strict control. Intercellular connections between cells in an *Anabaena* filament allow vegetative cells to provide fixed carbon to the heterocyst as an electron donor (for N_2 reduction by nitrogenase) in exchange for some of the NH_3 produced. However, the cell connections also allow for intercellular communication by regulatory molecules. In this regard, differentiating cells produce a small peptide called PatS that diffuses away from the developing heterocyst to form a gradient along the vegetative cells in the filament (Figure 7.17*b*). PatS is thought to inhibit differentiation in vegetative cells by preventing HetR from activating genes necessary for heterocyst formation. A second regulator called PatA, a response regulator analogous to the chemotaxis response regulator CheY ($\Leftrightarrow$ Figure 6.19), also participates in heterocyst pattern development. PatA promotes the activity of HetR, decreases the activity of PatS, and may also participate in cell division.

While other regulatory links in heterocyst formation remain active areas of study, the differentiation of vegetative cells to heterocysts in heterocystous cyanobacteria is a unique example of multicellular patterning and intercellular communication in prokaryotic cells.

MINIQUIZ

- Vegetative cells produce oxygen, while heterocysts do not. Why?
- What is the major transcriptional regulator that controls heterocyst formation?

7.9 Biofilm Formation

In Chapter 5 we briefly discussed the basic characteristic of bacterial biofilms—attached polysaccharide matrices containing embedded bacterial cells—along with their environmental and medical significance. In this section we focus on the molecular mechanisms that control biofilm development. Biofilm formation is a diverse process that is controlled in different ways. We focus here on two well-studied models of biofilm formation in gram-negative bacteria, those of *Pseudomonas aeruginosa* and *Vibrio cholerae*.

Steps to Biofilm Formation

Biofilm formation can be considered a type of developmental cycle that has four basic stages: (1) attachment, (2) colonization, (3) development, and (4) dispersal. Random collision of cells with a surface accounts for the initial cell attachment. Cell attachment is facilitated by structures such as flagella and pili or by proteins on the cell surface. Attachment of a cell to a surface (**Figure 7.18**) is a signal for the expression of biofilm-specific genes. The latter include genes encoding proteins that produce intercellular signaling molecules and extracellular polysaccharides that initiate matrix formation. Once committed to biofilm formation, a previously suspended (planktonic) cell typically loses its flagella and becomes nonmotile. However, biofilms are not static entities

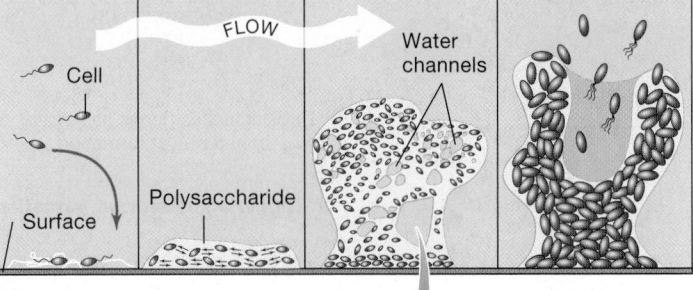

Attachment (adhesion of a few motile cells to a suitable solid surface)	Colonization (intercellular communication, growth, and polysaccharide formation)	Development (more growth and polysaccharide)	Active Dispersal (triggered by environmental factors such as nutrient availability)

(a)

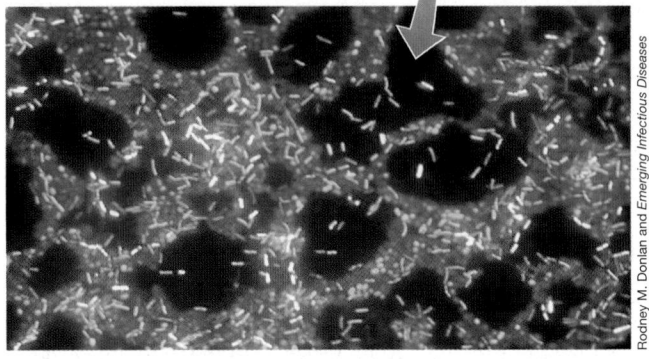

(b)

Figure 7.18 Biofilm formation. *(a)* Biofilms begin with the attachment of a few cells that then grow and communicate with other cells. The matrix is formed and becomes more extensive as the biofilm grows, eventually releasing cells. *(b)* Photomicrograph of a DAPI-stained biofilm that developed on a stainless steel pipe. Note the water channels.

and cells can be released from the biofilm through an active process of dispersal.

Several signals guide bacteria in transitioning from planktonic growth to life in a semisolid matrix. The actual switch from planktonic to biofilm growth in many bacteria is triggered by the cellular accumulation of the regulatory nucleotide *cyclic di-guanosine monophosphate* (c-di-GMP) (**Figure 7.19**). Although various nucleotides play important regulatory roles in all domains of life (⟳ Section 6.4), c-di-GMP is widely distributed only in *Bacteria*. The synthesis or degradation of c-di-GMP depends on both environmental and cellular cues, and its synthesis triggers a variety of physiological events. For example, c-di-GMP binds to proteins that reduce the activity of the flagellar motor, regulates cell surface proteins required for attachment, and mediates the biosynthesis of extracellular matrix polysaccharides of the biofilm.

Pseudomonas aeruginosa and Biofilms

Pseudomonas aeruginosa can form a tenacious biofilm (**Figure 7.20**) containing specific polysaccharides that subsequently increase its pathogenicity and prevent the penetration of antibiotics. *P. aeruginosa* is a classic opportunistic pathogen and from its primary reservoir in soil can infect the blood, lungs, urinary tract, ears, skin, and other

c-di-GMP

Figure 7.19 Molecular structure of the second messenger cyclic di-guanosine monophosphate. This is used as an intracellular signaling molecule by many bacteria to control specific physiological processes.

tissues of humans. The major symptoms of the genetic disease cystic fibrosis are caused by thick biofilms of *P. aeruginosa* that develop in the lungs, and the bacterium is a significant nosocomial (hospital-acquired) pathogen.

Besides the intracellular activities triggered by c-di-GMP, intercellular communication by quorum sensing (⟳ Section 6.8) is critical for the development and maintenance of *P. aeruginosa* biofilms. As *acyl homoserine lactones* (AHLs) accumulate, they signal to adjacent *P. aeruginosa* cells that the population of this species is enlarging. The production of AHL also triggers expression of a subset of the

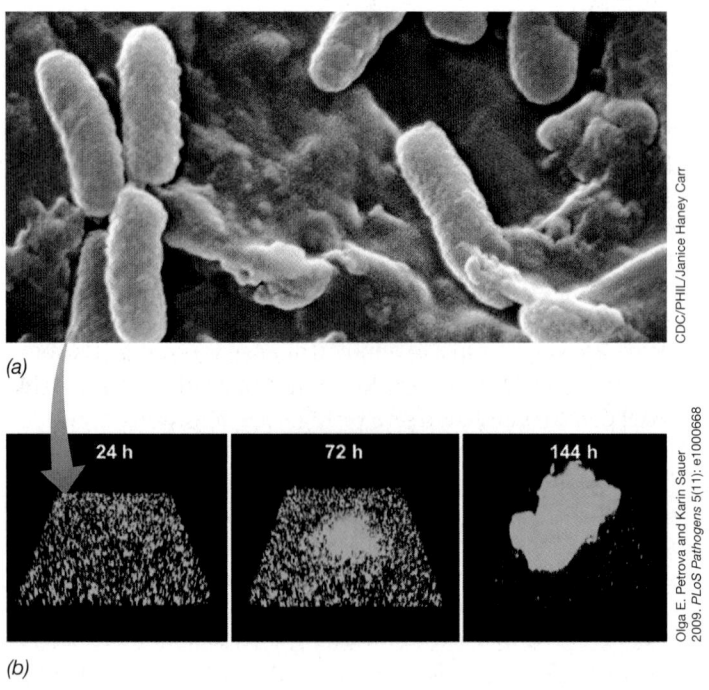

(a)

(b)

Figure 7.20 Biofilm formation in *Pseudomonas*. *(a)* Biofilms begin with the attachment of a few cells that then grow and communicate with other cells. The matrix is formed and becomes more extensive as the biofilm grows, eventually releasing cells. *(b)* Confocal scanning laser microscopy showing the progression of *P. aeruginosa* biofilm formation over a 144-h period. Cells are stained with the LIVE/DEAD viability stain, which stains live cells green. Each rectangular pattern of cells is about 0.2 mm wide. The mature biofilm is about 0.1 mm wide and 60 µm high. Data adapted from Petrova, O.E., and K. Sauer. 2009. A novel signaling network essential for regulating *Pseudomonas aeruginosa* biofilm development. *PLoS Pathogens* 5(11): e1000668.

genes necessary for biofilm formation including those that increase extracellular polysaccharide and c-di-GMP synthesis.

Elevated c-di-GMP levels initiate the production of extracellular polysaccharide, including Pel, which functions as both a primary scaffold for the microbial community and a mechanism for resisting antibiotics. c-di-GMP also leads to decreased flagellar function. Over time in nutrient-rich conditions, *P. aeruginosa* cells can develop mushroom-shaped microcolonies that can be more than 0.1 mm high and contain millions of cells. The final architecture of the biofilm is determined by multiple factors in addition to signaling molecules including nutritional factors and local flow environment.

In the biofilm-forming *Pseudomonas* species *P. fluorescens*, increases in c-di-GMP also promote biofilm formation. However, the biofilm machinery regulated by c-di-GMP in *P. fluorescens* is very different from that of *P. aeruginosa*. In *P. fluorescens*, changes in c-di-GMP levels affect secretion and cell surface localization of a large adhesion protein called LapA that helps stick the cell to the surface. For example, in response to low extracellular phosphate, *P. fluorescens* cells maintain a low c-di-GMP level that prevents localization of LapA to the outer membrane, thereby disabling the attachment mechanism required to initiate biofilm formation. If phosphate levels continue to fall within the biofilm, the associated reduction in c-di-GMP levels also results in the activation of a protease that cleaves LapA; this releases already attached cells and promotes their dispersal to explore for available nutrients in habitats nearby.

Vibrio cholerae and Biofilms

Vibrio cholerae, the causative agent of cholera (⊘ Section 32.3), also uses both inter- and intracellular signaling to control biofilm formation. While signaling by c-di-GMP activates the expression of genes for biofilm formation, quorum sensing acts in an opposite manner from that in *P. aeruginosa*. As *V. cholerae* cell density increases, the level of quorum signaling molecules also increases. The accumulation of these molecules triggers a regulatory cascade that ultimately results in the *repression* of biofilm formation genes and the *activation* of flagellar and virulence genes. Thus, in contrast to *P. aeruginosa*, biofilm formation in *V. cholerae* is triggered by low cell densities and repressed by high cell densities.

The ecological significance of this phenomenon is that biofilm formation is more likely to occur when *V. cholerae* is found in its natural marine environment where nutrients are typically scarce, thus leading to smaller populations than those found inside the close quarters of an intestinal cell where nutrients are more plentiful. Biofilm formation allows the *V. cholerae* cell to attach to marine surfaces such as plankton, crustaceans, and sediments for greater access to nutrients and protection from perturbations. The final step in biofilm formation, dispersal, ultimately aids in the transmission of *V. cholerae* to new host cells.

Because the biofilm lifestyle of many pathogenic bacteria often leads to either persister cells (Section 7.11) or antibiotic-resistant cells (or both), and because biofilms themselves function as barriers to the penetration of antimicrobial drugs, there is great medical interest in understanding the molecular biology of biofilm development. We continue our discussion of the ecology of biofilms and strategies to control their formation in Chapter 20.

III • Antibiotics and Microbial Growth

Throughout this chapter we have discussed how cells coordinate their molecular biology to optimize steps in microbial growth. But what about mechanisms that short-circuit these synchronized processes? Here we focus on how antibiotics target growth processes and the bacterial responses that can lead to resistance and persistence. In Chapter 28 we will focus on the clinical significance of antibiotics and consider their spectrum of activity and spread of resistance.

7.10 Antibiotic Targets and Antibiotic Resistance

Antibiotics are antimicrobial agents naturally produced by microorganisms, primarily certain bacteria and fungi. These agents are characterized by their ability to either kill or inhibit the growth of bacteria, and all target molecular processes essential to growth and survival. In this section we consider how antibiotics work and some key resistance mechanisms that bacteria have evolved to counter their effects.

Antibiotics That Target Major Molecular Processes

Because all steps within the central dogma (which describes the flow of genetic information; Chapter 4) are essential for growth, many antibiotics specifically target enzymes that catalyze DNA replication, RNA synthesis, and translation (**Figure 7.21a**). Quinolones such as ciprofloxacin target DNA gyrase in gram-negative bacteria and topoisomerase IV in gram-positive bacteria. Thus quinolones lead to cell death by interfering with DNA unwinding and replication (⊘ Section 4.1). Likewise, when transcription is inhibited in a growing cell, mRNA cannot be made and new protein synthesis is interrupted. The antibiotics rifampin and actinomycin prevent RNA synthesis by either blocking the RNA polymerase active site (rifampin) or blocking RNA elongation by binding to the major groove in DNA (⊘ Section 4.5).

Many antibiotics inhibit bacterial growth by targeting some aspect of protein synthesis (Figure 7.21a). Recall that ribosomes in *Bacteria* are 70S structures while eukaryotic ribosomes are 80S. The antibiotic puromycin contains a region that mimics the 3′ end of a tRNA, and this structural mimicry results in specific binding of the antibiotic to the A site in the 70S ribosome (⊘ Section 4.10); this induces chain termination and inhibits protein synthesis. Aminoglycoside antibiotics such as streptomycin specifically target the 16S rRNA of the 30S ribosome and result in the ribosome misreading mRNAs, thus leading to error-filled proteins that accumulate in the cell and ultimately inhibit growth.

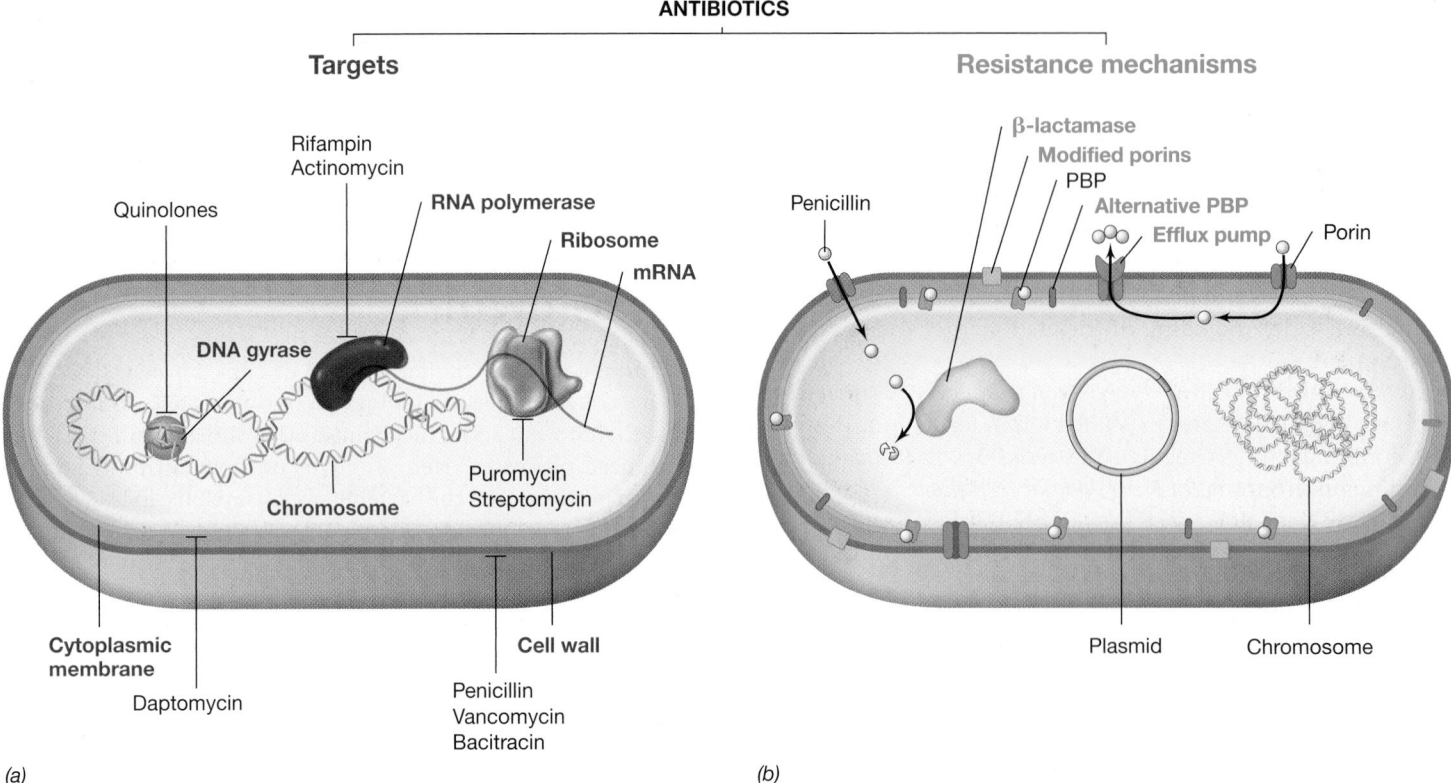

Figure 7.21 Antibiotics and antibiotic resistance. *(a)* Growth targets of select antibiotics. Antibiotic targets are in red bold. *(b)* Mechanisms of antibiotic resistance. Penicillin (β-lactam) is used to illustrate the mechanisms of modified target (modified porin), drug inactivation (β-lactamase), efflux pump, and metabolic bypass (alternative penicillin-binding protein). Genes for efflux pumps and β-lactamase enzymes can be plasmid or chromosomally encoded, while the alternative penicillin-binding protein is encoded only on the chromosome. PBP: Penicillin-binding protein. Mechanisms of antibiotic resistance are in green bold.

Antibiotics That Target the Cell Membrane and Wall

Other common antibiotic targets include the cell membrane, the cell wall, and specific metabolic processes (Figure 7.21*a*). Daptomycin is a lipopeptide produced by *Streptomyces* that specifically binds to phosphatidylglycerol residues of the bacterial cytoplasmic membrane (⇔ Section 2.3); this leads to pore formation and depolarization of the membrane, ultimately resulting in cell death. Some antibiotics target the gram-negative cell outer membrane. For example, polymyxins are cyclic peptides whose long hydrophobic tails specifically target the LPS layer and ultimately disrupt membrane structure, causing leakage and cell death.

Several antibiotics target the synthesis of peptidoglycan in bacteria such as the β-lactams penicillin, cephalosporin, and their derivatives. These antibiotics inhibit growth by interfering with proteins that catalyze transpeptidation (penicillin-binding proteins, Section 7.5); transpeptidation is the formation of cross-links between muramic acid residues that contribute to the structural strength of peptidoglycan (⇔ Section 2.4). Similarly, the antibiotic vancomycin inhibits peptidoglycan synthesis in gram-positive bacteria by binding to the pentapeptide of peptidoglycan precursors and preventing the formation of peptide interbridges by transpeptidases. The topical antibiotic bacitracin prevents peptidoglycan synthesis by binding to the peptidoglycan precursor transport system (bactoprenol, Section 7.5) and preventing new

peptidoglycan precursors from reaching the site of peptidoglycan synthesis. As autolysins continue to introduce small gaps in the existing peptidoglycan (Figure 7.14*a*), a shortage of precursors to patch the gaps weakens the cell wall and leads to cell lysis.

Many other antibiotics and related antimicrobials are known, some of which target other aspects of a bacterium's biology, such as particular metabolic reactions. But the bulk of clinically useful antibiotics strike one of the targets shown in Figure 7.21*a*. We finish this section with a look at how bacteria have countered the activity of antibiotics by evolving major mechanisms of antibiotic resistance.

Antibiotic Resistance: Spontaneous Mutations and Antibiotic Modification

If bacteria are to survive the onslaught of antibiotics produced by other microbes (or from their own antibiotics if they are an antibiotic-producing organism) they require resistance mechanisms of one sort or another. Resistance mechanisms are genetically encoded and fall into four classes: (1) modification of the drug target, (2) enzymatic inactivation, (3) removal from the cell via efflux pumps, and (4) metabolic bypasses (Figure 7.21*b*).

Random chromosomal mutations can lead to antibiotic resistance. For example, spontaneous mutants of *Escherichia coli* or other bacteria resistant to the antibiotic rifampin—which inhibits the activity of RNA polymerase—can be obtained by simply exposing a large cell population to the antibiotic. Under such conditions, spontaneous mutants that produce an altered RNA polymerase unaffected by rifampin are strongly selected for.

Resistance genes can also exist on a variety of mobile genetic elements (Chapter 4), and such genes can be readily transmitted between bacteria of the same or different species by horizontal gene flow (Chapter 11). Unlike a spontaneous mutation that affects the target of the antibiotic, many of these mobile resistance genes, especially ones transmitted by R plasmids, encode enzymes that inactivate the antibiotic by altering its structure, either through chemical modification or actual cleavage. As examples, β-lactamase binds to β-lactam-type antibiotics and cleaves a key ring structure in the molecule, and an acetylating enzyme adds acetyl groups to free hydroxy groups of chloramphenicol; once cleaved or chemically modified, the drugs can no longer bind to their target and microbial growth can continue.

Antibiotic Resistance: Efflux Pumps and Metabolic Bypasses

Another effective resistance strategy is to pump out antibiotics that have entered the cell. *Efflux pumps* are ubiquitous in gram-negative bacteria and work by transporting various molecules, including antibiotics, out of the cell (Figure 7.21*b*). Efflux lowers the intracellular concentration of an antibiotic and thus allows the cell to survive at higher external concentrations. Many efflux pumps act promiscuously and by transporting different classes of antibiotics outside of the cell, they contribute to the problem of multidrug resistance. The AcrAB-TolC efflux system of *E.coli* is one of the best-characterized efflux pumps and can pump out several antibiotics including rifampicin, chloramphenicol, and fluoroquinolones.

Growth as a biofilm (Section 7.9) leads to increased antibiotic resistance, a characteristic that makes infections caused by biofilm-forming bacteria difficult to treat. Although the exopolysaccharide matrix of a biofilm decreases the permeability of antibiotics, efflux pumps also play a major role. For example, genes encoding the AcrAB-TolC efflux pump in *E. coli* are upregulated when cells enter a biofilm growth mode. *Pseudomonas aeruginosa*, a classic biofilm former, encodes several multidrug efflux pumps that are also more active when cells grow in an attached state.

A final form of antibiotic resistance can occur when the target of the antibiotic is no longer essential to the cell's metabolism or survival. A classic example of this is *methicillin-resistant Staphylococcus aureus* (MRSA), the causative agent of a variety of serious, even life-threatening, infections. Methicillin is a β-lactam antibiotic that, like other penicillins, targets the activity of penicillin-binding proteins. However, unlike many penicillin derivatives, methicillin is resistant to β-lactamase cleavage. Although many strains of *S. aureus* are killed by methicillin, MRSA strains contain a 20- to 60-kilobase DNA chromosomal (or genomic) island (⇄ Section 9.7) called the *Staphylococcus chromosomal cassette for methicillin resistance* (*SCCmec*) that encodes an *alternative* penicillin-binding protein called MecA that is not recognized by methicillin or other β-lactams (Figure 7.21*b*).

Interestingly, MRSA strains synthesize MecA *only* in the presence of methicillin or other β-lactams. This regulation is due to the presence of the repressor protein MecI and the β-lactam sensor MecR1. In the absence of a β-lactam-type antibiotic, MecI binds to the operator of *mecA* and represses transcription. If the cell is treated with methicillin or other β-lactams, it triggers the MecR1 protease to degrade the MecI repressor. This induces expression of *mecA* when methicillin is present and thus the production of the alternative penicillin-binding protein allows MRSA strains to synthesize peptidoglycan in the presence of the antibiotic. The chromosomal island also possesses a transposable element (⇄ Section 4.2) that encodes resistance to the antibiotics erythromycin and spectinomycin. The island is thus likely subject to horizontal transfer to confer multidrug resistance.

MINIQUIZ

- Describe two targets of antibiotics and discuss why the drugs are effective.
- Some antibiotics target peptidoglycan synthesis. What is a molecular growth target of an antibiotic that inhibits peptidoglycan synthesis?
- Why are efflux pumps capable of conferring multidrug resistance?

7.11 Persistence and Dormancy

Besides being sensitive or resistant to antibiotics, a third growth response is possible—**persistence**. Persistence occurs when a population of antibiotic-sensitive bacteria produces rare cells that *transiently* become tolerant to multiple antibiotics (**Figure 7.22*a***). These *persisters*, as they are called, are genetically identical to their antibiotic-susceptible siblings and thus persistence is not inheritable or conferred by genetic transfer. Instead, persisters, which form at a rate of 10^{-6} to 10^{-4} in exponentially growing cultures, derive from a state of *dormancy* in which the cells are viable but do not grow.

Because antibiotics target active processes, the dormant state prevents the antibiotic from killing the cell. When antibiotic treatment is stopped, the dormant cells emerge from dormancy and grow. Persistence is believed to be the cause of recurring infections of *Mycobacterium tuberculosis* in those with chronic tuberculosis (⇄ Section 30.4) and of *Pseudomonas aeruginosa* in those suffering from the genetic disease cystic fibrosis.

How does a subpopulation of cells go dormant while genetically identical daughter cells do not? For a cell to become a persister and be able to later reverse the process, coordinated molecular events that affect cell growth must take place. The keys to this unusual phenomenon are chromosomally encoded toxin–antitoxin modules, the stringent response (⇄ Section 6.9), and phenotypic heterogeneity.

Toxin–Antitoxin Modules

Toxin–antitoxin (TA) modules are genetic loci that encode two components: a toxin whose production inhibits cell growth and an antitoxin that counteracts the activity of the toxin. TA modules are found in almost all *Bacteria* and in many *Archaea*. The identification of over 30 TA systems in *Escherichia coli* and

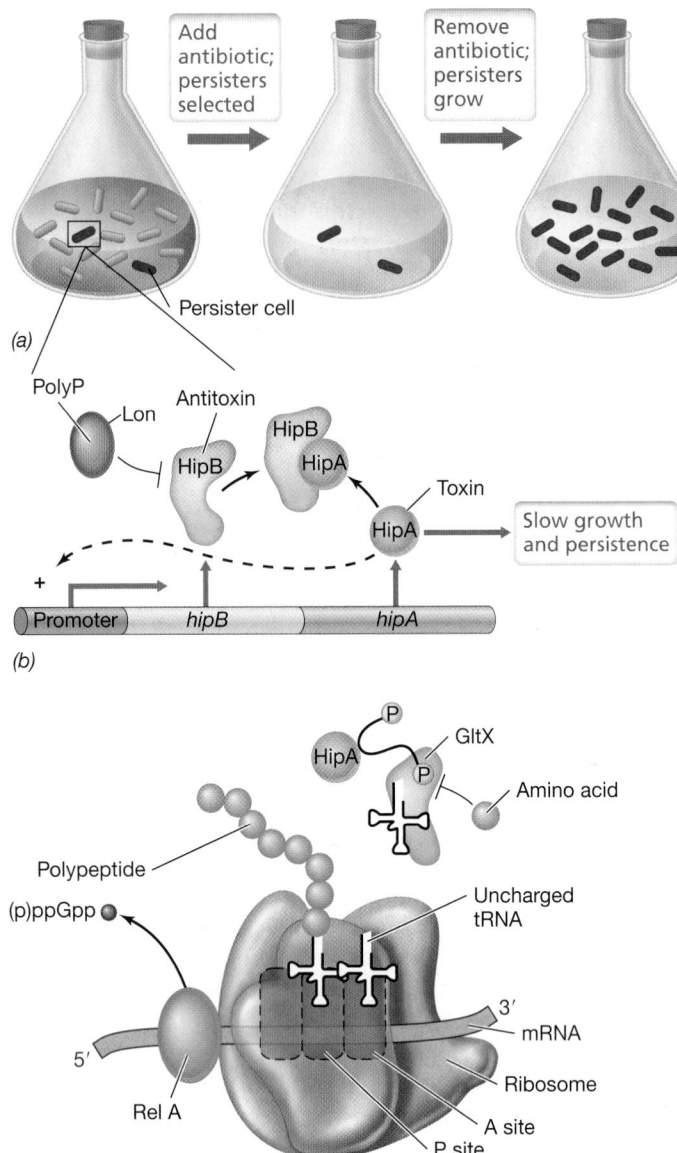

(a)

(b)

(c)

Figure 7.22 Persistence and the HipAB toxin–antitoxin module. *(a)* Antibiotic selection of persister cells and recovery from dormancy. *(b)* HipAB toxin–antitoxin module. During normal conditions, HipB sequesters the toxin HipA. The presence of Lon-PolyP leads to inactivation of the HipB antitoxin and free HipA toxin. HipA leads to slow growth and persistence and positively regulates the *hipAB* operon. *(c)* Mechanism of HipA toxin. HipA phosphorylates GltX, preventing the tRNA synthetase from charging amino acids. This leads to an uncharged tRNA entering the A site of the ribosome and the stimulation of RelA to produce (p)ppGpp.

60 TA systems in *M. tuberculosis* suggests that the TA modules play a role not only in normal physiology but also in pathogenicity. It is thought that toxin activity promotes cellular adaptation to ever-changing environments by slowing down cell growth to help ensure survival during stressful conditions.

The *hipAB* genes encode a TA module that has been shown to trigger persistence in *E. coli* (Figure 7.22*b*). In this module, HipA is a toxin that inhibits translation and HipB is an antitoxin that is susceptible to a protease called Lon. Under normal cellular conditions, the HipA toxin and HipB antitoxin form a stable complex that prevents

HipA from exerting its toxicity. However, if Lon is activated by its signal molecule polyphosphate (PolyP), the HipB antitoxin is degraded. Without the HipB antitoxin present to neutralize the HipA toxin, translation is inhibited and cell growth is arrested (Figure 7.22*b*).

How does the HipA toxin inhibit translation? HipA acts as a kinase (phosphorylating protein) targeting glutamyl-tRNA synthetase (GltX). If HipA is free to phosphorylate GltX, GltX can no longer charge its respective tRNAs with glutamine. This leads to uncharged tRNAs entering the A site of the ribosome (⇨ Section 4.10) and ultimately to stalling of the ribosome and activation of RelA (Figure 7.22*c*). Stalling, in turn, inhibits translation and thus protein synthesis. Free HipA also binds upstream of the *hipAB* operon, further activating transcription and thus increasing the concentration of cellular HipA (Figure 7.22*b*).

Steps to Dormancy

As we have seen, free HipA induces ribosome stalling. This stalling not only inhibits translation, but it also leads to the most important criteria for the development of persistence—production of the alarmone (p)ppGpp (guanosine tetraphosphate and guanosine pentaphosphate) by RelA and induction of the *stringent response pathway* (Figure 7.22*c*).

Recall that the stringent response leads to decreased rRNA and tRNA synthesis, and thus protein synthesis is inhibited (⇨ Section 6.9). The pathway also leads to decreased DNA replication and cell division. Thus, those cells in a population that have triggered the stringent response no longer actively grow and enter a state of dormancy. Although this cascade of events only occurs in a small subpopulation of cells, antibiotic treatment strongly selects and enriches for multidrug-tolerant persister cells. Once the antibiotic exposure has ended, the persisters can exit the stringent response pathway and resume producing antitoxin to neutralize the effect of the toxin. When this happens, protein synthesis returns to normal levels, allowing cells to grow.

Why does this cascade of events only occur in a *subpopulation* of cells without any sort of genetic direction? HipA-induced persistence occurs in cells that randomly produce higher amounts of the signal molecule PolyP, a phenomenon called *phenotypic heterogeneity*. While it is often assumed that cells within a clonal population are homogeneous, cell-to-cell differences in gene expression and in metabolite and protein content do occur. These differences can be the result of RNA polymerase producing a subset of mRNAs through random interactions with DNA or of unequal distribution of macromolecules between daughter cells during cell division. These cell-to-cell variations can have many outcomes, but a major one for clinical medicine is the persistence of difficult-to-eradicate pathogenic bacteria whose dormant state can falsely signal a cure, only to be followed by reinfection by the same pathogen at a later date.

MINIQUIZ

- Is persistence a heritable trait?
- What prevents the toxin component of TA modules from killing the cell under normal conditions?
- What occurs in the cell that frees HipA toxin?

Chapter Review

I • Bacterial Cell Division

7.1 Macromolecules such as proteins and nucleic acids can be visualized in living cells using fluorescent tagging and microscopy techniques. Super-resolution microscopy, the most advanced form of microscopy, can even resolve two individual molecules, generate 3D images, and track the movement of molecules throughout the cell.

> **Q** **What type of probes must be used for super-resolution microscopy and why?**

7.2 Before cell division occurs in prokaryotic cells, two complete copies of the genome must be present. While multiple replication forks can exist to decrease generation times, the process must be regulated. The nucleoid must also segregate to the poles of the cell to allow the septum to form.

> **Q** **What is the role of PopZ in segregating the *Caulobacter* daughter chromosomes?**

7.3 Cell division and chromosome replication are coordinately regulated, and the Fts proteins are keys to these processes. With the help of MinE, FtsZ defines the cell-division plane and helps assemble the divisome, the protein complex that orchestrates cell division.

> **Q** **What is the role of the penicillin-binding protein FtsI in cell division?**

7.4 MreB helps define cell shape, and in rod-shaped cells, MreB forms patchlike filaments that direct cell wall synthesis along the long axis of the cell. The protein crescentin plays an analogous role in *Caulobacter*, leading to formation of a curved cell. The eukaryotic proteins actin and tubulin involved in shape and cell division have prokaryotic counterparts.

> **Q** **What morphology do cells have that lack MreB or crescentin?**

7.5 During bacterial growth, new cell peptidoglycan is synthesized by the insertion of peptidoglycan precursors into the preexisting peptidoglycan fabric. Bactoprenol facilitates transport of these units through the cytoplasmic membrane. Transpeptidation completes the process of cell wall synthesis by cross-linking adjacent ribbons of peptidoglycan at muramic acid residues.

> **Q** **What are autolysins and why are they necessary?**

II • Regulation of Development in Model *Bacteria*

7.6 Sporulation in *Bacillus* during adverse conditions is triggered via a complex phosphotransfer relay system that monitors multiple aspects of the environment. The sporulation factor Spo0A then sets in motion a cascade of regulatory responses under the control of several alternative sigma factors.

> **Q** **What is a common trigger for sporulation in *Bacillus*?**

7.7 Differentiation in *Caulobacter* consists of the alternation between motile cells and those that are attached to surfaces. Three major regulatory proteins—CtrA, GcrA, and DnaA—act in succession to control the three phases of the cell cycle. Each in turn controls many other genes needed at specific times in the cell cycle.

> **Q** **How is the *Caulobacter* life cycle similar to the eukaryotic cell cycle?**

7.8 Heterocyst formation requires expression of the major regulatory protein HetR in the protoheterocysts. However, the protein must be inactivated in vegetative cells by diffusion of the PatS peptide along the filament.

> **Q** **What is meant by "patterning" during heterocyst formation?**

7.9 During biofilm formation, cells make contact with a surface using cell structures such as flagella and pili. Once cells are attached to a surface, both intra- and intercellular signaling molecules lead to the production of an extracellular matrix and colonization. Biofilms are not static entities, and the final step in their development is dispersal and release of cells.

> **Q** **What type of genes does c-di-GMP activate during biofilm formation?**

III • Antibiotics and Microbial Growth

7.10 Antibiotics are antimicrobial agents produced by microorganisms that target processes critical for growth such as DNA replication, transcription, protein synthesis, and cell wall synthesis. Specific mechanisms of antibiotic resistance include modification of the drug target, enzymatic inactivation, removal via efflux pumps, and metabolic bypass. These mechanisms are either chromosomally or plasmid encoded.

> **Q** **Why do antibiotics that target bacterial ribosomes not make humans sick?**

UNIT 2

7.11 Persistence is a dormant growth state in which cells are not susceptible to antibiotics. This lack of sensitivity is transient and thus not considered antibiotic resistance. Persistence is the result of toxin–antitoxin modules and the induction of the stringent response. Cells in the persistent state can resume growth once the antibiotic exposure has been stopped.

Q **What are the steps to inducing the stringent response during the development of persistence?**

Application Questions

1. If DnaA was not regulated in *Escherichia coli* and multiple rounds of replication were completed before cell division, what would be the consequence to the daughter cell and why? Would the resulting cell still be considered haploid?

2. Explain how cells exhibiting different phenotypes under the same conditions can possess the same genotype. Give examples in your explanation.

3. Describe how you would genetically design a superbug resistant to β-lactams, methicillin, streptomycin, daptomycin, and trimethoprim.

Chapter Glossary

Antibiotic an antimicrobial agent naturally produced by microorganisms

FtsZ a protein that forms a ring at the future site of septum formation in a prokaryotic cell

Green fluorescent protein (GFP) a protein that fluoresces green and is widely used in genetic analysis

Heterocysts specialized cells dedicated to nitrogen fixation

Persistence a dormant growth state in which a few members of a genetically identical population are transiently tolerant to multiple antibiotics

Reporter gene a gene whose product is easy to assay or detect

Super-resolution microscopy a form of light microscopy that employs special fluorescent molecules to resolve structures as small as 10–50 nm

Toxin–antitoxin module a genetic locus encoding a toxin that inhibits growth and an antitoxin that neutralizes the activity of the toxin

Viruses and Their Replication

microbiology**now**

Virophages: Viruses That Parasitize Other Viruses

The simplest definition of a virus is an infectious nucleic acid surrounded by protein. Viruses are also typically much smaller than bacteria. However, giant viruses comparable to some bacterial cells have been discovered. These "megaviruses" can actually be seen under the light microscope, as their diameters reach up to 0.75 μm, and because of their size, they clearly distort the line between a cell and a virus. Megaviruses infect protozoa such as *Amoeba* in diverse aquatic environments.

Virologists have made several exciting discoveries about megaviruses other than their size. For example, for megavirus replication, infected amoeba cells form cytoplasmic "factories" dedicated to virus production, and these factories can be as large as the cell's nucleus. A particularly remarkable discovery was the detection of tiny virus-like particles surrounding intracellular megaviruses. These structures contain DNA surrounded by capsid proteins, and so by definition, they should be viruses. However, the DNA in these particles does not encode proteins required for their replication. With this in mind, how do these virus-like particles reproduce?

Although they cannot replicate in amoeba cells by themselves, these tiny structures hijack the host's megavirus-producing factories for their own benefit. This leads to competition for virus replicating enzymes and the ultimate formation of defective megaviruses, which are unable to infect new host cells. Because of this parasitic activity toward megaviruses, these pesky little structures are called virophages in analogy to the term bacteriophage, which describes viruses that infect *Bacteria*. The transmission electron micrograph here shows active cell infection by Samba megaviruses isolated from a river in the Amazon, along with their associated virophages (arrow). The dark mass in the bottom right is a viral factory (VF), and the inset illustrates the formation of a defective Samba virus that is unable to infect new amoeba cells.

While it is unclear what role(s) virophages play in the biology of the amoeba cell, it is clear that the very existence of the virophage depends on its ability to parasitize the megavirus. Hence, the ability to inactivate megavirus replication and protect host cells from death may well be the selective force that maintains the virophage in this cozy arrangement of three different microbes.

Source: Campos, R.K., et al. 2014. Samba virus: A novel mimivirus from a giant rain forest, the Brazilian Amazon. *Virol. J. 11:* 95.

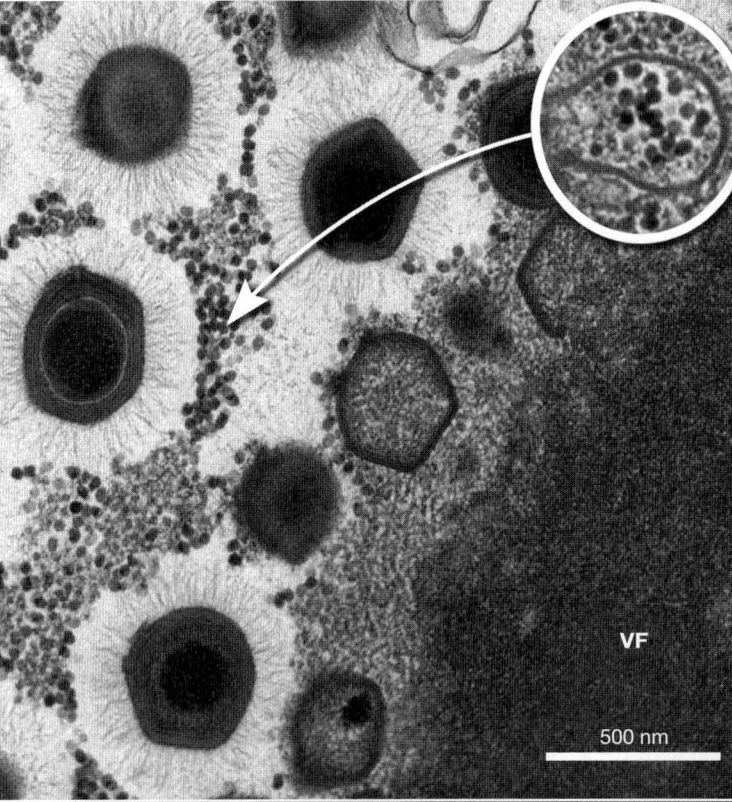

VF

500 nm

Throughout this unit we have focused on the growth of bacterial and archaeal cells. Here we shift gears and discuss how viruses "grow," or more accurately, replicate, and how this replication affects cells of all three domains of life.

A **virus** is a genetic element that can replicate only inside a living cell, called the **host cell**. Not considered living entities, viruses rely on the host cell for energy, metabolic intermediates, and protein synthesis, and so they are *obligate intracellular parasites*. However, viruses possess their own genomes and in this sense are independent of the host's genome.

Viruses infect both prokaryotic and eukaryotic cells and are responsible for many infectious diseases of humans and other organisms. The study of viruses is called *virology*, and this chapter covers the basic principles of this science. In Chapter 10 we consider the genomic and diversity aspects of viruses in detail.

I • The Nature of Viruses

8.1 What Is a Virus?

Although viruses are not cells, they nonetheless contain a genome that encodes those functions needed to replicate and they have an extracellular form, called the **virion**, which allows the virus to travel from one host cell to another. Viruses cannot replicate or reproduce unless the virion itself (or its genome, in the case of bacterial viruses) has gained entry into a suitable growing host cell, a process called *infection*.

Viral Components and Activities

The virion of any virus includes a protein shell, called the **capsid**, and the virus genome that the capsid contains. Most bacterial viruses are *naked*, with no further layers, whereas many animal viruses have an outer layer called the *envelope* that consists of a phospholipid bilayer (taken from the host cell membrane) and viral proteins (**Figure 8.1**). In enveloped viruses, the inner structure

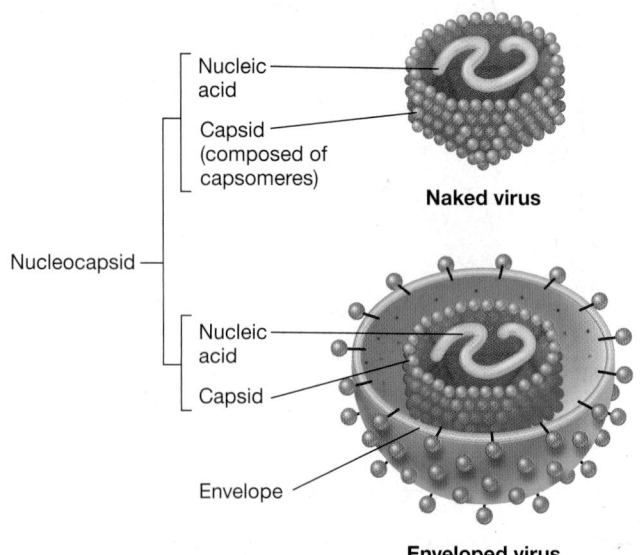

Figure 8.1 Comparison of naked and enveloped virus particles. The envelope originates from host cytoplasmic membrane.

of nucleic acid plus capsid protein is called the **nucleocapsid**. The virion protects the viral genome when the virus is outside the host cell, and proteins on the virion surface are important in attaching it to its host cell. The virion may also contain one or more virus-specific enzymes that play a role during infection and replication, as discussed later.

Once inside the host cell, a viral genome can orchestrate one of two quite different events. The virus may replicate and destroy the host in a **virulent** (**lytic**) infection. In a lytic infection, the virus redirects the host cell's metabolism from growth to support virus replication and the assembly of new virions. Eventually, new virions are released, and the process can repeat itself with new host cells. Alternatively, some viruses can cause a *lysogenic* infection; in this case, the host cell is not destroyed but is genetically altered because the viral genome becomes part of the host genome. These types of infection are discussed in detail later in the chapter.

Viral Genomes

All cells contain double-stranded DNA genomes. By contrast, viral genomes consist of either DNA *or* RNA and are further subdivided based on whether the genome is *single-stranded* or *double-stranded*. A very few highly unusual viruses use both DNA and RNA as genetic material, but at different stages of their life cycle (**Figure 8.2**).

Viral genomes can be either linear or circular, and single-stranded viral genomes may be of either the *plus sense* or *minus sense* in terms of their base sequence. Viral genomes of the plus configuration have the *exact same base sequence* as that of the viral mRNA that will be translated to form viral proteins. By contrast, viral genomes of the minus configuration are *complementary in base sequence* to viral mRNA. This interesting feature of viral genomes requires special genetic information flow processes, and we reserve our discussion of the details of these processes to Chapter 10.

Viral genomes are typically smaller than those of cells. The smallest bacterial genome known is about 139 kilobase pairs, encoding about 110 genes. Most viral genomes encode from a few up to about 350 genes. The smallest viral genomes are those of some small RNA viruses that infect animals. The genomes of these tiny viruses contain fewer than 2000 nucleotides and only two genes. A few very large viral genomes are known, such as the 1.25-Mbp DNA genome of a marine virus called *Megavirus*, which infects protozoans. RNA viruses typically have the smallest genomes and only DNA viruses have genomes encoding more than 40 genes.

Viruses can be classified on the basis of the hosts they infect as well as by their genome structure. Thus, we have bacterial viruses, archaeal viruses, animal viruses, plant viruses, protozoan viruses, and so on. Bacterial viruses are called **bacteriophages** (or simply *phage* for short) and have been intensively studied as model systems for the molecular biology and genetics of virus replication. In this chapter we will use bacteriophages many times to illustrate basic viral principles. Indeed, many of the key tenets of virology were discovered in studies of bacteriophages and subsequently applied to viruses of higher organisms. Because of their frequent medical importance, animal viruses have been extensively studied, whereas plant viruses, although of enormous importance to modern agriculture, have been less well studied.

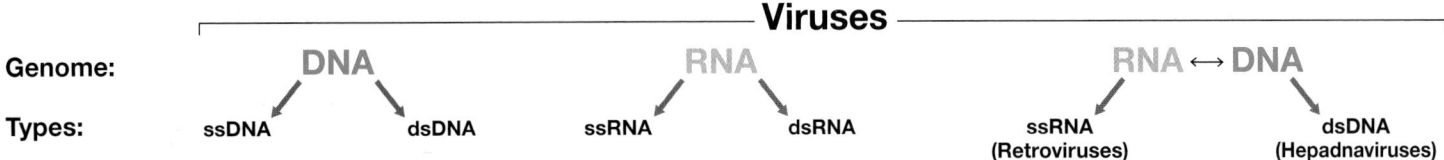

Viruses

Genome:	DNA		RNA		RNA ⟷ DNA	
Types:	ssDNA	dsDNA	ssRNA	dsRNA	ssRNA (Retroviruses)	dsDNA (Hepadnaviruses)

Figure 8.2 Viral genomes. The genomes of viruses can be either DNA or RNA, and some viruses use both at different stages in their replication cycle. However, only one type of genomic nucleic acid is found in the virion of any particular type of virus. Viral genomes can be single-stranded (ss) or double-stranded (ds) and circular or linear.

MINIQUIZ

- How does a virus differ from a cell?
- Why does a virus need a host cell?
- Compared with cells, what is unusual about viral genomes?

8.2 Structure of the Virion

Virions come in many shapes and sizes. Most viruses are smaller than prokaryotic cells, ranging in size from 0.02 to 0.3 μm (20–300 nanometers, nm). Smallpox virus, one of the larger viruses, is about 200 nm in diameter, which is about the size of the smallest known bacterial cells. Poliovirus, one of the smallest viruses, is only 28 nm in diameter, which is about the size of a ribosome, the cell's protein-synthesizing machine (Chapter 4).

Virion Structure

The structures of virions are quite diverse, varying widely in size, shape, and chemical composition (Chapter 10). The nucleic acid of a virion is always surrounded by its capsid (Figure 8.1). The capsid is composed of a number of individual protein molecules called **capsomeres** that are arranged in a precise and highly repetitive pattern around the nucleic acid.

The small size of most viral genomes restricts the number of distinct viral proteins that can be encoded. As a consequence, a few viruses have only a single kind of protein in their capsid. An example is the well-studied tobacco mosaic virus (TMV), which causes disease in tobacco, tomato, and related plants. TMV is a single-stranded RNA virus in which the 2130 copies of the simple capsomere protein are arranged in a helix with dimensions of 18 × 300 nm (**Figure 8.3**).

The information required for the proper folding and assembly of viral proteins into capsomeres and subsequently into capsids is often embedded within the amino acid sequence of the viral proteins themselves. When this is the case, virion assembly is a spontaneous process and is called *self-assembly*. However, some virus proteins and structures require assistance from host cell folding proteins for proper folding and assembly. For example, the capsid protein of bacteriophage lambda (Section 8.7) requires assistance from the *Escherichia coli* chaperonin GroE (⇨ Section 4.11) in order to fold into its active conformation.

Virus Symmetry

Virions are highly symmetric structures. When a symmetric structure is rotated around an axis, the same form is seen again after a certain number of degrees of rotation. Two kinds of symmetry are recognized in viruses, which correspond to the two primary viral shapes, rod and spherical. Rod-shaped viruses have *helical* symmetry while spherical viruses have *icosahedral* symmetry. A typical virus with helical symmetry is TMV (Figure 8.3). The lengths of helical viruses are determined by the length of the nucleic acid, and the width of the helical virion is determined by the size and packaging of the capsomeres.

Viruses with icosahedral symmetry contain 20 triangular faces and 12 vertices and are roughly spherical in shape (**Figure 8.4a**). Axes of symmetry divide the icosahedron into 5, 3, or 2 segments of identical size and shape (Figure 8.4b). Icosahedral symmetry is the most efficient arrangement of subunits in a closed shell because it requires the smallest number of capsomeres to build the shell. The simplest arrangement of capsomeres is 3 per triangular face, for a total of 60 capsomeres per virion. However, most viruses have more nucleic acid than can be packed into a shell made from 60 capsomeres and so viruses with some multiple of

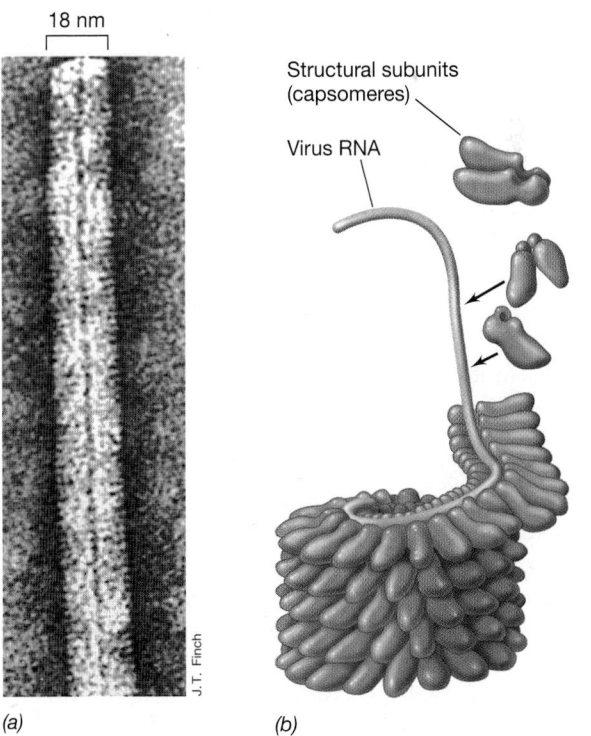

18 nm

Structural subunits (capsomeres)

Virus RNA

J.T. Finch

(a)   (b)

Figure 8.3 The arrangement of RNA and protein coat in a simple virus, tobacco mosaic virus. (a) High-resolution electron micrograph of a portion of the tobacco mosaic virus particle. (b) Cutaway showing structure of the virion. The RNA forms a helix surrounded by the protein subunits (capsomeres). The center of the virus particle is hollow.

UNIT 2

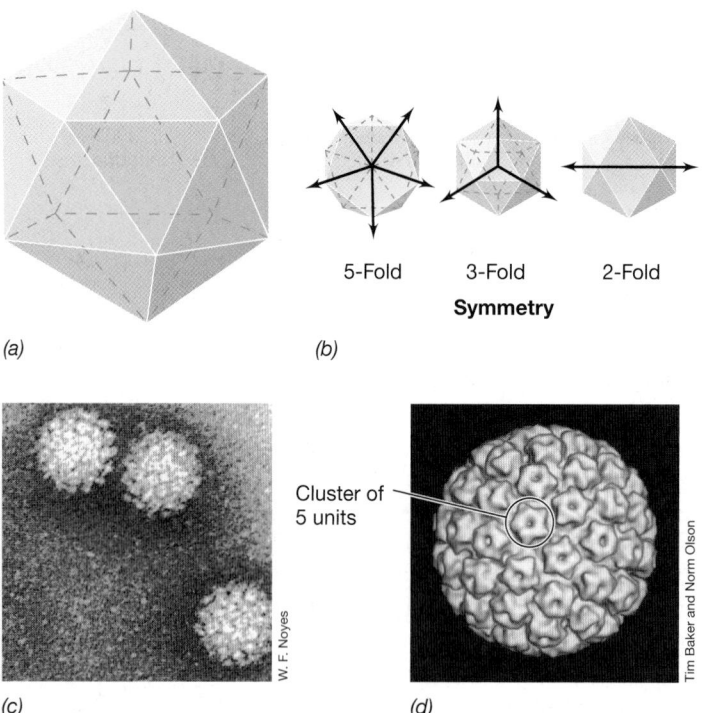

Figure 8.4 Icosahedral symmetry. *(a)* Model of an icosahedron. *(b)* Three views of an icosahedron showing 5-, 3-, or 2-fold symmetry. *(c)* Electron micrograph of human papillomavirus, a virus with icosahedral symmetry. The virion is about 55 nm in diameter. *(d)* Three-dimensional reconstruction of human papillomavirus; a virion contains 360 units arranged in 72 clusters of 5 each.

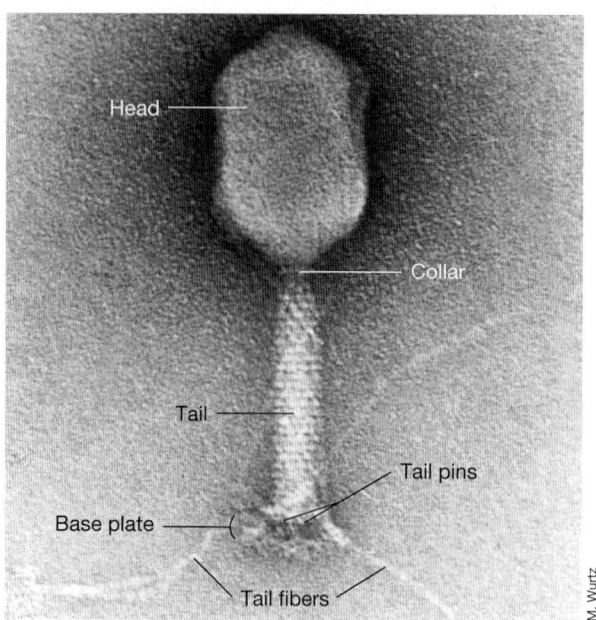

Figure 8.5 Structure of T4, a complex bacteriophage. Transmission electron micrograph of bacteriophage T4 of *Escherichia coli*. The tail components function in attachment of the virion to the host and injection of the nucleic acid (see Figure 8.12). The T4 head is about 85 nm in diameter.

60 capsomeres, such as 180, 240, or 360, are more common. The capsid of the human papillomavirus virus (Figure 8.4c), for example, consists of 360 capsomeres, with the capsomeres arranged into 72 clusters of 5 each (Figure 8.4d).

The structure of some viruses is highly complex, with the virion consisting of several parts each displaying its own shape and symmetry. The most structurally complex of all viruses are the head-plus-tail bacteriophages that infect *Escherichia coli*, such as phage T4. A T4 virion consists of an icosahedral head plus a helical tail (**Figure 8.5**). Some large viruses that infect eukaryotes are also structurally complex, although in ways quite distinct from the head-plus-tail bacteriophages. Mimivirus and vaccinia virus (see Figure 8.6d) are good examples and are discussed in more detail in Chapter 10.

Enveloped Viruses

Enveloped viruses have a lipoprotein membrane surrounding the nucleocapsid (**Figure 8.6**) and can have either RNA or DNA genomes. Most enveloped viruses (for example, Ebola, Figure 8.6a, b) use proteins on the virion's envelope to attach to and infect animal cells in which the cytoplasmic membrane is directly exposed to the environment. By contrast, plant and bacterial cells are surrounded by a cell wall outside the cytoplasmic membrane, and thus few examples of enveloped viruses are known in these organisms. Typically, the entire virion enters an animal cell during infection, with the envelope, if present, assisting in the infection process by fusing with the host membrane. Enveloped viruses also exit more easily from animal cells. As they pass out of the host cell, they are draped in cytoplasmic membrane material. The viral envelope consists

primarily of host cytoplasmic membrane, but some viral surface proteins (Figure 8.6a, b) become embedded in the envelope as the virus passes out of the cell.

The viral envelope is important in infection, as it is the component of the virion that makes contact with the host cell. The specificity of infection by enveloped viruses and some aspects of their penetration are thus controlled in part by the biochemistry of their envelopes. The virus-specific envelope proteins are critical for both attachment of the virion to the host cell during infection and for release of the virion from the host cell after replication.

Enzymes inside Virions

Viruses do not carry out metabolic processes and are thus metabolically inert. Nonetheless, some viruses carry enzymes in their virions that play important roles in infection. For example, some bacteriophages contain an enzyme that resembles lysozyme (↩ Section 2.4), which is used to make a small hole in the bacterium's peptidoglycan layer to allow nucleic acid from the virion to get into the host cytoplasm. A similar protein is produced in the later stages of infection to lyse the host cell and release new virions. Some animal viruses also contain enzymes that aid in their release from the host. For example, influenza virus (Figure 8.6c) has envelope proteins called *neuraminidases* that destroy glycoproteins and glycolipids of animal cell connective tissue, thus liberating the virions (↩ Section 10.9).

RNA viruses carry their own nucleic acid polymerases (RNA-dependent RNA polymerases called *RNA replicases*) that function to replicate the viral RNA genome and produce viral-specific mRNA. Such enzymes are necessary because DNA polymerase cannot make RNA and cells lack enzymes of any type that can make RNA from an RNA template. Retroviruses are unusual RNA animal viruses that replicate via DNA intermediates. Because making

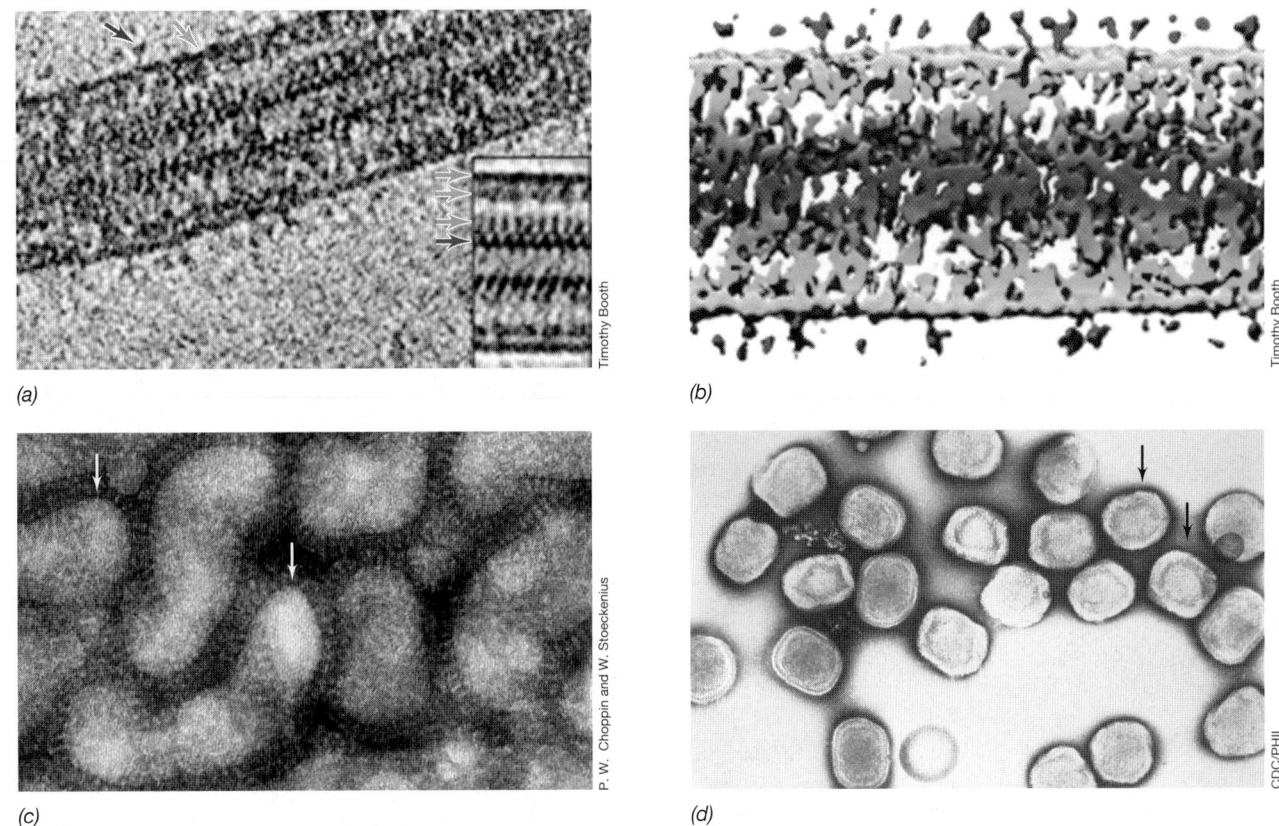

(a)

(b)

Timothy Booth

Timothy Booth

(c)

(d)

P. W. Choppin and W. Stoeckenius

CDC/PHIL

Figure 8.6 Enveloped viruses. (a) Cryo-electron micrograph of an Ebola virion. The virions are helical with a diameter of 80 nm. (b) Three-dimensional surface representation of an Ebola tomograph. Color-coding (also applies to arrows in part a) is as follows: red, spikes of envelope surface glycoproteins; orange, lipid envelope; green, membrane-associated proteins; blue/purple, nucleocapsid proteins. (c) Electron micrograph of influenza virus. The virions are about 80 nm in diameter, and can have many shapes. (d) Electron micrograph of vaccinia virus, an enveloped icosahedral pox virus about 350 nm wide. The arrows in both c and d point to the envelopes surrounding the nucleocapsids. Parts a and b modified from Beniac, D.R., Melito, P.L., deVarennes, S.L., Hiebert, S.L., Rabb, M.J., Lamboo, L.L., Jones, S.M., Booth, T.F. 2012. *PLoS ONE 7*(1): e29608.

DNA from an RNA template is another process cells cannot do, retroviral virions contain an RNA-dependent DNA polymerase called *reverse transcriptase* (Section 8.8). To sum up, although most viruses do not need to carry special enzymes in their virions, those that do absolutely require them for successful infection and replication.

MINIQUIZ

- Distinguish between a capsid and a capsomere. What is a common symmetry for spherical viruses?
- What is the difference between a naked virus and an enveloped virus?
- What kinds of enzymes can be found within the virions of RNA viruses? Why are they there?

8.3 Overview of the Virus Life Cycle

For a virus to replicate, it must induce a living host cell to synthesize all the essential components needed to make new virions. Because of these biosynthetic and energy requirements, dead host cells cannot replicate viruses. During an active viral infection, viral components are assembled into new virions that are released from the cell. Although replication steps are similar in most viruses, a major difference between viral infection of a *prokaryotic* cell and viral infection of a *eukaryotic* cell surrounds the initial step in infection. In cells of *Bacteria* and those *Archaea* in which the infection process has been studied, only the viral nucleic acid enters the host cell. By contrast, in plant and animal cells, the entire virion is taken up. Despite this key difference, the replication of bacterial viruses has been extremely well-studied and so we use them here as a general model of major viral replication events.

A cell that supports the complete replication cycle of a virus is said to be *permissive* for that virus. In a permissive host, the viral replication cycle can be divided into five steps (**Figure 8.7**):

1. *Attachment* (adsorption) of the virion to the host cell
2. *Penetration* (entry, injection) of the virion nucleic acid into the host cell
3. *Synthesis* of virus nucleic acid and protein by host cell machinery as redirected by the virus
4. *Assembly* of capsids and *packaging* of viral genomes into new virions
5. *Release* of new virions from the cell

The growth response during virus replication is illustrated in **Figure 8.8**. The response takes the form of a *one-step growth curve*, so named because a time course of virion numbers in the culture

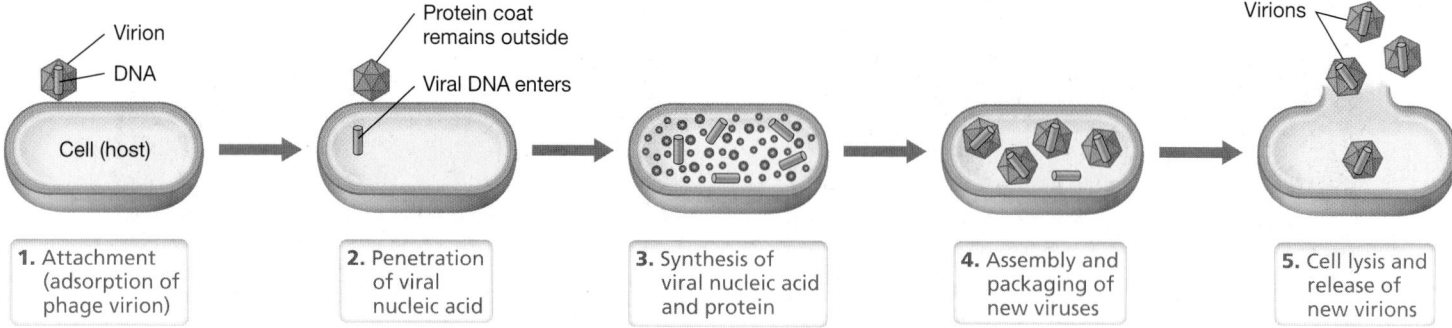

Figure 8.7 The replication cycle of a bacterial virus. The virions and cells are not drawn to scale. The burst size can be a hundred or more virions per host cell.

medium shows essentially no increase during the replication cycle until cells burst and release their newly synthesized virions. In the first few minutes after infection, the virus enters the *eclipse* phase, during which the viral genome and proteins will be replicated and translated, respectively. Once attached to a permissive host cell, a virion is no longer available to infect another cell. This is followed by the entry of viral nucleic acid into the host cell (Figure 8.7). If the infected cell breaks open at this point, the virion no longer exists as an infectious entity since the viral genome is no longer inside its capsid.

The *maturation* phase (Figure 8.8) begins as newly synthesized viral nucleic acid molecules become packaged inside their capsids. During the maturation phase, the number of infectious virions inside the host cell rises dramatically. However, the new virions still cannot be detected in the culture medium unless the cells are artificially lysed to release them. Because newly assembled virions are not yet present outside the cell, the eclipse and maturation periods together comprise the *latent period* of viral infection (Figure 8.8).

At the end of maturation, mature virions are released, either as a result of cell lysis or by budding or excretion, depending on the

virus. The number of virions released per cell, called the *burst size*, varies with the particular virus and the particular host cell, and can range from a few to a few thousand. The duration of the virus replication cycle also varies, from 20–60 min (in many bacterial viruses) to 8–40 h (in most animal viruses).

In Sections 8.5 and 8.6 we use a specific example to revisit these stages of the virus replication cycle and examine each in more detail.

--- **MINIQUIZ** ---

- What is packaged into capsids during maturation?
- Explain the term burst size.
- Why is the latent period so named?

8.4 Culturing, Detecting, and Counting Viruses

Host cells need to be growing in order for viruses to replicate in them. Pure cultures of bacterial hosts are either inoculated in liquid medium or spread as "lawns" on the surface of agar plates and then inoculated with a virus suspension. Animal viruses are cultivated in *tissue cultures*, which are cells obtained from an animal organ and grown in sterile glass or plastic vessels containing an appropriate culture medium (see Figure 8.10). Tissue culture media are often highly complex, containing a wide assortment of nutrients including blood serum and other highly nutritious substances to feed the animal cells as well as antimicrobial agents to prevent bacterial contamination.

Detecting and Counting Viruses: The Plaque Assay

A viral suspension can be quantified to estimate the number of infectious virions present per volume of fluid, a quantity called the **titer**. This is typically done using a *plaque assay*. When a virus infects host cells growing on a flat surface, a zone of cell lysis called a **plaque** forms and appears as a clear area in the lawn of host cells (**Figure 8.9**).

With bacteriophages, plaques may be obtained when virions are mixed into a small volume of molten agar containing host bacteria that is spread on the surface of an agar medium (Figure 8.9*a*).

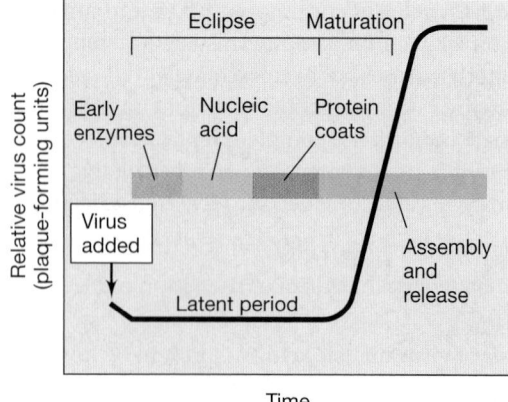

Figure 8.8 One-step growth curve of virus replication. Following adsorption, infectious virions cannot be detected in the growth medium, a phenomenon called *eclipse*. During the latent period, which includes the eclipse and early maturation phases, viral nucleic acid replicates and protein synthesis occurs. During the maturation period, virus nucleic acid and protein are assembled into mature virions and then released.

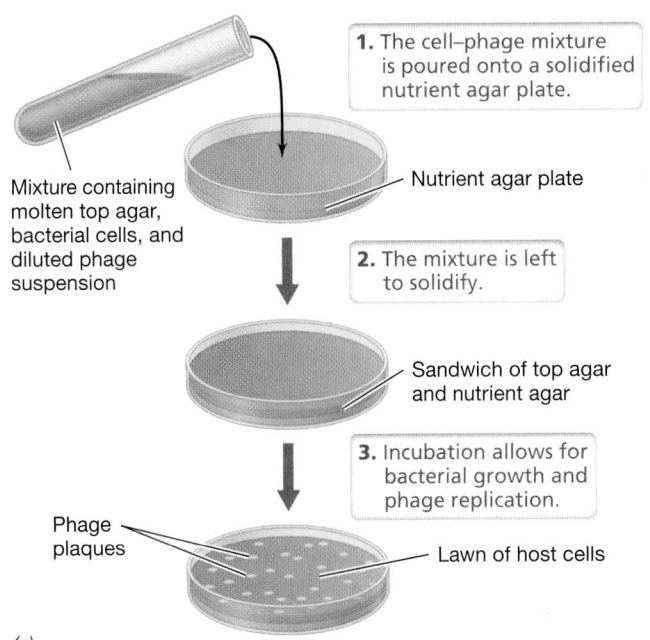

1. The cell–phage mixture is poured onto a solidified nutrient agar plate.

Mixture containing molten top agar, bacterial cells, and diluted phage suspension

Nutrient agar plate

2. The mixture is left to solidify.

Sandwich of top agar and nutrient agar

3. Incubation allows for bacterial growth and phage replication.

Phage plaques

Lawn of host cells

(a)

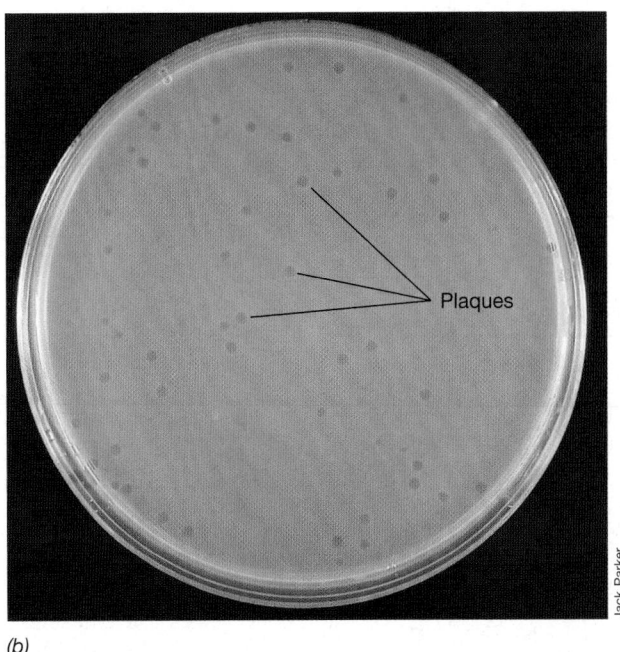

Plaques

Jack Parker

(b)

Figure 8.9 Quantification of bacterial virus by plaque assay. *(a)* "Top agar" containing a dilution of virions mixed with permissive host bacteria is poured over a plate of "bottom agar." Infected cells are lysed, forming plaques in the lawn. *(b)* Plaques (about 1–2 mm in diameter) formed by bacteriophage T4.

During incubation, the bacteria grow and form a turbid layer (lawn) that is visible to the naked eye. However, wherever a successful viral infection has occurred, cells are lysed, forming a plaque (Figure 8.9b). By counting the number of plaques, one can calculate the titer of the virus sample. The titer is typically expressed as the number of "plaque-forming units" per milliliter rather than an absolute viral number because of variations in plating efficiency, as described below. For replicating animal viruses, a tissue culture is grown and a diluted virus suspension overlaid upon it. As for bacterial viruses, plaques are revealed as cleared zones in the tissue culture cell layer, and from the number of plaques produced, an estimate of the virus titer can be made (**Figure 8.10**).

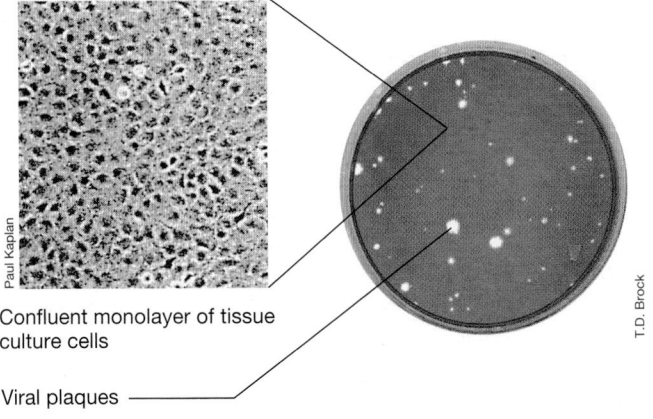

Confluent monolayer of tissue culture cells

Viral plaques

Paul Kaplan

T.D. Brock

Figure 8.10 Animal cell cultures and viral plaques. The animal cells support replication of the virus, and lysed cells result in plaques.

Plating Efficiency in Estimates of Viral Titers

The concept of *plating efficiency* is important in quantitative virology for all virus types. In any given viral preparation, the number of plaque-forming units is always lower than actual counts of viral particles made microscopically (using an electron microscope). This is because the efficiency with which virions infect host cells is rarely 100% and may often be considerably less. Virions that fail to infect may have assembled incompletely during the maturation process, may contain defective genomes, or may have suffered a spontaneous mutation that prevents them from attaching or otherwise properly replicating. Alternatively, a low plating efficiency may mean that viral growth conditions were not optimal or that some virions were damaged by handling or storage conditions.

Although plating efficiencies of bacterial viruses can often be higher than 50%, with many animal viruses it may be much lower, 0.1% or 1%. Knowledge of plating efficiency is useful in cultivating viruses because it allows the investigator to estimate what a titer needs to be to yield a certain number of plaques. If the titer is extremely low, the viral suspension may need to be concentrated by centrifugation or filtration before being used to infect host cells. This is especially true of animal viruses, as the costs of growing and maintaining tissue cultures can be significant.

— **MINIQUIZ** —

• What is meant by a viral titer?
• What is a plaque-forming unit?
• What is meant by the term plating efficiency?

II • The Viral Replication Cycle

Much of our understanding of lytic virus replication comes from the study of bacteriophages that infect *Escherichia coli*. Many RNA as well as DNA bacteriophages replicate in *E. coli* (**Table 8.1**). Here we choose one, bacteriophage T4, as our model for reviewing the individual stages of the virus life cycle (Figure 8.7) in more detail.

8.5 Attachment and Entry of Bacteriophage T4

The early steps in the life cycle of any bacteriophage are attachment to the surface of its host cell followed by penetration of the host cell outer layer(s) and entry of the viral genome into the cell.

Attachment

A major factor in host specificity of a virus is *attachment*. The virion itself has one or more proteins on its external surface that interact with specific components called *receptors* on the host cell surface. In the absence of its specific receptor, the virus cannot attach to the cell and hence cannot infect. Moreover, if the receptor is altered, for example by mutation, the host may become resistant to virus infection. The host range of a given virus is thus to a major extent determined by the presence of a suitable receptor that the virus can recognize and attach to.

Viral receptors are surface components of the host, such as proteins, carbohydrates, glycoproteins, lipids, or lipoproteins, or cell structures made from these macromolecules (**Figure 8.11**). The receptors carry out normal functions for the cell; for example, the receptor for phage T1 is an iron-uptake protein (Figure 8.11) and that for bacteriophage lambda functions in maltose uptake. Carbohydrates in the lipopolysaccharide (LPS) outer membrane of gram-negative bacteria are the receptors recognized by bacteriophage T4, a phage that binds to the LPS of *Escherichia coli* (Figure 8.11). Appendages that project from the cell surface, such as flagella and pili, are also common receptors for bacterial viruses. Small icosahedral viruses often bind to the side of these structures, whereas filamentous bacteriophages typically bind at the

tip, such as on the pilus (Figure 8.11). Regardless of the receptor used, however, once attachment has occurred, the stage is set for viral infection.

Penetration

Attachment of a virus to its host cell causes changes to both the virus and the host cell surface that result in penetration. Bacteriophages abandon their capsid outside the cell and only the viral genome reaches the cytoplasm. However, entry of the viral genome into a host cell only results in virus replication if the viral genome can be read. Consequently, for the replication of some viruses, for example RNA viruses, specific viral proteins must also enter the host cell along with the viral genome (Section 8.2).

The most intricate viral penetration mechanisms exist with the tailed bacteriophages. Bacteriophage T4 consists of an icosahedral head, within which the viral linear double-stranded DNA is folded, and a long, complex tail, which ends in a series of tail fibers and tail pins that contact the cell surface. Phage T4 virions first attach to *Escherichia coli* cells using their tail fibers (**Figure 8.12**). The ends of the tail fibers interact specifically with polysaccharides in the cell's LPS layer and then the tail fibers retract, allowing the tail itself to contact the cell wall via the tail pins. The activity of T4 lysozyme then forms a small pore in the peptidoglycan layer and the tail sheath contracts. When this occurs, T4 DNA enters the cytoplasm of the *E. coli* cell through a tail tube in a fashion resembling that of injection by a syringe. By contrast, the T4 capsid remains outside the cell (Figure 8.12). DNA inside bacteriophage heads is under high pressure, and because the interior of a bacterial cell is also under pressure from osmotic forces, the phage DNA injection process takes several minutes to complete.

TABLE 8.1 Some bacteriophages of *Escherichia coli*

Bacteriophage	Virion structure	Genome composition[a]	Genome structure	Size of genome[b]
MS2	Icosahedral	ssRNA	Linear	3,600
φX174	Icosahedral	ssDNA	Circular	5,400
M13, f1, and fd	Filamentous	ssDNA	Circular	6,400
Lambda	Head & tail	dsDNA	Linear	48,500
T7 and T3	Head & tail	dsDNA	Linear	40,000
T4	Head & tail	dsDNA	Linear	169,000
Mu	Head & tail	dsDNA	Linear	39,000

[a]ss, single-stranded; ds, double-stranded.
[b]In bases (ss genomes) or base pairs (ds genomes). These viral genomes have been sequenced and thus their lengths are known precisely. However, the sequence and length often vary slightly among different isolates of the same virus. Hence, the genome sizes listed here have been rounded off in all cases.

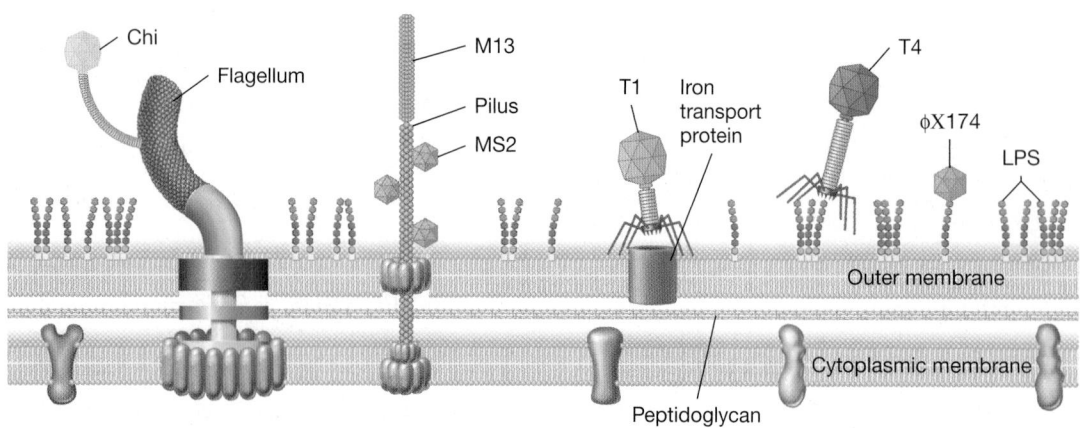

Figure 8.11 Bacteriophage receptors. Examples of the cell receptor sites used by different bacteriophages that infect *Escherichia coli*. All phages depicted except for MS2 are DNA phages.

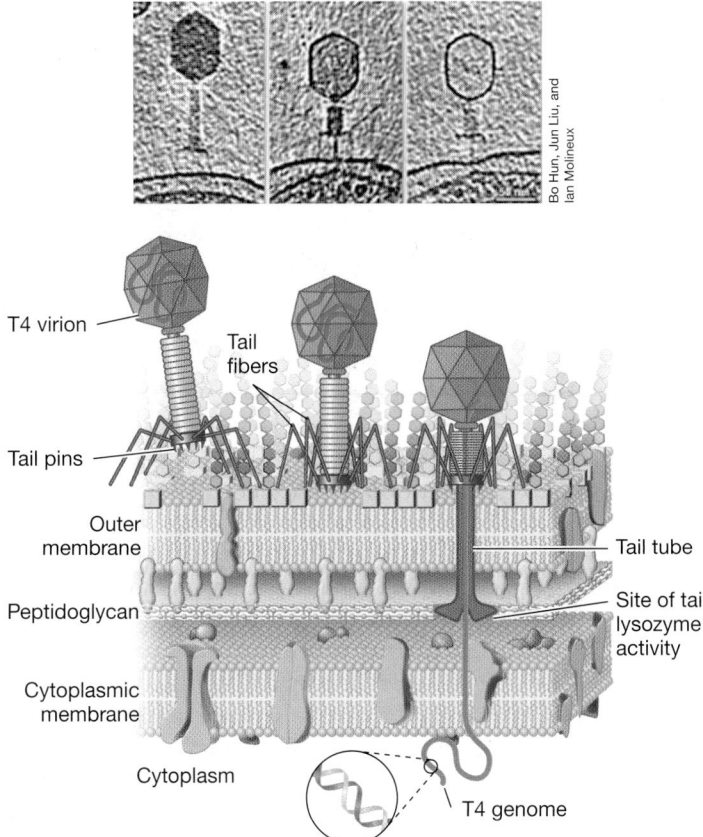

Bo Hun, Jun Liu, and Ian Molineux

T4 virion

Tail fibers

Tail pins

Outer membrane

Peptidoglycan

Cytoplasmic membrane

Cytoplasm

Tail tube

Site of tail lysozyme activity

T4 genome

Figure 8.12 Attachment and infection of an *Escherichia coli* cell by bacteriophage T4. The three transmission electron tomographs and the art beneath each depict (left to right) the initial attachment of a T4 virion to the cell outer membrane by tail fiber interactions with lipopolysaccharide (LPS); contact of the cell wall by the tail pins; and contraction of the tail sheath and injection of the T4 genome. The tail tube penetrates the outer membrane, and T4 lysozyme digests a small opening through the *E. coli* cell peptidoglycan layer.

Once a bacteriophage injects its genome into a host cell, a productive infection is not absolutely ensured. Below we consider some mechanisms employed by prokaryotic cells to protect against viral attack and avenues that subsequently evolved in bacteriophage to evade these processes.

Restriction and Modification

Although they lack the immune systems of animals (Chapters 26 and 27), *Bacteria* and *Archaea* possess several weapons against viral attack. Toxin–antitoxin modules (🔗 Section 7.11) and an antiviral system called CRISPR (🔗 Section 10.13) are two of these mechanisms. Additionally, *Bacteria* and *Archaea* can destroy double-stranded viral DNA through the activity of *restriction endonucleases,* enzymes that cleave foreign DNA at specific sites (🔗 Section 12.2). This process is called *restriction* and is a general host mechanism to prevent invasion by viral (or any other foreign) DNA. For such a system to be effective, however, the host must protect its own DNA from restriction enzyme attack. The host accomplishes this by *modification* of its DNA, typically by methylation of nucleotides at the sites where the restriction enzymes cut.

Restriction enzymes are specific for double-stranded DNA, and thus single-stranded DNA viruses and all RNA viruses are unaffected by restriction enzymes. Although host restriction systems confer significant protection from viral attack, some double-stranded DNA viruses have overcome host restriction by modifying their own DNA so it is no longer subject to restriction enzyme attack (**Figure 8.13a**). Many protective mechanisms are known, but in the *E. coli* bacteriophage T4 this is accomplished by substituting the base 5-*hydroxymethylcytosine* in place of cytosine in viral DNA. The hydroxyl group of this modified base is glucosylated, meaning that a molecule of glucose is added (Figure 8.13a); DNA with this modification is resistant to cleavage by all known *E. coli* restriction enzymes. By virtue of this viral protection mechanism, copies of the T4 genome are preserved until they are packaged later in the phage replication cycle and released by cell lysis (see next section) to attack uninfected cells of *E. coli.*

— MINIQUIZ —

- How does attachment contribute to virus–host specificity?
- Why does phage T4 need a lysozyme-like protein in order to infect its host, and what part of T4 enters the host cytoplasm?
- How does *Escherichia coli* try to protect itself from phage attack, and how does T4 protect itself from these weapons?

8.6 Replication of Bacteriophage T4

Drawing on what we already know about T4 phage attachment and penetration from the previous section, we now consider some unusual properties of the T4 genome and examine the steps in its replication cycle.

Genome Replication and Circular Permutation

Once a virus has infected a permissive host cell, the earliest events surround the synthesis of new copies of the viral genome. Because there are many types of viral genomes (Figure 8.2), there are many different schemes for virus genome replication (🔗 Section 10.1). In small DNA viruses, the cell's DNA polymerase is needed to replicate the viral genome. However, in more complex DNA viruses such as bacteriophage T4, the virus encodes its own DNA polymerase. Other proteins that function in viral DNA replication such as primases and helicases (🔗 Section 4.3) are also encoded by the T4 genome. In fact, T4 produces its own eight-protein DNA replisome complex (🔗 Section 4.4) to facilitate phage-specific genome synthesis.

Besides encoding its own replication machinery, the T4 genome has another unusual feature: In a population of T4 virions, although each copy of the genome contains the same set of genes, they are arranged in a different order. This is a phenomenon called *circular permutation,* which is a feature of many virus genomes. The term circular permutation is derived from the fact that DNA molecules that are circularly permuted appear to have been linearized by opening identical circular genomes at different locations. Circularly permuted genomes are also *terminally redundant,* meaning that some DNA sequences are duplicated on both ends of the DNA molecule as a result of the mechanism that generated them.

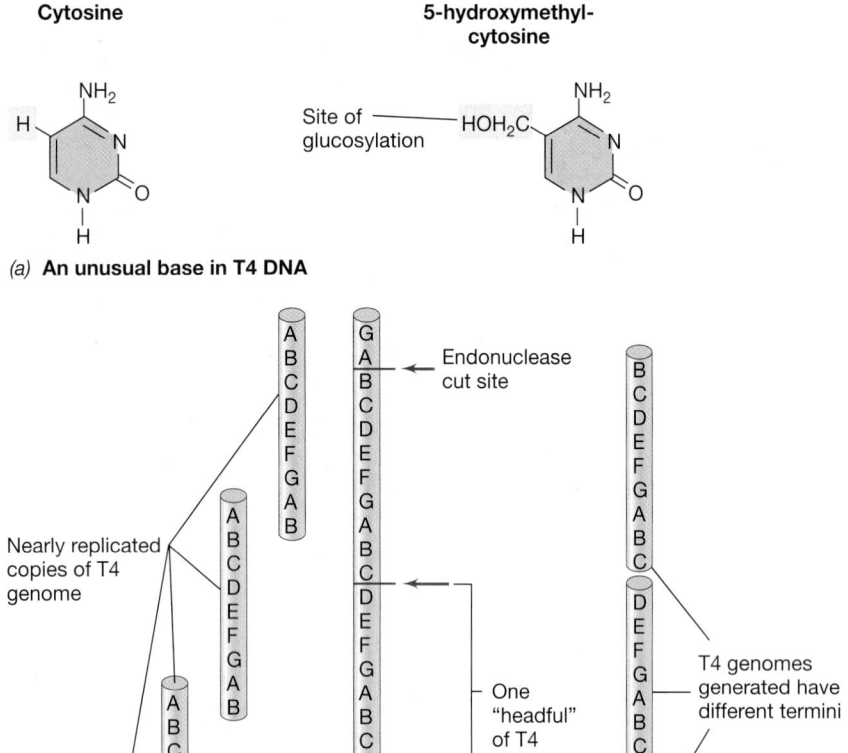

(a) **An unusual base in T4 DNA**

(b) **Circularly permuted T4 DNA**

Figure 8.13 Circular permutation and the unique DNA of bacteriophage T4. *(a)* The unique base 5-hydroxymethylcytosine in the DNA of bacteriophage T4. Once this base is glucosylated, the T4 DNA is resistant to restriction enzyme attack. *(b)* Generation of virus-length T4 DNA molecules with permuted sequences by an endonuclease that cuts off constant lengths of DNA from a concatemer regardless of their sequence.

The T4 genome is first replicated as a unit and then several genomic units are recombined end to end to form a long DNA molecule called a **concatemer** (Figure 8.13*b*). When the T4 DNA is packaged into capsids, the concatemer is not cut at a specific sequence; instead linear segments of DNA just long enough to fill a phage head are generated. This is called *headful packaging*, and is common among bacteriophages. However, because the T4 head holds slightly more than a genome length, the headful mechanism generates terminal repeats of about 3–6 kbp at each end of the DNA molecule (Figure 8.13*b*).

Transcription and Translation

Shortly after infection, T4 DNA is transcribed and translated, and the process of new virion synthesis begins. In less than half an hour, the process culminates in the release of new virions from the lysed cell. The major events are summarized in **Figure 8.14**.

Within a minute after T4 DNA enters the host cytoplasm, the synthesis of host DNA and RNA ceases and transcription of specific phage genes begins, thus inhibiting normal host growth. Translation of viral mRNA also begins quickly, and within 4 min of infection, phage DNA replication has already begun. The T4 genome encodes three major sets of proteins called **early proteins**, **middle proteins**, and **late proteins**, the terms referring to the general order of their appearance in the cell. Early proteins include enzymes for the synthesis and glucosylation of the unusual T4 base 5-hydroxymethylcytosine (Figure 8.13*a*), enzymes that function in the T4 replisome to produce copies of the phage-specific genome, and early proteins that modify host RNA polymerase. By contrast, middle and late proteins include additional RNA polymerase–modifying proteins, and virion structural and release proteins. These include, in particular, viral head and tail proteins and the enzymes required to liberate new virions from the cell (Figure 8.14).

The T4 genome does not encode its own RNA polymerase; instead, T4-specific proteins modify the specificity of the host RNA polymerase so that it recognizes only phage promoters (recall that promoters are the regions upstream of a structural gene where RNA polymerase binds to initiate transcription, ↩ Section 4.5). These modification proteins are encoded by T4 early genes and are transcribed by the host RNA polymerase. Host transcription is shut down shortly after this by a phage-encoded anti-sigma factor (↩ Section 6.15) that binds to the host RNA polymerase sigma factor and prevents it from recognizing promoters on host genes. This effectively switches the activity of host RNA polymerase from transcribing host genes to transcribing T4 genes. Later in the infection process, other phage proteins modify the host RNA polymerase so it now recognizes T4 middle gene promoters. Finally, transcription of T4 late genes begins, and this requires a new T4-encoded sigma factor that directs host RNA polymerase to promoters for these genes only. At this point, viral assembly can begin.

Packaging the T4 Genome and Virion Assembly and Release

The bacteriophage T4 DNA genome is forcibly pumped into a preassembled capsid using an energy-linked packaging motor. The motor components are encoded by viral genes, but host cell metabolism is needed to produce the proteins and supply the ATP required for the pumping process. The packaging process can be divided into three stages (**Figure 8.15*a***). First, precursors of the bacteriophage head called *proheads* are assembled but remain empty. Proheads contain temporary "scaffolding proteins" as well as head structural proteins. Second, a packaging motor

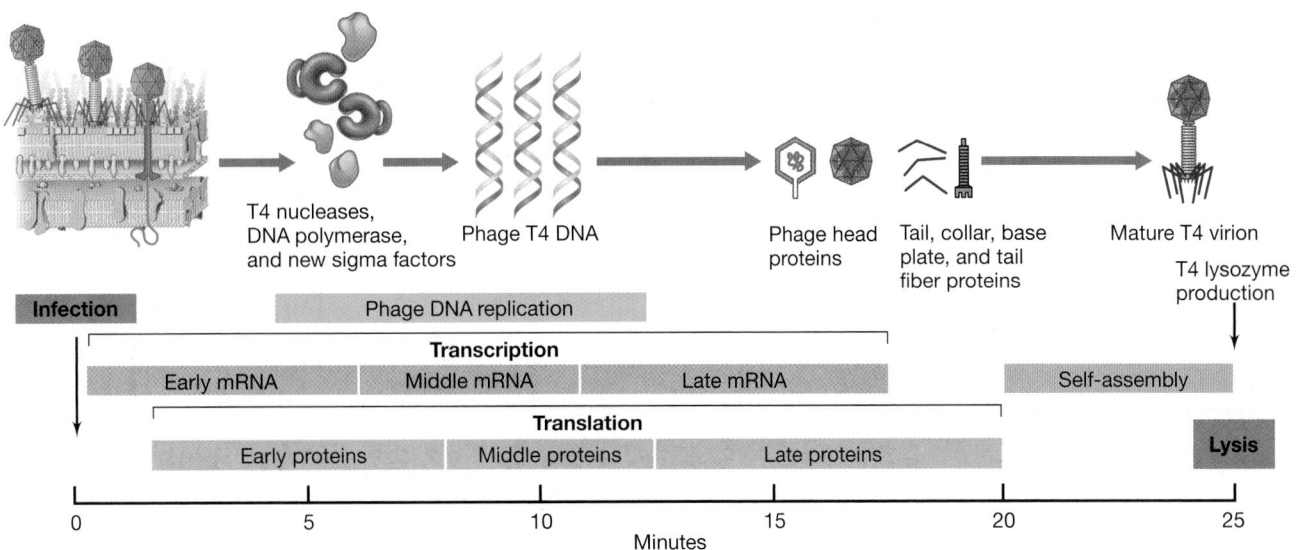

Figure 8.14 Time course of events in phage T4 infection. Following injection of DNA, early and middle mRNAs are produced that encode nucleases, T4 DNA polymerase, new phage-specific sigma factors, and other proteins needed for DNA replication. Late mRNAs encode virion structural proteins and T4 lysozyme, which are needed to lyse the cell and release new virions.

is assembled at the opening to the prohead (Figure 8.15b). The double-stranded linear T4 DNA genome (Figure 8.13b) is then pumped into the prohead under pressure using ATP as the driving force. The prohead expands when pressurized by the entering DNA and the scaffolding proteins are simultaneously discarded. Third, the packaging motor itself is discarded and the capsid head is sealed.

After the head has been filled, the T4 tail, tail fibers, and the other components of the virion are added, primarily by spontaneous reactions (self-assembly, Figures 8.14 and 8.15). The phage genome encodes a pair of very late enzymes that combine to breach the two major barriers to virion release: the host cytoplasmic membrane and peptidoglycan layer. Once these structures are compromised, the cell breaks open by osmotic lysis and the newly

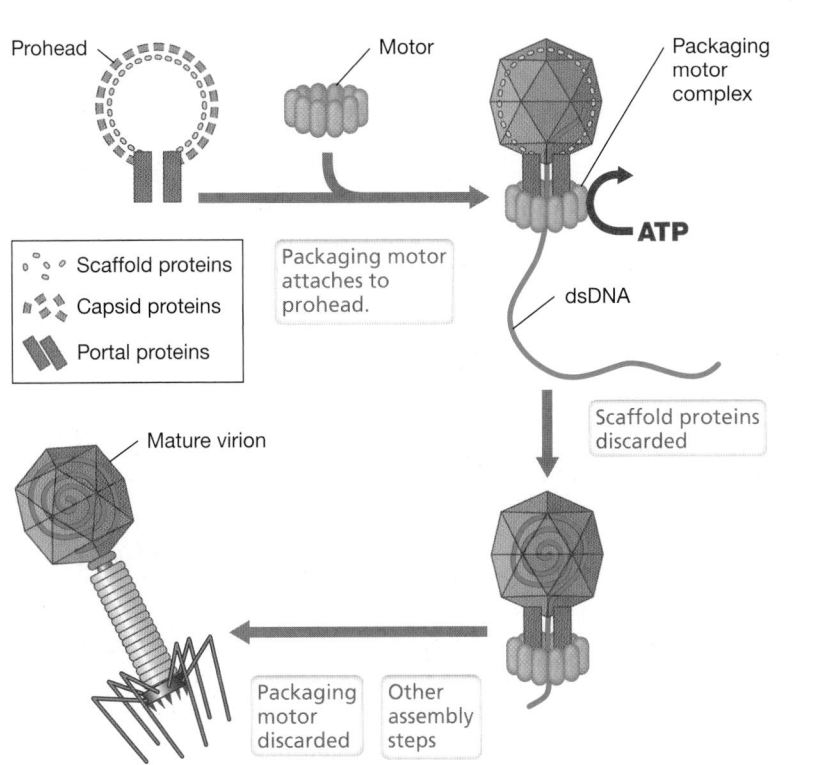

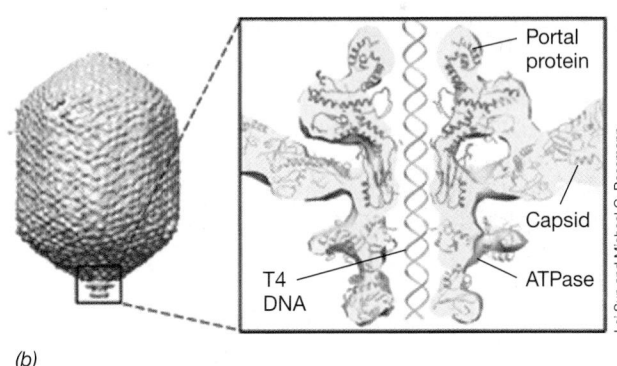

(b)

Figure 8.15 Packaging of DNA into a T4 phage head. *(a)* Proheads are assembled from capsid and portal proteins, both of which remain in the mature virion. As the head fills with DNA (driven by ATP hydrolysis), it expands and becomes more angular. Once the head is filled, the packaging motor detaches and the tail components are added. *(b)* Cryo-electron micrograph reconstruction of the T4 prohead and packaging motor. The blowup shows proteins within the motor and the prohead and their contact with DNA.

(a)

synthesized virions are released. After each replication cycle, which takes only about 25 min (Figure 8.14), over 100 new virions are released from each host cell (the *burst size*, Section 8.3), and these are now free to infect neighboring host cells.

MINIQUIZ

- What is a concatemer?
- Give one example each of T4 early, middle, and late proteins.
- What is required to package the T4 genome into its phage head?

8.7 Temperate Bacteriophages and Lysogeny

Bacteriophage T4 is a virulent virus and once infection begins, it always kills its host. However, some double-stranded DNA bacterial viruses, although capable of a virulent cycle, can also infect their host and establish a long-term stable relationship. These viruses are called **temperate viruses**.

Temperate viruses can enter into a state called **lysogeny**. In this state, most virus genes are not transcribed and instead, the virus genome is replicated in synchrony with the host chromosome and passed to daughter cells at cell division. A cell that harbors a temperate virus is therefore called a **lysogen**. While lysogen growth is controlled by its local environment and nutritional profile, the lysogenic state may confer new genetic properties—a condition called *lysogenic conversion*. We will see several examples in later chapters of pathogenic bacteria whose virulence (ability to cause disease) is at least in part linked to a lysogenic bacteriophage.

The Replication Cycle of a Temperate Phage

Two well-characterized temperate bacteriophages are lambda and P1. The life cycle of a temperate bacteriophage is shown in **Figure 8.16**. During lysogeny, the temperate virus genome is either integrated into the bacterial chromosome (lambda) or can exist in the cytoplasm as a plasmid (P1). In either case, the viral DNA, now called a **prophage**, replicates along with the host cell as long as the genes that activate the phage virulent pathway are repressed.

Maintenance of the lysogenic state is due to a phage-encoded *repressor protein*. Normally, low-level transcription of repressor genes and their subsequent translation maintains the repressor at a low level in the cell. However, if the phage repressor is inactivated or if its synthesis is in some way prevented, the prophage can be induced into the lytic stage. If induction occurs while the viral DNA is incorporated into the bacterial chromosome, the viral DNA is excised and phage genes are transcribed and translated; new virions are then produced, and the host cell is lysed (Figure 8.16). Various cell stress conditions, especially damage to host cell DNA, can induce a prophage to enter the lytic pathway. In contrast to this process, the viral "decision" to proceed to lysogeny or the lytic pathway upon initial viral infection is another matter altogether, and has been particularly well studied in bacteriophage lambda. We explore this story now.

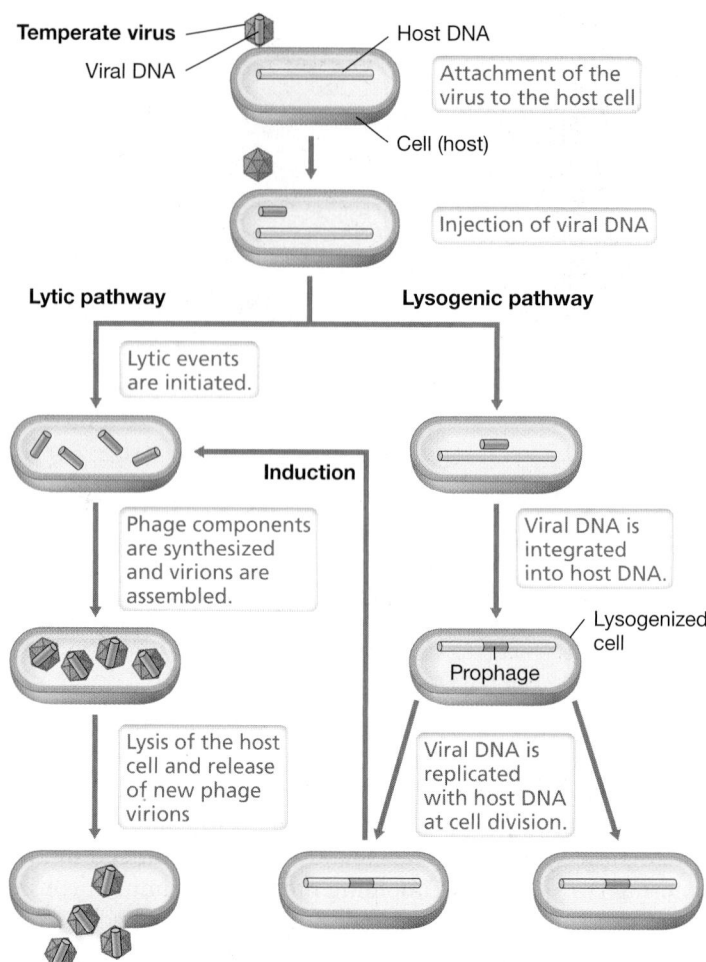

Figure 8.16 Consequences of infection by a temperate bacteriophage. The alternatives upon infection are replication and release of mature virions (lysis) or lysogeny, often by integration of the virus DNA into the host DNA, as shown here. The lysogen can be induced to produce mature virions and lyse.

Bacteriophage Lambda

Bacteriophage lambda, which infects *Escherichia coli*, is a double-stranded DNA virus with a head and tail (**Figure 8.17a**). At the 5′ end of each DNA strand of the linear lambda genome is a single-stranded region 12 nucleotides long. These single-stranded "cohesive" ends are complementary in base sequence; when lambda DNA enters the host cell, they base-pair to form the *cos* site and cyclize (circularize) the genome (Figure 8.17b).

If lambda enters the lytic pathway, long, linear concatemers of genomic DNA are synthesized by a mechanism called **rolling circle replication**. In this process, one strand in the circular lambda genome is nicked and is "rolled out" as a template for synthesis of the complementary strand (Figure 8.17c). The double-stranded concatemer is then cut into genome-sized lengths at the *cos* sites and the resulting genomes packaged into lambda phage heads. Once the tail has been added and mature lambda virions have been assembled (Figure 8.17a), cell lysis occurs and the virions are released. In its role as a lytic phage, lambda can also package a few chromosomal genes from its lysed host in newly synthesized virions and then transfer these to a second host cell,

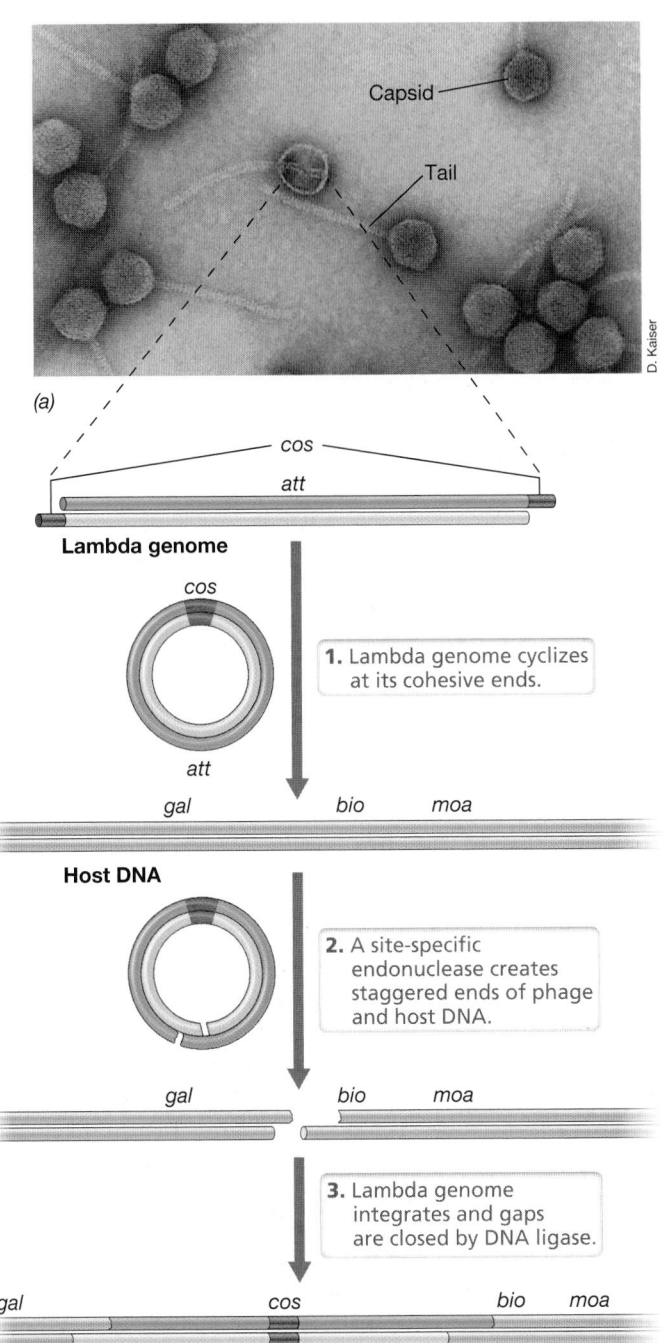

(a)

Figure 8.17 Bacteriophage lambda: virions, integration of viral DNA and rolling circle replication. *(a)* Transmission electron micrograph of phage lambda virions. The head of each virion is about 65 nm in diameter and contains linear dsDNA. *(b)* Lambda DNA integrates at specific attachment *(att)* sites on both the host and phage genomes. Host genes near *att* include *gal,* galactose utilization; *bio,* biotin synthesis; and *moa,* molybdenum cofactor synthesis. Lambda integrase is required, and specific pairing of the complementary ends results in integration of lambda DNA. *(c)* During rolling circle replication, as one strand (dark green) rolls out, it is both replicated at its opposite end and serves as a template for synthesis of the complementary strand.

a process called *transduction.* Transduction is an important means of horizontal gene transfer in nature and is also an important tool in bacterial genetics (⟳ Section 11.7).

Instead of the lytic pathway, if lambda takes the lysogenic route, its genome integrates into the *E. coli* chromosome. This requires a protein called *lambda integrase,* a phage-encoded enzyme that recognizes the phage and bacterial genome attachment sites (*att* in Figure 8.17*b*) and facilitates integration of the lambda genome. From this relatively stable state, certain events such as host DNA damage can initiate the lytic cycle once again. After such a trigger, a lambda excision protein excises the lambda genome from the host chromosome, transcription of lambda DNA begins, and lytic events unfold.

We now consider how these opposing processes of lysis and lysogeny are controlled upon initial infection of an *E. coli* cell by a phage lambda virion.

Lysis or Lysogeny: Regulation of the Lambda Lifestyle

Whether lysis or lysogeny occurs in a lambda infection depends in large part on the levels of two key repressor proteins that can accumulate in the cell following infection: the *lambda repressor,* also called the *cI protein,* and a second repressor called *Cro.* In a nutshell, the first repressor to accumulate will control the outcome of the infection.

If genes encoding the cI protein are rapidly transcribed following infection and cI accumulates, it represses the transcription of all other lambda-encoded genes, including Cro. When this happens, the lambda genome integrates into the host's genome and becomes a prophage (**Figure 8.18**). The host continues to grow, yielding more lysogens, until an event that triggers lysis is encountered. Cro, on the other hand, represses expression of a protein called cII whose function is to activate the synthesis of cI. Hence, following infection, if cI is present at insufficient levels to repress expression of phage-specific genes, Cro can accumulate in the cell; if this happens, lambda travels the lytic pathway.

Control of these alternative lifestyles—lysis or lysogeny—of lambda has been likened to a "genetic switch," where a defined series of events must occur for one pathway to be favored over the other. Although infection of an *E. coli* cell by a lambda virion typically results in the lytic cycle, as we have said, lytic events can be switched off if sufficient concentrations of cII are present to ensure adequate levels of cI (Figure 8.18). But how does this come about? Levels of protein cII are controlled by the relative activity of a protease in the cell that slowly degrades cII and by levels of yet another protein, cIII, whose function is to stabilize cII and protect it from protease attack. We thus have a cascade of regulatory

1. Lambda genome cyclizes at its cohesive ends.

2. A site-specific endonuclease creates staggered ends of phage and host DNA.

3. Lambda genome integrates and gaps are closed by DNA ligase.

(b) **Integration of lambda DNA into the host**

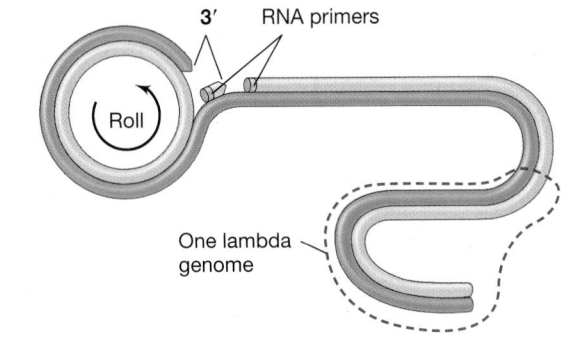

(c) **Rolling circle replication of lambda genome**

D. Kaiser

UNIT 2

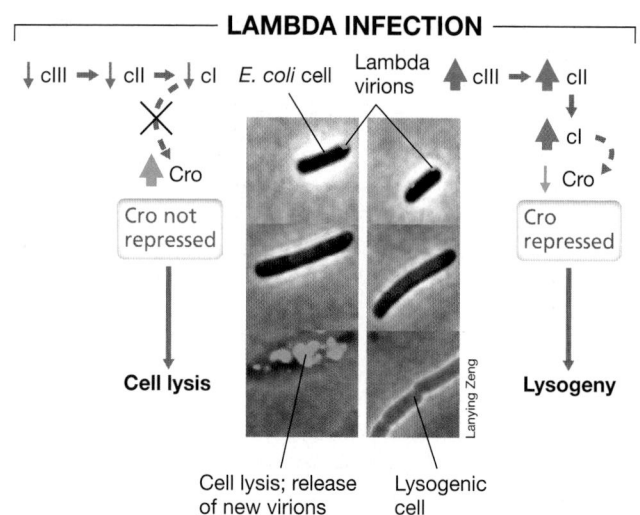

Figure 8.18 Regulation of lytic and lysogenic events in phage lambda. The photomicrographs show time courses of cells of *Escherichia coli* following a course of lytic (left panel, green) or lysogenic (right panel, red) events, as controlled by various repressors. The colors originate from genetically engineered lambda phage that trigger the production of specific fluorescent proteins when either lytic genes (green) or lysogenic genes (red) are expressed. Lytic cells are killed whereas *E. coli* lysogens continue to grow and divide.

TABLE 8.2 Representative viral diseases of humans

Disease	Virus	Genome DNA or RNA[a]	Size[b]
Cold sores/genital herpes	Herpes simplex	dsDNA	152,000
Smallpox	Variola major	dsDNA	190,000
Polio	Poliovirus	ssRNA (+)	7,500
Rabies	Rabies virus	ssRNA (−)	12,000
Influenza	Influenza A virus	ssRNA (−)	13,600
Measles	Measles virus	ssRNA (−)	15,900
Ebola hemorrhagic fever	Ebola virus	ssRNA (−)	19,000
Severe acute respiratory syndrome (SARS)	SARS virus	ssRNA (+)	29,800
Infant diarrhea	Rotavirus	dsRNA	18,600
Acquired immunodeficiency syndrome (AIDS)	Human immunodeficiency virus (HIV)	ssRNA/dsDNA (a retrovirus) (+)	9,700

[a]ss, single-stranded; ds, double-stranded. +, plus-strand virus; −, negative-strand virus (Section 8.1).
[b]In bases (ss genomes) or base pairs (ds genomes). These viral genomes have been sequenced and thus their lengths are known precisely. However, the sequence and length often vary slightly among different isolates of the same virus. Hence, the genome sizes listed here have been rounded off in all cases.

events here: cIII controls cII, which in turn controls cI. But even this is not the end of the story. Several other proteins not described here also play a role in the lambda lytic/lysogenic "decision," and hence the progress of a lambda infection is a highly intricate series of events.

Indeed, this tiny bacteriophage employs some of the most elaborate regulatory systems known in virology. One of the few lambda proteins expressed during lysogeny prevents the growing lysogen from entering dormancy during stressful conditions (⮌ Section 7.11). This regulatory mechanism helps ensure the continued spread of the lambda virus in a mixed population of *Escherichia coli* lysogens and nonlysogens.

--- MINIQUIZ ---

- What is a lysogen and what is a prophage?
- How does DNA replication in lambda differ from that of its host?
- What commits lambda to the lytic versus the lysogenic pathway?

8.8 An Overview of Animal Virus Infection

The major tenets of virology—presence of a capsid to carry the viral DNA or RNA genome, infection and takeover of host metabolic processes, and assembly and release from the cell—are universal, regardless of the nature of the host. Like bacterial viruses, animal viruses are classified by their genomes, with the majority of important human viral diseases caused by RNA viruses (Table 8.2). However, two key differences between bacterial and animal viruses are that (1) the entire virion of animal viruses (rather than just the nucleic acid) enters the host cell, and

(2) eukaryotic cells contain a nucleus, where many animal viruses replicate. We explore some aspects of animal viruses here.

Viral Infection of Animal Cells

While a large number of animal viruses exist (⮌ Section 10.14), most animal viruses that have been studied in detail are those that can replicate in cell cultures (Section 8.4 and Figure 8.10). To initiate infection, animal viruses also bind to specific host cell receptors. These virus receptors are typically animal cell surface proteins used in cell–cell contact or that function in the immune system. For example, the receptors for poliovirus and for HIV (the causative agent of AIDS) are normally used in intercellular communication between human cells. In multicellular organisms, cells in different tissues or organs often express different proteins on their cell surfaces. Consequently, viruses that infect animals often infect only certain tissues. For example, viruses that cause the common cold infect only cells of the upper respiratory tract.

Once an animal virus is bound to a receptor, its entry into a host cell generally occurs by fusion with the cytoplasmic membrane or by endocytosis (**Figure 8.19**). After entry into the cell, animal viruses must eventually lose their outer coat to deliver their genetic cargo to the cytosol. Some enveloped animal viruses are uncoated at the host cytoplasmic membrane, releasing the nucleocapsid into the cytoplasm, while naked and enveloped animal viruses that enter via endocytosis are uncoated in the host cytoplasm. If the viral genome is DNA, the genome passes through the nuclear membrane to the nucleus for replication. Conversely, the genomes of most RNA-based viruses are replicated or converted to DNA by viral enzymes within the nucleocapsid. We will focus shortly on the unique mode of replication of retroviruses, highly unusual RNA animal viruses with significant medical implications (Table 8.2).

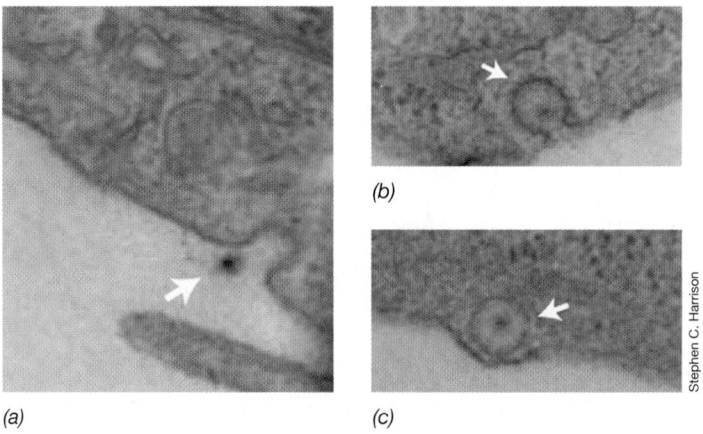

Stephen C. Harrison

(a)   (b)   (c)

Figure 8.19 Rotavirus cell entry. Electron micrographs of rotavirus virions illustrating the stages of cell entry. The arrows point to the virion. *(a)* Virion bound to the cell surface. *(b)* Engulfment of virion by the cell membrane. *(c)* Compartmentalization of the virion. Modified from Abdelhakim, A.H., Salgado, E.N., Fu, X., Pasham, M., Nicastro, D., Kirchhausen, T., and Harrison, S.C. 2014. *PLoS Pathogens 10:* e1004355.

Virion Assembly and Infection Outcomes

Virion assembly and morphogenesis occurs once viral mRNAs corresponding to the nucleocapsid have been translated by the host's machinery. After viral genome copies are packaged within their outer coat, many animal viruses must be enveloped (see Figure 8.22). This typically occurs when the virion exits the animal cell through budding or cell lysis. During this process the

virus may pick up part of the cell's cytoplasmic membrane and use it as part of the viral envelope.

Unlike a bacteriophage infection, in which one of only *two* outcomes—lysis or lysogeny—is possible depending on the virus, other events are possible in an animal virus infection. If an animal virus initially evades the immune system, animal viruses can catalyze at least four different outcomes (**Figure 8.20**). A *virulent infection* results in lysis of the host cell; this is the most common outcome. By contrast, in a *latent infection*, the viral DNA exists in the host's genome and virions are not produced; this leaves the host cell unharmed unless and until an event triggers the virulent pathway. With some enveloped animal viruses, release of virions, which occurs by a kind of budding process, may be slow, and the host cell may not be lysed (and thus not killed); instead it continues to grow and produce more virions. Such infections are called *persistent infections*. Finally, certain animal viruses can convert a normal cell into a tumor cell, a process called *transformation* (Figure 8.20).

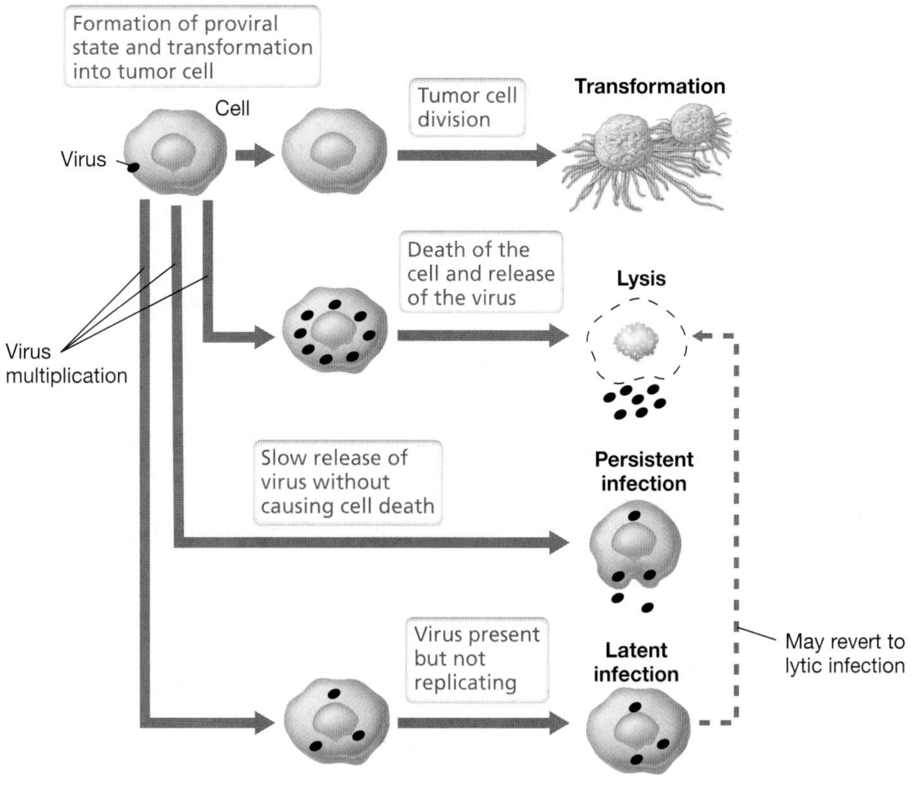

Retroviruses and Reverse Transcriptase

Retroviruses are structurally complex animal viruses that contain an RNA genome. However, unlike other RNA viruses, the genome is replicated inside the host cell by way of a *DNA* intermediate. The prefix *retro* means "backward," and the term *retrovirus* refers to the fact that these viruses transfer information from RNA to DNA (in contrast to genetic information flow in cells, which occurs from DNA to RNA). Retroviruses use the enzyme **reverse transcriptase** to carry out this unusual process. Retroviruses were the first viruses shown to cause cancer, and the human immunodeficiency virus (HIV) is a retrovirus that causes acquired immunodeficiency syndrome (AIDS).

Retroviruses are enveloped viruses and carry several enzymes within the virion (**Figure 8.21a**). These include *reverse transcriptase, integrase,* and a retroviral-specific *protease*. The genome of the retrovirus is unique and consists of two identical single-stranded RNAs of the plus sense (Section 8.1). The genome contains the genes *gag* (structural proteins), *pol* (reverse transcriptase and integrase), and *env* (envelope proteins) (Figure 8.21b). At each end of the retrovirus genome are repeated sequences that are essential for viral replication.

The replication of a retrovirus begins with the virion entering the host cell where the envelope is removed and reverse transcription begins in the nucleocapsid (**Figure 8.22**). A single strand of DNA is produced and then reverse transcriptase uses this as a template to make a complementary strand; double-stranded DNA is the final product. The dsDNA is released from the nucleocapsid and enters the host nucleus along with the integrase protein, and the integrase facilitates the incorporation of the retroviral DNA into the host genome.

Figure 8.20 Possible effects that animal viruses may have on cells they infect. Most animal viruses are lytic, and only a very few are known to cause cells to transform and become cancerous.

Formation of proviral state and transformation into tumor cell

Cell

Virus

Transformation

Tumor cell division

Virus multiplication

Death of the cell and release of the virus

Lysis

Slow release of virus without causing cell death

Persistent infection

Virus present but not replicating

Latent infection

May revert to lytic infection

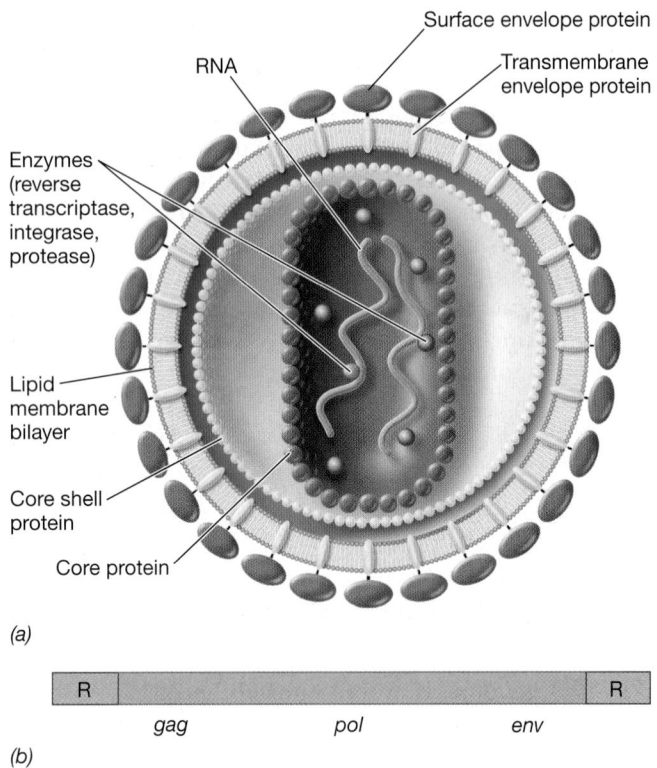

(a)

(b)

Figure 8.21 Retrovirus structure and function. *(a)* Structure of a retrovirus. *(b)* Genetic map of a typical retrovirus genome. Each end of the genomic RNA contains direct repeats (R).

The retroviral DNA is now a **provirus**. The provirus remains in the host genome indefinitely, and proviral DNA can be transcribed by the host RNA polymerase to form copies of the retroviral RNA genome and mRNA. Eventually, nucleocapsids are assembled that contain two copies of the retroviral RNA genome and are enveloped as they bud through the host cell cytoplasmic membrane (Figure 8.22). From here, the mature retrovirus virions are free to infect neighboring cells.

As exemplified by the retroviruses and some of the other viruses we have considered in this chapter, viruses are truly fascinating microbes. On the one hand, the growth (replication) of any virus is completely dependent on the growth and activities of its host cell; on the other hand, viruses can control the growth, survival, and genetic properties of their hosts. We revisit the viruses in Chapter 10 with a focus on viral genomics as the criterion by which we reveal the enormous genetic diversity of these important microbes. Although our discussion of viruses in this chapter has covered some of their remarkable characteristics, we have only scratched the surface.

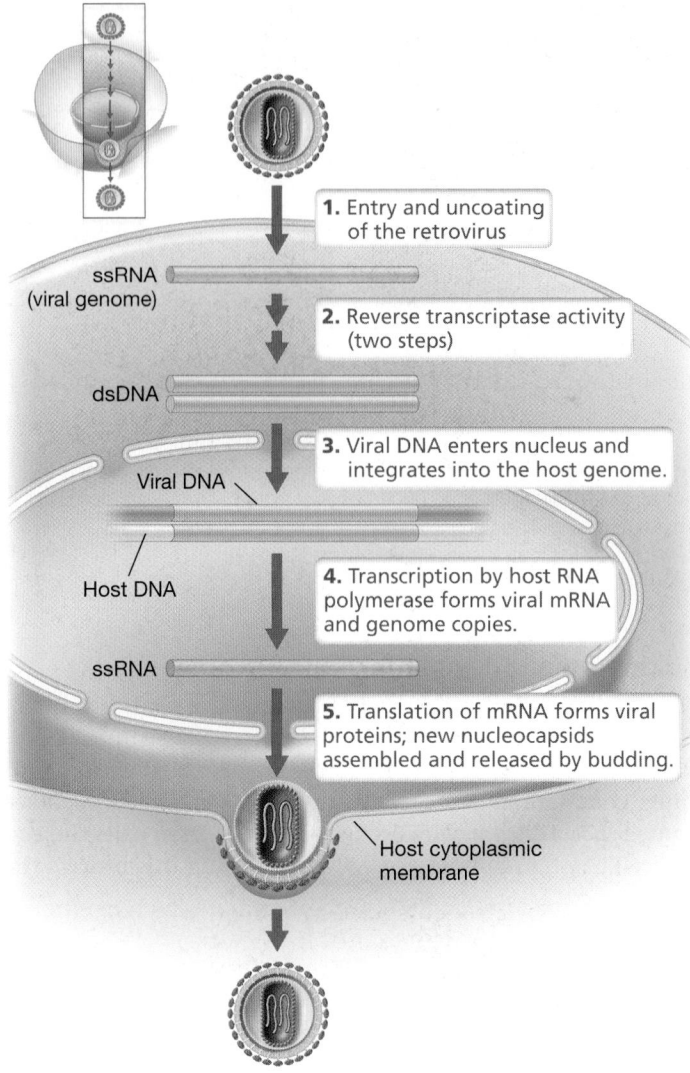

Figure 8.22 Replication of a retrovirus. The virion carries two identical copies of the RNA genome (orange). Reverse transcriptase, carried in the virion, makes single-stranded DNA from viral RNA and then double-stranded DNA that integrates into the host genome as a provirus. Transcription and translation of proviral genes leads to the production of new virions that are then released by budding.

--- **MINIQUIZ** ---

- Contrast the ways in which animal and bacterial viruses enter their hosts.
- What is the difference between a persistent and a latent viral infection?
- Why are retroviruses so named? What is required to carry out this process?

Chapter Review

I • The Nature of Viruses

8.1 A virus is an obligate intracellular parasite that requires a suitable host cell for replication. A virion is the extracellular form of a virus and contains either an RNA or a DNA genome inside a protein shell. Once the virus is inside the cell, the viral nucleic acid, and sometimes viral enzymes, redirects host metabolism to support virus replication. Viruses are classified by the characteristics of their genome and hosts. Bacteriophages infect bacterial cells.

Q **Once inside a host prokaryotic cell, what are the two types of infection that can be triggered by the virion genome?**

8.2 In the virion of a naked virus, only nucleic acid and protein are present; the entire unit is called the nucleocapsid. Enveloped viruses have one or more lipoprotein layers surrounding the nucleocapsid. The nucleocapsid is arranged in a symmetric fashion, with the icosahedron being a common morphology. Although virus particles are metabolically inert, one or more key enzymes are present within the virion in some viruses.

Q **Where does the envelope surrounding animal viruses originate?**

8.3 The virus replication cycle can be divided into five major stages: attachment (adsorption), penetration (uptake of the entire virion or injection of the nucleic acid only), protein and nucleic acid synthesis, assembly and packaging, and virion release.

Q **Why does a one-step growth curve differ in shape from that of a bacterial growth curve?**

8.4 Viruses can replicate only in their correct host cells. Bacterial viruses have proved useful as model systems because their host cells are easy to grow and manipulate in culture. Many animal viruses can be grown in cultured animal cells. Viruses can be quantified (titered) by a plaque assay. Plaques are clearings that develop on lawns of host cells, and in analogy to bacterial colonies, arise from the viral infection of a single cell.

Q **Describe the events that occur on an agar plate containing a bacterial lawn when a single bacteriophage particle causes the formation of a bacteriophage plaque.**

II • The Viral Replication Cycle

8.5 The attachment of a virion to a host cell is a highly specific process. Recognition proteins on the virus recognize specific receptors on the host cell. Sometimes the entire virion enters the host cell, whereas in other cases, as with most bacteriophages, only the viral genome enters. Host cells employ restriction enzymes in attempts to destroy viral and other foreign DNA, but T4 has chemically modified its DNA to make it resistant to such attack. Cells also modify their own DNA to protect it from their own restriction enzymes.

Q **What is required for a bacteriophage T4 virion to attach to an *Escherichia coli* cell?**

8.6 Bacteriophage T4 contains a double-stranded DNA genome that is both circularly permuted and terminally redundant. T4 encodes its own DNA polymerase and several other replication proteins. After a T4 virion penetrates a host cell, viral genes are expressed and regulated so as to redirect the host synthetic machinery to make viral nucleic acid and protein. Early viral genes encode viral genome replication events; middle and late viral genes encode structural proteins and capsid assembly. Once T4 components have been synthesized, new virions are made, primarily by self-assembly, and the virions are released after lysis of the host cell.

Q **Bacteriophage T4 lacks its own RNA polymerase. How do T4 genes get expressed or converted to mRNA? What host barriers must be broken before release of virions from the host cell?**

8.7 Some bacteriophages are temperate, meaning that they can initiate lytic events or integrate into the host genome as a prophage. Integration initiates a state called lysogeny in which the virus does not destroy the cell. A well-studied lysogenic virus of *Escherichia coli* is phage lambda; this phage uses an intricate regulatory system to govern whether the lytic or lysogenic state is initiated following infection.

Q **What enzyme is required to form a prophage, and what are the two main transcriptional repressors that control whether lambda proceeds to lysogeny or lysis?**

8.8 There are animal viruses with all known modes of viral genome replication. Many animal viruses are enveloped, picking up portions of host membrane as they leave the cell. Viral infection of animal host cells can result in cell lysis, but latent or persistent infections are also common, and a few animal viruses can cause cancer. Retroviruses like the AIDS virus are RNA viruses that employ the enzyme reverse transcriptase to replicate their RNA genome through a DNA intermediate. The DNA can integrate into the host chromosome where it can later be transcribed to yield viral mRNA and genomic RNA.

Q **Why can it be said that the retrovirus genome is unique in all of biology?**

Application Questions

1. What causes the viral plaques that appear on a bacterial lawn to stop growing larger?

2. The promoters on genes encoding early proteins in viruses like T4 have a different sequence than the promoters on genes encoding late proteins in the same virus. Explain how this benefits the virus.

3. Under some conditions, it is possible to obtain nucleic acid–free protein coats (capsids) of certain viruses. Under the electron microscope, these capsids look very similar to complete virions. What does this tell you about the role of the virus nucleic acid in the virus assembly process? Would you expect such particles to be infectious?

Chapter Glossary

Bacteriophage a virus that infects bacterial cells

Capsid the protein shell that surrounds the genome of a virus particle

Capsomere the subunit of a virus capsid

Concatemer two or more linear nucleic acid molecules joined covalently in tandem

Early protein a protein synthesized soon after virus infection and before replication of the virus genome

Enveloped in reference to a virus, having a lipoprotein membrane surrounding the virion

Host cell a cell inside which a virus replicates

Late protein a protein, typically a structural protein, synthesized late in virus infection

Lysogen a bacterium containing a prophage

Lysogeny a state in which the viral genome is replicated in step with the genome of the host

Lytic pathway the type of virus infection that leads to virus replication and destruction of the host cell

Middle protein a protein with either a structural or catalytic function synthesized after the early proteins in a virus infection

Nucleocapsid the complex of nucleic acid and capsid (shell) proteins of a virus

Plaque a zone of lysis or growth inhibition caused by virus infection of a bacterial lawn or other culture of sensitive host cells

Prophage the lysogenic form of a bacteriophage (see *provirus*)

Provirus the genome of a temperate or latent animal virus when it is replicating in step with the host chromosome

Retrovirus a virus whose RNA genome is replicated via a DNA intermediate

Reverse transcriptase the retroviral enzyme that can produce DNA from an RNA template

Rolling circle replication a DNA replication mechanism in which one strand is nicked and unrolled for use as a template to synthesize a complementary strand

Temperate virus a virus whose genome can replicate along with that of its host without causing cell death, in a state called lysogeny (bacterial viruses) or latency (animal viruses)

Titer the number of infectious virions per volume of fluid in a viral suspension

Virion the infectious virus particle; the viral genome surrounded by a protein coat and sometimes other layers

Virulent virus a virus that lyses or kills the host cell after infection

Virus a genetic element containing either RNA or DNA surrounded by a protein capsid and that replicates only inside host cells

Microbial Systems Biology

microbiology**now**

DNA Sequencing in the Palm of Your Hand

DNA sequencing technologies are revolutionizing microbiology at a remarkable pace. Innovations in next-generation sequencing have even tackled the issues of cost and portability. The world's first mobile nucleic acid sequencer—the MinION—is a palm-sized device that possesses 2000 tiny pore-containing proteins called nanopores. As single strands of nucleic acid travel through the nanopores, individual nucleotides are identified based on changes in electrical current. These current changes are relayed to a computer through a USB connection, which also powers the MinION. This miniature but mighty machine can display nucleic acid sequences from critical field samples in real time on a computer screen.

The utility of the MinION was clearly on display during the 2014–2015 Ebola virus hemorrhagic fever outbreak in West Africa. Scientists traveled to Guinea with three MinIONs in their luggage, a feat in itself as most DNA sequencers are too large and delicate to travel in baggage. Once in Guinea, scientists were able to survey the spread of different Ebola virus strains by analyzing unique nucleotide sequences present in each strain's genome. In as little as 48 hours after sample collection, Ebola virus genomes from 14 patients were determined using MinION sequencing. The photo here shows a researcher loading a patient's sample onto a MinION set up in a mobile field laboratory (inset).

Because the Ebola genome mutates on average every two weeks, the astonishing turnaround time provided by the MinION allowed epidemiologists to track geographical movements of different strains of the virus. This real-time analysis indicated that two major viral strains were the cause of Ebola persistence and that cross-border transmission between Sierra Leone and Guinea severely prolonged the outbreak. Traditional sequencing methods would not have supported such surveillance, as it requires weeks to obtain results after shipment of samples to remote laboratories.

While field biologists have envisioned a myriad of uses for the MinION, developers are currently attempting to modify it to operate from a smartphone instead of a computer. Also in its immediate future is outer space—NASA plans on testing the MinION on the International Space Station. And, because of its size, relatively low cost, and ease of use, the next frontier for the MinION will undoubtedly be the classroom.

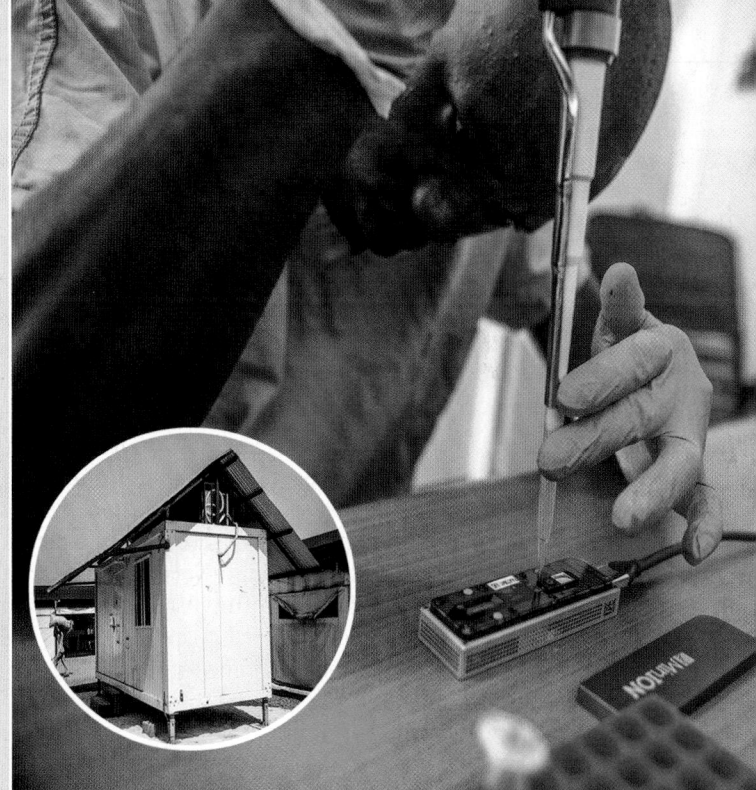

Source: Quick, J., et al. 2016. Real-time, portable genome sequencing for Ebola surveillance. *Nature 530:* 228–232. Photo credits: ©EMLab/Tommy Trenchard 2016.

Traditional approaches to studying microbes have focused on the analysis of individual biochemical pathways or molecular responses under specific conditions. While informative, this reductionist approach can only target a specific gene or subset of genes (or gene products) and fails to address the dynamic nature of organisms and how a network of biological molecules controls their behavior. By contrast, **systems biology** integrates different methodologies to yield an overview of an organism's response to its environment. Systems biology has been bolstered by the "*omics*" revolution—the ability to characterize and quantify large pools of biological molecules. Because the ability to store and analyze massive amounts of biological information by computer is essential to systems biology, the understanding of entire biological systems is evolving in parallel with computing power and storage and retrieval capabilities.

I • Genomics

The foundation of omics and systems biology lies in nucleic acid and protein *sequences*, characteristics ultimately controlled by the cell's genome. The **genome** is an organism's entire complement of genetic information, including genes that encode proteins, RNAs, and regulatory sequences, as well as any noncoding DNA that may be present. The genome sequence of an organism not only reveals its genes, but also yields important clues to how the organism functions. While new omics are coined with regularity, this chapter focuses on the major omics of systems biology—*genomics, transcriptomics, proteomics*, and *metabolomics*—and describes how these various pieces of the puzzle are integrated into microbial systems biology today (**Figure 9.1**).

The word **genomics** refers to the discipline of mapping, sequencing, analyzing, and comparing genomes. Here we review how genomes are sequenced and some techniques used to analyze these genomes and their gene content.

9.1 Introduction to Genomics

Advances in genomics rely heavily on improvements in molecular technology and computing power. The automation of DNA sequencing and the development of powerful computational tools for DNA and protein sequence analysis have reduced the cost and increased the speed at which genomes are analyzed. Thus the number of sequenced genomes has grown rapidly, with the major genomics bottleneck being the digestion of vast amounts of nucleic acid sequence data.

Genomics: Then and Now

The first genomes sequenced were those of small viruses over 40 years ago, and the first bacterial genome sequence was published in 1995. Today, DNA sequences from over 50,000 *Bacteria, Archaea*, and viruses, as well as datasets from metagenomic projects (Section 9.8), are available in public databases such as the *Genomes Online Database* (GOLD; see https://gold.jgi.doe.gov for an up-to-date list). With the goal of using genome sequences to advance systems and ecosystems biology, the United States Department of Energy's *Joint Genome Institute* (JGI) sponsors GOLD. **Table 9.1** lists some representative genomes from *Bacteria*

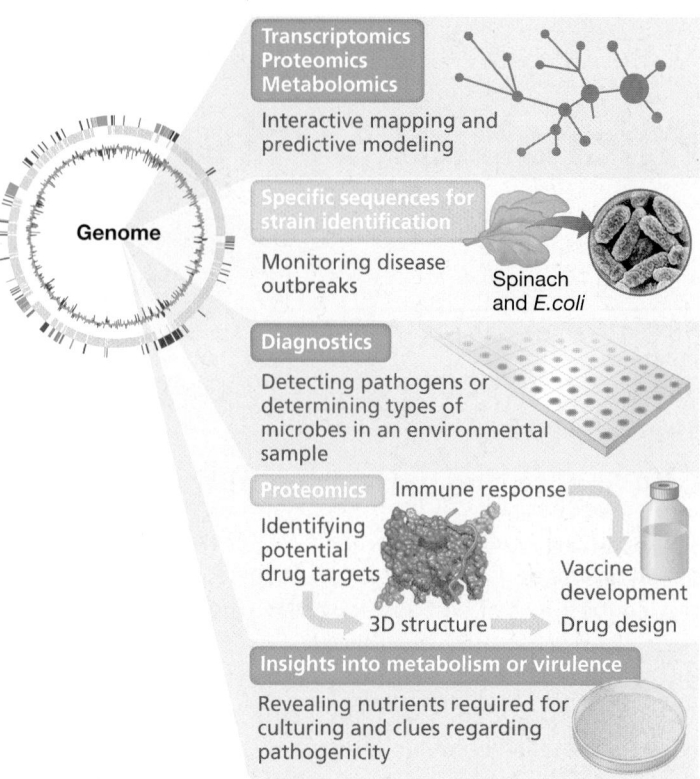

Figure 9.1 Utility of microbial genome sequences. A genome sequence allows for the development of omics approaches and tools for understanding, investigating, and monitoring microorganisms. It can also provide targets for drug and vaccine design.

and *Archaea*. The genomes of many eukaryotic organisms have also been sequenced, including the haploid human genome, which contains about 3.2 billion bp (~20,000 protein-encoding genes, **Figure 9.2**).

What Can Genomes Tell Us?

As we will discuss throughout this chapter, modern microbiology thrives on genome sequences; indeed, little in microbiology has been left untouched by genomic sequences. Microbial genome sequencing has discovered everything from genes encoding heat-stable enzymes in microbes that thrive in boiling water to genes that encode virulence factors in the most vicious pathogens. Genome sequencing has also been instrumental in developing microarrays for studying gene expression (Section 9.9), detecting horizontal transfer events (between microbes of different species, genera, and even kingdoms), monitoring and diagnosing disease outbreaks (based on the presence of "signature genes" of different pathogens), discovering CRISPRs (↩ Sections 10.13 and 11.12), understanding metabolic pathways, and discerning the growth requirements of microbes that have thus far defied laboratory culture.

The ability to sequence genomes has also been used to solve obscure medical mysteries. An excellent example is the genomics that revealed the causative agent of the "Black Death," which swept through Europe in the middle of the fourteenth century (**Figure 9.3a**). While it was believed that the Black Death was caused by a massive outbreak of bubonic plague, a typically fatal disease caused by the bacterium *Yersinia pestis* (↩ Section 31.7),

TABLE 9.1 Genomes of select species of *Bacteria* and *Archaea*[a]

Organism	Lifestyle[b]	Size (bp)	ORFs[c]	Comments
Bacteria				
Nasuia deltocephalicola	E	112,091	137	Degenerate sap-feeding insect endosymbiont
Tremblaya princeps	E	138,931	121	Degenerate mealybug endosymbiont
Hodgkinia cicadicola	E	143,795	169	Degenerate cicada endosymbiont
Buchnera aphidicola BCc	E	422,434	362	Aphid endosymbiont
Mycoplasma genitalium	P	580,070	470	Smallest nonsymbiotic bacterial genome
Borrelia burgdorferi	P	910,725	853	Spirochete, linear chromosome, causes Lyme disease
Rickettsia prowazekii	P	1,111,523	834	Obligate intracellular parasite, causes epidemic typhus
Treponema pallidum	P	1,138,006	1,041	Spirochete, causes syphilis
Methylophilaceae family, strain HTCC2181	FL	1,304,428	1,354	Marine methylotroph, smallest free-living genome
Thermotoga maritima	FL	1,860,725	1,877	Hyperthermophile
Deinococcus radiodurans	FL	3,284,156	2,185	Radiation resistant, multiple chromosomes
Bdellovibrio bacteriovorus	FL	3,782,950	3,584	Predator of other bacteria
Bacillus subtilis	FL	4,214,810	4,100	Gram-positive genetic model
Mycobacterium tuberculosis	P	4,411,529	3,924	Causes tuberculosis
Escherichia coli K-12	FL	4,639,221	4,288	Gram-negative genetic model
Escherichia coli O157:H7	FL	5,594,477	5,361	Enteropathogenic strain of *E. coli*
Pseudomonas aeruginosa	FL	6,264,403	5,570	Metabolically versatile opportunistic pathogen
Streptomyces coelicolor	FL	8,667,507	7,825	Linear chromosome, produces antibiotics
Bradyrhizobium japonicum	FL	9,105,828	8,317	Nitrogen fixation, nodulates soybeans
Sorangium cellulosum	FL	14,782,125	11,559	Forms multicellular fruiting bodies
Archaea				
Nanoarchaeum equitans	P	490,885	552	Smallest nonsymbiotic cellular genome
Methanocaldococcus jannaschii	FL	1,664,976	1,738	Methanogen, hyperthermophile
Pyrococcus horikoshii	FL	1,738,505	2,061	Hyperthermophile
Sulfolobus solfataricus	FL	2,992,245	2,977	Hyperthermophile, sulfur chemolithotroph
Haloarcula marismortui	FL	4,274,642	4,242	Extreme halophile, bacteriorhodopsin
Methanosarcina acetivorans	FL	5,751,000	4,252	Acetate using methanogen

[a]Information on prokaryotic genomes can be found at https://gold.jgi.doe.gov.
[b]E, endosymbiont; P, parasite; FL, free-living.
[c]Open reading frames. Genes encoding known proteins are included, as well as ORFs that could encode a protein greater than 100 amino acid residues. Smaller ORFs are not included unless they show similarity to a gene from another organism or unless the codon bias is typical of the organism being studied.

scientists could not be positive until they recovered and sequenced DNA samples from the teeth and bones of corpses of people known to have died from the Black Death. By comparisons of this ancient DNA with the genome of *Y. pestis*, the mystery behind this devastating medieval disease was unraveled: The Black Death was indeed bubonic plague.

Microbial genomics has also been used to identify new microbial phyla. For example, until recently, only three phyla of *Archaea* were known—*Euryarchaeota*, *Crenarchaeota*, and *Nanoarchaeota*. Because every cultured species was isolated from an extreme environment, many microbiologists concluded that *Archaea* were mainly extremophiles and that they did not inhabit oceans, lakes, and soil in significant numbers. However, based on environmental 16S rRNA gene sequencing, *Archaea* only marginally affiliated with *Crenarchaeota* were detected in marine and freshwater samples. Who were these organisms, and how were

they making a living? Subsequently, *Nitrosopumilus*, the first ammonia-oxidizing (nitrifying) archaeon known, was isolated (Figure 9.3b) (↺ Section 14.11). Using the powerful analytical tools of genomics, the genomes of two distinct ammonia-oxidizing *Archaea* were compared with those of all other *Archaea*. This genomic analysis clearly showed that these ammonia-oxidizing *Archaea* belonged in a new phylum, now called the *Thaumarchaeota* (↺ Section 17.5).

The above is just a taste of how genomics has impacted microbiology. Other relevant examples will appear regularly as you make your way through this book. The major message here is twofold: (1) We are clearly living in the era of microbial genomics, and (2) the genomics revolution has spawned a wealth of powerful tools to attack old problems in new ways. Indeed, in the past 40 years or so, microbiology as a science has leapt forward farther and faster than at any time in its history.

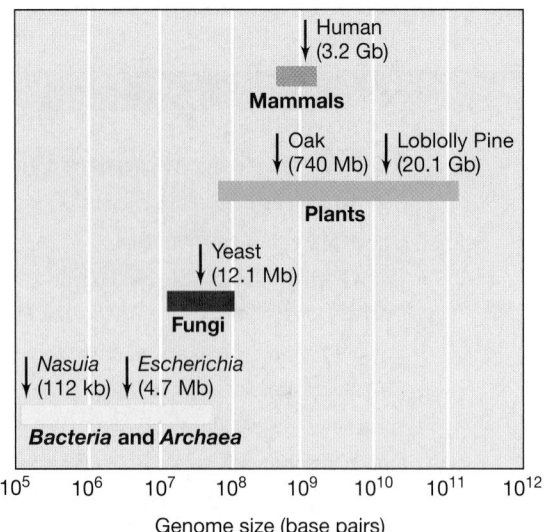

Figure 9.2 Genome sizes of microbial cells and higher organisms. Compare with viral genome sizes in Figure 10.1.

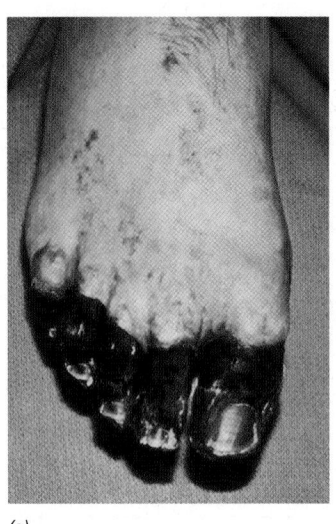
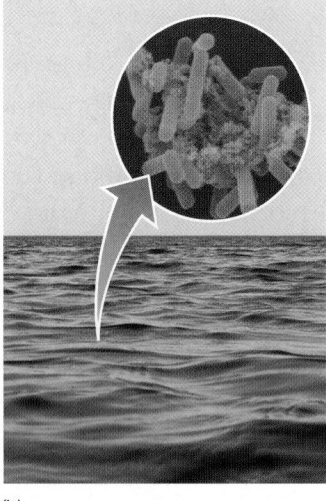

(a)   *(b)*

Figure 9.3 What genomes can tell us. *(a)* Genomics helped solve an ancient medical mystery surrounding plague. (The blackened skin on the toes of this modern plague victim originates from hemorrhaging due to systemic infection with *Yersinia pestis*.) *(b)* Genome sequencing was used to assign the marine ammonia-oxidizing archaeon *Nitrosopumilus* to a new phylum of *Archaea*, the *Thaumarchaeota*.

MINIQUIZ

- How many protein-encoding genes are in the human genome?
- List three examples of how genomics has led to major new discoveries in microbiology.

9.2 Sequencing and Annotating Genomes

In biology, the term **sequencing** refers to determining the precise order of subunits in a macromolecule. In the case of DNA (or RNA), the sequence is the *order* in which the nucleotides are aligned. DNA sequencing today forms the heart of the omics revolution and its technology is advancing so quickly that new

methods appear every year. Interestingly, however, despite the technological breakthroughs that have catapulted us into the omics age, some of the earliest sequencing methodologies—born of simple yet brilliant basic science—form the foundation of the latest methods today (see next subsection).

After sequencing and assembly of the gene fragments, the next step is *genome annotation*, the conversion of raw sequence data into a list of the genes and other functional sequences present in the genome. The term **bioinformatics** refers to the use of computers to store and analyze the sequences and structures of nucleic acids and proteins. Improved sequencing methods are now generating data faster than it can be properly analyzed. Thus, at present, annotation is the major "bottleneck" in genomics. Here we focus on the process of genome sequencing, assembly, and annotation.

DNA Sequencing

The first widely used method for sequencing DNA was the dideoxy method developed by the British scientist Fred Sanger, who won a Nobel Prize (his second) for this accomplishment. In the Sanger procedure the sequence is determined by making a copy of the original single-stranded DNA in a process similar to the polymerase chain reaction (PCR, ↩ Section 12.1). The secret behind the Sanger method was the addition of a mixture of normal deoxyribonucleotides (dNTPs) and small amounts of the corresponding *dideoxyribonucleotides* (ddNTPs), one for each of the four bases—adenine, guanine, cytosine, and thymine—to the mixture used to make the DNA copy (**Figure 9.4a**). The dideoxy analog is a specific *chain-terminator*; because it lacks a 3'-hydroxyl, the analog prevents further elongation of the chain after its insertion. Because ddNTPs insert randomly, DNA chains of varying length are produced in the synthesis reaction (Figure 9.4a). Sanger sequencing originally used radioactive labels, but automated systems were quickly developed that used a separate fluorescent label for each different ddNPT and that detected the DNA products (separated by passing through a sizing column) with a laser (Figure 9.4a).

Because the original Sanger method was dependent on primers binding to a known sequence and was limited to around 800 nucleotides per reaction, chromosomes or large DNA molecules could not be sequenced in a single reaction. Instead large DNA molecules had to be cut into smaller fragments and cloned into vectors for sequencing. This led to the development of new sequencing technologies, which appear now with such regularity that the term "next-generation sequencing" is commonly used to describe the latest and greatest in nucleic acid sequencing. For example, *pyrosequencing*, a *second-generation sequencing method* still widely used today, was developed to improve the process by employing the light-emitting enzyme luciferase to detect incorporation of dNTPs by emitting a pulse of light (Figure 9.4b). **Table 9.2** summarizes modern sequencing methods and illustrates how the cost of sequencing 1 megabase (Mbp, million base pairs) of DNA has dropped over 100,000-fold in the last 15 years.

Genome Assembly and Annotation

Regardless of which sequencing system is used, the sequences obtained must be *assembled* before they can be analyzed. Genome assembly consists of putting the fragments in the correct order and eliminating overlaps. Then, for assembled genomic sequences

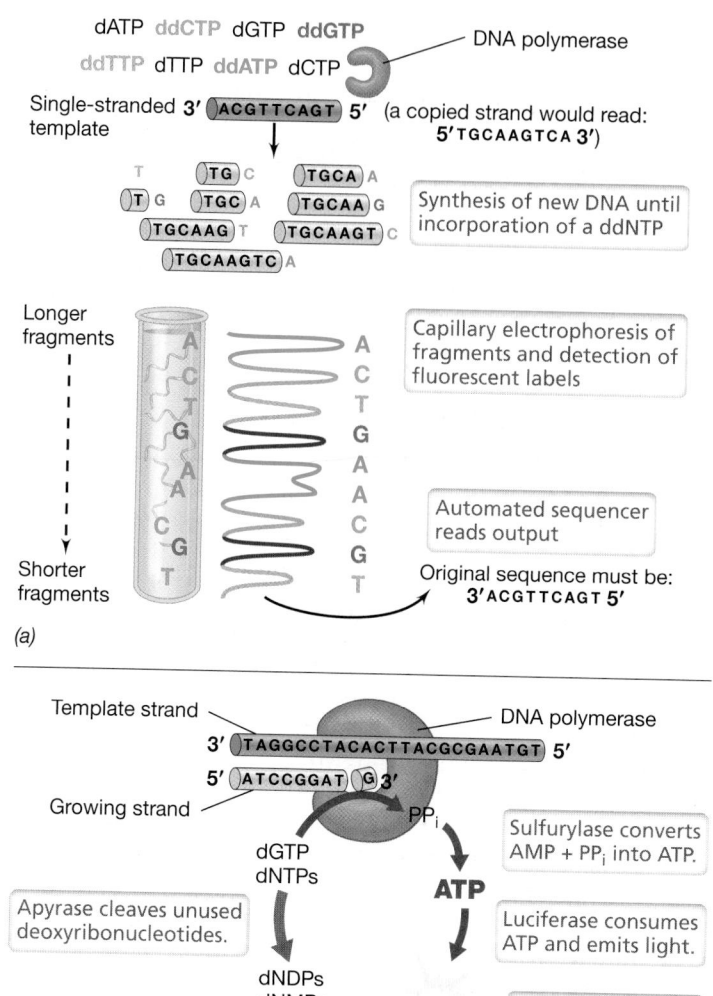

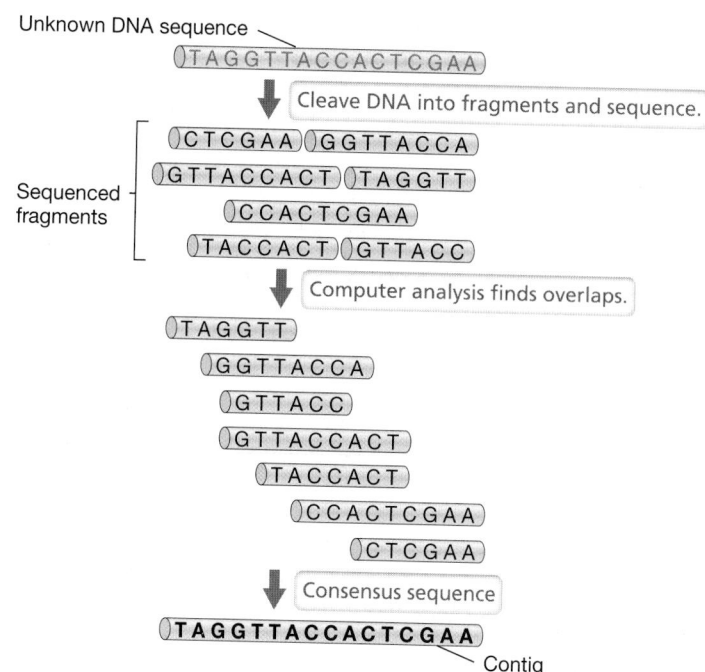

Figure 9.5 Computer assembly of DNA sequence. Most DNA sequencing methods generate vast numbers of short sequences (30 to several hundred bases) that must be assembled. The computer searches for overlaps in the short sequences and then arranges them to form contigs, or a consensus sequence.

Figure 9.4 DNA sequencing. *(a)* Sanger sequencing. When a polymerase incorporates a ddNTP during synthesis, the chain of DNA is terminated. The identity of the terminal ddNTP can be determined by capillary electrophoresis and fluorescence detection. *(b)* Pyrosequencing. Whenever a new dNTP is inserted into the growing strand of DNA (red arrows), pyrophosphate (PP$_i$) is released and is used to make ATP from AMP by the enzyme sulfurylase. The ATP is consumed by the enzyme luciferase, which releases light. Unused dNTPs are degraded by the enzyme apyrase (gray arrow).

to be useful, they must be *annotated* in order to identify genes and other functional regions. Many of the tasks surrounding genome assembly and annotation are highly computational. For genome assembly, a computer examines many short DNA fragments that have been sequenced and deduces their order by detecting all of the instances where two fragments of DNA possess overlapping sequence (**Figure 9.5**). These overlaps are used to merge sequencing reads into *contigs*, or contiguous consensus sequences. Individual contigs with overlapping ends are then aligned to form scaffolds (contigs as well as gaps) that are ultimately used to generate a map representing the complete genome.

From the genome map, the annotation process can begin. Because the genomes of *Bacteria* and *Archaea* possess very few intervening sequences (introns, ⮥ Section 4.6), their genomes essentially consist of a series of **open reading frames (ORFs)**

separated by short regulatory regions and transcriptional terminators. A *functional ORF* is one that actually encodes a protein (⮥ Section 4.9) and can be identified from a computer search of the sequence (**Figure 9.6**). Although any given cellular gene is always transcribed from one DNA strand, a gene can actually be

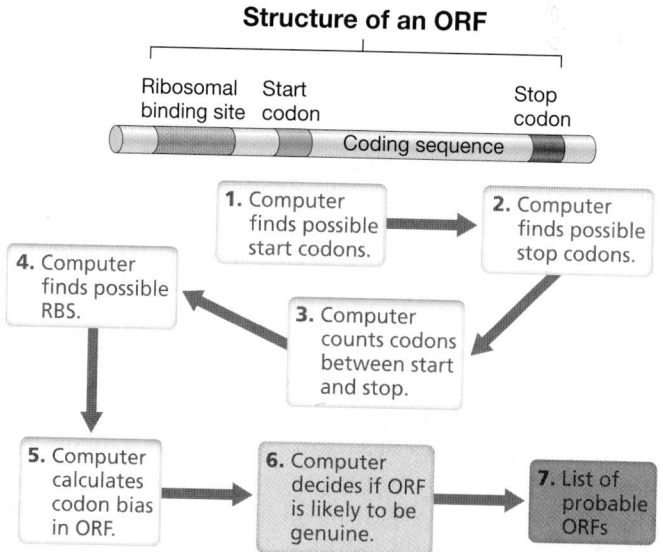

Figure 9.6 Computer identification of possible ORFs. The computer scans the DNA sequence looking first for start and stop codons. It then counts the number of codons in each uninterrupted reading frame and rejects those that are too short. The probability of a genuine ORF is made stronger if a likely ribosomal binding site (RBS) is found the correct distance in front of the reading frame. Codon bias calculations are used to test whether an ORF complies with the codon usage of the organism being examined.

TABLE 9.2 DNA sequencing methods

Generation	Method	Features
First generation	Sanger dideoxy method (radioactivity or fluorescence; DNA amplification)	Read length: 700–900 bases Used for the Human Genome Project
Second generation	454 Pyrosequencing (fluorescence; DNA amplification; massively parallel)	Read length: 400–500 bases Used to sequence genome of James Watson (completed 2007)
	Illumina/Solexa method (fluorescence; DNA amplification; massively parallel)	Read length: 50–100 bases Giant panda genome (2009; Beijing Genome Institute); Denisovan genome (2010)
	SOLiD method (fluorescence; DNA amplification; massively parallel)	Read length 50–100 bases
Third generation	HeliScope Single Molecule Sequencer (fluorescence; single molecule)	Read length: up to 55 bases Fossil DNA accuracy greatly improved
	Pacific Biosciences SMRT (fluorescence; single molecule; zero mode waveguide)	Read length: 2500–3000 bases
Fourth generation	Ion torrent (electronic—pH; DNA amplification)	Read length: 100–200 bases Sequenced genome of Intel cofounder Gordon Moore (originator of Moore's law), 2011
	Oxford nanopore (electronic—current; single molecule; real time)	Read length: thousands of bases Portable MinION unit is approximately the size of a flash drive

located on either strand and thus computer inspection of both strands of DNA is required.

Finding and Identifying ORFs

The first step in finding an ORF is to locate *start* and *stop* codons in the sequence (↩ Section 4.9 and Table 4.4). However, in-frame start and stop codons appear randomly with reasonable frequency; thus, further clues are needed. In *Bacteria*, translation begins at start codons located immediately downstream of a ribosome-binding sequence (Shine–Dalgarno site) on the mRNA (↩ Section 4.9). Thus, locating potential ribosome-binding sequences in addition to start and stop codons helps decide both whether an ORF is functional and which start codon is actually used. In addition, an ORF is more likely to be functional if its sequence is similar to those of ORFs in the genomes of other organisms (regardless of whether they encode known proteins) or if the ORF includes a sequence known to encode a protein functional domain. This is because proteins with similar functions in different cells tend to share a common evolutionary origin and thus share sequence and structural features (Section 9.5). A computer can search for sequence similarities in major databases such as GenBank (http://www.ncbi.nlm.nih.gov/Genbank/) using *BLAST* (*B*asic *L*ocal *A*lignment *S*earch *T*ool), an algorithm that can compare a nucleic acid or protein sequence with all other such sequences in the database.

Other issues must be considered in a genome annotation as well. For example, more than one codon exists for many of the 20 common amino acids (↩ Table 4.4), and some codons are used more frequently than others. The latter is known as **codon bias** (codon usage) and differs greatly between organisms. For example, **Table 9.3** shows the different usage of the six arginine codons in *Escherichia coli* compared to their usage in humans and fruit flies. If the codon bias in a given ORF differs greatly from the consensus for the organism containing it, that ORF may be nonfunctional or may be functional but obtained by horizontal gene transfer (Section 9.6).

Genomic Analyses: The Final Tally

No genome sequence project ends with 100% of the genome identified. In fact, this is one of the exciting findings of genomic analyses: Many genes in microbes almost certainly encode proteins whose function(s) remain unknown. Although there are differences among organisms, in most genomes the percentage of genes whose role can be clearly identified is approximately 70% of the total number of ORFs detected. Uncharacterized (or unknown) ORFs are said to encode *hypothetical proteins*, proteins that probably exist although their function is unknown. These ORFs have uninterrupted reading frames of reasonable length and the necessary start and stop codons and ribosome-binding site (Figure 9.6); however, the proteins they encode lack sufficient amino acid sequence homology with any known protein to be unambiguously identified. Some gene annotations can only assign a gene to a protein family or to a general function (such as "transport protein") without being more specific. Many of the unidentified genes in *E. coli* are thought to encode proteins that play a role in some unidentified regulatory process or are proteins required only for special nutritional or environmental conditions. A few may also function as "backups" of key enzymes.

TABLE 9.3 Examples of codon bias

	Usage of each arginine codon (%)		
Arginine codon[a]	**Escherichia coli**	**Fruit fly**	**Human**
AGA	1	10	22
AGG	1	6	23
CGA	4	8	10
CGC	39	49	22
CGG	4	9	14
CGU	49	18	9

[a]Arginine has six codons; see Table 4.6.

In addition to protein-encoding genes, some genes encode RNA molecules that are not translated. Such genes therefore lack start codons and may well have multiple stop codons within the gene. Some noncoding RNAs, such as tRNAs and rRNAs, are easy to detect because they are well characterized and are highly conserved. However, many noncoding regulatory RNA molecules (↩ Section 6.11) are conserved only in their three-dimensional structure, with little sequence homology. Thus transcriptomics, specifically RNA-Seq (Section 9.9), has become instrumental in identifying these noncoding genes.

With this general background in nucleic acid sequencing and the coding features of genomes, we move on to compare the nature of genomes in various microbial groups. We begin with the *Bacteria* and *Archaea* where thousands of genome sequences are available for comparative analyses.

MINIQUIZ

- What key molecules are essential for Sanger sequencing?
- What is an open reading frame (ORF)? What is a hypothetical protein?
- What is the major problem in identifying genes encoding nontranslated RNA?

9.3 Genome Size and Gene Content in *Bacteria* and *Archaea*

Once a genome has been assembled, *comparative genomics* using databases such as MicrobesOnline (http://www.microbesonline.org)—which contains nearly 4000 microbial genome sequences—can be used to probe its biological secrets. By using comparative genomics, it has been determined that genomes of *Bacteria* and *Archaea* show a strong correlation between genome size and open reading frame (ORF) content (**Figure 9.7**). Regardless of the organism, each megabase pair of DNA in a prokaryotic cell encodes about 1000 ORFs, and as the size of these genomes increases, the gene number also increases proportionally.

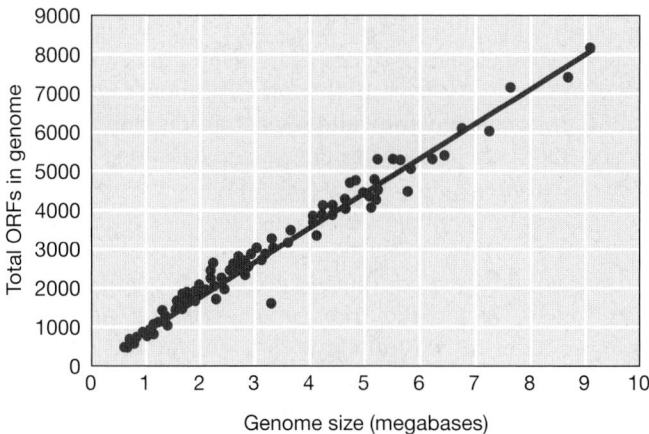

Figure 9.7 Correlation between genome size and ORF content in prokaryotic cells. Analyses of 115 completed genomes from species of both *Bacteria* and *Archaea*. Data from *Proc. Natl. Acad. Sci. (USA) 101*: 3160–3165 (2004).

This contrasts markedly with the genomes of eukaryotes, in which noncoding DNA (introns, ↩ Section 4.6) may constitute a large fraction of the genome, especially in organisms with large genomes (Figure 9.2).

Bacterial genomes range in size from the insect symbiont *Tremblaya princeps* with 121 protein-encoding genes to the soil-dwelling *Sorangium cellulosum* with nearly 100 times as many genes (Figure 9.2 and Table 9.1). While about 1300 genes has been the benchmark for the number of genes necessary for a cell to have a free-living existence, the recent discovery of free-living marine *Actinobacteria* containing approximately 800 genes has called this earlier estimate into question.

Small Genomes

The smallest cellular genomes belong to bacteria that are parasitic or endosymbiotic (cells that live inside other cells), with the insect symbionts *Tremblaya* (in mealybugs) and *Hodgkinia* (in cicadas) possessing some of the smallest genomes (around 140 kbp, Table 9.1 and **Figure 9.8**). The absolute smallest genome discovered thus far is that of *Nasuia deltocephalicola*, a sap-feeding insect symbiont whose genome is only 112 kbp. Because of their reduced

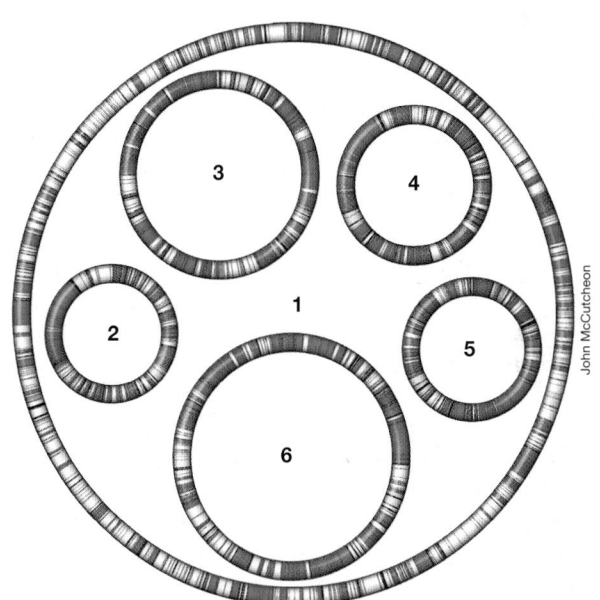

1. *Mycoplasma genitalium* (*Mollicutes*) 580.1 kbp GC: 31.7%	4. *Carsonella* (*Gammaproteobacteria*) 159.6 kbp GC: 16.6%
2. *Tremblaya* (*Betaproteobacteria*) 138.9 kbp GC: 58.8%	5. *Hodgkinia* (*Alphaproteobacteria*) 143.7 kbp GC: 58.4%
3. *Zinderia* (*Betaproteobacteria*) 208.5 kbp GC: 13.5%	6. *Sulcia* (*Bacteroidetes*) 245.5 kbp GC: 22.4%

Figure 9.8 Symbiont genomes. Five symbiont genomes are shown drawn to scale inside the circle representing the genome of a *Mycoplasma*. Blue: genes encoding genetic information processing; Red: genes encoding amino acid and vitamin biosyntheses; Yellow: rRNA genes; White: other genes; Gaps indicate noncoding DNA. Kbp, kilobase pairs. GC indicates percentage of nucleotides that are guanine or cytosine.

genome size, such symbionts are totally dependent on their insect host cells for survival and nutrients. In turn, the symbionts provide the insect with essential amino acids and other nutrients that the insect cannot synthesize.

With genomes of around 500 kbp, *Mycoplasma* (*Bacteria*) and *Nanoarchaeum equitans* (*Archaea*) have the smallest genomes among parasitic prokaryotic cells (Table 9.1). *N. equitans* is a hyperthermophile and a parasite of another hyperthermophile, the archaeon *Ignicoccus* (⇌ Section 17.6). *N. equitans* lacks virtually all genes that encode metabolic proteins and presumably depends on its host for most catabolic as well as anabolic functions. While some pathogens such as *Mycobacterium tuberculosis* have quite large genomes (4.4 Mbp), the genomes of most human pathogens, such as *Mycoplasma, Chlamydia*, and *Rickettsia*, are smaller than the largest known viral genome, that of *Pandoravirus* (2.5 Mbp, ⇌ Section 10.1).

Using *Mycoplasma*, which has around 500 genes, as a starting point, it has been estimated that around 250–300 genes are the minimum number possible for a viable cell. These estimates rely partly on comparisons with other small genomes. In addition, systematic mutagenesis has been performed to identify essential genes. For example, experiments with *Escherichia coli* and *Bacillus subtilis*, both of which have about 4000 genes, indicate that approximately 300–400 genes are essential depending on the growth conditions. However, in these experiments the bacteria were provided with many nutrients, allowing them to survive without many genes that encode biosynthetic functions. Most of the "essential genes" identified are present in other *Bacteria* as well and approximately 70% have also been found in *Archaea* and eukaryotes.

Large Genomes

Some *Bacteria* have genomes that are as large as those of some eukaryotic microbes. In fact, because eukaryotes tend to have significant amounts of noncoding DNA and bacteria do not, some bacterial genomes actually have more genes than microbial eukaryotes, despite having less DNA. For example, the genome of *Bradyrhizobium japonicum*, a bacterium that forms nitrogen-fixing root nodules on leguminous plants such as soybeans (⇌ Section 23.3), has 9.1 Mbp of DNA and 8300 ORFs, whereas the genome of the baker's yeast *Saccharomyces cerevisiae*, a eukaryote, has 12.1 Mbp of DNA and only 5800 ORFs (see Tables 9.1 and 9.5).

The largest bacterial genome known is that of *S. cellulosum*, a species of the gliding myxobacteria (⇌ Section 15.17). With just under 14.8 Mbp on a single circular chromosome, it has more DNA than several eukaryotes including yeast and the pathogenic protozoans *Cryptosporidium* and *Giardia* (see Table 9.5). The *S. cellulosum* genome is composed of roughly 10.5% noncoding DNA and 11,559 protein-encoding genes, making it over three times larger than the genome of *E. coli*. Interestingly, the *S. cellulosum* genome encodes 508 kinases (enzymes that phosphorylate other proteins to regulate their activity), which is over three times that of any other genome including those of eukaryotes. This suggests that the lifestyle of *S. cellulosum* is highly diverse and that its ecological success requires extensive regulation. In contrast to *Bacteria*, the

largest genomes found in species of *Archaea* thus far are only about 5 Mbp (Table 9.1).

Gene Content of Bacterial Genomes

The complement of genes in a particular organism reveals its capabilities. Conversely, genomes are molded by adaptation to particular lifestyles. Comparative analyses are useful when searching for genes that encode enzymes that probably exist because of the lifestyle of an organism, and in some cases these searches yield big surprises. For example, *Vampirovibrio chlorellavorus* has been reported to be a predatory bacterium that attacks its host, the green alga *Chlorella*, by surface attachment and ultimate ingestion of its cellular contents (thus the terms "vámpír" from Hungarian, meaning "blood sucker," and "vorus" from Latin, meaning "to devour," in the organism's name). Isolates of *V. chlorellavorus* existed only as 36-year-old freeze-dried samples that had not been successfully revived. However, by using advanced sequencing techniques, the genome of *V. chlorellavorus* was recovered, and surprisingly, genomic analyses indicated that *V. chlorellavorus* falls within the phylum *Cyanobacteria*, even though it lacks genes for photosynthesis.

Figure 9.9, reprinted from a scientific journal, is included here to give you an idea of why a microbe's genome should be sequenced and the amazing amount of information that can be gleaned from annotation, though the details are beyond the scope of this chapter. The figure summarizes some of the metabolic pathways and transport systems of *V. chlorellavorus* deduced from analysis of its genome. These include an electron transport chain for microaerobic growth, the ability to ferment, chemotaxis abilities, and the synthesis of 15 of the 20 essential amino acids. Comparative genomics also indicated that *V. chlorellavorus* used a conjugative type IV secretion system (⇌ Section 4.13) to attack its prey, the first discovery of this strategy in a predatory bacterium.

A functional analysis of genes and their activities in several bacteria is given in **Table 9.4**. Thus far, a distinct pattern of gene distribution in *Bacteria* has emerged. Metabolic genes are typically the most abundant class in bacterial genomes, although genes for protein synthesis overtake metabolic genes on a percentage basis as genome size decreases (Table 9.4 and **Figure 9.10**). Although many genes can be dispensed with, genes that encode the protein-synthesizing apparatus cannot. Thus, the *smaller* the genome, the greater the percentage of genes that encode translational processes. Conversely, the *larger* the genome, the more genes there are for transcriptional regulation and signal transduction.

Analyses of gene categories have also been done for several *Archaea*. On average, *Archaea* devote a higher percentage of their genomes to energy and coenzyme production than do *Bacteria* (this result is undoubtedly skewed a bit due to the large number of novel coenzymes produced by methanogenic *Archaea*, ⇌ Section 14.17). On the other hand, *Archaea* appear to contain fewer genes for carbohydrate metabolism and membrane functions (such as transport and membrane biosynthesis) than do *Bacteria*. However, this conclusion may also be skewed a bit because the corresponding pathways have been less studied in *Archaea* than in *Bacteria* and many of the relevant archaeal genes remain unidentified.

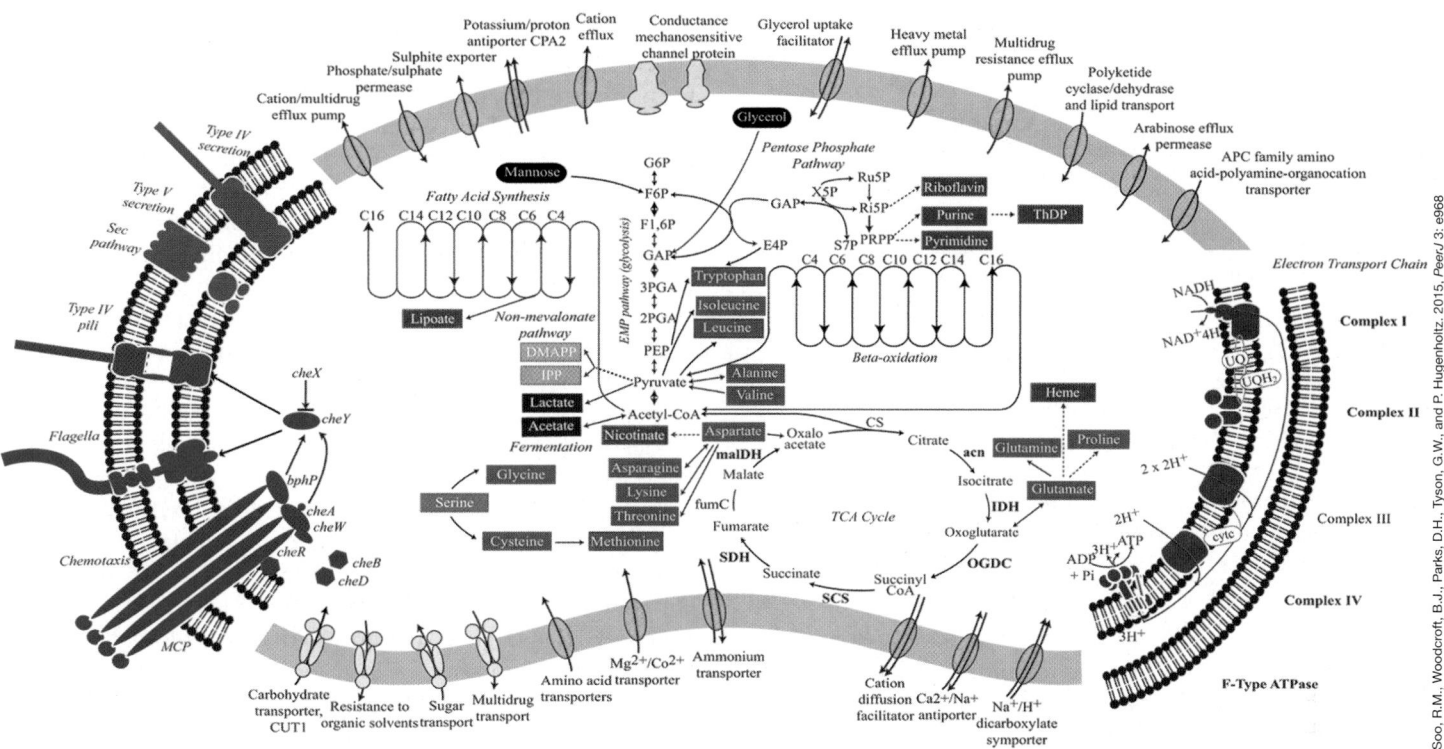

Soo, R.M., Woodcroft, B.J., Parks, D.H., Tyson, G.W., and P. Hugenholtz. 2015, *PeerJ 3*: e968

Figure 9.9 Functional and metabolic predictions for *Vampirovibrio chlorellavorus* based on genomic annotation. Although the details are beyond our discussion, the figure illustrates the power of genomic sequencing and annotation. Within the membrane, the following systems are highlighted: secretion (green), chemotaxis and movement (blue), electron transport (red), ATP-binding cassette transporters (yellow), and permeases/pumps/transporters (orange). Black ovals indicate substrates that enter the glycolysis pathway, while fermentation end-products are indicated as black rectangles. Colors of internal compounds correspond to the following: green (amino acids), red (cofactors and vitamins), purple (nucleotides), and orange (non-mevalonate pathway products). Note that genes for synthesis of serine (highlighted in blue) are not present, so presumably it is transported into the cell. Adapted from Soo, R.M., et al. 2015. *Peer J 3:* e968.

MINIQUIZ

- What lifestyle is typical of *Bacteria* and *Archaea* that contain fewer than 500 protein-encoding genes?
- Which is likely to have more genes, a species of *Bacteria* with 8 Mbp of DNA or a eukaryote with 10 Mbp? Explain.
- In prokaryotic cells with the largest genomes, which gene category contains the largest percentage of genes?

9.4 Organelle and Eukaryotic Microbial Genomes

Mitochondria and chloroplasts are eukaryotic cell organelles derived from endosymbiotic bacteria (↩ Sections 2.15 and 18.1), and thus share many fundamental traits with *Bacteria* to which they are phylogenetically related. The genomes of both organelles encode the machinery necessary for protein synthesis including

---

TABLE 9.4 Gene function in some genomes of *Bacteria*

	Percentage of genes		
Functional categories	Escherichia coli *(4.64 Mbp)*[a]	Haemophilus influenzae *(1.83 Mbp)*[a]	Mycoplasma genitalium *(0.58 Mbp)*[a]
Metabolism	21.0	19.0	14.6
Structure	5.5	4.7	3.6
Transport	10.0	7.0	7.3
Regulation	8.5	6.6	6.0
Translation	4.5	8.0	21.6
Transcription	1.3	1.5	2.6
Replication	2.7	4.9	6.8
Other, known	8.5	5.2	5.8
Unknown	38.1	43.0	32.0

[a]Chromosome size, in megabase pairs. Each organism listed contains only a single circular chromosome.

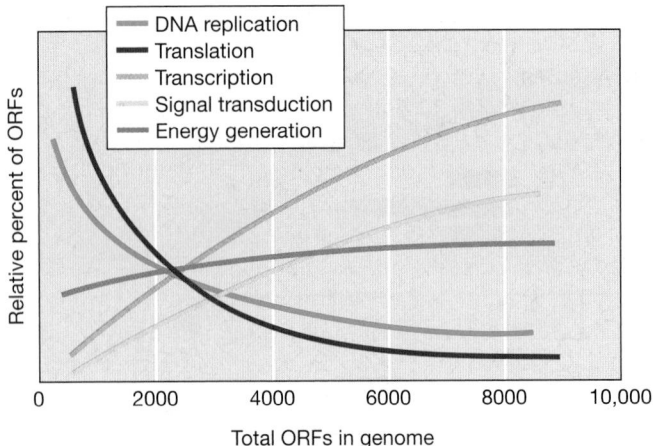

Figure 9.10 Functional category of genes as a percentage of the genome. The percentage of genes encoding products for translation or DNA replication is greater in organisms with small genomes, whereas the percentage of transcriptional regulatory genes is greater in organisms with large genomes.

ribosomes, transfer RNAs, and the other components necessary to drive translation. The genomes of several microbial eukaryotes have also been sequenced (**Table 9.5**), and their size varies widely (Figure 9.2). Certain single-celled protozoans, including the free-living ciliate *Paramecium* (40,000 genes) and the pathogen *Trichomonas* (60,000 genes), have significantly more genes than do humans (Table 9.5). In this section we focus on organellar genomes and the genomes of a few select microbial eukaryotes.

The Chloroplast Genome

Green plant cells contain chloroplasts, the organelles that perform photosynthesis (♻ Section 14.1). All known chloroplast genomes are circular DNA molecules, and each chloroplast contains several identical copies of the genome. The typical chloroplast genome is about 120–160 kbp and contains two inverted repeats of 6–76 kbp that each encode copies of the three rRNA genes (**Figure 9.11**). As might be expected, many chloroplast genes encode proteins for photosynthetic reactions and autotrophy. For example, the enzyme RubisCO catalyzes the first step in CO_2 fixation in the Calvin cycle (♻ Section 14.5). The *rbcL* gene encoding the large subunit of RubisCO is present on the chloroplast genome (see Figure 9.11), whereas the gene for the small subunit, *rbcS*, resides in the plant cell nucleus and its protein product must be imported from the cytoplasm into the chloroplast after synthesis.

The chloroplast genome also encodes tRNAs used in translation, several proteins used in transcription and translation, and some other proteins. Not all chloroplast proteins are encoded by the chloroplast genome; some are nuclear encoded. These are thought to be genes that migrated to the nucleus as the chloroplast evolved from an endosymbiotic cell into a photosynthetic organelle. Introns are common in chloroplast genes and are primarily of the self-splicing type (♻ Section 4.6).

Mitochondrial Genomes and Proteomes

Mitochondria are the eukaryotic cell's respiratory organelles and are present in all but a few eukaryotes (♻ Sections 2.15 and 18.1). Mitochondrial genomes primarily encode proteins for oxidative

TABLE 9.5 Some eukaryotic nuclear genomes[a]

Organism	Comments	Lifestyle[b]	Genome size (Mbp)	Haploid chromosomes	ORFs
Nucleomorph of *Bigelowiella natans*	Degenerate endosymbiotic nucleus	E	0.37	3	331
Encephalitozoon intestinalis	Smallest known eukaryotic genome, human pathogen	P	2.3	11	1,800
Cryptosporidium parvum	Parasitic protozoan	P	9.1	8	3,800
Plasmodium falciparum	Malignant malaria	P	23	14	5,300
Saccharomyces cerevisiae	Yeast, a model eukaryote	FL	13.4	16	5,800
Ostreococcus tauri	Marine green alga, smallest free-living eukaryote	FL	12.6	20	8,200
Aspergillus nidulans	Filamentous fungus	FL	30	8	9,500
Giardia intestinalis (also called *Giardia lamblia*)	Flagellated protozoan, causes acute gastroenteritis	P	12	5	9,700
Drosophila melanogaster	Fruit fly, model organism for genetic studies	FL	180	4	13,600
Caenorhabditis elegans	Roundworm, model for animal development	FL	97	6	19,100
Mus musculus	Mouse, a model mammal	FL	2,500	23	25,000
Homo sapiens	Human	FL	2,850	23	25,000
Arabidopsis thaliana	Model plant for genetics	FL	125	5	26,000
Paramecium tetraurelia	Ciliated protozoan	FL	72	>50	40,000
Pinus taeda	Loblolly pine tree	FL	20,000	19	50,000
Trichomonas vaginalis	Flagellated protozoan, human pathogen	P	160	6	60,000

[a]All data are for the haploid nuclear genomes of these organisms in megabase pairs. For most large genomes, both size and ORFs listed are best estimates due to large numbers of repetitive sequences and/or introns in the genomes.
[b]E, endosymbiont; P, parasite; FL, free-living.

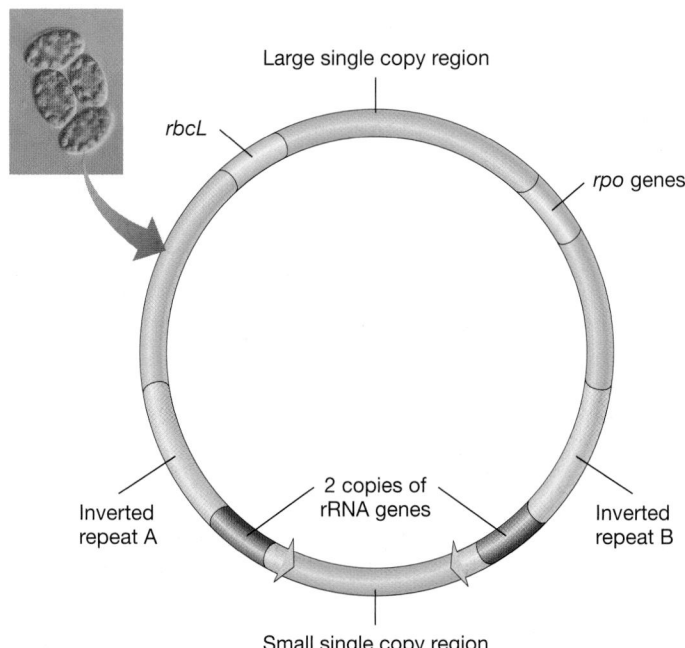

Figure 9.11 Map of a typical chloroplast genome. The inverted repeats each contain a copy of the three genes for rRNA (5S, 16S, and 23S). The large subunit of RubisCO is encoded by *rbcL* and the chloroplast RNA polymerase by *rpo* genes. Inset: Photo of four cells of the green alga *Makinoella* with chloroplasts clearly visible.

phosphorylation and, like chloroplast genomes, also encode proteins, rRNAs, and tRNAs for protein synthesis. However, most mitochondrial genomes encode far fewer proteins than those of chloroplasts. The largest mitochondrial genome known has only 62 protein-encoding genes, but others contain as few as three. The mitochondria of almost all mammals, including humans, encode only 13 proteins in addition to 22 tRNAs and 2 rRNAs. **Figure 9.12a** shows a map of the 16,569-bp human mitochondrial genome. While human mitochondrial genomes are circular, diverse arrangements exist in other organisms. For example, some mitochondrial genomes are linear, including those of certain algae, protozoans, and fungi. Finally, the mitochondria of many fungi and flowering plants contain, in addition to the mitochondrial genome, small circular or linear plasmids (⟳ Section 4.2).

Mitochondria require many more proteins than their genome encodes (in particular, proteins needed for translation), and thus many mitochondrial proteins are encoded by genes in the nucleus. The yeast mitochondrion contains as many as 800 different proteins in its proteome (all the proteins encoded by a genome; Section 9.10). However, only eight (~1%) of them are encoded by the yeast mitochondrial genome, the remaining proteins being encoded by nuclear genes (Figure 9.12b). However, the nuclear-encoded proteins required for translation and energy generation in mitochondria are more closely related to their counterparts in *Bacteria* than to those in the eukaryotic cytoplasm, consistent with both the evolutionary history of the mitochondrion and with a scenario—like that seen in the chloroplast—of genes having migrated from the original endosymbiont to the host cell nucleus.

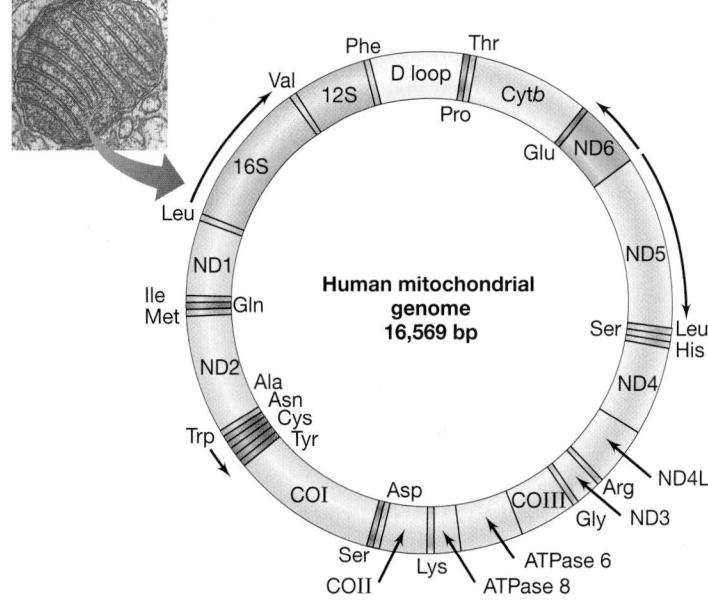

(a)

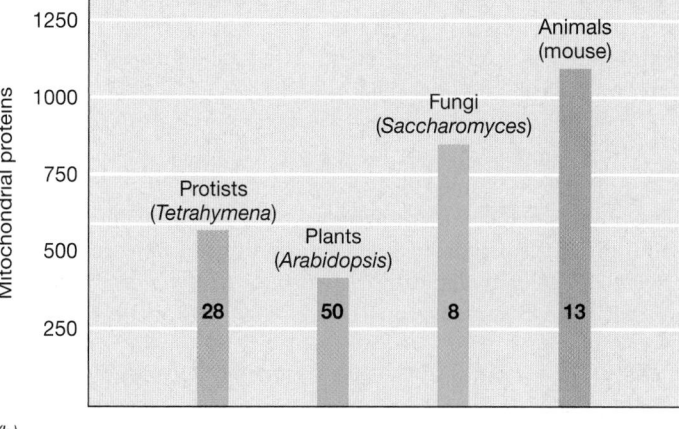

(b)

Figure 9.12 Map of the human mitochondrial genome and the mitochondrial proteome. *(a)* The genome encodes rRNAs, 22 tRNAs, and several proteins. Arrows show direction of transcription for genes of a given color, and the three-letter amino acid designations for tRNA genes are also shown. The 13 protein-encoding genes are in green. Cyt*b*, cytochrome *b*; ND1–6, components of the NADH dehydrogenase complex; COI–III, subunits of the cytochrome oxidase complex; ATPase 6 and 8, polypeptides of the mitochondrial ATPase complex. The two promoters are in the region called the D loop, which is also involved in DNA replication. Inset: Transmission electron micrograph of a mitochondrion (credit, D. W. Fawcett). *(b)* Mitochondrial proteomes. The numbers in each colored bar are the number of proteins encoded on the mitochondria of some model eukaryotes.

Genomes and Introns in Some Microbial Eukaryotes

Apart from the human pathogen *Trichomonas*, which contains almost three times more genes than human cells, parasitic eukaryotic microorganisms typically have relatively small genomes of 10–40 Mbp containing between 4000 and 11,000 genes. For example, *Trypanosoma brucei*, the agent of African sleeping sickness, has 11 chromosomes, 35 Mbp of DNA, and almost 11,000 genes. The four species of *Plasmodium* that infect humans (causing malaria, ↩ Section 33.5) have genomes ranging from 23 to 27 Mbp arranged in 14 chromosomes and about 5500 genes.

As in *Bacteria*, the smallest eukaryotic genome belongs to an endosymbiont. Known as a *nucleomorph*, it is the degenerate remains of a eukaryotic endosymbiont of a certain green alga that has acquired the ability to photosynthesize by secondary endosymbiosis (↩ Section 18.1). Nucleomorph genomes range from about 0.37 to 0.85 Mbp. The smallest genome in a *parasitic* eukaryote belongs to *Encephalitozoon intestinalis*, an intracellular pathogen of humans and other animals. *E. intestinalis* even lacks mitochondria, and although its haploid genome contains 11 chromosomes, the genome size is only 2.3 Mbp with approximately 1800 genes (Table 9.5); this is smaller than many bacterial genomes (Table 9.1).

The baker's yeast *Saccharomyces cerevisiae* is widely used as a model eukaryote and its genome contains 16 chromosomes (13.4 Mbp of DNA). Yeast has approximately 6000 ORFs, which is fewer than that of some genomes of *Bacteria* (Tables 9.1 and 9.5). How many of these yeast genes are actually essential? This question has been addressed by systematically inactivating each gene in turn with *knockout mutations* (mutations that completely inactivate genes, ↩ Section 12.4). Knockout mutations cannot normally be obtained in essential genes in a haploid organism. However, yeast can be grown in both diploid and haploid states (↩ Section 18.9). By generating knockout mutations in diploid cells and then investigating whether they can also exist in haploid cells, it is possible to determine whether a particular gene is essential for cell viability. Using knockout mutations, it has been shown that around 900 yeast ORFs (17% of its genome) are absolutely essential. Note that this number of essential genes is much greater than the approximately 300 genes (Section 9.3) estimated to be the minimal number required in a bacterial cell.

Being a eukaryote, the yeast genome contains introns (↩ Section 4.6). However, the total number of introns in the protein-encoding genes of yeast is a mere 225. Most yeast genes that contain introns have only a single small intron near the 5′ end of the gene. This situation differs greatly from that seen in more complex eukaryotes (**Figure 9.13**). For example, in the worm *Caenorhabditis elegans*, the average gene has five introns, and in the fruit fly *Drosophila*, the average gene has four. Introns are also common in the genes of plants, averaging around four per gene. The model flowering plant *Arabidopsis* averages five introns per gene, and over 75% of *Arabidopsis* genes have introns. In humans almost all protein-encoding genes have introns, and it is common for a single gene to have 10 or more. Moreover, introns in human genes are typically much longer than exons, the DNA that actually encodes proteins. Indeed, exons make up only about 1% of the human genome, whereas introns account for 24%. The remaining DNA is made up of repetitive sequences, noncoding RNA, and regulatory regions. Nevertheless, as we will see later, much of this DNA is indeed functional (Section 9.14).

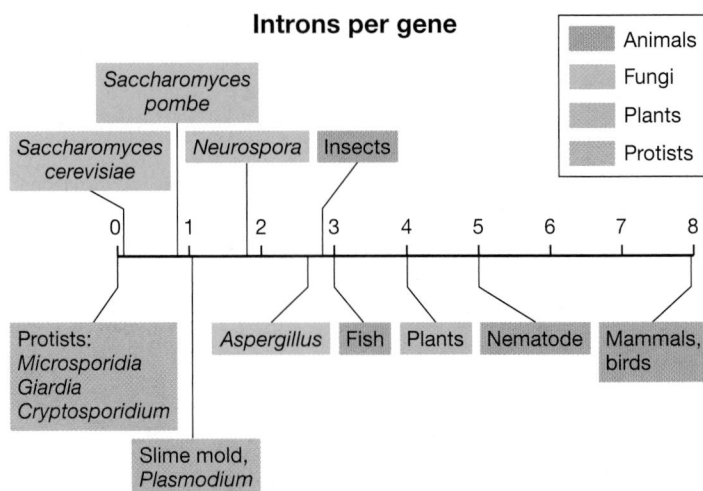

Figure 9.13 Intron frequency in the genes of different eukaryotes. The average number of introns per gene is shown for a range of eukaryotic organisms.

MINIQUIZ

- What is unusual about the genes that encode mitochondrial proteins?
- What do chloroplast genomes typically encode?
- What is unusual about the genome of the eukaryote *Encephalitozoon*?

II • The Evolution of Genomes

In addition to revealing how genes function and how organisms interact with their environments, comparative genomics can illuminate evolutionary relationships between organisms. Reconstructing evolutionary trees from genome sequences helps to distinguish between primitive and derived characteristics and can resolve ambiguities in phylogenetic trees based on analyses of a single gene, such as an rRNA gene (↩ Section 13.3). Genomics is also a link to understanding early life forms and may eventually help answer the most fundamental of all questions in biology: How did life originate?

9.5 Gene Families, Duplications, and Deletions

Genomes from both prokaryotic and eukaryotic cells often contain multiple copies of genes that are related in sequence due to shared evolutionary ancestry; such genes are called *homologous genes*, or **homologs**. Groups of gene homologs are called **gene families**. Not surprisingly, larger genomes tend to contain more individual members from a particular gene family than do smaller genomes. *Gene duplication* is thought to be a major driving force behind the evolution of gene families and the organisms that contain them.

Paralogs, Orthologs, and Gene Duplications

Comparative genomics shows that many genes have arisen by duplication of other genes. Such homologs may be subdivided,

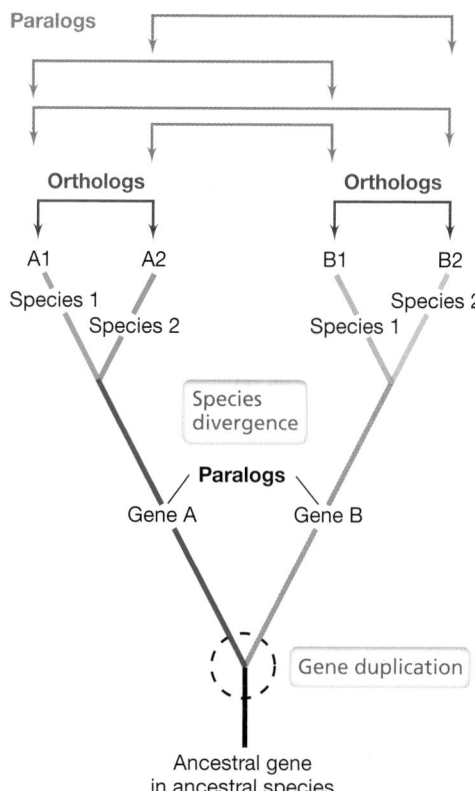

Figure 9.14 Orthologs and paralogs. This family tree depicts an ancestral gene that duplicated and diverged into two paralogous genes, A and B. Later, the ancestral species diverged into species 1 and species 2, both of which have genes for A and B (designated A1 and B1 and A2 and B2, respectively). Each such pair are paralogs. However, because species 1 and 2 are now separate species, A1 is an ortholog of A2 and B1 is an ortholog of B2.

depending on their origins. Genes whose similarity is the result of gene duplication at some time in the evolution of an organism are called **paralogs**. Genes found in one organism that are similar to genes in another organism because of descent from a common ancestor are called **orthologs** (**Figure 9.14**). An example of paralogous genes is evident in the genes that encode variant human lactate dehydrogenases (LDH), which recycle NAD⁺ through the conversion of pyruvate to lactate in tissues where oxygen is absent. These variants, called *isoenzymes*, are structurally distinct yet all highly related and carry out the same enzymatic reaction. By contrast, the corresponding LDH from the lactic acid bacterium *Lactobacillus* would be said to be orthologous to the human LDH isoenzymes. Thus, gene families contain both paralogs and orthologs.

If a segment of duplicated DNA is long enough to include an entire gene or group of genes, the organism with the duplication has multiple copies of these particular genes. After duplication, one of the duplicates is free to undergo spontaneous mutations while the other copy continues to supply the cell with the original function (**Figure 9.15a**). In this way, evolution can "experiment" with one copy of the gene. Such gene duplication events, followed by diversification of one copy, are thought to be the major events that fuel microbial evolution. Genomic analyses have revealed many examples of protein-encoding genes that were clearly derived from gene duplication. Figure 9.15b shows this for the enzyme RubisCO, a key enzyme of autotrophic metabolism (⇄ Section 14.5): An ancestral RubisCO gene gave rise to enzymes with different but related catalytic activities.

UNIT 3

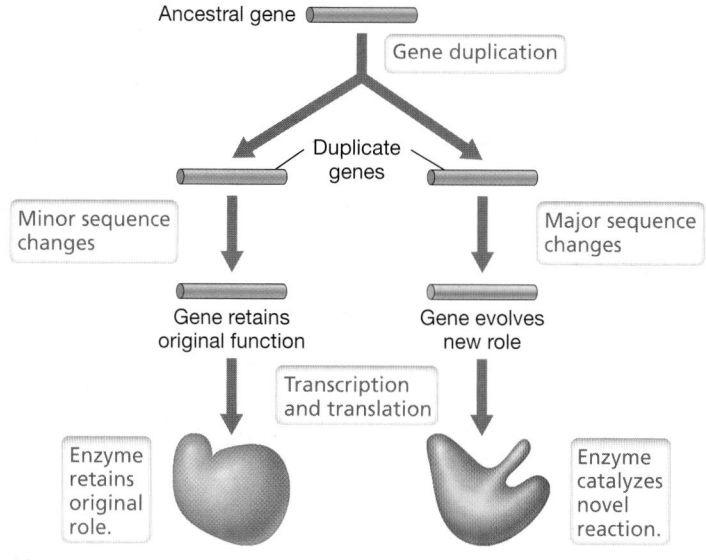

(a)

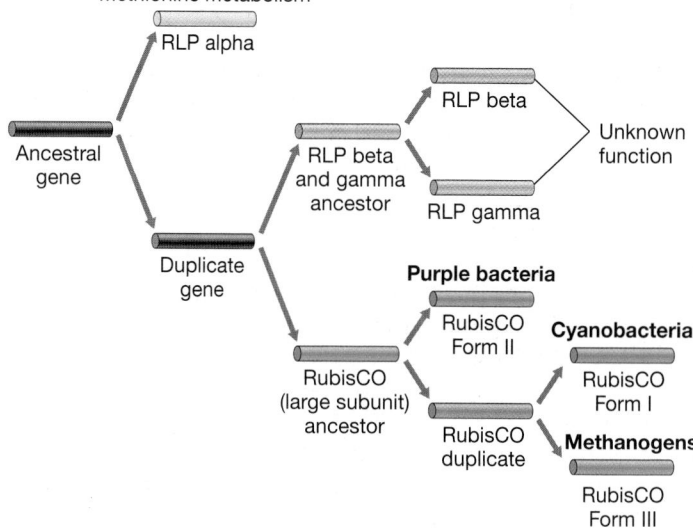

(b)

Figure 9.15 Evolution by gene duplication. (a) The principle of gene duplication. After duplication, the "spare" copy of a gene is free to evolve to encode a new function. (b) The RubisCO (*rbcL*) family of genes. The large subunit of the enzyme RubisCO that fixes CO₂ during photosynthesis is an ancestor of three closely related forms (I, II, and III) that all retain the original function (green bars). However, RubisCO is in turn derived from an ancestral gene (black bars) of unknown function that divided to produce a gene encoding an enzyme in methionine metabolism (yellow bar) and several genes whose function is still unknown (purple bars). RLP, RubisCO-like protein.

Entire Genome Duplications

Duplications of genetic material may include just a handful of bases, one or more genes, or even whole genomes. For example, comparison of the genomes of the yeast *Saccharomyces cerevisiae* and other fungi indicates that the ancestor of *S. cerevisiae* duplicated its entire genome. This was followed by extensive deletions that eliminated much of the duplicated genetic material. Analysis of the genome of the model plant *Arabidopsis* suggests that there were one or more whole genome duplications in the ancestor of the flowering plants, as well.

Some bacterial genomes show evidence of having once been duplicated. The distribution of duplicated genes and gene families in the genomes of *Bacteria* and *Archaea* suggest that several duplications have occurred. For example, the soil bacterium *Myxococcus* has a genome of 9.1 Mbp. This is approximately twice that of the genomes of its close relatives. However, while genomic analyses point to frequent small-scale gene duplications being rather common in prokaryotic cells, entire genome duplications appear to be rather rare. Conversely, in parasitic bacteria, frequent successive gene *deletions* have eliminated genes no longer needed for a parasitic lifestyle. This has been the driving force behind the unusually small genomes of many endosymbiotic bacteria. Insect endosymbionts have carried this theme to an extreme and show the smallest of all cellular genomes (Section 9.3, Table 9.1, and Figure 9.8).

MINIQUIZ

- What is a homologous gene?
- What is a gene family?
- Contrast gene paralogs with gene orthologs.

9.6 Horizontal Gene Transfer and the Mobilome

Genetic traits are transferred from one generation to the next by what's called a vertical process (from mother to daughter). However, vertical transfer in *Bacteria* and *Archaea* can be embellished by **horizontal gene transfer** (sometimes called *lateral gene transfer*), and this can complicate the analysis of genomes. Horizontal transfer refers to gene transfer from one cell to another by means other than the vertical process (**Figure 9.16**). In prokaryotic cells, at least three mechanisms of horizontal gene transfer are known: *transformation, transduction*, and *conjugation*, and these are discussed in detail in Chapter 11.

Detecting Horizontal Gene Flow

Horizontal gene transfers can be detected in genomes once the genes have been annotated (Section 9.2). The presence of genes that encode proteins typically found only in distantly related species is one signal that the genes originated from horizontal transfer. However, another clue to horizontally transferred genes is the presence of a stretch of DNA whose guanosine/cytosine (GC) content or codon bias (⬥ Section 4.9) differs significantly from that of the rest of the genome. Using these clues, many likely examples of horizontal transfer have been documented in the genomes of both *Bacteria* and *Archaea*. A classic example exists

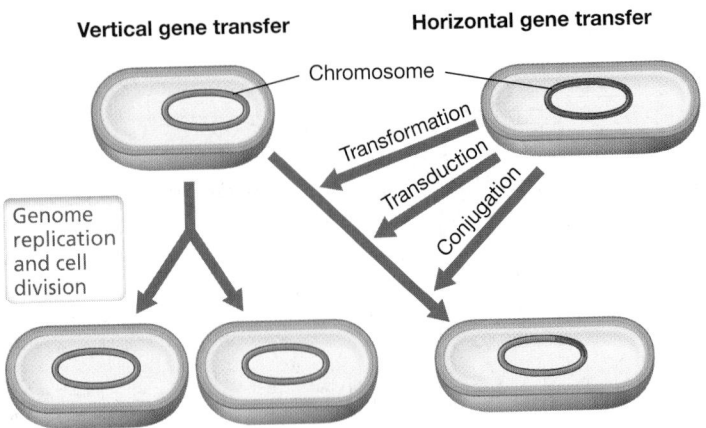

Figure 9.16 Vertical versus horizontal gene transfer. Vertical gene transfer occurs when cells divide. Horizontal gene transfer occurs when a donor cell contributes genes to a recipient cell. In *Bacteria* and *Archaea*, horizontal transfer occurs through one of three mechanisms: transformation, transduction, and conjugation.

with the thermophilic bacterium *Thermotoga maritima*, whose genome was shown to contain over 400 genes (greater than 20% of the entire genome) of archaeal origin. Of these genes, 81 were found in discrete clusters. The latter is a strong indication that the archaeal genes were transferred to *T. maritima* by horizontal gene flow from thermophilic *Archaea* that share its hot habitat.

For horizontal gene transfer to be readily detectable by comparative genomics, the phylogenetic difference between the organisms must be rather large. For example, several eukaryotic genes have been found in *Chlamydia*, a bacterial pathogen that causes both a sexually transmitted disease and an eye infection called trachoma in humans. In particular, two genes encoding histone H1-like proteins have been found in the *Chlamydia trachomatis* genome, suggesting horizontal transfer from a eukaryotic source, possibly even its human host. Note that this is conceptually the reverse of mitochondrial and chloroplast gene flow in which genes from the ancestor of the mitochondrion and the chloroplast were transferred to the eukaryotic nucleus (Section 9.4). In contrast to the *Chlamydia* findings, more subtle horizontal transfers, such as those between fairly closely related organisms, can be easily hidden and therefore missed during genome annotation.

Horizontally transferred genes typically encode metabolic functions distinct from the core molecular processes of DNA replication, transcription, and translation, and may account for observed similarities of metabolic genes in *Archaea* and *Bacteria*. In addition, there are several examples of virulence genes of pathogenic bacteria that have been transferred by horizontal means. It is clear that prokaryotic cells are actively exchanging genes in nature, and the process likely functions to "fine-tune" an organism's genome to a particular ecological situation or habitat. Nevertheless, it is necessary to be cautious when invoking horizontal gene transfer to explain the distribution of genes in a given organism. Homologs of the genes in question may be present in close relatives whose genome sequences are not yet available.

The Mobilome

Horizontal gene flow is facilitated by the **mobilome**, the sum total of all mobile genetic elements in a genome. The mobilome includes plasmids, prophages (integrated virus genomes), integrons, insertion sequences, and transposons (⏎ Section 11.11). Integrons are genetic elements that can capture gene cassettes (mobile DNA containing a recombination site) through the activity of an enzyme called integrase. All constituents of the mobilome also play important roles in genome evolution by shuffling genes between species.

Transposons are mobile genetic elements that move between different host DNA molecules, including chromosomes, plasmids, and viruses (**Figure 9.17**), by the activity of an enzyme called *transposase* (⏎ Section 11.11). In doing so, transposons may pick up and horizontally transfer genes encoding various characteristics, including resistance to antibiotics and production of toxins. Because of this tendency, transposable elements are a strong driver of genome evolution. However, transposons may also mediate a variety of large-scale chromosomal changes (Figure 9.17). Bacteria undergoing rapid evolutionary change often contain relatively large numbers of mobile elements, especially *insertion sequences*, simple transposable elements whose genes encode only transposition. Recombination among identical elements generates chromosomal rearrangements such as deletions, inversions, or translocations, and these provide a source of genomic diversity upon which natural selection can act. Thus, chromosomal rearrangements that accumulate in bacteria during stressful growth conditions are often flanked by repeats or insertion sequences.

Chromosomal rearrangements due to insertion sequences have apparently contributed to the evolution of several bacterial pathogens, increasing their pathogenic potential. In the genera *Bordetella, Yersinia,* and *Shigella*, the more highly pathogenic species show a much greater content of insertion sequences. For example, *Bordetella bronchiseptica*, which causes a bronchitis-like cough in dogs and other domestic animals, has a genome of 5.3 Mbp but carries no known insertion sequences. Its more pathogenic relative, *Bordetella pertussis*, the causative agent of whooping cough in humans (⏎ Section 30.3), has a smaller genome (4.1 Mbp) but has more than 260 insertion sequences. Comparison of these genomes suggests that the insertion sequences are responsible for major genome rearrangements, including the deletions that reduced the genome size in *B. pertussis*, possibly as a means for streamlining its genome to allow for more virulence factors to be encoded.

The mobilome, especially the horizontal transfer of transposable elements and prophages, is also responsible for the presence of *chromosomal islands* in certain strains. We turn our attention to these islands now and examine their relationship to the rest of the genome.

MINIQUIZ

- Which class of genes is rarely transferred horizontally? Why?
- List the major mechanisms by which horizontal gene transfer occurs in *Bacteria* and *Archaea*.
- How might transposons be especially important in the evolution of pathogenic bacteria?

9.7 Core Genome versus Pan Genome

One of the most important concepts to emerge from comparing the genome sequences of multiple strains of the same bacterial species is the distinction between the **pan genome** and the **core genome**. The *core genome* is that shared by all strains of a given species, whereas the *pan genome* of an organism includes the core plus genes not shared by all strains of that species. As we have seen, horizontal gene transfer of entire genetic elements such as plasmids, viruses, or transposable elements is widespread. Consequently, there may be major differences in the total amount of DNA and the suite of accessory capabilities (virulence, symbiosis, biodegradation, and the like) between strains of a single bacterial species. In other words, one could say that the core genome is typical of the species as a whole, whereas the other components that make up the pan genome, frequently including mobile elements, are restricted to particular strains within a species.

It is difficult to define the size of the pan genome precisely because it increases as the genomes of more strains of a species are sequenced. In some cases, such as the enteric bacteria *Escherichia coli* and *Salmonella enterica*, many different isolates have been found that carry a wide range of different mobilome elements. Consequently the pan genome is extremely large. **Figure 9.18** illustrates the pan genome for serovars (strains distinguished by their immunological properties) of the important human pathogen

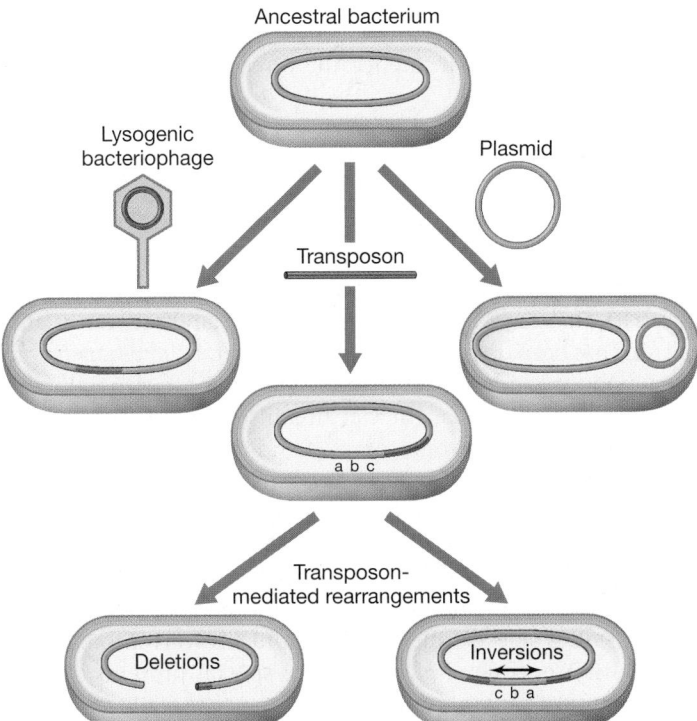

Figure 9.17 Mobile elements promote genome evolution. A variety of mobile genetic elements can move from one organism to another, thus adding genes to the genome of the recipient. The most common of these are plasmids, bacteriophages, and transposons. When a transposon moves, chromosomal rearrangements, such as deletions and inversions of DNA neighboring the transposon, may be mediated by the activity of the transposase.

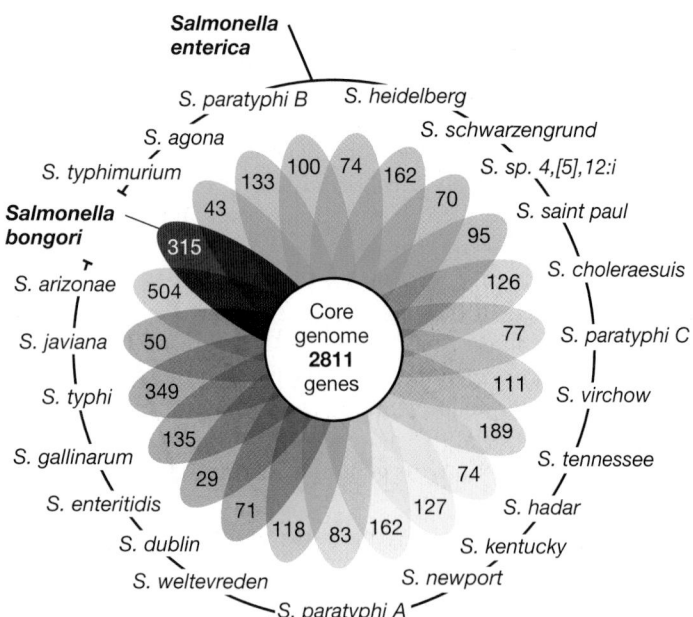

Figure 9.18 Flowerplot of the *Salmonella enterica* pan genome.
A "flowerplot" of gene families in serovars (strains) of the gram-negative pathogenic bacterium *Salmonella enterica* (the names surrounding the flowerplot are immunologically unique serovars [S.] of *S. enterica*). The figure presents the average number of gene families found in each genome as being unique to each serovar. *Salmonella bongori* is a species distinct from *S. enterica*. Serovar 4,[5],12:i has not yet been named. Data from Jacobsen, A., R.S. Hendriksen, F.M. Aaresturp, D.W. Ussery, and C. Friis. 2011. The *Salmonella enterica* pan-genome. *Microb Ecol 62*: 487–504.

S. enterica (salmonellosis) depicted in a "flowerplot" schematic. As is evident, although all strains contain at least 2811 genes, some contain several hundred more (Figure 9.18).

Chromosomal Islands

Comparison of the core and pan genomes of a particular species with their close relatives sometimes reveals extra blocks of genetic material that are part of the chromosome, rather than existing as plasmids or integrated viruses. These so-called **chromosomal islands**, or *genomic islands*, contain clusters of genes for specialized functions that are not essential for survival (**Figure 9.19**). Consequently, two strains of the same bacterial species may show significant differences in genome size. Chromosomal islands are presumed to be of "foreign" origin based on several lines of evidence. First, these extra genes are often flanked by inverted repeats, implying that the whole region was inserted into the chromosome by transposition (Section 9.6) at some point. Second, the base composition and codon bias (Table 9.3) in chromosomal islands often differ significantly from that of the genome proper. Third, chromosomal islands are found in some strains of a particular species but not in others.

Chromosomal islands in pathogenic bacteria have drawn the most attention because in many cases genes encoding disease-associated functions have been linked to an island. However, chromosomal islands are also known that encode the biodegradation of pollutants such as aromatic hydrocarbons and herbicides. In addition, many of the genes essential for the symbiotic relationship of species of the nitrogen-fixing bacterium *Rhizobium* with

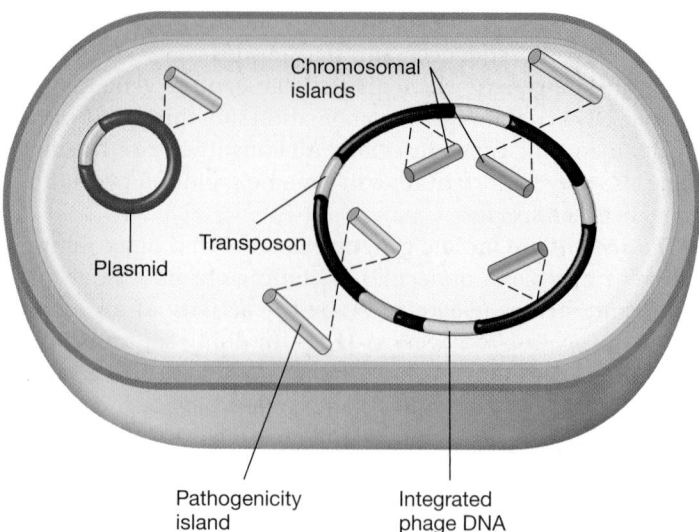

Figure 9.19 Possible genome insertions. The core genome is represented by the black regions of the chromosome and is present in all strains of a species. Each colored wedge indicates a single insertion. Where two wedges emerge from the same location, they represent alternative chromosomal islands that can insert at that site. However, only one insertion can be present at a given location. Plasmids, like the chromosome, may also have insertions.

the root nodules of plants (⇨ Section 23.3) are carried in chromosomal islands. Perhaps the most unique chromosomal island is the magnetosome island of the bacterium *Magnetospirillum*; this DNA fragment carries genes that encode the formation of magnetosomes, intracellular magnetic particles that orient the cell in a magnetic field and influence the direction of its movement (⇨ Section 2.8).

Some chromosomal islands carry a gene encoding an integrase enzyme, suggesting that the islands move within the genome in a manner similar to conjugative transposons (Section 9.6). Chromosomal islands are typically inserted into a gene for a tRNA; however, because the target site is duplicated upon insertion, an intact tRNA gene is regenerated during the insertion process. In a few cases, transfer of a whole chromosomal island between related bacteria has been demonstrated in the laboratory, and transfer can presumably occur by any of the mechanisms of horizontal transfer (Figure 9.17). It is thought that after insertion into the genome of a new host cell, chromosomal islands gradually accumulate mutations, and hence, over many generations, chromosomal islands tend to lose their ability to move.

Pathogenicity Islands and the Evolution of Virulence

Comparison of the genomes of pathogenic bacteria with those of their harmless or less virulent relatives often reveals chromosomal islands that encode *virulence factors*: special proteins, toxins, enzymes, or other molecules or structures that facilitate disease symptoms (Chapter 25). Some virulence genes are carried on lysogenic bacteriophages or plasmids (⇨ Sections 8.7 and 11.8); however, many others are clustered in chromosomal regions called **pathogenicity islands**.

The pathogenicity islands of uropathogenic strains of *E. coli* have been particularly well studied (**Figure 9.20**). Only a few strains

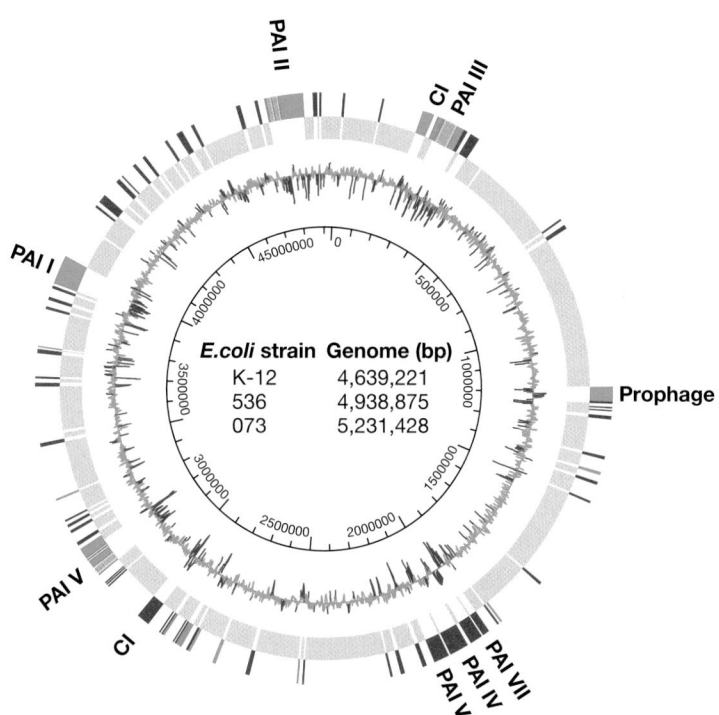

- What is the difference between core genome and pan genome?
- What is a chromosomal island and how can one be identified as being of foreign origin?
- What is a pathogenicity island and how does one move between bacterial species?

Figure 9.20 Pathogenicity islands in *Escherichia coli.* Genome comparisons of uropathogenic *E. coli* strain 536, uropathogenic strain 073, and the nonpathogenic strain K-12. The uropathogenic strains (urinary tract pathogens) contain pathogenicity islands and thus their chromosomes are larger than that of K-12. The inner circle represents nucleotide base pairs. The jagged circle shows the DNA GC distribution; regions where GC content varies dramatically from the genome average are in red. The outermost circle is a three-way genomic comparison: green, genes common to all strains; red, genes present in the pathogenic strains only; blue, genes found only in strain 536; orange, genes of strain 536 present in a different location in strain 073. Some very small inserts are deleted for clarity. PAI, pathogenicity island; CI, chromosomal island. Prophage, DNA from a temperate bacteriophage. Note the correlation between chromosomal islands and skewed GC content. Data adapted from *Proc. Natl. Acad. Sci. (USA) 103:* 12879–12884 (2006).

of *E. coli* cause bladder and kidney infections, but those that do contain pathogenicity islands that encode a variety of virulence factors including adhesins that facilitate binding to host tissues and a capsule that helps cells evade immune surveillance. For example, although *E. coli* strain K-12 (a harmless laboratory strain) and strain 073 (a urinary tract pathogen) share a core genome of some 4.6 Mbp, strain 073 contains 11% more DNA than strain K-12, much of which is devoted to its pathogenic lifestyle.

Small pathogenicity islands that encode a series of virulence factors are also present in certain strains of the gram-positive pathogenic bacterium *Staphylococcus aureus* (skin infections, boils, and the like) and can be moved between cells in temperate bacteriophages by transduction, a major mechanism of horizontal gene transfer (⮌ Section 11.7). The islands are smaller than the phage genome, and when the islands excise from the chromosome and replicate, they induce the formation of defective phage particles that carry the genes for the islands but are too small to carry the phage genome. In this way, when strains of *S. aureus* that lack the islands are infected, they are not killed by the phage but instead acquire the islands and become more potent pathogens.

III • Functional Omics

Despite the major effort required to generate an annotated genome sequence, the net result is simply a "list of parts." To understand how a cell *functions*, we need to know more than which genes are present. We must also understand (1) gene expression, (2) the function of gene products, (3) the activity of the proteins made, and (4) the metabolites produced during growth.

In analogy to the term "genome," the entire complement of RNA, proteins, or metabolites produced under a given set of conditions is known as the **transcriptome, translatome**, and **metabolome**, respectively. Adding the suffix "omic" denotes their corresponding areas of study. Table 9.6 summarizes some of the "omics" terminology used in microbiology today.

9.8 Metagenomics

Microbial communities contain many microbial species, many of which have never been cultured or formally identified. **Metagenomics** is the science that analyzes pooled DNA or RNA

TABLE 9.6 Some omics terminology		
DNA	**Genome** the total complement of genetic information of a cell or a virus	
	Metagenome the total genetic complement of all the cells present in a particular environment	
	Epigenome the total number of possible epigenetic changes	
	Methylome the total number of methylated sites on the DNA (whether epigenetic or not)	
	Mobilome the total number of mobile genetic elements in a cell	
RNA	**Transcriptome** the total RNA produced in an organism under a specific set of conditions	
Protein	**Proteome** the total set of proteins encoded by a genome; sometimes also used in place of *translatome*	
	Translatome the total set of proteins present under specified conditions	
	Interactome the total set of interactions between proteins (or other macromolecules)	
	Secretome the total set of proteins secreted by a cell	
Metabolites	**Metabolome** the total complement of small molecules and metabolic intermediates	
	Glycome the total complement of sugars and other carbohydrates	
Organisms	**Microbiome** the total complement of microorganisms in an environment (including those associated with a higher organism)	
	Virome the total complement of viruses in an environment	
	Mycobiome the total complement of fungi in a natural environment	

UNIT 3

from an environmental sample containing organisms that have not been isolated and identified (**Figure 9.21**). Just as the total gene content of an organism is its *genome*, so the total gene content of a microbial community is its **metagenome** (Table 9.6). In addition to metagenomic analyses based on DNA sequencing, analyses based on RNA or proteins (metatranscriptomics or metaproteomics, respectively) may be used to explore the patterns of gene expression in natural microbial communities. With today's molecular technology, these studies can even be done on individual cells (Section 9.12).

Examples of Metagenomic Studies

Several environments have been surveyed by large-scale metagenome sequencing projects. Extreme environments, such as highly acidic runoff waters from mining operations, tend to have low microbial species diversity. Consequently it has been possible to isolate community DNA (and metabolites, Section 9.11) and assemble much of it into nearly complete individual genomes. Conversely, complex environments such as fertile soils or aquatic environments are much more challenging, and complete genome assemblies here are much more difficult. Nonetheless, a surprising finding that has emerged from metagenomic studies thus far is that most genes recovered from natural habitats do not originate from cells but from viruses. This is discussed further in Chapter 10 where we consider the genomics and phylogeny of viruses.

Even if complete genomes cannot be assembled from environmental DNA, much useful information can be derived from metagenomic surveys. For example, environments can be analyzed for the presence and distribution of specific microbial groups. These vary greatly in relative abundance in different environments, and Figure 9.21*b* illustrates this for subgroups of *Proteobacteria* (a major

phylum of gram-negative *Bacteria*, Chapter 16) at a sampling site in the Pacific Ocean near the Hawaiian Islands. Light, oxygen, nutrients, and temperature all change with depth in a water column, and these factors can be correlated with proteobacterial subgroups to show which are most competitive at each depth (Figure 9.21*b*). One curious observation that has emerged from such metagenomic studies is that much cellular DNA in natural habitats does not reside in *living* cells. Around 50–60% of the DNA in the oceans is extracellular DNA present in deep-sea sediments. Presumably this was DNA deposited when dead organisms from the upper layers of the ocean sank to the bottom and lysed. Because nucleic acids are major reservoirs of phosphate, marine sediment DNA is thought to be a major component of the global phosphorus cycle.

Metagenomics and "Biome" Studies

The human body is estimated to contain about 10 trillion (10^{13}) cells, but each of us also carries around ten times more prokaryotic cells than human ones. This collection of prokaryotic cells is called the human *microbiome* (Chapter 24). Most of these organisms inhabit the large intestine, with the majority belonging to one of two phylogenetic groups of *Bacteria*, the *Bacteroidetes* and the *Firmicutes* (Chapter 16). A fascinating finding is that the composition of the gut microbiome correlates with obesity in both humans and experimental mouse models. The data show that the higher the proportion of *Firmicutes* (mostly species of *Clostridium* and relatives) in the gut, the more obese is the human or mouse. A suggested mechanism behind this finding is that fermentative species of *Firmicutes* convert more dietary fiber into fatty acids than can be absorbed by the host (⇔ Section 24.8). In this way, the obese host gets more organic carbon than the thinner host from the same amount of food.

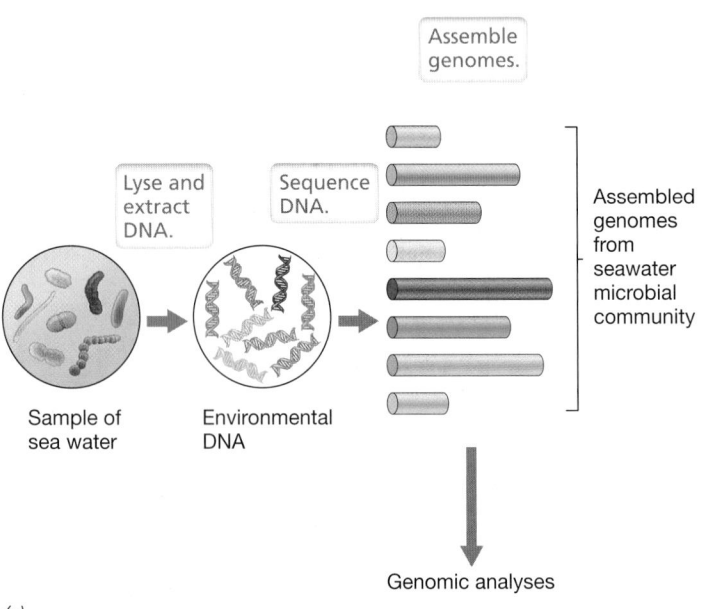

(a)

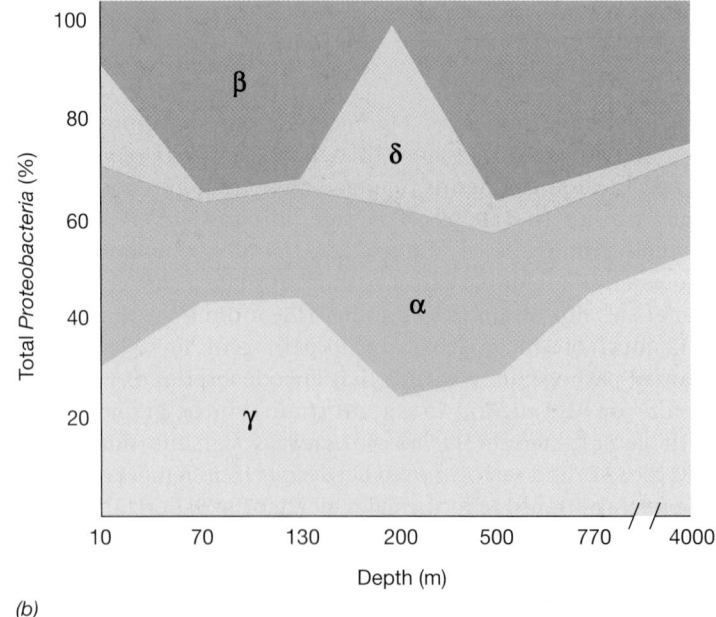

(b)

Figure 9.21 Metagenomics and the microbiome. *(a)* Isolation, sequencing, and identification of DNA from a sample of seawater. *(b) Proteobacteria* in the ocean. The distribution with depth of the major subgroups (alpha α, beta β, gamma γ, and delta δ) of *Proteobacteria* in the Pacific Ocean is shown. Many other types of bacteria are also present (not shown). Data adapted from Kembel, S.W., J.A. Eisen, K.S. Pollard, and J.L. Green. 2011. *PLoS One 6:* e23214.

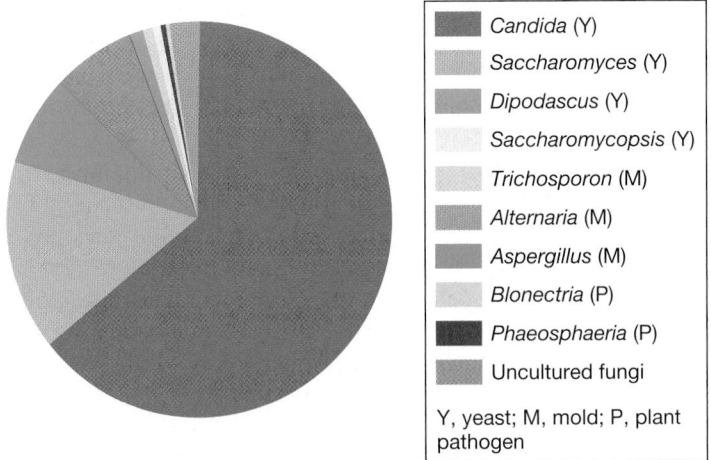

Figure 9.22 The mouse mycobiome. The data shown represent the relative amount of different fungal genera of the mouse intestine. The pie chart shows the most common fungi present are yeasts. Data adapted from Iliev, I.D., et al. *Science 336:* 1314–1317 (2012).

Recent surveys of the human and mouse gut microbiome have also revealed the rather surprising finding that over 60 species of fungi are present (**Figure 9.22**). These constitute the gut *mycobiome* (the prefix "myco" means fungal). Many fungi, typically non-pathogenic yeasts, inhabit the skin, the oral cavity, and virtually all moist surfaces on the human body. Many of these are common and generally harmless yeasts, such as *Saccharomyces*, *Cladosporium*, and most species of *Candida*. Most of these also are found in the gut, although some gut fungi—such as *Aspergillus* and *Trichosporon*—are potential serious pathogens. Moreover, although gut fungi constitute less than 1% of the microbiome, it is known that certain conditions such as inflammatory bowel disease and some cases of obesity correlate strongly with specific fungal populations. Thus metagenomics holds great promise for exploring possible connections between specific microbial populations and specific diseases in humans and other animals. Moreover, in cases where a clear cause-and-effect relationship is strongly suspected, metagenomics also holds great promise as a clinical tool for making medical diagnoses.

MINIQUIZ

- What is a metagenome? The mycobiome?
- How is a metagenome analyzed?

9.9 Gene Chips and Transcriptomics

Once a genome sequence is available, the sequence can be used to synthesize gene chip devices that can be used to detect specific microbes, determine genome differences between closely related strains of the same species (for example, the presence of chromosomal islands), identify sequences specifically bound by DNA-binding proteins, and measure gene expression (transcription). *Transcriptomics* refers to the global study of transcription and is done by monitoring the total RNA generated under a chosen

growth condition. In the case of genes whose role is still unknown, discovering the conditions under which they are transcribed may yield clues to their function. Two main approaches are used in transcriptomics: *microarrays* and *RNA-Seq*.

Microarrays and the DNA Gene Chip

Microarrays are small, solid supports to which genes or, more often, oligonucleotides corresponding to segments of genes are fixed and arrayed spatially in a known pattern; they are often called **gene chips** (Figure 9.23*a*). Microarrays measure the DNA or RNA that hybridizes to the DNA sequences on the chip. When DNA is denatured (that is, the two strands are separated), the single strands can form hybrid double-stranded molecules with other nucleic acid molecules by complementary or almost complementary base pairing (Figure 9.23*b*; ⮌ Section 12.1). This process is called *nucleic acid hybridization*, or **hybridization** for short, and is widely used in detecting, characterizing, and identifying segments of DNA or RNA. The single-stranded segments of nucleic acid, whose identity is already known, are called **nucleic acid probes** or, simply, *probes*. To detect hybridization to the probes, the nucleic acid added to the chip must be labeled with a fluorescent dye and then the hybridized chip is scanned with a laser fluorescence detector that measures which of the probes contain hybridized DNA (Figure 9.23*b*, *c*).

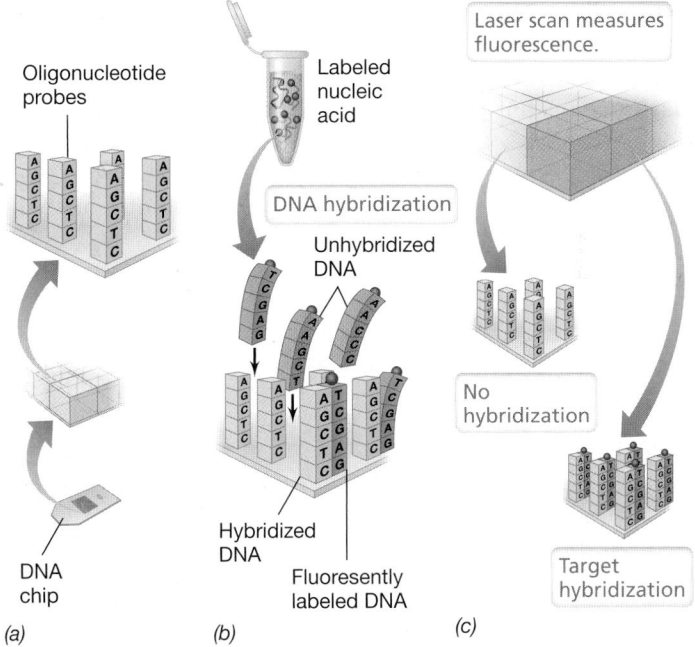

Figure 9.23 DNA chip design and application. *(a)* DNA chip design. Short single-stranded oligonucleotides (probes) corresponding to each gene in an organism or to diagnostic sequences corresponding to numerous organisms are synthesized and affixed at known locations to make a microarray. *(b)* Microarray hybridization. The presence of specific DNA or RNA (in the form of cDNA) is assayed by hybridizing fluorescently labeled samples (DNA or cDNA) to the DNA probes on the chip. Labeled DNA or cDNA will bind to the probes on the chip if they possess sequence complementarity. *(c)* Analysis of microarray hybridization. A scanning laser is used to identify regions of the chip where labeled nucleic acid has bound to the probes.

Gene chips are typically about 1 to 2 cm and are inserted into a plastic holder that can easily be manipulated (**Figure 9.24a**); each chip holds thousands of different DNA fragments. In practice, each gene is usually represented more than once in the array to increase reliability. Whole genome arrays contain DNA segments that cover the entire genome of an organism. For example, a chip that covers the entire human genome (Figure 9.24a) can analyze over 47,000 human transcripts and has room for 6500 additional oligonucleotides for use in clinical diagnostics.

Measuring Gene Expression and Other Uses of Gene Chips

In a gene expression microarray, the probes are designed and synthesized for each gene based on the genomic sequence. Once attached to the solid support, the DNA segments can be hybridized with labeled RNA from cells grown under specific conditions and scanned and analyzed by computer. Because mRNA levels are typically too low for use directly, the mRNA sequences are first amplified and converted into DNA using a modified version of the polymerase chain reaction (PCR) that converts RNA to *complementary DNA* (cDNA, ⮌ Section 12.1).

To monitor global gene expression, total RNA (or cDNA) from a test sample is hybridized to an array of oligonucleotides corresponding to the entire genome. Figure 9.24b shows part of a chip containing probes for over 6000 protein-encoding genes of the yeast *Saccharomyces cerevisiae*. After hybridizing yeast cDNA to the chip, a distinct hybridization pattern is observed, and the fluorescence and its intensity reveal both which genes were expressed and at what level (Figure 9.24b); these data yield the *transcriptome* of the yeast culture grown under specified conditions (Table 9.6). Using such analyses, gene expression under different growth conditions can be measured. For example, in yeast—which can grow by either fermentation or respiration—transcriptome analyses have shown that genes that control production of ethanol (a key fermentation product) are strongly repressed while genes encoding citric acid cycle functions (needed for aerobic growth) are strongly activated when the organism is shifted from anaerobic to aerobic conditions. Overall, over 700 genes are turned on and over 1000 turned off during this metabolic transition. In "shift" experiments of this type, the expression pattern of genes of unknown function is also revealed, and analysis of these expression patterns sometimes yields valuable clues to the cellular function of these unknown proteins.

Microarrays can also be used to identify specific microbes. For example, identification (ID) chips have been used in the food industry to detect DNA sequences unique to specific pathogens, such as the gastrointestinal pathogen *Escherichia coli* O157:H7, an occasional foodborne pathogen. In environmental work, microarrays called *PhyloChips* have been used to assess microbial diversity. These contain oligonucleotides complementary to the 16S rRNA of different bacterial species, a molecule widely used in microbial systematics (Chapter 13). After extraction of bulk DNA or RNA from an environment, the presence or absence of a given species can be assessed by the hybridization response on the chip (⮌ Section 19.7). Although ID chips and PhyloChips can be made highly specific, the inexpensive nature of DNA isolation, sequencing, and analysis have made metagenomic approaches to the identification of specific pathogens or phylogenetic groups in natural samples the preferred method of assessment.

RNA-Seq Analysis

RNA-Seq analysis is a transcriptomic method in which all the RNA molecules from a cell are converted to DNA (cDNA, ⮌ Section 12.1) and then sequenced. Provided that the genome sequence is available for comparison, RNA-Seq reveals both which genes were transcribed and how many RNA copies of each gene were made. Because RNA-Seq targets *all* transcripts, it is ideal for measuring the expression of mRNA, to identify long untranslated regions, and to discover noncoding RNAs. RNA-Seq requires high-throughput sequencing (second- or third-generation sequencing, Section 9.2) and is complicated a bit by the fact that the most abundant RNA in a cell is ribosomal RNA (rRNA). Nevertheless, methods are available to remove rRNA or enrich mRNA and primary transcripts from a total RNA pool. In addition, recent advancements in sequencing technology may allow sequencing without removing rRNA.

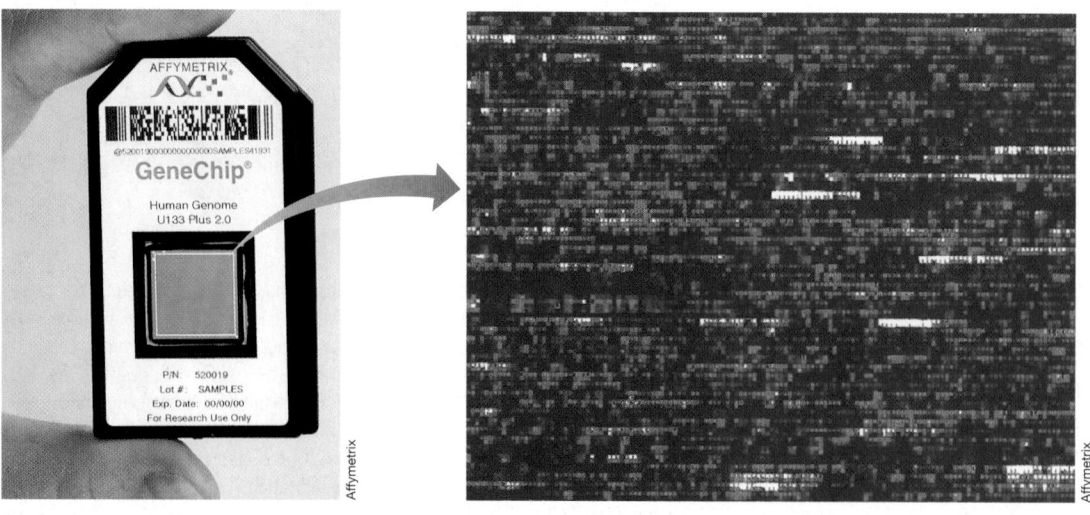

(a)   *(b)*

Figure 9.24 Using gene chips to assay gene expression. *(a)* The human genome chip contains over 47,000 gene fragments. Blowup from part *a* to part *b* indicates location of actual microarray. *(b)* A hybridized yeast chip shows fragments from a quarter of the genome of baker's yeast, *Saccharomyces cerevisiae*. Each gene is present in several copies and has been probed with fluorescently labeled cDNA (derived from mRNA) from yeast cells grown under a specific condition. The background of the chip is blue. Locations where the cDNA has hybridized are indicated by a gradation of colors up to a maximum number of hybridizations, which shows as white. Because the location of each gene on the chip is known, when the chip is scanned, it reveals which genes were expressed.

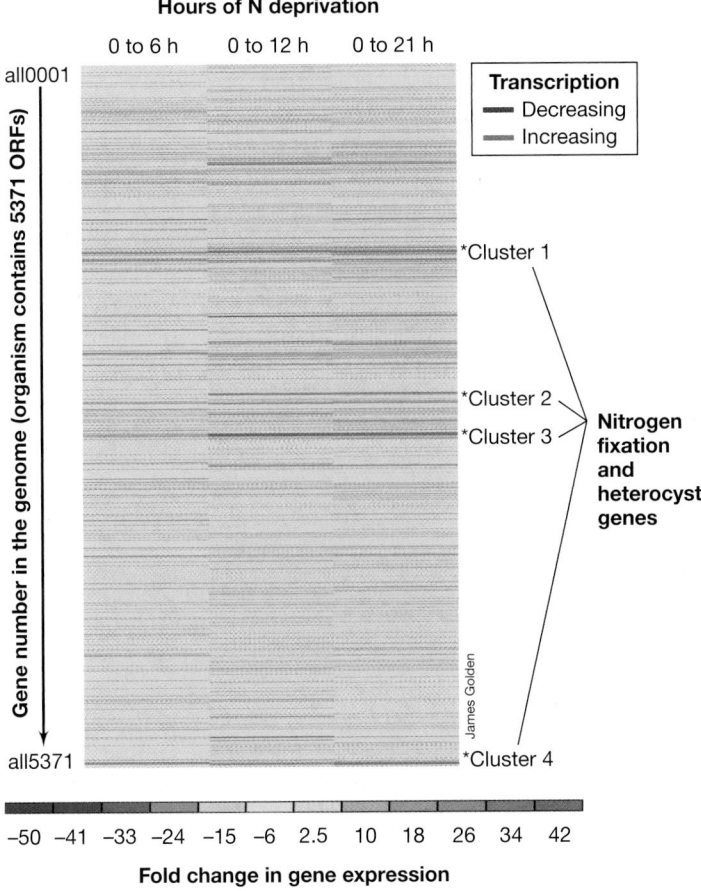

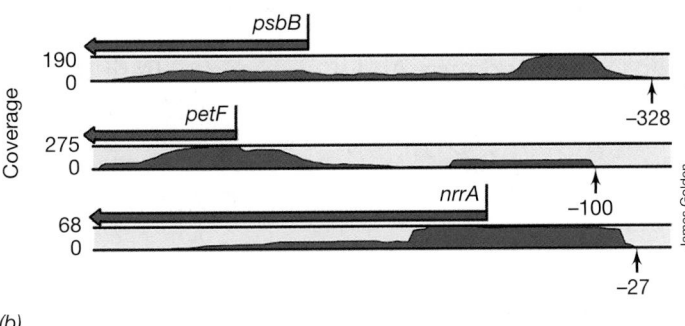

Figure 9.25 RNA-Seq analysis of the heterocyst-forming cyanobacterium *Anabaena* during nitrogen starvation. Cyanobacteria are oxygenic phototrophs (🔗 Section 14.4) and only some species, such as those that form heterocysts, can fix nitrogen under fully oxic conditions. *(a)* Heat map of gene expression 6, 12, and 21 h after cells were starved for fixed nitrogen. Genes that display increased expression are in red, whereas those that display decreased expression are in blue. Gene clusters 1–4 all encode proteins linked to nitrogen fixation. *(b)* Mapping of RNA-Seq reads. The arrows indicate the annotated open reading frames, and the plots underneath correspond to the number of sequencing reads detected for each chromosomal nucleotide position. Note that the negative numbers represent chromosomal positions upstream of the predicted start codon for the genes. The genes *psbB* and *petF* encode key proteins of photosynthesis, and *nrrA* encodes a protein that regulates heterocyst formation (the heterocyst is the site of N_2 fixation, 🔗 Section 14.6). Parts *a* and *b* modified from Flaherty, B.L., F. van Nieuwerburgh, S.R. Head, and J.W. Golden. 2011. *BMC Genomics 12:* 332.

RNA-Seq has overtaken microarray analysis as the method of choice for global studies of gene expression. The data from transcriptomic experiments can be presented in the form of a *heat map*, which uses different colors to show the level of gene expression. For example, **Figure 9.25a** identifies gene clusters that are upregulated (more intensely transcribed) during nitrogen deprivation in the heterocyst-forming *Anabaena*, a cyanobacterium that can use N_2 as its nitrogen source (nitrogen fixation, 🔗 Section 14.6). Gene clusters 1–4 represent increased expression of nitrogen fixation and heterocyst formation genes (heterocysts are the site of N_2 fixation) as the time of nitrogen deprivation increases (Figure 9.25a). RNA-Seq data can also be used for transcript mapping by plotting the sequencing reads against the genome annotation. Figure 9.25b illustrates the sequencing coverage at each base along the open reading frame regions for *psbB*, *petF*, and *nrrA*, genes that encode two key proteins needed for photosynthesis and a regulator protein for heterocyst formation in *Anabaena*, respectively (🔗 Section 7.8). These plots demonstrate long 5′ untranslated regions present in these transcripts, which may be associated with regulation.

Transcript abundance under different culture conditions can also be analyzed using RNA-seq data, as indicated by a comparison of cultures of a *Clostridium* species in exponential and stationary phase (**Figure 9.26**). Clostridia are gram-positive bacteria that produce endospores, the highly resistant and dormant stage of the cell's life cycle (🔗 Section 2.10). As one might predict, transcription of genes of the glycolytic pathway (the major means by which the organism makes ATP) is elevated during exponential growth, whereas expression of sporulation genes increases in stationary phase, when nutrients become limiting. RNA-Seq is also being used for microbial community analysis and can provide information on relative transcription levels when a genome sequence is not available for comparison. In this case the sequences detected must be identified by homology with sequences present in databanks.

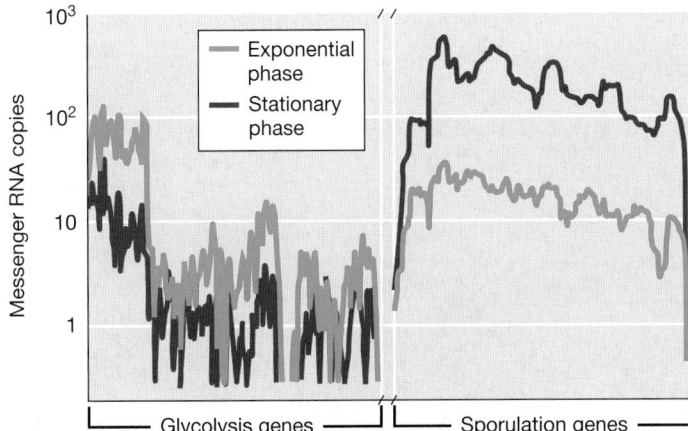

Figure 9.26 Transcriptomic analysis of sporulation genes in *Clostridium*. RNA-Seq analysis of cultures of a *Clostridium* species grown for 4.5 h (cells in exponential phase) or 14 h (cells in stationary phase). Two genomic regions are shown: (left) ~5.4-kb segment surrounding the *gap-pgk-tpi* glycolytic operon, and (right) ~1.2-kb segment surrounding the *cotJC-cotJB* sporulation operon. Production of endospores is triggered by nutrient starvation (🔗 Section 2.10). Data from Wang, Y., X. Li, Y. Mao, and H.P. Blaschek. 2011. *BMC Genomics 12:* 479–489.

UNIT 3

As discussed in Section 9.8, metagenomics is the genomic analysis of pooled DNA or RNA obtained from organisms in an environment. Metagenomic analysis using RNA-Seq has been exploited for culturing a bacterium from nature that had previously resisted laboratory culture. This was accomplished by using RNA-Seq to identify highly transcribed genes in the microbial community that contained the bacterium, followed by sequence analyses to identify the proteins encoded by the highly transcribed genes. These data allowed the researchers to deduce which nutrients the bacterium was likely to be using given the predicted enzymatic activities of these proteins. Culture media were then devised using this information as a guide and the previously uncultured bacterium was successfully cultured.

MINIQUIZ

- Why is it useful to survey expression of the entire genome under particular conditions?
- What do microarrays tell you that studying gene expression by assaying individual enzymes cannot?
- What technological advances does RNA-Seq depend on?

9.10 Proteomics and the Interactome

The genome-wide study of the structure, function, and activity of an organism's *proteins* is called **proteomics**. The number and types of proteins in a cell change in response to the cell's environment or to other factors, such as developmental cycles. As a result, the term **proteome** has unfortunately become ambiguous. In its wider sense, a proteome refers to *all* the proteins encoded by an organism's genome. In its narrower sense, however, it refers to those proteins present in a cell *at any given time*. The term *translatome* has been used to describe the latter; that is, it refers to every protein made by a cell under specific conditions.

Methods in Proteomics

Modern proteomics relies on some form of *mass spectrometry* to characterize the proteome, a method that can also be used to identify metabolites (Section 9.11). The mass of ^{12}C is defined as exactly 12 molecular mass units (daltons). However, the masses of other atoms, such as ^{14}N or ^{16}O, are not exact integers. Mass spectrometry using extremely high mass resolution techniques, which can distinguish between slightly different mass-to-charge ratios, allows the unambiguous determination of the molecular formula of any small molecule. Thus mass spectrometry can be used to identify several peptides in a sample. The amino acid sequence of these peptides can then be searched against the translation of a genome to identify the presence of specific proteins. To increase sensitivity, liquid chromatography is increasingly used to separate protein mixtures. In high-pressure liquid chromatography (HPLC), a protein sample is dissolved in a suitable liquid and forced under pressure through a special column that separates proteins by differences in their chemical properties, such as size, charge, or hydrophobicity. Fractions are collected at the end of the column, the proteins in each fraction are digested by proteases, and the peptides are identified by mass spectrometry.

MALDI (*matrix-assisted laser desorption ionization*) is an advanced version of mass spectrometry that does not require that the proteins be separated or digested. Instead the sample is affixed to a matrix and then ionized and vaporized by a laser (**Figure 9.27**). The ions generated are accelerated along the column toward the detector by an electric field. The time of flight (TOF) for each ion depends on its mass/charge ratio—the smaller this ratio, the faster the ion moves. The detector measures the TOF for each ion and the computer calculates the mass and hence the molecular formula. The combination of these two techniques is known as *MALDI-TOF*.

Utility of Proteomics: An Environmental Case Study

While proteomic analyses can be performed on pure cultures of specific microbes, *metaproteomics* of environmental samples are

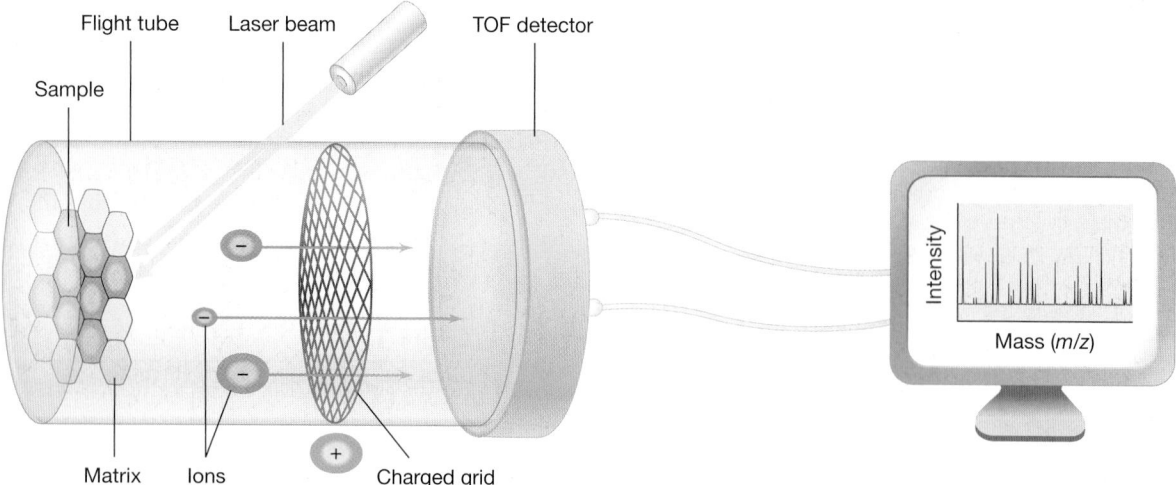

Figure 9.27 MALDI-TOF mass spectrometry. In matrix-assisted laser desorption ionization (MALDI) spectroscopy, the sample is ionized by a laser and the ions travel down the tube to the detector. The time of flight (TOF) depends on the mass/charge (*m/z*) ratio of the ion. The computer identifies the ions based on their time of flight, that is, the time it takes to reach the detector.

particularly insightful for revealing the collective metabolic potential of a microbial community. For example, permafrost covers an estimated 20% of the land on Earth and sequesters large amounts of organic carbon. Because climate change could lead to the release of the "greenhouse gases" CO_2 and CH_4 from permafrost, identifying the microbes inhabiting actively melting permafrost along with their metabolic potential is essential for climate predictions. Using a combination of metagenomics (Section 9.8) and metaproteomics, a revealing snapshot of the genes and proteins of the microbial community present in a permafrost meltwater bog has been obtained (**Figure 9.28**).

During the metagenomic analysis of the bog, DNA sequences were assembled and annotated into complete and partial genomes (Figure 9.28a). This analysis indicated that *Bacteria* of the phyla *Proteobacteria*, *Actinobacteria*, and *Chloroflexi* predominated, while methanogens (*Euryarchaeota*) were the dominant *Archaea*. From the large amount of sequence data obtained, the genomes of three novel (and as yet uncultured) methanogens could be assembled to the draft stage without any cultures of the organisms having been obtained. This indicated that unique methanogens resided in the permafrost—most likely species that function well in the cold (psychrophiles)—and therefore that the potential for increasing rates of methanogenesis as the permafrost melted was high. Using the annotated DNA sequences and the output from the mass spectrometry detection of peptides, the identity of proteins extracted from the bog sample was determined along with their microbial sources (Figure 9.28b). The proteomics detected an abundance of functionally diverse proteins including those that participate in cellular housekeeping, transport, and organic carbon respiration. In agreement with the metagenomics data, several proteins associated with C_1 metabolism, including those necessary for both making methane (methanogenesis) and oxidizing methane (methanotrophy) as well as oxidizing and reducing carbon monoxide, were also detected in the bog microbial community.

This combined metagenomic and metaproteomic snapshot of a major microbial community highlights the power of omics for resolving the "who" and "how" of complex microbial communities and for using the results to predict the response(s) of these communities to environmental changes. In the case of permafrost melting, the potential for the release of large amounts of CO_2 (due to increased respiration of organic carbon) and CH_4 (due to increased activities of methanogens) from permafrost as climate change progresses was clearly apparent (Figure 9.28).

The Interactome

By analogy with the terms genome and proteome, the **interactome** is the complete set of *interactions* among the macromolecules within a cell (**Figure 9.29**). Originally, the word *interactome* was applied only to interactions between proteins, many of which assemble into complexes. However, it is also possible to consider interactions between different classes of macromolecules, such as between protein and RNA (see the protein–RNA interactome in Figure 9.38).

Interactome data are typically expressed in the form of network diagrams, with each node representing a protein and the connecting lines representing the interactions. Diagrams of whole interactomes can be extremely complex (see Figure 9.35a) and thus more

focused interactomes, such as the motility protein network from the bacterium *Campylobacter jejuni* (Figure 9.29), are more instructive. This figure shows the core interactions between well-known components of the chemotaxis system (⟲ Sections 2.13 and 6.7), including all other proteins that are known to interact with these.

MINIQUIZ

- Why is the term "proteome" ambiguous, whereas the term "genome" is not?
- What are the most common experimental methods used to survey the proteome?
- What is the interactome?

9.11 Metabolomics

The metabolome is the complete set of *metabolic intermediates* and small molecules produced by an organism. Thus the metabolome reflects the enzymatic pathways encoded by the genome. While gene expression and the presence of corresponding proteins suggest the activity of specific pathways, the metabolomic data confirm that these potential reactions actually occurred in the cell in a given physiological state.

Advances in Metabolomic Techniques: NIMS

Metabolomics has lagged behind other omics due largely to the immense chemical diversity of small metabolites present in cells. This makes systematic metabolomic screening technically challenging. Early attempts used nuclear magnetic resonance (NMR) analysis of extracts from cells labeled with ^{13}C-glucose (^{13}C is a heavy isotope of carbon, most of which is ^{12}C). However, this method is limited in sensitivity, and the number of compounds that can be identified in a mixture simultaneously is too low for resolution of complete cell extracts.

While MALDI-TOF mass spectrometry (Section 9.10) can be used to detect and identify metabolites, *nanostructure-initiator mass spectrometry* (NIMS) is a more useful technique that can directly analyze biological samples without the need for special analytical preparation (**Figure 9.30**). Thus biofluids, tissues, or even individual cells can be analyzed. A laser is used to ionize the sample in NIMS, just as in MALDI-TOF, but the silicon-coated surface used in NIMS does not generate the background interference seen during ionization of a matrix by MALDI-TOF. This allows for the accurate identification of small metabolites present at low concentrations and increased spatial resolution during tissue imaging. Modifications to the initiator surface can also be made to detect notoriously difficult molecules such as structurally similar carbohydrates or steroids. These traits also make NIMS more sensitive than MALDI, which is illustrated by the ability of NIMS to detect drugs such as the heart drug propafenone in a single heart cell in the yoctomole (10^{-24}) range (Figure 9.30).

Utility of Metabolomics

Metabolome analysis has been particularly useful for the study of plant biochemistry, since plants produce several thousand different metabolites—more than most other types of organism.

UNIT 3

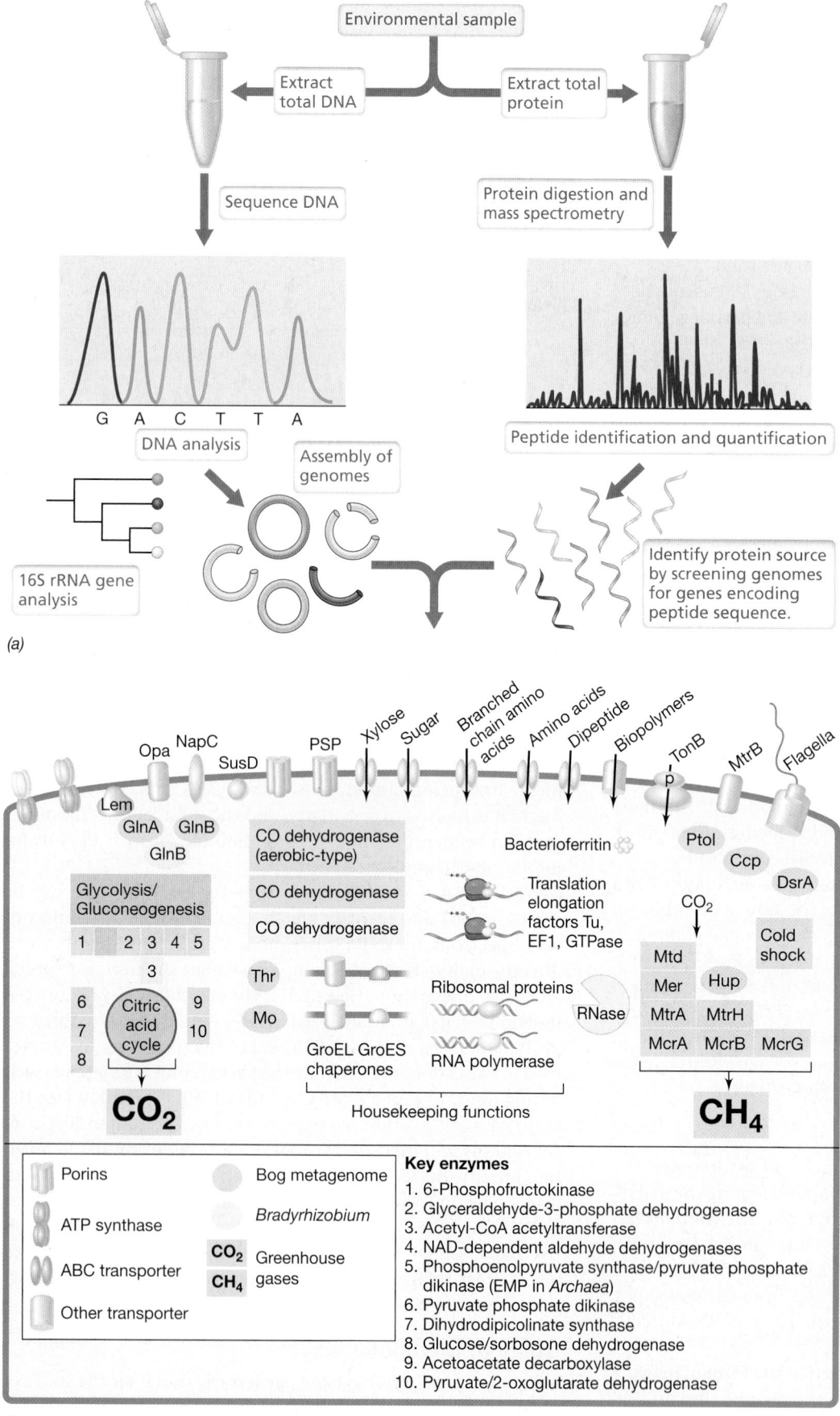

(a)

(b)

Figure 9.28 Meta-omics of permafrost. *(a)* Metagenomic and metaproteomic analysis of an environmental sample. Total DNA and protein are extracted from a sample and analyzed to identify microorganisms and their associated proteins. Following sequencing, the DNA is assembled into partial and complete individual genomes. The 16S rRNA gene can also be profiled to determine the phylogenetic affiliation of *Bacteria* and *Archaea* present in the sample. After identifying the digested proteins by mass spectrometry, their microbial source can be determined by searching for corresponding DNA sequences in the metagenomic data. *(b)* Visualization of a subset of the proteins identified from metagenomic and metaproteomic analysis of a bog sample. Ccp: cytochrome C peroxidase, DsrA: sulfite reductase subunit A, GlnA/GlnB: nitrogen storage, Hup: heterosulfide reductase, Lem: peptidoglycan-associated lipoprotein, Mtd/MtrA/MtrB/MtrH/McrA/McrB/McrG/Mer/Mo: methane metabolism, NapC: NapC/NirT cytochrome C, Opa: opacity protein, PSP: phosphate-selective porin, Ptol: periplasmic component of the Tol biopolymer transport system, Thr: thermosome, SusD: sulfate ABC transport system ATP-binding protein, TonB: periplasmic protein. Data adapted from Hultman, J., et al. 2015. *Nature 521:* 208–212.

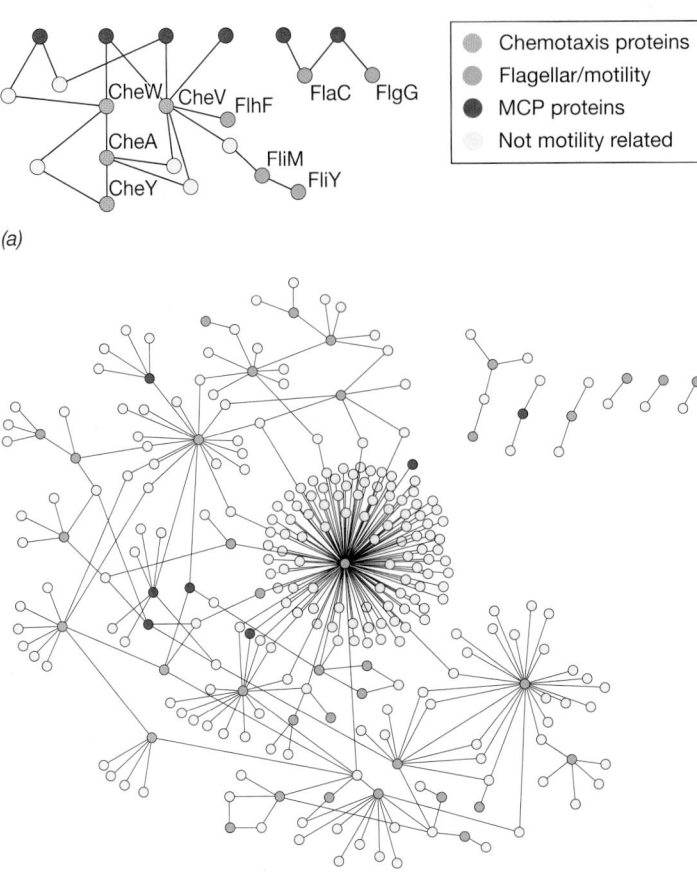

(a)

(b)

Figure 9.29 Motility protein interactome for *Campylobacter jejuni*. This network illustrates the way in which interactome data are depicted. *(a)* A subsection of the network highlighting the well-known proteins of the chemotaxis signal transduction pathway (CheW, CheA, and CheY) and their partners. MCP, methyl-accepting chemotaxis proteins (⟳ Section 6.7). *(b)* High-confidence interactions between all proteins known to have roles in motility. Note the six small networks that fall outside the single large network.

These compounds include many so-called *secondary metabolites*, chemicals such as scents, flavors, alkaloids, and pigments, many of which are commercially important. Metabolomic investigations have monitored the levels of several hundred metabolites in the model plant *Arabidopsis*, and significant changes were observed in the levels of many of these metabolites in response to changes in temperature, a hint that climate change will likely alter plant metabolism in major ways. Metabolomics can also be done on microbial cultures as well as natural microbial communities such as biofilms. For example, microbiologists have detected over 3500 different metabolites in a relatively simple microbial biofilm growing in extremely acidic (pH ~0.9) and heavy-metal-rich water draining from an abandoned mine site in northern California (USA). Many of these metabolites were suspected of being osmolytes and other protective molecules for combating the osmotic and other life stressors in this extreme environment.

Metabolomics has also been deployed to help characterize the human microbiome (Chapter 24). For instance, our skin epidermis is composed not only of cells but also of microbes that contribute to epithelial health. Thus metabolites corresponding to our skin

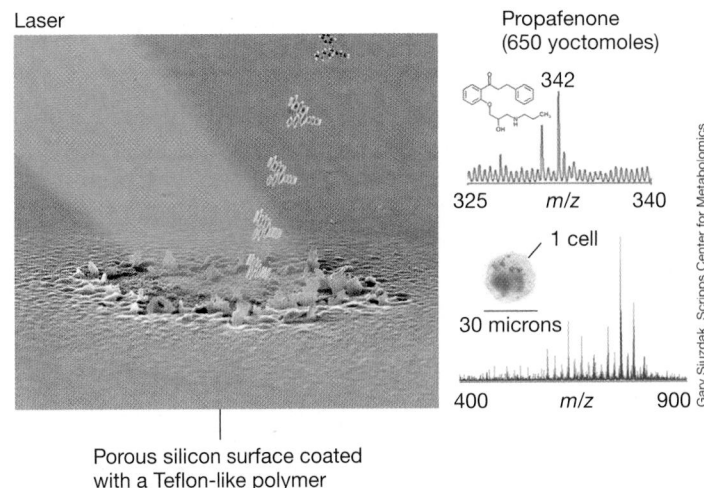

Figure 9.30 NIMS identification of metabolites. In nanostructure-initiator mass spectrometry (NIMS), a cell is placed on the silicon initiator surface and vaporized using a laser. The resulting ionized metabolites within a cell are then detected using mass spectrometry. Because NIMS lacks a matrix, it has extremely high sensitivity and resolution. Ionized metabolites are represented rising from the surface.

cells, to associated microorganisms, and to personal hygiene products are present. **Figure 9.31** shows results of a study on the pattern of metabolites and microbial diversity on the skin of two human subjects plotted simultaneously on a three-dimensional

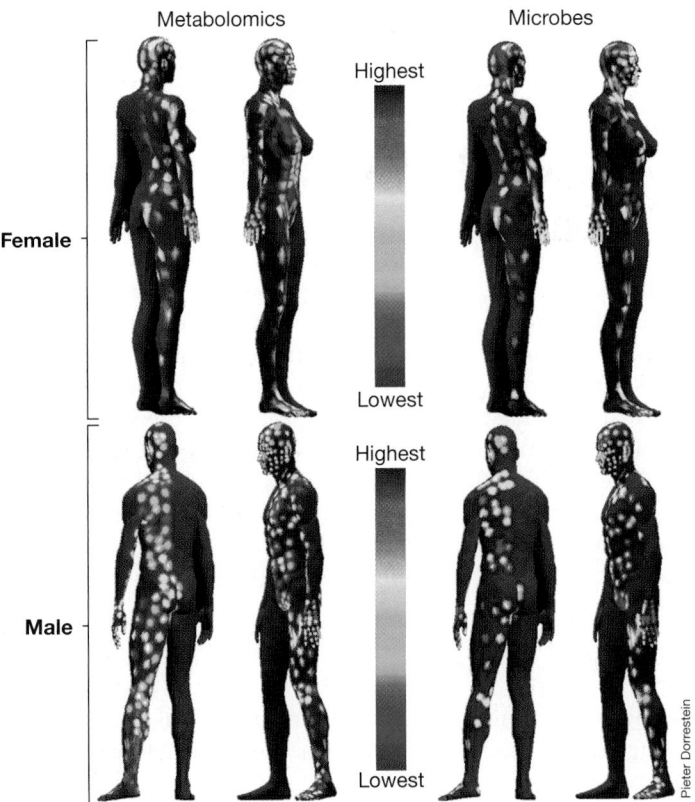

Figure 9.31 Metabolomic and metagenomic mapping of skin. Three-dimensional heat maps represent the diversity of metabolites and microorganisms detected on various areas of the skin on a male and a female subject. Red indicates the highest level of diversity and purple indicates the lowest level of diversity.

topographical heat map of the human body. One of the goals of this type of research is to determine if the presence of particular metabolites can be linked to the presence of particular microbial species. Such a picture of the human skin could offer microbiologists as well as medical clinicians a better understanding of the diversity and ecology of the skin microbiome and pave the way for the future use of studies of this type in the diagnosis of skin health or disease.

As expected, the omics study of the skin (Figure 9.31) revealed many microbes known from previous studies of the skin microbiome (↩ Section 24.5). But the study also discovered a diverse mixture of metabolites. While common chemicals produced by human skin cells such as triacylglycerides and diacylglycerides were readily detected, various metabolites resulting from microbial processing of these compounds were also detected. Overall, the level of metabolite diversity detected on specific areas of the body did not correlate well with microbial diversity. Instead, the complexity of the human skin metabolome significantly exceeded the diversity of the microbial profile, and this indicates that each species is likely producing several distinct metabolites. Interestingly, and somewhat surprisingly, the results also showed that personal hygiene products are the major source of metabolites on human skin.

Throughout this chapter we have discussed various omics and their applications as more or less individual entities. We now shift our focus to integrating multiple omics to better understand the entire organism—the major goal of systems biology.

MINIQUIZ

- What techniques are used to monitor the metabolome?
- What is a secondary metabolite?

IV · The Utility of Systems Biology

The basic strategy of systems biology is to generate comprehensive models for predicting the behavior or properties of an organism that were not obvious from pre-omics era observations. These are referred to as the *emergent properties* of the organism. Understanding an organism's emergent properties provides a deeper insight into its overall biology than can any single omics study by itself. The goal is to integrate the numerous omic datasets to create meaningful models in systems biology (**Figure 9.32**). We begin by seeing how all of these can come together in ecological studies of individual cells in a microbial community.

9.12 Single-Cell Genomics

Besides sequencing total environmental DNA as described for metagenomics (Section 9.8), the genomes of individual cells can also be sequenced—a technique called *single-cell genomics* (SCG) (**Figure 9.33**). This is now possible because of the ability to amplify tiny amounts of DNA. Single-cell genomics is critical for studying the metabolic potential of microorganisms in natural microbial communities. While metagenomic analysis of a microbial community can detect the presence of genes specific to certain

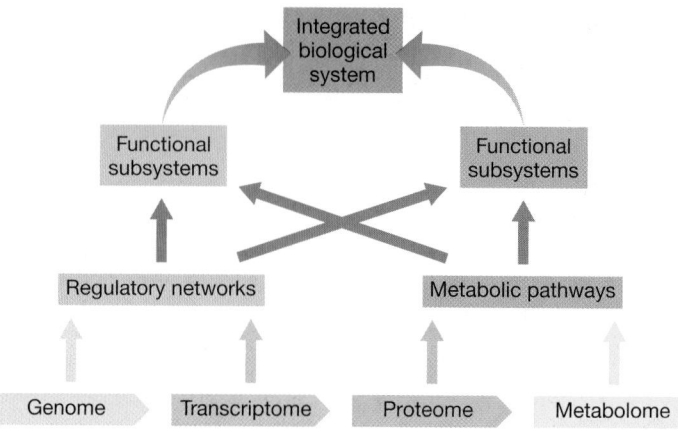

Figure 9.32 The components of systems biology. The results of various "omics" analyses are combined and successively integrated into higher-level views of the entire biology of an organism.

pathways, it is difficult to discern whether the pathways are contained within the same organism. Besides genome sequencing, transcriptome and proteome analyses can also be performed on individual cells, leading to a comprehensive omics study on one component of a microbial ecosystem.

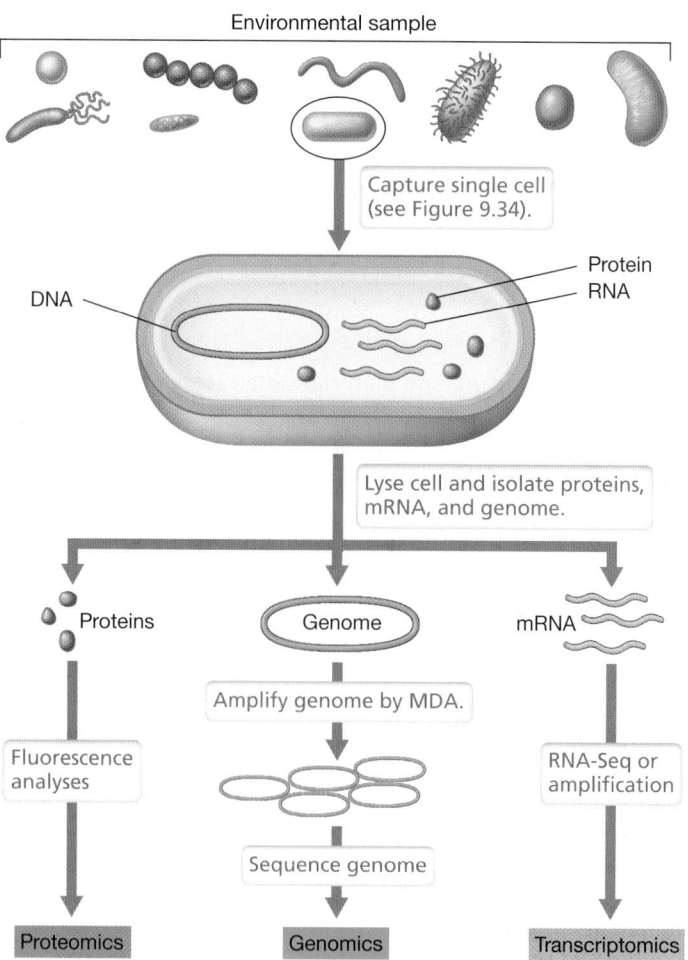

Figure 9.33 Single-cell genomics. A single cell isolated from an environmental sample can be the source of a diversified omics study.

Cell Isolation and Sample Preparation

The ability to isolate cells is obviously essential for single-cell genomics, and various physical techniques including dilution in microwells (🔗 Section 19.3), encapsulation, and *fluorescence-activated cell sorting* (FACS) have been used to do so. For encapsulation, the sample is diluted and added to sterile oil to form microdroplets. Approximately 30% of the resulting droplets from this technique will contain only one cell (**Figure 9.34**). Combining droplet encapsulation with FACS, which is able to optically detect single cells, allows droplets that do not contain a cell to be rejected. This method of single-cell isolation has also been shown effective in isolating single virions for omics analyses.

Sequencing DNA from single cells relies on a modified version of the polymerase chain reaction (PCR, 🔗 Section 12.1) called *multiple displacement amplification* (MDA) (🔗 Section 19.12 and Figure 19.37). This PCR technique uses a special viral DNA polymerase to amplify the femtogram (10^{-15} g) quantities of DNA present in a single bacterial cell into the micrograms (10^{-6} g) of DNA required for sequencing (a billionfold amplification). However, because of the sensitivity of MDA, contamination is one of its biggest problems. Contaminating DNA can originate from the sample itself or from the laboratory equipment and reagents. If great care is not taken in isolating the cell and amplifying its genome, contaminating DNA can make up half or more of the reaction products and create major problems for genome assembly and further genomic analyses. In addition to a cell's genome, its RNA can also be analyzed using RNA-Seq following amplification to form cDNA by a modified version of PCR (Section 9.9). Single-cell proteomic analyses are trickier than nucleic acid studies because amplification is not employed, but analyses using extremely sensitive fluorescence methods are available for this purpose (Figure 9.33).

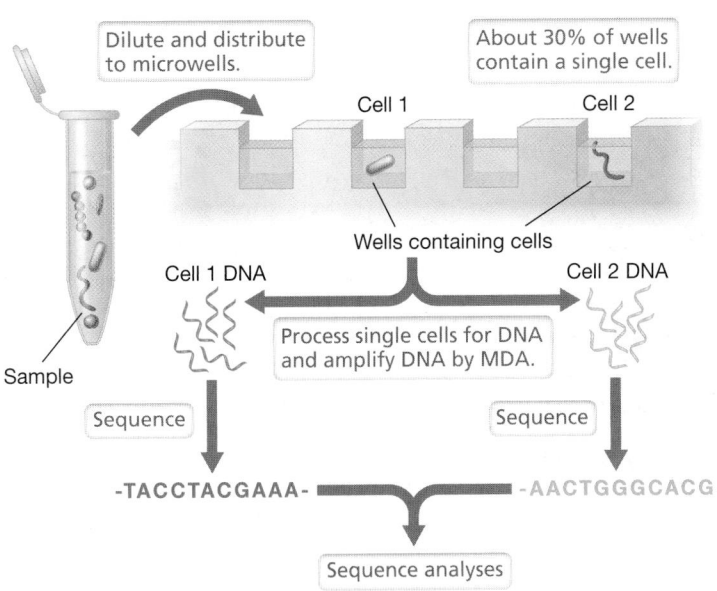

Figure 9.34 Isolation and sequencing of single cells. Droplets of a diluted sample are added to microfluidic wells prior to multiple displacement amplification for DNA sequencing.

Applications of Single-Cell Omics

Single-cell omics have the unique power to probe several facets of an organism's biology in an individual cell rather than on a cell population basis. Using SCG, metabolic genes present in an environment can be not only identified but assigned to particular species; this reveals which particular organisms are degrading which particular nutrients. For example, single-cell genomics has been used to analyze hydrocarbon degradation by bacteria in polluted environments, leading to a better understanding of which organisms are doing what in the overall process. Similarly, plasmids and viruses can be allocated to their correct host when the genome of a single cell is sequenced.

Single-cell genomics has also been used to explore the genomes of candidate phyla of *Bacteria* and *Archaea* that are detected in environmental 16S rRNA gene surveys but for which no laboratory cultures are available. These yet-to-be-cultivated microbes have been called *microbial dark matter*. One dark matter study applied SCG to over 200 uncultivated archaeal and bacterial cells from nine different environments. The results revealed several surprising findings including the following: Some archaeal RNA polymerase sigma factors are similar to their bacterial counterparts (🔗 Sections 4.5 and 4.6); some stop codons (🔗 Section 4.9 and Table 4.4) have been reassigned to incorporate actual tRNAs; the genomes of some bacterial cells encode an oxidoreductase enzyme previously seen only in eukaryotes; some *Archaea* contain genes encoding a bacterial-like stringent response (🔗 Section 6.9); and some *Archaea* produce an enzyme that functions in peptidoglycan synthesis (recall that peptidoglycan is a "signature molecule" of *Bacteria* and is not known from any *Archaea*, 🔗 Section 2.4).

Single-cell genomics is an excellent example of serendipity, the "pleasant surprise" of finding one thing while working on another. In this case, the serendipity occurred when methods designed with one goal in mind (the genomic analysis *of a population of cells*) were refined to probe the biology *of a single cell* in ways never before thought possible. Single-cell genomics is poised to complement the *Earth Microbiome Project*, a sequencing endeavor to archive the genome sequence of all cultured bacterial and archaeal type strains (http://www.earthmicrobiome.org/). For those species of *Bacteria* and *Archaea* that reside in the microbial dark matter, SCG offers a way to include these species in the archive while simultaneously revealing valuable clues that may help bring these organisms into laboratory culture.

MINIQUIZ

- How are single cells isolated from a mixed population?
- What must be done before minute amounts of DNA can be sequenced?

9.13 Integrating *Mycobacterium tuberculosis* Omics

Mycobacterium tuberculosis is a pathogen that infects one-third of the world's population and kills approximately 2 million people every year. Multidrug resistance and the ability to temporarily enter dormancy in response to stress (🔗 Section 7.11) are two characteristics that contribute to the persistence of *M. tuberculosis*. Thus the identification of new treatment methods is critical for combating *M. tuberculosis*. This challenge is being tackled using a systems biology approach for understanding how *M. tuberculosis*—an intracellular pathogen—adapts to the oxygen deficiency (hypoxia, a condition that develops in tuberculosis) in host cells and identifying potential drug targets for therapeutic design.

Tuberculosis Gene Expression and Regulatory Networks

As we discussed in Section 9.9, RNA-Seq can be used to characterize an organism's complete expression profile. A modification of this technique termed *dual RNA-Seq* can be used to simultaneously profile the transcriptomes of a pathogen *and* its host cell. This approach allows for the responses of both the pathogen and host to be captured during the infection process and is especially useful for understanding how *M. tuberculosis* evades the host's defense systems. Data from dual RNA-Seq and the integration of over 600 separate expression experiments have facilitated the construction of gene expression models. These models have identified the primary energy source of phagocytosed *M. tuberculosis* cells as host cholesterol, and they have shown that the amino acid aspartate produced by the host is used by *M. tuberculosis* not only as a nitrogen source but also as protection against reactive oxygen species (⊄ Section 5.14) produced by macrophages trying to kill *M. tuberculosis* cells.

The *M. tuberculosis* research has integrated transcriptome and other omics datasets including ChIP-Seq, a method where an antibody to a specific DNA-binding protein is used to trap the protein bound to its DNA, after which the DNA is removed and sequenced. From ChIP-Seq analyses, several transcription factors

and regulatory networks critical to the pathogenesis of *M. tuberculosis* have been identified. These include regulation by the major bacterial cell regulator LexA (⊄ Section 11.4) of certain genes for DNA damage response, and control by members of the DevR regulon of the entry of *M. tuberculosis* into dormancy (**Figure 9.35**). Additional potential drug targets have also been identified through the screening of over 1000 different mutant strains of *M. tuberculosis*. This analysis mapped 18 previously uncharacterized genes to the persistence response of *M. tuberculosis* (⊄ Section 7.11), illustrating the power of integrating omics and mutant analyses to identify important pathogenicity genes in *M. tuberculosis*.

Tuberculosis Proteomics and Metabolomics

Proteomics has been used to identify proteins essential to the *M. tuberculosis* hypoxia response. While expected proteins that combat reactive oxygen species such as superoxide dismutase were detected, so were toxin–antitoxin systems (⊄ Section 7.11) and proteins that participate in the biosynthesis of the unique lipoproteins of *M. tuberculosis*. Production of nitrate and nitrite transporters also increased as *M. tuberculosis* switched to anaerobic respiration (⊄ Section 3.12) for energy metabolism. Moreover, at least 160 different uncharacterized proteins containing the N-terminal amino acid sequence "Pro-Glu/Pro-Pro-Glu" were detected in the pathogen. If drugs could be developed that target this sequence, they might be effective treatments for tuberculosis.

Online fundraising and social media have even been combined for the common goal of identifying new drugs to treat tuberculosis. The *Connect to Decode* initiative (http://c2d.osdd.net) is composed of more than 800 researchers whose collective goal is to identify new tuberculosis drug targets by mapping the entire interactome of *M. tuberculosis*. By analyzing over 10,000 *M. tuberculosis*

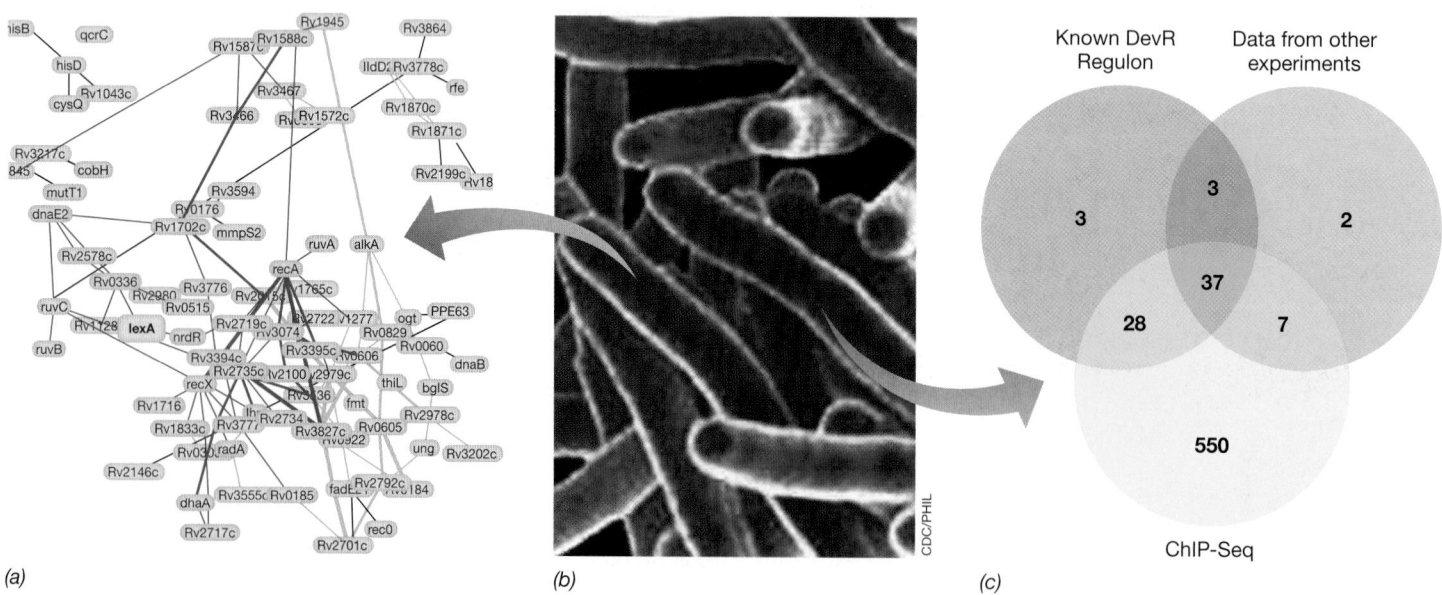

(a)   *(b)*   *(c)*

Figure 9.35 Genes controlled by the regulators LexA and DevR in *Mycobacterium tuberculosis*. *(a)* A snapshot of the interactome of some of the gene clusters that participate in DNA repair as identified from their expression pattern. Members of the LexA regulon are in red, while green connectors indicate interactions of other DNA repair genes. *(b)* Colorized scanning electron micrograph of cells of *M. tuberculosis*. *(c)* Venn diagram of results of various studies of the DevR regulon. Three methods of assessing genes in the regulon were employed and the number of genes identified from each detection method is indicated. A total of 662 genes were identified by ChIP-Seq analysis; of these, only 37 were identified by every method. Data adapted from van Dam, J.C.J., et al. 2014. *BMC Biology Systems 8:* 111.

experimental datasets, 87% of the proteins encoded by the genome have now been annotated. This is a vast improvement over the 52% of genes annotated in the original genome sequence. This collective effort has also resulted in a map of the complete *M. tuberculosis* interactome, with over 1400 proteins connected by over 2500 functional relationships (**Figure 9.36**). So far, 17 potential drug targets and their interactomes have been identified by this initiative. These targets lack homology to human proteins or to proteins of the human oral and gastrointestinal microbiome, thus increasing the probability of identifying drugs with the least side effects.

While tuberculosis metabolomics is in its infancy, studies comparing virulent versus attenuated *Mycobacterium* strains have identified over 1000 different lipids—many of them unique lipoproteins—that are likely quite important to the biology of *M. tuberculosis*. Thus, the *M. tuberculosis* lipid metabolism proteome may contain several potential therapeutic targets for tuberculosis. Finally, by profiling the *secretome* (an inventory of metabolites secreted) of *M. tuberculosis*, molecules unique to virulent strains (such as the nucleotide 1-tuberculosinyladenosine) have been discovered. This modified nucleotide is present in the urine of humans infected with *M. tuberculosis*, and its discovery illustrates the power of omics to link a particular pathogen to a specific molecule—be it a gene, protein, or metabolite—and yield new disease markers for use in clinical diagnostics. Moreover, if this modified nucleotide is essential to *M. tuberculosis*, drugs that interfere with its synthesis or activity are potential anti-tuberculosis drugs.

Figure 9.36 *Mycobacterium tuberculosis* **protein interactome.** Node color indicates proteins of the same category, connecting lines indicate interactions, and node size indicates relative number of interactions. Adapted from Vashisht, R., et al. 2012. *PLoS One 7*(7): e39809.

Vashisht, R. et al. 2012. PLoS One 7 (7): e39809

9.14 Systems Biology and Human Health

In 2003 the sequence of the 3-billion-base-pair human genome was completed and released in an international effort that cost nearly $3 billion. Scientists initially estimated that the human genome contained about 100,000 protein-encoding genes; however, we now know this number is much smaller, around 20,000. Since much more DNA is present in the human genome than the DNA that encodes proteins, what does the remaining DNA do? Is the bulk of the human genome simply "junk DNA"? Systems biology has attacked this question and revealed new and important information.

The Human Genome and ENCODE

To help unravel the mystery of the human genome and discover the role of the predicted genes in the human body, an international collaborative project termed the *Encyclopedia of DNA Elements* (ENCODE) project was initiated (www.encodeproject.org). The goal of this project is to create a catalog of functional elements in the human genome by measuring RNA expression (transcriptomics), identifying proteins that interact with DNA and RNA, and measuring DNA methylation to assess epigenetic interactions (changes to DNA that are not sequence based) in specific cells under different conditions. Thus far, ENCODE has revealed the surprising result that 80% of the human genome is *functional* in one cell type or another! If not encoding protein, this DNA functions as binding sites for proteins that influence gene activity, or functions as sites where chemical modifications lead to gene silencing, or encodes regulatory RNA. Hence, much of what was previously thought to be junk DNA in the human genome (stretches of DNA that lacked open reading frames) actually has a function, mainly in regulatory roles. ENCODE experiments have also been instrumental in revealing nucleotide polymorphisms (slight sequence differences in the same gene from different people) that correlate with certain genetic diseases. Thus, insights provided by ENCODE regarding how human DNA works are already bearing fruit for diagnostic medicine.

Personalized Omics Profiles and Medicine

How can the human genome and systems biology be applied to health? With the advent of next-generation sequencing, the cost to sequence a human genome has fallen under $1000, triggering an onslaught of human genome sequence data. By comparing genome sequences in over 2500 people from different continents, scientists have already identified over 88 million genome sites subject to variation. These genomic variants include single nucleotide differences, insertions and deletions, and rearrangements, each of which may turn out to be harmless, to be beneficial, to contribute to noninfectious "lifestyle" diseases such as obesity, diabetes, and heart disease, or to govern susceptibility to certain

UNIT 3

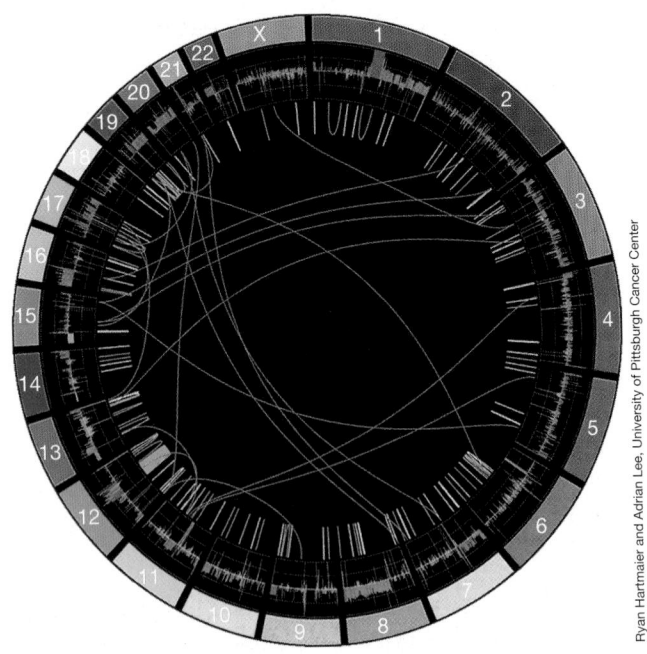

Ryan Hartmaier and Adrian Lee, University of Pittsburgh Cancer Center

Figure 9.37 Breast cancer tumor genome. Numbers on the outside ring represent individual chromosomes. The next ring represents increased (red) and decreased (green) copy number of DNA loci compared to a nontumor genome. The inner circle represents structural variations: red, translocations; blue, inversions; black, deletions; green, segmental duplications.

cancers. Understanding if and how genomic variants contribute to disease can be used to improve diagnostics, treatments, and prevention.

The genomics age has also benefited routine tumor genome sequencing. Data from these analyses are deposited in the Cancer Genome Atlas (http://cancergenome.nih.gov/) and have shown that each form of cancer contains a unique set of somatic (non-germ-line) mutations. For example, **Figure 9.37** illustrates copy number variations and genome rearrangements in a metastatic breast cancer sample. Patients at risk can have regions of specific genes sequenced (called *genotyping*) to determine if they are more susceptible to a certain cancer according to a comparison with these datasets. One example of this is genotyping the tumor suppressor encoding the genes *BRCA1* and *BRCA2*; certain mutations in these genes are known to increase the risk of developing breast cancer.

In addition to cancer diagnostics, omics has opened the era of *personalized medicine*, where genomic, transcriptomic, proteomic, metabolomic, and pharmacogenomic (omics responses to drugs) data are exploited to generate a snapshot of normal and disease states along with immune processes that occur in between. The first step in personalized medicine is the generation of an *i*ntegrative *P*ersonal *O*mics *P*rofile (iPOP). Tracking changes in this profile during healthy and disease states can aid in assessing medical risks and diagnosing and treating patients. The utility of personalized medicine can be illustrated by a study where blood samples taken from a male subject 20 times over a period of two years were subjected to proteomic and

metabolic profiling of about 5000 proteins and 4000 metabolites (**Figure 9.38**). The results suggested that the subject was at risk for coronary artery disease—which was not surprising based on his medical history and familial incidence—but also indicated that he was at high risk for type 2 diabetes. While this was a surprising finding considering his overall condition and familial health history, the prediction was based on profiling the subject's immune response to a viral infection. During these infections, both an increase in autoantibodies targeting an insulin receptor–binding protein and changes in gene expression related to the insulin response were detected. Just as this subject's personalized medicine profile predicted, he developed diabetes over the course of the two-year study (Figure 9.38). By tracking his iPOP, the onset of diabetes was revealed by the increasing production of RNAs, proteins, and metabolites, such as hemoglobin A1c and lauric acid, related to glucose metabolism (Figure 9.38). This case study illustrates the potential of iPOPs in monitoring the immune response and disease states. However, problems such as error rates, the analysis and storage of such large datasets, assessment of complicating factors

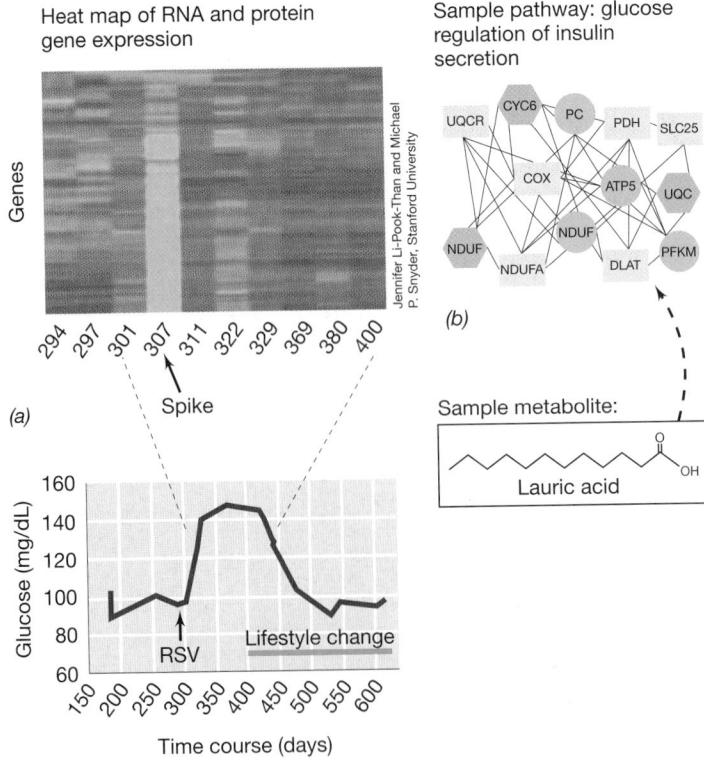

Figure 9.38 A personal integrated omics profile (iPOP). (*a*) Partial heat map of RNA and protein expression. Data plotted are from days 294–400 of the two-year study, with an increase in RNAs and proteins related to diabetes spiking on day 307. (*b*) Glucose regulation interactome. RNAs are indicated with blue circles, proteins with yellow squares, and RNA and its corresponding protein with green hexagons. A metabolite from the interactome is also shown. (*c*) Blood glucose levels during the time course of the study. The time of contracting a respiratory virus (RSV) is indicated as well as the time frame in which the patient made lifestyle and diet changes. Parts *b* and *c* adapted from Jennifer Li-Pook-Than and Michael P. Snyder, Stanford University.

(such as additional diseases during the course of the study, Figure 9.38), and ethical issues need to be addressed before iPOPs become common tools in clinical medicine.

Microbiology as well as biology in general has been forever changed by the dawn of genomics. Our journey through this chapter has only scratched the surface in describing how omics can be used to address previously intractable scientific questions. Much more is almost certainly in store, as major leaps forward in the omics field so far have typically been only one technological advancement away. In Chapter 10 we continue our genomics theme but change our focus from cells to viruses—the most abundant and genetically diverse microbes on Earth.

MINIQUIZ
- What is the ENCODE project?
- What are genomic variants?

MasteringMicrobiology® Visualize, explore, and think critically with Interactive Microbiology, MicroLab Tutors, MicroCareers case studies, and more. MasteringMicrobiology offers practice quizzes, helpful animations, and other study tools for lecture and lab to help you master microbiology.

Chapter Review

I • Genomics

9.1 Small viruses were the first organisms whose genomes were sequenced, but now many prokaryotic and eukaryotic cellular genomes have been sequenced.

Q What is one discovery resulting from the availability of a microbial genome?

9.2 DNA sequencing technology is advancing quickly. These advances have greatly increased the speed of DNA sequencing. Computer analysis of resulting sequencing data is also a vital part of genomics. Computational tools are used not only to annotate genomes but also to analyze sequences and the structures of biological macromolecules.

Q How can protein homology assist in genome annotation?

9.3 Sequenced genomes of *Bacteria* and *Archaea* range in size from 0.14 to 14.7 Mbp. The smallest are smaller than those of the largest viruses, whereas the largest have more genes than some eukaryotes. Gene content in prokaryotic cells is typically proportional to genome size. Many genes can be identified by their sequence similarity to genes found in other organisms. However, a significant percentage of sequenced genes are of unknown function.

Q What is the relationship between genome size and open reading frame content in genomes from prokaryotic cells?

9.4 Virtually all eukaryotic cells contain mitochondria, and in addition, plant cells contain chloroplasts. Although the genomes of organelles are independent of the nuclear genome, the organelles themselves are not. Many genes in the nucleus encode proteins required for organelle function. The complete genomic sequence of many microbial eukaryotes has also been determined, and the number of genes ranges from 1000 (less than many bacteria) to 60,000 (more than twice as many as humans).

Q Which genomes are larger, those of chloroplasts or those of mitochondria? How does your genome compare with that of yeast in overall size and gene number?

II • The Evolution of Genomes

9.5 Genomics can be used to study the evolutionary history of an organism. Organisms contain gene families, genes with related sequences. If these arose because of gene duplication, the genes are said to be paralogs; if they arose by speciation, they are called orthologs.

Q What is the major difference in how duplications have contributed to the evolution of the genomes of prokaryotic and eukaryotic cells?

9.6 Organisms may acquire genes from other organisms in their environment by horizontal gene transfer, and such a transfer may even cross phylogenetic domain boundaries. The mobilome, which includes transposons, integrons, and viruses, is important in genome evolution and often carries genes encoding antibiotic resistance or virulence factors.

Q Explain how horizontally transferred genes can be detected in a genome.

9.7 Comparison of the genomes of multiple strains of the same bacterial species shows a conserved component (the core genome) plus many variable genetic modules only present in certain strains of the species (combined with the core genome, this constitutes the pan genome). Many bacteria contain relatively large inserts of foreign origin known as chromosomal islands. These contain clusters of genes that encode specialized metabolic functions or pathogenesis and virulence factors (pathogenicity islands).

Q Explain how chromosomal islands might move between different bacterial hosts.

III • Functional Omics

9.8 Most microorganisms in the environment have never been cultured. Nonetheless, analysis of DNA samples has revealed enormous sequence diversity in most habitats. The concept of the metagenome embraces the total genetic content of all the organisms in a particular habitat.

> **Q** How do the human microbiome and mycobiome differ?

9.9 Microarrays consist of oligonucleotide probes corresponding to genes or gene fragments attached to a solid support in a known pattern; mRNA, cDNA, or DNA can then be labeled and hybridized to the gene chip to determine patterns of gene expression or the presence or absence of specific organisms. RNA-Seq combined with sequencing of cDNA can be used to profile the entire transcriptome of an organism.

> **Q** Besides gene expression, what else can be assayed using gene chips?

9.10 Proteomics is the analysis of all the proteins present in an organism. The ultimate aim of proteomics is to understand the structure, function, and regulation of these proteins. The interactome is the total set of interactions between macromolecules inside the cell.

> **Q** How does metaproteomics differ from proteomics?

9.11 Metabolomics profiles the complete set of metabolic intermediates produced by an organism. This analysis can determine active metabolic pathways and potential

cross-feeding (one organism supplies a nutrient for another organism) in community samples.

> **Q** Why is investigation of the metabolome lagging behind that of the proteome?

IV • The Utility of Systems Biology

9.12 With advances in molecular techniques, the genomes of single cells can be sequenced. Expression and protein profiles of single cells can also be determined. These techniques are instrumental for studying as yet uncultured microbes.

> **Q** How can single-cell genomics be used to address microbial dark matter?

9.13 By integrating multiple omics datasets in a systems biology approach, computer models predicting molecular activities and interactions in cells can be generated. For example, potential drug targets for the treatment of *Mycobacterium tuberculosis* have been identified using systems biology.

> **Q** How can systems biology be used to discover new diagnostic markers for disease?

9.14 Systems biology can also be applied to personal medicine. Besides detecting genetic variants, disease risks can be predicted by profiling a person's genome, transcriptome, proteome, and metabolome.

> **Q** What percentage of the human genome is now predicted to have functionality in at least one cell type?

Application Questions

1. Apart from genome size, what factors make complete assembly of a eukaryotic genome more difficult than assembly of a genome from a species of *Bacteria* or *Archaea*?

2. Describe how one might determine which proteins in *Escherichia coli* are repressed when a culture is shifted from a minimal medium (containing only a single carbon source) to a rich medium containing many amino acids, bases, and vitamins. How might one study which genes are expressed during each growth condition?

3. The gene encoding the beta subunit of RNA polymerase from *Escherichia coli* is said to be orthologous to the *rpoB* gene of *Bacillus subtilis*. What does that mean about the relationship between the two genes? What protein do you suppose the *rpoB* gene of *B. subtilis* encodes? The genes for the different sigma factors of *E. coli* are paralogous. What does that say about the relationship among these genes?

4. Describe how you could use systems biology to discover a new biologically produced antibiotic.

Chapter Glossary

Bioinformatics the use of computational tools to acquire, analyze, store, and access DNA and protein sequences

Chromosomal island a bacterial chromosome region of foreign origin that contains clustered genes for some extra property such as virulence or symbiosis

Codon bias the relative proportions of different codons encoding the same amino acid; it varies in different organisms. Same as codon usage

Core genome the part of a genome shared by all strains of a species

Gene chip small, solid supports to which genes or portions of genes are affixed and arrayed spatially in a known pattern (also called microarrays)

Gene family genes related in sequence to each other because of common evolutionary origin

Genome the total complement of genetic information of a cell or a virus

Genomics the discipline that maps, sequences, analyzes, and compares genomes

Homologs genes related in sequence to an extent that implies common genetic ancestry; includes both orthologs and paralogs

Horizontal gene transfer the transfer of genetic information between organisms as opposed to transfer from parent to offspring

Hybridization the joining of two single-stranded nucleic acid molecules by complementary base pairing to form a double-stranded hybrid DNA or DNA–RNA molecule

Interactome the total set of interactions between proteins (or other macromolecules) in an organism

Metabolome the total complement of small molecules and metabolic intermediates of a cell or organism

Metagenome the total genetic complement of all the cells present in a particular environment

Metagenomics the genomic analysis of pooled DNA or RNA from an environmental sample containing organisms that have not been isolated; same as environmental genomics

Microarray small, solid supports to which genes or portions of genes are affixed and arrayed spatially in a known pattern (also called gene chips)

Mobilome the mobile genetic elements in a genome

Nucleic acid probe a strand of nucleic acid that can be used to hybridize with a complementary strand of nucleic acid in a mixture; one of the two strands is labeled.

Open reading frame (ORF) a sequence of DNA or RNA that could be translated to give a polypeptide

Ortholog a gene in one organism that is similar to a gene in another organism because of descent from a common ancestor (see also *paralog*)

Pan genome the totality of the genes present in the different strains of a species

Paralog a gene whose similarity to one or more other genes in the same organism is the result of gene duplication (see also *ortholog*)

Pathogenicity island a bacterial chromosome region of foreign origin that contains clustered genes for virulence

Proteome (1) the total set of proteins encoded by a genome or (2) the total protein complement of an organism under a given set of conditions, also called the translatome

Proteomics the genome-wide study of the structure, function, and regulation of the proteins of an organism

Sequencing deducing the order of nucleotides in a DNA or RNA molecule by a series of chemical reactions

Systems biology the integration of data from genomics and other "omics" areas to build an overall picture of a biological system

Transcriptome the complement of all RNA produced in an organism under a specific set of conditions

Translatome the total set of proteins produced by an organism under a specific set of conditions

10

Viral Genomics, Diversity, and Ecology

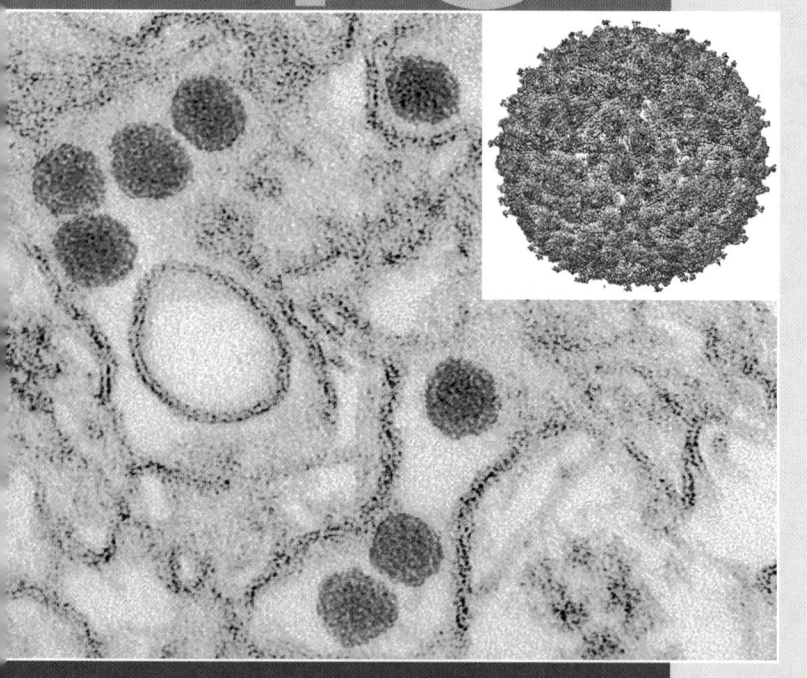

microbiology**now**

Viral Imaging to the Rescue: Structural Blueprint of Zika

Prior to 2016, Zika was known as a relatively benign mosquito-transmitted viral disease. Little research had been performed on the virus because most Zika infections caused only mild flulike symptoms lasting a few days. However, in recent outbreaks in South America, Central America, and Mexico, Zika infections in pregnant women have been linked to a birth defect called microcephaly, a severe brain malformation in which a baby is born with an abnormally small head. As the virus reached epidemic status, unexpected reports of Zika transmission from sexual contact emerged that resulted in the World Health Organization declaring the Zika virus "a public health emergency of international concern."

What can be done to combat the Zika epidemic? Zika is an enveloped icosahedral virus with a single-stranded plus-sense RNA genome. As a member of the *Flaviviridae* virus family, Zika is similar in many respects to the mosquito-transmitted dengue, yellow fever, and West Nile viruses. However, Zika differs from these other viruses in its ability to enter the central nervous system and potentially cross the placenta, and such capabilities likely set the stage for the virus to cause birth defects.

To better understand the Zika virus, scientists have zeroed in on the structure of the virion using advanced imaging techniques. The image here is a colorized transmission electron micrograph of Zika virions (blue, 40 nm in diameter) and a three-dimensional (3D) cryo-electron micrograph reconstruction of a mature Zika virion at near-atomic resolution (inset). The 3D analysis has revealed a striking difference between Zika and its close relatives—a unique carbohydrate group is associated with each of the 180 proteins that make up the icosahedral capsid of the Zika virus. Scientists believe that this molecule may be the key to Zika's ability to bind to nerve cells. If so, the structure could make an excellent antiviral target and offer opportunities for the development of both therapeutic and preventive drugs.

Besides answering questions regarding Zika's ability to attack nerve cells, the structural blueprint generated by these advanced microscopic techniques could be instrumental in future vaccine development, scientists say. Immediately, however, this work has provided a Zika virus fingerprint for its diagnosis and detection, and illustrates how basic science often contributes to improving human health.

 Source: Sirohi, D., et al. 2016. The 3.8 Å resolution cryo-EM structure of Zika virus. *Science 352:* 467–470.

Viruses infect all organisms and are the greatest repository of genetic diversity on Earth. In this chapter we explore viral diversity from both genomic and ecological perspectives and revisit and reinforce many of the basic concepts of virology developed in Chapter 8.

I • Viral Genomes and Evolution

Viruses have DNA or RNA genomes that can be either single-stranded or double-stranded (Chapter 8). Compared with cells, viral genomes can create some unusual challenges for genetic information flow. We begin our coverage by grouping viruses by their genome structure rather than by the hosts they infect, because viruses with the same genome structure face common problems in genetic information flow. We then consider which viruses infect cells in each of the domains of life and conclude with some hypotheses for how viruses may have first appeared and how viruses may have found a home on the universal tree of life.

10.1 Size and Structure of Viral Genomes

Viral genomes vary almost a thousandfold in size from smallest to largest. DNA viruses exist along this entire gradient from the tiny circovirus, whose 1.75-*kilobase* single-stranded genome pales in comparison to that of the recently discovered 2.5-*megabase*-pair double-stranded DNA genome of Pandoravirus (**Figure 10.1**). The genome of the latter is over twice that of the previously known largest virus (Mimivirus, see Figure 10.5*a*) and is larger than the genomes of several species of *Bacteria* and *Archaea* (⊘ Table 9.1). Pandoravirus infects certain marine amoebae, and with dimensions of 1 μm × 0.5 μm, is also larger than some bacterial cells (Figure 10.1).

RNA genomes, whether single- or double-stranded, are typically smaller than DNA viruses. Although some viral genomes are larger than those of some prokaryotic cells, genomes of *Bacteria* and *Archaea* are typically much larger than those of viruses (Figure 10.1), and genomes of eukaryotes are the largest of all. Viroids, naked infectious RNAs that cause certain plant diseases (Section 10.15), have the smallest genomes of all microbes (Figure 10.1).

Whether a viral genome is large or small, once a virus has infected its host, transcription of viral genes must occur and new copies of the viral genome must be made. Only later, once viral proteins begin to appear from the translation of viral transcripts, can viral assembly begin. For certain RNA viruses, the genome is also the mRNA. For most viruses, however, viral mRNA must first be made by transcription off of the DNA or RNA genome, and we now consider the variations on how this occurs.

The Baltimore Scheme: DNA Viruses

The American virologist David Baltimore, who shared with the American Howard Temin and Italian American Renato Dulbecco the Nobel Prize for Physiology or Medicine in 1975 for the discovery of retroviruses and their key enzyme, reverse transcriptase, developed a classification scheme for viruses. The scheme is based on the relationship of the viral genome to its mRNA and recognizes seven classes of viruses (**Figure 10.2**). By convention in virology, viral mRNA is always considered to be of the *plus*

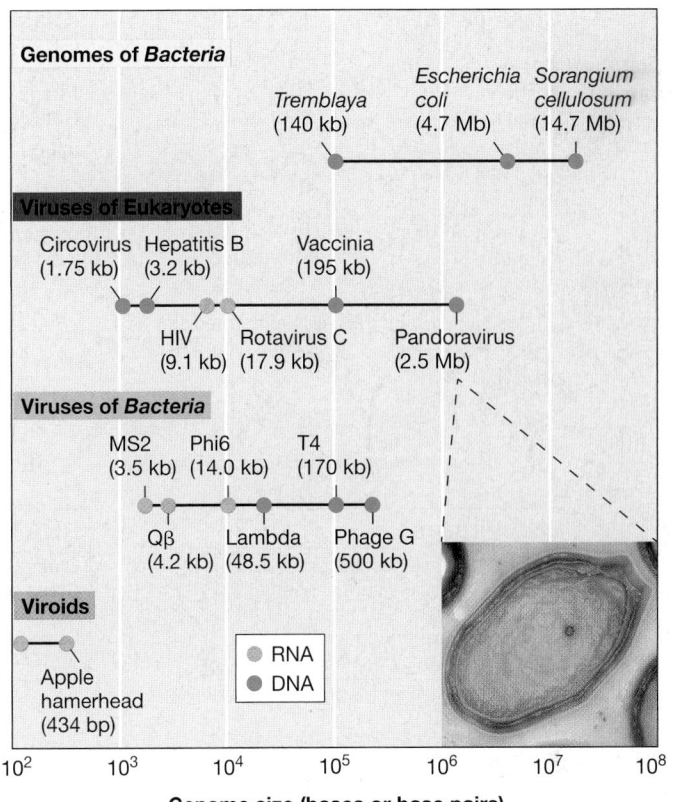

Figure 10.1 Comparative genomics. Genome sizes of select viroids, viruses, and prokaryotic cells. Inset: Micrograph of Pandoravirus, ~1 μm in length. Image courtesy of Chantal Abergel, IGS, UMR7256 CNRS-AMU. Bacteriophage phi6 and phage G infect *Pseudomonas* and *Bacillus* species, respectively; other bacterial viruses infect *Escherichia coli*.

configuration. Thus, to understand the molecular biology of a particular class of virus, one must know the nature of the viral genome and what steps are necessary to produce plus complementarity mRNA from it (Figure 10.2).

Double-stranded DNA viruses are in Baltimore class I. The mechanism of mRNA production and genome replication of class I viruses is the same as that used by the host cell, and we saw this with bacteriophage T4, a typical class I virus (⊘ Section 8.6). A virus containing a single-stranded genome may be either a **positive-strand** virus (also called a "plus-strand virus") or a **negative-strand** virus (also called a "minus-strand virus"). Class II viruses contain single-stranded plus-strand DNA genomes. Transcription of such a genome would yield a message of the minus sense. Therefore, before mRNA can be produced from class II viruses, a complementary DNA strand must first be made to form a double-stranded DNA intermediate; this is called the **replicative form**. The latter is used for transcription and as the source of new genome copies, the plus strand becoming the genome while the minus strand is discarded (Figure 10.2). With only one known exception, all single-stranded DNA viruses are positive-strand viruses.

The Baltimore Scheme: RNA Viruses

The production of mRNA and genome replication will obviously be different for RNA viruses than for DNA viruses. Cellular RNA

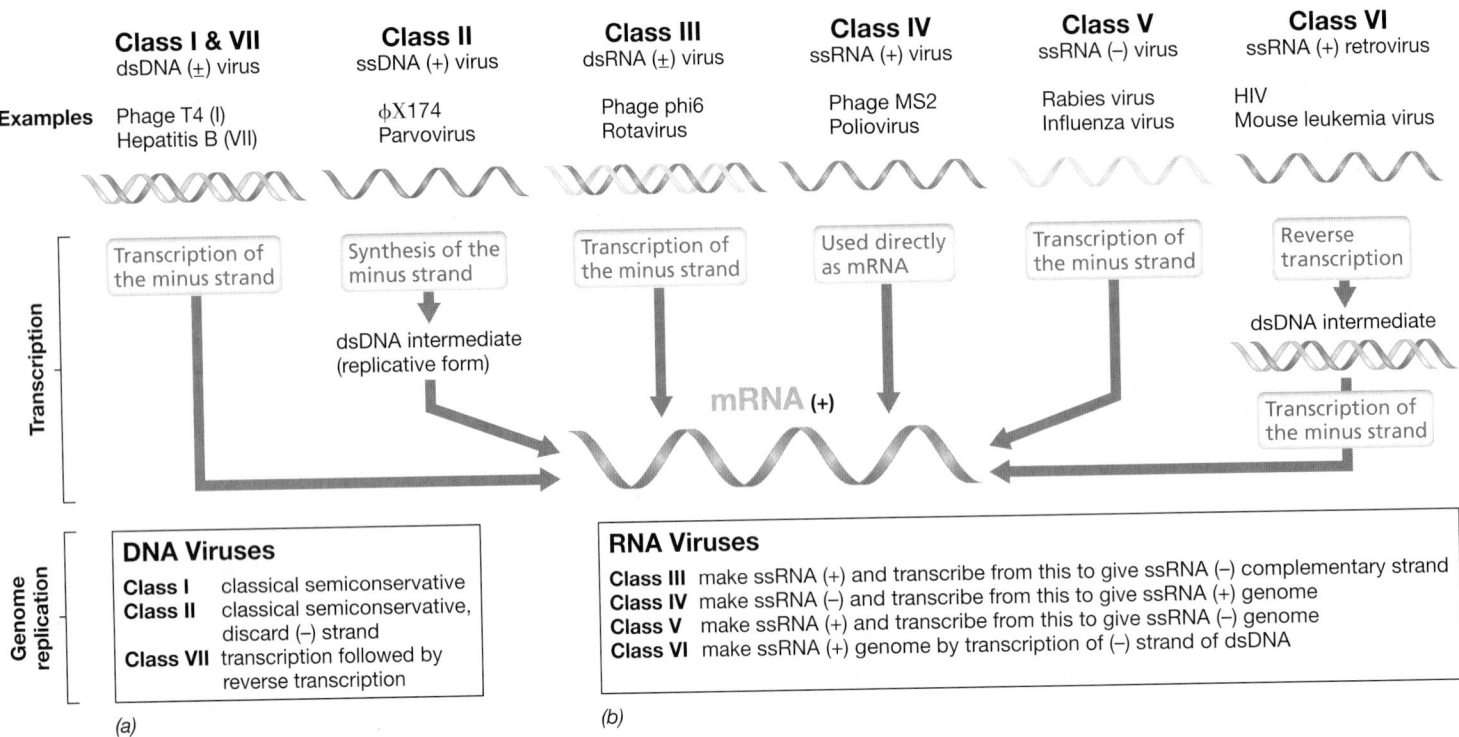

Figure 10.2 The Baltimore classification of viral genomes. Seven classes of viral genomes are known. The genomes can be either *(a)* DNA or *(b)* RNA, and either single-stranded (ss) or double-stranded (ds). With the exception of classes V and VI viruses, where the only known examples infect eukaryotic hosts, the top example listed is a bacterial virus and the bottom example an animal virus. The path each viral genome takes to form its mRNA and the strategy each uses for replication is shown.

polymerases do not catalyze the formation of RNA from an RNA template, but instead require a DNA template. Therefore, depending on the virus, RNA viruses must either carry in their virions or encode in their genomes an RNA-dependent RNA polymerase called *RNA replicase* (⌘ Section 8.2). With positive-strand RNA viruses (class IV), the genome is also mRNA. But for negative-strand RNA viruses (class V), RNA replicase must synthesize a plus strand of RNA off of the negative-strand template, and the plus strand is then used as mRNA. The plus strand is also used as a template to make more negative-strand genomes (Figure 10.2). RNA viruses of class III face a similar problem but start with double-stranded (+/−) RNA instead of only a positive or negative strand.

Retroviruses are animal viruses whose genomes consist of single-stranded RNA of the plus configuration but which replicate through a double-stranded DNA intermediate (class VI). The process of copying the information found in RNA into DNA is called **reverse transcription** and is catalyzed by an enzyme called *reverse transcriptase*. Human immunodeficiency virus, HIV, is a retrovirus. Finally, class VII viruses are those highly unusual viruses (hepatitis B virus is an example) whose genomes consist of double-stranded DNA but which replicate through an RNA intermediate. As we will see later, these viruses also use reverse transcriptase.

Hosts for Viruses of Each Baltimore Class

Only certain of the Baltimore classes of viruses are known to infect cells of a particular phylogenetic domain. For example, only two

classes of virus are known in *Archaea* and only four in *Bacteria*; only in animals do we find examples of all seven Baltimore classes of viruses (**Figure 10.3***a*). Examples of viruses from each of the Baltimore classes are drawn to scale in Figure 10.3*b*.

Double-stranded DNA viruses (class I) are the primary viruses infecting prokaryotic cells, while single-stranded plus-sense RNA viruses (class IV) are the major viral predators of eukaryotic cells (Figure 10.3*b*). As far as is known, fungi are only infected by RNA viruses of classes III and IV, whereas the vast majority of class I viruses that infect eukaryotes replicate in animal hosts rather than plants. By contrast, plants serve as hosts to many more class II viruses than do animals, whereas virtually all class V viruses (viruses with a single-strand minus-sense genome) infect animals rather than plants. And finally, retroviruses (class VI) are known only from animal hosts, while class VII viruses, which like retroviruses depend on reverse transcriptase to replicate their genome, are much more common in plants than in animals (Figure 10.3*a*). Although the reasons that different genomic classes of virus show specific host preferences is unclear, the fact that *Bacteria* and *Archaea* are hosts for only a relatively small group of viruses suggests that some viral classes may have evolved only later in the timeline of life when more complex eukaryotic hosts were available for infection. However, computational analyses to be discussed in Section 10.2 indicate that this hypothesis is probably incorrect, in that most viral groups, especially the RNA viruses, appear to be of ancient origin.

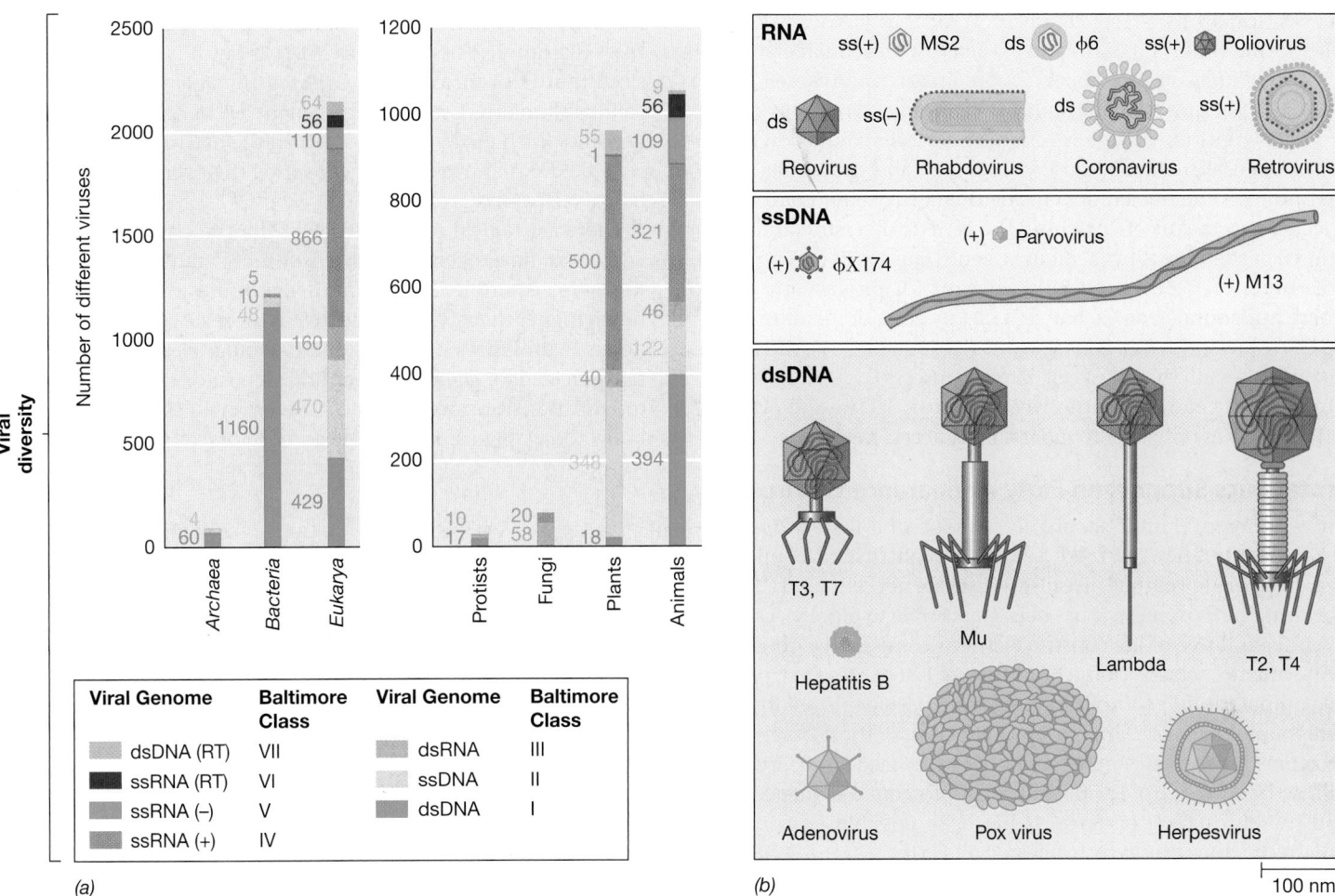

Figure 10.3 Viral hosts and viral diversity. *(a)* Virus host preferences by cellular domain (left graph) and in different major groups of eukaryotes (right graph). Data adapted from Nasir, A., and G. Caetano-Anollés. 2015. *Sci. Adv. 2015;1:* e1500527. *(b)* Drawings to scale of several viruses discussed in this chapter.

Viral Protein Synthesis

Once viral mRNA is made (Figure 10.2), viral proteins can be synthesized. In all viruses, these proteins can be grouped into two broad categories: (1) proteins synthesized soon after infection, called *early proteins*, and (2) proteins synthesized later in the infection, called *late proteins*. Both the timing and amount of viral protein synthesis is highly regulated. Early proteins are typically enzymes and are therefore synthesized in relatively small amounts. These include not only nucleic acid polymerases but also proteins that function to shut down host transcription and translation. By contrast, late proteins are typically structural components of the virion and other proteins that are not needed until virion assembly begins, and these are made in much larger amounts (⮑ Section 8.6).

Virus infection upsets the regulatory mechanisms of the host because there is a marked overproduction of viral nucleic acid and protein in the infected cell. Eventually, when the proper proportions of viral genome copies and virion structural components have been synthesized, new virions are assembled—typically spontaneously—and exit the host cell by either lysing and killing it or by a budding process in which the host cell may remain alive.

MINIQUIZ

- Distinguish between a positive-strand RNA virus and a negative-strand RNA virus.
- Contrast mRNA production in the two classes of single-stranded RNA viruses.
- What is unusual about genetic information flow in retroviruses?

10.2 Viral Evolution

When did viruses first appear on Earth and what is their relationship to cells? All known viruses require a host cell for their replication, and this leads to the natural conclusion that viruses evolved at some time *after* cells first appeared on Earth, about 4 billion years ago. Following this line of reasoning, viruses would be remnant cell components that evolved an ability to replicate with assistance from the cell. However, other hypotheses for the origin of viruses have been proposed, including that viruses are relics of the "RNA world," a period in evolution when RNA is hypothesized to have been the sole carrier of genetic information

UNIT 3

(⟲ Section 13.1 and see Figure 10.4), or that viruses were the "first forms of life" on Earth and existed in a precellular era.

Although *how* viruses appeared remains an unanswered question, so is the question of *why* viruses appeared. One likely driver of viral evolution was as a mechanism for cells to quickly move genes about in nature. Because viruses have an extracellular form that protects the nucleic acid inside them, they could have been selected as a means of enriching the genetic diversity (and thus fitness) of their hosts by facilitating gene transfers between them. This function seems especially relevant for prokaryotic cells, where horizontal gene exchange is clearly a major factor in their rapid evolution (⟲ Sections 9.6 and 13.3, Chapter 11). Although many viruses kill their host cell, latent viruses do not, and it is possible that the earliest viruses were primarily latent and evolved lytic capacities only later to more rapidly access new hosts.

Proteomics Support an Early Appearance of Viruses

An experimental analysis of the proteins of a wide variety of viruses has shed new light on how viruses might first have appeared and diversified over time. Because viruses are not cells (and thus do not contain ribosomes), it has been impossible to place viruses on the universal tree of life constructed from comparative ribosomal RNA sequences (⟲ Section 1.13 and Figure 1.36b). However, powerful computational methods have recently been deployed to compare the *proteomes* of a large group of viruses with those of cells (the proteome is the total complement of proteins made by a virus or a cell, ⟲ Section 9.10). From analyses of proteomic sequence data and protein folding patterns, it has been possible to gain new insights on how viruses first appeared on Earth.

Proteomics point to an origin of viruses from ancient cells that contained segmented RNA genomes and that existed before the last universal common ancestor (LUCA) of modern cells appeared (**Figure 10.4a**). This would have been the era of the "RNA World," a time preceding the appearance of DNA. Within this line of "virocells," strong evolutionary pressure for a reduction in both genomic and compartment size eventually eliminated the cellular nature of virocells altogether, leaving only a protein shell to protect the genome from damage. Such reduction would also have triggered a strict dependence in these emerging structures on archaeal, bacterial, or eukaryal cells for replication functions, a characteristic property of viruses. Proteomic analyses point to RNA viruses in general as being older than DNA viruses and, more specifically, to dsRNA viruses (Baltimore class III, Figure 10.2) as being the most ancient of all viruses. Interestingly, many different types of dsRNA viruses (Section 10.10) contain segmented genomes, a possible remnant from their virocell ancestors. Retroviruses also appear to be ancient and may have played a role in the transition from an RNA to a DNA world; we explore this possibility now.

The RNA to DNA Transition

If *RNA* viruses originated in the scenario just described (Figure 10.4a), how did *DNA* viruses arise? It is thought that some RNA viruses evolved DNA genomes as a mechanism to protect their genomes from cellular ribonucleases—cellular enzymes that destroy foreign RNA. Because DNA is not RNA, these viruses would have had to evolve their own DNA replication machinery to replicate their genomes. It is conceivable that an enzyme like reverse transcriptase

was a key to the conversion of RNA into DNA, just as it is in retroviruses (Baltimore class VI, Figure 10.2) today. It is further hypothesized that DNA viruses then infected the ancestors of the three cellular domains. Gradually, by genetic exchange with DNA viral genomes, each group of cells obtained the machinery necessary to replicate DNA and eventually converted their genomes from RNA-based to DNA-based chemistry.

There are logical reasons for why the transition from RNA to DNA may have occurred. DNA is a more stable molecule than RNA—for example, the spontaneous mutation rate of RNA is much higher than of DNA, and RNA is more susceptible to spontaneous hydrolysis—and this stability would over time have naturally selected for DNA as the genomic repository in cells. This RNA to DNA transition would then have initiated the DNA world we know today (Figure 10.4b). The absence of extant cells with RNA

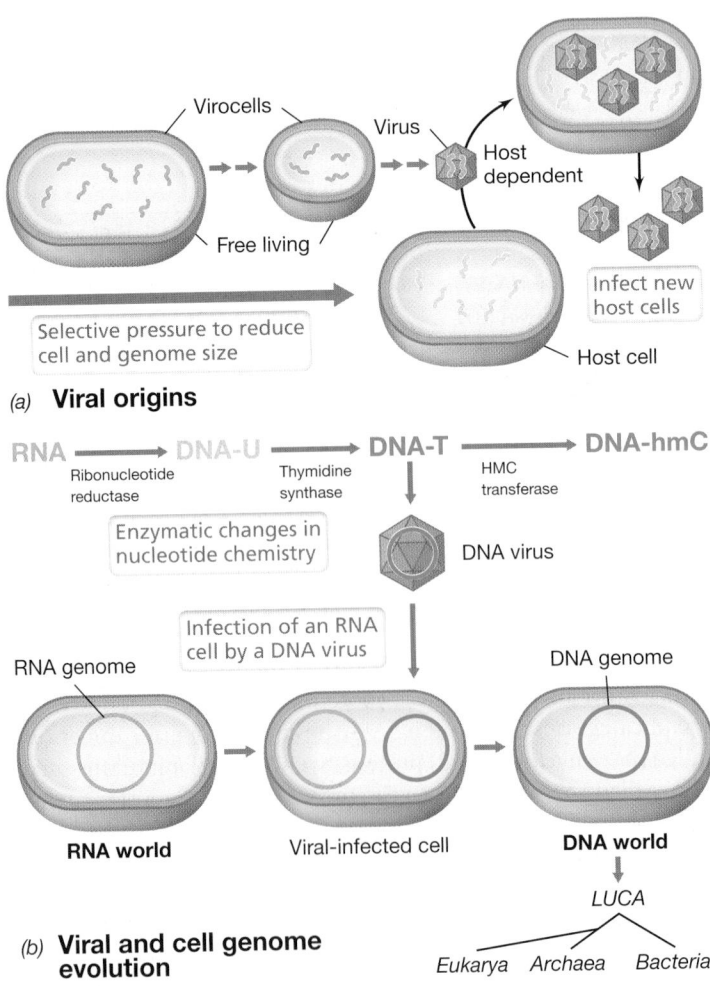

(a) **Viral origins**

(b) **Viral and cell genome evolution**

Figure 10.4 Viral origins and the role of viruses in the transition from an RNA world to a DNA world. *(a)* Viruses are thought to have arisen from primitive "virocells" that contained RNA genomes. Selection for reduced cell size and genomic demands led to the evolution of viruses. *(b)* The evolution of DNA-specific enzymes would have allowed RNA viruses to become DNA viruses. DNA-U, DNA with uracil (uracil is a base now found mainly in RNA); DNA-T, DNA with thymine (a base found in DNA but not RNA); DNA-hmC, DNA with 5-hydroxycytosine; DNA-U and DNA-hmC are DNA variants known from one virus or another. Infection of an RNA cell by a DNA virus could then have transferred DNA synthetic capacity to the cell, which led to DNA becoming the genomic repository of cells. LUCA, last universal common ancestor.

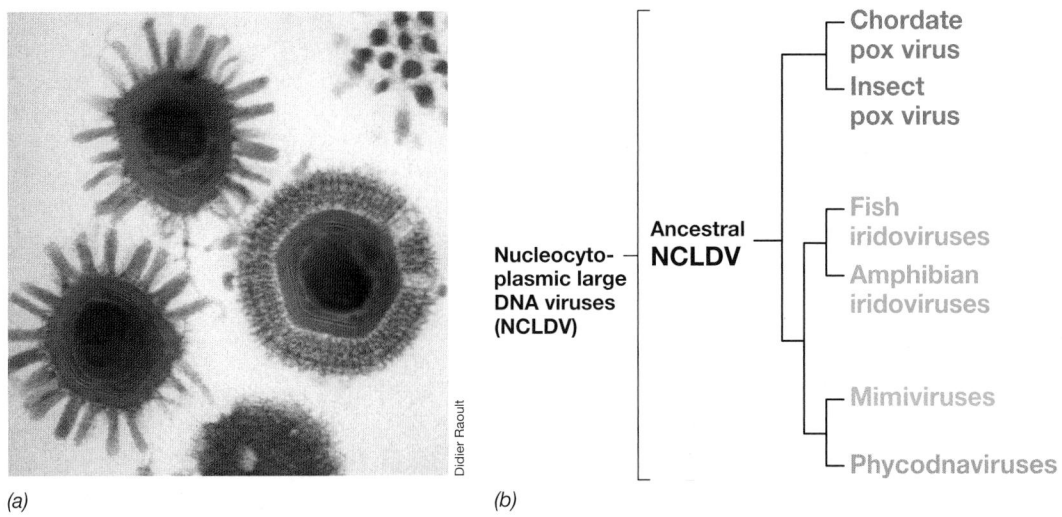

(a)   (b)

Figure 10.5 Phylogeny of nucleocytoplasmic large DNA viruses (NCLDV). *(a)* Transmission electron micrograph of Mimivirus, a member of the NCLDV group. A virion is about 0.75 μm in diameter. *(b)* Phylogeny of major groups of NCLDV based on comparative sequences of several proteins of DNA metabolism. See page 223 for additional coverage of large viruses.

genomes may be because such cells were never infected by DNA viruses and thus never evolved DNA genomes; Darwinian selection would have eventually driven these less fit cells to extinction. However, the fact that some RNA viruses still remain today may actually be a result of their high spontaneous mutation rate because it would allow them to stay one step ahead of evolving host defenses and more quickly adapt to available hosts.

Viral Phylogeny

Using the newly developed proteomic analyses just discussed, it has been possible to place viruses on a universal phylogenetic tree of life constructed from a combination of protein sequences and structural features rather than ribosomal RNA sequences (⟲ Section 1.13). As expected, such trees position viruses at the root of the tree—with RNA viruses preceding DNA viruses—and contain a long branch leading to the three domains of cellular life, with the latter branching out in much the same way as in trees based on ribosomal RNA. The branching order of viruses in the proteomics tree remains a bit fuzzy, but the analyses clearly show RNA viruses to lie basal to DNA viruses in agreement with hypotheses for how DNA viruses arose and how DNA replaced RNA in cellular genomes (Figure 10.4).

In only a few groups of viruses has it been possible to reliably trace phylogenies more precisely, and in these cases, trees have been assembled from sequences of a group of genes or proteins shared in common among the group. One such example is Mimivirus and its relatives, one of the larger known viruses (**Figure 10.5**). Mimivirus capsids are multilayered and icosahedral. The virion is surrounded by spikes and is nearly 0.75 μm in diameter, larger than some prokaryotic cells (Figure 10.5*a*). Mimivirus contains a 1.2-megabase-pair genome consisting of double-stranded DNA. Mimivirus infects the protozoan *Acanthamoeba* and belongs to a group of giant viruses with large genomes called *nucleocytoplasmic large DNA viruses* (NCLDV) (Figure 10.5*b*). The NCLDV comprise several virus families, including pox viruses (Section 10.6), iridoviruses, and certain plant viruses. These viruses share a set of highly homologous

proteins, most of which function in DNA metabolism. A phylogenetic tree of these viruses constructed from DNA sequences encoding these proteins shows how they have diverged from a common ancestor (Figure 10.5*b*). It is thus possible to track the phylogeny of particular viral groups with some confidence, but to do so, one needs to start with a group that is already known to share a number of properties in common.

It is clear that diversity in the viral world is enormous and that obtaining a detailed phylogenetic tree of all viruses will remain a challenge. The continual isolation of highly unusual new viruses makes this difficult task even more challenging. For example, only about 7% of the genes of Pandoravirus (Figure 10.1) have gene homologs in existing genomic databases. What this means is that over 90% of the genome of this giant virus will likely be new to biology—a striking example of what awaits discovery in the fascinating world of viruses.

MINIQUIZ

- How could viruses have accelerated the evolution of cells?
- Explain how viruses could have "invented" the genetic material found in all cells.
- Lacking ribosomes, how can viruses be placed on the universal tree of life?

II • DNA Viruses

Although DNA viruses likely appeared later in evolutionary times than RNA viruses (Section 10.2), DNA viruses today infect a wide variety of organisms, in particular, species of *Bacteria* and *Archaea*. In fact, the majority of viruses that infect prokaryotic cells are DNA viruses, mainly of the double-stranded variety (Figure 10.3). We examine several of these here along with some DNA viruses that infect eukaryotes, and keep our focus on the processes involved in transcription and genome replication in DNA viruses of different genomic makeup.

10.3 Single-Stranded DNA Bacteriophages: φX174 and M13

In this section we discuss two single-stranded DNA bacteriophages, φX174 and M13. Many single-stranded DNA plant and animal viruses are also known, and because these share with bacterial viruses the fact that their genomes are of the plus complementarity (Baltimore class II, Figure 10.2*a*), many molecular events are similar. Hence, our focus here will be on the phages.

Bacteriophage ΦX174

Bacteriophage ΦX174 contains a circular genome of 5386 nucleotides inside a tiny icosahedral virion, about 25 nm in diameter. Phage ΦX174 has only a few genes and shows the phenomenon of **overlapping genes**, a condition in which there is insufficient DNA to encode all viral-specific proteins unless parts of the genome are read more than once in different reading frames. For example, in the ΦX174 genome, gene B resides within gene A, and gene K resides within both genes A and C (**Figure 10.6a**). Genes D and E also overlap, gene E being contained completely within gene D. Also, the termination codon of gene D overlaps the initiation codon of gene J (Figure 10.6a).

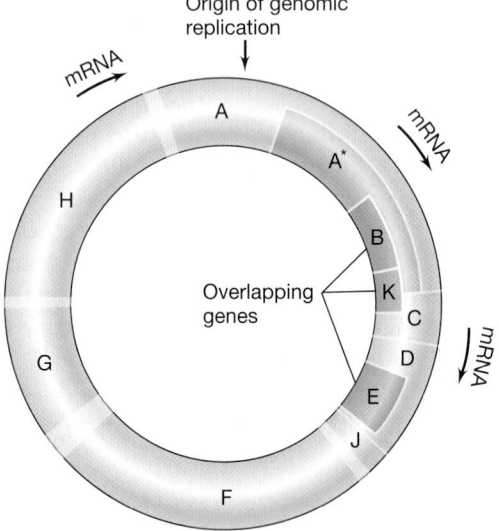

A   Replicative form DNA synthesis
A* Shutoff of host DNA synthesis
B   Formation of capsid precursors
C   DNA maturation
D   Capsid assembly
E   Host cell lysis
F   Major capsid protein
G   Major spike protein
H   Minor spike protein
J   DNA packaging protein
K   Function unknown

(a) Genetic map of ΦX174

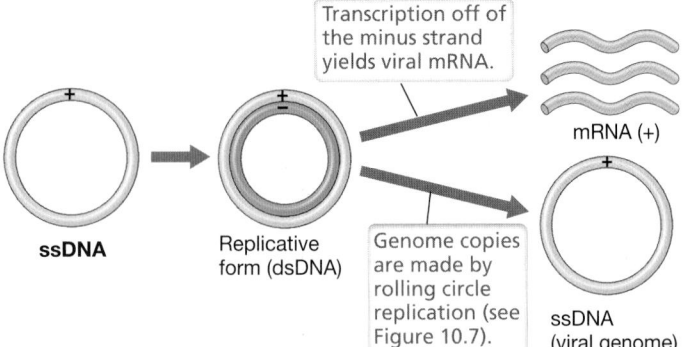

(b) Flow of events during ΦX174 replication

Figure 10.6 Bacteriophage ΦX174, a single-stranded DNA phage. *(a)* Genetic map. Note regions of gene overlap. Protein A* is formed using only part of the coding sequence of gene A by reinitiation of translation. The key indicates the functions of the proteins encoded by each gene. Unlabeled parts of the chromosome are regions of noncoding DNA. *(b)* Genetic information flow in ΦX174. Progeny single-stranded DNA is produced from the replicative form by rolling circle replication (see Figure 10.7).

The distinct gene products from overlapping genes are made by reinitiating transcription *in a different reading frame* within a gene to yield a second (and distinct) transcript. In addition to overlapping genes, a small protein in ΦX174 called A*, which functions to shut down host DNA synthesis, is synthesized by the reinitiation of *translation* (not transcription) within the mRNA for gene A. The A* protein is read from the same mRNA reading frame as A protein but has a different in-frame start codon and is thus a shorter protein.

Before a single-stranded DNA genome can be transcribed, a complementary strand of DNA must be synthesized, forming a double-stranded molecule called the replicative form. This can then be used as a source of both mRNA and genome copies. Upon infection of an *Escherichia coli* cell by ΦX174, the viral DNA is separated from the protein coat and the genome is converted into the replicative form by host enzymes. From this, several copies are made by semiconservative replication, and phage-specific transcripts are made by transcription off of the negative strand of the replicative form (Figure 10.6b). The replicative form is also the starting point for making copies of the phage genome by a mechanism we have already seen used in phage lambda (⇄ Section 8.7): **rolling circle replication** (Figure 10.7).

In the synthesis of the ΦX174 genome, the rolling circle facilitates the continuous production of positive strands from the replicative form. To do this, the positive strand of the latter is nicked and the 3′ end of the exposed DNA is used to prime synthesis of a new strand (Figure 10.7). Cutting of the plus strand is accomplished by the A protein (Figure 10.6a). Continued rotation of the circle leads to the synthesis of a linear ΦX174 genome. Note that rolling circle synthesis differs from semiconservative replication (⇄ Section 4.3) because only the negative strand serves as a template.

When the growing viral strand reaches unit length (5386 residues for ΦX174), the A protein cleaves it and then ligates the two ends of the newly synthesized single strand to give a single-stranded DNA circle. Ultimately, assembly of mature ΦX174 virions occurs and cell lysis follows. The E protein (Figure 10.6a) promotes cell lysis by inhibiting the activity of an enzyme in peptidoglycan synthesis (⇄ Section 7.5) in the host cell. Because of the resulting weakness in newly synthesized cell wall material, the host cell ruptures, releasing the phage virions.

Bacteriophage M13

Bacteriophage M13 is a filamentous virus with helical symmetry; the virion is long and thin and attaches to the pilus of its host cell (⇄ Section 8.5). Filamentous phages such as M13 have the unusual property of being released from the host cell without the cell undergoing lysis; infected cells continue to grow, and typical viral plaques (⇄ Section 8.4) are not observed. To facilitate the nonlytic release, M13 DNA is covered with coat proteins as it exits across the cell envelope. Four minor coat proteins cover the tips of the virion while the major coat protein covers the sides (**Figure 10.8**). Thus with M13, there is no intracellular accumulation of mature virions as occurs with typical lytic bacteriophages. Instead, these filamentous bacteriophages cause chronic infections.

Several features of phage M13 have made it useful as a cloning and DNA sequencing vehicle in the past. For example, many aspects of DNA replication in M13 are similar to those of ΦX174 and the genome is very small; this facilitates sequencing efforts.

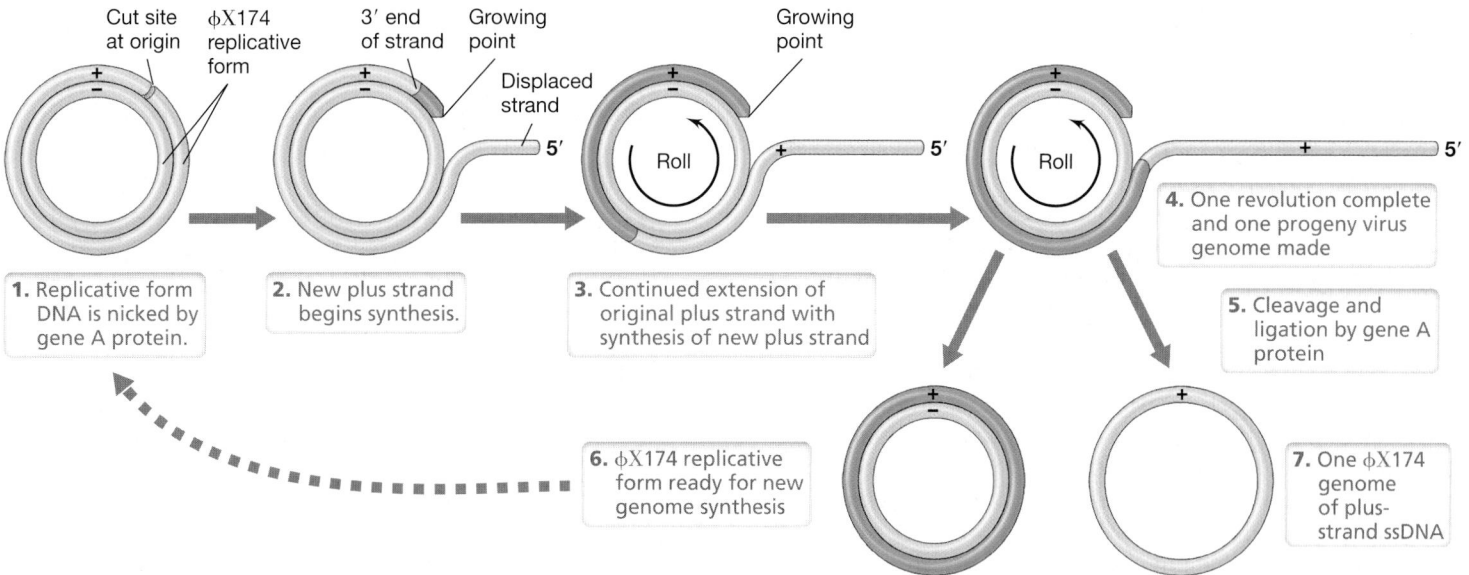

Figure 10.7 Rolling circle replication in phage φX174. Replication begins at the origin of the double-stranded replicative form with the cutting of the plus strand of DNA by gene A protein (both strands of DNA are shown in light green here for simplification). After one new progeny strand has been synthesized (one revolution of the circle), the gene A protein cleaves the new strand and ligates its two ends.

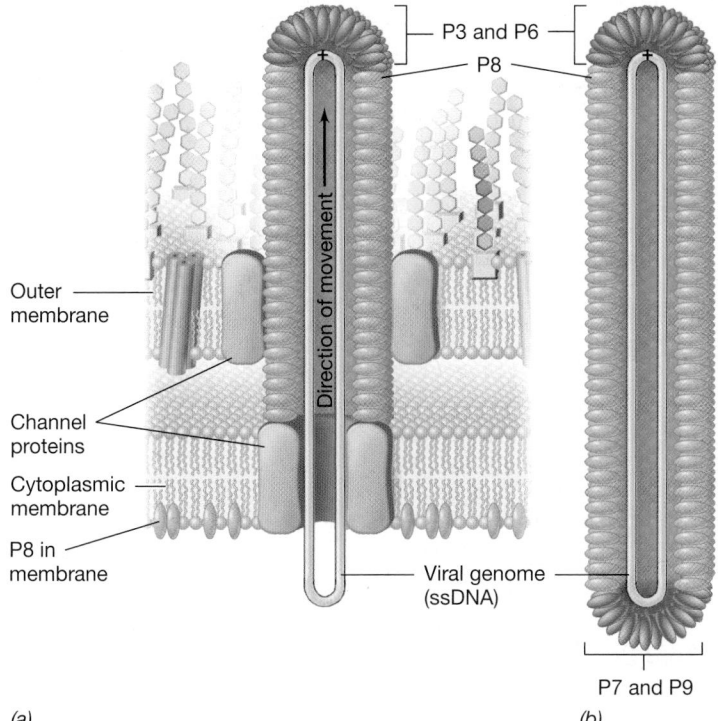

Figure 10.8 Release of phage M13. The virions of phage M13 exit infected cells without lysis. (a) Budding. The virus DNA crosses the cell envelope through a channel constructed from virus-encoded proteins. As this occurs, the DNA is coated with phage proteins that have been embedded in the cytoplasmic membrane. (b) Complete virion. The two ends of the virion are covered with small numbers of the minor coat proteins P3 and P6 (front end) or P7 and P9 (rear end). Because bacteriophage M13 is a single-stranded DNA phage, it was widely used in the past as a tool for molecular cloning and DNA sequencing (Chapters 9 and 12).

Second, a double-stranded form of genomic DNA essential for cloning purposes is produced naturally when M13 produces its replicative form. Third, as long as infected cells are kept growing, phage can be produced indefinitely, yielding a continuous source of the cloned DNA. These and other features of M13 made this phage a workhorse of the genetic engineering field for many years, although today M13 has been replaced for most genetic engineering tasks by a variety of even more convenient and useful tools.

MINIQUIZ

- Why is formation of the replicative form of φX174 necessary in order to make phage-specific mRNA?
- In the φX174 genome, describe the difference between how the gene B and gene A* proteins are made.
- How can M13 virions be released without killing the infected host cell?

10.4 Double-Stranded DNA Bacteriophages: T7 and Mu

The double-stranded DNA (dsDNA) (Baltimore class I, Figure 10.2) bacteriophages are among the best studied of all viruses, and we have already discussed two important ones, T4 and lambda, in Chapter 8. However, because of their importance in molecular biology, gene regulation, and genomics, we consider two more such viruses here, T7 and Mu, each of which has features distinct from those of T4 and lambda.

Bacteriophage T7

Bacteriophage T7 is a relatively small DNA virus that infects *Escherichia coli* and a few related enteric bacteria. The virion has an icosahedral head and a very short tail, and the T7 genome is a linear double-stranded DNA molecule of about 40 kilobase pairs.

When a T7 virion attaches to a host cell and the DNA is injected, early genes are quickly transcribed by host RNA polymerase and then translated. One of these early proteins inhibits the host restriction system, a mechanism for protecting the cell from foreign DNA (⇔ Section 8.5). This occurs very rapidly, as the T7 anti-restriction protein is made and becomes active before the entire T7 genome has entered the cell. Other early proteins include a T7 RNA polymerase and proteins that inhibit host RNA polymerase activity. T7 RNA polymerase recognizes only T7 gene promoters distributed along the T7 genome. This transcriptional strategy differs from that of phage T4 because T4 uses the host RNA polymerase throughout its replication cycle but modifies the host polymerase such that it recognizes only phage genes (⇔ Section 8.6).

Genome replication in T7 begins at an origin of replication within the molecule and proceeds bidirectionally from this point (**Figure 10.9a**). Phage T7 uses its own DNA polymerase, which is a composite protein including one polypeptide encoded by the phage and one by the host. As in phage T4, T7 DNA contains terminal repeats at both ends of the molecule and these are eventually used to form *concatemers* (Figure 10.9b). Continued replication and recombination leads to concatemers of considerable length, but ultimately a phage-encoded endonuclease cuts each concatemer at a specific site, resulting in the formation of linear DNA molecules with terminal repeats that are packaged into phage heads (Figure 10.9c). However, because T7 endonuclease cuts the concatemer at specific sequences, the DNA sequence in each T7 virion is identical. This differs from the situation in phage T4, where DNA concatemers are processed using a "headful mechanism" that generates circularly permuted genomes (⇔ Section 8.6).

Bacteriophage Mu

Like bacteriophage lambda (⇔ Section 8.7), bacteriophage Mu is a temperate phage but has the unusual property of replicating by *transposition*. Transposable elements are sequences of DNA that can move within their host genome from one location to another as discrete genetic units (⇔ Section 11.11); transposition is facilitated by an enzyme called **transposase**. Mu was so named because it generates *mu*tations when it integrates into the host cell chromosome, and thus it has been useful in bacterial genetics because it can generate mutants easily.

Bacteriophage Mu has an icosahedral head, a helical tail, and several tail fibers (**Figure 10.10a**). The genome of Mu consists of linear double-stranded DNA, and most Mu genes encode head and tail proteins, other replication factors such as the Mu transposase, and factors that affect host range. Host range is controlled by the kind of tail fibers that are made, with one type allowing only infection of *E. coli* while the other type allows the phage to infect several other enteric bacteria as well.

Figure 10.9 Replication of the bacteriophage T7 genome. *(a)* The linear, double-stranded DNA undergoes bidirectional replication, giving rise to intermediate "eye" and "Y" forms (for simplicity, both template strands are shown in light green and both newly synthesized strands in dark green). *(b)* Formation of concatemers by joining DNA molecules at their unreplicated terminal ends. *(c)* Production of mature viral DNA molecules from T7 concatemers by activity of the cutting enzyme, an endonuclease.

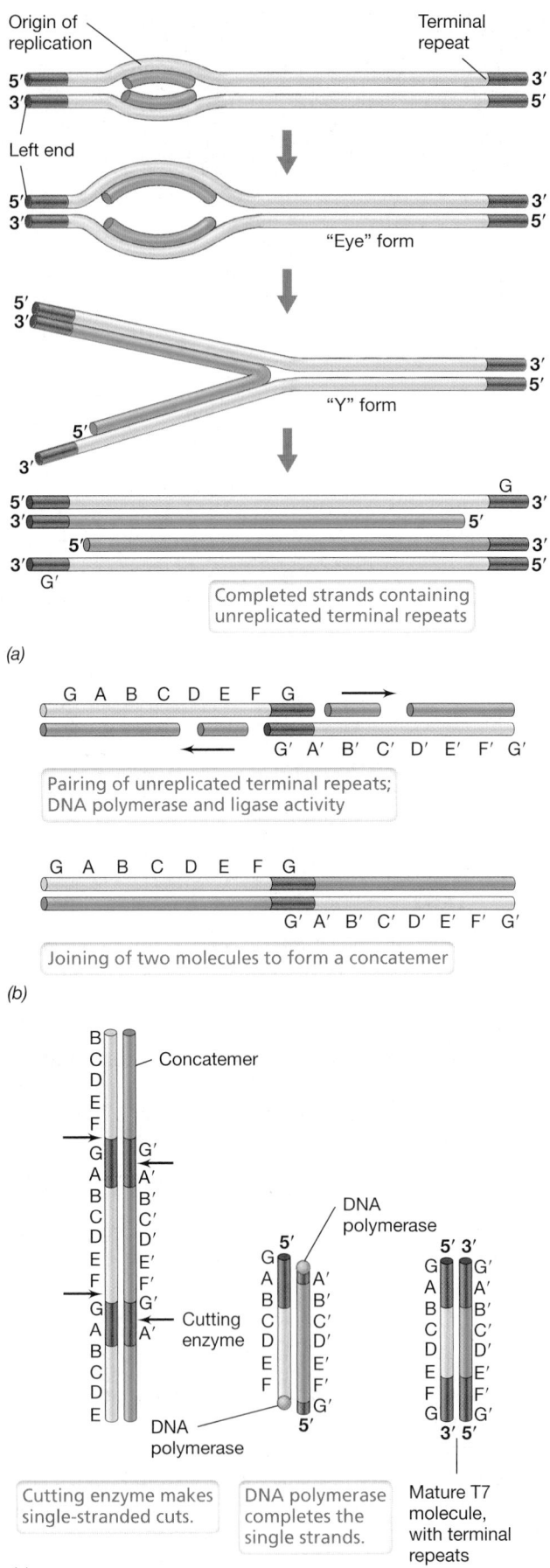

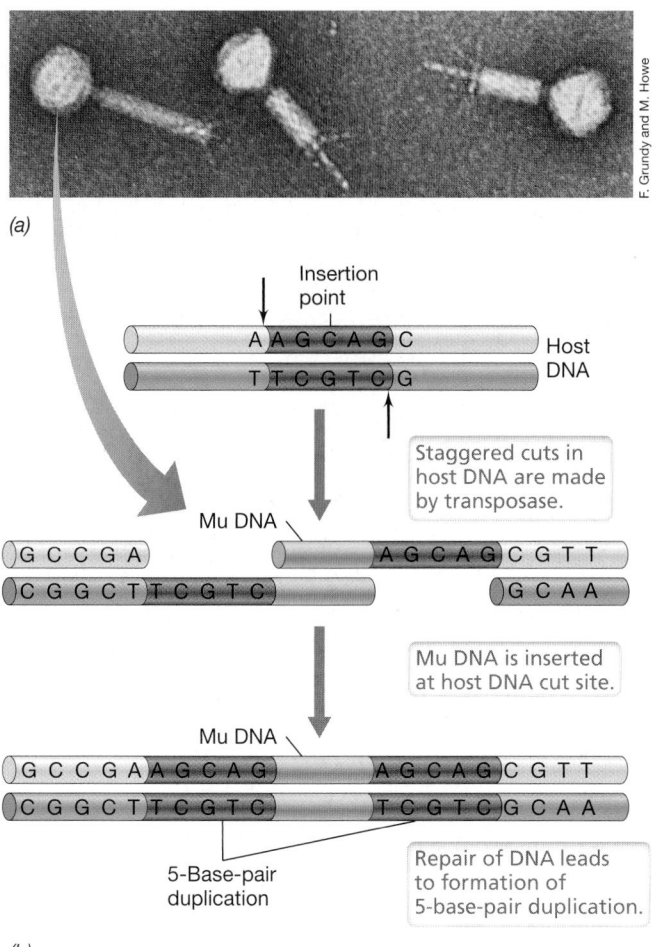

F. Grundy and M. Howe

Figure 10.10 Bacteriophage Mu. *(a)* Electron micrograph of virions of the double-stranded DNA phage Mu. *(b)* Integration of Mu into the host DNA, showing the generation of a 5-base-pair duplication of host DNA.

Phage Mu replicates in a completely different manner from all other bacteriophages because its genome is replicated as part of a larger DNA molecule (Figure 10.10*b*). Thus, integration of Mu DNA into the host genome is essential for both lytic and lysogenic development. Integration requires the activity of Mu transposase, and a 5-base-pair fragment of host DNA is duplicated at the target site where Mu DNA is integrated. This host DNA duplication arises because staggered cuts are made at the point in the host genome where Mu DNA is inserted. The resulting single-stranded segments are converted to the double-stranded form as part of the Mu integration process (Figure 10.10*b*).

Phage Mu can enter the lytic pathway upon initial infection if its repressor *is not* made; alternatively, Mu can form a lysogen if its repressor *is* made. In either case, Mu DNA is replicated by repeated transposition of Mu to multiple sites on the host genome. If the lytic cycle pathway is triggered, only the early genes of Mu are initially transcribed. Then, following expression of a Mu transcriptional activating protein, Mu head and tail proteins are synthesized. Following self-assembly, the cell is lysed and mature Mu virions are released. The lysogenic state in Mu requires that sufficient Mu repressor protein be present to prevent transcription of integrated Mu DNA.

MINIQUIZ
- In what major way does transcription of phage DNA differ in phages T4 and T7?
- What is unusual about the replication mechanism of the Mu genome?

10.5 Viruses of *Archaea*

Many bacteriophages and archaeal viruses have been isolated and characterized thus far. For *Bacteria*, these include both DNA and RNA phages, some with single-stranded and others with double-stranded genomes. However, all characterized archaeal viruses have DNA genomes, and with rare exception, double-stranded circular DNA genomes (Figure 10.3*a*).

DNA Archaeal Viruses

Several DNA viruses have been discovered whose hosts are species of *Archaea*, including representatives of both the *Euryarchaeota* and *Crenarchaeota* phyla (Chapter 17). Most viruses that infect species of *Euryarchaeota*, including both methanogenic and halophilic *Archaea*, are of the "head and tail" type, resembling the structurally complex phages that infect enteric bacteria, such as phage T4. One novel archaeal virus infects a halophile and is unusual because it is both enveloped and contains a single-stranded DNA genome. By contrast, all other characterized archaeal DNA viruses contain double-stranded and typically circular DNA genomes.

The most distinctive archaeal viruses infect hyperthermophilic *Crenarchaeota*. For example, the sulfur chemolithotroph *Sulfolobus* is host to several structurally unusual viruses. One such virus, called *SSV*, forms spindle-shaped virions that often cluster in rosettes (**Figure 10.11*a***). Such viruses are widespread in acidic hot springs worldwide. Virions of SSV contain a circular DNA genome of about 15 kilobase pairs. A second morphological type of *Sulfolobus* virus forms a rigid, helical rod-shaped structure (Figure 10.11*b*). Viruses in this class, nicknamed *SIFV*, contain linear DNA genomes about twice the size of that of SSV. Many variations on the spindle- and rod-shaped patterns have been seen in archaeal viral isolation studies, and a few species of *Crenarchaeota* are even infected by filamentous viruses.

A spindle-shaped virus that infects the hyperthermophile *Acidianus* displays a novel behavior. The virion, called *ATV*, contains a circular genome of about 68 kilobase pairs and is lemon-shaped when first released from the host cells. However, shortly after release from its lysed host cell, the virion produces long, thin tails, one at each end (Figure 10.11*d*). The tails are actually tubes, and as they form, the virion becomes thinner and its volume is reduced. Remarkably, this is the first example of virus development in the complete absence of host cell contact. It is thought that the extended tails of ATV help the virus in some way survive in its hot (85°C), acidic (pH 1.5) environment. This unusually shaped virus is also lysogenic, a property rarely seen in other archaeal viruses.

A spindle-shaped virus also infects the hyperthermophile *Pyrococcus* (*Euryarchaeota*). This virus, named *PAV1*, resembles SSV but is larger and contains a very short tail (Figure 10.11*c*). PAV1 has a small circular DNA genome and is released from host cells without cell lysis, probably by a budding mechanism similar to that of the *Escherichia coli* bacteriophage M13 (Section 10.3).

UNIT 3

UNIT 3

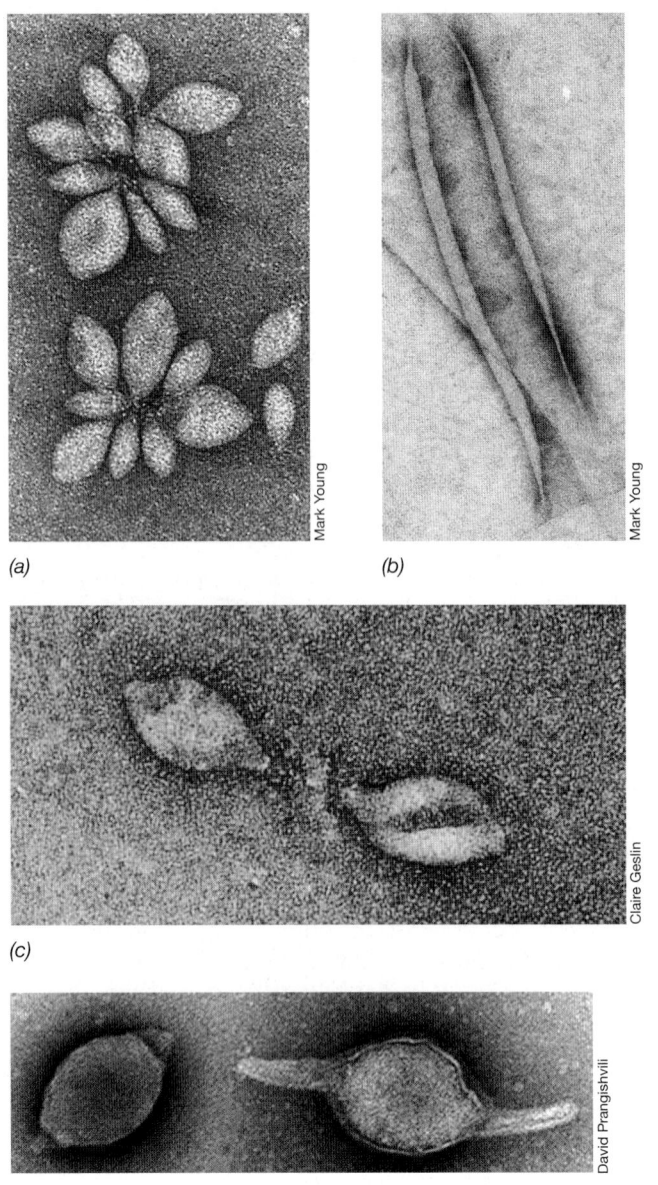

(a)

(b)

(c)

(d)

Figure 10.11 Archaeal viruses. Electron micrographs of viruses of *Crenarchaeota* (parts *a*, *b*, *d*), and a virus of a euryarchaeote *(c)*. *(a)* Spindle-shaped virus SSV1 that infects *Sulfolobus solfataricus* (virions are 40 × 80 nm). *(b)* Filamentous virus SIFV that infects *S. solfataricus* (virions are 50 × 900–1500 nm). *(c)* Spindle-shaped virus PAV1 that infects *Pyrococcus abyssi* (virions are 80 × 120 nm). *(d)* ATV, the virus that infects the hyperthermophile *Acidianus convivator*. When released from the cell the virions are lemon-shaped (left), but they proceed to grow appendages on both ends (right). ATV virions are about 100 nm in diameter.

Pyrococcus has a growth temperature optimum of 100°C and thus PAV1 virions must be extremely heat-stable. Despite their similar morphologies, genomic comparisons of PAV1 and SSV-type viruses show little sequence similarity, indicating that the two types of viruses do not have common evolutionary roots.

RNA Archaeal Viruses

Thus far, RNA viruses that can replicate in the laboratory on archaeal hosts are unknown, despite the fact that a variety of RNA viruses infect *Bacteria* and eukaryotes (Figure 10.3). Although no

Figure 10.12 An acidic Yellowstone hot spring and its archaeal viruses. Inset: Transmission electron micrograph of a mixture of archaeal viruses from the spring. Compare with Figure 10.11.

concrete examples of archaeal RNA viruses have emerged, environmental genomics have shown that they almost certainly exist. In some acidic hot springs of Yellowstone National Park that support large communities of *Crenarchaeota*, a large number of unusually shaped and structurally tough archaeal viruses have been discovered (**Figure 10.12**) and grown in the laboratory. Thus far, all of these have been DNA viruses. However, using the powerful tools of metagenomics (∂∂ Section 9.8), researchers studying these hot springs have discovered viral RNAs whose RNA sequences bear no resemblance to those of any known RNA viruses that infect *Bacteria*. Because these springs are too hot for eukaryotes and cell numbers of *Bacteria* are few, the unusual RNA is almost certainly from RNA archaeal viruses that are yet to be propagated in the laboratory.

Sequence analyses of the hot spring viral RNA show that it originated from single-stranded plus-sense RNA viruses (Baltimore class IV, Figure 10.2) (Section 10.8). These viral genomes also encode an RNA replicase—a hallmark of RNA viruses—and are likely to replicate by way of polyprotein formation, a replication mechanism employed by some class IV viruses of eukaryotes, such as poliovirus (Section 10.8). Replication steps of the putative RNA archaeal viruses, including important molecular details such as the extent to which viral (rather than host) polymerases participate in the replication process, are unclear and await laboratory cultivation of the viruses. However, now that scientists know that such viruses almost certainly exist, they can be on the lookout for them in viral enrichment and isolation studies.

— MINIQUIZ —

- What type of genome is seen in most archaeal viruses?
- Compared with other archaeal viruses, what are two unusual features of the virus that infects *Acidianus*?

10.6 Uniquely Replicating DNA Animal Viruses

Two groups of double-stranded DNA (Baltimore class I, Figure 10.2) animal viruses show unusual replication strategies: pox viruses and adenoviruses. Pox viruses are unique because all replication events, including DNA replication, occur in the host *cytoplasm* instead of the nucleus, and adenoviruses are unique because the replication of their genome proceeds in a leading fashion on *both* DNA template strands.

Pox Viruses

Pox viruses have been important historically as well as medically. Smallpox virus was the first virus to be studied in any detail and was the first virus for which a vaccine was developed (over 200 years ago the British physician Edward Jenner was the first to protect people from infection by smallpox virus by exposing them to the similar but much less virulent cowpox virus). Pox viruses are among the largest of all viruses, the brick-shaped vaccinia virions measuring almost 400 nm in diameter (**Figure 10.13**). Other pox viruses of importance are cowpox and vaccinia virus. Because it closely resembles the smallpox virus but is not pathogenic, vaccinia is used as a smallpox vaccine today and as a laboratory model for smallpox virus molecular biology.

The vaccinia virus genome consists of linear double-stranded DNA about 190 kilobase pairs in length and encoding about 250 genes. Following attachment, vaccinia virions are taken up into host cells and the nucleocapsids (Figure 10.13) are liberated in the cytoplasm; all replication events take place in the cytoplasm. Uncoating of the viral genome requires the activity of a viral protein that is synthesized after infection (the gene encoding this protein is transcribed by a viral RNA polymerase contained within the virion). In addition to this uncoating gene, a number of other viral genes are transcribed, including genes that encode a DNA polymerase that synthesizes copies of the viral genome. These are then incorporated into virions that accumulate in the cytoplasm, and the virions are released when the infected cell lyses.

Vaccinia virus has been genetically engineered to contain certain proteins from other viruses for use in recombinant vaccines (⇨ Section 12.8). A vaccine is a substance capable of eliciting an immune response in an animal that protects the animal from future infection with the same agent. Vaccinia virus causes no serious health effects in humans but elicits a strong immune response. Therefore, as a carrier of proteins from pathogenic viruses, vaccinia virus is a relatively safe and effective tool for stimulating an immune response against these pathogens. Success has been obtained with vaccinia virus vaccines against the viruses that cause influenza, rabies, herpes simplex type 1, and hepatitis B.

Adenoviruses

Adenoviruses are a group of small and naked icosahedral viruses (**Figure 10.14a**) that contain linear double-stranded DNA genomes. Adenoviruses are of minor health importance, causing mild respiratory infections in humans, but they have unique stature in virology because of the mechanism by which they replicate their genomes. Attached to the 5′ end of adenoviral genomic DNA is a protein called the adenoviral *terminal protein*, and it is essential for replication of the adenoviral genome. The complementary DNA strands also have inverted terminal repeats that play a role in the replication process (Figure 10.14b).

Following infection, the adenoviral nucleocapsid is released into the host cell nucleus, and transcription of the early genes proceeds by activity of the host RNA polymerase. Most early transcripts encode important replication proteins such as the terminal protein and a viral DNA polymerase. Replication of the adenoviral genome begins at either end of the DNA genome and the terminal protein facilitates this process because it contains a covalently bound cytosine that functions as a primer for DNA polymerase (Figure 10.14b). The products of this initial replication are a completed double-stranded viral genome and a single-stranded minus-sense DNA molecule. At this point, a unique replication event occurs. The single DNA strand cyclizes by means of its inverted terminal repeats, and a complementary (plus-sense) DNA strand is synthesized beginning from its 5′ end (Figure 10.14b). This mechanism is unique because double-stranded DNA is replicated *without the formation of a lagging strand*, as occurs in conventional semiconservative DNA replication (⇨ Section 4.3). Once sufficient copies of the adenoviral genome have formed and virion structural components accumulate in the host cell, mature adenoviral virions are assembled and released from the cell following lysis.

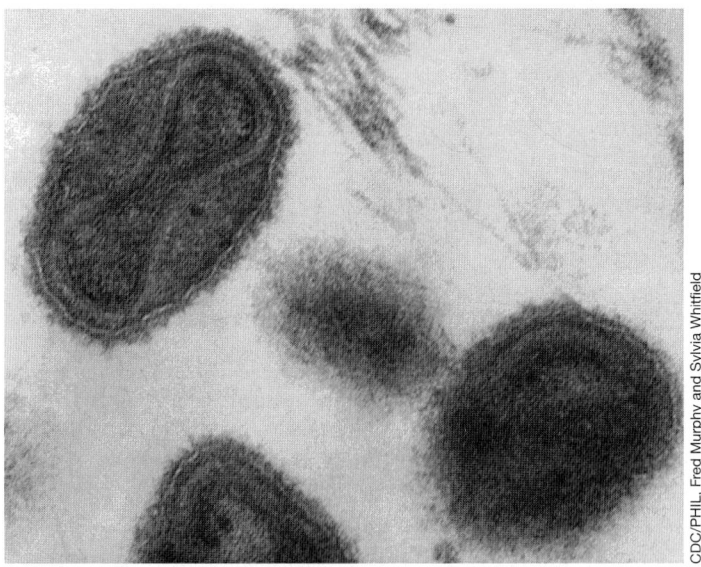

CDC/PHIL, Fred Murphy and Sylvia Whitfield

Figure 10.13 Smallpox virus. Transmission electron micrograph of a negatively stained thin section of smallpox virus virions. The virions are approximately 350 nm (0.35 μm) long. The dumbbell-shaped structure inside the virion is the nucleocapsid, which contains the double-stranded DNA genome. All replication functions for pox virus occur in the host cytoplasm.

— MINIQUIZ —
- What is unusual about genome replication in pox viruses?
- What is unusual about genome replication in adenoviruses?
- Why is the adenovirus terminal protein essential for replicating its genome?

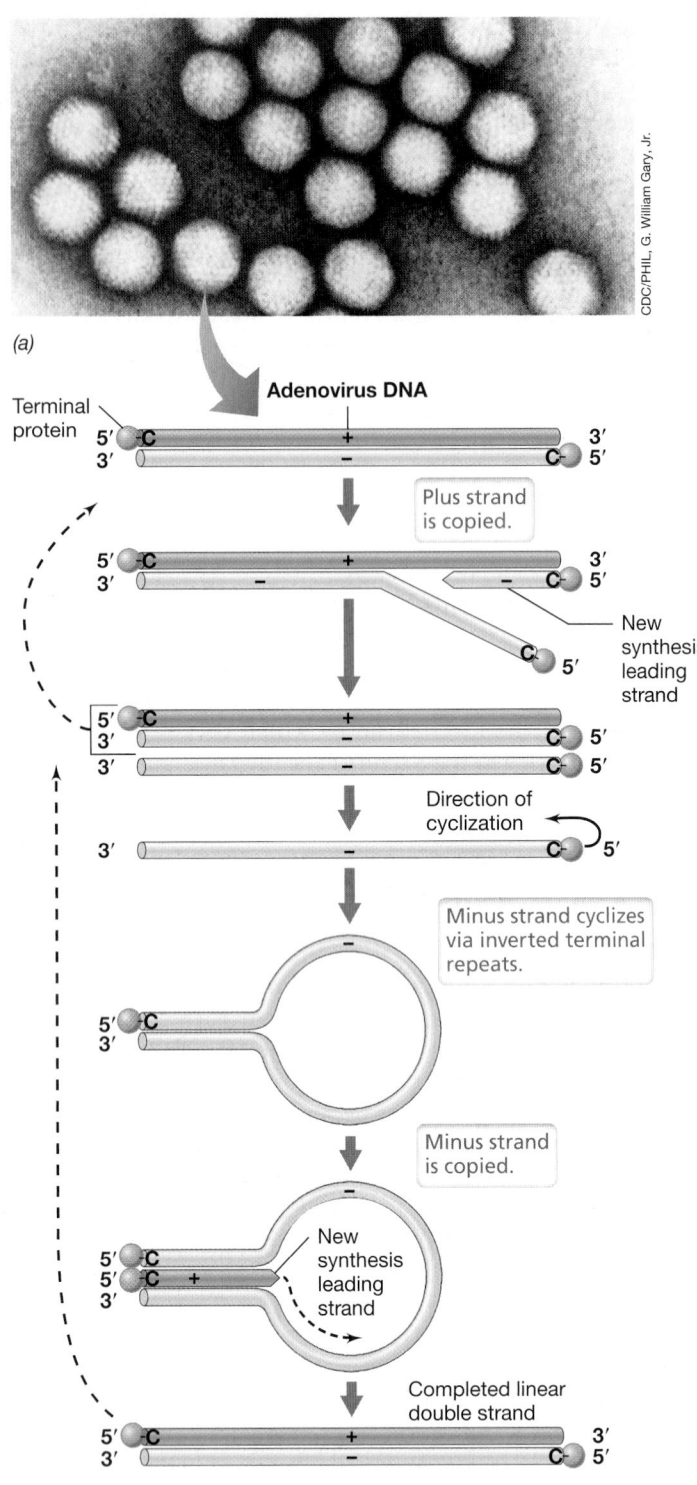

(a)

(b)

Figure 10.14 Adenoviruses and their genomic replication. *(a)* Transmission electron micrograph of adenoviral virions. Note the icosahedral structure. *(b)* Adenoviral genome replication. Because of loop formation (cyclization), there is no lagging strand; DNA synthesis is leading on both strands. A cytosine (**C**) is attached to the terminal protein. Adenoviruses are one of several classes of human viruses that cause upper respiratory infections such as the common cold (↩ Section 30.7). Rhinoviruses (single-stranded plus-sense RNA viruses, see Section 10.8) cause the vast majority of colds.

10.7 DNA Tumor Viruses

Besides catalyzing lytic events or becoming integrated into a genome in a latent state, some DNA animal viruses can induce cancers. These include viruses of the polyomavirus family and some herpesviruses, both of which contain double-stranded DNA genomes (Baltimore class II, Figure 10.2).

Polyomavirus SV40

Polyomavirus SV40 is a naked icosahedral virus that can cause tumors in small mammals, such as hamsters and rats. Its circular genome consists of double-stranded DNA (**Figure 10.15a**). The genome is too

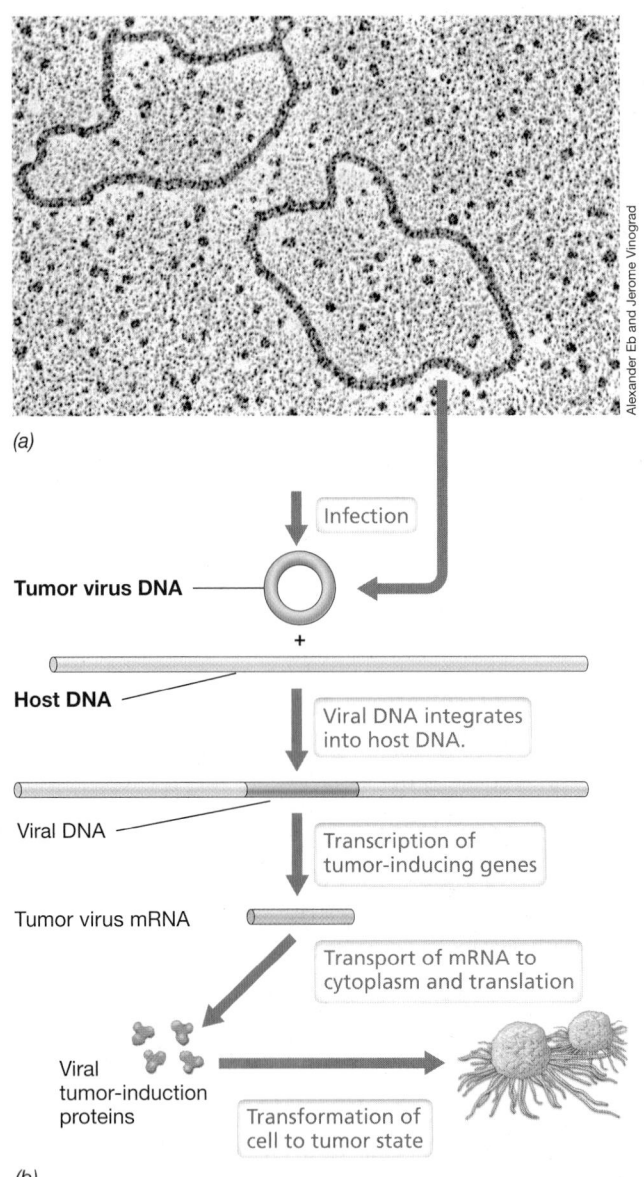

(a)

(b)

Figure 10.15 Polyomaviruses and tumor induction. *(a)* Transmission electron micrograph of relaxed (nonsupercoiled) circular DNA from a tumor virus. The contour length of each circle is about 1.5 μm. *(b)* Events in cell transformation by a polyomavirus such as SV40. Viral DNA becomes incorporated into the host genome. From there, viral genes encoding cell transformation events are transcribed and transported to the cytoplasm for translation.

small (5.2 kb) to encode its own DNA polymerase, so host DNA polymerases are used and SV40 DNA is replicated in a bidirectional fashion from a single origin of replication. Because of the small genomes of polyomaviruses, the strategy of overlapping genes, typical of many small bacteriophages (Sections 10.3 and 10.8), is also employed here. Transcription of the viral genome occurs in the nucleus and mRNAs are exported to the cytoplasm for protein synthesis. Eventually SV40 virion assembly occurs (in the nucleus) and the cell is lysed to release the new virions.

When SV40 infects a host cell, one of two outcomes can occur, depending on the host cell. In *permissive* hosts, virus infection results in the usual formation of new virions and the lysis of the host cell. In *nonpermissive* hosts, lytic events do not occur; instead, the viral DNA becomes integrated into host DNA, genetically altering the cells in the process (Figure 10.15b). Such cells can show loss of growth inhibition and become malignant, a process called *transformation* (⮑ Figure 8.20). As in certain tumor-causing retroviruses (Section 10.11), expression of specific SV40 genes is required to convert the cell to the transformed state. These tumor-inducing proteins bind to and inactivate host cell proteins that control cell division, and in this way, they promote uncontrolled cell development.

Herpesviruses

Herpesviruses are a large group of double-stranded DNA viruses that cause a variety of human diseases, including fever blisters (cold sores), venereal herpes, chicken pox, shingles, and infectious mononucleosis. An important group of herpesviruses cause cancer. For example, Epstein–Barr virus causes Burkitt's lymphoma, a tumor endemic in children of central Africa and New Guinea. A widespread herpesvirus is cytomegalovirus (CMV), present in nearly three-quarters of all adults in the United States over 40 years of age. For healthy individuals, infection with CMV comes with no apparent symptoms or long-term health consequences. However, CMV can cause pneumonia, retinitis (an eye condition), and certain gastrointestinal disorders, as well as serious disease or even death in immune-compromised individuals.

Herpesviruses can remain latent in the body for long periods of time and become active under conditions of stress or when the immune system is compromised. Herpesvirus virions are enveloped and can have many distinct structural layers over the icosahedral nucleocapsid (**Figure 10.16**). Following viral attachment, the host cytoplasmic membrane fuses with the virus envelope, and this releases the nucleocapsid into the cell. The nucleocapsid is transported to the nucleus, where the viral DNA is uncoated and three classes of mRNA are produced: *immediate early, delayed early,* and *late* (Figure 10.16). Immediate early mRNA encodes certain regulatory proteins that stimulate the synthesis of the delayed early proteins. Among the key proteins synthesized during the delayed early stage is a viral-specific DNA polymerase and a DNA-binding protein, both of which are needed for viral DNA replication. As for other viruses, late proteins are primarily viral structural proteins.

Herpesvirus DNA replication takes place in the nucleus. After infection, the herpesvirus genome circularizes and replicates by a

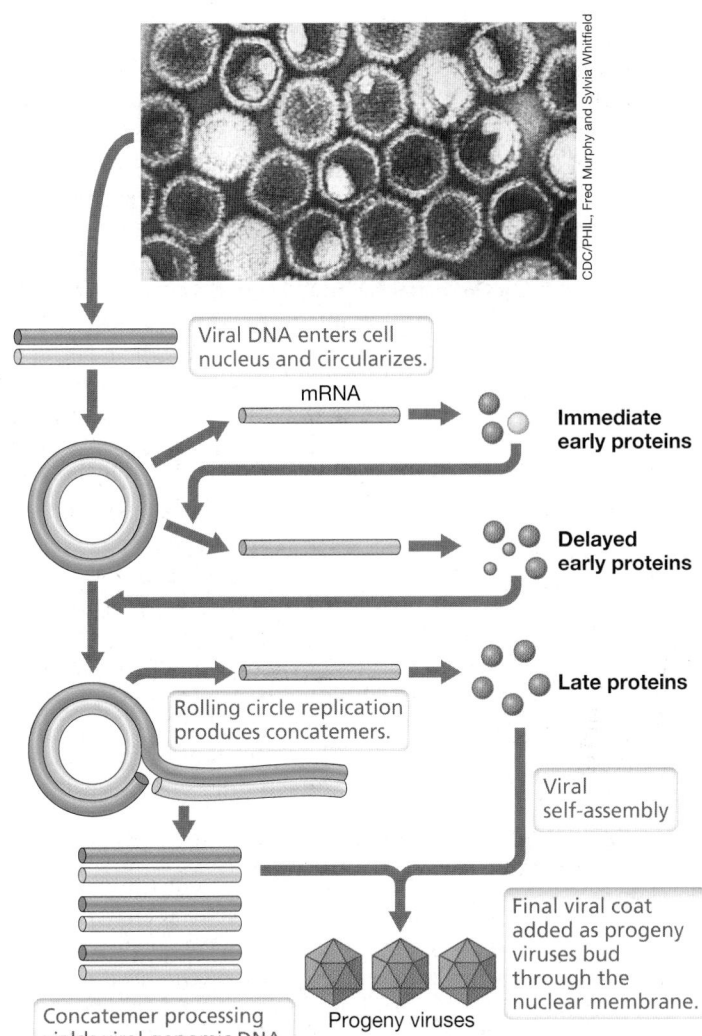

Figure 10.16 Herpesvirus. Flow of events in replication of herpes simplex virus starting from a transmission electron micrograph of herpes simplex virus (diameter of a single virion is about 150 nm). Although the viral genome is linear within the virion, it circularizes once inside the host.

rolling circle mechanism (Section 10.3). Long concatemers are formed that become processed into virus-length genomic DNA during the assembly process (Figure 10.16). Viral nucleocapsids are assembled in the nucleus, and the viral envelope is added during budding through the nuclear membrane. Mature herpesvirus virions are subsequently released through the endoplasmic reticulum to the outside of the cell. The assembly of herpesvirus virions thus differs from that of other enveloped viruses, which typically receive their envelope from the cytoplasmic membrane during exit from the cell.

MINIQUIZ

- What genomic feature does SV40 share with bacteriophage φX174?
- How can the outcome of an SV40 viral infection differ in permissive versus nonpermissive hosts?
- Name two common diseases caused by herpesviruses.

III • Viruses with RNA Genomes

We have seen that RNA viruses infect a multitude of hosts and were likely the first viruses to appear on Earth (Section 10.2). As in the foregoing sections that dealt with DNA viruses, we organize our coverage of RNA viruses here by genomic characteristics. RNA viruses make up Baltimore classes III–VI (Figure 10.2).

10.8 Positive-Strand RNA Viruses

Many viruses contain single-stranded RNA genomes of the plus sense and are therefore *positive-strand RNA viruses*. In these viruses, the sequence of the genome and the mRNA are the same (Figure 10.2). A number of positive-strand animal and bacterial viruses are known, so we restrict our coverage here to just a few well-studied cases. We begin with the tiny bacteriophage MS2.

Bacteriophage MS2

Bacteriophage MS2 is about 25 nm in diameter and has an icosahedral capsid. The virus infects cells of *Escherichia coli* by attaching to the cell's pilus (**Figure 10.17a**), a structure that normally functions in a form of horizontal gene exchange (conjugation) in bacteria. How MS2 RNA actually gets inside the *E. coli* cell from the pilus is unknown, but once it has, MS2 replication events begin quickly; the genetic map and major activities of this virus are shown in Figure 10.17b and c.

The MS2 genome is just 3.5 kb in size and encodes only four proteins, including the maturation protein, coat protein, lysis protein, and one subunit of **RNA replicase**, the enzyme that replicates the viral RNA. Interestingly, MS2 RNA replicase is a composite protein, with three subunits encoded by the host genome and one subunit by the viral genome. The gene encoding the MS2 lysis protein overlaps that encoding the coat protein and replicase subunit (Figure 10.17b). We have seen this phenomenon of *overlapping genes* before (Section 10.3) as a strategy for making small genomes encode more proteins.

Because the genome of phage MS2 is plus-sense RNA, it is translated directly upon entry into the cell by the host RNA polymerase. When RNA replicase is made, it begins synthesis of minus-sense RNA using plus strands as templates. As minus-sense RNA copies accumulate, more plus-sense RNA is made using the minus-sense strands as templates, and some of these are translated for continued synthesis of viral structural proteins.

Phage MS2 regulates synthesis of its proteins by controlling access of host ribosomes to translational start sites on its RNA. MS2 genomic RNA is folded into a complex secondary structure. Of the four AUG translational start sites (⟲ Section 4.9) on the MS2 RNA, the most accessible to the cell's translation machinery is that for the coat protein and replicase. Hence, translation begins at these sites very early following infection. However, as coat protein molecules accumulate, they bind to the RNA around the AUG start site for the replicase protein, effectively turning off synthesis of replicase. Although the gene for the maturation protein is at the 5′ end of the RNA, the extensive folding of the RNA limits access to the maturation protein translational start site, and consequently, only a few copies are synthesized. In this way, all MS2 proteins are made in the relative amounts needed for virus assembly. Ulti-

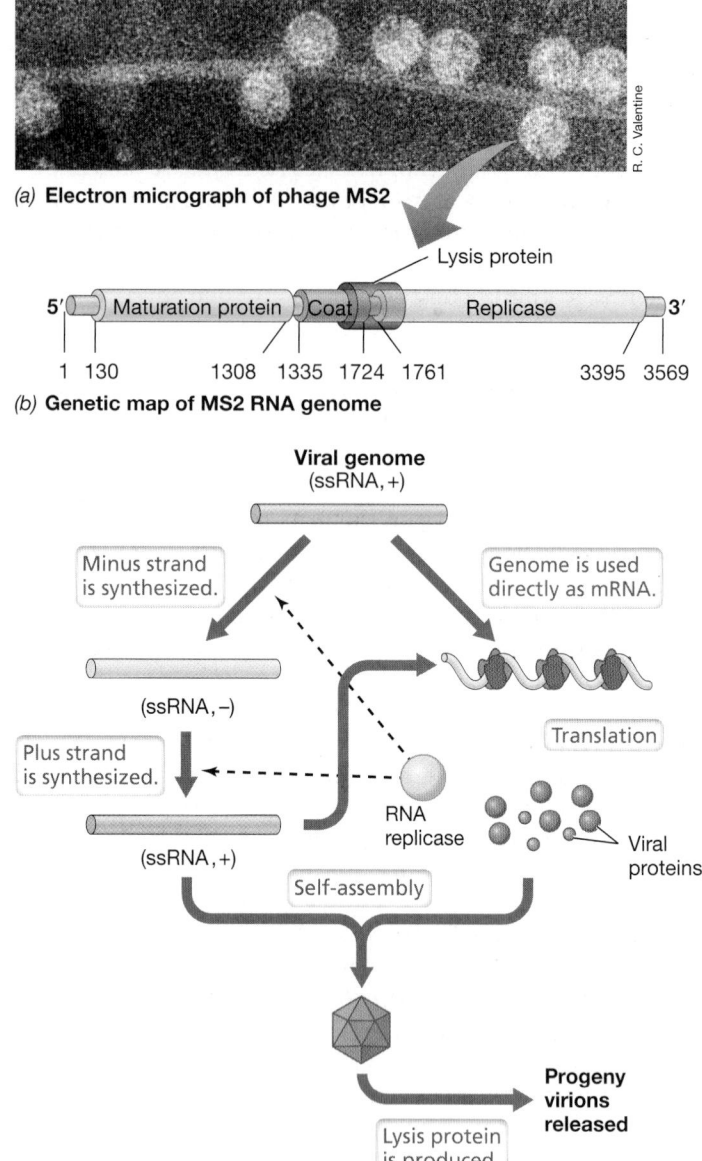

(a) **Electron micrograph of phage MS2**

R. C. Valentine

(b) **Genetic map of MS2 RNA genome**

(c) **Flow of events during viral multiplication**

Figure 10.17 A small RNA bacteriophage, MS2. *(a)* Transmission electron micrograph of the pilus of a cell of *Escherichia coli* showing virions of phage MS2 attached. *(b)* Genetic map of MS2. Note how the lysis protein gene overlaps with both the coat protein and replicase genes. The numbers refer to the nucleotide positions on the RNA, the entire genome being 3569 nucleotides in length. *(c)* Flow of events during MS2 replication.

mately, spontaneous assembly of MS2 virions begins, and the virions are released from the cell as a result of cell lysis.

Poliovirus

Several positive-strand RNA animal viruses cause disease in humans and other animals. These include poliovirus, the rhinoviruses that cause many cases of the common cold, the coronaviruses that cause respiratory syndromes, including severe acute respiratory syndrome (SARS), and the hepatitis A virus. We focus here on poliovirus and coronaviruses, both of which have linear RNA genomes.

Poliovirus is one of the smallest of all viruses with a 30-nm icosahedral structure containing the minimum 60 morphological units per virion (**Figure 10.18a, b**). At the 5′ terminus of the viral RNA is a protein, called the *VPg protein*, that is attached covalently to the genomic RNA, and at the 3′ terminus is a poly(A) tail (Figure 10.18*c*), a common feature of eukaryotic cell transcripts (🔗 Section 4.6). The poliovirus genome (about 7.4 kb) is also the mRNA, and the VPg protein facilitates binding of the RNA to host ribosomes. Translation yields a **polyprotein**, a single protein that self-cleaves into several smaller proteins including virion structural proteins. Other proteins generated from the polyprotein include the VPg protein, an RNA replicase responsible for synthesis of both minus-strand and plus-strand RNA, and a virus-encoded protease, which carries out the polyprotein cleavage (Figure 10.18*c*). This mechanism is called *post-translational cleavage* and is common in many animal viruses as well as animal cells.

Poliovirus replication occurs in the host cell cytoplasm. To initiate infection, the poliovirus virion attaches to a specific receptor on the surface of a sensitive cell and enters the cell. Once inside the cell, the virion is uncoated, and the genomic RNA is attached to ribosomes and translated to yield the polyprotein. Replication of viral RNA by the poliovirus RNA replicase begins shortly after infection. Both the positive and negative strands that are made pick up the VPg protein, which also functions as a primer for RNA synthesis (Figure 10.18*c*). Once poliovirus replication begins, host events are inhibited, and about 5 h after infection, cell lysis occurs with the release of new poliovirus virions.

Coronaviruses

Coronaviruses are single-stranded plus RNA viruses that, like poliovirus, replicate in the cytoplasm, but they differ from poliovirus in their larger size and details of replication. Coronaviruses cause respiratory infections in humans and other animals, including about 15% of common colds and SARS, an occasionally fatal infection of the lower respiratory tract in humans (🔗 Section 30.7).

Coronavirus virions are enveloped and contain club-shaped glycoprotein spikes on their surfaces (**Figure 10.19a**). These give the virus the appearance of having a "crown" (*corona* is Latin for crown). Coronavirus genomes are noteworthy because they are the largest of any known RNA viruses, about 30 kb. Because it is of the plus sense, the coronavirus genome can function directly in the cell as mRNA; however, most coronavirus proteins are not made by translating genomic RNA. Instead, only a portion of the genome is translated, in particular the region encoding RNA replicase (Figure 10.19*b*). The latter then uses the genomic RNA as a template to produce complementary negative strands from which several mRNAs are produced, and these mRNAs are translated to produce coronaviral proteins (Figure 10.19*b*). Full-length genomic RNA is also made off of the negative strands. New coronaviral virions are assembled within the Golgi complex, a major secretory organelle in eukaryotic cells (🔗 Section 2.16), and the fully assembled virions are released later from the cell surface.

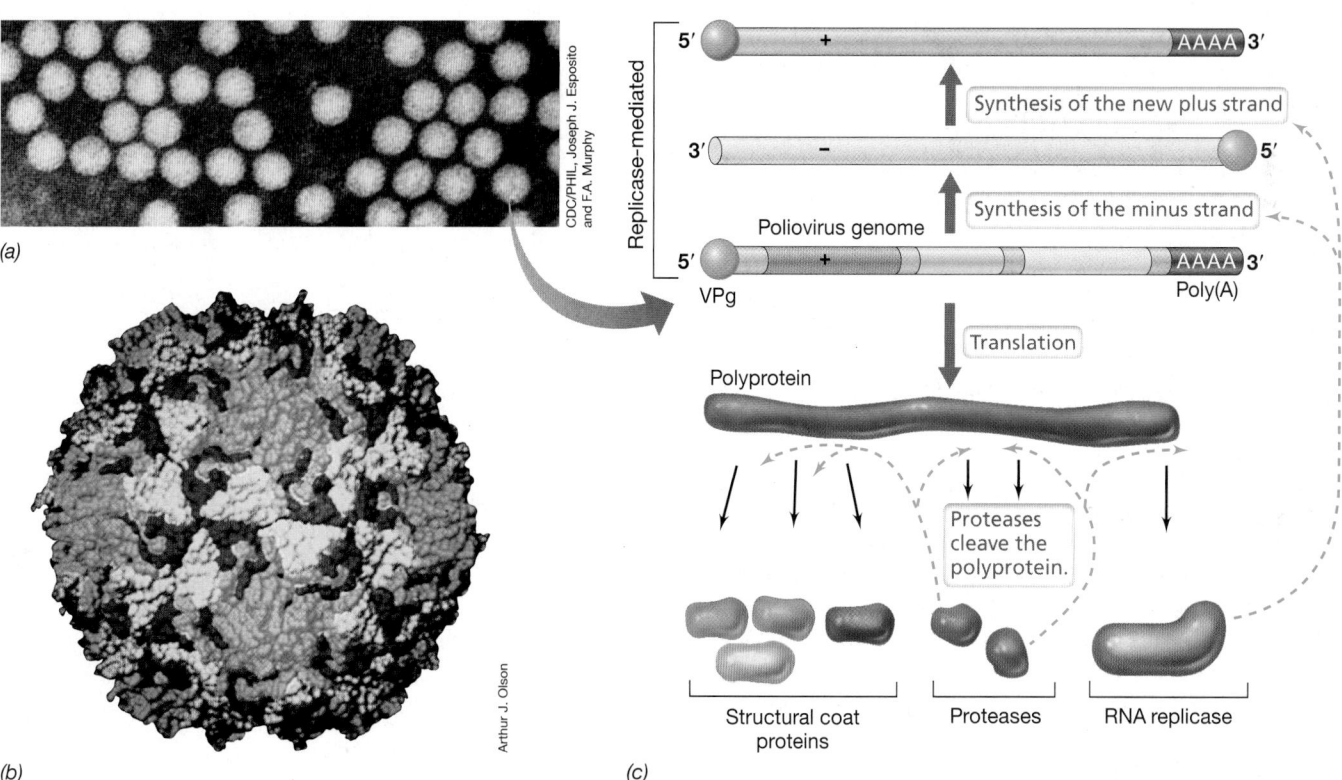

(a)

(b)

(c)

Figure 10.18 Poliovirus. *(a)* Transmission electron micrograph of poliovirus virions; a single virion is about 30 nm in diameter. *(b)* A computer model of a poliovirus virion. The various structural proteins are shown in distinct colors. *(c)* Genomic replication and formation of poliovirus proteins. Note the importance of the RNA replicase.

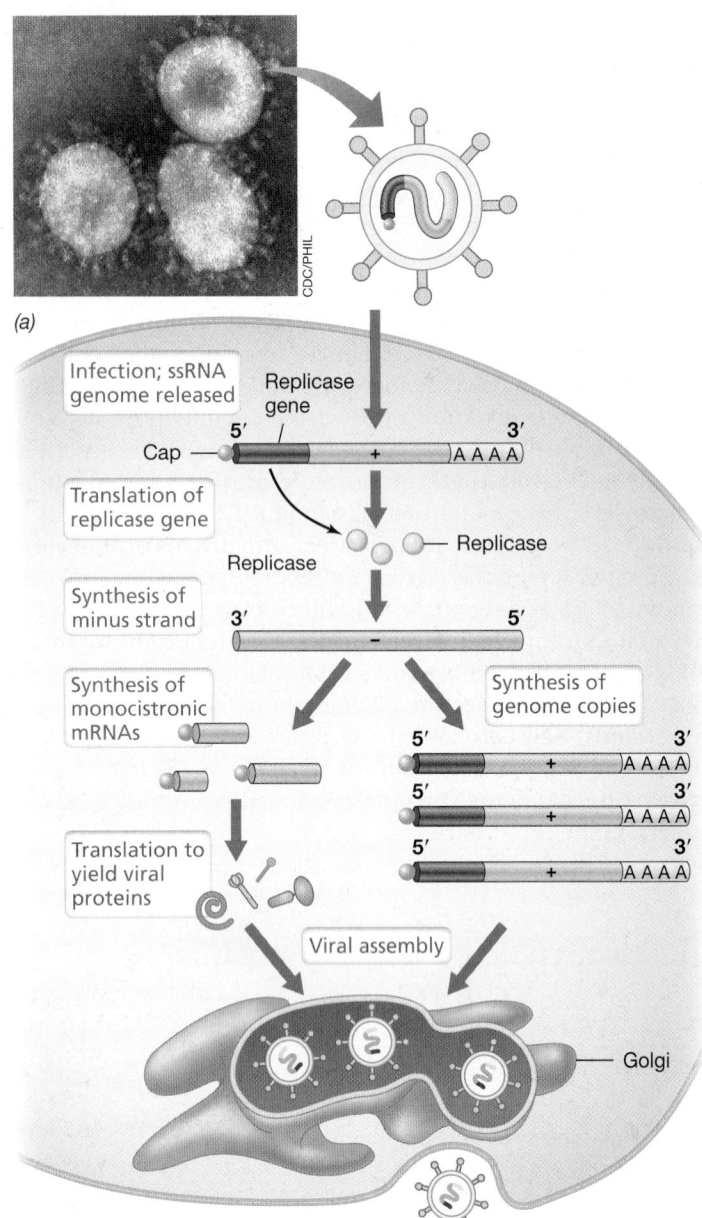

(a)

(b)

Figure 10.19 Coronaviruses. *(a)* Electron micrograph of a coronavirus; a virion is about 150 nm in diameter. *(b)* Steps in coronavirus replication. The mRNA encoding viral proteins is transcribed from the negative strand made by the RNA replicase using the viral genome as a template.

Coronavirus differs from poliovirus in terms of virion and genome size, lack of the VPg protein, and absence of polyprotein formation and cleavage. Nevertheless, their single-stranded plus-sense RNA genomes dictate that many other molecular events must occur in a similar way.

—————————— **MINIQUIZ** ——————————

• How can poliovirus RNA be synthesized in the cytoplasm whereas host RNA must be made in the nucleus?

• What is present in the poliovirus polyprotein?

• How are protein synthesis and genomic replication similar or different in poliovirus and the SARS virus?

10.9 Negative-Strand RNA Animal Viruses

A number of animal viruses have minus-sense RNA genomes (Baltimore class V, Figures 10.2 and 10.3). In contrast to the plus-strand viruses just considered, the genomes of these negative-strand RNA viruses are *complementary* in base sequence to the mRNA that is formed. We discuss here two important examples of negative-strand RNA viruses: rabies virus and influenza virus. There are no known negative-strand RNA bacteriophages or archaeal viruses.

Rabies Virus

Rabies virus, which causes the fatal neuroinflammatory disease rabies (⇔ Section 31.1), is a rhabdovirus, a name that refers to the characteristic shape of the virion (*rhabdos* is Greek for rod). Rhabdoviruses are commonly bullet-shaped (**Figure 10.20a**) and have an extensive and complex lipid envelope surrounding the helically

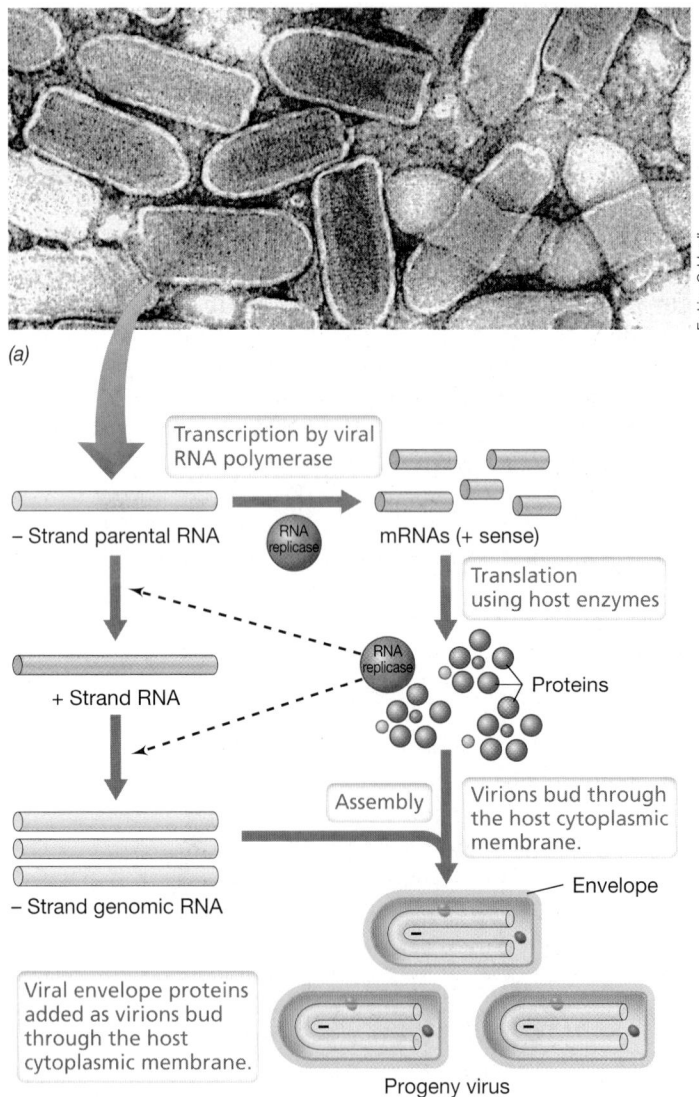

(a)

(b)

Figure 10.20 Negative-strand RNA viruses: Rhabdoviruses. *(a)* Transmission electron micrograph of vesicular stomatitis virus virions. A virion is about 65 nm in diameter. *(b)* Flow of events during replication of a negative-strand RNA virus. Note the importance of the viral-encoded RNA replicase.

symmetrical nucleocapsid. A rhabdovirus virion contains several enzymes that are essential for the infection process, including an RNA replicase. Unlike positive-strand viruses, a rhabdovirus genome cannot be directly translated but must first be transcribed by the replicase. This occurs in the cytoplasm and generates two classes of RNAs. The first is a series of mRNAs encoding each of the viral proteins, and the second is a complementary copy of the entire viral genome; the latter functions as a template for the synthesis of genomic RNA copies (Figure 10.20*b*).

Assembly of a rhabdovirus virion is a highly orchestrated process. Two different coat proteins are made, nucleocapsid and envelope. The nucleocapsid is formed first by assembly of nucleocapsid protein molecules around the viral RNA genome. The envelope proteins are glycoproteins and they migrate to the cytoplasmic membrane where they are inserted into the membrane. Nucleocapsids then migrate to areas on the cytoplasmic membrane where these virus-specific glycoproteins are embedded and bud through them, becoming coated by the glycoprotein-enriched cytoplasmic membrane in the process. The final result is the release of new virions that can infect neighboring cells.

Influenza Virus

Another group of negative-strand RNA viruses contains the important human pathogen *influenza virus*. Influenza virus has been well studied over many years, beginning with early work during the 1918 influenza pandemic that killed millions of people worldwide (⮂ Sections 29.8 and 30.8). Influenza virus is an enveloped virus in which the viral genome is present in the virion in a number of separate pieces, a condition called a *segmented genome*. In the case of influenza A virus, a common strain, the genome is segmented into eight linear single-stranded molecules ranging in size from 890 to 2341 nucleotides and totaling 13.5 kb. The nucleocapsid of the virus is of helical symmetry, about 6–9 nm in diameter and about 60 nm long, and is embedded in an envelope that has a number of virus-specific proteins as well as lipid derived from the host cytoplasmic membrane. Because of the way influenza virus buds as it leaves the cell, virions do not have a uniform shape and instead are pleomorphic (**Figure 10.21*a***).

Several proteins on the outside of the influenza virion envelope interact with the host cell surface. One of these is *hemagglutinin*. Hemagglutinin is highly immunogenic (capable of stimulating the immune system) and antibodies against it prevent the virus from infecting a cell. This is the mechanism by which immunity to influenza is brought about by immunization (⮂ Section 30.8). A second important influenza virus surface protein is the enzyme *neuraminidase* (Figure 10.21*b*). Neuraminidase breaks down sialic acid (a derivative of neuraminic acid) in the host cytoplasmic membrane. Neuraminidase functions primarily in virus assembly, destroying host membrane sialic acid that would otherwise block assembly or become incorporated into the virion. In addition to hemagglutinin and neuraminidase, influenza virions possess two other key enzymes. These include an RNA replicase, which converts the minus-strand genome into a plus strand, and an RNA endonuclease, which cuts the cap from host mRNAs (⮂ Section 4.6) and

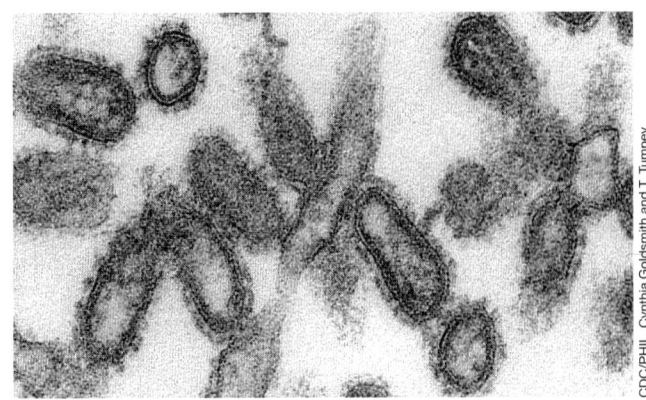

(a)

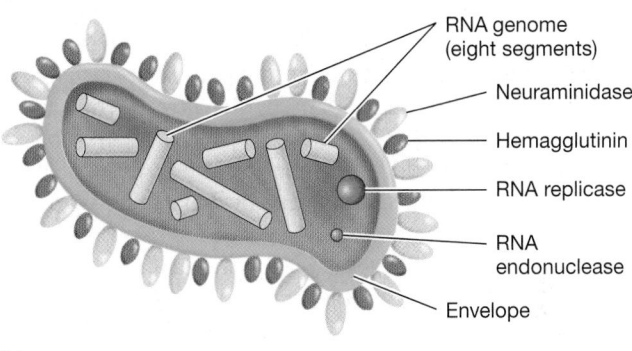

(b)

Figure 10.21 Influenza virus. *(a)* Transmission electron micrograph of thin sections of human influenza virus virions. *(b)* Some of the major components of the influenza virus, including the segmented genome.

uses them to cap viral mRNAs so they can be translated by the host translational machinery.

After the influenza virion enters the cell, the nucleocapsid separates from the envelope and migrates to the nucleus. Uncoating activates the virus RNA replicase and transcription begins. Ten proteins are encoded by the eight segments of the influenza virus genome; the mRNAs transcribed from six segments each encode a single protein, while the other two segments encode two proteins each. Some of the viral proteins are needed for influenza virus RNA replication, whereas others are structural proteins of the virion. The overall pattern of genomic RNA synthesis resembles that of the rhabdoviruses (Figure 10.20*b*), with full-length positive-strand RNA used as a template for making negative-strand genomic RNA. The complete enveloped virion forms by budding, as for the rhabdoviruses.

The segmented genome of the influenza virus has important practical consequences. Influenza virus exhibits a phenomenon called *antigenic shift* in which segments of the RNA genome from two different strains of the virus infecting the same cell are reassorted. This generates hybrid influenza virions that express unique surface proteins unrecognized by the immune system. Antigenic shift is thought to trigger major outbreaks of influenza because immunity to the new forms of the virus is essentially absent from the population. We discuss antigenic shift, and a related phenomenon called *antigenic drift*, in Section 30.8.

10.10 Double-Stranded RNA Viruses

Viruses with double-stranded RNA genomes (Baltimore class III, Figure 10.2) infect animals, plants, fungi, and a few bacteria. *Reoviruses* are an important family of animal viruses, and we focus on them here.

Rotavirus is a typical reovirus and is the most common cause of diarrhea in infants 6 to 24 months of age. Other reoviruses cause respiratory infections and some infect plants. Reovirus virions consist of a nucleocapsid 60–80 nm in diameter, surrounded by a double shell of icosahedral symmetry (**Figure 10.22a, b**). As we have seen with single-stranded RNA viruses, the virions of double-stranded RNA viruses must carry their own enzyme to synthesize their mRNA and replicate their RNA genomes. Like the influenza virus genome, the reoviral genome is segmented, in this case into 10–12 molecules of linear double-stranded RNA totaling 18 kb.

To initiate infection, a reovirus virion binds to a cellular receptor protein. The attached virus then enters the cell and is transported into lysosomes, where normally it would be destroyed (⇄ Section 2.16). However, only the outer coats of the virion are removed by the lysosome, revealing the nucleocapsid; the latter is released into the cytoplasm. This uncoating process activates the viral RNA replicase and initiates virus replication (Figure 10.22c).

Reovirus Replication

Reovirus replication occurs exclusively in the host cytoplasm but *within* the nucleocapsid itself (Figure 10.22c) because the host has enzymes that recognize double-stranded RNA as foreign and would destroy it. The plus strand of the reoviral genome is inactive as mRNA, and thus the first step in replication is the synthesis of plus-sense mRNA by the viral-encoded RNA replicase, using minus-strand RNA as a template; the nucleotide triphosphates necessary for RNA synthesis are supplied by the host (Figure 10.22c). The mRNAs are then capped with a methylated nucleotide (as is typical of eukaryotic mRNAs, ⇄ Section 4.6) by viral enzymes and exported from the nucleocapsid into the cytoplasm and translated by host ribosomes.

Most RNAs in the reovirus genome encode a single protein, although in a few cases the protein formed is cleaved to yield the final products. However, one of the reovirus mRNAs encodes two proteins but the RNA does not have to be processed in order to translate both of these. Instead, a ribosome occasionally "misses" the start codon for the first gene in this mRNA and travels on to the start codon of the second gene to begin translation. When this occurs, the second protein, needed in small amounts, is made but the first protein is not. This "molecular mistake" can be viewed as a primitive form of translational control that ensures that viral proteins are made in their proper amounts.

As viral proteins are formed in the host cytoplasm, they aggregate to form new nucleocapsids, trapping copies of RNA replicase inside

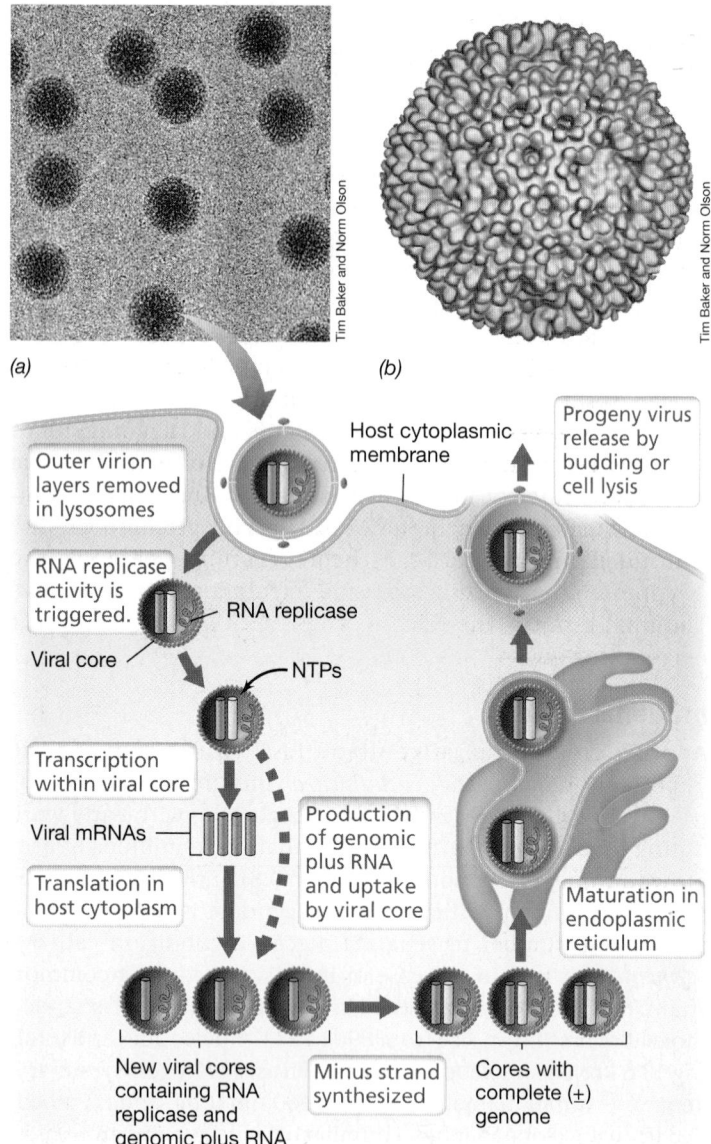

(a)

(b)

(c)

Figure 10.22 Double-stranded RNA viruses: The reoviruses. *(a)* Transmission electron micrograph showing reovirus virions (diameter, about 70 nm). *(b)* Three-dimensional computer reconstruction of a reovirus virion calculated from electron micrographs of frozen-hydrated virions. *(c)* The reovirus life cycle. All replication and transcription steps occur inside the nucleocapsids. NTPs, nucleotide triphosphates.

as they form (Figure 10.22c). Newly formed nucleocapsids then take up the correct complement of genomic (plus-strand) RNA fragments—probably by recognition of specific sequences on each fragment—and as each single-stranded RNA enters a newly formed nucleocapsid, a double-stranded form is produced from it by RNA replicase. Once genomic synthesis is complete, viral coat proteins are added in the host's endoplasmic reticulum, and the mature reoviral virions are released by budding or cell lysis (Figure 10.22c).

Reoviruses and RNA Replication

In addition to the unusual genome structure of reoviruses that forces them to employ special mechanisms to protect their double-stranded RNA genomes from cleavage by host ribonucleases

(Figure 10.22c), genomic replication in these viruses is also unique and differs in a fundamental way from that of cells and all other viruses.

Because the reovirus RNA genome is double-stranded, one might predict that reovirus replication would parallel that of organisms with double-stranded DNA genomes, but this is not the case. RNA replication in reoviruses is actually a *conservative* process rather than the well-known *semiconservative* process typical of cellular DNA replication and replication in viruses that contain double-stranded DNA genomes (Section 4.3 and Figure 10.2). This is because synthesis of reovirus mRNA occurs *only* off of the minus strand as a template in the infecting nucleocapsids, whereas synthesis of double-stranded genomic RNA from assimilated plus-strand RNA copies in progeny virions occurs *only* off of the plus strand as a template (Figure 10.22c). Hence, in addition to their unique double-stranded RNA genomes, reoviruses also display their unusual molecular biology by employing a unique genomic replication mechanism that is neither semiconservative nor rolling circle (Figure 10.7) in nature.

MINIQUIZ

- What does the reovirus genome consist of?
- How does reovirus genome replication resemble that of influenza virus, and how does it differ?
- Why must reoviral replication events occur within the nucleocapsid?

10.11 Viruses That Use Reverse Transcriptase

Two different classes of viruses use *reverse transcriptase*, and they differ in the type of nucleic acid in their genomes; retroviruses have *RNA* genomes while hepadnaviruses have *DNA* genomes (Baltimore classes VI and VII, respectively, Figure 10.2). Besides their unique molecular properties, both classes of viruses include important human pathogens, including HIV (a retrovirus) and hepatitis B (a hepadnavirus).

Retroviruses: Integration of Viral Genes into the Host Genome

Retroviruses have enveloped virions that contain two identical copies of the RNA genome (Figure 8.21a). The virion also contains several enzymes, including reverse transcriptase, and also a specific viral tRNA. Enzymes for retrovirus replication must be carried in the virion because although the retroviral genome is of the plus sense, it is not used directly as mRNA. Instead, the genome is converted to DNA by reverse transcriptase and integrated into the host genome. The DNA formed is a linear double-stranded molecule and is synthesized within the virion and then released to the cytoplasm. The major steps in reverse transcription are presented in **Figure 10.23**.

Reverse transcriptase has three enzymatic activities: (1) *reverse transcription* (to synthesize DNA from an RNA template), (2) *ribonuclease activity* (to degrade the RNA strand of an RNA:DNA hybrid), and (3) *DNA polymerase* (to make double-stranded

DNA from single-stranded DNA). Reverse transcriptase needs a primer for DNA synthesis and this is the function of the viral tRNA. Using this primer, nucleotides near the 5′ terminus of the RNA are reverse-transcribed into DNA. Once reverse transcription reaches the 5′ end of the RNA, the process stops. To copy the remaining RNA, a different mechanism comes into play. First, terminally redundant RNA sequences at the 5′ end of the molecule are removed by reverse transcriptase. This leads to the formation of a small, single-stranded DNA that is complementary to the RNA segment at the *other end* of the viral RNA. This short, single-stranded piece of DNA then hybridizes with the other end of the viral RNA molecule, where synthesis of DNA begins once again.

Continued reverse transcription leads to the formation of a double-stranded DNA molecule with long terminal repeats, and these assist in integration of the retroviral DNA into the host chromosome (Figure 10.23). For HIV, the chromosomal integration site is not random. Through the use of a special form of fluorescence microscopy, scientists have shown that HIV incorporates into chromosomal loci near the outer shell of the nucleus (photo inset, Figure 10.23). This location is likely favored due to the short life of the viral integrase. Recall from Section 8.8 that reverse transcription of a retrovirus occurs in the nucleocapsid. Thus the retroviral DNA must be integrated quickly into the host genome upon entry into the nucleus.

Retroviruses: Induction to Form New Retrovirus Virions

Once integrated, retroviral DNA becomes a permanent part of the host chromosome; the genes may be expressed or they may remain in a latent state indefinitely. However, if induced, retroviral DNA is transcribed by a cellular RNA polymerase to form RNA transcripts that can be either packaged into virions as genomic RNA or translated to yield retroviral proteins. Translation and processing of retroviral mRNAs is shown in **Figure 10.24**. All retroviruses have the genes *gag*, *pol*, and *env*, arranged in that order in their genomes (Figure 8.21). The *gag* gene at the 5′ end of the mRNA actually encodes several viral structural proteins. These are first synthesized as a single protein (polyprotein) that is subsequently processed by a protease which itself is part of the polyprotein. The structural proteins make up the capsid, and the protease is packaged in the virion.

Next, the *pol* gene is translated into a large polyprotein that also contains the *gag* proteins (Figure 10.24a). Compared to *gag* proteins, *pol* proteins are required in only small amounts. This regulation is achieved because *pol* protein synthesis requires the ribosome to either read through a stop codon at the end of the *gag* gene or switch to a different reading frame in this region. Both of these are rare events and can be considered a form of translational regulation. Once produced, the *pol* gene product is processed to yield *gag* proteins, reverse transcriptase, and integrase; the latter is the protein required for viral DNA integration into the host chromosome. For the *env* gene to be translated, the full-length mRNA is first processed to remove the *gag* and *pol* regions, and then the *env* product is made and immediately processed into two distinct envelope proteins by the viral-encoded protease (Figure 10.24b). Retroviral assembly occurs on the inner side of the host cytoplasmic membrane and virions are released across the membrane by budding (Figure 8.22).

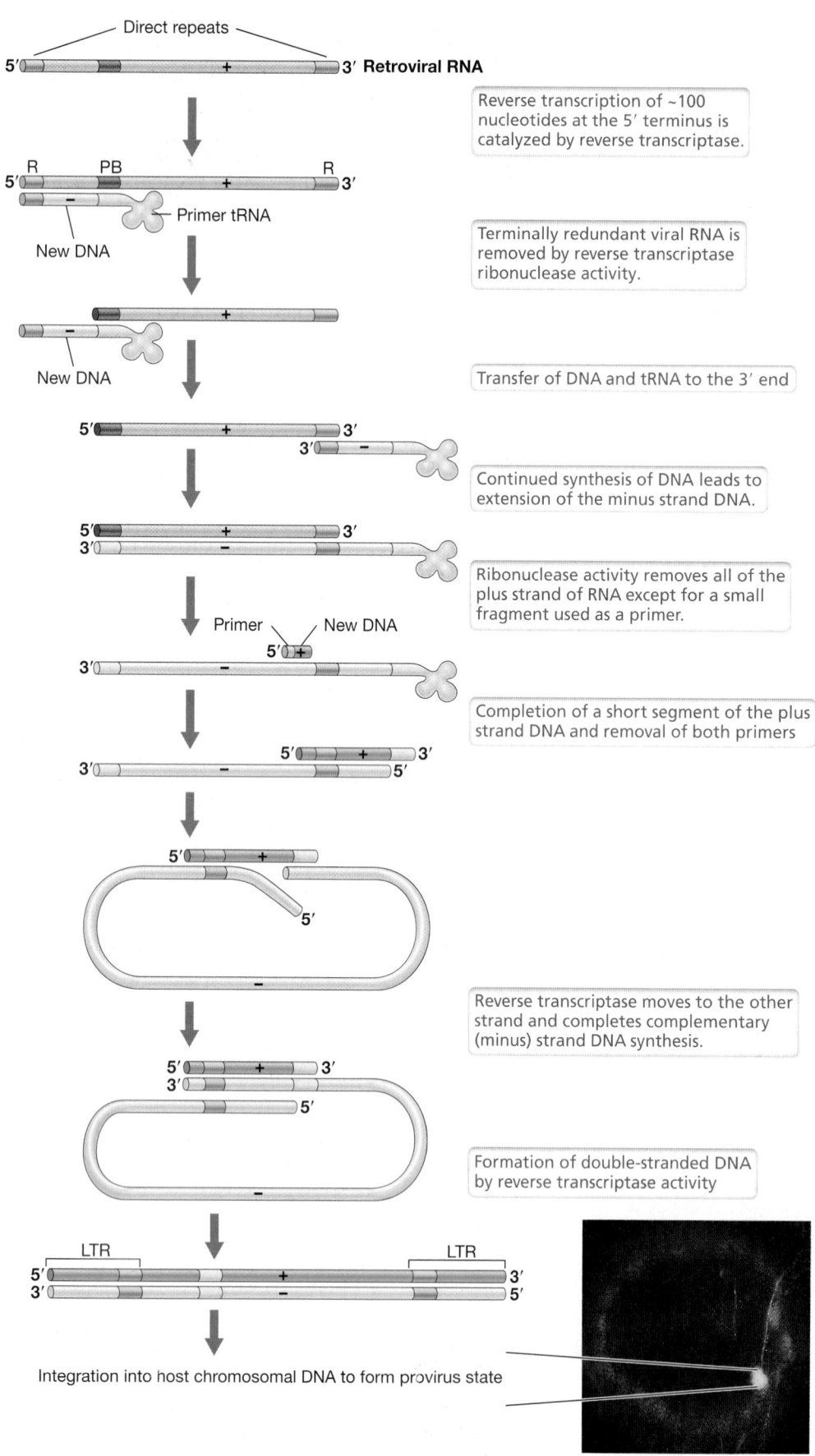

Direct repeats

5′ ───────────────────────── + ───────── 3′ **Retroviral RNA**

Reverse transcription of ~100 nucleotides at the 5′ terminus is catalyzed by reverse transcriptase.

R   PB   R
5′ ─────────────────────── + ─────────── 3′
New DNA   Primer tRNA
New DNA

Terminally redundant viral RNA is removed by reverse transcriptase ribonuclease activity.

New DNA

Transfer of DNA and tRNA to the 3′ end

5′ ─── + ─── 3′
                    3′ ─ − ─

Continued synthesis of DNA leads to extension of the minus strand DNA.

5′ ─── + ─── 3′
3′ ─── − ───

Ribonuclease activity removes all of the plus strand of RNA except for a small fragment used as a primer.

Primer   New DNA
5′ ─ + ─
3′ ─── − ───

Completion of a short segment of the plus strand DNA and removal of both primers

5′ ── + ── 3′
3′ ───── − ───── 5′

Reverse transcriptase moves to the other strand and completes complementary (minus) strand DNA synthesis.

5′ ── + ── 3′
3′ ─ − ─   5′

Formation of double-stranded DNA by reverse transcriptase activity

      LTR   LTR
5′ ─────── + ──────────── 3′
3′ ─────── − ──────────── 5′

Integration into host chromosomal DNA to form provirus state

With retroviruses we thus see a replication scheme that is perhaps the most complex of all known viruses. Despite the complexity of retroviruses, molecular studies of them suggest an ancient origin and possible central importance in the transition from self-replicating RNA life forms to the DNA world of cellular organisms. It is the signature enzyme of retroviruses—reverse transcriptase, the only enzyme known that can make DNA from RNA—that brings retroviruses into this evolutionary limelight, and so it is possible that all cells owe their very existence to this class of viruses (Section 10.2 and Figure 10.4).

Hepadnaviruses

In addition to the retroviruses a second class of unusual viruses employ the enzyme reverse transcriptase; these are the **hepadnaviruses**, such as human hepatitis B virus (**Figure 10.25a**). The tiny DNA genomes of hepadnaviruses are unusual because they are neither single-stranded nor double-stranded but *partially* double-stranded. Despite their small size (3–4 kilobase pairs), the hepadnavirus genomes encode several proteins by employing overlapping genes, a strategy we have seen before in very small viruses (Sections 10.3 and 10.8).

Besides the usual activities of reverse transcriptase we have just considered (Figure 10.23), hepadnaviral reverse transcriptase also functions as a protein primer for synthesis of one of its own DNA strands. In terms of its role in replication events, however, reverse transcriptase plays different roles in

Figure 10.23 Formation of double-stranded DNA from retrovirus single-stranded genomic RNA. The sequences labeled R on the RNA are direct repeats found at either end. The sequence labeled PB is where the primer (tRNA) binds. Note that DNA synthesis yields longer direct repeats on the DNA than were originally on the RNA. These are called long terminal repeats (LTRs). Inset: A special form of fluorescence microscopy allows visualization of HIV genome integration (green) into a chromosome region near the nuclear membrane (red) of a CD4 lymphocyte. Image courtesy of Marina Lusic, University Hospital, Heidelberg, Germany.

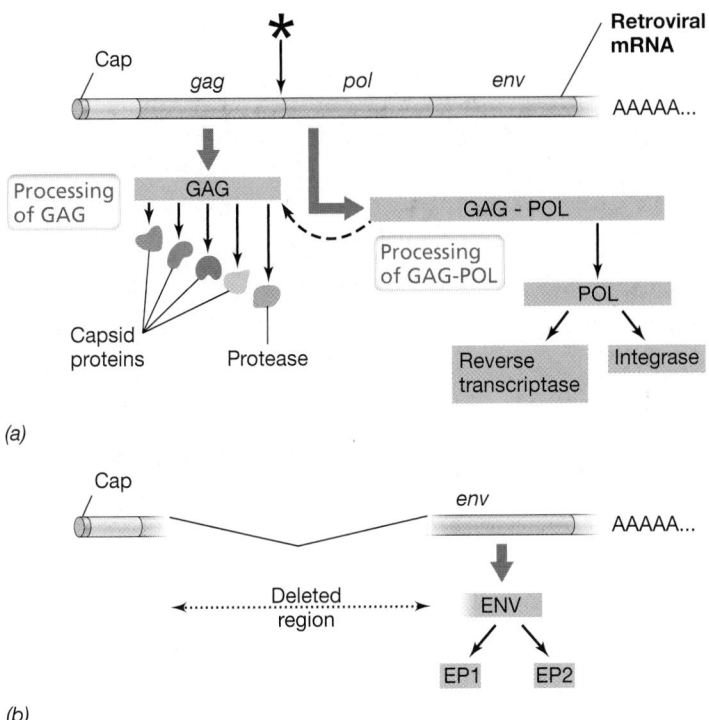

(a)

(b)

Figure 10.24 Translation of retrovirus mRNA and processing of the proteins. *(a)* Full-length retroviral mRNA encodes *gag*, *pol*, and *env*. The asterisk shows the site where a ribosome must read through a stop codon or do a precise shift of reading frame to synthesize the GAG-POL polyprotein. The thick gray arrows indicate translation, and the black arrows indicate protein-processing events. One of the *gag* gene products is a protease. The POL product is processed by this protease to yield reverse transcriptase and integrase, two key enzymes that catalyze retrovirus replication events (Figure 10.23). *(b)* The mRNA has been processed to remove most of the *gag-pol* region. This shortened message is translated to give the ENV polyprotein, which is cleaved into two envelope proteins (EP), EP1 and EP2.

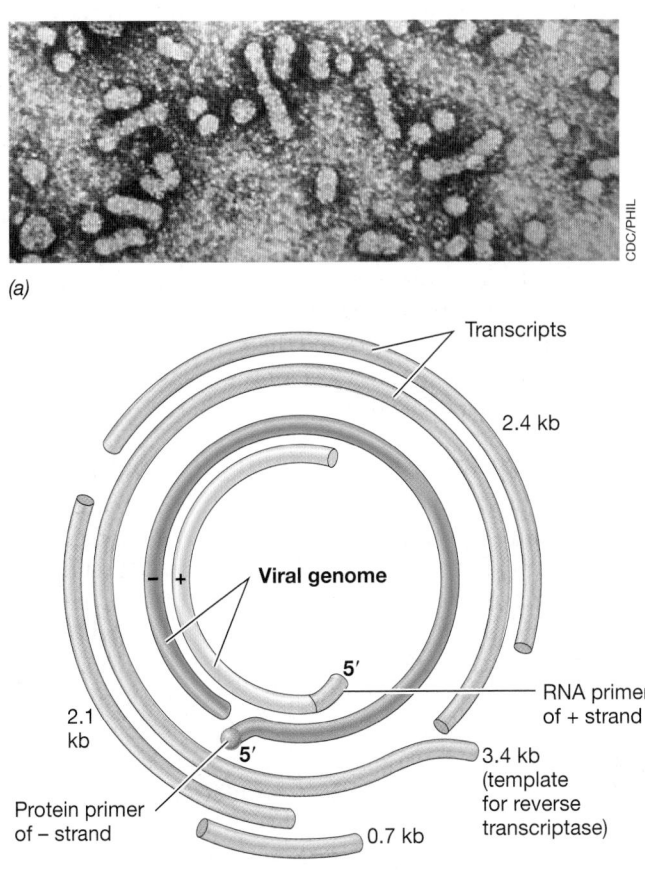

(a)

(b)

Figure 10.25 Hepadnaviruses. *(a)* Electron micrograph of hepatitis B virions. *(b)* Hepatitis B genome. The partially double-stranded genome is shown in the standard green colors. The sizes of the transcripts (orange) are also shown; all of the genes in the hepatitis B virus overlap. Reverse transcriptase produces the DNA genome from a single genome-length mRNA made by host RNA polymerase.

retroviral and hepadnaviral genome replication. In hepadnaviruses, the *DNA* genome is replicated through an *RNA* intermediate, whereas in retroviruses, the *RNA* genome is replicated through a *DNA* intermediate (Figures 10.23 and 10.25).

Upon infection, the hepadnavirus nucleocapsid enters the host nucleus where the partial genomic DNA strand is completed by the viral polymerase to form an entire double-stranded molecule. Transcription by host RNA polymerase yields four size classes of viral mRNAs (Figure 10.25*b*), which are subsequently translated to yield the hepadnaviral proteins. The largest of these transcripts is slightly larger than the viral genome and, together with reverse transcriptase, associates with viral proteins in the host cytoplasm to form genomes for new virions. Reverse transcriptase forms single-stranded DNA off of this large transcript inside the virion to form the minus-sense strand of the DNA genome and then uses this as a template to form a portion of the plus-sense strand, yielding the partially double-stranded genome characteristic of hepadnaviruses (Figure 10.25*b*). Once mature virions are produced, these associate with membranes in the endoplasmic reticulum and Golgi complex, from which they are exported across the cytoplasmic membrane by budding.

MINIQUIZ

- Why are protease inhibitors an effective treatment for human AIDS?
- Contrast the genomes of HIV and hepatitis B virus.
- How does the role of reverse transcriptase in the replication cycles of retroviruses and hepadnaviruses differ?

IV • Viral Ecology

Viruses can be found everywhere on and in Earth where cellular life is present (including on and in plants and animals) and are present in some environments in enormous numbers. The number of bacterial and archaeal cells on Earth is far greater than the total number of eukaryotic cells; estimates of total prokaryotic cell numbers are on the order of 10^{30}. However, the number of viruses is even greater than this, an estimated 10^{31} (10 nonillion)! Thus, one might expect that, despite their small size, viruses would play a major ecological role in nature. We consider some aspects of viral ecology here including a mechanism that protects *Bacteria* and *Archaea* from viral destruction (and countermeasures

UNIT 3

that have evolved in viruses) and explore the viral world that exists in and on the human body, the *human virome*.

10.12 The Bacterial and Archaeal Virosphere

The best estimates of the total number of cells of *Bacteria* and *Archaea* and their respective viruses have come from quantitative studies of seawater, although surveys of soil, freshwater, Earth's deep subsurface, and microbial mats, among many other habitats, are also teeming with microbes and the viruses that feed on them. We focus on seawater to give a feel for the numbers involved and how viruses interact with their prokaryotic hosts.

Bacteriophages and Archaeal Viruses in Seawater

There are about 10^6 prokaryotic cells/ml of seawater and approximately ten times as many viruses. It has been estimated that at least 5% and as many as 50% of the *Bacteria* in seawater are killed by bacteriophages each day, and most of the others are eaten by protozoa. For example, Syn5 is a bacteriophage that attacks and lyses *Synechococcus* species, who, along with their relative *Prochlorococcus*, are the major primary producers of the ocean and account for over 30% of the CO_2 fixed globally (↪ Section 20.10) (**Figure 10.26**). The cytoplasm released as a result of viral attack on cells of these species (Figure 10.26c) provides a significant amount of organic matter for other microbes in the ocean. Although viruses account for most of the total microbes present in seawater in terms of numbers, because of their very small sizes they constitute only about 5% of the total microbial biomass (**Figure 10.27**).

The most common bacteriophages in the oceans are head-and-tail bacteriophages containing double-stranded DNA genomes (Baltimore class I, Figure 10.2); by contrast, RNA-containing phages are relatively rare. As we have seen, lysogenic bacteriophages can integrate into the genomes of their bacterial hosts (↪ Section 8.7), and when they do, they can confer new properties on the cell. Moreover, some lytic phages facilitate the transfer of bacterial genes from one cell to another through the process of transduction, a major means of horizontal gene transfer in

which a virus ferries host genes between cells, for example by picking up host DNA and becoming a nonlytic *transducing particle* (↪ Section 11.7). As agents of transduction, bacteriophages are thought to have a major influence on bacterial evolution. For example, transferred genes may confer new metabolic or other beneficial properties on the recipient cells and allow them to successfully colonize new habitats.

A good example of phage gene transfer are the cyanophages that have been shown to transfer certain photosynthesis genes among strains of *Synechococcus* and *Prochlorococcus*. When these phages are released from their lysed host cells (Figure 10.26c), some of them incorporate host genes that encode photosystem (PS) II, one of the key components of oxygenic photosynthesis (↪ Section 14.4). When such a phage infects a new host cell, it provides the cell with genes encoding a modified PSII. It is hypothesized that a more diverse complement of PSII proteins improves both cell and phage fitness by allowing the host cell to better adapt to changing environmental

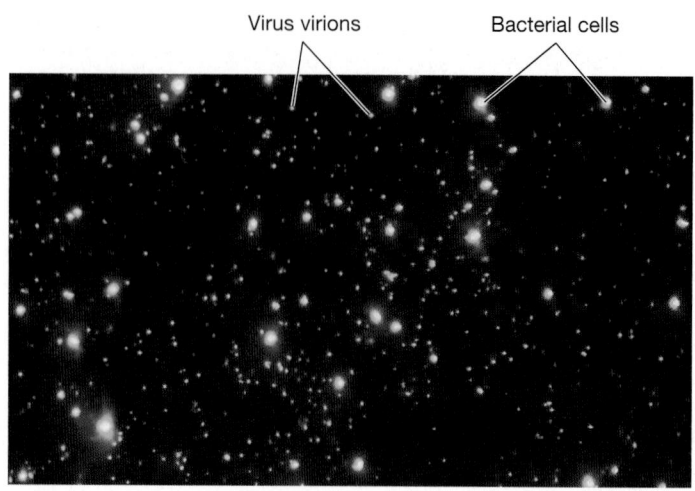

Figure 10.27 Viruses and bacteria in seawater. A fluorescence photomicrograph of seawater stained with the dye SYBR Green to reveal prokaryotic cells and viruses. Although viruses are too small to be seen with the light microscope, fluorescence from a stained virus is visible.

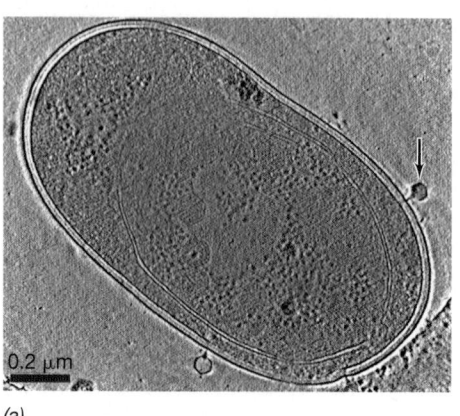 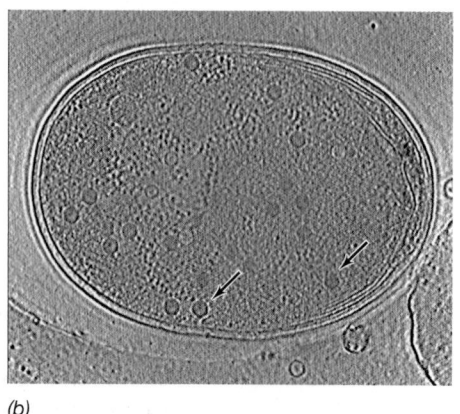 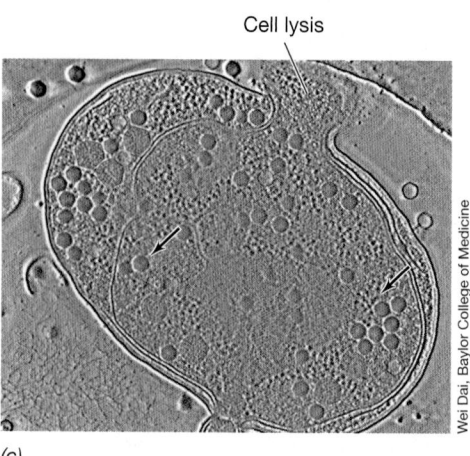

Figure 10.26 Cyanophage Syn5 infection of *Synechococcus*. Phase contrast–electron cryotomography sections of cells in various stages of infection. *(a)* Early. *(b)* Intermediate. *(c)* Late. Arrows point to phage virions.

conditions—for example, changes in light intensity or light quality—and in the process, producing more cyanophage.

Many *Archaea* are present in the oceans, and a major group of marine *Archaea* of ecological importance is the *Thaumarchaeota* (⇔ Section 17.5). These ammonia-oxidizing chemolithotrophs are capable of consuming the vanishingly low levels of ammonia present in planktonic (open ocean) waters. Although lytic archaeal viruses have yet to be isolated for this group, several species of the thaumarchaeotan genus *Nitrosopumilus* have been shown to harbor viral genomes within their chromosome (that is, the cells contain a *provirus*) whose genes suggest that the infecting virus was either of the head-and-tailed dsDNA bacteriophage type (Section 10.4) or was icosahedral-shaped, similar to the herpesviruses (Section 10.7) (Figure 10.3*b*). It is thus likely that at least some, and perhaps even many, of the viruses in seawater (Figure 10.27) infect marine *Archaea* instead of marine *Bacteria*. This is bolstered by the observation that virtually all known archaeal viruses contain double-stranded DNA genomes (Figure 10.3*a*), and these are the most commonly observed viral genomes in the oceans.

Survival Strategies and Metagenomics of Viruses in Nature

When hosts are plentiful in nature, it is thought that bacteriophages adopt the lytic lifestyle and thus large numbers of host cells are killed. By contrast, when host numbers are low, it may be difficult for viruses to find a new host cell, and under such circumstances, lysogeny would be favored if the virus is lysogenic (⇔ Section 8.7). Under these conditions, the virus would survive as a prophage until host numbers rebounded and a lytic lifestyle could once again be supported. This hypothesis is consistent with the observation that in the depths of the ocean where bacterial numbers are lower than in surface waters, around half the bacteria examined have been found to contain one or more lysogenic viruses. As far as is known, no single-stranded DNA viruses and no RNA viruses can enter a lysogenic state, and so how these viruses might survive periods of low host numbers is unknown.

Most of the genetic diversity on Earth resides in viruses, mostly bacteriophages. The *viral metagenome* is the sum total of all the virus genes in a particular environment. Several viral metagenomic studies have been undertaken, and they invariably show that immense viral diversity exists on Earth. For example, approximately 75% of the gene sequences found in viral metagenomic studies show no similarity to any other genes cataloged in viral or cellular gene databases. By comparison, surveys of bacterial metagenomes typically reveal approximately 10% unknown genes. Thus, most viruses still await discovery and most viral genes have unknown functions. This makes the study of the virosphere and viral diversity one of the most exciting areas of microbiology today.

MINIQUIZ

- What type of bacteriophages are most common in the oceans?
- How can bacteriophages affect bacterial evolution?
- What does the viral metagenome suggest about our understanding of viral diversity?

10.13 Viral Defense Mechanisms of *Bacteria* and *Archaea*

The viruses of *Bacteria* and *Archaea* can be viewed in two different ways: as deadly predators that cull cell populations, or as agents of diversity, enriching their hosts through gene transfers. Despite the importance of the latter, with viruses outnumbering their prokaryotic hosts by a factor of about 10, *Bacteria* and *Archaea* have evolved an arsenal of strategies to defend against their viral predators and we consider these here.

The Microbial Arms Race

The relationship between bacteriophages and their hosts is not static but instead is extremely dynamic. Although bacteria possess several weapons to battle phage attack, phages counter these weapons with weapons of their own, triggering a microbiological "arms race" for cell survival and viral propagation.

We previously discussed how *Bacteria* produce restriction endonucleases—proteins that target and destroy foreign DNA—and how in the case of bacteriophage T4 the virus avoids restriction enzyme surveillance by its *Escherichia coli* host by substituting the base 5-hydroxymethylcytosine for cytosine in its genome (⇔ Section 8.5). In response, some *E. coli* strains have evolved altered restriction enzyme systems that recognize this viral DNA modification and still degrade incoming T4 DNA, preventing infection. This host adaptation has then selected for T4 bacteriophages that modify their DNA in a different manner—by glycosylating (adding sugars) to specific DNA bases. Then, as one might expect in an expanding arms race, some strains of *E. coli* have evolved restriction systems that recognize glycosylated viral DNA and destroy it. As a countermeasure to this, some T4 strains have evolved a protein that inhibits these modified restriction enzymes. In a constant attempt to stay ahead, other *E. coli* strains have evolved endonucleases that are not inhibited by these bacteriophage proteins, and so it goes, back and forth as both predator and prey fight to survive and propagate in the face of "the enemy."

Alterations in viral receptor sites are also a common mechanism to avoid viral infection; for a virus to attach to a host cell, the virus must first recognize and attach to a cell receptor (⇔ Section 8.5 and Figure 8.11). To prevent viral infection, hosts can modify the structure of the cell receptor or protect the receptor by producing a shield, such as an outer cell surface capsule (⇔ Section 2.7). However, viruses can counter these shielding mechanisms either by a mutated viral receptor able to bind to its modified complement on the cell surface or by carrying enzymes within the capsid that can degrade the capsule. Some temperate bacteriophages (⇔ Section 8.7) ensure their self-preservation by hijacking the host-encoded toxin–antitoxin system (⇔ Section 7.11). They do this by way of a prophage-encoded protein that inactivates the host's antitoxin protein and replaces it with a phage antitoxin protein. Thus for the cell to avoid toxicity from its chromosomally encoded toxin (a protein that slows growth during stressful conditions in order to preserve resources but which could make cells uncompetitive in times of plenty if not controlled by its antitoxin), the cell is forced to retain the prophage and gain protection from its encoded antitoxin.

The Antiviral System of *Bacteria* and *Archaea*: CRISPR

A major antiviral defense of both *Bacteria* and *Archaea* is CRISPR, the *c*lustered *r*egularly *i*nterspaced *s*hort *p*alindromic *r*epeats found in the chromosomes of many species that help protect them from bacteriophage infection (ᴄ⊋ Section 11.12). CRISPR regions contain short repeats of *constant* DNA sequence alternating with short *variable* DNA sequences called *spacers* (**Figure 10.28** and ᴄ⊋ Figure 11.33). These spacers correspond to pieces of viral or other foreign DNA and function as a "memory bank" of past encounters with a virus in a manner analogous to how animals produce antibodies and long-lasting memory cells against an infecting virus (adaptive immunity, Chapter 27).

Besides the spacer regions, another essential component of the CRISPR system are the *Cas* (CRISPR-*as*sociated) *proteins*. These proteins possess endonuclease activity and both mediate the defense against foreign DNA and incorporate new spacer regions into the CRISPR region. When a virus attaches to a host cell and injects its DNA, the Cas proteins of a CRISPR region may recognize specific DNA sequences known as *p*rotospacer *a*djacent *m*otifs (PAMs) (Figure 10.28*a*). The Cas protein then cleaves the viral DNA at a locus near this PAM (termed the *protospacer*) and inserts the short DNA region into the CRISPR region of the chromosome, where it becomes a *spacer* (Figure 10.28*a*). The insertion of a spacer into the CRISPR region confers "genetic memory"

(referred to as *immunization*) on the cell and sets the stage for later encounters with the same virus.

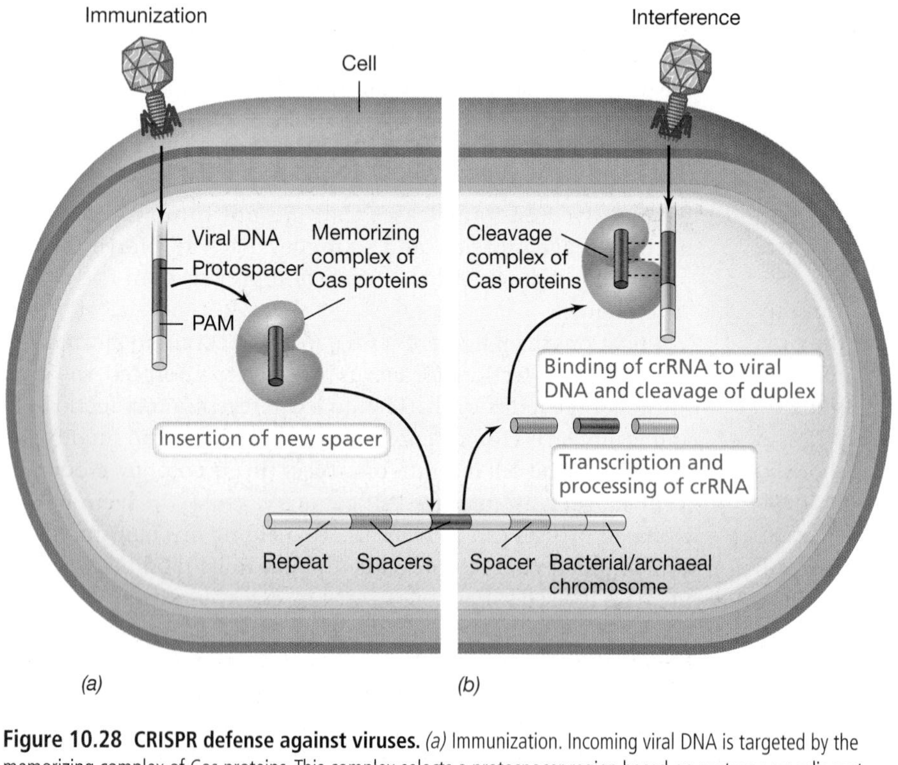

Immune Memory and Other Aspects of CRISPR

When an immunized cell encounters the same virus at a later date, the Cas proteins quickly destroy the incoming DNA in an RNA-dependent process. While the genomic CRISPR region does not contain open reading frames, it does have a promoter (ᴄ⊋ Section 4.5). The resulting transcript is considered the *pre-CRISPR RNA* (pre-crRNA) and contains an array of RNA sequences complementary to both the repeat and spacer regions. The Cas proteins then process the transcript into individual spacer RNAs by targeting the repeat regions (Figure 10.28*b*). These crRNAs then associate with Cas proteins within the cleavage complex and begin surveillance for complementary incoming viral DNAs. Any viral DNA:crRNA duplexes formed are cleaved by the endonuclease activity of Cas proteins and the invading DNA is degraded in a process called *interference* (Figure 10.28*b*). With part of its genome destroyed in this way, an invading virus cannot proceed to replicate and the infection (and threat to the cell) is thwarted.

One of the major conundrums of the CRISPR system is how a host can survive the *initial* viral invasion long enough to become immunized. For the CRISPR system to successfully prevent viral infection, a spacer corresponding to a region of the viral genome must already be present in the CRISPR locus. Immunization probably occurs when an incoming virus has been inactivated by environmental factors (such as ultraviolet radiation) or when the host's restriction enzyme system cleaves the invading DNA before a successful infection can be initiated.

Because of the dynamic nature of the phage–host interaction, viruses have evolved mechanisms to avoid CRISPR surveillance and destruction. These include mutation of the PAM regions recognized by the memorizing complex of Cas proteins and the production of proteins that inhibit activity of the cleavage complex of Cas proteins (Figure 10.28*b*). In addition, a *phage*-encoded (in contrast to *cell*-encoded) CRISPR system has been identified in the genome of a bacteriophage that infects *Vibrio cholerae*, the bacterium that causes cholera. The phage-encoded crRNAs target genes in *V. cholerae* that encode a defense system that prevents bacteriophage propagation; when these genes are inactivated, the cell's defense system is defunct, a rather elaborate example of the "arms race" between bacteria and their viruses.

While CRISPRs are essential for viral surveillance in *Bacteria* and *Archaea*, we discuss in later chapters how their mode of action is also beneficial for maintaining cellular genome integrity (ᴄ⊋ Section 11.12), and how the CRISPR/Cas system has been developed as a powerful tool for synthetic biology (ᴄ⊋ Section 12.12) .

Figure 10.28 CRISPR defense against viruses. *(a)* Immunization. Incoming viral DNA is targeted by the memorizing complex of Cas proteins. This complex selects a protospacer region based on protospacer adjacent motif (PAM) sequences located on the virus genome. Once the protospacer is excised from the viral genome, the memorizing complex inserts the protospacer into the CRISPR region of the chromosome, resulting in a unique spacer region. *(b)* Interference. The chromosomal CRISPR region is transcribed and processed into crRNAs that correspond to the individual spacer regions. Cas proteins bind to these crRNAs and search for complementary DNA. If a crRNA binds to the DNA of an invading virus, forming a crRNA:DNA duplex, the endonuclease activity of the cleavage complex is triggered and results in the degradation of incoming viral DNA.

MINIQUIZ

- Describe two ways that prokaryotic cells can avoid viral infection and how viruses may overcome these defenses.
- How does a prokaryotic cell become immunized against a specific virus?
- How do the crRNAs and Cas proteins protect the cell from invading viral DNA?

10.14 The Human Virome

In Chapter 8 we highlighted the infection process of animal viruses, and in Figures 10.2 and 10.3 we described the general morphology and genomic structure of common animal viruses. Animal viruses differ from the bacteriophages and archaeal viruses in that an animal cell takes up the *entire virion* instead of just the viral genome. There are also more than two lifestyles for animal viruses, including virulent infection, latent infection, persistent infection, and cellular transformation (↩ Figure 8.20). A healthy human is teeming with viruses—not just animal viruses but also bacteriophages and possibly even some plant viruses—and we now explore this remarkable suite of viruses living within us and on us.

The Human Body and the Virome

While profiling the viruses of the human body has been hampered by limitations in cell culturing, the power of metagenomics has not only allowed for the human *microbiome* to be characterized (Chapter 24), but also the human **virome**. The human virome encompasses the entire population of viruses present in and on the human body (**Figure 10.29**). And, as for the human microbiome, the human virome is both unique to an individual and relatively stable over long periods.

Depending on the individual, animal viruses of the human virome include those that cause severe diseases such as hepatitis (Section 10.11) and severe acute respiratory syndrome (SARS, a coronavirus, Section 10.8) as well as those that cause milder acute infections such as influenza (Section 10.9) and viruses of the common cold (rhinoviruses and coronaviruses, Section 10.8, and adenoviruses, Section 10.6). Other common animal viruses of the human virome cause latent infections, such as the herpesviruses (Section 10.7), in particular human cytomegalovirus, present in most human adults.

The viromes of healthy humans are dominated by viruses that contain DNA genomes. Areas of the human body in which the virome has been assessed are the nose, skin, mouth, and gastrointestinal tract (from fecal samples). Animal viruses commonly detected in these areas include the single-stranded DNA anelloviruses and circoviruses and the double-stranded DNA adenoviruses, polyomaviruses, and papillomaviruses (Figure 10.29a). Anelloviruses are nonenveloped viruses that establish persistent infections in the body early in life with no known connection to disease. Circoviruses possess some of the smallest of all viral genomes (Figure 10.1) and are commonly found in poultry and pigs, suggesting that their presence in the human gastrointestinal tract may originate from these food sources.

Adenoviruses are common respiratory viruses that have been detected in children with fevers, but are also found in the nose and upper respiratory tract of healthy humans. Similarly, polyomaviruses

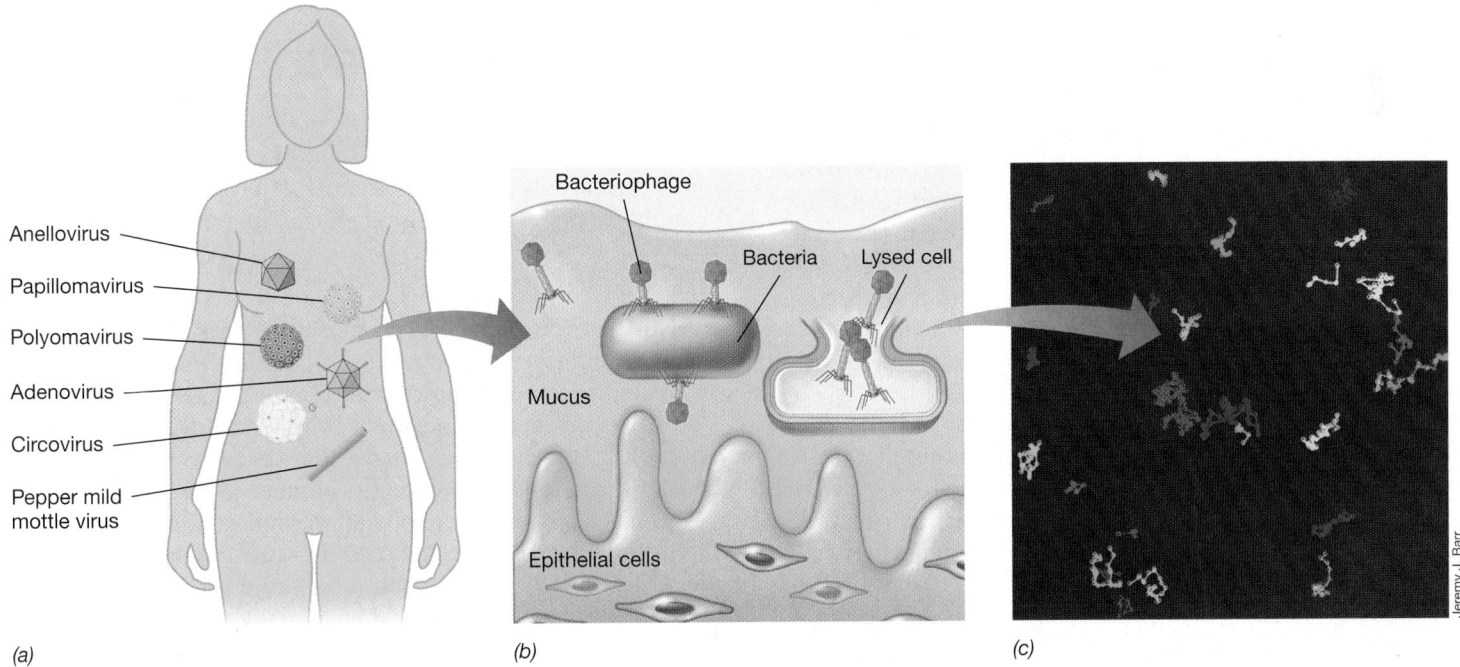

(a)   (b)   (c)

Jeremy J. Barr

Figure 10.29 The human virome. *(a)* Common viruses of eukaryotes found in the virome of a healthy human. *(b)* Bacteriophages within the mucosa of the human respiratory and gastrointestinal tracts protect epithelial cells from the invasion of bacterial pathogens. *(c)* Movement of T4 bacteriophages in mucus. Tracks represent movement of fluorescently labeled individual bacteriophage virions within a microfluidic chip containing 1% mucin to simulate a mucosal matrix.

are commonly found in healthy individuals, but can also cause a type of brain disorder called leukoencephalopathy as well as urinary tract infections in immunocompromised individuals. Several different papillomaviruses are also common in the human virome, especially in the skin and saliva. *Papillomaviruses* are nonenveloped double-stranded DNA viruses (Baltimore class I, Figure 10.2) that replicate on skin and in mucosal epithelia. Most papillomavirus infections are asymptomatic; however, the human papillomavirus (HPV) causes persistent infections that can develop into skin warts, and certain strains of HPV can cause premalignant lesions in the female reproductive tract that can lead to cervical cancer later in life.

Besides these common animal viruses, most human viromes also contain viruses that infect plants, such as the pepper mild mottle virus (Figure 10.29a). These viruses are undoubtedly transmitted to humans from foods and pass through the intestines. While the host specificity of these viruses is for plants, it is thought that their presence in humans may trigger inflammation, which may lead to some of the symptoms susceptible individuals have to spicy foods or certain plant products. Another possible interaction between the human virome and the immune system is the prevalence of *human endogenous retrovirus* (HERV) elements in human chromosomes; HERVs are remnants of retroviral genes that have been incorporated into and constitute 5–8% of the human genome. Although most HERVs are thought to be harmless, connections have been proposed between certain HERVs and the autoimmune disorders rheumatoid arthritis and systemic lupus erythematosus (⇄ Section 27.9), and with other afflictions such as inflammatory bowel (Crohn's) disease and multiple sclerosis.

Bacteriophages and the Human Virome

While every human body contains a unique mixture of different animal viruses, the most abundant viruses in all body sites are not animal viruses but instead are bacteriophages. The large intestine is the hotbed for such viruses, as this organ contains a roughly equal abundance of prokaryotic cells and viruses (about 10^9 of each per gram of feces). As with the animal viruses of the human virome, DNA bacteriophages dominate, and the majority of these viruses are thought to benefit bacterial cells by transferring genes encoding antibiotic resistance or enzymes for specialized metabolisms through the processes of transduction (⇄ Section 11.7) and lysogeny (⇄ Section 8.7). Genetic transfers from gut bacteriophages to their hosts likely help the gut microbiota adapt to changing nutrient conditions and stabilize it from the stresses of antibiotic treatment.

Bacteriophages of the human virome can also be a first line of defense against certain pathogens, especially within mucosal surfaces where bacteriophages accumulate. It has been estimated that 20 times more bacteriophages than bacteria exist in the mucosa of our lungs and intestines. The bacteriophages present in mucous linings are anchored to sugar residues produced by mucosal cells (Figure 10.29b). Here bacteriophages scavenge invading pathogens and kill them before they can cross the mucous barrier. Thus, phages within the mucous layer can be considered to have a symbiotic relationship with the human host and provide a form of host-independent immunity. Using fluorescently labeled T4 phages, movement of the phages through a model mucus layer can actually be tracked microscopically in the laboratory (Figure 10.29c).

Besides killing or conferring antibiotic resistance to bacteria within the microbiome, the bacteriophage component of the virome can also *enhance* the pathogenicity of certain bacteria. An example of this is the temperate (lysogenic, ⇄ Section 8.7) bacteriophage CTXϕ and its host, the bacterium *Vibrio cholerae* (cholera). It is the phage genome rather than the host genome that actually encodes the cholera toxin that induces disease symptoms (⇄ Section 32.3), and cells of *V. cholerae* that are not CTXϕ lysogens are nonpathogenic. This phage–bacterium link is also seen in the *toxin-coregulated pilus*, a structure essential for *V. cholerae* cells to attach to an intestinal cell. Genes encoding the pilus are part of a viral genome that is integrated into the host's genome (a prophage, ⇄ Section 8.7) and is only present in pathogenic strains of *V. cholerae*.

While the human microbiome has been extensively profiled (Chapter 24), these surveys have relied on targeting genes encoding ribosomal RNAs, highly conserved phylogenetic barometers present in all cells. However, because viruses do not possess a universal gene marker, characterization of the human virome has understandably lagged behind that of the cellular microbiome. Nevertheless, recognition of the potentially huge impact on human health by the virome coupled with the continually improving field of metagenomics should bring the human virome into clearer focus in the near future.

— MINIQUIZ —

- How does the virome differ from the microbiome?
- How do bacteriophages benefit the microbiome?
- How do bacteriophages benefit and harm human hosts?

V • Subviral Agents

We conclude our genomic tour of the viral world by considering two *subviral* agents: the viroids and the prions. These are infectious agents that resemble viruses but lack either nucleic acid (prions) or protein (viroids) and are thus not viruses.

10.15 Viroids

Viroids are infectious RNA molecules that lack a protein component. Viroids are small, circular, single-stranded RNA molecules that are the smallest known pathogens. They range in size from 246 to 399 nucleotides and show a considerable degree of sequence homology to each other, suggesting that they have common evolutionary roots. Viroids cause a number of important plant diseases and can have a severe agricultural impact (**Figure 10.30**). A few well-studied viroids include coconut cadang-cadang viroid (246 nucleotides) and potato spindle tuber viroid (359 nucleotides). No viroids are known that infect animals or microorganisms.

Viroid Structure and Function

The extracellular form of a viroid is naked RNA; there is no protein capsid. Although the viroid RNA is a single-stranded, covalently closed circle, its extensive secondary structure forms a hairpin-shaped double-stranded molecule with closed ends (**Figure 10.31**).

Figure 10.30 Viroids and plant diseases. Photograph of healthy tomato plant (left) and one infected with potato spindle tuber viroid (PSTV) (right). The host range of most viroids is quite restricted. However, PSTV infects tomatoes as well as potatoes, causing growth stunting, a flat top, and premature plant death.

This apparently makes the viroid sufficiently stable to exist outside the host cell. Because it lacks a capsid, a viroid does not use a receptor to enter the host cell. Instead, the viroid enters a plant cell through a wound, as from insect or other mechanical damage. Once inside, viroids move from cell to cell via plasmodesmata, which are the thin strands of cytoplasm that link plant cells (**Figure 10.32**).

Viroid RNAs do not encode proteins and thus the viroid is totally dependent on its host for replication. Plants have several RNA polymerases, one of which has RNA replicase activity, and this is the enzyme that replicates the viroid. The replication mechanism itself resembles the rolling circle mechanism used for genome synthesis by some small viruses (Sections 10.3 and 10.7). The result is a large RNA molecule containing many viroid units joined end to end. The viroid has ribozyme (catalytic RNA) activity and uses it to self-cleave the large RNA molecule, releasing individual viroids.

Viroid Disease

Viroid-infected plants can be symptomless or show symptoms that range from mild to lethal, depending on the viroid (Figure 10.30). Most disease symptoms are growth related, and it is believed that viroids mimic or in some way interfere with plant small regulatory RNAs. In fact, viroids could themselves be derived from regulatory RNAs that have evolved away from carrying out beneficial roles in the cell to inducing destructive events. Viroids are known to yield small interfering RNAs (siRNAs) as a by-product of replication. It has been proposed that these siRNAs may function by way of the RNA-interference silencing pathway to suppress

Figure 10.31 Viroid structure. Viroids consist of single-stranded circular RNA that forms a seemingly double-stranded structure by intra-strand base pairing.

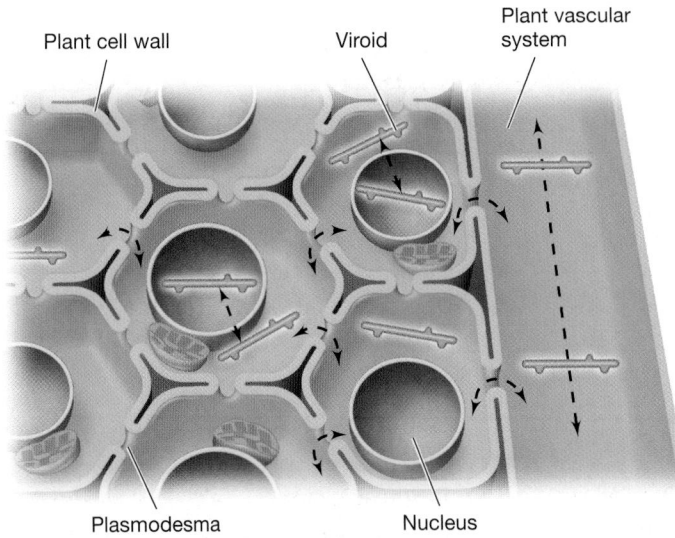

Figure 10.32 Viroid movement inside plants. After entry into a plant cell, viroids (orange) replicate either in the nucleus or the chloroplast. Viroids can move between plant cells via the plasmodesmata (thin threads of cytoplasm that penetrate the cell walls and connect plant cells). On a larger scale, viroids can also move around the plant via the plant vascular system.

the expression of plant genes that show some homology to viroid RNA, and in this way induce disease symptoms. This mechanism of regulation is similar to how some bacterial and archaeal regulatory small RNAs target mRNAs for degradation (Figure 6.27b).

— **MINIQUIZ** —

- If viroids are circular molecules, why are they depicted as hairpins?

- How might viroids cause disease in plants?

10.16 Prions

Prions represent the opposite extreme from that of viroids. Prions are infectious agents whose extracellular form consists entirely of protein. That is, *a prion lacks both DNA and RNA*. Prions cause several neurological diseases such as scrapie in sheep, bovine spongiform encephalopathy (BSE or "mad cow disease") in cattle, chronic wasting disease in deer and elk, and kuru and variant Creutzfeldt–Jakob disease in humans by catalyzing protein conformational changes that lead to protein clumping and accumulation. No prion diseases of plants are known, although prions have been detected in yeast. Collectively, animal prion diseases are known as *transmissible spongiform encephalopathies*.

Prion Proteins and the Prion Infectious Cycle

If prions lack nucleic acid, how is prion protein encoded? The answer to this conundrum is that the host cell itself encodes the prion. The host contains a gene, *Prnp* (*Prion* protein), which encodes the native form of the prion, known as PrP^C (Prion Protein Cellular). This is primarily found in the neurons of healthy animals, especially in the brain (**Figure 10.33a**). The pathogenic form of the prion protein is designated PrP^{Sc} (prion protein

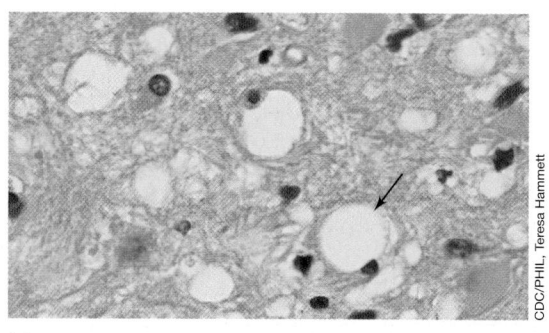

(a)

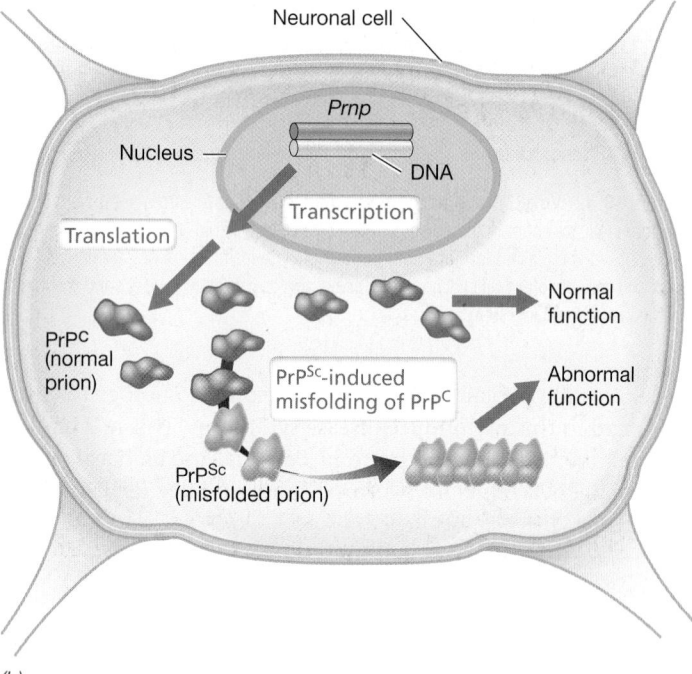

(b)

Figure 10.33 Prions. *(a)* Section through brain tissue of a human with variant Creutzfeldt–Jakob disease. Note the spongy nature of the tissue (clearings, arrow) where neural tissue has been lost. *(b)* Mechanism of prion misfolding. Neuronal cells produce the native form of the prion protein. The pathogenic form catalyzes the refolding of native prions into the pathogenic form. The pathogenic form is protease resistant, insoluble, and forms aggregates in neural cells. This eventually leads to destruction of neural tissues (see part *a*) and neurological symptoms.

*Sc*rapie) because the first prion disease to be discovered was that of scrapie in sheep. PrP^{Sc} is identical in amino acid sequence to PrP^{C} from the same animal species, but it has a different conformation. For example, native prion proteins are largely α-helical, whereas the pathogenic forms contain less α-helix and more β-sheet secondary structure. Prion proteins from different species of mammals are similar but not identical in amino acid sequence, and host range is linked in some way to protein sequence. For example, PrP^{Sc} from BSE-diseased cattle can infect humans, whereas PrP^{Sc} from scrapie-infected sheep apparently cannot.

When the PrP^{Sc} form enters a host cell that is expressing PrP^{C}, it promotes the conversion of PrP^{C} into the pathogenic form (Figure 10.33*b*). That is, the pathogenic prion "replicates" by converting preexisting native prions into the pathogenic form. As the pathogenic prions accumulate and aggregate, they form insoluble crystalline fibers referred to as amyloids in neural cells; this leads to disease symptoms including the destruction of brain and other nervous tissue (Figure 10.33*a*). PrP^{C} functions in the cell as a cytoplasmic membrane glycoprotein, and it has been shown that membrane attachment of pathogenic prions is necessary before disease symptoms commence. Mutant versions of PrP^{Sc} that can no longer attach to nerve cell cytoplasmic membranes still aggregate but no longer cause disease.

Besides the transmissible spongiform encephalopathies, amyloids are also associated with debilitating human diseases such as Alzheimer's, Huntington's, Parkinson's, and type 2 diabetes. Whether all of these are truly prion diseases is unclear. However, protein aggregation has been linked to these diseases, so the possibility remains that they are manifestations of prion infection.

Nonpathogenic Prions

Certain fungi have proteins that fit the prion definition of an inherited self-perpetuating change in protein structure, although these proteins do not cause noticeable disease. Instead they adapt the fungal cells to survive dynamic environmental conditions by conferring traits such as altered nutrient utilization, antibiotic resistance, and biofilm formation. In yeast, for example, the [URE3] prion is a protein that regulates the transcription of genes encoding certain nitrogen metabolism functions. The normal (soluble) form of this protein functions to repress genes encoding proteins that metabolize certain nitrogen sources. However, when the [URE3] prion accumulates, it forms insoluble aggregates and when this occurs, the normally repressed genes are derepressed and the nitrogen sources are metabolized.

Humans also have beneficial prions. MAVS is a human protein that is part of our innate immune system and has been shown to convert to a self-perpetuating prion-like form in cells that become infected with a virus. The aggregation of MAVS protein triggers the production of immune modulators called *interferons* (↩ Section 26.10). These proteins trigger the recruitment of phagocytic cells such as macrophages (cells that ingest and destroy pathogens, foreign particles, and cell debris) and other immune factors to the viral-infected cell, resulting in its destruction. While the conversion of MAVS to a prion-like form ultimately kills the cell, it prevents the virus from hijacking the cell's molecular machinery, which would allow the virus to replicate and lyse the host cell to infect adjacent cells.

MINIQUIZ

- On what basis can prions be differentiated from all other infectious agents?
- What is the difference between the native and pathogenic forms of the prion protein?
- How does a prion differ from a viroid? How does a prion differ from a virus?

MasteringMicrobiology® **Visualize, explore,** and **think critically** with Interactive Microbiology, MicroLab Tutors, MicroCareers case studies, and more. MasteringMicrobiology offers practice quizzes, helpful animations, and other study tools for lecture and lab to help you master microbiology.

Chapter Review

I • Viral Genomes and Evolution

10.1 Viral genomes can be single-stranded or double-stranded DNA or RNA and vary from a few to hundreds of kilobases in size. Viral mRNA is always of the plus configuration by definition, but single-stranded genomes can be of the plus or minus configuration. Viruses with RNA genomes must either carry an RNA replicase in their virions or encode this enzyme in their genomes in order to synthesize RNA from an RNA template.

Q **Describe the classes of viruses based on their genomic characteristics. For each class, describe how viral mRNA is made and how the viral genome is replicated.**

10.2 Viruses may have evolved as agents of gene transfer in cells, or they may be the remnants of "virocells" that contained RNA genomes and eventually streamlined their biology until they became dependent on a host cell for replication. RNA viruses are likely more ancient than DNA viruses, and the latter may have triggered the transition from RNA genomes to DNA genomes in cells.

Q **How would virocells have differed from the RNA viruses we know today, and how would they have been similar?**

II • DNA Viruses

10.3 Single-stranded DNA viruses contain DNA of the plus configuration, and a double-stranded replicative form is necessary for transcription and genome replication. The genome of the virus φX174 is so small that some of its genes overlap, and the genome replicates by a rolling circle mechanism. Some related viruses, such as M13, have filamentous virions that are released from the host cell without lysis.

Q **Describe how the genome of bacteriophage φX174 is transcribed and translated.**

10.4 The head-and-tail bacteriophage T7 contains a double-stranded DNA genome that encodes both early genes, transcribed by the host RNA polymerase, and late genes, transcribed by a virus-encoded RNA polymerase. Replication of the T7 genome employs T7 DNA polymerase and involves terminal repeats and concatemers. Bacteriophage Mu is a temperate virus that is also a transposable element. Mu replicates by transposition in the host chromosome.

Q **Why can it be said that transcription of the bacteriophage T7 genome requires two enzymes?**

Why is bacteriophage Mu mutagenic, and what features are necessary for Mu to insert into DNA?

10.5 Several double-stranded DNA viruses infect cells of *Archaea*, most of which inhabit extreme environments. Many of these genomes are circular, in contrast to the linear double-stranded DNA genomes of bacteriophages. Although many head-and-tail-type viruses are known, other archaeal viruses have an unusual spindle-shaped morphology.

Q **List three unusual features of the archaeal virus that infects *Acidianus* that distinguish it from bacteriophage T7.**

10.6 Pox viruses are large double-stranded DNA viruses that replicate entirely in the cytoplasm and are responsible for several human diseases, including smallpox. Adenoviruses are double-stranded DNA viruses whose genome replication employs protein primers and a mechanism that occurs without lagging-strand synthesis.

Q **Of all the double-stranded DNA animal viruses, pox viruses stand out concerning one unique aspect of their DNA replication process. What is this unique aspect, and how can this be accomplished without special DNA replication enzymes being packaged in the virion?**

10.7 Some double-stranded DNA viruses cause cancer in humans. SV40 is such a tumor virus and has a tiny genome containing overlapping genes. The virus can trigger cell transformation (tumor induction) from the activity of certain genes. Some herpesviruses also cause cancer, but most cause various human infectious diseases. Herpesviruses can maintain themselves in a latent state in the host indefinitely, initiating viral replication periodically.

Q **What is unusual about the envelope of the herpesvirus?**

III • Viruses with RNA Genomes

10.8 In single-stranded plus-sense RNA viruses, the genome is also the mRNA, and a negative strand is synthesized to produce more mRNA and genome copies. The tiny bacteriophage MS2 contains only four genes, one of which encodes a subunit of its RNA replicase. In poliovirus, the viral RNA is translated directly, producing a polyprotein that is cleaved into several small viral proteins. Coronaviruses are large RNA viruses that resemble poliovirus in some but not all of their replication features.

Q **What is the function of the VPg protein of poliovirus, and how can coronaviruses replicate without a VPg protein?**

UNIT 3

10.9 In negative-strand viruses, the virus RNA is not mRNA but must first be copied to form mRNA by RNA replicase present in the virion. The positive strand is the template for production of genome copies. Important pathogenic negative-strand viruses include rabies virus and influenza virus.

> Q **Rabies virus and poliovirus both have single-stranded RNA genomes, but only in poliovirus can the genome be translated directly. Explain.**

10.10 Reoviruses contain segmented linear double-stranded RNA genomes. Like negative-strand RNA viruses, reoviruses contain an RNA-dependent RNA polymerase within the virion. All replication events occur within newly forming virions.

> Q **Compare the reovirus genome to those of influenza virus and bacteriophage MS2. Why do reovirus replication events have to happen in the nucleocapsid?**

10.11 Some viruses employ reverse transcriptase, including retroviruses (HIV) and hepadnaviruses (hepatitis B). Retroviruses have single-stranded RNA genomes and use reverse transcriptase to make a DNA copy. Hepadnaviruses contain partially double-stranded DNA genomes and use reverse transcriptase to make a single strand of genomic DNA from a full-length complementary strand of RNA.

> Q **Why do both hepadnaviruses and retroviruses require reverse transcriptase when their genomes are double-stranded DNA and single-stranded RNA, respectively?**

IV • Viral Ecology

10.12 The number of viruses on Earth is greater than the number of cells by 10-fold. Most of the genetic diversity on Earth resides in virus genomes, most of which are still to be investigated. Viruses affect their host cells by either culling the host population or by carrying out horizontal gene transfer from one bacterial cell to another. In the oceans, both *Bacteria* and *Archaea* are likely to be infected with viruses.

> Q **How do viral numbers compare to those of bacteria in seawater?**

10.13 Prokaryotic cells employ various mechanisms to defend against viral infection. However, the fast mutation rate and genome plasticity of viruses allows them to adapt to these strategies. The CRISPR system is a type of immune system for *Bacteria* and *Archaea* that resists viral infection by recognizing and degrading incoming foreign DNA.

> Q **How does a prokaryotic cell survive initial viral infection to become immunized by way of CRISPR?**

10.14 The human virome is the entire population of viruses associated with the human body. Each individual possesses a unique and relatively stable virome. Animal viruses that cause asymptomatic infections are common in a healthy individual's virome. However, bacteriophages are the major constituent of the human virome. These phages can protect against pathogen invasion and also increase the fitness of the microbiome through the genetic transfer of beneficial genes.

> Q **How do bacterial viruses help prevent human diseases?**

V • Subviral Agents

10.15 Viroids are circular single-stranded RNA molecules that do not encode proteins and are dependent on host-encoded enzymes for replication. Unlike viruses, viroid RNA is not enclosed within a capsid, and all known viroids are plant pathogens.

> Q **What are the similarities and differences between viruses and viroids?**

10.16 Prions consist of protein but have no nucleic acid of any kind. Prions exist in two conformations, the native cellular form and the pathogenic form, which takes on a different protein structure. The pathogenic form "replicates" itself by converting native prion proteins, encoded by the host cell, into the pathogenic conformation.

> Q **What are the similarities and differences between prions and viruses?**

Application Questions

1. Not all proteins are made from the RNA genome of bacteriophage MS2 in the same amounts. Can you explain why? One of the proteins functions very much like a repressor, but it functions at the translational level. Which protein is it and how does it function?

2. Replication of both strands of DNA in adenoviruses occurs in a continuous (leading) fashion. How can this happen without violating the rule that DNA synthesis always occurs in a $5' \rightarrow 3'$ direction?

3. Imagine that you are a researcher at a pharmaceutical company charged with developing new drugs against human RNA viral pathogens. Describe at least two types of drugs you might pursue, what class of virus they would affect, and why you feel that the drugs would not harm the patient.

4. Reoviruses contain genomes that are unique in all of biology. Why? Why can't reovirus replication occur in the host cytoplasm? Contrast reovirus genomic replication events with those of a cell. Why can it be said that reovirus genome replication is not semiconservative even though the reovirus genome consists of complementary strands?

Chapter Glossary

Hepadnavirus a virus whose DNA genome replicates by way of an RNA intermediate

Negative strand a nucleic acid strand that has the opposite sense to (is complementary to) the mRNA

Overlapping genes two or more genes in which part or all of one gene is embedded in another gene

Polyprotein a large protein expressed from a single gene and subsequently cleaved to form several individual proteins

Positive strand a nucleic acid strand that has the same sense as the mRNA

Prion an infectious protein whose extracellular form lacks nucleic acid

Replicative form a double-stranded molecule that is an intermediate in the replication of viruses with single-stranded genomes

Retrovirus a virus whose RNA genome has a DNA intermediate as part of its replication cycle

Reverse transcription the process of copying genetic information found in RNA into DNA

RNA replicase an enzyme that can produce RNA from an RNA template

Rolling circle replication a mechanism, used by some plasmids and viruses, of replicating circular DNA, which starts by nicking and unrolling one strand. For a single-stranded genome, this is preceded by using the still-circular strand as a template for DNA synthesis; for a double-stranded genome, the unrolled strand is used as a template for DNA synthesis

Transposase an enzyme that catalyzes the insertion of DNA segments into other DNA molecules

Viroid an infectious RNA whose extracellular form lacks protein

Virome the entire population of viruses associated with the human body

11

Genetics of *Bacteria* and *Archaea*

microbiology**now**

Killing and Stealing: DNA Uptake by the Predator *Vibrio cholerae*

Vibrio cholerae, the causative agent of cholera, is a marine bacterium that can also flourish in the nutrient-rich human intestine. The pathogen reaches new human hosts by releasing a toxin that triggers diarrheal purges, resulting in transmission of the pathogen into new environments. Besides this virulent lifestyle, *V. cholerae* also competes for nutrients in marine environments by employing an arsenal of proteins to kill neighboring microbial cells. Using an elegant contractile structure known as the type VI secretion system (T6SS), predatory *V. cholerae* cells inject toxic molecules called effectors into prey cells. Cells lacking immunity to these effectors ultimately lyse and release their cellular contents.

While predation reduces competition and releases nutrients, microbiologists have recently observed another fascinating aspect of the "assassin" lifestyle of *V. cholerae*. Once prey cells have been killed using T6SS, *V. cholerae* predator cells can scavenge the DNA released from their victims using a special DNA-uptake system (natural competence). This system facilitates horizontal gene transfer, as it can incorporate DNA fragments containing up to 40 genes. If prey cell genes recombine with the *V. cholerae* chromosome, the *V. cholerae* population stands to benefit from the transformation event.

These images show predator cells killing and stealing DNA from defenseless prey cells. DNA-uptake proteins of the competence system are labeled with a red fluorescent tag and proteins of the T6SS with a green fluorescent tag so that predatory *V. cholerae* cells appear red with green "crossbows" inside. In contrast, prey lacking T6SS and immunity proteins appear as solid green cells. The panel on the left shows red predator cells with their green crossbows attacking prey cells. The white arrows point to dead prey cells, which circularize upon death. The panel on the right shows the same population of cells 30 minutes later. As the predator cells take up DNA from killed prey, the corresponding DNA-binding proteins localize for effective uptake (circled).

The antagonistic nature of *V. cholerae* provides this predatory bacterium with a mechanism for stealing valuable traits such as antibiotic resistance and virulence factors from its prey by exploiting horizontal gene transfer. Thus, this form of microbial warfare not only reduces competition for nutrients but also functions to increase the predator's genetic fitness.

Source: Borgeaud, S., L.C. Metzger, T. Scrignari, and M. Blokesch. 2015. The type VI secretion system of *Vibrio cholerae* fosters horizontal gene transfer. *Science 347:* 63–67.

In 1946 the microbiologist Joshua Lederberg made a groundbreaking discovery—like plants and animals, bacteria can also exchange genes! Lederberg's work showcasing genetic recombination in bacteria not only earned him a Nobel Prize but also helped launch the field of molecular biology and the use of bacteria to study how genes work in higher organisms such as animals.

Understanding the varied mechanisms by which *Bacteria* and *Archaea* exchange genes has helped tackle the conundrum of how these microbes can exhibit so much diversity while reproducing asexually. Gene exchange, along with genetic innovations that arise from random changes in a cell's genetic blueprint, confer selectable advantages that ultimately drive genetic diversity.

In this chapter we discuss mechanisms of genetic exchange in *Bacteria* and *Archaea*. We first describe how changes arise in the genome, and then we consider how *horizontal gene transfer* can move genes from one cell to another. While changes to the genome underlie microbial diversity and habitat adaptation, microorganisms also possess mechanisms to maintain genomic stability, and we end this chapter by considering these. Taken together, both genomic change and genomic stability are important to the evolution of an organism and its competitive success in nature.

I · Mutation

All organisms contain a specific sequence of nucleotide bases in their genome, their genetic blueprint. A **mutation** is a *heritable* change in the base sequence of that genome, that is, a change that is passed from the mother cell to progeny cells. Mutations can lead to changes in the properties of an organism; some mutations are beneficial, some are detrimental, but most are neutral and have no effect. Although the rate of spontaneous mutation is low (Section 11.3), the speed at which many prokaryotic cells divide and their characteristic exponential growth ensure that mutations accumulate in a population surprisingly fast. Moreover, whereas a single mutation typically brings about only a small change in a cell, genetic exchange often generates much larger change. Taken together, mutation and genetic exchange fuel the evolutionary process.

We begin by considering the molecular mechanism of mutation and the properties of mutant microorganisms.

11.1 Mutations and Mutants

The genomes of cells consist of double-stranded DNA. In viruses, by contrast, the genome may consist of double- or single-stranded DNA (or RNA) (Chapters 8 and 10). By convention, a strain of an organism or a virus isolated from nature is referred to as the **wild-type strain** and therefore contains the wild-type genome. A cell or virus derived from the wild type that carries a change in nucleotide sequence is called a **mutant**. A mutant by definition differs from the wild-type strain in its **genotype**, the nucleotide sequence of its genome. In addition, the observable properties of the mutant—its **phenotype**—may also be altered relative to its parent (**Figure 11.1**). This altered phenotype is called a *mutant phenotype*.

The term "wild-type" may be used to refer to an entire organism or just to the status of a particular gene that is under investigation.

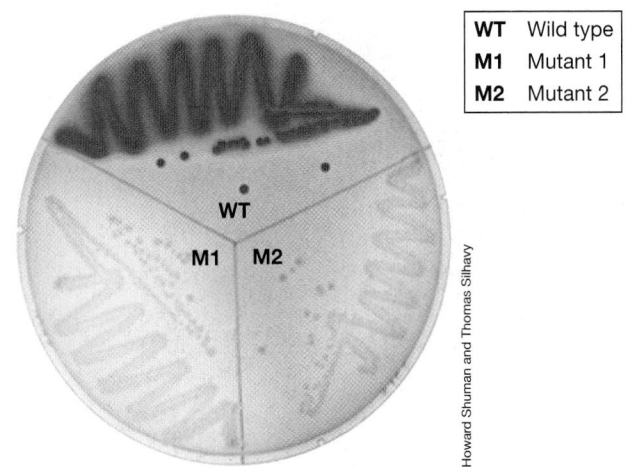

WT	Wild type
M1	Mutant 1
M2	Mutant 2

Figure 11.1 Wild-type versus mutant phenotype. Growth of wild-type *Escherichia coli* and maltose utilization mutants on a plate of MacConkey agar, a differential medium. The medium contains maltose as the carbon source and a pH indicator that turns red if maltose is fermented. Mutants 1 and 2 are unable to ferment maltose due to a deletion of the *malB* gene and a point mutation in the *malQ* gene, respectively.

Mutant derivatives can be obtained either directly from a wild-type strain or from another strain—referred to as a *parental strain*—previously derived from the wild type; for example, another mutant. Figure 11.1 shows a plate of MacConkey agar (a culture medium that contains a pH indicator that turns red if sugar is fermented) that shows the phenotypic difference between wild-type *Escherichia coli* and mutant derivatives in the sugar utilization pathway.

Depending on the mutation, a mutant strain may or may not differ in phenotype from its parent. By convention in bacterial genetics, the *genotype* of an organism is designated by three lowercase letters followed by a capital letter (all in italics) indicating a particular gene. For example, the *hisC* gene of *E. coli* encodes a protein called HisC that functions in biosynthesis of the amino acid histidine. Mutations in the *hisC* gene would be designated as *hisC1*, *hisC2*, and so on, the numbers referring to the order of isolation of the mutant strains. Each *hisC* mutation would be different, and each *hisC* mutation might affect the HisC protein in different ways.

The *phenotype* of an organism is designated by a capital letter followed by two lowercase letters, with either a plus or minus superscript to indicate the presence or absence of that property. For example, a His⁺ strain of *E. coli* is one that is capable of making its own histidine, whereas a His⁻ strain is not. The His⁻ strain would therefore require a histidine supplement for growth. A mutation in the *hisC* gene will lead to a His⁻ phenotype if it eliminates the function of the HisC protein.

Isolation of Mutants: Screening versus Selection

Virtually any characteristic of an organism can be changed by mutation. Some mutations are *selectable*, conferring some type of advantage on organisms possessing them, whereas others are nonselectable, even though they may lead to a very clear change in the phenotype of an organism. A selectable mutation confers a

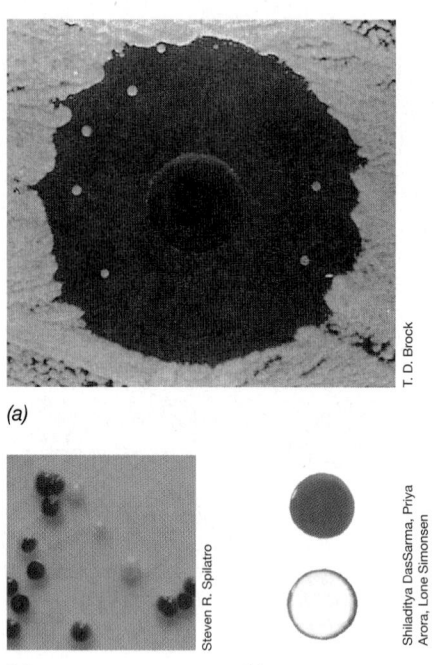

(a)

(b)   (c)

T. D. Brock

Steven R. Spilatro

Shiladitya DasSarma, Priya Arora, Lone Simonsen

Figure 11.2 Selectable and nonselectable mutations. *(a)* Development of antibiotic-resistant mutants, a type of easily selectable mutation, within the inhibition zone of an antibiotic assay disc. *(b)* Nonselectable mutations. UV-radiation-induced nonpigmented mutants of *Serratia marcescens*. The wild type has a dark red pigment. The white or colorless mutants make no pigment. *(c)* Colonies of mutants of a species of *Halobacterium*, a member of the *Archaea*. The wild-type colonies are white. The orangish-brown colonies are mutants that lack gas vesicles (⮕ Section 2.9). The gas vesicles scatter light and mask the color of the colony.

clear advantage on the mutant strain under certain environmental conditions, so the progeny of the mutant cell are able to outgrow and replace the parent. A good example of a selectable mutation is drug resistance: An antibiotic-resistant mutant can grow in the presence of an antibiotic that inhibits or kills the parent (**Figure 11.2a**) and is thus selected under these conditions. It is relatively easy to detect and isolate selectable mutants by choosing the appropriate environmental conditions. **Selection** is therefore an extremely powerful genetic tool, allowing the isolation of a single mutant from a population containing millions or even billions of parental cells.

An example of a nonselectable mutation is color loss in a pigmented organism (Figure 11.2*b, c*). Nonpigmented cells usually have neither an advantage nor a disadvantage over the pigmented parent cells when grown in the laboratory, although pigmented organisms may have a selective advantage in nature. We can detect nonselectable mutations only by examining large numbers of colonies and looking for the "different" ones, a process called **screening**. In microbial genetics, screening is typically a much more laborious and time-consuming process than is selection. Thus in a genetic experiment if selection is possible, it is almost always the preferred strategy.

Isolation of Nutritional Auxotrophs

Although screening is more tedious than selection, useful methods have been developed for screening large numbers of colonies for certain types of mutations. For instance, nutritionally defective mutants

can be detected by the technique of *replica plating* (**Figure 11.3**). A colony from a master plate can be transferred onto an agar plate lacking the nutrient by using a sterile loop, toothpick, or even a robotic arm. Parental colonies will grow normally, whereas those of the mutant will not. Thus, the inability of a colony to grow on medium lacking the nutrient signals that it is a mutant. The colony on the master plate corresponding to the vacant spot on the replica plate can then be picked, purified, and characterized.

A mutant strain with an additional nutritional requirement for growth is called an **auxotroph**, and the parental strain from which it was derived is called a *prototroph*. For instance, mutants of *E. coli* with a His⁻ phenotype are histidine auxotrophs, while the parental His⁺ strain from which the auxotroph was derived is the prototroph of such strains. As described earlier, many different mutations can lead to a strain showing a His⁻ phenotype, and thus an initial step in characterizing the genetics of a metabolic pathway (such as histidine biosynthesis) would be the isolation of several His⁻ strains followed by their comparative genetic analyses (Section 11.5).

Examples of common classes of mutants and the means by which they are detected are listed in **Table 11.1**.

> #### MINIQUIZ
> - Distinguish between a mutation and a mutant.
> - Distinguish between screening and selection.
> - How does an auxotroph differ from a prototroph?

11.2 Molecular Basis of Mutation

Mutations can be either spontaneous or induced. **Spontaneous mutations** are those that occur without external intervention, and most result from occasional errors in the pairing of bases by DNA polymerase during DNA replication. **Induced mutations**, by contrast, are those caused by agents in the environment and include mutations made deliberately by humans. Induced mutations can result from exposure to natural radiation (cosmic rays and so on) that alters the structure of bases in the DNA, or from a variety of chemicals that chemically modify DNA (Section 11.4).

Mutations that change only one base pair are called **point mutations** and occur when a single base-pair substitution takes place in the DNA. Many point mutations do not actually cause any phenotypic change, as discussed below. However, as for all mutations, any phenotypic change that results from a point mutation depends on exactly where in the genome the mutation occurs and the nature of the nucleotide change.

Base-Pair Substitutions: Missense, Nonsense, and Silent Mutations

If a point mutation is within the region of a gene that encodes a polypeptide, any change in the phenotype of the cell is most likely the result of a change in the amino acid sequence of that polypeptide. The error in the DNA is transcribed into mRNA, and the erroneous mRNA in turn is translated to yield a polypeptide. **Figure 11.4** shows the consequences of some base-pair substitutions.

In interpreting the results of a mutation, we must first recall that the genetic code is degenerate (⮕ Section 4.9 and Table 4.4).

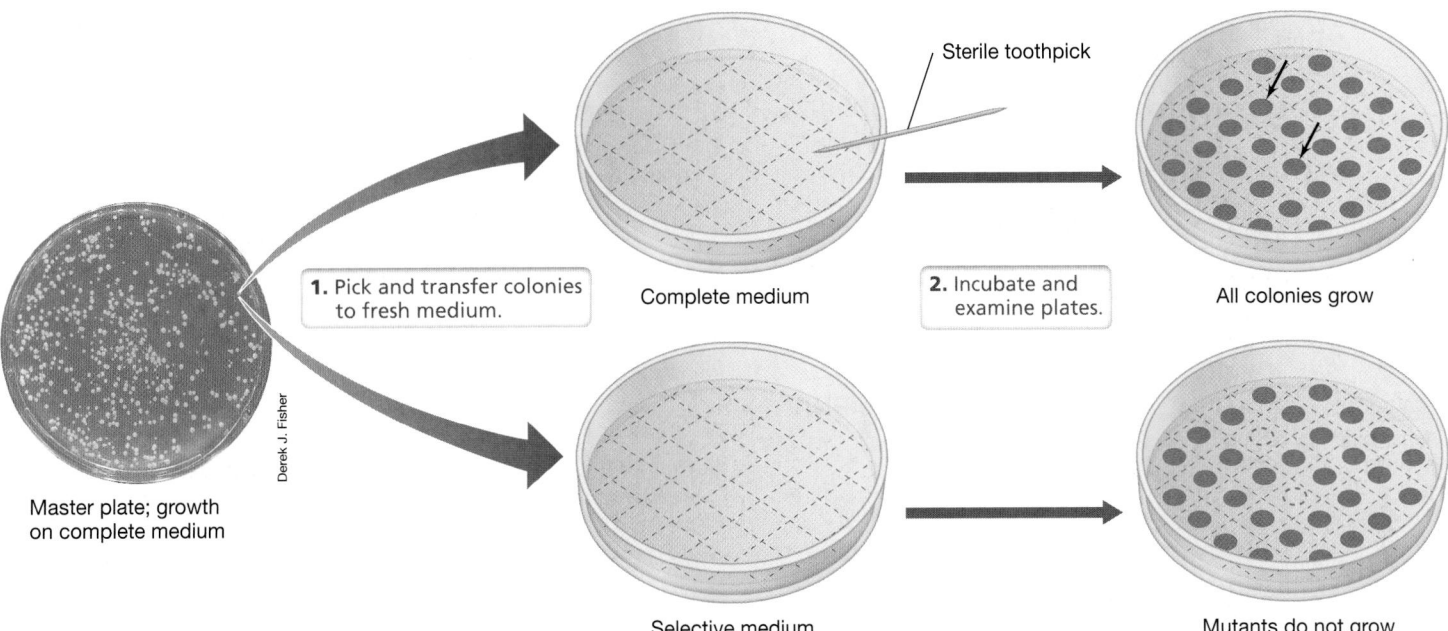

Figure 11.3 Screening for nutritional auxotrophs. The replica-plating method can be used for the detection of nutritional mutants. Colonies from the master plate are transferred using a sterile toothpick to a gridded plate containing different media for selection. The colonies not appearing on the selective medium are indicated with arrows. The selective medium lacked one nutrient (leucine) present in the master plate. Therefore, the colonies indicated with arrows on the master plate are leucine auxotrophs.

Consequently, not all mutations in the base sequence encoding a polypeptide will change the polypeptide. This is illustrated in Figure 11.4, which shows several possible results when the DNA that encodes a single tyrosine codon in a polypeptide is mutated. First, a change in the RNA from *UAC* to *UAU* would have no apparent effect because UAU is also a tyrosine codon. Although they do not affect the sequence of the encoded polypeptide, such changes in the DNA are considered one type of **silent mutation**, that is, a mutation that does not affect the phenotype of the cell. Note that

silent mutations in coding regions are almost always in the third base of the codon (arginine and leucine can also have silent mutations in the first position).

Changes in the first or second base of the codon more often lead to significant changes in the amino acid sequence of the polypeptide. For instance, a single base change from *UAC* to *AAC* (Figure 11.4) results in an amino acid change within the polypeptide from tyrosine to asparagine at a specific site. This is called a **missense mutation** because the informational "sense" (precise sequence

TABLE 11.1 Some examples of mutants

Phenotype	Nature of change	Detection of mutant
Auxotroph	Loss of enzyme in biosynthetic pathway	Inability to grow on medium lacking the nutrient
Temperature-sensitive	Alteration of an essential protein so it is more heat-sensitive	Inability to grow at a high temperature that normally supports growth
Cold-sensitive	Alteration of an essential protein so it is inactivated at low temperature	Inability to grow at a low temperature that normally supports growth
Drug-resistant	Detoxification of drug or alteration of drug target or permeability to drug	Growth on medium containing a normally inhibitory concentration of the drug
Rough colony	Loss or change in lipopolysaccharide layer	Granular, irregular colonies instead of smooth, glistening colonies
Nonencapsulated	Loss or modification of surface capsule	Small, rough colonies instead of larger, smooth colonies
Nonmotile	Loss of flagella or nonfunctional flagella	Compact instead of flat, spreading colonies; lack of motility by microscopy
Pigmentless	Loss of enzyme in biosynthetic pathway leading to loss of one or more pigments	Presence of different color or lack of color
Sugar fermentation	Loss of enzyme in degradative pathway	Lack of color change on agar containing sugar and a pH indicator
Virus-resistant	Loss of virus receptor	Growth in presence of large amounts of virus

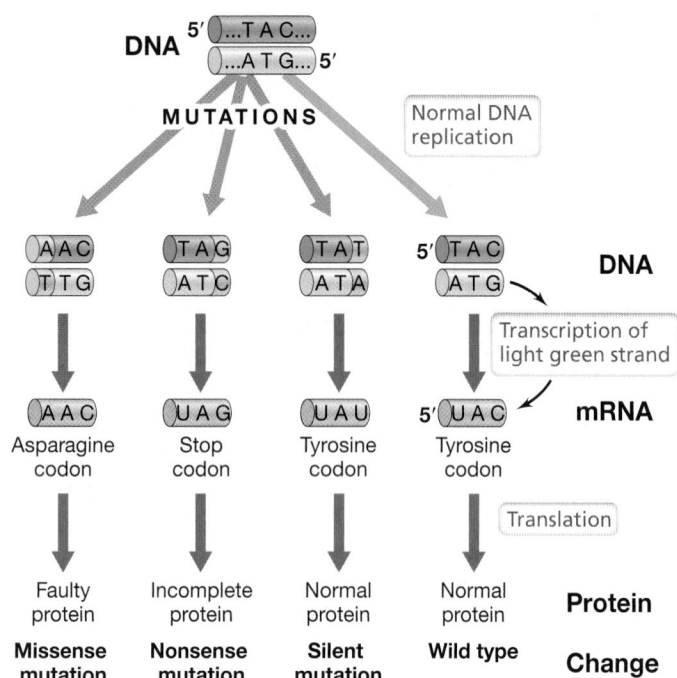

Figure 11.4 Possible effects of base-pair substitution in a gene encoding a protein. Three different protein products are possible from changes in the DNA for a single codon.

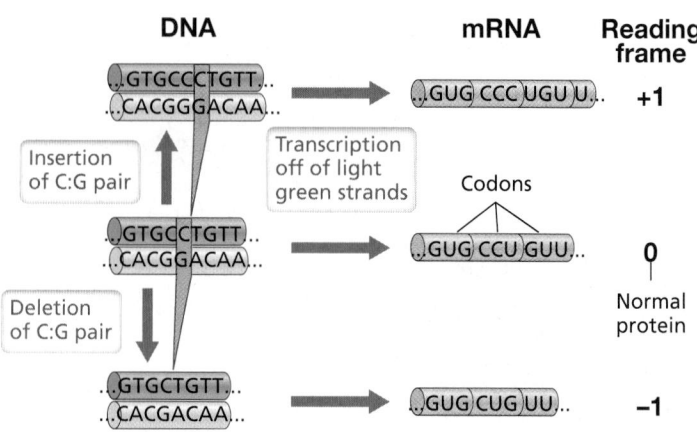

Figure 11.5 Shifts in the reading frame of mRNA caused by insertions or deletions. The reading frame in mRNA is established by the ribosome, which begins at the 5′ end (toward the left in the figure) and proceeds by units of three bases (codons). The normal reading frame is referred to as the 0 frame, that missing a base the −1 frame, and that with an extra base the +1 frame.

of amino acids) in the polypeptide has changed. If the change is at a critical location in the polypeptide chain, the protein could be inactive or have reduced activity. However, not all missense mutations necessarily lead to nonfunctional proteins. The outcome depends on where the substitution lies in the polypeptide chain and on how it affects protein folding and activity. For example, mutations in the active site of an enzyme are more likely to destroy activity than mutations in other regions of the protein.

Another possible outcome of a base-pair substitution is the formation of a nonsense (stop) codon. This results in premature termination of translation, leading to an incomplete polypeptide (Figure 11.4). Mutations of this type are called **nonsense mutations** because the change is from a sense (coding) codon to a nonsense codon (Table 4.4). Unless the nonsense mutation is very near the end of the gene, the product is considered *truncated* (incomplete). Truncated proteins are completely inactive or, at the very least, lack normal activity.

Other terms are occasionally used in microbial genetics to describe the precise type of base substitution in a point mutation. **Transitions** are mutations in which one purine base (A or G) is substituted for another purine, or one pyrimidine base (C or T) is substituted for another pyrimidine. **Transversions** are point mutations in which a purine base is substituted for a pyrimidine base, or vice versa.

Frameshifts and Other Insertions or Deletions

Because the genetic code is read from one end of the nucleic acid in consecutive blocks of three bases (codons), any deletion or insertion of a single base pair results in a shift in the reading frame. These **frameshift mutations** often have serious consequences.

Single base insertions or deletions change the primary sequence of the encoded polypeptide, typically in a major way (**Figure 11.5**). Such microinsertions or microdeletions can result from replication errors. Insertion or deletion of two base pairs also causes a frameshift. However, insertion or deletion of three base pairs does not cause a frameshift but does add or remove a codon; this results in the addition or deletion of a single amino acid in the polypeptide sequence. Although an amino acid addition or deletion may well be deleterious to protein function, it is usually not as bad as a frameshift, which scrambles the entire polypeptide sequence downstream of the mutation.

Insertions or deletions can also result in the gain or loss of hundreds or even thousands of base pairs. Such changes inevitably result in complete loss of gene function. Some deletions are so large that they may include several genes. If any of the deleted genes are essential, the mutation will be lethal. Such deletions cannot be restored through further mutations, but only through genetic recombination. Larger insertions and deletions may arise as a result of errors during genetic recombination (Section 11.5). In addition, many large insertion mutations are due to the insertion of specific identifiable DNA sequences called *transposable elements* (Section 11.11). The effect of transposable elements on the evolution of bacterial genomes was discussed in Section 9.6.

MINIQUIZ

- Do missense mutations occur in genes encoding tRNA? Why or why not?
- Why do frameshift mutations generally have more serious consequences than missense mutations?

11.3 Reversions and Mutation Rates

The rates at which different kinds of mutations occur vary widely. Some types of mutations occur so rarely that they are almost impossible to detect, whereas others occur so frequently that they present difficulties for an experimenter trying to maintain a genetically stable stock culture. Sometimes a second mutation can reverse the effect of an initial mutation. Furthermore, all organisms possess a variety of systems for DNA repair. Consequently, the observed mutation rate depends not only on the frequency of DNA changes but also on the efficiency of DNA repair.

Reversions (Back Mutations) and Suppressors

Point mutations are typically reversible, a process known as **reversion**. A revertant is a strain in which the original phenotype that was changed in the mutant is restored by a second mutation. Revertants can be of two types, same site or second site. In *same-site* revertants, the mutation that restores activity is at the same site as the original mutation. If the back mutation is not only at the same site but also restores the original sequence, it is called a *true* revertant.

In *second-site* revertants, the mutation is at a different site in the DNA. Second-site mutations can restore a wild-type phenotype if they function as *suppressor mutations*—mutations that compensate for the effect of the original mutation. Several classes of suppressor mutations are known. These include (1) a mutation somewhere else in the same gene that restores enzyme function, such as a second frameshift mutation near the first that restores the original reading frame; (2) a mutation in another gene that restores the function of the original mutated gene; and (3) a mutation in another gene that results in the production of an enzyme that can replace the nonfunctional one.

Suppressors can be best illustrated by mutations in tRNAs. Nonsense mutations can be suppressed by changing the anticodon sequence of a tRNA molecule so that it now recognizes a stop codon (**Figure 11.6**). Such an altered tRNA is called a *suppressor tRNA* and will insert the amino acid it carries at the stop codon that it now reads. Suppressor tRNA mutations would be lethal unless a cell has more than one tRNA for a particular codon. One tRNA may then be mutated into a suppressor, while the other performs the original function. Most cells have multiple tRNAs and so suppressor mutations are reasonably common, at least in microorganisms. Sometimes the amino acid inserted by the suppressor tRNA is identical to the original amino acid and the protein is fully active. In other cases, however, a different amino acid is inserted and only a partially active protein may be produced.

Mutation Rates

For most microorganisms, errors in DNA replication occur at a frequency of 10^{-6} to 10^{-7} per thousand bases during a single round of replication. A typical gene has about 1000 base pairs. Therefore, the frequency of a mutation *in a given gene* is also in the range of 10^{-6} to 10^{-7} per round of replication. For instance, in a bacterial culture having 10^8 cells/ml, there are likely to be a number of different mutants for any given gene in each milliliter of culture. Eukaryotes with very large genomes tend to have replication error rates about 10-fold lower than typical bacteria, whereas DNA viruses, especially those with very small genomes, may have error rates 100-fold to 1000-fold higher than those of cellular organisms. RNA viruses have even higher error rates due to less proofreading (↩ Section 4.4) and the lack of RNA repair mechanisms.

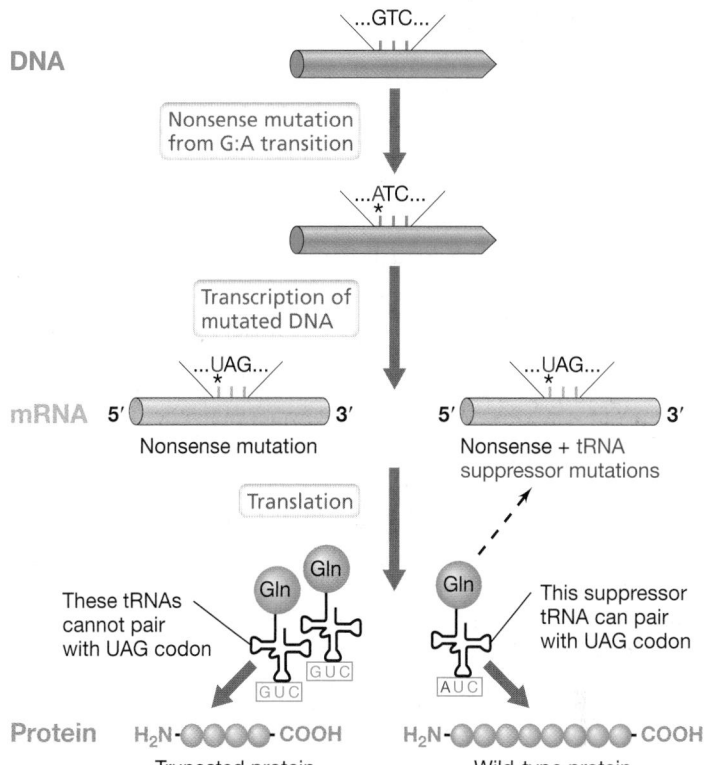

Figure 11.6 Suppression of nonsense mutations. Introduction of a nonsense mutation in a gene encoding a protein results in the incorporation of a stop codon (indicated by the *) in the corresponding mRNA. This single mutation leads to the production of a truncated polypeptide. The mutation is suppressed if a second mutation occurs in the anticodon of a tRNA, a tRNA charged with glutamine in this example, which allows the mutated tRNA or suppressor tRNA to bind to the nonsense codon.

Single base errors during DNA replication are more likely to lead to missense mutations than to nonsense mutations because most single base substitutions yield codons that encode other amino acids (↩ Table 4.4). The next most frequent type of codon change caused by a single base change leads to a silent mutation. This is because for the most part alternate codons for a given amino acid differ from each other by a single base change in the "silent" third position. A given codon can be changed to any of 27 other codons by a single base substitution, and on average, about two of these will be silent mutations, one a nonsense mutation, and the rest missense mutations.

Unless a mutation can be selected for, its experimental detection is difficult, and much of the skill of the microbial geneticist requires increasing the efficiency of mutation detection. This can be done most effectively by increasing the pool of mutations. As we see in the next section, it is possible to greatly increase the mutation rate by treatment with mutagenic agents. In addition, the mutation rate may change under certain circumstances, such as when cells are placed under high-stress conditions.

— MINIQUIZ —
- Why are suppressor tRNA mutations not lethal?
- Which class of mutation, missense or nonsense, is more common, and why?

11.4 Mutagenesis

The spontaneous rate of mutation is very low, but a variety of chemical, physical, and biological agents can increase the mutation rate and are therefore said to induce mutations. These agents are called **mutagens**, and we discuss some of the major categories of mutagens and their activities here.

Chemical Mutagens and Radiation

An overview of some of the major chemical mutagens and their modes of action is given in **Table 11.2**. Several classes of chemical mutagens exist. The *nucleotide base analogs* are molecules that resemble the purine and pyrimidine bases of DNA in structure yet display faulty base-pairing properties (**Figure 11.7**). If a base analog is incorporated into DNA in place of the natural base, the DNA may replicate normally most of the time. However, DNA replication errors occur at higher frequencies at these sites due to incorrect base pairing. The result is the incorporation of a mismatched base into the new strand of DNA and thus introduction of a mutation. During subsequent segregation of this strand in cell division, the mutation is revealed.

Other chemical mutagens induce *chemical modifications* in one base or another, resulting in faulty base pairing or related changes (Table 11.2). For example, alkylating agents (chemicals that react

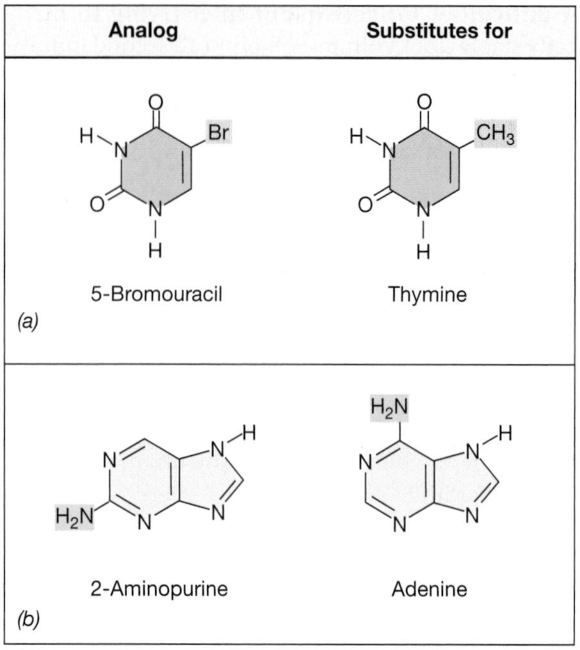

Figure 11.7 Nucleotide base analogs. Structure of two common nucleotide base analogs used to induce mutations and the normal nucleic acid bases for which they substitute. *(a)* 5-Bromouracil can base-pair with guanine, causing AT to GC substitutions. *(b)* 2-Aminopurine can base-pair with cytosine, causing AT to GC substitutions.

with amino, carboxyl, and hydroxyl groups by substituting them with alkyl groups) such as nitrosoguanidine are powerful mutagens and generally induce mutations at higher frequency than base analogs. Unlike base analogs, which have an effect only when incorporated during DNA replication, alkylating agents can introduce changes even in nonreplicating DNA. Both base analogs and alkylating agents tend to induce base-pair substitutions (Section 11.2).

Another group of chemical mutagens, the acridines, are planar molecules that function as *intercalating agents*. These mutagens become inserted between two DNA base pairs and push them apart. Then, during replication, this abnormal conformation can trigger single base insertions or deletions. Thus, acridines typically induce frameshift rather than point mutations (Section 11.2). Ethidium bromide, which is commonly used to detect DNA in gel electrophoresis, is also an intercalating agent and therefore a mutagen.

Nonionizing and *ionizing* radiation are two forms of electromagnetic radiation that are highly mutagenic (**Figure 11.8**). Ultraviolet (UV) radiation is widely used to generate mutations as the purine and pyrimidine bases of nucleic acids absorb UV radiation strongly (the absorption maximum for DNA and RNA is at 260 nm). The primary mutagenic effect is the production of *pyrimidine dimers*, in which two adjacent pyrimidine bases (cytosine or thymine) on the same strand of DNA become covalently bonded to one another. This either greatly impedes DNA polymerase activity or greatly increases the probability of DNA polymerase misreading the sequence at this point. Thus the killing of cells by UV radiation is due primarily to its effect on DNA. Conversely, ionizing radiation is more powerful than UV radiation and includes short-wavelength radiation such as X-rays, cosmic rays, and gamma rays

TABLE 11.2 Chemical and physical mutagens and their modes of action		
Agent	**Action**	**Result**
Base analogs		
5-Bromouracil	Incorporated like T; occasional faulty pairing with G	AT → GC and occasionally GC → AT
2-Aminopurine	Incorporated like A; faulty pairing with C	AT → GC and occasionally GC → AT
Chemicals reacting with DNA		
Nitrous acid (HNO$_2$)	Deaminates A and C	AT → GC and GC → AT
Hydroxylamine (NH$_2$OH)	Reacts with C	GC → AT
Alkylating agents		
Monofunctional (for example, ethyl methanesulfonate)	Puts methyl on G; faulty pairing with T	GC → AT
Bifunctional (for example, mitomycin, nitrogen mustards, nitrosoguanidine)	Cross-links DNA strands; faulty region excised by DNase	Both point mutations and deletions
Intercalating agents		
Acridines, ethidium bromide	Inserts between two base pairs	Microinsertions and microdeletions
Radiation		
Ultraviolet (UV)	Pyrimidine dimer formation	Repair may lead to error or deletion
Ionizing radiation (for example, X-rays)	Free-radical attack on DNA, breaking chain	Repair may lead to error or deletion

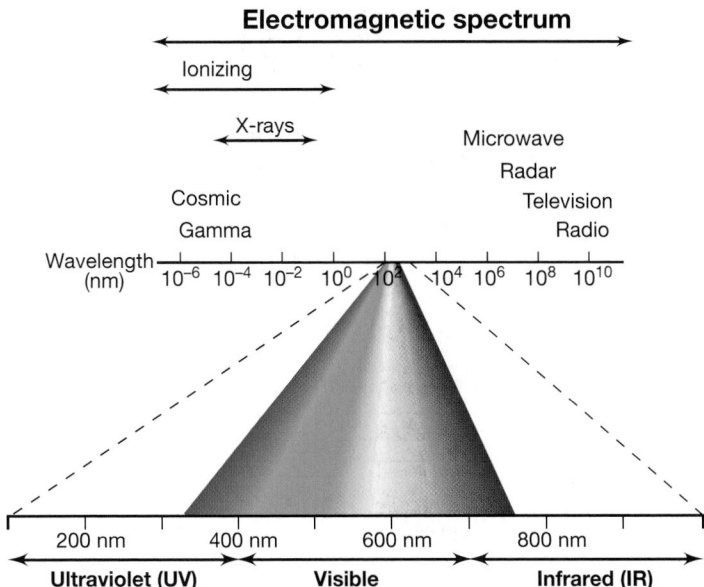

Electromagnetic spectrum

Figure 11.8 Wavelengths of radiation. Ultraviolet radiation consists of wavelengths just shorter than visible light. For any electromagnetic radiation, the shorter the wavelength, the higher the energy. DNA absorbs strongly at 260 nm.

(Figure 11.8). These rays cause water and other substances to ionize, resulting in the formation of free radicals such as the hydroxyl radical (OH·, ⮞ Section 5.14) that can damage macromolecules in the cell, including DNA. This causes double-stranded and single-stranded breaks that may lead to rearrangements or large deletions.

DNA Repair and the SOS System

By definition, a mutation is a *heritable* change in the genetic material. Therefore, if damaged DNA can be corrected before the cell divides, no mutation will occur. While cells have a variety of different DNA repair processes to correct mistakes (⮞ Section 4.4) or repair damage, some are error-prone and the repair process itself introduces the mutation. Some types of DNA damage, especially large-scale damage from highly mutagenic chemicals or large doses of radiation, may cause lesions that interfere with replication. If such lesions are not removed before replication occurs, DNA replication will stall and lethal breaks in the chromosome will result.

In *Bacteria*, stalled replication or major DNA damage activates the **SOS repair system**. The SOS system initiates a number of DNA repair processes, some of which are error-free. However, the SOS system also allows DNA repair to occur without a template, that is, with random incorporation of dNTPs. As might be expected, this results in many errors and hence many mutations. However, mutations may be less detrimental to cell survival than breaks in the chromosome, as mutations can often be corrected but chromosome breaks usually cannot.

In *Escherichia coli* the SOS repair system controls the transcription of approximately 40 genes located throughout the chromosome that participate in DNA damage tolerance and DNA repair (the SOS system thus forms a regulon, ⮞ Section 6.3). In DNA damage tolerance, DNA lesions remain in the DNA, but are

bypassed by specialized DNA polymerases that can move past DNA damage—a process known as *translesion synthesis*. Even if no template is available to allow insertion of the correct bases, it is less dangerous to cell survival in the long run to fill the gap than to let it remain. Consequently, translesion synthesis generates many errors. In *E. coli*, in which the process of mutagenesis has been studied in great detail, the two error-prone repair polymerases are DNA polymerase V, an enzyme encoded by the *umuCD* genes, and DNA polymerase IV, encoded by *dinB* (**Figure 11.9**). Both are induced as part of the SOS repair system.

The master regulators of the SOS system are the proteins LexA and RecA. LexA is a repressor that normally prevents expression of the SOS system. The RecA protein, which normally functions in genetic recombination (Section 11.5), is activated by the presence of DNA damage, in particular by the single-stranded DNA that results when replication stalls. The activated form of RecA then stimulates LexA to inactivate itself by self-cleavage. This leads to derepression of the SOS system and the coordinate expression of proteins that participate in DNA repair. Because some of the DNA repair mechanisms of the SOS system—such as DNA polymerases IV and V—are inherently error-prone, many mutations arise. However, once the original DNA damage has been repaired, the SOS regulon is repressed and further mutagenesis ceases.

— **MINIQUIZ** —

- How do mutagens cause mutations?
- What is meant by "error-prone" DNA repair?

II • Gene Transfer in *Bacteria*

Comparative genomic analyses of closely related microbes that exhibit different phenotypes have revealed distinct genome differences. Often these idiosyncratic differences result from *horizontal gene transfer*, the movement of genes between cells that are not direct descendants of one another (⮞ Section 9.6). Horizontal gene transfer allows cells to quickly acquire new characteristics and fuels metabolic diversity.

Three mechanisms of genetic exchange are known in bacteria: (1) *transformation*, in which free DNA released from one cell is taken up by another (Section 11.6); (2) *transduction*, in which DNA transfer is mediated by a virus (Section 11.7); and (3) *conjugation*, in which DNA transfer requires cell-to-cell contact and a conjugative plasmid in the donor cell (Sections 11.8 and 11.9). These processes are contrasted in **Figure 11.10**, and it should be noted that DNA transfer typically occurs in only one direction, *from donor to recipient*.

Before discussing the mechanisms of transfer, we consider the fate of transferred DNA. Regardless of how it was transferred, DNA that enters the cell by horizontal gene transfer faces three possible fates: (1) It may be degraded by the recipient cell's restriction enzymes or other DNA destruction systems (Section 11.12); (2) it may replicate by itself (but only if it possesses its own origin of replication, such as a plasmid or phage genome); or (3) it may recombine with the recipient cell's chromosome.

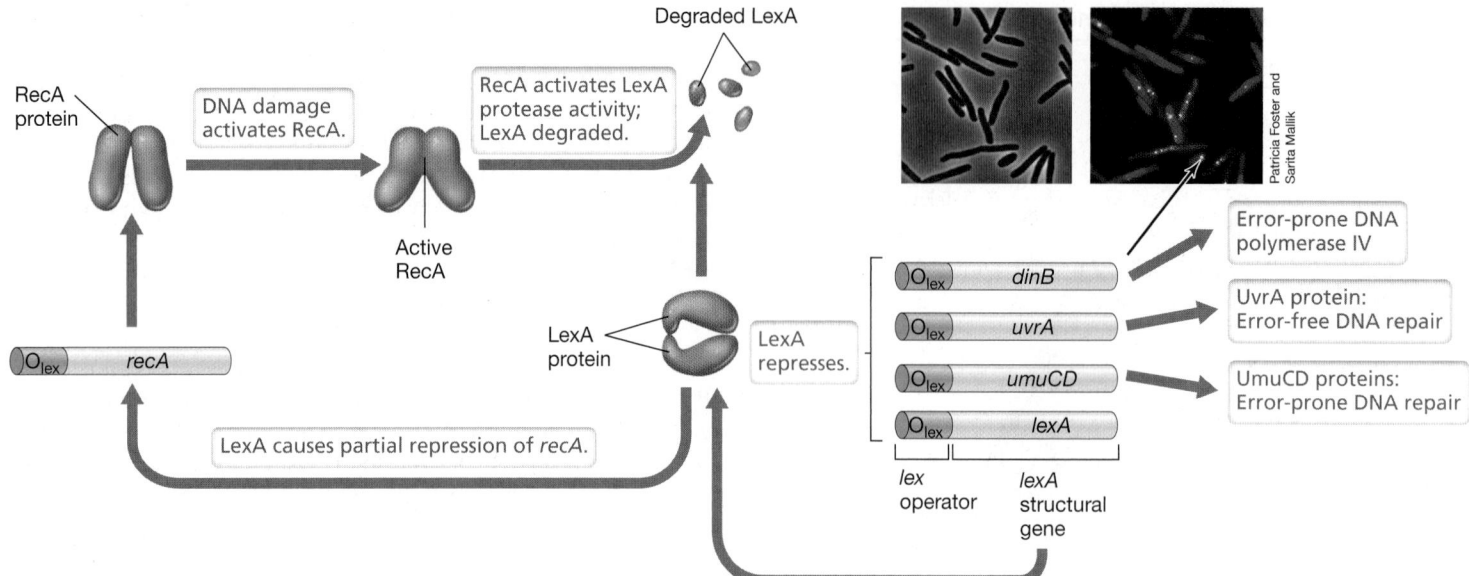

Figure 11.9 SOS response to DNA damage. DNA damage activates RecA protein, which in turn activates the protease activity of LexA, resulting in self-cleavage. LexA normally represses the activities of *recA*, the DNA repair genes *uvrA*, *umuCD* (the UmuCD proteins are part of DNA polymerase V), and *dinB*, which encodes DNA polymerase IV. However, repression is not complete. Some

RecA protein is produced even in the presence of LexA protein. When LexA is inactivated, DNA repair genes become highly active. Inset photos: Both photos show DNA polymerase IV localization to the nucleoid during the SOS response in *Escherichia coli*. Cells containing a fluorescently tagged DNA polymerase IV (DinB) were treated with an antibiotic to induce DNA damage.

Left: phase-contrast micrograph. Right: fluorescence micrograph showing cells stained with DAPI (blue) and DNA polymerase IV (yellow, in the nucleoid region). Expression of *dinB* requires not only the loss of LexA repression but also the protein RpoS, an RNA polymerase sigma factor whose synthesis is triggered by various stress responses.

11.5 Genetic Recombination

Recombination is the physical exchange of DNA between *genetic elements* (structures that carry genetic information). Here we focus on *homologous* recombination, a process that results in genetic exchange between homologous DNA sequences from two different sources. Homologous DNA sequences are those that have nearly the same sequence; therefore, bases can pair over an extended length of the two DNA molecules to facilitate exchange. This type of recombination drives the process of "crossing over" in classical genetics.

Molecular Events in Homologous Recombination

The RecA protein, previously mentioned in regard to the SOS repair system (Section 11.4 and Figure 11.9), is the key to homologous recombination. RecA is essential in nearly every homologous recombination pathway. RecA-like proteins have been identified in all *Bacteria* examined, as well as in the *Archaea* and most *Eukarya*.

A molecular mechanism for homologous recombination between two DNA molecules is shown in **Figure 11.11**. An enzyme that cuts DNA in the middle of a strand, called an *endonuclease*, begins the process by nicking one strand of the donor DNA molecule. This nicked strand is separated from the other strand by proteins with helicase activity; the resulting single-stranded segment binds single-strand binding protein (⮌ Section 4.3) and then RecA. This results in a complex that promotes base pairing with the complementary sequence in the recipient DNA molecule. Base pairing, in turn, displaces the other strand of the recipient DNA molecule (Figure 11.11) and is appropriately called *strand invasion*.

The base pairing of one strand from each of the two DNA molecules over long stretches generates recombination intermediates containing long **heteroduplex** regions, where each strand has originated from a different chromosome. The linked molecules are then resolved (separated) by enzymes that cut and rejoin the previously unbroken strands of both original DNA molecules. Depending on the orientation of the junction during resolution, two types of products, referred to as "patches" or "splices," are formed that differ in the conformation of the heteroduplex regions remaining after resolution (Figure 11.11).

Effect of Homologous Recombination on Genotype

For homologous recombination to generate new genotypes, the two homologous sequences must be related but genetically distinct. This is obviously the case in a diploid eukaryotic cell, which has two sets of chromosomes, one from each parent. In bacteria, genetically distinct but homologous DNA molecules are brought together in different ways. Genetic recombination in bacteria occurs after fragments of homologous DNA from a donor chromosome are transferred to a recipient cell by transformation, transduction, or conjugation. It is only after the transfer event, when the DNA fragment from the donor is in the recipient cell, that homologous recombination occurs.

For physical exchange of DNA segments to be detected, the cells resulting from recombination must be phenotypically different from both parents (**Figure 11.12**). Genetic crosses in bacteria usually depend on using recipient strains that lack some selectable character that the recombinants will gain. The recipient may be unable to grow on a particular medium or may exhibit a specific phenotype, while the genetic recombinants can grow on a particular

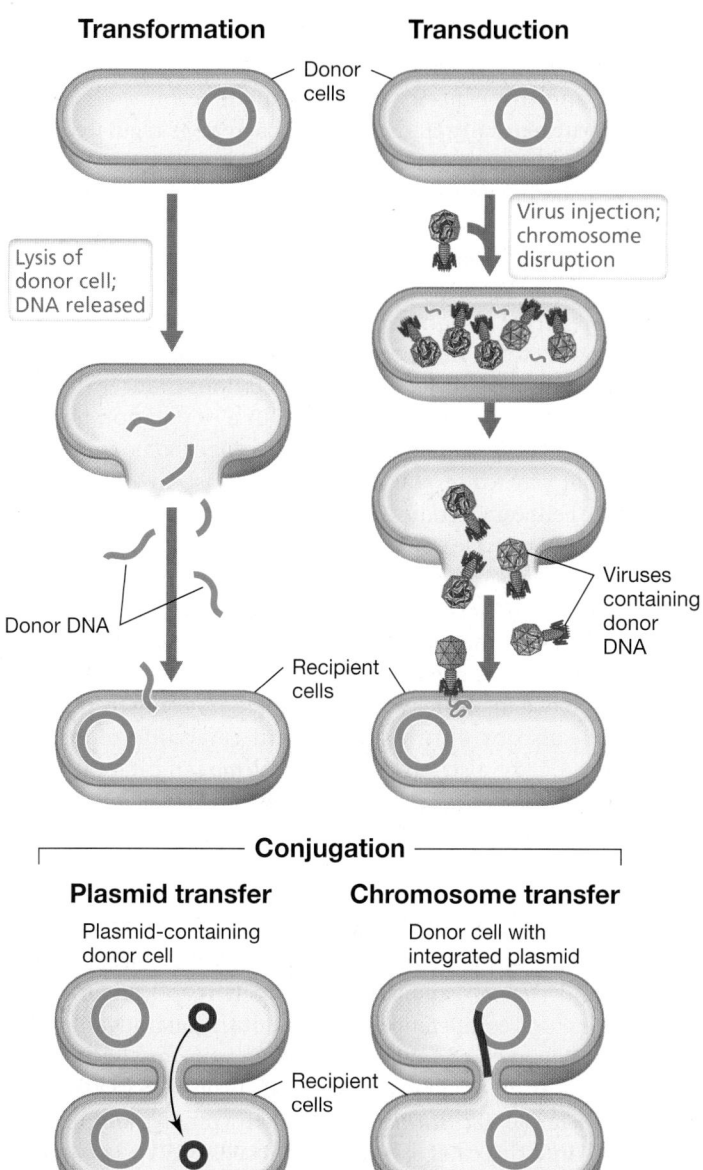

Figure 11.10 Processes by which DNA is transferred from donor to recipient bacterial cell. Just the initial steps in transfer are shown.

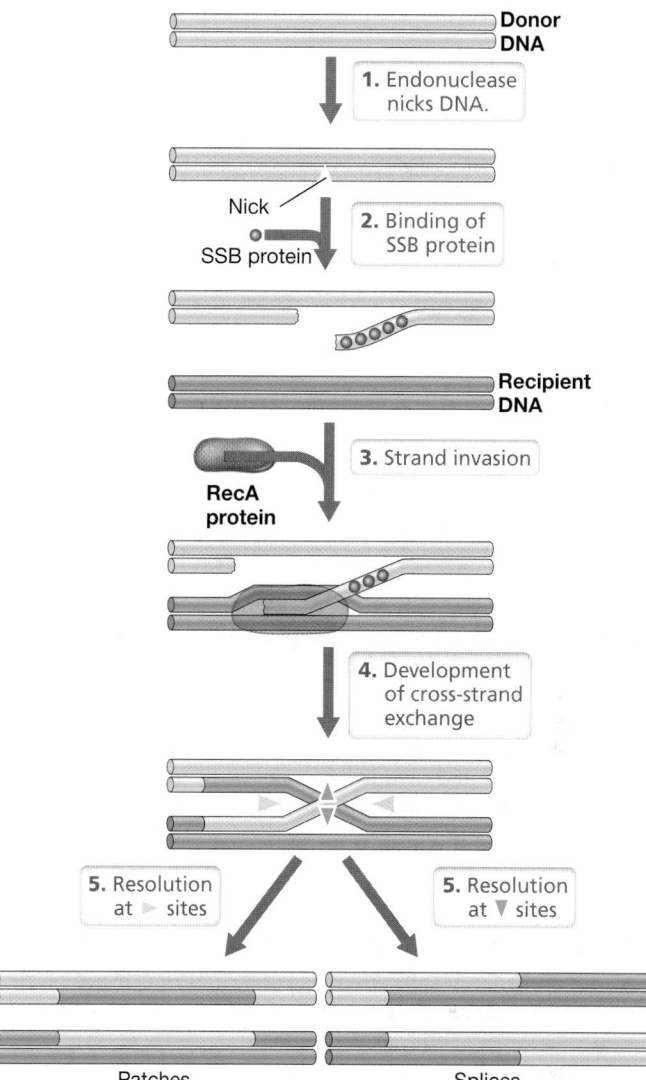

Figure 11.11 A simplified version of homologous recombination.
Homologous DNA molecules pair and exchange DNA segments. The mechanism requires breakage and reunion of paired segments. Two of the participating proteins, single-strand binding (SSB) protein and the RecA protein, are shown. The other participating proteins are not shown. The diagram is not to scale: Pairing may occur over hundreds or thousands of bases. Resolution occurs by cutting and rejoining the cross-linked DNA molecules. Note that there are two possible outcomes, patches or splices, depending on where strands are cut during the resolution process.

medium or exhibit a phenotype different from the recipient (Figures 11.1 and 11.12). Various kinds of selectable markers, such as drug resistance and nutritional requirements, were discussed in Section 11.1. The exceedingly great sensitivity of the selection process allows even a few recombinant cells to be detected in a large population of nonrecombinant cells, and thus selection is an important tool for the microbial geneticist.

Complementation

In all three methods of bacterial gene transfer, only a portion of the donor chromosome enters the recipient cell and thus transfer is just the first step; unless recombination takes place with the recipient chromosome, the donor DNA will be lost because it cannot replicate independently in the recipient. Nonetheless, it is possible to stably maintain a state of partial diploidy for use in

bacterial genetic analysis. A bacterial strain that carries *two copies* of any particular chromosomal segment is known as a partial diploid, or *merodiploid*. In general, one copy is present on the chromosome itself and the second copy on another genetic element, such as a plasmid or a bacteriophage.

Consequently, if the chromosomal copy of a gene is defective due to a mutation, it is possible to supply a functional (wild-type) copy of the gene on a plasmid or bacteriophage. For example, if one of the genes for biosynthesis of the amino acid tryptophan has a mutation resulting in a nonfunctional enzyme, this will yield a Trp⁻ phenotype. That is, the mutant strain will be a tryptophan auxotroph and must be supplied with tryptophan for growth. However, if a copy of the wild-type gene is introduced

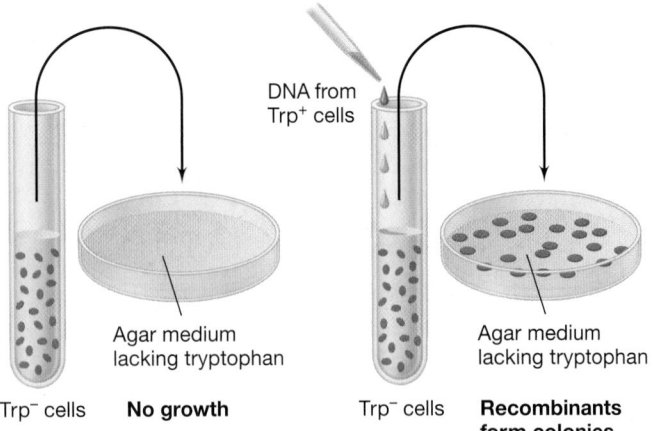

Figure 11.12 Using a selective medium to detect rare genetic recombinants. On the selective medium only the rare recombinants form colonies even though a very large population of bacteria was plated. Procedures such as this, which offer high resolution for genetic analyses, can ordinarily be used only with microorganisms. The type of genetic exchange being illustrated is transformation, but a similar outcome could result from any of the other forms of horizontal gene transfer.

into the same cell on a plasmid or viral genome, this gene will encode the necessary protein and, assuming the gene is transcribed and translated, will restore the wild-type phenotype. This process is called *complementation* because the wild-type gene is said to *complement* the mutation, in this case converting the Trp⁻ cell into Trp⁺ (Figure 11.12).

MINIQUIZ

- Which protein, found in virtually all cells, facilitates the pairing required for homologous recombination?
- Explain the fate of transferred chromosomal DNA if recombination does not occur in the recipient cell.
- What is a merodiploid, and what is genetic complementation?

11.6 Transformation

Transformation is a genetic transfer process by which *free DNA* is incorporated into a recipient cell and brings about genetic change. Several organisms are naturally transformable, including certain species of both gram-negative and gram-positive *Bacteria* and also some species of *Archaea* (Section 11.10). Because the DNA in prokaryotic cells is present as a large single molecule, when a cell is gently lysed, its DNA pours out.

Bacterial chromosomes break easily because of their extreme length (if linearized, the *Bacillus subtilis* chromosome would be 1700 μm long). Even after gentle extraction, the *B. subtilis* chromosome of 4.2 megabase pairs is converted to fragments of about 10 kilobase pairs each. Because an average gene contains about 1000 nucleotides, each of the fragments of *B. subtilis* DNA contains about ten genes. This is a typical transformable size. A single cell typically incorporates only one or at most a few DNA fragments, so only a small proportion of the genes of one cell can be transferred to another in a single transformation event.

Competence in Transformation

A cell that is able to take up DNA and be transformed is said to be *competent*, and this capacity is genetically determined. Competence in most naturally transformable bacteria is regulated, and special proteins play a role in the uptake and processing of DNA. These competence-specific proteins include a membrane-associated DNA-binding protein, a cell wall autolysin, and various nucleases. One pathway of natural competence in *B. subtilis*—an easily transformed species—is regulated by quorum sensing, a regulatory system that responds to cell density (⇄ Section 6.8). Cells produce and excrete a small peptide during growth, and the accumulation of this peptide to high concentrations induces the cells to become competent. But not all cells in a population become competent. In *Bacillus*, roughly 20% of the cells become competent and stay that way for several hours. By contrast, in *Streptococcus*, 100% of the cells can become competent, but only for a brief period during the growth cycle.

Multiple layers of regulation control natural competence in other bacteria. *Vibrio cholerae* (the causative agent of cholera) is naturally found in marine and freshwater environments associated with crustacean exoskeletons, which are composed of chitin. Competence in *V. cholerae* is controlled not only by quorum sensing but also by chitin sensing and catabolite repression (⇄ Section 6.4; see also page 306). *V. cholerae* can catabolize chitin (a polymer of *N*-acetylglucosamine and an abundant nutrient in the marine environment), and as cells aggregate on a chitin surface, they are in close proximity to one another and more likely to successfully exchange DNA (**Figure 11.13**).

High-efficiency, natural transformation is rare among *Bacteria*. For example, *Acinetobacter*, *Bacillus*, *Streptococcus*, *Haemophilus*, *Neisseria*, and *Thermus* are naturally competent and easy to transform. This natural competence provides a nutritional advantage, as free DNA is rich in carbon, nitrogen, and phosphorus. By contrast, many *Bacteria* are poorly transformed, if at all, under natural conditions. For example, *Escherichia coli* and many other gram-negative bacteria fall into this category. However, if cells of *E. coli* are treated with high concentrations of Ca^{2+} and then chilled, they become adequately competent. Cells treated in this manner take up double-stranded DNA, and therefore transformation of *E. coli* by plasmid DNA can be relatively efficient. This is important because getting DNA into *E. coli*—the workhorse of genetic engineering—is critical for biotechnology, as we will see in Chapter 12.

Electroporation is a physical technique that is used to get DNA into organisms that are difficult or impossible to transform, especially cells that contain thick cell walls. In electroporation, cells are mixed with DNA and then exposed to brief, high-voltage electrical pulses. This makes the cell envelope permeable and allows entry of the DNA. Electroporation works for getting free DNA into most types of cells, including *E. coli*, most other *Bacteria*, some species of *Archaea*, and even yeast and certain plant cells.

Uptake and Integration of DNA in Transformation

During natural transformation (**Figure 11.14**), competent bacteria reversibly bind DNA. Soon, however, the binding becomes irreversible. Competent cells bind much more DNA than do noncompetent cells—as much as 1000 times more. As noted earlier, the

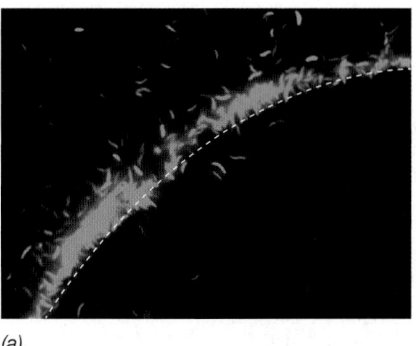

(a)

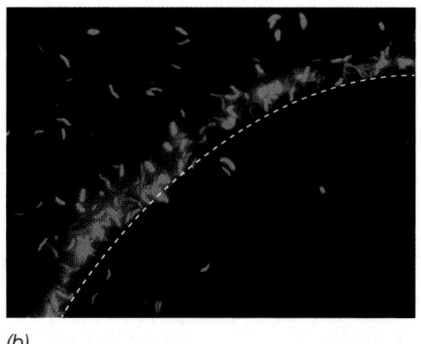

(b)

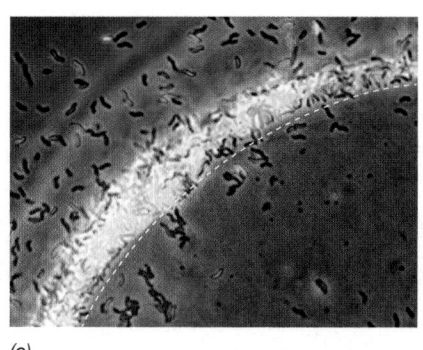

(c)

Figure 11.13 Regulation of natural competence in *Vibrio cholerae*. *V. cholerae* cells with fluorescent reporter genes linked to the promoters of competence genes were grown in the presence of chitin beads. White dashed line indicates edge of bead surface. (a) Cells with the *pilA* (pilus protein) promoter linked to a green fluorescent protein (GFP). (b) Cells with the *comEA* (DNA uptake proteins) promoter linked to a red fluorescent protein. (c) Merged image of parts *a* and *b* illustrating expression of competence genes in cells associated with the chitin bead. Cells of *V. cholerae* are about 0.5 μm wide and 1.5 μm long. Adapted from Lo Scrudato, M., and M. Blokesch. 2012. *PLoS Genetics* 8(6): e1002778. See page 306 for a different view of transformation in *V. cholerae*.

Melanie Blokesch

sizes of the transforming fragments are much smaller than that of the whole genome, and the fragments are further degraded during the uptake process. In *Streptococcus pneumoniae* (the cause of bacterial pneumonia) each cell can bind only about ten molecules of double-stranded DNA of 10–15 kbp each. However, as these fragments are taken up, they are converted into single-stranded pieces of about 8 kb, with the complementary strand being degraded. The DNA fragments in the mixture compete with each other for uptake and thus the probability of a transformant taking up DNA that confers an advantage or a selectable marker decreases.

During transformation, DNA is bound at the cell surface by a DNA-binding protein (Figure 11.14). In many species, this DNA-binding protein resembles a pilus (⟜ Section 2.7) that is able to pull the DNA into the periplasm of a gram-negative bacterium or through the thick cell wall of a gram-positive bacterium. Next, either the entire double-stranded fragment is taken up, or a nuclease degrades one strand and the remaining strand is taken up, depending on the organism. After uptake, a competence-specific protein binds the donor DNA. This protects the DNA from nuclease attack until it reaches the recipient's chromosome, where the

RecA protein takes over. The DNA is integrated into the genome of the recipient by recombination (Figures 11.11 and 11.14). The preceding applies only to small pieces of *linear* DNA. Many naturally transformable *Bacteria* are transformed only poorly by plasmid DNA because the plasmid must remain double-stranded and circular in order to replicate.

MINIQUIZ

- During transformation a cell usually incorporates only one or a few fragments of DNA. Explain.
- In genetic transformation, what is meant by the word competence?

11.7 Transduction

In **transduction**, a bacterial virus (bacteriophage) transfers DNA from one cell to another. Viruses can transfer host genes in two ways. In the first, called *generalized transduction*, DNA

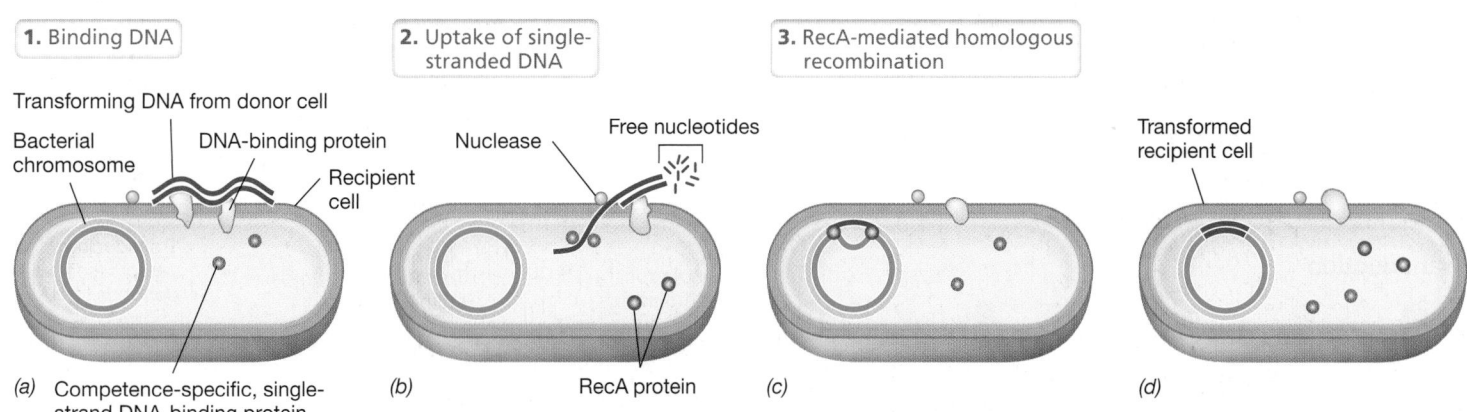

1. Binding DNA

2. Uptake of single-stranded DNA

3. RecA-mediated homologous recombination

Transforming DNA from donor cell

Bacterial chromosome   DNA-binding protein

Recipient cell

(a) Competence-specific, single-strand DNA-binding protein

Nuclease   Free nucleotides

(b)   RecA protein

(c)

Transformed recipient cell

(d)

Figure 11.14 Mechanism of transformation in a gram-positive bacterium. (a) Binding of double-stranded DNA by a membrane-bound DNA-binding protein. (b) Passage of one of the two strands into the cell while nuclease activity degrades the other strand. (c) The single strand of DNA in the cell is bound by specific proteins, and recombination with homologous regions of the bacterial chromosome is mediated by RecA protein. (d) Transformed cell.

derived from virtually any portion of the host genome is packaged inside the mature virion in place of the virus genome. In the second, called *specialized transduction*, DNA from a specific region of the host chromosome is integrated directly into the virus genome—usually replacing some of the virus genes. This occurs only with certain temperate viruses such as phage lambda (👁 Section 8.7).

In generalized transduction, the bacterial donor genes cannot replicate independently and are not part of a viral genome. Thus, unless the donor genes recombine with the recipient bacterial chromosome, they will be lost. In specialized transduction, homologous recombination may also occur. However, since the donor bacterial DNA is actually a part of a temperate phage genome, it may be integrated into the host chromosome during lysogeny (👁 Section 8.7).

Transduction occurs in a variety of *Bacteria*, including the genera *Desulfovibrio, Escherichia, Pseudomonas, Rhodococcus, Rhodobacter, Salmonella, Staphylococcus*, and *Xanthobacter*, as well as *Methanothermobacter thermautotrophicus*, a species of *Archaea*. Not all phages can transduce and not all bacteria are transducible, but with bacteriophages estimated to outnumber prokaryotic cells in nature by 10-fold, transduction likely plays an important role in gene transfer in the environment. Some examples of genes transferred by transducing bacteriophages include multiple-antibiotic-resistance genes among strains of *Salmonella enterica* (*typhimurium*), Shiga-like toxin genes in *Escherichia coli*, virulence factors in *Vibrio cholerae*, and genes encoding photosynthetic proteins in cyanobacteria (👁 Section 10.12).

Generalized Transduction

In generalized transduction, virtually any gene on the donor chromosome can be transferred to the recipient, forming a *transductant*. Generalized transduction was first discovered and extensively studied in the bacterium *S. enterica* with phage P22 and has also

been studied with phage P1 in *E. coli*. The mechanism of transduction is shown in **Figure 11.15**. When a bacterial cell is infected with a transducing phage, the lytic cycle may occur. However, during lytic infection, the enzymes responsible for packaging viral DNA into the bacteriophage sometimes package host DNA accidentally. The result is called a *transducing particle*. These cannot lead to a viral lytic infection because they contain no viral DNA, and are therefore said to be *defective*.

Upon lysis of the cell, transducing particles are released along with normal virions that contain the virus genome. When this lysate is used to infect a population of recipient cells, most of the cells are infected with a normal (lytic) virus. However, a small proportion of the population receives transducing particles that inject the DNA they packaged from the previous host bacterium. Although this DNA cannot replicate, it can recombine with the DNA (Section 11.5) of the new host (**Figure 11.16**). Because only a small proportion of the particles in the lysate are defective, and each of these contains only a small fragment of donor DNA, the probability of a given transducing particle containing a particular gene is quite low. Typically, only about 1 cell in 10^6 to 10^8 cells is transduced for any given gene.

Lysogeny and Specialized Transduction

Generalized transduction allows the transfer of any gene from one bacterium to another, but at a low frequency. In contrast, specialized transduction allows extremely efficient transfer but is selective and transfers only a small region of the bacterial chromosome. In the first case of specialized transduction to be discovered, genes for galactose catabolism were transduced by the temperate phage lambda of *E. coli*.

When lambda lysogenizes a host cell, the phage genome is integrated into the *E. coli* chromosome at a specific site (👁 Section 8.7). This site is next to the cluster of genes that encode the enzymes for galactose utilization. After insertion, viral DNA replication

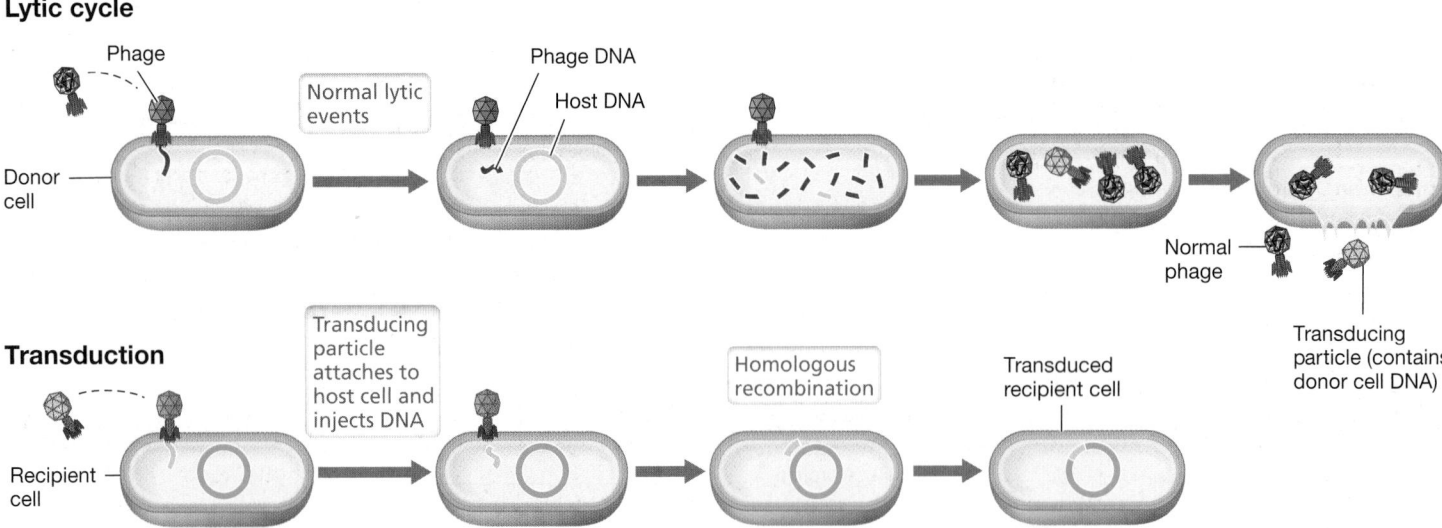

Figure 11.15 Generalized transduction. Note that "normal" virions contain phage genes, whereas a transducing particle contains host genes.

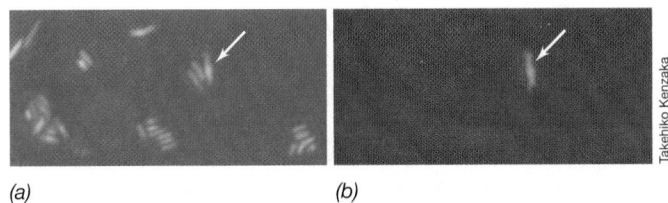

(a)   *(b)*

Takehiko Kenzaka

Figure 11.16 Visualization of generalized transduction. *Citrobacter freundii* cells were mixed with P1 bacteriophage carrying the β-lactamase (*bla*) gene for 10 minutes. *(a)* Fluorescence micrograph showing cells by DAPI staining. *(b)* Detection of a single *C. freundii* transductant containing the *bla* gene recombined into the genome using cycling primed in situ amplification–FISH, a modified version of fluorescence in situ hybridization (FISH, Ⳏ Section 19.5). Arrow indicates the cell that is transduced.

is under control of the bacterial host chromosome. Upon induction, the viral DNA separates from the host DNA by a process that is the reverse of integration (**Figure 11.17**). Usually the lambda DNA is excised precisely, but occasionally the phage genome is excised incorrectly. Some of the adjacent bacterial genes to one side of the prophage (for example, the galactose operon) are excised along with phage DNA. At the same time, some phage genes are left behind (Figure 11.17*b*). This transducing particle can subsequently transfer genes for galactose utilization to a recipient cell. This transfer can only be detected if a galactose-negative (Gal⁻) strain is infected with such a transducing particle and Gal⁺ transductants are selected.

For a lambda virion to be infectious, there is a limit to the amount of phage DNA that can be replaced with host DNA. Sufficient phage DNA must be retained to encode the phage protein coat and other phage proteins needed for lysis and lysogeny. However, if a helper phage is used together with a defective phage in a *mixed infection*, then far fewer phage-specific genes are needed in the defective phage. Only the *att* (attachment) region, the *cos* site (cohesive ends, for packaging), and the replication origin of the lambda genome (Ⳏ Figure 8.17*b*) are necessary.

Phage Conversion

Alteration of the phenotype of a host cell by lysogenization is called *phage conversion*. When a normal (that is, nondefective) temperate phage lysogenizes a cell and becomes a prophage, the cell becomes immune to further infection by the same type of phage. Such immunity may itself be regarded as a change in phenotype. However, other phenotypic changes unrelated to phage immunity are often observed in phage conversion of lysogenized cells.

Two cases of phage conversion have been especially well studied. One results in a change in structure of a polysaccharide on the cell surface of *S. enterica* (*anatum*) upon lysogenization with bacteriophage ε¹⁵. The second results in the conversion of non-toxin-producing strains of *Corynebacterium diphtheriae* (the bacterium that causes the disease diphtheria) to toxin-producing (pathogenic) strains following lysogeny with bacteriophage β (Ⳏ Section 30.3). In both cases, the genes responsible for the changes are an integral part of the phage genome and hence are automatically transferred to the cell upon phage infection and establishment of the lysogenic state.

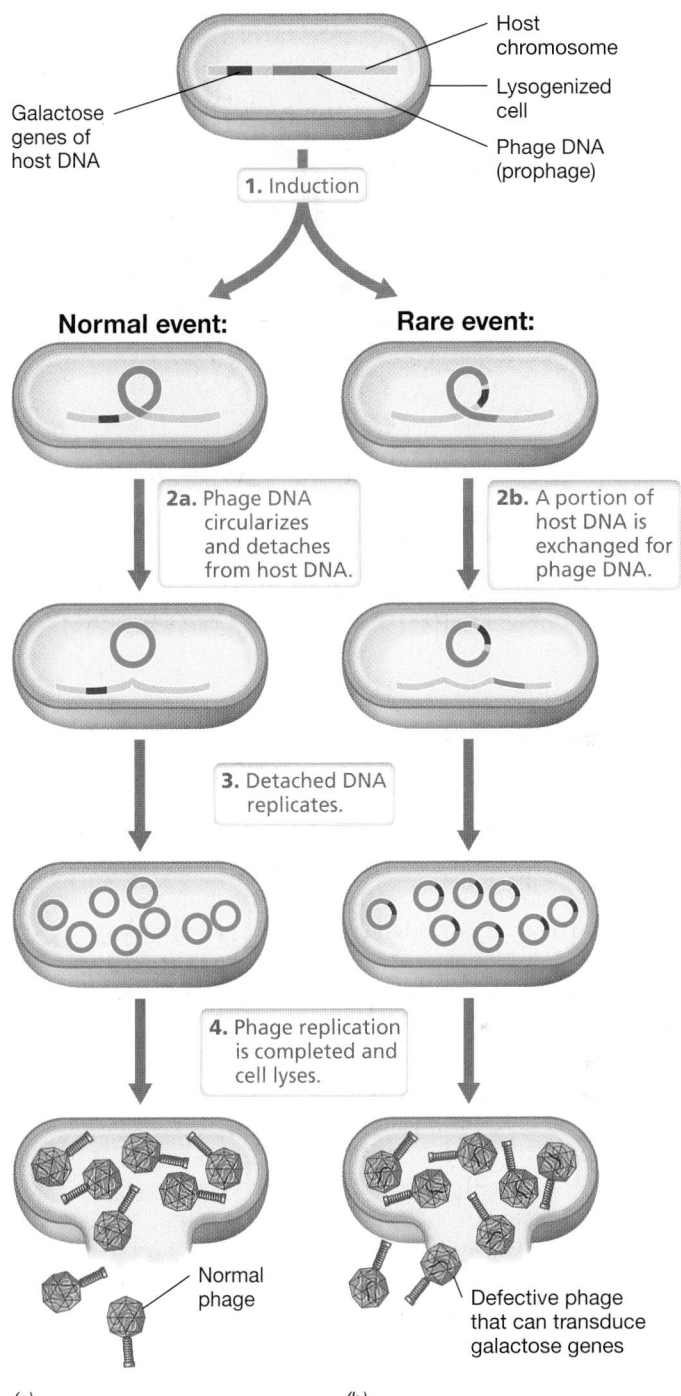

(a)   *(b)*

Figure 11.17 Specialized transduction. In an *Escherichia coli* cell containing a lambda prophage, *(a)* normal lytic events and *(b)* the production of particles transducing the galactose genes. Only a short region of the circular host chromosome is shown in the figure.

Lysogeny likely carries strong selective value for the host cell because it confers resistance to infection by viruses of the same type. Phage conversion may also be of considerable evolutionary significance because it results in genetic alteration of host cells. It has been found that many bacteria isolated from nature are natural lysogens, and thus it is likely that lysogeny is essential for survival of many host cells in nature.

UNIT 3

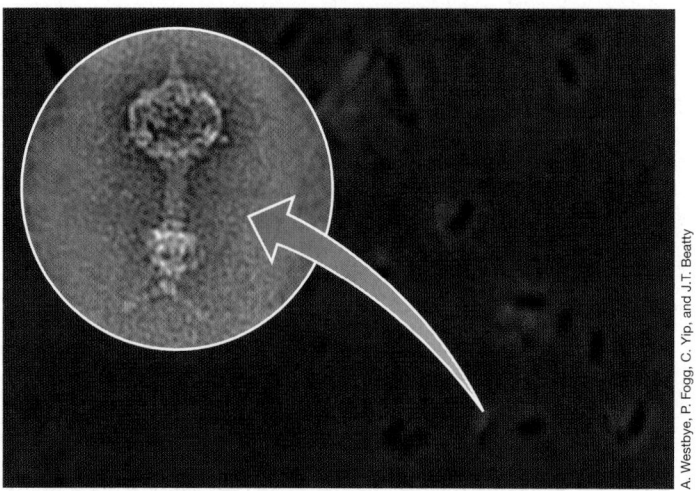

A. Westbye, P. Fogg, C. Yip, and J.T. Beatty

Figure 11.18 Gene transfer agents. Inset: Transmission electron micrograph of a gene transfer agent (GTA) isolated from *Rhodobacter capsulatus.* Visualization of a subset of *R. capsulatus* cells producing and releasing GTAs during stationary phase using a red fluorescent reporter gene linked to the promoter of an *R. capsulatus* gene essential for GTA production.

Gene Transfer Agents

DNA can also be transferred between prokaryotic cells by defective bacteriophages. These so-called *gene transfer agents* (*GTAs*) are the result of prokaryotic cells hijacking defective viruses and using them specifically for DNA exchange (**Figure 11.18**). GTAs resemble tiny tailed bacteriophages and contain random small pieces of *host* DNA. They are not considered true viruses because they do not contain genes encoding their own production and do not produce characteristic viral plaques. Instead, the genes encoding GTAs lie within the genome of the cell that produces them.

GTAs have been isolated from a wide range of *Bacteria* including the sulfate-reducing bacterium *Desulfovibrio desulfuricans* and a variety of anoxygenic purple bacteria and other *Alphaproteobacteria*, and also from certain methanogenic *Archaea*. GTAs seem to be particularly widely produced by marine *Bacteria*, especially purple bacteria such as *Roseovarius*. Exactly what triggers GTA synthesis has just begun to be studied and may vary in different species. However, by linking the promoter of a gene essential for GTA production to a reporter gene, microbial geneticists have determined that a subpopulation of cells in a culture of the anoxygenic phototrophic bacterium *Rhodobacter capsulatus* produce and release GTAs during stationary phase and nutrient fluctuations (Figure 11.18). This suggests that GTAs may have evolved as a mechanism for a cell to disperse its genes into the environment in a protected form before cell lysis released free DNA that could be quickly degraded.

While bacteriophages are considered the most abundant microbes on Earth, the percentage of these that are actually GTAs instead of viruses is unknown but could be significant. The fact that GTAs are produced by so many different species, do not result in cell lysis, and can transfer genes between bacteria that are taxonomically distinct, points to GTAs as major vehicles for gene flow between prokaryotic cells in nature. This could be especially true of open-ocean microbial communities, where

constant low nutrient levels might trigger GTA production as a means for cells to scavenge each other's genes for improving fitness and survival.

MINIQUIZ

- How does a transducing particle differ from an infectious bacteriophage?
- What is the major difference between generalized transduction and transformation?
- Why is phage conversion considered beneficial to host cells?

11.8 Conjugation

Conjugation is a form of horizontal gene transfer in both gram-negative and gram-positive bacteria that requires cell-to-cell contact (mating). Conjugation is a plasmid-encoded mechanism that can mediate DNA transfer between closely related cells or between more distantly related cells; for example, between cells of different genera. Conjugative plasmids use this mechanism to transfer copies of themselves and the genes they encode, such as those for antibiotic resistance, to new host cells.

The process of conjugation requires a *donor* cell, which contains the conjugative plasmid, and a *recipient* cell, which does not. In addition, genetic elements that cannot transfer themselves can sometimes be *mobilized* or transferred during conjugation. These other genetic elements can be other plasmids or the host chromosome itself. Indeed, conjugation was discovered because the F plasmid of *Escherichia coli* can mobilize the host chromosome (see Figure 11.24). Transfer mechanisms may differ depending on the participating plasmid, but most plasmids in gram-negative *Bacteria* employ a mechanism similar to that used by the F plasmid.

F Plasmid

The F plasmid (F stands for "fertility") is a circular DNA molecule of 99,159 bp. **Figure 11.19** shows a genetic map of the F plasmid. One region of the plasmid contains genes that regulate DNA replication. It also contains a number of transposable elements (Section 11.11) that allow the plasmid to integrate into the host chromosome. In addition, the F plasmid has a large region of DNA, the *tra* region, containing genes that encode transfer functions. Many genes in the *tra* region participate in mating pair formation, and most of these have to do with the synthesis of the sex pilus (⟜ Section 2.7) and a type IV secretion system (⟜ Section 4.13) to transfer the DNA. Only donor cells produce these pili. Different conjugative plasmids may have slightly different *tra* regions, and the pili may vary somewhat in structure. The F plasmid and its relatives encode F pili.

Pili allow specific pairing to take place between the donor and recipient cells. All conjugation in gram-negative *Bacteria* is thought to depend on cell pairing brought about by pili. The pilus makes specific contact with a receptor on the recipient cell and then is retracted by disassembling its subunits. This pulls the two cells together (**Figure 11.20a**). Following this process, donor and recipient cells remain in contact by binding coupling proteins located in the outer membrane of each cell (Figure 11.20b). DNA is then transferred from donor to recipient cell through this conjugation junction (Figure 11.20c).

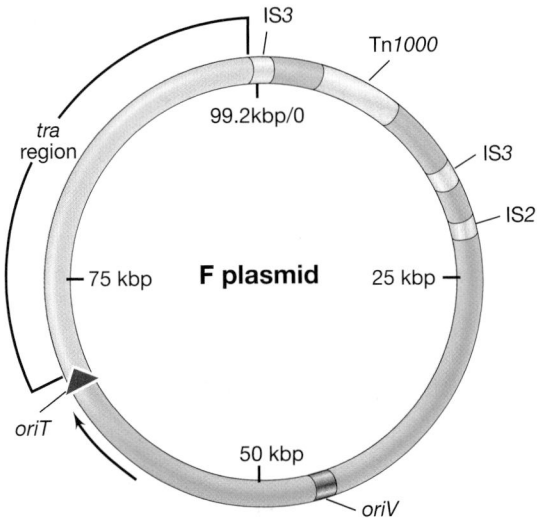

Figure 11.19 Genetic map of the F (fertility) plasmid of *Escherichia coli*. The numbers on the interior show the size in kilobase pairs (the exact size is 99,159 bp). The region in dark green at the bottom of the map contains genes primarily responsible for the replication and segregation of the F plasmid. The origin of vegetative replication is *oriV*. The light green *tra* region contains the genes needed for conjugative transfer. The origin of transfer during conjugation is *oriT*. The arrow indicates the direction of transfer (the *tra* region is transferred last). Insertion sequences are shown in yellow. These may recombine with identical elements on the bacterial chromosome, which leads to integration and the formation of different Hfr strains (Section 11.9).

Mechanism of DNA Transfer during Conjugation

DNA synthesis is necessary for DNA transfer by conjugation. This DNA is synthesized not by normal bidirectional replication (Section 4.4) but by **rolling circle replication**, a mechanism also used by some DNA viruses (Section 8.7) and shown in **Figure 11.21**. DNA transfer is triggered by cell-to-cell contact, at which time one strand of the circular plasmid DNA is nicked and is transferred to the recipient. The nicking enzyme required to initiate the process, TraI, is encoded by the *tra* operon of the F plasmid. TraI also has helicase activity and thus also unwinds the strand to be transferred. As this transfer occurs, DNA synthesis by the rolling circle mechanism replaces the transferred strand in the donor, while a complementary DNA strand is being made in the recipient. Therefore, at the end of the process, both donor and recipient possess complete plasmids. For transfer of the F plasmid, if an F-containing donor cell, which is designated F⁺, mates with a recipient cell lacking the plasmid (F⁻), the result is two F⁺ cells (Figure 11.21).

Transfer of plasmid DNA is efficient and rapid; under favorable conditions virtually every recipient cell that pairs with a donor acquires a plasmid. Transfer of the F plasmid (approximately 100 kbp) takes about 5 min. If plasmid genes can be expressed in the recipient, the recipient itself becomes a donor and can transfer the plasmid to other recipients. In this fashion, conjugative plasmids can spread rapidly among bacterial populations, behaving much like infectious agents. This is of major ecological significance because conjugative plasmids have been found in many *Bacteria* and some *Archaea* (Section 11.10), and a few plasmid-containing cells introduced into a population of potential recipients can convert the entire population into plasmid-bearing (and thus donating) cells in a short time.

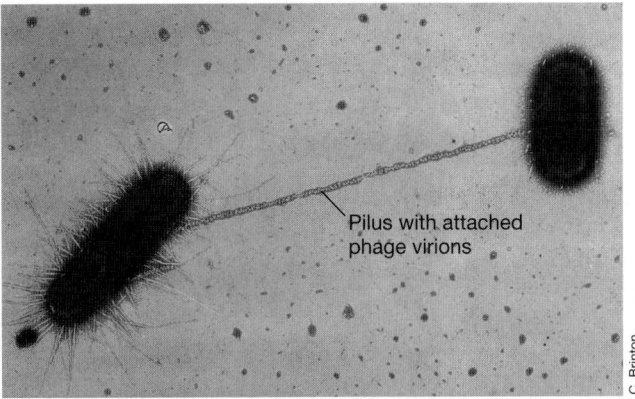

(a)

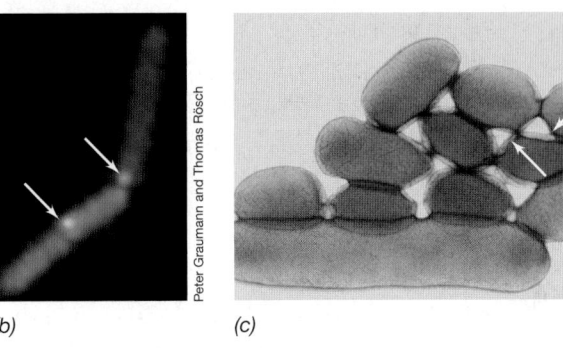

(b) (c)

Figure 11.20 Visualization of conjugation. *(a)* Formation of a mating pair. Direct contact between two conjugating *Escherichia coli* cells is first made via a pilus. The cells are then drawn together to form a mating pair by retraction of the pilus, which is achieved by depolymerization. Certain small phages (F-specific bacteriophages) use the sex pilus as a receptor and can be seen here attached to the pilus. *(b)* Coupling proteins near the cell membrane. These *Bacillus subtilis* cells contain the conjugative plasmid pLS20 encoding the VirD coupling protein linked to a fluorescent reporter gene. *(c)* Conjugation junctions. Negatively stained transmission electron micrograph of conjugation bridges between cells of *Yersinia pseudotuberculosis*. Arrows indicate connection sites. Adapted from Lesic, B., M. Zouine, M. Ducos-Galand, C. Huon, M-L. Rosso, M-C. Prévost, D. Mazel, and E. Carniel. 2012 *PLoS Genetics* 8(3): e1002529.

--- **MINIQUIZ** ---

• In conjugation, how are donor and recipient cells brought into contact with each other?

• Explain how rolling circle DNA replication allows both donor and recipient to end up with a complete copy of plasmids transferred by conjugation.

11.9 The Formation of Hfr Strains and Chromosome Mobilization

Chromosomal genes can be transferred by plasmid-mediated conjugation. As mentioned above, the F plasmid of *Escherichia coli* can, under certain circumstances, mobilize the chromosome for transfer during cell-to-cell contact. The F plasmid is actually an *episome*, a plasmid that can integrate into the host chromosome. When the F plasmid is integrated, chromosomal genes can be transferred along with the plasmid. Following genetic recombination between donor and recipient DNA, horizontal transfer of chromosomal genes by this mechanism can be extensive.

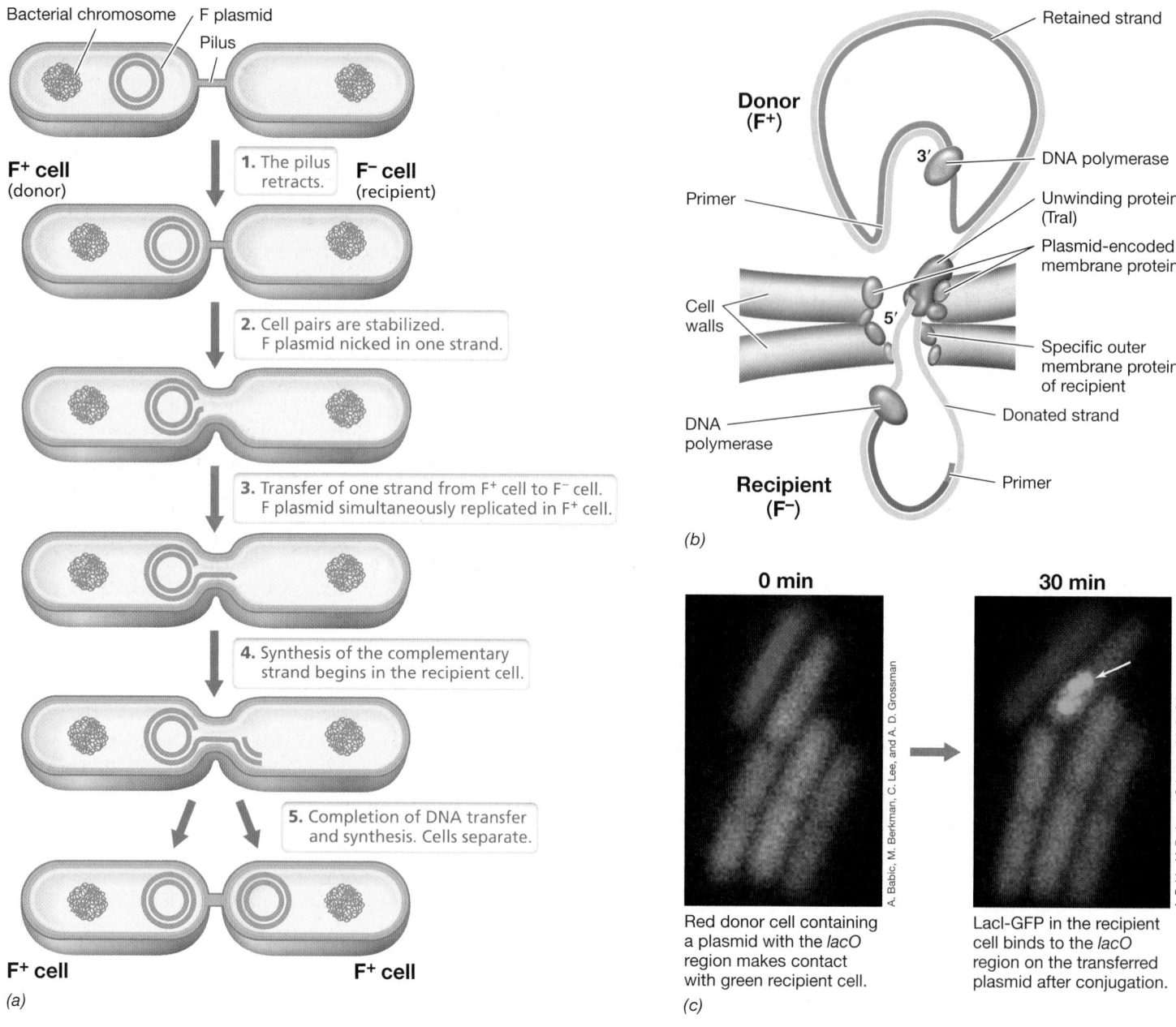

Figure 11.21 Transfer of plasmid DNA by conjugation. *(a)* The transfer of the F plasmid converts an F⁻ recipient cell into an F⁺ cell. Note the mechanism of rolling circle replication. *(b)* Details of the replication and transfer process. Note the large number of proteins needed for successful DNA transfer. *(c)* Visualization of DNA transfer by conjugation in *Bacillus subtilis* using fluorescence microscopy. The donor cell constitutively expresses a red fluorescent protein, while the recipient cells fluoresce green due to green fluorescent protein (GFP) fused to LacI (⊘ Figure 6.14). The DNA transferred from the donor contains a *lacO* operator region that binds LacI-GFP. Arrow indicates focal point in the recipient cell where LacI-GFP is bound to the *lacO* region obtained from conjugation.

Cells possessing a nonintegrated F plasmid are called F⁺, whereas those with an F plasmid integrated into the chromosome are called **Hfr cells** (for *high frequency of recombination*). This term refers to the high rates of genetic recombination between genes on the donor (Hfr) and recipient (F⁻) chromosomes. Both F⁺ and Hfr cells are donors, but unlike conjugation between an F⁺ and an F⁻, conjugation between an Hfr donor and an F⁻ leads to transfer of genes from the host chromosome. This is because the chromosome and plasmid now form a single molecule of DNA. Consequently, when rolling circle replication is initiated by the F plasmid, replication continues on into the chromosome. Thus, the chromosome is also replicated and parts of it get transferred. Hence, integration of a conjugative plasmid provides a mechanism for *mobilizing* a cell's genome.

Overall, the presence of the F plasmid results in three distinct changes in a cell: (1) the ability to synthesize the F pilus (Figure 11.20a), (2) the mobilization of DNA for transfer to another cell, and (3) the alteration of surface receptors so the cell can no longer be a recipient in conjugation and is therefore unable to take up a second copy of the F plasmid or any genetically related plasmids.

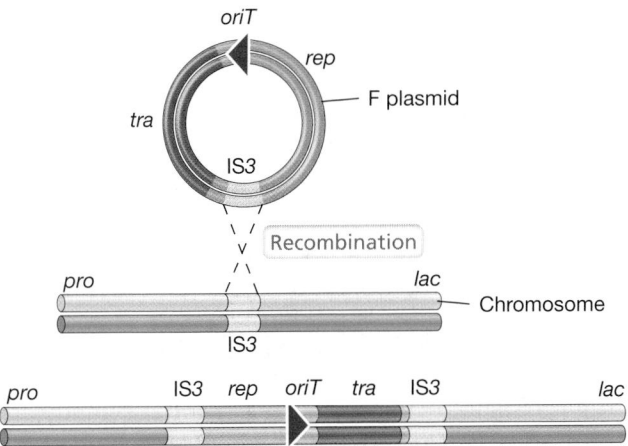

Figure 11.22 The formation of an Hfr strain. Integration of the F plasmid into the chromosome may occur at a variety of specific sites where IS elements are located. The example shown here is an IS*3* located between the chromosomal genes *pro* and *lac*. Some of the genes on the F plasmid are shown. The arrow indicates the origin of transfer, *oriT*, with the arrow as the leading end. Thus, in this Hfr, *pro* would be the first chromosomal gene to be transferred and *lac* would be among the last.

Integration of F Plasmid and Chromosome Mobilization

The F plasmid and the chromosome of *E. coli* both carry several copies of mobile genetic elements called *insertion sequences* (IS; Section 11.11). These provide regions of sequence homology between chromosomal and F plasmid DNA. Consequently, homologous recombination between an IS on the F plasmid and a corresponding IS on the chromosome results in integration of the F plasmid into the host chromosome (**Figure 11.22**). Once integrated, the plasmid no longer replicates independently, but the *tra* operon still functions normally and the strain synthesizes pili. When a recipient cell is encountered, conjugation is triggered just

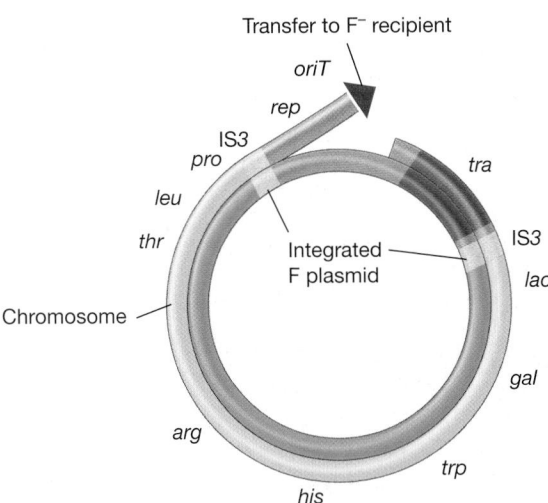

Figure 11.23 Transfer of chromosomal genes by an Hfr strain. The Hfr chromosome breaks at the origin of transfer within the integrated F plasmid. The transfer of DNA to the recipient begins at this point. DNA replicates during transfer as for a free F plasmid (Figure 11.21). This figure is not to scale; the inserted F plasmid is actually less than 3% of the size of the *Escherichia coli* chromosome.

as in an F⁺ cell, and DNA transfer is initiated at the *oriT* (origin of transfer) site. However, because the plasmid is now part of the chromosome, after part of the plasmid DNA is transferred, chromosomal genes begin to be transferred (**Figure 11.23**). As in the case of conjugation with just the F plasmid itself (Figure 11.21), chromosomal transfer also requires DNA replication.

Because the DNA strand usually breaks during transfer, only part of the donor chromosome is typically transferred. Consequently, the recipient does not become Hfr (or F⁺) because only part of the integrated F plasmid is transferred (**Figure 11.24**). However, after transfer, the Hfr strain remains genetically Hfr because it retains a copy of the integrated F plasmid. Following recombination, the recipient cell may express a new phenotype due to the incorporation of donor genes, but genetically it remains an F⁻ cell. Because a partial chromosome cannot replicate, for incoming donor DNA to survive, it must recombine with the recipient chromosome. As in transformation and transduction, genetic recombination between donor and recipient genes requires homologous recombination in the recipient cell.

Because several distinct insertion sequences are present on the *E. coli* chromosome, a number of different Hfr strains are possible. A given Hfr strain always donates genes in the same order, beginning at the same position. However, Hfr strains that differ in the

UNIT 3

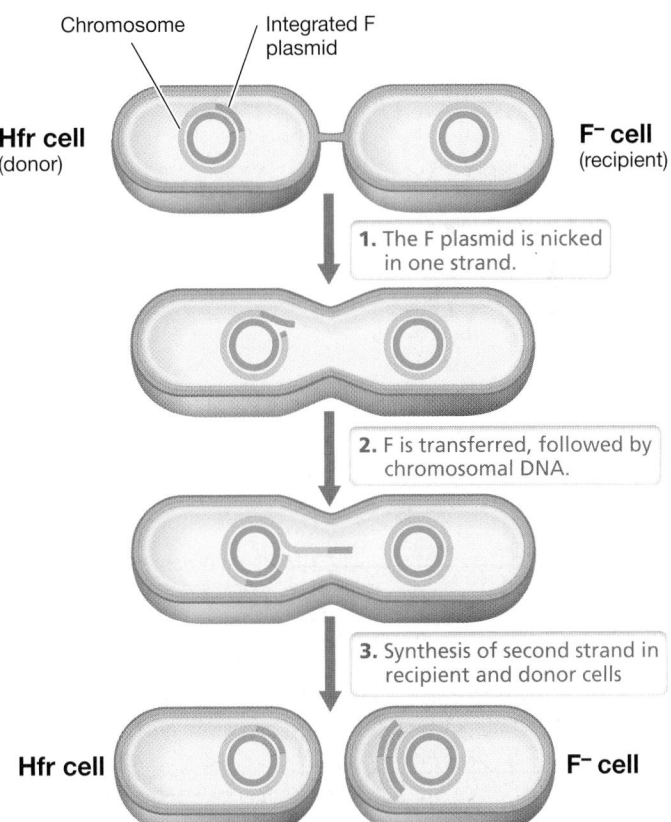

Figure 11.24 Transfer of chromosomal DNA by conjugation. Transfer of the integrated F plasmid from an Hfr strain results in the cotransfer of chromosomal DNA because it is linked to the plasmid. The steps in transfer are similar to those in Figure 11.21a. However, the recipient remains F⁻ and receives a linear fragment of donor chromosome attached to part of the F plasmid. For donor DNA to survive, it must be recombined into the recipient chromosome after transfer (not shown).

chromosomal integration site of the F plasmid transfer their genes in different orders (**Figure 11.25**). At some insertion sites, the F plasmid is integrated with its origin pointing in one direction, whereas at other sites the origin points in the opposite direction. The orientation of the F plasmid determines which chromosomal genes enter the recipient cell first and illustrate how virtually any gene on the chromosome can be mobilized by one Hfr configuration or another (Figure 11.25).

Transfer of Chromosomal Genes to the F Plasmid

Occasionally, integrated F plasmids may be excised from the chromosome. During excision, chromosomal genes may sometimes be incorporated into the liberated F plasmid. This can happen because both the F plasmid and the chromosome contain multiple identical insertion sequences where recombination can occur (Figure 11.22). F plasmids containing chromosomal genes are called *F'* (F-prime) *plasmids*. When F' plasmids promote conjugation, they transfer the chromosomal genes they carry at high frequency to the recipients. F'-mediated transfer resembles specialized transduction (Section 11.7) in that only a restricted group of chromosomal genes is transferred by any given F' plasmid. Transferring a known F' into a recipient allows one to establish diploids (two copies of each gene) for a limited region of the chromosome. Such partial diploids (merodiploids) are important for genetic complementation tests (Section 11.5).

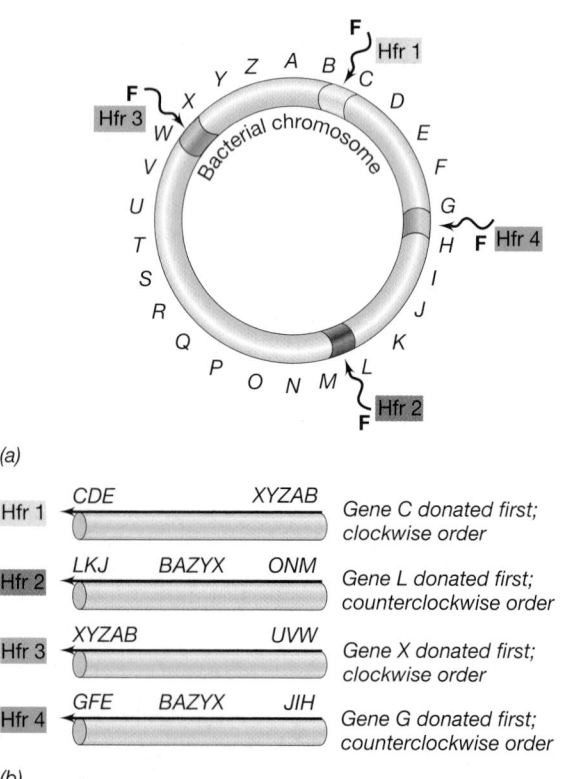

(a)

(b)

Figure 11.25 Formation of different Hfr strains. Different Hfr strains donate genes in different orders and from different origins. *(a)* F plasmids can be inserted into various insertion sequences on the bacterial chromosome, forming different Hfr strains. *(b)* Order of gene transfer for different Hfr strains. See Figures 11.22–11.24 for details of Hfr formation and DNA transfer.

MINIQUIZ

- In conjugation involving the F plasmid of *Escherichia coli*, how is the host chromosome mobilized?
- Why does an Hfr × F⁻ mating not yield two Hfr cells?
- At which sites in the chromosome can the F plasmid integrate?

III • Gene Transfer in *Archaea* and Other Genetic Events

Although studies of the genetics of *Archaea* lag behind genetics research in *Bacteria*, they are showing progress, along with archaeal versions of the tools necessary for detailed genetic analyses. In addition, some other genetic events in *Bacteria* reveal important genetic concepts even though they do not involve horizontal gene flow per se. We cover both of these topics here.

11.10 Horizontal Gene Transfer in *Archaea*

Although *Archaea* contain a single circular chromosome (**Figure 11.26**), as do most *Bacteria*, and genome analysis clearly shows that horizontal transfer of archaeal DNA also occurs in nature, laboratory-based gene transfer systems are not as well developed as those for *Bacteria*. Some problems here are of a practical nature including the fact that most well-studied *Archaea* are extremophiles, capable of growth only under conditions of high salt or high temperature (Chapter 17). The temperatures necessary to culture some hyperthermophiles, for example, will melt agar, and alternative materials are required to form solid media and obtain colonies.

Another problem is that most known antibiotics do not affect *Archaea*. For example, penicillins do not affect *Archaea* because their cell walls lack peptidoglycan. The choice of selectable markers for use in genetic crosses is therefore often limited. However, novobiocin (a DNA gyrase inhibitor) and mevinolin (an inhibitor of isoprenoid biosynthesis) have been used to inhibit growth of extreme halophiles, and puromycin and neomycin (both protein

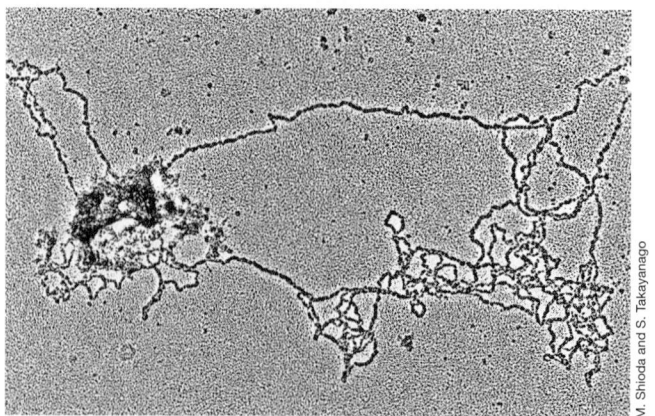

Figure 11.26 An archaeal chromosome, as shown in the electron microscope. The circular chromosome is from the hyperthermophile *Sulfolobus*, a member of the *Archaea*.

synthesis inhibitors) inhibit methanogens. Auxotrophic strains (Section 11.1) of a few *Archaea* have also been isolated for genetic selection purposes.

Examples of Archaeal Genetics

No single species of *Archaea* has become a model organism for archaeal genetics, although more genetic work has been done on select species of extreme halophiles (*Halobacterium, Haloferax*, ⮂ Section 17.1) than on any other *Archaea*. Various mechanisms of gene transfer have been found scattered among a range of *Archaea*. In addition, several plasmids have been isolated from *Archaea* and some have been used to construct cloning vectors, allowing genetic analysis through gene cloning and sequencing rather than through traditional genetic crosses. Transposon mutagenesis (Section 11.11) has been well developed in certain methanogens, including species of *Methanococcus* and *Methanosarcina*, and other tools such as shuttle vectors and other in vitro methods of genetic analysis have been developed for study of the unique biochemistry of the methanogens (⮂ Sections 14.17 and 17.2).

Transformation occurs in the methanogen *Methanococcus voltae* and the hyperthermophiles *Thermococcus kodakarensis* and *Pyrococcus furiosus*, all of which are naturally competent (Section 11.6). *Thermococcus* species can also exchange plasmids through the budding of their cell envelope, a process that results in DNA-containing membrane vesicles. Other conditions for transformation work reasonably well in several *Archaea* although the details vary from organism to organism. One approach requires removal of divalent metal ions, which in turn results in the partial disassembly of the glycoprotein cell wall layer that surrounds many archaeal cells (S-layer, ⮂ Section 2.6), and this allows access to transforming DNA. However, *Archaea* with rigid cell walls have proven difficult to transform, although electroporation (Section 11.6) sometimes works. One exception is in *Methanosarcina* species, organisms with a thick polysaccharide cell wall, for which high-efficiency transformation systems have been developed that employ DNA-loaded lipid preparations (liposomes) that traverse the cell wall to deliver DNA into the cell.

Although viruses that infect *Archaea* are plentiful, transduction is extremely rare. Only one archaeal virus, which infects the thermophilic methanogen *Methanothermobacter thermautotrophicus*, has been shown to transduce the genes of its host. Unfortunately the low burst size (about six phages liberated per cell) makes it impractical to use this system for gene transfer. Gene transfer agents (Section 11.7) have been found in one species of methanogen but do not appear to be widespread in *Archaea*.

Conjugation in *Archaea*

Different types of conjugation have been detected in *Archaea*. Some strains of the thermophilic and acidophilic *Sulfolobus solfataricus* (⮂ Section 17.9) contain plasmids that promote conjugation between two cells in a manner similar to that seen in *Bacteria*. In this process, cell pairing is independent of pili formation and DNA transfer is unidirectional. However, most of the genes encoding these functions in *S. solfataricus* seem to have little similarity to those in gram-negative *Bacteria*. The exception is a gene similar to *traG* from the F plasmid, whose protein product participates in stabilizing mating pairs. It thus seems likely

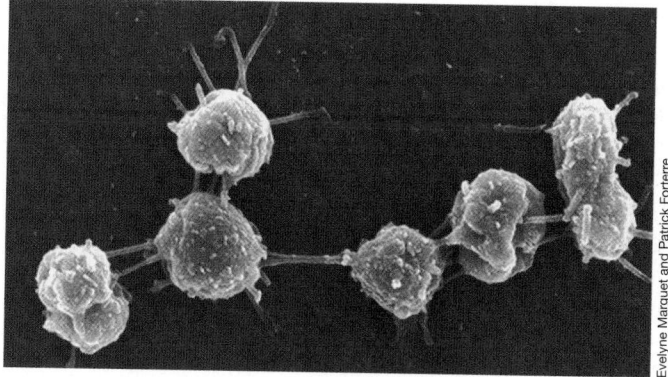

Figure 11.27 Nanotubes and *Thermococcus*. Scanning electron micrograph showing nanotubes linking cells of *Thermococcus* sp. 5-4. A single *Thermococcus* cell is about 1 μm in diameter.

that the actual mechanism of conjugation in *Archaea* is quite different from that in *Bacteria*.

Most species of *Sulfolobus* can also exchange DNA without the participation of a fertility plasmid. Such exchange is dependent on cell aggregation, a process that requires pili whose synthesis is triggered by UV radiation. While the exact mechanism of this unusual DNA exchange is unknown, these pili bring cells of the same species together through recognition of S-layer glycosylation patterns present on the cell walls.

Other *Archaea* form specialized structures between cells that allow for genetic exchange, and the formation of these structures is also independent of a fertility plasmid. For example, **Figure 11.27** illustrates the formation of DNA-transferring nanotubes between cells of a *Thermococcus* species. Interestingly, unlike the other means of horizontal gene transfer discussed in this chapter, the *Thermococcus* nanotubes allow for *bidirectional* transfer of DNA. Thus, with both cells in an exchange able to function as recipients, the *Thermococcus* nanotubes likely facilitate very dynamic gene flow. Similarly, some halobacteria also form cytoplasmic bridges between mating cells that are used for DNA transfer. Although the nanotube and cytoplasmic bridge systems are not yet in routine use, they may well be useful for developing more facile archaeal genetic transfer systems in the future.

MINIQUIZ

- Why is it usually more difficult to select recombinants with *Archaea* than with *Bacteria*?
- Why do penicillins not kill species of *Archaea*?

11.11 Mobile DNA: Transposable Elements

As we have seen, molecules of DNA may move from one cell to another, but to a geneticist, the phrase "mobile DNA" has a special meaning. Mobile DNA refers to discrete segments of DNA that move as units from one location to another *within* other DNA molecules.

Although the DNA of certain viruses can be inserted into and excised from the genome of the host cell (⮂ Section 8.7), most mobile DNA consists of **transposable elements**. These are stretches of DNA that can move from one site to another. However, transposable elements are always found inserted into another DNA molecule such as a plasmid, a chromosome, or a viral genome. Transposable elements do not possess their own origin of replication. Instead, they are replicated when the host DNA molecule into which they are inserted is replicated.

Transposable elements move by *transposition*, a process that is important in both genome rearrangement and genetic analysis. Transposable elements are abundant and widespread in nature and can be found in the genomes of all three domains of life as well as in many viruses and plasmids, suggesting that the elements offer a selective advantage by accelerating genome rearrangement.

The two major types of transposable elements in *Bacteria* are *insertion sequences* (IS) and *transposons*. Both elements have two important features in common: They carry genes encoding *transposase*, the enzyme necessary for transposition, and they have short inverted terminal repeats at their ends that are also needed for transposition (the ends of transposable elements are not free but are continuous with the host DNA molecule into which the transposable element has inserted). **Figure 11.28** shows genetic maps of two well-studied transposable elements: the insertion element IS*2* and the transposon Tn*5*.

Insertion Sequences and Transposons

Insertion sequences (IS) are the simplest type of transposable element. They are short DNA segments, about 1000 nucleotides long, and typically contain inverted repeats of 10–50 base pairs. Each different IS has a specific number of base pairs in its terminal repeats, and the only protein encoded is the transposase. Several hundred distinct IS elements have been characterized. IS elements are found in the chromosomes and plasmids of both *Bacteria* and *Archaea*, as well as in certain bacteriophages. Individual strains of the same bacterial species vary in the number and location of the IS elements they harbor. For instance, the genome of one strain of *Escherichia coli* has five copies of IS*2* and five copies of IS*3*. Many plasmids, such as the F plasmid, also carry IS elements. Indeed,

integration of the F plasmid into the *E. coli* chromosome is facilitated by recombination between identical IS elements on the F plasmid and the chromosome (Section 11.9 and Figure 11.22).

Transposons are larger than IS elements but have the same two essential components: inverted repeats at both ends and a gene that encodes transposase (Figure 11.28*b*). The transposase recognizes the inverted repeats and moves the segment of DNA flanked by them from one site to another. Consequently, any DNA that lies between the two inverted repeats is moved and is, in effect, part of the transposon. Genes included inside transposons vary widely. Some of these genes, such as antibiotic resistance genes, confer important new properties on the organism. Because antibiotic resistance is both important and easy to detect, most well-studied transposons contain antibiotic resistance genes as selectable markers. Examples include transposon Tn*5*, which encodes kanamycin resistance (Figure 11.28*b*) and Tn*10*, which encodes tetracycline resistance.

Because any genes lying between the inverted repeats become part of a transposon, it is possible to get hybrid transposons that display complex behavior. For example, conjugative transposons contain *tra* genes and can move between bacterial species by conjugation as well as transpose from place to place within a single bacterial genome. Even more complex is bacteriophage Mu, which is both a virus and a transposon (↩ Section 10.4). In this case a complete virus genome is contained within a transposon.

Mechanisms of Transposition

Both the inverted repeats (located at the ends of transposable elements) and transposase are essential for transposition. The transposase recognizes, cuts, and ligates the DNA during transposition. When a transposable element is inserted into target DNA, a short sequence in the target DNA at the site of integration is duplicated during the insertion process (**Figure 11.29**). The duplication arises because single-stranded DNA breaks are made by the transposase. The transposable element is then attached to the single-stranded ends that have been generated. Finally, enzymes of the host cell repair the single-strand portions, which results in the duplication.

Two mechanisms of transposition are known: conservative and replicative (**Figure 11.30**). In *conservative* transposition, as occurs with the transposon Tn*5* (Figure 11.28*b*), the transposon is excised

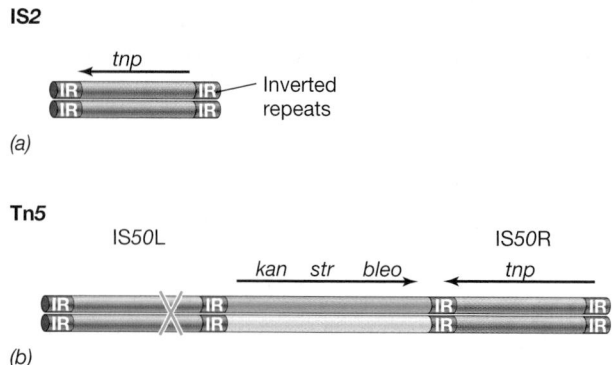

Figure 11.28 Maps of the transposable elements IS*2* and Tn*5*. The arrows above the maps show the direction of transcription of any genes on the elements. The gene encoding the transposase is *tnp*. (*a*) IS*2* is an insertion sequence of 1327 bp with inverted repeats of 41 bp at its ends. (*b*) Tn*5* is a composite transposon of 5.7 kbp containing the insertion sequences IS*50* L and IS*50* R at its left and right ends, respectively. IS*50* L is not capable of independent transposition because there is a nonsense mutation, marked by a blue cross, in its transposase gene. The genes *kan*, *str*, and *bleo* confer resistance to the antibiotics kanamycin (and neomycin), streptomycin, and bleomycin.

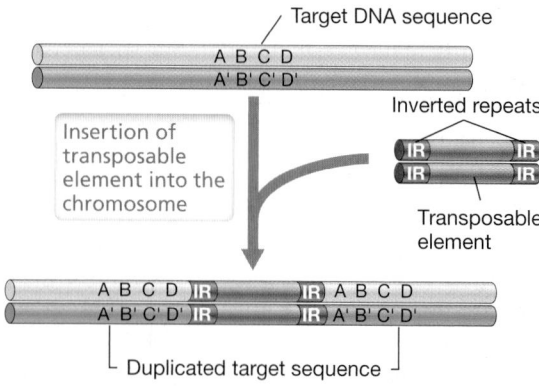

Figure 11.29 Transposition. Insertion of a transposable element generates a duplication of the target sequence. Note the presence of inverted repeats (IR) at the ends of the transposable element.

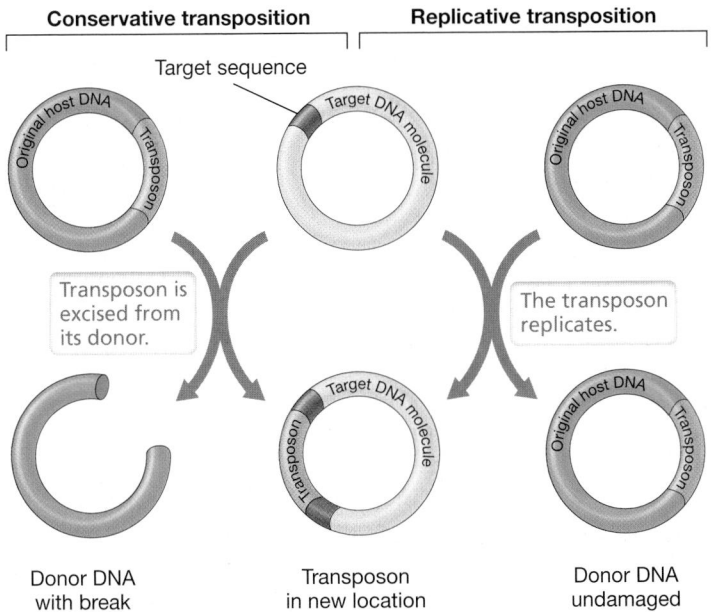

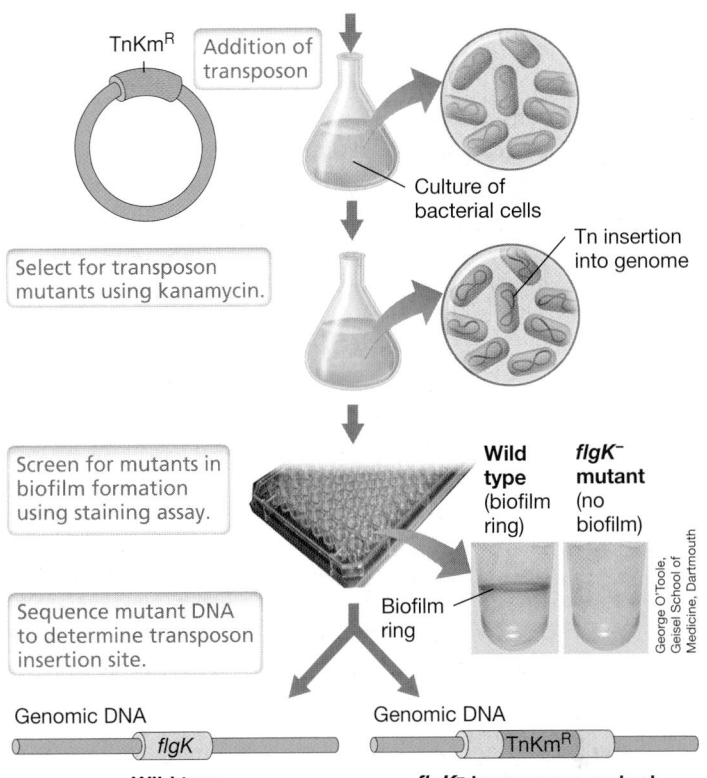

Figure 11.30 Two mechanisms of transposition. Donor DNA (carrying the transposon) is shown in green, and recipient DNA carrying the target sequence is shown in yellow. In both conservative and replicative transposition, the transposase inserts the transposon (purple) into the target site (red) on the recipient DNA. During this process, the target site is duplicated. In conservative transposition, the donor DNA is left with a double-stranded break at the previous location of the transposon. In contrast, after replicative transposition, both donor and recipient DNA possess a copy of the transposon.

from one location and is reinserted at a second location. The copy number of a conservative transposon therefore remains at one. By contrast, during *replicative* transposition, a new copy of the transposon is produced and is inserted at the second location. Thus, after a replicative transposition event, one copy of the transposon remains at the original site, while a second copy is incorporated at the new site.

Utility of Transposon Mutagenesis

When a transposon inserts itself within a gene, the DNA sequence in the gene is altered and a mutation occurs (**Figure 11.31**). Mutations due to transposon insertion do occur naturally. However, laboratory use of transposons has been a powerful genetic tool to create a library of bacterial mutants. To do this, transposons carrying antibiotic resistance genes are used. The transposon is introduced into the target cell on a plasmid that cannot replicate in that particular host using conjugation or transformation as transfer mechanisms. Consequently, antibiotic-resistant colonies will mostly be due to insertion of the transposon into the bacterial genome.

Figure 11.32 Utility of transposon mutagenesis. A transposon (Tn) conferring kanamycin resistance (Km^R) is added to a culture of wild-type *Pseudomonas aeruginosa*. After selection for integration of the transposon using the antibiotic kanamycin, the mutants are screened for biofilm formation in a microtiter plate. Cultures that produce biofilm adhere to the microtiter plate (ring in left tube of inset photo) and the resulting biofilm can be stained with crystal violet. Primers specific to the Tn can be used to determine the Tn insertion site and the identity of the interrupted gene. *flgK* encodes a protein in the flagellar hook.

Because bacterial genomes contain relatively little noncoding DNA, most transposon insertions will occur in genes that encode proteins. This technique can be used to determine the function of a novel gene if a screening method is available. For example, if a transposon inserts into a gene encoding a product required for biofilm formation (biofilms are colonies of microbes encased in an adhesive and attached to a surface, ⟿ Section 5.1), the transposon mutant will no longer grow in the biofilm mode. Then, further genetic analyses can be performed to reveal which gene the transposon has disrupted. **Figure 11.32** illustrates the use of transposon mutagenesis to study flagella structure and function in a notorious biofilm-producing bacterium, *Pseudomonas aeruginosa* (⟿ Section 7.9). In this research, transposon mutagenesis combined

Figure 11.31 Transposon mutagenesis. The transposon moves into the middle of gene 2. Gene 2 is now disrupted by the transposon and is inactivated. Gene A from the transposon is now expressed from the chromosome.

with a special staining technique was used to identify the flagellar hook protein FlgK (⮌ Section 2.11) as a protein required for biofilm formation.

Two transposons widely used for mutagenesis of *E. coli* and related bacteria are Tn*5* (Figure 11.28*b*) and Tn*10*. Many *Bacteria*, a few *Archaea*, and the yeast *Saccharomyces cerevisiae* have all been mutagenized using genetically engineered transposons. More recently, transposons have even been used to isolate mutations in animals, including mice.

MINIQUIZ

- Which features do insertion sequences and transposons have in common?
- What is the significance of the terminal inverted repeats of transposons?
- How can transposons be used in bacterial genetics?

11.12 Preserving Genomic Integrity: CRISPR Interference

Bacteria and *Archaea* not only produce restriction endonucleases (⮌ Section 8.5) that function to destroy incoming foreign DNA, they also have an RNA-based defense system to destroy invading DNA from viral infections and some horizontally transferred genes. This prokaryotic "immune system"—called CRISPR—was previously described as a major means for *Bacteria* and *Archaea* to evade viral destruction (⮌ Section 10.13). But CRISPR also helps the cell maintain the stability and integrity of its genome by destroying certain plasmids and other genes obtained from horizontal transfers.

CRISPR Mechanism

The CRISPR region on the bacterial chromosome is essentially a memory bank of incoming nucleic acid sequences used for surveillance against foreign DNA. It consists of many different segments of foreign DNA called *spacers* alternating with identical repeated sequences (**Figure 11.33**). The spacer sequences correspond to pieces of foreign DNA that have previously invaded the cell. Once the spacers are recombined into the CRISPR region, the system provides resistance to any incoming DNA (and sometimes RNA) that contains the same or very closely related sequences to those in individual spacer regions.

The key to the CRISPR region's ability to prevent horizontal transfer of some incoming DNAs is the transcription of a long RNA molecule that is then cleaved in the middle of each of the repeated sequences by the nuclease activity of CRISPR-*as*sociated (Cas) proteins. This converts the long RNA molecule into spacer segments of small RNAs called CRISPR RNAs (crRNAs). If one of these crRNAs base-pairs with an invading nucleic acid, then the foreign DNA or RNA duplex is destroyed by the nuclease activity of other Cas

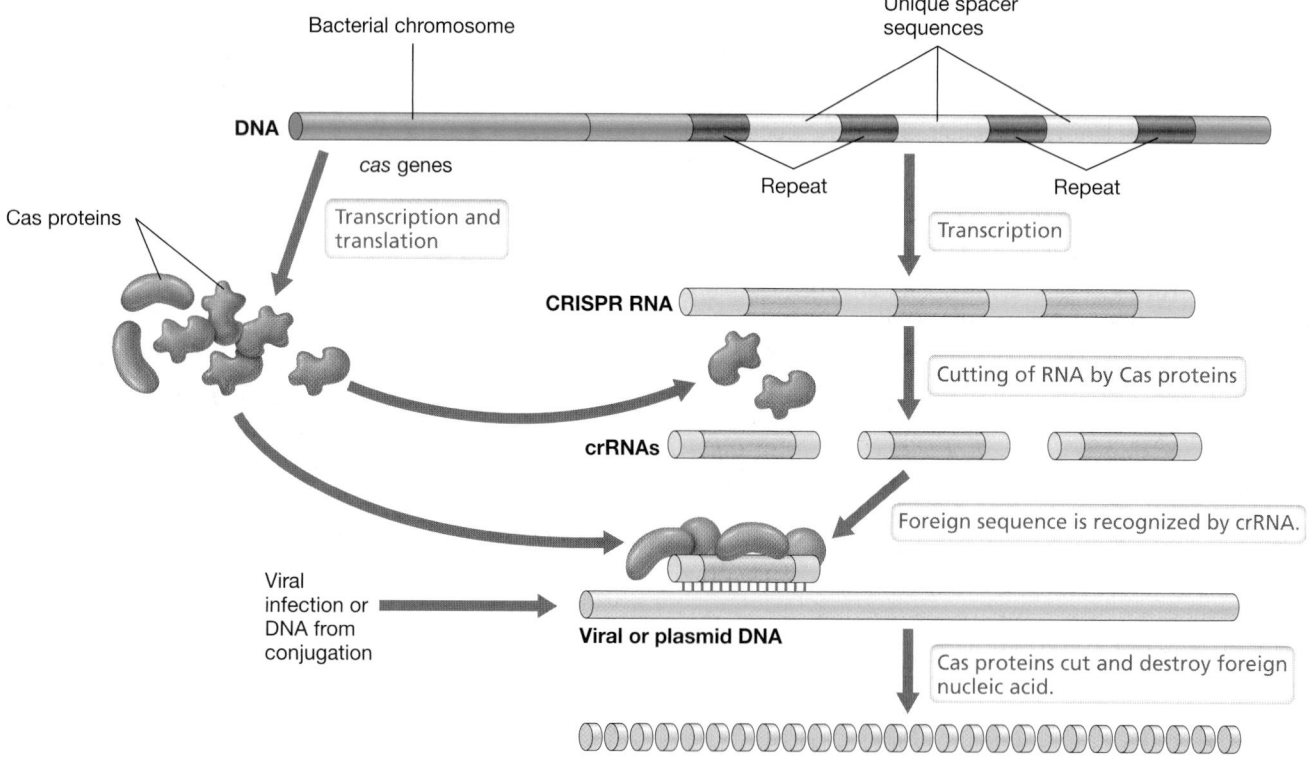

Figure 11.33 Operation of the CRISPR system. The CRISPR region on the bacterial chromosome is transcribed into a long RNA molecule that is then processed into segments by some of the Cas proteins. Each spacer segment corresponds to previous encounters with incoming foreign nucleic acid. If one of these short CRISPR RNA (cRNA) molecules (corresponding to a spacer) recognizes and base pairs with incoming nucleic acid from transduction or conjugation, other Cas proteins destroy the foreign nucleic acid.

proteins (Figure 11.33). This destruction prevents the foreign DNA from being replicated or recombined into the genome and thus keeps the genome from accumulating random bits of foreign DNA that could compromise genome integrity.

Distribution of CRISPR

The CRISPR system is widely distributed in both *Archaea* and *Bacteria*. Approximately 90% of the sequenced genomes of *Archaea* and 70% of those of *Bacteria* possess a CRISPR system. The utility of the system was first demonstrated in the dairy industry where starter cultures used for milk fermentation were susceptible to rampant bacteriophage infection. However, a strain of *Streptococcus thermophilus* was found to be resistant to virulent bacteriophage, and the difference between this *S. thermophilus* strain and those susceptible to viral infection was the spacers within its CRISPR region.

While it is unknown why some viruses and other foreign DNAs possess the initial recognition sequences (PAMs, Section 10.13)

for spacer incorporation by the CRISPR system and others do not, laboratory experiments have shown that bacteriophages can overcome recognition by the Cas proteins and crRNAs by modifying their genome through mutation. This illustrates how the CRISPR system—central as it may be to preventing attacks on a cell's genome—does have a drawback in its reliance on recognizing previously encountered DNA sequences that can continue to mutate in their host and then reappear in the future in a form not recognized by CRISPR.

In Chapter 12 we discuss the CRISPR system in a totally different context as a powerful tool in biotechnology for editing genomes and creating recombinant organisms.

— MINIQUIZ —
- Why is the CRISPR system considered a prokaryotic "immune system"?
- What do the spacers within the CRISPR region correspond to?

MasteringMicrobiology®　**Visualize, explore,** and **think critically** with Interactive Microbiology, MicroLab Tutors, MicroCareers case studies, and more. MasteringMicrobiology offers practice quizzes, helpful animations, and other study tools for lecture and lab to help you master microbiology.

Chapter Review

I • Mutation

11.1 Mutation is a heritable change in the nucleotide sequence of the genome and may lead to a change in phenotype. Selectable mutations are those that give the mutant a growth advantage under certain environmental conditions and are especially useful in genetic research. If selection is not possible, mutants must be identified by screening.

Q Write a one-sentence definition of the term "genotype." Do the same for "phenotype." Does the phenotype of an organism automatically change when a change in genotype occurs? Why or why not? Can phenotype change without a change in genotype? In both cases, give examples to support your answer.

11.2 Mutations, either spontaneous or induced, are in the base sequence of the nucleic acid in a genome. A point mutation is due to a single base-pair change. In a nonsense mutation, the codon becomes a stop codon and an incomplete polypeptide is made. Deletions and insertions cause more dramatic changes in the DNA, including frameshift mutations that often result in complete loss of gene function.

Q What are silent mutations? From your knowledge of the genetic code, why do you think most silent mutations affect the third position in a codon?

11.3 Different types of mutations occur at different frequencies. For a typical bacterium, mutation rates of

10^{-6} to 10^{-7} per kilobase pair are generally seen. Although RNA and DNA polymerases make errors at about the same rate, RNA genomes typically accumulate mutations at much higher frequencies than DNA genomes.

Q What is the difference between same-site and second-site revertants?

11.4 Mutagens are chemical, physical, or biological agents that increase the mutation rate. Mutagens can alter DNA in many different ways. However, alterations in DNA are not mutations unless they are inherited. Some DNA damage can lead to cell death if not repaired, and both error-prone and high-fidelity DNA repair systems exist.

Q Give an example of one biological, one chemical, and one physical mutagen and describe the mechanism by which each causes a mutation.

II • Gene Transfer in *Bacteria*

11.5 Homologous recombination occurs when closely related DNA sequences from two distinct genetic elements are combined together in a single element. Recombination is an important evolutionary process, and cells have specific mechanisms for ensuring that recombination takes place.

Q What are heteroduplex regions of DNA and what process leads to their formation?

11.6 Certain bacteria exhibit competence, a state in which cells are able to take up free DNA released by other bacteria.

Incorporation of donor DNA into a recipient cell requires the activity of single-strand binding protein, RecA protein, and several other enzymes. Only competent cells are transformable.

> **Q** **Explain why recipient cells do not successfully take up plasmids during natural transformation.**

11.7 Transduction is the transfer of host genes from one bacterium to another by a bacterial virus. In generalized transduction, defective virus particles randomly incorporate fragments of the cell's chromosomal DNA, but the transducing efficiency is low. In specialized transduction, the DNA of a temperate virus excises incorrectly and takes adjacent host genes along with it; the transducing efficiency here may be very high.

> **Q** **Explain how a generalized transducing particle differs from a specialized transducing particle.**

11.8 Conjugation is a mechanism of DNA transfer in *Bacteria* and *Archaea* that requires cell-to-cell contact. Conjugation is controlled by genes carried by certain plasmids (such as the F plasmid) and requires transfer of the plasmid from a donor cell to a recipient cell. Plasmid DNA transfer requires replication using the rolling circle mechanism.

> **Q** **What is a sex pilus and which cell type, F⁻ or F⁺, would produce this structure?**

11.9 The donor cell chromosome can be mobilized for transfer to a recipient cell. This requires an F plasmid to integrate into the chromosome to form the Hfr phenotype. Because transfer of the host chromosome is rarely complete, recipient cells rarely become F⁺. F′ plasmids are previously integrated F plasmids that have excised and captured some chromosomal genes.

> **Q** **What is a merodiploid and how does an F′ plasmid yield a merodiploid?**

III · Gene Transfer in *Archaea* and Other Genetic Events

11.10 Archaeal research lags behind bacterial research in the development of systems for gene transfer. Many antibiotics are ineffective against *Archaea*, making it difficult to select recombinants effectively. The unusual growth conditions needed by many *Archaea* also make genetic experimentation challenging. Nevertheless, the genetic transfer systems of *Bacteria*—transformation, transduction, and conjugation—are all known in *Archaea*.

> **Q** **Explain one type of conjugation in *Archaea* and how it differs from F-plasmid-mediated conjugation.**

11.11 Transposons and insertion sequences are genetic elements that can move from one location on a host DNA molecule to another by transposition. Transposition can be either replicative or conservative. Transposons often carry genes encoding antibiotic resistance and can be used to identify the functions of unknown genes.

> **Q** **What are the major differences between insertion sequences and transposons?**

11.12 The clustered regularly interspaced short palindromic repeat (CRISPR) system is an RNA-based mechanism of protecting the genome of *Bacteria* and *Archaea* from invading DNA resulting from viral infection or conjugation. If small RNA molecules resulting from the spacer regions of the CRISPR region bind to incoming complementary DNA, Cas proteins destroy the nucleic acid duplex.

> **Q** **Explain why incoming DNA recognized by a short RNA molecule expressed from the CRISPR region cannot be completely foreign to the cell.**

Application Questions

1. A constitutive mutant is a strain that continuously makes a protein that is inducible in the wild type. Describe two ways in which a change in a DNA molecule could lead to the emergence of a constitutive mutant. How could these two types of constitutive mutants be distinguished genetically?

2. Although a large number of mutagenic chemicals are known, none is known that induces mutations in only a single gene (gene-specific mutagenesis). From what you know about mutagens, explain why it is unlikely that a gene-specific chemical mutagen will be found. How then is site-specific mutagenesis accomplished?

3. Why is it difficult in a single experiment to transfer a large number of genes to a recipient cell using transformation or transduction?

4. Transposable elements cause mutations when inserted within a gene. These elements disrupt the continuity of a gene. Introns also disrupt the continuity of a gene, yet the gene is still functional. Explain why the presence of an intron in a gene does not inactivate that gene but insertion of a transposable element does.

Chapter Glossary

Auxotroph an organism that has developed an additional nutritional requirement compared with the wild type, often as a result of mutation

Conjugation the transfer of genes from one prokaryotic cell to another by a mechanism requiring cell-to-cell contact

Frameshift mutation a mutation in which insertion or deletion of nucleotides changes the groups of three bases in which the genetic code is read within an mRNA, usually resulting in a faulty product

Genotype the complete genetic makeup of an organism; the complete description of a cell's genetic information

Heteroduplex a DNA double helix composed of single strands from two different DNA molecules

Hfr cell a cell with the F plasmid integrated into the chromosome

Induced mutation a mutation caused by external agents such as mutagenic chemicals or radiation

Insertion sequence (IS) the simplest type of transposable element, which carries only genes that participate in transposition

Missense mutation a mutation in which a single codon is altered so that one amino acid in a protein is replaced with a different amino acid

Mutagen an agent that causes mutation

Mutant an organism whose genome carries a mutation

Mutation a heritable change in the base sequence of the genome of an organism

Nonsense mutation a mutation in which the codon for an amino acid is changed to a stop codon

Phenotype the observable characteristics of an organism

Point mutation a mutation that involves a single base pair

Recombination a resorting or rearrangement of DNA fragments resulting in a new sequence

Reversion an alteration in DNA that reverses the effects of a prior mutation

Rolling circle replication a mechanism of replicating double-stranded circular DNA that starts by nicking and unrolling one strand and using the other (still circular) strand as a template for DNA synthesis

Screening a procedure that permits the identification of organisms by phenotype or genotype, but does not inhibit or enhance the growth of particular phenotypes or genotypes

Selection placing organisms under conditions that favor or inhibit the growth of those with a particular phenotype or genotype

Silent mutation a change in DNA sequence that has no effect on the phenotype

SOS repair system a DNA repair system activated by DNA damage

Spontaneous mutation a mutation that occurs "naturally" without the help of mutagenic chemicals or radiation

Transduction the transfer of host cell genes from one cell to another by a virus

Transformation the transfer of bacterial genes through the uptake of free DNA from the environment

Transition a mutation in which a pyrimidine base is replaced by another pyrimidine or a purine is replaced by another purine

Transposable element a genetic element able to move (transpose) from one site to another on host DNA molecules

Transposon a type of transposable element that carries genes in addition to those required for transposition

Transversion a mutation in which a pyrimidine base is replaced by a purine or vice versa

Wild-type strain a bacterial strain isolated from nature or one used as a parent in a genetics investigation

12

Biotechnology and Synthetic Biology

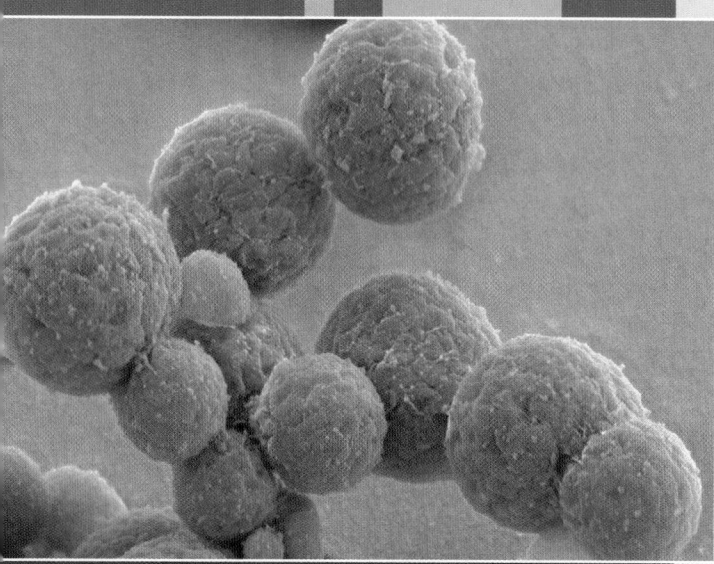

microbiology**now**

Creation of a New Life Form: Design of a Minimal Cell

A cell's genome is its blueprint for life. However, what is the bare minimum number of genes needed to sustain a free-living cell? This is a question that microbiologists at the J. Craig Venter Institute (JCVI) have attempted to answer ever since they sequenced the genomes of several *Mycoplasma* species in the 1990s. Because *Mycoplasma* species are parasitic bacteria, their genomes are already reduced in size and hence provide an excellent foundation for creating a "minimal cell." However, little did the scientists at JCVI suspect that it would take 20 years to satisfy their scientific curiosity!

Instead of beginning by genetically manipulating a *Mycoplasma* species, microbiologists at JCVI wanted to have more control. To begin unraveling the genetic requirements for life, they first generated a synthetic self-replicating *Mycoplasma* (described in this chapter). The genome of this pioneering synthetic life form was synthesized from scratch based on its known genome sequence. The synthetic cell did not possess a "designer genome," or even a minimal one; it simply contained its own genome but one completely constructed in the laboratory. This breakthrough in synthetic biology provided the technology needed for microbiologists to create designer genomes.

Using comparative genomics and prior knowledge about specific gene sequences, microbiologists at JCVI continued their work by designing and synthesizing several minimal genomes that they hypothesized would sustain life. To their dismay, none of these resulted in a viable cell. So instead, they generated modules of DNA corresponding to a *Mycoplasma* genome and sewed different combinations together to form synthetic genomes. Once viable cells were obtained from transplanting these genomes, nonessential genes from the smallest genome were identified by transposon mutagenesis. After removing these unnecessary genes, a synthetic minimal cell coined JCVI-syn3.0 was created (see photo). This autonomous life form possesses a 531-kilobase genome encoding 473 genes; JCVI-syn3.0 thus contains a genome smaller than any other free-living cell.

While this work showcases the amazing advancements in synthetic biology and the potential for creating designer cells with novel functions, a surprising mystery surrounds this minimal cell: The roles for almost a third of JCVI-syn3.0's genes remain unknown, highlighting how much we still need to learn about the genetic foundation of a living cell.

Source: Hutchison, C.A. 3rd, et al. 2016. Design and synthesis of a minimal bacterial genome. *Science 351(6280):* aad6253. Photo provided by Clyde Hutchison and J. Craig Venter, JCVI and Thomas Deerinck and Mark Ellisman, NCMIR.

Industrial microbiology uses microbes on a large scale to produce desired products such as enzymes, foods, and beverages. These microbes are usually not genetically modified. Instead, naturally overproducing strains are isolated from wild-type strains and used for industrial purposes. In contrast, biotechnology uses genetically modified microorganisms to produce high-value products that the organisms do not naturally produce. In this chapter we discuss the basic techniques of genetic engineering that underlie biotechnology, in particular those used to clone, alter, and express genes efficiently in host organisms. We also explore how genetic engineering and biotechnology can be used for industrial, medical, and agricultural applications and introduce the exciting new field of synthetic biology.

I • Tools of the Genetic Engineer

Performing genetics *in vivo* (in living organisms) has many limitations that can be overcome by manipulating DNA *in vitro* (in a test tube). **Genetic engineering** refers to the use of in vitro techniques to alter genes in the laboratory. Such altered genes may be reinserted into the original source organism or into some other host organism. Expression of a gene from one organism in a different host organism is called **heterologous expression**.

Genetic engineering requires that specific DNA be isolated, purified, and further manipulated. We begin by considering some of the basic tools of the genetic engineer, including amplification of DNA, the separation of nucleic acids by electrophoresis, nucleic acid hybridization, and molecular cloning. We also describe methods for expressing foreign genes in bacteria and targeted mutagenesis.

12.1 Manipulating DNA: PCR and Nucleic Acid Hybridization

The first objective of genetic engineering is to isolate copies of specific genes in pure form, and the key method for doing so is the **polymerase chain reaction (PCR)** (**Figure 12.1**). Simply put, the polymerase chain reaction is DNA replication in vitro, as segments of target DNA are multiplied by up to a billionfold in the process of *amplification*. During each round of amplification, the amount of DNA *doubles*, leading to an exponential increase in the target DNA. Using an automated PCR machine called a *thermocycler*, a large amount of amplified DNA can be produced from only a few molecules of target DNA. In some cases it is desirable to quantify the initial amount of target DNA, and a variation on PCR called *quantitative PCR (qPCR)* is used for this purpose (⮌ Figures 28.23 and 28.24). A second variation on the original PCR technique allows for amplification of RNA (following its conversion to DNA, as discussed later in this section).

PCR and Polymerases

PCR requires DNA polymerase, the enzyme that naturally copies DNA molecules (⮌ Section 4.3), and artificially synthesized oligonucleotide primers (Section 12.4) made of DNA (rather than RNA like the primers used by cells) to synthesize DNA. PCR does not actually copy whole DNA molecules but amplifies stretches of up to a few thousand base pairs (the *target*) from within a

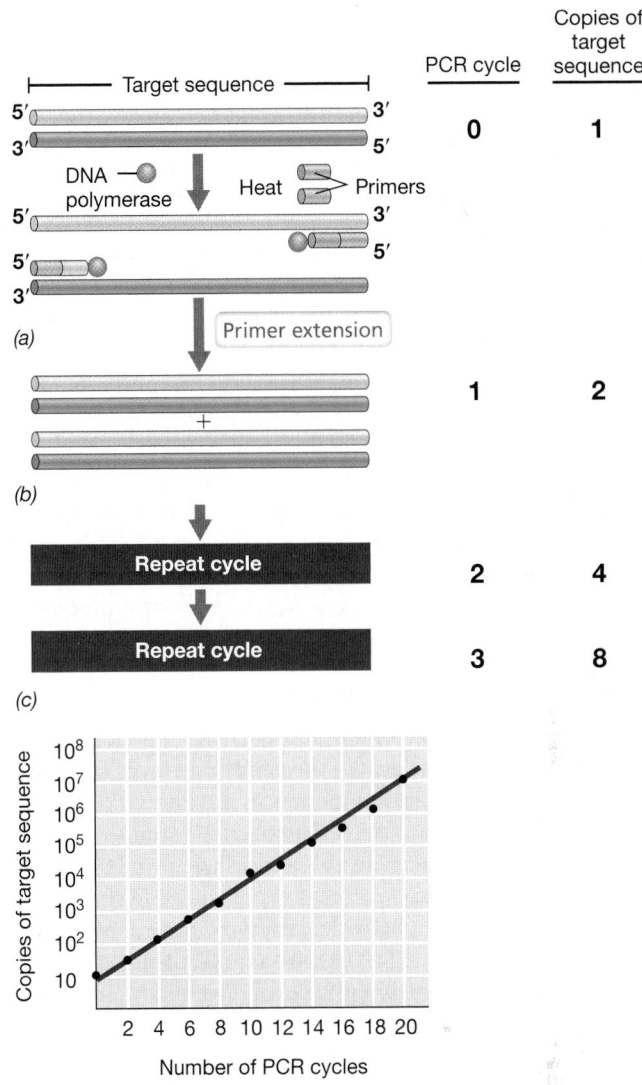

Figure 12.1 The polymerase chain reaction (PCR). The PCR amplifies specific DNA sequences. *(a)* Target DNA is heated to separate the strands, and a large excess of two oligonucleotide primers, one complementary to each strand, is added along with DNA polymerase. *(b)* Following primer annealing, primer extension yields a copy of the original double-stranded DNA. *(c)* Two additional PCR cycles yield four and eight copies, respectively, of the original DNA sequence. *(d)* Effect of running 20 PCR cycles on a DNA preparation originally containing ten copies of a target gene. Note that the plot is semilogarithmic.

larger DNA molecule (the *template*) during the following steps (Figure 12.1):

1. Template DNA is denatured by heating and then two DNA oligonucleotide primers flanking the target DNA on each strand are added in excess. This ensures that most template strands anneal to a primer, and not to each other, as the mixture cools (Figure 12.1*a*).

2. DNA polymerase then extends the primers using the original DNA as the template (Figure 12.1*b*).

3. After an appropriate incubation period, the mixture is heated again to separate the strands, but now the target gene is present in twice the original amount. The mixture is then cooled to

allow the primers to hybridize with complementary regions of newly synthesized and original DNA, and the process is repeated (Figure 12.1*c*). In practice, 20–30 cycles are typically run, yielding a 10^6-fold to 10^9-fold increase in the target sequence (Figure 12.1*d*).

Because high temperatures are used to denature the double-stranded copies of DNA in vitro, use of a thermostable DNA polymerase is critical. *Taq polymerase*, a DNA polymerase isolated from the thermophilic hot spring bacterium *Thermus aquaticus* (🔗 Section 16.20), is stable to 95°C and thus is unaffected by the denaturation step employed in the PCR (Figure 12.1). A DNA polymerase from *Pyrococcus furiosus*, a hyperthermophilic species of *Archaea* with a growth temperature optimum of 100°C (🔗 Section 17.4), is called *Pfu polymerase* and is even more thermostable than *Taq* polymerase. Moreover, unlike *Taq* polymerase, *Pfu* polymerase has proofreading activity (🔗 Section 4.4), making it especially useful when high accuracy is crucial. To supply the demand for thermostable DNA polymerases, the genes encoding these enzymes have been cloned into *Escherichia coli*, allowing the enzymes to be produced commercially in large quantities.

PCR Applications and RT-PCR

PCR is extremely valuable for obtaining DNA for gene cloning or for sequencing purposes because the gene or genes of interest can easily be amplified if flanking sequences are known. PCR is also used routinely in comparative or phylogenetic studies to amplify genes from various sources. In these cases the primers are made commercially to base-pair with regions of the gene that are conserved in sequence across a wide variety of organisms. Because small ribosomal subunit (SSU) rRNA—a molecule used for phylogenetic analyses—has both highly conserved and highly variable regions (🔗 Section 13.7 and Figure 13.15), primers specific for the SSU rRNA gene from different taxonomic groups can be used to survey habitats for their microbial contents (🔗 Section 19.6). Also, because it is so sensitive, PCR can be used to amplify very small quantities of DNA. For example, PCR has been used to amplify DNA from sources as varied as mummified human remains, fossilized plants and animals, and even single microbial cells (🔗 Section 9.12). It is also a common tool of medical diagnostics in clinical microbiology laboratories (🔗 Section 28.8) and is widely used in forensic science to attach an identity to evidence from a crime scene such as blood, semen, or tissue samples.

An important extension of the standard PCR procedure is *reverse transcription PCR* (RT-PCR), used to make *DNA* from an *mRNA* template (**Figure 12.2**). This procedure can be used to detect if a gene is expressed or to produce an intron-free eukaryotic gene for expression in bacteria (Section 12.3). RT-PCR uses the retroviral enzyme *reverse transcriptase* to convert RNA into **complementary DNA (cDNA)** (🔗 Section 10.11). Figure 12.2 illustrates how reverse transcriptase makes a single strand of cDNA using RNA as a template. To make cDNA, a primer complementary to the 3′ end of the target RNA is used by the enzyme reverse transcriptase to initiate RNA synthesis. If the template is eukaryotic mRNA, a primer complementary to the poly(A) tail (🔗 Section 4.6) of the mRNA can be used. The activity of reverse transcriptase results in a hybrid nucleic acid molecule containing both DNA and RNA. RNase H,

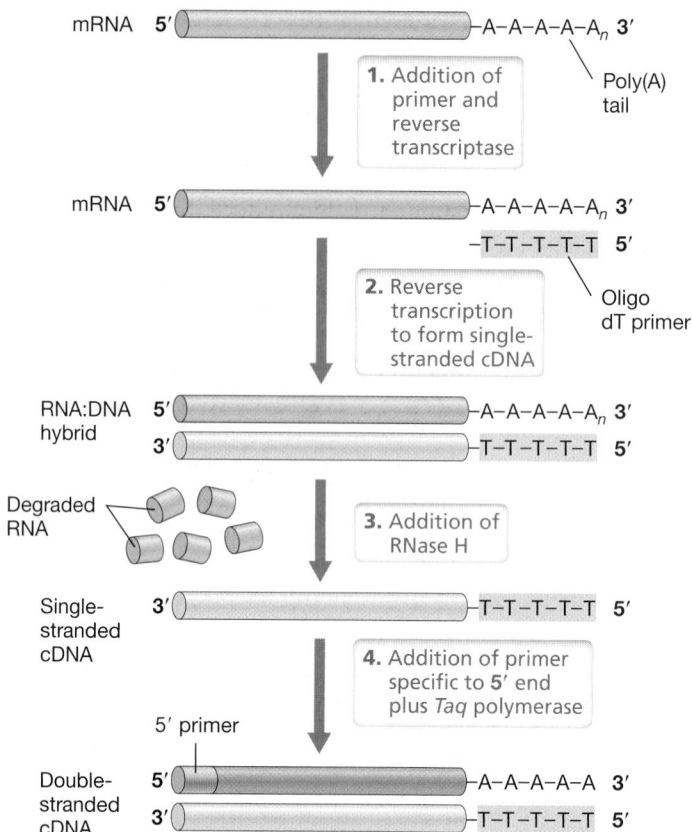

Figure 12.2 Reverse transcription PCR. Steps in the synthesis of cDNA from a eukaryotic mRNA. Reverse transcriptase synthesizes a hybrid molecule containing both RNA and DNA using the mRNA as a template and oligo-dT primer as a substrate. Next, the enzyme RNase H hydrolyzes the RNA portion of the hybrid molecule, yielding a single-stranded molecule of complementary DNA (cDNA). Following the addition of a primer complementary to the 5′ end of the cDNA, *Taq* polymerase produces a double-stranded cDNA.

a ribonuclease specific for the hybrid molecule, hydrolyzes the RNA, leaving the single-stranded cDNA as template for standard PCR using an additional primer complementary to the 5′ end.

Gel Electrophoresis and Nucleic Acid Hybridization

To verify that amplification of a nucleic acid was successful and for other nucleic acid manipulation steps, DNA or RNA fragments can be separated from each other by **gel electrophoresis**, a technique that employs an agarose gel to separate nucleic acid fragments based on differences in their size and charge (**Figure 12.3*a***). When an electrical current is applied, nucleic acids move through the gel toward the positive electrode because of their negatively charged phosphate groups, and small molecules migrate more rapidly than large molecules. After the gel has been run for a time sufficient to separate the molecules, the gel is stained with *ethidium bromide*, a compound that binds to nucleic acids and makes them fluoresce (Figure 12.3*b*). To determine the size of the DNA or RNA of interest, the migration is compared to a standard sample consisting of nucleic acid fragments of known sizes, called a *ladder*. DNA fragments can then be purified from gels and used for a variety of purposes, such as cloning or hybridization.

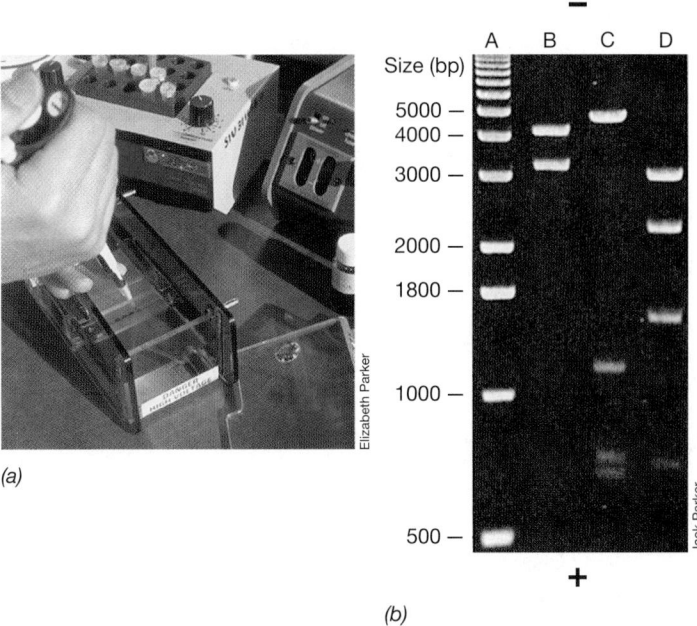

(a)

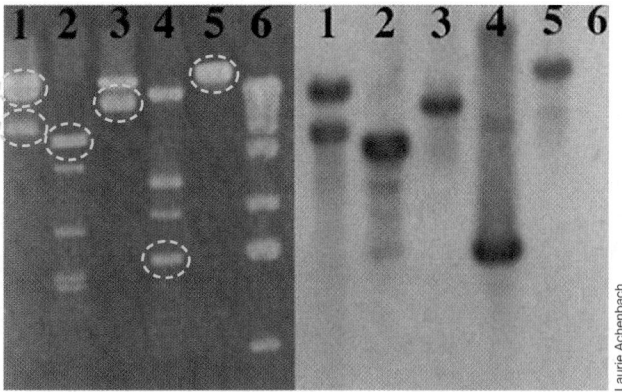

(a) **Southern blot**

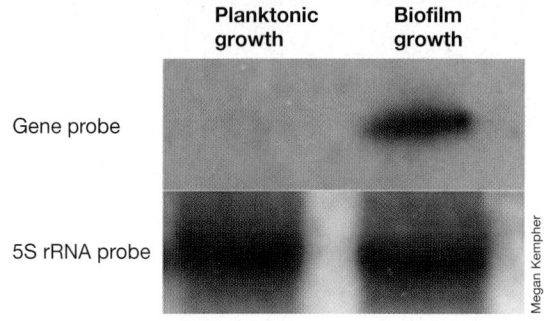

(b) **Northern blot**

Figure 12.3 Agarose gel electrophoresis of DNA. *(a)* DNA samples are loaded into wells in a submerged agarose gel. *(b)* A photograph of a stained agarose gel. The DNA was loaded into wells toward the top of the gel (negative pole) as shown, and the positive electrode is at the bottom. The standard sample in lane A (DNA ladder) has fragments of known size that may be used to determine the sizes of the fragments in the other lanes. Bands stain less intensely at the bottom of the gel because the fragments are smaller, and thus there is less DNA to stain.

Figure 12.4 Nucleic acid hybridization. *(a)* Southern blotting. (Left panel) Purified molecules of DNA from several different plasmids were treated with restriction enzymes and then subjected to agarose gel electrophoresis. (Right panel) Blot of the DNA gel shown to the left. After blotting, DNA in the gel was hybridized to a radioactive probe. The positions of the bands were visualized by X-ray autoradiography. Note that only some of the DNA fragments (circled in yellow in the left panel) have sequences complementary to the labeled probe. Lane 6 contained DNA used as a size marker and none of the bands hybridized to the probe. *(b)* Northern blotting. (Top panel) Hybridization and detection of a radioactive gene-specific probe to a blot of total RNA. The probe only bound to RNA from biofilm-grown cells, indicating that the target gene is not expressed during planktonic (suspended) growth. (Bottom panel) Hybridization and detection of a radioactive probe corresponding to the 5S rRNA to the same blot. The signal intensity indicates that equal amounts of RNA from each sample were loaded into the gel.

When DNA is denatured (that is, the two strands are separated), the single strands can be used to form hybrid double-stranded molecules with other single-stranded DNA (or RNA) molecules by complementary base pairing (⇄ Section 4.1) in a process called *nucleic acid hybridization*, or **hybridization** for short. Hybridization is widely used in detecting, characterizing, and identifying segments of DNA and RNA. Single-stranded nucleic acids whose identity is already known and that are used in hybridization are called **nucleic acid probes**, or simply *probes*. To allow detection, probes are made radioactive or are labeled with chemicals that are colored or yield fluorescent products (⇄ Section 19.5), and by varying the hybridization conditions, the "stringency" of the hybridization can be adjusted such that complementary base pairing is somewhat flexible or, alternatively, must be nearly exact.

Hybridization is useful for finding related sequences in different genomes or other genetic elements and to determine if a gene is expressed into an RNA transcript. In *Southern blotting*, probes of known sequence are hybridized to target DNA fragments that have been separated by gel electrophoresis. The hybridization procedure in which DNA is the target sequence in the gel, and RNA or DNA is the probe, is called a **Southern blot**. By contrast, a **Northern blot** uses RNA as the target sequence and DNA or RNA as the probe to detect gene expression. In both techniques, the nucleic acid fragments must be in a single-stranded form and are transferred to a synthetic membrane. The membrane is then exposed to the labeled probe. If the probe is complementary to any of the fragments, hybrids form, and the probe attaches to the membrane at the locations of the complementary fragments. **Figure 12.4** shows

how a Southern blot can be used to identify fragments of DNA containing sequences that hybridize to the probe and how the intensity of a signal on a Northern blot gives a rough estimate of mRNA abundance from the target gene.

Nucleic acid hybridization has many other uses. Hybridization is the basis of the fluorescence in situ hybridization (FISH) technique (⇄ Section 19.5) (**Figure 12.5**), where fluorescent probes are used to target specific DNA (or RNA) sequences in cells. This approach allows the identification of pathogens in clinical samples or bacteria of interest in environmental samples. For example, Figure 12.5 demonstrates the simultaneous use of eight different oligonucleotide probes in combinations to distinguish between 28 different strains of *E. coli* whose SSU rRNA sequences varied only slightly from strain to strain. The variations in color give a visual indication of the specificity and power of nucleic acid probes. Hybridization is also important in various "omics," in particular transcriptomics and metatranscriptomics, where genome-wide gene expression can be monitored in pure cultures

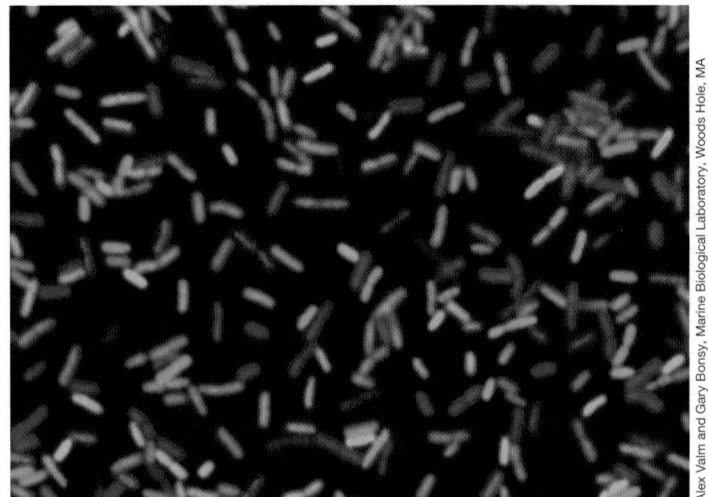

Figure 12.5 Fluorescence spectral image of 28 differently labeled strains of *Escherichia coli*. Cells were labeled with combinations of fluorophore-conjugated oligonucleotides that are complementary to *E. coli* 16S rRNA.

and natural populations, respectively, using microarray technology (⟺ Section 9.9).

MINIQUIZ

- Why is a primer needed at each end of the DNA segment being amplified by PCR?
- How does RT-PCR differ from traditional PCR?
- What are some applications of nucleic acid hybridization in molecular biology?

12.2 Molecular Cloning

The movement of desired genes from their original source to a small and manipulable genetic element (the **vector**) is called **molecular cloning**. Molecular cloning results in **recombinant DNA**, a molecule containing DNA from different sources. Once cloned, the gene(s) of interest can be manipulated, and when the recombinant vector is placed in an appropriate host, the cloned DNA is replicated, providing the foundation for much of genetic engineering.

An Overview of Gene Cloning

Following isolation of the source DNA, the major steps in gene cloning are (1) inserting the DNA into a cloning vector (**Figure 12.6**), and (2) inserting the vector into a host. The source DNA can be a gene or genes amplified by the polymerase chain reaction (Section 12.1), DNA synthesized from an RNA template by reverse transcriptase (Section 12.1), or even completely synthetic DNA made in vitro (Section 12.4). Cloning vectors are small, independently replicating genetic elements that can both carry and replicate cloned DNA segments (see Figure 12.8). Cloning vectors are typically designed to allow insertion of foreign DNA at a *restriction site* (Figure 12.6). *Restriction endonucleases*, or **restriction enzymes** for short, recognize specific base sequences (restriction sites) within DNA and cut

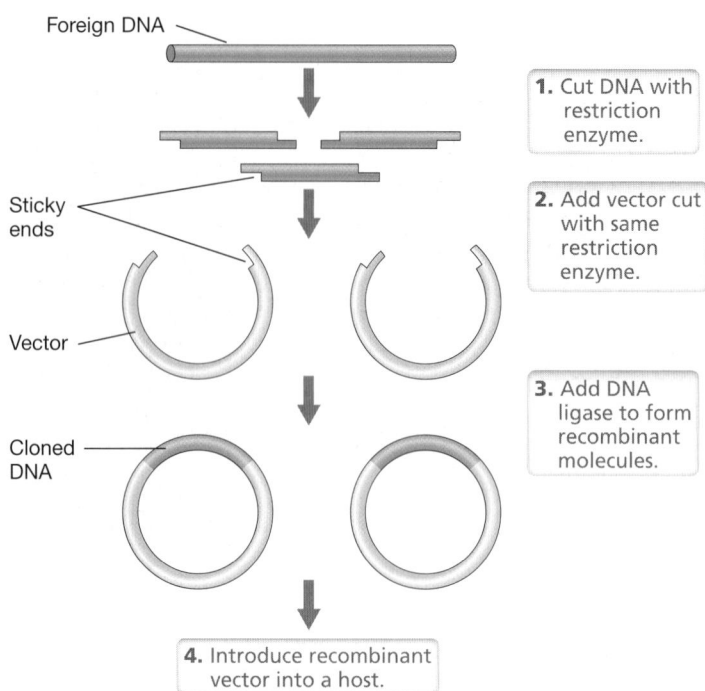

Figure 12.6 Major steps in gene cloning. By cutting the foreign DNA and the vector DNA with the same restriction enzyme, complementary sticky ends are generated that allow foreign DNA to be inserted into the vector.

the phosphodiester backbone, resulting in double-stranded breaks (**Figure 12.7**). The recognition sequences are typically inverted repeats and are called *palindromes*.

Restriction enzymes with different sequence specificities are widespread among *Bacteria*, where they help protect cells from attack by viral DNA. The cell is protected from its own restriction enzyme(s) by chemical modification (typically by methylation) of one of the bases in any potential restriction sites that exist in its genome. The restriction enzyme *Eco*RI makes staggered cuts,

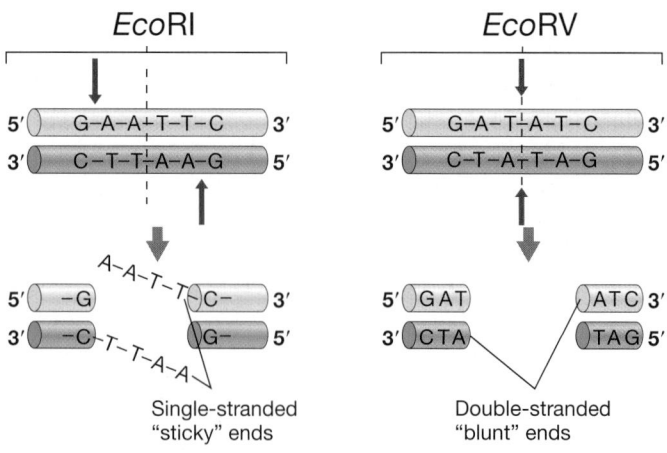

Figure 12.7 Restriction and modification of DNA. Sequences of DNA recognized by the restriction endonucleases *Eco*RI and *Eco*RV. The red arrows indicate the bonds cleaved by the enzyme, and the dashed line indicates the axis of symmetry of the sequence. After cutting DNA with these restriction enzymes, note the single-stranded "sticky" ends generated by *Eco*RI versus the "blunt" ends generated by *Eco*RV.

leaving short, single-stranded overhangs called "sticky" ends at the termini of the two fragments. Other restriction enzymes such as *Eco*RV cut both strands of the DNA directly opposite each other, resulting in blunt ends (Figure 12.7). If the source DNA and the vector are both cut with the *same* restriction enzyme that yields complementary sticky ends, the two molecules can be joined (annealed) using *DNA ligase*, an enzyme that covalently links the strands of the vector and the source DNA. If the source DNA is PCR generated, DNA ligase is used to join the amplified DNA to specialized vectors (see Figure 12.9a).

In the final step of gene cloning, recombinant DNA molecules are introduced into suitable host organisms where they can replicate. But in practice, this often yields a mixture of *recombinant* constructs, where only some of the cells contain the desired cloned gene. To identify a host colony containing the correct recombinant DNA, one can select host cells expressing a vector-encoded marker such as antibiotic resistance. Colonies can then be screened for recombinant vectors by looking for the inactivation of a vector gene due to insertion of foreign DNA (see Figure 12.8).

Cloning Vectors

Several types of cloning vectors exist, including viruses, cosmids, and artificial chromosomes, and their use is dependent on the size of the DNA fragment to be cloned and the host in which the vector will be inserted. Plasmids are widely used cloning vectors, and the plasmid pUC19 (**Figure 12.8**) is a good example. This plasmid possesses an ampicillin resistance gene for selection and a blue–white color-screening system to select for recombinants. It also contains a short segment of artificial DNA containing cut sites for many different restriction enzymes, called a *multiple cloning site* (*MCS*), inserted into the *lacZ* gene encoding the lactose-degrading enzyme β-galactosidase (⇄ Section 6.4 and Figure 6.14). The presence of the short MCS does not inactivate *lacZ*, and cut sites for restriction enzymes present in the MCS are absent from the rest of the vector.

The use of pUC19 in gene cloning is shown in Figure 12.8. A suitable restriction enzyme with a cut site within the MCS is chosen, and both the vector and the foreign DNA to be cloned are cut with this enzyme. The vector is linearized, and segments of the foreign DNA are inserted into the open cut site and ligated into position with the enzyme DNA ligase. This insertion disrupts the *lacZ* gene—a phenomenon called *insertional inactivation*—and is used to detect the presence of foreign DNA within the vector or recombinant vector. After DNA ligation, the resulting plasmids are transformed into cells of *Escherichia coli* and the cells plated on media containing both ampicillin (to select for cells containing the plasmid) and a lactose analog called *X-gal*, to detect β-galactosidase activity. X-gal, which is colorless, can be cleaved by β-galactosidase to generate a blue product. Thus, cells containing the vector *without* cloned DNA form blue colonies (that is, β-galactosidase is active), whereas cells containing the vector *with* an insert of cloned DNA do not form β-galactosidase and are therefore white and are the focus of further analyses.

Plasmids developed specifically for cloning DNA products synthesized by *Taq* polymerase in a polymerase chain reaction (PCR; Section 12.1) have also been designed (**Figure 12.9a**). The enzymatic activity of *Taq* polymerase adds a template-independent

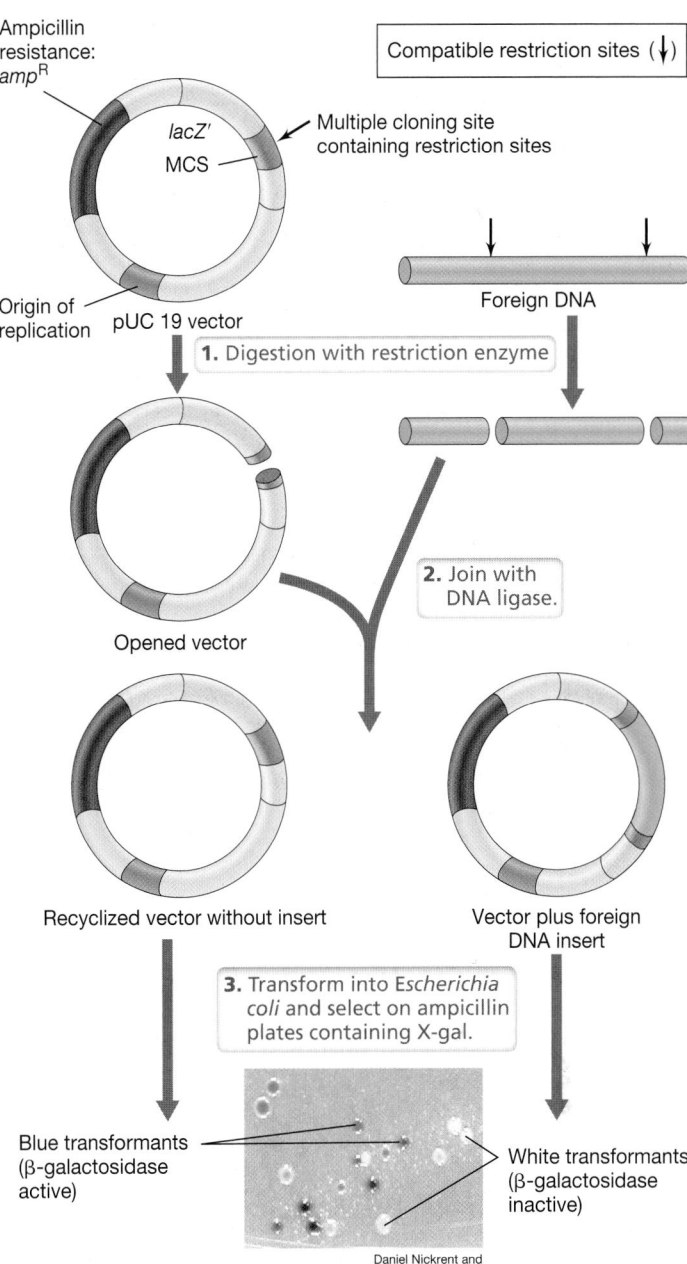

Figure 12.8 Cloning into the plasmid vector pUC19. Essential features include an ampicillin resistance marker and the multiple cloning site (MCS) with multiple restriction enzyme cut sites. The cloning vector and foreign DNA are cut with compatible restriction enzymes at positions indicated by the arrows. Insertion of DNA within the MCS inactivates β-galactosidase, allowing blue–white screening for the presence of the insert. The photo on the bottom shows colonies of *Escherichia coli* on an X-gal plate. The enzyme β-galactosidase can cleave the normally colorless X-gal to form a blue product.

adenosine residue to the 3′ ends of its products. Linearized vectors are commercially available that contain overhanging thymidine residues that allow for base pairing with the *Taq* PCR product and subsequent ligation using DNA ligase (Figure 12.9a). For cloning genes into the yeast *Saccharomyces cerevisiae*, **yeast artificial chromosomes (YACs)** are frequently used (Figure 12.9b). YACs are linear vectors that replicate in yeast like normal chromosomes but have sites where very large fragments of DNA can be inserted.

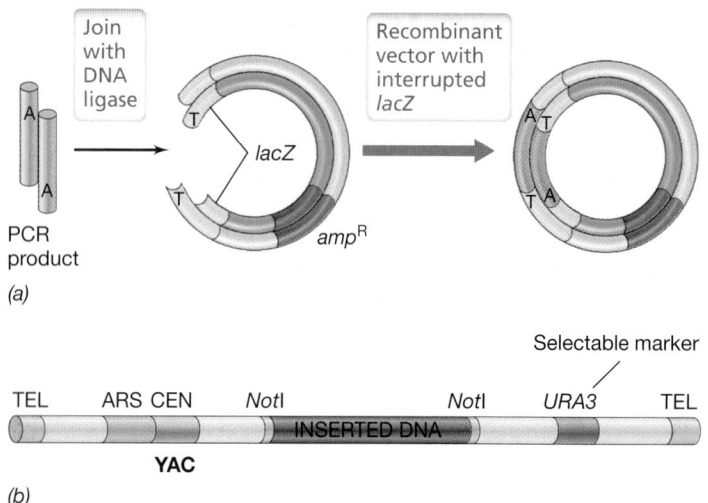

(a)

(b)

Figure 12.9 Specialized vectors. *(a)* PCR vector. The linearized cloning vector contains overhanging thymidine residues that base-pair with the adenosine residues present on the 3′ ends of *Taq*-polymerase-generated PCR. Ligation of the two pieces of DNA yields a circular plasmid containing an interrupted *lacZ*. *(b)* A yeast artificial chromosome (YAC) containing foreign DNA. The foreign DNA was cloned into the vector at a *Not*I restriction site. The telomeres are labeled TEL and the centromere CEN. The origin of replication is labeled ARS (for *autonomous replication sequence*). The *URA3* gene is used for selection. The host into which the clone is transformed has a mutation in *URA3* and requires uracil for growth (Ura⁻). Host cells containing this YAC become Ura⁺. The diagram is not to scale; vector DNA is only 10 kbp whereas cloned DNA can be up to 800 kbp.

To function like normal eukaryotic chromosomes, YACs have an origin of DNA replication, telomeres for replicating DNA at the ends of the chromosome, and a centromere for segregation during mitosis. YACs also contain a cloning site and a gene for selection following transformation into the host (Figure 12.9*b*).

Hosts for Cloning Vectors

The most useful hosts for cloning are microorganisms that are easy to grow and transform with engineered DNA. They must also be genetically stable in culture and have the appropriate enzymes to allow replication of the vector. It is also helpful if considerable background information on the host and a wealth of tools for its genetic manipulation exist. Hosts that meet these conditions include the bacteria *E. coli* and *Bacillus subtilis*, and the yeast *S. cerevisiae* (**Figure 12.10**). Complete genome sequences are available for all of these organisms, and they are widely used as cloning hosts.

Although *E. coli* is found in the human intestine and some wild-type strains are potentially harmful (⇔ Section 32.11), several modified *E. coli* strains have been developed specifically for cloning purposes. However, if cloned gene *expression* is desired, the outer membrane of this gram-negative bacterium (⇔ Section 2.5) can hinder protein secretion. This issue can be overcome using the gram-positive bacterium *B. subtilis* as a cloning host (Figure 12.10). Cloning of DNA from eukaryotic sources into eukaryotic rather than prokaryotic cells is often done since eukaryotic hosts already possess the complex RNA and post-translational processing systems required for the production of eukaryotic proteins (⇔ Section 4.6). Because it is easy to grow and manipulate, the

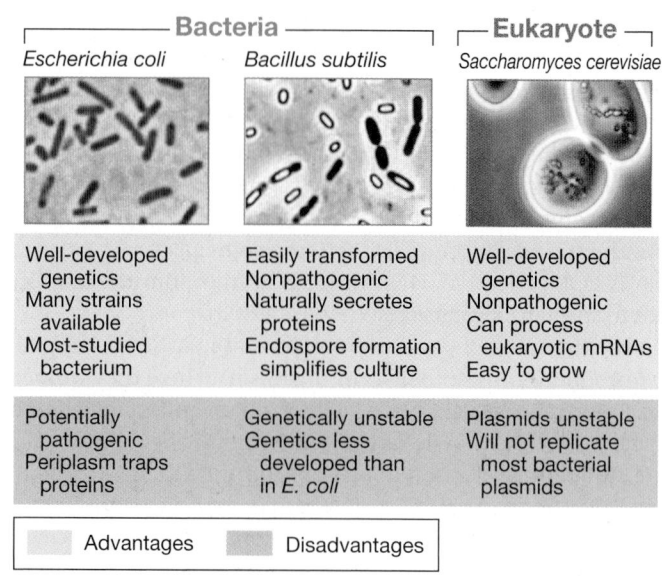

Figure 12.10 Hosts for molecular cloning. A summary of the advantages and disadvantages of some common cloning hosts.

workhorse for cloning in eukaryotic cells is the yeast *S. cerevisiae*. However, some cloning applications require the use of plant tissues, insect cell lines, or cultured mammalian cells. Regardless of eukaryotic host type, it is necessary to get cloned DNA into the host, and several methods including transfection (see Figure 12.20), microinjection, and electroporation (⇔ Section 11.6) can be used.

MINIQUIZ

- What is the purpose of molecular cloning?
- What is a multiple cloning site, and what is insertional inactivation?
- When would it be beneficial to use a eukaryotic host for molecular cloning?

12.3 Expressing Foreign Genes in *Bacteria*

Once genes are cloned, they can be transcribed and translated (expressed) to produce their encoded proteins. Obstacles to the expression of genes from mammalian or other eukaryotic sources include the following: (1) The genes must be placed under control of a bacterial promoter; (2) any introns (⇔ Section 4.6) must be removed; (3) codon usage (codon bias, ⇔ Section 4.9) may require edits to gene sequences; and (4) many eukaryotic proteins require host modification after translation to yield the active form and bacteria cannot perform most such modifications. Here we consider solutions to these challenges.

Transcription and Translation of Cloned Genes Using Expression Vectors

Expression vectors are designed to allow the experimenter to control the expression of cloned genes. However, the native promoter of a cloned gene may work poorly in a new host, and the overproduction of foreign proteins may also damage the host cell. Therefore, it is important to regulate the expression of cloned genes. Typically, the regulation is at the transcriptional level, and in practice, high levels of transcription require strong promoters that bind RNA polymerase efficiently (↩ Section 4.5). An example of this is the use of the bacteriophage T7 promoter and T7 RNA polymerase to regulate gene expression. When T7 infects *Escherichia coli*, it encodes its own RNA polymerase that recognizes only T7 promoters (↩ Section 10.4). In T7 expression vectors, cloned genes are placed under control of the T7 promoter. To achieve this, the gene for T7 RNA polymerase must also be present in the cell under the control of an easily regulated system, such as *lac* (↩ Section 6.2) (**Figure 12.11**). This is usually done by integrating the gene for T7 RNA polymerase with a *lac* promoter into the chromosome of a specialized host strain.

The BL21 series of *E. coli* host strains are especially designed to work with the pET series of T7 expression vectors (Figure 12.11). The cloned genes are expressed shortly after T7 RNA polymerase transcription has been switched on by a *lac* inducer, such as the chemical IPTG (↩ Section 6.2). Because it recognizes only T7 promoters, the T7 RNA polymerase transcribes only the cloned genes. The T7 RNA polymerase is so highly active that it uses most of the RNA precursors, thereby limiting transcription to the cloned genes. Consequently, host genes that require host RNA polymerase are for the most part not transcribed and thus the cells stop growing; translation in such cells then yields primarily the protein of interest. The T7 control system is thus very effective for generating large amounts of a specific protein.

Expression vectors must also be designed to ensure that the mRNA produced is efficiently translated. To synthesize protein from an mRNA molecule, it is essential for the ribosomes to bind at the correct site and begin reading in the correct frame. In bacteria this is accomplished by having a ribosome-binding site (RBS, ↩ Section 4.9) and a nearby start codon on the mRNA. Bacterial RBSs are not found in eukaryotic genes and must be engineered into the vector if high levels of expression of the eukaryotic gene are to be obtained.

Other adjustments to a cloned gene may be necessary to ensure high-efficiency translation. For example, *codon usage* can be an obstacle. Codon usage is related to the concentration of the appropriate tRNA in the cell (↩ Section 9.2 and Table 9.3). Because of the redundancy of the genetic code, more than one tRNA exists for most amino acids (↩ Section 4.9). Therefore, if a cloned gene has a codon usage pattern distinct from that of its expression host, it will probably be translated inefficiently in that host. Site-directed mutagenesis (Section 12.4) can then be used to change selected codons in the gene, making it more amenable to the codon usage pattern of the host.

Cloning the Gene via mRNA or Artificial Synthesis

If a cloned gene contains introns, as eukaryotic genes typically do (↩ Section 4.6), the correct protein product will not be made in a bacterial host unless modifications are made. This can be done via mRNA. In a typical mammalian cell, less than 5% of the total RNA is mRNA. However, eukaryotic mRNA is unique because of the poly(A) tails found at the 3′ end (↩ Section 4.6), and this makes it easy to isolate, even though it is of low abundance. If a cell extract is passed over a chromatographic column containing strands of poly(T) linked to a cellulose support, most of the mRNA separates from other RNAs by sticking to the support by specific pairing of As and Ts. The RNA is then released from the column by a low-salt buffer, which gives a preparation greatly enriched in mRNA.

Once mRNA has been isolated, the genetic information is converted into complementary DNA (cDNA) by RT-PCR as was illustrated in Figure 12.2. This double-stranded cDNA contains the coding sequence but lacks introns (**Figure 12.12**), and thus it can be inserted into a plasmid or other vector for cloning. However, because the cDNA contains only coding sequences, it lacks a promoter and other upstream regulatory sequences necessary for expression. Thus expression vectors containing bacterial promoters and ribosome-binding sites are used to obtain high-level expression of genes cloned in this way (see Figure 12.13).

For small proteins it is possible to artificially synthesize the entire gene (Section 12.4). Many mammalian proteins such as high-value peptide hormones are made by protease cleavage of large precursor molecules. Thus, in order to produce a short peptide such as insulin in its active form, construction and cloning of an artificial gene that encodes just the final hormone rather than the larger precursor protein from which it was derived may have several advantages. Constructed genes are naturally free of introns and thus the mRNA does not need processing. Also, promoters and other regulatory sequences can be inserted into the gene upstream of the coding sequences, and codon bias (↩ Sections 4.9 and 9.2) can be adjusted to best suit the expression host.

Protein Stability and Purification

The synthesis of a protein in a new host may spawn additional problems. For example, some proteins are susceptible to degradation by protease enzymes and others may be toxic to their host.

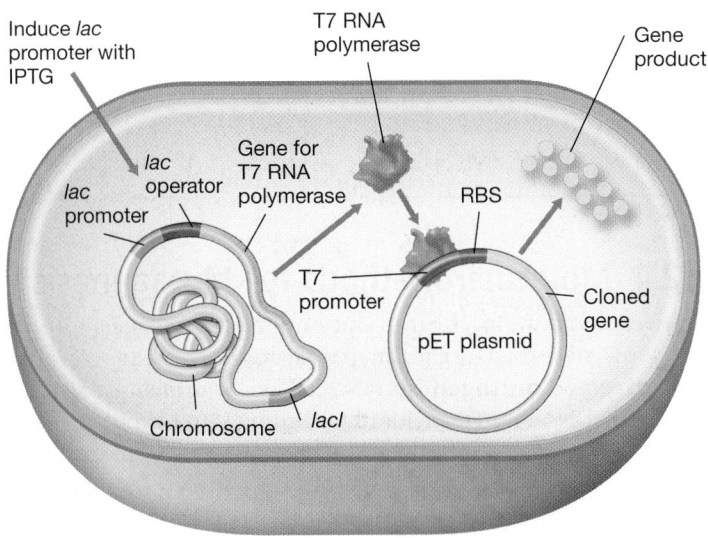

Figure 12.11 The T7 expression system. The gene for T7 RNA polymerase is in a gene fusion under control of the *lac* promoter and is inserted into the chromosome of a special host strain of *Escherichia coli* (BL21). Addition of IPTG induces the *lac* promoter, causing expression of T7 RNA polymerase. This transcribes the cloned gene, which is under control of the T7 promoter and is carried by the pET plasmid. RBS, ribosome-binding site.

UNIT 3

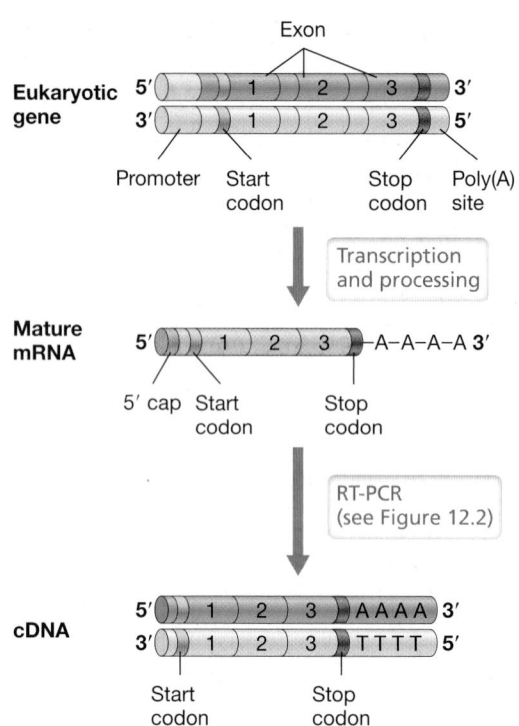

Figure 12.12 Complementary DNA (cDNA). Steps illustrating the synthesis of an intron-lacking cDNA corresponding to a eukaryotic gene generated by reverse transcription PCR (RT-PCR).

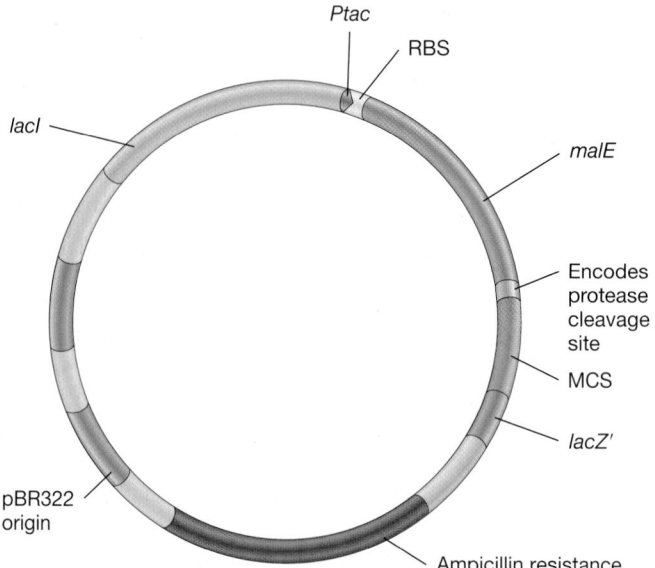

Figure 12.13 An expression vector for gene fusions. The gene to be cloned is inserted into the multiple cloning site (MCS) so it is in frame with the *malE* gene, which encodes maltose-binding protein. The insertion inactivates the gene for the alpha fragment of *lacZ*, which encodes β-galactosidase. The fused gene is under control of the hybrid *tac* promoter (*Ptac*) and an *Escherichia coli* ribosome-binding site (RBS). The plasmid also contains the *lacI* gene, which encodes the *lac* repressor. Therefore, an inducer must be added to turn on the *tac* promoter. The plasmid contains a gene conferring ampicillin resistance on its host.

Also, when proteins are massively overproduced, they sometimes aggregate into insoluble inclusions. Although inclusions are relatively easy to purify, the protein they contain is often difficult to solubilize and may be partially denatured. Protein purification can be simplified if the target protein is made as a *fusion protein* along with a carrier protein encoded by the vector. To do this, the two genes are fused to yield a single coding sequence. A short segment that is recognized and cleaved by a commercially available protease is included between them. After transcription and translation, a single protein is made that is purified by methods designed for the carrier protein. The fusion protein is then cleaved by the protease to release the target protein from the carrier protein. Fusion proteins simplify purification of the target protein because a carrier protein is chosen that will not form inclusions and is easy to purify.

Several vectors are available to generate fusion proteins, and **Figure 12.13** shows an example of a fusion vector that is also an expression vector. In this example, the carrier protein is the *E. coli* maltose-binding protein (encoded by *malE*, Figure 12.13), a protein that is easily purified by methods based on its high affinity for maltose. Once purified, the two portions of the fusion protein are separated by protease or chemical treatment. One other advantage of making a fusion protein is that the carrier protein can be chosen to contain the bacterial *signal sequence*, a peptide rich in hydrophobic amino acids that enables transport of the protein across the cytoplasmic membrane (⊂⊃ Section 4.12). This makes possible a bacterial expression system that not only *makes* and *secretes* mammalian proteins, but also allows for the heterologously expressed protein to be separated from all of the other

proteins secreted by the cell using binding resins specific for the maltose-binding protein. Thus, carrier proteins can be used to save time, money, and effort in obtaining a desired product.

— MINIQUIZ —

- How can the bacteriophage T7 promoter be used to control expression of a eukaryotic gene in *Escherichia coli*?

- What major advantage does cloning mammalian genes from mRNA or using synthetic genes have over PCR amplification and cloning of the native gene?

- How is a fusion protein made?

12.4 Molecular Methods for Mutagenesis

Conventional mutagens introduce mutations *at random* in the DNA of the intact organism (⊂⊃ Section 11.4). In contrast, **site-directed mutagenesis** (also called *in vitro mutagenesis*) uses synthetic DNA plus DNA cloning techniques to introduce mutations into genes *at precisely determined sites*. In addition to changing one or just a few bases, mutations may also be engineered by inserting large segments of DNA at precisely determined locations.

Site-Directed Mutagenesis

Site-directed mutagenesis requires that short sequences of DNA (*oligonucleotides*) of precise sequence be available, and these are chemically synthesized; primers or probes for use in the polymerase chain reaction and hybridization (Section 12.1) are also made in this way. Oligonucleotides of 12–40 bases are inexpensive and commercially available, and oligonucleotides of over 100 bases in length can be made if necessary. Site-directed mutagenesis then allows any base pair in a specific gene to be changed. When the mutated gene is expressed, a protein with an altered amino acid sequence will be produced. Site-directed mutagenesis can thus be used to manipulate proteins to test the functional importance of specific amino acids.

One procedure for site-directed mutagenesis is illustrated in **Figure 12.14**. A cloned target gene is denatured to form single-stranded DNA and then allowed to hybridize with the mutagenized oligonucleotide containing a one-base mismatch. After extension by DNA polymerase, the complementary DNA strand formed will contain the mismatch. After transformation of the vector into host cells followed by its semiconservative replication and subsequent cell division, one daughter cell will carry the mutation while the other will be wild type. Progeny bacteria are then screened for those carrying the mutation.

Site-directed mutagenesis may also be carried out using PCR. In this case, the short DNA oligonucleotide with the required mutation is used as a PCR primer. The mutation-carrying primer is designed to anneal to the target with the mismatch in the middle and must have enough matching nucleotides on both sides for binding to be stable during the PCR reaction. The mutant primer is then paired with a normal primer, and when the PCR reaction amplifies the target DNA, it incorporates the mutation(s) into the final amplified product.

Site-directed mutagenesis has many applications. The technique has been widely used by enzymologists to change a specific amino acid in the active site of an enzyme to see how the modified enzyme compares with the wild-type enzyme. In such experiments, the vector encoding the mutant enzyme is inserted into a mutant host strain unable to make the original enzyme. Consequently, the activity measured is due to the mutant version of the enzyme alone. Using in vitro mutagenesis, enzymologists can link virtually any aspect of an enzyme's activity—catalysis, resistance, susceptibility to chemical or physical agents, interactions with other proteins—to specific amino acids in the enzyme. In genetic engineering, site-directed mutagenesis has been used to improve the properties of specific proteins, and we discuss some examples in Section 12.6.

Cassette Mutagenesis and Gene Disruption

To make more than a few base-pair changes or replace sections of a gene of interest, synthetic fragments called **DNA cassettes** (or cartridges) can be used to mutate DNA in a process known as **cassette mutagenesis**. These cassettes can be synthesized using the polymerase chain reaction or by direct DNA synthesis. The cassette can then replace sections of the DNA of interest using restriction sites. However, if sites for the appropriate restriction enzyme are not present at the required location, they can be inserted by site-directed mutagenesis (Figure 12.14). Cassettes used to replace sections of genes are typically the same size as the wild-type DNA fragments they replace.

Another type of cassette mutagenesis is called **gene disruption**. In this technique, cassettes are inserted into the middle of a gene, thus disrupting the coding sequence (**Figure 12.15**). Cassettes used for making insertion mutations can be almost any size and can even carry an entire gene. To facilitate selection, cassettes that encode antibiotic resistance are commonly used. For example, a DNA cassette containing a gene conferring kanamycin resistance is inserted at a restriction site in a cloned gene. The vector carrying the disrupted gene is then converted from a circular to a linear form by cutting it with a different restriction enzyme. Finally, the linear DNA is transformed into the host and kanamycin resistance is selected. The linear plasmid cannot replicate, and so resistant cells arise mostly by homologous recombination (⮌ Section 11.5) between the mutated gene on the plasmid and the wild-type gene on the chromosome (Figure 12.15).

When a cassette is inserted, the cells not only gain antibiotic resistance, but they also *lose the function of the gene* into which the cassette is inserted. Such mutations are called *knockout mutations* and are widely used in biology. Knockouts are similar to insertion mutations made by transposons (⮌ Section 11.11), but here the experimenter chooses which gene will be mutated. Knockout mutations in haploid organisms yield viable cells only if the disrupted gene is nonessential. Thus, gene knockouts are commonly used for determining whether a gene of interest is essential.

UNIT 3

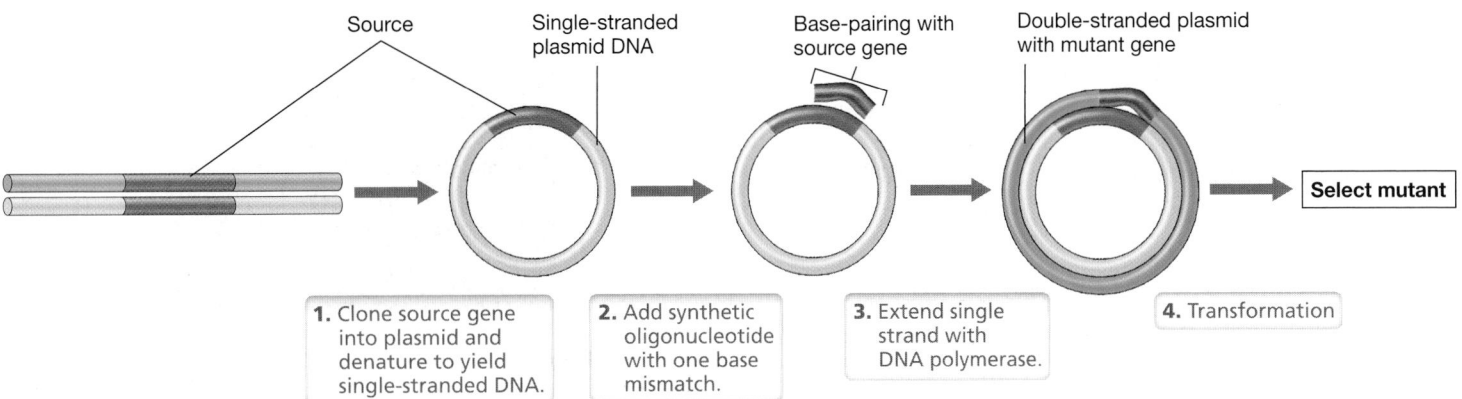

Figure 12.14 Site-directed mutagenesis using synthetic DNA. Short synthetic oligonucleotides hybridized to the cloned gene may be used to generate mutations. Cloning the source DNA into a plasmid followed by denaturation yields the single-stranded DNA needed for site-directed mutagenesis to work.

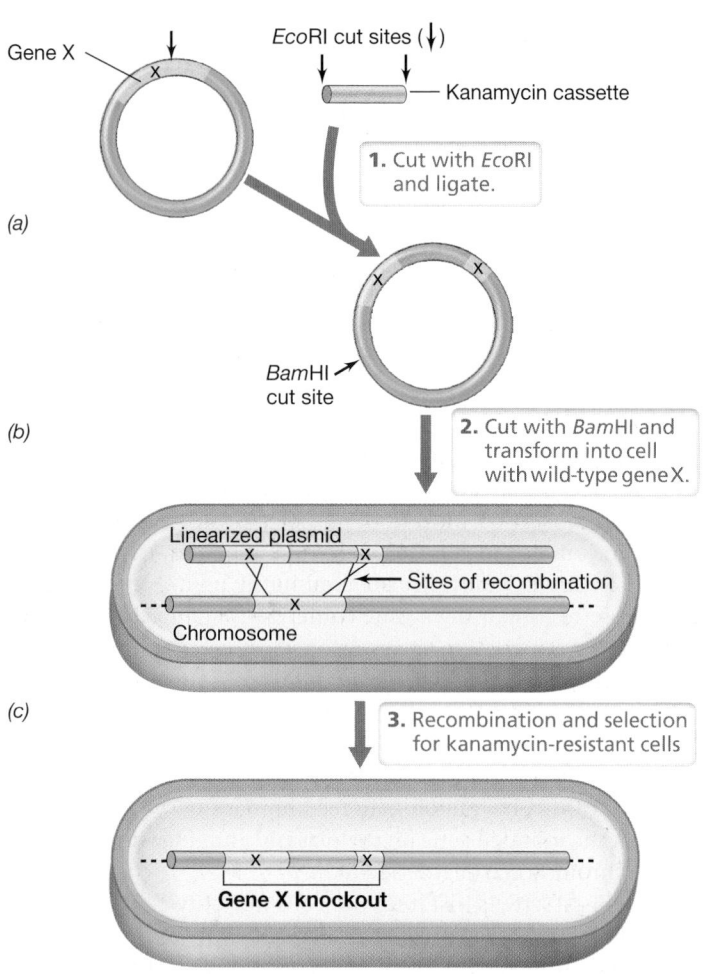

(a)

(b)

(c)

(d)

Figure 12.15 Gene disruption by cassette mutagenesis. (a) A cloned wild-type copy of gene X, carried on a plasmid, and a kanamycin cassette are cut with *Eco*RI and mixed. (b) The cut plasmid and the cassette are ligated, creating a plasmid with the kanamycin cassette as an insertion mutation within gene X. This new plasmid is cut with *Bam*HI and transformed into a cell. (c) The transformed cell contains the linearized plasmid with a disrupted gene X and its own chromosome with a wild-type copy of the gene. (d) In some cells, homologous recombination occurs between the wild-type and mutant forms of gene X. Cells that can grow in the presence of kanamycin have only a single, disrupted copy of gene X.

MINIQUIZ

- How can site-directed mutagenesis be useful to enzymologists?
- What is used to alter more than a few base pairs in a gene of interest?
- What are knockout mutations?

12.5 Reporter Genes and Gene Fusions

DNA manipulation has revolutionized the study of gene regulation, and *gene fusions* have been a major tool for studying regulatory events. In a reporter gene fusion, a coding sequence from one source (the *reporter*) is fused to a regulatory region from another source to form a hybrid gene. Regulation of gene expression is then studied by assaying for the product of the reporter as a function of different conditions sensed by the regulator.

Reporter Genes

The key property of a **reporter gene** is that it encodes a protein that is easy to detect and assay. Reporter genes are used for a variety of purposes. They may be used to report the presence or absence of a particular genetic element (such as a plasmid) or DNA inserted within a vector. They can also be fused to other genes or to the promoter of other genes so that gene expression can be studied (⊘ Figure 7.2).

The first widely used reporter gene was *lacZ* from *Escherichia coli*, a gene that encodes the enzyme β-*galactosidase*, required for lactose catabolism (⊘ Section 6.2). Cells expressing β-galactosidase can be detected easily by the color of their colonies on indicator plates that contain the artificial substrate X-gal (5-bromo-4-chloro-3-indolyl-β-D-galactopyranoside); X-gal is cleaved by β-galactosidase to yield a blue color (see Figure 12.8).

The **green fluorescent protein (GFP)** is widely used as a reporter (**Figure 12.16**). Although the gene for GFP was originally cloned from the jellyfish *Aequorea victoria*, GFP may be expressed in most cells as it is stable and causes little or no disruption of host cell metabolism. If expression of a cloned gene is linked to the expression of GFP, the latter signals (reports) that the cloned gene has also been expressed (Figure 12.16). Since the advent of GFP, many similar but differently colored fluorescent proteins have been developed as reporters (⊘ Section 7.1).

Gene Fusions

Gene fusions are genetic constructs that consist of segments from two different genes. If the promoter that controls a coding sequence is removed, the coding sequence can be fused to a different regulator to place the gene under the control of a different promoter. Alternatively, the promoter region can be fused to a gene

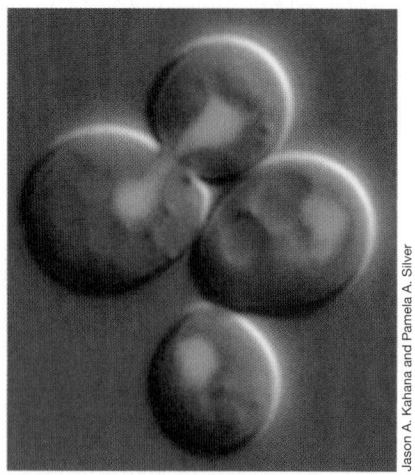

Figure 12.16 Green fluorescent protein (GFP). GFP can be used as a tag for protein localization in vivo. In this example, the gene encoding Pho2, a DNA-binding protein from the yeast *Saccharomyces cerevisiae*, was fused to the gene encoding GFP and photographed by fluorescence microscopy. The recombinant gene was transformed into budding yeast cells. These expressed the fluorescent fusion protein localized in the nucleus.

whose product is easy to assay. There are two different types of gene fusions. In **operon fusions**, a coding sequence that retains its own translational start site and signals is fused to the transcriptional signals of another gene. In **protein fusions**, genes that encode two different proteins are fused together so that they share the same transcriptional and translational start and stop signals. Following translation, protein fusions yield a single hybrid polypeptide (Section 12.3).

Gene fusions are often used in studying gene regulation, especially if measuring the levels of the natural gene product is difficult, expensive, or time consuming. The regulatory region of the gene of interest is fused to the coding sequence for a reporter gene, such as that for β-galactosidase or GFP. The reporter is then made under the conditions that would trigger expression of the target gene (**Figure 12.17**). The expression of the reporter is assayed under a variety of conditions to determine how the gene of interest is regulated (Chapter 6). *Transcriptional control* is assayed by fusing the transcriptional start signals of the gene of interest to a reporter gene, whereas *translational control* is assayed by fusing translational start signals of a gene of interest to a reporter gene under the control of a known promoter.

Gene fusions may also be used to test for the effects of regulatory genes. Mutations that affect regulatory genes are introduced into cells carrying gene fusions, and expression is measured and compared to cells lacking the regulatory mutations. This allows the rapid screening of multiple regulatory genes that are suspected of controlling the target gene. Besides the use of fusions to monitor for the presence or expression of a gene, proteins that are easily purified can also be fused to proteins of interest to aid in purification of the latter (Section 12.3).

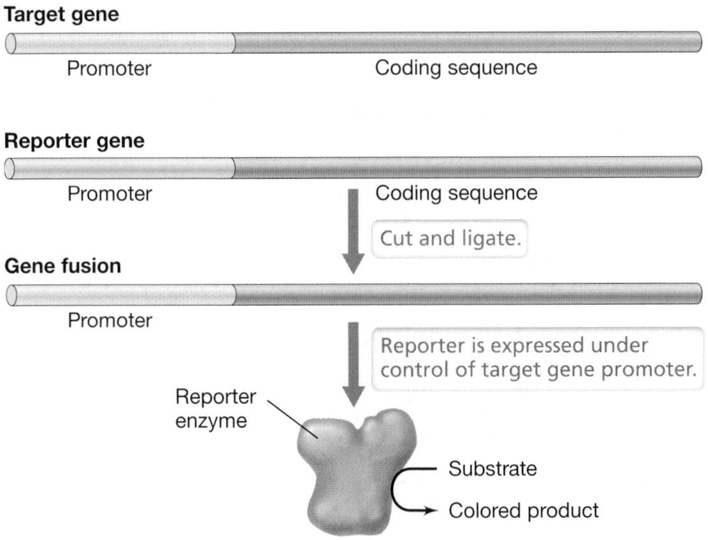

Figure 12.17 Construction and use of gene fusions. The promoter of the target gene is fused to the reporter coding sequence. Consequently, the reporter gene is expressed under those conditions where the target gene would normally be expressed. The reporter shown here is an enzyme (such as β-galactosidase) that converts a substrate to a colored product that is easy to detect. This approach greatly facilitates the investigation of regulatory mechanisms.

MINIQUIZ

- What is a reporter gene? The product of which reporter gene yields a green color?
- Why are gene fusions useful in studying gene regulation?

II • Making Products from Genetically Engineered Microbes: Biotechnology

Genetic engineering can transform microorganisms into tiny factories for the production of valuable products including fuels, chemicals, drugs, and human hormones, such as insulin. This is the science of **biotechnology**. Up to this point we have only considered the techniques used for manipulating, cloning, and expressing DNA. We now consider how these techniques are applied in biotechnology to produce valuable proteins and genetically modified plants, animals, vaccines, and metabolic pathways.

UNIT 3

12.6 Somatotropin and Other Mammalian Proteins

One of the most economically profitable areas of biotechnology has been the production of human proteins. Many mammalian proteins have high pharmaceutical value but are typically present in very low amounts in normal tissue, and it is therefore extremely costly to purify them. Even if the protein can be produced in cell culture, this is much more expensive and difficult than growing microbial cultures that produce the protein in high yield. Therefore, the biotechnology industry has developed genetically engineered microorganisms to produce many different mammalian proteins.

Somatotropin

Although insulin was the first human protein to be produced by bacteria, the genetic engineering required was complicated because insulin consists of two short polypeptides held together by disulfide bonds. A more straightforward example is *human somatotropin* (growth hormone), which consists of a single polypeptide encoded by a single gene; a deficiency of somatotropin in the body results in hereditary dwarfism. Because the human somatotropin gene has been successfully cloned and expressed in bacteria, children showing stunted growth can be treated with *recombinant human somatotropin* to correct this. However, some forms of dwarfism are caused by a lack of the somatotropin receptor, and in such cases, administration of somatotropin has no effect.

The human somatotropin gene was cloned as complementary DNA (cDNA) from mRNA as described in Section 12.3 (see Figure 12.18). The cDNA was then expressed in a bacterial expression vector. The main problem with producing relatively short polypeptide hormones such as somatotropin is their susceptibility to protease digestion, but this problem was overcome by using bacterial host strains lacking key protease enzymes. Today recombinant human growth hormone taken by injection is marketed under several brand names in the United States and has successfully treated thousands of children afflicted with any of several different syndromes that result in short stature. Recombinant somatotropin has also been used to treat some cases of tissue atrophy in adults. However, use in adults is not a common practice, and growth hormone is banned by the International Olympic Committee and by some professional sports leagues for its alleged performance-enhancing capabilities.

Recombinant bovine somatotropin (rBST) is used in the dairy industry (**Figure 12.18**). Injection of rBST into cows does not make them grow larger but instead stimulates milk production. This is because somatotropin has two binding sites; one is the somatotropin receptor that stimulates growth while the other is the prolactin receptor that promotes milk production. Thus, cows treated with rBST produce more milk. However, when human somatotropin is used to treat short stature conditions, it is desirable to avoid any side effects from the hormone's prolactin activity. To alleviate this problem, site-directed mutagenesis (Section 12.4) of the human somatotropin gene was used to alter those amino acids of somatotropin that bind to the prolactin receptor, thus ensuring that the hormone would only target growth. As this example shows, it is possible not only to make genuine human hormones but also to alter their specificity and activity to make them better pharmaceuticals.

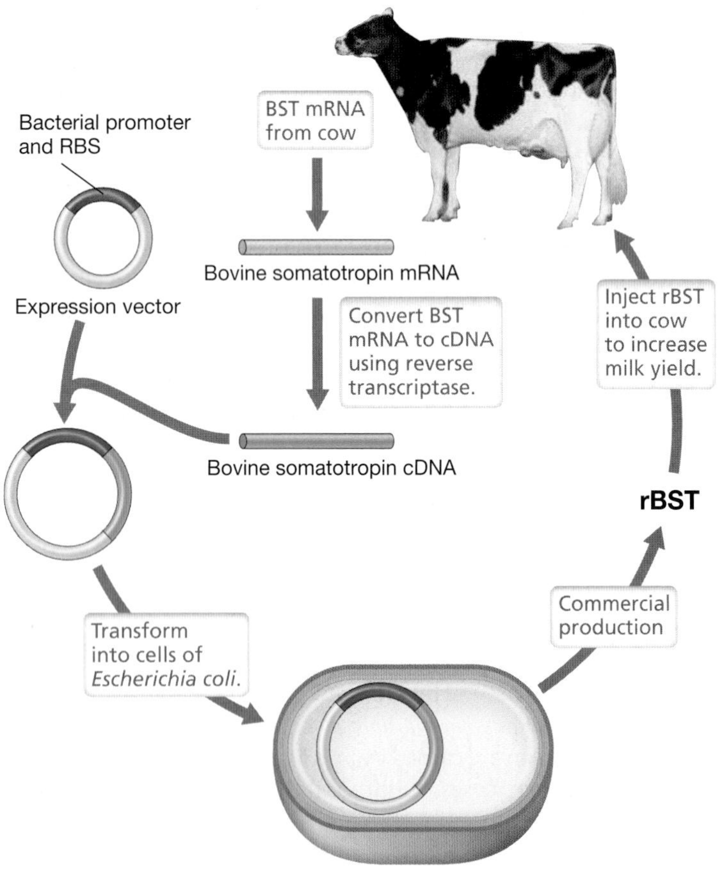

Bacterial promoter and RBS
BST mRNA from cow
Bovine somatotropin mRNA
Expression vector
Convert BST mRNA to cDNA using reverse transcriptase.
Inject rBST into cow to increase milk yield.
Bovine somatotropin cDNA
rBST
Commercial production
Transform into cells of *Escherichia coli*.

Figure 12.18 Cloning and expression of bovine somatotropin. The mRNA for bovine somatotropin (BST) is obtained from a cow, and the mRNA is converted to cDNA by reverse transcriptase. The cDNA version of the somatotropin gene is then cloned into a bacterial expression vector that has a bacterial promoter and ribosome-binding site (RBS). The construct is transformed into cells of *Escherichia coli*, and recombinant bovine somatotropin (rBST) is produced. Milk production increases in cows treated with rBST.

Other Mammalian Proteins

Many other mammalian proteins are produced today by genetic engineering (**Table 12.1**). These include, in particular, an assortment of hormones and proteins for blood clotting and other blood processes. For example, *tissue plasminogen activator* (TPA) is a protein that dissolves blood clots in the bloodstream that may form in the final stages of the healing process. TPA is primarily used in heart patients or others suffering from poor circulation to prevent the development of clots that can be life-threatening. Heart disease is a leading cause of death in many developed countries, especially in the United States, so microbially produced TPA is in high demand.

In contrast to TPA, the blood clotting factors VII, VIII, and IX are critically important for the *formation* of blood clots. Hemophiliacs suffer from a deficiency of one or more clotting factors and can

TABLE 12.1 **A few human medical products made by genetic engineering**

Product	Function
Blood proteins	
Erythropoietin	Treats certain types of anemia
Factors VII, VIII, IX	Promotes blood clotting
Tissue plasminogen activator	Dissolves blood clots
Urokinase	Promotes blood clotting
Human hormones	
Epidermal growth factor	Wound healing
Follicle-stimulating hormone	Treatment of reproductive disorders
Insulin	Treatment of diabetes
Nerve growth factor	Treatment of degenerative neurological disorders and stroke
Relaxin	Facilitates childbirth
Somatotropin (growth hormone)	Treatment of some growth abnormalities
Immune modulators	
α-Interferon	Antiviral, antitumor agent
β-Interferon	Treatment of multiple sclerosis
Colony-stimulating factor	Treatment of infections and cancer
Interleukin-2	Treatment of certain cancers
Lysozyme	Anti-inflammatory
Tumor necrosis factor	Antitumor agent, potential treatment of arthritis
Replacement enzymes	
β-Glucocerebrosidase	Treatment of Gaucher disease, an inherited neurological disease
Therapeutic enzymes	
Human DNase I	Treatment of cystic fibrosis
Alginate lyase	Treatment of cystic fibrosis

therefore be treated with microbially produced clotting factors. In the past hemophiliacs have been treated with clotting factor extracts from pooled human blood, some of which was contaminated with viruses such as HIV and hepatitis C, putting hemophiliacs at high risk for contracting AIDS, hepatitis, or liver cancer. Recombinant clotting factors have eliminated this problem.

Some mammalian proteins made by genetic engineering are enzymes rather than hormones (Table 12.1). For instance, *human DNase I* is used to treat the buildup of DNA-containing mucus in the lungs of patients with cystic fibrosis. The mucus forms because cystic fibrosis is often accompanied by life-threatening lung infections by the bacterium *Pseudomonas aeruginosa*. The bacterial cells form biofilms (⟨⟩ Sections 7.9 and 20.4) within the lungs that make drug treatment difficult. DNA is released when the bacteria lyse, and this fuels mucus formation, making it difficult to breathe. DNase digests the DNA and greatly decreases the viscosity of the mucus.

MINIQUIZ

- What is the advantage of using genetic engineering to make insulin?
- What are the major problems when manufacturing proteins in bacteria?
- Explain how an enzyme can be useful in treating a bacterial infection, such as that which occurs with cystic fibrosis.

12.7 Transgenic Organisms in Agriculture and Aquaculture

Genetic improvement of plants and animals by traditional selection and breeding has a long history, but recombinant DNA technology has led to revolutionary changes. Although the genetic engineering of higher organisms is not truly microbiology, much of the DNA manipulation is carried out using bacteria and their plasmids and genes. Hence, we consider the genetic manipulation

of plants and animals here with a focus on the microbiology that supported it.

Because genetically engineered plants or animals contain a gene from another organism—called a *transgene*—they are **transgenic organisms**. The public knows these as **genetically modified organisms (GMOs)**. Strictly speaking, the term *genetically modified* refers to genetically engineered organisms whether or not they contain foreign DNA. In this section we discuss how foreign genes are inserted into plant and fish genomes and how transgenic organisms may be used.

The Ti Plasmid and Transgenic Plants

While recombinant DNA can be transformed into plant cells by electroporation or transfection (see Figure 12.20), the **Ti plasmid** from the gram-negative bacterium *Agrobacterium tumefaciens*, a plant pathogen, can be used to transfer DNA directly into the cells of certain plants. This plasmid is responsible for *A. tumefaciens* virulence and encodes genes that mobilize DNA for transfer to the plant, which as a result contracts crown gall disease (⟨⟩ Section 23.5). The segment of the Ti plasmid DNA that is actually transferred to the plant is called **T-DNA**. The sequences at the ends of the T-DNA are essential for transfer, and the foreign DNA to be transferred must reside between these ends.

One common Ti-vector system that has been used for the transfer of genes to plants is a two-plasmid system called a *binary vector*, which consists of a cloning vector plus a helper plasmid. The cloning vector contains the two ends of the T-DNA flanking a multiple cloning site, two origins of replication so that it can replicate in both *Escherichia coli* (the host for cloning) and *A. tumefaciens*, and two antibiotic resistance markers, one for selection in plants and the other for selection in bacteria. The foreign DNA is inserted into the vector, which is transformed into *E. coli* and then moved to *A. tumefaciens* by conjugation (**Figure 12.19**).

This cloning vector lacks the genes needed to transfer T-DNA to a plant. However, when placed in an *A. tumefaciens* cell that contains a suitable helper plasmid, the T-DNA can be transferred to a plant. The "disarmed" helper plasmid, called *D-Ti*, contains the

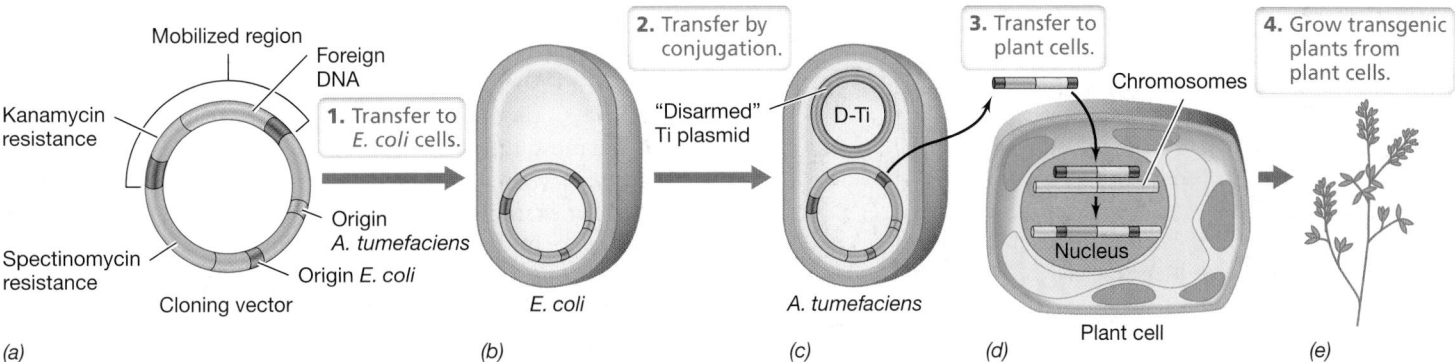

Figure 12.19 Production of transgenic plants using a binary vector system in *Agrobacterium tumefaciens*. *(a)* Plant cloning vector containing ends of T-DNA (red), foreign DNA, origins of replication, and resistance markers. *(b)* The vector is put into cells of *Escherichia coli* for cloning and then *(c)* transferred to *A. tumefaciens* by conjugation. The resident Ti plasmid (D-Ti) has been genetically engineered to remove key pathogenesis genes. *(d)* D-Ti can still mobilize the T-DNA region of the vector for transfer to plant cells grown in tissue culture. *(e)* From the recombinant plant cell, a whole plant can be grown. Details of Ti plasmid transfer from bacterium to plant are shown in Figure 23.25.

virulence (*vir*) region of the Ti plasmid but lacks the T-DNA. It also lacks the genes that initiate disease but supplies all the functions needed to transfer the T-DNA from the cloning vector. The cloned DNA and the kanamycin resistance marker of the vector are mobilized by D-Ti and transferred into a plant cell where they enter the nucleus (Figure 12.19*d*). Following integration into a plant chromosome, the foreign DNA can be expressed and confer new properties on the plant.

A number of transgenic plants have been produced using the Ti plasmid of *A. tumefaciens*. The Ti system works well with broadleaf plants (dicots), including crops such as tomato, potato, tobacco, soybean, alfalfa, and cotton. It has also been used to produce transgenic trees, such as walnut and apple. The Ti system does not work with plants from the grass family (monocots, including the important crop plant corn), but other methods of introducing DNA, such as transfection by microprojectile bombardment with a particle gun (**Figure 12.20**), have been used successfully for them.

Herbicide- and Insect-Resistant Plants

Major areas targeted for genetic improvement in plants include herbicide, insect, and microbial disease resistance, as well as improved product quality. The main genetically modified (GM) crops today are soybeans, corn, cotton, and canola. Almost all the GM soybeans and canola planted were herbicide resistant, whereas the corn and cotton were herbicide resistant or insect resistant, or both.

Herbicide resistance is genetically engineered into a crop plant to protect it from herbicides applied to kill weeds. Many herbicides inhibit a key plant enzyme or protein necessary for growth. For example, the herbicide *glyphosate* (Roundup™, made by

Figure 12.21 Transgenic plants: herbicide resistance. The photograph shows a portion of a field of soybeans that has been treated with Roundup™, a glyphosate-based herbicide manufactured by Monsanto Company (St. Louis, Missouri, USA). The remnants of plants on the right are normal soybeans; the plants on the left have been genetically engineered to be glyphosate resistant.

Monsanto) kills plants by inhibiting an enzyme necessary for making aromatic amino acids. Some bacteria contain an equivalent enzyme and are also killed by glyphosate. However, mutant bacteria were selected that were resistant to glyphosate and contained a resistant form of the enzyme. The gene encoding this resistant enzyme from *A. tumefaciens* was cloned, modified for expression in plants, and transferred into important crop plants, such as soybeans. When sprayed with glyphosate, plants containing the bacterial gene are not killed (**Figure 12.21**). Thus glyphosate can be used to kill weeds that compete for water and nutrients with the growing crop plants. Herbicide-resistant soybeans are now widely planted in the United States.

Transgenic plants resistant to damage by certain insects have been produced by genetic engineering (**Figure 12.22**). One widely used approach is based on introducing genes encoding the toxic proteins of the gram-positive, endospore-forming bacterium *Bacillus thuringiensis* into plants. As it sporulates, *B. thuringiensis* produces a crystalline protein called *Bt toxin* (⟳ Section 16.8) that is toxic to moth and butterfly larvae. Many variants of Bt toxin exist that are specific for different insects. Certain strains of *B. thuringiensis* produce additional proteins toxic to beetle and fly larvae and mosquitoes.

The Bt transgene is normally inserted directly into the plant genome. For example, a natural Bt toxin gene was cloned into a plasmid vector under control of a chloroplast ribosomal RNA promoter and then transfected into tobacco plant chloroplasts by microprojectile bombardment (Figure 12.20). This yielded transgenic plants that expressed Bt toxin at levels that were extremely toxic to larvae of several insect species. Binding Bt triggers a change in the toxin's conformation that disrupts the insect digestive system, causing death. Bt toxin is harmless to mammals (including humans) because any toxin ingested is destroyed in the stomach and the specific Bt receptors in the insect intestine are absent from the intestines of other groups of organisms.

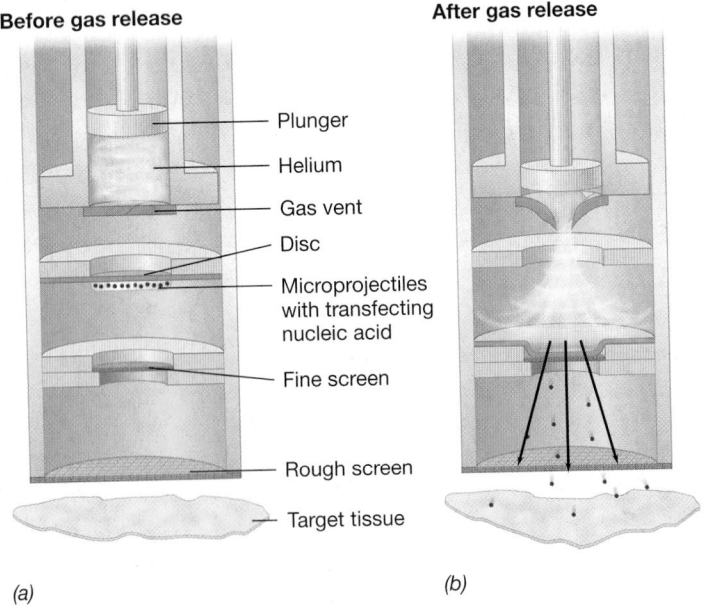

Before gas release

After gas release

- Plunger
- Helium
- Gas vent
- Disc
- Microprojectiles with transfecting nucleic acid
- Fine screen
- Rough screen
- Target tissue

(a)　　　　　*(b)*

Figure 12.20 DNA gun for transfection of eukaryotic cells. The inner workings of the gun show how metal pellets coated with nucleic acids (microprojectiles) are projected at target cells. *(a)* Before firing and *(b)* after firing. A shock wave due to gas release throws the disc carrying the microprojectiles against the fine screen. The microprojectiles continue on into the target tissue.

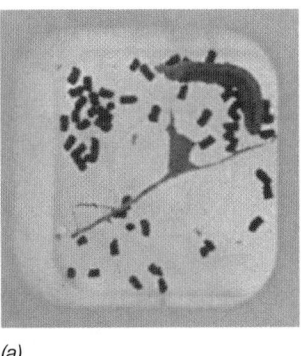 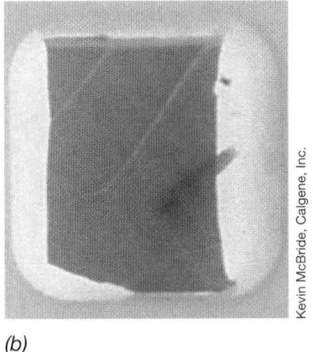

(a) *(b)*

Figure 12.22 Transgenic plants: insect resistance. The results of an assay to determine the effect of beet armyworm larvae on tobacco leaves. *(a)* Leaf from wild-type plant. *(b)* Leaf from transgenic plant that expresses Bt toxin in its chloroplasts.

Transgenic Fish

Many foreign genes have been incorporated and expressed in laboratory research animals and in commercially important animals. The genetic engineering uses microinjection to deliver cloned genes to fertilized eggs; genetic recombination then incorporates the foreign DNA into the genomes of the eggs. More recently, farm animals and fish have been genetically modified to improve yields.

An interesting practical example of a transgenic animal is the *AquAdvantage* salmon developed by AquaBounty Technologies (**Figure 12.23**). These transgenic salmon do not grow to be larger than normal salmon but simply reach market size much faster—18 months versus 3 years. The gene for growth hormone in native salmon is activated by light. Consequently, salmon grow rapidly only during the summer months. In the genetically engineered salmon, the growth promoter for the growth hormone gene was replaced with the promoter from another fish that grows at a more or less constant rate all year round. The result was salmon that make growth hormone continuously and thus grow faster. Such transgenic salmon can be grown commercially in aquaculture operations and harvested more quickly than with non-GMO farm-raised salmon.

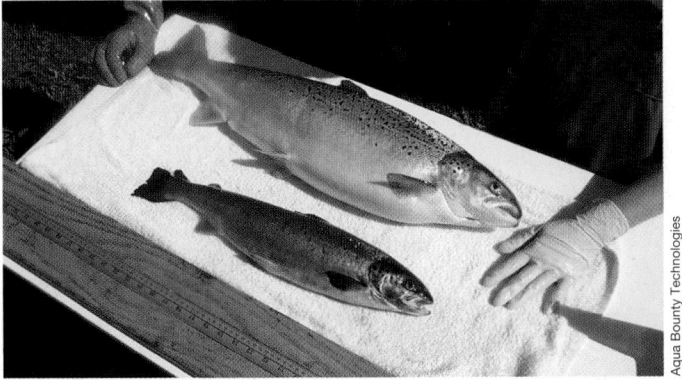

Figure 12.23 Fast-growing transgenic salmon. The *AquAdvantage*™ salmon (top) was engineered by AquaBounty Technologies (Maynard, Massachusetts, USA). The transgenic and control fish are both 18 months old but weigh 4.5 kg and 1.2 kg, respectively.

In 1995 AquaBounty applied to the U.S. Food and Drug Administration for approval to distribute the fast-growing salmon. After two decades of debate regarding the potential risks of consuming genetically modified fish, final approval occurred in 2015. Thus the AquAdvantage salmon is the first genetically engineered animal to be heading to the supermarket and is licensed to be sold without any GMO label.

— MINIQUIZ —

- What is a transgenic plant?
- Give an example of a genetically modified plant and describe how its modification benefits agriculture.
- How have transgenic salmon been engineered to reach market size faster?

12.8 Engineered Vaccines and Therapeutics

Genetic engineering is used to manufacture certain vaccines and medical therapeutics. Vaccines are substances that elicit immunity to a disease when injected into an animal (⇄ Section 27.2). Moreover, many pathogenic bacteria cause disease through their ability to infiltrate cells and release virulence factors such as toxins and destructive enzymes. Through genetic engineering, some of these activities have been harnessed to specifically target cancer cells. We consider both of these "medical miracles"—vaccines and engineered pathogens—here.

Recombinant Vaccines, Vaccinia Virus, and Subunit Vaccines

Genetic engineering can modify a pathogen by deleting genes that encode virulence factors (⇄ Section 25.3) while retaining those whose products elicit an immune response. This yields a recombinant and infective (but attenuated) vaccine. Conversely, one can add genes from a pathogenic virus to the genome of a relatively harmless virus, called a *carrier virus*. Such vaccines are called **vector vaccines** and induce immunity to the pathogenic virus. Indeed, one can even combine the two approaches by disarming one pathogen and adding back to it immunity-inducing genes from a second pathogen. This yields a **polyvalent vaccine**, a vaccine that immunizes against two different diseases at the same time.

Vaccinia virus (⇄ Section 10.6) is widely used to prepare recombinant vaccines for human use; however, cloning into vaccinia requires a selective marker, which is provided by the gene encoding the enzyme thymidine kinase. Vaccinia virus contains a gene encoding thymidine kinase, an enzyme that converts thymidine into thymidine triphosphate. However, this enzyme also converts the base analog 5-bromodeoxyuridine (BrdU) into a nucleotide that is incorporated into DNA, causing a lethal reaction. Thus, cells that express either host- or virus-encoded thymidine kinase are killed when treated with BrdU.

Genes to be put into vaccinia virus are first inserted into an *Escherichia coli* plasmid that contains part of the vaccinia thymidine kinase (*tdk*) gene (**Figure 12.24**). The foreign DNA is inserted into the *tdk* gene, which is therefore disrupted. This recombinant

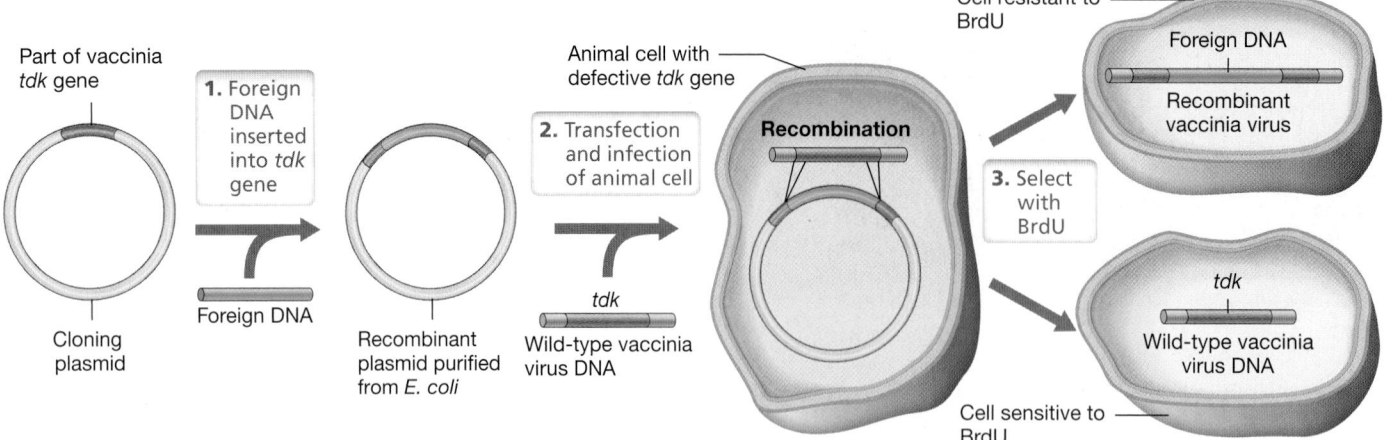

Figure 12.24 Production of recombinant vaccinia virus. Foreign DNA is inserted into a short segment of the thymidine kinase gene (*tdk*) from vaccinia virus carried on a plasmid. Following replication of this plasmid in *Escherichia coli*, both the recombinant plasmid and wild-type vaccinia virus are put into the same animal host cell to promote recombination. The animal cells are treated with 5-bromodeoxyuridine (BrdU), which kills only cells with an active thymidine kinase. Only recombinant vaccinia viruses whose *tdk* gene is inactivated by insertion of foreign DNA survive.

plasmid is then transformed into animal cells whose own *tdk* genes have been inactivated. These cells are also infected with wild-type vaccinia virus. The two versions of the *tdk* gene—one on the plasmid and the other on the virus—then recombine. Some viruses gain a disrupted *tdk* gene plus its foreign insert (Figure 12.24). Cells infected by wild-type vaccinia virus (with an active thymidine kinase) are killed by BDU. By contrast, cells infected by *recombinant* vaccinia virus (with a disrupted *tdk* gene) grow long enough to yield a new generation of virions (Figure 12.24). The protocol thus selects for viruses whose *tdk* gene contains a cloned insert of foreign DNA. Vaccinia viruses can also be engineered to carry genes from multiple viruses, forming polyvalent vaccines. Currently, several vaccinia vector vaccines have been developed and licensed for veterinary use, including one for rabies, while many other vaccinia vaccines are at the clinical trial stage.

Subunit vaccines, vaccines that contain only a specific protein or two from a pathogen, are also produced by recombinant means. For a pathogenic virus, the gene encoding its coat protein is often the best vaccine candidate because coat proteins are typically highly immunogenic. Subunit vaccines are popular because large amounts of immunogenic proteins are produced that can be administered at high dosage without the risk that exists with attenuated or killed-cell vaccines that may inadvertently contain viable pathogen cells or viruses. However, some subunit vaccines, such as that prepared against a surface protein of human hepatitis B, require that the immunogenic proteins be glycosylated by the host before they are immunologically active. To solve this problem, the subunit hepatitis B vaccine was produced in a eukaryotic host (yeast), which generated the glycosylated and immunologically active form of the vaccine.

Pathogens and Antibodies as Engineered Anticancer Therapeutics

While many cancers are treatable with radiation and chemotherapy, how to specifically target the drugs or radiation to tumor cells has been a long-standing problem, and biotechnology may

have a solution. *Listeria monocytogenes* is a pathogenic bacterium that causes listeriosis, a serious foodborne illness (ᴄ⊃ Section 32.13). *L. monocytogenes* grows within human cells, which allows wild-type strains to evade the immune system. By contrast, weakly pathogenic recombinant strains of *L. monocytogenes* can be cleared by the immune system of healthy cells but not by tumor cells. This observation hinted that weakened strains of *L. monocytogenes* might be turned into anticancer vehicles to deliver toxic drugs or radioisotopes specifically to tumor cells. This was accomplished by coupling the radionuclide [188]rhenium to a recombinant strain of *L. monocytogenes*. In experiments with mice, this therapeutic strain infected and multiplied in pancreatic tumor cells (**Figure 12.25**) without harming normal pancreatic cells.

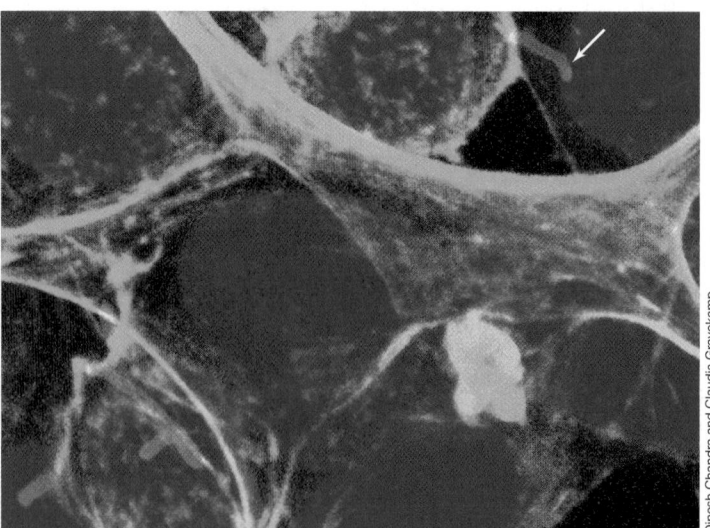

Figure 12.25 Therapeutic *Listeria*. *L. monocytogenes* cells (pink) linked to the radionuclide [188]rhenium enter and multiply inside mouse pancreatic tumor cells (blue, cell nuclei; green, cytoplasm) that have spread from the primary tumor. Radiation from the [188]rhenium slowly kills the tumor cells.

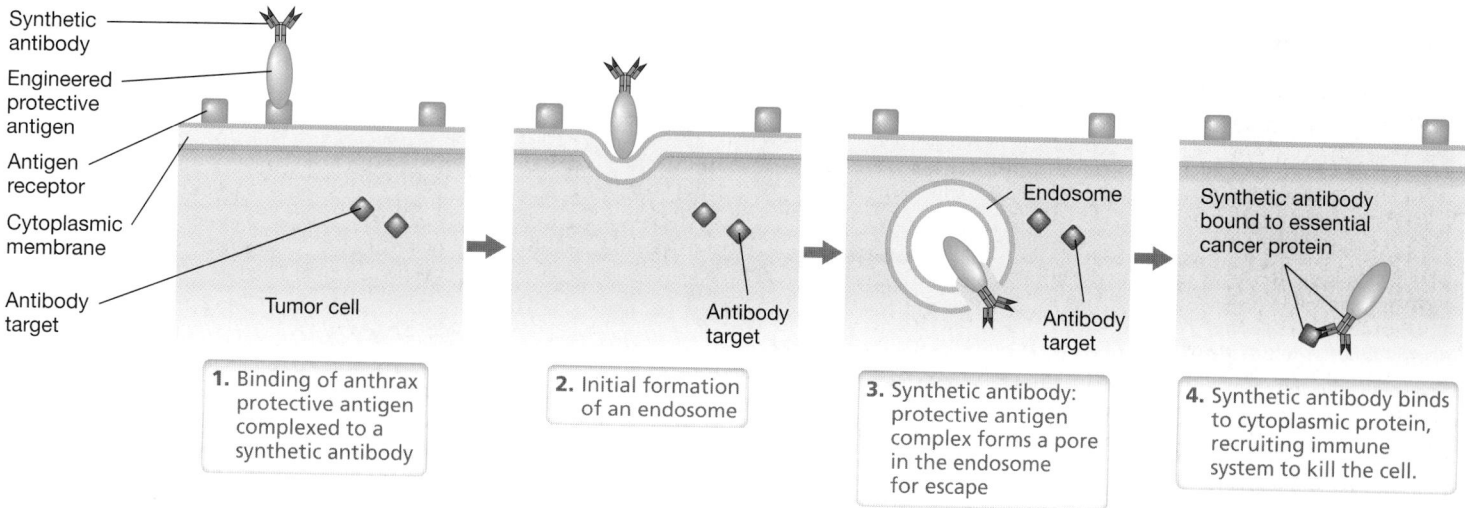

Figure 12.26 Engineered anthrax toxin. The protective antigen component of the anthrax toxin is engineered to carry a synthetic antibody. This engineered protective antigen specifically binds to a cell receptor on target cancer cells. After binding to the receptor, the engineered complex is taken up into the cell through an endosome. Following release into the cytoplasm, the synthetic antibody binds to an essential cellular protein, triggering cell death through the host's immune response. Anthrax toxin is discussed in more detail in Sections 29.9 and 31.8.

Another mechanism for treating cancer is by using antibodies, proteins produced by the immune system to attack foreign substances (⮌ Section 27.3). It has been found that the binding of antibodies to specific targets inside cancer cells can trigger the host's immune system to kill the cancer cell. However, antibodies do not freely enter cells and thus a transport mechanism has been genetically engineered using a toxin produced by *Bacillus anthracis*, the bacterium that causes anthrax (⮌ Sections 29.9 and 31.8) (**Figure 12.26**). Anthrax toxin contains three components: edema factor, lethal factor, and protective antigen; the latter is essential for carrying the toxic edema and lethal factors into the cell.

Using genetic engineering, scientists have modified the *B. anthracis* protective antigen to carry a *synthetic anticancer antibody* instead of the toxic edema and lethal factors. Injection of the modified and now harmless toxin results in the protective antigen recognizing and binding to a receptor on the outside of the cancer cell (Figure 12.26). The protective antigen: antibody complex is then taken up into the cancer cell through the formation of an endosome. Once the antibody is released from the endosome into the cytoplasm, it specifically binds a protein essential to tumor viability. This binding then triggers the cell's immune system, which recognizes the antibody: cellular protein complex and kills the cell (Figure 12.26). Any antigen:antibody complex incorporated by normal cells is harmless because both the toxic portions of the toxin and the specific cancer proteins targeted by the antibody are missing in normal cells.

Stimulating the immune system to fight cancer may well turn out to be a mechanism in a whole new line of anticancer therapies. The novel system devised here—combining antitumor antibodies with a delivery system crafted from a highly toxic bacterium—shows both the power and the promise of genetic engineering to accelerate the war on cancer.

MINIQUIZ

- Explain why recombinant vaccines might be safer than some vaccines produced by traditional methods.
- What are the important differences among a recombinant live attenuated vaccine, a vector vaccine, and a subunit vaccine?
- What feature of some pathogenic bacteria make them attractive for engineered cancer treatments?

12.9 Mining Genomes and Engineering Pathways

Complex environments, such as fertile soil, contain vast numbers of uncultured microbes and their genes that are ripe for harvesting by genetic engineering. In addition, microbial metabolic pathways can be altered, embellished, or otherwise modified to change their characteristics and improve their efficiency. We explore how genetic engineering can both mine environmental genomes and alter metabolic pathways here.

Environmental Gene Mining

Just as the total gene content of an organism is its *genome*, the collective genomes of an environment are its *metagenome* (⮌ Sections 9.8 and 19.8). *Gene mining* is the process of identifying and isolating potentially useful genes from the environment without the need to culture the organisms that contained them. In gene mining, DNA (or RNA) isolated directly from environmental samples is cloned into suitable vectors to construct a *metagenomic library* (**Figure 12.27**). If RNA is isolated, it must first be converted to cDNA by reverse transcriptase (Figure 12.2).

Screening of environmental metagenomic libraries has identified novel genes that encode enzymes that can degrade various pollutants and enzymes that make novel antibiotics. Retrieval of gene

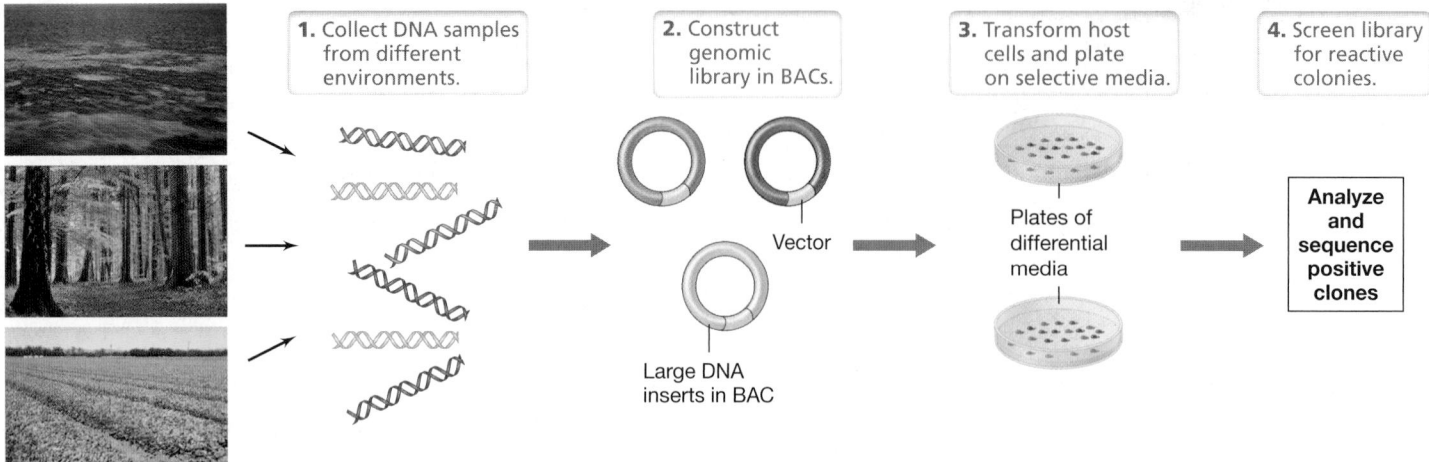

Figure 12.27 Metagenomic search for useful genes in the environment. DNA samples are obtained from different environments, such as seawater, forest soil, and agricultural soil. A metagenomic library is constructed using bacterial artificial chromosomes (BACs) and screened for genes of interest. Possibly useful clones are analyzed further.

clusters encoding entire metabolic pathways—such as for antibiotic synthesis—requires vectors such as **bacterial artificial chromosomes (BACs)**. BACs are similar to plasmids except that they can carry large inserts of DNA. BACs are especially useful for screening samples from rich environments, such as soil, where vast numbers of unknown genomes are present and correspondingly large numbers of genes are available to screen (Figure 12.27).

Several lipases, chitinases, esterases, and other degradative enzymes with novel substrate ranges and other properties have been isolated by this approach, and such enzymes have many industrial applications. Enzymes with improved resistance to industrial production conditions, such as high temperature, high or low pH, and oxidizing conditions, are especially valuable and desirable. Metagenomics can also target products with a particular combination of properties, such as a heat-stable lipase. Lipases hydrolyze fats, but their industrial production and use often requires that they remain active at high temperatures. To isolate a thermostable lipase, a metagenomic library was prepared from a hot spring sample and the DNA transformed into cells of *Escherichia coli*. Recombinant colonies expressing lipase activity were then selected and analyses indicated that certain of them remained active at 90°C. The gene encoding the heat-stable lipase was then introduced into an expression vector for commercial production of the enzyme.

By metagenomic mining of extreme environments, several useful heat- and acid-stable enzymes have been isolated for cleaning food-processing equipment in the food industry (**Figure 12.28**). To prevent foodborne infections, food-processing equipment must be rigorously cleaned, and cleaning protocols typically employ rigorous acid and base treatments, detergents, and sanitizers, all of which consume large amounts of chemicals and generate large volumes of wastewater that must be treated. By contrast, cleaning equipment with enzymes that function optimally near the boiling point of water in dilute acids (Figure 12.28) requires fewer chemicals and less water and more effectively removes microbial biofilms than do standard cleaning practices.

Pathway Engineering: Indigo Synthesis

Pathway engineering is the process of assembling a new or improved biochemical pathway using genes from one or more organisms. Engineered microbes are used to make alcohols, solvents, food additives, dyes, antibiotics, and many other products. They may also be used to degrade agricultural waste, pollutants, herbicides, and other toxic or undesirable materials. Here we discuss improving or modifying *existing pathways*, and later we explore the use of synthetic biology to create *entirely new* pathways (Section 12.11).

An interesting example of pathway engineering is the production of indigo by *E. coli* (**Figure 12.29**). Indigo is an important dye used for treating wool and cotton; blue jeans, for example, are made of cotton dyed with indigo. Although indigo can be synthesized chemically, the heavy demand for indigo by the textile industry has spawned new approaches for its synthesis, including a biotechnological approach using pathway engineering.

Figure 12.28 Application of CinderBio hyperstable enzymes for the cleaning of creamery equipment. These heat-stable enzymes clean industrial food-processing equipment as well as or better than traditional cleaning methods and do not generate large amounts of toxic wastewater.

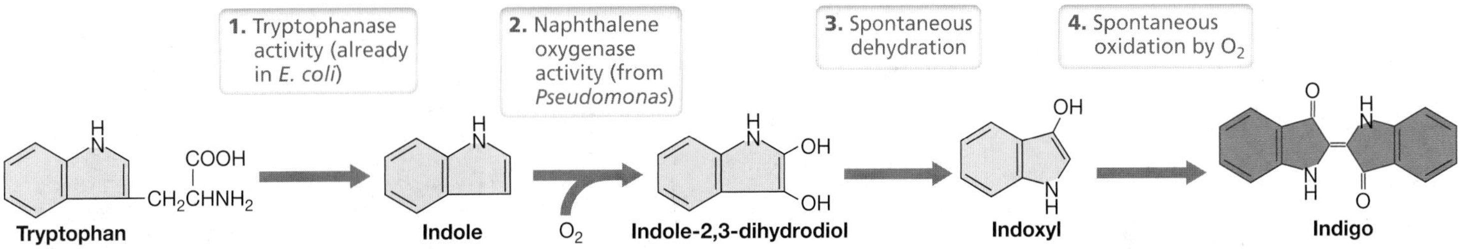

Figure 12.29 Engineered pathway for production of the dye indigo. *Escherichia coli* naturally expresses tryptophanase, which converts tryptophan into indole. Naphthalene oxygenase (originally from *Pseudomonas*) converts indole to dihydroxy-indole, which spontaneously dehydrates to indoxyl. Upon exposure to air, indoxyl dimerizes spontaneously to form indigo.

Because the structure of indigo is very similar to that of the aromatic hydrocarbon naphthalene, enzymes that oxygenate naphthalene also oxidize indole to its dihydroxy derivative; the latter oxidizes spontaneously in air to yield indigo, a bright blue pigment. Enzymes for oxygenating naphthalene are encoded by several plasmids found in *Pseudomonas* and other soil bacteria. When genes from such plasmids were cloned into *E. coli*, the cells turned blue because they had incorporated the genes encoding the enzyme naphthalene oxygenase.

The indigo pathway consists of four steps, two enzymatic and two spontaneous (Figure 12.29). *E. coli* naturally synthesizes the enzyme tryptophanase that carries out the first of these steps, the conversion of tryptophan to indole. In the engineered *E. coli*, a second step converts indole to the product that converts to indigo spontaneously (Figure 12.29). For indigo production, tryptophan must be supplied to the recombinant *E. coli* cells, and for commercial application, this was accomplished by affixing cells to a solid support in a bioreactor and then continuously trickling a tryptophan solution obtained from waste protein sources over the cells. If the tryptophan solution is recirculated over the cells several times, indigo levels steadily increase until the dye can be harvested.

Although bioproduction of indigo is clearly possible, it is currently difficult to compete with the chemical production of more than 20 kilotons per year. Indeed, some of the major challenges of pathway engineering are controlling the metabolic pathway and producing the desired compound at the yields necessary to be cost effective.

--- MINIQUIZ ---
- Explain why metagenomic cloning gives large numbers of novel genes.
- What types of environments are often sampled to prospect for industrial enzymes and why?
- How was *Escherichia coli* modified to produce indigo?

12.10 Engineering Biofuels

With the global supply of fossil fuels limited and environmental concerns growing about how climate change will affect our planet, renewable energy sources such as biologically produced fuels—*biofuels*—are in demand. The major biofuels in use today are ethanol, biodiesel, hydrogen, and methane. Select microorganisms can produce biofuels; however, the yield from wild-type organisms is often hindered by toxic by-products and missing enzymes for critical steps. Thus, to enhance the production of biofuels, microorganisms have been genetically modified to optimize production. Here we discuss how genetic engineering has allowed for the use of alternative biofuel feedstocks, how enzyme replacement can yield new biofuels, and how phototrophic microorganisms can be harnessed as biofuel factories.

Bacterial Conversion of Switchgrass to Ethanol

Over 14 billion gallons of ethanol are produced per year in the United States from the fermentation of corn sugar by yeast (**Figure 12.30a**). However, because corn requires considerable cost and energy inputs to grow and harvest and is a major food source for both humans and domesticated animals, alternative nonedible and low-resource-input plant materials are more desirable biofuel feedstocks. Much attention has been focused on fast-growing grasses such as *switchgrass* (*Panicum virgatum*) (Figure 12.30b) as a source of cellulose for the production of ethanol. However, switchgrass cellulose is integrated with other plant polymers such as hemicellulose and lignin and requires high temperature, chemical, and enzymatic pretreatment to break the polymers down to fermentable sugars.

By leveraging both microbial diversity and genetic engineering, a bacterium has been discovered and genetically modified not only to break down switchgrass cellulose to fermentable sugars but also to ferment this sugar to ethanol. *Caldicellulosiruptor*, a gram-positive anaerobic and thermophilic bacterium, naturally produces a cellulase enzyme that can convert cellulose and hemicellulose to glucose. Unique proteins called *tāpirins* that extrude from the outer layer of the bacterium's peptidoglycan allow the cell to directly bind to raw switchgrass during the conversion process (Figure 12.30c). While *Caldicellulosiruptor bescii* grows optimally at 80°C and can ferment sugars, it naturally yields mostly acetate, lactate, and hydrogen as fermentation products. To directly convert switchgrass to ethanol, genetic engineers altered the terminal steps of the *C. bescii* glycolytic pathway (⬗ Figure 3.14) by replacing genes encoding lactate dehydrogenase and other acidic fermentation products with a bifunctional acetaldehyde/alcohol dehydrogenase from another thermophile, *Clostridium thermocellum*. This shifted the fermentation in *C. bescii* from mainly acidic products to 70% ethanol.

UNIT 3

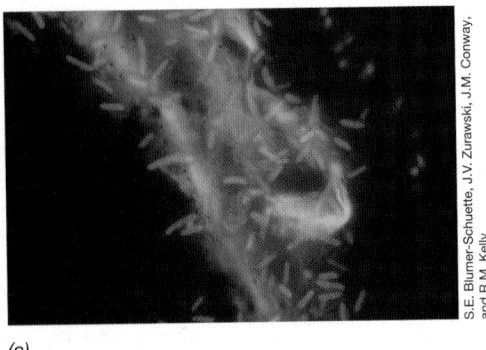

(a)   (b)   (c)

Figure 12.30 Biofuels. *(a)* A bioethanol plant in Nebraska (USA). In the plant, glucose from cornstarch is fermented by yeast to ethanol plus CO_2. The large tank in the foreground is the ethanol storage tank, and the pipes in the background are for distilling the alcohol from the fermentation broth. *(b)* Switchgrass, a source of cellulose as a feedstock for ethanol production. *(c)* Fluorescence photomicrograph of acridine orange–stained cells of the thermophilic and cellulolytic bacterium *Caldicellulosiruptor kronotskyensis* growing on switchgrass (a single cell measures about 0.6 μm × 3 μm). The green color represents the switchgrass.

Use of thermophilic microorganisms for biofuel production has several advantages such as reduced risk of contamination with mesophiles and improved substrate solubility. It also makes the collection of any volatile products easier. For example, separating small amounts of desired products from large amounts of growth media (such as collecting ethanol during yeast fermentation of corn sugar) requires significant energy inputs to cool the bioreactors for growth of the organism and then later to heat and distill off the ethanol (Figure 12.30*a*). By contrast, since *C. bescii* grows optimally at 80°C—just above the boiling point of ethanol—commercial production of alcohol from cellulose by this genetically engineered bacterium requires little or no cooling and saves energy by the continuous emission of the desired product, ethanol.

Engineered Alkenes and Alkanes

Petroleum contains a mixture of hydrocarbons of varying chain length. Propane (C_3H_8), produced from natural gas processing and petroleum refining, is a widely used heating and cooking fuel and a key fuel for agricultural applications. Using genetic engineering, scientists have modified strains of *Escherichia coli* to convert glucose into propane and some other petroleum hydrocarbons.

Hydrocarbon production in *E. coli* begins with the synthesis of a fatty aldehyde. This was done by heterologously expressing in *E. coli* the *Photorhabdus luminescens luxCED* genes, which encode enzymes that reduce fatty acids to their corresponding aldehydes **(Figure 12.31)**. The activity of these fatty acid reductase, synthetase, and transferase enzymes yields a fatty aldehyde that can be converted to hydrocarbons by the enzyme aldehyde decarbonylase. However, *E. coli* lacks this enzyme as well. To overcome this limitation, genetic engineers cloned the aldehyde decarbonylase gene from the cyanobacterium *Nostoc punctiforme* into *E. coli*, allowing the engineered *E. coli* to convert linear fatty acids added to its growth medium into linear hydrocarbons (Figure 12.31).

Because branched-chain hydrocarbons yield higher octane numbers (which is good for gasoline engine performance), scientists expanded this engineered pathway by heterologously expressing enzymes that participate in the first step of the fatty acid elongation cycle (⥀ Figure 3.30) from *Bacillus subtilis*. This allowed the engineered *E. coli* to use more branched-chain fatty acids as starting substrates and by doing so generate higher-octane fuels.

Microalgae and Biodiesel

Microalgae are unicellular phototrophic eukaryotes that produce an abundance of bioactive compounds including lipids, fatty

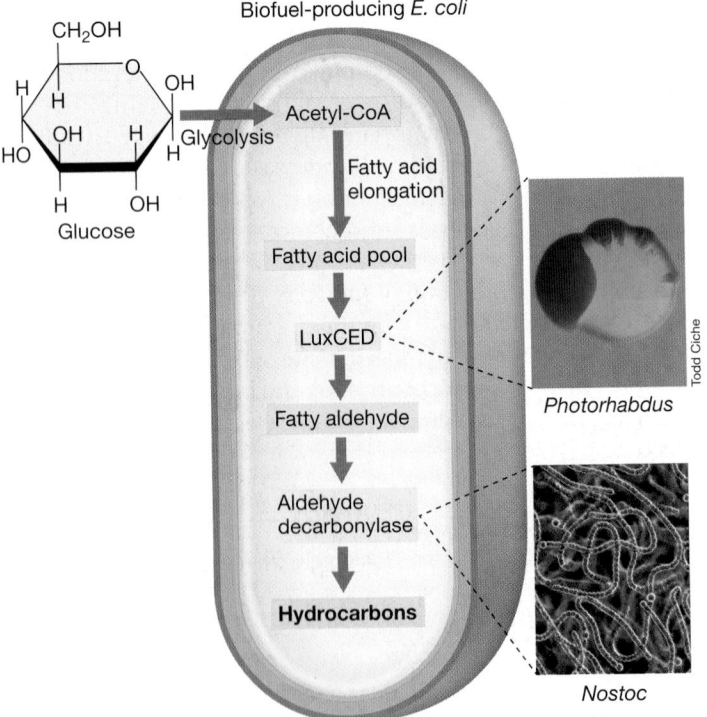

Figure 12.31 Hydrocarbon-producing *Escherichia coli*. *E. coli* naturally produces fatty acids from acetyl-CoA. By engineering a strain to express the LuxC (fatty acid reductase), LuxE (fatty acid synthetase), and LuxD (fatty acid transferase) proteins from the fluorescent bacterium *Photorhabdus luminescens* (inset: fluorescent colony), a fatty aldehyde intermediate is produced. This fatty aldehyde can then be converted to a hydrocarbon if the same strain has also been engineered to express the aldehyde decarbonylase enzyme from the cyanobacterium *Nostoc punctiforme* (inset photo).

acids, and carotenoids. These products are made using only sunlight, CO_2, a few minerals, and water. Microalgae of interest to biotechnology include the green algal genera *Chlorella* and *Chlamydomonas*. These organisms produce significant amounts of storage lipids known as *triacylglycerides* (TAG), substances that can be chemically treated to yield *biodiesel*, a fuel for use in the diesel engines present in many trucks and heavy transport vehicles.

Improving TAG synthesis in the microalgal biosynthetic pathway required some genetic engineering breakthroughs, and **Figure 12.32a** illustrates the successful design of vectors that allow proteins to be targeted to the nucleus or chloroplast of cells of *Chlamydomonas*; other vectors have been developed that allow the mitochondrion or the endoplasmic reticulum to be targeted. Organelle targeting is facilitated through the fusion of *signal sequences* (⇔ Section 4.12) to the protein of interest. A vector for the simultaneous expression of

two foreign genes in separate cellular locations has also been generated (Figure 12.32b). These targeting vectors are critical for manipulating compartmentalized TAG biosynthetic activities such as control of gene expression by transcription factors (encoded by the nuclear genome), ATP-producing enzymes (encoded by the mitochondrial genome), and enzymes for initial fatty acid synthesis reactions (encoded by the chloroplast genome). Translation of proteins by ribosomes on the endoplasmic reticulum is also important for protein secretion.

Using microalgal triacylglycerides as a feedstock for biodiesel increases the possibilities for biofuel production, and the fact that the process is driven by the energy of sunlight makes it an environmentally attractive process. However, the major obstacle facing all biofuel production schemes—especially those that require sunlight—is the expense of the equipment and engineering necessary to scale production up to levels necessary to compete with the petroleum industry. Currently, about 80 million barrels of oil a day arrive on the highly volatile world energy market, and until oil supplies diminish to the point where a significant price hike is encountered, any biofuel will have a difficult time competing.

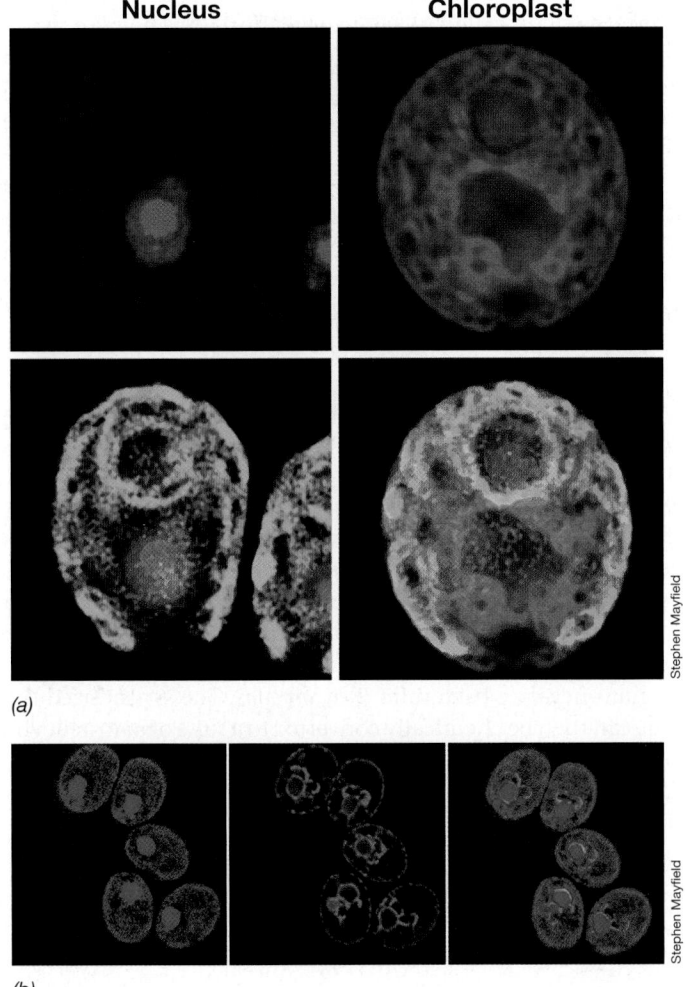

Nucleus **Chloroplast**

(a)

Stephen Mayfield

(b)

Stephen Mayfield

Figure 12.32 Genetic tools for engineering microalgae. *(a)* Fluorescent micrographs showing the expression of reporter genes in targeted regions of cells of *Chlamydomonas*. Top panel: A reporter gene encoding a red fluorescent protein that targets the nucleus (left) and chloroplasts (right). Bottom panel: both images overlaid with a green fluorescent stain for the chloroplast. *(b)* Dual expression of two reporter genes to different cellular locations using a single vector. A gene encoding a blue fluorescent protein is localized to the nucleus, while a gene encoding a red fluorescent protein is targeted to the endoplasmic reticulum in cells of *Chlamydomonas*. Adapted from Rasala, B.A., S-S. Chao, M. Pier, D.J. Barrera, and S.P. Mayfield. 2014 *PLoS ONE 9*(4): e94028.

MINIQUIZ

- How has *Caldicellulosiruptor* been modified to produce ethanol directly from switchgrass?
- What is the advantage of using thermophiles to produce biofuels?
- What has been the limiting factor in engineering microalgae to produce greater amounts of lipids?

III • Synthetic Biology and Genome Editing

The term *synthetic biology* refers to the use of genetic engineering to create novel biological systems out of available biological parts, often taken from several different organisms. These biological parts (promoters, enhancers, operators, riboswitches, regulatory proteins, enzyme domains, signal receivers, etc.) have been termed *biobricks*. Synthetic biology links these biobricks together in various combinations to form modules capable of generating complex behaviors. While we will discuss some amazing examples of synthetic biology (including the formation of synthetic cells) in Section 12.11, college students are also trying their hand at synthetic biology. An undergraduate competition called the *International Genetically Engineered Machine* (iGEM; http://igem.org/) occurs annually worldwide. Teams in this competition have used synthetic biology to engineer products ranging from biodegradable Styrofoam-like material to a strain of *Escherichia coli* that can protect honeybees from parasites.

Another powerful new and rapidly developing technology allows the precise editing of genomes in living cells, a technique that has revolutionized biotechnology. In Section 12.12 we explore how the microbial immune system has been leveraged to edit any genome and the remarkable applications of this genome editing technology.

UNIT 3

12.11 From Synthetic Metabolic Pathways to Synthetic Cells

A major focus of synthetic biology thus far has been the construction or modification of metabolic pathways. By using various enzyme and regulator biobricks, synthetic biologists can construct artificial pathways to convert cheap and abundant substrates into high-value products. Such products are often expensive because purifying them from their original source, typically plants, is costly. While genetically modified organisms (GMOs) are used for their production, no foreign DNA is present in the products.

Engineering a Major Food Product

Vanillin, one of the most popular flavoring agents in the world, is a secondary metabolite extracted from the seedpods (vanilla beans) of orchids of the genus *Vanilla*. Natural vanillin is expensive because of the slow growth of orchids, their relatively low output, and the high production costs of cultivating and harvesting the beans. By analyzing the natural metabolic pathway for the production of vanillin, genetic engineers have synthesized strains of *Escherichia coli* and yeast that can synthesize the flavoring agent from glucose. Vanillin synthesis in *E. coli* requires five heterologously expressed enzymes, and once the necessary biobricks were incorporated into the bacterium, it produced vanillin identical in structure and taste to naturally produced vanillin. (However, some argue that naturally produced vanilla contains additional compounds extracted from the vanilla beans that contribute to its unique taste.)

"Synbio vanillin," as the *E. coli* product has been called, is commercially available as an inexpensive source of vanillin and has been used primarily for flavoring ice cream and baked goods. As the second most expensive spice in the world (behind saffron), natural vanilla is an especially costly ingredient for bulk use such as in ice cream. Synbio vanillin is a good example of how synthetic biology can be used to convert inexpensive feedstocks (corn sugar) into high-value products.

Synthetic Pharmaceuticals: Artemisinin and Malaria

Pharmaceuticals are often derived from natural products; aspirin, for example, was originally obtained from willow bark. While aspirin is now chemically synthesized, not all pharmaceuticals can be economically synthesized. One example is the antimalarial drug *artemisinin*. Malaria, which is caused by protozoans of the genus *Plasmodium*, is transmitted by mosquitoes and infects nearly 500 million people each year, primarily in tropical and subtropical countries. While various traditional antimalarials are in use, the parasite has evolved resistance to many of these and thus new antimalarial drugs are constantly needed. Artemisinin is an alternative antimalarial that is produced in limited amounts by the cultivated sweet wormwood plant (*Artemisia annua*). To ensure availability of the drug, the *Semi-Synthetic Artemisinin Project* was initiated with the goal of engineering a microorganism for the synthesis of artemisinic acid, used for production of artemisinin (**Figure 12.33**).

A. annua naturally produces artemisinic acid by converting acetyl-CoA to farnesyl diphosphate (FPP; a 15-carbon intermediate) using the mevalonate biosynthetic pathway. FPP is then oxidized to artemisinic acid and dihydroartemisinic acid through an intermediate called *amorphadiene*. Initially, the plan was to have *E. coli* produce artemisinic acid through fermentation. Synthetic biologists made numerous attempts to engineer *E. coli* with the correct biobricks to produce artemisinic acid through permutations of the natural pathway, inhibiting natural competing enzymes, mutating genes for codon optimization (Section 12.3), and modifying fermentation conditions, but no strategy emerged that could yield the amorphadiene intermediate at sufficient levels. However, by transferring the necessary metabolic pathway biobricks to a modified strain of the baker's yeast *Saccharomyces cerevisiae* along with metabolic adjustments to divert carbon flux toward the final product, synthetic biologists designed a yeast strain that produces large amounts of artemisinic acid that can then be chemically converted to artemisinin (Figure 12.33).

Synthetic biology has also been successful at synthesizing powerful painkilling drugs. For example, genetic engineers rewired yeast to produce the chemical *thebaine*, a precursor to morphine and hydrocodone, using glucose as feedstock. This was accomplished by the heterologous expression in yeast of 21 different genes that originated from sources as diverse as plants, a rat, and the gram-negative bacterium *Pseudomonas*. The synthesized thebaine can then be chemically converted to a suite of pain-relieving drugs and marketed by pharmaceutical companies.

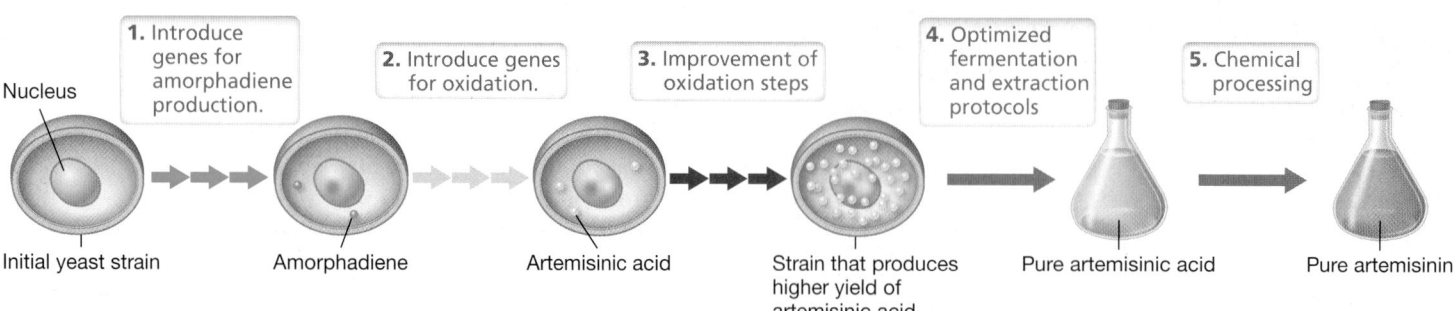

Figure 12.33 Artemisinin synthesis through synthetic biology. A summary of the engineering steps used to modify a *Saccharomyces cerevisiae* yeast strain to produce artemisinic acid. Colored arrows represent expression of genes from the plant *Artemisia annua* and other genetic modifications.

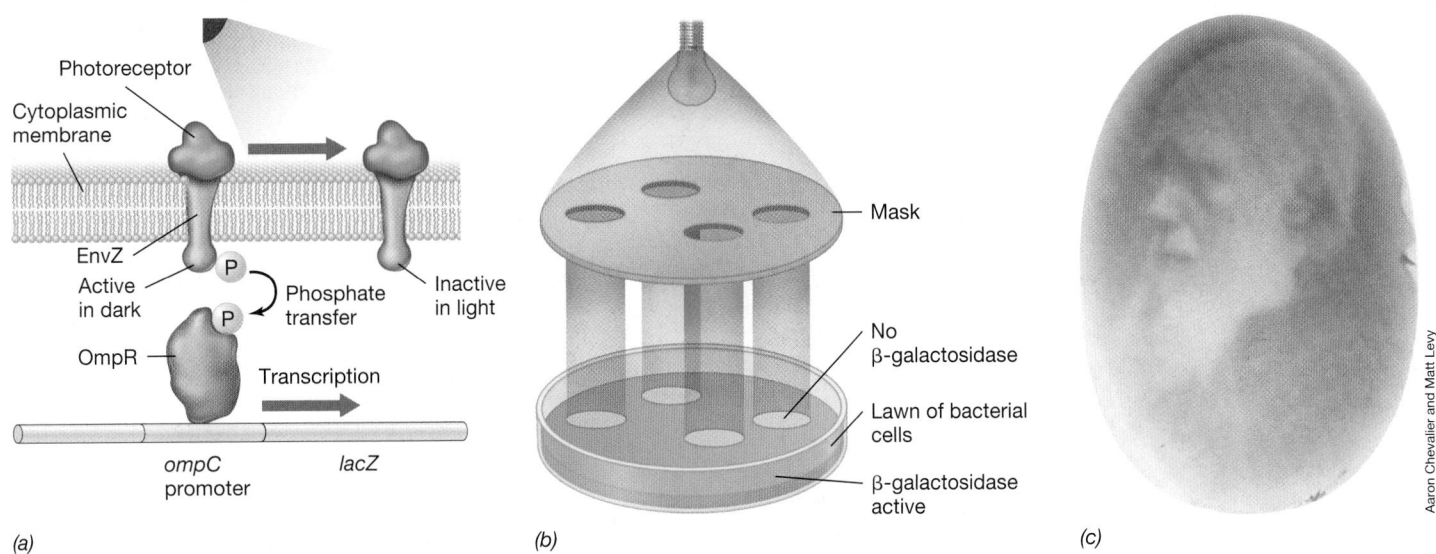

Figure 12.34 Bacterial photography. *(a)* Light-detecting *Escherichia coli* cells were genetically engineered using components from cyanobacteria and *E. coli* itself. Red light inhibits phosphate (P) transfer to the DNA-binding protein OmpR; phosphorylated OmpR is required to activate *lacZ* transcription (*lacZ* encodes β-galactosidase). *(b)* Setup for making a bacterial photograph. The opaque portions of the mask correspond to zones where β-galactosidase is active and thus to the dark regions of the final image. *(c)* A bacterial photograph of a portrait of Charles Darwin.

Photographic *Escherichia coli*

An early excursion into the world of synthetic biology is the use of genetically modified *E. coli* to produce photographs. The engineered bacteria are grown as a lawn on agar plates, and when an image is projected onto the lawn, unilluminated bacteria make a dark pigment while illuminated bacteria do not. The result is a primitive photograph of the projected image (**Figure 12.34**).

Construction of the photographic *E. coli* required synthetic biologists to create three key biobricks: (1) a light detector and signaling module; (2) a pathway to convert heme (already present in *E. coli*) into the photoreceptor pigment phycocyanobilin (an accessory light-harvesting pigment of cyanobacteria, ⊄ Section 14.2); and (3) an enzyme encoded by a gene whose transcription can be switched on and off to make the dark pigment (Figure 12.34*a*). The photoreceptor is a fusion protein in which the sensing half is the light-detecting part of the phytochrome protein from the cyanobacterium *Synechocystis*. This required phycocyanobilin, which is not naturally made by *E. coli*, hence the need to install the biobrick that contained the pathway to make phycocyanobilin.

The other half of the fusion protein is the signal transmission domain of the EnvZ sensor protein from *E. coli*. EnvZ is part of a two-component regulatory system, its partner being OmpR (⊄ Section 6.6). Normally, EnvZ activates the DNA-binding protein OmpR, and the latter in turn activates target genes by binding to the promoter. The hybrid protein was designed to activate OmpR in the dark but not in the light. This is because phosphorylation of OmpR is required for activation, and red light converts the sensor to a state in which phosphorylation is inhibited. Consequently, the target gene is *off* in the light and *on* in the dark. When a mask is placed over the Petri plate containing a lawn of the engineered *E. coli* cells (Figure 12.34*b*), cells in the dark make a pigment that cells in the light do not, and in this way a "photograph" of the masked image develops (Figure 12.34*c*).

The pigment made by the *E. coli* cells results from the activity of the lactose-degrading enzyme β-galactosidase, naturally present in *E. coli*. The target gene, *lacZ*, encodes this enzyme. In the dark, *lacZ* is expressed and β-galactosidase is made. The enzyme cleaves the lactose analog X-gal (Section 12.2) present in the growth medium to release galactose and a black dye. In the light, the *lacZ* gene is not expressed, no β-galactosidase is made, and so no dye is released. Contrast in the photograph is controlled by how much light cells see, which is governed by the nature of the mask that is used (Figure 12.34*c*).

Although bacterial photographs can hardly compare with digital photographs, the knowledge gained from assembling the biobricks required for bacterial photographs—work that is now many years old—helped form a foundation for the deployment of synthetic approaches in more complex biological systems, including entire cells, to which we turn now.

Synthetic Cells

The synthesis of an entire cell from scratch can be considered the pinnacle of synthetic biology (see page 332 for more on this). This has not happened as of 2017, but a related feat was announced in 2010: A group of synthetic biologists from California (USA) had produced a "synthetic" bacterium. However, the organism produced was not the result of assembling various biobricks to form a living organism—true de novo cell synthesis—but instead was the product of the artificial construction of a bacterial genome from a known genome sequence and the insertion of this synthetic genome into a different bacterial species to yield viable cells (**Figure 12.35**).

This feat was accomplished by artificially synthesizing a 1.08-million-base-pair (Mbp) genome based on the genome sequence of the bacterium *Mycoplasma mycoides*. This circular chromosome was pieced together from linear fragments of DNA containing homologous ends and the homologous recombination

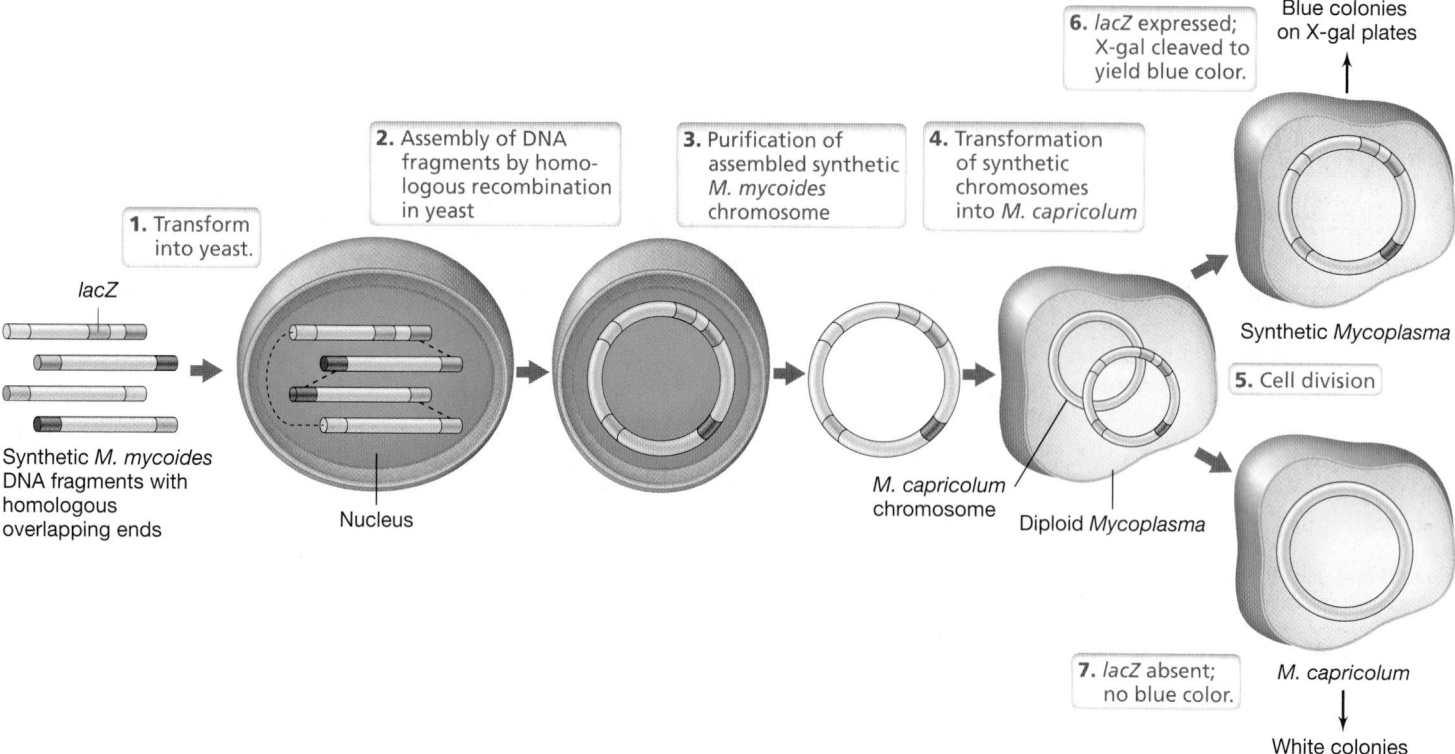

Figure 12.35 Formation of a synthetic *Mycoplasma* cell. DNA fragments corresponding to the desired *M. mycoides* chromosome are synthesized with overlapping ends (represented by the same colors). These DNA fragments are linked together using the homologous recombination machinery of *Saccharomyces cerevisiae*, with sites of homologous recombination indicated by lines. Once assembled, the synthetic chromosome is transformed into a cell of *M. capricolum*. This results in a cell with two separate chromosomes. After cell division, one cell possesses the synthetic chromosome (considered the synthetic *M. mycoides*) and the other remains a wild-type *M. capricolum*. The cell containing the synthetic chromosome was identified because it contained the reporter gene *lacZ* and thus produced blue colonies (see Figure 12.8). For more coverage of synthetic cells including research to define the minimalist genome, see page 332.

complex of yeast cells. The fully assembled synthetic chromosome was then purified and transformed into a cell of *Mycoplasma capricolum* (Figure 12.35). The synthetic chromosome contained a reporter gene, *lacZ*, absent from the *M. capricolum* genome, which caused colonies containing the synthetic chromosome to turn blue (see Figure 12.8); this was needed to distinguish *M. capricolum* cells containing the *M. mycoides* genome from those containing the *M. capricolum* genome after cell division (Figure 12.35). When cells in the blue colonies were examined, they showed all of the properties of the original *M. mycoides* cell.

Although this synthetic *M. mycoides* was not constructed solely from biobricks (the *M. capricolum* host cell contained ribosomes, various enzymes, and other important cytoplasmic components), the experiment did prove that an entire genome could be transplanted from one species to another. And just as importantly, the experiment offered a glimpse of the possibilities that lie on the horizon for synthetic biology. The possibilities are indeed endless, and the benefits of this science for humans and our planet will likely be significant.

MINIQUIZ

- What are biobricks?
- What organism has been genetically modified to produce precursors to the drugs artemisinin and morphine?
- How was *Escherichia coli* modified to produce a photograph?

12.12 Genome Editing and CRISPRs

Earlier in this book we discussed *c*lustered *r*egularly *i*nterspaced *s*hort *p*alindromic *r*epeat (CRISPR) systems and their role in protecting *Bacteria* and *Archaea* from foreign DNA and maintaining genome integrity (⇔ Sections 10.13 and 11.12). Microbiologists studying a CRISPR system called *CRISPR/Cas9* (Cas9 refers to CRISPR *a*ssociated protein 9) in the bacterium *Streptococcus pyogenes* discovered that the system could also recognize and cleave a specific DNA sequence within other cells and that foreign DNA could be inserted into the cut site.

Optimization of the CRISPR/Cas9 system has revolutionized biotechnology by providing the most powerful and precise tool yet for altering eukaryotic genomes in living cells. Indeed, *genome editing*, as it has come to be known, has been used successfully to edit the genomes of plants, animal embryos, and human cell lines. Here we explore how this system works and its benefits to synthetic biology.

Sequence Targeting by the Cas9 Protein

As illustrated in Figure 10.28, CRISPR systems possess Cas proteins that function as endonucleases when guided to a piece of nucleic acid by the complementary binding of CRISPR RNAs (crRNAs). By designing a synthetic RNA molecule that both recruits the *Streptococcus* Cas9 protein and binds to the desired target DNA sequence, genetic engineers have harnessed the power of the CRISPR–Cas system to cut specific DNA sequences in the genome of virtually any cell. At the cut site, the DNA can either be ligated (yielding a gene deletion) or used for inserting new DNA (**Figure 12.36a, b**).

The synthetic RNA molecule used for gene editing is called a *synthetic guide RNA* (sgRNA), and the Cas9 protein from *S. pyogenes* binding both the sgRNA and target DNA can be visualized in Figure 12.36c. For complete Cas9 endonuclease activity, a short *proto-spacer adjacent motif* (PAM; see Figure 10.28) must also occur on the target DNA. Without this PAM sequence, the Cas9 protein will bind to the region where the sgRNA binds, but will not cut. When the Cas9 protein does cut DNA, its two endonuclease domains (Figure 12.36c) cooperate to cut both DNA strands, and this generates double-stranded breaks (Figure 12.36a, b). Depending on the target cell, various methods of delivering the CRISPR system can be used. Genes corresponding to the designed sgRNA and Cas9 protein are often cloned into a plasmid under regulatory control of a strong promoter. Alternatively, the sgRNA and mRNA corresponding to the Cas9 protein can be generated in vitro. In either case, the materials are injected directly into target cells to trigger the gene editing process.

The presence of a PAM sequence close to the desired cleavage site is often the only limitation to cutting a specific DNA sequence. However, the PAM sequence is just three nucleotides in length and thus occurs with frequency in most genomes. CRISPR systems from other bacteria also recognize different PAMs, so alternative Cas proteins may be employed if needed. To delete a region of DNA, two target cleavage sites flanking the DNA sequence to be deleted must be identified and corresponding sgRNAs designed (Figure 12.36b). By contrast, only one cleavage site and corresponding sgRNA are needed to insert DNA (Figure 12.36a). To edit a region of a chromosome, sgRNAs are designed to bind to target DNA. This binding stimulates the Cas9 protein to cleave the target site if a PAM sequence is nearby (Figure 12.36a, b).

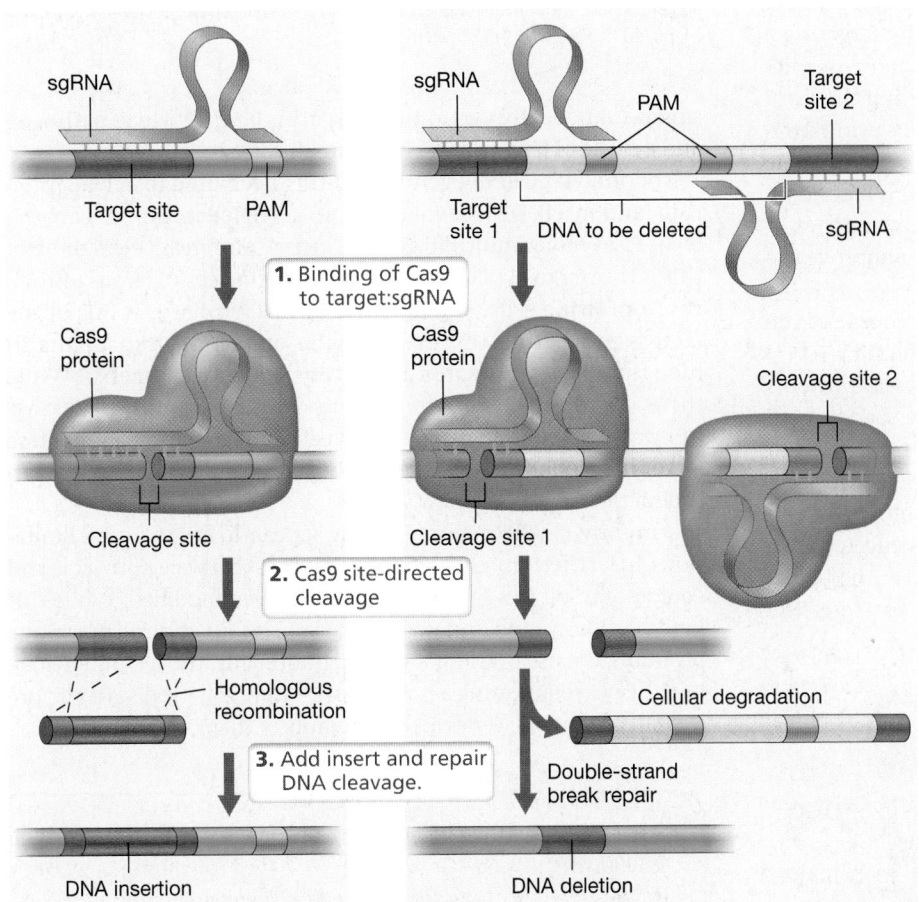

(a) **Genome insertion**   (b) **Genome deletion**

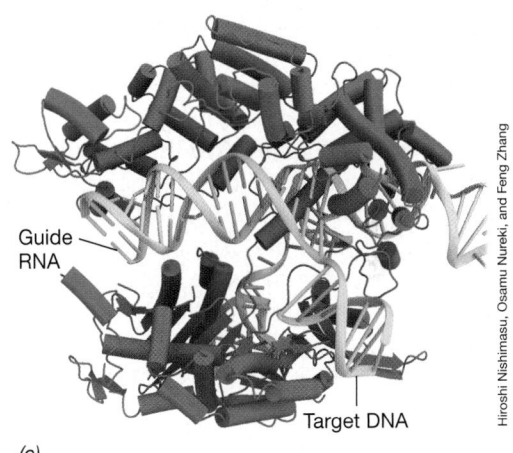

(c)

Figure 12.36 CRISPR/Cas9 genome editing.
sgRNA represents the synthetic guide RNA, while PAM represents a protospacer adjacent motif. Note that each genome target site must possess a PAM sequence for DNA cleavage to occur. (a) Insertion of foreign DNA into a targeted site of the genome. An sgRNA is synthesized to bind to a single target site on the genome through complementarity. This binding of the sgRNA to the DNA stimulates the Cas9 protein to cleave the genome at the target site. Foreign DNA with ends homologous to the cleavage site can be incorporated into the cut site through homologous recombination. This results in a genomic insertion. (b) Deletion of a genomic region. Two separate target sites flanking the DNA to be deleted are selected. After the design, addition, and binding of sgRNAs corresponding to these regions, Cas9 protein-dependent DNA cleavage occurs. This results in a double-stranded break in the target chromosome and a free piece of DNA. The double-strand break is then ligated by the cell's DNA double-strand break repair pathway, while the free piece of genomic DNA is degraded. This results in a genomic deletion. (c) Crystal structure of the *Streptococcus pyogenes* Cas9 protein. The target DNA is shown in green and the sgRNA in orange.

While a Cas9 protein and sgRNA can be used to cut DNA at specific sites (Figure 12.36), how is new DNA *inserted* at the cleavage site and how does the DNA get ligated back together? These tasks are accomplished by harnessing the cell's own DNA repair machinery. If a piece of DNA containing sequences with homology to the cut site is added to the system, homologous recombination will be used to incorporate the DNA (⮯ Section 11.5), yielding a genomic insertion (Figure 12.36a). If the goal is only to delete the chromosomal region between two cut sites, the nonhomologous double-strand DNA break repair pathway will be employed to ligate the DNA following the deletion event (Figure 12.36b).

CRISPR Editing in Practice

Applications of CRISPR genome editing have seen a meteoric rise since its discovery in 2013. By designing a Cas9-encoding gene to possess codons optimized for the organism of interest (Section 12.3) and engineering sgRNA to target the gene of interest, almost any DNA can be edited. While the genomes of rice, sorghum, and wheat have been edited using CRISPR, the CRISPR/Cas9 system has also been employed in tomato plants (a dicot) by targeting a gene region in which a mutation leads to leaves that are needle-like or wiry (**Figure 12.37**). In this study, DNA encoding the Cas9 protein and sgRNA was introduced into the plant cells using *Agrobacterium* and the Ti plasmid system (Figure 12.19). Because the resulting mutations were stable and heritable, CRISPR genome editing is poised for testing the functions of unknown genes in dicots and may possibly be used in the near future to improve the nutritional and other properties of fruits and vegetables.

While a Cas9 system has been used to excise the genome of the retrovirus HIV (the causative agent of AIDS, ⮯ Section 10.11) from the genome of infected human cells in vitro, Cas proteins from other bacteria can be used to target HIV or other specific viral RNAs. For example, the Csy4 CRISPR-associated protein from *Pseudomonas* is an endoribonuclease that processes the long CRISPR transcript (⮯ Figure 11.33). The Csy4 protein has been modified to recognize and destroy free HIV RNA in infected cells (**Figure 12.38**). Similarly, the CRISPR/Cas9 system from the bacterium *Francisella novicida* has been reprogrammed to target the

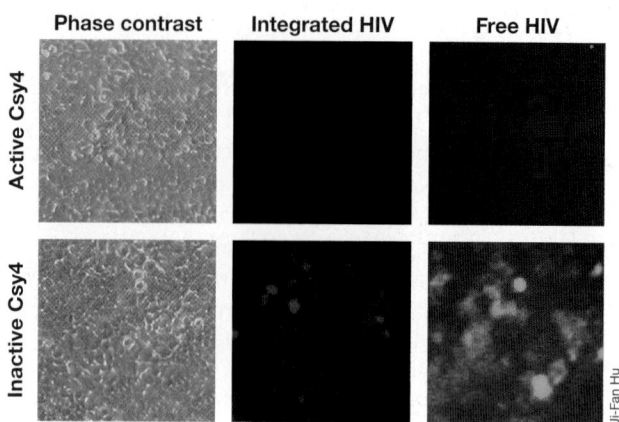

Figure 12.38 CRISPR-mediated inhibition of HIV infection. The results of using the *Pseudomonas* Csy4 endoribonuclease CRISPR protein to target RNA-based HIV in infected human embryonic kidney cells. Red indicates HIV provirus integration into infected cells, while free HIV virions are indicated in green. Top panel: expression of an HIV–Csy4 targeting vector. Bottom panel: expression of a mutated vector containing a nonfunctional Csy4 protein. Adapted from Guo, R., H. Wang, J. Cui, G. Wang, W. Li, and J-F Hu. 2015. *PLoS ONE 10*(10): e0141335.

single-stranded RNA genome of hepatitis C virus (a liver pathogen and a cause of liver cancer) in a human cell line.

Not only has CRISPR genome editing been used to delete, interrupt, and insert DNA sequences into a single location, it can also be used to target multiple genetic loci. An impressive example of this is the removal of 62 copies of the porcine endogenous retrovirus from swine cells. The presence of this retrovirus is one of the factors preventing the use of swine organs for transplants in humans (swine are anatomically very similar to humans). While there are other swine proteins that provoke an immune response in humans, genetic engineers envision editing away all of these factors to produce immune-friendly pig embryos for human organ production in the next few years!

Currently, CRISPR genome editing appears to have very few limitations. In fact, fertility clinic human embryos that were nonviable and could not result in a live birth have even been modified. While this landmark accomplishment has raised serious ethical questions regarding the use of CRISPR editing in humans, the technique may be the key to eradicating a host of devastating genetic diseases before a baby carrying the genes for one or more of them is born.

MINIQUIZ

- What is it about the Cas9 protein that makes it an efficient DNA editing tool?
- What is the role of the sgRNA in genome editing?
- How is recombinant DNA inserted into a genome using CRISPR editing?

12.13 Biocontainment of Genetically Modified Organisms

Throughout this chapter we have focused on genetically engineering microbes as factories for the synthesis of high-value products. While these applications of genetic engineering are clearly

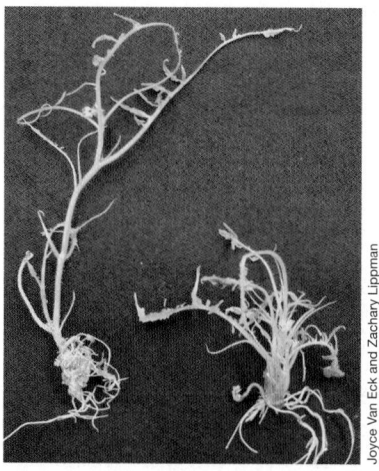

Figure 12.37 CRISPR editing of tomato genome. Interruption of the tomato SlAGO7 gene encoding an argonaute homolog results in plants that have wiry or spindly leaves (left) compared to a normal tomato plant (right).

beneficial, environmental concerns remain that genetically modified organisms (GMOs) may spread their modified genes to wild-type populations with adverse consequences. How can synthetic biology help solve this problem?

Early Containment Schemes

Through the years, various schemes have been proposed for containing GMOs. For example, both the use of auxotrophic strains and the induction of genes encoding self-toxins have been attempted with GMOs, but for both strategies, successful and reliable implementation has been a problem. Recall that an auxotroph is a mutant derivative of a microbial species that has a nutritional requirement; without the nutrient, the auxotroph cannot grow (꒰ Section 11.1), and this is the theory behind using auxotrophs to contain GMOs. However, in nature, auxotrophic strains can often survive by cross-feeding off the metabolites of other organisms, and the possibility always remains that an auxotrophic GMO could revert back to the wild type by back mutation and lose its nutritional dependencies.

In Section 7.11 we saw how certain bacteria produce a growth-inhibiting toxin that slows cell growth to help ensure survival of the population under stressful conditions. This has also been explored as a mechanism for biocontainment. That is, if a GMO were to escape confinement within a bioreactor (where all growth conditions are ideal for production of the desired product), the toxin would kick in and trigger the dormant state, from which the escaped GMO could not recover. However, in nature, where a cell has to compete not only with cells of its own kind but also with other microbial species, survival strongly selects for toxin gene mutations; if such a mutation occurred quickly, the toxin system could be disarmed and the GMO might survive.

Neither the auxotroph nor the toxin approaches adequately solve the problem of GMO containment. Moreover, neither of these mechanisms address the possibility of engineered DNA being released through industrial waste streams and finding its way into other microorganisms by horizontal gene transfer (Chapter 11). Thus, to more thoroughly and safely address the containment problem, genetic engineers have tapped into synthetic biology itself to devise novel methods of controlling GMOs in the environment, and we consider one now.

Biocontainment by Recoding the GMO Genome

A novel approach to prevent genetically modified bacteria from surviving outside of their bioreactor is to recode the genome of the GMO so the bacterium *can only grow if supplied with a synthetic amino acid* (**Figure 12.39**). This involves rewiring the organism's genetic code and translational machinery (Chapter 4) to synthesize proteins containing the synthetic amino acid. To accomplish this significant feat, synthetic biologists first replaced all the TAG (UAG on the RNA) stop codons associated with open reading frames on the *Escherichia coli* chromosome with the TAA stop codon. They also deleted the gene for release factor 1 that terminates translation when the ribosome encounters a UAG on the mRNA. Because this manipulation placed an alternative stop codon at the end of genes that had TAG codons, proteins of the correct length are still produced during translation and the recoded cell grows normally. This genetic manipulation

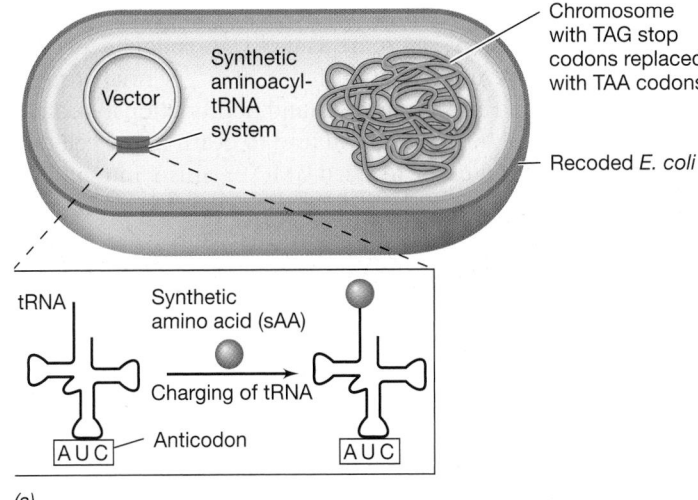

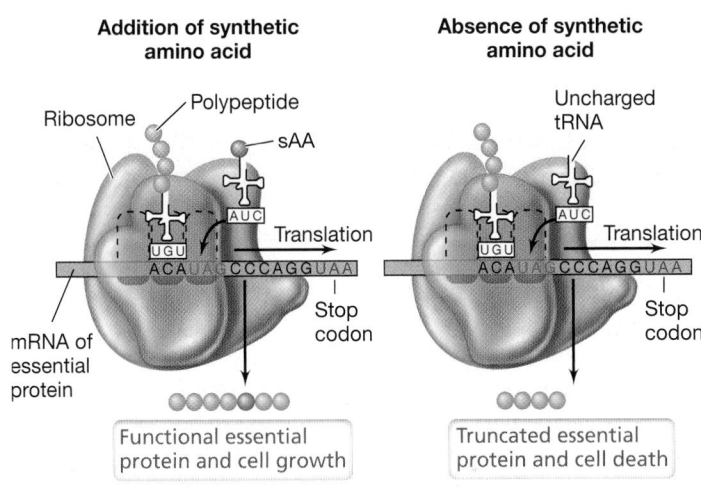

Figure 12.39 Recoding and control of genetically modified *Escherichia coli*. *(a)* The chromosome of an *E. coli* cell is genetically modified (recoded) to replace all TAG stop codons with TAA stop codons. The recoded *E. coli* stably maintains a vector expressing a tRNA with an AUC anticodon and an aminoacyl-tRNA synthetase (꒰ Figure 4.33) that charges the tRNA with a synthetic amino acid (sAA). *(b)* Control of cell growth by inserting a UAG codon into mRNAs of essential proteins. If the synthetic amino acid is added to the growth medium, the recoded *E. coli* will translate functional essential proteins. If the synthetic amino acid is not present, an uncharged tRNA will bind to UAG codons engineered into essential proteins. This results in truncated essential proteins and ultimately cell death.

also freed up the UAG codon to be reassigned another translational function.

To engineer dependence on a synthetic amino acid, genes for an aminoacyl-tRNA synthetase (꒰ Section 4.8) that recognizes the synthetic amino acid (sAA) and a corresponding tRNA with an AUC anticodon were expressed from a vector (Figure 12.39a). This resulted in tRNAs with the AUC anticodon carrying the sAA. A set of essential genes was then modified to contain the TAG codon in positions where incorporation of the sAA would not affect protein activity. Thus for the recoded bacterium to translate mRNAs possessing the UAG codon, the cell must be fed the artificial amino acid (Figure 12.39a). If the growth environment does not have the sAA (as would be the case if the GMO escaped to nature), uncharged tRNAs will enter into the ribosome when a UAG codon is encountered; if this occurs during the translation of an essential protein, translation will stall and a truncated (and nonfunctional) protein will result (Figure 12.39b). This will lead to death of the cell and puts ultimate control of the recoded GMO organism into human hands. Because the TAG codon was placed in three essential genes, it is extremely unlikely that sufficient mutations could arise to remove the GMO's dependence on the presence of the sAA.

While we have focused on how CRISPRs can be used to genetically modify organisms, synthetic biologists are even tinkering with the genome editing system itself to have it specifically clip out target DNA such as recombinant genes under specific conditions. Continued advances in synthetic biology and the control of GMOs not only will allow for more widespread use of these organisms for synthesizing desired products and therapeutics in carefully controlled production settings, but may also trigger the more extensive use of GMOs to solve the urgent problems that remain in medicine, agriculture, and the environment.

MINIQUIZ

- Why is the use of auxotrophy not a good method for controlling the growth of a genetically modified organism?
- How can a tRNA be engineered to encode for a synthetic amino acid?
- Why is it unlikely that GMOs recoded to depend on a synthetic amino acid will mutate to no longer depend on the exogenously supplied synthetic amino acid?

MasteringMicrobiology®   **Visualize, explore,** and **think critically** with Interactive Microbiology, MicroLab Tutors, MicroCareers case studies, and more. MasteringMicrobiology offers practice quizzes, helpful animations, and other study tools for lecture and lab to help you master microbiology.

Chapter Review

I • Tools of the Genetic Engineer

12.1 The polymerase chain reaction is a procedure for amplifying DNA in vitro and employs heat-stable DNA polymerases. This amplified DNA is often used for cloning purposes and can be visualized by gel electrophoresis. Complementary nucleic acid sequences may be detected by hybridization.

Q **Describe the basic principles of gene amplification using the polymerase chain reaction (PCR). How have thermophilic and hyperthermophilic prokaryotes simplified the use of PCR?**

12.2 The isolation of a specific gene or region of a chromosome by molecular cloning is done using a cloning vector. Plasmids are useful cloning vectors because they are easy to isolate and purify and are often able to multiply to high copy numbers in bacterial cells. The choice of a cloning host depends on the final application. In many cases the host can be a prokaryote, but in others, it is essential that the host be a eukaryote.

Q **How does the insertional inactivation of β-galactosidase allow the presence of foreign DNA in a plasmid vector such as pUC19 to be detected?**

12.3 Many cloned genes are not expressed efficiently in a foreign host. Expression vectors have been developed that both increase transcription of the cloned gene and control the level of transcription. To achieve very high levels of expression of eukaryotic genes in prokaryotes, the expressed gene must be free of introns. This can be accomplished by synthesizing cDNA from the mature mRNA encoding the protein of interest or by making an entirely synthetic gene. Protein fusions are often used to stabilize or solubilize the cloned protein.

Q **What is the significance of reverse transcriptase in the cloning of animal genes for expression in bacteria?**

12.4 Synthetic DNA molecules of desired sequence can be made in vitro and used to construct a mutated gene directly or to change specific base pairs within a gene by site-directed mutagenesis. Also, genes can be disrupted by inserting DNA fragments, called cassettes, into them, generating knockout mutants.

Q **What does site-directed mutagenesis allow you to do that normal mutagenesis does not?**

12.5 Reporter genes are genes whose products are easy to assay or detect. They are used to simplify and increase the speed of genetic analysis. In gene fusions, segments from two different genes, one of which is usually a reporter gene, are spliced together.

Q **Describe two widely used reporter genes.**

II • Making Products from Genetically Engineered Microbes: Biotechnology

12.6 The first human protein made commercially using engineered bacteria was human insulin. Recombinant bovine somatotropin is widely used in the United States to increase milk yield in dairy cows.

Q **What classes of mammalian proteins are produced by biotechnology? How are the genes for such proteins obtained?**

12.7 Genetic engineering can make plants resistant to disease and improve product quality. The Ti plasmid of the bacterium *Agrobacterium tumefaciens* can transfer DNA into plant cells. Genetically engineered commercial plants are called genetically modified organisms (GMOs).

> **Q** What is the Ti plasmid and how has it been of use in genetic engineering?

12.8 Many recombinant vaccines have been produced or are under development. These include live recombinant, vector, and subunit vaccines. Properties associated with pathogens can be used to develop cancer-treating therapeutics.

> **Q** What is a subunit vaccine and why are subunit vaccines considered a safer way of conferring immunity to viral pathogens than attenuated virus vaccines?

12.9 Genes for useful products may be cloned directly from DNA or RNA in environmental samples without first isolating the organisms that carry them. In pathway engineering, genes that encode the enzymes for a metabolic pathway are assembled. These genes may come from one or more organisms, but the engineering must achieve regulation of the coordinated sequence of expression required in the pathway.

> **Q** How has metagenomics been used to find novel useful products?

12.10 While select microorganisms can produce biofuels in small amounts, others can be modified to produce various biofuels through pathway engineering. This modification often requires genes from multiple microorganisms. New tools for genetically manipulating microalgae have also been developed to facilitate biofuel production.

> **Q** What do microalgae produce that can be chemically converted to biodiesel?

III • Synthetic Biology and Genome Editing

12.11 Instead of modifying or improving a single existing pathway, synthetic biology focuses on engineering novel biological systems by linking known biological components together in various combinations. These modifications can result in the production of high-value products.

> **Q** How does synthetic biology differ from engineering of *Escherichia coli* to produce indigo?

12.12 Not only can CRISPR systems be used as a prokaryotic immune system, they can also be modified to edit the genomes of eukaryotes.

> **Q** How has the CRISPR editing technology been applied to targeting virus-infected eukaryotic cells?

12.13 With the advancements in genetic engineering, methods for controlling genetically modified organisms are imperative. One promising method is the use of synthetic biology to recode organisms for dependence on the presence of synthetic amino acids.

> **Q** What are some mechanisms for controlling a genetically modified organism other than making it dependent on synthetic amino acids?

Application Questions

1. Suppose you have just determined the DNA base sequence for an especially strong promoter in *Escherichia coli* and you are interested in incorporating this sequence into an expression vector. Describe the steps you would use. What precautions are necessary to be sure that this promoter actually works as expected in its new location?

2. Many genetic systems use the *lacZ* gene encoding β-galactosidase as a reporter. What advantages or problems would there be if (a) luciferase or (b) green fluorescent protein were used instead of β-galactosidase as reporters?

3. You have just discovered a protein in mice that may be an effective cure for cancer, but it is present only in tiny amounts. Describe the steps you would use to produce this protein in therapeutic amounts. Which host would you want to clone the gene into and why? Which host would you use to express the protein in and why?

4. Describe how you could recode *Escherichia coli* to produce novel proteins containing more than the standard 22 amino acids.

Chapter Glossary

Bacterial artificial chromosome (BAC) a circular artificial chromosome with a bacterial origin of replication

Biotechnology the use of organisms, typically genetically altered, in industrial, medical, or agricultural applications

Cassette mutagenesis creating mutations by the insertion of a DNA cassette

Complementary DNA (cDNA) DNA made from an RNA template during the reverse transcription PCR (RT-PCR) procedure

DNA cassette an artificially designed segment of DNA that usually carries a gene for resistance to an antibiotic or some other convenient marker and is flanked by convenient restriction sites

Expression vector a cloning vector that contains the necessary regulatory sequences to allow transcription and translation of cloned genes

Gel electrophoresis a technique for separation of nucleic acid molecules by passing an electric current through a gel made of agarose or polyacrylamide

Gene disruption (also called gene knockout) the inactivation of a gene by insertion of a DNA fragment that interrupts the coding sequence

Gene fusion a structure created by joining together segments of two separate genes, in particular when the regulatory region of one gene is joined to the coding region of a reporter gene

Genetically modified organism (GMO) an organism whose genome has been altered using genetic engineering; the abbreviation GM is also used in terms such as GM crops and GM foods

Genetic engineering the use of in vitro techniques in the isolation, alteration, and expression of DNA or RNA and in the development of genetically modified organisms

Green fluorescent protein (GFP) a protein that glows green and is widely used in genetic analysis

Heterologous expression transcription and translation of a gene or genes from one organism in a different organism

Hybridization the formation of a double helix by the base pairing of single strands of DNA or RNA from two different sources

Molecular cloning the isolation and incorporation of a fragment of DNA into a vector where it can be replicated

Northern blot a hybridization procedure where RNA is the target and DNA or RNA is the probe

Nucleic acid probe a strand of nucleic acid that can be labeled and used to hybridize to a complementary molecule from a mixture of other nucleic acids

Operon fusion a gene fusion in which a coding sequence that retains its own translational signals is fused to the transcriptional signals of another gene

Pathway engineering the assembly of a new or improved biochemical pathway using genes from one or more organisms

Polymerase chain reaction (PCR) the artificial amplification of a DNA sequence by repeated cycles of strand separation and replication

Polyvalent vaccine a vaccine that immunizes against more than one disease

Protein fusion a gene fusion in which two coding sequences are fused so that they share the same transcriptional and translational start sites

Recombinant DNA a DNA molecule containing DNA originating from two or more sources

Reporter gene a gene used in genetic analysis because the product it encodes is easy to detect

Restriction enzyme an enzyme that recognizes a specific DNA sequence and then cuts the DNA; also known as a restriction endonuclease

Site-directed mutagenesis construction in vitro of a gene with a specific mutation

Southern blot a hybridization procedure where DNA is the target and RNA or DNA is the probe

Subunit vaccine vaccine that contains only a specific protein or two from a pathogen

T-DNA the segment of the *Agrobacterium tumefaciens* Ti plasmid that is transferred into plant cells

Ti plasmid a plasmid in *Agrobacterium tumefaciens* capable of transferring genes from bacteria to plants

Transgenic organism a plant or an animal with foreign DNA inserted into its genome

Vector (as in cloning vector) a self-replicating DNA molecule that is used to carry cloned genes or other DNA segments for genetic engineering

Vector vaccine a vaccine made by inserting genes from a pathogenic virus into a relatively harmless carrier virus

Yeast artificial chromosome (YAC) an artificial chromosome with a yeast origin of replication and a centromere sequence

Microbial Evolution and Systematics

microbiology now

Lokiarchaeota and the Origin of *Eukarya*

The domain *Eukarya* includes plants, animals, fungi, and a tremendous diversity of microorganisms. The plants, animals, and fungi are relative newcomers on the scene, as their evolutionary origins occurred some 400–600 million years ago. In contrast, the first eukaryotic microbes originated well over a billion years ago. The evolutionary origin of the eukaryotic cell remains enigmatic and we still do not know when or how the domain *Eukarya* was formed.

Genomic analyses clearly reveal *Eukarya* to be genetic chimeras. Eukaryotic genomes contain a mixture of genes that originated either within the *Bacteria* or within the *Archaea*, as well as many genes that are unique to *Eukarya*. Most evidence suggests that *Eukarya* share an ancestor with the domain *Archaea*, but *Eukarya* contain numerous "signature genes" not found in the *Archaea*. These unique eukaryotic genes encode proteins associated with the distinctive cell biology of *Eukarya* and were likely essential for the origins of multicellularity, a property widespread in the eukaryotic world.

The recent discovery of *Lokiarchaeota*—a new phylum of *Archaea*—has provided fresh insights into the origin of *Eukarya*. *Lokiarchaeota* were discovered through metagenomic analyses of microbial communities that inhabit deep marine sediments near a hydrothermal vent system known as Loki's Castle (see photo), located along the Mid-Atlantic Ridge between Greenland and Norway. Remarkably, the genomes of *Lokiarchaeota* contain a number of eukaryotic signature genes, and in particular, genes associated with membrane remodeling and the development of a cytoskeleton. The presence of a cytoskeleton and the ability to remodel intracellular membranes would have facilitated membrane invagination in primitive eukaryotic cells, and this would have allowed bacterial endosymbionts to be acquired and provided new nutritional strategies, such as phagocytosis.

These results suggest that features uniquely associated with the eukaryotic cell may actually have their origins in the domain *Archaea*. The discovery of the *Lokiarchaeota* also indicates that, rather than emerging as a sister group to the *Archaea*, the earliest eukaryotic cells emerged from within the *Archaea* following the endosymbiotic acquisition of the bacteria that gave rise to the eukaryotic cell's respiratory organelle, the mitochondrion. Hence, the first steps toward the origins of cellular complexity may have occurred within the domain *Archaea*.

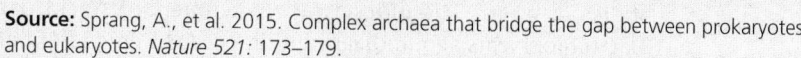 **Source:** Sprang, A., et al. 2015. Complex archaea that bridge the gap between prokaryotes and eukaryotes. *Nature 521:* 173–179.

Evolution is a topic that pervades all of biology, and microbiology in particular because microorganisms were the first life forms on Earth. In this chapter we tackle the topic of evolution and describe experimental methods for unraveling evolutionary relationships. We will also see how these powerful genetic methods along with phenotypic observations underlie the robust systematics of the microbial world that will unfold in the following five chapters.

I • Early Earth and the Origin and Diversification of Life

In these first sections, we consider the possible conditions under which life arose, the earliest evidence for cellular life, and its divergence into three evolutionary lineages: *Bacteria*, *Archaea*, and *Eukarya*. Although much about these events and processes remains speculative, geological and molecular evidence has combined to build a plausible scenario for the earliest events in the **evolution** of life and for the fundamental impacts that microbes have had on the history of our Earth.

13.1 Formation and Early History of Earth

The Earth of 4 billion years ago would be foreign and inhospitable to human eyes, but this sterile wasteland of blasted rock and boiling seas was the incubator from which all life sprang. Few fossils exist to tell the story of the early Earth, and most of what we know is inferred from chemical and isotopic analyses of ancient rocks and minerals. The story of life begins not long after the dawn of our solar system with the formation of Earth itself.

Origin of Earth

Earth formed about 4.5 billion years ago (**Figure 13.1**), based on analyses of slowly decaying radioactive isotopes. Our planet and the other planets of our solar system arose from materials making up a disc-shaped nebular cloud of dust and gases released by the supernova of a massive old star. As a new star—our sun—formed within this cloud, it began to compact, undergo nuclear fusion, and release large amounts of energy in the form of heat and light. Materials left in the nebular cloud began to clump and fuse due to collisions and gravitational attractions, forming tiny accretions that gradually grew larger to form clumps that eventually coalesced into planets. Energy released in this process heated the emerging Earth as it formed, as did energy released by radioactive decay within the condensing materials, forming a planet Earth of fiery hot magma. As Earth cooled over time, a metallic core, rocky mantle, and a lower-density, thin surface crust formed.

The inhospitable conditions of early Earth, characterized by a molten surface under intense bombardment by asteroids and other objects from space, are thought to have persisted for over 500 million years. Water on Earth originated from volcanic outgassing of the planet's interior and from innumerable collisions with icy comets and asteroids. Given Earth's heat at the time, water would have been present only as water vapor. The intense heat and the absence of liquid water indicate that early Earth was certainly a sterile planet. No rocks dating to the origin of Earth have yet been discovered, presumably because they have undergone geological metamorphosis. However, ancient crystals of the

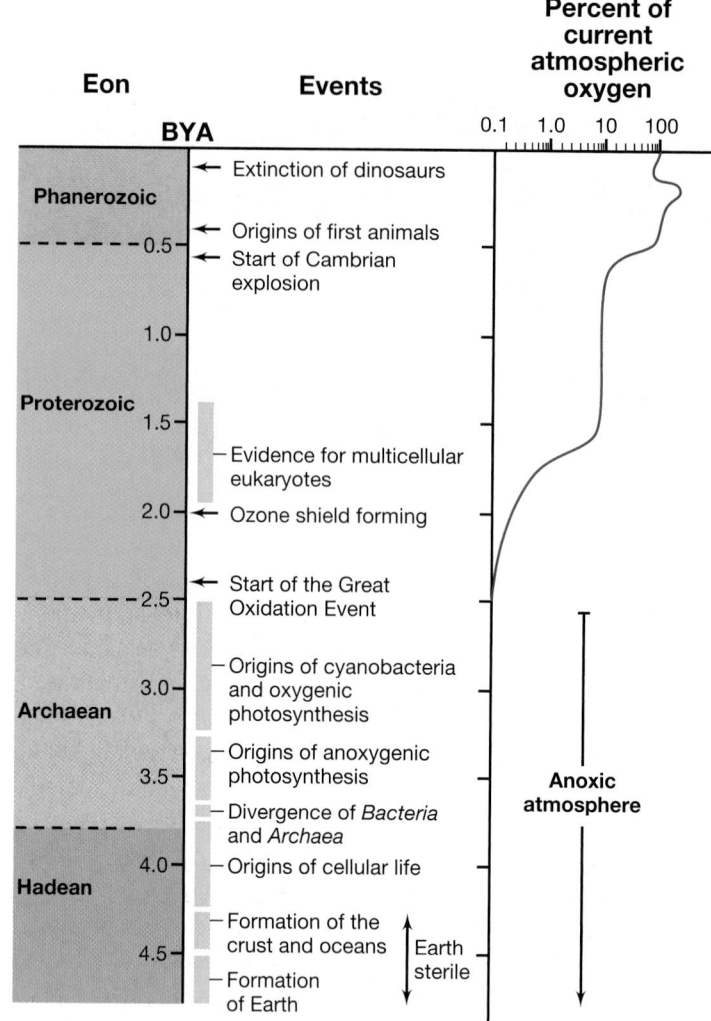

Figure 13.1 Major landmarks in biological evolution, Earth's changing geochemistry, and microbial metabolic diversification. The oldest date for the origin of life is fixed by the time of Earth's origin, and the minimum time for the origin of oxygenic photosynthesis is fixed by the Great Oxidation Event, about 2.4 billion years ago (BYA). Note how the oxygenation of the atmosphere from cyanobacterial metabolism was a gradual process, occurring over a period of about 2 billion years. Compare this figure with the introduction to the antiquity of life on Earth shown in Figure 1.5.

mineral zircon ($ZrSiO_4$) have been discovered that were formed on the early Earth, and these materials have given us a glimpse of conditions on Earth prior to the origin of life.

Liquid water is a requirement for life, and its presence on Earth made possible the origin of life. Analyses of ancient zircon crystals, including impurities trapped in these crystals and their oxygen isotope ratios (we discuss the use of isotopic analyses as indications of living processes in Section 19.10) indicate the presence of solid crust and liquid water as early as 4.3 billion years ago (Figure 13.1). Furthermore, graphite inclusions discovered within ancient zircon minerals have carbon isotope ratios that suggest biogenic origin, providing evidence for life on Earth 4.1 billion years ago. Among the oldest surviving sedimentary rocks are those found in southwestern Greenland. Sedimentary rocks are formed within bodies of water. The oldest known sedimentary rocks date to 3.86 billion years ago, indicating that oceans were present at the

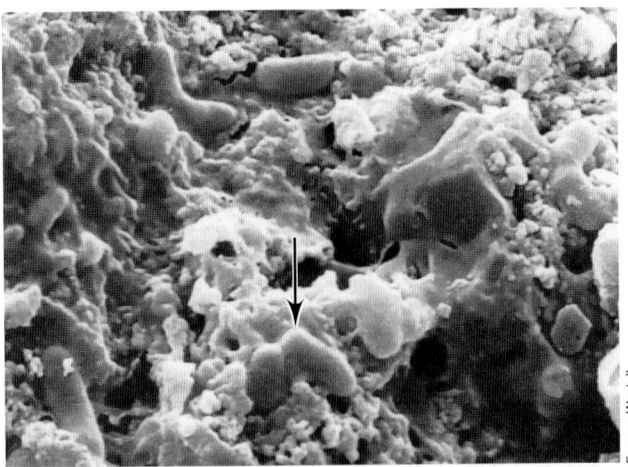

Figure 13.2 Ancient microbial life. Scanning electron micrograph of microfossil bacteria from 3.45-billion-year-old rocks of the Barberton Greenstone Belt, South Africa. Note the rod-shaped bacteria (arrow) attached to particles of mineral matter. The cells are about 0.7 μm in diameter.

time these rocks formed. These rocks contain fossilized remains of what appear to be cells (**Figure 13.2**), and carbon with isotopic ratios that provide further evidence for ancient microbial life.

Origin of Cellular Life

The origin of life on Earth remains the greatest of mysteries, obscured by the depths of time. Few rocks survive unaltered to testify about this period of Earth's history. Experimental evidence indicates that organic molecules such as RNA nucleotides, amino acids, and lipids can form spontaneously under conditions that were present on the early Earth, providing the preconditions needed for the first living systems. However, conditions on Earth's surface more than 4 billion years ago, in particular the extremely hot temperatures and levels of ultraviolet radiation, were likely hostile to the origin of life as we know it.

One hypothesis holds that life may have originated at hydrothermal systems on the ocean floor (**Figure 13.3**). Deep on the ocean floor, conditions would have been less hostile and more stable than on Earth's surface. A steady and abundant supply of energy in the form of reduced inorganic compounds—hydrogen (H_2) and hydrogen sulfide (H_2S), for example—would have been available at these hydrothermal systems. The unique geochemistry of these sites can support the abiotic production of molecules critical for the emergence of life. For example, molecules such as amino acids, lipids, sugars, and nucleotide bases can all be formed under conditions found at certain hydrothermal systems. Furthermore, mineral structures that form in these systems can produce compartmentalized structures that may have been necessary for conserving energy prior to the emergence of biological membranes. Whether on the seafloor or elsewhere, some form of prebiotic chemistry must have facilitated the development of the first self-replicating systems, the precursors to cellular life.

Molecules of RNA were likely a central component of the first self-replicating systems and it is possible that life began in an *RNA world* (**Figure 13.4**). RNA is a component of certain essential cofactors and molecules found in all cells (such as ATP, NADH, and coenzyme A); it can bind small molecules (such as ATP, other

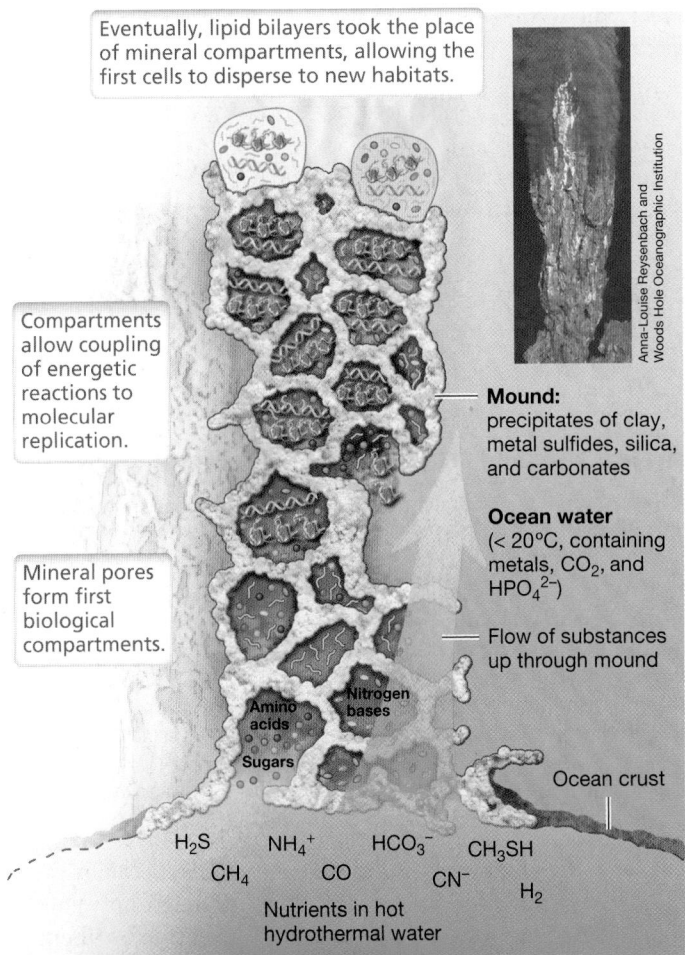

Eventually, lipid bilayers took the place of mineral compartments, allowing the first cells to disperse to new habitats.

Compartments allow coupling of energetic reactions to molecular replication.

Mineral pores form first biological compartments.

Mound: precipitates of clay, metal sulfides, silica, and carbonates

Ocean water (< 20°C, containing metals, CO_2, and HPO_4^{2-})

Flow of substances up through mound

Amino acids

Nitrogen bases

Sugars

Ocean crust

H_2S NH_4^+ HCO_3^- CH_3SH

CH_4 CO CN^- H_2

Nutrients in hot hydrothermal water

Figure 13.3 Submarine mounds and their possible link to the origin of life. Model of the interior of a hydrothermal mound with hypothesized transitions from prebiotic chemistry to cellular life depicted. Inset: photo of an actual hydrothermal mound. Hot, mineral-rich hydrothermal fluid mixes with cooler, more oxidized ocean water, forming precipitates of Fe and S compounds, clays, silicates, and carbonates. Mineral precipitates form pores that could have served as energy-rich compartments that facilitated the evolution of precellular forms of life.

nucleotides, and amino acids); it can possess catalytic activity; and it is known to catalyze protein synthesis through the activities of rRNA, tRNA, and mRNA (⊄ Section 4.10). It is thus possible to imagine that certain RNA molecules once had the ability to catalyze their own synthesis; these earliest forms of life may have had little or no need for DNA or protein. Indeed, comparative genomics studies of viruses suggest that the earliest viruses might have evolved from primitive cell-like structures that contained RNA (rather than DNA) genomes (⊄ Section 10.2).

Eventually, proteins replaced the catalytic role of RNAs in very primitive cells and at some point DNA, a molecule that is inherently more stable than RNA and therefore a better repository of genetic (coding) information, assumed the role of genome and template for all RNA synthesis (Figure 13.4). The earliest cellular forms of life likely possessed elements of this three-part system of DNA, RNA, and protein, in addition to a membrane system capable of conserving energy (see Figure 13.5). The *last universal common ancestor* (*LUCA*) of all extant life likely existed at 3.8–3.7 billion

UNIT 4

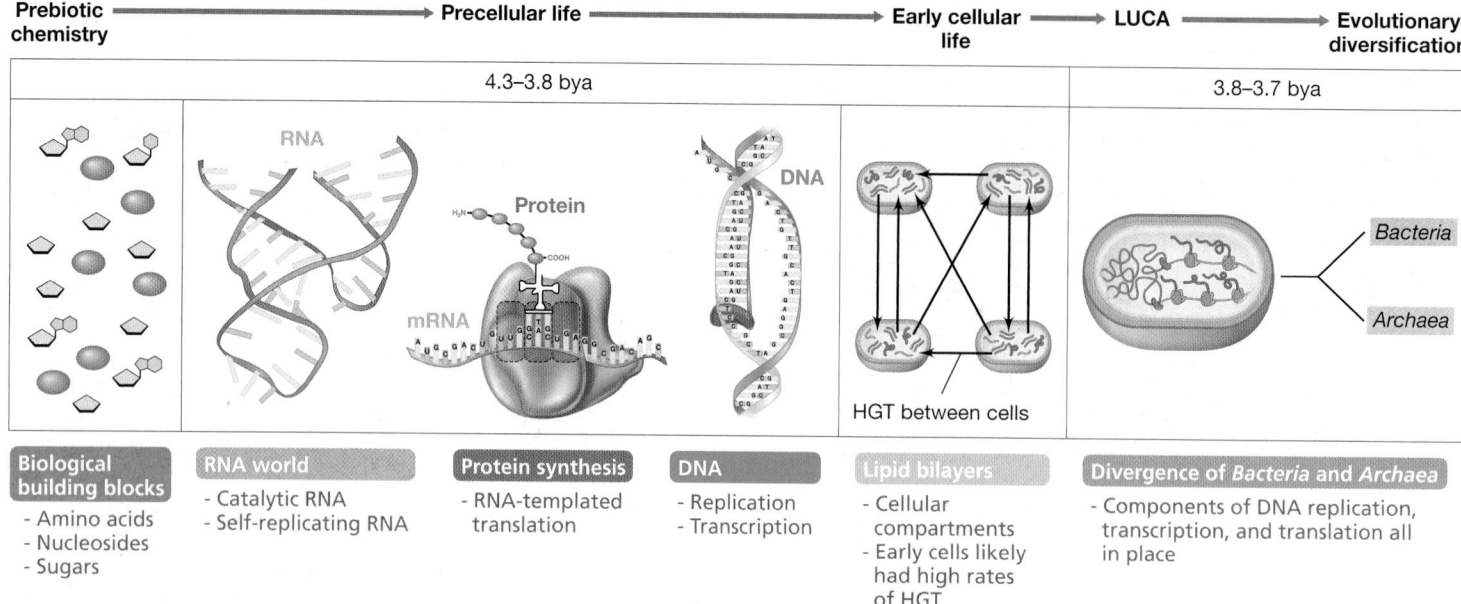

Figure 13.4 Events hypothesized to precede the origin of cellular life. The earliest self-replicating biological systems may have been based on catalytic RNA. At some point RNA enzymes evolved the capability to synthesize proteins, and proteins became the main catalytic molecules. Conversion from RNA- to DNA-based genomes required the evolution of DNA and RNA polymerases. The lipid bilayer is the site of electron transport and the evolution of this structure was likely important for energy conservation, in addition to containing and protecting biomolecules. The last universal common ancestor (LUCA), which preceded the divergence of *Bacteria* and *Archaea* (see Figure 13.9), was a cellular organism that had a lipid bilayer and used DNA, RNA, and protein. Horizontal gene transfer (HGT) may have allowed rapid transfer of beneficial genes among early forms of life.

years ago, the point at which *Bacteria* and *Archaea* became distinct and diverged and life began to diversify into the forms we recognize today. One can envision a time of intensive biochemical innovation and experimentation in which much of the structural and functional machinery of these earliest self-replicating systems evolved and was refined by natural selection.

Metabolic Diversification: Consequences for Earth's Biosphere

Following the origin of cells, microbial life experienced a long period of metabolic diversification, exploiting the various resources available on Earth. For much of Earth's history the planet, including all of its oceans, was anoxic (Figure 13.1). Thus, the energy-generating metabolism of primitive cells would have been exclusively anaerobic. During this era CO_2 may have been the major source of carbon for cells (autotrophy, ⇔ Section 14.5) because abiotic sources of organic carbon would quickly have become limiting as life became more and more abundant on Earth. Likewise, abiotic sources of fixed nitrogen would have become limiting on the early Earth, and microbes evolved the ability to use atmospheric N_2 as a source of nitrogen (nitrogen fixation, ⇔ Section 14.6) as early as 3.2 billion years ago, as indicated by isotopic ratios of nitrogen found in ancient sedimentary rocks. The capacity for both autotrophy and nitrogen fixation remain widespread among microorganisms today.

It is widely thought that H_2 was a major fuel for energy metabolism of early cells. This hypothesis is also supported by the tree of life (see Figure 13.9), in that virtually all of the earliest branching organisms in the *Bacteria* and *Archaea* use H_2 as an electron donor in energy metabolism and are also autotrophs. Elemental sulfur (S^0) may have been one of the earliest electron acceptors, as the reduction of S^0 to yield H_2S is exergonic and would likely have required relatively few enzymes (**Figure 13.5**). Moreover, because of the abundance of H_2 and sulfur compounds on early Earth, this scheme

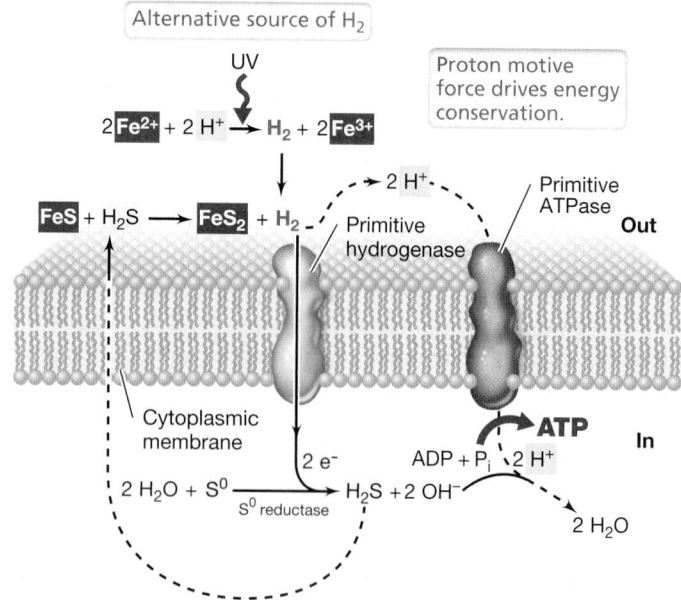

Figure 13.5 A possible energy-conserving scheme for primitive cells. Formation of pyrite (FeS_2) leads to H_2 production and S^0 reduction, which fuels a primitive ATPase. Note how H_2S plays only a catalytic role; the net substrates would be FeS and S^0. Also note how few different proteins would be required. $\Delta G^{0\prime} = -42$ kJ for the reaction $FeS + H_2S \rightarrow FeS_2 + H_2$.

would have provided cells with a nearly limitless supply of energy. Other early microorganisms may have used H_2 and CO_2 to produce acetate (⇔ Section 14.16) or methane (⇔ Section 14.17). These early forms of chemolithotrophic metabolism driven by H_2 would likely have supported the production of large amounts of organic compounds from autotrophic CO_2 fixation. Over time, these organic materials would have accumulated and could have provided the conditions needed for the evolution of new chemoorganotrophic bacteria with diverse metabolic strategies to conserve energy from the oxidation of organic compounds.

Eventually Earth became a highly oxic planet, with its characteristic high levels of O_2 that we breathe today (Figure 13.1). This key geochemical transition in Earth's history was catalyzed by microbes and we explore this topic now.

MINIQUIZ

- What characteristics would have made the surface of Earth inhospitable to the formation of life 4.5 billion years ago?

- How do we know when oceans were first present on Earth? Why is the presence of oceans significant to the origins and diversification of life?

- What lines of reasoning support the hypothesis that the first self-replicating systems were based on RNA molecules?

13.2 Photosynthesis and the Oxidation of Earth

The evolution of photosynthesis was a biological breakthrough that revolutionized the chemistry of Earth. Phototrophic organisms use energy from the sun to oxidize molecules such as H_2S, S^0, or H_2O and to synthesize complex organic molecules from carbon dioxide or simple organic molecules (⇔ Section 14.5). Over time, the products of photosynthesis accumulated in the biosphere, stimulating the further diversification of microbial life. Earth's first phototrophs were anoxygenic (cells that do not produce O_2, ⇔ Sections 14.3 and 15.4–15.7), but from these evolved the *Cyanobacteria*, the earliest O_2-producing (oxygenic phototrophs) (Figure 13.1, ⇔ Section 15.3).

Fossilized microbial formations called **stromatolites** can be found in rocks that are 3.5 billion years old, providing the earliest conclusive evidence of life on Earth (**Figure 13.6**). Stromatolites, or "layered rocks," are formed when certain kinds of microbial mats cause the deposition of carbonate or silicate minerals that promote fossilization (we discuss microbial mats in Section 20.5). Stromatolites were diverse and common on Earth between 2.8 and 1 billion years ago, but declined dramatically in abundance over the last billion years. Stromatolites are largely gone from Earth today, and yet modern examples of this ancient microbial ecosystem can still be found in certain shallow marine basins (Figure 13.6c, e) or in hot springs (Figure 13.6d). Phototrophic bacteria, such as cyanobacteria (⇔ Section 15.3) and the green nonsulfur bacterium *Chloroflexus* (⇔ Section 15.7), play a central role in the formation of modern stromatolites. Ancient stromatolites contain microfossils that appear remarkably similar to modern species of phototrophic bacteria (**Figure 13.7a**). Hence, the earliest phototrophic organisms

(a)   (b)

(c)

(d)   (e)

Figure 13.6 Ancient and modern stromatolites. (a) The oldest known stromatolite, found in a rock about 3.5 billion years old from the Warrawoona Group in Western Australia. Shown is a vertical section through the laminated structure preserved in the rock. Arrows point to the laminated layers. (b) Stromatolites of conical shape from 1.6-billion-year-old dolomite rock from northern Australia. (c) Modern stromatolites, Darby Island, Bahamas. The large stromatolite in the foreground is about 1 m in diameter. (d) Modern stromatolites composed of thermophilic cyanobacteria growing in a thermal pool in Yellowstone National Park, USA. Each structure is about 2 cm high. (e) Modern stromatolites from Shark Bay, Australia. Individual structures are 0.5–1 m in diameter.

may have evolved in the *Bacteria* more than 3.5 billion years ago, giving rise to the stromatolites we observe in the fossil record.

The first phototrophs were anoxygenic, using electron donors such as H_2S for CO_2 fixation and generating elemental sulfur (S^0) as a waste product (⇔ Section 14.3). The ability to use solar radiation as an energy source allowed phototrophs to diversify extensively. By 2.5–3.3 billion years ago, the cyanobacterial lineage evolved a photosystem capable of oxygenic photosynthesis (⇔ Section 14.4) in which H_2O supplanted H_2S as the reductant for CO_2, thereby generating O_2 as a waste product. The origin of oxygenic photosynthesis and the rise of O_2 in Earth's atmosphere caused the greatest change ever in the history of our biosphere and set the stage for the evolution of even newer forms of life that evolved to exploit the energy available from O_2 respiration. We look at the evidence for and consequences of this great oxidation event now.

The Rise of Oxygen: Banded Iron Formations

In the absence of O_2, most of Earth's iron would have been present in reduced forms (Fe^0 and Fe^{2+}) and large amounts of iron would have been dissolved in Earth's anoxic oceans. Molecular and chemical evidence indicates that oxygenic photosynthesis first appeared on Earth several hundred million years before significant levels of O_2 appeared

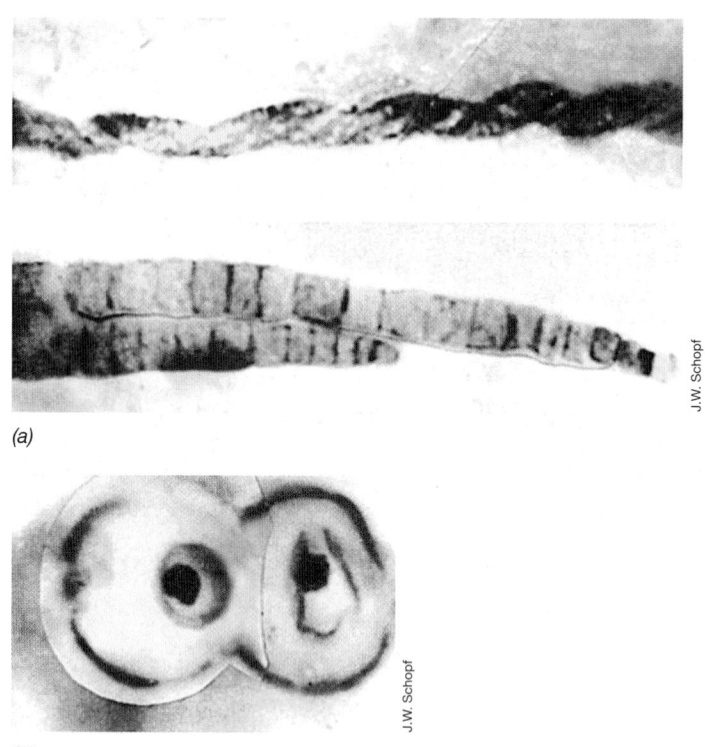

J.W. Schopf

(a)

J.W. Schopf

(b)

Figure 13.7 More recent fossil bacteria and eukaryotes. *(a)* One-billion-year-old microfossils from central Australia that resemble modern filamentous cyanobacteria. Cell diameters, 5–7 μm. *(b)* Microfossils of eukaryotic cells from the same rock formation. The cellular structure is similar to that of certain modern green algae, such as *Chlorella* species. Cell diameter, about 15 μm. Color was added to make cell form more apparent.

in the atmosphere. However, the O_2 that cyanobacteria produced could not accumulate in the atmosphere because it reacted spontaneously with the reduced iron minerals in the oceans to make iron oxides. By 2.4 billion years ago, O_2 levels had risen to one part per million, a tiny amount by present-day standards, but enough to initiate what has come to be called the *Great Oxidation Event* (Figure 13.1).

The metabolism of cyanobacteria yielded O_2 that oxidized reduced minerals containing Fe^{2+} to iron oxides containing Fe^{3+}. These iron oxide minerals became a prominent marker in the geological record. Iron oxides are poorly soluble in water and would have precipitated in the oceans, raining down onto the seafloor and forming sedimentary structures we see today as **banded iron formations** (Figure 13.8), laminated sedimentary rocks formed in deposits of iron- and silica-rich materials. Much of the iron in rocks of Precambrian origin (>0.5 billion years ago, see Figure 13.1) exists in such banded formations, and today these minerals are mined as a major source of iron ore. During this span of Earth history, lasting more than 1.5 billion years, the presence of precipitating iron minerals could have caused the oceans to appear brown, or black, or even red rather than the blue color we know today. Only after the abundant Fe^{2+} on Earth was oxidized could O_2 begin to accumulate in the atmosphere, and not until 600–900 million years ago did atmospheric O_2 reach present-day levels (~21%, Figure 13.1).

As O_2 accumulated on Earth, the atmosphere gradually changed from anoxic to oxic (Figure 13.1). Species of *Bacteria* and *Archaea* unable to adapt to this change were increasingly restricted to anoxic habitats because of the toxicity of O_2 and because it chemically

John M. Hayes

Figure 13.8 Banded iron formations. An exposed cliff made of sedimentary rock about 10 m in height in Western Australia contains layers of iron oxides (arrows) interspersed with layers containing iron silicates and other silica materials. The iron oxides contain iron in the ferric (Fe^{3+}) form produced from ferrous iron (Fe^{2+}) primarily by the oxygen released by cyanobacterial photosynthesis.

oxidized the reduced substances upon which their metabolisms depend. However, the oxic atmosphere also created conditions for the evolution of various new metabolic schemes, such as sulfide oxidation, nitrification, and the various other aerobic chemolithotrophic processes (Chapter 14). Microbes that evolved the capacity to respire O_2 gained a tremendous energetic advantage because of the high reduction potential of the O_2/H_2O couple (⮌ Section 3.6), and with more energy at their disposal, aerobes could reproduce far more rapidly than anaerobes.

The Ozone Shield

An important consequence of O_2 for the evolution of life was the formation of *ozone* (O_3). The sun bathes Earth in intense amounts of ultraviolet (UV) radiation, which is lethal to cells and can cause severe DNA damage (⮌ Section 11.4). When O_2 is subject to UV radiation from the sun, it is converted to ozone, which strongly absorbs UV radiation in wavelengths up to 300 nm. The conversion of O_2 to O_3 creates an *ozone shield*, a barrier that protects the surface of Earth from much of the UV radiation from the sun. Prior to the generation of the ozone shield, the punishing UV irradiation from the sun would have made Earth's surface fairly inhospitable for life, restricting life to environments that provided protection from UV radiation, such as in the oceans or in the subsurface. However, as Earth developed an ozone shield, organisms could range over the terrestrial surface of Earth, exploiting new habitats and evolving ever-greater diversity. Figure 13.1 summarizes some landmarks in biological evolution and Earth's geochemistry as Earth transitioned from an anoxic to a highly oxic planet.

--- **MINIQUIZ** ---

- Why is the origin of cyanobacteria considered a critical step in evolution?
- What caused the development of banded iron formations?
- What lines of evidence indicate that microbial life was present on Earth 3.5 billion years ago?

13.3 Living Fossils: DNA Records the History of Life

The evolutionary origins of microorganisms remained a mystery until it was discovered that certain molecular sequences are a record of evolutionary history (⇌ Section 1.13). Mutations occur at random and accumulate over time, causing heritable changes in the sequence of DNA (Chapter 11); this ultimately results in evolution. Organisms that share a recent ancestor have similar DNA sequences, and organisms that are more distantly related have more dissimilar DNA sequences. Hence, we can reconstruct the evolutionary history, or **phylogeny**, of a set of related DNA sequences by analyzing similarities in their nucleotide sequences (Section 13.7).

Carl Woese and the Tree of Life

Carl Woese was the first to construct a **universal tree of life** that he inferred from nucleotide sequence similarity in the **ribosomal RNA (rRNA)** genes of diverse organisms (⇌ Section 1.13).

The universal tree of life (**Figure 13.9**) is a genealogy of life on Earth. It depicts the evolutionary history of all cells and clearly reveals the three-domain concept in which all cells can be classified as *Bacteria*, *Archaea*, or *Eukarya*. The root of the universal tree represents a point in time when all extant life on Earth shared a common ancestor, the last universal common ancestor, LUCA (Figures 13.4 and 13.9). The universal tree of life shows that the first living things were microorganisms, and that microbes were the dominant life form for most of the history of life on Earth.

Genomics (Chapter 9) supports the three-domain concept through phylogenetic analysis of most genes central to cellular function. For example, at least 60 genes (including rRNA genes) are shared by nearly all extant cells and these genes must have been present in the universal ancestor. Most of these genes encode core functions in transcription, translation, and DNA replication. Across these conserved genes, those of *Eukarya* show greater sequence similarity to those of *Archaea* than to those of *Bacteria*. These and other data support the conclusion that the *Bacteria* and *Archaea* diverged prior to the origin of eukaryotic cells (Figure 13.9).

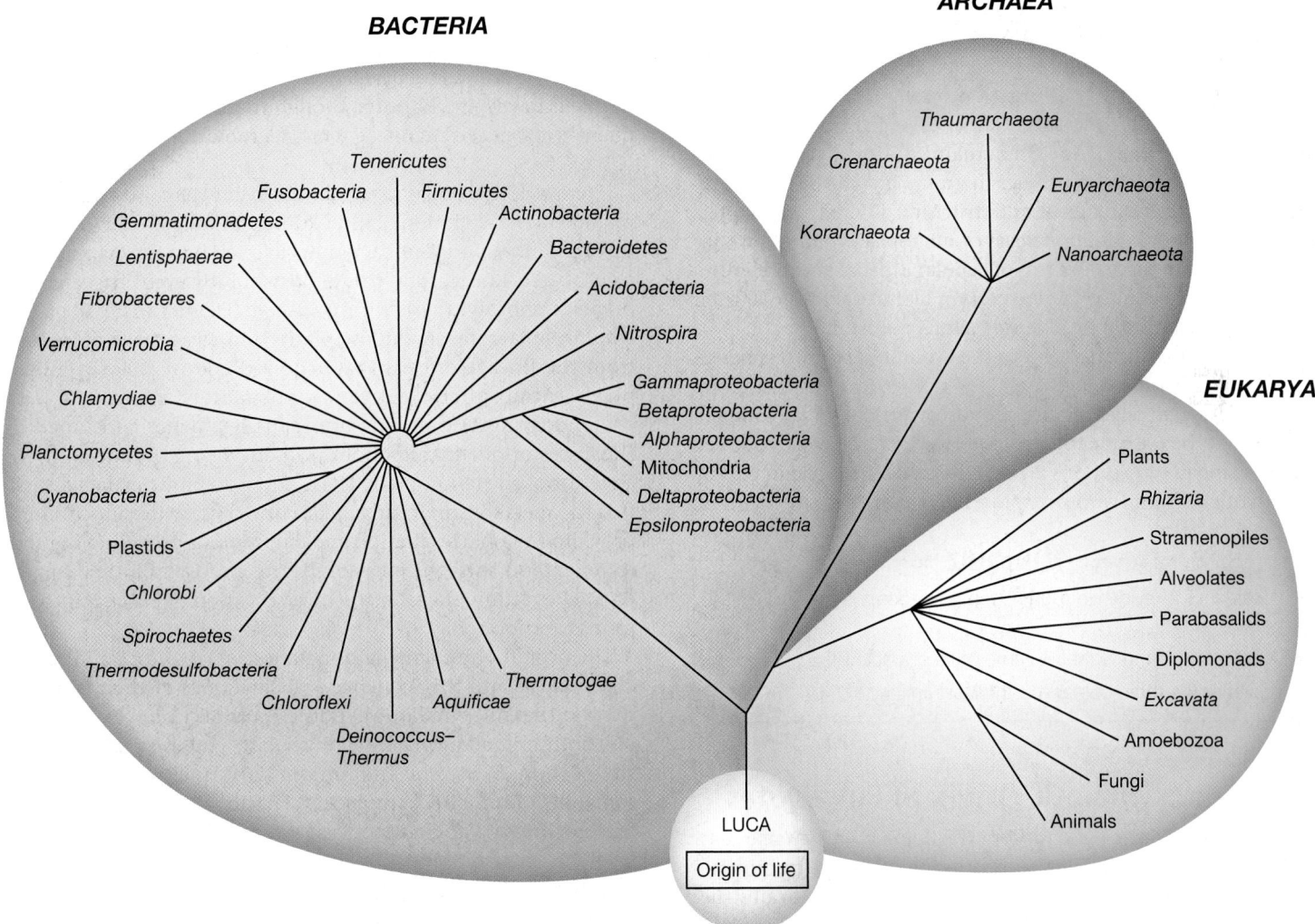

Figure 13.9 Universal phylogenetic tree as supported by comparative SSU rRNA gene sequence analysis. Only a few key organisms or lineages are shown in each domain. The branch lengths in this tree are arbitrary and nodes have been collapsed to reflect phylogenetic uncertainty. At least 84 phyla of *Bacteria* have now been identified, although many of these have not yet been cultured. LUCA, last universal common ancestor (Figure 13.4).

Hence, phylogenetic analyses of genome sequences suggest that LUCA had a prokaryotic cell structure along with a DNA-based genome and the ability to transcribe genes and synthesize proteins (Figure 13.9).

Other Influences Affecting Phylogeny

The manner in which the three domains were established remains a topic of debate. It is clear that the three domains represent the major evolutionary cell lineages that exist on Earth, and it is also clear that *Bacteria* and *Archaea* diverged prior to the divergence of *Eukarya* from *Archaea*. However, there are many examples of genes shared by any two of the three domains. One hypothesis is that early in the history of life, before the primary domains had diverged, **horizontal gene transfer** (⮌ Chapter 11) was extensive. During this time, any genes that provided a strong benefit may have been transferred rapidly among early forms of life (Figure 13.4). As the domains continued to diverge over time, barriers to unrestricted horizontal gene transfer likely evolved in order to maintain genomic stability (⮌ Section 11.12). As a result, the previously promiscuous cell population began to slowly sort out into the primary lines of evolutionary descent (Figures 13.4 and 13.9).

It seems likely that the domains *Bacteria* and *Archaea* had already diverged by about 3.7 billion years ago (Figure 13.1). Following this, there was a further bifurcation, perhaps 1.2 to 2.7 billion years ago, at which time the *Eukarya* diverged from *Archaea* to form a distinct domain. As each domain continued to evolve, certain traits became fixed within each group, giving rise to a multitude of genetic, physiological, and structural differences (Chapters 14–18). After nearly 4 billion years of microbial evolution, we see the grand result: three domains of cellular life that are each evolutionarily distinct and yet share certain features that reveal their common descent from a universal cellular ancestor.

Before we turn our attention to the evolutionary process per se, we focus briefly on the *Eukarya*, a phylogenetically distinct group whose eukaryotic cell structure contrasts with the prokaryotic cell structure of the *Bacteria* and *Archaea*. However, the eukaryotic cell we know today (⮌ Sections 2.14–2.16) has been shaped in part by prokaryotic cells, and we now explore how this might have occurred.

MINIQUIZ

- What kinds of evidence support the three-domain concept of life?
- What is LUCA and what are some of its characteristics?
- Which of the three domains is the least ancient?

13.4 Endosymbiotic Origin of Eukaryotes

The divergence of the *Eukarya* from the domain *Archaea* marked a major milestone in cellular evolution as the origin of a membrane-enclosed nucleus and organelles gave rise to eukaryotic cell structure. Here we consider the origin of the *Eukarya* and show how eukaryotes are genetic chimeras containing genes from at least two different phylogenetic domains. A recent hypothesis for the origin of eukaryotes from within the *Archaea* was presented on page 363.

Endosymbiosis

As Earth became more oxic, organelle-containing eukaryotic microorganisms arose, and the rise in O_2 likely had a major impact on their evolution. While the exact origins of the eukaryotic cell remain unclear, the oldest microfossils that have recognizable nuclei are about 2 billion years old. Multicellular and increasingly complex microfossils of algae are evident from 1.9 to 1.4 billion years ago (Figure 13.7b). By 0.6 billion years ago, with O_2 near present-day levels, large multicellular organisms, the Ediacaran fauna, were present in the sea (Figure 13.1). In a relatively short time, multicellular eukaryotes diversified into the ancestors of modern-day algae, plants, fungi, and animals.

A well-supported explanation for the origin of organelles in the eukaryotic cell is the **endosymbiotic hypothesis** (Figure 13.10). This hypothesis, for which supporting evidence today is so strong that its basic tenets can be considered more like a theory than a hypothesis, states that the mitochondrion of modern-day eukaryotes arose from the stable incorporation of a bacterium capable of aerobic respiration into the cytoplasm of early eukaryotic cells. Endosymbiotic mitochondria were beneficial to these early eukaryotic cells because they increased the cell's respiratory capacity, and thus these early mitochondria-containing cells proceeded to become the ancestors of all living *Eukarya*. Virtually all eukaryotic cells have mitochondria, though these structures were subsequently lost in certain lineages of anaerobic microbial eukaryotes (Chapter 18).

There was a second endosymbiotic event that also had a major impact on the evolution of life. Chloroplasts arose from the stable incorporation of a cyanobacterium-like cell into the cytoplasm of a eukaryotic lineage, and this endosymbiotic event triggered the origin of photosynthesis within *Eukarya* (Figure 13.10). All phototrophic eukaryotes, including plants and algae, have descended from the lineage of cells that acquired endosymbiotic chloroplasts. Atmospheric oxygen is intimately associated with the endosymbiotic origins of organelles, being consumed by the ancestor of the mitochondrion and being produced by the ancestor of the chloroplast. These endosymbiotic events diversified the metabolism of early eukaryotic cells, with mitochondria providing aerobic respiration and chloroplasts providing the ability to exploit sunlight for energy. The endosymbiotic origin of organelles set the stage for the diversification of *Eukarya* into the forms we know today.

The overall physiology and metabolism of mitochondria and chloroplasts and the sequence and structures of their genomes support the endosymbiotic hypothesis (⮌ Section 9.4). For example, both mitochondria and chloroplasts contain ribosomes the size of those in *Bacteria* and *Archaea* (70S), including a 16S ribosomal RNA (**16S rRNA**) molecule. The 16S rRNA gene sequences of mitochondria and chloroplasts are also characteristic of those from *Bacteria*, and sequence analyses place the ancestor of mitochondria in the **phylum** *Alphaproteobacteria*, and the ancestor of chloroplasts in the phylum *Cyanobacteria* (Figure 13.9). Moreover, the same antibiotics that inhibit ribosome function in free-living *Bacteria* inhibit ribosome function in these organelles. Mitochondria and chloroplasts also contain small amounts of DNA arranged in a covalently closed, circular form, which is typical of *Bacteria*,

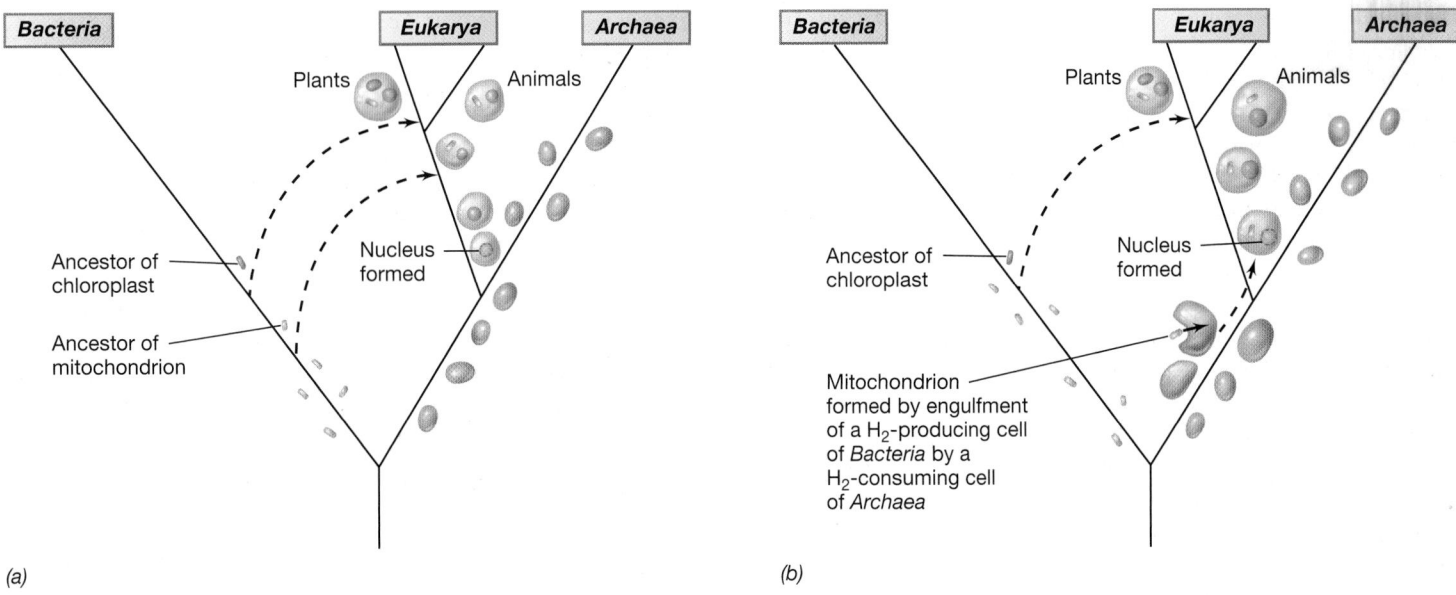

(a)   *(b)*

Figure 13.10 Endosymbiotic models for the origin of the eukaryotic cell. *(a)* The serial endosymbiosis hypothesis proposes that the eukaryotic ancestor diverged from the archaeal line and possessed a nucleus and other features of eukaryotic cells prior to endosymbiosis with the bacterial ancestor of the mitochondrion. A later endosymbiosis with the cyanobacterial ancestor of the chloroplast gave rise to the eukaryotic ancestor of all plants and all other photosynthetic eukaryotes. *(b)* The hydrogen hypothesis (which is a version of the symbiogenesis hypothesis) proposes that the eukaryotic cell evolved from a symbiotic relationship between H_2-producing cells of *Bacteria* and H_2-consuming cells of *Archaea*. The bacterial partner was eventually engulfed by its archaeal partner and evolved over time into the mitochondrion. The nucleus and other features of the eukaryotic cell evolved after establishment of endosymbiosis. A later endosymbiosis with the cyanobacterial ancestor of the chloroplast gave rise to the eukaryotic ancestor of all plants and all other photosynthetic eukaryotes. Note the position of the mitochondrion and plastids (chloroplasts are a type of plastid) on the universal phylogenetic tree (Figure 13.9).

and the phylogeny of these gene sequences indicates a bacterial ancestry. Indeed, these and many other telltale signs of *Bacteria* are present in organelles from modern eukaryotic cells.

Formation of the Eukaryotic Cell

The exact pathway by which the eukaryotic cell emerged remains a major unresolved question in evolution; however, it seems clear that the modern eukaryotic cell genome is a genetic chimera, made up of genes from both *Bacteria* and *Archaea*. In eukaryotes, most of the genes that encode information-processing machinery resemble those of *Archaea*, while most metabolic genes resemble those of *Bacteria*. For example, eukaryotic cells share with *Archaea* molecular features of transcription and translation (Chapter 4), while features they share with *Bacteria* include their ester-linked membrane lipids (Chapter 2) and their glycolytic pathway (Chapter 3). As we have seen, there is strong support for the endosymbiotic origin of mitochondria and chloroplasts from *Bacteria*, and it is clear that certain genes from these endosymbionts were transferred to the cell nucleus (⇔ Section 9.4).

Two major hypotheses have been put forward to explain the formation of the eukaryotic cell (Figure 13.10). In the *serial endosymbiosis hypothesis*, eukaryotes arose as a nucleus-bearing cell line that split from the *Archaea* and later acquired mitochondria and chloroplasts by endosymbiosis (Figure 13.10*a*). Endosymbiosis is posited to have occurred when this cell line engulfed a bacterial cell that, rather than being destroyed, managed to survive and replicate as an endosymbiont within the cytoplasm of the host cell. According to this hypothesis, eukaryotic genes that resemble those of *Bacteria* were acquired in gene transfers from the

endosymbiont to the nuclear genome. A major problem with this hypothesis is that it does not easily account for the fact that *Bacteria* and *Eukarya* have structurally similar membrane lipids, in contrast to the lipids of *Archaea* (⇔ Section 2.6).

The second hypothesis, called the *symbiogenesis hypothesis*, proposes that the eukaryotic cell arose from a symbiotic relationship between cells of *Bacteria* and *Archaea* that ultimately resulted in engulfment of the bacterial partner to form mitochondria. One version of this hypothesis is the *hydrogen hypothesis*, which proposes that the eukaryotic cell arose from an association between a H_2-producing species of *Bacteria* and a H_2-consuming species of *Archaea* (Figure 13.10*b*). In the hydrogen hypothesis, the bacterial cell was a facultative aerobe and a chemoorganotroph that could grow by aerobic respiration or by syntrophic production of H_2 (⇔ Section 14.23). The archaeal partner was an anaerobic chemolithotroph that required H_2 for growth. These symbiotic partners could have coevolved for an extended period prior to bacterial engulfment and formation of the mitochondria. The nucleus arose after endosymbiosis as genes for lipid biosynthesis were transferred from the symbiont to the host. The transfer of these bacterial genes allowed the host to synthesize lipids containing fatty acids, lipids that may have been conducive to the formation of internal membranes, such as the nuclear membrane system (⇔ Section 2.14).

The origin of the nucleus was critical to the evolution of the eukaryotic cell, yet it remains unclear whether the nucleus appeared before or after the endosymbiotic origin of mitochondria. One hypothesis for the origin of the nucleus is that its formation is associated with the evolution of RNA processing in

UNIT 4

Eukarya. In contrast to the genes of *Bacteria* and *Archaea*, eukaryotic genes frequently contain introns, which must be removed prior to translation (⟳ Section 4.6). For proper gene expression to occur in *Eukarya*, RNA splicing must occur prior to translation. Eukaryotic cells have a molecular complex called the *spliceosome* that performs RNA splicing and operates in the cell nucleus. Hence, the nuclear membrane may have evolved in *Eukarya* as a mechanism to separate spliceosomes in the nucleus from ribosomes in the cytoplasm.

In the next section we trace the evolutionary path of both eukaryotic and prokaryotic cells in detail. Analyses of molecular evolution provide direct evidence of the evolutionary history of cells, leading to the modern "tree of life."

MINIQUIZ

- What evidence supports the idea that the mitochondrion and chloroplast were once free-living members of the domain *Bacteria*?

- In what ways are modern eukaryotes a combination of attributes of *Bacteria* and *Archaea*?

- Describe the different hypotheses for the formation of the eukaryotic cell.

II • Microbial Evolution

While many of the basic principles of evolution are conserved across all domains of life, certain aspects of microbial evolution are uncommon in plants and animals. For example, *Bacteria* and *Archaea* are generally haploid and asexual, they have several mechanisms for horizontal gene transfer that result in the asymmetrical exchange of genetic material uncoupled from reproduction, and their genomes can be remarkably heterogeneous and highly dynamic. We now consider the processes that cause the diversification of microbial lineages and how these forces affect the evolution of microbial genomes.

13.5 The Evolutionary Process

In its simplest form, evolution is a change in **allele** frequencies in a population of organisms over time resulting in descent with modification. Alleles are alternate versions of a given gene. New alleles arise as a result of mutation and recombination, and changes in allele frequencies can occur through a variety of processes. We see here how these simple mechanisms give rise to the origin and divergence of microbial species.

Origins of Genetic Diversity

As we have seen, **mutations** are random changes in DNA sequence that accumulate over time, and they are a fundamental source of the natural variation that drives the evolutionary process. Most mutations are neutral or deleterious, though some can be beneficial. Mutations take several forms including *substitutions, deletions, insertions*, and *duplications* (Chapter 11). Duplication events produce a redundant copy of a gene that can be modified by further mutation without losing the function encoded by the

original gene (⟳ Section 9.5). Hence, duplications allow for the diversification of gene function.

Recombination is a process by which segments of DNA are broken and rejoined to create new combinations of genetic material (⟳ Section 11.5). Recombination can cause reassortment of genetic material already present in a genome and is also required for the integration into the genome of DNA acquired through horizontal gene transfer. Recombination can be broadly classified as either *homologous* or *nonhomologous*. Homologous recombination requires short segments of highly similar DNA sequence flanking the region of DNA being transferred (⟳ Section 11.5). By contrast, nonhomologous recombination is mediated by several mechanisms that share in common the fact that they do not require high levels of sequence similarity to initiate successful DNA integration.

Selection and Genetic Drift

New alleles result when mutation and recombination cause variation in gene sequences. Evolution occurs when different alleles change in frequency in a population over a span of many generations. Evolutionary biologists have described many different mechanisms that can govern this evolutionary process, but chief among them are the forces of *selection* and *genetic drift*.

Selection is defined on the basis of **fitness**, the ability of an organism to produce progeny and contribute to the genetic makeup of future generations. Most mutations are *neutral* with respect to fitness and they have no effect on the cell, as a result of the degeneracy of the genetic code (⟳ Section 4.9). These mutations generally accumulate in DNA over time. Some mutations are *deleterious*; these decrease the fitness of an organism by disrupting gene function. Deleterious mutations are generally purged from populations over time by natural selection. Some mutations can be *beneficial*, increasing the fitness of an organism, and these mutations are favored by natural selection, increasing in frequency in a population over time. An example of a beneficial mutation would be a mutation that induces antibiotic resistance in a pathogenic bacterium infecting a person undergoing antibiotic therapy. It is important to remember that all mutations occur by chance; the selective nature of the environment does not *cause* adaptive mutations but simply *selects* for the growth and reproduction of those organisms that have incurred mutations that provide a fitness advantage.

While Darwin proposed natural selection as the mechanism by which gene frequencies change over time, evolutionary change can occur through mechanisms other than selection. A chief example is **genetic drift** (Figure 13.11), a random process that can cause gene frequencies to change over time, resulting in evolution in the absence of natural selection. Genetic drift occurs because some members of a population will have more offspring than others simply as a result of chance; over time these chance events can result in evolutionary change in the absence of selection. Genetic drift is most powerful in small populations and in populations that experience frequent "bottleneck" events. The latter occur when a population experiences a severe reduction in population size followed by regrowth from the cells that remain. For example, genetic drift can be very important in the evolution of pathogens because each new infection is caused by a small number of cells colonizing a new host. Hence, pathogen populations can change rapidly as a result of random genetic drift (Figure 13.11).

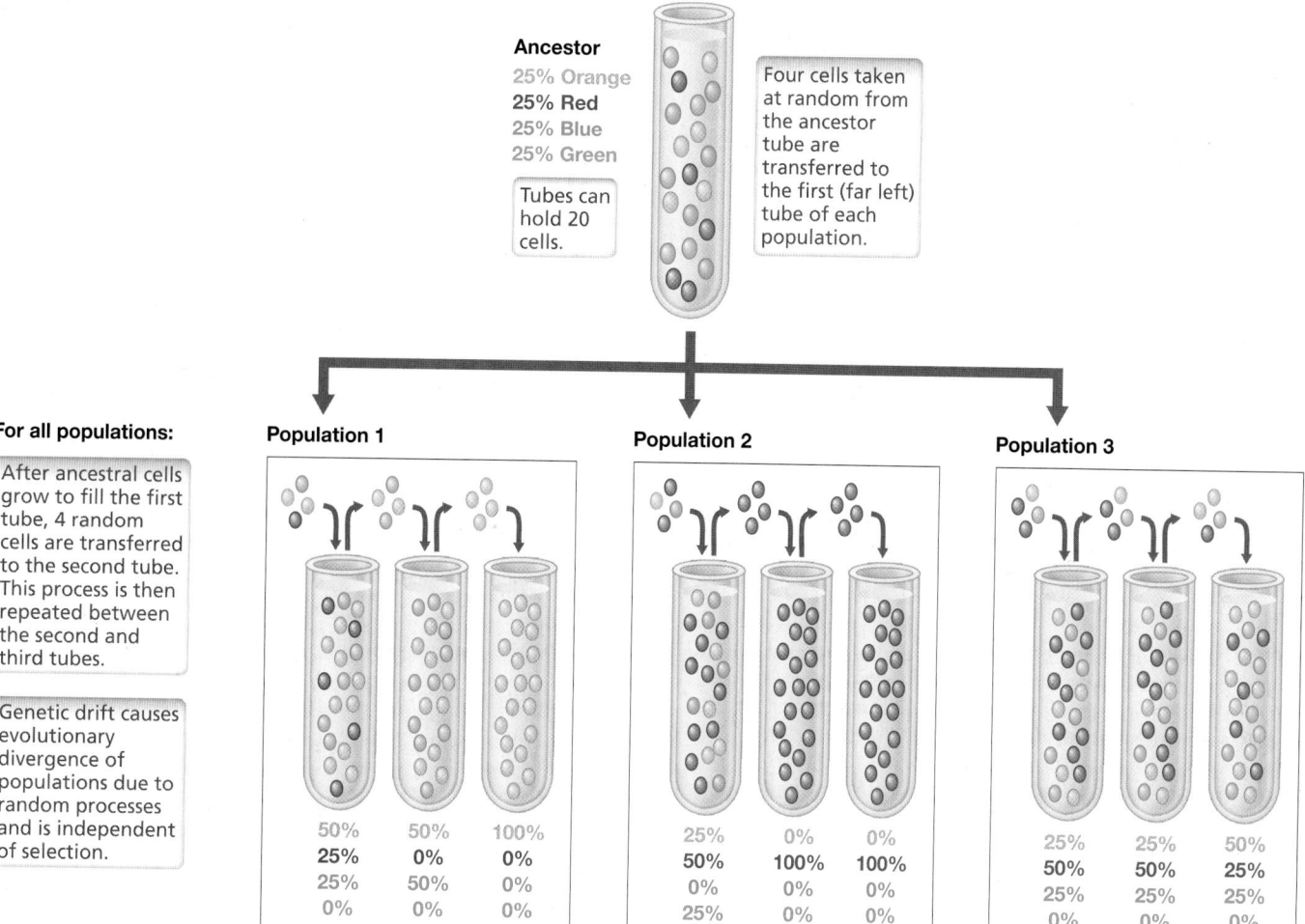

Ancestor

25% Orange
25% Red
25% Blue
25% Green

Tubes can hold 20 cells.

Four cells taken at random from the ancestor tube are transferred to the first (far left) tube of each population.

For all populations:

After ancestral cells grow to fill the first tube, 4 random cells are transferred to the second tube. This process is then repeated between the second and third tubes.

Genetic drift causes evolutionary divergence of populations due to random processes and is independent of selection.

Population 1

50%	50%	100%
25%	**0%**	**0%**
25%	50%	0%
0%	0%	0%

Population 2

25%	0%	0%
50%	**100%**	**100%**
0%	0%	0%
25%	0%	0%

Population 3

25%	25%	50%
50%	**50%**	**25%**
25%	25%	25%
0%	0%	0%

Figure 13.11 Genetic drift. Genetic drift is a random process that can cause gene frequencies in a population to change over time, causing evolution without natural selection. In this example, a population containing four different bacterial genotypes (indicated by different colors), each at equal frequency, is present in the ancestor tube. Four cells at random are then transferred to each of three new tubes and the cells allowed to grow to fill each tube. There is no difference in fitness between the cells and so they grow equally. Cells taken at random are then transferred in two successive rounds. Striking differences in genotype frequencies between the populations are observed after only three rounds of transfers.

New Traits Can Evolve Quickly in Microorganisms

A change in the environment or the introduction of cells to a new environment can cause rapid evolutionary changes in microbial populations. Microorganisms typically form large populations and can reproduce quickly, producing a new generation in as few as 20 minutes for some species, and thus evolutionary events in microbial populations can often be observed in the laboratory on relatively short time scales. The heritable variation already present in a population provides the raw material upon which natural selection acts following such a change in the selective environment. Here we consider two examples of rapid evolutionary change in bacteria, one involving the rapid loss of a characteristic trait in *Rhodobacter*, and one involving the acquisition of a new trait in *Escherichia coli*.

Rhodobacter is a phototrophic purple bacterium that carries out anoxygenic photosynthesis (⇔ Section 14.3) in illuminated anoxic environments in nature. When cultured anaerobically in either the light or the dark, the cells synthesize bacteriochlorophyll and carotenoids because it is the absence of O_2, not the presence of light, that triggers pigment synthesis. In the light these pigments participate in photosynthetic reactions that lead to ATP synthesis, but in darkness, the pigments provide no benefit to the cell. Random mutations occasionally generate *Rhodobacter* cells that produce either reduced levels of photopigments or no photopigments at all. In nature, the ability to carry out photosynthesis is of significant value and such pigment mutants cannot compete with wild-type cells. However, when cultured in constant darkness, wild-type cells have no such advantage. In darkness, mutants that produce reduced levels of photopigments gain the advantage because they do not waste energy and resources synthesizing photopigments that provide them no benefit. As a result, photopigment mutants quickly take over a *Rhodobacter* culture that is maintained in darkness for several generations because the mutants quickly become the fittest organisms in the population and therefore enjoy the greatest reproductive success (**Figure 13.12**). Mutations affecting photosynthesis occur at the same rate in the light as in the dark, but in the light the selection for the wild-type phototrophic

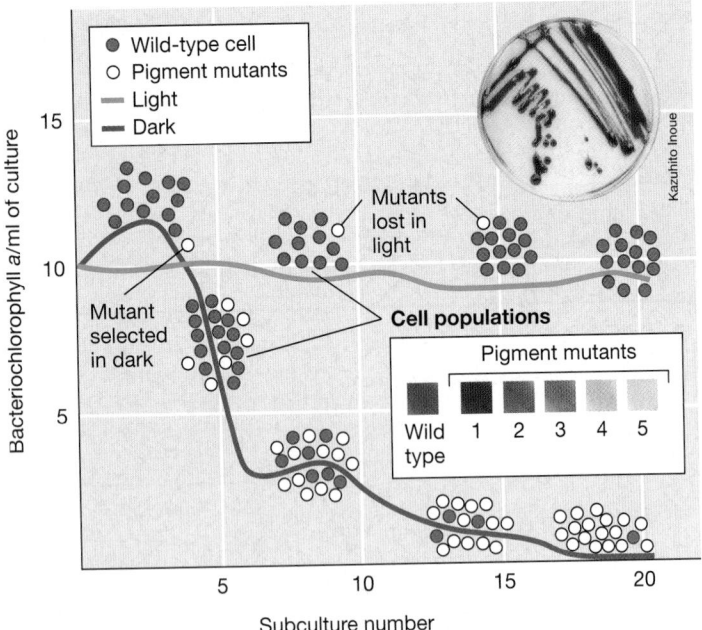

Figure 13.12 Survival of the fittest and natural selection in a population of phototrophic purple bacteria. Serial subculture of the purple bacterium *Rhodobacter capsulatus* in the light (green line) provides a benefit to wild-type phototrophic strains and they outcompete nonphototrophic mutants. In the dark (blue line), however, phototrophy provides no benefits and nonphototrophic mutants quickly outcompete wild-type cells still making bacteriochlorophyll and carotenoids. Photos: top, plate culture showing colonies of phototrophic cells of *R. capsulatus*; bottom, close-up photos showing the color of colonies of wild type and five pigment mutants (1–5) obtained during serial dark subculture. Wild-type cells are reddish-brown from their assortment of carotenoid pigments. The color of mutant colonies reflects the absence (or reduced synthesis) of one or more carotenoids. Mutant strain 5 lacked bacteriochlorophyll and was no longer able to grow phototrophically. Mutant strains 1–4 could grow phototrophically but at reduced growth rates from the wild type. Data adapted from Madigan, M.T., et al. 1982. *J. Bacteriol. 150:* 1422–1429.

phenotype is so strong that photopigment mutants are quickly lost from the population.

Experimental evolution is a growing field of study enabled by the rapid growth of bacterial populations and the ability to preserve bacteria indefinitely by freezing. The latter makes it possible to maintain a living "fossil record" of ancestral organisms that can be thawed later and compared to evolved strains. For example, the *E. coli* long-term evolution experiment (LTEE), which has been running since 1988, has tracked the evolution of 12 parallel lines of *E. coli* through more than 50,000 generations. The *E. coli* LTEE cultures have been grown aerobically on a minimal medium with glucose as a sole source of carbon and energy. *E. coli* is typically propagated in a rich medium that contains an excess of all the nutrients cells need to grow, and so the minimal glucose medium used in the LTEE represents a new adaptive environment in which *E. coli* can evolve over time.

In the LTEE, both the ancestor and the evolved lines were genetically engineered to contain a neutral marker that made their colonies either red or white. The marker made it possible to measure the fitness of evolved strains relative to the ancestor by competing them against one another (**Figure 13.13a**). Genome sequencing during the experiment revealed that mutations accumulated randomly over time in the evolved lines. However, the relative

fitness of the evolved lines on minimal glucose medium increased dramatically over the first 500 generations as a result of selection acting on mutations beneficial in this new environment (Figure 13.13*b*). The fitness of the evolved lines continued to increase,

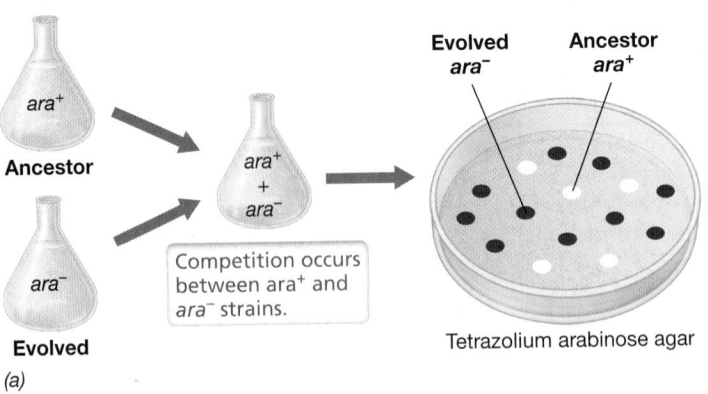

(a)

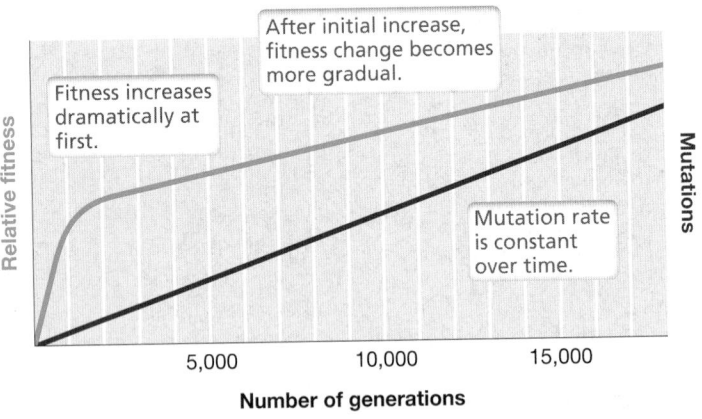

(b)

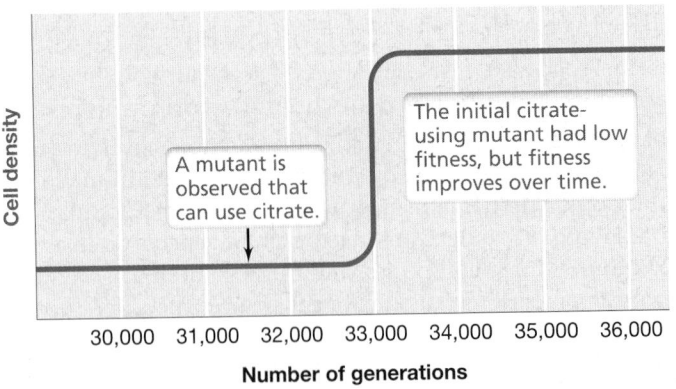

(c)

Figure 13.13 Long-term evolution of *Escherichia coli*. *(a)* In the *E. coli* long-term evolution experiment (LTEE), ancestral and derived lines differ in a mutation that affects their ability to use arabinose, allowing them to be differentiated by their colony color when grown on tetrazolium arabinose agar. *(b)* Competition experiments between evolved and ancestral strains show that relative fitness in minimal glucose media increases dramatically for evolved lines. *(c)* The ability to use citrate aerobically evolved in one of 12 LTEE lines. Cells growing on minimal glucose typically grow to low cell density, but the ability to use both glucose and citrate allowed the mutant cell line to reach significantly higher cell densities. Relative fitness is a measure of the growth rate of the evolved strain to that of the ancestral strain.

albeit at a reduced rate, as a result of further selection over the course of the experiment. Most remarkably, after 31,500 generations, one of the evolved lines obtained the ability to use citrate as an energy source (Figure 13.13c). Citrate was present as a pH buffer in the growth medium and was not considered a potential carbon source for *E. coli* because the inability to grow aerobically on citrate is a diagnostic trait for *E. coli*. However, the random accumulation of mutations in this one evolved line modified preexisting genes in such a way as to allow for the evolution of a new adaptive trait. The diverged strains can now exploit a new resource that was unavailable to the ancestral population. Since they can now catabolize both citrate and glucose, these cells grow to higher cell density than the ancestor (Figure 13.13c). The fact that only one of the 12 parallel lines evolved the ability to grow on citrate demonstrates the chance nature of evolution.

The transitions shown in these experiments remind us of how quickly evolutionary pressures can shift even major properties (such as metabolic strategies) of a microbial cell population. In the case of *Rhodobacter*, a mutation that is deleterious in the wild provides a selective advantage when the organism is grown in the laboratory in a continuously dark environment. Under this new condition, evolution causes *Rhodobacter* to lose unneeded metabolic machinery. In the case of *E. coli*, the accumulation of random mutations allows for the accumulation of genetic diversity in a population. Billions of different mutations were sampled by the population over thousands of generations and some rare combination of mutations, by chance, gave the cells the ability to exploit citrate as a resource. Natural variation caused by chance mutation generated a new trait, the ability to use citrate, and since the environment in which the cells were grown happened to contain citrate, this mutation provided a selective advantage to those cells. In the absence of citrate, these mutations would still occur at the same rate. However, in the absence of a selective benefit, cells able to use citrate would likely disappear from the population over time.

Speciation of Microorganisms Can Take a Long Time

Microbial species can possess a wide variety of individuals with diverse traits. As we discussed above, microorganisms can evolve new traits with remarkable speed, and as a result, microbial species can be genetically and phenotypically diverse. Sequence changes can be used as a **molecular clock** to estimate the time since two lineages have diverged. Major assumptions of the molecular clock approach are that nucleotide changes accumulate in a sequence in proportion to time, that such changes are generally neutral and do not interfere with gene function, and that they are random.

Molecular clock estimates are most reliable when they can be calibrated with evidence from the geological record. Molecular clock estimates can be calibrated using obligate bacterial symbionts of insects (⇆ Section 23.6) for which the insect host provides a suitable fossil record to date evolutionary events. From such calculations it is possible to estimate the divergence of microbial species based on dissimilarity in their gene sequences. For example, molecular clock estimates indicate that two well-characterized strains of *E. coli*, the harmless strain K-12 and the foodborne pathogenic strain O157:H7 (⇆ Section 32.11), diverged about 4.5 million years ago. Likewise, it is estimated that the closely related species *E. coli* and *Salmonella enterica* (*typhimurium*), which

have 2.8% dissimilarity in their 16S rRNA genes, last shared a common ancestor some 100–140 million years ago. Hence, although microbes can evolve new traits rapidly, microbial speciation appears to take a very long time.

MINIQUIZ

- What are the different processes that give rise to genetic variation?
- What is the difference between selection and genetic drift and how do they promote evolutionary change?
- In the experiment of Figure 13.12, why did the dark cell population lose its pigments?

13.6 The Evolution of Microbial Genomes

The dynamic nature of microbial genomes was revealed in dramatic fashion when the first genomes were sequenced from multiple strains of a single species. Genome sequencing of *Escherichia coli* strain K-12 and two pathogenic strains showed that only 39% of their genes were shared among all three genomes (**Figure 13.14**)!

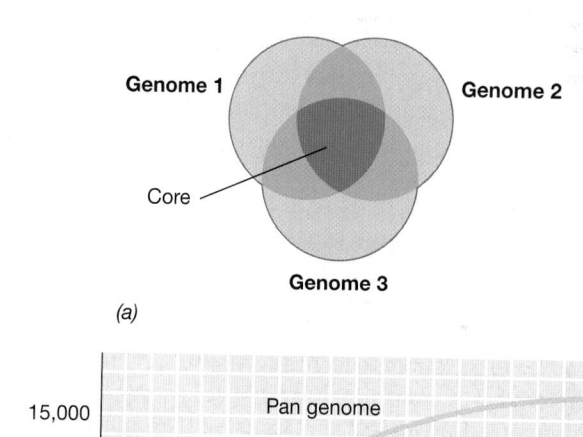

(a)

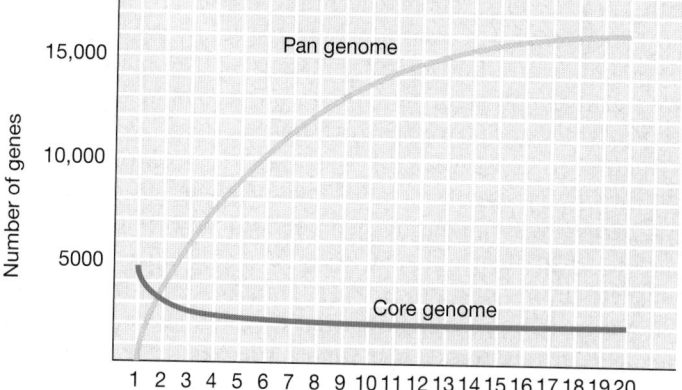

(b)

Figure 13.14 The core and pan genome concept. Microbial genomes are dynamic and heterogeneous. The first three genomes sequenced from different strains of *Escherichia coli* were found to have only 39% of their genes in common. The core genome is considered the set of genes that are shared by all members of a species (darkest green in part a), while the pan genome is the totality of all genes found in the different strains of a species (all the genes in genomes 1–3 in part a). The size of the core and pan genome can vary between species. In *E. coli* the core genome is composed of approximately 1976 genes (b). The size of the pan genome in *E. coli* is not fixed, as each different strain has a unique complement of genes acquired from horizontal gene exchange. Data adapted from Touchon, M., et al. 2009. *PLoS Genetics* 5:(1) e1000344.

The three genomes varied in size by more than a million base pairs in length and each contained a unique and diverse complement of genes acquired through horizontal gene transfer (⟳ Figure 9.20).

Genomes of many microbial species have now been examined in this way and the studies have revealed that genes from different strains of a single microbial species can be placed into two classes: the **core genome**, genes shared by all strains of a species, and the **pan genome**, the core genome plus genes that are not shared by all strains of a species and which are often acquired through horizontal gene transfer (see Figure 13.21). We previously introduced the core/pan genome concept (⟳ Section 9.7) and here we consider the forces that drive these patterns of genome evolution.

The Dynamic Nature of the *Escherichia coli* Genome

Escherichia coli genomes have on average 4721 genes, with individual strains having as few as 4068 or as many as 5379 total genes. The core genome consists of only 1976 genes present in all strains, accounting for less than half the genes present in the average *E. coli* genome. The size of the core genome can be expected to decrease as the evolutionary distance of strains increases. Taking this to its extreme, only about 60 genes are predicted to be universally present in all species of *Bacteria* and *Archaea* (Section 13.3).

The number of unique genes observed in *E. coli*, and the size of its pan genome, continues to increase indefinitely as each new strain of *E. coli* has its genome sequenced. For example, a survey of genomes from 20 *E. coli* strains revealed a total of 17,838 unique genes (Figure 13.14*b*). Subtracting the contribution from the core genome, this indicates more than 15,862 genes not shared by all strains. A great many of these genes have clearly been obtained from horizontal gene transfers rather than through vertical patterns of inheritance (from mother cell to daughter cell). Patterns of gene exchange appear to be governed by phylogenetic distance, with rates of gene exchange declining as the phylogenetic distance increases between genomes. In the core genome of *E. coli*, for example, most horizontal gene transfer takes place between close relatives and occurs by homologous recombination of DNA segments of 50 to 500 base pairs in length.

Genome analysis reveals that the core and pan genome concept is a general feature of microbial genomes, though the relative number of genes present in each pool can vary widely between species. The dramatic change in genome size and gene content between strains of a single species indicates that microbial genomes are highly dynamic; that is, genomes can shrink or enlarge relatively quickly over time. Moreover, the existence of a pan genome suggests that prokaryotic cells regularly sample genes from other microbes in their environment through horizontal gene transfers. Variations between genomes then arise through the forces of mutation and recombination, with the evolutionary dynamics of genomes being governed by selection and genetic drift.

Gene Deletions in Microbial Genomes

Gene deletions play an important role in microbial genome dynamics (see Explore the Microbial World, "The Black Queen Hypothesis"). Deletions occur with far greater frequency than insertions in microbial genomes, and this bias toward deletions is the force that maintains the small size of many microbial genomes.

Selection is the main force that counters the effect of deletions, preserving those genes that provide a fitness benefit to the cell.

It is common in prokaryotic cells for nonessential and nonfunctional materials to be deleted over evolutionary time (Figure 13.12), which is why the genomes of *Bacteria* and *Archaea* are tightly packed with genes and contain relatively few noncoding sequences (⟳ Section 9.3). Most genes acquired by horizontal gene transfer, like most mutations in general, can be expected to be neutral or deleterious to the cell. Hence, it is likely that of the new genes acquired from the environment, those that do not convey a fitness benefit are deleted from the genome over time. In addition, genetic drift (Figure 13.11) can promote deletion events when population sizes are small or when populations pass through a bottleneck. Deletions are also thought to be a driving force for the tiny genomes observed in many obligate intracellular symbionts and bacterial pathogens (⟳ Sections 9.3 and 23.6). The genomes of these bacteria can afford to be streamlined because several of the key metabolites each needs are provided to them by their host organisms.

— MINIQUIZ —

- What is the difference between the core and pan genomes of a given species?
- What kind of recombination might have the greatest impact on the core genome?
- What effects do deletions have on the evolution of microbial genomes?

III • Microbial Phylogeny and Systematics

S **ystematics** is the study of the diversity of organisms and their relationships. It links phylogeny with **taxonomy**, the science in which organisms are characterized, named, and classified according to defined criteria.

Bacterial taxonomy has changed substantially in the past few decades, embracing a combination of methods for the identification of bacteria and description of new species. This *polyphasic approach* to taxonomy uses three kinds of methods—*phenotypic, genotypic,* and *phylogenetic*—for the identification and description of bacteria. Phenotypic analysis examines the morphological, metabolic, physiological, and chemical characteristics of the cell. Genotypic analysis considers characteristics of the genome. These two kinds of analysis categorize organisms based on similarities. They are complemented by phylogenetic analysis, which seeks to place organisms within an evolutionary framework using molecular sequence data. We begin by exploring how molecular sequences can provide phylogenetic insight.

13.7 Molecular Phylogeny: Making Sense of Molecular Sequences

Molecular sequences provide a record of past evolutionary events and can be used to build **phylogenetic trees**, which are diagrams that depict evolutionary history. All cells contain DNA as

THE BLACK QUEEN HYPOTHESIS

It is a common misconception that evolution inevitably causes organisms to increase in complexity over time. In reality, evolution is both a give and a take proposition. Fitness changes are completely dependent on the environment, and fitness in some environments may actually be improved by a loss, rather than a gain, of specific genes.

The *Black Queen hypothesis*[1] posits a mechanism and a rationale for this loss of function whose end result is the evolution of mutual dependence in microbial communities. The term *Black Queen* refers to the card game Hearts in which there are two winning strategies. One winning strategy is to avoid getting stuck with the queen of spades. In this strategy each player seeks to lose as many contests ("tricks") as possible so as not to be forced to collect the black queen. The second winning strategy is to "shoot the moon" by collecting all of the trump cards including the black queen. In its microbial context, the Black Queen hypothesis embraces these card game strategies by proposing that some organisms optimize fitness (i.e., "win") by the selective loss of specific genes while others optimize fitness by keeping them all.

The Black Queen hypothesis proposes that certain microbial genes encode extracellular products, such as metabolites or enzymes, which can be used by all or most members of the community. If an organism remains in the community, then selection will be relaxed on genes that encode the synthesis of products that are provided by other members of the community. The presence of such shared products in the community renders genes with similar functions nonessential for some community members (**Figure 1**). The mutation bias toward deletions can then cause these genes to be lost from the genome (see Section 13.6).

The fitness of organisms that lose functions and develop dependencies will actually increase in the community since these organisms no longer bear the costs of production. Such organisms will remain competitive as long as they remain within the community, but they may be unable to grow if separated from the community in which they coevolved. In this way, mutual dependencies accumulate within microbial communities over time. The Black Queen hypothesis also explains the not-uncommon observation that some microorganisms can only be grown in the laboratory in coculture with one or more other species from their environment.

In contrast to the gene loss strategy, organisms that preserve all essential functions (those that shoot the moon in the Hearts analogy) bear the costs of maintaining all gene functions, which puts them at a disadvantage to mutually dependent competitors when competing in the native community. However, cells that maintain their ability to grow independently still have a winning strategy because, unlike their mutually dependent competitors, they retain the option of dispersing to new habitats and growing outside of the native community.

Finally, in addition to describing how microbial community interdependencies might come about, the Black Queen hypothesis also reminds us of how interwoven microbial communities actually are. We will see in later chapters that several molecular tools are available to unwind this complexity and reveal both the diversity of the community and its genetic and metabolic potential.

[1]Morris, J.J., R.E. Lenski, and E.R. Zinser. 2013. The Black Queen hypothesis: Evolution of dependencies through adaptive gene loss. *mBio 3:* e00036-13.

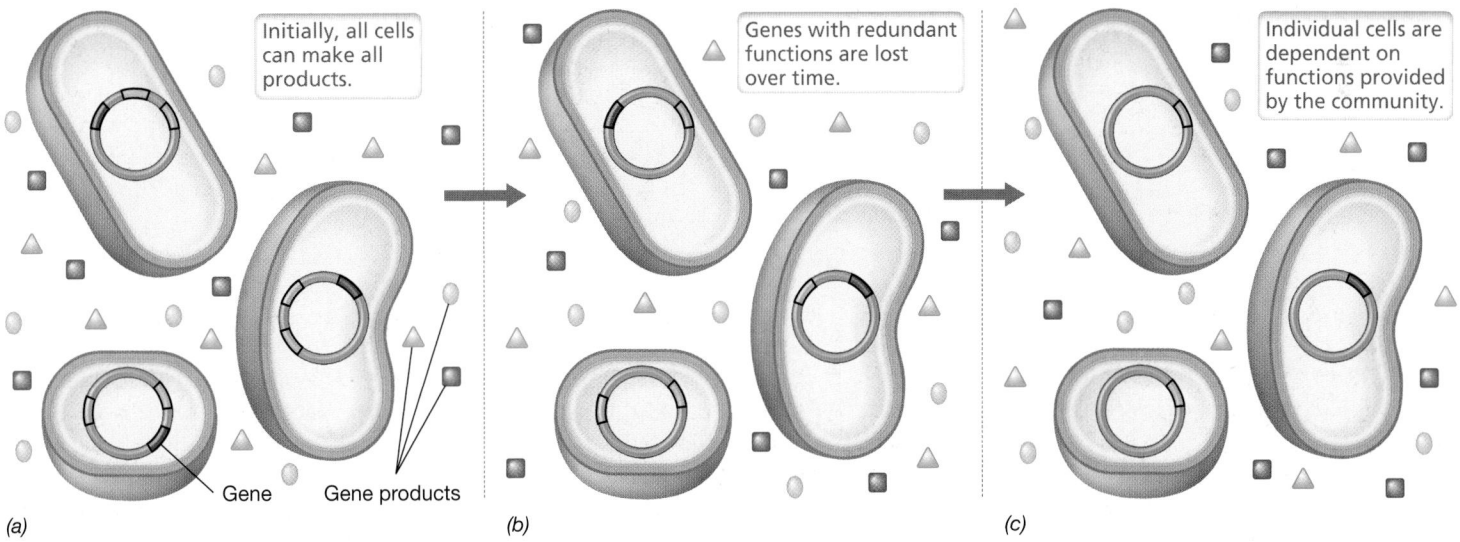

Initially, all cells can make all products.

Genes with redundant functions are lost over time.

Individual cells are dependent on functions provided by the community.

Gene   Gene products

(a)   (b)   (c)

Figure 1 The Black Queen hypothesis and the evolution of dependence in microbial communities.
(a) Three species in a community each have three different genes that make extracellular products that benefit the whole community (a gene and its product are shown in the same color). *(b)* Over time, random mutations cause functions to be lost from the genomes. *(c)* As long as some members of the community continue to make each product, there will be no fitness cost when a single species loses a single gene. Over time, the three species thus become mutually dependent.

their genetic material, and mutations accumulate in DNA sequences over time. These mutations occur at random and they are a major source of the natural variation that makes evolution possible (Section 13.5). Hence, the difference in nucleotide sequence between the DNA of any two organisms will be a function of the number of mutations that have accumulated since they shared a common ancestor. As a result, differences in DNA sequences can be used to infer evolutionary relationships. In this section we will learn how DNA sequences are used in the phylogenetic analysis of microbial life.

Obtaining DNA Sequences

The analysis of molecular phylogeny begins by determining the sequence of macromolecules such as DNA, RNA, or protein. Here we will focus on analysis of DNA sequence, which is used widely for determining phylogenetic relationships between organisms.

Obtaining DNA from a microorganism is relatively easy if the organism can be cultivated in isolation in the laboratory. In this case, genomic DNA is isolated and either the genome can be sequenced directly, or the genomic DNA can be used to amplify one or more specific genes, using the polymerase chain reaction (PCR, ⟷ Section 12.1). Advances in DNA sequencing technology (⟷ Section 9.2) have made genome sequencing a standard tool employed in analyses of microbial phylogeny. However, sequence analysis of small subunit (SSU) ribosomal RNA (rRNA) genes, which encode the rRNA molecule found within the small subunit of the ribosome (⟷ Section 4.10), remains a cornerstone of molecular phylogeny in microbiology because **SSU rRNA** genes are highly conserved, present in all cellular organisms, and easily sequenced and analyzed.

PCR primers can be designed to target any region of DNA from any organism. Standard primers exist for many highly conserved genes, such as the SSU rRNA gene (**Figure 13.15**). Primers for the SSU

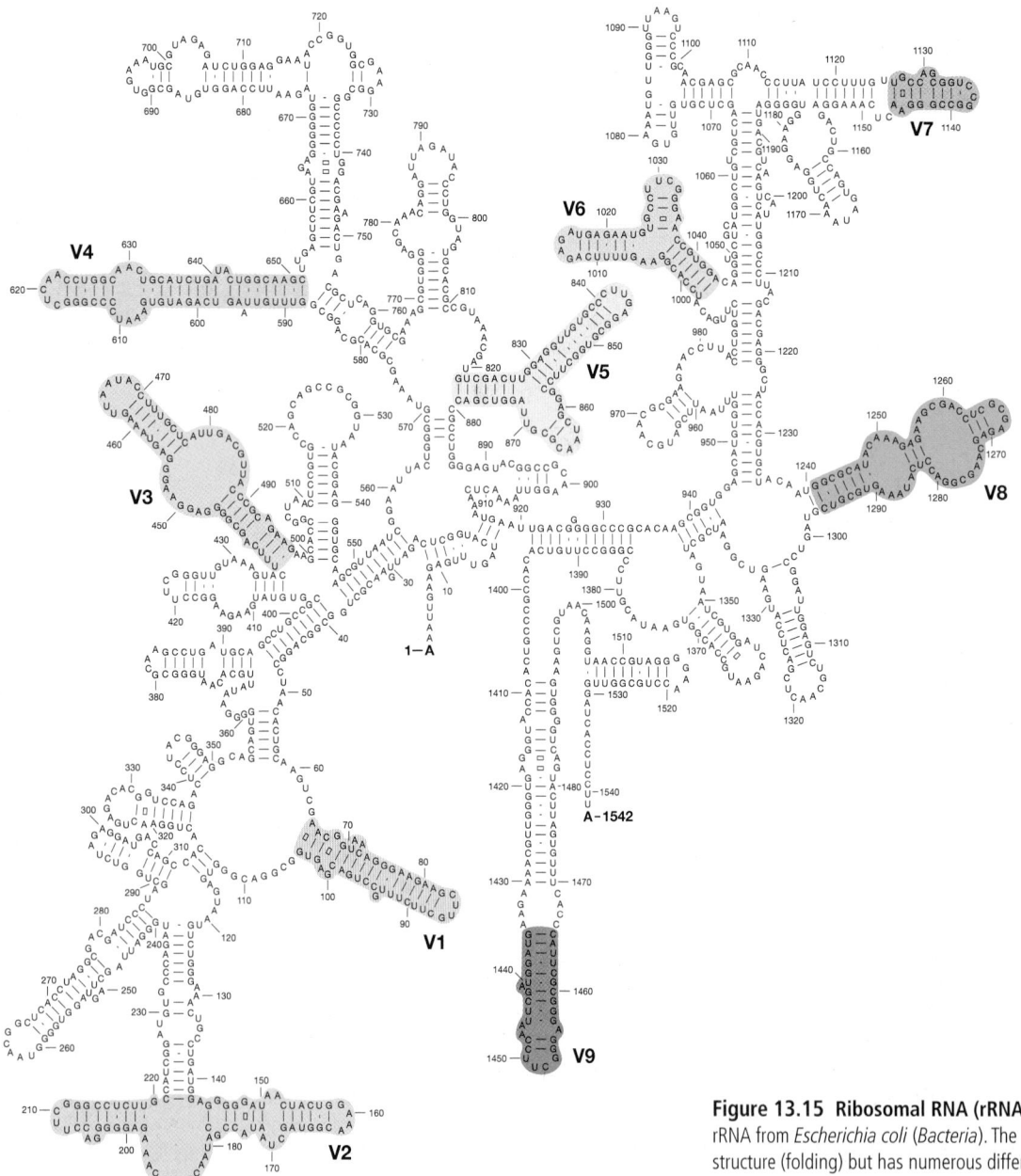

Figure 13.15 Ribosomal RNA (rRNA). Primary and secondary structure of 16S rRNA from *Escherichia coli* (*Bacteria*). The 16S rRNA from *Archaea* is similar in secondary structure (folding) but has numerous differences in primary structure (sequence). The molecule is composed of conserved and variable regions (V1–V9). The approximate positions of the variable regions are indicated in color.

rRNA gene can have different levels of phylogenetic specificity to target discrete species, genera, phyla, or domains, and there are even "universal" primers that will amplify the SSU rRNA gene from any organism. PCR products are visualized by agarose gel electrophoresis, excised from the gel, extracted and purified from the agarose, and then sequenced, often using the same oligonucleotides as primers for the sequencing reactions (**Figure 13.16**). Instead of starting with a microbial culture, it is also possible to amplify SSU rRNA genes from DNA extracted directly from an environmental sample or to sequence environmental DNA without amplification using a metagenomic approach (⇄ Sections 9.8 and 19.8). These latter approaches are widely used to characterize microorganisms that are difficult to grow in laboratory culture. Once sequences are obtained, they must be aligned and analyzed, issues we turn to now.

Sequence Alignment

Phylogeny can be inferred only from genes that have **homology**, that is, *genes that have been inherited from a common ancestor.* Thus homology is a binary trait; sequences are either homologous or they are not. The concept of homology is often confused with that of sequence similarity. The latter is a continuous trait defined as a percentage of nucleotide positions shared between any two sequences. Sequence similarity is used to infer homology, but a similarity value can be calculated between any two sequences regardless of their function or evolutionary relationship. Thus, the terms similarity and homology are not interchangeable. Genes that have homology can be either **orthologs**, if they have the same function and originate from a single ancestral gene in a common ancestor, or **paralogs**, if they have evolved to have different functions as a result of gene duplication (⇄ Section 9.5). Phylogenetic analyses typically focus on analysis of orthologous genes that have similar function.

Phylogenetic analyses estimate evolutionary changes from the number of sequence differences across a set of homologous nucleotide positions. Some mutations introduce nucleotide insertions or deletions, and these cause gene sequences to differ in length, making it necessary to *align* nucleotide positions prior to phylogenetic analysis of gene sequences. The purpose of **sequence alignment** is to add gaps to molecular sequences in order to establish positional homology, that is, to be sure that each position in the sequence was inherited from a common ancestor of all organisms under consideration (**Figure 13.17**). Proper sequence alignment is critical to phylogenetic analysis because the assignment of mismatches and gaps caused by deletions is in effect an explicit hypothesis of how the sequences have diverged from a common ancestral sequence.

Phylogenetic Trees: Composition and Construction

A phylogenetic tree is a diagram that depicts the evolutionary history of an organism and bears some resemblance to a family tree. Most microbes have not left fossils and so their ancestors are unknown, but ancestral relationships can be inferred from the DNA sequences of organisms that are alive today. Organisms that share a recent ancestor are likely to share characteristics, and thus phylogenetic trees allow us to make hypotheses about an organism's characteristics. Phylogenetic trees are also of great use in taxonomy and species identification, as we will discuss later (Section 13.9).

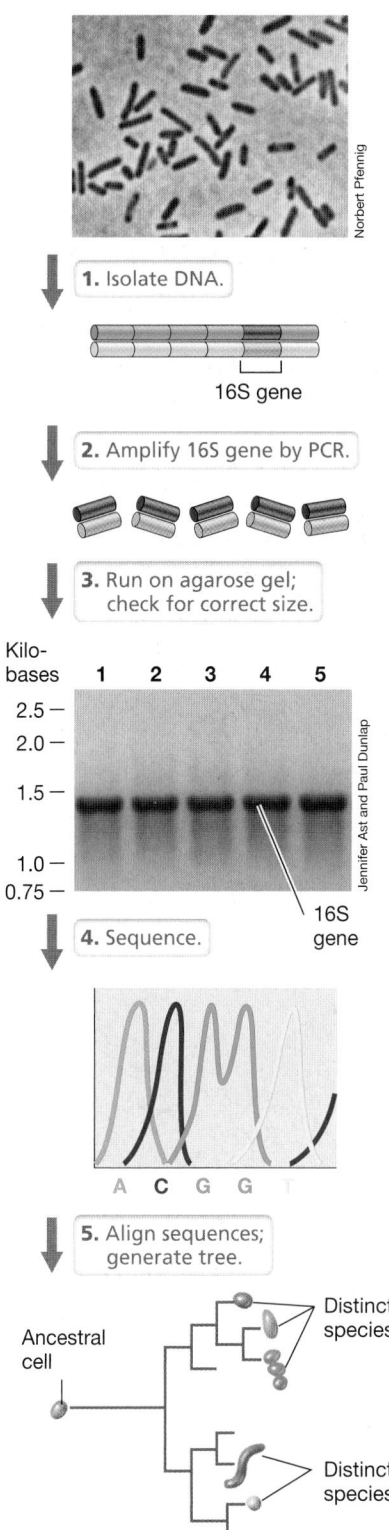

Figure 13.16 PCR amplification of the 16S rRNA gene. Following DNA isolation, primers complementary to the ends of the 16S rRNA (see Figure 13.15) are used to PCR-amplify the 16S rRNA gene from the genomic DNA of five different unknown bacterial strains, and the products are run on an agarose gel (second photo). The bands of amplified DNA are approximately 1465 nucleotides in length. Positions of DNA kilobase size markers are indicated at the left. Excision from the gel and purification of these PCR products is followed by sequencing and analysis to identify the bacteria.

Sequences before alignment

Sequence differences

	1	2	3
1 GGA CCT AAA TTT ATA CCC	1	–	–	–
2 GGA AAA GGG CCC AAA CGC	2	11	–	–
3 GGA GGG CCT TTT ATA CCC	3	6	11	–

Sequences after alignment

	1	2	3
1 GGA – – – – – – CCT AAA TTT ATA CCC	1	–	–	–
2 GGA AAA GGG CCC – – – – – – AAA CGC	2	3	–	–
3 GGA – – – GGG CCT – – – TTT ATA CCC	3	0	3	–

(a)   *(b)*

Figure 13.17 Alignment of DNA sequences. *(a)* Sequences for a hypothetical region of a gene are shown for three species before alignment and after alignment. A sequence alignment should display homologous positions in vertical columns. Sequence alignment is achieved by adding gaps, indicated by hyphens, to maximize local sequence similarity between the species in the alignment. *(b)* The distance matrices show the number of sequence differences that would be inferred for each species pair both before and after alignment.

A phylogenetic tree is composed of *nodes* and *branches* (**Figure 13.18**). The tips of the branches in a phylogenetic tree represent species that exist today. Phylogenetic trees can be constructed that are either *rooted trees* or *unrooted trees*. Rooted trees show the position of the

ancestor of all organisms being examined. Unrooted trees depict the relative relationships among the organisms under study but do not provide evidence of the most ancestral node in the tree. The nodes represent a past stage of evolution where an ancestor diverged into two new lineages. The branch length represents the number of changes that have occurred along that branch. In a phylogenetic tree, only the position of nodes and the branch lengths are informative; rotation around nodes has no effect on the tree's topology (Figure 13.18*b*).

There is only one correct phylogenetic tree that most accurately depicts the evolutionary history of a group of gene sequences, but inferring this tree from sequence data can be a challenging task. The complexity of the problem is revealed by considering the total number of different trees that can be formed from a random set of sequences. For example, only three possible trees can be drawn for any four arbitrary sequences. But if one doubles to eight the number of sequences, now 10,395 trees are possible. This complexity continues to expand exponentially such that 2×10^{182} different trees can be drawn to represent 100 arbitrary sequences. Phylogenetic analysis uses molecular sequence data in an attempt to identify the one correct tree that accurately represents the evolutionary history of a set of sequences.

A variety of methods are available for inferring phylogenetic trees from molecular sequence data. The structure of a phylogenetic tree is inferred by applying either an *algorithm* or some set of

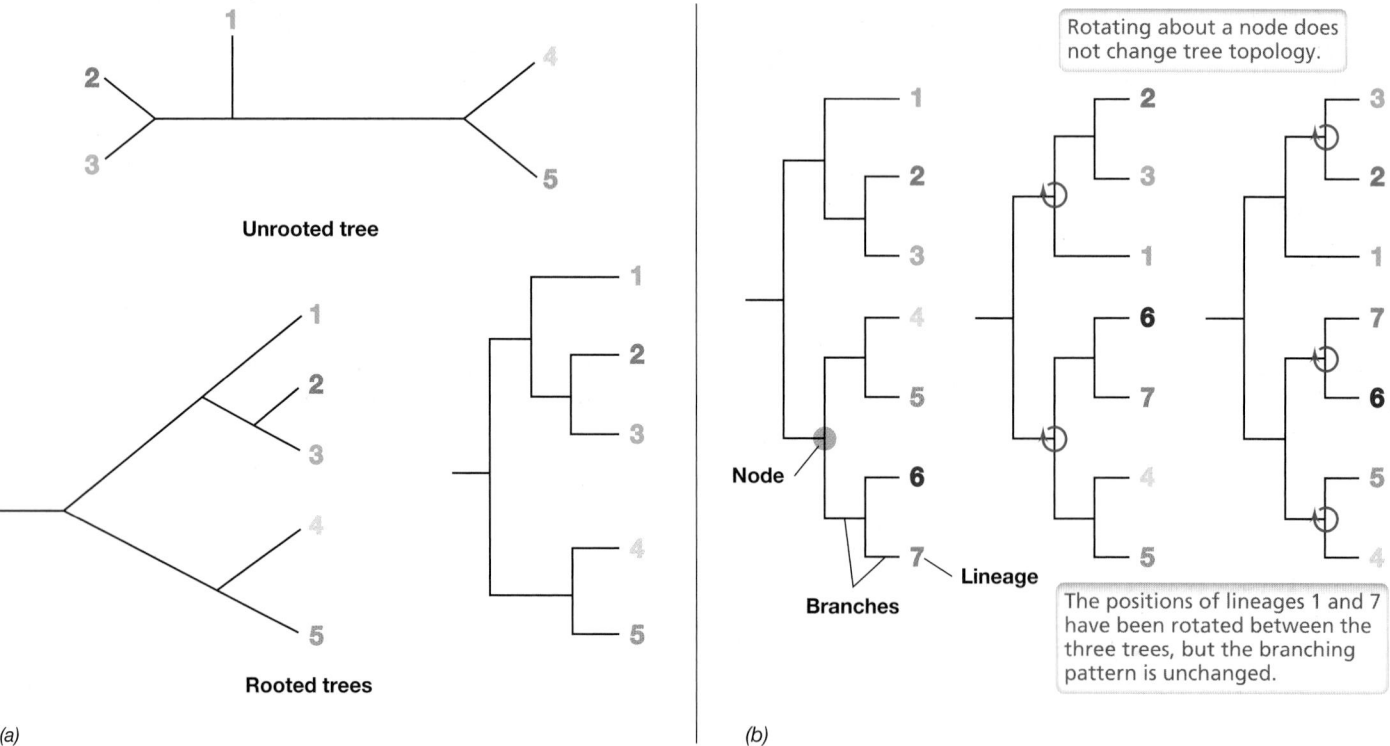

Figure 13.18 Phylogenetic trees and their interpretation. *(a)* Unrooted and rooted examples of phylogenetic trees. The tips of the branches are species (or strains) and the nodes are ancestors. Ancestral relationships are revealed by the branching order in rooted trees. *(b)* Three equivalent versions of the same phylogenetic tree are shown. The only difference between the trees is that their nodes have been rotated at the points indicated by red arrows. The vertical position of species is different between the trees but the pattern of ancestry (the nodes shared by each species) remains unchanged.

optimality criteria. An algorithm is a programmed series of steps designed to construct a single tree (**Figure 13.19**). Algorithms used to build phylogenetic trees include the *unweighted pair group method with arithmetic mean* (UPGMA) and the *neighbor-joining method.* Alternatively, phylogenetic methods that employ optimality criteria include *parsimony, maximum likelihood,* and *Bayesian* analyses. These latter methods evaluate many possible trees and select the one tree that has the best optimality score, that is, they select the tree that best fits the sequence data given a discrete model of molecular evolution. Optimality scores are calculated on the basis of evolutionary models that describe how molecular sequences change over time. For example, evolutionary models can account for variation in substitution rates and base frequencies between sequence positions.

Limitations of Phylogenetic Trees

Molecular phylogeny provides powerful insights into evolutionary history, but it is important to consider the limitations of building and interpreting phylogenetic trees. For example, it can be difficult to choose the true tree based on available sequence data if several different trees fit the data equally well. *Bootstrapping,* a statistical method in which information is resampled at random, is an approach used to deal with uncertainty in phylogenetic trees. Bootstrap values indicate the *percentage of the time* that a given node in a phylogenetic tree is supported by the sequence data. High bootstrap values indicate that a node in the tree is likely to be correct, while low bootstrap values indicate that the placement of a node cannot be accurately determined given the available data.

Homoplasy, also known as *convergent evolution,* occurs when organisms share a trait that was not inherited from a common ancestor. An example is the evolution of wings in insects and birds. These traits evolved separately and do not indicate that a winged ancestor was shared among insects and birds. Homoplasy occurs in molecular sequences as well, when similar sequence positions result from recurrent mutation rather than inheritance from a common ancestor (**Figure 13.20**). The problem of homoplasy in

Recurrent mutation can erase evolutionary information, causing sequence differences to underestimate true distances.

(a)

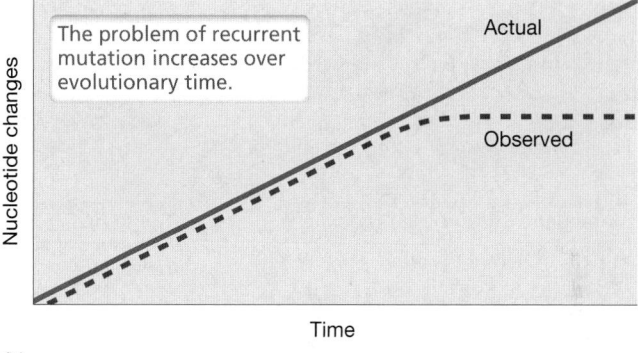

(b)

Figure 13.20 The problem of homoplasy due to recurrent mutation. It is possible for recurrent mutation to obscure the true number of mutations that have occurred since a pair of sequences have shared a common ancestor. *(a)* Two series of mutations during the evolution of a gene sequence are compared. On the left side, the number of mutations is equal to the number observed between species 1 and 4. However, if there is recurrent mutation (right side), the number of mutations observed between species 1 and 4 can be less than the number that actually occurred. *(b)* The likelihood of recurrent mutation increases as more and more mutations accumulate over time.

molecular phylogeny then increases in proportion to evolutionary time (Figure 13.20*b*). As a result of homoplasy, the reconstruction of accurate phylogenetic trees gets more difficult when sequence divergence between organisms is very high.

The prevalence of horizontal gene transfer (Chapter 11) also creates complications when considering the evolutionary history of microorganisms. When the sequence of a gene is used to infer the phylogeny of an organism, it must be assumed that the gene is inherited in a *vertical* fashion—from mother to daughter—throughout the evolutionary history of the organism. The *horizontal* exchange of genes between unrelated organisms violates this assumption (**Figure 13.21**). Hence, it is important to consider the difference between a *gene phylogeny,* which depicts the evolutionary history of an individual gene, and an *organismal phylogeny,* which depicts the evolutionary history of the cell.

In general, genes encoding SSU rRNAs appear to be transferred horizontally at very low frequencies, and rRNA gene phylogenies agree largely with those prepared from other genes that encode genetic informational functions. Thus, SSU rRNA gene sequences are generally considered to provide an accurate record of organismal phylogeny. Nevertheless, many microbial genomes contain genes that have been acquired by horizontal gene transfer at some point in

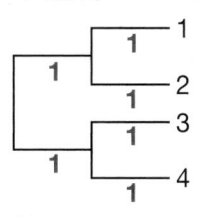

		1	2	3	4
1	A C T G A C	1	–			
2	A C T C A T	2	2	–		
3	A C A T G G	3	4	4	–	
4	A C A A G A	4	4	4	2	–

(a)   *(b)*   *(c)*

1. The first step in making a tree is to align sequences.
2. A distance matrix is calculated from the number of sequence differences.
3. The tree is constructed by adding nodes to join lineages that have the fewest differences.

Figure 13.19 Building phylogenetic trees. The number of nucleotide differences between gene sequences can be used to build a phylogenetic tree. In the sequence alignment *(a)* we can count the number of differences between each pair of sequences to build a distance matrix *(b)*. This distance matrix can be used to build a tree *(c)* where the cumulative lengths of the horizontal branches (labeled with a red "1") between any two species in the tree are proportional to the number of nucleotide differences between these species.

UNIT 4

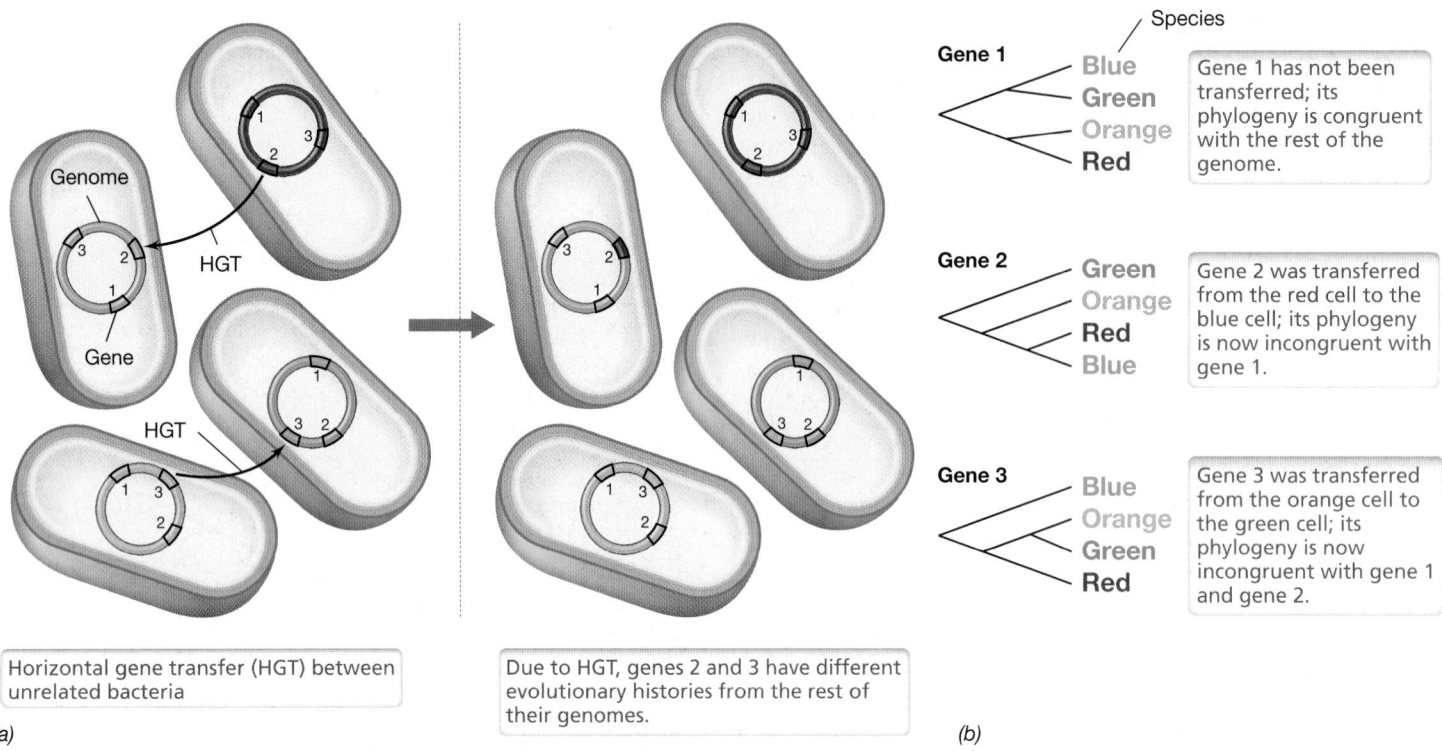

Figure 13.21 Horizontal gene transfer. The horizontal transfer of a gene will cause it to have a different evolutionary history from the rest of the genome. *(a)* Genes are transferred horizontally between distantly related microorganisms. Colors are used to match microorganisms with their genomes. *(b)* As a result of the horizontal transfer events in part *a*, different phylogenetic trees for gene 1, gene 2, and gene 3 are obtained. Only the gene tree for gene 1, which was not transferred, remains congruent with the organismal phylogeny.

their evolutionary history, and this has important implications for microbial evolution (Section 13.6 and Chapter 11).

MINIQUIZ

- How are DNA sequences obtained for phylogenetic analysis?
- What does a phylogenetic tree depict?
- Why is sequence alignment critical to phylogenetic analysis?

13.8 The Species Concept in Microbiology

Species are the fundamental units of biological diversity, and how we distinguish and classify species in microbiology greatly affects our ability to explain and assess the diversity of the microbial world. At present, there is no universally accepted concept of a microbial species. Microbial systematics combines phenotypic, genotypic, and sequence-based phylogenetic data within a framework of standards and guidelines for describing and identifying microorganisms in a taxonomic framework, but the issue of what actually constitutes a *species* remains controversial. However, a working definition of a microbial species has been developed and widely used, and we consider this here.

A Phylogenetic Species Concept for *Bacteria* and *Archaea*

From a taxonomic standpoint, all members of a species should be genetically and phenotypically cohesive, and their traits should be distinct from those described for other species. In addition, a species should be **monophyletic**, that is, the strains composing the species should all share a recent common ancestor to the exclusion of other species. The working definition of a microbial species tries to incorporate these principles and is best described as a *phylogenetic species concept*. The phylogenetic species concept defines a microbial species pragmatically as a group of strains that share certain characteristic traits and which are genetically cohesive and share a unique recent common ancestor. This species concept requires that a majority of genes in the species have congruent phylogenies and share a recent common ancestor. The phylogenetic species concept is not based on an evolutionary model of speciation, and thus species described in this way do not necessarily reflect meaningful units in terms of ecological or evolutionary processes. The phylogenetic species concept was developed to facilitate taxonomy, and species justifications derived from this concept are based largely on the expert judgment of taxonomists.

Under the phylogenetic species concept, species of *Bacteria* and *Archaea* are defined operationally as a group of strains sharing a high degree of similarity in many traits and sharing a recent common ancestor for their SSU rRNA genes. Species characterization employs a polyphasic approach that considers a range of different traits in making taxonomic judgments. Traits currently considered most important for identifying species include genomic similarity based on DNA hybridization and comparisons of SSU rRNA sequences.

The degree of **DNA–DNA hybridization** between the genomes of two organisms provides a measure of their genomic similarity (⇄ Section 12.1). In a hybridization experiment (**Figure 13.22**), *probe DNA* obtained from one organism is labeled with a fluorescent or radioactive label, sheared into small pieces, and heated to separate the two DNA strands. The probe is then added to single-stranded and sheared *target DNA* from a second organism and the mixture cooled to allow the DNA strands to reanneal. The genomic similarity between the two organisms is calculated as a percentage of probe hybridized to target relative to a control (probe DNA hybridized to target DNA from the same organism).

A value of 70% or less genomic hybridization and a difference in SSU rRNA gene sequence of 3% or more between two organisms is taken as evidence that the two are distinct species. Experimental data suggest that these criteria are valid, reliable, and consistent in identifying new microbial species for taxonomic

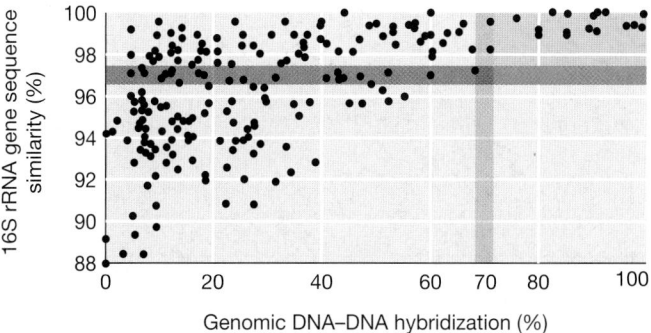

Figure 13.23 Relationship between 16S rRNA gene sequence similarity and genomic DNA–DNA hybridization for pairs of organisms. Pairs of microorganisms are compared on the basis of their 16S rRNA similarities and DNA–DNA hybridization values. Points in the upper right region represent pairs of strains that share greater than 97% 16S rRNA gene sequence similarity and 70% genomic hybridization values, and thus are likely members of the same species. Adapted from Rosselló-Mora, R., and R. Amann. 2001. *FEMS Microbiol. Revs. 25*: 39–67, and Stackebrandt, E., and J. Ebers. 2006. *Microbiology Today. 11:* 153–155.

purposes (**Figure 13.23**). On the basis of this phylogenetic species concept, over 10,000 species of *Bacteria* and *Archaea* have been formally recognized. The criteria that should be used to define a genus, the next highest taxon (see Table 13.2), is more a matter of judgment, but discrete genera typically have greater than 5% dissimilarity in their SSU rRNA gene sequences. There are no consensus criteria for defining taxonomic ranks above the level of genera.

How Many Species of *Bacteria* and *Archaea* Exist?

The result of nearly 4 billion years of evolution is the microbial world we see today (Figure 13.9). Microbial taxonomists agree that no firm estimate of the number of bacterial and archaeal species can be given at present, in part because of uncertainty about what defines a species in these domains. However, they also agree that in the final analysis, this number will be very large. The diversity of bacterial and archaeal species on Earth is unquestionably higher than that of all plant and animal species combined, and their total species numbers of *Bacteria* and *Archaea* are likely several orders of magnitude higher than the 10,000 species that have already been characterized.

Every environment on Earth contains a diverse community of microorganisms. For example, analyses of environmental SSU rRNA gene sequences (⇄ Section 19.6) using the phylogenetic species concept indicate that over 10,000 different species can coexist in a single gram of soil! Since 1977 more than 3.3 million SSU rRNA sequences have been generated and used to characterize the vast diversity of the microbial world. The Ribosomal Database Project (RDP; http://rdp.cme.msu.edu) contains an ever-growing collection of these sequences and provides computational programs for their analysis and for the construction of phylogenetic trees. While we cannot yet know the biodiversity of microbial life, we do know that nearly all plant and animal species have microbiomes (Chapter 23) that contain countless numbers of unique microbes. Thus, microorganisms are not only the oldest but also unquestionably the most diverse forms of life on our planet. We will see this diversity in action over the next five chapters.

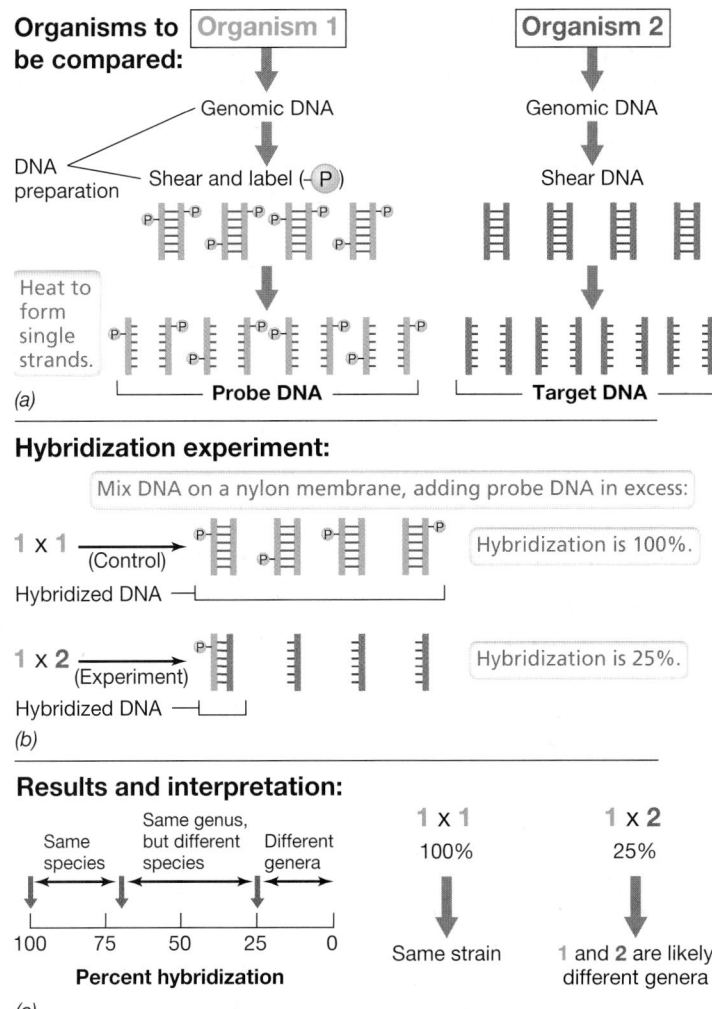

Figure 13.22 Genomic hybridization as a taxonomic tool. *(a)* Genomic DNA is isolated from the organisms to be compared and then sheared and denatured. Probe DNA is prepared from organism 1 by shearing, denaturing, and labeling the DNA (shown here as radioactive phosphate). *(b)* Sheared single-stranded target DNA from each genome is immobilized on a membrane and then hybridized with the labeled probe DNA from organism 1. Radioactivity in the hybridized DNA is measured. *(c)* Radioactivity in the control (organism 1 DNA hybridizing to itself) is taken as the 100% hybridization value.

UNIT 4

13.9 Taxonomic Methods in Systematics

A *polyphasic approach*, that is, an approach that uses many different methods in combination, is used to identify and name species of *Bacteria* and *Archaea* in accordance with the currently accepted phylogenetic species concept. In this section we describe methods commonly used for characterizing microbial, and primarily prokaryotic, species.

Gene Sequence Analyses

As we have described, gene sequences are commonly determined from PCR-amplified fragments of DNA, and the sequences are analyzed using phylogenetic analyses (Section 13.7). However, SSU rRNA gene sequences are highly conserved, and while they provide valuable phylogenetic information, they are not always useful for distinguishing closely related species. By contrast, other highly conserved genes, such as *recA*, which encodes a recombinase

protein (⇄ Sections 11.4 and 11.5), and *gyrB*, which encodes DNA gyrase (⇄ Section 4.1), can be useful for distinguishing bacteria at the species level. The DNA sequences of protein-encoding genes accumulate mutations more rapidly than rRNA genes; for this reason, sequences from such genes can often distinguish prokaryotic species that cannot be resolved by rRNA gene sequence analyses alone (**Figure 13.24**).

Multilocus Sequence Typing

Multilocus sequence typing (MLST) is a method in which several different "housekeeping" genes from several related organisms are sequenced and the sequences are used collectively to distinguish the organisms. Housekeeping genes encode essential functions in cells and are always located on the chromosome rather than on a plasmid. For each gene, an approximately 450-base-pair sequence is amplified and then sequenced. The alleles of each gene (variants that differ by at least one nucleotide) are each assigned a number. The strain being studied is then assigned an allelic profile, or multilocus sequence type, consisting of a series of numbers representing its particular combination of alleles (**Figure 13.25**). In MLST, strains with identical sequences for a given gene have the same allele number for that gene, and two strains with identical sequences for all the genes have the same allelic profile (and would be considered identical by this method). The relatedness between each allelic profile is expressed in a dendrogram of genetic distances that vary from 0 (strains are identical) to 1 (strains are only distantly related, if at all).

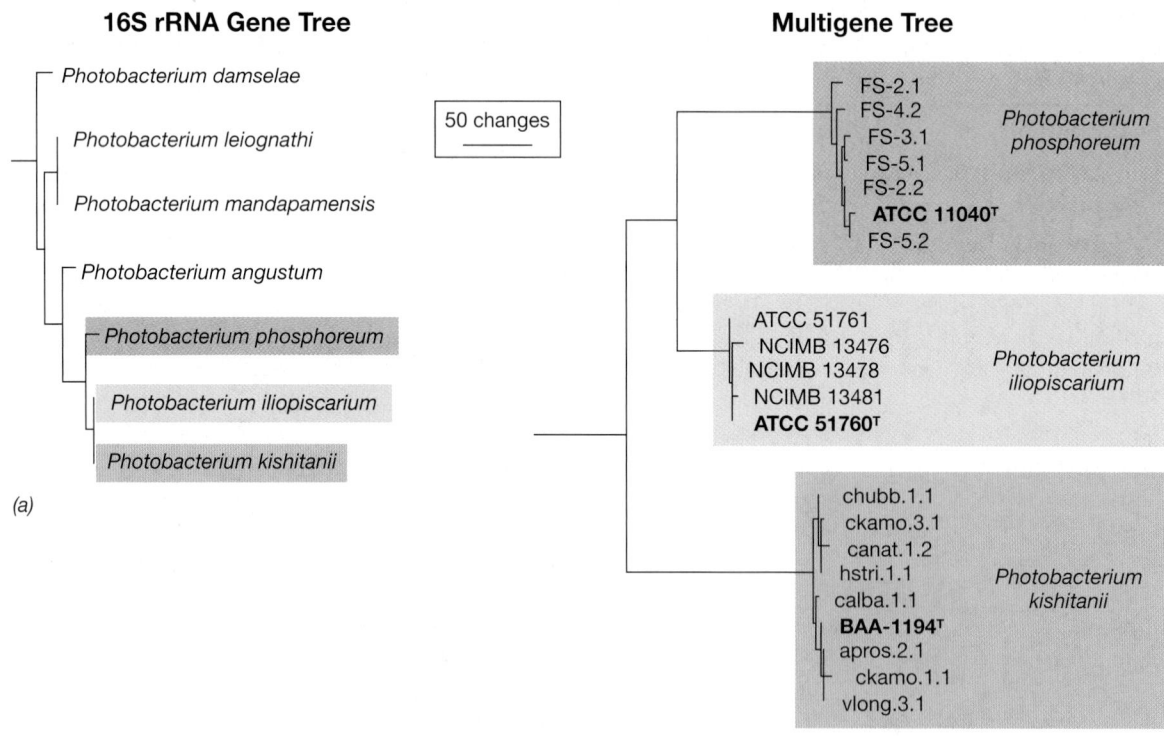

Figure 13.24 Multigene phylogenetic analysis. A phylogeny is shown for species in the genus *Photobacterium*. *(a)* 16S rRNA gene tree, showing the species to be poorly resolved. *(b)* Multigene analysis based on combined analysis of the 16S rRNA gene and *gyrB* and

luxABFE genes in 21 isolates from three *Photobacterium* species. Multigene analysis clearly resolves the strains into three distinct phylogenetic species, *P. phosphoreum*, *P. iliopiscarium*, and *P. kishitanii*. The scale bar indicates the branch length equal to a total of 50 nucleotide changes.

The type strain (Section 13.10) of each species is listed in bold. (All abbreviations are part of strain designations.) Phylogenetic analyses courtesy of Tory Hendy and Paul V. Dunlap, University of Michigan.

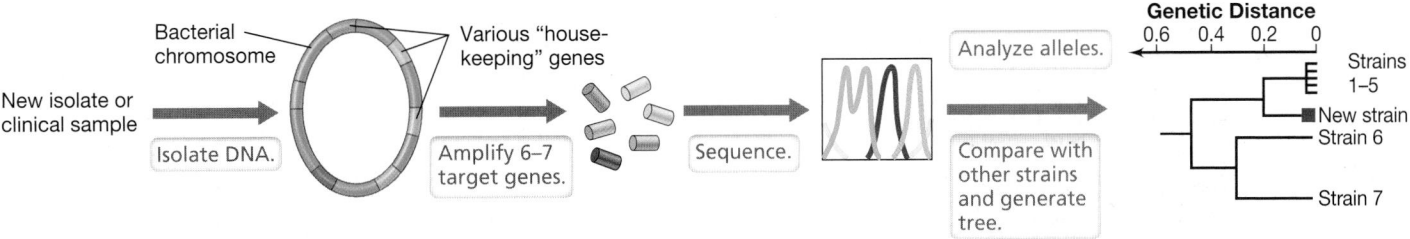

Figure 13.25 Multilocus sequence typing. Steps in MLST leading to a similarity phenogram are shown. Strains 1–5 are virtually identical, whereas strains 6 and 7 are distinct from one another and from both strains 1–5 and the new strain.

MLST has sufficient resolving power to distinguish among even very closely related strains of a given species. In practice, strains can be discriminated on the basis of a single nucleotide change in just one of the analyzed genes. MLST has found its greatest use in clinical microbiology, where it has been used to differentiate strains of various pathogens. This is important because some strains within a species—*Escherichia coli* K-12, for example—may be harmless, whereas others, such as *E. coli* strain O157:H7, can cause serious and even fatal infections (🔗 Section 32.11). MLST is also widely used in epidemiological studies to track a virulent strain of a bacterial pathogen as it moves through a population and in environmental studies to define the geographic distributions of strains.

Genome Fingerprinting

Genome fingerprinting is a rapid approach for evaluating polymorphisms between strains of a species. The fingerprints are generally fragments of DNA generated from individual genes or whole genomes. **Ribotyping** is a form of genome fingerprinting based on the localization of SSU rRNA genes on genome fragments. In this method, genomic DNA from an organism is digested by a restriction enzyme (🔗 Section 12.2) and the fragments are separated by gel electrophoresis, transferred to a nylon membrane, and labeled with an SSU rRNA gene probe (**Figure 13.26**). Different microbial species can have different numbers of rRNA operons, ranging from 1 to 15, and the number of rRNA operons present in a microbial genome is a conserved feature of all strains of a species. In addition, changes in genome sequence between strains can cause the endonuclease enzyme to cut in different locations, producing variation in the lengths of the restriction fragments that are visualized. Hence, the size and number of bands detected

generates a specific pattern, a kind of genome fingerprint called a *ribotype*, and this pattern can be compared with patterns of reference organisms in a computer database.

The ribotype of a particular organism can be unique and diagnostic, allowing rapid identification of different species and even different strains of a species. For these reasons, ribotyping has found many applications in clinical diagnostics and the microbial analyses of food, water, and beverages. Other common genome fingerprinting methods include *repetitive extragenic palindromic PCR* (*rep-PCR*) and *amplified fragment length polymorphism* (*AFLP*). The rep-PCR method is based on the presence of highly conserved repetitive DNA elements interspersed randomly around the bacterial chromosome. The number and positions of these elements differ between strains of a species. Oligonucleotide primers designed to be complementary to these elements enable PCR amplification of genomic fragments found between the repeated elements. These PCR products can be visualized using gel electrophoresis to reveal a pattern of bands that can be used as a fingerprint (**Figure 13.27**). AFLP is based on the digestion of genomic DNA with one or two restriction enzymes and selective PCR amplification of the resulting fragments, which are then separated by agarose gel electrophoresis. Strain-specific banding patterns similar to those of rep-PCR or other DNA fingerprinting methods are generated, with the large number of bands giving a high degree of discrimination between strains within a species.

Multigene and Whole Genome Analyses

The use of multiple genes and entire genomes for the identification and description of bacteria is becoming increasingly common as DNA sequencing capacities improve and costs decline (🔗 Section 9.2). A wide range of sequence analyses can be performed on entire genomes, providing insights into microbial physiology and microbial evolution. These analyses have also provided important insights into the large role that horizontal gene exchange has played in microbial evolution and on the highly dynamic nature of microbial genomes (Section 13.6).

Shared *orthologs* (homologous genes that share the same function, Section 13.7) can be aligned and examined using phylogenetic methods to determine the **average nucleotide identity** of these genes. Different microbial species typically have less than 95% average nucleotide identity for their shared orthologous genes. Comparative analyses of gene content (presence or absence of genes), *synteny* (the order of genes in the genome), and genome GC content provide further insights into relationships between

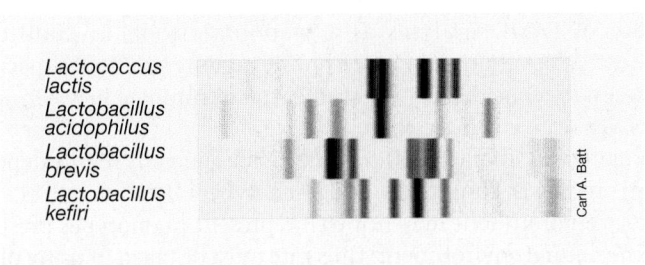

Figure 13.26 Ribotyping. Ribotype results for four different lactic acid bacteria. DNA was taken from a strain of each bacterium, digested into fragments by restriction enzymes, separated by gel electrophoresis, and then probed with a 16S rRNA gene probe. Variations in position and intensity of the bands are important in identification.

UNIT 4

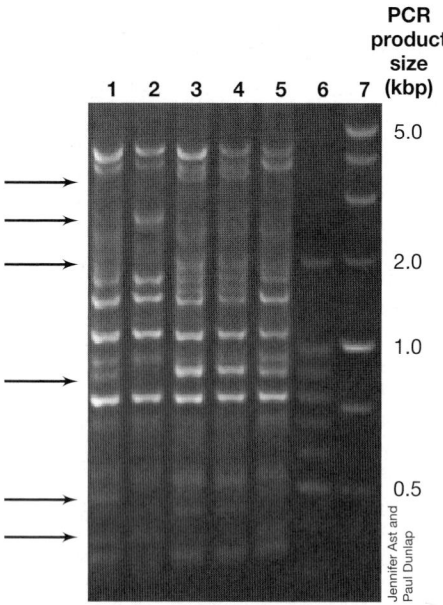

Figure 13.27 DNA fingerprinting with rep-PCR. Genomic DNAs from five strains (1–5) of a single species of bacteria were PCR-amplified using specific primers called *rep* (*rep*etitive *e*xtragenic *p*alindromic). The PCR products were separated in an agarose gel on the basis of size to generate DNA fingerprints. Arrows indicate some of the differing bands. Lanes 6 and 7 are 100-bp and 1-kbp DNA size markers, respectively, used for estimating sizes of the DNA fragments.

strains. Entire genome sequences can also be used for metabolic reconstruction and characterization of a cell's genetic capacities. Several methods in comparative genomics and population genomics (Chapter 9) have been developed for use in systematic analyses.

Phenotypic Analysis

The observable characteristics—the **phenotype**—of a bacterium provide many traits that can be used to differentiate species. Typically, several phenotypic traits are determined routinely when describing a new microorganism. These phenotypic results are then compared to the phenotypes of organisms that have been described previously. The phenotypic traits that are determined will depend on the type of organism being described. For example, in applied situations, such as clinical diagnostic microbiology, where timely identification is important, a well-defined subset of traits can be tested to rapidly discriminate between different types of microorganisms. Table 13.1 lists general categories and examples of some phenotypic traits used in identifications and species descriptions, and we examine one of these traits here.

The types and proportions of fatty acids present in cytoplasmic membrane lipids and the outer membrane lipids of gram-negative bacteria are phenotypic traits often used in taxonomic analyses. The technique for identifying these fatty acids has been nicknamed **FAME**, for *f*atty *a*cid *m*ethyl *e*ster, and is in widespread use in clinical, public health, and food- and water-inspection laboratories where pathogens routinely must be identified. The fatty acid composition of *Bacteria* varies from species to species in chain length and in the presence or absence

TABLE 13.1 Some phenotypic characteristics of taxonomic value

Category	Characteristics
Morphology	Colony morphology; Gram reaction; cell size and shape; pattern of flagellation; presence of spores, inclusion bodies (e.g., PHB,[a] glycogen, or polyphosphate granules, gas vesicles, magnetosomes); capsules, S-layers, or slime layers; stalks or appendages; fruiting body formation
Motility	Nonmotile; gliding motility; swimming (flagellar) motility; swarming; motile by gas vesicles
Metabolism	Mechanism of energy conservation (phototroph, chemoorganotroph, chemolithotroph); utilization of individual carbon, nitrogen, or sulfur compounds; fermentation of sugars; nitrogen fixation; growth factor requirements
Physiology	Temperature, pH, and salt ranges for growth; response to oxygen (aerobic, facultative, anaerobic); presence of catalase or oxidase; production of extracellular enzymes
Cell lipid chemistry	Fatty acids;[b] polar lipids; respiratory quinones
Cell wall chemistry	Presence or absence of peptidoglycan; amino acid composition of cross-links; presence or absence of cross-link interbridge
Other traits	Pigments; luminescence; antibiotic sensitivity; serotype; production of unique compounds, for example, antibiotics

[a]PHB, poly-β-hydroxybutyric acid (↪ Section 2.8).
[b]Figure 13.28.

of double bonds, rings, branched chains, or hydroxy groups (**Figure 13.28**). In FAME analyses, fatty acids extracted from cells grown in culture under standardized conditions are analyzed by gas chromatography. A chromatogram showing the types and amounts of fatty acids from the unknown bacterium is then compared with a database containing the fatty acid profiles of thousands of reference bacteria grown under the same conditions.

Fatty acid profiles of an organism, like many other phenotypic traits, can vary as a function of temperature, growth phase (exponential versus stationary), and growth medium. Hence, for comparative results to be valid, it is necessary to grow the unknown organism on a specific medium and at a specific temperature. For many organisms this is impossible, of course, limiting the applicability of FAME analyses as a taxonomic tool. In addition, the extent of variation in FAME profiles among strains of a species, a necessary consideration in studies to discriminate between species, is a work in progress.

Phenotypic characteristics of strains are generally highly dependent on growth conditions, and phenotypes observed in the laboratory environment may not well represent phenotypes present in the natural environment; thus care must be taken in using phenotypic characteristics in systematic analyses. In addition, the value of different phenotypic characteristics for reaching a firm systematics conclusion can vary with the taxonomic groups being examined.

Classes of Fatty Acids in *Bacteria*

Class/Example **Structure of example**

I. *Saturated:*
tetradecanoic acid

II. *Unsaturated:*
omega-*7-cis*
hexadecanoic acid

III. *Cyclopropane:*
cis-7,8-methylene
hexadecanoic acid

IV. *Branched:*
13-methyltetradecanoic acid

V. *Hydroxy:*
3-hydroxytetradecanoic acid

(a)

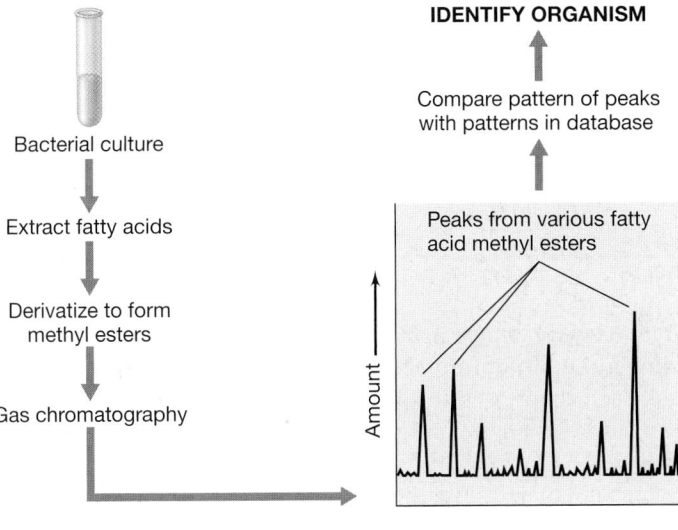

IDENTIFY ORGANISM

Compare pattern of peaks
with patterns in database

Peaks from various fatty
acid methyl esters

Bacterial culture

Extract fatty acids

Derivatize to form
methyl esters

Gas chromatography

(b)

Figure 13.28 Fatty acid methyl ester (FAME) analysis in bacterial identification. *(a)* Classes of fatty acids in *Bacteria*. Only a single example is given of each class; however, more than 200 structurally distinct fatty acids are known from bacterial sources. A methyl ester contains a methyl group (CH_3) in place of the proton on the carboxylic acid group (COOH) of the fatty acid. *(b)* Procedure. Each peak from the gas chromatograph is due to one particular fatty acid methyl ester, and the peak height is proportional to the amount.

MINIQUIZ
- What class of genes is used in MLST analyses?
- How is ribotyping different from rep-PCR?
- What is FAME analysis?

13.10 Classification and Nomenclature

We conclude our treatment of microbial evolution and systematics with a brief description of how *Bacteria* and *Archaea* are classified and named—the science of *taxonomy*. Information is also presented on culture collections, which are repositories for scientific deposition of live microbial cultures; on some key taxonomic resources available for microbiology; and on the procedures for naming new microbial species. The formal description of a new microbial species and the deposition of cultures into a culture collection form an important foundation of the systematics of prokaryotic species.

Taxonomy and Describing New Species

Classification is the organization of organisms into progressively more inclusive groups on the basis of either phenotypic similarity or evolutionary relationship. A species is made up of one to several strains, and similar species are grouped into genera (singular, genus). Similar genera are grouped into families, families into orders, orders into classes, up to the domain, the highest-level taxon based on a collection of phenotypic and genotypic information. This hierarchical scheme is illustrated in Table 13.2.

Nomenclature is the actual naming of organisms and follows the **binomial system** of nomenclature devised by the Swedish medical doctor and botanist Carl Linnaeus and used throughout biology; organisms are given *genus names* and *species epithets*. The names are Latin or Latinized Greek derivations, often descriptive of some key property of the organism, and are printed in *italics*. By classifying organisms into groups and naming them, we order the natural microbial world and make it possible to communicate effectively about all aspects of particular organisms, including their behavior, ecology, physiology, pathogenesis, and evolutionary relationships. The creation of new taxa of *Bacteria* and *Archaea* must follow the rules described in the *International Code of Nomenclature of Bacteria* (the Bacteriological Code). This source presents the formal framework by which *Bacteria* and *Archaea* are to be officially named and the procedures by which existing names can be changed, for example, when new data warrant taxonomic rearrangements.

When a new isolate of *Bacteria* or *Archaea* is isolated from nature and thought to be unique, a decision must be made as to whether it is sufficiently different from established taxa to be described as a new taxon. To achieve formal validation of taxonomic standing as a new genus or species, a detailed description of the organism's characteristics and distinguishing traits, along with its proposed name, must be published, and viable cultures of the organism must be deposited in at least two international culture collections (Table 13.3). The manuscript describing and naming a new taxon undergoes peer review before publication. A major vehicle for the description of new taxa is the *International Journal of Systematic and Evolutionary Microbiology* (*IJSEM*), the official publication of record for the taxonomy and classification of *Bacteria, Archaea*, and microbial eukaryotes. In each issue, the *IJSEM* also publishes an approved list of newly validated names. By providing validation of newly proposed names, publication in *IJSEM* paves the way for their inclusion in taxonomic reference sources. Two websites provide listings of valid, approved bacterial names: List of Prokaryotic Names with Standing in Nomenclature (http://www.bacterio.net), and Prokaryotic Nomenclature Up-to-Date (http://www.dsmz.de).

It is possible to use molecular and genomic techniques (Chapters 9 and 19) to characterize the phenotypic and genotypic characteristics of microorganisms without the need for the

UNIT 4

TABLE 13.2 Taxonomic hierarchy for the purple sulfur bacterium *Allochromatium warmingii*

Taxon	Name	Properties	Confirmed by
Domain	*Bacteria*	Bacterial cells; rRNA gene sequences typical of *Bacteria*	Microscopy; 16S rRNA gene sequence analysis; presence of unique biomarkers, for example, peptidoglycan
Phylum	*Proteobacteria*	rRNA gene sequence typical of *Proteobacteria*	16S rRNA gene sequence analysis
Class	*Gammaproteobacteria*	Gram-negative bacteria; rRNA sequence typical of *Gammaproteobacteria*	Gram-staining, microscopy
Order	*Chromatiales*	Phototrophic purple bacteria	Characteristic pigments (∂ Figures 14.2, 14.3, and 14.9)
Family	*Chromatiaceae*	Purple sulfur bacteria	Ability to oxidize H_2S and store S^0 within cells; microscopic observation of S^0 (see photo); 16S rRNA gene sequence
Genus	*Allochromatium*	Rod-shaped purple sulfur bacteria; <95% 16S rRNA gene sequence identity with other genera	Microscopy (see photo)
Species	*warmingii*	Cells 3.5–4.0 µm × 5–11 µm; storage of sulfur mainly in poles of cell (see photo); <97% 16S rRNA gene sequence identity with other species	Cell size measured microscopically with a micrometer; observation of polar position of S^0 globules in cells (see photo); 16S rRNA gene sequence

Sulfur (S^0) globules

Cells of *A. warmingii*

Norbert Pfennig

cultivation of an isolated strain. However, in the absence of an isolate that can be deposited in two international culture collections, it is not possible to validly name a new species of microorganism under the Bacteriological Code. However, when an organism is well characterized but not yet cultured or not yet obtained in pure culture, a provisional taxonomic name can be applied. The moniker *Candidatus* is appended to candidate taxonomic ranks. For example, "*Candidatus* Pelagibacter ubique" is a globally widespread and well-characterized marine bacterium that is difficult to grow in laboratory media (∂ Section 19.3) and so has not been formally named under the Bacteriological Code. By contrast, the bacterium "*Candidatus* Heliomonas lunata" can be grown in laboratory culture but the culture is not pure so this bacterium also retains *Candidatus* status.

The International Committee on the Systematics of Prokaryotes (ICSP) oversees the nomenclature and taxonomy of *Bacteria* and *Archaea*. The ICSP also oversees the publication of *IJSEM* and the *International Code of Nomenclature of Bacteria* and gives guidance to several subcommittees that meet to establish and revise standards for the description of new species in the different groups of *Bacteria* and *Archaea*.

Bergey's Manual and The Prokaryotes

Because taxonomy is to some degree a matter of scientific judgment, there is no "official" classification of *Bacteria* and *Archaea*. However, the classification system most widely accepted by microbiologists is that of *Bergey's Manual of Systematic Bacteriology*, a major taxonomic treatment of *Bacteria* and *Archaea*. Widely used, *Bergey's Manual* has served microbiologists since 1923 as a compendium of information on all recognized prokaryotic species. Each chapter, written by experts, contains tables, photos, figures, and other systematic information useful for identification purposes.

A second major source describing the physiology, ecology, phylogeny, enrichment, isolation, and cultivation of *Bacteria* and *Archaea* is *The Prokaryotes*. This work is available online by subscription through university libraries. Collectively, *Bergey's Manual* and *The Prokaryotes* offer microbiologists both the concepts and the details of the biology of *Bacteria* and *Archaea* as we know them today and are the primary resources for microbiologists characterizing newly isolated organisms.

Culture Collections

National microbial culture collections (Table 13.3) are an important foundation of microbial systematics. These permanent collections catalog and store microorganisms and provide cultures upon request (for a fee) to researchers in academia, medicine, and industry. The collections play an important role in protecting microbial biodiversity, just as museums do in preserving plant and animal specimens for future study. However, unlike museums, which maintain collections of chemically preserved or dried, *dead* specimens, microbial culture collections store microorganisms as *viable cultures*, typically frozen or in a freeze-dried state. These storage methods maintain the cells indefinitely in a living state and prevent genetic changes that might occur if the organisms were continually subcultured.

TABLE 13.3 **Some international microbial culture collections**

Collection	Name	Location	Web address
ATCC	American Type Culture Collection	Manassas, Virginia	http://www.atcc.org
BCCM/LMG	Belgian Coordinated Collection of Microorganisms	Ghent, Belgium	http://bccm.belspo.be
CIP	Collection of the Institut Pasteur	Paris, France	http://www.pasteur.fr
CBS	Centraalbureau voor Schimmelcultures	Utrecht, The Netherlands	http://www.cbs.knaw.nl
DSMZ	Deutsche Sammlung von Mikroorganismen und Zellkulturen	Braunschweig, Germany	http://www.dsmz.de
JCM	Japan Collection of Microorganisms	Saitama, Japan	http://jcm.brc.riken.jp/
NCCB	Netherlands Culture Collection of Bacteria	Utrecht, The Netherlands	http://www.cbs.knaw.nl
NCIMB	National Collection of Industrial, Marine and Food Bacteria	Aberdeen, Scotland	http://www.ncimb.com
NRRL	United States Department of Agriculture, Agricultural Research Service Culture Collection	Peoria, Illinois	http://nrrl.ncaur.usda.gov

A related and key role of culture collections is as repositories for *type strains*. When a new species of bacteria is described in a scientific journal, a strain is designated as the **type strain** and this strain serves as the nomenclatural type of the taxon for future taxonomic comparison with other strains of that species (see Figure 13.24 for a visual representation of this). Deposition of this type strain in the national culture collections of at least two countries—thereby making the strain publicly available internationally—is a prerequisite for validation of the new species name. Some of the large national culture collections are listed in Table 13.3. Their websites contain searchable databases of strain holdings together with information on the environmental sources of strains and their descriptions.

— **MINIQUIZ** —

- What roles do culture collections play in microbial systematics?
- What is the *IJSEM* and what taxonomic function does it fulfill?
- Why might viable cell cultures be of more use in microbial taxonomy than preserved specimens?

MasteringMicrobiology®   **Visualize**, **explore**, and **think critically** with Interactive Microbiology, MicroLab Tutors, MicroCareers case studies, and more. MasteringMicrobiology offers practice quizzes, helpful animations, and other study tools for lecture and lab to help you master microbiology.

Chapter Review

I • Early Earth and the Origin and Diversification of Life

13.1 Planet Earth is about 4.5 billion years old. After its formation, Earth was hot and sterile. Gradual cooling of the Earth allowed formation of liquid water, a requirement for the origin of life. The earliest evidence of life comes from isotopic analysis of ancient sedimentary rocks and zircon minerals indicating that life was present on Earth 3.86–4.1 billion years ago.

Q **What is LUCA, and what is a plausible explanation for the origin of cellular life?**

13.2 In rocks 3.5 billion years old or younger, microbial formations called stromatolites are abundant and show extensive microbial diversification. The evolution of oxygenic photosynthesis caused O_2 to accumulate 2.4 billion years ago, eventually leading to the formation of banded iron formations, the ozone shield, and an oxygenated atmosphere, which set the stage for rapid diversification of metabolic types and the evolution of multicellularity.

Q **Why was the origin of cyanobacteria of such importance to the evolution of life on Earth?**

13.3 Ribosomal RNA genes have been used to construct a universal tree of life revealing that life on Earth evolved in three major directions, forming the domains *Bacteria*, *Archaea*, and *Eukarya*. The universal tree of life shows that the domains *Bacteria* and *Archaea* diverged billions of years ago, that *Eukarya* split from *Archaea* later in the history of life, and that complex multicellular eukaryotes only began to diverge within the last 600 million years.

Q **What evidence supports the classification of life into three domains?**

13.4 The eukaryotic cell developed from endosymbiotic events. The modern eukaryotic cell is a chimera with genes and characteristics from both *Bacteria* and *Archaea*. SSU rRNA sequence analyses indicate the ancestors of mitochondria are found in the phylum *Proteobacteria* and those of chloroplasts are found in the phylum *Cyanobacteria*.

Q What is the endosymbiotic hypothesis for the origin of mitochondria and chloroplasts? What evidence supports this hypothesis?

II · Microbial Evolution

13.5 Evolution is defined as a change in allele frequencies in a population of organisms over time resulting in descent with modification. New alleles are created through the processes of mutation and recombination. Mutations occur at random and most mutations are neutral or deleterious, but some are beneficial. Natural selection and genetic drift are two mechanisms that cause allele frequencies to change in a population over time.

Q What is fitness? To what degree does fitness depend on the environment in which organisms live?

13.6 Microbial genomes are dynamic, and genome size and gene content can vary considerably between strains of a species. The core genome is defined as the set of all genes shared by a species, while the pan genome is defined as the core genome plus genes whose presence varies among strains of a species.

Q What are some processes that influence the content of the pan genome?

III · Microbial Phylogeny and Systematics

13.7 Molecular sequences accumulate random mutations over time, and molecular phylogenetic analysis examines differences in molecular sequences to determine the evolutionary history of life. A phylogenetic tree is a diagram that depicts the evolutionary history of a set of genes or organisms.

Q What is the difference between a gene tree and an organismal tree?

13.8 At present, species in *Bacteria* and *Archaea* are defined operationally based on shared genetic and phenotypic traits. The dynamic nature of microbial genomes and the abundance of genes acquired through horizontal gene transfer have raised questions about the nature of microbial species.

Q What is the "species problem" and why is the concept of microbial species difficult to resolve?

13.9 Systematics is the study of the diversity and relationships of living organisms. Polyphasic taxonomy is based on phenotypic, genotypic, and phylogenetic information. Bacterial species can be distinguished genotypically on the basis of DNA–DNA hybridization, DNA fingerprinting, MLST, or multigene or whole genome analyses. Phenotypic traits useful in taxonomy include morphology, motility, metabolism, and cell chemistry, especially lipid analyses.

Q What level of SSU rRNA gene dissimilarity, DNA–DNA hybridization, and whole genome average nucleotide identity would be required to support the designation of a new species of microorganism?

13.10 Nomenclature in microbiology follows the binomial system used in all of biology. Formal recognition of a new species of *Bacteria* or *Archaea* requires depositing a sample of the organism in culture collections and publishing the new species name and description.

Q Is it possible to provide a formal name for a microorganism that has not been cultivated in isolation? What kind of name might be used if a microorganism is well characterized but cannot yet be cultivated in isolation?

Application Questions

1. Compare and contrast the physical and chemical conditions on Earth at the time life first arose with conditions today. From a physiological standpoint, discuss at least two reasons why *animals* could not have existed on early Earth. In what ways has microbial metabolism altered Earth's biosphere? How might life on Earth be different if oxygenic photosynthesis had not evolved?

2. For the following sequences, construct the phylogenetic tree that best depicts their evolutionary relationships.

   Taxon 1: GTTCCCTTA
   Taxon 2: GTTCGGTAT
   Taxon 3: GAAAAACCCTAT
   Taxon 4: CTTCCCTTT
   Taxon 5: GTAAAACCCGAT

3. Imagine that you have been given several bacterial strains from various countries around the world and that all the strains are thought to cause the same gastrointestinal disease and to be genetically identical. Upon carrying out a DNA fingerprint analysis of the strains, you find that four different strain types are present. What methods could you use to test whether the different strains are actually members of the same species?

4. Imagine that you have discovered a new form of microbial life, one that appears to represent a fourth domain. How would you go about characterizing the new organism and determining if it actually is evolutionarily distinct from *Bacteria, Archaea*, and *Eukarya*?

Chapter Glossary

Allele a sequence variant of a given gene

Archaea a domain of life consisting of microorganisms that have prokaryotic cell structure and are distinct from *Bacteria*

Average nucleotide identity a genome-wide measure of genetic similarity between two microorganisms based on aligning all orthologous gene pairs across the two genomes and determining the percentage of matching nucleotides (see also *orthologs*)

Bacteria a domain of life consisting of microorganisms that have prokaryotic cell structure and are distinct from *Archaea*

Banded iron formation iron oxide–rich ancient sedimentary rocks containing zones of oxidized iron (Fe^{3+}) formed by oxidation of Fe^{2+} by O_2 produced by cyanobacteria

Binomial system the system devised by the Swedish scientist Carl Linnaeus for naming living organisms in which an organism is given a genus name and a species epithet

Core genome those genes found in common in the genomes of all strains of a species

DNA–DNA hybridization the experimental determination of genomic similarity by measuring the extent of hybridization of DNA from one organism with that of another

Endosymbiotic hypothesis the idea that a chemoorganotrophic bacterium and a cyanobacterium were stably incorporated into another cell type to give rise, respectively, to the mitochondria and chloroplasts of modern-day eukaryotes

Eukarya a domain of life consisting of organisms that have eukaryotic cell structure; includes algae, protists, fungi, slime molds, plants, and animals

Evolution a change in allele frequencies in a population of organisms over time, with new alleles arising due to mutation and recombination, resulting in descent with modification

FAME fatty acid methyl ester; a technique for identifying microorganisms by their fatty acids

Fitness the capacity of an organism to survive and reproduce as compared to that of competing organisms

Genetic drift a process that results in a change in allele frequencies in a population as a result of random changes in the number of offspring from each individual over time

Homology having shared ancestry

Homoplasy when two organisms have the same trait as a result of recurrent mutation or convergent evolution

Horizontal gene transfer the asymmetrical and unidirectional transfer of DNA from one cell to another

Molecular clock a DNA sequence, such as a gene for rRNA, that can be used as a comparative temporal measure of evolutionary divergence

Monophyletic in phylogeny, a group descended from a single shared ancestor

Multilocus sequence typing (MLST) a taxonomic tool for classifying organisms on the basis of gene sequence variations in several housekeeping genes

Mutation a heritable change in DNA sequence

Orthologs genes that are homologous (that is, they are descended from a common ancestor) and share the same function (see also *paralog*)

Pan genome the totality of the genes present in the different strains of a species

Paralogs genes that are homologous (that is, they are descended from a common ancestor) but which have diverged to have different functions; typically, paralogous genes are the result of gene duplication (see also *ortholog*)

Phenotype the physical and chemical characteristics of an organism that can be observed or measured

Phylogenetic tree a diagram that depicts the evolutionary history of an organism; consists of nodes and branches

Phylogeny evolutionary history

Phylum a major lineage of cells in one of the three domains of life

Recombination resorting or rearrange-ment of DNA fragments resulting in a new sequence

Ribosomal RNA (rRNA) RNA molecules found in the small and large subunits of the ribosome

Ribotyping a means of identifying microorganisms from analysis of DNA fragments generated from restriction enzyme digestion of the genes encoding their ribosomal RNA

Selection in an evolutionary context, a process that results in a change in allele frequencies in a population when individuals that are favored in a given environment are able to produce more offspring and make a greater contribution to the gene content of future generations

Sequence alignment the insertion of gaps into a set of molecular sequences organized in rows so that homologous positions are organized in vertical columns. Alignment is necessary prior to phylogenetic analysis because deletion and insertion mutations cause variations in the length of molecular sequences

16S rRNA the type of SSU rRNA found in *Bacteria* and *Archaea*; its eukaryotic counterpart is 18S rRNA (see also *SSU rRNA*)

Species defined in microbiology as a collection of strains that all share the same major properties and differ in one or more significant properties from other collections of strains; defined phylogenetically as a monophyletic, exclusive group based on DNA sequence

SSU rRNA (small subunit rRNA) the rRNA molecule found in the small subunit of the ribosome, comprised of 16S rRNA, which occurs in the 30S ribosomal subunit of *Bacteria* and *Archaea*, or its orthologous counterpart 18S rRNA, which occurs in the 40S ribosomal subunit of *Eukarya*. SSU rRNA genes are conserved in all forms of cellular life and this gene is often used in phylogenetic analysis of microorganisms

Stromatolite a laminated microbial mat which can become fossilized and is typically built from layers of filamentous microorganisms

Systematics the study of the diversity of organisms and their relationships; includes taxonomy and phylogeny

Taxonomy the science of identification, classification, and nomenclature

Type strain the strain chosen to represent the nomenclatural type of a species; type strains are deposited in microbial culture collections and are used to represent the typical characteristics of their species

Universal tree of life a phylogenetic tree that shows the positions of representatives of all domains of cellular life

UNIT 4

14

Metabolic Diversity of Microorganisms

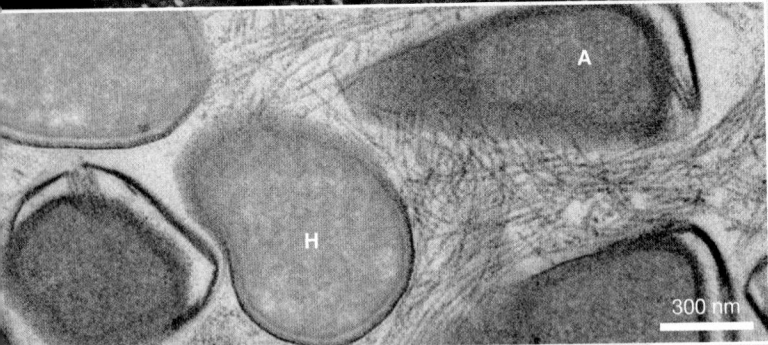

300 nm

microbiology now

Microbes That Plug into the Matrix

Living things conserve energy by moving electrons around. Most microbes transfer electrons to soluble electron acceptors, such as O_2 or NO_3^- that have been taken up by the cell. Some microbes, such as iron-reducing bacteria, transfer electrons to insoluble electron acceptors, such as metallic minerals. Recent discoveries, however, show that certain microbes are actually able to conserve energy by plugging into other cells. Such direct interspecies electron transfer was recently discovered at the heart of an exceptional symbiosis between anaerobic methanotrophic *Archaea* (ANME) and sulfate-reducing bacteria (SRB).

Methane is a powerful greenhouse gas that is abundant in the deep sea as methane hydrates. Methane seeps occur where these hydrates meet ocean water, creating enormous potential for methane to bubble into the atmosphere. However, this methane never reaches the atmosphere because it is consumed by methane-eating *Archaea* that live in marine sediments. ANME live within aggregates that also contain SRB, their symbiotic partners (see upper photo of part of an aggregate where ANME-1 [red] and SRB HotSeep-1 [green] are identified using fluorescent probes and confocal laser microscopy). ANME catalyze the anaerobic oxidation of methane, but in the absence of a suitable electron acceptor, this process alone does not conserve energy. Here is where the SRB partners help out. The SRB take electrons from ANME and use them to reduce sulfate, thereby conserving enough energy to support both symbiotic partners.

The mechanism that underlies this symbiotic association has long been a mystery. In most syntrophic partnerships, H_2 is the "currency" of electron transfer, being produced by one partner and consumed by the other. However, H_2 is not exchanged between ANME and SRB. Instead, ANME and SRB are connected by "nanowires" up to 1000 nm long (see lower photo where ANME-1 [A], SRB HotSeep-1 [H], and their nanowires are visible in an electron micrograph). These structures are produced by the SRB and connect to ANME cells. Both partners have large, multiheme cytochrome c proteins, which enable them to donate (ANME) or accept (SRB) electrons via the nanowires. Hence, the organisms in this intimate symbiosis survive by plugging into an extracellular matrix of tiny wirelike structures that allow electrons to be transferred between species.

 Source: Wegener, G., et al. 2015. Intercellular wiring enables electron transfer between methanotrophic archaea and bacteria. *Nature 526:* 587–590.

A major theme of microbiology is the great *phylogenetic diversity* of microbial life on Earth. We got a taste of this in the last chapter and will explore microbial diversity in detail in the following four chapters. In this chapter we focus on the *metabolic diversity* of microorganisms, with special emphasis on the processes and mechanisms that underlie this diversity. We will then return to the organisms themselves and unveil the phylogenetic breadth of the microbial world in the context of metabolic diversity.

I • Phototrophy

Phototrophy—the use of light energy—is widespread in the microbial world. In this first part of the chapter, we examine the properties and energy-conserving strategies of phototrophic microorganisms and see how these support a lifestyle based on the use of CO_2 as carbon source.

14.1 Photosynthesis and Chlorophylls

The most important biological process on Earth is **photosynthesis**, the conversion of light energy to chemical energy. Organisms that carry out photosynthesis are called **phototrophs**. Photosynthetic organisms are also **autotrophs**, capable of growing with CO_2 as the sole carbon source. Energy from light is used in the reduction of CO_2 to organic compounds (*photoautotrophy*). Some phototrophs can also use organic carbon as their carbon source; this lifestyle is called *photoheterotrophy*.

Photosynthesis originated within the *Bacteria*, and a wide diversity of bacterial species can harvest energy from light. No less than six different photosynthetic systems have evolved within *Bacteria*, and these various systems are found in the *Heliobacteria*, *Acidobacteria*, green sulfur bacteria, purple bacteria, filamentous anoxygenic phototrophs (green nonsulfur bacteria), and *Cyanobacteria*. Ultimately, photosynthesis also evolved within the *Eukarya* as a result of the endosymbiotic origin of chloroplasts from cyanobacterial relatives

(⮌ Section 13.4). These photosynthetic systems all differ in characteristic ways but they all reveal similar underlying principles, which we review in the sections that follow.

Photoautotrophy is comprised of two distinct sets of reactions that operate in parallel: (1) *light reactions* that produce ATP and (2) light-independent *dark reactions* that reduce CO_2 to cell material for autotrophic growth. We will discuss the light reactions in Sections 14.3 and 14.4 and the light-independent dark reactions in Section 14.5. Reduction of CO_2 by the light-independent dark reactions requires both energy, in the form of ATP, and electrons, in the form of NADH (or NADPH). The generation of NADH (or NADPH) from reduction of $NAD(P)^+$ requires an electron donor supplied from the environment. Water (H_2O) is the electron donor for photosynthesis in green plants, algae, and cyanobacteria. By contrast, phototrophic bacteria such as the *Heliobacteria*, *Acidobacteria*, green sulfur bacteria, purple bacteria, and filamentous anoxygenic phototrophs can use diverse electron donors but cannot use water. For example, in green and purple sulfur bacteria the donor could be a reduced sulfur compound such as hydrogen sulfide (H_2S), or even molecular hydrogen (H_2).

The oxidation of H_2O produces molecular oxygen (O_2) as a waste product; because of this, the photosynthetic process in cyanobacteria (and chloroplasts) is called **oxygenic photosynthesis**. However, in all other phototrophic bacteria O_2 is *not* produced, and thus the process is called **anoxygenic photosynthesis** (**Figure 14.1**). Oxygen produced by cyanobacteria billions of years ago converted Earth from an anoxic to an oxic world and set the stage for an explosion of eukaryotic microbial diversity that eventually gave rise to plants and animals.

Photosynthesis requires light-sensitive pigments, the *chlorophylls*—present in plants, algae, and the cyanobacteria—and *bacteriochlorophylls*, present in anoxygenic phototrophs. Absorption of light energy by chlorophylls and bacteriochlorophylls begins the process of photosynthetic energy conversion, and the net result is chemical energy, ATP.

UNIT 4

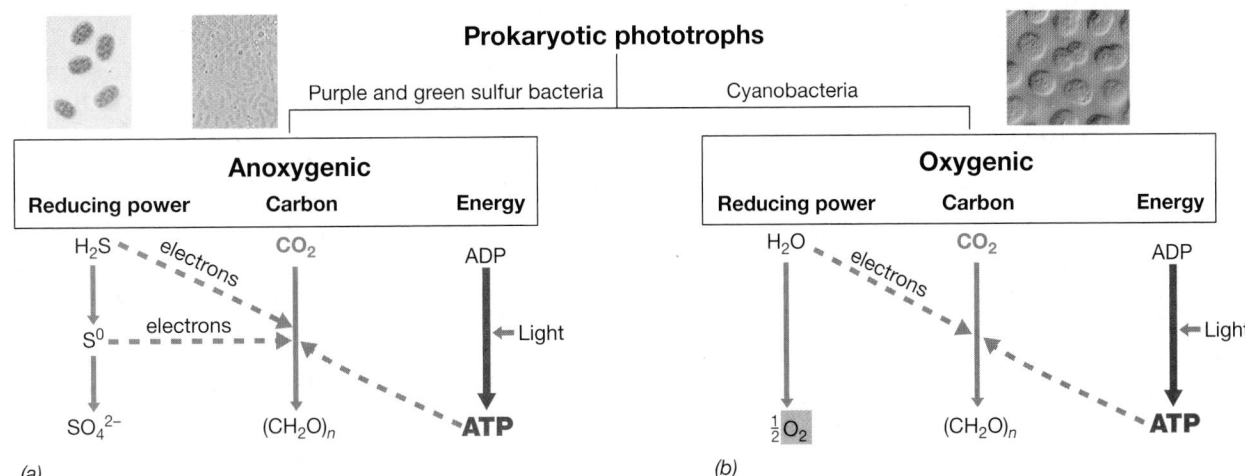

Figure 14.1 Patterns of photosynthesis. Energy and reducing power synthesis in *(a)* anoxygenic and *(b)* oxygenic phototrophs. Note that oxygenic phototrophs produce O_2, while anoxygenic phototrophs do not. Insets: Left, light photomicrographs of cells of a purple sulfur bacterium (*Chromatium*, cells 5 μm in diameter) and a green sulfur bacterium (*Chlorobium*, cells 0.9 μm in diameter). Note the sulfur globules inside or outside the cells produced from the oxidation of H_2S. Right, interference-contrast photomicrograph of cells of a coccoid-shaped cyanobacterium.

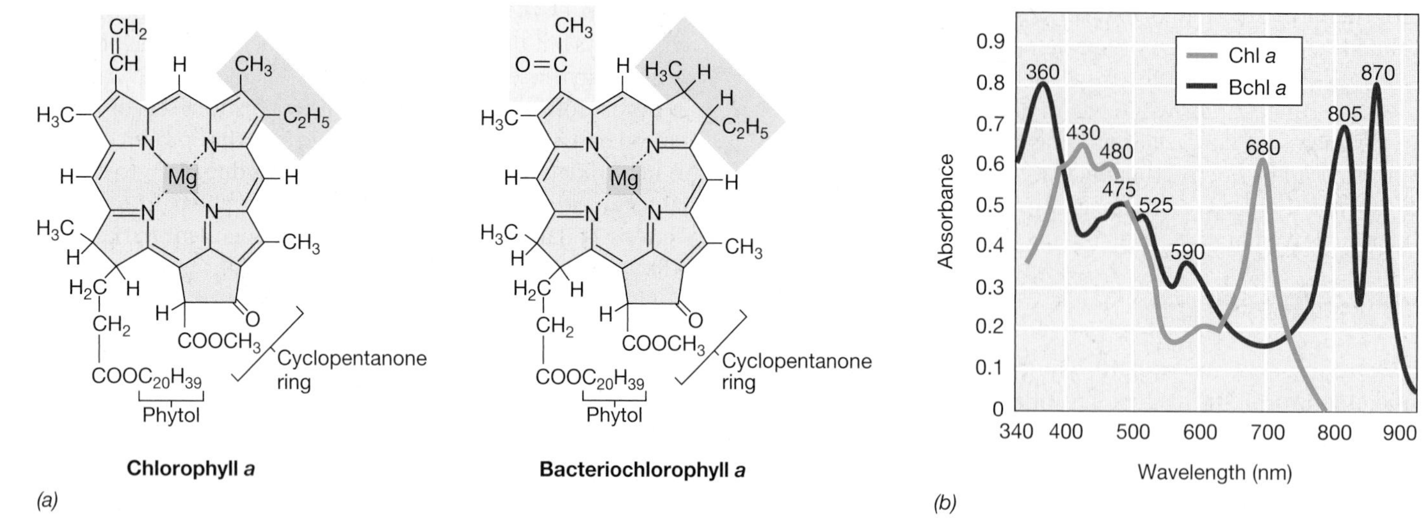

Figure 14.2 Structures and spectra of chlorophyll *a* and bacteriochlorophyll *a*. *(a)* The two molecules are identical except for those portions contrasted in yellow and green. *(b)* Absorption spectrum (green curve) of cells of the green alga *Chlamydomonas*. The peaks at 680 and 430 nm are due to chlorophyll *a*, and the peak at 480 nm is due to carotenoids. Absorption spectrum (red curve) of cells of the phototrophic purple bacterium *Rhodopseudomonas palustris*. Peaks at 870, 805, 590, and 360 nm are due to bacteriochlorophyll *a*, and peaks at 525 and 475 nm are due to carotenoids.

Chlorophyll and Bacteriochlorophyll

Chlorophyll and **bacteriochlorophyll** are tetrapyrroles that are related to the parent structure of the cytochromes. But unlike cytochromes, chlorophylls contain *magnesium* instead of *iron* at the center of the ring. Chlorophylls also contain specific substituents on the tetrapyrrole ring and a hydrophobic alcohol that helps anchor the chlorophyll into photosynthetic membranes. The structure of chlorophyll *a*, the principal chlorophyll of oxygenic phototrophs, is shown in **Figure 14.2a**. Chlorophyll *a* is green because it *absorbs* red and blue light and *transmits* green light; its absorption spectrum shows strong absorbance near 680 nm and 430 nm (Figure 14.2b). Several structurally distinct chlorophylls are known, each distinguished by its unique *absorption spectrum*. Cyanobacteria contain chlorophyll *a* (a few species contain chlorophyll *d*), while their relatives the prochlorophytes contain chlorophylls *a* and *b*.

Anoxygenic phototrophs produce one or more bacteriochlorophylls. Bacteriochlorophyll *a* (**Figure 14.3**), present in most purple bacteria (ⅎ Sections 15.4 and 15.5), absorbs maximally between 800 and 925 nm (different purple bacteria synthesize slightly different photocomplexes, and the absorption maxima of bacteriochlorophyll *a* in any given organism depend to some degree on how proteins in the photocomplexes are arranged in the photosynthetic membrane; see Figure 14.6). Other bacteriochlorophylls, whose distribution runs along phylogenetic lines, absorb in other regions of the visible and near infrared spectrum (Figure 14.3).

The existence of different forms of chlorophyll or bacteriochlorophyll that absorb light of different wavelengths allows phototrophs to make better use of the available energy in the electromagnetic spectrum. By employing different pigments with distinct absorption properties, different phototrophs can coexist in the same habitat, each absorbing wavelengths of light that others cannot. Thus, pigment diversity has *ecological* significance for the successful coexistence of different phototrophs in the same habitat.

Reaction Centers and Antenna Pigments

In oxygenic phototrophs and in purple anoxygenic phototrophs, chlorophyll/bacteriochlorophyll molecules do not exist freely in the cell but are attached to proteins and housed within membranes to form *photocomplexes* consisting of anywhere from 50 to 300 chlorophyll/bacteriochlorophyll molecules. A small number of these pigment molecules are present within photosynthetic **reaction centers** (**Figure 14.4**), the complex macromolecular structures that participate directly in the reactions that lead to energy conservation. Photosynthetic reaction centers are surrounded by larger numbers of light-harvesting chlorophylls/bacteriochlorophylls. These so-called **antenna pigments** (also called *light-harvesting pigments*) function to absorb light and funnel some of the energy to the reaction center (Figure 14.4). At the low light intensities that are often found in nature, this arrangement for concentrating energy allows reaction centers to receive light energy that would otherwise be missed.

Photosynthetic Membranes, Chloroplasts, and Chlorosomes

The chlorophyll pigments and all the other components of the light-gathering apparatus exist within membranes in the cell. The location of these photosynthetic membranes differs between prokaryotic and eukaryotic phototrophs. In eukaryotic phototrophs, photosynthesis takes place in intracellular organelles, the *chloroplasts*, which contain sheetlike photosynthetic membrane systems

Pigment/Absorption maxima (in vivo)	R1	R2	R3	R4	R5	R6	R7
Bchl a (purple bacteria)/ 805, 830–890 nm	$-\overset{\displaystyle O}{\overset{\|}{C}}-CH_3$	$-CH_3{}^a$	$-CH_2-CH_3$	$-CH_3$	$-\overset{\displaystyle O}{\overset{\|}{C}}-O-CH_3$	P/Gg^b	–H
Bchl b (purple bacteria)/ 835–850, 1020–1040 nm	$-\overset{\displaystyle O}{\overset{\|}{C}}-CH_3$	$-CH_3{}^c$	$=\overset{\displaystyle H}{\underset{}{C}}-CH_3$	$-CH_3$	$-\overset{\displaystyle O}{\overset{\|}{C}}-O-CH_3$	P	–H
Bchl c (green sulfur bacteria)/745–755 nm	$-\overset{\displaystyle H}{\underset{\displaystyle OH}{C}}-CH_3$	$-CH_3$	$-C_2H_5$ $-C_3H_7{}^d$ $-C_4H_9$	$-C_2H_5$ $-CH_3$	–H	F	$-CH_3$
Bchl c_s (green nonsulfur bacteria)/740 nm	$-\overset{\displaystyle H}{\underset{\displaystyle OH}{C}}-CH_3$	$-CH_3$	$-C_2H_5$	$-CH_3$	–H	S	$-CH_3$
Bchl d (green sulfur bacteria)/705–740 nm	$-\overset{\displaystyle H}{\underset{\displaystyle OH}{C}}-CH_3$	$-CH_3$	$-C_2H_5$ $-C_3H_7$ $-C_4H_9$	$-C_2H_5$ $-CH_3$	–H	F	–H
Bchl e (green sulfur bacteria)/719–726 nm	$-\overset{\displaystyle H}{\underset{\displaystyle OH}{C}}-CH_3$	$-\overset{\displaystyle O}{\overset{\|}{C}}-H$	$-C_2H_5$ $-C_3H_7$ $-C_4H_9$	$-C_2H_5$	–H	F	$-CH_3$
Bchl g (heliobacteria)/ 670, 788 nm	$-\overset{\displaystyle H}{\underset{}{C}}=CH_2$	$-CH_3{}^a$	$-C_2H_5$	$-CH_3$	$-\overset{\displaystyle O}{\overset{\|}{C}}-O-CH_3$	F	–H

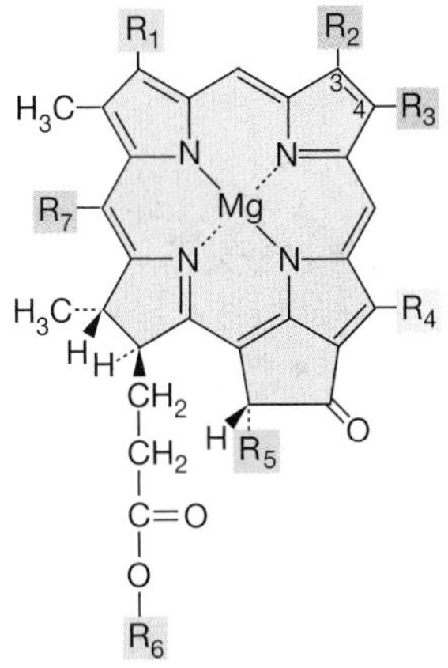

^{a}No double bond between C_3 and C_4; additional H atoms are in positions C_3 and C_4.

^{b}P, Phytyl ester ($C_{20}H_{39}O-$); F, farnesyl ester ($C_{15}H_{25}O-$); Gg, geranylgeraniol ester ($C_{10}H_{17}O-$); S, stearyl alcohol ($C_{18}H_{37}O-$).

^{c}No double bond between C_3 and C_4; an additional H atom is in position C_3.

^{d}Bacteriochlorophylls c, d, and e consist of isomeric mixtures with the different substituents on R_3 as shown.

Figure 14.3 Structure of all known bacteriochlorophylls (Bchl). The different substituents present in the positions R_1 to R_7 in the structure at the right are listed. Absorption properties can be determined by suspending intact cells of a phototroph in a viscous liquid such as 60% sucrose (this reduces light scattering and smooths out spectra) and running absorption spectra as shown in Figure 14.2b. In vivo absorption maxima are the physiologically relevant absorption peaks. The spectrum of bacteriochlorophylls extracted from cells and dissolved in organic solvents is often quite different.

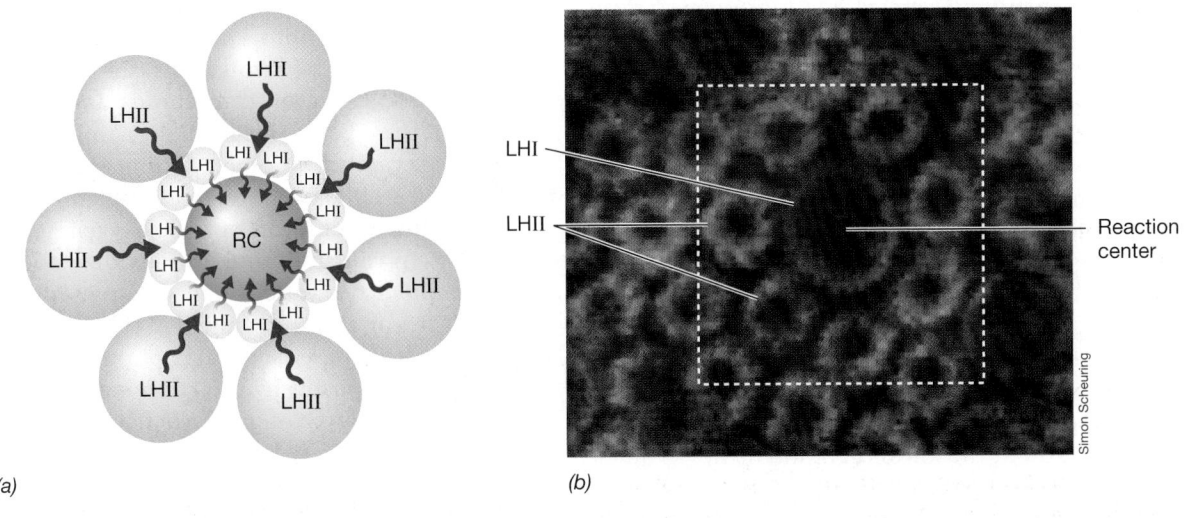

(a)   (b)

Simon Scheuring

LHI

LHII

Reaction center

Figure 14.4 Arrangement of light-harvesting chlorophylls/bacteriochlorophylls and reaction centers within a photosynthetic membrane. (a) Light energy absorbed by light-harvesting (LH) molecules (light green) is transferred to the reaction centers (dark green, RC) where photosynthetic electron transport reactions begin. Pigment molecules are secured within the membrane by specific pigment-binding proteins. Compare this figure to Figure 14.12b. (b) Atomic force micrograph of photocomplexes of the purple bacterium *Phaeospirillum molischianum*. This organism has two types of light-harvesting complexes, LHI and LHII. LHII complexes transfer energy to LHI complexes, and these transfer energy to the reaction center (see Figure 14.11b).

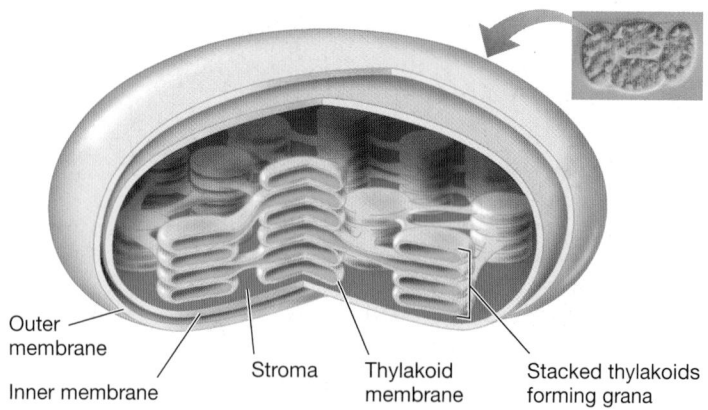

Outer membrane
Inner membrane
Stroma
Thylakoid membrane
Stacked thylakoids forming grana

Figure 14.5 The chloroplast. Details of chloroplast structure, showing how the convolutions of the thylakoid membranes define an inner space called the stroma and form membrane stacks called grana. Inset: Photomicrograph of cells of the green alga *Makinoella*. Each of the four cells in a cluster contains several chloroplasts.

called **thylakoids** (**Figure 14.5**); stacks of thylakoids within the chloroplast form *grana*. The thylakoids are arranged so that the chloroplast is divided into two regions, the matrix space that surrounds the thylakoids, called the *stroma*, and the inner space within the thylakoid array, called the *lumen*. This arrangement makes possible the generation of a light-driven proton motive force that is used to synthesize ATP (Section 14.4).

Chloroplasts are absent from prokaryotic phototrophs. In purple bacteria, the photosynthetic pigments are integrated into internal membrane systems that arise from invagination of the cytoplasmic membrane. Membrane vesicles called *chromatophores* or membrane stacks called *lamellae* are common membrane arrangements in purple bacteria (**Figure 14.6**). In cyanobacteria, photosynthetic pigments reside in lamellar membranes (see Figure 14.10) also called *thylakoids* because of their resemblance to the thylakoids in the chloroplasts of algae (Figure 14.5).

The ultimate structure for capturing energy from low light intensities is the **chlorosome** (**Figure 14.7**). Chlorosomes are present in the anoxygenic green sulfur bacteria (*Chlorobium*, Figure 14.1 and ⮌ Section 15.6), filamentous anoxygenic phototrophs (green nonsulfur bacteria; *Chloroflexus*, ⮌ Section 15.7), and photosynthetic *Acidobacteria* (*Chloracidobacterium*, ⮌ Section 15.8). Chlorosomes function as giant antenna systems, but unlike the antennae of purple bacteria or cyanobacteria, bacteriochlorophyll molecules in the chlorosome are not attached to proteins. Chlorosomes contain bacteriochlorophyll *c*, *d*, or *e* (Figure 14.3) arranged in dense arrays running along the long axis of the structure. Light energy absorbed by these antenna pigments is transferred to bacteriochlorophyll *a* in the reaction center in the cytoplasmic membrane through a small protein called the *FMO protein* (Figure 14.7).

Green bacteria can grow at the lowest light intensities of all known phototrophs and are often found in the deepest waters of lakes, inland seas, and other anoxic aquatic habitats where light levels are too low to support other phototrophs. Green nonsulfur bacteria are major components of microbial mats, thick

biofilms that form in hot springs and highly saline environments (⮌ Section 20.5). Microbial mats experience a steep light gradient, with light levels even a few millimeters into the mat approaching darkness. Hence, chlorosomes allow green nonsulfur bacteria to grow phototrophically with only the minimal light intensities available.

MINIQUIZ

- What is the fundamental difference between an oxygenic and an anoxygenic phototroph?
- What is the purpose of chlorophyll and bacteriochlorophyll molecules? In what ways do they resemble cytochromes and in what ways to they differ?
- Why can phototrophic green bacteria grow at light intensities that will not support purple bacteria?

14.2 Carotenoids and Phycobilins

Although chlorophyll/bacteriochlorophyll is required for photosynthesis, phototrophic organisms contain other pigments as well. These pigments include, in particular, the *carotenoids* and *phycobilins*.

Carotenoids

The most widespread accessory pigments in phototrophs are the **carotenoids**. Carotenoids are hydrophobic pigments that are firmly embedded in the photosynthetic membrane. **Figure 14.8** shows the structure of a common carotenoid, β-carotene. Carotenoids are typically yellow, red, brown, or green and absorb light in the blue region of the spectrum. The major carotenoids of anoxygenic phototrophs are shown in **Figure 14.9**. Because they tend to mask the color of bacteriochlorophylls, carotenoids are responsible for the brilliant colors of red, purple, pink, green, yellow, or brown that are observed in different species of anoxygenic phototrophs (⮌ Figure 15.12).

Carotenoids are closely associated with chlorophyll or bacteriochlorophyll in photosynthetic complexes, and some of the energy absorbed by carotenoids can be transferred to the reaction center. However, carotenoids function primarily as photoprotective agents. Bright light can be harmful to cells because it can catalyze photooxidation reactions that can produce toxic forms of oxygen, such as singlet oxygen ($^1O_2$). Like superoxide and other forms of toxic oxygen (⮌ Section 5.14), singlet oxygen can spontaneously oxidize photocomplexes, rendering them nonfunctional. Carotenoids quench toxic oxygen species by absorbing much of this harmful light and in this way prevent these dangerous photooxidations. Because phototrophic organisms by their very nature must live in the light, the photoprotection conferred by carotenoids is clearly advantageous.

Phycobiliproteins and Phycobilisomes

Cyanobacteria and the chloroplasts of red algae (which are descendants of cyanobacteria, ⮌ Section 18.1) contain pigments called

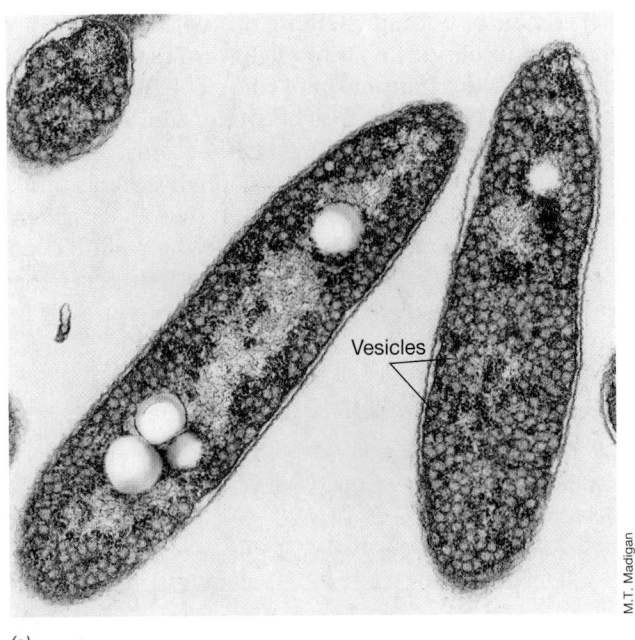

(a)

M.T. Madigan

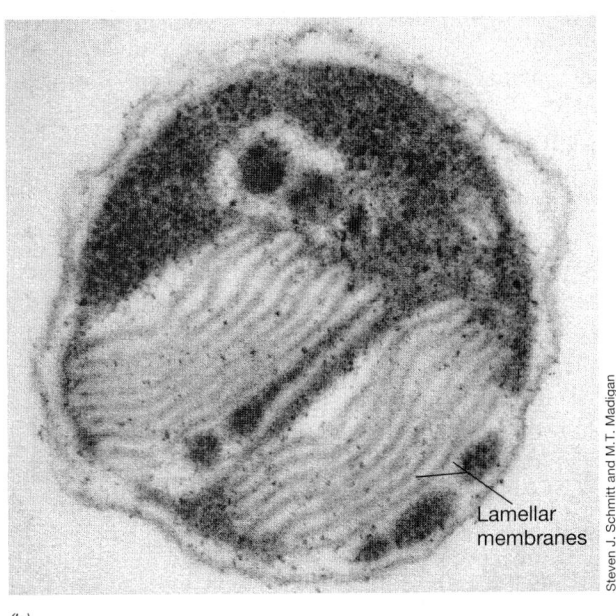

(b)

Steven J. Schmitt and M.T. Madigan

Lamellar
membranes

Figure 14.6 Membranes in anoxygenic phototrophs. *(a)* Chromatophores.
Section through a cell of the purple bacterium *Rhodobacter* showing vesicular
photosynthetic membranes. The vesicles are continuous with and arise by invagination of
the cytoplasmic membrane. A cell is about 1 μm wide. *(b)* Lamellar membranes in the
purple bacterium *Ectothiorhodospira*. A cell is about 1.5 μm wide. These membranes are
also continuous with and arise from invagination of the cytoplasmic membrane, but
instead of forming vesicles, they form membrane stacks.

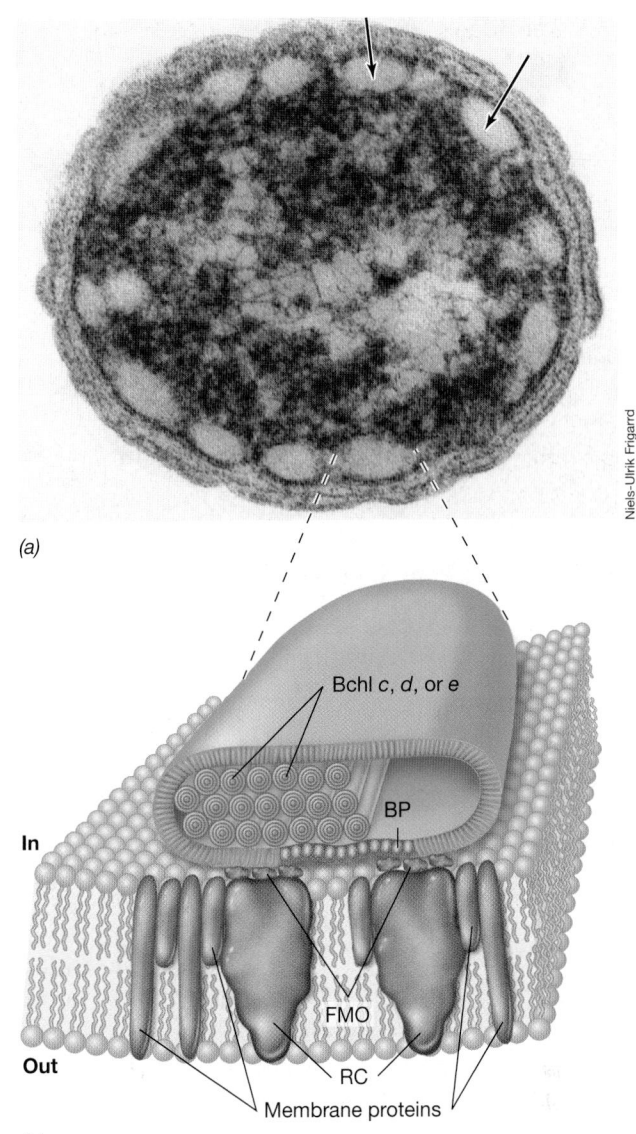

(a)

Niels-Ulrik Frigaard

In

Bchl *c*, *d*, or *e*

BP

Out

FMO

RC

Membrane proteins

(b)

Figure 14.7 The chlorosome of green sulfur and green nonsulfur bacteria.
(a) Transmission electron micrograph of a cross section of a cell of the green sulfur
bacterium *Chlorobaculum tepidum*. Note the chlorosomes (arrows). *(b)* Model of
chlorosome structure. The chlorosome (green) lies appressed to the inside surface of the
cytoplasmic membrane. Antenna bacteriochlorophyll (Bchl) molecules are arranged in
tubelike arrays inside the chlorosome, and energy is transferred from these to reaction
center (RC) Bchl *a* in the cytoplasmic membrane through a protein called FMO. Base
plate (BP) proteins function as connectors between the chlorosome and the cytoplasmic
membrane.

phycobiliproteins, which are the main light-harvesting systems of these phototrophs. Phycobiliproteins consist of red or
blue-green linear tetrapyrroles, called *bilins*, bound to proteins,
and give cyanobacteria and red algae their characteristic colors
(**Figure 14.10**). The red phycobiliprotein, called *phycoerythrin*,
absorbs most strongly at wavelengths around 550 nm, whereas the
blue phycobiliprotein, *phycocyanin* (Figure 14.10*b*), absorbs most

H₃C CH₃   CH₃   CH₃   H₃C
                                              CH₃
              CH₃   CH₃   H₃C CH₃
        CH₃

Figure 14.8 Structure of β-carotene, a typical carotenoid. The conjugated
double-bond system is highlighted in orange.

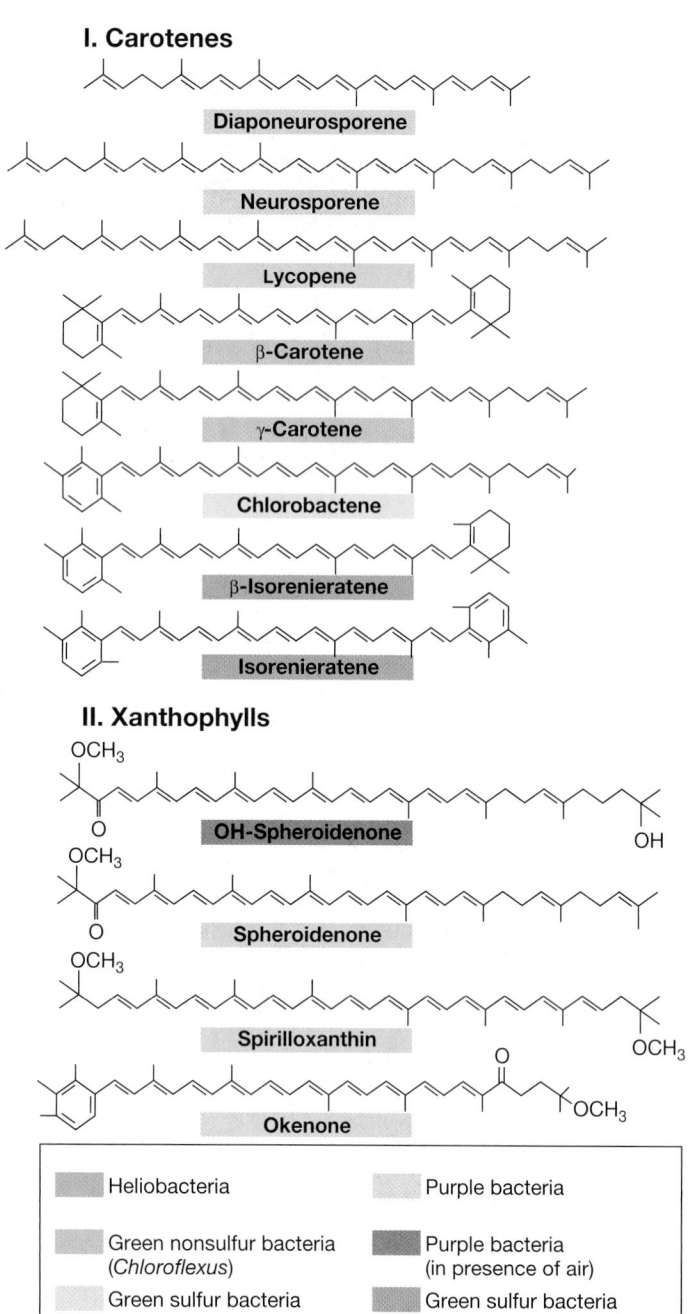

Figure 14.9 Structures of some common carotenoids found in anoxygenic phototrophs. Carotenes are hydrocarbon carotenoids, and xanthophylls are oxygenated carotenoids. Compare the structure of β-carotene shown in Figure 14.8 with how it is drawn here. For simplicity in the structures shown here, methyl (CH₃) groups are designated by bond only.

strongly at 620 nm. A third phycobiliprotein, called *allophycocyanin*, absorbs at about 650 nm.

Phycobiliproteins assemble into aggregates called **phycobilisomes** that attach to cyanobacterial thylakoids (Figure 14.10c). Phycobilisomes are arranged such that the allophycocyanin molecules are in direct contact with the photosynthetic membrane. Allophycocyanin is surrounded by phycocyanin or phycoerythrin (or both, depending on the organism). Phycocyanin and phycoerythrin absorb light of shorter wavelengths (higher energy) and transfer some energy to allophycocyanin, which is positioned closest to the reaction center chlorophyll and transfers energy to it (Figure 14.10b). Thus, in a fashion similar to how antenna bacteriochlorophyll systems function in anoxygenic phototrophs (Figure 14.4), energy transfer proceeds "downhill" from phycobilisomes to the reaction center. Phycobilisomes thus facilitate energy transfer to cyanobacterial reaction centers, allowing cyanobacteria to grow at lower light intensities than would otherwise be possible.

MINIQUIZ

- In which phototrophs are carotenoids found? Phycobiliproteins?
- How does the structure of a phycobilin compare with that of a chlorophyll?
- Phycocyanin is blue-green. What color of light does it absorb?

14.3 Anoxygenic Photosynthesis

In the photosynthetic light reactions, electrons travel through an electron transport chain whose components are arranged in a photosynthetic membrane in order of their increasingly more electropositive reduction potential (E_0'). This generates a proton motive force that drives ATP synthesis. Key parts of this process include photosynthetic reaction centers and photosynthetic membranes (Section 14.1).

Photosynthetic reaction centers are complex macromolecular structures localized within photosynthetic membranes. They are composed of multiple protein subunits and cofactors (including chlorophylls or bacteriochlorophylls), and they interact with both antenna pigments and components of the electron transport chain. Photosynthetic pigments funnel light energy to the reaction center to excite a special pair of chlorophylls (or bacteriochlorophylls), thereby generating high-potential electrons that can be donated to subsequent electron transport reactions. Several different types of reaction centers have been described but all fall into one of two classes; they are either of a *quinone type* (Q-type) or *iron–sulfur type* (FeS-type) depending on the electron acceptor in the reaction center.

Electron Flow in Purple Bacteria

Purple bacteria use a Q-type reaction center which contains three polypeptides, designated L, M, and H. These proteins, along with a molecule of cytochrome *c*, are firmly embedded in the photosynthetic membrane (Figure 14.6) and wind through the membrane several times (**Figure 14.11**). The L, M, and H polypeptides bind two molecules of bacteriochlorophyll *a*, called the *special pair*, two additional bacteriochlorophyll *a* molecules that function in photosynthetic electron flow, two molecules of bacteriopheophytin *a* (bacteriochlorophyll *a* minus its magnesium atom), two molecules of quinone (⊃ Section 3.10), and one carotenoid molecule (Figure 14.11).

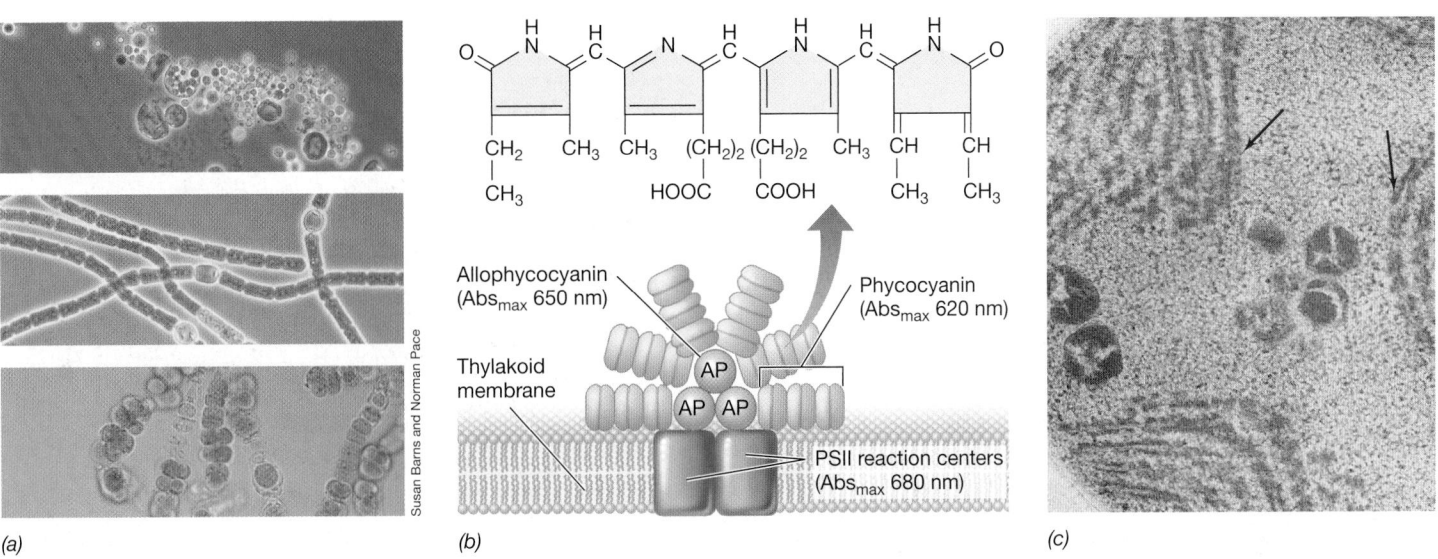

(a)   (b)   (c)

Figure 14.10 Phycobiliproteins and phycobilisomes. (a) Light photomicrographs of cells of the cyanobacteria (top to bottom) *Dermocarpa*, *Anabaena*, and *Fischerella*, showing the typical blue-green color of cells due to phycobiliproteins. (b) Structure of phycocyanin (top) and a phycobilisome. Phycocyanin absorbs at higher energies (shorter wavelengths) than allophycocyanin. Chlorophyll *a* absorbs at longer wavelengths (lower energies) than allophycocyanin. Energy flow is thus phycocyanin → allophycocyanin → chlorophyll *a* of PSII. (c) Electron micrograph of a thin section of the cyanobacterium *Synechocystis*. Note the darkly staining ball-like phycobilisomes (arrows) attached to the lamellar membranes.

Photosynthetic light reactions begin when light energy absorbed by the antenna systems is transferred to the special pair of bacteriochlorophyll *a* molecules (Figure 14.11*a*). This excites the special pair, converting it from a relatively weak to a very strong electron donor (very electronegative E_0', ⮌ Section 3.6). Once this strong donor has been produced, the remaining steps in photosynthetic electron flow are highly reminiscent of those we have seen before in respiration (⮌ Section 3.10 and Figure 3.22); that is, electrons flow through a membrane from carriers of low E_0' to those of high E_0', generating a proton motive force in the process (**Figure 14.12**).

Before excitation, the purple bacterial reaction center, which is called *P870*, has an E_0' of about +0.5 V; after excitation, it has a potential of about −1.0 V (Figure 14.12*a*). An excited electron

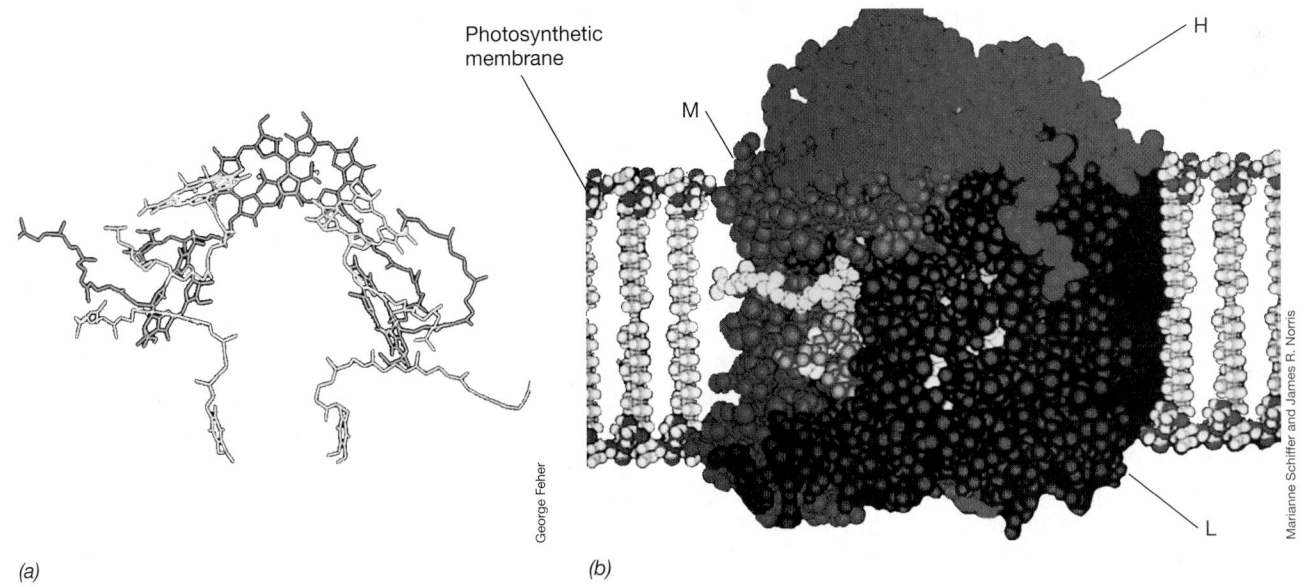

(a)   (b)

Figure 14.11 Structure of the reaction center of a purple phototrophic bacterium. (a) Arrangement of pigment molecules in the reaction center. The "special pair" of bacteriochlorophyll molecules (orange) overlap and occur at the top of the reaction center structural diagram; adjacent to and below the special pair in the diagram are a pair of accessory bacteriochlorophylls (light yellow). Bacteriopheophytin molecules (blue) are arranged below the bacteriochlorophylls, and quinones (dark yellow) are present at the bottom of the structural model. Compare the structure of these molecules in the reaction center to their role in electron transfer (see Figure 14.12). (b) Molecular model of the protein structure of the reaction center. The pigments described in part *a* are bound to membranes by protein H (blue), protein M (red), and protein L (green). The reaction center pigment–protein complex is integrated into the lipid bilayer.

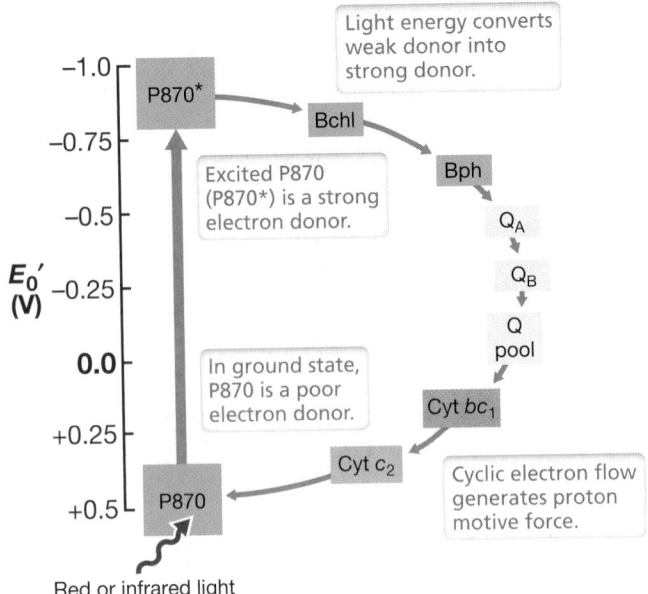

(a) **Electron flow in anoxygenic photosynthesis**

(b) **Arrangement of protein complexes in the purple bacterium reaction center**

Figure 14.12 Electron flow in anoxygenic photosynthesis in a purple bacterium. *(a)* Schematic of electron flow in a purple bacterium. Bchl, bacteriochlorophyll; Bph, bacteriopheophytin; Q_A, Q_B, intermediate quinones; Q pool, quinone pool in membrane; Cyt, cytochrome. *(b)* Arrangement of protein complexes in the purple bacterium reaction center leading to proton motive force (photophosphorylation) by ATPase. Two protons are translocated for every electron that passes from the reaction center to Cyt bc_1. LH, light-harvesting bacteriochlorophyll complexes; RC, reaction center; Bph, bacteriopheophytin; Q, quinone; FeS, iron–sulfur protein; bc_1, cytochrome bc_1 complex; c_2, cytochrome c_2. For a description of ATPase function, see Section 3.11.

within P870 proceeds to reduce a molecule of bacteriochlorophyll *a* within the reaction center (Figures 14.11*a* and 14.12*a*). This transition takes place incredibly fast, taking only about three-trillionths (3×10^{-12}) of a second. Once reduced, bacteriochlorophyll *a* proceeds to reduce bacteriopheophytin *a*, and the latter reduces quinone molecules within the membrane (Figure 14.12). These transitions are also very fast, taking less than one-billionth of a second. From the quinone, electrons are transported through the membrane more slowly (on a millisecond scale) through a series of iron–sulfur proteins and cytochromes (Figure 14.12), eventually returning to the reaction center.

Figure 14.12*b* shows electron flow within the actual context of the photosynthetic membrane. Key electron transport proteins include many that also participate in respiratory electron flow (⟨⟩ Figure 3.22)—cytochrome bc_1 and cytochrome c_2, in particular (Figure 14.12). Cytochrome c_2 is a periplasmic cytochrome (recall that the periplasm is the region between the cytoplasmic membrane and the outer membrane in gram-negative bacteria, ⟨⟩ Section 2.5) that functions as an electron shuttle between the membrane-bound bc_1 complex and the reaction center (Figure 14.12*b*). Electron flow is completed when cytochrome c_2 donates an electron to the special pair to return it to its original ground-state reduction potential. The reaction center can then absorb new light energy and repeat the process.

ATP is synthesized during photosynthetic electron flow from the activity of ATPase that couples the proton motive force to ATP synthesis (⟨⟩ Section 3.11). This mechanism of ATP synthesis is called **photophosphorylation**, specifically *cyclic photophosphorylation*, because electrons move within a closed loop. However, unlike respiration, where there is a net consumption of electrons, in cyclic photophosphorylation there is no net input or consumption of electrons; electrons simply travel a circuitous route, returning from whence they came (Figure 14.12).

Generation of Reducing Power

For a purple bacterium to grow as a photoautotroph, the formation of ATP is not enough. Reducing power (NADH) is also necessary to reduce CO_2 to cell material. Reducing power for purple bacteria can come from many sources, in particular reduced sulfur compounds such as H_2S. When H_2S is the electron donor in purple sulfur bacteria, globules of S^0 are stored inside the cells (Figure 14.1). When S^0 is formed, electrons end up in the "quinone pool" (Figure 14.12). However, the E_0' of quinone (about 0 V) is insufficiently electronegative to reduce NAD^+ (-0.32 V). Hence, electrons from the quinone pool must be forced backwards (against the electrochemical gradient) to reduce NAD^+ to NADH (see Figure 14.13). This energy-requiring process, called **reverse electron transport**, is driven by the energy of the proton motive force. We will see later that reverse electron flow is also the mechanism by which chemolithotrophs obtain reducing power for CO_2 fixation; in many of these cases, the electrons come from electron donors of quite positive E_0' (Sections 14.7–14.15). However, as we will see shortly, reverse electron flow is not necessary in other anoxygenic phototrophs.

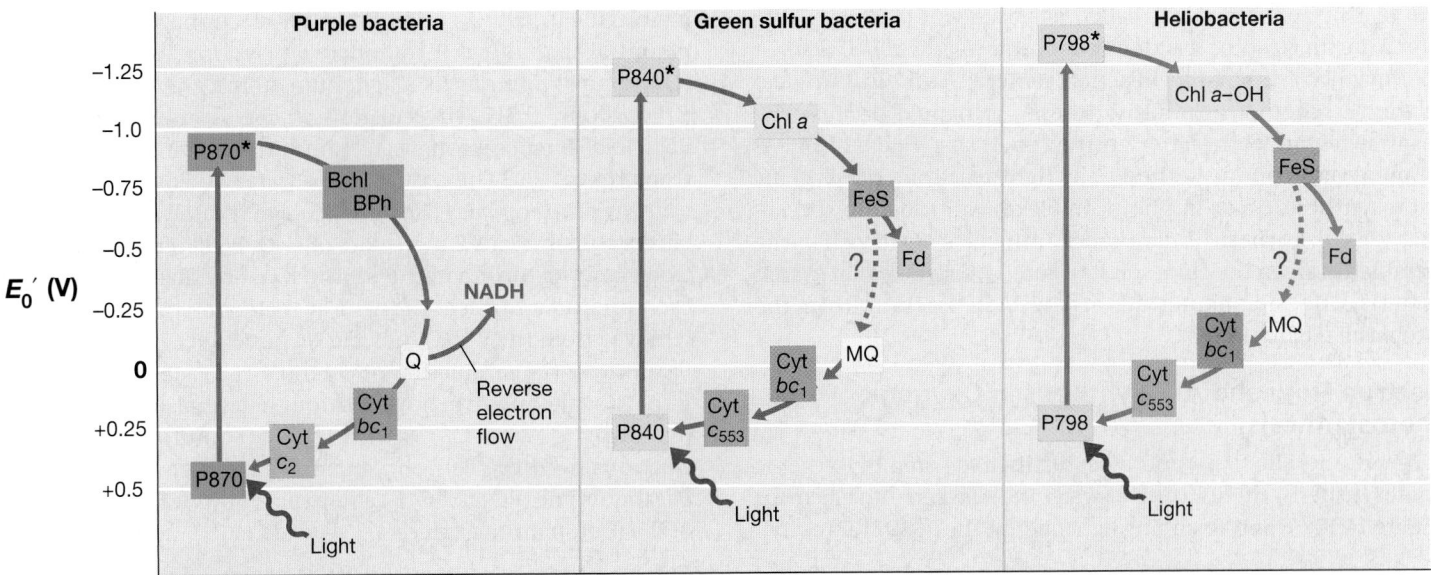

Figure 14.13 A comparison of electron flow in purple bacteria, green sulfur bacteria, and *Heliobacteria*. Reverse electron flow in purple bacteria is necessary to produce NADH because the primary acceptor (quinone, Q) is more positive in potential than the $NAD^+/NADH$ couple. In green sulfur bacteria and *Heliobacteria*, ferredoxin (Fd), whose E_0' is more negative than that of NADH, is produced by light-driven reactions for reducing power needs. Cyclic electron flow in green sulfur bacteria and *Heliobacteria* would require electron transfer from the FeS-type photosystem to the menaquinone pool, but evidence for this mechanism is limited, suggesting noncyclic electron flow in these phototrophs. Bchl, bacteriochlorophyll; BPh, bacteriopheophytin; Q, quinone; MQ, menaquinone. P870 and P840 are reaction centers of purple and green bacteria, respectively, and consist of Bchl *a*. The reaction center of *Heliobacteria* (P798) contains Bchl *g*, and the reaction center of *Chloroflexus* is of the purple bacterial type. Note that forms of chlorophyll *a* are present in the reaction centers of green bacteria and heliobacteria.

Photosynthetic Electron Flow in Other Anoxygenic Phototrophs

Thus far we have focused on electron flow in purple bacteria. Although analogous membrane-associated reactions drive photophosphorylation in other anoxygenic phototrophs, there are significant differences in the details. Both filamentous anoxygenic phototrophs and purple bacteria employ structurally similar Q-type reaction centers, but by contrast the green sulfur bacteria, *Acidobacteria*, and *Heliobacteria* all employ FeS-type reaction centers and this is reflected in differences in electron flow.

Figure 14.13 contrasts photosynthetic electron flow in purple bacteria, green sulfur bacteria, and the *Heliobacteria*. Note that in green sulfur bacteria and *Heliobacteria* the excited state of the reaction center bacteriochlorophylls is significantly *more* electronegative than in purple bacteria and that actual chlorophyll *a* (green sulfur bacteria) or a structurally modified form of chlorophyll *a* (hydroxychlorophyll *a* in *Heliobacteria*) is present in the reaction center. Thus, unlike in purple bacteria, where the first stable acceptor molecule (quinone) has an E_0' of about 0 V (Figure 14.12*a*), the acceptors in green sulfur bacteria and *Heliobacteria* are FeS-proteins that have a much more electronegative E_0' than does NADH. Hence, reverse electron flow is unnecessary in green sulfur bacteria or *Heliobacteria*. The highly electronegative electrons generated by the FeS-type reaction centers are ultimately transferred to a protein called *ferredoxin* ($E_0' = -0.4$ V). In green sulfur bacteria ferredoxin is the direct electron donor for CO_2 fixation (Section 14.5), and electrons from ferredoxin can also pass to ferredoxin–NAD oxidoreductase for the production of NADH.

It remains unclear whether electron transfer in green sulfur bacteria and *Heliobacteria* is cyclic or noncyclic. It has been proposed that the FeS-type reaction centers in these phototrophs can transfer electrons directly to menaquinone, thereby generating a proton motive force resulting in cyclic photophosphorylation as seen in purple bacteria (Figure 14.13). However, little evidence for cyclic photophosphorylation has been observed in green sulfur bacteria and *Heliobacteria*. Alternatively, these phototrophs may employ noncyclic electron flow whereby electrons from external electron donors, such as H_2S, enter at the level of the menaquinone pool. These electrons would be transferred through the reaction center and then to ferredoxin where they would ultimately be channeled into biosynthetic reactions.

MINIQUIZ

- What parallels exist in the processes of photophosphorylation and oxidative phosphorylation?
- What is reverse electron flow and why is it necessary? Which phototrophs need to use reverse electron flow?
- What is the difference between cyclic and noncyclic photophosphorylation?

14.4 Oxygenic Photosynthesis

In contrast to photosynthetic electron flow in *anoxygenic* phototrophs, which have either FeS-type or Q-type photosynthetic reaction centers, *oxygenic* phototrophs have both types of reaction centers. In oxygenic phototrophs, electrons flow through two distinct photosystems called *photosystem I* (*PSI*, or *P700*), which has an FeS-type reaction center, and *photosystem II* (*PSII*, or *P680*), which has a Q-type reaction center. PSI and PSII interact in the "Z scheme" of photosynthesis, so named because the pathway resembles the letter Z turned on its side (**Figure 14.14**). As in anoxygenic photosynthesis, the light reactions in oxygenic photosynthesis occur in photocomplexes embedded in specialized photosynthetic membranes. In eukaryotic cells, the membranes are in the chloroplast (Figure 14.5), whereas in cyanobacteria, the membranes are arranged in stacks within the cytoplasm (Figure 14.10c).

Electron Flow and ATP Synthesis in Oxygenic Photosynthesis

PSII performs the first—and most distinctive—step in oxygenic photosynthesis, the splitting of water into oxygen and electrons (Figure 14.14). Upon absorbing light energy, the P680 chlorophyll

a molecule in PSII is excited to a very electronegative reduction potential that allows it to donate an electron to pheophytin *a* (chlorophyll *a* minus its magnesium atom), a molecule with an E_0' of about −0.5 V. This creates a charge separation that causes P680 to become so strongly electropositive that it can accept electrons from H_2O. The oxidation of water by PSII occurs at the *water-oxidizing complex* (**Figure 14.15**) and is catalyzed by a Mn_4Ca cluster, which binds 2 molecules of H_2O. P680 removes one electron from the Mn_4Ca cluster of the water-oxidizing complex for each photon absorbed. In this way 4 electrons are sequentially removed from the 2 H_2O molecules bound to the Mn_4Ca cluster, resulting in the production of O_2 and 4 H^+. Each electron transferred to pheophytin travels through several other proteins, including Q_A and Q_B, within the PSII photocomplex. Two electrons from the PSII photocomplex are then used to reduce plastoquinone (PQ) to PQH_2, a step that allows for the generation of the proton motive force.

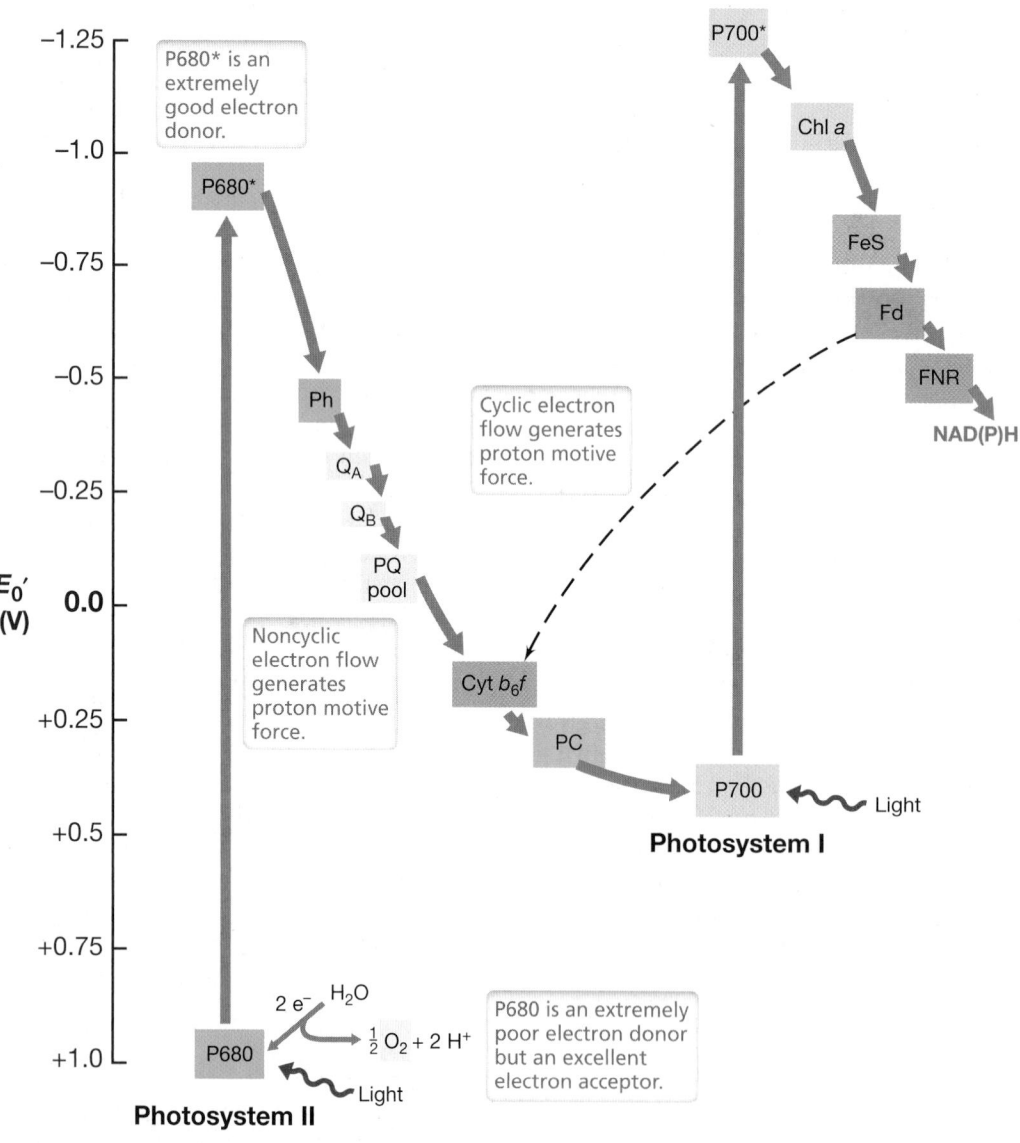

Figure 14.14 Electron flow in oxygenic photosynthesis, the "Z" scheme. Electrons flow through two photosystems, PSI and PSII. Ph, pheophytin; PQ, plastoquinone; Chl, chlorophyll; Cyt, cytochrome; PC, plastocyanin; FeS, nonheme iron–sulfur protein; Fd, ferredoxin; FNR, ferredoxin–NADP oxidoreductase; P680 and P700 are the reaction center chlorophylls of PSII and PSI, respectively. Compare with Figure 14.12a.

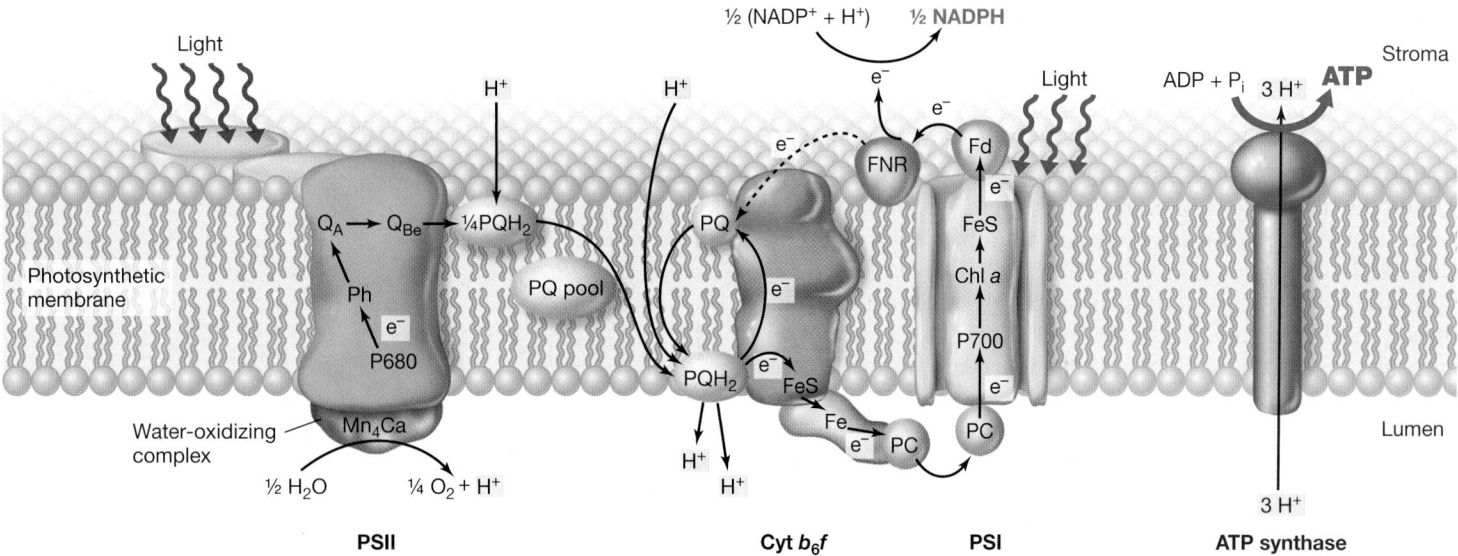

Figure 14.15 Electron transport in oxygenic photosynthesis. Photosystem II (PSII) is activated by photons, causing H_2O to be oxidized on the Mn_4Ca cluster of the water-oxidizing complex. Electrons are transferred from PSII to the plastoquinone pool (PQ/PQH_2). Protons are exchanged across the membrane when plastoquinone is oxidized by cytochrome b_6f. Electrons are then transferred to plastocyanin (PC), which carries them to photosystem I (PSI). Upon activation by light, PSI reduces ferredoxin (Fd), with sequential reduction of ferredoxin: $NADP^+$ oxidoreductase (FNR), and then $NADP^+$. The ATP and NADPH produced by the light reactions are used in CO_2 fixation by the Calvin cycle (see Section 14.5). Cyclic photophosphorylation occurs when FNR donates electrons to cytochrome b_6f instead of to $NADP^+$. During cyclic photophosphorylation, more ATP and less NADPH are produced than during noncyclic photophosphorylation.

The proton motive force is generated in oxygenic photosynthesis by electron transport through quinones and cytochromes of increasingly positive reduction potential. These electron transport reactions are similar to those encountered during our discussion of aerobic respiration (⇌ Section 3.11). Electrons from PQH_2 are transferred through cytochrome b_6f and through a copper-containing protein called *plastocyanin* before being donated to the PSI reaction center (Figure 14.15). The absorption of light by P700 of PSI allows it to accept electrons donated from plastocyanin. Electrons travel through several intermediates in PSI terminating with the reduction of $NADP^+$ to NADPH (Figure 14.14). Two protons are generated for each water molecule that is split by PSII and four protons are translocated across the membrane for every two electrons transferred through the electron transport chain, resulting in a total of 12 protons translocated for every molecule of O_2 produced. This proton motive force is then used by ATP synthase to produce ATP.

Oxygenic photosynthesis results in *noncyclic photophosphorylation* because electrons do not cycle back to reduce the oxidized P680, but instead are used in the reduction of $NADP^+$. However, when the cell requires less NADPH, oxygenic phototrophs can perform *cyclic photophosphorylation*. This occurs when, instead of reducing $NADP^+$, electrons from PSI are returned to the electron transport chain that connects PSII to PSI. When this happens, the recycled electrons can be used to generate a proton motive force that supports additional ATP synthesis (dashed line in Figures 14.14 and 14.15).

Anoxygenic Photosynthesis in Oxygenic Phototrophs

Photosystems I and II normally function in tandem in oxygenic photosynthesis. However, if PSII activity is blocked, some oxygenic phototrophs can perform photosynthesis using only PSI. Under these conditions, cyclic photophosphorylation (Figure 14.14) occurs exclusively, and reducing power for CO_2 reduction comes from sources other than water. In effect, this is anoxygenic photosynthesis occurring in oxygenic phototrophs.

Many cyanobacteria can use H_2S as an electron donor under these conditions and many green algae can use H_2. When H_2S is used, it is oxidized to elemental sulfur (S^0), and sulfur granules similar to those produced by green sulfur bacteria (Figure 14.1) are deposited outside the cyanobacterial cells. **Figure 14.16** shows this in the filamentous cyanobacterium *Oscillatoria limnetica*. This organism lives in anoxic salt ponds where it oxidizes sulfide and carries out anoxygenic photosynthesis along with green and purple bacteria.

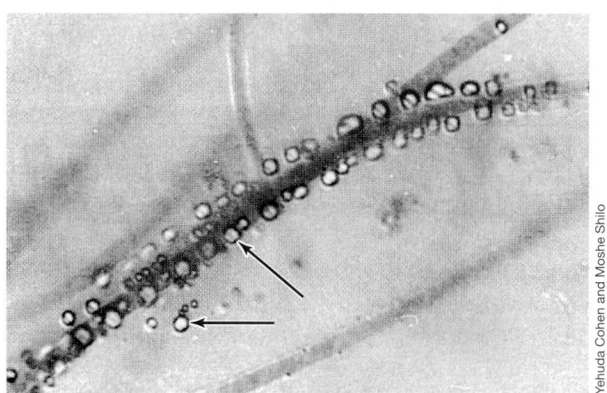

Figure 14.16 Oxidation of H_2S by *Oscillatoria limnetica*. Note the globules of S^0 (arrows), the oxidation product of H_2S, formed outside the cells. *O. limnetica* carries out oxygenic photosynthesis, but cells revert to the anoxygenic process in the presence of H_2S.

From an evolutionary standpoint, the process of cyclic photophosphorylation in both oxygenic and anoxygenic phototrophs is one of many indications of their close relationship. Further evidence of evolutionary relationships among phototrophs can be found in the fact that the structure of the purple bacterial and green nonsulfur photosynthetic reaction centers resembles that of PSII, whereas the structure of the reaction centers of green sulfur bacteria and heliobacteria resembles that of PSI.

Because the evidence is strong that purple and green bacteria preceded cyanobacteria on Earth by perhaps as many as 0.5 billion years (⇌ Section 13.2), it is clear that anoxygenic photosynthesis was the first form of photosynthesis on Earth. The key evolutionary inventions of cyanobacteria were to connect the two forms of reaction centers (as PSI and PSII) and evolve the capacity to use H_2O as a photosynthetic electron donor. The latter was a seminal event in Earth's history since it not only oxygenated the planet but also allowed photoautotrophs to tap an inexhaustible supply of electrons.

MINIQUIZ

- Differentiate between cyclic and noncyclic electron flow in oxygenic photosynthesis.
- What is the key role of light energy in the initial step of the photosynthetic light reactions?
- What evidence is there that anoxygenic and oxygenic photosynthesis are related processes?

II • Autotrophy and N_2 Fixation

All cells require a source of carbon and nitrogen to form cell biomass (⇌ Section 3.1). The atmosphere contains a large reservoir of inorganic carbon, as CO_2, and nitrogen, as N_2. However, these gases must be chemically reduced before they can be assimilated into cell material. The reductive processes associated with the assimilation of CO_2 and N_2 are called CO_2 *fixation* and

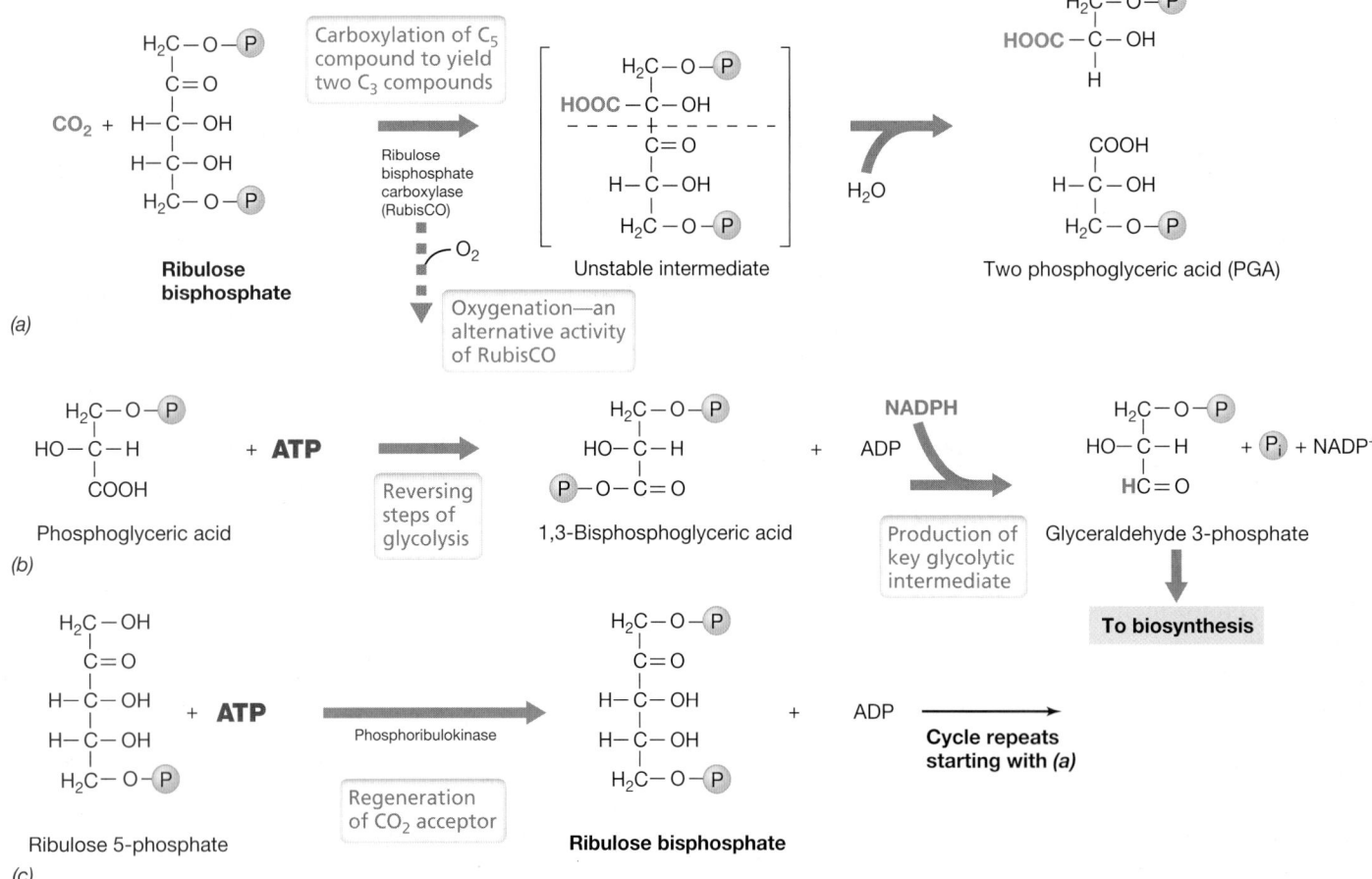

Figure 14.17 Key reactions of the Calvin cycle. *(a)* Reaction of the enzyme ribulose bisphosphate carboxylase. *(b)* Steps in the conversion of 3-phosphoglyceric acid (PGA) to glyceraldehyde 3-phosphate. Note that both ATP and NADPH are required. *(c)* Conversion of ribulose 5-phosphate to the CO_2 acceptor molecule ribulose 1,5-bisphosphate by the enzyme phosphoribulokinase.

N_2 *fixation*, respectively. Both of these pathways require substantial amounts of energy from the cell in the form of ATP and reducing power. These processes evolved very early in the history of life and are widely distributed among species of *Bacteria* and *Archaea*.

14.5 Autotrophic Pathways

Autotrophy is the process by which an energy-poor and highly oxidized form of carbon—CO_2—is reduced and assimilated into cell material. Many microbes are autotrophic, including virtually all phototrophs and chemolithotrophs. In photoautotrophic organisms, the assimilatory reactions responsible for CO_2 fixation are often called the light-independent *dark reactions* of photosynthesis, because they are not inhibited by the absence of light (see Section 14.1).

In oxygenic photosynthesis, CO_2 is reduced to the level of glyceraldehyde 3-phosphate by the **Calvin cycle**. Although the Calvin cycle is the most widespread and important pathway of CO_2 fixation in the biosphere, many autotrophic *Bacteria* and *Archaea* have evolved alternative pathways for fixing CO_2. These alternative autotrophic pathways all ultimately reduce CO_2 to the level of acetyl-coenzyme A (acetyl-CoA), a central metabolite that feeds into all major biosynthetic pathways (⇌ Section 3.9). We begin our discussion of autotrophy by focusing on the Calvin cycle.

The Calvin Cycle

The Calvin cycle is present in purple bacteria, cyanobacteria, algae, green plants, most chemolithotrophic *Bacteria*, and a few *Archaea*. The cycle requires CO_2, a CO_2-acceptor molecule, NADPH, ATP, and two key enzymes, *ribulose bisphosphate carboxylase* and *phosphoribulokinase*. The first step in the Calvin cycle is catalyzed by the enzyme ribulose bisphosphate carboxylase, **RubisCO** for short. RubisCO catalyzes the formation of two molecules of 3-phosphoglyceric acid (PGA) from ribulose bisphosphate and CO_2 as shown in **Figure 14.17a**. The PGA is then phosphorylated and reduced to a key intermediate of glycolysis, glyceraldehyde 3-phosphate. From this, glucose can be formed by reversal of the early steps in glycolysis (⇌ Figure 3.14).

Instead of focusing on the incorporation of a single molecule of CO_2, it is easiest to consider Calvin cycle reactions based on the incorporation of 6 molecules of CO_2, as this is what is required to make one hexose ($C_6H_{12}O_6$). For RubisCO to incorporate 6 molecules of CO_2, 6 molecules of ribulose bisphosphate (total, 30 carbons) are required; carboxylation of these yields 12 molecules of PGA (total, 36 carbon atoms) (**Figure 14.18**). These then form the carbon skeletons for the eventual synthesis of 6 molecules of ribulose bisphosphate (total, 30 carbons) plus one hexose (6 carbons) for cell biosynthesis. A series of biochemical rearrangements between various sugars follow, resulting in 6 molecules of ribulose 5-phosphate (30 carbons). The final step in the Calvin cycle is the phosphorylation of each of these by the enzyme phosphoribulokinase (Figures 14.17c and 14.18) to regenerate 6 molecules of the acceptor molecule, ribulose bisphosphate. All totaled, *12 NADPH and 18 ATP* are required to synthesize one glucose from 6 CO_2 by the Calvin cycle.

Many Calvin cycle autotrophs produce polyhedral cell inclusions called **carboxysomes**. These inclusions, about 100 nm in

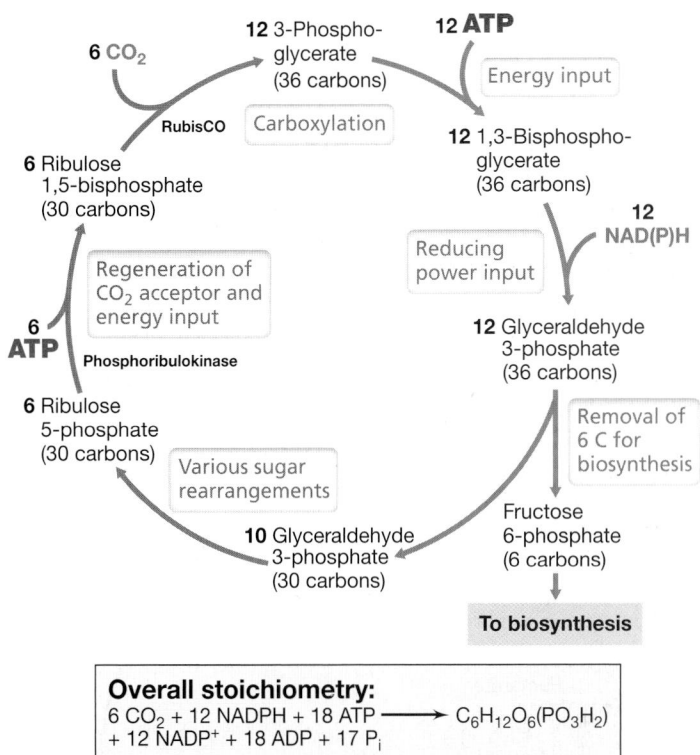

Overall stoichiometry:
$$6\ CO_2 + 12\ NADPH + 18\ ATP \longrightarrow C_6H_{12}O_6(PO_3H_2)$$
$$+ 12\ NADP^+ + 18\ ADP + 17\ P_i$$

Figure 14.18 The Calvin cycle. Shown is the production of one hexose molecule from CO_2. For each six molecules of CO_2 incorporated, one fructose 6-phosphate is produced. In phototrophs, ATP comes from photophosphorylation and NAD(P)H from light or reverse electron flow.

diameter, are surrounded by a thin, protein shell and consist of a crystalline array of RubisCO (**Figure 14.19**); about 250 RubisCO molecules are present per carboxysome. Carboxysomes function to improve the efficiency of RubisCO. Though CO_2 is the actual substrate for RubisCO, inorganic carbon is first incorporated into the cell as bicarbonate (HCO_3^-); the latter diffuses into carboxysomes and is converted to CO_2 by the enzyme *carbonic anhydrase*. CO_2 cannot escape the carboxysome and can be maintained at high concentration therein, and this increases the efficiency of CO_2 fixation. RubisCO can react with either CO_2 or O_2, and O_2 competes with CO_2 for access to the enzyme (Figure 14.17a), lowering the efficiency of CO_2 fixation. However, the protein

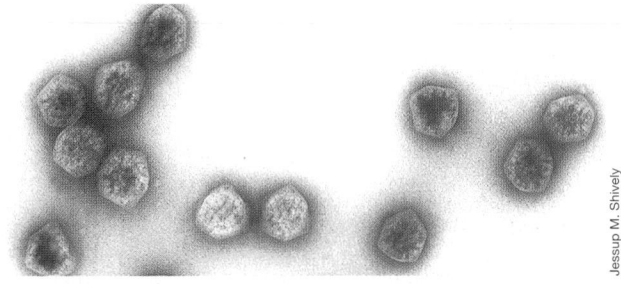

Figure 14.19 Crystalline Calvin cycle enzymes: Carboxysomes. Electron micrograph of carboxysomes purified from the chemolithotrophic sulfur oxidizer *Halothiobacillus neapolitanus*. The structures are about 100 nm in diameter. Carboxysomes are present in a wide variety of obligately autotrophic aerobic *Bacteria*.

shell of the carboxysome protects RubisCO from O_2 and maintains CO_2 at the high levels necessary for efficient CO_2 fixation. Plants lack carboxysomes and RubisCO in plants is instead concentrated in the stroma of chloroplasts (Figure 14.5). The stroma does not provide protection from O_2, however, and so the efficiency of CO_2 fixation in plants declines at high rates of O_2 production.

The Reverse Citric Acid Cycle

Not all phototrophic organisms rely on the Calvin cycle for CO_2 fixation. The **reverse citric acid cycle** (also called the *reductive TCA cycle*) is a pathway of CO_2 fixation that is used by green sulfur bacteria such as *Chlorobium* (Figure 14.1). In the reverse citric acid cycle, CO_2 is reduced by a reversal of steps in the citric acid cycle

($\rightleftarrows$ Section 3.9) (**Figure 14.20a**). The reverse citric acid cycle is more efficient than the Calvin cycle, requiring only 4 NADH, 2 reduced ferredoxins, and 10 ATP to synthesize one glucose from 6 CO_2 in contrast to the 18 ATP required by the Calvin cycle.

As the name implies, most of the reactions of the reverse citric acid cycle are catalyzed by reverse reactions of enzymes of the citric acid cycle. However, the cycle requires the activity of several unique enzymes. These include in particular the enzymes α-ketoglutarate synthase and pyruvate synthase, which catalyze the reductive fixation of CO_2 using electrons supplied by reduced ferredoxin. Ferredoxin is an iron–sulfur protein of very electronegative E_0', about −0.4 V, and is produced in the light reactions of green sulfur bacteria (Figure 14.13). These two ferredoxin-linked reactions are (1) the carboxylation of succinyl-CoA to α-ketoglutarate, and

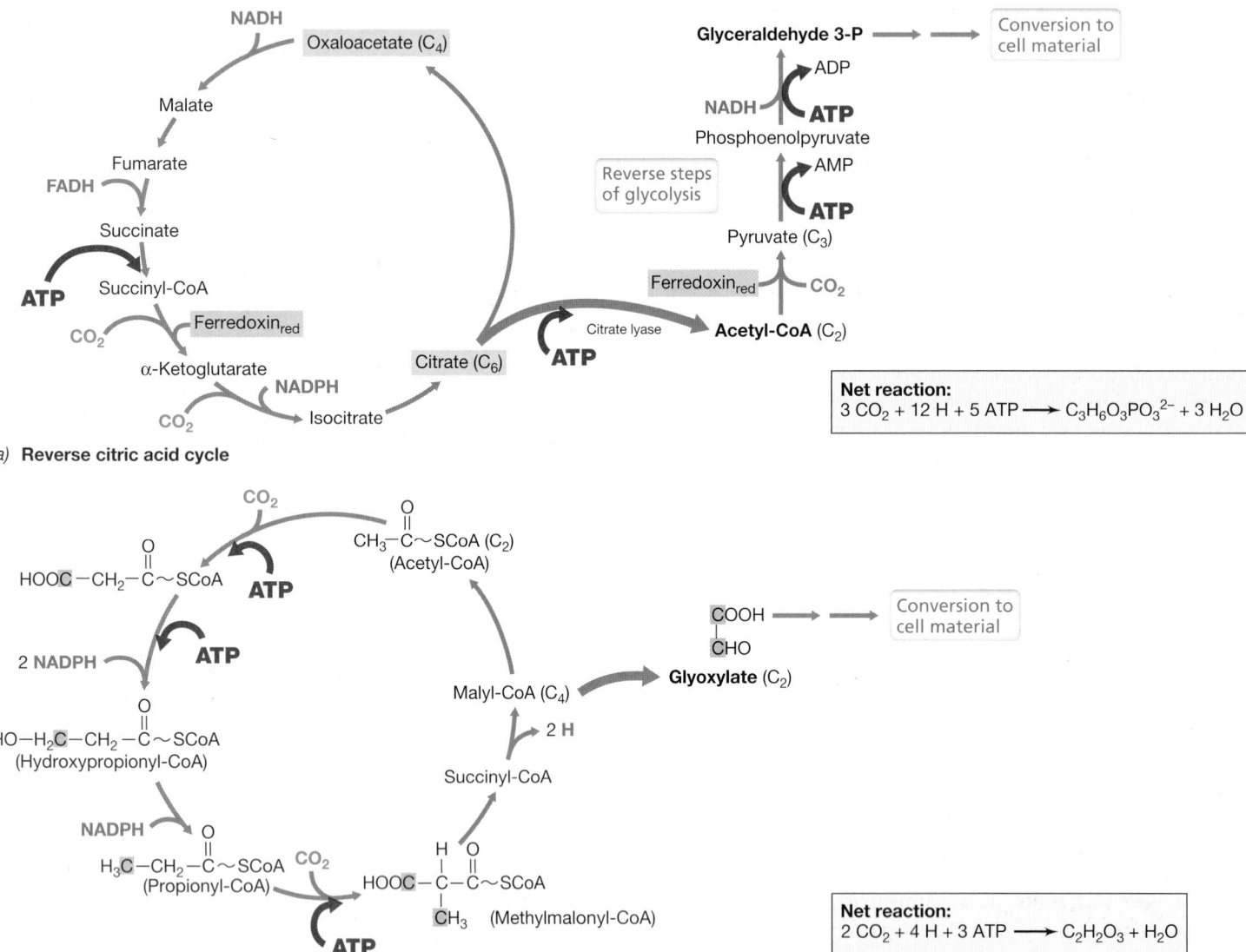

(a) **Reverse citric acid cycle**

Net reaction:
$3 CO_2 + 12 H + 5 ATP \longrightarrow C_3H_6O_3PO_3^{2-} + 3 H_2O$

(b) **Hydroxypropionate pathway**

Net reaction:
$2 CO_2 + 4 H + 3 ATP \longrightarrow C_2H_2O_3 + H_2O$

Figure 14.20 Unique autotrophic pathways in phototrophic green bacteria. *(a)* The reverse citric acid cycle is the mechanism of CO_2 fixation in green sulfur bacteria. Ferredoxin$_{red}$ indicates carboxylation reactions requiring reduced ferredoxin (2 H each). Starting from oxaloacetate, each turn of the cycle results in three molecules of CO_2 being incorporated and pyruvate as the product. *(b)* The hydroxypropionate pathway is the autotrophic pathway in the green nonsulfur bacterium *Chloroflexus*. Acetyl-CoA is carboxylated twice to yield methylmalonyl-CoA. This intermediate is rearranged to yield a new acetyl-CoA acceptor molecule and a molecule of glyoxylate, which is converted to cell material.

(2) the carboxylation of acetyl-CoA to pyruvate (Figure 14.20a). The reverse citric acid cycle also replaces the enzyme *citrate synthase* from the citric acid cycle (⥱ Figure 3.16) with the enzyme *citrate lyase* (an ATP-dependent enzyme that cleaves citrate into acetyl-CoA and oxaloacetate), and the enzyme *succinate dehydrogenase* from the citric acid cycle by *fumarate reductase* in the reverse cycle (Figure 14.20a).

The reverse citric acid cycle operates in certain nonphototrophic autotrophs as well. For example, the hyperthermophilic chemolithotrophs *Thermoproteus* and *Sulfolobus* (*Archaea*; ⥱ Section 17.9) and *Aquifex* (*Bacteria*; ⥱ Section 16.19) use the reverse citric acid cycle, as do certain mesophilic sulfur chemolithotrophic *Bacteria*, such as *Sulfurimonas*. Thus, this pathway, originally discovered in green sulfur bacteria, is likely distributed among several groups of autotrophic microbes.

Other Pathways of CO_2 Fixation

In addition to the Calvin cycle and the reverse citric acid cycle, at least four other pathways of CO_2 fixation are known. The filamentous anoxygenic phototroph *Chloroflexus* (⥱ Section 15.7) grows autotrophically with either H_2 or H_2S as electron donor. However, neither the Calvin cycle nor the reverse citric acid cycle operates in this organism. Instead, two molecules of CO_2 are reduced to glyoxylate by the **3-hydroxypropionate bi-cycle**. This cycle is so named because hydroxypropionate, a three-carbon compound, is a key intermediate and it couples two cycles, one that fixes two molecules of bicarbonate into one molecule of glyoxylate and a second that adds a third molecule of bicarbonate to ultimately yield pyruvate, which can then be funneled into biosynthetic reactions (Figure 14.20b). The 3-hydroxypropionate bi-cycle is somewhat less efficient than the reverse citric acid cycle, requiring 16 ATP per glucose generated.

In phototrophic bacteria, the 3-hydroxypropionate bi-cycle is found in *Chloroflexus*, thought to be one of the earliest phototrophs on Earth. This suggests that the hydroxypropionate pathway may have been one of the earliest mechanisms, if not *the* earliest, for autotrophy in anoxygenic phototrophs. In addition to *Chloroflexus*, the hydroxypropionate bi-cycle operates in several hyperthermophilic *Archaea*, including *Metallosphaera*, *Acidianus*, and *Sulfolobus*. These are all chemolithotrophs that lie near the base of the phylogenetic tree of *Archaea* (Chapter 17). The evolutionary roots of the hydroxypropionate pathway may thus be very deep, and it is possible that this pathway was nature's first attempt at autotrophy.

Other routes of CO_2 fixation include the **3-hydroxypropionate/4-hydroxybutyrate cycle** and the **dicarboxylate/4-hydroxybutyrate cycle**. These pathways of CO_2 fixation are found among diverse autotrophic species of *Archaea*. These cycles each include two connected pathways in which bicarbonate and/or CO_2 is converted into acetyl-CoA for use in biosynthesis. These pathways are named for their key intermediates, which include 3-hydroxypropionate, 4-hydroxybutyrate, and C_4 dicarboxylic acids.

The final pathway of CO_2 fixation is the **reductive acetyl-coenzyme A pathway**. This pathway is found in obligate anaerobes including methanogenic *Archaea*, diverse acetogens, and *Planctomyces* that carry out the anammox reaction (all of these are described in this chapter). The reductive acetyl-coenzyme A pathway is the most efficient of all the CO_2 fixation pathways and requires only 6–8 molecules of ATP per 6 molecules of CO_2 fixed. It is

also the only pathway of CO_2 fixation that can be coupled directly to energy conservation and we will consider it in detail in Section 14.16.

MINIQUIZ

- What reaction(s) does the enzyme RubisCO carry out?
- How much NADPH and ATP is required to make one hexose molecule by the Calvin cycle?
- Contrast autotrophy in the following phototrophs: cyanobacteria; purple and green sulfur bacteria; *Chloroflexus*.

14.6 Nitrogen Fixation

In addition to carbon, cells need a significant amount of nitrogen to synthesize proteins, nucleic acids, and many other organic molecules. Most microbes obtain this nitrogen from "fixed" forms of N in their environment, such as ammonia (NH_3) or nitrate (NO_3^-). However, many *Bacteria* and *Archaea* can form ammonia from gaseous dinitrogen (N_2), a process called **nitrogen fixation**. The ammonia produced is then assimilated into organic form. The ability to fix nitrogen frees an organism from a dependence on fixed nitrogen and confers a significant ecological advantage when fixed nitrogen is limiting. The process of nitrogen fixation is also of enormous agricultural importance, as it supports the nitrogen needs of key crops, such as soybeans and alfalfa.

Only certain species of *Bacteria* and *Archaea* can fix nitrogen, and a list of some important nitrogen-fixing organisms is given in **Table 14.1**. Some nitrogen-fixing bacteria are *free-living* and carry

TABLE 14.1 Some nitrogen-fixing organisms[a]

Free-living aerobes		
Chemoorganotrophs	**Phototrophs**	**Chemolithotrophs**
Azotobacter	Cyanobacteria (e.g.,	*Alcaligenes*
Azomonas	*Anabaena*,	*Acidithiobacillus*
Azospirillum	*Nostoc*,	
Klebsiella[b]	*Gloeothece*,	
Methylomonas	*Aphanizomenon*)	
Free-living anaerobes		
Chemoorganotrophs	**Phototrophs**	**Chemolithotrophs**[c]
Clostridium	Purple bacteria (e.g.,	*Methanosarcina*
Desulfotomaculum	*Chromatium*,	*Methanocaldococcus*
	Methanococcus,	
	Rhodobacter)	
	Green sulfur bacteria	
	(e.g., *Chlorobium*)	
	Heliobacteria	
Symbiotic		
With leguminous plants	**With nonleguminous plants**	
Soybeans, peas, clover, etc. with *Rhizobium*, *Bradyrhizobium*, *Sinorhizobium*	Alder, bayberry, autumn olive, many other bushy plants, with the actinomycete *Frankia*	

[a]Only some common genera are listed in each category; many other genera are known.
[b]Nitrogen fixation occurs only under anoxic conditions.
[c]All are *Archaea*.

out the process completely independently. By contrast, others are *symbiotic* and fix nitrogen only in association with certain plants (↩ Section 23.3). However, in symbiotic nitrogen fixation, it is the bacterium, not the plant, that fixes N_2; no eukaryotic organisms are known to fix nitrogen.

Nitrogenase

Nitrogen fixation is catalyzed by an enzyme complex called **nitrogenase**. Nitrogenase consists of two proteins, *dinitrogenase* and *dinitrogenase reductase*. Both proteins contain iron, and dinitrogenase contains molybdenum as well. The iron and molybdenum in dinitrogenase are part of the enzyme cofactor called the *iron–molybdenum cofactor* (*FeMo-co*), and reduction of N_2 occurs at this site. The composition of FeMo-co is $MoFe_7S_8•$homocitrate (**Figure 14.21**). Two "alternative" nitrogenases are known that lack molybdenum. These contain either vanadium (V) plus iron or iron-only in their cofactors and are made by certain nitrogen-fixing bacteria when molybdenum is limiting in their environment (↩ Section 15.12).

Nitrogen fixation is inhibited by oxygen (O_2) because dinitrogenase reductase is irreversibly inactivated by O_2. Nevertheless, many nitrogen-fixing bacteria are obligate aerobes. In these organisms, nitrogenase is protected from oxygen inactivation by a combination of the rapid removal of O_2 by respiration and the production of O_2-retarding slime layers (**Figure 14.22**). In heterocystous cyanobacteria, nitrogenase is protected by its localization in a differentiated cell called a *heterocyst* (Figure 14.22c; ↩ Section 15.3). Inside the heterocyst, conditions are anoxic, while in neighboring vegetative cells, conditions are just the opposite because oxygenic photosynthesis is occurring. Oxygen production is shut down in the heterocyst, thus protecting it as a dedicated site for N_2 fixation (↩ Section 7.8).

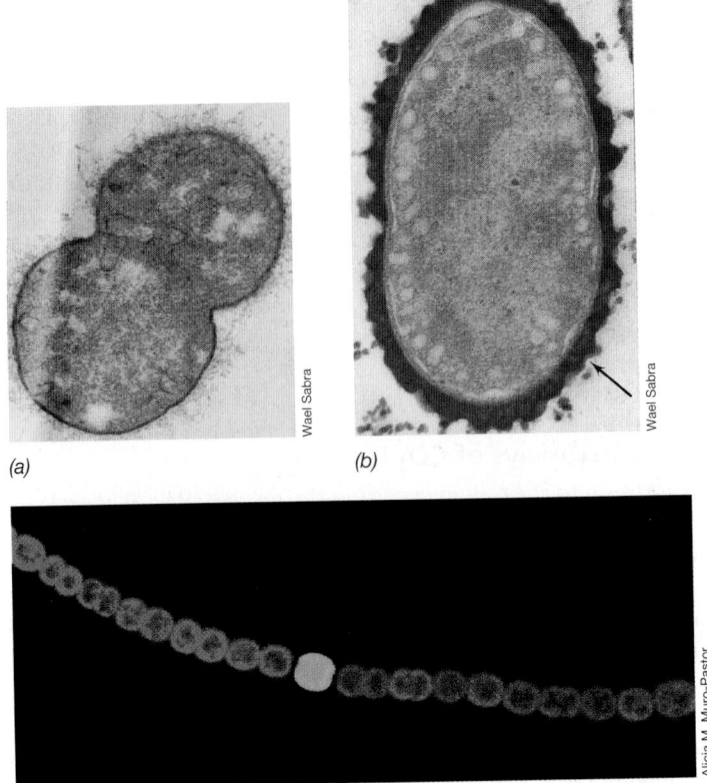

(a)   *(b)*

(c)

Figure 14.22 Two ways of protecting nitrogenase from O₂. Induction of slime formation by O_2 is demonstrated by comparing transmission electron micrographs of *(a)* nitrogen-fixing *Azotobacter vinelandii* cells grown with 2.5% O_2 and showing very little slime with *(b)* *A. vinelandii* cells grown in air (21% O_2) and showing an extensive darkly staining slime layer (arrow). The slime retards diffusion of O_2 into the cell, thus preventing nitrogenase inactivation by O_2. A single cell of *A. vinelandii* is about 2 μm in diameter. *(c)* Fluorescence photomicrograph of cells of the filamentous cyanobacterium *Anabaena* showing a single heterocyst (green). The heterocyst is a differentiated cell that specializes in nitrogen fixation and protects nitrogenase from O_2 inactivation.

Electron Flow in Nitrogen Fixation

Owing to the stability of the triple bond in N_2, its activation and reduction is very energy demanding. Six electrons are needed to reduce N_2 to NH_3, and the successive reduction steps occur directly on nitrogenase with no free intermediates accumulating (**Figure 14.23**). Although only *six* electrons are necessary to reduce N_2 to two NH_3, *eight* electrons are actually consumed in the process, two electrons being lost as H_2 for each mole of N_2 reduced. For unknown reasons, H_2 evolution is an obligatory step in nitrogen fixation and occurs in the first round of the nitrogenase reduction cycle. Following this, N_2 is reduced in successive steps, and ammonia is the released product (Figure 14.23).

The sequence of electron transfer in nitrogenase is as follows: electron donor → dinitrogenase reductase → dinitrogenase → N_2. The electrons for N_2 reduction are transferred to dinitrogenase reductase from the low-potential iron–sulfur proteins ferredoxin or flavodoxin (↩ Section 3.10). In addition to electrons, ATP is required for nitrogen fixation. ATP binds to dinitrogenase reductase, and, following its hydrolysis to ADP, lowers

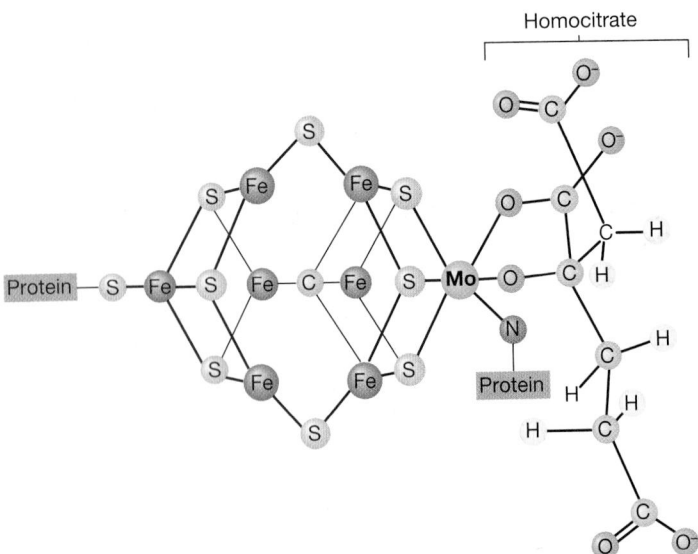

Homocitrate

Figure 14.21 FeMo-co, the iron–molybdenum cofactor from nitrogenase. On the left is the Fe_7S_8 cube that binds to Mo along with O atoms from homocitrate (right, all O atoms shown in purple) and N and S atoms from dinitrogenase.

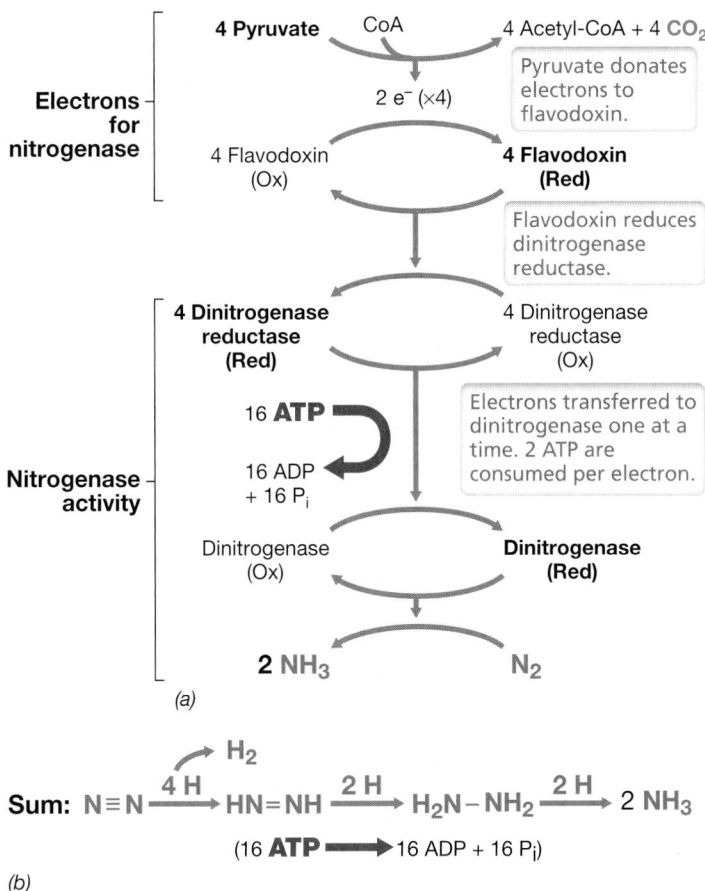

(a)

(b)

$$\text{Sum: } N\equiv N \xrightarrow{\text{4 H}} HN=NH \xrightarrow{\text{2 H}} H_2N-NH_2 \xrightarrow{\text{2 H}} 2\ NH_3$$

$$(16\ \textbf{ATP} \longrightarrow 16\ ADP + 16\ P_i)$$

Figure 14.23 Biological nitrogen fixation by nitrogenase. The nitrogenase complex is composed of dinitrogenase and dinitrogenase reductase. Electrons from reduced ferredoxin are used to reduce dinitrogenase reductase and these electrons in turn reduce dinitrogenase at the expense of ATP. Ditrogenase ultimately donates these electrons to N_2 at the active site of the enzyme, resulting in the formation of 2 NH_3. The iron-molybdenum cofactor (FeMo-co, Figure 14.21) is part of dinitrogenase.

the reduction potential of the protein. This allows dinitrogenase reductase to interact with and reduce dinitrogenase. Electrons are transferred from dinitrogenase reductase to dinitrogenase one at a time, and each cycle of reduction requires two ATP. Thus a total of 16 ATP are required for the reduction of N_2 to 2 NH_3 (Figure 14.23).

Assaying Nitrogenase: Acetylene Reduction

Nitrogenases are not entirely specific for N_2 and also reduce other triply bonded compounds, such as acetylene (HC≡CH). The reduction of acetylene by nitrogenase is only a two-electron process, and *ethylene* ($H_2C=CH_2$) is the final product. However, the reduction of acetylene to ethylene provides a simple, sensitive, and rapid method for measuring nitrogenase activity (**Figure 14.24**). This technique, known as the *acetylene reduction assay*, is widely used in microbiology to detect and quantify nitrogen fixation.

Although the reduction of acetylene is taken as strong proof of N_2 fixation, definitive proof requires an isotope of nitrogen, $^{15}N_2$, as a tracer. If a culture or natural sample is enriched with $^{15}N_2$ and incubated, the production of $^{15}NH_3$ is firm evidence of nitrogen fixation. Nevertheless, acetylene reduction is a more rapid and sensitive method for measuring N_2 fixation and can easily be used in laboratory studies of pure cultures or ecological studies of nitrogen-fixing bacteria directly in their habitat. To do this, a sample, which may be soil, water, or a culture, is incubated in a vessel with HC≡CH and the gas phase is later analyzed by gas chromatography for the production of $H_2C=CH_2$ (Figure 14.24).

MINIQUIZ

- Write a balanced equation for the reaction catalyzed by nitrogenase.
- What is FeMo-co and what does it do?
- How is acetylene useful in studies of nitrogen fixation?

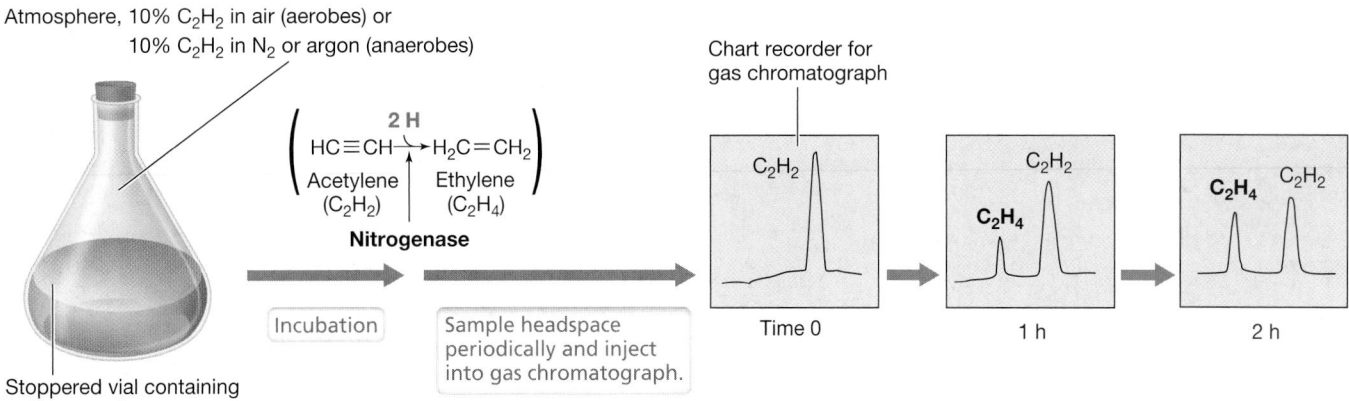

Figure 14.24 The acetylene reduction assay of nitrogenase activity in nitrogen-fixing bacteria. The results show no ethylene (C_2H_4) at time 0 but increasing production of C_2H_4 as the assay proceeds. As C_2H_4 is produced, a corresponding amount of C_2H_2 is consumed.

III • Respiratory Processes Defined by Electron Donor

As we learned in Chapter 3, energy is conserved in respiration by redox reactions that transfer electrons from an initial electron donor to a final electron acceptor. A tremendous diversity of respirations exist and the microbes that carry them out are typically characterized by the nature of their electron donors and/or electron acceptors. Some of these microbes oxidize organic compounds (*chemoorganotrophs*), whereas others oxidize inorganic compounds (*chemolithotrophs*) and obtain their carbon from CO_2. We first review some basic bioenergetics that apply to all forms of respiration.

14.7 Principles of Respiration

In all forms of respiration, low-potential electron donors (those that are more electronegative) are oxidized and the resulting electrons are driven through electron transport chains by their affinity for a high-potential electron acceptor (those that are more electropositive); the latter are ultimately reduced. Electron transport chains include components such as cytochromes, quinones, and iron–sulfur proteins (Section 3.10); collectively, these proteins harness the flow of electrons to generate a proton motive force. Finally, ATP synthesis is driven by an ion gradient through the activity of ATP synthase (Section 3.11).

Energetics of Respiration

Recall from Chapter 3 that the tendency of a substance to donate or accept electrons is defined by its reduction potential (E_0') and that reactions in which electrons are transferred are called *redox reactions*. Respiration can be understood as the coupling of two redox half reactions. Redox half reactions are reversible, and in a coupled reaction, the more electronegative (lower reduction potential) half reaction will proceed as an oxidation (the electron donor) while the more electropositive (higher reduction potential) half reaction will proceed as a reduction (the electron acceptor). The farther apart the two half reactions are in terms of the E_0' of their redox couples, the greater the amount of energy released (Figure 3.10). Based on this simple principle, a wide diversity of organic or inorganic electron donors can be coupled to terminal electron acceptors in the various forms of respiration (**Figure 14.25** and **Table 14.2**). These reactions can support growth provided that sufficient energy is released for the production of ATP (the energy-rich phosphate bond of ATP has a free energy of −31.8 kJ/mol).

Aerobic and Anaerobic Respiration

Respiration can occur under both oxic and anoxic conditions. *Anaerobic respirations* are those that have electron acceptors other than oxygen. Anaerobic respirations are distinct from fermentations because fermentations do not require an external electron acceptor and they generate ATP as a result of substrate-level phosphorylation, whereas respirations generate ATP by harnessing an ion motive force. From Figure 14.25 we can see that almost any half reaction can serve as an electron acceptor provided it is coupled with a sufficiently electronegative electron donor (that is, an electron donor with a lower reduction potential). The energy released from the oxidation of an electron donor using O_2 as electron acceptor is greater than if the same compound is oxidized with an alternate electron acceptor (Figure 3.10). These energy differences are dictated by the reduction potentials of each acceptor (Figure 14.25).

Because the O_2/H_2O couple is most electropositive, more energy is available from aerobic respiration than from anaerobic respiration. This means that for a given electron donor, aerobic organisms will always be able to conserve more energy—and will therefore outcompete—anaerobic organisms, and this is why aerobic respiration has been the dominant form of respiration on Earth ever since the Great Oxidation Event (Section 13.2). However, because oxygen is such a good electron acceptor, and because it is poorly soluble in water, it can be rapidly consumed; hence, anoxic habitats remain widespread in nature and are habitats for anaerobic microbes.

TABLE 14.2 Energy yields from the oxidation of various inorganic electron donors[a]

Electron donor	Chemolithotrophic reaction	Group of chemolithotrophs	E_0' of couple (V)	$\Delta G^{0'}$ (kJ/reaction)	Number of electrons/ reaction	$\Delta G^{0'}$ (kJ/2 e⁻)
Phosphite[b]	$4\ HPO_3^{2-} + SO_4^{2-} + H^+ \rightarrow 4\ HPO_4^{2-} + HS^-$	Phosphite bacteria	−0.69	−364	8	−91
Hydrogen[b]	$H_2 + \frac{1}{2}O_2 \rightarrow H_2O$	Hydrogen bacteria	−0.42	−237.2	2	−237.2
Sulfide[b]	$HS^- + H^+ + \frac{1}{2}O_2 \rightarrow S^0 + H_2O$	Sulfur bacteria	−0.27	−209.4	2	−209.4
Sulfur[b]	$S^0 + 1\frac{1}{2}O_2 + H_2O \rightarrow SO_4^{2-} + 2\ H^+$	Sulfur bacteria	−0.20	−587.1	6	−195.7
Ammonium[c]	$NH_4^+ + 1\frac{1}{2}O_2 \rightarrow NO_2^- + 2\ H^+ + H_2O$	Nitrifying bacteria	+0.34	−274.7	6	−91.6
Nitrite[b]	$NO_2^- + \frac{1}{2}O_2 \rightarrow NO_3^-$	Nitrifying bacteria	+0.43	−74.1	2	−74.1
Ferrous iron[b]	$Fe^{2+} + H^+ + \frac{1}{4}O_2 \rightarrow Fe^{3+} + \frac{1}{2}H_2O$	Iron bacteria	+0.77	−32.9	1	−65.8

[a]Data are from G_f^0 values in Table 3.2 (or references therein) and Figure 3.10, and from bioenergetics calculations as described in Sections 3.4 and 3.6; E_0' values for Fe^{2+} are for pH 2, and others are for pH 7. At pH 7, the E_0' for the Fe^{3+}/Fe^{2+} couple is about +0.2 V.
[b]Except for phosphite, all reactions are shown coupled to O_2 as electron acceptor. The only known phosphite oxidizer couples to SO_4^{2-} as electron acceptor. H_2 and most sulfur compounds can be oxidized anaerobically using one or more electron acceptors, and Fe^{2+} can be oxidized at neutral pH with NO_3^- as electron acceptor. For other chemolithotrophic reactions of sulfur compounds, see Table 14.3.
[c]Ammonium can also be oxidized with NO_2^- as electron acceptor (anammox, Section 14.12).

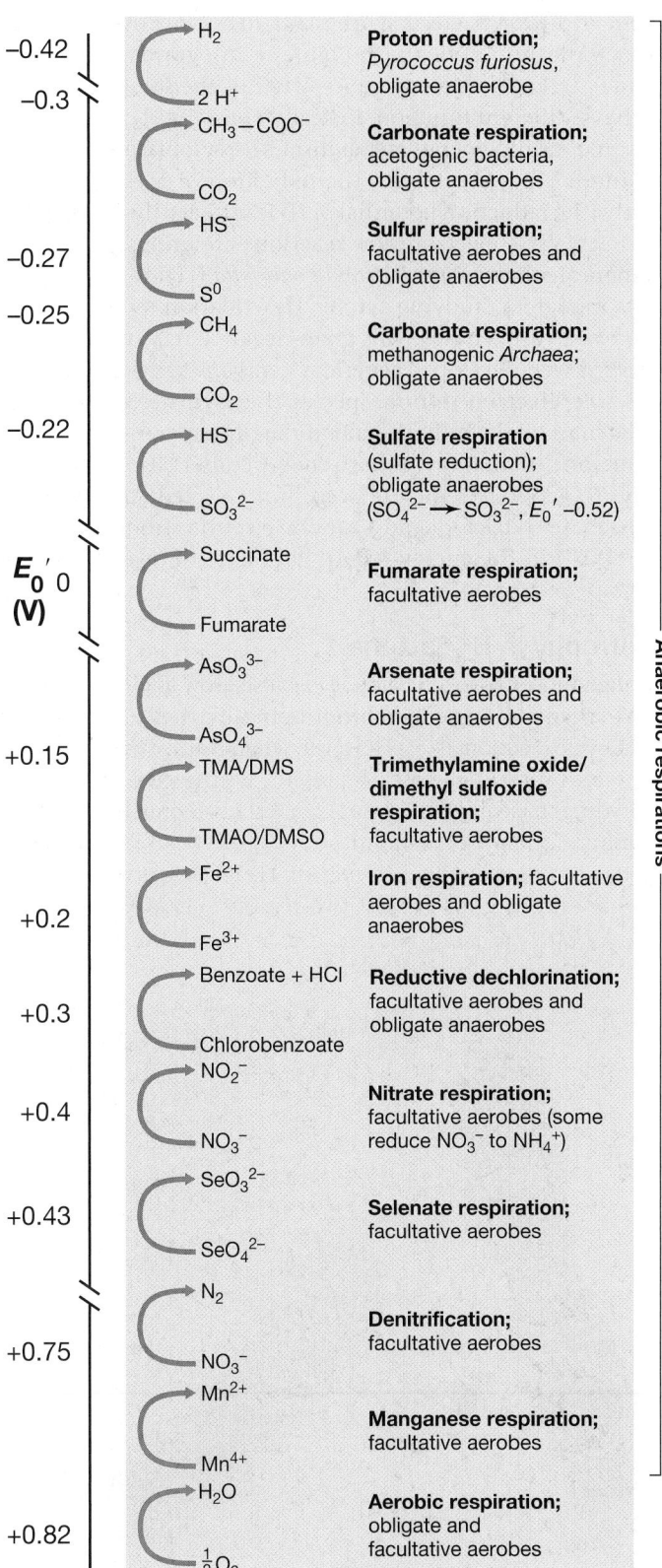

Figure 14.25 Major forms of anaerobic respiration. The redox couples are arranged in order from most electronegative E_0' (top) to most electropositive E_0' (bottom) assuming neutral pH. See Figure 3.10 to compare how the energy yields of some of these anaerobic respirations vary. The E_0' of the Fe^{3+}/Fe^{2+} couple at pH 2 is +0.77 V.

An organism's relationship to oxygen often defines its relationship to other electron acceptors. Organisms that are facultative aerobes (⊂⊃ Section 5.14) can switch to alternative electron acceptors when oxygen is limiting, but will switch back to using oxygen as soon as it becomes available. Alternative electron acceptors used by facultative organisms include those that are fairly near the O_2/H_2O couple, such as many inorganic and organic compounds, and metals such as Fe^{3+} and Mn^{4+} (Figure 14.25). Microorganisms that use more electronegative electron acceptors often employ enzymes that are inhibited by oxygen and hence are typically obligate anaerobes. Organisms that use electronegative electron acceptors such as sulfate (SO_4^{2-}), elemental sulfur (S^0), and carbon dioxide (CO_2) are thus locked into an anaerobic lifestyle.

Assimilative and Dissimilative Reductions

Biosynthetic reactions, such as CO_2 fixation, require both ATP and reducing power. Reducing power in the cell is typically in the form of NADH ($E_0' = -0.32$ V), though some obligate anaerobes also require reducing power in the form of reduced ferredoxin (Fd^{2-}_{red}, $E_0' = -0.37$ to -0.5 V). In the case of chemoorganotrophs, NADH is readily generated during the oxidation of organic molecules (⊂⊃ Sections 3.8 and 3.9). Reducing power is used to reduce inorganic compounds such as NO_3^-, SO_4^{2-}, and CO_2 so that they can be used as sources of N, S, and C in new cell material. The end products of such reductions are the amino groups (—NH_2) of amino acids and other nitrogenous substances, the sulfhydryl groups (—SH) of several sulfur-containing compounds in the cell, and the organic carbon found in all cell constituents, respectively. When NO_3^-, SO_4^{2-}, or CO_2 is reduced for these purposes, it is said to be *assimilated*, and the reduction process is called *assimilative* reduction. Assimilative metabolism is conceptually and physiologically quite different from the reduction of NO_3^-, SO_4^{2-}, and CO_2 during energy conservation in anaerobic respiration. To distinguish these two kinds of reductions, use of these compounds as electron acceptors for energy purposes is called *dissimilative* reduction.

Assimilative and dissimilative metabolisms differ markedly. In assimilative metabolism, *energy is consumed*, and so only enough of the compound (NO_3^-, SO_4^{2-}, or CO_2) is reduced to satisfy the needs for biosynthesis; the products of reduction are then converted to cell material in the form of macromolecules and other biomolecules. By contrast, in dissimilative metabolism, *energy is conserved*, a large amount of the electron acceptor is reduced, and the reduced product remains a small molecule (N_2, H_2S, or CH_4, for example), which is then excreted from the cell. Most microbes can perform a variety of assimilative reductions, whereas dissimilative reductions are only characteristic of organisms carrying out anaerobic respirations.

--- **MINIQUIZ** ---

- In a coupled reaction, how can you tell the electron donor half reaction from the electron acceptor half reaction?
- How does aerobic respiration differ from anaerobic respiration, and why does aerobic respiration repress anaerobic respiration?
- Describe the major differences between assimilative and dissimilative reductions.

14.8 Hydrogen (H₂) Oxidation

Chemolithotrophs are microbes that conserve energy from the oxidation of inorganic electron donors (↩ Section 3.12). Most chemolithotrophs are also autotrophs. However, a few chemolithotrophs lack autotrophic capacities and grow as **mixotrophs**, meaning that they use their inorganic electron donor for energy conservation but assimilate organic carbon as their carbon source.

A simple consideration of bioenergetics tell us which kinds of chemolithotrophs should be expected in nature (Table 14.2), and the reactions of these highly diverse microbes form the heart of the major nutrient cycles (Chapter 21). We begin with perhaps the simplest of all chemolithotrophs, the hydrogen bacteria. Hydrogen (H₂) is a common product of microbial metabolism, especially of some fermentations (Sections 14.19–14.23), and the classical "hydrogen bacteria" respire H₂ aerobically, forming water and ATP as the final products.

Hydrogenase and the Energetics of H₂ Oxidation

Synthesis of ATP during H₂ oxidation by O₂ is the result of electron transport reactions that generate a proton motive force. The overall reaction

$$H_2 + \tfrac{1}{2}O_2 \rightarrow H_2O \qquad \Delta G^{0\prime} = -237 \text{ kJ}$$

is highly exergonic and can be coupled to the synthesis of ATP. In this reaction, which is catalyzed by the enzyme **hydrogenase**, the electrons from H₂ are initially transferred to a quinone acceptor. From there electrons travel through a series of cytochromes to generate a proton motive force and eventually reduce O₂ to water (**Figure 14.26a**).

Some hydrogen bacteria synthesize two distinct hydrogenases, one cytoplasmic and one membrane-integrated. The latter enzyme participates in energetics, whereas the soluble hydrogenase has a different function. Instead of binding H₂ for use as an electron donor in energy metabolism, the cytoplasmic hydrogenase binds H₂ and catalyzes the reduction of NAD⁺ to NADH directly (the reduction potential of H₂ is sufficiently electronegative that reverse electron flow reactions are unnecessary). The gammaproteobacterium *Ralstonia eutropha* (Figure 14.26b) has been a model for studying aerobic H₂ oxidation by species that make two hydrogenases. This gram-negative relative of *Pseudomonas* has well-developed genetic systems and grows robustly on H₂ as sole electron donor. Species that synthesize only one hydrogenase make only the membrane-integrated form of the enzyme, and it functions in both energy conservation and autotrophy. To generate reducing power for CO₂ reduction, single-hydrogenase H₂ bacteria must back up electrons from quinone to form NADH in the energy-requiring process of reverse electron transport (Section 14.3).

Autotrophy in H₂ Bacteria

Although most hydrogen bacteria can also grow as chemoorganotrophs, when growing chemolithotrophically, they fix CO₂ by the Calvin cycle (Section 14.5). However, when readily usable organic compounds such as glucose are present, synthesis of Calvin cycle and hydrogenase enzymes by H₂ bacteria is repressed. Thus, H₂ bacteria are *facultative* chemolithotrophs. This flexibility undoubtedly has ecological value. In nature, H₂ levels in oxic environments are fleeting for at least two reasons: (1) Most biological

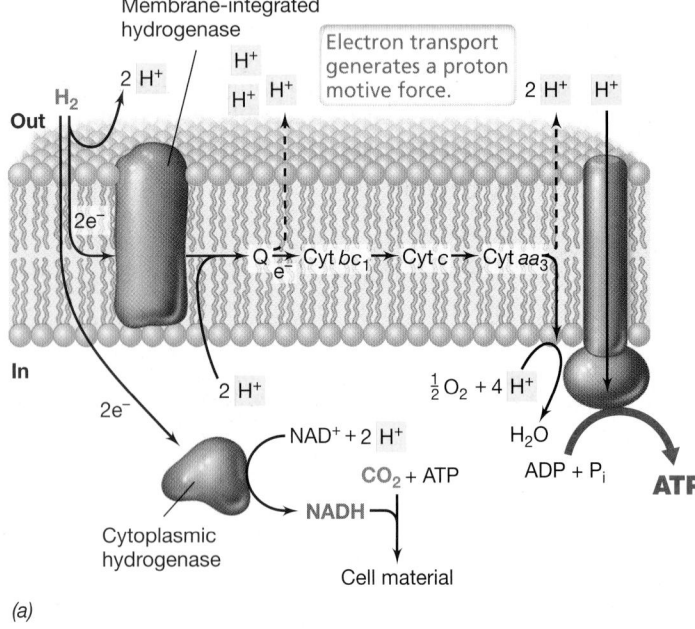

(a)

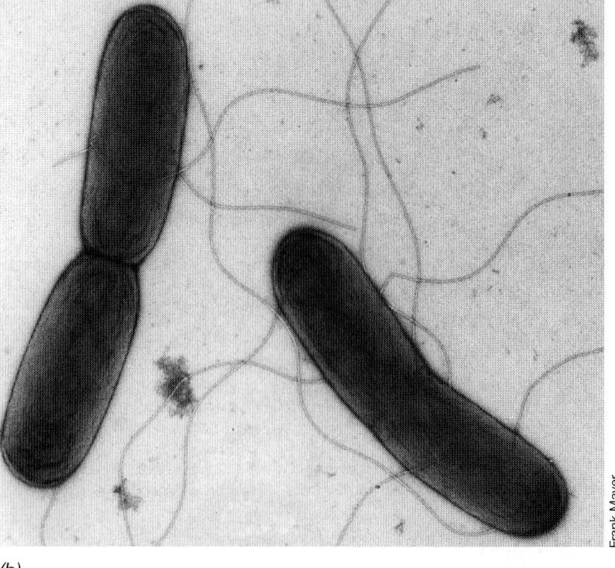

(b)

Figure 14.26 Bioenergetics and function of the two hydrogenases of aerobic hydrogen bacteria. *(a)* In *Ralstonia eutropha*, two hydrogenases are present; the membrane-bound hydrogenase participates in energetics, whereas the cytoplasmic hydrogenase makes NADH for the Calvin cycle. Some hydrogen bacteria have only the membrane-bound hydrogenase, and in these organisms reducing power is synthesized by reverse electron flow from Q back to NAD⁺ to form NADH. Cyt, cytochrome; Q, quinone. *(b)* Transmission electron micrograph of negatively stained cells of the hydrogen-oxidizing chemolithotroph *Ralstonia eutropha*. A cell is about 0.6 μm in diameter and contains several flagella.

H_2 production is the result of fermentations, which are anoxic processes, and (2) H_2 can be utilized by several different anaerobic *Bacteria* and *Archaea* and thus is exhausted before it reaches oxic regions of a habitat. Hence, aerobic hydrogen bacteria must have a backup metabolism to H_2 oxidation, and in nature they likely shift between chemoorganotrophic and chemolithotrophic lifestyles as nutrients in their habitats allow. Moreover, many aerobic H_2 bacteria grow best microaerobically and are probably most competitive as H_2 bacteria in oxic–anoxic interfaces where H_2 may be in greater and more continuous supply than in fully oxic habitats.

MINIQUIZ

- What enzyme is required for hydrogen bacteria to grow as H_2 chemolithotrophs?
- Why is reverse electron flow unnecessary in H_2 bacteria that contain two hydrogenases?

14.9 Oxidation of Sulfur Compounds

Many reduced sulfur compounds can be electron donors for the colorless sulfur bacteria, called *colorless* to distinguish them from the pigmented green and purple bacteria discussed earlier in this chapter (Figure 14.1 and Section 14.3). Historically, the concept of chemolithotrophy emerged in the late nineteenth century from studies of the sulfur bacteria by the Russian microbiologist Sergei Winogradsky (⇔ Section 1.11) and was a radically new idea at the time. However, as our understanding of metabolic diversity has improved, it has become clear that chemolithotrophy, and in particular sulfur chemolithotrophy, is a major metabolic lifestyle of many *Bacteria* and *Archaea*.

Energetics of Sulfur Oxidation

The most common sulfur compounds used as electron donors are hydrogen sulfide (H_2S), elemental sulfur (S^0), and thiosulfate ($S_2O_3^{2-}$); sulfite (SO_3^{2-}) can also be oxidized (Table 14.2 and Table 14.3). In most cases, the final oxidation product is sulfate (SO_4^{2-}). Sulfide oxidation occurs in stages, with the first oxidation step yielding elemental sulfur, S^0. Some sulfide-oxidizing bacteria, such as *Beggiatoa*, deposit this elemental sulfur inside the cell (**Figure 14.27a**), where the sulfur exists as a potential energy (electron) reserve. When the supply of sulfide has been depleted, additional energy can then be conserved from the

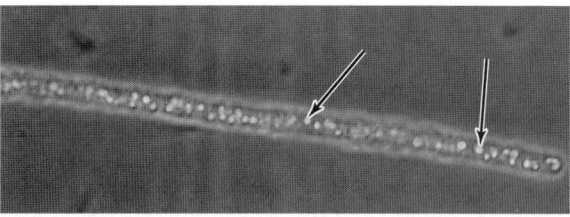

(a)

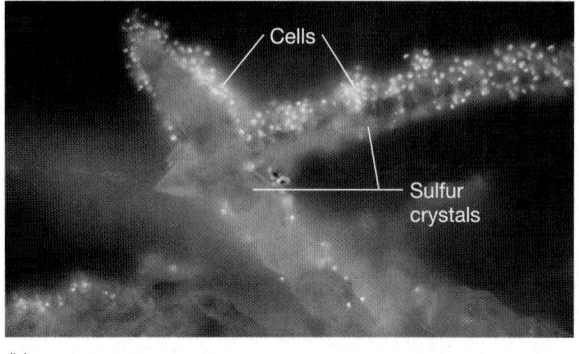

(b)

Figure 14.27 Sulfur bacteria. *(a)* Internal sulfur granules in *Beggiatoa* (arrows). *(b)* Attachment of cells of the sulfur-oxidizing archaeon *Sulfolobus acidocaldarius* to a crystal of elemental sulfur. Cells are visualized by fluorescence microscopy after being stained with the dye acridine orange. The sulfur crystal does not fluoresce.

oxidation of sulfur to sulfate. When S^0 is present externally, the organism must attach itself to the sulfur particle because elemental sulfur is rather insoluble (Figure 14.27b).

One product of the oxidation of reduced sulfur compounds is protons (Tables 14.2 and 14.3). Consequently, one result of sulfur chemolithotrophy is acidification of the environment. Because of this, many sulfur bacteria have evolved to be acid-tolerant or even acidophilic. *Acidithiobacillus thiooxidans*, for example, grows best at a pH between 2 and 3.

Biochemistry of Sulfur Oxidation: The Sox System

There are diverse pathways for conserving energy from the oxidation of sulfur compounds. One of the best characterized is the *Sox* (for *sulfur oxidation*) *system* (**Figure 14.28**), which has been described in *Paracoccus pantotrophus*. The Sox system contains over 15 genes encoding various cytochromes and other proteins necessary for the oxidation of reduced sulfur compounds directly to sulfate. Elements of the Sox system are found in diverse sulfur chemolithotrophs and also in some phototrophic sulfur bacteria,

TABLE 14.3 Comparison of the energetics of oxidation of some common reduced sulfur compounds

Chemolithotrophic reaction	Electrons	Stoichiometry[a]	Energetics (kJ/electron)[a]
Sulfide to sulfate	8	$H_2S + 2 O_2 \rightarrow SO_4^{2-} + 2 H^+$	$\Delta G^{0\prime} = -798.2$ kJ/reaction (-99.75 kJ/e$^-$)
Sulfite to sulfate	2	$SO_3^{2-} + \frac{1}{2}O_2 \rightarrow SO_4^{2-}$	$\Delta G^{0\prime} = -258$ kJ/reaction (-129 kJ/e$^-$)
Thiosulfate to sulfate	8	$S_2O_3^{2-} + H_2O + 2 O_2 \rightarrow 2 SO_4^{2-} + 2 H^+$	$\Delta G^{0\prime} = -818.3$ kJ/reaction (-102 kJ/e$^-$)

[a]All reactions are balanced, both atomically and electrically. See Table 3.2 and Sections 3.4 and 3.6 for details of calculations. For the reaction and energetics of the oxidation of sulfide to sulfur and sulfur to sulfate, see Table 14.2.

organisms that oxidize sulfide to obtain reducing power for CO_2 fixation rather than for energy conservation. The fact that this biochemical system is distributed among bacteria that oxidize sulfide for very different reasons is a good indication that the genes that encode Sox have been transferred between species by horizontal gene flow ($\leftrightarrow$ Section 9.6 and Chapter 11).

There are four key proteins in the Sox system: *SoxXA, SoxYZ, SoxB,* and *SoxCD*. All of these proteins are present in the periplasm. The pathway begins when the enzyme SoxXA forms a heterodisulfide bond between the sulfur compound to be oxidized (which can be HS^-, S^0, or $S_2O_3{}^{2-}$) and the carrier protein, SoxYZ (Figure 14.28). The sulfur compound remains bound to the carrier throughout the pathway, being ultimately released as sulfate through the activity of SoxB. The enzyme SoxCD (sulfur dehydrogenase) is the key enzyme that mediates the removal of 6 electrons from the sulfur compound bound to the carrier (Figure 14.28). Electrons from the Sox system are funneled into the electron transport chain (see later), while the protons generated in the periplasm are released to and acidify the external environment.

Other Aspects of Chemolithotrophic Sulfur Oxidation

Sulfur-oxidizing microbes that store sulfur granules (see Figure 14.27a) also use components of the Sox system but lack the key enzyme sulfur dehydrogenase (SoxCD). In the absence of SoxCD, a sulfur atom bound to SoxYZ is added to a growing sulfur granule in the periplasm (Figure 14.28). The sulfur in the granule can be reductively activated and transported to the cytoplasm where it is eventually oxidized to sulfite ($SO_3{}^{2-}$) by the reverse activity of DsrAB (an enzyme homologous to the enzyme sulfite reductase found in sulfate-reducing bacteria, Section 14.14). The sulfite is then oxidized to sulfate plus two electrons through one of two different pathways. The most widespread system employs the reverse

Sox/Dsr Systems

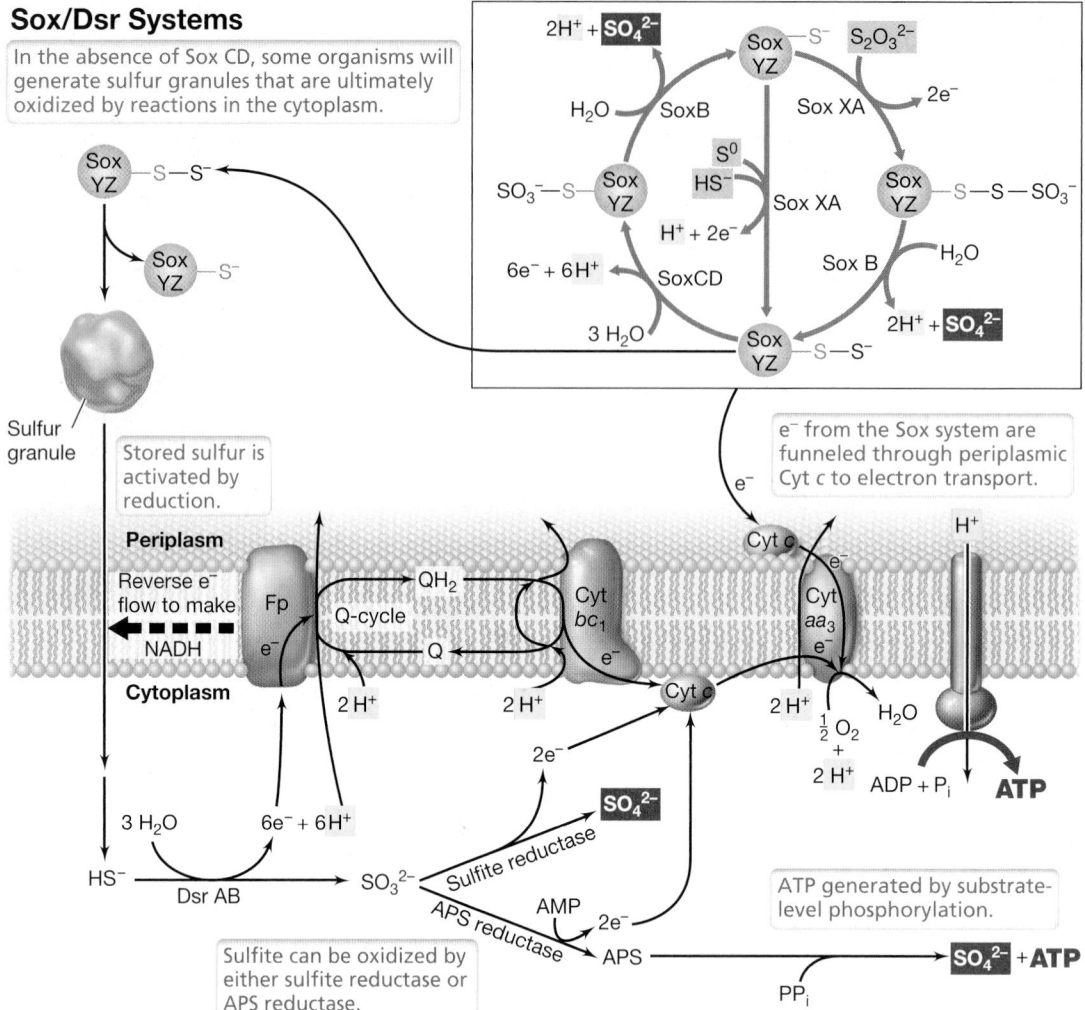

Figure 14.28 **Oxidation of reduced sulfur compounds by sulfur chemolithotrophs.** There are several different pathways for conserving energy through the oxidation of sulfide (H_2S), thiosulfate ($S_2O_3{}^-$), and elemental sulfur (S^0). In the Sox system (sulfur oxidation), SoxXA attaches a reduced sulfur compound to the carrier protein SoxYZ. The protein SoxCD, sulfur dehydrogenase, catalyzes removal of 6 e^- from the bound sulfur atom, and is a key enzyme for bacteria that use the complete Sox system for sulfur oxidation (such as *Paracoccus pantotrophus*). Sulfate ($SO_4{}^{2-}$) is released by the action of SoxB. In contrast, bacteria that form sulfur granules, such as *Beggiatoa* (Figure 14.27a), lack SoxCD and instead oxidize sulfur compounds using the enzymes DsrAB, dissimilatory sulfite reductase, and APS reductase (see Section 14.14). In sulfur oxidation, these enzymes are run backwards to oxidize sulfur compounds. In certain sulfur oxidizers, APS reductase is replaced by sulfite reductase. Reactions of the Sox cycle take place in the periplasm and electrons enter the electron transport chain through the activity of a periplasmic *c*-type cytochrome (Cyt *c*), while reactions of Sox/Dsr systems take place instead in the cytoplasm and electrons can enter electron transport at either the level of flavoproteins (Fp) or *c*-type cytochromes.

activity of the cytoplasmic enzyme *sulfite reductase*. This enzyme oxidizes sulfite and transfers the electrons to the electron transport chain. By contrast, some sulfur chemolithotrophs oxidize $SO_3{}^{2-}$ to $SO_4{}^{2-}$ via a reversal of the activity of the enzyme adenosine phosphosulfate reductase (an enzyme essential for the metabolism of sulfate-reducing bacteria, see Section 14.14 and Figure 14.37). The oxidation of $SO_3{}^{2-}$ to $SO_4{}^{2-}$ yields an energy-rich phosphate bond by substrate-level phosphorylation when AMP is converted to ATP (Figure 14.28).

Electrons from the oxidation of reduced sulfur compounds eventually reach the electron transport chain, as shown in Figure 14.28. Though the exact details remain unknown, electrons are likely to enter at the flavoprotein or cytochrome *c* ($E_0' = +0.3$ V) levels and are transported through the chain to O_2, generating a proton motive force that triggers ATP synthase activity (Figure 14.28). Electrons for CO_2 fixation come from reverse electron transport (Section 14.3), eventually yielding NADH, and autotrophy is driven by reactions of the Calvin cycle or some other autotrophic pathway (Section 14.5). Although the sulfur chemolithotrophs are primarily an aerobic group, some species can grow by anaerobic respiration using nitrate as an electron acceptor. The sulfur bacterium *Thiobacillus denitrificans* is a classic example, reducing nitrate to dinitrogen gas (the process of denitrification, Section 14.13).

MINIQUIZ

- How many electrons are available from the oxidation of H_2S if S^0 or SO_4^{2-} is the final product?
- In terms of intermediates, how does the Sox system differ from other sulfide-oxidizing systems?

14.10 Iron (Fe^{2+}) Oxidation

The aerobic oxidation of ferrous iron (Fe^{2+}) to ferric iron (Fe^{3+}) supports growth of the chemolithotrophic "iron bacteria" (⮌ Section 15.15). At acidic pH, only a small amount of energy is available from this reaction (Table 14.2), and for this reason the iron bacteria must oxidize large amounts of iron in order to produce only tiny amounts of cell material. The ferric iron produced becomes hydrated to form insoluble ferric hydroxide ($Fe^{3+} + 3 H_2O \rightarrow Fe(OH)_3 + 3 H^+$) and other iron precipitates in aquatic environments, and this drives down the pH (**Figure 14.29**). This inevitable chemical reaction probably explains why many iron-oxidizing bacteria have evolved to be strongly acidophilic.

Iron-Oxidizing Bacteria

The best-known iron bacteria, *Acidithiobacillus ferrooxidans* and *Leptospirillum ferrooxidans*, can both grow autotrophically using ferrous iron (Figure 14.29) as electron donor at pH values as low as 1; growth is optimal at pH 2–3. These bacteria are common in acid-polluted environments such as coal-mining runoff waters (Figure 14.29a). *Ferroplasma*, a species of *Archaea*, is an extremely acidophilic iron oxidizer and can grow at pH values below 0 (⮌ Section 17.3). We discuss the role of all of these organisms in acid mine pollution and mineral oxidation in Sections 21.4, 22.1, and 22.2.

At neutral pH, Fe^{2+} spontaneously oxidizes to Fe^{3+}, so opportunities for the iron bacteria in neutral habitats are restricted to locations where Fe^{2+} is transitioning from anoxic to oxic conditions. For example, anoxic groundwater often contains dissolved Fe^{2+}, and when it is released, as in iron-rich spring water, it becomes exposed to O_2. At such interfaces, iron bacteria oxidize Fe^{2+} to Fe^{3+} before it oxidizes spontaneously. *Gallionella ferruginea*, *Sphaerotilus natans*, and *Leptothrix discophora* are examples of bacteria that live at these interfaces. They are typically seen mixed in with the characteristic ferric iron deposits they form (⮌ Figures 15.36 and 21.14).

Energy from Iron Oxidation

The bioenergetics of ferrous iron oxidation by *Acidithiobacillus ferrooxidans* and other acidophilic iron oxidizers are of considerable

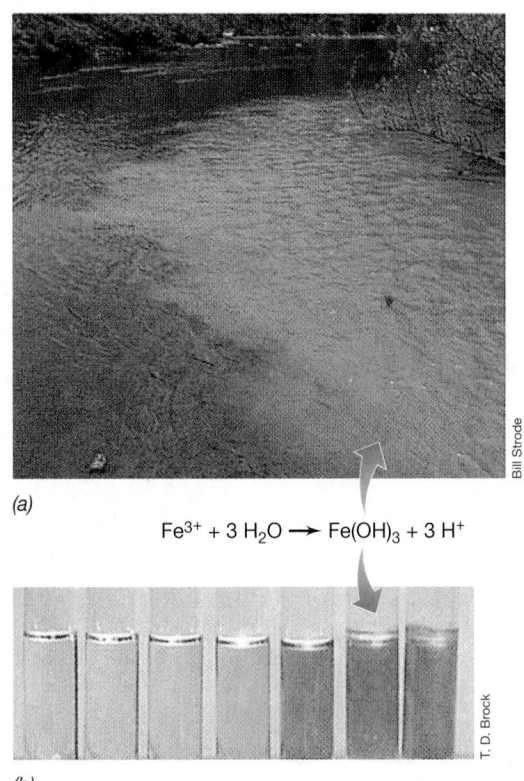

(a)

$$Fe^{3+} + 3\ H_2O \longrightarrow Fe(OH)_3 + 3\ H^+$$

(b)

Figure 14.29 Iron-oxidizing bacteria. *(a)* Acid mine drainage, showing the confluence of a normal river and a creek draining a coal-mining area. At low pH values, Fe^{2+} does not oxidize spontaneously in air, but *Acidithiobacillus ferrooxidans* carries out the oxidation; insoluble $Fe(OH)_3$ and complex ferric salts precipitate. *(b)* Cultures of *A. ferrooxidans*. Shown is a dilution series, with no growth in the tube on the left and increasing amounts of growth from left to right. Growth is evident from the production of $Fe(OH)_3$.

interest because of the very electropositive reduction potential of the Fe^{3+}/Fe^{2+} couple at acidic pH ($E_0' = +0.77$ V at pH 2). The respiratory chain of *A. ferrooxidans* contains cytochromes of the *c* and aa_3 types and a periplasmic copper-containing protein called *rusticyanin* (**Figure 14.30**). There is also an iron-oxidizing protein located in the outer membrane of this gram-negative bacterium.

Because the reduction potential of the Fe^{3+}/Fe^{2+} couple is so high, steps in electron transport to oxygen ($\frac{1}{2} O_2/H_2O$, $E_0' = +0.82$ V) can obviously be few. Iron oxidation begins in the outer membrane where the organism contacts either soluble Fe^{2+} or insoluble ferrous iron minerals. Fe^{2+} is oxidized to Fe^{3+}, a one-electron transition (Table 14.2), by an outer membrane cytochrome *c* that transfers electrons into the periplasm where rusticyanin ($E_0' = +0.68$ V) is the electron acceptor. This thermodynamically slightly unfavorable reaction is thought to be pulled forward by the immediate consumption of Fe^{3+} in $Fe(OH)_3$ formation (Figure 14.30). Rusticyanin then reduces a periplasmic cytochrome *c*, which transfers electrons to cytochrome aa_3, and it is the latter protein that reduces O_2 to H_2O; ATP is synthesized by ATPase in the usual fashion (Figure 14.30).

The nature of the proton motive force in *A. ferrooxidans* is of interest. In a highly acidic environment, a large gradient of protons already exists across the *A. ferrooxidans* cytoplasmic membrane (the periplasm is pH 1–2, whereas the cytoplasm is

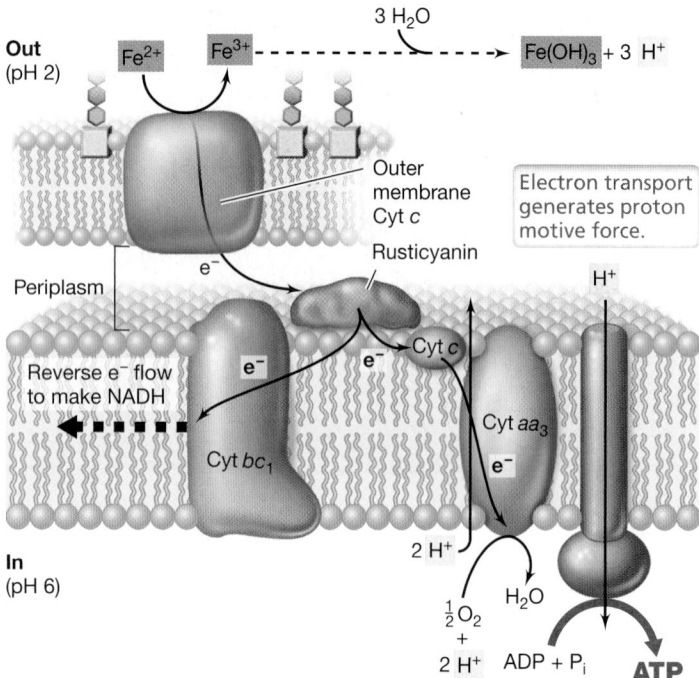

Figure 14.30 Electron flow during Fe²⁺ oxidation by the acidophile *Acidithiobacillus ferrooxidans*. The periplasmic copper-containing protein rusticyanin receives electrons from Fe^{2+} oxidized by a *c*-type cytochrome located in the outer membrane. From here, electrons travel a short electron transport chain, resulting in the reduction of O_2 to H_2O. Reducing power comes from reverse electron flow. Note the steep pH gradient across the membrane.

pH 5.5–6, Figure 14.30). Although one might think that with this gradient *A. ferrooxidans* could make ATP at no energetic cost, this is not the case; the organism cannot make ATP from this preformed proton motive force in the absence of an electron donor. This is because H^+ ions that enter the cytoplasm via ATPase must be consumed in order to maintain the internal pH within acceptable limits. Proton consumption occurs during the reduction of O_2 in the electron transport chain and this reaction requires electrons; the latter come from the oxidation of Fe^{2+} to Fe^{3+} (Figure 14.30).

Autotrophy in *A. ferrooxidans* is supported by the Calvin cycle (Section 14.5), and because of the high potential of the electron donor, much energy must be consumed in reverse electron flow reactions to obtain the reducing power (NADH) necessary to drive CO_2 fixation. NADH is formed by reduction of NAD^+ by electrons obtained from Fe^{2+} that are forced backwards through cytochrome bc_1 and the quinone pool at the expense of the proton motive force (Figure 14.30).

The relatively poor energetic yield from ferrous iron oxidation coupled with the large energetic demands of the Calvin cycle (Figure 14.18) means that *A. ferrooxidans* must oxidize large amounts of Fe^{2+} to produce even a very small amount of cell material. Thus, in environments where acidophilic iron-oxidizing bacteria thrive, their presence is signaled not by the formation of high cell numbers but by the presence of the extensive ferric iron precipitates they have generated (Figure 14.29). We consider the ecology of iron bacteria in Chapters 21 and 22.

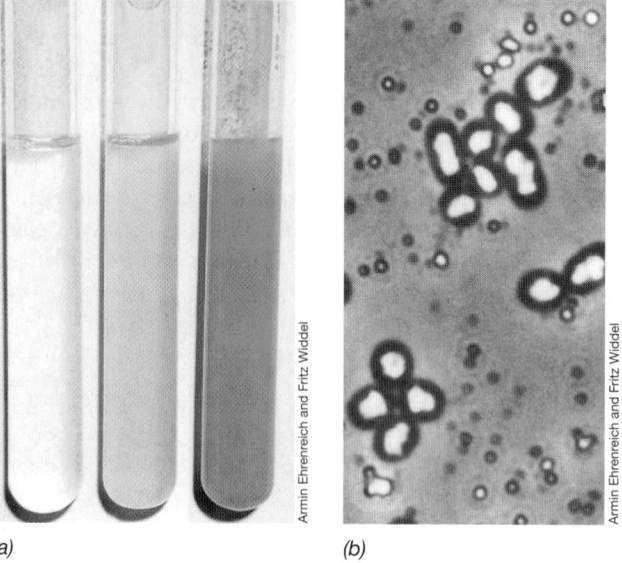

Figure 14.31 Fe²⁺ oxidation by anoxygenic phototrophic bacteria. *(a)* Oxidation in anoxic tube cultures. Left to right: Sterile medium, inoculated medium, a growing culture showing $Fe(OH)_3$. *(b)* Phase-contrast photomicrograph of an Fe^{2+}-oxidizing purple bacterium. The bright refractile areas within cells are gas vesicles. The granules outside the cells are iron precipitates. This organism is phylogenetically related to the purple sulfur bacterium *Chromatium*.

Ferrous Iron Oxidation under Anoxic Conditions

Ferrous iron can be oxidized under *anoxic* conditions by certain chemolithotrophs and phototrophic purple and green bacteria (**Figure 14.31**). In these cases, Fe^{2+} is used either as an electron donor in energy metabolism (chemolithotrophs) and/or as a reductant for CO_2 fixation (phototrophs). An important point to consider here is that at neutral pH where these organisms thrive, the E_0' of the Fe^{3+}/Fe^{2+} couple is significantly more electronegative than at acidic pH (+0.2 V versus +0.77 V, respectively, Figure 14.25). Hence, electrons from Fe^{2+} can reduce cytochrome *c* to initiate electron transport reactions. For chemolithotrophs, the electron acceptor is nitrate (NO_3^-), with either nitrite (NO_2^-) or dinitrogen gas (N_2) being the final product of this anaerobic respiration. For Fe^{2+}-oxidizing purple and green bacteria, either soluble Fe^{2+} or iron sulfide (FeS) can be used as electron donor. With FeS, both Fe^{2+} and S^{2-} are oxidized, Fe^{2+} to Fe^{3+} (one electron) and HS^- to SO_4^{2-} (eight electrons).

MINIQUIZ

- Why is only a very small amount of energy available from the oxidation of Fe^{2+} to Fe^{3+} at acidic pH?
- What is the function of rusticyanin and where is it found in the cell?
- How can Fe^{2+} be oxidized under anoxic conditions?

14.11 Nitrification

The reduced inorganic nitrogen compounds ammonia (NH_3) and nitrite (NO_2^-) are oxidized aerobically by the chemolithotrophic *nitrifying bacteria* in the process of **nitrification**

(⥰ Section 15.13). Nitrifying bacteria are widely distributed in soils, water, wastewaters, and the oceans. Nitrification consists of two different sets of reactions; the first set of reactions catalyze oxidation of ammonia to nitrite, and the second set catalyze oxidation of nitrite to nitrate (NO_3^-). Most nitrifying microbes are only able to catalyze one set of these reactions. For example, *Bacteria* such as *Nitrosomonas* and *Archaea* such as *Nitrosopumilus* oxidize NH_3 only to nitrite, and we call these organisms *ammonia oxidizers*. The full nitrification pathway is ultimately completed when other *Bacteria* such as *Nitrobacter* oxidize NO_2^- to NO_3^-, and we call these organisms *nitrite oxidizers*. As far as is known, only certain bacteria in the genus *Nitrospira* can catalyze both sets of reactions, oxidizing NH_3 all the way to NO_3^-.

Bioenergetics and Enzymology of Ammonia and Nitrite Oxidation

The bioenergetics of nitrification is based on the same principles that govern other chemolithotrophic reactions: Electrons from reduced inorganic substrates (in this case, reduced nitrogen compounds) enter an electron transport chain, and electron transport reactions establish a proton motive force that drives ATP synthesis. The complete oxidation of NH_3 to NO_3^- involves an eight-electron transfer, and the electron donors for the nitrifying bacteria are not particularly strong. The E_0' of the NO_2^-/NH_3 couple (the first step in the oxidation of NH_3) is +0.34 V, and the E_0' of the NO_3^-/NO_2^- couple is even more positive, about +0.43 V. By necessity, these reduction potentials force the nitrifying bacteria to donate electrons to rather high-potential electron acceptors, and this of course limits the amount of energy that can be conserved (Section 14.7).

Several key enzymes participate in the oxidation of reduced nitrogen compounds. In ammonia-oxidizing bacteria such as *Nitrosomonas*, NH_3 is oxidized by *ammonia monooxygenase* (monooxygenases are discussed in Section 14.24), producing hydroxylamine (NH_2OH) and H_2O (**Figure 14.32**). A second key enzyme, *hydroxylamine oxidoreductase*, then oxidizes NH_2OH to NO_2^-, removing four electrons in the process. Ammonia monooxygenase is an integral membrane protein, whereas hydroxylamine oxidoreductase is periplasmic (Figure 14.32). In the reaction carried out by ammonia monooxygenase,

$$NH_3 + O_2 + 2\,H^+ + 2\,e^- \rightarrow NH_2OH + H_2O$$

two electrons and protons are needed to reduce one molecule of oxygen (O_2) to H_2O. These electrons originate from the oxidation of hydroxylamine and are supplied to ammonia monooxygenase from hydroxylamine oxidoreductase via cytochrome *c* and ubiquinone (Figure 14.32). Thus, for every *four* electrons generated from the oxidation of NH_3 to NO_2^-, only *two* actually reach cytochrome aa_3, the terminal oxidase that interacts with O_2 to form H_2O (Figure 14.32).

Nitrite-oxidizing bacteria such as *Nitrobacter* oxidize NO_2^- to NO_3^- by the enzyme *nitrite oxidoreductase*, with electrons traveling a very short electron transport chain (because of the high potential of the NO_3^-/NO_2^- couple) to the terminal oxidase (**Figure 14.33**). Cytochromes of the *a* and *c* types are present in the

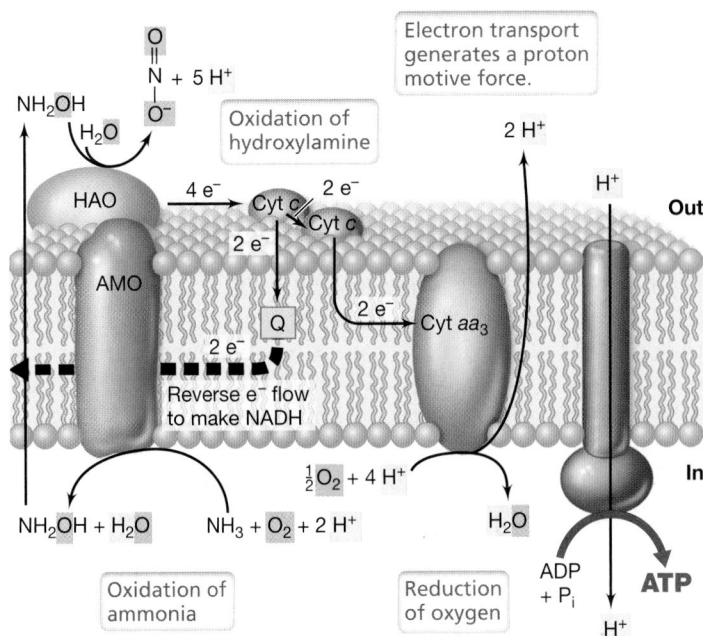

Figure 14.32 Oxidation of NH_3 and electron flow in ammonia-oxidizing bacteria. The reactants and the products of this reaction series are highlighted. The cytochrome *c* (Cyt *c*) in the periplasm is a different form of Cyt *c* than that in the membrane. AMO, ammonia monooxygenase; HAO, hydroxylamine oxidoreductase; Q, ubiquinone.

electron transport chain of nitrite oxidizers, and the activity of cytochrome aa_3 generates a proton motive force (Figure 14.33). As is the case with the iron bacteria (Section 14.10), only small amounts of energy are available from nitrite oxidation. Hence, minimal amounts of cell material are obtained even though large amounts of nitrite may be oxidized.

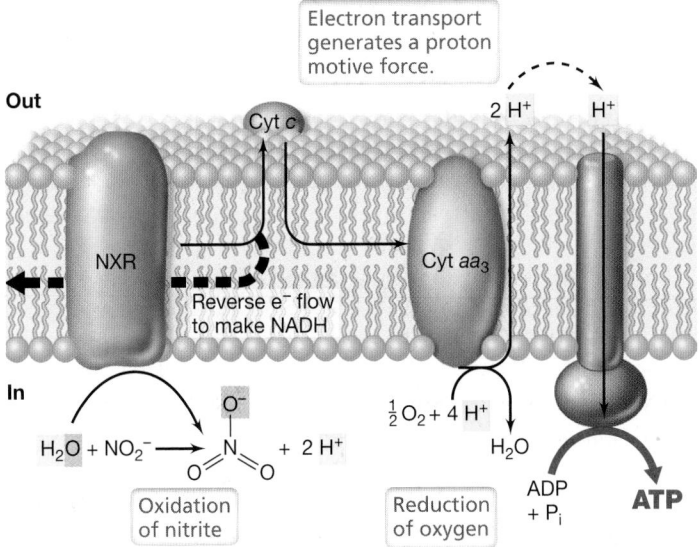

Figure 14.33 Oxidation of NO_2^- to NO_3^- by nitrifying bacteria. The reactants and products of this reaction series are highlighted to show the reaction clearly. NXR, nitrite oxidoreductase.

Carbon Metabolism and Ecology of Nitrifying Bacteria

Like sulfur- and iron-oxidizing chemolithotrophs (Sections 14.9 and 14.10), aerobic nitrifying *Bacteria* employ the Calvin cycle for CO_2 fixation. The ATP and reducing power requirements of the Calvin cycle place additional burdens on an energy-generating system that already has a relatively low yield (NADH to drive the Calvin cycle in nitrifiers is formed by reverse electron flow, Figures 14.32 and 14.33). The energetic constraints are particularly severe for nitrite oxidizers, and it is perhaps for this reason that most of these organisms have alternative energy-conserving mechanisms, being able to grow chemoorganotrophically on glucose and a few other organic substrates. By contrast, species of ammonia-oxidizing bacteria are either obligate chemolithotrophs or mixotrophs (Section 14.8). Autotrophy in ammonia-oxidizing *Archaea* is supported by a variation of the hydroxypropionate cycle (Section 14.5).

Nitrifying microbes play key ecological roles in the nitrogen cycle, converting ammonia into nitrate, a key plant nutrient. Nitrifiers are also important in sewage and wastewater treatment, removing toxic amines and ammonia and releasing less toxic nitrogen compounds (↩ Sections 22.6 and 22.7). Nitrifiers play a similar role in the water column of lakes, where ammonia produced in the sediments from the decomposition of organic nitrogenous compounds is oxidized to nitrate, a more usable fixed nitrogen source for algae and cyanobacteria.

MINIQUIZ

- What are the substrates for the enzyme ammonia monooxygenase?
- What is the difference between ammonia oxidation and nitrite oxidation and in what types of organisms are these reactions found?

14.12 Anaerobic Ammonia Oxidation (Anammox)

Although the ammonia-oxidizing microbes just discussed are strict *aerobes*, NH_3 can also be oxidized under anoxic conditions. This process is called **anammox** (for *an*aerobic *amm*onia *ox*idation) and is catalyzed by an unusual group of obligately anaerobic *Bacteria*.

Ammonia is oxidized in the anammox reaction using NO_2^- as the electron acceptor to yield N_2:

$$NH_4^+ + NO_2^- \rightarrow N_2 + 2 H_2O \qquad \Delta G^{0\prime} = -357\,kJ$$

A major anammox organism, *Brocadia anammoxidans*, is a species of the *Planctomycetes* phylum of *Bacteria* (↩ Section 16.16). *Planctomycetes* are unusual *Bacteria* in that their cytoplasm can contain membrane-enclosed compartments of various types (**Figure 14.34**). In cells of *B. anammoxidans*, this compartment is the *anammoxosome*, and it is within this structure that the anammox reaction occurs (Figure 14.34c). In addition to *Brocadia*, several other genera of anammox bacteria are known, including *Kuenenia, Anammoxoglobus, Jettenia,* and *Scalindua,* all of which are related to *Brocadia* and also contain anammoxosomes. Like

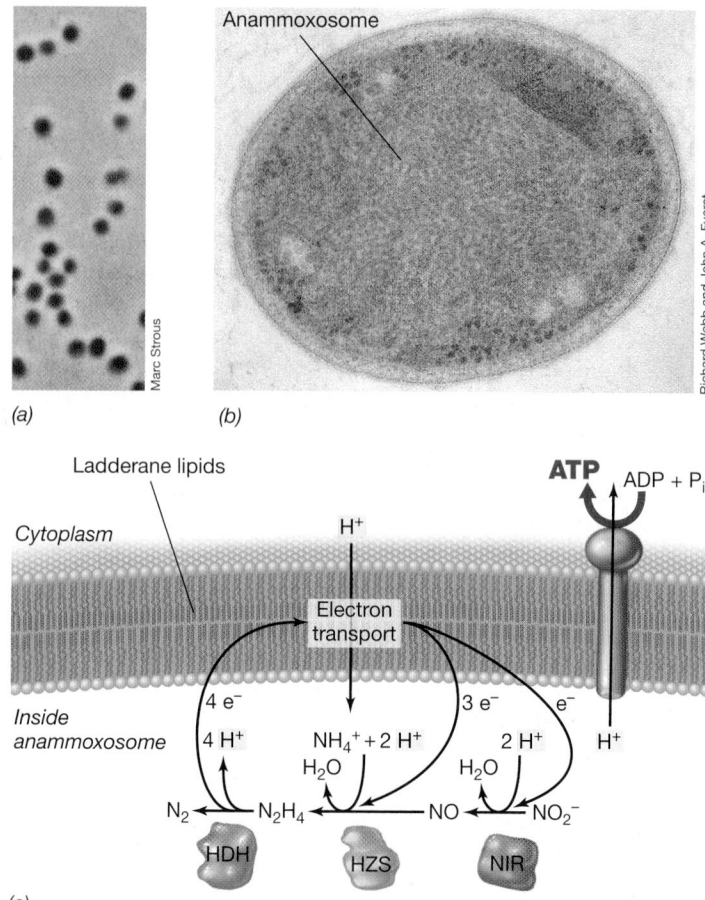

(a) (b)

(c)

Figure 14.34 Anammox. *(a)* Phase-contrast photomicrograph of cells of *Brocadia anammoxidans*. A single cell is about 1 μm in diameter. *(b)* Transmission electron micrograph of a cell; note the membrane-enclosed compartments including the large fibrillar anammoxosome. *(c)* Reactions in the anammoxosome. NIR, nitrite reductase, HZS, hydrazine synthase; HDH, hydrazine dehydrogenase.

aerobic ammonia oxidizers, anammox bacteria are also autotrophs, but they do not fix CO_2 using the pathways employed by aerobic ammonia oxidizers. Instead, anammox bacteria fix CO_2 by way of the reductive acetyl-CoA pathway, an autotrophic pathway widespread among some obligately anaerobic autotrophic *Bacteria* and *Archaea* (see Section 14.16).

The Anammoxosome and Its Reactions

The anammoxosome is a unit membrane–enclosed structure (Figure 14.34b) and in this respect is technically an organelle in the eukaryotic sense of the term. Lipids that form the anammoxosome membrane are not the typical lipids of *Bacteria* but instead consist of fatty acids constructed of multiple cyclobutane (C_4) rings bonded to glycerol by both ester and ether bonds. These *ladderane lipids*, as they are called, aggregate in the membrane to form an unusually dense membrane structure that prevents diffusion of substances from the anammoxosome into the cytoplasm.

The sturdy anammoxosome membrane is required to protect the cell from toxic intermediates produced during anammox reactions. These include, in particular, the compound *hydrazine* (N_2H_4), a very strong reductant. In the anammox reaction, NO_2^- is first reduced to nitric oxide (NO) by nitrite reductase, and then NO reacts with ammonium (NH_4^+) to yield N_2H_4 by activity of the enzyme hydrazine synthase (Figure 14.34c). N_2H_4 is then oxidized to N_2 plus electrons by the enzyme hydrazine dehydrogenase, and the electrons are funneled into the electron transport chain where they are used to reduce nitrite and nitric oxide earlier in the pathway. In this way, anammox generates a *cyclical* series of electron transfer reactions in order to generate a proton motive force; ATP is formed from the latter by ATPases in the anammoxosome membrane (Figure 14.34c).

Reducing power for CO_2 fixation by anammox bacteria is derived from reverse electron transport, but because electron transfer reactions are cyclic, the electrons needed for reverse electron transport derive from an independent set of reactions that oxidize nitrite to nitrate by a nitrite oxidoreductase, a reaction also present in *Nitrobacter* (Figure 14.33). Interestingly, then, nitrite serves two different purposes for anammox bacteria: The *reduction* of nitrite is required to generate ATP by chemiosmosis, and the *oxidation* of nitrite is required to generate reducing power for CO_2 fixation.

Ecology of Anammox

In nature, the source of NO_2^- for the anammox reaction is presumably aerobic ammonia-oxidizing *Bacteria* and *Archaea*. These organisms coexist with anammox bacteria in ammonia-rich habitats such as sewage and other wastewaters. The suspended particles that form in these habitats contain both oxic and anoxic zones in which ammonia oxidizers of different physiologies can coexist in close association. In mixed laboratory cultures, high levels of oxygen inhibit anammox and favor classic nitrification, and thus it is likely that in nature, the fraction of ammonia oxidation catalyzed by anammox bacteria is governed by the concentration of O_2 in the habitat.

From an environmental standpoint, anammox is a very beneficial process in the treatment of wastewaters. The anoxic removal of NH_3 and amines by the formation of N_2 (Figure 14.34c) helps reduce the input of fixed nitrogen from wastewater treatment effluents that flow into rivers and streams, thereby maintaining higher water quality than would otherwise be possible. Also, marine anammox bacteria are likely responsible for the large amount of NH_3 that is known to disappear during mineralization processes in anoxic marine sediments. At least some ammonia-rich freshwater lake sediments also support anammox, and thus it appears that anammox can occur in any anoxic environment in which NH_3 and NO_2^- coexist.

MINIQUIZ

- What are the electron donor and acceptor in the anammox process?
- What does electron transport in anammox bacteria have in common with electron transport in purple sulfur bacteria?
- Compare CO_2 fixation in anammox bacteria and purple sulfur bacteria. What characteristics do these processes share and how are they different?

IV • Respiratory Processes Defined by Electron Acceptor

We examined the process of aerobic respiration in Chapter 3 and reviewed the bioenergetics of respiration in Section 14.7. Here we consider the details of **anaerobic respiration** in the many variations it is found in the microbial world. A wide variety of compounds function as electron acceptors in anaerobic respirations (Figure 14.25), and each acceptor is typically linked to a specific group or groups of microbes. We begin with a common form of anaerobic respiration in which nitrate functions as electron acceptor.

14.13 Nitrate Reduction and Denitrification

Inorganic nitrogen compounds are some of the most common electron acceptors in anaerobic respiration. Table 14.4 summarizes the relevant forms of inorganic nitrogen with their oxidation states. One of the most common alternative electron acceptors for dissimilative purposes is nitrate (NO_3^-), which can be reduced with two electrons to nitrite (NO_2^-), or reduced further to nitric oxide (NO), nitrous oxide (N_2O), or dinitrogen (N_2). Because NO, N_2O, and N_2 are all gases, they can be lost from the environment, and their biological production is called **denitrification** (Figure 14.35).

Some nitrate reducers, for example *Escherichia coli*, are not true denitrifiers, but only carry out the first step (nitrate to nitrite) in the process. Moreover, some organisms can reduce NO_2^- to ammonia (NH_3) in a dissimilative process. But it is the production of gaseous products—*denitrification*—that is of greatest global significance because it consumes fixed nitrogen and produces some polluting gases.

Denitrifying Microorganisms and Their Ecological Activities

Many denitrifying *Bacteria* are phylogenetically *Proteobacteria* and physiologically facultative aerobes. *Pseudomonas* species, for example, are typically strong denitrifiers. Aerobic respiration occurs when O_2 is present, even if NO_3^- is also present in the medium. Many denitrifying bacteria also reduce other electron acceptors anaerobically, such as Fe^{3+} and certain organic electron acceptors (Figure 14.25), and some denitrifiers can even ferment. Thus, denitrifying bacteria are metabolically diverse in terms

TABLE 14.4 Oxidation states of key nitrogen compounds

Compound	Oxidation state of N atom
Organic N (—NH_2)	−3
Ammonia (NH_3)	−3
Nitrogen gas (N_2)	0
Nitrous oxide (N_2O)	+1 (average per N)
Nitric oxide (NO)	+2
Nitrite (NO_2^-)	+3
Nitrogen dioxide (NO_2)	+4
Nitrate (NO_3^-)	+5

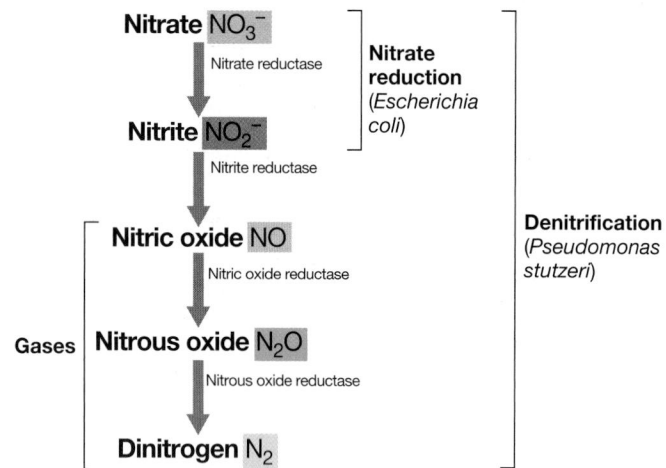

Figure 14.35 Steps in the dissimilative reduction of nitrate. Some organisms can carry out only the first step. All enzymes involved are derepressed by anoxic conditions. Also, some bacteria are known that can reduce NO_3^- to NH_4^+ in dissimilative metabolism. Note that colors used here match those used in Figure 14.36.

of alternative energy-generating mechanisms. Some species of *Archaea* can grow anaerobically by nitrate reduction to nitrite, and several archaeans can also denitrify. Interestingly, at least one eukaryote has also been shown to be a denitrifier. The protist *Globobulimina pseudospinescens*, a shelled amoeba (a foraminiferan, ↩ Section 18.6), can denitrify and likely employs this form of metabolism in its habitat, anoxic marine sediments.

Denitrification is a process with significant ecological ramifications. For agricultural purposes, denitrification is a detrimental process, as it removes nitrate—often intentionally added by grain and other crop farmers as potassium nitrate fertilizer—from the soil. Gaseous products of denitrification other than N_2 (N_2O and NO) are also of significant environmental concern. N_2O is a strong greenhouse gas (contributing to climate change) and can also be converted to NO by sunlight; NO reacts with and consumes ozone (O_3) in the upper atmosphere to form NO_2^-. When it rains, NO_2^- returns to Earth as nitrous acid (HNO_2) in *acid rain*. In contrast to these environmentally harmful processes, for a desirable process like sewage treatment, denitrification (as well as anammox, see earlier) is beneficial because it removes fixed nitrogen, a major trigger of algal growth if a nitrate-rich sewage effluent is released into rivers or lakes (↩ Sections 20.8, 22.6, and 22.7).

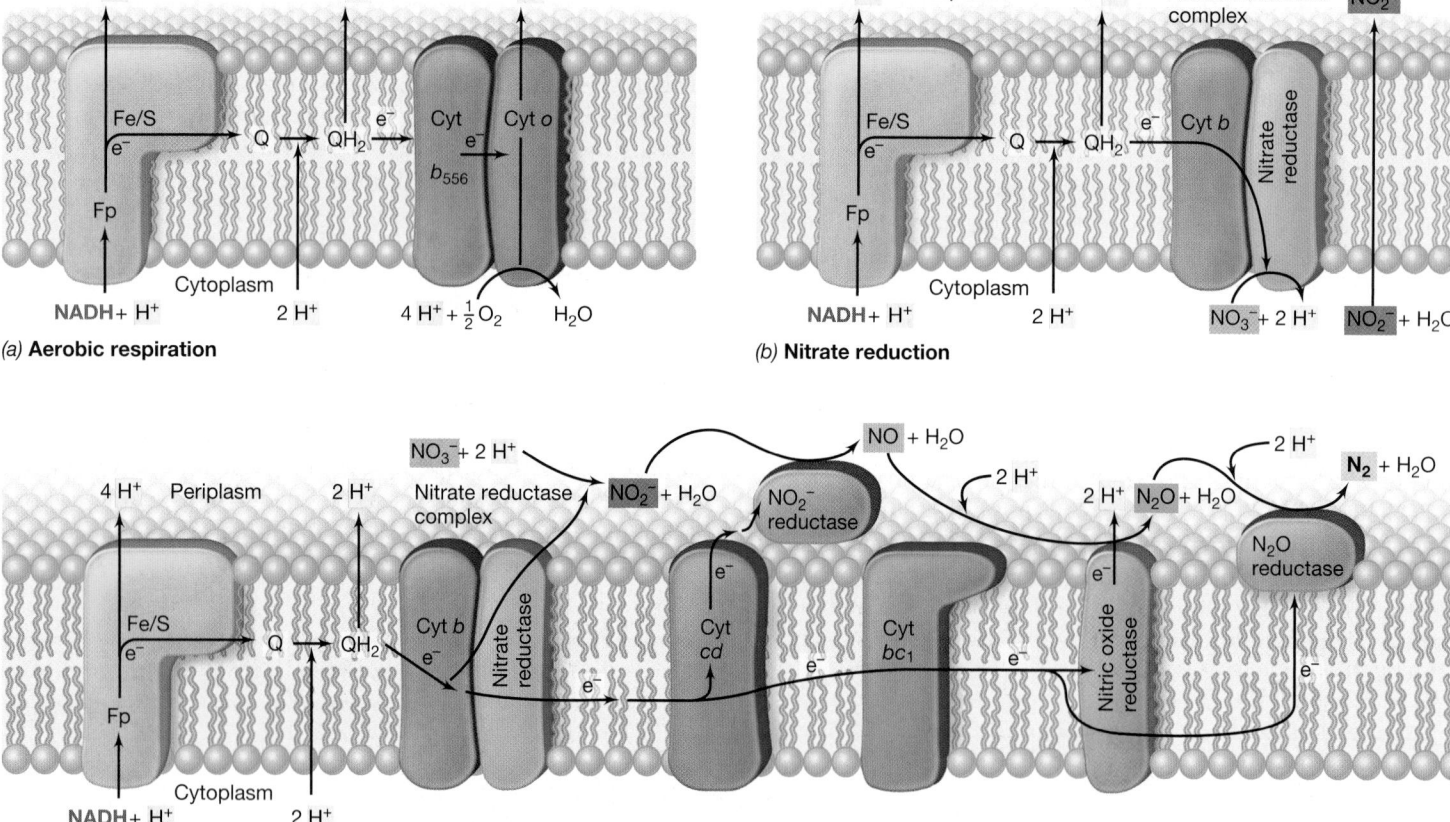

(a) **Aerobic respiration**

(b) **Nitrate reduction**

(c) **Denitrification**

Figure 14.36 Respiration and nitrate-based anaerobic respiration. Electron transport processes in the membrane of *Escherichia coli* when *(a)* O_2 or *(b)* NO_3^- is used as an electron acceptor and NADH is the electron donor. Fp, flavoprotein; Q, ubiquinone. Under high-oxygen conditions, the sequence of carriers is Cyt b_{556} → Cyt o → O_2. However, under low-oxygen conditions (not shown), the sequence is Cyt b_{568} → Cyt d → O_2. Note how more protons are translocated per two electrons oxidized aerobically during electron transport reactions than anaerobically with NO_3^- as electron acceptor, because the aerobic terminal oxidase (Cyt o) pumps two protons. *(c)* Scheme for electron transport in membranes of *Pseudomonas stutzeri* during denitrification. Nitrate and nitric oxide reductases are integral membrane proteins, whereas nitrite and nitrous oxide reductases are periplasmic enzymes.

Biochemistry of Dissimilative Nitrate Reduction

The electron transport pathways of aerobic respiration, nitrate respiration, and denitrification are compared in **Figure 14.36**. The enzyme that catalyzes the first step of dissimilative nitrate reduction is *nitrate reductase*, a molybdenum-containing, membrane-integrated enzyme whose synthesis is repressed by O_2. All subsequent enzymes of the pathway are coordinately regulated and thus also repressed by O_2. But, in addition to anoxic conditions, nitrate must also be present before these enzymes are fully expressed.

The biochemistry of dissimilative nitrate reduction has been studied in detail in *E. coli*, in which NO_3^- is reduced only to NO_2^-, and *Paracoccus denitrificans* and *Pseudomonas stutzeri*, in which denitrification occurs. The *E. coli* nitrate reductase accepts electrons from a *b*-type cytochrome, and a comparison of the electron transport chains in aerobic versus nitrate-respiring cells of *E. coli* is shown in Figure 14.35*a, b*. Because of the reduction potential of the NO_3^-/NO_2^- couple (+0.43 V), fewer protons are pumped during nitrate reduction than in aerobic respiration (O_2/H_2O, +0.82 V). In *P. denitrificans* and *P. stutzeri*, nitrogen oxides are formed from NO_2^- by the enzymes nitrite reductase, nitric oxide reductase, and nitrous oxide reductase. NO and N_2O are gaseous intermediates that are free to escape from the cell, and N_2O in particular is a major product of denitrification, though these intermediates are often reduced all the way to N_2. During these electron transport reactions, a proton motive force is established (Figure 14.36*c*), and ATPase couples this to the synthesis of ATP.

MINIQUIZ

- For *Escherichia coli*, why is more energy released in aerobic respiration than during NO_3^- reduction?
- How do the products of NO_3^- reduction differ between *E. coli* and *Pseudomonas*?
- Where is the dissimilative nitrate reductase found in the cell? What unusual metal does it contain?

14.14 Sulfate and Sulfur Reduction

Several inorganic sulfur compounds are important electron acceptors in anaerobic respiration, and Table 14.5 lists the oxidation states of the key compounds. Sulfate (SO_4^{2-}), the most oxidized form of sulfur, is reduced by the *sulfate-reducing bacteria*, a highly diverse group of obligately anaerobic bacteria widely distributed in nature. The end product of sulfate reduction is hydrogen sulfide, H_2S, an important natural product that participates in many biogeochemical processes (⇄ Sections 21.4, 22.11, and 22.12). Species in the genus *Desulfovibrio*, in particular *D. desulfuricans*, have been widely studied, and the general properties of sulfate-reducing bacteria are discussed in Section 15.9.

As with nitrate (Section 14.13), it is necessary to distinguish between assimilative and dissimilative sulfate metabolism. Most microbes can incorporate sulfate for biosynthetic purposes to make cysteine, methionine, and many other organosulfur compounds; this is *assimilative* sulfate metabolism. By contrast, the ability to use sulfate as an electron acceptor for energy conservation requires its

TABLE 14.5 Sulfur compounds for sulfate reduction

Compound	Oxidation state of S atom
Organic S (R—SH)	−2
Sulfide (H_2S)	−2
Elemental sulfur (S^0)	0
Thiosulfate (—$S-SO_3^{2-}$)	−2/+6
Sulfur dioxide (SO_2)	+4
Sulfite (SO_3^{2-})	+4
Sulfate (SO_4^{2-})	+6

large-scale reduction and is restricted to the sulfate-reducing bacteria. H_2S is produced on a very large scale by these organisms and is excreted from the cell, free to be oxidized by air, used by other organisms, or combined with metals to form metal sulfides.

Biochemistry and Energetics of Sulfate Reduction

As was shown in Figure 14.25, SO_4^{2-} is an energetically much less favorable electron acceptor than is O_2 or NO_3^-. However, sufficient free energy to make ATP is available from sulfate reduction when an electron donor is oxidized that yields NADH or FADH. Hydrogen (H_2) is used by virtually all species, whereas the use of organic electron donors is more restricted. For example, lactate and pyruvate are widely used by species found in freshwater anoxic environments while acetate and longer-chain fatty acids are widely used by marine sulfate-reducing bacteria. Many morphological and physiological types of sulfate-reducing bacteria are known, and with the exception of *Archaeoglobus* (⇄ Section 17.4), a genus of *Archaea*, all known sulfate reducers are *Bacteria* (⇄ Section 15.9).

The reduction of SO_4^{2-} to H_2S requires eight electrons and proceeds through a number of intermediate stages. The reduction of SO_4^{2-} requires that it first be *activated* in a reaction requiring ATP. The enzyme ATP sulfurylase catalyzes the attachment of SO_4^{2-} to a phosphate of ATP, forming *adenosine phosphosulfate* (*APS*) as shown in **Figure 14.37*a***. Activation raises the extremely electronegative E_0' of the SO_4^{2-}/SO_3^{2-} couple (−0.52 V) to near 0 V, making reduction of the sulfate moiety possible with electron donors such as NADH (−0.32 V).

In *dissimilative* sulfate reduction, the SO_4^{2-} in APS is reduced directly to sulfite (SO_3^{2-}) by the enzyme APS reductase with the release of AMP. In *assimilative* reduction, by contrast, a second phosphate is added to APS to form *phosphoadenosine phosphosulfate* (*PAPS*) (Figure 14.37*a*), and only then is the SO_4^{2-} reduced. However, in both cases the product of sulfate reduction is sulfite (SO_3^{2-}), and once SO_3^{2-} is formed, it is reduced to H_2S by activity of the enzyme sulfite reductase (Figure 14.37*b*).

During dissimilative sulfate reduction, electron transport reactions generate a proton motive force and this drives ATP synthesis by ATPase. A major electron carrier in this process is *cytochrome c_3*, a periplasmic low-potential cytochrome (**Figure 14.38**). Cytochrome c_3 accepts electrons from a periplasmic hydrogenase and transfers these electrons to a membrane-associated protein complex. This complex, called *Hmc*, carries the electrons across the

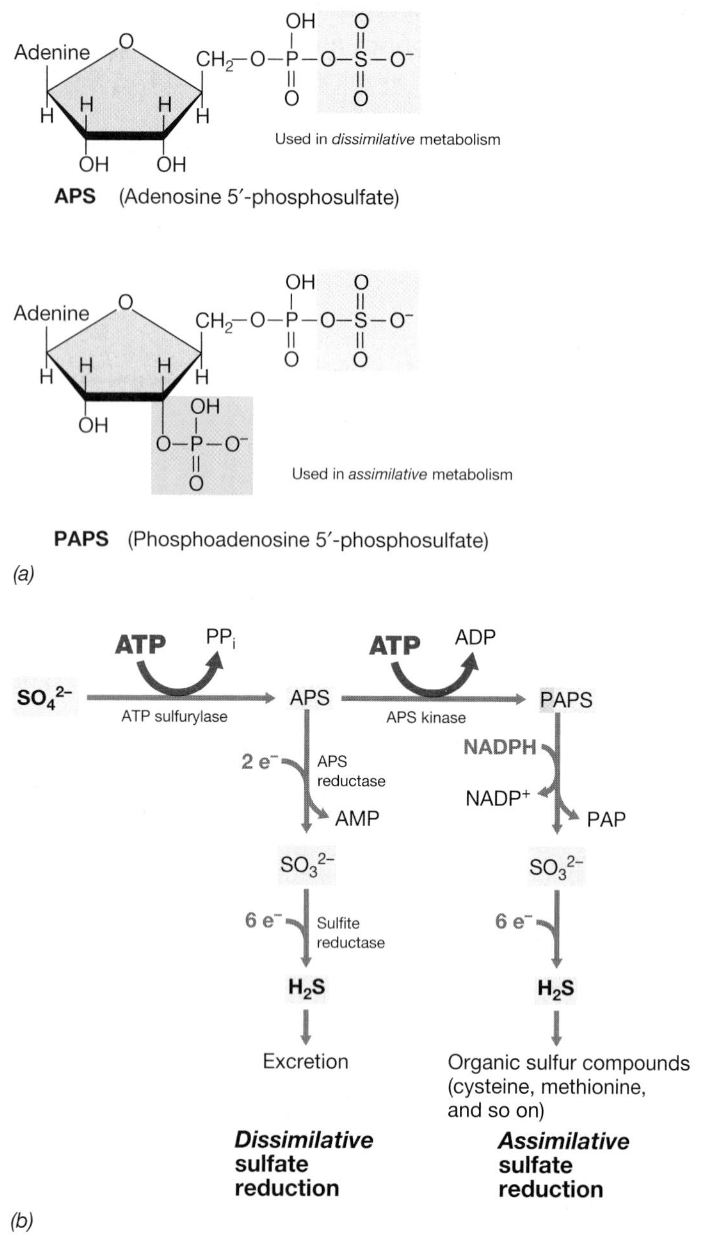

APS (Adenosine 5′-phosphosulfate)

Used in *dissimilative* metabolism

PAPS (Phosphoadenosine 5′-phosphosulfate)

Used in *assimilative* metabolism

(a)

SO_4^{2-} → (ATP, PP_i, ATP sulfurylase) → APS → (ATP, ADP, APS kinase) → PAPS

APS → (2 e⁻, APS reductase) → AMP → SO_3^{2-} → (6 e⁻, Sulfite reductase) → **H₂S** → Excretion

PAPS → (NADPH, NADP⁺, PAP) → SO_3^{2-} → (6 e⁻) → **H₂S** → Organic sulfur compounds (cysteine, methionine, and so on)

Dissimilative sulfate reduction

Assimilative sulfate reduction

(b)

Figure 14.37 Biochemistry of sulfate reduction: Activated sulfate. *(a)* Two forms of active sulfate can be made, adenosine 5′-phosphosulfate (APS) and phosphoadenosine 5′-phosphosulfate (PAPS). Both are derivatives of adenosine diphosphate (ADP), with the second phosphate of ADP being replaced by SO_4^{2-}. *(b)* Schemes of assimilative and dissimilative sulfate reduction.

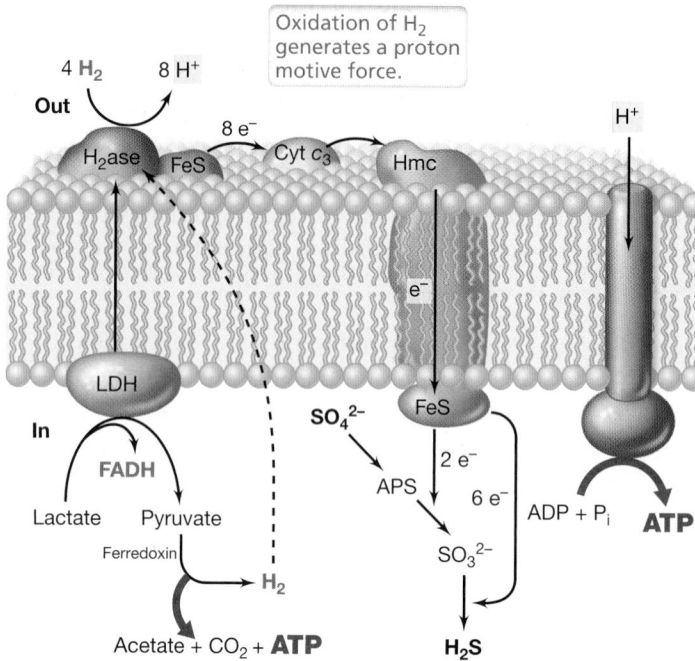

Figure 14.38 Electron transport and energy conservation in sulfate-reducing bacteria. In addition to external H_2, H_2 originating from the catabolism of organic compounds such as lactate and pyruvate can fuel hydrogenase. The enzymes hydrogenase (H_2ase), cytochrome (cyt) c_3, and a cytochrome complex (Hmc) are periplasmic proteins. A separate protein shuttles electrons across the cytoplasmic membrane from Hmc to a cytoplasmic iron–sulfur protein (FeS) that supplies electrons to APS reductase (forming SO_3^{2-}) and sulfite reductase (forming H_2S, Figure 14.37*b*). LDH, lactate dehydrogenase.

(Figure 14.38). A net of one ATP is produced for each SO_4^{2-} reduced to HS⁻ by H_2, and the reaction is

$$4 H_2 + SO_4^{2-} + H^+ \rightarrow HS^- + 4 H_2O \qquad \Delta G^{0'} = -152\,kJ$$

When lactate or pyruvate is the electron donor, ATP is produced not only from the proton motive force but also by substrate-level phosphorylation during the oxidation of pyruvate to acetate plus CO_2 (Figure 14.38).

Marine but not freshwater species of sulfate-reducing bacteria can couple sulfate reduction and the oxidation of acetate (and longer-chain fatty acids) to CO_2:

$$CH_3COO^- + SO_4^{2-} + 3 H^+ \rightarrow 2 CO_2 + H_2S + 2 H_2O$$

$$\Delta G^{0'} = -57.5\,kJ$$

The mechanism for acetate oxidation in most of these species is the *acetyl-CoA pathway*, a series of reversible reactions used by many anaerobes for acetate synthesis or acetate oxidation (Section 14.16). A few sulfate-reducing bacteria can also grow autotrophically with H_2. Under these conditions, the organisms use the acetyl-CoA pathway for making acetate as a carbon source. Such species can be cultured in a completely organic-free medium containing only mineral salts, sulfate, CO_2, and H_2.

Special Metabolisms of Sulfate-Reducing Bacteria

Certain species of sulfate-reducing bacteria can catalyze unusual reactions not characteristic of all species. These include *disproportionation, phosphite oxidation*, and *sulfur reduction*.

cytoplasmic membrane and transfers them to APS reductase and sulfite reductase, cytoplasmic enzymes that generate sulfite and sulfide, respectively (Figure 14.38).

The enzyme hydrogenase plays a central role in sulfate reduction whether *Desulfovibrio* is growing on H_2, per se, or on an organic compound such as lactate. This is because lactate is converted through pyruvate to acetate (much of the latter is either excreted or assimilated into cell material because *Desulfovibrio* cannot oxidize acetate to CO_2) with the production of H_2. This H_2 crosses the cytoplasmic membrane and is oxidized by the periplasmic hydrogenase to electrons, which are fed back into the system, and protons, which establish the proton motive force

Disproportionation is a process in which one molecule of a substance is oxidized while a second molecule is reduced, ultimately forming two different products. For example, *Desulfovibrio sulfodismutans* can disproportionate thiosulfate ($S-SO_3^{2-}$) as follows:

$$S-SO_3^{2-} + H_2O \rightarrow H_2S + SO_4^{2-} \quad \Delta G^{0\prime} = -21.9 \text{ kJ/reaction}$$

Note that in this reaction, the right-hand sulfur atom of $S-SO_3^{2-}$ is oxidized (forming SO_4^{2-}), while the left-hand atom is reduced (forming H_2S). The free energy available from the oxidation of thiosulfate by *D. sulfodismutans* is insufficient to couple to substrate-level phosphorylation and so instead is coupled to a proton "pump" that uses the minimal energy available in the reaction to establish a proton motive force. Other reduced sulfur compounds such as sulfite (SO_3^{2-}) and sulfur (S^0) can also be disproportionated. These forms of sulfur metabolism allow sulfate-reducing bacteria to recover energy from sulfur intermediates produced from the oxidation of H_2S by sulfur chemolithotrophs that coexist with them in nature and also from intermediates generated in their own metabolism during SO_4^{2-} reduction (Figure 14.37*b*).

At least one sulfate-reducing bacterium can couple phosphite (HPO_3^-) oxidation to SO_4^{2-} reduction. This chemolithotrophic reaction yields phosphate and sulfide:

$$4 HPO_3^- + SO_4^{2-} + H^+ \rightarrow 4 HPO_4^{2-} + HS^- \quad \Delta G^{0\prime} = -364 \text{ kJ}$$

This bacterium, *Desulfotignum phosphitoxidans*, is an autotroph and also a strict anaerobe, which by necessity it must be because phosphite spontaneously oxidizes in air. The natural sources of phosphite are likely to be organophosphorous compounds called *phosphonates* that are generated from the anoxic degradation of nucleic acids, phospholipids, and other cellular sources. Along with sulfur disproportionation (also a chemolithotrophic process) and H_2 utilization, phosphite oxidation underscores the diversity of chemolithotrophic reactions carried out by sulfate-reducing bacteria.

Sulfur Reduction

Besides sulfate, most sulfate-reducing bacteria can also conserve energy from the reduction of elemental sulfur to sulfide ($S^0 + 2 H \rightarrow H_2S$). In addition, however, a variety of non-sulfate-reducing microbes can also reduce sulfur in anaerobic respiration. These are the *sulfur reducers*, a large group of *Bacteria* and *Archaea* that coexist with sulfate-reducing bacteria in anoxic, sulfur-rich habitats in nature.

The electrons for sulfur reduction come from H_2 or any of a number of organic compounds. For example, *Desulfuromonas acetoxidans* can oxidize acetate or ethanol to CO_2 coupled to the reduction of S^0 to H_2S. Sulfur reducers lack the capacity to activate sulfate to APS (Figure 14.37), and presumably this is what excludes them from using SO_4^{2-} as an electron acceptor. *Desulfuromonas* contains several cytochromes, including an analog of cytochrome c_3, a key electron carrier in sulfate-reducing bacteria. In culture some sulfur reducers including *Desulfuromonas* can also use Fe^{3+} as an electron acceptor in place of sulfur, but sulfur is probably the major electron acceptor used in nature. Indeed, it is the reduction of oxidized sulfur compounds and the production of H_2S that connects the sulfur- and sulfate-reducing bacteria in an ecological sense.

14.15 Other Electron Acceptors

In addition to the electron acceptors for anaerobic respiration discussed thus far, several metals, metalloids, and halogenated and unhalogenated organic compounds are important electron acceptors for bacteria in nature (Figure 14.25). In addition to these, even protons can be used by a very few strict anaerobes. We consider these forms of anaerobic respiration here.

Metal and Metalloid Reduction

Several metals and metalloids can be reduced in anaerobic respirations. Ferric iron (Fe^{3+}) and manganic ion (Mn^{4+}) are the most important metals reduced. The reduction potential of the Fe^{3+}/Fe^{2+} couple is +0.2 V (at pH 7), and that of the Mn^{4+}/Mn^{2+} couple is +0.8 V; thus, several electron donors can couple to Fe^{3+} and Mn^{4+} reduction. In these reactions, electrons typically travel from the donor through an electron transport chain that generates a proton motive force and terminates in a metal reductase system, reducing Fe^{3+} to Fe^{2+} or Mn^{4+} to Mn^{2+}. The gram-negative bacteria *Shewanella* and *Geobacter* are major species here. Other inorganic substances can function as electron acceptors for anaerobic respiration, including the metalloids selenium, tellurium, and arsenic, and various oxidized chlorine compounds (some of these are shown in Figure 14.25). Several chlorate and perchlorate-reducing bacteria have also been isolated and are likely responsible for the removal of these toxic compounds from nature; the typical end product of these reactions is chloride (Cl^-).

The sulfate-reducing bacterium *Desulfotomaculum* can reduce both AsO_4^{3-} to AsO_3^{3-} and sulfate to sulfide (Figure 14.25), precipitating the yellow mineral orpiment (As_2S_3) in the process (**Figure 14.39**). This is an example of *biomineralization*, the formation of a mineral by bacterial activity. As_2S_3 formation also functions as a means of detoxifying what would otherwise be a toxic compound (arsenic), and thus such microbial activities may have practical applications for the cleanup of arsenic-containing toxic wastes and groundwater.

Organic Electron Acceptors

Several organic compounds can be electron acceptors in anaerobic respirations. Of those listed in Figure 14.25, the compound that has been most extensively studied is *fumarate*, a citric acid cycle intermediate (↪ Figure 3.16), which is reduced to succinate. The role of fumarate as an electron acceptor for anaerobic respiration derives from the fact that the fumarate/succinate couple has a reduction potential near 0 V, which allows coupling of fumarate reduction to the oxidation of

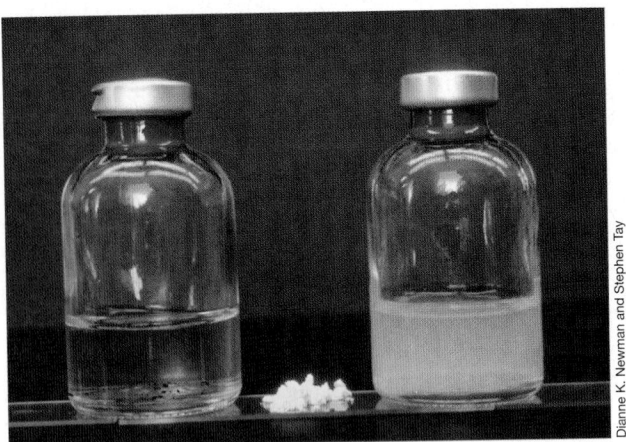

Figure 14.39 Biomineralization during arsenate reduction by the sulfate-reducing bacterium *Desulfotomaculum auripigmentum*. Left, appearance of culture bottle after inoculation. Right, following growth for 2 weeks and biomineralization of arsenic trisulfide, As_2S_3. Center, synthetic sample of As_2S_3.

NADH, FADH, or H_2. Many facultatively aerobic bacteria can grow anaerobically on fumarate as electron acceptor, including *Escherichia coli*.

Trimethylamine oxide (TMAO) and dimethyl sulfoxide (DMSO) (Figure 14.25) are important organic electron acceptors. TMAO is a product of marine fish, and several bacteria can reduce it to trimethylamine (TMA), which has a strong odor and flavor (the odor of spoiled seafood is due primarily to TMA produced by bacterial action). Dimethyl sulfoxide (DMSO), which is reduced to dimethyl sulfide (DMS), is a common natural product and is found in both marine and freshwater environments. The reduction potentials of the TMAO/TMA and DMSO/DMS couples are about the same (near +0.15 V, Figure 14.25) and so the electron transport chains that terminate with TMAO or DMSO reductases typically contain cytochromes of the *b* type.

Several halogenated organic compounds function as electron acceptors in **reductive dechlorination** (also called *dehalorespiration*). For example, the sulfate-reducing bacterium *Desulfomonile* grows anaerobically with H_2 or organic electron donors and chlorobenzoate as an electron acceptor that is reduced to benzoate and hydrochloric acid (HCl):

$$C_7H_4O_2Cl^- + 2\,H \rightarrow C_7H_5O_2^- + HCl$$

Several other anaerobic bacteria can reductively dechlorinate, and some are even restricted to chlorinated compounds as electron acceptors for anaerobic respiration. For example, *Dehalobacter* and *Dehalococcoides* oxidize H_2 and reduce tetrachloroethylene to dichloroethylene and ethene, respectively. *Dehalococcoides* can also reduce polychlorinated biphenyls (PCBs). PCBs are widespread organic pollutants that contaminate freshwater environments, where they accumulate in fish and other aquatic life. However, removal of the chlorine groups from these molecules greatly reduces their toxicity and hence reductive dechlorination is not only a form of anaerobic respiration but also an environmentally significant process of bioremediation.

Proton Reduction

Perhaps the simplest of all anaerobic respirations is one carried out by the hyperthermophile *Pyrococcus furiosus*. *P. furiosus* is a species of *Archaea* that grows optimally at 100°C (Chapter 17) on sugars and small peptides as electron donors and protons as electron acceptors. This is possible because of a unique biochemical feature of the glycolytic pathway of *P. furiosus*.

During glycolysis, the oxidation of glyceraldehyde 3-phosphate forms 1,3-bisphosphoglyceric acid, an intermediate with two energy-rich phosphate bonds; this compound is then converted to 3-phosphoglyceric acid plus ATP (⇌ Figure 3.14). However, in *P. furiosus* this glycolytic step is bypassed and instead, 3-phosphoglyceric acid is formed directly from glyceraldehyde 3-phosphate (**Figure 14.40**). This prevents *P. furiosus* from making ATP by substrate-level phosphorylation at this step, but this problem is compensated for by the fact that glyceraldehyde 3-phosphate oxidation is coupled to the production of *ferredoxin* rather than NADH; ferredoxin has a more negative E_0' (−0.42 V) than does $NAD^+/NADH$ (−0.32 V). This highly negative E_0' allows for the coupling of ferredoxin oxidation to the reduction of 2 H^+ to H_2, and this reaction pumps a proton across the membrane (Figure 14.40). Proton pumping by hydrogenase is analogous to proton pumping by terminal electron carriers in other respirations. Additional ATP is produced by *P. furiosus* by substrate-level phosphorylations in the conversions of phosphoenolpyruvate to pyruvate and acetyl-CoA to acetate (Figure 14.40).

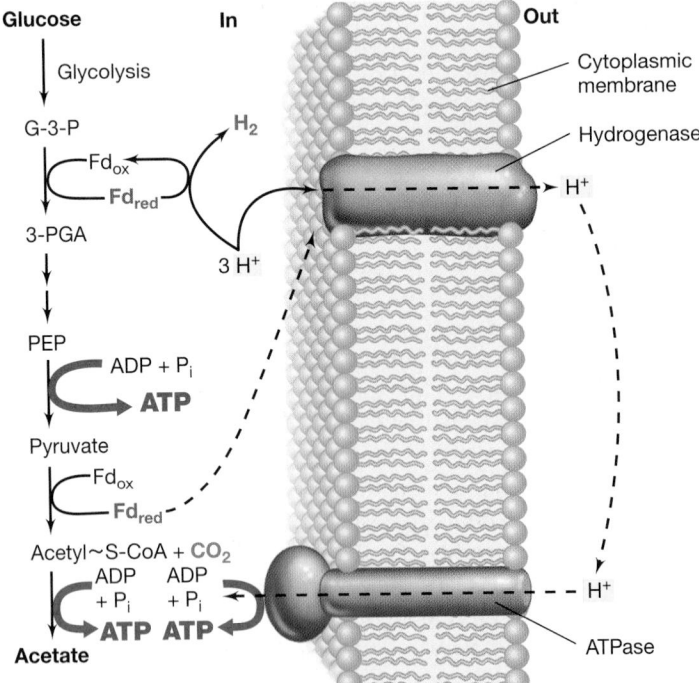

Figure 14.40 Modified glycolysis and proton reduction in anaerobic respiration in the hyperthermophile *Pyrococcus furiosus*. Hydrogen (H_2) production is linked to H^+ pumping by a hydrogenase that receives electrons from reduced ferredoxin (Fd_{red}). All intermediates from G-3-P downward in the pathway are present in two copies. Compare this figure with classical glycolysis in Figure 3.14. G-3-P, glyceraldehyde 3-phosphate; 3-PGA, 3-phosphoglycerate; PEP, phosphoenolpyruvate.

V • One-Carbon (C_1) Metabolism

Carbon dioxide (CO_2) and methane (CH_4) are abundant in many anoxic habitats, and a wide diversity of microbes have evolved metabolic pathways that conserve energy from either the reduction of CO_2 or the oxidation of CH_4. A number of the enzymatic reactions in the metabolism of one-carbon compounds (C_1 metabolism) are unique to this kind of metabolism. In this section we consider the metabolism of organisms that perform C_1 metabolism of one sort or the other, highlighting the major similarities and differences.

14.16 Acetogenesis

Two major groups of strictly anaerobic microbes use CO_2 as an electron acceptor for energy conservation. One of these is the *acetogens*, and we discuss them here. The other group, the *methanogens*, are considered in the next section. Hydrogen (H_2) is a major electron donor for both of these organisms, and an overview of their energy metabolism, **acetogenesis** and **methanogenesis**, is shown in **Figure 14.41**. Both processes are linked to ion pumps, of either protons (H^+) or sodium ions (Na^+), as the mechanism of energy conservation, and these pumps fuel ATP synthases in the membrane. The pathway of acetogenesis also conserves energy in a substrate-level phosphorylation reaction.

Organisms and Pathway

Acetogens carry out the reaction

$$4 H_2 + H^+ + 2 HCO_3^- \rightarrow CH_3COO^- + 4 H_2O \qquad \Delta G^{0\prime} = -105 \, kJ$$

In addition to H_2, electron donors for acetogenesis include various C_1 compounds such as methanol, several methoxylated aromatic compounds, sugars, organic and amino acids, alcohols, and

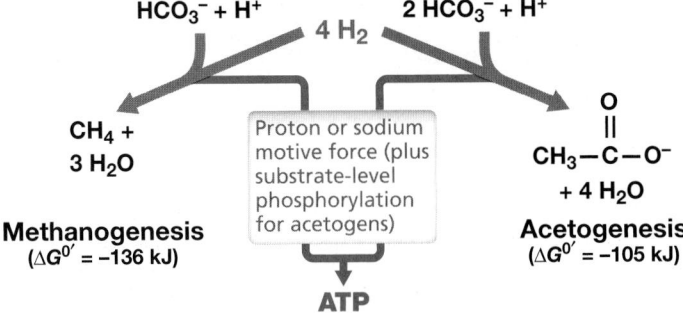

Figure 14.41 The contrasting processes of methanogenesis and acetogenesis. Note the difference in free energy released in the reactions.

certain nitrogen bases, depending on the organism. Many acetogens can also reduce nitrate (NO_3^-) and thiosulfate ($S_2O_3^{2-}$) in dissimilative metabolisms. However, CO_2 reduction is the major reaction of ecological relevance.

A major unifying thread among acetogens is the pathway of CO_2 reduction. Acetogens reduce CO_2 to acetate by the reductive acetyl-CoA pathway (also called the *Wood–Ljungdahl pathway*), the major pathway in obligate anaerobes for the production of acetate (see Figure 14.42). The reactions of the reductive acetyl-CoA pathway are reversible, and some microbes reverse this pathway to oxidize acetate. **Table 14.6** lists the groups that either produce or oxidize acetate by way of the acetyl-CoA pathway.

Acetogens such as *Acetobacterium woodii* and *Clostridium aceticum* can grow either chemoorganotrophically by fermentation of sugars (reaction A) or chemolithotrophically and autotrophically through the reduction of CO_2 to acetate with H_2 as electron donor (reaction B). In either case, the sole product is *acetate*:

(A) $C_6H_{12}O_6 \rightarrow 3 CH_3COO^- + 3 H^+$

(B) $2 HCO_3^- + 4 H_2 + H^+ \rightarrow CH_3COO^- + 4 H_2O$

When growing on glucose, acetogens use glycolysis (⟸ Figure 3.14) to oxidize the glucose into two molecules of pyruvate and two molecules of NADH. The pyruvate is then further oxidized to produce two molecules of acetate:

(C) $2 Pyruvate^- \rightarrow 2 acetate^- + 2 CO_2 + 2 NADH$

The CO_2 generated in reaction (C) is then used as a terminal electron acceptor in the reductive acetyl-CoA pathway. The NADH generated during glycolysis and pyruvate oxidation is used as an

TABLE 14.6 **Organisms employing the acetyl-CoA pathway**
I. Pathway drives acetate synthesis for energy purposes
Acetoanaerobium noterae
Acetobacterium woodii
Acetobacterium wieringae
Acetogenium kivui
Acetitomaculum ruminis
Clostridium aceticum
Clostridium formicaceticum
Clostridium ljungdahlii
Moorella thermoacetica
Desulfotomaculum orientis
Sporomusa paucivorans
Eubacterium limosum (also produces butyrate)
Treponema primitia (from termite hindguts)
II. Pathway drives acetate synthesis for cell biosynthesis
Acetogens
Methanogens
Sulfate-reducing bacteria
III. Pathway drives acetate oxidation for energy purposes
Reaction: Acetate + H^+ + 2 $H_2O \rightarrow$ 2 CO_2 + 8 H
Group II sulfate reducers (other than *Desulfobacter*)
Reaction: Acetate + $H^+ \rightarrow CO_2 + CH_4$
Acetotrophic methanogens (*Methanosarcina*, *Methanosaeta*)

electron donor in CO_2 reduction. Starting from pyruvate, then, the overall production of acetate can be written as

$$2\,\text{Pyruvate}^- + 4\,H \rightarrow 3\,\text{acetate}^- + H^+$$

Most acetogenic bacteria that produce acetate in energy metabolism are gram-positive *Bacteria*, and many are species of the genera *Clostridium* or *Acetobacterium* (Table 14.6). A few other gram-positive and many different gram-negative *Bacteria* and *Archaea* use the reductive acetyl-CoA pathway for autotrophic purposes, reducing CO_2 to acetate as a source of cell carbon; these include autotrophic sulfate-reducing bacteria (Section 14.14), anammox bacteria (Section 14.12), and methanogens (Section 14.17). And finally, some microbes run the acetyl-CoA pathway in the *reverse direction* as a means of oxidizing acetate to CO_2; these include acetate-utilizing methanogens and sulfate-reducing bacteria. The acetyl-CoA pathway is thus a metabolically highly versatile series of reactions.

The Reductive Acetyl-CoA Pathway and Energy Conservation in Acetogenesis

Unlike other autotrophic pathways (Section 14.5), the reductive acetyl-CoA pathway of CO_2 fixation is not a cycle. Instead,

it catalyzes the reduction of CO_2 along two linear pathways, with one molecule of CO_2 being reduced to the methyl group of acetate (the methyl branch of the pathway) and the other to the carbonyl group of acetate (the carbonyl branch of the pathway). These two C_1 units are then combined to form acetyl-CoA (**Figure 14.42**).

A key enzyme of the acetyl-CoA pathway is *carbon monoxide* (*CO*) *dehydrogenase*. CO dehydrogenase contains Ni, Zn, and Fe as cofactors and catalyzes the reaction

$$CO_2 + H_2 \rightarrow CO + H_2O$$

The CO produced by CO dehydrogenase ends up as the *carbonyl* carbon of acetate (Figure 14.42). The methyl group of acetate originates from the reduction of CO_2 by a series of reactions in which the coenzyme *tetrahydrofolate* plays a major role (Figure 14.42). The methyl group is then transferred from tetrahydrofolate to a cobalt- and iron-containing *corrinoid iron–sulfur protein* (CoFeSP) coenzyme. In the final step of the pathway, the methyl group is combined with CO by the activity of both CO dehydrogenase and acetyl-CoA synthase to form acetyl-CoA. Conversion of

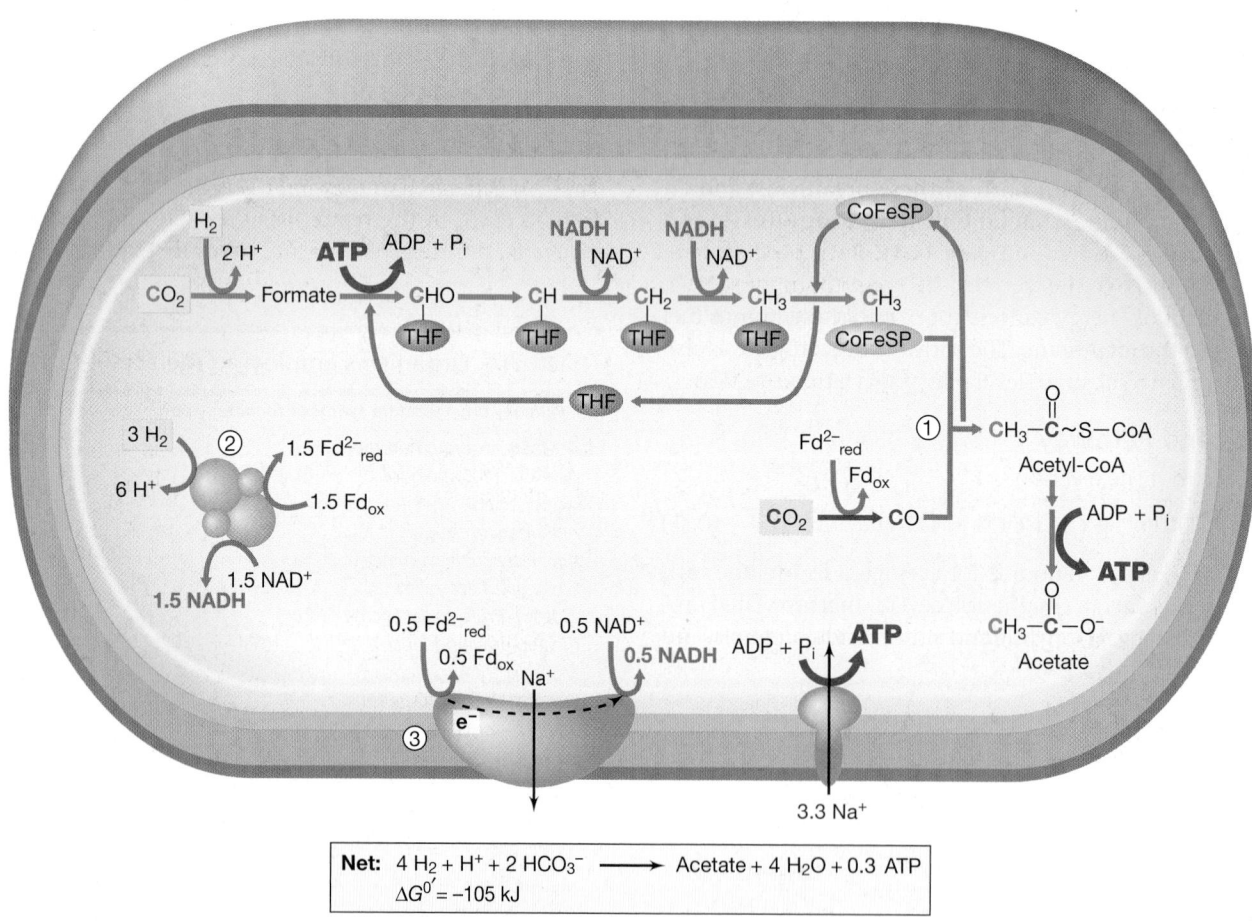

Net: $4\,H_2 + H^+ + 2\,HCO_3^- \longrightarrow$ Acetate $+ 4\,H_2O + 0.3\,$ATP
$\Delta G^{0\prime} = -105\,$kJ

Figure 14.42 Reactions of acetogenesis from H_2 and CO_2 in *Acetobacterium woodii*. The reductive acetyl-CoA pathway is used to reduce CO_2 to acetate. The pathway has a methyl branch (blue arrows) and a carbonyl branch (red arrows). Carbon monoxide from the carbonyl branch and a methyl group from the methyl branch are combined into acetyl-CoA by carbon monoxide dehydrogenase (1), a key enzyme for the pathway. Note that the reductive acetyl-CoA pathway does not conserve energy. An electron-bifurcating hydrogenase (2) (see also Figure 14.43) is used to reduce ferredoxin (Fd^{2-}_{red}), and energy is conserved at the Rnf complex (3), which generates a Na^+ motive force. THF, tetrahydrofolate; CoFeSP, Co/Fe-containing corrinoid iron–sulfur protein.

acetyl-CoA to acetate is the last step in the pathway, generating one ATP by substrate-level phosphorylation (Figure 14.42, see Table 14.7). However, this ATP is consumed in the first step of the acetyl-CoA pathway. Considering this, how then do acetogens get their ATP?

Acetogens conserve energy by the generation of an ion motive force. In *Acetobacterium woodii* the ion motive force is generated by activity of the *Rnf complex* (or in some acetogens a related complex called the *Ech complex*); these enzymes use reduced ferredoxin (Fd^{2-}_{red}) as electron donor and NAD^+ as electron acceptor. The Rnf complex pumps one Na^+ across the membrane for each electron exchanged, thereby generating a Na^+ motive force, which can be used to make ATP using a Na^+-dependent ATP synthase (Figure 14.42). Alternatively, in other acetogens such as *Clostridium ljungdahlii*, the Rnf complex instead pumps H^+ instead of Na^+ and has a typical H^+-dependent ATP synthase. Ultimately, only 0.3 ATP are produced for every 4 H_2 and 2 CO_2 that are consumed by the acetogens, making these organisms minimalists in terms of the energy that can be conserved from their metabolism.

Flavin-Based Electron Bifurcation

The reduced ferredoxin (Fd^{2-}_{red}) that donates electrons to the Rnf cluster has a reduction potential of −0.45 V, but H_2 has a reduction potential of −0.414 V. How is it possible for acetogens to generate an ion motive force when the reduction of ferredoxin by H_2 is thermodynamically unfavorable? The answer is a process called *electron bifurcation* (**Figure 14.43**).

Electron bifurcation is an application of a basic concept we considered in Chapter 3: Endergonic reactions can be driven

forward by coupling them to exergonic reactions. In electron bifurcation the endergonic reduction of Fd^{2-}_{red} by a hydrogenase is coupled to the exergonic reduction of $NAD^+/NADH$ ($E_0' = -0.32$ V) by the same hydrogenase. The electron-bifurcating hydrogenase contains a flavin coenzyme. The flavin accepts two electrons at a time and donates one of these electrons to a higher-potential electron acceptor (NAD^+) in order to drive the unfavorable reduction of ferredoxin by the other electron (Figure 14.43). Hence, a total of 2 H_2 must be oxidized to generate one Fd^{2-}_{red} and one NADH.

Flavin-based electron bifurcation is used by many obligate anaerobes to generate the Fd^{2-}_{red} that they need. In particular, we will see shortly that flavin-based electron bifurcation is essential to energy conservation in methanogens that lack cytochromes.

--- **MINIQUIZ** ---

- What is the purpose of CO dehydrogenase?
- If acetogens conserve energy using the Rnf complex, then what is the purpose of the reductive CoA pathway?
- What is electron bifurcation and what role does it play in acetogens?

14.17 Methanogenesis

The biological production of methane—*methanogenesis*—is catalyzed by a group of strictly anaerobic *Archaea* called the **methanogens**. These organisms are present in freshwater sediments (**Figure 14.44**), sewage sludge digesters (⇌ Section 22.6) and other bioreactors, and the intestines of warm-blooded animals, including humans. The reduction of CO_2 by H_2 to form methane

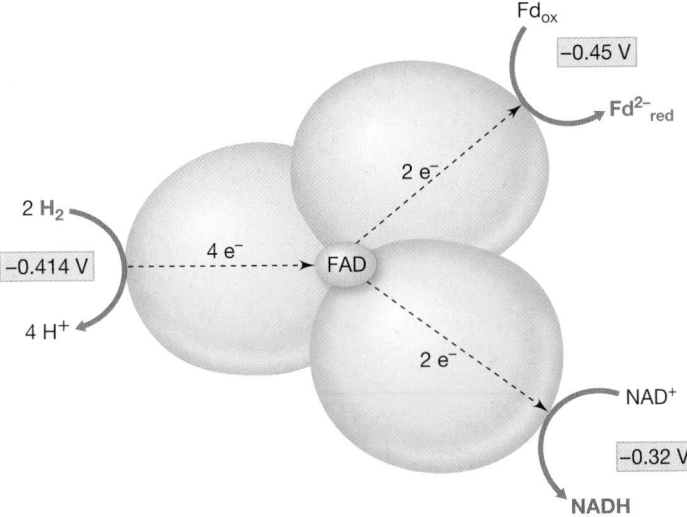

Figure 14.43 The reaction scheme for flavin-based electron bifurcation. Many obligate anaerobes require reduced ferredoxin as an electron donor (see Figure 14.42), but lack external electron donors that are sufficiently electronegative to reduce ferredoxin. In flavin-based electron bifurcation, two electrons from an electron donor (such as H_2) are transferred to a flavin (FAD), and one electron is used to reduce a favorable electron acceptor (such as NAD^+), making it possible to drive the second electron to an unfavorable electron acceptor (such as Fd_{ox}). A total of 4 electrons from H_2 are bifurcated to produce reduced ferredoxin (Fd^{2-}_{red}) and NADH.

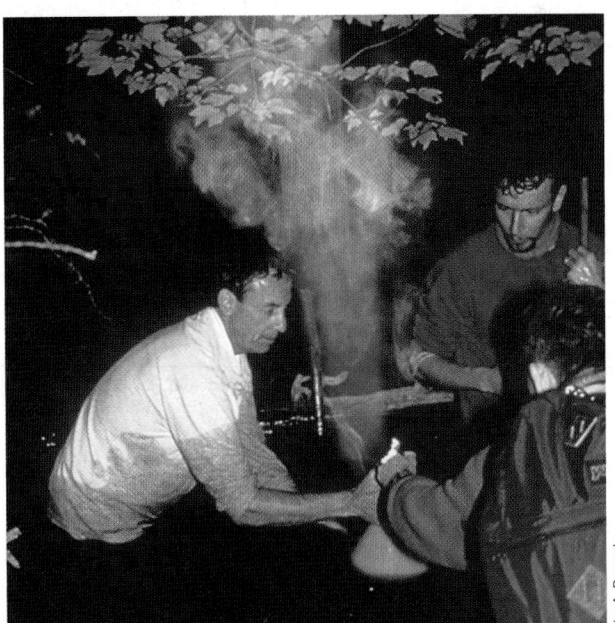

Figure 14.44 Methanogenesis. Methane is collected in a funnel from swamp sediments where it was produced by methanogens and then ignited in a demonstration experiment at Woods Hole, Massachusetts (USA).

UNIT 4

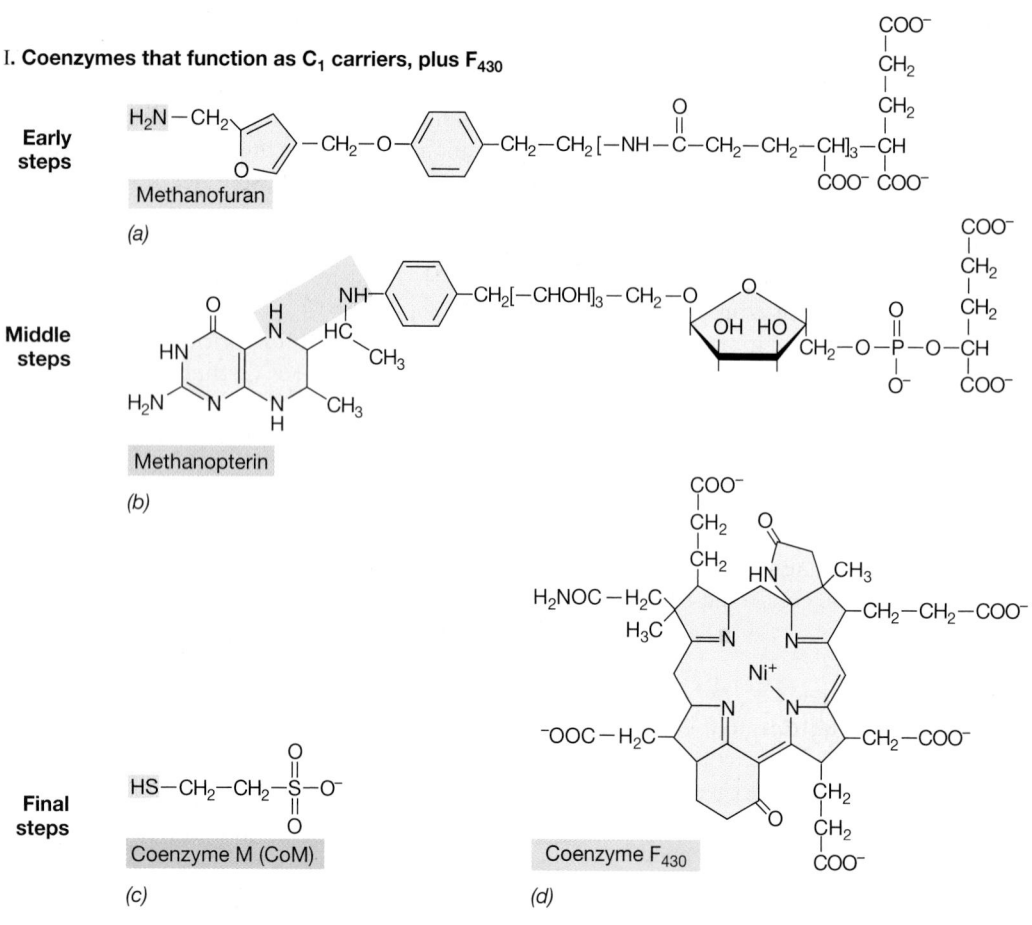

Figure 14.45 Coenzymes of methanogenesis. The atoms shaded in brown or yellow are the sites of oxidation–reduction reactions (brown in F_{420} and CoB) or the position to which the C_1 moiety is attached during the reduction of CO_2 to CH_4 (dark yellow in methanofuran, methanopterin, and coenzyme M). The same colors used to highlight a particular coenzyme (CoB is orange, for example) are also used in Figures 14.47 and 14.48 to follow the reactions in each figure. Coenzyme F_{430} participates in the terminal step of methanogenesis catalyzed by the enzyme methyl reductase, with the methyl group binding to Ni^+ in F_{430} prior to its reduction to CH_4.

(CH_4) is a major pathway of methanogenesis and is a form of anaerobic respiration. We consider the basic properties, phylogeny, and taxonomy of the methanogens in Section 17.2. Here we focus on their bioenergetics and unique biochemistry.

C_1 Carriers in Methanogenesis

Methanogenesis from CO_2 requires eight electrons, and these electrons are added two at a time. This leads to intermediary oxidation states of the carbon atom from +4 (CO_2) to −4 (CH_4). Several novel coenzymes participate in methanogenesis and can be divided into two classes: (1) those that carry the C_1 unit along its path of enzymatic reduction (C_1 *carriers*) and (2) those that donate electrons (*redox coenzymes*) (**Figure 14.45**). We consider the C_1 carriers first.

The coenzyme *methanofuran* is required for the first step of methanogenesis. Methanofuran contains the five-membered furan ring and an amino nitrogen atom that binds CO_2 (Figure 14.45*a*). *Methanopterin* (Figure 14.45*b*) is a methanogenic coenzyme that resembles the vitamin folic acid and plays a role analogous to that of tetrahydrofolate (a coenzyme that participates in C_1 transformations; see Figure 14.42) by carrying the C_1 unit in the intermediate steps of CO_2 reduction to CH_4. *Coenzyme M (CoM)* (Figure 14.45*c*) is required for the terminal step of methanogenesis, the reduction of the methyl group (CH_3) to CH_4. Although not a C_1 carrier, the nickel (Ni^{2+})-containing tetrapyrrole *coenzyme F_{430}* (Figure 14.45*d*) also participates in the terminal step of methanogenesis as part of the methyl reductase enzyme complex (discussed later).

Redox Coenzymes

The coenzymes *F_{420}* and *7-mercaptoheptanoylthreonine phosphate* (also called coenzyme B, CoB) are electron donors in methanogenesis. Coenzyme F_{420} (Figure 14.45*e*) is a flavin derivative, structurally resembling the flavin coenzyme FMN (↩ Figure 3.18). F_{420} participates in methanogenesis as the electron donor in several steps of CO_2 reduction (see Figure 14.47). Coenzyme F_{420} takes its name from the fact that its oxidized form absorbs light at 420 nm and fluoresces blue-green. Such fluorescence is useful for the microscopic identification of a methanogen (**Figure 14.46**). CoB is required for the terminal step of methanogenesis catalyzed by the *methyl reductase enzyme complex*. As shown in Figure 14.45*f*, the structure of CoB resembles the vitamin pantothenic acid, which is part of acetyl-CoA (↩ Figure 3.13).

Methanogenesis from CO_2 + H_2

Electrons for the reduction of CO_2 to CH_4 typically come from H_2, but a few other substrates can also supply the electrons in some methanogens. **Figure 14.47** shows the steps in CO_2 reduction by H_2:

1. CO_2 is activated by a methanofuran-containing enzyme and reduced to the formyl level. The immediate electron donor is ferredoxin, a strong reductant with a reduction potential (E_0') near −0.4 V.

2. The formyl group is transferred from methanofuran to an enzyme containing methanopterin (MP in Figure 14.47). It is subsequently dehydrated and reduced in two separate steps (total of 4 H) to the methylene and methyl levels. The immediate electron donor here is reduced F_{420}.

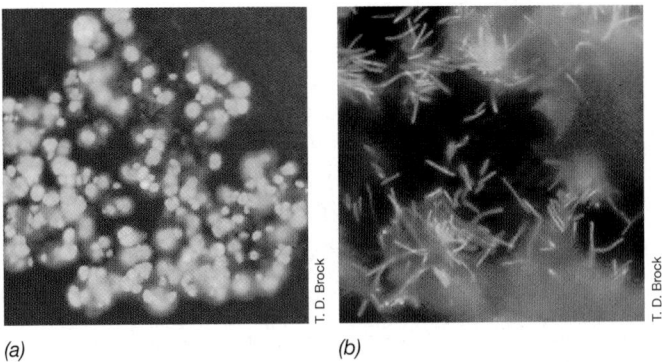

Figure 14.46 Fluorescence due to the methanogenic coenzyme F_{420}. *(a)* Autofluorescence in cells of the methanogen *Methanosarcina barkeri* due to the presence of the unique electron carrier F_{420}. A single cell is about 1.7 μm in diameter. The organisms were made visible by excitation with blue light in a fluorescence microscope. *(b)* F_{420} fluorescence in cells of the methanogen *Methanobacterium formicicum*. A single cell is about 0.6 μm in diameter.

3. The methyl group is transferred from methanopterin to an enzyme containing CoM by the enzyme methyl transferase. This reaction is highly exergonic and linked to the pumping of Na^+ across the membrane from inside to outside the cell.

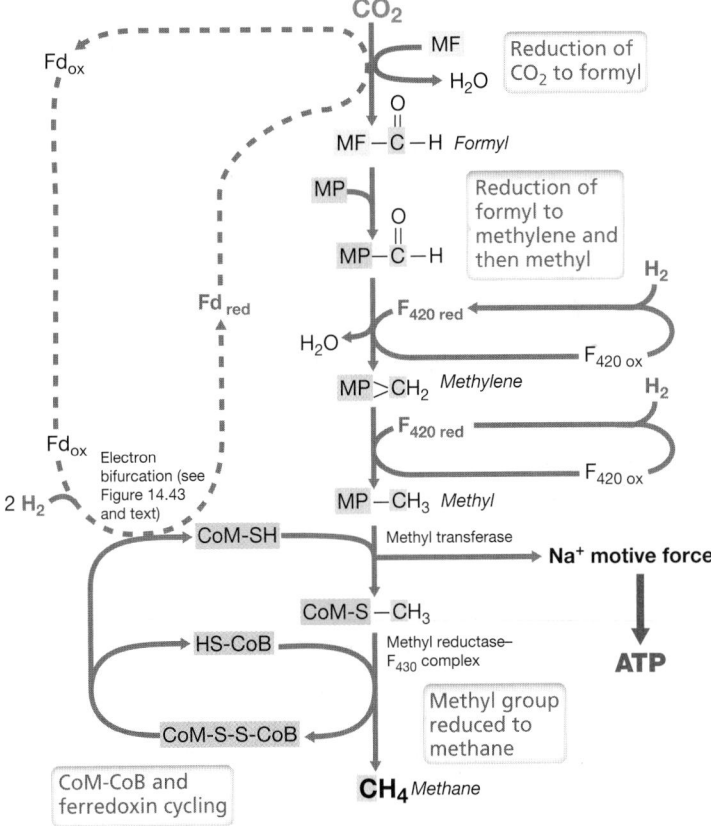

Figure 14.47 Methanogenesis from CO_2 plus H_2. The carbon atom reduced is highlighted in green, and the source of electrons is highlighted in brown. See Figure 14.45 for the structures of the coenzymes. MF, methanofuran; MP, methanopterin; CoM, coenzyme M; $F_{420\ red}$, reduced coenzyme F_{420}; F_{430}, coenzyme F_{430}; Fd, ferredoxin; CoB, coenzyme B.

4. Methyl-CoM is reduced to methane by methyl reductase. In this reaction, F_{430} and CoB are required. Coenzyme F_{430} removes the CH_3 group from CH_3–CoM, forming a Ni^+–CH_3 complex. This complex is reduced by CoB, generating CH_4 and a disulfide complex of CoM and CoB (CoM-S—S-CoB).

5. Free CoM and CoB are regenerated by the reduction of CoM-S—S-CoB ($E_0' = 0.14$) through a flavin-based electron-bifurcation reaction (see Figure 14.43) with H_2 (−0.414 V) as the electron donor. The high reduction potential of CoM-S—S-CoB is used to drive electrons to reduce ferredoxin ($E_0' = -0.45$ V). The Fd^{2-}_{red} is then used for CO_2 reduction in the first step of the pathway (Figure 14.47).

Methanogenesis from Methyl Compounds and Acetate

We will learn in Section 17.2 that methanogens can form CH_4 from certain methylated compounds such as methanol and acetate, as well as from $H_2 + CO_2$. Methanogens that reduce methylated compounds, such as *Methanosarcina*, typically have cytochromes, a feature that distinguishes them from methanogens that use only $H_2 + CO_2$. Methanol is catabolized by donating methyl groups to an enzyme containing a corrinoid coenzyme to form CH_3–corrinoid. Corrinoids are the parent structures of compounds such as vitamin B_{12} and contain a porphyrin-like ring with a central cobalt atom. The

CH_3–corrinoid complex then transfers the methyl group to CoM, yielding CH_3–CoM (**Figure 14.48a**) from which methane is formed in the same way as in the terminal step of CO_2 reduction. If H_2 is unavailable to drive the terminal step, some of the methanol must be oxidized to CO_2 to yield electrons for this purpose. This occurs by reversal of steps in methanogenesis (Figures 14.47 and 14.48a).

When acetate is the substrate for methanogenesis, it is first activated to acetyl-CoA, which interacts with CO dehydrogenase from the acetyl-CoA pathway (Section 14.16 and Figure 14.42). The methyl group of acetate is then transferred to the corrinoid enzyme to yield CH_3–corrinoid, and from there it follows the CoM-mediated terminal step of methanogenesis. Simultaneously, the CO group is oxidized to yield CO_2 and electrons (Figure 14.48b).

Autotrophy

Autotrophy in methanogens is supported by the reductive acetyl-CoA pathway (Section 14.16). As we have just seen, parts of this pathway are already integrated into the catabolism of methanol and acetate by methanogens (Figure 14.48). However, methanogens lack the tetrahydrofolate-driven series of reactions of the acetyl-CoA pathway that lead to the production of a methyl group (Figure 14.42). But this is not a problem because methanogens either derive methyl groups directly from their electron donors (Figure 14.48) or make methyl groups during methanogenesis from $H_2 + CO_2$ (Figure 14.47). Thus methanogens have access to

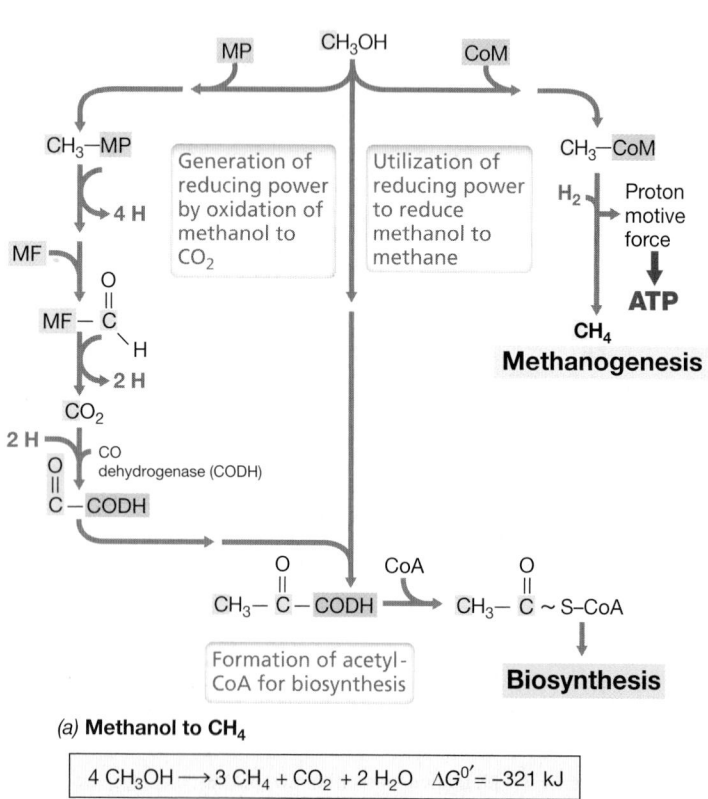

(a) **Methanol to CH₄**

$$4\ CH_3OH \longrightarrow 3\ CH_4 + CO_2 + 2\ H_2O \quad \Delta G^{0'} = -321\ kJ$$

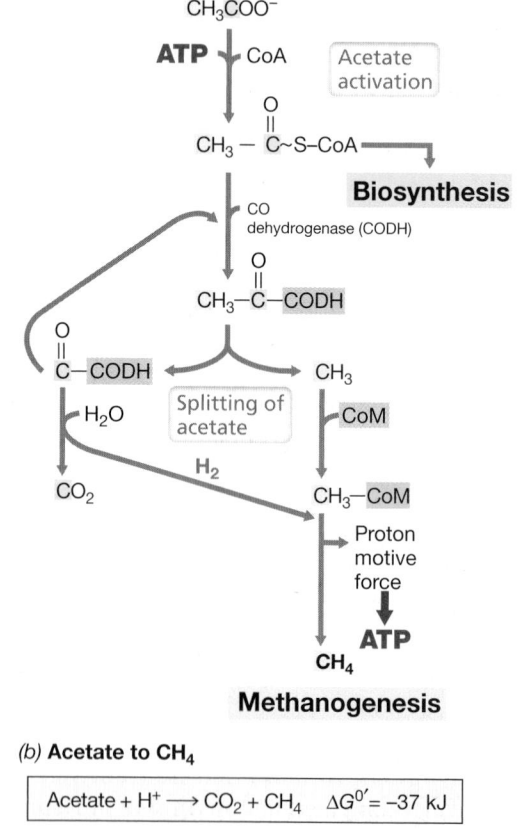

(b) **Acetate to CH₄**

$$Acetate + H^+ \longrightarrow CO_2 + CH_4 \quad \Delta G^{0'} = -37\ kJ$$

Figure 14.48 Methanogenesis from methanol and acetate. Both reaction series contain parts of the acetyl-CoA pathway. *(a)* For growth on methanol (CH_3OH), most CH_3OH carbon is converted to CH_4, and a smaller amount is converted to either CO_2 or, via formation of acetyl-CoA, is assimilated into cell material. *(b)* Acetate is split into CH_4 and CO_2. Abbreviations and color-coding are as in Figures 14.45 and 14.47; CODH, carbon monoxide dehydrogenase.

abundant methyl groups, and the removal of a portion for biosynthesis is of little consequence. The carbonyl group of the acetate produced during autotrophic growth of methanogens is derived from the activity of carbon monoxide dehydrogenase, and the terminal step in acetate synthesis is as described for acetogens (Section 14.16 and Figure 14.42).

Energy Conservation in Methanogenesis

Under standard conditions, the free energy of methanogenesis from $H_2 + CO_2$ is −131 kJ/mol. Energy conservation in methanogenesis occurs at the expense of a proton or sodium motive force, depending on the substrate used; substrate-level phosphorylation (⟲ Sections 3.8 and 14.19) does not occur. When methane is formed from $H_2 + CO_2$, ATP is produced from the sodium motive force generated during methyl transfer from MP to CoM by the enzyme methyl transferase (Figure 14.47). This energized state of the membrane then drives the synthesis of ATP, probably by way of an H^+-linked ATPase following conversion of the sodium motive force into a proton motive force by exchange of Na^+ for H^+ across the membrane. The ATP yield per CH_4 produced is about 0.5.

In some methanogens, such as *Methanosarcina*, a nutritionally versatile organism that can make methane from acetate or methanol as well as from $CO_2 + H_2$, a different mechanism of energy conservation occurs from acetate or methanol, since the methyl transferase reaction cannot be coupled to the generation of a sodium motive force under these conditions. Instead, in acetate- and methanol-grown cells energy conservation is linked to the terminal step in methanogenesis, the methyl reductase step (Figures 14.47, 14.48, and **Figure 14.49**). In this reaction, the interaction of CoB with CH_3–CoM and methyl reductase forms CH_4 and a heterodisulfide product, CoM-S—S-CoB. The latter is reduced by H_2 to regenerate CoM-SH and CoB-SH (Figure 14.49). This reduction, carried out by the enzyme *heterodisulfide reductase*, is exergonic and is coupled to the pumping of H^+ across the membrane (Figure 14.49). Electrons from H_2 flow to the heterodisulfide reductase through a membrane-associated electron carrier called *methanophenazine*. This compound is reduced by F_{420} and subsequently oxidized by a *b*-type cytochrome; the latter is the electron donor to the heterodisulfide reductase (Figure 14.49). Cytochromes and methanophenazine are absent in methanogens that can use only $H_2 + CO_2$ for methanogenesis, and $H_2 + CO_2$ methanogens instead regenerate CoM-SH and CoB-SH by using an electron-bifurcation reaction (Figure 14.43).

In methanogens we thus see at least two mechanisms for energy conservation: (1) a proton motive force linked to the methyl reductase reaction and used to drive ATP synthesis in acetate- or methanol-grown cells, and (2) a sodium motive force (which is likely converted to a proton motive force) during methanogenesis from $H_2 + CO_2$.

(a)

(b)

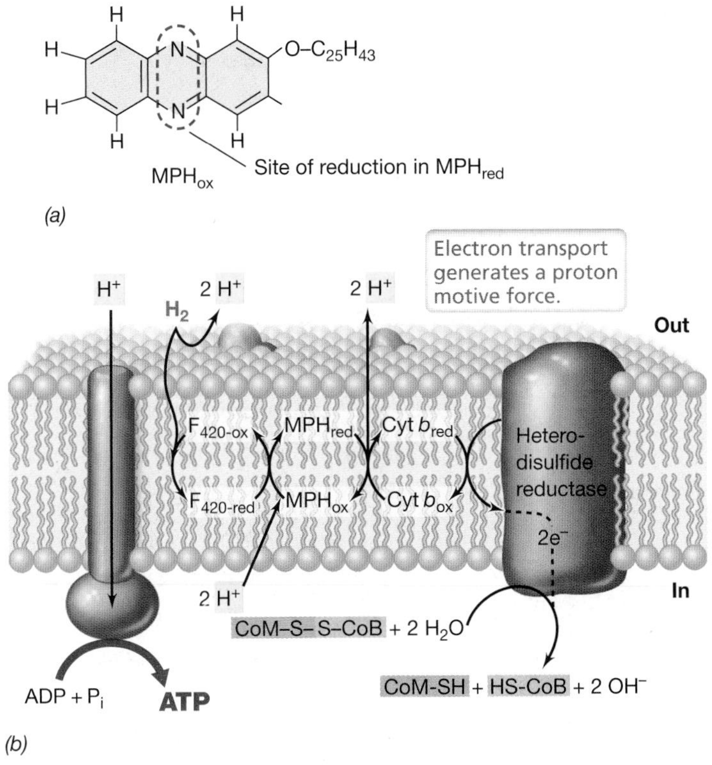

Figure 14.49 Energy conservation in methanogenesis from methanol or acetate. *(a)* Structure of methanophenazine (MPH in part *b*), an electron carrier in the electron transport chain leading to ATP synthesis; the central ring of the molecule can be alternately reduced and oxidized. *(b)* Steps in electron transport. Electrons originating from H_2 reduce F_{420} and then methanophenazine. The latter, through a cytochrome of the *b* type, reduces heterodisulfide reductase with the extrusion of H^+ to the outside of the membrane. In the final step, heterodisulfide reductase reduces CoM-S—S-CoB to HS-CoM and HS-CoB. See Figure 14.45 for the structures of CoM and CoB.

MINIQUIZ

- Which coenzymes function as C_1 carriers in methanogenesis? Which function as electron donors?
- In methanogens growing on $H_2 + CO_2$, how is carbon obtained for cell biosynthesis?
- How is ATP made in methanogenesis when the substrates are $H_2 + CO_2$? Acetate?

14.18 Methanotrophy

Methane (CH_4) and many other C_1 organic compounds can be catabolized both aerobically and anaerobically. In this section, we consider the oxidation of these compounds by **methylotrophs**, organisms that use organic compounds that lack C—C bonds as electron donors and carbon sources. The oxidations of CH_4 and methanol (CH_3OH) have been the best-studied reactions, and we focus here on the oxidation of CH_4 as an example of a methylotrophic lifestyle.

Aerobic Methane Oxidation

The steps in CH_4 oxidation to CO_2 can be summarized as

$$CH_4 \rightarrow CH_3OH \rightarrow CH_2O \rightarrow HCOO^- \rightarrow CO_2$$

Not all methylotrophs can use methane. **Methanotrophs** are those methylotrophs that can use CH_4, and methanotrophy has been especially well studied in the gram-negative bacterium *Methylococcus capsulatus*. Methanotrophs assimilate either all or one-half of their cell carbon (depending on the pathway used) from the C_1 compound formaldehyde (CH_2O).

The initial step in the *aerobic* oxidation of CH_4 is catalyzed by the enzyme *methane monooxygenase* (MMO). Monooxygenases

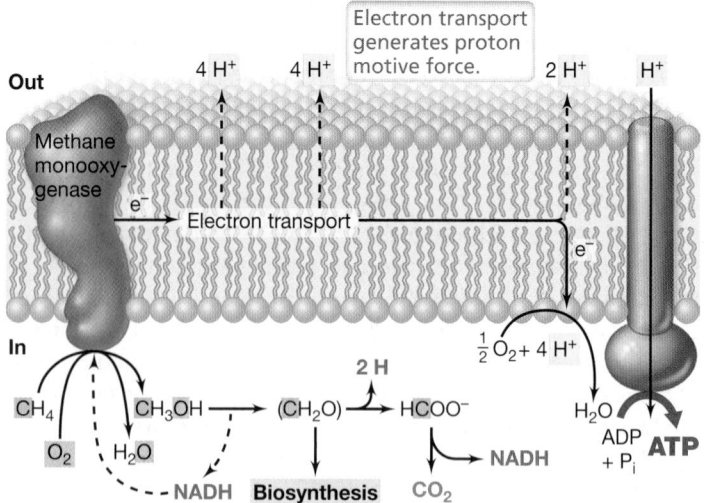

Figure 14.50 Oxidation of methane by methanotrophic bacteria. CH_4 is oxidized to CH_3OH by the membrane-integrated enzyme methane monooxygenase (MMO). A proton motive force is established from electron flow in the membrane, and this fuels ATPase. Note how carbon for biosynthesis comes from CH_2O.

incorporate one oxygen atom from O_2 into a carbon compound (see Section 14.24 and Figure 14.64a). *M. capsulatus* contains two MMOs, one cytoplasmic (soluble MMO, sMMO) and the other membrane-integrated (particulate MMO, pMMO). In the MMO reaction, an atom of oxygen is introduced into CH_4, forming CH_3OH, and the second atom of O is reduced to form H_2O (**Figure 14.50**). CH_3OH is oxidized by an alcohol dehydrogenase, yielding formaldehyde (CH_2O) and NADH, and the CH_2O is either oxidized to CO_2 or used to make new cell material.

C_1 Assimilation by Aerobic Methanotrophs

At least two distinct pathways exist for the incorporation of C_1 units into cell material in methanotrophs. The **serine pathway** is outlined in **Figure 14.51a**. In this pathway, acetyl-CoA is synthesized from one molecule of CH_2O (produced from the oxidation of CH_3OH, Figure 14.50) and one molecule of CO_2. The serine pathway requires reducing power and energy in the form of two molecules each of NADH and ATP, respectively, for each acetyl-CoA synthesized. The serine pathway employs a number of enzymes of the citric acid cycle and one enzyme, *serine transhydroxymethylase*, unique to the pathway (Figure 14.51a).

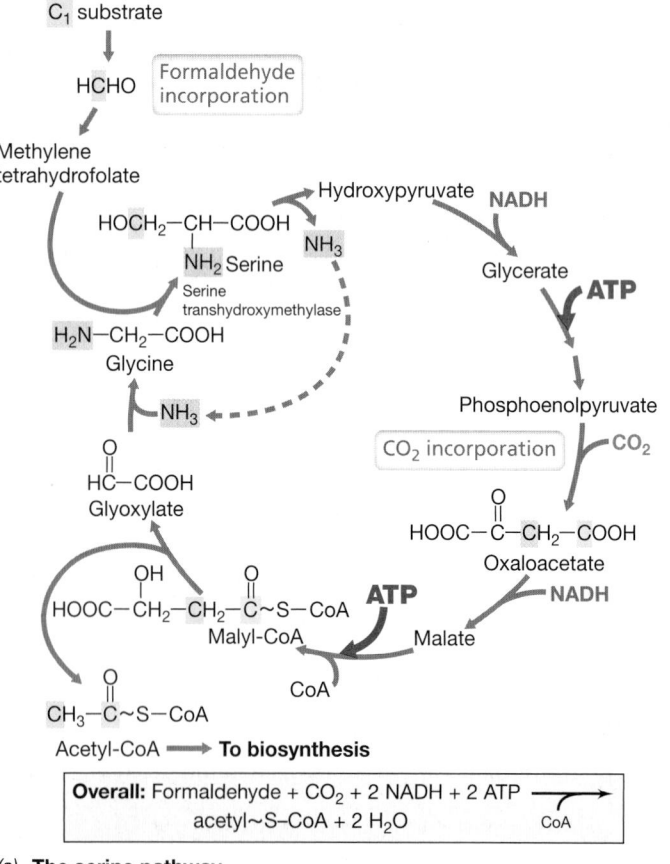

Overall: Formaldehyde + CO_2 + 2 NADH + 2 ATP ⟶ acetyl~S–CoA + 2 H_2O + CoA

(a) **The serine pathway**

Overall: 3 Formaldehyde + ATP ⟶ glyceraldehyde-3-P

(b) **The ribulose monophosphate pathway**

Figure 14.51 The serine and ribulose monophosphate pathways for the assimilation of C_1 units into cell material by methylotrophic bacteria. *(a)* Serine pathway. The product, acetyl-CoA, is used as the starting point for making new cell material. The key enzyme of the pathway is serine transhydroxymethylase. *(b)* Ribulose monophosphate pathway. Three molecules of CH_2O are required, with the product being glyceraldehyde 3-phosphate. The key enzyme of this pathway is hexulosephosphate synthase. The sugar rearrangements require enzymes of the pentose phosphate pathway (⟿ Figure 3.26).

An alternative pathway for C_1 incorporation is the **ribulose monophosphate pathway** (Figure 14.51*b*). This pathway is more energy efficient than the serine pathway because *all* of the carbon for cell material is derived from CH_2O. Because CH_2O is at the same oxidation level as cell material, no reducing power is needed for its incorporation. Hence, all of the NADH from the oxidation of methane can be oxidized in the electron transport chain.

The ribulose monophosphate pathway consumes one molecule of ATP for each molecule of glyceraldehyde 3-phosphate (G-3-P) synthesized (Figure 14.51*b*); two G-3-Ps can then be converted into glucose by reversal of the glycolytic pathway ($\Rightarrow$ Figure 3.14). The enzymes *hexulosephosphate synthase*, which condenses one molecule of formaldehyde with one molecule of ribulose 5-phosphate, and *hexulose 6-P isomerase* (Figure 14.51*b*) are unique to the ribulose monophosphate pathway. The remaining enzymes of this pathway are enzymes of intermediary metabolism widely distributed in bacteria.

Anaerobic Oxidation of Methane (AOM)

The *anaerobic* oxidation of methane uses a variety of enzymes and cofactors common to other forms of C_1 metabolism; however, anaerobic methanotrophy shows some novel features, as well. Methane can be oxidized anaerobically by an association (called a *consortium*) of two organisms, a sulfate-reducing bacterium (SRB) plus a species of *Archaea* phylogenetically related to methanogens. These consortia thrive in anoxic marine sediments and are responsible for oxidizing more than 90% of the methane produced there. The components of the consortium coexist in spatially structured aggregates (**Figure 14.52**). The archaeal component, called ANME (*anaerobic methanotroph*), of which there are several different types, oxidizes CH_4 as an electron donor. Electrons from methane oxidation are then transferred to the sulfate reducer, which uses them to reduce SO_4^{2-} to H_2S (Figure 14.52*b*).

ANME *Archaea* oxidize CH_4 to CO_2 by reversing the steps of methanogenesis (Figure 14.47). This process is endergonic but is made possible by the SRB partner organism, which consumes electrons from ANME, thereby making the oxidation CH_4 to CO_2 energetically favorable. Remarkably, electrons are transferred between the ANME and SRB partners by *direct electron transfer* (Figure 14.52*b*). Cells of ANME make electrically conductive multiheme cytochromes that span their outer cell layer and transfer electrons from the cytoplasmic membrane to the outside of the cell. The SRB have similar large electrically conductive cytochromes as well as pili ($\Rightarrow$ Section 2.7) that serve as electrically conductive "nanowires." These pili can be more than 1 μm in length and electrically connect the cells within aggregates to facilitate the two microbes' metabolisms (see page 392 for more on this). The presence of direct electron transfer may explain why some ANME can also use insoluble metals such as iron (Fe^{3+}) and manganese oxides (Mn^{4+}) as terminal electron acceptors (Section 14.15) in CH_4 oxidation.

AOM is not limited to consortia of ANME and SRB. ANME *Archaea* include *Methanoperedens nitroreducens*, which uses nitrate as a terminal electron acceptor for the anaerobic oxidation of CH_4. This organism couples reverse methanogenesis to the reduction of NO_3^- to NO_2^-. *Methanoperedens nitroreducens* can be found in consortia with denitrifying bacteria that then use NO_2^- as an electron acceptor. These consortia are active in anoxic environments where CH_4 and NO_3^- coexist, such as certain freshwater sediments.

Intra-Aerobic Methanotrophy

The methanotrophic denitrifying bacterium *Methylomirabilis oxyfera* is an obligate anaerobe that catalyzes AOM linked to NO_2^- as an electron acceptor. *M. oxyfera* can grow on CH_4 in pure culture and has a highly unusual polygonal morphology (**Figure 14.53**). Analysis of the *M. oxyfera* genome reveals all of the genes required for the *aerobic* oxidation of CH_4 to CO_2. *M. oxyfera* is also able to reduce NO_2^- to N_2 and it has most of the genes required for denitrification although it lacks nitric oxide reductase and nitrous oxide reductase (Section 14.13 and Figure 14.36). This leads to the interesting question of why an *anaerobic* bacterium would use *aerobic* pathways for CH_4 oxidation.

The answer to this conundrum is that *M. oxyfera* reduces nitrite in a novel way. The organism reduces NO_2^- to nitric oxide (NO) like a normal denitrifying bacterium would, but then *M. oxyfera* does something remarkable: The organism generates O_2 through the reaction $2\,NO \rightarrow N_2 + O_2$ and proceeds to use this O_2 as the electron acceptor for CH_4 oxidation. Because *M. oxyfera* produces its own O_2, its methanotrophic metabolism has been termed *intra-aerobic methanotrophy*. As it turns out, O_2 is toxic for the anaerobic *M. oxyfera*. However, if the O_2 is consumed (by its reduction to H_2O

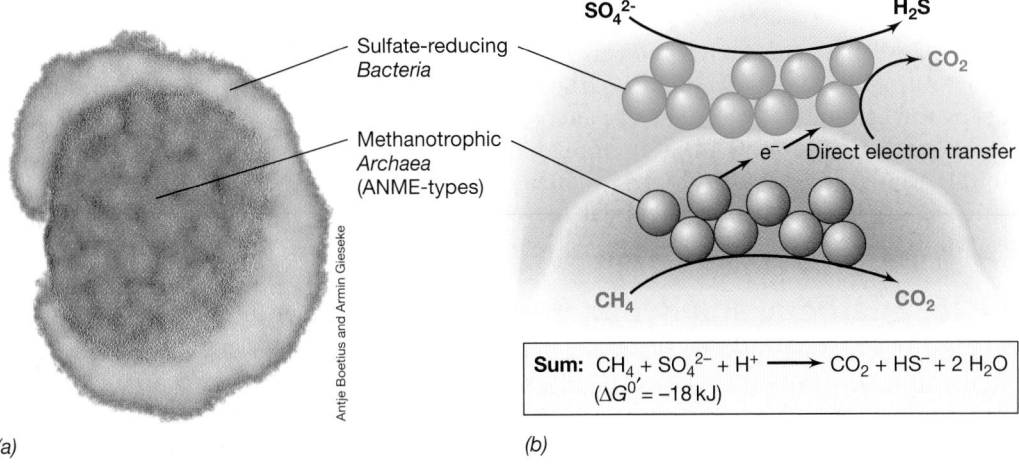

(a)   *(b)*

Antje Boetius and Armin Gieseke

Sum: $CH_4 + SO_4^{2-} + H^+ \longrightarrow CO_2 + HS^- + 2\,H_2O$
$(\Delta G^{0'} = -18\,kJ)$

Figure 14.52 Anaerobic oxidation of methane. *(a)* Methane-oxidizing cell aggregates from marine sediments. The aggregates contain methanotrophic *Archaea* (red) surrounded by sulfate-reducing bacteria (green). Each cell type has been stained by a different FISH probe ($\Rightarrow$ Section 19.5). The aggregate is about 30 μm in diameter. *(b)* Mechanism for the cooperative degradation of CH_4. An organic compound or some other carrier of reducing power transfers electrons from methanotroph to sulfate reducer. ANME, anaerobic methanotroph. See pages 137 and 392 for more on the ANME consortium.

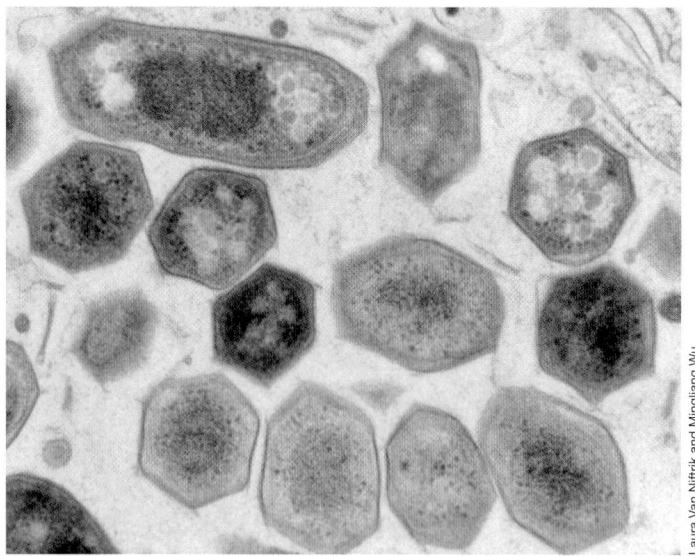

Figure 14.53 The cell morphology of *Methylomirabilis oxyfera*. The denitrifying methanotroph *M. oxyfera* has a unique polygonal morphology as revealed in this transmission electron micrograph of cells from a microbial community grown in a bioreactor. The scale bar is 0.5 μm.

Laura Van Niftrik and Mingliang Wu

using electrons from the oxidation of CH_4) as soon as it is produced, O_2 never accumulates, and the organism's environment remains anoxic.

MINIQUIZ

- When using CH_4 as electron donor, why is *Methylococcus capsulatus* an obligate aerobe?
- In which two ways does the ribulose monophosphate pathway save energy over reactions of the serine pathway?
- What is unique about methanotrophy in *Methylomirabilis oxyfera*?

VI • Fermentation

Thus far we have considered phototrophy and many types of aerobic and anaerobic respiration; these processes are unified by the fact that they all conserve energy at the expense of an ion gradient (H^+ or Na^+) that fuels oxidative phosphorylation. Here we turn our focus to *fermentations*, metabolisms in which energy conservation occurs at the expense of other reactions.

14.19 Energetic and Redox Considerations

If terminal electron acceptors, such as sulfate (SO_4^{2-}), nitrate (NO_3^-), and ferric iron (Fe^{3+}), are absent from anoxic habitats, then organic compounds are catabolized by **fermentation**. Recall from Chapter 3 that we emphasized how redox balance is achieved in fermentations by having the substrate serve as both electron donor and electron acceptor and that ATP is synthesized by *substrate-level phosphorylation*. We pick up on these two essential features of fermentation here (**Figure 14.54**).

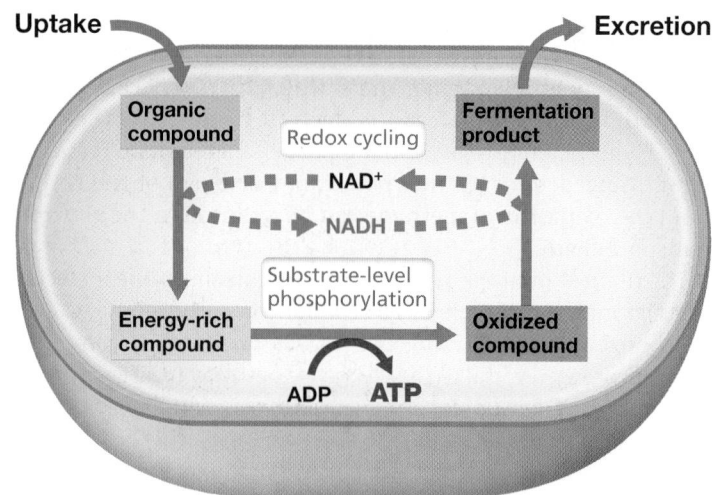

Figure 14.54 The essentials of fermentation. The fermentation product is excreted from the cell, and only a relatively small amount of the original organic compound is used for biosynthesis.

Energy-Rich Compounds and Substrate-Level Phosphorylation

Energy can be conserved by substrate-level phosphorylation from many different compounds. However, central to an understanding of substrate-level phosphorylation is the concept of *energy-rich compounds*. These are organic compounds that contain an energy-rich phosphate bond or a molecule of coenzyme A. The bond is "energy-rich" because its hydrolysis is highly exergonic. **Table 14.7** lists some energy-rich compounds formed during metabolism; the hydrolysis of most of these yields sufficient free energy to be coupled to ATP synthesis ($\Delta G^{0\prime} = -31.8$ kJ/mol). If an organism can form one of these compounds during fermentative metabolism, it can make ATP by transferring the phosphate bond from the energy-rich compound to ADP to form ATP—substrate-level phosphorylation.

Redox Balance and H_2 and Acetate Production

In any fermentation there must be atomic and redox balance. That is, the total number of each type of atom and electrons in the products of the reaction must balance those in the reactants (the substrates). Redox balance is achieved in fermentations by the excretion from the cell of *fermentation products*, reduced substances such as acids or alcohols that are produced as end products of the catabolism of the original fermentable substance (Figure 14.54).

In several fermentations, redox balance is facilitated by the production of molecular hydrogen (H_2). The production of H_2 is associated with the activity of the iron–sulfur protein *ferredoxin*, a very low-potential electron carrier, and is catalyzed by the enzyme *hydrogenase*. H_2 can also be produced from the C_1 fatty acid formate (**Figure 14.55**). Although the H_2 can no longer be used by the fermenter and is thus excreted, H_2 is a very powerful electron donor and can be oxidized by many different *Bacteria* and *Archaea*. Indeed, with its very electronegative $E_0\prime$ (making it suitable as an electron donor for any form of respiration), H_2 is never wasted in microbial ecosystems.

TABLE 14.7 Energy-rich compounds that can couple to substrate-level phosphorylation[a]

Compound	Free energy of hydrolysis, $\Delta G^{0\prime}$ (kJ/mol)[b]
Acetyl-CoA	−35.7
Propionyl-CoA	−35.6
Butyryl-CoA	−35.6
Caproyl-CoA	−35.6
Succinyl-CoA	−35.1
Acetyl phosphate	−44.8
Butyryl phosphate	−44.8
1,3-Bisphosphoglycerate	−51.9
Carbamyl phosphate	−39.3
Phosphoenolpyruvate	−51.6
Adenosine phosphosulfate (APS)	−88
N^{10}-Formyltetrahydrofolate	−23.4
Energy of hydrolysis of ATP (ATP → ADP + P_i)	−31.8

[a]Data from Thauer, R.K., K. Jungermann, and K. Decker. 1977. Energy conservation in chemotrophic anaerobic bacteria. *Bacteriol. Rev. 41:* 100–180.
[b]The $\Delta G^{0\prime}$ values shown here are for "standard conditions," which are not necessarily those of cells. Including heat loss, the energy costs of making an ATP are more like 60 kJ than 32 kJ, and the energy of hydrolysis of the energy-rich compounds shown here is thus likely higher. But for simplicity and comparative purposes, the values in this table will be taken as the actual energy released per reaction.

Many anaerobic bacteria produce acetate or other fatty acids as a major or minor fermentation product. The production of these is energy conserving because it offers the organism the opportunity to make ATP by substrate-level phosphorylation. The key intermediate generated is the coenzyme-A derivative of each fatty acid, since these are energy-rich compounds (Table 14.7). For example, acetyl-CoA can be converted to acetyl phosphate (Figure 14.55) and the phosphate group subsequently transferred to ADP, yielding ATP. Fatty acid production is common in fermentations and if the fatty acid is metabolized through a CoA intermediate, the potential for ATP synthesis by substrate-level phosphorylation is a possibility.

With these foundational principles of fermentative bioenergetics firmly in hand, we explore the metabolic diversity of

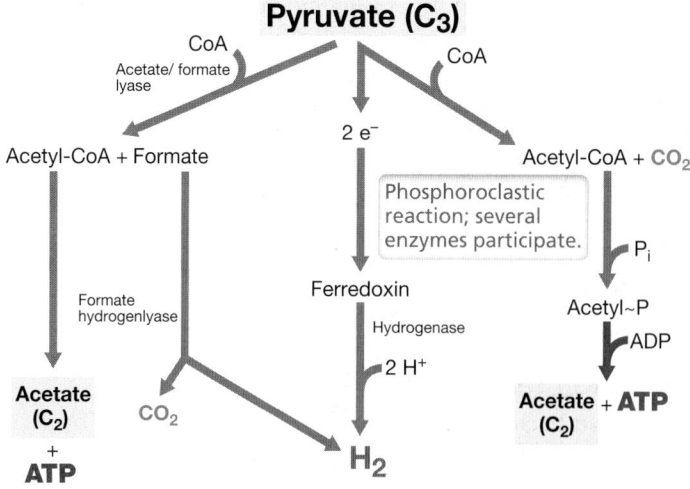

Figure 14.55 Production of H_2 and acetate from pyruvate. At least two mechanisms are known, one that produces H_2 directly and the other that makes formate as an intermediate. When acetate is produced, ATP synthesis is possible (Table 14.7).

fermentations beginning with species that produce acidic fermentation products, common and widespread bacteria in most anoxic environments.

MINIQUIZ

- Why is H_2 produced during many types of fermentation?
- Why is acetate formation in fermentation energetically beneficial to the cell?

14.20 Lactic and Mixed-Acid Fermentations

Fermentations are classified by either the substrate fermented or the products formed. Table 14.8 lists some major fermentations classified on the basis of the products formed, such as alcohol, lactic acid, propionic acid, mixed acid, butyric acid, and acetate. Other fermentations are classified by the substrate

TABLE 14.8 Common fermentations and their energetics and example organisms

Type	Reaction	Energy yield ($\Delta G^{0\prime}$, kj/mol)	Organisms
Alcoholic	Hexose → 2 ethanol + 2 CO_2	−239	Yeast, *Zymomonas*
Homolactic	Hexose → 2 lactate⁻ + 2 H⁺	−196	*Streptococcus*, some *Lactobacillus*
Heterolactic	Hexose → lactate⁻ + ethanol + CO_2 + H⁺	−216	*Leuconostoc*, some *Lactobacillus*
Propionic acid	3 Lactate⁻ → 2 propionate⁻ + acetate⁻ + CO_2 + H_2O	−170	*Propionibacterium*, *Clostridium propionicum*
Mixed acid	Hexose → ethanol + 2,3-butanediol + succinate²⁻ + lactate⁻ + acetate⁻ + formate⁻ + H_2 + CO_2	Depends on product ratio	Enteric bacteria including *Escherichia*, *Salmonella*, *Shigella*, *Klebsiella*, *Enterobacter*
Butyric acid	Hexose → butyrate⁻ + 2 H_2 + 2 CO_2 + H⁺	−264	*Clostridium butyricum*
Butanol	2 Hexose → butanol + acetone + 5 CO_2 + 4 H_2	−468	*Clostridium acetobutylicum*
Caproate/ Butyrate	6 Ethanol + 3 acetate⁻ → 3 butyrate⁻ + caproate⁻ + 2 H_2 + 4 H_2O + H⁺	−183	*Clostridium kluyveri*
Acetogenic	Fructose → 3 acetate⁻ + 3 H⁺	−276	*Clostridium aceticum*

fermented rather than the fermentation product; for instance, amino acid, purine/pyrimidine, or the succinate fermentation. Some anaerobes even ferment aromatic compounds and other unusual substrates (Table 14.9). Clearly, a wide variety of organic compounds can be fermented, and in a few cases, only a very restricted group of anaerobes can carry out the fermentation. Many of these are metabolic specialists, having evolved the capacity to ferment a substrate not catabolized by other bacteria (Table 14.9).

We begin with two very common fermentations of sugars in which lactic acid is the sole or major product.

Lactic Acid Fermentation

Lactic acid bacteria are gram-positive nonsporulating bacteria that produce lactic acid as a major or sole fermentation product from the fermentation of sugars (↩ Section 16.6). Two fermentative patterns are observed. One, called **homofermentative**,

TABLE 14.9 Some unusual bacterial fermentations

Type	Reaction	Organisms
Acetylene	$2\ C_2H_2 + 3\ H_2O \rightarrow$ ethanol $+$ acetate$^-$ $+$ H$^+$	*Pelobacter acetylenicus*
Glycerol	4 Glycerol $+ 2\ HCO_3^- \rightarrow$ 7 acetate$^-$ $+ 5\ H^+ + 4\ H_2O$	*Acetobacterium* spp.
Phloroglucinol (aromatic)	$C_6H_6O_3 + 3\ H_2O \rightarrow 3$ acetate$^-$ $+ 3\ H^+$	*Pelobacter massiliensis* *Pelobacter acidigallici*
Putrescine	$10\ C_4H_{12}N_2 + 26\ H_2O \rightarrow$ 6 acetate$^-$ $+ 7$ butyrate$^-$ $+ 20\ NH_4^+ + 16\ H_2 + 13\ H^+$	Unclassified gram-positive nonsporulating anaerobes
Citrate	Citrate^{3-} $+ 2\ H_2O \rightarrow$ formate$^-$ $+ 2$ acetate$^-$ $+ HCO_3^- + H^+$	*Bacteroides* spp.
Benzoate (aromatic)	2 Benzoate$^- \rightarrow$ cyclohexane carboxylate$^-$ $+ 3$ acetate$^-$ $+ HCO_3^- + 3\ H^+$	*Syntrophus aciditrophicus*

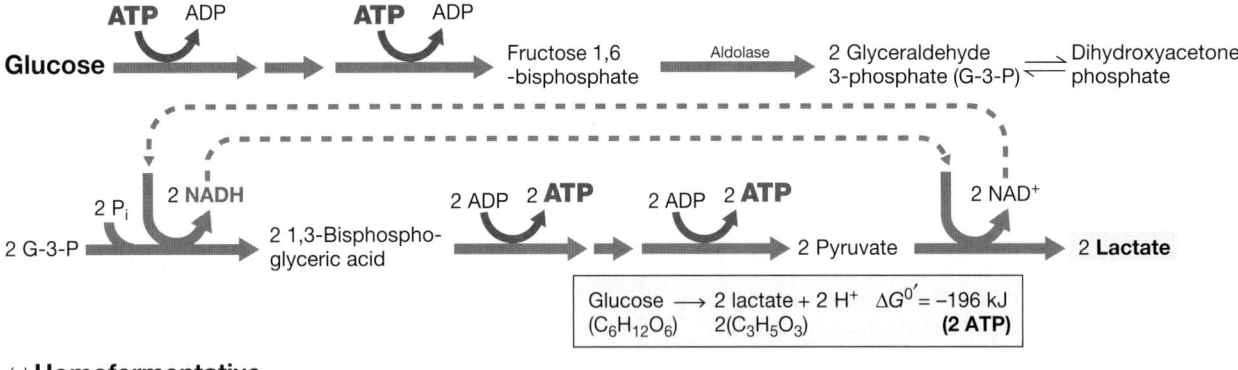

(a) **Homofermentative**

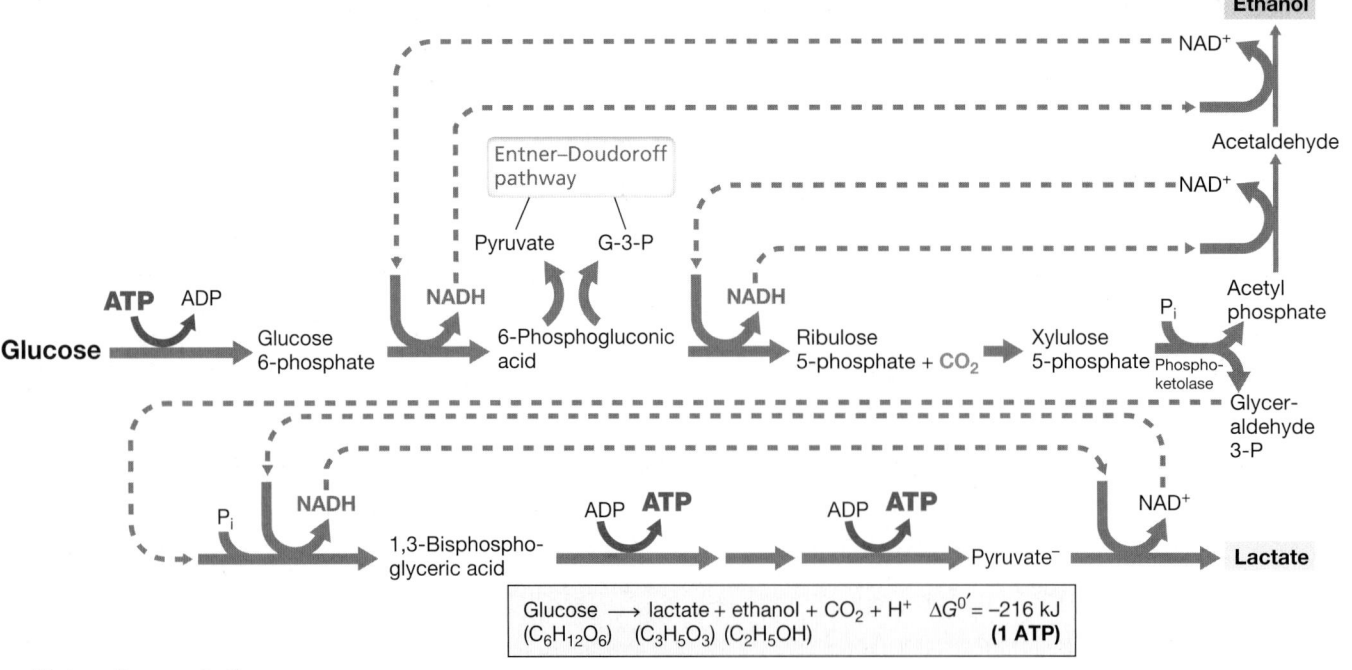

(b) **Heterofermentative**

Figure 14.56 The fermentation of glucose in *(a)* homofermentative and *(b)* heterofermentative lactic acid bacteria. Note that no ATP is made in reactions leading to ethanol formation in heterofermentative organisms.

yields a single fermentation product, lactic acid. The other, called **heterofermentative**, yields products in addition to lactate, mainly ethanol plus CO_2.

Figure 14.56 summarizes pathways for the fermentation of glucose by homofermentative and heterofermentative lactic acid bacteria. The differences observed can be traced to the presence or absence of the enzyme *aldolase*, a key enzyme of glycolysis (⮌ Figure 3.14). Homofermentative lactic acid bacteria contain aldolase and produce *two* molecules of lactate from glucose by the glycolytic pathway (Figure 14.56a). Heterofermenters lack aldolase and thus cannot break down fructose bisphosphate to triose phosphate. Instead, they oxidize glucose 6-phosphate to 6-phosphogluconate and then decarboxylate this to pentose phosphate. The latter compound is then converted to triose phosphate and acetyl phosphate by the key enzyme *phosphoketolase* (Figure 14.56b). The early steps in catabolism by heterofermentative lactic acid bacteria are those of the pentose phosphate pathway (⮌ Figure 3.26).

In heterofermenters, triose phosphate is converted to lactic acid with the production of ATP (Figure 14.56b). However, to achieve redox balance the acetyl phosphate produced is used as an electron acceptor and is reduced by NADH (generated during the production of pentose phosphate) to ethanol. This occurs without ATP synthesis because the energy-rich CoA bond is lost during ethanol formation. Because of this, heterofermenters produce only *one* ATP/glucose instead of the *two* ATP/glucose produced by homofermenters. In addition, because heterofermenters decarboxylate 6-phosphogluconate, they produce CO_2 as a fermentation product; homofermenters do not produce CO_2. Thus an easy way to differentiate a homofermenter from a heterofermenter is to observe for the production of CO_2 in laboratory cultures.

Entner–Doudoroff Pathway

The initial stages of glucose fermentation (⮌ Section 3.8) often rely on glycolysis or a variant of the glycolytic pathway called the *Entner–Doudoroff pathway*. In the Entner–Doudoroff pathway, glucose 6-phosphate is oxidized to 6-phosphogluconic acid and NADPH; the 6-phosphogluconic acid is dehydrated and split into pyruvate and glyceraldehyde 3-phosphate (G-3-P), a key intermediate of the glycolytic pathway. G-3-P is then catabolized as in glycolysis, generating NADH and two ATP, and used as an electron acceptor to balance redox reactions (Figure 14.56a).

Because pyruvate is formed directly in the Entner–Doudoroff pathway and cannot yield ATP as can G-3-P (Figure 14.56a), the Entner–Doudoroff pathway yields only half the ATP of the glycolytic pathway. Organisms using the Entner–Doudoroff pathway therefore share this physiological characteristic with heterofermentative lactic acid bacteria (Figure 14.56b). *Zymomonas*, an obligately fermentative gram-negative bacterium, and *Pseudomonas*, a strictly respiratory bacterium (⮌ Section 16.4), are major genera that employ the Entner–Doudoroff pathway for glucose catabolism.

Mixed-Acid Fermentations

In *mixed-acid fermentations* (Table 14.8), characteristic of enteric bacteria (⮌ Section 16.3), three different acids—*acetic, lactic,* and *succinic*—are formed from the fermentation of glucose or other sugars that can be converted into glucose. Ethanol, CO_2, and H_2 are also typically formed as fermentation products. Glycolysis is the pathway used by mixed-acid fermenters, such as *Escherichia coli*, and we detailed the steps in that pathway in Figure 3.14.

Some enteric bacteria produce acidic products in lower amounts than *E. coli* and balance redox in their fermentations by producing larger amounts of neutral products. One key neutral product is the four-carbon alcohol *butanediol*. In this variation of the mixed-acid fermentation, the main products observed are butanediol, ethanol, CO_2, and H_2 (**Figure 14.57**). In the mixed-acid fermentation of *E. coli*, equal amounts of CO_2 and H_2 are produced, whereas in a

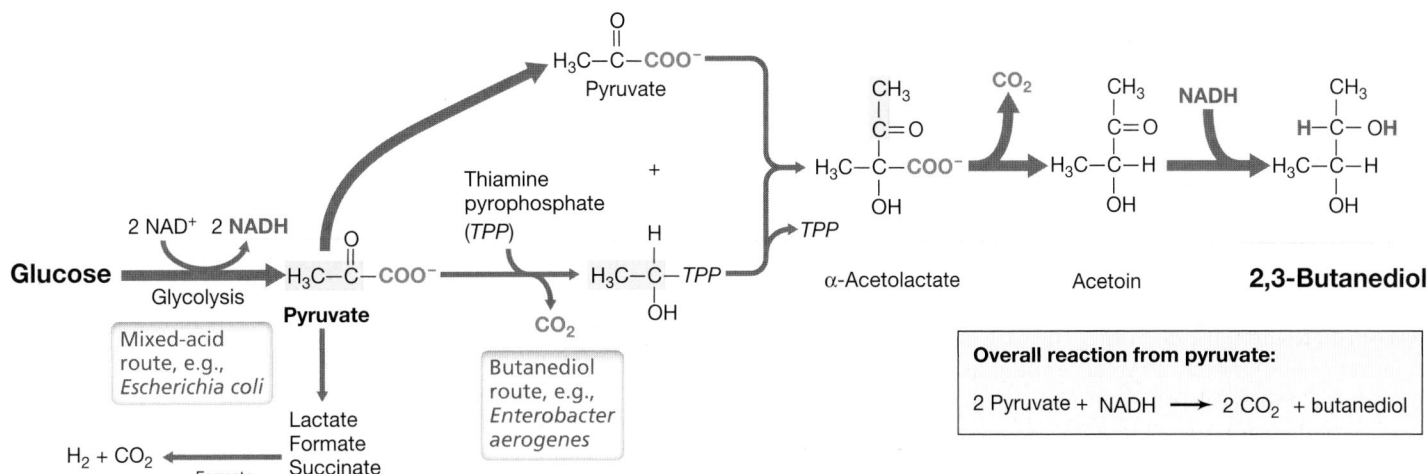

Figure 14.57 Butanediol production and mixed-acid fermentations. Note how only one NADH but two molecules of pyruvate are used to make one butanediol. This leads to redox imbalance and the production of more ethanol by butanediol producers than by mixed-acid fermenters.

butanediol fermentation, considerably more CO_2 than H_2 is produced. This is because mixed-acid fermenters produce CO_2 only from formic acid by means of the enzyme *formate hydrogenlyase* (Figure 14.55):

$$HCOOH \rightarrow H_2 + CO_2$$

By contrast, butanediol producers, such as *Enterobacter aerogenes*, produce CO_2 and H_2 from formic acid but also produce two additional molecules of CO_2 during the formation of each molecule of butanediol. However, because butanediol production consumes only one-half of the NADH generated in glycolysis, more ethanol is produced by these organisms than by non-butanediol fermenters in order to achieve redox balance (Figure 14.57).

MINIQUIZ

- How can homo- and heterofermentative metabolism be differentiated in pure cultures of lactic acid bacteria?
- Butanediol production leads to greater ethanol production than in the mixed-acid fermentation of *Escherichia coli*. Why?

14.21 Clostridial and Propionate Fermentations

Species of the genus *Clostridium* are obligately fermentative anaerobes (⇨ Section 16.7). Different clostridia ferment sugars, amino acids, purines and pyrimidines, and a few other compounds. In all cases ATP synthesis is linked to substrate-level phosphorylations either in the glycolytic pathway or from the hydrolysis of a CoA intermediate (Table 14.7). We begin with sugar-fermenting (called *saccharolytic*) clostridia.

Sugar Fermentation by *Clostridium* Species

A number of clostridia ferment sugars, producing *butyric acid* as a major fermentation product. Some species also produce the neutral products acetone and butanol; *Clostridium acetobutylicum* is a classic example of this pattern. The biochemical steps in the formation of butyric acid and neutral products from sugars are shown in **Figure 14.58**.

In saccharolytic clostridia, glucose is converted to pyruvate and NADH via the glycolytic pathway, and pyruvate is split to yield acetyl-CoA, CO_2, and H_2 (through ferredoxin) by the phosphoroclastic reaction (Figure 14.55). Most of the acetyl-CoA is then reduced to butyrate or other fermentation products using NADH derived from glycolytic reactions as electron donor. The actual products observed are influenced by the duration and the conditions of the fermentation. During the early stages of the butyric fermentation, butyrate and a small amount of acetate and ethanol are produced. But as the pH of the medium drops, acid production decreases and acetone and butanol begin to appear. If the pH of the medium is kept neutral by buffering, there is very little formation of acetone and butanol; instead, butyric acid production continues, and this is for a good reason.

When *C. acetobutylicum* synthesizes butyrate, extra ATP is produced (Figure 14.58) and so the organism will continue to make butyrate unless conditions become overly acidic. However,

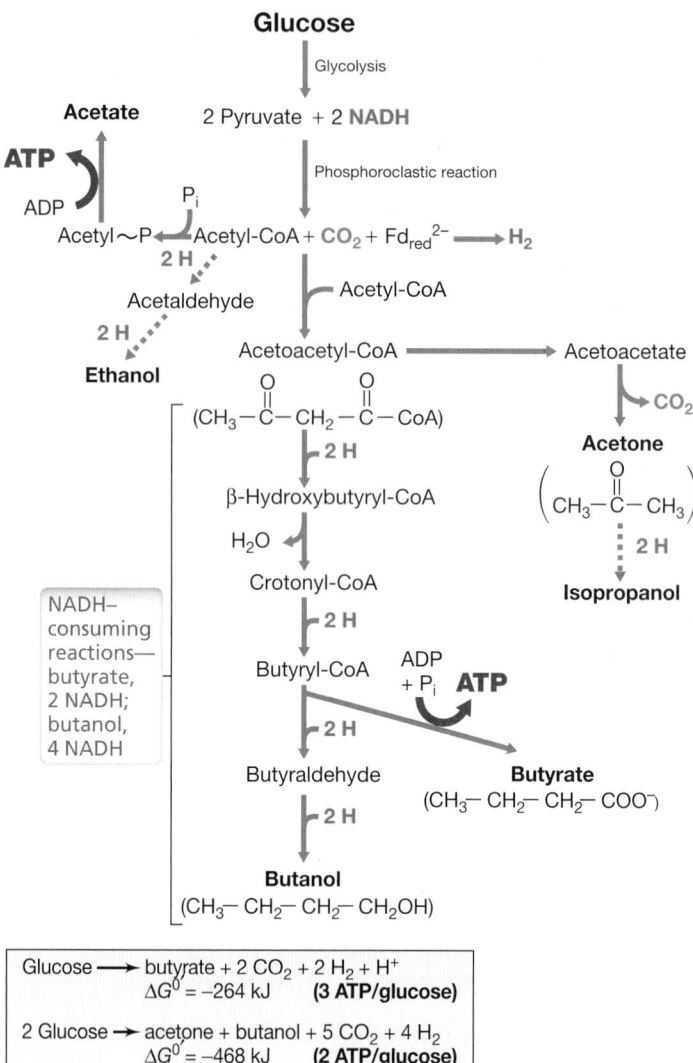

Figure 14.58 The butyric acid and butanol/acetone fermentation. All fermentation products from glucose are shown in bold (dashed lines indicate minor products). Note how the production of acetate and butyrate lead to additional ATP by substrate-level phosphorylation. By contrast, formation of butanol and acetone reduces the ATP yield because the butyryl-CoA to butyrate step is bypassed. 2 H, NADH; Fd_{red}^{2-}, reduced ferredoxin.

C. acetobutylicum is acid-sensitive, and if the pH drops below about pH 5, genes encoding enzymes that make neutral products are derepressed and the fermentation shifts to acetone/butanol production. Interestingly, the production of butanol is a consequence of the production of acetone. For each acetone that is made, two NADH produced during glycolysis are not reoxidized as they would be if butyrate were produced. To achieve redox balance, the cell then uses butyrate as an electron acceptor with butanol being the final fermentation product (Figure 14.58). Previously excreted butyrate can also be reincorporated by the cell and reduced to butanol and then excreted again. Although neutral product formation helps *C. acetobutylicum* keep its environment from becoming too acidic, there is an energetic price to pay for this. In producing butanol, the cell loses the opportunity to convert butyryl-CoA to butyrate and gain ATP (Figure 14.58).

Amino Acid Fermentation by *Clostridium* Species and the Stickland Reaction

Some *Clostridium* species ferment amino acids. These are the *proteolytic* clostridia, organisms that degrade proteins released from dead organisms. Some of these, such as the animal pathogen *Clostridium tetani* (tetanus), are strictly proteolytic, while other species are both saccharolytic and proteolytic.

Depending on the species, some proteolytic clostridia ferment individual amino acids, typically glutamate, glycine, alanine, cysteine, histidine, serine, or threonine. The biochemistry behind these fermentations is quite complex, but the metabolic strategy is simple. In virtually all cases, the amino acids are catabolized in such a way as to eventually yield a fatty acid–CoA derivative, typically acetyl (C_2), butyryl (C_4), or caproyl (C_6). From these, ATP is produced by substrate-level phosphorylation (Table 14.7). Other typical products of amino acid fermentation include ammonia (NH_3) and CO_2.

Some clostridia ferment only an amino acid *pair*. In this situation one amino acid functions as the electron donor and is oxidized, whereas the other amino acid is the electron acceptor and is reduced. This *coupled* amino acid fermentation is called a **Stickland reaction**, named for the scientist who discovered it. For example, *Clostridium sporogenes* ferments glycine and alanine, and in this reaction, alanine is the electron donor and glycine is the electron acceptor (**Figure 14.59**). The products of the Stickland reaction are invariably NH_3, CO_2, and a carboxylic acid with one fewer carbons than the amino acid that was oxidized (Figure 14.59).

Many of the products of amino acid fermentation by clostridia are foul-smelling substances, and the odor that results from putrefaction is mainly a result of clostridial activity. In addition to fatty acids, other odoriferous compounds produced include hydrogen sulfide (H_2S), methyl mercaptan (CH_3SH, derived from sulfur-containing amino acids), cadaverine (from lysine), putrescine (from ornithine, see Table 14.9), and NH_3. Purines and pyrimidines, released from the degradation of nucleic acids, lead to many of the same fermentation products and yield ATP by substrate-level phosphorylation from the hydrolysis of fatty acid–CoA derivatives (Table 14.7) produced in their respective fermentative pathways.

Clostridium kluyveri Fermentation

Another species of *Clostridium* also ferments a mixture of substrates in which one is the donor and one is the acceptor, as in the Stickland reaction. However, this organism, *C. kluyveri*, ferments not amino acids but instead *ethanol plus acetate*. In this fermentation, ethanol is the electron donor and acetate is the electron acceptor. The overall reaction is shown in Table 14.8.

The ATP yield in the caproate/butyrate fermentation is low, 1 ATP/6 ethanol fermented. However, *C. kluyveri* has a selective advantage over all other fermenters in its apparently unique ability to oxidize a highly reduced fermentation product of other anaerobes (ethanol) and couple it to the reduction of another common fermentation product (acetate), reducing the latter to longer-chain fatty acids, reactions that consume NADH. The single ATP produced per 6 ethanol oxidized comes from substrate-level phosphorylation during conversion of a fatty acid–CoA derivative formed during the fermentation. The fermentation of *C. kluyveri* is an example of a **secondary fermentation**, which can be viewed as a fermentation of fermentation products. We see another example of this next.

Propionic Acid Fermentation

The gram-positive bacterium *Propionibacterium* and some related bacteria produce *propionic acid* as a major fermentation product from either glucose or lactate. Lactate, a fermentation product of the lactic acid bacteria (Section 14.20), is probably the major substrate for propionic acid bacteria in nature, where these two groups live in close association. *Propionibacterium* is an important agent in the ripening of Emmental (Swiss) cheese, which gets its unique bitter and nutty taste from the propionic and acetic acids produced, and the CO_2 produced during the fermentation forms bubbles that leave the characteristic holes (eyes) in the cheese.

Figure 14.60 shows the reactions leading from lactate to propionate. When glucose is the starting substrate, it is first catabolized to pyruvate by the glycolytic pathway. Then pyruvate, produced either from glucose or from the oxidation of lactate, is converted to acetate plus CO_2 or carboxylated to form methylmalonyl-CoA; the latter is converted into oxaloacetate and, eventually, propionyl-CoA (Figure 14.60). Propionyl-CoA reacts with succinate in a step catalyzed by the enzyme CoA transferase, producing succinyl-CoA and propionate. This results in a lost opportunity for ATP production from propionyl-CoA (Table 14.7) but avoids the energetic costs of having to activate succinate with ATP to form succinyl-CoA. The succinyl-CoA is

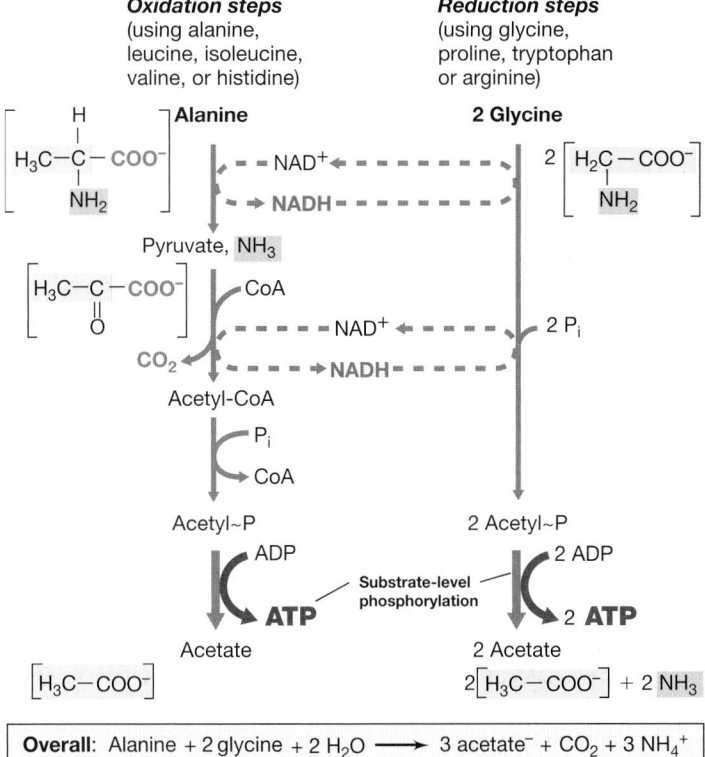

Figure 14.59 The Stickland reaction. This example shows the co-catabolism of the amino acids alanine and glycine. The structures of key substrates, intermediates, and products are shown in brackets to allow the chemistry of the reaction to be followed. Note how in the reaction shown, alanine is the electron donor and glycine is the electron acceptor.

UNIT 4

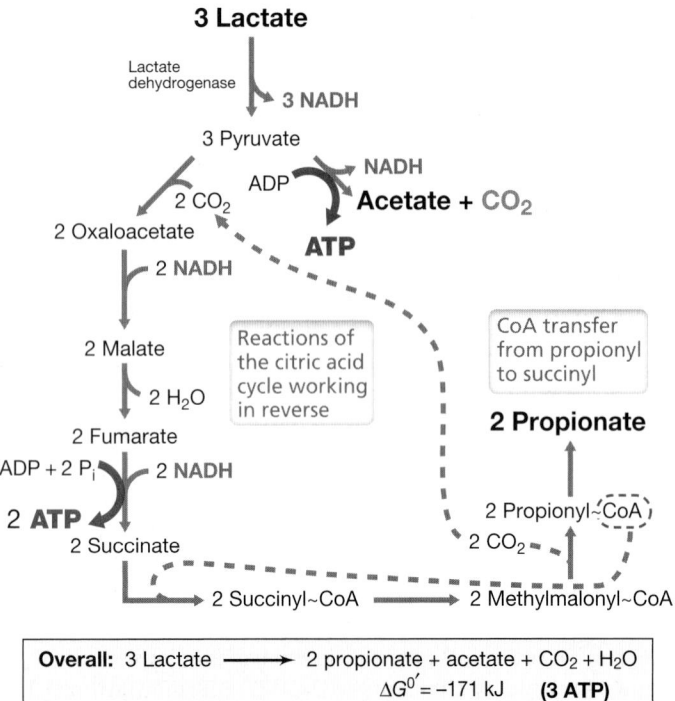

Overall: 3 Lactate $\longrightarrow$ 2 propionate + acetate + CO_2 + H_2O
$\Delta G^{0'} = -171$ kJ   **(3 ATP)**

Figure 14.60 The propionic acid fermentation of *Propionibacterium*. Products are shown in bold. The four NADH made from the oxidation of three lactate are reoxidized in the reduction of oxaloacetate and fumarate, and the CoA group from propionyl-CoA is exchanged with succinate during the formation of propionate.

then isomerized to methylmalonyl-CoA and the cycle is complete; propionate is formed and CO_2 regenerated (Figure 14.60).

NADH is oxidized in the steps between oxaloacetate and succinate. The reduction of fumarate to succinate (Figure 14.60) is linked to electron transport reactions and the formation of a proton motive force; this yields one ATP by oxidative phosphorylation. The propionate pathway also converts some lactate to acetate plus CO_2, which allows for additional ATP to be made by substrate-level phosphorylation (Figure 14.60). Thus, in the propionate fermentation, both substrate-level *and* oxidative phosphorylation occur.

Propionate is also formed in the fermentation of succinate by the bacterium *Propionigenium*, but by a completely different mechanism than that described here for *Propionibacterium*. *Propionigenium*, to be considered next, is phylogenetically and ecologically unrelated to *Propionibacterium*, but aspects of its energy metabolism are of considerable interest from the standpoint of metabolic diversity and the energetic limits to life.

MINIQUIZ

- Compare the mechanisms for energy conservation in *Clostridium acetobutylicum* and *Propionibacterium*.
- What type of substrates are fermented by saccharolytic clostridia? By proteolytic clostridia?
- What are the substrates for the *Clostridium kluyveri* fermentation? In nature, where do these come from?

14.22 Fermentations That Lack Substrate-Level Phosphorylation

Certain fermentations yield insufficient energy to synthesize ATP by substrate-level phosphorylation (that is, less than −32 kJ, Table 14.7), yet still support anaerobic growth without added electron acceptors. In these cases, catabolism of the compound is linked to ion pumps that establish an ion gradient across the cytoplasmic membrane. Examples of these include the fermentation of succinate by *Propionigenium modestum* and the fermentation of oxalate by *Oxalobacter formigenes*.

Propionigenium modestum

Propionigenium modestum was first isolated in anoxic enrichment cultures lacking electron acceptors and fed succinate as an electron donor. *Propionigenium* inhabits marine and freshwater sediments and can also be isolated from the human oral cavity. The organism is a gram-negative short rod and, phylogenetically, is a species of *Fusobacteria*. During studies of the physiology of *P. modestum*, it was shown to require sodium chloride (NaCl) for growth and to catabolize succinate under strictly anoxic conditions:

$$\text{Succinate}^{2-} + H_2O \rightarrow \text{propionate}^- + HCO_3^- \quad \Delta G^{0'} = -20.5 \text{ kJ}$$

This decarboxylation releases insufficient free energy to support ATP synthesis by substrate-level phosphorylation (Table 14.7) but sufficient free energy to pump a sodium ion (Na^+) from the cytoplasm to the periplasm across the cytoplasmic membrane. Energy conservation in *Propionigenium* is then linked to the resulting *sodium motive force*; a sodium-translocating (instead of proton-translocating) ATPase exists in the membrane of this organism that uses the sodium motive force to drive ATP synthesis (**Figure 14.61a**).

In a related decarboxylation reaction, the bacterium *Malonomonas* decarboxylates the C_3 dicarboxylic acid malonate, forming acetate plus CO_2. As for *Propionigenium*, energy metabolism in *Malonomonas* is linked to Na^+ and a sodium-driven ATPase. But the free energy available from malonate fermentation by *Malonomonas* (−17.4 kJ) is even less than that of succinate fermentation by *P. modestum*. *Sporomusa*, an endospore-forming bacterium and also an acetogen (Section 14.16), is also capable of fermenting malonate, as are a few other *Bacteria*.

Oxalobacter formigenes

Oxalobacter formigenes is a bacterium present in the intestinal tract of animals, including humans. It catabolizes the C_2 dicarboxylic acid oxalate, producing formate plus CO_2. Oxalate degradation by *O. formigenes* is thought to be important in the human colon for preventing the accumulation of oxalate, a substance that can form calcium oxalate kidney stones. *O. formigenes* is a gram-negative strict anaerobe that carries out the following reaction:

$$\text{Oxalate}^{2-} + H_2O \rightarrow \text{formate}^- + HCO_3^- \quad \Delta G^{0'} = -26.7 \text{ kJ}$$

As in the catabolism of succinate by *P. modestum*, insufficient energy is available from this reaction to drive ATP synthesis by substrate-level phosphorylation (Table 14.7). However, the reaction supports growth of the organism because the decarboxylation

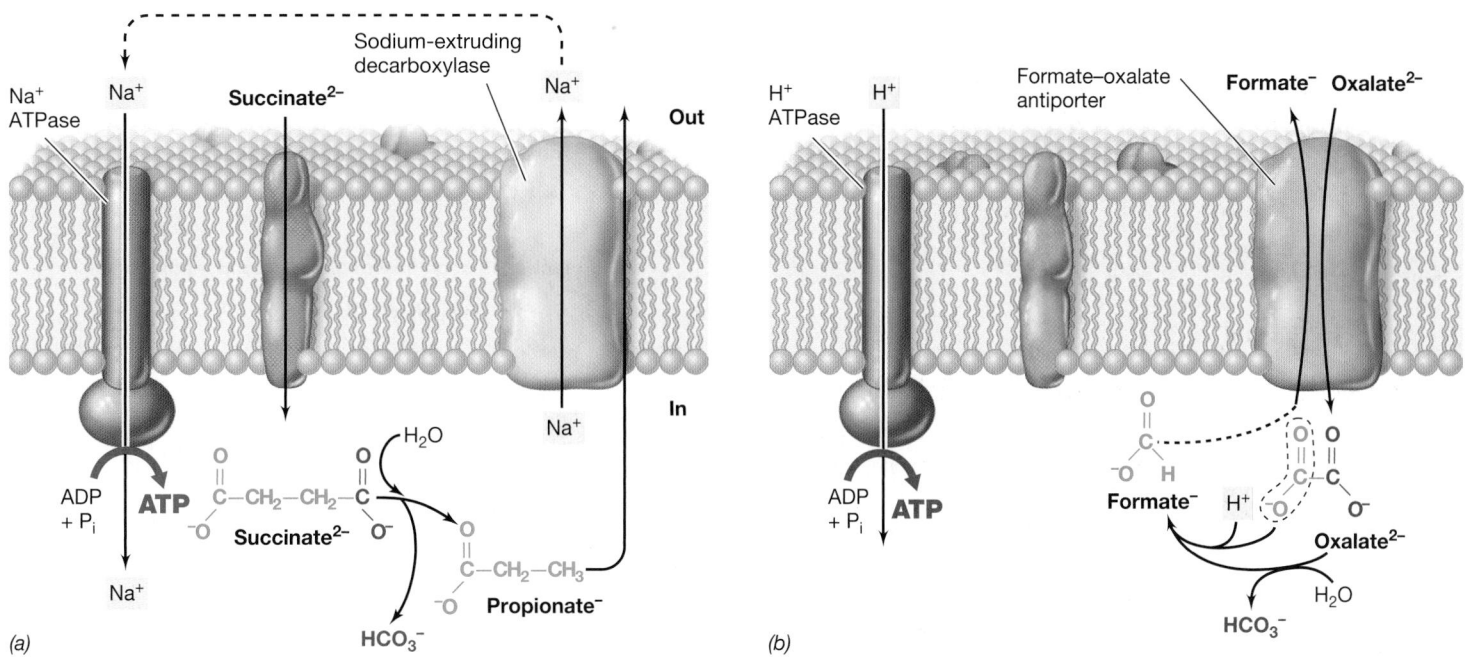

Figure 14.61 The unique fermentations of succinate and oxalate. *(a)* Succinate fermentation by *Propionigenium modestum.* Sodium export is linked to the energy released by succinate decarboxylation, and a sodium-translocating ATPase produces ATP. *(b)* Oxalate fermentation by *Oxalobacter formigenes.* Oxalate import and formate export by a formate–oxalate antiporter (⮂ Figure 3.3) consume cytoplasmic protons. ATP synthesis is linked to a proton-driven ATPase. All substrates and products are shown in bold.

of oxalate is exergonic and forms formate, which is excreted from the cell. This is because the internal consumption of protons during the oxidation of oxalate and production of formate is, in effect, a proton pump; a divalent molecule (oxalate) enters the cell while a univalent molecule (formate) is excreted. The continued exchange of oxalate for formate establishes a proton motive force that is coupled to ATP synthesis by the proton-translocating ATPase in the membrane (Figure 14.61*b*).

What Can Be Learned from the Decarboxylations of Succinate and Oxalate?

The unique aspect of decarboxylation-type fermentations is that ATP is made without substrate-level phosphorylation *or* oxidative phosphorylation fueled by electron transport. Instead, ATP synthesis is driven by ion pumps coupled to the small amount of energy released from the decarboxylation reaction. Organisms such as *Propionigenium, Malonomonas,* or *Oxalobacter* thus offer an important lesson in microbial bioenergetics: ATP synthesis from reactions that yield less than −32 kJ is still possible if the reaction can be coupled to an ion pump.

At a minimum, then, an energy-conserving reaction must yield sufficient free energy to pump at least one ion. This energy requirement is estimated to be near −12 kJ. Reactions that release less free energy than this should not be able to drive ion pumps and should therefore not be potential energy-conserving reactions. However, as we will see in the next section, bacteria are known that push this theoretical limit even lower, and their energetics, consequently, are still incompletely understood. These are the syntrophs, bacteria living on the energetic margin of existence.

MINIQUIZ

- Why does *Propionigenium modestum* require sodium for growth?
- Of what benefit is the organism *Oxalobacter* to human health?
- How can a fermentation that yields insufficient free energy to make an ATP still support growth?

14.23 Syntrophy

There are many examples in microbiology of **syntrophy**, a situation in which two different microbes cooperate to degrade a substance that neither can degrade alone. Most syntrophic reactions are secondary fermentations in which organisms ferment the fermentation products of other anaerobes. We will see in Chapter 21 how syntrophy is often a key step in the anoxic catabolism that leads to the production of methane in nature. Here we consider the microbiology and energetic aspects of syntrophy.

H₂ Consumption in Syntrophy: The Metabolic Link

Table 14.10 lists some major groups of syntrophs and the compounds they degrade. Many organic compounds can be degraded syntrophically, including even aromatic and aliphatic hydrocarbons. But the major compounds of interest in syntrophic environments are fatty acids and alcohols. The heart of syntrophic reactions is *interspecies H₂ transfer—H₂ production* by one partner, the syntroph, linked to H₂ *consumption* by the other. The H₂ consumer can be any one of a number of physiologically distinct organisms: denitrifying bacteria, ferric iron–reducing bacteria, sulfate-reducing bacteria, acetogens, or methanogens, groups we have already considered.

UNIT 4

TABLE 14.10 Properties of major syntrophic bacteria[a]

Genus	Number of known species	Phylogeny[b]	Substrates fermented in coculture[c]
Syntrophobacter	4	*Deltaproteobacteria*	Propionate (C₃), lactate; some alcohols
Syntrophomonas	8	*Firmicutes*	C_4–C_{18} saturated/unsaturated fatty acids; some alcohols
Pelotomaculum	5	*Firmicutes*	Propionate, lactate, several alcohols; some aromatic compounds
Syntrophus	3	*Deltaproteobacteria*	Benzoate and several related aromatic compounds; some fatty acids and alcohols

[a]All syntrophs are obligate anaerobes.
[b]See Chapters 15 and 16.
[c]Not all species can use all substrates listed.

As an example of syntrophy, consider the fermentation of ethanol to acetate plus H_2 by the syntroph *Pelotomaculum* coupled to the production of methane (**Figure 14.62**). As can be seen, the syntroph carries out a reaction whose standard free-energy change ($\Delta G^{0\prime}$) is positive. Hence, in pure culture, the organism will not grow. However, the H_2 produced by *Pelotomaculum* can be used as an electron donor by a methanogen to produce methane, an exergonic reaction. When the two reactions are summed, the overall reaction is exergonic (Figure 14.62), and when *Pelotomaculum* and a methanogen are cultured together (cocultured), both organisms grow luxuriously.

A second example of syntrophy is the oxidation of a fatty acid such as butyrate to acetate plus H_2 by the fatty acid–oxidizing bacterium *Syntrophomonas* (**Figure 14.63a**):

$$\text{Butyrate}^- + 2\,H_2O \rightarrow 2\,\text{acetate}^- + H^+ + 2\,H_2 \quad \Delta G^{0\prime} = +48.2\,\text{kJ}$$

The free-energy change of this reaction is even more unfavorable than that of ethanol oxidation (Figure 14.63a), and in pure culture *Syntrophomonas* will obviously not grow on butyrate. However, as with ethanol fermentation by *Pelotomaculum*, if the H_2 produced by *Syntrophomonas* is consumed by a partner organism, *Syntrophomonas* will grow on butyrate in coculture with the H_2-consuming partner. How does this occur?

Energetics of H_2 Transfer

In a syntrophic relationship, the removal of H_2 by a partner organism shifts the equilibrium of the entire reaction and pulls it in the direction of product formation; this can greatly affect the energetics of the reaction. Recall from our consideration of the principles of free energy (⇨ Section 3.4) that the concentration of reactants and products in a reaction can have a major effect on the reaction's energetics. This is usually not the case for most fermentation

Ethanol fermentation carried out by the syntroph:

$$2\,CH_3CH_2OH + 2\,H_2O \rightarrow 4\,H_2 + 2\,CH_3COO^- + 2\,H^+$$
$$\Delta G^{0\prime} = +19.4\,\text{kJ/reaction}$$

Methanogenesis carried out by the methanogen:

$$4\,H_2 + CO_2 \rightarrow CH_4 + 2\,H_2O$$
$$\Delta G^{0\prime} = -130.7\,\text{kJ/reaction}$$

Coupled reaction in coculture of syntroph and methanogen:

$$2\,CH_3CH_2OH + CO_2 \rightarrow CH_4 + 2\,CH_3COO^- + 2\,H^+$$
$$\Delta G^{0\prime} = -111.3\,\text{kJ/reaction}$$

(a) Reactions

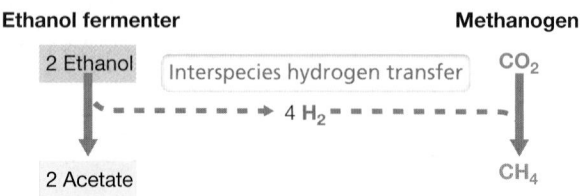

(b) Overview of syntrophic transfer of H_2

Figure 14.62 Syntrophy: Interspecies H_2 transfer. Shown is the fermentation of ethanol to methane and acetate by syntrophic association of an ethanol-oxidizing syntroph and a H_2-consuming partner (in this case, a methanogen). *(a)* Reactions involved. The two organisms share the energy released in the coupled reaction. *(b)* Overview of the syntrophic transfer of H_2.

products because they are not consumed to extremely low levels. H_2, by contrast, is an exception and can be consumed to levels near zero; at these tiny H_2 concentrations, the energetics of a reaction whose product is H_2 is dramatically affected.

For convenience, the $\Delta G^{0\prime}$ of a reaction is calculated on the basis of *standard conditions*—one molar concentration of products and reactants (⇨ Section 3.4). By contrast, the related term ΔG is calculated on the basis of the *actual concentrations* of products and reactants present. At near-zero levels of H_2, the energetics of the oxidation of ethanol or fatty acids to acetate plus H_2, reactions that are endergonic under standard conditions, become exergonic. For example, if the concentration of H_2 is kept extremely low from consumption by the partner organism, ΔG for the oxidation of butyrate by *Syntrophomonas* yields -18 kJ (Figure 14.63a).

While H_2 transfer is characteristic of many syntrophic associations, particularly those associated with fermentation, it is also possible to have syntrophic associations in which *only electrons* are transferred. In direct electron transfer, such as the anoxic consumption of CH_4 by the ANME–sulfate reducer consortium (Section 14.18 and Figure 14.52), we learned that some respiratory syntrophs can use electrically conductive proteins (large multiheme cytochromes and "nanowires") to transfer electrons directly between cells separated by significant distances. In such syntrophic reactions, direct electron transfer does not depend on diffusion rates or H_2 concentrations (see also page 392).

Energetics in Syntrophs

Energy conservation in syntrophs is grounded in both substrate-level and oxidative phosphorylations. From biochemical studies

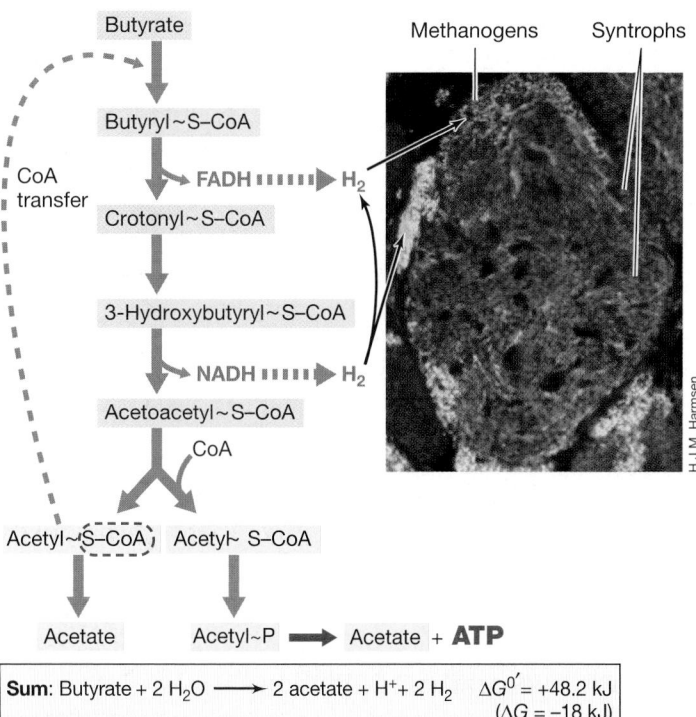

Sum: Butyrate + 2 H$_2$O $\longrightarrow$ 2 acetate + H$^+$ + 2 H$_2$ $\Delta G^{0\prime}$ = +48.2 kJ
(ΔG = −18 kJ)

(a) **Syntrophic culture**

1. Crotonate oxidation:

$$CH_3HC{=}CH{-}\overset{\overset{\displaystyle O}{\|}}{C}\diagdown_{O^-} + 2\,H_2O \longrightarrow \begin{array}{l}\text{2 acetate}\\ \text{+ H}_2 \text{ + H}^+\end{array}$$

2. Crotonate reduction:

$$CH_3HC{=}CH{-}\overset{\overset{\displaystyle O}{\|}}{C}\diagdown_{O^-} + H_2 \longrightarrow \boxed{\text{butyrate}}$$

Proton motive force

Sum: 2 Crotonate + 2 H$_2$O $\longrightarrow$ 2 acetate + butyrate + H$^+$ $\Delta G^{0\prime}$ = −352 kJ

(b) **Pure culture**

Figure 14.63 Energetics of growth of *Syntrophomonas* in syntrophic culture and in pure culture. *(a)* In syntrophic culture, growth requires a H$_2$-consuming organism, such as a methanogen. H$_2$ production is driven by reverse electron flow because the $E_0{}'$ values of the FADH and NADH couples are more electropositive than that of 2 H$^+$/H$_2$. *(b)* In pure culture, energy conservation is linked to anaerobic respiration with crotonate reduction to butyrate. Inset: photomicrograph of cells of a fatty acid–degrading syntrophic bacterium (red) in association with a methanogen (green-yellow).

of syntrophic butyrate catabolism, substrate-level phosphorylation has been shown to occur during the conversion of acetyl-CoA to acetate (Figure 14.63*a*) although the −18 kJ of energy released (ΔG) is in theory insufficient for this. However, the energy released is sufficient to produce *a fraction* of an ATP, so it is possible that in some way *Syntrophomonas* can couple two or more rounds of butyrate oxidation to the synthesis of one ATP by substrate-level phosphorylation.

Besides a syntrophic lifestyle, many syntrophs can also carry out anaerobic respirations in pure culture by the disproportionation of unsaturated fatty acids (disproportionation is a process in which one molecule of a substrate is oxidized while another is reduced). For example, crotonate, an intermediate in syntrophic butyrate metabolism (Figure 14.63*a*), supports growth of pure

cultures of *Syntrophomonas*. Under these conditions some of the crotonate is oxidized to acetate and some is reduced to butyrate (Figure 14.63*b*). Because crotonate reduction by *Syntrophomonas* is coupled to the formation of a proton motive force, as occurs in other anaerobic respirations that employ organic electron acceptors (such as fumarate reduction to succinate, Section 14.15), it is possible that some step(s) in syntrophic metabolism generate a proton motive force as well. Pumping protons or some other ion would almost certainly be required for benzoate- and propionate-fermenting syntrophs, whose free energy yield (ΔG) is vanishingly low, only about −5 kJ per reaction.

Regardless of how ATP is made during syntrophic growth, an additional energetic problem burdens syntrophs. During syntrophic metabolism, syntrophs produce H$_2$ ($E_0{}' = -0.42$ V) from more electropositive electron donors such as FADH ($E_0{}' = -0.22$ V) and NADH ($E_0{}' = -0.32$ V), generated during fatty acid oxidation reactions (Figure 14.63*a*); it is unlikely that this occurs without an energy input. Thus, some fraction of the meager ATP generated by *Syntrophomonas* during syntrophic growth is probably consumed to drive reverse electron flow reactions (Section 14.3) to produce H$_2$. Combining this energy drain with the inherently poor energetic yields of syntrophic reactions, it should be obvious that syntrophic bacteria thrive on a very marginal energy economy.

Ecology of Syntrophs

Ecologically, syntrophic bacteria are key links in the anoxic steps of the carbon cycle ($\Rightarrow$ Section 21.1). Syntrophs consume highly reduced fermentation products and release a key product, H$_2$, for anaerobic respirations. Without syntrophs, a bottleneck would develop in anoxic environments in which electron acceptors (other than CO$_2$) were limiting. By contrast, when conditions are oxic or alternative electron acceptors are abundant, syntrophic relationships are unnecessary. For example, if O$_2$ or NO$_3{}^-$ is available as an electron acceptor, the energetics of the respiration of a fatty acid or an alcohol is so favorable that syntrophic relationships are unnecessary. Thus, syntrophy is characteristic of anoxic catabolism in which primarily methanogenesis or acetogenesis are the terminal processes in the ecosystem. Methanogenesis is a major process in anoxic wastewater biodegradation, and microbiological studies of sludge granules that form in such systems have shown the close physical relationship that develops between H$_2$ producer and H$_2$ consumer in such habitats (Figure 14.63*a* inset).

— **MINIQUIZ** —

- Give an example of interspecies H$_2$ transfer. Why can it be said that both organisms benefit from this process?
- Why can a pure culture of *Syntrophomonas* grow on crotonate but not butyrate?

VII • Hydrocarbon Metabolism

Hydrocarbons, molecules that contain only carbon and hydrogen atoms, are widely used by microbes as electron donors, and we wrap up our coverage of metabolic diversity with a consideration of this process. Unlike methane, hydrocarbons containing two or more carbons typically have to be oxygenated before they can be catabolized. We first consider the aerobic catabolism of aliphatic and aromatic hydrocarbons, where this oxygenation involves O_2. However, we then proceed with a consideration of anoxic hydrocarbon metabolism, a situation where oxygenation of the hydrocarbon is still necessary, but where O_2 obviously plays no role.

14.24 Aerobic Hydrocarbon Metabolism

We previously discussed the role of molecular oxygen (O_2) as an *electron acceptor* in energy-generating reactions. By contrast, O_2 also plays an important role as a *reactant* in the catabolism of hydrocarbons, and oxygenase enzymes are key players in the process.

Oxygenases and Aliphatic Hydrocarbon Oxidation

Oxygenases are enzymes that catalyze the incorporation of O_2 into organic compounds and in some cases, inorganic compounds (Section 14.11). There are two classes of oxygenases: *dioxygenases*, which catalyze the incorporation of *both atoms* of O_2 into the molecule, and *monooxygenases*, which catalyze the incorporation of *only one* of the two oxygen atoms of O_2 into an organic compound with the second atom of O_2 being reduced to H_2O. For most monooxygenases, the required electron donor is NADH or NADPH.

In the initial oxidation step of a saturated aliphatic hydrocarbon, one of the atoms of O_2 is incorporated, typically at a terminal carbon atom. This reaction is catalyzed by a monooxygenase, and a typical reaction sequence is shown in **Figure 14.64a**. The end product of the reaction sequence is a fatty acid of the same length as the original hydrocarbon. The fatty acid is then oxidized by *beta-oxidation*, a series of reactions in which two carbons of the fatty acid are split off at a time (Figure 14.64b). During beta-oxidation, NADH is formed and is oxidized in the electron transport chain for energy conservation purposes. A single round of beta-oxidation releases acetyl-CoA plus a new fatty acid that is two carbon atoms shorter than the original fatty acid. The process of beta-oxidation is then repeated, and another acetyl-CoA molecule is released. The acetyl-CoA formed by beta-oxidation is either oxidized through the citric acid cycle (⮌ Figure 3.16) or used to make new cell material. With the exception of how the hydrocarbon is oxygenated, much of the biochemistry of anoxic hydrocarbon catabolism is the same as that shown for aerobic catabolism (Figure 14.64), with beta-oxidation reactions being of prime importance in both cases.

Aromatic Hydrocarbon Oxidation

Many aromatic hydrocarbons can also be used as electron donors aerobically by microorganisms. The metabolism of these

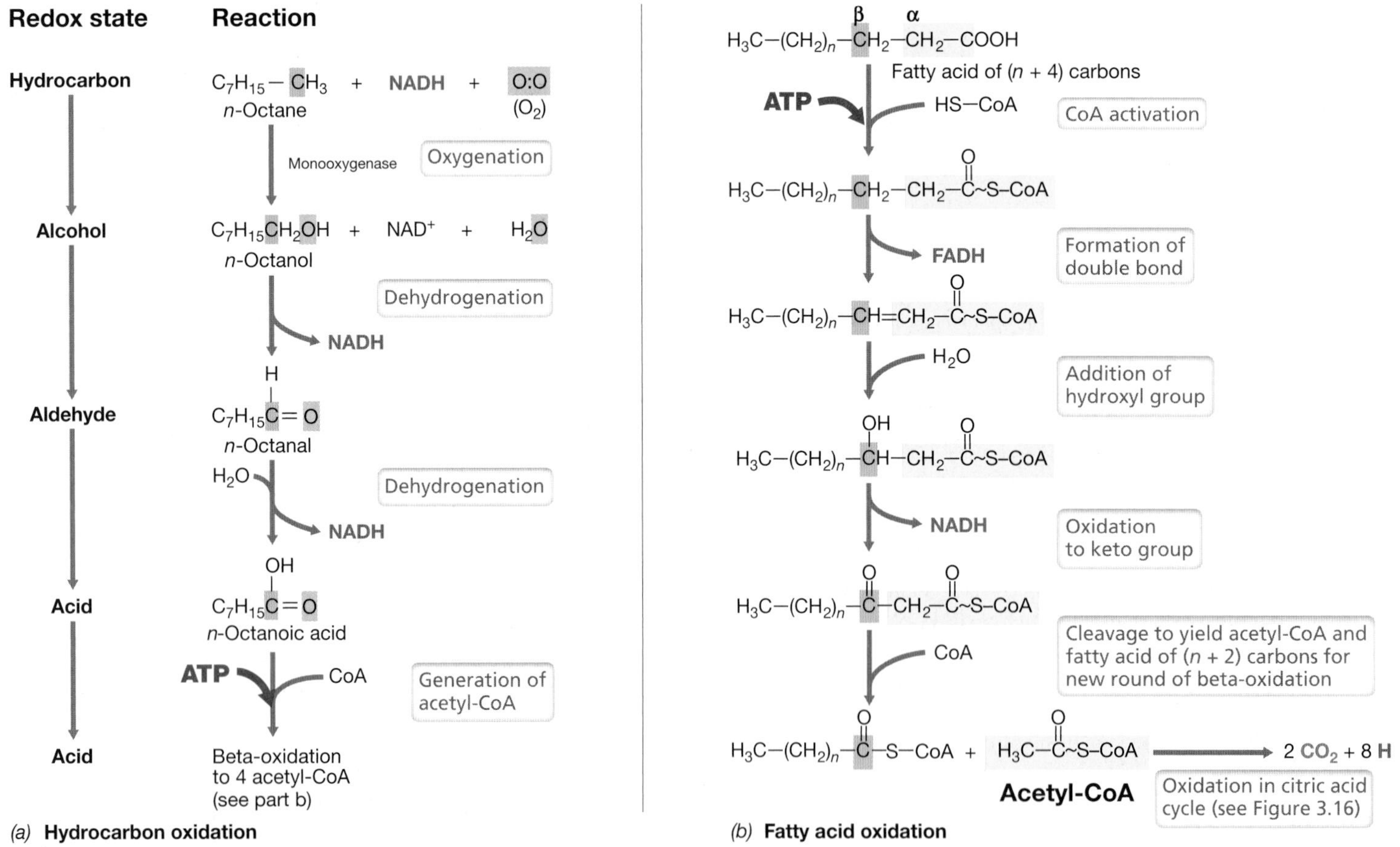

(a) **Hydrocarbon oxidation**

(b) **Fatty acid oxidation**

Figure 14.64 Monooxygenase activity and beta-oxidation. *(a)* Steps in the oxidation of an aliphatic hydrocarbon, the first of which is catalyzed by a monooxygenase. *(b)* Fatty acid oxidation by beta-oxidation leads to the successive formation of acetyl-CoA.

compounds, some of which contain multiple rings, such as naphthalene or biphenyls, typically has as its initial stage the formation of catechol or a structurally related compound via catalysis by oxygenase enzymes, as shown in **Figure 14.65**. Once catechol is formed it can be cleaved and further degraded into compounds that can enter the citric acid cycle, such as succinate, acetyl-CoA, and pyruvate.

Several steps in the aerobic catabolism of aromatic hydrocarbons require oxygenases. Figure 14.65a–c shows four different oxygenase-catalyzed reactions, one using a monooxygenase, two using a ring-cleaving dioxygenase, and one using a ring-hydroxylating dioxygenase. As in aerobic aliphatic hydrocarbon catabolism (Figure 14.64), aromatic compounds, whether single- or multi-ringed, are typically oxidized completely to CO_2, with electrons entering an electron transport chain or used to make new cell material.

MINIQUIZ

- How do monooxygenases differ in function from dioxygenases?
- What is the final product of catabolism of a hydrocarbon?
- What is meant by the term beta-oxidation?

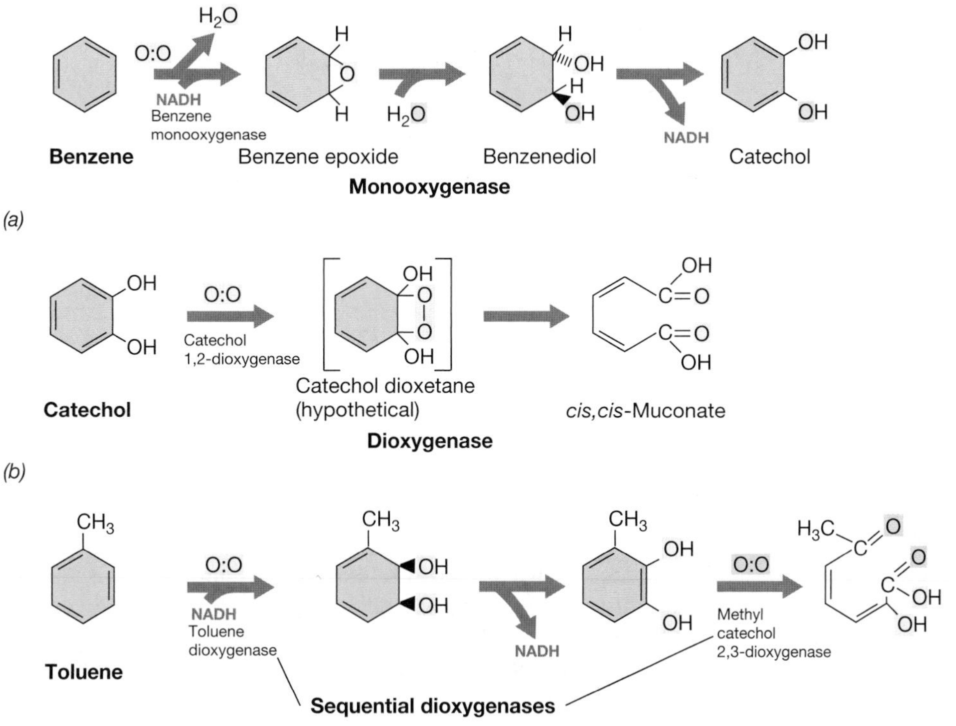

14.25 Anaerobic Hydrocarbon Metabolism

Although aerobic hydrocarbon oxidation is a major process in nature, anaerobic hydrocarbon oxidation linked to the reduction of nitrate, sulfate, or ferric iron as electron acceptors in anaerobic respirations is also possible. And, as for the aerobic process, both aliphatic and aromatic hydrocarbons can be degraded anaerobically.

Aliphatic Hydrocarbons

Aliphatic hydrocarbons are straight-chain saturated or unsaturated organic compounds, and many are substrates for denitrifying and sulfate-reducing bacteria. Saturated aliphatic hydrocarbons as long as C_{20} support growth, although shorter-chain hydrocarbons are more soluble and therefore more readily catabolized. The mechanism of anoxic hydrocarbon degradation has been well studied for hexane (C_6H_{14}) metabolism in denitrifying bacteria (NO_3^- as electron acceptor). However, the mechanism appears to be the same for the anoxic catabolism of longer-chain hydrocarbons and for anoxic hydrocarbon oxidation linked to other electron acceptors, and so we focus on the hexane/nitrate system here.

In anoxic hexane metabolism, hexane is modified on carbon atom 2 by attachment of a molecule of *fumarate*, a C_4 intermediate of the citric acid cycle (Figure 3.16), forming the compound *1-methylpentylsuccinate* (**Figure 14.66a**). The enzymatic addition of fumarate to hexane effectively oxygenates the hexane and allows the molecule to be further catabolized anaerobically. Following the addition of coenzyme A, a series of reactions occurs that includes beta-oxidation (Figure 14.64b) and regeneration of fumarate. The electrons released during beta-oxidation generate a proton motive force and are consumed in nitrate or sulfate reduction (Sections 14.13 and 14.14, respectively).

Aromatic Hydrocarbons

Aromatic hydrocarbons can be degraded anaerobically by some nitrate, ferric iron, and sulfate-reducing bacteria. For anoxic catabolism of the aromatic hydrocarbon toluene, oxygen needs to be added to the compound to begin catabolism, and this occurs by the addition of fumarate, as in aliphatic hydrocarbon catabolism (Figure 14.66a). The reaction series eventually yields benzoyl-CoA, which is further degraded by ring reduction (Figure 14.66b). Benzene (C_6H_6) can also be catabolized anaerobically, likely by a similar mechanism. Multi-ringed aromatic hydrocarbons such as naphthalene ($C_{10}H_8$) can be degraded by certain sulfate-reducing and denitrifying bacteria. In contrast to other hydrocarbons, the oxygenation of multi-ringed hydrocarbons occurs by the addition of CO_2 to the ring to form a carboxylic acid derivative rather than by fumarate

(a)

(b)

(c)

Figure 14.65 Roles of oxygenases in catabolism of aromatic compounds. Monooxygenases introduce one atom of oxygen from O_2 into a substrate, whereas diooxygenases introduce both atoms of oxygen. *(a)* Hydroxylation of benzene to catechol by a monooxygenase in which NADH is an electron donor. *(b)* Cleavage of catechol to *cis, cis*-muconate by an intradiol ring-cleavage dioxygenase. *(c)* The activities of a ring-hydroxylating dioxygenase and an extradiol ring-cleavage dioxygenase in the degradation of toluene. The oxygen atoms that each enzyme introduces are distinguished by different colors. Compare aerobic toluene catabolism to anoxic toluene catabolism shown in Figure 14.66b.

UNIT 4

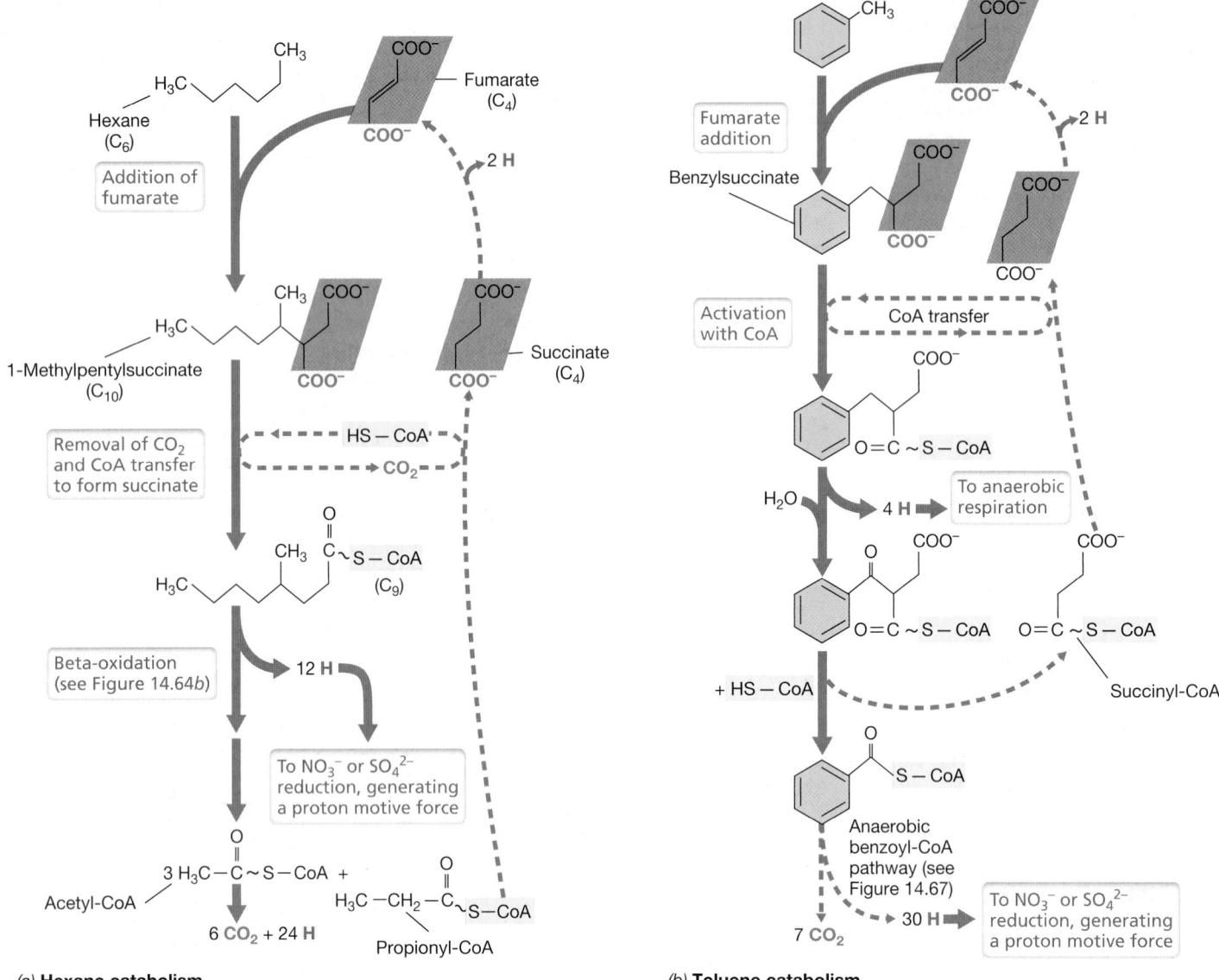

(a) **Hexane catabolism**

(b) **Toluene catabolism**

Figure 14.66 Anoxic catabolism of two hydrocarbons. *(a)* In anoxic catabolism of the aliphatic hydrocarbon hexane, the addition of fumarate provides the oxygen atoms necessary to form a fatty acid derivative that can be catabolized by beta-oxidation (see Figure 14.64) to yield acetyl-CoA. Electrons (H) generated from hexane catabolism are used to reduce sulfate or nitrate in anaerobic respirations. *(b)* Fumarate addition during the anoxic catabolism of the aromatic hydrocarbon toluene forms benzylsuccinate.

addition. But this carboxylation reaction serves the same purpose as oxygenase reactions (Figures 14.64*a* and 14.65) or the addition of fumarate (Figure 14.66); an O atom becomes part of the hydrocarbon and facilitates its catabolism.

Many bacteria can catabolize certain aromatic hydrocarbons anaerobically, including even fermentative and phototrophic bacteria. However, except for toluene, only aromatic compounds that already contain an O atom are degraded, typically by a common mechanism. In contrast to aerobic catabolism that occurs by way of ring *oxidation* (Figure 14.65), anaerobic catabolism proceeds by ring *reduction*. The anaerobic degradation of aromatic hydrocarbons is often facilitated by their conversion into benzoate followed by aromatic ring cleavage and benzoate catabolism

performed by the *benzoyl-CoA pathway* (**Figure 14.67**). Benzoate catabolism in this pathway begins by forming the coenzyme A derivative followed by ring cleavage to yield fatty or dicarboxylic acids that can be further catabolized to intermediates of the citric acid cycle (Figure 14.67).

As we reach the end of our survey of metabolic diversity, we need not keep in mind all of the metabolic details, for they are numerous and formidable, but instead, we need to see the "metabolic big picture" formed by the major themes of this chapter: phototrophy and chemolithotrophy, CO_2 and N_2 fixations, respirations by electron donor or electron acceptor, C_1 metabolisms, fermentations, and hydrocarbon metabolism. These overarching principles will guide us through the next four

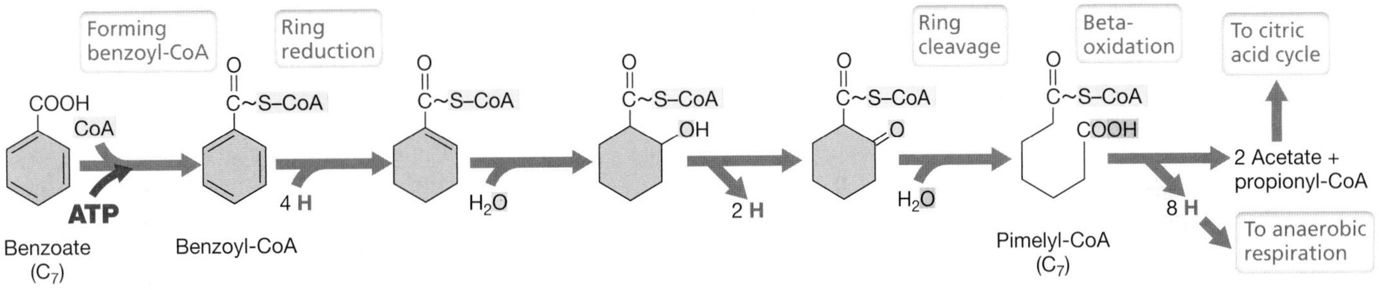

Figure 14.67 Anoxic degradation of benzoate by the benzoyl-CoA pathway. This pathway operates in the purple phototrophic bacterium *Rhodopseudomonas palustris* and many other facultative bacteria, both phototrophic and chemotrophic. Note that all intermediates of the pathway are bound to coenzyme A. The acetate produced is further catabolized in the citric acid cycle.

chapters where our focus will be on the diversity of the microbes themselves rather than their metabolisms. Chapters 15–17 in particular will cover many of the *Bacteria* and *Archaea* whose metabolisms we have described here, and we will witness there the many cases in which metabolic diversity and prokaryotic diversity are inextricably linked.

MINIQUIZ

- What is the benzoyl-CoA pathway, and how might it participate in the anaerobic degradation of toluene?
- How is hexane oxygenated during anoxic catabolism?

MasteringMicrobiology®   **Visualize, explore,** and **think critically** with Interactive Microbiology, MicroLab Tutors, MicroCareer case studies, and more. MasteringMicrobiology offers practice quizzes, helpful animations, and other study tools for lecture and lab to help you master microbiology.

Chapter Review

I • Phototrophy

14.1 In photosynthesis, ATP is generated from light and then consumed in the reduction of CO_2. Two forms of photosynthesis are known: oxygenic, where O_2 is produced (for example, in cyanobacteria), and anoxygenic, where it is not (for example, in purple and green bacteria). Photosynthetic reaction centers and photosynthetic pigments reside in membranes where the light reactions of photosynthesis are carried out.

Q **What are the functions of light-harvesting and reaction center chlorophylls?**

14.2 Accessory pigments including carotenoids and phycobilins absorb light and transfer the energy to reaction center chlorophyll, thus broadening the wavelengths of light usable in photosynthesis. Carotenoids also play an important photoprotective role in preventing photooxidative damage to cells.

Q **What accessory pigments are present in phototrophs, and what are their functions?**

14.3 Electron transport reactions occur in the photosynthetic reaction center of anoxygenic phototrophs, forming a proton motive force and ATP.

Q **What is reverse electron transport and why is it necessary in purple sulfur bacteria?**

14.4 In oxygenic photosynthesis, H_2O donates electrons to drive CO_2 fixation, and O_2 is a by-product. There are two separate but interconnected photosystems in oxygenic phototrophs, PSI and PSII, whereas anoxygenic phototrophs contain a single photosystem.

Q **How does the reduction potential (E_0') of chlorophyll *a* in PSI and PSII differ? Why must the reduction potential of PSII chlorophyll *a* be so highly electropositive?**

II • Autotrophy and N_2 Fixation

14.5 Autotrophy is supported in most phototrophic and chemolithotrophic bacteria by the Calvin cycle, in which the enzyme RubisCO plays a key role. The reverse citric acid and hydroxypropionate cycles are autotrophic pathways in green sulfur and green nonsulfur bacteria, respectively.

Q **What is a carboxysome, and what is its role in CO_2 fixation? Plants do not have carboxysomes; how might this affect the efficiency with which they fix CO_2?**

14.6 The reduction of N_2 to NH_3 is called nitrogen fixation and is catalyzed by the enzyme nitrogenase. Nitrogenase can be assayed using the triply bonded compound acetylene as a surrogate of N_2; nitrogenase reduces acetylene to ethylene.

Q **How might the ability to fix nitrogen help a bacterium be more competitive in its environment?**

III • Respiratory Processes Defined by Electron Donor

14.7 During respiration, energy is conserved by chemiosmotic ion gradients generated by the transport of electrons through the electron transport chain, and ATP is synthesized by ATP synthase. Anaerobic respiration yields less energy than aerobic respiration but can proceed in environments where O_2 is absent but alternative electron acceptors are present.

> **Q** Why is NO_3^- a better electron acceptor for anaerobic respiration than is SO_4^{2-}?

14.8 The chemolithotrophic hydrogen bacteria use H_2 as an electron donor, reducing O_2 to H_2O. The enzyme hydrogenase is required to oxidize H_2, and H_2 also supplies reducing power for the fixation of CO_2 in these autotrophs.

> **Q** Which inorganic electron donors are used by the organisms *Ralstonia*, *Thiobacillus*, and *Acidithiobacillus*?

14.9 Reduced sulfur compounds such as H_2S, $S_2O_3^{2-}$, and S^0 are electron donors for energy conservation in sulfur chemolithotrophs. Electrons from these substances enter electron transport chains, yielding a proton motive force. Sulfur chemolithotrophs are also autotrophs and fix CO_2 by the Calvin cycle.

> **Q** Compare and contrast the utilization of H_2S by a purple phototrophic bacterium and by a colorless sulfur bacterium such as *Beggiatoa*. What role does H_2S play in the metabolism of each organism?

14.10 Chemolithotrophic iron bacteria oxidize Fe^{2+} as an electron donor. Most iron bacteria grow at acidic pH and are often associated with acidic pollution from mineral and coal mining. A few chemolithotrophic and phototrophic bacteria can oxidize Fe^{2+} to Fe^{3+} anaerobically.

> **Q** Why is it necessary for iron-oxidizing bacteria to have cytochromes in their outer membranes even though electron transport and energy conservation takes place at the cytoplasmic membrane?

14.11 The ammonia-oxidizing *Bacteria* and *Archaea* produce nitrite from ammonia, which is then oxidized by nitrite-oxidizing *Bacteria* to nitrate.

> **Q** What is the key enzyme that mediates ammonia oxidation, and how does this enzyme function in the cell?

14.12 Anoxic ammonia oxidation (anammox) consumes both ammonia and nitrite, forming N_2. The anammox reaction occurs within a membrane-enclosed compartment called the anammoxosome.

> **Q** Contrast classical nitrification with anammox in terms of oxygen requirements, organisms involved, and the need for monooxygenases.

IV • Respiratory Processes Defined by Electron Acceptor

14.13 Nitrate is a common electron acceptor in anaerobic respiration. Nitrate reduction is catalyzed by the enzyme nitrate reductase, reducing NO_3^- to NO_2^-. Many bacteria that use NO_3^- in anaerobic respiration produce gaseous nitrogen compounds (NO, N_2O, or N_2) as final end products of reduction (denitrification).

> **Q** What is the difference in nitrate respiration by *Escherichia coli* and *Pseudomonas*?

14.14 Sulfate-reducing bacteria are obligately anaerobic bacteria that reduce SO_4^{2-} to H_2S in a process in which SO_4^{2-} must first be activated to adenosine phosphosulfate (APS). Disproportionation is an additional energy-yielding strategy for certain species. Some organisms, such as *Desulfuromonas*, cannot reduce SO_4^{2-} but can reduce S^0 to H_2S.

> **Q** Why is the enzyme hydrogenase useful to *Desulfovibrio* even when it is not grown on H_2 as electron donor?

14.15 Besides inorganic nitrogen and sulfur compounds and CO_2, several other substances can function as electron acceptors for anaerobic respiration. These include Fe^{3+}, Mn^{4+}, fumarate, certain organic and chlorinated organic compounds, and even protons.

> **Q** Compare and contrast ferric iron reduction with reductive dechlorination in terms of (1) product of the reduction and (2) environmental significance.

V • One-Carbon (C_1) Metabolism

14.16 Acetogens are strict anaerobes that reduce CO_2 to acetate, usually with H_2 as electron donor. The mechanism of acetate formation is the acetyl-CoA pathway, a pathway widely distributed in obligate anaerobes for either autotrophic purposes or acetate oxidation.

> **Q** Compare and contrast acetogens with methanogens in terms of (1) substrates and products of their energy metabolism, (2) ability to use organic compounds as electron donors in energy metabolism, and (3) phylogeny.

14.17 Methanogenesis is the production of CH_4 from $CO_2 + H_2$ or from acetate or methanol by strictly anaerobic methanogenic *Archaea*. Several unique coenzymes are required for methanogenesis, and energy conservation is linked to either a proton motive or sodium motive force.

> **Q** What are the major differences in the conservation of energy by methanogens that only use $H_2 + CO_2$ versus those that can use methyl compounds?

14.18 Methanotrophy is the use of CH_4 as both carbon source and electron donor, and the enzyme methane monooxygenase is a key enzyme in the aerobic catabolism of methane. In methanotrophs, C_1 units are assimilated into cell material as formaldehyde or formaldehyde plus CO_2 by the ribulose monophosphate or serine pathways, respectively. The anaerobic oxidation of methane can be performed by ANME *Archaea*, which are related to methanogens, but which conserve energy by running the pathway for methanogenesis in reverse. The anaerobe *Methylomirabilis oxyfera* performs intra-aerobic methane oxidation in which O_2 is formed and then consumed.

> **Q** How does a methano*troph* differ from a methano*gen*? Which pathway for C_1 assimilation found in aerobic methanotrophs is most energetically efficient and why? In what ways does *M. oxyfera* resemble denitrifiers and aerobic methanotrophs, and in what important ways does it differ?

VI • Fermentation

14.19 In the absence of external electron acceptors, organic compounds can be catabolized anaerobically only by fermentation. Most fermentations require that an energy-rich organic compound be formed that can yield ATP by substrate-level phosphorylation. Redox balance is achieved by the production of fermentation products.

> **Q** Define the term substrate-level phosphorylation: How does it differ from oxidative phosphorylation? What compound(s) do fermentative bacteria need to synthesize in order to make ATP by substrate-level phosphorylation?

14.20 The lactic acid fermentation is carried out by homofermentative species, where lactate is the sole product, and heterofermentative species, where lactate, ethanol, and CO_2 are produced. The mixed-acid fermentation typical of enteric bacteria yields various acids plus neutral products (ethanol, butanediol), depending on the organism.

> **Q** What are the major fermentation products of *Lactobacillus* and *Escherichia*?

14.21 Clostridia ferment sugars, amino acids, and other organic compounds, with butyric acid being a major product. Butyrate production allows for an additional ATP to be produced. *Propionibacterium* produces propionate, acetate, and CO_2 in a secondary fermentation of lactate.

> **Q** How does *Propionibacterium* use secondary fermentation to conserve energy?

14.22 Energy conservation in *Propionigenium*, *Oxalobacter*, and *Malonomonas* is linked to decarboxylation reactions that pump Na^+ or H^+ across the membrane; ATPases use the energy in the ion gradient to form ATP. The reactions catalyzed by these organisms yield insufficient free energy to make ATP by substrate-level phosphorylation.

> **Q** Give an example of a fermentation that does not employ substrate-level phosphorylation.

14.23 In syntrophy, two organisms cooperate to degrade a compound that neither can degrade alone. In this process H_2 produced by one organism is consumed by the partner. H_2 consumption affects the energetics of the reaction carried out by the H_2 producer, allowing it to make ATP where it otherwise could not.

> **Q** Why is syntrophy also called "interspecies H_2 transfer"? When does syntrophy not involve H_2 transfer?

VII • Hydrocarbon Metabolism

14.24 In addition to its role as a terminal electron acceptor, O_2 can also be a substrate. In aerobic metabolism, oxygenases introduce atoms of oxygen from O_2 into hydrocarbons. Once oxygenated, aliphatic hydrocarbons can be further degraded by beta-oxidation and aromatic hydrocarbons by ring splitting and oxidation.

> **Q** How do monooxygenases differ from dioxygenases in terms of the reactions they catalyze? Why are oxygenases necessary for the aerobic catabolism of hydrocarbons?

14.25 Hydrocarbons can be oxidized under anoxic conditions following addition of the dicarboxylic acid fumarate. Aromatic compounds are catabolized anaerobically by ring reduction and cleavage to form intermediates that can be catabolized in the citric acid cycle.

> **Q** How do denitrifying and sulfate-reducing bacteria degrade hydrocarbons anaerobically and without oxygenases?

Application Questions

1. The growth rate of the phototrophic purple bacterium *Rhodobacter* is about twice as fast when the organism is grown *phototrophically* in a medium containing malate as the carbon source as when it is grown with CO_2 as the carbon source (with H_2 as the electron donor). Discuss the reasons why this is true, and list the nutritional class in which we would place *Rhodobacter* when growing under each of the two different conditions.

2. Although physiologically distinct, aerobic chemolithotrophs and chemoorganotrophs share a number of features with respect to the production of ATP. Discuss these common features along with reasons why the growth yield (grams of cells per mole of substrate consumed) of a chemoorganotroph respiring glucose is so much higher than for a chemolithotroph respiring sulfur.

3. A fatty acid such as butyrate cannot be fermented in pure culture, although its anaerobic catabolism under other conditions occurs readily. How do these conditions differ, and why does the latter allow for butyrate catabolism? How then can butyrate be fermented in mixed culture?

4. When methane is made from CO_2 (plus H_2) or from methanol (in the absence of H_2), various steps in the metabolic pathways are shared in common. Compare and contrast methanogenesis from these two substrates, highlighting the similarities and differences in the processes.

Chapter Glossary

Acetogenesis energy metabolism in which acetate is produced from either H_2 plus CO_2 or from organic compounds

Anaerobic respiration use of an electron acceptor other than O_2 in an electron transport–based oxidation leading to a proton motive force

Anammox anoxic ammonia oxidation

Anoxygenic photosynthesis photosynthesis in which O_2 is not produced

Antenna pigments light-harvesting chlorophylls or bacteriochlorophylls in photocomplexes that funnel energy to the reaction center

Autotroph an organism that uses CO_2 as its sole carbon source

Bacteriochlorophyll the chlorophyll pigment of anoxygenic phototrophs

Calvin cycle the biochemical pathway for CO_2 fixation in many autotrophic organisms

Carboxysomes crystalline inclusions of RubisCO

Carotenoid a hydrophobic accessory pigment present along with chlorophyll in photosynthetic membranes

Chlorophyll a light-sensitive, Mg-containing porphyrin of phototrophic organisms that initiates the process of photophosphorylation

Chlorosome a cigar-shaped structure present in the periphery of cells of green sulfur and green nonsulfur bacteria and containing the antenna bacteriochlorophylls (*c*, *d*, or *e*)

Denitrification anaerobic respiration in which NO_3^- or NO_2^- is reduced to nitrogen gases, primarily N_2

Dicarboxylate/4-hydroxybutyrate cycle an autotrophic pathway found in certain *Archaea*

Fermentation anaerobic catabolism of an organic compound in which the compound serves as both an electron donor and an electron acceptor and in which ATP is usually produced by substrate-level phosphorylation

Heterofermentative producing a mixture of products, typically lactate, ethanol, and CO_2, from the fermentation of glucose

Homofermentative producing only lactic acid from the fermentation of glucose

Hydrogenase an enzyme, widely distributed in anaerobic microorganisms, capable of oxidizing or evolving H_2

3-Hydroxypropionate bi-cycle an autotrophic pathway found in *Chloroflexus* and a few *Archaea*

3-Hydroxypropionate/4-hydroxybutyrate cycle an autotrophic pathway found in certain *Archaea*

Methanogen a methane-producing member of the *Archaea*

Methanogenesis the biological production of CH_4

Methanotroph an organism that oxidizes CH_4

Methylotroph an organism capable of growth on compounds containing no C—C bonds; some methylotrophs are methanotrophic

Mixotroph an organism in which an inorganic compound serves as the electron donor in energy metabolism and organic compounds serve as the carbon source

Nitrification the microbial oxidation of ammonia to nitrate

Nitrogenase the enzyme complex required to reduce N_2 to NH_3 in biological nitrogen fixation

Nitrogen fixation the reduction of N_2 to NH_3 by the enzyme nitrogenase

Oxygenase an enzyme that catalyzes the incorporation of oxygen from O_2 into organic or inorganic compounds

Oxygenic photosynthesis photosynthesis carried out by cyanobacteria and green plants in which O_2 is evolved

Photophosphorylation the production of ATP in photosynthesis

Photosynthesis the series of reactions in which ATP is synthesized by light-driven reactions and CO_2 is fixed into cell material

Phototroph an organism that uses light as an energy source

Phycobiliprotein the antenna pigment complex in cyanobacteria that contains phycocyanin and allophycocyanin or phycoerythrin coupled to proteins

Phycobilisome an aggregate of phycobiliproteins

Reaction center a photosynthetic complex containing chlorophyll or bacteriochlorophyll and several other components; the initial electron transfer reactions of photosynthetic electron flow occur here

Reductive acetyl-coenzyme A (acetyl-CoA) pathway a pathway used for acetogenesis, autotrophic CO_2 fixation, and acetate oxidation (when run in the reverse direction) widespread in obligate anaerobes including methanogens, acetogens, and sulfate-reducing bacteria

Reductive dechlorination an anaerobic respiration in which a chlorinated organic compound is used as an electron acceptor, usually with the release of Cl^-

Reverse citric acid cycle a mechanism for autotrophy in green sulfur bacteria and a few other autotrophic *Bacteria*, and also in some *Archaea*

Reverse electron transport the energy-dependent movement of electrons against the thermodynamic gradient to form a strong reductant from a weaker electron donor

Ribulose monophosphate pathway a reaction series in certain methylotrophs in which formaldehyde is assimilated into cell material using ribulose monophosphate as the C_1 acceptor molecule

RubisCO the acronym for ribulose bisphosphate carboxylase, a key enzyme of the Calvin cycle

Secondary fermentation a fermentation in which the substrates are the fermentation products of other organisms

Serine pathway a reaction series in certain methylotrophs in which CH_2O plus CO_2 are assimilated into cell material by way of the amino acid serine

Stickland reaction the fermentation of an amino acid pair

Syntrophy a process whereby two or more microorganisms cooperate to degrade a substance neither can degrade alone

Thylakoids membrane stacks in cyanobacteria or in the chloroplast of eukaryotic phototrophs

Functional Diversity of Microorganisms

microbiology**now**

New Discoveries Have Redefined the Global Nitrogen Cycle

Nitrification, the aerobic oxidation of ammonia to nitrate, is a critical component of the global nitrogen cycle and is mediated exclusively by microorganisms. The Russian microbiologist Sergei Winogradsky was the first to describe the microbial basis of nitrification when he isolated chemolithotrophic bacteria that oxidized ammonia to nitrite over a century ago. Since then, the process of nitrification has been considered a two-step process that requires two physiological classes of chemolithotrophs: (1) microbes that oxidize ammonia to nitrite, and (2) microbes that oxidize nitrite to nitrate. Ammonia-oxidizing *Bacteria* and *Archaea* typically associate closely with nitrite-oxidizing *Bacteria*, and it has long been thought that nitrification absolutely required the coordinated activities of both of these functional groups. That is, only by working together could these microbes catalyze the overall nitrification process.

Recent discoveries continue to redefine our understanding of the nitrogen cycle, and sometimes these discoveries come from unexpected sources, such as a microbial biofilm found growing in an oil well. This biofilm was used to inoculate an enrichment culture designed to grow nitrifying bacteria. Over time the culture was found to oxidize ammonia all the way to nitrate, but the microbes in this culture did not match any known nitrifying organisms. Since the bacteria could not be grown in isolation, metagenomic sequencing was used to reconstruct the genomes of the species present. One of the dominant genomes belonged to the phylum *Nitrospira* (the photo shows clusters of fluorescently stained *Nitrospira* cells [yellow] in a 3D reconstruction of a biofilm observed by confocal laser microscopy). Bacteria from this phylum were already known to oxidize nitrite to nitrate (the second step of nitrification), but this microbe was different. It also contained the genes required for ammonia oxidation, most notably the gene encoding the key enzyme ammonia monooxygenase.

This novel bacterium, named "*Candidatus* Nitrospira inopinata," is the first microbe shown to perform both steps of nitrification. Now that they have been identified, unique gene sequences from these complete ammonia-oxidizing bacteria—nicknamed "comammox" bacteria for their dual metabolic activities—have been found in freshwater, soil, and engineered environments worldwide. The discovery of comammox bacteria once again emphasizes the fact that new microbial metabolisms are typically just "one culture away."

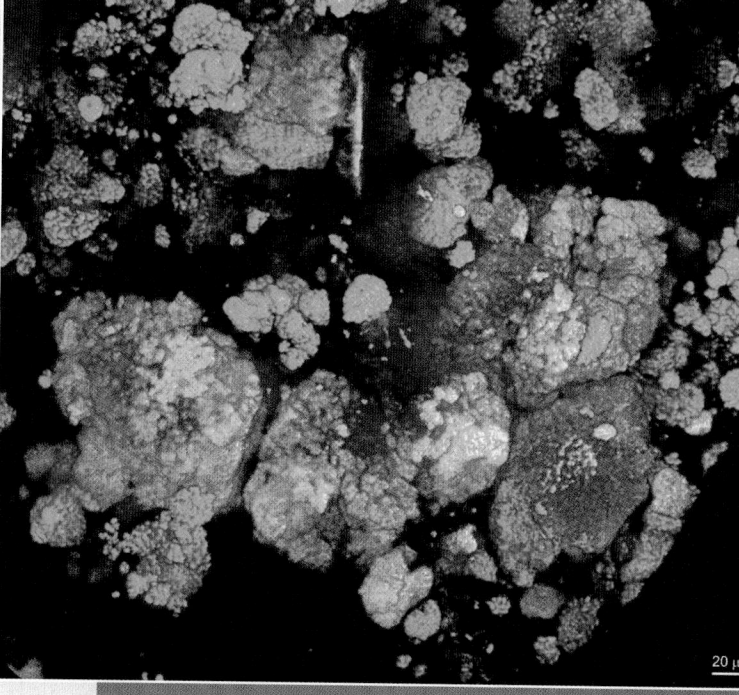

20 μm

 **Source:** Daims, H., et al. 2015. Complete nitrification by *Nitrospira* bacteria. *Nature* 528: 504–509.

I • Functional Diversity as a Concept

Microbial diversity can be understood in terms of both phylogenetic diversity and functional diversity. We begin by defining and contrasting the concepts of phylogenetic diversity and functional diversity.

15.1 Making Sense of Microbial Diversity

Phylogenetic diversity is the component of microbial diversity that deals with evolutionary relationships between microorganisms. Most fundamentally, phylogenetic diversity deals with the diversity of evolutionary lineages such as phyla, genera, and species. At its broadest, phylogenetic diversity encompasses the genetic and genomic diversity of evolutionary lineages and so can be defined on the basis of either genes or organisms (⇨ Section 13.7). Most commonly, though, phylogenetic diversity is defined on the basis of ribosomal RNA gene phylogeny, which is thought to reflect the phylogenetic history of the entire organism (⇨ Section 13.7). Phylogenetic diversity is the overarching theme of our coverage of microbial diversity in Chapters 16–18.

Functional diversity is the component of microbial diversity that deals with diversity in form and function as it relates to microbial physiology and ecology. It is useful to consider microbial diversity in terms of functional groupings because organisms with common traits and common genes often share physiological characteristics and have similar ecological roles. In many cases, functional traits align with phylogenetic groups (for example, with the organisms described in Sections 15.3, 15.4, 15.6, 15.7, 15.19). Microbial functional diversity, however, often does not correspond with phylogenetic diversity as defined by the 16S ribosomal RNA gene. We will see many examples in this chapter where functional traits are widely distributed among the *Bacteria* and *Archaea* (**Figure 15.1**).

At least three reasons can account for why a functional trait is shared between divergent organisms with dissimilar 16S ribosomal RNA gene sequences. The first is *gene loss*, a situation where a trait present in the common ancestor of several lineages is subsequently lost in some lineages but retained in others that over evolutionary time became quite divergent. The second is **convergent evolution**, in which a trait has evolved independently in two or more lineages and is not encoded by homologous genes shared by these lineages. The third is **horizontal gene transfer**

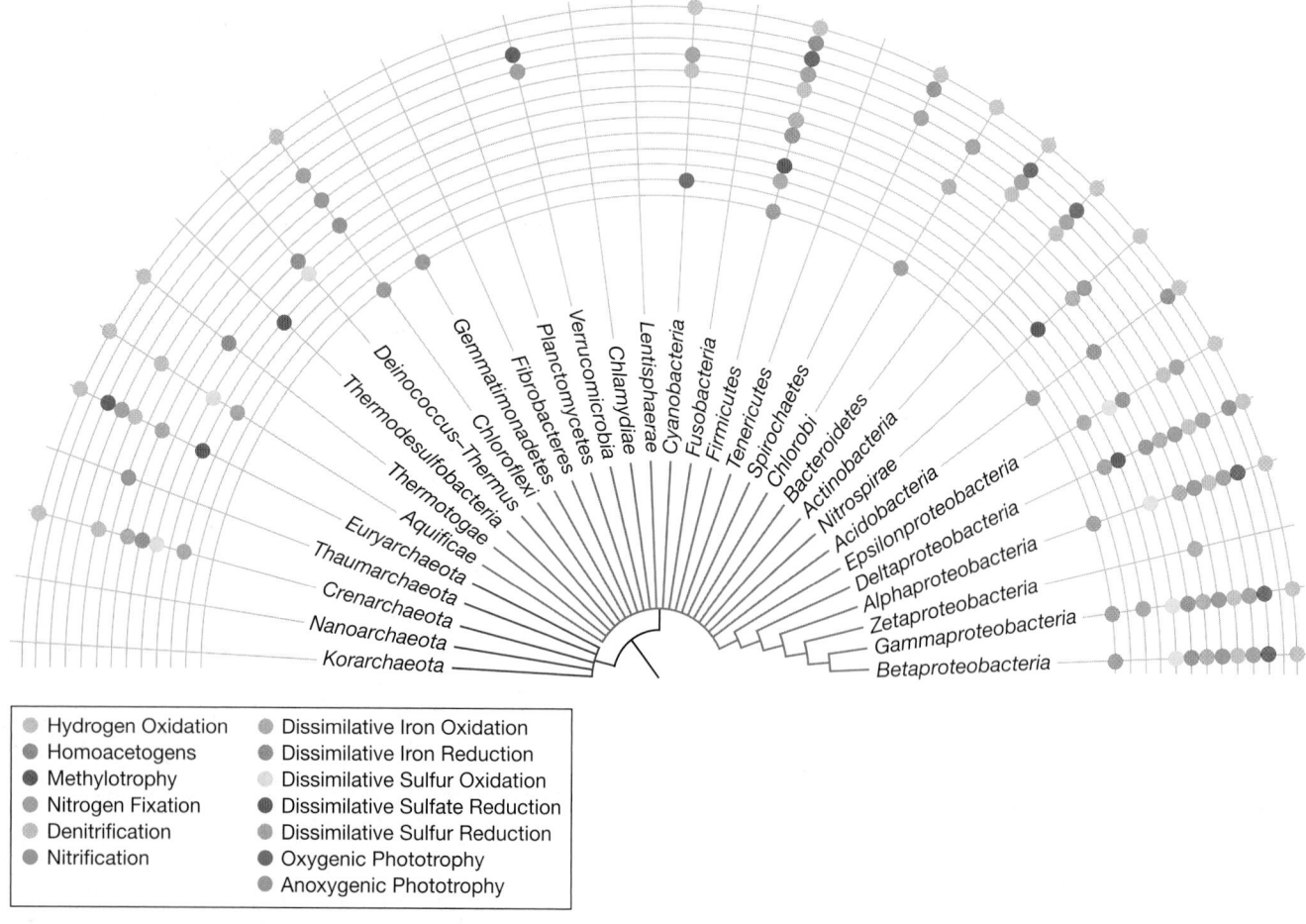

Figure 15.1 Major functional traits mapped across major phyla of *Bacteria* and *Archaea*.
The dendrogram shows relationships between microbial phyla as inferred by analysis of 16S ribosomal RNA gene sequences. Blue branches are used to denote phyla of *Bacteria* and red branches phyla of *Archaea*. Colored circles indicate phyla that contain at least one species with a functional trait indicated in the color key.

(⇄ Sections 9.6 and 13.7), a situation where genes that confer a particular trait are homologous and have been exchanged between distantly related lineages.

Functional diversity can be further defined in terms of physiological diversity, ecological diversity, and morphological diversity. *Physiological diversity* relates to the functions and activities of microorganisms. Physiological diversity is most commonly described in terms of microbial metabolism and cellular biochemistry (Chapter 14). *Ecological diversity* relates to relationships between organisms and their environments. Organisms with similar physiological characteristics can have different ecological strategies (Section 15.11). Causes and consequences of ecological diversity will also be considered when we consider the science of microbial ecology in Chapters 19 and 20. *Morphological diversity* relates to the outward appearance of an organism, as cell shape and cellular structures often have ecological significance for microorganisms (Sections 15.19–15.22). In some cases, the morphology of a group is so distinctive that the group is essentially defined by this property, for example, with the spirochetes (Section 15.19).

The concepts of physiological, ecological, and morphological diversity are often intertwined. The examples provided in this chapter are meant to be illustrative and not exhaustive, and we will consider other organisms with important ecological functions in Chapters 16–18 and 20–23.

MINIQUIZ

- Why is it necessary to consider microbial diversity in terms of phylogenetic diversity and functional diversity?
- What are three reasons that functional traits might not correspond with distinct phylogenetic groups as defined by 16S ribosomal RNA gene sequences?

II • Diversity of Phototrophic *Bacteria*

In this section we consider the diversity of phototrophic microorganisms, those microbes that conserve energy from light. We will see that phototrophy is widespread within the domain *Bacteria* and that several distinct types of phototrophs can be defined on the basis of their physiological traits.

15.2 Overview of Phototrophic Bacteria

The ability to conserve energy from light evolved early in the history of life, when the Earth was anoxic (⇄ Section 13.2). Photosynthesis originated within the *Bacteria*, and the first phototrophic organisms were *anoxygenic phototrophs*, organisms that do not generate O_2 as a product of photosynthesis (⇄ Section 14.3). Instead of H_2O, these early phototrophs likely used H_2, ferrous iron (Fe^{2+}), or H_2S as the electron donor for photosynthesis. Anoxygenic photosynthesis is present in six bacterial phyla: the *Proteobacteria, Chlorobi, Chloroflexi, Firmicutes, Acidobacteria*, and *Gemmatimonadetes*. Oxygenic photosynthesis, by contrast, is known only

within the *Cyanobacteria* (Figure 15.1). There is extensive metabolic diversity among the anoxygenic phototrophs, which are found in a wide range of habitats. It is clear that horizontal gene exchange has had a major impact on the evolution of photosynthesis and on the distribution of photosynthetic genes across the phylogenetic tree of *Bacteria*.

Phototrophic bacteria have several common features. All phototrophic bacteria use chlorophyll-like pigments and various accessory pigments to harvest energy from light and transfer this energy to a membrane-bound reaction center where it is used to drive electron transfer reactions that ultimately result in the production of ATP (⇄ Sections 14.1–14.4). There are two different types of photosynthetic reaction centers: type I reaction centers (FeS-type), which are found in photosystem I of oxygenic phototrophs, and type II reaction centers (quinone-type, or Q-type), which are found in photosystem II of oxygenic phototrophs (⇄ Sections 14.1–14.4). Both types of reaction centers are present in *Cyanobacteria* (⇄ Section 14.4), whereas only one type or the other is present in anoxygenic phototrophs. In some cases photosynthetic pigments are found in the cytoplasmic membrane, but often they are present in intracellular photosynthetic membrane systems that originate from invaginations of the cytoplasmic membrane. These internal membranes allow phototrophic bacteria to increase the amount of pigment they contain for better use of light of low intensities.

Many phototrophic bacteria couple light energy to carbon fixation through a variety of different mechanisms (⇄ Section 14.5), but not all phototrophs fix CO_2; some instead either prefer or require organic sources of carbon to support growth. We will see that many of the characteristics of phototrophic bacteria, including their membrane systems and photosynthetic pigments, have evolved as a result of niche adaptation for the light environment.

MINIQUIZ

- What form of photosynthesis was most likely the first to appear on Earth?

15.3 *Cyanobacteria*

KEY GENERA: *Prochlorococcus, Crocosphaera, Synechococcus, Trichodesmium, Oscillatoria, Anabaena*

Cyanobacteria comprise a large, morphologically and ecologically heterogeneous group of oxygenic, phototrophic *Bacteria*. As we saw in Section 13.2, these organisms were the first oxygen-evolving phototrophic organisms on Earth, and over billions of years converted the once anoxic atmosphere of Earth to the oxygenated atmosphere we see today.

Phylogeny and Classification of *Cyanobacteria*

The morphological diversity of the *Cyanobacteria* is impressive (**Figure 15.2**). Both unicellular and filamentous forms are known, and there is considerable variation within these morphological types. Cyanobacterial cells range in size from 0.5 μm in diameter to cells as large as 100 μm in diameter. *Cyanobacteria* can be divided into five morphological groups: (1) *Chroococcales* are unicellular,

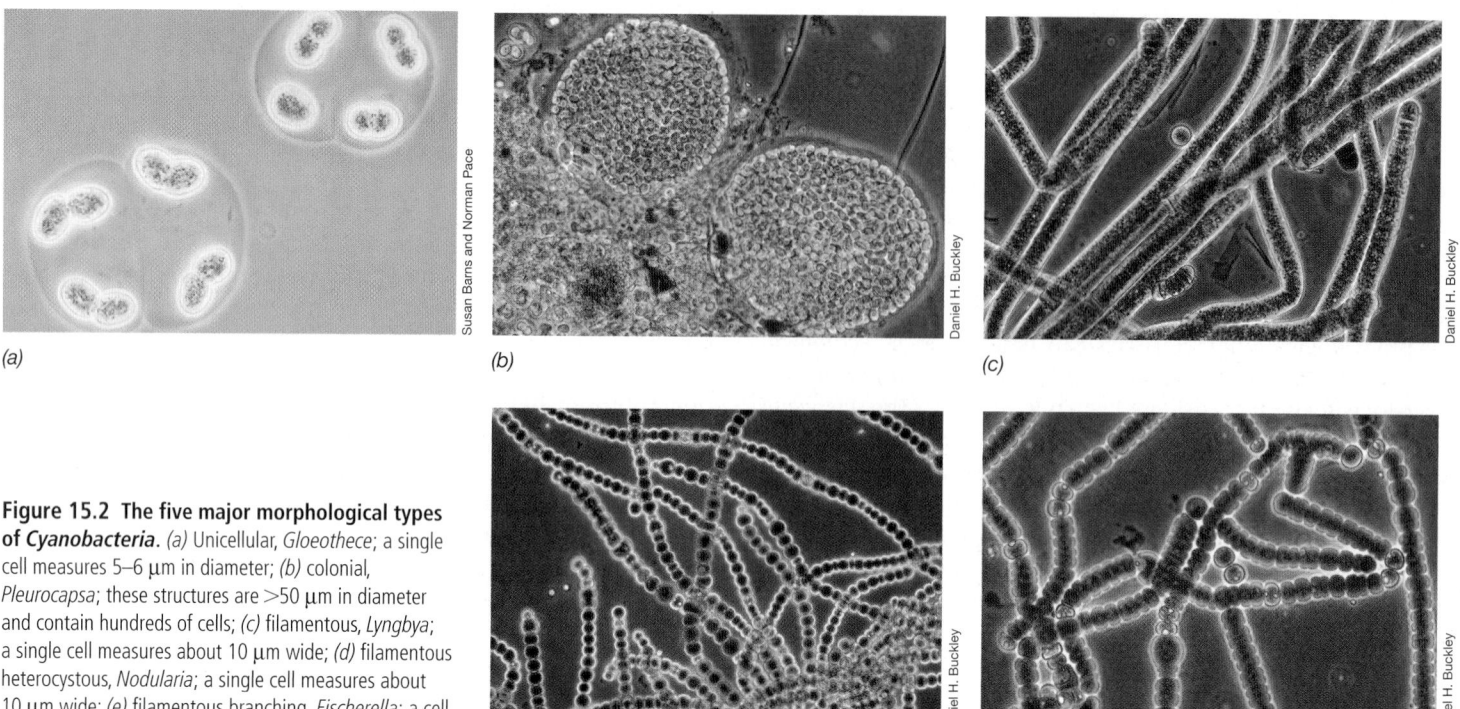

Figure 15.2 The five major morphological types of *Cyanobacteria*. *(a)* Unicellular, *Gloeothece*; a single cell measures 5–6 μm in diameter; *(b)* colonial, *Pleurocapsa*; these structures are >50 μm in diameter and contain hundreds of cells; *(c)* filamentous, *Lyngbya*; a single cell measures about 10 μm wide; *(d)* filamentous heterocystous, *Nodularia*; a single cell measures about 10 μm wide; *(e)* filamentous branching, *Fischerella*; a cell is about 10 μm wide. See how morphological diversity relates to phylogenetic diversity in Figure 15.3.

dividing by binary fission (Figure 15.2*a*); (2) *Pleurocapsales* are unicellular, dividing by multiple fission (colonial) (Figure 15.2*b*); (3) *Oscillatoriales* are filamentous nonheterocystous forms (Figure 15.2*c*); (4) *Nostocales* are filamentous, divide along a single axis, and are capable of cellular differentiation (Figure 15.2*d*); and (5) *Stigonematales* are morphologically similar to *Nostocales* except that cells divide in multiple planes, forming branching filaments (Figure 15.2*e*). Finally, the **prochlorophytes** are a lineage of unique unicellular *Cyanobacteria* once thought to be distinct but now classified within the *Chroococcales*.

Some of the major morphological classifications of *Cyanobacteria* correspond to coherent phylogenetic groups, but others do not (**Figure 15.3**). Species of *Pleurocapsales* form a coherent group within the cyanobacteria, indicating that reproduction by multiple fission arose only once in the evolutionary history of *Cyanobacteria* (Figure 15.3). Likewise, species of the *Nostocales* and *Stigonematales* share a common ancestor and form a coherent phylogenetic group indicating a single origin of cellular differentiation within the *Cyanobacteria* (Figure 15.3). All *Stigonematales* share a single ancestor within the clade composed of *Nostocales* and *Stigonematales*, indicating that the capacity to form branching filaments arose only once within the lineage of *Cyanobacteria* capable of cellular differentiation (Figure 15.3). In contrast, unicellular and simple filamentous *Cyanobacteria* (*Chroococcales* and *Oscillatoriales*, respectively) are dispersed in the cyanobacterial phylogeny, and these morphological groups do not represent coherent evolutionary lineages (Figure 15.3).

Physiology and Photosynthetic Membranes

Cyanobacteria are oxygenic phototrophs and therefore have both FeS-type and Q-type photosystems. All species are able to fix CO_2 by the Calvin cycle, many can fix N_2, and most can synthesize their own vitamins. Cells harvest energy from light and fix CO_2 during the day. During the night, cells generate energy by fermentation or aerobic respiration of carbon storage products such as glycogen. While CO_2 is the predominant source of carbon for most species, some *Cyanobacteria* can assimilate simple organic compounds such as glucose and acetate if light is present, a process called *photoheterotrophy*. A few *Cyanobacteria*, mainly filamentous species, can also grow in the dark on glucose or sucrose, using the sugar as both carbon and energy source. Finally, when sulfide concentrations are high, some *Cyanobacteria* are able to switch from oxygenic photosynthesis to anoxygenic photosynthesis using hydrogen sulfide rather than water as electron donor for photosynthesis (⇆ Figure 14.16).

Cyanobacteria have specialized membrane systems called *thylakoids* that increase the ability of cells to harvest light energy (⇆ Figure 14.10). The cell wall of cyanobacteria contains peptidoglycan and is structurally similar to that of other gram-negative bacteria. Photosynthesis takes place in the thylakoid membrane, a complex and multilayered photosynthetic membrane system containing photopigments and proteins that mediate photosynthesis (⇆ Sections 14.1 and 14.2). In most unicellular *Cyanobacteria*, the thylakoid membranes are arranged in regular concentric circles around the periphery of the cytoplasm (**Figure 15.4**). *Cyanobacteria* produce chlorophyll *a*, and most also have characteristic pigments called **phycobilins** (⇆ Figure 14.10), which function as accessory pigments in photosynthesis. One class of phycobilins, *phycocyanins*, are blue and, together with the green chlorophyll *a*, are responsible for the blue-green color of most cyanobacteria. Some *Cyanobacteria* produce *phycoerythrin*, a red phycobilin, and species producing phycoerythrin are red or brown. Photopigments are

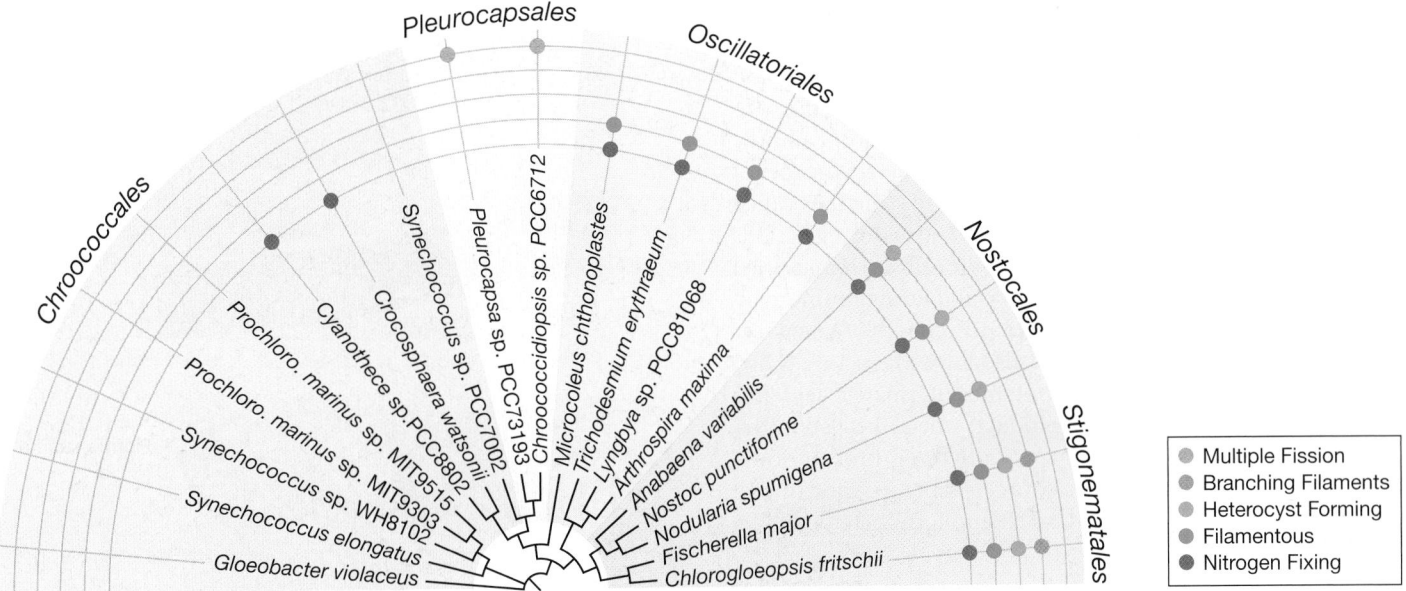

Figure 15.3 Taxonomically informative traits mapped onto the phylogeny of *Cyanobacteria*. The dendrogram depicts phylogenetic relationships inferred from analysis of conserved protein families in cyanobacterial genomes. Colored circles are used to indicate species traits as indicated by the key. Color shading is used to indicate taxonomic groupings. "*Prochloro.*" is used to indicate *Prochlorococcus*, which is a distinct group within the *Chroococcales*. Note that the *Chroococcales* and *Oscillatoriales* are not monophyletic in origin, meaning that these traits have arisen independently on multiple occasions in the phylogeny.

fluorescent and emit light when visualized using a fluorescence microscope; chlorophyll *a*, for example, fluoresces bright red (**Figure 15.5**). Prochlorophytes, such as *Prochlorococcus* and *Prochloron*, are unique among *Cyanobacteria* in that all members of this group contain chlorophyll *a* and *b* but do not contain phycobilins.

Motility and Cellular Structures

Cyanobacteria possess several mechanisms for motility. Many cyanobacteria exhibit gliding motility (⇌ Section 2.12). Gliding occurs only when a cell or filament is in contact with a solid surface or with another cell or filament. In some *Cyanobacteria*, gliding is not a simple translational movement but is accompanied by rotations, reversals, and flexing of filaments. Most gliding species exhibit directional movement toward light (phototaxis), and chemotaxis (⇌ Section 2.13) may occur as well. *Synechococcus* exhibits

an unusual form of swimming motility that does not require flagella or any other extracellular organelle. The cell surface of *Synechococcus* has specialized proteins that provide direct thrust by way of a mechanism that has yet to be resolved. Gas vesicles (⇌ Section 2.9) are also found in a variety of aquatic *Cyanobacteria* and are important in positioning cells in the water column. The function of gas vesicles is to regulate cell buoyancy such that cells can remain in a position in the water column where light intensity is optimal for photosynthesis.

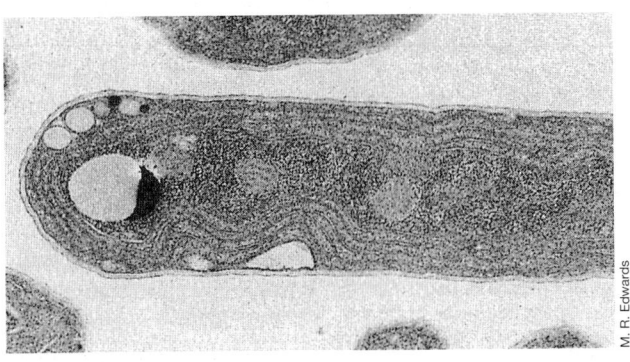

Figure 15.4 Thylakoids in *Cyanobacteria*. Electron micrograph of a thin section of the cyanobacterium *Synechococcus lividus*. A cell is about 5 μm in diameter. Note thylakoid membranes running parallel to the cell wall.

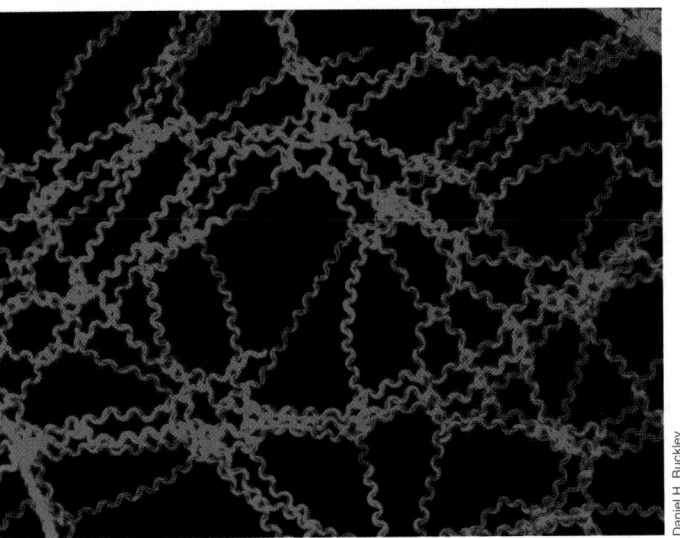

Figure 15.5 Phycocyanin fluorescence in *Cyanobacteria*. Fluorescence micrograph of *Spirulina*. Filaments consist of chains of helical cells with each cell approximately 5 μm wide.

Cyanobacteria are able to form a variety of structures associated with energy storage, reproduction, and survival. Many *Cyanobacteria* produce extensive mucilaginous envelopes, or sheaths, that bind groups of cells or filaments together (Figure 15.2a). Some filamentous cyanobacteria form *hormogonia* (**Figure 15.6**), short, motile filaments that break off from longer filaments to facilitate dispersal in times of stress. Some species also form resting structures called *akinetes* (Figure 15.6c), which protect the organism during periods of darkness, desiccation, or cold. Akinetes are cells with thickened outer walls. When conditions improve, akinetes germinate by breaking down their outer wall and initiating growth of a new vegetative filament. Many *Cyanobacteria* also form a structure called *cyanophycin*. This structure is a copolymer of aspartic acid and arginine and is a nitrogen storage product; when nitrogen in the environment becomes deficient, cyanophycin is broken down and used as a cellular nitrogen source. Many species of the *Nostocales* and *Stigonematales* are also able to form heterocysts, as discussed next.

Heterocysts and Nitrogen Fixation

Many *Cyanobacteria* are capable of nitrogen fixation (Figure 15.3). The nitrogenase enzyme, however, is inhibited by oxygen and thus nitrogen fixation cannot occur along with oxygenic photosynthesis (⟗ Section 14.6). *Cyanobacteria* have evolved several regulatory mechanisms for separating nitrogenase activity from photosynthesis (⟗ Section 7.8). For example, many unicellular *Cyanobacteria*, such as *Cyanothece* and *Crocosphaera* (**Figure 15.7a**), fix nitrogen only at night when photosynthesis does not occur. In contrast, the filamentous cyanobacteria *Trichodesmium* (Figure 15.7b) fixes nitrogen only during the day through a mechanism that remains somewhat unclear, but appears to require transient suppression of photosynthetic activity within filaments. Finally, many filamentous *Cyanobacteria* of the *Nostocales* and *Stigonematales* facilitate nitrogen fixation by forming specialized cells called *heterocysts*, either on the ends of filaments (**Figure 15.8a, b**) or along the filament (Figure 15.8c, d).

Heterocysts arise from differentiation of vegetative cells and are the sites of nitrogen fixation in heterocystous *Cyanobacteria*. Heterocysts are surrounded by a thickened cell wall that slows the diffusion of O_2 into the cell and permits nitrogenase activity to occur in an anoxic environment. Heterocysts lack photosystem II, the oxygen-evolving photosystem that generates reducing power from H_2O (⟗ Section 14.4), and thus do not fluoresce as strongly as vegetative cells (Figure 15.8). Without photosystem II, heterocysts are unable to fix CO_2 and thus lack the necessary electron donor (pyruvate) for nitrogen fixation. However, heterocysts have intercellular connections with adjacent vegetative cells that allow for mutual exchange of materials between these cells. Fixed carbon is imported by the heterocyst from adjacent vegetative cells, and this is oxidized to yield electrons for nitrogen fixation. The products of photosynthesis move from vegetative cells to heterocysts, and fixed nitrogen moves from heterocysts to vegetative cells (⟗ Figure 7.17).

Ecology of *Cyanobacteria*

Cyanobacteria are of central importance to the productivity of the oceans. Small unicellular *Cyanobacteria*, such as *Synechococcus*

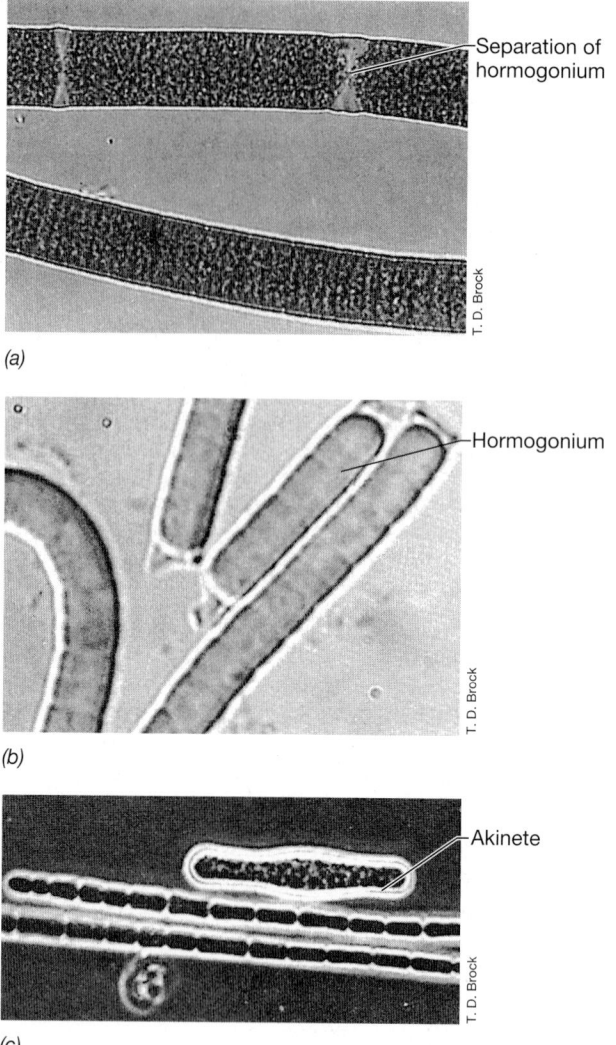

(a)

(b)

(c)

Figure 15.6 Structural differentiation in filamentous *Cyanobacteria.* *(a)* Initial stage of hormogonium formation in *Oscillatoria*. Notice the empty spaces where the hormogonium is separating from the filament. *(b)* Hormogonium of a smaller *Oscillatoria* species. Notice that the cells at both ends are rounded. Cells are about 10 μm wide. Differential interference contrast microscopy. *(c)* Akinete (resting spore) of *Anabaena* in a phase-contrast micrograph, cells about 5 μm wide.

and *Prochlorococcus* (⟗ Section 20.10), are the most abundant phototrophs in the oceans. Together these organisms contribute 80% of marine photosynthesis and 35% of all photosynthetic activity on Earth.

Cyanobacterial nitrogen fixation represents the dominant input of new nitrogen into vast segments of Earth's oceans, particularly in oligotrophic tropical and subtropical waters. Marine nitrogen fixation is dominated by two groups of *Cyanobacteria*, the unicellular species, such as *Crocosphaera*, and the filamentous *Trichodesmium*. *Crocosphaera* (Figure 15.7a) and relatives dominate nitrogen fixation in most of the Pacific Ocean and are widespread in tropical and subtropical habitats. *Trichodesmium* is the dominant nitrogen-fixer in the North Atlantic Ocean and parts of the Pacific where dissolved iron concentrations are elevated. *Trichodesmium* forms macroscopically visible tufts of filaments (Figure 15.7b) and relies on gas vesicles to remain suspended in the photic

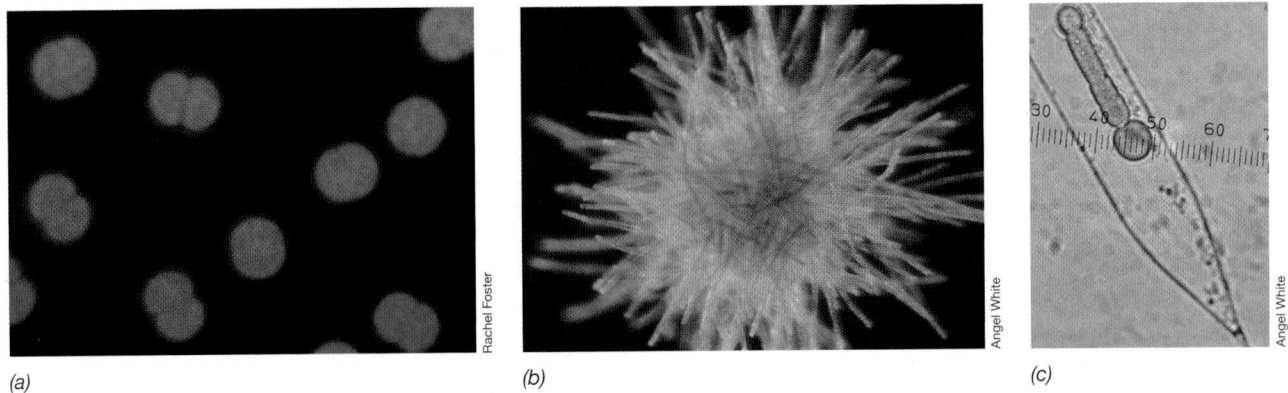

(a)   *(b)*   *(c)*

Figure 15.7 Marine *Cyanobacteria* that fix N₂. *(a)* Unicellular *Crocosphaera*-like cells in the process of dividing; cells are approximately 5 μm diameter. *(b)* Colonial "tuft" of *Trichodesmium*. The tuft is composed of many attached, undifferentiated, unbranched filaments and has a diameter of approximately 100 μm. *(c)* A diatom containing the cyanobacterial symbiont *Richelia* (scale in micrometers). The *Richelia* symbiont is an unbranched filament with a terminal heterocyst; cells are about 5 μm wide.

zone where it is often observed in dense masses of cells called *blooms*. In addition, other marine nitrogen-fixers including species of *Calothrix* and *Richelia* form symbiotic associations with diatoms (Figure 15.7*c*); these symbiotic associations are often observed in tropical and subtropical oceans. Finally, heterocystous cyanobacteria such as *Nodularia* (Figure 15.2*d*) and *Anabaena* can sometimes dominate nitrogen fixation in cold waters of the Northern Hemisphere and are often observed in the Baltic Sea.

Cyanobacteria are also widely found in terrestrial and freshwater environments. In general, they are more tolerant of environmental extremes, particularly extremes of desiccation, than are eukaryotic algae. *Cyanobacteria* are often the dominant or sole oxygenic phototrophic organisms in hot springs, saline lakes, desert soils, and other extreme environments. In some of these environments, cyanobacterial mats of variable thickness may form (⮌ Figure 20.7). Freshwater lakes, especially those rich in inorganic nutrients, often develop blooms of *Cyanobacteria*, especially in late summer when temperatures are warmest (⮌ Figures 20.1 and 20.17). A few *Cyanobacteria* are symbionts of liverworts, ferns, and cycads, and a number are phototrophic

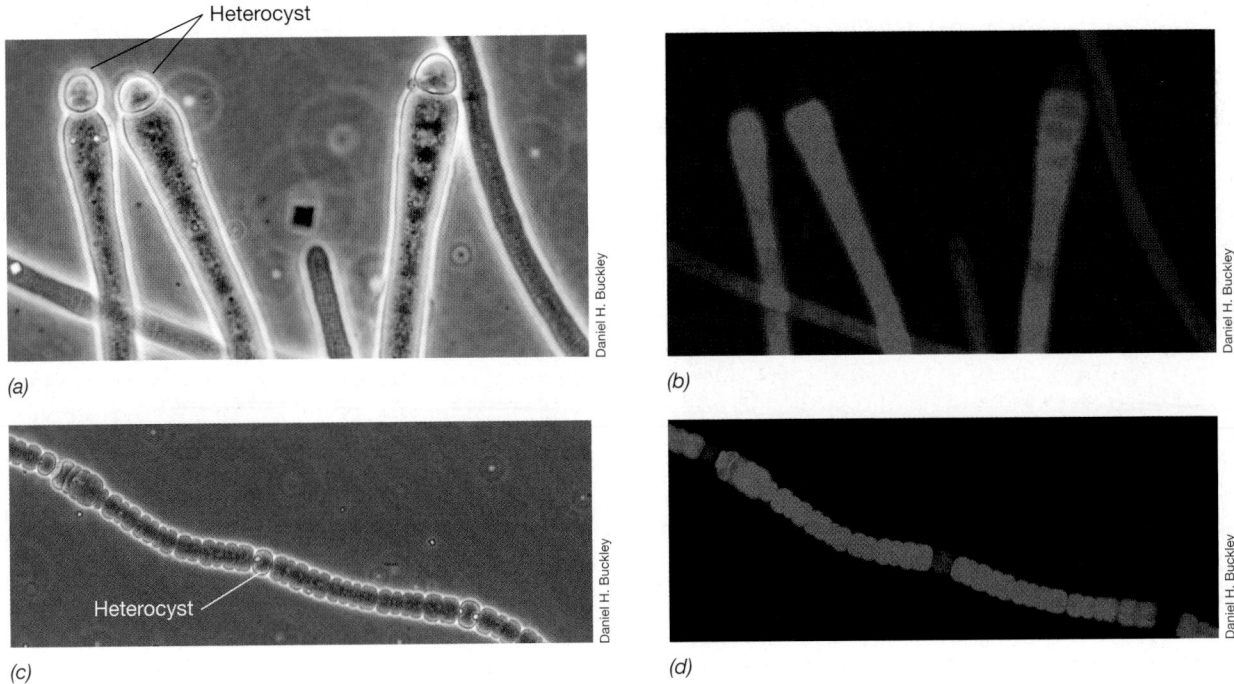

Heterocyst

(a)   *(b)*

Heterocyst

(c)   *(d)*

Figure 15.8 Heterocysts. Differentiation of heterocysts causes the loss of photopigments and inability to carry out photosynthesis. *(a)* Phase-contrast micrograph of *Calothrix* with terminal heterocysts. *(b)* Fluorescence micrograph of the same *Calothrix* filaments; cells are about 10 μm wide. *(c)* Phase-contrast micrograph of *Fischerella*. *(d)* Fluorescence micrograph of the same *Fischerella* filaments; cells are about 10 μm wide. See how heterocyst formation is regulated at the genetic level in the well-studied cyanobacterium *Anabaena* in Figure 7.17.

UNIT 4

components of lichens, a symbiosis between a phototroph and a fungus (🔗 Section 23.1).

Several metabolic products of *Cyanobacteria* are of considerable practical importance. Some *Cyanobacteria* produce potent neurotoxins, and toxic blooms may form when massive accumulations of *Cyanobacteria* develop. Animals ingesting water containing these toxic products may be killed. Many *Cyanobacteria* are also responsible for the production of earthy odors and flavors in some freshwater, and if such waters are used as drinking water sources, aesthetic problems may arise. The major compound produced is geosmin, a substance also produced by many actinomycetes (🔗 Section 16.12).

MINIQUIZ

- What are the differentiating properties of the five major morphological groups of *Cyanobacteria*?
- What is a heterocyst and what is its function?

15.4 Purple Sulfur Bacteria

KEY GENERA: *Chromatium, Ectothiorhodospira*

Purple sulfur bacteria are anoxygenic phototrophs that use hydrogen sulfide (H_2S) as an electron donor for photosynthesis (🔗 Figure 14.1). Purple sulfur bacteria are a phylogenetically coherent group found within the order *Chromatiales* in the *Gammaproteobacteria*.

Purple sulfur bacteria are generally found in illuminated anoxic zones where H_2S is present. Such habitats occur commonly in lakes, marine sediments, and "sulfur springs," where H_2S produced geochemically or biologically can support the growth of purple sulfur bacteria (**Figure 15.9**). Purple sulfur bacteria are also commonly found in microbial mats (🔗 Section 20.5) and in salt marsh sediments. The characteristic color of purple sulfur bacteria comes from their carotenoids, accessory pigments involved in

light harvesting (🔗 Section 14.2). These bacteria use a Q-type photosystem (🔗 Figure 14.12), contain either bacteriochlorophyll *a* or *b*, and carry out CO_2 fixation by the Calvin cycle (🔗 Section 14.5).

During autotrophic growth of purple sulfur bacteria, H_2S is oxidized to elemental sulfur (S^0), which is deposited as sulfur granules (**Figure 15.10**). When sulfide is limiting, the sulfur is used as an electron donor for photosynthesis, resulting in the oxidation of S^0 to sulfate (SO_4^{2-}). Many purple sulfur bacteria can also use other reduced sulfur compounds as photosynthetic electron donors; for example, thiosulfate ($S_2O_3^{2-}$) is commonly used to grow laboratory cultures.

The purple sulfur bacteria form two families: the *Chromatiaceae* and the *Ectothiorhodospiraceae*. Species of the two families are readily distinguished by the location of sulfur granules and by their photosynthetic membranes. *Chromatiaceae*, including the genera *Chromatium* and *Thiocapsa*, store S^0 granules *inside* their cells (in the periplasmic space) and have vesicular intracellular photosynthetic membrane systems (**Figure 15.11b**). These organisms are common in stratified lakes containing sulfide and in the anoxic sediments of salt marshes. *Ectothiorhodospiraceae*, including the two main genera *Ectothiorhodospira* and *Halorhodospira*, oxidize H_2S to S^0 that is deposited *outside* the cell (Figure 15.10d) and have lamellar intracellular photosynthetic membrane systems (Figure 15.11a). These genera are also interesting because many species are extremely halophilic (salt loving) or alkaliphilic (alkalinity loving) and are among the most extreme in these characteristics of all known *Bacteria*. These organisms are typically found in saline lakes, soda lakes, and salterns, where abundant levels of SO_4^{2-} support sulfate-reducing bacteria (🔗 Section 21.4 and Section 15.9), the organisms that produce H_2S.

Purple sulfur bacteria are often observed in high density in meromictic (permanently stratified) lakes. Meromictic lakes form layers because they have denser (usually saline) water on the bottom and less dense (usually freshwater) water nearer the surface.

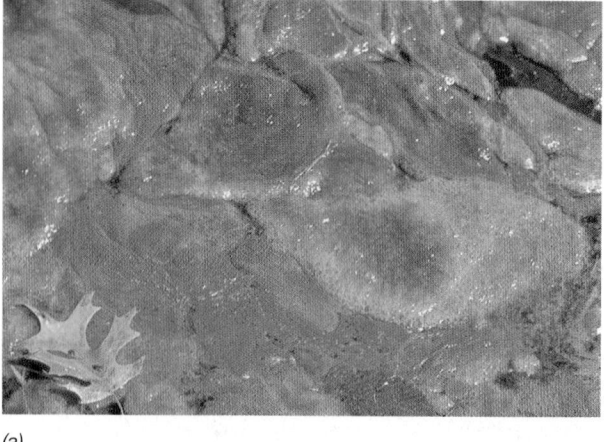
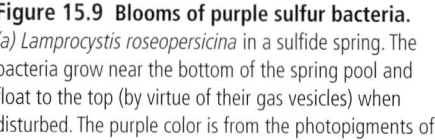

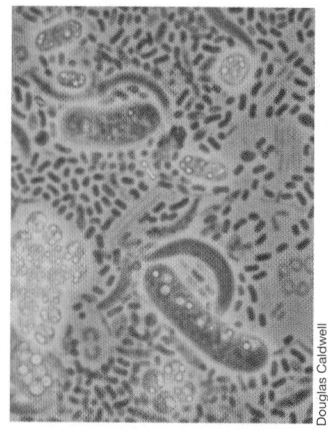

(a)   (b)   (c)

Figure 15.9 Blooms of purple sulfur bacteria. *(a) Lamprocystis roseopersicina* in a sulfide spring. The bacteria grow near the bottom of the spring pool and float to the top (by virtue of their gas vesicles) when disturbed. The purple color is from the photopigments of the purple sulfur bacteria and the green color is from cells of the alga *Spirogyra*. *(b)* Sample of water from a depth of 7 m in Lake Mahoney, British Columbia; the major phototroph is the purple sulfur bacterium *Amoebobacter purpureus*. *(c)* Phase-contrast photomicrograph of layers of purple sulfur bacteria from a small, stratified lake in Michigan. The purple sulfur bacteria include *Chromatium* species (large rods) and *Thiocystis* (small cocci). The small green-colored rods are green sulfur bacteria such as *Chlorobium* (🔗 Section 15.6).

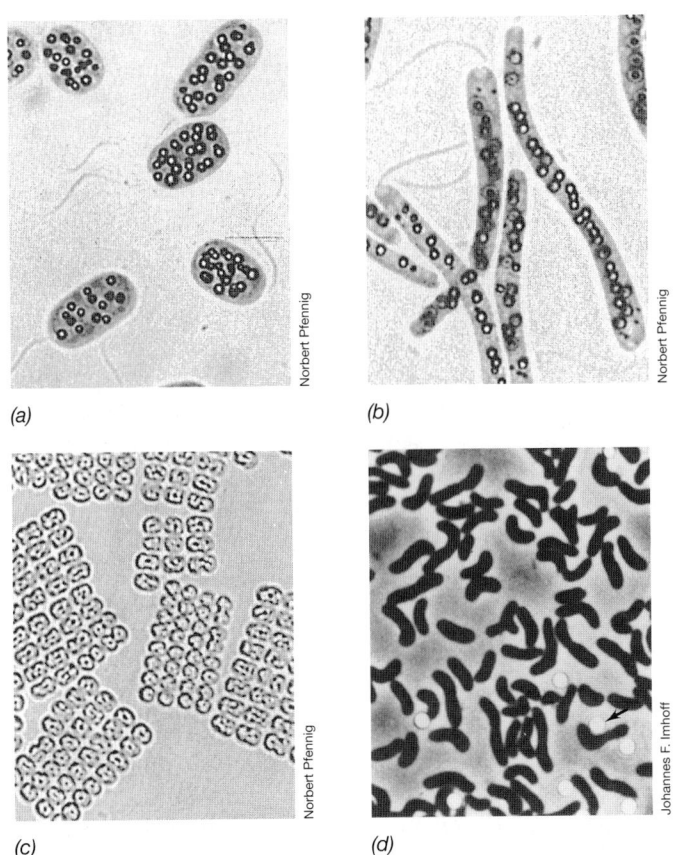

(a)

(b)

(c)

(d)

Norbert Pfennig

Norbert Pfennig

Norbert Pfennig

Johannes F. Imhoff

Figure 15.10 Bright-field and phase-contrast photomicrographs of purple sulfur bacteria. *(a) Chromatium okenii;* cells are about 5 μm wide. Note the globules of elemental sulfur inside the cells. *(b) Thiospirillum jenense,* a very large, polarly flagellated spiral; cells are about 30 μm long. Note the sulfur globules. *(c) Thiopedia rosea;* cells are about 1.5 μm wide. *(d)* Phase-contrast micrograph of cells of *Ectothiorhodospira mobilis;* cells are about 0.8 μm wide. Note external sulfur globules (arrow).

If sufficient sulfate is present to support sulfate reduction, sulfide is produced in the sediments and diffuses upward into the anoxic bottom waters. The presence of sulfide and light in the anoxic layers of the lake allow purple sulfur bacteria to form dense cell masses (Figure 15.9b), usually in association with green phototrophic bacteria.

─────── **MINIQUIZ** ───────

- What is the source of the purple color from which the purple sulfur bacteria get their name?
- Where would you expect to find purple sulfur bacteria in nature?

15.5 Purple Nonsulfur Bacteria and Aerobic Anoxygenic Phototrophs

Purple Nonsulfur Bacteria

KEY GENERA: *Rhodospirillum, Rhodoferax, Rhodobacter*

The **purple nonsulfur bacteria** are the most metabolically versatile of all microbes. Despite their name, they are not always

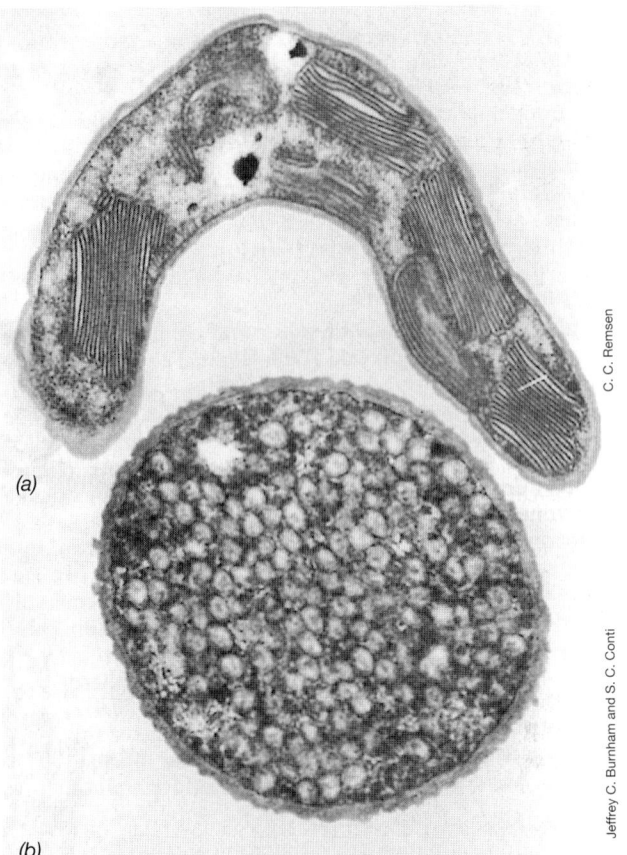

(a)

(b)

C. C. Remsen

Jeffrey C. Burnham and S. C. Conti

Figure 15.11 Membrane systems of phototrophic purple bacteria as revealed by transmission electron microscopy. *(a) Ectothiorhodospira mobilis,* showing the photosynthetic membranes in flat sheets (lamellae). *(b) Allochromatium vinosum,* showing the membranes as individual, spherical vesicles.

purple; these organisms synthesize an array of carotenoids (⮌ Section 14.2) that can lend them a variety of spectacular colors (**Figure 15.12**). Together, these pigments give purple bacteria their colors, usually purple, red, or orange. Purple nonsulfur bacteria are typically photoheterotrophs (a condition where light is the energy source and an organic compound is the carbon source), and species are able to use a wide range of carbon sources and electron donors for photosynthesis, including organic acids, amino acids, alcohols, sugars, and even aromatic compounds like benzoate or toluene. Like purple sulfur bacteria, purple nonsulfur bacteria use a Q-type photosystem, and contain either bacteriochlorophyll *a* or *b*. The purple nonsulfur bacteria are phylogenetically and morphologically diverse (**Figure 15.13**) and reside within the *Alphaproteobacteria* (e.g., *Rhodospirillum, Rhodobacter, Rhodopseudomonas*) or *Betaproteobacteria* (e.g., *Rubrivivax, Rhodoferax*).

Purple nonsulfur bacteria can conserve energy through a variety of metabolic processes. For example, some species can grow photoautotrophically using H_2, low levels of H_2S, or even ferrous iron (Fe^{2+}) as the electron donor for photosynthesis with CO_2 fixation carried out by the Calvin cycle. Most species are also able to grow in darkness by using aerobic respiration of organic or even some inorganic compounds; synthesis of the photosynthetic machinery is typically repressed by O_2. Finally, some species can

UNIT 4

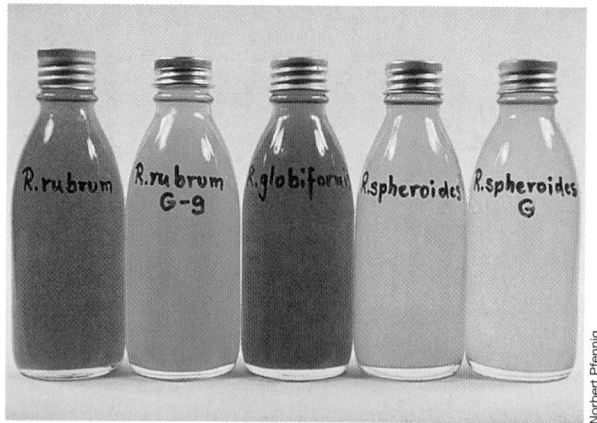

Figure 15.12 Photograph of liquid cultures of phototrophic purple bacteria showing the color of species with various carotenoid pigments. All species contain bacteriochlorophyll *a*. The blue culture is a carotenoidless mutant strain of *Rhodospirillum rubrum* showing that bacteriochlorophyll *a* is actually blue. The bottle on the far right (*Rhodobacter sphaeroides* strain G) lacks one of the carotenoids of the wild type and thus is less red and more green.

also grow by fermentation or anaerobic respiration using a variety of electron donors and acceptors.

Enrichment and isolation of purple nonsulfur bacteria is easy using a mineral salts medium supplemented with an organic acid as carbon source. Such media, inoculated with a mud, lake water, or sewage sample and incubated anaerobically in the light, invariably select for purple nonsulfur bacteria. Enrichment cultures can be made even more selective by omitting fixed nitrogen sources (for example, ammonia) or organic nitrogen sources (for example, yeast extract or peptone) from the medium and supplying a gaseous headspace of N_2. Virtually all purple nonsulfur bacteria can fix N_2 ($\leftrightarrows$ Section 14.6) and will thrive under such conditions, rapidly outcompeting other bacteria.

Aerobic Anoxygenic Phototrophs

KEY GENERA: *Roseobacter, Erythrobacter*

The **aerobic anoxygenic phototrophs** are obligatory aerobic heterotrophs that use light as a supplemental source of energy to support growth. Like purple nonsulfur bacteria, aerobic anoxygenic phototrophs are phylogenetically diverse and are *Alphaproteobacteria* or *Betaproteobacteria*. The primary physiological difference with the purple nonsulfur bacteria is that aerobic anoxygenic phototrophs are strict heterotrophs and employ anoxygenic photosynthesis only under *oxic* conditions as a supplemental source of energy. Aerobic anoxygenic phototrophs contain bacteriochlorophyll *a* and a Q-type photosystem, but are unable to fix CO_2 and thus rely on organic forms of carbon as their carbon source. Carotenoids of various types lend colors of yellow, orange, or pink to cultures.

Aerobic anoxygenic phototrophs are only able to photosynthesize when grown on a day/night cycle. Under these conditions, bacteriochlorophyll *a* is made only in the dark and then used to conserve energy by photophosphorylation when the light returns. Aerobic anoxygenic phototrophs can account for as much as a quarter of the microbial community inhabiting coastal marine waters and 5% of gross photosynthesis in such systems

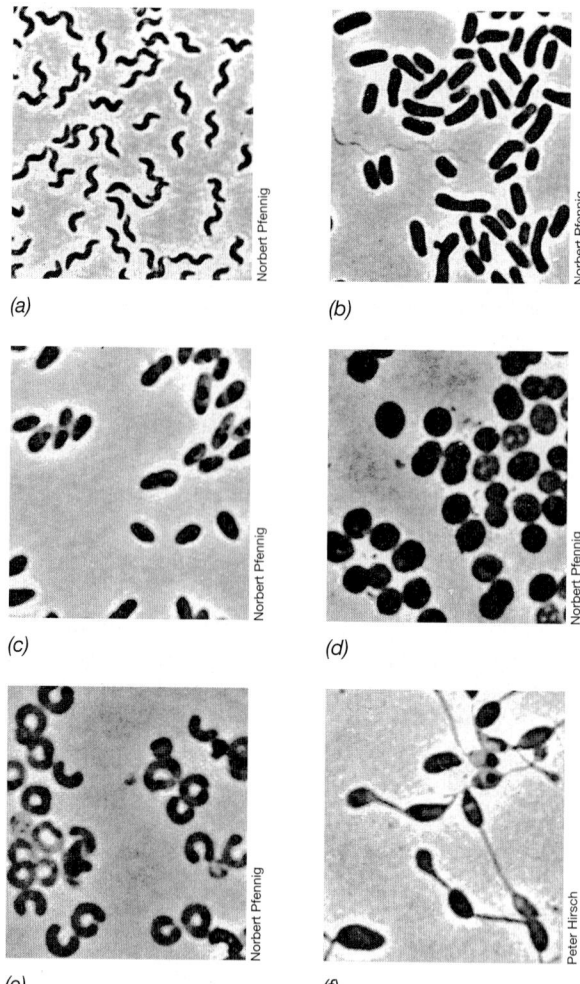

Figure 15.13 Representatives of several genera of purple nonsulfur bacteria. *(a)* Phaeospirillum fulvum*; cells are about 3 μm long. *(b)* Rhodoblastus acidophilus*; cells are about 4 μm long. *(c)* Rhodobacter sphaeroides*; cells are about 1.5 μm wide. *(d)* Rhodopila globiformis*; cells are about 1.6 μm wide. *(e)* Rhodocyclus purpureus*; cells are about 0.7 μm in diameter. *(f)* Rhodomicrobium vannielii*; cells are about 1.2 μm wide.

($\leftrightarrows$ Section 20.10). Common genera found in coastal marine habitats include *Roseobacter* and *Erythrobacter*.

> **MINIQUIZ**
> - What are some similarities between purple nonsulfur bacteria and aerobic anoxygenic phototrophs? What are the differences between these two groups?
> - Where would you expect to find aerobic anoxygenic phototrophs?

15.6 Green Sulfur Bacteria

KEY GENERA: *Chlorobium, Chlorobaculum,* "*Chlorochromatium*"

Green sulfur bacteria are a phylogenetically coherent group of anoxygenic phototrophs that forms the phylum *Chlorobi*. Green sulfur bacteria have little metabolic versatility and they are typically nonmotile and strictly anaerobic anoxygenic phototrophic bacteria. The group is also morphologically restricted and includes primarily short to long rods (**Figure 15.14**).

Like purple sulfur bacteria, green sulfur bacteria oxidize hydrogen sulfide (H_2S) as an electron donor for autotrophic growth, oxidizing it first to sulfur (S^0) and then to sulfate (SO_4^{2-}). But unlike most purple sulfur bacteria, the S^0 produced by green sulfur bacteria is deposited only outside the cell (Figure 15.14*a*). Autotrophy is supported not by the reactions of the Calvin cycle, as in purple bacteria, but instead by a reversal of steps in the citric acid cycle (↺ Section 14.5 and Figure 14.20*a*), a unique means of autotrophy in phototrophic bacteria.

Pigments and Ecology

Green sulfur bacteria contain bacteriochlorophyll *c, d,* or *e* and house these pigments in unique structures called **chlorosomes** (**Figure 15.15**). A small amount of bacteriochlorophyll *a* is present in the reaction center and FMO protein, the latter of which connects the chlorosome to the cytoplasmic membrane (↺ Figure 14.7*b*). Chlorosomes are oblong bacteriochlorophyll-rich bodies bounded by a thin, nonunit membrane and attached to the cytoplasmic membrane in the periphery of the cell (Figure 15.15 and ↺ Figure 14.7). Chlorosomes function to funnel energy into the photosystem, and this eventually leads to ATP synthesis. Unlike purple anoxygenic phototrophs, green sulfur bacteria use an FeS-type photosystem. Both green- and brown-colored species of green sulfur bacteria are known, the brown-colored species containing bacteriochlorophyll *e* and carotenoids that turn dense cell suspensions brown (**Figure 15.16**).

Like purple sulfur bacteria (Section 15.4), green sulfur bacteria live in anoxic, sulfidic, illuminated aquatic environments. However, the chlorosome is a very efficient light-harvesting structure, which allows green sulfur bacteria to grow at light intensities much lower than those required by other phototrophs. Green sulfur bacteria also tend to have a greater tolerance of H_2S than do other anoxygenic phototrophs. As a result, green sulfur bacteria are typically found at the greatest depths of all phototrophic microorganisms in lakes or microbial mats, where light intensities are low and H_2S levels the highest.

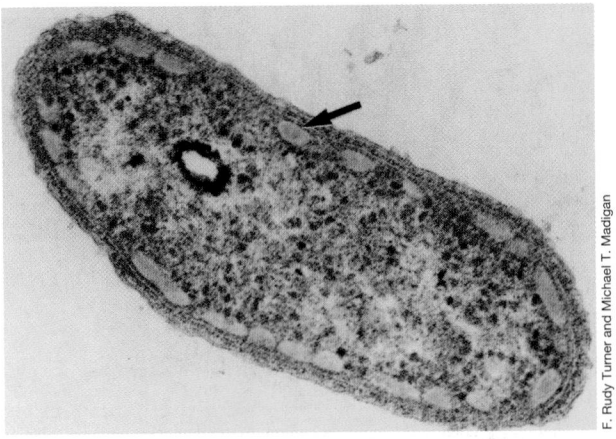

Figure 15.15 The thermophilic green sulfur bacterium *Chlorobaculum tepidum*. Transmission electron micrograph. Note chlorosomes (arrow) in the cell periphery. A cell is about 0.7 μm wide.

As an example, a species of green sulfur bacteria isolated from a deep-sea hydrothermal vent (↺ Section 20.14) was found to be growing phototrophically on the weak glow of infrared radiation emitted from the geothermally heated rock. One species, *Chlorobaculum tepidum* (Figure 15.15), is thermophilic and forms dense microbial mats in high-sulfide hot springs. *C. tepidum* also grows rapidly and is amenable to genetic manipulation by both conjugation and transformation (Chapter 11). Because of these features, *C. tepidum* has become the model organism for studying the molecular biology of green sulfur bacteria.

Green Sulfur Bacteria Consortia

Certain species of green sulfur bacteria form an intimate two-membered association, called a **consortium**, with a chemoorganotrophic bacterium. In the consortium, each organism benefits, and thus a variety of such consortia containing different phototrophic and chemotrophic components probably exist in nature. The phototrophic component, called the *epibiont*, is physically attached to the nonphototrophic central

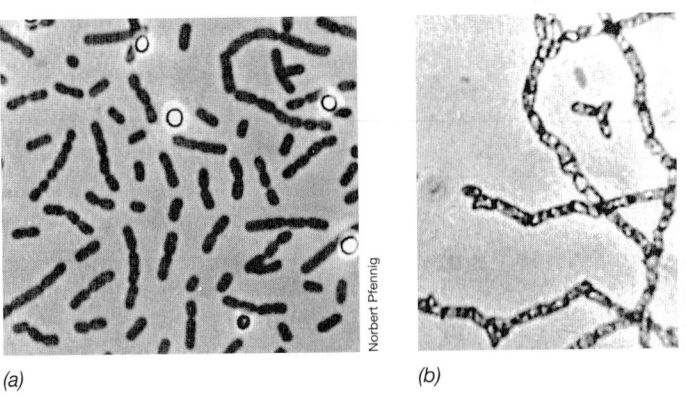

Figure 15.14 Phototrophic green sulfur bacteria. *(a) Chlorobium limicola*; cells are about 0.8 μm wide. Note the spherical sulfur granules deposited extracellularly. *(b) Chlorobium clathratiforme*, a bacterium forming a three-dimensional network; cells are about 0.8 μm wide.

Figure 15.16 Green and brown chlorobia. Tube cultures of *(a) Chlorobaculum tepidum* and *(b) Chlorobaculum phaeobacteroides*. Cells of *C. tepidum* contain bacteriochlorophyll *c* and green carotenoids, and cells of *C. phaeobacteroides* contain bacteriochlorophyll *e* and isorenieratene, a brown carotenoid. The structures of bacteriochlorophylls *c* and *e* and green bacteria carotenoids were shown in Figures 14.3 and 14.9.

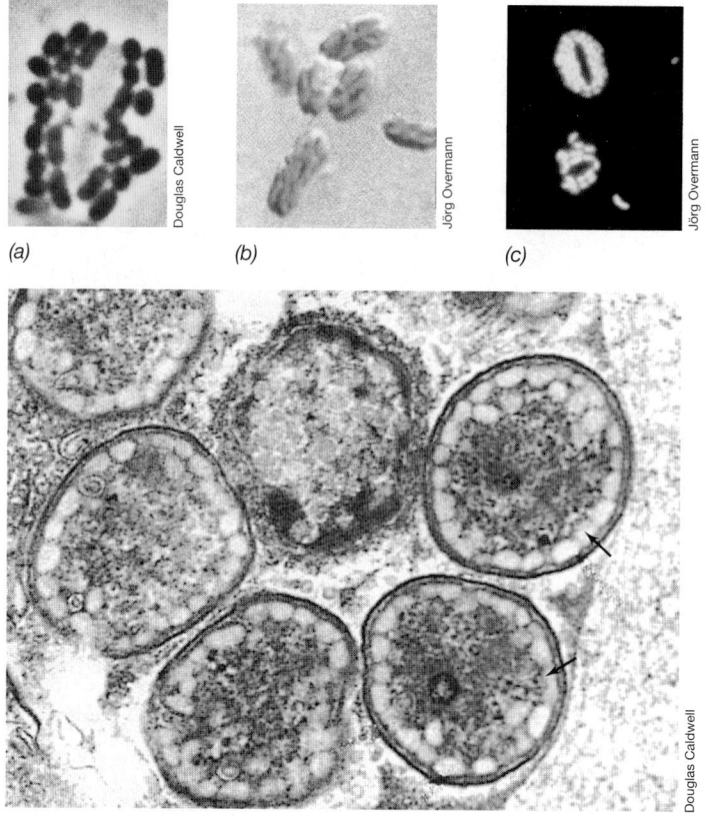

(a) (b) (c)

(d)

Figure 15.17 **"*Chlorochromatium aggregatum*."** Consortia of green sulfur bacteria and a chemoorganotroph. *(a)* In a phase-contrast micrograph, the nonphototrophic central organism is lighter in color than the pigmented phototrophic bacteria. *(b)* Green carotenoids lend their color to the phototrophs in a differential interference contrast micrograph. *(c)* A fluorescence micrograph shows the cells stained with a fluorescent probe specific for green sulfur bacteria. *(d)* Transmission electron micrograph of a cross section through a single consortium; note the chlorosomes (arrows) in the epibionts. The entire consortium is about 3 μm in diameter.

cell (**Figure 15.17**) and communicates with it in various ways (⇌ Section 23.2).

The name "*Chlorochromatium aggregatum*" (not a formal name because this is a mixed culture) has been used to describe a commonly observed green-colored consortium that is green because the epibionts are green sulfur bacteria that contain green-colored carotenoids (Figure 15.17*b*). Evidence that the epibionts are indeed green sulfur bacteria comes from pigment analyses, the presence of chlorosomes (Figure 15.17*d*), and phylogenetic staining (Figure 15.17*c*). A structurally similar consortium called "*Pelochromatium roseum*" is brown because its epibionts produce brown-colored carotenoids (⇌ Figures 23.3 and 23.4). We examine the symbiotic nature of the *Chlorochromatium* consortium in more detail in Section 23.2.

— MINIQUIZ —

- Which pigments are present in the chlorosome?
- What evidence exists that the epibionts of green bacterial consortia are truly green sulfur bacteria?

15.7 Green Nonsulfur Bacteria

KEY GENERA: *Chloroflexus, Heliothrix, Roseiflexus*

Green nonsulfur bacteria, which are also called *filamentous anoxygenic phototrophs*, are anoxygenic phototrophs of the phylum *Chloroflexi*. The latter contains several distinct lineages, one of which, the class *Chloroflexi*, contains green nonsulfur bacteria. The remainder of the phylum contains metabolically diverse organisms including both aerobic and anaerobic chemoorganotrophs as well as the *Dehalococcoidetes*, a group of dehalogenating bacteria that use halogenated organic compounds as electron acceptors in anaerobic respiration (⇌ Section 14.15). Analyses of 16S ribosomal RNA sequences from environmental samples (⇌ Section 19.6) indicate that species of the phylum *Chloroflexi* are widespread and that most species in the phylum have yet to be cultivated in isolation; thus the metabolic diversity of this phylum remains poorly characterized.

All cultured representatives of the green nonsulfur bacteria are filamentous bacteria that are capable of gliding motility. *Chloroflexus*, one of the most studied of the green nonsulfur bacteria, forms thick microbial mats in neutral to alkaline hot springs along with thermophilic cyanobacteria (**Figure 15.18**; ⇌ Figure 20.7*b*). Green nonsulfur bacteria grow best as photoheterotrophs using simple carbon sources as electron donors for photosynthesis. However, growth also occurs photoautotrophically using H_2 or H_2S as electron donors for photosynthesis. The 3-hydroxypropionate bi-cycle, a pathway of CO_2 incorporation unique to only a few *Bacteria* and *Archaea*, supports autotrophic growth (⇌ Section 14.5). Most green nonsulfur bacteria also grow well in the dark by aerobic respiration of a wide variety of carbon sources.

The photosynthetic features of the green nonsulfur bacteria form a "hybrid" between those of both green sulfur bacteria (Section 15.6) and purple phototrophic bacteria (Sections 15.4, 15.5). Green nonsulfur bacteria have reaction centers that contain bacteriochlorophyll *a* and chlorosomes that contain bacteriochlorophyll *c* (Figure 15.15) and in this way are similar to green sulfur bacteria. However, in contrast to green sulfur bacteria, green nonsulfur bacteria contain a Q-type photosynthetic reaction center and in this respect resemble purple bacteria.

Other *Chloroflexi*

In addition to *Chloroflexus*, other phototrophic green nonsulfur bacteria include the thermophile *Heliothrix* and the large-celled mesophiles *Oscillochloris* (Figure 15.18*b*) and *Chloronema* (Figure 15.18*c*). *Oscillochloris* and *Chloronema* form rather large cells, 2–5 μm wide and up to several hundred micrometers long (Figure 15.18*c*). Species of both genera inhabit freshwater lakes containing H_2S. *Roseiflexus* and *Heliothrix* are similar to *Chloroflexus* in their filamentous morphology and thermophilic lifestyle, but differ in a major photosynthetic property. *Roseiflexus* and *Heliothrix* lack bacteriochlorophyll *c* and chlorosomes and thus more closely resemble purple phototrophic bacteria (Sections 15.4, 15.5) than *Chloroflexus*. This can be seen in cultures of *Roseiflexus* that are yellow-orange instead of green from their extensive carotenoid pigments and lack of bacteriochlorophyll *c* (Figure 15.18*d*).

Thermomicrobium is a nonphototrophic genus of *Chloroflexi* and a strictly aerobic, gram-negative rod, growing optimally in

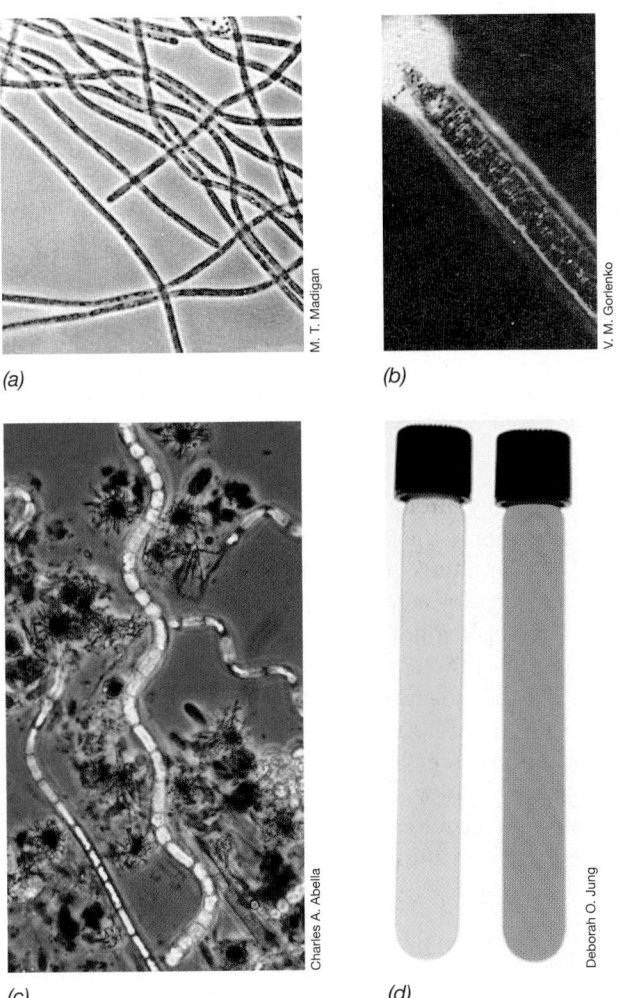

(a)   (b)

(c)   (d)

Figure 15.18 Green nonsulfur bacteria. (a) Phase-contrast micrograph of the anoxygenic phototroph *Chloroflexus aurantiacus*; cells are about 1 µm in diameter. (b) Phase-contrast micrograph of the large phototroph *Oscillochloris*; cells are about 5 µm wide. The brightly contrasting material on the top is a holdfast, used for attachment. (c) Phase-contrast micrograph of filaments of a *Chloronema* species; the cells are wavy filaments and about 2.5 µm in diameter. (d) Tube cultures of *C. aurantiacus* (right) and *Roseiflexus* (left). *Roseiflexus* is yellow because it lacks bacteriochlorophyll *c* and chlorosomes.

complex media at 75°C. Besides its phylogenetic properties, *Thermomicrobium* is also of interest because of its membrane lipids (**Figure 15.19**). Recall that the lipids of *Bacteria* and *Eukarya* contain fatty acids esterified to *glycerol* (⮑ Section 2.3). By contrast, the lipids of *Thermomicrobium* are formed on *1,2-dialcohols* instead of glycerol, and have neither ester *nor* ether linkages (Figure 15.19; ⮑ Section 2.3). In addition, cells of *Thermomicrobium* contain only small amounts of peptidoglycan, and the cell wall is composed primarily of protein.

MINIQUIZ

- In what ways do *Chloroflexus* and *Roseiflexus* resemble *Chlorobium*? *Rhodobacter*?

- What is unique about *Thermomicrobium*?

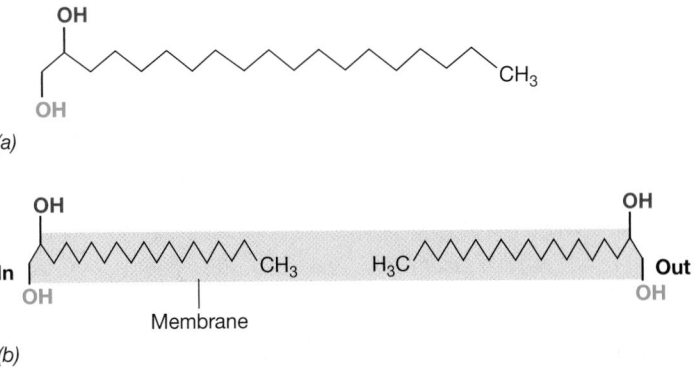

(a)

(b)

Figure 15.19 The unusual lipids of *Thermomicrobium*. (a) Membrane lipids from *Thermomicrobium roseum* contain long-chain diols like the one shown here (1,2-nonadecanediol). Note that unlike the lipids of other *Bacteria* or of *Archaea*, neither ester- nor ether-linked side chains are present. (b) To form a bilayer membrane, dialcohol molecules oppose each other at the methyl groups, and the —OH groups are the inner and outer hydrophilic surfaces. Small amounts of the diols have fatty acids esterified to the secondary —OH group (shown in red), whereas the primary —OH group (shown in green) can bond a hydrophilic molecule like phosphate.

15.8 Other Phototrophic *Bacteria*

KEY GENERA: *Heliobacterium, Chloracidobacterium*

Heliobacteria

Heliobacteria are a phylogenetically coherent group of phototrophic gram-positive *Bacteria* found within the phylum *Firmicutes*. The heliobacteria are anoxygenic phototrophs that have an FeS-type photosystem and that produce a unique pigment, bacteriochlorophyll *g* (⮑ Figure 14.3). Heliobacteria grow photoheterotrophically using a narrow range of organic compounds including pyruvate, lactate, acetate, or butyrate, and the group contains five genera: *Heliobacterium, Heliophilum, Heliorestis, Heliomonas,* and *Heliobacillus*. All known heliobacteria form rod-shaped or filamentous cells (**Figure 15.20**), although *Heliophilum* is unusual because its cells form into bundles (Figure 15.20b) that are motile as a unit.

Heliobacteria are strict anaerobes, but in addition to phototrophic growth, they can grow chemotrophically in darkness by pyruvate fermentation (as can many clostridia, close relatives of the heliobacteria). Heliobacteria produce endospores, the highly resistant structures produced by certain gram-positive bacteria (⮑ Section 2.10). Like the endospores of *Bacillus* or *Clostridium* species, the endospores of heliobacteria (Figure 15.20c) contain elevated calcium (Ca^{2+}) levels and the signature molecule of the endospore, *dipicolinic acid*. Heliobacteria reside in soil, especially paddy (rice) field soils, where their nitrogen fixation activities (⮑ Section 14.6) may benefit rice productivity. A large diversity of heliobacteria have also been found in highly alkaline environments, such as soda lakes and surrounding alkaline soils.

Phototrophic *Acidobacteria*

A novel group of anoxygenic phototrophs has been discovered growing in photosynthetic microbial mats of certain thermal springs in Yellowstone National Park. *Chloracidobacterium thermophilum* is a thermophilic oxygen-tolerant anoxygenic

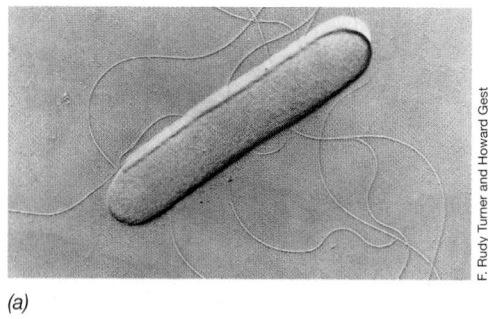

(a)

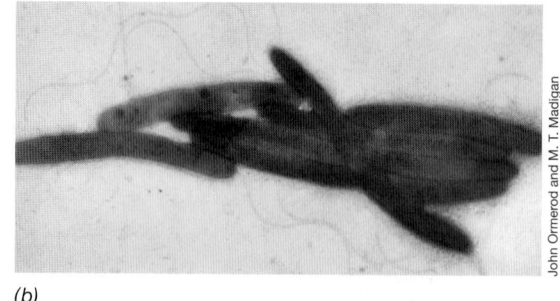

(b)

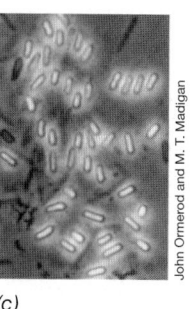

(c)

Figure 15.20 Cells and endospores of heliobacteria. *(a)* Electron micrograph of *Heliobacillus mobilis*, a peritrichously flagellated species. *(b)* *Heliophilum fasciatum* cell bundles as observed by electron microscopy. *(c)* Phase-contrast micrograph of endospores from *Heliobacterium gestii*. Most heliobacteria cells are about 1–2 μm in diameter.

phototroph of the phylum *Acidobacteria* (⟜ Section 16.21). Similar to green sulfur bacteria, *C. thermophilum* produces bacteriochlorophyll *a* and *c*, the latter in chlorosomes (**Figure 15.21**), and uses an FeS-type photosystem. However, unlike green sulfur bacteria, *C. thermophilum* can also grow aerobically, as is true for the aerobic anoxygenic phototrophs (Section 15.5). In terms of its carbon metabolism, *C. thermophilum* is a photoheterotroph that uses short-chain fatty acids as carbon sources, but unlike green sulfur or green nonsulfur bacteria, it is incapable of autotrophy.

Phototrophic *Gemmatimonadetes*

Another novel group of anoxygenic phototrophs has been discovered in a freshwater lake in the western Gobi desert (China and Mongolia). *Gemmatimonas phototrophica* is an aerobic facultative photoheterotroph of the phylum *Gemmatimonadetes*. It gains

most of its energy through the aerobic respiration of organic compounds, whether in the light or in the dark. However, in the light, *G. phototrophica* uses photophosphorylation to supplement energy generated by aerobic respiration. *G. phototrophica* cannot grow as an obligate phototroph, it cannot fix CO_2, and it cannot grow anaerobically. *G. phototrophica* contains a photosynthetic gene cluster that resembles those of aerobic anoxygenic phototrophs (Section 15.5), and it produces bacteriochlorophyll *a* and a Q-type reaction center, both of which are characteristic properties of purple bacteria (⟜ Section 14.3). It thus seems likely that *G. phototrophica* acquired its photosynthetic gene cluster and the ability to perform photophosphorylation as the result of an ancient horizontal gene transfer event.

MINIQUIZ

- What types of anoxygenic phototrophs contain chlorosomes?
- What kind of phototrophic bacteria make endospores?

Chlorosome

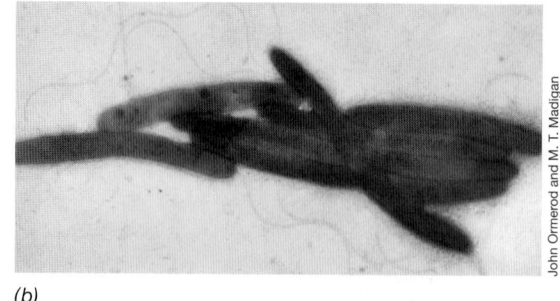

(a)   *(b)*

Figure 15.21 Chlorosomes in *Chloracidobacterium thermophilum*, a phototrophic member of the phylum *Acidobacteria*. *(a)* Electron micrograph of *C. thermophilum* showing chlorosomes. *(b)* Fluorescence photomicrograph of *C. thermophilum*. The red color is the fluorescence of bacteriochlorophyll *c* present in chlorosomes. A cell of *C. thermophilum* is about 0.8 μm wide.

III • Microbial Diversity in the Sulfur Cycle

Sulfur metabolism may have fueled the earliest forms of life on our planet (⟜ Section 13.1), and the sulfur cycle (⟜ Section 21.4) continues to support an enormous diversity of microorganisms. In this section we consider the diversity of organisms capable of *dissimilative sulfur metabolism*; that is, organisms that conserve energy through the oxidation or reduction of sulfur compounds (⟜ Sections 14.9 and 14.14).

The remarkable diversity of *Bacteria* and *Archaea* capable of dissimilative sulfur metabolism is in part a function of the chemical diversity in which sulfur occurs in the biosphere. Sulfur has eight oxidation states that range from its most oxidized form, sulfate (SO_4^{2-}, oxidation state of +6), to thiosulfate ($S_2O_3^{2-}$, oxidation state of +2), to elemental sulfur (S^0, oxidation state of 0), and finally to hydrogen sulfide (H_2S, oxidation state of −2), its most reduced form. In addition, sulfur compounds can take on diverse chemical forms including inorganic sulfur compounds, organosulfur compounds, and metal sulfides.

In this part of the chapter we will focus on the diversity of **dissimilative sulfate-reducers, dissimilative sulfur-reducers**, and **dissimilative sulfur-oxidizers**. Anoxygenic phototrophs, such as the purple and green sulfur bacteria discussed in Sections 15.4–15.6, are also important links in the sulfur cycle. However, here we restrict our focus to chemotrophic dissimilative sulfur metabolisms.

15.9 Dissimilative Sulfate-Reducers

KEY GENERA: *Desulfovibrio, Desulfobacter*

Sulfate-reducing bacteria gain energy by coupling the oxidation of H_2 or organic compounds to the reduction of SO_4^{2-} (anaerobic respiration). There are more than 30 known genera of sulfate reducers found across five phyla of *Bacteria* and *Archaea* (**Figure 15.22**). Most sulfate reducers reside in the *Deltaproteobacteria*, though sulfate reducers are also found in the *Firmicutes* (e.g., *Desulfotomaculum* and *Desulfosporosinus*), *Thermodesulfobacteria* (e.g., *Thermodesulfobacterium)*, and *Nitrospirae* (e.g., *Thermodesulfovibrio*). Sulfate reduction also occurs in *Archaeoglobus*, a genus of the archaeal phylum *Euryarchaeota*.

Physiology of Sulfate-Reducing Bacteria

Sulfate-reducing bacteria are morphologically and biochemically diverse. The biochemistry of sulfate reduction was discussed in Section 14.14, so here we consider some of the more general physiological properties of this group. Sulfate reducers are generally obligate anaerobes, and strict anoxic techniques must be used in their cultivation (**Figure 15.23***g*).

Sulfate reducers use H_2 or organic compounds as electron donors for growth, and the range of organics used is fairly broad. Lactate and pyruvate are almost universally used, and many species also oxidize short-chain alcohols (ethanol, propanol, and butanol) as electron donors. Some species, such as *Desulfosarcina and Desulfonema*, grow chemolithotrophically and autotrophically with H_2 as an electron donor, SO_4^{2-} as an electron acceptor, and CO_2 as the sole carbon source. A few sulfate reducers can oxidize hydrocarbons as electron donors (Section 14.25).

There are two physiological types of dissimilative sulfate-reducers, the *complete oxidizers*, which can oxidize acetate and other fatty acids completely to CO_2, and the *incomplete oxidizers*, which are unable to oxidize acetate to CO_2. The latter group includes the best studied of the sulfate-reducing bacteria, *Desulfovibrio* (Figure 15.23*a*), along with *Desulfomonas, Desulfotomaculum*, and *Desulfobulbus* (Figure 15.23*c*). The acetate oxidizers include *Desulfobacter* (Figure 15.23*d*), *Desulfococcus, Desulfosarcina* (Figure 15.23*e*), and *Desulfonema* (Figure 15.23*b*), among many others. These bacteria specialize in the oxidation of fatty acids (in particular acetate) to CO_2, and reducing SO_4^{2-} to H_2S. These two physiological groups are not phylogenetically coherent but instead are distributed widely across the phylogeny of sulfate-reducing bacteria (Figure 15.22).

Some sulfate-reducing bacteria can exploit alternative metabolic pathways. In addition to SO_4^{2-} or S^0, some sulfate reducers can also reduce nitrate and sulfonates (such as isethionate, $HO-CH_2-CH_2-SO_3^-$). Certain organic compounds can also be fermented by sulfate-reducing bacteria. The most common of these is pyruvate, which is fermented by way of the phosphoroclastic reaction to acetate, CO_2, and H_2 (Figure 14.55). Moreover, although generally obligate anaerobes, a few sulfate-reducing bacteria are quite O_2-tolerant (primarily strains that coexist with O_2-producing cyanobacteria in microbial mats). At least one species, *Desulfovibrio oxyclinae*, can actually grow with O_2 as the electron acceptor under microaerophilic conditions.

Ecology of Sulfate-Reducing Bacteria

Sulfate reducers are widespread in aquatic and terrestrial environments that contain SO_4^{2-} and become anoxic as a result of microbial decomposition. Sulfate reducers are abundant in marine sediments, and the H_2S they generate is responsible for the pungent smell (like that of rotten eggs) often encountered in decaying vegetation near coastal ecosystems. *Desulfotomaculum*, phylogenetically a species of *Firmicutes* (gram-positive *Bacteria*), consists of endospore-forming rods found primarily in soil. Growth and reduction of SO_4^{2-} by *Desulfotomaculum* in certain

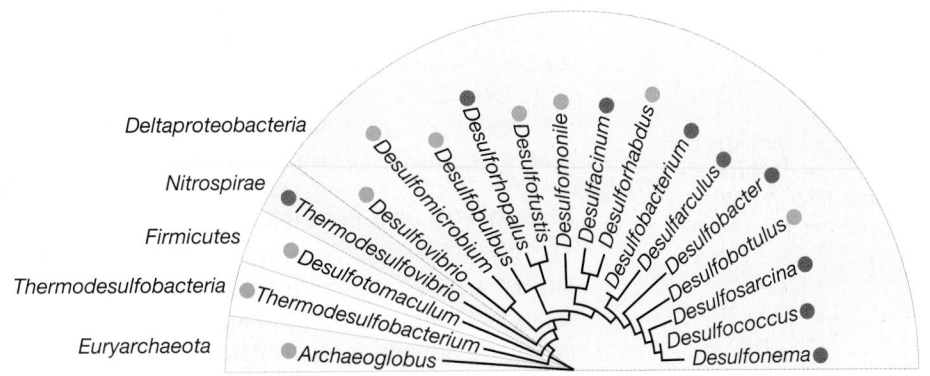

Figure 15.22 Dissimilative sulfate-reducers. The dendrogram depicts phylogenetic relationships among some genera of sulfate reducers as inferred by analysis of their 16S ribosomal RNA gene sequences. Color shading is used to differentiate the five main phyla that contain genera of sulfate reducers. Colored circles indicate whether species are complete oxidizers, which are able to oxidize acetate to CO_2, or incomplete oxidizers, which cannot oxidize acetate. The physiology of sulfate-reducing bacteria is considered in Section 14.14, and their role in the sulfur cycle in Section 21.4.

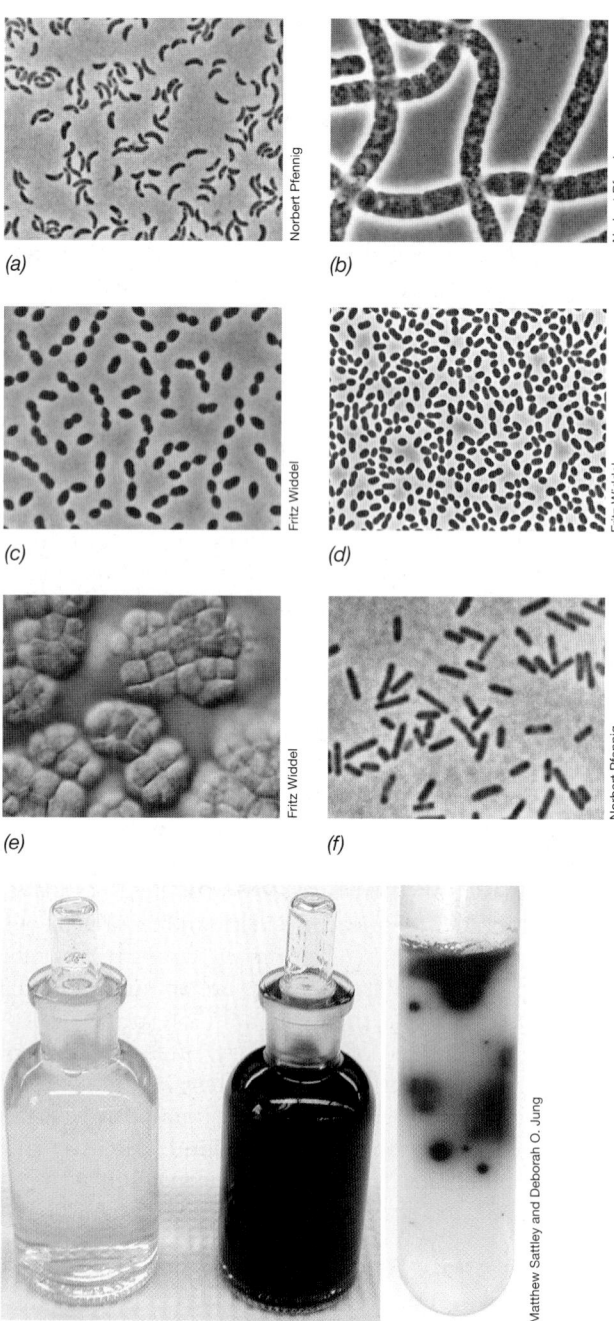

(a)

(b)

(c)

(d)

(e)

(f)

(g)

Figure 15.23 Representative sulfate-reducing and sulfur-reducing bacteria. *(a) Desulfovibrio desulfuricans*; cell diameter about 0.7 μm. *(b) Desulfonema limicola*; cell diameter 3 μm. *(c) Desulfobulbus propionicus*; cell diameter about 1.2 μm. *(d) Desulfobacter postgatei*; cell diameter about 1.5 μm. *(e) Desulfosarcina variabilis*; cell diameter about 1.25 μm. *(f) Desulfuromonas acetoxidans*; cell diameter about 0.6 μm. *(g)* Enrichment culture of sulfate-reducing bacteria. Left, sterile medium; center, a positive enrichment showing black FeS; right, colonies of sulfate-reducing bacteria in a dilution tube (⮌ Sections 19.1 and 19.2). Photos *a–d* and *f* are phase-contrast photomicrographs; part *e* is an interference contrast micrograph.

canned foods leads to a type of spoilage called *sulfide stinker*. Species of *Thermodesulfobacterium*, *Thermodesulfovibrio*, and *Archaeoglobus* (an archaeon) are all thermophilic and found in geothermally heated environments such as hot springs, hydrothermal

vents, and oil reserves. The remaining genera of sulfate reducers are indigenous to anoxic marine and freshwater environments and can occasionally be isolated from the mammalian gut.

The enrichment of *Desulfovibrio* species is straightforward in an anoxic lactate–sulfate medium containing ferrous iron (Fe^{2+}). A reducing agent, such as thioglycolate or ascorbate, is required to achieve a low reduction potential (E_0') in the medium. When sulfate-reducing bacteria grow, the H_2S they form combines with the ferrous iron to form black, insoluble ferrous sulfide (Figure 15.23*g*). Purification can be accomplished by diluting the culture in molten agar tubes (⮌ Section 19.2 and Figure 19.3*b*). Upon solidification, individual cells of sulfate-reducing bacteria become distributed throughout the agar and grow to form black colonies (Figure 15.23*g*) that can be removed aseptically to yield pure cultures.

MINIQUIZ

- What are the typical electron donors used by dissimilative sulfate-reducers?
- What bacterial phyla are known to contain dissimilative sulfate-reducers?

15.10 Dissimilative Sulfur-Reducers

KEY GENERA: *Desulfuromonas, Wolinella, Sulfolobus*

Here we consider the dissimilative *sulfur*-reducers, microorganisms that are able to use the respiratory reduction of S^0 to conserve energy. Dissimilative sulfur-reducing bacteria can reduce S^0 and other oxidized forms of sulfur (such as SO_3^{2-}) to H_2S but are unable to reduce SO_4^{2-}. There are more than 25 genera of dissimilative sulfur-reducers spread across five bacterial and archaeal phyla (Figure 15.1).

Most sulfur-reducing bacteria are *Proteobacteria*, primarily *Deltaproteobacteria* (e.g., *Desulfuromonas, Pelobacter, Desulfurella, Geobacter*), with some genera residing in the *Epsilonproteobacteria* (e.g., *Wolinella* and *Sulfurospirillum*) and *Gammaproteobacteria* (e.g., *Shewanella* and *Pseudomonas mendocina*). Other sulfur-reducing bacteria are species of *Firmicutes* (e.g., *Desulfitobacterium* and *Ammonifex*), *Aquificae* (e.g., *Desulfurobacterium* and *Aquifex*), *Synergistetes* (e.g., *Dethiosulfovibrio*), or *Deferribacteres* (e.g., *Geovibrio*). The sulfur-reducing *Archaea*—of which there are many—are all genera of the phylum *Crenarchaeota* (e.g., *Acidianus, Sulfolobus, Pyrodictium*, and *Thermodiscus*).

Physiology and Ecology of Sulfur-Reducing Bacteria

The physiology of sulfur reducers is more diverse than that of sulfate reducers. Most sulfur reducers are obligate anaerobes, but facultatively aerobic species are also common. Sulfur reducers are often able to reduce electron acceptors such as nitrate, ferrous iron, or thiosulfate as alternatives to S^0. Like sulfate reducers (Section 15.9), the physiology of sulfur reducers is characterized by whether they completely oxidize acetate and other fatty acids to CO_2. Species of *Desulfuromonas* (Figure 15.23*f*) are complete oxidizers that grow anaerobically by coupling the oxidation of acetate, succinate, ethanol, or propanol to the reduction of S^0.

In contrast, *Sulfospirillum* and *Wolinella* are incomplete oxidizers and cannot use acetate as an electron donor. *Sulfospirillum* can reduce S^0 using either H_2 or formate as electron donor.

Dissimilative sulfur-reducing bacteria reside in many of the same habitats as dissimilative sulfate-reducing bacteria and often form associations with bacteria that oxidize H_2S to S^0, such as green sulfur bacteria (Section 15.6). The S^0 produced from H_2S oxidation is then reduced back to H_2S during metabolism of the sulfur reducer, completing an anoxic sulfur cycle (⟳ Section 21.4).

MINIQUIZ

- What are the typical electron donors used by dissimilative sulfur-reducers?
- What bacterial phyla contain dissimilative sulfur-reducers?

15.11 Dissimilative Sulfur-Oxidizers

KEY GENERA: *Thiobacillus, Achromatium, Beggiatoa*

Dissimilative sulfur-oxidizers are **chemolithotrophs** that oxidize reduced sulfur compounds such as H_2S, S^0, thiosulfate, or thiocyanate (^-SCN) as electron donors in energy conservation, typically with O_2 as electron acceptor. These organisms are common in environments such as marine sediments, sulfur springs, and hydrothermal systems where H_2S produced by sulfate- or sulfur-reducing bacteria (Sections 15.9, 15.10), or abiotically by geothermal reactions, is released into oxygenated waters (**Figure 15.24**). The sulfur oxidizers are found in three phyla of *Bacteria* (*Proteobacteria, Aquificae, Deinococcus–Thermus*) and one of *Archaea* (*Crenarchaeota*) (Figure 15.1). Most sulfur-oxidizing bacteria are *Beta-* (*Thiobacillus*), *Gamma-* (*Achromatium, Beggiatoa*), or *Epsilonproteobacteria* (*Thiovulum, Sulfurimonas*).

Physiological Diversity of Sulfur-Oxidizing Bacteria

The morphological and physiological diversity of sulfur oxidizers is significant. Cells can be less than 1 micrometer in diameter (e.g., *Sulfurimonas denitrificans*) or as large as 750 micrometers in diameter (e.g., *Thiomargarita namibiensis*). Most sulfur oxidizers are obligate aerobes; however, species of *Thiomargarita* and *Sulfurimonas* can also reduce NO_3^- in denitrification (⟳ Section 14.13 and Section 15.13). Many species oxidize H_2S to elemental sulfur (S^0), which they deposit as either intracellular or extracellular granules for later use as an electron donor (⟳ Figure 14.27) if H_2S becomes limiting.

Some sulfur chemolithotrophs are *obligate chemolithotrophs*, locked into a lifestyle of using inorganic instead of organic compounds as electron donors. When growing in this fashion, they are also autotrophs, converting CO_2 into cell material by reactions of the Calvin cycle. **Carboxysomes** are often present in cells of obligate chemolithotrophs (**Figure 15.25a**). These structures contain high levels of Calvin cycle enzymes and probably increase the rate at which these organisms fix CO_2 (⟳ Section 14.5).

Other sulfur chemolithotrophs are *facultative chemolithotrophs*, facultative in the sense that they can grow *either* chemolithotrophically (and thus, also as autotrophs) *or* chemoorganotrophically. Most species of *Beggiatoa* can obtain energy from the

(a)

(b)

Figure 15.24 Habitats of sulfur oxidizers. *(a)* A sulfidic artesian spring in Florida (USA). The outside of the spring is coated with a mat of *Thiothrix* (see Figure 15.26b). The mat is about 1.5 m in diameter. *(b)* Hydrothermal chimneys at Cathedral Hill in the Guaymas Basin (Mexico), 2000-m depth. Sulfide-rich waters vent from the chimneys, which are covered by mats composed of orange, white, and yellow cells of *Beggiatoa*.

oxidation of inorganic sulfur compounds but lack enzymes of the Calvin cycle. They thus require organic compounds as carbon sources. Organisms that use a mix of carbon and energy sources, for example those that simultaneously assimilate carbon from both CO_2 and organic sources, are called **mixotrophs**.

Thiobacillus and *Achromatium*

The genus *Thiobacillus* and related genera include several gram-negative, rod-shaped *Betaproteobacteria* (Figure 15.25a) that are the best studied of the sulfur chemolithotrophs. The oxidation of H_2S, S^0, or thiosulfate by *Thiobacillus* generates sulfuric acid (H_2SO_4), and thus thiobacilli are often acidophilic. One highly acidophilic species, *Acidithiobacillus ferrooxidans*, can also grow chemolithotrophically by the oxidation of Fe^{2+} and is a major biological agent for the oxidation of this metal. Iron pyrite (FeS_2) is a major natural source of ferrous iron as well as of sulfide. The oxidation of FeS_2, especially in mining operations, can be both beneficial (because leaching of the ore releases the iron from the sulfide mineral) and ecologically disastrous (the environment can become acidic and contaminated with toxic metals such as aluminum, cadmium, and lead) (⟳ Sections 22.1 and 22.2).

Achromatium is a spherical sulfur-oxidizing chemolithotroph that is common in freshwater sediments of neutral pH containing H_2S. Cells of *Achromatium* are large cocci that can have diameters

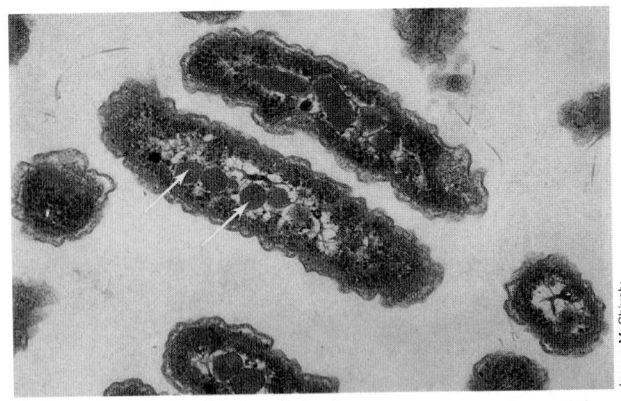

(a)

Jessup M. Shively

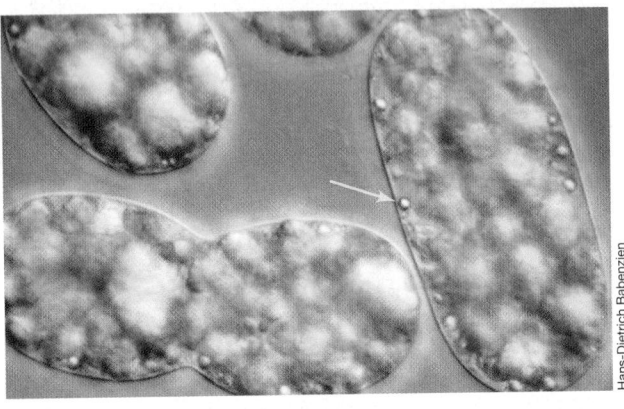

(b)

Hans-Dietrich Babenzien

Figure 15.25 Nonfilamentous sulfur chemolithotrophs. *(a)* Transmission electron micrograph of cells of the chemolithotrophic sulfur-oxidizer *Halothiobacillus neapolitanus*. A single cell is about 0.5 μm in diameter. Note the polyhedral bodies (carboxysomes) distributed throughout the cell (arrows) (↩ Figure 14.19). *(b) Achromatium*. Cells photographed by differential interference contrast microscopy. The small globular structures near the periphery of the cells (arrow) are elemental sulfur, and the large granules are calcium carbonate. A single *Achromatium* cell is about 25 μm in diameter.

of 10–100 μm (Figure 15.25*b*). *Achromatium* is a species of *Gammaproteobacteria* and is specifically related to purple sulfur bacteria, such as its phototrophic counterpart *Chromatium* (Section 15.4 and Figure 15.10*a*). Like *Chromatium*, cells of *Achromatium* store S^0 internally (Figure 15.25*b*); the granules later disappear as S^0 is oxidized to SO_4^{2-}. Cells of *Achromatium* also store large granules of calcite ($CaCO_3$) (Figure 15.25*b*), possibly as a carbon source (in the form of CO_2) for autotrophic growth. The physiology of chemolithotrophic sulfur-oxidizers is discussed in Section 14.9.

Ecological Diversity and Strategies of Sulfide-Oxidizing Bacteria

Aerobic sulfide-oxidizers provide a case study that demonstrates the degree of ecological diversification that can occur among microbes that share the same basic metabolic features. The chemical oxidation of H_2S to H_2SO_4 is spontaneous and rapid in the presence of O_2. Hence, aerobic H_2S-oxidizers have evolved diverse ecological strategies that allow them to metabolize two molecules that otherwise react with each other spontaneously. We consider

here six different strategies used by aerobic sulfide-oxidizers to cope with the chemical instability of H_2S in the presence of O_2.

1. *Thiothrix* is a filamentous sulfur chemolithotroph that forms filaments that group together at their ends by way of a holdfast to form cell arrangements called *rosettes* (**Figure 15.26**). The ecological strategy of *Thiothrix* is to use its holdfast to position itself in high-flow environments downstream from a source of H_2S. Such environments are common near sulfur springs and in creeks draining sulfidic salt marshes where abundant H_2S is produced and carried away in waters rich with O_2 (Figure 15.26*a*). Physiologically, *Thiothrix* is an obligately aerobic mixotroph, and in this and most other respects it resembles *Beggiatoa*.

2. *Beggiatoa* are filamentous, gliding, sulfur-oxidizing bacteria that are usually large in both diameter and length, consisting of many short cells attached end to end (**Figure 15.27***a*). Filaments can flex and twist so that many filaments become intertwined to form a complex tuft. *Beggiatoa* is found primarily in microbial mats, sediments, sulfur springs, and hot springs. The ecological strategy of *Beggiatoa* is to use gliding motility to position itself at the point where H_2S and O_2 co-occur in an environment. For example, *Beggiatoa* in microbial mats can move vertically by as much as several centimeters per day in response to cyanobacterial O_2 production, moving up to obtain O_2 when photosynthesis ceases at night and down during the day when cyanobacterial O_2 production at the mat surface causes H_2S to be found deeper in the mat.

3. The genus *Thiomargarita* contains some of the largest bacteria yet observed, with diameters that can be as much as 0.75 millimeter (**Figure 15.28**). *Thiomargarita* is nonmotile, and its ecological strategy is to separate in time the oxidation of H_2S from the reduction of O_2. To accomplish this, *Thiomargarita* contains a giant vacuole (Figure 15.28*b*) that it fills with high concentrations of nitrate (NO_3^-). This vacuole can fill almost the entire volume of the cell. Cells live in sulfide-rich marine sediments that are mixed occasionally with O_2-rich waters, such as that in salt marshes and in ocean upwelling zones. When buried in sediment, cells oxidize H_2S to S^0 anaerobically by reducing NO_3^- stored in the vacuole to ammonium (NH_4^+). They then store the S^0 as intracellular granules (Figure 15.28*a*). When turbulent waters mix the cells into the water column where H_2S is lacking, they switch to the aerobic oxidation of stored S^0. The energy they gain from S^0 oxidation is used to refill their vacuole with NO_3^- from the water column so they will be able to survive the next period of anoxia.

4. *Thioploca* are large filamentous bacteria that use a strategy similar to that of *Thiomargarita*. *Thioploca* also have intracellular S^0 granules and large vacuoles filled with NO_3^- (Figure 15.27*b*). However, filaments of *Thioploca* are motile by gliding and they occur in large sheaths that can be filled with many parallel filaments (Figure 15.27*b*). Sheaths are arranged vertically in the sediments and filaments glide up and down in the sheaths, going down to anaerobically respire H_2S using stored NO_3^- as electron acceptor and going up to aerobically respire S^0 and to refill their vacuoles with NO_3^- (↩ Figure 20.8).

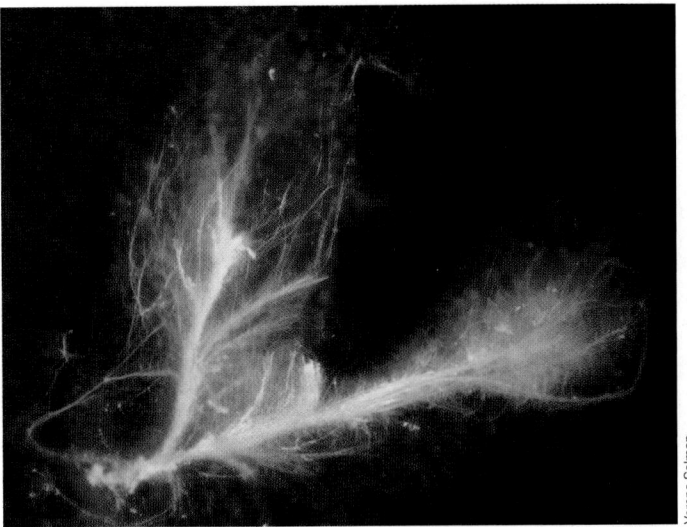

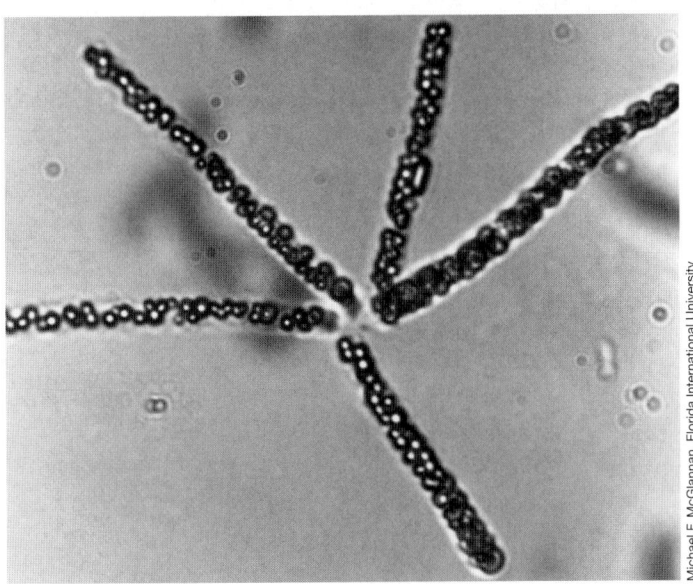

(a)

(b)

Figure 15.26 *Thiothrix.* *(a)* Filaments of *Thiothrix* attached to plant material found in the outwash stream of a sulfidic cave in Frasassi, Italy. From the plant branch point, the longest branch is about 4 mm long. *(b)* Phase-contrast photomicrograph of a rosette of cells of *Thiothrix* isolated from the sulfide-containing artesian spring shown in Figure 15.24a. Note the internal sulfur globules produced from the oxidation of sulfide. Each filament is about 4 μm in diameter.

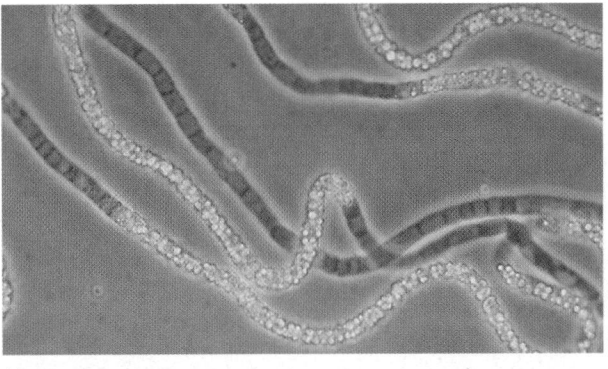

(a)

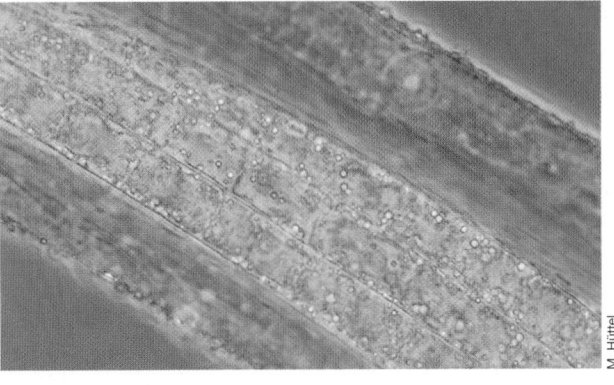

(b)

Figure 15.27 Filamentous sulfur-oxidizing bacteria. *(a)* Phase-contrast photomicrograph of a *Beggiatoa* species isolated from a sewage treatment plant. Note the abundant elemental sulfur granules in some of the cells. *(b)* Cells of a large marine *Thioploca* species. Cells contain sulfur granules (yellow) and are about 40–50 μm wide.

5. *Thiovulum* are found in freshwater and marine habitats in which sulfide-rich muds interface with oxic zones (**Figure 15.29**). *Thiovulum* cells are fairly large (10–20 μm) cocci, and when motile, they swim at exceptionally high speed, perhaps the fastest of all known bacteria (~0.6 mm/sec). The ecological strategy of *Thiovulum* is to actually control the flow of nutrients to cells. *Thiovulum* cells secrete a slime that links cells together in a veil-like structure that can be centimeters in diameter (Figure 15.29a). The veils, composed of many *Thiovulum* cells, are formed over a source of H_2S. Cells have long flagella that attach to the veil and to solid surfaces. Since the terminal end of the flagellum is attached and immobile, flagellar rotation causes cells to rotate along their flagellar axis. The simultaneous unidirectional rotation of all of the *Thiovulum* cells in the veil creates a flow of water through the veil, allowing the cells to generate and regulate the gradients of H_2S and O_2 they require to generate energy.

6. The final ecological strategy of sulfur chemolithotrophs is for the organism to form a symbiotic association with a eukaryote. There are diverse symbiotic associations in which the host provides a mechanism for regulating H_2S and O_2 levels and the sulfide-oxidizing symbiont fixes CO_2 and provides a source of carbon and energy to the host. The best example is the tube worm *Riftia*, which contains sulfide-oxidizing endosymbionts and lives at deep-sea hydrothermal vents (⇨ Section 23.9). A variety of other such symbiotic associations are present at hydrothermal vent ecosystems, including symbionts living in the gill tissue of the giant clam *Calyptogena magnifica* and on the surface of the yeti crab, which farms sulfide-oxidizing bacteria by waving its claws over sulfide-rich vent fluid. Symbioses involving invertebrates are also common in the sulfide-rich marine sediments of shallow coastal systems. For example, bivalves in the family *Solemyidae* burrow into sulfide-rich sediments and pump sulfide- and oxygen-rich water over gills that contain sulfide-oxidizing bacteria.

UNIT 4

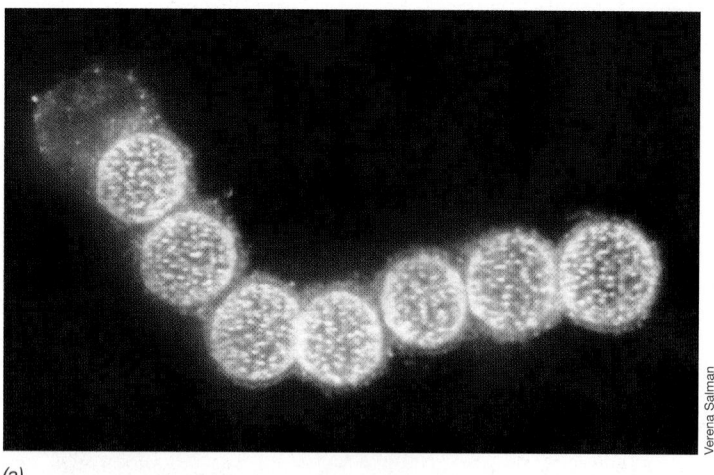

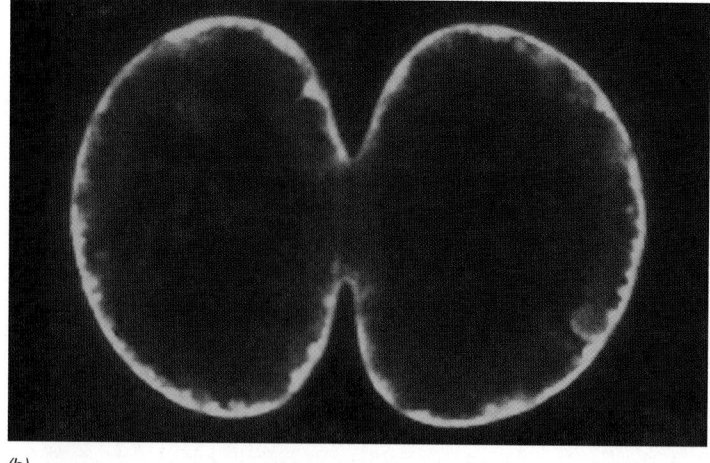

(a)

(b)

Figure 15.28 The giant sulfide-oxidizing bacterium, *Thiomargarita*. *(a) Thiomargarita namibiensis* recovered from the Namibian upwelling (off the Namibian coast, southwest Africa). Cells are about 100 μm in diameter. *(b)* Dividing cells of vacuole-containing sulfide-oxidizers recovered from the same location. The fluorescence micrograph shows ribosomes of *Thiomargarita* stained with a fluorescent nucleic acid probe. Ribosomes are found in the cytoplasm, which is present as a thin layer along the outer edge of the cells. The cytoplasm is squeezed between the cell wall and the large central vacuole, which appears dark in the image. Cells are about 50 μm wide.

From these examples it should be clear how ecological diversity drives bacteria that carry out the same energy metabolism—in this case sulfide oxidation—to best exploit the different environments they inhabit. In each case, the goal of the organism is the same, to obtain the electron donor and acceptor it needs. But also in each case, the strategy to accomplish this is unique and the best fit to both the properties of the organism and the habitat it exploits.

MINIQUIZ

- Describe the energy and carbon metabolism of *Thiobacillus* in terms of how ATP and new cell material are made.
- What are some ecological strategies that sulfur oxidizers use to compete with the spontaneous (chemical) oxidation of H$_2$S?

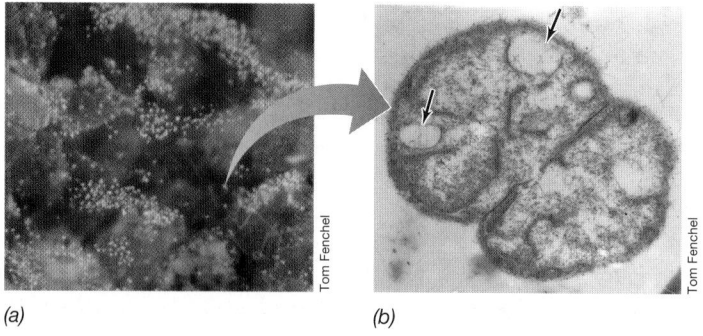

(a)

(b)

Figure 15.29 The sulfur-oxidizing *Thiovulum*. *(a)* Macrophotograph of cells of *Thiovulum* (yellow dots) that formed a thin veil in marine sand containing H$_2$S (large, irregular structures are sand grains). The *Thiovulum* veil is employed as a strategy for regulating nutrient flow, in particular for obtaining H$_2$S and O$_2$ for energy needs. *(b)* Transmission electron micrograph of a dividing cell of *Thiovulum*. Sulfur (S^0) globules are shown with arrows. Single cells of *Thiovulum* are typically 10–20 μm in diameter.

IV • Microbial Diversity in the Nitrogen Cycle

All forms of life must assimilate nitrogen for growth and thus all organisms must catalyze certain nitrogen transformations. The *Bacteria* and *Archaea*, however, are the only domains in which representatives exist that can conserve energy from the transformation of inorganic nitrogen species. In this section we will consider the diversity of three physiological groups of bacteria that participate in the nitrogen cycle: *diazotrophs, nitrifiers,* and *denitrifiers*. The physiology of these groups was considered in Sections 14.6, 14.11, and 14.13. We start our tour of microbial diversity in the nitrogen cycle by considering those microbes that reduce atmospheric nitrogen: the nitrogen fixers.

15.12 Diversity of Nitrogen Fixers

KEY GENERA: *Mesorhizobium, Desulfovibrio, Azotobacter*

Diazotrophs are microorganisms that fix dinitrogen gas (N$_2$) into NH$_3$ that can be assimilated as a source of nitrogen for cells. Nitrogen fixation is an assimilative process and requires ATP and the enzyme nitrogenase ($\Leftrightarrow$ Section 14.6). Diazotrophs typically fix N$_2$ only when other forms of N are absent, and nitrogenase expression is inhibited when NH$_3$ is available to cells. Nitrogenase is irreversibly inhibited by O$_2$ and this is one cause of ecological diversification among diazotrophs; we will see that different organisms have evolved different solutions to protecting nitrogenase from O$_2$.

Nitrogen fixation is widespread among *Bacteria* and is also found in a few *Archaea*, and it is thought that the last universal common ancestor ($\Leftrightarrow$ Section 13.1) possessed a primitive nitrogenase. The *nifH* gene encodes the dinitrogenase reductase component of nitrogenase ($\Leftrightarrow$ Section 14.6) and can be used as a measure of diazotroph diversity. More than 30,000 unique *nifH* gene sequences have been described spanning nine bacterial phyla and one archaeal phylum (Figure 15.1). The phylogenetic distribution of nitrogenase in the tree of life has been influenced strongly by horizontal gene exchange. As a result, the phylogeny of *nifH* is largely inconsistent with the 16S ribosomal RNA gene phylogeny (**Figure 15.30**). We consider here the diversity of both symbiotic and free-living diazotrophic *Bacteria*.

Symbiotic Diazotrophs

Diazotrophs form several symbiotic relationships with plants, animals, and fungi. These relationships are generally defined by the host providing a hospitable environment, including a source of carbon and energy and a system for regulating oxygen concentrations, and the microbial symbiont providing in return a supply of fixed nitrogen to the host.

The symbiosis between rhizobia and leguminous plants is one of the best-characterized nitrogen-fixing symbiotic associations (♻ Section 23.3). Root-nodule-forming bacteria are *Alphaproteobacteria* (e.g., *Mesorhizobium, Bradyrhizobium, Sinorhizobium*), *Betaproteobacteria* (e.g., *Burkholderia*), or *Actinobacteria* (e.g., *Frankia*). Other genera of symbiotic diazotrophs are found in association with shipworms (*Teredinibacter*), termite guts (*Treponema*) (♻ Section 23.7), endomycorrhizal fungi (*Glomeribacter*) (♻ Sections 18.11 and 23.4), and several fungi, algae, and plants (*Cyanobacteria*) (♻ Sections 23.1 and 23.3). These different symbioses have evolved independently multiple times as a result of convergent evolution (Figure 15.30).

Free-Living Diazotrophs

Free-living diazotrophs need a mechanism for protecting nitrogenase from oxygen (♻ Sections 14.6 and 7.8). The simplest solution to this problem is to grow only in anoxic environments. The origin of nitrogen fixation predates the origin of oxygenic photosynthesis and thus the first nitrogen-fixing organisms were free-living anaerobes. Obligately anaerobic free-living diazotrophs are common in anoxic environments including marine and freshwater sediments and microbial mats. Obligately anaerobic free-living diazotrophs are found in the bacterial phyla *Firmicutes* (e.g., *Clostridium*), *Chloroflexi* (e.g., *Oscillochloris*), *Chlorobi* (e.g., *Chlorobium*), *Spirochaetes* (e.g., *Spirochaeta*), and *Proteobacteria* (e.g., *Desulfovibrio, Chromatium*) and in the archaeal phylum *Euryarchaeota* (e.g., *Methanosarcina*). *Desulfovibrio* occur in anoxic salt marsh sediments dominated by *Spartina* grass, and their N$_2$ fixation is an important nitrogen source to plants that live in this ecosystem.

Other simple mechanisms for protecting nitrogenase from oxygen include fixing N$_2$ only at times when oxygen is absent or present in low concentration. For example, facultative aerobes will often fix N$_2$ only while growing anaerobically (e.g., *Klebsiella*). Some aerobic nitrogen-fixers are *microaerophiles*; these organisms fix nitrogen only in environments where oxygen is present at low concentration (typically less than 2%). However, some organisms have evolved more complex mechanisms for protecting nitrogenase from oxygen and are able to grow in the presence of air.

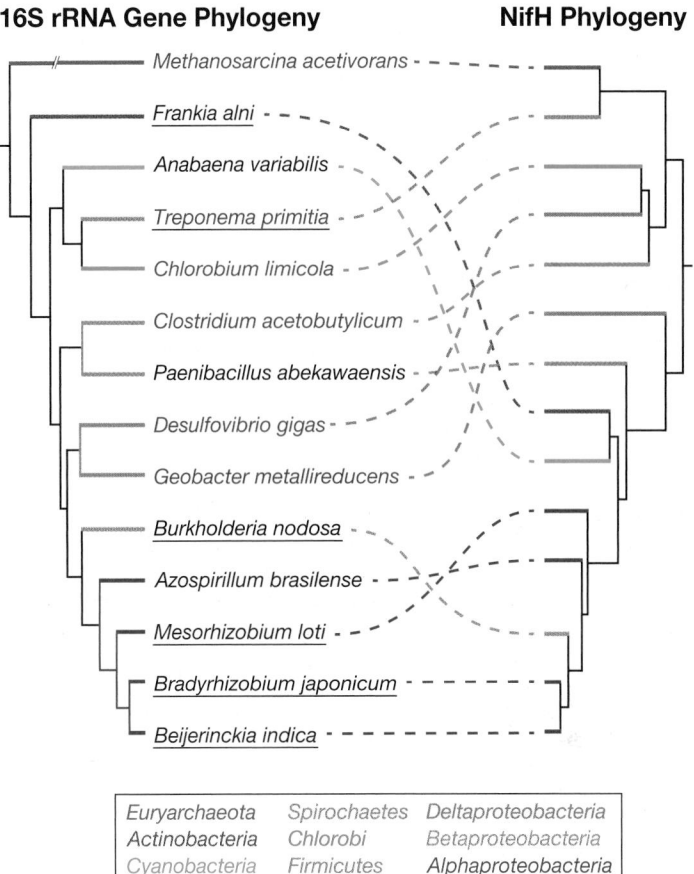

16S rRNA Gene Phylogeny   **NifH Phylogeny**

Euryarchaeota	Spirochaetes	Deltaproteobacteria
Actinobacteria	Chlorobi	Betaproteobacteria
Cyanobacteria	Firmicutes	Alphaproteobacteria

Figure 15.30 Relationships among diazotrophic (nitrogen-fixing) bacteria as inferred from 16S ribosomal RNA gene sequences and NifH amino acid sequences. Branches in each tree are colored to indicate phyla. The dashed lines indicate branches shared between the two trees. The incongruence between the two trees has resulted from multiple horizontal transfer events of the *nifH* gene. Red text denotes obligate anaerobes and underlined text indicates species that form symbioses with *Eukarya*.

Obligately aerobic free-living diazotrophs include the *Cyanobacteria*, which have evolved a variety of mechanisms of protecting nitrogenase from oxygen (Section 15.3), as well as a variety of unicellular free-living chemoorganotrophic bacteria. Obligately aerobic free-living diazotrophs include *Azotobacter, Azospirillum*, and *Beijerinckia*. *Azotobacter* cells are large rods or cocci with diameters of 2–4 μm or more. When they are growing on N$_2$ as a nitrogen source, extensive capsules or slime layers are typically produced (**Figure 15.31** and ♻ Figures 2.17 and 14.22a, b). It is thought that the high respiratory rate characteristic of *Azotobacter* cells and the abundant capsular slime they produce help protect nitrogenase from O$_2$. *Azotobacter* is able to grow on many different carbohydrates, alcohols, and organic acids, and metabolism is strictly oxidative.

Azotobacter can form resting structures called *cysts* (**Figure 15.32b**). Like bacterial endospores, *Azotobacter* cysts show negligible endogenous respiration and are resistant to desiccation, mechanical disintegration, and ultraviolet and ionizing radiation. In contrast to endospores, however, cysts are not very

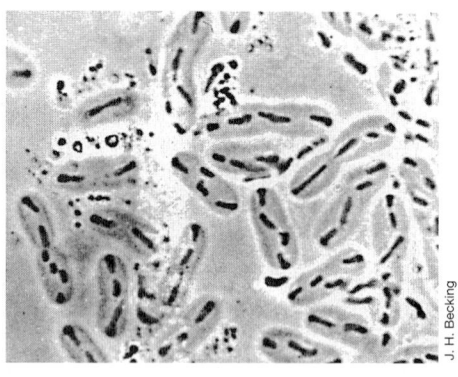

J. H. Becking

(a)

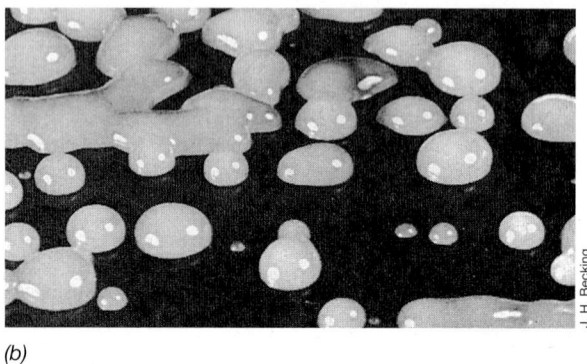

J. H. Becking

(b)

Figure 15.31 Examples of slime production by free-living N₂-fixing bacteria. (a) Cells of *Derxia gummosa* encased in slime. Cells are about 1–1.2 μm wide. (b) Colonies of *Beijerinckia* species growing on a carbohydrate-containing medium. Note the raised, glistening appearance of the colonies due to abundant capsular slime.

heat resistant, and they are not completely dormant because they rapidly oxidize carbon sources if supplied.

Azotobacter and Alternative Nitrogenases

We considered the important process of biological N₂ fixation in Section 14.6 and discussed the central importance of the metals molybdenum (Mo) and iron (Fe) to the enzyme nitrogenase.

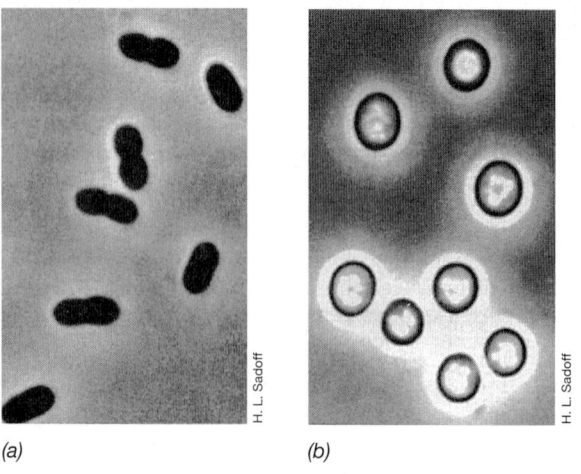

H. L. Sadoff

(a)

H. L. Sadoff

(b)

Figure 15.32 *Azotobacter vinelandii.* (a) Vegetative cells and (b) cysts visualized by phase-contrast microscopy. A cell measures about 2 μm in diameter and a cyst about 3 μm.

The species *Azotobacter chroococcum* was the first nitrogen-fixing bacterium shown to grow on N₂ in the absence of molybdenum. This is because either of two "alternative nitrogenases" are formed when Mo limitation prevents the MoFe nitrogenase from being synthesized. These nitrogenases are less efficient than the MoFe nitrogenase and contain either vanadium (V) or Fe in place of Mo. The three different types of nitrogenase (MoFe, VFe, and FeFe) are encoded by paralogous genes and likely arose as the result of gene duplications (⇄ Sections 9.5 and 13.7). Subsequent investigations of other nitrogen-fixing bacteria have shown that these genetically distinct "backup" nitrogenases are widely distributed among nitrogen-fixing microbes, in particular in the *Cyanobacteria* and *Archaea*.

MINIQUIZ

- What mechanisms do free-living diazotrophs use to protect nitrogenase from oxygen?
- Where might you expect to find nitrogen-fixing bacteria?

15.13 Diversity of Nitrifiers and Denitrifiers

Microorganisms that grow by the anaerobic respiration of inorganic nitrogen (NO_3^-, NO_2^-) to the gaseous products NO, N_2O, and N_2 are called **denitrifiers** (⇄ Section 14.13). These organisms are typically facultative aerobes and chemoorganotrophs that use organic carbon as both carbon source and electron donor.

Microorganisms able to grow chemolithotrophically at the expense of reduced inorganic nitrogen compounds (NH_3, NO_2^-) are called **nitrifiers** (Figure 15.33) (⇄ Section 14.11). These organisms are typically obligate aerobes that can also grow autotrophically; most species fix CO_2 by the Calvin cycle. A few species have also been shown to grow mixotrophically by assimilating organic carbon in addition to CO_2.

Physiology of Nitrifying *Bacteria* and *Archaea*

Nitrification often results from the sequential activities of two physiological groups of organisms, the *ammonia oxidizers* (which oxidize NH_3 to nitrite, NO_2^-) (Figure 15.33a), and the *nitrite oxidizers*, the actual nitrate-producing microorganisms, which oxidize NO_2^- to NO_3^- (Figure 15.33b). Ammonia oxidizers typically have genus names beginning in *Nitroso-*, whereas genus names of nitrate producers begin with *Nitro-*. However, certain microbes within the genus *Nitrospira* are able to carry out both ammonia oxidation and nitrite oxidation, and are therefore able to oxidize ammonia all the way to nitrate (see page 451 for more on these bacteria).

Many species of nitrifiers have internal membrane stacks (Figure 15.33) that closely resemble the photosynthetic membranes found in their close phylogenetic relatives, the purple phototrophic bacteria (Section 15.4) and the methane-oxidizing (methanotrophic) bacteria (Section 15.16). The membranes are the location of key enzymes in nitrification: *ammonia monooxygenase*, which oxidizes NH_3 to hydroxylamine (NH_2OH), and *nitrite oxidoreductase*, which oxidizes NO_2^- to NO_3^- (⇄ Section 14.11).

Enrichment cultures of nitrifying bacteria can be achieved using mineral salts media containing NH_3 or NO_2^- as electron

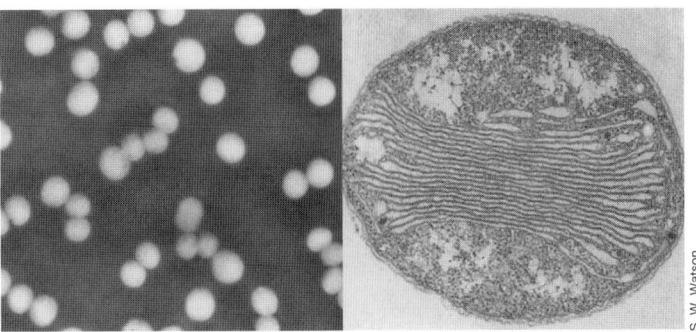

Reaction: $NH_3 + 1\frac{1}{2}\,O_2 \longrightarrow NO_2^- + H^+ + H_2O$

(a)

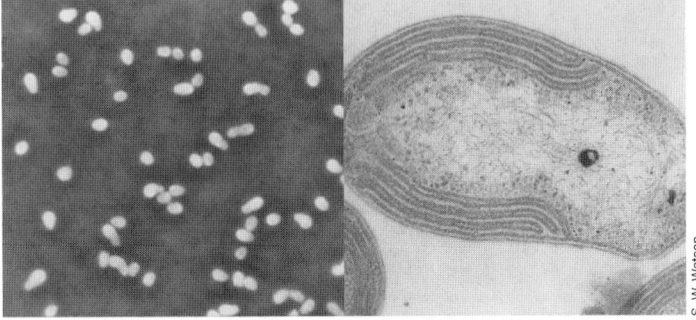

Reaction: $NO_2^- + \frac{1}{2}\,O_2 \longrightarrow NO_3^-$

(b)

Figure 15.33 Nitrifying bacteria. *(a)* Phase-contrast photomicrograph (left) and electron micrograph (right) of the ammonia-oxidizing bacterium *Nitrosococcus oceani*. A single cell is about 2 μm in diameter. *(b)* Phase-contrast photomicrograph (left) and electron micrograph (right) of the nitrite-oxidizing bacterium *Nitrobacter winogradskyi*. A cell is about 0.7 μm in diameter. Beneath each panel is the chemolithotrophic reaction that each organism catalyzes. The distinct internal membranes of each species are sites of key enzymes of nitrification.

donors and bicarbonate (HCO_3^-) as the sole carbon source. Because these organisms produce very little ATP from their electron donors ($\rightleftharpoons$ Section 14.11), visible turbidity may not develop in cultures even after extensive nitrification has occurred. An easy means of monitoring growth is thus to assay for the production of NO_2^- (with NH_3 as electron donor) or NO_3^- (with NO_2^- as electron donor).

Nitrifying *Bacteria* and *Archaea*: Ammonia Oxidizers

KEY GENERA: *Nitrosomonas, Nitrosospira, Nitrosopumilus*
Ammonia oxidizers are found in the *Beta-* (e.g., *Nitrosomonas, Nitrosospira, Nitrosolobus, Nitrosovibrio*) and *Gammaproteobacteria* (*Nitrosococcus*), in the phylum *Nitrospirae*, and in the archaeal phylum *Thaumarchaeota* (*Nitrosopumilus, Nitrosocaldus, Nitrosoarchaeum, Nitrososphaera*).

Ammonia oxidizers are widespread in soil and water. Bacterial ammonia-oxidizers are present in highest numbers in habitats where NH_3 is abundant, such as sites with extensive protein decomposition (ammonification), and also in sewage treatment facilities ($\rightleftharpoons$ Sections 22.6 and 22.7). Nitrifying bacteria develop especially well in lakes and streams that receive inputs of sewage or other wastewaters because these are frequently high in NH_3. *Nitrosomonas* is often observed in the activated sludge present in

aerobic wastewater treatment facilities. Bacterial ammonia-oxidizers are also common in soils (e.g., *Nitrosospira, Nitrosovibrio*) and in the oceans (e.g., *Nitrosococcus*).

Archaeal ammonia-oxidizers ($\rightleftharpoons$ Section 17.5) appear to be most common in habitats where NH_3 is present in low concentration. These organisms are thought to be the dominant ammonia-oxidizers in the oceans where ammonia levels are very low ($\rightleftharpoons$ Sections 20.9 and 20.11). Archaeal ammonia-oxidizers are also common in soils, and in some soils they outnumber bacterial ammonia-oxidizers by several orders of magnitude. The availability of NH_3 relative to NH_4^+ declines with pH, and thus acid soils (pH < 6.5), which are common, may favor organisms able to grow at low NH_3 concentration.

Nitrifying Bacteria: Nitrite Oxidizers

KEY GENERA: *Nitrospira, Nitrobacter*
Nitrite oxidizers are found in the classes *Alpha-* (*Nitrobacter*), *Beta-* (*Nitrotoga*), *Gamma-* (*Nitrococcus*), and *Deltaproteobacteria* (*Nitrospina*), as well as in the phylum *Nitrospirae* (genus *Nitrospira*) ($\rightleftharpoons$ Section 16.21).

Like nitrite-oxidizing *Proteobacteria*, *Nitrospira* oxidizes nitrite (NO_2^-) to nitrate (NO_3^-) and grows autotrophically (**Figure 15.34**). However, *Nitrospira* lacks the extensive internal membranes found in species of nitrifying *Proteobacteria*. Nevertheless, *Nitrospira* inhabits many of the same environments as nitrite-oxidizing *Proteobacteria* such as *Nitrobacter*, so it has been suggested that its capacity for NO_2^- oxidation may have been acquired by horizontal gene flow from nitrifying *Proteobacteria* (or vice versa). As we know, this mechanism for acquiring physiological traits has been widely exploited in the bacterial world ($\rightleftharpoons$ Chapter 11 and Section 13.7). However, environmental surveys for the presence of nitrifying bacteria in nature have shown *Nitrospira* to be much more abundant than *Nitrobacter*; thus most of the NO_2^- oxidized in natural environments is probably due to the activities of *Nitrospira*.

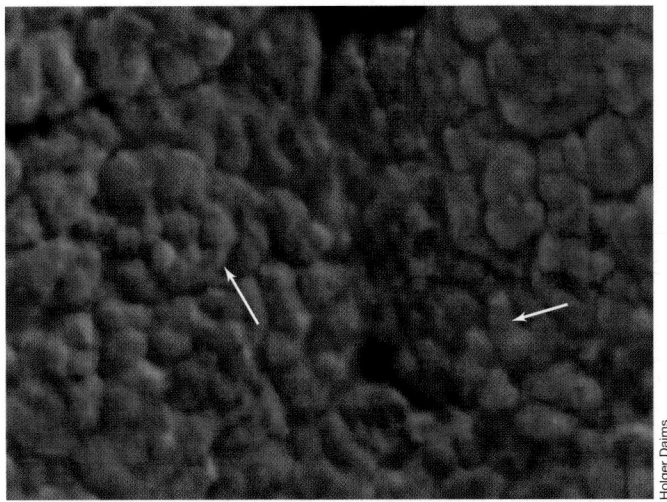

Figure 15.34 The nitrifying bacterium *Nitrospira*. An aggregate of *Nitrospira* cells enriched from activated sludge from a wastewater treatment facility. Individual cells are curved (arrows) and group into tetrads in the aggregate. A single cell of *Nitrospira* is about 0.3×1–2 μm (see page 451 for a metabolically unique *Nitrospira*).

UNIT 4

Denitrifying *Bacteria* and *Archaea*

KEY GENERA: *Paracoccus, Pseudomonas*

Denitrifiers are capable of growth by the anaerobic respiration of NO_3^- or NO_2^- to the gaseous products NO, N_2O, and N_2 (⊃ Section 14.13). Nearly all denitrifiers are chemoorganotrophs that use organic carbon as both carbon source and electron donor. Exceptions include the denitrifying sulfur-oxidizers discussed in Section 15.11. Denitrifiers are typically facultative aerobes and in nearly all cases will grow preferentially as aerobes if O_2 is present. Denitrifiers are of great importance in agricultural soils where they cause the loss of nitrogen fertilizers and the production of N_2O, which is a dominant component of greenhouse gases produced by agricultural soils (⊃ Section 21.8).

Denitrifiers are phylogenetically and metabolically diverse and include two archaeal phyla and six bacterial phyla, including five classes of *Proteobacteria* (Figure 15.1). One of the best-characterized denitrifiers is *Paracoccus denitrificans* (*Alphaproteobacteria*). Denitrification of NO_3^- to N_2 requires several key enzymatic steps (⊃ Section 14.13), and the genes that encode these enzymes are present throughout the tree of life, indicating the strong influence of horizontal gene exchange. However, many nitrate reducers possess only part of the denitrification pathway and are thus unable to reduce NO_3^- completely to N_2, producing final products such as NO_2^-, NO, or N_2O.

MINIQUIZ

- Under what conditions would you expect microorganisms to grow as a result of denitrification?
- Which traits are shared among ammonia oxidizers and nitrite oxidizers?

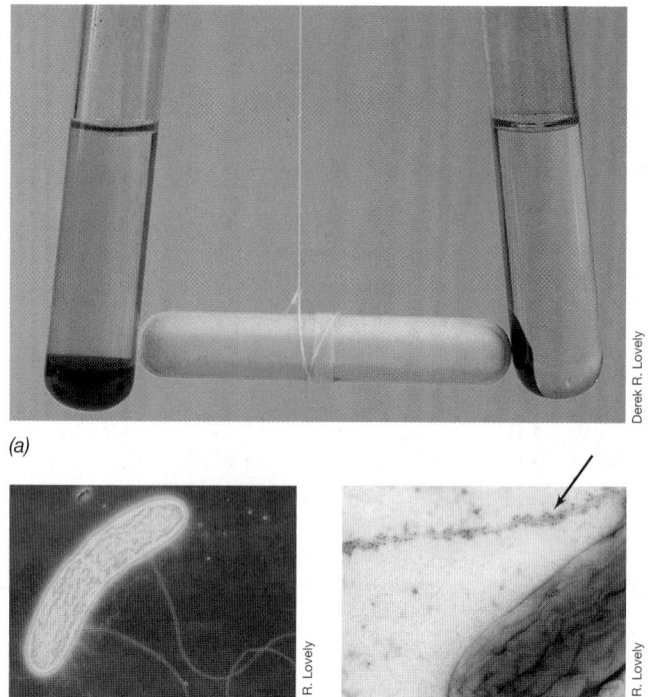

(a)

(b)   (c)

Figure 15.35 The dissimilative iron-reducing bacterium *Geobacter*. *(a)* The uninoculated tube (left) contains an anoxic medium that includes acetate and ferrihydrite, a poorly magnetic iron oxide. Following growth of *Geobacter* (right tube) the ferrihydrite is reduced to magnetite, which is magnetic. *(b)* Transmission electron micrograph of *Geobacter sulfurreducens* showing flagella and pili. The cell is about 0.7×3.5 μm. *(c)* Transmission electron micrograph of *G. sulfurreducens* showing immunogold labeling of cytochrome OmcS on the pili (arrow).

V • Other Distinctive Functional Groupings of Microorganisms

We continue our focus on functional groups whose physiological and ecological traits span different phyla as a result of convergent evolution or horizontal gene transfer. From a physiological standpoint, all groups here are chemotrophs—either chemolithotrophs or chemoorganotrophs—that contribute to specific steps in the carbon cycle or that metabolize hydrogen or metals.

15.14 Dissimilative Iron-Reducers

KEY GENERA: *Geobacter, Shewanella*

Dissimilative iron-reducers couple the reduction of oxidized metals or metalloids to cellular growth. These organisms need to overcome the fundamental obstacle of using an insoluble solid material as an electron acceptor in respiration. A variety of microorganisms are able to enzymatically reduce metals as a consequence of either fermentation reactions or sulfur or sulfate reduction, but such organisms do not conserve energy from metal reduction. In contrast, dissimilative iron-reducers carry out metal respiration by coupling the oxidation of H_2 or organic compounds to the reduction of ferric iron (Fe^{3+}) (**Figure 15.35a**) or manganese (Mn^{6+}).

Dissimilative iron-reducers are phylogenetically diverse (Figure 15.1). Iron-reducing bacterial genera are found in the *Proteobacteria* (*Geobacter, Shewanella*), *Acidobacteria* (*Geothrix*), *Deferribacteres* (*Geovibrio*), *Deinococcus–Thermus* (*Thermus*), *Thermotogae* (*Thermotoga*), and *Firmicutes* (*Bacillus, Thiobacillus*), while archaeal genera are found in the *Crenarchaeota* (*Pyrobaculum*). Iron respiration likely evolved early in the history of life and its wide distribution may be due to its presence in the universal ancestor coupled with subsequent gene loss in some lineages and horizontal gene transfer to others.

Physiology

Dissimilative iron-reducers specialize in using insoluble external electron acceptors, and these organisms are typically extremely versatile at anaerobic respiration. Dissimilative iron-reducers are unusual in that they possess outer membrane cytochromes that facilitate electron transfer with insoluble minerals. Most species are able to use either iron oxides or manganese oxides as electron acceptors, and various species are also able to use nitrate, fumarate, and oxidized inorganic sulfur, cobalt, chromium, uranium, tellurium, selenium, arsenic, and humic compounds (⊃ Section 14.15). Most genera of iron-reducing bacteria are obligate anaerobes, but some, such as *Shewanella* and relatives, are facultative aerobes. Electron donors are typically organic compounds such as fatty acids, alcohols, sugars, and in certain cases, even aromatic compounds. Many species are also able to use H_2 as an electron donor, but they are generally unable to grow autotrophically, requiring a source of organic carbon to support growth.

The family *Geobacteraceae* in the *Deltaproteobacteria* contains four genera of dissimilative iron-reducing bacteria (*Geobacter, Desulfuromonas, Desulfuromusa, Pelobacter*) that aptly demonstrate the physiological diversity of the obligately anaerobic metal reducers. *Geobacter, Desulfuromonas,* and *Desulfuromusa* can all use acetate as an electron donor as well as a diversity of other small organics, and they oxidize these substrates completely to CO_2. These genera typically specialize in anaerobic respiration. *Geobacter* in particular can use a wide range of electron donors and acceptors. *Geobacter* produce pili (Figure 15.35*b*) that contain cytochromes (Figure 15.35*c*), and these pili facilitate electron transfer to the surface of iron oxide minerals. *Pelobacter*, in contrast, are primarily fermentative organisms having a more limited respiratory capacity. For example, *Pelobacter carbinolicus* can only use lactate as the electron donor and can only use ferric iron or S^0 as the electron acceptors. *Pelobacter* species are unable to oxidize their carbon substrates completely to CO_2.

Shewanella and its relatives *Ferrimonas* and *Aeromonas* in the *Gammaproteobacteria* are facultative aerobes and will grow aerobically when O_2 is available. *Shewanella* are able to use a wide diversity of electron donors and acceptors in addition to ferric iron and manganese. However, like *Pelobacter* species, they are unable to oxidize their carbon substrates completely to CO_2 and are unable to oxidize acetate as an electron donor for anaerobic respiration.

Ecology

Dissimilative iron-reducers are common in anoxic freshwater and marine sediments. These organisms are thought to play an important role in organic matter oxidation in many anoxic habitats. Dissimilative iron-reducers are also common in the deep subsurface, found in shallow aquifers as well as in the deep subsurface environment (⊂⊃ Section 20.7). In addition, several thermophilic and hyperthermophilic iron-reducing species are known (e.g., *Thermus, Thermotoga*) and are often found in hot springs and other geothermally heated systems, including the deep subsurface.

MINIQUIZ

- In what phylogenetic groups are *Geobacter* and *Shewanella* found?
- Which genera of dissimilative iron-reducers contain facultative aerobes?

15.15 Dissimilative Iron-Oxidizers

KEY GENERA: *Acidithiobacillus, Gallionella*

The ability to couple the oxidation of ferrous iron (Fe^{2+}) to cell growth is widespread in the tree of life and thought to be a trait that evolved early in Earth's history. Genera capable of using ferrous iron as an electron donor to support growth are spread across five bacterial and two archaeal phyla (Figure 15.1).

Aerobic iron-oxidizer diversity and distribution are influenced strongly by pH and O_2. Ferrous iron oxidizes spontaneously to form insoluble precipitates in the presence of O_2 at neutral to alkaline pH (pH > 7) but is stable either under anoxic conditions or aerobically at acidic pH (pH < 4). Iron oxidizers can be divided into four functional groups on the basis of their physiology: acidophilic aerobic iron-oxidizers, neutrophilic aerobic iron-oxidizers, anaerobic chemotrophic iron-oxidizers, and anaerobic phototrophic iron-oxidizers.

Acidophilic Aerobic Iron-Oxidizing *Bacteria*

The growth of iron-oxidizing bacteria is favored in iron-rich acidic environments where soluble ferrous iron is present. Aerobic iron-oxidizers are often abundant in acid mine drainage generated from abandoned coal or iron mines or from mine tailings (⊂⊃ Sections 22.1 and 22.2). Acidophilic aerobic iron-oxidizers also inhabit iron-rich acidic springs in volcanic areas. In these environments, sulfur is often present along with ferrous iron, and many acidophilic aerobic iron-oxidizers are able to oxidize both elemental sulfur and ferrous iron. Species can be either autotrophic or heterotrophic, and commonly observed genera include *Acidithiobacillus* (*Gammaproteobacteria*), *Leptospirillum* (*Nitrospirae*), and *Ferroplasma* (*Euryarchaeota*). Other acidophilic aerobic iron-oxidizers can be found in the *Actinobacteria* and *Firmicutes*.

Neutrophilic Aerobic Iron-Oxidizing *Bacteria*

Neutrophilic aerobic iron-oxidizers are organisms adapted to a specialized niche (⊂⊃ Section 14.10). This is because ferrous iron is relatively insoluble at neutral pH and its chemical oxidation is spontaneous and rapid in the presence of air. Furthermore, at neutral pH, iron oxidation at the cell surface causes the formation of an iron oxide crust that can effectively entomb growing cells. Neutrophilic aerobic iron-oxidizers therefore thrive where iron-rich anoxic waters are exposed to air. Such habitats are common near wetlands or soils where anoxic groundwater forms a spring, but iron oxidizers also inhabit the rhizosphere of wetland plants and certain submarine hydrothermal systems.

Few genera of neutrophilic aerobic iron-oxidizers have been described and they all belong to the *Proteobacteria*. Those species found in freshwater habitats belong to a set of closely related genera in the *Betaprotobacteria*, while species found in marine habitats belong to the *Zetaproteobacteria*. The metabolism of these organisms is fairly narrow. Species are typically microaerophiles and obligate chemolithotrophs, though in certain cases mixotrophy has been observed. The genera *Leptothrix* and *Sphaerotilus* are exceptions (Section 15.21). *Leptothrix* and *Sphaerotilus* are common in freshwater environments containing neutrophilic aerobic iron-oxidizers. They catalyze the oxidation of both iron and manganese but do not appear to conserve energy from these reactions, conserving energy instead from the oxidation of organic matter.

Characteristic species of neutrophilic aerobic iron-oxidizers are found in the genus *Gallionella* (freshwater) and the marine genus *Mariprofundus* (marine). Species of *Gallionella* and *Mariprofundus* each form a twisted stalklike structure containing $Fe(OH)_3$ from the oxidation of ferrous iron (**Figure 15.36**). The iron-encrusted stalk contains an organic matrix on which $Fe(OH)_3$ accumulates as it is excreted from the cell surface. Stalk formation is presumably an adaptation that prevents cells from becoming entombed in an iron oxide crust.

Gallionella is common in the waters draining bogs, iron springs, and other habitats where ferrous iron is present. *Mariprofundus*

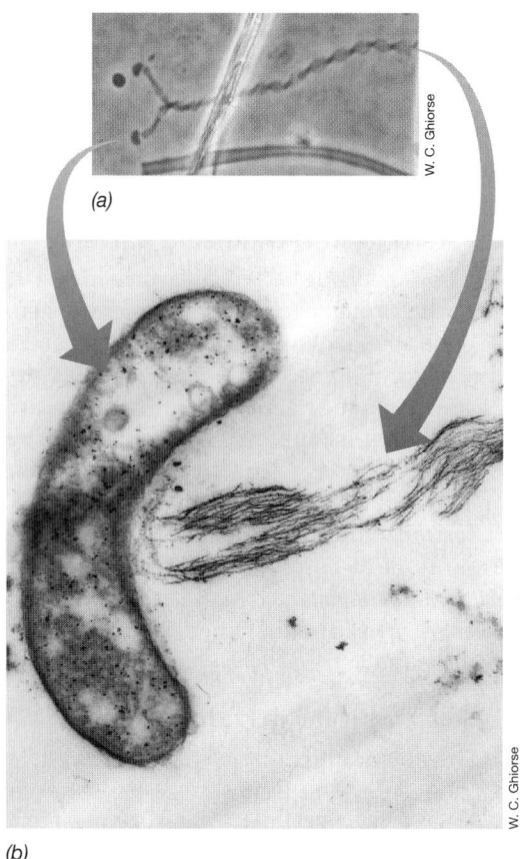

W. C. Ghiorse

(a)

W. C. Ghiorse

(b)

Figure 15.36 The neutrophilic ferrous iron-oxidizer *Gallionella ferruginea*, from an iron seep near Ithaca, New York. *(a)* Photomicrograph of two bean-shaped cells with stalks that combine to form one twisted mass. *(b)* Transmission electron micrograph of a thin section of a *Gallionella* cell with stalk. Cells are about 0.6 μm wide.

was first isolated from Lōʻihi Seamount, a submarine volcano near Hawaii (⇨ Section 21.5 and Figure 21.14). *Gallionella* and *Mariprofundus* are both autotrophic chemolithotrophs containing enzymes of the Calvin cycle (⇨ Section 14.5).

Anaerobic Iron-Oxidizing *Bacteria*

Anaerobic ferrous iron oxidation can be mediated by both chemotrophic and phototrophic bacteria. These groups are common in anoxic sediments and wetlands. Anoxic conditions promote the solubility of ferrous iron across a wide range of pH and so, unlike the aerobic iron-oxidizing bacteria, growth of anaerobic iron-oxidizers is not strictly limited to neutral pH. These groups contain organisms that are metabolically diverse and able to grow by using a variety of different electron donors and acceptors.

Phototrophic iron oxidation occurs in select species of purple nonsulfur bacteria of the *Alphaproteobacteria* (e.g., *Rhodopseudomonas palustris*), select species of purple sulfur bacteria of the *Gammaproteobacteria* (⇨ Figure 14.31), and select species of green sulfur bacteria found in the *Chlorobi* (*Chlorobium ferrooxidans*). In all cases ferrous iron is one of several compounds that these organisms can use as an electron donor in photosynthesis.

Anaerobic chemotrophic iron-oxidizers couple the oxidation of ferrous iron to nitrate reduction, producing either NO_2^- or

nitrogen gases (denitrification). These organisms are *Alpha-, Beta-, Gamma-,* or *Deltaproteobacteria*, and most are also able to use various organic electron donors in nitrate reduction; many can also grow aerobically. The bacterial genera *Acidovorax, Aquabacterium,* and *Marinobacter* all contain anaerobic iron-oxidizers. While most species are mixotrophs when growing with ferrous iron as electron donor, species such as *Marinobacter aquaeolei* and *Thiobacillus denitrificans* are able to grow autotrophically as iron-oxidizing chemolithotrophs.

MINIQUIZ

- What habitat characteristics govern the diversity and distribution of iron oxidizers?
- How do aerobic neutrophilic iron-oxidizers keep their cells from becoming entombed in a crust of iron?

15.16 Methanotrophs and Methylotrophs

Methylotrophs are organisms that grow using organic compounds lacking C—C bonds as electron donors in energy metabolism and as carbon sources. Methylotrophy occurs in the bacterial phyla *Proteobacteria, Firmicutes, Actinobacteria, Bacteroidetes, Verrucomicrobia,* and in the archaeal phylum *Euryarchaeota* (Figure 15.1). **Methanotrophs** are a subset of methylotrophs defined by their ability to use methane as a substrate for growth (⇨ Section 14.18).

Aerobic methylotrophs are common in soil and aquatic environments where O_2 is present. Anaerobic methylotrophs are common in anoxic environments, particularly in marine sediments. Many anaerobic methylotrophs are methanogenic *Archaea*. In addition, a consortium of methanogenic *Archaea* and sulfate-reducing bacteria combine to oxidize methane from gas hydrates found in deep-sea sediments (⇨ Section 14.18). We consider here only the aerobic methylotrophs.

Aerobic Facultative Methylotrophs

KEY GENERA: *Hyphomicrobium, Methylobacterium*
Aerobic facultative methylotrophs are unable to use methane but can use many other methylated compounds. They are species of *Alpha-, Beta-,* and *Gammaproteobacteria, Actinobacteria,* and *Firmicutes*. Facultative methylotrophs are metabolically diverse and, in addition to methylated substrates, most species can grow aerobically using other organic compounds, such as organic acids, ethanol, and sugars. When growing as methylotrophs, most species can grow aerobically with methanol and some can also metabolize methylated amines, methylated sulfur compounds, and halomethanes. Most are obligate aerobes, though some species are capable of denitrification.

The genus *Hyphomicrobium* provides an example of the metabolic versatility of the aerobic facultative methylotrophs. Certain species of *Hyphomicrobium* can grow as aerobic methylotrophs using methanol, methylamine, or dimethyl sulfide. Species of *Hyphomicrobium* can also grow as anaerobic methylotrophs using methanol as an electron donor coupled to denitrification. Finally, *Hyphomicrobium* can grow aerobically on a range of C_2 and C_4 compounds.

TABLE 15.1 Some characteristics of methanotrophic *Bacteria*

Organism	Morphology	Phylogenetic group[a]	Internal membranes[b]	Carbon assimilation pathway[c]	N$_2$ fixation
Methylomonas	Rod	*Gamma*	I	Ribulose monophosphate	No
Methylomicrobium	Rod	*Gamma*	I	Ribulose monophosphate	No
Methylobacter	Coccus to ellipsoid	*Gamma*	I	Ribulose monophosphate	No
Methylococcus	Coccus	*Gamma*	I	Ribulose monophosphate and Calvin cycle	Yes
Methylosinus	Rod or vibrioid	*Alpha*	II	Serine	Yes
Methylocystis	Rod	*Alpha*	II	Serine	Yes
Methylocella[d]	Rod	*Alpha*	II	Serine	Yes
Methylacidiphilum[d]	Rod	*Verrucomicrobiaceae*[d]	Membrane vesicles	Serine and Calvin cycle	Yes

[a]All except for *Methylacidiphilum* are *Proteobacteria*.

[b]Internal membranes: type I, bundles of disc-shaped vesicles distributed throughout the organism; type II, paired membranes running along the periphery of the cell. See Figure 15.37.

[c]See Figure 14.51.

[d]Acidophiles. For the properties of *Verrucomicrobiaceae*, see Section 16.17.

Aerobic Methanotrophs

KEY GENERA: *Methylomonas, Methylosinus*

Aerobic methanotrophs are methylotrophs that can use methane as an electron donor and typically can use it as a carbon source as well. **Table 15.1** gives a taxonomic overview of the methanotrophs. Most methanotrophs are *Proteobacteria* and are classified into two major groups based on their internal cell structure, phylogeny, and carbon assimilation pathway. *Type I methanotrophs* assimilate one-carbon compounds via the ribulose monophosphate cycle and are *Gammaproteobacteria*. By contrast, *type II methanotrophs* assimilate C$_1$ intermediates via the serine pathway and are *Alphaproteobacteria* (Table 15.1). We discussed the biochemical details of these pathways in Section 14.18. Most methanotrophs are metabolically specialized for aerobic growth on methane, though some can grow on either methane or methanol. Methanotrophs are typically obligate methylotrophs; however, the methanotrophic genus *Methylocella* contains species that can also grow on acetate or organic acids such as pyruvate and succinate.

In addition to the proteobacterial methanotrophs described above, the phylum *Verrucomicrobia* contains the bacterium *Methylacidiphilum*. Genome analysis has shown that species of *Methylacidiphilum* lack key enzymes of both the ribulose monophosphate and serine pathways. Instead, *Methylacidiphilum* uses the Calvin cycle to assimilate carbon from CO$_2$.

Physiology

Methanotrophs possess a key enzyme, *methane monooxygenase*, which catalyzes the incorporation of an atom of oxygen from O$_2$ into CH$_4$, forming methanol (CH$_3$OH, ↩ Section 14.18). The requirement for O$_2$ as a reactant in the initial oxygenation of CH$_4$ explains why these methanotrophs are obligate aerobes. Methane monooxygenase is located in extensive internal membrane systems that are the site of methane oxidation. Membranes in type I methanotrophs are arranged as bundles of disc-shaped vesicles distributed throughout the cell (**Figure 15.37b**). Type II species possess paired membranes running

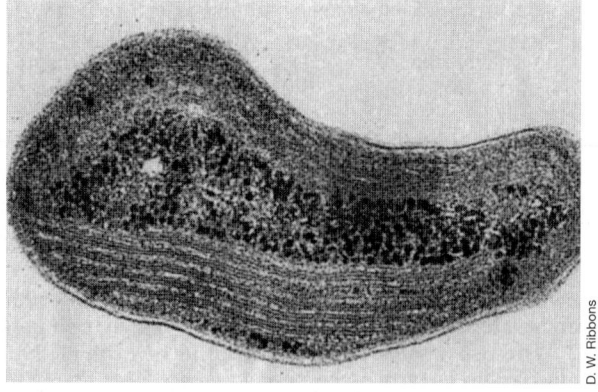

(a)

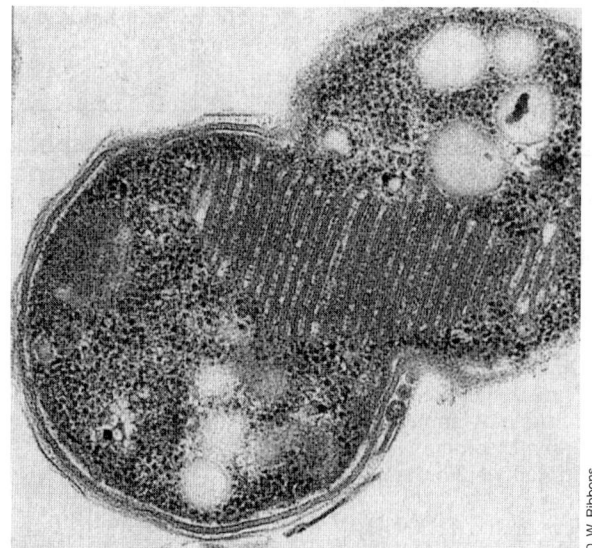

(b)

Figure 15.37 Methanotrophs. *(a)* Electron micrograph of a cell of *Methylosinus*, illustrating a type II membrane system. Cells are about 0.6 μm in diameter. *(b)* Electron micrograph of a cell of *Methylococcus capsulatus*, illustrating a type I membrane system. Cells are about 1 μm in diameter. Compare with Figure 15.33.

UNIT 4

along the periphery of the cell (Figure 15.37a). Verrucomicrobial methanotrophs possess membrane vesicles. Methylotrophs unable to use methane lack these internal membrane arrays.

Methanotrophs are virtually unique among bacteria in possessing relatively large amounts of sterols. Sterols are rigid planar molecules found in the cytoplasmic and other membranes of eukaryotes but are absent from most bacteria. Sterols may be an essential part of the complex internal membrane system for methane oxidation (see Figure 15.37). The only other group of bacteria in which sterols are widely distributed is the mycoplasmas, bacteria that lack cell walls and thus probably require a tougher cytoplasmic membrane ($\Rightarrow$ Section 16.9). Many methylotrophs contain various carotenoid pigments and high levels of cytochromes in their membranes, and these features often render colonies of aerobic methylotrophs pink.

Ecology

Aerobic methylotrophs are found in the open ocean, soils, in association with plant roots and leaf surfaces, and at the oxic interface of many anoxic environments. Methanol is produced during the breakdown of plant pectin and this is likely an important substrate for methylotrophs in terrestrial ecosystems. In addition, soils contain aerobic methanotrophs that consume atmospheric methane and are an important biological sink for atmospheric methane. Aerobic methanotrophs are also common at the oxic interface of anoxic environments found in lakes, sediments, and wetlands where methanogens provide a constant source of methane. These methanotrophs play an important role in the global carbon cycle by oxidizing CH_4 and converting it into cell material and CO_2 before it reaches the atmosphere (CH_4 is a strong greenhouse gas).

Methanotrophs also form a variety of symbioses with eukaryotic organisms. For example, some marine mussels live in the vicinity of hydrocarbon seeps on the seafloor, places where CH_4 is released in substantial amounts. Methanotrophic symbionts reside within the animal's gill tissue (**Figure 15.38**), which ensures effective gas exchange with seawater. Assimilated CH_4 is distributed throughout the animal by the excretion of organic compounds by the methanotrophs. These methanotrophic symbioses are therefore conceptually similar to those that develop between sulfide-oxidizing chemolithotrophs and hydrothermal vent tube worms and giant clams ($\Rightarrow$ Section 23.9).

Methylomirabilis oxyfera is a methanotroph isolated from anoxic waters in the Black Sea, and was the first isolate obtained from the unique bacterial phylum NC-10. *M. oxyfera* is an obligate anaerobe; however, it uses the O_2-dependent enzyme of aerobic methanotrophs (methane monooxygenase) to oxidize methane to CO_2. *M. oxyfera* accomplishes this by reducing nitrite to nitric oxide (NO), which is then dismutated to N_2 and O_2 (2 NO $\rightarrow N_2 + O_2$). The O_2 produced by this pathway is then consumed by methane monooxygenase during the oxidation of CH_4 ($\Rightarrow$ Section 14.18). Like the methanotroph *Methylacidiphilum*, *M. oxyfera* assimilates C_1 units as CO_2, probably by the Calvin cycle.

--- **MINIQUIZ** ---

- What is the difference between a methanotroph and a methylotroph?
- What is unique about the methanotroph *Methylomirabilis*?

15.17 Microbial Predators

KEY GENERA: *Bdellovibrio, Myxococcus*

Some bacteria are predators that consume other bacteria. Known bacterial predators reside among several classes of *Proteobacteria* and in the *Bacteroidetes* and *Cyanobacteria*. Several different

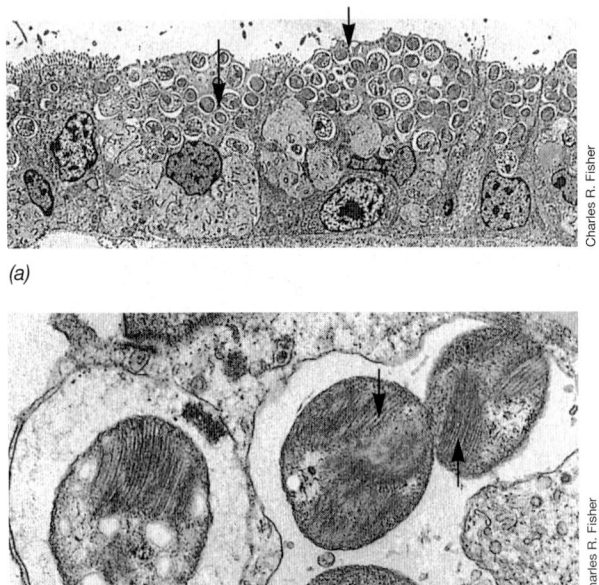

(a)

(b)

Charles R. Fisher

Figure 15.38 Methanotrophic symbionts of marine mussels. *(a)* Electron micrograph of a thin section at low magnification of gill tissue from a marine mussel living near hydrocarbon seeps in the Gulf of Mexico. Note the symbiotic methanotrophs (arrows) in the tissues. *(b)* High-magnification view of gill tissue showing methanotrophs with type I membrane bundles (arrows). Cells of the methanotrophs are about 1 μm in diameter. Compare with Figure 15.37b.

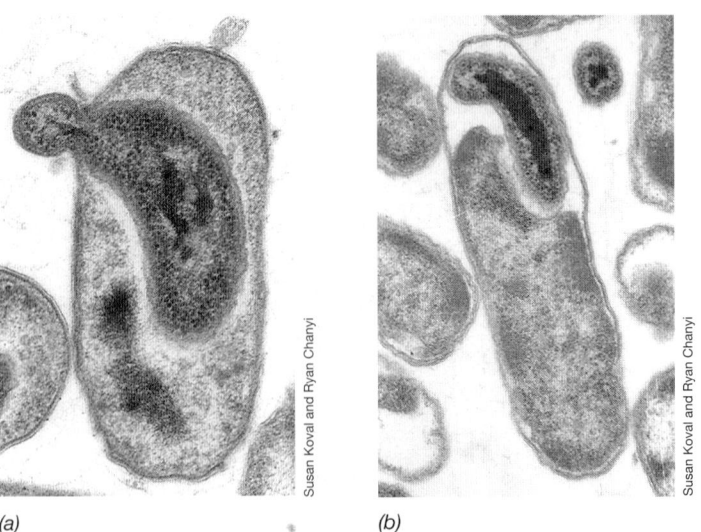

(a)

(b)

Susan Koval and Ryan Chanyi

Figure 15.39 Attack on a prey cell by *Bdellovibrio*. Thin-section electron micrographs of *Bdellovibrio* attacking a cell of *Delftia acidovorans*. *(a)* Entry of the predator cell. *(b)* *Bdellovibrio* cell inside the host. The *Bdellovibrio* cell is enclosed in the bdelloplast and replicates in the periplasmic space. A *Bdellovibrio* cell measures about 0.3 μm in diameter.

Figure 15.40 Developmental cycle of the bacterial predator *Bdellovibrio bacteriovorus*. *(a)* Electron micrograph of a cell of *Bdellovibrio bacteriovorus*; note the thick flagellum. A cell is 0.3 μm wide. *(b)* Events in predation. Following primary contact with a gram-negative bacterium, the highly motile *Bdellovibrio* cell attaches to and penetrates into the prey periplasmic space. Once inside the periplasmic space, *Bdellovibrio* cells elongate and within 4 h progeny cells are released. The number of progeny cells released varies with the size of the prey; 5–6 bdellovibrios are released from *Escherichia coli* and 20–30 for a larger prey cell, such as *Aquaspirillum*.

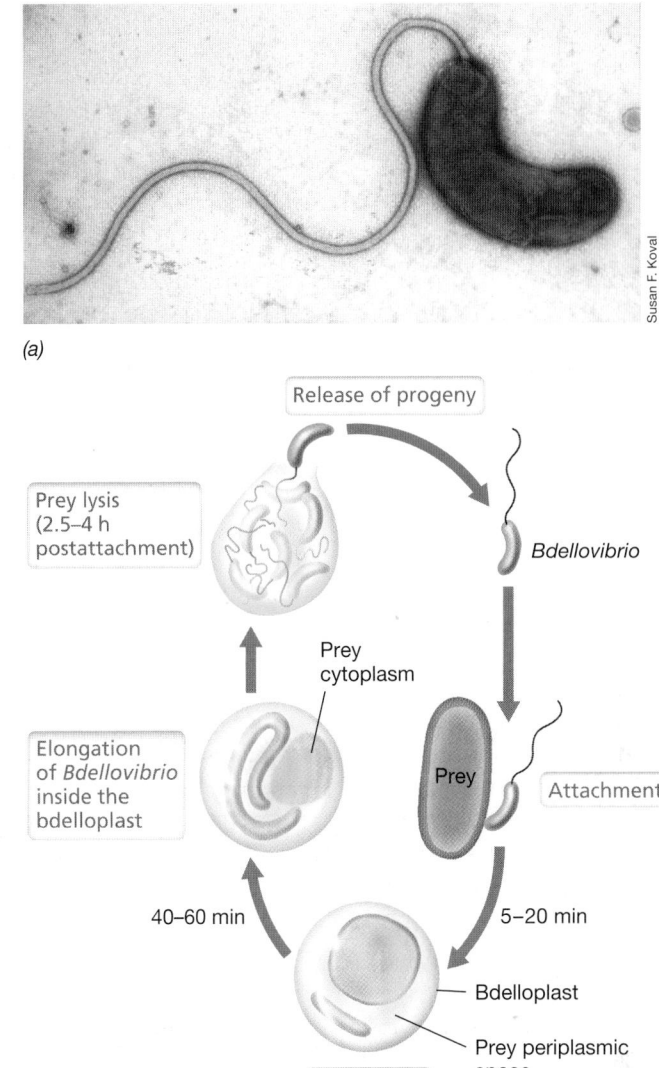

methods of predation have been observed. Some predators, such as *Vampirococcus* (phylogeny unknown), *Micavibrio* (*Alphaproteobacteria*), and the green algal predator *Vampirovibrio* (related to *Cyanobacteria*, ⇄ Section 9.3 and Figure 9.9) are *epibiotic predators*; they attach to the surface of their prey and acquire nutrients from its cytoplasm or periplasm. Other predators, such as *Daptobacter* (*Epsilonproteobacteria*), are *cytoplasmic predators*, as they invade their host cells and replicate in the cytoplasm, consuming their prey from the inside out. *Bdellovibrio* have a similar lifestyle as *periplasmic predators*; they invade and replicate within the periplasmic space of their prey cells. Finally, predators such as *Lysobacter* (*Gammaproteobacteria*) and *Myxococcus* (*Deltaproteobacteria*) are *social predators*. These gliding bacteria use swarming behavior to find prey, which they lyse and feed upon collectively. *Bdellovibrio* and *Myxococcus* are the most thoroughly described genera of bacterial predators.

Bdellovibrio

Bdellovibrio are small, highly motile and curved bacteria that prey on other bacteria, using the cytoplasmic constituents of their hosts as nutrients (*bdello* is a prefix meaning "leech"). After attachment of a *Bdellovibrio* cell to its prey, the predator penetrates the cell wall of the prey and replicates in the periplasmic space, eventually forming a spherical structure called a *bdelloplast*. Two stages of penetration are shown in electron micrographs in **Figure 15.39** and diagrammatically in **Figure 15.40**. A wide variety of gram-negative prey bacteria can be attacked by *Bdellovibrio*, but gram-positive cells are not attacked.

Bdellovibrio is an obligate aerobe, obtaining its energy from the oxidation of amino acids and acetate. In addition, *Bdellovibrio* assimilates nucleotides, fatty acids, peptides, and even some intact

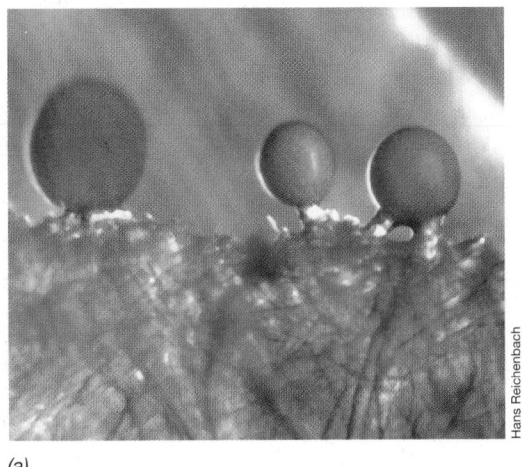
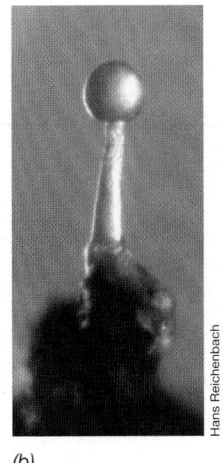
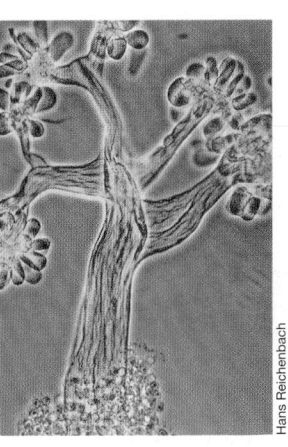

Figure 15.41 Fruiting bodies of three species of fruiting myxobacteria. *(a) Myxococcus fulvus* (125 μm high). *(b) Myxococcus stipitatus* (170 μm high). *(c) Chondromyces crocatus* (560 μm high).

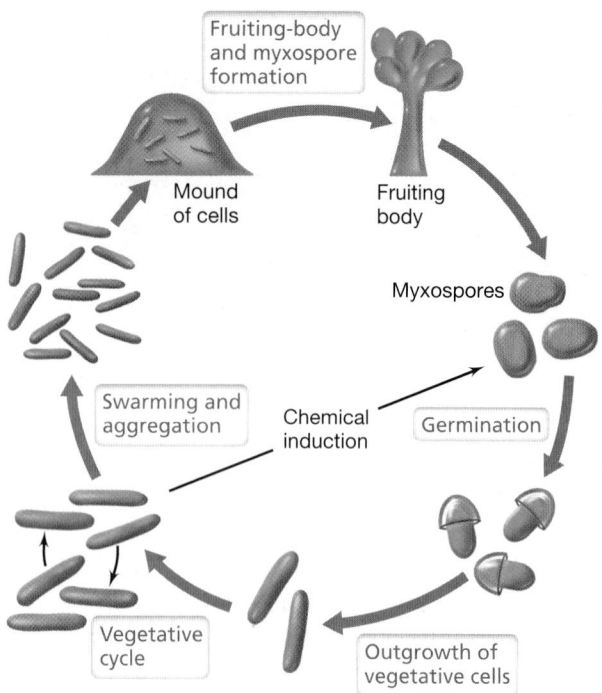

Figure 15.42 Life cycle of *Myxococcus xanthus*. Aggregation assembles vegetative cells that then undergo fruiting body formation, within which some vegetative cells undergo morphogenesis to form resting cells called myxospores. The myxospores germinate under favorable nutritional and physical conditions to yield vegetative cells.

proteins directly from its host without first hydrolyzing them. Prey-independent derivatives of *Bdellovibrio* can be isolated and grown on complex media, however, showing that predation is not an obligatory lifestyle.

Phylogenetically, bdellovibrios are *Deltaproteobacteria*, and they are widespread in aquatic habitats. Procedures for their isolation are similar to those used to isolate bacterial viruses (⮂ Section 8.4). Prey bacteria are spread on the surface of an agar plate forming a lawn, and the surface is inoculated with a small amount of soil suspension that has been filtered through a membrane filter; the filter retains most bacteria but allows the small *Bdellovibrio* cells to pass. On incubation of the agar plate, plaques analogous to

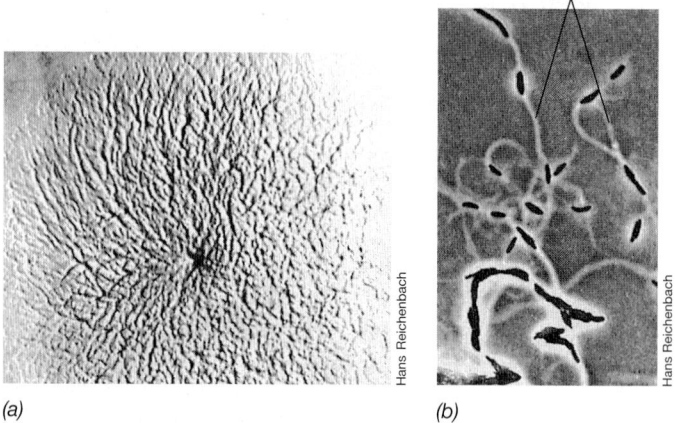

Figure 15.44 Swarming in *Myxococcus*. (a) Photomicrograph of a swarming colony (5-mm radius) of *Myxococcus xanthus* on agar. *M. xanthus* has been used as a model for developmental events in myxobacteria. (b) Single cells of *Myxococcus fulvus* from an actively gliding culture, showing the characteristic slime trails on the agar. A cell of *M. fulvus* is about 0.8 μm in diameter.

bacteriophage plaques (⮂ Figure 8.9*b*) are formed at locations where *Bdellovibrio* cells are multiplying. Pure cultures of *Bdellovibrio* can then be isolated from these plaques. *Bdellovibrio* are widely distributed, as cultures have been obtained from many soils and from sewage.

Myxobacteria

Myxobacteria exhibit the most complex behavioral patterns of all known bacteria. The life cycle of myxobacteria results in the formation of multicellular structures called *fruiting bodies*. The fruiting bodies are often strikingly colored and morphologically elaborate (**Figure 15.41**), and these can often be seen with a hand lens on moist pieces of decaying wood or plant material. The fruiting myxobacteria are classified on morphological grounds using characteristics of the vegetative cells, the myxospores, and fruiting body structure.

The life cycle of a typical myxobacterium is shown in **Figure 15.42**. The vegetative cells of the myxobacteria are simple,

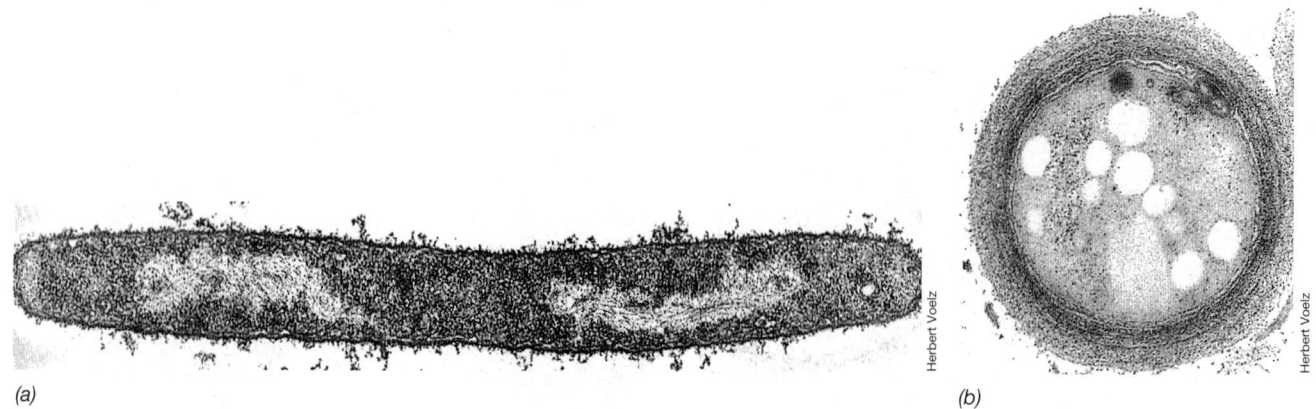

Figure 15.43 *Myxococcus*. (a) Electron micrograph of a thin section of a vegetative cell of *Myxococcus xanthus*. A cell measures about 0.75 μm wide. (b) Myxospore of *M. xanthus*, showing the multilayered outer wall. Myxospores measure about 2 μm in diameter.

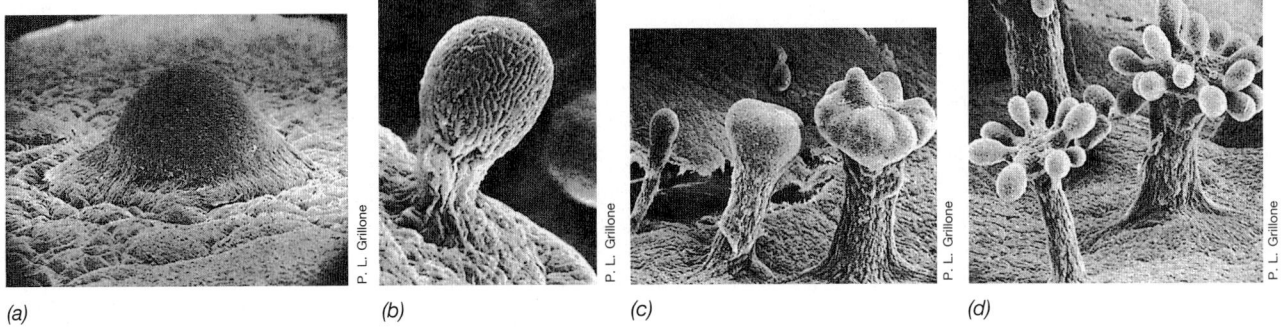

(a)   *(b)*   *(c)*   *(d)*

P. L. Grillone

Figure 15.45 Scanning electron micrographs of fruiting body formation in *Chondromyces crocatus*. *(a)* Early stage, showing aggregation and mound formation. *(b)* Initial stage of stalk formation. Slime formation in the head has not yet begun and so the cells that compose the head are still visible. *(c)* Three stages in head formation. Note that the diameter of the stalk also increases. *(d)* Mature fruiting bodies. The entire fruiting structure is about 600 μm in height (compare with Figure 15.41*c*).

nonflagellated, gram-negative rods (**Figure 15.43**) that glide across surfaces and obtain their nutrients primarily by using extracellular enzymes to lyse other bacteria and use the released nutrients. A vegetative cell excretes slime, and as it moves across a solid surface, it leaves behind a slime trail (**Figure 15.44**). The vegetative cells form a swarm that exhibits self-organizing behavior, and this allows them to behave as a single coordinated entity in response to environmental cues.

Upon nutrient exhaustion, vegetative cells of myxobacteria begin to migrate toward each other, aggregating together in mounds or heaps (**Figure 15.45**). Aggregation is likely mediated by chemotactic or quorum-sensing responses (⊄ Sections 6.7 and 6.8). As the cell masses become higher, they begin to differentiate into fruiting bodies (**Figure 15.46**) containing *myxospores*. Myxospores are specialized cells that are resistant to drying, ultraviolet radiation, and heat, but the degree of heat resistance is much less than that of the bacterial endospore (⊄ Section 2.10). Fruiting bodies can be simple, consisting of masses of myxospores embedded in slime, or complex, consisting of a stalk and heads (Figure 15.46). The fruiting body stalk is composed of slime within which a few cells are trapped. The majority of the cells migrate to the fruiting body head, where they undergo differentiation into myxospores (Figure 15.42).

Hans Reichenbach

David White

Figure 15.46 The myxobacterium *Stigmatella aurantiaca*. Scanning electron micrograph of a fruiting body growing on a piece of wood. Note the individual cells visible in each fruiting body. Inset: Phase-contrast photomicrograph of a single fruiting body about 150 μm high. The color is due to the production of glucosylated carotenoid pigments.

MINIQUIZ

• What environmental conditions trigger fruiting body formation in myxobacteria?

• What are the different ways in which species of *Myxococcus* and *Bdellovibrio* kill their prey?

15.18 Microbial Bioluminescence

KEY GENERA: *Vibrio, Aliivibrio, Photobacterium*

Several species of bacteria can emit light, a process called **bioluminescence** (**Figure 15.47**). Most bioluminescent bacteria are classified in the genera *Photobacterium, Aliivibrio,* and *Vibrio,* but a few species reside in *Shewanella,* a genus of primarily marine bacteria, and in *Photorhabdus,* a genus of terrestrial bacteria (all *Gammaproteobacteria*).

Most bioluminescent bacteria inhabit the marine environment, and some species colonize specialized *light organs* of certain marine fishes and squids, producing light that the animal uses for signaling, avoiding predators, and attracting prey (Figure 15.47 *c–f* and ⊄ Section 23.8). When living symbiotically in light organs of fish and squids, or saprophytically, for example on the skin of a dead fish, or parasitically in the body of a crustacean, luminous bacteria can be recognized by the light they produce.

UNIT 4

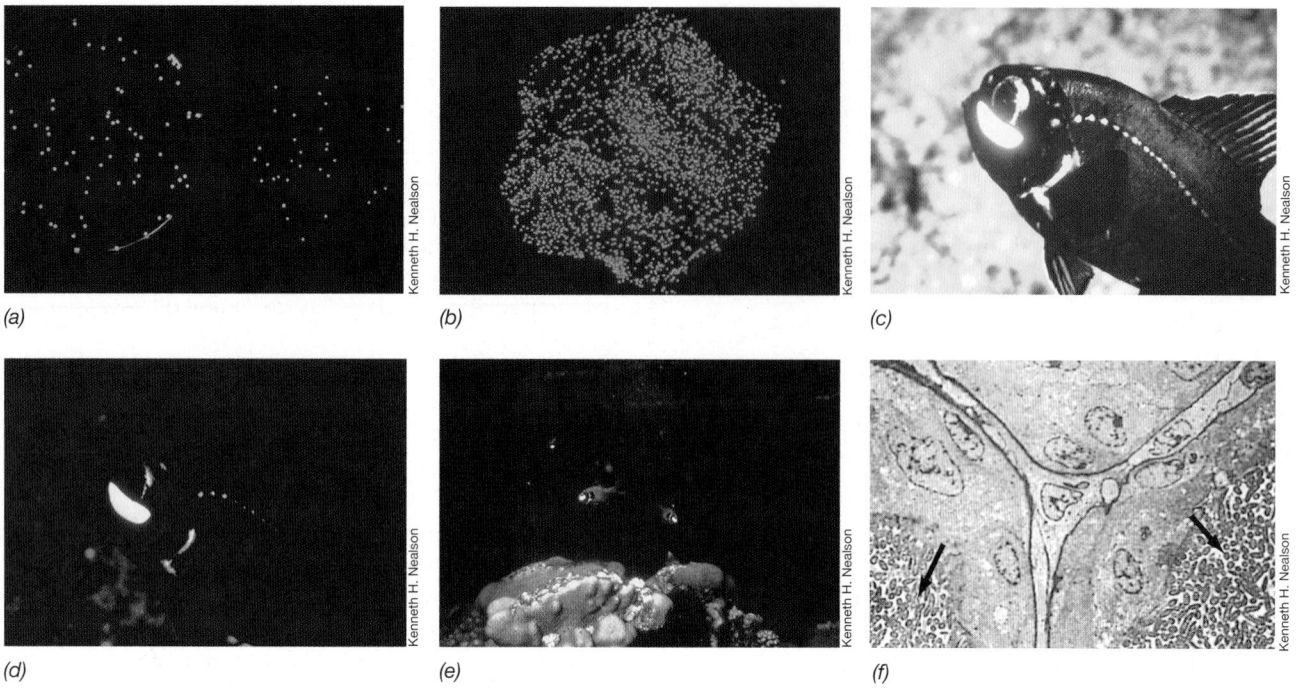

Figure 15.47 Bioluminescent bacteria and their role as light organ symbionts in the flashlight fish. *(a)* Two Petri plates of luminous bacteria photographed by their own light. Note the different colors. Left, *Aliivibrio fischeri* strain MJ-1, blue light, and right, strain Y-1, green light. *(b)* Colonies of *Photobacterium phosphoreum* photographed by their own light (⇄ Figure 1.2). *(c)* The flashlight fish *Photoblepharon palpebratus*; the bright area is the light organ containing bioluminescent bacteria. *(d)* Same fish photographed by its own light. *(e)* Underwater photograph taken at night of *P. palpebratus*. *(f)* Electron micrograph of a thin section through the light-emitting organ of *P. palpebratus* showing the dense array of bioluminescent bacteria (arrows).

Mechanism and Ecology of Bioluminescence

Although *Photobacterium*, *Aliivibrio*, and *Vibrio* isolates are facultative aerobes, they are bioluminescent only when O_2 is present. Luminescence in bacteria requires the genes *luxCDABE* (⇄ Section 6.8) and is catalyzed by the enzyme *luciferase*, which uses O_2, a long-chain aliphatic aldehyde (RCHO) such as tetradecanal, and reduced flavin mononucleotide ($FMNH_2$) as substrates:

$$FMNH_2 + O_2 + RCHO \xrightarrow{\text{Luciferase}} FMN + RCOOH + H_2O + light$$

The light-generating system constitutes a metabolic route for shunting electrons from $FMNH_2$ to O_2 directly, without employing other electron carriers such as quinones and cytochromes.

Luminescence in many luminous bacteria only occurs at high population density. The enzyme luciferase and other proteins of the bacterial luminescence system exhibit a population density–responsive induction, called **autoinduction**, in which transcription of the *luxCDABE* genes is controlled by a regulatory protein, LuxR, and an inducer molecule, acyl homoserine lactone (AHL, ⇄ Section 6.8 and Figure 6.20). During growth, cells produce AHL, which can rapidly cross the cytoplasmic membrane in either direction, diffusing in and out of cells. Under conditions in which a high local population density of cells of a given species is attained, as in a test tube, a colony on a plate, or in the light organ of a fish or squid (⇄ Section 23.8), AHL accumulates. Only when it reaches a certain concentration in the cell is AHL bound by LuxR,

forming a complex that activates transcription of *luxCDABE*; cells then become luminous (Figure 15.47*b*, ⇄ Figure 1.2). This gene regulatory mechanism is also called *quorum sensing* because of the population density–dependent nature of the phenomenon (⇄ Section 6.8).

The strategy of population-density-responsive induction of luminescence ensures that luminescence develops only when population densities are high enough to allow the light produced to be visible to animals. The bacterial light can then attract animals to feed on the luminous material, thereby bringing the bacteria into the animal's nutrient-rich gut for further growth. Alternatively, the luminous material may function as a light source in symbiotic light organ associations.

Quorum sensing is a form of regulation that has also been found in many different nonluminous bacteria, including several animal and plant pathogens. Quorum sensing in these bacteria controls activities such as the production of extracellular enzymes and expression of virulence factors for which a high population density is beneficial if the bacteria are to have a biological effect.

MINIQUIZ

- What substrates and enzyme are required for an organism such as *Aliivibrio* to emit visible light?
- What is quorum sensing and how does it control bioluminescence?

VI • Morphologically Diverse *Bacteria*

15.19 Spirochetes

KEY GENERA: *Spirochaeta, Treponema, Cristispira, Leptospira, Borrelia*

Spirochetes are morphologically unique bacteria found only within the bacterial phylum *Spirochaetes*. Spirochetes are gram-negative, motile, tightly coiled *Bacteria*, typically slender and flexuous in shape (**Figure 15.48**). Spirochetes are widespread in aquatic sediments and in animals. Some cause diseases, including syphilis, an important human sexually transmitted disease. Spirochetes are classified into eight genera (**Table 15.2**) primarily on the basis of habitat, pathogenicity, phylogeny, and morphological and physiological characteristics.

Spirochetes have an unusual mode of motility conveyed by their unusual morphology. Spirochetes contain *endoflagella*, which resemble normal flagella but are found in the cell periplasm (**Figure 15.49**). The endoflagella are anchored at the cell poles and extend back along the length of the cell. Both the endoflagella and the protoplasmic cylinder are surrounded by a flexible membrane called the *outer sheath* (Figure 15.49*b*). Endoflagella rotate, as do typical bacterial flagella. However, when both endoflagella rotate in the same direction, the protoplasmic cylinder rotates in the opposite direction, placing torsion on the cell (Figure 15.49*b*). This torsion causes the spirochete cell to flex, resulting in a corkscrew-like motion that allows cells to burrow through viscous materials or tissues.

Spirochetes are often confused with spirilla. **Spirilla** are helically curved rod-shaped cells, usually motile by means of polar flagella (**Figure 15.50**). The word *spirillum* refers to a general cell shape that is widespread among *Bacteria* and *Archaea*. The number of helical turns in a single spirillum may vary from less than one complete turn (in which case the organism looks like a vibrio) to many turns. In addition, spirilla that divide terminally, such as the cyanobacterium *Spirulina* (Figure 15.5), can form long helical filaments that superficially resemble spirochetes. Spirilla, however, lack the outer sheath, endoflagella, and corkscrew-like motility of spirochetes. In addition, spirilla are typically fairly rigid cells while spirochetes are highly flexible and quite thin (<0.5 μm).

Spirochaeta and *Cristispira*

The genus *Spirochaeta* includes free-living, anaerobic, and facultatively aerobic spirochetes. These organisms, of which several species are known, are common in aquatic environments such as freshwater and sediments, and also in the oceans. *Spirochaeta plicatilis* (Figure 15.48*b*) is a large spirochete found in sulfidic freshwater and marine habitats. The 20 or so endoflagella inserted at each pole of *S. plicatilis* are arranged in a bundle that winds around the coiled protoplasmic cylinder. Another species, *Spirochaeta stenostrepta* (Figure 15.48*a*), is an obligate anaerobe commonly found in H_2S-rich black muds. It ferments sugars to ethanol, acetate, lactate, CO_2, and H_2.

Cristispira (**Figure 15.51**) is a unique spirochete found in nature only in the crystalline style of certain molluscs, such as clams and oysters. The crystalline style is a flexible, semisolid rod seated in a sac and rotated against a hard surface of the digestive tract, thereby mixing and grinding the small particles of food taken in by the animal. *Cristispira* lives in both freshwater and marine molluscs, but not all species of molluscs possess this organism. Unfortunately, *Cristispira* has not been cultured, and so the

UNIT 4

TABLE 15.2 **Genera of spirochetes and their characteristics**

Genus	Dimensions (μm)	General characteristics	Number of endoflagella	Habitat	Diseases
Cristispira	30–150 × 0.5–3.0	3–10 complete coils; bundle of endoflagella visible by phase-contrast microscopy	~100	Digestive tract of molluscs; has not been cultured	None known
Spirochaeta	5–250 × 0.2–0.75	Anaerobic or facultatively aerobic; tightly or loosely coiled	2–40	Aquatic, free-living, freshwater and marine	None known
Treponema	5–15 × 0.1–0.4	Microaerophilic or anaerobic; helical or flattened coil amplitude up to 0.5 μm	2–32	Commensal or parasitic in humans, other animals	Syphilis, yaws, swine dysentery, pinta
Borrelia	8–30 × 0.2–0.5	Microaerophilic; 5–7 coils of approximately 1 μm amplitude	7–20	Humans and other mammals, arthropods	Relapsing fever, Lyme disease, ovine and bovine borreliosis
Leptospira	6–20 × 0.1	Aerobic, tightly coiled, with bent or hooked ends; requires long-chain fatty acids	2	Free-living or parasitic in humans, other mammals	Leptospirosis
Leptonema	6–20 × 0.1	Aerobic; does not require long-chain fatty acids	2	Free-living	None known
Brachyspira	7–10 × 0.35–0.45	Anaerobe	8–28	Intestine of warm-blooded animals	Causes diarrhea in chickens and swine
Brevinema	4–5 × 0.2–0.3	Microaerophile; forms deep branch in spirochete lineage as assessed by 16S rRNA sequence analysis	2	Blood and tissue of mice and shrews	Infectious for laboratory mice

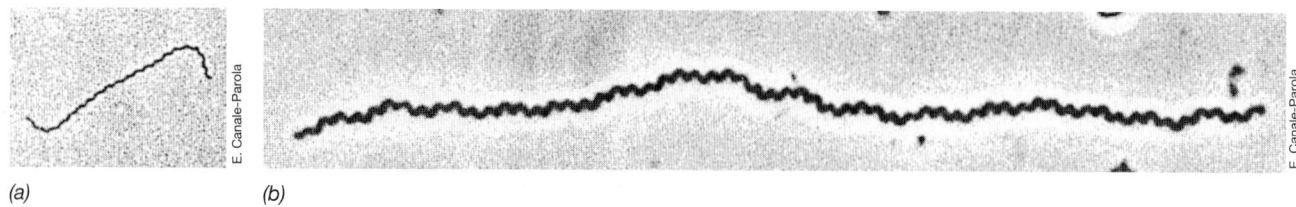

(a) *(b)*

Figure 15.48 Morphology of spirochetes. Two spirochetes at the same magnification, showing the wide size range in the group. *(a) Spirochaeta stenostrepta,* by phase-contrast microscopy. A single cell is 0.25 μm in diameter. *(b) Spirochaeta plicatilis.* A single cell is 0.75 μm in diameter and can be up to 250 μm (0.25 mm) in length.

physiological rationale for its restriction to this unique habitat is unknown.

Treponema and Borrelia

Anaerobic or microaerophilic host-associated spirochetes that are commensals or pathogens of humans and animals reside in the genus *Treponema*. *T. pallidum*, the causal agent of syphilis (⇔ Section 30.13), is the best-known species of *Treponema*. It differs in morphology from other spirochetes in that the *Treponema* cell is not helical but flat and wavy. The *T. pallidum* cell is remarkably thin, measuring only 0.2 μm in diameter. Because of this, dark-field microscopy has long been used to examine exudates from suspected syphilitic lesions (⇔ Figure 30.37).

Other species of *Treponema* are also often found as commensals in humans and other animals. For example, *Treponema denticola* is common in the human oral cavity and is associated with gum disease. It ferments amino acids such as cysteine and serine, forming acetate as the major fermentation acid, as well as CO$_2$, NH$_3$, and H$_2$S. Spirochetes are also common in the rumen, the digestive forestomach of ruminant animals (⇔ Section 23.13). For instance, *Treponema saccharophilum* (**Figure 15.52a**) is a large, pectinolytic spirochete found in the bovine rumen where it ferments pectin, starch, inulin, and other plant polysaccharides. *Treponema primitia* can be found in the hindgut of certain termites. In the termite gut, fermentation of cellulose causes production of H$_2$ and CO$_2$. *T. primitia* is

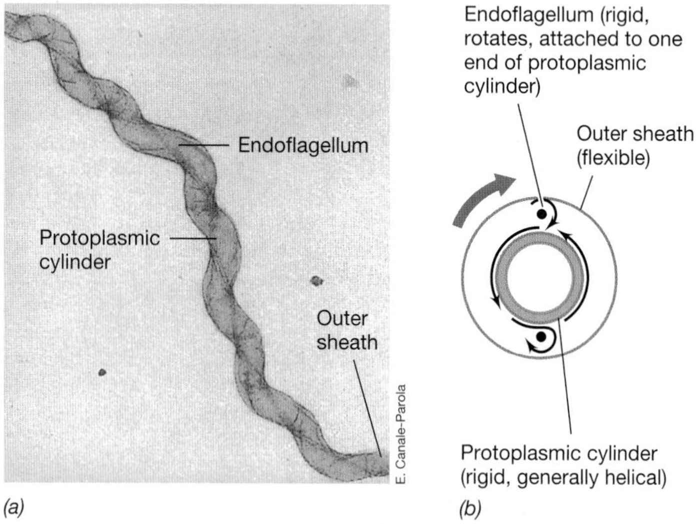

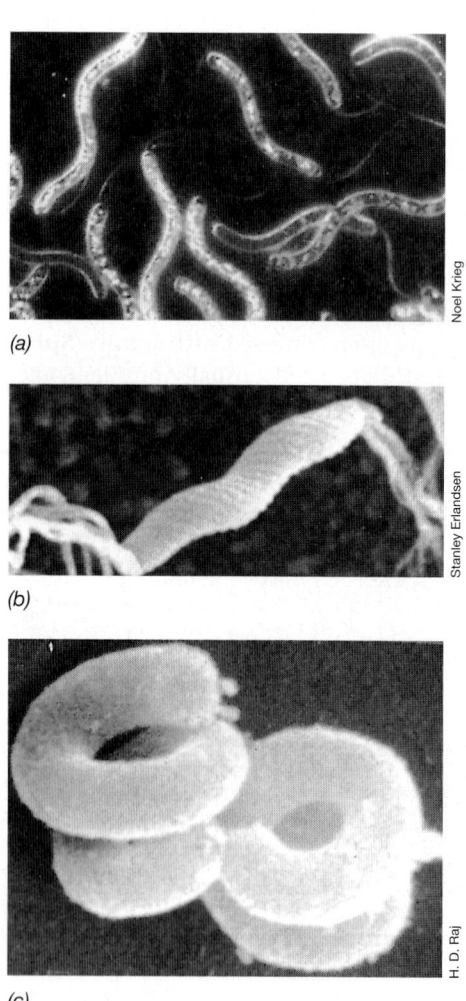

(a)

(b)

(c)

Figure 15.50 Spirilla. *(a) Spirillum volutans,* visualized by dark-field microscopy, showing flagellar bundles and volutin (polyphosphate) granules. Cells are about 1.5 × 25 μm. *(b)* Scanning electron micrograph of an intestinal spirillum. Note the polar flagellar tufts and the spiral structure of the cell surface. *(c)* Scanning electron micrograph of cells of *Ancylobacter aquaticus*. Cells are about 0.5 μm in diameter.

(a)

Endoflagellum —

Protoplasmic cylinder —

Outer sheath —

Endoflagellum (rigid, rotates, attached to one end of protoplasmic cylinder)

Outer sheath (flexible)

Protoplasmic cylinder (rigid, generally helical)

(b)

Figure 15.49 Motility in spirochetes. *(a)* Electron micrograph of a negatively stained cell of *Spirochaeta zuelzerae,* showing the position of the endoflagellum; the cell is about 0.3 μm in diameter. *(b)* Diagram of a spirochete cell, showing the arrangement of the protoplasmic cylinder, endoflagella, and external sheath, and how rotation of the endoflagellum generates rotation of both the protoplasmic cylinder and the external sheath.

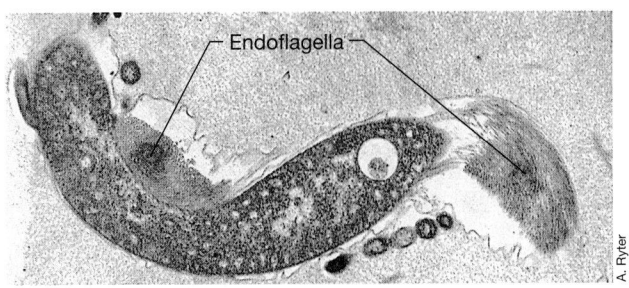

Figure 15.51 *Cristispira.* Electron micrograph of a thin section of a cell of *Cristispira.* This large spirochete is about 2 µm in diameter. Notice the numerous endoflagella.

an acetogen (⇨ Section 14.16) that grows on H_2 plus CO_2, forming acetate, which is an important component of the insect's nutrition. *Treponema azotonutricium* is also found in the termite hindgut and is capable of nitrogen fixation (⇨ Section 14.6).

The majority of species of *Borrelia* are animal or human pathogens. *Borrelia burgdorferi* (Figure 15.52*b*) is the causative agent of the tickborne *Lyme disease*, which infects humans and animals (⇨ Section 31.4). *B. burgdorferi* is also of interest because it is one of the few known bacteria that has a linear (as opposed to a circular) chromosome (⇨ Sections 4.2 and 9.3). Other species of *Borrelia* are primarily of veterinary importance, causing diseases in cattle, sheep, horses, and birds. In most cases, the bacterium is transmitted to the animal host from the bite of a tick.

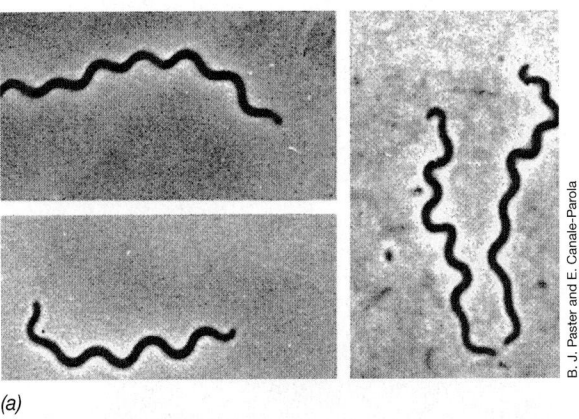

(a)

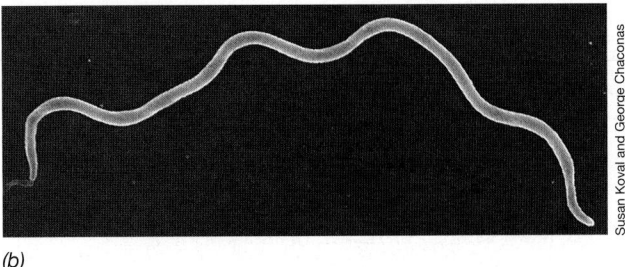

(b)

Figure 15.52 *Treponema* and *Borrelia.* (a) Phase-contrast micrographs of *Treponema saccharophilum*, a large pectinolytic spirochete from the bovine rumen. A cell measures about 0.4 µm in diameter. Left, regularly coiled cells; right, irregularly coiled cells. (b) Scanning electron micrograph of a cell of *Borrelia burgdorferi*, the causative agent of Lyme disease.

Leptospira and *Leptonema*

The genera *Leptospira* and *Leptonema* contain strictly aerobic spirochetes that oxidize long-chain fatty acids (for example, the C_{18} fatty acid oleic acid) as electron donors and carbon sources. With few exceptions, these are the only substrates utilized for growth. Leptospiras are thin, finely coiled, and usually bent at each end into a semicircular hook. At present, several species are recognized in this group, some free-living and many parasitic. Two major species of *Leptospira* are *L. interrogans* (parasitic) and *L. biflexa* (free-living). Strains of *L. interrogans* are parasitic for humans and animals. Rodents are the natural hosts of most leptospiras, although dogs and pigs are also important carriers of certain strains.

In humans the most common leptospiral syndrome is *leptospirosis*, a disorder in which the organism localizes in the kidneys and can cause renal failure or even death. Leptospiras ordinarily enter the body through the mucous membranes or through breaks in the skin during contact with an infected animal. After a transient multiplication in various parts of the body, the organism localizes in the kidneys and liver, causing nephritis and jaundice. Domestic animals such as dogs are vaccinated against leptospirosis with a killed virulent strain in the combined distemper–leptospira–hepatitis vaccine.

MINIQUIZ

- What are the major differences between spirochetes and spirilla?
- Name two diseases of humans caused by spirochetes.

15.20 Budding and Prosthecate/Stalked Microorganisms

KEY GENERA: *Hyphomicrobium, Caulobacter*

The growth of most bacteria is coupled to cell division by the well-known process of binary fission (⇨ Section 5.1 and Figure 5.1). In this section, we consider organisms that grow and divide in different ways, including budding and the formation of appendages. Budding and appendaged species often have life cycles that are distinct among bacteria.

Budding Division

As we learned in Section 5.1, budding bacteria divide as a result of unequal cell growth. Cell division in stalked and budding bacteria forms a totally new daughter cell, with the mother cell retaining its original identity (⇨ Figure 5.3). In contrast, binary fission produces two equivalent cells.

A fundamental difference between budding bacteria and bacteria that divide by binary fission is the formation of new cell wall material from a single point (polar growth) rather than throughout the whole cell (intercalary growth) as in binary fission (Chapter 7). Several genera not normally considered to be budding bacteria show polar growth without differentiation of cell size (⇨ Figure 5.3). An important consequence of polar growth is that internal structures, such as membrane complexes, are not partitioned in the cell division process and must be formed de novo. However, this has an advantage in that more complex internal structures can

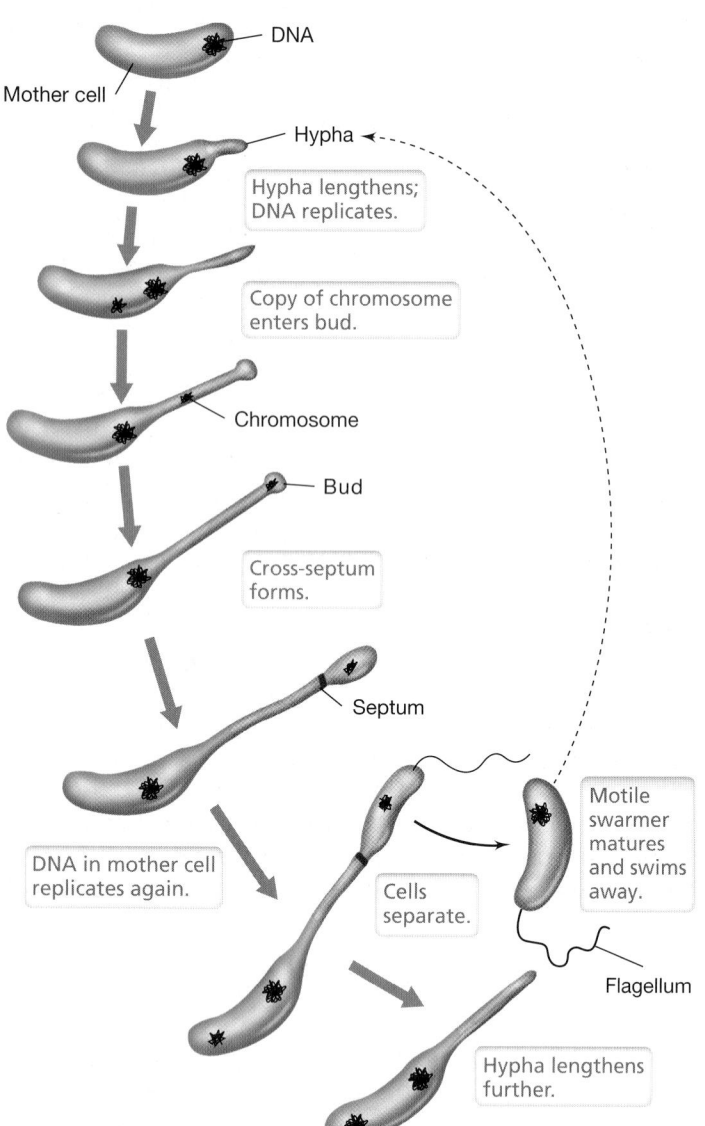

Figure 15.53 Stages in the *Hyphomicrobium* cell cycle. The single chromosome of *Hyphomicrobium* is circular.

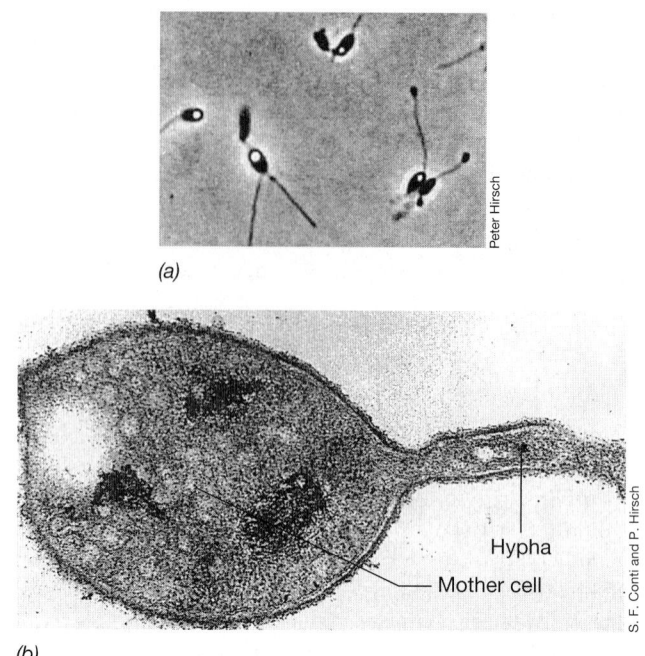

Figure 15.54 Morphology of *Hyphomicrobium*. *(a)* Phase-contrast micrograph of cells of *Hyphomicrobium*. Cells are about 0.7 μm wide. *(b)* Electron micrograph of a thin section of a single *Hyphomicrobium* cell. The hypha is about 0.2 μm wide.

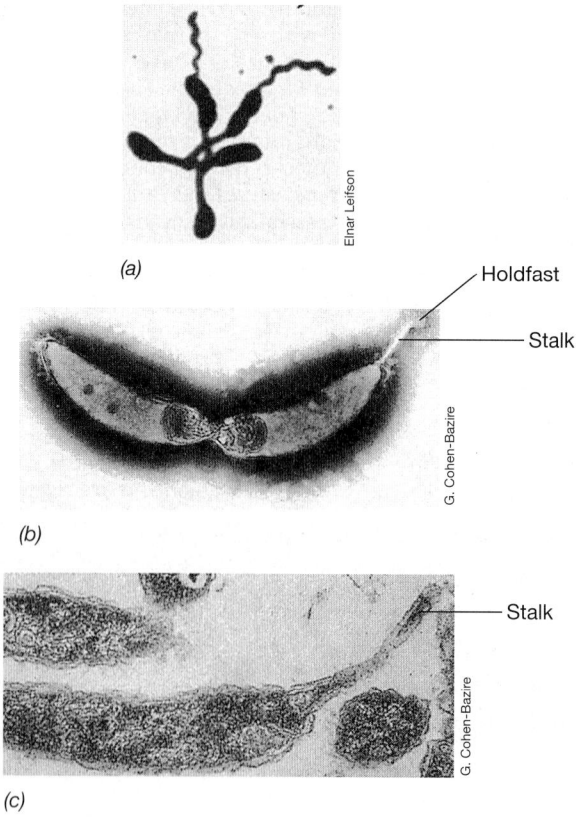

Figure 15.55 Stalked bacteria. *(a)* A *Caulobacter* rosette. A single cell is about 0.5 μm wide. The five cells are attached by their stalks, which are also prosthecae. Two of the cells have divided, and the daughter cells have formed flagella. *(b)* Negatively stained preparation of a *Caulobacter* cell in division. *(c)* A thin section of *Caulobacter* showing that cytoplasm is present in the stalk. Parts *b* and *c* are transmission electron micrographs.

be formed in budding cells than in cells that divide by binary fission, since the latter cells would have to partition these structures between the two daughter cells. Not coincidentally, many budding bacteria, particularly phototrophic and chemolithotrophic species, contain extensive internal membrane systems.

Budding Bacteria: *Hyphomicrobium*

Two well-studied budding bacteria are closely related *Alphaproteobacteria*: *Hyphomicrobium* (**Figure 15.53**), which is chemoorganotrophic, and *Rhodomicrobium*, which is phototrophic. These organisms release buds from the ends of long, thin hyphae. The hypha is a direct cellular extension and contains cell wall, cytoplasmic membrane, and ribosomes, and can contain DNA.

Figure 15.53 shows the life cycle of *Hyphomicrobium*. The mother cell, which is often attached by its base to a solid substrate, forms a thin outgrowth that lengthens to become a hypha. At the end of the hypha, a bud forms. This bud enlarges, forms a flagellum, breaks loose from the mother cell, and swims away. Later, the daughter cell loses its flagellum and after a period of maturation forms a hypha and buds. More buds can also form at the hyphal tip of the mother cell, leading to arrays of cells connected by hyphae. In some cases, a bud begins to form directly from the mother cell without the intervening formation of a hypha, whereas in other cases a single cell forms hyphae from each end (**Figure 15.54**). Nucleoid replication events occur before the bud emerges, and then once a bud has formed, a copy of the chromosome moves down the hypha and into the bud. A cross-septum then forms, separating the still-developing bud from the hypha and mother cell (Figure 15.54).

Physiologically, *Hyphomicrobium* is a methylotrophic bacterium (⇄ Sections 14.18 and 15.16), and it is widespread in freshwater, marine, and terrestrial habitats. Preferred carbon sources are methanol (CH_3OH), methylamine (CH_3NH_2), formaldehyde (CH_2O), and formate ($HCOO^-$). A fairly specific enrichment procedure for *Hyphomicrobium* is to use CH_3OH as an electron donor with nitrate (NO_3^-) as an electron acceptor in a dilute medium incubated under anoxic conditions. The only rapidly growing denitrifying bacterium known that uses CH_3OH as an electron donor is *Hyphomicrobium*, and so this procedure can select this organism out of a wide variety of environments.

Prosthecate and Stalked *Bacteria*

A variety of bacteria are able to produce cytoplasmic extrusions including *stalks* (**Figure 15.55**), *hyphae*, and *appendages* (**Table 15.3**). Extrusions of these kinds, which are smaller in diameter than the mature cell and contain cytoplasm and a cell wall, are collectively called **prosthecae** (**Figure 15.56**). Prosthecae allow organisms to attach to particulate matter, plant material, or other microorganisms in aquatic habitats. In addition, prosthecae can be used to increase the ratio of surface area to cell volume. Recall that the high surface-to-volume ratio of prokaryotic cells in general confers an increased ability to take up nutrients and expel wastes (⇄ Section 2.2). The unusual morphology of appendaged bacteria (Figure 15.56) carries this theme to an extreme, and may be an evolutionary adaptation to life in oligotrophic (nutrient-poor) waters where these organisms are most commonly found.

Prosthecae may also function to reduce cell sinking. Because these organisms are aquatic and their metabolism is typically aerobic, prosthecae may keep cells from sinking into anoxic zones in their aquatic environments where they would be unable to respire. Some prosthecate bacteria produce gas vesicles (⇄ Section 2.9) (Table 15.3), which would also help prevent sinking.

Caulobacter

Two common stalked bacteria are *Caulobacter* (Figure 15.55) and *Gallionella* (Figure 15.36). The former is a chemoorganotroph that produces a cytoplasm-filled stalk, that is, a prostheca, while the latter is a chemolithotrophic iron-oxidizing bacterium whose stalk is composed of ferric hydroxide [$Fe(OH)_3$] (Section 15.15). *Caulobacter* cells are often seen on surfaces in aquatic environments with the stalks of several cells attached to form *rosettes* (Figure 15.55a). At the end of the stalk

TABLE 15.3 Characteristics of major genera of stalked, appendaged (prosthecate), and budding *Bacteria*

Characteristics	Genus	Phylogenetic group[a]
Stalked bacteria		
Stalk an extension of the cytoplasm and involved in cell division	*Caulobacter*	Alpha
Stalked, fusiform-shaped cells	*Prosthecobacter*	Verrucomicrobiaceae[b]
Stalked, but stalk is an excretory product not containing cytoplasm:		
Stalk depositing iron, cell vibrioid	*Gallionella*	Beta
Laterally excreted gelatinous stalk not depositing iron	*Nevskia*	Gamma
Appendaged (prosthecate) bacteria		
Single or double prosthecae	*Asticcacaulis*	Alpha
Multiple prosthecae:		
Short prosthecae, multiply by fission, some with gas vesicles	*Prosthecomicrobium*	Alpha
Flat, star-shaped cells, some with gas vesicles	*Stella*	Alpha
Long prosthecae, multiply by budding, some with gas vesicles	*Ancalomicrobium*	Alpha
Budding bacteria		
Phototrophic, produce hyphae	*Rhodomicrobium*	Alpha
Phototrophic, budding without hyphae	*Rhodopseudomonas*	Alpha
Chemoorganotrophic, rod-shaped cells	*Blastobacter*	Alpha
Chemoorganotrophic, buds on tips of slender hyphae:		
Single hypha from parent cell	*Hyphomicrobium*	Alpha
Multiple hyphae from parent cell	*Pedomicrobium*	Alpha

[a]All but *Prosthecobacter* are *Proteobacteria*.
[b]See Section 16.17.

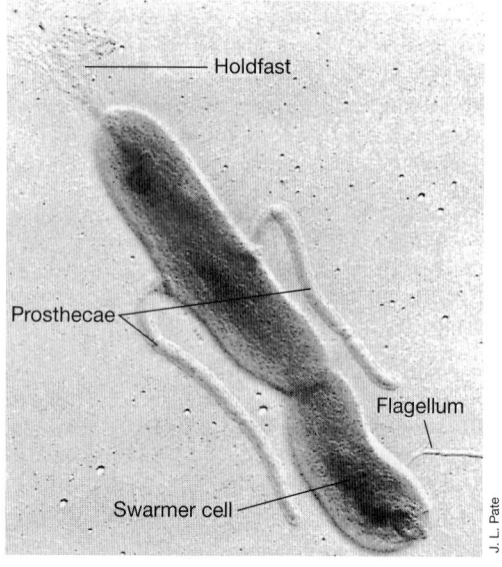

(a)

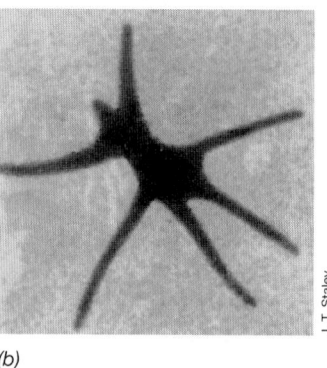

(b)

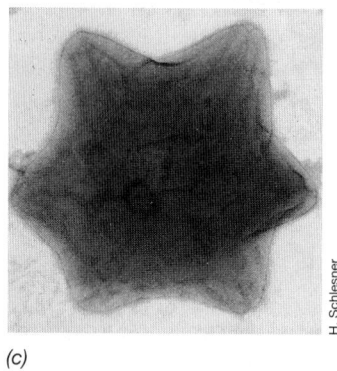

(c)

Figure 15.56 Prosthecate bacteria. *(a)* Electron micrograph of a shadow-cast preparation of *Asticcacaulis biprosthecum*, illustrating the location and arrangement of the prosthecae, the holdfast, and a swarmer cell. The swarmer cell breaks away from the mother cell and begins a new cell cycle. Cells are about 0.6 μm wide. *(b)* Negatively stained electron micrograph of a cell of *Ancalomicrobium adetum*. The prosthecae are bounded by the cell wall, contain cytoplasm, and are about 0.2 μm in diameter. *(c)* Electron micrograph of the star-shaped prosthecate bacterium *Stella*. Cells are about 0.8 μm in diameter.

is a structure called a *holdfast* by which the stalk anchors the cell to a surface.

The *Caulobacter* cell division cycle (**Figure 15.57**; ⮌ Section 7.7 and Figure 7.16) is unique because cells undergo unequal binary fission. A stalked cell of *Caulobacter* divides by elongation of the cell followed by binary fission, and a single flagellum forms at the pole opposite the stalk. The flagellated cell so formed, called a *swarmer*, separates from the nonflagellated mother cell and eventually attaches to a new surface, forming a new stalk at the flagellated pole; the flagellum is then lost. Stalk formation is a necessary precursor of cell division and is coordinated with DNA synthesis (Figure 15.57). The cell division cycle in *Caulobacter* is thus more complex than simple binary fission or budding division because the stalked and swarmer cells are structurally different and the growth cycle must include both forms.

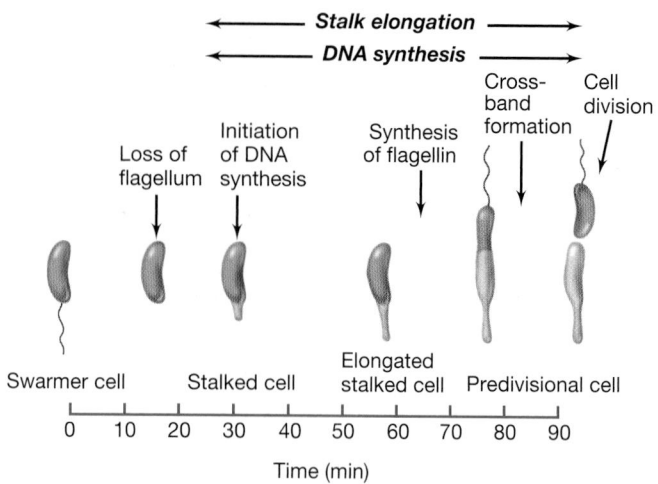

Figure 15.57 Growth of *Caulobacter*. Stages in the *Caulobacter* cell cycle, beginning with a swarmer cell. Compare with Figure 7.16.

— MINIQUIZ —

• How does budding division differ from binary fission? How does binary fission differ from the division process in *Caulobacter*?

• What advantage might a prosthecate organism have in a very nutrient-poor environment?

15.21 Sheathed Microorganisms

KEY GENERA: *Sphaerotilus, Leptothrix*

Bacteria in many phyla form sheaths made of polysaccharide or protein that encase one or many cells. Sheaths often function to bind cells together into long multicellular filaments (Sections 15.3 and 15.11). *Sphaerotilus* and *Leptothrix* are filamentous bacteria that grow within a sheath and have a unique life cycle. Under favorable conditions, the cells grow vegetatively, leading to the formation of long, cell-packed sheaths. Flagellated swarmer cells form within the sheath under unfavorable growth conditions, and the swarmer cells break out and are dispersed to new environments, leaving behind the empty sheath.

Sphaerotilus and *Leptothrix* are common in freshwater habitats that are rich in organic matter, such as wastewaters and polluted streams. Because they are typically found in flowing waters, they are also abundant in trickling filters and activated sludge digesters in sewage treatment plants (⮌ Section 22.6). In habitats in which reduced iron (Fe^{2+}) or manganese (Mn^{2+}) is present, the sheaths may become coated with ferric hydroxide [$Fe(OH)_3$] or manganese oxides from the oxidation of these metals.

Leptothrix

The ability of *Sphaerotilus* and *Leptothrix* to precipitate iron oxides on their sheaths is well established, and when sheaths become iron encrusted, as occurs in iron-rich waters, they can frequently be seen microscopically (**Figure 15.58**). Iron precipitates form when ferrous iron (Fe^{2+}), chelated to organic materials such as humic or tannic acids, is oxidized. These chemoorganotrophic bacteria use the organic materials as a carbon or energy source and, when no longer chelated, the ferrous iron becomes oxidized and precipitates

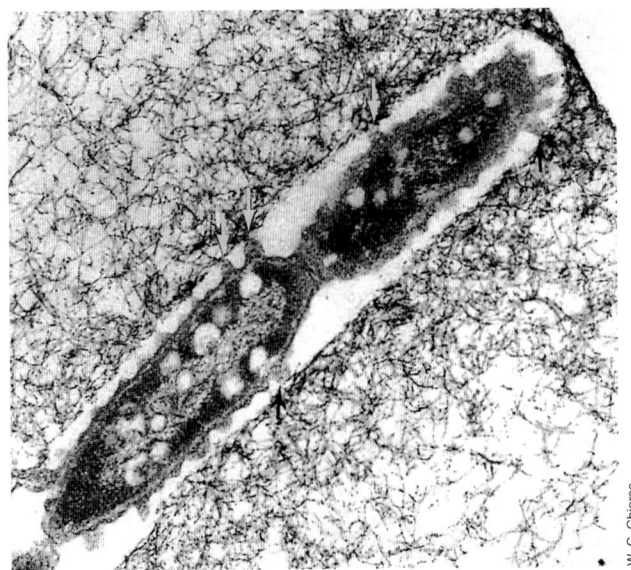

Figure 15.58 *Leptothrix* **and iron precipitation.** Transmission electron micrograph of a thin section of *Leptothrix* growing in a ferromanganese film in a swamp in Ithaca, New York. A single cell measures about 0.9 μm in diameter. Note the protuberances of the cell envelope that contact the sheath (arrows).

on the sheath. Iron oxidation is fortuitous and though these organisms are closely related to dissimilative iron-oxidizers (Section 15.15), the organism does not gain energy from iron oxidation. In a similar way, *Leptothrix* can also oxidize manganese.

Sphaerotilus

The *Sphaerotilus* filament is composed of a chain of rod-shaped cells enclosed in a closely fitting sheath. This thin, transparent structure is difficult to see when it is filled with cells, but when it is partially empty, the sheath can more easily be resolved (**Figure 15.59a**). Individual cells are 1–2 × 3–8 μm in dimensions and stain gram-negatively. The cells within the sheath (Figure 15.59b) divide by binary fission, and the new cells synthesize new sheath material at the tips of the filaments. Eventually, motile swarmer cells are liberated from the sheaths (Figure 15.59c) and then migrate, attach to a solid surface, and begin to grow, with each swarmer being the forerunner of a new filament. The sheath, which is devoid of peptidoglycan, consists of protein and polysaccharide.

Sphaerotilus species are nutritionally versatile and use simple organic compounds as carbon and energy sources; one species can grow mixotrophically with thiosulfate as electron donor. Befitting its habitat in flowing waters, *Sphaerotilus* is an obligate aerobe. Large masses (blooms) of *Sphaerotilus* often occur in the fall of the year in streams and brooks when leaf litter causes a temporary increase in the organic content of the water. In addition, its filaments are the main component of a microbial complex that wastewater engineers call "sewage fungus," a filamentous slime found on the rocks in streams receiving sewage pollution. In activated sludge of sewage treatment plants (⇌ Section 22.6), *Sphaerotilus* is often responsible for a condition called *bulking*, where the tangled masses of *Sphaerotilus* filaments so increase the bulk of the sludge that it remains suspended and does not settle as it should. This has a negative effect on the oxidation of organic matter and the

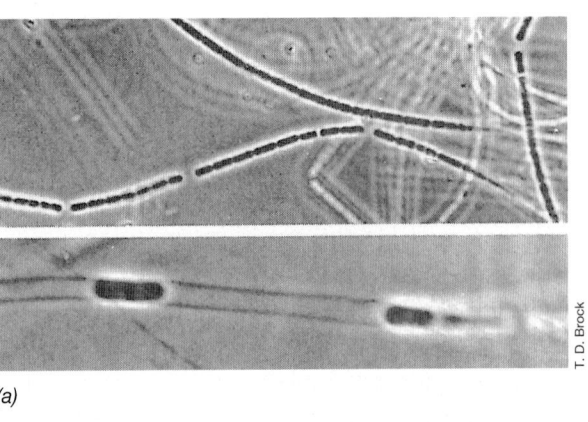

(a)

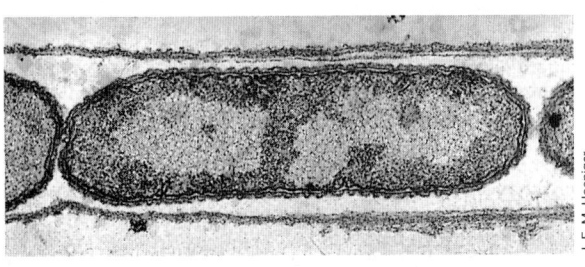

(b)

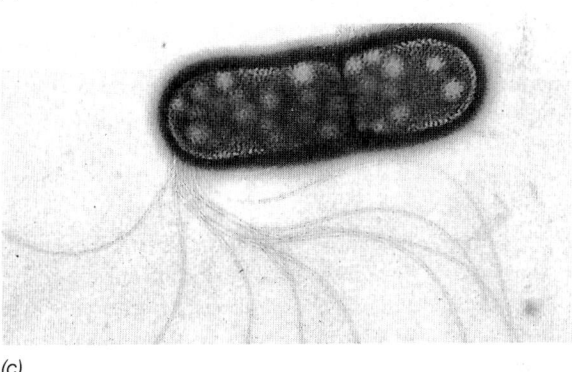

(c)

Figure 15.59 *Sphaerotilus natans.* A single cell is about 2 μm wide. *(a)* Phase-contrast photomicrographs of material collected from a polluted stream. Active growth stage (above) and swarmer cells leaving the sheath. *(b)* Electron micrograph of a thin section through a filament, clearly showing the sheath. *(c)* Electron micrograph of a negatively stained swarmer cell. Notice the polar flagellar tuft.

recycling of inorganic nutrients and leads to treatment plant discharges with high nitrogen and carbon loads.

— **MINIQUIZ** —

- Describe how a sheathed bacterium such as *Sphaerotilus* grows.
- List two metals that are oxidized by sheathed bacteria.

15.22 Magnetic Microbes

KEY GENERA: *Magnetospirillum*

In a magnetic field, magnetic bacteria demonstrate a dramatic directed movement called *magnetotaxis*. Within these cells are structures called *magnetosomes*, which consist of chains of magnetic particles made of magnetite (Fe_3O_4) or greigite (Fe_3S_4)

(⇨ Section 2.8 and Figure 2.24). Magnetosomes are localized within invaginations of the cell membrane that are organized in a linear conformation by a protein scaffold. Magnetic bacteria orient along the north–south magnetic moment of a magnetic field, aligning parallel to the field lines in much the same manner as a compass needle. Magnetic bacteria are typically microaerophilic or anaerobic and are most often found near the oxic–anoxic interface in sediments or stratified lakes. The magnetosomes of aerobic species typically contain the mineral magnetite while those of anaerobes contain exclusively greigite.

Although the ecological role of bacterial magnets is unclear, the ability to orient in a magnetic field may be of selective advantage in maintaining these organisms in zones of low O_2 concentration. Generally, the concentration of O_2 decreases with depth through sediments or the water column of stratified lakes. Since Earth is spherical, its magnetic field lines have a strong vertical component in the Northern and Southern Hemispheres. Thus, bacteria that orient along these field lines can preferentially swim down and away from O_2. The magnetosome functions like a compass needle to "point" the bacterium in the right direction; rotation of the flagellum, by contrast, is controlled by a chemotactic response to O_2 (⇨ Section 2.13).

Magnetic bacteria display one of two magnetic polarities depending on the orientation of magnetosomes within the cell. Cells in the Northern Hemisphere have the north-seeking pole of their magnetosomes forward with respect to their flagella and thus move in a northward direction (which in the Northern Hemisphere is downward). Cells in the Southern Hemisphere have the opposite polarity and move southward.

Most of the magnetic bacteria that have been described are species of *Alphaproteobacteria*, but species have also been observed in the *Gammaproteobacteria*, the *Deltaproteobacteria*, and the *Nitrospira* group. One of the best-characterized species is *Magnetospirillum magnetotacticum* (**Figure 15.60**), which is a chemoorganotrophic microaerophile that can also grow anaerobically by reducing NO_3^- or N_2O. In contrast, the species *Desulfovibrio magneticus* is a sulfate reducer and an obligate anaerobe. In addition, magnetosomes have been observed in a few species of sulfur oxidizers and purple nonsulfur bacteria. Multicellular magnetotactic bacteria are also known. These are *Deltaproteobacteria* that form multicellular aggregates of 10–20 cells organized as a hollow sphere. While

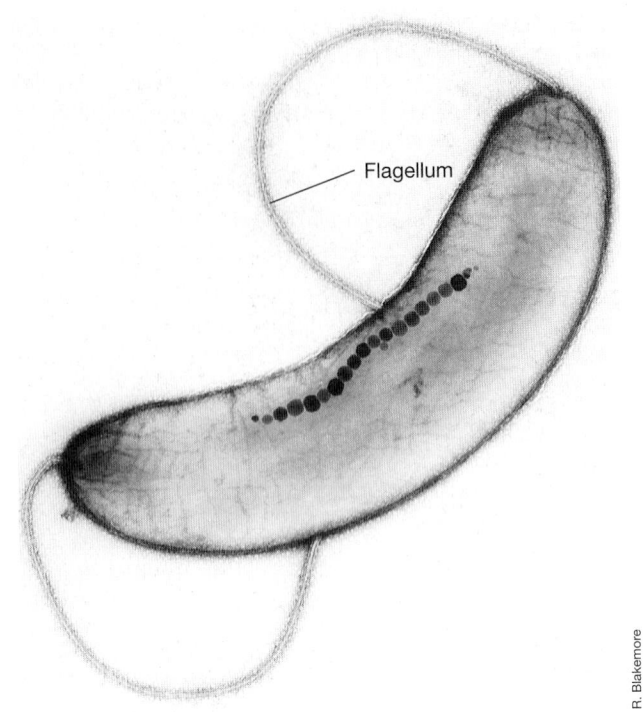

Flagellum

R. Blakemore

Figure 15.60 A magnetotactic spirillum. Electron micrograph of a single cell of *Magnetospirillum magnetotacticum*; a cell measures 0.3×2 µm. The cell contains particles of magnetosomes made of Fe_3O_4 arranged in a chain.

multicellular magnetotactic bacteria are obligate anaerobes, the basis of their metabolism has not yet been determined.

We transition now from viewing *Bacteria* from a functional diversity perspective to consider some other important phyla in a phylogenetic perspective in Chapter 16. We then conclude our coverage of prokaryotic microbes with Chapter 17 dedicated to the *Archaea*.

MINIQUIZ

- What benefit do magnetic bacteria accrue from having magnetosomes?
- Would you expect to find greigite or magnetite in the magnetosomes of *Desulfovibrio magneticus*?

MasteringMicrobiology®   **Visualize, explore,** and **think critically** with Interactive Microbiology, MicroLab Tutors, MicroCareer case studies, and more. MasteringMicrobiology offers practice quizzes, helpful animations, and other study tools for lecture and lab to help you master microbiology.

Chapter Review

I • Functional Diversity as a Concept

15.1 Phylogenetic diversity is that component of microbial diversity that deals with evolutionary relationships between microorganisms. By contrast, functional diversity deals with diversity in form and function as it relates to microbial physiology and ecology. Incongruence between phylogeny and the functional traits of microorganisms can result from patterns of gene loss, horizontal gene transfer, and/or convergent evolution.

Q **What is convergent evolution and how is it different from horizontal gene transfer?**

II • Diversity of Phototrophic *Bacteria*

15.2 Anoxygenic phototrophs, which do not produce oxygen, were the first phototrophic organisms to evolve. The evolution of photosynthesis has been affected strongly by patterns of horizontal gene transfer.

> **Q** Which bacterial phyla contain phototrophs?

15.3 *Cyanobacteria* is the only bacterial phylum that contains oxygenic phototrophs. All species of cyanobacteria can fix CO_2 and many can fix N_2, making these organisms important primary producers in many ecosystems.

> **Q** How are prochlorophytes, such as *Prochlorococcus*, different from other cyanobacteria?

15.4 Purple sulfur bacteria are anoxygenic phototrophic *Gammaproteobacteria*. Purple sulfur bacteria use H_2S and S^0 as electron donors and fix CO_2 by the Calvin cycle. These phototrophs have bacteriochlorophylls *a* or *b* and use a Q-type reaction center.

> **Q** Compare and contrast the metabolism, morphology, and phylogeny of purple sulfur and purple nonsulfur bacteria.

15.5 Purple nonsulfur bacteria are anoxygenic phototrophic *Alpha-* and *Betaproteobacteria*. Purple nonsulfur bacteria are metabolically diverse, growing best as photoheterotrophs, and can also grow in darkness. These phototrophs have bacteriochlorophylls *a* or *b* and use type II photosystems with Q-type reaction centers. Aerobic anoxygenic phototrophs have type II photosystems but only possess bacteriochlorophyll *a*.

> **Q** Compare and contrast the metabolism of purple nonsulfur bacteria and aerobic anoxygenic phototrophs.

15.6 Green sulfur bacteria are anoxygenic phototrophs of the phylum *Chlorobi*. Green sulfur bacteria use H_2S or S^0 as electron donors and fix CO_2 by the reverse citric acid cycle. These phototrophs contain bacteriochlorophylls *c*, *d*, or *e* localized in chlorosomes and bacteriochlorophyll *a* localized in their FeS-type photosynthetic reaction centers.

> **Q** Compare and contrast the metabolism, morphology, and phylogeny of green sulfur and purple sulfur bacteria.

15.7 Green nonsulfur bacteria, also known as filamentous anoxygenic phototrophs, are anoxygenic phototrophs of the phylum *Chloroflexi* and grow best as photoheterotrophs. These phototrophs contain bacteriochlorophyll *c* in chlorosomes and bacteriochlorophyll *a* in their Q-type photosynthetic reaction centers.

> **Q** What traits do green nonsulfur bacteria share with green sulfur bacteria and purple sulfur bacteria?

15.8 Heliobacteria are anoxygenic phototrophic *Firmicutes* that grow as photoheterotrophs or in darkness as chemotrophs. Heliobacteria produce bacteriochlorophyll *g* and have FeS-type reaction centers. *Chloracidobacterium thermophilum* is an anoxygenic phototrophic acidobacterium that grows photoheterotrophically, possesses bacteriochlorophyll *a* and *c* as well as chlorosomes, and has an FeS-type reaction center.

> **Q** In what ways is *Chloracidobacterium thermophilum* similar to green sulfur bacteria, and in what ways is it different?

III • Microbial Diversity in the Sulfur Cycle

15.9 Dissimilative sulfate-reducers are obligate anaerobes that grow by reducing SO_4^{2-} with H_2 or simple organic compounds as electron donors. Most sulfate reducers are *Deltaproteobacteria*.

> **Q** With respect to sulfate-reducing bacteria, what is the difference between complete and incomplete oxidizers?

15.10 Dissimilative sulfur-reducers are metabolically and phylogenetically diverse organisms that grow by reducing S^0 and other oxidized sulfur compounds (other than SO_4^{2-}) as electron acceptors.

> **Q** In what ways are sulfur-reducing bacteria different from sulfate-reducing bacteria, and in what ways are they similar?

15.11 Sulfur chemolithotrophs, most of which are species of *Proteobacteria*, oxidize H_2S and other reduced sulfur compounds as electron donors with O_2 or NO_3^- as electron acceptors and use either CO_2 or organic compounds as carbon sources. Sulfur chemolithotrophs use a variety of ecological strategies to conserve energy from H_2S and O_2, substances that otherwise react together spontaneously.

> **Q** What are some ecological strategies that aerobic sulfide-oxidizers use to compete with the chemical oxidation of H_2S by atmospheric O_2?

IV • Microbial Diversity in the Nitrogen Cycle

15.12 Diazotrophs are bacteria that assimilate N_2 through activity of the enzyme nitrogenase. Diazotrophs are metabolically and phylogenetically diverse and employ various adaptations to protect nitrogenase from oxygen inactivation.

> **Q** What are some ways that diazotrophs protect nitrogenase from O_2?

15.13 Nitrifying bacteria are aerobic chemolithotrophs that oxidize NH_3 to NO_2^- (genus prefix *Nitroso-*) or NO_2^- to NO_3^- (genus prefix *Nitro-*). Ammonia oxidizers are *Proteobacteria* or *Thaumarchaeota*, while nitrite oxidizers are *Proteobacteria* or *Nitrospirae*. Denitrifiers are metabolically and phylogenetically diverse facultative aerobes and chemoorganotrophs that reduce NO_3^- to the gaseous products NO, N_2O, and N_2.

> **Q** Compare and contrast the nitrogen metabolism of nitrifiers with that of denitrifiers.

V • Other Distinctive Functional Groupings of Microorganisms

15.14 Dissimilative iron-reducers reduce insoluble electron acceptors in anaerobic respirations. Most species can grow anaerobically by reducing ferric iron using H_2 or simple organic compounds as electron donor. The best-characterized genera include *Geobacter*, which contains exclusively obligate anaerobes, and *Shewanella*, which contains facultative aerobes.

> **Q** In what ways are the dissimilative iron-reducing bacteria *Shewanella* and *Geobacter* similar, and in what ways are they different?

15.15 Dissimilative iron-oxidizers conserve energy from the aerobic oxidation of ferrous iron. These microbes use several ecological strategies to cope with the chemical instability of ferrous iron in oxic habitats at neutral pH. Iron oxidizers are found in four physiological groups: aerobic acidophiles, aerobic neutrophiles, anaerobic chemotrophs, and anaerobic phototrophs.

> **Q** Which group of dissimilative iron-oxidizers is the least diverse and in what way is this related to oxygen and pH?

15.16 Methylotrophs grow on organic compounds that lack carbon–carbon bonds. Some methylotrophs are also methanotrophs, organisms able to catabolize methane. Most methanotrophs are *Proteobacteria* that contain extensive internal membranes and incorporate carbon by either the serine or ribulose monophosphate pathways.

> **Q** What are the differences between type I and type II methanotrophs?

15.17 Bacterial predators such as *Bdellovibrio* and *Myxococcus* consume other microorganisms. Myxobacteria have a complex developmental cycle that involves the formation of fruiting bodies that contain myxospores.

> **Q** Compare and contrast the life cycle of *Myxococcus* with that of *Bdellovibrio*.

15.18 *Vibrio*, *Aliivibrio*, and *Photobacterium* species are marine bacteria, some of which are pathogenic and bioluminescent.

Bioluminescence, catalyzed by the enzyme luciferase, is controlled by a quorum-sensing mechanism that ensures that light is not emitted until a large cell population has been attained.

> **Q** Describe the manner in which cell density regulates light production in luminescent bacteria.

VI • Morphologically Diverse *Bacteria*

15.19 The phylum *Spirochaetes* contains helically shaped bacteria that show a novel form of motility that allows them to "corkscrew" through viscous materials. These organisms are common in anoxic habitats and are the cause of many well-known human diseases, such as syphilis.

> **Q** Contrast the motility of spirochetes with that of spirilla.

15.20 Prosthecate bacteria, such as *Hyphomicrobium*, *Caulobacter*, and *Gallionella*, are appendaged cells that form stalks or prosthecae used for attachment or nutrient absorption, and are primarily aquatic. Some prosthecate bacteria, such as *Hyphomicrobium*, have a complex life cycle in which new cells form by budding from hyphae.

> **Q** Contrast the life cycle of *Hyphomicrobium* with that of *Caulobacter*.

15.21 Sheathed bacteria are filamentous *Proteobacteria* in which individual cells form chains within an outer layer called the sheath. *Sphaerotilus* and *Leptothrix* are major genera of sheathed bacteria and can oxidize metals, such as Fe^{2+} and Mn^{2+}.

> **Q** In what environment might you expect to find *Leptothrix*?

15.22 Magnetosomes are specialized magnetic structures present in magnetotactic bacteria. Magnetosomes orient cells along the magnetic field lines of Earth, and this allows cells to use their normal chemotactic response to move vertically in a directed fashion in sediments or stratified aquatic systems.

> **Q** In what way does a magnetosome contribute to the fitness of microaerophilic bacteria in sediments?

Application Questions

1. Describe a key physiological feature of the following *Bacteria* that would differentiate each from the others: *Acetobacter*, *Methylococcus*, *Azotobacter*, *Photobacterium*, *Desulfovibrio*, and *Spirillum*.

2. Describe the metabolism for each of the following *Bacteria* and state whether the organism is an aerobe or an anaerobe: *Thiobacillus*, *Nitrosomonas*, *Methylomonas*, *Pseudomonas*, *Acetobacter*, and *Gallionella*.

3. Using an example from each of the morphologically diverse groups of *Bacteria* (Sections 15.19–15.22), describe how you could distinguish them from each other using only microscopy. How do the habitats of your example organisms differ from each other? Could you find any of these organisms in or on the human body? Despite their ability to oxidize inorganic electron donors, why are *Sphaerotilus* and *Leptothrix* not considered chemolithotrophs?

Chapter Glossary

Aerobic anoxygenic phototroph an organism that is an aerobic heterotroph that uses anoxygenic photosynthesis as a supplemental source of energy

Autoinduction a gene regulatory mechanism involving small, diffusible signal molecules that are produced in larger amounts as population size increases

Bioluminescence the enzymatic production of visible light by living organisms

Carboxysome a polyhedral cellular inclusion of crystalline ribulose bisphosphate carboxylase (RubisCO), the key enzyme of the Calvin cycle

Chemolithotroph an organism able to oxidize inorganic compounds (such as H_2, Fe^{2+}, S^0, or NH_4^+) as energy sources (electron donors)

Chlorosome a cigar-shaped structure bounded by a nonunit membrane and containing the light-harvesting bacteriochlorophyll (*c, d,* or *e*) in green sulfur bacteria and *Chloroflexus*

Consortium a two- or more-membered association of bacteria, usually living in an intimate symbiotic fashion

Convergent evolution a circumstance where a trait or set of traits that are similar in form and/or function between two organisms are not inherited from a shared ancestor (that is, traits that are similar but not homologous)

Cyanobacteria prokaryotic oxygenic phototrophs containing chlorophyll *a* and phycobilins

Denitrifier an organism that carries out anaerobic respiration with NO_3^- or NO_2^-, reducing it to the gaseous products NO, N_2O, and N_2

Diazotroph an organism that can assimilate N_2 into biomass by activity of the enzyme nitrogenase

Dissimilative sulfate-reducer an anaerobic microorganism that conserves energy through the reduction of SO_4^{2-}

Dissimilative sulfur-oxidizer a microorganism that gains energy for growth through oxidation of reduced sulfur compounds

Dissimilative sulfur-reducer an anaerobic or facultatively aerobic microorganism that conserves energy through the reduction of S^0 but cannot reduce SO_4^{2-}

Functional diversity the component of biological diversity that deals with the forms and functions of organisms as they relate to differences in physiology and ecology

Green nonsulfur bacteria anoxygenic phototrophs containing chlorosomes, Q-type photosynthetic reaction centers, bacteriochlorophylls *a* and *c* as light-harvesting chlorophylls, and typically growing best as photoheterotrophs

Green sulfur bacteria anoxygenic phototrophs containing chlorosomes, FeS-type photosynthetic reaction centers, bacteriochlorophylls *c, d,* or *e* as antenna bacteriochlorophylls, and typically growing with H_2S as an electron donor

Heliobacteria anoxygenic phototrophs containing bacteriochlorophyll *g* and FeS-type reaction centers

Horizontal gene transfer a unidirectional transfer of genes between unrelated organisms; can cause homologous genes to be dispersed in a phylogeny

Methanotroph an organism capable of oxidizing methane (CH_4) as an electron donor in energy metabolism

Methylotroph an organism capable of oxidizing organic compounds that do not contain carbon–carbon bonds; if able to oxidize CH_4, also a methanotroph

Mixotroph an organism that conserves energy from the oxidation of inorganic compounds but requires organic compounds as a carbon source

Nitrifier a chemolithotroph capable of carrying out the oxidation of NH_3 or NO_2^-

Phycobilin a protein containing the pigment phycocyanin or phycoerythrin that functions as a photosynthetic accessory pigment in cyanobacteria

Prochlorophyte a bacterial oxygenic phototroph that contains chlorophylls *a* and *b* but lacks phycobilins

Prosthecae extrusions of cytoplasm, often forming distinct appendages, bounded by the cell wall

Purple nonsulfur bacteria a group of phototrophic bacteria that contain bacteriochlorophyll *a* or *b* and Q-type reaction centers, and that grow best as photoheterotrophs

Purple sulfur bacteria a group of phototrophic bacteria that contain bacteriochlorophylls *a* or *b*, Q-type reaction center, and can oxidize H_2S as photosynthetic electron donor

Spirilla (singular, spirillum) spiral-shaped cells

Spirochete a slender, tightly coiled, gram-negative bacterium of the phylum *Spirochaetes* characterized by possession of endoflagella used for motility

16

Diversity of *Bacteria*

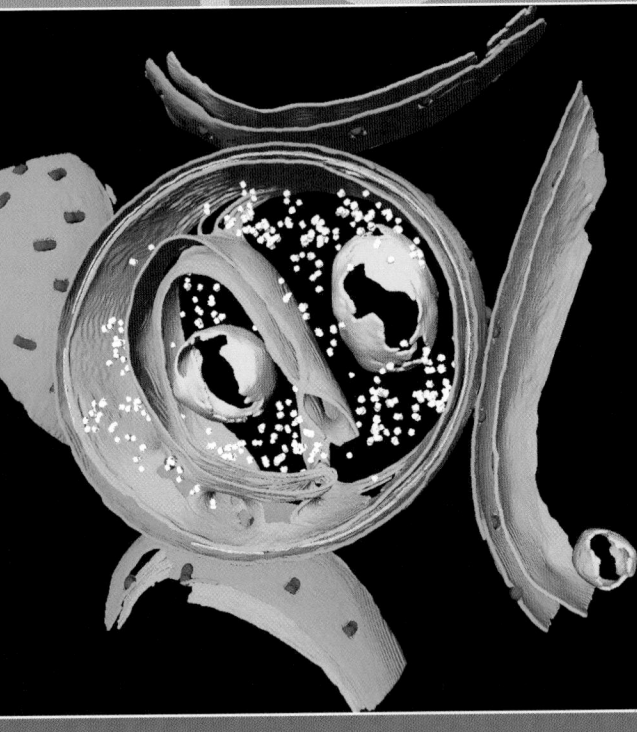

microbiology**now**

The Mystery of the Missing Peptidoglycan

Planctomycetes have many enigmatic features. Unlike most *Bacteria* they lack the protein FtsZ, which is required for binary fission; they divide by budding much like yeast; and they appear capable of endocytosis, a feature uniquely associated with eukaryotic cells. What is more, *Planctomycetes* were long thought to lack peptidoglycan, and transmission electron micrographs (TEM) suggested the presence of a membrane-enclosed nucleus in some species. Even more incredible, some *Planctomycetes* actually have membrane-enclosed organelles (anammoxosomes). This unique mix of prokaryotic and eukaryotic features has for years led many microbiologists to argue that *Planctomycetes* are a missing link in cellular evolution. However, the mysteries of the missing peptidoglycan and the nature of the "nucleus"—like many other microbial mysteries—were ultimately solved by advances in microbiological techniques.

Cryo-electron tomography (CET) is revolutionizing our understanding of cellular structures. CET generates detailed three-dimensional images with nanometer-scale resolution. For example, CET scans of *Planctomycetes* revealed that what appeared to be a nuclear membrane in two-dimensional transmission electron micrographs was actually a system of deep invaginations of the cytoplasmic membrane. Two-dimensional slices through this three-dimensional system of membranes gave the false appearance of a nuclear membrane. Furthermore, these images, such as the CET scan of *Planctopirus limnophila* shown here, clearly reveal that cells of *Planctomycetes* have a gram-negative cell envelope composed of an outer membrane (green), a peptidoglycan layer (orange), and the cytoplasmic membrane (blue).

The *Planctopirus* cytoplasmic membrane is heavily invaginated, and when viewed in three dimensions, these invaginations can be seen to wrap around both the DNA containing nucleoids (yellow) and ribosomes (white) present in the cytoplasm. Genomic analyses of *Planctomycetes* revealed that key genes required for the biosynthesis of peptidoglycan were present in the genome, and biochemical experiments confirmed that the major constituents of peptidoglycan (*N*-acetylglucosamine, *N*-acetylmuramic acid, and 2,6-diaminopimelic acid) were present.

For every mystery solved in science, another is often revealed. While we now know that cells of *Planctomycetes* contain both peptidoglycan and a gram-negative cell envelope, an understanding of the remarkable membrane invaginations that appear in CET scans of cells of *Planctomycetes* and the unusual crateriform structures (magenta) that stud the surface of their outer membranes remains elusive and beyond our current understanding of prokaryotic cell biology.

 Source: Jeske, O., et al. 2015. *Planctomycetes* do possess a peptidoglycan cell wall. *Nature Communications 6:* 7116.

In the last chapter we examined the *functional* diversity of microorganisms. In this and the next two chapters we shift our focus to the *phylogenetic* diversity of microorganisms. We discussed the difference between functional diversity and phylogenetic diversity in Section 15.1. In this chapter we examine the major lineages of *Bacteria* (**Figure 16.1a**) and we focus on the *Archaea* and microbial *Eukarya* in Chapters 17 and 18, respectively.

Including phyla of *Bacteria* known only from 16S ribosomal RNA (rRNA) gene sequences retrieved from the environment (⮑ Section 19.6), over 80 phyla can be distinguished. However, fewer than half of these contain species that have been characterized in laboratory culture (Figure 16.1b). Remarkably, more than 90% of characterized genera and species of *Bacteria* originate in only four phyla: *Proteobacteria, Actinobacteria, Firmicutes,* and *Bacteroidetes* (Figure 16.1b).

With more than ten thousand species of bacteria described, we obviously cannot consider them all. Therefore, using phylogenetic trees to focus our discussion, we will explore some of the best-known species from a broad diversity of phyla. In this chapter we

will consider species from more than 20 bacterial phyla, focusing on those with the largest numbers of characterized species. We begin our tour of the *Bacteria* with the phylum *Proteobacteria*, a hotbed of cultured species in this domain.

I • *Proteobacteria*

The **Proteobacteria** are by far the largest and most metabolically diverse phylum of *Bacteria* (**Figure 16.2**). More than a third of characterized species of *Bacteria* originate within this group (Figure 16.1b), and *Proteobacteria* constitute the majority of known bacteria of medical, industrial, and agricultural significance.

As a group, the *Proteobacteria* are all gram-negative bacteria. They show an exceptionally wide diversity of energy-generating mechanisms, with chemolithotrophic, chemoorganotrophic, and phototrophic species (Figure 16.2). Indeed, we have already encountered diverse representatives of this group in Chapters 14 and 15 where we considered the tremendous metabolic and functional diversity of the *Proteobacteria*. The *Proteobacteria* are equally

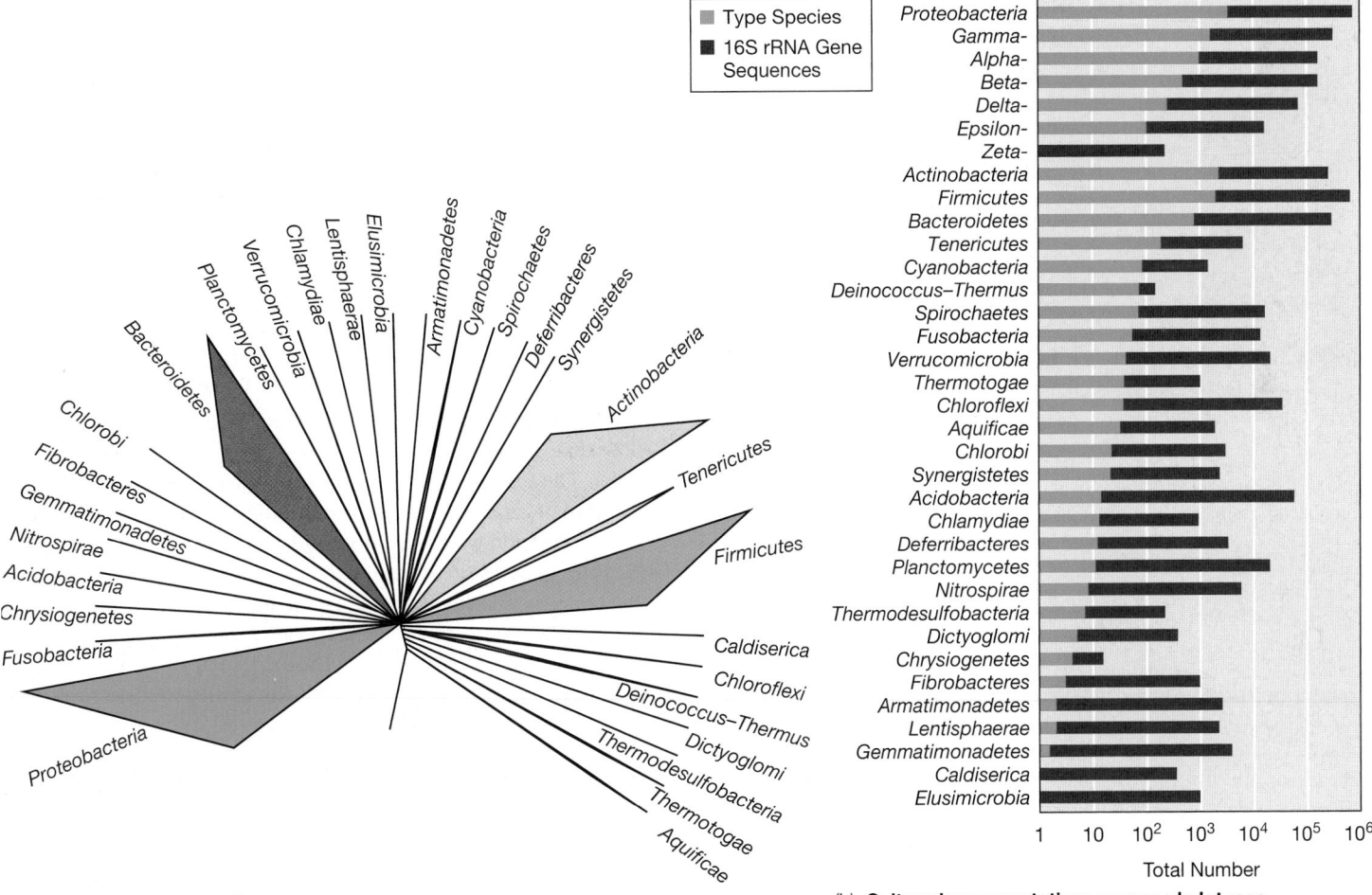

(a) **Major phyla of *Bacteria***

(b) **Cultured representatives versus phylotypes**

Figure 16.1 Some major phyla of *Bacteria* based on 16S ribosomal RNA gene sequence comparisons. (a) Depicted are the major phyla of *Bacteria* that have cultivated species. Analyses of 16S rRNA gene sequences from natural environments suggest there are more than 80 bacterial phyla. (b) Numbers of cultured and characterized species (green bars) and known 16S rRNA gene sequences (phylotypes, red bars) for each of the 29 major bacterial phyla that have at least one characterized species in pure culture. Also shown are related data for the different classes of *Proteobacteria*. Differences between the size of the red and green bars indicate the degree to which members of each group are common in natural environments but difficult to cultivate in isolation. Note that the abscissa is a log scale.

UNIT 4

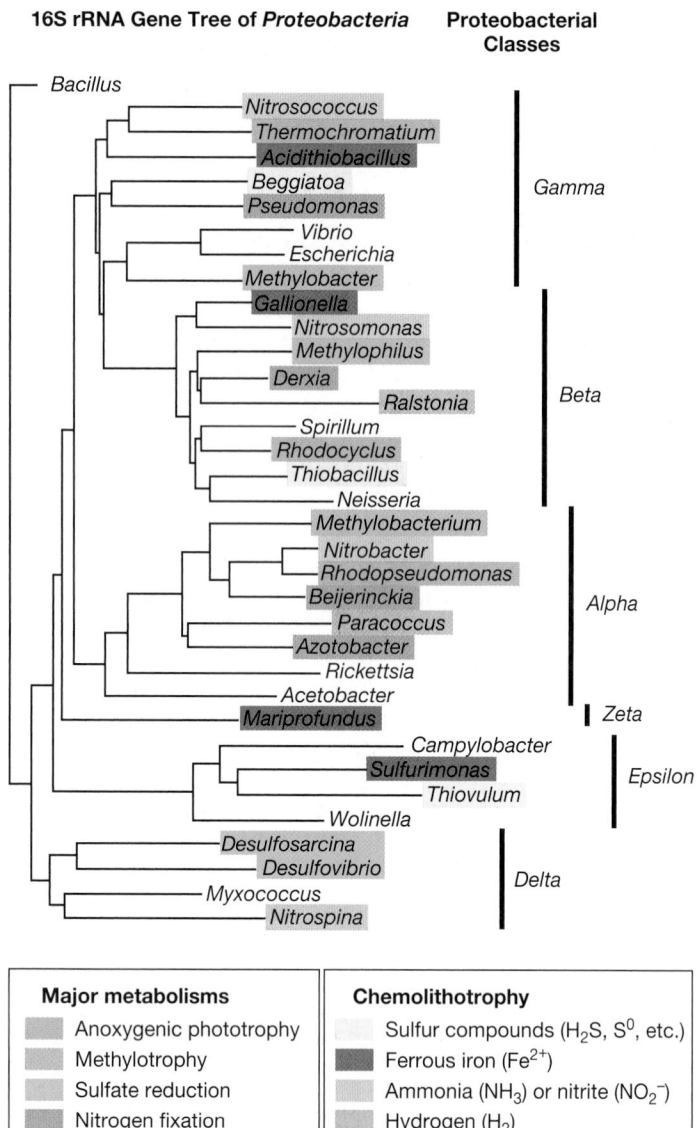

16S rRNA Gene Tree of *Proteobacteria*

Proteobacterial Classes

Major metabolisms
- Anoxygenic phototrophy
- Methylotrophy
- Sulfate reduction
- Nitrogen fixation

Chemolithotrophy
- Sulfur compounds (H_2S, S^0, etc.)
- Ferrous iron (Fe^{2+})
- Ammonia (NH_3) or nitrite (NO_2^-)
- Hydrogen (H_2)

Figure 16.2 Phylogenetic tree and metabolic links of some key genera of *Proteobacteria*. Phylogeny of representative genera of *Proteobacteria* as revealed by analysis of 16S rRNA gene sequences. Note how identical metabolisms are often distributed in phylogenetically distinct genera, suggesting that horizontal gene flow has been extensive in the *Proteobacteria*. Some organisms listed may have multiple properties; for example, some sulfur chemolithotrophs are also iron or hydrogen chemolithotrophs, and several of the organisms listed can fix nitrogen. Phylogenetic analyses were performed and the phylogenetic tree constructed by Marie Asao, Ohio State University.

diverse in terms of their relationship to oxygen (O_2), with anaerobic, microaerophilic, and facultatively aerobic species known. Morphologically, they also exhibit a wide range of cell shapes, including straight and curved rods, cocci, spirilla, and filamentous, budding, and appendaged forms.

As we proceed through this chapter, we will need to recall the descending hierarchy of microbial systematics: domain (*Bacteria* or *Archaea*)→phylum→class→order→family→genus→species. We will use these terms frequently in this chapter. Based on 16S rRNA gene sequences, the phylum *Proteobacteria* can be divided into six classes: *Alpha-, Beta-, Gamma-, Delta-, Epsilon-,* and

Zetaproteobacteria. Each class contains many genera with the exception of the *Zetaproteobacteria*, which is composed of a single known species, the marine iron-oxidizing bacterium *Mariprofundus ferrooxydans* (🔗 Section 15.15).

Despite the phylogenetic breadth of the *Proteobacteria*, species in different classes often have similar metabolisms. For example, phototrophy and methylotrophy occur in three different classes of *Proteobacteria*, and nitrifying bacteria span four classes of *Proteobacteria* (🔗 Figure 15.1). This suggests that horizontal gene flow (Chapter 11 and 🔗 Section 13.7) has played a major role in shaping the metabolic diversity of the *Proteobacteria*. The sharing of metabolic traits in the different classes of *Proteobacteria* is also a good reminder that phenotype and phylogeny often provide different views of prokaryotic diversity (🔗 Section 15.1).

16.1 *Alphaproteobacteria*

With about one thousand described species, the *Alphaproteobacteria* are the second largest class of *Proteobacteria* (Figure 16.1*b*). The *Alphaproteobacteria* contain extensive functional diversity (Figure 16.2, 🔗 Figure 15.1), and many genera in this group have already been considered in Chapter 14. Most species are obligate aerobes or facultative aerobes and many are **oligotrophic**, preferring to grow in environments that have low nutrient concentration. A total of 10 orders have been described within the *Alphaproteobacteria*, but the vast majority of species fall within the *Rhizobiales, Rickettsiales, Rhodobacterales, Rhodospirillales, Caulobacterales,* and *Sphingomonadales* (**Figure 16.3, Table 16.1**).

Rhizobiales

KEY GENERA: *Bartonella, Methylobacterium, Pelagibacter, Rhizobium, Agrobacterium*

The *Rhizobiales* (Figure 16.3) are the largest and most metabolically diverse order of *Alphaproteobacteria* and contain phototrophs (e.g., *Rhodopseudomonas*), chemolithotrophs (e.g., *Nitrobacter*), symbionts (e.g., rhizobia), free-living nitrogen-fixing bacteria (e.g., *Beijerinckia*), a few pathogens of plants and animals, and diverse chemoorganotrophs. The group gets its name from the *rhizobia*, a *polyphyletic* collection of genera that form root nodules and fix nitrogen in symbiotic association with leguminous plants (🔗 Section 23.00).

Among the *Rhizobiales* are nine genera that contain rhizobia: *Bradyrhizobium, Ochrobactrum, Azorhizobium, Devosia, Methylobacterium, Mesorhizobium, Phyllobacterium, Sinorhizobium,* and *Rhizobium*. These are typically chemoorganotrophs and obligate aerobes, and the genes that convey the ability to form root nodules have clearly been distributed among these genera by horizontal gene transfer. Each rhizobial genus has a distinct range of plant hosts that can be colonized (🔗 Table 23.1). Rhizobia can be isolated by crushing nodules and spreading their contents on nutrient-rich solid media; colonies typically produce copious amounts of exopolysaccharide slime (**Figure 16.4**).

The organism *Agrobacterium tumefaciens* (also called *Rhizobium radiobacter*) is closely related to root nodule *Rhizobium* species but is a plant pathogen that causes crown gall disease (🔗 Section 23.5). *A. tumefaciens* is unable to form root nodules, and the genes that encode gall formation are unrelated to those that mediate nodule formation.

Alphaproteobacteria

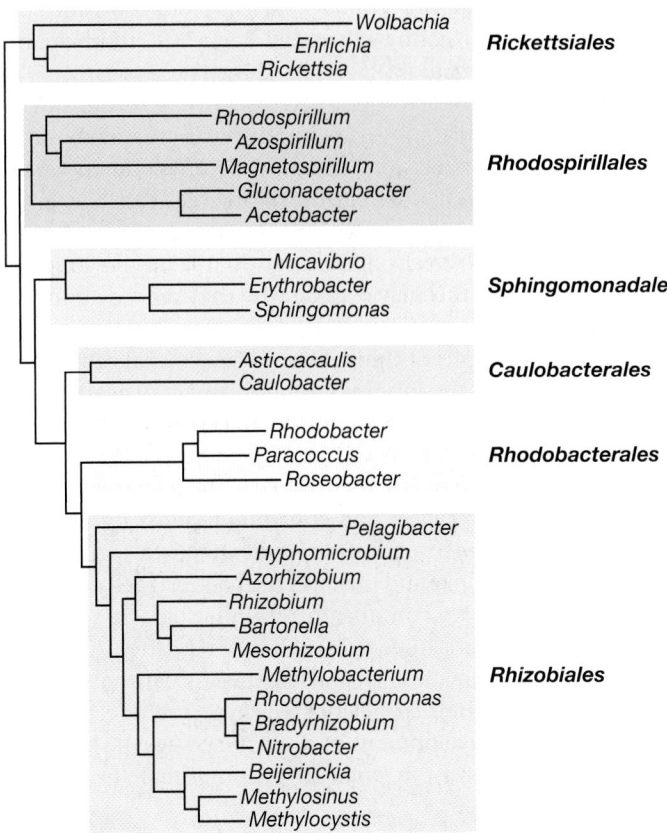

Figure 16.3 Major orders of *Proteobacteria* in the class *Alphaproteobacteria*. The phylogenetic tree was constructed using 16S rRNA gene sequences from representative genera of *Alphaproteobacteria*. Order names are shown in bold.

TABLE 16.1 Notable genera of *Alphaproteobacteria*

Family	Genus	Notable characteristics
Caulobacterales	Caulobacter	Asymmetric cell division and formation of prosthecae
Rickettsiales	Rickettsia	Obligate intracellular parasites, transmitted by arthropods
	Wolbachia	Live within arthropods and affect their reproduction
Rhizobiales	Bartonella	Obligate intracellular parasites, transmitted by arthropods
	Bradyrhizobium	Form root nodules with soybean and other legumes
	Brucella	Facultative intracellular parasites of animals, zoonotic pathogen
	Hyphomicrobium	Stalked cells, metabolically versatile
	Mesorhizobium	Form root nodules with bird's-foot trefoil and other legumes
	Methylobacterium	Methylotroph found on plants and in soil
	Nitrobacter	Nitrifying bacterium that oxidizes NO_2^- to NO_3^-
	Pelagibacter	Oligotrophic chemoorganotroph; high abundance in ocean surface
	Rhodopseudomonas	Metabolically versatile purple nonsulfur bacterium
Rhodobacterales	Paracoccus	Species used as a model for studying denitrification
	Rhodobacter	Metabolically versatile purple nonsulfur bacteria
	Roseobacter	Aerobic anoxygenic phototroph
Rhodospirillales	Acetobacter	Used industrially for producing acetic acid
	Azospirillum	Obligately aerobic diazotroph
	Gluconobacter	Used industrially for producing acetic acid
	Magnetospirillum	Magnetotactic bacterium
Sphingomonadales	Sphingomonas	Aerobic degradation of aromatic organics, biodegradation
	Zymomonas	Ferments sugars into ethanol, potential for biofuel production

The genus *Methylobacterium* is one of the largest in the *Rhizobiales*. These species are often called "pink-pigmented facultative methylotrophs" (⇌ Section 15.16) because of the pink color of their colonies and their good growth on methanol. Species are commonly found on the surface of plants and in soils and freshwater systems. These organisms are also commonly encountered in toilets and baths where their growth on shower curtains, caulk, and in toilet bowls results in the formation of pink-pigmented biofilms. Species of *Methylobacterium* are readily isolated by pressing the surface of a plant leaf onto an agar Petri plate containing methanol as the sole source of carbon.

Bartonella is another notable genus of *Rhizobiales*. These organisms, once classified with the *Rickettsiales*, are intracellular pathogens of humans. Species of *Bartonella* can cause a variety of diseases in humans and other vertebrate animals. *Bartonella quintana* is the causative agent of *trench fever*, a disease that decimated troops in World War I. Other species of *Bartonella* can cause bartonellosis, cat scratch disease, and a variety of inflammatory diseases. Disease transmission is mediated by arthropod vectors including fleas, lice, and sand flies (Chapter 31). Species of *Bartonella* are fastidious and difficult to cultivate, and isolation is most commonly achieved using blood agar. When growing in tissue culture, cells of *Bartonella* grow on the outside surface of the eukaryotic host cells rather than within the cytoplasm or the nucleus.

Finally, the genus *Pelagibacter* also belongs to the *Rhizobiales*. *Pelagibacter ubique* is an oligotroph and an obligately aerobic chemoorganotroph that inhabits the photic zone of Earth's oceans. This organism can make up 25% of the bacterial cells found at the ocean's surface, and its numbers can reach 50% of cells in temperate

UNIT 4

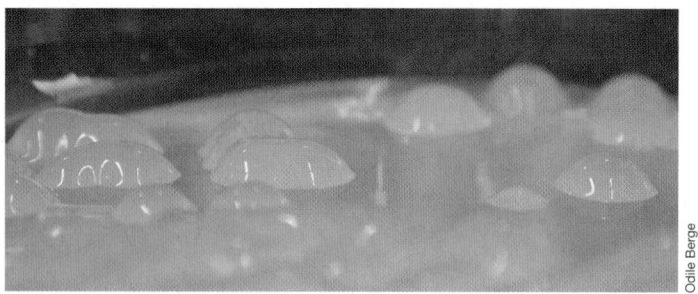

Figure 16.4 Colonies of *Rhizobium mongolense*. Colonies of rhizobia often produce copious exopolysaccharide slime. These colonies of *Rhizobium mongolense* were grown on a medium low in nitrogen with sucrose as carbon source.

waters in the summer; as a consequence, *Pelagibacter ubique* is likely the most abundant bacterial species on Earth (⇐⇒ Section 20.11).

Rickettsiales

KEY GENERA: *Rickettsia, Wolbachia*

Rickettsiales (Figure 16.3) are all obligate intracellular parasites or mutualists of animals. Species in this order have not yet been cultivated in the absence of host cells (**Figure 16.5**) and must be grown in chicken eggs or in host cell tissue culture. Typically, *Rickettsiales* are closely associated with arthropods. Those genera that cause disease such as *Rickettsia* and *Ehrlichia* are transmitted by arthropod bites; other genera such as *Wolbachia* are obligate parasites or mutualists of insects and other arthropods.

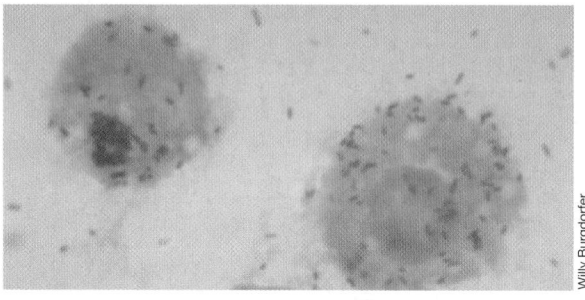

(a)

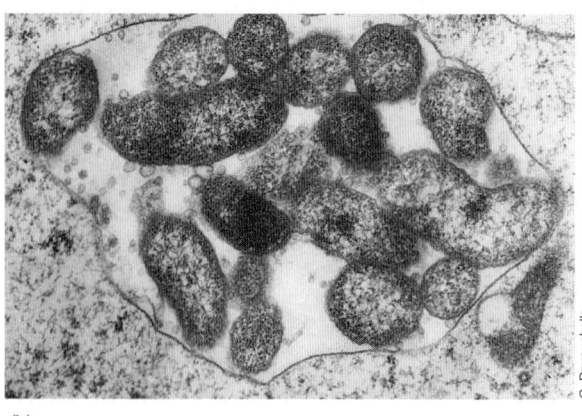

(b)

Figure 16.5 Rickettsias growing within host cells. *(a) Rickettsia rickettsii* in tissue culture. Cells are about 0.3 μm in diameter. *(b)* Electron micrograph of cells of *Rickettsiella popilliae* within a blood cell of its host, the beetle *Melolontha melolontha*. The bacteria grow inside a vacuole within the host cell.

Species of the genus *Rickettsia* are the causative agents of several human diseases, including typhus (*Rickettsia prowazekii*) and spotted fever rickettsiosis, commonly called Rocky Mountain spotted fever (*Rickettsia rickettsii*) (⇐⇒ Section 31.3). These organisms are closely associated with arthropod vectors and can be transmitted by ticks, fleas, lice, and mites. Most rickettsias are metabolically specialized, able to oxidize only the amino acids glutamate or glutamine and unable to oxidize glucose or organic acids. Rickettsias are unable to synthesize certain metabolites and must instead obtain them from host cells. Rickettsias do not survive long outside their hosts, and this may explain why they must be transmitted from animal to animal by arthropod vectors.

Electron micrographs of thin sections of rickettsial cells show a typical prokaryotic morphology including a cell wall (Figure 16.5*b*). The penetration of a host cell by a rickettsial cell is an active process, requiring both host and parasite to be viable and metabolically active. Once inside the host cell, the bacteria multiply primarily in the cytoplasm and continue replicating until the host cell is loaded with parasites (Figure 16.5; ⇐⇒ Figure 31.6). The host cell then bursts and liberates the bacterial cells.

The genus *Wolbachia* contains intracellular parasites of many insects (**Figure 16.6**), a huge group that constitutes 70% of all known arthropod species. *Wolbachia* species can have any of several effects on their insect hosts. These include inducing parthenogenesis (development of unfertilized eggs), the killing of males, and feminization (the conversion of male insects into females).

Wolbachia pipientis is the best-studied species in the genus. Cells of *W. pipientis* colonize the insect egg (Figure 16.6), where they multiply in vacuoles of host cells surrounded by a membrane of host origin. Cells of *W. pipientis* are passed from an infected female to her offspring through this egg infection. *Wolbachia*-induced parthenogenesis occurs in a number of species of wasps. In these insects, males normally arise from unfertilized eggs (which contain only one set of chromosomes), while females arise from fertilized eggs (which contain two sets of chromosomes). However, in unfertilized eggs infected with

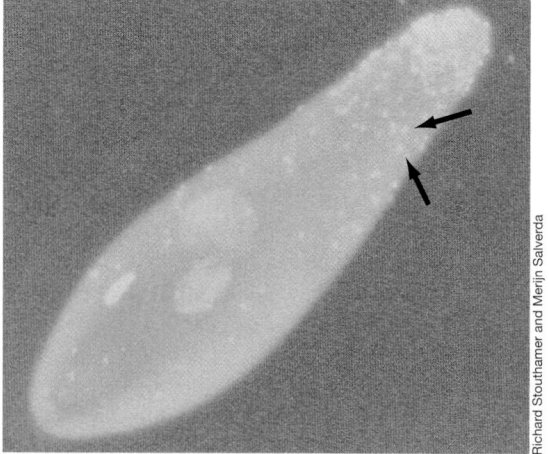

Figure 16.6 *Wolbachia*. Photomicrograph of a DAPI-stained egg of the parasitoid wasp *Trichogramma kaykai* infected with *Wolbachia pipientis*, which induces parthenogenesis. The *W. pipientis* cells are primarily located in the narrow end of the egg (arrows).

Wolbachia, the organism somehow triggers a doubling of the chromosome number, thus yielding only females. Predictably, if female insects are fed antibiotics that kill *Wolbachia*, parthenogenesis ceases.

Other Groups of *Alphaproteobacteria*

KEY GENERA: *Rhodobacter, Acetobacter, Caulobacter, and Sphingomonas*

The orders *Rhodobacterales* and *Rhodospirillales* (Figure 16.3) contain metabolically diverse organisms that have been discussed previously, including purple nonsulfur bacteria (*Rhodobacter* and *Rhodospirillum*, ↩ Section 15.5), aerobic anoxygenic phototrophs (*Roseobacter*, ↩ Section 15.5), nitrogen-fixing bacteria (*Azospirillum*, ↩ Section 15.12), denitrifiers (*Paracoccus*, ↩ Section 15.13), methylotrophs (*Methylobacterium*, ↩ Section 15.16), and magnetotactic bacteria (*Magnetospirillum*, ↩ Section 15.22), among others.

The *Caulobacterales* are typically oligotrophic and strictly aerobic chemoorganotrophs. Species typically form prosthecae or stalks (↩ Section 15.20), and many species display asymmetric forms of cell division. The characteristic genus is *Caulobacter*, which has a characteristic life cycle that we have discussed previously (↩ Sections 7.7 and 15.20).

The *Sphingomonadales* include diverse aerobic and facultatively aerobic chemoorganotrophs as well as species of aerobic anoxygenic phototrophs (*Erythrobacter*) and a few obligate anaerobes. The characteristic genus is *Sphingomonas*, which consists of obligately aerobic and nutritionally versatile species. Sphingomonads are widespread in aquatic and terrestrial environments and are notable for their ability to metabolize a wide range of organic compounds including many aromatic compounds that are common environmental contaminants (e.g., toluene, nonylphenol, dibenzo-*p*-dioxin, naphthalene, and anthracene, among others). As a consequence, sphingomonads have been widely studied as potential agents of bioremediation (↩ Section 22.4). These organisms are typically easy to cultivate and grow well on a variety of complex culture media.

MINIQUIZ

- What are some ways in which *Wolbachia* species can affect insects?
- What organisms might form the pink scum you find on the edge of a bathtub? How might you try to cultivate these organisms?

16.2 *Betaproteobacteria*

With about 500 described species, the *Betaproteobacteria* are the third largest class of *Proteobacteria* (Figure 16.1). The *Betaproteobacteria* contain an immense amount of functional diversity (Figure 16.2 and ↩ Figure 15.1), and many species in this group have already been considered in Chapter 15. A total of six orders of *Betaproteobacteria* have many characterized species: *Burkholderiales, Hydrogenophilales, Methylophilales, Neisseriales, Nitrosomonadales*, and *Rhodocyclales* (**Figure 16.7**), and we focus on these here.

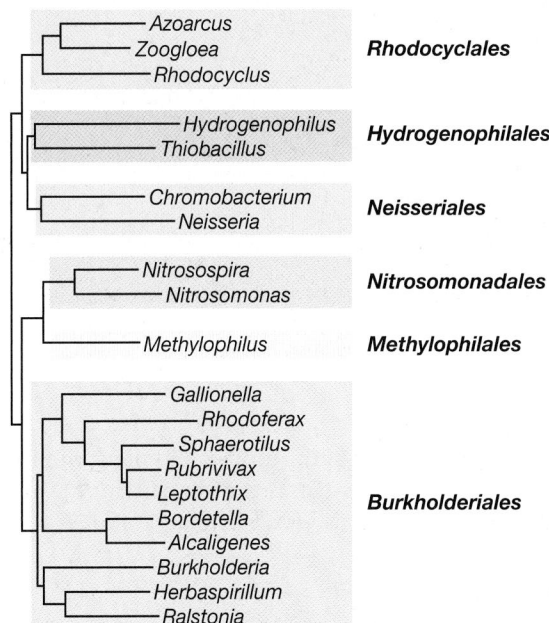

Betaproteobacteria

Figure 16.7 Major orders of *Proteobacteria* in the class *Betaproteobacteria*. The phylogenetic tree was constructed using 16S rRNA gene sequences from representative genera of *Betaproteobacteria*. Order names are shown in bold.

Burkholderiales

KEY GENERA: *Burkholderia*

The *Burkholderiales* contain species with a wide range of metabolic and ecological characteristics. Species include strictly aerobic, facultatively aerobic, and obligately anaerobic chemoorganotrophs, anoxygenic phototrophs, obligate and facultative chemolithotrophs, free-living nitrogen fixers, and pathogens of plants, animals, and humans.

Burkholderia is the type genus for the *Burkholderiales*. The genus *Burkholderia* includes diverse species of chemoorganotrophs with strictly respiratory metabolism. All species can grow aerobically, some also grow anaerobically with nitrate as the electron acceptor, and many strains are able to fix N_2. The metabolic versatility of *Burkholderia* species with respect to organic compounds, and aromatic compounds in particular, has led to interest in their use in bioremediation (↩ Section 22.5). Certain strains of *Burkholderia* have also been shown to promote plant growth. However, many species are potentially pathogenic for plant or animals. One of the best known of the pathogenic species is *Burkholderia cepacia*.

B. cepacia is primarily a soil bacterium but also an opportunistic pathogen (**Figure 16.8**). Often found in the rhizosphere of plants, *B. cepacia* can produce both anti-fungal and anti-nematodal compounds, and thus its ability to colonize plant roots can provide disease protection and promote plant growth. However, *B. cepacia* is also known as a plant pathogen in certain circumstances, and it is a major cause of soft rot in onions. *B. cepacia* has also emerged as an opportunistic hospital-acquired infection in humans, as it is a hardy organism that is difficult to eradicate from the clinical setting. *B. cepacia* can form secondary lung infections in patients who are immunocompromised or have pneumonia or cystic fibrosis.

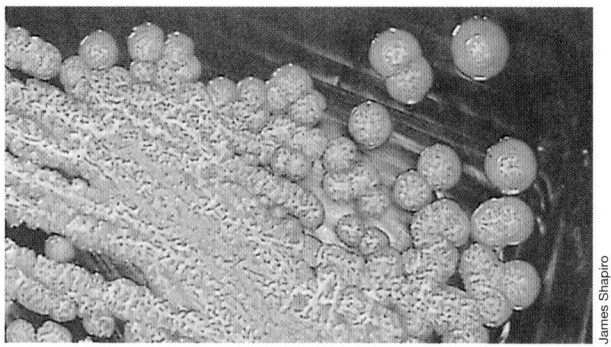

Figure 16.8 Colonies of *Burkholderia*. Photograph of colonies of *Burkholderia cepacia* on an agar plate.

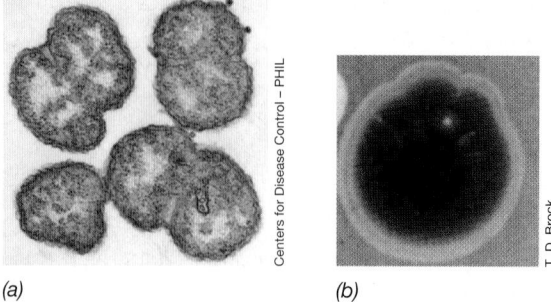

(a)   *(b)*

Figure 16.9 *Neisseria* and *Chromobacterium*. *(a)* Transmission electron micrograph of cells of *Neisseria gonorrhoeae* showing the typical diplococcus cell arrangements. *(b)* A large colony of *Chromobacterium violaceum*.

Its ability to form biofilms in the lung and its natural resistance to many antibiotics has made this organism particularly dangerous for patients with cystic fibrosis (⮑ Sections 7.9 and 20.4).

Rhodocyclales

KEY GENERA: *Rhodocyclus, Zoogloea*

Like the *Burkholderiales*, the order *Rhodocyclales* contains species with diverse metabolic and ecological characteristics. The type genus for the *Rhodocyclales* is *Rhodocyclus*, a purple nonsulfur bacterium (⮑ Section 15.5). Like most purple nonsulfur bacteria, *Rhodocyclus* species grow best as photoheterotrophs but most can also grow as photoautotrophs with H_2 as electron acceptor. Species can also grow by respiration in darkness, but they are typically found in illuminated anoxic environments where organic matter is present.

Zoogloea is another important genus of the *Rhodocyclales*. *Zoogloea* species are aerobic chemoorganotrophs that are distinctive for producing a thick gelatinous capsule that binds cells together into a complex matrix with branching, fingerlike projections. This gelatinous matrix can cause *flocculation*, the formation of macroscopic particles that settle out of solution. *Zoogloea ramigera* is of particular importance in aerobic wastewater treatment (⮑ Section 22.6), where it degrades much of the organic carbon in the waste stream and promotes flocculation and settling, crucial steps in water purification.

Neisseriales

KEY GENERA: *Chromobacterium, Neisseria*

The order *Neisseriales* contains at least 29 genera of diverse chemoorganotrophs. The best-characterized species are in the genera *Neisseria* and *Chromobacterium*. Species of *Neisseria* are commonly isolated from animals, and some of them are pathogenic. *Neisseria* species are always cocci (**Figure 16.9a**). Some *Neisseria* are free-living saprophytes and reside in the oral cavity and other moist areas on the animal body. Others are serious pathogens, such as *Neisseria meningitidis*, which can cause a potentially fatal inflammation of the membranes lining the brain (meningitis, ⮑ Section 30.5). We discuss the clinical microbiology of *Neisseria gonorrhoeae*—the causative agent of the sexually transmitted disease gonorrhea—in Section 28.3, and the pathogenesis of gonorrhea itself in Section 30.13.

Chromobacterium is a close phylogenetic relative of *Neisseria* but is rod-shaped in morphology. The best-known *Chromobacterium* species is *C. violaceum*, a purple-pigmented organism (Figure 16.9b) found in soil and water and occasionally in pus-forming wounds of humans and other animals. *C. violaceum* and a few other chromobacteria produce the purple pigment *violacein* (Figure 16.9b), a water-insoluble pigment with both antimicrobial and antioxidant properties. *Chromobacterium* is a facultative aerobe, growing fermentatively on sugars and aerobically on various carbon sources.

Hydrogenophilales, Methylophilales, and Nitrosomonadales

KEY GENERA: *Hydrogenophilus, Thiobacillus, Methylophilus, Nitrosomonas*

These three orders contain organisms that have fairly specialized metabolic capabilities including chemolithotrophs and methylotrophs; most species are obligate aerobes and many are autotrophic. *Hydrogenophilus thermoluteolus* is an obligate aerobe that can grow as a chemolithotroph using H_2 as an electron donor for respiration (⮑ Section 14.8) and the Calvin cycle to fix CO_2. This species is a facultative chemolithotroph, and can also grow as a chemoorganotroph on simple carbon sources. *Thiobacillus* is another important genus of *Hydrogenophilales*. Species of *Thiobacillus* can be chemoorganotrophs or chemolithotrophs. Chemolithotrophic species of *Thiobacillus* are sulfur bacteria (⮑ Sections 14.9 and 15.11) that oxidize reduced sulfur compounds as electron donors and grow by aerobic respiration or denitrification (⮑ Sections 14.13 and 15.13). Species of *Thiobacillus* can also fix CO_2 using the Calvin cycle and are commonly found in soils, sulfur springs, marine habitats, and other locales where reduced sulfur compounds are available.

The *Methylophilales* and *Nitrosomonadales* contain metabolically specialized organisms. *Methylophilus* species are obligate and facultative methylotrophs (⮑ Section 14.18) that grow on methanol and other C_1 compounds, but not on CH_4. Facultative species can grow as chemoorganotrophs through aerobic respiration of simple sugars. The order *Nitrosomonadales* contains obligately chemolithotrophic ammonia-oxidizing bacteria, the key genera being *Nitrosomonas* and *Nitrosospira* (⮑ Section 15.13).

16.3 *Gammaproteobacteria: Enterobacteriales*

KEY GENERA: *Enterobacter, Escherichia, Klebsiella, Proteus, Salmonella, Serratia, Shigella*

The *Gammaproteobacteria* are the largest and most diverse class of *Proteobacteria*, containing nearly half of all characterized species in the phylum. The class contains more than 1500 characterized species among its 15 orders (**Figure 16.10**, Figure 16.1*b*). Its species have diverse metabolic and ecological characteristics (Figure 16.2 and ♁ Figure 15.1) and include many well-known human pathogens. Species can be phototrophic (including the purple sulfur bacteria, ♁ Section 15.4), chemoorganotrophic, or chemolithotrophic, and can have either respiratory or fermentative metabolisms. Members of this group often develop quickly in laboratory media and can be isolated from a wide diversity of habitats. In this section we consider the *Enterobacteriales*, one of the largest and best-known orders within the *Gammaproteobacteria*.

The *Enterobacteriales*, commonly called the **enteric bacteria**, comprise a relatively homogeneous phylogenetic group within the *Gammaproteobacteria* and consist of facultatively aerobic, gram-negative, nonsporulating rods that are either nonmotile or motile by peritrichous flagella (**Figure 16.11**). The *oxidase test* and the *catalase test* are common assays used to characterize bacteria (♁ Section 28.3), and these tests can be used to discriminate enteric bacteria from many other *Gammaproteobacteria*. The oxidase test is an assay for the presence of cytochrome *c* oxidase, an enzyme present in many respiring bacteria. The catalase test assays for the enzyme catalase, which detoxifies hydrogen peroxide and is commonly found in bacteria able to grow in the presence of oxygen (♁ Section 5.14 and Figure 5.29). Enteric bacteria are oxidase-negative and catalase-positive. They also produce acid from glucose and reduce nitrate but only to nitrite. Enteric bacteria have relatively simple nutritional requirements and ferment sugars to a variety of end products.

Among the enteric bacteria are many species pathogenic to humans, other animals, or plants, as well as other species of industrial importance. *Escherichia coli*, the best known of all organisms, is the classic enteric bacterium. Because of the medical importance of many enteric bacteria, an extremely large number have been characterized, and numerous genera and species have been defined, largely for ease in identification purposes in clinical microbiology. However, because enteric bacteria are genetically very closely related, their positive identification often presents considerable difficulty. In clinical laboratories, identification is typically based on the combined analysis of a large number of diagnostic tests carried out using miniaturized rapid diagnostic media kits along with immunological and genomic analyses to identify signature proteins or genes of particular species (Chapter 28).

Fermentation Patterns in Enteric Bacteria

One major taxonomic characteristic separating the various genera of enteric bacteria is the type and proportion of fermentation products generated from the fermentation of glucose. Two broad

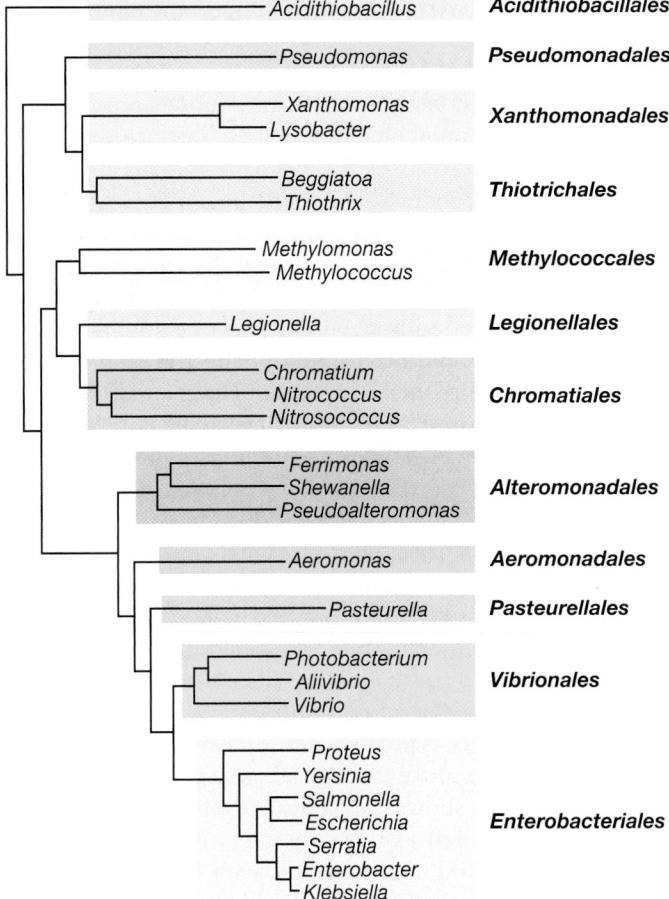

Gammaproteobacteria

Genus	Order
Acidithiobacillus	**Acidithiobacillales**
Pseudomonas	**Pseudomonadales**
Xanthomonas / Lysobacter	**Xanthomonadales**
Beggiatoa / Thiothrix	**Thiotrichales**
Methylomonas / Methylococcus	**Methylococcales**
Legionella	**Legionellales**
Chromatium / Nitrococcus / Nitrosococcus	**Chromatiales**
Ferrimonas / Shewanella / Pseudoalteromonas	**Alteromonadales**
Aeromonas	**Aeromonadales**
Pasteurella	**Pasteurellales**
Photobacterium / Aliivibrio / Vibrio	**Vibrionales**
Proteus / Yersinia / Salmonella / Escherichia / Serratia / Enterobacter / Klebsiella	**Enterobacteriales**

Figure 16.10 Major orders of *Proteobacteria* in the class *Gammaproteobacteria*. The phylogenetic tree was constructed using 16S rRNA gene sequences from representative genera of *Gammaproteobacteria*. Order names are shown in bold.

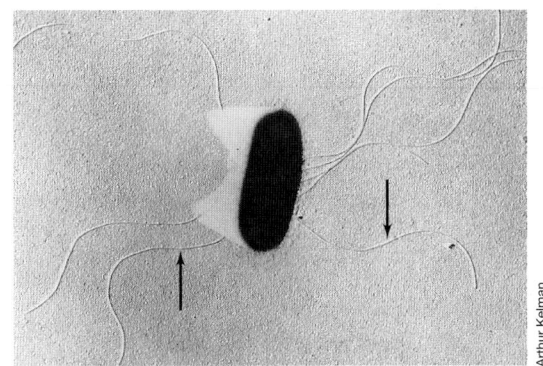

Arthur Kelman

Figure 16.11 A butanediol-producing enteric bacterium. Electron micrograph of a shadow-cast preparation of a cell of the butanediol-producing bacterium *Erwinia carotovora*. The cell is about 0.8 μm wide. Note the peritrichously arranged flagella (arrows), typical of enteric bacteria.

patterns are recognized, the *mixed-acid fermentation* and the *2,3-butanediol fermentation* (**Figure 16.12**).

In the mixed-acid fermentation, three acids are formed in significant amounts: acetic, lactic, and succinic. Ethanol, CO_2, and H_2 are also formed, but not butanediol. In the butanediol fermentation, smaller amounts of acids are formed, and butanediol, ethanol, CO_2, and H_2 are the main products (Figure 14.57). As a result of mixed-acid fermentation, equal amounts of CO_2 and H_2 are produced, whereas in the butanediol fermentation, considerably more CO_2 than H_2 is produced. This is because mixed-acid fermenters produce CO_2 only from formic acid by means of the enzyme formate hydrogenlyase:

$$HCOOH \rightarrow H_2 + CO_2$$

This reaction results in equal amounts of CO_2 and H_2. The butanediol fermenters also produce CO_2 and H_2 from formic acid, but they produce two additional molecules of CO_2 during the formation of each molecule of butanediol (Figure 16.12*b*). Butanediol fermentation is characteristic of *Enterobacter, Klebsiella, Erwinia*, and *Serratia*, whereas mixed-acid fermentation is observed in *Escherichia, Salmonella, Shigella, Citrobacter, Proteus*, and *Yersinia*.

Mixed-Acid Fermenters: *Escherichia, Salmonella, Shigella,* and *Proteus*

Species of *Escherichia* are almost universal inhabitants of the intestinal tract of humans and other warm-blooded animals, although they are by no means the dominant organisms in this habitat. *Escherichia* may play a nutritional role in the intestinal tract by synthesizing vitamins, particularly vitamin K. As a facultative aerobe, this organism probably also helps consume O_2, thus rendering the large intestine anoxic. Wild-type *Escherichia* strains rarely show any growth-factor requirements and are able to grow on a wide variety of carbon and energy sources such as sugars, amino acids, organic acids, and so on.

Some strains of *Escherichia* are pathogenic and have been implicated in diarrheal diseases, especially in infants; diarrheal diseases are a major public health problem in developing countries (Sections 32.1, 32.3, 32.10, and 32.11). *Escherichia* is also a major cause of urinary tract infections in women. Enteropathogenic *E. coli* strains are becoming more frequently implicated in gastrointestinal infections and generalized fevers. Some strains, such as enterohemorrhagic *E. coli*, an important representative of which is strain O157:H7, can cause sporadic outbreaks of severe foodborne disease. Infection occurs primarily through consumption of contaminated foods, such as raw or undercooked ground beef, unpasteurized milk, or contaminated water. In a small percentage of cases, *E. coli* O157:H7 causes a life-threatening complication related to its production of a very potent enterotoxin.

Salmonella and *Escherichia* are quite closely related. However, in contrast to *Escherichia*, species of *Salmonella* are almost always pathogenic, either to humans or to other warm-blooded animals (*Salmonella* is also found in the intestines of cold-blooded animals, such as turtles and lizards). In humans the most common diseases caused by salmonellas are typhoid fever and gastroenteritis (Sections 32.5 and 32.10). The shigellas are also genetically very closely related to *Escherichia*. Genomic analyses strongly suggest that *Shigella* and *Escherichia* have exchanged a significant number of genes by horizontal gene flow. In contrast to most *Escherichia*, however, species of *Shigella* are typically pathogenic to humans, causing a rather severe gastroenteritis called *bacillary dysentery*. *Shigella dysenteriae*, transmitted by food- and waterborne routes, is a good example of this. The bacterium, which contains endotoxin, invades intestinal epithelial cells, where it excretes a neurotoxin that causes acute gastrointestinal distress.

The genus *Proteus* typically contains highly motile cells (**Figure 16.13**) that produce the enzyme *urease*. Unlike *Salmonella* and *Shigella*, *Proteus* shows only a distant relationship to *E. coli*. *Proteus* is a frequent cause of urinary tract infections in humans and probably benefits in this regard from its ready ability to degrade urea by urease. Because of the rapid motility of *Proteus* cells, colonies growing on agar plates often exhibit a characteristic *swarming* phenotype (Figure 16.13*b*). Cells at the edge of the growing colony are more rapidly motile than those in the center of the colony. The former move a short distance away from the colony in

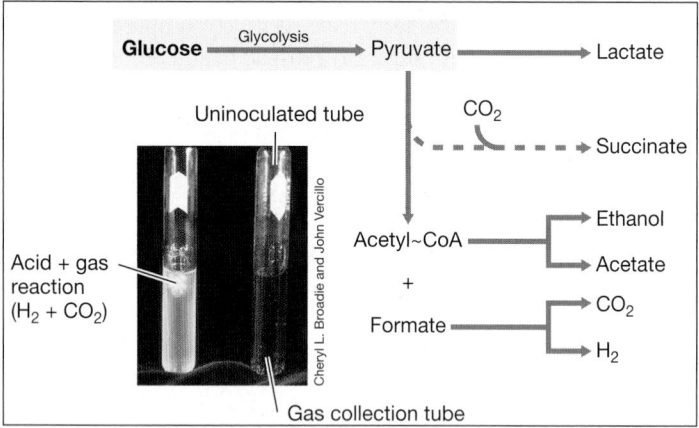

(a) **Mixed-acid fermentation** (for example, *Escherichia coli*)

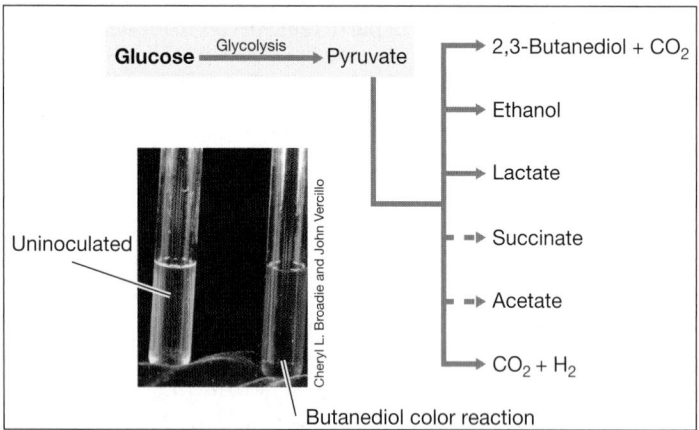

(b) **Butanediol fermentation** (for example, *Enterobacter aerogenes*)

Figure 16.12 Enteric fermentations. Distinction between *(a)* mixed-acid and *(b)* butanediol fermentation in enteric bacteria (Figure 14.57). The solid arrows indicate reactions leading to major products. Dashed arrows indicate minor products. *(a)* The photo shows the production of acid (yellow) and gas (in the inverted Durham tube) in a culture of *Escherichia coli* carrying out a mixed-acid fermentation (purple tube was uninoculated). *(b)* The photo shows the pink-red color in the Voges–Proskauer (VP) test, which indicates butanediol production, following growth of *Enterobacter aerogenes*. The left (yellow) tube was not inoculated. Note that the mixed-acid fermentation produces less CO_2 but more acid products from glucose than does the butanediol fermentation.

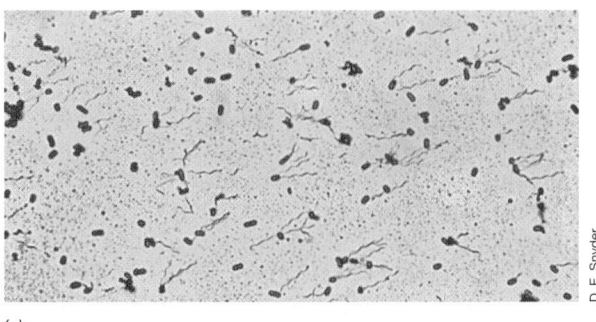

(a)

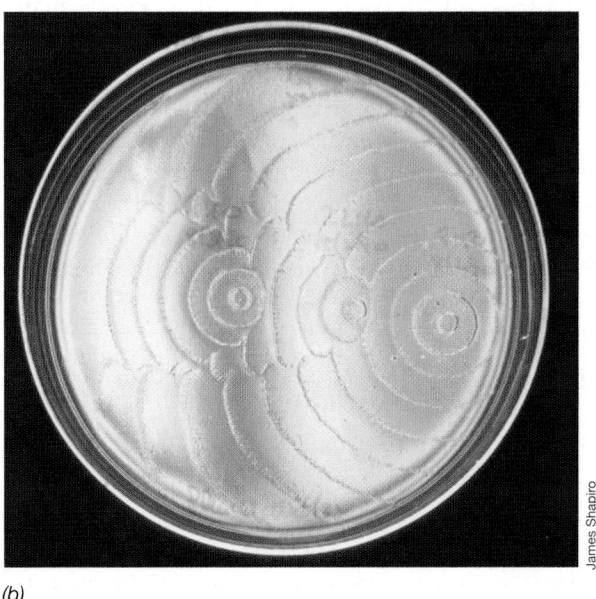

(b)

Figure 16.13 Swarming in *Proteus*. (a) Cells of *Proteus mirabilis* stained with a flagella stain; the peritrichous flagella of each cell form into a bundle to rotate in synchrony. (b) Photo of a swarming colony of *Proteus vulgaris*. Note the concentric rings.

a mass and then undergo a reduction in motility, settle down, and divide, forming a new population of motile cells that again swarm. As a result, the mature colony appears as a series of concentric rings, with higher concentrations of cells alternating with lower concentrations (Figure 16.13b).

Butanediol Fermenters: *Enterobacter, Klebsiella,* and *Serratia*

The butanediol fermenters are genetically more closely related to each other than to the mixed-acid fermenters, a finding that is in agreement with the observed physiological differences (Figure 16.12). *Enterobacter aerogenes* is a common species in water and sewage as well as the intestinal tract of warm-blooded animals and is an occasional cause of urinary tract infections. One species of *Klebsiella, K. pneumoniae*, occasionally causes pneumonia in humans, but klebsiellas are most commonly found in soil and water. Most *Klebsiella* strains also fix nitrogen (⟜ Sections 14.6 and 15.12), a property not characteristic of other enteric bacteria.

The genus *Serratia* forms a series of red pyrrole-containing pigments called *prodigiosins* (**Figure 16.14**). Prodigiosin is produced in stationary phase as a secondary metabolite and is of interest

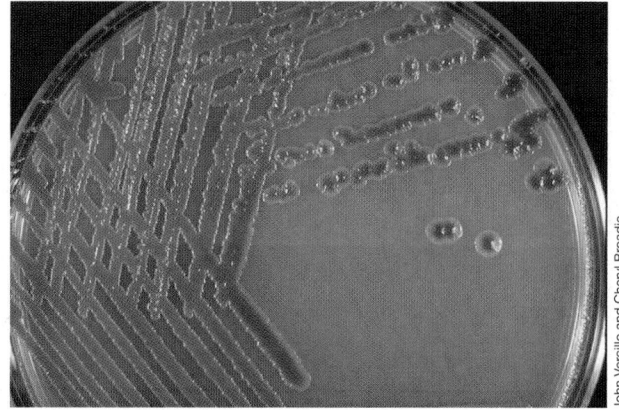

Figure 16.14 Colonies of *Serratia marcescens*. The orange-red pigmentation is due to the pyrrole-containing pigment prodigiosin.

because it contains the pyrrole ring also found in the pigments for energy transfer: porphyrins, chlorophylls, bacteriochlorophylls, and phycobilins (⟜ Sections 14.1–14.3). However, it is unclear if prodigiosin plays any role in energy transfer, and its exact function is unknown. Species of *Serratia* can be isolated from water and soil as well as from the gut of various insects and vertebrates and occasionally from the intestines of humans. *Serratia marcescens* is also a human pathogen that can cause infections in many body sites. It has been implicated in infections caused by some invasive medical procedures and is an occasional contaminant in intravenous fluids.

MINIQUIZ

- What is a mixed-acid fermentation, and of what significance is this trait to enteric bacteria?
- What characteristics would you use to distinguish between *Escherichia coli* and *Klebsiella pneumoniae*?

16.4 *Gammaproteobacteria: Pseudomonadales* and *Vibrionales*

KEY GENERA: *Aliivibrio, Pseudomonas, Vibrio*

The phylogenetic and metabolic diversity of the *Gammaproteobacteria* makes it difficult to select the many notable species in this class of *Proteobacteria*. We focus here on the *Pseudomonadales* and *Vibrionales*, since these groups (along with the *Enterobacteriales*) represent three of the most abundant and most commonly encountered orders of *Gammaproteobacteria* (Figure 16.10).

Pseudomonadales

The *Pseudomonadales* contain exclusively chemoorganotrophs that carry out respiratory metabolisms. All species can grow as aerobes and are typically oxidase- and catalase-positive, but some are also capable of anaerobic respiration with nitrate as the electron acceptor. Most species are able to use a wide diversity of organic compounds as sources of carbon and energy for growth. These organisms are ubiquitous in soil and aquatic systems, and many species cause diseases of plants and animals, including humans.

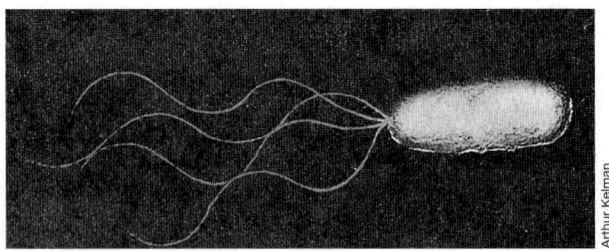

Figure 16.15 Cell morphology of pseudomonads. Shadow-cast transmission electron micrograph of a *Pseudomonas* cell. The cell measures about 1 μm in diameter.

The term **pseudomonad** is often used to describe any gram-negative, polarly flagellated, aerobic rod that is able to use diverse carbon sources. Pseudomonads can be found in several different groups of *Proteobacteria*, but here we consider only those organisms in the order *Pseudomonadales*. The type genus for this order is *Pseudomonas*.

Several species of *Pseudomonas* are pathogenic. Among these, *Pseudomonas aeruginosa* (**Figure 16.15**) is frequently associated with infections of the urinary and respiratory tracts in humans. *P. aeruginosa* is not an obligate pathogen. Instead, the organism is an opportunist, initiating infections in individuals with weakened immune systems. *P. aeruginosa* is a common cause of hospital-acquired (nosocomial) infections from catheterizations, tracheostomies, lumbar punctures, and intravenous infusions, and often emerges in patients given prolonged treatment with immunosuppressive agents. *P. aeruginosa* is also a common pathogen in patients receiving treatment for severe burns or other traumatic skin damage and in people suffering from cystic fibrosis. In addition to localized infections, *P. aeruginosa* can also cause systemic infections, usually in individuals who have experienced extensive skin damage.

P. aeruginosa is naturally resistant to many widely used antibiotics, so treatment of infections is often difficult. Resistance is typically due to a resistance transfer plasmid (R plasmid) (⬡ Sections 4.2 and 28.12), which is a plasmid whose genes encode proteins that detoxify various antibiotics or pump them out of the cell. Polymyxin, an antibiotic not ordinarily used in human therapy because of its toxicity, is effective against *P. aeruginosa* and is used in critical medical situations.

Certain species of *Pseudomonas*, such as *Pseudomonas syringae*, are well-known plant pathogens (phytopathogens). Phytopathogens frequently inhabit nonhost plants (in which disease symptoms are inapparent) and from there become transmitted to host plants and initiate infection. Disease symptoms vary considerably, depending on the particular phytopathogen and host plant. The pathogen releases plant toxins, lytic enzymes, plant growth factors, and other substances that destroy or distort plant tissue, releasing nutrients for use by the bacterium. In many cases the disease symptoms help identify the phytopathogen. Thus, *Pseudomonas syringae* is typically isolated from leaves showing chlorotic (yellowing) lesions, whereas *Pseudomonas marginalis*, a "soft-rot" pathogen, infects stems and shoots but rarely leaves.

Vibrionales

The *Vibrionales* contain facultatively aerobic rods and curved rods that employ a fermentative metabolism. One key difference between the *Vibrio* group and enteric bacteria is that *Vibrio* are oxidase-positive whereas enteric bacteria are oxidase-negative. Although *Pseudomonas* species are also oxidase-positive, they are not fermentative and so are clearly distinct from *Vibrio* species. The best-known genera in this group are *Vibrio*, *Aliivibrio*, and *Photobacterium*, which contain several species that are bioluminescent (⬡ Section 15.18).

Most vibrios and related bacteria are aquatic, being found in marine, brackish, or freshwater habitats. *Vibrio cholerae* is the cause of the disease cholera in humans (⬡ Sections 29.8 and 32.3); the organism does not normally cause disease in other hosts. Cholera is one of the most common human infectious diseases in developing countries and is transmitted almost exclusively via water.

Vibrio parahaemolyticus inhabits the marine environment and is a major cause of gastroenteritis in Japan, where raw fish is widely consumed; the organism has also been implicated in outbreaks of gastroenteritis in other parts of the world, including the United States. *V. parahaemolyticus* can be isolated from seawater itself or from shellfish and crustaceans, and its primary habitat is probably marine animals, with humans being an accidental host.

MINIQUIZ

- What species of *Pseudomonas* is a common cause of lung infection in cystic fibrosis patients?
- What major characteristic could be used to differentiate strains of *Pseudomonas* from those of *Vibrio*?

16.5 Deltaproteobacteria and Epsilonproteobacteria

These classes of *Proteobacteria* contain fewer species and less functional diversity than we have encountered in the *Alpha-, Beta-,* and *Gammaproteobacteria* (Figure 16.2 and ⬡ Figure 15.1). The *Deltaproteobacteria* are primarily sulfate- and sulfur-reducing bacteria (⬡ Sections 14.14, 15.9, and 15.10), dissimilative iron-reducers (⬡ Section 15.14), and bacterial predators (⬡ Section 15.17). *Epsilonproteobacteria*, by contrast, contain many species that oxidize the H$_2$S produced by the sulfate and sulfur reducers. The final class of *Proteobacteria*, the *Zetaproteobacteria*, contains only one characterized species (the iron oxidizer *Mariprofundus ferrooxydans*) and was considered earlier (⬡ Section 15.15).

Deltaproteobacteria

KEY GENERA: *Bdellovibrio, Myxococcus, Desulfovibrio, Geobacter, Syntrophobacter*

Eight orders have been characterized within the *Deltaproteobacteria* (**Figure 16.16**). The *Myxococcales* and *Bdellovibrionales* contain notable genera of bacterial predators (⬡ Section 15.17). In contrast, the *Desulfuromonadales* contains diverse species of metal- and sulfur-reducing genera such as *Geobacter* (⬡ Sections 15.10 and 15.14). Indeed, like the *Desulfuromonadales*, many genera from the *Deltaproteobacteria* are associated with the reduction of sulfur compounds.

The largest and most common order containing sulfate reducers is the *Desulfovibrionales*. These organisms are readily cultivated

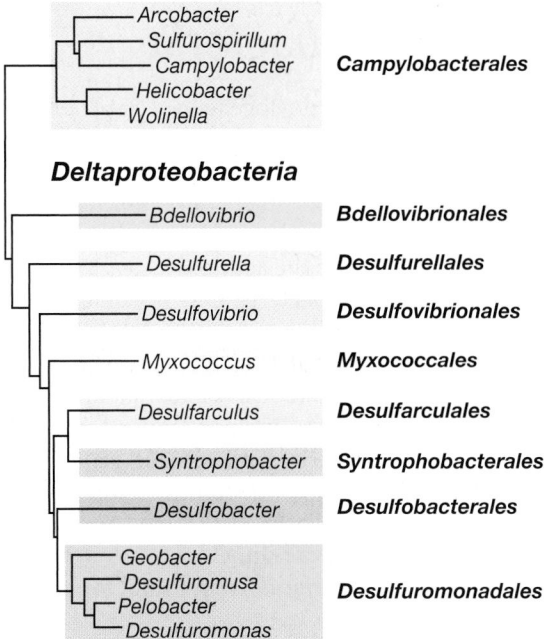

Epsilonproteobacteria

Arcobacter
Sulfurospirillum
Campylobacter   **Campylobacterales**
Helicobacter
Wolinella

Deltaproteobacteria

Bdellovibrio   **Bdellovibrionales**

Desulfurella   **Desulfurellales**

Desulfovibrio   **Desulfovibrionales**

Myxococcus   **Myxococcales**

Desulfarculus   **Desulfarculales**

Syntrophobacter   **Syntrophobacterales**

Desulfobacter   **Desulfobacterales**

Geobacter
Desulfuromusa
Pelobacter   **Desulfuromonadales**
Desulfuromonas

Figure 16.16 Major orders of *Proteobacteria* in the classes *Deltaproteobacteria* and *Epsilonproteobacteria*. The phylogenetic tree was constructed using 16S rRNA gene sequences from representative genera in the *Delta-* and *Epsilonproteobacteria*. Order names are shown in bold.

from marine sediments and nutrient-rich anoxic environments that contain sulfate. Species of *Desulfovibrionales* are typically incomplete oxidizers (cp Section 15.9). All use sulfate as the terminal electron acceptor, and all require small organic compounds such as lactate as a source of carbon and energy for growth. Species within the orders *Desulfobacterales* and *Desulfarculales* also typically reduce sulfate; however, in contrast to the *Desulfovibrionales*, these species can be complete or incomplete acetate oxidizers

(cp Section 15.9). In addition to sulfate, some species in these three orders can also reduce sulfite, thiosulfate, or nitrate, and some are capable of certain fermentations.

The final order containing sulfate reducers is the *Syntrophobacterales*. Some but not all species of the *Syntrophobacterales* are able to reduce sulfate. In nature, however, species of *Syntrophobacterales* primarily interact with H_2-consuming bacteria in a metabolic partnership called *syntrophy* (cp Section 14.23). For example, syntrophic species such as *Syntrophobacter wolinii* oxidize propionate, producing acetate, CO_2, and H_2. However, such growth is only possible when a H_2-consuming partner is present. If sulfate is present, *S. wolinii* can grow as a sulfate reducer without the need for a partner. *S. wolinii* can also grow without a partner organism by fermenting pyruvate, fumarate, or malate.

Epsilonproteobacteria

KEY GENERA: *Campylobacter, Helicobacter*

The *Epsilonproteobacteria* (Figure 16.16) were initially defined by only a few pathogenic bacteria; in particular, by species of *Campylobacter* and *Helicobacter*. However, environmental studies of marine and terrestrial microbial habitats have shown that a diversity of *Epsilonproteobacteria* exist in nature, and their numbers and metabolic capabilities suggest they play important ecological roles (Table 16.2). Species of *Epsilonproteobacteria* are especially abundant at oxic–anoxic interfaces in sulfur-rich environments, and play major roles in the oxidation of sulfur compounds in nature.

Campylobacter and Helicobacter

These two genera of *Epsilonproteobacteria* share a number of characteristics. *Campylobacter* and *Helicobacter* species are gram-negative, oxidase- and catalase-positive, motile spirilla, and most species are pathogenic to humans or other animals (Table 16.2). These organisms are also microaerophilic (cp Section 5.14) and must therefore be cultured from clinical specimens at low (3–15%) O_2 and high (3–10%) CO_2.

TABLE 16.2 Characteristics of key genera of *Epsilonproteobacteria*

Genus	Habitat	Descriptive characters	Physiology and metabolism
Campylobacter	Reproductive organs, oral cavity, and intestinal tract of humans and other animals; pathogenic	Slender, spirally curved rods; corkscrew-like motility by single polar flagellum	Microaerophilic; chemoorganotrophic
Arcobacter	Diverse habitats (freshwater, sewage, saline environments, animal reproductive tract, plants); some species pathogenic for humans and other animals	Slender, curved rods; motile by single polar flagellum	Microaerophilic; aerotolerant or aerobic; chemoorganotrophic; oxidation of sulfide to elemental sulfur (S^0) by some species; nitrogen fixation in one species
Helicobacter	Intestinal tract and oral cavity of humans and other animals; pathogenic	Rods to tight spiral; some species with tightly coiled periplasmic fibers	Microaerophilic, chemoorganotrophic; produce high levels of urease (nitrogen assimilation)
Sulfurospirillum	Freshwater and marine habitats containing sulfur	Vibrioid to spiral-shaped cells; motile by polar flagella	Microaerophilic; reduces elemental sulfur (S^0)
Thiovulum	Freshwater and marine habitats containing sulfur; not yet in pure culture (cp Figure 15.29)	Cells contain orthorhombic S^0 granules; rapid motility by peritrichous flagella	Microaerophilic; chemolithotrophic, oxidizing H_2S
Wolinella	Bovine rumen	Rapidly motile by polar flagellum; single species known: *W. succinogenes*	Anaerobe; anaerobic respiration using fumarate, nitrate, or other compounds as terminal electron acceptor, and with H_2 or formate as electron donor

UNIT 4

Campylobacter species, over a dozen of which have been described, cause acute gastroenteritis that typically results in a bloody diarrhea. Pathogenesis is due to several factors, including an enterotoxin that is related to cholera toxin. *Helicobacter pylori*, also a pathogen, causes both chronic and acute gastritis, leading to the formation of peptic ulcers. We consider these diseases, including their modes of transmission and clinical symptoms, in more detail in Section 30.10.

Sulfurospirillum and *Wolinella*

Species of *Sulfurospirillum*, a *Campylobacter* relative, are nonpathogenic, free-living microaerophiles found in freshwater and marine habitats (Table 16.2). These bacteria also carry out anaerobic respirations using elemental sulfur (S^0), selenate, or arsenate as electron acceptors (⇌ Section 14.15).

Wolinella is an anaerobic bacterium isolated from the bovine rumen (Table 16.2; ⇌ Section 23.13). Unlike other *Epsilonproteobacteria*, the single known species, *W. succinogenes*, grows best as an anaerobe and can catalyze anaerobic respirations using fumarate or nitrate as electron acceptors with H_2 or formate as electron donors. Although *W. succinogenes* has thus far been found only in the rumen, its genome shows significant homologies to both the *Campylobacter* and *Helicobacter* genomes and contains additional genes that encode nitrogen fixation, extensive cell signaling mechanisms, and virtually complete metabolic pathways, absent from closely related genomes. This suggests that *Wolinella* inhabits diverse environments outside of the rumen.

Environmental *Epsilonproteobacteria*

In addition to cultured representatives of the genera mentioned above, and many additional species and genera not considered here, there are large groups within this class that are known only from 16S ribosomal RNA gene sequences obtained from the environment (⇌ Section 19.6). Through environmental sequencing studies and ongoing cultivation efforts, species of *Epsilonproteobacteria* are now becoming recognized as ubiquitous in marine and terrestrial environments where sulfur-cycling activities are ongoing, particularly in deep-sea hydrothermal vent habitats where sulfide-rich and oxygenated waters mix (⇌ Section 20.14). Also, living attached to the surface of animals such as the tube worm *Alvinella* and the shrimp *Rimicaris* that reside near hydrothermal vents, a large variety of uncultured *Epsilonproteobacteria* may, through their sulfur metabolism, detoxify H_2S that would otherwise be deleterious to their animal hosts, allowing the animals to thrive in a chemically hostile environment (⇌ Section 23.9). Further exploration of the phylogeny, metabolic activities, and ecological roles of *Epsilonproteobacteria* will likely uncover exciting new aspects of prokaryotic diversity.

MINIQUIZ

- What four metabolic traits are most common in species of *Deltaproteobacteria*?
- Why is *Wolinella* physiologically unusual among the *Epsilonproteobacteria*?

II • *Firmicutes, Tenericutes, and Actinobacteria*

We continue our tour of phylogenetic bacterial diversity with the gram-positive bacteria of the phyla *Actinobacteria* and *Firmicutes*, and the closely related phylum *Tenericutes* (**Figure 16.17**). These three phyla contain nearly half of all characterized species of *Bacteria* (Figure 16.1*b*).

The *Actinobacteria* include the actinomycetes, a huge group of primarily filamentous soil bacteria. One distinguishing feature of the *Actinobacteria* is that their genomic DNA typically has a high GC content, and as a result they are also called the **high GC gram-positive bacteria**. The *Tenericutes* include cells that lack a cell wall, and the *Firmicutes* include the endospore-forming bacteria, lactic acid bacteria, and several other groups. In contrast to the *Actinobacteria*, the genomes of *Firmicutes* generally have a low GC content, and as a result, they are also called the **low GC gram-positive bacteria**.

We begin by examining *Firmicutes* that do not form endospores.

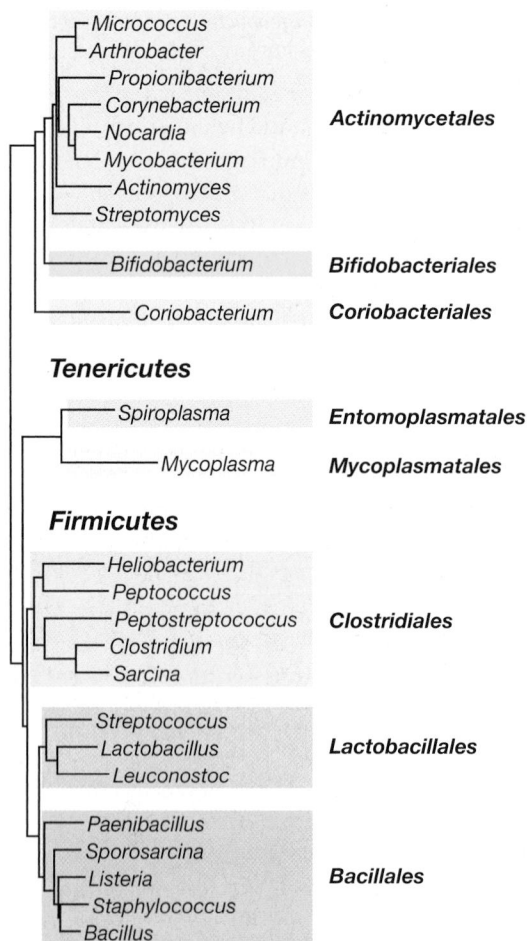

Figure 16.17 Major orders of gram-positive bacteria and relatives. The phylogenetic tree was constructed from 16S rRNA gene sequences of representative genera of *Actinobacteria, Firmicutes,* and *Tenericutes*. Order names are shown in bold.

16.6 *Firmicutes: Lactobacillales*

KEY GENERA: *Lactobacillus, Streptococcus*

The order *Lactobacillales* contains the **lactic acid bacteria**, fermentative organisms that produce lactic acid as a major end product of metabolism. These organisms are used widely in food production and preservation. Lactic acid bacteria are nonsporulating, oxidase- and catalase-negative rods or cocci that show an exclusively fermentative metabolism. All lactic acid bacteria produce lactic acid as a major or sole fermentation product. Members of this group lack porphyrins and cytochromes, do not carry out oxidative phosphorylation, and hence obtain energy only by substrate-level phosphorylation. Unlike many anaerobes, however, most lactic acid bacteria are not sensitive to oxygen (O_2) and can grow in its presence; thus they are called *aerotolerant anaerobes.*

Most lactic acid bacteria obtain energy only from the metabolism of sugars and therefore are usually restricted to habitats in which sugars are present. They typically have limited biosynthetic abilities, and their complex nutritional requirements include needs for amino acids, vitamins, purines, and pyrimidines (for example, ⮌ Table 5.1 for *Leuconostoc mesenteroides*). One important difference between subgroups of the lactic acid bacteria lies in the pattern of products formed from the fermentation of sugars. One group, called **homofermentative**, produces a single fermentation product, *lactic acid*. The other group, called **heterofermentative**, produces other products, mainly ethanol and CO_2, as well as lactate (⮌ Section 14.20 provides additional coverage of homofermentative and heterofermentative pathways).

Lactobacillus

Lactobacilli are typically rod-shaped and grow in chains, varying from long and slender to short, bent rods (**Figure 16.18**), and most are homofermentative. Lactobacilli are common in dairy products, and some strains are used in the preparation of fermented milk products. For instance, *Lactobacillus acidophilus* (Figure 16.18*a*) is used in the production of acidophilus milk; *Lactobacillus delbrueckii* (Figure 16.18*c*) is used in the preparation of yogurt; and other species are used in the production of sauerkraut, silage, and pickles (⮌ Section 32.6).

Lactobacilli are typically more resistant to acidic conditions than are other lactic acid bacteria and are able to grow well at pH values as low as 4. Because of this, they can be selectively enriched from dairy products and fermenting plant material on acidic carbohydrate-containing media. The acid resistance of the lactobacilli enables them to continue growing during natural lactic fermentations, even when the pH value has dropped too low for other lactic acid bacteria to grow. The lactobacilli are therefore typically responsible for the final stages of most lactic acid fermentations. They are rarely, if ever, pathogenic.

Streptococcus and Other Cocci

The genera *Lactococcus* and *Streptococcus* (**Figure 16.19**) contain homofermentative species of coccoid-shaped lactic acid bacteria with quite distinct habitats and activities that are of considerable practical importance to humans. Some species are pathogenic to humans and animals (⮌ Section 30.2). *Streptococcus* species (Figure 16.19*a*) have a characteristic cell morphology of cocci in chains or tetrads

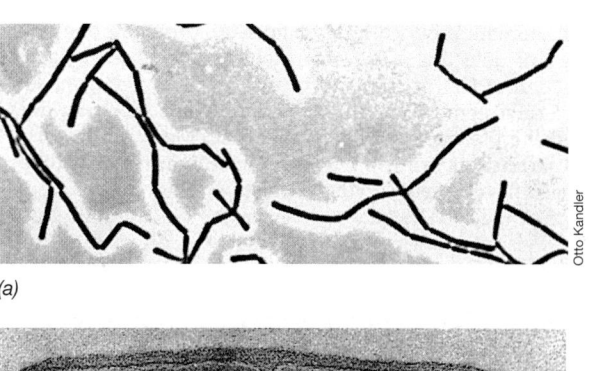

(a)

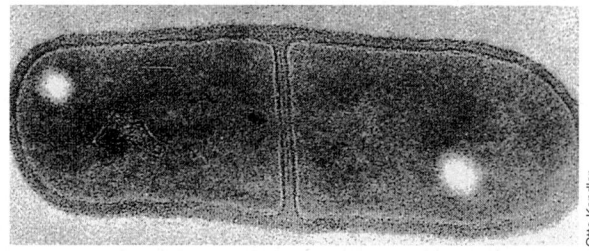

(b)

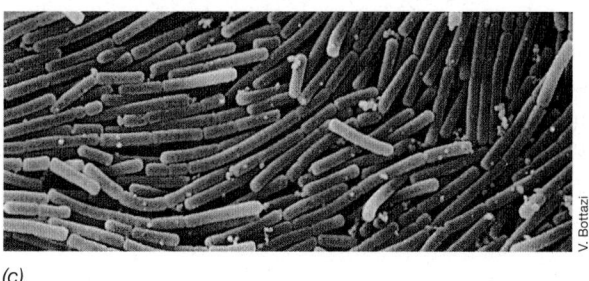

(c)

Figure 16.18 *Lactobacillus* species. *(a) Lactobacillus acidophilus,* phase-contrast. Cells are about 0.75 μm wide. *(b) Lactobacillus brevis,* transmission electron micrograph. Cells measure about 0.8 × 2 μm. *(c) Lactobacillus delbrueckii,* scanning electron micrograph. Cells are about 0.7 μm in diameter.

and so are readily resolved from the rod-shaped lactobacilli. As producers of lactic acid, other streptococci play important roles in the production of buttermilk, silage, and other fermented products (⮌ Section 32.6), and certain species play a major role in the formation of dental caries (⮌ Sections 24.3 and 25.2).

There are several other genera of homofermentative cocci. The genus *Lactococcus* (Figure 16.19*b*) contains those streptococci of

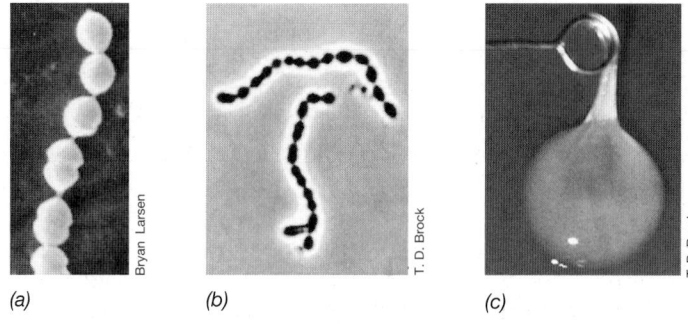

(a)   *(b)*   *(c)*

Figure 16.19 Gram-positive cocci. *(a) Streptococcus* sp., scanning electron micrograph. *(b) Lactococcus lactis,* phase-contrast micrograph. Cells in both photos are 0.5–1 μm in diameter. *(c)* Colony of *Leuconostoc mesenteroides* showing the extensive dextran slime produced by cells grown on sucrose.

dairy significance, whereas the genus *Enterococcus* includes streptococci that are primarily of fecal origin and can be human pathogens. Species of the genera *Peptococcus* and *Peptostreptococcus* are obligate anaerobes that ferment proteins rather than sugars.

Streptococci have been divided into two groups of related species: the *pyogenes subgroup*, characterized by *Streptococcus pyogenes*, the cause of strep throat (⮑ Section 30.2), and the *viridans subgroup*, characterized by *Streptococcus mutans*, the major cause of dental caries (⮑ Sections 24.3 and 25.2). Hemolysis on blood agar is of considerable importance in the subdivision of the genus into species. For example, species that produce the virulence factors streptolysin O or S form colonies surrounded by a large zone of complete red blood cell hemolysis when plated on blood agar, a condition called *β-hemolysis* (⮑ Figure 25.16a). β-hemolysis is diagnostic for streptococci in the pyogenes subgroup. In contrast, streptococci in the viridans subgroup cause incomplete hemolysis on blood agar, a condition that leads to greening of the agar under colonies. Streptococci are also divided into immunological groups (designated by the letters A, B, C, F, G), based on the presence of specific carbohydrate antigens (antigens are substances that elicit an immune response). Those β-hemolytic streptococci found in humans usually contain the group A antigen, whereas enterococci contain the group D antigen.

Heterofermentative lactococci reside in the genus *Leuconostoc*. Strains of *Leuconostoc* also produce the flavoring ingredients diacetyl and acetoin from the catabolism of citrate; they have been used as starter cultures in dairy fermentations. Some strains of *Leuconostoc* produce large amounts of glucose or fructose polysaccharide slimes, especially when cultured on sucrose as the carbon and energy source (Figure 16.19c), and some of these polymers have found medical use as plasma extenders in blood transfusions.

MINIQUIZ

- How do heterofermentative and homofermentative bacteria differ physiologically?
- How can *Streptococcus pyogenes* be distinguished from *Streptococcus mutans*?

16.7 *Firmicutes:* Nonsporulating *Bacillales* and *Clostridiales*

KEY GENERA: *Listeria, Staphylococcus, Sarcina*

Firmicutes that form endospores reside in the orders *Bacillales* and *Clostridiales*. However, numerous *Bacillales* and *Clostridiales* are unable to form endospores, and we consider some of these here.

Listeria

The order *Bacillales* typically contains aerobic and facultatively aerobic chemoorganotrophs. Members of this group are widespread and particularly common in soils. For example, *Listeria* is found widely in soils and is an opportunistic pathogen and a common cause of foodborne illness. *Listeria* are gram-positive, catalase-positive, rod-shaped, facultatively aerobic chemoorganotrophs. Although several species of *Listeria* are known, the

species *Listeria monocytogenes* is most noteworthy because it causes a major foodborne illness, *listeriosis* (⮑ Section 32.13). The organism is transmitted in contaminated, usually ready-to-eat foods such as cheese and sausages, and can cause anything from a mild illness to a fatal form of meningitis. Species of *Listeria* often grow well at low temperatures, allowing growth in refrigerated foods.

Staphylococcus

Staphylococcus (**Figure 16.20**) is a facultative aerobe that shows a typical respiratory metabolism but can also grow fermentatively. Cells typically grow in clusters and produce acid from glucose both aerobically and anaerobically. *Staphylococcus* species are catalase-positive, and this permits their distinction from *Streptococcus* and some other genera of lactic acid bacteria. Staphylococci are relatively resistant to reduced water potential and tolerate drying and high salt (NaCl) fairly well. Their ability to grow in media containing salt provides a selective means for isolation. For example, if an appropriate inoculum such as a skin swab, dry soil, or room dust is spread on a rich-medium agar plate containing 7.5% NaCl and the plate is incubated aerobically, gram-positive cocci often form the predominant colonies. Many species are pigmented, and this provides an additional aid in selecting gram-positive cocci.

Staphylococci are common commensals and parasites of humans and animals, and they occasionally cause serious infections. In humans, there are two major species, *Staphylococcus epidermidis*, a nonpigmented, nonpathogenic organism usually found on the skin or mucous membranes, and *Staphylococcus aureus* (Figure 16.20), a yellow-pigmented species that is most commonly associated with pathological conditions including boils, pimples, pneumonia, osteomyelitis, meningitis, and arthritis. Some *S. aureus* strains are resistant to multiple antibiotics (so-called *MRSA* strains) and are fierce pathogens that can cause extensive tissue damage (⮑ Figure 30.29). We discuss the pathogenesis of MRSA and other strains of *S. aureus* and staphylococcal diseases in Sections 24.5, 28.2, and 30.9.

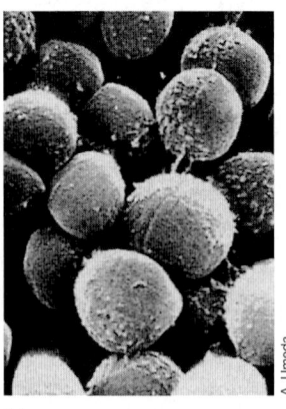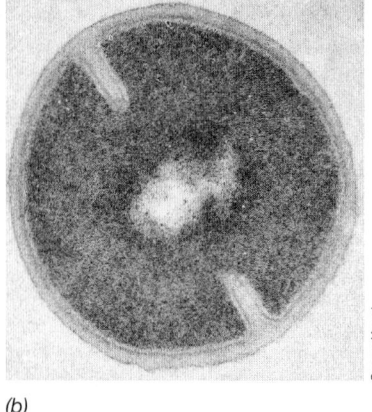

(a)   *(b)*

Figure 16.20 *Staphylococcus.* *(a)* Scanning electron micrograph of typical *Staphylococcus aureus* cells, showing the irregular arrangement of the cell clusters. Individual cells are about 0.8 µm in diameter. *(b)* Transmission electron micrograph of a dividing cell of *S. aureus*. Note the thick gram-positive cell wall.

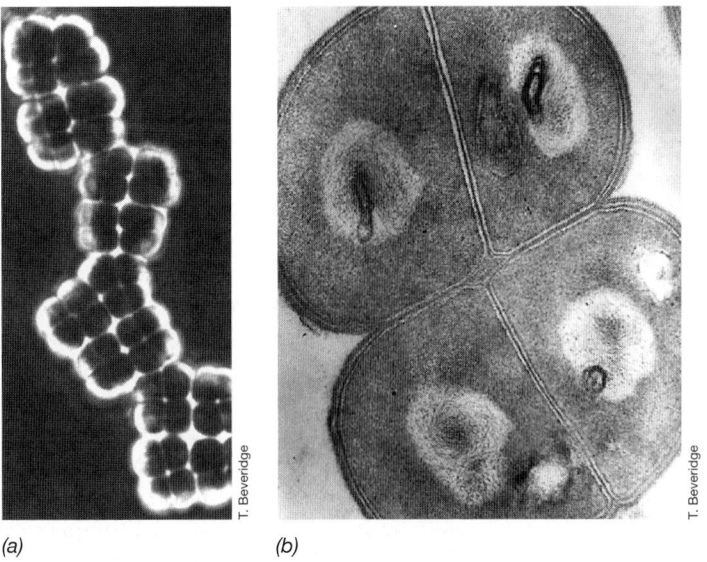

(a)   *(b)*

Figure 16.21 *Sarcina*. *(a)* Phase-contrast photomicrograph of cells of a typical gram-positive coccus *Sarcina*. A single cell is about 2 μm in diameter. *(b)* Electron micrograph of a thin section from *Sarcina ventriculi*. The outermost layer of the cell consists of cellulose.

Sarcina

The genus *Sarcina* groups obligate anaerobes that are catalase-negative within the order *Clostridiales*. *Sarcina* species divide in three perpendicular planes to yield packets of eight or more cells and are notable for this morphology (**Figure 16.21**). *Sarcina* are also extremely acid-tolerant, being able to ferment sugars and grow in environments at a pH as low as 2. Cells of one species, *Sarcina ventriculi*, contain a thick, fibrous layer of cellulose surrounding the cell wall (Figure 16.21*b*). The cellulose layers of adjacent cells become attached, and this functions as a cementing material to hold together packets of *S. ventriculi* cells.

Sarcina species can be isolated from soil, mud, feces, and stomach contents. Because of its extreme acid tolerance, *S. ventriculi* is one of only a few bacteria that can inhabit and grow in the stomach of humans and other monogastric animals. Rapid growth of *S. ventriculi* is observed in the stomach of humans suffering from certain gastrointestinal disorders, such as pyloric ulcerations. These pathological conditions retard the flow of food to the intestine and often require surgery to correct.

MINIQUIZ
- How could species of *Staphylococcus* be differentiated from *Streptococcus*?
- What characteristics differentiate *Sarcina* from *Staphylococcus*?

16.8 *Firmicutes*: Sporulating *Bacillales* and *Clostridiales*

KEY GENERA: *Bacillus, Clostridium, Sporosarcina*

All endospore-forming bacteria are gram-positive species of *Bacillales* or *Clostridiales*. The ability to form endospores evolved only once in a common ancestor of the *Bacillales, Clostridiales*,

and *Lactobacillales* (Figure 16.17). However, many *Bacillales* and *Clostridiales* and the entire order *Lactobacillales* are unable to form endospores. The capacity to make endospores requires many genes (⮔ Sections 2.10 and 7.6) and has not been acquired by horizontal gene transfer. It thus appears that the phylogenetic distribution of endospores has seen many cases where the capacity to form endospores has been lost during the course of evolution.

Endospore-forming bacteria are distinguished on the basis of cell morphology, shape and cellular position of the endospore (**Figure 16.22**), relationship to O_2, and energy metabolism. The two genera about which most is known are *Bacillus*, species of which are aerobic or facultatively aerobic, and *Clostridium*, which contains species that are obligately anaerobic and fermentative. All endospore-forming bacteria are ecologically related because they are found in nature primarily in soil. Even those species that are pathogenic to humans or other animals are primarily saprophytic soil organisms and infect animals only incidentally. Indeed, the ability to produce endospores should be advantageous for a soil microorganism because soil is a highly variable environment in terms of nutrient levels, temperature, and water activity.

Endospore-forming bacteria can be selectively isolated from soil, food, dust, and other materials by heating the sample to 80°C for 10 min, a treatment that effectively kills vegetative cells while any endospores present remain viable. Streaking such heat-treated samples on plates of the appropriate medium and incubating either aerobically or anaerobically selectively yields species of *Bacillus* or *Clostridium*, respectively.

Bacillus and *Paenibacillus*

Species of *Bacillus* and *Paenibacillus* grow well on defined media containing any of a number of carbon sources. Many bacilli produce extracellular hydrolytic enzymes that break down complex polymers such as polysaccharides, nucleic acids, and lipids, permitting the organisms to use these products as carbon sources and electron donors. Many bacilli produce antibiotics, including bacitracin, polymyxin, tyrocidin, gramicidin, and circulin. In most cases the antibiotics are released when the culture enters the stationary phase of growth and is committed to sporulation.

Several bacilli, most notably *Paenibacillus popilliae* and *Bacillus thuringiensis*, produce toxic insecticidal proteins. *P. popilliae* causes a fatal condition called milky disease in Japanese beetle larvae and larvae of closely related beetles of the family *Scarabaeidae*. *B. thuringiensis* causes a fatal disease of many different groups of insects. Both of these insect pathogens form a crystalline protein during sporulation called the *parasporal body*, which is deposited within the sporangium but outside the endospore proper (**Figure 16.23**). In *B. thuringiensis*, the parasporal body is a protoxin that is converted to a toxin in the insect gut. The toxin binds to specific receptors in the intestinal epithelial cells of certain insects and induces pore formation that causes leakage of the host cell cytoplasm followed by lysis. Diverse strains of *B. thuringiensis* can make different types of toxin that have specificity for different groups of insects. Endospore preparations derived from *B. thuringiensis* and *P. popilliae* are commercially available as biological insecticides.

The *cry* genes that encode crystal proteins have been isolated from several *B. thuringiensis* strains. The genes for the *B. thuringiensis* crystal protein (known commercially as "Bt toxin") have been

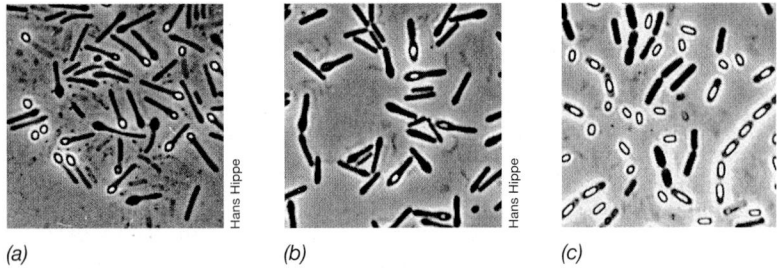

(a)   (b)   (c)

Figure 16.22 *Clostridium* **species and endospore location.** *(a) Clostridium cadaveris,* terminal endospores. Cells are about 0.9 μm wide. *(b) Clostridium sporogenes,* subterminal endospores. Cells are about 1 μm wide. *(c) Clostridium bifermentans,* central endospores. Cells are about 1.2 μm wide. All are phase-contrast micrographs.

introduced into genetically modified crops (e.g., maize, soybeans, and cotton) to render the plants resistant to insects. These genetically modified "Bt crops" are used widely around the world. Genetically altered Bt toxins have also been developed by genetic engineering to help increase toxicity and reduce resistance (↩ Section 12.7).

Clostridium

Clostridia lack a respiratory chain, and so unlike *Bacillus* species, they obtain ATP by substrate-level phosphorylation. Many anaerobic energy-yielding mechanisms are known in the clostridia (↩ Section 14.21). Indeed, the separation of the genus *Clostridium* into subgroups is based primarily on these properties and on the fermentable substrate used. A number of clostridia are *saccharolytic* and ferment sugars, producing butyric acid as a major end product. Some of these also produce acetone and butanol, such as *Clostridium pasteurianum*, which is also a vigorous nitrogen-fixing bacterium (↩ Section 14.6).

One group of clostridia including the species *C. thermocellum*, *C. cellulolyticum*, and *C. cellulovorans* ferments cellulose with the formation of acids and alcohols. These species are likely the major organisms decomposing cellulose in anoxic environments such as the rumen and sediments. Cellulolytic clostridia possess *cellulosomes*, a complex multienzyme structure found on the outer surface of the cell wall. The cellulosome binds insoluble cellulose and degrades it into soluble products that are transported into the

cytoplasm and metabolized by the cell. This cellulosome mechanism is common to bacteria that are able to degrade cellulose anaerobically.

Another group of clostridia are *proteolytic* and conserve energy from the fermentation of amino acids. Some species ferment individual amino acids, but others ferment only amino acid pairs. The products of amino acid fermentation are typically acetate, butyrate, CO_2, and H_2. The coupled catabolism of an amino acid pair is called a *Stickland reaction*; for example, *Clostridium sporogenes* ferments glycine plus alanine. In the Stickland reaction, one amino acid functions as the electron donor and is oxidized, whereas the other is the electron acceptor and is reduced (↩ Figure 14.59). Many of the products of amino acid fermentation by clostridia are foul-smelling substances, and the odor that results from putrefaction is mainly the result of clostridial action. In addition to butyric acid, other odoriferous compounds produced are isobutyric acid, isovaleric acid, caproic acid, hydrogen sulfide, methyl mercaptan (from sulfur amino acids), cadaverine (from lysine), putrescine (from ornithine), and ammonia.

The main habitat of clostridia is the soil, where they live primarily in "pockets" made anoxic by facultative or obligately aerobic bacteria. In addition, a number of clostridia inhabit the anoxic environment of the mammalian intestinal tract. Several clostridia are capable of causing severe diseases in humans, as will be discussed in Sections 24.8, 31.9, and 32.9. For example, botulism is caused by *Clostridium botulinum*, tetanus by *Clostridium tetani*, and gas gangrene by *Clostridium perfringens* and a number of other clostridia, both sugar and amino acid fermenters. These pathogenic clostridia seem in no way unusual metabolically but are distinct in that they produce specific toxins or, in those causing gas gangrene, a group of toxins. *C. perfringens* and related species can also cause gastroenteritis in humans and domestic animals (↩ Section 32.9), and botulism outbreaks are not uncommon in birds such as ducks and a variety of other animals.

Sporosarcina

The genus *Sporosarcina* (**Figure 16.24**) is unique among endospore formers because cells are cocci instead of rods. *Sporosarcina* consists of strictly aerobic spherical to oval cells that divide in two or three perpendicular planes to form tetrads or packets of eight or more cells. The major species is *Sporosarcina ureae*. This bacterium can be enriched from soil by plating dilutions of a pasteurized soil sample on alkaline nutrient agar supplemented with 8% urea and incubating in air. Most soil bacteria are strongly inhibited by as little as 2% urea. However, *S. ureae* tolerates this, catabolizing urea to CO_2 and ammonia (NH_3), which dramatically raises the pH. *S. ureae* is remarkably alkaline-tolerant and can be grown in media up to pH 10, and this feature can be used to advantage in its enrichment from soil.

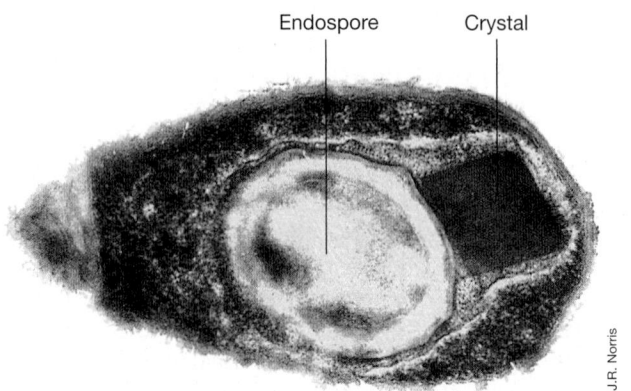

Endospore   Crystal

Figure 16.23 The toxic parasporal crystal in the insect pathogen *Bacillus thuringiensis.* Electron micrograph of a thin section of a sporulating cell. The crystalline protein (Bt toxin) is toxic to certain insects by causing lysis of their intestinal cells.

MINIQUIZ

• What is the major physiological distinction between *Bacillus* and *Clostridium* species?

• What is the crystalline protein made by *Bacillus thuringiensis* and what is its significance to agriculture?

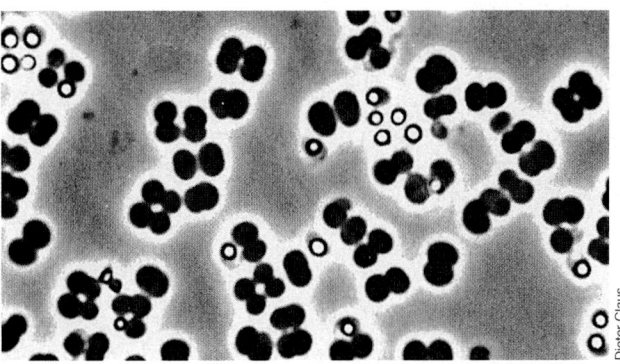

Figure 16.24 *Sporosarcina ureae.* Phase-contrast micrograph. A single cell is about 2 μm wide. Note bright refractile endospores. Most cell packets contain eight cells.

16.9 *Tenericutes:* The Mycoplasmas

KEY GENERA: *Mycoplasma, Spiroplasma*

The *Tenericutes*, which contain the single class *Mollicutes*, are bacteria that lack cell walls (*mollis* is Latin for "soft") and are some of the smallest organisms known. This group is often called the *mycoplasmas* because *Mycoplasma*, a notable genus containing human pathogens, is the best-characterized genus in the phylum (**Table 16.3**).

Although they do not stain gram-positively (because they lack cell walls), mycoplasmas are phylogenetically related to the *Firmicutes*. Mycoplasmas typically live in close association with animal and plant hosts and this may eliminate the need for a gram-positive cell wall. These organisms also have small genomes (ranging in size from 0.6 to 2.2 megabase pairs [Mbp]), a characteristic common in obligate symbionts (Sections 9.3 and 23.6).

Properties of Mycoplasmas

The absence of cell walls in mycoplasmas has been confirmed by electron microscopy and chemical analyses, which show that peptidoglycan is absent. Mycoplasmas resemble protoplasts (bacteria treated to remove their cell walls), but they are more resistant to osmotic lysis and are able to survive conditions under which protoplasts lyse. This ability to resist osmotic lysis is at least partially determined by the presence of sterols, which make the cytoplasmic membrane of mycoplasmas more stable than that of other bacteria. Indeed, some mycoplasmas require sterols in their growth media, and this sterol requirement can aid in the classification of mycoplasmas (Table 16.3).

In addition to sterols, certain mycoplasmas contain compounds called *lipoglycans* (Table 16.3). Lipoglycans are long-chain heteropolysaccharides covalently linked to membrane lipids and embedded in the cytoplasmic membrane of many mycoplasmas. Lipoglycans in some ways resemble the lipopolysaccharides in the outer membrane of gram-negative bacteria, except that they lack the lipid A backbone (Section 2.5). Lipoglycans function to help stabilize the cytoplasmic membrane and have also been identified as facilitating attachment of mycoplasmas to cell surface receptors of animal cells.

Growth of Mycoplasmas

Mycoplasmas can be grown in the laboratory and are small and pleomorphic cells. A single culture may exhibit small coccoid

TABLE 16.3 *Major characteristics of mycoplasmas*

Genus	Properties	Genome size (megabase pairs)	Presence of lipoglycans
Require sterols			
Mycoplasma	Many pathogenic; facultative aerobe (Figure 16.25)	0.66–1.35	+
Anaeroplasma	May or may not require sterols; obligate anaerobes; degrade starch, producing acetic, lactic, and formic acids plus ethanol and CO_2; found in the bovine and ovine rumen	1.5–1.6	+
Spiroplasma	Spiral to corkscrew-shaped cells; associated with various phytopathogenic (plant disease) conditions; facultative aerobe	0.94–2.2	–
Ureaplasma	Coccoid cells; occasional clusters and short chains; growth optimal at pH 6; strong urease reaction; associated with certain urinary tract infections in humans; microaerophile	0.75	–
Entomoplasma	Facultative aerobe; associated with insects and plants	0.79–1.14	Unknown
Do not require sterols			
Acholeplasma	Facultative aerobe	1.5	+
Asteroleplasma	Obligate anaerobe; isolated from the bovine or ovine rumen	1.5	+
Mesoplasma	Phylogenetically and ecologically related to *Entomoplasma*; facultative aerobe	0.87–1.1	Unknown

elements; larger, swollen forms; and filamentous forms, often highly branched (**Figure 16.25**). The small coccoid elements (0.2–0.3 μm in size) are among the smallest of free-living cells (Section 2.2). The mode of growth of mycoplasmas differs in liquid and agar cultures. On agar the organisms tend to grow so that they become embedded in the medium. These colonies show a characteristic "fried-egg" appearance consisting of a dense central core that penetrates downward into the agar, surrounded by a circular spreading area that is lighter in color (**Figure 16.26**). As would be expected of cells lacking cell walls, growth of *Mollicutes* is not inhibited by antibiotics that inhibit cell wall synthesis. However,

UNIT 4

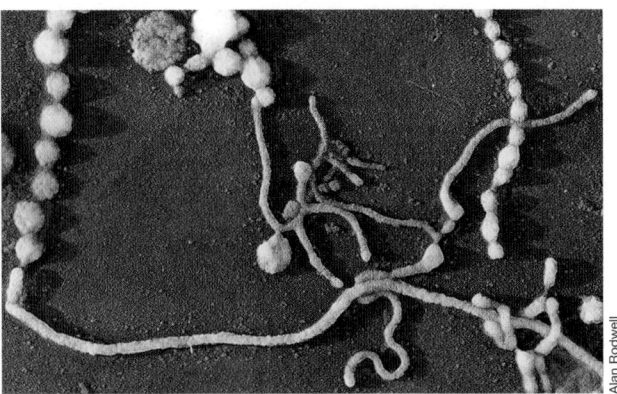

Figure 16.25 *Mycoplasma mycoides*. Metal-shadowed transmission electron micrograph. Note the coccoid and hyphae-like elements. The average diameter of cells in chains is about 0.5 μm.

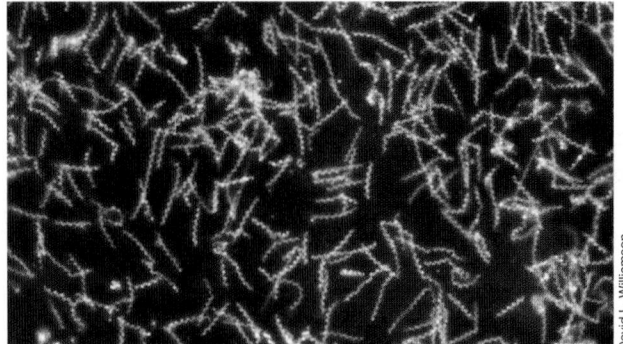

Figure 16.27 "Sex-ratio" spiroplasma from the hemolymph of the fly *Drosophila pseudoobscura*. Dark-field micrograph. Female flies infected with the sex-ratio spiroplasma bear only female progeny. Cells are about 0.15 μm in diameter.

mycoplasmas are as sensitive as most *Bacteria* to antibiotics whose targets are other than the cell wall.

Media for the culture of mycoplasmas are typically quite complex. For many species, growth is poor or absent even in complex yeast extract–peptone–beef heart infusion media. Fresh serum or ascitic fluid (peritoneal fluid) is needed as well to provide unsaturated fatty acids and sterols. Some mycoplasmas can be cultivated on relatively simple culture media, however, and even defined media have been developed for some species. Most mycoplasmas use carbohydrates as carbon and energy sources and require vitamins, amino acids, purines, and pyrimidines as growth factors. The energy metabolism of mycoplasmas is variable; some species are strictly aerobic, whereas others are facultative aerobes or obligate anaerobes (Table 16.3).

Spiroplasma

The genus *Spiroplasma* consists of helical or spiral-shaped *Mollicutes*. Amazingly, although they lack a cell wall and flagella, spiroplasmas are motile by means of a rotary (screw) motion or a slow undulation. Intracellular fibrils that are thought to play a role in motility have been demonstrated. The organism has been isolated from ticks, the hemolymph (**Figure 16.27**) and gut of insects, vascular plant fluids and insects that feed on these fluids, and the surfaces of flowers and other plant parts. For example, *Spiroplasma citri* has been isolated from the leaves of citrus plants, where it causes a disease called citrus stubborn disease, and from corn plants suffering from corn stunt disease. A number of other

mycoplasma-like organisms have been detected in diseased plants by electron microscopy, which indicates that a large group of plant-associated *Mollicutes* may exist. Some species of *Spiroplasma* are known that cause insect diseases, such as honeybee spiroplasmosis and lethargy disease of the beetle *Melolontha*.

— MINIQUIZ —

- Why do mycoplasmas need to have stronger cytoplasmic membranes than other bacteria?
- Motile spiroplasmas cannot contain a normal bacterial flagellum; why?

16.10 *Actinobacteria:* Coryneform and Propionic Acid Bacteria

KEY GENERA: *Arthrobacter, Corynebacterium, Propionibacterium*

The other major group of gram-positive bacteria is the *Actinobacteria*, which form their own phylum within the *Bacteria*. The *Actinobacteria* contain rod-shaped to filamentous and primarily aerobic bacteria that are common inhabitants of soil and plant materials. For the most part they are harmless commensals, species of *Mycobacterium* (for example, *Mycobacterium tuberculosis*) being notable exceptions. Some are of great economic value in either the production of antibiotics or certain fermented dairy products. While there are nine orders of *Actinobacteria*, the vast majority of species belong to the order *Actinomycetales* (Figure 16.17). We consider here the coryneform bacteria, species of *Actinomycetales* that have an unusual method of cell division, and the propionic acid bacteria, important agents in the ripening of Swiss cheese.

Coryneform Bacteria

Coryneform bacteria are gram-positive, aerobic, nonmotile, rod-shaped organisms that form irregular-shaped, club-shaped, or V-shaped cell arrangements during growth. V-shaped cells arise as a result of an abrupt movement that occurs just after cell division, a process called *snapping division* (**Figure 16.28**). Snapping division occurs because the cell wall consists of two layers; only the inner layer participates in cross-wall formation, and so after the cross-wall is formed, the two daughter cells remain attached by the

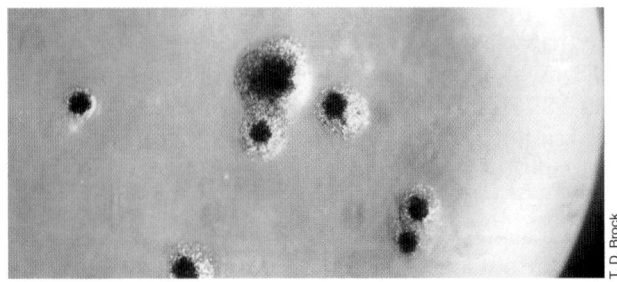

Figure 16.26 Colonies of a *Mycoplasma* species on agar. Note the typical "fried-egg" appearance. The colonies are about 0.5 mm in diameter.

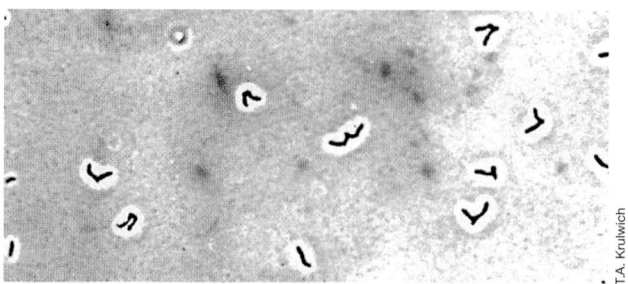

Figure 16.28 Snapping division in *Arthrobacter*. Phase-contrast micrograph of characteristic V-shaped cell groups in *Arthrobacter crystallopoietes* resulting from snapping division. Cells are about 0.9 μm in diameter.

outer layer of the cell wall. Localized rupture of this outer layer on only one side of the cell results in a bending of the two cells away from the ruptured side (**Figure 16.29**) and thus development of V-shaped forms.

The main genera of coryneform bacteria are *Corynebacterium* and *Arthrobacter*. The genus *Corynebacterium* consists of an extremely diverse group of bacteria, including animal and plant pathogens and saprophytes. Some species, such as *Corynebacterium diphtheriae*, are pathogenic (diphtheria, ⮂ Section 30.3). The genus *Arthrobacter*, consisting primarily of soil organisms, is distinguished from *Corynebacterium* on the basis of a developmental cycle involving conversion from rod to coccus and back to rod again (**Figure 16.30**). However, some coryneform bacteria are pleomorphic and form coccoid cells during growth, and so the distinction between the two genera on the basis of life cycle is not absolute. The *Corynebacterium* cell frequently has a swollen end, so it has a club-shaped appearance, whereas *Arthrobacter* species are less commonly club-shaped.

Along with the *Acidobacteria* (Section 16.21), species of *Arthrobacter* are among the most common of all soil bacteria. They are remarkably resistant to desiccation and starvation, despite the fact that they do not form spores or other resting cells. Arthrobacters are a heterogeneous group that have considerable nutritional versatility, and strains have been isolated that decompose herbicides, caffeine, nicotine, phenols, and other unusual organic compounds.

Propionic Acid Bacteria

The **propionic acid bacteria** (genus *Propionibacterium*) were first discovered in Swiss (Emmentaler) cheese, where their fermentative production of CO_2 produces the characteristic holes and the propionic acid they produce is at least partly responsible for

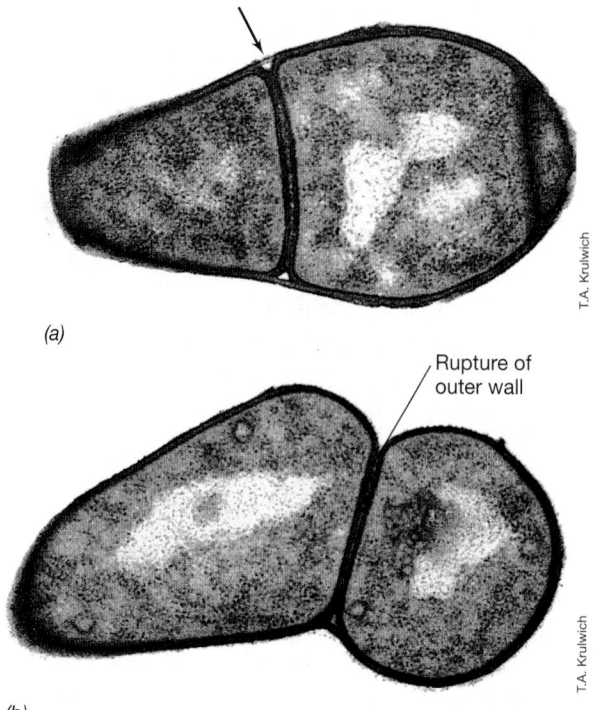

Figure 16.29 Cell division in *Arthrobacter*. Transmission electron micrograph of cell division in *Arthrobacter crystallopoietes,* illustrating how snapping division and V-shaped cell groups arise. *(a)* Before rupture of the outer cell wall layer (arrow). *(b)* After rupture of the outer layer on one side. Cells are 0.9–1 μm in diameter.

the unique flavor of the cheese. The bacteria in this group are gram-positive anaerobes that ferment lactic acid, carbohydrates, and polyhydroxy alcohols, producing primarily propionic acid, acetic acid, and CO_2 (⮂ Section 14.21).

The fermentation of lactate is of interest because lactate itself is a fermentation product of many bacteria (Section 16.6). The starter culture in Swiss cheese manufacture consists of a mixture of homofermentative streptococci and lactobacilli, plus propionic acid bacteria. The homofermentative organisms carry out the initial fermentation of lactose to lactic acid during formation of the curd (protein and fat). After the curd has been drained, the propionic acid bacteria develop rapidly. The eyes (or holes) characteristic of Swiss cheese are formed by the accumulation of CO_2, the gas diffusing through the curd and gathering at weak points. The propionic acid bacteria are thus able to obtain energy anaerobically

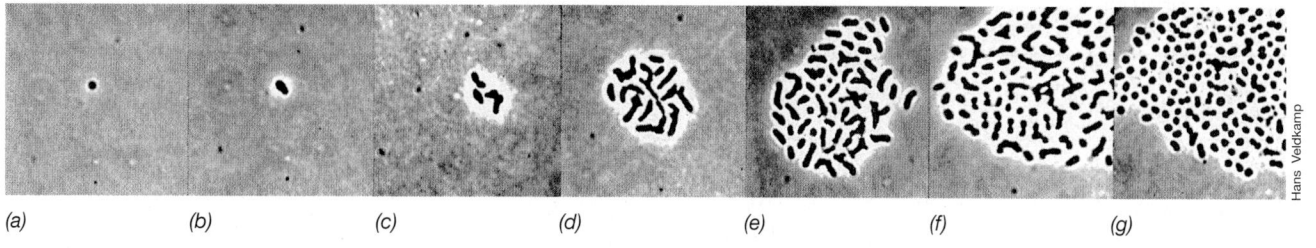

Figure 16.30 Stages in the life cycle of *Arthrobacter globiformis* as observed in slide culture.
(a) Single coccoid element; *(b–e)* conversion to rod and growth of a microcolony consisting predominantly of rods; *(f–g)* conversion of rods to coccoid forms. Cells are about 0.9 μm in diameter.

from a product that other bacteria have produced by fermentation. This metabolic strategy is called a *secondary fermentation*.

Propionate is also formed in the fermentation of succinate by the bacterium *Propionigenium*. This organism is phylogenetically and ecologically unrelated to *Propionibacterium*, but energetic aspects of its fermentation are of considerable interest. We discussed the mechanism of the *Propionigenium* fermentation in Section 14.22.

MINIQUIZ

- What is snapping division and what organism exhibits it?
- What organism is involved in the production of Swiss cheese, and what products does it make that help to flavor the cheese and make the holes?

16.11 *Actinobacteria: Mycobacterium*

KEY GENUS: *Mycobacterium*

Mycobacteria are common in soils and most are harmless, but the genus *Mycobacterium* contains several notable human pathogens, chief among them *Mycobacterium tuberculosis*, the cause of tuberculosis (c⊃ Section 30.4). Species are rod-shaped bacteria that at some stage of their growth cycle possess the distinctive staining property called **acid-fastness**. This property is due to the presence of unique lipids called *mycolic acids*, found only in species of the genus *Mycobacterium*, on the surface of the mycobacterial cell. Mycolic acids are a group of complex branched-chain hydroxylated lipids (**Figure 16.31a**) covalently bound to peptidoglycan in the cell wall; the complex gives the cell surface a waxy, hydrophobic consistency.

Because of their waxy surface, mycobacteria do not stain well with Gram stain. A mixture of the red dye basic fuchsin and phenol is used in the acid-fast (Ziehl–Neelsen) stain. The stain is driven into the cells by slow heating, and the role of the phenol is to enhance penetration of the fuchsin into the lipids. After washing in distilled water, the preparation is decolorized with acid alcohol and counterstained with methylene blue. Cells of acid-fast organisms stain red (Figure 16.31 inset), whereas the background and non-acid-fast organisms appear blue (c⊃ Figure 30.15a).

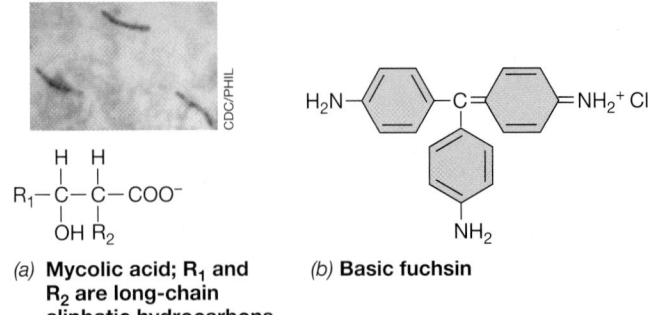

(a) **Mycolic acid; R$_1$ and R$_2$ are long-chain aliphatic hydrocarbons**

(b) **Basic fuchsin**

Figure 16.31 Acid-fast staining. Structure of *(a)* mycolic acid and *(b)* basic fuchsin, the dye used in the acid-fast stain. The fuchsin dye combines with mycolic acids in the cell wall via ionic bonds between COO$^-$ and NH$_2$$^+$. Inset: Acid-fast stain of cells (red) of *Mycobacterium tuberculosis* present in a sputum sample from a tuberculosis patient.

Mycobacteria are somewhat pleomorphic and may undergo branching or even filamentous growth. However, in contrast to the filaments of the actinomycetes (Section 16.12), the filaments of the mycobacteria do not form a true mycelium. Mycobacteria can be separated into two major groups: slow-growing species (e.g., *M. tuberculosis*, *M. avium*, *M. bovis*, and *M. gordonae*) and fast-growing species (e.g., *M. smegmatis*, *M. phlei*, *M. chelonae*, *M. parafortuitum*). *Mycobacterium tuberculosis* is a typical slow grower, and visible colonies are produced from dilute inoculum only after days to weeks of incubation. When growing on solid media, mycobacteria form tight, compact, often wrinkled colonies (**Figure 16.32**). This colony morphology is probably due to the high lipid content and hydrophobic nature of the cell surface, which facilitates cells sticking together.

For the most part, mycobacteria have relatively simple nutritional requirements. Most species can grow aerobically in a simple mineral salts medium with ammonium as the nitrogen source and glycerol or acetate as the sole carbon source and electron donor. Growth of *M. tuberculosis* is more difficult and is stimulated by lipids and fatty acids. The virulence of *M. tuberculosis* cultures has been correlated with the formation of long, cordlike structures (Figure 16.32b) that form as a result of side-to-side aggregation and intertwining of long chains of bacteria. Growth in cords reflects the presence of a characteristic glycolipid, the *cord factor*, on the cell surface (**Figure 16.33**).

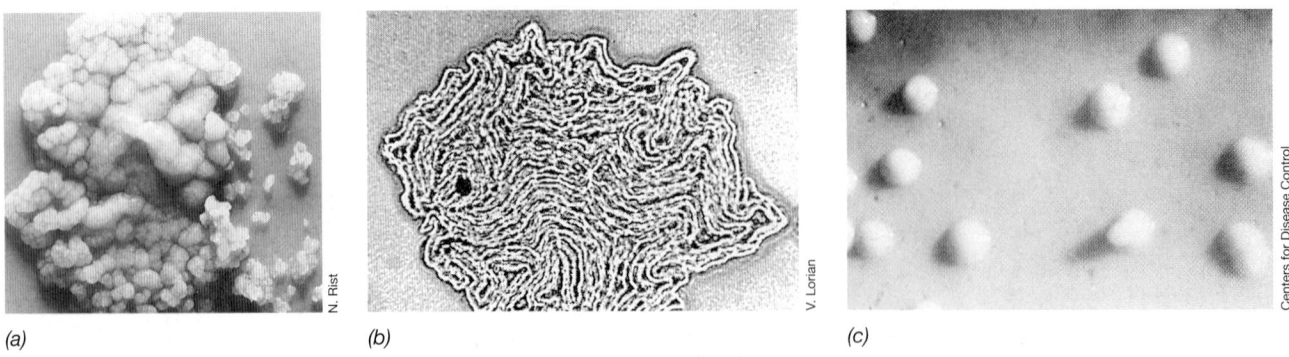

(a)   *(b)*   *(c)*

Figure 16.32 Characteristic colony morphology of mycobacteria. *(a)* *Mycobacterium tuberculosis*, showing the compact, wrinkled appearance of the colony. The colony is about 7 mm in diameter. *(b)* A colony of virulent *M. tuberculosis* at an early stage, showing the characteristic cordlike growth. Individual cells are about 0.5 μm in diameter. (See also the historic drawings of *M. tuberculosis* cells made by Robert Koch, c⊃ Figure 1.31). *(c)* Colonies of *Mycobacterium avium* from a strain of this organism isolated as an opportunistic pathogen from an AIDS patient.

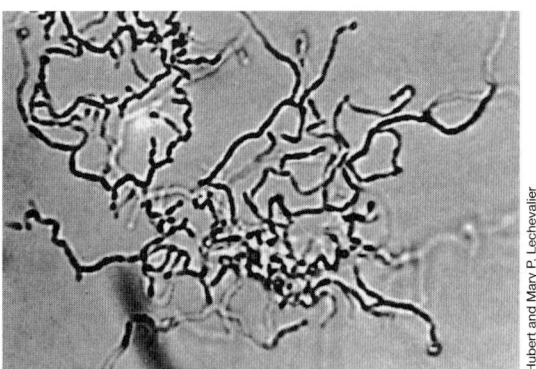

$$CH_2O-\overset{\displaystyle O}{\overset{\|}{C}}-\overset{\displaystyle H}{\underset{|}{\overset{|}{C}}}-\overset{\displaystyle OH}{\underset{|}{\overset{|}{C}}}-C_{60}H_{120}(OH)$$

Figure 16.33 Structure of cord factor, a mycobacterial glycolipid: 6,6′-di-O-mycolyl trehalose. The two identical long-chain dialcohol groups are shown in purple. Inset: Photomicrograph of acid-fast stained cells of *Mycobacterium tuberculosis* (Figure 16.31) that have formed cords.

The pathogenesis of tuberculosis, along with the related mycobacterial disease leprosy, is discussed in Section 30.4.

Some mycobacteria produce yellow carotenoid pigments (Figure 16.32*c*), and pigmentation can aid in identification. Mycobacteria can either be nonpigmented (e.g., *M. tuberculosis, M. bovis, M. smegmatis, M. chelonae*); or can form pigment only when cultured in light, a property called *photochromogenesis* (e.g., *M. parafortuitum*); or can form pigment even when cultured in the dark, a property called *scotochromogenesis* (e.g., *M. gordonae, M. phlei*). Photochromogenesis is triggered by the blue region of the visible spectrum and is characterized by the photoinduction of one of the early enzymes in carotenoid biosynthesis. As with other carotenoid-containing bacteria, it is likely that carotenoids protect mycobacteria against oxidative damage from singlet oxygen (⟢ Section 5.14).

MINIQUIZ

- What is mycolic acid, and what properties does this substance confer on mycobacteria?

16.12 Filamentous *Actinobacteria: Streptomyces* and Relatives

KEY GENERA: *Streptomyces, Actinomyces, Nocardia*

The **actinomycetes** are a large group of phylogenetically related, filamentous and aerobic gram-positive *Bacteria* common in soils. Many actinomycetes have a characteristic developmental cycle that culminates in the production of desiccation-resistant spores. Filaments elongate from their ends and form branching *hyphae*. Hyphal growth results in a network of filaments called a *mycelium* (**Figure 16.34**), analogous to that formed by filamentous fungi (⟢ Section 18.8). When nutrients are depleted, the mycelium forms aerial hyphae that differentiate into spores that allow for survival and dispersal. We focus here on the genus *Streptomyces*, the most important genus in this group.

Figure 16.34 *Nocardia*. A young colony of an actinomycete of the genus *Nocardia*, showing typical filamentous cellular structure (mycelium). Each filament is about 0.8–1 μm in diameter.

Streptomyces

Over 500 species of *Streptomyces* are recognized. *Streptomyces* filaments are typically 0.5–1.0 μm in diameter and of indefinite length, and often lack cross-walls in the vegetative phase. *Streptomyces* grow at the tips of the filaments and may branch often. Thus, the vegetative phase consists of a complex, tightly woven matrix, resulting in a compact, convoluted mycelium and subsequent colony. As the colony ages, characteristic aerial filaments called *sporophores* are formed, which project above the surface of the colony and give rise to spores (**Figure 16.35**).

(a)

(b)

Figure 16.35 Spore-bearing structures of actinomycetes. Phase-contrast micrographs. Compare these photos with the art in Figure 16.37. *(a) Streptomyces,* a monoverticillate type. *(b) Streptomyces,* a closed spiral type. Filaments are about 0.8 μm wide in both types.

UNIT 4

Growth phase
Tip curls
Partitioning of tip
Cell walls thicken and constrict
Spores mature

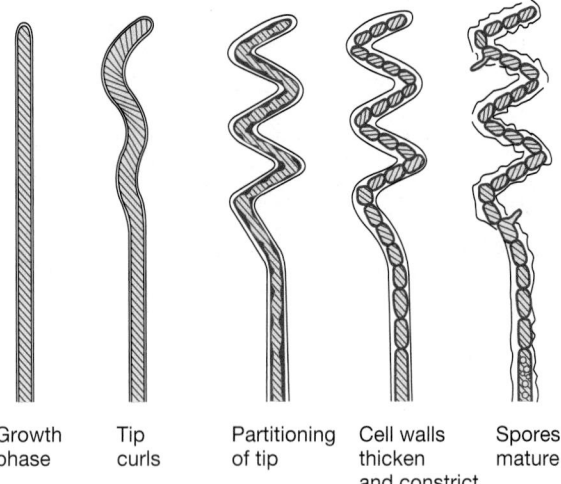

Figure 16.36 Spore formation in *Streptomyces*. Diagram of stages in the conversion of an aerial hypha (sporophore) into spores (conidia).

Streptomyces spores, called *conidia*, are quite distinct from the endospores of *Bacillus* and *Clostridium*. Unlike the elaborate cellular differentiation that leads to the formation of an endospore, conidia are produced by the formation of cross-walls in the multinucleate sporophores followed by separation of the individual cells directly into spores (**Figure 16.36**). Differences in the shape and arrangement of aerial filaments and spore-bearing structures of various species are among the fundamental features used in classifying the *Streptomyces* species (**Figure 16.37**). The conidia and sporophores are often pigmented and contribute a characteristic color to the mature colony (**Figure 16.38**). The dusty appearance of the mature colony, its compact nature, and its color make detection of *Streptomyces* colonies on agar plates relatively easy (Figure 16.38*b*).

Ecology and Isolation of *Streptomyces*

Although a few streptomycetes are aquatic, they are primarily soil organisms. In fact, the characteristic earthy odor of soil is caused by the production by streptomycetes of a series of complex metabolites all called *geosmin*. Alkaline to neutral soils are more favorable for the development of *Streptomyces* than are acid soils. Moreover, higher numbers of *Streptomyces* are found in well-drained soils (such as sandy loams or soils covering limestone), where conditions are more likely to be aerobic, than in waterlogged soils, which quickly become anoxic.

Isolation of *Streptomyces* from soil is relatively easy: A suspension of soil in sterile water is diluted and spread on selective agar medium, and the plates are incubated aerobically at 25°C (Figure 16.38). Media selective for *Streptomyces* contain mineral salts plus polymeric substances such as starch or casein as organic nutrients. Streptomycetes typically produce extracellular hydrolytic enzymes that permit utilization of polysaccharides (starch, cellulose, and hemicellulose), proteins, and fats, and some

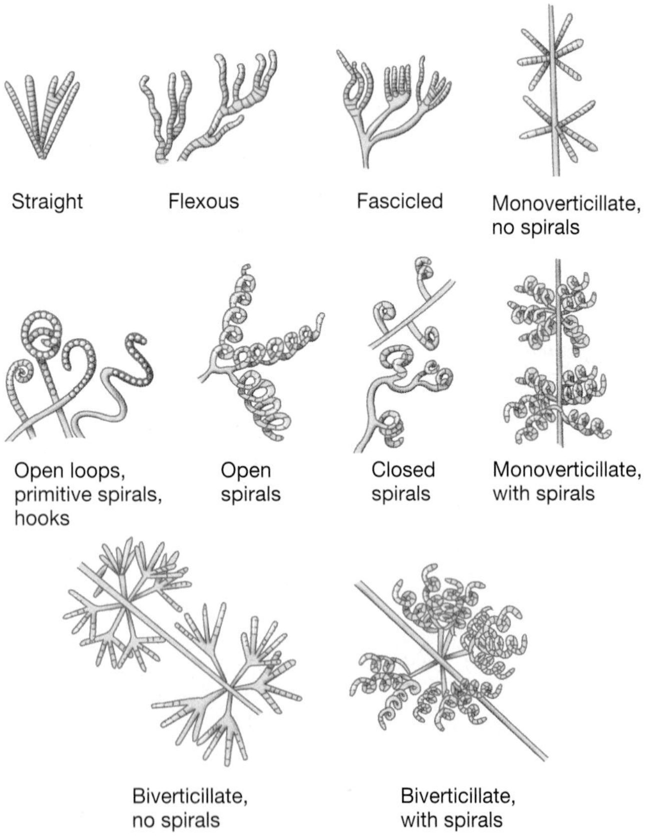

Straight
Flexous
Fascicled
Monoverticillate, no spirals
Open loops, primitive spirals, hooks
Open spirals
Closed spirals
Monoverticillate, with spirals
Biverticillate, no spirals
Biverticillate, with spirals

Figure 16.37 Morphologies of spore-bearing structures in the streptomycetes. A given species of *Streptomyces* produces only one morphological type of spore-bearing structure. The term "verticillate" means "whorls."

M. T. Madigan
(a)
David A. Hopwood
(b)

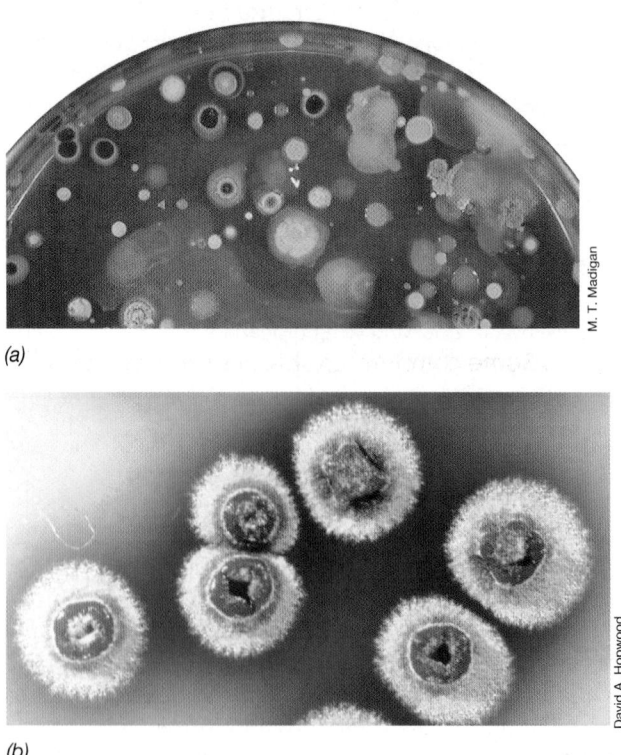

Figure 16.38 Streptomycetes. *(a)* Colonies of *Streptomyces* and other soil bacteria derived from spreading a soil dilution on a casein–starch agar plate. The *Streptomyces* colonies are of various colors (several black *Streptomyces* colonies are near the top of the plate) but can easily be identified by their opaque, rough, nonspreading morphology. *(b)* Close-up photo of colonies of *Streptomyces coelicolor*.

strains can use hydrocarbons, lignin, tannin, and other polymers. After incubation for 5–7 days in air, the plates are examined for the presence of the characteristic *Streptomyces* colonies (Figure 16.38), and spores from colonies can be restreaked to isolate pure cultures.

Antibiotics of *Streptomyces*

Perhaps the most striking physiological property of the streptomycetes is the extent to which they produce *antibiotics* (Table 16.4). Evidence for antibiotic production is often seen on the agar plates used in their initial isolation: Adjacent colonies of other bacteria show zones of inhibition (**Figure 16.39***a*).

About 50% of all *Streptomyces* isolated have been found to be antibiotic producers. Over 500 distinct antibiotics are produced by streptomycetes and many more are suspected; most of these have been identified chemically. Some species produce more than one antibiotic, and often the several antibiotics produced by one organism are chemically unrelated. Although an antibiotic-producing organism is resistant to its own antibiotics, it usually remains sensitive to antibiotics produced by other streptomycetes. Many genes are required to encode the enzymes for antibiotic synthesis, and because of this, the genomes of *Streptomyces* species are typically quite large (8 Mbp and larger; ⇄ Table 9.1). More than 60 streptomycete antibiotics have been used in human and veterinary medicine, and some of the most commonly used are listed in Table 16.4.

Ironically, despite the extensive research done on antibiotic-producing streptomycetes by the antibiotic industry and the fact that *Streptomyces* antibiotics are a multibillion-dollar-a-year industry, the ecology of *Streptomyces* remains poorly understood. The interactions of these organisms with other bacteria and the ecological rationale for antibiotic production remains an important topic about which we know very little. One hypothesis for why *Streptomyces* species produce antibiotics is that antibiotic production, which is linked to sporulation (a process itself triggered by nutrient depletion), might be a mechanism to inhibit the growth of other organisms competing with *Streptomyces* cells for limiting

nutrients. This would allow the *Streptomyces* to complete the sporulation process and form a dormant structure that would increase their chances of survival.

MINIQUIZ

- Contrast spores and sporulation in *Streptomyces* and *Bacillus* species.
- Why might antibiotic production be of advantage to streptomycetes?

III • *Bacteroidetes*

The phylum *Bacteroidetes* contains more than 700 characterized species spread across four orders: *Bacteroidales, Cytophagales, Flavobacteriales,* and *Sphingobacteriales* (**Figure 16.40**). The *Bacteroidetes* are gram-negative nonsporulating rods; species are typically saccharolytic and can be aerobic or fermentative, including obligate aerobes, facultative aerobes, and obligate anaerobes. Gliding motility (⇄ Section 2.12) is widespread in the phylum, though many species are nonmotile and a few are motile by flagella. The genus *Bacteroides* has been particularly well studied as these organisms are a major component of the microbial community in the human gut.

16.13 *Bacteroidales*

KEY GENUS: *Bacteroides*

The order *Bacteroidales* primarily contains obligately anaerobic fermentative species. The type genus is *Bacteroides,* which contains species that are saccharolytic, fermenting sugars or proteins (depending on the species) to acetate and succinate as major fermentation products. *Bacteroides* are normally commensals, found in the intestinal tract of humans and other animals. In fact, *Bacteroides* species are the numerically dominant bacteria in the human large intestine, where measurements have shown that

---

TABLE 16.4 Some common antibiotics synthesized by species of *Streptomyces* and related *Actinobacteria*

Chemical class	Common name	Produced by	Active against[a]
Aminoglycosides	Streptomycin	*S. griseus*[b]	Most gram-negative *Bacteria*
	Spectinomycin	*Streptomyces* spp.	*Mycobacterium tuberculosis,* penicillinase-producing *Neisseria gonorrhoeae*
	Neomycin	*S. fradiae*	Broad spectrum, usually used in topical applications because of toxicity
Tetracyclines	Tetracycline	*S. aureofaciens*	Broad spectrum, gram-positive and gram-negative *Bacteria,* rickettsias and chlamydias, *Mycoplasma*
	Chlortetracycline	*S. aureofaciens*	As for tetracycline
Macrolides	Erythromycin	*Saccharopolyspora erythraea*	Most gram-positive *Bacteria,* frequently used in place of penicillin; *Legionella*
	Clindamycin	*S. lincolnensis*	Effective against obligate anaerobes, especially *Bacteroides fragilis,* the major cause of anaerobic peritoneal infections
Polyenes	Nystatin	*S. noursei*	Fungi, especially *Candida* (a yeast) infections
	Amphotericin B	*S. nodosus*	Fungi
None	Chloramphenicol	*S. venezuelae*	Broad spectrum; drug of choice for typhoid fever

[a]Most antibiotics are effective against several different *Bacteria.* The entries in this column refer to the common clinical application of a given antibiotic. The structures and mode of action of many of these antibiotics are discussed in Sections 28.10–28.12.
[b]All species names beginning with an "*S.*" are species of *Streptomyces.*

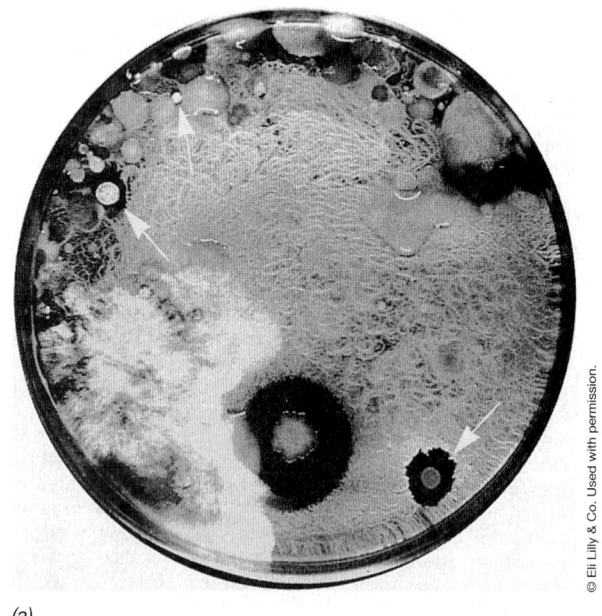

(a)

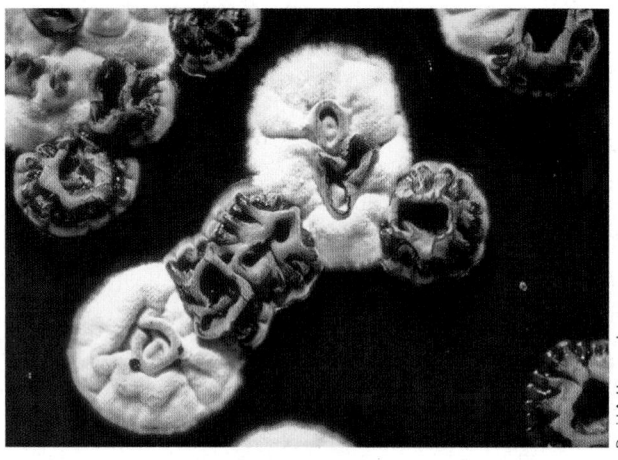

(b)

Figure 16.39 Antibiotics from *Streptomyces*. *(a)* Antibiotic action of soil microorganisms on a crowded agar plate. The smaller colonies surrounded by inhibition zones (arrows) are streptomycetes; the larger, spreading colonies are *Bacillus* species, some of which are also producing antibiotics. *(b)* The red-colored antibiotic undecylprodigiosin is being excreted by colonies of *S. coelicolor*.

about 10^{11} prokaryotic cells are present per gram of wet feces (➶ Section 24.2). However, species of *Bacteroides* can occasionally be pathogens and are the most important anaerobic bacteria associated with human infections such as *bacteremia* (bacteria in the blood).

Bacteroides thetaiotaomicron is one of the most prominent species of *Bacteroides* found in the lumen of the large intestine. *B. thetaiotaomicron* specializes in the degradation of complex polysaccharides. A majority of its genome is devoted to making enzymes that degrade polysaccharides. The diversity and number of genes for carbohydrate metabolism found in its genome far exceeds those found in any other bacterial species. *B. thetaiotaomicron* produces many enzymes that are not encoded by the human genome and thus it vastly increases the diversity of plant polymers that can be degraded in the human digestive tract.

Bacteroidetes

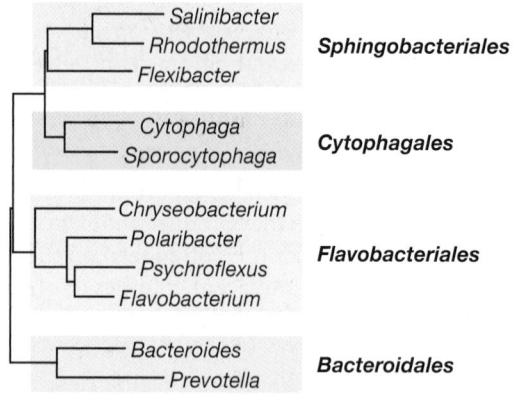

Figure 16.40 Major orders of *Bacteroidetes*. The phylogenetic tree was constructed from 16S rRNA gene sequences of representative genera of *Bacteroidetes*. Order names are shown in bold.

Species of *Bacteroides* are unusual in that they are one of the few groups of bacteria to synthesize a special type of lipid called *sphingolipid* (**Figure 16.41**), a collection of lipids characterized by the long-chain amino alcohol sphingosine in place of glycerol in the lipid backbone. Sphingolipids such as sphingomyelin, cerebrosides, and gangliosides are common in mammalian tissues, especially in the brain and other nervous tissues, but are rare in most bacteria. The production of sphingolipids can be found in a number of other genera in the phylum *Bacteroidetes* including *Flectobacillus, Prevotella, Porphyromonas*, and *Sphingobacterium*.

— **MINIQUIZ** —

• What is the role of *Bacteroides thetaiotaomicron* in the human gut?

16.14 *Cytophagales, Flavobacteriales, and Sphingobacteriales*

KEY GENERA: *Cytophaga, Flavobacterium, Flexibacter*

Cytophagales

The order *Cytophagales* (Figure 16.40) contains almost exclusively obligate aerobes, though some species have limited fermentative

(a)

(b)

Figure 16.41 Sphingolipids. Comparison of *(a)* glycerol with *(b)* sphingosine. In sphingolipids, characteristic of *Bacteroides* species, sphingosine is the esterifying alcohol; a fatty acid is bonded by peptide linkage through the N atom (shown in red), and the terminal —OH group (shown in green) can be any of a number of compounds including phosphatidylcholine (sphingomyelin) or various sugars (cerebrosides and gangliosides).

capabilities. Cells are typically long, slender, gram-negative rods, often containing pointed ends, and move by gliding (**Figure 16.42**). Cytophagas specialize in the degradation of complex polysaccharides. They are widespread in toxic soils and freshwaters, where they probably account for much of the bacterial cellulose digestion. Cellulose decomposers can easily be isolated by placing small crumbs of soil on pieces of cellulose filter paper laid on the surface of mineral salts agar. The bacteria attach to and digest the cellulose fibers, forming spreading colonies (Figure 16.42*b*).

Cellulose degradation by cytophagas can proceed by two different mechanisms. The typical mechanism is the free cellulase mechanism in which cells secrete extracellular enzymes called *exoenzymes* that degrade insoluble cellulose outside of the cell. A complex mixture of enzymes is secreted including *processive endocellulases*, which cleave *internal* β-1,4 glycosidic bonds, and *processive exocellulases*, which cleave *terminal* β-1,4 glucosidic bonds, releasing cellobiose. These exoenzymes degrade insoluble cellulose into soluble polysaccharides and disaccharides that can be readily assimilated by cells. *Cytophaga hutchinsonii* does not produce processive cellulases, and its degradation of cellulose likely requires physical contact of cellulose fibers with cellulase enzymes located on the outer surface of its cell wall.

The genus *Cytophaga* contains species that can degrade not only cellulose (Figure 16.42*c*) but also agar (Figure 16.42*a*) and chitin. In pure culture *Cytophaga* can be grown on agar containing embedded cellulose fibers (Figure 16.42*b*). The related genus *Sporocytophaga* is similar to *Cytophaga* in morphology and physiology, but the cells form resting spherical structures called *microcysts* (Figure 16.42*d*), similar to those produced by some fruiting myxobacteria (⮡ Section 15.17).

Several species of *Cytophaga* are fish pathogens and can cause serious problems in the cultivated fish industry. Two of the most important diseases are *columnaris disease*, caused by *Cytophaga columnaris*, and *cold-water disease*, caused by *Cytophaga psychrophila*. Both diseases preferentially affect stressed fish, such as those living in waters receiving pollutant discharges or living in high-density confinement situations such as fish hatcheries and aquaculture facilities. Infected fish show tissue destruction, frequently around the gills, probably from proteolytic activities of the *Cytophaga* pathogen.

Flavobacteriales and *Sphingobacteriales*

Flavobacteriales and *Sphingobacteriales* (Figure 16.40) typically contain aerobic and facultatively aerobic chemoorganotrophs. Like most *Bacteroidetes*, these organisms are gram-negative rods, and are saccharolytic with many species motile by gliding. Species are found widely in soils and in aquatic habitats, where they typically degrade complex polysaccharides.

Flavobacteriales can be particularly abundant in marine waters including aquatic systems in polar environments. *Flavobacterium* species are primarily found in aquatic habitats, both freshwater and marine, as well as in foods and food-processing plants. Most species are obligate aerobes, though some species are able to reduce nitrate in an anaerobic respiration. Flavobacteria frequently produce yellow pigments and are generally saccharophilic; most can also degrade starch and proteins. Flavobacteria are rarely pathogenic; however, one species, *Flavobacterium meningosepticum*,

(a)

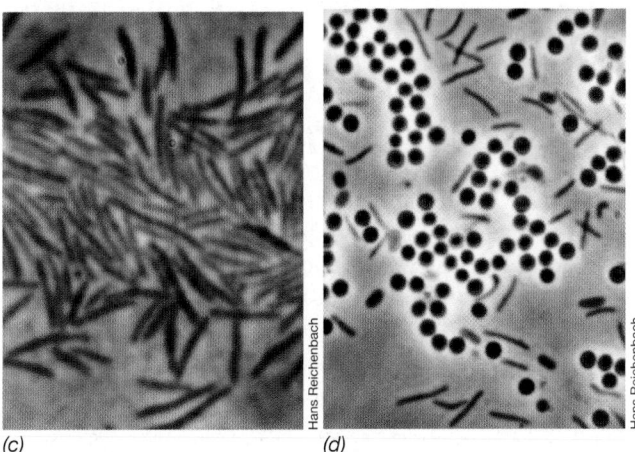

(b)

(c)   *(d)*

Figure 16.42 *Cytophaga* and *Sporocytophaga*. *(a)* Streak of an agarolytic marine *Cytophaga* hydrolyzing agar in a Petri dish. *(b)* Colonies of *Sporocytophaga* growing on cellulose. Note the clearing zones (arrows) where the cellulose has been degraded. *(c)* Phase-contrast micrograph of cells of *Cytophaga hutchinsonii* grown on cellulose filter paper (cells are about 1.5 μm in diameter). *(d)* Phase-contrast micrograph of the rod-shaped cells and spherical microcysts of *Sporocytophaga myxococcoides* (cells are about 0.5 μm and microcysts about 1.5 μm in diameter). Although *Sporocytophaga* microcysts are only slightly more heat-tolerant than vegetative cells, they are extremely resistant to desiccation and thus help the organism survive dry periods in soil. The genera *Cytophaga* and *Sporocytophaga* form a major clade within the phylum *Bacteriodetes* (Figure 16.40).

has been implicated in cases of infant meningitis, and several fish pathogens are also known.

Some *Flavobacteriales* are psychrophilic or psychrotolerant (↩ Section 5.10). These include, in particular, the genera *Polaribacter* and *Psychroflexus*, organisms commonly isolated from cold environments, especially permanently cold environments such as polar waters and sea ice. Many related genera are also capable of good growth below 20°C and can thus be agents of food spoilage. None are pathogenic.

Sphingobacteriales are phenotypically similar to many *Flavobacteriales*. In terms of physiology, species of *Sphingobacteriales* are generally able to degrade a greater breadth of complex polysaccharides than are *Flavobacteriales*, and in this regard they resemble species of *Cytophagales*. The genus *Flexibacter* is typical of many genera of *Sphingobacteriales*. Species of *Flexibacter* differ from those of *Cytophaga* in that they usually require complex media for good growth and are unable to degrade cellulose. Cells of some *Flexibacter* species also undergo changes in cell morphology from long, gliding, threadlike filaments lacking cross-walls to short, nonmotile rods. Many flexibacteria are pigmented due to carotenoids located in their cytoplasmic membrane, or from related pigments called *flexirubins*, located in the cell's outer membrane. *Flexibacter* species are common in soil and freshwaters where they degrade polysaccharides, and none have been identified as pathogens.

— MINIQUIZ —

- Describe a method for isolating *Cytophaga* species from nature.
- What characteristics are shared between the genera *Cytophaga* and *Bacteroides*, and in what ways do they differ?

IV • *Chlamydiae, Planctomycetes, and Verrucomicrobia*

The phyla *Chlamydiae, Planctomycetes*, and *Verrucomicrobia* share an ancestor and are more closely related to each other than to other bacterial phyla (**Figure 16.43**). These three groups contain organisms that can be found in a variety of habitats including soils, aquatic systems, and in association with eukaryotic hosts. We first consider the chlamydias, a group of small gram-negative bacteria that cause some serious human and animal diseases.

16.15 *Chlamydiae*

KEY GENERA: *Chlamydia, Chlamydophila, Parachlamydia*

The phylum *Chlamydiae* contains a single order, the *Chlamydiales*. The entire phylum consists of obligate intracellular parasites of eukaryotes. Though the species that are human pathogens have been characterized in most detail, the phylum contains diverse species that interact with a wide variety of eukaryotic hosts. Species are typically very small cocci, approximately 0.5 μm in diameter, and display a distinctive developmental cycle. Like many obligate parasites and symbionts, the genomes of *Chlamydiae* are typically reduced, ranging in size from 0.55–1 Mbp (↩ Section 9.3).

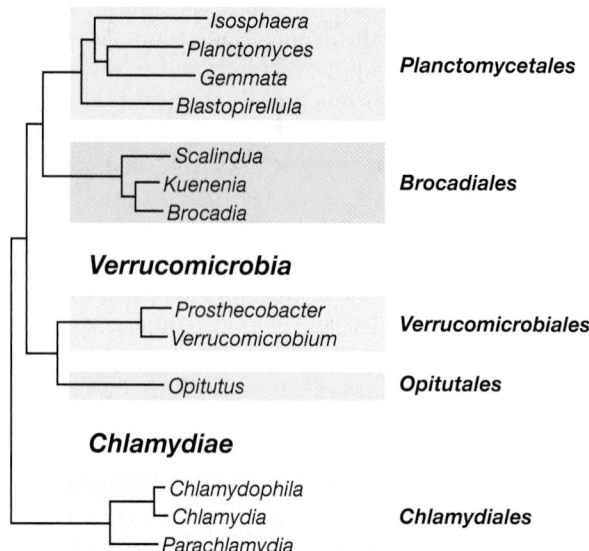

Figure 16.43 Major orders of *Chlamydiae, Planctomycetes*, and *Verrucomicrobia*. The phylogenetic tree was constructed from 16S rRNA gene sequences of representative genera of *Chlamydiae, Planctomycetes*, and *Verrucomicrobia*. Order names are shown in bold.

Life Cycle of *Chlamydiae*

All species of *Chlamydiae* demonstrate a unique chlamydial life cycle (**Figure 16.44**). Two types of cells are seen in the life cycle: (1) a small, dense cell, called an *elementary body*, which is relatively resistant to drying and is the means of dispersal, and (2) a larger, less dense cell, called a *reticulate body*, which divides by binary fission and is the vegetative form.

Elementary bodies are nonmultiplying cells specialized for infectious transmission. By contrast, reticulate bodies are noninfectious forms that function only to multiply inside host cells to form a large inoculum for transmission. Unlike the rickettsias, the chlamydias are not transmitted by arthropods but are primarily airborne invaders of the respiratory system—hence the significance of resistance to drying of the elementary bodies. A dividing reticulate body can be seen in **Figure 16.45**. After a number of cell divisions, these vegetative cells are converted into elementary bodies that are released when the host cell disintegrates (Figure 16.44*b*) and can then infect other nearby host cells. Generation times of 2–3 h have been measured for reticulate bodies, considerably faster than times found for the rickettsias (Section 16.1).

Notable Genera of *Chlamydiae*

Chlamydiae are particularly well adapted to invading and colonizing eukaryotic cells, and different species can infect a diverse array of eukaryotic hosts. The species *Parachlamydia acanthamoebae* infects free-living amoebae, particularly amoebae in the genus *Acanthamoeba*. *Parachlamydia* demonstrates the typical chlamydial life cycle during infection of amoebae (Figure 16.44). Most species of *Chlamydiae* can multiply or survive within free-living amoebae, and these hosts may be important for the survival and dispersal of *Chlamydiae* in nature. A diversity of 16S rRNA gene sequences from *Chlamydiae* can be detected in natural environments, suggesting

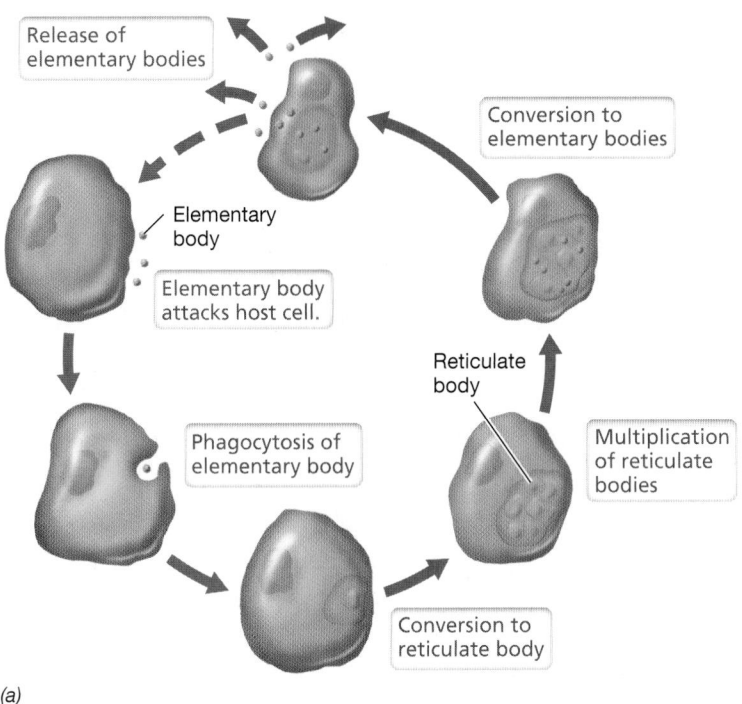

(a)

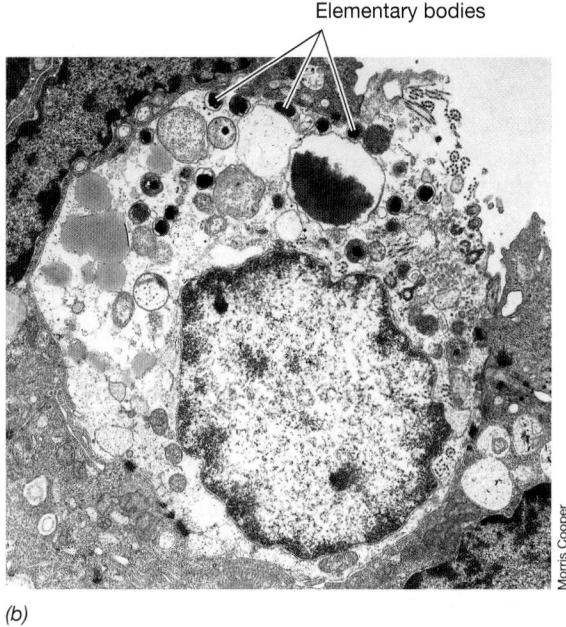

Elementary bodies

Morris Cooper

(b)

Figure 16.44 The infection cycle of a chlamydia. *(a)* Schematic diagram of the cycle: The entire cycle takes about 48 h. *(b)* Human chlamydial infection. Elementary bodies (~0.3 μm in diameter) are the infectious form and reticulate bodies (~1 μm in diameter) are the multiplying form. An infected fallopian tube cell is bursting, releasing mature elementary bodies.

that these organisms are widespread and that many of their natural hosts have yet to be identified. While free-living amoebae are the natural hosts for *P. acanthamoebae*, this species can also infect humans, although only weakly compared with *Chlamydiae* whose natural hosts are human.

The best-studied human pathogens are found in the genera *Chlamydia* and *Chlamydophila*. Several species are recognized within these genera: *Chlamydophila psittaci*, the causative agent of the disease psittacosis; *Chlamydia trachomatis*, the causative agent of trachoma and a variety of other human diseases; and *Chlamydophila pneumoniae*, the cause of some respiratory syndromes. Psittacosis is an epidemic disease of birds that is occasionally transmitted to humans and causes pneumonia-like symptoms. Trachoma, a debilitating disease of the eye characterized by vascularization and scarring of the cornea, is the leading cause of blindness in humans. Other strains of *C. trachomatis* infect the genitourinary tract, and chlamydial infections are currently one of the leading sexually transmitted diseases (⮂ Section 30.14).

Molecular and Metabolic Properties

The chlamydias are among the most biochemically limited of all known *Bacteria*. Indeed, their genomes, approximately 1 Mbp in size, appear to be even more biosynthetically limited than those of the rickettsias, the other group of obligate intracellular parasites known among the *Bacteria* (Section 16.1). Interestingly, the *C. trachomatis* genome lacks a gene encoding the protein FtsZ, a key protein in septum formation during cell division (⮂ Section 7.3) and thought to be indispensable for growth of all prokaryotic cells. Also, some genes in *C. trachomatis* are distinctly *eukaryotic*, suggesting horizontal transfer from host to bacterium; these genes may encode functions that facilitate the pathogenic lifestyle of *C. trachomatis* (⮂ Section 30.14). In sum, the chlamydias appear to have evolved an efficient and effective survival strategy including parasitizing the resources of the host and producing resistant cell forms for transmission.

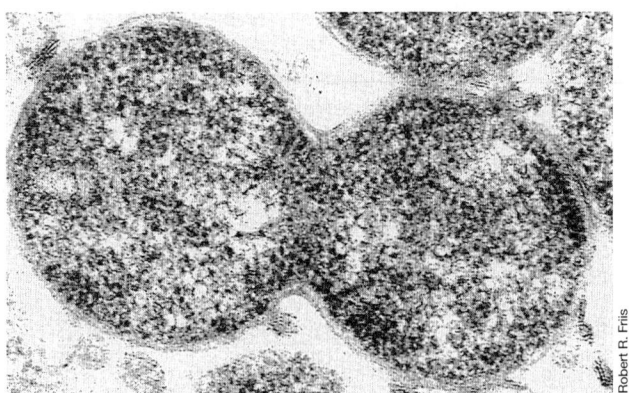

Robert R. Friis

Figure 16.45 *Chlamydia*. Thin-section electron micrograph of a dividing reticulate body of *Chlamydophila psittaci* within a mouse tissue-culture cell. A single chlamydial cell is about 1 μm in diameter.

--- **MINIQUIZ** ---

- How are *Chlamydia* and *Mycoplasma* (Section 16.9) similar? How are they different?

- What is the difference between an elementary body and a reticulate body?

16.16 *Planctomycetes*

KEY GENERA: *Planctomyces, Blastopirellula, Gemmata, Brocadia*

The phylum *Planctomycetes* contains several morphologically unique bacteria found primarily in two orders, *Planctomycetales* and *Brocadiales* (Figure 16.43).

Planctomycetes are gram-negative bacteria and many divide by budding. They often have stalks or appendages and their cells arranged in rosettes. *Planctomycetes* are unusual among bacteria because they can have an S-layer in their cell envelope (⊂⊃ Section 2.6). Another remarkable feature of *Planctomycetes* is that they often contain intracellular compartments that resemble the organelles of eukaryotes.

Compartmentalization in *Planctomycetes*

We learned in Section 1.2 of the major structural differences between prokaryotic and eukaryotic cells. In particular, eukaryotes have a membrane-enclosed nucleus whereas in *Bacteria* and *Archaea*, DNA supercoils and compacts to form the nucleoid present in the cytoplasm. However, *Planctomycetes* are unique among all known bacteria in that they show evidence of cell compartmentalization.

All *Planctomycetes* produce a structure enclosed by a nonunit membrane and called a *pirellulosome*; this structure contains the nucleoid, ribosomes, and other necessary cytoplasmic components. But in some *Planctomycetes*, for example, in the bacterium *Gemmata* (**Figure 16.46**), the nucleoid itself is surrounded by invaginations of the cell membrane. DNA in *Gemmata* remains in a covalently closed, circular, and supercoiled form, typical of bacteria (⊂⊃ Section 4.2), but it is highly condensed and remains

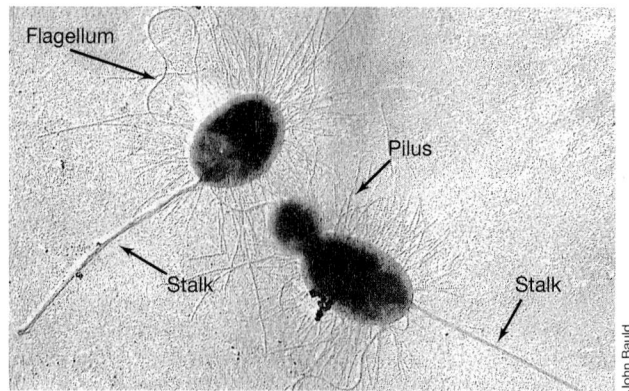

Figure 16.47 *Planctomyces maris*. Metal-shadowed transmission electron micrograph. A single cell is about 1–1.5 μm long. Note the fibrillar nature of the stalk. Pili are also abundant. Note also the flagella (curly appendages) on each cell and the bud that is developing from the nonstalked pole of one cell.

partitioned from the remaining cytoplasm by a true unit membrane (Figure 16.46) (see page 494 for more on this).

Another interesting compartment is the *anammoxosome*, found in species of the *Brocadiales* including *Brocadia anammoxidans*. This bacterium catalyzes the anaerobic oxidation of ammonia (NH_3) within the anammoxosome structure. The anammoxosome membrane is composed of unique lipids that form a tight seal, protecting cytoplasmic components from toxic intermediates produced during the anaerobic oxidation of ammonia (⊂⊃ Section 14.12).

Planctomyces

Planctomyces is the best-characterized genus in the *Planctomycetes*. In Section 15.20 we considered the stalked proteobacterium *Caulobacter*. *Planctomyces* is also a stalked bacterium (**Figure 16.47**). However, unlike *Caulobacter*, the stalk of *Planctomyces* consists of protein and does not contain a cell wall or cytoplasm (compare Figure 16.47 with Figure 15.56). The *Planctomyces* stalk presumably functions in attachment, but it is a much narrower and finer structure than the prosthecal stalk of *Caulobacter*.

Like *Caulobacter* (⊂⊃ Figures 7.16 and 15.56), *Planctomyces* is a budding bacterium that displays a life cycle. Its motile swarmer cells attach to a surface, grow a stalk from the attachment point, and generate a new cell from the opposite pole by budding. This daughter cell produces a flagellum, breaks away from the attached mother cell, and begins the cycle anew. Physiologically, *Planctomyces* species are facultatively aerobic chemoorganotrophs, growing either by fermentation or respiration of sugars.

The habitat of *Planctomyces* is primarily aquatic, both freshwater and marine, and the genus *Isosphaera* is a filamentous, gliding hot spring bacterium. The isolation of *Planctomyces* and relatives, like that of *Caulobacter*, requires dilute media.

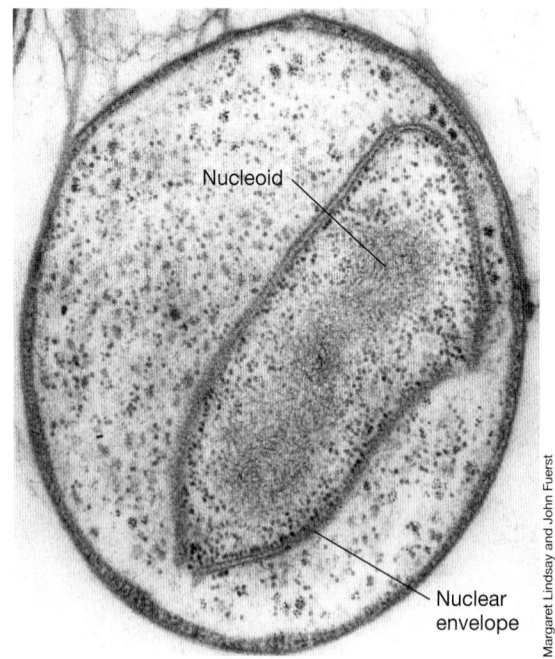

Figure 16.46 *Gemmata*: a nucleated bacterium. Thin-section electron micrograph of a cell of *Gemmata obscuriglobus* showing the nucleoid surrounded by a "nuclear envelope" (see page 494). The cell is about 1.5 μm in diameter.

MINIQUIZ

- How does the stalk of *Planctomyces* differ from the stalk of *Caulobacter*?
- What is unusual about the bacterium *Gemmata*?

16.17 *Verrucomicrobia*

KEY GENERA: *Verrucomicrobium, Prosthecobacter*

The phylum *Verrucomicrobia* contains at least four orders with characterized species, but most are found within the order *Verrucomicrobiales* (Figure 16.43). Species of *Verrucomicrobia* are aerobic or facultatively aerobic bacteria capable of fermenting sugars. An exception is the genus *Methylacidiphilum*, which contains aerobic methanotrophs (↩ Section 15.16). In addition, some *Verrucomicrobia* form symbiotic associations with protists. *Verrucomicrobia* are widespread in nature, inhabiting freshwater and marine environments as well as forest and agricultural soils. The *Verrucomicrobia* can have membrane-bound intracellular structures similar to those found in the *Planctomycetes*. The *Verrucomicrobia* typically form cytoplasmic appendages called *prosthecae* (↩ Section 15.20). *Verrucomicrobia* share with other prosthecate bacteria the presence of peptidoglycan in their cell walls and in this way are clearly distinct from *Planctomycetes*.

The genera *Verrucomicrobium* and *Prosthecobacter* produce two to several prosthecae per cell (**Figure 16.48**). Unlike cells of *Caulobacter* (↩ Figures 7.16 and 15.56), which contain a single prostheca and produce flagellated and nonprosthecate swarmer cells, *Verrucomicrobium* and *Prosthecobacter* divide symmetrically, and both mother and daughter cells contain prosthecae at the time of cell division. The genus name *Verrucomicrobium* derives from Greek roots meaning "warty," which is an appropriate description of cells of *Verrucomicrobium spinosum* with their multiple projecting prosthecae (Figure 16.48).

Species of the genus *Prosthecobacter* contain two genes that show significant homology to the genes that encode tubulin in eukaryotic cells. Tubulin is the key protein that makes up the cytoskeleton of eukaryotic cells (↩ Section 2.16). Although the important cell division protein FtsZ (↩ Section 7.3) is also a tubulin homolog, the *Prosthecobacter* proteins are structurally more similar to eukaryotic tubulin than is FtsZ. The role of the tubulin proteins in *Prosthecobacter* is unknown since a eukaryotic-like cytoskeleton has not been observed in these organisms.

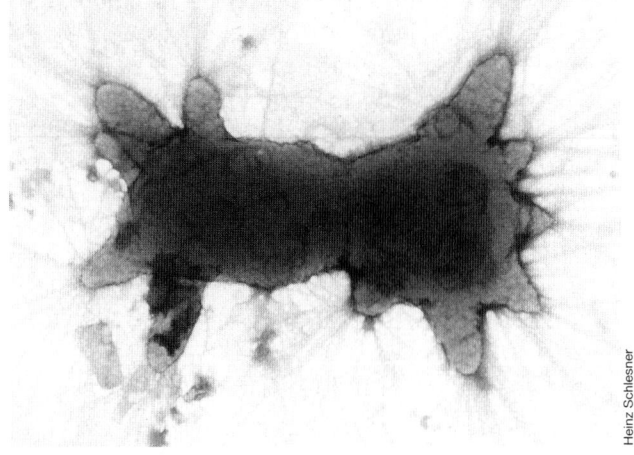

Figure 16.48 *Verrucomicrobium spinosum*. Negatively stained transmission electron micrograph. Note the wartlike prosthecae. A cell is about 1 μm in diameter.

— MINIQUIZ —
- Describe two ways that *Verrucomicrobia* differ from *Planctomycetes*.

V • Hyperthermophilic *Bacteria*

Three phyla of hyperthermophilic bacteria cluster deep in the phylogenetic tree of *Bacteria*, near the root (Figure 16.1). Each group consists of one or two major genera, and a key physiological feature of most species is *hyperthermophily*—optimal growth at temperatures above 80°C (↩ Section 5.11). We begin with *Thermotoga* and *Thermodesulfobacterium*, each representative of its own lineage.

16.18 *Thermotogae* and *Thermodesulfobacteria*

KEY GENERA: *Thermotoga, Thermodesulfobacterium*

Thermotoga species are rod-shaped hyperthermophiles that form a sheathlike envelope (called a *toga*; thus the genus name) (**Figure 16.49a**), stain gram-negatively, and are nonsporulating. *Thermotoga* species are fermentative anaerobes, catabolizing sugars or starch and producing lactate, acetate, CO_2, and H_2 as fermentation products. The organisms can also grow by anaerobic respiration using H_2 as an electron donor and ferric iron as an electron acceptor. Species of *Thermotoga* have been isolated from terrestrial hot springs as well as marine hydrothermal vents.

Despite being bacterial, the genome of *Thermotoga* contains many genes that show strong homology to genes from hyperthermophilic *Archaea*. In fact, over 20% of the genes of *Thermotoga*

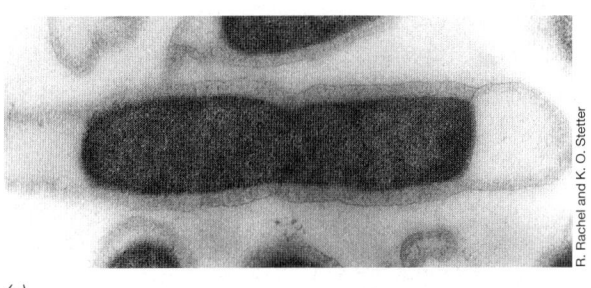
(a)

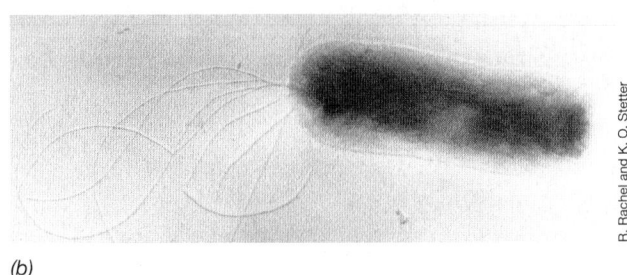
(b)

Figure 16.49 Hyperthermophilic *Bacteria*. Electron micrographs of two hyperthermophiles: *(a) Thermotoga maritima*—temperature optimum, 80°C. Note the outer covering, the toga. *(b) Aquifex pyrophilus*—temperature optimum, 85°C. Cells of *Thermotoga* measure 0.6 × 3.5 μm; cells of *Aquifex* measure 0.5 × 2.5 μm.

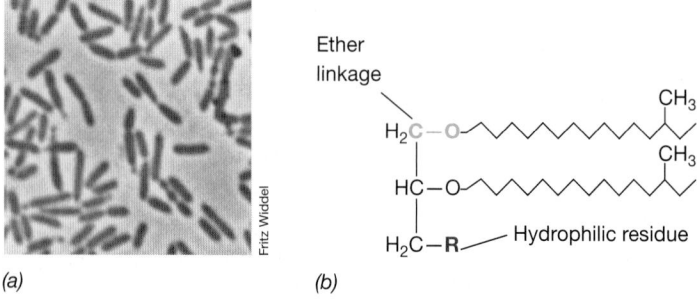

(a)   *(b)*

Figure 16.50 *Thermodesulfobacterium*. *(a)* Phase-contrast micrograph of cells of *Thermodesulfobacterium thermophilum*. *(b)* Structure of one of the lipids of *Thermodesulfobacterium mobile*. Note that although the two hydrophobic side chains are ether-linked, they are not phytanyl units, as in *Archaea*. The designation "R" is for a hydrophilic residue, such as a phosphate group.

probably originated from *Archaea* by horizontal gene transfers (⟳ Sections 9.6 and 13.7). Although a few archaea-like genes have been identified in the genomes of other *Bacteria* and vice versa, only in *Thermotoga* has such large-scale horizontal transfer of genes between domains been detected thus far.

Thermodesulfobacterium (**Figure 16.50**) is a thermophilic sulfate-reducing bacterium, positioned on the phylogenetic tree in a separate phylum after *Thermotoga* and *Aquifex* (Figure 16.1*a*). *Thermodesulfobacterium* is a strict anaerobe that uses compounds such as lactate, pyruvate, and ethanol (but not acetate) as electron donors, as do sulfate-reducing bacteria such as *Desulfovibrio* (⟳ Section 15.9), reducing SO_4^{2-} to H_2S.

An unusual biochemical feature of *Thermodesulfobacterium* is the production of *ether-linked lipids*. Recall that such lipids are a hallmark of the *Archaea* and that a polyisoprenoid C_{20} hydrocarbon (phytanyl) replaces fatty acids as the side chains in archaeal lipids (⟳ Section 2.3). However, the ether-linked lipids in *Thermodesulfobacterium* are unusual because the glycerol side chains are not phytanyl groups, as they are in *Archaea*, but instead are composed of unique C_{17} hydrocarbons along with some fatty acids (Figure 16.50*b*). Thus we see in *Thermodesulfobacterium* both a deep phylogenetic lineage (Figure 16.1) and a lipid profile that combines features of both the *Archaea* and the *Bacteria*. However, a few other *Bacteria* have also been found to contain ether-linked lipids, and thus these lipids may be more common among *Bacteria* than previously thought.

MINIQUIZ

- What is unique about the genome of *Thermotoga* and the lipids of *Thermodesulfobacterium*?

16.19 *Aquificae*

KEY GENERA: *Aquifex, Thermocrinis*

The genus *Aquifex* (Figure 16.49*b*) is an obligately chemolithotrophic and autotrophic hyperthermophile and is the most thermophilic of all known *Bacteria*. Various *Aquifex* species utilize H_2, sulfur (S^0), or thiosulfate ($S_2O_3^{2-}$) as electron donors and O_2 or nitrate (NO_3^-) as electron acceptors, and grow at temperatures up

to 95°C. *Aquifex* can tolerate only very low O_2 concentrations (microaerophilic), and is unable to oxidize all tested organic compounds. *Hydrogenobacter*, a relative of *Aquifex*, shows most of the same properties as *Aquifex*, but is an obligate aerobe.

Aquifex and Autotrophy

Autotrophy in *Aquifex* occurs by way of the reverse citric acid cycle, a series of reactions previously detected only in green sulfur bacteria (⟳ Sections 14.5 and 15.6) within the domain *Bacteria*. The genome sequence of *Aquifex aeolicus* reveals that an entirely chemolithotrophic and autotrophic lifestyle is encoded by a genome of only 1.55 Mbp (one-third the size of the *Escherichia coli* genome). The discovery that so many hyperthermophilic species of *Archaea* and *Bacteria*, like *Aquifex*, are H_2 chemolithotrophs, coupled with the finding that they branch as very early lineages on their respective phylogenetic trees (Figure 16.1*a*), suggests that H_2 was a key electron donor for energy metabolism in primitive organisms that appeared on early Earth (⟳ Sections 13.1 and 17.13).

Thermocrinis

Thermocrinis (**Figure 16.51**) is a relative of *Aquifex* and *Hydrogenobacter*. This bacterium grows optimally at 80°C as a chemolithotroph oxidizing H_2, $S_2O_3^{2-}$, or S^0 as electron donors, with O_2 as electron acceptor. *Thermocrinis ruber*, the only known species, grows in the outflow of certain hot springs in Yellowstone National Park (Figure 16.51*a*) where it forms pink "streamers" consisting of a filamentous form of the cells attached to siliceous sinter (Figure 16.51*b*). In static culture, cells of *T. ruber* grow as individual rod-shaped cells (Figure 16.51*c*). However, when cultured in a flowing system in which growth medium is trickled over a solid glass surface to which cells can attach, *Thermocrinis* assumes the streamer morphology it forms in its constantly flowing habitat in nature.

T. ruber is of historical significance in microbiology because it was one of the organisms discovered in the 1960s by Thomas Brock, a pioneer in the field of thermal microbiology and an author of the first seven editions of this textbook. The discovery by Brock that the pink streamers (Figure 16.51*b*) contained protein and nucleic acids clearly indicated that they were living organisms and not just mineral debris. Moreover, the presence of streamers in 80–90°C hot spring outflow waters but not those of lower temperatures supported Brock's hypothesis that these organisms actually *required* heat for growth and were therefore likely to be present in even boiling or superheated waters. Both of these conclusions were subsequently supported by the discovery by Brock and other microbiologists of dozens of genera of hyperthermophilic *Bacteria* and *Archaea* inhabiting hot springs, hydrothermal vents, and other thermal environments. More coverage of hyperthermophiles can be found in Section 5.11 and Chapter 17.

MINIQUIZ

- Of what evolutionary significance is the fact that organisms in the *Aquifex* lineage are both hyperthermophilic and H_2 chemolithotrophs?

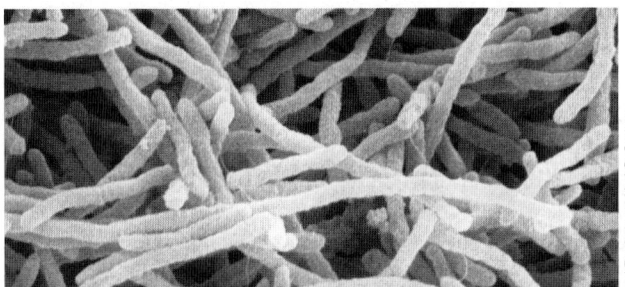

Figure 16.51 *Thermocrinis*. *(a)* Octopus Spring, Yellowstone National Park (USA). The source water of this alkaline and siliceous hot spring is 92°C. *(b)* Cells of *Thermocrinis ruber* growing as filamentous streamers (arrow) attached to siliceous sinter in the outflow (85°C) of Octopus Spring. *(c)* Scanning electron micrograph of rod-shaped cells of *T. ruber* grown on a silicon-coated cover glass. A single cell of *T. ruber* is about 0.4 μm in diameter and from 1 to 3 μm long.

VI • Other *Bacteria*

Thus far in this chapter we have focused on phyla that have many described species (Figure 16.1). Beyond these mainstream bacterial phyla are many others that have but one or at most a handful of characterized species (Figure 16.1*b*). In addition, many more phyla are known only from community sampling of 16S rRNA genes from nature (⇌ Section 19.6). We cannot cover them all. So in this final part of the chapter we consider one phylum that has been well studied and then summarize some other phyla that are emerging into the mainstream of microbial diversity.

16.20 *Deinococcus–Thermus*

KEY GENERA: *Deinococcus, Thermus*

The *deinococci* group contains only a few characterized genera in two orders, the *Deinococcales* and the *Thermales*. Members of this phylum are typically aerobic chemoorganotrophs that metabolize sugars, amino and organic acids, or various complex mixtures. Though deinococci stain gram-positively, they have a gram-negative cell wall structure (**Figure 16.52**) made up of several layers, including an outer membrane, which is characteristic of gram-negative bacteria (⇌ Section 2.5). However, unlike the outer membrane of bacteria such as *Escherichia coli*, the outer membrane of deinococci lack lipid A. Deinococci also contain an unusual form of peptidoglycan in which ornithine replaces diaminopimelic acid in the *N*-acetylmuramic acid cross-links (⇌ Section 2.4).

Species of *Thermales* are typically thermophiles or hyperthermophiles and the type genus is *Thermus*. *Thermus aquaticus*, discovered in a Yellowstone National Park hot spring in the mid-1960s by Thomas Brock (Section 16.19), has been a model organism for studying life at high temperatures. *T. aquaticus* has subsequently been isolated from many geothermal systems, and is the source of *Taq* DNA polymerase. Because it is so heat-stable, *Taq* polymerase allowed the polymerase chain reaction (PCR) technique for amplifying DNA to be fully automated (⇌ Section 12.1), an advance that has revolutionized all of biology.

Radiation Resistance of *Deinococcus radiodurans*

Species of *Deinococcales* have the unusual property of being extremely radiation resistant, and *Deinococcus radiodurans* is the best-studied species in this regard. Most deinococci are red or pink due to carotenoids, and many are highly resistant to both radiation and desiccation. Resistance to ultraviolet (UV) radiation can be used to advantage in isolating deinococci. These remarkable organisms can be selectively isolated from soil, ground meat, dust, and filtered air following exposure of the sample to intense UV (or even gamma) radiation and plating on a rich medium containing tryptone and yeast extract. For example, *D. radiodurans* cells survive exposure to 15,000 grays (Gy) of ionizing radiation (1 Gy = 100 rad). This is sufficient to shatter the organism's chromosome into hundreds of fragments (by contrast, a human can be killed by exposure to less than 10 Gy) (⇌ Section 5.16).

In addition to impressive radiation resistance, *D. radiodurans* is resistant to the effects of many mutagenic agents. The only chemical mutagens that seem to work on *D. radiodurans* are agents such as nitrosoguanidine, which induces deletions in DNA. Deletions are apparently not repaired as efficiently as point mutations in this organism, and mutants of *D. radiodurans* can be isolated in this way.

DNA Repair in *Deinococcus radiodurans*

Studies of *D. radiodurans* have shown that it is highly efficient in repairing damaged DNA. Several different DNA repair enzymes exist in *D. radiodurans*. In addition to the DNA repair enzyme RecA (⇌ Sections 11.4 and 11.5), several RecA-independent DNA systems exist in *D. radiodurans* that can repair breaks in single- or

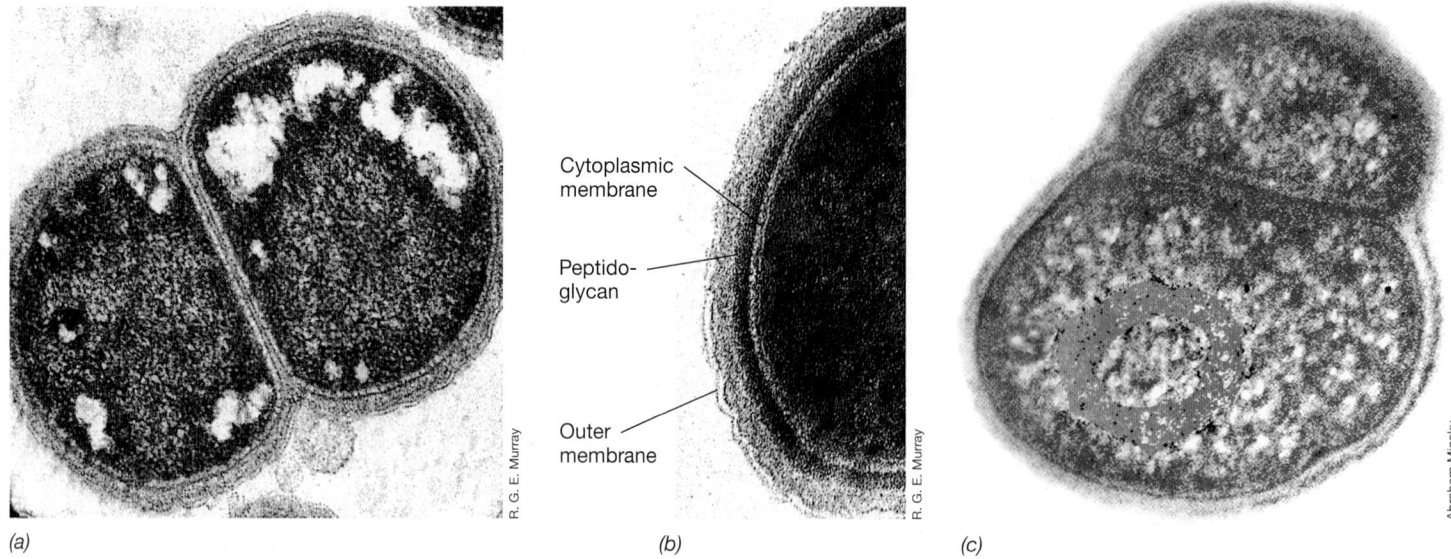

Figure 16.52 The radiation-resistant coccus *Deinococcus radiodurans*. An individual cell is about 2.5 μm in diameter. *(a)* Transmission electron micrograph of *D. radiodurans*. Note the outer membrane layer. *(b)* High-magnification micrograph of wall layer. *(c)* Transmission electron micrograph of cells of *D. radiodurans* colored to show the toroidal morphology of the nucleoid (green).

double-stranded DNA, and excise and repair misincorporated bases. In fact, repair processes are so effective that the chromosome can even be reassembled from a fragmented state.

It is also thought that the unique arrangement of DNA in *D. radiodurans* cells plays a role in radiation resistance. Cells of *D. radiodurans* always exist as pairs or tetrads (Figure 16.52*a*). Instead of scattering DNA within the cell as in a typical nucleoid, DNA in *D. radiodurans* is ordered into a toroidal (coiled, or stack of rings) structure (Figure 16.52*c*). Repair is then facilitated by the fusion of nucleoids from adjacent compartments, because their toroidal structure provides a platform for homologous recombination. From this extensive recombination, a single repaired chromosome emerges, and the cell containing this chromosome can then grow and divide.

> **─── MINIQUIZ ───**
>
> • Describe a commercial application of *Thermus aquaticus*.
>
> • Describe an unusual biological feature of *Deinococcus radiodurans*.

16.21 Other Notable Phyla of *Bacteria*

The basic properties of seven other phyla of *Bacteria* are discussed briefly below. Although most of these have few cultured representatives (Figure 16.1*b*), many may well be of considerable ecological importance. If so, future research on their culture and ecological activities will supply the necessary proof. Until then, we paint a picture of these phyla with a broad brush to summarize their major characteristics in a general way.

Acidobacteria

Acidobacteria are widespread in the environment as revealed by analyses of 16S rRNA genes retrieved from environmental samples

(Figure 16.1*b*). *Acidobacteria* are abundant in soils, particularly acid soils (pH < 6.0) where they often comprise a majority of some soil communities. *Acidobacteria* also inhabit freshwater, hot spring microbial mats, wastewater treatment reactors, and sewage sludge. There is evidence for as many as 25 major subgroups within the *Acidobacteria*, indicating substantial phylogenetic and metabolic diversity of the species in this phylum. Their abundance, widespread distribution, and likely metabolic diversity indicate they play important ecological roles, especially in soil. Unfortunately, while *Acidobacteria* are widespread in the environment, they have proven difficult to cultivate; as a result, few species have been isolated (Figure 16.1*b*) and only a handful of genera have been described.

The few species of *Acidobacteria* that have been characterized are metabolically diverse, including both chemoorganotrophs and photoheterotrophs as well as obligate aerobes and obligately fermentative anaerobes. Three species of *Acidobacteria* have been well characterized, *Acidobacterium capsulatum, Geothrix fermentans*, and *Holophaga foetida*; all are gram-negative chemoorganotrophs. *A. capsulatum* is an acidophilic, encapsulated, obligately aerobic bacterium isolated from acid mine drainage; it utilizes various sugars and organic acids. *G. fermentans*, a strict anaerobe, oxidizes simple organic acids (acetate, propionate, lactate, fumarate) to CO_2 coupled to the reduction of ferric iron as electron acceptor (dissimilative iron reduction, ⟲ Section 15.14), and can also ferment citrate to acetate plus succinate as products. *H. foetida* is a strictly anaerobic homoacetogen (⟲ Section 14.16) that grows by degrading methylated aromatic compounds to acetate. Some *Acidobacteria* degrade polymers such as cellulose and chitin, and at least one genus, *Chloracidobacterium*, is phototrophic (⟲ Section 15.8).

Nitrospirae, Deferribacteres, and Chrysiogenetes

The phylum *Nitrospirae* is named for the genus *Nitrospira*, a chemolithotroph that oxidizes nitrite to nitrate and grows autotrophically ($\Rightarrow$ Section 15.13), as do species of the proteo-bacterium *Nitrobacter* ($\Rightarrow$ Section 15.13). *Nitrospira* inhabits many of the same environments as *Nitrobacter*. However, environmental surveys have shown that *Nitrospira* is much more abundant than *Nitrobacter* in nature, and thus most of the nitrite oxidized in nitrogen-rich environments such as wastewater treatment plants and ammonia-rich soils is probably due to *Nitrospira*. Some species of *Nitrospira* found widely in soils have also been observed to contain the complete pathway for nitrification, oxidizing ammonia all the way to nitrate ($\Rightarrow$ Sections 14.11 and 15.13). Other key *Nitrospirae* include *Leptospirillum*, an aerobic, acidophilic, iron-oxidizing chemolithotroph ($\Rightarrow$ Section 15.15) common in acid mine drainage associated with the mining of coal and iron ($\Rightarrow$ Section 22.2).

The phyla *Deferribacteres* and *Chrysiogenetes* (Figure 16.1) contain anaerobic chemoorganotrophs that display considerable metabolic diversity with respect to the electron acceptors used in anaerobic respirations (Chapter 14 and $\Rightarrow$ Figure 14.25). Most, though not all, species are able to grow through anaerobic respiration of nitrate to nitrite or ammonium. The *Deferribacteres* group is named for the genus *Deferribacter*, a thermophilic dissimilative ferric iron-reducer ($\Rightarrow$ Sections 14.15 and 15.14) that can also reduce nitrate and metal oxides. *Geovibrio* is a related genus that can also grow using elemental sulfur as an electron acceptor ($\Rightarrow$ Section 15.10). The bacterium *Chrysiogenes arsenatis* and its relatives are notable for the ability to couple the oxidation of acetate and a few other organic compounds to the reduction of arsenate as a terminal electron acceptor, reducing it to arsenite. In addition to arsenate, many species of *Chrysiogenetes* can reduce selenate, nitrite, nitrate, thiosulfate, and elemental sulfur in anaerobic respirations ($\Rightarrow$ Sections 14.14 and 14.15).

Synergistetes, Fusobacteria, Fibrobacteres

The phyla *Synergistetes, Fusobacteria,* and *Fibrobacteres* contain relatively few characterized species (Figure 16.1*b*), but those that have been cultured employ fermentative metabolisms. Species in these groups are often associated with the gastrointestinal tracts of animals and some have been associated with human disease.

Synergistetes are gram-negative nonsporulating rods found in association with animals and in anoxic environments in terrestrial and marine systems. Described species are typically obligate anaerobes that degrade proteins and are capable of fermenting amino acids. In animals they are most often found in the gastrointestinal tract; for example, *Synergistes jonesii* inhabits the rumen ($\Rightarrow$ Section 23.13). In humans, species of *Synergistetes* have been associated with certain soft tissue wounds and abscesses, dental plaque, and periodontal conditions.

Fusobacteria are gram-negative nonsporulating rods found in sediments and the gastrointestinal systems and oral cavities of animals. *Fusobacteria* are obligate anaerobes that ferment carbohydrates, peptides, and amino acids. Species of the genus *Fusobacterium* are common components of the human microbiome where they colonize mucous membranes. Different species can be found in the oral cavity, the gastrointestinal tract, and the vagina (Chapter 24). *Fusobacterium nucleatum* is often found in gingival crevices in the human oral cavity. Some fusobacteria may be human pathogens and *F. nucleatum* is often present in patients suffering from periodontal disease.

While 16S ribosomal RNA genes from *Fibrobacteres* can be recovered from a wide range of habitats, the only characterized species have come from either the rumen or gastrointestinal tracts of animals. The genus *Fibrobacter* contains gram-negative fermentative strict anaerobes. However, unlike most *Fusobacteria* and *Synergistetes*, species of *Fibrobacter* are unable to ferment proteins or amino acids and specialize instead in the fermentation of carbohydrates, including cellulose. In the rumen, cellulose is the major source of energy, and in such environments it supports not only cellulolytic bacteria such as *Fibrobacter* but many noncellulolytic anaerobes that use glucose released during cellulose degradation.

— MINIQUIZ —

- What is a major habitat for many species of *Acidobacteria*?
- How do *Nitrospira* and *Deferribacter* differ in terms of lifestyle and metabolism?
- What metabolic characteristics are shared by most *Synergistetes, Fusobacteria,* and *Fibrobacteres,* and what disease in humans has been correlated with the presence of *Synergistetes* and *Fusobacteria*?

MasteringMicrobiology®

Visualize, explore, and **think critically** with Interactive Microbiology, MicroLab Tutors, MicroCareers case studies, and more. MasteringMicrobiology offers practice quizzes, helpful animations, and other study tools for lecture and lab to help you master microbiology.

Chapter Review

I • *Proteobacteria*

16.1 The *Alphaproteobacteria* are the second largest class of *Proteobacteria* and metabolically diverse. Key genera are: *Rhizobium, Rickettsia, Rhodobacter,* and *Caulobacter.*

> **Q** Which genera of *Alphaproteobacteria* are known to form nitrogen-fixing nodules in plants?

16.2 The *Betaproteobacteria* are the third largest class of *Proteobacteria* and metabolically diverse. Key genera are *Burkholderia, Rhodocyclus, Neisseria,* and *Nitrosomonas.*

> **Q** What are the six major orders within the *Betaproteobacteria*?

16.3 The *Gammaproteobacteria* are the largest and most diverse class of *Proteobacteria* and contain many human pathogens. The *Enterobacteriales*, or enteric bacteria, are the most heavily studied of all bacteria. Key genera are *Escherichia* and *Salmonella*.

> **Q** What is the catalase test? What catalase reaction would you expect from an obligate aerobe? What reaction would you expect from an obligate anaerobe?

16.4 The *Pseudomonadales* and *Vibrionales* are among the most common *Gammaproteobacteria*. Key genera are *Pseudomonas* and *Vibrio*.

> **Q** What morphological and physiological features distinguish *Escherichia* from *Vibrio*?

16.5 The *Deltaproteobacteria* and *Epsilonproteobacteria* are smaller and less metabolically diverse classes of *Proteobacteria*. Key genera of *Deltaproteobacteria* are *Myxococcus, Desulfovibrio*, and *Geobacter*. Key genera of *Epsilonproteobacteria* are *Campylobacter* and *Helicobacter*.

> **Q** What types of metabolism are typically associated with members of the *Deltaproteobacteria*?

II • *Firmicutes, Tenericutes,* and *Actinobacteria*

16.6 Lactic acid bacteria such as *Lactobacillus* and *Streptococcus* produce lactate as the primary end product of fermentation, and they have many roles in food production and preservation. The *Firmicutes* are one of the two main phyla of gram-positive bacteria.

> **Q** What is the difference between homofermentative and heterofermentative lactic acid bacteria, and what genera are commonly associated with each of these types?

16.7 Many genera of *Firmicutes* in the orders *Bacillales* and *Clostridiales*, including *Staphylococcus, Listeria*, and *Sarcina*, are unable to form endospores.

> **Q** What characteristics of *Listeria* make it a frequent cause of foodborne illness?

16.8 Production of endospores is a hallmark of the key genera *Bacillus* and *Clostridium* and is only found in the phylum *Firmicutes*.

> **Q** What is a good strategy for isolating spore-forming bacteria from an environmental sample?

16.9 The phylum *Tenericutes* contains the mycoplasmas, organisms that lack cell walls and have very small genomes. Many species are pathogenic for humans, other animals, and plants. The key genus is *Mycoplasma*.

> **Q** What two phyla are most closely related to the *Tenericutes*?

16.10 *Actinobacteria* are the second major phylum of gram-positive bacteria. *Corynebacterium* and *Arthrobacter* are common gram-positive soil bacteria. *Propionibacterium* ferments lactate to propionate and is the key agent responsible for the unique flavor and texture of Swiss cheese.

> **Q** In what sort of environment would you expect to find large numbers of *Actinobacteria*?

16.11 Species of *Actinobacteria* in the genus *Mycobacterium* are mainly harmless soil saprophytes, but *Mycobacterium tuberculosis* causes the disease tuberculosis.

> **Q** How does the cell wall of *Mycobacterium* influence its reaction to the Gram stain and the acid-fast stain?

16.12 The streptomycetes are a large group of filamentous, gram-positive bacteria that form spores at the end of aerial filaments and are found in the phylum *Actinobacteria*. Many clinically useful antibiotics such as tetracycline and neomycin have come from *Streptomyces* species.

> **Q** How are the spores of streptomycetes different from endospores?

III • *Bacteroidetes*

16.13 The phylum *Bacteroidetes* includes gram-negative rods that do not form spores, many of which have gliding motility. Most species in the order *Bacteroidales* are obligate anaerobes that ferment carbohydrates in anoxic environments. The genus *Bacteroides* contains species that are common in the gastrointestinal tract of animals.

> **Q** What species of *Bacteroidetes* is most abundant in the human gastrointestinal tract, and what role does this organism play in the human gut?

16.14 The *Cytophagales* and *Flavobacteriales* are orders in the *Bacteroidetes* that include aerobic bacteria able to degrade complex polysaccharides such as cellulose. These bacteria are important in organic matter decomposition.

> **Q** What is an exoenzyme and why are these types of organisms important in the degradation of cellulose?

IV • *Chlamydiae, Planctomycetes,* and *Verrucomicrobia*

16.15 The phylum *Chlamydiae* includes small obligate intracellular parasites that are adept at invading eukaryotic cells. Many species cause various diseases in humans and other animals.

> **Q** Describe the infection cycle of *Chlamydia*.

16.16 The *Planctomycetes* are a group of stalked, budding bacteria that form intracellular compartments of various types, and show extensive invaginations of the cytoplasmic membrane.

> **Q** What are two types of intracellular compartments that have been observed within the *Planctomycetes*?

16.17 Species of *Verrucomicrobia* are distinguished by the multiple prosthecae on their cells and their unique phylogeny.

> **Q** Verruca is a word that means "wart." How do you think the *Verrucomicrobia* got their name?

V • Hyperthermophilic *Bacteria*

16.18 *Thermotogae* and *Thermodesulfobacteria* form two deeply branching phyla within the *Bacteria*. These hyperthermophilic bacteria have proven that extensive horizontal gene transfer has occurred from *Archaea* to *Bacteria* (*Thermotoga*) and that ether-linked lipids are not limited to the *Archaea* (*Thermodesulfobacterium*).

> **Q** What major physiological property unites species of *Thermotoga, Aquifex,* and *Thermocrinis*?

16.19 The *Aquificae* phylum contains a group of hyperthermophilic, H_2-oxidizing bacteria that form the earliest branch on the tree of the domain *Bacteria*.

> **Q** In what environment might you observe *Thermocrinis ruber,* and what role did this organism play in the discovery of hyperthermophiles?

VI • Other *Bacteria*

16.20 *Deinococcus* and *Thermus* are the major genera in a distinct phylum of *Bacteria*. *Thermus* is the source of the key enzyme in automated PCR, whereas *Deinococcus* is the most radiation-resistant bacterium known, exceeding even endospores in this regard.

> **Q** What are some of the remarkable properties that allow *Deinococcus* to survive exposure to massive doses of radiation?

16.21 *Acidobacteria* are widespread in many environments, especially soils, and show various physiologies. The genus *Nitrospira* includes nitrite-oxidizing bacteria, while species of *Deferribacteres* and *Chrysiogenetes* specialize in various forms of anaerobic respiration. Species of *Synergistetes, Fusobacteria,* and *Fibrobacteres* are fermentative anaerobes that inhabit the gastrointestinal tract and other anoxic niches in animals.

> **Q** What are four ways in which different species of *Acidobacteria* have been shown to generate energy?

Application Questions

1. Enteric bacteria, lactic acid bacteria, and propionic acid bacteria have distinctive metabolic traits that can be used to characterize and identify these organisms. Describe the metabolic characteristics of these organisms, name a genus that belongs to each group, and indicate in what way these organisms can be differentiated.

2. Microorganisms can have a variety of different relationships with oxygen. Describe the terms used to characterize a cell's response to oxygen, and give an example from this chapter of an organism that can be described by each of these terms.

Chapter Glossary

Acid-fastness a property of *Mycobacterium* species in which cells stained with the dye basic fuchsin resist decolorization with acidic alcohol

Actinomycetes a term used to refer to aerobic filamentous bacteria in the phylum *Actinobacteria*

Coryneform bacteria gram-positive, aerobic, nonmotile, rod-shaped organisms with the characteristic of forming irregular-shaped, club-shaped, or V-shaped cell arrangements, typical of several genera of unicellular *Actinobacteria*

Enteric bacteria a large group of gram-negative, rod-shaped *Bacteria* characterized by a facultatively aerobic metabolism and commonly found in the intestines of animals

Heterofermentative in reference to lactic acid bacteria, capable of making more than one fermentation product

High GC gram-positive bacteria a term that refers to bacteria in the *Actinobacteria*

Homofermentative in reference to lactic acid bacteria, producing only lactic acid as a fermentation product

Lactic acid bacteria fermentative bacteria that produce lactic acid, are found in the *Firmicutes*, and are important in the production and preservation of many foods

Low GC gram-positive bacteria a term that refers to bacteria in the *Firmicutes*

Oligotrophic a term that refers to organisms that grow best under low-nutrient conditions

Propionic acid bacteria gram-positive fermentative bacteria that generate propionate as a fermentation end product and are important in the production of cheese

Proteobacteria the largest and most metabolically diverse phylum of bacteria

Pseudomonad a term used to refer to any gram-negative, polarly flagellated, aerobic rod able to use a diverse suite of carbon sources

Diversity of *Archaea*

microbiology**now**

The *Archaea* Just Under Your Feet

The domain *Archaea* is named for the Archaean eon, the period of geological history when life first spread across Earth. In the Archaean, high temperatures and an atmosphere thick in toxic gases enveloped Earth. *Archaea* were once thought to be remnants of this forgotten age since most isolates had been obtained from extreme environments such as volcanic systems or salt ponds. However, in the microbial world, things are not always as they seem.

Our knowledge of microbial diversity has changed dramatically in recent years. Now we can use molecular techniques to characterize an organism's DNA without needing to cultivate it first in the laboratory. One of the earliest discoveries made using molecular techniques was that *Archaea* are not restricted to extreme environments; in fact, they are plentiful in our oceans and soils. The vast majority of these *Archaea* belong to the phylum *Thaumarchaeota*, a diverse group of microbes that account for up to 20% of prokaryotic cells in the oceans and 1% of all microbes in soils. But while the discovery of this novel phylum relied on molecular techniques, puzzling out their purpose required cultivation of strains that could be analyzed in the laboratory.

Thaumarchaeota are ammonia oxidizers and critically important to the biosphere, being major players in the global nitrogen cycle. Prior to the discovery of *Thaumarchaeota*—for over 100 years in fact—microbiologists thought that ammonia oxidation was catalyzed only by *Bacteria*. The marine genus *Nitrosopumilus* was the first of the *Thaumarchaeota* to be isolated, and now *Nitrososphaera viennensis* (see photo) is the first species to be described from soil. *N. viennensis* was isolated from a backyard garden using clues derived from cultivation-independent molecular analyses. Cells of *N. viennensis* form irregular coccoids, and the organism is both mesophilic and neutrophilic. Physiologically, *N. viennensis* is a mixotroph (an organism that can fix CO_2 but which grows best with organic matter present) and conserves energy from either ammonia or urea as chemolithotrophic electron donors.

N. viennensis and related soil *Archaea* carry out nitrification in soils worldwide, a fact that reminds us that when we open our eyes to the microbial world we find that major discoveries are often "just underfoot."

Source: Stieglmeier, M., et al. 2014. *Nitrososphaera viennensis* gen. nov., sp. nov., an aerobic and mesophilic, ammonia-oxidizing archaeon from soil and a member of the archaeal phylum *Thaumarchaeota*. *IJSEM* 64: 2738.

The domain *Archaea* was once thought to contain ancient forms of microbial life that only survived in extreme habitats having environmental conditions similar to those of Earth during the Archaean age (from ~4.0 billion to 2.5 billion years before the present, ♂ Figure 13.1). We now know, however, that *Archaea* compose a considerable fraction of living matter in the biosphere and perform important biogeochemical reactions in soil, the oceans, wetlands, and even in the guts of animals. In this chapter we will learn about the enormous phylogenetic and physiological diversity found within the domain *Archaea*.

The *Archaea* are composed of five phyla: *Euryarchaeota, Crenarchaeota, Thaumarchaeota, Korarchaeota*, and *Nanoarchaeota* (**Figure 17.1**). The exact ancestry and number of archaeal phyla remains a contentious issue, and it is likely that many other phyla of *Archaea* remain to be discovered and described. Given the relationship of *Archaea* and *Eukarya* (♂ Section 13.4), the exploration of genomic diversity within *Archaea* is essential to understanding the evolutionary origins of eukaryotic cells. Many *Archaea* are quite difficult to cultivate in isolation, however, and the use of metagenomics (♂ Section 9.8) and single-cell genome sequencing (♂ Section 9.12) continues to provide remarkable new insights into their phylogenetic and physiological diversity.

While all *Archaea* share certain features, this domain encompasses considerable physiological diversity. Common traits of all *Archaea* include ether-linked lipids, a lack of peptidoglycan in cell walls (Chapter 2), and structurally complex RNA polymerases that resemble those of *Eukarya* (♂ Figure 4.20). Despite these shared features, *Archaea* display diverse metabolisms including various forms of chemoorganotrophy and chemolithotrophy that employ aerobic respiration, anaerobic respiration, or fermentation for energy conservation. For example, chemolithotrophy is well established in

the *Archaea*, with H_2 being a common electron donor (Section 17.13) for species in many phyla and with ammonia oxidation widespread among species of *Thaumarchaeota*. Anaerobic respiration, especially forms employing elemental sulfur (S^0) as electron acceptor, is also prevalent among the *Archaea*, especially in *Crenarchaeota*. And finally, aerobic respiration also occurs widely in *Thaumarchaeota*, among some groups of *Euryarchaeota*, and among certain species of *Crenarchaeota*.

In addition, certain physiological capabilities are uniquely found within the domain *Archaea*. Methane production, for example, is a unique characteristic of **methanogens**, *Euryarchaeota* that conserve energy from the production of methane (♂ Section 14.17). *Methanogenesis* is a globally important process that is uniquely archaeal (♂ Sections 14.17, 21.1, and 21.2). *Archaea* are also well known for containing many examples of **extremophiles**, including **hyperthermophiles** (organisms with growth temperature optima above 80°C), as well as halophiles, acidophiles, and psychrophiles (Chapter 5).

With this brief background and the phylogeny of *Archaea* firmly in mind (Figure 17.1), we now consider the organismal diversity of this fascinating domain of life.

I • *Euryarchaeota*

Euryarchaeota comprise a large and physiologically diverse group of *Archaea*. This phylum includes methanogens as well as many genera of extremely halophilic (salt-loving) *Archaea*. As a study in physiological contrasts, these two groups are remarkable: Methanogens are the strictest of anaerobes while extreme halophiles are primarily obligate aerobes. Other groups of *Euryarchaeota* include the hyperthermophiles *Thermococcus* and

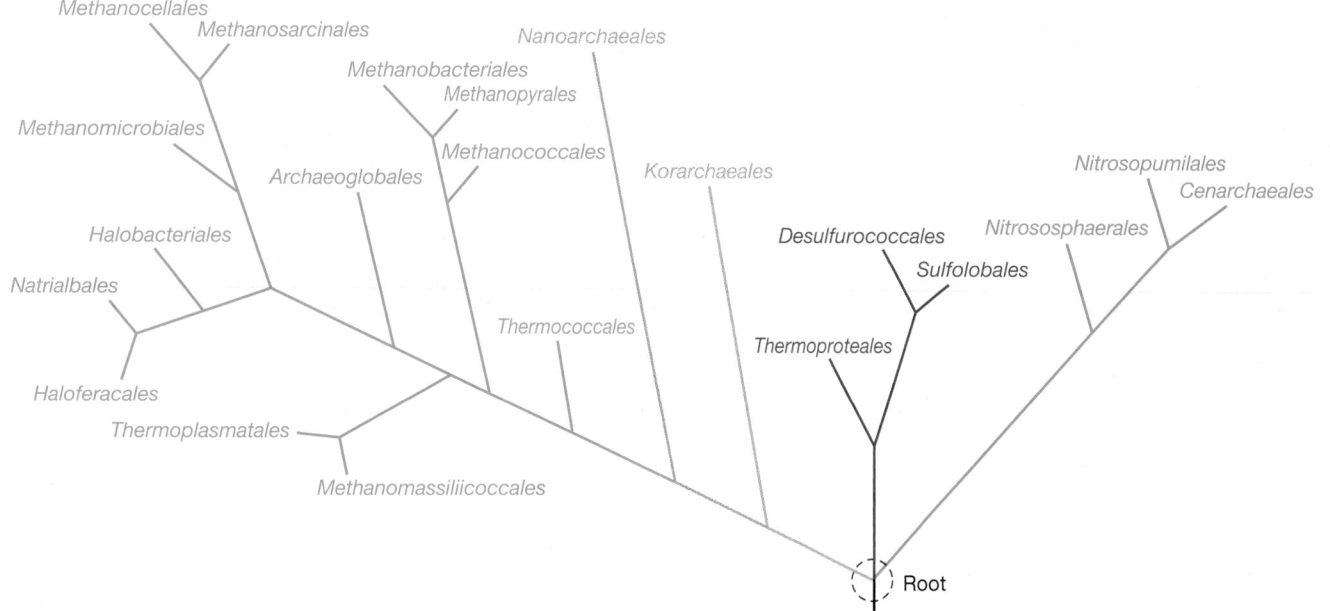

Figure 17.1 Schematic representation of the phylogeny of the major taxonomic orders within the domain *Archaea*. Each of the five archaeal phyla and their major orders are indicated in a different color.

Pyrococcus, the hyperthermophilic methanogen *Methanopyrus*, and the cell wall–less *Thermoplasma*, an organism phenotypically similar to the mycoplasmas (⇔ Section 16.9). We begin our review of *Euryarchaeota* by reviewing the extremely halophilic *Archaea*.

17.1 Extremely Halophilic *Archaea*

KEY GENERA: *Halobacterium, Haloferax, Natronobacterium*

Extremely halophilic *Archaea*, often just called the "haloarchaea," are a diverse group that inhabits environments high in salt. These include naturally salty environments, such as solar salt evaporation ponds and salt lakes, and artificial saline habitats such as the surfaces of heavily salted foods, for example, certain fish and meats. Such salty habitats are called *hypersaline* (**Figure 17.2**). The term **extreme halophile** is used to indicate that these organisms are not only halophilic, but that their requirement for salt is very high, in some cases at levels near saturation (⇔ Figure 5.24).

An organism is considered an extreme halophile if it requires 1.5 M (about 9%) or more sodium chloride (NaCl) for growth. Most species of extreme halophiles require 2–4 M NaCl (12–23%) for optimal growth and can grow at salinities as high as 5.5 M NaCl (32%, the limit of saturation for NaCl), although some species grow very slowly at this salinity. Some phylogenetic relatives of extremely halophilic *Archaea*, for example species of *Haloferax* and *Natronobacterium*, are able to grow at much lower salinities, such as at or near that of seawater (about 2.5% NaCl).

Hypersaline Environments: Chemistry and Productivity

Hypersaline habitats are common throughout the world, but extremely hypersaline habitats are rare. Most such environments are in hot, dry areas of the world. Salt lakes can vary considerably in ionic composition. The predominant ions in a hypersaline lake depend on the surrounding topography, geology, and general climatic conditions.

(a) *(b)*

(c) *(d)*

Figure 17.2 Hypersaline habitats for halophilic *Archaea*. These organisms not only tolerate salt but *require* salt, and typically in large amounts. *(a)* The north arm of Great Salt Lake, Utah, a hypersaline lake in which the ratio of ions is similar to that in seawater, but in which absolute concentrations of ions are several times that of seawater. The green color is primarily from cells of cyanobacteria and green algae. *(b)* Aerial view near San Francisco Bay, California, of a series of seawater evaporating ponds where solar salt is prepared. The red-purple color is predominantly due to bacterioruberins and bacteriorhodopsin in cells of haloarchaea. *(c)* Lake Hamara, Wadi El Natroun, Egypt. A bloom of pigmented haloalkaliphiles is growing in this pH 10 soda lake. Note the deposits of trona (NaHCO$_3$·Na$_2$CO$_3$·2 H$_2$O) around the edge of the lake. *(d)* Scanning electron micrograph of halophilic bacteria including square *Archaea* present in a saltern in Spain.

Great Salt Lake in Utah (USA) (Figure 17.2*a*), for example, is essentially concentrated seawater. In this hypersaline lake the relative proportions of the various ions [e.g., sodium (Na^+), chloride (Cl^-), and sulfate (SO_4^{2-})] are those of seawater, although the overall concentration of ions is much higher. In addition, the pH of this hypersaline lake is slightly alkaline.

Soda lakes, in contrast, are highly alkaline hypersaline environments. The water chemistry of soda lakes resembles that of hypersaline lakes such as Great Salt Lake, but because high levels of carbonate minerals are also present in the surrounding strata, the pH of soda lakes is quite high. Waters of pH 10–12 are not uncommon in these environments (Figure 17.2*c*). In addition, calcium (Ca^{2+}) and Mg^{2+} are virtually absent from soda lakes because they precipitate out at high pH and carbonate concentrations.

The diverse chemistries of hypersaline habitats have selected for a large diversity of halophilic microorganisms. Some organisms are unique to one environment while others are widespread. Moreover, despite their extreme conditions, salt lakes can be highly productive ecosystems (the word *productive* here means high levels of autotrophic CO_2 fixation). *Archaea* are not the only microorganisms present. The eukaryotic alga *Dunaliella* (⇨ Figure 18.33*a*) is the major, if not the sole, oxygenic phototroph in most salt lakes. In highly alkaline soda lakes where *Dunaliella* is absent, anoxygenic phototrophic purple bacteria of the genera *Ectothiorhodospira* and *Halorhodospira* (⇨ Section 15.4) predominate. Organic matter originating from primary production by oxygenic or anoxygenic phototrophs sets the stage for growth of haloarchaea, which are chemoorganotrophic organisms. In addition, a few extremely halophilic chemoorganotrophic *Bacteria*, such as *Halanaerobium*, *Halobacteroides*, and *Salinibacter*, thrive in such environments.

Marine salterns are also habitats for extreme halophiles. Marine salterns are enclosed basins filled with seawater left to evaporate, eventually yielding solar sea salt (Figure 17.2*b, d*). As salterns approach the minimum salinity limits for haloarchaea, the waters turn a reddish purple color due to the massive growth—called a *bloom*—of cells (the red coloration apparent in Figure 17.2*b* and *c* is due to carotenoids and other pigments to be discussed later). Morphologically unusual *Archaea* are often present in salterns, including species with a square or cup-shaped morphology (Figure 17.2*d*). Extreme halophiles are also present in highly salted foods, such as certain types of sausages, marine fish, and salted pork.

Taxonomy and Physiology of Extremely Halophilic *Archaea*

The extremely halophilic *Archaea* are found within the orders *Halobacteriales, Natrialbales*, and *Haloferacales*, which share a common ancestor within the *Euryarchaeota* (Figure 17.1). These three orders constitute the haloarchaea, but they are also sometimes called "halobacteria" because the genus *Halobacterium* (**Figure 17.3**), the best-studied representative of the extreme halophiles, was discovered and characterized prior to the discovery of *Archaea*. Many genera of *Natrialbales*, including *Natronobacterium, Natronomonas*, and their relatives, differ from other extreme halophiles in being extremely alkaliphilic as well as halophilic. As befits their soda lake habitat (Figure 17.2*c*), these natronobacteria grow optimally at low Mg^{2+} concentrations and alkaline pH (9–11).

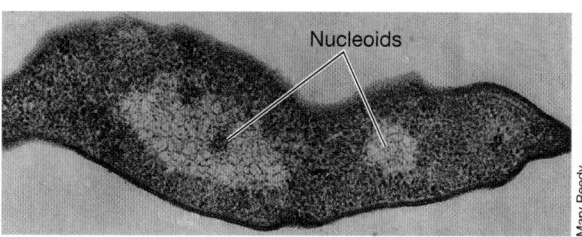

(a)

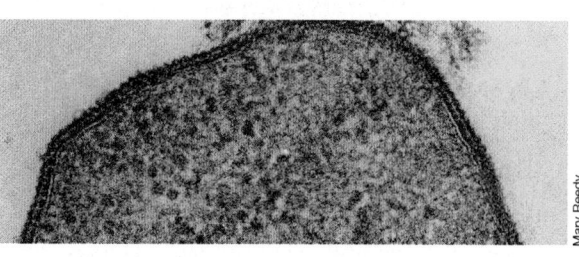

(b)

Figure 17.3 Electron micrographs of thin sections of the extreme halophile *Halobacterium salinarum*. A cell is about 0.8 μm in diameter. *(a)* Longitudinal section of a dividing cell showing the nucleoids. *(b)* High-magnification electron micrograph showing the glycoprotein subunit structure of the cell wall.

Haloarchaea stain gram-negatively, reproduce by binary fission, and do not form resting stages or spores. Cells of the various cultured genera are rod-shaped, cocci, or cup-shaped, but even cells that form squares are known (Figure 17.2*d*). Cells of *Haloquadratum* are square and only about 0.1 μm thick. *Haloquadratum* also forms gas vesicles (⇨ Section 2.9) that allow cells to float in its salty hypersaline habitat, probably as a means to be in contact with air since most extreme halophiles are obligate aerobes. Many other extremely halophilic *Archaea* also produce gas vesicles. A few strains of extreme halophiles are weakly motile by *archaella*, the archaeal analog of bacterial flagella, that rotate to propel the cell forward (⇨ Section 2.11), but most halophiles lack archaella. The genomes of *Halobacterium* and *Halococcus* are unusual in that large plasmids containing up to 30% of the total cellular DNA are present and the GC base ratio of these plasmids (near 60% GC) differs significantly from that of chromosomal DNA (66–68% GC). Plasmids from extreme halophiles are among the largest naturally occurring plasmids known.

Most haloarchaea use amino acids or organic acids as electron donors aerobically and require a number of growth factors such as vitamins for optimal growth. A few haloarchaea oxidize carbohydrates aerobically, but this capacity is rare; sugar fermentation does not occur. Electron transport chains containing cytochromes of the *a*, *b*, and *c* types are present in *Halobacterium*, and energy is conserved via a proton motive force arising from electron transport. Some haloarchaea can grow anaerobically, as growth by anaerobic respiration (⇨ Section 14.7) linked to the reduction of nitrate or fumarate has been demonstrated in certain species.

Water Balance in Extreme Halophiles

Extremely halophilic *Archaea* require large amounts of NaCl for growth. Detailed salinity studies of *Halobacterium* have shown that the requirement for Na^+ cannot be satisfied by any other ion, even the chemically related ion potassium (K⁺). However, cells of *Halobacterium* need *both* Na⁺ and K⁺ for growth because each plays an important role in maintaining osmotic balance.

As we learned in Section 5.13, microbes must withstand the osmotic forces they face in their habitats. To do so in a high-solute environment such as the salt-rich habitats of *Halobacterium*, organisms must either accumulate or synthesize solutes intracellularly. These solutes are called **compatible solutes**. These compounds counteract the tendency of the cell to become dehydrated under conditions of high osmotic strength by placing the cell in positive water balance with its surroundings. Cells of *Halobacterium*, however, do not synthesize or accumulate organic compounds but instead pump large amounts of K⁺ from the environment into the cytoplasm. This ensures that the concentration of K⁺ *inside* the cell is even greater than the concentration of Na⁺ *outside* the cell (Table 17.1). This ionic condition maintains positive water balance.

The *Halobacterium* cell wall (Figure 17.3*b*) is composed of glycoprotein and is stabilized by Na⁺. Sodium ions bind to the outer surface of the *Halobacterium* wall and are absolutely essential for maintaining cellular integrity. When insufficient Na⁺ is present, the cell wall breaks apart and the cell lyses. This is a consequence of the exceptionally high content of the *acidic* (negatively charged) amino acids aspartate and glutamate in the glycoprotein of the *Halobacterium* cell wall. The negative charge on the carboxyl group of these amino acids is bound to Na⁺; when Na⁺ is diluted away, the negatively charged parts of the proteins tend to repel each other, leading to cell lysis.

Halophilic Cytoplasmic Components

Like cell wall proteins, cytoplasmic proteins of *Halobacterium* are highly acidic, but it is K⁺, not Na⁺, that is required for activity. This makes sense because K⁺ is the predominant cation in the cytoplasm of cells of *Halobacterium* (Table 17.1). Besides having a high acidic amino acid composition, halobacterial cytoplasmic proteins typically contain lower levels of hydrophobic amino acids and lysine, a positively charged (basic) amino acid, than proteins of nonhalophiles. This is also to be expected because in a highly ionic cytoplasm, more polar proteins would tend to remain in solution whereas less polar proteins would tend to cluster and perhaps lose activity. The ribosomes of *Halobacterium* also require high KCl levels for stability, whereas ribosomes of nonhalophiles have no KCl requirement.

Extremely halophilic *Archaea* are thus well adapted to life in a highly ionic environment. Cellular components exposed to the external environment require high Na⁺ for stability, whereas cytoplasmic components require high K⁺. With the exception of a few extremely halophilic species of *Bacteria* that also use KCl as a compatible solute, in no other group of prokaryotic cells do we find this unique requirement for such high amounts of specific cations.

Bacteriorhodopsin and Light-Mediated ATP Synthesis in Haloarchaea

Certain species of haloarchaea can catalyze a light-driven synthesis of ATP. This form of phototrophy is not linked to CO_2 fixation, and does not require chlorophyll pigments, and so it is not photosynthesis in the traditional sense. However, other light-sensitive pigments are present, including red and orange carotenoids—primarily C_{50} pigments called *bacterioruberins*—and inducible pigments involved in energy conservation; we discuss these pigments here.

Under conditions of low aeration, *Halobacterium salinarum* and some other haloarchaea synthesize a protein called **bacteriorhodopsin** and insert it into their cytoplasmic membranes. Bacteriorhodopsin is so named because of its structural and functional similarity to rhodopsin, the visual pigment of the eye. Conjugated to bacteriorhodopsin is a molecule of retinal, a carotenoid-like molecule that can absorb light energy and pump a proton across the cytoplasmic membrane. The retinal gives bacteriorhodopsin a purple hue. Thus cells of *Halobacterium* that are switched from growth under high-aeration conditions to oxygen-limiting growth conditions (a trigger of bacteriorhodopsin synthesis) gradually change color from orange-red to purple-red as they synthesize bacteriorhodopsin and insert it into their cytoplasmic membranes.

Bacteriorhodopsin absorbs green light around 570 nm. Following absorption, the retinal of bacteriorhodopsin, which normally exists in a *trans* configuration (Ret$_T$), becomes excited and converts to the *cis* (Ret$_C$) form (**Figure 17.4**). This transformation is coupled to the translocation of a proton across the cytoplasmic membrane. The retinal molecule then decays to the *trans* isomer along with the uptake of a proton from the cytoplasm, and this completes the cycle. The proton pump is then ready to repeat the cycle (Figure 17.4). As protons accumulate on the outer surface of the membrane,

TABLE 17.1 Concentration of ions in cells of *Halobacterium salinarum*[a]		
Ion	Concentration in medium (M)	Concentration in cells (M)
Na⁺	4.0	1.4
K⁺	0.032	4.6
Mg²⁺	0.13	0.12
Cl⁻	4.0	3.6

[a]Data from Christian, J.H.B, and Waltho, J.A. *Biochim. Biophys. Acta 65*: 506–508 (1962).

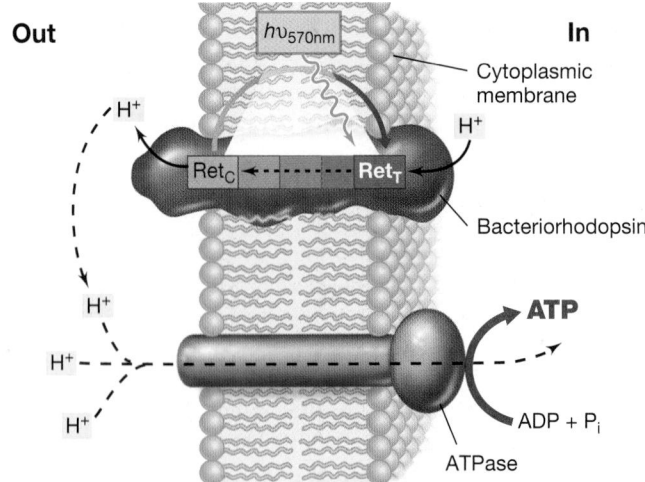

Figure 17.4 Model for the mechanism of bacteriorhodopsin. Light of 570 nm (hv_{570nm}) converts the protonated retinal of bacteriorhodopsin from the *trans* form (Ret$_T$) to the *cis* form (Ret$_C$), along with translocation of a proton to the outer surface of the cytoplasmic membrane, thus establishing a proton motive force. ATPase activity is driven by the proton motive force.

a proton motive force is generated that is coupled to ATP synthesis through the activity of a proton-translocating ATPase (Figure 17.4) (⇄ Section 3.11).

Bacteriorhodopsin-mediated ATP production in *H. salinarum* supports slow growth of this organism under anoxic conditions. The light-stimulated proton pump of *H. salinarum* also functions to pump Na^+ out of the cell by activity of a Na^+–H^+ antiport system and to drive the uptake of nutrients, including the K^+ needed for osmotic balance. Amino acid uptake by *H. salinarum* is indirectly driven by light as well, because amino acids are cotransported into the cell with Na^+ by an amino acid–Na^+ symporter (⇄ Section 3.2); removal of Na^+ from the cell occurs by way of the light-driven Na^+–H^+ antiporter.

Other Rhodopsins

Besides bacteriorhodopsin, at least three other rhodopsins are present in the cytoplasmic membrane of cells of *H. salinarum*. **Halorhodopsin** is a light-driven chloride (Cl^-) pump that brings Cl^- into the cell as the anion for K^+. The retinal of halorhodopsin binds Cl^- and transports it into the cell. Two other light sensors, called *sensory rhodopsins*, are present in *H. salinarum*. These light sensors control phototaxis (movement toward light, ⇄ Section 2.13) by the organism. Through the interaction of a cascade of proteins similar to those in chemotaxis (⇄ Section 6.7), sensory rhodopsins affect archaellar rotation, moving cells of *H. salinarum* toward light where bacteriorhodopsin can function to make ATP (Figure 17.4).

We will learn when we consider marine microbiology (⇄ Sections 20.10 and 20.11) that diverse species of chemoorganotrophic *Bacteria* that inhabit the upper layers of the ocean contain bacteriorhodopsin-like proteins called *proteorhodopsins*. As far as is known, proteorhodopsin functions like bacteriorhodopsin except that different spectral forms exist, each form being tuned to the absorption of its own specific wavelengths of light. Although the energy generated from proteorhodopsin alone is insufficient to sustain growth, these marine bacteria use proteorhodopsin as a supplement to the ATP they generate from respiration. Proteorhodopsin as a mechanism for energy conservation in marine bacteria makes good ecological sense because levels of dissolved organic matter in the open oceans are typically very low, and thus a strictly chemoorganotrophic lifestyle would be difficult.

> **MINIQUIZ**
>
> - Since cells of *Halobacterium* require high levels of Na^+ for growth, why is this not true for the organism's cytoplasmic enzymes?
> - What benefit does bacteriorhodopsin confer on *Halobacterium salinarum*?

17.2 Methanogenic *Archaea*

KEY GENERA: *Methanobacterium, Methanocaldococcus, Methanosarcina, Methanopyrus*

Many *Euryarchaeota* are methanogens, microorganisms that produce methane (CH_4) as an integral part of their energy metabolism (methane production is called *methanogenesis*). In Section 14.17 we considered the unique biochemistry of methanogenesis. Later, we will learn how methanogenesis is a major component of the global carbon cycle, serving as the terminal step in the biodegradation of organic matter in many anoxic habitats (⇄ Sections 21.1 and 21.2). Methanogens are important in a wide range of anoxic habitats including freshwater sediments, wetlands, rice paddies, wastewater treatment plants, geothermal systems, the subsurface of the Earth's crust, and within the guts of many animals.

Diversity and Physiology of Methanogens

Methanogens occur in at least seven orders, including *Methanobacteriales, Methanococcales, Methanopyrales, Methanomassiliicoccales, Methanomicrobiales, Methanocellales*, and *Methanosarcinales* (Figure 17.1). Methanogens exhibit considerable morphological and physiological diversity (**Figure 17.5** and see Table 17.2). As can be seen in Figure 17.1, methanogens are spread widely within the *Euryarchaeota* and do not represent a single coherent phylogenetic group. We have already been introduced to the factors that cause inconsistency between phylogenetic diversity and functional diversity (⇄ Section 15.1). In the case of methanogenesis it is most likely that the ability to reduce CO_2 to CH_4 evolved only once within the *Euryarchaeota*, and that gene loss caused lineages such as the haloarchaea and *Thermoplasmatales* to lose the capacity to produce methane. We will also see that *Archaeoglobus* (Section 17.4) still retains some of the genes encoding methanogenesis and can actually produce methane under certain growth conditions.

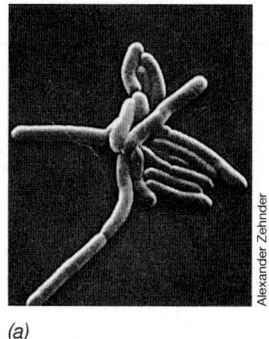 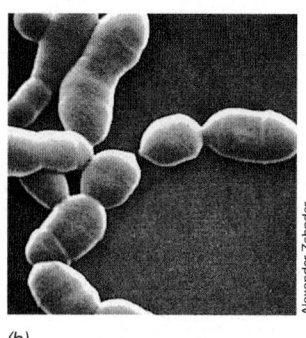 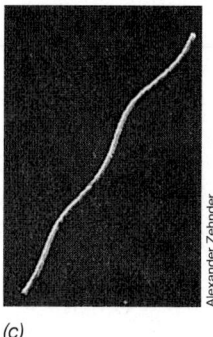 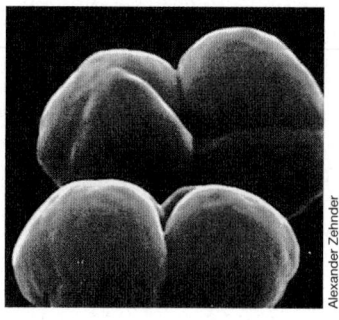

(a) (b) (c) (d)

Figure 17.5 Scanning electron micrographs of cells of diverse species of methanogenic *Archaea*.
(a) Methanobrevibacter ruminantium. A cell is about 0.7 μm in diameter. *(b) Methanobrevibacter arboriphilus.* A cell is about 1 μm in diameter. *(c) Methanospirillum hungatei.* A cell is about 0.4 μm in diameter. *(d) Methanosarcina barkeri.* A cell is about 1.7 μm wide.

UNIT 4

Methanogens have diverse physiological characteristics, but they are united by their ability to produce methane and by their intolerance to oxygen. Several cofactors required by methanogenic pathways are inhibited by oxygen, and hence all methanogens are obligate anaerobes. As a result, strict anoxic techniques are necessary for their cultivation in isolation. While most methanogens are mesophilic and live in non-extreme environments, a diversity of species have been described including some that grow optimally at very high (see Figure 17.7) or very low temperatures, at very high salt concentrations, or at extremes of pH.

Methanogenesis can be performed through three different pathways (Table 17.2): methanogenesis by CO_2 reduction ($\Leftrightarrow$ Figure 14.47), methylotrophic methanogenesis ($\Leftrightarrow$ Figure 14.48a), and acetoclastic methanogenesis ($\Leftrightarrow$ Figure 14.48b). Each of these pathways relies on *coenzyme M* with methane ultimately produced by reduction of methyl-CoM to methane ($\Leftrightarrow$ Section 14.17). There is a limited set of substrates that methanogens can convert into methane through these pathways (Table 17.2). Interestingly, these substrates do not include such common compounds as glucose, organic acids, or fatty acids (other than acetate and pyruvate). Methanogens often form syntrophic associations with fermentative anaerobes ($\Leftrightarrow$ Sections 14.23 and 21.2). In this way the fermentative organisms degrade a wide range of organic carbon molecules into H_2, CO_2, and acetate, which are ultimately used as substrates for methanogenesis.

The three methanogenic pathways are found in different phylogenetic groups of methanogens (Table 17.3). Methanogenesis by CO_2 reduction is found widely across the known diversity of methanogens, but not all methanogens can produce methane by CO_2 reduction. Notably, many species in the *Methanosarcinales*, species of *Methanosphaera*, and species in the *Methanomassiliicoccales* have lost the ability to produce methane by CO_2 reduction (Table 17.3). The CO_2 reduction pathway can also be used by some species to produce methane from formate, or carbon monoxide. In addition, while most species use H_2 to reduce CO_2, some can also reduce CO_2 by using electrons from pyruvate or certain alcohols (Tables 17.2 and 17.3).

The acetoclastic and methylotrophic pathways of methanogenesis are found mainly within the *Methanosarcinales* (Table 17.3). Methylated substrates include methanol (CH_3OH) and many others (Table 17.2). Many species in the *Methanosarcinales* will actually oxidize one molecule of CH_3OH to CO_2 to generate

---

TABLE 17.3 Characteristics of some methanogenic *Archaea*[a]

Order/Genus	Substrates for methanogenesis[b]
Methanobacteriales	
Methanobacterium	$H_2 + CO_2$, formate
Methanobrevibacter	$H_2 + CO_2$, formate
Methanosphaera	$H_2 +$ methanol
Methanothermus	$H_2 + CO_2$
Methanothermobacter	$H_2 + CO_2$, formate, CO
Methanococcales	
Methanococcus	$H_2 + CO_2$, pyruvate + CO_2, formate
Methanothermococcus	$H_2 + CO_2$, formate
Methanocaldococcus	$H_2 + CO_2$
Methanotorris	$H_2 + CO_2$
Methanomicrobiales	
Methanomicrobium	$H_2 + CO_2$, formate
Methanogenium	$H_2 + CO_2$, formate
Methanospirillum	$H_2 + CO_2$, formate
Methanoplanus	$H_2 + CO_2$, formate
Methanocorpusculum	$H_2 + CO_2$, $H_2 +$ alcohols, formate
Methanoculleus	$H_2 + CO_2$, $H_2 +$ alcohols, formate
Methanofollis	$H_2 + CO_2$, formate
Methanolacinia	$H_2 + CO_2$, $H_2 +$ alcohols
Methanosarcinales	
Methanosarcina	$H_2 + CO_2$, methanol, methylamines, acetate, CO
Methanolobus	Methanol, methylamines
Methanohalobium	Methanol, methylamines
Methanococcoides	Methanol, methylamines
Methanohalophilus	Methanol, methylamines, methyl sulfides
Methanosaeta	Acetate
Methanosalsum	Methanol, methylamines, dimethyl sulfide
Methanimicrococcus	$H_2 +$ methanol, $H_2 +$ methylamines
Methanopyrales	
Methanopyrus	$H_2 + CO_2$
Methanocellales	
Methanocella	$H_2 + CO_2$
Methanomassiliicoccales	
Methanomassiliicoccus	$H_2 +$ methanol
Methanomethylophilus	$H_2 +$ methanol

[a]Taxonomic orders are listed in bold. An order is a taxonomic rank that consists of several families; families consist of several genera ($\Leftrightarrow$ Table 13.2).
[b]Methylamines can include the substrates methylamine ($CH_3NH_3^+$), dimethylamine (($CH_3)_2NH_2^+$), and trimethylamine (($CH_3)_3NH^+$); methyl sulfides can include dimethyl sulfide (($CH_3)_2S$) and methyl mercaptan (CH_3SH).

TABLE 17.2 The three methanogenic pathways and their substrates

I. **CO_2 reduction pathway**
   Carbon dioxide, CO_2 (with electrons derived from H_2, certain alcohols, or pyruvate)
   Formate, $HCOO^-$
   Carbon monoxide, CO

II. **Methylotrophic pathway**
   Methanol, CH_3OH
   Methylamine, $CH_3NH_3^+$
   Dimethylamine, $(CH_3)_2NH_2^+$
   Trimethylamine, $(CH_3)_3NH^+$
   Methyl mercaptan, CH_3SH
   Dimethyl sulfide, $(CH_3)_2S$

III. **Acetoclastic pathway**
   Acetate, CH_3COO^-
   Pyruvate, CH_3COCOO^-

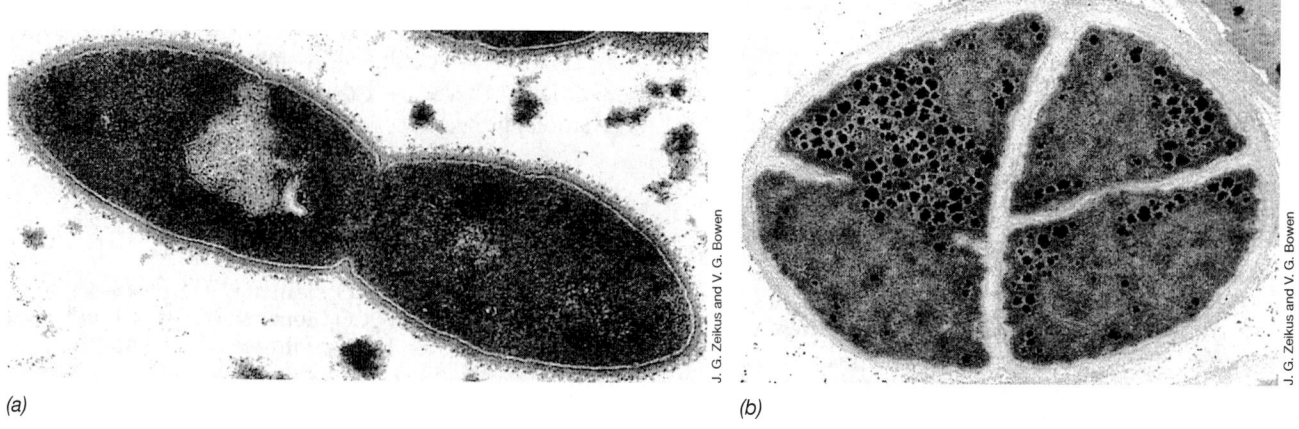

(a)

(b)

J. G. Zeikus and V. G. Bowen

Figure 17.6 Transmission electron micrographs of thin sections of methanogenic *Archaea*.
(a) Methanobrevibacter ruminantium. A cell is 0.7 μm in diameter. *(b) Methanosarcina barkeri,* showing the thick cell wall and the manner of cell segmentation and cross-wall formation. A cell is 1.7 μm in diameter.

the electrons needed to reduce three molecules of CH_3OH to CH_4 and H_2O (⥊ Figure 14.48*a*). However, other methylotrophic species lack this ability (Table 17.3) and require an external electron donor such as H_2 to reduce methylated compounds to CH_4. Only a few methanogens have been shown to use acetate as a substrate and they are all members of the *Methanosarcinales* (Table 17.3). Methanogenesis from acetate is a dominant source of methane production in a wide range of environments (⥊ Section 21.2).

Methanogens show a diversity of cell wall chemistries. These include the pseudomurein walls of *Methanobacterium* species and relatives (**Figure 17.6*a***), walls composed of methanochondroitin (so named because of its structural resemblance to chondroitin, the connective tissue polymer of vertebrate animals) in *Methanosarcina* and relatives (Figure 17.6*b*), the protein or glycoprotein walls of *Methanocaldococcus* (**Figure 17.7*a***) and *Methanoplanus* species, respectively, and the S-layer walls of *Methanospirillum* (Figure 17.5*c*) (⥊ Section 2.6).

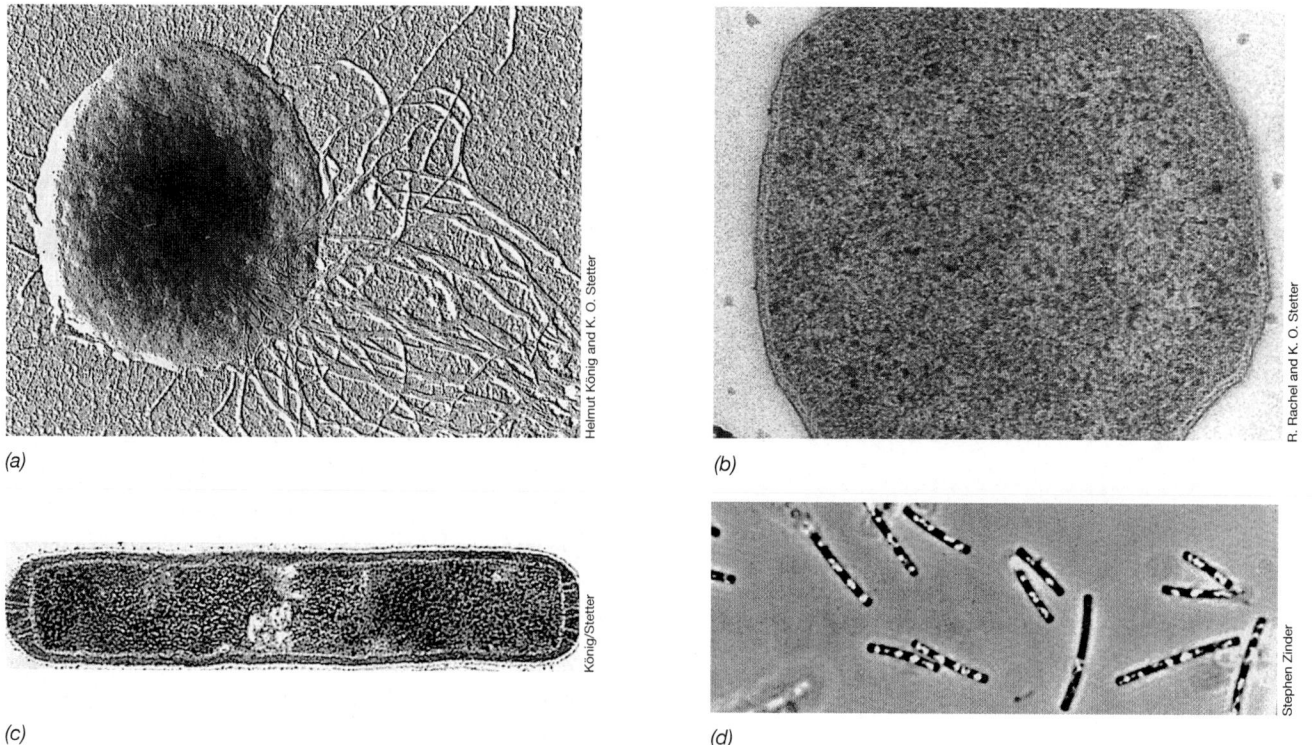

(a)

(b)

(c)

(d)

Helmut König and K. O. Stetter

R. Rachel and K. O. Stetter

König/Stetter

Stephen Zinder

Figure 17.7 Hyperthermophilic and thermophilic methanogens. *(a) Methanocaldococcus jannaschii* (temperature optimum, 85°C), shadowed preparation electron micrograph. A cell is about 1 μm in diameter. *(b) Methanotorris igneus* (temperature optimum, 88°C), thin section. A cell is about 1 μm in diameter. *(c) Methanothermus fervidus* (temperature optimum, 88°C), thin-sectioned electron micrograph. A cell is about 0.4 μm in diameter. *(d) Methanosaeta thermophila* (temperature optimum, 60°C), phase-contrast micrograph. A cell is about 1 μm in diameter. The refractile bodies inside the cells are gas vesicles.

UNIT 4

Methanocaldococcus jannaschii as a Model Methanogen

The genomes of the hyperthermophilic methanogen *Methanocaldococcus jannaschii* (Figure 17.7*a*) and many other methanogens have been sequenced. The 1.66-megabase-pair (Mbp) circular genome of *M. jannaschii*, an organism that has been used as a model for the molecular study of methanogenesis and archaeal motility, contains about 1700 genes, and genes encoding enzymes of methanogenesis and several other key cell functions have been identified. Interestingly, the majority of *M. jannaschii* genes encoding functions such as central metabolic pathways and cell division are similar to those in *Bacteria*. By contrast, most of the *M. jannaschii* genes encoding core molecular processes such as transcription and translation more closely resemble those of eukaryotes. These findings reflect the various traits shared by organisms in the three cellular domains and are consistent with our understanding of how the three domains of cells evolved, as discussed in Chapter 13. However, analyses of the *M. jannaschii* genome also show that fully 40% of its genes have no counterparts in genes from either of the other domains. Some of these are genes that encode the enzymes needed for methanogenesis, of course, but many others likely encode novel cellular functions absent from cells in the other domains or may encode redundant functions carried out by classes of enzymes distinct from those found in *Bacteria* and *Eukarya*.

Methanopyrus, a Hyperthermophilic Methanogen

Methanopyrus (**Figure 17.8**), the only genus in the order *Methanopyrales*, is a rod-shaped hyperthermophilic methanogen that shares phenotypic properties with both the hyperthermophiles (see Section 17.13) and the methanogens. *Methanopyrus* was isolated from hot sediments near submarine hydrothermal vents and from the walls of "black smoker" hydrothermal vent chimneys (Section 17.10; ⇨ Section 20.14). *Methanopyrus*

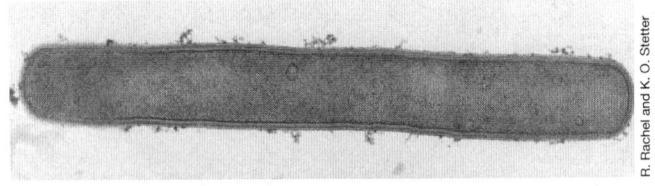

(a)

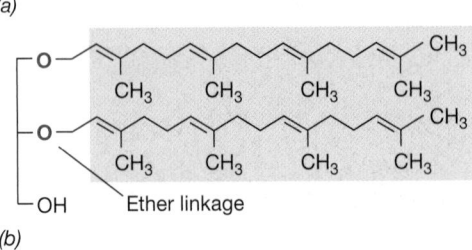

(b)

R. Rachel and K. O. Stetter

Figure 17.8 *Methanopyrus*. *Methanopyrus* grows optimally at 100°C and can make CH_4 only from $CO_2 + H_2$. (a) Electron micrograph of a cell of *Methanopyrus kandleri*, the most thermophilic of all known organisms (upper temperature limit, 122°C). This cell measures 0.5×8 μm. (b) Structure of the novel lipid of *M. kandleri*. This is the normal ether-linked lipid of the *Archaea* except that the side chains are an unsaturated form of phytanyl (geranylgeraniol).

produces CH_4 only from $H_2 + CO_2$ and grows rapidly for an autotrophic organism (generation time <1 h at 100°C). In special pressurized vessels, growth of one strain of *Methanopyrus* has been recorded at 122°C, the highest temperature yet shown to support microbial growth (Sections 17.11 and 17.12).

Methanopyrus was once thought to be one of the most ancestral lineages of *Archaea*, but genome sequence analysis reveals it to be a relative to the *Methanobacteriales* and *Methanococcales* (Figure 17.1). *Methanopyrus* also has a pseudomurein cell wall, a trait that it shares with species of *Methanobacteriales*.

In addition to its remarkable tolerance to high temperature, *Methanopyrus* is also unusual because it contains membrane lipids found in no other known organism. Recall that in the lipids of *Archaea*, the glycerol side chains contain **phytanyl** rather than fatty acids bonded in ether linkage to the glycerol (⇨ Section 2.3). In *Methanopyrus*, this ether-linked lipid is an *unsaturated* form of the otherwise saturated biphytanyl tetraethers found in all other hyperthermophilic *Archaea* (Figure 17.8*b*). These unusual lipids may help stabilize the cytoplasmic membrane of *Methanopyrus* at its unusually high growth temperatures.

MINIQUIZ

- What are the three major pathways of methanogenesis and in what phylogenetic groups are they found?
- What physiological and structural features distinguish the *Methanosarcinales* from other methanogens?

17.3 *Thermoplasmatales*

KEY GENERA: *Thermoplasma, Picrophilus, Ferroplasma*

A phylogenetically distinct line of *Archaea* contains thermophilic and extremely acidophilic genera: *Thermoplasma, Ferroplasma,* and *Picrophilus*. These organisms are among the most acidophilic of all known microbes, with *Picrophilus* being capable of growth even below pH 0. Most are thermophilic as well. These genera also form their own taxonomic order within the *Euryarchaeota*, the *Thermoplasmatales* (Figure 17.1). We begin with a description of the mycoplasma-like organisms *Thermoplasma* and *Ferroplasma*.

Archaea Lacking Cell Walls

Thermoplasma and *Ferroplasma* lack cell walls, and in this respect they resemble the mycoplasmas (⇨ Section 16.9). *Thermoplasma* (**Figure 17.9**) is a chemoorganotroph that grows optimally at 55°C and pH 2 in complex media. Two species of *Thermoplasma* have been described, *Thermoplasma acidophilum* and *Thermoplasma volcanium*. Species of *Thermoplasma* are facultative aerobes, growing either aerobically or anaerobically by sulfur respiration (⇨ Section 14.14). Most strains of *T. acidophilum* have been obtained from self-heating coal refuse piles. Coal refuse contains coal fragments, pyrite (FeS_2), and other organic materials extracted from coal. When dumped into piles in surface mining operations, coal refuse heats as a result of microbial metabolism bringing it to combustion temperature (**Figure 17.10**). This sets the stage for growth of *Thermoplasma*, which likely metabolizes organic

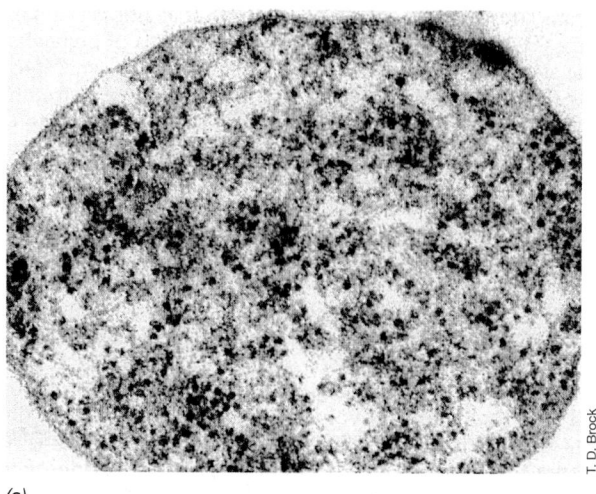

(a)

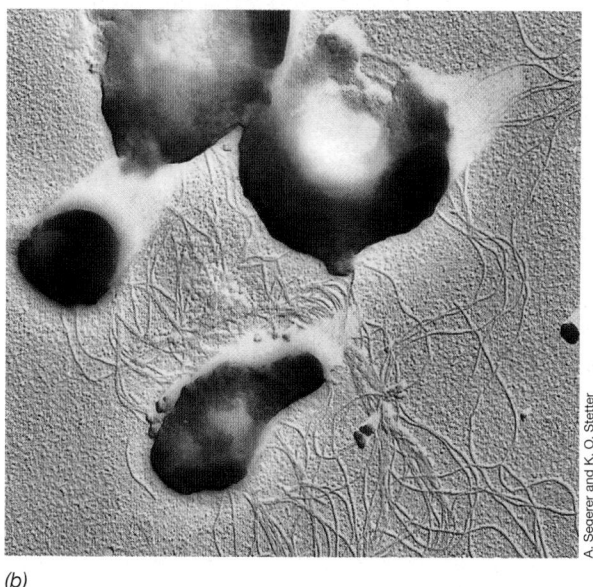

(b)

Figure 17.9 *Thermoplasma* species. *(a) Thermoplasma acidophilum*, an acidophilic and thermophilic mycoplasma-like archaeon; electron micrograph of a thin section. The diameter of cells varies from 0.2 to 5 μm. The cell shown is about 1 μm in diameter. *(b)* Shadowed preparation of cells of *Thermoplasma volcanium* isolated from hot springs. Cells are 1–2 μm in diameter. Notice the abundant archaella and irregular cell morphology.

compounds leached from the hot coal refuse. The second species, *T. volcanium*, has been isolated in hot acidic soils throughout the world and is highly motile by multiple archaella (Figure 17.9*b*).

To survive the osmotic stresses of life without a cell wall and to withstand the dual environmental extremes of low pH and high temperature, *Thermoplasma* has evolved a unique cytoplasmic membrane structure. The membrane contains a lipopolysaccharide-like material called *lipoglycan*. This substance consists of glycolipids containing sugars such as mannose and glucose (**Figure 17.11**) and these glycolipids form a tetraether lipid monolayer membrane. The hydrophobic core of this glycolipid consists of a biphytanyl (⊃ Section 2.3). In *Thermoplasma* and similar organisms such as *Sulfolobus*, this basic biphytanyl structure can be modified to include one to four cyclopentane rings, with the number of rings

Figure 17.10 A typical self-heating coal refuse pile, habitat of *Thermoplasma*. The pile, containing coal debris, pyrite, and other microbial substrates, self-heats as a result of microbial metabolism.

tending to increase in proportion to the temperature of the environment. These glycolipids constitute a major fraction of the total lipids of *Thermoplasma*. The membrane also contains glycoproteins and glycophospholipids, but not sterols. These molecules render the *Thermoplasma* membrane stable to hot, acidic conditions.

Like mycoplasmas (⊃ Section 16.9), *Thermoplasma* contains a relatively small genome (1.5 Mbp). In addition, *Thermoplasma* DNA is complexed with a highly basic DNA-binding protein that organizes the DNA into globular particles resembling the nucleosomes of eukaryotic cells. This protein is homologous to the histone-like DNA-binding protein HU of *Bacteria*, which plays an important role in organization of the DNA in the cell. In contrast, several other *Euryarchaeota* contain basic proteins homologous to the DNA-binding histone proteins of eukaryotic cells.

Ferroplasma

Ferroplasma is a chemolithotrophic relative of *Thermoplasma*. *Ferroplasma* is a strong acidophile; however, it is not a thermophile, as it grows optimally at 35°C. *Ferroplasma* oxidizes ferrous iron (Fe^{2+}) to ferric iron (Fe^{3+}) to obtain energy (this reaction generates acid, Figure 17.18*d*) and uses CO_2 as its carbon source (autotrophy). *Ferroplasma* grows in mine tailings containing

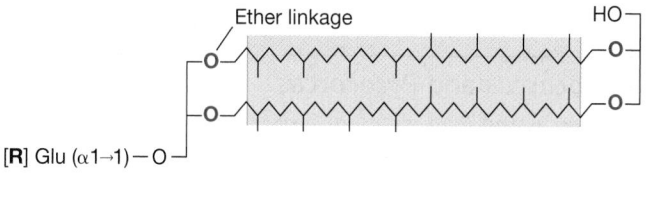

R = Man (α1 → 2) Man (α1 → 4) Man (α1 → 3)

Figure 17.11 Structure of a tetraether glycolipid of *Thermoplasma acidophilum*. The dominant glycolipids of *T. acidophilum* have two polar head groups that are connected by ether linkages to a hydrophobic core. This structure causes them to form a thermostable lipid monolayer. One or both of the polar head groups typically contains a mono- or oligosaccharide that can be composed of glucose (Glu), mannose (Man), or other sugars. The hydrophobic core consists of caldarchaeol (shown), which can be modified to include one to four cyclopentane rings (not shown). The lipid shown contains an oligosaccharide linked to a single polar head group.

pyrite, which is its energy source. The extreme acidophily of *Ferroplasma* allows it to drive the pH of its habitat down to extremely acidic values. After moderate acidity is generated from Fe^{2+} oxidation by acidophilic organisms such as *Acidithiobacillus ferrooxidans* and *Leptospirillum ferrooxidans* (⇔ Section 21.5), *Ferroplasma* becomes active and subsequently generates the very low pH values typical of acid mine drainage. Acidic waters at pH 0 can be generated by the activities of *Ferroplasma*.

Picrophilus

A phylogenetic relative of *Thermoplasma* and *Ferroplasma* is *Picrophilus*. Although *Thermoplasma* and *Ferroplasma* are extreme acidophiles, *Picrophilus* is even more so, growing optimally at pH 0.7 and capable of growth at pH values lower than 0. *Picrophilus* also has a cell wall (an S-layer; ⇔ Section 2.6) and a much lower DNA GC base ratio than does *Thermoplasma* or *Ferroplasma*. Although phylogenetically related, *Thermoplasma*, *Ferroplasma*, and *Picrophilus* have quite distinct genomes. Two species of *Picrophilus* have been isolated from acidic Japanese solfataras, and like *Thermoplasma*, both grow heterotrophically on complex media.

The physiology of *Picrophilus* is of interest as a model for extreme acid tolerance. Studies of its cytoplasmic membrane point to an unusual arrangement of lipids that forms a highly acid-impermeable membrane at very low pH. By contrast, at moderate acidities such as pH 4, the membranes of cells of *Picrophilus* become leaky and disintegrate. Obviously, this organism has evolved to survive only in highly acidic habitats.

MINIQUIZ

- In what ways are *Thermoplasma* and *Picrophilus* similar? In what ways do they differ?
- How does *Thermoplasma* strengthen its cytoplasmic membrane to survive without a cell wall?

17.4 *Thermococcales* and *Archaeoglobales*

KEY GENERA: *Thermococcus, Pyrococcus, Archaeoglobus, Ferroglobus*

A few euryarchaeotes thrive in thermal environments and some are hyperthermophiles. We consider here the *Thermococcales* and *Archaeoglobales*, which are two orders of *Euryarchaeota* that contain hyperthermophilic species (Figure 17.1).

Thermococcus and *Pyrococcus*

Thermococcus and *Pyrococcus* are genera within the order *Thermococcales*. *Thermococcus* is a spherical hyperthermophilic euryarchaeote indigenous to anoxic thermal waters in various locations throughout the world. The spherical cells contain a tuft of polar archaella and are thus highly motile (**Figure 17.12a**). *Thermococcus* is an obligately anaerobic chemoorganotroph that metabolizes proteins and other complex organic mixtures (including some sugars) with elemental sulfur (S^0) as electron acceptor at temperatures from 55 to 95°C.

Pyrococcus is morphologically similar to *Thermococcus* (Figure 17.12b). *Pyrococcus* differs from *Thermococcus* primarily by its

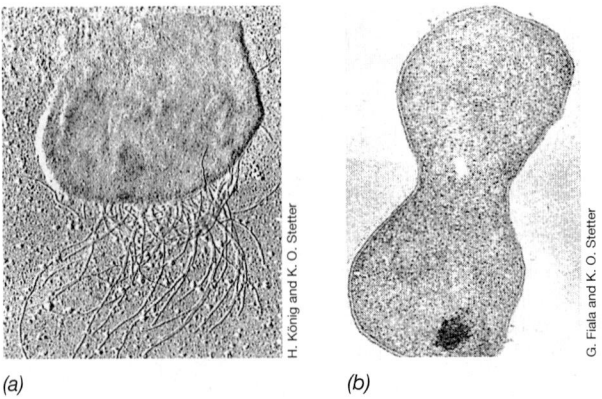

(a)   (b)

Figure 17.12 Spherical hyperthermophilic *Euryarchaeota* from submarine volcanic areas. *(a) Thermococcus celer*; electron micrograph of shadowed cells (note tuft of archaella). *(b)* Dividing cell of *Pyrococcus furiosus*; electron micrograph of thin section. Cells of both organisms are about 0.8 μm in diameter.

higher temperature requirements; *Pyrococcus* grows between 70 and 106°C with an optimum of 100°C. *Thermococcus* and *Pyrococcus* are also metabolically quite similar. Proteins, starch, or maltose are oxidized as electron donors, and S^0 is the terminal electron acceptor and is reduced to hydrogen sulfide (H_2S). Both *Thermococcus* and *Pyrococcus* form H_2S when S^0 is present, but form H_2 when S^0 is absent (see Table 17.4).

Archaeoglobus and *Ferroglobus*

Archaeoglobus was isolated from hot marine sediments near hydrothermal vents. In its metabolism, *Archaeoglobus* couples the oxidation of H_2, lactate, pyruvate, glucose, or complex organic compounds to the reduction of SO_4^{2-} to H_2S. Cells of *Archaeoglobus* are irregular cocci (**Figure 17.13a**) and grow optimally at 83°C.

Archaeoglobus and methanogens share some characteristics. We learned in Section 14.17 about the unique biochemistry of

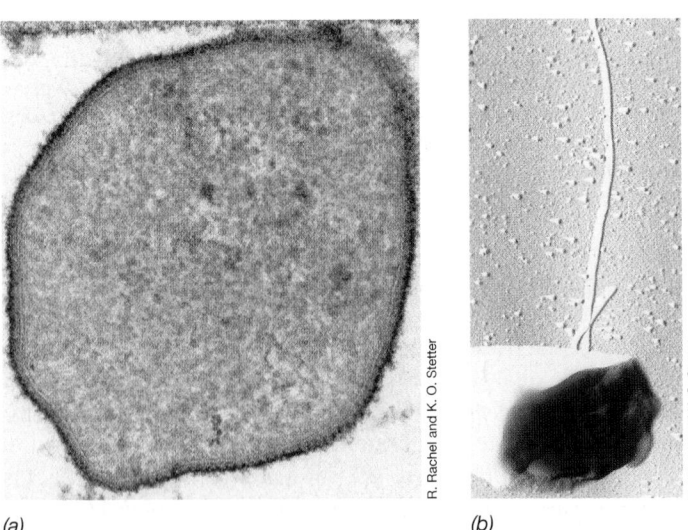

(a)   (b)

Figure 17.13 *Archaeoglobales*. *(a)* Transmission electron micrograph of the sulfate-reducing hyperthermophile *Archaeoglobus fulgidus*. The cell measures 0.7 μm in diameter. *(b)* Freeze-etched electron micrograph of *Ferroglobus placidus*, a ferrous iron–oxidizing, nitrate-reducing hyperthermophile. The cell measures about 0.8 μm in diameter.

methanogenesis. Briefly, this process requires a series of novel coenzymes, and with rare exceptions, these coenzymes have only been found in methanogens. *Archaeoglobus*, however, also contains many of these coenzymes, and cultures of this organism actually produce small amounts of CH_4. In addition, the genome of *Archaeoglobus*, which contains about 2400 genes, shares many genes in common with methanogens (Section 17.2). It seems likely that the ancestor of *Archaeoglobus* was a methanogen that lost many of the genes required for methanogenesis. Furthermore, genome analysis suggests that the ancestor of *Archaeoglobus* acquired genes for sulfate reduction as a result of horizontal gene transfer from sulfate-reducing bacteria within the *Deltaproteobacteria* (⇝ Section 15.9).

Ferroglobus (Figure 17.13b) is related to *Archaeoglobus* but is not a sulfate reducer. Instead, *Ferroglobus* is an iron-oxidizing chemolithotroph, conserving energy from the oxidation of Fe^{2+} to Fe^{3+} coupled to the reduction of nitrate (NO_3^-) to nitrite (NO_2^-) (see Table 17.4). *Ferroglobus* grows autotrophically and can also use H_2 or H_2S as electron donors in its energy metabolism. *Ferroglobus* was isolated from a shallow marine hydrothermal vent and grows optimally at 85°C.

Ferroglobus is interesting for several reasons, but especially for its ability to oxidize Fe^{2+} to Fe^{3+} under anoxic conditions. This process might help explain the origin of some Fe^{3+} found in ancient rocks dated to before the predicted appearance of cyanobacteria on Earth (⇝ Section 13.2). With organisms like *Ferroglobus*, it would have been possible for Fe^{2+} oxidation to proceed without the need for molecular oxygen (O_2) as an electron acceptor. The metabolism of *Ferroglobus* thus has implications for dating the origin of cyanobacteria and the subsequent oxygenation of Earth. Certain anoxygenic phototrophic bacteria can also oxidize Fe^{2+} under anoxic conditions (⇝ Section 14.10), and so several anaerobic routes to ancient Fe^{3+} are possible. This makes it difficult to estimate when cyanobacteria first appeared on Earth and to what degree nonphototrophic organisms helped trigger the Great Oxidation Event (⇝ Figure 13.1).

MINIQUIZ

- How do *Thermococcus* and *Pyrococcus* make ATP?
- Compare the energy-yielding metabolisms of *Archaeoglobus* and *Ferroglobus*.

II • *Thaumarchaeota, Nanoarchaeota, and Korarchaeota*

Our understanding of *Archaea* has been revolutionized by the development of molecular phylogeny (⇝ Sections 1.13 and 13.7) and methods for studying microorganisms without the need for cultivation in laboratory cultures (⇝ Sections 19.4–19.8). The *Thaumarchaeota, Nanoarchaeota,* and *Korarchaeota* were all discovered and characterized initially with the aid of techniques for analysis of 16S ribosomal RNA genes. From these initial efforts,

species representing each of these phyla were subsequently isolated or at least grown in enrichment cultures. We begin our consideration of these unusual phyla with the *Thaumarchaeota*.

17.5 *Thaumarchaeota* and Nitrification in *Archaea*

KEY GENERA: *Nitrosopumilus, Nitrososphaera*

Early surveys of 16S ribosomal RNA genes from open-ocean microbial communities resulted in the shocking conclusion that *Archaea* were abundant and widespread in the oceans. At the time, the archaeal domain was considered to contain only extremophiles and obligate anaerobes, and their presence in oxygen-rich temperate and even polar oceanic environments was something of a mystery. Even more remarkable, these novel *Archaea* were widespread and common in soils all over the world (see page 530).

Phylogenetic analyses of 16S ribosomal RNA gene sequences initially suggested that this novel group of *Archaea* was a deeply divergent lineage of the *Crenarchaeota*, a group of hyperthermophilic *Archaea* (Section 17.8). It was only after genome sequence analysis of the marine nitrifier *Nitrosopumilus maritimus* that it became clear that the **Thaumarchaeota** constitute a unique phylum of *Archaea* and that they diverged from the primary line of archaeal descent prior to the divergence of *Crenarchaeota* and *Euryarchaeota* (Figure 17.1).

Physiological Characteristics of *Thaumarchaeota*

The physiology of *Thaumarchaeota* remained a mystery until the isolation of *Nitrosopumilus maritimus* (**Figure 17.14**). *N. maritimus* grows chemolithotrophically by aerobically oxidizing ammonia (NH_3) to nitrite (NO_2^-), the first step in nitrification (⇝ Sections 14.11, 15.13, and 21.3). This organism uses CO_2 as its sole carbon source (autotrophy), as do nitrifying *Bacteria* (⇝ Section 15.13). However, unlike ammonia-oxidizing *Bacteria* such as *Nitrosomonas*, *N. maritimus* is adapted to life under extreme nutrient limitation, as would befit an organism indigenous to open ocean waters. *N. maritimus* can grow at NH_3 concentrations that are a hundred times lower than those required by bacterial nitrifiers, and their growth is actually inhibited at the higher NH_3 concentrations required to support growth of nitrifying species of *Bacteria*.

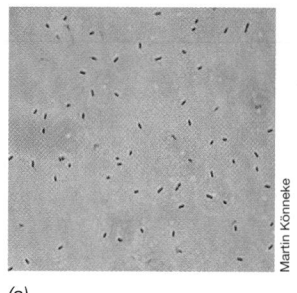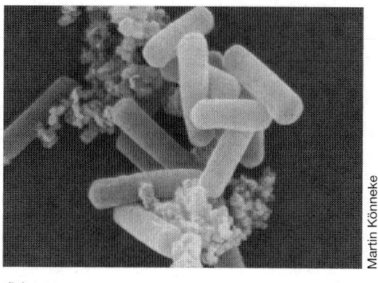

(a) *(b)*

Figure 17.14 *Nitrosopumilus maritimus*, a nitrifying species of *Archaea*. This organism can oxidize NH_3 present at the very low amounts typical of marine environments. *(a)* Phase-contrast photomicrograph. *(b)* Scanning electron micrograph. A single cell of *N. maritimus* is about 0.2 μm in diameter.

Several species of *Thaumarchaeota* have been isolated and characterized, revealing a number of properties common to this group. Species have been isolated from habitats including the oceans, marine sediment, an estuary, soil, and hot springs. All existing isolates are chemolithotrophic ammonia-oxidizers (↩ Section 14.11), and most species, like *N. maritimus*, are able to grow at very low concentrations of NH_3. The membranes of all *Thaumarchaeota* also have a unique lipid called *crenarchaeol* (↩ Figure 2.6c), a compound limited to species of this phylum. In addition, autotrophy in *Thaumarchaeota* is supported by the 3-hydroxypropionate/4-hydroxybutyrate cycle, a finding that further distinguishes archaeal nitrifiers from nitrifying *Bacteria* that employ the Calvin cycle for CO_2 fixation (↩ Section 14.5). The 3-hydroxypropionate/4-hydroxybutyrate cycle also allows for the assimilation of organic carbon, and some archaeal nitrifiers have been shown to assimilate pyruvate during mixotrophic growth. Growth temperatures of *Thaumarchaeota* vary widely, as some species thrive in polar seas while others inhabit hot spring environments up to about 75°C.

Nitrososphaera viennensis, which represents a lineage of *Thaumarchaeota* found widely in soils (see page 530), can grow at a wide range of NH_3 concentrations. Like marine species of *Thaumarchaeota*, *N. viennensis* can grow at low concentrations of NH_3, but *N. viennensis* can also tolerate high levels (up to 10 mM) of ammonium at neutral pH. Hence, *N. viennensis*, and other archaeal nitrifiers, may be active in soils that have fairly high levels of ammonia, and in these environments they may compete directly with bacterial nitrifiers. In addition, several species of *Thaumarchaeota*, including *N. viennensis*, possess urease activity. *N. viennensis* can grow with urea, which is hydrolyzed to ammonia and subsequently used as an electron donor for energy conservation.

Environmental Distribution of *Thaumarchaeota*

Thaumarchaeota are ubiquitous in soils (see page 530) and found throughout the marine water column from the equator to the polar seas. Indeed, these *Archaea* are one of the most abundant and widespread phyla on Earth, and in surveys of soil or marine samples, thaumarchaea are often found to be the dominant group of *Archaea*. With the use of fluorescent phylogenetic probes (FISH, ↩ Section 19.5), thaumarchaea have been detected in oxic marine waters worldwide; they thrive even in waters and sea ice near Antarctica (**Figure 17.15**). Marine species are planktonic (suspended freely or attached to suspended particles in the water column, Figure 17.15b) and present in significant numbers (~10^4/ml) in waters that are both nutrient-poor and very cold (0–4°C in seawater and below 0°C in sea ice). Marine thaumarchaea are found throughout the oceans and may constitute 20% of picoplankton (very small prokaryotic cells) worldwide. They are particularly abundant in the deep ocean where they can also account for up to 40% of picoplankton (↩ Section 20.11). The NH_3 concentration in marine waters is often at the threshold for archaeal nitrification, suggesting that *Thaumarchaeota* may play a major role in controlling the levels of NH_3 in the oceans.

Thaumarchaea are also common in soils, comprising as much as 1–2% of the total ribosomal RNA in soil microbial communities and, in some soils, outnumbering nitrifying *Bacteria* by 1000-fold. They are found in soils across a wide range of pH from 3.5 to 8.7. While widely present in soils, thaumarchaea may be particularly

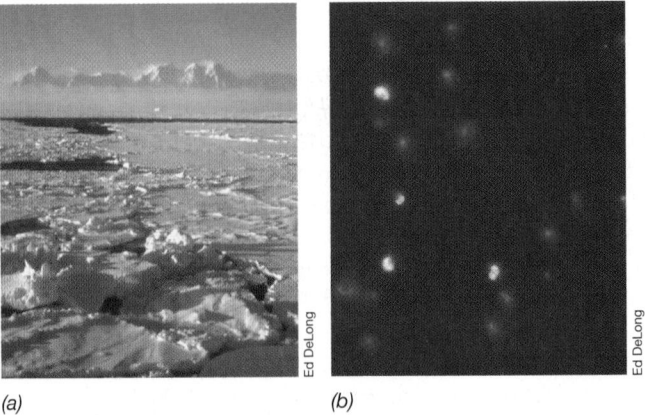

Figure 17.15 Cold-dwelling *Thaumarchaeota*. *(a)* Photo of the Antarctic Peninsula taken from shipboard. The frigid waters that lie under the surface ice shown here are habitats for cold-dwelling *Thaumarchaeota*. *(b)* Fluorescence photomicrograph of seawater treated with a fluorescent probe (↩ Section 19.5) specific for species of *Thaumarchaeota* (green cells). Blue cells are stained with DAPI, a fluorescent DNA stain that stains all cells.

important in acid soils (pH < 5.5), which make up more than 30% of all soils. Nitrifiers oxidize NH_3, but at low pH, NH_4^+ predominates and is thus unavailable for nitrification. While nitrification occurs in acid soils, and often at high rates, bacterial nitrifiers have not been observed to grow below pH 6.3. In contrast, the thaumarchaeotal species *Nitrosotalea devanaterra*, isolated from acidic agricultural soil, grows optimally at pH 4–5. The ability of thaumarchaea to grow at low NH_3 concentrations explains how they can be successful in acidic soils where free NH_3 is present in very low concentration.

MINIQUIZ

- How does the organism *Nitrosopumilus maritimus* conserve energy and obtain carbon?
- In what environments might you expect to find species of *Thaumarchaeota*?

17.6 *Nanoarchaeota* and the "Hospitable Fireball"

KEY GENUS: *Nanoarchaeum*

The **Nanoarchaeota** are represented by a single isolated species, the highly unusual *Nanoarchaeum equitans* (**Figure 17.16**). *N. equitans* is one of the smallest cellular organisms known and has the smallest genome among species of *Archaea* (0.49 Mb). The coccoid cells of *N. equitans* are very small, about 0.4 μm in diameter, and have only about 1% of the volume of an *Escherichia coli* cell. They cannot grow in pure culture and replicate only when attached to the surface of their host organism, *Ignicoccus hospitalis* (Section 17.10), a hyperthermophilic species of *Crenarchaeota* whose name means "the hospitable fireball." *N. equitans* grows to 10 or more cells per *Ignicoccus* cell and lives an apparently parasitic lifestyle, making it the only known archaeal symbiont. Indeed, in agreement with its lifestyle, the species epithet *equitans* means "riding," as in "riding the fireball."

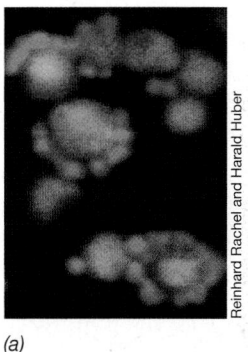
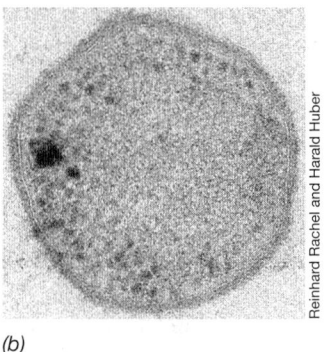

(a)   (b)

Figure 17.16 *Nanoarchaeum equitans.* *(a)* Fluorescence micrograph of cells of *N. equitans* (red) attached to cells of *Ignicoccus* (green). Cells were stained by FISH (↩ Section 19.5) using specific nucleic acid probes targeted to each organism. *(b)* Transmission electron micrograph of a thin section of a cell of *N. equitans*. Note the distinct cell wall. Cells of *N. equitans* are about 0.4 μm in diameter.

Nanoarchaeum and Its Host

N. equitans and its host *Ignicoccus* were first isolated from a submarine hydrothermal vent (↩ Section 20.14) off the coast of Iceland. However, environmental sampling of 16S ribosomal RNA genes (↩ Section 19.6) indicates that organisms phylogenetically similar to *N. equitans* exist in other submarine hydrothermal vents and in terrestrial hot springs, so *Archaea* of this kind are probably distributed worldwide in suitable hot habitats. Like its host *Ignicoccus, N. equitans* grows at temperatures from 70 to 98°C and optimally at 90°C.

The metabolism of *Nanoarchaeum* is not fully understood, but it likely depends on its host for many metabolic functions. *Ignicoccus* is an autotroph that uses H_2 as an electron donor and S^0 as an electron acceptor and so probably supplies *N. equitans* with organic carbon. *N. equitans* is incapable of metabolizing H_2 and S^0 for energy, and whether it generates ATP from substances obtained from *Ignicoccus* or obtains its ATP directly from its host is unknown. The appearance of *N. equitans* cells is typical of *Archaea*, with a cell wall consisting of an S-layer (↩ Section 2.6) that overlays what appears to be a periplasmic space (Figure 17.16*b*).

Although the sequence of its 16S ribosomal RNA gene clearly places *N. equitans* in the domain *Archaea*, the sequence differs at many sites from 16S ribosomal RNA gene sequences of other *Archaea*, even in regions of the molecule that are highly conserved. These differences initially led to the conclusion that *N. equitans* was an early-branching lineage of *Archaea*. However, more detailed phylogenetic analyses of genes encoding ribosomal proteins suggest that the divergence of *N. equitans* occurred at about the time that the *Euryarchaeota* were formed (Figure 17.1). Some analyses even suggest that *N. equitans* may be a species of *Euryarchaeota*. However, genomic analyses show the organism to lack several genes that encode information processing and cell division in *Euryarchaeota*. Conclusive phylogenetic placement of the *Nanoarchaeota* will ultimately require the discovery of more species from this group.

The *N. equitans* Genome

The sequence of the *N. equitans* genome provides insight into this organism's obligately parasitic lifestyle. Its single, circular genome is only 490,885 nucleotides long, one of the smallest cellular genomes

yet sequenced (↩ Table 9.1). Genes for several important metabolic functions are missing from the *N. equitans* genome, including those for the biosynthesis of amino acids, nucleotides, coenzymes, and lipids. Also missing are genes encoding proteins for widely distributed catabolic pathways, such as glycolysis. Presumably, all of these functions are carried out for *N. equitans* by its *Ignicoccus* host, with transfer of needed substances from *Ignicoccus* to the attached *N. equitans* cells. *N. equitans* also lacks some of the genes necessary to encode ATPase, and this indicates that it may not synthesize a functional ATPase. If true, this would be the only cellular organism that lacks ATPase. If no ATPase is present and substrate-level phosphorylation does not occur (because of the lack of glycolytic enzymes), then *N. equitans* would be dependent on *Ignicoccus* for energy as well as carbon.

With so many genes missing, which genes remain in the *N. equitans* genome? *N. equitans* contains genes encoding the key enzymes for DNA replication, transcription, and translation as well as genes for DNA repair enzymes. In addition to its small size, the genome of *N. equitans* is also among the most gene dense of any organism known as over 95% of the *N. equitans* chromosome encodes proteins—a value higher than that of most all other prokaryotic cells (↩ Section 9.3).

MINIQUIZ

• Which aspects of the biology of *Nanoarchaeum equitans* make it especially interesting from an evolutionary point of view?

• Why can it be said that *N. equitans* is both a carbon and an energy parasite?

17.7 *Korarchaeota* and the "Secret Filament"

KEY GENUS: *Korarchaeum*

Ribosomal RNA sequences of **Korarchaeota** have been observed in a range of geothermal habitats, both submarine and terrestrial. However, *Korarchaeum cryptofilum*, whose name means "the secret filament of youth," is the only characterized species in the phylum *Korarchaeota*.

First observed as a 16S ribosomal RNA gene phylotype recovered from the hot spring named Obsidian Pool in Yellowstone National Park, USA, *K. cryptofilum* has yet to be grown in pure culture. However, its genome sequence has been determined from metagenomic analyses (see ↩ Sections 9.8 and 19.8) of an enrichment culture. *K. cryptofilum* is an obligately anaerobic chemoorganotroph and a hyperthermophile, growing at 85°C. Cells are long, thin (<0.2-μm diameter) filaments of variable length (**Figure 17.17***a–c*), with most filaments being around 15 μm long but some reaching as much as 100 μm. Filaments of *K. cryptofilum* have a tough paracrystalline S-layer (Figure 17.17*d*), which maintains cell integrity in its extremely hot habitat.

Though *K. cryptofilum* cannot be grown in isolation, its genome sequence provides clues about its lifestyle. *K. cryptofilum* lacks the ability to perform anaerobic respiration (with the possible exception of proton reduction, ↩ Section 14.15) and lives a fermentative lifestyle. Similar to other archaeal hyperthermophiles, *K. cryptofilum* grows by fermentation of peptides or amino acids (see Table 17.4).

UNIT 4

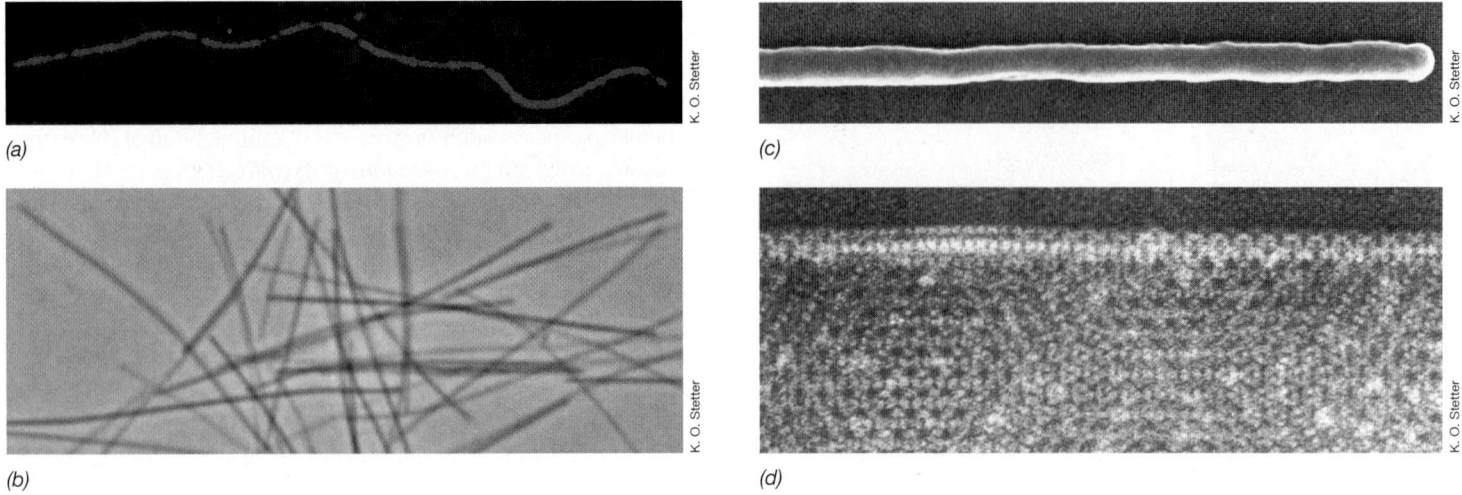

Figure 17.17 *Korarchaeum cryptofilum*. *(a)* Fluorescence in situ hybridization (FISH) was used to identify the morphology of *Korarchaeota* growing in an enrichment culture at 85°C. *(b)* Phase-contrast image of filaments of *K. cryptofilum*. *(c)* Scanning electron micrograph of a *K. cryptofilum* filament. *(d)* Transmission electron micrograph of the surface of a *K. cryptofilum* filament showing the paracrystalline S-layer (⊘ Section 2.6). Filaments of *K. cryptofilum* are about 0.17 μm wide and 15 μm long.

K. cryptofilum lacks many core genes in biosynthesis including the ability to synthesize purines, coenzyme A, and several essential cofactors. Presumably *K. cryptofilum* obtains these essential components from its environment. The inability of *K. cryptofilum* to synthesize molecules essential for its own growth may be explained by the evolution of mutual dependence as described by the Black Queen hypothesis (⊘ Explore the Microbial World, Chapter 13). This dependence on other members of the hot spring microbial community may explain why *K. cryptofilum* has proven difficult to obtain in pure culture.

As with the *Nanoarchaeota*, there is some uncertainty about the phylogenetic position of the *Korarchaeota*. The genome of *K. cryptofilum* includes some gene families that share affinity with *Euryarchaeota* and others that share affinity with *Crenarchaeota*. For example, phylogenetic analyses of ribosomal proteins, RNA polymerase subunits, and ribosomal RNA genes indicate affinity between *Crenarchaeota* and *Korarchaeota*. In contrast, genes for cell division, tRNA maturation, and DNA replication and repair indicate affinity between *Euryarchaeota* and *Korarchaeota*. The unique genetic composition of *K. cryptofilum* supports its placement near the base of the archaeal radiation (Figure 17.1), and future work on this interesting archaeon should clarify its actual phylogenetic position.

MINIQUIZ

- What is the most likely reason that *Korarchaeum cryptofilum* has been difficult to isolate in pure culture?

III • *Crenarchaeota*

Among *Archaea* in laboratory culture, the **Crenarchaeota** are mostly hyperthermophiles and include species growing optimally above the boiling point of water. Many hyperthermophiles are chemolithotrophic autotrophs, and because no phototrophs can survive such temperatures, these organisms are the sole primary producers in these habitats.

17.8 Habitats and Energy Metabolism

Most hyperthermophilic *Archaea* have been isolated from geothermally heated soils or waters containing S^0 and H_2S, and most species metabolize sulfur in one way or another. In terrestrial environments, sulfur-rich springs, boiling mud, and soils may have temperatures up to 100°C and are mildly to extremely acidic owing to the production of sulfuric acid (H_2SO_4) from the biological oxidation of H_2S and S^0 (⊘ Sections 14.9 and 21.4). Such hot, sulfur-rich environments, called **solfataras**, are found throughout the world (**Figure 17.18**), including Italy, Iceland, New Zealand, and Yellowstone National Park in Wyoming (USA). Depending on the surrounding geology, solfataras can be mildly acidic to slightly alkaline (pH 5–8) or extremely acidic, with pH values below 1. Hyperthermophilic crenarchaeotes have been obtained from all of these environments, but most inhabit neutral or weakly acidic thermal habitats.

Hyperthermophilic *Crenarchaeota* also inhabit undersea hot springs called **hydrothermal vents**. We discuss the geology and microbiology of these habitats in Section 20.14. Here we only note that submarine waters can be much hotter than surface waters because the water is under hydrostatic pressure. Indeed, all hyperthermophiles with growth temperature optima above 100°C originate from submarine sources. These sources include shallow (2–10 m depth) vents such as those off the coast of Vulcano, Italy, as well as deep (2000–4000 m depth) vents near ocean-spreading centers (see Figure 17.24). Deep hydrothermal vents are the hottest habitats so far known to yield viable life forms.

With a few exceptions, hyperthermophilic *Crenarchaeota* are obligate anaerobes. Their energy-yielding metabolism is either chemoorganotrophic or chemolithotrophic (or both, for example, in *Sulfolobus*) and is dependent on diverse electron donors and

(a)

(b)

(c)

(d)

Figure 17.18 Terrestrial habitats of hyperthermophilic *Archaea:* Yellowstone National Park.
(a) A typical solfatara; steam rich in H_2S rises to the surface. *(b)* Sulfur-rich hot spring, a habitat containing dense populations of *Sulfolobus*. The acidity in solfataras and sulfur springs comes from the oxidation of H_2S and S^0 to H_2SO_4 (sulfuric acid) by *Sulfolobus* and related sulfur-oxidizing microbes. *(c)* A typical neutral pH boiling spring, Imperial Geyser. Many different species of hyperthermophilic *Archaea* may reside in such a habitat. *(d)* An acidic iron-rich geothermal spring, another *Sulfolobus* habitat; here the oxidation of Fe^{2+} to Fe^{3+} generates acidity.

acceptors. Fermentation is rare and most bioenergetic strategies employ anaerobic respiration (Table 17.4). Energy is conserved during these respiratory processes by the same general mechanism widespread in *Bacteria*: electron transfer within the cytoplasmic membrane leading to the formation of a proton motive force from which ATP is made by way of proton-translocating ATPases (Section 3.11).

Many hyperthermophilic crenarchaeotes can grow chemolithotrophically under anoxic conditions, with H_2 as the electron donor and S^0 or NO_3^- as the electron acceptor; a few can also oxidize H_2 aerobically (Table 17.4). H_2 respiration with ferric iron (Fe^{3+}) as electron acceptor also occurs in several hyperthermophiles. Other chemolithotrophic lifestyles include the oxidation of S^0 and Fe^{2+} aerobically or Fe^{2+} anaerobically with NO_3^- as the acceptor (Table 17.4). Only one sulfate-reducing hyperthermophile is known (the euryarchaeote *Archaeoglobus*, Section 17.4). The only bioenergetic option apparently impossible is photosynthesis, a means of energy conservation that is apparently limited to temperatures below 74°C (see Figure 17.28).

MINIQUIZ
- What form of energy metabolism is widespread among hyperthermophiles?
- How might the temperature and pH tolerance of a hyperthermophile living in a solfatara differ from that of a hyperthermophile living in a hydrothermal vent?

17.9 *Crenarchaeota* from Terrestrial Volcanic Habitats

KEY GENERA: *Sulfolobus, Acidianus, Thermoproteus, Pyrobaculum*
Terrestrial volcanic habitats can have temperatures as high as 100°C and are thus suitable for hyperthermophilic *Archaea*. Two phylogenetically related organisms isolated from these environments are *Sulfolobus* and *Acidianus*. These genera form the heart of an order called the *Sulfolobales* (Table 17.5). In addition, *Sulfolobus* has been a model organism for molecular biology studies of *Archaea*.

TABLE 17.4 **Energy-yielding reactions of hyperthermophilic *Archaea* by nutritional class**

Energy-yielding reaction	Metabolic type[a]	Example genera[b]
Chemoorganotrophic		
Organic compound + $S^0 \rightarrow H_2S + CO_2$	AnR	Thermoproteus, Thermococcus, Desulfurococcus, Thermofilum, Pyrococcus
Organic compound + $SO_4^{2-} \rightarrow H_2S + CO_2$	AnR	Archaeoglobus
Organic compound + $O_2 \rightarrow H_2O + CO_2$	AeR	Sulfolobus
Organic compound $\rightarrow CO_2 + H_2$ + fatty acids	AnR	Staphylothermus, Pyrodictium
Organic compound + $Fe^{3+} \rightarrow CO_2 + Fe^{2+}$	AnR	Pyrodictium
Organic compound + $NO_3^- \rightarrow CO_2 + N_2$	AnR	Pyrobaculum
Pyruvate $\rightarrow CO_2 + H_2$ + acetate	AnR	Pyrococcus
Peptides $\rightarrow CO_2$ + acetate + butanol	F	Hyperthermus, Korarchaeum
Chemolithotrophic		
$H_2 + S^0 \rightarrow H_2S$	AnR	Acidianus, Pyrodictium, Thermoproteus, Stygiolobus, Ignicoccus
$H_2 + NO_3^- \rightarrow NO_2^- + H_2O$ (NO_2^- is reduced to N_2 by some species)	AnR	Pyrobaculum
$4 H_2 + NO_3^- + H^+ \rightarrow NH_4^+ + 2 H_2O + OH^-$	AnR	Pyrolobus
$H_2 + 2 Fe^{3+} \rightarrow 2 Fe^{2+} + 2 H^+$	AnR	Pyrobaculum, Pyrodictium, Archaeoglobus
$2 H_2 + O_2 \rightarrow 2 H_2O$	AeR	Acidianus, Sulfolobus, Pyrobaculum
$2 S^0 + 3 O_2 + 2 H_2O \rightarrow 2 H_2SO_4$	AeR	Sulfolobus, Acidianus
$2 FeS_2 + 7 O_2 + 2 H_2O \rightarrow 2 FeSO_4 + 2 H_2SO_4$	AeR	Sulfolobus, Acidianus, Metallosphaera
$2 FeCO_3 + NO_3^- + 6 H_2O \rightarrow 2 Fe(OH)_3 + NO_2^- + 2 HCO_3^- + 2 H^+ + H_2O$	AnR	Ferroglobus
$4 H_2 + SO_4^{2-} + 2 H^+ \rightarrow 4 H_2O + H_2S$	AnR	Archaeoglobus
$4 H_2 + CO_2 \rightarrow CH_4 + 2 H_2O$	AnR	Methanopyrus, Methanocaldococcus, Methanothermus

[a] AnR, anaerobic respiration; AeR, aerobic respiration; F, fermentation.
[b] Most are *Crenarchaeota*; ⟳ Figure 17.1.

Sulfolobales

Sulfolobus grows in sulfur-rich acidic thermal areas (Figure 17.18) at temperatures up to 90°C and at pH values of 1–5. *Sulfolobus* is an aerobic chemolithotroph that oxidizes H_2S or S^0 to H_2SO_4 and fixes CO_2 as a carbon source. *Sulfolobus* can also grow chemoorganotrophically. Cells of *Sulfolobus* are more or less spherical but contain distinct lobes (**Figure 17.19a**). Cells adhere tightly to sulfur crystals, where they can be seen with a microscope after staining with fluorescent dyes (⟳ Figure 14.27b). Besides the aerobic respiration of sulfur or organic compounds, *Sulfolobus* can also oxidize Fe^{2+} to Fe^{3+}, and this ability has been harnessed for the high-temperature leaching of iron and copper ores (⟳ Sections 21.5 and 22.1).

A facultative aerobe resembling *Sulfolobus* also lives in acidic solfataric springs. This organism, *Acidianus* (Figure 17.19b), differs from *Sulfolobus* most clearly by its ability to grow using S^0 both anaerobically as well as aerobically. Under aerobic conditions the organism uses S^0 as an electron *donor*, oxidizing S^0 to H_2SO_4, with O_2 as an electron acceptor. Anaerobically, *Acidianus* uses S^0 as an electron *acceptor* with H_2 as an electron donor, forming H_2S as the reduced product. Thus, the metabolic fate of S^0 in cultures of *Acidianus* depends on the presence or absence of O_2. Like *Sulfolobus*, *Acidianus* is roughly spherical in shape but is not as lobed (Figure 17.19b). It grows at temperatures from 65°C up to a maximum of 95°C, with an optimum of about 90°C. As a group, then, the *Sulfolobales* contain the most thermophilic of all highly acidophilic *Archaea*.

Thermoproteales

Key genera within the *Thermoproteales* are *Thermoproteus*, *Thermofilum*, and *Pyrobaculum*. The genera *Thermoproteus* and *Thermofilum* consist of rod-shaped cells that inhabit neutral or slightly acidic hot springs. Cells of *Thermoproteus* are rigid rods about 0.5 μm in diameter and highly variable in length, ranging from short cells of 1–2 μm (**Figure 17.20a**) up to filaments 70–80 μm long. Filaments of *Thermofilum* are thinner, some as little as 0.17–0.35 μm wide, with filament lengths ranging up to 100 μm (Figure 17.20b).

Both *Thermoproteus* and *Thermofilum* are strict anaerobes that carry out an S^0-based anaerobic respiration (Table 17.4). Most *Thermoproteus* isolates can grow chemolithotrophically on H_2 or chemoorganotrophically on complex carbon substrates such as yeast extract, small peptides, starch, glucose, ethanol, malate, fumarate, or formate (Table 17.4). *Pyrobaculum* (Figure 17.20c) is a rod-shaped hyperthermophile but is physiologically distinct from other *Thermoproteales* in that some species of *Pyrobaculum* can respire aerobically. However, *Pyrobaculum* can also grow by anaerobic respiration with NO_3^-, Fe^{3+}, or S^0 as electron acceptors and H_2 as an electron donor (that is, they can grow chemolithotrophically and autotrophically). Other species of *Pyrobaculum* can grow anaerobically on organic electron donors, reducing S^0 to H_2S. The growth temperature optimum of *Pyrobaculum* is 100°C, and species of this organism have been isolated from terrestrial hot springs and from hydrothermal vents.

TABLE 17.5 **Properties of some hyperthermophilic *Crenarchaeota***

Order/Genus[a]	Morphology	Relationship to O_2[b]	Temperature			Optimum pH
			Minimum	Optimum	Maximum	
Sulfolobales						
Sulfolobus	Lobed coccus	Ae	55	75	87	2–3
Acidianus	Coccus	Fac	60	88	95	2
Metallosphaera	Coccus	Ae	50	75	80	2
Stygiolobus	Lobed coccus	An	57	80	89	3
Sulfurisphaera	Coccus	Fac	63	84	92	2
Sulfurococcus	Coccus	Ae	40	75	85	2.5
Thermoproteales						
Thermoproteus	Rod	An	60	88	96	6
Thermofilum	Rod	An	70	88	95	5.5
Pyrobaculum	Rod	Fac	74	100	102	6
Caldivirga	Rod	An	60	85	92	4
Thermocladium	Rod	An	60	75	80	4.2
Desulfurococcales						
Desulfurococcus	Coccus	An	70	85	95	6
Aeropyrum	Coccus	Ae	70	95	100	7
Staphylothermus	Cocci in clusters	An	65	92	98	6–7
Pyrodictium	Disc-shaped with filaments	An	82	105	110	6
Pyrolobus	Lobed coccus	Fac	90	106	113	5.5
Thermodiscus	Disc-shaped	An	75	90	98	5.5
Ignicoccus	Irregular coccus	An	65	90	103	5
Hyperthermus	Irregular coccus	An	75	102	108	7
Stetteria	Coccus	An	68	95	102	6
Sulfophobococcus	Disc-shaped	An	70	85	95	7.5
Thermosphaera	Coccus	An	67	85	90	7
Strain 121[c]	Coccus	An	85	106	121	7

[a]The group names ending in "ales" are order names.
[b]Ae, aerobe; An, anaerobe; Fac, facultative.
[c]Also known by the unofficial taxonomic name of "*Geogemma barossii*."

— MINIQUIZ —

- What are the similarities and differences between *Sulfolobus* and *Pyrobaculum*?
- Among *Thermoproteales*, what is unusual about the metabolism of *Pyrobaculum*?

17.10 *Crenarchaeota* from Submarine Volcanic Habitats

KEY GENERA: *Pyrodictium, Pyrolobus, Ignicoccus, Staphylothermus*

We now consider the microbiology of submarine volcanic habitats, homes to the most thermophilic of all known *Archaea*. These habitats include both shallow-water thermal springs and deep-sea hydrothermal vents. We discuss the geology of these fascinating microbial habitats in Section 20.14 and the interesting animal communities that develop there in Section 23.9. The organisms to be described here constitute an order of *Archaea* called the *Desulfurococcales* (Table 17.5).

Pyrodictium and *Pyrolobus*

Pyrodictium and *Pyrolobus* are examples of microorganisms whose growth temperature optimum lies above 100°C; the optimum for *Pyrodictium* is 105°C and for *Pyrolobus* is 106°C. Cells of *Pyrodictium* are irregularly disc-shaped and grow in culture in a mycelium-like layer attached to crystals of S^0. The cell mass consists of a network of fibers to which individual cells are attached (**Figure 17.21a**). The fibers are hollow and consist of protein arranged in a fashion similar to that of bacterial flagella (Section 2.11). However, the filaments do not function in motility but instead as organs of attachment. The cell walls of *Pyrodictium* are composed of glycoprotein. Physiologically, *Pyrodictium* is a strict anaerobe that grows chemolithotrophically on H_2 as an electron donor and S^0 as an electron acceptor (Section 14.14) or chemoorganotrophically on complex mixtures of organic compounds (Table 17.4).

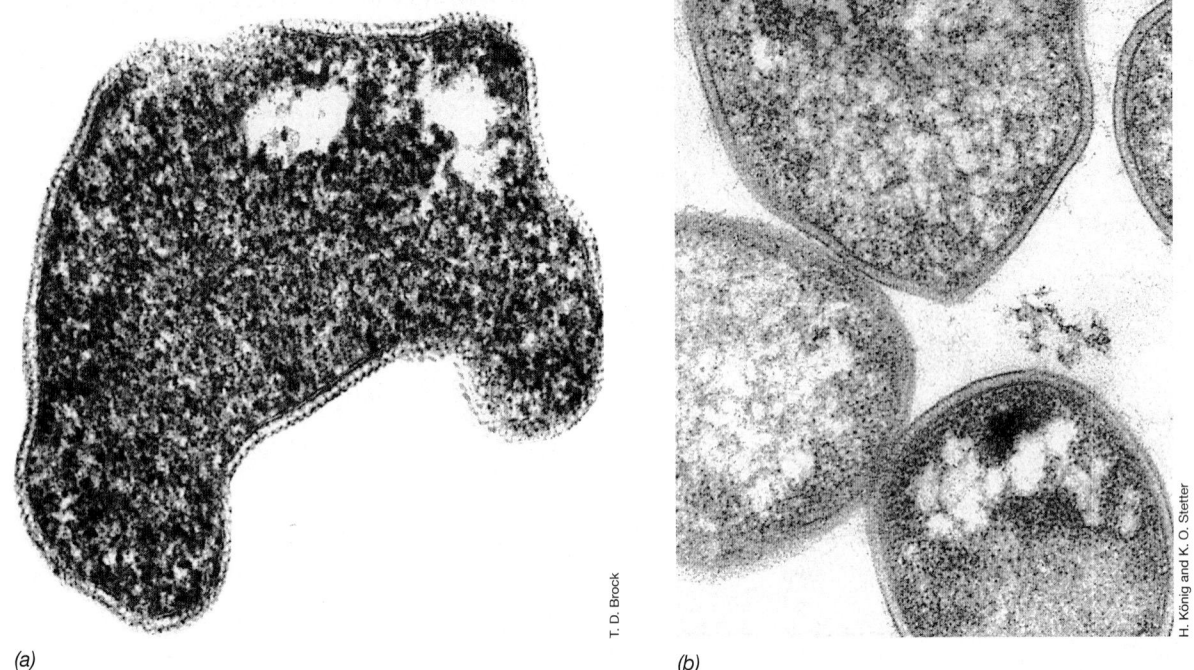

(a)

T. D. Brock

(b)

H. König and K. O. Stetter

Figure 17.19 Acidophilic hyperthermophilic *Archaea*, the *Sulfolobales*. *(a) Sulfolobus acidocaldarius;* electron micrograph of a thin section. *(b) Acidianus infernus*; electron micrograph of a thin section. Cells of both organisms vary from 0.8 to 2 μm in diameter. *Sulfolobales* typically show temperature optima below 90°C (Table 17.5).

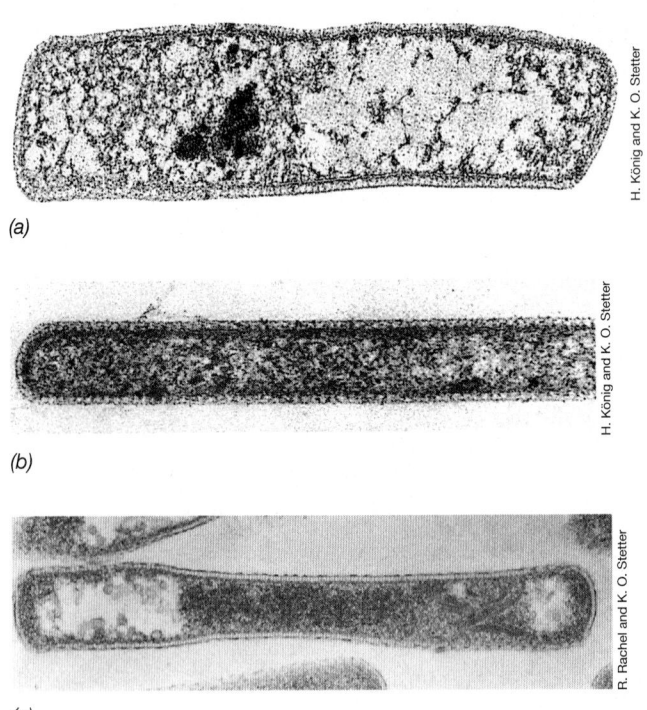

(a)

H. König and K. O. Stetter

(b)

H. König and K. O. Stetter

(c)

R. Rachel and K. O. Stetter

Figure 17.20 Rod-shaped hyperthermophilic *Archaea*, the *Thermoproteales*. *(a) Thermoproteus neutrophilus*; electron micrograph of a thin section. A cell is about 0.5 μm in diameter. *(b) Thermofilum librum*. A cell is about 0.25 μm in diameter. *(c) Pyrobaculum aerophilum*. Transmission electron micrograph of a thin section; a cell measures 0.5×3.5 μm. Although the temperature optimum of *P. aerophilum* is 100°C, optima for other *Thermoproteales* are all below 90°C (Table 17.5).

Pyrolobus fumarii (Figure 17.21*c*) is one of the most thermophilic of the hyperthermophiles. Its growth temperature maximum is 113°C (Table 17.5). *P. fumarii* lives in the walls of "black smoker" hydrothermal vent chimneys (⊘ Section 20.14 and Figures 20.37, 20.38, and 20.40) where its autotrophic abilities contribute organic carbon to this otherwise inorganic environment. *P. fumarii* cells are coccoid-shaped (Figure 17.21*c*), and the cell wall is composed of protein. The organism is an obligate H_2 chemolithotroph, growing by the oxidation of H_2 coupled to the reduction of NO_3^- to ammonium (NH_4^+), thiosulfate ($S_2O_3^{2-}$) to H_2S, or very low concentrations of O_2 to H_2O. Besides its extremely thermophilic nature, *P. fumarii* can withstand temperatures substantially above its growth temperature maximum. For example, cultures of *P. fumarii* survive autoclaving (121°C) for 1 h, a condition that even bacterial endospores (⊘ Section 2.10) cannot withstand.

Another organism in this group shares with *Pyrolobus* a growth temperature optimum of 106°C. However, "Strain 121," as this organism has been called, actually shows weak growth at 121°C, and cells remain viable for 2 h at 130°C. Only *Methanopyrus*, a hyperthermophilic methanogen, can grow at a higher temperature (122°C, Section 17.2). Strain 121 consists of coccoid cells with archaella (Figure 17.21*d*); the organism is also a strict anaerobe and grows chemolithotrophically and autotrophically with Fe^{3+} as electron acceptor and formate or H_2 as electron donors. It is thus clear that the *Pyrodictium/Pyrolobus* group collectively contains the most hyperthermophilic examples of all known microbes.

Desulfurococcus and *Ignicoccus*

Other notable members of the *Desulfurococcales* include *Desulfurococcus*, the genus for which the order is named (**Figure 17.22a**), and

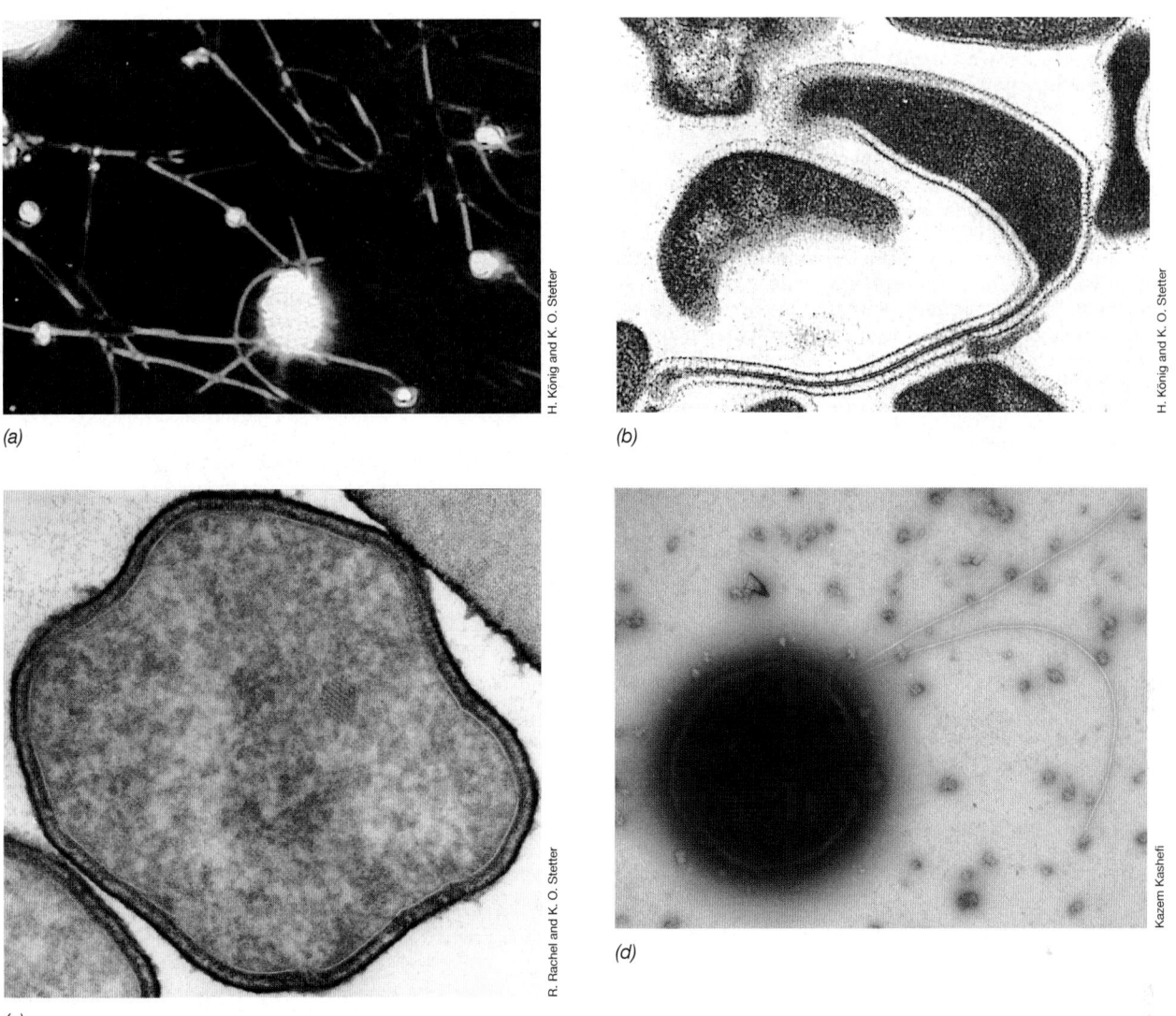

(a)

(b)

H. König and K. O. Stetter

(c)

R. Rachel and K. O. Stetter

(d)

Kazem Kashefi

Figure 17.21 *Desulfurococcales* with growth temperature optima >100°C. *(a) Pyrodictium occultum* (growth temperature optimum, 105°C), dark-field micrograph. *(b)* Thin-section electron micrograph of *P. occultum.* Cells are highly variable in diameter from 0.3 to 2.5 μm. *(c)* Thin section of a cell of *Pyrolobus fumarii,* one of the most thermophilic organisms ever described (growth temperature optimum, 106°C); a cell is about 1.4 μm in diameter. *(d)* Negative stain of a cell of "Strain 121," capable of growth at 121°C; a cell is about 1 μm wide. Although the *Desulfurococcales* contain the greatest number of hyperthermophiles capable of growth above 100°C, the most thermophilic of all known *Archaea* is actually a euryarchaeote, *Methanopyrus* (Section 17.2).

Ignicoccus. Desulfurococcus is a strictly anaerobic S⁰-reducing organism like *Pyrodictium,* but differs from this organism in its phylogeny and the fact that it is much less thermophilic, growing optimally at about 85°C.

Ignicoccus grows optimally at 90°C, and its energy metabolism is based on H_2 as an electron donor and S^0 as an electron acceptor, as is that of so many hyperthermophilic *Archaea* (Table 17.4). Some *Ignicoccus* species are hosts to the small parasitic archaeon *Nanoarchaeum equitans* (Section 17.6). *Ignicoccus* (Figure 17.22*b*) has a novel cell structure that lacks an S-layer and possesses a unique *outer cellular membrane.* This outer cellular membrane is distinct in several ways from the outer membrane of gram-negative *Bacteria* (⇄ Section 2.5). Most notably, the outer cellular membrane of *Ignicoccus* contains ATPase and is the site of energy conservation. *Ignicoccus* also contains an inner cellular membrane that contains

the cytoplasm and the enzymes responsible for biosynthesis and information processing. In this way neither the outer nor inner cellular membrane satisfies the typical definition of a cytoplasmic membrane (⇄ Section 2.3).

Between the inner and outer cellular membranes of *Ignicoccus* is a large *intermediate compartment* that is analogous to the periplasm of gram-negative *Bacteria* but is much larger in volume, representing some two to three times the volume of the cytoplasm (Figure 17.22*b*). The periplasm of *Ignicoccus* also contains membrane-bound vesicles (Figure 17.22*b*) that may function in exporting substances outside the cell. In this way, the cell structure of *Ignicoccus* resembles that of *Eukarya.* Hence, *Ignicoccus* has been proposed to be a modern descendant of the ancestral cell type that gave rise to the origin of eukaryotic cells (⇄ Figure 13.9).

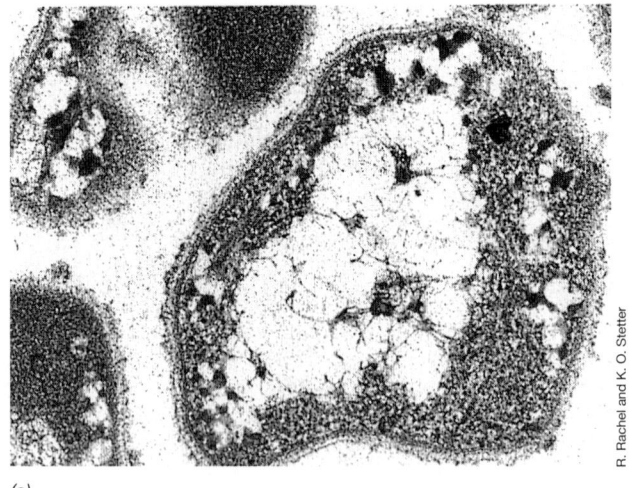

(a)

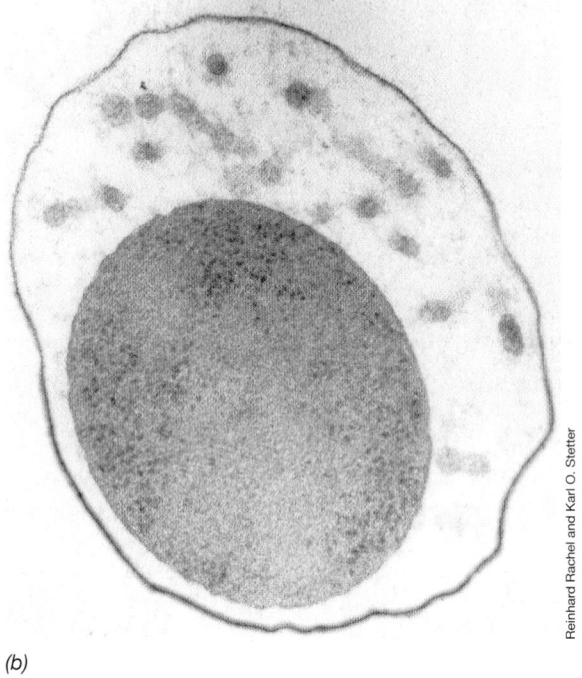

(b)

Figure 17.22 *Desulfurococcales* **with growth temperature optima <100°C.** *(a)* Thin section of a cell of *Desulfurococcus saccharovorans*; a cell is 0.7 μm in diameter. *(b)* Thin section of a cell of *Ignicoccus islandicus*; the cell proper is surrounded by an extremely large periplasmic compartment. The cell itself measures about 1 μm in diameter and the cell plus periplasm measures 1.4 μm.

Staphylothermus

A morphologically unusual member of the order *Desulfurococcales* is the genus *Staphylothermus* (**Figure 17.23**). Cells of *Staphylothermus* are spherical, about 1 μm in diameter, and form aggregates of up to 100 cells, much like its morphological counterpart among the *Bacteria, Staphylococcus* (cp Figures 16.20 and 30.28*a*). Unlike many hyperthermophiles, *Staphylothermus* is not a chemolithotroph but instead a chemoorganotroph, growing optimally at 92°C. Energy is obtained from the fermentation of peptides, producing the fatty acids acetate and isovalerate as fermentation products (Table 17.4).

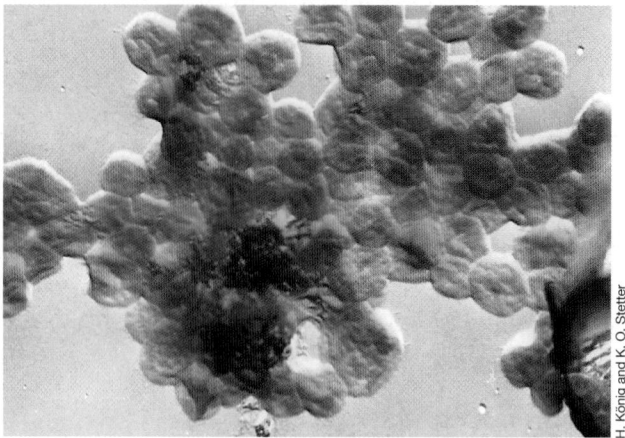

Figure 17.23 **The hyperthermophile** *Staphylothermus marinus.* Electron micrograph of shadowed cells. A single cell is about 1 μm in diameter.

Isolates of *Staphylothermus* have been obtained from both shallow marine hydrothermal vents and very hot black smokers (see Figure 17.24; cp Section 20.14). This organism is apparently widely distributed in submarine thermal areas, where it is likely to play a significant role in consuming proteins released from dead organisms.

--- **MINIQUIZ** ---

- What can we conclude about the *Pyrodictium/Pyrolobus* group in terms of life at high temperature?
- What unusual structural features are present in *Ignicoccus* and *Staphylothermus*?

IV • Evolution and Life at High Temperature

Most of the hyperthermophiles discovered so far are species of *Archaea* and some grow near to what may be the upper temperature limit for life. Here we consider the major factors that likely define the upper temperature limit for life and the biological adaptations of hyperthermophiles that permit them to exist at the exceptionally high temperatures of 100°C and higher. We end with a discussion of the importance of hydrogen (H₂) metabolism to the biology of hyperthermophiles.

17.11 An Upper Temperature Limit for Microbial Life

Habitats that contain liquid water—a prerequisite for cellular life—and that have temperatures higher than 100°C are only found where geothermally heated water flows out of vents or rifts in the ocean floor (cp Figures 13.3 and 20.37–20.40). The hydrostatic pressure that overlies the water keeps it from boiling, allowing it to reach temperatures of up to about 400°C in vents at several thousand meters' depth. In contrast, terrestrial hot springs can boil and therefore only attain temperatures near 100°C. It is not surprising, then, that hydrothermal vents have been rich sources of hyperthermophilic *Archaea* with growth temperature optima above 100°C (Table 17.5).

Black smokers emit hydrothermal vent fluid at 250–350°C or higher. Metallic mounds or more upright structures called *chimneys* form from the metal sulfides that precipitate out of the hot fluid as it mixes with the surrounding, much cooler seawater (**Figure 17.24**). As far as is known, the superheated vent water itself is sterile. However, hyperthermophiles thrive in mounds or smoker chimney walls where temperatures are compatible with their survival and growth (co Figure 20.40). By studying structures such as these, we can address the question, "What is the upper temperature for microbial (and presumably all forms of) life?"

What Is the Upper Temperature Limit for Life?

How high a temperature can hyperthermophiles withstand? Over the past several decades, the known upper temperature limit for life has been pushed higher and higher with the isolation and characterization of new species of thermophiles and hyperthermophiles (**Figure 17.25**). For some time the record holder was *Pyrolobus fumarii* (Figure 17.21c), with its upper temperature limit for growth of 113°C. The current record holder, *Methanopyrus* (Section 17.2 and Figure 17.8), however, has pushed the limit somewhat higher, with the ability to grow at 122°C and to survive substantial periods at even higher temperatures. Given the trend over the past several years (Figure 17.25), one can predict that *Archaea* even more hyperthermophilic than *Methanopyrus* may inhabit hydrothermal environments but have yet to be isolated. Indeed, many experts predict that the upper temperature limit for *Archaea* is likely to exceed 140°C, perhaps even 150°C, and that the maximum temperature allowing survival but not growth is even hotter yet.

Figure 17.24 Hydrothermal vents. Hydrothermal mound from the Rainbow vent field, Mid-Atlantic Ridge hydrothermal system. The hydrothermal fluid emitting from the two short chimneys is >300°C.

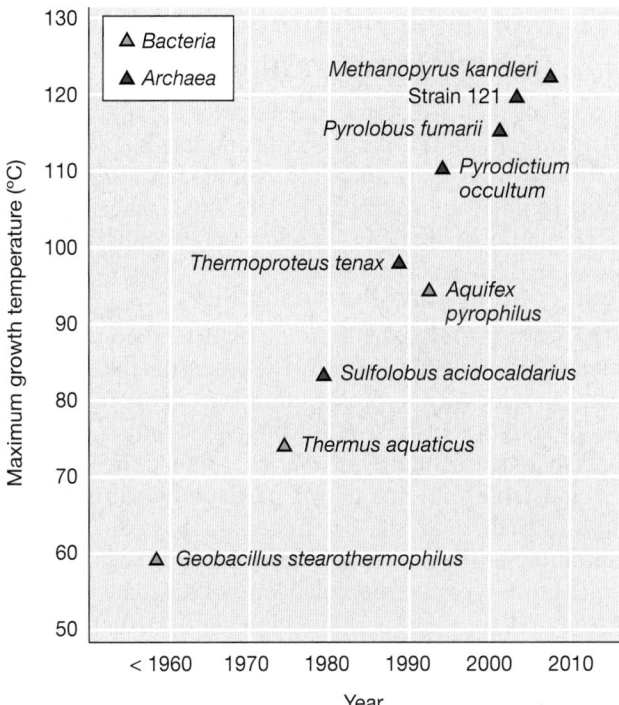

Figure 17.25 Thermophilic and hyperthermophilic *Bacteria* and *Archaea*. The graph gives the species that were, in turn, the record holders for growing at the highest temperature, from before 1960 to the present.

Biochemical Problems at Supercritical Temperatures

Whatever the upper temperature limit is for life, it is likely to be defined by one or more biochemical challenges that evolution has been unable to solve. There is obviously an upper limit, but we do not yet know what it is. Water samples taken directly from superheated (>250°C) hydrothermal vent discharges are devoid of measurable biochemical markers (DNA, RNA, and protein) that would signal life as we know it, while vents emitting water at temperatures below about 150°C yield evidence of macromolecules. These results are consistent with laboratory experiments on the stability of key biomolecules. For example, ATP is degraded almost instantly at 150°C. Thus, above 150°C, any life forms would have to deal with the heat lability of a molecule that is, as far as is known, universally distributed in cells. As a caveat, however, the stability of small molecules such as ATP may be significantly greater under cytoplasmic conditions of high levels of dissolved solutes than in pure solutions tested in the laboratory. Nevertheless, if life forms exist at temperatures above 150°C, they must be unique in many ways, either using a suite of novel small molecules absent from cells as we know them, or deploying special protection systems that maintain small molecules in a stable state such that biochemistry can proceed.

--- MINIQUIZ ---

- Where are the hottest potential microbial habitats located on Earth?

- Why would it be impossible for organisms to grow at 200 or 300°C?

17.12 Molecular Adaptations to Life at High Temperature

Because all cellular structures and activities are affected by heat, hyperthermophiles are likely to exhibit multiple adaptations to the exceptionally high temperatures of their habitats. Here we briefly examine some adaptations employed by hyperthermophiles to protect their proteins and nucleic acids at high temperatures.

Protein Folding and Thermostability

Because most proteins denature at high temperatures, much research has been done to identify the properties of thermostable proteins. Protein thermostability derives from the folding of the molecule itself, not because of the presence of any special amino acids. Perhaps surprisingly, however, the amino acid composition of thermostable proteins is not particularly unusual except perhaps in their slight bias for increased levels of amino acids that promote alpha-helical secondary structures. In fact, many enzymes from hyperthermophiles contain the same major structural features in both primary and higher-order structure (⟂ Section 4.7) as their heat-labile counterparts from organisms that grow best at much lower temperatures.

Thermostable proteins typically do display some structural features that likely improve their thermostability. These include having highly hydrophobic cores, which decrease the tendency of the protein to unfold in an ionic environment, and more ionic interactions on the protein surfaces, which also help hold the protein together and work against unfolding. Ultimately, it is the *folding* of the protein that most affects its heat stability, and noncovalent ionic bonds called *salt bridges* on a protein's surface likely play a major role in maintaining the biologically active structure. But, as previously stated, many of these changes are possible with only minimal changes in primary structure (amino acid sequence), as seen when thermostable and heat-labile forms of the same protein are compared.

Chaperones: Assisting Proteins to Remain in Their Native State

Earlier we discussed a class of proteins called *chaperones* (heat shock proteins; ⟂ Section 4.11) that function to refold partially denatured proteins. Hyperthermophilic *Archaea* have special classes of chaperones that function only at the highest growth temperatures. In cells of *Pyrodictium abyssi* (**Figure 17.26**), for example, a major chaperone is the protein complex called the **thermosome**. This complex keeps other proteins properly folded and functional at high temperature, helping cells survive even at temperatures above their maximal growth temperature. Cells of *P. abyssi* grown near its maximum temperature (110°C) contain high levels of the thermosome. Possibly because of this, the cells can remain viable following a heat shock, such as a 1-h treatment in an autoclave (121°C). In cells experiencing such a treatment and then returned to the optimum temperature, the thermosome, which is itself quite heat resistant, is thought to refold sufficient copies of key denatured proteins that *P. abyssi* can once again begin to grow and divide. Thus, as a result of chaperone activity, the upper temperature limit at which many hyperthermophiles can *survive* is higher than the upper temperature at which they can *grow*. The "safety

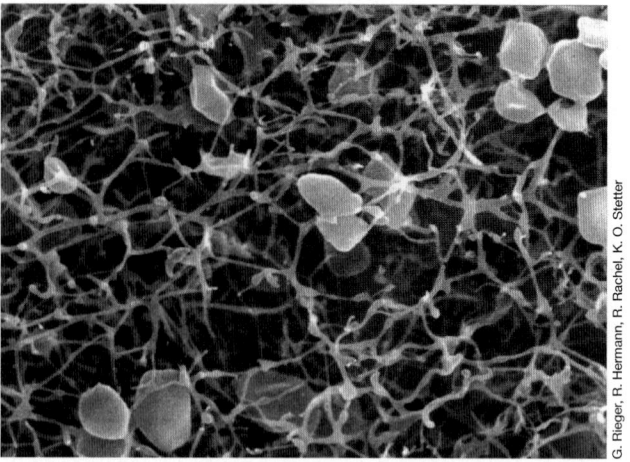

Figure 17.26 *Pyrodictium abyssi*, **scanning electron micrograph.** *Pyrodictium* has been studied as a model of macromolecular stability at high temperatures. Cells are enmeshed in a sticky glycoprotein matrix that binds them together.

net" of chaperone activity probably ensures that cells in nature that briefly experience temperatures above their growth temperature maximum are not killed by the exposure.

DNA Stability: Solutes, Reverse Gyrase, and DNA-Binding Proteins

What keeps DNA from melting at high temperatures? Various mechanisms are known to contribute. One such mechanism increases cellular solute levels, in particular potassium (K^+) or compatible organic compounds. For example, the cytoplasm of the hyperthermophilic methanogen *Methanopyrus* (Section 17.2) contains greater than 1 M potassium cyclic 2,3-diphosphoglycerate. This solute prevents chemical damage to DNA, such as depurination or depyrimidization (loss of a nucleotide base through hydrolysis of the glycosidic bond) from high temperatures, events that can lead to mutation (⟂ Section 11.2). This compound and other compatible solutes, such as potassium di-*myo*-inositol phosphate, which protects against osmotic stress, and the polyamines putrescine and spermidine, which stabilize both ribosomes and nucleic acids at high temperature, help maintain key cellular macromolecules in hyperthermophiles in their active forms.

A unique protein found *only* in hyperthermophiles is responsible for DNA stability in these organisms. All hyperthermophiles produce a special DNA topoisomerase called **reverse DNA gyrase**. This enzyme introduces positive supercoils into the DNA of hyperthermophiles (in contrast to the negative supercoils introduced by DNA gyrase present in *Bacteria* and most *Archaea*; ⟂ Section 4.1). Positive supercoiling stabilizes DNA to heat and thereby prevents the DNA helix from spontaneously unwinding. The noticeable absence of reverse DNA gyrase in organisms whose growth temperature optima lie below 80°C strongly suggests a specific role for this enzyme in DNA stability at high temperatures.

Species of *Euryarchaeota* also contain highly basic (positively charged) DNA-binding proteins that are remarkably similar in amino acid sequence and folding properties to the core histones of the *Eukarya* (⟂ Figure 2.46). Archaeal histones from the hyperthermophilic methanogen *Methanothermus fervidus* (Figure 17.7c)

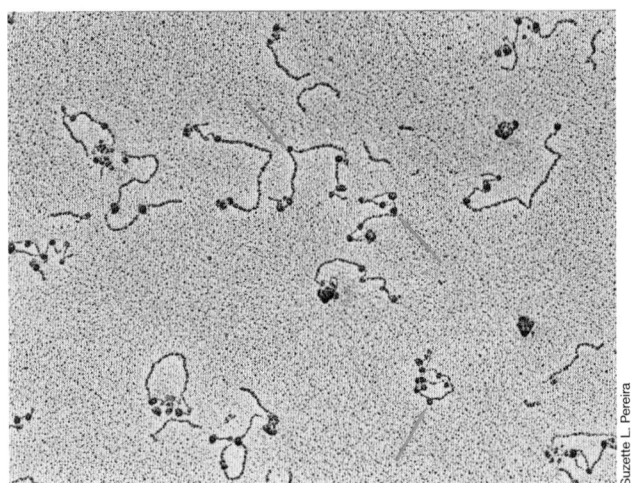

Figure 17.27 Archaeal histones and nucleosomes. Electron micrograph of linearized plasmid DNA wrapped around copies of archaeal histone Hmf (from the hyperthermophilic methanogen *Methanothermus fervidus*) to form the roughly spherical, darkly stained nucleosome structures (arrows). Compare this micrograph with an artist's depiction of the histones and nucleosomes of *Eukarya* shown in Figure 2.46*b*.

have been particularly well studied. These proteins wind and compact DNA into nucleosome-like structures (**Figure 17.27**) and maintain the DNA in a double-stranded form at very high temperatures. Archaeal histones are found in most *Euryarchaeota*, including extremely halophilic *Archaea*, such as *Halobacterium*. However, because the extreme halophiles are not thermophiles, archaeal histones may have other functions besides DNA stability, in particular in assisting in gene expression by opening the helix to allow transcriptional proteins to bind.

Lipid and Ribosomal RNA Stability

How have the lipids and the protein-synthesizing machinery of hyperthermophiles adjusted to high temperatures? Virtually all hyperthermophilic *Archaea* synthesize lipids of the biphytanyl tetraether type (⇨ Section 2.3). These lipids are naturally heat resistant because the phytanyl units forming each half of the membrane structure are covalently bonded to one another; this yields a *lipid monolayer* membrane instead of the normal lipid bilayer (⇨ Figure 2.6). This structure resists the tendency of heat to pull apart a lipid bilayer constructed of fatty acid or phytanyl side chains that are not covalently bonded.

A final point on molecular adaptations to life at high temperatures is the base composition of ribosomal RNAs. Ribosomal RNAs are key structural and functional components of the ribosome, the cell's protein-synthesizing apparatus (⇨ Section 4.10). Hyperthermophilic species of both *Bacteria* and *Archaea* show as much as a 15% greater proportion of GC base pairs in their small ribosomal subunit RNAs compared with organisms that grow at lower temperatures. GC base pairs form three hydrogen bonds compared to the two of AU base pairs (⇨ Figure 4.1*c*), and thus the higher GC content of the ribosomal RNAs should confer greater thermal stability on the ribosomes of these organisms and this should assist protein synthesis at high temperatures. By contrast to ribosomal RNAs, the GC content of genomic DNA of hyperthermophiles is often rather low, which suggests that the thermal stability

of ribosomal RNA might be an especially significant factor for life under hyperthermophilic conditions.

— MINIQUIZ —

- How do hyperthermophiles keep proteins and DNA from being destroyed by high heat?
- How are the lipids and ribosomes of hyperthermophiles protected from heat denaturation?

17.13 Hyperthermophilic *Archaea*, H₂, and Microbial Evolution

When cellular life first arose on Earth nearly 4 billion years ago, temperatures almost certainly were far hotter than they are today. Thus, for hundreds of millions of years, Earth may have been suitable only for hyperthermophiles. Given the discussion above on the temperature limits to life, it has been hypothesized that biological molecules, biochemical processes, and the first cells arose on Earth around hydrothermal springs and vents on the seafloor as they cooled to temperatures compatible with biological molecules (⇨ Section 13.1 and Figures 13.3 and 13.4). The phylogeny of modern hyperthermophiles (Figure 17.1), as well as the similarities in their habitats and metabolism to those of early cells on Earth, suggests that hyperthermophiles may be the closest remaining descendants of ancient cells and are a living window into the biology of ancient microbial life.

Hyperthermophilic Habitats and H₂ as an Energy Source

The oxidation of H_2 linked to the reduction of Fe^{3+}, S^0, NO_3^-, or, rarely, O_2 is a widespread form of energy metabolism in hyperthermophiles (Table 17.4 and **Figure 17.28**). This, coupled with the likelihood that these hyperthermophiles best characterize early Earth phenotypes, points to the important role H_2 has played in the evolution of microbial life. Hydrogen metabolism may have evolved in primitive organisms because of the ready availability of H_2 and suitable inorganic electron acceptors in their primordial environments, but also because a H_2-based energy economy requires relatively few proteins (⇨ Figure 13.5). As chemolithotrophs, these organisms may have obtained all of their carbon from CO_2 or might have assimilated available organic compounds directly for biosynthetic needs. Either way it is likely that the oxidation of H_2 was the energetic driving force for maintaining life processes.

If one compares microbial energy conservation mechanisms as a function of temperature from data of cultured *Bacteria* and *Archaea*, only chemolithotrophic organisms are known at the hottest temperatures (Figure 17.28). Chemoorganotrophy occurs up to at least 110°C, as this is the upper temperature limit for growth of *Pyrodictium occultum*, an organism that can conserve energy and grow by fermentation and by chemolithotrophic growth on H_2 with S^0 as electron acceptor (Table 17.4). Photosynthesis is apparently the least heat-tolerant of all bioenergetic processes, with no hyperthermophilic representatives known and an upper temperature limit of 73°C. This is consistent with the conclusion that anoxygenic photosynthesis first appeared on Earth some

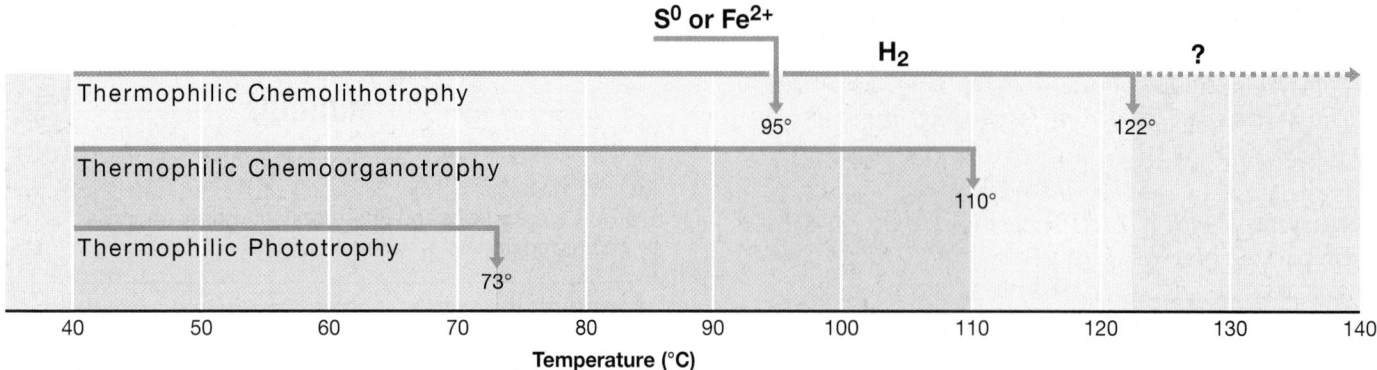

Figure 17.28 Upper temperature limits for energy metabolism. The record holder for phototrophy is *Synechococcus lividus* (*Bacteria*, cyanobacteria); for chemoorganotrophy, *Pyrodictium occultum* (*Archaea*); for chemolithotrophy with S^0 as electron donor, *Acidianus infernus* (*Archaea*); for chemolithotrophy with Fe^{2+} as electron donor, *Ferroglobus placidus* (*Archaea*); and for chemolithotrophy with H_2 as electron donor, *Methanopyrus kandleri* (*Archaea*, 122°C).

hundreds of millions of years after the first life forms are thought to have appeared (↩ Figure 13.1).

Comparisons of bioenergetic options as a function of temperature (Figure 17.28) point to the H_2-oxidizing hyperthermophilic *Archaea* and *Bacteria* as the most likely extant examples of Earth's earliest cellular life forms. More so than any other microbes, these organisms retain the metabolic and physiological traits one would predict to be necessary for existence on a hot early Earth.

— **MINIQUIZ** —

- What phylogenetic and physiological evidence suggests that today's hyperthermophiles are the closest living links to Earth's earliest cells?
- Which mechanism of energy conservation is least heat tolerant?
- Which chemolithotrophic lifestyle seems best suited to life at the highest temperatures?

MasteringMicrobiology®

Visualize, **explore**, and **think critically** with Interactive Microbiology, MicroLab Tutors, MicroCareers case studies, and more. MasteringMicrobiology offers practice quizzes, helpful animations, and other study tools for lecture and lab to help you master microbiology.

Chapter Review

I • *Euryarchaeota*

17.1 Extremely halophilic *Archaea* require large amounts of NaCl for growth and accumulate large levels of KCl in their cytoplasm as a compatible solute. These salts affect cell wall stability and enzyme activity. The light-mediated proton pump bacteriorhodopsin helps extreme halophiles make ATP.

Q **Contrast the roles of bacteriorhodopsin, halorhodopsin, and sensory rhodopsin in *Halobacterium salinarum*.**

17.2 Methanogenic *Archaea* are strict anaerobes whose metabolism is linked to the production of CH_4. Methane can be produced by CO_2 reduction by H_2, from methyl substrates such as CH_3OH, or from acetate.

Q **What is the most likely evolutionary explanation for the observation that methanogens are found widely within the *Euryarchaeota* while related lineages, including the haloarchaea, lack the ability to make methane?**

17.3 *Thermoplasma, Ferroplasma,* and *Picrophilus* are extremely acidophilic thermophiles that form their own phylogenetic family of *Archaea*. Cells of *Thermoplasma* and *Ferroplasma* lack cell walls, resembling the mycoplasmas in this regard.

Q **What two major physiological features unify species of *Thermoplasmatales*? Why does this allow some of them to successfully colonize coal refuse piles?**

17.4 *Archaeoglobus* and *Ferroglobus* are related anaerobic *Archaea* that carry out different anaerobic respirations. *Archaeoglobus* is a sulfate reducer and *Ferroglobus* is a nitrate reducer that oxidizes ferrous iron.

Q **What is physiologically unique about *Archaeoglobus*?**

II • *Thaumarchaeota, Nanoarchaeota, and Korarchaeota*

17.5 *Thaumarchaeota* are widespread and abundant in soils and marine environments. All cultivated species of thaumarchaea are autotrophic ammonia-oxidizers and these organisms are important in the global nitrogen cycle.

> **Q** What is physiologically unusual about the thaumarchaeotal species *Nitrosopumilus maritimus*?

17.6 *Nanoarchaeum equitans* is a hyperthermophile that forms its own phylum, the *Nanoarchaeota*, and is a parasite of the crenarchaeote *Ignicoccus*. *N. equitans* has a tiny, highly compact genome and depends on *Ignicoccus* for most of its cellular needs, including both carbon and energy.

> **Q** How is *Nanoarchaeum* similar to other *Archaea*? How does it differ?

17.7 *Korarchaeum cryptofilum* forms its own phylum, the *Korarchaeota*, and is a hyperthermophile that lacks important biosynthetic pathways, obtaining key building blocks from its environment. *K. cryptofilum* has some genes that are similar to *Euryarchaeota* and other genes that are similar to *Crenarchaeota*.

> **Q** Why is it difficult to determine the phylogenetic placement of *Nanoarchaeota* and *Korarchaeota*?

III • *Crenarchaeota*

17.8 A wide variety of chemoorganotrophic and chemolithotrophic energy metabolisms have been found in hyperthermophilic *Crenarchaeota*, including fermentation and anaerobic respirations. Strictly autotrophic lifestyles are common but photosynthesis is absent.

> **Q** What forms of energy metabolism are present in *Crenarchaeota*? What form is not present?

17.9 Hyperthermophilic *Crenarchaeota* thrive in terrestrial hot springs of various chemistries. These include in particular organisms such as *Sulfolobus*, *Acidianus*, *Thermoproteus*, and *Pyrobaculum*.

> **Q** What is unusual about the metabolism of S^0 by *Acidianus*?

17.10 In deep-sea hydrothermal systems, *Crenarchaeota* such as *Pyrolobus*, *Pyrodictium*, *Ignicoccus*, and *Staphylothermus* thrive. With the exception of the methanogen *Methanopyrus* (*Euryarchaeota*), species of these genera grow at the highest temperatures of all *Archaea*, in many cases well above the boiling point of water.

> **Q** What is unusual about the organism *Pyrolobus fumarii*?

IV • Evolution and Life at High Temperature

17.11 Life as we know it is probably limited to temperatures below 150°C. Key small molecules, such as ATP, are quickly destroyed above this temperature.

> **Q** What organism is the current record holder for the upper temperature limit for growth?

17.12 Macromolecules in hyperthermophiles are protected from heat denaturation by their heat-stable folding patterns (proteins), solutes and binding proteins (DNA), unique monolayer membrane architecture (lipids), and the high GC content of their ribosomal RNAs.

> **Q** What is reverse DNA gyrase and why is it important to hyperthermophiles?

17.13 Hydrogen metabolism is likely to have been the driving force behind the energetics of the earliest cells on Earth. Chemolithotrophic metabolisms based on H_2 as an electron donor are found in the most heat-tolerant of all known *Bacteria* and *Archaea*.

> **Q** Why might H_2 metabolism have evolved as a mechanism for energy conservation in the earliest organisms on Earth?

Application Questions

1. Using the phylogenetic tree in Figure 17.1 as a guide, discuss what indicates that bacteriorhodopsin may have been a late evolutionary invention and that anaerobic respiration with S^0 as electron acceptor might have been an early evolutionary invention.

2. Defend or refute the following statement: The upper temperature limit to life is unrelated to the stability of proteins or nucleic acids.

Chapter Glossary

Bacteriorhodopsin a protein containing retinal that is found in the membranes of certain extremely halophilic *Archaea* and that is involved in light-mediated ATP synthesis

Compatible solute an organic or inorganic substance that is accumulated in the cytoplasm of a halophilic organism and maintains osmotic pressure

Crenarchaeota a phylum of *Archaea* that contains hyperthermophilic organisms

Euryarchaeota a phylum of *Archaea* that contains primarily methanogens, extreme halophiles, *Thermoplasma*, and some marine hyperthermophiles

Extreme halophile an organism whose growth is dependent on large concentrations (generally 9% or more) of NaCl

Extremophile an organism whose growth is dependent on extremes of temperature, salinity, pH, pressure, or radiation, which are generally inhospitable to most forms of life

Halorhodopsin a light-driven chloride pump that accumulates Cl⁻ within the cytoplasm

Hydrothermal vent a deep-sea hot spring emitting warm (~20°C) to superheated (>300°C) water

Hyperthermophile an organism with a growth temperature optimum of 80°C or greater

Korarchaeota a phylum of *Archaea* that contains the hyperthermophile *Korarchaeum cryptophilum*

Methanogen a CH_4-producing organism

Nanoarchaeota a phylum of *Archaea* that contains the hyperthermophilic parasite *Nanoarchaeum equitans*

Phytanyl a branched-chain hydrocarbon containing 20 carbon atoms and commonly found in the lipids of *Archaea*

Reverse DNA gyrase a protein universally present in hyperthermophiles that introduces positive supercoils into circular DNA

Solfatara a hot, sulfur-rich, generally acidic environment commonly inhabited by hyperthermophilic *Archaea*

Thaumarchaeota a phylum of *Archaea* that contains widespread species capable of aerobic ammonia oxidization

Thermosome a heat shock (chaperone) protein complex that functions to refold partially heat-denatured proteins in hyperthermophiles

Diversity of Microbial *Eukarya*

18

microbiology**now**

Arbuscular Mycorrhizal Fungi: Intimate, Unseen, and Powerful

Fungi are common eukaryotic microbes in soil and their role in decomposition is of critical importance to the health of terrestrial ecosystems. Perhaps just as important, though less well recognized, is the role that fungi play in promoting plant productivity.

Mycorrhizae are fungi that form symbiotic associations with plant roots. These fungi acquire carbon and energy from their plant host. In exchange, the mycorrhizal fungi use their extensive mycelial networks to acquire mineral nutrients from the soil and then pass these nutrients along to their plant partners. Arbuscular mycorrhizal fungi (AMF) are one of the most important groups of mycorrhizae. AMF are found within the division *Glomeromycota*, an ancient fungal lineage that diverged early in the history of the fungal kingdom. AMF symbioses are both ancient and intimate. Indeed, fossil evidence suggests that the earliest terrestrial plants formed symbioses with AMF. More than two-thirds of plant species today can form AMF symbioses, including most flowering plants and many important crop species. And they do this for good reason: AMF can increase plant photosynthesis by 20%, a substantial increase in plant fitness.

AMF are obligate biotrophs, symbionts that require a live host. AMF exist as spores in the soil and then germinate and infect their hosts. Upon encountering a plant root, AMF form hyphae that penetrate the root epithelium. The fungi then colonize root cortical cells where they develop elaborately branched structures called arbuscules; the arbuscules are enveloped by a plant membrane that controls nutrient exchange between the plant and the fungus. Arbuscules of the fungus *Rhizophagus irregularis* are shown here within *Medicago truncatula* (barrel clover) cells where they express a green fluorescent marker in the plant cell membrane (the image shows a section of root colonized by AMF with an inset that shows a single cortical cell, scale bar 10 μm).

The 153-megabase genome of *R. irregularis* has been sequenced, revealing at least 23,561 genes. And, as one would expect considering the ecology of this fungus, *R. irregularis* genes encoding nutrient uptake systems are particularly highly expressed. Analyses of AMF, enabled by genomic techniques, may eventually yield discoveries that have tangible effects on agricultural productivity and shed light on the evolutionary origins of land plants.

Source: Tisserant, E., et al. 2013. Genome of an arbuscular mycorrhizal fungus provides insight into the oldest plant symbiosis. *PNAS 110:* 20117–20122.

I • Organelles and Phylogeny of Microbial *Eukarya*

In this chapter we consider the phylogeny and diversity of microbial eukaryotes. A tremendous diversity of microorganisms can be found within the domain *Eukarya*. Indeed, eukaryotic microorganisms are far more phylogenetically diverse than their macroscopic relatives. The majority of microbial eukaryotes are protists. **Protists** are *single-celled* eukaryotic microorganisms. While protists are found widely within the *Eukarya*, microbial eukaryotes can also be colonial or multicellular. Many of the microbial eukaryotes that have been discovered have unusual characteristics and lifestyles and we have only begun to describe their extensive diversity.

The exact evolutionary origin of the eukaryotic cell remains a mystery (↩ Section 13.4). It is now clear that the last eukaryotic common ancestor was a single-celled microorganism closely related to *Archaea*. This microorganism had certain features now shared by all eukaryotic cells, including a nucleus, a cytoskeleton, spliceosomes, a genome with spliceosomal introns (that is, those introns processed by spliceosomes), and mitochondria (↩ Sections 2.14–2.16 and 4.6). Once eukaryotic cell structure had evolved, it was remarkably successful and made possible the evolution of diverse microbial lineages in addition to complex multicellular organisms such as plants and animals.

It is clear that endosymbiosis has played a major role in the origin and diversification of *Eukarya*. It is thus fitting to begin our coverage of microbial eukaryotes by reviewing the properties of their organelles—the mitochondrion and the chloroplast—derived from endosymbiosis. The evolutionary history of these energy powerhouses is distinct from that of the eukaryotic cell itself (↩ Sections 2.15 and 13.4). However, once the mitochondrion and chloroplast were established as characteristic features of eukaryotic cells, they exercised a foundational role in the further evolution of *Eukarya*.

18.1 Endosymbioses and the Eukaryotic Cell

Initial speculation on the link between organelles and bacteria goes back over a century and was based on the fact that microscopically, mitochondria and chloroplasts "looked like" bacteria. Through the years this idea slowly gathered experimental support to yield the current view that mitochondria and chloroplasts are ancestors of respiratory or phototrophic *Bacteria*, respectively, that established residence inside another cell type to provide ATP in exchange for a safe and stable existence. This is the **endosymbiotic hypothesis** (↩ Section 13.4) and is a major tenet of modern biology.

Support for the Endosymbiotic Hypothesis

Several lines of evidence support the endosymbiotic hypothesis:

1. **Mitochondria and chloroplasts contain DNA.** Although most mitochondrial and chloroplast proteins are encoded by nuclear DNA, a few are encoded by a small genome residing

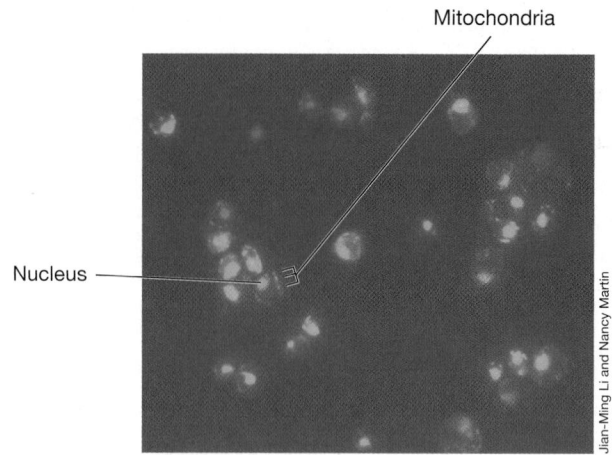

Figure 18.1 Organellar DNA. Cells of the yeast *Saccharomyces cerevisiae* have been stained with the fluorescent dye DAPI that binds to DNA. Each mitochondrion has two to four circular chromosomes that stain blue with the dye.

within the organelle itself (↩ Section 9.4). These include proteins of the respiratory chain (mitochondrion) and photosynthetic apparatus (chloroplast) as well as ribosomal RNAs and transfer RNAs. Most mitochondrial DNA and all chloroplast DNA is of a covalently closed circular form like that of most *Bacteria* (↩ Sections 1.2, 4.2, and 9.3). Organellar DNA can be visualized in eukaryotic cells with special staining methods (**Figure 18.1**).

2. **The eukaryotic nucleus contains genes derived from *Bacteria*.** Genomic sequences of eukaryotic cells (↩ Section 9.4) have clearly shown that several nuclear genes encode functions specific to mitochondria and chloroplasts. Moreover, because these gene sequences more closely resemble those of *Bacteria* than those of *Archaea* or *Eukarya*, it is concluded that these genes were translocated to the nucleus from ancestors of mitochondria and chloroplasts during the transition from engulfed cells to dedicated organelles.

3. **Organellar ribosomes and their phylogeny.** Ribosomes are either 80S in size, typical of the cytoplasm of eukaryotic cells, or 70S, typical of *Bacteria* and *Archaea* (↩ Section 4.10). Mitochondria and chloroplasts contain 70S ribosomes, and phylogenetic analyses of their ribosomal RNA gene sequences (Chapter 13) along with genomic studies of organellar DNA (↩ Section 9.4) show unequivocally that these structures were originally *Bacteria*.

4. **Antibiotic specificity.** Several antibiotics (for example, streptomycin) kill or inhibit *Bacteria* by interrupting the protein synthesis functions of 70S ribosomes. These antibiotics also inhibit protein synthesis in mitochondria and chloroplasts.

5. **Hydrogenosomes.** Hydrogenosomes are membrane-enclosed organelles found in certain anaerobic eukaryotes that lack mitochondria. They supply the cell with ATP from fermentative reactions (↩ Figure 2.49). Like mitochondria, hydrogenosomes also contain their own DNA and ribosomes, and phylogenetic analyses of hydrogenosome ribosomal RNA have revealed their connection to *Bacteria*.

Secondary Endosymbioses

The mitochondrion, chloroplast, and hydrogenosome are structures that originated from *primary* endosymbiosis events. That is, these structures are derived from cells of *Bacteria*. Primary endosymbioses gave rise to the chloroplast in the common ancestor of green algae, red algae, and plants (**Figure 18.2** and see Figure 18.3). However, following this primary event, several unrelated groups of nonphototrophic microbial eukaryotes also acquired chloroplasts but by *secondary* rather than primary endosymbioses. The secondary events occurred when entire green algal or red algal cells were engulfed and their chloroplasts stably retained, thereby making the engulfing cell phototrophic.

Secondary endosymbioses of green algae account for the presence of chloroplasts in euglenids and chlorarachniophytes, while alveolates (ciliates, apicomplexans, and dinoflagellates) and stramenopiles obtained their chloroplasts through secondary endosymbioses with red algae (Figure 18.2 and see Figure 18.3).

The ancestral red algal chloroplasts were apparently lost from some lineages, such as the ciliates, or became greatly reduced in size in others, such as the apicomplexans, where only molecular traces of chloroplasts remain. In some other organisms, such as the dinoflagellates, the red algal chloroplast was apparently replaced altogether with a chloroplast from different algae, including green algae.

These many examples of endosymbiotic events underscore the importance of endosymbiosis in the evolution and diversification of microbial eukaryotes. It is unlikely that primary endosymbiotic events occurred only once in evolutionary history—after all, trial and error is the essence of evolution—and secondary endosymbioses almost certainly occurred quite commonly (Figure 18.2). Even today there are many examples of nonphototrophic protists that engulf phototrophic protists, and the entrapped phototrophs carry out photosynthesis for extended periods (⟲ Section 23.11). Indeed, it appears that endosymbioses are a common and ongoing occurrence in the eukaryotic world.

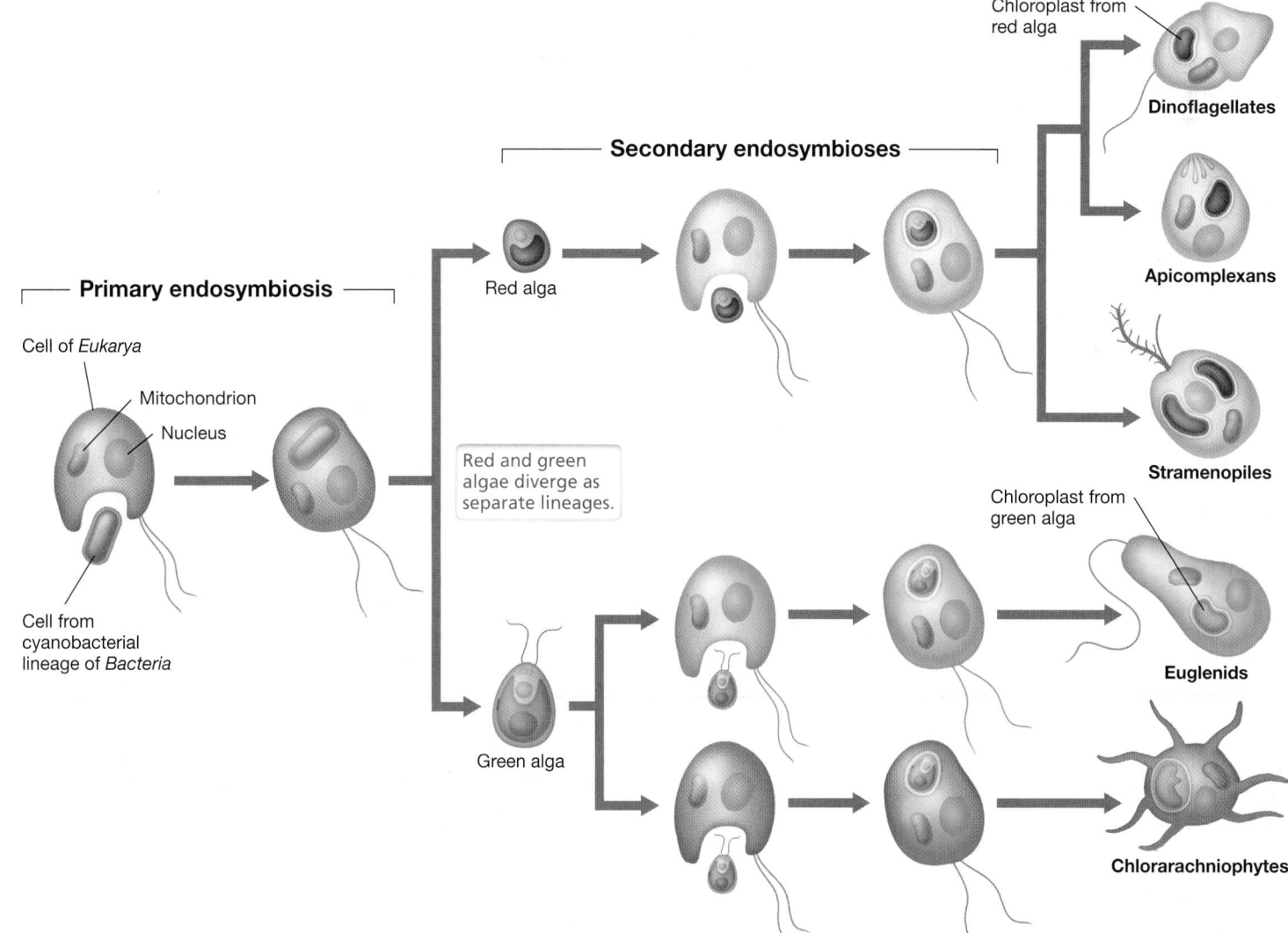

Figure 18.2 Endosymbioses. Following primary endosymbiotic association(s) leading to the mitochondrion, primary endosymbioses with phototrophic *Bacteria* led to the red and green algae. Secondary symbioses of green and red algae spread the property of photosynthesis to many independent lineages of protists.

— **MINIQUIZ** —

- What is the endosymbiotic hypothesis?
- Summarize the molecular evidence that supports the relationship of organelles to *Bacteria*.
- Distinguish between primary and secondary endosymbiosis.

18.2 Phylogenetic Lineages of *Eukarya*

Biologists agree that the eukaryotic cell is a genetic chimera. The main part of the eukaryotic cell including its cytoplasm (and likely its nucleus) can be traced to the *Eukarya* domain, whereas its energy-producing organelles—the mitochondria and chloroplasts—contain their own DNA and are clearly derived from *Bacteria* (Section 18.1). Many genes of bacterial and archaeal origin can be found in the nucleus of eukaryotic cells (◁ Section 9.4), suggesting that extensive horizontal gene transfer took place early in the evolution of eukaryotic cells, and many of these gene transfer events are a direct consequence of primary endosymbiosis.

It appears likely that a major phylogenetic radiation took place early in eukaryote evolution, possibly triggered by the endosymbiotic acquisition of mitochondria. All extant *Eukarya* contain mitochondria, structures homologous to mitochondria (for example, hydrogenosomes), or some genetic trace of these structures. The mitochondrion would have provided the early eukaryotic cell with dramatic new metabolic capabilities. What promoted this primary endosymbiotic event is unknown, but quite possibly it was the accumulation of O_2 in the atmosphere from cyanobacterial photosynthesis (◁ Figure 13.2). The ready availability of O_2 would have selected for bacterial cells that could carry out aerobic respiration and favored those eukaryotes that could stably incorporate them and employ them as energy organelles. Somewhat later in evolutionary time, the ancestor of the chloroplast was acquired in another primary endosymbiotic event, with further eukaryotic phototrophic diversity unfolding later through secondary endosymbioses (Section 18.1).

Eukaryotic Evolution: The Big Picture

Although phylogenies based on ribosomal RNA gene sequences (Chapter 13) confirm the three domains of life—*Bacteria, Archaea,* and *Eukarya*—our picture of eukaryotic evolution has changed dramatically with the incorporation of data from whole genome sequences. There are currently six recognized *supergroups* of *Eukarya* (a supergroup is not an official taxon but essentially equivalent to a kingdom in taxonomic hierarchy): *Archaeplastida, Rhizaria, Chromalveolata (Stramenopiles* and *Alveolata), Excavata, Amoebozoa,* and *Opisthokonta* (**Figure 18.3**).

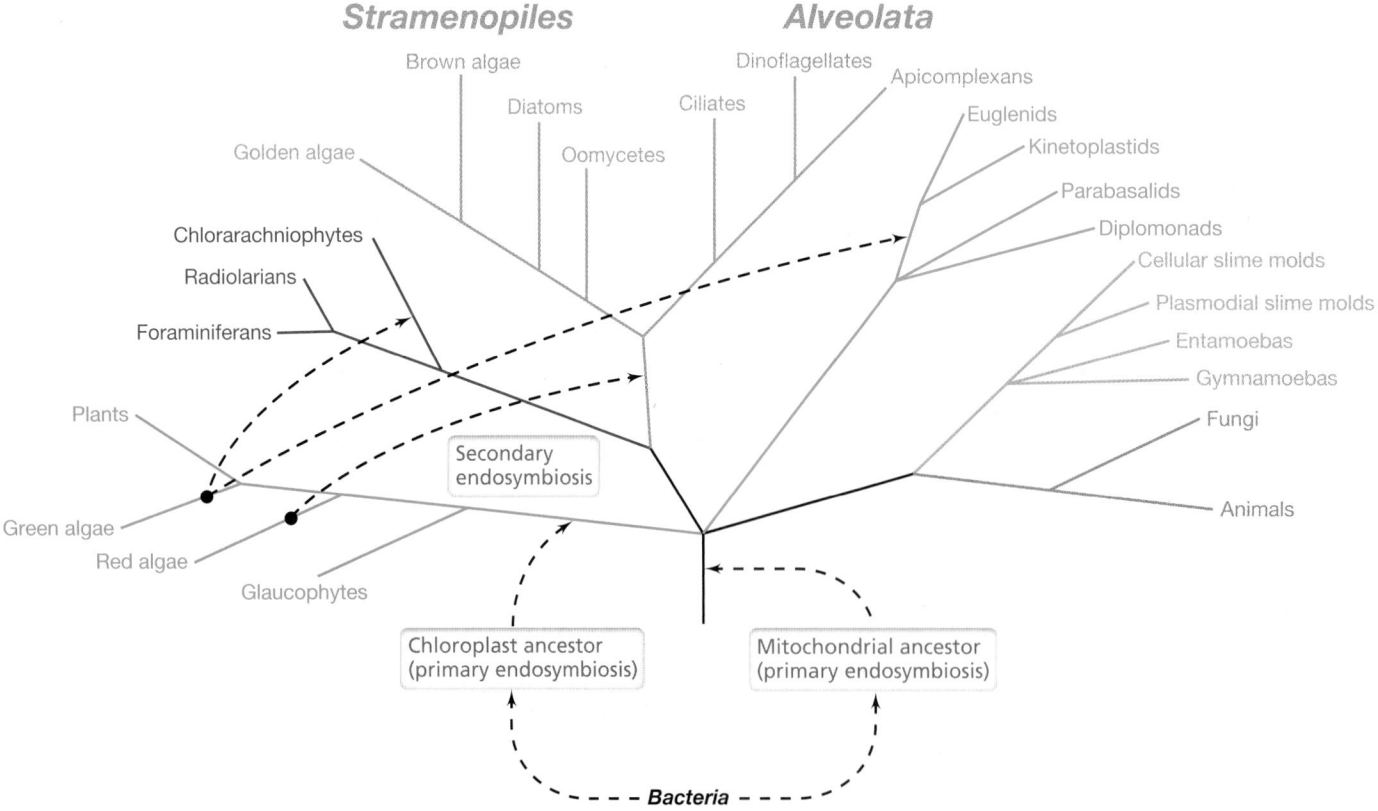

Figure 18.3 Phylogenetic tree of *Eukarya*. The tree diagram provides a schematic view of phylogenetic relationships for six of the best characterized supergroups within the *Eukarya*. Colors are used to identify the supergroups in the tree. Dashed lines indicate both primary endosymbioses of mitochondria and chloroplasts, and the secondary endosymbioses of red and green algae. Protists are common in all eukaryotic lineages except the *Opisthokonta* (*Fungi* and *Animalia*) and the plants.

The *Archaeplastida* include the entire plant kingdom as well as all red and green algae; these photosynthetic lineages resulted from the primary endosymbiosis of the ancestor of all chloroplasts. The *Stramenopiles, Alveolata,* and *Rhizaria* include highly diverse protists including both heterotrophic and phototrophic species. These three groups share an ancestor and they are classified together into a large phylogenetic cluster known as *SAR* (an acronym for *Stramenopiles, Alveolata,* and *Rhizaria*). Where it occurs, phototrophy within these lineages was acquired as a result of secondary endosymbiosis (Figure 18.3 and Section 18.1). The *Excavata* include diverse heterotrophic protists, many of which are anaerobic, and some of which have acquired phototrophy as a result of secondary endosymbiosis (Figure 18.3 and Section 18.1). The *Amoebozoa* include many forms of amoebae and slime molds, though cells with an amoeboid morphology occur in many other lineages of *Eukarya* as well. Finally, the *Opisthokonta* include the well-known kingdoms *Fungi* and *Animalia*.

The phylogenetic tree shown in Figure 18.3 should not be considered the final word on eukaryotic evolution. New genome sequences continue to be determined, new organisms continue to be discovered, and new aspects of eukaryotic biology continue to be revealed. With each new discovery we shed more light on eukaryotic phylogeny and it is likely that our understanding of eukaryotic phylogeny will continue to improve as our knowledge grows.

Phylogenetic Insights on Endosymbiosis

As we have seen, endosymbiosis has clearly been an important aspect of eukaryotic evolution, and the acquisition of the mitochondrion by primitive *Eukarya* was central to the evolutionary success of this domain. However, there are some parasitic microbial eukaryotes such as *Giardia* and *Microsporidia* that lack mitochondria. The *Microsporidia* were once thought to be ancient members of the *Eukarya*, descended from an ancestor of *Eukarya* that lacked mitochondria. We now know, however, that these amitochondriate eukaryotes are descended from eukaryotic ancestors that once had mitochondria (see the position of *Microsporidia* in Figure 18.23) but then lost them for some reason, perhaps while transitioning to an anaerobic lifestyle such as in *Giardia, Entamoeba,* and several other parasitic eukaryotes. However, amitochondriate eukaryotes typically retain a few genes of mitochondrial origin, and these molecular leftovers are strong evidence that the organisms once had mitochondria.

The tree of *Eukarya* (Figure 18.3) also shows how secondary endosymbioses account for the origin of chloroplasts in some unicellular phototrophic eukaryotes. Following primary endosymbiosis of the cyanobacterial ancestor of chloroplasts by early mitochondrion-containing eukaryotes, these now phototrophic eukaryotes diverged into red and green algae. Then, in secondary endosymbioses, ancestors of the euglenids, kinetoplastids, and chlorarachniophytes engulfed green algae while ancestors of the alveolates and stramenopiles engulfed red algae (Section 18.1). These secondary endosymbioses account for the great phylogenetic diversity of phototrophic eukaryotes, which we explore later in this chapter.

MINIQUIZ

- What does the endosymbiotic hypothesis propose?
- How does the composite tree of eukaryotes differ from the ribosomal RNA–based tree?
- How does secondary endosymbiosis help explain the diversity of phototrophic eukaryotes?

II • Protists

With the big picture of eukaryotic cell phylogeny in mind, we proceed to examine the major groups of eukaryotic microorganisms. We begin with protists other than the green and red algae. Protists include both phototrophic and nonphototrophic microbial eukaryotes. These organisms are widely distributed in nature, exhibit a wide range of morphologies, and show great phylogenetic diversity. Indeed, protists are common in all eukaryotic lineages except plants, fungi, and animals (Figure 18.3); thus they represent much of the diversity found in the domain *Eukarya*.

18.3 *Excavata*

KEY GENERA: *Giardia, Trichomonas, Trypanosoma, Euglena*

Excavata encompasses diverse protists including both chemoorganotrophs and phototrophs, with some species being anaerobic. We begin with the diplomonads and parabasalids, flagellated protists that lack mitochondria and chloroplasts. These microbes live in anoxic habitats, such as animal intestines, either symbiotically or as parasites, and conserve energy from fermentation. Some diplomonads cause serious and common diseases in fish, domestic animals, and humans, and one parabasalid causes a major sexually transmitted disease of humans. Both groups share a relatively recent common ancestor before they diverged to form separate phylogenetic lineages (Figure 18.3).

Diplomonads

Diplomonads (**Figure 18.4a**) characteristically contain two nuclei of equal size, and also contain mitosomes, much reduced mitochondria lacking electron transport proteins and enzymes of the citric acid cycle. The diplomonad *Giardia* has a relatively small genome for a eukaryote, about 12 megabase pairs (Mbp). The genome is also quite compact, contains few introns, and lacks genes for many metabolic pathways, including the citric acid cycle (⇨ Figure 3.16). These characteristics likely account for the organism's parasitic and anaerobic lifestyle. *Giardia intestinalis* (Figure 18.4a), also known as *Giardia lamblia*, causes giardiasis, one of the most common waterborne diarrheal diseases in the United States. We examine the disease giardiasis in Section 33.4.

Parabasalids

Parabasalids contain a *parabasal body* that, among other functions, gives structural support to the cell's Golgi complex. These anaerobic microbial eukaryotes lack mitochondria but contain hydrogenosomes (⇨ Section 2.15). Parabasalids live in the intestinal and urogenital tract of vertebrates and invertebrates as parasites or as commensal symbionts (⇨ Section 33.4).

UNIT 4

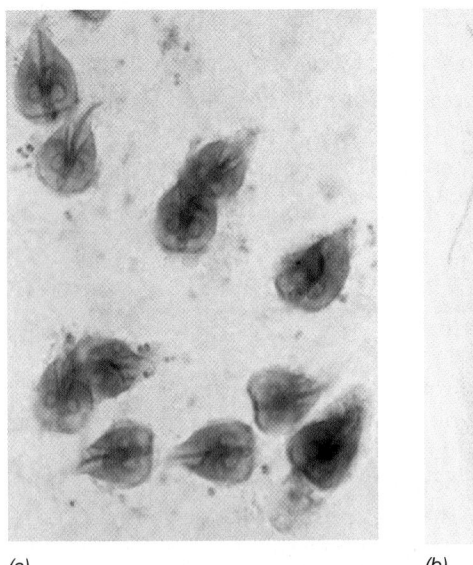
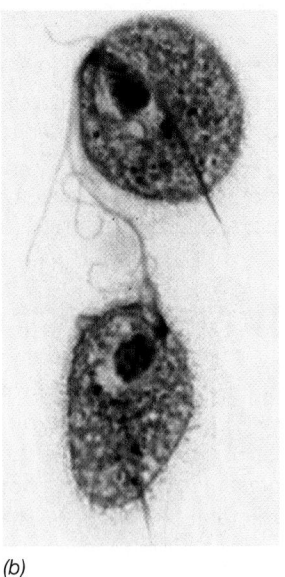

(a)   *(b)*

Figure 18.4 Diplomonads and parabasalids. *(a)* Light photomicrograph of cells of *Giardia intestinalis*, a typical diplomonad. Note the dual nuclei. Cells are about 10 μm wide. *(b)* Light photomicrograph of cells of the parabasalid *Trichomonas vaginalis*. Cells are about 6 μm wide. The spearlike structure (axostyle) is used to attach the cell to urogenital tissues.

The parabasalid *Trichomonas vaginalis* is motile by a tuft of flagella (Figure 18.4*b*) and causes a widespread sexually transmitted disease in humans.

The genomes of parabasalids are unique among eukaryotes in that most of them lack introns, the noncoding sequences characteristic of eukaryotic genes (⊖ Sections 4.6 and 9.4). In addition, the genome of *T. vaginalis* is surprisingly huge for a parasitic organism, about 160 Mbp, and shows evidence of genes acquired from bacteria by horizontal gene transfer. Much of the genome of *T. vaginalis* contains repetitive DNA sequences and transposable elements (⊖ Section 11.11), which has made genomic analyses difficult. But *Trichomonas* is still thought to contain nearly 60,000 genes, about twice that of the human genome and near the upper limit observed thus far for eukaryotic genomes.

Kinetoplastids

Kinetoplastids are a well-studied group of *Excavata* and are named for the presence of the *kinetoplast*, a mass of DNA present in their single, large mitochondrion. Kinetoplastids live primarily in aquatic habitats, where they feed on bacteria. Some species, however, are parasites of animals and cause serious diseases in humans and vertebrate animals. Cells of *Trypanosoma*, a genus infecting humans, are small, about 20 μm long, thin, and crescent-shaped. Trypanosomes have a single flagellum that originates in a basal body and folds back laterally across the cell where it is enclosed by a flap of cytoplasmic membrane (**Figure 18.5**). Both the flagellum and the membrane participate in propelling the organism, making effective movement possible even in viscous liquids, such as blood, where pathogenic trypanosomes are often found.

Trypanosoma brucei (Figure 18.5) causes *African sleeping sickness*, a chronic and usually fatal human disease. The parasite lives and grows primarily in the bloodstream, but in the later stages of the

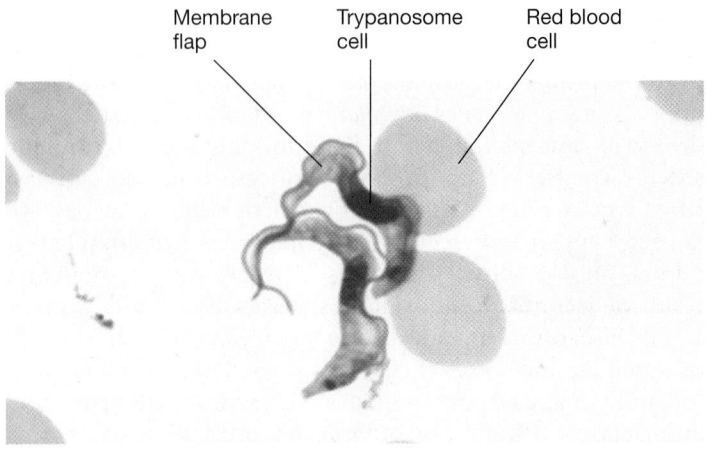

Membrane flap   Trypanosome cell   Red blood cell

Figure 18.5 Trypanosomes. Photomicrograph of the flagellated kinetoplastids *Trypanosoma brucei*, the causative agent of African sleeping sickness. Blood smear preparation. The cells of *T. brucei* are about 3 μm wide.

disease it invades the central nervous system, causing an inflammation of the brain and spinal cord that is responsible for the characteristic neurological symptoms of the disease. The parasite is transmitted from host to host by the tsetse fly, *Glossina* spp., a bloodsucking fly found only in certain parts of Africa. After moving from the human to the fly in a blood meal, the parasite proliferates in the intestinal tract of the fly and invades the insect's salivary glands and mouthparts, from which it is transferred to a new human host by a fly bite (⊖ Section 33.6).

Other kinetoplastids that are human parasites include *Trypanosoma cruzi*, the causative agent of *Chagas disease*, and *Leishmania* species, the causative agents of cutaneous and systemic *leishmaniasis*. Chagas disease is spread by the bite of a blood-feeding insect called the "kissing bug." The disease is usually self-limiting, but it can become chronic and lead to a fatal infection. Leishmaniasis is a disease of tropical and subtropical regions transmitted to humans and other mammals by a bite from the sand fly. This potentially fatal disease can be localized to the skin surrounding a fly bite or can infect the spleen and liver and cause systemic infection. Both Chagas disease and leishmaniasis are covered in more detail in Section 33.6.

Euglenids

Another well-studied group of *Excavata* are the euglenids (**Figure 18.6**). Unlike the kinetoplastids, these motile microbial eukaryotes are nonpathogenic and are both chemotrophic *and* phototrophic. Most euglenids contain two flagella, dorsal and ventral, and their active motility allows the organisms to access both illuminated and dark habitats in their environment to support their alternate nutritional lifestyles.

Euglenids live exclusively in aquatic habitats, both freshwater and marine, and contain chloroplasts, which support phototrophic growth (Figure 18.6). In darkness, however, cells of *Euglena*, a typical euglenid, can lose their chloroplasts and exist as chemoorganotrophs. Many euglenids can also feed on bacterial cells via **phagocytosis**, a process of surrounding a particle with a portion of their flexible cytoplasmic membrane to engulf the particle and bring it into the cell where it is digested.

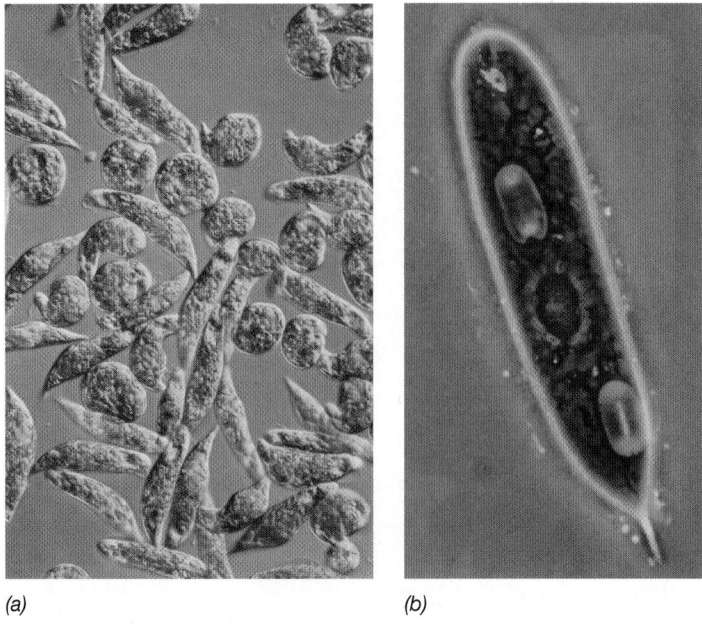

(a)   (b)

Figure 18.6 *Euglena*, a euglenid. *(a)* This phototrophic protist, like other euglenids, is not pathogenic. A cell is about 15 μm wide. *(b)* High-magnification view.

MINIQUIZ

- Contrast the two nutritional options for *Euglena*.
- What diseases are caused by *Trypanosoma cruzi, Leishmania*, and *Giardia*?
- How do diplomonads obtain energy?

18.4 *Alveolata*

KEY GENERA: *Gonyaulax, Plasmodium, Paramecium*

The alveolates as a group are characterized by their *alveoli*, cytoplasmic sacs located just under the cytoplasmic membrane. Although the function of alveoli is unknown, they may help the cell maintain osmotic balance by controlling water influx and efflux, and in the dinoflagellates, they may function as armor plates (see Figure 18.9). Three phylogenetically distinct, although related, kinds of alveolates are known: the *ciliates*, which use cilia for motility; the *dinoflagellates*, which are motile by means of a flagellum; and the *apicomplexans*, which are parasites of humans and other animals (Figure 18.3).

Ciliates

Ciliates possess *cilia* (**Figure 18.7**) at some stage of their life cycle. Cilia are structures that function in motility and may cover the cell or form tufts or rows, depending on the species. Probably the best-known and most widely distributed ciliates are those of the genus *Paramecium* (Figure 18.7). Like many other ciliates, *Paramecium* uses cilia not only for motility but also to obtain food by ingesting particulate materials such as bacterial cells through a distinctive funnel-shaped oral groove. Cilia that line the oral groove move material down the groove to the cell mouth, also called the *gullet* (Figure 18.7*b*). Once in the gullet, the material is enclosed in a

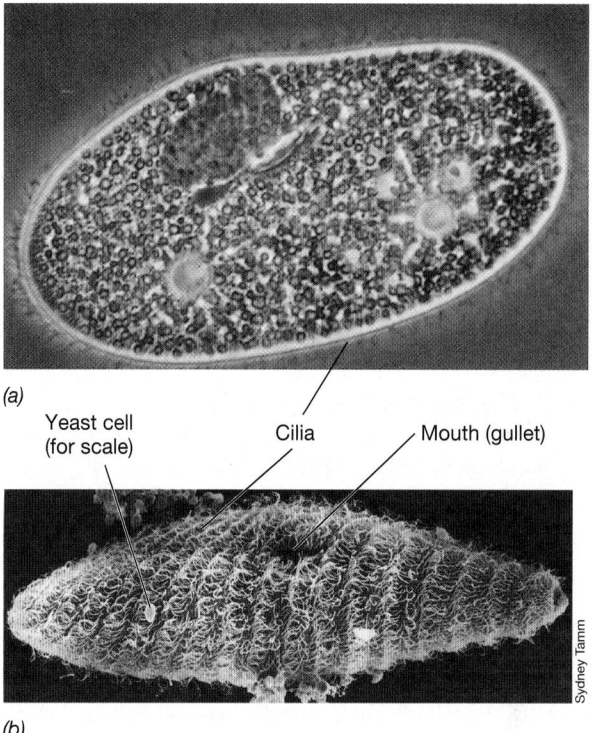

(a)

Yeast cell (for scale)   Cilia   Mouth (gullet)

(b)

Figure 18.7 *Paramecium*, a ciliated protist. *(a)* Phase-contrast photomicrograph. *(b)* Scanning electron micrograph. Note the cilia in both micrographs. A single *Paramecium* cell is about 60 μm in diameter.

vacuole by phagocytosis. Digestive enzymes secreted into the vacuole then break down the material as a source of nutrients.

Ciliates are unique among protists in having two kinds of nuclei, *micronuclei* and *macronuclei*. Genes in the macronucleus regulate basic cellular functions, such as growth and feeding, whereas those of the micronucleus are involved in sexual reproduction, which occurs through a partial fusion of two *Paramecium* cells and exchange of micronuclei. The genome of *Paramecium* is huge, with macronuclear genes numbering about 40,000, nearly twice that of humans.

Many *Paramecium* species (as well as many other protists) are hosts for endosymbiotic *Bacteria, Archaea*, or eukaryotes, the latter usually green algae. These organisms may play a nutritional role, synthesizing vitamins or other growth factors used by the host cell. Several anaerobic ciliated protists contain endosymbionts. For example, ciliated protists in the termite hindgut contain endosymbiotic methanogens (*Archaea*) that consume H_2 plus CO_2 to form methane (CH_4). Ciliates themselves can also be symbiotic: Obligately anaerobic ciliates are present in the rumen, the forestomach of ruminant animals, and play an important role in the digestive and fermentative processes of the animal (⟳ Section 23.13).

In contrast to symbioses, some ciliates are animal parasites, although this lifestyle is less common in ciliates than in some other groups of protists. The species *Balantidium coli* (**Figure 18.8**), for example, is primarily an intestinal parasite of domestic animals but occasionally infects the intestinal tract of humans, producing dysentery-like symptoms. Cells of *B. coli* form cysts (Figure 18.8) that promote disease transmission in infected food or water.

Sydney Tamm

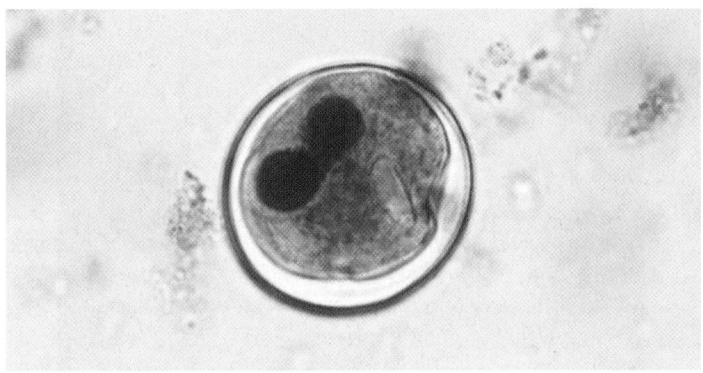

Figure 18.8 *Balantidium coli*, **a ciliated protist that causes a dysentery-like disease in humans.** The dark blue–stained lobed structure in this *B. coli* cyst obtained from swine intestine is a dividing macronucleus. The cell is about 50 μm wide.

Dinoflagellates

Dinoflagellates are a diverse group of marine and freshwater phototrophic alveolates (**Figure 18.9**) that acquired the capacity to photosynthesize through secondary endosymbioses (Figures 18.2 and 18.3). Flagella encircling the cell impart spinning movements that give dinoflagellates their name (*dinos* is Greek for "whirling"). Dinoflagellates have two flagella of different lengths and with different points of insertion into the cell, transverse and longitudinal. The transverse flagellum is attached laterally, whereas the longitudinal flagellum originates from the lateral groove of the cell and extends lengthwise (see Figure 18.10*b*). Some dinoflagellates are free-living, whereas others live a symbiotic existence with animals that form coral reefs, obtaining a sheltered and protected habitat in exchange for supplying phototrophically fixed carbon as a food source for the reef. A number of free-living species are capable of bioluminescence and emit light when disturbed at night; this bioluminescence results in a "sparkling" effect that can often be observed in coastal seas and bioluminescent bays.

Several species of dinoflagellates are toxic. For example, dense suspensions of *Gonyaulax* cells, called "red tides" (**Figure 18.10***a*) because of the red-colored pigments of this organism, can form in warm and typically polluted coastal waters. Such blooms are often

Figure 18.9 **The marine dinoflagellate** *Ornithocercus magnificus* **(an alveolate).** The cell proper is the globular central structure; the attached ornate structures are called *lists*. A cell is about 30 μm wide.

(a)

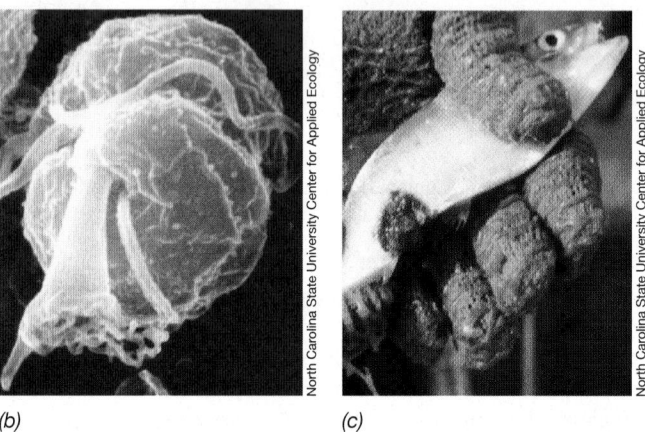

(b)   *(c)*

Figure 18.10 **Toxic dinoflagellates (alveolates).** *(a)* Photograph of a "red tide" caused by massive growth of toxin-producing dinoflagellates such as *Gonyaulax*. The toxin is excreted into the water and also accumulates in shellfish that feed on the dinoflagellates. *(b)* Scanning electron micrograph of a toxic spore of *Pfiesteria piscicida*; the structure is about 12 μm wide. *(c)* A fish killed by *P. piscicida*; note the lesions of decaying flesh.

associated with fish kills and poisoning in humans following consumption of mussels that have accumulated *Gonyaulax* through filter feeding. Toxicity results from a neurotoxin that can cause a condition called *paralytic shellfish poisoning* in humans and some marine animals, such as sea otters. Symptoms include numbness of the lips, dizziness, and difficulty breathing; in severe cases, death can result from respiratory failure. *Pfiesteria* is another toxic dinoflagellate. Toxic spores of *Pfiesteria piscicida* (Figure 18.10*b*) infect fish and eventually kill them by way of neurotoxins that affect movement and destroy skin. Lesions form on areas of the fish, allowing opportunistic bacterial pathogens to grow (Figure 18.10*c*). Symptoms of human toxemia from *Pfiesteria* poisoning include skin rashes and respiratory problems.

Apicomplexans

Apicomplexans are nonphototrophic obligate parasites that cause severe human diseases such as malaria (*Plasmodium* species, **Figure 18.11***a*), toxoplasmosis (*Toxoplasma*, Figure 18.11*b*), and coccidiosis (*Eimeria*). These organisms are characterized by nonmotile adult stages, and nutrients are taken up in soluble form across the cytoplasmic membrane as in bacteria and fungi.

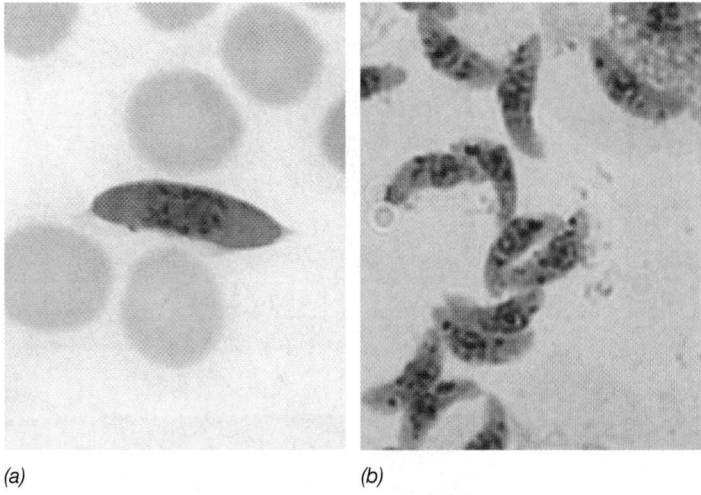

(a)   *(b)*

Figure 18.11 Apicomplexans. *(a)* A gametocyte of *Plasmodium falciparum* in a blood smear. The gametocyte is the stage in the malarial parasite life cycle that infects the mosquito vector. *(b)* Sporozoites of *Toxoplasma gondii*.

Apicomplexans produce structures called *sporozoites* (Figure 18.11*b*), which function in transmission of the parasite to a new host, and the name apicomplexan derives from the presence at one apex of the sporozoite of a complex of organelles that penetrate host cells. Apicomplexans also contain *apicoplasts*. These are degenerate chloroplasts that lack pigments and photosynthetic capacity but contain a few of their own genes. Apicoplasts catalyze fatty acid, isoprenoid, and heme biosyntheses, and export their products to the cytoplasm. It is hypothesized that apicoplasts are derived from red algal cells engulfed by apicomplexans in a secondary endosymbiosis (Figures 18.2 and 18.3). Over time, the chloroplast of the red algal cell degenerated to play a nonphototrophic role in the apicomplexan cell.

Both vertebrates and invertebrates can be hosts for apicomplexans. In some cases, an alternation of hosts takes place, with some stages of the life cycle linked to one host and some to another. Important apicomplexans are the coccidia, which are typically parasites of birds, and species of *Plasmodium* (malarial parasites) (Figure 18.11*a*). We reserve detailed discussion of malaria—a disease that throughout the course of history has killed more humans than any other disease—for Section 33.5.

MINIQUIZ

- How does the organism *Paramecium* move?
- What health problem is associated with the organism *Gonyaulax*?
- What are apicoplasts, which organisms have them, and which functions do they carry out?

18.5 *Stramenopiles*

KEY GENERA: *Phytophthora, Nitzschia, Ochromonas, Macrocystis*

The *Stramenopiles* include both chemoorganotrophic and phototrophic microorganisms as well as macroorganisms. Members of this group bear flagella with many short, hairlike extensions (Figure 18.2), and this morphological feature gives the group its name (from Latin *stramen* for "straw" and *pilus* for "hair"). The diatoms, oomycetes, golden algae, and brown algae are the major groups of *Stramenopiles* (Figure 18.3).

Diatoms

Diatoms include over 200 genera of unicellular, phototrophic, microbial eukaryotes, and are major components of the planktonic (suspended) phytoplankton microbial community in marine and freshwaters. Diatoms characteristically produce a cell wall made of silica to which protein and polysaccharide are added. The wall, which protects the cell against predation, exhibits widely different shapes in different species and can be highly ornate (**Figure 18.12**). The external structure formed by this wall, called a *frustule*, often remains after the cell dies and the organic materials have disappeared. Diatom frustules typically show morphological symmetry, including *pinnate symmetry* (having similar parts arranged on opposite sides of an axis, as in the common diatom *Nitzschia*, Figure 18.12*b*), and *radial symmetry*, as in the marine diatoms *Thalassiosira* and *Asterolampra* (Figure 18.12*c, d*). Because the diatom frustules, which are composed mainly of silica, are resistant to decay, these structures can remain intact for long periods of time and often sink and remain in the sediments for millions of years. Diatom frustules constitute some of the best-known unicellular eukaryotic fossils, and from dating of frustule samples, it has been shown that diatoms first appeared on Earth relatively recently, about 200 million years ago.

Oomycetes

The oomycetes, also called *water molds*, were previously grouped with fungi based on their filamentous growth and the presence of **coenocytic** (that is, multinucleate) hyphae, morphological traits characteristic of fungi (Section 18.8). Phylogenetically, however, the oomycetes are distant from fungi and are closely related to other *Stramenopiles* (Figure 18.3). Oomycetes differ from fungi in other fundamental ways, as well. For example, the cell walls of oomycetes are typically made of cellulose instead of the chitin cell walls of fungi, and the water molds have flagellated cells, which are lacking in all but a few fungi. Nonetheless, oomycetes are ecologically similar to fungi in that they grow as a mass of hyphae decomposing dead plant and animal material in aquatic habitats.

Oomycetes have had a major impact on human society, as many species are plant pathogens (phytopathogens). The oomycete *Phytophthora infestans*, which causes late blight disease of potatoes, contributed to massive famines in Ireland in the mid-nineteenth century. The famines led to the death of a million Irish and triggered great waves of Irish immigration to North America. Other major phytopathogens include *Pythium*, a common pathogen of greenhouse seedlings, and *Albugo*, which causes "white rusts" on several agricultural crops.

Golden Algae and Brown Algae

Along with the diatoms, golden and brown algae form major lineages of *Stramenopiles*. Golden algae, also called *chrysophytes*, are primarily unicellular marine and freshwater phototrophs. Some

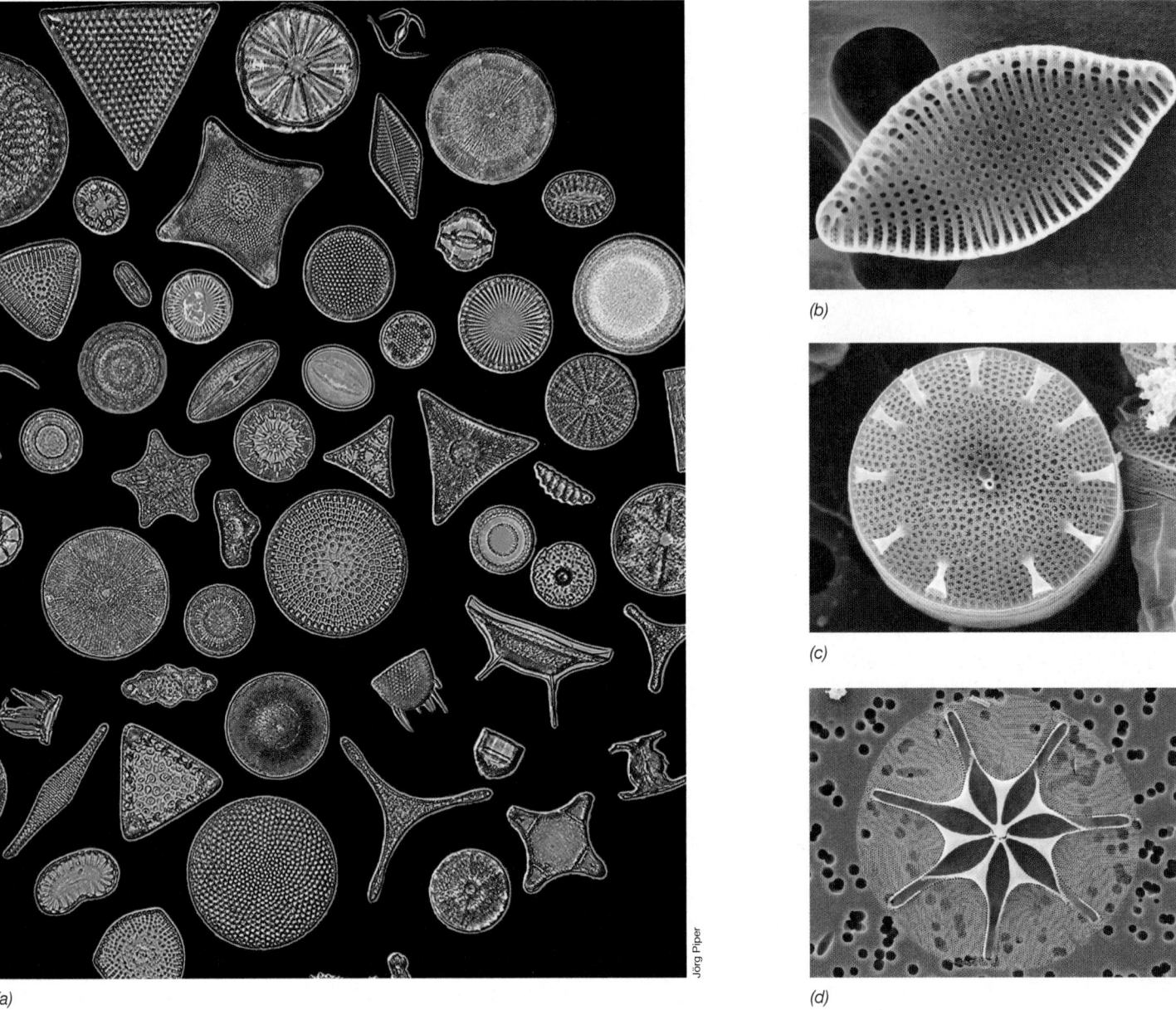

Figure 18.12 Diatom frustules. *(a)* Dark-field photomicrograph of a collage of frustules from different diatom species showing various forms of symmetry. *(b–d)* Scanning electron micrographs of diatom frustules showing pinnate (part *b*) or radial (parts *c, d*) symmetry. Diatoms vary considerably in size from very small species about 5 μm wide to larger species up to 200 μm wide.

species are chemoorganotrophs and feed either by phagocytosis or by transporting soluble organic compounds across the cytoplasmic membrane. Some golden algae, such as *Dinobryon* (**Figure 18.13***a*), found in freshwater, are colonial. However, most golden algae are unicellular and motile by the activity of two flagella of unequal length.

Golden algae are so named because of their golden-brown color (Figure 18.13*a, c*). This is due to chloroplast pigments dominated by the brown-colored carotenoid fucoxanthin. The major chlorophyll pigment in golden algae is chlorophyll *c* rather than chlorophyll *a*, and they lack the phycobiliproteins present in red algal chloroplasts (Section 18.14). Cells of the unicellular golden alga *Ochromonas*,

the best-studied genus of this group, have only one or two chloroplasts (Figure 18.13*c*).

Brown algae are primarily marine and are multicellular and typically macroscopic. No unicellular brown algae are known. The kelps, such as the giant kelp *Macrocystis* (Figure 18.13*b*), which can grow up to 50 m in length, are perhaps the most widespread of brown algae. *Fucus*, another common seaweed of intertidal regions, can grow up to 2 m. As their name implies, brown algae are brown or green-brown in color depending on how much of the carotenoid pigment fucoxanthin they produce. Most marine "seaweeds" are brown algae and their rapid growth, especially in cold marine waters, can cause nuisance odor problems when they wash ashore and decay.

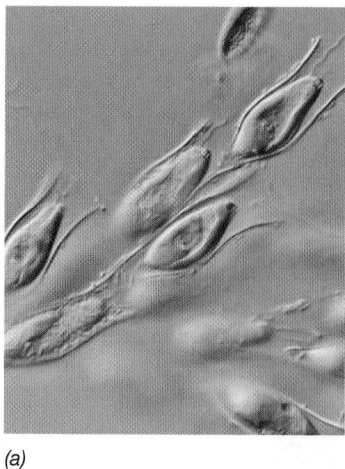

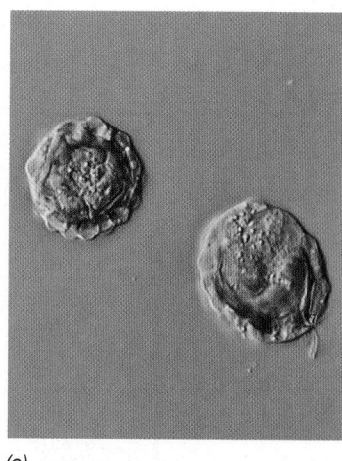

(a) (b) (c)

Figure 18.13 Golden and brown algae. *(a) Dinobryon*, a golden alga (family *Chrysophyceae*) that forms branched colonies. *(b) Macrocystis*, a marine kelp belonging to the brown algae (family *Phaeophyceae*). *(c) Ochromonas*, a unicellular chrysophyte. The golden or brown color of the chloroplasts of these algae is due to the pigment fucoxanthin.

MINIQUIZ

- What structure of diatoms accounts for their excellent fossil record?
- In what ways do oomycetes differ from and resemble fungi?
- Which chlorophyll pigment is found in golden and brown algae?

18.6 *Rhizaria*

Rhizaria form a diverse group of protists that includes the *Chlorarachniophyta*, *Radiolaria*, and *Foraminifera* (Figure 18.3). They are distinguished from other protists by their threadlike cytoplasmic extrusions (pseudopodia) by which they move and feed. Some *Rhizaria* have amoeboid morphology and were once mistakenly classified as *amoebae* because of this and their pseudopodia, but it is now known that many phylogenetically diverse organisms employ pseudopodia for motility and feeding purposes.

Chlorarachniophyta

Chlorarachniophytes are freshwater and marine amoeba-like phototrophs that develop a flagellum for dispersal. The group's acquisition of green algal chloroplasts is a prime example of a secondary endosymbiosis (Figure 18.2) and shows how extensively this process has molded several phylogenetically distinct lineages of microbial eukaryotes (Figure 18.3).

Chloroplasts typically have two membranes, derived from the inner and outer membranes of *Cyanobacteria*, which have a gram-negative cell envelope (Section 2.5). Chloroplasts in chlorarachniophytes, however, have four membranes. In addition, they have **nucleomorphs** tucked in between the two sets of chloroplast membranes. Nucleomorphs are remnant nuclei left over from when the algal endosymbionts were acquired during secondary endosymbiosis (Section 18.1). Both the nucleomorphs and the extra membranes were derived from this algal endosymbiont (Figure 18.2).

It is a testament to endosymbiosis to consider that chlorarachniophytes have merged a minimum of five different genomes: the host genome, the host mitochondrial genome, the algal endosymbiont genome (which has become the nucleomorph), the algal endosymbiont mitochondrial genome, and the algal endosymbiont chloroplast genome! Most nonessential or duplicate genes have been deleted over time and many genes from the mitochondria and chloroplasts have been transferred to the host genome. As a result, the majority of genes in a chlorarachniophyte are present in the nucleus of the host cell. The nucleomorph itself is greatly reduced in size relative to its ancestor and over time may be lost completely from the chloroplast.

Foraminifera

In contrast to chlorarachniophytes, foraminiferans are exclusively marine microbes and form shell-like structures called *tests*, which have distinctive characteristics and are often quite ornate (**Figure 18.14a**). Tests are typically made of organic materials reinforced with calcium carbonate. The test is not firmly attached to the cell, and the amoeba-like cell may extend partway out of the test during feeding. However, because of the weight of the test, the cell usually sinks to the bottom of the water column, and it is thought that the organisms feed on dissolved organic matter and particulate deposits, primarily bacteria, other protists, and the remains of dead organisms near the sediments. Foraminiferan cells can also host a variety of algae that form endosymbiotic relationships with the protist and supply it with organic carbon, probably in exchange for inorganic nutrients derived from the breakdown of dead organisms. Phototrophs are found primarily in planktonic foraminifera that remain suspended in the water column to provide their endosymbionts with sufficient sunlight.

Foraminiferan tests (Figure 18.14a) are relatively resistant to decay and are readily fossilized. These buried and preserved tests are quite useful to geologists. Because particular taxa of foraminifera are typically associated with particular strata in the geological record, fossilized foraminiferan tests in samples obtained from exploratory wells are used by oil industry paleontologists as a means to date and assess the petroleum potential of a given drill site.

UNIT 4

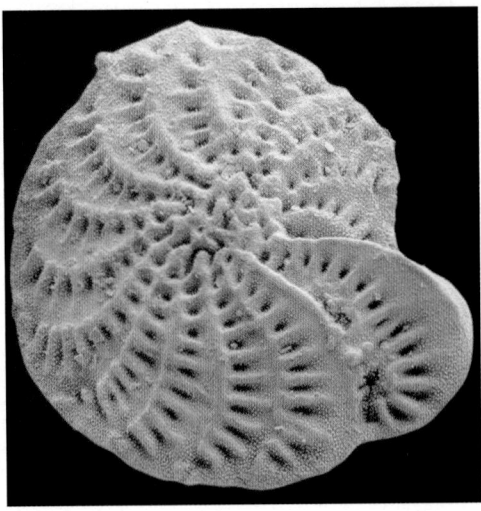

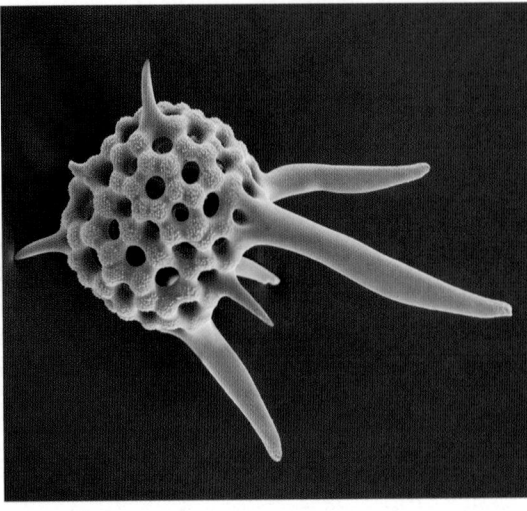

(a) *(b)*

Figure 18.14 Foraminifera and radiolaria. *(a)* A foraminiferan. Note the ornate and multilobed test. The test is about 1 mm wide. *(b)* A spiked radiolarian of the *Nassellaria* group. A test is about 150 μm wide. Both *a* and *b* are colorized scanning electron micrographs.

Radiolaria

Radiolarians are chemoorganotrophic and mostly planktonic marine eukaryotes that reside in the upper 100 m of the ocean where they consume bacteria and particulate organic matter. Some species associate with algae that take on a symbiotic (but not endosymbiotic) role and supply nutrients to the radiolarian.

The name "radiolarian" comes from the radial symmetry of their tests, transparent or translucent mineral skeletons made of silica in one fused piece (Figure 18.14*b*). Along with the accumulation of lipid droplets and large cytoplasmic vacuoles, the needle-like pseudopodia of radiolarians probably help keep the organisms from sinking in their mainly open ocean (planktonic) habitats. However, when cells eventually die, their tests settle to the ocean floor and can build up over time into thick layers of slowly decaying cell material.

MINIQUIZ

- What structure distinguishes rhizaria from all other protists?
- How are chlorarachniophytes thought to have acquired the ability to photosynthesize?

18.7 *Amoebozoa*

KEY GENERA: *Amoeba, Entamoeba, Physarum, Dictyostelium*

The *Amoebozoa* are a large group of terrestrial and aquatic protists that use lobe-shaped pseudopodia for movement and feeding, in contrast to the threadlike pseudopodia of rhizaria. The major groups of *Amoebozoa* are the *gymnamoebas*, the *entamoebas*, and the *plasmodial* and *cellular slime molds*. Phylogenetically, the *Amoebozoa* diverged from a lineage that eventually led to the fungi and animals (Figure 18.3).

Gymnamoebas and Entamoebas

The gymnamoebas are free-living protists that inhabit aquatic and soil environments. They use pseudopodia to move by a process called *amoeboid movement* (**Figure 18.15**) and feed by phagocytosis on bacteria, other protists, and particulate organic materials. Amoeboid movement results from streaming of the cytoplasm as it flows forward at the less contracted and viscous cell tip, taking the path of least resistance. Cytoplasmic streaming is facilitated by microfilaments (∞ Section 2.16), which exist in a thin layer just beneath the cytoplasmic membrane. *Amoeba* (Figure 18.15) is a common organism in pond waters, with species varying in size from 15 μm in diameter (clearly microscopic) to over 750 μm (visible with the naked eye).

In contrast to gymnamoebas, the entamoebas are parasites of vertebrates and invertebrates. Their usual habitat is the oral cavity or intestinal tract of animals. The anaerobe *Entamoeba histolytica* is pathogenic in humans and causes amebic dysentery, an ulceration of the intestinal tract that results in bloody diarrhea. This parasite forms cysts that are transmitted from person to person by fecal contamination of water, food, and eating utensils. In Section 33.3 we discuss the etiology and pathogenesis of amebic dysentery, an important cause of death from intestinal parasites in humans.

Slime Molds

The **slime molds** were previously grouped with fungi since they undergo similar life cycles and produce fruiting bodies with spores for dispersal. As protists, however, slime molds are motile and can move across a solid surface fairly quickly (see Figures 18.16–18.18). Slime molds are divided into two groups: *plasmodial slime molds*

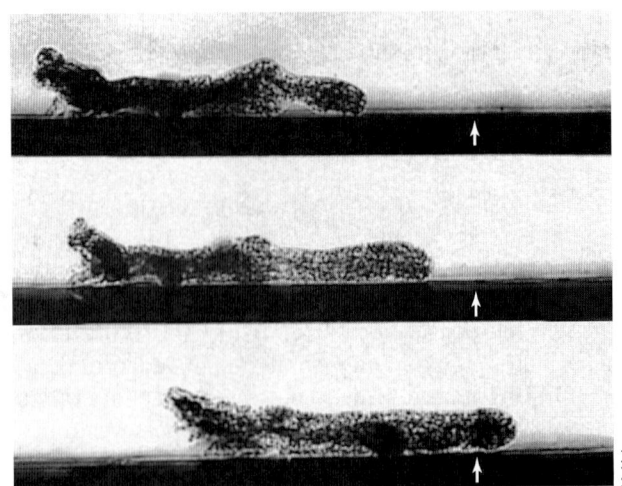

M. Haberey

Figure 18.15 Time-lapse view of the amoebozoan *Amoeba proteus*. The time interval from top to bottom is about 6 sec. The arrows point to a fixed spot on the surface. A single cell is about 80 μm wide.

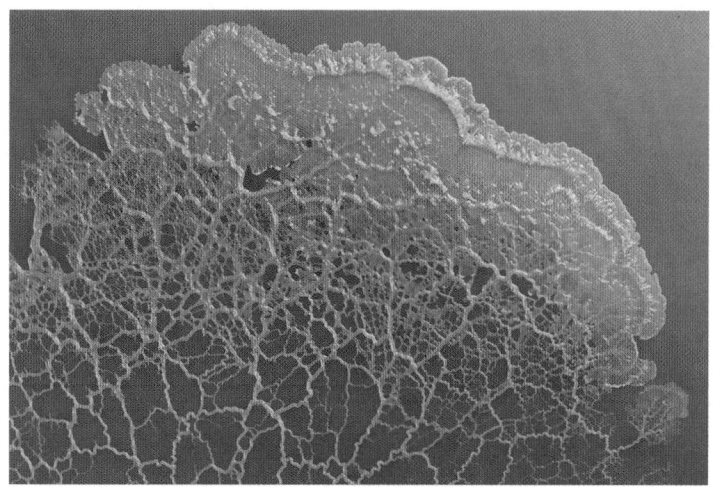

Figure 18.16 Slime mold. The plasmodial slime mold *Physarum* growing on an agar surface. The plasmodium is about 5 cm long and 3.5 cm wide.

(also called *acellular slime molds*), whose vegetative forms are masses of protoplasm of indefinite size and shape called plasmodia (**Figure 18.16**), and *cellular slime molds*, whose vegetative forms are single amoebae. Slime molds live primarily on decaying plant matter, such as leaf litter, logs, and soil, where they consume other microorganisms, especially bacteria. Slime molds can maintain themselves in

a vegetative state for long periods but eventually form differentiated sporelike structures that can remain dormant and then germinate later to once again generate the active amoeboid state.

Plasmodial slime molds, such as *Physarum*, exist in the vegetative phase as an expanding single mass of protoplasm called the *plasmodium* that contains many diploid nuclei (Figure 18.16). The plasmodium is actively motile by amoeboid movement, and from this phase, a sporangium containing haploid spores can be produced; when conditions are favorable, the spores germinate to yield haploid flagellated swarm cells. The fusion of two swarm cells then regenerates a diploid plasmodium.

In contrast to their plasmodial relatives, cellular slime molds are individual haploid cells and form diploids only under certain conditions. The well-studied cellular slime mold *Dictyostelium discoideum* undergoes an asexual life cycle in which vegetative cells aggregate, migrate as a cell mass, and eventually produce fruiting bodies in which cells differentiate and form spores (**Figures 18.17** and **18.18**). When cells of *Dictyostelium* are starved, they aggregate and form a pseudoplasmodium; in this stage cells lose their individuality, but do not fuse. Aggregation is triggered by the production of cyclic adenosine monophosphate (cAMP). The first cells of *Dictyostelium* that produce this compound attract neighboring cells and eventually aggregate into motile masses of cells called *slugs*. Fruiting body formation begins when the slug becomes stationary and vertically oriented. The emerging structure differentiates into

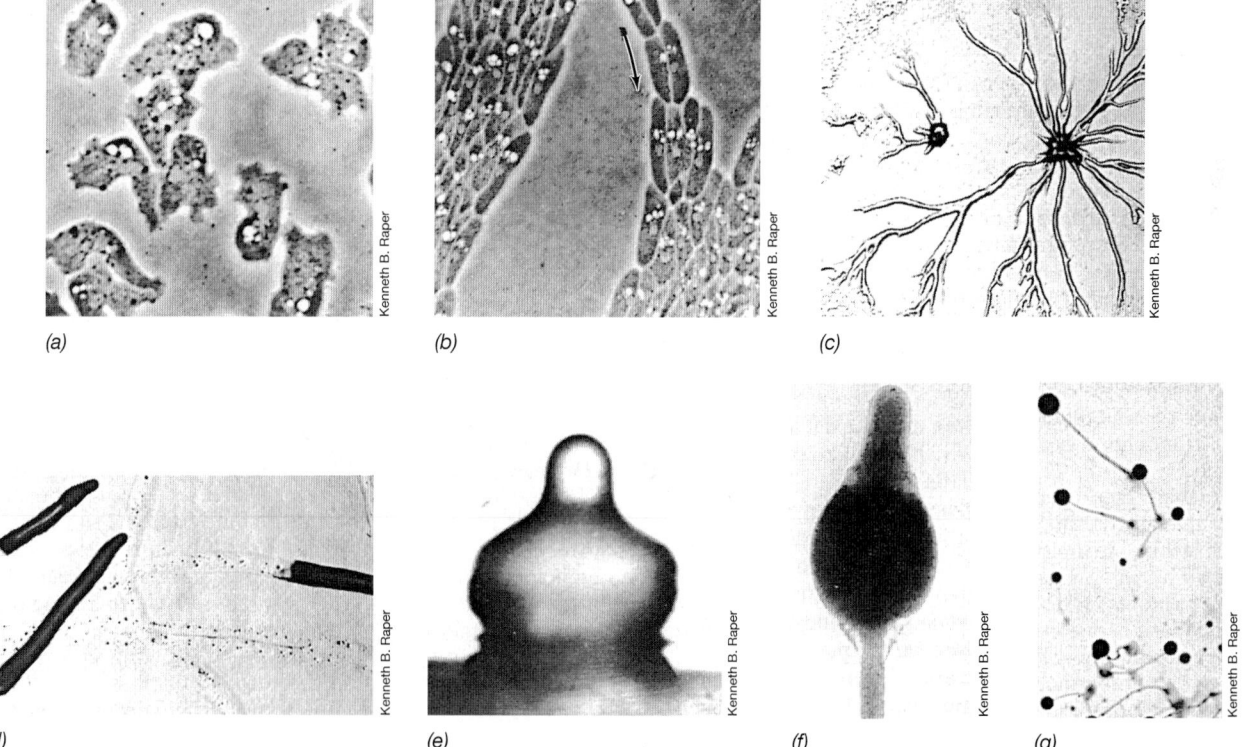

Figure 18.17 Photomicrographs of various stages in the life cycle of the cellular slime mold ***Dictyostelium discoideum.*** *(a)* Amoebae in preaggregation stage. *(b)* Aggregating amoebae. Amoebae are about 300 μm in diameter. *(c)* Low-power view of aggregating amoebae. *(d)* Migrating pseudoplasmodia (slugs) moving on an agar surface and leaving trails of slime behind. *(e, f)* Early stages of fruiting body. *(g)* Mature fruiting bodies. Figure 18.18 shows the sizes of these structures. *Dictyostelium* has long served as a model for development in multicellular organisms, and its genome of 12,500 genes is about half that of the human genome.

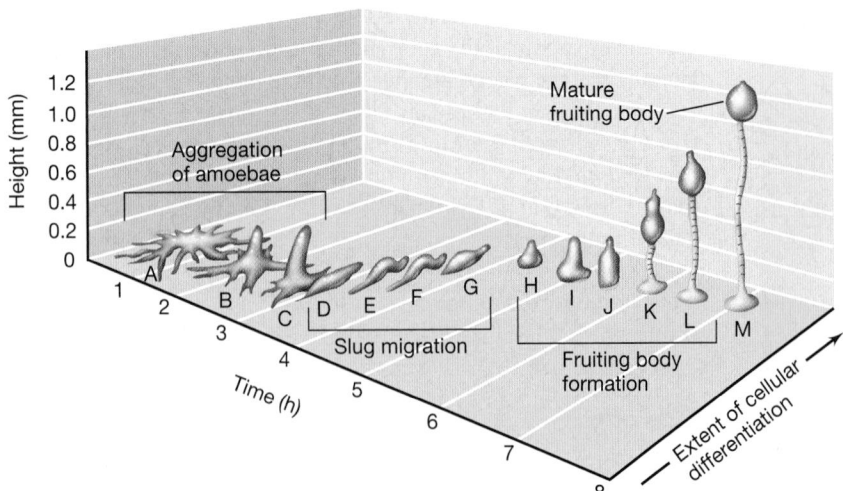

Figure 18.18 Stages in fruiting body formation in the cellular slime mold *Dictyostelium discoideum*. *(A–C)* Aggregation of amoebae. *(D–G)* Migration of the slug formed from aggregated amoebae. *(H–I)* Culmination of migration and formation of the fruiting body. *(M)* Mature fruiting body composed of stalk and head. Cells from the rear of the slug form the head and become spores. *Dictyostelium* also undergoes sexual reproduction (not shown) when two amoebae fuse to form a macrocyst; the fused nuclei in the macrocyst return to the haploid stage when meiosis forms new vegetative amoebae.

a stalk and a head. The stalk cells form cellulose, which provides the rigidity of the stalk, and the head cells differentiate into spores. Eventually, spores are released and dispersed, with each spore forming a new amoeba (Figures 18.17 and 18.18).

In addition to this asexual process, *Dictyostelium* can produce sexual spores. These form when two amoebae in an aggregate fuse to form a single giant amoeba. A thick cellulose wall develops around this cell to form a structure called the *macrocyst*, and this can remain dormant for long periods. Eventually, the diploid nucleus undergoes meiosis to form haploid nuclei that become integrated into new amoebae that can once again initiate the asexual cycle.

MINIQUIZ

- How can amoebozoans be distinguished from rhizaria?
- Compare and contrast the lifestyles of gymnamoebas and entamoebas.
- Describe the major steps in the life cycle of *Dictyostelium discoideum*.

III • Fungi

The **Fungi** are a large, diverse, and widespread group of organisms that includes such well-known groups as the *molds, mushrooms*, and *yeasts*. Approximately 100,000 fungal species have been described, and as many as 1.5 million species may exist. Fungi form a phylogenetic cluster distinct from other protists and are the microbial group most closely related to animals (Figure 18.3).

Most fungal species are microscopic and terrestrial. They inhabit soil or dead plant matter and play crucial roles in the mineralization of organic carbon. A large number of fungal species are plant

pathogens, and a few cause diseases of animals, including humans. Certain species of fungi also establish symbiotic associations with plants, facilitating the plant's acquisition of minerals from soil, and many fungi benefit humans through fermentation and the synthesis of antibiotics.

18.8 Fungal Physiology, Structure, and Symbioses

In this section we describe some general features of fungi, including their physiology, cell structure, and the symbiotic associations they develop with plants and animals. In the following section we examine fungal reproduction and phylogeny.

Nutrition, Physiology, and Ecology

Fungi are chemoorganotrophs—typically displaying simple nutritional requirements—and most are aerobic. Fungi feed by secreting extracellular enzymes that digest polymeric materials, such as polysaccharides or proteins, into monomers that are assimilated as sources of carbon and energy. As decomposers, fungi digest dead animal and plant materials. As parasites of plants or animals, fungi use the same mode of nutrition but take up nutrients from the living cells of the plants and animals they invade rather than from dead organic materials.

A major ecological activity of fungi, especially the basidiomycetes, is the decomposition of wood, paper, cloth, and other products derived from these natural sources. Lignin, a complex polymer in which the building blocks are phenolic compounds, is an important constituent of woody plants, and in association with cellulose it confers rigidity on them. Lignin is decomposed in nature almost exclusively through the activities of certain basidiomycetes called *wood-rotting fungi*. Two types of wood rot are known: *brown rot*, in which the cellulose is attacked preferentially and the lignin left unmetabolized, and *white rot*, in which both cellulose and lignin are decomposed. The white rot fungi are of major ecological importance because they play such a key role in decomposing woody materials in forests.

Fungal Morphology, Spores, and Cell Walls

Most fungi are multicellular, forming a network of filaments called *hyphae* (singular, hypha) from which asexual spores are produced (**Figure 18.19**). Hyphae are tubular cell walls that surround the cytoplasmic membrane. Fungal hyphae are often septate, with cross-walls dividing each hypha into separate cells. In some cases, however, the vegetative cell of a fungal hypha contains more than one nucleus, and hundreds of nuclei can form as a result of repeated nuclear divisions without the formation of cross-walls, a condition called *coenocytic*. Each hyphal filament grows mainly at the tip by extension of the terminal cell (Figure 18.19).

Hyphae typically grow together across and above a surface to form a compact, macroscopically visible tuft called a *mycelium* (**Figure 18.20a**). From the mycelium, aerial hyphae reach up into the air above the surface, and spores called **conidia** are formed

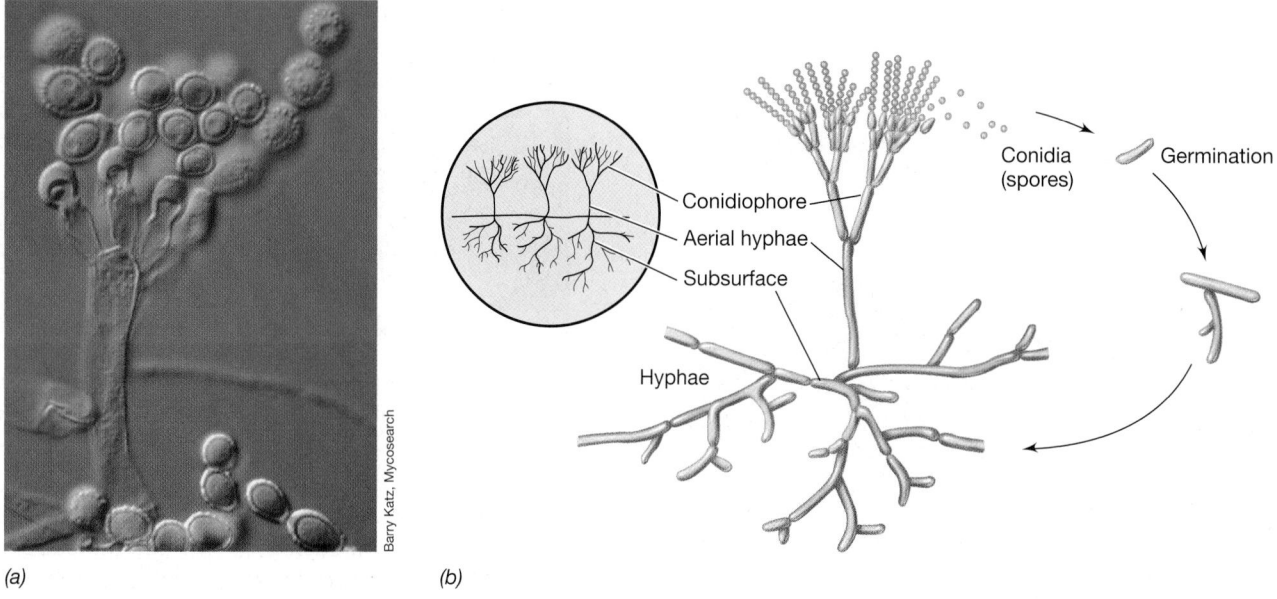

(a)   *(b)*

Figure 18.19 Fungal structure and growth. *(a)* Photomicrograph of a typical mold. Spherical structures at the ends of aerial hyphae are asexual spores (conidia). *(b)* Diagram of a mold life cycle. The conidia can be dispersed by either wind or animals and are about 2 μm wide.

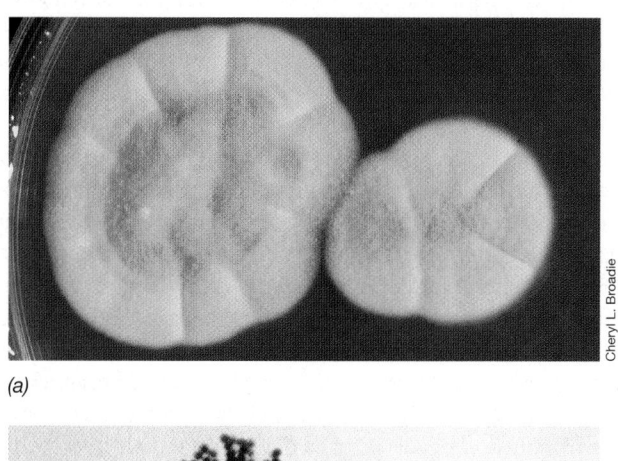

(a)

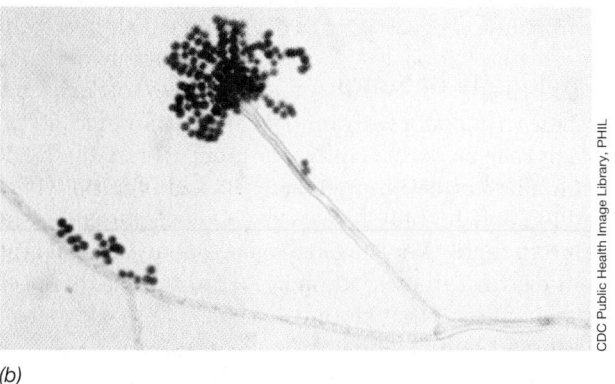

(b)

Figure 18.20 Hyphal fungi (molds). *(a)* Colonies of an *Aspergillus* species (ascomycete), growing on an agar plate. Note the masses of filamentous cells (mycelia) and asexual spores that give the colonies a dusty, matted appearance. *(b)* Conidiophore and conidia of *Aspergillus fumigatus* (see Figure 18.19*b*). The conidiophore is about 300 μm long and the conidia here are about 3 μm wide. These cells were stained to improve contrast. Besides being a common saprophyte, *Aspergillus* can be pathogenic, causing serious pulmonary and occasional systemic infections in humans and some domestic animals. Cancer patients and others with weakened immune systems are particularly susceptible to aspergillosis.

on their tips (Figure 18.20*b*). Conidia are asexual spores and they are often pigmented black, green, red, yellow, or brown (Figure 18.20). Conidia give the mycelium a dusty appearance (Figure 18.20*a*) and function to disperse the fungus to new habitats. Some fungi form macroscopic reproductive structures called *fruiting bodies* (**mushrooms** or puffballs, for example), in which millions of spores are produced that can be dispersed by wind, water, or animals (**Figure 18.21**). In contrast to mycelial fungi, some fungi grow as single cells; these are the **yeasts**.

Most fungal cell walls consist of **chitin**, a polymer of *N*-acetyl-glucosamine. Chitin is arranged in the walls in microfibrillar bundles, as is cellulose in plant cell walls, to form a thick, tough wall structure. Other polysaccharides such as mannans and galactosans, or even cellulose itself, replace or supplement chitin in some fungal cell walls. Fungal cell walls are typically 80–90% polysaccharide, with only small amounts of proteins, lipids, polyphosphates, and inorganic ions making up the wall-cementing matrix.

Symbioses and Pathogenesis

Most plants are dependent on certain fungi to facilitate their uptake of minerals from soil. These fungi form symbiotic associations with the plant roots called *mycorrhizae* (the word means, literally, "fungus roots"). Mycorrhizal fungi establish close physical contact with the roots and help the plant obtain phosphate and other minerals and also water from the soil. In return, the fungi obtain nutrients such as sugars from the plant root (⟳ Figure 23.21). There are two kinds of mycorrhizal associations. One, *ectomycorrhizae*, typically forms between basidiomycetes (Section 18.13) and the roots of woody plants, while the second, *endomycorrhizae*, forms between glomeromycete fungi (Section 18.11) and many nonwoody plants. Some fungi also form associations with cyanobacteria or green algae. These are the *lichens*, the colorful and crusty growths often seen on the surfaces of trees and rocks.

UNIT 4

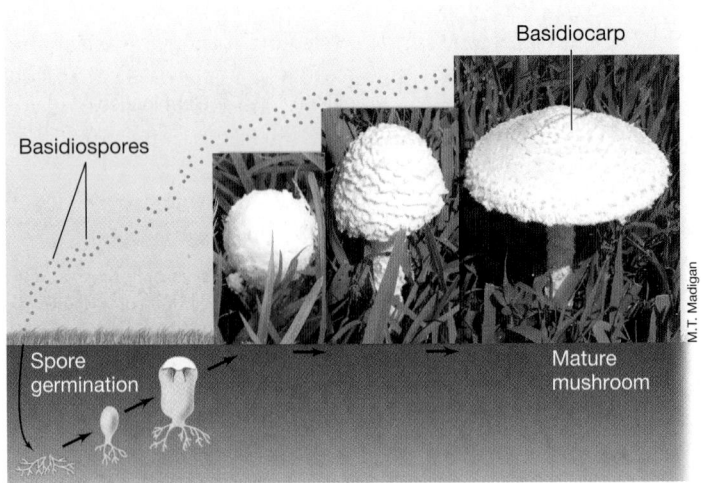

Figure 18.21 Mushroom life cycle. Mushrooms typically develop underground and then emerge on the surface rather suddenly (usually overnight), triggered by an influx of moisture. Photos of stages in formation of a common lawn mushroom (see also Section 18.13).

We explore the biology of lichens and mycorrhizae in more detail in Sections 23.1 and 23.4, respectively.

Fungi can invade and cause disease in plants and animals. Fungal plant pathogens cause widespread crop and plant damage worldwide, and fruit and grain crops in particular suffer significant yearly losses as a result of fungal infection. Human fungal diseases, called *mycoses*, range from relatively minor and easily cured conditions, such as athlete's foot and jock itch, to serious, life-threatening systemic mycoses, such as histoplasmosis. We consider some major fungal diseases of humans in Chapter 33.

MINIQUIZ

- What are conidia? How does a conidium differ from a hypha? From a mycelium?
- What is chitin and where is it present in fungi?
- Distinguish between mycorrhizae and lichens.

18.9 Fungal Reproduction and Phylogeny

Fungi reproduce by *asexual* means in one of three ways: (1) by the growth and spread of hyphal filaments; (2) by the asexual production of spores (conidia; Figures 18.19 and 18.20); or (3) by simple cell division, as in budding yeasts (**Figure 18.22**). Most fungi also form sexual spores, typically as part of an elaborate life cycle. Some fungi, such as the well-known mold *Penicillium* (the source of the antibiotic penicillin), were long thought to lack a sexual stage and reproduce only by way of conidia. But it has now been shown that *Penicillium* (and probably all fungi of its taxonomic class, the *Deuteromycetes*) go through a sexual stage in their life cycles.

Sexual Spores of Fungi

Some fungi produce spores as a result of sexual reproduction. The spores develop from the fusion of either unicellular gametes or

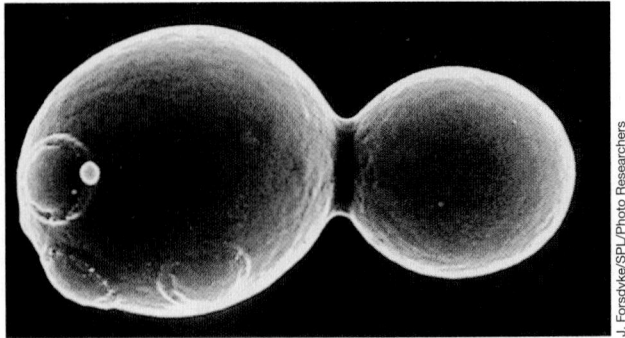

Figure 18.22 The common baker's and brewer's yeast *Saccharomyces cerevisiae* (*Ascomycota*). In this colorized scanning electron micrograph, note the budding division and scars from previous buds. A single cell is about 6 μm in diameter.

specialized hyphae called *gametangia*. Alternatively, sexual spores can originate from the fusion of two haploid cells to yield a diploid cell; this then undergoes meiosis and mitosis to yield individual haploid spores.

Depending on the group, different types of sexual spores are produced. Spores formed within an enclosed sac (ascus) are called *ascospores*. Many yeasts produce ascospores, and we consider sporulation in the common baker's yeast *Saccharomyces cerevisiae* in Section 18.12. Sexual spores produced on the ends of a club-shaped structure (basidium) are *basidiospores* (Figure 18.21 and see Figure 18.30c). *Zygospores*, produced by zygomycetous fungi such as the common bread mold *Rhizopus* (Section 18.11), are macroscopically visible structures that result from the fusion of hyphae and genetic exchange. Eventually the zygospore matures and produces asexual spores that are dispersed by air and germinate to form new fungal mycelia. Chytrid fungi produce sexual spores, motile by eukaryotic flagella, called *zoospores*.

Sexual spores of fungi are typically resistant to drying, heating, freezing, and some chemical agents. However, neither sexual nor asexual spores of fungi are as resistant to heat as bacterial endospores (⟷ Section 2.10). Both asexual and sexual spores of fungi can germinate and develop into a new hypha.

The Phylogeny of Fungi

Fungi share an ancestor with animals, and are more closely related to animals than any other eukaryotic group (Figure 18.3). The last common ancestor of all fungi likely existed sometime between 450 million and 1.5 billion years ago. One of the earliest fungal lineages is thought to be the *Chytridiomycota*, an unusual group of motile fungi that produce zoospores. Thus the lack of flagella in most fungi indicates that motility is a characteristic that has been lost at various times in different fungal lineages.

Some of the major groups of fungi are shown in the evolutionary tree in **Figure 18.23**. The phylogeny shown in this figure includes several distinct fungal groups: the *Microsporidia*, *Chytridiomycota*, *Zygomycota*, *Glomeromycota*, *Ascomycota*, and *Basidiomycota*. The vast majority of described fungal species belong to the *Ascomycota* and *Basidiomycota*. The *Ascomycota* are a large and diverse group of fungi that includes the yeasts, such as *Saccharomyces* (Figure 18.22), and molds, such as *Aspergillus* (Figure 18.20). The *Basidiomycota* include fungi that form mushrooms (Figure 18.21 and see Figure 18.30), as well as many

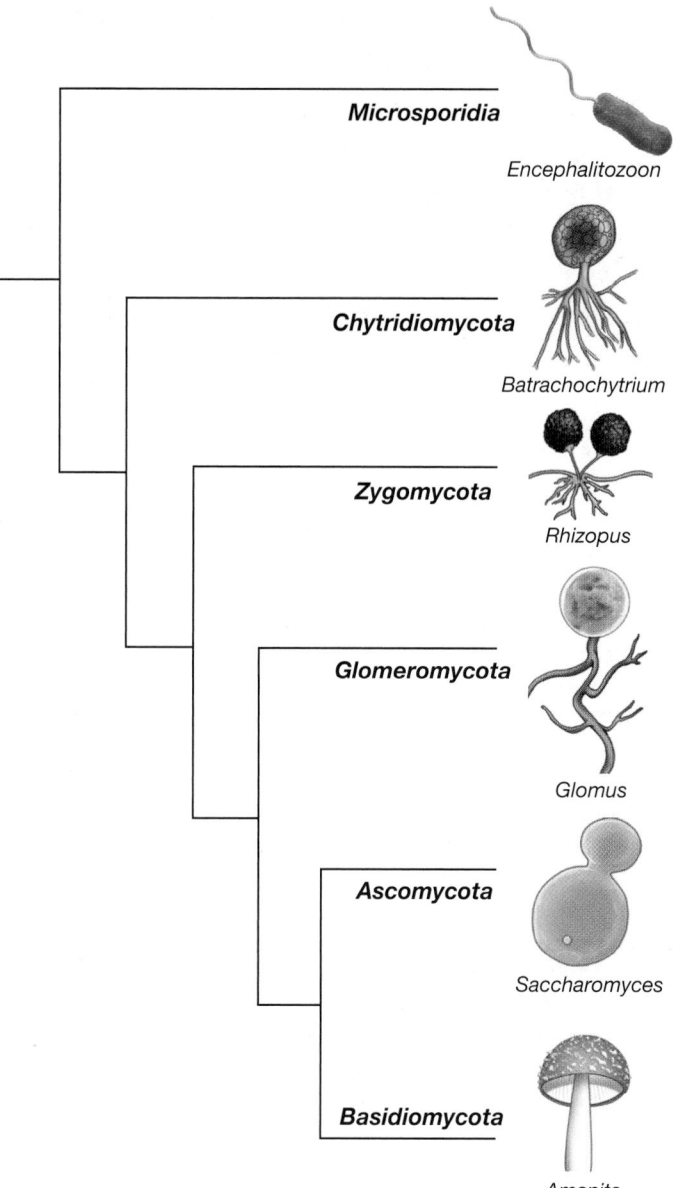

Figure 18.23 Phylogeny of fungi. This schematic phylogenetic tree depicts the relationships among the major groups (phyla) of fungi. A typical genus is listed for each group and depicted in the tree.

important plant pathogens such as rusts and smuts. While a tremendous diversity of fungal species have already been cultured and described, phylogenetic analyses of fungal DNA sequences recovered from environmental samples (⟳ Section 19.6) indicate that more than 90% of fungal species remain to be discovered. It is thus clear that we have much to learn about the biology and phylogeny of the fungi.

— **MINIQUIZ** —

- Why is the mold *Penicillium* economically important?
- What are the major differences between ascospores and conidia?
- To what major group of macroorganisms are fungi most closely related?

18.10 *Microsporidia* and *Chytridiomycota*

KEY GENERA: *Allomyces, Batrachochytrium, Encephalitozoon*

The *Microsporidia* and *Chytridiomycota* are ancient phylogenetic groups of parasitic or saprophytic fungi. Microsporidia species are obligate parasites of a wide variety of animal hosts including humans, whereas the *Chytridiomycota* are primarily aquatic fungi whose species are either parasites or saprophytes.

Microsporidia

Microsporidia are tiny (2–5 μm) unicellular parasites of animals and protists. Based on 18S ribosomal RNA gene sequencing and their lack of mitochondria, *Microsporidia* were once thought to form a very early-branching lineage of *Eukarya*. However, composite gene and protein sequencing has shown the microsporidians are fungi closely related to the *Chytridiomycota* (Figure 18.23). There remains debate as to whether the *Microsporidia* should be included as one of the most deeply divergent lineages within the fungi or whether they should instead be classified as a distinct lineage closely related to fungi.

Microsporidia have adapted to a parasitic lifestyle. They exist as spores when outside of the host. When near a host cell they extend a helical polar tubule that penetrates the host cytoplasmic membrane. The spore then injects its *sporoplasm* into the host cell. The sporoplasm replicates within the cytoplasm of the host cell, forming new spores as it completes its life cycle. The cell membrane of the host is disrupted and the spores are released in the surrounding environment, free to infect a new cell.

Like most obligate parasites, microsporidia have undergone significant genome reduction, losing many features that would allow them to live outside of a host cell. The microsporidium *Encephalitozoon* (**Figure 18.24a**), for example, lacks not only mitochondria and hydrogenosomes but also a Golgi complex (another key eukaryotic cell structure, ⟳ Section 2.16). Moreover, *Encephalitozoon* contains a very small genome of only 2.9 Mbp and contains only about 2000 genes (this is 1.5 Mbp and 2600 genes *smaller* than that of the bacterium *Escherichia coli*). The *Encephalitozoon* genome lacks genes for major metabolic pathways, such as the citric acid cycle (⟳ Section 3.9), meaning that this pathogen must depend on its host for even the most basic of metabolites and metabolic processes.

In humans, *Encephalitozoon* causes chronic debilitating diseases of the intestine, lung, eye, muscle, and some internal organs but is uncommon among healthy adults with normal immune systems. However, microsporidial diseases have appeared with increasing frequency in immune-compromised individuals, such as those with HIV/AIDS or those on long-term administration of immune-suppressing drugs, such as those who have had organ transplants.

Chytridiomycota

Chytridiomycota, or *chytrids*, are the earliest diverging lineage of fungi (Figure 18.23), and their name refers to the structure of the fruiting body, which contains their *zoospores* (Section 18.9). These spores are unusual among fungal spores in being flagellated and motile, and are ideal for dispersal of these organisms in the aquatic environments, mostly freshwater and moist soils, where they are commonly found.

Many species of chytrids are known and some exist as single cells, whereas others form colonies with hyphae. They include both free-living forms that degrade organic material, such as *Allomyces*, and parasites of animals, plants, and protists.

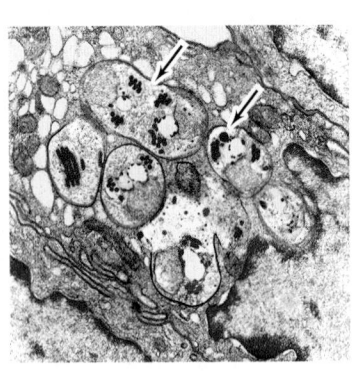

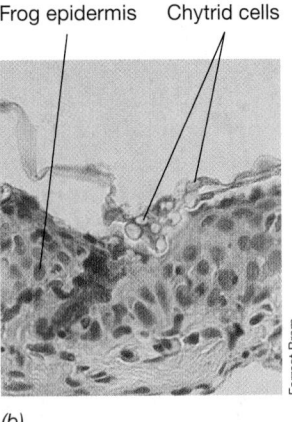

Frog epidermis   Chytrid cells

Forrest Brem

(a)   (b)

Figure 18.24 *Microsporidia* and *Chytridiomycota*. *(a)* Colorized transmission electron micrograph of thin sections of cells of the microsporidium *Encephalitozoon intestinalis* (arrows) growing in human intestinal cells. *(b)* Cells of the chytrid *Batrachochytrium dendrobatidis* stained pink growing on the surface of frog epidermis.

The chytrid *Batrachochytrium dendrobatidis* causes chytridiomycosis of frogs (Figure 18.24*b*), a condition in which the organism infects the frog's epidermal layers, leading to a loss of ions across the membrane and osmotic imbalance. Chytrids have been implicated in the massive die-off of frogs and some other amphibians worldwide, probably in response to increases in global temperatures that have stimulated chytrid proliferation and to increased animal susceptibility due to habitat loss and aquatic pollution.

Unresolved aspects of the phylogeny of chytrids suggest that this group is not monophyletic. That is, some organisms currently classified as chytrids may actually be more closely related to species of other fungal groups, such as the *Zygomycota*, to which we turn next. As is true for the protists, much about the evolution of the chytrids and other groups of fungi remains to be learned.

MINIQUIZ

- What animal group has been most affected by chytrids?
- What are some features of *Microsporidia* that distinguish them from chytrids?

18.11 *Zygomycota* and *Glomeromycota*

KEY GENERA: *Rhizopus, Glomus*

We consider two groups of fungi here, the *Zygomycota*, known primarily for their role in food spoilage, and the *Glomeromycota*, important fungi in certain mycorrhizal associations. *Zygomycota* are commonly found in soil and on decaying plant material, whereas *Glomeromycota* form symbiotic relationships with plant roots. All of these fungi are coenocytic (multinucleate), and a unifying feature is the formation of sexual spores called *zygospores* (Section 18.9).

Zygomycota

The common black bread mold *Rhizopus nigricans* (**Figure 18.25**) is a widespread zygomycete. This organism undergoes a complex life cycle that includes both asexual and sexual reproduction. In the asexual phase the mycelia form sporangia within which haploid spores are produced. Once released, spores disperse and eventually

(a)   (b)

Figure 18.25 *Zygomycota*. *(a)* Moldy bread from growth of the zygomycete *Rhizopus nigricans*. *(b)* Stained mycelium of *Rhizopus* showing the black aerial sporangia containing asexual spores.

germinate, giving rise to vegetative mycelia. In the sexual phase, mycelial gametangia of different mating types (analogous to male and female, see Section 18.12) fuse to yield a cell with two nuclei called a *zygosporangium*, which can remain dormant and resist dryness and other unfavorable conditions. When conditions are favorable, the different haploid nuclei fuse to form a diploid nucleus followed by meiosis to yield haploid spores. As in the asexual phase, the release of the spores, in this case genetically nonidentical spores, disperses the organism for vegetative hyphal growth.

Most species of *Rhizopus* and related zygomycetes are harmless saprophytes whose airborne spores land and form spreading colonies on stale bread (Figure 18.25*a*) and various moist surfaces in the home or on walls and crevices in buildings where moisture is trapped. However, some species are human pathogens. If inhaled in sufficient amounts, spores of pathogenic *Rhizopus* species can cause serious infections of the lungs, sinuses, eyes, nose, and mouth leading to swelling of facial features, asthma-like symptoms, and even fatal systemic fungal infections if the initial infection is not promptly treated.

Glomeromycota

The *Glomeromycota* are a relatively small and unique group of obligately symbiotic fungi in which all known species form associations with plants called *endomycorrhizae* (Section 18.8 and ↩ Section 23.4). As many as 80% or more of land plant species form these associations in which the fungal hyphae enter the plant cell and aid the plant's acquisition of phosphate from the soil in return for fixed carbon from the plant. Most of these *Glomeromycota* are *arbuscular mycorrhizae*, fungi that form structures called arbuscules that penetrate cells of their plant host and specialize in nutrient exchange (↩ Section 23.4). As plant symbionts, glomeromycetes are thought to have played a pivotal role in the ability of early vascular plants to colonize land (see page 557).

As far as is known, glomeromycetes reproduce only asexually and are mostly coenocytic in their hyphal morphology. Asexual spores of *Glomus* (Figure 18.23), a major genus of endomycorrhizae, are collected from the roots of cultivated plants and used as an agricultural inoculant to promote vigorous symbiotic associations between plant and fungus. This natural approach to plant fertilization is a widespread practice in small, sustainable farming operations and has been shown to increase both the growth and nutrient content of tomato, pepper, squash, bean, and several other small fruit and vegetable plants.

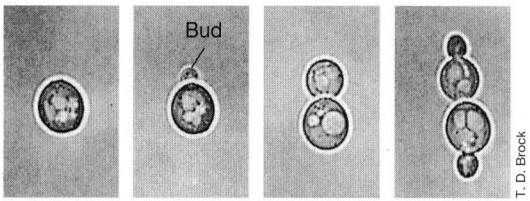

Figure 18.26 Growth by budding division in *Saccharomyces cerevisiae*. A time-lapse series of phase-contrast micrographs shows the budding division process starting from a single cell. Note the pronounced nucleus. A single cell of *S. cerevisiae* is about 6 μm in diameter.

— MINIQUIZ —

- Contrast the habitats of *Zygomycota* and *Glomeromycota*.
- How does the fungus *Glomus* aid the acquisition of nutrients by plants?

18.12 *Ascomycota*

KEY GENERA: *Saccharomyces, Candida, Aspergillus*

The *Ascomycota*, also called the ascomycetes, are the largest and most diverse group of fungi and they range from single-celled species, such as the baker's yeast *Saccharomyces* (**Figure 18.26** and Figure 18.22), to species that grow as filaments, such as the common mold *Aspergillus* (Figure 18.20). Ascomycetes are found in aquatic and terrestrial environments and take their name from the production of *asci* (singular, ascus), cells in which two haploid nuclei from different mating types fuse to form a diploid nucleus that eventually undergoes meiosis to form haploid ascospores. In addition to ascospores, ascomycetes reproduce asexually by the production of conidia that form at the tips of specialized hyphae called *conidiophores* (Figure 18.20). Both saprophytic and pathogenic yeasts, such as *Candida albicans*, are common in nature. We focus here on the yeast *Saccharomyces* as a model ascomycete.

Saccharomyces cerevisiae

The cells of *Saccharomyces* and other single-celled ascomycetes are spherical, oval, or cylindrical, and cell division typically takes place by budding. In the budding process, a new cell forms as a small outgrowth of the old cell; the bud gradually enlarges and then separates from the parent cell (Figures 18.22 and 18.26).

Yeast cells are typically much larger than bacterial cells and can be distinguished from bacteria microscopically by their larger size and by the obvious presence of internal cell structures, such as the nucleus or cytoplasmic vacuoles (Figure 18.26). Yeasts flourish in sugar-rich habitats such as fruits, flowers, and the bark of trees. Yeasts are typically facultative aerobes, growing aerobically as well as by fermentation. Several yeasts live symbiotically with animals, especially insects, and a few species are pathogenic for animals and humans (↩ Sections 33.1 and 33.2). The most important commercial yeasts are the baker's and brewer's yeasts, which are species of *Saccharomyces*. The yeast *S. cerevisiae* has been studied as a model eukaryote for many years and was the first eukaryote to have its genome completely sequenced (↩ Section 9.4).

Mating Types and Sexual Reproduction in *Saccharomyces*

The yeast *Saccharomyces* can reproduce by sexual means in which two cells fuse. Within the fused cell, called a *zygote*, meiosis occurs and ascospores are eventually formed. The life cycle of *S. cerevisiae* is depicted in **Figure 18.27**. *S. cerevisiae* can grow vegetatively as either haploid or diploid cells. *S. cerevisiae* forms two different types of haploid cells called *mating types* designated α (alpha) and

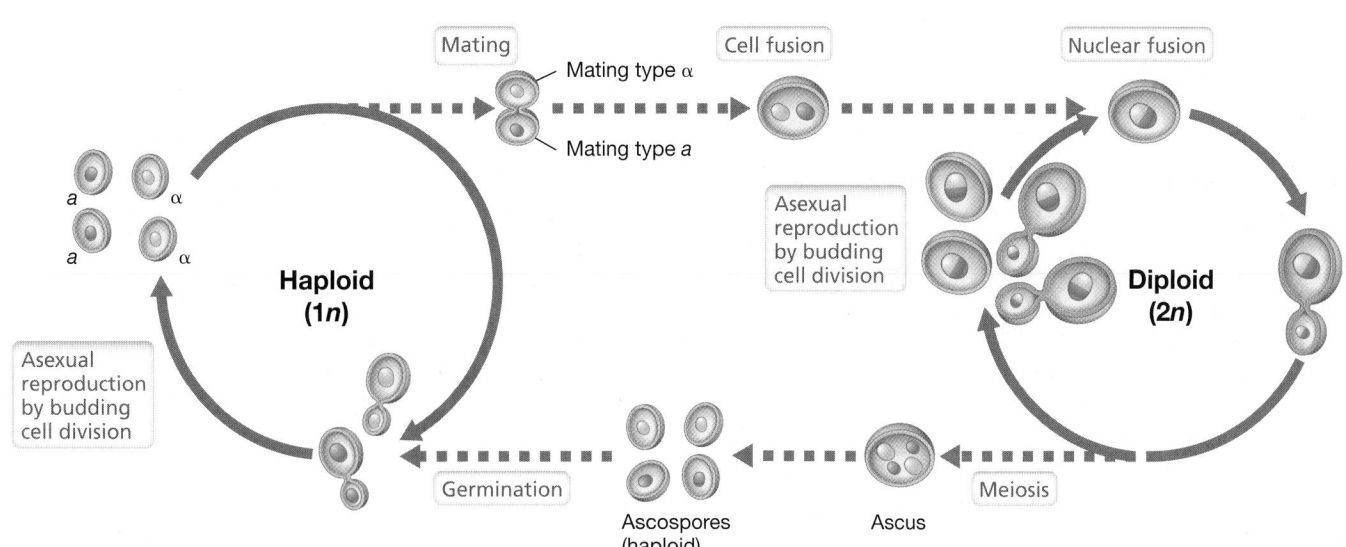

Figure 18.27 Life cycle of a typical ascomycete yeast, *Saccharomyces cerevisiae*. Cells can grow vegetatively for long periods as haploid cells or as diploid cells before life cycle events (dashed lines) generate the alternate genetic form.

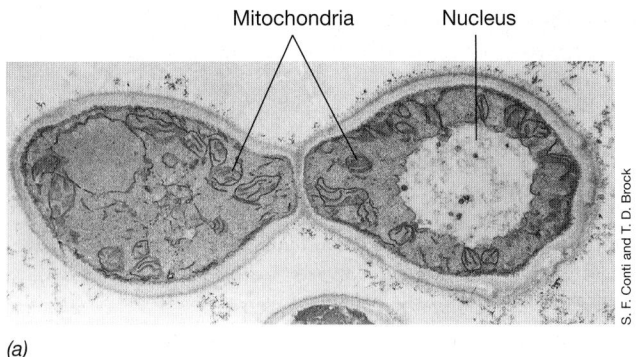

Mitochondria   Nucleus

(a)

S. F. Conti and T. D. Brock

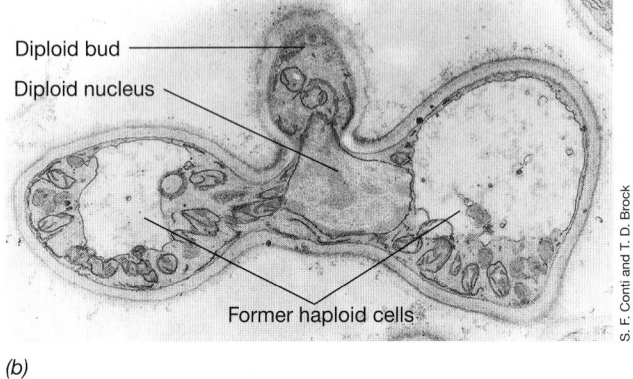

Diploid bud

Diploid nucleus

Former haploid cells

(b)

S. F. Conti and T. D. Brock

Figure 18.28 Electron micrographs of mating in the ascomycete yeast *Hansenula wingei.* (a) Two cells have fused at the point of contact. (b) Late stage of mating. The nuclei of the two cells have fused, and a diploid bud has formed at a right angle to the mating cells. This bud becomes the progenitor of a diploid cell line. A cell of *Hansenula* is about 10 μm in diameter.

a (encoded by genes α and *a*); these are analogous to male and female gametes. The α and *a* genes regulate the production of the peptide hormones α *factor* or *a factor*, which are excreted by yeast cells during mating. The hormones bind to cells of the opposite mating type and bring about changes in their cell surfaces that enable the cells to fuse; once mating has occurred, the nuclei fuse, forming a diploid zygote (**Figure 18.28**). The zygote undergoes vegetative growth by budding, but under starvation conditions it undergoes meiosis and generates ascospores (Figure 18.27).

Haploid strains of *S. cerevisiae* are genetically predisposed to be either *a* or α but are able to switch their mating type. This switch occurs when the active mating-type gene is replaced with one of two otherwise "silent" genes, as shown in **Figure 18.29**. There is a single location on one of the *S. cerevisiae* chromosomes called the *MAT* (for *mating type*) locus, at which either gene *a* or gene α can be inserted. At this locus, the *MAT* promoter controls transcription of whichever gene is present. If gene *a* is at that locus, then the cell is mating-type *a*, whereas if gene α is at that locus, the cell is mating-type α. Elsewhere in the yeast genome are copies of genes *a* and α that are not expressed, and these are the source of the inserted gene. In the switch (Figure 18.29), the appropriate gene, *a* or α, is copied from its silent site and inserted into the *MAT* location, replacing the gene already present. The old mating-type gene is excised and discarded, and the new gene is inserted. Whichever gene is inserted in the *MAT* locus is the one that will govern the

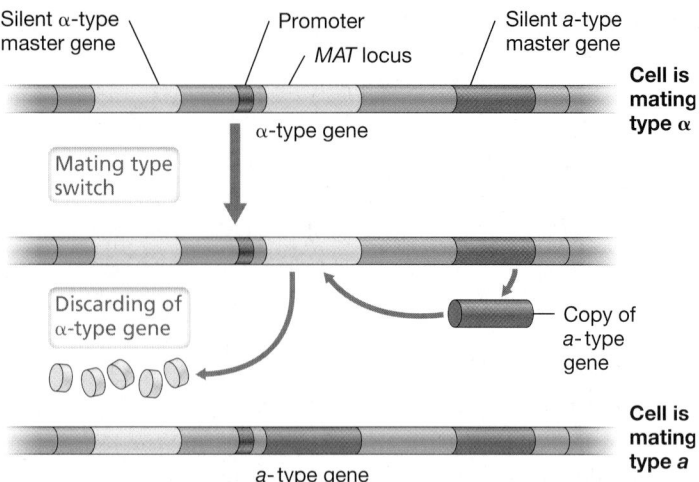

Figure 18.29 The cassette mechanism that switches an ascomycete yeast from mating type α to *a*. The cassette inserted at the *MAT* locus determines the mating type. The process shown is reversible, so type *a* can also revert to type α.

mating type of the strain. It is thus possible for cells from a pure culture of *S. cerevisiae* derived from a single cell to mate, following a mating-type switch in one or more cells in the culture.

--- MINIQUIZ ---

- Are ascospores haploid or diploid cells?
- Explain how a *single* haploid cell of *Saccharomyces* can eventually yield a diploid cell.

18.13 *Basidiomycota*

KEY GENERA: *Agaricus, Amanita*

Basidiomycota are a large group of fungi, with over 30,000 species described. Many are the commonly recognized mushrooms and toadstools, some of which are edible, such as the commercially grown mushroom *Agaricus*. Others, such as the mushroom *Amanita* (**Figure 18.30a**), are highly poisonous. Other basidiomycetes include puffballs, smuts, rusts, and an important human fungal pathogen, *Cryptococcus* (⇌ Sections 33.1 and 33.2). The defining characteristic of the *Basidiomycota* is the *basidium* (plural, *basidia*), a structure in which haploid basidiospores are formed by meiosis. The basidium, a word that means "little pedestal" (Figure 18.30c), gives the group its name.

Mushroom Development

During most of its existence, a mushroom fungus lives as a simple haploid mycelium, growing vegetatively in soil, leaf litter, or decaying logs. It is the sexual reproductive phase of basidiomycetes that produces the visible mushroom structure (Figures 18.21 and 18.30). In this process, mycelia of different mating types fuse, and the faster growth of the dikaryotic (two nuclei per cell) mycelium formed from that fusion overgrows and crowds out the parental haploid mycelia. Then, when environmental conditions are favorable, usually following periods of wet and cool weather, the dikaryotic mycelium develops rapidly into the fruiting body.

(a)

(b)

Gills

USDA

Basidiospore

Basidium

(c)

Figure 18.30 Mushrooms. *(a) Amanita*, a highly poisonous mushroom. *(b)* Gills on the underside of the mushroom fruiting body contain the spore-bearing basidia. *(c)* Light micrograph of basidia and basidiospores from the mushroom *Coprinus*.

The mushroom fruiting body, called a *basidiocarp*, begins as a mycelium that differentiates into a small button-shaped structure underground that then expands into the full-grown basidiocarp that we see aboveground, the mushroom (Figures 18.21 and 18.30). The dikaryotic basidia are borne on the underside of the basidiocarp on flat plates called *gills*, which are attached to the cap of the mushroom (Figure 18.30*b, c*). The basidia then undergo a fusion of the two nuclei, forming basidia with diploid nuclei. The two rounds of meiotic division generate four haploid nuclei in the basidia, and each of the nuclei becomes a basidiospore. The genetically distinct basidiospores can then be dispersed by wind to new habitats to begin the cycle again, germinating under favorable conditions and growing as haploid mycelia (Figure 18.21).

Pathogenic Basidiomycetes

The smuts and rusts are plant pathogenic basidiomycetes. Smuts are pathogens of cereal grains and other crops, and the genus *Ustilago* contains several species whose hosts include corn, sugarcane, wheat, and several other grains. The fungus targets the reproductive system of the plant, and in corn smut causes the developing kernels to form massive tumorlike kernels that give the cob a swollen and burned appearance.

Rust fungi attack rapidly growing plant tissues such as emerging shoots, leaves, and fruits. Rust fungi of the genus *Puccinia* can infect any of a wide variety of plants whereas other rust fungi are more host-restricted. Some common rusts include stem rust of wheat and white pine blister rust, a disease of several pine species that can trigger branch dieback and even death of the entire tree. Pine trees weakened by rust fungi are typically more susceptible to insect attack such as from infestations of the pine bark beetle; this highly destructive pest has destroyed major stands of various pine species in the western United States and central Europe in recent years. Hence, even if the rust fungus itself is not lethal, the damage it does to the overall vigor of the tree can set the stage for lethal attack by a secondary agent.

—————————— **MINIQUIZ** ——————————

- What are some distinguishing features of basidiospores and of zoospores (Section 18.9)?
- Are basidiospores haploid or diploid?

IV • *Archaeplastida*

We conclude our tour of eukaryotic microbial diversity with the **algae**. The kingdom *Archaeplastida* includes both the red and green algae as well as land plants. As we have previously discussed, only the red and green algae originated from primary endosymbiotic events, whereas other protists containing chloroplasts were the result of secondary endosymbioses (Figures 18.2 and 18.3). Here we focus on the red and green algae, a large and diverse group of eukaryotic organisms that contain chlorophyll and carry out oxygenic photosynthesis.

18.14 Red Algae

KEY GENERA: *Polysiphonia, Cyanidium, Galdiera*

The red algae, also called *rhodophytes*, mainly inhabit the marine environment, but a few species are found in freshwater and terrestrial habitats. Both unicellular and multicellular species are known, and some of the latter are macroscopic.

Basic Properties

Red algae are phototrophic and contain chlorophyll *a*; their chloroplasts lack chlorophyll *b* but contain phycobiliproteins, the major light-harvesting pigments of the cyanobacteria (⟳ Section 14.2). The reddish color of many red algae (**Figure 18.31**) results from phycoerythrin, an accessory pigment that masks the green color of chlorophyll. This pigment is present along with phycocyanin and allophycocyanin in structures called *phycobilisomes*, the light-harvesting (antenna) components of cyanobacteria. At greater depths in aquatic habitats, where less light penetrates, cells compensate by producing more phycoerythrin and are a darker red, whereas shallow-dwelling species often have less phycoerythrin and can be green in color (see Figure 18.32).

Most species of red algae are multicellular and lack flagella. Some are considered seaweeds and are the source of agar, the

UNIT 4

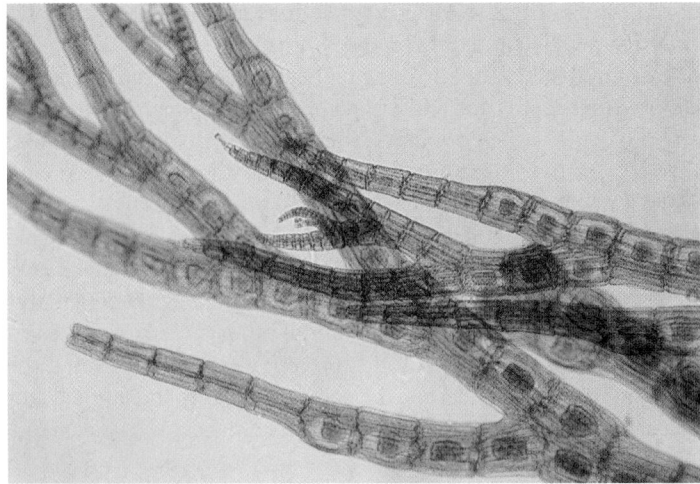

Figure 18.31 *Polysiphonia*, **a filamentous marine red alga.** Light micrograph. *Polysiphonia* grows attached to the surfaces of marine plants. Cells are about 150 μm wide.

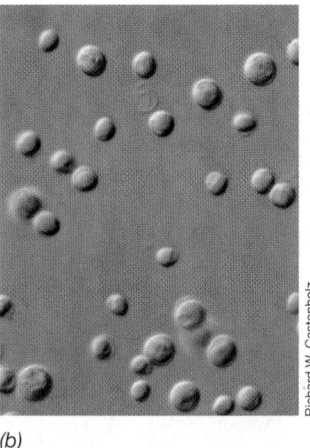

Christine Oesterhelt and Gerald Schönknecht

Richard W. Castenholz

(a)   (b)

Figure 18.32 *Galdieria*, **a unicellular red alga.** *(a) Galdieria* inhabits acidic and sulfidic hot springs where it often grows attached to mineral debris as shown here (arrow). *(b)* Cells of *Galdieria* are about 25 μm in diameter and are more blue-green than red in color because the phototroph contains mainly phycocyanin rather than phycoerythrin as its phycobilin (🔗 Section 14.2).

solidifying agent used in bacteriological culture media, and carrageenans, thickening and stabilizing agents used in the food industry. Other species of red algae, such as the genus *Porphyra*, are harvested, dried, and used as a wrap in sushi. Different species of red algae are filamentous, leafy, or, if they deposit calcium carbonate, *coralline* (coral-like) in morphology. Coralline red algae play an important role in the development of coral reefs and help strengthen reefs against wave damage (🔗 Section 23.11).

Polysiphonia (Figure 18.31) is a genus of filamentous and branched red algae found worldwide in marine environments. The organism is found primarily near shore where the cells grow attached to rocks, other algae, and man-made surfaces such as jetties, concrete abutments, retaining walls, and docks. Nearly 200 species of this genus are recognized, and the organism undergoes a complex life cycle in which an alternation of generations occurs. In this cycle, haploid male and female gametes released from a diploid multicellular organism develop into haploid male and female multicellular organisms. The male alga then releases haploid "sperm" cells that fuse with a specialized reproductive structure on the female alga to yield a diploid zygote; the latter develops a multicellular form and, following meiosis, releases male and female gametes to complete the life cycle.

Cyanidium, Galdieria, and Relatives

In addition to multicellular red algae like *Polysiphonia* (Figure 18.31), unicellular species are also known. One such group, members of the *Cyanidiales* that includes the genera *Cyanidium, Cyanidioschyzon*, and *Galdieria* (**Figure 18.32**), live in hot, acidic and metal-rich hot springs at temperatures from 30 to 60°C and acidic pH (0.5 to 4.0); under these extreme conditions, no other phototrophic microorganisms (including anoxygenic phototrophs) can exist. The unicellular red algae are unusual in other ways as well. For example, cells of *Cyanidioschyzon merolae* are unusually small (1–2 μm in diameter) for eukaryotes, and the genome of this species, approximately 16.5 Mbp, is one of the smallest genomes known for a phototrophic eukaryote.

Molecular analyses of the *Galdieria* genome have revealed the surprising result that this *eukaryotic* phototroph contains at least 75 genes acquired by horizontal transfer from various *prokaryotic* sources. Some of the key genes transferred encode protections against salt stress and metal toxicity and the construction of a toughened cytoplasmic membrane to withstand the heat and acidity of the *Galdieria* habitat (Figure 18.32*a*). The ability of *Galdieria* to grow in darkness—an unusual feature for an alga—has also been linked to horizontally transferred genes, in particular to genes encoding transport systems for various organic compounds. Although these genetic transfers are unusual in the sense that donor and recipient belong to different phylogenetic domains, they once again underscore the importance of horizontal gene transfer in molding microbial genomes (🔗 Sections 9.6 and 13.6).

--- **MINIQUIZ** ---

- What traits link cyanobacteria and red algae?
- What physiological properties would be necessary for *Galdieria* to live in its habitat?

18.15 Green Algae

KEY GENERA: *Chlamydomonas, Volvox*

The green algae, also called *chlorophytes*, have chloroplasts containing chlorophylls *a* and *b*, which give them their characteristic green color, but they lack phycobiliproteins and so do not develop the red or blue-green colors of red algae (Figures 18.31 and 18.32). In the composition of their photosynthetic pigments, green algae are similar to plants and are closely related to plants phylogenetically. There are two main groups of green algae, the *chlorophytes*, examples of which are the microscopic *Chlamydomonas* and *Dunaliella* (**Figure 18.33***a*), and the charophyceans such as *Chara* (Figure 18.33*b*), macroscopic organisms that often resemble land plants and are actually most closely related to land plants.

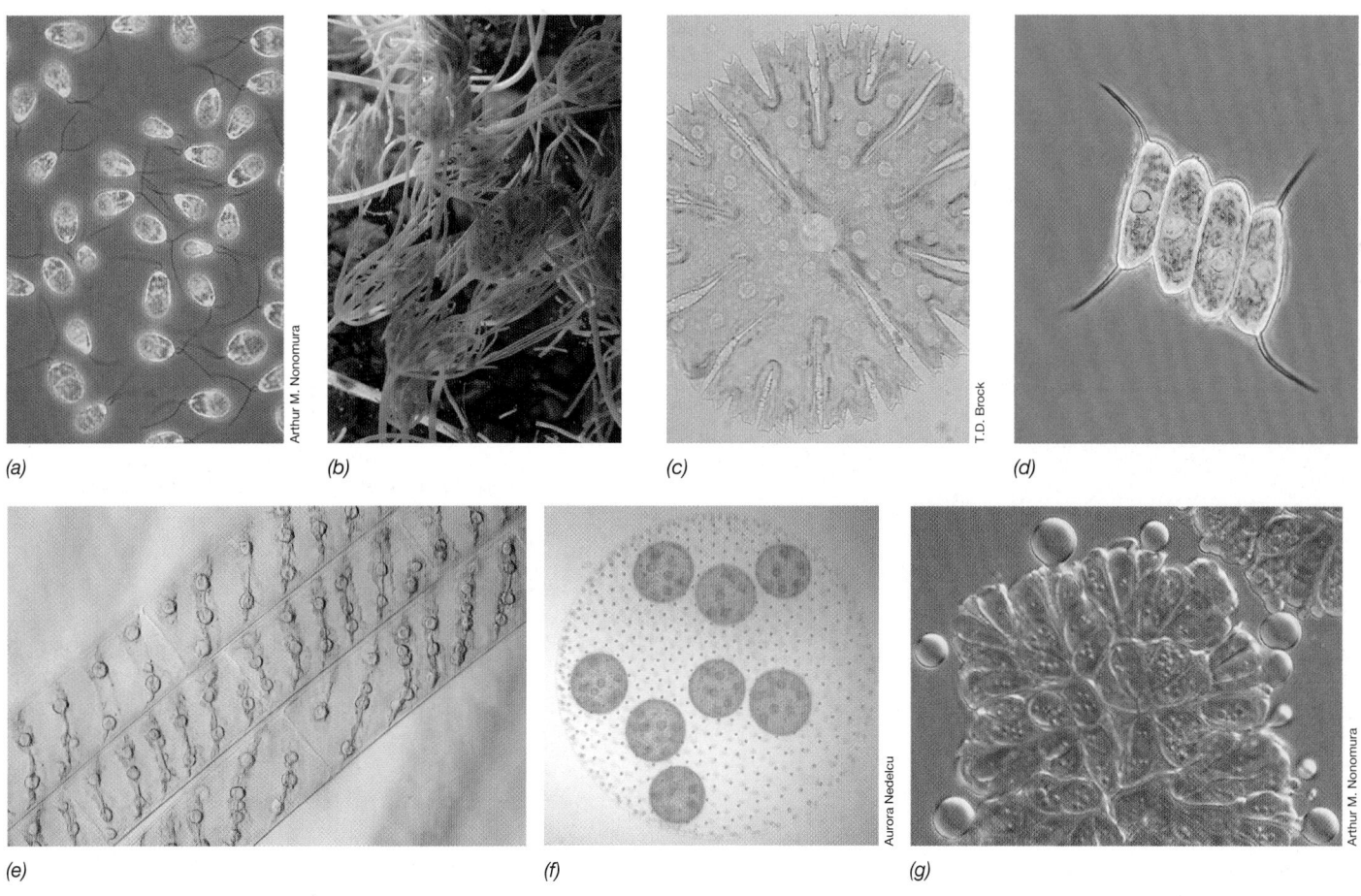

Figure 18.33 Green algae. *(a)* A single-celled, flagellated green alga, *Dunaliella*. A cell is about 5 μm wide. *(b)* The plantlike green alga *Chara*. *(c) Micrasterias*. This single multilobed cell is about 100 μm wide. *(d) Scenedesmus*, showing packets of four cells each. *(e) Spirogyra*, a filamentous alga with cells about 20 μm wide. Note the green spiral-shaped chloroplasts. *(f) Volvox carteri* colony with eight daughter colonies. *(g)* The petroleum-producing green alga *Botryococcus braunii*. Note the excreted oil droplets surrounding the cell.

Most green algae inhabit freshwater while others are found in moist soil or growing in snow, to which they impart a pink color (⮌ Figure 5.20). Other green algae live as symbionts in lichens (⮌ Section 23.1). The morphology of chlorophytes ranges from unicellular (Figure 18.33*a, c*) to filamentous, with individual cells arranged end to end (Figure 18.33*e*), to **colonial**, as aggregates of cells (Figure 18.33*f*). Even multicellular species exist, an example of which is the seaweed *Ulva*. Most green algae have a complex life cycle, with both sexual and asexual reproductive stages.

Very Small Green Algae and Colonial Green Algae

One of the smallest eukaryotes known is the green alga *Ostreococcus tauri*, a common unicellular species of marine phytoplankton (⮌ Section 20.10 and Figure 20.24*b*). Cells of *O. tauri* have a diameter of approximately 2 μm, and the organism contains the smallest genome of any known phototrophic eukaryote, approximately 12.6 Mbp. *Ostreococcus* has thus provided a model organism for research into the evolution of genome reduction and specialization in eukaryotes.

At the colonial level of organization in green algae is *Volvox* (Figure 18.33*f*). This alga forms colonies composed of several hundred flagellated cells, some of which are motile and primarily carry out photosynthesis, while others specialize in reproduction. Cells in a *Volvox* colony are interconnected by thin strands of cytoplasm that allow the entire colony to swim in a coordinated fashion. *Volvox* has been a long-term model for research on the genetic mechanisms controlling multicellularity and the distribution of functions among cells in multicellular organisms.

Some colonial green algae have potential as sources of biofuels. For example, the colonial green alga *Botryococcus braunii* excretes long-chain (C_{30}–C_{36}) hydrocarbons that have the consistency of crude oil (Figure 18.33*g*). About 30% of the *B. braunii* cell dry weight consists of this petroleum, and there has been heightened interest in using this and other oil-producing algae as renewable sources of petroleum. Evidence from biomarker studies have shown that some known petroleum reserves originated from green algae such as *B. braunii* that settled in lakebeds in ancient times. Hence, if the scale-up challenges for commercial algal petroleum production could ever be met, it is possible that some fraction of the world's oil supply could someday come from photosynthesis by green algae.

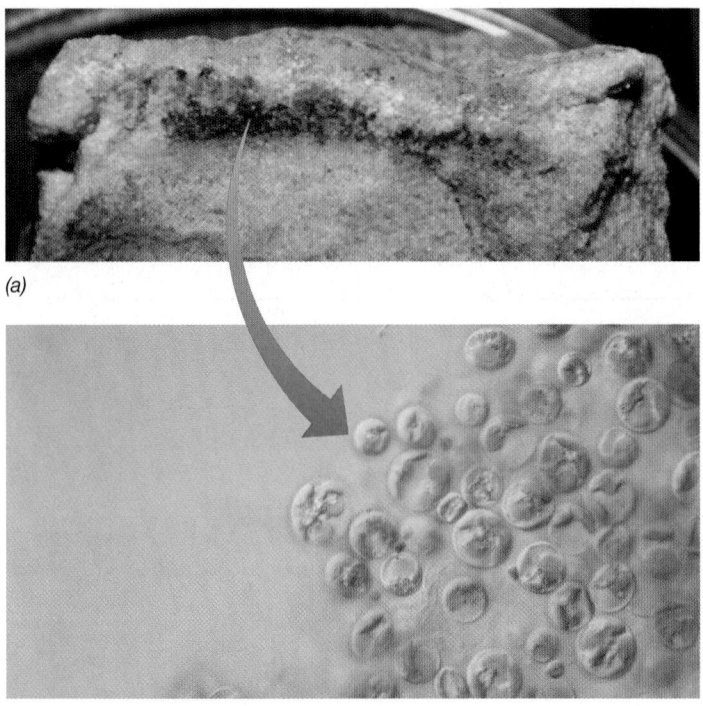

(a)

(b)

Figure 18.34 Endolithic phototrophs. *(a)* Photograph of a limestone rock from the McMurdo Dry Valleys region of Antarctica broken open to show the layer of endolithic green algae. *(b)* Light micrograph of cells of the green alga *Trebouxia*, a widespread endolithic alga in Antarctica.

Endolithic Phototrophs

Some green algae grow inside rocks. These *endolithic* (*endo* means "inside") phototrophs inhabit porous rocks, such as those containing quartz, and are typically found in layers near the rock surface (**Figure 18.34a**). Endolithic phototrophic communities are most common in dry environments such as deserts or cold, dry environments such as Antarctica. For example, in the McMurdo Dry Valleys of Antarctica, where temperatures and humidity are extremely low (⏧ Figure 5.19*d*, *e*), life within a rock has its advantages. Rocks in these harsh environments are heated by the sun, and water from snowmelt can be absorbed and retained for relatively long periods, supplying moisture needed for growth. Moreover, water absorbed by a porous rock makes the rock more transparent, thus funneling more light to the algal layers.

A wide variety of phototrophs can form endolithic communities, including cyanobacteria and various green algae (Figure 18.34*b*). In addition to being free-living phototrophs, green algae and cyanobacteria coexist with fungi in endolithic lichen communities (⏧ Section 23.1 for discussion of the lichen symbiosis). Metabolism and growth of these internal rock microbial communities slowly weathers the rock, allowing gaps to develop where water can enter, freeze and thaw, and eventually crack the rock, producing new habitats for microbial colonization. The decomposing rock also forms a crude soil that can support development of plant and animal communities in environments where conditions (temperature, moisture, and so on) permit.

MINIQUIZ

- What phototrophic properties link green algae and plants?
- What is unusual about the green algae *Ostreococcus*, *Volvox*, and *Botryococcus*?
- What are endolithic phototrophs?

MasteringMicrobiology® **Visualize**, **explore**, and **think critically** with Interactive Microbiology, MicroLab Tutors, MicroCareers case studies, and more. MasteringMicrobiology offers practice quizzes, helpful animations, and other study tools for lecture and lab to help you master microbiology.

Chapter Review

I • Organelles and Phylogeny of Microbial *Eukarya*

18.1 Key metabolic organelles of eukaryotes are the chloroplast, which functions in photosynthesis, and the mitochondrion or hydrogenosome, which function in respiration or fermentation. These organelles were originally *Bacteria* that established permanent residence inside other cells (endosymbiosis).

> **Q** **Distinguish between a primary and a secondary endosymbiosis. Which groups of protists are derived from which form of endosymbiosis?**

18.2 Ribosomal RNA gene sequences do not yield as reliable a phylogenetic tree of the *Eukarya* as do other genes and proteins. The modern, multigene tree of eukaryotes shows a major radiation of eukaryotic diversity emerging at some time following symbiotic events that led to the mitochondrion.

> **Q** **What are the six major supergroups within the *Eukarya*?**

II • Protists

18.3 Diplomonads such as *Giardia* are unicellular, flagellated, nonphototrophic protists. Parabasalids such as *Trichomonas* are human pathogens and contain huge genomes that lack introns. Euglenids and kinetoplastids are unicellular, flagellated protists. Some are phototrophic. This group includes some important human pathogens, such as *Trypanosoma*, and some well-studied nonpathogens, such as *Euglena*.

> **Q** What morphological feature unites kinetoplastids and euglenids?

18.4 Three groups make up the alveolates: ciliates, dinoflagellates, and apicomplexans. Most ciliates and dinoflagellates are free-living organisms, whereas apicomplexans are obligate parasites of animals.

> **Q** What organism causes "red tides," and why is this organism toxic?

18.5 *Stramenopiles* are protists that bear a flagellum with fine, hairlike extensions. They include oomycetes, diatoms, and brown and golden algae.

> **Q** In terms of their photosynthetic pigments, how are brown and golden algae similar?

18.6 The *Rhizaria* include diverse protists such as the phototrophic chlorarachniophytes and foraminiferans, as well as radiolarians, which are chemoorganotrophs.

> **Q** What morphological trait unite the *Rhizaria* and distinguish them from other protists?

18.7 *Amoebozoa* are protists that use pseudopodia for movement and feeding. Within *Amoebozoa* are gymnamoebas, entamoebas, and slime molds. Plasmodial slime molds form masses of motile protoplasm, whereas cellular slime molds are individual cells that aggregate to form fruiting bodies from which spores are released.

> **Q** Explain the major differences between the slime molds *Dictyostelium* and *Physarum*.

III • Fungi

18.8 Fungi include the molds, mushrooms, and yeasts. Other than phylogeny, fungi primarily differ from protists by their rigid cell wall, production of spores, and lack of motility.

> **Q** What is the major difference between a mold and a yeast?

18.9 A variety of sexual spores are produced by fungi, including ascospores, basidiospores, and zygospores. From a phylogenetic standpoint, fungi are the closest relatives of animals.

> **Q** List the different types of sexual spores of fungi. Are conidia sexual or asexual spores?

18.10 *Chytridiomycota* and *Microsporidia* are basal to all other known fungal groups in the fungal phylogeny. Some chytrids are amphibian pathogens, while microsporidians are parasites of animals.

> **Q** In what way do chytrids and microsporidians differ from other fungi?

18.11 *Zygomycota* form coenocytic hyphae and undergo both asexual and sexual reproduction, and the common bread mold *Rhizopus* is a good example. *Glomeromycota* are fungi that form endomycorrhizal associations with plants.

> **Q** What is the major feature of the ecology of glomeromycetes?

18.12 The *Ascomycota* are a large and diverse group of mostly saprophytic fungi. Some, such as *Candida albicans*, can be pathogenic in humans. There are two mating types in the yeast *Saccharomyces cerevisiae*, and yeast cells can convert from one type to the other by a genetic switch mechanism.

> **Q** How is the mating type of a yeast cell determined?

18.13 *Basidiomycota* include the mushrooms, puffballs, smuts, and rusts. Basidiomycetes undergo both vegetative reproduction as haploid mycelia and sexual reproduction via fusion of mating types and formation of haploid basidiospores.

> **Q** What morphological feature unites the *Basidiomycota*, and where is this feature found?

IV • *Archaeplastida*

18.14 Red algae are mostly marine and range from unicellular to multicellular. Their reddish color is due to the pigment phycoerythrin, a key cyanobacterial pigment, present in their chloroplast.

> **Q** In what kinds of habitats would one likely find red algae?

18.15 Green algae are common in aquatic environments and can be unicellular, filamentous, colonial, or multicellular. A unicellular green alga, *Ostreococcus*, has the smallest genome known for a phototrophic eukaryote, while the green alga *Volvox* is a model colonial phototroph.

> **Q** What traits link green algae and plants?

Application Questions

1. Explain why the process of endosymbiosis can be viewed as both an ancient event and a more recent event. What advantages could endosymbiosis give to both the endosymbiont and the host?

2. Summarize the evidence for endosymbiosis. How could the endosymbiotic hypothesis have originated before the era of molecular biology? How has molecular biology supported the theory?

3. Considering all of the groups of microbes covered in this chapter, which of them seem most similar to prokaryotic cells in cell structure? Discuss at least two lines of evidence that show that these microbes are not prokaryotic cells.

Chapter Glossary

Algae an informal term for phototrophic eukaryotes other than plants; algae are polyphyletic, being found in diverse groups of *Eukarya* as a result of secondary endosymbioses

Chitin a polymer of *N*-acetylglucosamine commonly found in the cell walls of fungi

Ciliate any protist characterized in part by rapid motility driven by numerous short appendages called cilia

Coenocytic the presence of multiple nuclei in fungal hyphae without septa

Colonial the growth form of certain protists and green algae in which several cells live together and cooperate for feeding, motility, or reproduction; an early form of multicellularity

Conidia the asexual spores of fungi

Endosymbiotic hypothesis the concept that a respiratory bacterium and a cyanobacterium were stably incorporated into another cell type to yield the mitochondria and chloroplasts, respectively, of eukaryotic cells

Fungi nonphototrophic eukaryotic microorganisms with rigid cell walls

Mushroom the aboveground fruiting body, or basidiocarp, of basidiomycete fungi

Nucleomorphs remnant nuclei found in certain algae; derived from an ancestral algal endosymbiont and associated with chloroplasts acquired by secondary endosymbiosis

Phagocytosis a mechanism for ingesting particulate material in which a portion of the cytoplasmic membrane surrounds the particle and brings it into the cell

Protist an informal term used to describe any unicellular eukaryotic microorganism whether heterotrophic or photosynthetic

Secondary endosymbiosis the endosymbiotic acquisition by a mitochondrion-containing eukaryotic cell of a red or green algal cell, which itself contains a chloroplast derived from primary endosymbiosis

Slime mold a nonphototrophic protist that lacks cell walls and that aggregates to form fruiting structures (cellular slime molds) or masses of protoplasm (acellular slime molds)

Yeast the single-celled growth form of various fungi

UNIT 4

Taking the Measure of Microbial Systems

microbiology**now**

The Vineyard Microbiome Revealed by Next-Generation Sequencing Technology

It has been recognized for millennia that the quality of wine produced from the same variety of grape can vary dramatically depending on where the grapevines are grown. The French refer to the influence of location on wine characteristics as *terroir* (from the French *terre*, meaning "land"). The influence of locale on wine quality is the primary basis for defining the appellation boundaries for vineyards within major growing regions. The Burgundy region in France, for example, has dozens of such appellations. It has long been considered that *terroir* is a function of climate and the physical and chemical characteristics of the local soil. However, a recent study using next-generation sequencing of 16S rRNA genes to determine the diversity and relative abundance of different microorganisms on grape plants and in soils from major wine-growing areas in the United States and France has revealed that *terroir* may result from a combination of microbial and abiotic factors.

Distinct microbial communities are imprinted on the grapes from different winegrowing regions, even for the same grape variety. Next-generation sequencing technology can unravel these microbial communities and define their species composition and abundance. Modern sequencing efforts have shown that the soil microbial community can vary significantly between appellations and that soil is the major source of the microbiota found on grapevine leaves and fruit. This suggests that *terroir* is influenced not only by soil physics and chemistry and the prevailing climate, but also by the local soil microbial community. Apart from a direct contribution soil microbes might make in wine fermentation by influencing the flavor, color, and quality of the final product, differences in the soil microbial community may also affect the physiology of the plant, for example, by controlling the production of specific plant compounds that alter grape chemistry.

These initial applications of next-generation sequencing to the study of *terroir* highlight the growing importance of modern microbial ecology in agriculture. As our understanding of plant–soil–microbial interactions develops, a soil microbial census could well become a valuable tool for improving soil quality and crop performance by ensuring that each crop plant has the best possible microbial associates.

Source: Zarraonaindia, I., et al. 2015. The soil microbiome influences grapevine-associated microbiota. *MBio:* 6 e02527-14.

We now begin a new unit devoted to microorganisms in their natural habitats. We learned in Chapter 1 that *microbial communities* consist of cell populations living in association with other populations in nature. The science of **microbial ecology** is focused on how microbial populations assemble to form communities and how these communities interact with each other and their environments.

The major components of microbial ecology are *biodiversity* and *microbial activity*. To study biodiversity, microbial ecologists must identify and quantify microorganisms in their habitats. Knowing how to do this is often helpful for isolating organisms of interest as well, which is another goal of microbial ecology. To study microbial activity, microbial ecologists must measure the metabolic processes that microorganisms carry out in their habitats. In this chapter we consider modern methods for assessing microbial diversity and activity. Chapter 20 will outline the basic principles of microbial ecology and examine the types of environments that microorganisms inhabit. Chapters 21–24 will complete our coverage of microbial ecology by exploring nutrient cycles, applied microbiology, and the role microbes play in symbiotic associations with other life forms, including humans.

We begin with the microbial ecologist's toolbox, which includes a collection of powerful tools for dissecting the structure and function of microbial communities in relation to their natural habitats.

I • Culture-Dependent Analyses of Microbial Communities

The vast majority of microorganisms, more than 99% of all species by most estimates, have never been grown in laboratory cultures. Recognition of this fact, based on molecular diversity surveys (Sections 19.4–19.8) of various microbial habitats, has stimulated the development of new methods for isolating microbes from nature in order to establish pure cultures. Even though a host of sophisticated methods are available for studying microbes in their native environments, culturing a microorganism remains the only way to fully characterize its properties and predict its impact on its environment.

In the first part of this chapter we cover the enrichment approach, a time-honored and useful method for isolating microorganisms from nature but one with limitations. Enrichment is based on culturing in a selective growth medium, and thus the tools and methods used in this approach are considered *culture-dependent* analyses. As we will see, considerable progress has been made in culturing the more elusive microorganisms in natural populations by using robotics and associated microfabrication technology to establish large numbers of enrichment cultures that can be monitored simultaneously. In the second and third parts of this chapter we consider *culture-independent* analyses, techniques that can tell us much about the structure and function of microbial communities in the absence of actual laboratory cultures. In the final part of this chapter, we consider methods for measuring microbial activities in nature and linking them to specific organisms. Collectively, these methods allow the microbial ecologist to ask both "who's there" and "what are they doing."

19.1 Enrichment Culture Microbiology

For an **enrichment culture**, a medium and a set of incubation conditions are established that are *selective* for the desired organism and *counterselective* for undesired organisms. Effective enrichment cultures duplicate as closely as possible the resources and conditions of a particular ecological niche. Hundreds of different enrichment strategies have been devised, and **Tables 19.1** and **19.2** summarize some simple and direct ones.

Inocula

Successful enrichment requires an appropriate inoculum containing the organism of interest. Thus, the making of an enrichment culture begins with collecting a sample from the appropriate habitat to serve as the inoculum (Tables 19.1 and 19.2). Enrichment cultures are established by placing the inoculum into selective media and incubating under specific conditions. In this way, many common microbes can be isolated. For example, the great Dutch microbiologist Martinus Beijerinck, who conceptualized the enrichment culture technique (⇌ Section 1.11), used enrichment cultures to isolate the nitrogen-fixing bacterium *Azotobacter* (**Figure 19.1**). Because *Azotobacter* is a rapidly growing bacterium capable of N_2 fixation in air (⇌ Sections 14.6 and 15.12), enrichment using media devoid of fixed nitrogen, such as ammonia or nitrate, and

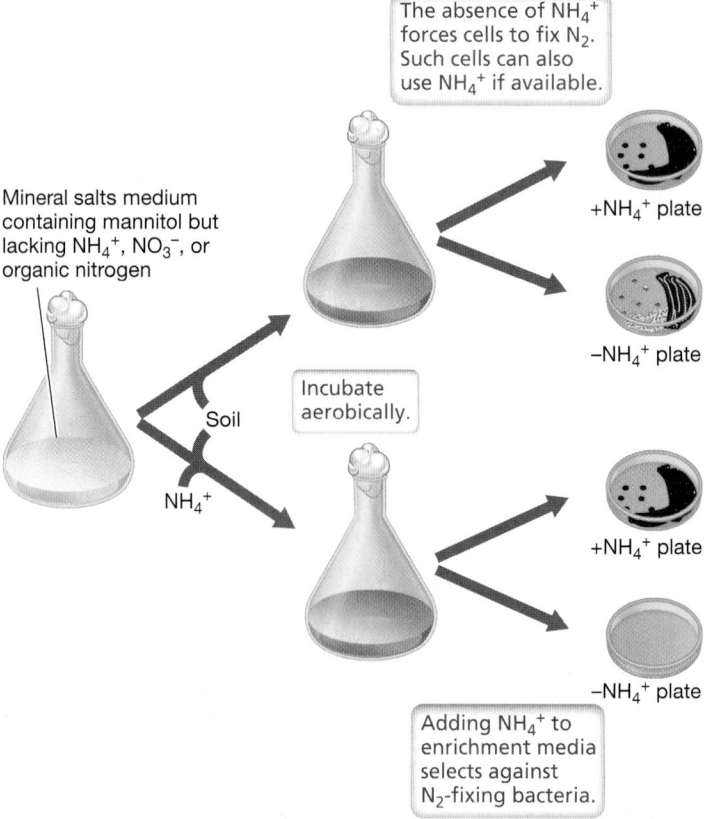

Figure 19.1 The isolation of *Azotobacter*. Selection for aerobic nitrogen-fixing bacteria usually results in the isolation of *Azotobacter* or its relatives. The selective basis of the enrichment is the absence of fixed nitrogen (NH_4^+ in this case) in the culture medium in the upper flask. Thus the medium *selects* from the microbial community those species that can fix N_2 aerobically, of which *Azotobacter* is one of the most rapidly growing. See Section 1.11 and Figure 1.33 for more on the historical importance of *Azotobacter*.

TABLE 19.1 Some enrichment culture methods for phototrophic and chemolithotrophic bacteria

Light-phototrophic bacteria: main C source, CO_2		
Incubation condition	*Organisms enriched*	*Inoculum*
Incubation in air		
N_2 as nitrogen source	Cyanobacteria	Pond or lake water; sulfide-rich muds; stagnant water; raw sewage; moist, decomposing leaf litter; moist soil exposed to light
NO_3^- as nitrogen source, 55°C	Thermophilic cyanobacteria	Hot spring microbial mat
Anoxic incubation		
H_2 or organic acids; N_2 as sole nitrogen source	Purple nonsulfur bacteria, heliobacteria	Same as above plus hypolimnetic lake water (⌐⊃ Section 20.8); pasteurized soil (heliobacteria); microbial mats for thermophilic species
H_2S as electron donor	Purple and green sulfur bacteria	
Fe^{2+}, NO_2^- as electron donor	Purple bacteria	

Dark-chemolithotrophic bacteria: main C source, CO_2 (medium must lack organic C)			
Electron donor	*Electron acceptor*	*Organisms enriched*	*Inoculum*
Incubation in air: aerobic respiration			
NH_4^+	O_2	Ammonia-oxidizing *Bacteria* (*Nitrosomonas*) or *Archaea* (*Nitrosopumilus*)	Soil, mud, sewage effluent, seawater
NO_2^-	O_2	Nitrite-oxidizing bacteria (*Nitrobacter, Nitrospira*)	
H_2	O_2	Hydrogen bacteria (various genera)	
H_2S, S^0, $S_2O_3^{2-}$	O_2	*Thiobacillus* spp.	
Fe^{2+}, low pH	O_2	*Acidithiobacillus ferrooxidans*	
Anoxic incubation			
S^0, $S_2O_3^{2-}$	NO_3^-	*Thiobacillus denitrificans*	Mud, lake sediments, soil
H_2	NO_3^-	*Paracoccus denitrificans*	
Fe^{2+}, neutral pH	NO_3^-	*Acidovorax* and various other gram-negative autotrophic bacteria	

incubation in air selects strongly for this bacterium and its close relatives. Non-nitrogen-fixing bacteria and anaerobic nitrogen-fixing bacteria are counterselected in this technique.

Enrichment Culture Outcomes

For success with enrichment cultures, attention to both the culture medium and the incubation conditions is important. That is, the *resources* (nutrients) and *conditions* (temperature, pH, osmotic considerations, aerobic or anaerobic, and the like) must closely mimic those of the habitat to offer the best chance of obtaining the organism of interest (⌐⊃ Table 20.1).

Some enrichment cultures yield nothing. This may be because the organism capable of growing under the enrichment conditions specified is absent from the habitat. Alternatively, even though the organism of interest exists in the habitat sampled, the resources and conditions employed in the enrichment may simply be incompatible with its growth. Thus enrichment cultures can yield a firm *positive* conclusion (that is, that an organism with certain capacities exists in a particular environment because it was enriched) but never a firm *negative* conclusion (that such an organism is not present because the enrichment failed). Moreover, the isolation of the desired organism from an enrichment culture says nothing about the ecological importance or abundance of the organism in its habitat. A positive enrichment proves only that

the organism was present in the sample, and in practice, this can result from as few as a single viable cell.

The Winogradsky Column

The **Winogradsky column** is an artificial microbial ecosystem and a long-term source of various bacteria for enrichment cultures. Winogradsky columns have been used to isolate phototrophic purple and green bacteria, sulfate-reducing bacteria, and many other anaerobes. Named for the famous Russian microbiologist Sergei Winogradsky (⌐⊃ Section 1.11), the column was first used by Winogradsky in the late nineteenth century in his classic studies of soil microorganisms.

A Winogradsky column is prepared by filling a glass cylinder about half full with organically rich, preferably sulfidic mud into which carbon substrates have been mixed. The substrates determine which organisms are enriched. Fermentative substrates, such as glucose, that can lead to acidic conditions and excessive gas formation (which can create gas pockets that disrupt the enrichment and let in air) are avoided. The mud is supplemented with small amounts of calcium carbonate ($CaCO_3$) as a buffer and gypsum ($CaSO_4$) as a source of sulfate. The mud is packed tightly in the cylinder, taking care to avoid trapping air, and then covered with lake, pond, or ditch water (or seawater if it is a marine column). The top of the cylinder is covered to prevent evaporation, and the container is placed near a window that receives diffuse sunlight for a period of months.

UNIT 5

TABLE 19.2 Some enrichment culture methods for chemoorganotrophic and strictly anaerobic bacteria[a]

Electron donor (and nitrogen source)	Electron acceptor	Typical organisms enriched	Inoculum
Incubation in air: aerobic respiration			
Lactate + NH_4^+	O_2	*Pseudomonas fluorescens*	Soil, mud; lake sediments; decaying vegetation; pasteurize inoculum (80°C for 15 min) for all *Bacillus* enrichments
Benzoate + NH_4^+	O_2	*Pseudomonas fluorescens*	
Starch + NH_4^+	O_2	*Bacillus polymyxa*, other *Bacillus* spp.	
Ethanol (4%) + 1% yeast extract, pH 6.0	O_2	*Acetobacter, Gluconobacter*	
Urea (5%) + 1% yeast extract	O_2	*Sporosarcina ureae*	
Hydrocarbons (e.g., mineral oil, gasoline, toluene) + NH_4^+	O_2	*Mycobacterium, Nocardia, Pseudomonas*	
Cellulose + NH_4^+	O_2	*Cytophaga, Sporocytophaga*	
Mannitol or benzoate, N_2 as N source	O_2	*Azotobacter*	
$CH_4 + NO_3^-$	O_2	*Methylobacter, Methylomicrobium*	Lake sediments, thermocline (⇨ Section 20.8) of stratified lake
Anoxic incubation: anaerobic respiration			
Organic acids	NO_3^-	*Pseudomonas* (denitrifying species)	Soil, mud; lake sediments
Yeast extract	NO_3^-	*Bacillus* (denitrifying species)	
Organic acids	SO_4^{2-}	*Desulfovibrio, Desulfotomaculum*	
Acetate, propionate, butyrate	SO_4^{2-}	Fatty acid-oxidizing sulfate reducers	As above; or sewage digester sludge; rumen contents; marine sediments
Acetate, ethanol	S^0	*Desulfuromonas*	
Acetate	Fe^{3+}	*Geobacter, Geospirillum*	
Acetate	ClO_3^-	Various chlorate-reducing bacteria	
H_2	CO_2	Methanogens (chemolithotrophic species only), homoacetogens	Mud, sediments, sewage sludge
CH_3OH	CO_2	*Methanosarcina barkeri*	
CH_3NH_2 or CH_3OH	NO_3^-	*Hyphomicrobium*	
Hydrocarbons	SO_4^{2-} or NO_3^-	Anoxic hydrocarbon-degrading bacteria	Freshwater or marine sediments
Acetate + H_2 + NH_4^+	Tetrachloroethene (PCE)	*Dehalococcoides* spp.	PCE-polluted groundwater
Anoxic incubation: fermentation			
Glutamate or histidine	No exogenous electron acceptors added	*Clostridium tetanomorphum* or other proteolytic *Clostridium* species	Mud, lake sediments; rotting plant or animal material; dairy products (lactic and propionic acid bacteria); rumen or intestinal contents (enteric bacteria); sewage sludge; soil; pasteurize inoculum for *Clostridium* enrichments
Starch + NH_4^+	None	*Clostridium* spp.	
Starch + N_2 as N source	None	*Clostridium pasteurianum*	
Lactate + yeast extract	None	*Veillonella* spp.	
Glucose or lactose + NH_4^+	None	*Escherichia, Enterobacter*, other fermentative organisms	
Glucose + yeast extract (pH 5)	None	Lactic acid bacteria (*Lactobacillus*)	
Lactate + yeast extract	None	Propionic acid bacteria	
Succinate + NaCl	None	*Propionigenium*	
Oxalate	None	*Oxalobacter*	
Acetylene	None	*Pelobacter* and other acetylene fermenters	

[a]All media must contain an assortment of mineral salts including N, P, S, Mg^{2+}, Mn^{2+}, Fe^{2+}, Ca^{2+}, and other trace elements (⇨ Sections 3.1–3.2). Certain organisms may have requirements for vitamins or other growth factors. This table is meant as an overview of enrichment methods and does not speak to the effect incubation temperature might have in isolating thermophilic (high temperature), hyperthermophilic (very high temperature), and psychrophilic (low temperature) species, or the effect that extremes of pH or salinity might have, assuming an appropriate inoculum was available. Some enrichment substrates are naturally more specific than others. For example, glucose is quite nonspecific as an enrichment substrate compared with benzoate or methanol.

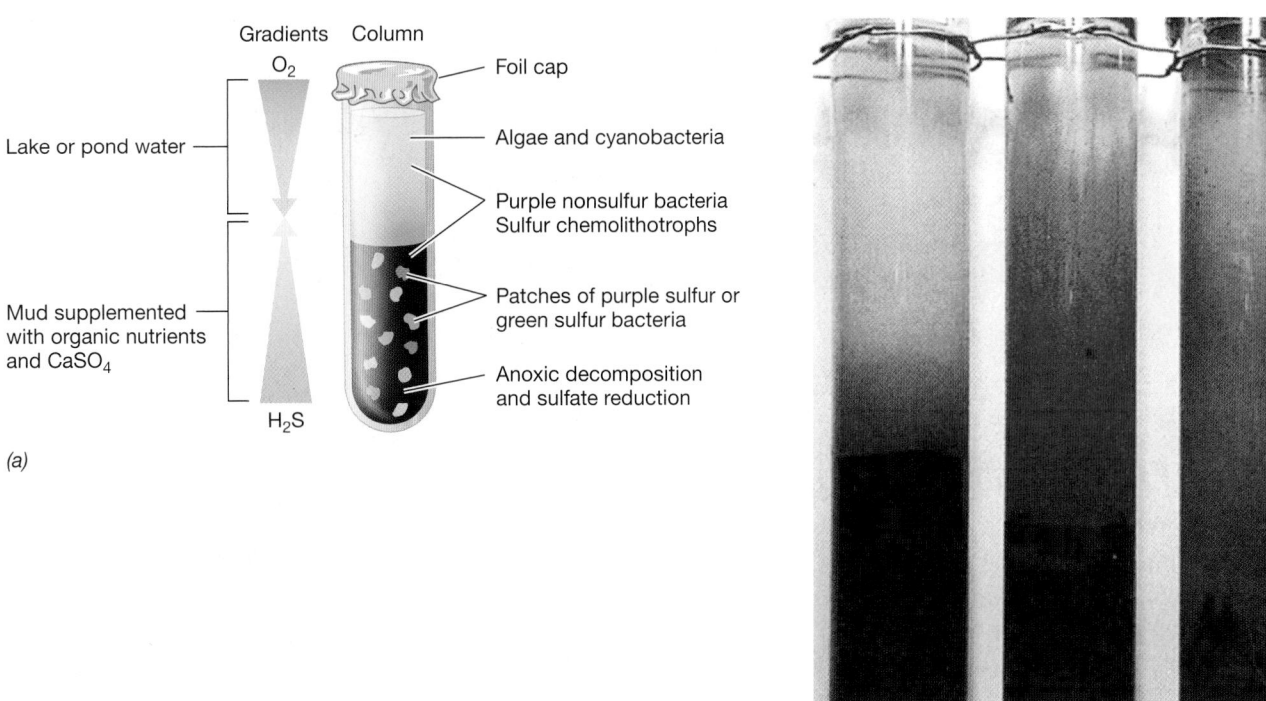

Figure 19.2 The Winogradsky column. *(a)* Schematic view of a typical column used to enrich phototrophic bacteria. The column is incubated in a location that receives subdued sunlight. Anoxic decomposition leading to SO_4^{2-} reduction creates the gradient of H_2S. *(b)* Photo of Winogradsky columns that have remained anoxic up to the top; each column had a bloom of a different phototrophic bacterium. Left to right: *Thiospirillum jenense*, *Chromatium okenii*, both of which are purple sulfur bacteria, and *Chlorobium limicola* (green sulfur bacterium).

In a typical Winogradsky column, a diverse community of microbes develops (**Figure 19.2a**). Algae and cyanobacteria develop quickly in the upper portions of the water column; by producing O_2 these organisms help to keep this zone of the column oxic much as they do in the upper zones of a lake. Fermentative processes in the mud lead to the production of organic acids, alcohols, and H_2, suitable substrates for sulfate-reducing bacteria (⟳ Section 14.14). Hydrogen sulfide (H_2S) from the sulfate reducers triggers the development of purple and green sulfur bacteria (anoxygenic phototrophs, ⟳ Sections 14.3 and 15.4–15.8) that use sulfide as a photosynthetic electron donor. These organisms typically grow in patches in the mud on the sides of the column but may bloom in the water itself if oxygenic phototrophs are scarce (Figure 19.2b). The pigmented cells of the anoxygenic phototrophs can be sampled with a pipette for microscopy, isolation, and characterization (Table 19.1).

Winogradsky columns have been used to enrich both aerobic and anaerobic *Bacteria* and *Archaea*. Besides supplying a ready source of inocula for enrichment cultures, columns can also be supplemented with a specific compound to enrich an organism in the inoculum that can degrade it. Once a crude enrichment has been established in the column, culture media can be inoculated for the isolation of pure cultures, as discussed in Section 19.2.

Enrichment Bias

Although the enrichment culture technique is quite useful and still widely practiced, there exists a bias, and sometimes a very severe bias, in the outcome of enrichments. This bias is typically most profound in liquid enrichment cultures where the most rapidly growing organism(s) for the chosen set of conditions dominate. However, using molecular techniques to be described later, we now know that the most rapidly growing organisms in laboratory cultures are often only minor components of the microbial community rather than the most abundant and ecologically relevant organisms carrying out the process of interest. This could be for several reasons including the fact that the levels of resources available in laboratory cultures are typically much higher than those in nature, and the conditions in the natural habitat, including both the types and proportions of different organisms present as well as the physical and chemical conditions, are nearly impossible to reproduce and maintain for long periods in laboratory cultures.

This problem of **enrichment bias** can be demonstrated by comparing the results obtained in dilution cultures (Section 19.2) with classical liquid enrichment. Dilution of an inoculum followed by liquid enrichment or plating often yields different organisms than liquid enrichments established with the same but undiluted inocula. It is thought that dilution of the inoculum eliminates quantitatively insignificant but rapidly growing "weed" species, allowing development of organisms that are more abundant in the community but slower growing. Dilution of the inoculum is thus a common practice in enrichment culture microbiology today. As discussed below, the problem of overgrowth by "weed" species can also be circumvented by physical isolation of the desired organism before introducing it into a growth medium. This can be accomplished by dilution and a variety of classical isolation procedures that we turn to in the next section. However, more recently, sophisticated methods have been developed to physically isolate single cells of interest (or a single type of cells) and place them in a growth medium that is free of undesired cells. We consider these techniques in Section 19.3.

MINIQUIZ

- Describe the enrichment strategy behind Beijerinck's isolation of *Azotobacter*.
- Why is sulfate (SO_4^{2-}) added to a Winogradsky column?
- What is enrichment bias? How does dilution reduce enrichment bias?

19.2 Classical Procedures for Isolating Microbes

Once a positive enrichment culture has been obtained, the next step is typically to attempt to get the enriched organism in *pure culture*—one containing a single kind of microorganism. Pure cultures are valuable because genomes can be quickly dissected and experiments done under controlled laboratory conditions to clearly define the physiology of the isolate. Pure cultures have been studied since the days of Robert Koch (⇔ Section 1.10) and we considered some of these methods earlier (⇔ Section 5.5).

Agar Dilution Tubes and the Most-Probable-Number Technique

Common isolation procedures include the streak plate, agar dilution, and liquid dilution. For organisms that form colonies on agar plates, the streak plate is quick, easy, and the method of choice (**Figure 19.3**); if a well-isolated colony is selected and restreaked several successive times, a pure culture is usually obtained. With proper incubation facilities (for example, anoxic jars or anoxic chambers for anaerobes, ⇔ Section 5.14), it is possible to purify both aerobes and anaerobes on agar plates by the streak plate method.

In the agar dilution tube method, a mixed culture is diluted in tubes of molten agar medium, resulting in colonies embedded in the agar. This method is useful for purifying anaerobic organisms such as phototrophic sulfur bacteria and sulfate-reducing bacteria from samples taken from Winogradsky columns or other sources. A culture is purified by successive dilutions of cell suspensions in tubes of molten agar medium (Figure 19.3, ⇔ Figure 15.23*g*). Repeating this procedure using a colony from the highest-dilution tube as inoculum for a new set of dilutions usually yields pure cultures. A related procedure called the *roll tube method* uses tubes containing a thin layer of agar on their inner surface. The agar can then be streaked for isolated colonies. Because the tubes can be flushed with an oxygen-free gas during streaking, the roll tube method is primarily used for the isolation of anaerobic microbes.

Another purification procedure is the serial dilution of an inoculum in a liquid medium until the final tube in the series shows no growth. When a 10-fold serial dilution is used, for example, the last tube showing growth should have originated from ten or fewer cells. Besides being a method for obtaining pure cultures, serial dilution techniques are widely used to estimate viable cell numbers in the **most-probable-number (MPN) technique**

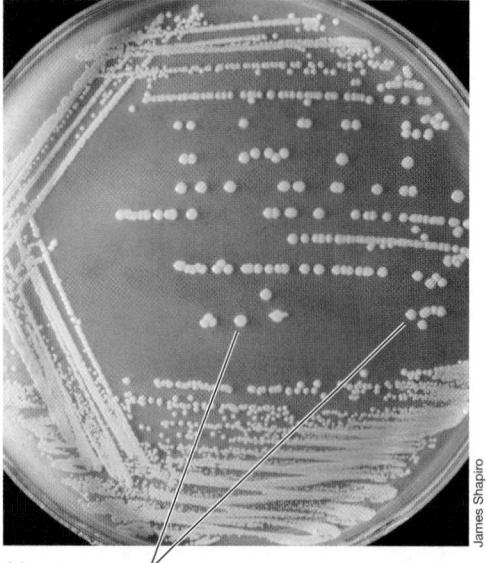

(a)

Colonies   Paraffin–mineral oil seal

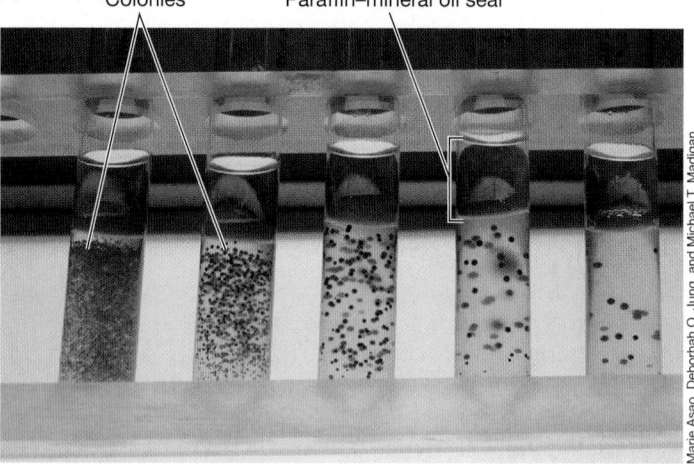

(b)

Figure 19.3 Pure culture methods. *(a)* Organisms that form distinct colonies on plates are usually easy to purify. *(b)* Colonies of phototrophic purple bacteria in agar dilution tubes; the molten agar was cooled to approximately 45°C before inoculation. A dilution series was established from left to right, eventually yielding well-isolated colonies. The tubes were sealed with a 1:1 mixture of sterile paraffin and mineral oil to maintain anaerobiosis.

(**Figure 19.4**). MPN methods have been used for estimating the numbers of microorganisms in foods, wastewater, and other samples in which cell numbers need to be assessed routinely. An MPN count of a natural sample can be done using highly selective media and incubation conditions to target one or a small group of organisms or a particular pathogen. Alternatively, a count can be done using complex media to get a general estimate of viable cell numbers (but see Section 5.7 for a caveat that applies to such estimates). Use of several replicate tubes at each dilution improves accuracy of the final MPN obtained.

Criteria for Culture Purity

Regardless of the methods used to purify a culture, once a putative pure culture has been obtained, it is essential to verify its purity. This is typically done through a combination of (1) microscopy, (2) observation of colony characteristics on plates or in dilution

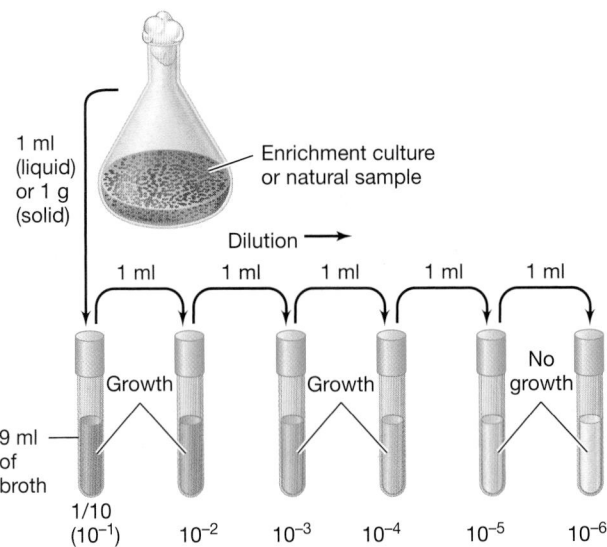

Figure 19.4 Procedure for a most-probable-number (MPN) analysis. Growth in the 10^{-4} but not the 10^{-5} dilution means that cell numbers were at least 10^4 cells/ml in the sample used for inoculation. Because particle-attached microorganisms can skew numbers significantly, gentle methods to disassociate microorganisms from particles are often used prior to dilution. In addition, each dilution tube is mixed thoroughly before removing a sample for the next dilution.

tubes, and (3) tests of the culture for growth in other media. In the latter, it is important to test the culture for growth in media and under growth conditions in which the desired organism is predicted to grow poorly or not at all but in which contaminants will grow vigorously. In the final analysis, the microscopic observation of a single morphological type of cell that displays uniform staining characteristics (for example, in a Gram stain) coupled with uniform colony characteristics and the absence of contamination in growth tests with various culture media is strong evidence that a culture is a pure (*axenic*) culture.

Certain molecular methods described in this chapter for characterizing natural microbial communities can also be applied to the verification of culture purity. However, these techniques are complementary and do not substitute for the more fundamental observations of culture characteristics and cellular morphology.

MINIQUIZ

- What is a pure culture and why is obtaining one useful in microbial ecology?
- How does the agar dilution method differ from streaking to obtain isolated colonies?

19.3 Selective Single-Cell Isolation: Laser Tweezers, Flow Cytometry, Microfluidics, and High-Throughput Methods

The problem of enrichment bias has fueled the development of new methods for culturing microbes from nature. These advancements have emerged from the understanding that every microbe has a *fundamental niche* and a *realized niche*.

The **fundamental niche** refers to the range of environments in which a species will be sustained when it is not resource-limited, such as may result from competition with other species. By contrast, the **realized niche** refers to the range of natural environments supporting a species when it is confronted with factors such as resource limitation, predation, and competition from other species.

Establishing laboratory conditions that fall within the fundamental niche may be sufficient to support an organism once it is in pure culture but may fail to selectively enrich the same organism from a natural sample. Because the realized niche of most microorganisms is unknown, there has been an increasing emphasis on developing methods that physically isolate single cells into separate compartments free from competition with other microbes. These include both manual and robotic methods that function to sort individual cells from an environmental sample, and we consider these methods now.

Laser Tweezers and Flow Cytometry

Laser tweezers consist of an inverted light microscope equipped with a strongly focused infrared laser and a micromanipulation device. Trapping a single cell is possible because the laser beam creates a force that pushes down on a microbial cell (or other small object) and holds it in place (**Figure 19.5a**). Then when the laser beam is moved, the trapped cell moves along with it. If a mixed sample is in a capillary tube, a single cell can be optically trapped

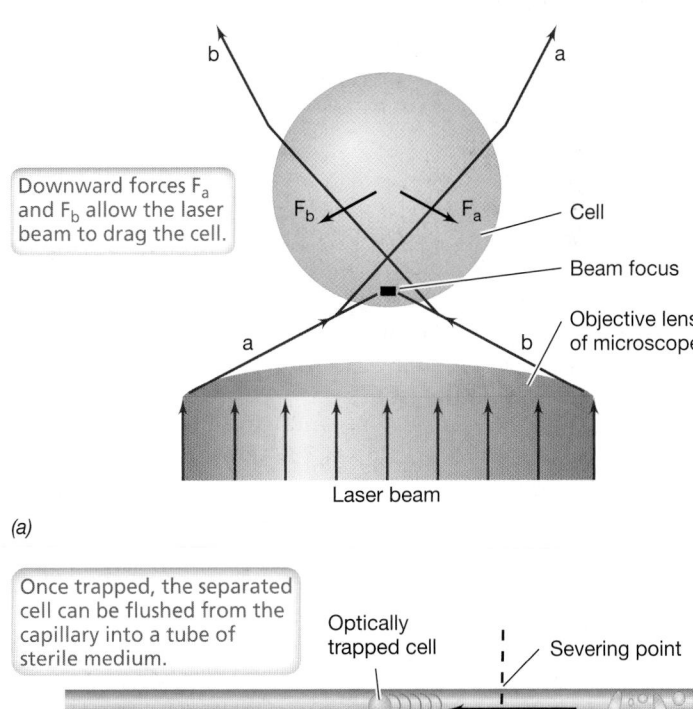

Figure 19.5 The laser tweezers for the isolation of single cells. (a) Mechanism by which individual cells can be isolated. (b) Once a cell has been isolated in a capillary tube, it can be tested for subsequent growth in pure culture.

and moved away from contaminating organisms (Figure 19.5*b*). The cell can then be isolated by breaking the tube at a point between the cell and the contaminants and flushing the cell into a small tube of sterile medium. Laser tweezers, when coupled with staining techniques that identify particular organisms (Sections 19.4 and 19.5), can be used to select organisms of interest from a mixture for purification and further laboratory study.

Flow cytometry is a technique for counting and examining a mixture of cells by suspending them in a stream of fluid and passing them through an electronic detector that sorts them according to defined criteria; for example, by cell size, shape, or fluorescent properties. This ability makes cell sorting useful not only for isolating single cells but also for enriching a particular cell type from a mixture. Cell sorters can deposit individual cells into wells of a microtiter plate where each well contains the same growth medium or a slightly different growth medium. Because the growth requirements of some organisms include organic compounds and metabolites produced by other organisms that share their environment, addition of filter-sterilized source water (for aquatic organisms) or soil water extract (for soil organisms) can be used to supplement the media tested. Each well in the microtiter plate can then be monitored for growth or some other property either manually or using robotic methods (high-throughput culture, see next subsection). We explore the mechanism and uses of flow cytometry in more detail in Section 19.12 (see Figures 19.36 and 19.37).

High-Throughput Culture and Microfluidic Devices

Continuing innovations in single-cell isolation methodology have spawned **high-throughput culturing methods** and related methods for use on an even smaller scale. High-throughput methods require dilution (or cell sorting) of a sample to yield a single cell in each well of a microtiter plate (**Figure 19.6**). From there, each well is robotically monitored over time for cell growth or a specific target gene. High-throughput methods allow the experimenter to test many alternative growth conditions simultaneously in an attempt to replicate the realized niche or, alternatively, to allow the organism to occupy its fundamental niche by relieving it from competition. Microtiter wells that are positive for growth or a target gene of interest identify the acceptable resources and conditions for growth of a particular microbe and supply valuable clues for the design of laboratory culture media to obtain its growth in pure culture.

High-throughput cultivation has shown increasing success in isolating unique bacteria. For example, high-throughput methods were used for the isolation of one of the most abundant bacteria on Earth, the small marine planktonic bacterium *Pelagibacter ubique* (Figure 19.6). This bacterium thrives on the very dilute pool of dissolved organic matter present in the open oceans and eluded classical enrichment methods for years. But with high-throughput technology, this ecologically important bacterium was brought into laboratory culture where its biology could be studied in more detail.

Microfluidic devices carry the high-throughput concept even further by using microfabrication technology to combine channels and wells for fluid transfer and collection on a miniaturized platform. One such device is less than 10 centimeters long yet holds 3200 nanoliter-sized wells, each well functioning as a small culture vessel (**Figure 19.7**). An environmental sample is introduced into the microfluidic device such that each well receives a single cell. Different medium formulations can be tested, and the media supplemented with a small amount of filter sterilized water or soil extract collected from the sampled niche (these additions may stimulate growth by providing trace nutrients missing in the culture medium).

Both growth and target genes can be assessed in each well of the microfluidic device; growth is assessed by direct microscopic examination of a well under the microscope. A variation on this technique employs a microchamber device modified such that each of the tiny chambers is separated from the external environment by a membrane that traps the microbes but allows soluble nutrients to diffuse in and out.

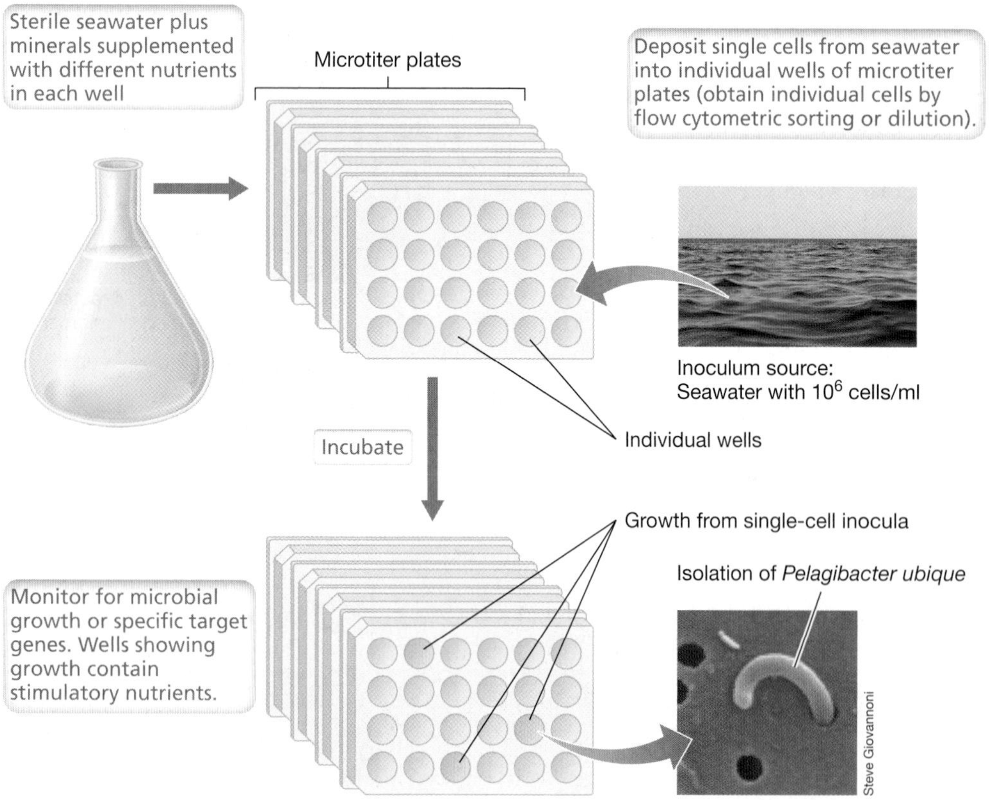

Sterile seawater plus minerals supplemented with different nutrients in each well

Microtiter plates

Deposit single cells from seawater into individual wells of microtiter plates (obtain individual cells by flow cytometric sorting or dilution).

Inoculum source: Seawater with 10^6 cells/ml

Individual wells

Incubate

Growth from single-cell inocula

Isolation of *Pelagibacter ubique*

Monitor for microbial growth or specific target genes. Wells showing growth contain stimulatory nutrients.

Steve Giovannoni

Figure 19.6 Methodological pipeline for high-throughput cultivation of previously uncultured microorganisms. The method shown here was used to isolate the marine bacterium *Pelagibacter ubique*. Following the addition of filter-sterilized seawater and low nutrient concentrations to the individual wells, pure cultures of *Pelagibacter* and other novel marine *Bacteria* were obtained. *Pelagibacter* is the most abundant bacterium in the open oceans (↪ Section 20.11).

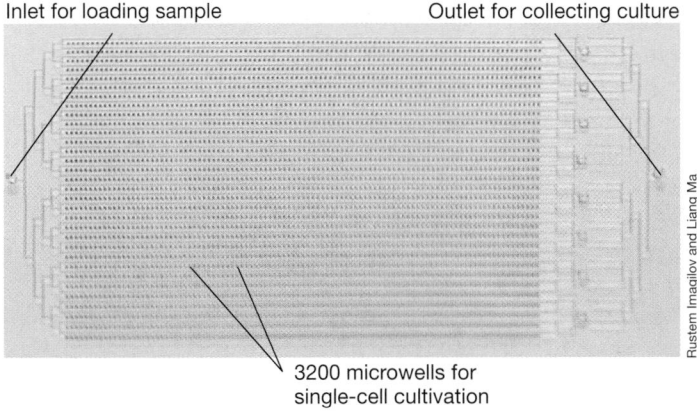

Inlet for loading sample   Outlet for collecting culture

3200 microwells for
single-cell cultivation

Rustem Imagilov and Liang Ma

Figure 19.7 Microfluidic platform for cultivation. An environmental inoculum is suspended in a cultivation medium and loaded onto this microfluidic device, enabling confinement of as many as 3200 single cells in nanoliter wells to promote the growth of microcolonies. Following different periods of incubation, cultured populations are collected at the outlet and further grown under conditions demonstrated to support growth on the microfluidic device. The device is about 7 cm wide.

Following the introduction of a single cell into each chamber, the device is placed back into the environment from which the inoculum was obtained. Then, after incubation for a month or more, microbes that initiate growth only when incubated under the conditions and resources present in their habitats can often be isolated and subsequently propagated in the laboratory.

Although technology is rapidly advancing the art of isolating new microbes, patience is still needed in any cultivation effort, as the discovery of slow-growing or dormant organisms may require months of incubation. Also, many microbes in nature are likely adapted to extremely low nutrient concentrations and may be inhibited by levels of nutrients used to grow organisms commonly studied in the laboratory. Both high-throughput and microfluidic methods overcome these problems by their ability to separate individual cells from other cells that may release inhibitory materials and by surveying a nearly limitless variety of nutrient conditions. Currently, these methods offer the best opportunity for culturing the most interesting (and likely ecologically relevant) microorganisms from nature.

MINIQUIZ

- How might you isolate a morphologically unique bacterium present in an enrichment culture in relatively low numbers?
- What is meant by "high-throughput" in culturing microorganisms? How has it benefited microbiology?

II • Culture-Independent Microscopic Analyses of Microbial Communities

Microbial ecologists quantify cells in a microbial habitat to estimate relative abundances of different species. Cell stains are necessary to obtain these types of data, and we detail these

methods here. Organisms in natural environments can also be detected by assaying their genes. Genes encoding either ribosomal RNA (rRNA, ⟳ Section 13.7) or enzymes that support a specific physiology are the usual targets in these studies. *Environmental genomics* is a method for assessing the entire gene complement of a habitat, revealing both the biodiversity and metabolic capabilities of the microbial community at the same time, and we consider this exploding field of microbial ecology in Section 19.8.

19.4 General Staining Methods

Several general staining methods are suitable for quantifying microorganisms in natural samples. Although these methods do not reveal the physiology or phylogeny of the cells, they are nonetheless reliable and widely used by microbial ecologists for measuring total cell numbers. One method also allows cell viability to be assessed.

Fluorescent Staining with Dyes That Bind Nucleic Acids

Fluorescent dyes can be used to stain microorganisms from virtually any microbial habitat. **DAPI** (4′, 6-diamidino-2-phenylindole) is a popular stain for this purpose, as is the dye **acridine orange**. There is also increasing use of *SYBR Green I*, a dye that confers very bright fluorescence to all microorganisms, including viruses. These stains bind to DNA and are strongly fluorescent when exposed to ultraviolet (UV) radiation (DAPI absorption maximum, 400 nm; acridine orange absorption maximum, 500 nm; SYBR Green I absorption maximum, 497 nm), making the microbial cells in the sample readily visible and easy to enumerate. Cells stained with DAPI fluoresce blue, cells stained with acridine orange fluoresce orange or greenish-orange, and cells stained with SYBR Green I fluoresce green (**Figure 19.8**).

Dyes that stain DNA are widely used for the enumeration of microorganisms in environmental, food, and clinical samples. Depending on the sample, background staining is occasionally a problem with fluorescent stains, but because these dyes specifically stain nucleic acids, they are for the most part nonreactive with inert matter. Thus, for many samples, from soil as well as aquatic sources, they can give a reasonable estimate of the cell numbers present. Staining with the brightly fluorescent SYBR Green I also provides excellent enumeration of aquatic virus populations (⟳ Section 20.11). For dilute aquatic samples, cells can be stained following collection on a membrane surface by filtration.

DNA staining is a nonspecific process; *all* microorganisms in a sample are stained. Although this may at first seem desirable, it is not necessarily so. For example, DAPI and acridine orange fail to differentiate between living and dead cells or between different species of microorganisms, so they cannot be used to assess cell viability or to track specific microorganisms in an environment.

Viability Staining

Viability staining differentiates live cells from dead ones. Hence, viability stains yield both abundance and viability data at the same time. The basis of differentiating between live and dead cells lies with whether a cell's cytoplasmic membrane is intact. Two dyes that fluoresce green and red are added to a sample; the green-fluorescing dye penetrates all cells, viable or not, whereas the red

UNIT 5

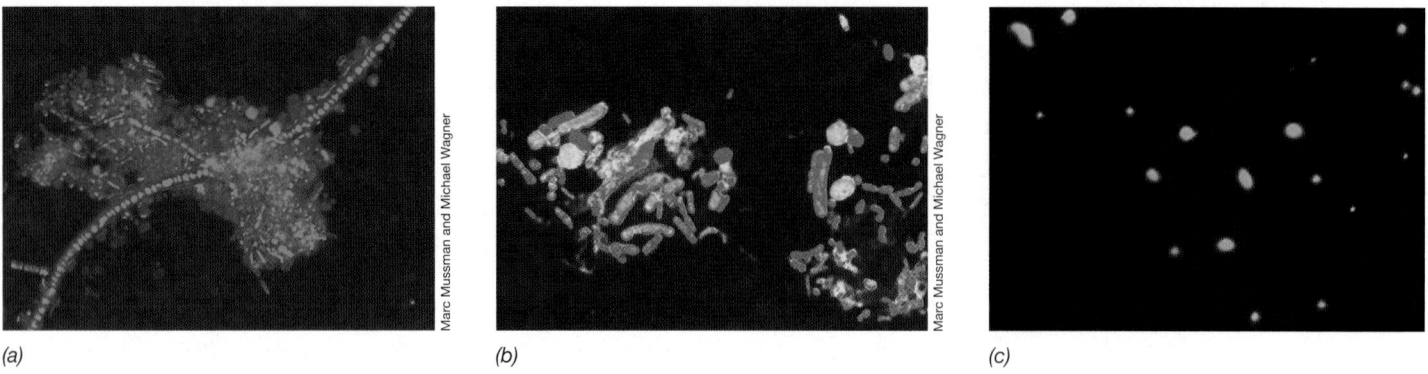

(a)   *(b)*   *(c)*

Figure 19.8 Nonspecific fluorescent stains. *(a)* DAPI and *(b)* acridine orange staining showing microbial communities inhabiting activated sludge in a municipal wastewater treatment plant. With acridine orange, cells containing low RNA levels stain green. *(c)* SYBR Green-stained sample of Puget Sound (Washington, USA) surface water showing green-fluorescing bacterial cells. The large cells near the center of the field are 0.8–1.0 μm in diameter.

dye, which contains the chemical propidium iodide, penetrates only those cells whose cytoplasmic membrane is no longer intact and that are therefore dead. Thus, when viewed microscopically, green cells are scored as alive and red cells as dead, yielding an instant assessment of both abundance and viability (**Figure 19.9**).

Although useful for research that uses laboratory cultures, the live/dead staining method is not suitable for use in the direct microscopic examination of samples from many natural habitats because of problems with nonspecific staining of background materials. However, procedures have been developed to overcome this problem in analyses of aquatic environments; a water sample is filtered and the filters are stained with the live/dead stain and examined microscopically. Thus in aquatic microbiology, live/dead staining is often used to measure the viability of cell populations in the water column of lakes or oceans, or in the flowing waters of streams, rivers, and other aquatic environments.

Fluorescent Proteins as Cell Tags and Reporter Genes

Bacterial cells can be altered by genetic engineering to make them autofluorescent. As discussed earlier, a gene encoding the

green fluorescent protein (GFP) can be inserted into the genome of virtually any cultured bacterium (⤇ Sections 7.1 and 12.5). When the gene encoding the GFP (*gfp*) is expressed, cells fluoresce green when observed with ultraviolet microscopy (**Figure 19.10**). Although GFP is not useful for the study of natural populations of microorganisms (because these cells lack the GFP gene), GFP-tagged cells can be introduced into an environment, such as plant roots, and then tracked over time by microscopy. Using this method, microbial ecologists can study competition between the native microbiota and a GFP-tagged introduced strain and can assess the effect of perturbations of an environment on the survivability of the introduced strain.

The gene *gfp* and those encoding other **fluorescent proteins** have also been used extensively in laboratory cultures of various bacteria and in controlled environments as *reporter genes*. When the gene is fused with an operon under the control of a specific regulatory protein, transcription can be studied by using fluorescence as the indicator (a "reporter") of activity. That is, when genes containing the fused fluorescent protein gene are transcribed and translated, both the protein of interest and the fluorescent protein are made, and cells fluoresce the characteristic color (⤇ Figures 7.2 and 12.16). For example, expression of *gfp* was used to demonstrate that colonization of alfalfa roots by *Sinorhizobium meliloti* (legume–root nodule symbiosis, ⤇ Section 23.3) is promoted by sugars and dicarboxylic acids released by the plant (Figure 19.10*b, c*). The photophysical properties of GFP and other fluorescent proteins isolated from different marine invertebrates (jellyfish, corals, anemones) have since been altered through mutation to yield a broad palette of fluorescent proteins of varying spectral properties (Figure 19.10*a*), offering the experimenter the capability to monitor several species simultaneously.

One drawback to the use of GFP is that to become fluorescent it requires O_2, and thus it is not suitable for tracking cells introduced into strictly anoxic habitats. However, flavin-based fluorescent proteins that do not require O_2 are available to overcome this limitation. These proteins are derived from bacterial and plant photosensory flavoproteins and are more thermally stable than the GFP, making them useful for tracking mildly thermophilic species.

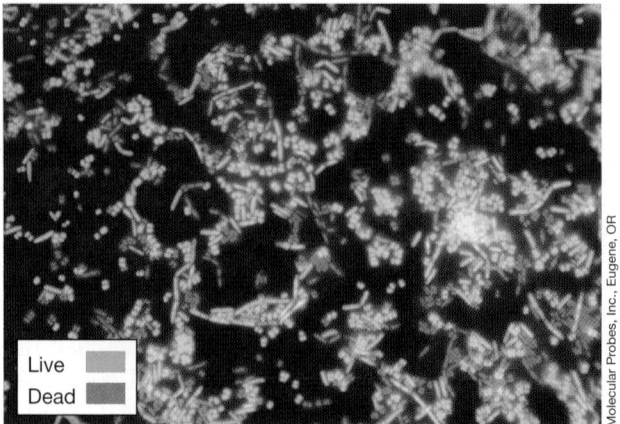

Live
Dead

Figure 19.9 Viability staining. Live (green) and dead (red) cells of *Micrococcus luteus* (cocci) and *Bacillus cereus* (rods) stained by the LIVE/DEAD BacLight Bacterial Viability Stain.

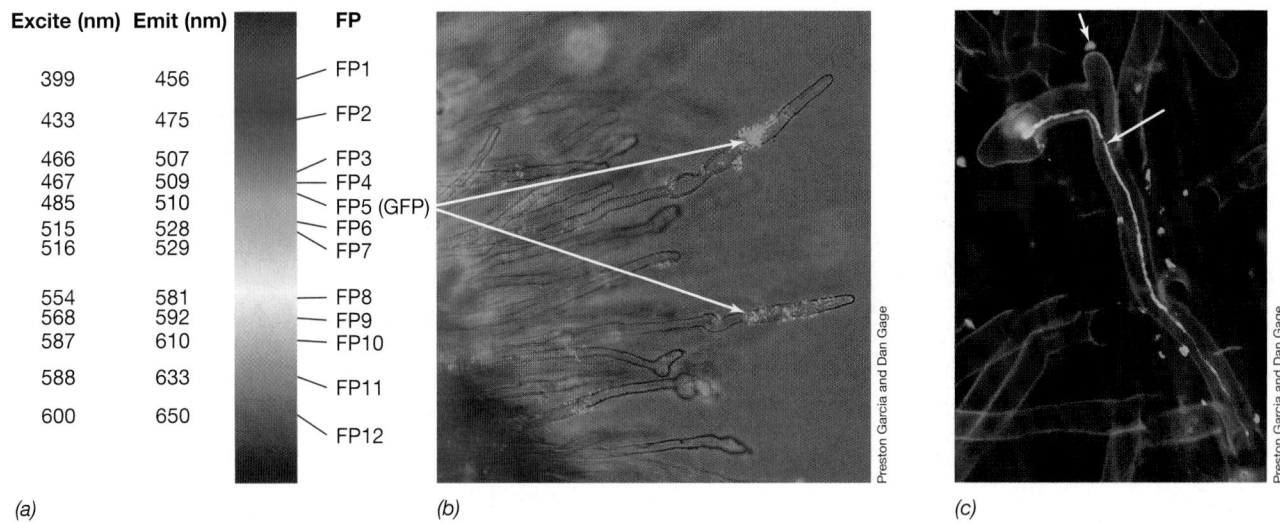

Excite (nm)	Emit (nm)	FP
399	456	FP1
433	475	FP2
466	507	FP3
467	509	FP4
485	510	FP5 (GFP)
515	528	FP6
516	529	FP7
554	581	FP8
568	592	FP9
587	610	FP10
588	633	FP11
600	650	FP12

(a)   *(b)*   *(c)*

Figure 19.10 Fluorescent protein reporters. *(a)* Twelve different fluorescent proteins (FP1–FP12) are known that have distinct excitation (Excite) and emission (Emit) properties. *(b)* Cells of *Sinorhizobium meliloti* (arrows) carrying a plasmid with an α-galactoside-inducible promoter fused to the GFP (FP5); the cells are on clover seedling roots. Green fluorescence indicates that α-galactosides are released and available to support the growth of this bacterium. *(c)* *S. meliloti* cells (arrows) carrying a plasmid with a succinate-inducible promoter fused to GFP; green fluorescence indicates that succinate or other C_4 dicarboxylic acids have been secreted by the plant root hairs.

In addition to these fluorescent tags for tracking microbes, phylogenetic stains (Section 19.5) are widely used for identifying microbes and a wide variety of new fluorescent "super-resolution" microscopy techniques are available for tracking individual molecules within a microbial cell (↩ Section 7.1). Thus, fluorescence technology has come a long way since the days when only DAPI and acridine orange were available for visualizing microbial cells in nature.

Limitations of Microscopy

The microscope is an essential tool for exploring microbial diversity and for enumerating and identifying microorganisms in natural samples. However, microscopy alone does not suffice for the study of microbial diversity. Prokaryotic cells vary greatly in size (↩ Section 2.2 and Table 2.1). Very small cells can be a major problem and can go totally unnoticed, and some cells are near the limits of resolution of the light microscope. Such cells can easily be overlooked in the examination of natural samples, especially if the sample contains high levels of particulate matter or high numbers of large cells. Also, it is often difficult to differentiate live cells from dead cells or cells in general from certain inert materials in natural samples. However, the biggest limitation with the microscopic methods we have discussed thus far is that none of them reveal the phylogenetic diversity of the microorganisms in the habitat under study.

We will see in the next section and get a preview here (**Figure 19.11**) of powerful staining methods that can reveal the *phylogeny* of organisms observed in a natural sample. These methods have revolutionized microbial ecology and have helped microbiologists overcome the major limitation of the light microscope in microbial ecology: identifying from a phylogenetic perspective cells observed in a microscopic field. These methods have also taught microbial ecologists an important lesson—when observing unstained or nonspecifically stained natural populations of microorganisms under the microscope, one must remember that the sample almost certainly contains a genetically diverse community, even if many cells "look" the same (Figure 19.11). The simple shapes of bacteria conceal their remarkable diversity.

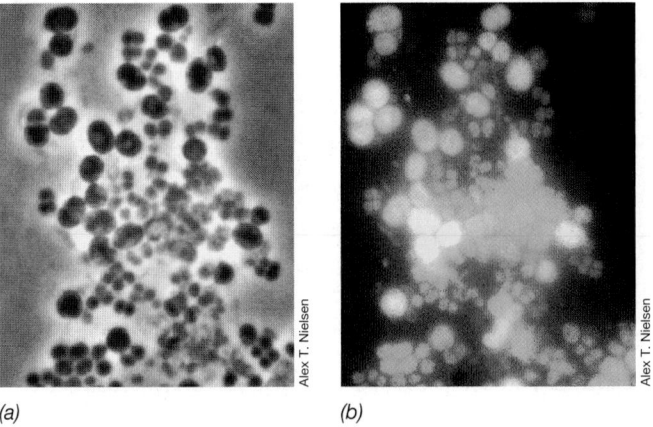

(a)   *(b)*

Figure 19.11 Morphology and genetic diversity. The photomicrographs shown here, produced by *(a)* phase contrast and *(b)* a technique called phylogenetic FISH (Section 19.5), are of the same field of cells. Although the large oval cells are of a rather unusual size for prokaryotic cells and all look similar by phase-contrast microscopy, the phylogenetic stains reveal that there are two genetically distinct types (one stains yellow and one stains blue). Both cell types are about 2.25 μm in diameter. The smaller, green cells in pairs or clusters are about 1 μm in diameter.

MINIQUIZ

- How does viability staining differ from stains like DAPI?
- What types of environments limit the application of GFP?
- Why is it incorrect to say that the GFP is a "staining" method?

UNIT 5

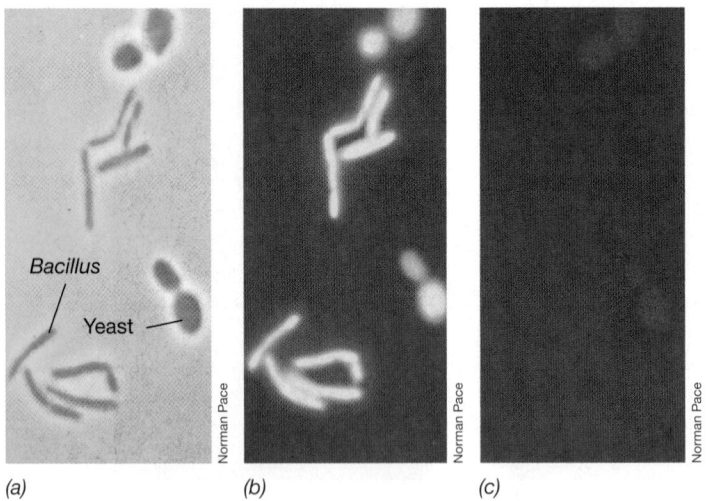

(a)   *(b)*   *(c)*

Figure 19.12 Fluorescently labeled rRNA probes: Phylogenetic stains.
(a) Phase-contrast photomicrograph of cells of *Bacillus megaterium* (rod, *Bacteria*) and the yeast *Saccharomyces cerevisiae* (oval cells, *Eukarya*). *(b)* Same field; cells stained with a yellow-green universal rRNA probe (this probe hybridizes with rRNA from organisms of any phylogenetic domain). *(c)* Same field; cells stained with a eukaryal probe (only cells of *S. cerevisiae* react). Cells of *B. megaterium* are about 1.5 μm in diameter and cells of *S. cerevisiae* are about 6 μm in diameter.

19.5 Fluorescence In Situ Hybridization (FISH)

Because of their great specificity, nucleic acid probes are powerful tools for identifying and quantifying microorganisms. Recall that a **nucleic acid probe** is a DNA or RNA oligonucleotide complementary to a sequence in a target gene or RNA; when the probe and the target come together they hybridize (c⊃ Section 12.1). Nucleic acid probes can be made fluorescent by attaching fluorescent dyes to them. The fluorescent probes can often be used to identify organisms that contain a nucleic acid sequence complementary to the probe. This technique is called **fluorescence in situ hybridization (FISH)**, and different applications are described here, including methods that target phylogeny (**Figure 19.12**) or gene expression (see Figure 19.14).

Phylogenetic Identification Using FISH

Phylogenetic FISH stains are fluorescing oligonucleotides complementary in base sequence to sequences in ribosomal RNA (16S or 23S rRNA in *Bacteria* and *Archaea* or 18S or 28S rRNA in eukaryotes). Phylogenetic stains penetrate cells without lysing them and hybridize with rRNA directly in the ribosomes. The number of fluorescent probes bound to a cell reflects the number of its ribosomes. As single microbial cells can contain tens of thousands of ribosomes, strong signals can be achieved. Because ribosomes are scattered throughout the cell in most prokaryotic cells, the entire cell becomes fluorescent (Figures 19.11*b* and 19.12).

By targeting sites in the rRNA sequence that are variable between different organisms, phylogenetic stains can be designed to be very specific and react with only one species or a handful of related microbial species. Alternatively, by targeting conserved sequences in the rRNA they can be made more general and react with, for example, all cells of a given phylogenetic domain. Using FISH, an investigator can identify or track an organism of interest or a

domain of interest in a natural sample. For example, if one wishes to determine the percentage of a given microbial population that are *Archaea*, an archaea-specific phylogenetic stain may be used in combination with DAPI (Section 19.4) to assess *Archaea* and total numbers, respectively, and a percentage could then be derived by calculation.

FISH technology can also employ multiple phylogenetic probes. With a suite of probes, each designed to react with a particular organism or group and each containing its own fluorescent dye, FISH can image multiple taxa in a habitat in a single experiment (**Figure 19.13**). If FISH is combined with confocal microscopy (c⊃ Section 1.7), it is possible to explore microbial populations with depth, as, for example, in a biofilm (c⊃ Section 20.4). In addition to microbial ecology, FISH is also an important tool in the food industry and in clinical diagnostics for the microscopic detection of specific pathogens in food products or clinical specimens.

CARD-FISH

Besides characterizing the abundance of different taxa in a habitat, FISH can be used to measure *gene expression* in organisms in a natural sample. Because the target in this case is messenger RNA (mRNA), a form of RNA that is much less abundant in cells than is rRNA, standard FISH techniques cannot be applied. Instead, the signal (fluorescence) must be amplified. A FISH method that enhances the signal is called *catalyzed reporter deposition FISH* (*CARD-FISH*).

In CARD-FISH the specific nucleic acid probe contains a molecule of the enzyme peroxidase conjugated to it instead of a fluorescent dye. After there has been time for hybridization, the preparation is treated with a fluorescently labeled soluble compound called *tyramide*, which is a substrate for peroxidase. Within cells containing the nucleic acid probe, the tyramide is converted by the activity of peroxidase into a very reactive intermediate that

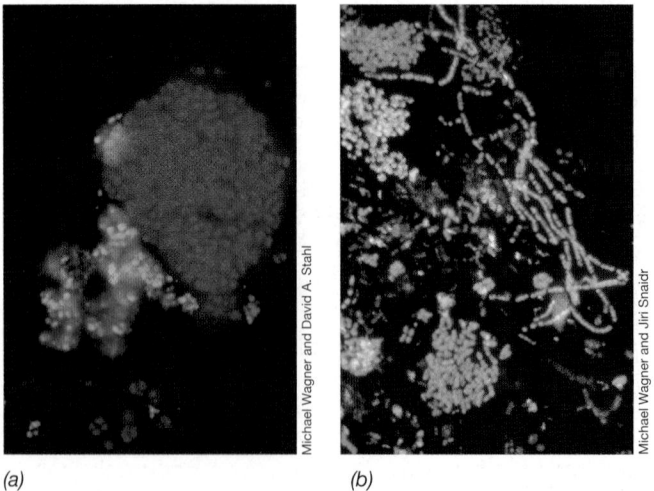

(a)   *(b)*

Figure 19.13 FISH analysis of activated sludge from a wastewater treatment plant. *(a)* Nitrifying bacteria. Red, ammonia-oxidizing bacteria; green, nitrite-oxidizing bacteria. *(b)* Confocal laser scanning micrograph of a sewage sludge sample treated with three phylogenetic FISH probes, each containing a fluorescent dye (green, red, or blue) that identifies a particular group of *Proteobacteria*. Green-, red-, or blue-stained cells reacted with only a single probe; other cells reacted with two (turquoise, yellow, purple) or three (white) probes.

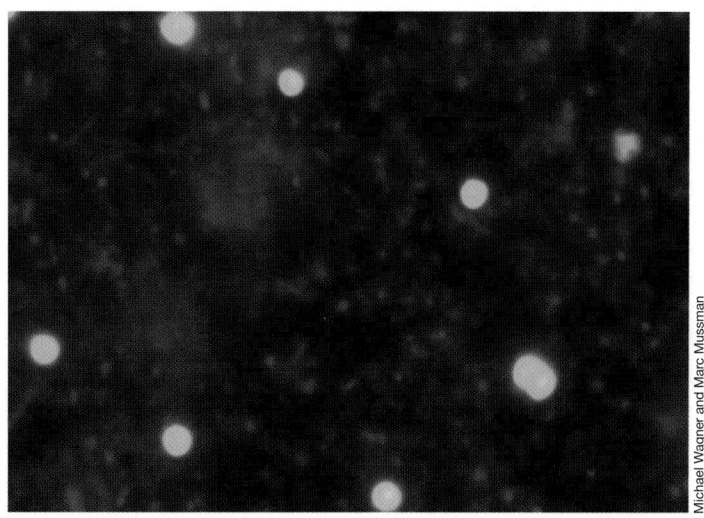

Figure 19.14 Catalyzed reporter deposition FISH (CARD-FISH) labeling of *Archaea*. Archaeal cells in this preparation fluoresce intensely (green) relative to DAPI-stained cells (blue).

covalently binds to adjacent proteins; this amplifies the signal sufficiently to be detected by fluorescence microscopy (**Figure 19.14**). Each molecule of peroxidase activates many molecules of tyramide so that even mRNAs present at very low abundance can be visualized.

Besides detecting mRNA, CARD-FISH is also useful in phylogenetic studies of microbes that may be growing very slowly, such as organisms inhabiting the open oceans where cold temperatures and low nutrient concentrations limit growth rates (Figure 19.14). Because such cells have few ribosomes compared with more actively growing cells, standard FISH often yields only a weak signal.

MINIQUIZ
- What structure in the cell is the target for fluorescent probes in phylogenetic FISH?
- FISH and CARD-FISH can be used to reveal different things about cells in nature. Explain.

III • Culture-Independent Genetic Analyses of Microbial Communities

Microbial biodiversity studies often forgo isolating organisms or even quantifying or identifying them microscopically. Instead, *specific genes* are used as measures of biodiversity and metabolic capacities. Some genes are unique to particular organisms. Detection of such a gene in an environmental sample implies that the organism is present. The major techniques employed in this type of microbial community analysis are the polymerase chain reaction (PCR), DNA fragment analysis by gel electrophoresis (DGGE, T-RFLP, ARISA) or molecular cloning, and DNA sequencing and analysis. In addition, entire genomes extracted from cells present in an environmental sample can be analyzed as a measure of the biodiversity of microbial communities.

19.6 PCR Methods of Microbial Community Analysis

We discussed the principle of the polymerase chain reaction (PCR) in Section 12.1. Recall the major steps involved: (1) Two nucleic acid primers are hybridized to a complementary sequence in a target gene; (2) DNA polymerase copies the target gene; and (3) multiple copies of the target gene are made by repeated melting of complementary strands, hybridization of primers, and new synthesis (Figure 12.1). From a single copy of a gene, several million copies can be made for subsequent analyses. PCR finds wide applications in microbial ecology.

PCR and Microbial Community Analysis

Which genes are suitable as target genes for microbial community analyses? Because genes encoding the small subunit ribosomal (SSU) rRNAs are phylogenetically informative and techniques for their analysis well developed (Section 13.7), they are widely used in community analyses. Moreover, because rRNA genes are universal and contain several regions of high sequence conservation, it is possible to amplify them from all organisms using only a few different PCR primers, even though the organisms may be phylogenetically distantly related. In addition to rRNA genes, genes that encode enzymes for metabolic functions unique to a specific organism or group of related organisms can be the target genes (Table 19.3).

Genes such as those encoding rRNAs that have retained ancestral function while changing in sequence over time as species have diverged are called *orthologs* (Sections 9.5 and 13.7). Organisms that share the same or very closely related orthologous genes are called a **phylotype**. In microbial ecology, the phylotype concept is primarily used to provide a natural (phylogenetic) framework for describing the microbial diversity of a given habitat, regardless of whether the identified phylotypes are cultured organisms or not. Thus, the word *phylotype* is widely used to describe the microbial diversity of a habitat based solely on nucleic acid sequences. It is only when additional physiological and genetic information becomes available, typically after the organism is brought into laboratory culture (Sections 19.2 and 19.3), that proposing a genus and species name for a phylotype becomes possible.

In a typical community analysis experiment, total DNA is isolated from a microbial habitat (**Figure 19.15**). Commercially available kits that yield high-purity DNA from soil or other complex habitats are available for this purpose. The DNA obtained is a mixture of genomic DNA from all of the microorganisms that were in the sample from the habitat (Figure 19.15). From this mixture, PCR is used to amplify the target gene and make multiple copies of each variant (phylotype) of the target gene. If RNA is isolated instead of DNA (to detect those genes being transcribed), the RNA can be converted into complementary DNA (cDNA) by the enzyme reverse transcriptase (Section 10.11) and the cDNA subjected to PCR as for isolated DNA. However, regardless of whether DNA or RNA is originally isolated, the different phylotypes need to be sorted out following the PCR step before they can be sequenced.

UNIT 5

TABLE 19.3 **Genes commonly used for evaluating specific microbial processes in the environment using PCR**

Metabolic process[a]	Target gene	Encoded enzyme
Denitrification	narG	Nitrate reductase
	nirK, nirS	Nitrite reductase
	norB	Nitric oxide reductase
	nosZ	Nitrous oxide reductase
Nitrogen fixation	nifH	Nitrogenase
Nitrification	amoA	Ammonia monooxygenase
Methane oxidation	pmoA	Methane monooxygenase
Sulfate reduction	apsA	Adenosine phosphosulfate reductase
	dsrAB	Sulfite reductase
Methane production	mcrA	Methyl coenzyme M reductase
Degradation of petroleum compounds	nahA	Naphthalene dioxygenase
	alkB	Alkane hydroxylase
Anoxygenic photosynthesis	pufM	M subunit of photosynthetic reaction center

[a]All of these metabolic processes are discussed in Chapter 14 and Section 3.12.

Sorting can be accomplished using one of three different methods: (1) physical separation by gel electrophoresis (⇄ Section 12.1), (2) clone library construction (⇄ Sections 12.2 and 12.9), and (3) next-generation sequencing technology (⇄ Section 9.2). We consider these methods now.

Denaturing Gradient Gel Electrophoresis: Separating Very Similar Genes

One method to resolve phylotypes is **denaturing gradient gel electrophoresis (DGGE)**, which separates genes of the same *size* that differ in their melting (denaturing) profile because of differences in their *base sequence* (**Figure 19.16a, b**). DGGE employs a gradient of a DNA denaturant, typically a mixture of urea and formamide. When a double-stranded DNA fragment moving through the gel reaches a region containing sufficient denaturant, the strands begin to "melt"; at this point, their migration stops (Figures 19.15 and 19.16b). Differences in base sequence cause differences in the melting properties of DNA. Thus, the different bands observed in a DGGE gel are phylotypes that can differ in base sequence significantly or by as little as a single base change.

Once DGGE has been performed, the individual bands are excised and sequenced (Figure 19.15). With 16S rRNA as the target gene, for example, the DGGE pattern immediately reveals the number of phylotypes (distinct 16S rRNA genes) present in a habitat (Figure 19.16c). The method provides an excellent mechanism to quickly evaluate temporal and spatial shifts in microbial community structure (Figure 19.16c). If PCR primers specific for genes other than 16S rRNA are used, such as a metabolic gene (Table 19.3), the variants of this specific gene that exist in the sample can also be assessed. Thus, although the number of bands on a DGGE gel is an overview of the biodiversity in a habitat (Figure 19.16c), sequence analysis is still required for identification and to infer phylogenetic relationship.

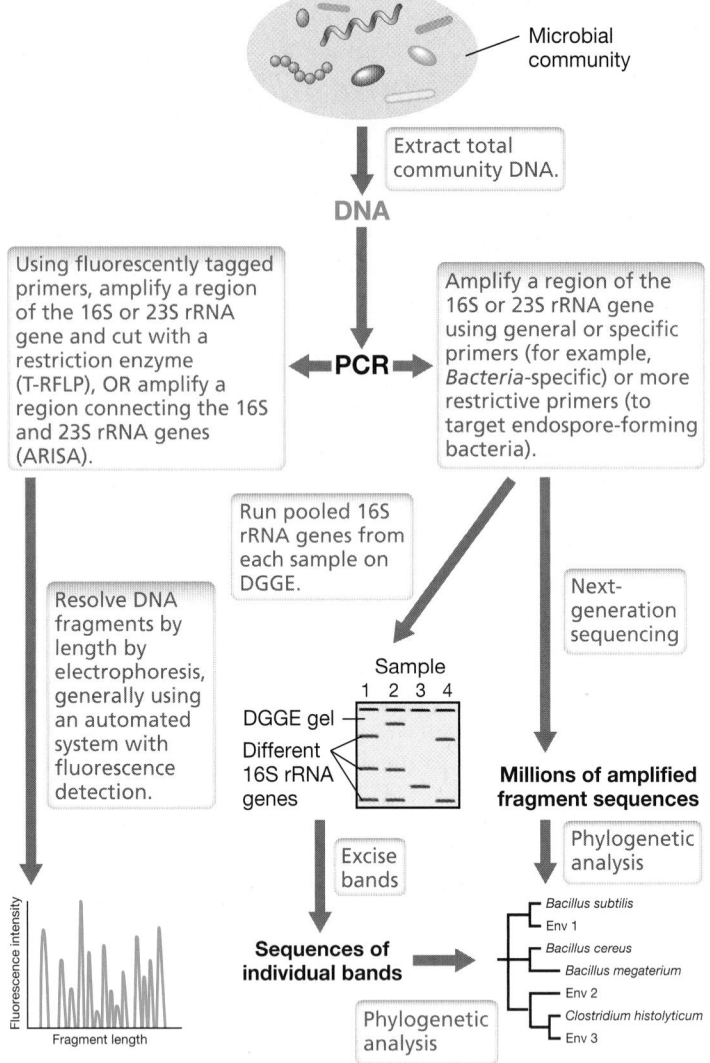

Figure 19.15 Steps in single-gene biodiversity analysis of a microbial community. From total community DNA, 16S rRNA genes are amplified using primers that target only *Firmicutes*, a group of gram-positive *Bacteria* that includes the endospore-forming genera *Bacillus* and *Clostridium*. The 16S rRNA gene products obtained from PCR are then either separated by DGGE or sequenced directly by next-generation sequencing; from the sequence data, a phylogenetic tree is generated. "Env" indicates an environmental sequence (phylotype). In T-RFLP analyses, the number of peaks indicates the number of phylotypes.

T-RFLP and ARISA

A rapid method of microbial community analysis is *terminal restriction fragment length polymorphism* (*T-RFLP*). In this method a target gene (usually an rRNA gene) is amplified by PCR from community DNA using a primer set in which one of the primers is end-labeled with a fluorescent dye. The PCR products are then treated with a restriction enzyme (⇄ Section 12.2) that cuts the DNA at specific sequences. This generates a series of DNA fragments of varying length, the number of which depends on how many restriction cut sites are present in the DNA. The fluorescently labeled terminal fragments are then separated by size on an automated DNA sequencer that detects fluorescent fragments (thus, only the terminal dye-labeled fragments are detected). The pattern obtained

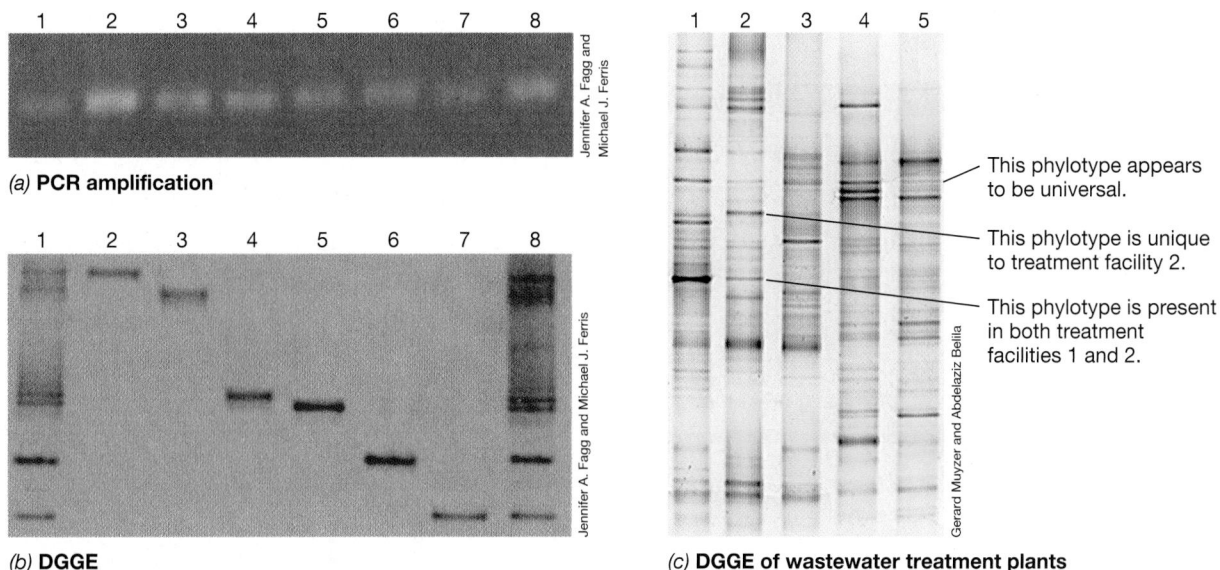

(a) **PCR amplification**

(b) **DGGE**

(c) **DGGE of wastewater treatment plants**

This phylotype appears to be universal.

This phylotype is unique to treatment facility 2.

This phylotype is present in both treatment facilities 1 and 2.

Figure 19.16 PCR and DGGE gels. Bulk DNA was isolated from a microbial community and amplified by PCR using primers for 16S rRNA genes of *Bacteria* (*a*, lanes 1 and 8). Six bands later resolved by DGGE (*b*, lanes 2–7) were excised and reamplified and each gave a single band at the same location on the PCR gel (*a*, lanes 2–7). However, by DGGE analysis, each band migrated to a different location on the gel (*b*, lanes 2–7). Note that all bands migrate to the same location in the nondenaturing PCR gel because they are all of the same size, but they migrate to different locations on the DGGE gel because they have different sequences. *(c)* DGGE profiles of microbial communities from different wastewater treatment facilities amplified using primers for the 16S rRNA genes of *Bacteria*.

shows the rRNA sequence variation and general abundance of different sequence types (fragment fluorescence intensity) in the microbial community sampled (Figure 19.15).

DGGE and T-RFLP both measure *single-gene diversity*, but in different ways. The pattern of bands on a DGGE gel reflects the number of same-length sequence variants of a single gene (Figure 19.16), whereas the pattern of bands on a T-RFLP gel reflects variants differing in DNA sequence of a single gene as measured by differences in restriction enzyme cut sites. The information obtained from a T-RFLP analysis, in addition to providing insight into the diversity and population abundances of a microbial community, can also be used to infer phylogeny. Diagnostic information for each fragment includes knowledge of sequences near both ends (primer sequence and restriction enzyme cut site), knowledge that a second restriction site does not exist within the fragment, and fragment length. Using specialized software, this information can be used to search for matching 16S rRNA sequences in public databases. Although this is of some predictive value, closely related sequences are often not differentiated by these criteria. Thus, T-RFLP generally underestimates the diversity within a microbial community.

A technique related to T-RFLP that provides more detailed analysis of microbial communities is *automated ribosomal intergenic spacer analysis* (ARISA), which exploits the proximity of the 16S rRNA and 23S rRNA genes in the genomes of *Bacteria* and *Archaea*. The DNA separating these two genes, called the *internal transcribed spacer* (ITS) region, differs in length among species and often also differs in length among the multiple rRNA operons of a single species (**Figure 19.17a**). The PCR primers for ARISA are complementary to conserved sequences in the 16S and 23S rRNA genes that flank the spacer region. Amplification (Figure 19.17b) and analysis (Figure 19.17c) are conducted as described for T-RFLP, resulting in a

complex pattern of bands that can be used for community analysis. However, ARISA differs from T-RFLP in that ARISA does not require a restriction enzyme digestion following PCR amplification. The word "automated" in the ARISA acronym refers to the use of a DNA sequencer that automatically identifies and assigns sizes to each dye-labeled fragment (Figure 19.17c), as can also be done in T-RFLP analyses. ARISA has received greatest application in the study of microbial community dynamics by monitoring, for example, changes in the presence and relative abundance of a specific community member through time and space.

Diversity Studies Using Clone Libraries or Next-Generation Sequencing

Early molecular microbial diversity research relied on the construction of clone libraries to separate individual amplified DNA molecules (*amplicons*); each clone in the library contained a unique sequence that was then used as a template for sequence determination (⮂ Sections 9.2 and 12.2). Figure 19.16a shows that a 16S rRNA gene amplicon mixture appears as a single band when run on a nondenaturing gel. However, because the amplified target gene came from a *mixture* of different cells, the different phylotypes in the single band need to be sorted out before they are sequenced. Today this is done by next-generation sequencing systems (⮂ Section 9.2) rather than by DGGE or by cloning.

Next-generation sequencing does not require a cloning step, as individual DNA fragments are separated and amplified on the sequencing device itself; thus, PCR products can be used directly for sequencing. Since millions of amplification reactions are then conducted simultaneously, the total number of sequencing reads vastly exceeds what is possible by sequencing individual clones obtained in a clone library one at a time (**Figure 19.18**). This tremendous volume of sequence data allows for what has been called

UNIT 5

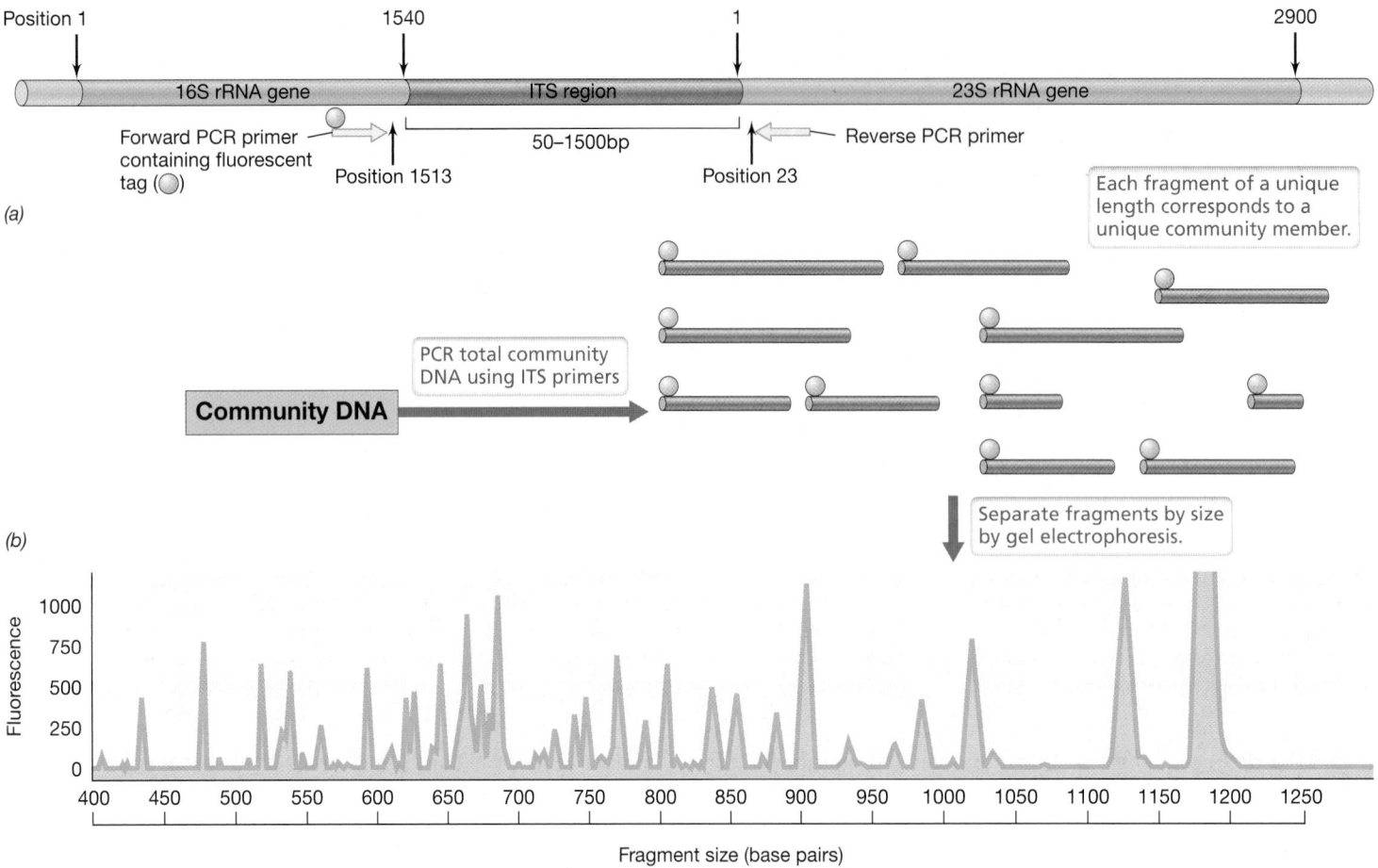

Figure 19.17 Automated ribosomal intergenic spacer analysis (ARISA). *(a)* Structure of rRNA operon spanning the 16S rRNA gene (positions 1–1540), an internal transcribed spacer (ITS) region of variable length, and the 23S rRNA gene (positions 1–2900). The PCR primers, one labeled with a fluorescent dye, are complementary to conserved sequences near the ITS region. *(b)* Amplified DNA fragments of different lengths, each corresponding to a community member. *(c)* Fragment analysis determined by an automated DNA sequencer. The peaks, which correspond to different ITS regions, can be identified by cloning and sequencing the amplified products.

deep sequence analyses, meaning that minor phylotypes that were possibly missed by the more limited and expensive clone library method are now revealed (Figure 19.18*b*). For example, if a particular phylotype were present at only 0.01% in a library of cloned sequences, it could take one thousand clones or more to ever detect it. By contrast, next-generation sequencing would detect this low-abundance phylotype along with its more abundant neighbors. This collection of minor phylotypes, which represent a substantial fraction of total diversity but only a minor component of total organism abundance in most environments, has been called the *rare biosphere* (Figure 19.18).

Results of PCR Phylogenetic Analyses

Phylogenetic analyses of microbial communities have yielded surprising results. For example, using the gene encoding 16S rRNA as the target, analyses of natural microbial communities typically show that many phylogenetically distinct *Bacteria* and *Archaea* (phylotypes) are present whose rRNA gene sequences differ from those of all known laboratory cultures (Figure 19.15). Moreover,

using quantitative PCR (⇄ Section 12.1), a variant of PCR that allows each phylotype to be quantified as well as amplified, it has often been observed that the most abundant phylotypes in a natural microbial community are ones that have thus far defied laboratory culture. These sobering results make it clear that our knowledge of microbial diversity from enrichment cultures is far from complete and that enrichment bias (Section 19.1) is a serious problem in culture-dependent biodiversity studies. Obviously, much work remains to put our abilities to culture microbes on a par with our existing abilities to detect and identify them in nature.

--- **MINIQUIZ** ---

- What could you conclude from PCR/DGGE analysis of a sample that yielded one band by PCR and one band by DGGE? One band by PCR and four bands by DGGE?

- What surprising finding has come out of many molecular studies of natural habitats using 16S rRNA as the target gene?

- How has next-generation sequencing technology altered our understanding of microbial community diversity?

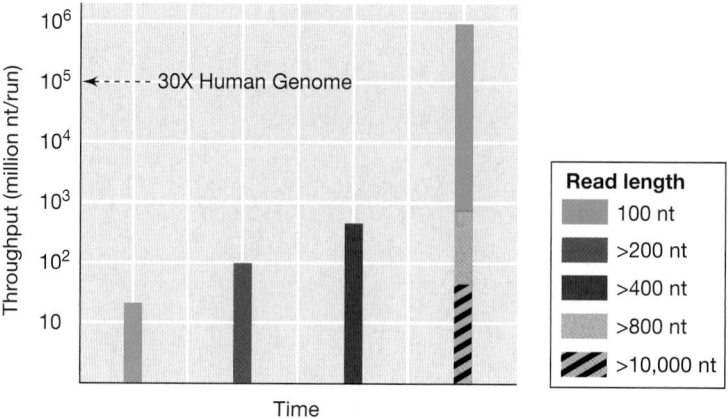

(a)

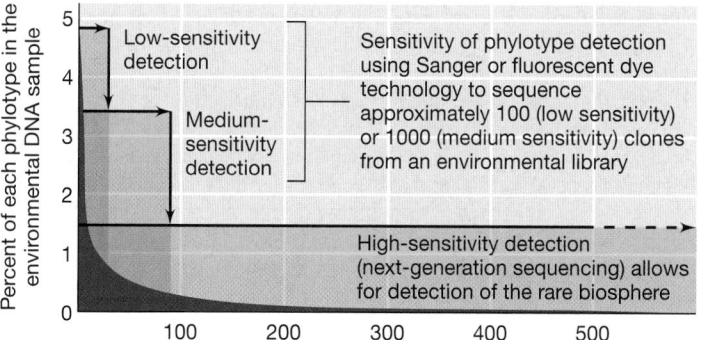

(b)

Figure 19.18 Community diversity analyses using next-generation sequencing technology. *(a)* Current sequencing platforms (⇄ Section 9.2) have the capacity to generate 10^{12} nucleotides (nt) of sequence in a single sequencing run (requiring a week or less), with individual read lengths varying from 100 to 800 nucleotides. The three segments in the rightmost bar show that technologies generating longer reads have lower throughput per sequencing run. *(b)* This enormous sequencing capacity revealed many unique phylotypes that were not detected using DGGE or clone library sequencing. Fewer than 100 unique phylotypes would be detected by Sanger (first-generation) sequencing of 1000 clones in a library of 16S rRNA gene PCR amplicons. Jed Fuhrman is acknowledged for input to part *b*.

19.7 Microarrays for Analysis of Microbial Phylogenetic and Functional Diversity

We previously considered the use of DNA chips—a type of *microarray*—for assessing overall gene expression in a microbial pure culture (⇄ Section 9.9). Specific microarrays can also be constructed for rapid gene-based analyses of biodiversity and the functional potential of natural microbial communities. Microarrays designed for biodiversity studies, called *PhyloChips*, have been developed for screening microbial communities for specific phylogenetic groups. Another type of microarray, the *GeoChip*, has been designed to detect genes encoding metabolic functions of biogeochemical significance, such as genes encoding proteins required for sulfate respiration, ammonia oxidation, denitrification, or nitrogen fixation (Table 19.3).

PhyloChips and GeoChips

PhyloChips are constructed by affixing rRNA probes or rRNA gene–targeted oligonucleotide probes to the chip surface in a known pattern, and several thousand different probes can be added to a single PhyloChip. As an example, consider a Phylo-Chip designed to assess the diversity of sulfate-reducing bacteria (⇄ Section 15.9) in a sulfidic natural environment, such as marine sediments (**Figure 19.19**). Oligonucleotides complementary to specific sequences in the 16S rRNA genes of all known sulfate-reducing bacteria (over 100 species) would be affixed to the chip. Then, following the isolation of total community DNA from the sample and PCR amplification and fluorescence labeling of the 16S rRNA genes, the environmental DNA would be hybridized to the probes on the PhyloChip. The species of sulfate reducers present in the sample would then be determined by detecting which probes hybridized sample DNA (Figure 19.19). Alternatively, rRNA could be extracted from the microbial community, labeled with a fluorescent dye, and hybridized directly to the PhyloChip without an amplification step.

In contrast to the PhyloChip, the GeoChip (**Figure 19.20**) targets *functional* genes rather than phylogenetic genes. However, because genes encoding enzymes of similar function can vary significantly in primary sequence, these *functional gene microarrays*, as they are called, must contain many thousands of probes in order to achieve reasonable coverage of natural diversity. Even then, such arrays may only sample a fraction of the actual functional diversity in a habitat. The most recent version of the Geo-Chip functional gene microarray contains over 160,000 probes covering more than 1400 gene families related to carbon, nitrogen, and sulfur cycling. By detecting which of the probes contain hybridized DNA from the natural sample, a rapid

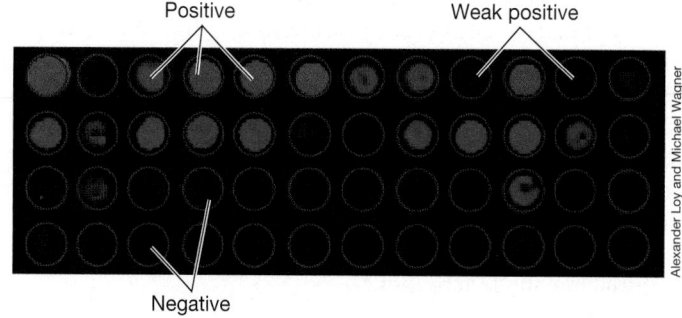

Figure 19.19 PhyloChip analysis of sulfate-reducing bacteria diversity. Each spot on the microarray shown has an oligonucleotide complementary to a sequence in the 16S rRNA of a different species of sulfate-reducing bacteria. After the microarray is hybridized with 16S rRNA genes PCR-amplified from a microbial community and then fluorescently labeled, the presence or absence of each species is signaled by fluorescence (positive or weak positive) or nonfluorescence (negative), respectively.

UNIT 5

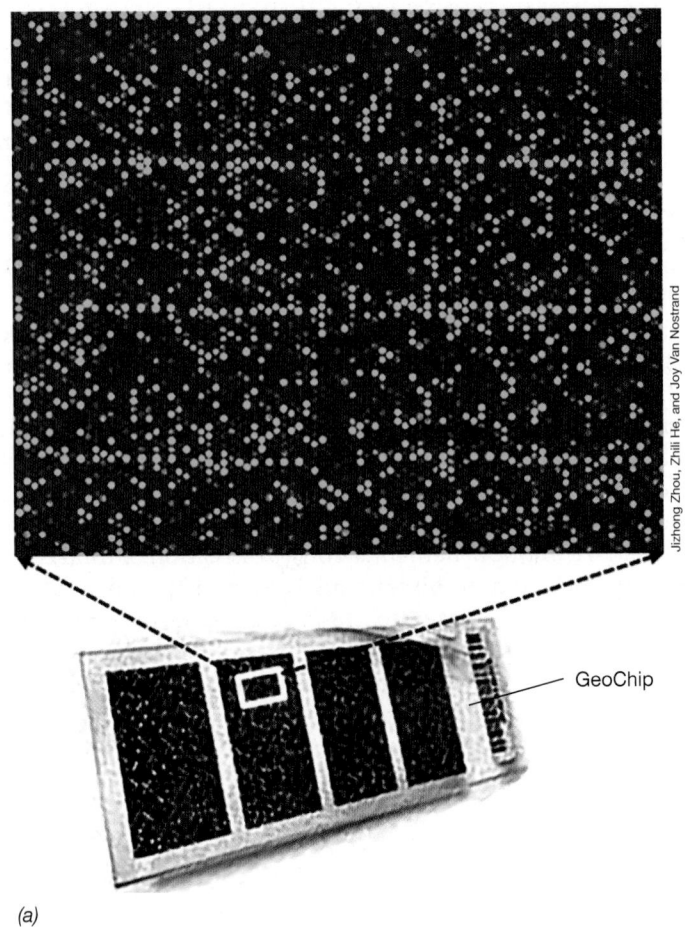

Functional category	Genes families	Total probes	Database gene coverage (%)
Carbon cycling	149	26922	49
Nitrogen cycling	32	6493	52
Sulfur cycling	27	4739	64
Phosphorus	7	3260	52
Metal homeostasis	121	43432	47
Viruses	115	2857	55
Other	81	10380	42
Organic remediation	104	11591	41
Virulence	639	21152	45
Secondary metabolism	68	4032	56
Electron transfer	15	797	65
Stress response	89	26306	33
TOTAL	**1447**	**161,961**	**44**

(b)

(a)

GeoChip

Jizhong Zhou, Zhili He, and Joy Van Nostrand

Figure 19.20 GeoChip analysis of functional gene diversity. The current version of the GeoChip contains over 160,000 probes covering more than 365,000 gene sequences in public databases, encompassing most major biogeochemical processes. The image shows green fluorescence of varying intensity (approximating gene abundance) following hybridization of fluorescent dye-labeled environmental DNA to the individual probes in one region of the high-density array of probes. The red spots correspond to repeated applications of a known amount of a reference DNA standard. A red-dye-labeled probe complementary to the reference standard was added to the environmental DNA prior to hybridization. Red fluorescence of equal intensity among the reference standard spots confirms that hybridization was uniform throughout the array.

appraisal of the potential metabolisms operating in a particular habitat can be obtained.

Advantages and Caveats of Environmental Microarrays

PhyloChips and GeoChips circumvent many of the time-consuming steps of molecular microbial ecology—PCR, DGGE, cloning, and sequencing (Figure 19.15). However, an important caveat to keep in mind is the possibility of *nonspecific hybridization*. That is, gene variants that are closely related in sequence may not be resolved because of overlapping hybridization patterns. Moreover, totally unrelated genes may yield false positive results if they are sufficiently complementary in base sequence to hybridize to the probe. And finally, unlike nucleic acid sequencing, whose costs keep plummeting yearly, designing and manufacturing gene chips are not inexpensive activities. Nevertheless, functional gene arrays offer another important tool for the culture-independent assessment of microbial biodiversity and potential metabolic activities.

MINIQUIZ

- What is a PhyloChip and what can it tell you? How does a PhyloChip differ from a GeoChip?
- What are the advantages and disadvantages of microarray technology compared to sequencing PCR products?

19.8 Environmental Genomics and Related Methods

A more encompassing approach to the molecular study of microbial communities is **environmental genomics**, also called **metagenomics**. Before the metagenomics era, microbial community analyses typically focused on the diversity of a *single* gene in an environmental sample. By contrast, in environmental genomics, *all* genes in a given microbial community can be sampled, and if the experiment is properly designed, the information obtained can support a much deeper understanding of the structure and function of the community than can single-gene analyses.

Metagenomics and Reconstructing Environmental Genomes

The goal of a metagenomics study today is to use next-generation DNA sequencing (⊘ Section 9.2) to identify as many genes as possible from an environmental DNA sample and determine the phylogeny of the organism(s) to which the genes belong. Although complete and finished genomes are often not the goal of metagenomics, there is increasing interest in assembling individual genomes from large metagenomic datasets to at least the draft stage. Rather than simply generating a list of all genes present in an environment, these nearly complete genomes can better connect functional and phylogenetic aspects of a microbial habitat.

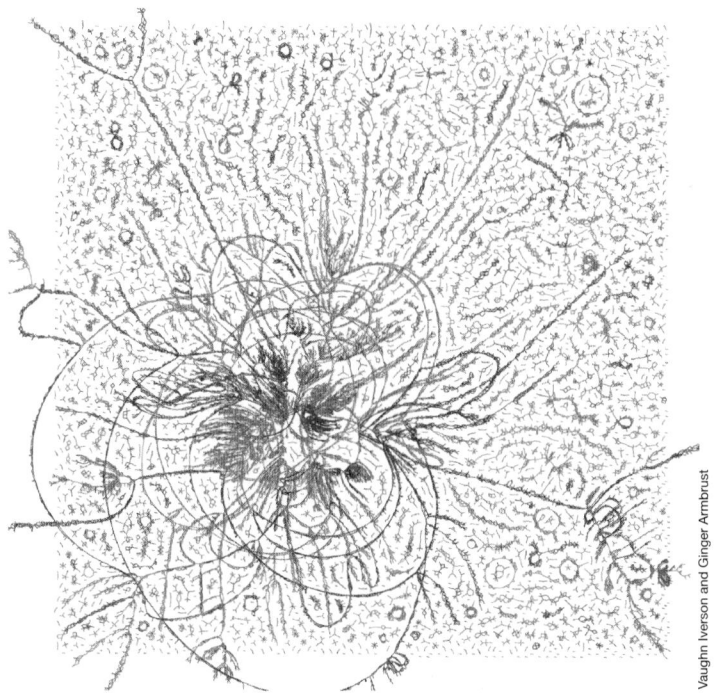

Vaughn Iverson and Ginger Armbrust

Figure 19.21 Genome assembly from a coastal marine metagenome consisting of 58.5 *billion* nucleotides of sequence. This "connection graph" is intended as a visual representation of the complexity and abundance of partial and complete genomes assembled from the water sample. The long strands, colored by differences in the percentage of guanine plus cytosine content, correspond to prokaryotic genomes and the small circular strands are most likely from viruses or plasmids.

An example of this is shown in the "connection graph" of **Figure 19.21** that depicts an assembly of genomes from a coastal marine water sample. A total of 58.5 *billion* nucleotides in the metagenome were used to stitch together these complete and near-complete genomes. Such massive metagenomic undertakings often reveal links between physiologies and phylogenies not obtainable in the absence of reconstructed genomes.

A problem with genome assembly from a mixture of environmental DNA sequence reads, however, is that the genomes obtained are unlikely to be either complete or clonal, instead being composed of fragments of DNA from closely related strains of a species (**Figure 19.22**). This has proven to be a major problem in the assembly of soil microbial genomes from metagenomic data. A single gram of fertile soil contains about 10^{12} bacterial and archaeal genes and 10^9 genomes; complete coverage of these genes is not yet feasible with available technology, even with the sequencing of 300 billion nucleotides in one soil study. Single-cell genomics (⊂⊃ Section 9.12) may ultimately overcome this problem (see also Section 19.12), but by definition it only provides information from a single microbial cell.

Of critical importance in evaluating a genome reconstructed from metagenomic DNA is to assess whether it contains all of the genes required by a cell (for example, all necessary tRNA and rRNA genes and genes encoding essential proteins such as DNA and RNA polymerases) and is therefore a legitimate genome candidate. In addition, an assessment of the relative abundance of genes encoding specific functions is equally valuable, since abundance

changes suggest interactions among species or a common response to a particular environmental variable. For example, if a high number of genes were recovered in the pathway for nitrogen fixation, this would suggest that the environment sampled was limited in NH_4^+, NO_3^-, and other forms of fixed nitrogen, thus selecting for nitrogen-fixing bacteria. Figure 19.22 contrasts the environmental genomic approach with single-gene analysis of microbial communities.

Although metagenomics can reveal much about a microbial habitat, there are many things metagenomics cannot tell us about environmental microbial communities. Currently no methods are available for translating metagenomic sequence data into fundamental physiological information about the microbial community, such as the maximum specific growth rate of different species, their saturation constants for nutrients, their optimum, minimum, and maximum pH or temperature for growth, or their speed of recovery from starvation. Moreover, any metabolisms never before encountered are unlikely to be deduced from nucleotide sequence data alone. These realities once again underscore why it is important to culture new microbes from nature; at present, there is simply no substitute for culture-based characterization to define many critical aspects of a microbe's functional biology.

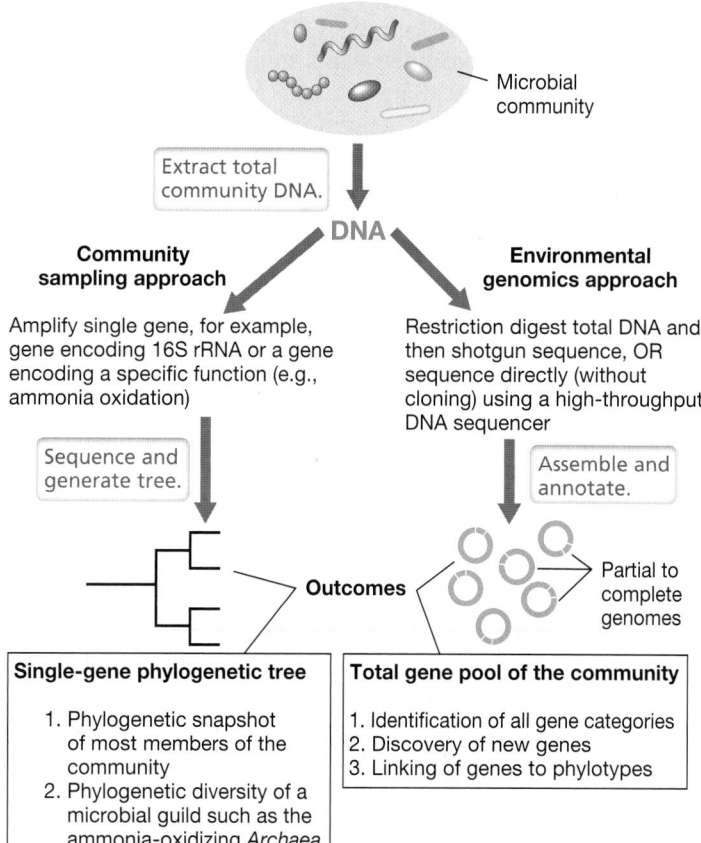

Figure 19.22 Single-gene versus environmental genomic approaches to microbial community analysis. In the environmental genomic approach, all community DNA is sequenced, but the assembled genomes may not all be complete. Total gene recovery is variable and depends on several factors including the complexity of the habitat and the amount of sequence determined. Recovery is typically better when diversity is low and sequence redundancy is high.

UNIT 5

Some Examples of Environmental Genomics

Environmental genomics can detect both new genes in known organisms and known genes in new organisms. A large number of interesting microbial communities have been probed using early metagenomic tools. In an early study of bacterial and archaeal diversity in the Sargasso Sea (a low-nutrient region of the Atlantic Ocean near Bermuda), over one billion nucleotides were sequenced and from this over 1800 bacterial and archaeal species were detected, including 148 previously unknown phylotypes and many novel genes. Many of these species had previously been missed using rRNA community analyses that employed PCR and cloning or PCR and DGGE (Section 19.6). Genes that fail to amplify, of course, remain undetected in community analyses, and cloning efficiency is far from 100%. Metagenomics sidesteps these problems by sequencing DNA directly without the need to amplify it or resolve different phylotypes before sequencing.

The Sargasso Sea metagenome study revealed several novel findings such as the presence of ammonia-oxidizing genes in archaeal genomes, a result that led to the discovery of a new group of *Archaea*, the *Thaumarchaeota* ($\mathcal{C}\wp$ Section 17.5). Moreover, genes encoding *proteorhodopsin*, a light-sensitive proton pump present in certain *Proteobacteria* and related to bacteriorhodopsin of extreme halophiles ($\mathcal{C}\wp$ Section 17.1), were found in the genomes of several new phylogenetic lineages of *Bacteria*. However, despite this major sequencing undertaking, much was missed. This is because 1 milliliter of seawater contains approximately 5 *trillion* base pairs (bp) of bacterial genomic DNA and would therefore require 5000 times the sequencing effort just to cover each base pair once on average! Hence, even with current technology—which can generate over a trillion bp of sequence in a few days (Figure 19.18)—no one natural environment has yet been sequenced completely.

Genomic/metagenomic approaches have also revealed variations in genes associated with a single phylotype; that is, in strains that contain identical, or nearly identical, rRNA genes. For example, in studies of *Prochlorococcus*, the most abundant cyanobacterium (oxygenic phototroph) in the ocean ($\mathcal{C}\wp$ Section 15.3), comparison of the genome sequences of cultured strains with *Prochlorococcus* genes obtained from metagenomic analyses of ocean water identified extensive regions shared between the cultured and environmental populations (**Figure 19.23**). This high level of gene conservation confirms that the organisms in culture are typical of environmental populations. In addition, however, these analyses also identified several highly variable regions in which the genomes of cultured strains differed significantly from those of environmental populations. These variable regions were clustered in the genome as *genomic islands* (chromosomal islands, $\mathcal{C}\wp$ Section 9.7), and likely encode functions that control the growth response of particular *Prochlorococcus* populations to environmental variables such as temperature or light quality and intensity.

Metatranscriptomics and Metaproteomics

As we discussed in Chapter 9, the genomics era has spawned several additional "omics," in particular, *metatranscriptomics* and *metaproteomics*. **Metatranscriptomics** is analogous to metagenomics but analyzes the sequences of community *RNA* rather than community *DNA*. The isolated RNA is converted into cDNA by reverse transcription ($\mathcal{C}\wp$ Section 10.11) before sequencing and analysis as for DNA. Whereas metagenomics describes the functional capacities of the community (for example, the relative abundance of specific genes), metatranscriptomics reveals which genes in the community *are actually being expressed* and the relative level of that expression, at a specific time and place. Because the expression of most genes in *Bacteria* and *Archaea* is controlled at the level of transcription (Chapter 6), mRNA abundance can be taken as a census of individual gene expression levels. Thus, gene transcript abundance determined for an entire community can be used to infer the operation of major

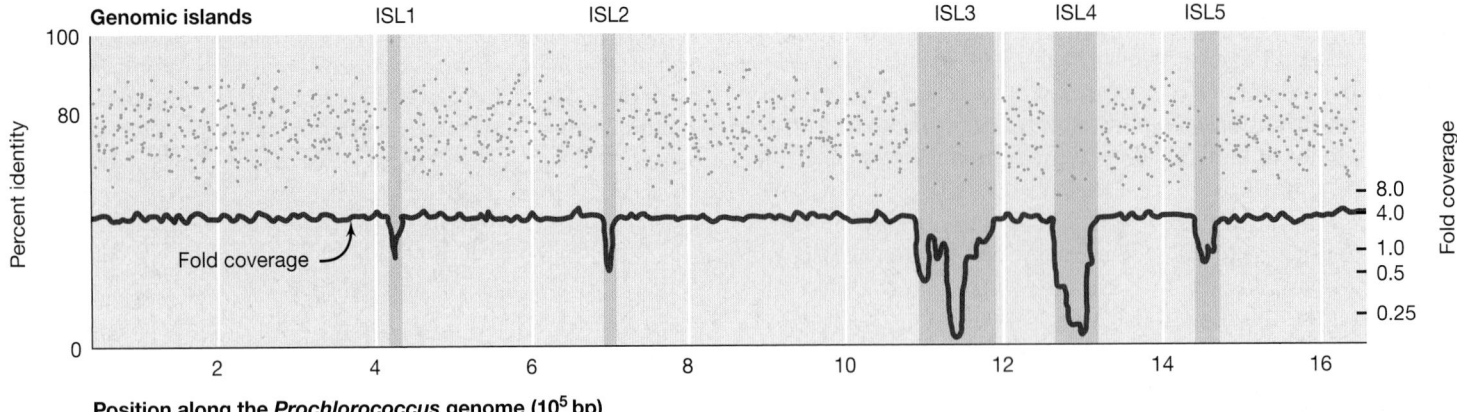

Figure 19.23 Metagenomic analysis. Sequences (represented as green dots) from the Sargasso Sea metagenome that align to the genome sequence of a cultured *Prochlorococcus*, showing regions where the cultured strain has genes of high similarity (high % identity) with sequences in the metagenome, and other regions (shaded) where it lacks genes in common (genomic islands, ISL1–ISL5). Since the DNA sequence contained within the genomic islands is thought to encode niche-specific functions, the cultured strain would likely not exhibit the same environmental distribution as strains containing all the island genes. Fold coverage is a measure of how completely the various regions in the *Prochlorococcus* genome are accounted for by similar sequences in the metagenome.

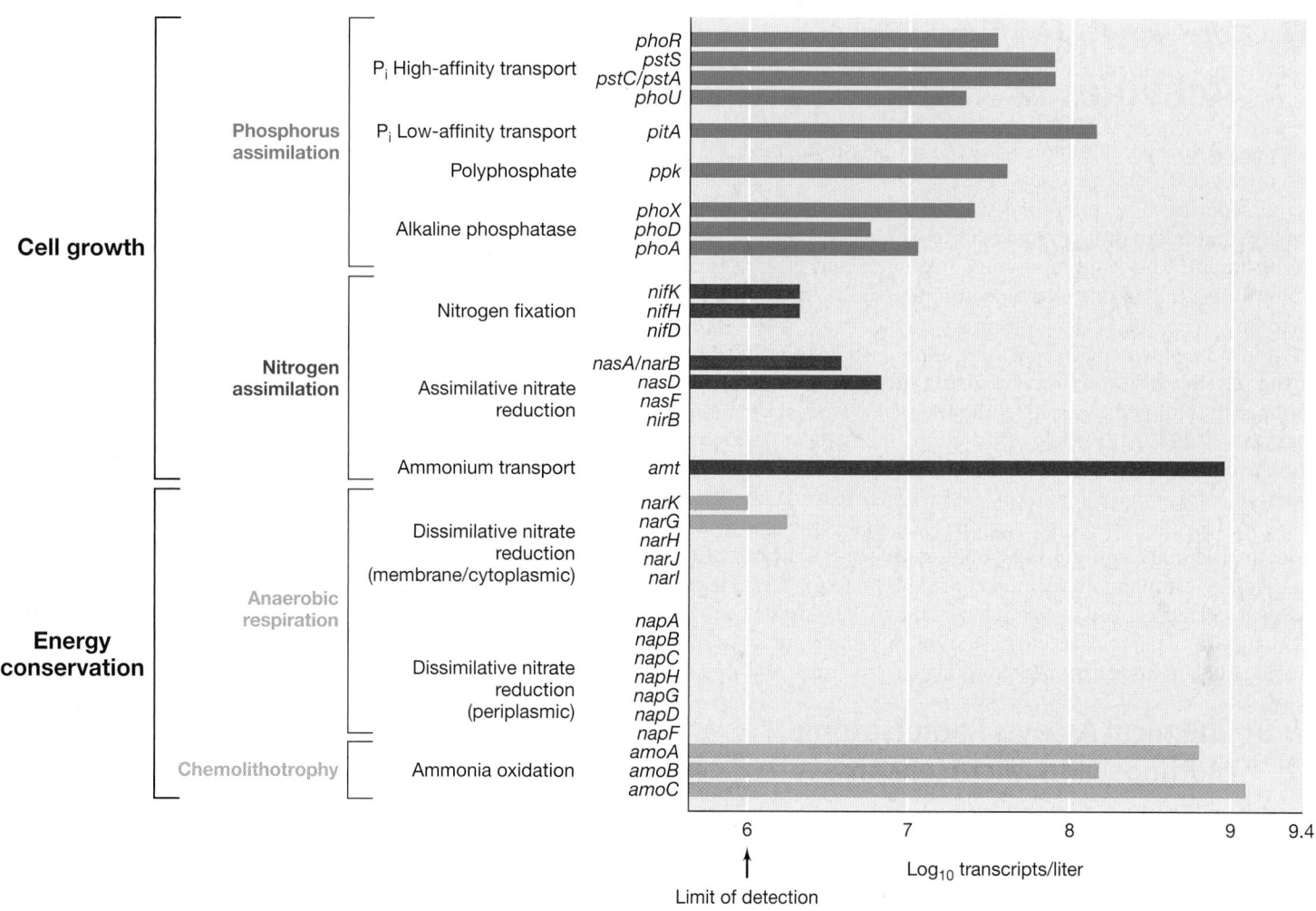

Figure 19.24 Metatranscriptomic analysis of coastal marine surface waters. Expression of genes for key steps in the N and P cycle in a seawater sample determined by sequencing environmental mRNA. These data showed that the microbial community was using both inorganic (high expression of P transporters) and organic (alkaline phosphatase) forms of phosphate (PO_4^{3-}). Low levels of transcripts for genes required for NO_3^- assimilation contrasted with the high expression of genes for NH_3 transport and chemolithotrophic NH_3 oxidation. Also, as expected for oxic marine surface waters, there was little expression of genes for NO_3^- respiration. Data courtesy of Mary Ann Moran, University of Georgia Marine Sciences.

metabolic processes catalyzed by that community at the time of sampling (**Figure 19.24**).

Metaproteomics, the measure of the diversity and abundance of different *proteins* in a community, is an even more direct measure of cell function than is metatranscriptomics. This is because different mRNAs have different half-lives and efficiencies of translation, and thus will not all yield the same number of protein copies. However, metaproteomics is more of a technical challenge than is either metagenomics or metatranscriptomics. Protein identification, usually by mass spectrometric characterization of peptides released from enzymatic digestion of the total protein pool using specific proteases (⟷ Section 9.10), relies on naturally available material since it is not possible to amplify protein sequences as one does using PCR to amplify nucleic acids for sequencing. Protein identification also requires at least partial separation of the individual peptides in order to reduce complexity of the sample as well as a reference genome or metagenome to identify potential coding sequences. As a consequence, as a tool in microbial ecology, metaproteomics has thus far been mainly restricted to the qualitative characterization of rather simple microbial communities, such as those in some extreme environments, or to the characterization of only very abundant proteins in more complex microbial communities.

MINIQUIZ

- What is a metaproteome, and how does it differ from a metagenome and from a metatranscriptome?

- How do environmental genomic approaches differ from environmental single-gene analyses, such as that based on 16S rRNA gene analysis for microbial community characterization?

- How can the most metabolically active cell populations in a community be identified using environmental omics methods?

IV • Measuring Microbial Activities in Nature

Up until now our emphasis has been on measuring microbial phylogenetic and genomic *diversity*. We wrap up this chapter by considering how microbial ecologists measure microbial *activity;* that is, what microorganisms are actually *doing* in their environment. The techniques we consider include the use of radioisotopes, microsensors, stable isotopes, and several genomic methods.

Activity measurements in a natural sample are *collective* estimates of the physiological reactions occurring in the entire microbial community, although several techniques to be discussed later (see Sections 19.10–19.12) allow for a more targeted assessment of physiological activity. Activity measurements reveal both the types and rates of the major metabolic reactions occurring in a habitat, and the various techniques can be used alone or in combination in microbial community analyses. In conjunction with biodiversity estimates and gene expression analyses, these help define the structure and function of the microbial ecosystem, the ultimate goal of microbial ecology. Activity measurements can also provide valuable information for the design of enrichment cultures.

19.9 Chemical Assays, Radioisotopic Methods, and Microsensors

In many studies, direct chemical measurements of microbial reactions are sufficient for assessing microbial activity in an environment. For example, the fate of lactate oxidation by sulfate-reducing bacteria in a sediment sample can be tracked easily. If sulfate-reducing bacteria are present and active in a sediment sample, then lactate added to the sediment will be consumed and SO_4^{2-} will be reduced to H_2S. Since lactate, SO_4^{2-}, and S^{2-} can all be measured with fairly high sensitivity using simple chemical assays, the transformations of these substances relative to one another in a sample can easily be followed (**Figure 19.25a**).

Radioisotopes

When very high sensitivity is required, or turnover rates need to be determined, or the fate of portions of a molecule needs to be followed, *radioisotopes* are more useful than strictly chemical assays. For instance, if measuring photoautotrophy is the goal, the light-dependent uptake of radioactive carbon dioxide ($^{14}CO_2$) into microbial cells can be measured (Figure 19.25b). If SO_4^{2-} reduction is of interest, the rate of conversion of $^{35}SO_4^{2-}$ to $H_2{}^{35}S$ can be assessed (Figure 19.25c). Heterotrophic activities can be measured by tracking the release of $^{14}CO_2$ from ^{14}C-labeled organic compounds (Figure 19.25d), and so on.

Both isotopic and chemical methods are widely used in microbial ecology. To be valid, however, these must employ proper controls because some isotopic transformations might be due to abiotic processes. The *killed cell control* is the key control in such experiments. That is, it is essential to show that the transformation being measured stops when chemical agents or heat treatments that kill microorganisms are applied to the sample. Formalin at a final concentration of 4% is commonly used as a

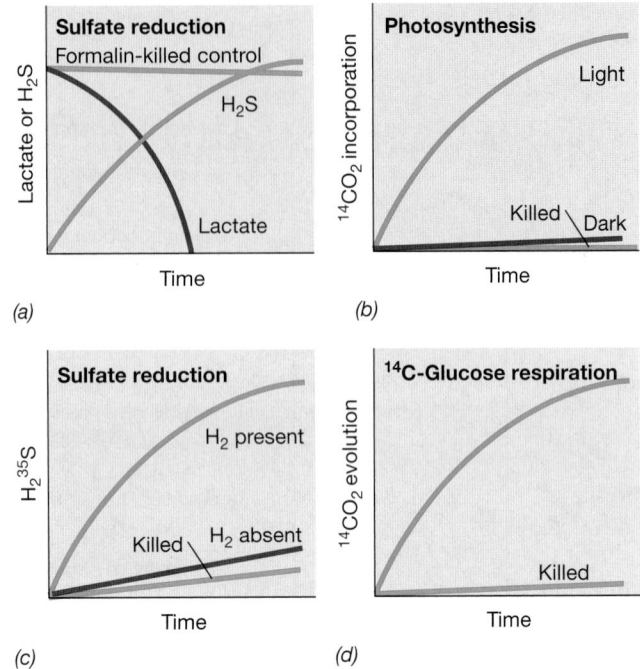

Figure 19.25 Microbial activity measurements. *(a)* Chemical measurements of lactate and H_2S transformations during SO_4^{2-} reduction. Radioisotopic measurements: *(b)* photosynthesis measured with $^{14}CO_2$; *(c)* SO_4^{2-} reduction measured with $^{35}SO_4^{2-}$; *(d)* production of $^{14}CO_2$ from ^{14}C-glucose.

chemical sterilant in microbial ecology studies. This kills all cells, and transformations that are insensitive to the presence of 4% formalin can be ascribed to abiotic processes (Figure 19.25a).

Microsensors

Microsensors in the form of glass needles containing a sensing mechanism at the tip have been used to study the activity of microorganisms in nature. Microsensors have been constructed that measure many chemical species including pH, O_2, NO_2^-, NO_3^-, nitrous oxide (N_2O), CO_2, H_2, and H_2S. As the name *microsensor* implies, these devices are very small, their tips ranging in diameter from 2 to 100 μm (**Figure 19.26**). The sensors are carefully inserted into the habitat in small increments to follow microbial activities over very short distances.

Microsensors have many applications. For example, O_2 concentrations in microbial mats (⊄⊅ Figure 20.7c), aquatic sediments, or soil particles (⊄⊅ Figure 20.3) can be very accurately measured over extremely fine intervals using microsensors. A micromanipulator is used to insert the sensors gradually through the sample such that measurements can be taken every 50–100 μm (**Figure 19.27**). Using a bank of microsensors, each sensitive to a different chemical, simultaneous measurements of several transformations in a habitat can be made.

Microbial processes in the sea are extensively studied because they have a profound impact on nutrient cycles and the overall health of the planet. As it is difficult to reproduce in the laboratory the conditions found at great depths, it is useful to use microsensors on robotic devices to analyze microbial activities on the seafloor. **Figure 19.28** shows deployment of an instrument "lander" equipped with various microsensors so that the distribution of

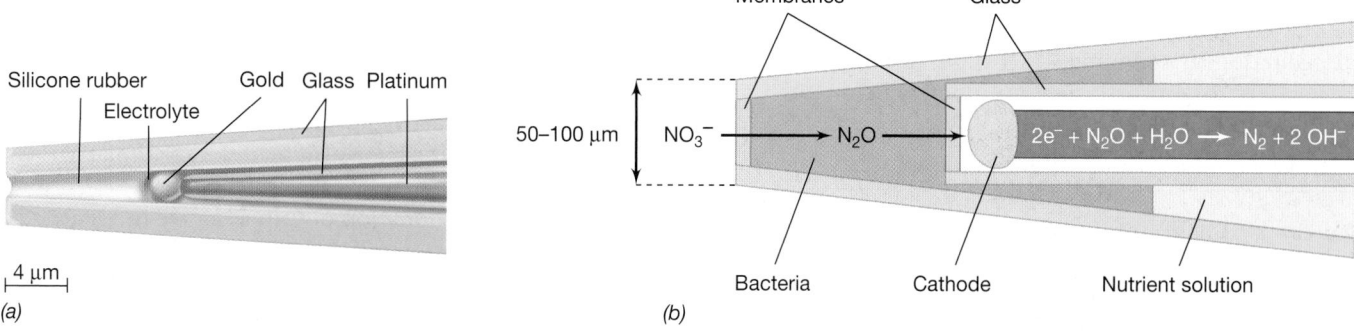

Figure 19.26 Microsensors. *(a)* Schematic drawing of an oxygen (O_2) microsensor. Oxygen diffuses through the silicone membrane in the microsensor tip and reacts with electrons on the gold surface of the cathode, forming hydroxide ions (OH^-); the latter generates a current proportional to the O_2 concentration in the sample. Note the scale of the electrode. *(b)* Biological microsensor for the detection of nitrate (NO_3^-). Bacteria immobilized at the sensor tip denitrify NO_3^- or NO_2^- to N_2O, which is detected by electrochemical reduction to N_2 at the cathode. Based on drawings by Niels Peter Revsbech.

chemicals in the sediment can be analyzed and compared with that in overlying ocean water.

One of the biologically most important chemical species in the oceans is NO_3^-, but electrochemical sensors cannot measure NO_3^- in seawater, as the high concentrations of salts interfere. To circumvent this problem, a "living" microsensor was designed that contains bacteria within its tip that reduce NO_3^- (or NO_2^-) to N_2O. The N_2O produced by the bacteria is then detected following its abiotic reduction to N_2 at the cathode of the microsensor (Figure 19.26*b*); this provides an electrical impulse signaling

the presence of NO_3^-. In the oxic layer of marine sediments, NO_3^- is produced from the oxidation of NH_4^+ (nitrification, ⮌ Section 14.11), so there is often a peak of NO_3^- in the sediment

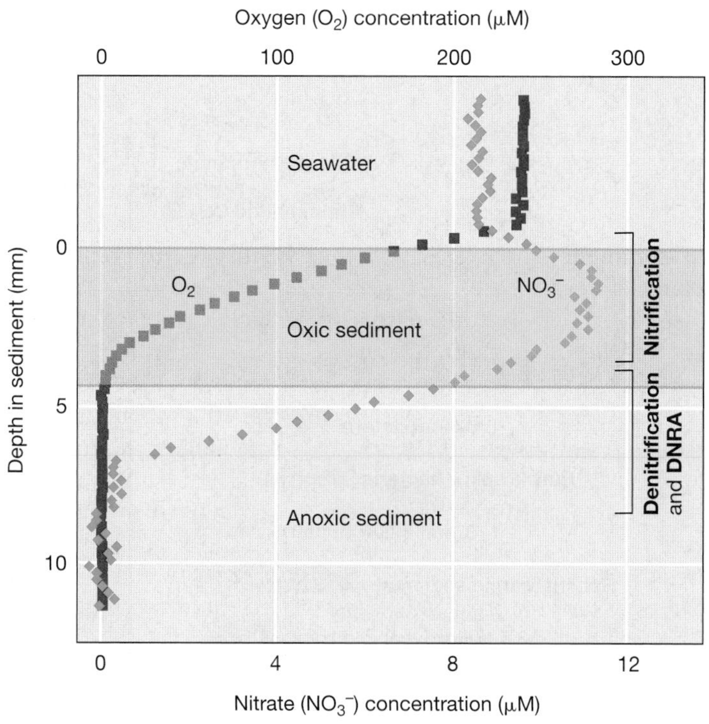

Figure 19.27 Depth profiles of O_2 and NO_3^-. Data obtained using the lander shown in Figure 19.28 equipped with microelectrode sensors for remote chemical characterization of deep-sea sediments. Note the zones of nitrification and denitrification. DRNA, dissimilative reduction of NO_3^- to NH_4^+. Based on data and drawings by Niels Peter Revsbech.

Figure 19.28 Deployment of a deep-sea lander. The lander is equipped with a bank of microsensors (arrow) to measure distribution of chemicals in marine sediments.

surface layer (Figure 19.27). In the deeper, anoxic layers of the sediment, NO_3^- is consumed by denitrification and dissimilative nitrate reduction to ammonia (DRNA) (⇌ Section 14.13), and NO_3^- therefore disappears a few millimeters below the oxic–anoxic interface (Figure 19.27).

19.10 Stable Isotopes and Stable Isotope Probing

Many of the chemical elements have more than one isotope, which differ in their number of neutrons. Certain isotopes are unstable and break down as a result of radioactive decay. Others, called *stable isotopes*, are not radioactive, but are metabolized differently by microorganisms and can be used to study microbial transformations in nature. There are two methods in which stable isotopes can yield information on microbial activities, *isotopic fractionation* and *stable isotope probing*.

Isotopic Fractionation

The two elements most useful for stable isotope studies in microbial ecology are carbon (C) and sulfur (S), although the heavy isotope of nitrogen, ^{15}N, is also widely used. Carbon (C) exists in nature primarily as ^{12}C, but about 5% exists as ^{13}C. Likewise, S with its four stable isotopes exists primarily as ^{32}S. Some S is found as ^{34}S, and very small amounts as ^{33}S and ^{36}S. The relative abundance of these isotopes changes when certain C or S compounds are metabolized by microorganisms because the enzymes that act on these compounds typically favor the *lighter* isotope of C or S. That is, relative to the lighter isotope, the heavier isotope is discriminated against when both are metabolized by an enzyme (**Figure 19.29**).

For example, when CO_2 is fixed into cell material by an autotrophic organism, the cellular C becomes *enriched* in ^{12}C and *depleted* in ^{13}C, relative to an inorganic carbon standard of known isotopic composition. Likewise, the S atom in H_2S produced from the bacterial reduction of SO_4^{2-} is isotopically lighter than H_2S that has formed geochemically. These discriminations are called **isotopic fractionations** (Figure 19.29) and are typically the result of biological activities. Thus this technique can be used as a measure of whether or not a particular transformation has been catalyzed by microorganisms.

The isotopic fractionation of C in a sample is calculated as the extent of ^{13}C depletion relative to a standard having an isotopic composition of geological origin. The standard for C isotope analysis is rocks from a Cretaceous (65- to 150-million-year-old) limestone formation (the Pee Dee belemnite). Because the magnitude of fractionation is usually very small, depletion is calculated as "per mil" (‰, or parts per thousand) and reported as the $\delta^{13}C$ (pronounced "delta C 13") of a sample using the following formula:

$$\delta^{13}C = \frac{(^{13}C/^{12}C\text{ sample}) - (^{13}C/^{12}C\text{ standard})}{(^{13}C/^{12}C\text{ standard})} \times 1000\text{‰}$$

The same formula is used to calculate the fractionation of S isotopes, in this case using iron sulfide (FeS) mineral from the Canyon Diablo meteorite as the standard:

$$\delta^{34}S = \frac{(^{34}S/^{32}S\text{ sample}) - (^{34}S/^{32}S\text{ standard})}{(^{34}S/^{32}S\text{ standard})} \times 1000\text{‰}$$

Use of Isotopic Fractionation in Microbial Ecology

The isotopic composition of a material can reveal its biological or geological past. For example, plant material and petroleum (which is derived from plant material) have similar isotopic compositions (**Figure 19.30**). Carbon from both plants and petroleum is isotopically

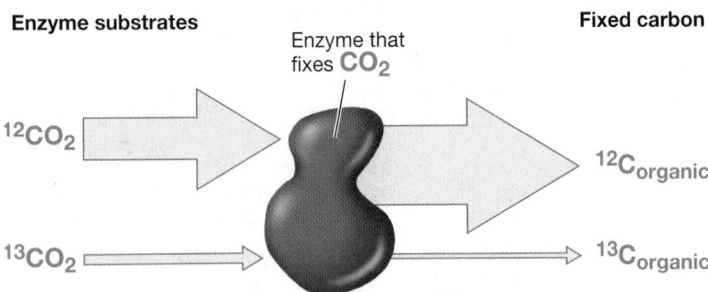

Figure 19.29 Mechanism of isotopic fractionation with C as an example. Enzymes that fix CO_2 preferentially fix the lighter isotope (^{12}C). This results in fixed carbon being enriched in ^{12}C and depleted in ^{13}C relative to substrate CO_2. The size of the arrows indicates the relative abundance of each isotope of carbon.

Enzyme substrates   Fixed carbon

Enzyme that fixes CO_2

$^{12}CO_2$ → $^{12}C_{organic}$

$^{13}CO_2$ → $^{13}C_{organic}$

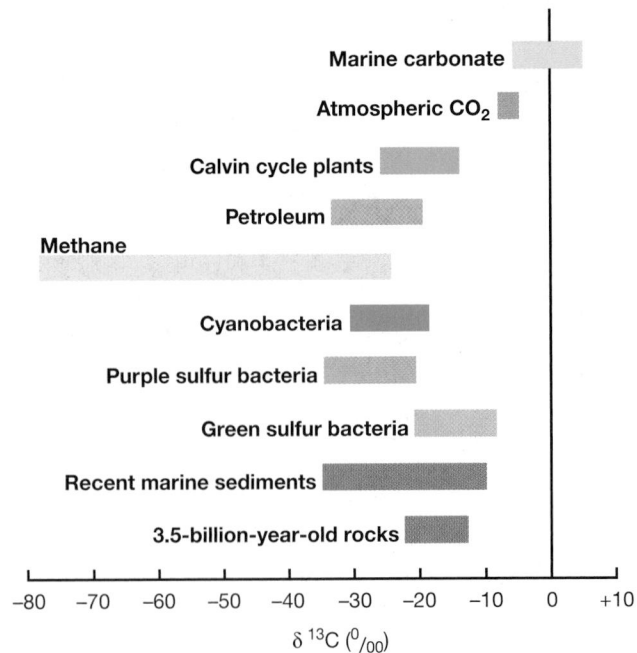

Figure 19.30 Isotopic geochemistry of ^{13}C and ^{12}C. Note that C fixed by autotrophic organisms is enriched in ^{12}C and depleted in ^{13}C. Methane formed from the reduction of CO_2 with H_2 by methanogenic *Archaea* shows extreme isotopic fractionation.

Marine carbonate
Atmospheric CO_2
Calvin cycle plants
Petroleum
Methane
Cyanobacteria
Purple sulfur bacteria
Green sulfur bacteria
Recent marine sediments
3.5-billion-year-old rocks

−80 −70 −60 −50 −40 −30 −20 −10 0 +10
$\delta^{13}C$ (‰)

lighter than the CO_2 from which it was formed because the biochemical pathway used to fix CO_2 discriminated against $^{13}CO_2$ (Figures 19.29 and 19.30). Moreover, methane (CH_4) produced by methanogenic *Archaea* (⮂ Section 17.2) is isotopically extremely light, indicating that methanogens discriminate strongly against $^{13}CO_2$ when they reduce CO_2 to CH_4 (⮂ Section 14.17). By contrast, carbon in isotopically heavier marine carbonates is clearly of geological origin (Figure 19.30).

Because of the differences in the proportion of ^{12}C and ^{13}C in carbon of biological versus geological origin, the $^{13}C/^{12}C$ ratio of rocks of different ages has been used as evidence for or against past biological activity in Earth's ancient environments. Organic C in rocks as old as 3.5 billion years shows evidence of isotopic fractionation (Figure 19.30), supporting the idea that autotrophic life existed at this time. Indeed, we now believe that the first life on Earth appeared somewhat before this, about 3.8–3.9 billion years ago (⮂ Sections 1.3 and 13.1).

The activity of sulfate-reducing bacteria is easy to recognize from their fractionation of stable S isotopes in sulfides (**Figure 19.31**). As compared with an H_2S standard, sedimentary H_2S is highly enriched in ^{32}S (depleted in ^{34}S, Figure 19.31). Fractionation during sulfate reduction allows one to identify biologically produced S and has been widely used to trace the activities of sulfur-cycling *Bacteria* and *Archaea* through geological time. Sulfur isotopic analyses have also been used as evidence for the lack of life on the Moon. For example, the data in Figure 19.31 show that the isotopic composition of sulfides in lunar rocks closely approximates that of the H_2S standard, which represents primordial Earth, and differs from that of microbially produced H_2S.

Stable Isotope Probing

Beyond stable isotope fractionation, an alternative stable isotope method called **stable isotope probing (SIP)** can be used to identify an organism or organisms carrying out the transformation of a nutrient labeled with a specific stable heavy isotope, such as ^{13}C or ^{15}N or even ^{18}O (the lighter, more common isotopes of these elements are ^{12}C, ^{14}N, and ^{16}O, respectively). The idea behind SIP is that the label will be selectively incorporated only into the cellular material of organisms actively metabolizing the nutrient.

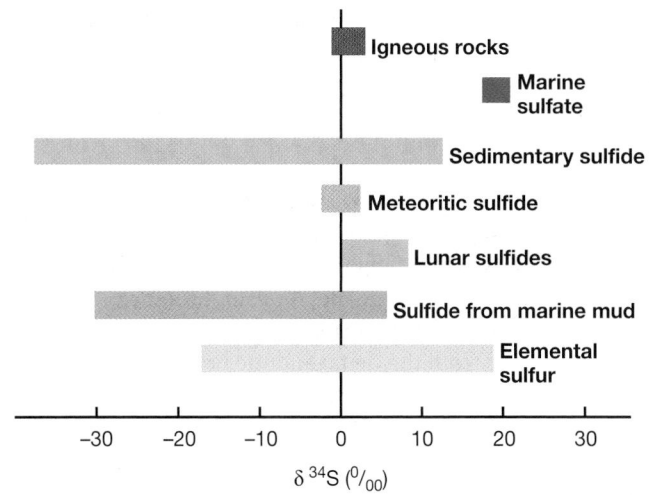

Figure 19.31 Isotopic geochemistry of ^{34}S and ^{32}S. Note that H_2S and S^0 of biogenic origin are enriched in ^{32}S and depleted in ^{34}S.

Then, following the isolation and sequencing of isotopically labeled DNA, the organisms carrying out the transformation can be identified.

Stable isotope probing can be used to ask general to more specific questions. For example, if ^{13}C-labeled benzoate were added to a sediment sample and the sample incubated for an appropriate period, the ^{13}C label would end up in the DNA of the organism (or organisms) that metabolized the benzoate (**Figure 19.32**). Thus, although all DNA from the sample would be isolated, the ^{13}C-DNA is heavier, albeit only slightly heavier, than ^{12}C-DNA, and this difference is sufficient to separate the DNAs by a special type of centrifugation technique (Figure 19.32). Once the ^{13}C-DNA is isolated, it can be analyzed for phylogenetic genes or metabolic genes to yield genomic information about the benzoate degrader(s) in the sample. If instead of an organic compound, the experimental study focused on nitrogen fixation (conversion of N_2 to cell nitrogen, ⮂ Section 14.6), $^{15}N_2$ could be supplied to a sample. When nitrogen fixers in the sample incorporate this, they will produce

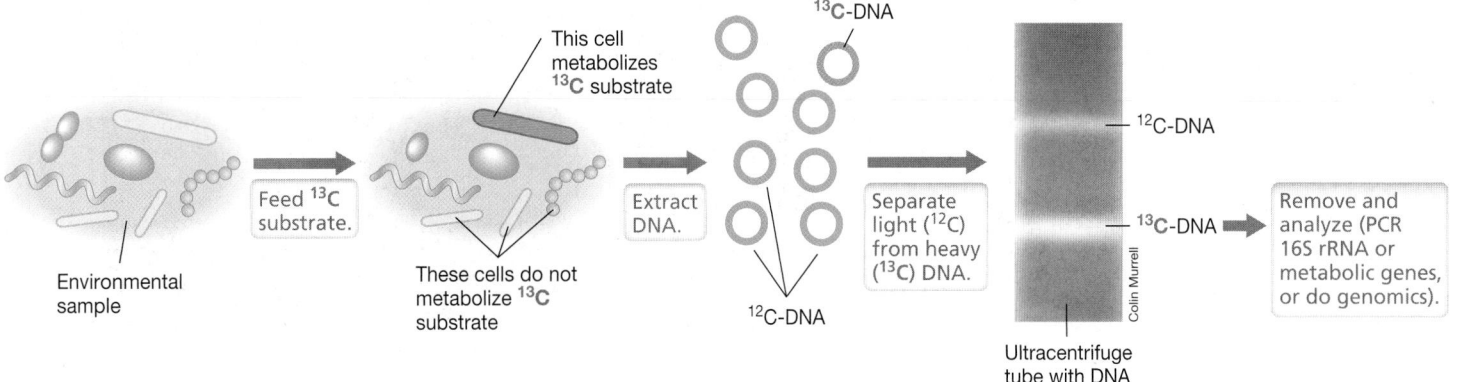

Figure 19.32 Stable isotope probing. The microbial community in an environmental sample is fed a ^{13}C substrate. Organisms that can metabolize the substrate produce ^{13}C-DNA as they grow and divide; ^{13}C-DNA can be separated from lighter ^{12}C-DNA by density gradient centrifugation (photo). The isolated DNA is then subjected to specific gene analysis or entire genomic analysis.

slightly heavier DNA than organisms that cannot fix nitrogen, and the heavier DNA can be isolated and analyzed for genes of interest.

SIP can also be used in combination with metagenomics to pinpoint organisms carrying out a specific metabolism (from SIP results) in the context of all the other species and metabolisms present in the sample (as revealed by metagenomic results). Besides a phylogenetic "hit" from the SIP results, additional genomic analyses of the labeled DNA could reveal functional genes required for the specific metabolism. Moreover, the phylogenetic and functional results from the SIP experiment could be further confirmed from the metagenomic profile.

MINIQUIZ

- What is the simplest explanation for why lunar sulfides are isotopically similar to those of the primordial Earth?

- What is the expected isotopic composition of carbon in methanotrophs (bacteria that consume CH_4)?

- How might exchange of metabolites among members of a microbial community complicate interpretation of an SIP experiment?

19.11 Linking Functions to Specific Organisms

The isotopic methods described thus far used samples containing large numbers of cells to infer that specific metabolisms were occurring within a community or in particular species within the community. These methods give an overview of community activities but do not reveal the contribution of individual cells. To do this, new methods have been developed that can measure the activity and the elemental and isotopic composition of single cells. These are powerful methods for connecting cells of a specific

microbial population with a specific activity or ecological niche, but in most cases, the phylogeny of the organisms of interest must be known such that the necessary FISH probes (Section 19.5) can be developed.

Single-Cell Metabolisms Imaged by Secondary Ion Mass Spectrometry (SIMS)

Secondary ion mass spectrometry (*SIMS*) is based on the detection of ions released from a sample placed under a focused high-energy primary ion beam, for example, of cesium (Cs^+); from the data generated, the elemental and isotopic composition of released materials can be obtained. When the primary ion beam impacts the sample, most chemical bonds are broken and atoms or polyatomic fragments are ejected from a very thin layer (1–2 nm) of the surface as either neutral or charged particles (secondary ions), a process called *sputtering*. These secondary ions are directed to a mass spectrometer, an instrument that can determine their mass-to-charge ratio.

NanoSIMS instruments are SIMS devices designed to yield information on single cells. The instrument is equipped with Cs^+ and O_2 primary beam sources with a resolution of 50 nm for the Cs^+ ion beam and 200 nm for the O_2 beam. The O_2 beam generates positive secondary ions and is used to analyze metals (e.g., Fe, Na, Mg) while the Cs^+ beam generates negative secondary ions for the analysis of major cellular elements (C, N, P, S, O, H) and halogens. The NanoSIMS instrument also records where on the specimen the ion beam is directed such that a two-dimensional image of the distribution of specific ions on the sample surface is obtained. In addition, by focusing the ion beam on the same spot during repeated cycles of sputtering, material can be slowly burned away to expose deeper regions of the sample. This high-resolution SIMS analysis is where the term *NanoSIMS* got its name. NanoSIMS instruments have multiple detectors that provide for the simultaneous analysis of ions of different mass-to-charge ratios originating from the same sample location (**Figure 19.33a**).

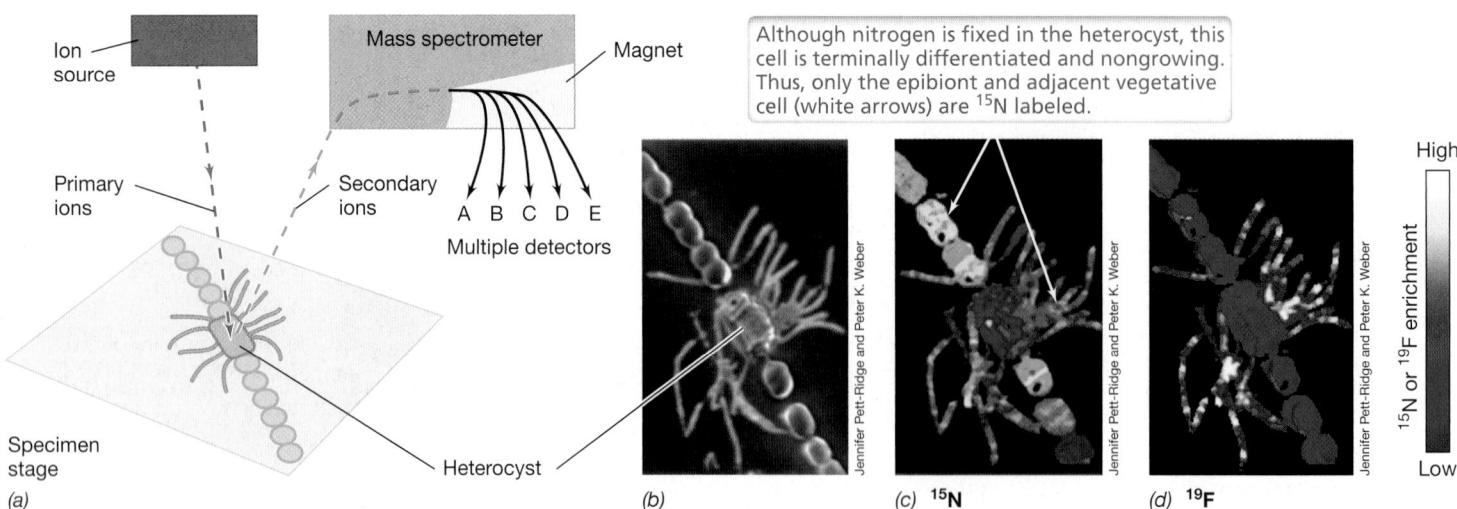

Figure 19.33 NanoSIMS technology. *(a)* Schematic of NanoSIMS operation showing the beams of primary (red) and secondary (blue) ions and five different detectors, each of which identifies ions of a different mass-to-charge ratio. *(b–d)* Demonstration of interspecies nutrient transfer from a filamentous cyanobacterium (*Anabaena*) to a *Rhizobium* species attached to the cyanobacterial heterocyst. The coculture was incubated with $^{15}N_2$, and the transfer of ^{15}N-labeled compounds from *Anabaena* to *Rhizobium* was imaged using a combination of EL-FISH and NanoSIMS. *(b)* Total ^{12}C abundance shown by gray tones. *(c)* ^{15}N enrichment. *(d)* ^{19}F abundance conferred by a probe that hybridizes only to the attached rhizobial cells (EL-FISH).

When NanoSIMS is combined with FISH (Section 19.5) in a technology called *FISH-SIMS*, the incorporation of different elements, natural isotopes, or isotope-labeled substrates can be tracked into individual cells of specific cell populations. A variation on the FISH-SIMS method that simplifies the identification of cells scanned by NanoSIMS uses probe-conferred deposition of a halide (Br, Fl, I), either through direct incorporation of the halide into an oligonucleotide probe or by using a halide-containing tyramide substrate (see CARD-FISH, Section 19.5). Halogens possess a high ionization yield compared with other elements and are thus easy to detect and are typically of low natural abundance. Using this technology, one of the NanoSIMS detectors is dedicated to identifying cells to which the probe has hybridized (Figure 19.33*d*) by halogen ionization while the remaining detectors are used for assessing elemental composition (Figure 19.33*c*).

Because of its excellent spatial resolution, NanoSIMS technology is being increasingly used to examine metabolite transfer among single cells of interacting microorganisms. For example, labeling with $^{15}N_2$ followed by NanoSIMS was used to demonstrate the transfer of N_2 fixed by filamentous cyanobacteria to attached heterotrophic bacteria (Figure 19.33*c*). Labeling with $^{15}NH_4$ and ^{13}C-labeled CO_2 or organic substrates is also being used to explore the assimilation of key nutrients and the transfer of metabolites among microbial species in both aquatic and soil microbial communities.

Raman Microspectroscopy

Raman microspectroscopy can be used to characterize the molecular and isotopic composition of single cells by nondestructive illumination with monochromatic light generated by a laser. Raman is a form of spectroscopy that measures light scattering and can yield both qualitative and quantitative results. Compositional analysis is based on photon scattering after interaction with different cellular components. Although most of the scattered photons have the same energy as the incident photons, a small fraction of scattered photons are shifted in wavelength (relative to the incident wavelength) to either a longer wavelength (a phenomenon known as *Stokes Raman scattering*) or shorter wavelength (*anti-Stokes scattering*).

Raman spectrometers separate the more abundant Stokes scattered photons for analysis and, when combined with confocal microscopy (Section 1.7), can generate a compositional spectrum of a single microbial cell (**Figure 19.34**). Although the spectrum is complex, several compounds and molecules have characteristic peaks that can identify specific cell types, physiological states, or metabolic activities following incorporation of compounds labeled with stable isotopes.

Major advantages of Raman microspectroscopy include the following: It is nondestructive and can be used on living cells; water does not cause interference; it can be combined with FISH; and cells of interest can be further manipulated, for example, through capture by laser tweezers (Section 19.3). In the example of Raman shown in Figure 19.34, the incorporation of phenylalanine (an amino acid) by an obligate chlamydial symbiont of an amoeba has been tracked following addition of phenylalanine

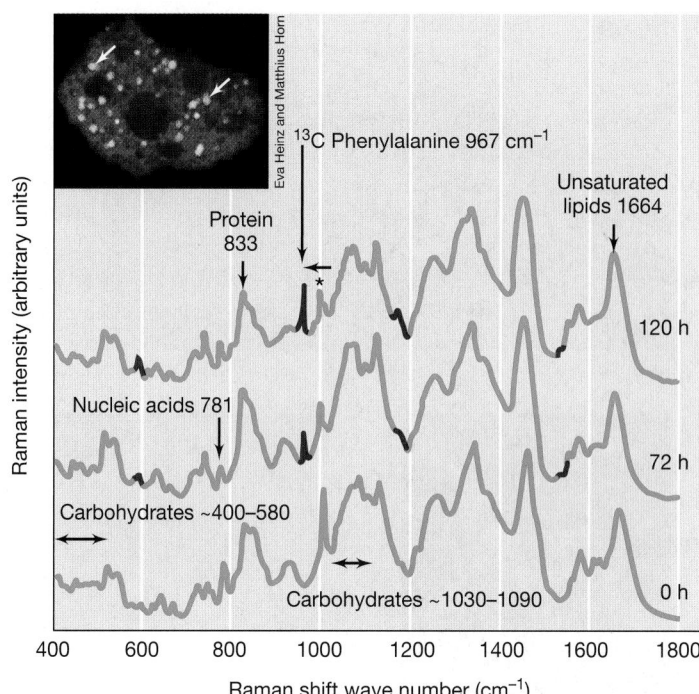

Figure 19.34 Raman microspectroscopic analysis of single cells. A cell of *Acanthamoeba* containing FISH-stained chlamydial symbionts (inset photo, arrows point to blue chlamydial cells) and Raman spectroscopy of isolated chlamydial cells. After incubation for 72 and 120 h in medium containing ^{13}C phenylalanine, symbionts were released by lysis of the amoeba and their Raman spectra recorded. Peaks diagnostic for labeled phenylalanine are shown in red. Since the Raman wave number of ^{13}C phenylalanine at 967 cm^{-1} is well resolved from other spectral features, the ratio of 1003 cm^{-1} (unlabeled phenylalanine, at *) to 967 cm^{-1} (labeled phenylalanine) peak areas corresponds to the relative amount of labeled phenylalanine incorporated by single cells (note how this increases with time). Other red peaks correspond to spectral peaks specific for different chemical bonds in ^{13}C phenylalanine. Also shown are selected peaks and their wave numbers for other cellular components of the symbiont.

labeled with ^{15}N and ^{13}C to the amoeba growth medium. A less specific but still useful application of Raman is based on the incorporation of heavy protons from deuterated water (^2H_2O). Since water is a reactant in many biochemical reactions, metabolically active microorganisms that incorporate the heavy hydrogen atom can be identified by Raman microspectroscopy for further processing, including characterization by single-cell genomics (Section 19.12).

Radioisotopes in Combination with FISH: Microautoradiography-FISH

Radioisotopes are used as measures of microbial activity in a microscopic technique called **microautoradiography (MAR)**. In this method, cells from a microbial community are exposed to a substrate containing a radioisotope, such as an organic compound or CO_2. Heterotrophs take up the radioactive organic compounds and autotrophs take up the radioactive CO_2. Following incubation in the substrate, cells are affixed to a slide and the slide is dipped in photographic emulsion. While the slide is left in darkness for a period, radioactive decay from the incorporated substrate induces formation of silver grains in the emulsion; these appear as black

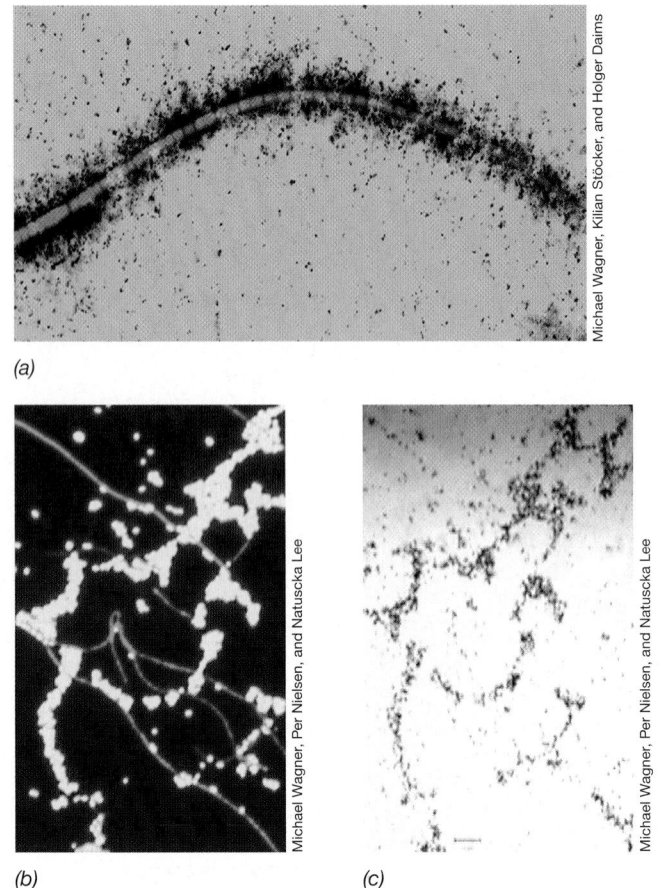

(a)

(b)   *(c)*

Michael Wagner, Kilian Stöcker, and Holger Daims

Michael Wagner, Per Nielsen, and Natuscka Lee

Michael Wagner, Per Nielsen, and Natuscka Lee

Figure 19.35 MAR-FISH. Fluorescence in situ hybridization (FISH) combined with microautoradiography (MAR). *(a)* An uncultured filamentous cell belonging to the *Gammaproteobacteria* (as revealed by FISH) is shown to be an autotroph (as revealed by MAR-measured uptake of $^{14}CO_2$). *(b)* Uptake of ^{14}C-glucose by a mixed culture of *Escherichia coli* (yellow cells) and *Herpetosiphon aurantiacus* (filamentous green cells). *(c)* MAR of the same field of cells shown in part *b*. The radioactivity of incorporated glucose exposes the film and shows that glucose was assimilated mainly by cells of *E. coli*.

dots above and around the cells. **Figure 19.35a** shows a MAR experiment in which an autotrophic cell has taken up $^{14}CO_2$.

Microautoradiography can be done simultaneously with FISH (Section 19.5) in **MAR-FISH**, a powerful technique that combines identification with activity measurements. MAR-FISH allows a microbial ecologist to determine (by MAR) which organisms in a natural sample are metabolizing a particular radiolabeled substance while at the same time identifying these organisms (by FISH) (Figure 19.35). MAR-FISH thus goes a step beyond phylogenetic identification by revealing physiological information about the organisms, as is also true of NanoSIMS. Such data are useful not only for understanding the activity of the microbial ecosystem but also for guiding enrichment cultures. For example, knowledge of the phylogeny and morphology of an organism metabolizing a particular substrate in a natural sample can be used to design an enrichment protocol to isolate the organism. In addition, MAR-FISH results can be quantified by counting the silver grains as a measure of the amount of substrate consumed by single cells, allowing the activity distribution in a community to be described. The technique is limited only by the availability of suitable radioactive isotopes.

For example, although C-labeled substrates work well, it is not feasible to track N incorporation using MAR-FISH because the radioactive isotope ^{13}N has a very short half-life. However, it is feasible to track N incorporation using the nonradioactive stable isotope ^{15}N with NanoSIMS, as we saw earlier (Figure 19.33).

MINIQUIZ

- How could NanoSIMS be used to identify a nitrogen-fixing bacterium?
- Why is Raman microspectroscopy suited for the selective isolation of microorganisms and NanoSIMS is not?
- How does MAR-FISH link microbial diversity and activity?

19.12 Linking Genes and Cellular Properties to Individual Cells

We have seen in the previous section how the combination of FISH with MAR or FISH with NanoSIMS allows for analyses of both microbial diversity and activity. Coupled with advanced DNA sequencing methods that can determine a genome sequence from the DNA contained in a single cell (⇔ Section 9.12), these techniques are at the cutting edge of microbial ecology today. Improvements in single-cell DNA sequencing technology, combined with high-throughput analysis and isolation of single cells by flow cytometry, now allow for gene identification and selected physiological analyses (e.g., size and intrinsic fluorescence) to be performed on selected populations and single cells in the environment.

Flow Cytometry and Multiparametric Analyses

Because of the large population sizes of natural microbial communities, methods that rely on microscopy can examine only a very small part of a whole community. It is difficult to assess cell numbers by counting cells microscopically, and this problem is compounded if populations are present in low numbers. However, flow cytometry (Section 19.3) offers an alternative to more labor-intensive microscopic methods.

Flow cytometers can examine specific cell parameters such as size, shape, or fluorescent properties as the cells pass through a detector at rates of many thousands of cells per second (**Figure 19.36**). Fluorescence may be intrinsic (for example, chlorophyll fluorescence of phototrophic microorganisms); or it may be conferred by DNA staining, or by differential staining of live versus dead cells (vital stains), or by fluorescent DNA probes (FISH), all methods discussed in this chapter.

A major advantage of flow cytometry is the ability to carry out *multiparametric analyses*, that is, the capacity to combine multiple parameters in the analysis of a microbiological sample or to sort cells in order to find a specific population. A good example of this was the discovery in the late 1980s of a novel and abundant community of marine cyanobacteria, all species of the genus *Prochlorococcus*. *Prochlorococcus* cells are smaller and have different fluorescent properties than another common marine cyanobacterium, *Synechococcus*. Based on differences in size and fluorescence, flow cytometry resolved these two populations and *Prochlorococcus* was subsequently shown to be the predominant

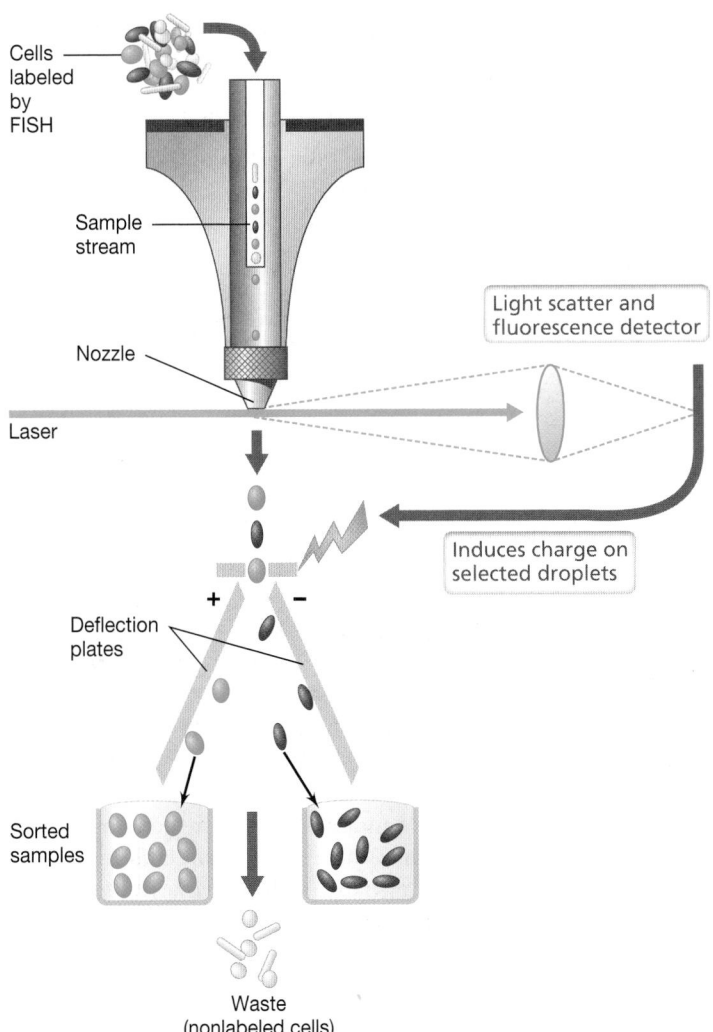

Figure 19.36 Flow cytometric cell sorting. As the fluid stream exits the nozzle, it is broken into droplets containing no more than a single cell. Droplets containing desired cell types (detected by fluorescence or light scatter) are charged and collected by redirection into collection tubes or microtiter plates by positively or negatively charged deflection plates.

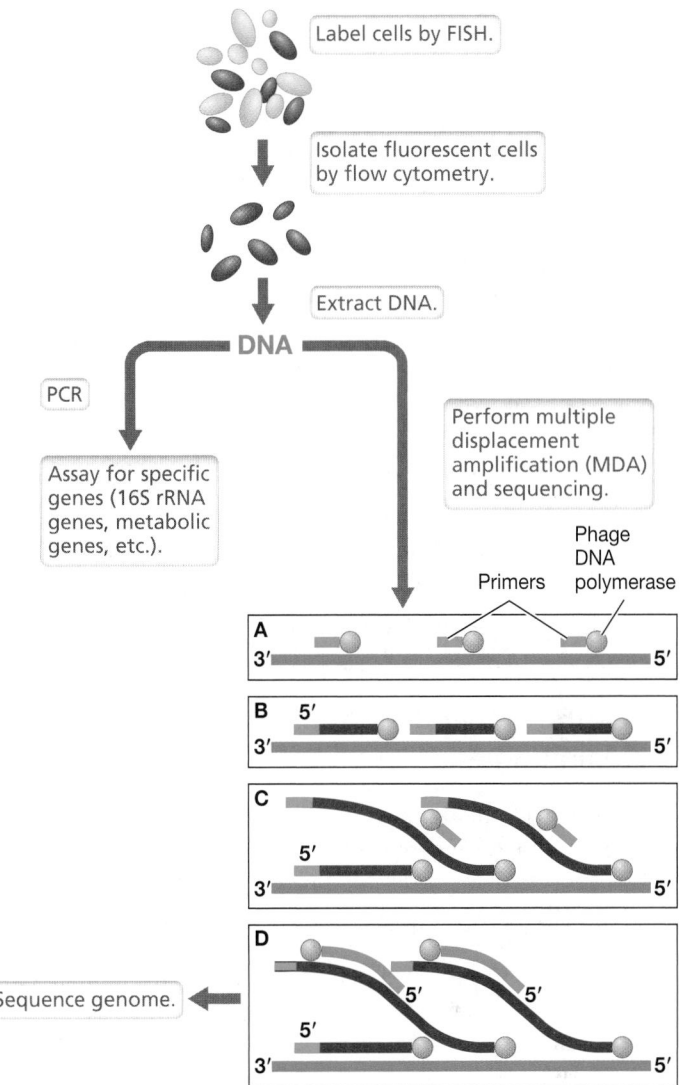

Figure 19.37 Genetic analysis of sorted cells. DNA is recovered from a specific population of cells following FISH labeling and flow cytometric sorting (Figure 19.36). DNA is characterized by PCR amplification and sequencing of specific genes, or by amplification of the entire genome by multiple displacement amplification (MDA) followed by sequencing. For MDA, an amount of DNA sufficient for full genome sequence determination is produced using short DNAs of random sequence as primers (A) to initiate genome replication by a bacteriophage DNA polymerase. The bacteriophage polymerase copies DNA from multiple points in the genome and also displaces newly synthesized DNA (B, C), thereby freeing additional DNA for primer annealing and (D) initiation of polymerization.

oxygenic phototroph in ocean waters between 40°S and 40°N latitude, reaching concentrations greater than 10^5 cells/ml. Metagenomics has also been used to identify the unique genomic features of different natural populations of *Prochlorococcus* (Figure 19.23). These findings have led to the conclusion that *Prochlorococcus* is the most abundant phototrophic organism on Earth. We discuss the biology of *Prochlorococcus* in more detail in Section 20.10.

Single-Cell Genomics

A major stumbling block in a PCR-based gene recovery method is the requirement that a specific gene that will react with the primers used in the amplification be identified prior to analysis. Newer methods of DNA amplification now provide an alternative method for associating specific genes with a specific organism without the problems and biases associated with PCR. These methods employ *single-cell genomics* (⊄⊃ Section 9.12), one of the most recent tools to enter the microbial ecologist's toolbox.

Multiple displacement amplification (MDA) (Figure 19.37) is key to single-cell genomics because it can amplify chromosomal DNA from a single cell isolated from a natural environment using a cell sorting technique, such as flow cytometry (Figure 19.36). MDA uses a specific bacteriophage DNA polymerase to initiate replication of cell DNA at random points in the chromosome, displacing the complementary strand as each polymerase molecule synthesizes new DNA. The phage polymerase has strong strand displacement activity, resulting in the synthesis of numerous high-molecular-weight DNA products. The number of genome copies produced by amplification is sufficient to determine the complete, or nearly complete, genome sequence using next-generation sequencing systems. In this way, both phylogenetic and metabolic functions can be inferred from the genome sequence and PCR is not required.

As could be predicted, MDA requires stringent control over purity to eliminate contaminating DNA, but when combined with high-throughput DNA sequencing methods, MDA provides a powerful tool for linking specific metabolic functions to individual cells that have never been grown in laboratory culture. Information about the metabolic capacities of these uncultured organisms can then be used to develop strategies to recover them by either classical enrichment culture and isolation methods (Sections 19.1 and 19.2) or by any of the several single-cell isolation culturing

techniques now available to tease out individual cells and get them growing in the laboratory (Section 19.3).

— **MINIQUIZ** —

- How can stable isotope probing reveal the identity of an organism that carries out a particular process?
- What key method is required to do genomics on a single cell?
- Compared with microscopy, what are the advantages and disadvantages of flow cytometry for characterizing a microbial community?

MasteringMicrobiology®   **Visualize**, **explore**, and **think critically** with Interactive Microbiology, MicroLab Tutors, MicroCareers case studies, and more. MasteringMicrobiology offers practice quizzes, helpful animations, and other study tools for lecture and lab to help you master microbiology.

Chapter Review

I • Culture-Dependent Analyses of Microbial Communities

19.1 The enrichment culture technique is a means of obtaining microorganisms from natural samples. Successful enrichment and isolation prove that an organism of a specific metabolic type was present in the sample, but do not indicate its ecological importance or abundance. Enrichments following dilution of the sample often yield different organisms than enrichments with undiluted samples.

> **Q** Why do the results of a direct enrichment of an environmental sample and enrichment following dilution of the sample often differ with respect to the types of populations recovered?

19.2 Once a successful enrichment culture has been established, pure cultures can often be obtained by conventional microbiological procedures, including streak plates, agar dilution, and liquid dilution methods.

> **Q** What criteria serve to demonstrate that a culture of a previously undescribed microorganism is pure?

19.3 Several methods are available to isolate and culture single cells. Laser tweezers allow one to isolate a cell from a microscope field and move it away from contaminants. Flow cytometric sorting combined with high-throughput culturing technology allow for isolated cells to be cultured in a large variety of culture media simultaneously to identify the resources and conditions best suited to the growth of the isolated cell.

> **Q** What feature of high-throughput culturing relieves the human demands that would otherwise be required to set up and monitor huge numbers of individual cultures?

II • Culture-Independent Microscopic Analyses of Microbial Communities

19.4 DAPI, acridine orange, and SYBR Green are general stains for quantifying microorganisms in natural samples. Some stains can differentiate live versus dead cells. The GFP makes cells autofluorescent and is a means for tracking cells introduced into the environment and reporting gene expression. In natural samples, morphologically identical cells may actually be genetically distinct.

> **Q** What limits the application of GFP for general studies of microbial activity and distribution in the environment?

19.5 FISH methods have combined the power of nucleic acid probes with fluorescent dyes and are thus highly specific in their staining properties. FISH methods include phylogenetic stains and CARD-FISH.

> **Q** Why is CARD-FISH more suitable than FISH for characterizing very slowly growing microorganisms in the environment?

III • Culture-Independent Genetic Analyses of Microbial Communities

19.6 PCR can be used to amplify specific target genes such as rRNA genes or key metabolic genes for subsequent analysis of community structure and potential functions. DGGE and T-RFLP can identify the different variants of these genes among the species in a community. Application of ARISA is limited to amplification of the internal transcribed spacer region separating the 16S and 23S rRNA genes.

Q **Which method, ARISA or T-RFLP, would provide more detail about microbial community complexity? Why?**

19.7 Microarrays comprised of thousands of DNA probes are used to screen communities for specific phylogenetic groups as well as genes encoding key biochemical processes.

Q **Why might a microarray be superior to using high-throughput sequencing to identify a rare population member in a complex microbial community? What are the advantages and limitations of FISH and PhyloChips for analysis of microbial communities?**

19.8 Environmental genomics (metagenomics) is based on cloning (except in direct sequencing), sequencing, and analysis of the collective genomes of the organisms present in a microbial community. Metatranscriptomics and metaproteomics are offshoots of metagenomics whose focus is mRNA and proteins, respectively.

Q **Give an example of how environmental genomics has discovered a known metabolism in a new organism. Why is it difficult to use environmental genomics to discover new biochemical properties?**

IV • Measuring Microbial Activities in Nature

19.9 The activity of microorganisms in natural samples can be assessed very sensitively using radioisotopes or microsensors, or both. The measurements obtained give the net activity of the microbial community.

Q **What are the major advantages of radioisotopic methods in the study of microbial ecology? What type of controls (discuss at least two) would you include in a radioisotopic experiment to show $^{14}CO_2$ incorporation by phototrophic bacteria or to show $^{35}SO_4^{2-}$ reduction by sulfate-reducing bacteria?**

19.10 Natural isotopic composition, the result of isotopic fractionation by enzymes that discriminate against the heavier form of an element, can reveal the biological origin and/or biochemical mechanisms involved in the formation of various substances. Stable isotope probing (SIP) uses compounds labeled with isotopes not naturally abundant to identify microorganisms metabolizing and assimilating the compound added to a community.

Q **Will autotrophic organisms contain more or less ^{12}C in their organic compounds than was present in the CO_2 that fed them? Why would SIP using $^{15}NO_3^-$ not be useful for identifying bacteria carrying out nitrate respiration?**

19.11 A variety of advanced technologies such as NanoSIMS, MAR-FISH, and Raman microspectroscopy make it possible to examine metabolic activity, gene content, and gene expression of single cells in natural microbial communities. NanoSIMS employs secondary ion mass spectrometry technology while MAR-FISH combines the uptake of radiolabeled substrates (MAR) along with phylogenetic identification (FISH). Raman microspectroscopy is a nondestructive method (retaining cell viability) for identifying metabolically active microorganisms in the environment.

Q **What can MAR-FISH tell you that FISH alone cannot? How might you combine SIP and NanoSIMS to identify novel methane-consuming cells in a natural community?**

19.12 Flow cytometry combined with cell sorting can rapidly evaluate many thousands of single cells in a natural environment for basic cellular properties (size, shape) or gene content (using specific fluorescent probes). Single-cell genomics incorporates methods for analyzing the genome of individual cells isolated from a natural microbial community, for example by flow cytometry.

Q **How would you use cytometric cell sorting to evaluate genome sequence variation among a minor population of marine bacteria?**

Application Questions

1. Design an experiment for measuring the activity of sulfur-oxidizing bacteria in soil. If only certain species of the sulfur oxidizers present were metabolically active, how could you tell this? How would you prove that your activity measurement was due to biological activity?

2. You wish to know whether *Archaea* exist in a lake water sample but are unsuccessful in culturing any. Using techniques described in this chapter, how could you determine whether *Archaea* exist in the sample, and if they do, what proportion of the cells in the lake sample are *Archaea*?

3. Design an experiment to solve the following problem: Determine the rate of methanogenesis ($CO_2 + 4H_2 \rightarrow CH_4 + 2H_2O$) in anoxic lake sediments and whether or not it is H_2-limited. Also, determine the morphology of the dominant methanogen (recall that these are *Archaea*, ⇄ Section 17.2). Finally, calculate what percentage the dominant methanogen is of the total archaeal and total prokaryotic populations in the sediments. Remember to specify necessary controls.

4. Design a SIP experiment that would allow you to determine which organisms in a lake water sample were capable of oxidizing the hydrocarbon hexane (C_6H_{14}). Assume that four different species could do this. How would you combine SIP with other molecular analyses to identify these four species?

Chapter Glossary

Acridine orange a nonspecific fluorescent dye used to stain DNA in microbial cells in a natural sample

DAPI a nonspecific fluorescent dye that stains DNA in microbial cells; used to obtain total cell numbers in natural samples

Denaturing gradient gel electrophoresis (DGGE) an electrophoretic technique capable of separating nucleic acid fragments of the same size that differ in base sequence

Enrichment bias a problem with enrichment cultures in which "weed" species tend to dominate in the enrichment, often to the exclusion of the most abundant or ecologically significant organisms in the inoculum

Enrichment culture a culture that employs highly selective laboratory methods for obtaining microorganisms from natural samples

Environmental genomics (metagenomics) the use of genomic methods (sequencing and analyzing genomes) to characterize natural microbial communities

Flow cytometry a technique for counting and examining microscopic particles by suspending them in a stream of fluid and passing them by an electronic detection device

Fluorescence in situ hybridization (FISH) a method employing a fluorescent dye covalently bonded to a specific nucleic acid probe for identifying or tracking organisms in the environment

Fluorescent protein any of a large group of proteins that fluoresce different colors, including the green fluorescent protein, for tracking genetically modified organisms and determining conditions that induce the expression of specific genes

Fundamental niche the range of environments in which a species will be sustained when it is not resource-limited, such as may result from competition with other species

Green fluorescent protein (GFP) a protein that glows green and is widely used in genetic analysis

High-throughput culturing methods the use of microtiter plates whose wells contain various culture media that can be inoculated with single cells whose growth or target gene content is measured robotically

Isotopic fractionation the discrimination by enzymes against the heavier isotope of the various isotopes of C or S, leading to enrichment of the lighter isotopes

Laser tweezers a device for obtaining pure cultures by optically trapping a single cell with a laser beam and moving it away from surrounding cells into sterile growth medium

MAR-FISH a technique that combines identification of microorganisms with measurement of metabolic activities

Metaproteomics the measurement of whole-community protein expression using mass spectrometry to assign peptides to the amino acid sequences encoded by unique genes

Metatranscriptomics the measurement of whole-community gene expression using RNA sequencing

Microautoradiography (MAR) the measurement of the uptake of radioactive substrates by visually observing the cells in an exposed photographic emulsion

Microbial ecology the study of the interaction of microorganisms with each other and their environment

Microfluidic devices miniaturized systems for fluid handling that are increasingly used for high-throughput culturing of microorganisms

Microsensor a small glass sensor or electrode for measuring pH or specific compounds such as O_2, H_2S, or NO_3^- that can be immersed into a microbial habitat at microscale intervals

Most-probable-number (MPN) technique the serial dilution of a natural sample to determine the highest dilution yielding growth

Multiple displacement amplification (MDA) a method to generate multiple copies of chromosomal DNA from a single organism

Nucleic acid probe an oligonucleotide, usually 10–20 bases in length, complementary in base sequence to a nucleic acid sequence in a target gene or RNA

Phylotype one or more organisms with the same or related sequences of a phylogenetic marker gene

Realized niche the range of natural environments supporting a species when that organism is confronted with factors such as resource limitation, predation, and competition from other species

Stable isotope probing (SIP) a method for characterizing an organism that incorporates a particular substrate by supplying the substrate in ^{13}C or ^{15}N form and then isolating heavy isotope–enriched DNA and analyzing the genes

Winogradsky column a glass column packed with mud and overlaid with water to mimic an aquatic environment, in which various bacteria develop over a period of months

Microbial Ecosystems

microbiology now

Microbes of the Abyss

The biology of deep ocean regions, the abyssal zone (below 4000 m) and hadal zone (below 6000 m), has stirred the imagination for centuries. These regions remain among the least explored of Earth's biosphere because their study requires very specialized equipment. Only a few remotely operated vehicles (ROVs), autonomous underwater vehicles, and human-occupied submersibles can dive to the deepest marine waters, which are at depths over 10,000 m; pressures at such depths are around 16,000 pounds per square inch (110,000 kilopascals). For example, the Japanese ROV Kaiko (40% scale model in photo) retrieved sediment samples from the Mariana Trench (Pacific Ocean, depth of 10,900 m) (see Kaiko in action in Figure 20.32), but unfortunately was lost in a later dive when its tethering cable snapped.

Animals and microbes that inhabit the deep sea experience extreme pressure, low levels of nutrients, and near-freezing temperatures. The microbiology of these regions was first explored decades ago using high-pressure cultivation methods to isolate the first pressure-loving bacteria (piezophiles). Cultivation was complemented by limited culture-independent molecular surveys of diversity based on 16S rRNA gene sequencing. However, recent advances in metagenomics and high-throughput sequencing have revolutionized studies of deep-sea microbes and their adaptive strategies, such as alterations in membrane and protein structure that allow growth under such extreme conditions.

Diversity studies have shown that both free-living and particle-attached archaeal and bacterial piezophiles are abundant in the abyssal and hadal zones and that most have not yet been cultured. For example, *Bacteria* related to the phyla *Marinimicrobia* and *Gemmatimonadetes* are enriched at these great depths. Since no member of the *Marinimicrobia* has been isolated and the few *Gemmatimonadetes* available are from soils or wastewater, a more complete understanding of the unique properties of these hadal *Bacteria* awaits their future cultivation. Ammonia-oxidizing *Archaea* related to *Nitrosopumilus*, an organism common in upper marine waters, are the most abundant free-living *Archaea* in hadal waters, highlighting their global dominance in both marine and terrestrial environments.

The presence of these novel *Archaea* and *Bacteria* at great depths points to a unique hadal microbiology that we are only now beginning to understand. Future ROV descents such as those pioneered by Kaiko should bring deep-sea microbiology into better focus.

Source: Tarn, J., et al. 2016. Identification of free-living and particle-associated microbial communities present in hadal regions of the Mariana Trench. *Front. Microbiol. 7:* Article 665.

Microorganisms do not live alone in nature but instead interact with other organisms and with their environment. In so doing, microorganisms carry out many essential activities that support all life on Earth. In this chapter we explore some of the major habitats of microorganisms; these include soil, freshwater, and the oceans. In addition to these, microbes have also established more specific, and often very intimate, associations with plants and animals. We examine a few examples of such microbial partnerships and symbioses in Chapter 23.

I • Microbial Ecology

We begin with a broad overview of the science of microbial ecology, including ways that organisms interact with each other and their environments and the difference between species *diversity* and species *abundance*. These basic ecological concepts pervade this and the next three chapters.

20.1 General Ecological Concepts

The distribution of microorganisms in nature resembles that of macroorganisms in the sense that a given species resides in certain places but not others; that is, everything is not everywhere. Also, environments differ in their abilities to support diverse microbial populations, from the highly diverse microbial world of undisturbed fertile soil to the rather restricted world of some highly extreme environments.

Ecosystems and Habitats

An **ecosystem** is a dynamic complex of plant, animal, and microbial communities and their abiotic surroundings, all of which interact as a functional unit. An ecosystem contains many different **habitats**, parts of the ecosystem best suited to one or a few populations. Although microorganisms are present in any habitat containing plants and animals, many microbial habitats are unsuitable for plants and animals. For example, microorganisms are ubiquitous on Earth's surface and even deep within it; they inhabit boiling hot springs and solid ice, acidic environments near pH 0, saturated brines, environments contaminated with radionuclides and heavy metals, and the interior of porous rocks that contain only traces of water. Therefore, some ecosystems are mostly or even exclusively microbial.

Collectively, microorganisms show great metabolic diversity and are the primary catalysts of nutrient cycles in nature (Chapter 21). The *types* of microbial activities possible in an ecosystem are a function of the species present, their population sizes, and the physiological state of the microorganisms in each habitat. By contrast, the *rates* of microbial activities in an ecosystem are controlled by the nutrients and growth conditions that prevail. Depending on several factors, microbial activities in an ecosystem can have minimal or profound impacts and can diminish or enhance the activities of both the microorganisms themselves and the macroorganisms that may coexist with them.

Species Diversity in Microbial Habitats

A group of microorganisms of the same species that reside in the same place at the same time constitutes a microbial **population** and may be descendants of a single cell. A microbial population differs from a microbial **community**. A community consists of populations of one species living in association with populations of one or more other species. The microbial species that reside in a certain habitat are those best able to grow with the nutrients and conditions that prevail there.

The diversity of microbial species in a community can be expressed in two ways. One is **species richness**, the total *number* of different species present. Identifying cells is, of course, basic to determining microbial species richness, but this need not require their isolation and culture. Species richness may also be expressed in molecular terms by the diversity of phylotypes (for example ribosomal RNA genes, ⟳ Section 19.6) observed in a given community. **Species abundance**, by contrast, is the *proportion* of each species in the community (compare Figure 20.1*b* and *c*). Species richness and abundance can change quickly over a short time as shown by the change in abundance of cyanobacteria in a lake receiving nutrient-rich agricultural runoff (**Figure 20.1a**). One goal of microbial ecology is to understand species richness and abundance in microbial communities along with the community's

(a)

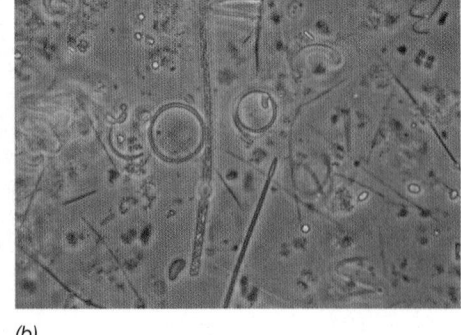
(b)

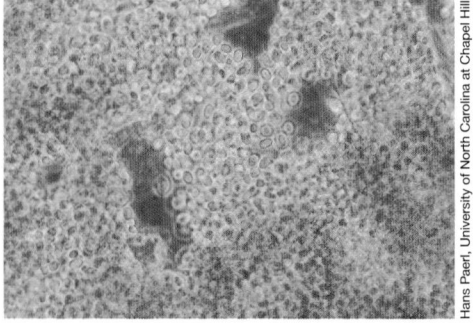
(c)

Hans Paerl, University of North Carolina at Chapel Hill

Figure 20.1 Microbial species diversity: Richness versus abundance. *(a)* Collecting samples from Lake Taihu, China, following a bloom of the cyanobacterium *Microcystis*. *(b)* High species richness in St. John's River, Florida, shown by microscopy of planktonic microorganisms including cyanobacteria, diatoms, green algae, flagellates, and bacteria. *(c)* Shift of St. John's River community to low richness but high abundance following a bloom of the cyanobacterium *Microcystis*.

TABLE 20.1 Resources and conditions that govern microbial growth in nature

Resources
Carbon (organic, CO_2)
Nitrogen (organic, inorganic)
Other macronutrients (S, P, K, Mg)
Micronutrients (Fe, Mn, Co, Cu, Zn, Mn, Ni)
O_2 and other electron acceptors (NO_3^-, SO_4^{2-}, Fe^{3+})
Inorganic electron donors (H_2, H_2S, Fe^{2+}, NH_4^+, NO_2^-)

Conditions
Temperature: cold → warm → hot
Water potential: dry → moist → wet
pH: 0 → 7 → 14
O_2: oxic → microoxic → anoxic
Light: bright light → dim light → dark
Osmotic conditions: freshwater → marine → hypersaline

associated activities and the abiotic environment. Once all of these factors are known, microbial ecologists can model the ecosystem by perturbing it in some way and observing whether predicted changes match experimental results.

The microbial species richness and abundance of a community are functions of the conditions that prevail and the kinds and amounts of nutrients available in the habitat. Table 20.1 lists common nutrients and conditions relevant to microbial growth. In some microbial habitats, such as undisturbed organic-rich soils, high species richness is common (see Figure 20.12), with most species present at only moderate abundance. Nutrients in such a habitat are of many different types, and this helps select for high species richness. In other habitats, such as some extreme environments, species richness is often very low and abundance of one or a few species very high. This is because the physical and chemical (physicochemical) in the environment exclude all but a handful of species, and key nutrients are present at such high levels that the highly adapted species can grow to high cell densities. Bacteria that catalyze acid mine runoff from the oxidation of iron are a good example. These organisms thrive in highly acidic, iron-rich but organic-poor waters, where the acidic conditions and the dearth of organic carbon limit species richness. However, the elevated levels of ferrous iron (Fe^{2+}) present, which is oxidized to Fe^{3+} in energy-yielding reactions (⇔ Section 14.10), fuel high species abundance. We examine the activities of acidophilic iron-oxidizing microorganisms in Sections 21.5 and 22.1.

--- **MINIQUIZ** ---

- What is the difference between species richness and species abundance?
- How does an ecosystem differ from a habitat?
- What are the characteristics of a microbial population?
- How does a microbial population differ from a microbial community?

20.2 Ecosystem Service: Biogeochemistry and Nutrient Cycles

In any ecosystem whose resources and growth conditions are suitable, microorganisms will grow to form populations. Metabolically similar microbial populations that exploit the same resources in a similar way are called **guilds**. A habitat that is shared by a guild and supplies the resources and conditions the cells require for growth is called a **niche**. Sets of guilds form microbial communities (**Figure 20.2**). Microbial communities interact with macroorganisms and abiotic factors in the ecosystem in a way that defines the workings of that ecosystem.

Energy Inputs to the Ecosystem

Energy enters ecosystems as sunlight, organic carbon, and reduced inorganic substances. Light is used by phototrophs to make ATP and synthesize new organic matter (Figure 20.2). In addition to carbon (C), new organic matter contains nitrogen (N), sulfur (S), phosphorus (P), iron (Fe), and the other elements of life (⇔ Section 3.1). This newly synthesized organic material along with organic matter that enters the ecosystem from the outside (called *allochthonous* organic matter) fuels the catabolic activities of chemoorganotrophic organisms. These activities oxidize the organic matter to CO_2 by respiration or ferment it to various reduced substances. If chemolithotrophs are present and metabolically active in the ecosystem, they can conserve energy from the oxidation of inorganic electron donors, such as H_2, Fe^{2+}, S^0, or NH_3 (Chapters 14 and 15),

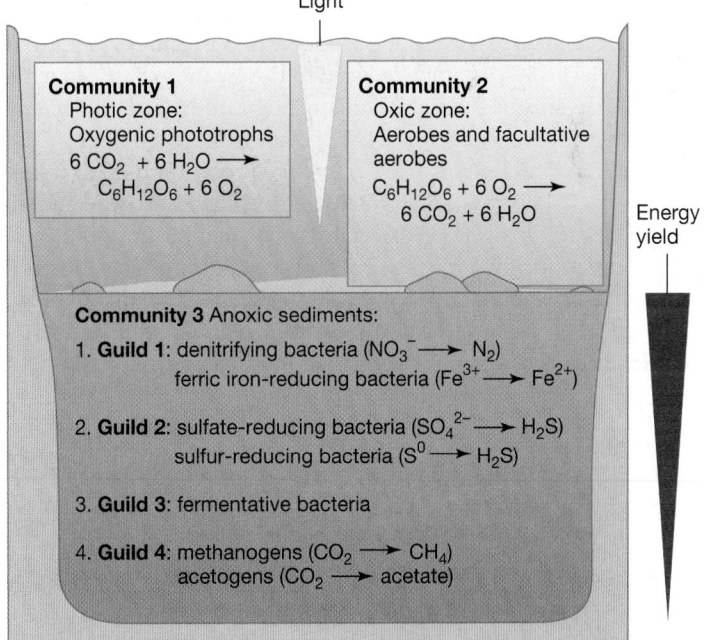

Figure 20.2 Populations, guilds, and communities. Microbial communities consist of populations of cells of different species. A freshwater lake ecosystem would likely have the communities shown here. The reduction of NO_3^-, Fe^{3+}, SO_4^{2-}, S^0, and CO_2 are examples of anaerobic respirations. The region of greatest activity for each of the different respiratory processes would differ with depth in the sediment. As more energetically favorable electron acceptors are depleted by microbial activity near the surface, less favorable reactions occur deeper in the sediment.

and contribute new organic matter through their autotrophic activities (Figure 20.2).

Biogeochemical Cycling

Microorganisms play an essential role in cycling elements, in particular C, N, S, and Fe, between their different chemical forms. The study of these transformations is part of **biogeochemistry**, an interdisciplinary science that includes biology, geology, and chemistry. Figure 20.2 shows how the activities of different guilds of microorganisms influence the chemistry of one environment, a lake ecosystem. The sequence of changing chemistry with increasing depth in the sediments corresponds to the layers of different microbial guilds. The location of each guild in the ecosystem is primarily determined by the availability of electron donors and acceptors, both of which tend to decrease with increasing depth in the sediments.

A *biogeochemical cycle* defines the transformations of an element that are catalyzed by either biological or chemical means (or both). Many different microorganisms participate in biogeochemical cycling reactions, and in many cases, microorganisms are the *only* biological agents capable of regenerating forms of the elements needed by other organisms, particularly plants. Thus, biogeochemical cycles are often also *nutrient cycles*, reactions that generate important nutrients for other organisms.

Most biogeochemical cycles proceed by oxidation–reduction reactions as the element moves through the ecosystem and are often tightly *coupled*, with transformations in one cycle affecting one or more other cycles. For example, hydrogen sulfide (H_2S) is oxidized by phototrophic and chemolithotrophic microbes to sulfur (S^0) and sulfate (SO_4^{2-}), the latter being a key nutrient for plants. Phototrophs and chemolithotrophs are also autotrophs, and thus affect the carbon cycle by producing new organic carbon from CO_2. However, SO_4^{2-} can be reduced to H_2S by the sulfate-reducing bacteria, organisms that consume organic carbon, and this reduction closes the biogeochemical sulfur cycle while regenerating CO_2. The cycling of nitrogen is also a microbial process and is key to the regeneration of forms of nitrogen usable by plants and other organisms. The nitrogen cycle is driven by both chemolithotrophic and chemoorganotrophic bacteria, organisms that produce and consume organic carbon, respectively. We considered the microbiology of biogeochemical cycles and their coupled nature in Chapters 14 and 15 and will revisit this theme in more detail in Chapter 21.

MINIQUIZ

- How does a microbial guild differ from a microbial community?
- What is a biogeochemical cycle? Give an example based on sulfur. Why are biogeochemical cycles also called nutrient cycles?

II • The Microbial Environment

Microorganisms define the limits of life throughout aquatic and terrestrial environments on our planet. Specific conditions required by a particular organism or group of organisms may be subject to rapid change as a result of inputs to and outputs from their habitat or as a result of microbial activities or physical disturbances. Thus, within one environment there can be multiple habitats, some relatively stable and others changing rapidly over time and space.

20.3 Environments and Microenvironments

Besides living in the common habitats of soil and water, microorganisms thrive in extreme environments and also reside on and within the cells of other organisms. The intimate associations developed between microorganisms and other organisms will be presented in Chapters 23 and 24. Here we focus on terrestrial and aquatic microbial habitats.

The Microorganism, Niches, and the Microenvironment

The habitat in which a microbial community resides is governed by physicochemical conditions that are determined in part by the metabolic activities of the community. For example, the organic material used by one species may have been a metabolic by-product of a second species. Another example is oxygen (O_2), which can become limiting if biological consumption exceeds the rate at which it is supplied.

Because microbes are very small, they directly experience only a tiny local environment; this small space is called their **microenvironment**. For example, for a typical 3-μm rod-shaped bacterium, a distance of 3 mm is equivalent to that which a human would experience over a distance of 2 km! As a consequence of the smallness of microorganisms, the variable metabolic activities of nearby microbes, and the changes in physicochemical conditions over short intervals of time and distance, numerous microenvironments can exist within a given habitat. The conditions supporting growth within a microenvironment correspond to the general requirements for growth we considered in Chapter 5.

Ecological theory states that for every organism there exists at least one niche, the *realized niche* (also called the *prime niche*), where it will be most successful. The organism dominates the realized niche but may also inhabit other niches; in other niches it is less ecologically successful than in its realized niche but it may still be able to compete. The full range of environmental conditions under which an organism can exist is called its *fundamental niche* (we considered the realized and fundamental niche in the context of enrichment culture and isolation in Sections 19.1–19.3). The word "niche" should not be confused with the word "microenvironment" because the microenvironment describes conditions at a specific location and can change rapidly. In other words, the general conditions that describe a specific niche may be transient at many places in a microenvironment.

Another important consequence of microbes being so small is that diffusion often determines the availability of resources. Consider, for example, the distribution of an important microbial nutrient such as O_2 in a soil particle. Microsensors (👁 Section 19.9) can be used to measure oxygen concentrations throughout small soil particles. As shown in the data from an actual microsensor experiment (**Figure 20.3**), soil particles are not homogeneous in terms of their O_2 content but instead contain many adjacent microenvironments. The outer layer of the soil particle may be fully oxic (21% O_2) while the center, only a very short distance away (in human terms, but of course a great distance from a microbial standpoint), may be anoxic (O_2-free). The microorganisms near the outer edges

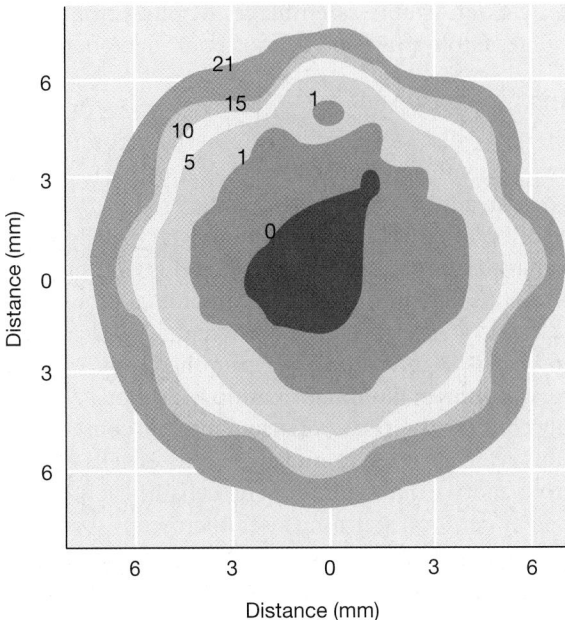

Figure 20.3 Oxygen microenvironments. Contour map of O_2 concentrations in a small soil particle as determined by a microsensor ($\iff$ Section 19.9). The axes show the dimensions of the particle. The numbers on the contours are percentages of O_2 concentration (air is 21% O_2). Each zone can be considered a different microenvironment.

consume all of the O_2 before it can diffuse to the center of the particle. Thus, anaerobic organisms could thrive near the center of the particle, microaerophiles (aerobes that require very low oxygen levels) farther out, and obligately aerobic organisms in the outermost region of the particle. Facultatively aerobic bacteria (organisms that can grow either aerobically or anaerobically) could be distributed throughout the particle ($\iff$ Section 5.14). Nutrient transfer is particularly important in thick assemblages of cells, such as biofilms and microbial mats, and we explore this in Section 20.4.

Physicochemical conditions in a microenvironment are subject to rapid change in both time and space. For example, the O_2 concentrations shown in the soil particle in Figure 20.3 represent "instantaneous" values. Measurements taken in the same particle following a period of intense microbial respiration or disturbance due to wind, rain, or disruption by soil animals could differ dramatically from those shown. During such events certain populations may temporarily dominate the activities in the soil particle and grow to high numbers, while others remain dormant or nearly so. However, if the microenvironments shown in Figure 20.3 are eventually reestablished, the various microbial activities characteristic of different regions of the soil particle will eventually return as well.

Nutrient Levels and Growth Rates

Resources (Table 20.1) typically enter an ecosystem intermittently. A large pulse of nutrients—for example, an input of leaf litter or the carcass of a dead animal—may be followed by a period of nutrient deprivation. Because of this, microorganisms in nature often face a "feast-or-famine" existence. It is thus common for them to produce storage polymers as reserve materials when resources are abundant and draw upon these reserves in periods of starvation. Examples of storage materials are poly-β-hydroxyalkanoates, polysaccharides, and polyphosphate ($\iff$ Section 2.8).

Extended periods of exponential microbial growth in nature are probably rare. Microorganisms typically grow in spurts, linked closely to the availability and types of resources. Because all relevant physicochemical conditions in nature are rarely optimal for microbial growth at the same time, growth rates of microorganisms in nature are usually well below the maximum growth rates recorded in the laboratory. For instance, the generation time of *Escherichia coli* in the intestinal tract of a healthy adult eating at regular intervals is about 12 h (two doublings per day), whereas in pure culture it can grow much faster, with a minimum generation time of about 20 min under optimal conditions. In addition, research-based estimates indicate that most cultured soil bacteria typically grow in nature at less than 1% of the maximal growth rate measured in the laboratory.

These slower growth rates in nature than in laboratory culture reflect the facts that (1) resources and growth conditions (Table 20.1) are frequently suboptimal; (2) the distribution of nutrients throughout the microbial habitat is not uniform; and (3) except in rare instances, microorganisms in nature grow in mixed populations rather than pure culture. An organism that grows rapidly in pure culture may grow much slower in a natural environment where it must compete with other organisms that may be better suited to the resources and growth conditions available.

Microbial Competition and Cooperation

Competition among microorganisms for resources in a habitat may be intense, with the outcome dependent on several factors, including rates of nutrient uptake, inherent metabolic rates, and ultimately, growth rates. A typical habitat contains a mixture of different species (Figures 20.1 and 20.2), with the density of each population dependent on how closely its niche resembles its realized niche.

Some microbes work together to carry out transformations that neither can accomplish alone—a process called *syntrophy*—and these microbial partnerships are particularly important for anoxic carbon cycling ($\iff$ Sections 14.23 and 21.2). Metabolic cooperation can also be seen in the activities of organisms that carry out *complementary* metabolisms. For example, we have previously considered metabolic transformations that are carried out by two distinct groups of organisms, such as those of the nitrifying *Bacteria* and *Archaea* ($\iff$ Sections 14.11, 15.13, and 17.5). Together, these nitrifiers oxidize ammonia (NH_3) to nitrate (NO_3^-). Because nitrite (NO_2^-), the product of ammonia-oxidizing nitrifiers, is the substrate for the nitrite-oxidizing bacteria, the two groups of organisms often live in nature in tight association within their habitats ($\iff$ Figure 19.13).

Winogradsky discovered nitrification in 1890 ($\iff$ Section 1.11), but until lately no single organism capable of complete ammonia oxidation was known. However, the recent discovery of *Nitrospira* species (*Bacteria*) whose genomes encode enzyme systems for both ammonia and nitrite oxidation shows that a single species can catalyze both oxidations, a process called *comammox*. With the use of molecular tools (Chapter 19) to survey various habitats for comammox bacteria, related organisms have been identified in wetlands, riverbeds, aquifers and lake sediments, and wastewater treatment systems. However, no marine comammox species have been identified, and thus the overall significance of comammox organisms in the global nitrogen cycle is unclear at this point.

UNIT 5

━━━ **MINIQUIZ** ━━━
- What characteristics define the realized niche of a particular microorganism?
- Why can many different physiological groups of organisms live in a single habitat?

20.4 Surfaces and Biofilms

Surfaces are important microbial habitats, typically offering greater access to nutrients, protection from predation and physicochemical disturbances, and a means for cells to remain in a favorable habitat, modify the habitat from their own activities, and not be washed away. Moreover, flow across a colonized surface increases transport of nutrients to the surface, providing more resources than are available to planktonic cells (cells that live a floating existence) in the same environment. A surface may also be provided by another organism or by a nutrient such as a particle of organic matter. For example, plant roots become heavily colonized by soil bacteria living on organic exudates from the plant, as revealed when fluorescent stains are used (**Figure 20.4a**).

Virtually any natural or artificial surface exposed to microorganisms will be colonized. For example, microscope slides have been used as experimental surfaces to which organisms can attach and grow. A slide can be immersed in a microbial habitat, left for a period of time, and then retrieved and examined microscopically (Figure 20.4b). Clusters of a few cells that develop from a single colonizing cell—called *microcolonies*—form readily on such surfaces, much as they do on natural surfaces in nature. Periodic microscopic examination of immersed slides has been used to measure growth rates of attached organisms in nature.

Surface colonization may be sparse, consisting only of microcolonies not visible to the naked eye, or may consist of so many cells that microbial accumulation becomes visible as, for example, in a stagnant toilet bowl. Surface growth can be particularly problematic in a hospital setting where microbial colonization of indwelling devices such as catheters and intravenous lines can cause serious infection. In a few extreme environments that lack small animal grazers (for example, hot springs), microbial accumulation on a surface can be many centimeters in thickness. Called **microbial mats** (Section 20.5), such accumulations often contain highly complex yet very stable assemblages of phototrophic, chemolithotrophic, autotrophic, and heterotrophic microbes.

Biofilms

As bacterial cells grow on surfaces they commonly form **biofilms**—assemblages of bacterial cells attached to a surface and enclosed in an adhesive matrix that is the product of excretion by cells and cell death (**Figure 20.5**). The matrix is typically a mixture of polysaccharides, proteins, and nucleic acids that bind the cells together. Biofilms trap nutrients for microbial growth and help prevent the detachment of cells on dynamic surfaces, such as in flowing systems (Figure 20.5c). We examined some of the genetic regulatory features of biofilm formation in Section 7.9 and so here primarily consider their ecological and medical consequences.

Biofilms typically contain multiple layers of cells embedded in the porous matrix material, and the cells in each layer can be examined by confocal scanning laser microscopy (⇔ Section 1.7; Figure 20.5b). Biofilms may contain one or two species or, more commonly, many species of bacteria. The biofilms that form on tooth and soft surfaces of the mouth, for example, contain between 100 and 200 different phylotypes (⇔ Section 19.6), including species of both *Bacteria* and *Archaea*; in total, the human mouth is a habitat for approximately 700 phylotypes (⇔ Sections 24.3 and 25.2). Biofilms are thus functional and growing microbial communities and not just cells trapped in a sticky matrix.

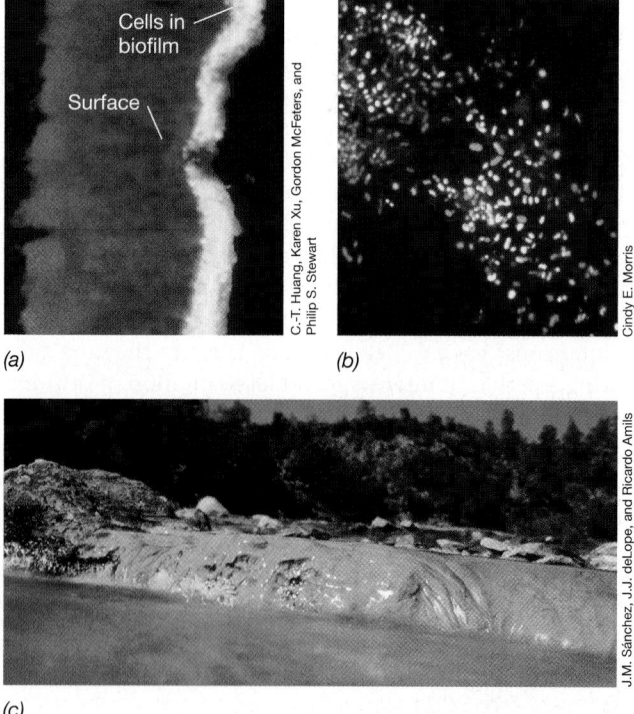

(a)

(b)

(c)

Figure 20.5 Examples of microbial biofilms. (a) A cross-sectional view of an experimental biofilm made up of cells of *Pseudomonas aeruginosa*. The yellow layer (about 15 μm in depth) contains cells and is stained by a reaction showing activity of the enzyme alkaline phosphatase. (b) Confocal scanning laser microscopy of a natural biofilm (top view) on a leaf surface. The color of the cells indicates their depth in the biofilm: red, surface; green, 9-μm depth; blue, 18-μm depth. (c) A biofilm of iron-oxidizing bacteria attached to rocks in the Rio Tinto, Spain. As Fe^{2+}-rich water passes over and through the biofilm, the organisms oxidize Fe^{2+} to Fe^{3+}.

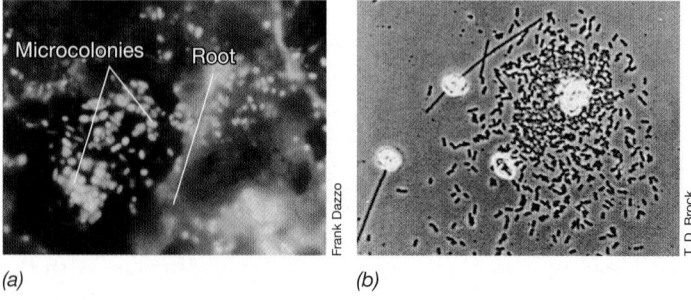

(a)

(b)

Figure 20.4 Microorganisms on surfaces. (a) Fluorescence photomicrograph of a natural microbial community living on plant roots in soil and stained with acridine orange. Note microcolony development. (b) Bacterial microcolonies developing on a microscope slide that was immersed in a river. The bright particles are mineral matter. The short, rod-shaped cells are about 3 μm long.

Wherever submerged surfaces are present in natural environments, biofilm growth is almost always more extensive and diverse than the planktonic growth in the liquid that surrounds the surface. Biofilms differ from planktonic communities in supporting critical transport and transfer processes, which generally control growth in biofilm environments. For example, if consumption of O_2 by populations near the surface exceeds diffusion of O_2 into deeper regions of the biofilm, the deeper regions will become anoxic, opening up new niches for colonization by obligate anaerobes or facultative aerobes. This is similar to the depletion of O_2 in the interior of a soil particle that was depicted in Figure 20.3.

One of the most clinically and industrially relevant properties of biofilm microbial communities is their inherent tolerance to antibiotics and other antimicrobial chemicals. A given species growing in a biofilm can be up to 1000 times more tolerant of an antimicrobial substance than planktonic cells of the same species. Reasons for this greater tolerance include slower growth rates in biofilms, reduced penetration of antimicrobial substances through the extracellular matrix, and the expression of genes that increase tolerance to stress. This tolerance of antimicrobial substances may explain why biofilms are responsible for many untreatable or difficult-to-treat chronic infections and are also hard to eradicate in industrial systems, such as wastewater plants, where surface growth (fouling) by microbes may impair important processes.

Pseudomonas aeruginosa and Cystic Fibrosis

Pseudomonas aeruginosa is a notorious biofilm former (**Figure 20.6**). In this bacterium (and many other *Bacteria*), elevated levels of the small molecule cyclic di-guanosine monophosphate (c-di-GMP, ⮌ Figure 7.19) initiate the production of extracellular polysaccharide, decrease flagellar function, and prepare cells for cell–cell and cell–surface interactions. Over time, in nutrient-rich conditions, *P. aeruginosa* cells can develop mushroom-shaped microcolonies that can be over 0.1 mm high and contain millions of cells enmeshed in a sticky polysaccharide matrix (Figure 20.6). The final architecture of the biofilm is determined by multiple factors in addition to signaling molecules, including nutritional factors and local flow environment (⮌ Section 7.9).

P. aeruginosa biofilms also form in human lungs in those with the genetic disease *cystic fibrosis*. Once in the biofilm state, *P. aeruginosa* is difficult to treat with antibiotics and the biofilm helps the bacteria persist in individuals with this disease. Like most biofilms, that which develops in the lungs of cystic fibrosis patients is composed of more than one bacterial species. So, in addition to *intra*species signaling, *inter*species signaling likely contributes to both initiating and maintaining the cystic fibrosis biofilm as well as biofilms of other types.

Why Bacteria Form Biofilms

At least three reasons have been proposed for the formation of biofilms. First, biofilms are a means of microbial self-defense that increase survival. Biofilms resist physical forces that could otherwise remove cells only weakly attached to a surface. Biofilms also resist phagocytosis by protozoa and cells of the immune system, and retard the penetration of toxic molecules such as antibiotics. These advantages improve the chances for survival of cells in the biofilm. Second, biofilm formation allows cells to remain in a favorable niche. Biofilms attached to nutrient-rich surfaces, such as animal tissues, or to surfaces in flowing systems (Figure 20.5c) fix bacterial cells in locations where nutrients may be more abundant or are constantly being replenished. Third, biofilms form because they allow bacterial cells to live in close association with each other. This facilitates cell-to-cell communication, offers more opportunities for nutrient and genetic exchange, and in general increases chances for survival.

Biofilms form on virtually any surface capable of supporting bacterial growth, and this suggests that biofilms are the "default" growth mode for bacteria in natural habitats, environments that typically differ dramatically in nutrient levels from the rich liquid culture media used in the laboratory. If this is true, planktonic growth may be the atypical growth mode and the norm only for those bacteria adapted to life at extremely low nutrient concentrations (discussed in Sections 20.7, 20.9, and 20.11).

Common Biofilms and Their Control

Biofilms have significant implications in human medicine and commerce. In the body, bacterial cells within a biofilm are protected from attack by the immune system, and antibiotics and other antimicrobial agents sometimes fail to penetrate the biofilm. Besides cystic fibrosis, biofilms have been implicated in several medical and dental conditions, including periodontal disease, chronic wounds, kidney stones, tuberculosis, Legionnaires' disease, and *Staphylococcus* infections (⮌ Figure 5.4a). Medical implants are ideal surfaces for biofilm development. These include both short-term devices, such as urinary catheters, as well as long-term implants, such as artificial joints. It is estimated that 10 million people a year in the United States experience biofilm infections from implants or intrusive medical procedures.

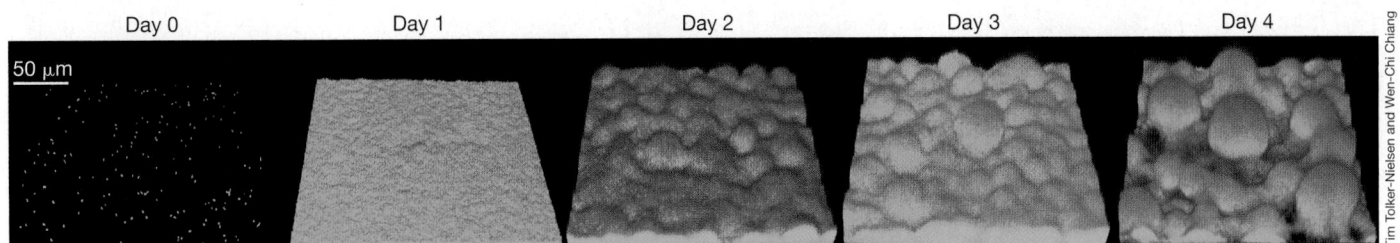

Figure 20.6 *Pseudomonas aeruginosa* **biofilm development.** Confocal scanning laser micrographs of a developing *Pseudomonas aeruginosa* biofilm in a flow cell continuously irrigated with nutrient-rich medium. *P. aeruginosa* cells first attach to the glass surface (day 0), then rapidly grow and move on the surface to cover the entire surface (day 1); by day 4 mushroom-shaped microcolonies over 0.1 mm high have developed.

Biofilms explain why routine oral hygiene is so important for maintaining dental health. Dental plaque is a typical biofilm and contains acid-producing bacteria responsible for dental caries (c₂ Sections 24.9 and 25.2 and Figures 25.7 and 25.8).

Biofilms can slow the flow of water, oil, or other liquids through pipelines and can accelerate corrosion of the pipes themselves. Biofilms also initiate the degradation of submerged objects, such as structural components of offshore oil platforms, boats, and shoreline installations. The safety of drinking water may be compromised by biofilms that develop in water distribution pipes, many of which in the United States are nearly 100 years old (c₂ Section 22.9). Water-pipe biofilms mostly contain harmless microbes, but if pathogens successfully colonize a biofilm, standard chlorination practices may fail to kill them. Periodic releases of pathogenic cells can then lead to outbreaks of disease. For example, it is thought that *Vibrio cholerae*, the causative agent of cholera (c₂ Section 32.3), may be propagated in this manner.

Biofilm control is big business, and thus far, only a limited number of tools exist to fight biofilms. Collectively, industries commit huge financial resources to treating pipes and other surfaces to keep them free of biofilms. New antimicrobial agents that can penetrate biofilms, as well as drugs that eliminate biofilm formation by interfering with intercellular communication, are being developed. A class of chemicals called *furanones*, for example, has shown promise as biofilm preventives on abiotic surfaces.

MINIQUIZ

- Why might a biofilm be a good habitat for bacterial cells living in a flowing system?
- Give an example of a medically relevant biofilm that forms in virtually all healthy humans.
- How is it possible for both aerobes and obligate anaerobes to coexist in the same biofilm?

20.5 Microbial Mats

Microbial mats are among the most visibly conspicuous of microbial communities and can be thought of as extremely thick biofilms. Supported by phototrophic or chemolithotrophic bacteria, these layered microbial communities can be several centimeters thick (**Figure 20.7a, b**). The layers are composed of species of different microbial guilds whose activities are governed by light availability and other resources (Table 20.1). The combination of microbial metabolism and nutrient transport controlled by diffusion results in steep concentration gradients of different microbial nutrients and metabolites, creating unique niches at different depth intervals in the mats. The most abundant and versatile phototrophic mat builders are filamentous cyanobacteria, which are oxygenic phototrophs and many of which tolerate extreme environmental conditions. For example, some species of cyanobacteria grow in waters as hot as 73°C or as cold as 0°C, and others tolerate salinities in excess of 12% and pH values as high as 10.

Cyanobacterial Mats

Cyanobacterial mats (Figure 20.7a, b) are complete microbial ecosystems, containing large numbers of **primary producers** (cyanobacteria and other phototrophic bacteria) that use light energy to synthesize new organic material from CO_2. These along with populations of consumers in the mat community mediate all key nutrient cycles.

Microbial mats have existed for over 3.5 billion years (c₂ Section 13.1) but are found today only in aquatic environments where environmental stresses such as high temperatures or high salt concentrations restrict grazing by small animals and insects. Well-studied microbial mats are found in hypersaline solar evaporation basins; such basins have either formed naturally, such as Solar Lake (Sinai, Egypt), or have been constructed for the recovery of sea salt (Figure 20.7a). Because microbial mats are restricted to extreme environments, most are found in remote

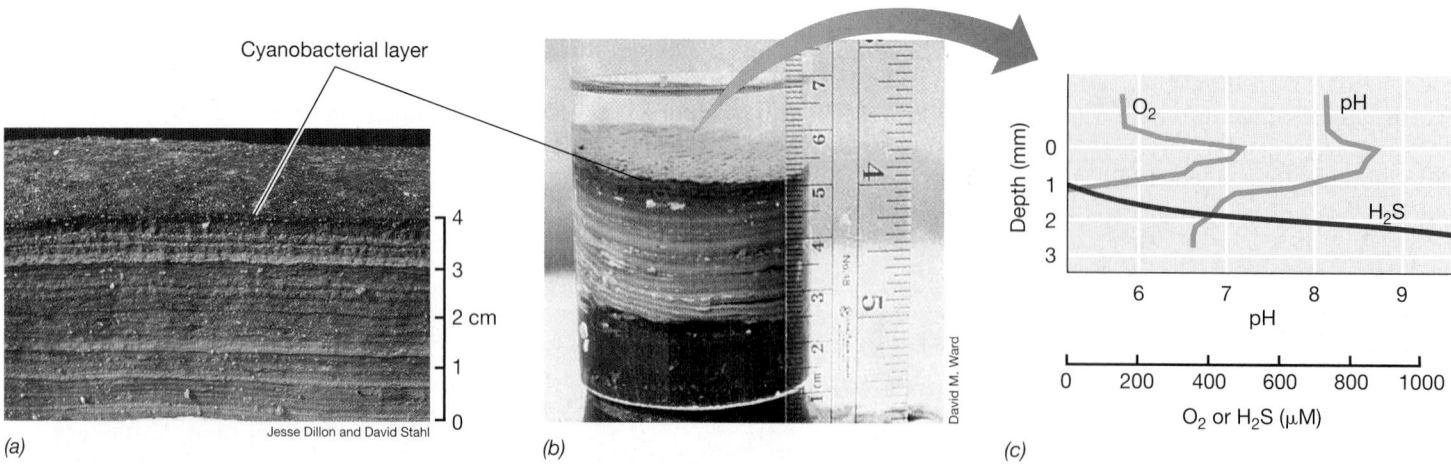

Figure 20.7 Microbial mats. *(a)* Mat specimen collected from the bottom of a hypersaline pond at Guerrero Negro, Baja California (Mexico). Most of the bottom of this shallow pond is covered with mats built by the major primary producer, the filamentous cyanobacterium *Microcoleus chthonoplastes*. *(b)* Microbial mat core from an alkaline Yellowstone National Park (USA) hot spring. The upper (green) layer contains mainly cyanobacteria, while the reddish layers contain anoxygenic phototrophic bacteria. *(c)* Daylight oxygen (O_2), H_2S, and pH profiles through a hot spring mat core such as that shown in part *b*.

locations and many are not readily accessible to study. In contrast, however, the cyanobacterial mats that colonize the outflow channels of hot springs in Yellowstone National Park (USA), Iceland, and many other thermal regions in the world are easily accessible and have been widely studied (Figure 20.7b, c).

The chemical and biological structure of a microbial mat can change dramatically during a 24-h period (called a *diel cycle*) as a consequence of changing light intensity. Using microsensors (⮌ Section 19.9) it is possible to measure pH, H_2S, and O_2 repeatedly over a diel cycle in zones in the mat separated vertically by only a few micrometers. During the day, there is intense oxygen production in the cyanobacterial surface layer of microbial mats and active sulfate reduction throughout the lower regions. Near the zone where O_2 and H_2S begin to mix, intense metabolic activity by phototrophic and chemolithotrophic sulfur bacteria may consume these substrates rapidly over very short vertical distances. Detecting the rate of these changes reveals the zones of greatest microbial activity (Figure 20.7c). These gradients disappear at night when photosynthesis stops and the entire mat turns anoxic and H_2S accumulates. Some mat organisms rely on motility to follow the shifting chemical gradients. For example, sulfur-oxidizing filamentous phototrophic bacteria such as *Chloroflexus* and *Roseiflexus* (⮌ Section 15.7) follow the up-and-down movement of the O_2–H_2S interface on a diel basis.

Chemolithotrophic Mats

The most common types of chemolithotrophic mats are composed of filamentous sulfur-oxidizing bacteria, such as *Beggiatoa* and *Thioploca* species, which grow on marine sediment surfaces at the interface between O_2 supplied from the overlying water and H_2S produced by sulfate-reducing bacteria living in the sediment. In these dark habitats, photosynthesis cannot occur and so the bacteria oxidize H_2S to support energy conservation and autotrophic reactions (⮌ Sections 14.9 and 15.11).

Chemolithotrophic mats composed of sulfur-oxidizing *Thioploca* species on sediments of the Chilean and Peruvian continental shelf are thought to be the most extensive microbial mats of any type on Earth (**Figure 20.8**). *Thioploca* has developed a

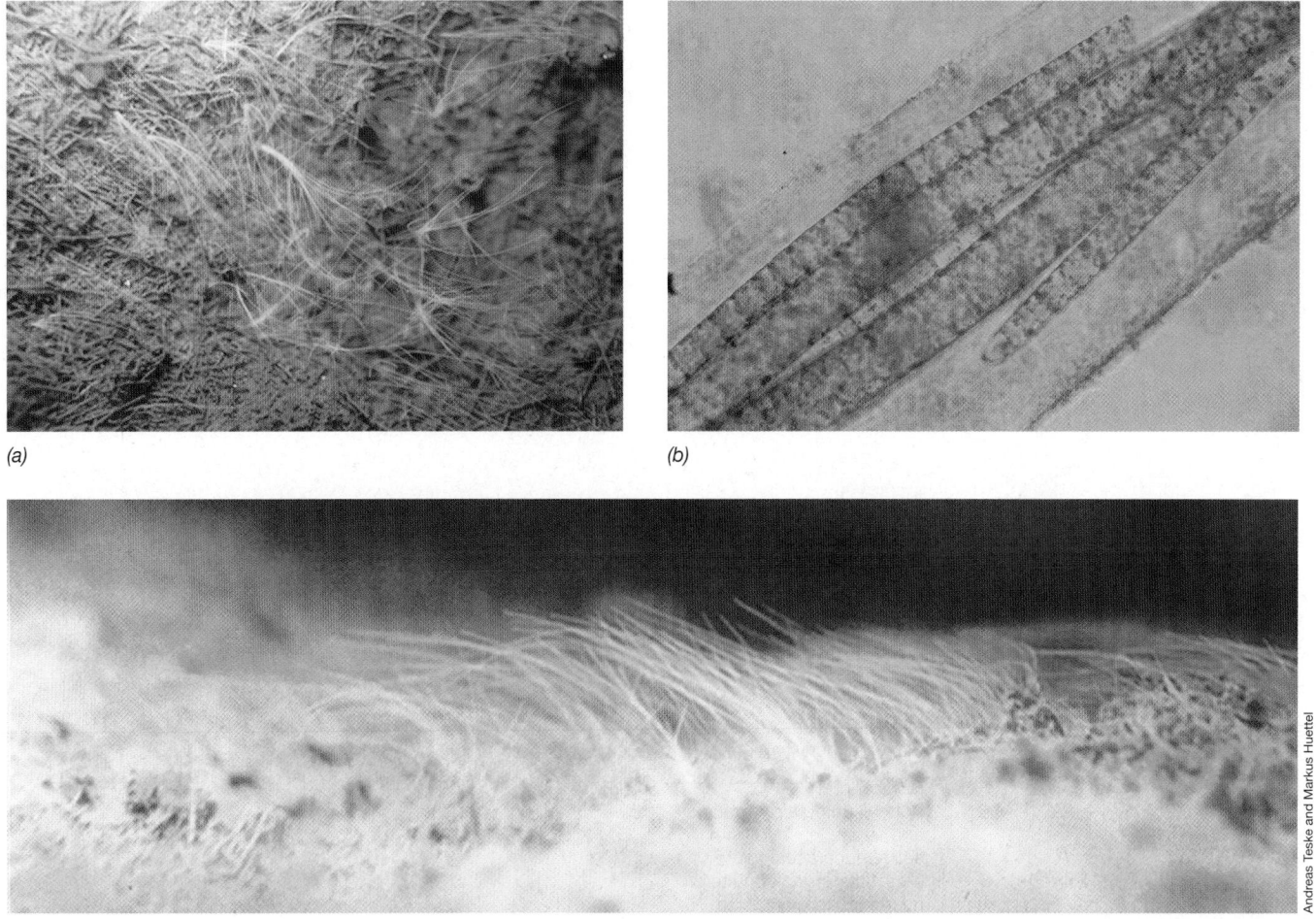

(a)

(b)

(c)

Figure 20.8 *Thioploca* **mats.** *(a, c)* Filaments of the large sulfur-oxidizing chemolithotroph *Thioploca* extend into the water above the sediment (87 m depth) in the Bay of Concepción off the Chilean coast. *(b) Thioploca* form bundles of 10 to 20 filaments (trichomes) held together by a gelatinous sheath, each bundle approximately 1.5 mm in diameter and 10–15 cm in length. Two species of *Thioploca* commonly inhabit the same bundle: *T. chileae*, about 20 μm in diameter, and *T. araucae*, about 40 μm in diameter. Individual trichomes glide independently within the sheaths and can extend up to 3 cm into the water.

Andreas Teske and Markus Huettel

UNIT 5

remarkable strategy to bridge spatially separated resources. These mat bacteria contain large internal vacuoles that store high concentrations of nitrate (NO_3^-) as an electron acceptor to support the anaerobic respiration of H_2S. Much like a scuba diver filling tanks with oxygen to dive into the water, cells of *Thioploca* migrate up to the sediment surface (Figure 20.8*a*, *b*) to charge internal vacuoles with NO_3^- from the water column and then return ("dive") into the anoxic sediment (gliding at speeds of 3–5 mm per hour) to use their stored NO_3^- as an electron acceptor for H_2S oxidation.

The physical and biological structures of both biofilms and microbial mats are determined by metabolic interactions among the microbes within them and the diffusion of nutrients. Thus, as biofilms form on a surface they become increasingly more complex, and in so doing generate new niches for organisms of differing physiologies. This diversity reaches a maximum in mature microbial mats (Figure 20.7*a*, *b*), as molecular community sampling (Section 19.6) has shown these structures to be among the most complex microbial communities yet discovered.

MINIQUIZ

- What is a microbial mat and what major nutrient changes occur in mats during a diel cycle?

- How would motile aerobic bacteria in a microbial mat respond to changing O_2 concentrations over a diel cycle?

III • Terrestrial Environments

Extensive microbial habitats exist in two terrestrial environments on Earth that are similar in lacking sunlight, being periodically or permanently anoxic, and sharing several other physicochemical conditions in common. These two habitats are soils and water enclosed in soils and bedrock. We cover these microbial habitats in the next two sections, and in each case we begin with the abiotic part of the environment and conclude with a discussion of the microbial communities that live there.

20.6 Soils

The word *soil* refers to the loose outer material of Earth's surface, a layer distinct from the bedrock that lies underneath (**Figure 20.9**). Soil develops over long periods through complex interactions among the parent geological materials (rock, sand, glacial drift materials, and so on), the topography, climate, and the presence and activities of living organisms.

Soils can be divided into two broad groups: *Mineral soils* are derived from the weathering of rock and other inorganic materials, and *organic soils* are derived from sedimentation in bogs and marshes. Most soils are a mixture of these two basic types. Although mineral soils, which are the primary focus of this section, predominate in most terrestrial environments, there is increasing interest in the role that organic soils play in carbon storage. A detailed understanding of carbon storage (sinks) and sources (such as release of CO_2) is of great relevance to the science of climate change. The carbon cycle is a major focus of Chapter 21.

Soil Composition and Formation

Vegetated soils have at least four components. These include (1) inorganic mineral matter, typically 40% or so of the soil volume; (2) organic matter, usually about 5%; (3) air and water, roughly 50%; and (4) microorganisms and macroorganisms, about 5%. Particles of various sizes are present in soil. Soil scientists classify soil particles on the basis of size: Those in the

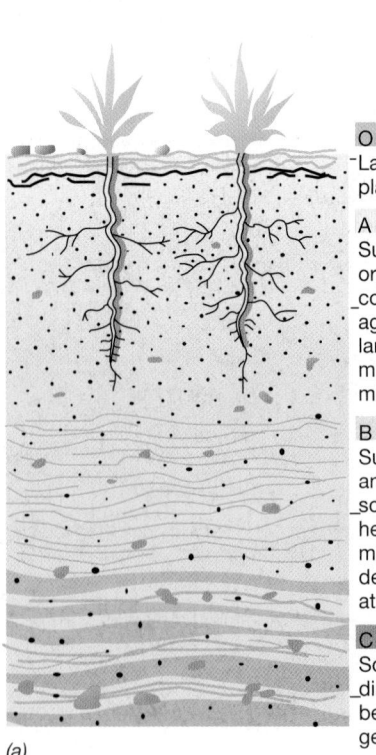

O horizon —Layer of undecomposed plant materials

A horizon — Surface soil (high in organic matter, dark in color, is tilled for agriculture; plants and large numbers of microorganisms grow here; microbial activity high)

B horizon — Subsoil (minerals, humus, and so on, leached from soil surface accumulate here; little organic matter; microbial activity detectable but lower than at A horizon)

C horizon — Soil base (develops directly from underlying bedrock; microbial activity generally very low)

(a)

(b)

Figure 20.9 Soil. *(a)* Profile of a mature soil. The soil horizons are zones defined by soil scientists. *(b)* Photo of a soil profile, showing O, A, and B horizons. This soil from Carbondale, Illinois (USA), is rich in clay and very compact. Such soils are not as well drained as those rich in sand. Note the clear color delineation between the organic-rich A horizon and the less-organic-rich B horizon.

range of 0.1–2 mm in diameter are called *sand*, those between 0.002 and 0.1 mm *silt*, and those less than 0.002 mm *clay*. Different textural classes of soil are then given names such as "sandy clay" or "silty clay" based on the percentages of sand, silt, and clay they contain. A soil in which no one particle size dominates is called a *loam*.

Physical, chemical, and biological processes all contribute to the formation of soil. An examination of almost any exposed rock reveals the presence of algae, lichens, or mosses. These organisms are phototrophic and produce organic matter, which supports the growth of chemoorganotrophic bacteria and fungi. More complex chemoorganotrophic communities composed of *Bacteria*, *Archaea*, and eukaryotes then develop as the extent of the earlier colonizing organisms increases. Carbon dioxide produced during respiration becomes dissolved in water to form carbonic acid (H_2CO_3), which slowly dissolves the rock, especially rocks containing limestone ($CaCO_3$). In addition, many chemoorganotrophs excrete organic acids, which also promote the dissolution of rock into smaller particles.

Freezing, thawing, and other physical processes assist in soil formation by forming cracks in the rocks. As the particles generated combine with organic matter, a crude soil forms in these crevices, providing sites needed for pioneering plants to become established. The plant roots penetrate farther into the crevices, further fragmenting the rock; the excretions of the roots promote development in the **rhizosphere** (the soil that surrounds plant roots and receives plant secretions) of high microbial cell abundance (Figure 20.4a). When the plants die, their remains are added to the soil and become nutrients for more extensive microbial development. Minerals are rendered soluble, and as water percolates, it carries some of these substances deeper into the soil.

As weathering proceeds, the soil increases in depth and becomes able to support the development of larger plants and small trees. Soil animals such as earthworms colonize the soil and play an important role in keeping the upper layers of the soil mixed and aerated. Eventually, the movement of materials downward results in the formation of soil layers, called a *soil profile* (Figure 20.9). The rate of development of a typical soil profile depends on climatic and other factors, but it can take hundreds to thousands of years.

Water Availability: Vegetated and Dryland Soils as Microbial Habitats

The limiting nutrients in soils are often inorganic nutrients such as phosphorus and nitrogen, key components of several classes of macromolecules. Another major factor affecting microbial activity in soil is the availability of water, and we have previously emphasized the importance of water for microbial growth (⌨ Section 5.13).

Water is a highly variable component of soil, and a soil's water content depends on soil composition, rainfall, drainage, and plant cover. Water is held in the soil in two ways—by adsorption onto surfaces or as free water in thin sheets or films between soil particles (**Figure 20.10**). The water present in soils has materials

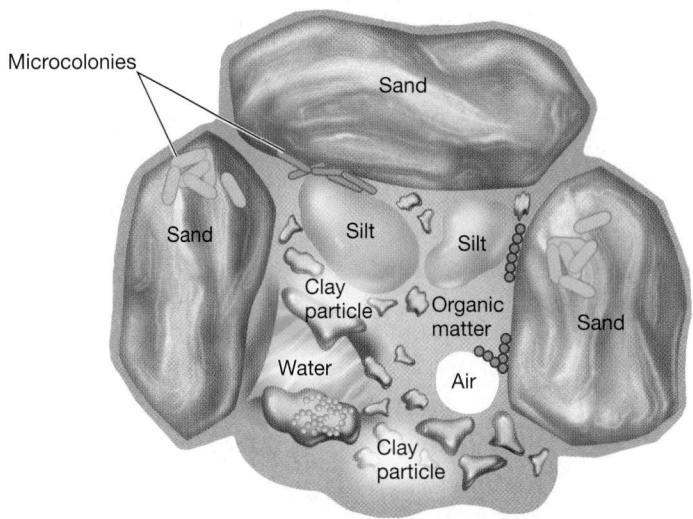

Figure 20.10 A soil microbial habitat. Very few microorganisms are free in the soil solution; most of them reside in microcolonies attached to the soil particles. Note the relative size differences among sand, clay, and silt particles.

dissolved in it, and the mixture is called the *soil solution*. In well-drained soils, air penetrates readily, and the oxygen concentration of the soil solution can be high, similar to that of the soil surface. In waterlogged soils, however, the only oxygen present is that dissolved in water, and this can be rapidly consumed by the resident microbiota. Such soils then become anoxic, and, as described for freshwater environments (Section 20.8), show profound changes in their biological activities. There is also water in the larger channels in soil, where bulk flow is important for rapid transport of microorganisms and their substrates and products.

Arid Soils

The greatest microbial activity in soils is in the organic-rich surface layers in and around the rhizosphere (Figure 20.4a). However, some soils are so dry that plant coverage is greatly limited and only special microbial communities can thrive. These are *arid soils*, and approximately 35% of Earth's landmass is permanently or seasonally arid. Aridity can be defined by the *aridity index*, expressed as the ratio of precipitation to potential evapotranspiration (P/PET). Evapotranspiration is the sum of water loss through evaporation and plant transpiration. A region is deemed arid if there is a P/PET of less than 1; that is, water entering through precipitation (and fog and dew) is less than that lost through evapotranspiration.

Arid soils are among the most extreme environments on Earth, with temperature highs in excess of 60°C and lows of −24°C, high insolation (exposure to solar rays), and low water activity. Although arid regions are typically nearly devoid of leafy plants, they sustain important microbial communities that assemble in and stabilize soil near the surface and reside within and on the surfaces of rocks. The dominant microorganisms present in these carbon-limited environments are cyanobacteria, with lesser numbers of green algae, fungi, heterotrophic bacteria, lichens, and mosses.

(a)

(b)

(c)

Figure 20.11 Biological soil crust (BSC). *(a)* BSC on the Colorado Plateau shown adjacent to lighter disturbed soils. *(b, c)* Scanning electron micrographs of filamentous cyanobacteria (*Microcoleus* species) that bind sand grains together with their sheath material. The sand grains in part *b* are about 100 μm in diameter and the filaments in part *c* about 5 μm in diameter.

Dryland microbial habitats include *biological soil crusts* (BSCs) (**Figure 20.11**), ventral surfaces of translucent stones (*hypolithic* colonists), exposed rock surfaces (*epilithic* colonists), and the interior pore spaces, cracks, and fissures of rocks (*endolithic* colonists). The soil crusts are dominated by cyanobacterial *Microcoleus* species (Figure 20.11*b, c*), whereas coccoid *Chroococcidiopsis* species are the predominant endolithic population. The rock colonists play an important role in weathering and soil formation as described above; here we primarily consider the BSC communities.

The BSC serves a critical function in soil stabilization of desert ecosystems. Stabilization is critical because of the very slow rate of desert soil formation (<1 cm per 1000 years). Here, the filamentous cyanobacteria (*Microcoleus*) and fungi provide soil cohesion, which is further stabilized aboveground by lichens and mosses when present. Importantly, this microbial network functions to eliminate soil erosion from wind and water. The BSCs are major determinants of water infiltration and influence local hydrological cycles and water availability to vegetation. Remarkably, when moisture and temperature conditions are optimal, the photosynthetic rates of BSC are comparable to those of vascular plant leaves. Cyanobacteria and other nitrogen-fixing bacteria (⊘ Sections 7.8, 14.6, and 15.3) provide nitrogen, and much of the fixed nitrogen is released immediately and made available to other soil biota.

The disruption of BSCs is a major contributor to *desertification*, a process exacerbated by climate change and human activities. Dust storms resulting from BSC destruction reduce soil fertility, and when heavy dust is deposited on nearby snowfields it accelerates melt and evapotranspiration rates, thereby reducing freshwater inputs to rivers. Once compromised, soil crusts have recovery times varying from 15 to 50 years. Given the expansive terrestrial presence of BSCs, their importance to human and ecosystem function, and the projected increase in aridity associated with climate change, a better understanding of BSC formation and the rehabilitation of compromised BSCs is important for a healthy planet Earth.

A Phylogenetic Snapshot of Soil Bacterial and Archaeal Diversity

As we saw in Figure 20.3, even a single soil particle can contain many different microenvironments and can thus support the growth of several physiological types of microorganisms. To examine soil particles directly for microbes, fluorescence microscopes are often used, the organisms in the soil having been previously stained with a fluorescent dye. To visualize a specific microorganism in a soil particle, fluorescent staining, such as with fluorescent gene probes (⊘ Sections 19.4, 19.5), can also be used. Microorganisms can also be observed on soil surfaces directly by scanning electron microscopy (Figure 20.11*b, c*).

We learned in Chapter 19 that sequence analyses of 16S ribosomal RNA (rRNA) genes obtained from the environment are commonly used as a measure of bacterial and archaeal diversity (⊘ Section 19.6). As yet, no natural communities have been so thoroughly characterized by these techniques that *all* resident species have been identified. However, within limits, the method is widely considered to be a valid measure of microbial diversity and avoids the more serious problems of enrichment bias that plague culture-dependent diversity studies (⊘ Section 19.1). Here and in later sections of this chapter we present a "phylogenetic snapshot" of major microbial habitats, with the goal of emphasizing trends and patterns rather than absolute details.

Molecular community sampling of a typical vegetated surface has shown typically *thousands* of different species of *Bacteria* and *Archaea* in a single gram of soil, likely reflecting the numerous microenvironments present there. A "species" is operationally defined here as a 16S rRNA gene sequence obtained from a microbial community that differs from all other sequences by more than 3% (⟳ Section 13.8). Such an environmental sequence is called a *phylotype*. Besides very large species numbers, soil microbial diversity studies have also showed that diversity varies with soil type and geographical location. For example, analysis of an Alaska forest soil, an Oklahoma prairie soil, and a Minnesota farm soil (all sites in the USA) revealed approximately 5000, 3700, and 2000 different phylotypes, respectively. The Alaska and Minnesota soils showed similar distributions at the phylum level of taxonomy (for example, *Proteobacteria*, *Acidobacteria*, *Bacteroidetes*, *Actinobacteria*,

Verrucomicrobia, and *Planctomycetes*) but shared only about 20% of their species in common. This indicates that although the *proportions* of the dominant phyla in different soils are relatively constant, the *actual species present* within a phylum may vary considerably between different soils. In addition, lower bacterial diversity was observed in the farm soil than the Alaska soil, an indication that modern intensive agricultural practices that rely heavily on fertilization, low plant diversity, and the chemical suppression of unwanted plants and animals negatively affect bacterial diversity.

Figure 20.12 shows the general composition of prokaryotic soil communities based on pooled 16S rRNA sequence data taken

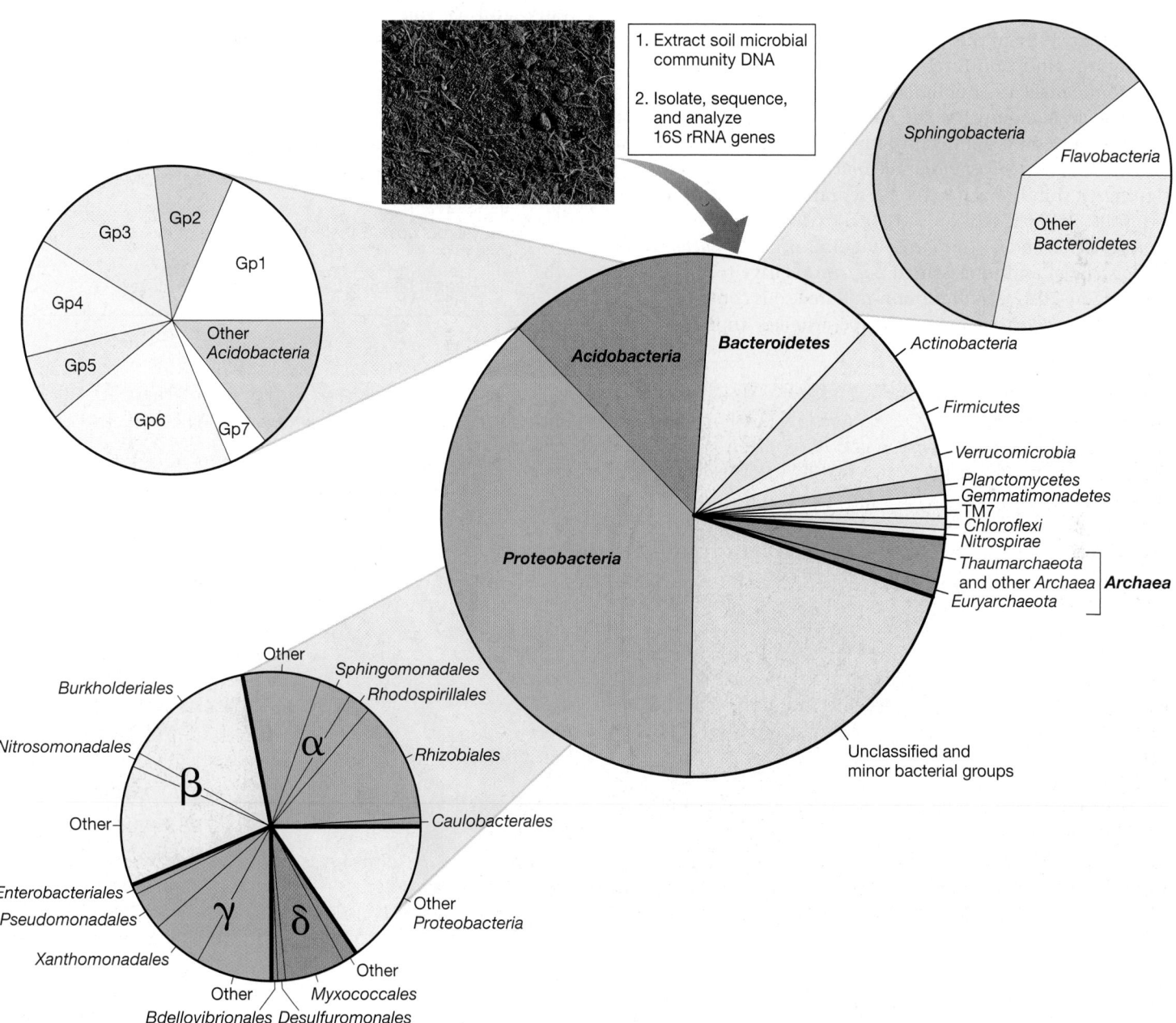

Figure 20.12 Soil bacterial and archaeal diversity. The results are pooled analyses from several studies of the 16S rRNA gene content of soil environments. Many of these groups are covered in Chapters 15 and 16 (*Bacteria*) or 17 (*Archaea*). For *Proteobacteria*, *Acidobacteria*, and *Bacteroidetes*, major subgroups are indicated (Gp, group). Note high species richness as indicated by the large proportion of the total community composed of unclassified and minor bacterial groups. Also note the relatively low proportion of the total prokaryotic soil diversity represented by *Archaea* and that many soil *Archaea* are not close relatives of known *Euryarchaeota* or *Crenarchaeota*. Data assembled and analyzed by Nicolas Pinel.

from several soils. As can be seen, *Proteobacteria* (Chapters 15 and 16) make up nearly half of the total phylotypes recovered, with all major subgroups except for *Epsilonproteobacteria* well represented. *Acidobacteria* and *Bacteroidetes* are also abundant groups; *Actinobacteria* and *Firmicutes* are less so. In addition to these, a major proportion of soil phylotypes are unclassified species or members of minor bacterial groups. This underscores the high bacterial diversity typical of soil ecosystems. In contrast to *Bacteria*, the diversity of *Archaea* in soil is minimal, with relatively few sequences within each major phylum of *Archaea* (*Euryarchaeota, Thaumarchaeota,* and *Crenarchaeota*) represented. However, since there have been fewer selective surveys of archaeal diversity in soils, their diversity may be greater than now recognized.

A similar study to that shown in Figure 20.12 but performed on hydrocarbon-polluted soil showed that the general taxonomic makeup of polluted and unpolluted soils is similar: *Proteobacteria* comprise the largest fraction in both soil types, followed by significant representation of *Acidobacteria, Bacteroidetes, Actinobacteria,* and *Firmicutes*. However, there was a significant shift in fractional representation of these taxa in the two soils. Polluted soils are enriched in *Actinobacteria, Gammaproteobacteria,* and *Euryarchaeota* but diminished in *Bacteroidetes, Acidobacteria,* and unclassified *Bacteria* relative to unpolluted soils (Figure 20.12). Hydrocarbon-polluted soils contained a single dominant *Bacteroidetes* phylotype, whereas unpolluted soils

contained several phylotypes of *Bacteroidetes* (Figure 20.12). Notably, *Thaumarchaeota* are absent from all surveys of hydrocarbon-polluted soils, suggesting that hydrocarbon pollutants suppress ammonia-oxidizing *Thaumarchaeota* (*Archaea,* ⟳ Section 17.5).

Although the *functional* significance of the observed diversity of microbial communities in polluted versus unpolluted soils is unknown, the shifts observed signal that the two soils will likely differ in their capacity to process carbon and nitrogen and to carry out other important nutrient cycling events. However, despite this lack of a functional connection, different 16S rRNA gene surveys of soils agree on two things: (1) undisturbed, unpolluted soils support very high prokaryotic diversity, and (2) soil perturbations trigger measurable shifts in community composition toward species that are more competitive in the disturbed soil and are accompanied by an overall reduction in diversity.

MINIQUIZ

- Which phylum of *Bacteria* dominates bacterial diversity in vegetated soil?
- What factors govern the extent and type of microbial activity in soils?
- Which region of soil is the most microbially active?

(a)

(b)

Figure 20.13 Sampling the deep subsurface. *(a)* Sampling hot (55°C) fissure water from a depth of 3000 m in the Tau Tona South African gold mine. *(b)* Drilling to 600 m in Allendale, South Carolina (USA), for the U.S. Department of Energy (DOE) Deep Subsurface Microbiology Program. Subsurface microbiology is both expensive and challenging from the standpoint of obtaining uncontaminated samples from deep underground. However, it is clear that some very interesting *Bacteria* and *Archaea* inhabit earth's deep subsurface (see Figures 20.14 and 20.15).

20.7 The Terrestrial Subsurface

In the soils and rocks of Earth's subsurface, there is water. This underground water, called *groundwater*, is a vast but little-explored microbial habitat. As recently as three decades ago most microbiologists were of the opinion that significant microbial numbers were limited to the top 100 m or so of Earth's crust. However, from research made possible by the development of improved drilling and aseptic sampling technology, it is now known that microbial life extends down at least 3 *kilometers* into the Earth in regions containing trapped water. The microbiology of relatively shallow groundwater is quite similar to the microbiology of soils. However, microorganisms in deep subsurface waters exist at temperatures that can exceed 50°C and in anoxic and nutrient-depleted surroundings; thus, their microbial diversity is distinct from that of soil.

Bacteria in the Deep Subsurface

Subsurface microbiology initially focused on relatively shallow and easily accessible aquifer systems, revealing diverse populations of *Archaea* and *Bacteria* and a limited presence of protozoa and fungi. An *aquifer* is an underground layer of water-bearing permeable material, such as fractured rock or gravel. Microorganisms in aquifers are metabolically active and greatly influence the chemistry of groundwater. For example, the presence of ferrous iron (Fe^{2+}) in groundwater is largely attributable to the activity of microorganisms such as *Geobacter* that reduce ferric iron (Fe^{3+}) as an electron acceptor (↻ Section 14.15).

Research on the deep microbial biosphere has been facilitated by mining and drilling operations that expose water in fractured rock at great depths. For example, samples collected from a nearly 3-km-deep gold-mining operation in South Africa (**Figure 20.13a**) revealed chemolithotrophic and autotrophic *Bacteria* and *Archaea*. DNA extracted from fissure water deep in this mine showed that a H_2-oxidizing, sulfate-reducing bacterium was virtually the only bacterium present. Genome analysis of the organism, as yet uncultured but given the provisional name *Desulforudis audaxviator*, indicated that it should be thermophilic and should be capable of autotrophic growth using H_2 as the electron donor for anaerobic respiration and CO_2 fixation. In addition, the organism contained genes encoding a nitrogen fixation system (↻ Section 14.6), meaning that it should be able to live in an anoxic environment on a diet of a few minerals, CO_2, SO_4^{2-}, N_2, and H_2.

D. audaxviator would be well suited to long-term isolation in the deep subsurface, as would other autotrophic and nitrogen-fixing bacteria that could use H_2 as electron donor. Possible subsurface sources of H_2 for this include the radiolysis of water by uranium, thorium, and other radioactive elements, and geochemical processes such as the release of H_2 from the oxidation of iron silicate minerals in aquifers. H_2 can satisfy the needs of bacteria that carry out many different types of bacterial anaerobic respirations, including sulfate reduction, acetogenesis, and ferric iron reduction (Chapter 14), and examples of all these physiologies have been identified from genomic analyses of subsurface materials. Hence, *Bacteria* capable of these physiologies

undoubtedly inhabit the subsurface microbial ecosystem along with *D. audaxviator*.

Archaea in the Deep Subsurface

Many novel lineages of *Archaea* appear to be adapted to the extremely nutrient-limited environments of the terrestrial subsurface and deep marine sediments (Section 20.13). In addition to archaeal species that affiliate with phyla having cultivated representatives (*Euryarchaeota, Crenarchaeota, Thaumarchaeota*, Chapter 17), novel phyla so far identified only through PCR-based and metagenomic surveys (↻ Sections 19.6 and 19.8) include the *Aigarchaeota* and *Bathyarchaeota* (**Figure 20.14**). These surveys have also revealed a remarkable diversity of extremely small *Archaea* having cells 0.15–1.2 μm in diameter, containing small genomes (~500–1000 genes), and living in the subsurface and also in a variety of other nutrient-depleted environments

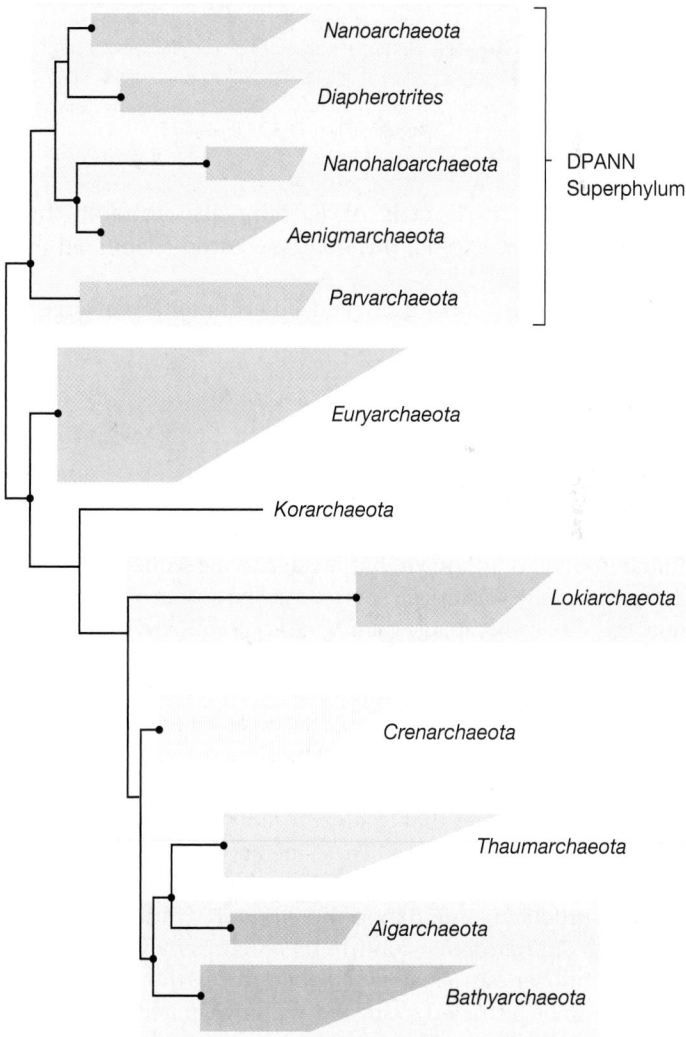

Figure 20.14 Diversity and abundance of subsurface *Archaea*. Phylogeny of archaeal clades found predominantly in the terrestrial and marine deep subsurfaces (this is a consensus tree based on 16S rRNA gene sequences and other conserved marker genes). The novel phyla previously unknown until studies of the deep subsurface are highlighted in light blue. The *Lokiarchaeota* may be the closest known relatives of eukaryotic cells (see page 363 for more on the *Lokiarchaeota*–eukaryote connection).

UNIT 5

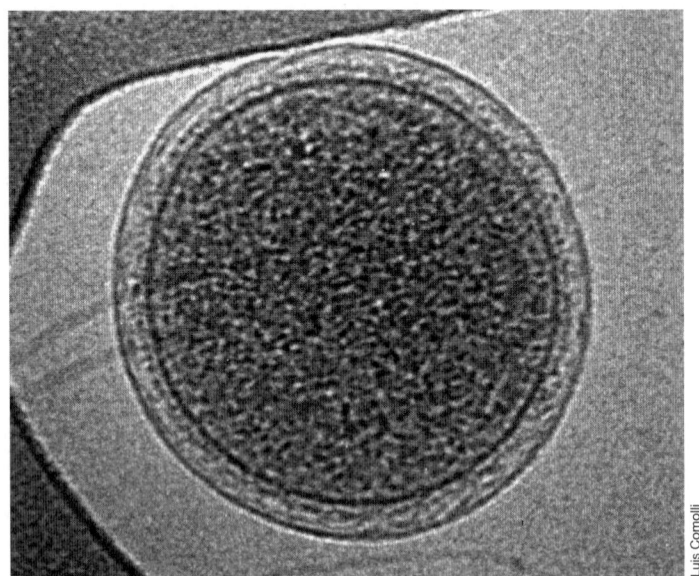

Figure 20.15 Small *Archaea*. Electron micrographic section of a cell of a small species of *Archaea* inhabiting acid mine drainage (↩ Section 22.2). The small *Archaea* found in this acidic environment affiliate with the *Diapherotrites* and *Parvarchaeota* phyla (Figure 20.14). This cell is approximately 0.4 μm in diameter.

(**Figure 20.15**); small cells of *Bacteria* also inhabit these environments (see Explore the Microbial World, "Tiny Cells," in Chapter 2).

The tiny *Archaea* form a single deep evolutionary divergence within the *Archaea* that encompass multiple phyla (a "superphylum," Figure 20.14). The first described member of this superphylum (called DPANN as an acronym from the beginning initials of the five phyla within this lineage) was the thermophile *Nanoarchaeum equitans*, a small parasitic species of limited metabolic capacity that grows in obligate physical association with a host archaeon, *Ignicoccus* (↩ Section 17.6). Whether other tiny *Archaea* will show a similar lifestyle is unknown, but metagenomic sequences suggest that some of the subsurface species may be autotrophs and thus more metabolically capable than *Nanoarchaeum*. Nevertheless, any physiology inferred solely from genome sequences is only a hypothesis that must be confirmed with studies of cultured species.

At least two major metagenomic surprises emerged from these deep subsurface studies. First was the discovery of genomes from subsurface *Bathyarchaeota* (Figure 20.14) that contained genes encoding the enzymes that catalyze reactions of methanogenesis and the acetyl-CoA pathway. These metabolisms are the mechanisms of energy conservation and autotrophy, respectively, in methane-producing (methanogenic) *Archaea*. If future laboratory cultures of *Bathyarchaeota* confirm these results they would show that methanogenesis, a major metabolic process thought for some time now to be restricted to species in a single phylum of *Archaea* (*Euryarchaeota*, Figure 20.14 and Chapter 17), is actually more broadly distributed within this domain.

Another remarkable finding from the deep biosphere was the discovery of an archaeal clade given the name *Lokiarchaeota* (Figure 20.14). It has long been known that *Archaea* and *Eukarya* diverged at some point in the distant past (↩ Section 13.3 and Figure 13.9), but the closest archaeal ancestors to the *Eukarya* had

thus far remained unclear. However, based on metagenomic analyses, the *Lokiarchaeota* are hypothesized to be the lineage from which the eukaryotic domain emerged. This phylogenetic connection was supported by criteria other than just ribosomal RNA gene sequences, including the fact that *Lokiarchaeota* genomes encode a eukaryotic-type cytoskeleton that include proteins such as actin (↩ Section 2.16) as well as several eukaryotic-like proteins with specific membrane functions and the potential for taking up substances by engulfment or membrane invagination. Thus, analysis of the deep subsurface biosphere has not only greatly expanded the known phylogenetic diversity of *Archaea*, but has possibly revealed new information about the origin of eukaryotes.

Growth Rates and the Future of Subsurface Microbiology

Bacterial numbers in uncontaminated groundwater vary by several orders of magnitude (10^2–10^8 per ml), reflecting limited nutrient availability, mostly in the form of dissolved organic carbon. Measured and estimated generation times for deep subsurface bacteria vary by many orders of magnitude, from days to centuries, as determined by the physicochemical environment, the physiology of the resident populations, and nutrient availability. For example, microorganisms appear to be attached to surfaces or within biofilms in the nutrient-depleted subsurface, but it is unknown whether these are genetically and physiologically distinct from microbes in planktonic populations and the extent to which they exploit different metabolic strategies.

The many unanswered questions in subsurface microbiology have encouraged the establishment of permanent science laboratories at great depths in the Earth. For example, the Sanford Underground Research Facility in Lead, South Dakota (USA) (2400 m deep), is supported by government and private agencies for research in physics, geology, and microbiology. The Integrated Ocean Drilling Program, an international effort, has probed for microbial populations at great depths below the seafloor. Results thus far have shown *Archaea* and *Bacteria* as far down as 2000 m below the seafloor (Section 20.13) and in rocks more than 100 *million* years old. Although this may sound ancient, such ages are actually relatively young compared with viable bacteria that have been recovered from salt crystals nearly a half *billion* years old. Obviously, bacterial cells can remain viable for enormously long periods of time.

--- **MINIQUIZ** ---

- Why are possible sources of biologically available energy in the terrestrial subsurface?
- What environmental factors determine the abundance and type of cells in the deep subsurface?
- What information obtained from the *Lokiarchaeota* metagenome associate it with the origin of eukaryotes?

IV • Aquatic Environments

Freshwater and marine environments differ in many ways including salinity, average temperature, depth, and nutrient content, but both provide many excellent habitats for microorganisms.

In this part of the chapter we focus first on freshwater microbial habitats. We then consider two marine environments: (1) coastal and ocean waters, and (2) the deep sea. Much new information is emerging about marine microorganisms from studies using the molecular tools of microbial ecology, especially genetic stains, microbial community sampling, and metagenomics (Chapter 19).

20.8 Freshwaters

Freshwater environments vary significantly in the resources and conditions (Table 20.1) available for microbial growth because some lakes and streams are isolated and nearly pristine while others are highly polluted from agricultural, industrial, or residential runoff. Both oxygen-producing and oxygen-consuming organisms are present in aquatic environments, and the balance between photosynthesis and respiration (Figure 20.2) controls the natural cycles of oxygen, carbon, and other nutrients (nitrogen, phosphorus, metals).

Among microorganisms, oxygenic phototrophs include the algae and cyanobacteria. These can either be *planktonic* (floating) and distributed throughout the water columns of lakes, sometimes accumulating in large numbers at a particular depth, or *benthic*, meaning they are attached to the bottom or sides of a lake or stream. Oxygenic phototrophs, which obtain their energy from light and use water as an electron donor to reduce CO_2 to organic matter (Chapter 14), are the main primary producers in freshwater aquatic ecosystems.

The activity and diversity of chemoorganotrophic aquatic microbial communities depend to a major extent on primary production, in particular its rates and temporal and spatial distributions. Oxygenic phototrophs produce new organic material as well as O_2. If primary production rates are very high, the resultant excessive organic matter production can lead to bottom-water O_2 depletion from respiration and the development of anoxic conditions. This in turn stimulates anaerobic metabolisms such as anaerobic respirations and fermentations (Chapter 14). Like oxygenic phototrophs, anoxygenic phototrophs can also fix CO_2 into organic material. But these organisms use reduced substances other than water, such as H_2S or H_2, as electron donors in photosynthesis (⇨ Section 14.3). Organic matter produced by anoxygenic phototrophs can also support and enhance respiration, accelerating the spread of anoxia.

Oxygen Relationships in Freshwater Environments

The biological and nutrient structure of lakes is greatly influenced by seasonal changes in physical gradients of temperature and salinity. In many lakes in temperate climates the water column becomes stratified, separated into layers of differing physical and chemical characteristics that constitute a **stratified water column**. During the summer, warmer and less dense surface layers, called the **epilimnion**, are separated from the colder and denser bottom layers (the **hypolimnion**). The *thermocline* is the transition zone from epilimnion to hypolimnion (**Figure 20.16**).

In the late fall and early winter, lake surface waters become colder and thus more dense than the bottom layers. This, combined with wind-driven mixing, causes the cooled surface water to sink and the lake to "turn over," mixing surface and bottom waters and their nutrients. The separation of a relatively well-mixed surface layer from a relatively static bottom layer limits the transfer of nutrients between layers until fall turnover once again mixes the water column.

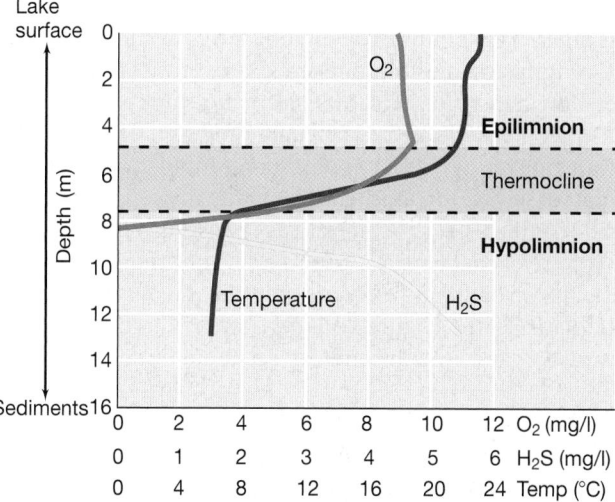

Figure 20.16 Development of anoxic conditions in a temperate lake due to summer stratification. The colder bottom waters are more dense and contain H_2S from bacterial sulfate reduction. The thermocline is the zone of rapid temperature change. As surface waters cool in the fall and early winter, they reach the temperature and density of hypolimnetic waters and sink, displacing bottom waters and effecting lake turnover. Data are from a small freshwater lake in northern Wisconsin (USA).

During periods of stratification, transfer between surface and bottom waters is controlled not by mixing but by the much slower process of diffusion. As a result, bottom waters can experience seasonal periods of either low or no dissolved O_2. Although O_2 is one of the most plentiful gases in the atmosphere (21% of air), it has relatively limited solubility in water, and in a large body of water its exchange with the atmosphere is slow. Whether a body of water actually becomes O_2-depleted depends on several factors, including the amount of organic matter present and the degree of mixing of the water column. Organic matter that is not consumed in surface layers sinks to the depths and is decomposed by anaerobes (Figure 20.2). Lakes may contain high levels of dissolved organic matter because inorganic nutrients that run off the surrounding land can trigger algal and cyanobacterial blooms; these organisms typically excrete various organic compounds and also release complex organic compounds when they die and decay. The combination of water body stratification during early summer, high organic loading, and limited O_2 transfer results in O_2 depletion of the bottom waters (Figure 20.16), making them unsuitable for aerobic organisms such as plants and animals.

The annual turnover cycle allows the bottom waters of a lake to pass from oxic to anoxic and back to oxic. Microbial activity and community composition is altered with these changes in oxygen content, but other factors that accompany fall turnover of the water column, especially changes in temperature and nutrient levels, govern microbial diversity and activity as well. If organic matter is sparse, as it is in pristine lakes or in the open ocean, there may be insufficient substrate available for chemoorganotrophs to consume all the oxygen. The microorganisms that dominate such environments are typically **oligotrophs**, organisms adapted to growth under very dilute conditions (Section 20.11). Alternatively, where currents are strong or there is turbulence because of wind mixing, the water column may be well mixed, and consequently oxygen may be transferred to the deeper layers.

Oxygen levels in rivers and streams are also of interest, especially those that receive inputs of organic matter from urban, agricultural, or industrial pollution. Even in a river well mixed by rapid water flow and turbulence, large organic inputs can lead to a marked oxygen deficit from bacterial respiration (**Figure 20.17a**). As the water moves away from a point source input, for example, from an input of sewage, organic matter is gradually consumed, and the oxygen content returns to previous levels. As in lakes, nutrient inputs to rivers and streams from sewage or other pollutants can trigger massive blooms of cyanobacteria and algae (Figure 20.1) and aquatic plants (Figure 20.17b), thereby diminishing overall water quality and growth conditions for aquatic animals.

Biochemical Oxygen Demand

The microbial oxygen-consuming capacity of a body of water is called its **biochemical oxygen demand (BOD)**. The BOD of water is determined by taking a sample, aerating it well to saturate the water with dissolved O_2, placing it in a sealed bottle, incubating it in the dark (usually for 5 days at 20°C), and determining the residual oxygen in the water at the end of incubation. A BOD determination gives a measure of the amount of organic material in the water that can be oxidized by the microbes present in the water. As a lake or river recovers from an input of organic matter or from excessive primary production, the initially high BOD becomes lower and is accompanied by a corresponding increase in

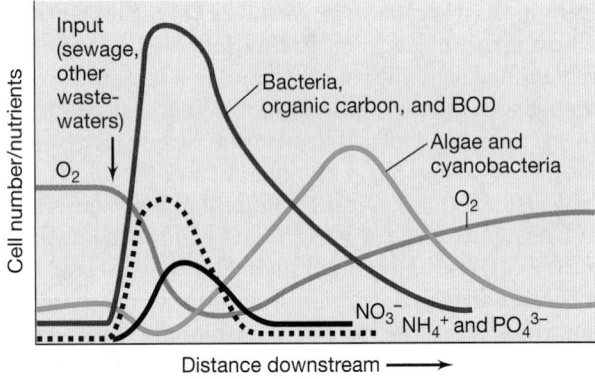

(a)

(b)

Figure 20.17 Effect of the input of organic-rich wastewaters into aquatic systems. *(a)* In a river, bacterial numbers increase and O_2 levels decrease with a spike of organic matter. The rise in algae and cyanobacteria is a response to inorganic nutrients, especially PO_4^{3-}. *(b)* Photo of a eutrophic (nutrient-rich) lake, Lake Mendota, Madison, Wisconsin (USA), showing algae, cyanobacteria, and aquatic plants that bloom in response to nutrients from agricultural runoff. (See also Figure 20.1.)

dissolved oxygen in the ecosystem (Figure 20.17a). Another related measure of the organic material in a body of water is the *chemical oxygen demand (COD)*. This determination uses a strong oxidizing agent, such as acidic potassium dichromate, to oxidize the organic matter to CO_2; the amount of organic matter present is proportional to the amount of dichromate consumed. COD is often used as a rapid measure of water quality and of its potential BOD.

We thus see that in freshwaters the oxygen and carbon cycles are linked, with the levels of organic carbon and oxygen being inversely related. Although oxygenic photosynthesis produces O_2, the corresponding production of organic matter leads to O_2 deficiencies. Anoxic aquatic environments, which are typically rich in organic material, are the end result of respiratory processes that remove dissolved oxygen from the ecosystem, leaving the remaining organic material to be mineralized by organisms employing the anaerobic energy metabolisms we discussed in Chapter 14. It is also important to recognize the importance of storms, floods, and droughts in determining delivery, transport, and cycling of organic matter and inorganic nutrients in freshwater systems, including streams, rivers, lakes, and reservoirs. These less predictable changes also affect microbial productivity, diversity, distribution, and interactions in freshwater systems.

A Phylogenetic Snapshot of Freshwater Prokaryotic Diversity

The importance of *Bacteria* and *Archaea* in lakes, streams, and rivers to the production, regeneration, and mobilization of nutrients is well recognized. However, only more recently have molecular methods been used to identify the participating microbial populations, their interactions, and seasonal patterns. As we saw for studies of soil diversity (Section 20.6), 16S ribosomal RNA gene sequencing is used as a culture-independent method to identify and quantify microbial phylotypes (Section 19.6). Since most molecular studies of freshwater systems have focused on lakes, the emerging picture of lake community structure is examined here.

Figure 20.18 shows the major prokaryotic groups that inhabit lake surface samples (the epilimnion). Five major bacterial phyla are routinely observed: *Proteobacteria*, *Actinobacteria*, *Bacteroidetes*, *Cyanobacteria*, and *Verrucomicrobia*. *Archaea* affiliated with *Euryarchaeota*, *Crenarchaeota*, and *Thaumarchaeota* are also present. This phylum-level composition shares features in common with the ocean, where *Proteobacteria* and *Bacteroidetes* also comprise the greater part of diversity (see Figure 20.29). However, compared with lakes, in the oceans the diversity of *Betaproteobacteria* is lower and *Gammaproteobacteria* and *Alphaproteobacteria* are the more diverse subgroups of the *Proteobacteria* (see Figure 20.29).

A functional interpretation of lake prokaryotic community structure is constrained by the limited availability of cultured representatives. Freshwater *Thaumarchaeota* affiliate with known ammonia-oxidizing species and can be inferred to oxidize ammonia, but the metabolic features of freshwater *Euryarchaeota* are not yet known. *Actinobacteria* are chemoorganotrophic bacteria that in lakes may be responsible for the breakdown of nucleic acid and proteins. In addition, metagenomic analyses (Section 19.8) have shown that at least some *Actinobacteria* contain genes related to those that encode bacteriorhodopsin, a membrane-integrated protein that converts light energy into ATP (Section 20.11 and

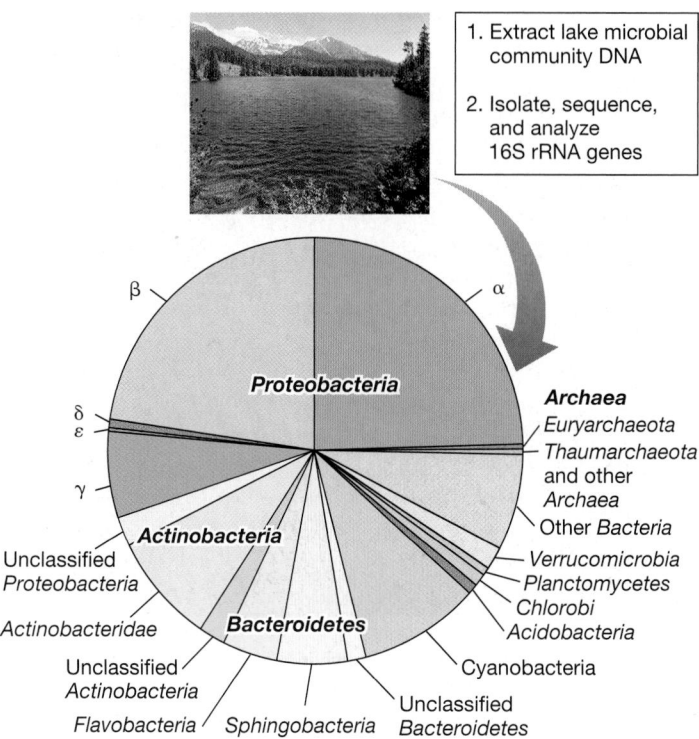

1. Extract lake microbial community DNA

2. Isolate, sequence, and analyze 16S rRNA genes

Figure 20.18 Freshwater lake bacterial and archaeal diversity. Distribution of 16S ribosomal RNA gene sequences by phylum determined from analysis of a collective dataset of 16S genes detected in the epilimnion of several freshwater lakes. Data assembled and analyzed by Nicolas Pinel.

⊂⊃ Section 17.1) in extremely halophilic *Archaea*. The *Actinobacteria* analog is called *actinorhodopsin*. Hence some *Actinobacteria* may be able to harvest light as an energy source.

Bacteroidetes are well represented in lake ecosystems. These organisms are known for their significant metabolic diversity and are likely to be important in lakes in the degradation of various biopolymers and humic materials. The abundant *Betaproteobacteria* tend to be fast-growing species that can respond quickly to pulses of organic nutrients, whereas the *Alphaproteobacteria* are more competitive under conditions of low availability of organic nutrients; this likely accounts for their prevalence in the oligotrophic open ocean (see Figure 20.29).

Taken as a whole, high prokaryotic diversity in freshwater lakes (Figure 20.18) reflects the dynamic character of these habitats. Lakes typically receive seasonally variable inputs of endogenous and exogenous nutrients, a pattern that sustains a phylogenetically and metabolically complex community of *Bacteria* and *Archaea*.

MINIQUIZ

- What is a primary producer? In a freshwater lake, would primary producers more likely reside in the epilimnion or the hypolimnion, and why?
- Will addition of organic matter to a water sample increase or decrease its BOD?
- What factors might account for the prokaryotic diversity of freshwater lakes?

20.9 The Marine Environment: Phototrophs and Oxygen Relationships

With the exception of oxygen, nutrient levels in the open ocean (the *pelagic zone*) are often very low compared with many freshwater environments. This is especially true of key inorganic nutrients for phototrophic organisms, such as nitrogen, phosphorus, and iron. In addition, water temperatures in the oceans are cooler and more constant seasonally than those of most freshwater lakes. The activity of marine phototrophs is limited by these factors, and thus total microbial cell numbers are typically about 10-fold lower in the oceans than in freshwater environments ($\sim 10^6$/ml versus 10^7/ml, respectively). These are average numbers, and studies of marine prokaryotic diversity are just beginning to reveal recurrent temporal patterns of diversity and abundance.

The Bermuda Atlantic Time-Series Study (BATS) has a history of continuous biogeochemical monitoring of ocean waters since the mid-1950s, and is now incorporating molecular analyses of microbial population structure. BATS has revealed three seasonal microbial communities in ocean waters: (1) the community corresponding to the spring surface water bloom (featuring small eukaryotic algae, marine actinobacteria, and two groups of *Alphaproteobacteria*); (2) the summertime community in the upper water column associated with water column stratification (featuring *Pelagibacter*, *Puniceispirillum*, and two groups of *Gammaproteobacteria*); and (3) the deeper, more stable community (featuring *Nitrosopumilus*, representatives of the SAR11 group [see Figure 20.29] with which the genus *Pelagibacter* affiliates, a group of *Deltaproteobacteria*, and species of two additional groups related to the *Chloroflexi* and *Fibrobacter*). A complex and as yet poorly understood interplay of seasonal changes in physicochemical and biotic conditions likely controls these microbial communities that wax and wane in these recurring annual cycles.

Most microbes in marine waters have very small cells, a typical characteristic of organisms living in nutrient-poor environments. Smallness is an adaptive feature for nutrient-limited microorganisms in that it requires less energy for cellular maintenance. The trade-off is that a greater number of transport enzymes relative to cell volume are needed for organisms to acquire nutrients from very dilute (oligotrophic) than from nutrient-rich (eutrophic) aquatic environments. For example, ammonia-oxidizing *Archaea* (*Nitrosopumilus*, ⊂⊃ Section 17.5) are the dominant chemolithotrophs in pelagic waters and have very high-affinity transport systems for acquiring the ammonia they need as an electron donor in energy metabolism.

In pelagic waters there is a lower return of nutrients from the bottom waters than in freshwater lakes, and thus lower average primary productivity. However, because the oceans are so large, the collective carbon dioxide sequestration and oxygen production from oxygenic photosynthesis in the oceans are major factors in Earth's carbon balance. Salinity is more or less constant in the pelagic zone but is more variable in coastal areas. Terrestrial inputs, retention of nutrients, and upwelling of nutrient-rich waters combine to support higher populations of phototrophic microorganisms in near-shore waters than in pelagic waters (**Figure 20.19**); the more productive near-shore waters in turn support higher densities of chemoorganotrophic bacteria and aquatic animals, such as fish and shellfish.

In shallow marine waters such as marine bays and inlets, eutrophication resulting from nutrient inputs can actually lead to the waters becoming intermittently anoxic from the removal of O_2 by respiration and the production of H_2S by sulfate-reducing bacteria (**Figure 20.20**). An extensive region (6000–7000 square miles) of oxygen depletion in the Gulf of Mexico is associated with high loads of nitrogen and phosphorus carried in by the Mississippi River from agricultural runoff in the Mississippi Valley. This region, called the *Gulf of Mexico Dead Zone*, contributes to the loss and impairment of fish and benthic sea life that sustain major seafood industries in this region. The Gulf of Mexico experiences other ecological problems as well, as we examine now.

The Deepwater Horizon Catastrophe

In addition to the chronic degradation of the Gulf of Mexico ecology through agricultural runoff, increased offshore oil drilling also poses significant environmental risk. A major catastrophe for the Gulf of Mexico was the April 2010 explosion and sinking of the Deepwater Horizon offshore drilling platform; failure to control well pressure resulted in the rupture of the wellhead at a depth of 1.5 km and the release of over 4 million barrels of oil before the well was capped three months later (**Figure 20.21**). This largest marine oil spill ever was unique in

Figure 20.19 Distribution of chlorophyll in the western North Atlantic Ocean as recorded by satellite. The east coast of the United States from the Carolinas to northern Maine is shown in dotted outline. Areas rich in phototrophic plankton are shown in red (>1 mg chlorophyll/m³); blue and purple areas have lower chlorophyll concentrations (<0.01 mg/m³). Note the high primary productivity of coastal areas and the Great Lakes.

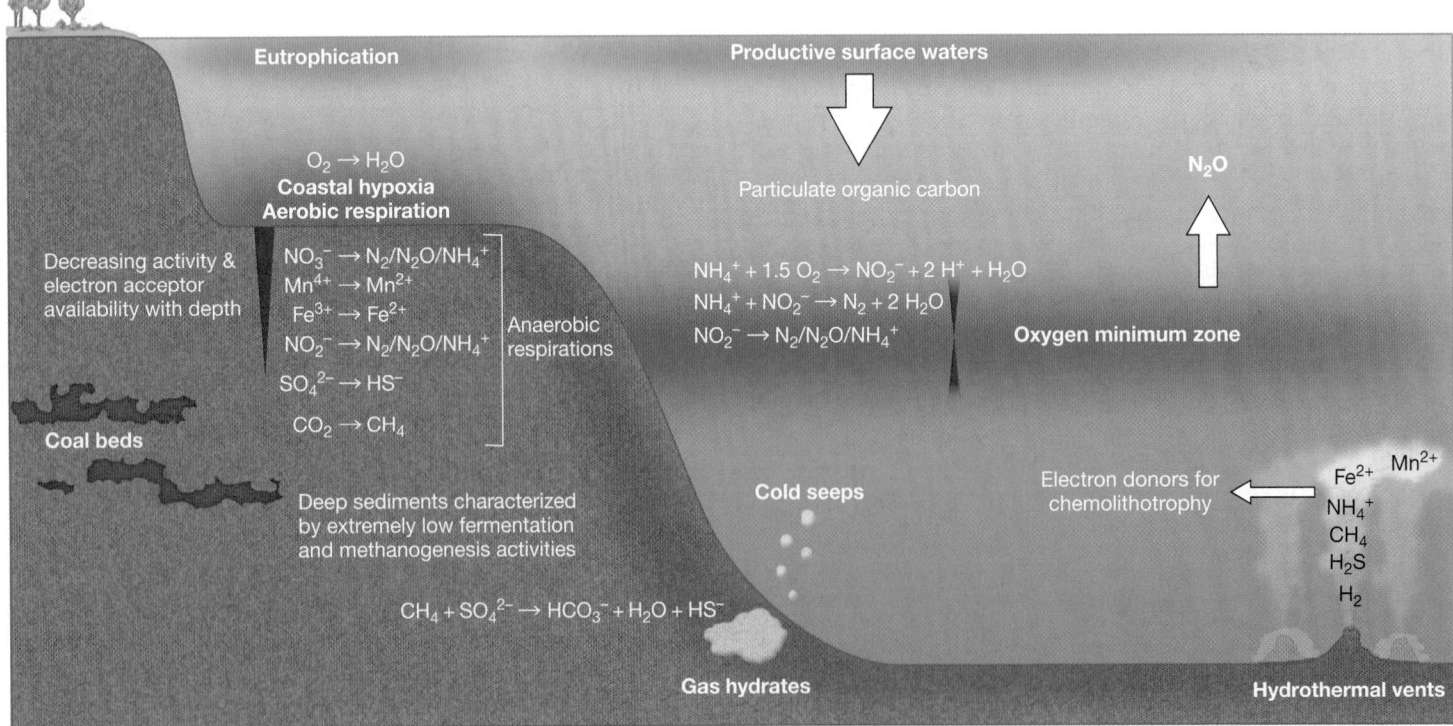

Figure 20.20 Diversity of marine systems and associated microbial metabolic processes. Decreasing electron acceptor availability with depth into the sediment or with increasing distance into an oxygen minimum zone is indicated by red wedges. Sulfate becomes limiting only at greater depths in marine sediments. The indicated metabolic diversity is covered in Chapter 14.

(a)

(b)

Figure 20.21 The Deepwater Horizon oil spill in the Gulf of Mexico. *(a)* Inferno resulting from the wellhead blowout. *(b)* NASA Terra satellite image taken on May 24, 2010, of the Gulf of Mexico near New Orleans, Louisiana. A large plume of oil was released at about 1500 m depth, some of which reached the surface where sunlight reflects off of the oil slick (arrows).

that most of the oil was released as a plume at great depths in the water column. Typically, marine oil spills contaminate primarily the surface waters, resulting in rapid volatilization and loss to the atmosphere of low-molecular-weight oil components such as naphthalene, ethylbenzene, toluene, and xylene. By contrast, the Deepwater Horizon spill released both low-molecular-weight components *and* natural gas (methane, ethane, propane) deep into the water column. These components comprised about 35% of the hydrocarbon plume that extended across many miles of the Gulf from the surface to depths greater than 800 m (Figure 20.21*b*).

The microbial response to hydrocarbon contamination was tracked over several months using both culture-based and molecular methods, including 16S ribosomal RNA gene and metagenomic sequencing, and PhyloChip and GeoChip microarray analyses (⇄ Sections 19.6–19.8). These methods showed that the initial microbial response to the spill (May and June 2010) was a bloom of hydrocarbon-degrading *Gammaproteobacteria* species related to genera in the *Oceanospirillales* group, and of the genera *Colwellia* and *Cycloclasticus*. Increased numbers of *Colwellia* and *Oceanospirillales* species were attributed to their use of gaseous hydrocarbons, since both grew rapidly when ethane or propane was added to enrichment cultures (⇄ Section 19.1). *Colwellia* species also contributed to the degradation of a variety of other hydrocarbons, as indicated by their growth in crude oil enrichment cultures lacking natural gas and by stable isotope probing experiments (⇄ Section 19.10) showing their incorporation of ^{13}C benzene. Although there remains considerable uncertainty about the fate of all the hydrocarbons released during the Deepwater Horizon spill, it appears that the early stimulation of a bloom of hydrocarbon-degrading bacteria by the more easily degraded, soluble, low-molecular-weight components helped reduce the environmental impact of this immense oil spill.

Oxygen Minimum Zones

Another feature of the marine water column is **oxygen minimum zones (OMZs)**, regions of oxygen-depleted waters at intermediate depths, typically in waters between 100 and 1000 m, that extend over wide expanses of the open and coastal ocean (Figure 20.20). These oxygen-depleted regions arise when the respiratory demand for oxygen exceeds oxygen availability, and they are associated with nutrient-rich, highly productive regions. In this way they are similar to the depletion of oxygen caused by agricultural runoff in coastal zones, such as that contributing to the Gulf of Mexico Dead Zone. However, OMZs predate human activity and originate naturally in regions of high surface production and little mixing with oxygen-rich water.

The oxygen saturation values of the largest of the OMZs in the eastern Pacific off the coast of Peru are less than 10% of that at the surface. Oxygen levels at certain depth intervals in the OMZs of the Bay of Bengal and Arabian Sea approach or reach zero. Because of this, OMZs have been recognized as significant sinks for the loss of fixed nitrogen through denitrification (Figure 20.20, ⇄ Section 14.13) and anammox processes (Figure 20.20, ⇄ Section 14.12). In addition to contributing to a significant fraction of the 50% loss of fixed nitrogen from the oceans, these regions are also a source of nitrous oxide (N_2O), a potent greenhouse gas (⇄ Section 21.8) of which approximately one-third is emitted from the oceans. Reduced sulfur can also be an important electron donor for denitrification by sulfide-oxidizing bacteria in the OMZ, the sulfide coming from microbial sulfate reduction (⇄ Sections 14.14 and 15.9). At times of exceptionally high surface water productivity, the accumulation and release of sulfide from OMZ regions has been implicated in massive fish kills.

Ongoing studies of OMZs have shown that these regions of oxygen depletion are expanding, and that their recent expansion is almost certainly associated with global warming. As the oceans absorb more heat, warming of the surface waters increases stratification of near-surface waters and reduces oxygen transfer through mixing to deeper regions. Expansion of the OMZs will favor anaerobic microbial processes at the expense of the aerobic processes that sustain critical oceanic food webs. These changes may further affect atmospheric chemistry by increasing the release of N_2O and will negatively impact marine food webs by reducing levels of fixed nitrogen. Expansion of OMZs might also increase the frequency of toxic sulfidic waters. Ultimately, these changes are expected to impact commercial fisheries.

MINIQUIZ

- What did the Deepwater Horizon spill tell us about how mixed hydrocarbons are degraded in nature?

- What is an oxygen minimum zone and why is expansion of these zones a problem for marine and global ecology?

20.10 Major Marine Phototrophs

The oceans contain large numbers of phototrophic microorganisms, including both prokaryotic and eukaryotic oxygenic phototrophs as well as significant numbers of a special group of purple (anoxygenic) phototrophs. We consider these organisms here as a prelude to exploring the marine prokaryotic world in general in Section 20.11.

Primary Productivity: *Prochlorococcus*

Much of the primary productivity in the open oceans, even at significant depths, comes from photosynthesis by prochlorophytes, tiny bacterial phototrophs that phylogenetically affiliate with the cyanobacteria (Section 15.3); **prochlorophytes** contain chlorophylls *a* and *b* but do not contain phycobilins. The organism *Prochlorococcus* is a particularly important primary producer in the marine environment (**Figure 20.22**). Because *Prochlorococcus* lacks phycobilins, the accessory pigments of the cyanobacteria (Section 14.2), dense suspensions of *Prochlorococcus* cells are olive green (as are green algae) rather than the blue-green color of cyanobacteria (compare Figures 20.1c and 20.22).

Prochlorococcus accounts for up to half of the photosynthetic biomass and primary production in the tropical and subtropical regions of the world's oceans, reaching cell densities of 10^5/ml. A number of strains of *Prochlorococcus* have now been identified in culture, and each inhabits its own depth range in pelagic waters. The different *Prochlorococcus* strains are considered distinct *ecotypes*, genetic variants of a species that differ physiologically and therefore occupy slightly different niches. For example, different *Prochlorococcus* ecotypes photosynthesize at different light intensities (high-light versus low-light ecotypes) and use different inorganic and organic nitrogen and phosphorus sources. *Prochlorococcus* is thus distributed in both surface waters and deeper waters to depths of 200 m, and when an oxygen minimum zone (Section 20.9) is present, *Prochlorococcus* extends into the upper regions of this zone (Figure 20.20). This is near the bottom of the photic zone where light intensities are very low (see Figure 20.26). Genome sequences of about a dozen *Prochlorococcus* strains in

Figure 20.22 *Prochlorococcus*, the most abundant oxygenic phototroph in the oceans. A bottle of *Prochlorococcus* showing the olive green color of the cells containing chlorophylls *a* and *b*. Inset: FISH-stained cells of *Prochlorococcus* in a marine water sample (Section 19.6 and Figure 19.23).

culture revealed that although each contains about 2000 genes, only about 1100 genes are shared by all strains. Each presumptive ecotype contains approximately 200 unique genes, which likely have adaptive significance for growth in the realized niche of that ecotype. This was illustrated in Chapter 19 where we compared the genome of a single cultured *Prochlorococcus* ecotype to metagenome sequences obtained from pelagic waters (Section 19.8 and Figure 19.23).

Single-cell genomic studies (Sections 9.12 and 19.12) of natural *Prochlorococcus* populations have refined our understanding of genetic diversity and the relationship between environmental conditions and the fitness of individual genotypes. These analyses revealed a high degree of fine-scale genetic diversity within the high-light (HL) and low-light (LL) ecotypes. Each ecotype is actually composed of hundreds of subpopulations, each of which is united by a common set of core genes that encode functions that control the interaction of the organism with its environment (e.g., transport functions, oxidative stress responses, and cell surface structure) (**Figure 20.23b**). However, each subpopulation also contains a small set of genes termed "flexible" genes (ranging from 4 to 14 genes per characterized genome) that vary within that subpopulation. The flexible genes are packaged as "cassettes" localized to highly variable regions (genomic islands) within the *Prochlorococcus* genome(Sections 9.7, 19.8, and Figure 19.23). Variation in flexible gene content contributes to exceptionally high microdiversity within natural populations of *Prochlorococcus*, pointing to tremendous versatility in their adaptive response to new niche opportunities. The importance of genotypic variability in sustaining large numbers of this major marine phototroph is shown by seasonal shifts in different genotypes in response to changing light intensity, nutrient availability, and predator populations (Figure 20.23a).

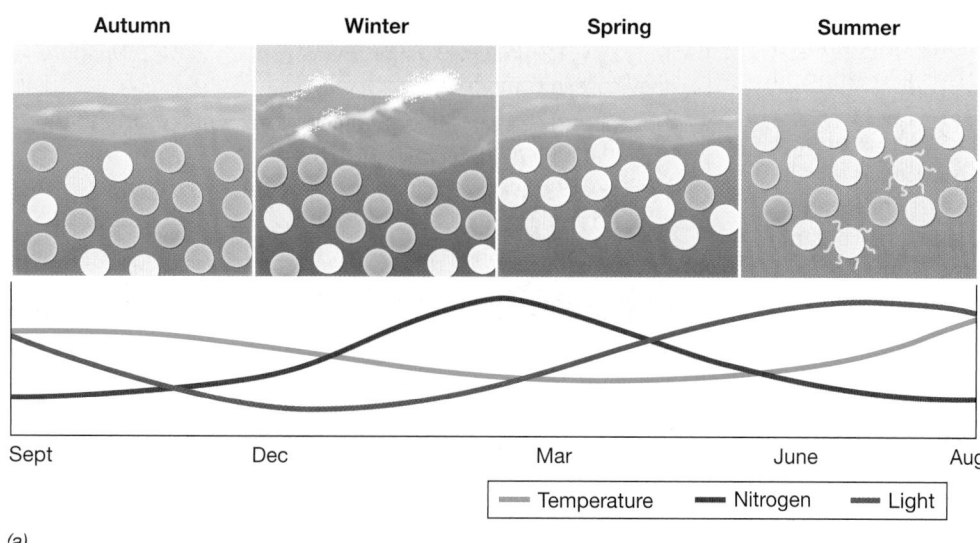

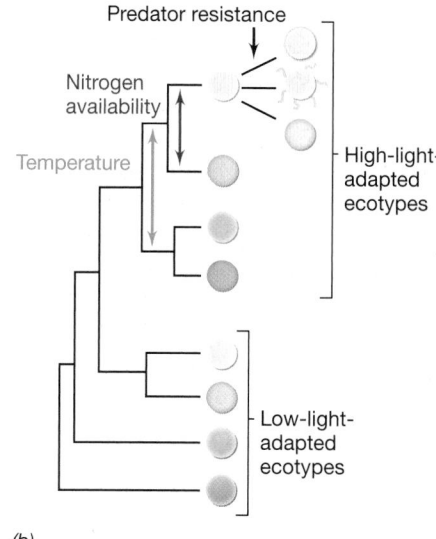

(a)

(b)

Figure 20.23 Seasonal variation of *Prochlorococcus* ecotypes in marine surface waters. *(a)* Single-cell genome sequencing (⟳ Sections 9.12 and 19.12) has shown that the abundance of different genotypes, presumptive ecotypes (represented by cell color and surface features), correlate with seasonal changes in temperature, light, nutrients, and predators (grazers and phage). *(b)* Adaptations to different physical conditions (light and temperature) partition at higher taxonomic levels than do adaptations to different nutrient conditions (e.g., nitrogen availability) and resistance to specific predators, which show closer genetic relationships among ecotypes.

Other Pelagic Oxygenic Phototrophs

In tropical and subtropical oceans, the planktonic filamentous marine cyanobacterium *Trichodesmium* (**Figure 20.24a**) is a widespread and occasionally abundant phototroph. Cells of *Trichodesmium* form puffs (colonies) of filaments. Each puff can contain many hundreds of individual filaments, each filament composed of 20–200 cells. In the Caribbean Sea surface waters, colonies of *Trichodesmium* can approach 100/m³. *Trichodesmium* is a nitrogen-fixing cyanobacterium, and the production of fixed nitrogen by this organism is thought to be an important link in the marine nitrogen cycle. *Trichodesmium* contains phycobilins, absent from prochlorophytes, and thus differs from these organisms in its absorption properties (⟳ Section 14.2).

Very small phototrophic eukaryotes also inhabit coastal and pelagic waters, and some of these are among the smallest eukaryotic cells known. Three common genera—*Bathycoccus, Micromonas*, and *Ostreococcus*—contain only one mitochondrion and one chloroplast per cell. These genera are now assigned to the *Prasinophyceae*, a family of green algae that diverged early from other lineages of green algae (⟳ Section 18.15). Cells of *Ostreococcus* are cocci that measure only about 0.7 µm in diameter (Figure 20.24b), which is even smaller than a cell of *Escherichia coli*.

Although cells of *Ostreococcus* and *Prochlorococcus* are of roughly the same dimensions and they are both oxygenic phototrophs, their genomes are distinct. The genome of *Ostreococcus* is 12.6 Mbp (distributed over 20 chromosomes), which is more than seven times the size of the *Prochlorococcus* genome. Even though this is large relative to cyanobacteria, the *Ostreococcus* genome is very gene dense, containing about 8000 genes,

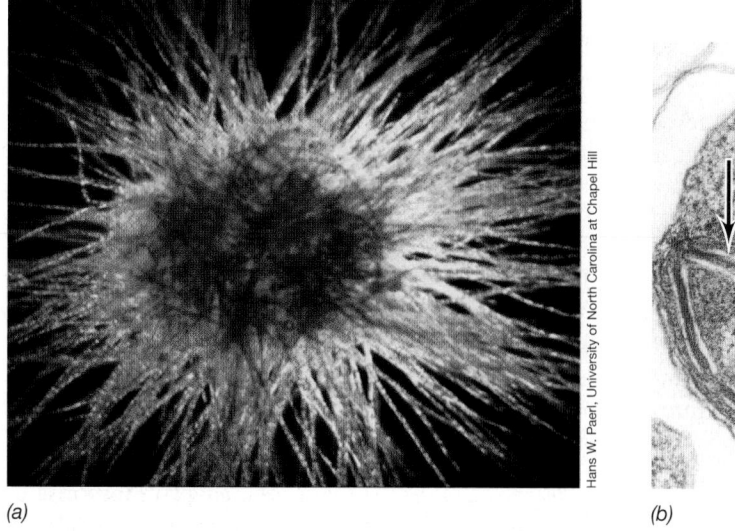

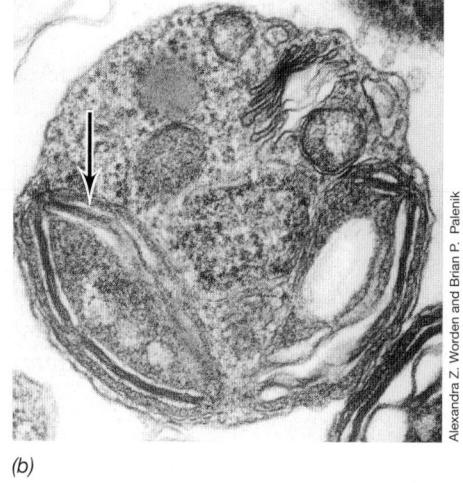

(a)   (b)

Figure 20.24 *Trichodesmium* and *Ostreococcus*. *(a)* Light photomicrograph of a puff of cells of the nitrogen-fixing cyanobacterium *Trichodesmium*. The filaments in the puff are chains of cells, each of which is about 6 µm in diameter. *(b)* Transmission electron micrograph of a cell of *Ostreococcus*, a small green alga found primarily in marine coastal waters. The arrow points to the chloroplast. An *Ostreococcus* cell is about 0.7 µm in diameter.

and thought to be near the minimum genome size of a free-living photosynthetic eukaryote. As a reference, the genome of a common plant, Japanese rice (*Oryza sativa* subsp. *japonica*), is 420 Mbp and contains about 50,000 genes.

In many marine waters, other small eukaryotic cells are present at about 10^4/ml. Although many of these are *Ostreococcus* or relatives, some are chemoorganotrophs and some are phototrophs unrelated to *Ostreococcus* that incorporate small amounts of organic matter to supplement their primarily phototrophic lifestyle.

Aerobic Anoxygenic Phototrophs

Besides *oxygenic* phototrophs, *anoxygenic* phototrophs are also present in coastal and pelagic marine waters. Like purple anoxygenic phototrophs, these organisms contain bacteriochlorophyll *a* (⟜ Sections 14.1, 14.3, 15.4, and 15.5). However, unlike classical purple bacteria that carry out photosynthesis only under *anoxic* conditions, these anoxygenic phototrophs carry out photosynthetic light reactions only under *oxic* conditions.

Aerobic anoxygenic phototrophs include bacteria such as *Erythrobacter, Roseobacter*, and *Citromicrobium* (**Figure 20.25**), all genera of *Alphaproteobacteria*. Aerobic anoxygenic phototrophs synthesize ATP by photophosphorylation when oxygen is present (which is all of the time in oxic pelagic waters), but they are unable to grow autotrophically and thus rely on organic carbon for their carbon sources (a nutritional condition called *photoheterotrophy*). These organisms thus use the ATP produced by photophosphorylation to supplement their otherwise chemoorganotrophic metabolism.

Surveys have shown that a great diversity of aerobic anoxygenic phototrophs exist in marine waters, especially near-shore waters. Oligotrophic and highly oxic freshwater lakes are also habitats for these interesting phototrophic bacteria. The physiology of aerobic anoxygenic phototrophs is thus ideal for their illuminated and highly oxic habitats.

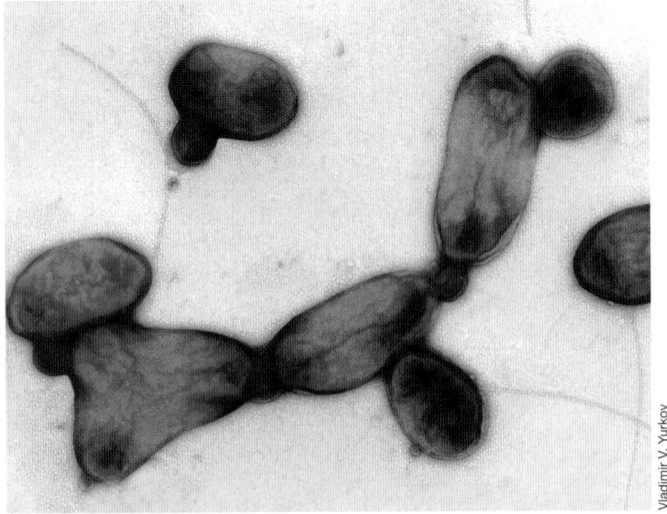

Figure 20.25 Aerobic anoxygenic phototrophic bacteria. Transmission electron micrograph of negatively stained cells of *Citromicrobium*. Cells of this marine, aerobic anoxygenic phototroph produce bacteriochlorophyll *a* only under oxic conditions and divide by both budding and binary fission, yielding morphologically unusual and irregular-shaped cells.

Vladimir V. Yurkov

20.11 Pelagic *Bacteria, Archaea*, and Viruses

Despite vanishingly low nutrient levels, significant numbers of *Bacteria* and *Archaea* live a planktonic existence in pelagic marine waters. Of these, one species in particular has garnered significant attention, a bacterium named *Pelagibacter*.

Distribution and Activity of *Archaea* and *Bacteria* in Pelagic Waters

The abundance of prokaryotic cells in the open oceans decreases with depth. In surface waters, cell numbers average about 10^6/ml. Below 1000 m, however, total cell numbers fall to between 10^3 and 10^5/ml. The distribution of *Bacteria* and *Archaea* with depth has been tracked in pelagic waters using fluorescence in situ hybridization (FISH) technology (⟜ Section 19.5).

Species of *Bacteria* tend to predominate in waters above 1000 m, although cells of *Bacteria* and *Archaea* are found in near-equal abundance in deeper waters (**Figure 20.26**). Deep-water *Archaea* are almost exclusively species of *Thaumarchaeota* (⟜ Section 17.5). Many or perhaps even most are ammonia-oxidizing chemolithotrophs (⟜ Sections 14.11 and 17.5) that play an important role in coupling the marine carbon and nitrogen cycles (Chapter 21). Extrapolating from the data in Figure 20.26, it is estimated that 1.3×10^{28} and 3.1×10^{28} cells of *Archaea* and *Bacteria*, respectively, exist in the world's oceans. This means that the oceans contain the largest microbial biomass on Earth's surface.

Pelagic *Bacteria* and *Archaea* are ecologically important because they consume dissolved organic carbon in the oceans, one of the largest pools of usable organic carbon on Earth. These small and free-living planktonic microbes consume about half the total oceanic organic carbon produced from photosynthesis and are responsible for about half of all marine respiration and nutrient regeneration. Planktonic marine microbes thus return organic matter to the marine food web that would otherwise be lost because of the inability of larger marine organisms to take up such diluted organic nutrients. This so-called "secondary production" is balanced by cell losses from grazing protists and from virus attack (see Figure 20.28), leading to a near-steady state in which bacterial abundance in the open ocean remains roughly constant over time. But importantly, secondary production both recycles nutrients and allows some of the dissolved organic carbon in seawater to reach larger organisms, including fish, because protists are passed up the food web by the feeding activities of larger organisms.

Pelagibacter: The Most Abundant Bacterium

Small planktonic chemoorganotrophic bacteria inhabit pelagic marine waters in numbers of 10^5–10^6 cells/ml. The most abundant

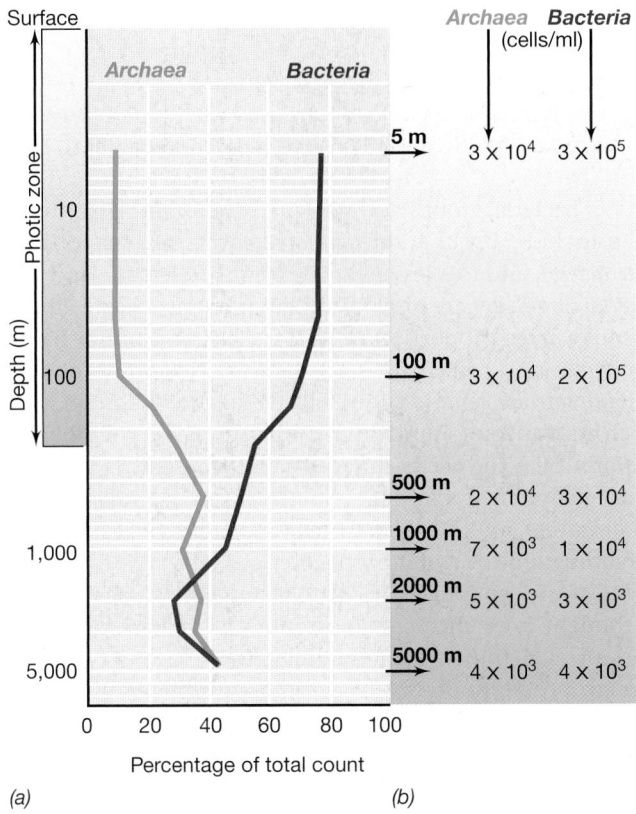

	Archaea	Bacteria
	(cells/ml)	
5 m	3×10^4	3×10^5
100 m	3×10^4	2×10^5
500 m	2×10^4	3×10^4
1000 m	7×10^3	1×10^4
2000 m	5×10^3	3×10^3
5000 m	4×10^3	4×10^3

(a)

(b)

Figure 20.26 Distribution of *Archaea* and *Bacteria* in North Pacific Ocean water. *(a)* Percentage of *Archaea* and *Bacteria* with depth. *(b)* Absolute numbers per milliliter of *Archaea* and *Bacteria* with depth in the open ocean.

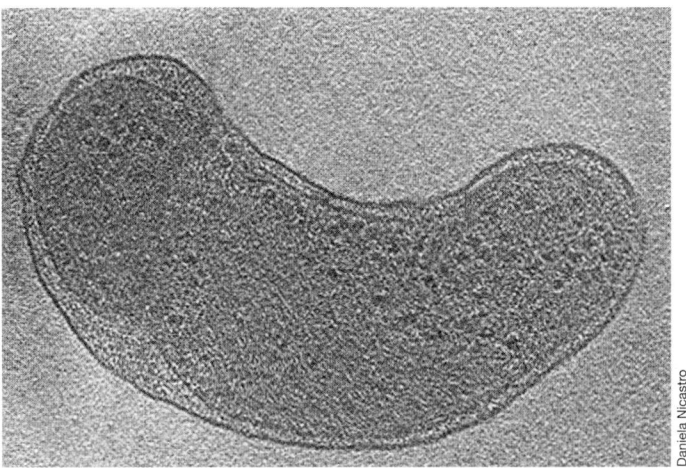

Figure 20.27 *Pelagibacter*, the most abundant bacterium in the ocean. Electron micrograph taken by electron tomography, a technique that introduces a three-dimensional effect. A single cell of *Pelagibacter* is about 0.2 µm in diameter.

of these are members of the "SAR11 group" within the *Alphaproteobacteria*, which includes the genus *Pelagibacter*. Environmental metagenomic studies (⇄ Sections 9.8 and 19.8) and cell counts done using FISH (⇄ Section 19.5) have revealed a great abundance of SAR11 group organisms in pelagic waters. The total oceanic population of this group is estimated to be about 2.4×10^{28} cells, making it the most successful microbial group, as reflected by abundance, on the planet. *Pelagibacter* is an oligotroph, an organism that grows best at very low concentrations of nutrients and grows in laboratory culture only up to the densities it is found in nature.

What makes *Pelagibacter* so successful in the open oceans? In part, its success is related to small size. Cells of *Pelagibacter* are small rods with a diameter of only 0.2–0.5 µm, near the limits of resolution of the light microscope (**Figure 20.27**), and a volume of 0.01 µm³. The resulting high surface-to-volume ratio (⇄ Section 2.2) facilitates nutrient transport, increasing substrate concentration and processing rates within the cell. Proteomic analyses (⇄ Section 9.10) have also revealed a high abundance of periplasmic substrate-binding proteins for soluble nutrients such as phosphate, amino acids, and sugars in *Pelagibacter*. Another adaptive feature of *Pelagibacter* is its fairly small genome (1.3 Mbp). Consistent with the proteome analysis, the genome encodes an unusually high number of ABC-type transport systems—transporters that have an extremely high affinity for their substrates (⇄ Section 3.2)—and other

enzymes useful for an oligotrophic lifestyle. The *Pelagibacter* genome is also highly "streamlined," with intergenic spacings averaging only 3 base pairs; such a highly compact genome reduces the cost of replication.

The *Pelagibacter* genome also contains genes encoding a form of the visual pigment rhodopsin that can convert light energy into ATP. In Section 17.1 we discussed the now well-studied molecule *bacteriorhodopsin*, a light-activated protein complex present in the extreme halophile *Halobacterium* (*Archaea*); bacteriorhodopsin functions in ATP synthesis as a simple light-driven proton pump (⇄ Figure 17.4). The form of rhodopsin in *Pelagibacter* and other pelagic *Bacteria* is structurally similar to bacteriorhodopsin and has been called **proteorhodopsin** ("proteo" referring to *Proteobacteria*). Although proteorhodopsin was first discovered in species of *Proteobacteria*, it is actually fairly widely distributed in *Bacteria*, including many *Gamma-* and *Alphaproteobacteria*, *Bacteroidetes*, and *Actinobacteria*, and it has also been found in nonhalophilic species of *Archaea*, such as species of the marine *Euryarchaeota*. The different variants of proteorhodopsins in marine microbes have absorption properties that reflect changing spectral properties of light at increasing depths in the water column, with near-surface variants absorbing green light and those at greater depths absorbing blue light.

Proteorhodopsin-containing marine bacteria survive starvation better in the light than in the dark. This shows that energy-starved cells use light-mediated ATP production to compensate for energy unavailable from carbon respiration when organic carbon levels are low. Proteorhodopsins are thought to exist in ~80% of bacteria in some marine waters and are thus a widespread strategy to supplement the energy metabolism of marine microbes such that they need not rely solely on scarce organic carbon for their energy needs.

Marine Viruses

In the oceans, viruses are more abundant than cellular microorganisms, often numbering over 10^7 virions/ml in typical seawater (⇄ Section 10.12). In coastal waters, where bacterial cell

UNIT 5

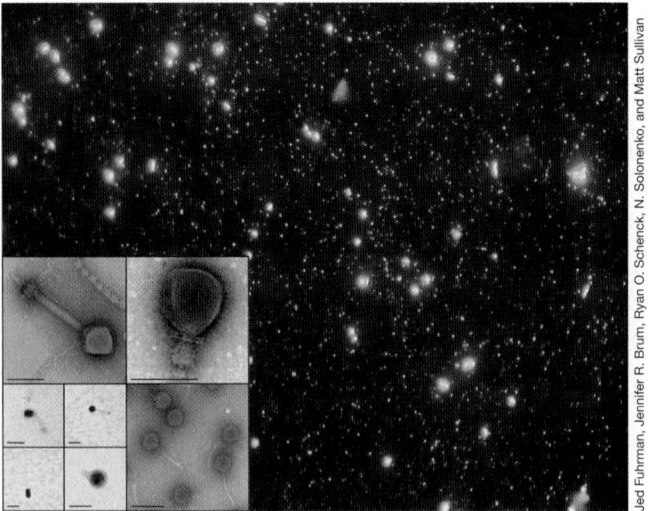

Figure 20.28 Viruses in seawater. A water sample collected on a 0.02-μm filter is stained with SYBR Green and viewed by epifluorescence microscopy. The tiny green dots are viruses while the larger, brighter dots are prokaryotic cells about 0.5 μm in diameter. Viruses are typically 10 times more abundant than the total of prokaryotic cells in seawater. Inset transmission electron micrographs show various marine bacterial viruses (scale bars, 100 nm in all images).

numbers are higher than in the oceans, viral numbers are also higher, as many as 10^8 virions/ml. Most of these viruses are bacteriophages, which infect species of *Bacteria*, and archaeal viruses, which infect species of *Archaea*. The number of virions in seawater is about 10-fold greater than average prokaryotic cell numbers, suggesting that viruses are actively infecting their hosts, replicating, and being released into seawater (**Figure 20.28**). Only a small fraction of released viruses (an average of one per burst) successfully infects a new host, and most are inactivated or destroyed by sunlight and hydrolytic enzymes. In these ways, the entire viral population is replaced in periods of only a few days or weeks. We considered the diversity of bacterial and archaeal viruses in Chapter 10.

Along with feeding by protists, marine virus infections probably help to maintain prokaryotic cells at the levels that are observed, but viruses may also have other important ecosystem functions. These include facilitating genetic exchange between prokaryotic cells and allowing for lysogeny, the state in which a virus genome integrates within the cellular genome; lysogeny can confer new genetic properties on the cell (⇨ Sections 8.7 and 11.7). For example, the discovery that some of the viruses that infect *Prochlorococcus*, the most abundant oxygenic phototroph in the oceans (Figure 20.22 and Section 20.10), contain genes that encode proteins for photosynthesis indicates that even key metabolic properties may be encoded by viruses. Although the genetic diversity of marine viruses is just now being recognized, it is thought that the diversity of marine viral genomes could surpass even that of all prokaryotic cells, making the oceans a hotbed of genetic diversity.

A Phylogenetic Snapshot of Marine Bacterial and Archaeal Diversity

Several studies have attempted to characterize the diversity of planktonic marine *Bacteria* and *Archaea* by analysis of 16S ribosomal RNA genes obtained from seawater. The existence of abundant alphaproteobacterial populations to which *Pelagibacter* is affiliated was first revealed by such 16S rRNA sequence analysis. Mesophilic *Archaea* related to *Nitrosopumilus maritimus* (⇨ Section 17.5) were discovered using similar methods.

Major bacterial groups now recognized as abundant in the open ocean include *Alpha-* and *Gammaproteobacteria*, cyanobacteria, *Bacteroidetes*, and to a lesser extent, *Betaproteobacteria* and *Actinobacteria*; *Firmicutes* are only minor components (**Figure 20.29**). As for soil, a large proportion of unclassified and minor bacterial groups are also present in seawater. A major group of marine *Gammaproteobacteria* is the yet to be cultured "SAR86 group," which accounts for approximately 10% of the total prokaryotic community in the ocean surface layer. Representing the *Archaea* in pelagic waters is a rather restricted diversity of *Euryarchaeota*, *Crenarchaeota*, and *Thaumarchaeota*, most of which have not yet been brought into laboratory culture.

With the exception of the cyanobacteria, most marine *Bacteria* are thought to be chemoorganotrophs adapted to extremely low nutrient availability, some augmenting energy conservation through proteorhodopsin or aerobic anoxygenic phototrophy (Section 20.10). The discovery of the chemolithotroph *Nitrosopumilus* suggested the possibility that many marine *Archaea* specialize in ammonia oxidation, although heterotrophic species likely exist as well. "Dilution culture" methods employing very dilute culture media have been successful in bringing some pelagic microbes into culture (⇨ Section 19.2). It appears that most of these organisms have evolved to grow only at very low nutrient concentrations, so it is either difficult or impossible to culture them to higher cell densities. Cell densities of marine oligotrophs in laboratory cultures are similar to those in their natural environments (10^5–10^6/ml), which renders many of the common tools for measuring cell growth (turbidity, microscopic counts) useless on samples that are not first concentrated. Nevertheless, there have been notable successes with dilution culturing of marine bacteria, and the aforementioned bacterium *Pelagibacter* is a good example (⇨ Figure 19.6).

--- MINIQUIZ ---

- What is proteorhodopsin and why is it so named? Why might proteorhodopsin make a bacterium such as *Pelagibacter* more competitive in its habitat?

- How do numbers of pelagic prokaryotic cells and viruses compare?

- Which phylum and subgroups of *Bacteria* dominate pelagic marine waters?

- Why are dilute culture media used for isolating pelagic microbes?

20.12 The Deep Sea

Light penetrates no farther than about 300 m in pelagic waters; as has been mentioned, this illuminated region is called the *photic zone* (Figure 20.26). Beneath the photic zone, down to a depth

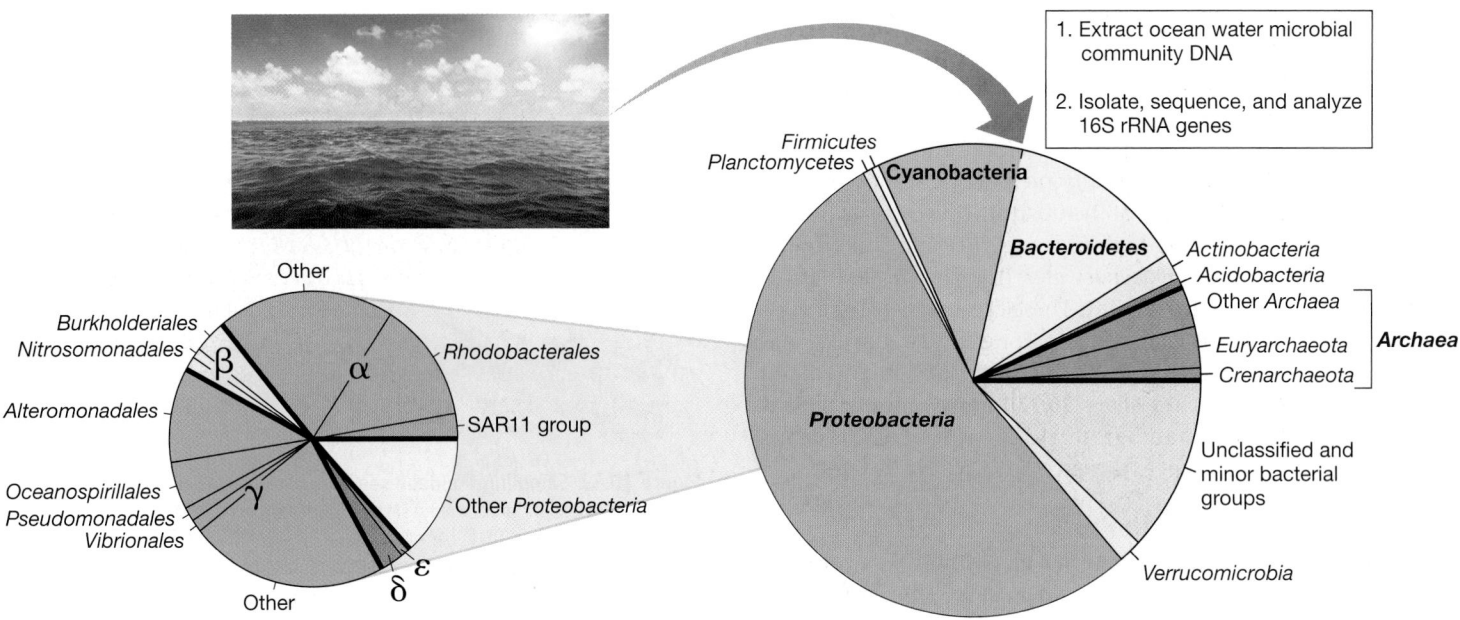

Figure 20.29 Ocean bacterial and archaeal diversity. The results are pooled analyses of 25,975 16S ribosomal RNA sequences from several studies of pelagic ocean waters. Many of these groups are covered in Chapters 15 and 16 (*Bacteria*) or 17 (*Archaea*). For *Proteobacteria*, major subgroups are indicated. Note the high proportion of cyanobacterial and *Gammaproteobacteria* sequences. Data assembled and analyzed by Nicolas Pinel. Compare the prokaryotic diversity of seawater with that of freshwater shown in Figure 20.18.

of about 1000 m, there is still considerable biological activity. However, water at depths greater than 1000 m is, by comparison, much less biologically active and is known as the *deep sea*. Greater than 75% of all ocean water is deep-sea water, lying primarily at depths between 1000 and 6000 m. The deepest waters in the oceans lie below 10,000 m. However, because depressions this deep are very rare, the waters in them make up only a very small proportion of all pelagic waters.

Conditions in the Deep Sea

Organisms that inhabit the deep sea face three major environmental extremes: (1) low temperature, (2) high pressure, and (3) low nutrient levels. In addition, deep-sea waters are completely dark such that photosynthesis is impossible. Thus, microbes that inhabit the deep sea must be chemotrophic and able to grow under high pressure and oligotrophic conditions in the cold.

Below depths of about 100 m, ocean water temperatures stay constant at 2–3°C. We discussed the responses of microorganisms to changes in temperature in Sections 5.9–5.11. As would be expected, bacteria isolated from marine waters below 100 m are psychrophilic (cold-loving) or at least psychrotolerant. Deep-sea microbes must also be able to withstand the enormous hydrostatic pressures associated with great depths. Pressure increases by 1 atm for every 10 m of depth in a water column. Thus, an organism growing at a depth of 5000 m must be able to withstand pressures of 500 atm. We will see that microorganisms in general are remarkably tolerant of high hydrostatic pressures; many species can withstand pressures of 500 atm, and some species can withstand far more than this. Moreover, from studies of deep-sea microbial diversity performed thus far, some unusual microbes call this extreme environment home (see page 617).

Piezotolerant and Piezophilic *Bacteria* and *Archaea*

Different physiological responses to pressure are observed in different deep-sea microorganisms. Some organisms simply tolerate high hydrostatic pressure, but do not grow optimally under such pressure; these organisms are **piezotolerant** (Figure 20.30). By contrast, others actually *grow best* under elevated hydrostatic pressure; these are called **piezophiles**. Organisms isolated from surface waters down to about 3000 m are typically piezotolerant. In piezotolerant organisms, higher metabolic rates are observed

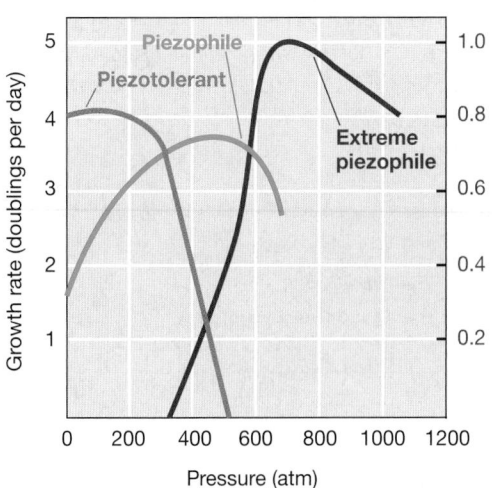

Figure 20.30 Growth of piezotolerant, piezophilic, and extremely piezophilic bacteria. Compare the slower growth rate of the extreme piezophile (right ordinate) with the growth rate of the piezotolerant and piezophilic bacteria (left ordinate), and note the inability of the extreme piezophile to grow at low pressures.

at 1 atm than at 300 atm, although growth rates at the two pressures may be similar (Figure 20.30). However, piezotolerant isolates typically do not grow at pressures greater than about 500 atm (Figure 20.30).

By contrast, cultures derived from samples taken at greater depths, 4000–6000 m, are typically piezophilic, growing optimally at pressures of around 300–400 atm. However, although piezophiles grow best under high pressure, they can still grow at 1 atm (Figure 20.30). In even deeper waters (for example, 10,000 m), **extreme piezophiles** are present. These organisms require very high pressure for growth (**Figure 20.31**). For example, the extreme piezophile *Moritella*, isolated from the Mariana Trench (Pacific Ocean, >10,000-m depth) (**Figure 20.32**), grows optimally at a pressure of 700–800 atm and grows nearly as well at 1035 atm (Figure 20.31), the pressure it experiences in its natural habitat.

Molecular Effects of High Pressure

High pressure affects cellular physiology and biochemistry in many ways. In general, pressure decreases the ability of the subunits of multi-subunit proteins to interact. Thus, large protein complexes in extreme piezophiles must interact in such a way as to minimize pressure-related effects. Protein synthesis, DNA synthesis, and nutrient transport are sensitive to high pressure. Piezophilic bacteria grown under high pressure have a higher proportion of unsaturated fatty acids in their cytoplasmic membranes than when grown at 1 atm. Unsaturated fatty acids allow membranes to remain functional and keep from gelling at high pressures or at low temperatures. The rather slow growth rates of extreme piezophiles such as *Moritella* compared with other marine bacteria (Figure 20.30) are likely due to the combined effects of pressure and low temperature; low temperature slows down

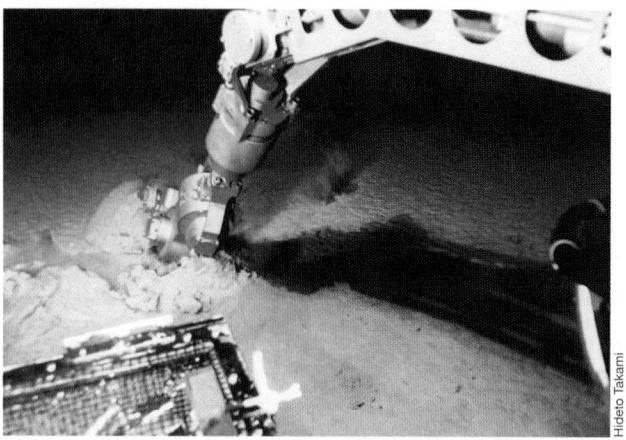

Figure 20.32 Sampling the deep sea. The unmanned Japanese submersible *Kaiko* collecting a sediment sample on the seafloor of the Mariana Trench off the Philippines at a depth of 10,897 m. The tubes of sediment are used for enrichment and isolation of piezophilic bacteria, such as the extreme piezophile (*Moritella*) isolated from this seafloor.

the reaction rates of enzymes, directly affecting cell growth (ᴄⴢ Sections 5.9 and 5.10).

Studies of gene expression and adaptive features contributing to growth at high pressure have required special pressurized incubation devices (**Figure 20.33**). These studies have shown that when a gram-negative piezophile is grown under high pressure, a specific outer membrane protein called OmpH (outer *m*embrane *p*rotein H) is present that is absent from cells grown at 1 atm. OmpH is a type of porin. Porins are proteins that form channels through which molecules diffuse into the periplasm (ᴄⴢ Section 2.5). Presumably, the porin made by cells grown at 1 atm cannot function properly at high pressure and thus a different porin must be synthesized. Interestingly, pressure controls transcription of *ompH*, the gene encoding OmpH. In characterized gram-negative piezophiles, a pressure-sensitive membrane protein complex is present that monitors pressure and triggers transcription of *ompH* only when conditions of high pressure warrant it. Transcriptomic analyses (ᴄⴢ Section 9.9) indicate that even relatively modest changes in hydrostatic pressure alter the expression of a large number of genes in piezophiles, so it is likely that many other pressure-monitoring proteins exist in these organisms.

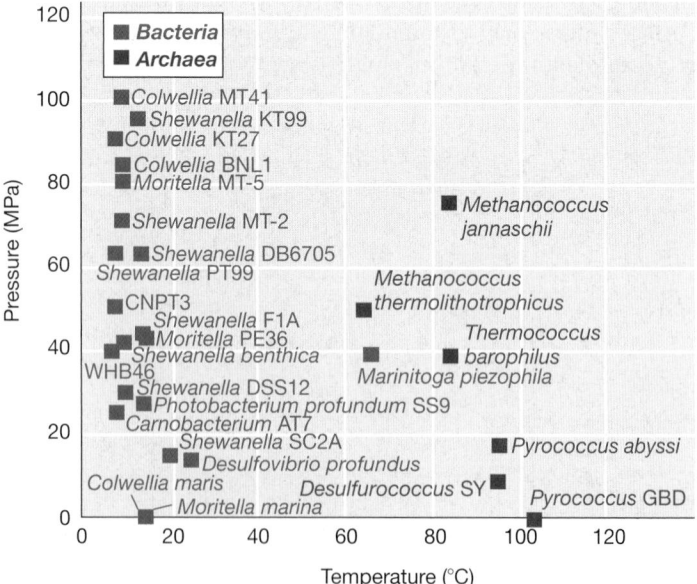

Figure 20.31 Pressure and temperature optima for cultured bacterial and archaeal piezophiles. Pressure is in pascals (Pa), the SI unit for pressure. One megapascal (MPa) corresponds to approximately 10 atm. Note that different species of the same genus can have vastly different pressure optima. Data assembled by Doug Bartlett.

MINIQUIZ

- How does pressure change with depth in a water column?
- What molecular adaptations are found in piezophiles that allow them to grow optimally under high pressure?

20.13 Deep-Sea Sediments

In addition to deep-sea *waters*, another vast and mostly unexplored microbial biosphere exists below the seafloor in the deep-sea *sediments*. Drilling expeditions to explore far below the ocean seafloor have revealed both archaeal and bacterial populations

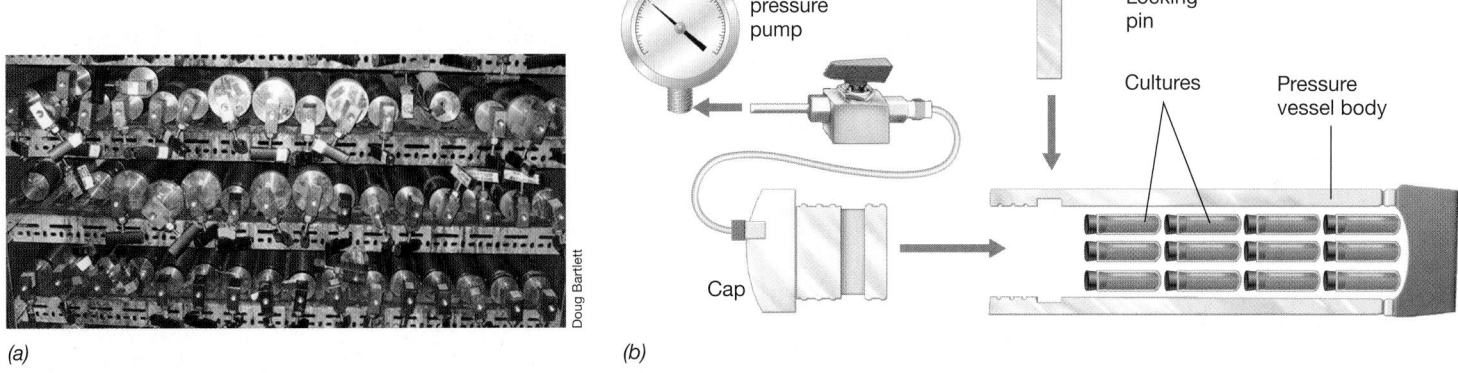

(a)   (b)

Figure 20.33 Pressure cells for growing piezophiles under elevated pressure. (a) Photo of several pressure cells incubating in a cold room (4°C). (b) Schematic design of a pressure cell. These vessels are designed to maintain pressures of 1000 atm. Illustration based on drawing by Doug Bartlett.

at depths greater than 2000 m (**Figure 20.34**). Most studies thus far have focused on relatively organic-rich sediments along continental margins. Here, cell numbers typically decrease from about 10^9 cells/g in surface sediment to about 10^6 cells/g at depths as great as 1000 m below the seafloor. In these coastal sediments, sulfate-reducing bacteria and other anaerobes deplete sulfate and other electron acceptors within a few meters of the sediment–water interface (Figure 20.20). This depletion of electron acceptors and organic matter with increasing depth constrains energy available to the deep subsurface microbial communities, accounting for the major decrease in cell numbers with depth.

Cell Numbers in Deep-Sea Sediments

The better-studied continental margins and shelf sediments are not representative of most of the ocean floor, about 90% of which is at greater than 2000 m depth and associated with marine waters of low productivity and therefore of much lower organic matter content (Figure 20.20). In the absence of significant transport of organic material to the sediment surface, sulfate and other electron acceptors may permeate all the way through the sediments to the underlying bedrock. However, because of the dearth of organic matter, cell numbers in these sediments are several orders of magnitude lower than in organic-rich sediments, ranging from about

(a)

(b)

Figure 20.34 Drilling deep-sea sediments. (a) Deep-sea drilling vessel the JOIDES *Resolution*. Inset: Red dot indicates the location of sediment sampling in the Peru Basin. (b) Sediment cores recovered from the Peru Basin at 4800 m depth. Cores were split lengthwise to allow subsampling for molecular characterization. See Section 20.5 and Figure 20.8 for discussion of sulfide-oxidizing microbial mats that grow on the sediment surface off the Chilean and Peruvian coasts.

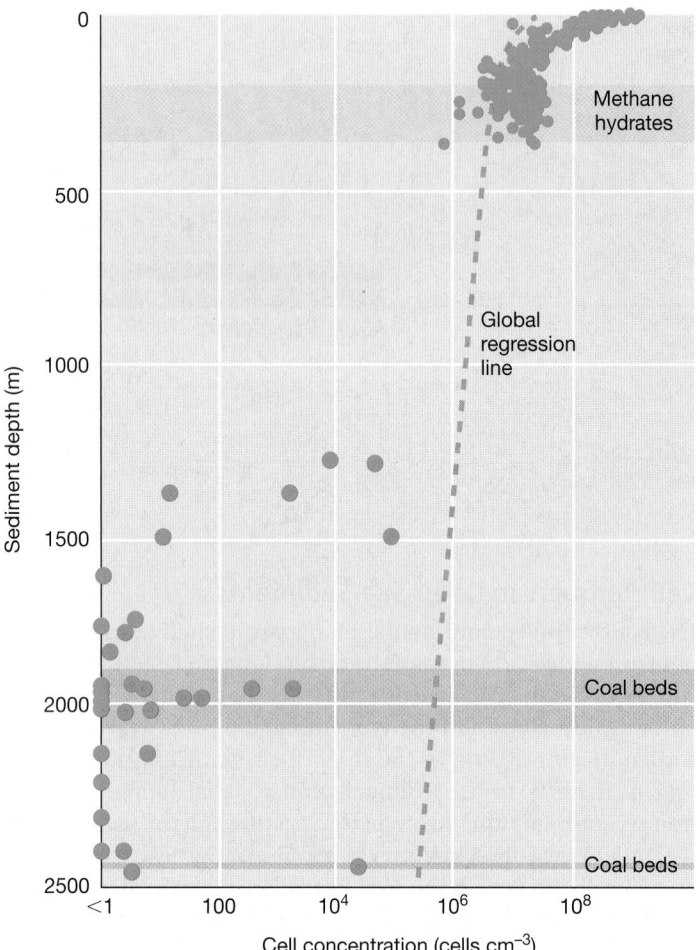

Figure 20.35 Microbial cell numbers in deep-sea sediments. Depth-related cell abundance from the analysis of one coastal sediment (filled circles) and the global average for all sampled ocean sediments (dashed regression line) based on cell count data.

10^6 cells/gram at the surface to fewer than 10^3 cells/gram at depths of a few hundred meters (**Figure 20.35**). Cell counts in all sediments generally decline significantly with depth, reflecting lower organic carbon availability and quality in older deep sediment material. Despite these low numbers, because deep-sea sediments are so vast, it has been estimated that a total of $\sim 5.4 \times 10^{29}$ prokaryotic cells exist in deep-sea sediments ($\sim 4 \times 10^{15}$ g), a number similar to the sum total in all the oceans.

Figure 20.35 also shows that cell numbers can deviate from the global regression as a function of spotty distributions of resources with depth at any particular drill site. In the example shown in Figure 20.35, cell numbers in coastal plain sediments are elevated near methane seeps (Figure 20.20), a region associated with active anaerobic methane oxidation (Section 14.25), and in coal bed deposits (Figures 20.20 and 20.35), where, in addition to coal, associated organic materials boost electron donor availability.

A Phylogenetic Snapshot of Marine Sediment Prokaryotic Diversity

Marine sediment communities have been explored only to a limited extent because of the difficulty and expense of obtaining

uncontaminated drilling cores from great depths (Figure 20.34). However, analyses of 16S ribosomal RNA gene sequences obtained by using PCR methods (Section 19.6) on deep core samples have clearly established that novel *Archaea* make up a large fraction of the archaeal diversity and that bacterial diversity in both deep and shallow marine sediments is dominated by *Proteobacteria* (**Figure 20.36**). This abundance of *Proteobacteria* is also true of all the other habitats explored by culture-independent techniques (Figures 20.12, 20.18, and 20.29, and see Figure 20.41).

Within marine sediment *Proteobacteria*, phylotypes associated with sulfate-reducing bacteria such as the *Desulfobacterales* are quite common (Figure 20.36), a fact that is not surprising considering that sulfate reduction is the major form of anaerobic respiration in marine sediments (Sections 14.14 and 15.9). *Bacteroidetes* and the unclassified/minor groups are also well represented in shallow marine sediments. Although major players in marine waters, cyanobacteria compose just a tiny proportion of the total cell population in the permanently dark and anoxic sediments and probably represent cells that have reached the sediments after attaching to a particle or dead animal that eventually sank.

In addition to a variety of *Bacteria*, novel phyla of *Archaea* unrelated to cultured representatives are widespread in the deep subsurface. Genome sequences of uncultured *Bathyarchaeota* recovered from both terrestrial and marine sediments (Figure 20.14) revealed a physiological capacity to degrade and assimilate protein and a capacity for some species to degrade carbohydrates. Other *Bathyarchaeaota* appear to carry out methanogenesis and are autotrophs, likely existing on the small amounts of H_2 produced from geochemical reactions (Section 20.7) in deep sediments.

Energy Limitation and Microbial Life below the Seafloor

How microbes in the deepest marine sediments survive in their nutrient-depleted environment remains unclear, but it is likely that they employ many of the strategies we have seen in marine pelagic microbes including small cell size (Figure 20.15) and small, compact genomes. Sequencing of 16S ribosomal RNA genes selectively amplified by PCR (Section 19.6) using DNA extracted from deep drilling cores, as well as from more limited metagenomic surveys, has identified relatively few sequences related to the classical sulfate-reducing bacteria (Section 15.9) or methanogenic and known methane-oxidizing *Archaea* (Sections 14.17 and 17.2), organisms that by contrast are quite common in surface sediments. Thus, from a nutritional standpoint, how do buried cells thrive under conditions of extreme energy limitation?

In the marine environment, easily degraded organic material is removed by microbial respiration in the water column and anaerobic respiration in surficial sediments (the sediments at and near the sediment surface), leaving behind a dilute pool of less readily degradable organic material that slowly trickles down into deeper sediments. Microbes that inhabit the deep sediments presumably utilize this low-quality organic material along with cell necromass (cellular components released following cell death) as electron donors in energy metabolism. However, since representative organisms have yet to be isolated from deep sediments, our understanding of the supporting energy-conserving metabolisms is sketchy and can only be inferred from a few partial genome

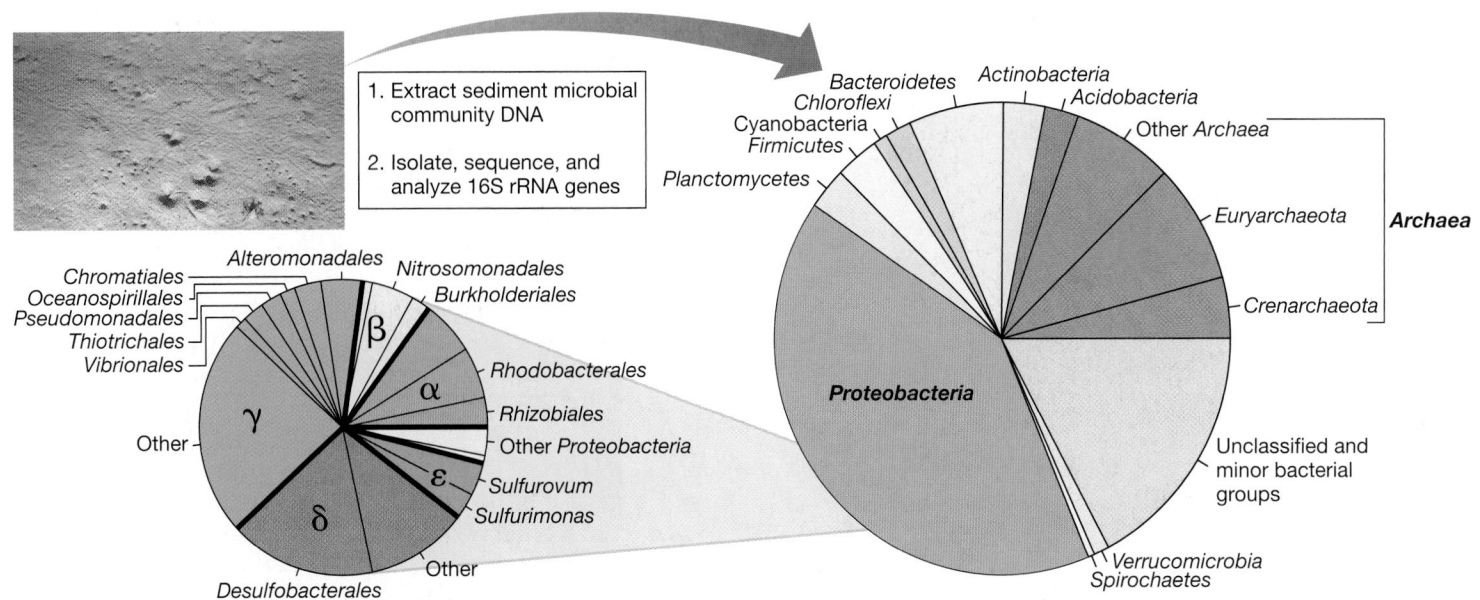

1. Extract sediment microbial community DNA

2. Isolate, sequence, and analyze 16S rRNA genes

Figure 20.36 Marine sediment bacterial and archaeal diversity. The results are pooled analyses of 13,360 16S ribosomal RNA gene sequences from several studies of shallow and deep marine sediments. Many of the groups indicated are covered in Chapters 15 and 16 (*Bacteria*) or 17 (*Archaea*). For *Proteobacteria*, major subgroups are indicated. Note the high proportion of archaeal sequences and of *Gamma-*, *Delta-*, and *Epsilonproteobacteria*. Data assembled and analyzed by Nicolas Pinel. Compare the prokaryotic diversity of marine sediments with that of open ocean water shown in Figure 20.29.

sequences assembled from metagenomes (⟨⟩ Sections 9.8 and 19.8) or obtained from single cells isolated directly from sediments, using single-cell genomic technology (⟨⟩ Sections 9.12 and 19.12). Various known and yet-to-be-discovered forms of anaerobic respiration and fermentation are the most likely metabolic candidates for deep seafloor metabolisms along with other metabolic options not readily obvious from the conditions known to exist in deep marine sediments.

Since sediment microorganisms largely control the fate of carbon in this vast subsurface reservoir of organic matter, the discovery of novel *Archaea* (Figure 20.14) living at the thermodynamic edge of life has given microbial ecologists a new perspective on carbon cycling in marine sediments. However, the extent to which abiotic processes may also contribute energy sources to these buried microbes is unclear but could be significant. For example, the higher temperatures found at greater depths may promote the alteration of organic material, releasing methane and other hydrocarbons, H_2, acetate, and CO_2 that may diffuse upwards to nourish microbes inhabiting the deep sediment biosphere. Thus there is clearly much more to learn about the microbial ecology of Earth's deep subsurface, both in terrestrial and marine sediment contexts.

MINIQUIZ

- Give two reasons why sulfate-reducing bacteria are common in shallow marine sediments.
- How and why do the numbers of bacterial and archaeal cells vary with depth in marine deep sediments?
- What alternative sources of energy are suggested to nourish microorganisms in extremely deep marine sediments?

20.14 Hydrothermal Vents

Although we have thus far described the deep sea and deep-sea sediments as a remote, low-temperature, high-pressure, oligotrophic environment suitable only for slow-growing piezotolerant and piezophilic microorganisms, there are some amazing exceptions. Thriving animal and microbial communities are found clustered in and around deep-sea hot springs throughout the world. These hot springs are located at depths from less than 1000 m to greater than 4000 m in regions of the seafloor where volcanic magma and hot rock have caused the floor to rift apart at crustal spreading centers (Figure 20.20 and **Figure 20.37**), or where iron and magnesium minerals associated with ancient rocks react with seawater and generate heat. Seawater seeping into these dynamic cracking regions of the crust reacts with hot rock, resulting in hot springs saturated with inorganic chemicals and dissolved gases. Collectively, these types of underwater hot springs are called **hydrothermal vents** (Figure 20.20). We discuss several remarkable symbiotic associations between hydrothermal vent–associated animals and microorganisms in Chapter 23. Here we consider the vent environment as a habitat for free-living microbes.

Types of Vents

Volcanic hydrothermal systems are typically either warm (~5 to >50°C), diffuse vents or very hot vents that emit hydrothermal fluids at 270 to >400°C. The gently flowing, warm, diffuse fluids are emitted from cracks in the seafloor and the exterior walls of hydrothermal chimneys. The fluids originate from the mixing of cold seawater with hot hydrothermal fluids in subsurface regions of the sediments. Hot vents, called *black smokers*, form upright sulfide edifices called *chimneys* that can be less than 1 m to over 30 m in height.

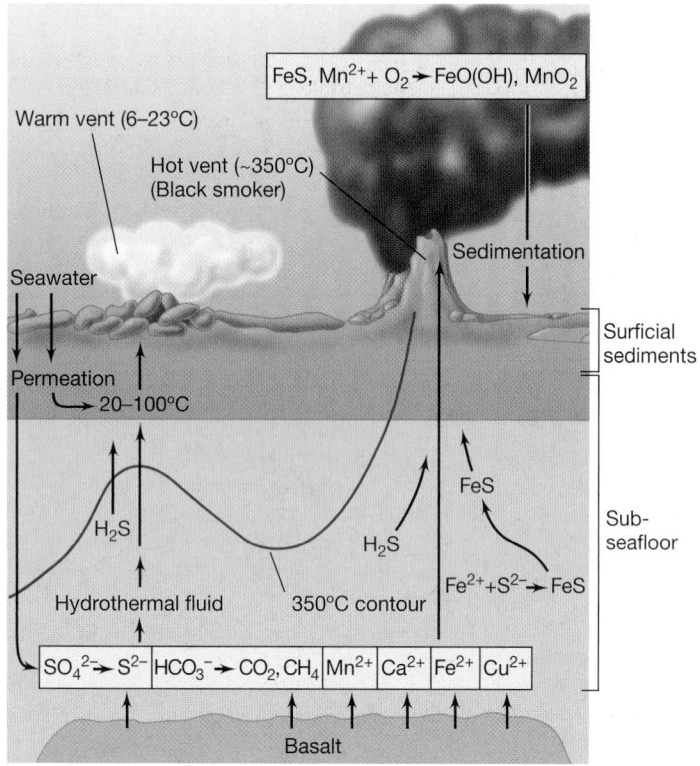

Figure 20.37 Hydrothermal vents. Schematic showing geological formations and major inorganic chemicals and minerals that are emitted from warm vents and black smokers. In warm vents, the hot hydrothermal fluid is cooled by cold 2–3°C seawater permeating the sediments. In black smokers, hot hydrothermal fluid near 350°C reaches the seafloor directly. The term "surficial" is a geological term pertaining to Earth's surface.

Chimneys form when acidic hydrothermal fluids rich in dissolved metals and magmatic gases are suddenly mixed with cold, oxygenated seawater. The rapid mixing causes fine-grained metal sulfide minerals such as pyrite and sphalerite to precipitate out, forming dark, buoyant plumes that rise above the seafloor (**Figure 20.38**).

A quite different type of hydrothermal vent environment is the "Lost City" formation located in the mid-Atlantic Ocean. Lost City is formed from the exposure of minerals associated with ocean crust 1–2 million years old that was once deep beneath the seafloor. Geological faults in these slow-spreading systems exposed magnesium and iron-rich rocks called *peridotites* at the seafloor. Chemical reactions of seawater and newly exposed peridotite are highly exothermic, generating heat and also driving the pH up to as high as pH 11. Extremely high levels of H_2, CH_4, and other low-molecular-weight hydrocarbons are also present in the hot (200°C) hydrothermal fluids. In contrast to the acidic volcanic black smoker systems (Figure 20.38), which are relatively transient, mixing of these alkaline fluids with seawater results in the formation of calcium carbonate (limestone) chimneys (**Figure 20.39**) that can reach up to 60 m in height and be active for 100,000 years or more.

Bacteria and Archaea in Hydrothermal Vents

Bacteria displaying chemolithotrophic metabolisms dominate hydrothermal vent microbial ecosystems (Figure 20.20).

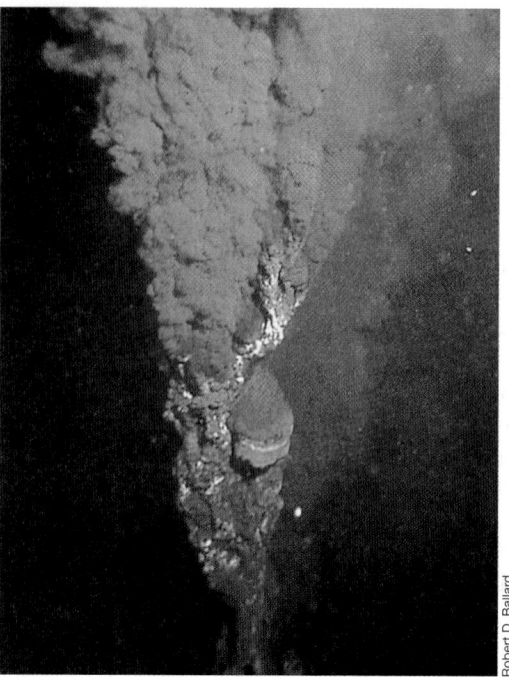

Figure 20.38 A hydrothermal vent black smoker emitting sulfide- and mineral-rich water at temperatures of 350°C. The walls of the black smoker chimneys display a steep temperature gradient and contain several types of *Bacteria* and *Archaea*.

Figure 20.39 Massive carbonate chimney formation at Lost City peridotite-hosted vent system. Microbial colonization of freshly exposed mineral surfaces was studied by placing sterile mineral fragments in the green-topped device positioned over an actively venting area of the chimney. The diameter of the cylindrical collection device is approximately 10 cm.

Sulfidic vents support sulfur bacteria, whereas vents that emit other inorganic electron donors support nitrifying, hydrogen-oxidizing, iron- and manganese-oxidizing, or methylotrophic bacteria, the latter presumably growing on the CH_4 and carbon monoxide (CO) emitted from the vents. **Table 20.2** summarizes the inorganic electron donors and electron acceptors that are thought to play a role in chemolithotrophic metabolisms at hydrothermal vents. All of these metabolisms were discussed in Chapter 14.

Although microbes cannot survive in the superheated hydrothermal fluids of black smokers, thermophilic and hyperthermophilic organisms do thrive in the *gradients* that form as the superheated water mixes with cold seawater. For example, the walls of smoker chimneys are teeming with hyperthermophiles such as *Methanopyrus,* a species of *Archaea* that oxidizes H_2 and makes CH_4 (⟜ Section 17.2). Phylogenetic FISH staining (⟜ Section 19.5) has detected cells of both *Bacteria* and *Archaea* in smoker chimney walls (**Figure 20.40**). The most thermophilic of all known sulfur-reducing microbes, species of *Pyrolobus* and *Pyrodictium* (Chapter 17), were isolated from black smoker chimney walls. In contrast to the significant microbial diversity in volcanic vent chimney walls, the carbonate chimney walls of the Lost City vents are comprised primarily of methanogens of the genus *Methanosarcina* and are nourished by the H_2-rich fluids that permeate the porous chimney walls.

When smokers plug up from mineral debris, hyperthermophiles presumably drift away to colonize active smokers and somehow become integrated into the growing chimney wall. Surprisingly, although they require very high temperatures for growth, hyperthermophiles are remarkably tolerant of cold temperatures and oxygen. Thus, transport of cells from one vent site to another in cold oxic seawater apparently is not a problem.

A Phylogenetic Snapshot of Hydrothermal Vent Prokaryotic Diversity

Using the powerful tools developed for microbial community sampling (⟜ Section 19.6), studies of prokaryotic diversity near volcanic hydrothermal vents have revealed an enormous diversity

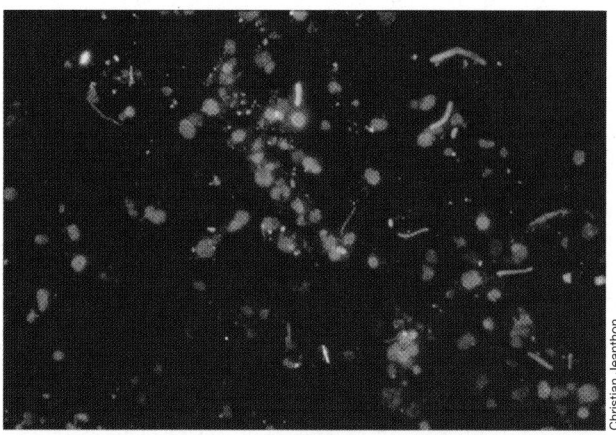

Figure 20.40 Phylogenetic FISH staining of black smoker chimney material. Taken from the Snake Pit vent field in the Mid-Atlantic Ridge, depth of 3500 m. A green fluorescing dye was conjugated to a probe that reacts with the 16S rRNA of all *Bacteria* and a red dye to a 16S rRNA probe for *Archaea*. The hydrothermal fluid going through the center of this chimney was at 300°C.

of *Bacteria*. These 16S rRNA gene sequence surveys include both warm and hot vents. Hydrothermal vent microbial communities are dominated by *Proteobacteria*, in particular *Epsilonproteobacteria* (⟜ Section 16.5; **Figure 20.41**). *Alpha-, Delta-,* and *Gammaproteobacteria* are also abundant, whereas *Betaproteobacteria* are much less so. Many *Epsilon-* and *Gammaproteobacteria* oxidize sulfide and sulfur as electron donors with either O_2 or nitrate (NO_3^-) as electron acceptors.

As shown in the detailed diagram of *Proteobacteria* in Figure 20.41, vent *Epsilonproteobacteria* phylotypes most closely match those of chemolithotrophic sulfur bacteria such as *Sulfurimonas, Arcobacter, Sulfurovum,* and *Sulfurospirillum*. These bacteria oxidize reduced sulfur compounds as electron donors (⟜ Sections 14.9 and 15.11), and such a physiology is consistent with their presence near vent fluids charged with sulfur and sulfide. In addition, most *Deltaproteobacteria* specialize in anaerobic metabolisms using oxidized sulfur compounds as electron acceptors.

In contrast to *Bacteria*, the diversity of volcanic hydrothermal vent *Archaea* is quite limited. Estimates of the number of unique phylotypes indicate that the diversity of *Bacteria* near hydrothermal vents is about 10 times that of *Archaea*. However, *Archaea* are prevalent in samples recovered from the walls of hot vent chimneys (Figure 20.40). Most of the *Archaea* detected near hydrothermal vents are either methanogens (⟜ Section 17.2) or species of marine *Crenarchaeota* and *Euryarchaeota* (Chapter 17). With the exception of the ammonia-oxidizing *Nitrosopumilus* (*Thaumarchaeota*, ⟜ Section 17.5), organisms in these groups remain uncultured and their physiologies poorly understood.

TABLE 20.2 Chemolithotrophic *Bacteria* and *Archaea* present near deep-sea hydrothermal vents[a]

Chemolithotroph	Electron donor	Electron acceptor	Product from donor
Sulfur-oxidizing	HS^-, S^0, $S_2O_3^{2-}$	O_2, NO_3^-	S^0, SO_4^{2-}
Nitrifying	NH_4^+, NO_2^-	O_2	NO_2^-, NO_3^-
Sulfate-reducing	H_2	S^0, SO_4^{2-}	H_2S
Methanogenic	H_2	CO_2	CH_4
Hydrogen-oxidizing	H_2	O_2, NO_3^-	H_2O
Iron- and manganese-oxidizing	Fe^{2+}, Mn^{2+}	O_2	Fe^{3+}, Mn^{4+}
Methylotrophic	CH_4, CO	O_2	CO_2

[a]See Chapter 14 for detailed discussions of these metabolisms and Chapters 15–17 for further coverage of each group of organisms.

— MINIQUIZ —

• How does a warm hydrothermal vent differ from a black smoker, both chemically and physically?

• Why is 350°C water emitted from a black smoker not boiling?

• Which phylum of *Bacteria* and which subgroups of this phylum dominate hydrothermal vent ecosystems, and why?

UNIT 5

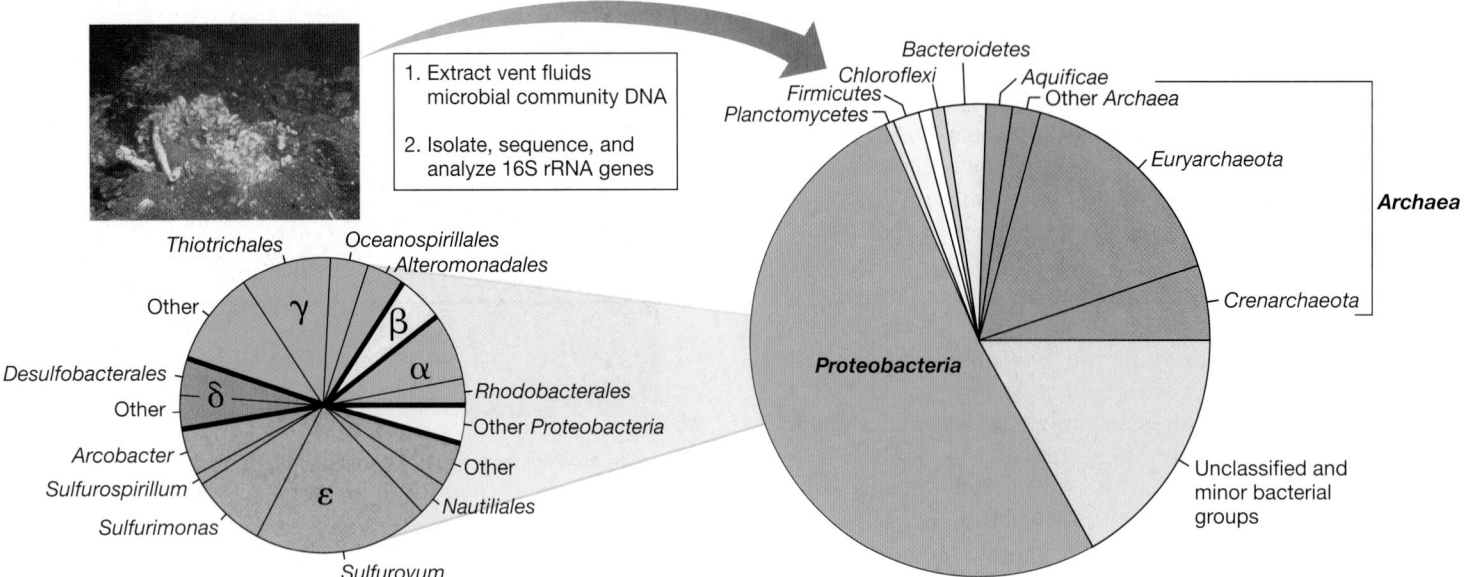

Figure 20.41 Hydrothermal vent bacterial and archaeal diversity. The results are pooled analyses from several studies of the 16S rRNA gene content of warm and hot hydrothermal vents. Many of these groups are covered in Chapters 15 and 16 (*Bacteria*) or 17 (*Archaea*). For *Proteobacteria*, major subgroups are indicated. Note the high proportion of *Archaea* and of *Epsilonproteobacteria*. The physiology of many of these organisms is summarized in Table 20.2. Data assembled and analyzed by Nicolas Pinel.

MasteringMicrobiology®

Visualize, **explore**, and **think critically** with Interactive Microbiology, MicroLab Tutors, MicroCareers case studies, and more. MasteringMicrobiology offers practice quizzes, helpful animations, and other study tools for lecture and lab to help you master microbiology.

Chapter Review

I • Microbial Ecology

20.1 Ecosystems consist of organisms, their environments, and all of the interactions among the organisms and environments. The organisms are members of populations and communities and are adapted to habitats. Species richness and abundance are aspects of species diversity in a community and an ecosystem.

Q **What types of environmental disturbance would likely reduce species richness in an aquatic habitat? What types of environments would support high species diversity?**

20.2 Microbial communities consist of guilds of metabolically similar organisms. Microorganisms play major roles in energy transformations and biogeochemical processes that result in the recycling of elements essential to living systems.

Q **In what forms does potential energy enter a microbial ecosystem? Which energy classes of microorganisms can exploit each?**

II • The Microbial Environment

20.3 The niche for a microorganism consists of the specific assortment of biotic and abiotic factors within a microenvironment in which that microorganism can be competitive. Microorganisms in nature often live a feast-or-famine existence such that only the best-adapted species reach high population density in a given niche. Cooperation among microorganisms is also important in many microbial interrelationships.

Q **Why is it possible to isolate both obligately anaerobic and obligately aerobic bacteria from the same soil sample?**

20.4 When surfaces are available, bacteria grow in attached masses of cells called biofilms. Biofilm formation confers several protective advantages on cells. Biofilms can have significant medical and economic effects on humans when unwanted biofilms develop on inert as well as living surfaces.

Q **The surface of a rock in a flowing stream will often contain a biofilm. What specific advantages could be conferred on bacteria growing in a biofilm compared with growth within the flowing stream?**

UNIT 5

20.5 Microbial mats can be phototrophic or chemolithotrophic. Phototrophic cyanobacterial mats are thick biofilms consisting of microbial cells and trapped particulate materials. They are widespread in hypersaline or thermal waters where grazing animals are excluded by salt or temperature from feeding on the mat cells. Sulfur-oxidizing chemolithotrophs form extensive mat communities on marine sediment surfaces below the photic zone at the interface between O_2 supplied from the overlying water and H_2S produced by sulfate-reducing bacteria.

Q What is the importance of gliding motility in the formation of marine chemolithotrophic mat systems?

III • Terrestrial Environments

20.6 Soils are complex microbial habitats with numerous microenvironments and niches. Microorganisms are present in the soil primarily attached to soil particles. The most important factors influencing microbial activity in soil are the availability of water and nutrients. However, in very arid soils microorganisms play important roles in stabilizing soil structure.

Q In what soil horizon are microbial numbers and activities the highest, and why?

20.7 The deep subsurface is a significant microbial habitat, most likely sustaining chemolithotrophic populations that can live on a diet of a few minerals, CO_2, SO_4^{2-}, N_2, and H_2. Hydrogen is thought to be continually produced by interaction of water with iron minerals or by the radiolysis of water. Novel phyla of small-celled *Archaea* are present in the deep subsurface of both terrestrial and marine sediment environments.

Q Studies of the deep subsurface have also identified heterotrophic populations. What types and sources of nutrients are available to those organisms?

IV • Aquatic Environments

20.8 In freshwater aquatic ecosystems, phototrophic microorganisms are the main primary producers. Most of the organic matter produced is consumed by bacteria, which can lead to depletion of oxygen in the environment. The BOD of a body of water indicates its relative content of organic matter that can be biologically oxidized.

Q How and in what way does an input of organic matter, such as sewage, affect the oxygen content of a river or stream?

20.9 Pelagic marine waters are more nutrient deficient than most freshwaters, yet substantial numbers of microbes inhabit the oceans. However, in some highly productive and expansive oceanic regions, oxygen can be drawn down to low or unmeasurable levels at depths between 100 and 1000 m; oxygen-depleted waters at these depths are called oxygen minimum zones.

Q Why is release of sulfide from oxygen minimum zones infrequent, occurring at times of exceptionally high surface water productivity?

20.10 The major microbial oxygenic phototrophs in the open oceans include the bacterium *Prochlorococcus* and the alga *Ostreococcus*; both of these phototrophs are small microorganisms. Marine anoxygenic phototrophs include *Roseobacter* and its relatives, the aerobic phototrophic purple bacteria.

Q How has single-cell genomics contributed to new understanding of factors controlling the diversity of *Prochlorococcus* in the ocean?

20.11 Species of *Bacteria* tend to predominate in marine surface waters, whereas in deeper waters *Archaea* make up a larger fraction of the microbial community. Many pelagic *Bacteria* use light to make ATP by rhodopsin-driven proton pumps. Viruses outnumber microbial cells by over an order of magnitude in marine waters.

Q Many pelagic *Bacteria* can use light energy but are not considered "phototrophs" in the same sense as cyanobacteria or purple bacteria. Explain.

20.12 The deep sea is a cold, dark habitat where hydrostatic pressure is high and nutrient levels are low. Piezophiles grow best under pressure but do not require pressure, whereas extreme piezophiles require high pressure, typically several hundred atmospheres, for growth.

Q What property is shared by piezotolerant, piezophilic, and extremely piezophilic microorganisms?

20.13 Deep-sea sediments show decreasing nutrient levels with depth and thus the microbes that reside there experience constant near-starvation conditions. Although not yet cultured, novel phyla of *Archaea* inhabit deep marine sediments (as well as the terrestrial deep subsurface), and these organisms likely survive by scavenging the trace levels of both organic matter and electron acceptors present there.

Q What are sources of organic matter in deep-sea sediments that could be used to support growth?

20.14 Hydrothermal vents are deep-sea hot springs where either volcanic activity or unusual chemistry generates fluids containing large amounts of inorganic electron donors that can be used by chemolithotrophic bacteria.

Q Would you expect to find the same types of microorganisms associated with black smoker and carbonate vent systems? Explain.

Application Questions

1. Imagine a sewage plant that is releasing sewage containing high levels of ammonia and phosphate and very low levels of organic carbon. Which types of microbial blooms might be triggered by this sewage? How would the graphs of oxygen near and beyond the plant's release point differ from the graph shown in Figure 20.17a?

2. Keeping in mind that the open-ocean waters are highly oxic, predict the possible metabolic lifestyles of open-ocean *Archaea* and *Bacteria*. Why might rhodopsin-like pigments be more abundant in one group of organisms than in the other?

3. Global warming has been suggested to result in reduced transfer of oxygen to deeper waters in the ocean (Section 20.9). How might global warming also result in reduced nutrient availability to planktonic species in marine surface waters?

Chapter Glossary

Biochemical oxygen demand (BOD) the microbial oxygen-consuming properties of a water sample

Biofilm colonies of microbial cells encased in a porous organic matrix and attached to a surface

Biogeochemistry the study of biologically mediated chemical transformations in the environment

Community two or more cell populations coexisting in a certain area at a given time

Ecosystem a dynamic complex of organisms and their physical environment interacting as a functional unit

Epilimnion the warmer and less dense surface waters of a stratified lake

Extreme piezophile an organism requiring several hundred atmospheres of pressure for growth

Guild metabolically similar microbial populations that exploit the same resources in a similar way

Habitat part of the ecosystem best suited to one or a few populations

Hydrothermal vents warm or hot water–emitting springs associated with crustal spreading centers on the seafloor

Hypolimnion the colder, denser, and often anoxic bottom waters of a stratified lake

Microbial mat a thick, layered, diverse community nourished either by light in a hypersaline or an extremely hot aquatic environment, in which cyanobacteria are essential; or by chemolithotrophs growing on the surface of sulfide-rich marine sediments

Microenvironment a micrometer-scale space surrounding a microbial cell or group of cells

Niche in ecological theory, the biotic and abiotic characteristics of the microenvironment that contribute to an organism's competitive success

Oligotroph an organism that grows only or grows best at very low levels of nutrients

Oxygen minimum zone (OMZ) an oxygen-depleted region of intermediate depth in the marine water column

Piezophile an organism that grows best under a hydrostatic pressure greater than 1 atm

Piezotolerant able to grow under elevated hydrostatic pressures but growing best at 1 atm

Population a group of organisms of the same species in the same place at the same time

Primary producer an organism that synthesizes new organic material from CO_2 and obtains energy from light or from oxidation of inorganic compounds

Prochlorophyte a bacterial oxygenic phototroph that contains chlorophylls *a* and *b* and lacks phycobiliproteins

Proteorhodopsin a light-sensitive protein present in some pelagic *Bacteria* that fuels a proton pump that yields ATP

Rhizosphere the region immediately adjacent to plant roots

Species abundance the proportion of each species in a community

Species richness the total number of different species present in a community

Stratified water column a body of water separated into layers having distinct physical and chemical characteristics

Nutrient Cycles

The Big Thaw and the Microbiology of Climate Change

As Earth warms in response to the accumulation of greenhouse gases, attention has focused on high-latitude regions where temperatures are rising twice as fast as the global average. As temperatures rise, normally frozen ground (the permafrost subsurface layer in tundra, photo on left) thaws, making vast stores of organic carbon available for microbial decomposition. Accumulated over thousands of years from past plant and animal life, the pool of carbon in the permafrost zone is estimated to be 1300–1600 billion tons. This is twice as much carbon as is currently in the atmosphere, and predicting the rate and extent of microbial conversion of permafrost organic carbon to greenhouse gases (primarily CO_2 and CH_4) is essential in developing future climate scenarios.

It has been estimated that 5–15% of permafrost carbon will decompose in this century, releasing primarily CO_2 at a rate unlikely to cause abrupt climate change in the near future. However, many uncertainties remain with this estimate, including how localized permafrost thawing might accelerate the rate at which carbon is made available for microbial decomposition. Abrupt thawing forms surface depressions that accumulate snow and water (right photo), and this creates saturated and anoxic conditions that favor microbial production of methane, a potent greenhouse gas.

A recent study highlighted the importance of incorporating vegetation change in modeling permafrost thawing. For example, the disappearance of a dominant tundra shrub species (the dwarf birch, which covers the surface in the wide-angle photo on the left and is indicated with an arrow in the close-up photo on the right) left grasses and sedges as the primary ground cover and resulted in localized thawing and formation of saturated soils within 6 years. Increased surface warming from the loss of shade provided by the low-lying shrub triggered a positive feedback loop: Depressions formed that trapped snow and water, thereby accelerating thawing. This transition also shifted the microbial community in the disturbed tundra from that of a methane sink to a methane source.

Thus, although overall permafrost carbon emissions may only rise gradually as temperatures rise, localized pockets of intense microbial activity must also be considered if reliable estimates of total greenhouse gas emissions from the massive store of organic carbon in permafrost regions are to be achieved.

Source: Nauta, A.L., et al. 2016. Permafrost collapse after shrub removal shifts tundra ecosystem to a methane source. *Nature Climate Change 5:* 67–70.

I • Carbon, Nitrogen, and Sulfur Cycles

The key nutrients for life are cycled by both microorganisms and macroorganisms, but for any given nutrient, it is microbial activities that dominate. Understanding how microbial nutrient cycles work is important because the cycles and their many feedback loops are essential for plant agriculture and the overall health of sustainable plant life.

We begin our coverage of nutrient cycles with the carbon cycle. Major areas of interest here are the magnitude of carbon reservoirs on Earth, the rates of carbon cycling within and between reservoirs, and the coupling of the carbon cycle to other nutrient cycles. We emphasize the gases *carbon dioxide* (CO_2) and *methane* (CH_4) as major components of the carbon cycle and of human impacts on the global ecosystem. We return to the carbon cycle in Section 21.8 to consider how human activities are affecting this critical nutrient cycle.

21.1 The Carbon Cycle

On a global basis, carbon (C) cycles as CO_2 through all of Earth's major carbon reservoirs: the atmosphere, the land, the oceans, freshwaters, sediments and rocks, and biomass (**Figure 21.1**). As we have already seen for freshwater environments, the carbon and oxygen cycles are intimately linked (↪ Section 20.8). All nutrient cycles link in some way to the carbon cycle, but the nitrogen (N) cycle links particularly strongly because, other than water (H_2O), C and N make up the bulk of living organisms (↪ Section 3.1 and see Figure 21.5).

Carbon Reservoirs

By far the largest C reservoir on Earth is the sediments and rocks of Earth's crust (Figure 21.1), but the rate at which sediments and rocks decompose and carbon is removed as CO_2 is so slow that flux out of this reservoir is insignificant on a human time scale. A large amount of C is found in land plants. This is the organic C of forests, grasslands, and agricultural crops—the major sites of phototrophic CO_2 fixation. However, more C is present in dead organic material, called **humus**, than in living organisms. Humus is a complex mixture of organic materials that have resisted rapid decomposition and is derived primarily from dead plants and microorganisms. Some humic substances are quite recalcitrant, with a decomposition time of several decades, but certain other humic components decompose much more rapidly.

The most rapid means of transfer of C is via the atmosphere. Carbon dioxide is removed from the atmosphere primarily by photosynthesis of land plants and marine microorganisms and is returned to the atmosphere by respiration of animals and chemoorganotrophic microorganisms (Figure 21.1). The single most important contribution of CO_2 to the atmosphere is by microbial decomposition of dead organic material. However, since the Industrial Revolution, human activities have increased atmospheric CO_2 levels by nearly 40%, primarily from the combustion of fossil fuels. This rise in CO_2, a major *greenhouse gas*, has triggered a period of steadily increasing global temperatures called **global warming** (see Section 21.8 and Figure 21.19). Although the consequences of global warming on microbial nutrient cycling are currently unpredictable, everything we know about the physiology of microorganisms tells us that microbial activities in nature will change in response to higher temperatures. Whether these responses will be favorable or unfavorable to higher organisms (including humans) is a major area of active research today (Section 21.8).

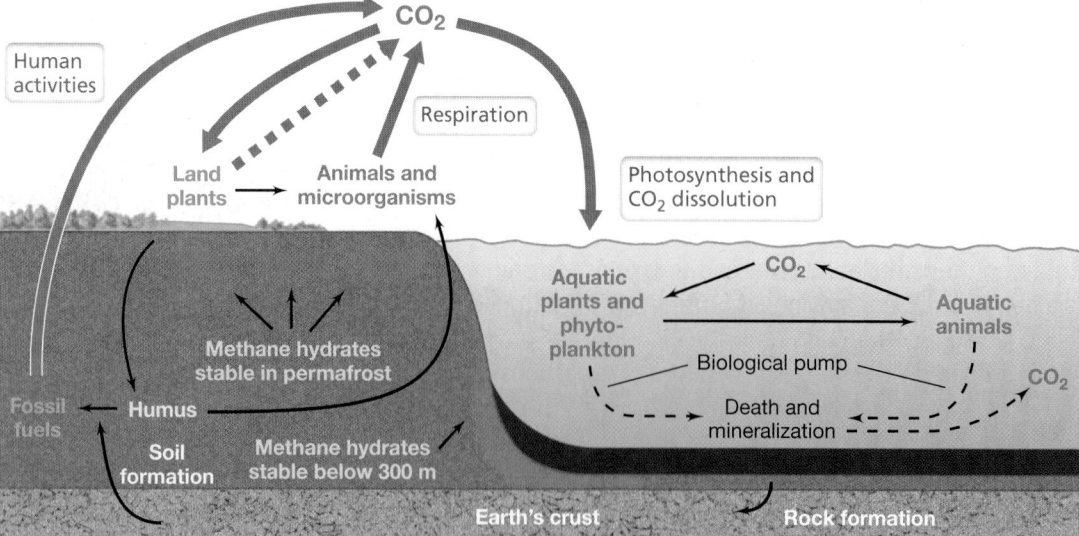

Major Carbon Reservoirs on Earth

Reservoir	Percent of Total[a]
Rocks and sediments	99.5[b]
Oceans	0.05
Methane hydrates	0.014
Fossil fuels	0.006
Terrestrial biosphere	0.003
Aquatic biosphere	0.000002

[a]Total carbon, 76×10^{15} tons
[b]80% inorganic

Figure 21.1 The carbon cycle. The carbon and oxygen cycles are closely connected, as oxygenic photosynthesis both removes CO_2 and produces O_2, and respiration both produces CO_2 and removes O_2. The accompanying table shows that the greatest reservoir of carbon on Earth is in rocks and sediments, and most of this is in inorganic form as carbonates.

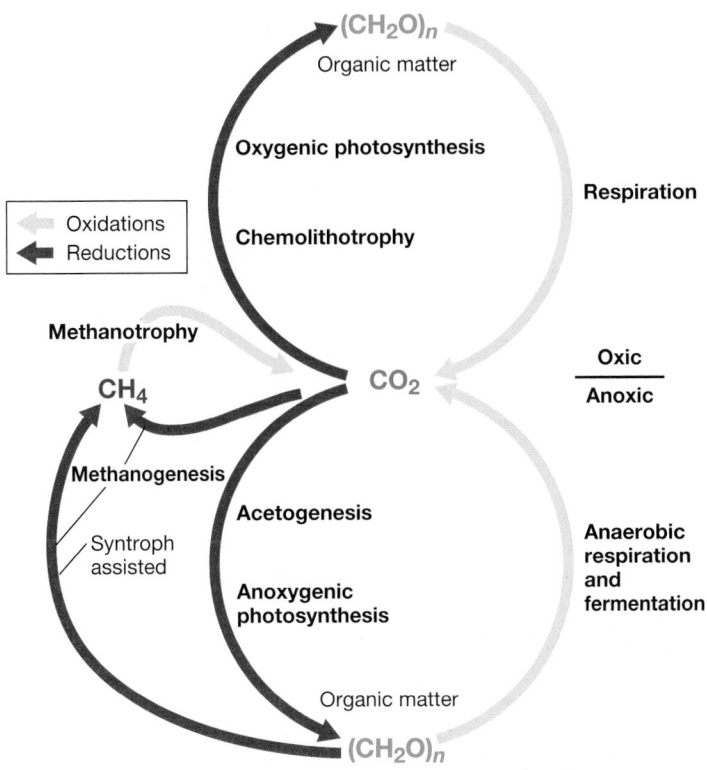

Figure 21.2 Redox cycle for carbon. The diagram contrasts autotrophic processes (CO_2 → organic compounds) and heterotrophic processes (organic compounds → CO_2).

Photosynthesis and Decomposition

New organic compounds are biologically synthesized on Earth only by CO_2 fixation by phototrophs and chemolithotrophs. Most organic compounds originate in photosynthesis and thus phototrophic organisms are the foundation of the carbon cycle (Figure 21.1). However, phototrophic organisms are abundant in nature only in habitats where light is available. The deep sea, deep terrestrial subsurface, and other permanently dark habitats are devoid of indigenous phototrophs. There are two groups of oxygenic phototrophic organisms: *plants* and *microorganisms*. Plants are the dominant phototrophic organisms of terrestrial environments, whereas phototrophic microorganisms dominate in aquatic environments.

The redox cycle for C (**Figure 21.2**) begins with photosynthetic CO_2 fixation, driven by the energy of light:

$$CO_2 + H_2O \rightarrow (CH_2O) + O_2$$

CH_2O represents organic matter at the oxidation–reduction level of cell material. Phototrophic organisms also carry out respiration, both in the light and the dark. The overall equation for respiration is the reverse of oxygenic photosynthesis:

$$(CH_2O) + O_2 \rightarrow CO_2 + H_2O$$

For organic matter to accumulate, the rate of photosynthesis must exceed the rate of respiration. In this way, autotrophic organisms build biomass from CO_2, and then this biomass in one way or another supplies the C heterotrophic organisms need. Anoxygenic phototrophs and chemolithotrophs also

produce excess organic compounds, but in most environments the contributions of these organisms to the net accumulation of organic matter are minor compared to the inputs of oxygenic phototrophs. This is because the reductant used by oxygenic phototrophs, H_2O (⇄ Section 14.1), is in virtually unlimited supply.

Organic compounds are degraded biologically to CH_4 and CO_2 (Figure 21.2). Carbon dioxide, most of which is of microbial origin, is produced by aerobic and anaerobic respirations (⇄ Section 14.7). Methane is produced in anoxic environments by *methanogens* from the reduction of CO_2 with hydrogen (H_2) or from the splitting of acetate into CH_4 and CO_2 (⇄ Section 14.17). However, any naturally occurring organic compound can eventually be converted to CH_4 from the cooperative activities of methanogens and various fermentative bacteria, as we will see in Section 21.2. Methane produced in anoxic habitats is insoluble and most often diffuses rapidly to oxic environments where it is either released to the atmosphere or oxidized to CO_2 by *methanotrophs* (Figure 21.2). Hence, most of the C in organic compounds eventually returns to CO_2, and the links in the carbon cycle are closed.

Methane Hydrates

Although present in the atmosphere at levels lower than even CO_2, CH_4 is a potent greenhouse gas that is over 20 times more effective in trapping heat than is CO_2. Some CH_4 enters the atmosphere from methanogenic production, but not all biologically produced CH_4 is immediately consumed or released to the atmosphere. Huge amounts of CH_4 derived primarily from past microbial activities are trapped underground or under marine sediments as *methane hydrates* (Figure 21.1 and **Figure 21.3**), molecules of frozen CH_4. Methane hydrates form when sufficient CH_4 is present in environments of high pressure and low temperature such as beneath the permafrost in the Arctic and in marine sediments (Figure 21.1). These deposits can be up to several hundred meters thick and are estimated to contain 700–10,000 petagrams (1 petagram = 10^{15} g) of CH_4. This exceeds other known CH_4 reserves on Earth by several orders of magnitude.

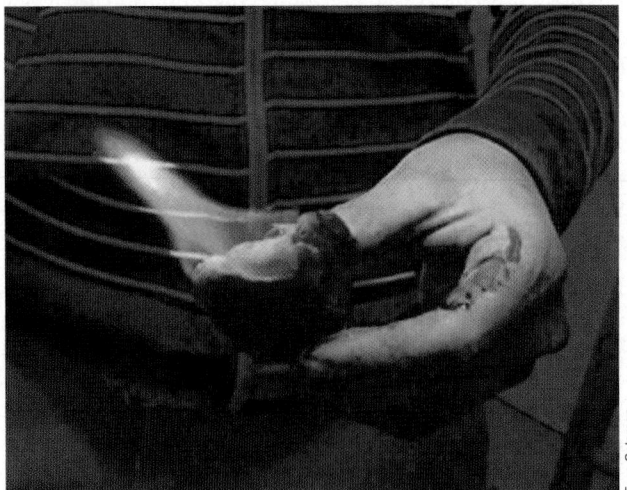

Figure 21.3 Burning methane hydrate. Frozen methane ice retrieved from marine sediments is ignited.

UNIT 5

Methane hydrates are highly dynamic, absorbing and releasing CH_4 in response to changes in pressure, temperature (**Figure 21.4**), and fluid movement. Methane hydrates also fuel deep-water ecosystems, called *cold seeps* (⮂ Figure 20.20). Here, the slow release of CH_4 from seafloor hydrates nourishes not only anaerobic methane-oxidizing *Archaea* (⮂ Section 14.25), but also animal communities that contain aerobic methane-oxidizing endosymbionts that oxidize CH_4 and release organic matter to the animals (⮂ Section 23.9). Anaerobic oxidation of CH_4 is coupled to the reduction of sulfate (SO_4^{2-}), nitrate (NO_3^-), and oxides of iron and manganese [e.g., $FeO(OH)$], dampening release of free methane.

Although deep oceanic methane hydrates are stabilized by high pressure, hydrates in shallower coastal sediments are much more sensitive to small changes in temperature (Figure 21.4). Hydrates in shallow sediments are at the margin of what is called the *gas hydrate stability zone* (GHSZ). In these marginal regions, relatively small changes in temperature can destabilize the hydrates. For example, on-site monitoring of hydrate deposits in a marine coastal plain showed them to be sensitive to 1–2°C seasonal changes in bottom-water temperature. During periods of seasonally elevated water temperature, methane was observed to bubble freely (a phenomenon called methane flares) from the sediments (Figure 21.4). In addition to the release of methane from marine hydrates, as permafrost melts, its huge reserve of organic matter could be catabolized by microbes, leading to the formation of yet more methane (Section 21.8).

Carbon Balances and Coupled Cycles

Although it is convenient to consider carbon cycling as a series of reactions separate from those in other nutrient cycles, in reality, all nutrient cycles are *coupled cycles*; major changes in one cycle affect the functioning of others. But certain cycles, such as the carbon and nitrogen cycles (**Figure 21.5**), are extremely closely coupled because of the large amount of C and N in living organisms. The rate of primary productivity (CO_2 fixation) is controlled by several factors, in particular by the magnitude of photosynthetic biomass and by available N, often a limiting nutrient. Thus, large-scale reductions in biomass, for instance by widespread deforestation, reduce rates of primary productivity and increase levels of CO_2. High levels of organic C stimulate microbial nitrogen fixation ($N_2 \rightarrow NH_3$) and this in turn adds more fixed N to the pool for primary producers; low levels of organic C have just the opposite effect (Figure 21.5). High levels of ammonia (NH_3) stimulate

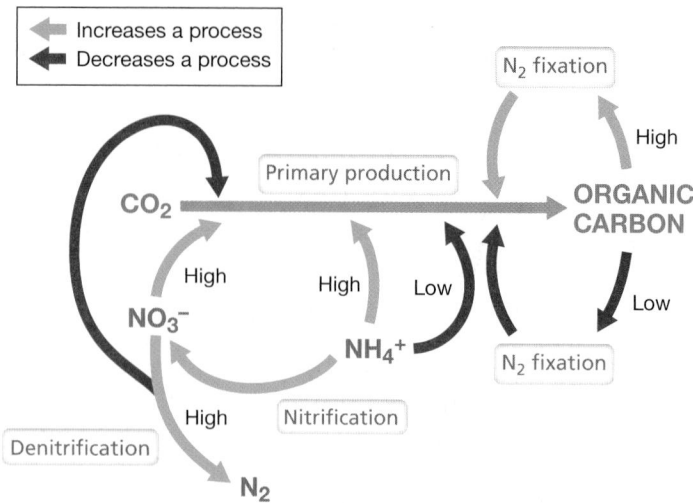

Figure 21.5 Coupled cycles. All nutrient cycles are interconnected, but the carbon and nitrogen cycles are intimately coupled. In the carbon cycle, CO_2 supplies the C for carbon compounds. The N cycle, shown in more detail in Figure 21.8, supplies N for many biological compounds.

primary production and nitrification, but inhibit nitrogen fixation. High levels of nitrate (NO_3^-), an excellent N source for plants and aquatic phototrophs, stimulate primary production but also increase the rate of denitrification; the latter removes fixed forms of N from the environment and feeds back in a negative way on primary production (Figure 21.5).

This simple example illustrates how nutrient cycles are anything but isolated entities. Instead, they are coupled systems that must maintain a delicate balance of inputs and outputs. Thus, one could expect the C and N cycles to respond to large inputs in specific components (for example, through inputs of CO_2 or nitrogen fertilizers) in ways that are not always beneficial to the biosphere and that can have unintended consequences (Section 21.8).

MINIQUIZ

- How is new organic matter made in nature?
- In what ways are oxygenic photosynthesis and respiration related?
- What is a methane hydrate?

21.2 Syntrophy and Methanogenesis

Most organic compounds are oxidized in nature by *aerobic* microbial processes. However, because oxygen (O_2) is a poorly soluble gas and is actively consumed when available, much organic carbon still ends up in anoxic environments. Methanogenesis, the biological production of CH_4, is a major process in anoxic habitats and is catalyzed by a large group of *Archaea*, the *methanogens*, which are strict anaerobes. We discussed the biochemistry of methanogenesis in Section 14.17 and other aspects of methanogens in Section 17.2.

Most methanogens can use CO_2 as a terminal electron acceptor in anaerobic respiration, reducing it to CH_4 with H_2 as electron

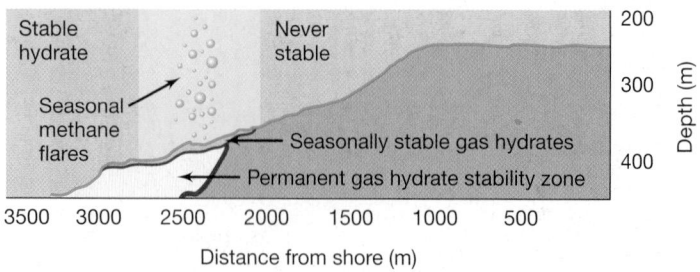

Figure 21.4 Seasonal flares of methane bubbling from methane hydrates. Methane hydrates in shallow coastal sediments are sensitive to seasonal changes in bottom water temperature. Flares of methane bubbles are observed when water temperatures warm by as little as 1–2°C.

donor. Only a very few other substrates, chiefly acetate, are directly converted to CH_4 by methanogens. To convert most organic compounds to CH_4, methanogens must team up with partner organisms called *syntrophs* that function to supply them with precursors for methanogenesis.

Anoxic Decomposition and Syntrophy

In Section 14.23 we discussed the biochemistry of **syntrophy**, a process in which two or more organisms cooperate in the anaerobic degradation of organic compounds. Here we consider the interactions of syntrophic bacteria with their partner organisms and their significance for the carbon cycle. Our focus will be anoxic freshwater sediments and anoxic wastewater treatment, both of which are major sources of CH_4.

Polysaccharides, proteins, lipids, and nucleic acids from dead organisms find their way into anoxic habitats, where they are catabolized. The monomers released by hydrolysis of these polymers become major electron donors for energy metabolism. For the breakdown of a typical polysaccharide such as cellulose, the process begins with *cellulolytic* bacteria (**Figure 21.6**). These organisms hydrolyze cellulose into glucose, which is catabolized by fermentative organisms to short-chain fatty acids (acetate, propionate, and butyrate), alcohols such as ethanol and butanol, and the gases H_2 and CO_2. Hydrogen (H_2) and acetate are consumed by methanogens directly, but the bulk of the carbon remains in the form of fatty acids and alcohols; these cannot be directly catabolized by methanogens and require the activities of syntrophic bacteria (⇄ Section 14.23; Figure 21.6).

Syntrophic bacteria are *secondary* fermenters because they ferment the products of the primary fermenters, yielding H_2, CO_2, and acetate as products. For example, *Syntrophomonas wolfei* oxidizes C_4 to C_8 fatty acids, yielding acetate, CO_2 (if the fatty acid was C_5 or C_7), and H_2 (Table 21.1 and Figure 21.6). Other species of *Syntrophomonas* use fatty acids up to C_{18} in length, including some unsaturated fatty acids. *Syntrophobacter wolinii* specializes in propionate (C_3) fermentation, generating acetate, CO_2, and H_2,

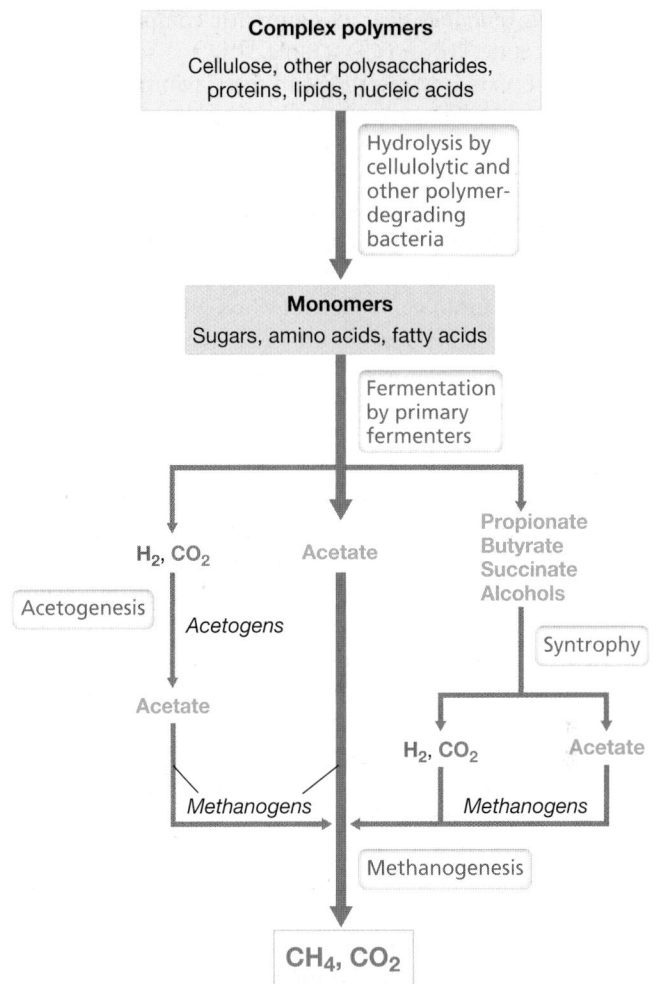

Figure 21.6 Anoxic decomposition. In anoxic decomposition, various groups of fermentative anaerobes cooperate in the conversion of complex organic materials to CH_4 and CO_2. This pattern holds for environments in which sulfate-reducing bacteria play only a minor role; for example, in freshwater lake sediments, sewage sludge bioreactors, or the rumen.

TABLE 21.1 Major reactions in the anoxic conversion of organic compounds to methane[a]

		Free energy change (kJ/reaction)	
Reaction type	Reaction	$\Delta G^{0\prime}$ [b]	ΔG^c
Fermentation of glucose to acetate, H_2, and CO_2	Glucose $+ 4\ H_2O \rightarrow 2$ acetate$^- + 2\ HCO_3^- + 4\ H^+ + 4\ H_2$	−207	−319
Fermentation of glucose to butyrate, CO_2, and H_2	Glucose $+ 2\ H_2O \rightarrow$ butyrate$^- + 2\ HCO_3^- + 2\ H_2 + 3\ H^+$	−135	−284
Fermentation of butyrate to acetate and H_2	Butyrate$^- + 2\ H_2O \rightarrow 2$ acetate$^- + H^+ + 2\ H_2$	+48.2	−17.6
Fermentation of propionate to acetate, CO_2, and H_2	Propionate$^- + 3\ H_2O \rightarrow$ acetate$^- + HCO_3^- + H^+ + H_2$	+76.2	−5.5
Fermentation of ethanol to acetate and H_2	2 Ethanol $+ 2\ H_2O \rightarrow 2$ acetate$^- + 4\ H_2 + 2\ H^+$	+19.4	−37
Fermentation of benzoate to acetate, CO_2, and H_2	Benzoate$^- + 7\ H_2O \rightarrow 3$ acetate$^- + 3\ H^+ + HCO_3^- + 3\ H_2$	+70.1	−18
Methanogenesis from $H_2 + CO_2$	$4\ H_2 + HCO_3^- + H^+ \rightarrow CH_4 + 3\ H_2O$	−136	−3.2
Methanogenesis from acetate	Acetate$^- + H_2O \rightarrow CH_4 + HCO_3^-$	−31	−24.7
Acetogenesis from $H_2 + CO_2$	$4\ H_2 + 2\ HCO_3^- + H^+ \rightarrow$ acetate$^- + 4\ H_2O$	−105	−7.1

[a]Data adapted from Zinder, S. 1984. Microbiology of anaerobic conversion of organic wastes to methane: Recent developments. *Am. Soc. Microbiol. 50:* 294–298.
[b]Standard conditions: solutes, 1 M; gases, 1 atm; 25°C.
[c]Concentrations of reactants in typical anoxic freshwater ecosystems: fatty acids, 1 mM; HCO_3^-, 20 mM; glucose, 10 μM; CH_4, 0.6 atm; H_2, 10^{-4} atm. Data from G_f^0 values in Table 3.2 (or references therein) and Figure 3.10, and bioenergetics calculations as described in Sections 3.4 and 3.6.

and *Syntrophus gentianae* degrades aromatic compounds such as benzoate to acetate, H_2, and CO_2 (Table 21.1).

Despite rather extensive metabolic diversity, syntrophs are unable to carry out any of these reactions in pure culture. Instead, they depend on a H_2-consuming partner organism because of the unusual bioenergetics linked to the syntrophic process. As described in Section 14.23, H_2 consumption by a partner organism is absolutely essential for growth of syntrophic bacteria (in the absence of other electron acceptors), and the association of H_2 producer and H_2 consumer can be very intimate. In fact, it is thought that H_2 transfer in some syntrophic associations may not be directly via H_2 transfer but through *direct conduction*, where electrons are transferred between species using electrically conductive wire-like structures (see Explore the Microbial World, "Microbially Wired," later in this chapter). But no matter the mechanisms, it is the transfer of H_2 (or electrons) that makes the syntrophic association work. How is this so?

When the reactions listed in Table 21.1 for the fermentation of butyrate, propionate, ethanol, or benzoate are written with all reactants at standard conditions (solutes, 1 M; gases, 1 atm; 25°C), the reactions yield free-energy changes ($\Delta G^{0'}$, ⌘ Section 3.4) that are positive in arithmetic sign; that is, the reactions *require* rather than *release* energy. But the consumption of H_2 dramatically affects the energetics, making the reaction favorable and allowing energy to be conserved. This can be seen in Table 21.1, where the ΔG values (free-energy change measured under actual conditions in the habitat) are negative in arithmetic sign if H_2 concentrations are kept near zero by a H_2-consuming partner organism. This allows the syntrophic bacterium to conserve a small amount of energy that is used to produce ATP.

The final products of the syntrophic partnership are CO_2 and CH_4 (Figure 21.6), and any naturally occurring organic compound that enters a methanogenic habitat will eventually be converted to these products. This includes even complex aromatic and aliphatic hydrocarbons. Additional organisms other than those shown in Figure 21.6 may participate in such degradations, but eventually fatty acids and alcohols will be generated and they will be converted to methanogenic substrates by syntrophs. Acetate produced by syntrophs (as well as by the activities of acetogenic bacteria, Figure 21.6 and ⌘ Section 14.16) is a direct methanogenic substrate and is converted to CO_2 and CH_4 by various methanogens.

Methanogenic Symbionts and Acetogens in Termites

A variety of anaerobic protists that thrive under strictly anoxic conditions, including ciliates and flagellates, play a major role in the carbon cycle. Methanogenic *Archaea* live within some of these protist cells as H_2-consuming endosymbionts. For example, methanogens are present *within* cells of trichomonal protists inhabiting the termite hindgut (**Figure 21.7**) where methanogenesis and acetogenesis are major metabolic processes. Methanogenic symbionts of protists are species of the genera *Methanobacterium* or *Methanobrevibacter* (⌘ Section 17.2).

In the termite hindgut, endosymbiotic methanogens along with acetogenic bacteria are thought to benefit their protist hosts by consuming H_2 generated from glucose fermentation by cellulolytic protists. The acetogens are not endosymbionts but instead reside in the termite hindgut itself, consuming H_2 from primary fermenters and reducing CO_2 to make acetate. Unlike methanogens,

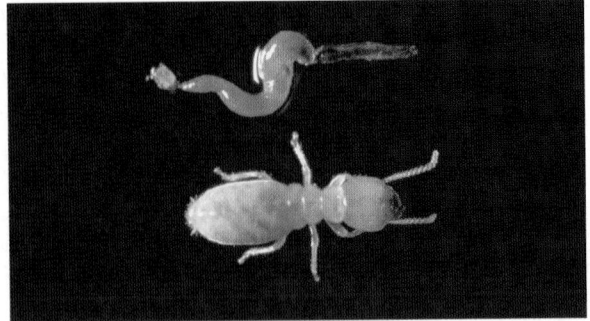

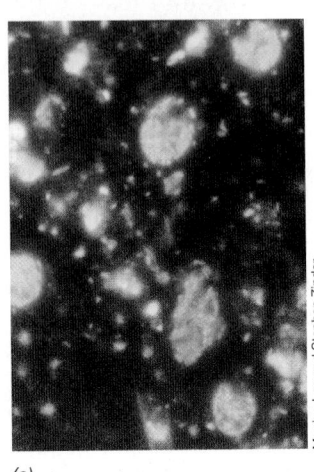

Figure 21.7 Termites and their carbon metabolism. *(a)* A subterranean termite worker larva shown beneath a hindgut dissected from another worker. The animal is about 0.5 cm long. Two views of the same microscopic field show termite hindgut protists photographed by *(b)* phase-contrast and *(c)* epifluorescence. Endosymbiotic methanogens in the protist cells fluoresce blue-green because of the methanogenic coenzyme F_{420} (compare with ⌘ Figure 14.46). The average diameter of the protist cells is 15–20 μm.

acetogens can ferment glucose directly to acetate. Acetogens can also ferment methoxylated aromatic compounds to acetate. This is especially important in the termite hindgut because termites live on wood, which contains lignin, a complex polymer of methoxylated aromatic compounds. The acetate produced by acetogens in the termite hindgut is consumed by the insect as its primary carbon and energy source. Microbial symbioses in the termite hindgut are discussed in more detail in Section 23.7.

—— **MINIQUIZ** ——

- Why does *Syntrophomonas* need a partner organism in order to ferment fatty acids or alcohols?
- What kinds of organisms are used in coculture with *Syntrophomonas*?
- What is the final product of acetogenesis?

21.3 The Nitrogen Cycle

Nitrogen is an essential element for life (⌘ Section 3.1) and exists in a number of oxidation states. We have discussed four major microbial N transformations thus far: nitrification, denitrification,

Key Processes and Microbes in the Nitrogen Cycle

Processes	Example organisms
Nitrification ($NH_4^+ \rightarrow NO_3^-$)	
$NH_4^+ \rightarrow NO_3^-$	Comammox (*Nitrospira* species)
$NH_4^+ \rightarrow NO_2^-$	*Nitrosomonas, Nitrosopumilus* (*Archaea*)
$NO_2^- \rightarrow NO_3^-$	*Nitrobacter*
Denitrification ($NO_3^- \rightarrow N_2$)	*Bacillus, Paracoccus,*
	Pseudomonas
N_2 Fixation ($N_2 + 8 H \rightarrow NH_3 + H_2$)	
Free-living	
Aerobic	*Azotobacter*
	Cyanobacteria
Anaerobic	*Clostridium*, purple and
	green phototrophic bacteria
	Methanobacterium (*Archaea*)
Symbiotic	*Rhizobium*
	Bradyrhizobium
	Frankia
Ammonification (organic-N $\rightarrow NH_4^+$)	
	Many organisms can do this
Anammox ($NO_2^- + NH_3 \rightarrow 2 N_2$)	*Brocadia*

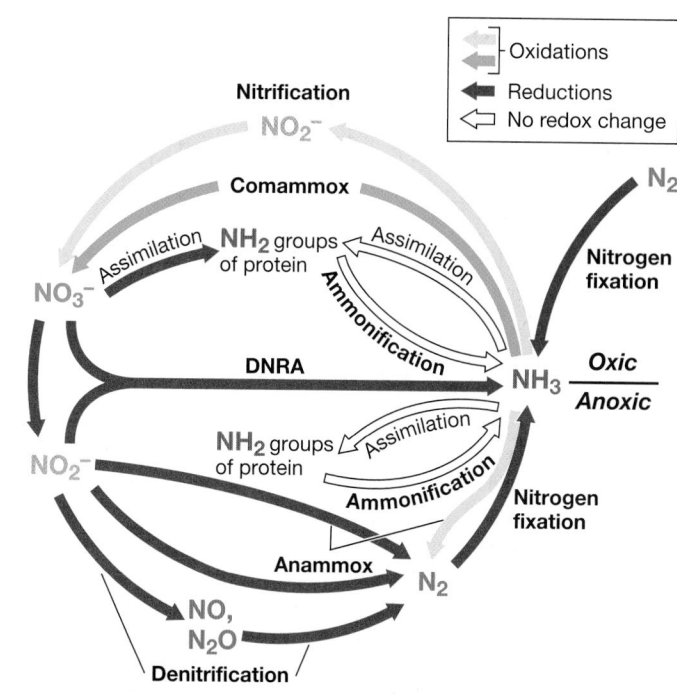

Figure 21.8 Redox cycle for nitrogen. The actual anammox reaction is $NH_4^+ + NO_2^- \rightarrow N_2 + 2 H_2O$ (↩ Figure 14.34). DNRA, dissimilative reduction of nitrate to ammonia.

anammox, and nitrogen fixation (Chapter 14). These and other key N transformations are summarized in the redox cycle shown in **Figure 21.8**.

Nitrogen Fixation and Denitrification

Nitrogen gas (N_2) is the most stable form of N and is a major reservoir for N on Earth. However, only a relatively small number of *Bacteria* and *Archaea* are able to use N_2 as a cellular N source by the process of *nitrogen fixation* ($N_2 + 8 H \rightarrow 2 NH_3 + H_2$) (↩ Section 14.6). The N recycled on Earth is mostly already "fixed N"; that is, N in combination with other elements, such as in ammonia (NH_3) or nitrate (NO_3^-). In many environments, however, the short supply of fixed N puts a premium on biological nitrogen fixation, and in these habitats, nitrogen-fixing bacteria flourish.

We discussed the role of NO_3^- as an alternative electron acceptor in anaerobic respiration in Section 14.13. Under most conditions, the end product of NO_3^- reduction is N_2, nitric oxide (NO), or nitrous oxide (N_2O). The reduction of NO_3^- to these gaseous N compounds, called **denitrification** (Figure 21.8), is the main means by which N_2 and N_2O are formed biologically. On the one hand, denitrification is a detrimental process. For example, if agricultural fields fertilized with NO_3^- fertilizer become waterlogged following heavy rains, anoxic conditions can develop and denitrification can be extensive; this removes fixed N from the soil. On the other hand, denitrification can aid in wastewater treatment (↩ Section 22.7). By converting NO_3^- in wastewater to volatile forms of N, denitrification minimizes fixed N in discharge waters and subsequent algal growth triggered by the nitrate influx.

The production of N_2O and NO by denitrification can have other environmental consequences. Nitrous oxide can be photochemically oxidized to NO in the atmosphere. Nitric oxide reacts with ozone (O_3) in the upper atmosphere to form nitrite (NO_2^-), and this

returns to Earth as nitric acid (HNO_2). In addition, N_2O is a very potent greenhouse gas. Although N_2O molecules persist on average only about 100 years because of their reactivity, on a per weight basis, the contribution of N_2O to warming is about 300 times that of CO_2. Thus, denitrification contributes to global warming; to O_3 destruction, which increases passage of ultraviolet radiation to the surface of Earth; and to acid rain, which increases acidity of soils. Increases in soil acidity can change microbial community structure and function and, ultimately, soil fertility, impacting both plant diversity and agricultural yields of crop plants.

Ammonification and Ammonia Fluxes

Ammonia is released during the decomposition of organic N compounds such as amino acids and nucleotides, a process called *ammonification* (Figure 21.8). Another process contributing to the generation of NH_3 is the respiratory reduction of NO_3^- to NH_3, called *dissimilative nitrate reduction to ammonia* (DNRA, Figure 21.8). DNRA dominates NO_3^- and nitrite (NO_2^-) reduction in reductant-rich anoxic environments, such as highly organic marine sediments and the human gastrointestinal tract. It is thought that nitrate-reducing bacteria exploit this pathway primarily when NO_3^- is limiting because DNRA consumes eight electrons compared with the four and five electrons consumed when NO_3^- is reduced to N_2O or N_2, respectively.

At neutral pH, NH_3 exists as ammonium (NH_4^+). Much of the NH_4^+ released by aerobic decomposition in soils is rapidly recycled and converted to amino acids in plants and microorganisms. However, because NH_3 is volatile, some of it can be lost from alkaline soils by vaporization, and there are major losses of NH_3 to the atmosphere in areas with dense animal populations (for example, cattle feedlots). On a global basis, however, NH_3 constitutes only about 15% of the N released to the atmosphere, the rest being primarily N_2 or N_2O from denitrification.

UNIT 5

Nitrification and Anammox

Nitrification, the oxidation of NH_3 to NO_3^-, is a major process in well-drained oxic soils at neutral pH, and is carried out by the nitrifying *Bacteria* and *Archaea* (Figure 21.8). Whereas denitrification *consumes* NO_3^-, nitrification *produces* NO_3^-. If materials high in NH_3, such as manure or sewage, are added to soils, the rate of nitrification increases.

Nitrification was long considered an obligatory two-step process in which some species oxidize NH_3 to NO_2^- and then other species oxidize NO_2^- to NO_3^-. Many species of both *Bacteria* and *Archaea* can oxidize NH_3 ($\rightleftharpoons$ Sections 14.11, 15.13, 17.5), whereas thus far, only species of *Bacteria* are known that oxidize NO_2^-. Archaeal nitrifiers generally greatly outnumber their bacterial counterparts in marine and most terrestrial systems and likely control rates of NH_3 oxidation in nature. However, novel *Nitrospira* species were recently discovered that have the capacity to oxidize ammonia completely to nitrate. Although the environmental significance of these organisms (termed *comammox* bacteria) has yet to be established, their discovery has overturned a century of conventional wisdom.

Although NO_3^- is readily assimilated by plants, it is very soluble and therefore is rapidly leached or denitrified from waterlogged soils; consequently, nitrification is not beneficial for plant agriculture. Ammonium, on the other hand, is positively charged and strongly adsorbed to negatively charged soils. Anhydrous NH_3 is therefore used extensively as an agricultural fertilizer, but to prevent its conversion to NO_3^-, chemicals are added to the NH_3 to inhibit nitrification. One common inhibitor is a pyridine compound called *nitrapyrin* (2-chloro-6-trichloromethylpyridine). Nitrapyrin specifically inhibits the *first* step in nitrification, the oxidation of NH_3 to NO_2^-. However, this effectively inhibits both steps in nitrification because the second step, $NO_2^- \rightarrow NO_3^-$, depends on the first ($\rightleftharpoons$ Section 14.11). The addition of nitrapyrin to anhydrous NH_3 has greatly increased the efficiency of crop fertilization and has helped prevent pollution of waterways by NO_3^- leached from nitrified soils.

Ammonia can be oxidized under anoxic conditions in a process called *anammox*. The anammox bacteria affiliate with five genera (*Brocadia, Kuenenia, Scalindua, Anammoxoglobus,* and *Jettenia*) in a single phylogenetically cohesive family (*Brocadiaceae*) within the *Planctomycetes* ($\rightleftharpoons$ Section 16.16). Since anammox bacteria have yet to be isolated in pure culture, the genus and species names are prefaced by the term "Candidatus" to indicate this tentative taxonomic status ($\rightleftharpoons$ Section 13.10). In the anammox reaction, NH_3 is oxidized anaerobically with NO_2^- as the electron acceptor, forming N_2 as the final product (Figure 21.8), which is released to the atmosphere. Although a major process in sewage and in anoxic marine basins and sediments, anammox is not a significant process in well-drained (oxic) soils. The microbiology and biochemistry of anammox was discussed in Section 14.12.

─── **MINIQUIZ** ───

- What is nitrogen fixation and why is it important to the nitrogen cycle?
- How do the processes of nitrification and denitrification differ? How do nitrification and anammox differ?
- How does the compound nitrapyrin benefit both agriculture and the environment?

21.4 The Sulfur Cycle

Microbial transformations of sulfur (S) are even more complex than those of N because of the large number of oxidation states of S and the fact that several transformations of S also occur spontaneously (abiotically). Chemolithotrophic S oxidation and sulfate (SO_4^{2-}) reduction were covered in Sections 14.9, 14.14, 15.9, and 15.11. The redox cycle for microbial S transformations is shown in **Figure 21.9**.

Although a number of oxidation states of S are possible, only three are significant in nature, -2 (sulfhydryl, R–SH, and sulfide, HS^-), 0 (elemental sulfur, S^0), and $+6$ (sulfate, SO_4^{2-}). The bulk of S on Earth resides in sediments and rocks in the form of sulfate minerals, primarily gypsum ($CaSO_4$), and sulfide minerals (pyrite, FeS_2), but the oceans constitute the most significant reservoir of SO_4^{2-} in the biosphere. A significant amount of S, in particular sulfur dioxide (SO_2, a gas), enters the S cycle from human activities, primarily the burning of fossil fuels.

Hydrogen Sulfide and Sulfate Reduction

A major volatile S gas is hydrogen sulfide (H_2S). Hydrogen sulfide is produced from bacterial sulfate reduction ($SO_4^{2-} + 4 H_2 \rightarrow H_2S + 2 H_2O + 2 OH^-$) (Figure 21.9) or is geochemically produced and emitted from sulfide springs and volcanoes. Although H_2S is volatile, different forms exist depending on pH: H_2S predominates below pH 7 and the nonvolatile HS^- and S^{2-} predominate above pH 7. Collectively, H_2S, HS^-, and S^{2-} are referred to as "sulfide."

Sulfate-reducing bacteria are a large and highly diverse group ($\rightleftharpoons$ Sections 14.14 and 15.9) and are widespread in nature. However, in anoxic habitats such as freshwater sediments and many soils, sulfate reduction is limited by SO_4^{2-} availability. Moreover, because organic electron donors (or H_2, which is a product of the fermentation of organic compounds) are needed to support sulfate reduction, it only occurs where significant amounts of organic material are present.

In marine sediments, the rate of sulfate reduction is typically carbon-limited and can be greatly increased by an influx of organic matter. This is important because the disposal of sewage or garbage in the oceans or coastal regions can trigger sulfate reduction. Hydrogen sulfide is toxic to many plants and animals and therefore its formation is potentially detrimental; sulfide is toxic because it combines with the iron of cytochromes and blocks respiration. Sulfide is commonly detoxified in nature by combination with iron, forming the insoluble minerals FeS (pyrrhotite) and FeS_2 (pyrite). The black color of sulfidic sediments or sulfate-reducing bacterial cultures is due to these metal sulfide minerals ($\rightleftharpoons$ Figure 15.23g).

Sulfide and Elemental Sulfur Oxidation–Reduction

Under oxic conditions, sulfide rapidly oxidizes spontaneously at neutral pH. Sulfur-oxidizing chemolithotrophic bacteria, most of which are aerobes ($\rightleftharpoons$ Sections 14.9 and 15.11), can catalyze the oxidation of sulfide. However, because of the rather rapid spontaneous reaction, microbial sulfide oxidation is significant only in areas where H_2S emerging from anoxic environments meets air. Where light is available, there can be anoxic oxidation of sulfide,

Key Processes and Microbes in the Sulfur Cycle

Process	Example organisms
Sulfide/sulfur oxidation ($H_2S \longrightarrow S^0 \longrightarrow SO_4^{2-}$)	
Aerobic	Sulfur chemolithotrophs (*Thiobacillus, Beggiatoa*, many others)
Anaerobic	Purple and green phototrophic bacteria, some chemolithotrophs
Sulfate reduction (anaerobic) ($SO_4^{2-} \longrightarrow H_2S$)	*Desulfovibrio, Desulfobacter Archaeoglobus* (*Archaea*)
Sulfur reduction (anaerobic) ($S^0 \longrightarrow H_2S$)	*Desulfuromonas*, many hyperthermophilic *Archaea*
Sulfur disproportionation ($S_2O_3^{2-} \longrightarrow H_2S + SO_4^{2-}$)	*Desulfovibrio*, and others
Organic sulfur compound oxidation or reduction ($CH_3SH \longrightarrow CO_2 + H_2S$) ($DMSO \longrightarrow DMS$)	Many organisms can do this
Desulfurylation (organic–$S \longrightarrow H_2S$)	Many organisms can do this

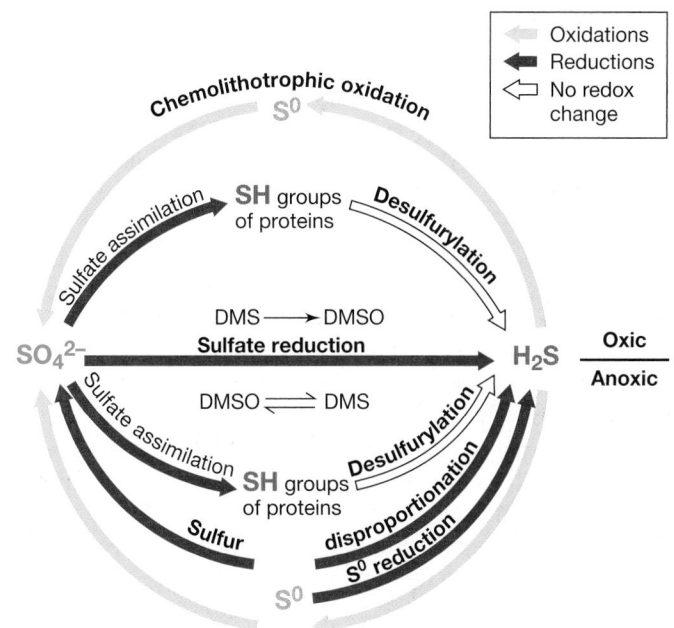

Figure 21.9 Redox cycle for sulfur. DMS, dimethyl sulfide; DMSO, dimethyl sulfoxide.

catalyzed by the phototrophic purple and green sulfur bacteria (⇔ Sections 14.3, 15.4, and 15.6).

S^0 is chemically stable but is readily oxidized by sulfur-oxidizing chemolithotrophic bacteria such as *Thiobacillus* and *Acidithiobacillus*. Because S^0 is insoluble, the bacteria that oxidize it must attach to the S^0 crystals to obtain their substrate (⇔ Figure 14.27). The oxidation of S^0 forms sulfuric acid (H_2SO_4), and thus S^0 oxidation characteristically lowers the pH in the environment, sometimes drastically. For this reason, small amounts of S^0 can be added to alkaline soils as an inexpensive and natural way to lower the pH, relying on the ubiquitous sulfur chemolithotrophs to carry out the acidification process.

S^0 can be reduced as well as oxidized. The reduction of S^0 to sulfide (a form of anaerobic respiration) is a major ecological process of some *Bacteria* and hyperthermophilic *Archaea* (⇔ Section 15.10 and Chapter 17). Although sulfate-reducing bacteria can also reduce S^0, most S^0 reduction occurs by the activities of the physiologically specialized sulfur reducers, organisms that are incapable of SO_4^{2-} reduction (⇔ Section 15.10). The habitats of the sulfur reducers are generally those of the sulfate reducers, so from an ecological standpoint, the two groups form a metabolic guild unified by their formation of H_2S.

Organic Sulfur Compounds

In addition to *inorganic* forms of S, several *organic* S compounds are also cycled in nature. Many of these foul-smelling compounds are highly volatile and can thus enter the atmosphere. The most abundant organic S compound in nature is *dimethyl sulfide* (CH_3—S—CH_3); it is produced primarily in marine environments as a degradation product of dimethylsulfoniopropionate [$(CH_3)_2S^+CH_2CH_2COO^-$], a major osmoregulatory solute in marine algae (⇔ Section 5.13). This compound can be used as a carbon source and electron donor by microorganisms and is catabolized to CH_3—S—CH_3 and acrylate (CH_2=$CHCOO^-$). The latter, a derivative of the fatty acid propionate, is used to support growth.

Dimethyl sulfide released to the atmosphere undergoes photochemical oxidation to methanesulfonate (CH_3SO_3), SO_2, and SO_4^{2-}. By contrast, CH_3—S—CH_3 produced in anoxic habitats can be microbially transformed in at least three ways: (1) by methanogenesis (yielding CH_4 and H_2S), (2) as an electron donor for photosynthetic CO_2 fixation in phototrophic purple bacteria (yielding dimethyl sulfoxide, DMSO), and (3) as an electron donor in energy metabolism in certain chemoorganotrophs and chemolithotrophs (also yielding DMSO). DMSO can be an electron acceptor for anaerobic respiration (⇔ Section 14.15), producing CH_3—S—CH_3. Many other organic S compounds affect the global sulfur cycle, including methanethiol (CH_3SH), dimethyl disulfide (H_3C—S—S—CH_3), and carbon disulfide (CS_2), but on a global basis, CH_3—S—CH_3 is the most significant.

--- **MINIQUIZ** ---

- Is H_2S a substrate or a product of the sulfate-reducing bacteria? Of the chemolithotrophic sulfur bacteria?
- Why does the bacterial oxidation of sulfur result in a pH drop?
- What organic sulfur compound is most abundant in nature?

II • Other Nutrient Cycles

I n Part II of this chapter we explore the interactions of microorganisms with metals—in particular iron and manganese—and with some nonmetals whose microbial transformations are of major global significance.

21.5 The Iron and Manganese Cycles

Iron (Fe) is one of the most abundant elements in Earth's crust. On the surface of Earth, Fe exists naturally in two oxidation states, ferrous [Fe^{2+}, also Fe(II)] and ferric [Fe^{3+}, also Fe(III)]. A third oxidation state, Fe^0, is abundant in Earth's core and is also a major product of human activities from the smelting of iron ores to form cast iron.

In nature, iron cycles primarily between the Fe^{2+} and Fe^{3+} forms, and these redox transitions are one-electron oxidations and reductions. Ferric iron is reduced both chemically and as a form of anaerobic respiration, and Fe^{2+} is oxidized both chemically and as a form of chemolithotrophic metabolism (**Figure 21.10**). Manganese (Mn), although present at 5- to 10-fold lesser abundance than Fe in the near-surface environment, is another redox-active metal of microbiological significance, existing primarily in two oxidation states (Mn^{2+} and Mn^{4+}, see Figure 21.11).

A key feature of the iron and manganese cycles is the different solubilities of these metals in their oxidized versus reduced forms. Reduced iron (Fe^{2+}) and manganese (Mn^{2+}) are soluble. In contrast, oxidized minerals of iron such as iron oxide-hydroxides [e.g., $Fe(OH)_3$, FeOOH, and Fe_2O_3] and manganese oxide (MnO_2) are insoluble and tend to settle out in aquatic environments. As a consequence, these strong oxidants can comprise several percent by weight of marine and freshwater sediments, making them among the most abundant of potential electron acceptors in many anoxic systems (see Figure 21.11).

Bacterial Reduction of Iron and Manganese Oxides

Some *Bacteria* and *Archaea* can use Fe^{3+} as an electron acceptor in anaerobic respiration (⮌ Section 14.15). These organisms also commonly have the capacity to use Mn^{4+} as an electron acceptor, and some have the capacity to reduce oxidized uranium (⮌ Section 22.3).

Ferric iron and manganese oxide reduction is common in waterlogged soils, bogs, and anoxic lake sediments (**Figure 21.11**). When soluble reduced iron and manganese reach oxic regions, for example, through diffusion from anoxic regions of sediments, they are oxidized chemically [e.g., $Fe^{2+} + \frac{1}{4} O_2 + 2\frac{1}{2} H_2O \rightarrow Fe(OH)_3 + 2 H^+$] or microbiologically (Figure 21.10). The chemical oxidation of Fe^{2+} is very rapid at near-neutral pH. Although the spontaneous oxidation of Mn^{2+} is very slow at neutral pH, the rate of oxidation can be increased up to five orders of magnitude by a variety of manganese-oxidizing bacteria and even fungi. The oxidized metal oxides and hydroxides then precipitate, returning the oxidized metals to the sediments where they can again serve as electron acceptors, completing the cycle.

Oxidized iron (Fe^{3+}) and manganese (Mn^{4+}) are chemically very reactive. Phosphate is trapped as insoluble ferric phosphate precipitates. Chemical oxidation of refractory organic compounds by Mn^{4+} oxide may yield more available sources of carbon for microbial growth. Other metals [e.g., copper (Cu), cadmium (Cd), cobalt (Co), lead (Pb), and arsenic (As)] form insoluble complexes with the iron and manganese oxides. When these oxides are subsequently reduced, the bound phosphate is liberated along with the soluble forms of these metals.

In recent years it has been recognized that the surfaces and appendages of cells of bacteria that interact with iron and manganese oxides, such as *Geobacter*, are electrically conductive, functioning as "nanowires" to move electrons around in microbial habitats. This movement of electrons is a form of electricity, and the process may eventually have commercial applications for power generation (see Explore the Microbial World, "Microbially Wired"). Humic substances (Section 21.1) can also facilitate microbial metal reduction. Since some constituents of humics

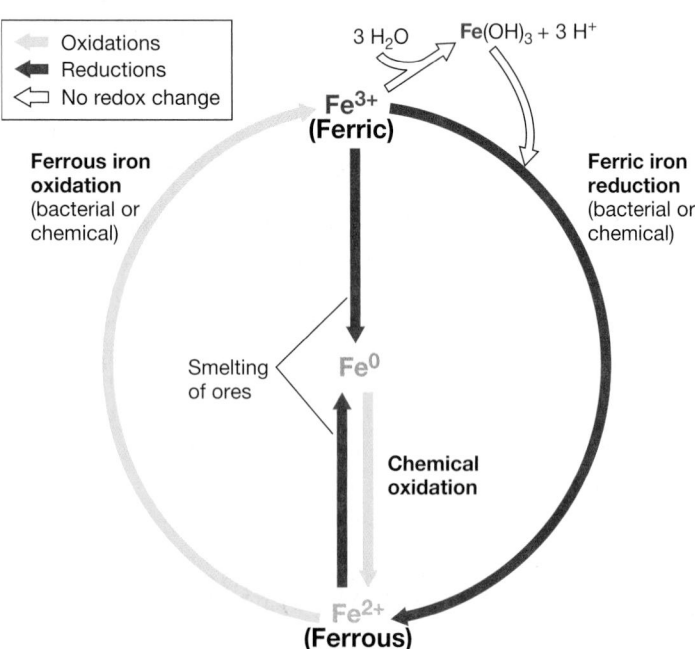

Figure 21.10 Redox cycle for iron. The major forms of iron in nature are Fe^{2+} and Fe^{3+}. Fe^0 is primarily a product of smelting of iron ores. Fe^{3+} forms various minerals such as ferric hydroxide, $Fe(OH)_3$.

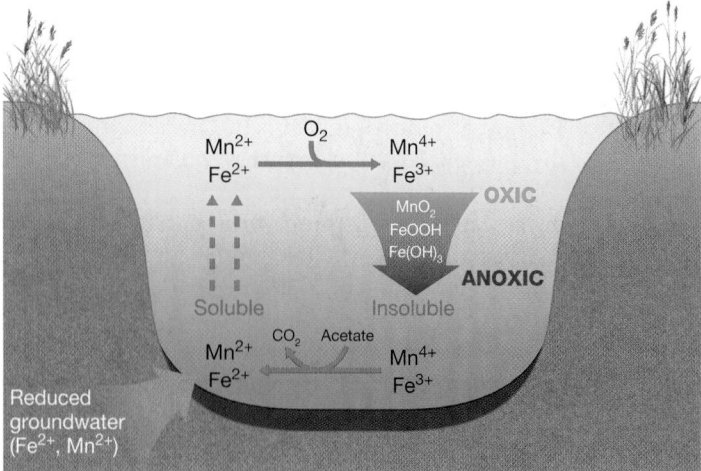

Figure 21.11 Iron and manganese redox cycling in a typical freshwater system. Iron and manganese oxides in sediments are used as electron acceptors by metal-reducing bacteria. The resulting reduced forms are soluble and diffuse into the oxic regions of the sediment or water column, where they are oxidized microbially or chemically. Precipitation of the insoluble oxidized metals then returns the metals to the sediments, completing the redox cycle.

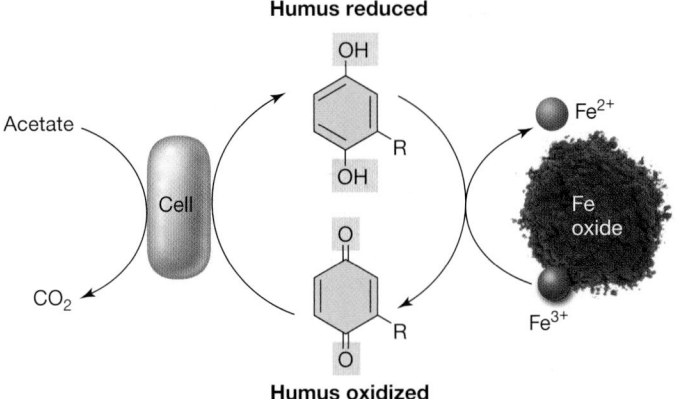

Humus reduced

Humus oxidized

Figure 21.12 Role of humic substances in humus as an electron shuttle in microbial metal reduction. Quinone-like functional groups in humus are reduced by acetate-oxidizing bacteria. The reduced humus then donates electrons to metal oxides, releasing reduced soluble iron (Fe^{2+}) and oxidized humus. The cycle continues as oxidized humus is again reduced by the bacteria.

Figure 21.13 Oxidation of ferrous iron (Fe^{2+}). A microbial mat growing in the Rio Tinto, Spain. The mat consists of acidophilic green algae (eukaryotes) and various iron-oxidizing chemolithotrophic bacteria. The Rio Tinto has a pH of about 2 and contains high levels of dissolved metals, in particular Fe^{2+}. The red-brown precipitates consist of $Fe(OH)_3$ and other ferric minerals.

can alternate between oxidized and reduced forms, they can function to shuttle electrons from the bacterium to the reduction of the iron or manganese oxides (**Figure 21.12**).

Microbial Oxidation of Reduced Iron and Manganese

At neutral pH, ferrous iron (Fe^{2+}) is rapidly oxidized abiotically in oxic environments. In contrast, at *acidic* pH (pH < 4), Fe^{2+} is not oxidized spontaneously. Thus, much of the research on microbial oxidation of iron is focused on acidic, ferrous-iron-rich habitats, where acidophilic chemolithotrophs such as *Acidithiobacillus ferrooxidans* and *Leptospirillum ferrooxidans* oxidize Fe^{2+} to Fe^{3+} (**Figure 21.13**).

The oxidation of Fe^{2+} to Fe^{3+} yields a single electron; consequently very little energy can be conserved (⇆ Sections 14.10 and 15.15) and so these bacteria must oxidize large amounts of Fe^{2+} in order to grow. In such environments, even a relatively small population of cells can precipitate a large amount of iron minerals. Although O_2 is the most environmentally significant electron acceptor, Fe^{2+} oxidation can also be coupled to NO_3^- reduction by some anaerobic microbes (⇆ Sections 14.10 and 15.15) and Fe^{2+} functions as an electron donor in photosynthesis for some anoxygenic phototrophs (⇆ Sections 14.10, 15.2, and 15.5). Even though the oxidation of Mn^{2+} to Mn^{4+} is also potentially energetically favorable for growth, and many microorganisms catalyze Mn^{2+} oxidation, as yet no organism has been conclusively shown to derive energy from the oxidation of reduced manganese.

Since the abiotic oxidation of reduced iron is rapid at near-neutral pH, iron-oxidizing bacteria that inhabit nonacidic environments are restricted to a very narrow redox region in which ferrous iron–rich water impinges on oxygenated water (Figure 21.11). These microoxic habitats include freshwater and coastal sediments, slow-moving streams, ferrous iron–rich waters from springs, and hydrothermal vents (**Figure 21.14**). For example, when ferrous iron–rich groundwaters are exposed to air, Fe^{2+} is oxidized at the interface of these two zones by iron-oxidizing bacteria such as *Leptothrix* and *Gallionella* (Figure 21.14*b, c, d*; ⇆ Sections 14.10

and 15.15). Thus, their physiology dictates that they maintain a position within a narrow environment of low levels of O_2 and high levels of reduced metals. How these organisms secure and maintain a position within such a narrow range of abiotic conditions is not well understood, but the sheath and stalk structures typically found in iron oxidizers may assist in their proper positioning or attachment (Figure 21.14*b, d, f*; ⇆ Figures 15.36 and 15.58).

As we have seen, organisms that reduce insoluble metal oxides can use extracellular conductors for electron transfer, such as electrically conductive pili or cell-surface-associated cytochromes. However, a similar problem exists for organisms that oxidize metals: Insoluble metal oxides are the product of metal oxidation, and the organism must ensure that these insoluble oxides are deposited external to the cell. Thus, organisms that oxidize Fe^{2+} or Mn^{2+} use surface-associated electron transfer proteins to ensure that metals are oxidized *outside* the cytoplasm. Cytochromes participate in both iron reduction and iron oxidation, and genomes of metal-oxidizing *Gallionella* and *Sideroxydans* species contain genes encoding periplasmic *c*-type cytochromes that resemble those encoding proteins known to reduce metal oxides in *Shewanella*, suggesting that mechanistically similar electron transfer pathways are likely used for both the reduction and oxidation of extracellular metals.

Although possibly sharing similar electron transfer mechanisms, metal-oxidizing bacteria are confronted with another problem—their metabolism could soon encase the cell in an iron oxide shell. To prevent this, metal oxidizers produce extracellular organic material that captures metal oxides and deposits it some distance away from the cell. Some metal oxidizers, such as *Gallionella*, produce extended organic stalks that become encrusted with metal oxides away from the cell (Figure 21.14*d*; see also Figure 15.36). An alternative strategy is used by *Leptothrix* species. These bacteria produce an organic sheath surrounding the cells that becomes encrusted with metal oxides (Figure 21.14*b* and

UNIT 5

MICROBIALLY WIRED

Regardless of the electron acceptor they use, when bacteria respire, they carry out oxidations and reductions that generate electricity. They do this when they oxidize an organic or inorganic electron donor and separate electrons from protons during electron transport. The electrons eventually reduce some electron acceptor and the protons generate the proton motive force.

In any form of respiration, electron disposal is necessary for energy conservation. When the electron acceptor is oxygen (O_2), nitrate (NO_3^-), or many of the other soluble substances used by bacteria as electron acceptors (⇌ Sections 14.13–14.17), the final product diffuses away from the cell. Many bacteria reduce ferric iron (Fe^{3+}) as an electron acceptor under anoxic conditions, including the bacterium *Geobacter sulfurreducens* (**Figure 1a**). However, in contrast to soluble electron acceptors, Fe^{3+} is typically present in nature as an insoluble mineral, such as an iron oxide (Figure 21.12), and thus the reduction of Fe^{3+} occurs *outside* the cell. Under such conditions, the ferric iron functions as an electrical anode, and the bacterial cell facilitates transfer of electrons from the electron donor to the anode.[1]

Research has shown that *Geobacter* forms direct electrical connections with insoluble materials that can either accept or donate electrons. Electron transfer involves cytochromes localized along the length of pili that are generally 10–20 micrometers long (⇌ Figure 15.35c). These electrically conductive structures function as electrical nanowires, much as copper wire does in a household electrical circuit. Being conductive structures, nanowires can form direct electrical connections with insoluble materials that either accept or donate electrons, or alternatively, nanowires can form connections between cells. In this way, electrons obtained by *Geobacter* from the oxidation of organic electron donors or from H_2 can be shuttled to a suitable electron acceptor.

Surprisingly, bacterial electron shuttling can occur over rather large spatial distances, much larger than the cell itself. For example, in microsensor studies (⇌ Section 19.9) of hydrogen sulfide (H_2S) oxidation in anoxic marine sediments (sulfide is the product of sulfate-reducing bacteria), the oxidation of H_2S deep in the sediments released electrons that reduced O_2 at the sediment water interface, some 2.5 cm away (**Figure 1c**). The electrical conductors in the sediment are filamentous bacteria (**Figure 1b**) affiliated with the *Desulfobulbaceae* family of sulfate-reducing bacteria (⇌ Section 15.9). Although phylogenetically affiliated with *sulfate reducers*, the filamentous bacteria actually function as *sulfide oxidizers*, using O_2 as the terminal electron acceptor.[2]

The surface of the filamentous bacterial cells has ridges running along its entire length. Microscopically these ridges appear much like cables, each microbial filament surrounded by 15 to 17 structures 400–700 nm in width that run continuously along the length of the filament (**Figure 1b**). These structures are implicated in electron transfer from the sulfide oxidized at one end of the filament to the reduction of O_2 near the sediment surface at the other end of the filament. Although reminiscent of *Geobacter* nanowires, the mechanism for electron transfer over such large distances is unknown.

In nature, electrical communication between bacterial cells may be a major way by which electrons generated from microbial metabolism in anoxic habitats are shuttled to oxic regions. Moreover, research on the microbiology of the process indicates that this microbial electricity could be harnessed in the form of microbial "fuel cells" that could oxidize toxic and waste carbon compounds in anoxic environments, with the resulting electrons coupled to power generation. In such a scheme, bacteria would be exploited to function as catalysts for diverting electrons from electron donors directly to artificial anodes, with the resulting electrical current being siphoned off to supply a portion of human power needs.

[1] Lovley, D.R. 2006. Bug juice: Harvesting electricity with microorganisms. *Nat. Rev. Microbiol. 4:* 497–508.
[2] Pfeffer, C., et al. 2012. Filamentous bacteria transport electrons over centimetre distances. *Nature 491:* 218–221.

⇌ Figure 15.58). In this case the cells can move out of the sheath, leaving the metal oxide crust behind.

Although not all metal oxidizers produce such morphologically conspicuous structures, it is thought that most, if not all, metal oxidizers are forced to produce some form of extracellular organic material in order to sequester the insoluble product of their energy metabolism. In addition, the incorporation of this organic matter into the metal oxides likely alters the physical and chemical properties of the minerals themselves.

— MINIQUIZ —

- In what oxidation state is Fe in $Fe(OH)_3$? In FeS? How is $Fe(OH)_3$ formed?
- Why does biological Fe^{2+} oxidation under oxic conditions occur mainly at acidic pH?
- Why is excreted organic matter important to many iron oxidizers?

21.6 The Phosphorus, Calcium, and Silica Cycles

Many other chemical elements undergo microbial cycling and we focus on three key ones here—phosphorus (P), calcium (Ca), and silica (Si). The cycling of these elements is important in aquatic environments, particularly in the oceans, which are major reservoirs of Ca and Si and where large amounts of Ca and Si are incorporated into the exoskeletons of certain microbes. However, unlike the C, N, and S cycles, in the Ca and Si cycles there are no redox changes or gaseous forms that can escape and alter Earth's atmospheric chemistry, and only recently have different redox states of phosphorus been discovered to be biogeochemically significant. However, as we will see, keeping these cycles in balance—especially that of Ca—is important for maintaining sustainable life on Earth.

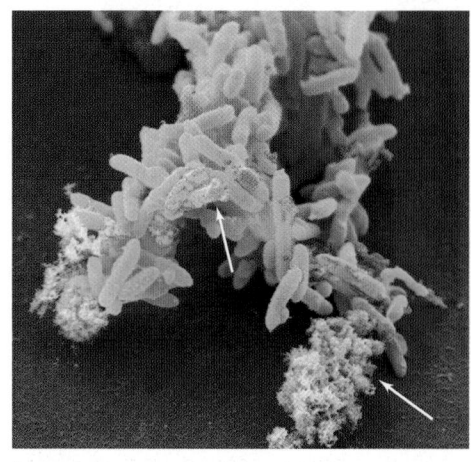

(a)

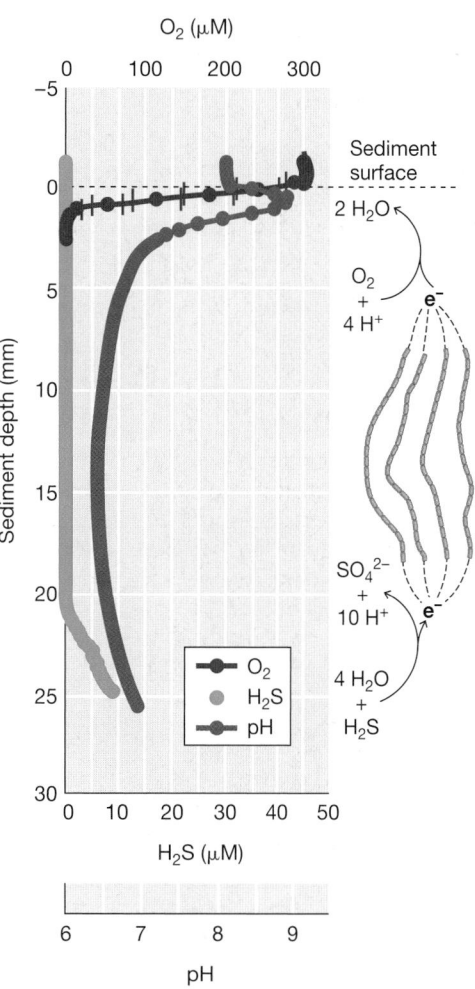

O_2 (µM)

Sediment depth (mm)

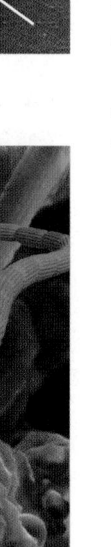

Sediment surface

$2 H_2O$

O_2 + $4 H^+$

e^-

SO_4^{2-} + $10 H^+$

e^-

$4 H_2O$ + H_2S

- O_2
- H_2S
- pH

H_2S (µM)

pH

(b)

(c)

Figure 1 (a) Cells of *Geobacter* attached to ferric iron precipitates (arrows) reduce Fe^{3+} to Fe^{2+}. (b) Three-dimensional rendering of cabled filamentous sulfide-oxidizers, with inset transmission electron microscopic cross-section displaying presumptive electrically conductive filaments surrounding the cell perimeter. (c) Microsensor profiles of sediments colonized by cabled filamentous bacteria, showing wide separation of depths where sulfide oxidation and oxygen reduction occur. The profiles are explained as follows: Oxygen (O_2) is quickly consumed by bacterial respiration in the upper regions of the sediment, whereas H_2S, produced in anoxic regions, only accumulates deeper in the sediments. The pH is more acidic lower in the sediments because sulfide oxidation yields protons. As O_2 is consumed near the sediment surface, protons are also consumed and the pH rises.

Phosphorus

Phosphorus exists in nature primarily as organic and inorganic phosphates. Phosphorus reservoirs include phosphate-containing minerals in rocks, dissolved phosphates in freshwaters and marine waters, and the nucleic acids and phospholipids of living organisms. Although P has multiple oxidation states, most environmental phosphates are at the +5 oxidation state (for example, inorganic phosphate, HPO_4^{2-}). In nature P cycles through living organisms (as cellular P), waters and soils (as inorganic and organic P), and Earth's crust (as inorganic P). P is typically the limiting nutrient for photosynthesis in freshwaters, which receive it from the weathering of rocks.

In the oceans, a fraction of dissolved P is organic, in the form of phosphate esters and *phosphonates*. Phosphonates are organophosphate compounds that contain a direct bond between the P and C atoms. In this form the P atom is more reduced (+3 oxidation state) than in phosphate (+5 oxidation state). Phosphonates are produced by certain microorganisms and comprise about a quarter of the organic P pool in nature. For many organisms, phosphonates are a less available source of P than is HPO_4^{2-} because the organisms lack the enzymes required to degrade phosphonates. Such organisms can be phosphorus-limited even when sufficient P is present as phosphonates.

Phosphonates and reduced inorganic forms of P—phosphite ($H_2PO_3^-$, +3 oxidation state) and hypophosphite ($H_2PO_2^-$, +1 oxidation state)—are rapidly cycled in the marine environment by producers and consumers. This previously unrecognized P cycle is now thought to be important for organisms that live in P-depleted environments and have the capacity to make or consume these alternative forms of phosphorus. Why marine organisms produce so much of these reduced phosphorus species is unknown. However, the fact that some marine microorganisms produce methylphosphonate may have solved another metabolic conundrum. The degradation of methylphosphonate

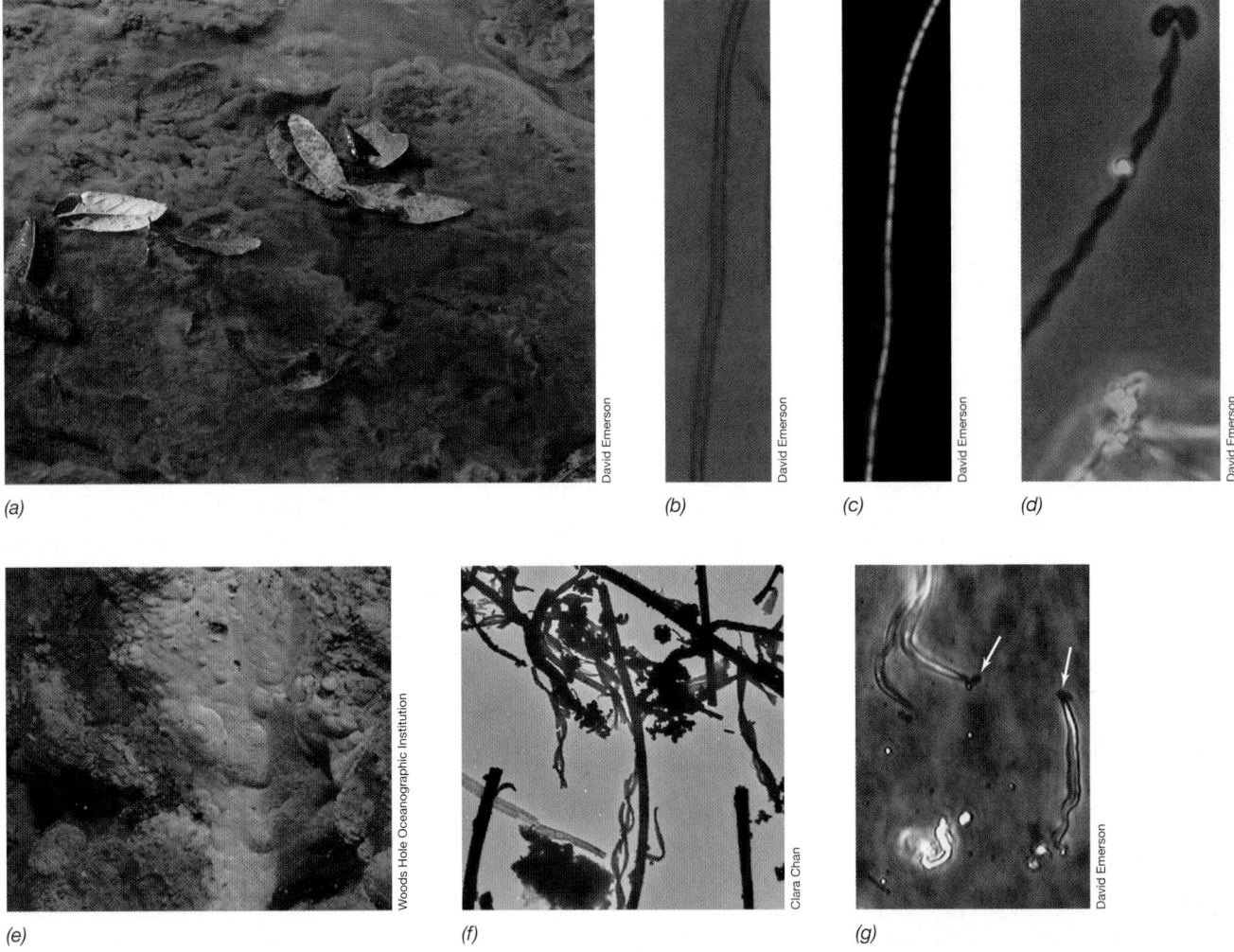

Figure 21.14 Fe-oxidizing microbial mats.
(a) Freshwater microbial mat in a slow-moving stream where Fe^{2+}-enriched groundwater is mixing with oxygenated surface water, triggering growth of Fe^{2+}-oxidizing bacteria and precipitation of iron oxides. *(b–d)* Fe-oxidizing *Bacteria. (b, c)* Phase-contrast and epifluoresence photomicrographs of the sheath-forming Fe-oxidizer *Leptothrix ochracea* (the sheath is approximately 2 μm wide). *(d)* The stalk-forming Fe^{2+}-oxidizer *Gallionella ferruginea* showing bean-shaped cells in the process of cell division at the end of the iron oxide–encrusted stalk (each bean-shaped cell is about 2 μm long). *(e)* An iron-oxidizing mat at a deep-sea hydrothermal vent (1000-meter depth) at Lōʻihi Seamount. *(f)* TEM image of biogenic oxides produced at Lōʻihi; note the variety of helical stalks and tubular sheathlike filaments (the filaments vary from 2 to 4 μm wide). *(g)* Phase-contrast photomicrograph of marine Fe^{2+}-oxidizers growing at the ends of iron oxide filaments (cells denoted by arrows) from an experimental incubation at Lōʻihi (the filaments are approximately 2 μm wide).

$(CH_4O_3P^-)$ by some marine microorganisms—a process that liberates CH_4—may explain the previously puzzling observation that relatively high levels of CH_4 are present in the oxygenated surface waters of the ocean (methanogenic *Archaea* are strict anaerobes; ⇄ Section 17.2).

Calcium

The major global reservoirs of calcium (Ca) are calcareous rocks and the oceans. In the oceans, where dissolved Ca exists as Ca^{2+}, calcium cycling is a highly dynamic process although the concentration of Ca^{2+} in seawater remains constant at about 10 mM. Several marine eukaryotic phototrophic microorganisms take up Ca^{2+} to form their calcareous exoskeletons; these include in particular the *coccolithophores* and *foraminifera* (**Figure 21.15**; ⇄ Section 18.6). The calcium-cycling activities of these planktonic phototrophs are also tightly coupled with inorganic components of the carbon cycle.

The precipitation of calcium carbonate $(CaCO_3)$ to form the shells of calcareous phytoplankton controls both CO_2 flux into ocean surface water and inorganic C transport into deep ocean water and the sediments. Moreover, the formation of $CaCO_3$ both depletes surface dissolved bicarbonate (HCO_3^-) and increases the level of dissolved CO_2 (Figure 21.15*c*). The latter reduces the influx of atmospheric CO_2 into surface ocean waters and this helps maintain the slightly alkaline pH of the oceans. When these calcareous organisms die and sink toward the sediments, Ca^{2+} and inorganic and organic C are transported to the deep ocean from which they are only slowly released over long periods.

The formation of $CaCO_3$ exoskeletons brings into play a delicate balance between Ca^{2+} and C and is a process sensitive to changes in atmospheric CO_2 levels. This is because increased levels of atmospheric CO_2 increase the formation of carbonic acid (H_2CO_3), and as this dissociates to form HCO_3^- and H^+, $CaCO_3$ dissolves and

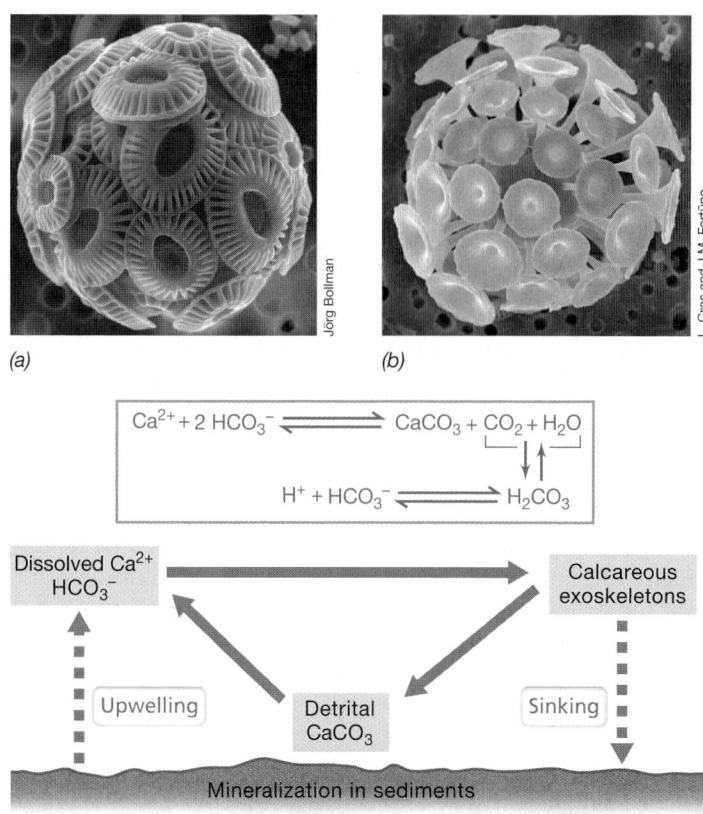

(a)   (b)

$$Ca^{2+} + 2\ HCO_3^- \rightleftharpoons CaCO_3 + CO_2 + H_2O$$

$$H^+ + HCO_3^- \rightleftharpoons H_2CO_3$$

Dissolved Ca^{2+} HCO_3^- → Calcareous exoskeletons

Upwelling   Detrital $CaCO_3$ Sinking

Mineralization in sediments

(c)

Figure 21.15 The marine calcium cycle. Scanning electron micrographs of cells of the calcareous phytoplankton *(a) Emiliania huxleyi* and *(b) Discosphaera tubifera.* The exoskeletons of these coccolithophores are made of calcium carbonate ($CaCO_3$). A cell of *Emiliana* is about 8 μm wide and a cell of *Discosphaera* is about 12 μm wide. *(c)* The marine calcium cycle; dynamic pools of Ca^{2+} are shaded in green. Detrital $CaCO_3$ is that in fecal pellets and other organic matter from dead organisms. Note how H_2CO_3 formation lowers ocean pH when it dissolves to form H^+ and HCO_3^-.

seawater pH decreases (Figure 21.15*c*). Greater ocean acidity resulting from rising atmospheric CO_2 is predicted to reduce the rate of formation of calcareous shells, which will likely have effects on other microbial nutrient cycles and plant and animal communities (Section 21.8).

Silica

The marine Si cycle is controlled primarily by unicellular eukaryotes (diatoms, silicoflagellates, and radiolarians) that build ornate external cell skeletons called *frustules* (**Figure 21.16***a*) (⟲ Sections 18.5 and 18.6). These structures are not constructed of $CaCO_3$ as in the coccolithophores, but of opal (SiO_2), whose formation begins with the uptake by the cell of dissolved silicic acid (Figure 21.16*b*).

Diatoms are rapidly growing phototrophic eukaryotes and often dominate blooms of phytoplankton in coastal and open ocean waters. However, unlike other major phytoplankton groups, diatoms require Si and can become silica-limited when blooms develop. Also, because of their large size, diatom cells tend to sink faster than other organic particles, and in this way, they contribute significantly to the return of Si and C to deeper ocean waters. The transport of organic material produced through primary

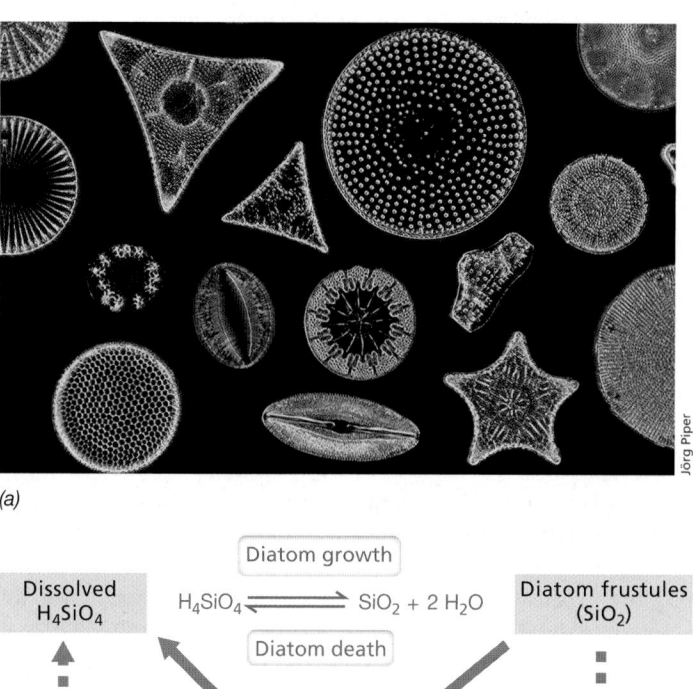

(a)

Diatom growth

Dissolved H_4SiO_4 $H_4SiO_4 \rightleftharpoons SiO_2 + 2\ H_2O$ Diatom frustules (SiO_2)

Diatom death

Upwelling   Detrital SiO_2 Sinking

Mineralization in sediments

(b)

Figure 21.16 The marine silica cycle. *(a)* Dark-field photomicrograph of a collection of diatom shells (frustules). The frustules are made of SiO_2. *(b)* The marine silica cycle; dynamic pools of Si are shaded in green.

production in near-surface waters to deeper ocean waters, primarily by sinking particles, is called the *biological pump* and is an important aspect of the carbon cycle in terms of carbon burial and mineralization in marine environments (Figure 21.1).

In addition to the major nutrient requirements of any phototrophic organism (CO_2, N, P, Fe), diatoms require sufficient dissolved Si, and in nature this originates primarily from Si released from the skeletons of dead diatoms (Figure 21.16*b*). Although Si is released fairly rapidly following cell death, during periods of high diatom production in relatively shallow waters, a significant fraction of dissolved Si can be buried in sediments and remain there for millions of years. This has consequences for continued diatom growth and their phototrophic consumption of dissolved CO_2 from ocean waters. The flux of CO_2 into and out of ocean water affects its pH (Figure 21.15*c*), and through this link, the Si and C cycles are coupled in similar fashion to what we have seen with the Ca and C cycles.

--- **MINIQUIZ** ---

- How does the formation of $CaCO_3$ skeletons by calcareous phytoplankton retard CO_2 uptake and help maintain ocean water pH?

- How might Si depletion in the photic zone influence the biological pump?

UNIT 5

III • Humans and Nutrient Cycling

Humans have a profound impact on microbial nutrient cycles by adding and removing specific components of the cycles in large amounts. Here we consider human inputs of three major species: mercury (Hg), CO_2 and other atmospheric gases, and various fixed N compounds. These compounds either cause toxicity problems (Hg) or affect planet Earth in globally significant ways (gases and N compounds). We begin with the very toxic metal Hg, which is transformed by bacteria in many different ways.

21.7 Mercury Transformations

Mercury is not a biological nutrient (↩ Figure 3.1), but microbial transformations of various mercuric compounds help to detoxify some of the most toxic forms. Mercury is a widely used industrial product, especially in the electronics industry. Mercury is also an active ingredient in many pesticides, a pollutant from the chemical and mining industries and from the combustion of coal and municipal wastes, and a common contaminant of aquatic ecosystems and wetlands. Because of its propensity to concentrate in living tissues, Hg is of considerable environmental importance. The major form of Hg in the atmosphere is elemental mercury (Hg^0), which is volatile and is oxidized to mercuric ion (Hg^{2+}) photochemically. Most Hg thus enters aquatic environments as Hg^{2+} (**Figure 21.17**).

Microbial Redox Cycle for Mercury

Mercuric ion readily adsorbs to particulate matter, and when deposited in anoxic environments, such as lake or marine sediments, Hg^{2+} can be metabolized from there by anaerobic microbes. Microbial activity methylates Hg, yielding *methylmercury* (CH_3Hg^+) (Figure 21.17), and this can be further methylated to form *dimethylmercury* (CH_3—Hg—CH_3). Methylmercury and dimethylmercury are extremely toxic to animals because they are readily absorbed through the skin and are potent neurotoxins. But in addition, CH_3Hg^+ is soluble and can be concentrated in the food chain, primarily in the muscle tissues of fish, and its concentration increases with each trophic level (a process called *biomagnification*), causing a threat to humans whose diets rely on fish. Methylmercury is about 100 times more toxic than Hg^0 or Hg^{2+}, and its accumulation in the aquatic food chain seems to be particularly acute in freshwater lakes and marine coastal waters where enhanced levels of CH_3Hg^+ have been detected in fish caught for human consumption. Mercuric compounds can cause liver and kidney damage in humans and other animals.

Methylation was long associated with sulfate-reducing and iron-reducing bacteria, but only recently have the contributing enzyme systems been identified. Methylation by sulfate-reducing bacteria requires two genes, *hgcA* (encoding a putative methyltransferase corrinoid protein) and *hgcB*, encoding a putative [4Fe-4S] ferredoxin (Figure 21.17). In the proposed reaction sequence (Figure 21.17), a methyl group is transferred from the methylated HgcA protein to

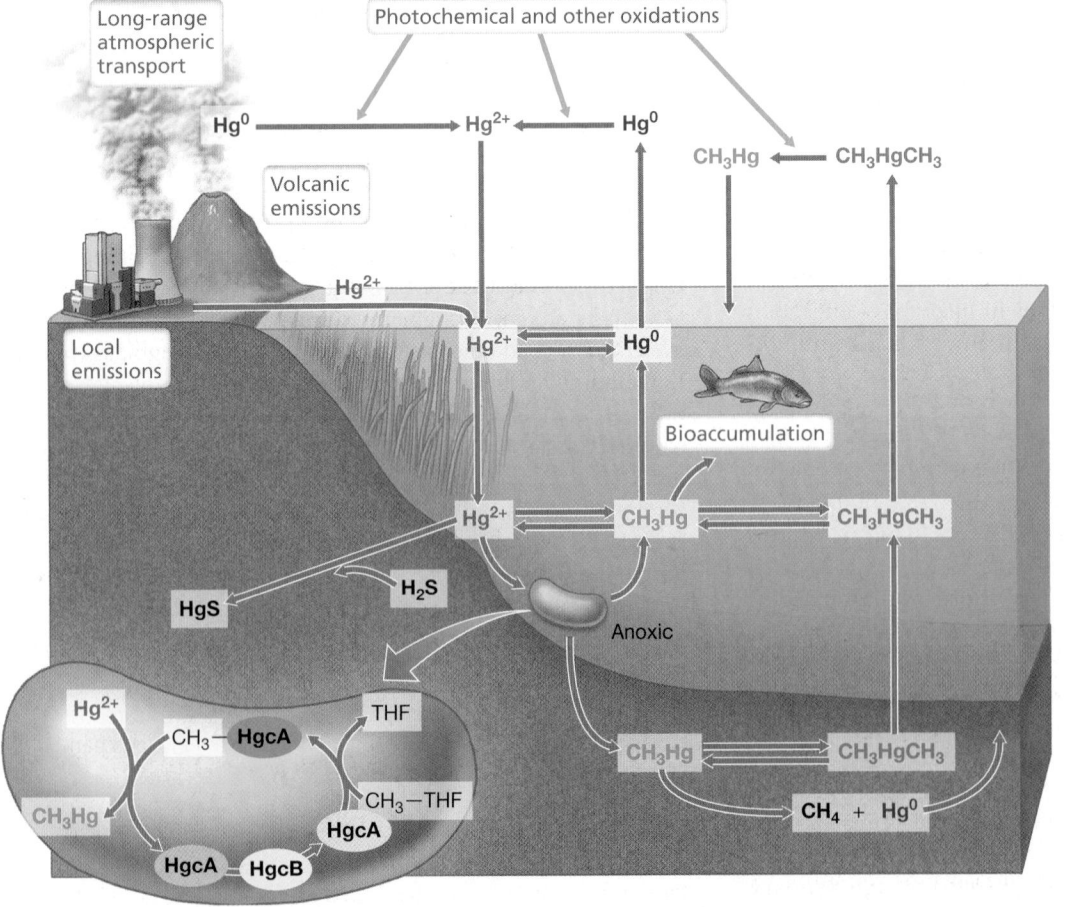

Figure 21.17 Biogeochemical cycling of mercury. The major reservoirs of mercury are water and sediments. Mercury in water can be concentrated in animal tissues; it can be precipitated as HgS from sediments. The volatile forms of mercury are Hg^0 and CH_3HgCH_3. The enlarged bacterial cell shows the enzyme system responsible for mercury methylation by sulfate-reducing bacteria (Section 21.7). Common forms of mercury are shown by different colors. THF, tetrahydrofolate. HgcA and HgcB are proteins that function to methylate mercury.

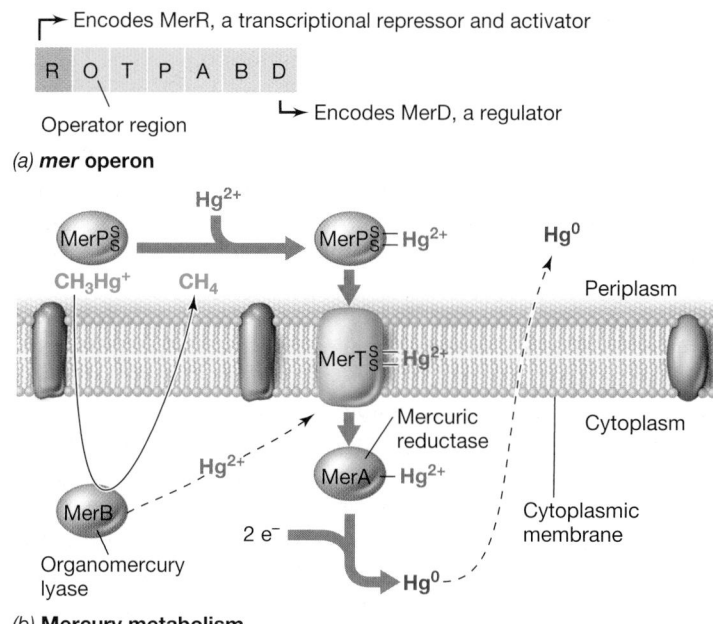

(a) *mer* operon

(b) **Mercury metabolism**

Figure 21.18 Mechanism of mercury transformations and resistance.
(a) The *mer* operon. MerR can function as either a repressor (in the absence of Hg^{2+}) or a transcriptional activator (in the presence of Hg^{2+}). *(b)* Transport and reduction of Hg^{2+} and CH_3Hg^+; the Hg^{2+} is bound by cysteine residues in the MerP and MerT proteins. MerA is the enzyme mercuric reductase and MerB is organomercury lyase.

inorganic Hg^{2+}. The HgcB protein then regenerates the reduced form of HgcA required for accepting the methyl group from methyltetrahydrofolate (THF). Identification of the genes encoding these enzymes provided the opportunity to survey their distribution in available genomic and metagenomic sequences, and this has revealed that genes for mercury methylation are widely distributed in both *Bacteria* and methanogenic *Archaea*. The discovery of mercury methylation genes not only paves the way for developing genetic tools for surveying environments for the potential to methylate mercury, but will contribute to our understanding of the physiological significance of methylation (hypothesized to be used for detoxification). It also allows for the identification of environmental variables that control gene expression and could possibly provide insight into the observation that mercury methylation also occurs in oxic marine surface waters.

Several other microbial Hg transformations occur in anoxic sediments, including reactions catalyzed by sulfate-reducing bacteria ($H_2S + Hg^{2+} \rightarrow HgS$) and methanogens ($CH_3Hg^+ \rightarrow CH_4 + Hg^0$) (Figure 21.17). The solubility of mercuric sulfide (HgS) is very low, so in sulfidic sediments, most Hg exists as HgS. But upon aeration, HgS can be oxidized to Hg^{2+} and SO_4^{2-} by metal-oxidizing bacteria (Section 21.5), and the Hg^{2+} is eventually converted to CH_3Hg^+. However, it is not the Hg in HgS that is oxidized here, but instead the *sulfide*, probably by organisms related to *Acidithiobacillus* (⟲ Section 15.11).

Mercury Resistance

At sufficiently high concentrations, Hg^{2+} and CH_3Hg^+ can be toxic to microorganisms as well as to macroorganisms. However, several gram-positive and gram-negative bacteria convert toxic forms of

Hg to nontoxic or less toxic forms. These mercury-resistant bacteria employ the enzyme *organomercury lyase* to degrade the highly toxic CH_3Hg^+ to Hg^{2+} and methane (CH_4), and the NADPH (or NADH)-linked enzyme *mercuric reductase* to reduce Hg^{2+} to Hg^0, which is volatile and thus mobile (**Figure 21.18**).

In many mercury-resistant bacteria, genes encoding Hg resistance reside on plasmids or transposons (⟲ Sections 4.2 and 11.11). These *mer* genes are arranged in an operon under control of the regulatory protein MerR, which can function as either a repressor or an activator of transcription (⟲ Sections 6.2 and 6.3), depending on Hg availability. In the absence of Hg^{2+}, MerR functions as a *repressor* and binds to the operator region of the *mer* operon, thus preventing transcription of the structural genes, *merTPABD*. However, when Hg^{2+} is present, it forms a complex with MerR, which then binds to the *mer* operon and functions as an *activator* of transcription of *mer* structural genes (Figure 21.18).

MerP is a periplasmic mercuric ion–binding protein; it binds Hg^{2+} and transfers it to the membrane transport protein MerT, which interacts with mercuric reductase (MerA) to reduce Hg^{2+} to Hg^0 (Figure 21.18b). Thus, Hg^{2+} is not released into the cytoplasm and the final result is the release of Hg^0 from the cell. Mercuric ion produced from the activity of MerB is trapped by MerT and reduced by MerA, again releasing Hg^0 (Figure 21.18b). In this way, Hg^{2+} and CH_3Hg^+ are converted to the relatively nontoxic Hg^0.

─────────────── **MINIQUIZ** ───────────────

• What forms of mercury are most toxic to organisms?

• How is mercury methylated by microbes?

• How is mercury detoxified by bacteria?

21.8 Human Impacts on the Carbon and Nitrogen Cycles

Human activities have major effects on the carbon and nitrogen cycles, and these effects have significance for the health of our planet in general. The period of marked human influence on these nutrient cycles began with the Industrial Revolution and is informally termed the *Anthropocene*, a new geological epoch. Although the greatest human impacts have been on the release of CO_2 through the burning of fossil fuels (oil, gas, and coal) and from extensive and ongoing deforestation, human activity has also profoundly affected the nitrogen cycle. We discussed earlier the close coupling of the carbon and nitrogen cycles (Section 21.1), and here we consider some of the projected biogeochemical consequences of human alteration of these two critical nutrient cycles.

CO$_2$, Other Trace Gases, and Global Warming

Atmospheric CO_2 levels have increased approximately 40% since the beginning of the Industrial Revolution in the 1800s, and are now higher than at any time in the last 800,000 years. Carbon dioxide is one of several *trace gases* (primarily water vapor, CO_2, CH_4, and N_2O) which comprise less than 0.5% of the atmosphere but contribute significantly to terrestrial and atmospheric warming due to the *greenhouse effect*, the ability of these gases to trap the infrared radiation emitted by the Earth. Atmospheric concentrations of all of these trace gases are rapidly increasing as a consequence of human activities (**Figure 21.19**). The change in global warming potential resulting from the addition of these gases to the atmosphere is expressed as **radiative forcing**, defined as the difference between sunlight energy absorbed by Earth and energy radiated back to space, measured in watts per square meter of Earth's surface (W/m^{-2}) (Figure 21.19*d*). Radiative forcing is used to calculate an Annual Greenhouse Gas Index (AGGI), defined as the ratio of the total direct radiative forcing due to long-lived greenhouse gases for a given year to that which was present in 1990 (the baseline year for the Kyoto Protocol for controlling greenhouse gas emissions).

In 2014 the AGGI was 1.36 (an increase in radiative forcing of 36% since 1990). Climate models incorporate both the radiative forcing values and the atmospheric lifetimes of the major greenhouse gases. Although methane and N_2O have greater radiative forcing per unit molecule than CO_2 (58- and 206-fold, respectively), they are at much lower concentrations and therefore contribute less to warming (Figure 21.19*d*). CO_2 is the major contributor to the AGGI in terms of both amount and rate of increase. The atmospheric lifetime of a molecule of N_2O is around 150 years. In contrast, there is more uncertainty about the atmospheric lifetimes of methane and CO_2, now estimated at about 10 and 120 years, respectively. The long lifetimes of N_2O and CO_2 mean that once these gases are added to the atmosphere they will impact Earth's climate for centuries.

Also shown in Figure 21.19*d* is radiative forcing attributed to chlorofluorocarbons (CFCs). There is no natural source of CFCs. These compounds were chemically synthesized for use as refrigerants, aerosol propellants, and cleaning solvents. Following the discovery that they were destroying stratospheric ozone, their manufacture was effectively suspended through an international agreement, the Montreal Protocol on substances that deplete the ozone layer, which entered into force on January 1, 1989. As a result, levels of major CFCs are now constant or declining. Although their long atmospheric lifetimes mean that some CFCs will remain in the atmosphere for over 100 years, the Montreal Protocol offers some hope that nations can come together to address shared environmental threats.

CO_2 and Its Effects on Aquatic Microbial Ecosystems

The increase in atmospheric CO_2 concentration, measured across a global network of sampling stations (Figure 21.19*a*), is currently about 2 parts per million per year. This increase would be much more rapid were it not for the high solubility of CO_2 in water, which produces carbonic acid; much anthropogenic CO_2 thus dissolves in the oceans (Figures 21.1 and 21.15*c*). The surface waters of the oceans have taken up an estimated 500 billion tons of CO_2 from the atmosphere out of a total of 1300 billion tons of total anthropogenic emissions, thus modulating the greenhouse effect somewhat. The increase in average Earth air temperature, estimated to have increased 0.75°C in the twentieth century and projected to increase by anywhere from 1.1 to 6.4°C in the twenty-first century, would also have been more rapid without the buffering influence of the oceans. Since three orders of magnitude more energy is required to raise the temperature of a cubic meter of water than a cubic meter of air, it can be calculated that over 80% of the heat retained on Earth as a result of the greenhouse effect thus far has actually entered the ocean.

Although there is considerable uncertainty about the consequences of ocean warming and CO_2 consumption on Earth's biological systems, there is agreement on how these changes will affect biogeochemistry. Warmer ocean surface waters are more buoyant (because of their lower density) than are deeper waters. Thus, as occurs seasonally in lakes (⇄ Section 20.8), the oceans will become more stratified with future global warming. Stratification tends to slow the transfer of nutrients from deeper waters that are needed to nourish phototrophic microbes at the base of the food web in surface waters. This in turn reduces ocean productivity and export of a portion of that production to the deeper ocean through sedimentation (the biological pump, Figure 21.1).

Ocean warming is also contributing to the expansion of oxygen minimum zones (OMZs), regions of naturally occurring low O_2 concentration in subsurface waters between 100 and 1000 m in depth (⇄ Section 20.9). OMZs are a consequence of both the reduced solubility of O_2 in warmer water and the increasing stratification associated with surface warming, which reduces mixing of

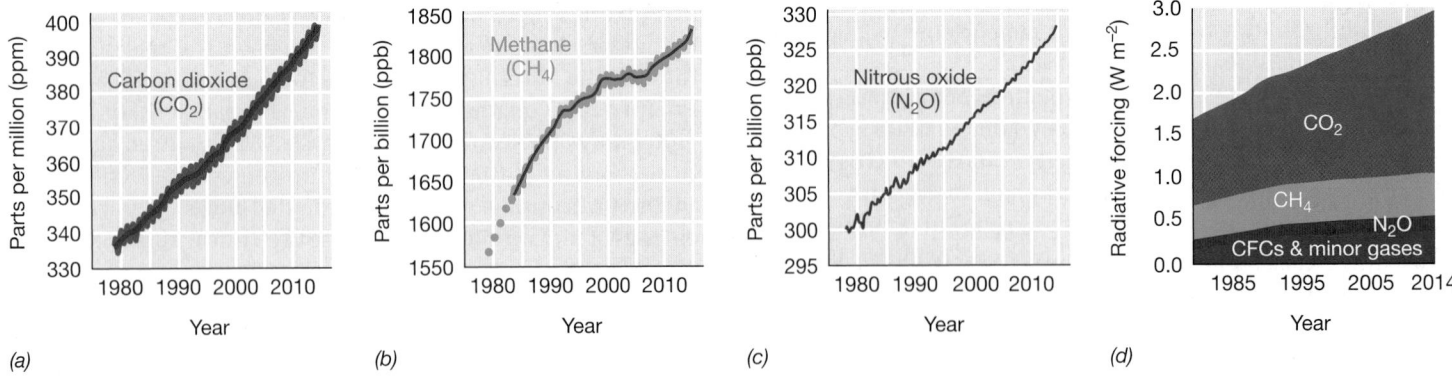

(a)   *(b)*   *(c)*   *(d)*

Figure 21.19 Past 35-year increases in greenhouse gases and associated radiative forcing. Global average values for *(a)* CO_2, *(b)* CH_4, *(c)* N_2O. Increases in CO_2 have averaged 1.4 ppm per year before 1995 and 2 ppm per year thereafter. *(d)* Radiative forcing (see text for definition) from chlorofluorocarbons (CFCs) is now stable or declining as a result of the Montreal Protocol banning production of substances that deplete the ozone layer. These data are continuously collected by the Global Monitoring Division of the National Oceanic and Atmospheric Association/Earth System Research Laboratory.

surface and subsurface waters. Animals will be excluded from the expanding OMZs whereas anaerobic microbial processes, such as denitrification and anammox, that directly influence the nitrogen cycle and production of the greenhouse gas N_2O, will be enhanced.

Acidification of the ocean resulting from the ongoing dissolution of anthropogenic CO_2 has reduced ocean pH by 0.1 pH units since the beginning of the Industrial Revolution and may further reduce the pH by 0.3–0.4 units by the year 2100. The ongoing reduction in carbonate (CO_3^{2-}) concentration, a consequence of increasing acidification, is expected to be detrimental to marine calcifiers (organisms synthesizing $CaCO_3$ shells or skeletons, Figure 21.15). Since the concentration of Ca in seawater is relatively constant, continued reduction in CO_3^{2-} will ultimately reach a point where the dissolution of existing $CaCO_3$ is chemically favored, ultimately releasing more dissolved CO_2 (Figure 21.15), which reduces the capacity of the oceans to absorb more atmospheric CO_2.

Although the biological response to ocean acidification is unknown, it is likely that coral reef ecosystems, a major component of the marine biosphere (↶ Section 23.11), will cease to occur naturally on Earth if CO_2 emissions continue at their present rate (Figure 21.19*a*). Calcification in foraminifera (↶ Section 18.6) will likely be impaired significantly by ocean acidification, as will calcification in coccolithophores (Figure 21.15). Over periods of a century or so, the invasion of anthropogenic CO_2 into the deep ocean will ultimately result in a significant reduction in the levels of $CaCO_3$ sequestered there, and this will likely affect the carbon cycle in major but as yet unpredictable ways.

Methane and Global Warming

Increases in atmospheric methane have contributed to about one-fifth of the increase in radiative forcing since 1750. About two-thirds of this increase is related to industrial activities (e.g., coal mining, natural gas wells, pipelines, and fracking). Major natural sources of methane are wetlands, ruminants, thawing permafrost, and methane hydrates (Figures 21.1, 21.3, and 21.4). Although increases in atmospheric methane slowed during the 1990s, and were nearly constant between 1999 and 2006, strong growth resumed in 2007 (Figure 21.19*b*). Analysis of the stable isotopic composition of new atmospheric methane (↶ Section 19.10) suggests that recent growth is from increased release from tropical wetlands. Wetlands are now the world's largest natural source of atmospheric methane, and increased methane release has been attributed to recent changes in tropical climate.

Ongoing warming will also increase atmospheric inputs of methane by destabilizing coastal methane hydrates (Figure 21.4) and by the melting of permafrost (see page 651 for more on this problem). That is, there is *positive climate feedback* linked to methane release. Permafrost contains about 50% of the global soil carbon, and thawing will lead to a loss of soil carbon in the form of methane and carbon dioxide. The positive climate feedback of these new greenhouse gas emissions is unknown and may depend on the microbial community composition controlling organic matter decomposition. For example, recent studies in the Arctic have documented changes in methanogenic community structure (from H_2-oxidizing to partly acetate-oxidizing methanogens) as a function of changes in vegetation patterns and permafrost

melting. Thus, a refined understanding of the microbial sources of methane in a warming Arctic will likely be essential to developing predictive models of further climate change.

An additional positive climate feedback is associated with the rapid loss of summer sea ice in the Arctic Ocean, reduced by more than 11% since 1979. Instead of reflecting back most sunlight as ice does, the sea ice melt has opened a new expanse of dark open water to absorb the sun's energy. This extra energy input, and associated flux of moisture and heat to the Arctic atmosphere, is contributing to a strong local positive feedback called the *Arctic amplification*. Surface temperatures in the Arctic are rising twice as fast as at lower latitudes, and will influence methane release from hydrates and permafrost. Reduction of the temperature differential between the Arctic and global midlatitudes may also be contributing to recent changes in weather patterns. The reduced temperature differential is associated with weakening of west-to-east winds and a slowing of the jet stream, causing the jet stream to "meander" deeper south. In turn, these meanders have been linked to the increasing occurrence of extreme weather events in midlatitude regions.

Anthropogenic Effects on the Nitrogen Cycle

Anthropogenic impacts on the microbial ecology of the nitrogen cycle are as profound as those on the carbon cycle. The yearly industrial production of nitrogenous fertilizers through the Haber–Bosch process, which combines $N_2 + H_2$ to form NH_3 under high temperature and pressure, is now comparable to the amount of fixed nitrogen entering the biosphere through biological nitrogen fixation, a key link in the nitrogen cycle (Section 21.3). Most of the industrially produced N is applied to farmland, but a significant fraction runs off to the oceans and contributes to coastal eutrophication (↶ Section 20.9). Large amounts are also lost as gaseous nitrogen compounds (N_2, N_2O, and NO), primarily from nitrification of NH_3 and denitrification of NO_3^- (Section 21.3). N_2O emission is now increasing at a rate of 0.2–0.3% per year (Figure 21.19*c*). Agriculture—including the microbial breakdown of manure and urine—contributes about 80% of N_2O emissions in the United States. Lesser amounts come from motor vehicle emissions, and industrial production of fertilizers and nitrogen-based polymers.

Transport of N from industrial and agricultural centers through the atmosphere fertilizes both terrestrial and marine systems. Atmospheric deposition of industrially sourced fixed nitrogen to the oceans is now about the same as that which enters through biological nitrogen fixation. The ecological consequences of this fertilization are a major unknown. On the one hand, if deposition suppresses microbial nitrogen fixation, this would to some degree mitigate the fertilization effect. On the other hand, a greater supply of both CO_2 and iron (caused by greater deposition of dust from areas of increasing desertification, ↶ Section 20.6) along with increased N depositions could enhance primary production, since iron is also often a limiting nutrient. Either way, major effects on the carbon cycle should be expected from human inputs in the nitrogen cycle.

Although changes in Earth's biosphere from human intervention in microbial nutrient cycles are a certainty, precisely what these changes will be is less clear. However, because major nutrient cycles are closely coupled (Section 21.1 and Figure 21.5), it is likely that the significant changes already in play in the carbon and nitrogen cycles will have feedback effects on other nutrient cycles as well. Collectively, these events could upset the balance and interrelationships of Earth's nutrient cycles in general—cycles driven by the activities of microbial communities keenly attuned to their environments—and have significant (and likely negative) consequences for plants, animals, and humans.

— **MINIQUIZ** —

- What is the greenhouse effect and what causes it?
- What is the fate of most nitrogen used in agricultural applications?
- Why are the OMZs expanding and what are the likely impacts on nutrient cycles?

MasteringMicrobiology®

Visualize, **explore**, and **think critically** with Interactive Microbiology, MicroLab Tutors, MicroCareers case studies, and more. MasteringMicrobiology offers practice quizzes, helpful animations, and other study tools for lecture and lab to help you master microbiology.

Chapter Review

I • Carbon, Nitrogen, and Sulfur Cycles

21.1 The oxygen and carbon cycles are interconnected through the complementary activities of autotrophic and heterotrophic organisms. Microbial decomposition is the single largest source of CO_2 released to the atmosphere.

Q **What are methane hydrates, and why are these deposits of concern to climate scientists?**

21.2 Under anoxic conditions, organic matter is degraded to CH_4 and CO_2. Methane is formed primarily from the reduction of CO_2 by H_2 and from acetate, both supplied by syntrophic bacteria; these organisms depend on H_2 consumption as the basis of their energetics.

Q **How can organisms such as *Syntrophobacter* and *Syntrophomonas* grow when their metabolism is based on thermodynamically unfavorable reactions? How does coculture of these syntrophs with certain other bacteria allow them to grow?**

21.3 The principal form of nitrogen on Earth is N_2, which can be used as a N source only by nitrogen-fixing bacteria. Ammonia produced by nitrogen fixation or by ammonification can be assimilated into organic matter or oxidized to NO_3^-. Denitrification and anammox cause major losses of fixed nitrogen from the biosphere.

Q **Compare and contrast the processes of nitrification and denitrification in terms of the organisms involved, the environmental conditions that favor each process, and the changes in nutrient availability that accompany each process.**

21.4 Bacteria play major roles in both the oxidative and reductive sides of the sulfur cycle. Sulfur- and sulfide-oxidizing bacteria produce SO_4^{2-}, whereas sulfate-reducing bacteria consume SO_4^{2-}, producing H_2S. Because sulfide is toxic and reacts with various metals, SO_4^{2-} reduction is an important biogeochemical process.

Dimethyl sulfide is the major organic sulfur compound of ecological significance in nature.

Q **If sulfur chemolithotrophs had never evolved, would there be a problem in the microbial cycling of sulfur compounds? What impact would coastal marine eutrophication have on microbial sulfur transformations?**

II • Other Nutrient Cycles

21.5 Iron and manganese exist naturally in two oxidation states: Fe^{2+}/Fe^{3+} and Mn^{2+}/Mn^{4+}. Bacteria reduce the oxidized metals in anoxic environments and oxidize the reduced forms primarily in oxic environments. At neutral pH, bacteria compete with abiotic oxidation in the presence of O_2.

Q **Why are most iron-oxidizing chemolithotrophs obligate aerobes, and why are the better-studied iron oxidizers acidophilic?**

21.6 P, Ca, and Si are elements cycled by microbial activities, primarily in aquatic environments. Calcium and silica play important roles in the biogeochemistry of the oceans as components of the exoskeletons of coccolithophores and diatoms, respectively.

Q **In what ways are Ca and Si cycling in ocean waters similar, and in what ways do they differ? How do the calcium and silica cycles couple to the carbon cycle?**

III • Humans and Nutrient Cycling

21.7 A major toxic form of Hg in nature is CH_3Hg^+, which can yield Hg^{2+}, which is reduced by bacteria to Hg^0. Genes conferring resistance to the toxicity of Hg, such as those that encode enzymes that can detoxify or pump out the metal, often reside on plasmids or transposons.

Q **How are Hg^{2+} and CH_3Hg^+ detoxified by the *mer* system?**

21.8 Anthropogenic inputs of CO_2 and reactive nitrogen are impacting major nutrient cycles. Although some consequences are reasonably well understood, including expansion of OMZs and impaired growth of calcareous organisms, the long-term changes to the nutrient cycles that sustain Earth's biosphere are not well understood.

Q **Discuss two different ways that microorganisms contribute to positive feedbacks in climate change.**

Application Questions

1. Compare and contrast the carbon, sulfur, and nitrogen cycles in terms of the physiologies of the organisms that participate in the cycle. Which physiologies are part of one cycle but not another?

2. ^{14}C-labeled cellulose is added to a vial containing some anoxic freshwater lake sediments and sealed under anoxic conditions. A few hours later, $^{14}CH_4$ appears in the vial. Discuss what has happened to yield such a result.

3. Carbon can be sequestered in the ocean in a variety of forms. Discuss the different forms, their biological sources, and how global warming will influence them.

Chapter Glossary

Denitrification the biological reduction of nitrate (NO_3^-) to gaseous N compounds

Global warming the predicted and ongoing warming of the atmosphere and oceans attributed to anthropogenic release of greenhouse gases, primarily CO_2, that trap infrared radiation emitted by Earth

Humus dead organic matter, some of which functions as electron shuttles for the microbial reduction of metal oxides

Radiative forcing the difference between sunlight energy absorbed by Earth and energy radiated back to space

Syntrophy the cooperation of two or more microorganisms to degrade anaerobically a substance neither can degrade alone

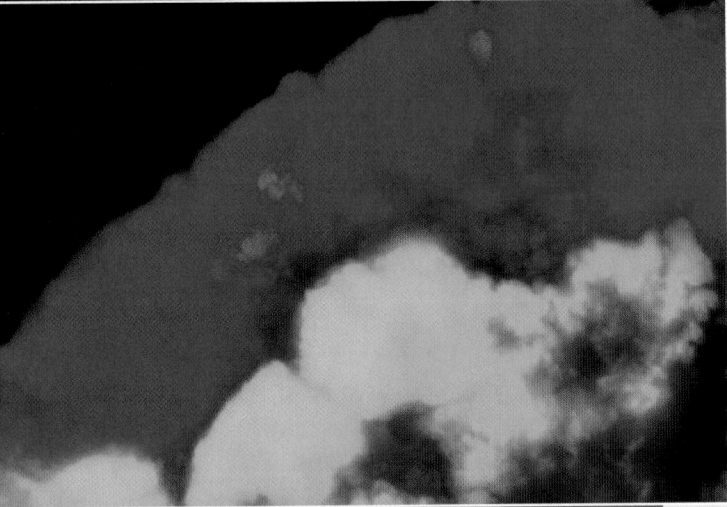

22

Microbiology of the Built Environment

microbiologynow

After the Toilet Flushes

Few give much thought to the fate of water leaving the toilet and the importance of sanitation technology in reducing the spread of infectious disease. However, sewage treatment technology is a critical part of our built environment without which large urban centers would simply not be possible.

Most wastewater treatment plants rely on a technology developed over a century ago called *activated sludge*, a process in which microbial degradation of organic material in sewage is promoted by vigorously mixing sewage in large, well-aerated reactor tanks. The resulting flocculated microbial biomass (sludge) is allowed to settle by gravity in large settling tanks, and the treated water is returned to the environment or directed to additional treatment processes. Both pathogenic microbes and most organic materials are removed in the activated sludge process.

Human population growth and demands for a more sustainable society have prompted the development of wastewater treatment systems that are simpler, more energy efficient, and require less space. A new wastewater treatment process called *aerobic granular sludge technology* (AGS) relies on microbes selected to grow as compact 1- to 2-millimeter granules (upper photo), rather than as flocs. The dense granules settle ten times faster than flocs (eliminating the need for large settling tanks), function in smaller reactors (a reduction of up to 80% in the wastewater built environment is possible), and generate a smaller volume of sludge needing disposal.

The slow diffusion of oxygen into the granules results in the formation of layered structures (lower photo, thin section of a FISH-stained granule) in which aerobes localize to the outer layer and anaerobes to the interior, and this makes the single reactor system ideal for the complete removal of nitrogen from wastewater. For example, aerobic ammonia-oxidizers (blue) in the outer layers of each AGS granule supply nitrite to anaerobic denitrifying microbes found mostly in the interior (green and red). The combined activities of these two physiological groups convert NH_3—an excellent microbial N source—into N_2, and thus when the treated effluent is released into the environment, it will not trigger massive microbial growth. The lower operational cost and complexity of AGS technology also make it ideal for use in small communities and developing countries.

 Source: van Loosdrecht, M.C.M., and Brdjanovic, D. 2014. Anticipating the next century of wastewater treatment. *Science 344:* 1452–1453.

In this chapter we address the microbiology of "built" systems. These include the infrastructure for drinking water and wastewater distribution and treatment, gas and oil transmission, building materials, private and public spaces, and environments modified for mineral extraction or for the cleanup of pollutants. By their very nature, built systems create new microbial habitats, and these promote both desired and undesired microbial activities. From the standpoint of microbial ecology, these activities are simply the natural result of microbes exploiting resources provided to them.

Examples of built systems designed for desirable microbial activities include the construction of biological reactors for the treatment of wastewater and the stimulation of microbial activity in aquifers to clean up environmental pollutants. A notable example of an unwanted activity is microbially influenced corrosion of the pipelines used for transmission of wastewater, drinking water, and oil. Essential infrastructure costing several billion dollars is lost every year to microbially influenced corrosion. For example, the American Association of Civil Engineers estimates that between now and the year 2050, 30% of the drinking water distribution system in the United States will need to be replaced at an annual cost of about $11 billion.

I • Mineral Recovery and Acid Mine Drainage

The biogeochemical capacities of microorganisms seem almost limitless, and it is often said that microorganisms are "Earth's greatest chemists." The activities of these great little chemists have been exploited in many ways. Here we consider how microbial activities help extract valuable metals from low-grade ores.

22.1 Mining with Microorganisms

One of the most common forms of iron in nature is **pyrite** (FeS_2), which is often present in coal and in metal ores. Sulfide (HS^-) also forms insoluble minerals with many metals, and many ores mined as sources of these metals are sulfide ores. If the concentration of metal in the ore is low, it may be economically feasible to mine the ore only if the desired metals are first concentrated by **microbial leaching (Figure 22.1)**. The promotion of acid production and dissolution of FeS_2 by acidophilic bacteria such as *Acidithiobacillus ferrooxidans* is used to leach the metal ores in large-scale mining operations. Leaching is especially useful for copper ores because copper sulfate ($CuSO_4$), formed during the oxidation of copper sulfide ores, is very water-soluble. Indeed, approximately a quarter of all copper mined worldwide is obtained by microbial leaching.

The Leaching Process

The susceptibility to oxidation varies among minerals, and those minerals that are most readily oxidized are most amenable to microbial leaching. Thus, iron and copper sulfide ores such as pyrrhotite (FeS) and covellite (CuS) are readily leached, whereas lead and molybdenum ores are much less so. In microbial leaching, low-grade ore is dumped in a large pile called the *leach dump* and a dilute sulfuric acid solution at pH 2 is percolated down through the pile (Figure 22.1). The liquid emerging from the

Figure 22.1 The leaching of low-grade copper ores using iron-oxidizing bacteria. *(a)* A typical leaching dump. The low-grade ore has been crushed and dumped in such a way that the surface area exposed is as high as possible. Pipes distribute the acidic leach water over the surface of the pile. The acidic water slowly percolates through the pile and exits at the bottom. *(b)* Effluent from a copper leaching dump. The acidic water is very rich in Cu^{2+}. *(c)* Recovery of copper as metallic copper (Cu^0) by passage of the Cu^{2+}-rich water over metallic iron in a long flume. *(d)* A small pile of metallic copper removed from the flume, ready for further purification.

(a)   (b)   (c)   (d)

UNIT 5

bottom of the pile (Figure 22.1*b*) is rich in dissolved metals and is transported to a precipitation plant (Figure 22.1*c*) where the desired metal is precipitated and purified (Figure 22.1*d*). The liquid is then pumped back to the top of the pile and the cycle repeated. As needed, acid is added to maintain an acidic pH.

We illustrate microbial leaching of copper with the common copper ore CuS, in which copper exists as Cu^{2+}. *A. ferrooxidans* oxidizes the sulfide in CuS to SO_4^{2-}, releasing Cu^{2+} as shown in **Figure 22.2**. However, this reaction can also occur spontaneously. Indeed, the key reaction in copper leaching is actually not the bacterial oxidation of sulfide in CuS but the spontaneous oxidation of sulfide by ferric iron (Fe^{3+}) generated from the bacterial oxidation of ferrous iron (Fe^{2+}) (Figure 22.2). In any copper ore, FeS_2 is also present, and its oxidation by bacteria leads to the formation of Fe^{3+}. The spontaneous reaction of CuS with Fe^{3+} proceeds in the absence of O_2 and forms Cu^{2+} plus Fe^{2+}; importantly for efficiency of the leaching process, this reaction can take place deep in the leach dump where conditions are anoxic (Figure 22.2).

Metal Recovery

The precipitation plant is where the Cu^{2+} from the leaching solution is recovered (Figure 22.1*c, d*). Shredded scrap iron (a source of elemental iron, Fe^0) is added to the precipitation pond to recover copper from the leach liquid by the chemical reaction shown in the

lower part of Figure 22.2. This results in a Fe^{2+}-rich liquid that is pumped to a shallow oxidation pond where iron-oxidizing chemolithotrophs oxidize the Fe^{2+} to Fe^{3+}. This now ferric-iron-rich acidic liquid is pumped to the top of the pile and the Fe^{3+} is used to oxidize more CuS (Figure 22.1). The entire CuS leaching operation is thus driven by the oxidation of Fe^{2+} to Fe^{3+} by iron-oxidizing bacteria.

Temperatures rise in a leaching dump and this leads to shifts in the iron-oxidizing microbial community. *A. ferrooxidans* is a mesophile, and when heat generated by microbial activities raises temperatures above about 30°C inside a leach dump, this bacterium is outcompeted by mildly thermophilic iron-oxidizing chemolithotrophic *Bacteria* such as *Leptospirillum ferrooxidans* and *Sulfobacillus*. At even higher temperatures (60–80°C), hyperthermophilic *Archaea* such as *Sulfolobus* (👉 Section 17.9) predominate in the leach dump.

Other Microbial Leaching Processes: Uranium and Gold

Bacteria are also used in the leaching of uranium (U) and gold (Au) ores. In uranium leaching, *A. ferrooxidans* oxidizes U^{4+} to U^{6+} with O_2 as an electron acceptor. However, U leaching depends more on the abiotic oxidation of U^{4+} by Fe^{3+} with *A. ferrooxidans* contributing to the process mainly through the reoxidation of Fe^{2+} to Fe^{3+}, as in copper leaching (Figure 22.2). The reaction observed is as follows:

$$UO_2 + Fe_2(SO_4)_3 \rightarrow UO_2SO_4 + 2\,FeSO_4$$
$$(U^{4+})\;(Fe^{3+})\qquad\quad (U^{6+})\qquad (Fe^{2+})$$

Unlike UO_2, the uranyl sulfate (UO_2SO_4) formed is highly soluble and is concentrated by other processes.

Gold is typically present in nature in deposits associated with minerals containing arsenic (As) and FeS_2. *A. ferrooxidans* and related bacteria can leach the arsenopyrite minerals, releasing the trapped Au:

$$2\,FeAsS[\,Au\,] + 7\,O_2 + 2\,H_2O + H_2SO_4 \rightarrow Fe_2(SO_4)_3$$
$$+\;2\,H_3AsO_4 + [\,Au\,]$$

The Au is then complexed with cyanide (CN^-) by traditional gold-mining methods. Unlike copper leaching, which is done in a huge dump (Figure 22.1*a*), gold leaching is done in small bioreactor tanks (**Figure 22.3**), where more than 95% of the trapped Au can be released. Moreover, the potentially toxic As and CN^- residues from the mining process are removed in the gold-leaching bioreactor. Arsenic is removed as a ferric precipitate, and CN^- is removed by its bacterial oxidation to CO_2 plus urea in later stages of the Au recovery process. Small-scale microbial-bioreactor leaching has thus become popular as an alternative to the environmentally devastating gold-mining techniques that leave a toxic trail of As and CN^- at the extraction site. Pilot processes are also being developed for bioreactor leaching of zinc, lead, and nickel ores.

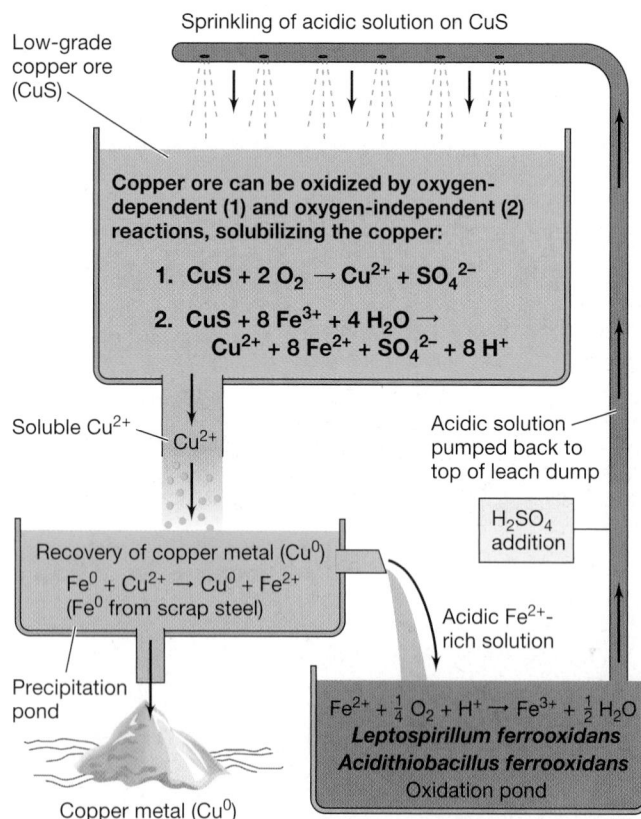

Sprinkling of acidic solution on CuS

Low-grade copper ore (CuS)

Copper ore can be oxidized by oxygen-dependent (1) and oxygen-independent (2) reactions, solubilizing the copper:

1. CuS + 2 O_2 → Cu^{2+} + SO_4^{2-}

2. CuS + 8 Fe^{3+} + 4 H_2O →
    Cu^{2+} + 8 Fe^{2+} + SO_4^{2-} + 8 H^+

Soluble Cu^{2+} — Cu^{2+}

Acidic solution pumped back to top of leach dump

H_2SO_4 addition

Recovery of copper metal (Cu^0)
$Fe^0 + Cu^{2+} \rightarrow Cu^0 + Fe^{2+}$
(Fe^0 from scrap steel)

Acidic Fe^{2+}-rich solution

Precipitation pond

$Fe^{2+} + \frac{1}{4} O_2 + H^+ \rightarrow Fe^{3+} + \frac{1}{2} H_2O$
Leptospirillum ferrooxidans
Acidithiobacillus ferrooxidans
Oxidation pond

Copper metal (Cu^0)

Figure 22.2 Arrangement of a leaching pile and reactions in the microbial leaching of copper sulfide minerals to yield metallic copper. Reaction 1 occurs both biologically and chemically. Reaction 2 is strictly chemical and is the most important reaction in copper-leaching processes. For reaction 2 to proceed, it is essential that the Fe^{2+} produced from the oxidation of sulfide in CuS to sulfate be oxidized back to Fe^{3+} by iron chemolithotrophs (see chemistry in the oxidation pond).

MINIQUIZ

- What is required to oxidize CuS under anaerobic conditions?
- What key role does *Acidithiobacillus ferrooxidans* play in the copper leaching process?

Figure 22.3 Gold bioleaching. Gold leaching tanks in Ghana (Africa). Within the tanks, a mixture of *Acidithiobacillus ferrooxidans*, *Acidithiobacillus thiooxidans*, and *Leptospirillum ferrooxidans* solubilizes the pyrite/arsenic mineral containing trapped gold, which releases the gold.

22.2 Acid Mine Drainage

Although microbial leaching has tremendous value in mining operations, the same process has contributed to extensive environmental destruction where mining operations improperly handle or dispose of pyrite-containing coal and mineral deposits. Bacterial and spontaneous oxidation of sulfide minerals is the major cause of **acid mine drainage**, an environmental problem worldwide caused by surface mining operations. As described for the oxidation of copper sulfides promoted in microbial mining (Section 22.1), the oxidation of FeS_2 is a combination of chemically and bacterially catalyzed reactions, and two electron acceptors participate in the process: O_2 and Fe^{3+}. When FeS_2 is first exposed in a mining operation (**Figure 22.4b**), a slow chemical reaction with O_2 begins (Figure 22.4c). This reaction, called the *initiator reaction*, leads to the oxidation of HS^- to SO_4^{2-} and the development of acidic conditions as Fe^{2+} is released. *Acidithiobacillus ferrooxidans* and *Leptospirillum ferrooxidans* then oxidize Fe^{2+} to Fe^{3+}, and the Fe^{3+} formed under these acidic conditions, being soluble, reacts spontaneously with more FeS_2 and oxidizes the HS^- to sulfuric acid (H_2SO_4), which immediately dissociates into SO_4^{2-} and H^+:

$$FeS_2 + 14\,Fe^{3+} + 8\,H_2O \rightarrow 15\,Fe^{2+} + 2\,SO_4^{2-} + 16\,H^+$$

Again, the bacteria oxidize Fe^{2+} to Fe^{3+}, and this Fe^{3+} reacts with more FeS_2. Thus, there is a progressive, rapidly increasing rate at which FeS_2 is oxidized, called the *propagation cycle* (Figure 22.4c). Under natural conditions some of the Fe^{2+} generated by the bacteria leaches away and is subsequently carried by anoxic groundwater into surrounding streams. However, bacterial or spontaneous oxidation of Fe^{2+} then takes place in the aerated streams, and because O_2 is present, the insoluble $Fe(OH)_3$ is formed.

As we have seen (Figure 22.4c), the breakdown of FeS_2 ultimately leads to the formation of H_2SO_4 and Fe^{2+}; in waters in which these products have formed, pH values can be lower than 1. Mixing of acidic mine waters into rivers (**Figure 22.5**) and lakes seriously degrades water quality because both the acid and the dissolved metals (in addition to iron, there is aluminum, and heavy metals such as cadmium and lead) are toxic to aquatic organisms.

The O_2 requirement for the oxidation of Fe^{2+} to Fe^{3+} explains how acid mine drainage develops. As long as the pyritic material is not mined, FeS_2 cannot be oxidized because O_2, water, and the bacteria cannot reach it. However, when a mineral or coal seam is exposed

(a)   *(b)*

(c)

Figure 22.4 Coal and pyrite. *(a)* Coal from the Black Mesa formation in northern Arizona (USA); the gold-colored spherical discs (about 1 mm in diameter) are particles of pyrite (FeS_2). *(b)* A coal seam in a surface coal-mining operation. Exposing the coal to oxygen and moisture stimulates the activities of iron-oxidizing bacteria growing on the pyrite in the coal. *(c)* Reactions in pyrite degradation. The primarily abiotic initiator reaction sets the stage for the primarily bacterial oxidation of Fe^{2+} to Fe^{3+}. The Fe^{3+} attacks and oxidizes FeS_2 abiotically in the propagation cycle.

(Figure 22.4b), O_2 and water are introduced, making both spontaneous and bacterial oxidation of FeS_2 possible. The acid formed can then leach into surrounding aquatic systems (Figure 22.5).

Figure 22.5 Acid mine drainage from a surface coal-mining operation. The yellowish-red color is due to the precipitated iron oxides in the drainage (see Figure 22.4c for the reactions in acid mine drainage).

Where acid mine drainage is extensive and Fe^{2+} levels high, a strongly acidophilic species of *Archaea, Ferroplasma*, is often present. This aerobic iron-oxidizing organism is capable of growth at pH 0 and at temperatures up to 50°C. Cells of *Ferroplasma* lack a cell wall and are phylogenetically related to *Thermoplasma*, also a cell-wall-lacking and strongly acidophilic (but chemoorganotrophic) member of the *Archaea* (⇨ Section 17.3).

— MINIQUIZ —

- In what oxidation state is iron in the mineral $Fe(OH)_3$? In FeS? How is $Fe(OH)_3$ formed?

- Natural pyritic deposits, such as underground coal seams, do not contribute to acid mine drainage; why not?

II • Bioremediation

The term **bioremediation** refers to the microbial cleanup of oil, toxic chemicals, or other environmental pollutants, usually by stimulating the activities of indigenous microorganisms in some way. These pollutants include both natural materials, such as petroleum products, and **xenobiotic** chemicals, synthetic chemicals not produced by organisms in nature.

Although bioremediation of many toxic substances has been proposed, most successes have been in cleaning up spills of crude oil or the leakage of hydrocarbons from bulk storage tanks. More recently, the targeted destruction of chlorinated environmental pollutants, including commonly used solvents and pesticides, has become more amenable to bioremediation as a result of a better understanding of associated microbiology. There has also been increasing success in the bioremediation of uranium-contaminated environments, many of which are the legacy of poorly regulated past mining of uranium for nuclear fuel and weapons. We begin here with a consideration of this very toxic pollutant.

22.3 Bioremediation of Uranium-Contaminated Environments

Major classes of inorganic pollutants are metals and radionuclides that cannot be destroyed, but only altered in chemical form. Often the extent of environmental pollution is so great that physical removal of the contaminated material is impossible. Thus, *containment* is the only real option, and a common goal in the bioremediation of inorganic pollutants is to change their mobility, making them less likely to move with groundwater and so contaminate surrounding environments. Here we consider how the radioactive element uranium can be contained by the activities of bacteria.

Bioremediation of Uranium

Uranium contamination of groundwater has occurred at sites in the United States and elsewhere where uranium ores have been processed or stored (**Figure 22.6**), and the movement of radioactive materials offsite via groundwater is a threat to environmental and human health. Because the contamination is often widespread, making mechanical methods of recovery very expensive, microbiologists have joined forces with engineers to develop biological treatments that exploit the ability of some bacteria to reduce U^{6+} to U^{4+}. Uranium as U^{6+} is soluble, whereas U^{4+} forms an immobile uranium mineral called *uraninite*, thus limiting the movement of U into groundwater and potential contact with humans and other animals.

Bacterial Transformations of Uranium

The major strategy for immobilizing uranium has been to use bacteria to change the oxidation state of U in major uranium contaminants to a form that will stabilize the element. In this regard, *Bacteria*, including metal-reducing *Shewanella* and *Geobacter* species (⇨ Section 15.14) and sulfate-reducing *Desulfovibrio* species (⇨ Section 15.9), couple the oxidation of organic matter and H_2 to the reduction of U^{6+} to U^{4+}.

Field studies in which organic electron donors have been injected into uranium-contaminated aquifers to stimulate U^{6+} reduction have shown that this approach can lower U levels to below the U.S. Environmental Protection Agency's drinking water standard of 0.126 μM. However, even though uraninite is stable under reducing conditions, if conditions become oxic, it reoxidizes. Thus, much ongoing uranium bioremediation research is focused on questions of whether microbially reduced uranium is stable if the composition of the

Figure 22.6 Uranium bioremediation. An experimental plot at a United States Department of Energy uranium-contaminated site. Organic carbon (acetate) is being infused into the site (see inset photo) and travels in groundwater in the direction of the arrow shown in the main photo. Acetate is an electron donor for reduction of U^{6+} to U^{4+}, which immobilizes the uranium.

(a)   (b)   (c)

Figure 22.7 Environmental consequences of large oil spills and the effect of bioremediation. *(a)* A contaminated beach along the coast of Alaska containing oil from the *Exxon Valdez* spill of 1989. *(b)* The rectangular plot (arrow) was treated with inorganic nutrients to stimulate bioremediation of spilled oil by microorganisms, whereas areas above and to the left were untreated. *(c)* Oil spilled into the Mediterranean Sea from the Jiyeh (Lebanon) power plant that flowed to the port of Byblos during the 2006 war in Lebanon.

microbial community changes or if oxidants, such as O_2, NO_3^-, and Fe^{3+}, are introduced via groundwater. This is obviously an important question because uraninite stability must be targeted for the long term in order to account for the long half-life of nuclear decay of uranium.

MINIQUIZ

- Which reaction, oxidation or reduction, is key to uranium bioremediation?
- Why is immobilization a good strategy for dealing with uranium pollution?

22.4 Bioremediation of Organic Pollutants: Hydrocarbons

Organic pollutants, unlike inorganic pollutants, can generally be completely degraded by microorganisms, eventually to CO_2. This is true of petroleum released in oil spills (**Figure 22.7**), which can be attacked by many different microorganisms. These organisms have been exposed to complex mixtures of hydrocarbons through natural oil seeps for millennia and thus have evolved the catabolic machinery necessary to degrade this naturally occurring pollutant. In contrast, xenobiotic pollutants tend to be more persistent and are degraded by more specialized groups of microorganisms. In this section we focus on hydrocarbons and in the next section on xenobiotics.

Petroleum and Hydrocarbon Bioremediation

Petroleum is a rich source of organic matter, and because of this, microorganisms readily attack hydrocarbons when petroleum is pumped to Earth's surface and comes into contact with air and moisture. Under some circumstances, such as in bulk petroleum storage tanks, microbial growth is undesirable. However, in oil spills, biodegradation is desirable and can be promoted by the addition of inorganic nutrients to balance the huge influx of organic carbon from the oil (Figure 22.7).

The biochemistry of hydrocarbon catabolism was covered in Sections 14.24 and 14.25. Both oxic and anoxic biodegradation is possible. We emphasized that under oxic conditions, oxygenase enzymes play an important role in introducing oxygen atoms into the hydrocarbon. Our discussion here will focus on *aerobic* processes, because it is only when O_2 is present that oxygenase enzymes can function and hydrocarbon bioremediation can be effective in a relatively short time.

Diverse bacteria, fungi, and a few green algae can oxidize petroleum products aerobically. Small-scale oil pollution of aquatic and terrestrial ecosystems from human as well as natural activities is common. Oil-oxidizing microorganisms develop rapidly on oil films and slicks, and hydrocarbon oxidation is most extensive if the temperature is warm enough and supplies of inorganic nutrients (primarily N and P) are sufficient. Moreover, because oil is insoluble in water and is less dense, it floats to the surface and forms slicks. There, hydrocarbon-degrading bacteria attach to the oil droplets (**Figure 22.8**) and eventually decompose the oil and disperse the slick. Certain oil-degrading bacteria are specialist species; for example, the bacterium *Alcanivorax borkumensis* grows only on hydrocarbons, fatty acids, or pyruvate. This organism produces surfactant chemicals that help break up the oil and solubilize it. Once solubilized, the oil can be incorporated more readily and catabolized as an electron donor and carbon source.

In large surface oil spills such as those shown in Figure 22.7, volatile hydrocarbons, both aliphatic and aromatic, evaporate quickly without bioremediation, leaving nonvolatile components

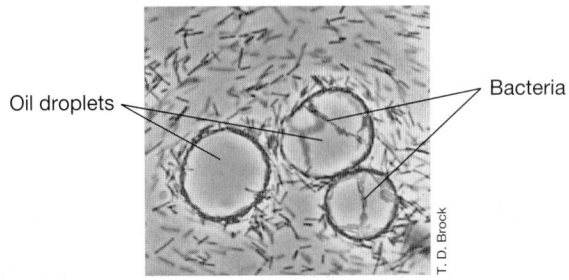

Figure 22.8 Hydrocarbon-oxidizing bacteria in association with oil droplets. The bacteria are concentrated in large numbers at the oil–water interface, but are actually not within the droplet itself.

for cleanup crews and microorganisms to tackle. Microorganisms consume oil by oxidizing it to CO_2. When bioremediation activities are promoted by inorganic nutrient application, oil-oxidizing bacteria typically develop quickly after an oil spill (Figure 22.7*b*), and under ideal conditions, 80% or more of the nonvolatile oil components can be oxidized within one year. However, certain oil fractions, such as those containing branched-chain and polycyclic hydrocarbons, are not preferred microbial substrates and remain in the environment much longer. Spilled oil that finds its way into sediments is even more slowly degraded and can have a significant long-term impact on fisheries that depend on unpolluted waters for productive yields.

A notable exception to the more common surface spill of oil was the 2010 sinking of the Deepwater Horizon offshore drilling platform in the Gulf of Mexico, resulting in the rupture of the wellhead at a depth of 1.5 km and release of over 4 million barrels (635 million liters) of oil into the deep ocean (⇌ Section 20.9 and Figure 20.21). About 35% of the resulting hydrocarbon plume was comprised of low-molecular-weight components and natural gas (methane, ethane, propane). The availability of these more easily degraded oil components is thought to have accelerated the natural degradation process by stimulating the development of a large bloom of bacteria having the capacity to oxidize both the easily degraded and more recalcitrant hydrocarbon components. It remains uncertain whether the industry decision to promote dispersal of the oil (which was intended to increase the oil's surface area and bioavailability) by injecting thousands of gallons of chemical dispersants directly into the plume actually accelerated microbial degradation. Regardless, although some legacy of this major oil spill remains, much of the oil did disappear from a combination of volatilization and microbial activities.

Degradation of Stored Hydrocarbons

Interfaces where oil and water meet often form on a large scale. Besides water that separates from crude petroleum during storage and transport, moisture can condense inside bulk fuel storage tanks (**Figure 22.9**) where there are leaks. This water eventually accumulates in a layer beneath the petroleum. Gasoline and crude oil storage tanks are thus potential habitats for hydrocarbon-oxidizing microorganisms. If sufficient sulfate (SO_4^{2-}) is present in the oil, as it often is in crude oils, sulfate-reducing bacteria can grow in the

tanks, consuming hydrocarbons under anoxic conditions (⇌ Sections 14.25 and 15.9). The sulfide (H_2S) produced is highly corrosive and causes pitting and subsequent leakage of the tanks along with souring of the fuel. Aerobic degradation of stored fuel components is less of a problem because the storage tanks are sealed and the fuel itself contains little dissolved O_2.

MINIQUIZ

- Why do petroleum-degrading bacteria need to attach to the *surface* of oil droplets?
- What is unique about the physiology of the bacterium *Alcanivorax*?

22.5 Bioremediation of Organic Pollutants: Pesticides and Plastics

Unlike hydrocarbons, many chemicals that humans put into the environment have never been there before. These are the xenobiotics, and we consider their microbial degradation here.

Pesticide Catabolism

Xenobiotics include pesticides, polychlorinated biphenyls (PCBs), munitions, dyes, and chlorinated solvents, among many other chemicals. Some xenobiotics differ chemically in such major ways from anything organisms have experienced in nature that they

DDT, dichlorodiphenyltrichloroethane (an organochlorine)

Malathion, mercaptosuccinic acid diethyl ester (an organophosphate)

2,4-D, 2,4-dichlorophenoxy-acetic acid

Site of additional Cl for 2,4,5,-T

Atrazine, 2-chloro-4-ethylamino-6-isopropylaminotriazine

Monuron, 3-(4-chlorophenyl)-1,1-dimethylurea (a substituted urea)

Chlorinated biphenyl (PCB), shown is 2,3,4,2′,4′,5′-hexachlorobiphenyl

Trichloroethylene

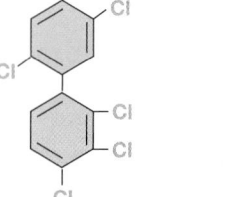

Figure 22.10 Examples of xenobiotic compounds. Although none of these compounds exist naturally, microorganisms exist that can break them down.

Figure 22.9 Bulk petroleum storage tanks. Fuel tanks often support microbial growth at oil–water interfaces.

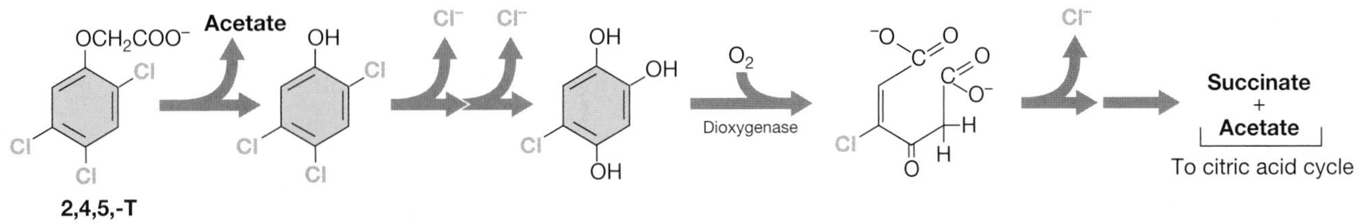

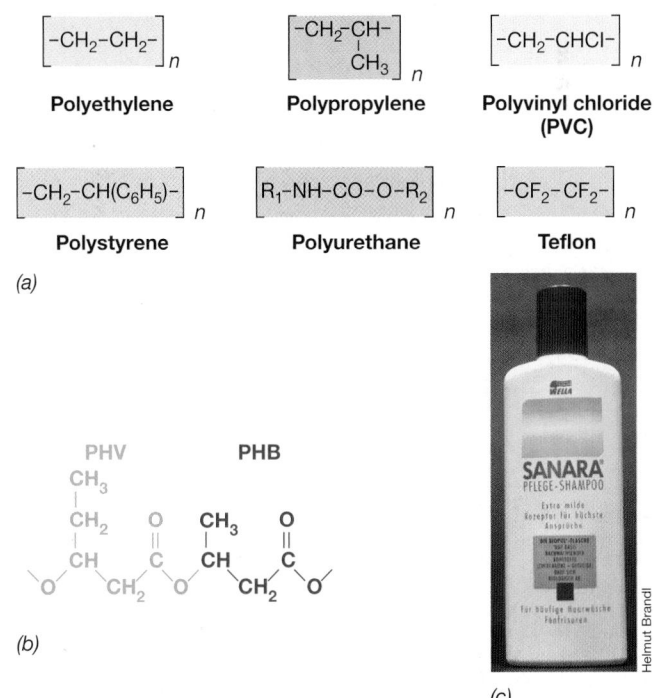

Figure 22.11 Biodegradation of the herbicide 2,4,5-T. Pathway of aerobic 2,4,5-T biodegradation; note the importance of a dioxygenase enzyme (⮀ Section 14.24) in the degradation process.

biodegrade extremely slowly, if at all. Other xenobiotics are structurally related to one or more natural compounds and can sometimes be degraded slowly by enzymes that normally degrade the structurally related natural compounds. We focus here on pesticide bioremediation.

Over 1000 pesticides have been marketed worldwide for pest control purposes. Pesticides include *herbicides, insecticides*, and *fungicides*. Pesticides display a wide variety of chemistries, and include chlorinated, aromatic, and nitrogen- and phosphorus-containing compounds (**Figure 22.10**). Some of these substances can be used as carbon and energy sources by microorganisms, whereas others are utilized only poorly or not at all. Highly chlorinated compounds are typically the pesticides most resistant to microbial attack. However, related compounds may differ remarkably in their degradability. For example, chlorinated compounds such as DDT persist relatively unaltered for years in soils, whereas chlorinated compounds such as 2,4-D are significantly degraded in just a few weeks.

Environmental factors, such as temperature, pH, aeration, and organic content of the soil, influence the rate of pesticide decomposition, and some pesticides can disappear from soils nonbiologically by volatilization, leaching, or spontaneous chemical breakdown. In addition, some pesticides are degraded only when other organic material is present that can be used as the primary energy source, a phenomenon called *cometabolism*. In most cases, pesticides that are cometabolized are only partially degraded, generating new xenobiotic compounds that may be even more toxic or difficult to degrade than the original compound. Thus, from an environmental standpoint, cometabolism of a pesticide is not always a good thing.

Dechlorination

Many xenobiotics are chlorinated compounds and their degradation proceeds through *dechlorination*. For example, the bacterium *Burkholderia* dechlorinates the pesticide 2,4,5-T aerobically, releasing chloride ion (Cl^-) in the process (**Figure 22.11**); this reaction is catalyzed by oxygenase enzymes (⮀ Section 14.24). Following dechlorination, a dioxygenase enzyme breaks the aromatic ring to yield compounds that can enter the citric acid cycle and yield energy.

Although the aerobic breakdown of chlorinated xenobiotics is undoubtedly ecologically important, **reductive dechlorination** may be even more so because of the rapidity with which anoxic conditions develop in polluted microbial habitats. We previously described reductive dechlorination as a form of anaerobic respiration in which chlorinated organic compounds such as chlorobenzoate ($C_7H_4O_2Cl^-$) are terminal electron acceptors and when reduced, release chloride (Cl^-), a nontoxic substance (⮀ Section 14.15).

Many compounds can be reductively dechlorinated including dichloro-, trichloro-, and tetrachloro- (perchloro-) ethylene,

chloroform, dichloromethane, and polychlorinated biphenyls (Figure 22.10). In addition, several brominated and fluorinated organic compounds can be dehalogenated in analogous fashion. Many of these chlorinated or halogenated compounds are highly toxic and some have been linked to cancer (particularly trichloroethylene). Some of these compounds, such as PCBs, have been widely used as insulators in electrical transformers and enter anoxic environments from slow leakage of the transformer or from leaking storage containers. Eventually these compounds end up in groundwater or sediment, where they are among the most common contaminants detected in the United States. There is therefore great interest in reductive dechlorination as a bioremediation strategy for their removal from anoxic environments.

Plastics

Plastics are classic examples of xenobiotics, and the plastics industry worldwide produces about 300 million tons of plastic per year, almost half of which are discarded rather than recycled. Plastics are polymers of various chemistries (**Figure 22.12a**).

$$\left[-CH_2-CH_2- \right]_n$$
Polyethylene

$$\left[\begin{array}{c} -CH_2-CH- \\ \quad\quad CH_3 \end{array} \right]_n$$
Polypropylene

$$\left[-CH_2-CHCl- \right]_n$$
Polyvinyl chloride (PVC)

$$\left[-CH_2-CH(C_6H_5)- \right]_n$$
Polystyrene

$$\left[R_1-NH-CO-O-R_2 \right]_n$$
Polyurethane

$$\left[-CF_2-CF_2- \right]_n$$
Teflon

(a)

(b)

(c)

Figure 22.12 Synthetic and microbial plastics. *(a)* The monomeric structure of several synthetic plastics. *(b)* Structure of the copolymer of poly-β-hydroxybutyrate (PHB) and poly-β-hydroxyvalerate (PHV). *(c)* A brand of shampoo previously marketed in Germany and packaged in a bottle made of the PHB/PHV copolymer.

UNIT 5

Many plastics remain essentially unaltered for long periods in landfills, refuse dumps, and as litter in the environment. As much as 9 million metric tons of plastic per year enters the marine environment, and this is of particular environmental concern. Weathering of plastic debris in the ocean causes fragmentation into small particles that small marine invertebrates can ingest, possibly disrupting important marine food webs. This problem has fueled the search for biodegradable alternatives called **microbial plastics** as replacements for some synthetic plastics.

Polyhydroxyalkanoates (PHAs) are a common bacterial storage polymer (Section 2.8), and these readily biodegradable polymers have many of the desirable properties of xenobiotic plastics. PHAs can be biosynthesized in various chemical forms, each with its own unique physical properties (stiffness, shear and impact strength, and the like). A PHA *copolymer* containing equal amounts of poly-β-hydroxybutyrate and poly-β-hydroxyvalerate (Figure 22.12*b*) has been marketed in Europe as a container for personal care products and has had the greatest success as a plastic substitute thus far (Figure 22.12*c*). However, because synthetic plastics are currently less expensive than microbial plastics, synthetic petroleum-based plastics make up virtually the entire plastics market today.

The bacterium *Ralstonia eutropha* has been used as a model organism for the commercial production of PHAs. This genetically manipulable and metabolically diverse bacterium produces PHAs in high yield, and specific copolymers can be obtained by simple nutritional modifications. Nevertheless, the microbial plastics industry is burdened by the reality that the best substrates for PHA biosyntheses are glucose and related organic compounds, substances obtained from corn or other crops. And even when the price of oil is high, plant products cannot compete with oil as feedstocks for the plastics industry.

MINIQUIZ

- Why might the addition of inorganic nutrients stimulate oil degradation whereas the addition of glucose would not?
- What is reductive dechlorination and how does it differ from the reactions shown in Figure 22.11?
- What main advantage do microbial plastics have over synthetic plastics?

III • Wastewater and Drinking Water Treatment

Water is the most important potential common source of infectious diseases and a potential source for chemically induced intoxications (Chapter 32). This is because a single water source often serves large numbers of people, as, for example, in large cities. Everyone in these cities must use the available water, and contaminated water has the potential to spread disease to all exposed individuals. Similarly, appropriate treatment of wastewater is essential for maintaining environmental quality and for reducing the spread of disease. Thus, the microbiology of water,

water transport systems, and water treatment are of the utmost importance to public health.

The outbreak of cholera in Haiti following the 2010 earthquake is a reminder of the importance of well-maintained waste and drinking water treatment systems in ensuring public health (Sections 29.8 and 32.3). Here we examine systems built for the chemical and biological treatment of water and the transmission systems used for delivering treated water to consumers. We also examine the human health significance of the microbial communities that develop within the pipes of municipal water distribution systems and premise plumbing.

22.6 Primary and Secondary Wastewater Treatment

Wastewater is domestic sewage or liquid industrial waste that cannot be discarded in untreated form into lakes or streams because of public health, economic, environmental, and aesthetic considerations. Wastewater treatment employs physical and chemical methods as well as industrial-scale use of microorganisms. Wastewater enters a treatment plant and, following treatment, the **effluent water**—treated wastewater discharged from the wastewater treatment facility—is suitable for release into surface waters such as lakes and streams or to drinking water purification facilities (**Figure 22.13**).

Wastewater and Sewage

Wastewater from domestic sewage or industrial sources cannot be discarded in untreated form into lakes or streams. **Sewage** is liquid effluent contaminated with human or animal fecal materials. Wastewater may also contain potentially harmful inorganic and organic compounds as well as pathogenic microorganisms. Wastewater treatment can use physical, chemical, and biological (microbiological) processes to remove or neutralize contaminants.

On average, each person in the United States uses 100–200 gallons (380–760 liters) of water every day for washing, cooking, drinking, and sanitation. Wastewater collected from these activities must be treated to remove contaminants before it can be released into surface waters. About 16,000 publicly owned treatment works (POTW) operate in the United States. Most POTWs are fairly small, treating 1 million gallons (3.8 million liters) or less of wastewater per day. Collectively, however, these plants treat about 32 billion gallons (121 billion liters) of wastewater daily. Wastewater plants are usually constructed to handle both domestic and industrial wastes. Domestic wastewater is made up of sewage, "gray water" (the water resulting from washing, bathing, and cooking), and wastewater from small-scale food processing in homes and restaurants.

Industrial wastewater includes liquid discharged from the petrochemical, pesticide, food and dairy, plastics, pulp and paper, pharmaceutical, and metallurgical industries. Industrial wastewater may contain toxic substances; some manufacturing and processing plants are required by the U.S. Environmental Protection Agency (EPA) to pretreat toxic or heavily contaminated discharges before they enter POTWs. Pretreatment may

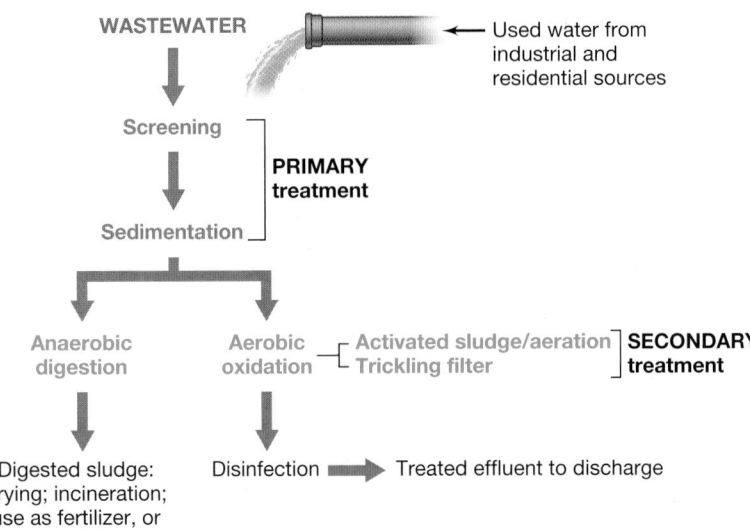

Figure 22.13 Wastewater treatment processes. Effective water treatment plants use the primary and secondary treatment methods shown here. Tertiary treatment may also be used to reduce nutrient levels in waters released to the environment, reducing biochemical oxygen demand (BOD), nitrogen, and phosphorus to very low to undetectable levels.

involve mechanical processes in which large debris is removed. Some wastewaters are pretreated biologically or chemically to remove highly toxic substances such as cyanide; heavy metals such as arsenic, lead, and mercury; or organic materials such as acrylamide, atrazine (a herbicide), and benzene. These substances are converted to less toxic forms by treatment with chemicals or microorganisms capable of neutralizing, oxidizing, precipitating, or volatilizing these wastes. The pretreated wastewater can then be released to the POTW.

Wastewater Treatment and Biochemical Oxygen Demand

The goal of a wastewater treatment facility is to reduce organic and inorganic materials in wastewater to a level that no longer supports microbial growth and to eliminate other potentially toxic materials. The efficiency of treatment is expressed in terms of a reduction in the **biochemical oxygen demand (BOD)**, the relative amount of dissolved oxygen consumed by microorganisms to completely oxidize all organic and inorganic matter in a water sample (⟳ Section 20.8). High levels of organic and inorganic materials in the wastewater result in a high BOD.

Typical values for domestic wastewater, including sewage, are approximately 200 BOD units. For industrial wastewater from sources such as dairy plants, the values can be as high as 1500 BOD units. An efficient wastewater treatment facility reduces BOD levels to less than 5 BOD units in the final treated water. Wastewater facilities are designed to treat both low-BOD sewage and high-BOD industrial wastes.

Treatment is a multistep operation employing a number of independent physical and biological processes (Figure 22.13). *Primary*, *secondary*, and sometimes additional treatments are

employed to reduce biological and chemical contamination in the wastewater, and each higher level of treatment employs more complex technologies.

Primary Wastewater Treatment

Primary wastewater treatment uses only physical separation methods to separate solid and particulate organic and inorganic materials from wastewater. Wastewater entering the treatment plant is passed through a series of grates and screens that remove large objects. The effluent is allowed to settle for a few hours. Solids settle to the bottom of the separation reservoir and the effluent is drawn off to be discharged or for further treatment (**Figure 22.14**).

Municipalities that provide only primary treatment, as is true for the city of Victoria (British Columbia, Canada), discharge extremely polluted water with high BOD into adjacent waterways; high levels of soluble and suspended organic matter and other nutrients remain in water following primary treatment. These nutrients trigger undesirable microbial growth, further reducing water quality. Most treatment plants employ secondary and even *tertiary* (Section 22.7) treatments to reduce the organic content of the wastewater before release to natural waterways. Secondary treatment processes use both aerobic and anaerobic microbial digestion to further reduce organic nutrients in wastewater.

Secondary Aerobic Wastewater Treatment

Secondary aerobic wastewater treatment uses oxidative degradation reactions carried out by microorganisms under *oxic* conditions to treat wastewater containing low levels of organic materials (**Figure 22.15a, b**). In general, wastewaters that originate from residential sources can be treated efficiently using only aerobic treatment. Several aerobic degradative processes can be used for wastewater treatment; *activated sludge* methods are the most common (Figure 22.15a, b). Here, wastewater is continuously mixed and aerated in large tanks. Slime-forming aerobic bacteria, including *Zoogloea ramigera* and others, grow and form aggregated

Figure 22.14 Primary treatment of wastewater. Wastewater is pumped into the reservoir (left) where solids settle. As the water level rises, the water spills through the grates to successively lower levels. Water at the lowest level, now virtually free of solids, enters the spillway (arrow) and is pumped to a secondary treatment facility.

(a)

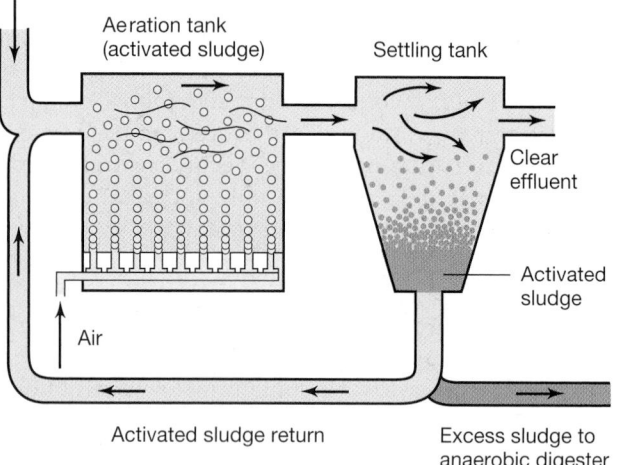

Wastewater from primary treatment

Aeration tank (activated sludge)

Settling tank

Clear effluent

Activated sludge

Air

Activated sludge return

Excess sludge to anaerobic digester

(b)

(c)

Figure 22.15 Secondary aerobic wastewater treatment processes. Parts *a* and *b* show the activated sludge method. *(a)* Aeration tank of an activated sludge installation in a metropolitan wastewater treatment plant. The tank is 30 m long, 10 m wide, and 5 m deep. *(b)* Wastewater flow through an activated sludge installation. Recirculation of activated sludge to the aeration tank introduces microorganisms responsible for oxidative degradation of the organic components of the wastewater. *(c)* Trickling filter method. The booms rotate, distributing wastewater slowly and evenly on the rock bed. The rocks are 10–15 cm in diameter and the bed is 2 m deep.

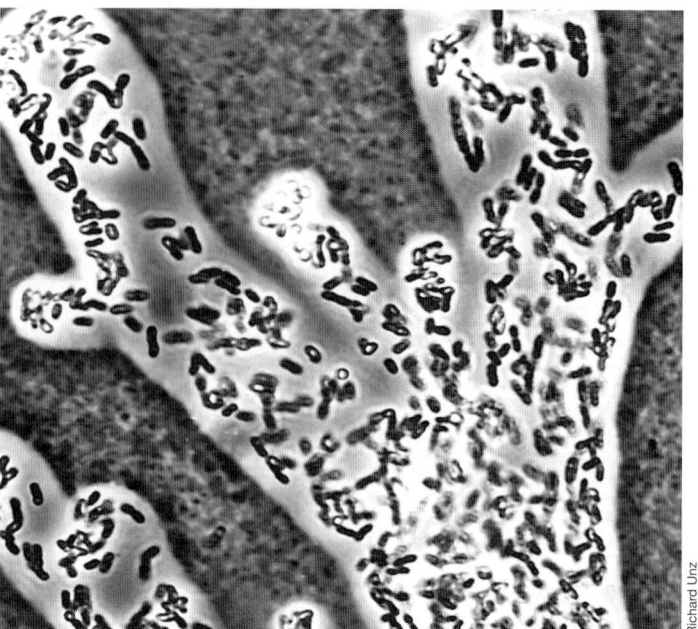

Figure 22.16 A wastewater floc formed by the bacterium *Zoogloea ramigera*. Floc formed in the activated sludge process consists of a large number of small, rod-shaped cells of *Z. ramigera* surrounded by a polysaccharide slime layer, arranged in characteristic fingerlike projections in this negative stain with India ink.

masses called flocs (**Figure 22.16**). The biology of *Zoogloea* is discussed in Section 16.2. Protists, small animals, filamentous bacteria, and fungi attach to the flocs. Oxidation of organic matter occurs on the floc as it is agitated and exposed to air. The aerated effluent containing the flocs is pumped into a holding tank or clarifier where the flocs settle. Some of the floc material (called activated sludge) is then returned to the aerator as inoculum for new wastewater, and the rest is pumped to an anaerobic sludge digester (see Figure 22.17), or removed, dried, and burned, or is used for fertilizer.

Wastewater normally stays in an activated sludge tank for 5–10 hours, a time too short for complete oxidation of all organic matter. However, during this time much of the soluble organic matter is adsorbed to the floc and incorporated by the microbial cells. The BOD of the liquid effluent is considerably reduced (up to 95%) when compared to the incoming wastewater; most of the material with high BOD is now in the settled flocs. The flocs can then be transferred to the anoxic sludge digester for conversion to CO_2 and CH_4.

Activated sludge treatment performance is sometimes diminished by the overgrowth of certain filamentous microbes (commonly members of the *Actinobacteria*, ↩ Section 16.12) that cause slow settling of the flocs, a problem called *sludge bulking*. This persistent problem, as well as the large reactor volumes required for activated sludge treatment, has prompted the development of new types of reactors that promote the growth of dense microbial aggregates, called *granular sludge*. The individual dense microbial granules (of several mm size) in granular sludge have both high metabolic activity and good gravity settling properties, greatly reducing the size and energy costs of a treatment facility. Although much of this work is still laboratory based, there should be a relatively rapid transfer of this technology to practice (see page 672).

The *trickling filter* is an alternative method of aerobic secondary treatment (Figure 22.15c). A trickling filter is a bed of crushed rocks, about 2 m thick. Wastewater is sprayed on top of the rocks and slowly passes through the bed. The organic material in the wastewater adsorbs to the rocks, and microorganisms grow and form biofilms on the large, exposed rock surfaces. The complete mineralization of organic matter to CO_2, ammonia, nitrate, sulfate, and phosphate takes place in the extensive microbial biofilm that develops on the rocks.

Most treatment plants chlorinate the effluent after secondary treatment to further reduce the possibility of biological contamination. The treated effluent can then be discharged into streams or lakes. In the eastern United States, many wastewater treatment facilities use ultraviolet (UV) radiation to disinfect effluent water. Ozone (O_3), a strong oxidizing agent that is an effective bactericide and viricide (⇄ Sections 5.15–5.17), is also used for wastewater disinfection in some treatment plants in the United States.

Secondary or Tertiary Anaerobic Treatment

Anaerobic treatment involves a series of catabolic reactions carried out by various *Bacteria* and *Archaea* under *anoxic* conditions. Anaerobic treatment is typically used to treat wastewater containing large quantities of insoluble organic matter (and therefore having a very high BOD) such as fiber and cellulose waste from food and dairy plants. It is also used for additional treatment of the sludge originating from secondary aerobic wastewater treatment (Figure 22.15b). In that case, the additional step is called **tertiary treatment**, defined as any treatment process in which unit operations are added for the further processing of the secondary treatment effluent or solids.

The anaerobic degradation process is carried out in large, enclosed tanks called *sludge digesters* or *bioreactors* (**Figure 22.17**). The process requires the collective activities of many different microbes and the major reactions are summarized in Figure 22.17c. First, anaerobes use polysaccharidases, proteases, and lipases to digest suspended solids and large macromolecules into soluble components. These soluble components are then fermented to yield a mixture of fatty acids, H_2, and CO_2; the fatty acids are further fermented by the cooperative actions of syntrophic bacteria (⇄ Sections 14.23 and 21.2) to produce acetate, CO_2, and H_2. These products are then used as substrates by methanogenic *Archaea* (⇄ Sections 17.2 and 21.2), fermenting acetate to produce methane (CH_4) and CO_2, the major products of anoxic sewage treatment (Figure 22.17c). The CH_4 is burned off or used as fuel to heat and power the wastewater treatment plant.

MINIQUIZ

- What is biochemical oxygen demand (BOD), and why is its reduction important in wastewater treatment?
- How do primary and secondary wastewater treatment methods differ?
- Other than treated water, what are the final products of wastewater treatment? How might these end products be used?

22.7 Advanced Wastewater Treatment

Advanced wastewater treatment is any process designed to produce an effluent of higher quality than normally achieved by secondary treatment. This includes tertiary treatment, physical–chemical treatment, or combined biological–physical treatment. Typical goals of advanced treatment include additional removal of organic matter and suspended solids, removal of key inorganic nutrients required for microbial growth (including ammonia, nitrate, nitrite, phosphorus, and dissolved organic carbon), and degradation of any potentially toxic materials. Advanced water treatment is the most complete method of treating sewage but has not been widely adopted because of the costs associated with such complete nutrient removal. Here we examine biological removal of phosphorus, nitrogen, and trace contaminants, three areas of advanced treatment of increasing importance to wastewater treatment.

Biological Phosphorus Removal

Conventional secondary biological treatment removes only about 20% of phosphorus from wastewater, necessitating additional

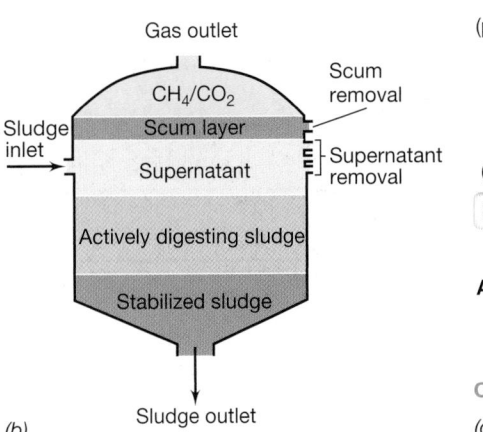
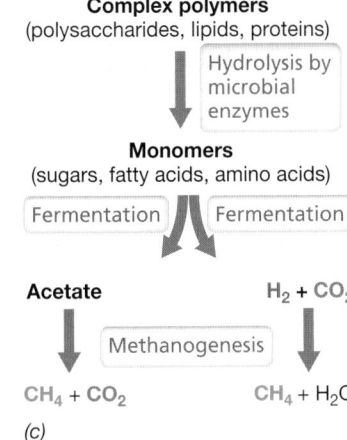

Figure 22.17 Anaerobic treatment. *(a)* Anaerobic sludge digester. Only the top of the tank is shown; the remainder is underground. *(b)* Inner workings of a sludge digester. *(c)* Major microbial processes in anaerobic sludge digestion. Methane (CH_4) and carbon dioxide (CO_2) are the major products of anaerobic biodegradation.

chemical or biological treatment. Chemical precipitation is the most commonly used process, removing up to 90% of the influent phosphorus. Removal is accomplished by the addition of either Fe or Al as chloride or sulfate salts, with Fe^{2+} or Fe^{3+} salts more commonly used. At near-neutral pH, the Fe^{3+} forms insoluble ferric phosphate ($FePO_4$) or ferric hydroxide-phosphate complexes. These then precipitate and are removed as sludge.

The chemical precipitation process results in up to 95% more sludge, contributing to additional disposal problems. As an alternative, tertiary treatment that encourages the growth of phosphorus-accumulating bacteria can also remove up to 90% of phosphorus, a process called *enhanced biological phosphorus removal* (*EBPR*). Here the waste stream is processed by sequential passage through anaerobic and aerobic bioreactors (**Figure 22.18**). In the anaerobic reactor, *phosphorus-accumulating organisms* (*PAOs*) use energy available from stored polyphosphate to assimilate short-chain fatty acids, and produce intracellular polyhydroxyalkanoates (PHAs) (Figure 22.18*a*; ⇌ Section 2.8); as this occurs, soluble orthophosphate (PO_4^{3-}) is released. During the following aerobic phase of treatment, the stored PHA is metabolized, providing energy and carbon for new cell growth. The energy is used to form intracellular polyphosphate, removing orthophosphate from solution (Figure 22.18*a*). The new biomass (sludge) with high polyphosphate content is then collected for phosphorus removal (Figure 22.18*b*).

The EBPR process sometimes fails as a result of the overgrowth of competing microbial populations, commonly microorganisms that accumulate glycogen as opposed to phosphorus, thus rendering the process less efficient. Hence, better control of the process will require improved understanding of the ecology and physiology of the PAOs. Recent progress in this area has been made with identification of one of the principal PAOs, the appropriately named bacterium *Accumulibacter phosphatis*. *A. phosphatis* is part of a clade of related phosphorus-accumulating *Betaproteobacteria* (⇌ Section 16.2) that have been identified in different EBPR systems. Although no pure cultures are yet available, laboratory reactor systems enriched in these organisms are now providing insight into operating conditions necessary for stable operation of the EBPR.

Biological Nitrogen Removal: The Conventional Process

The strict regulatory limitations on release of nitrogen as ammonia or nitrate/nitrite, also called *reactive nitrogen* (*Nr*), from wastewater treatment facilities reflects their adverse health effects on human and aquatic life and contribution to eutrophication of receiving water bodies (⇌ Section 20.8). Thus, there is increasing use of tertiary treatment to remove remaining Nr from wastewater following secondary treatment or anaerobic sludge digestion (Figure 22.15).

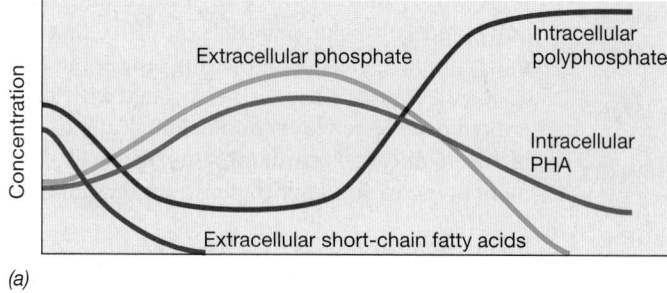

(a)

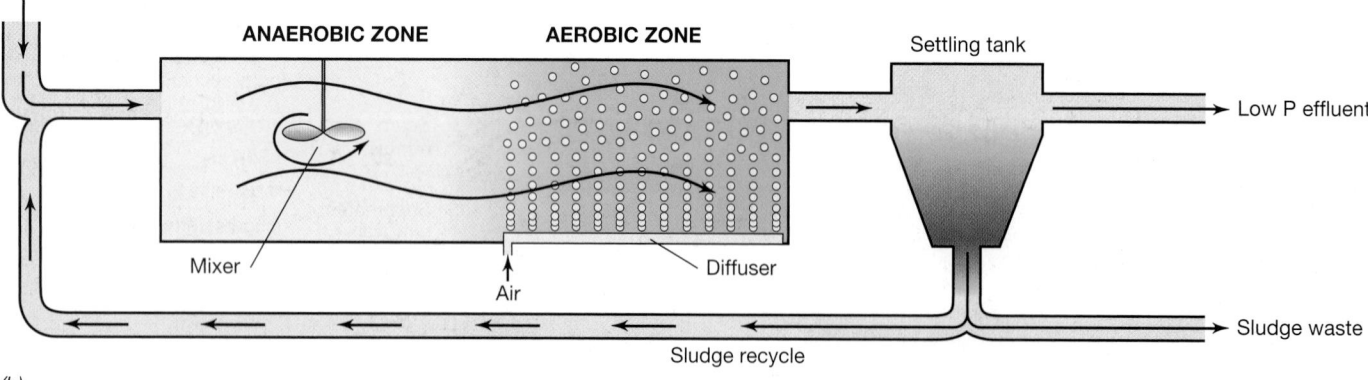

(b)

Figure 22.18 Enhanced biological phosphorus removal process. *(a)* Carbon and phosphate transitions during the treatment process. *(b)* Wastewater processing. During the passage of wastewater through the reactor system, the microbial community transitions from anaerobic to aerobic growth. In the anaerobic zone, short-chain fatty acids are taken up and internal stores of polyphosphate (polyP) are released as extracellular orthophosphate. In the aerobic zone, the extracellular phosphate is reassimilated as polyP and the intracellular stores of polyhydroxyalkanoates (PHAs) are metabolized. High-phosphorus sludge is harvested for disposal.

Conventional ("classical") treatment converts the Nr to its inert atmospheric form (N_2) using a combination of nitrification and denitrification (**Figure 22.19a**). Nitrification (⟺ Section 14.11) is first used to convert ammonia to nitrate followed by anaerobic conversion of nitrate to N_2 (and some N_2O) by denitrification (⟺ Section 14.13). In the following reactions, organic carbon is represented by chemical oxygen demand (COD, ⟺ Section 20.8), the mass of oxygen that is reduced by a specific amount of organic matter; for example, a COD of 4 g reduces 4 g of O_2 (0.125 moles) to water:

$$\text{Ammonia oxidation: } NH_4^+ + 1.5\,O_2 \rightarrow NO_2^- + H_2O + 2\,H^+$$

$$\text{Nitrite oxidation: } NO_2^- + 0.5\,O_2 \rightarrow NO_3^-$$

$$\text{Denitrification: } NO_3^- + 40\,g\,COD + H^+ \rightarrow 0.5\,N_2$$
$$+ 15\,g\,biomass$$

$$\text{Combined: } NH_4^+ + 2\,O_2 + 40\,g\,COD \rightarrow 0.5\,N_2 + H_2O + H^+$$
$$+ 15\,g\,biomass$$

Although the primarily autotrophic ammonia- and nitrite-oxidizing microbes require little or no reduced carbon for growth, a carbon source is required for the subsequent microbial reduction of nitrate to N_2 by denitrification. This may require supplementation with additional carbon for wastewaters having a low ratio of organic carbon to reactive nitrogen (C/N). Supplementation is commonly accomplished by adding a relatively inexpensive carbon source, such as methanol. However, because a typical treatment plant treats well over a million gallons (3.8 million liters) of wastewater a day, organic carbon addition can add significant costs, and this has fueled the development of less costly advanced treatment processes.

Biological Nitrogen Removal: Partial Nitrification–Denitrification and Anammox

The cost of Nr removal can be reduced through advanced treatment processes that limit the activities of nitrite-oxidizing bacteria. This leaves more of the oxidized NH_4^+ in the form of NO_2^- than NO_3^- and thus less organic matter is needed to denitrify:

$$\text{Ammonia oxidation: } NH_4^+ + 1.5\,O_2 \rightarrow NO_2^- + H_2O + 2\,H^+$$

$$\text{Denitrification: } NO_2^- + 24\,g\,COD + H^+ \rightarrow 0.5\,N_2$$
$$+ 9\,g\,biomass$$

$$\text{Combined: } NH_4^+ + 1.5\,O_2 + 24\,g\,COD \rightarrow 0.5\,N_2 + H_2O + H^+$$
$$+ 9\,g\,biomass$$

By limiting the oxidation of ammonia to nitrite in the wastewater, the requirement for carbon is reduced by ~40%, oxygen by ~25%, and biomass by ~40%. This makes nitrogen removal from low C/N wastewater less costly. The only complication to this treatment strategy is the need to suppress nitrite-oxidizing bacteria (NOB), since they generally have higher substrate utilization rates than the ammonia-oxidizing bacteria (AOB). Achieving cost-effective control of partial ammonia oxidation is therefore key to advanced nitrogen removal treatment systems.

Suppression of nitrite oxidation has been achieved by adjusting growth conditions that differentially affect AOB and NOB, primarily those of pH and temperature. AOB grow faster than NOB at temperatures above room temperature, having a specific growth rate (⟺ Section 5.4) approximately two times that of NOB at 35°C. For example, one widely used process relies on washing out NOB from a reactor operated at 30–40°C

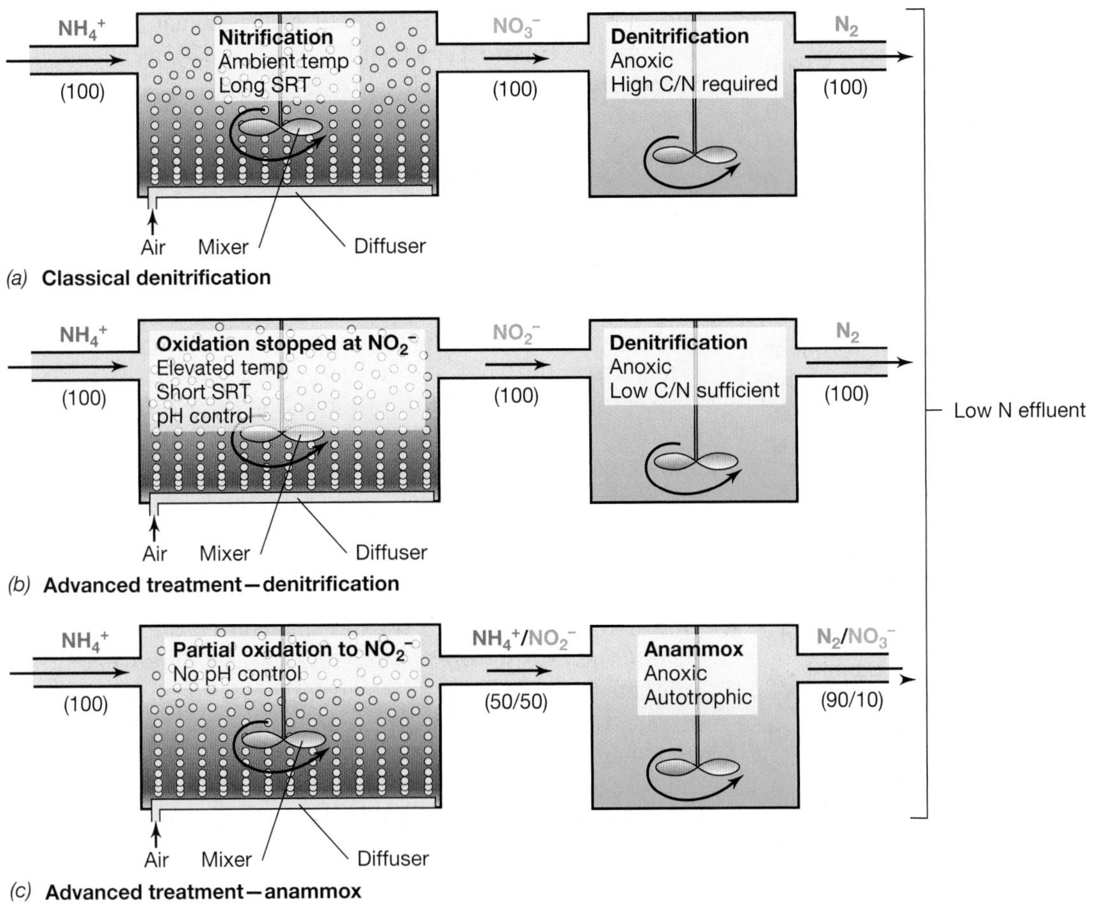

(a) **Classical denitrification**

(b) **Advanced treatment—denitrification**

(c) **Advanced treatment—anammox**

Figure 22.19 Alternative treatment processes for nitrogen removal from wastewater. Boxes show the conceptual—not full-scale operational—design of biological reactors and operating conditions that promote the indicated processes. (a) Classic denitrification process. (b, c) Advanced treatment processes for more efficient and economical removal of nitrogen. Numbers in parentheses correspond to percentages of each form of nitrogen entering and leaving a reactor. Since NO_2^- is also used as a source of electrons for autotrophic carbon fixation by anammox bacteria, the product of its oxidation (NO_3^-) is also generated in the anammox reactor. SRT, Sludge retention time.

UNIT 5

and a short sludge retention time of 1–1.5 days (Figure 22.19*b*). The retention time can be controlled by the rate the reactor is fed wastewater or by allowing sludge to settle during a period without mixing before withdrawing the treated effluent. Control of reactor pH provides another way to achieve partial nitrification, as an alkaline pH (7.5–8.5) favors the growth of AOB over NOB (Figure 22.19*b*).

Anammox bacteria (⮌ Section 14.12) are now increasingly used to remove Nr from concentrated ammonia streams of low C/N, such as is found in anaerobic sludge digester liquor and animal production wastewater. The following reaction includes the production of cellular biomass during growth of anammox bacteria:

$$\text{Anammox: } NH_4^+ + 1.32\,NO_2^- + 0.066\,HCO_3^- + 0.13\,H^+ \rightarrow$$
$$1.02\,N_2 + 0.26\,NO_3^- + 2.03\,H_2O + 0.066\,C_{biomass}$$

Since anammox bacteria are chemolithotrophic autotrophs and also anaerobes, they do not depend on organic carbon as an electron donor and O_2 is not needed in the reaction. Anammox also reduces the biomass generated and the emission of greenhouse gases (primarily CO_2 and N_2O, ⮌ Section 21.8). When pH is not controlled in the ammonia-oxidation process, the primary reactor will generate a nearly equal molar ratio of ammonia and nitrite. This is because after about half the ammonia is oxidized, the resulting drop in pH prevents further ammonia oxidation (Figure 22.19*c*). The N-rich effluent is then treated in a separate reactor that separates partial nitrification in the first reactor from anammox in the second (Figure 22.19*c*).

Contaminants of Emerging Concern

As one can imagine, a wide variety of chemicals end up in wastewaters. Until recently, studies of the environmental fate of chemicals in wastewaters have focused primarily on priority pollutants, including heavily used agricultural products and chemicals that demonstrate acute toxicity or carcinogenicity. However, it is now clear that new bioactive pollutants are entering the environment that will pose new challenges for microbial bioremediation, and most of these enter the environment in wastewaters. These pollutants include pharmaceuticals, active ingredients in personal care products, fragrances, household products, sunscreens, prescription and other medications, and many other unusual or xenobiotic molecules.

Unlike pesticides, these "new" pollutants are more or less continuously discharged into the environment primarily through release of treated or untreated sewage and, because of this, they do not need to persist to have environmental effects. For example, it is known that low levels of synthetic estrogen compounds, excreted in the urine of women taking birth control pills and eventually discharged in active form from wastewater treatment plants, can activate estrogen response genes in aquatic animals such as fish and contribute to the feminization of males.

Wastewater treatment plants were originally designed to handle natural materials, primarily human and industrial wastes, but now there is a growing interest in carefully researching the design of a new generation of treatment facilities to stimulate bioremediation of these emerging contaminants. Because these contaminants are often present in very low concentrations and are often new classes of xenobiotic chemicals, they may not actually support microbial growth but rather may be degraded only by cometabolism (Section 22.5) or by highly specialized species. We can therefore expect that the bioremediation of emerging contaminants will be an active area of microbiological research and environmental public policy in coming years.

— MINIQUIZ —

- What are the advantages of enhanced biological phosphorus removal (EBPR) relative to traditional chemical removal of phosphorus? Are there any disadvantages?
- Why is the incomplete oxidation of ammonia useful in tertiary treatment of wastewater? What advantages does the anammox process have over classical wastewater treatments for N removal?
- Give an example of an "emerging" contaminant.

22.8 Drinking Water Purification and Stabilization

Wastewater treated by secondary methods can usually be discharged into rivers and streams. However, such water is not **potable** (safe for human consumption). The production of potable water requires further treatment to remove potential pathogens, eliminate taste and odor, reduce nuisance chemicals such as iron and manganese, and decrease **turbidity**, which is a measure of suspended solids. **Suspended solids** are small particles of solid pollutants that resist separation by ordinary physical means.

Intestinal infections due to waterborne pathogens are still common, even in developed countries (⮌ Section 32.1), and some estimates indicate that waterborne diseases impact the health of several million people each year in the United States alone. Water treatment practices, however, have significantly improved access to safe water, starting with public works projects coupled with the application and development of water microbiology in the early twentieth century.

A century ago, water purification in the United States was limited to *filtration* to reduce turbidity, and this resulted in high rates of waterborne disease. Although filtration significantly decreased the microbial load of water, many microorganisms still passed through the filters. However, around 1913, **chlorination** using Cl_2 came into use as a disinfectant for large water supplies. Chlorine gas was an effective and inexpensive general disinfectant for drinking water, and its use quickly reduced the incidence of waterborne disease (⮌ Section 32.2). Major improvements in public health in the United States were largely due to the adoption of water filtration and disinfection treatment procedures. Public works engineering and microbiology working hand in hand were thus the major contributors to the dramatic advances in public health in the United States and other developed countries in the twentieth century.

Physical and Chemical Purification

A typical city installation for drinking water treatment is shown in **Figure 22.20a**. Figure 22.20*b* shows the process that purifies **raw water** (also called **untreated water**) that flows through the treatment plant. Raw water is first pumped from the source, in this

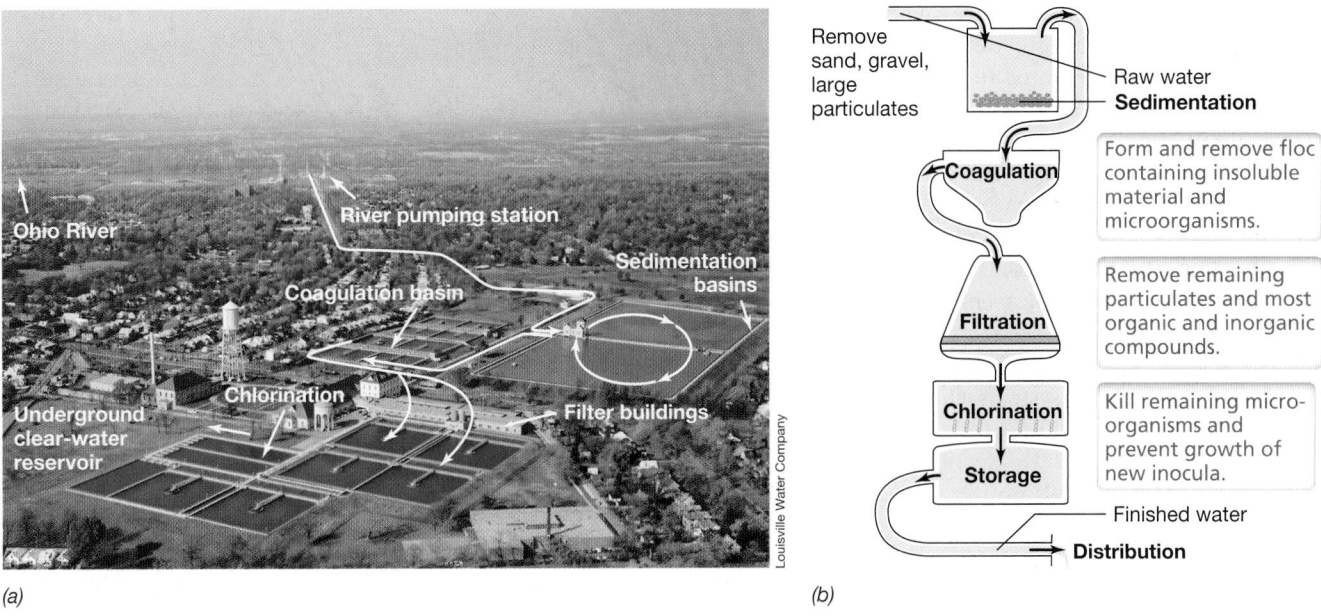

Figure 22.20 Water purification plant. *(a)* Aerial view of a water treatment plant in Louisville, Kentucky, USA. The arrows indicate direction of flow of water through the plant. *(b)* Schematic overview of a typical community water purification system.

case a river, to a sedimentation basin where anionic polymers, alum (aluminum sulfate), and chlorine are added. **Sediment**, including soil, sand, mineral particles, and other large particles, settles out. The sediment-free water is then pumped to a **clarifier** or coagulation basin, which is a large holding tank where **coagulation** takes place. The alum and anionic polymers form large particles from the much smaller suspended solids. After mixing, the particles continue to interact, forming large, aggregated masses, a process called **flocculation**. The large, aggregated particles (floc) settle out by gravity, trapping microorganisms and adsorbing suspended organic matter and sediment.

After coagulation, flocculation, and sedimentation, the clarified water undergoes **filtration** through a series of filters designed to remove organic and inorganic solutes, as well as remaining suspended particles and microorganisms. The filters typically consist of thick layers of sand, activated charcoal, and ion exchangers. After this step and the previous purification steps, the filtered water is free of particulate matter, most organic and inorganic chemicals, and nearly all microorganisms.

Disinfection

Clarified, filtered water must be disinfected before it is released to the supply system as pure, potable **finished water**. **Primary disinfection** is the introduction of sufficient disinfectant into clarified, filtered water to kill existing microorganisms and inhibit further microbial growth. Chlorination is the most common method of primary disinfection. In sufficient doses, chlorine kills most microorganisms within 30 minutes. A few pathogenic protists such as *Cryptosporidium*, however, are not easily killed by chlorine treatment (⇌ Section 33.4). In addition to killing microorganisms, chlorine oxidizes and effectively neutralizes many organic compounds. Since most taste- and odor-producing

chemicals are organic compounds, chlorination thus improves water taste and smell as well as disinfecting.

Chlorine is added to water either from a concentrated solution of sodium hypochlorite or calcium hypochlorite, or as chlorine gas from pressurized tanks. Chlorine gas is commonly used in large water treatment plants because it is most amenable to automatic control. When dissolved in water, chlorine gas is extremely volatile and dissipates within hours from treated water. To maintain adequate levels of chlorine for primary disinfection, many municipal water treatment plants introduce ammonia gas with the chlorine to form the more stable, nonvolatile chlorine-containing compound **chloramine**, $HOCl + NH_3 \rightarrow NH_2Cl + H_2O$.

Chlorine is consumed when it reacts with organic materials. Therefore, sufficient quantities of chlorine must be added to finished water containing organic materials so that a small amount, called the *chlorine residual*, remains. The chlorine residual reacts to kill any remaining microorganisms. The water plant operator performs chlorine analyses on the treated water to determine the level of chlorine to be added for **secondary disinfection**, the maintenance of sufficient chlorine residual or other disinfectant residual in the water distribution system to inhibit microbial growth. A chlorine residual level of 0.2–0.6 mg/liter is suitable for most water supplies. After chlorine treatment, the now potable water is pumped to storage tanks from which it flows by gravity or pumps through a **distribution system** of storage tanks and supply lines to the consumer. Residual chlorine levels inhibit growth of bacteria in the finished water prior to the water reaching the consumer. It does not protect against catastrophic system failures such as a broken pipe in the distribution system.

Ultraviolet (UV) radiation is also used as an effective means of disinfection. As we discussed in Section 5.16, UV radiation is used to treat secondarily treated effluent from water treatment plants.

In Europe, UV irradiation is commonly used for drinking water applications, and it is increasingly used in the United States. For disinfection, UV light is generated from mercury vapor lamps. Their major energy output is at 253.7 nm, a wavelength that is bactericidal and may also kill cysts and oocysts of protists such as *Giardia* and *Cryptosporidium*, important eukaryotic pathogens in water (↩ Section 33.4). Viruses, however, are more resistant.

UV radiation has several advantages over chemical disinfection procedures like chlorination. First, UV irradiation is a physical process that introduces no chemicals into the water. Second, UV radiation–generating equipment can be used in existing flow systems. Third, no disinfection by-products are formed with UV disinfection. Especially in smaller systems where finished water is not pumped long distances or held for long periods (reducing the need for residual chlorine), UV disinfection may be preferable to reduce dependence on chlorination.

MINIQUIZ

- What specific purposes do sedimentation, coagulation, filtration, and disinfection accomplish in the drinking water treatment process?
- What general procedures are used to reduce microbial numbers (microbial load) in water supplies?
- What are the advantages of UV disinfection versus, or as a complement to, chemical disinfection with chlorine?

22.9 Water Distribution Systems

Once drinking water leaves the treatment facility, the water often travels through many miles of municipal and premise distribution pipes from the facility to the consumer (**Figure 22.21**). In addition to taste and odor problems often associated with source water, the long transit and residence times may also contribute to undesirable taste and odors from biological and chemical processes. Although undesirable, taste and odor alone usually do not signal a health threat. However, water distribution systems may also promote the growth of obligate or opportunistic pathogens (↩ Section 25.4), sequester and protect pathogens, or select for more pathogenic and resistant forms of microorganisms. Even though drinking water–associated disease often goes unreported, in the United States alone in 2009 and 2010, 33 disease outbreaks associated with drinking water affected over 1000 persons and were linked to nine deaths.

The Microbiology of Municipal Water Distribution Systems

Microbial growth in drinking water distribution systems can be eliminated only through complete nutrient removal (elimination of growth substrates originating from the source water and from distribution system structural materials) or by maintaining appropriate residual chlorine levels throughout the distribution system. In reality, neither of these is attainable. Growth is unavoidable as a consequence of reduction in chlorine concentration with increasing distance from the point of production together with the tendency for microorganisms to form biofilms on the pipe

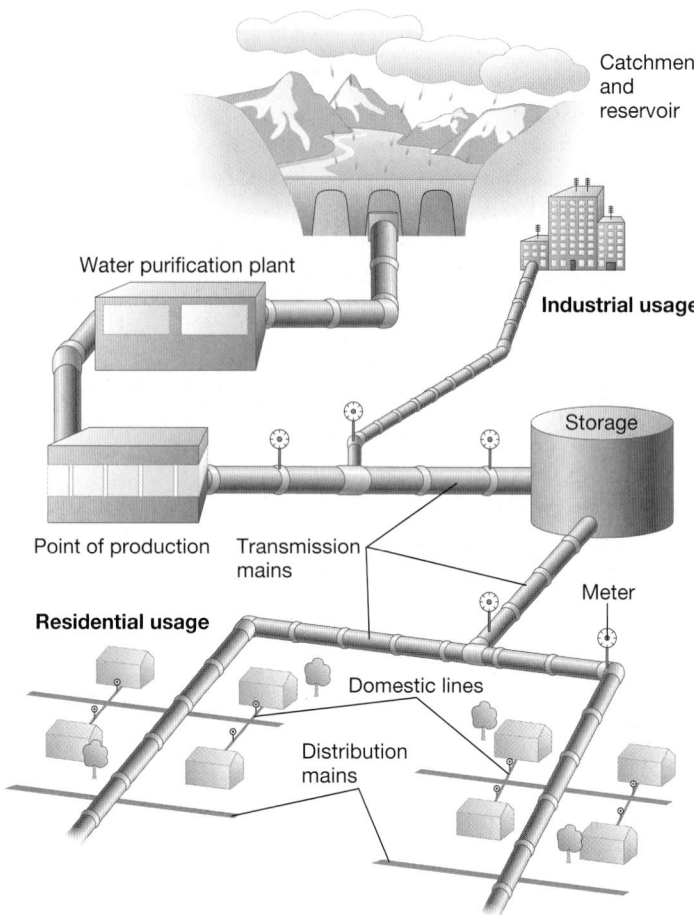

Figure 22.21 Drinking water distribution system. A municipal distribution system includes a surface reservoir, water purification plant, distribution mains, and domestic lines that encompass many miles of pipes in a typical community.

walls. Microorganisms in biofilms are more resistant to disinfection (↩ Sections 7.9 and 20.4) and significant microbial accumulation is found in all distribution systems, over 90% of which is in the form of biofilms that coat the pipe walls.

Only recently have culture-independent molecular techniques, including 16S rRNA sequence analysis (↩ Section 19.6), begun to fully resolve the species that commonly colonize water distribution pipes. Although these studies are showing that pathogenic species are rare, some opportunistic pathogens (↩ Section 25.4) are present and can infect susceptible humans, including infants and the elderly or individuals with compromised immune systems. Opportunistic pathogens that have been found in water distribution systems include (1) nontuberculous mycobacteria (including *Mycobacterium avium, M. intracellulare, M. kansasii,* and *M. fortuitum*) associated with many thousands of clinical cases each year in the United States; (2) *Legionella pneumophila* (the causative agent of Legionnaires' disease, ↩ Section 32.4); (3) *Pseudomonas aeruginosa* (which can infect the eyes, ears, skin, and lungs); and (4) opportunistic protozoan pathogens such as *Naegleria* and *Acanthamoeba* (↩ Section 33.3) that can cause keratitis and encephalitis.

Because infection by these and other opportunistic pathogens is often of unclear origin and much waterborne disease goes unreported, the significance of water distribution systems as a source (or reservoir) for pathogenic microorganisms is unclear. However, because of the potential large-scale health risk, the issue of pathogens in drinking water has been receiving much greater attention in recent years, including the use of molecular microbial ecology (Chapter 19) to investigate the problem.

Water distribution systems also support numerous grazing protists that subsist by consuming bacteria. For example, as many as 300 amoebae/cm^2 have been observed in some water distribution systems. Bacteria that survive and replicate following ingestion by these protists are potentially also less susceptible to clearance by the mammalian immune system. The best example of this is *Legionella*, an opportunistic pathogen that has emerged as a relatively new public health risk because of its ability to establish residence and replicate in protists inhabiting water-handling systems (**Figure 22.22**), including premise plumbing, shower heads, and air-conditioning systems. The basic cellular mechanisms *Legionella* uses to gain entry and replicate in a broad variety of protists (including *Acanthamoeba, Hartmannella, Naegleria*, and *Tetrahymena*) also allow it to more easily infect human cells. It has even been suggested that protists have been the driving force in the evolution of pathogenic *Legionella*. Opportunistic pathogens now recognized to have the ability to survive or grow within protists include *Legionella, Pseudomonas*, and *Mycobacterium* species.

The Microbiology of Premise Water Distribution Systems

One of the best-recognized microbiological concerns with premise water is *L. pneumophila* (⇔ Section 32.4). This pathogen multiplies in premise water systems at temperatures between 20 and 46°C. It survives for months in drinking water and its survival is augmented by the presence of other bacteria and protozoa—in which intracellular growth is possible (Figure 22.22)—and also through sequestration in biofilms. Temperatures greater than 50°C lead to a decrease in numbers, and temperatures greater than 60°C result in rapid elimination (cell death). Thus, to

prevent growth of *L. pneumophila*, premise water must be kept below 20°C or above 50°C from storage units to the tap.

Nontuberculous mycobacteria (including the species *Mycobacterium avium, M. intracellulare, M. kansasii*, and *M. fortuitum*) are also more resistant to chlorine disinfection and protozoal grazing and are enriched in showerheads receiving municipal water that still maintains a chlorine residual. As yet the significance of showers as a reservoir of opportunistic pathogens is unknown. However, the increasing frequency of showering as opposed to bathing, and possible aerosolization of opportunistic pathogens through showering, has prompted additional research in this area. The general picture that is emerging is that changes in treatment processes and the architecture of water distribution systems, coupled with the aging condition of some systems, can compromise human health (see Figure 22.23*a*).

MINIQUIZ

- Trace the treatment of water through a drinking water treatment plant, from the inlet to the final distribution point (faucet).

- What features of a premise water distribution system might encourage the growth of *Legionella*? Suppress growth?

IV • Indoor Microbiology and Microbially Influenced Corrosion

Although one might think that being indoors protects a person from the microbial world, nothing could be farther from the truth. Indoor air, both in private dwellings and public places, as well as surfaces of the structures themselves, can be teeming with microbes. In addition, billions of dollars' worth of metal, stone, and concrete infrastructure in dwellings and buried pipes is lost every year from corrosion catalyzed by microbial activities. Microbial metabolism accelerates corrosion through alteration of pH or redox, production of corrosive metabolites, and creation of corrosive microenvironments in biofilms. In the final part of this chapter we examine the microbiology of indoor microbial ecosystems and a few cases in which the microbial contribution to corrosion is relatively well understood.

22.10 The Microbiology of Homes and Public Spaces

Humans in urban environments spend the majority of their life indoors. They share this indoor environment with a microbiota that inhabit the air, dust, surfaces, and ventilation and water systems (**Figure 22.23**). The health effects of indoor microbial exposure may be positive or negative (Figure 22.23*a*). Pathogens, such as antibiotic-resistant *Staphylococcus aureus*, may be elevated in the indoor environment as a consequence of shedding from human skin. In contrast, the increased incidence of allergies and autoimmune disorders in children in developed countries has

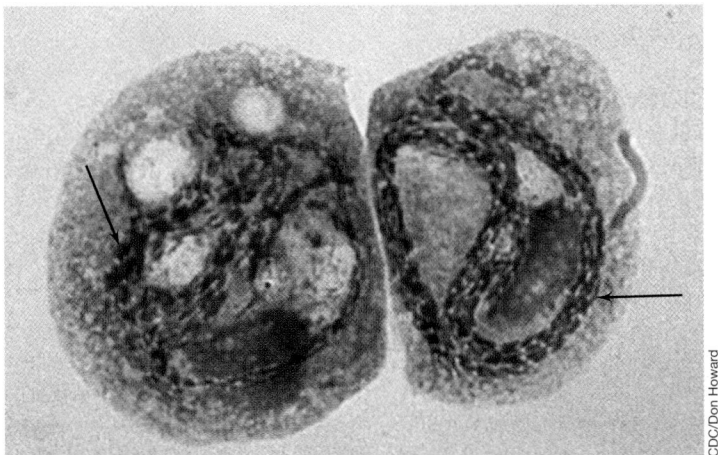

Figure 22.22 Protists as reservoirs of *Legionella*. Two cells of the protist *Tetrahymena* contain chains of the rod-shaped pathogen *Legionella pneumophila* (arrows). In premise water systems, protists can persist and be reservoirs of bacterial pathogens. *L. pneumophila* and legionellosis are discussed in Section 32.4.

(a)

(b)

Figure 22.23 Sources of airborne and surface-associated microorganisms in the built environment. *(a)* Sources of microorganisms in a typical household, including surfaces, humans, pets, plumbing systems, and outdoor air. The colors correspond to microorganisms or spores that may be beneficial or not harmful (green) or potentially detrimental (red) to human health. *(b)* Air particle collector (arrow) deployed in the New York City subway system for surveying the diversity of airborne bacteria by 16S rRNA gene sequencing (⮂ Section 19.6).

been in part attributed to the "too-clean" indoor environments that reduce exposure to microbes important for "training" the immune system early in life. Thus, a microbially depleted indoor environment may actually be detrimental to human health.

In addition to living with microbes in our home environment, what microbial exposures does a person experience in major public spaces, such as subway systems, the supermarket, or even the classroom? These questions can now be addressed using powerful culture-independent molecular methods to quantify microbial diversity and abundance patterns using 16S (or 18S) rRNA gene sequencing of DNA isolated from samples collected from different parts of the indoor environment (⮂ Section 19.6).

Microbiology of the Indoor Air in Private Dwellings

Although dedicated studies of premise microbiology are as yet limited, some general trends are emerging. A study of dust collected from upper trims of inside and outside doors of over a thousand homes throughout the USA revealed distinct indoor and outdoor microbial communities composed of fungi and bacteria. The outdoor fungi closely resemble the outdoor fungal populations found in different geographical regions, whereas the indoor bacterial communities more strongly reflect the type and number of occupants, including pets. The overall indoor diversity of bacteria and fungi is greater than that found outdoors, reflecting a mixing of indoor and outdoor sources. A small number of fungal species are more abundant inside the house, including common molds such as *Aspergillus, Penicillium, Alternaria,* and *Fusarium.* The incidence of these fungal populations increases with the age of the dwelling and whether or not a basement is present (basements are often damp and this increases mold abundance).

Bacteria that are characteristic of the human skin (gram-positive bacteria such as *Staphylococcus, Streptococcus, Corynebacterium,* and *Propionibacterium*), feces (*Bacteroides, Faecalibacterium, Ruminococcus*), or the vagina (*Lactobacillus, Bifidobacterium, Lactococcus*) are much more commonly found inside a house than

outdoors. The increased incidence of species of the bacteria *Prevotella, Porphyromonas, Moraxella,* and *Bacteroides* is typically associated with the presence of domestic animals—in particular dogs or cats—in a home. In fact, it is possible to predict with near certainty whether a home has these pets based only on a molecular analysis of household microbiota. Certain skin-associated taxa such as *Corynebacterium* and *Dermabacter* are elevated in homes with more men than women, possibly reflecting the known tendency for men to shed more skin than do women.

In addition to the surveys of dust described above, molecular surveys of home surfaces show that the microbial communities of human hands, noses, and bare feet (⮂ Sections 9.11 and 24.5, and Figure 9.31) closely resemble the organisms found on household surfaces, including floors, light switches, countertops, and door knobs (Figure 22.23). Moreover, the surface-associated microbiota of a home is highly predictive of a specific family, and the composition of the microbial community has been found to shift within days of a change in home occupants.

Microbiology of Public Places

Public buildings, office spaces, and transit systems are another important component of the built environment. The heavily trafficked New York City municipal subway system alone moves over 1.5 billion passengers a year. Similar to homes, subways and offices contain a mixture of airborne microorganisms sourced from humans and the outdoor air. Because air exchange in a subway system must be extensive, most airborne microorganisms in subway systems are typical of those found outdoors. In addition, however, about 5% of the microbial population in subway air is composed of microbes that normally reside on the feet, hands, arms, and heads of humans (⮂ Section 24.5); these are most likely shed from the more exposed areas of the subway ridership.

Indoor plumbing of public and private buildings is another well-recognized point of microbial exposure. Each flush of a toilet generates over 100,000 small (<5-μm) aerosol particles. Since aerosolized bacteria and viruses can remain viable for hours after they deposit on bathroom surfaces, flushing is a potential mechanism of enteric pathogen transmission as well as a means of transmitting harmless saprophytes from person to person. Transmission of human diseases caused by direct contact (such as sexually transmitted diseases) is not a major issue with bathroom fixtures because pathogens such as *Neisseria gonorrhoeae* (gonorrhea) and *Treponema pallidum* (syphilis) are very sensitive to drying.

However, because of the enormous numbers of bacteria shed in feces, transmission of enteric bacteria by bathroom aerosols is a distinct possibility.

As ongoing studies of indoor microbiology continue to reveal the types and origins of microorganisms we spend much of our day in contact with, we can also anticipate important future changes in construction that will enhance beneficial exposures and limit detrimental exposures to microorganisms in the built environment. At this point in our understanding, indoor air, per se, seems relatively innocuous. However, and depending greatly on the degree of cleanliness of the structure, certain areas in the home and public buildings can be more dangerous than others.

MINIQUIZ

- How can a microbial inventory reveal information about the presence or absence of household pets?
- Which room(s) in a private dwelling are potentially the most dangerous from a microbiology perspective, and why?

22.11 Microbially Influenced Corrosion of Metals

Iron is the most commonly used metal in the built environment. On a global basis, millions of miles of water, gas, and oil distribution pipelines are made of metal, and their corrosion contributes to the greatest loss of infrastructure in the built environment. Corrosion of iron by oxygen in air is thought to be solely an electrochemical process. However, much critical iron-containing infrastructure is buried or submerged, restricting exposure to oxygen. At near-neutral pH, in the absence of oxygen, corrosion of iron and steel is significantly accelerated by **microbially influenced corrosion (MIC)**. Microbial groups implicated in MIC include sulfate-reducing bacteria (⮠ Sections 14.14 and 15.9), ferric-iron-reducing bacteria (⮠ Sections 15.14 and 21.5),

ferrous-iron-oxidizing bacteria (⮠ Sections 14.10, 15.15, and 21.5), and methanogens (⮠ Sections 14.17, 17.2, and 21.2).

Metal Corrosion by Sulfate-Reducing Bacteria

Metal structures submerged in the marine environment and pipelines used for transmission of low-grade oil are particularly subject to MIC through the activities of sulfate-reducing bacteria. Corrosion by sulfate-reducing bacteria is partly attributable to the chemically corrosive nature of hydrogen sulfide (H_2S), the product of their metabolism. Crude oils containing more than about 0.5% sulfur by weight are called "sour" and may be naturally corrosive because of the H_2S that is present. In oil fields near the ocean, such as in the Middle East and Alaska, seawater is injected to maintain reservoir pressure and force oil into the producing well. Since seawater contains nearly 30 mM sulfate, an undesirable consequence of injection is further souring by stimulating the growth of sulfate-reducing bacteria.

A strategy now used by the petroleum industry to control souring is inclusion of nitrate (NO_3^-) in the injection water, stimulating the growth of nitrate-reducing bacteria. Since nitrate respiration is energetically more favorable than sulfate respiration (⮠ Sections 14.13 and 20.2), the nitrate reducers outcompete sulfate reducers for usable organic electron donors in the oil. Nitrate also stimulates the growth of sulfide-oxidizing, nitrate-reducing chemolithotrophs (⮠ Sections 14.9 and 15.11), thereby reversing souring by removing the sulfide.

Mechanisms of Metal Corrosion

At least two mechanisms have been described by which sulfate reducers could corrode iron. In the first mechanism, H_2 consumption by the sulfate reducer accelerates electrochemical pitting of the iron surface (**Figure 22.24a**). This model is based on the capacity of many sulfate reducers to use hydrogen (H_2) as an electron donor, thereby accelerating the energetically favorable but kinetically slow H_2 production originating from the chemical oxidation of iron ($Fe^0 + 2\,H^+ \rightarrow Fe^{2+} + H_2$). The overall stoichiometry of this reaction

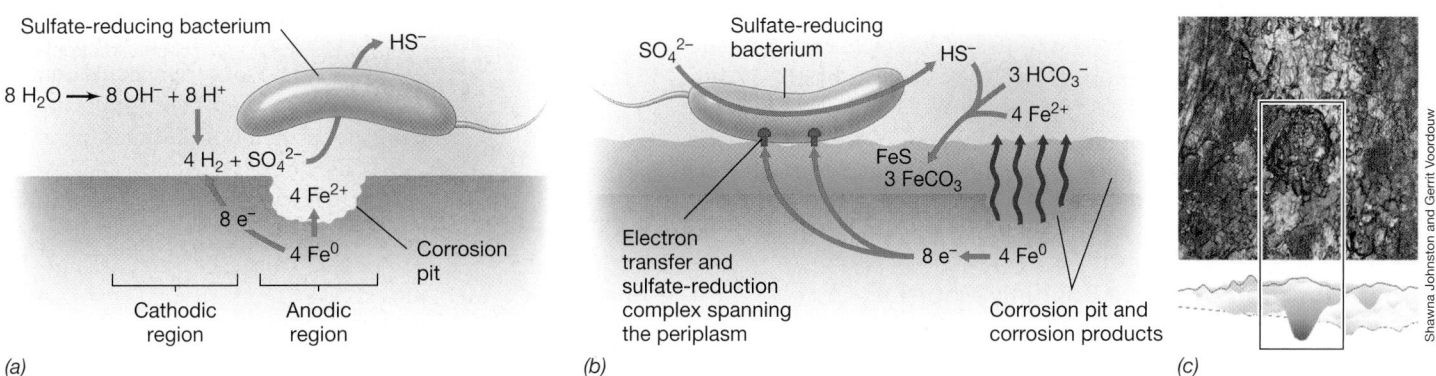

Figure 22.24 Corrosion of iron by sulfate-reducing bacteria. Two models for the activities of sulfate-reducing bacteria in metal corrosion. *(a)* Accelerating oxidation of metallic iron by consuming H_2 produced abiotically by proton reduction at the metal surface. *(b)* Direct electron transfer from the metal using electron-conductive outer cell wall structures connecting to an electron transfer system spanning the periplasm. *(c)* Top: photo of a model iron surface undergoing sulfidic corrosion. Bottom: scan of a side view of the metal surface in the photo revealing the areas where corrosion and pitting of the metal surface has occurred.

shows that Fe^{2+} formed from pitting reacts with sulfide from sulfate reduction and that the reaction is energetically favorable:

$$4\,Fe^0 + SO_4^{2-} + 3\,HCO_3^- + 5\,H^+ \rightarrow FeS + 3\,FeCO_3 + 4\,H_2O$$
$$(\Delta G^{0\prime} = -925\ kJ)$$

This mechanism, although feasible, has been questioned because H_2 formation from the iron surface at neutral pH is an intrinsic bottleneck, controlled by the limited availability of protons required for the reactions generating H_2.

Detailed electrochemical studies have shown that some sulfate-reducing bacteria, such as *Desulfopila corrodens*, have the capacity to take up electrons directly from the metal (Fe^0, Figure 22.24*b*). In this mechanism, the sulfate reducers attached to the metal surface engage in direct (cathodic) electron uptake from the metal through an electroconductive sulfidic corrosion layer (Figure 22.24*b*). A similar ability to take up electrons directly from Fe^0 has been observed for a *Methanobacterium* species that produces methane (CH_4) rather than sulfide from growth on Fe^0. The direct electron uptake model also suggests that associated with the cell surface are redox-active proteins, or other conductive structures, that conduct electrons from the corrosion layer to the cell. This represents yet another of a growing number of examples of the microbial use of conductive cellular structures for the oxidation or reduction of insoluble electron acceptors or electron donors, respectively (⟿ Sections 15.14 and 21.5, and see Explore the Microbial World, "Microbially Wired," in Chapter 21).

MINIQUIZ

- How does a nitrate addition prevent sulfide souring of crude oil?
- Why is accelerated microbial corrosion of iron metal thought to require a direct interaction between the sulfate reducers and the metal surface?

22.12 Biodeterioration of Stone and Concrete

In the same way that microorganisms contribute to soil formation through the dissolution of mineral and rock surfaces by their physical and metabolic activities (⟿ Section 20.6), buildings or other structures composed of natural stone or concrete are also subject to microbial colonization that may contribute to a slow loss of structural integrity through the microbes' metabolic activities. This degradative process is called *biodeterioration*.

Biodeterioration of Stone Building Materials

Microbial colonization of natural and structural stone building material is ubiquitous. Microbes can colonize the surface and penetrate several millimeters into rocky material depending on its physical characteristics (e.g., surface roughness, porosity, light penetration). Microbes can also grow on and within the facades of buildings constructed of limestone, sandstone, granite, basalt, and soapstone. These "within stone," or *endolithic*, communities are phylogenetically diverse, comprised of chemoorganotrophic and chemolithotrophic *Bacteria* and *Archaea*, microbial eukaryotes

including fungi and algae, and cyanobacteria. The cyanobacteria and algae primarily nourish the community, living in close or symbiotic association with other microbial members. For example, endolithic fungi have been observed to enclose the phototrophs in lichen-like associations (⟿ Figure 23.2).

Although not generally included in discussions of "extreme environments," life on and within stone building materials requires adaptation to multiple extreme conditions, including intense solar radiation, desiccation, temperature and moisture fluctuations, and lack of nutrients. Protection from solar radiation is conferred by production of UV-absorbing pigments (for example, melanin, mycosporines, and carotenoids) by fungi and other community members. The fungi also play a central role in this process of slow biodeterioration through the production of oxalic acid, which dissolves and mobilizes mineral constituents of the stone. Mineral dissolution and mobilization provides the communities with nutrients and increases habitability by enlarging pore spaces within the stone and thereby accelerating deterioration.

Crown Corrosion of Wastewater Distribution Systems

A very rapid form of microbial biodeterioration is observed in the **crown corrosion** of concrete sewer tiles, a process leading ultimately to the collapse of the pipe. Crown corrosion is a consequence of interactions between sulfate-reducing bacteria (⟿ Sections 14.14 and 15.9) and chemolithotrophic sulfur-oxidizing bacteria (⟿ Sections 14.9 and 15.11) in these underground wastewater transmission systems (**Figure 22.25**).

The first step in crown corrosion is the reduction of sulfate in the sewage to H_2S by sulfate-reducing bacteria, using primarily organic electron donors available in the waste stream water for

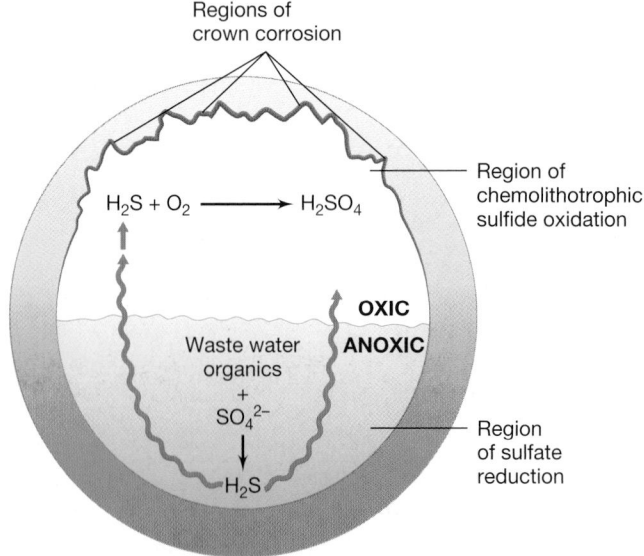

Figure 22.25 Crown corrosion of concrete sewer pipes. Corrosion is the result of a microbial sulfur cycle that develops within the transmission pipe. Sulfate-reducing bacteria consume organic material in the anoxic wastewater, producing H_2S. The latter is oxidized by sulfur-oxidizing chemolithotrophic bacteria that attach to the oxic upper (crown) pipe surface, accelerating corrosion from the production of H_2SO_4 (sulfuric acid).

sulfate reduction. The H_2S is then released into the headspace of the pipe where conditions are oxic. The sulfide, or partially oxidized intermediates such as thiosulfate or sulfur, is then oxidized by neutrophilic thiobacilli such as *Thiobacillus thioparus* (Section 15.11). As the pH drops to 4–5 with continued microbial production of sulfuric acid, acidophilic sulfur-oxidizing species such as *Acidithiobacillus thiooxidans* displace the neutrophilic species. Destruction and ultimate structural failure of the concrete results from the reaction of sulfuric acid with the free lime in the concrete, producing $CaSO_4·2H_2O$ (gypsum) that penetrates into the concrete. The gypsum then reacts with calcium aluminate in the concrete, leading to the production of the calcium aluminum sulfate mineral ettringite $[(CaO)_3·(Al_2O_3)·(CaSO_4)_3·(32H_2O)]$, which by increasing internal pressure contributes to cracking and further acceleration of the corrosion process.

A series of steps and microbial ecology similar to that of crown corrosion contributes to the corrosion of concrete holding tanks and cooling towers, particularly those in or near the marine environment where sulfate levels are typically high. In the United States alone such corrosion costs billions of dollars a year for replacement structures and control of the progressing corrosion.

— **MINIQUIZ** —

- How does the production of oxalic acid by fungi contribute to the deterioration of stone building materials?
- Prior to better regulatory control of metal release into domestic wastewater systems, crown corrosion of sewer tiles was less of a problem. Why?

MasteringMicrobiology®

Visualize, **explore**, and **think critically** with Interactive Microbiology, MicroLab Tutors, MicroCareers case studies, and more. MasteringMicrobiology offers practice quizzes, helpful animations, and other study tools for lecture and lab to help you master microbiology.

Chapter Review

I • Mineral Recovery and Acid Mine Drainage

22.1 The capacity of bacteria to oxidize Fe^{2+} aerobically at acidic pH is used to mine metals, principally copper-, uranium-, and gold-containing low-grade ores, through a process called microbial leaching. Bacterial oxidation of Fe^{2+} to Fe^{3+} is the key reaction in most microbial leaching processes because Fe^{3+} can oxidize sulfide ores, liberating extractable metals in the ores under either oxic or anoxic conditions.

> **Q** **Which crucial step in the oxidation of copper ores is carried out by *Acidithiobacillus ferrooxidans*? How is copper recovered from copper solutions produced by leaching?**

22.2 Spontaneous microbial oxidation of ferrous iron in pyritic ore or coal that has been exposed to air and water, such as occurs during some coal-mining operations, causes a type of pollution called acid mine drainage.

> **Q** **Which *Bacteria* and *Archaea* play a major role in acid mine drainage? Why do they carry out the reactions that they do? Why is air necessary for this process?**

II • Bioremediation

22.3 Although an inorganic pollutant such as uranium cannot be destroyed, containment is possible by reducing its mobility. For example, metal-reducing microorganisms in a region of uranium contamination can be stimulated to reduce U^{6+} to U^{4+}, forming the immobile uranium mineral *uraninite* that does not move into the groundwater.

> **Q** **What could thwart microbial bioremediation of a site that contains buried nuclear weapons that are leaking uranium?**

22.4 Hydrocarbons are excellent carbon sources and electron donors for bacteria and are readily oxidized when O_2 is available. Hydrocarbon-oxidizing bacteria bioremediate spilled oil, and their activities can be assisted by addition of inorganic nutrients.

> **Q** **What physical and chemical conditions are necessary for the rapid microbial degradation of oil in aquatic environments? Design an experiment that would allow you to test which conditions optimized the oil oxidation process.**

22.5 Some xenobiotics (chemicals new to nature) persist, whereas others are readily degraded, depending on their chemistries. Dechlorination is a major means of detoxifying xenobiotics that reach anoxic environments. With the exception of readily degradable microbial plastics, recalcitrant synthetic plastics are major environmental concerns.

> **Q** **Why are some bacterial transformations of xenobiotics possible only through cometabolism?**

III • Wastewater and Drinking Water Treatment

22.6 Sewage and industrial wastewater treatment reduces the BOD of wastewater. Primary, secondary, and tertiary wastewater treatment employs physical, biological, and physicochemical processes. After secondary or tertiary treatment, effluent water has significantly reduced BOD and is suitable for release into the environment.

> **Q** **Trace the treatment of wastewater in a typical plant from incoming water to release. What is the overall reduction in the BOD for typical household wastewater? What is the overall reduction in the BOD for typical industrial wastewater?**

22.7 Advanced wastewater treatment, such as enhanced biological phosphorus removal, is used to improve the quality of the treated wastewater. Of increasing concern are pharmaceuticals and ingredients in personal care products that are not degraded by conventional treatment systems and that can have adverse environmental effects even at very low concentrations.

> **Q** Why is advanced wastewater treatment desirable from an environmental point of view?

22.8 Drinking water purification plants employ industrial-scale physical and chemical systems that remove or neutralize biological, inorganic, and organic contaminants from natural, community, and industrial sources. Water purification plants employ clarification, filtration, and chlorination processes to produce potable water.

> **Q** Identify (stepwise) the process of purifying drinking water. What important contaminants are targeted by each step in the process?

22.9 The many miles of pipes for municipal drinking water distribution systems and premise plumbing have created new microbial habitats. Most microbes here are associated with the pipe walls as biofilms, resulting in a community that is more resistant to chlorine and that can sustain or sequester opportunistic pathogenic bacteria, such as *Mycobacterium, Legionella*, and *Pseudomonas*. The ability of some of these to grow within protist cells may increase their pathogenicity.

> **Q** What features of municipal and premise water distribution systems might contribute to a microbial health hazard? Why might showering increase your exposure to opportunistic pathogens?

IV • Indoor Microbiology and Microbially Influenced Corrosion

22.10 The indoor air and surfaces of dwellings and other buildings contain a diversity of mostly harmless saprophytic microbes that are for the most part a reflection of the humans and animals that reside there. However, the microbiota of certain parts of the indoor built environment, such as bathrooms and toilets, may contain enteric pathogens from aerosols generated there.

> **Q** From the perspective of a child's health, can a home be "too clean"? Explain.

22.11 Corrosion of metal structures exposed to the environment can be accelerated by microbial activity during microbially influenced corrosion. Structures in or near seawater are particularly prone to corrosion as a consequence of the direct and indirect activities of sulfate-reducing bacteria.

> **Q** What are the two general models for the acceleration of corrosion by sulfate-reducing bacteria?

22.12 Microbial contribution to the structural degradation of stone and concrete is called biodeterioration. Complex microbial communities colonize the stone and produce substances that dissolve and mobilize its mineral constituents. Crown corrosion of concrete sewer lines results from the concerted activities of sulfate-reducing and sulfur-oxidizing bacteria growing within the wastewater and the headspace of sewer pipes, respectively. The resulting sulfuric acid is primarily responsible for the destruction of the concrete.

> **Q** How does the involvement of sulfate-reducing bacteria differ for metal and crown corrosion?

Application Questions

1. Acid mine drainage is in part a chemical process and in part a biological process. Discuss the chemistry and microbiology that lead up to acid mine drainage and point out the key reactions that are biological. What ways can you think of to prevent acid mine drainage? How might you prevent further generation of acid drainage?

2. Why is reduction of BOD in wastewater a primary goal of wastewater treatment? What are the consequences of releasing wastewater with a high BOD into local water sources such as lakes or streams?

3. Discuss the microbial ecology contributing to crown corrosion of concrete sewer lines. In consideration of this ecology, what intervention strategies might be useful in reducing or eliminating corrosion?

Chapter Glossary

Acid mine drainage acidic water containing H_2SO_4 derived from the microbial and spontaneous oxidation of iron sulfide minerals released by coal mining

Anaerobic treatment degradative and fermentative reactions carried out by microorganisms under anoxic conditions to treat sludge solids or wastewater containing high levels of insoluble organic materials

Biochemical oxygen demand (BOD) the relative amount of dissolved oxygen consumed by microorganisms for complete oxidation of bioavailable organic and inorganic material in a water sample

Bioremediation the cleanup of oil, toxic chemicals, and other pollutants by organisms, usually microorganisms

Chloramine a disinfectant chemical manufactured on-site by combining chlorine and ammonia at precise ratios

Chlorination disinfecting water with Cl_2 at a sufficiently high concentration that a residual level is maintained throughout the distribution system

Clarifier a reservoir in which suspended solids in raw water are coagulated and removed through precipitation

Coagulation the formation of large, insoluble particles from much smaller, colloidal particles by the addition of aluminum sulfate and anionic polymers

Crown corrosion the destruction of the upper half, or crown, of concrete wastewater pipes by sulfuric acid produced through the concerted activities of sulfate-reducing and sulfur-oxidizing bacteria

Distribution system water pipes, storage reservoirs, tanks, and other equipment used to deliver drinking water to consumers or store it before delivery

Effluent water treated wastewater discharged from a wastewater treatment facility

Filtration the removal of suspended particles from water by passing it through one or more permeable membranes or media (e.g., sand, anthracite, or diatomaceous earth) and ion exchangers

Finished water water delivered to the distribution system after treatment

Flocculation the water treatment process after coagulation that uses gentle stirring to cause suspended particles to form larger, aggregated masses (flocs)

Microbial leaching the extraction of valuable metals such as copper from sulfide ores by microbial activities

Microbial plastics polymers consisting of microbially produced (and thus biodegradable) substances, such as polyhydroxyalkanoates

Microbially influenced corrosion (MIC) the contribution of microbial metabolic activities to accelerating the corrosion of metal and concrete structures

Potable drinkable; safe for human consumption

Primary disinfection the introduction of sufficient chlorine or other disinfectant into clarified, filtered water to kill existing microorganisms and inhibit further microbial growth

Primary wastewater treatment physical separation of wastewater contaminants, usually by separation and settling

Pyrite a common iron-containing ore, FeS_2

Raw water surface water or groundwater that has not been treated in any way (also called untreated water)

Reductive dechlorination an anaerobic respiration in which a chlorinated organic compound is used as an electron acceptor, usually with the release of Cl^-

Secondary aerobic wastewater treatment oxidative reactions carried out by microorganisms under aerobic conditions to treat wastewater containing low levels of organic materials

Secondary disinfection the maintenance of sufficient chlorine or other disinfectant residual in the water distribution system to inhibit microbial growth

Sediment soil, sand, minerals, and other large particles found in raw water

Sewage liquid effluents contaminated with human or animal fecal material

Suspended solid a small particle of solid pollutant that resists separation by ordinary physical means

Tertiary treatment any treatment process in which unit operations are added for the further processing of the secondary treatment effluent or solids

Turbidity a measurement of suspended solids in water

Untreated water surface water or groundwater that has not been treated in any way (also called raw water)

Wastewater liquid derived from domestic sewage or industrial sources, which cannot be discarded in untreated form into lakes or streams

Xenobiotic a synthetic compound not produced by organisms in nature

23

Microbial Symbioses with Microbes, Plants, and Animals

microbiology**now**

The Inner Life of Bees

Bees and other animal pollinators provide vital ecosystem services. Thus, the loss of nearly 60% of honeybee colonies in the USA since 1947 is a major concern. Several factors are thought to contribute to this decline including fragmentation of habitats, pesticides, pollution, land use changes, and climate change. The decline of bee populations has also drawn attention to the importance of the bee gut microbiome for combating pathogens and environmental stressors. For example, when the European bumblebee is experimentally deprived of its gut microbiota, the insect is much more susceptible to infection with a common protozoan parasite.

The gut communities of bumblebees and honeybees (photo) are surprisingly simple and consist of only five dominant and culturable bacterial species. These include two gram-negative *Proteobacteria* (*Snodgrassella alvi* and *Gilliamella apicola*), two *Lactobacillus* species, and a *Bifidobacterium* species. *S. alvi* and *G. apicola* pack the lumen of the honeybee gut (see transmission electron micrograph). Although different strains of each honeybee gut microbe are closely related in 16S rRNA gene sequence, their genomes have diverged significantly through coevolution with bees over millions of years. As a result, strains have become host-specific. For example, *S. alvi* isolated from honeybees will not colonize bumblebees, and vice versa.

Genome sequence analyses have also identified metabolic interdependencies among microbes and the bee. For instance, *S. alvi* oxidizes the fermentation products of *G. apicola*, and *G. apicola* has genes for the degradation of pectin, a polysaccharide in the cell wall of pollen grains. Because bees do not produce pectinase and pollen is a key component of their diet, bee gut microbiota are essential for utilization of this important food source. Some bee gut microbes also carry genes for the utilization of uncommon sugars such as arabinose and raffinose that are indigestible and potentially toxic to the bee.

The relative simplicity of the bee gut microbial community has made this insect an excellent model for animal–microbe associations in general and the microbiology of bee health in particular. Bees are major pollinators of fruits and other plants important in the human diet. Thus, studies of the bee microbiome highlight both the environmental and economic importance of developing a better understanding of the "microbial life" of bees.

Source: Kwong, W.K., P. Engel, H. Koch, and N.A. Moran. 2014. Genomics and host specialization of honey bee and bumble bee gut symbionts. *Proc. Natl. Acad. Sci. USA 111:* 11509–11514.

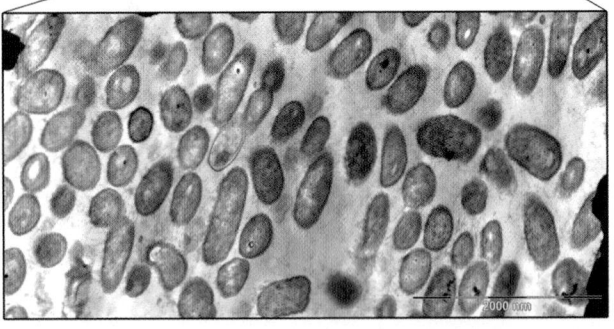

In this chapter we consider relationships of microorganisms with other microorganisms or with macroorganisms—prolonged and intimate relationships of a type called **symbiosis**, a word that means "living together." In Chapter 24 we examine microbial symbioses with humans.

Microorganisms that live within or on plants and animals are grouped according to how they affect their hosts. *Parasites* are microorganisms that benefit at some expense to the host, *pathogens* actually cause a disease in the host, *commensals* have no discernible effect on the host, and *mutualists* are beneficial to the host. In one way or another, all microbial symbioses benefit the microorganism. In this chapter we focus on **mutualisms**—relationships in which both partners benefit. We view the microorganisms as intimate evolutionary partners that influence both the evolution and physiology of their hosts. Most mutualistic symbioses of microbes with plants or animals had their origins many millions of years ago and have evolved to benefit the physiology of both partners, a process called **coevolution**. Over time, the changes in both partners may be so extensive that the symbiosis becomes obligate—neither the microbe nor the host can survive in the absence of the other.

I • Symbioses between Microorganisms

Many microbial species have intimate and beneficial associations with other microbial species. Direct microscopic observations of natural samples show that many microbes are not solitary entities but are associated with other microbes on surfaces or as suspended aggregates of cells. In most cases, the advantages conferred by an association are unknown. Because microbial ecologists have recognized that *communities* of interacting microbial populations—not individual organisms—control critical environmental processes, research to discover the nature of strictly microbial symbioses has increased. We present here two microbial mutualisms where the advantages to both partners are clear.

23.1 Lichens

Lichens are readily visible, leafy or encrusting microbial symbioses often found growing on bare rocks, tree trunks, house roofs, and bare soils—surfaces where other organisms typically do not grow (**Figure 23.1**). A lichen is a mutualistic association between two dominant microorganisms, a fungus, usually an ascomycete (⊂⊃ Section 18.12), and either an alga or a cyanobacterium. The alga or cyanobacterium is the phototrophic partner and produces organic matter that feeds the fungus. The fungus, unable to carry out photosynthesis, provides a firm anchor within which the phototrophic partner can grow, protected from erosion by rain or wind. Cells of the phototroph are embedded in defined layers or clumps among cells of the fungus (**Figure 23.2**). The characteristic morphology of any given lichen is primarily determined by the fungus, and many fungi (more than 18,000 named species) are able to form lichen associations. Diversity among the phototrophs is much lower, and thus many different kinds of lichens have the same phototrophic partner. Many cyanobacteria that partner in

Figure 23.1 Lichens. *(a)* A lichen growing on a branch of a dead tree. *(b)* Lichens coating the surface of a large rock.

lichens are nitrogen-fixing species, organisms such as *Anabaena* or *Nostoc* (⊂⊃ Sections 14.6 and 15.3).

The fungus clearly benefits from associating with the phototroph in the lichen symbiosis, but how does the phototroph benefit besides just having a sturdy substratum? *Lichen acids*, complex organic compounds secreted by the fungus, promote the dissolution and chelation of inorganic nutrients from the rock or other surface that are needed by the phototroph. Another role of the fungus is to protect the phototroph from drying; most of the habitats

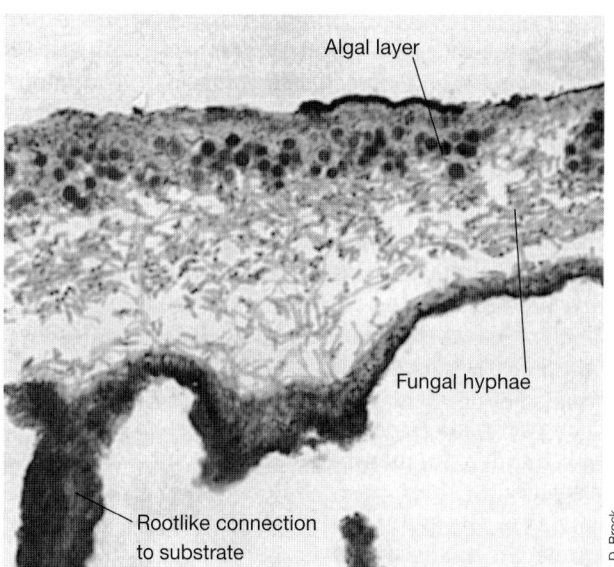

Figure 23.2 Lichen structure. Photomicrograph of a cross section through a lichen. The algal layer is positioned within the lichen structure near the top so as to receive the most sunlight. The fungal partners form the structural layers (cortex) positioned above and below the algal layer.

in which lichens live are dry, and fungi are, in general, better able to tolerate dry conditions than are the phototrophs. The fungus actually facilitates the uptake of water and sequesters some for the phototroph. Lichens typically grow quite slowly. For example, a lichen 2 cm in diameter growing on the surface of a rock may be several years old. Lichen growth varies from 1 mm or less per year to over 3 cm per year, depending on the organisms composing the symbiosis and the amount of rainfall and sunlight received.

The general view that lichens are simple two-partner assemblages has been challenged by recent culture-independent studies (Chapter 19) showing that lichens also host a bacterial and archaeal microbiota that may benefit the association by supplying additional nutrients such as vitamin B_{12}, protection from toxic compounds, and metabolites that have antimicrobial activity. Another remarkable recent discovery based on culture-independent analyses overturns the accepted understanding that only a single fungal species is a partner in this symbiosis. The cortex of many lichens, formed by fungal structural tissue bounding the phototroph layer (Figure 23.2), is now known to be composed of a basidiomycete yeast in addition to the known ascomycete (Chapter 18). This previously unrecognized yeast partner may account for the inability to reconstitute natural lichen thalli in axenic conditions from only the ascomycete and phototroph. These examples demonstrate how recent advances in culture-independent methods, despite being in an early stage, are bringing new understanding even to well-studied symbioses.

MINIQUIZ

- What two microbes form a partnership in the lichen symbiosis? What are the benefits to both partners?
- Besides organic compounds, of what benefit to the fungus is a mutualism with *Anabaena*?

23.2 "*Chlorochromatium aggregatum*"

Microbial mutualisms called **consortia** form in freshwater environments. A commonly observed consortium develops between nonmotile green sulfur bacteria (phototrophs that are colored either green or brown) and certain motile, nonphototrophic bacteria. These consortia are found worldwide in stratified sulfidic freshwater lakes and can account for up to 90% of the green sulfur bacteria present and nearly 70% of the bacterial biomass in these lakes. The basis of the mutualism of these consortia is in the phototrophic production of organic matter by the green sulfur bacterium and the motility and organic matter consumption of the chemotrophic partner organism. Each consortium has been given a genus and species name, but since these names do not denote true species (because they are not a single organism), the names are enclosed in quotation marks. We examined the general biology of these consortia in Section 15.6.

Nature of the Consortium

The morphology of a green sulfur bacterial consortium depends upon its species composition. The consortium generally consists of 13–69 green sulfur bacteria, called *epibionts*, surrounding and

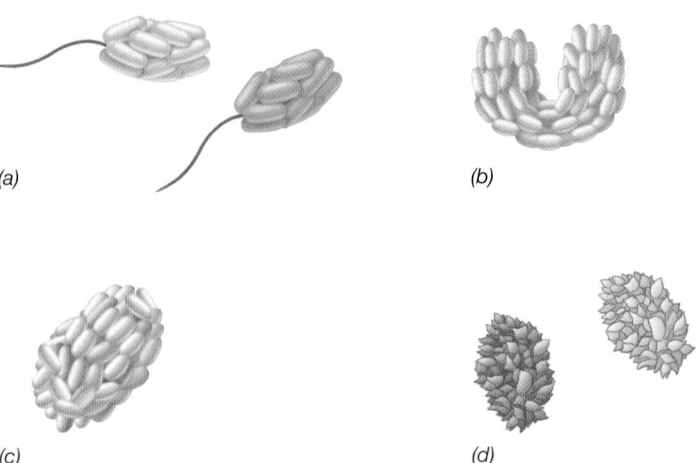

Figure 23.3 Drawings of some motile phototrophic consortia found in freshwater lakes. Green epibionts: *(a)* "*Chlorochromatium aggregatum*," *(b)* "*C. glebulum*," *(c)* "*C. magnum*," *(d)* "*C. lunatum*." Brown epibionts: *(a)* "*Pelochromatium roseum*," *(d)* "*P. selenoides*." The epibionts are 0.5–0.6 μm in diameter. Adapted from Overmann, J., and H. van Gemerden. 2000. *FEMS Microbiol. Rev. 24:* 591–599.

attached to a central, colorless, flagellated, rod-shaped bacterium (**Figure 23.3**). Several distinct motile phototrophic consortia have been recognized based on the color, morphology, and presence or absence of gas vesicles (Section 2.9) of the epibionts. For example, in "*Chlorochromatium aggregatum*" the central bacterium is surrounded by rod-shaped green bacteria, whereas in "*Pelochromatium roseum*" the epibiont is brown. The consortium "*Chlorochromatium glebulum*" is bent and includes green epibionts that contain gas vesicles (Figure 23.3).

Green sulfur bacteria are obligately anaerobic phototrophs that form a distinct phylum (*Chlorobi*, Section 15.6). The green and brown species differ in the types of bacteriochlorophyll and carotenoids they contain. Both green and brown species are found in stratified lakes where light penetrates to depths at which the water contains hydrogen sulfide (H_2S), the primary electron donor for photosynthetic CO_2 fixation by the phototroph. In stratified lakes, the motile consortia reposition rapidly to remain in regions where conditions are most favorable for photosynthesis in the constantly changing gradients of light, oxygen, and sulfide that occur throughout the course of a day. Water samples collected from depths where these conditions are most favorable are enriched in this morphologically conspicuous consortium (**Figure 23.4**). The consortia show dark aversion (scotophobotaxis, Section 2.13) and positive chemotaxis toward sulfide.

Some free-living green sulfur bacteria, such as *Pelodictyon* (*Chlorobium*) *phaeoclathratiforme*, produce gas vesicles that regulate buoyancy and vertical position in the water column. However, the time they require for repositioning in the water column is from one to several days, which is not fast enough for tracking the more rapidly changing gradients. By contrast, motile consortia move up and down in the water column fast enough to follow the gradients of light and sulfide as they change on a diel basis.

Although green bacterial consortia were discovered almost a century ago, only with the advent of molecular techniques and newer culture methods has it become possible to study certain aspects of

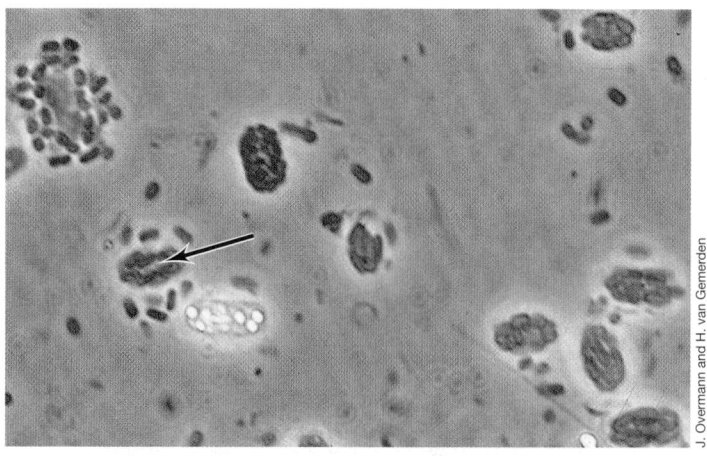

J. Overmann and H. van Gemerden

Figure 23.4 Phase-contrast micrograph of "*Pelochromatium roseum*" from Lake Dagow (Brandenburg, Germany). The preparation was compressed between a coverslip and microscope slide to reveal the central rod-shaped bacterium (arrow). A single consortium is about 3.5 µm in diameter. Used with permission from J. Overmann and H. van Gemerden. 2000. *FEMS Microbiol. Rev. 24*: 591–599.

these remarkable associations. Sequencing of 16S ribosomal RNA (rRNA) genes revealed a significant biogeography of epibionts in lakes of Europe and the United States. *Biogeography* is the study of the geographic distribution of organisms, in this case, the genetically distinct phototrophic consortia in different lakes. Epibionts in neighboring lakes have identical 16S rRNA gene sequences, whereas the sequences of morphologically similar epibionts in widely separated lakes differ. Phylogenetic analysis has shown that the mechanisms of cell–cell recognition responsible for stable morphology have evolved between particular epibionts and their central bacterium.

Phylogeny and Metabolism of a Consortium

The epibiont of "*Chlorochromatium aggregatum*" has been isolated and grown in pure culture. Although this green sulfur bacterium, named *Chlorobium chlorochromatii*, can be grown in pure culture, no naturally free-living variant has been observed, supporting the view that in nature, a symbiotic lifestyle is obligate for epibionts. The central bacterium of "*Chlorochromatium aggregatum*" belongs to the *Betaproteobacteria* (⟿ Section 16.2). Interestingly, this bacterium requires α-ketoglutarate, an intermediate of the citric acid cycle (⟿ Section 3.9), and this is presumably supplied to it by the epibiont. However, the central cell only assimilates fixed carbon in the presence of light and sulfide—conditions in which the epibionts are active and can transfer nutrients to the central bacterium. Genomic analysis of the central bacterium of one consortium revealed massive gene loss, indicating that this organism is unable to grow independently of the green sulfur bacterium.

Recent studies comparing the transcriptome and proteome (⟿ Sections 9.9 and 9.10) of *C. chlorochromatii* growing alone or in association with the central rod bacterium have identified some features specifically related to the symbiosis. Approximately 50 proteins are unique to the symbiotic state. Most of approximately 350 differentially regulated genes are repressed when the organism is symbiotically associated, whereas only 19 genes are more highly expressed. Many of the more highly expressed genes encode proteins of amino acid metabolism and nitrogen regulation.

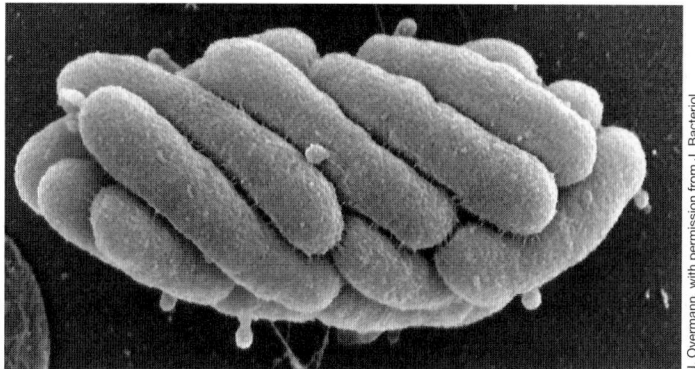

(a)

J. Overmann, with permission from J. Bacteriol

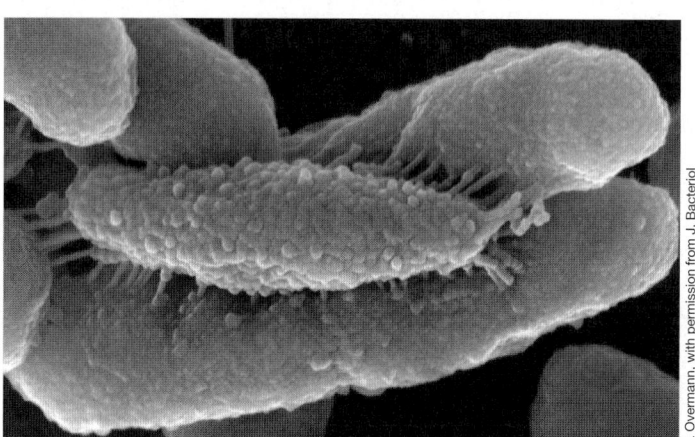

(b)

J. Overmann, with permission from J. Bacteriol

Figure 23.5 Scanning electron micrographs of "*Chlorochromatium aggregatum*." *(a) Chlorobium chlorochromatii* epibionts tightly clustered around a flagellated central bacterium. *(b)* The central bacterium exhibits numerous protrusions of its outer membrane that make intimate contact with the epibionts, possibly fusing the periplasms of the two gram-negative organisms. Cells of the epibiont are about 0.6 µm in diameter. Used with permission from G. Wanner, *et al.* 2008. *J. Bacteriol. 190*: 3721–3730.

These include the enzyme glutamate synthase (⟿ Section 3.14) and an ABC transporter of branched amino acids, suggesting that the metabolic coupling between the epibiont and central rod bacterium involves the exchange of amino acids.

Although it is not yet known whether the central bacterium transfers any organic compounds to the epibiont, this hypothesis can be tested now that the genome sequence of the central bacterium is known. Scanning electron microscopy of the consortium (**Figure 23.5**) has revealed that tubular extensions of the central bacterium's periplasm (⟿ Section 2.5) cover much of its surface and appear to fuse with the periplasm of the epibiont. If the two bacterial partners actually share a common periplasmic space, this would facilitate the transfer of nutrients from phototroph to chemotroph. The fact that the central bacterium is unable to grow without its phototrophic partner (while the phototroph can be grown in pure culture), and that organic compounds are only assimilated by the chemotroph in the light, is strong evidence that nutrients flow from the phototroph to feed the chemotroph and that the chemotroph is obligatorily dependent on its phototrophic partner.

II • Plants as Microbial Habitats

Plants interact closely with microbes through their roots and leaf surfaces and even more intimately within their vascular tissue and cells. Most mutualisms between plants and microorganisms increase nutrient availability to the plants or defend them against pathogens. We consider three examples in the next three sections: (1) a mutualism where the nature of the symbiosis is understood in exquisite detail (root nodules), (2) a mutualism in which plants expand and interconnect their root system through association with a fungus (mycorrhizae), and (3) a symbiosis that is harmful to the plant (crown gall disease).

23.3 The Legume–Root Nodule Symbiosis

A plant–bacterial mutualism of great importance to humans is that of leguminous plants and nitrogen-fixing bacteria. *Legumes* are flowering plants that bear their seeds in pods and include such agriculturally important members as soybeans, clover, alfalfa, beans, and peas. These crops are key commodities for the food and agricultural industries, and the ability of legumes to grow without added nitrogen saves farmers millions of dollars in fertilizer costs yearly and reduces the polluting effects of fertilizer runoff.

The partners in a symbiosis are called *symbionts*, and most nitrogen-fixing bacterial symbionts of plants are collectively called *rhizobia*, derived from the name of a major genus, *Rhizobium*. Species of rhizobia are *Alpha-* or *Betaproteobacteria* (⇨ Sections 16.1 and 16.2) (**Figure 23.6**) that can grow freely in soil or infect leguminous plants and establish a symbiotic relationship. The same genus (or even species) of legume can contain both rhizobial and

Figure 23.7 Soybean root nodules. The nodules developed from infection by *Bradyrhizobium japonicum*. The main stem of this soybean plant is about 0.5 cm in diameter.

nonrhizobial strains. Infection of legume roots by rhizobia leads to the formation of **root nodules** (**Figure 23.7**) in which the bacteria fix gaseous nitrogen (N_2) (⇨ Section 14.6). Nitrogen fixation in root nodules accounts for a fourth of the N_2 fixed annually on Earth and is of enormous agricultural importance, as it increases the fixed nitrogen content of soil. Nodulated legumes can grow well on unfertilized bare soils that are nitrogen deficient, while other plants grow only poorly on them (**Figure 23.8**).

Leghemoglobin and Cross-Inoculation Groups

In the absence of its bacterial symbiont, a legume cannot fix N_2. Rhizobia, on the other hand, can fix N_2 when grown in pure culture under microaerophilic conditions (a low-oxygen environment is

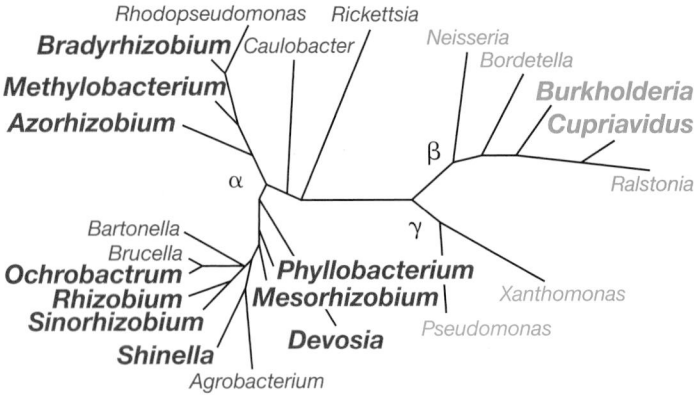

Figure 23.6 Phylogeny of rhizobial (names in boldface) and related genera inferred from analysis of 16S rRNA gene sequences. More than 70 species of rhizobia are found in 12 genera of *Alpha-* and *Betaproteobacteria*.

Figure 23.8 Effect of nodulation on plant growth. A field of unnodulated (left) and nodulated (right) soybean plants growing in nitrogen-poor soil. The yellow color is typical of chlorosis, the result of nitrogen starvation.

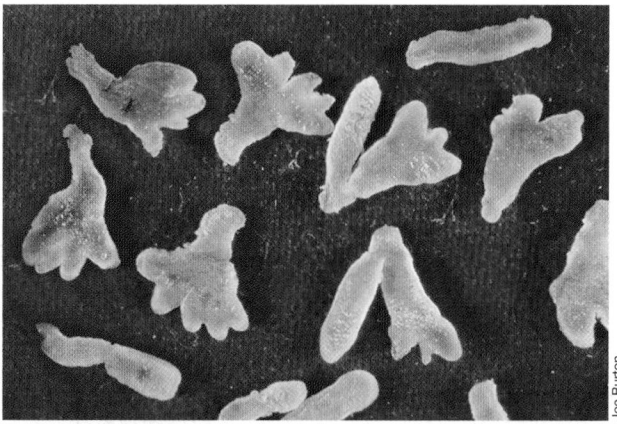

Figure 23.9 Root nodule structure. Sections of root nodules from the legume *Coronilla varia*, showing the reddish pigment leghemoglobin.

Joe Burton

TABLE 23.1 **Major cross-inoculation groups of leguminous plants**

Host plant	Nodulated by
Pea	*Rhizobium leguminosarum* biovar *viciae*[a]
Bean	*Rhizobium leguminosarum* biovar *phaseoli*[a]
Bean	*Rhizobium tropici*
Lotus	*Mesorhizobium loti*
Clover	*Rhizobium leguminosarum* biovar *trifolii*[a]
Alfalfa	*Sinorhizobium meliloti*
Soybean	*Bradyrhizobium japonicum*
Soybean	*Bradyrhizobium elkanii*
Soybean	*Sinorhizobium fredii*
Sesbania rostrata (a tropical legume)	*Azorhizobium caulinodans*

[a]Several varieties (biovars) of *Rhizobium leguminosarum* exist, each capable of nodulating a different legume.

necessary because the key nitrogen-fixing enzyme, called *nitrogenase*, is inactivated by high levels of O_2, ⮌ Section 14.6). In the nodule, O_2 levels are precisely controlled by the O_2-binding protein **leghemoglobin**. Production of this iron-containing protein in healthy N_2-fixing nodules (**Figure 23.9**) is induced through the interaction of the plant and bacterial partners. Leghemoglobin functions as an "oxygen buffer," cycling between the oxidized (Fe^{3+}) and reduced (Fe^{2+}) forms of iron to supply sufficient O_2 for bacterial respiration while keeping unbound O_2 within the nodule low. The ratio of leghemoglobin-bound O_2 to free O_2 in the root nodule is thus maintained on the order of 10,000:1.

There is a marked specificity between the species of legume and rhizobium that can establish a symbiosis. A particular rhizobial species is able to infect certain species of legumes but not others. A group of related legumes that can be infected by a particular rhizobial species is called a *cross-inoculation group*. Each group consists of all the legume species that will develop nodules when inoculated with rhizobia obtained from any other legume of the group (**Table 23.1**). If legumes are inoculated with the correct rhizobial strain, leghemoglobin-rich, N_2-fixing nodules develop on their roots (Figures 23.7–23.9).

Steps in Root Nodule Formation

How root nodules form is well understood for most rhizobia (**Figure 23.10**). The steps are as follows:

1. Recognition of the correct partner by both plant and bacterium and attachment of the bacterium to the root hairs

2. Secretion of oligosaccharide signaling molecules (Nod factors) by the bacterium

3. Bacterial invasion of the root hair

4. Movement of bacteria to the main root by way of the infection thread

5. Formation of modified bacterial cells (bacteroids) within the plant cells, development of the N_2-fixing state, and continued plant and bacterial cell division forming the mature root nodule

Another mechanism of nodule formation that does not require Nod factors is used by some species of phototrophic rhizobia. This mechanism has yet to be fully elucidated, but appears to require

UNIT 5

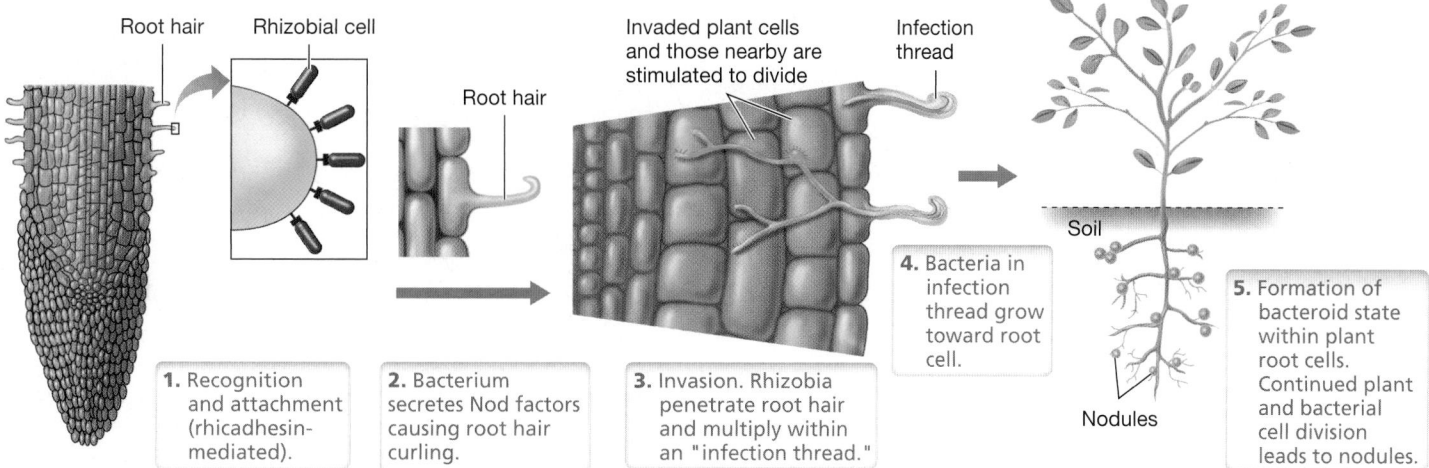

Root hair   Rhizobial cell

Root hair

Invaded plant cells and those nearby are stimulated to divide

Infection thread

Soil

4. Bacteria in infection thread grow toward root cell.

5. Formation of bacteroid state within plant root cells. Continued plant and bacterial cell division leads to nodules.

Nodules

1. Recognition and attachment (rhicadhesin-mediated).

2. Bacterium secretes Nod factors causing root hair curling.

3. Invasion. Rhizobia penetrate root hair and multiply within an "infection thread."

Figure 23.10 Steps in the formation of a root nodule in a legume infected by *Rhizobium*. Formation of the bacteroid state is a prerequisite for nitrogen fixation. The time course of nodulation events from infection to effective nodule is about 1 month for soybeans. See Figure 23.15 for physiological activities in the nodule.

the bacterial production of *cytokinins*. Cytokinins are plant hormones derived from adenine or phenylurea necessary for cell growth and differentiation.

Attachment and Infection

The roots of leguminous plants secrete organic compounds that stimulate the growth of a diverse rhizosphere microbial community. If rhizobia of the correct cross-inoculation group are in the soil, they will form large populations and eventually attach to the root hairs that extend from the roots of the plant (Figure 23.10). An adhesion protein called *rhicadhesin* is present on the cell surfaces of rhizobia. Other substances, such as carbohydrate-containing proteins called *lectins* and specific receptors in the plant cytoplasmic membrane, also play roles in plant–bacterium attachment.

After attaching, a rhizobial cell penetrates into the root hair, which curls in response to substances (Nod factors) secreted by the bacterium. The bacterium then induces formation by the plant of a cellulosic tube, called the **infection thread** (**Figure 23.11a**), which spreads down the root hair. Root cells adjacent to the root hairs subsequently become infected by rhizobia, and plant cells divide. Continued plant cell division forms the tumorlike nodule (Figure 23.11a) consisting of plant cells filled with bacteroids (discussed below and Figure 23.11b). A different mechanism of infection is used by some rhizobia adapted to aquatic or semiaquatic tropical legumes (see Figure 23.16). These rhizobia enter the plant at the loose cellular junctions of roots emerging perpendicular from an established root (*lateral roots*). Following entry into the plant, some of the rhizobia develop infection threads, whereas others do not.

Bacteroids

The rhizobia multiply rapidly within the plant cells and become transformed into swollen, misshapen, and branched cells called **bacteroids**. A microcolony of bacteroids becomes surrounded by portions of the plant cytoplasmic membrane to form a structure called the *symbiosome* (Figure 23.11c), and only after the symbiosome forms does N₂ fixation begin. Nitrogen-fixing nodules can

be detected experimentally by the reduction of acetylene to ethylene (⇄ Section 14.6). When the plant dies, the nodule deteriorates, releasing bacteroids into the soil. Although bacteroids are incapable of division, a small number of dormant rhizobial cells are always present in the nodule. These now proliferate, using some of the products of the deteriorating nodule as nutrients. The bacteria can then initiate infection the next growing season or maintain a free-living existence in the soil.

Nodule Formation: *nod* Genes, Nod Proteins, and Nod Factors

Rhizobial genes that direct the steps in nodulation of a legume are called *nod genes*. It is thought that the ability to form nodules has independently emerged multiple times through the horizontal transfer of such genes as *nod* and *nif* that are located on plasmids or transferable regions of chromosomal DNA. In *Rhizobium leguminosarum* biovar *viciae*, which nodulates peas, ten *nod* genes have been identified. The *nodABC* genes encode proteins that produce oligosaccharides called **Nod factors**; these induce root hair curling and trigger cell division in the pea plant, eventually leading to formation of the nodule (see Figure 23.15 for a description of root nodule biochemistry).

Nod factors are lipochitin oligosaccharides to which various substituents are bonded (**Figure 23.12**) that function as primary rhizobial signaling molecules triggering legumes to develop new plant organs: root nodules that host the bacteria as nitrogen-fixing bacteroids (Figure 23.11). Resolving the details of the signaling pathway triggered by Nod binding to cell surface receptors (NFR1 and NFR2) and leading to the induction of organogenesis (nodule formation) is an active area of research (**Figure 23.13**). Interestingly, many elements of the signaling pathway leading to nodulation are also used by the mycorrhizal fungi for infection of plant roots (Figure 23.13 and Section 23.4).

Which plants a given rhizobial species can infect is in part determined by the structure of the Nod factor it produces. Besides the *nodABC* genes, which are universal and whose

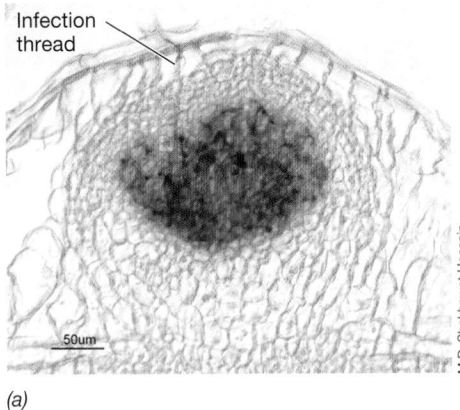

(a)

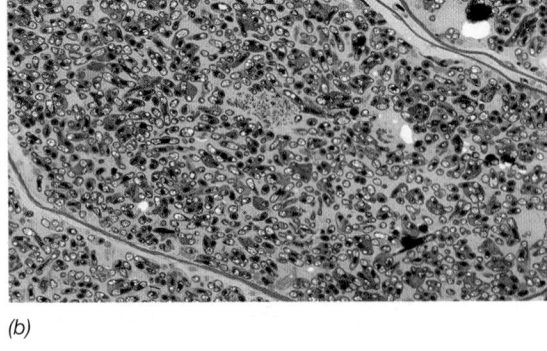

(b)

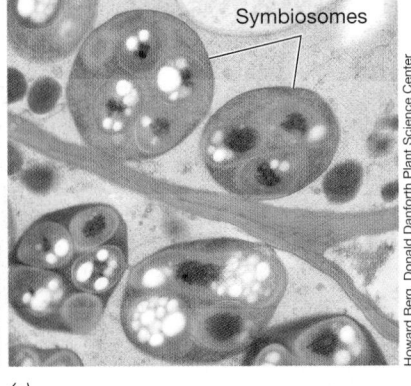

(c)

Figure 23.11 The infection thread and formation of root nodules. *(a)* Light micrograph of an early-stage nodule from a legume (*Lotus japonicus*) infected with a rhizobium strain containing a chromosomal copy of the *lacZ* gene. The nodule was sectioned and bacterial distribution (blue) determined using an activity stain (X-gal) that turns blue when cleaved by the enzyme β-galactosidase (⇄ Section 12.2 and Figure 12.8). An infection thread, consisting of a cellulosic tube through which bacteria move to root cells, is clearly visible extending from the surface to the interior. *(b)* Transmission electron micrograph through a soybean (*Glycine max*) nodule infected with *Bradyrhizobium japonicum*, showing bacteroid-filled plant cells. The plant cell is approximately 50 μm long. *(c)* Higher-magnification micrograph showing individual symbiosomes, each filled with several bacteroids. The clear areas in each bacteroid are the storage polymer poly-β-hydroxybutyrate (⇄ Section 2.8). Bacteroids are about 2 μm long.

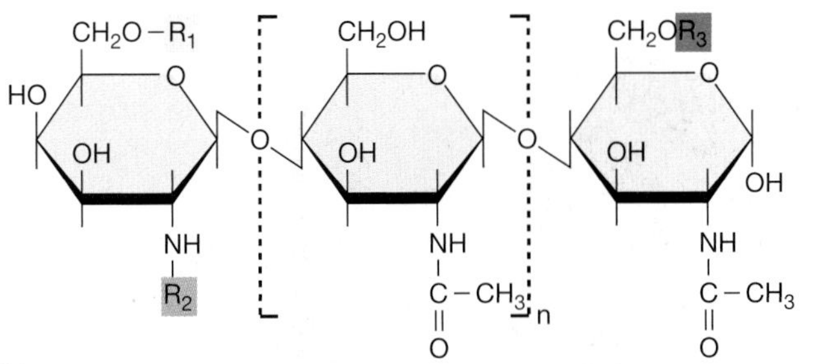

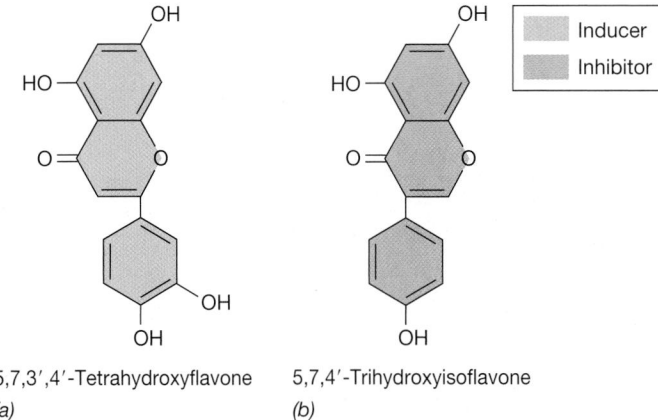

(a)

Rhizobial or AM fungus species	R₁	R₂	R₃
Sinorhizobium meliloti (alfalfa)	Ac	C16:2 or C16:3	SO₃H
Rhizobium leguminosarum biovar *viciae* (pea)	Ac	C18:1 or C18:4	H or Ac
Glomus intraradices (many agricultural crops)	H	C16 or C16:1 or C16:2 or C18 or C18:1Δ9Z	H or SO₃H

(b)

Figure 23.12 Nod and Myc factors. *(a)* General structure of Nod factors produced by rhizobia species (*Sinorhizobium meliloti* and *Rhizobium leguminosarum* biovar *viciae*) and the Myc factor produced by *Glomus intraradices*, an arbuscular mycorrhizal (AM) fungus (Section 23.4). The central hexose unit can repeat up to three times for different Nod factors, and repeat either two or three times for the different Myc factors. *(b)* Table of the structural differences (R₁, R₂, R₃) that define the precise signaling factors of each species. C16:1, C16:2, and C16:3, palmitic acid with either one, two, or three double bonds, respectively; C18:1, oleic acid with one double bond; C18:1Δ9Z, the *trans* isomer of oleic acid with one double bond at the 9th C–C bond; C18:4, oleic acid with four double bonds; Ac, acetyl.

5,7,3′,4′-Tetrahydroxyflavone

5,7,4′-Trihydroxyisoflavone

(a)

(b)

Figure 23.14 Plant flavonoids and nodulation. Structures of flavonoid molecules that are *(a)* an inducer of *nod* gene expression and *(b)* an inhibitor of *nod* gene expression in *Rhizobium leguminosarum* biovar *viciae*. Note the similarities in the structures of the two molecules. The common name of the structure shown in part *a* is *luteolin*, and it is a flavone derivative. The structure in part *b* is called *genistein*, an isoflavone derivative.

the rhizobia–legume symbioses is controlled by the chemistry of the flavonoids secreted by each species of legume.

Biochemistry of Root Nodules

As discussed in Section 14.6, N₂ fixation requires the enzyme *nitrogenase*. Nitrogenase from bacteroids shows the same biochemical properties as the enzyme from free-living N₂-fixing bacteria, including O₂ sensitivity and the ability to reduce acetylene as well as N₂. Bacteroids are dependent on the plant for the electron donor for N₂ fixation. The major organic compounds transported across

products synthesize the Nod backbone, each cross-inoculation group contains *nod* genes that encode proteins that chemically modify the Nod factor backbone to form its species-specific molecule (Figure 23.12). In *R. leguminosarum* biovar *viciae*, *nodD* encodes the regulatory protein NodD, which controls transcription of other *nod* genes. After interacting with inducer molecules, NodD *promotes* transcription and is thus a *positive* regulatory protein (⮌ Section 6.3). NodD inducers are plant flavonoids, organic molecules that are widely secreted by plants. Some flavonoids that are structurally very closely related to *nodD* inducers in *R. leguminosarum* biovar *viciae* inhibit *nod* gene expression in other rhizobial species (**Figure 23.14**). This indicates that part of the specificity observed between plant and bacterium in

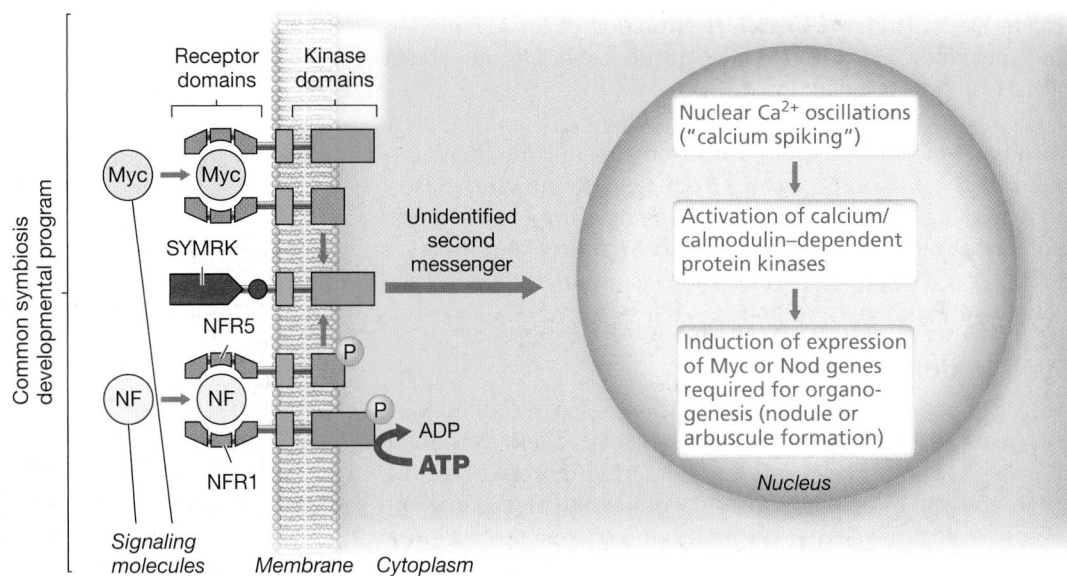

Figure 23.13 Nod and Myc signaling pathways in root nodule and mycorrhizal arbuscule formation. Nod factor (NF) signaling involves at least three membrane-associated receptors (NFR1, NFR5, and SYMRK) that together initiate nodulation via protein phosphorylation. NFR1 and SYMRK have active kinase domains (blue), whereas NFR5 kinase is inactive. The direct binding of NF to a complex of NFR1 and NFR5 at the plant cell cytoplasmic membrane initiates signal transduction by activation of the NFR1 kinase. The resulting phosphorylation of the NFR cytoplasmic domains triggers events leading to formation of the infection thread. Signal transduction to the SYMRK by the NFR1–NFR5–Nod factor complex (or by an unidentified receptor for the Myc factor) is part of a conserved symbiosis program in which induction of calcium signaling in the plant cell nucleoplasm triggers gene expression changes and production of plant growth hormones (cytokinins) required for nodule or arbuscule formation. See Section 23.4 for a discussion of mycorrhizae.

Plant cytoplasm

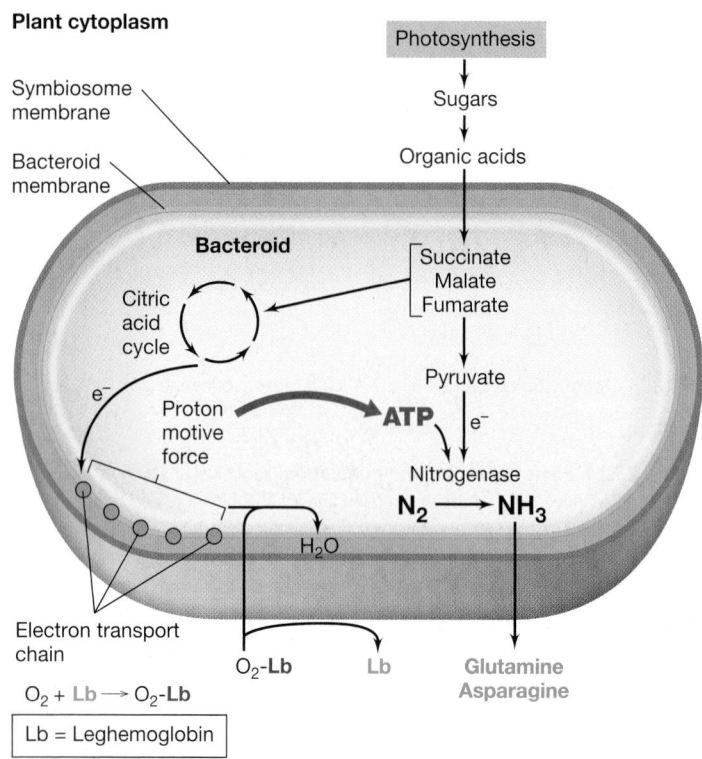

Figure 23.15 The root nodule bacteroid. Schematic diagram of major metabolic reactions and nutrient exchanges in the bacteroid. The symbiosome is a collection of bacteroids surrounded by a membrane originating from the plant (see Figure 23.11c).

the symbiosome membrane (Figure 23.11c) and into the bacteroid proper are citric acid cycle intermediates, in particular, the C_4 organic acids *succinate, malate,* and *fumarate* (Figure 3.16) (**Figure 23.15**). These are used as electron donors for ATP production and, following conversion to pyruvate, as the ultimate source of electrons for the reduction of N_2.

The product of N_2 fixation is ammonia (NH_3), and the plant assimilates most of this NH_3 by forming organic nitrogen compounds. The NH_3-assimilating enzyme glutamine synthetase is present in high levels in the plant cell cytoplasm and can convert glutamate and NH_3 into glutamine (Section 3.14). This and a few other organic nitrogen compounds transport bacterially fixed nitrogen throughout the plant.

Stem-Nodulating Rhizobia

Although most leguminous plants form N_2-fixing nodules on their *roots*, a few legume species bear nodules on their *stems*. Stem-nodulated leguminous plants are widespread in tropical regions where soils are often nitrogen deficient because of leaching and intense biological activity. The best-studied system is the tropical aquatic legume *Sesbania*, which is nodulated by the bacterium *Azorhizobium caulinodans* (**Figure 23.16**). Stem nodules typically form in the submerged portion of the stems or just above the water level. The general sequence of events by which stem nodules form in *Sesbania* resembles that of root nodules: attachment, formation of an infection thread, and bacteroid formation.

Some stem-nodulating rhizobia produce bacteriochlorophyll *a* and thus have the potential to carry out anoxygenic photosynthesis (Section 14.3). Bacteriochlorophyll-containing rhizobia, called

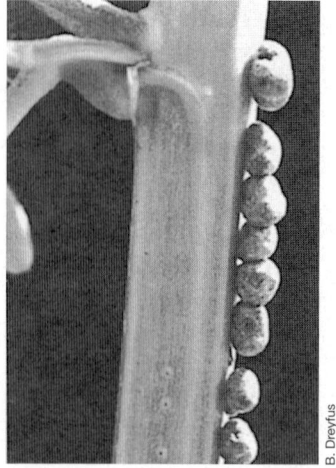

Figure 23.16 Stem nodules formed by stem-nodulating *Azorhizobium*. The right side of this stem of the tropical legume *Sesbania rostrata* was inoculated with *Azorhizobium caulinodans*, but the left side was not.

photosynthetic *Bradyrhizobium*, are widespread in nature, particularly in association with tropical legumes. In these species, light energy converted to chemical energy (ATP) in photosynthesis probably supplies part of the energy source needed by the bacterium to support N_2 fixation.

Nonlegume N_2-Fixing Symbioses: *Azolla–Anabaena* and *Alnus–Frankia*

Various nonleguminous plants form N_2-fixing symbioses with bacteria other than rhizobia. For example, the water fern *Azolla* harbors within small pores of its fronds a species of heterocystous N_2-fixing cyanobacteria (Section 15.3) called *Anabaena azollae* (**Figure 23.17**). *Azolla* has been used for centuries to enrich Asian rice paddies with fixed nitrogen. Before planting rice, the farmer allows the surface of the rice paddy to become densely covered with *Azolla*. As the rice plants grow, they eventually crowd out the *Azolla*, causing its death and the release of its nitrogen, which is

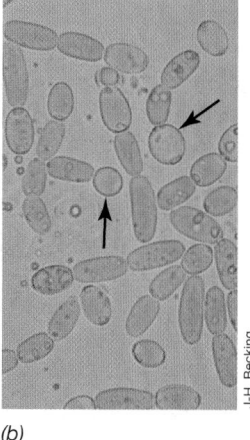

(a)   *(b)*

Figure 23.17 *Azolla–Anabaena* symbiosis. *(a)* Intact association showing a single plant of *Azolla pinnata*. The diameter of the plant is approximately 1 cm. *(b)* Cyanobacterial symbiont *Anabaena azollae* as observed in crushed leaves of *A. pinnata*. Single cells of *A. azollae* are about 5 μm wide. Vegetative cells are oblong; the spherical heterocysts (lighter color, arrows) are differentiated for nitrogen fixation.

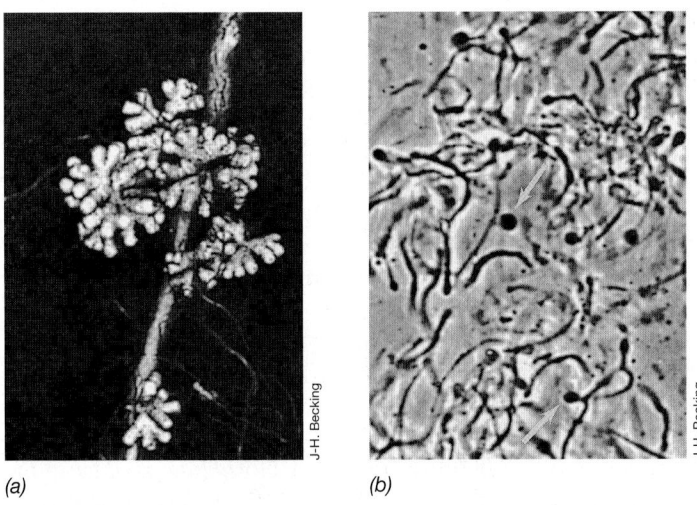

(a)   (b)

J.-H. Becking

Figure 23.18 *Frankia* nodules and *Frankia* cells. *(a)* Root nodules of the common alder *Alnus glutinosa*. *(b) Frankia* culture purified from nodules of *Comptonia peregrina*. Note vesicles (arrows) on the tips of hyphal filaments.

assimilated by the rice plants. By repeating this process each growing season, rice farmers can obtain high yields of rice without applying nitrogenous fertilizers.

The alder tree (genus *Alnus*) has N_2-fixing root nodules (**Figure 23.18a**) that harbor filamentous, N_2-fixing actinomycetes of the genus *Frankia*. When assayed in cell extracts the nitrogenase of *Frankia* is sensitive to O_2, but cells of *Frankia* fix N_2 at full oxygen tensions. This is because *Frankia* protects its nitrogenase from O_2 by localizing the enzyme in terminal swellings on the cells called *vesicles* (Figure 23.18b). The vesicles contain thick walls that retard O_2 diffusion, thus maintaining the O_2 tension within vesicles at levels compatible with nitrogenase activity. In this regard, *Frankia* vesicles resemble the heterocysts produced by some filamentous cyanobacteria as localized sites of N_2 fixation (↩ Section 15.3).

Alder is a characteristic pioneer tree able to colonize nutrient-poor soils, probably because of its ability to enter into a symbiotic N_2-fixing relationship with *Frankia*. A number of other small or bushy, woody plants are nodulated by *Frankia*. As is the case for the rhizobial symbionts of leguminous plants, a single strain of *Frankia* can form nodules on several different species of plants.

— **MINIQUIZ** —

- How do rhizobial root nodules benefit a plant?
- What are Nod factors and what do they do?
- What is a bacteroid and what occurs within it? What is the function of leghemoglobin?
- What are the major similarities and differences between rhizobia and *Frankia*?

23.4 Mycorrhizae

Mycorrhizae are mutualisms between plant roots and fungi in which nutrients are transferred in both directions. The fungus transfers inorganic nutrients—in particular, phosphorus and nitrogen—from the soil to the plant, and the plant in turn transfers primarily carbohydrates to the fungus. These mutualisms are harnessed in agricultural applications. From fungal spores produced in culture or from root scrapings of infected plants, soil inoculants are produced that enhance plant growth.

Classes of Mycorrhizae

There are two classes of mycorrhizae. In *ectomycorrhizae*, fungal cells form an extensive sheath (*fungal mantle*) around the outside of the root with only a slight penetration into the root cellular structure (**Figure 23.19**). In *endomycorrhizae*, a part of the fungus

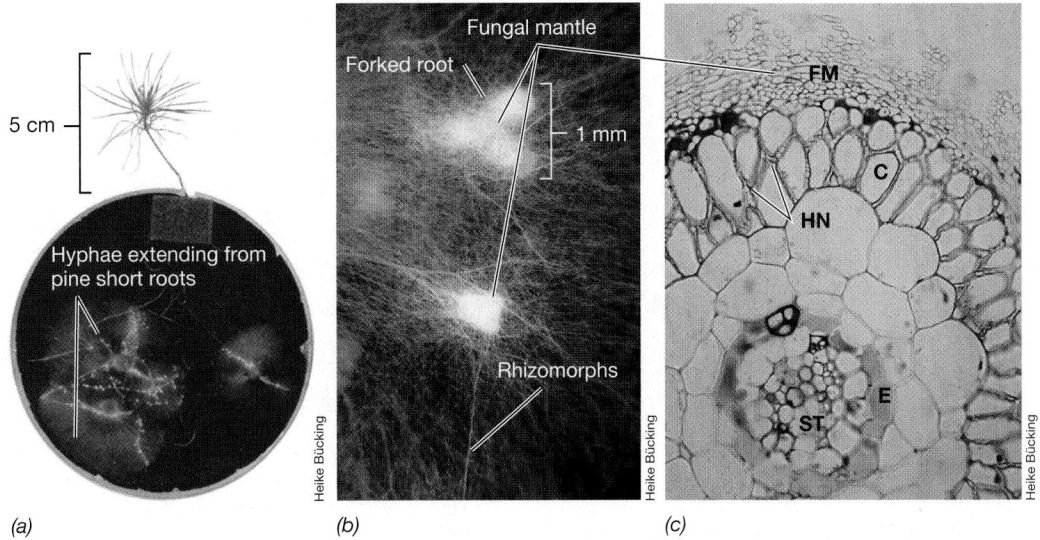

(a)   (b)   (c)

Heike Bücking

Figure 23.19 Ectomycorrhizal colonization of pine and beech tree roots. *(a)* Colonization of pine (*Pinus sylvestris*) roots with the ectomycorrhizal fungus *Suillus bovinus*. *(b)* Ectomycorrhizal root tips of pine roots enclosed by a mantle of ectomycorrhizal fungi (white) and associated hyphae extending into the soil matrix. The ectomycorrhizal fungus *Suillus bovinus* also forms rhizomorphs, hyphal aggregations that are involved in the long-distance transport from the soil to the mycorrhizal root. *(c)* Cross section of a beech fine root showing the fungal mantle (FM) and Hartig net (HN) within the root cortex (C). The Hartig net is the location of nutrient exchange between the plant and fungus. Also shown are root vascular tissue (root stele, ST) and endodermis (E).

UNIT 5

becomes deeply embedded within cells comprising the root tissue. Ectomycorrhizae are found mainly on the roots of forest trees, especially conifers, beeches, and oaks, and are most highly developed in boreal and temperate forests. In such forests, almost every root of every tree is mycorrhizal. The root system of a mycorrhizal tree such as a pine (genus *Pinus*) is composed of both long and short roots. The short roots, which are characteristically dichotomously branched in *Pinus* (Figure 23.19b), show typical fungal colonization, and long roots are also frequently colonized.

Ectomycorrhizal hyphae extending from the fungal mantle and penetrating between the epidermal and cortical cells form the *Hartig net* (Figure 23.19c). This network is where nutrient exchange between the fungus and the host plant occurs. Most mycorrhizal fungi do not catabolize cellulose and other leaf litter polymers. Instead, they catabolize simple carbohydrates and typically have one or more vitamin requirements. They obtain their carbon from root secretions and obtain inorganic minerals from the soil. Mycorrhizal fungi are rarely found in nature except in association with roots, and most are probably obligate symbionts.

Despite the close symbiotic association between fungus and root, a single species of tree can form multiple mycorrhizal associations. One pine species can associate with over 40 species of fungi. This relative lack of host specificity allows ectomycorrhizal mycelia to interconnect trees, providing linkages for transfer of carbon and other nutrients between trees of the same or different species. Nutrient transfer from well-illuminated overstory plants to shaded trees is thought to help equalize resource availability, subsidizing young trees and increasing biodiversity by promoting the coexistence of different species.

Arbuscular Mycorrhizae

Although ectomycorrhizal fungi play a significant role in the ecology of forests, there is a greater diversity of endomycorrhizae. Most are *arbuscular mycorrhizae* (*AM*) that comprise a phylogenetically distinct fungal division, the *Glomeromycota* (⬡ Section 18.11), of which all or most species are obligate plant mutualists (the word *arbuscular* means "little tree"). AM colonize 70–90% of all terrestrial plants, including most grassland species and many crop species. The association between plants and the *Glomeromycota* is thought to be the ancestral type of mycorrhizae, established about 450 million years ago and an important evolutionary step in the successful invasion of dry land by terrestrial plants (see page 557).

It is now known that AM fungi produce lipochitin oligosaccharide signaling factors (**Myc factors**) very closely related to the Nod factors in the rhizobium–legume symbiosis (Section 23.3), and Myc factors initiate formation of the mycorrhizal state (Figures 23.12 and 23.13). Root colonization by an AM fungus begins with germination of a soilborne spore, producing a branched germination mycelium that recognizes the host plant through reciprocal chemical signaling. Spore germination and mycelial branching is induced by *strigolactones*, plant hormones released by the roots that also play a key role in plant development. When a plant is nutrient-limited, this hormone acts to repress aboveground plant growth (suppressing formation of secondary shoots) and to stimulate the growth of the root system, enhancing the production of lateral roots and root hairs. These developmental changes help the plant secure nutrients before using them later for aboveground growth.

The Myc factor produced by the AM fungal mycelium signals the plant to initiate the developmental process (Figure 23.13). The fungus then forms a contact structure called the *hyphopodium* with root epidermal cells (**Figure 23.20c**). Penetrating hyphae extend into the plant from each hyphopodium, usually forging an intracellular path through epidermal and outer cortical cell layers of the root before forming dichotomously branched or coiled hyphal structures called **arbuscules** within plant inner cortex cells, near vascular tissues. However, the arbuscular hyphae remain separated from plant protoplasm by an extensive plant cytoplasmic membrane that forms a region called the *apoplast* (**Figure 23.21**), and this structure functions to increase the surface area of contact between plant and fungus. Inorganic nitrogen and phosphorus are then "mined" from the soil by the fungi, converted to arginine and polyphosphate, and translocated through the hyphae to the plant (Figure 23.21).

Myc factors are very similar to the rhizobial Nod factors and only relatively minor modifications of the chitin backbone structure confer specificity (Figure 23.12). It is now suspected that the basic signaling and developmental systems used in the legume–root nodule symbiosis (Section 23.3), which arose about 60 million years ago, first evolved in the

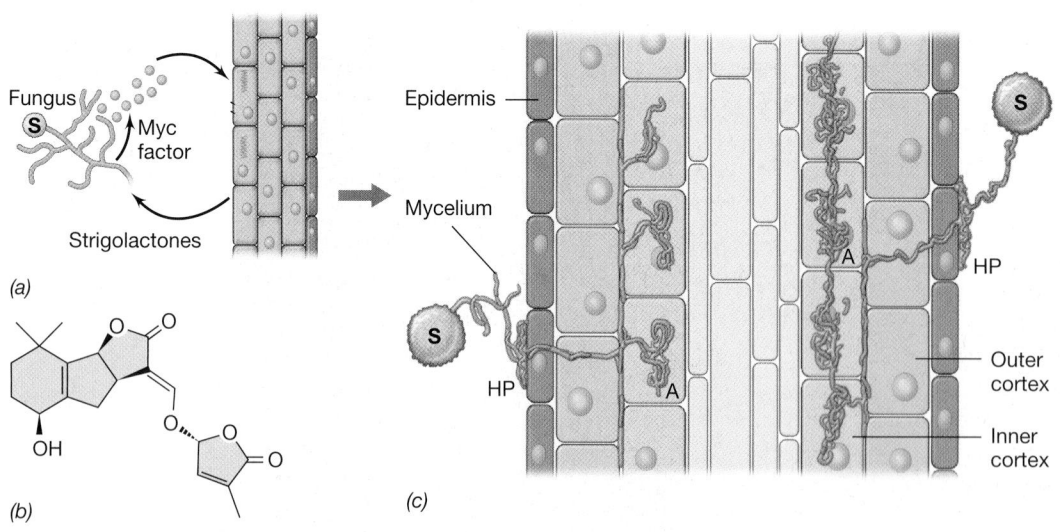

Figure 23.20 Arbuscular mycorrhizal root colonization. *(a)* A spore (S) recognizes the presence of a nearby root by sensing the presence of strigolactones released by the plant root. *(b)* The structure of strigol, one type of strigolactone. Strigolactones stimulate spore germination and mycelial branching. Myc factor (see Figure 23.12) produced by the growing fungal mycelium then initiates the infection process. *(c)* The mycelium forms an attachment structure called the hyphopodium (HP) and then enters the inner cortex region of the root by penetrating epidermal cells and cells of the outer cortex. Arbuscules (dichotomously branched invaginations, A) are formed by mycelia spreading either *inter*cellularly (left) or *intra*cellularly (right).

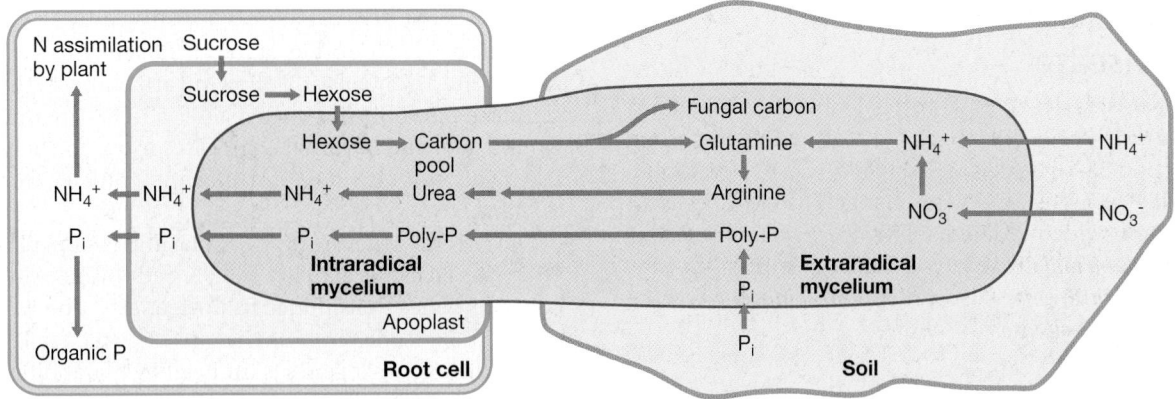

Figure 23.21 Pathways of N, P, and C exchange between plant and arbuscular mycorrhizal fungi. Inorganic nitrogen (NH_4^+ and NO_3^-) and phosphorus (P_i) mined from the soil by the extraradical (soil-associated) mycelia are translocated to the plant as arginine and polyphosphate (poly-P) through the mycelial network and delivered to the plant at the intraradical (plant cell–associated) mycelium. Ammonia and phosphate are regenerated in the intraradical mycelium for transfer to the plant cell. In exchange for the N and P, the plant provides organic carbon to the fungus.

much more ancient AM fungi–plant symbiosis. Apparently the AM fungal system was recruited and adapted for the legume–root nodule symbiosis (Figure 23.13).

Although the arbuscular mycorrhizae are a much more ancient and widely distributed microorganism–plant symbiosis, understanding of their signaling and developmental program has been much slower to develop because the AM fungi cannot be maintained in pure culture. AM fungi are obligately *biotropic* (meaning that they obtain their nutrients only from living cells of their symbiotic partner), and unlike rhizobia they have no supporting genetic system as has been exploited to help unravel the complex developmental steps leading to legume–root nodule formation.

Benefits for the Plant

The beneficial effect of the mycorrhizal fungus on the plant is best observed in poor soils where plants that are mycorrhizal thrive, but nonmycorrhizal ones do not. For example, if trees planted in prairie soils, which ordinarily lack a suitable fungal inoculum, are artificially inoculated at the time of planting, they grow much more rapidly than uninoculated trees (**Figure 23.22**). The mycorrhizal plant can absorb nutrients from its environment more efficiently (Figure 23.21) and thus has a competitive advantage. This improved nutrient absorption is due to the greater surface area provided by the fungal mycelium. For example, in the pine seedling shown in Figure 23.19*a* and *b*, the ectomycorrhizal fungal mycelium makes up the overwhelming part of the absorptive capacity of the plant root system. The mycorrhizal plant is better able to function physiologically and compete successfully in a species-rich plant community, and the fungus benefits from a steady supply of organic nutrients.

In addition to helping plants absorb nutrients, mycorrhizae also play a significant role in supporting plant diversity. Field experiments have clearly shown a positive correlation between the abundance and diversity of mycorrhizae in a soil and the extent of the plant diversity that develops in it. However, although most mycorrhizae are true mutalistic symbioses, there are also parasitic mycorrhizae. In these less common mycorrhizal symbioses,

either the plant parasitizes the fungus or, in some cases, the fungus parasitizes the plant. Obviously, there is much more to learn about fungal–plant symbioses, and such discoveries will be important not only in agriculture but also in facilitating the restoration of damaged ecosystems.

— **MINIQUIZ** —

- How do endomycorrhizae differ from ectomycorrhizae?
- What features of mycorrhizal fungi might have assisted in colonization of dry land by plants?
- How do mycorrhizal fungi promote plant diversity?

Figure 23.22 Effect of mycorrhizal fungi on plant growth. Six-month-old seedlings of Monterey pine (*Pinus radiata*) growing in pots containing prairie soil: left, nonmycorrhizal; right, mycorrhizal.

23.5 *Agrobacterium* and Crown Gall Disease

Some microorganisms develop parasitic symbioses with plants. The genus *Agrobacterium*, a relative of the root nodule bacterium *Rhizobium* (Section 23.3), is such an organism, causing the formation of tumorlike growths on diverse plants. The two species of *Agrobacterium* most widely studied are *Agrobacterium tumefaciens* (also called *Rhizobium radiobacter*), which causes *crown gall disease*, and *Agrobacterium rhizogenes* (also called *Rhizobium rhizogenes*), which causes *hairy root disease*.

The Ti Plasmid

Although wounded plants often form a benign accumulation of tissue called a *callus*, the growth in crown gall disease (**Figure 23.23**) is different in that it is uncontrolled growth, resembling an animal tumor. *A. tumefaciens* cells induce tumor formation only if they contain a large plasmid called the **Ti plasmid** (Ti for *t*umor *i*nducing). In *A. rhizogenes*, a similar plasmid called the *Ri plasmid* is necessary for induction of hairy root disease. Following infection, a part of the Ti plasmid called the *transferred DNA* (T-DNA) is integrated into the plant's genome. T-DNA carries the genes for tumor formation and also for the synthesis of a number of modified amino acids called *opines*. Octopine [N^2-(1,3-dicarboxyethyl)-L-arginine] and *nopaline* [N^2-(1,3-dicarboxypropyl)-L-arginine] are two common opines. Opines are produced by plant cells transformed by T-DNA and are a source of carbon and nitrogen, and sometimes phosphate, for the parasitic *A. tumefaciens* cells. These nutrients are the benefits for the bacterial symbiont.

Recognition and T-DNA Transfer

To initiate the tumorous state, *A. tumefaciens* cells attach to a wound site on the plant. Following attachment, the synthesis of cellulose microfibrils by the bacteria helps anchor them to the wound site, and bacterial aggregates form on the plant cell surface. This sets the stage for plasmid transfer from bacterium to plant.

The general structure of the Ti plasmid is shown in **Figure 23.24**. Only the T-DNA is actually transferred to the plant. The T-DNA contains genes that induce tumorigenesis. The *vir* genes on the Ti plasmid encode proteins that are essential for T-DNA transfer. Transcription of *vir* genes is induced by metabolites synthesized by wounded plant tissues. Examples of inducers include the phenolic compounds acetosyringone and ferulate. The transmissibility genes on the Ti plasmid (Figure 23.24) allow the plasmid to be transferred by conjugation from one bacterial cell to another.

The *vir* genes are the key to T-DNA transfer. The *virA* gene encodes a protein kinase (VirA) that interacts with inducer molecules and then phosphorylates the product of the *virG* gene (**Figure 23.25**). VirG is activated by phosphorylation and functions to activate other *vir* genes. The product of the *virD* gene (VirD) has endonuclease activity and nicks DNA in the Ti plasmid in a region adjacent to the T-DNA. The product of the *virE* gene is a DNA-binding protein that binds the single strand of T-DNA in the plant cell to protect it from destruction by nucleases. The *virB* operon encodes 11 different proteins that form a type IV secretion system (⇦ Section 4.13) for single-strand T-DNA and protein transfer between bacterium and plant (Figure 23.25) and thus resembles bacterial conjugation (⇦ Section 11.8). Laboratory studies of *A. tumefaciens* have shown that it can transfer T-DNA into many types of eukaryotic cells, including fungi, algae, protists, and even human cell lines.

Once inside the plant cell, T-DNA becomes inserted into the genome of the plant. Tumorigenesis (*onc*) genes on the Ti plasmid (Figure 23.24) encode enzymes for plant hormone production and

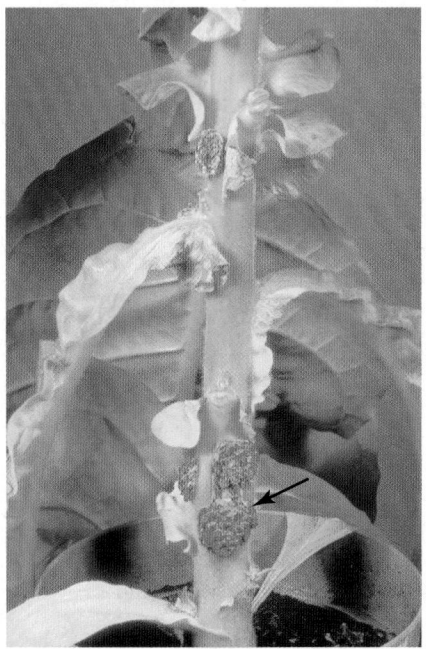

Figure 23.23 Crown gall. Photograph of a crown gall tumor (arrow) on a tobacco plant caused by *Agrobacterium tumefaciens*. The disease usually does not kill the plant but may weaken it and make it more susceptible to drought and diseases.

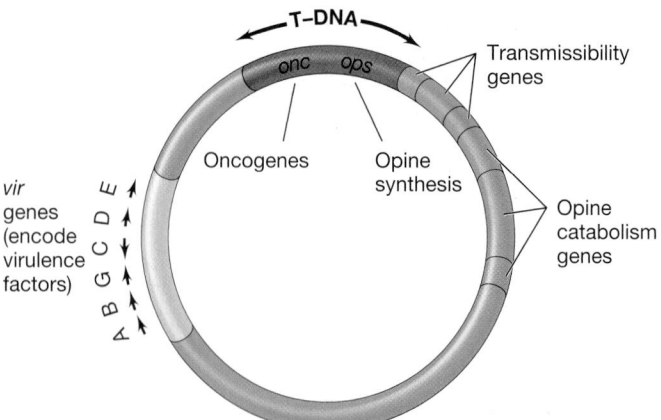

Figure 23.24 Structure of the Ti plasmid of *Agrobacterium tumefaciens*. T-DNA is the region transferred to the plant. Arrows indicate the direction of transcription of each gene. The entire Ti plasmid is about 200 kilobase pairs (kbp) of DNA and the T-DNA is about 20 kbp.

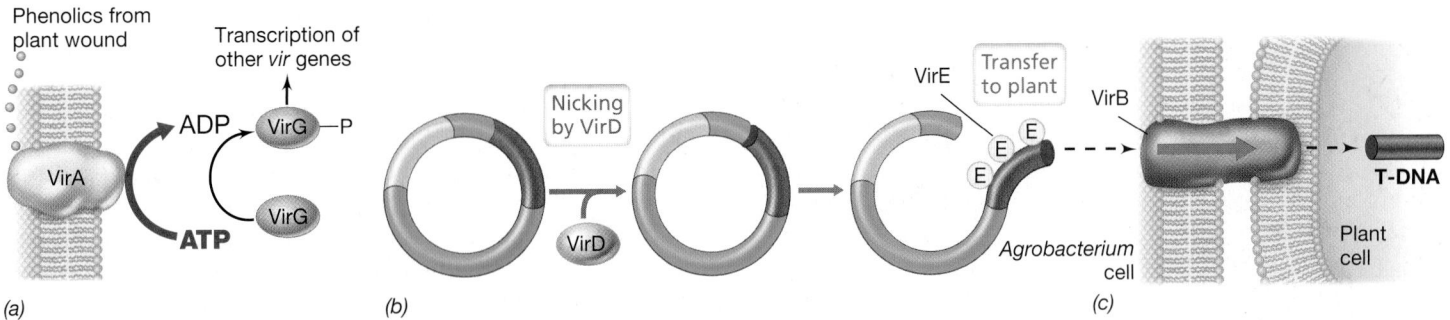

Figure 23.25 Mechanism of transfer of T-DNA to the plant cell by *Agrobacterium tumefaciens*.
(a) VirA activates VirG by phosphorylation and VirG activates transcription of other *vir* genes. *(b)* VirD is an endonuclease that nicks the Ti plasmid, exposing the T-DNA. *(c)* VirB functions as a conjugation bridge between the *A. tumefaciens* cell and the plant cell, and VirE is a single-strand binding protein that assists in T-DNA transfer. Plant DNA polymerase produces the complementary strand to the transferred single strand of T-DNA.

at least one key enzyme of opine biosynthesis. Expression of these genes leads to tumor formation and opine production. The Ri plasmid responsible for hairy root disease also contains *onc* genes. However, in this case the genes confer increased auxin responsiveness on the plant, and this promotes overproduction of root tissue and the symptoms of the disease. The Ri plasmid also encodes several opine biosynthetic enzymes.

Genetic Engineering with the Ti Plasmid

From the standpoint of microbiology and plant pathology, crown gall disease and hairy root disease both require intimate interactions that lead to genetic exchange from bacterium to plant. In other words, tumor induction in these diseases is the result of a natural plant-transformation system. Because of this, recent interest in the Ti–crown gall system has shifted away from the disease itself toward applications of this natural genetic exchange process in plant genetic engineering and biotechnology.

Several modified Ti plasmids that lack disease genes but that can still transfer DNA to plants have been developed by genetic engineering. These have been used for the construction of genetically modified (transgenic) plants. Many transgenic plants have been constructed thus far, including crop plants carrying genes for resistance to herbicides, insect attack, and drought. We discuss the use of the Ti plasmid as a vector in plant biotechnology in Section 12.7.

— MINIQUIZ —

- What are opines and whom do they benefit?
- How do the *vir* genes differ from T-DNA in the Ti plasmid?
- How has an understanding of crown gall disease benefited plant agriculture?

III • Insects as Microbial Habitats

Insects are the most abundant class of animals living today, with over 1 million species known. As many as 20% of all insects are thought to support symbiotic microbes in a mutually beneficial way. The symbioses contribute to the insects' ecological success by providing them either nutritional advantages or protection from parasites (see page 696 for an example of how the gut microbiota of bees benefits them in these ways). Some symbionts are found on insects' outer surfaces or in their digestive tracts. *Endosymbionts* are intracellular bacteria and are typically localized at specialized organs within the insect.

23.6 Heritable Symbionts of Insects

How symbionts are transferred from one generation to the next determines how a mutualism functions and how stable it is. Microbial symbionts can either be acquired by a host from an environmental reservoir (*horizontal* transmission) or be transferred directly from the parent to the next generation (*heritable* or *vertical* transmission). The mode of symbiont transmission is related to the specificity and persistence of an association. In general, less specificity is associated with horizontal transmission. In this section we focus only on mutualisms in which the microbial symbiont has no free-living form; that is, the symbionts are transmitted in a vertical fashion.

Types of Heritable Symbionts

All known heritable symbionts of insects lack a free-living replicative stage. Thus, they are *obligate* symbionts. However, although these bacteria require the host for replication, not all hosts are dependent upon the symbiont. Relative to host dependence, heritable symbionts are either *primary* or *secondary symbionts*. Primary symbionts are required for host reproduction. They are restricted to a specialized region called the **bacteriome** present in several insect groups; within the bacteriome the bacterial cells reside in specialized cells called **bacteriocytes**. Secondary symbionts are not required for host reproduction. Unlike primary symbionts, secondary symbionts are not always present in every individual of a species and are not restricted to particular host tissues.

Secondary symbionts are broadly distributed among insect groups. Like pathogens, they invade different cell types and may live extracellularly within the insect's *hemolymph* (the fluid bathing the body cavity). In insects with bacteriomes, secondary symbionts can invade the bacteriocytes, co-residing with or sometimes

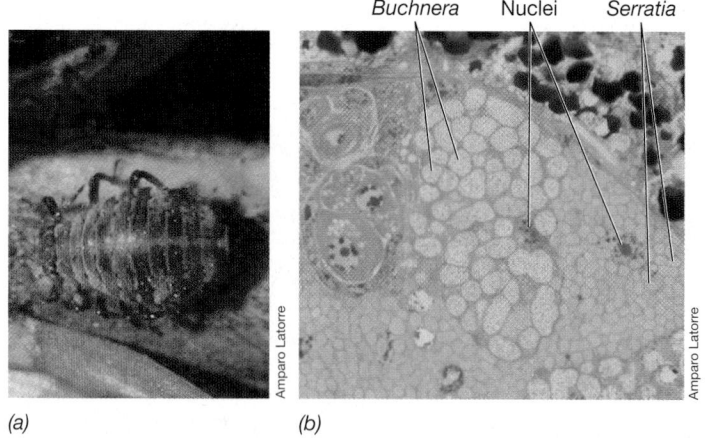

Buchnera   Nuclei   Serratia

(a)   (b)

Figure 23.26 Primary and secondary symbionts of an aphid. *(a)* The cedar aphid *Cinara cedri*, a model organism for studies of symbioses. *(b)* Transmission electron micrograph of the bacteriome of *C. cedri* showing two bacteriocytes. Packed within each bacteriocyte are cells of *Buchnera aphidicola* (the primary symbiont) or *Serratia symbiotica*, the smaller, secondary symbiont. The nucleus of each bacteriocyte is identified. The bacteriocyte containing *Buchnera* cells is about 40 μm wide.

displacing the primary symbionts (**Figure 23.26b**). However, in order to persist in the insect host, the secondary symbiont must confer some advantage, such as a nutritional advantage or protection from environmental stresses such as heat. For example, whiteflies infected with *Rickettsia* bacteria (⮌ Section 16.1) produce offspring at about twice the rate of uninfected flies, and more offspring survive to adulthood. Secondary symbionts may also provide protection against invasion by pathogens or predators. A *Spiroplasma* species (⮌ Section 16.9), which was first observed in *Drosophila neotestacea* in the 1980s, provides protection against a parasitic nematode worm. In most instances the basis for increased fitness or protection is unknown, but in one case a toxin encoded by a lysogenic bacteriophage (⮌ Section 8.7) carried by the symbiont is known to confer protection on the insect from infection by a parasitic wasp.

There are heritable parasitic symbionts that manipulate the host's reproductive system, increasing the frequency of female progeny (sex-ratio skewing, ⮌ Figure 16.27). Because most heritable symbionts are transmitted maternally, the suppression of male progeny serves to expand the number of infected individuals and increase the rate of spread through an insect population. Since symbiont-conferred functions can spread rapidly within a population, acquisition of symbiont-encoded traits provides a mechanism for much more rapid adaptation than is possible through mutations in insect genes. *Rickettsia* infection of the whitefly population provides one example of how rapidly symbiont-conferred traits can spread through a population. Only 1% of whiteflies in the southwestern United States were infected with *Rickettsia* in 2000. In 2006, 97% of flies were infected. In another example, a strain of *Wolbachia* (⮌ Section 16.1) swept through populations of *Drosophila simulans* in California in only 3 years.

An important applied benefit of improved basic understanding of insect symbionts is the increased use of symbionts in insect pest management and the control of vectorborne diseases, such as malaria and filariasis in humans (⮌ Sections 33.5 and 33.7).

For example, symbiotic *Wolbachia*, which are reproduction manipulators, are widely distributed among insect species (possibly infecting as many as 60–70% of all insect species). The sperm of *Wolbachia*-infected males can sterilize uninfected females. Although the mechanism for sterilization is not fully understood, the phenomenon is being tested as a means to suppress disease transmission. Release of a large number of *Wolbachia*-infected male *Culex quinquefasciatus* mosquitoes, the vector of the filarial nematode causing elephantiasis (⮌ Section 33.7), in Myanmar (Burma) effectively eliminated the local mosquito population.

In some cases, the presence of the symbiont decreases insect transmission of disease. For instance, *Aedes aegypti* mosquitoes infected with *Wolbachia* are less likely to transmit the virus causing dengue fever (⮌ Section 31.5). However, in some other cases, the presence of the symbiont *increases* disease transmission. For example, whiteflies infected with *Hamiltonella* bacteria (a symbiont affiliated with the *Enterobacteriaceae*) are more likely than uninfected flies to transmit tomato yellow leaf curl virus.

Nutritional Significance of Obligate Intracellular Symbionts of Insects

The association of bacteria and insects has allowed many insects to use food resources that are rich in some nutrients, but poor in others. To achieve adequate nutrition, some insects exploit the metabolic potential of their symbionts. For instance, aphids feed on the carbohydrate-rich but otherwise nutrient-poor sap of phloem vessels in plants. Early on it was suspected that obligate symbionts might benefit the insect by providing nutrients not provided by their primary diet, and this is now known to be true.

Molecular analyses have shown that most families of aphids harbor the bacterium *Buchnera* in their bacteriomes. The role of *Buchnera* in host nutrition was first indicated by experiments using defined diets to examine the nutrient requirements of aphids. Compared with infected controls, symbiont-free aphids required a diet containing all amino acids that are either lacking or rare in phloem sap. Subsequent genomic studies documented the presence in *Buchnera* of genes encoding the biosynthesis of nine amino acids missing from the sap. There are also examples of synergy between host and symbiont where the synthesis of certain amino acids becomes a joint venture. For example, *Buchnera* lacks the enzyme needed for the last step in leucine biosynthesis, but the necessary gene is present in the aphid's genome. Presumably, this enzyme is made by the aphid and participates in the leucine biosynthetic pathway along with the bacterial enzymes.

A secondary symbiont can also contribute to a joint venture. For example, the *Buchnera* symbiont of the cedar aphid is unable to supply tryptophan to the aphid. Two genes in the tryptophan biosynthetic pathway are present in *Buchnera*, but the remaining genes for the pathway are located on the chromosome of a secondary endosymbiont (Figure 23.26). Thus, different parts of a required metabolic pathway can be encoded by different endosymbionts present in the same insect. The fungus-cultivating ants provide yet another example of a complex symbiosis that has formed between an insect and multiple microorganisms (see Explore the Microbial World, "The Multiple Microbial Symbionts of Fungus-Cultivating Ants").

THE MULTIPLE MICROBIAL SYMBIONTS OF FUNGUS-CULTIVATING ANTS

The attine ants are an example of an elaborate symbiotic association between multiple microbial species and insect. These ants have established an obligate mutualism with a fungus they cultivate in fungal gardens for food, using small leaf fragments to mulch these gardens. A close symbiotic relationship between ant and fungus was first indicated by the observation that one specific fungus was cultivated by each ant lineage. The ants and their mutualistic fungi can be divided into five agricultural systems, each requiring distinct lineages of ants and fungi. Ants grouped in the "lower attine agriculture" system form associations with specific groups of fungi they capture from the environment. By contrast, the "higher attine agriculture" group cultivates fungi that apparently are no longer capable of existing apart from the ant mutualism.

In addition to the close mutualistic relationship between ant species and the specific fungus they cultivate, this symbiosis is now known to include four other microbial symbionts: a small fungus that is parasitic on the garden fungus, nitrogen-fixing bacteria (⊘ Section 14.6) associated with the garden fungus, an actinobacterium that antagonizes the parasitic fungus, and a black yeast that interferes with the actinobacterium.

The fungus is vertically transmitted between ant generations by colony-founding queens. The queen collects a pellet of fungus prior to her mating flight, storing it in a pouch in her oral cavity. After mating, she uses the fungus pellet to establish a new nest and fungus garden (**Figure 1a**). Nitrogen-fixing *Klebsiella* and *Pantoea* species associated with the fungus enrich the nutritional quality of the garden by adding new nitrogen to the

Figure 1 **Attine ants.** *(a)* Queen and worker ants in their fungal garden. *(b)* Mutualism with *Actinobacteria* can cover much of the exoskeleton of workers (white areas).

nitrogen-poor leaf growth substrate. A single leaf-cutter ant colony may contribute as much as 1.8 kg of fixed nitrogen per year. This new nitrogen benefits the ant colony and also results in higher overall plant diversity near leaf-cutter colonies.

However, the garden is at risk of being destroyed by a parasitic fungus of the genus *Escovopsis*. To repel the parasitic microfungus, the ant has formed another symbiotic association with an actinobacterium (genus *Pseudonocardia*) that appears as a "waxy bloom" growing on the cuticle of the ant (**Figure 1b**). These bacteria, housed in specialized cuticular modifications on the ant's body, secrete secondary metabolites that inhibit the growth of *Escovopsis*. The *Pseudonocardia* likely receive nourishment from the ant in glandular secretions through pores localized

in regions of cuticular modification. Comparative genomic sequencing has revealed good congruence between the phylogenies of the ants, fungal cultivars, *Escovopsis*, and *Pseudonocardia*, pointing to very specific interactions among microbes and ants in this complex symbiosis.

The fourth and final microorganism identified in this symbiosis is a yeast that grows in the same cuticular regions colonized by the bacterium *Pseudonocardia*. This black-pigmented yeast interferes with chemical protection of the garden by stealing nutrients from the *Pseudonocardia*, thereby indirectly reducing its ability to suppress *Escovopsis* growth. The attine ant–microbial symbiosis is thus a complex maze of interactions between a macroorganism—the ant—and two groups of microbes, fungi and bacteria.

Mealybugs (*Planococcus citri*) present one of the most unusual examples of a partnership between two symbionts infecting the same insect. Mealybugs have two stable bacterial symbionts, "*Candidatus* Tremblaya princeps" (a *Betaproteobacterium*) and "*Candidatus* Moranella endobia" (a *Gammaproteobacterium*) (the term "*Candidatus*" means that these organisms are not yet in pure culture, ⊘ Section 13.10). These symbionts cooperate in providing their host with essential amino acids missing in its diet, as is true for the symbionts of many sap-feeding insects. However, the Moranella bacterium actually lives *inside of*

Tremblaya, the only known example of a bacterium-within-a-bacterium symbiosis. The highly reduced Tremblaya genome has lost all genes for tRNA synthetases, an essential function supplied either by the host or by the Moranella residing within the cytoplasm of Tremblaya.

Genome Reduction and Gene Transfer Events

Common features of primary symbionts are extreme genome reduction (⊘ Table 9.1), high adenine plus thymine content, and accelerated rates of mutation. Genomes of most insect symbionts

TABLE 23.2 Genome features of some endosymbiotic *Bacteria* of animals[a]

Host	Symbiont (genus)	Genome size (Mbp)	G+C (%)	Genes
Aphid	Heterotroph (*Buchnera*)	0.42–0.62	20–26	362–574
Tsetse fly	Heterotroph (*Wigglesworthia*)	0.70	22	617
Carpenter ant	Heterotroph (*Blochmannia*)	0.71–0.79	27–30	583–610
Sharpshooter	Heterotroph (*Sulcia*)	0.25	22	227
Mealybug	Heterotroph ("*Candidatus* Moranella endobia" Gammaproteobacteria)	0.54	43.5	406
Mealybug	Heterotroph ("*Candidatus* Tremblaya princeps" Betaproteobacteria)	0.14	58.8	121
Clam (*Calyptogena okutanii*)	Sulfur oxidizer (unnamed)	1.0	32	975
Clam (*Calyptogena magnifica*)	Sulfur oxidizer (*Ruthia*)	1.2	34	1248
Tube worm (*Riftia pachyptila*)	Sulfur oxidizer (unnamed)	3.3[b]	NA	NA

[a]All listed symbionts are obligately associated with their hosts, with the exception of the symbiont of *Riftia*, which also has a free-living stage. For a comparison with the genomes of free-living *Bacteria*, see Table 9.1.
[b]The free-living sulfur-oxidizing bacterium *Thiomicrospira crunogena* has a genome significantly smaller (2.4 Mb) than this symbiont.

fall within a range from 0.14 to 0.80 megabase pairs (Mbp) and 16.5 to 33% G+C (**Table 23.2**). The 0.14-Mbp (140 kilobase-pair) genome of "*Candidatus* Tremblaya princeps" is among the smallest genome known for any cell. In contrast, the genomes of related free-living bacteria range from 2 to 8 Mbp with a base composition closer to 50% G+C. Two common types of spontaneous mutation, cytosine deamination and the oxidation of guanosine, if not repaired, change a GC pair to an AT pair (⇌ Section 11.4). Symbionts with reduced genomes have fewer DNA repair enzymes (⇌ Section 11.4), and this likely facilitates a shift over time to genomes of lower G+C content.

The streamlined genomes of insect symbionts have lost genes from most functional categories (Chapter 9) and tend to retain only genes required for host fitness and essential molecular processes, such as translation, replication, and transcription. Genome reduction implies that the symbionts are reliant on the host for many functions no longer encoded in the symbiont genome (⇌ Section 9.4). For example, in many cases genes needed for the biosynthesis of cell wall components are missing, including lipid A and peptidoglycan, suggesting that the host supplies these functions or that the structures are not required to form stable cells within the bacteriocyte.

There is an interesting genomic contrast between primary symbionts and typical disease-causing bacteria (pathogens). While primary symbionts tend to lose genes encoding proteins required in *catabolic* pathways, pathogenic bacteria typically retain these, but lose genes for *anabolic* pathways. This reflects their differing relationships with the hosts; the insect symbiont provides the host with essential biosynthetic nutrients while the pathogen obtains important biosynthetic nutrients from the host.

Because genome sequences for a large number of insects and their symbionts are now appearing, microbiologists can begin to evaluate the frequency of gene transfer between them. Horizontal gene transfer is the movement of genetic information across normal mating barriers (Chapters 11 and 13). Although early research demonstrated that DNA of *Wolbachia* bacteria has been transferred to the nuclear genomes of their insect and nematode hosts, inspection of other insect mutualisms for which both host and symbiont genome sequences have become available (e.g., aphid and body louse) indicate that DNA transfer is very rare. This suggests that horizontal transfer is highly variable for reasons yet to be determined.

MINIQUIZ

- What factors stabilize the presence of a secondary insect symbiont?
- What are the consequences of symbiont genome reduction?
- How could it be determined if a symbiont and its host have experienced a long period of coevolution?

23.7 Termites

Microorganisms are primarily responsible for the degradation of wood and cellulose in natural environments. However, the activities of free-living microbial species have been exploited by certain groups of insects that have established microbial symbioses in order to digest lignocellulosic materials. Like the rumen of herbivorous animals (see Section 23.13), the insect gut provides a protective niche for microbial symbionts, and in return, the insect gains access to nutrients derived from an otherwise indigestible carbon source. Termites are among the most abundant representatives of this type of symbiotic alliance.

Termite Natural History and Biochemistry

Microbial symbionts in termites decompose the greater part of cellulose (74–99%) and hemicellulose (65–87%) in the plant material termites ingest. In contrast to the insect examples discussed in the previous section, most termites do not harbor *intracellular* bacteria. Instead, the symbiotic bacteria are present in digestive organs (guts) as in the case of mammals. Termite diets include lignocellulosic plant materials (either intact or at various stages of decay), dung, and soil organic matter (humus). About two-thirds of the terrestrial environment supports one or more termite species, with the greatest representation in tropical and subtropical regions, where termites may constitute as much as 10% of all animal biomass and 95% of soil insect biomass. In savannas, their numbers sometimes exceed 4000/m², and their biomass density (1–10 g/m²) may be higher than that of grazing mammalian herbivores.

Termites are categorized as higher or lower based on their phylogeny, and this classification correlates with different symbiotic strategies. The posterior alimentary tract of *higher* termites (family *Termitidae*, comprising about three-fourths of termite species) contains a dense and diverse community of mostly anaerobic bacteria, including cellulolytic species. In contrast, the *lower* termites harbor diverse populations of both anaerobic bacteria and cellulolytic

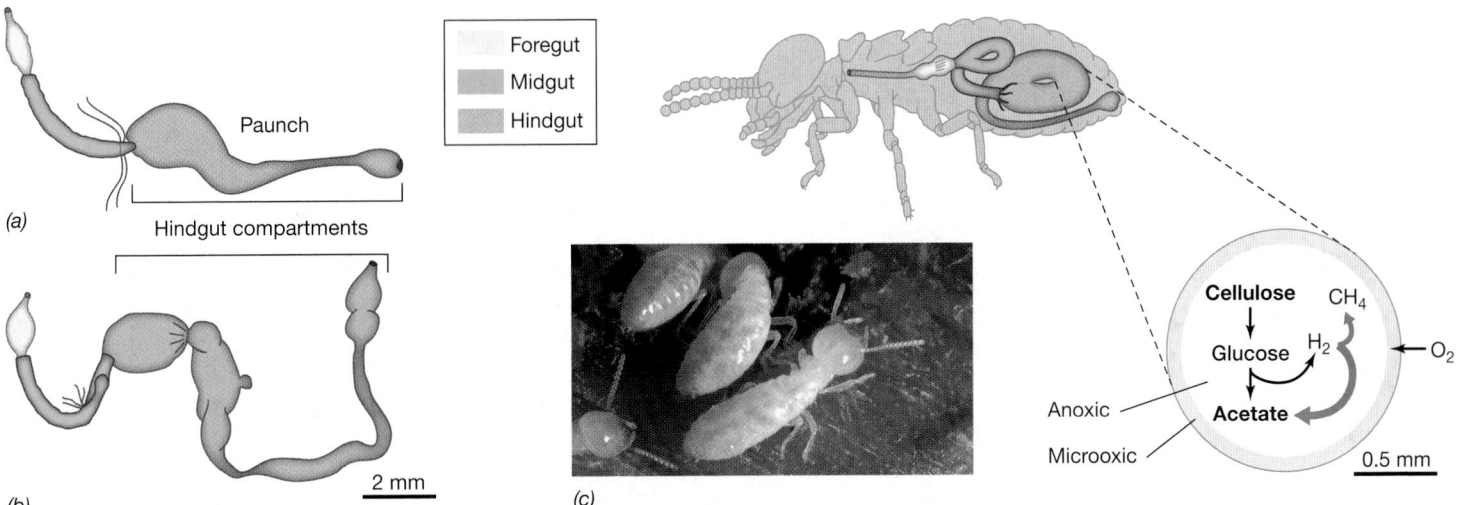

Figure 23.27 Termite gut anatomy and function. Gut architecture of lower *(a)* and higher *(b)* termites, showing the foregut, midgut, and differing complexity of the hindgut compartments. *(c)* Photo of workers, gut architecture, and biochemical activities of the lower termite *Coptotermes formosanus*. Microbial fermentation products, particularly acetate, are assimilated by the termite. Hydrogen produced by fermentation is consumed primarily by CO_2-reducing acetogens, with a smaller amount going to hydrogenotrophic methanogens. Acetogenesis and methanogenesis are discussed in Sections 14.16 and 14.17, respectively.

protists. Bacteria of lower termites participate little or not at all in cellulose digestion; only the protists phagocytize and degrade the wood particles ingested by the termites. The termite itself produces cellulases in its salivary glands or the midgut epithelium, but the relative contributions of microbial and termite enzymes to lignocellulosic breakdown are unclear.

The termite gut consists of a foregut (including the crop and muscular gizzard), a tubular midgut (site of secretion of digestive enzymes and absorption of soluble nutrients), and a relatively large hindgut of about 1 microliter volume (**Figure 23.27**). In lower termites, the hindgut consists primarily of a single chamber, the *paunch* (Figure 23.27a). The hindgut of most higher termites is more complex, being divided into several compartments (Figure 23.27b). For both higher and lower termites, the hindgut harbors a dense and diverse microbial community and is a major site of nutrient absorption. Acetate and other organic acids are produced during microbial fermentation of carbohydrate in the hindgut, and these products are primary carbon and energy sources for the termite (Figure 23.27c). High O_2 consumption by bacteria near the gut wall keeps the interior of the hindgut anoxic. However, microsensor measurements (Section 19.9) have shown that O_2 can penetrate up to 200 μm into the gut before it is completely removed by microbial respiration (Figure 23.27c). Thus, this tiny gut compartment offers distinct microbial niches with respect to O_2 and can support diverse microbial activities.

Bacterial Diversity and Lignocellulose Digestion in Higher Termites

In termites of different genera, the microbial gut communities differ significantly. Analysis of 16S rRNA gene sequences from hindgut contents of three genera (*Nasutitermes, Reticulitermes*, and *Microcerotermes*) of higher termites revealed a high diversity of microbial species from 12 phyla of *Bacteria*, but few *Archaea* (**Figure 23.28**).

Spirochetes of the genus *Treponema* (Section 15.19) dominated, with a lesser contribution from thus far uncultured organisms distantly related to the phylum *Fibrobacteres* (Section 16.21), a group also present in the rumen (Figure 23.40). Metagenomic analysis (Section 9.8) of the *Nasutitermes* hindgut microbial community has revealed bacterial genes encoding glycosyl hydrolases that hydrolyze cellulose and hemicelluloses. These metagenomic data clearly implicate spirochetes and *Fibrobacteres* in the digestion of lignocellulose, although the corresponding cellulolytic bacteria have not yet been isolated from the higher termites. At every molting of an individual termite, gut symbionts are lost, yet there is good conservation of the gut community within each termite species. Stable horizontal transmission of gut symbionts likely occurs as a result of the intimate social behavior and close contact characteristic of termites.

Acetogenesis and Nitrogen Fixation in the Termite Gut

Genes encoding enzymes of the acetyl-CoA pathway are highly represented in the spirochetes of the *Nasutitermes* hindgut, consistent with their function as the major CO_2-reducing acetogens (Section 14.16). The termite gut microbial communities have long been recognized as important to host nitrogen metabolism, providing new fixed nitrogen through nitrogen fixation (Section 14.6) and helping to conserve nitrogen by recycling excretory nitrogen back to the insect for biosynthesis. Consistent with this, metagenomic analyses reveal that many bacteria, including *Fibrobacteres* and treponeme spirochetes, contain genes encoding nitrogenase, the enzyme required to fix N_2.

From a simple energetic viewpoint, methanogenesis from H_2 and CO_2 is more favorable than acetogenesis from the same substrates (-34 kJ/mol of H_2 versus -26 kJ/mol of H_2, respectively), and thus methanogens should have a competitive advantage in all habitats in which the two processes compete (Sections 14.16–14.17).

UNIT 5

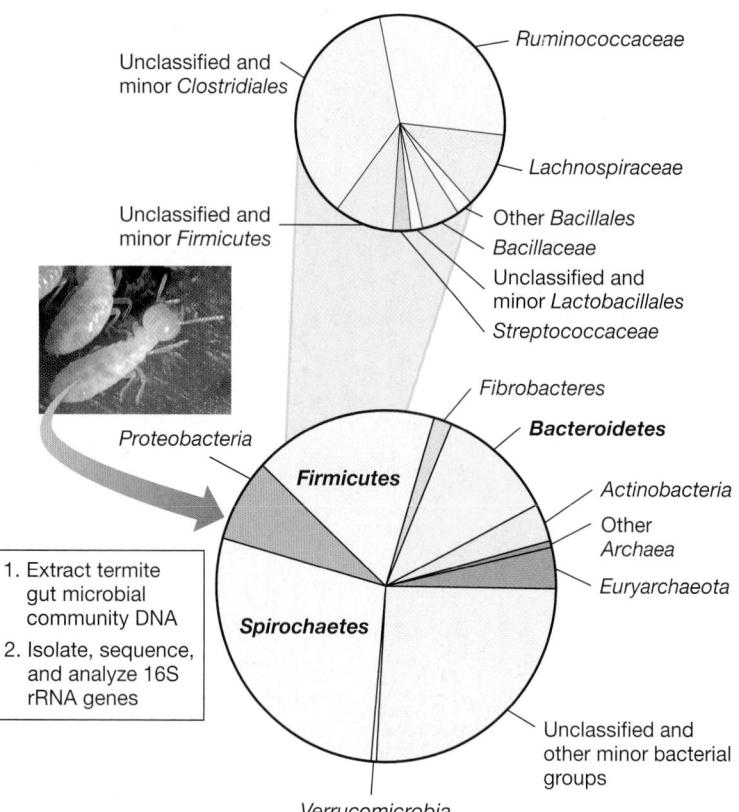

Figure 23.28 Microbial composition of termite hindgut inferred from 16S rRNA sequences. The results are pooled analyses of 5075 sequences from amplified or metagenomic sequencing studies of three genera of wood-feeding higher termites, *Nasutitermes, Reticulitermes,* and *Microcerotermes.* The data provide information primarily of diversity, not relative abundance. Data assembled and analyzed by Nicolas Pinel.

However, in termites they do not. There are at least two reasons for this. First, unlike methanogens, acetogens are able to use other substrates such as sugars or methyl groups from lignin degradation as electron donors for energy metabolism. Second, termite acetogens (which seem to consist mostly of spirochetes) can for some reason better colonize the H_2-rich termite gut center, whereas methanogens are largely restricted to the gut wall. On the gut wall, methanogens are located downstream of the H_2 gradient and thus receive only a fraction of the H_2 flux. In addition, the wall likely contains higher O_2 tensions, which may negatively affect the physiology of methanogens. So, despite the fact that termites are methanogenic, producing up to 150 teragrams of CH_4 per year on a global basis (1 teragram = 10^{12} grams), carbon and electron flow in the termite gut favor acetogenesis in this interesting anoxic microbial habitat.

MINIQUIZ

- How are anoxic conditions maintained in the termite hindgut?
- Why does reductive acetogenesis predominate over methanogenesis in many termites?
- Which group of morphologically unusual bacteria, absent from molecular surveys of prokaryotic cells in the rumen, seem to dominate activities in the termite hindgut?

IV • Other Invertebrates as Microbial Habitats

Thus far in this chapter we have discussed how certain macroorganisms that live in terrestrial environments provide habitats for microbial symbionts. We now consider some additional symbioses occurring in both terrestrial and aquatic environments. Aquatic environments—especially marine environments—impose different constraints on symbioses and offer different opportunities and challenges for the evolution of symbioses between macroorganisms and microorganisms. Nevertheless, microbial symbioses with marine animals, especially with invertebrates, are common. By finding habitats in marine invertebrates, microorganisms establish a safe residence in a nutritionally rich environment. The invertebrates benefit, too, as exemplified by the squid and the hydrothermal vent animal symbioses we cover here.

23.8 Hawaiian Bobtail Squid

The Hawaiian bobtail squid, *Euprymna scolopes,* is a small marine invertebrate (**Figure 23.29a**) that harbors a large population of the bioluminescent gram-negative gammaproteobacterium *Aliivibrio fischeri* (🔗 Section 16.4) in a light organ located on its ventral side. Squid and bacterium are partners in a mutualism. The bacteria emit light that resembles moonlight penetrating marine waters, and this is thought to camouflage the squid from predators that strike from beneath. Several other species of *Euprymna* inhabit marine waters near Japan and Australia and in the Mediterranean, and these contain *Aliivibrio* symbionts as well.

The Squid–*Aliivibrio* System as a Model Symbiosis

Many features of the *E. scolopes–A. fischeri* symbiosis have made it an important model for studies of animal–bacterial symbioses. These include the facts that the animals can be grown in the laboratory and that there is only a single bacterial species in the symbiosis in contrast to the huge number of species in symbioses such as those of the termite (Figure 23.28) or the mammalian large intestine (Chapter 24). In addition, the symbiosis is not an essential one; both the squid and its bacterial partner can be cultured apart from each other in the laboratory. This allows juvenile squid to be grown without bacterial symbionts and then experimentally colonized. Experiments can be done to study specificity in the symbiosis, the number of bacterial cells needed to initiate an infection, the capacity of genetically defined mutants of *A. fischeri* to initiate infection of the squid, and many other aspects of the relationship. Moreover, because the genome of *A. fischeri* has been sequenced, the powerful techniques of microbial genomics may be employed.

Establishing the Squid–*Aliivibrio* Symbiosis

Juvenile squid just hatched from eggs do not contain cells of *A. fischeri.* Thus, transmission of bacterial cells to juvenile squid is a horizontal (environmental) rather than a vertical (parent to offspring) event. Almost immediately after juveniles emerge from eggs, cells of *A. fischeri* in surrounding seawater begin to colonize them, entering through ciliated ducts that end in the immature light organ. Amazingly, the light organ becomes colonized specifically with

(a)

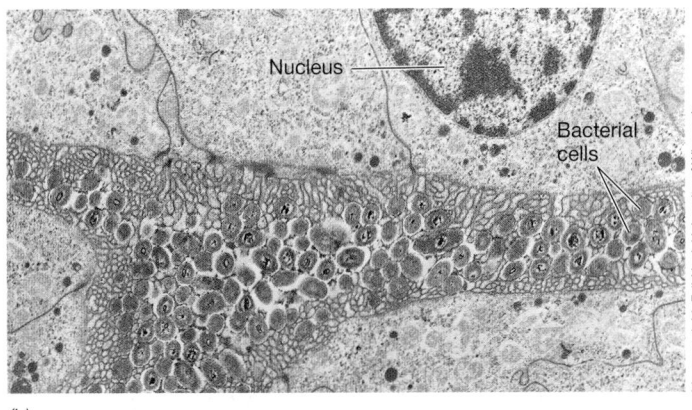

(b)

Figure 23.29 Squid–*Aliivibrio* symbiosis. *(a)* An adult Hawaiian bobtail squid, *Euprymna scolopes*, is about 4 cm long. (b) Thin-sectioned transmission electron micrograph through the *E. scolopes* light organ shows a dense population of bioluminescent *Aliivibrio fischeri* cells.

A. fischeri and not with any of the many other species of gram-negative bacteria present in the seawater. Even if large numbers of other species of bioluminescent bacteria are offered to juvenile squid along with low numbers of *A. fischeri*, only *A. fischeri* establishes residence in the light organ. This implies that the animal in some way recognizes and accepts *A. fischeri* cells and excludes those of other species.

The squid–*Aliivibrio* symbiosis develops in several stages. Contact of the squid with any bacterial cells triggers recognition in a very general way. Upon contact with peptidoglycan (a component of the cell wall of *Bacteria*, ⇄ Section 2.4), the young squid secretes mucus from its developing light organ. The mucus is the first layer of specificity in the symbiosis, as it makes gram-negative but not gram-positive bacteria aggregate. Within the aggregates of gram-negative cells that may contain only low numbers of *A. fischeri*, this bacterium outcompetes the other gram-negative bacteria to form a monoculture. Formation of the monoculture is aided by antimicrobial substances including the gas nitric oxide (NO, see below) in the mucus outside the light organ and is established quickly, within 2 h of a juvenile's hatching from an egg. The highly motile *A. fischeri* cells present in the aggregate migrate up the ducts and into the light organ tissues. Once there, they lose their flagella, become nonmotile, divide to form dense populations (Figure 23.29*b*), and trigger developmental events that lead to maturation of the host light organ. The light organ in a mature *E. scolopes* contains between 10^8 and 10^9 *A. fischeri* cells.

Colonization of *A. fischeri* by the squid is assisted by nitric oxide (NO). Nitric oxide is a well-known defense response of animal cells to attack by bacterial pathogens; the gas is a strong oxidant and causes sufficient oxidative damage to bacterial cells to kill them (⇄ Section 26.7). Nitric oxide produced by the squid is incorporated into the aggregates in the mucus and is present in the light organ itself. As *A. fischeri* colonizes the light organ, NO levels diminish rapidly. It appears that cells of *A. fischeri* can tolerate exposure to NO and consume it through the activity of NO-inactivating enzymes. The inability of other gram-negative bacteria in the aggregates to detoxify NO helps explain the sudden enrichment of *A. fischeri* in the ducts even before the actual colonization of the light organ. Then, after establishment, continued production of NO in the light organ prevents colonization by other bacterial species.

Propagating the Symbiosis

The squid matures into an adult in about 2 months and then lives a strictly nocturnal existence in which it feeds mostly on small crustaceans. During the day, the animal buries itself and remains quiescent in the sand. Each morning at dawn the squid nearly empties its light organ of *A. fischeri* cells and begins to grow a new population of the bacterium. The bacterial cells grow rapidly in the light organ; by midafternoon, the structure contains the dense populations of *A. fischeri* cells required for the production of visible light. The actual emission of light requires a certain density of cells and is controlled by the regulatory mechanism called *quorum sensing* (⇄ Section 6.8). The daily expulsion of bacterial cells is thought to be a mechanism for seeding the environment with cells of the bacterial symbionts. This, of course, increases the chances that the next generation of juvenile squid will be colonized.

A. fischeri grows much faster in the light organ than in the open ocean, presumably because it is supplied with nutrients by the squid. Thus *A. fischeri* benefits from the symbiosis by having an alternative habitat to seawater in which rapid growth and dense populations are possible. Isolation studies have shown that *A. fischeri* is not a particularly abundant marine bacterium. Daily expulsion of *A. fischeri* cells from the light organ increases the bacterium's numbers in the microbial community. Thus, the symbiotic relationship of the bacterium with the squid probably helps maintain larger *A. fischeri* populations than would exist if all cells were free-living. Because the competitive success of a microbial species is to some degree a function of population size (⇄ Section 20.1), this boost in cell numbers may confer an important ecological advantage on *A. fischeri* in its marine habitat.

— **MINIQUIZ** —

- Of what value is the squid–*Aliivibrio* symbiosis to the squid? To the bacterium?

- What features of the squid–*Aliivibrio* symbiosis make it an ideal model for studying animal–bacterial symbioses?

23.9 Marine Invertebrates at Hydrothermal Vents and Cold Seeps

Diverse invertebrate communities develop near undersea hot springs called *hydrothermal vents*. We covered the geochemistry and microbiology of hydrothermal vents and cold seeps of natural gas in Sections 20.14 and 21.1. Here we focus on hydrothermal vent animals and their microbial symbionts.

Macroinvertebrates, including tube worms over 2 m in length and large clams and mussels, are present near these vents (**Figure 23.30**). Photosynthesis cannot support these invertebrate communities because they exist below the photic zone. However, hydrothermal fluids contain large amounts of reduced inorganic materials, including H_2S, Mn^{2+}, H_2, and CO (carbon monoxide), and some vents contain high levels of ammonium (NH_4^+) instead of H_2S; all of these are good electron donors for chemolithotrophs, *Bacteria* and *Archaea* that use inorganic compounds as electron donors and fix CO_2 as their carbon source (Chapter 14). Thus, hydrothermal vent invertebrates are able to exist in permanent darkness because they receive nourishment through a symbiotic association with these autotrophic bacteria.

(a)

Dudley Foster, Woods Hole Oceanographic Institution

(b)

Carl Wirsen, Woods Hole Oceanographic Institution

Figure 23.30 Invertebrates living near deep-sea thermal vents. *(a)* *Riftia* (tube worms, phylum *Annelida*), showing the sheath (white) and plume (red) of the worm bodies. *(b)* Mussel bed in vicinity of a warm vent. Note yellow deposition of elemental sulfur from the oxidation of H_2S emitted from the vents.

Tube Worms, Mussels, and Giant Clams

Hydrothermal vent–associated animals either feed directly on cells of free-living chemolithotrophs or have formed tight symbiotic associations with them. Mutualistic chemolithotrophs are either tightly attached to the animal surface (that is, as *epibionts*) or actually live within the animal tissues, supplying organic compounds to the animals in exchange for a safe residence and ready access to the electron donors needed for their energy metabolism. For example, the 2-m-long tube worms (Figure 23.30*a*) lack a mouth, gut, and anus, but contain an organ consisting primarily of spongy tissue called the *trophosome*. This structure, which constitutes half the worm's weight, is filled with sulfur granules and large populations of spherical sulfur-oxidizing bacteria (**Figure 23.31**). Bacterial cells taken from trophosome tissue show activity of enzymes of the Calvin cycle, a major pathway for autotrophy (⇔ Section 14.5), but interestingly, they also contain enzymes of the reverse citric acid cycle, a second autotrophic pathway (⇔ Section 14.5). In addition, they show a suite of sulfur-oxidizing enzymes necessary to obtain energy from reduced sulfur compounds (⇔ Sections 14.9 and 15.11). The tube worms are thus nourished by organic compounds produced from CO_2 and secreted by the sulfur chemolithotrophs.

Along with tube worms, giant clams and mussels (Figure 23.30*b*) are also common near hydrothermal vents, and sulfur-oxidizing bacterial symbionts have been found in the gill tissues of these animals. Phylogenetic analyses have shown that each individual animal harbors one or more different strains of bacterial symbiont and that a variety of species of bacterial symbionts inhabit different species of vent animal. With the exception of the bacterial symbiont of *Riftia* (tube worms), which also has a free-living stage (Table 23.2), none of the bacterial symbionts of hydrothermal vent animals have yet been obtained in laboratory culture, even though they are fairly closely related to free-living sulfur chemolithotrophs (⇔ Sections 14.9, 15.11, and 16.5).

The red plume of the tube worm (Figure 23.30*a*) is rich in blood vessels and is used to trap and transport inorganic substrates to the

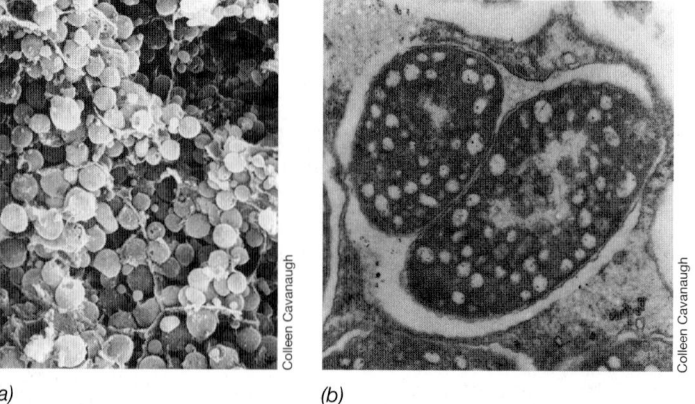

(a)

Colleen Cavanaugh

(b)

Colleen Cavanaugh

Figure 23.31 Chemolithotrophic sulfur-oxidizing bacteria associated with the trophosome tissue of tube worms from hydrothermal vents. *(a)* Scanning electron micrograph of trophosome tissue showing spherical chemolithotrophic sulfur-oxidizing bacteria. Cells are 3–5 μm in diameter. *(b)* Transmission electron micrograph of bacteria in sectioned trophosome tissue. The cells are frequently enclosed in pairs by an outer membrane of unknown origin. Reprinted with permission from *Science 213:* 340–342 (1981), © AAAS.

bacterial symbionts. Tube worms contain unusual hemoglobins that bind H_2S and O_2; these are then transported to the trophosome where they are released to the bacterial symbionts. The CO_2 content of tube-worm blood is also high, about 25 mM, and presumably this is released in the trophosome as a carbon source for the symbionts. In addition, stable isotope analyses (Section 19.10) of elemental sulfur from the trophosome have shown that its $^{34}S/^{32}S$ composition is the same as that of the sulfide emitted from the vent. This ratio is distinct from that of seawater sulfate and is further proof that geothermal sulfide is actually entering the worm in large amounts.

Other marine invertebrates have coevolved bacterial symbioses that supply their nutrition as well (**Table 23.3**). For example, methanotrophic (CH_4-consuming) symbionts are present in giant clams that live near cold seeps of natural gas at relatively shallow depths in the Gulf of Mexico. Although not autotrophs (CH_4 is an organic compound), the methanotrophs do provide nutrition to the clams; the methanotrophs use CH_4 as their electron donor and carbon source and secrete organic carbon to the clams. Molecular hydrogen (H_2) is used as an electron donor by the mussel *Bathymodiolus puteoserpentis*, the most abundant macrofauna in vent fields associated with the peridotite-hosted vent systems of the Mid-Atlantic Ridge (Section 20.14). These systems release extremely high levels of H_2 and CH_4, with measured H_2 concentrations as high as 19 mM. This mussel was previously shown to live in a dual symbiosis with methane-oxidizing bacteria and chemolithotrophic sulfur-oxidizing bacteria localized to the gill tissue. Remarkably, the sulfur-oxidizing symbiont of *B. puteoserpentis* also has the capacity to use H_2 as an electron donor, making this mussel one of the most versatile of vent macrofauna.

Genomics and Hydrothermal Vent Symbioses

Genome sequencing is revealing additional features of the metabolic interaction and coevolution of marine invertebrates and their bacterial symbionts. The genome sequence of the gill endosymbiont of the giant vent clam *Calyptogena magnifica* offers direct evidence for carbon fixation via the Calvin cycle; the genome encodes the key enzymes of the Calvin cycle, ribulose bisphosphate carboxylase (RubisCO) and phosphoribulokinase (Section 14.5), and contains genes encoding key sulfur oxidation processes. The genome of this symbiont also encodes the biosynthesis of most vitamins and cofactors and all 20 amino acids needed to support the host. However, because few substrate-specific transporters are encoded by the symbiont genome, it is suspected that the clam actually digests symbiont cells for nutrition, as do mussels (Table 23.3).

Like the obligate symbionts of insects, most symbionts of marine invertebrates have small genomes (Table 23.2), indicating reduced function and an obligate association with their host. The bacterial symbiont of the giant tube worm *Riftia pachyptila* is an exception, having a genome larger than some free-living sulfur-oxidizing chemolithotrophs (Table 23.2). The *R. pachyptila* symbiont is acquired by uninfected juvenile animals from the environment (horizontal transmission), and its larger genome is likely important for survival as a free-living bacterium.

— MINIQUIZ —

- How do giant tube worms receive their nutrition?
- What are the similarities of the obligate symbioses of insects and hydrothermal vent invertebrates?
- What factors determine the genome size of the symbionts of marine invertebrates?

TABLE 23.3 **Marine animals with chemolithotrophic or methanotrophic endosymbiotic *Bacteria***

Host phylum (genus or order)	Common name	Habitat	Symbiont metabolic type
Porifera (*Demospongiae*)	Sponge	Seeps	Methanotrophs
Platyhelminthes (*Catenulida*)	Flatworm	Shallow water	Sulfur chemolithotrophs
Nematoda (*Monhysterida*)	Mouthless nematode	Shallow water	Sulfur chemolithotrophs
Mollusca (*Solemya*, *Lucina*)	Clam	Vents, seeps, shallow water	Sulfur chemolithotrophs
Mollusca (*Calyptogena*)	Clam	Vents, seeps, whale falls[a]	Sulfur chemolithotrophs
Mollusca (*Bathymodiolus*)	Mussel	Vents, seeps, whale and wood falls[a]	Sulfur and H_2 chemolithotrophs, methanotrophs
Mollusca (*Alviniconcha*)	Snail	Vents	Sulfur chemolithotrophs
Annelida (*Riftia*)	Tube worm	Vents, seeps, whale and wood falls[a]	Sulfur chemolithotrophs

[a]Whale and wood falls are sunken whale carcasses and wood, respectively.

23.10 Entomopathogenic Nematodes

Here we consider two families of nematodes (the *Heterorhabditidae* and *Steinernematidae*) that are obligate pathogens of insects; together, they constitute a group of entomopathogenic (insect-killing) nematodes that are globally distributed and have a wide range of insect hosts. The basis for their insect lethality is a specific association between the species of nematode and its bacterial symbionts that produce a variety of insecticidal compounds.

Specificity of Entomopathogenic Nematodes for Their Symbionts and Insect Hosts

Entomopathogenic nematodes have been extensively studied as biocontrol agents, providing an alternative to broad-spectrum chemical pesticides in the control of insect pests. In addition to alleviating possible human health effects of chemical pesticides, biologically based insect control strategies can be highly specific, targeting only the pest species and not other, co-resident insects. This helps maintain the natural insect biodiversity in environments that require pest species management, most commonly in agricultural settings. Apart from their demonstrated utility in pest management, this symbiosis has also provided a powerful model system for developing a detailed understanding of the evolution, physiology, and genetics of beneficial host–microbe interactions.

UNIT 5

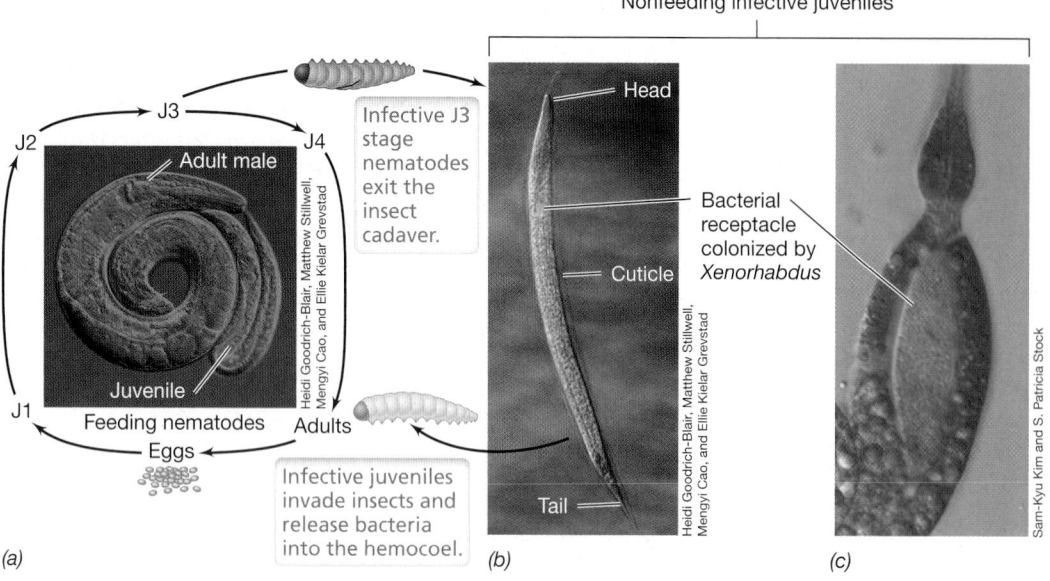

Figure 23.32   Nematode–*Xenorhabdus* life cycle. *(a)* In the presence of nutrients released in the infected insect, the nematodes develop through four juvenile stages (J1–J4), molting between stages, to egg-laying adults. When nutrients become limiting, the nematodes develop into an alternative nonfeeding J3 stage (known as the infective juvenile) that closes off most of its intestine and forms a receptacle in the gut where symbiotic bacteria localize. The infective J3 stage nematodes are then released into the environment to infect other insects. *(b)* *Xenorhabdus* localized to the bacterial receptacle of the infective juvenile. *(c)* Bacteria-filled receptacle (br) imaged by differential interference contrast microscopy (⇌ Section 1.6). Nematodes shown in *a* and *b* were genetically engineered to express the green fluorescent protein (⇌ Section 7.1) for imaging by fluorescence microscopy. Nematodes are approximately 50 µm in diameter.

Species of the gram-negative bacteria *Photorhabdus* and *Xenorhabdus* are the primary bacterial symbionts of the entomopathogenic nematodes. Comparative 16S rRNA sequence analysis (⇌ Section 13.7) has shown very high specificity of association between bacterial and nematode species. Species of *Photorhabdus* specifically associate with nematode species within the *Heterorhabditidae*, whereas *Xenorhabdus* species selectively associate with *Steinernematidae* nematodes. Each *Steinernema* species is thought to associate specifically with only one *Xenorhabdus* species. However, a single *Xenorhabdus* species may be associated with more than one nematode species. In turn, the nematodes comprising these two families can be separated into several phylogenetic groups that correspond to the phylogenetic relationships of their insect hosts. Together these data point to a history of cospeciation (coevolution) between nematodes and their bacterial symbionts that has resulted in the emergence of entomopathogenic nematodes having different insect host specificities.

The Nematode Life Cycle and Lethality

All entomopathogenic nematodes have a similar life cycle in which only one stage, the nonfeeding third-stage infective juvenile, survives outside the host (**Figure 23.32**). The long association between nematode and bacterial symbiont has also resulted in modification of the anterior part of the infective juvenile nematode's intestine. This region forms a discrete chamber in the infective juvenile, called the *bacterial receptacle* (Figure 23.32*b* and *c*). This vessel becomes filled with bacteria by growth of the symbiont during maturation of the infective juvenile, protecting the symbiont for the next infective cycle.

When the infective juvenile encounters an appropriate insect host, it invades the insect's hemocoel (the body cavity containing the bloodlike hemolymph of the insect's circulatory system) through a natural opening such as the mouth or anus. The bacteria are then released from the bacterial receptacle into the hemolymph and rapidly multiply. They are able to thwart the insect's natural immune system in part by producing a variety of hemolysins, toxins, and digestive enzymes (e.g., proteases, lipases, chitinases) that promote the release of nutrients from host tissues. The bacterial symbionts also produce antibiotics to inhibit competitive colonization by other microbes. Both the multiplying bacteria and digested host tissues nourish the multiplying nematodes, and this results in a slow insect death, taking anywhere from 1 to 22 days. As the nutrients in the host become depleted, the adult nematodes produce new infective nonfeeding juveniles that are adapted to withstand the outside environment (Figure 23.32*a*), and the life cycle repeats.

MINIQUIZ

- What evidence suggest that the nematodes and their bacterial symbionts have coevolved?
- What prevents other bacteria from colonizing the dead insect and competing with the nematode and *Xenorhabdus* for nutrients?

23.11 Reef-Building Corals

Coral reef ecosystems are the products of mutualistic associations between microscopic algae and simple marine animals. The extensive ecosystems associated with the worldwide distribution of these mutualisms support tens of thousands of species.

Phototrophic Symbioses with Animals

We began this chapter exploring lichens—the mutualism between a fungus and a phototrophic partner, either an alga or cyanobacterium. Like lichen fungi, some animals establish mutualistic associations with algae or cyanobacteria (**Table 23.4**). The animals in most of these associations are in phyla that display very simple body plans, for example, the *Porifera* (sponges) and *Cnidaria* (corals, sea anemones, and hydroids). These mutualistic animal–bacterial associations live in clear tropical waters where nutrients for the

TABLE 23.4 Symbioses between animals and phototrophic symbionts

Host	Common name	Symbionts[a]
Porifera	Sponge	Cyanobacteria, *Chlorella*, *Symbiodinium*
Cnidaria	Coral, sea anemone	*Symbiodinium*, *Chlorella*
Platyhelminthes	Flatworm	Diatoms, primitive chlorophytes
Mollusca	Snail, clam	*Symbiodinium*, *Chlorella*
Ascidia	Sea squirt	Cyanobacteria

[a]Cyanobacteria are *Bacteria*; all others are eukaryotic phototrophs.

animals are scarce, and the animal body typically has a large surface area relative to its volume and is thus well suited for capturing light.

There are only a few instances of algae forming associations with more complex animals, such as those in the phyla *Platyhelminthes* (flatworms), *Mollusca* (snails and clams), and *Urochordata* (sea squirts). In these cases either the animal has a suitable surface-to-volume ratio or has evolved specific light-gathering surfaces. The unicellular phototrophic symbionts are phylogenetically diverse and include cyanobacteria (⊆ Section 15.3), and red and green algae, diatoms, and dinoflagellates (Chapter 18). Most common are the green algae *Chlorella* (associating with sponges and freshwater hydras), cyanobacteria (associating with marine sponges), and species of the dinoflagellate genus *Symbiodinium*. Dinoflagellates and other alveolates comprise eight genera and around 2000 extant species (⊆ Section 18.4). Although dinoflagellate mutualisms are common, most are between species of *Symbiodinium* and marine invertebrates or protists. We focus here on the symbiotic association between the dinoflagellate *Symbiodinium* and the stony coral cnidarians.

The mutualisms between the cnidarian stony corals (order *Scleractinia*) and *Symbiodinium* are among the most spectacular and ecologically significant animal–phototroph associations (**Figure 23.33**). Together the corals and dinoflagellates form the trophic and structural foundation of the coral reef ecosystem. The cnidarians possess a very simple two-tissue-layer body plan (ectoderm and gastroderm) and harbor the dinoflagellate symbiont intracellularly in membrane-bound vesicles (symbiosomes) within cells of the inner (gastrodermal) tissue layer (Figure 23.33c). The coral symbiosomes are analogous to the bacteroid-filled vesicles that develop in plant cells of the legume root nodules (Section 23.3). The coral skeleton is an extremely efficient light-gathering structure that greatly enhances light harvesting by the *Symbiodinium*. The algae receive key inorganic nutrients from host metabolism and pass photosynthetically produced organic compounds to the corals. This mutualism has allowed coral reefs to develop in large expanses of nutrient-poor ocean waters.

Transmission, Specificity, and Benefits of the *Symbiodinium*–Coral Association

Reef-building corals reproduce sexually by releasing gametes into the seawater (broadcast spawning). A male and a female gamete fuse to form a free-swimming larva that later settles on a surface, where it may initiate a new coral colony. Algal symbionts are typically present in the egg before it is released from the parent (vertical transmission), although free-living *Symbiodinium* cells can also be ingested by juvenile corals (horizontal transmission). A developing coral that ingests dinoflagellates digests all of them except the particular *Symbiodinium* of its mutualism. After establishing an association, the coral controls the growth of *Symbiodinium* via chemical signaling and, following each cell division, each *Symbiodinium* daughter cell is allocated to a new symbiosome.

Both partners in the cnidarian–dinoflagellate mutualism have evolved adaptations for nutritional exchange. The dinoflagellates donate most of their photosynthetically fixed carbon (in the form of small molecules such as sugars, glycerol, and amino acids) to the cnidarian in exchange for inorganic nitrogen, phosphorus, and inorganic carbon from the host. The cnidarian is thought to control division of *Symbiodinium* through nitrogen limitation. Moreover, in addition to providing protection and inorganic nutrients, the calcium carbonate skeleton of corals is one of the most efficient collectors of solar radiation in nature, amplifying the incident light field for the symbionts by as much as fivefold; this benefits the symbiont in carrying out photosynthesis under a light-absorbing water column.

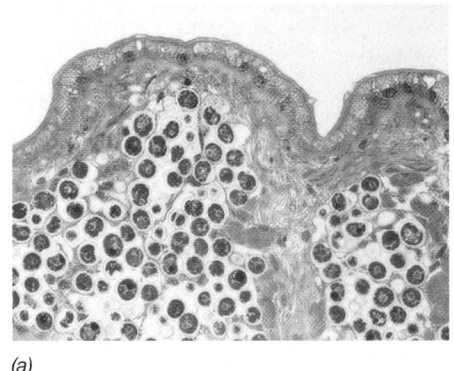
(a)

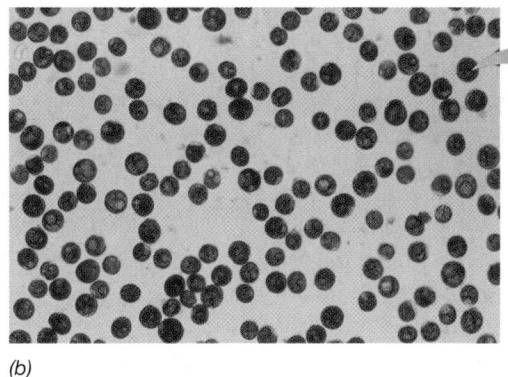
(b)

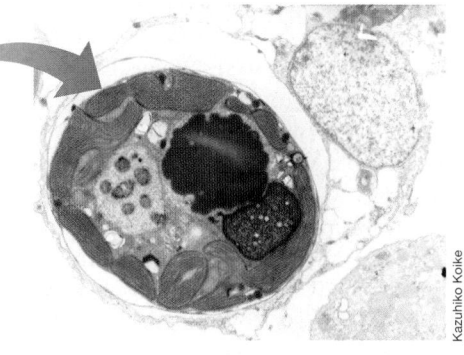
(c)

Figure 23.33 *Symbiodinium* symbiont of marine invertebrates. *(a)* Thin-section micrograph of *Symbiodinium* in the mantle tissue of a giant clam. *(b) Symbiodinium* cells recovered from a soft coral. *(c)* Transmission electron micrograph of a *Symbiodinium* cell within a vacuole of a cell of the stony coral *Ctenactis echinata*. The *Symbiodinium* cell is about 10 μm in diameter.

UNIT 5

(a)

(b)

Figure 23.34 Coral bleaching. *(a)* Two colonies of the brain coral *Colpophyllia natans*. The coral on the left is a healthy brown color, whereas the coral on the right is fully bleached. *(b)* A large colony of partially bleached mountainous star coral (*Orbicella faveolata*).

Coral Bleaching—The Risk of Harboring a Phototrophic Symbiont in a Changing World

Many of the extensive coral reef systems in the oceans worldwide are now threatened with extinction, primarily as a consequence of human activities. Ongoing loss of these beautiful and productive ecosystems is thought to be the result of elevated atmospheric CO_2; namely, increased sea surface temperature, rising sea levels, and ocean acidification (Sections 21.6 and 21.8). Coastal development also threatens reef systems, contributing to pollution from sewage discharge, eutrophication from nutrient runoff, and overfishing. These environmental changes are contributing to high mortality through disease, loss of coral structure from reduced calcification caused by acidification, and bleaching. Healthy corals harbor millions of cells of *Symbiodinium* per square centimeter of tissue. Coral bleaching is the loss of color from host tissues caused by the lysis of these symbionts, revealing the underlying white limestone skeleton (**Figure 23.34**).

Coral reefs live close to their optimum temperature, and it is the synergistic effect of increased sea surface temperature and irradiance that causes massive bleaching. Elevated temperature and high irradiance impair the photosynthetic apparatus of the dinoflagellates, resulting in the production of reactive oxygen species (for example, singlet oxygen and superoxide, Section 5.14) that cause damage to both host and symbiont. Bleaching is thought to be caused by a protective immune response of the host that destroys compromised symbionts. Increases in sea surface temperatures as small as 0.5–1.5°C above the local maximum, if sustained for several weeks, can induce rapid coral bleaching. A significant decrease in temperature below the optimum range for coral growth can have a similar effect. Thermal stress, accentuated by seasonal increases in electromagnetic radiation of ultraviolet and some visible wavelengths, has resulted in bleaching of huge expanses of coral reefs.

Although coral reefs are clearly threatened, there is much uncertainty in projecting their future. The more ominous projections, based on projected increases in sea temperature, point to a collapse of Indian Ocean coral reef systems within only a few years and a possible global collapse of coral reefs by the middle of this century. Moreover, the Great Barrier Reef off the northeast coast of Australia in the Coral Sea—the world's largest coral reef system—has experienced significant coral bleaching in recent years; nearly half the reef has been bleached. However, these observations and future projections lack much basic knowledge on the vulnerability of individual coral species and the adaptive capacity of individual coral–symbiont mutualisms. For example,

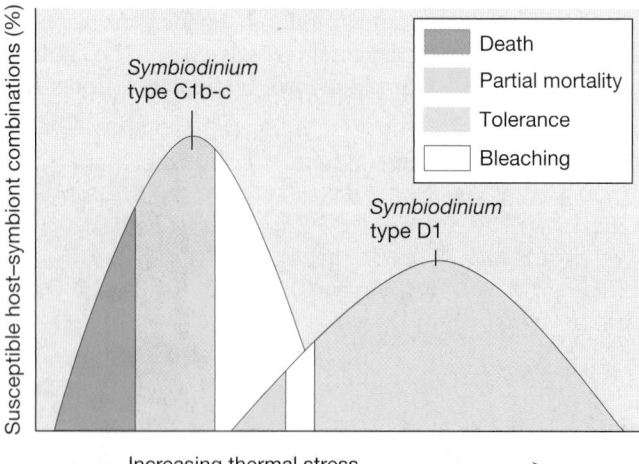

Figure 23.35 Differential stress tolerance of a coral species associated with different *Symbiodinium* phylotypes. *Pocillopora* corals symbiotically associated with *Symbiodinium* type C1b-c were much more sensitive to a thermal stress event than the same coral species associated with *Symbiodinium* type D1. The more tolerant *Symbiodinium* type D1–*Pocillopora* association suffered very low mortality. The response also suggested additional genetic variation within each *Symbiodinium* type, since the two mutualisms displayed a range of sensitivity to increasing thermal stress.

thermal tolerance is in part conferred by the species or strain of *Symbiodinium*, and following a bleaching event the mutualism can shift to a more thermally tolerant symbiont (**Figure 23.35**). Nevertheless, coral bleaching stands as additional evidence that climate change is upon us and that microbially based ecosystems are responding.

Molecular results have indicated that there are over 150 different *Symbiodinium* phylotypes, each possibly representing a distinct species with different stress tolerance. Both symbiont swapping and symbiont switching have been proposed as the underlying mechanism for shifting between symbionts. In switching, the symbiont is taken up from a water column population. In swapping, the shift results from differential growth of a genetic variant already associated with the coral, but in very low numbers, thereby swapping places with the previously dominant mutualist following the bleaching event. Most studies indicate that swapping is the more common adaptive mechanism, but uncertainty remains. Because the type of symbiont influences the ability of the coral to adapt to stresses associated with climate change, a more complete understanding of the alternative mechanisms of adaptive response, including possible symbiont switching, is essential to predicting the future health of corals, their symbionts, and the reefs they build.

MINIQUIZ

- What gives corals their spectacular colors?
- What are the two mechanisms of *Symbiodinium* transfer to developing corals?
- What are the major environmental factors contributing to coral bleaching?

V • Mammalian Gut Systems as Microbial Habitats

The evolution of animals has been shaped in part by a long history of symbiotic associations with microorganisms and includes vertebrates as well as the invertebrate systems we just explored. We end this chapter by considering microbial symbioses with mammals other than humans. In the next chapter, we focus specifically on the microbiota of humans. Microbes inhabit all sites on mammalian bodies, but the greatest diversity and density of microbes are found in the mammalian gut, and so we center our discussion there.

23.12 Alternative Mammalian Gut Systems

Some mammals are *herbivores*, consuming only plant materials, whereas others are *carnivores*, eating primarily the flesh of other animals. *Omnivores* eat both plants and animals. As **Figure 23.36** indicates, closely related mammals have evolved adaptations for differing diets. Notice that mammals of different lineages independently evolved the herbivorous lifestyle, mostly during the Jurassic period, an era in Earth's history of roughly 60 million years beginning about 200 million years ago.

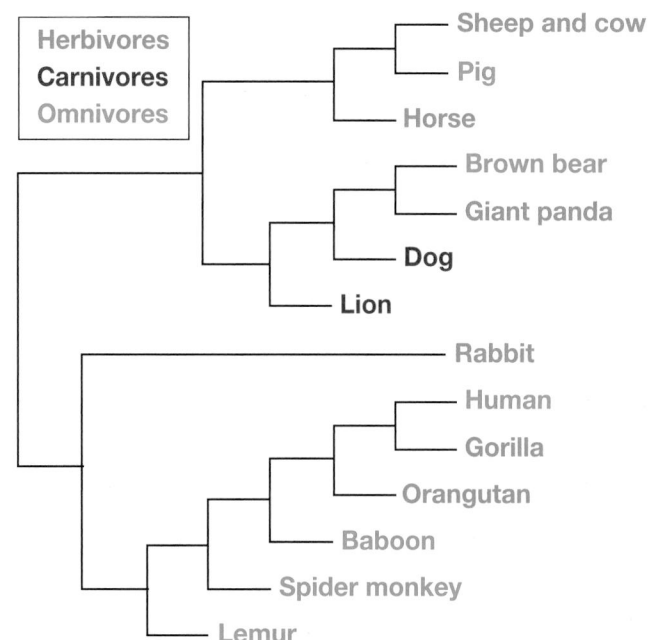

Figure 23.36 Phylogenetic tree showing multiple origins of herbivory among mammals. Some of the herbivores listed are foregut fermenters, while others are hindgut fermenters (see Figure 23.37). Instead of animal flesh, some mammalian carnivores eat only insects (the *insectivores*, such as bats), or fish (the *piscivores*, such as the river otter).

The massive evolutionary radiation of mammals during the Jurassic led to the evolution of several feeding strategies. Most mammalian species evolved gut structures that foster mutualistic associations with microorganisms. As anatomical differences evolved, microbial fermentation remained important or essential in mammalian digestion. *Monogastric* mammals, such as humans, have a single compartment, the stomach, positioned before the intestine. Such animals may get a substantial part of their energy requirement from microbial fermentation of otherwise indigestible foods, while herbivores are totally dependent on such fermentations.

Plant Substrates

Microbial associations with various mammalian species led to the capacity to catabolize plant fiber, the structural component of plant cell walls. Fiber is composed primarily of insoluble polysaccharides of which cellulose is the most abundant component. Mammals—and indeed virtually all animals—lack the enzymes necessary to digest cellulose and certain other plant polysaccharides. However, many microbes have genes encoding the glycoside hydrolases and polysaccharide lyases required to decompose these polysaccharides.

As the most abundant organic compound on Earth and one composed exclusively of glucose, *cellulose* offers a rich source of carbon and energy for animals that can digest it. The two primary traits that evolved to support herbivory are (1) an enlarged anoxic fermentation chamber for holding ingested plant material and (2) an extended retention time—the time that ingested material remains in the gut. A longer retention time allows for a longer association of microorganisms with the ingested material and thus a more complete degradation of the plant polymers.

Foregut fermenters Examples: Ruminants (photo 1), colobine monkeys, macropod marsupials, hoatzin (photo 2)

1.

2.

Hindgut fermenters Examples: Cecal animals (photos 3 and 4), primates, some rodents, some reptiles

3.

4.

Figure 23.37 Variations on vertebrate gut architecture. All vertebrates have a small intestine, but vary in other gut structures. Most host absorption of dietary nutrients occurs in the small intestine, whereas microbial fermentation can occur in the forestomach, cecum, or large intestine (colon). Foregut fermentation is found in four major clades of mammals and one avian species (the hoatzin). Hindgut fermentation, either in the cecum or large intestine/colon, is common to many clades of mammals (including humans), birds, and reptiles. Compare with Figure 23.36.

Labels on diagram: Foregut fermentation chamber; Acidic stomach; Small intestine; Large intestine (colon); Cecum; Hindgut fermentation chambers

Foregut versus Hindgut Fermenters

Two digestive patterns have evolved in herbivorous mammals. In herbivores with a *foregut* fermentation, the microbial fermentation chamber *precedes* the small intestine. This gut architecture originated independently in ruminants, colobine monkeys, sloths, and macropod marsupials (**Figure 23.37**). These all share the common feature that ingested nutrients are degraded by the gut microbiota *before* reaching the acidic stomach and small intestine. We examine the digestive processes of ruminants, as examples of foregut fermenters, in the next section.

Horses and rabbits are herbivorous mammals, but they are not foregut fermenters. Instead, these animals are *hindgut* fermenters. They have only one stomach but use an organ called the *cecum*, a digestive organ located between the small and large intestines, as their fermentation vessel (Figure 23.37). The cecum contains fiber- and cellulose-digesting (cellulolytic) microorganisms. Mammals, such as the rabbit, that rely primarily on microbial breakdown of plant fiber in the cecum are called *cecal fermenters*. In other hindgut fermenters, both the cecum and colon are major sites of fiber breakdown by microorganisms.

Anatomical differences among monogastric mammals, foregut fermenters, and hindgut fermenters are summarized in Figure 23.37.

Nutritionally, foregut fermenters have an advantage over hindgut fermenters in that the cellulolytic microbial community of the foregut eventually passes through an acidic stomach. As this occurs, most microbial cells are killed by the acidity and become a protein source for the animal. By contrast, in animals such as horses and rabbits, the remains of the cellulolytic community pass out of the animal in the feces because of its position posterior to the acidic stomach.

MINIQUIZ

- How do animals with foregut and hindgut fermentation differ in recovery of nutrients from plants?
- How does retention time affect microbial digestion of food in a gut compartment?

23.13 The Rumen and Ruminant Animals

A very successful group of foregut fermenters are *ruminants*, herbivorous mammals that possess a special digestive organ, the **rumen**, within which cellulose and other plant polysaccharides are digested by microorganisms. Some of the most important

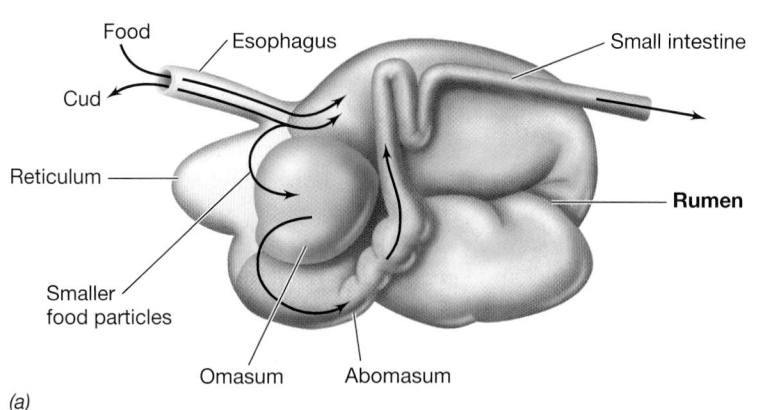

Figure 23.38 The rumen. *(a)* Schematic diagram of the rumen and gastrointestinal system of a cow. Food travels from the esophagus into the reticulo-rumen, consisting of the reticulum and rumen. Cud is regurgitated and chewed until food particles are small enough to pass from the reticulum into the omasum, abomasum, and intestines, in that order. The abomasum is an acidic vessel, analogous to the stomach of monogastric animals like pigs and humans. *(b)* Photo of a fistulated Holstein cow. The fistula, shown unplugged, is a sampling port that allows access to the rumen.

domesticated animals—cows, sheep, and goats—are ruminants. Camels, buffalo, deer, reindeer, caribou, and elk are also ruminants. Indeed, ruminants are Earth's dominant herbivores. Because the human food economy depends to a great extent on ruminant animals, rumen microbiology is of considerable economic significance and importance.

Rumen Anatomy and Activity

Unique features of the rumen as a site of cellulose digestion are its relatively large size (capable of holding 100–150 liters in a cow, 6 liters in a sheep) and its position in the gastrointestinal system *before* the acidic stomach. The rumen's warm and constant temperature (39°C), narrow pH range (5.5–7, depending on when the animal was last fed), and anoxic environment are also important factors in overall rumen function. **Figure 23.38***a* shows the relationship of the rumen to other parts of the ruminant digestive system. The digestive processes and microbiology of the rumen have been well studied, in part because it is possible to remove samples for analysis by way of a sampling port, called a *fistula*, implanted into the rumen of a cow (Figure 23.38*b*) or a sheep.

After a cow swallows its food, the food enters the first chamber of the four-compartment stomach, the reticulum. Partially digested plant materials flow freely between the rumen and reticulum, sometimes referred to together as the *reticulo-rumen*. The main function of the reticulum is to collect smaller food particles and move them to the omasum. Larger food particles (called *cud*) are regurgitated, chewed, mixed with saliva containing bicarbonate, and returned to the reticulo-rumen where they are digested by rumen bacteria. Solids may remain in the rumen for more than a day during digestion. Eventually, small and more thoroughly digested food particles are passed to the omasum and from there to the abomasum, an organ similar to a true, acidic stomach. In the abomasum, chemical digestive processes begin that continue in the small and large intestine.

Microbial Fermentation in the Rumen

Food remains in the rumen for 20–50 h depending on the feeding schedule and other factors. During this relatively long retention time, cellulolytic microorganisms hydrolyze cellulose, which frees

glucose. The glucose then undergoes bacterial fermentation with the production of **volatile fatty acids (VFAs)**, primarily *acetic, propionic,* and *butyric* acids, and the gases carbon dioxide (CO_2) and methane (CH_4) (**Figure 23.39**). The VFAs pass through the

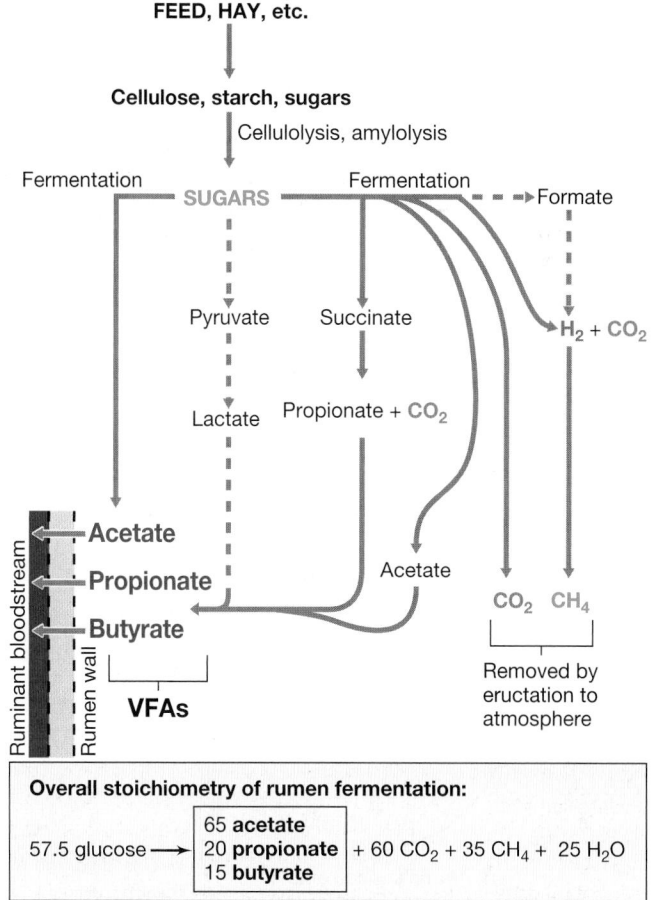

Figure 23.39 Biochemical reactions in the rumen. The major pathways are solid lines; dashed lines indicate minor pathways. Approximate steady-state rumen levels of volatile fatty acids (VFAs) are acetate, 60 mM; propionate, 20 mM; butyrate, 10 mM.

UNIT 5

rumen wall into the bloodstream and are oxidized by the animal as its main source of energy. The gaseous fermentation products CO_2 and CH_4 are released by eructation (belching).

The rumen contains enormous numbers of bacteria (10^{10}–10^{11} cells/g of rumen contents). Most of the bacteria adhere tightly to food particles. These particles proceed through the gastrointestinal tract of the animal where they undergo further digestive processes similar to those of nonruminant animals. Bacterial cells that digested plant fiber in the rumen are themselves digested in the acidic abomasum. Because bacteria living in the rumen biosynthesize amino acids and vitamins, the digested bacterial cells are a major source of protein and vitamins for the animal.

Rumen Bacteria

Anaerobic bacteria dominate in the rumen because it is a strictly anoxic compartment, and some anaerobic microbial eukaryotes are also present. Cellulose is converted to fatty acids, CO_2, and CH_4 (Figure 23.39) in a multistep microbial food chain, with several different anaerobes participating in the process. Recent estimates of rumen microbial diversity from analysis of 16S rRNA gene sequences suggest that the typical rumen contains 300–400 bacterial "species" (defined as "operational taxonomic units" sharing less than 97% sequence identity, ⮌ Section 13.8) (**Figure 23.40**). This is more than 10 times higher than culture-based diversity estimates. Molecular surveys show that species of *Firmicutes* and *Bacteroidetes* dominate the *Bacteria* in the rumen, while methanogens make up virtually the entire archaeal population (Figure 23.40).

A number of rumen anaerobes have been cultured and their physiology characterized (**Table 23.5**). Several different rumen

bacteria hydrolyze cellulose to sugars and ferment the sugars to VFAs. *Fibrobacter succinogenes* and *Ruminococcus albus* are the two most abundant cellulolytic rumen anaerobes. Although both organisms produce cellulases, *Fibrobacter*, a gram-negative bacterium, produces enzymes localized to its outer membrane. *Ruminococcus*, a gram-positive bacterium (and therefore lacking an outer membrane) produces a cellulose-degrading protein complex stabilized by scaffold proteins and bound to its cell wall. Thus, cells of both *Fibrobacter* and *Ruminococcus* need to bind to cellulose particles in order to degrade them.

If a ruminant is gradually switched from cellulose to a diet high in starch (grain, for instance), the starch-digesting bacteria *Ruminobacter amylophilus* and *Succinomonas amylolytica* grow to high numbers in the rumen. On a low-starch diet these organisms are typically minor constituents. If an animal is fed legume hay, which is high in pectin, a complex polysaccharide containing both hexose and pentose sugars, then the pectin-digesting bacterium *Lachnospira multipara* (Table 23.5) becomes an abundant member of the rumen microbial community. Some of the fermentation products of these rumen bacteria are used as energy sources by secondary fermenters in the rumen. For example, succinate is fermented to propionate plus CO_2 (Figure 23.39) by the bacterium *Schwartzia*, and lactate is fermented to acetate and other fatty acids by *Selenomonas* and *Megasphaera* (Table 23.5). Hydrogen (H_2) produced in the rumen by fermentative processes never accumulates because it is quickly consumed by methanogens for the reduction of CO_2 to CH_4. H_2 removal facilitates greater fermentative activity because H_2 accumulation negatively affects the energetics of fermentative reactions that produce H_2 (⮌ Section 14.23).

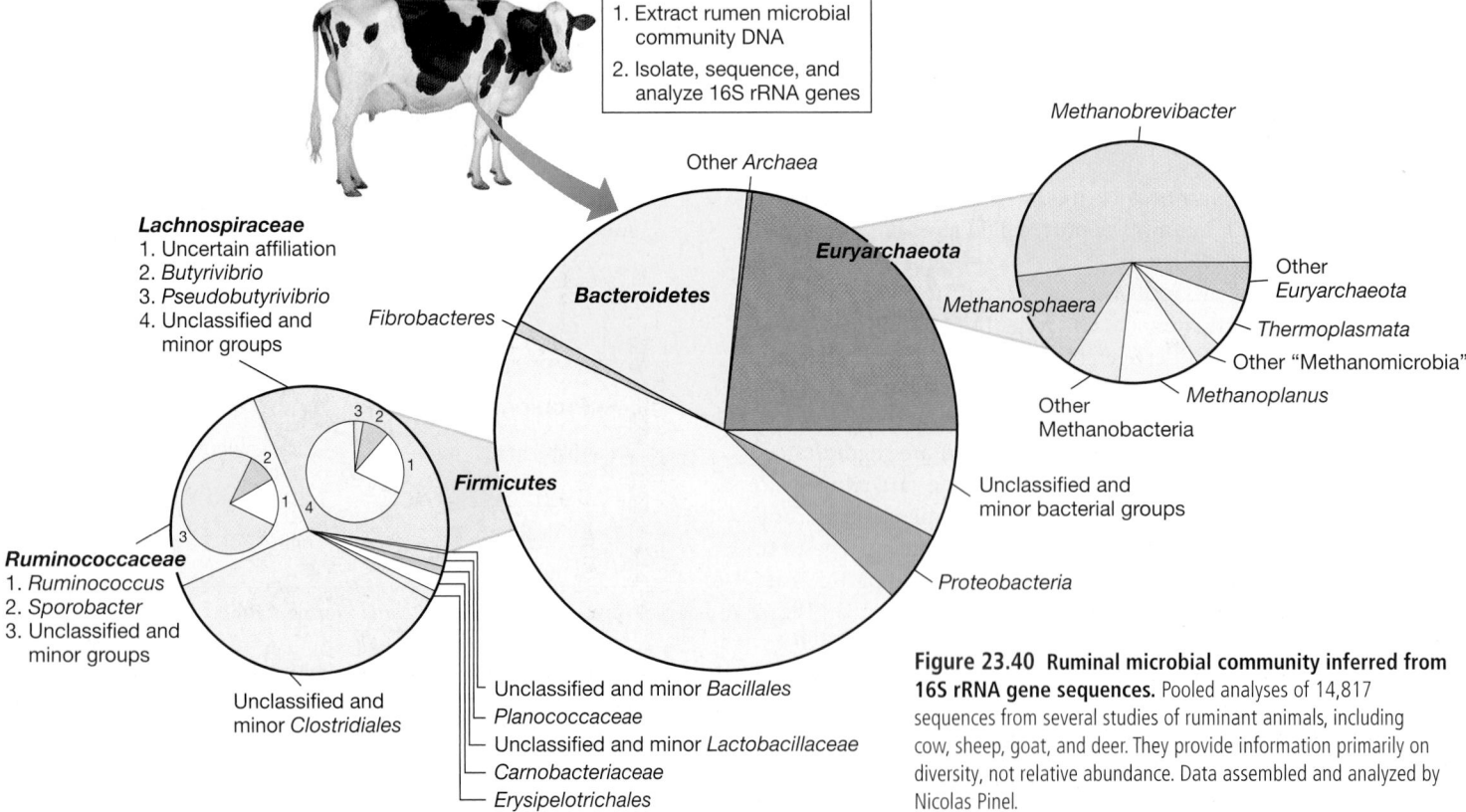

Figure 23.40 Ruminal microbial community inferred from 16S rRNA gene sequences. Pooled analyses of 14,817 sequences from several studies of ruminant animals, including cow, sheep, goat, and deer. They provide information primarily on diversity, not relative abundance. Data assembled and analyzed by Nicolas Pinel.

TABLE 23.5 Characteristics of some rumen *Bacteria* and *Archaea*

Organism[a]	Morphology	Fermentation products
Cellulose decomposers		
Gram-negative		
Fibrobacter succinogenes[b]	Rod	Succinate, acetate, formate
Butyrivibrio fibrisolvens[c]	Curved rod	Acetate, formate, lactate, butyrate, H_2, CO_2
Gram-positive		
Ruminococcus albus[c]	Coccus	Acetate, formate, H_2, CO_2
"*Clostridium lochheadii*"	Rod (endospores)	Acetate, formate, butyrate, H_2, CO_2
Starch decomposers		
Gram-negative		
Prevotella ruminicola[d]	Rod	Formate, acetate, succinate
Ruminobacter amylophilus	Rod	Formate, acetate, succinate
Selenomonas ruminantium	Curved rod	Acetate, propionate, lactate
Succinomonas amylolytica	Oval	Acetate, propionate, succinate
Gram-positive		
Streptococcus bovis	Coccus	Lactate
Lactate decomposers		
Gram-negative		
Selenomonas ruminantium subsp. *lactilytica*	Curved rod	Acetate, succinate
Megasphaera elsdenii	Coccus	Acetate, propionate, butyrate, valerate, caproate, H_2, CO_2
Succinate decomposer		
Gram-negative		
Schwartzia succinovorans	Rod	Propionate, CO_2
Pectin decomposer		
Gram-positive		
Lachnospira multipara	Curved rod	Acetate, formate, lactate, H_2, CO_2
Methanogens		
Methanobrevibacter ruminantium	Rod	CH_4 (from $H_2 + CO_2$ or formate)
Methanomicrobium mobile	Rod	CH_4 (from $H_2 + CO_2$ or formate)

[a]Except for the methanogens, which are *Archaea*, all organisms listed are *Bacteria*.
[b]These species also degrade xylan, a major plant cell wall polysaccharide.
[c]Also degrades starch.
[d]Also ferments amino acids, producing NH_3. Several other rumen bacteria ferment amino acids as well, including *Peptostreptococcus anaerobius* and *Clostridium sticklandii*.

Dangerous Changes in the Rumen Microbial Community

Significant changes in the microbial composition of the rumen can cause illness or even death of the animal. For example, if a cow is changed abruptly from forage to a grain diet, the gram-positive bacterium *Streptococcus bovis* grows rapidly in the rumen. The normal level of *S. bovis*, about 10^7 cells/g, is an insignificant fraction of total rumen bacterial numbers. But if large amounts of grain are fed abruptly, numbers of *S. bovis* can quickly rise to over 10^{10} cells/g and dominate the rumen microbial community. This occurs because grasses contain mainly cellulose, which does not support growth of *S. bovis*, while grain contains high levels of starch, on which *S. bovis* grows rapidly.

Because *S. bovis* is a lactic acid bacterium (⟳ Sections 14.20 and 16.6), large populations are capable of producing large amounts of its fermentation product, lactic acid. Lactic acid is a much stronger acid than the VFAs produced during normal rumen function. Lactate production thus acidifies the rumen below its lower functional limit of about pH 5.5, thereby disrupting the activities of normal rumen bacteria. Rumen acidification, a condition called *acidosis*, causes inflammation of the rumen epithelium, and severe acidosis can cause hemorrhaging in the rumen, acidification of the blood, and death of the animal.

Despite the activities of *S. bovis*, ruminants such as cattle can be fed a diet exclusively of grain. However, to avoid acidosis, they must be switched from forage to grain *gradually* over a period of many days. The slow introduction of starch selects for VFA-producing, starch-degrading bacteria (Table 23.5) instead of *S. bovis*, and thus normal rumen functions continue and the animal remains healthy.

Protective Changes in the Rumen Microbial Community

The overgrowth of *S. bovis* is an example of how a single microbial species can have a deleterious effect on animal health. There is also at least one well-studied example of how a single bacterial species can *enhance* the health of ruminant animals; in this case, animals fed the tropical legume *Leucaena leucocephala*. This plant has a very high nutritional value, but contains an amino acid–like compound called *mimosine* that is converted to toxic 3-hydroxy-4(1H)-pyridone (3,4-DHP) and 2,3-dihydroxypyridine (2,3-DHP) by rumen microorganisms (**Figure 23.41**). The observation that ruminants in

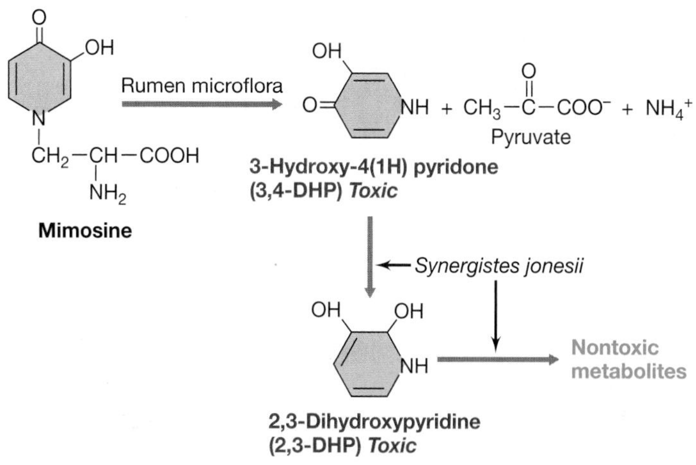

Figure 23.41 Conversion of mimosine to toxic pyridine and pyridone metabolites by ruminal microorganisms. Mimosine is converted to toxic 3,4-DHP by normal ruminal microbiota. *Synergistes jonesii* converts 3,4-DHP to nontoxic metabolites through a 2,3-DHP intermediate, preventing buildup of toxic metabolites of mimosine.

Hawaii, but not Australia, could feed on *Leucaena* without toxic effect led investigators to hypothesize that further metabolism of DHP by bacteria present in Hawaiian ruminants alleviated DHP toxicity. This was subsequently confirmed by the isolation of the bacterium *Synergistes jonesii*, a unique anaerobe related to the *Deferribacter* group (⟵ Section 16.21) and not closely related to any other rumen bacteria. Inoculation of Australian ruminants with cells of *S. jonesii* conferred resistance to mimosine by-products, allowing the animals to feed on *Leucaena* without ill effect.

The success of this single-organism modification of the rumen microbial community has encouraged further studies of this sort, including genetic engineering of bacteria to improve their ability to utilize available nutrients or to detoxify toxic substances. A notable success has been inoculation of the rumen of sheep with genetically engineered cells of *Butyrivibrio fibrisolvens* (Table 23.5) containing a gene encoding the enzyme fluoroacetate dehalogenase; this successfully prevented fluoroacetate poisoning of sheep fed plants containing high levels of this highly toxic inhibitor of the citric acid cycle.

Rumen Protists and Fungi

In addition to huge populations of *Bacteria* and *Archaea*, the rumen has characteristic populations of ciliated protists (Chapter 18) present at a density of about 10^6 cells/ml. Many of these protists are obligate anaerobes, a physiology that is rare among eukaryotes. Although these protists are not essential for rumen fermentation, they contribute to the overall process. In fact, some protists are able to hydrolyze cellulose and starch and ferment glucose with the production of the same VFAs formed by cellulose-fermenting bacteria (Figure 23.39 and Table 23.5). Rumen protists also consume rumen bacteria and smaller rumen protists and are likely to play a role in controlling bacterial densities in the rumen. An interesting commensal interaction has been observed between rumen protists that produce VFAs and H_2 as products and methanogenic bacteria that consume the H_2, producing CH_4. Because their cells autofluoresce (⟵ Section 14.17), methanogens are easily observed in rumen fluid bound to the surface of H_2-producing protists.

Anaerobic fungi also inhabit the rumen and play a role in its digestive processes. Rumen fungi are typically species that alternate between a flagellated and a thallus form, and studies with pure cultures have shown that they can ferment cellulose to VFAs. *Neocallimastix*, for example, is an obligately anaerobic fungus that ferments glucose to formate, acetate, lactate, ethanol, CO_2, and H_2. Although a eukaryote, this fungus lacks mitochondria and cytochromes and thus lives an obligately fermentative existence. However, *Neocallimastix* cells contain a redox organelle called the *hydrogenosome*; this mitochondrial analog evolves H_2 and has otherwise been found only in certain anaerobic protists (⟵ Section 2.15).

Rumen fungi play an important role in the degradation of polysaccharides other than cellulose, including a partial solubilization of lignin (the strengthening agent in the cell walls of woody plants), hemicellulose (a derivative of cellulose that contains pentoses and other sugars), and pectin.

We now move on to Chapter 24 where the microbial world that populates humans will be center stage and considered in some detail.

— **MINIQUIZ** —

- What physical and chemical conditions prevail in the rumen?
- What are VFAs and of what value are they to the ruminant?
- Why is the metabolism of *Streptococcus bovis* of special concern for ruminant nutrition?

MasteringMicrobiology®　**Visualize**, **explore**, and **think critically** with Interactive Microbiology, MicroLab Tutors, MicroCareers case studies, and more. MasteringMicrobiology offers practice quizzes, helpful animations, and other study tools for lecture and lab to help you master microbiology.

Chapter Review

I • Symbioses between Microorganisms

23.1 Lichens are a mutualistic association between one or two species of fungi and an oxygenic phototroph, either an alga or a cyanobacterium.

> **Q** Describe some similarities and differences between the lichen and coral symbioses (Section 23.11).

23.2 The consortium "*Chlorochromatium aggregatum*" is a mutualism between a phototrophic green sulfur bacterium and a motile heterotrophic bacterium. Mutual benefit is based on the phototroph supplying organic matter to the heterotroph in exchange for motility that permits rapid repositioning in stratified lakes to obtain optimal light and nutrients.

> **Q** What mechanisms do the consortia use to orient at the appropriate depth in the water column? What have genomic studies revealed about the nature of the *Chlorochromatium* mutualism?

II • Plants as Microbial Habitats

23.3 One of the most agriculturally important plant–microbial symbioses is that between legumes and nitrogen-fixing bacteria. The bacteria induce the formation of root nodules within which nitrogen fixation occurs. The plant provides the energy needed by the root nodule bacteria, and the bacteria provide fixed nitrogen for the plant.

> **Q** Describe the steps in the development of root nodules on a leguminous plant. What is the nature of

the recognition between plant and bacterium and how do Nod factors help control this? How does this compare with recognition in the *Agrobacterium*–plant system (Section 23.5)?

23.4 Mycorrhizae are mutualistic associations between fungi and the roots of plants that allow the plant to extend its root system via intimate interaction with an extensive network of fungal mycelia. Both ectomycorrhizae and endomycorrhizae are known. The mycelial network provides the plant with essential inorganic nutrients, and the plant, in turn, supplies organic compounds to the fungus.

> **Q** **How do mycorrhizae improve the growth of trees? In what way(s) are the root nodule and mycorrhizal symbioses similar? In what major way do they differ?**

23.5 The crown gall bacterium *Agrobacterium* enters into a unique relationship with plants. Part of the Ti plasmid in the bacterium can be transferred into the genome of the plant, initiating crown gall disease. The Ti plasmid has also been used for the genetic engineering of crop plants.

> **Q** **Compare and contrast the production of a plant tumor by *Agrobacterium tumefaciens* and a root nodule by a *Rhizobium* species. In what ways are these structures similar? In what ways are they different? Of what importance are plasmids to the development of both structures?**

III • Insects as Microbial Habitats

23.6 A large proportion of insects have established obligate mutualisms with bacteria, the basis of the mutualism often being bacterial biosynthesis of nutrients such as amino acids that are absent from the food the insect feeds on. Long-established obligate mutualisms are marked by extreme genome reduction of the symbiont, with retention of only those genes essential for the mutualism.

> **Q** **How is it possible for aphids to feed only on the carbohydrate-rich but nutrient-poor sap of phloem vessels in plants? Why do symbionts that are transmitted horizontally show less genome reduction, as opposed to the significant genome reduction observed in heritable symbionts?**

23.7 Termites associate symbiotically with bacteria and protists capable of digesting plant cell walls. The unique termite gut configuration and the hindgut microbial community composed largely of cellulolytic bacteria and protists and acetogenic bacteria result in high levels of acetate, the primary source of carbon and energy for the termite.

> **Q** **How do the microbial communities of higher and lower termite guts differ in composition and degradation of cellulose?**

IV • Other Invertebrates as Microbial Habitats

23.8 A light-emitting organ on the underside of the Hawaiian bobtail squid provides a habitat for bioluminescent cells of the bacterium *Aliivibrio fischeri*. From the mutualism in the light organ, the squid gains protection from predators while the bacterium benefits from a habitat in which it grows quickly and contributes cells to its free-living population.

> **Q** **How is the correct bacterial symbiont selected in the squid–*Aliivibrio* symbiosis?**

23.9 Most invertebrates living on the seafloor near regions receiving hydrothermal fluids have established obligate mutualisms with chemolithotrophic bacteria. These mutualisms are nutritional, allowing the invertebrates to thrive in an environment enriched in reduced inorganic materials, such as H_2S, that are abundant in vent fluids. The invertebrates provide the symbionts an ideal nutritional environment in exchange for organic nutrients.

> **Q** **Why is the genome of the tube worm symbiont thought to be so much larger than the genomes of insect symbionts?**

23.10 Entomopathogenic nematodes have established symbiotic associations with species of *Photorhabdus* and *Xenorhabdus* bacteria. Following invasion of the insect by the nematode, bacterial symbionts are released into the insect's hemolymph and multiply rapidly by thwarting the insect's immune system, killing the insect by the release of toxins and digestive enzymes. When nutrients are depleted, the nematodes then transition to a nonfeeding juvenile form, harboring the symbiont in a specialized receptacle that will go on to infect other insects.

> **Q** **Why are entomopathogenic nematodes so attractive for the biocontrol of insect pest species?**

23.11 The mutualism between the dinoflagellate *Symbiodinium* and the stony corals produces the extensive worldwide coral reef ecosystems that sustain a tremendous diversity of marine life. Coral bleaching caused by climate change threatens these ecosystems.

> **Q** **How does the body plan of corals influence their ability to symbiotically associate with *Symbiodinium*?**

V • Mammalian Gut Systems as Microbial Habitats

23.12 Microbial fermentation is important for digestion in all mammals. Several microbial mutualisms have evolved in different mammals that allow for the digestion of different types of food. Herbivores derive almost all of their carbon and energy from plant fiber.

> **Q** **What are the major benefits and the disadvantages of a rumen system? How does a cecal animal compare with a ruminant?**

23.13 The rumen, the digestive organ of ruminant animals, specializes in cellulose digestion, which is carried out by microorganisms. Bacteria, protists, and fungi in the rumen produce volatile fatty acids that provide energy for the ruminant. Rumen microorganisms synthesize vitamins and amino acids and are also a major source of protein—all used by the ruminant.

Q Give an example of a single microbial species contributing to herbivore nutrition. What is an example of a single microbial species that can contribute to herbivore pathology?

Application Questions

1. Imagine that you have discovered a new animal that consumes only grass in its diet. You suspect it to be a ruminant and have available a specimen for anatomical inspection. If this animal is a ruminant, describe the position and basic components of the digestive tract you would expect to find and any key microorganisms and substances you might look for. What metabolic types of microorganisms or specific genes would you predict would be present?

2. Why would you be very surprised to find the exact same microbial symbionts inhabiting a lichen and the rumen of a cow? Consider both the physical and chemical conditions of the habitat and the requirement for specific microbes to be present in order for the symbiosis to be successful.

Chapter Glossary

Arbuscule branched or coiled hyphal structure within cells of the inner cortex of plants with a mycorrhizal infection

Bacteriocyte a specialized insect cell in which bacterial symbionts reside

Bacteriome a specialized region in several insect groups that contains insect bacteriocyte cells packed with bacterial symbionts

Bacteroid the misshapen cells of rhizobia inside a leguminous plant root nodule; can fix N_2

Coevolution evolution that proceeds jointly in a pair of intimately associated species owing to the effects each has on the other

Consortium a mutualism between bacteria, for example, a phototrophic green sulfur bacterium and a motile nonphototrophic bacterium

Infection thread in the formation of root nodules, a cellulosic tube through which *Rhizobium* cells can travel to reach and infect root cells

Leghemoglobin an O_2-binding protein found in root nodules

Lichen one or two species of fungi and an alga (or cyanobacterium) living in symbiotic association

Mutualism a symbiosis in which both partners benefit

Myc factors lipochitin oligosaccharides produced by mycorrhizal fungi to initiate symbiosis with a plant

Mycorrhizae a symbiotic association between a fungus and the roots of a plant

Nod factors lipochitin oligosaccharides produced by root nodule bacteria that help initiate the plant–bacterial symbiosis

Root nodule a tumorlike growth on plant roots that contains symbiotic nitrogen-fixing bacteria

Rumen the major fermentation vessel in the multichambered gut of ruminant animals, where most cellulose digestion occurs

Symbiosis an intimate relationship between organisms, often developed through prolonged association and coevolution

Ti plasmid a conjugative plasmid in the bacterium *Agrobacterium tumefaciens* that can transfer genes into plants

Volatile fatty acids (VFAs) the major fatty acids (acetate, propionate, and butyrate) produced during fermentation in the rumen

Microbial Symbioses with Humans

microbiology**now**

Frozen in Time: The Iceman Microbiome

Humans and their microbial associates—collectively called the *human microbiome*—have coevolved for millennia. As we will see in this chapter, the human microbiome influences a person's health, disease, and predisposition to disease. Among our intimate microbial associates, the pathogenic bacterium *Helicobacter pylori* is known to have developed a close relationship with humans in the distant past and to have coevolved with humans. *H. pylori* colonizes the stomachs of about half the human race. Although this bacterium generally does not cause overt disease, it is a major risk factor for the development of ulcers and stomach cancer. Moreover, because *H. pylori* is transmitted primarily by contact within families, the distribution of genetic variants of this bacterium may yield clues to past human migrations.

Unraveling the details of the *H. pylori* ancestry is complicated by the ability of different strains of this bacterium to recombine their genetic information. Because the DNA of various strains has mixed over long periods, the reconstruction of population movement inferred from genome sequences of modern *H. pylori* strains is incomplete. One of the biggest unanswered questions was the origin of strains now common among modern Europeans, which appear to be hybrids of strains originating in Asia and Africa. Unfortunately, the sequence data did not point to a reliable time interval in which that mingling of human populations occurred—an important period of human migration that was estimated to have occurred 10,000–50,000 years ago.

This estimate has now been greatly refined following the remarkable discovery of a well-preserved 5300-year-old European Copper Age mummy frozen in the Italian Alps. Using the newest methods for DNA sequencing, it was possible to reconstruct the genome of *H. pylori* preserved in the stomach of the "Iceman" (see photo), the corpse discovered when melting ice revealed the human remains on the side of a mountain. The Iceman *H. pylori* genome sequence turned out to be an almost pure representative of the Asian population, which means this *H. pylori* strain was present in Europe before hybridization of African and Asian strains produced the modern European variant. Thus, by employing historical biogeography, we now know this important period of human migration was much more recent than previously thought.

Source: Maixner, F., et al. 2016. The 5300-year-old *Helicobacter pylori* genome of the Iceman. *Science 351:* 162–165.

we one, or are we many? This somewhat philosophical question has been asked about the functional significance of microorganisms colonizing our bodies. In their totality, this massive assemblage of microorganisms is called the **human microbiome**. A **microbiome** can be defined as a functional collection of different microbes in a particular environmental system. The term **microbiota**, by contrast, is generally used in reference to the types of organisms present in an environmental habitat. The human microbiome is comprised of different microbiota that colonize different habitats of the body. For example, the microbiota colonizing the skin is different from that of the gut, but they are all part of the human microbiome.

With this chapter we begin a new unit focused on microbe–human interactions and the immune system. The vast majority of microbes in and on the human body contribute in one way or another to the health of the individual; relatively few cause harm. In this chapter we explore the human microbiome with a focus on the composition of the **normal microbiota** of various body systems and the growing appreciation for how these important microbes maintain health and occasionally trigger disease.

I • Structure and Function of the Healthy Adult Human Microbiome

We begin with an overview of the human microbiome and then consider the microbiota present in specific regions of the body.

24.1 Overview of the Human Microbiome

The sites of the human body inhabited by microorganisms include the mouth, nasal cavities, throat, stomach, intestines, urogenital tracts, and skin (**Figure 24.1**). It is estimated that the microbes in the human microbiome number between 10^{13} and 10^{14} cells, which is roughly the same to ten times the total number of human cells in a single person. Together, the human body as the **host** and its associated microbes are increasingly recognized to constitute a *host–microbiome supraorganism*. For example, the gut microbial community in the healthy human was once considered to consist of microorganisms that were merely commensals, but we now know that this community is critical to development of the immune system, overall health later in life, and predisposition to disease.

Future Benefits of Knowing the Human Microbiome

The clinical benefits of knowing a person's microbiome seem promising and include the development of biomarkers for predicting predisposition to specific diseases, the design of therapies targeting selected microbial species in particular body sites, personalized drug therapies, and tailor-made probiotics (Section 24.11). However, a caveat must also be introduced. It should be emphasized that we are only at the early stages of resolving the many interactions between our body and the microbes we host. Sorting out the nature and activities of the human microbiome is an

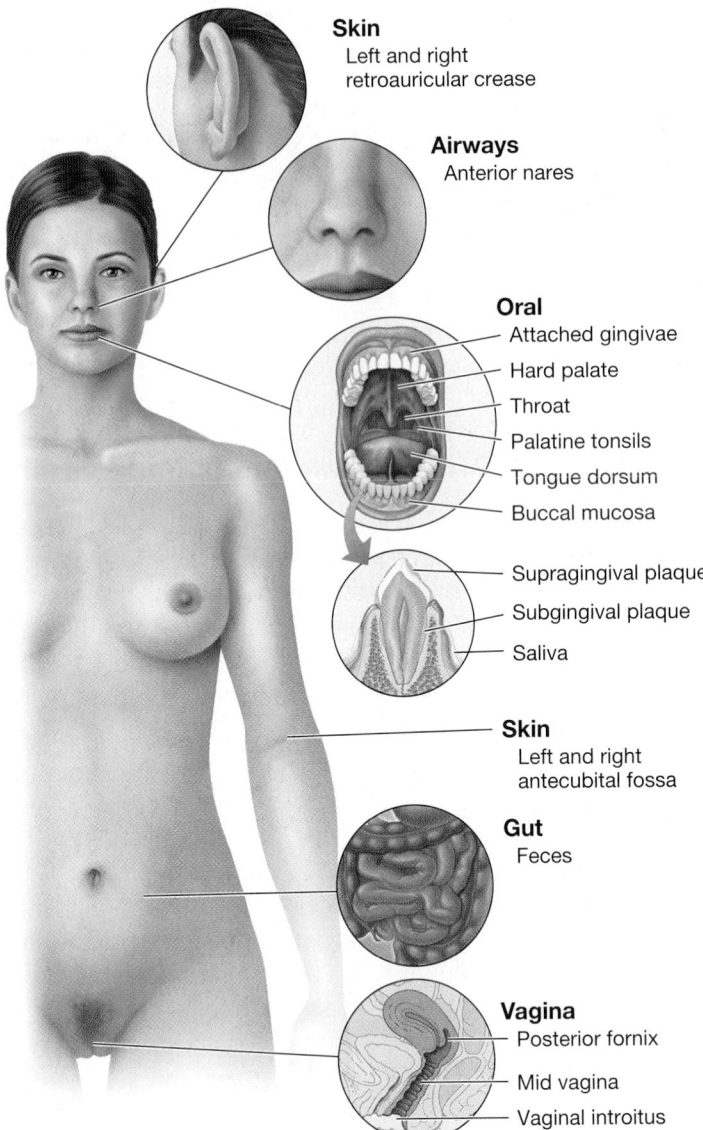

Figure 24.1 Microbial habitats of the human body. Primary body sites characterized in ongoing studies of the human microbiome (see Table 24.1).

extremely complex problem because not only is a person affected by his or her microbiota but that microbiota is also affected by the person's activities, health, and diet. Thus, cause and effect relationships are often not immediately obvious and can sometimes be difficult to sort out.

Microbiome studies to date have revealed an incredible diversity of microbial life in and on the human body, and have signaled several likely connections between the microbial composition at a body site and health status. However, at this point microbial associations with health or disease states are for the most part only correlations, and a causal relationship between microbiome and the health status of the host has been well established in only a few instances.

Experimental Protocols and Body Target Sites

Because the vast majority of microorganisms cannot yet be readily cultured or enumerated using growth-dependent approaches, it was the development of advanced nucleic acid sequencing methods (Chapters 9 and 19) that first offered the means to initiate a robust census of microbial diversity and abundance patterns in humans. The megasequencing era has also allowed microbiologists to compare the microbiomes of different individuals and to begin to resolve the dynamic relationship between the microbiome and host history, genetics, health status, and diet.

The focus on molecular sequences in microbiome studies does not diminish the importance of cultivation in the study of the human microbiome. Many hundreds of microorganisms have been isolated from the human body. The physiological and molecular characterization of those isolates has provided and will continue to provide essential understanding of their functional significance in human health and disease. Now that sequence-based surveys show that many human-associated microorganisms have not been isolated, there is also concerted effort to bring those microorganisms into culture. In addition, the development of appropriate culture conditions for isolation is being guided by metagenomic sequencing (⮌ Sections 9.8 and 19.8), which provides insights into the nutritional requirements of the uncultured microorganisms. Thus, the fusion of molecular and culture-based analyses will be critical to a full understanding of the functional significance of the human microbiome, as highlighted throughout this chapter.

The microbial diversity of the human microbiome has been mapped for the most part by surveys using 16S rRNA gene sequencing (⮌ Sections 13.7 and 19.6) and selected metagenomic analyses (⮌ Sections 9.8 and 19.8) as major tools in several national and international microbiome research efforts (Table 24.1). Initial studies examined the microbial diversity in hundreds of healthy volunteers, collecting 1518 total samples from various body sites (Figure 24.1) and analyzing them for their microbial species composition. These culture-independent assessments of microbial diversity typically use the definition of a microbial species given

earlier in this book (in which 97% or greater 16S rRNA gene sequence identity defines a single species cluster, ⮌ Section 13.8). However, some studies have used different species definitions, and for this reason, along with the fact that different methodologies (for example, different DNA isolation methods) have been used, the various analyses are not all directly comparable and should be considered only an approximate characterization of these complex microbial systems. Nevertheless, collectively, these studies have shown on the one hand that the microbial diversity between individuals is so great that no one microbial species is present in greatest abundance in all individuals, but that on the other hand, particular microbial groups typically dominate. Similarities in microbial diversity between individuals are more evident at higher bacterial taxonomic levels (such as phyla) and when certain genes are linked to particular body sites.

The general microbial composition of the four major human body sites (skin, oral, urogenital, and gastrointestinal), as defined by higher taxonomic levels, is shown in **Figure 24.2**. The sections to follow will examine each of these sites at a higher taxonomic resolution to more fully describe the huge diversity of microbes inhabiting the human body. Later sections address the functional significance of the microbiome and how it might be rationally modified for health benefit. These ongoing studies—including relationships to disease, ethnicity, and diet—are coordinated under the International Human Genome Consortium. Some of the major questions posed by these integrated projects include the following: (1) Do individuals share a core human microbiome? (2) Is there a correlation between the composition of microbiota colonizing a body site and host genotype? (3) Do differences in the human microbiome correlate with differences in human health? (4) Are differences in the relative abundance of specific bacterial populations important to either health or disease?

We start with a focus on the human gut, the human body site most heavily colonized by microbes.

TABLE 24.1 Major human microbiome research programs

Research program	Participating countries	Programmatic objectives
MetaGenoPolis	France	Demonstrate the impact of the human gut microbiota on health and disease using metagenomics technology
International Human Microbiome Standards	European Commission	Optimize methods for the assessment of the effects of the gut microbiome on human health through the standardization of procedures and protocols
Korean Twin Cohort Project	Korea	Characterize microbiota associated with epithelial tissue in a twin cohort study group, with the goal of identifying targets for early disease diagnosis and prevention
NIH Human Microbiome Project (HMP)	USA	Characterize the microbes that live in and on the human body, and assess the ability to demonstrate correlations of changes of the human microbiome with health
Canadian Human Microbiome Initiative	Canada	Characterize the microorganisms colonizing the human body. Evaluate their relationship to health and examine compositional changes associated with chronic disease
NIH Jumpstart Program	USA	Generate the complete genome sequences of 200 bacterial strains isolated from the human body; recruit donors for securing samples from five body regions, and perform 16S rRNA and metagenomic sequence analysis of the sampled body regions
Integrative Human Microbiome Project	USA	Crowdsourcing model to secure fecal samples for 16S rRNA sequence analysis

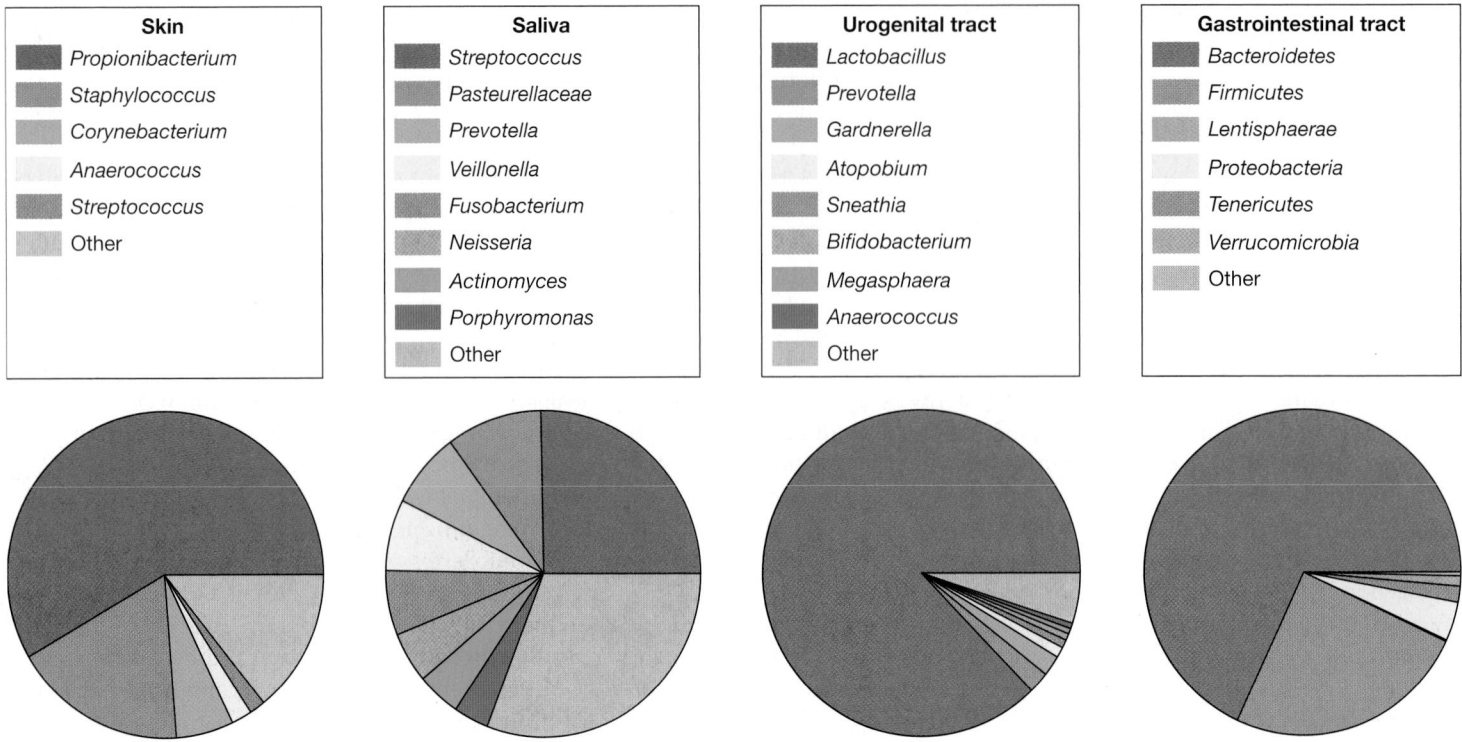

Figure 24.2 Overview of major microbial populations in the body sites sampled by human microbiome projects. Diversity among body sites was evaluated by 16S rRNA gene sequence analysis (↩ Section 13.7). Note that each body site tends to be dominated by one population type: skin by *Propionibacterium* species; oral by *Streptococcus* species; urogenital by *Lactobacillus* species; and gastrointestinal by *Bacteroides* species.

MINIQUIZ

- Which major body sites are heavily colonized by microbes?
- What methods have been used to assess the human microbiome?
- Why might knowing our microbiome and how it functions be useful?

24.2 Gastrointestinal Microbiota

In the previous chapter we reviewed how gut architecture differs among herbivores, carnivores, and omnivores (↩ Section 23.12). Here we examine the microorganisms throughout the entire gastrointestinal tract as well as their functions and special properties. The guts of herbivores include, in addition to the stomach, other compartments needed to foster microbial fermentation of large amounts of ingested plant material. In contrast, *monogastric* mammals, such as omnivorous humans, have only the stomach positioned before the intestines. The human gastrointestinal tract consists of the stomach, small intestine, and large intestine (**Figure 24.3**), and is responsible for the digestion of food and the absorption of nutrients; many important nutrients are also produced by the indigenous microbiota.

Starting with the stomach, the human digestive tract is a long folded tube of nutrients mixed with microbes, primarily species of *Bacteria*. The host and its gut microorganisms share the easily digestible nutrients. The nutrients are moved through this tube, encountering ever-changing microbial communities and abundance (Figure 24.3). The gastrointestinal tract has about 400 m^2 of surface area and is home to a total of about 10^{13} microbial cells. In the human duodenum, ingested food passed down from the stomach is blended with bile, bicarbonate, and digestive enzymes. About 1–4 h after ingestion, food reaches the gut (the large intestine, with near-neutral pH) and total bacterial numbers have increased from about 10^4/g (in the stomach) to 10^8/g (in the small intestine) to 10^{11}–10^{12} per g (in the large intestine)(Figure 24.3).

The Stomach and Small Intestine

The stomach was long thought to be either sterile or only minimally populated until the discovery in 1983 of *Helicobacter pylori* colonizing the stomachs of about 50% of the world's human population. This discovery prompted a closer inspection of the microbiology of the human stomach through both cultivation and 16S rRNA sequence analyses. The stomach is now recognized as the home of a vibrant bacterial community with hundreds of phylotypes distributed between the gastric lumen (pH 1–2) and the mucus layer of the wall (pH 6–7). Although low pH prevents bacterial overgrowth, the healthy human stomach holds a core microbiome that is distinct from the transient passage of ingested oral populations and is dominated by *Bacteroidetes* (*Prevotella*), *Firmicutes* (*Streptococcus, Veillonella, Lactobacillus*), *Actinobacteria* (*Rothia, Propionibacterium*), *Fusobacteria*, and *Proteobacteria* (*Haemophilus, Methylobacterium*).

UNIT 6

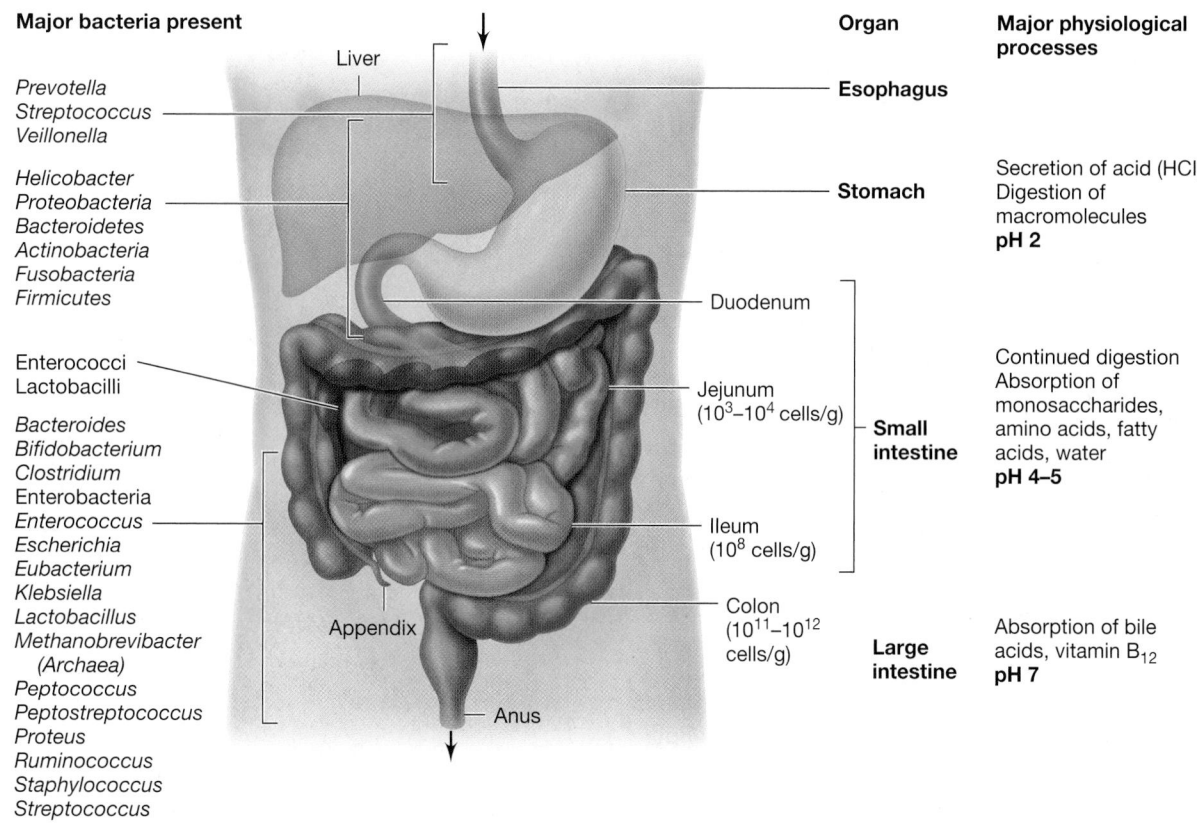

Major bacteria present

Liver

Prevotella
Streptococcus
Veillonella

Helicobacter
Proteobacteria
Bacteroidetes
Actinobacteria
Fusobacteria
Firmicutes

Enterococci
Lactobacilli

Bacteroides
Bifidobacterium
Clostridium
Enterobacteria
Enterococcus
Escherichia
Eubacterium
Klebsiella
Lactobacillus
Methanobrevibacter
   (Archaea)
Peptococcus
Peptostreptococcus
Proteus
Ruminococcus
Staphylococcus
Streptococcus

Appendix

Anus

Duodenum

Jejunum
(10^3–10^4 cells/g)

Ileum
(10^8 cells/g)

Colon
(10^{11}–10^{12}
cells/g)

Organ

Esophagus

Stomach

**Small
intestine**

**Large
intestine**

**Major physiological
processes**

Secretion of acid (HCl)
Digestion of
macromolecules
pH 2

Continued digestion
Absorption of
monosaccharides,
amino acids, fatty
acids, water
pH 4–5

Absorption of bile
acids, vitamin B$_{12}$
pH 7

Figure 24.3 The human gastrointestinal tract and major members of its microbiota. These taxa are representative of microorganisms found in healthy adults. Not every individual harbors all of these microorganisms.

Firmicutes, Bacteroidetes, and *Actinobacteria* dominate gastric fluid samples, whereas *Firmicutes* and *Proteobacteria* are more abundant in gastric mucosal samples. When present, *H. pylori* accounts for the vast majority of stomach microbial biomass. This pathogenic proteobacterium is typically transmitted orally between family members and can persist for decades in the gastric mucosa (see page 729). About 20% of *H. pylori*-infected individuals will suffer from upper gastrointestinal tract symptoms during their lifetime (⌘ Section 30.10). However, even when infection is asymptomatic, chronic inflammation is considered a major risk factor for the development of ulcers and gastric malignancies. As such, *H. pylori* was recognized as a "definite carcinogen" (group 1) by the World Health Organization in 1994.

Distal to the stomach, the intestinal tract consists of the small intestine and the large intestine, each of which is divided into different anatomical segments and supports a characteristic microbiota (Figure 24.3). The small intestine has three distinct environments in the *duodenum* and the *ileum*, which are connected by the *jejunum*. The duodenum, adjacent to the stomach, is fairly acidic and its normal microbiota resembles that of the stomach. From the duodenum to the ileum, the pH gradually becomes less acidic and bacterial numbers increase. In the lower ileum, cell numbers of 10^5–10^7/gram of intestinal contents are common, even though the environment becomes progressively more anoxic. Fusiform (spindle-shaped) anaerobic bacteria are typically present, attached to the intestinal wall at one end (**Figure 24.4**). Whereas the colonic microbiota discussed in the following section is largely

driven by the efficient degradation of complex indigestible carbohydrates, the microbiota residing in the small intestine must compete with the host for rapid uptake and conversion of small carbohydrates, and consists of species adapted to rapid changes in overall nutrient availability.

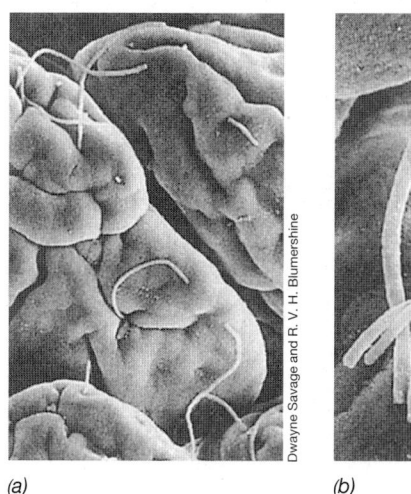
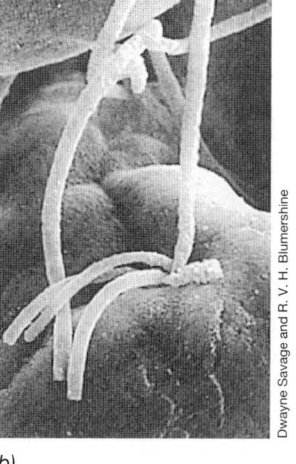

(a) *(b)*

Dwayne Savage and R. V. H. Blumershine

Figure 24.4 Microbiota in the small intestine. Scanning electron micrographs of the microbial community on epithelial cells in the mouse ileum. *(a)* An overview at low magnification shows long, filamentous fusiform bacteria on epithelial cells in the mouse ileum. *(b)* Higher magnification shows several filaments attached at a single depression. The attachment is at the end of the filaments. Individual cells are 10–15 µm long.

UNIT 6

The Large Intestine

The ileum empties into the *cecum*, the pouch that is considered the beginning of the large intestine. The *colon* makes up the rest of the large intestine. In the colon, *Bacteria* are present in enormous numbers and large numbers of *Archaea* (primarily methanogens) can be present, too. The colon is essentially an in vivo fermentation vessel, with the microbiota using nutrients derived from the digestion of food (Figure 24.3). Facultative aerobes such as *Escherichia coli* are present but in smaller numbers than other bacteria. The facultative aerobes consume any remaining oxygen, rendering the large intestine strictly anoxic. Anoxia promotes growth of obligate anaerobes such as *Clostridium* and *Bacteroides* species.

The total number of obligate anaerobes in the colon is enormous. Bacteria are by far the dominant population—cell counts of 10^{10} to 10^{11} bacterial cells/gram in distal gut and fecal contents are normal—with *Bacteroidetes* and gram-positive *Bacteria* accounting for greater than 99% of all prokaryotic cells. *Archaea* comprise a small fraction of the gut microbiota and are represented by 0.05–1% of total gene counts in gut content metagenomes (♂ Sections 9.8 and 19.8); these *Archaea* consist only of methanogens. The hydrogenotrophic *Methanobrevibacter smithii* is the most abundant and often the exclusive archaeal population, with the methanol-reducing *Methanosphaera stadtmanae* occasionally present but in much lower abundance. Eukaryotic microbes are also only a minor part of the human gut community (<1%) and are represented primarily by the yeasts *Candida albicans* and *Candida rugosa*, which are considered part of the normal gut microbiota. Protists are absent in the gastrointestinal tract of healthy humans, but they can cause gastrointestinal infections if ingested in contaminated food or water (Chapter 32). **Figure 24.5** gives a molecular snapshot of bacterial diversity in the human colon as determined by 16S rRNA gene sequence analysis of feces. At this finer taxonomic resolution, it becomes possible to see the dominance of two major families of *Firmicutes* (*Lachnospiraceae* and *Ruminococcaceae*) at this body site (Figure 24.5). Species of these two families of anaerobic bacteria are important in the digestion of polysaccharide polymers in plant fiber, such as cellulose and pectin; these are depolymerized and the sugars fermented in the large intestine.

During the passage of food through the gastrointestinal tract, water is absorbed from the digested material, which gradually becomes more concentrated and is converted to feces. Bacteria compose about one-third of the weight of fecal matter. Organisms living in the lumen of the large intestine are continuously displaced downward by the flow of material, and bacteria that are lost are continuously replaced by new growth, similar to a continuous culture system (♂ Section 5.4). The time needed for passage of material through the human gastrointestinal tract is about 24 h, and the growth rate of bacteria in the lumen is one to two doublings per day.

A person sheds a total of about 10^{13} bacterial cells each day in feces. Most of those organisms are restricted to the lumen of the large intestine (**Figure 24.6**). Production of **mucin** (a thick liquid secretion containing water-soluble proteins and glycoproteins) by *goblet cells* (a specialized class of epithelial cells) in

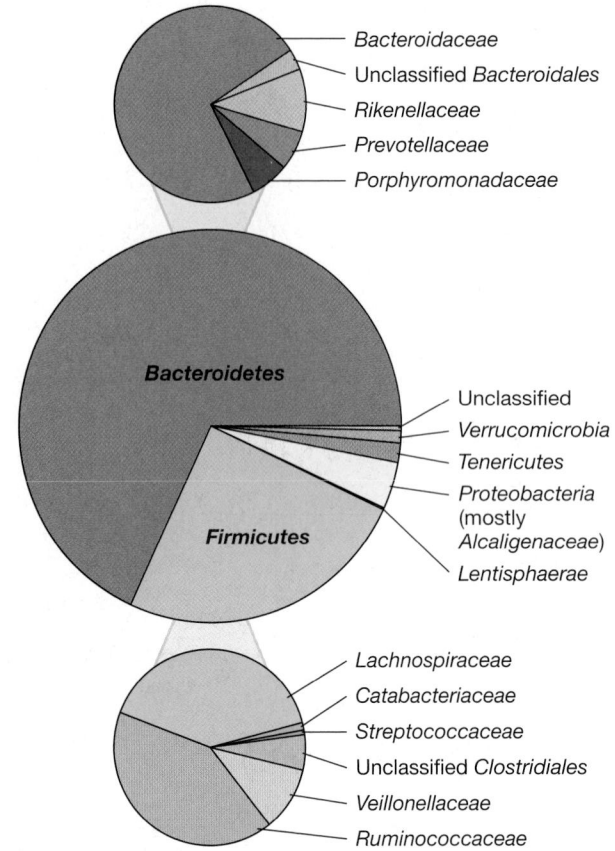

Figure 24.5 Bacterial diversity of feces. The results are pooled analyses of approximately 1 million sequences (from 184 samples) of the V1–V3 region of the 16S ribosomal RNA gene (♂ Figure 13.15). Members of the *Bacteroidetes* and *Firmicutes* dominate the normal microbiota of the large intestine. Many of these groups are covered in Chapters 15 and 16 (*Bacteria*). Data assembled and analyzed by Nícolas Pinel.

the intestinal epithelium forms a protective layer (inner mucus layer) immediately adjacent to the intestinal epithelium. This inner mucus layer, unlike the outer mucus layer, is rarely colonized by bacteria (Figure 24.6). Goblet cells also produce various antimicrobial peptides that help prevent microbial contact with the underlying epithelium.

Summary of the Gut Microbiota: The Two Major Components

The human gut microbial community is composed of only a few major phyla and shows a species composition distinct from that of any free-living microbial communities. Although we may think of *Escherichia coli* as a significant gut bacterium, the entire phylum *Gammaproteobacteria* (to which *E. coli* belongs; ♂ Section 16.3) makes up <1% of all gut bacteria. The vast majority (~98%) of all human gut phylotypes fall into one of three major bacterial phyla: *Firmicutes, Bacteroidetes,* and *Proteobacteria* (Figure 24.5). *Bacteroidetes* and *Firmicutes* dominate but can vary dramatically in relative abundance among individuals. Surveys of the human gut have shown the gut composition to vary from >90% *Bacteroidetes* to >90% *Firmicutes,* and the ratio of these two groups in a given individual may influence aspects of their health, such as leanness versus obesity (Section 24.8).

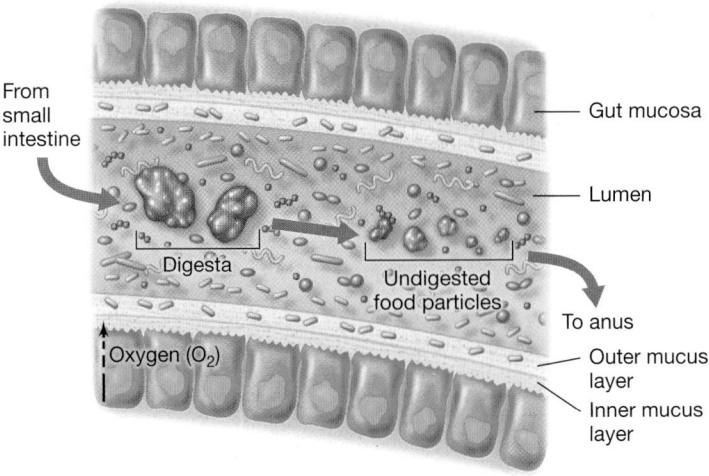

Figure 24.6 Different microenvironments in the large intestine. The inner mucin layer, produced by and contacting the gut mucosa, is partly oxygenated but generally free of bacteria. The sparsely populated outer mucus layer is adjacent to the heavily colonized anoxic lumen, which receives undigested food particles from the small intestine.

In contrast to the rather limited *phylum-level* diversity, the *species* diversity in the mammalian gut is enormous. For instance, a single study of diversity in human fecal samples (based on the analysis of millions of 16S rRNA sequences from a large sample of people) identified between 3500 and 35,000 bacterial "species." Which end of this large range is more accurate depends mainly on whether the 16S rRNA similarity threshold for defining a species (⇌ Section 13.8) is set at the 97% sequence identity cutoff or is set more stringently (~98–99% identity), as some microbiologists have proposed. By the less stringent criteria, *any one individual* harbors a total of about 160 species in their large intestine. *Archaea* (represented by a phylotype closely related to the methanogen *Methanobrevibacter smithii*), yeasts, fungi, and protists are either absent or compose only a minor part of the human gut community (compare this with the rumen, ⇌ Section 23.13).

Gut Enterotypes

Not surprisingly, comparative studies have also shown that humans share more bacterial genera with each other than with other species of mammals. This suggests that the mammalian gut microbiota may be "fine-tuned" to each mammalian species. Interestingly, however, although there is high variability from person to person in gut community composition, any particular individual's community is relatively stable over long periods. Also, ongoing metagenomic sequencing studies hint at the existence of a limited number of distinct, well-balanced general types of gut microbial communities that may reflect early life experiences, such as breast-feeding versus formula feeding.

Three general gut communities (called *enterotypes)* have been described, differing primarily by the enrichment of one microbial group in each enterotype. Enterotype 1 is enriched in *Bacteroides*, enterotype 2 is enriched in *Prevotella*, and enterotype 3 is enriched in *Ruminococcus*. The association of an individual with a particular

enterotype seems to transcend national borders, nutrition, and ethnicity. Moreover, metabolic pathway reconstructions (based on annotation of metagenomic gene sequences, Chapter 9) suggest that each enterotype is functionally distinct; for example, they differ in their capacity for vitamin production. These results also suggest that enterotype influences an individual's response to diet and drug therapy and may contribute to health or disease status in general. As a note of caution, other analyses of these data suggest that enterotypes may not be fully distinct but instead parts of a continuous variation in population structure. Nevertheless, a better understanding of the strengths and weaknesses of each enterotype could introduce exciting new concepts and practices into the field of clinical medicine (for example, fecal transplants, Section 24.10).

Products of Intestinal Microbiota and "Educating" the Immune System

Some products of gut microbial metabolism are relatively simple compounds, such as the volatile fatty acids generated by microbial fermentation of plant material, as also occurs in the rumen of herbivores (⇌ Figure 23.39). Other products generated by the activities of fermentative bacteria and methanogens include H_2, CO_2, and CH_4 (similar to fermentation products of the rumen) and several other substances and nutrients (**Table 24.2**). Gut microbes also produce vitamins B_{12} and K. These essential vitamins are not synthesized by humans (and vitamin B_{12} is not present in plants) but are made by the gut microbiota and absorbed from the colon. In addition, steroids, produced in the liver and released into the intestine from the gallbladder as bile acids, are modified in the intestine by the microbiota; the modified bioactive steroid compounds are then absorbed from the gut. Gut microbes also function in amino acid metabolism (Table 24.2). Humans are unable to make 9 of the 20 amino acids; these "essential" amino acids—such as lysine—can be obtained in our diet but are also produced and secreted by certain gut microbes.

Many other microbial metabolites or transformation products that can be generated in the gut have significant influence on host physiology. These include post-translationally modified peptides

TABLE 24.2 Biochemical/metabolic contributions of intestinal microorganisms

Process	Product or enzyme
Vitamin synthesis	Thiamine, riboflavin, pyridoxine, B_{12}, K
Amino acid synthesis[a]	Asparagine, glutamate, methionine, tryptophan, lysine, and others
Gas production	CO_2, CH_4, H_2
Odor production	H_2S, NH_3, amines, indole, skatole, butyric acid
Organic acid production	Acetic, propionic, butyric acids
Glycosidase reactions	β-Glucuronidase, β-galactosidase, β-glucosidase, α-glucosidase, α-galactosidase
Steroid metabolism (bile acids)	Esterified, dehydroxylated, oxidized, or reduced steroids

[a]Capacity for amino acid biosynthesis inferred from the identification of biochemical pathways encoded in gut metagenomic sequences (⇌ Sections 9.8 and 19.8).

such as the lantibiotics and bacteriocins (**Table 24.3**), substances that function to help secure colonization of the producing organism by inhibiting organisms closely related to the producer. Gut bacteria can also synthesize high levels (>100 mg/day in some cases) of metabolites derived from the reduction of amino acids. These include substances such as *tryptamine*, a tryptophan metabolite thought to function as a biogenic neurotransmitter that signals the enteric nervous system, and *4-ethylphenylsulfate*, shown to affect behavior in mice (Table 24.3). These examples suggest that an animal's microbiome and its nervous system—called the *gut–brain axis*—are likely connected (see Explore the Microbial World, "The Gut–Brain Axis").

It is known that the immune system does not properly develop in the absence of microbial stimulation and that early life exposure to a variety of microorganisms is essential for developing tolerance to beneficial microorganisms (our normal microbiota) and recognizing pathogens as foreign. For example, the consequences of excessive hygiene in an infant's development may be a poorly trained immune system that is more likely to attack beneficial organisms with an inflammatory response. Such an immune status is thought to promote autoimmune conditions such as allergies, asthma, and inflammatory bowel diseases later in life.

Much of our understanding of how the immune system is trained to accept the normal microbiota comes from the study of mice (Section 24.6). For example, a key factor contributing to the successful colonization of *Bacteroides fragilis*, part of the normal microbiota of the mouse gut, is production of a "symbiosis factor" called *polysaccharide A* (Table 24.3). Polysaccharide A is a diffusible oligosaccharide derived from the *B. fragilis* outer membrane that in some way signals the host's immune system to promote the tolerance needed for successful colonization by this bacterium. Besides promoting its own colonization, *B. fragilis* has also been shown to protect mice from colitis caused by a pathogenic bacterium, *Helicobacter hepaticus*. In experimental animals colonized with *B. fragilis* mutants unable to express polysaccharide A, *H. hepaticus* colonizes the gut and elicits inflammatory bowel disease, unlike the results for a control group of normal animals.

These are just a few examples of the complex "dialogue" that takes place between the host and its normal gut microbiota early in life, a dialogue that is critical to the health of the host. Developing a truly mechanistic understanding of these events will be essential to promoting health and preventing diseases related to imbalances in the microbiome of humans. The identification of common and variably distributed biologically active compounds produced by the human microbiota is an emerging research area of great significance. Indeed, this information will be critical to predicting health outcomes and developing appropriate therapies for pathologies now attributed to imbalances or other abnormalities in the human microbiome.

> **MINIQUIZ**
> - How does the general metabolism of microorganisms colonizing the small and large intestines differ and why?
> - What is an enterotype?

24.3 Oral Cavity and Airways

Mucous membranes throughout the body support the growth of a normal microbiota that prevents infection by pathogenic microorganisms. Mucous membranes consist of epithelial cells, tightly packed cells that interface with the external environment, and are found throughout the body, lining the urogenital, respiratory, and gastrointestinal tracts (Figure 24.6). Mucous membranes secrete *mucin*, forming the mucus layer that retains moisture and inhibits microbial attachment; invaders are usually swept away by physical processes like swallowing or sneezing, but some microorganisms adhere to the epithelial surface and colonize. Here we discuss two mucosal environments and their resident microbes. In Chapter 25 we explore the mechanism of specific attachment of microbes to mucous membranes leading to growth and development of the normal microbiota or, alternatively, to microbial disease.

Oral Microbes

Saliva contains microbial nutrients, but it is not a good microbial growth medium because the nutrients are present in low concentration and saliva contains antibacterial substances. These include *lysozyme*, an enzyme that cleaves glycosidic linkages in peptidoglycan of the bacterial cell wall, weakening the wall and causing cell lysis (↩ Section 2.4). Another enzyme, *lactoperoxidase*, found in both milk and saliva, kills bacteria by a reaction that generates singlet oxygen (a toxic form of oxygen, ↩ Section 5.14). Despite

TABLE 24.3 **Small bioactive molecules produced by bacteria in the large intestine**

Class	Compound	Example producer	Activity
RiPP[a] (lantibiotic)	Ruminococcin A	*Ruminococcus gnavus*	Antibiotic
RiPP[a] (bacteriocin)	Ruminococcin C	*Ruminococcus gnavus*	Antibiotic
Amino acid metabolite	Indolepropionic acid	*Clostridium sporogenes*	Protective anti-oxidant
Amino acid metabolite	4-Ethylphenylsulfate	Undefined	Neuromodulatory
Amino acid metabolite	Tryptamine	*Ruminococcus gnavus*	Neurotransmitter
Volatile fatty acid	Propionic acid	*Bacteroides* spp.	Immunomodulatory[b]
Oligosaccharide	Polysaccharide A	*B. fragilis*	Immunomodulatory[b]

[a]Ribosomally synthesized and post-translationally modified peptides.
[b]These small molecules promote colonization by normal microbiota.

THE GUT–BRAIN AXIS

Interactions between the gut microbiota and the host brain and general nervous system (called the *gut–brain axis*) have been gaining attention because of possible contributions to behavioral disorders. We know that there is a clear relationship between gut dysbiosis and pathologies such as inflammatory bowel disease and obesity (Section 24.8). However, gut dysbiosis is also associated with autism, a general term for a spectrum of behavioral disorders (together called *autism spectrum disorder*) that can emerge in the first 36 months of life, causing substantial impairments in social interaction and communication and marked by repetitive behaviors and unusual interests.

Human autism has been attributed to a combination of genetic and environmental factors. An environmental risk factor for a child developing autism and other behavioral disorders is maternal immune activation (MIA), which involves elevated levels of inflammatory factors in the blood, placenta, and amniotic fluid during pregnancy and can be caused, for example, by viral infection. Although the relationship between MIA during pregnancy and impaired development of the fetal central nervous system is complex, mouse models suggest that some features of autism can be alleviated by correcting associated defects in the gut microbial community.

The offspring of mice in which MIA has been induced display autism-like behaviors, such as communication defects and anxiety (**Figure 1a, b**). Those behavioral changes are associated with a loss of intestinal integrity, a shift in the composition of gut clostridial and bacteroidal populations reminiscent of changes observed in autistic humans, and a 46-fold increase in serum levels of the chemical *4-ethylphenylsulfate* (**Figure 1d**). This compound is made from the amino acid tyrosine by certain gut microbiota. When administered alone, it induces anxiety-like behavior in normal mice.[1] However, a remarkable finding is the ability to ameliorate defects in communication, anxiety, 4-ethylphenylsulfate levels, and gut permeability by feeding the affected mice a human gut commensal bacterium, either *Bacteroides fragilis* or *Bacteroides thetaiotaomicron* (**Figure 1c**). It is thought that these organisms displace the 4-ethylphenylsulfate producers and return gut chemistry to normal.

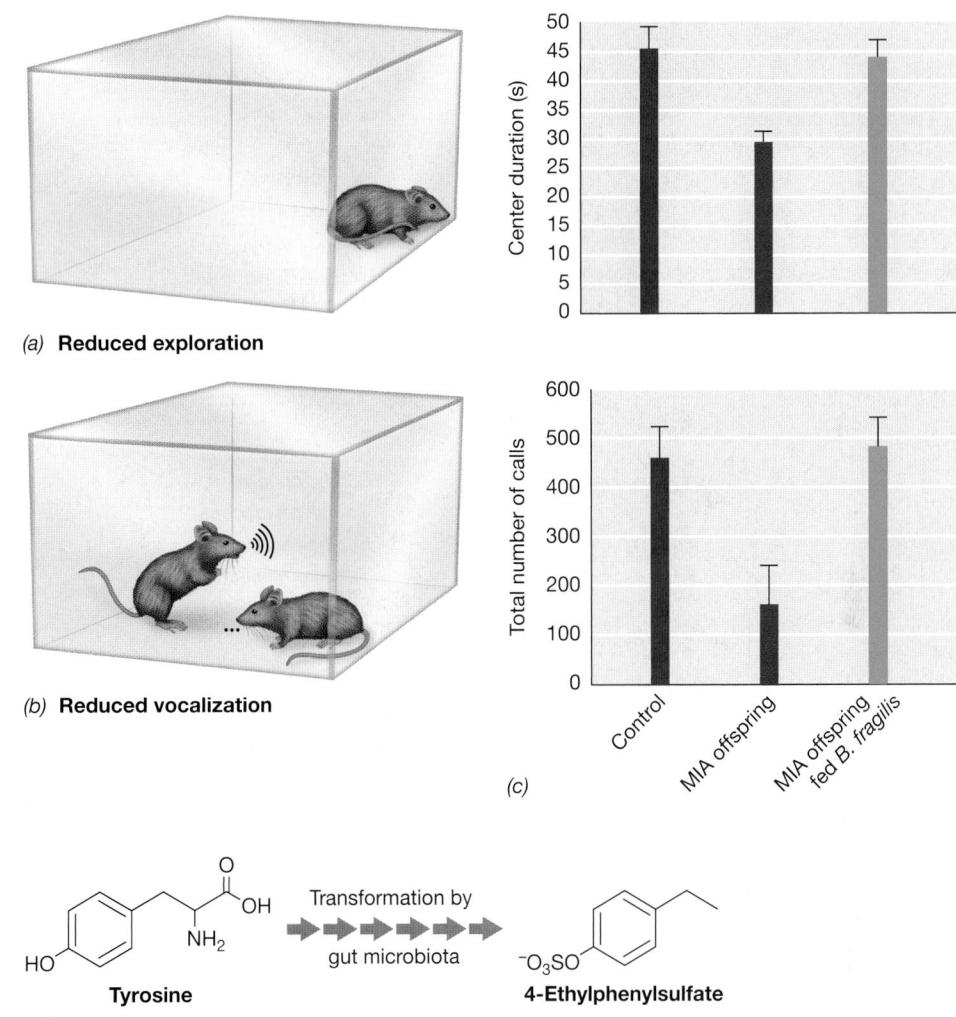

(a) **Reduced exploration**

(b) **Reduced vocalization**

(c)

(d)

Tyrosine → Transformation by gut microbiota → **4-Ethylphenylsulfate**

Figure 1 Influence of gut microbiota on behavior. Mice offspring of mothers with maternal immune activation (MIA) show autistic-like behavior, marked by (a) a fear of exploring the center of the test field and (b) reduced vocalization. (c) Feeding the affected offspring the human commensal bacterium *Bacteroides fragilis* ameliorates these behavioral abnormalities. (d) Certain gut microbiota can convert the amino acid tryptophan into 4-ethylphenylsulfate, the neuroactive compound thought to trigger the mouse autistic-like behaviors shown in *a* and *b*. Modified from Hsiao, E.Y., et al. 2013. *Cell 155*: 1451–1463.

These studies offer an example of the importance of the gut microbiota–brain connection in behavior and point to the possibility of using a rational modification of the gut community, such as might be possible with targeted probiotic therapy (Section 24.11), to alleviate symptoms in human neurodevelopmental disorders such as autism. Furthermore, amelioration of aberrant behavior by reducing serum levels of the gut microbiota–derived metabolite 4-ethylphenylsulfate to normal levels suggests that other neurodevelopmental illnesses may also be linked to the accumulation of microbial metabolites in serum by an unbalanced gut microbiota. However, one must be cautious about overinterpreting this study because the science thus far is only at the early stages of associating the gut microbiota with behavioral disorders.

[1]Hsiao, E.Y., S.W. McBride, S. Hsien, et al. 2013. Microbiota modulate behavioral and physiological abnormalities associated with neurodevelopmental disorders. *Cell 155*: 1451–1463.

UNIT 6

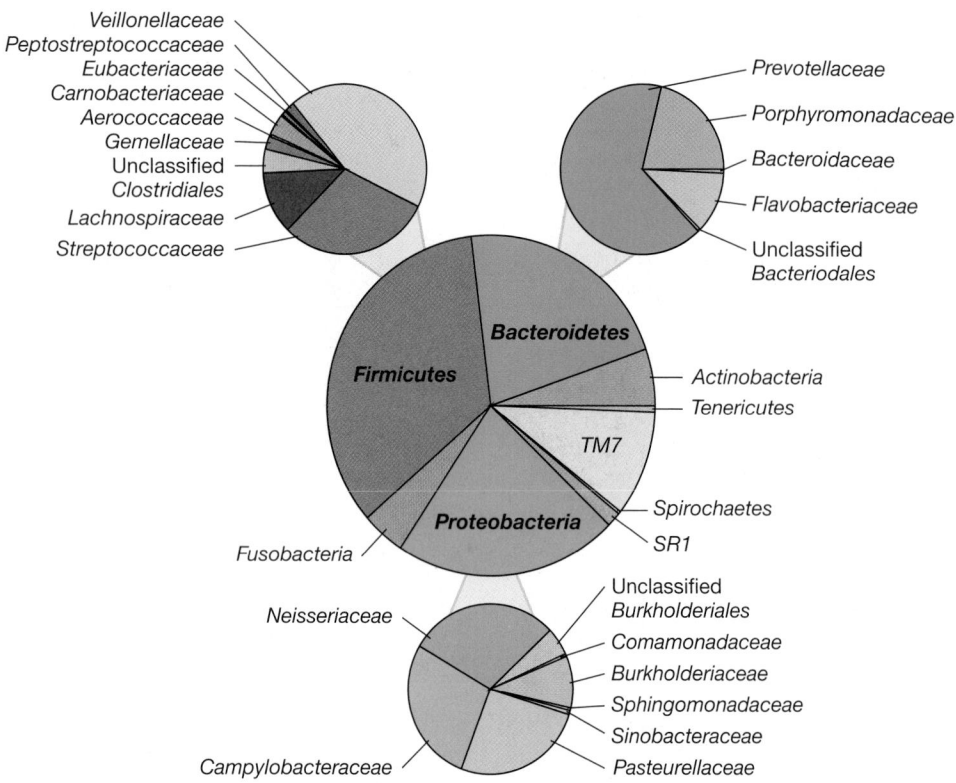

Figure 24.7 Bacterial diversity of saliva. The results are pooled analyses of approximately 750,000 sequences (from 159 samples) of the V1–V3 region of the 16S ribosomal RNA gene (⮌ Figure 13.15). Note the lower fraction of anaerobes affiliated with *Fusobacteria* and *Spirochaetes* relative to the distribution of taxa observed in subgingival plaque (Figure 24.9). Many of these groups are covered in Chapters 15 and 16 (*Bacteria*). Data assembled and analyzed by Nícolas Pinel.

the activity of these antibacterial substances, food particles and cell debris provide high concentrations of nutrients near surfaces such as the teeth and gums, creating favorable conditions for extensive local microbial growth and sometimes contributing to tissue damage and disease.

The oral cavity provides a variety of microbial habitats, each colonized by species that grow primarily as biofilms (⮌ Sections 7.9 and 20.4). The microbiota found in saliva consists of microorganisms shedding from multiple sites within the oral cavity and provides an overview of oral microbial diversity (**Figure 24.7**). The oral microbiome is essentially as diverse as the gut, but humans share greater proportions of common taxa for the mouth than for the gut. Abundant bacterial genera in the oral cavity include *Streptococcus*, *Haemophilus*, *Veillonella*, *Actinomyces*, and *Fusobacterium* (see page 757).

As for all microbial communities reexamined by molecular methods, 16S rRNA-based sequence surveys of the oral cavity have shown that the culture-based methods of the past have provided a very incomplete census of microbial diversity. At least 750 species of aerobic and anaerobic microbes, including a minor representation of methanogenic *Archaea* and yeast, are known to reside in the oral cavity, distributed among teeth, tissue surfaces, and saliva. Most of these microorganisms have facultatively aerobic metabolisms, but some, such as *Bacteroidetes*, are obligately anaerobic and some have strictly aerobic metabolisms, such as the *Neisseria*, *Acinetobacter*, and *Moraxella* genera in the *Proteobacteria* phylum. The most abundant genera in the oral cavity are *Firmicutes*;

Veillonella parvula, an obligate anaerobe, is the most abundant single species and *Streptococcus* is the most abundant genus in the mouth, comprising about 25% of cells found in some individuals. The related *Firmicutes* genera *Abiotrophia* (a member of *Aerococcaceae*), *Gemella* (a member of *Gemellaceae*), and *Granulicatella* (a member of *Carnobacteriaceae*) are also common; species from these genera were among the 10 taxa most frequently detected. Other genera are present in much lower numbers, with only 17 taxa each contributing more than 1% of the oral microbiome. As is the case for the skin microbiota (Section 24.5), not all bacterial taxa are present or similarly distributed in all individuals.

Oral Microenvironments and Their Microbiota

Bacteria found in the mouth during the first year of life (when teeth are absent) are predominantly aerotolerant anaerobes such as streptococci and lactobacilli, and a few aerobes. When the teeth appear, the newly created surfaces are rapidly colonized by anaerobes that are specifically adapted to growth in biofilms on the surfaces of the teeth and in the gingival crevices (**Figure 24.8**). The primary colonizers of clean tooth surfaces are species of *Streptococcus*; obligate anaerobes such as *Veillonella* and *Fusobacterium* colonize habitats below the gum line. Most of these organisms

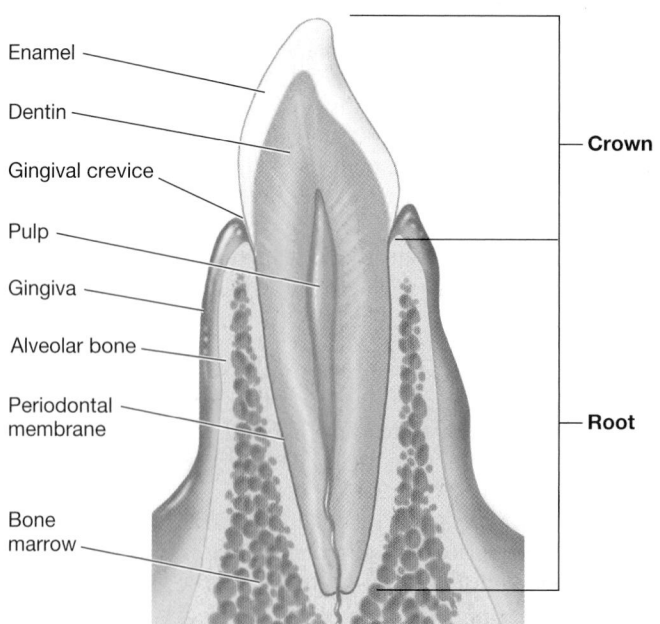

Figure 24.8 Section through a tooth. The diagram shows the tooth architecture and the surrounding tissues that anchor the tooth in the gum.

contribute to the health of the host by keeping pathogenic bacteria in check and preventing them from adhering to mucosal surfaces. Tooth decay, gum inflammation, and periodontal disease are among the most visible manifestations of a breakdown in these generally stable mutualisms. We discuss the microorganisms associated with the hard tooth surface and their contribution to the formation of dental plaque and dental caries in Section 25.2 (↩ Figures 25.7 and 25.8).

There is significant site variation in the diversity and specificity of colonization by different oral bacterial species. For example, there are differences in abundant microbial taxa associated with subgingival plaque (**Figure 24.9**) compared with that found in saliva (Figure 24.7). Different species of *Corynebacterium* demonstrate significant site specificity. For instance, *Corynebacterium matruchotii* is almost exclusively found in the supragingival plaque, whereas *Corynebacterium argentoratense* occurs mostly in the saliva. *Lautropia mirabilis* selectively colonizes the supragingival plaque, whereas the spirochete *Treponema socranskii* is found mostly in subgingival plaque, presumably because this site provides the low-oxygen environment needed for microaerobic growth of this bacterium. The distribution of *Firmicutes, Proteobacteria*, and *Bacteroidetes* is similar between the oropharynx (see Figure 24.10) and saliva. The hard palate (roof of the mouth) harbors a much lower diversity of microbiota than does the gingival plaque, probably as a result of constant shedding of epithelial cells and the shear forces associated with chewing and swallowing.

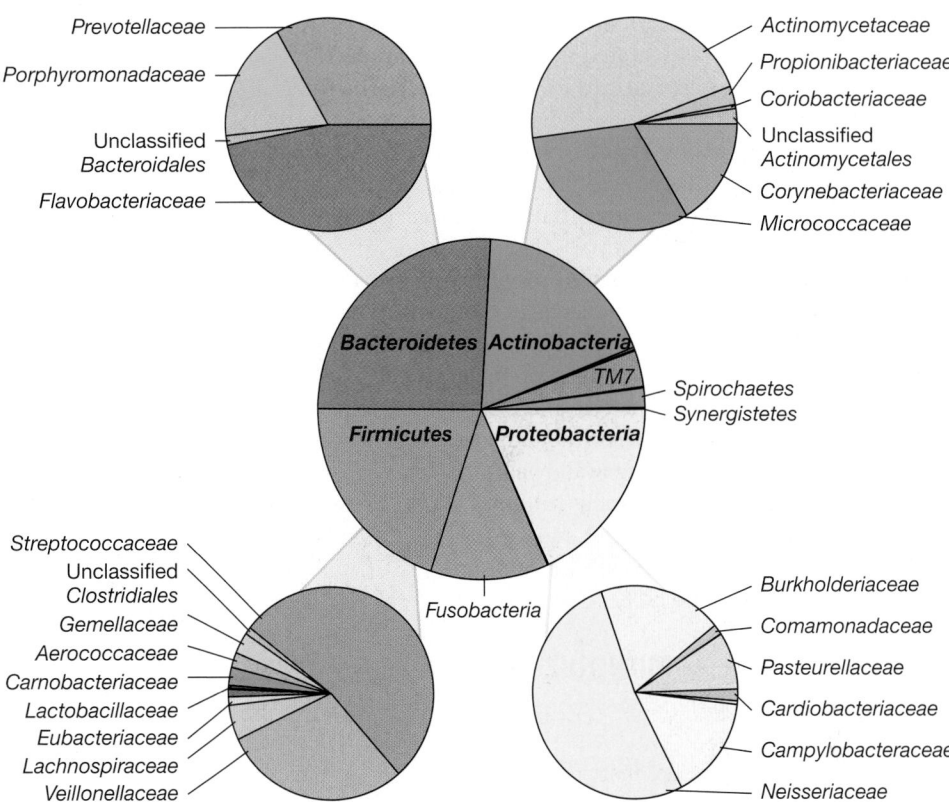

Figure 24.9 Bacterial diversity of subgingival plaque. The results are pooled analyses of approximately one million sequences (from 183 samples) of the V1–V3 region of the 16S ribosomal RNA gene (↩ Figure 13.15). Compare the fractional distribution of different bacterial taxa with that observed in saliva (Figure 24.7), noting the higher representation of anaerobic *Fusobacteria* and *Spirochaete* populations in the oxygen-limited gingival crevice. Many of these groups are covered in Chapters 15 and 16 (*Bacteria*). Data assembled and analyzed by Nicolas Pinel.

Microenvironments of the Respiratory Tract

The anatomy of the respiratory tract is shown in **Figure 24.10** and Figure 24.1. In the **upper respiratory tract** (including the throat/tonsils, nasopharynx, oral cavity, oropharynx, and larynx), microbes live in areas bathed with secretions from the mucous membranes. Bacteria continually enter the upper respiratory tract from the air during breathing, but most are trapped in the mucus of the nasal and oral passages and expelled with nasal secretions, or swallowed and then killed in the stomach. However, a few microbes colonize respiratory mucosal surfaces in all individuals; those most commonly present are species of staphylococci, streptococci, diphtheroid bacilli, and gram-negative cocci.

Occasionally, potential pathogens such as *Staphylococcus aureus* and *Streptococcus pneumoniae* are part of the normal microbiota in the nasopharynx of healthy individuals. These individuals are *carriers* of the pathogens but do not normally develop disease, presumably because the other resident microorganisms compete successfully for nutritional and metabolic resources and limit pathogen attachment, colonization, or activities. The innate

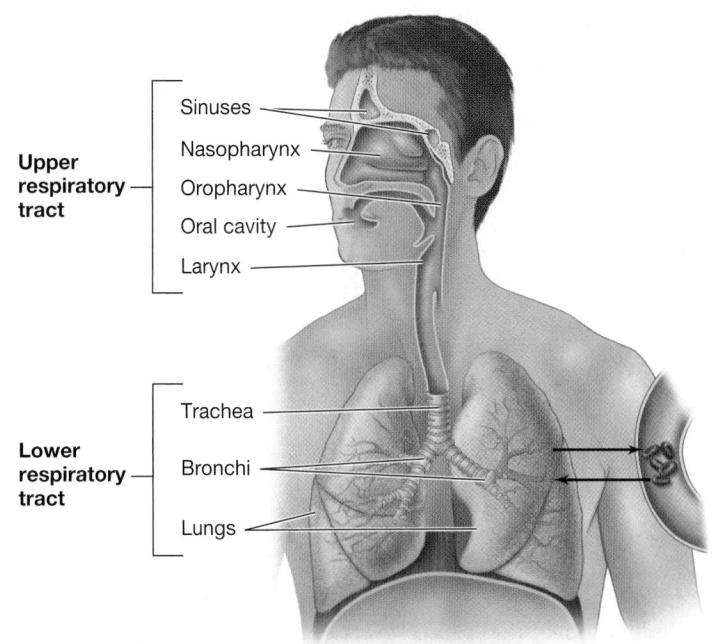

Figure 24.10 The respiratory tract. In healthy individuals the upper respiratory tract has a large variety and number of microorganisms. By contrast, the lower respiratory tract in a healthy person has few if any microorganisms.

immune system (Chapter 26) and components of the adaptive immune system such as secreted antibodies (Chapter 27) are particularly active at mucosal surfaces and inhibit growth and invasion by potential pathogens.

The **lower respiratory tract** (trachea, bronchi, and lungs, Figure 24.10) has no resident microbiota in healthy adults, despite the large number of organisms potentially able to reach this region during normal breathing. Dust particles, which are fairly large, settle in the upper respiratory tract. As the air passes into the lower respiratory tract, the flow rate decreases, and organisms settle onto the walls of the respiratory passages. The walls of the entire respiratory tract are lined with ciliated epithelial cells, and the cilia, beating upward, push bacteria and other particulate matter toward the upper respiratory tract where they are then expelled in saliva and nasal secretions or are swallowed. Only particles smaller than about 10 μm in diameter reach the lungs. Nevertheless, these include some pathogenic microbes, most notably certain bacteria or viruses that cause pneumonia (inflammation of the lungs, ⊃ Sections 30.1, 30.2, and 30.8).

MINIQUIZ

- Compare the microbial microenvironments in the oral cavity in newborns and adults.
- Identify the major microbes that predominate in the adult oral cavity by taxa and metabolic requirements.
- Why is the lower respiratory tract typically microbe-free?

24.4 Urogenital Tracts and Their Microbes

In the urogenital tracts of healthy adults (**Figure 24.11**), the kidneys and bladder are sterile; however, epithelial cells lining the distal urethra are colonized by facultatively aerobic gram-negative *Bacteria*. Potential pathogens such as *Escherichia coli* and *Proteus mirabilis*, normally present in small numbers in the body or in the local environment, can multiply in the urethra and cause disease if conditions such as changes in pH occur. Such organisms are a frequent cause of urinary tract infections, especially in females. *Proteus* can be especially notorious as a urinary tract pathogen. This bacterium is a strong urease producer; it generates ammonia from urea and uses the ammonia as a nitrogen source. However, ammonia also causes urine pH to become quite alkaline, and this can trigger other urinary tract conditions such as the formation of kidney stones.

The vagina of the adult female is weakly acidic (pH~5) and contains significant amounts of glycogen. *Lactobacillus acidophilus*, a resident organism in the vagina, ferments the polysaccharide glycogen, producing lactic acid that maintains a local acidic environment (Figure 24.11*b*). Other organisms, such as species of the yeasts *Torulopsis* and *Candida*, various streptococci, and *E. coli*, may also be present. Before puberty, *L. acidophilus* is absent, the female vagina is neutral in pH and does not produce glycogen, and the microbiota consists predominantly of staphylococci, streptococci, diphtheroids, and *E. coli*. After menopause, glycogen production ceases, the pH rises, and the microbiota again resembles that found before puberty.

Female

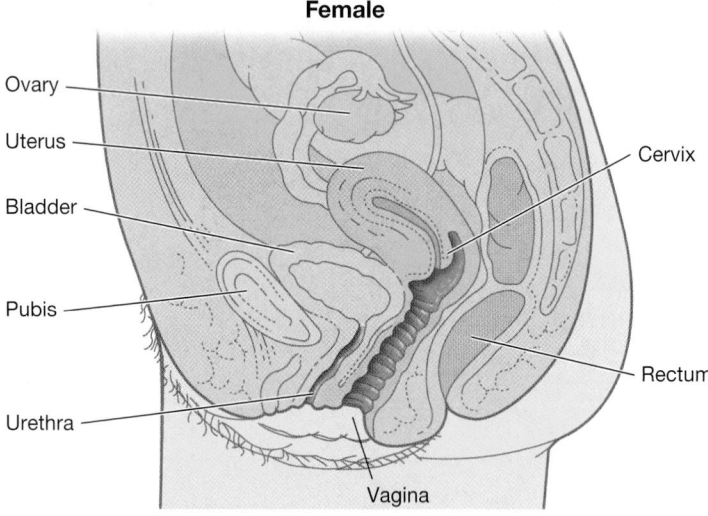

(a)

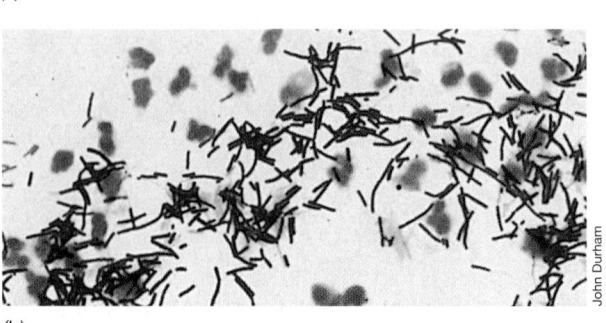

(b)

John Durham

Male

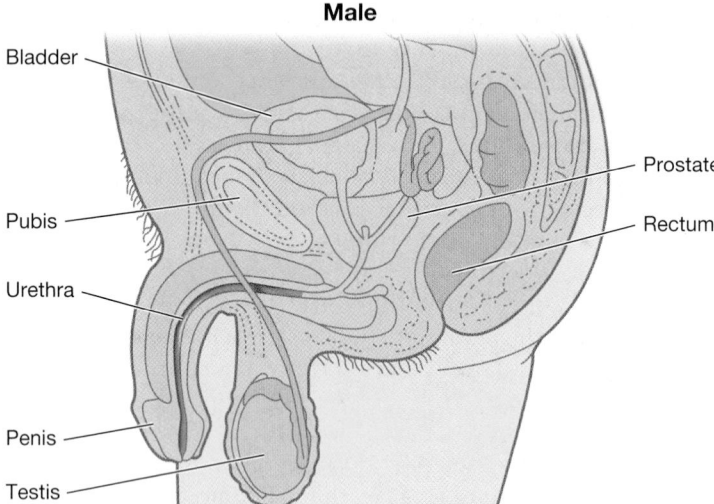

Figure 24.11 Microbial growth in the urogenital tracts. *(a)* The urogenital tracts of the human female and male, showing regions (red) where microorganisms often grow. The upper regions of the urogenital tracts of both males and females are sterile in healthy individuals. *(b)* Gram stain of *Lactobacillus acidophilus*, the predominant organism in the vagina of women between the onset of puberty and the end of menopause. The individual gram-positive rods are 3–4 μm long. Species of *Lactobacillus* are anaerobic bacteria that ferment glucose and other sugars primarily to lactic acid as a fermentation product (⊃ Section 16.6).

Culture-independent 16S rRNA sequence analyses have confirmed earlier culture-based observations that the vaginal microbial community is less complex at the genus level than the oral or gut communities, that a healthy vaginal microenvironment is dominated by lactobacilli (Figures 24.1 and 24.11*b*), and that *vaginosis* (major changes in the balance of microbes in the vagina) is characterized by increased bacterial diversity, elevated pH, and a vaginal discharge. But even in the healthy adult female, vaginal microbiota are more diverse than suggested by culture alone (**Figure 24.12**). For example, one molecular study identified 112 genera of *Bacteria* in the vagina. These analyses have also shown that multiple community types constitute a normal vaginal microbiome, but that these normal states can vary in their stability (see Figure 24.22). There appear to be at least five types of "normal vaginal communities" containing different compositions of *Lactobacillus* spp. Four types are defined by dominance by one of *L. crispatus, L. iners, L. reuteri,* or *L. jensenii* (see Figure 24.22), while the fifth is a more heterogeneous type characterized by higher overall diversity and a greater proportion of other strict anaerobes relative to the lactobacilli. Although all vaginal community types are associated with an acidic pH, pH varies with community type. The *L. crispatus* type shows the lowest average pH (~4.0), whereas the heterogeneous type shows the highest average pH (~5.3).

Thus, unlike the picture of microbial diversity we have seen in other body sites or products (Figures 24.2, 24.5, 24.7, and 24.9), the microbiota in the vagina of the healthy female is dominated by lactobacilli (Figure 24.12). In contrast to the vagina, studies of the penis microbiota are fewer, but the general picture shows that the bacterial diversity of the penis is typical of those in the vagina, the patterns being especially so in sexual partners. However, the microbiota on the circumcised versus the uncircumcised penis can be quite different, and bacterial abundance on the uncircumcised penis is typically much greater as well.

— **MINIQUIZ** —

- What is the importance of vaginal *Lactobacillus* in healthy adult women?
- What variable feature of the vagina is most closely associated with different community compositions of *Lactobacillus* species?

24.5 The Skin and Its Microbes

The skin is a complex human organ functioning primarily to prevent loss of moisture and restrict the entry of pathogens. An average adult human has about two square meters (2 m^2) of skin surface that varies greatly in chemical composition and moisture content. Skin also provides an environment for part of the human microbiome. The skin microbiota consists of a rich community of microorganisms that associates intimately with the host's hormonal, nervous, and immunological systems.

There are approximately 1 million resident bacteria per square centimeter of skin, for a total of about 10^{10} skin microorganisms covering the average adult. Although these numbers are much lower than the oral and gut communities, molecular analyses have shown that the skin harbors a diverse microbial community of bacteria and fungi (primarily yeast) that vary significantly with location on the body as a function of the diversity of habitats. These habitats consist of microenvironments of varying temperature, pH, moisture, sebum content (sebum is the oily secretions of the sebaceous glands), and surface characteristics. One distinct set of microenvironments includes moist skin areas such as the inside of the nostril, the armpit, and the umbilicus. Moist skin is separated by only a few centimeters from dry microenvironments such as the forearms and the palms of the hands. A third microenvironment consists of areas with high concentrations of sebaceous glands such as those by the side of

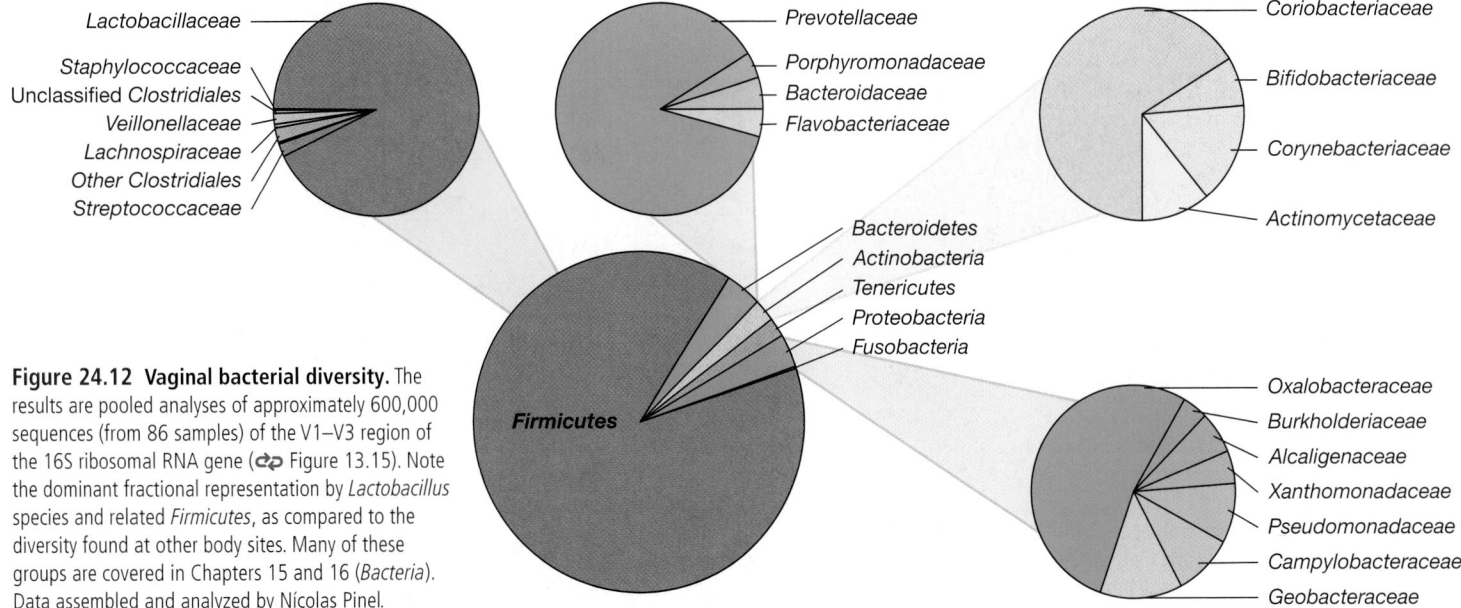

Figure 24.12 Vaginal bacterial diversity. The results are pooled analyses of approximately 600,000 sequences (from 86 samples) of the V1–V3 region of the 16S ribosomal RNA gene (⌘ Figure 13.15). Note the dominant fractional representation by *Lactobacillus* species and related *Firmicutes*, as compared to the diversity found at other body sites. Many of these groups are covered in Chapters 15 and 16 (*Bacteria*). Data assembled and analyzed by Nícolas Pinel.

Lactobacillaceae
Staphylococcaceae
Unclassified *Clostridiales*
Veillonellaceae
Lachnospiraceae
Other *Clostridiales*
Streptococcaceae

Prevotellaceae
Porphyromonadaceae
Bacteroidaceae
Flavobacteriaceae

Coriobacteriaceae
Bifidobacteriaceae
Corynebacteriaceae
Actinomycetaceae

Bacteroidetes
Actinobacteria
Tenericutes
Proteobacteria
Fusobacteria

Firmicutes

Oxalobacteraceae
Burkholderiaceae
Alcaligenaceae
Xanthomonadaceae
Pseudomonadaceae
Campylobacteraceae
Geobacteraceae

UNIT 6

the nose, the back of the scalp, and the upper chest and back. In addition to these site-specific differences, sweat is high in salt and other antimicrobial substances such as free fatty acids and antimicrobial peptides and thus plays a role in controlling diversity.

Microbial Diversity of Skin Microenvironments

A 16S rRNA sequencing comparison of 20 diverse skin sites categorized as moist, dry, or oily revealed tremendous diversity and variation among sites and individuals, but also showed some common patterns (**Figure 24.13a**). Collectively, nearly 20 bacterial phyla were detected, but most phylotypes affiliated with one of four groups: *Actinobacteria* (~52%), *Firmicutes* (~24%), *Proteobacteria* (~16%), and *Bacteroidetes* (~6%). Over 200 different genera were identified, but species of three genera, *Corynebacterium* and *Propionibacterium* (both *Actinobacteria*) and *Staphylococcus* (*Firmicutes*) typically dominated the observed phylotypes (Figure 24.13b).

Each skin microenvironment showed a unique microbiota. Moist sites are dominated by corynebacteria and staphylococci, while drier sites support a mixed population dominated by *Betaproteobacteria*, corynebacteria, and *Flavobacteriales*. Species of *Propionibacterium* predominate in sebaceous areas (Figure 24.13b). For example, colonization of the follicular sebaceous gland system by *Propionibacterium acnes* is promoted by its ability to hydrolyze triglycerides present in sebum, resulting in release of free fatty acids that promote adherence of this bacterium, which sometimes causes disease (acne, Section 24.9).

A higher-resolution molecular diversity study of 400 different body sites of one male and one female subject revealed 850 distinct species (using 97% 16S rRNA sequence identity as the species cutoff). The results of a high-resolution analysis of a single dry skin site (the inside of the elbow) are shown in **Figure 24.14**. As shown by the more general studies (Figure 24.13a), the most common microbial phyla were *Actinobacteria, Firmicutes, Proteobacteria*, and *Bacteroidetes*, but in this work, the breakdown of each phylum into family level taxa shows a significant hidden diversity within each major group. Although staphylococci, propionibacteria, and *Betaproteobacteria* dominate, many other groups are also present (Figure 24.14).

The abundance of specific bacterial taxa can also be visualized in a "heat map" diagram (Section 9.11 and Figure 9.31) showing the major locations of different taxa on the skin. An example is shown in **Figure 24.15** where it can be seen that *Propionibacterium* tends to localize on sebaceous regions (head, face, upper back, and upper chest), whereas species of *Staphylococcus* and *Corynebacterium* are more prevalent on less exposed regions, such as the groin, under arm, and toe web—areas higher in temperature and moisture content (Figure 24.15a–c).

Other Aspects of the Skin Microbiome

Eukaryotic microbes and *Archaea* are also present on the skin. The yeast *Malassezia* is the most common fungus found on the skin, and at least five different species of this yeast are typically present on the skin of healthy individuals. In an individual with a weakened immune system, for example someone who has HIV/AIDS or whose normal microbiota has been compromised, *Candida* and other potentially pathogenic fungi can also colonize the skin and cause serious (even fatal) infections. Fungal pathogens are discussed in Chapter 33. A remarkable finding emerging from 16S rRNA gene surveys of skin is that ammonia-oxidizing *Archaea* (Section 17.5) can comprise as much as 4% of the skin microbiota in some individuals, presumably sustained by ammonia present in the sweat of more physically active individuals.

Environmental and host factors influence the composition of the normal skin microbiota. For example, the *weather* may cause an increase in skin temperature and moisture, which increases

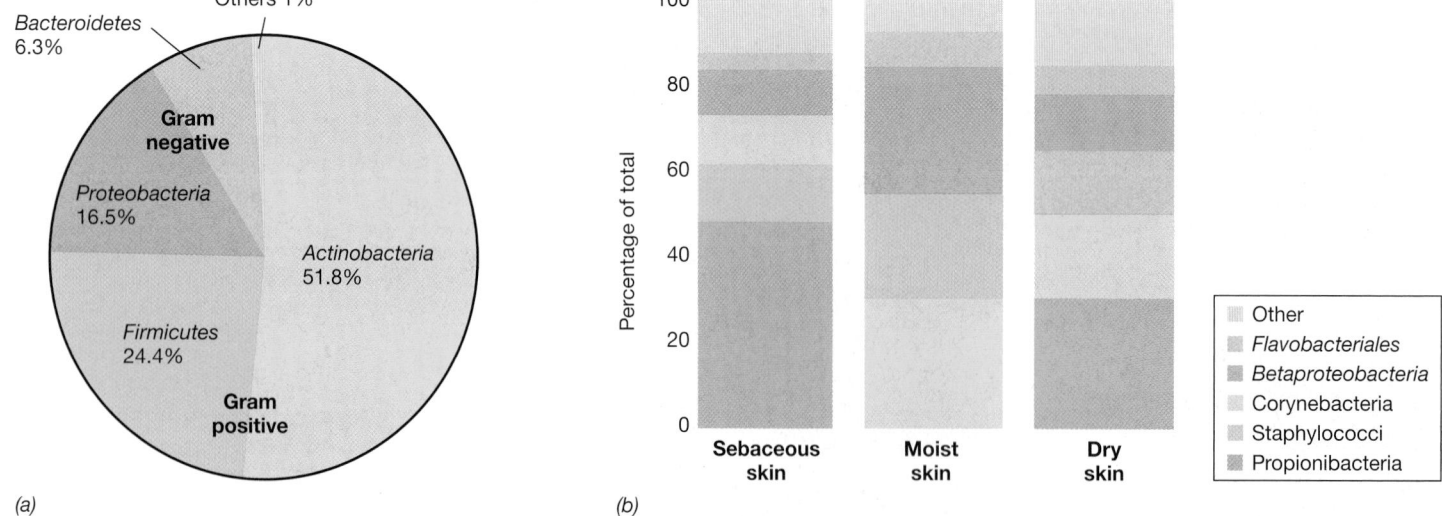

(a)   (b)

Figure 24.13 Normal skin microbiota. *(a)* Analysis of the skin microbiome from 10 healthy human volunteers detected 19 bacterial phyla. Four phyla were predominant. *(b)* Composite populations of *Bacteria* from the same volunteers, divided according to sebaceous, moist, and dry skin microenvironments. Data are adapted from Grice et al., 2009, *Science 324:* 1190.

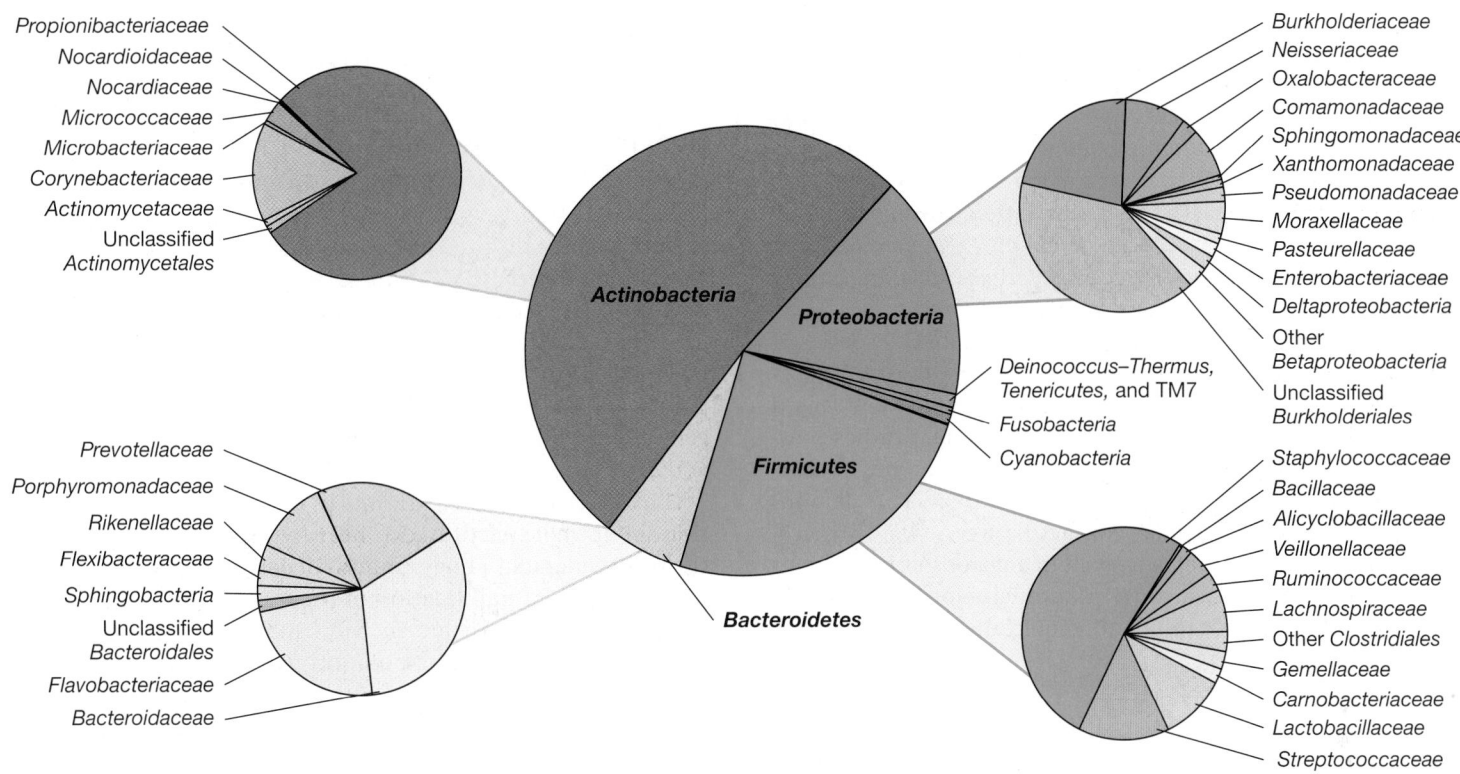

Figure 24.14 Skin bacterial diversity at the inside of the elbow (antecubital fossa; see Figure 24.1). Note that *Propionibacterium* species comprise the vast majority of the *Actinobacteria*, and that *Staphyloccocus* species dominate skin-associated *Firmicutes*. The results are pooled analyses of approximately 80,000 sequences (from 123 samples) of the V1–V3 region of the 16S ribosomal RNA gene (⟳ Figure 13.15). Many of these groups are covered in Chapters 15 and 16 (*Bacteria*). Data assembled and analyzed by Nícolas Pinel.

the abundance of the skin microbiota. The *age* of the host also has an effect; young children have a more varied skin microbiota and carry more potentially pathogenic gram-negative *Bacteria* than do adults. *Personal hygiene* also greatly influences the resident skin microbiota; individuals with poor hygiene typically

have higher microbial population densities on their skin. And finally, many microorganisms that would otherwise colonize skin cannot survive there simply because of its low moisture content and presence of antimicrobial fatty acids. Thus, the skin is a natural barrier to microbial colonization (⟳ Figure 26.2)

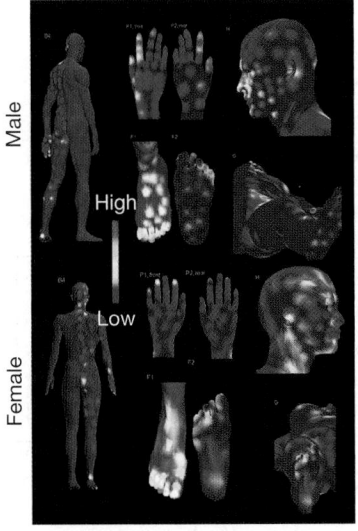

(a) **Staphylococcus**

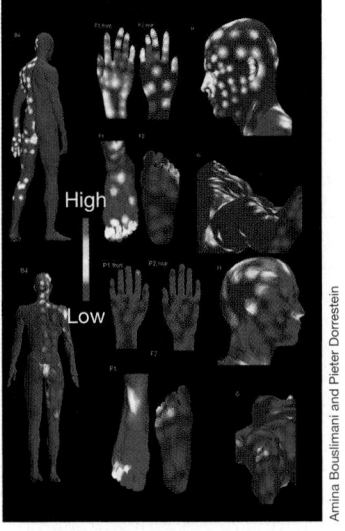

(b) **Propionibacterium**

(c) **Corynebacterium**

Male

Female

High

Low

High

Low

High

Low

Amina Bouslimani and Pieter Dorrestein

Figure 24.15 Distributions of *Staphylococcus*, *Propionibacterium*, and *Corynebacterium* on the human skin. Heat maps of microbial species distributions at 400 different body sites inferred from 16S rRNA gene sequences from the skin of male (top) and female (bottom) subjects (in scale bar, red indicates higher relative abundance and blue lower abundance). *Propionibacterium* species tend to localize on sebaceous regions (head, face, upper back, and upper chest); *Corynebacterium* are most common on the head, groin, and toes; and *Staphylococcus* are most abundant on the foot.

UNIT 6

while at the same time supporting a diverse array of normal microbiota.

As yet, there have been no significant longitudinal studies of early colonization and succession of the skin microbiome as has been conducted for the gut microbiome (Section 24.7). However, it is well recognized that there is a major transition of the skin microbiome associated with sexual maturation. The skin of young children is dominated by *Streptococcus* spp., *Betaproteobacteria*, and *Gammaproteobacteria*. By contrast, these taxa are largely absent from the skin of postadolescent young adults; in this group the skin is characteristically dominated by species of *Propionibacterium* and *Corynebacterium*, as we have seen (Figures 24.13 and 24.15). And finally, encounters with man's best friend can contribute to the human skin microbiota. Studies have shown that adult dog owners have more skin microbes in common with their own dogs than with other dogs. This demonstrates that close and regular contact between two distinctly different species of animal can result in a major sharing of their microbiomes. In contrast to the skin microbiota, microbes in the mouths and guts of canines differ quite distinctly from those of their owners.

MINIQUIZ

- Compare the populations of microorganisms in the three major skin microenvironments.
- Describe the properties of microorganisms that grow well on the skin.

II • From Birth to Death: Development of the Human Microbiome

At birth, a baby is exposed to both maternal microorganisms and microbes present in the local environment. These early encounters determine the composition of microorganisms that first colonize different body sites. In this part of the chapter we focus primarily on colonization of the gut, the body site harboring the largest part of the human microbiome, and consider factors that govern early colonization and subsequent successional events of a community now recognized to be critical to general health and the education of the immune system.

24.6 Human Study Groups and Animal Models

Establishing relationships between the composition of a person's microbiome and its contribution to health and disease has been a difficult exercise because of complications surrounding sampling, the uncertainty about the contribution of the host's genetic background, and the limitations of controlling diet and other contributing lifestyle factors. Nevertheless, several human study groups and animal model experiments have revealed some of the basic principles and given microbiome studies a starting point.

Human Microbiome Study Groups

Most functional understanding of the human microbiome has been based on surveys of selected study groups (Table 24.1). For example, one of the most ambitious early studies was the Human Microbiome Project (HMP) funded by the U.S. National Institutes of Health. This study collected samples from 242 individuals (all American medical students in good health), sampling different body sites (15 to 18 sites depending on the sex) at one to three time points, and then evaluated bacterial diversity based on 16S rRNA gene sequencing and limited metagenomic analyses (Figures 24.1 and 24.2).

The objective of the HMP was to develop baseline information about what constituted a "healthy" microbiome. Although the HMP generated a huge amount of data, the study group represented only a small fraction of human diversity. This limitation was clearly revealed in the Global Gut Project, which examined three distinct populations (U.S. citizens, Malawians, and a group of indigenous peoples from Venezuela) and different age groups within those populations. The gut microbiomes of the two non-U.S. populations were distinct from those of the U.S. individuals, showing that the HMP greatly underestimated the potential for variation in gut microbiomes across nationalities. These studies also suffered from a lack of appropriate metadata; for example, detailed information about dietary habits (vegetarian, vegan, omnivorous) and how much fiber or protein was ingested daily. Because there is still very limited understanding of the influence of specific environmental variables such as lifestyle, diet, gender, and genetics on microbiome structure and function, new and ongoing studies (e.g., the American Gut Project) are obtaining these valuable metadata at the time of sampling. The overarching goal is to develop robust correlations between the microbiome, host genetics, diet, lifestyle, health, and pathologies thought to have a microbial connection, in particular, heart disease, cancer, stroke, diabetes, and obesity.

The Mouse Model

Even if appropriate metadata are available, a major limitation of the human microbiome studies to date is establishing causality, something that is only possible with highly controlled animal studies. Hence the mouse has become the major model animal system for linking cause and effect in the gut microbiome.

Although the mouse and human digestive system have some significant differences (**Figure 24.16**), the mouse (and to a more limited extent, the rat) has been the workhorse of experimental microbiome studies. Compared to humans, mice have a relatively larger colon and cecum, which are needed to extract nutrients derived primarily from the fermentation of plant materials. For example, the average ratio of the length of the murine small intestine to its colon is about 2.5 whereas in humans it is about 7. In mice, fermentation of plant material is

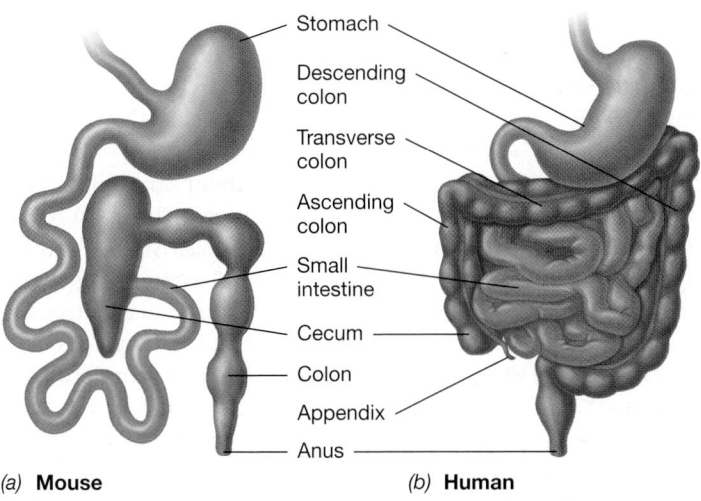

(a) **Mouse**   *(b)* **Human**

Figure 24.16 Anatomy of the mouse and human intestinal tracts. The mouse *(a)* and human *(b)* intestinal tracts have significant structural differences that are associated with differences in the composition of the microbiota of mice and humans despite similarities in their physiologies.

Stomach
Descending colon
Transverse colon
Ascending colon
Small intestine
Cecum
Colon
Appendix
Anus

compartmentalized in the cecum, while in humans this fermentation takes place in the colon. Nevertheless, mice have several experimental advantages, including the availability of well-defined genetic lines, low maintenance costs, and short life cycle. This has allowed researchers to explore the importance of the host genetic background through selective gene knockouts, the manipulation of gut microbiota composition using germ-free mice (raised free of all microbes), tests of the influence of a strictly controlled diet, assessing the consequences of antibiotic treatment, and exploring the transfer of physiological traits through fecal transplantation. These studies have clearly associated the composition of the mouse microbiome to different pathologies, including obesity and inflammatory bowel disease, as we will discuss in Section 24.8.

Unfortunately, because of anatomical differences (Figure 24.16), mouse studies are not directly applicable to humans, and these anatomical differences may in part account for observed differences in the relative abundance of dominant bacterial genera in mouse and human guts. For example, the genera *Prevotella, Faecalibacterium,* and *Ruminococcus* are in high abundance in the human gut, whereas the genera *Lactobacillus, Alistipes,* and *Turicibacter* are more abundant in the mouse gut. Thus, although the mouse model gives us much experimental latitude not available in gut microbiome studies of humans, actual results from the two systems are not directly comparable. However, despite these differences in bacterial composition, experimenters have glimpsed much useful information from the study of animal models that has accelerated our understanding of the human gut microbiome (for example, see Section 24.8 and Figure 24.20).

We now follow development of the human gut microbiota from birth through adulthood, comparing and contrasting the major organisms seen in healthy humans. An overview of the mature and highly stable gut microbiota was presented in Figures 24.3 and 24.5.

24.7 Colonization, Succession, and Stability of the Gut Microbiota

Colonization of an initially sterile gut begins immediately after birth; a succession of microbial populations replaces each other in turn until a relatively stable, adult microbial community is established. The source of early colonizers is not clear, although some species are clearly transmitted from mother to infant through the birth canal. The infant gut community is dominated by bifidobacteria—fermentative anaerobes of the bacterial class *Actinobacteria* (⊃ Section 16.10)—and does not reach an adultlike composition until about age 3. There are also major changes in the gut community in the aged. Indeed, recent studies have correlated frailty in the elderly with two major microbial factors: (1) an overall decrease in gut bacterial diversity, and (2) reduced abundance of *Firmicutes* and increased abundance of *Bacteroides*.

Microbial Activities in the First Year of Life

During the first year of life the newborn's relatively simple community evolves into a more complex and adultlike composition. The early microbial colonizers are an important source of amino acids and vitamins. Microbial genes encoding the synthesis of vitamins K_2 (menaquinone), B_6 (pyridoxal), and B_7 (biotin) are elevated in the newborn's microbiota. As the gut microbiome matures, there is a greater prevalence of microbial genes encoding synthesis of the vitamins thiamine (B_1), pantothenate (B_5), and cobalamin (B_{12}). The presence of bacterial genera such as *Enterococcus* and *Escherichia*, both facultative microbes, in the newborn gut is also indicative of a more aerobic state of the early gut system and a greater role for the citric acid cycle and respiration in microbial energy production in the neonate.

Major factors controlling the early assembly of the gut microbiome following birth is whether birth was vaginal or by cesarean section (C-section) and whether initial nutrition came from breast milk or from formula. A vaginally delivered infant is colonized by a gut microbiota similar to that of the mother, suggesting direct transfer from mother to neonate during passage through the birth canal and subsequent intimate contact with the mother. In contrast, the gut microbiota of a child delivered by C-section is significantly different from that of the mother. In an analysis of approximately 100 newborns, of the 187 taxonomically annotated 16S rRNA gene sequence types present in vaginally delivered newborns, 135 (72%) were found in their own mothers, including species of *Escherichia, Bifidobacterium, Enterococcus, Bacteroides,* and *Bilophila*. By contrast, only 41% of the species matched those of the mother in the gut of the C-section neonate. The C-section neonate microbiome tends to be enriched in groups such as

UNIT 6

Enterobacter hormaechei, Haemophilus parainfluenzae, Staphylococcus saprophyticus, S. aureus, Streptococcus australis, and *Veillonella dispar*, indicating that skin and oral microbes, as well as environmental populations, were the first colonizers. *Bacteroides* species were either less prevalent or totally missing in the infants delivered by C-section.

In 4-month-old infants delivered vaginally, the gut microbiome is defined by *Bifidobacterium, Lactobacillus, Collinsella, Granulicatella*, and *Veillonella*, reflecting reduced oxygen availability and increased production and utilization of lactose associated with a diet comprised primarily of milk. These populations are enriched in genes for carbohydrate uptake, and genes encoding lactose-specific transporters are most abundant in the 4-month-old infant. At 4 months there is also a clear difference between children who have received exclusively breast milk compared with those given formula milk. Breast-fed infants have increased levels of taxa commonly used as probiotics (Section 24.11), including *Lactobacillus johnsonii, L. gasseri, L. paracasei, L. casei*, and *Bifidobacterium longum*. The enrichment of *Bifidobacterium* species, in particular *B. longum*, is related to the composition of human milk. Breast milk contains a complex mixture of unusual oligosaccharides that most gut microbes and humans are unable to digest (**Figure 24.17**). However, these sugars are metabolized by *B. longum* growing in the infant's gut. In addition, since the structures of human milk oligosaccharides mimic carbohydrates lining the infant gut, they also function to suppress infection by pathogenic bacteria by blocking the receptors on the pathogens' cells required for attachment.

The abundance of *B. longum* in breast-fed infants also leads to the production of short-chain fatty acids; this creates an environment favoring the growth of commensal normal microbiota important for "educating" the immune system. In contrast, 4-month-old formula-fed infants tend to have elevated numbers of *Clostridium difficile*—a potentially serious pathogen—*Granulicatella adiacens, Citrobacter* spp., *Enterobacter cloacae*, and *Bilophila wadsworthia* instead of large numbers of *B. longum*. The gut of children that remain on breast milk at 12 months continues to be dominated by *Bifidobacterium* and *Lactobacillus* as well as genera of

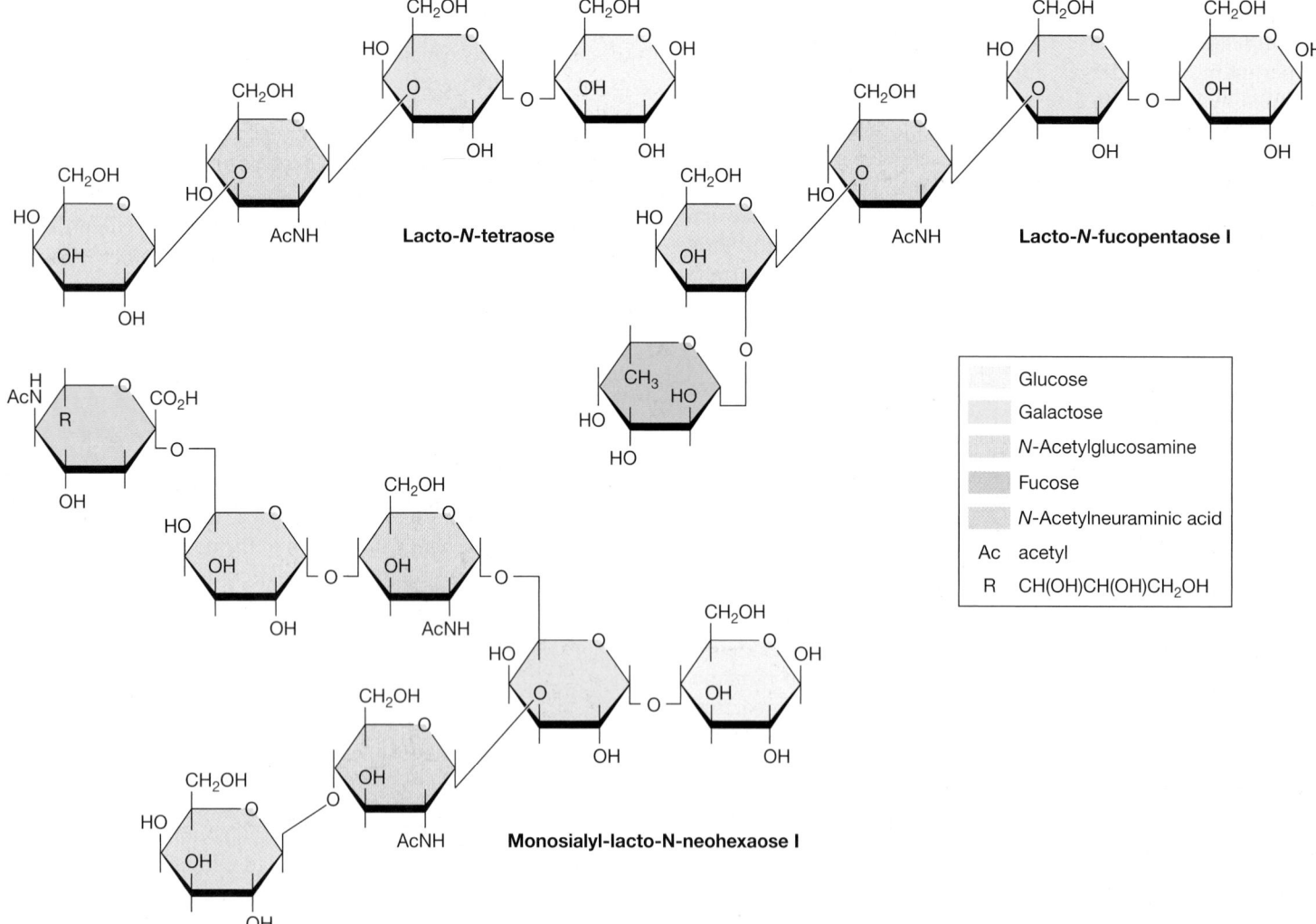

Figure 24.17 Examples of oligosaccharides found in human breast milk. These short polysaccharides selectively nourish and enrich for desirable microbial populations, such as *Bifidobacterium* and *Lactobacillus* species, in the infant's developing gut microbial community.

bacteria typical of a less mature (younger) gut community (e.g., *Collinsella*, *Megasphaera*, and *Veillonella*). Thus, in addition to providing nutrition to the newborn, mother's milk selects for a specific group of earlier colonizing normal microbiota important to the overall health of the child and proper development of its immune system.

The difference in the gut microbiomes of babies born vaginally or by C-section is significantly less by 12 months of age, but the C-section gut microbial community at 12 months remains distinctive in being more heterogeneous in composition than that of infants born vaginally. Termination of breast-feeding results in a shift of the community toward a more adultlike composition, as reflected by enrichment in *Bacteroides*, *Bilophila*, *Roseburia*, *Clostridium*, *Anaerostipes*, and *Eikenella*. Many of these later-arriving genera are efficient degraders of dietary fibers and complex carbohydrates, consistent with a transition to more solid foods following the cessation of breast-feeding; they also produce short-chain fatty acids. The increased abundance of *Bacteroides thetaiotaomicron*, a bacterium encoding a diverse set of glycan-degrading enzymes, is specifically associated with the increase in pectin in solid foods. But it is the cessation of breast-feeding rather than the introduction of solid foods, per se, that actually triggers the development of an adultlike gut microbiota. The time of development of a fully adultlike gut microbial community varies among study groups, but the gut microbiome of children generally reaches an adultlike composition when they are 1–3 years old.

Stability of the Adult Microbiome and Transitions with Age

As stated earlier, the 16S rRNA sequence-based census of microbial diversity in adult human fecal material (Figure 24.5) has identified between 3500 and 35,000 microbial species, depending on whether a 97% or more stringent 98–99% identity threshold is used to define a species (Section 24.2). The number of species in any given individual is much lower, fewer than 200 unique species. However, an individual's gut microbiota is relatively stable through time. Longitudinal studies over a 5-year period combining metagenomic and 16S rRNA sequence analyses to achieve precise resolution have shown that 70% of an individual's unique species persist over a one-year sampling period, and that samples taken 4 years later show only a few additional changes in species composition.

The most stable species are typically affiliated with the *Bacteroidetes* and *Actinobacteria*. Although *Actinobacteria*, such as *Bifidobacterium* spp., comprise a very small part of the adult gut microbiota compared to the *Bacteroidetes*, specific species tend to associate with individuals over long periods of time. Species of *Firmicutes* and *Proteobacteria* appear to be significantly less stable members of the gut community. Studies extrapolating gut stability over longer time periods suggest that the majority of species comprising an individual's gut microbiome constitute a stable core that persists for his or her entire adult life. These studies have also shown that species of an individual's core gut microbiome tend to be shared among family members, and likely were acquired very early in life. Thus, the early colonizers—including microorganisms acquired from parents or siblings—may in large part determine the metabolic and immunological character of the

adult microbiome. In other words, early life experiences that govern or influence microbial colonization may be an important factor in adult health and predisposition to disease.

As we age, so does our microbiome. The most well-established change is the age-related alteration in the relative proportions of *Firmicutes* and *Bacteroidetes*, with an increasing proportion of *Bacteroidetes* seen in the elderly as compared with the higher proportions of *Firmicutes* in young adults. Also observed with age are significant decreases in bifidobacteria and certain clostridial species. There is also the observation that old age–related inability to perform routine daily activities, a measure of frailty, is correlated with decreased diversity in the gut microbiota. Consistent with this observation, elderly individuals living at home or in supportive communities have a more diverse gut community compared with those living in residential care facilities, such as nursing homes. Thus, maintaining a diverse gut microbiota is likely one of many links to health for the aged and is possibly a positive effector of longevity.

With this overview of changes in the human microbiome with time, we move on to consider some startling discoveries about human diseases linked to unfavorable changes in the human gut microbiota. We then end this chapter by looking at some therapies for dealing with these problems.

MINIQUIZ

- What factors contribute to early colonization of the newborn's gut community?
- How do microorganisms in the infant and adult gut community contribute differently to vitamin and amino acid requirements?
- What factor(s) are most important in the transition from an immature to a mature gut microbial community?

III • Disorders Attributed to the Human Microbiome

Alteration of the structure and activity of the human microbiome is associated with a variety of pathologies, including obesity, type 2 (non-insulin-dependent) diabetes, asthma, atopic dermatitis, liver disease, colorectal cancer, kidney stones, psoriasis, tooth decay (caries), and periodontitis. As we learn more about the relationship between the human microbiome and health and disease, therapeutic intervention may well be possible. This might include promoting the growth of protective beneficial bacteria, inhibiting the growth of specific microbes (or specific assemblages of microbes) that compromise health, fecal transplants to introduce a desirable gut microbiota, and developing appropriate behavioral interventions, such as diet modification.

24.8 Disorders Attributed to the Gut Microbiota

Two well-studied examples of links between the gut microbiota and clinical disease are inflammatory bowel disease and obesity. Both conditions show strong links to alterations in the gut microbiota.

Inflammatory Bowel Disease

The gut microbiota plays important roles in shaping the immune system during infancy, during which time the commensal microbiota and their products interact with immune cells to initiate and maintain host tolerance. The failure to develop tolerance to the normal microbiota early in life is associated with different immune-mediated diseases, including allergies and chronic inflammation of the gut such as *inflammatory bowel disease* (IBD).

It is widely accepted that IBD is not caused by a specific pathogenic microbe but rather an imbalance between the immune system and the normal gut microbiota. The observation that antibiotic use in early life increases the risk of IBD points to the importance of the development of a normal gut microbiota in "educating" the immune system to differentiate between the normal microbiota and invading pathogens. For example, relative to healthy children, children with IBD have higher levels of species of *Veillonella*, *Prevotella*, *Lactobacillus*, and *Parasporobacterium* and lower levels of species of *Bifidobacterium* and *Verrucomicrobium*. This type of disruption of the homeostasis between the gut microbiota and the host is called **dysbiosis**.

There is also some evidence that once developed, IBD is transmissible. Fostering or co-caging healthy mice with IBD-predisposed mice was sufficient to cause IBD development in the healthy mice and was correlated with the transfer of the enteric bacterial species *Klebsiella pneumoniae* and *Proteus mirabilis* from the IBD mice to the healthy mice. Metagenomic analyses of healthy subjects and patients with IBD shows that the gut microbiota of IBD patients shares fewer genes in common with healthy subjects, relative to the number of genes shared among healthy subjects. The microbial community of IBD patients also tends to have significantly reduced functional capacity, as reflected by a reduction in the number of nonredundant (functionally unique) genes relative to subjects that do not have IBD (**Figure 24.18**).

However, as for the relationship between the gut microbiome and obesity, to be considered next, the causes of IBD and its possible transmission are not well understood. IBD may follow the disruption of mucosal barrier integrity (a condition called *leaky gut*) by a gut pathogen or toxic insult, permitting commensal bacteria to interact with and activate the adaptive immune system (Chapter 27); this stimulates the proliferation and differentiation of T cells into commensal-specific effector cells that can persist in the intestine long after resolution of the infection. For example, the IBD syndromes of *Crohn's disease* and *ulcerative colitis* are known to be associated with a T cell response to intestinal commensal bacteria.

There is also a strong correlation between IBD and diet. With the typical Western diet rich in animal protein, carbohydrates are first fermented in the upper colon (Figure 24.3). As digesta move further along the colon, protein and amino acids are then fermented, generating potentially harmful metabolites such as ammonia, phenols, amines, and H_2S that have been implicated in promoting IBD as well as playing a role in colon cancer. Animal studies have shown that these compounds can promote the development of a leaky gut and gut inflammation. In contrast, a diet rich in plant-based foods (high fiber) appears to inhibit development of these

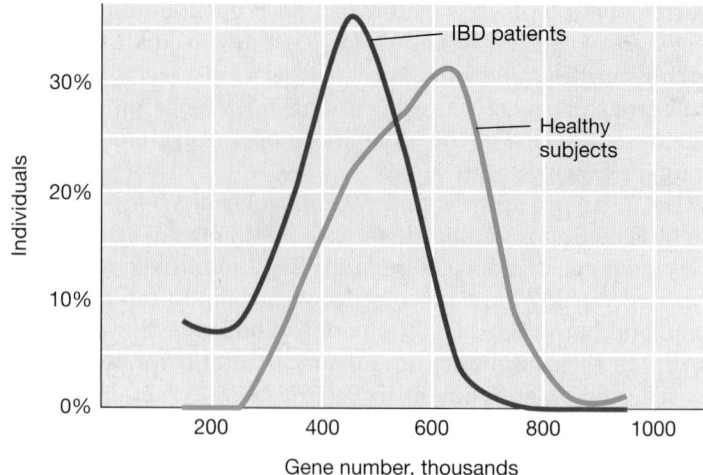

Figure 24.18 Reduced functional capacity of the gut microbiome of patients with inflammatory bowel disease. Metagenomic analysis of human gut microbiota in healthy subjects and patients with inflammatory bowel disease (IBD) revealed a tendency toward fewer nonredundant bacterial genes in patients with IBD.

pathologies, highlighting the importance of diet in maintaining a primarily carbohydrate-based fermentation in the healthy gut. Finally, IBD does not always affect identical twins; this observation argues against a strong genetic connection to IBD and supports the hypothesis that the environment and diet are the major triggers of this intestinal condition.

The Role of the Gut Microbiota in Obesity: Mouse Models

Obesity is a significant health risk that contributes to secondary health issues such as high blood pressure, cardiovascular disease, and diabetes. Gut microorganisms likely play a part in human obesity, although the mechanisms are unclear. However, relatively minor changes in gut energy metabolism can have significant long-term effects on the accumulation of body fat. A small but persistent difference (+12 kcal/day) would result in a greater than 0.45 kg (~1 pound) gain in fat per year. This is the average weight increase experienced by people in the United States from ages 25 to 55.

Initial evidence linking the gut microbiota to host fat accumulation came from studies using germ-free mice. In these experiments, normal mice had 40% more total body fat than those raised under germ-free conditions, although both mouse populations were fed the same amount and type of food. After germ-free mice were inoculated with cecal material from a normal mouse, they developed a gut microbiota and their total body fat increased, although there had been no changes in food intake or energy expenditure. Studies of experimental colonization of germ-free mice with individual microbial species or microbial communities have demonstrated that colonization triggers the expression of *host* genes for glucose uptake and lipid absorption and transport in the ileum. This also indicates that there may be a link between gut microbial composition and the ability of the host to harvest energy from its diet, ultimately contributing to obesity.

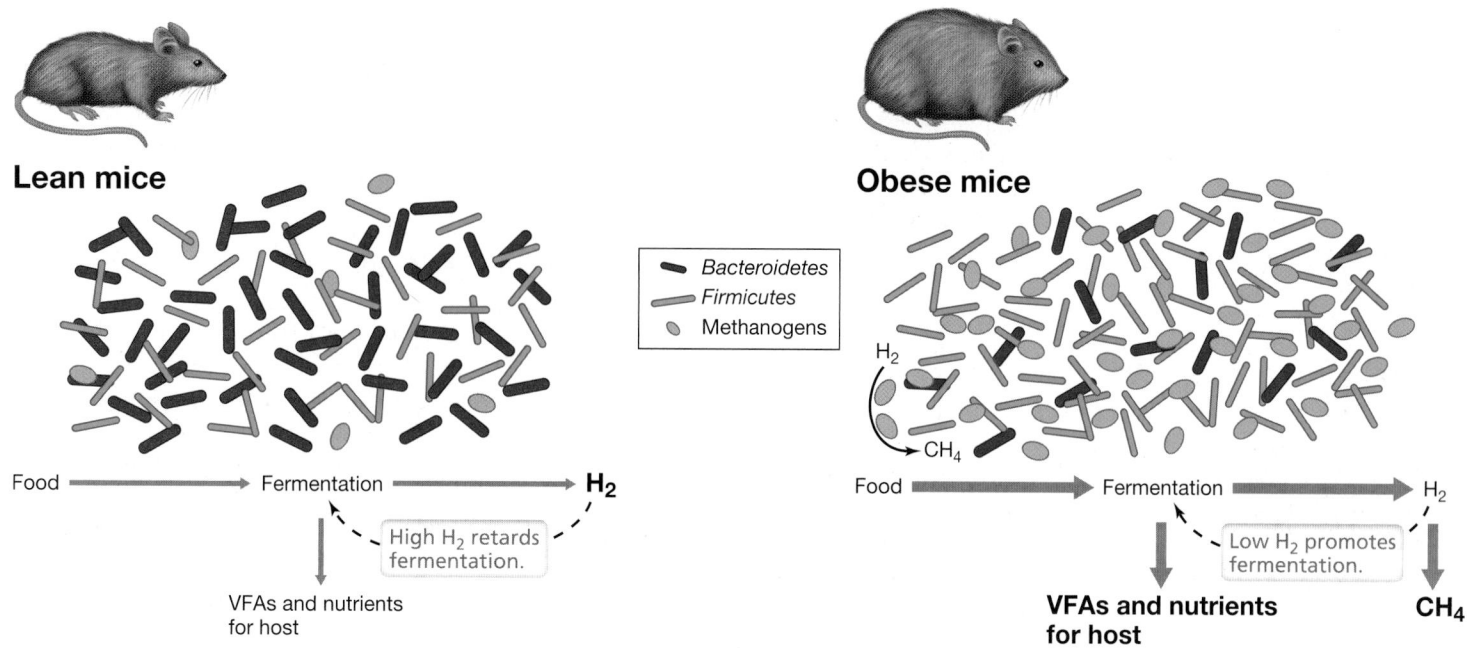

Figure 24.19 Differences in gut microbial communities between lean and obese mice. Obese mice have more methanogens, a 50% reduction in *Bacteroidetes*, and a proportional phylum-wide increase in *Firmicutes*. Nutrient production from fermentation is higher in obese mice due to removal of H_2 by methanogens.

One of the main activities of the intestinal microbiota is to break down and ferment dietary fibers into **volatile fatty acids (VFAs)**, including acetate, propionate, and butyrate. The host absorbs these acids, and humans obtain about 10% of their daily energy requirements from them. Mice that are genetically obese have microbial gut communities that differ from those of normal mice, with 50% fewer *Bacteroidetes*, a proportional increase in *Firmicutes*, and a greater number of methanogenic *Archaea* (**Figure 24.19**). Methanogens are thought to increase the efficiency of microbial conversion of fermentable substrates by consuming molecular hydrogen (H_2), as mentioned for fermentation in the rumen ($\rightleftarrows$ Section 23.13). The working hypothesis is that H_2 removal stimulates fermentation, making more fermentation products available for absorption by the host and thus contributing to obesity.

The importance of the gut community to a predisposition to obesity in mice has also been shown by *fecal transplant studies*, transplanting a small sample of the gut contents from a set of paired human twins (one obese and one lean) into germ-free mice (see Section 24.10 for more coverage of fecal transplants). Although the recipient mice were fed identical high-fiber diets, the mice receiving fecal contents from the obese twin gained significantly more weight than the mice receiving fecal contents from the lean twin (**Figure 24.20**). This is direct experimental evidence that a lean or obese body type can be altered by changes in the gut microbiota, even when the microbiota originates from a different species. Put another way, specific but widely distributed gut microbes may exist that can in some way control an animal's metabolism to yield a lean or obese body type.

The Gut Microbiota and Human Obesity

Animal model inferences have been more difficult to demonstrate with human subjects, since strict control of diet and host genotype is not feasible and gut microbiota manipulation is more difficult to achieve. Nevertheless, studies of humans, while not strictly confirming the *Bacteroidetes–Firmicutes* relationship established in mice, have shown that obese humans are more likely to harbor species of *Prevotella* (a genus of *Bacteroidetes*) and methanogenic *Archaea* than are lean humans, suggesting that the mouse model (Figures 24.19 and 24.20) is likely applicable to humans. In humans, the methanogens are proposed to remove H_2 produced by *Prevotella*, facilitating fermentation by *Prevotella* and increasing nutrients to the host. This model is also supported by the study of germ-free mice colonized with *Bacteroides thetaiotaomicron* (having a metabolism similar to *Prevotella*) and the methanogen *Methanobrevibacter smithii*. Relative to controls containing just one of these species, co-colonized mice have a higher number of total gut bacteria, higher acetate levels in the intestinal lumen and blood, and greater body fat.

This relatively simple explanation for the cause of obesity is insufficient to fully explain microbial contributions to obesity. For example, the gut microbiota in lean mice produce greater amounts of the fatty acids propionate and butyrate and actually digest *more* of the plant fiber than do the microbiota of obese mice. Similarly, although a fiber-rich diet increases the amount of material available for fermentation in the human large intestine, such a diet also reduces the risk of obesity. This may in part be a consequence of volatile fatty acids binding to free fatty acid receptors in the gut and triggering the production of hormones associated with feelings of a full stomach (satiety). In humans, obesity is also associated with a variety of metabolic complications, including low-grade inflammation, hypertension, glucose intolerance, and diabetes, and these factors have not been adequately addressed in the simple microbiota models of obesity proposed to date.

An influence of host physiology on the gut community is also suggested by changes during pregnancy in humans. The period

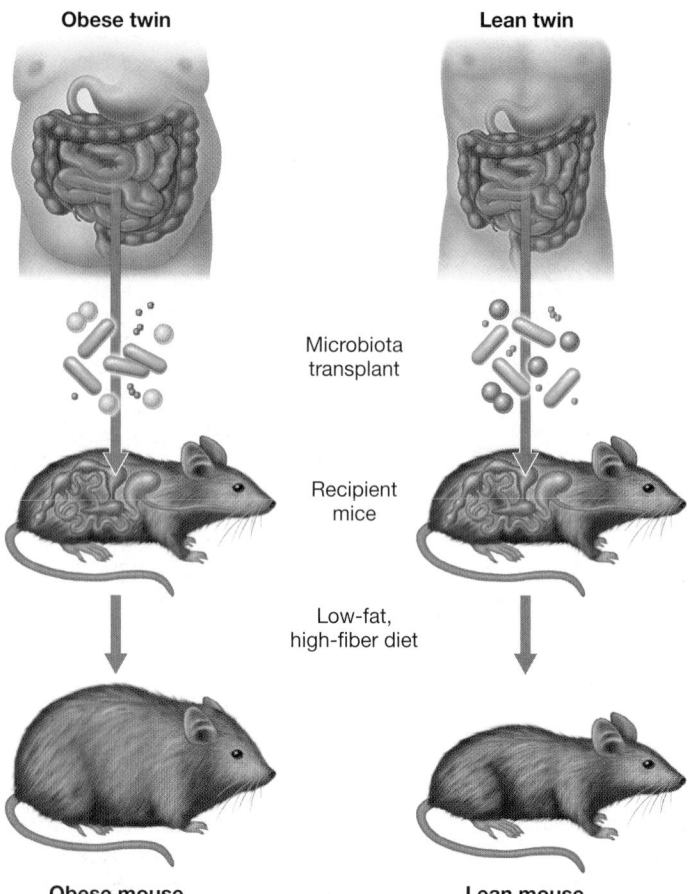

Figure 24.20 Transfer of an obese condition by fecal transplant. Transplanting fecal material from the gut contents of a paired identical human twin study group (one twin was obese and the other lean) to germ-free mice showed that the obese twin microbiota made the mouse obese. Conversely, transfer of gut contents from the lean twin did not contribute to an obese phenotype. Adapted from Ridaura, V.K., et al. *Science 341:* DOI:10.1126/science.1241214.

between the first and third trimesters of pregnancy is associated with a decrease in gut microbial diversity and enrichment in the gut community of species of *Proteobacteria* and *Actinobacteria*. These changes are associated with the increased body fat and insulin insensitivity that develop later in gestation. A simple interpretation of these findings is that a pregnant woman's body in some way manipulates her gut microbiome as part of its preparations for a greater demand on stored energy reserves. If true, this once again underscores the complex interplay between the gut microbiota and multiple host variables (genetic, physiological, behavioral, and environmental) that may be contributing factors in the development of human obesity and associated metabolic disorders.

— **MINIQUIZ** —

- What is dysbiosis? How might this condition lead to inflammatory bowel disease?
- What is the mechanism by which higher numbers of gut *Firmicutes* are thought to be linked to obesity?
- Why are the gut microbiota results obtained in mice more difficult to confirm through human studies?

24.9 Disorders Attributed to the Oral, Skin, and Vaginal Microbiota

Imbalances in the normal microbiota have been linked to human health issues other than those affecting the gut. These include in particular diseases of the oral cavity, skin, and vagina, and we consider some of these important disease centers here.

Dental Caries and Periodontitis

Dental caries and periodontal diseases are among the most common chronic human maladies. We consider dental caries in Chapter 25, where we examine how bacteria attach to solid surfaces. The mechanisms by which oral bacteria stick to the teeth and gums has been used as a major model system for bacterial attachment to solid surfaces. The attached cells eventually form a diverse microbial community in *dental plaque* and this leads to dental disease from the generation of lactic acid from fermentation (⟜ Section 25.2 and Figures 25.7 and 25.8). Although *Streptococcus mutans* is the major pathogen associated with caries, colonization of tooth surfaces (**Figure 24.21**) with organisms other than *S. mutans* can also lead to

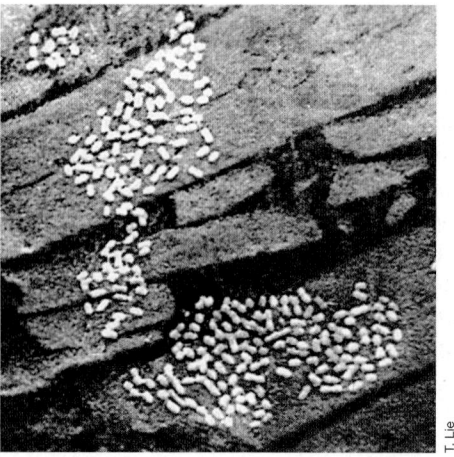

(a)

(b)

Figure 24.21 Colonization of tooth surfaces. *(a)* The colonies are growing on a model tooth surface inserted into the mouth for 6 h. *(b)* Higher magnification of the preparation in part *a*. Note the diverse morphology of the organisms present and the slime layer (arrows) holding the organisms together.

caries. In the absence of *S. mutans*, other acidogenic and acid-tolerant species can initiate caries.

Species of the bacterial genera *Streptococcus, Granulicatella,* and *Actinomyces* are typically elevated in children who develop severe early dental caries, whereas caries-free children have a higher relative abundance of *Aestuariimicrobium*. In older children, caries is associated with decreasing bacterial diversity and increasing prevalence of *Porphyromonas* and *Prevotella* species. Similarly, the dental plaque of healthy subjects is more complex than those with caries. Chronic periodontitis (inflammation and loss of tissue and bone that supports the teeth) is also associated with decreased microbial diversity, although no single bacterial phylotype seems to be directly associated with disease.

Collectively, these observations suggest that no single pathogen leads to the formation of dental caries or periodontal disease, but rather that community-wide changes in microbial composition trigger the diseases. Periodontal disease is particularly serious because it is thought to contribute to several debilitating systemic conditions, including cardiovascular disease, diabetes, pneumonia, and arthritis. Although the etiology of these systemic pathologies is unclear, they appear to be associated at least in part with dysbiosis of the oral cavity similar to that observed in dysbiosis of the gut microbiota in inflammatory bowel disease (Section 24.8).

Acne Vulgaris

The interrelationship of species of the bacterial genus *Propionibacterium* with the innate and adaptive arms of immunity (Chapters 26 and 27, respectively) has been examined extensively because of the association of this bacterium with *acne vulgaris*, usually just called *acne*, a cutaneous inflammatory disease of the skin that affects more than 85% of adolescents worldwide. However, the suspected causative agent of acne, *P. acnes*, is also a member—and a dominant member—of the normal cutaneous microbiota (Figures 24.2, and 24.13–24.15) and thus is also present in individuals without inflammatory disease.

Although *Propionibacterium* species have the capacity to elicit an inflammatory response, this alone is insufficient to demonstrate a causal association with disease. A better understanding of the cause(s) of acne will likely emerge from a better understanding of variation among strains of *P. acnes*. Notably, recent experiments using metagenomic and multilocus sequence typing (⟳ Section 13.9) suggest that specific strains of *P. acnes* may trigger the acne condition, whereas others are associated with skin health. Because of the scarcity of nutrients available for the skin microbiota, even minor differences in nutritional requirements between strains may favor the development of one or another *P. acnes* strain. If true, this might lead to therapies for acne using probiotics or prebiotics (see Section 24.11) that foster the growth of harmless strains. For example, a lotion could be applied to the skin that contains a nonpathogenic strain or a nutrient selective for nonpathogenic strains as a means of protecting the skin from colonization by pathogenic strains.

Vaginal Conditions

Dysbiosis of the normal vaginal microbiota (Section 24.4 and Figure 24.12) is associated with either an inflammatory infection (*vaginitis*) or a less severe clinical form (*vaginosis*) that may be

asymptomatic or associated with malodor and discharges. Common types of vaginitis result from the overgrowth of species of the yeast *Candida* (candidiasis) or from infection by the sexually transmitted protozoan *Trichomonas vaginalis*, both discussed in Chapter 33. Diagnosis of the less highly inflammatory vaginosis in a clinical research lab is commonly based on a microscopic scoring of a Gram-stained vaginal smear (*Nugent score*). This ten-point scoring system scores a vaginal swab as normal (0–3), intermediate (4–7), or vaginosis (7–10) based on a *decrease* in gram-positive rods (presumptive lactobacilli, see Figure 24.11*b*) and an *increase* in two morphotypes of gram-variable rods as compared to the normal range of relative abundances.

Culture-independent molecular methods (Chapter 19) are now increasingly being employed in place of the Nugent system to more precisely define vaginal communities associated with health, disease, and predisposition to disease, including neonatal infections, miscarriage, pre-term birth, and increased susceptibility to HIV and sexually transmitted infections. Longitudinal studies of individual women over multiple months have revealed that the vaginal microbial communities of some women are highly dynamic, changing in composition over relatively short time periods, whereas the communities of others are relatively stable (**Figure 24.22**).

Most vaginal microbial communities appear to be stable, displaying *resilience* (the capacity to return to the pre-disturbance community structure following disruption, such as menstruation) or *resistance* (the capacity to resist community disruption in response to disturbance). For example, a low pH–tolerant, *Lactobacillus crispatus*–dominated vaginal community is relatively stable

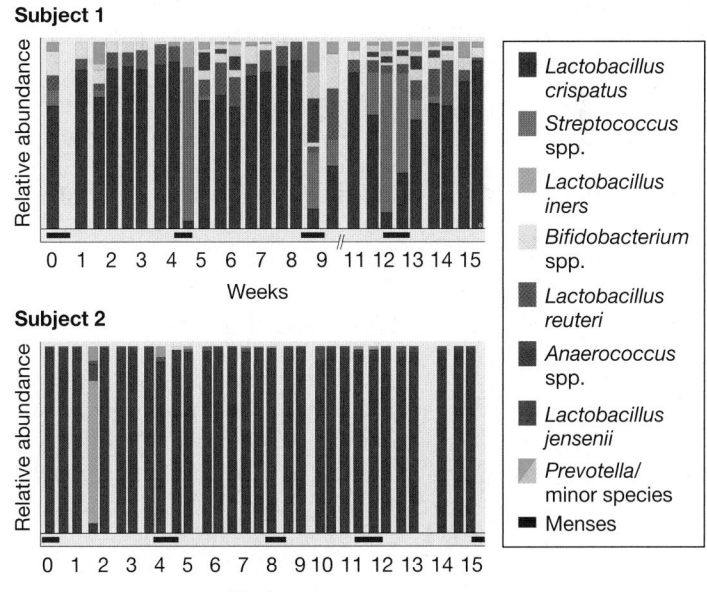

Figure 24.22 Resilience of the normal vaginal microbiota in two subjects. Both vaginal communities were dominated by *Lactobacillus crispatus* and demonstrated resilience to perturbation, most notably the disruption associated with menstruation (indicated by horizontal red bars) in subject 1. In addition to the major population types identified in these figures, populations detected in minor amounts at different times included the following: for subject 1, species of *Alloscardovia, Escherichia, Peptostreptococcus, Finegoldia, Prevotella,* and *Anaerococcus*; and for subject 2, *L. jensenii, Staphylococcus, L. gasseri, Corynebacterium, Clostridium,* and *L. vaginalis*.

(Figure 24.22), showing either resilience (subject 1) or resistance (subject 2). By contrast, other vaginal communities are more prone to converting to a vaginosis form. For example, the healthy *L. crispatus*–dominated community can convert to a healthy *L. iners*–dominated community before returning to *L. crispatus* (Figure 24.22b). However, although both the *L. crispatus* and *L. iners* communities are associated with vaginal health, an *L. iners*–dominated community is more likely to transition to vaginosis. Vaginal bacterial diversity also waxes and wanes over shorter time periods, such as a woman's menstrual cycle. Higher bacterial diversity is observed during days of menstruation than in intervening days, with diversity following a cycle inversely related to that of the estrogen hormone estradiol.

As for most ongoing human microbiome research, molecular microbial community analyses of the vagina have revealed possible connections between the microbial diversity of the vagina, vaginal health, and the susceptibility to disease. But clear and direct links have not yet emerged. We do not yet fully understand how the vaginal community is established and maintained or how vaginal dysbiosis develops and resolves. However, if some community types are associated with clinical symptoms or are associated with higher risk for vaginal disorders, then early intervention may be important to maintaining women's health and the health of a mother and baby in particular.

MINIQUIZ

- What observations indicate that dental carries are not due solely to *Streptococcus mutans*?
- Why might it be possible to have high abundance of *Propionibacterium acnes* without developing acne vulgaris?
- What are some clinical advantages of a community-structure-based evaluation of vaginal health?

IV • Modulation of the Human Microbiome

One important basic-science goal of human microbiome studies is to understand how the microbial composition and activity of human-associated microbes promotes health or predisposes the body to a variety of health disorders. An important practical goal is to then use this knowledge to improve the health and fitness of humans. Although at this point we have only a sketchy understanding of the relationships between microbes and humans, there is a strong indication that in at least some cases significant health benefits are associated with altering the human microbiome.

24.10 Antibiotics and the Human Microbiome

Antibiotics are naturally produced antimicrobial substances whose efficacy varies against different bacteria (Chapter 28). However, when an antibiotic is taken orally it kills or inhibits to at least some extent the normal microbiota as well as the targeted

pathogen(s). Antibiotic treatment can lead to a significant loss of the gut normal microbiota (Section 24.2). When antibiotic therapy ends, the normal intestinal microbiota is usually, but not always, reestablished spontaneously in adults.

Use of antibiotics during the first few months of life is a particular problem for the developing normal microbiota. This disruption can affect normal development of the immune system and predispose the infant to later autoimmune disorders, such as IBD and allergies (Section 24.8). Recent research has also shown that early disruption of the gut microbiota influences host energy metabolism. For example, antibiotic exposure during the first 6 months of life is associated with increased weight gain in infants between 10 and 38 months of age compared to those not receiving antibiotics. With growing concerns about the frequency of childhood obesity leading to adult obesity along with the growing number of childhood autoimmune disorders, it is clearly important to avoid, if possible, disruption of the normal development of the human gut microbiota early in life.

Clostridium difficile Infections

Sometimes antibiotic-resistant opportunistic pathogens become established in young children or the elderly following treatment with antibiotics that disrupts the normal microbiota. A particularly problematic complication of antibiotic therapy is infection with toxigenic *Clostridium difficile* and the inability to resolve infections with repeated follow-on administrations of antibiotics (**Figure 24.23**). There is a strong association between antimicrobial therapy and the subsequent development of *C. difficile* infection, and the risk of infection is greater if *C. difficile* is resistant to the antimicrobial agents used in therapy.

Clostridium difficile was first described in 1935 as part of the normal intestinal microbiota of healthy neonates. Its role in diarrhea was first described in 1978, and the marked increase in hospital-acquired diarrhea since 2003 is attributed in part to the emergence of extremely virulent and toxigenic *C. difficile* strains. In those infected, symptoms vary from a mild diarrhea to severe abdominal pain and fever. The most severe complications are inflammatory lesions and bowel perforation; as a result of these, septic shock (⟳ Section 25.2) and death are possible.

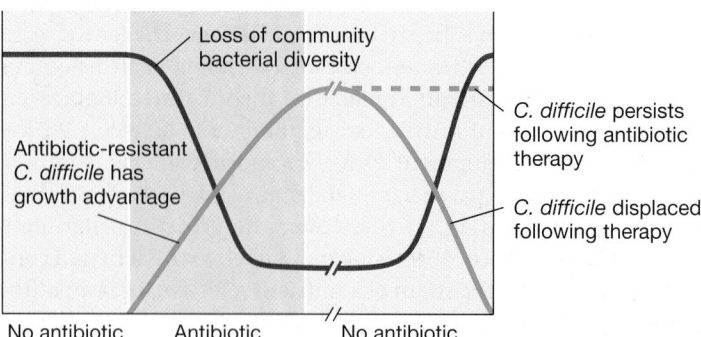

Figure 24.23 Antibiotic treatment increases the risk of *Clostridium difficile* infection. During antibiotic therapy the displacement of sensitive gut populations results in a significant loss of community diversity, allowing antibiotic-resistant populations such as toxigenic *C. difficile* to increase in abundance. When antibiotic therapy ends, *C. difficile* may be displaced (solid green line) by reestablishment of a normal gut microbiota, or alternatively, it may persist (dotted line), causing intestinal disease.

Although elderly hospitalized patients receiving antibiotics are still the main risk group, there is increasing incidence of *C. difficile* infection in younger populations having no previous contact either with a hospital environment or antibiotics. Until recently, there was no effective means to treat patients that were nonresponsive to antibiotic treatment. If medical management with intravenous fluids and antibiotics was not effective, then surgical removal of the colon was a last resort life-saving measure. However, the prognosis for recovery from a *C. difficile* infection today has been dramatically improved by a novel nondrug and nonsurgical therapy: the infusion of fecal material from the gut contents of a healthy donor into the gut of a *C. difficile*–infected patient—more commonly called a **fecal transplant**. In most cases such treatment has been shown to restore a healthy colon in patients suffering recurring *C. difficile* infections. We examine this procedure now.

Fecal Transplants

A changing paradigm in the treatment of inflammatory bowel disease and *C. difficile* infections is the use of a fecal transplant. The goal of a fecal transplant is to reintroduce normal microbiota into the gut of someone with an intestinal disease of bacterial origin and have the transplanted microbial community become established and exclude the disease-causing organisms. Although it has taken the medical community some time to adopt this therapy, the dramatic improvement in the rate of resolution from *C. difficile* infection by fecal transplant from a healthy donor has prompted widespread adoption of this procedure.

As we have seen, once established, *C. difficile* is a particularly difficult pathogen to eradicate (Figure 24.23). However, the percentage of otherwise nonresponsive patients cured of *C. difficile* infection without relapse is nearly 90% with fecal transplant therapy versus only about 25% with standard antibiotic treatment. It has also been shown that fecal transplants can alleviate some forms of *metabolic syndrome*, a condition characterized by elevated blood pressure and glucose levels, excess belly fat, and abnormal cholesterol levels, that is a frequent precursor of type 2 (insulin-nonresponsive) diabetes. Metabolic syndrome patients receiving a fecal transplant from a lean donor had significantly increased insulin sensitivity, and also showed increased fecal levels of butyrate, greater overall gut microbial diversity, and increased abundance of gut bacteria related to the butyrate-producing bacterium *Roseburia intestinalis*, thought to be a beneficial microbe.

Fecal transplants require that the donor be healthy and free of infection. And, because feces are considered a body fluid, screening and testing of the donor is essential. Potential donors must complete a screening questionnaire similar to that required for donating blood, and any prospective donors who have risk factors for HIV (AIDS) or hepatitis infections or any history of gastrointestinal disease, autoimmune disease, or cancer are screened out. Prospective fecal donors must also undergo blood tests for a suite of pathogens and their feces are tested for fecal bacterial pathogens and parasites. If everything checks out, samples of their feces are frozen in small vials and then thawed and used when needed. Fecal transplant recipients typically receive the transplant via a saline solution during colonoscopy to ensure that the transplanted feces actually reach the upper colon.

Although a fecal transplant may seem somewhat crude and an essentially uncontrolled or nonprecise therapy, studies are accumulating that show that the appropriate manipulation of the gut microbiota can have significant health benefits and that gut microbiota transplants can be maintained for long periods if not indefinitely. More generally, the success of fecal transplants has shown that manipulation of the human microbiome for improved health is feasible and that with greater understanding of these complex communities, more targeted intervention may be possible, as we consider in the next section.

— MINIQUIZ —

- Why do healthy adults usually not contract *Clostridium difficile* infections?
- Why is it unsafe for a fecal transplant recipient to receive feces from an unscreened donor?

24.11 Probiotics and Prebiotics

We close this chapter with a brief discussion of how ingestion of living microbial cultures (*probiotics*) or particular nutrients (*prebiotics*), generally derived from plants, could be used to treat disease and promote health by modulating the gut microbial community.

Probiotics

The United Nations Food and Agriculture Organization and the World Health Organization define a **probiotic** as a "live microorganism which, when administered in adequate amounts, confer a health benefit on the host." Species of *Bifidobacterium* and *Lactobacillus* bacteria are the most commonly discussed probiotics. They are usually delivered to the gastrointestinal system by ingestion of a fermented milk product, such as yogurt (**Figure 24.24**), with the hope of suppressing gut or urogenital disturbances.

Most physicians and scientists agree that ingested probiotic foodstuffs probably have limited therapeutic value. Nevertheless, human microbiome research has spurred a renaissance in probiotics research and investment and specifically for the development of effective and more targeted probiotics. The remarkable success of fecal transplants (Section 24.10) points to the therapeutic

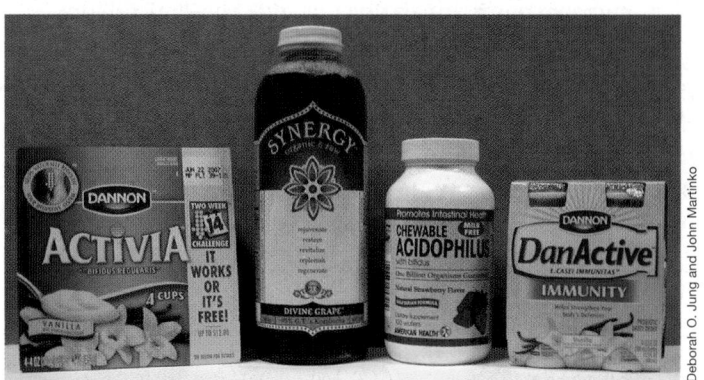

Figure 24.24 Examples of some probiotic foods and supplements widely available worldwide. Some probiotic products also contain prebiotics (specific polysaccharides obtained from various plants).

potential of rational modification of the gut microbiome. However, although fecal transplants have proven very effective, there is minimal understanding of the mechanisms behind their successes and also why they do not work in a limited number of cases.

Molecular analyses of *C. difficile* infections have highlighted the importance of a mechanistic understanding for the development of effective probiotic therapies. Resistance to *C. difficile* infection following antibiotic therapy (Figure 24.23) in both mice and humans is correlated with the presence of a second *Clostridium* species, *C. scindens*. Administration of this species to mice infected with *C. difficile* has been shown to be an effective probiotic therapy; suppression has been linked to the ability of *C. scindens* to convert cholic acid (found in bile) into deoxycholic acid, a growth inhibitor of *C. difficile*. In similar studies, recovery from diarrhea caused by *Vibrio cholerae* (cholera, ↩ Section 32.3) has been associated with an increased abundance of the bacterium *Ruminococcus obeum*. This bacterium produces a quorum-sensing autoinducer (AI-2)(↩ Section 6.8) that suppresses expression of *V. cholerae* colonization factors and effectively prevents the cholera bacterium from establishing an infection. In both of these cases this level of understanding is expected to foster the development of more ecologically based, specifically targeted, and scientifically proven probiotic therapies than can be hoped for from general and nutritionally based approaches (Figure 24.24).

Prebiotics

Probiotics should not be confused with *prebiotics*. The prebiotic approach to developing a healthy normal microbiota promotes the ingestion of certain plant nutrients as microbial growth stimulants with the idea that they will nurture particular bacterial species in the gut known to be associated with a healthy colon. Prebiotics are typically carbohydrates that are indigestible by the body but are excellent carbon and energy sources for certain fermentative gut bacteria. Fructooligosaccharides, polymers of fructose present in many vegetables, are thought to stimulate desirable gut microbes, and the complex polysaccharide *inulin* is widely promoted commercially as a major prebiotic.

Some probiotic formulations (such as certain yogurts) contain prebiotics as well. As has occurred with probiotics, dramatic health benefits have been claimed in some prebiotic studies. But at present, clear benefits of prebiotics for restoring or maintaining a healthy colonic microbiota await carefully controlled and quantitative scientific studies.

— MINIQUIZ —

- What is a probiotic? How does it differ from a prebiotic?
- Why do you think fecal transplants have stimulated renewed research into the development of probiotic and prebiotic therapies?

MasteringMicrobiology®

Visualize, **explore**, and **think critically** with Interactive Microbiology, MicroLab Tutors, MicroCareers case studies, and more. MasteringMicrobiology offers practice quizzes, helpful animations, and other study tools for lecture and lab to help you master microbiology.

Chapter Review

I • Structure and Function of the Healthy Adult Human Microbiome

24.1 The human body is colonized by a diverse assemblage of microorganisms in numbers that equal or exceed the total number of cells in the human body. Different body sites provide distinctive habitats that influence the types of microorganisms inhabiting each site. These microbiota together constitute the human microbiome, now recognized to be intimately associated with human health and disease.

Q **What are some of the possible future benefits of studies of the human microbiome?**

24.2 The human gastrointestinal tract is composed of many segments whose nutritional and pH characteristics differ dramatically and which support quite distinct bacterial populations.

Q **How do microbial diversity and abundance vary along the length of the gastrointestinal tract?**

24.3 Distinctive groups of microorganisms colonize the mucous membranes and hard surfaces of the mouth, and are constantly shed in saliva. The different microhabitats of the mouth (hard and soft surfaces, gingival crevice) sustain a high diversity of both aerobic and anaerobic populations.

Q **Where would you expect to find the highest numbers of spirochetes in the mouth?**

24.4 With the exception of the vagina and distal urethra, the male and female urogenital tracts are sterile. The healthy vaginal microbial community is dominated by *Lactobacillus* species, which suppress the growth of other populations by lowering the vaginal pH to about 5 by fermenting glycogen to lactic acid.

Q **What factor most affects the vaginal microbial community at the onset of puberty?**

24.5 The skin is a complex human organ covering about 2 square meters of the body and hosting about 10^{10} microorganisms in the healthy adult. Microbial community composition varies among the diversity of skin sites, generally categorized as moist, dry, or oily. The most common skin microorganisms are bacteria of the phyla *Actinobacteria* and *Firmicutes*.

Q **What characteristic of *Propionibacterium* species accounts for their colonization of the sebaceous gland system?**

II • From Birth to Death: Development of the Human Microbiome

24.6 Our current understanding of relationships between the human microbiome and health status is derived from complementary studies of select human study groups and experimental manipulation of animal models, in particular the mouse. Human studies are limited by incomplete control of lifestyle and genetic factors influencing the microbiome. Animal studies are limited by differences from humans in microbiology, anatomy, and physiology.

Q **What are the major anatomical differences between mouse and human gastrointestinal systems, and how might those differences influence microbial composition?**

24.7 The newborn gut microbial community evolves into a more complex adultlike community over the first 2–3 years of life. Initial colonization is influenced by vaginal or C-section delivery and by feeding of breast milk or formula. The unique oligosaccharides in human breast milk suppress pathogen colonization and select for beneficial microorganisms such as *Bifidobacterium* species. Changes in the gut microbial community with age have been associated with frailty in the elderly.

Q **When a child is transitioned from breast milk to solid food, what are the associated changes in the gut microbial community and what physiological differences do those changes reflect?**

III • Disorders Attributed to the Human Microbiome

24.8 The two major disorders associated with changes in the human gut microbial community are inflammatory bowel disease and obesity. Inflammatory bowel disease is associated with dysbiosis, a breakdown in the normal interactions between the human host and its intestinal microbiota reflected by reduced diversity of microbial genes in the diseased state. Obesity is directly associated with the microbial community composition of the gut, which influences host energy recovery, but obesity is likely not determined by a single contributing microbial or host factor.

Q **In what way are methanogenic *Archaea* implicated in obesity?**

24.9 Dental caries and periodontitis are diseases of the mouth caused by the combined activities of multiple microbial species, not a single pathogen. Strain variation among *Propionibacterium* species is associated with ability to cause acne, a common cutaneous inflammatory disease of adolescents. The vaginal community is dominated by *Lactobacillus* species that limit colonization by other organisms through lowering pH by production of lactic acid. Although generally very resilient to perturbation, the vaginal community transitions to a more diverse community during menstruation and during incidents of vaginitis or vaginosis.

Q **Why might a therapy based on colonization of the skin by selected *Propionibacterium* strains be effective in preventing the development of acne?**

IV • Modulation of the Human Microbiome

24.10 Antibiotic therapy may predispose an individual to infection by *Clostridium difficile*, a toxigenic bacterium that causes severe intestinal disease, by reducing competition from the normal microbiota of the gut. Infections that cannot be resolved by standard or repeated antibiotic therapy have been shown to respond to fecal transplants, which introduce fecal material from a healthy donor into the gut of the infected individual.

Q **Why has the success of fecal transplants in treating *C. difficile* infection encouraged the development of ecologically based therapies for other disorders associated with the human microbiome?**

24.11 Probiotics are live microorganisms which, when administered in adequate amounts, confer a health benefit on the host. Although the probiotic health benefits of *Lactobacillus* species are commonly promoted, there is little clinical evidence of their efficacy. Nonetheless, new understanding of the mechanisms by which organisms such as *Clostridium scindens* suppress pathogen colonization show that targeted probiotics can likely be developed to prevent or treat diseases, such as that caused by *C. difficile* infection. Prebiotics are nutritional supplements thought to promote the growth of beneficial gut microbes.

Q **What is the mechanism by which *C. scindens* suppresses colonization by *C. difficile*?**

Application Question

1. You are told that you must be placed on a high dose of a broad-spectrum antibiotic to treat a serious infection. You are concerned that this therapy will seriously disrupt your intestinal microbiota, possibly leading to intestinal disease such as caused by *Clostridium difficile*. In advance of the treatment, what might you do to ensure that your normal intestinal microbiota is restored after extended antibiotic therapy? Discuss two ways this restoration could be accomplished, one that would return your exact microbiota and one that would return some representative species.

Chapter Glossary

Dysbiosis an alteration or imbalance of an individual's microbiome relative to the normal, healthy state, primarily observed in the microbiota of the digestive tract or the skin

Fecal transplant the transfer of microbiota from the colon of one individual into the colon of another

Host an organism that can harbor pathogenic or beneficial (micro) organisms

Human microbiome the total microbial content in and on the human body

Lower respiratory tract the trachea, bronchi, and lungs

Microbiome a functional collection of different microbes in an environmental system such as the human body

Microbiota the types of organisms present in an environmental habitat, such as the human skin or the human gastrointestinal tract

Mucin a secretion from specialized epithelial cells containing water-soluble glycoproteins and proteins, forming the mucus that retains moisture and impedes microbial invasion on mucosal surfaces

Normal microbiota microorganisms that are usually found associated with healthy body tissue

Probiotic a live microorganism that, when administered in adequate amounts, confers a health benefit on the host

Upper respiratory tract the nasopharynx, oral cavity, and throat

Volatile fatty acids (VFAs) the major fatty acids (acetate, propionate, and butyrate) produced during fermentation in the large intestine of monogastric animals and the rumen or cecum of herbivores

Microbial Infection and Pathogenesis

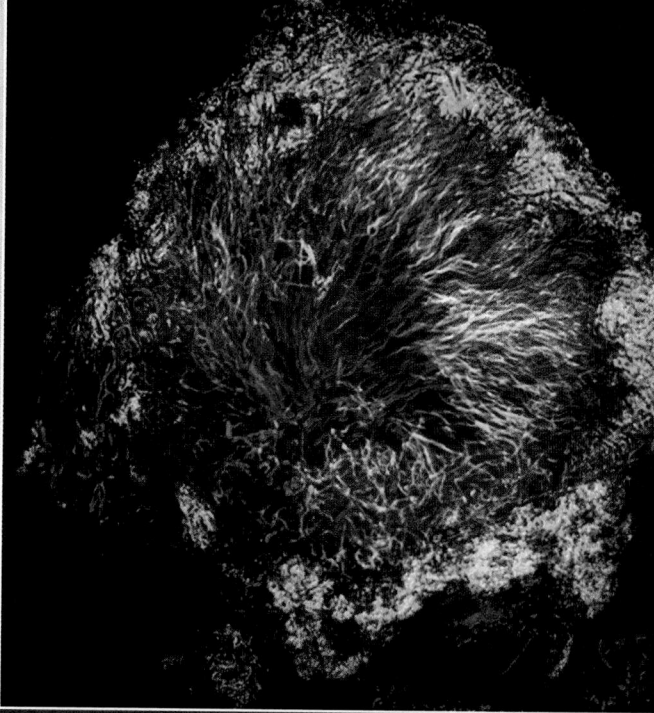

microbiology**now**

The Microbial Community That Thrives on Your Teeth

Few people have such superb oral hygiene that they lack dental plaque, the microbial biofilm that forms on and between teeth and along or below the gumline. If not removed regularly, dental plaque invariably leads to dental caries (cavities), the condition in which portions of tooth enamel and dentin break down from the onslaught of bacterial activities. Dental plaque and dental caries develop from the natural tendency of oral bacteria such as *Streptococcus mutans* and its close relative *S. sobrinus* to attach firmly to the teeth and gums and ferment sucrose (table sugar) to lactic acid, which attacks the teeth and slowly rots them away.

Until recently, dental plaque was thought to consist largely of the aforementioned streptococci. Both species could easily be isolated from dental plaque and both light and electron microscopy typically showed large numbers of cocci in chains, a hallmark of the genus *Streptococcus*. But a recent molecular ecology study of the microbial diversity of dental plaque revealed that this material is composed of more than just streptococci and develops in a precisely structured way.

The photo here is a light micrograph of a section through human dental plaque stained by fluorescence in situ hybridization (FISH). Different oligonucleotides, each specific for a different major phylum of *Bacteria* and containing a distinct fluorescent dye, were allowed to hybridize to the ribosomal RNA in cells in the plaque and then observed by fluorescence microscopy. Surprisingly, instead of seeing primarily streptococci, the researchers saw a diverse and highly organized microbial community. The micrograph shows streptococci (stained green) located primarily at the periphery of the plaque beyond several other bacteria that combine to form a scaffold emerging from the tooth surface. These include *Corynebacterium* (purple), *Capnocytophaga* (red), *Fusobacterium* (yellow), *Leptotrichia* (blue-green), and *Haemophilus* (orange), among others. A major conclusion that emerged from this study was that the scaffolding microbes likely function to position the streptococci out into the oral cavity where sucrose should be more available.

New views of old problems often reveal surprising results. In the case of dental plaque, FISH technology has revealed a whole new microbial world in a habitat previously thought to be dominated by only two species of well-characterized bacteria.

 Source: Mark Welch, J.L., et al. 2016. Biogeography of a human oral microbiome at the micron scale. *Proc. Natl. Acad. Sci. (USA) 113*: doi: 10.1073/pnas.1522149113.

I • Human–Microbial Interactions

Humans are exposed to microorganisms of all sorts in their environment. Whether one is walking outdoors, sitting indoors, or participating in any type of physical activity, environmental microbes and humans interact. The human body is also a natural home to enormous numbers of microorganisms, as we saw in Chapter 24. Most of these are harmless, and only a very small percentage cause disease. However, those that do, along with pathogens that are not part of the normal human microbiota, possess specific traits that underlie their pathogenic lifestyles. A major focus of this chapter will be a consideration of these specific traits and how they trigger the diseased state.

25.1 Microbial Adherence

In the world of infectious diseases, the term **infection** is used to imply the growth of microorganisms on or in the host, whereas the term **disease** is reserved for actual tissue damage or injury that impairs host function. If a **pathogen** gains access to the specific tissues it infects, disease will occur only if it first adheres to those tissues, multiplies to yield many cells or viral particles, and then proceeds to damage tissues (or the entire organism) by the release of toxic or invasive substances (**Figure 25.1**). Adherence is the first step, and although adherence is *required* to initiate disease, it is not *sufficient* to initiate disease because the host has many innate defenses that can thwart infection; we consider these in Chapter 26.

Adherence Molecules

Pathogens typically adhere to epithelial cells through specific interactions between molecules on the pathogen and molecules on the host tissues. In addition, pathogens may adhere to each other, forming biofilms (⮂ Sections 5.1 and 20.4), with the biofilm itself adhering to specific tissues. In medical microbiology, **adherence** is the enhanced ability of a microorganism to attach to a cell or a surface. Pathogens gain access to host tissues by way of a *portal of entry* of one sort or another. These include mucous membranes, the skin surface, or under mucous membranes or the skin during penetration of these sites from puncture wounds, insect bites, cuts, or other abrasions. The portal of entry may be critical for the establishment of an infection because a pathogen

that gains access to incompatible tissues is typically ineffective. For example, if cells of the bacterium *Streptococcus pneumoniae* are swallowed, they will be killed by the strong acidity of the stomach, whereas if the same cells reach the respiratory tract, they could trigger a fatal case of pneumonia.

Receptor molecules coating the surfaces of both the pathogen and cells of its host are often critical for adhering the pathogen to host tissues. Specific receptors can be important for the binding of any type of pathogenic microbe including bacteria, viruses, and parasites (**Figure 25.2**). Pathogen receptors have evolved to bind specifically to complementary molecules on the host cell cytoplasmic membrane, and the complementary nature of the pathogen and host cell receptors alerts the pathogen that it has arrived on a suitable infection site. Receptors on the pathogen surface are called **adhesins** and are composed of glycoprotein or lipoprotein covalently bound to the outer layer of the cell (Figure 25.2*a*). Host cell receptors are typically glycoproteins or complex membrane lipids such as gangliosides or globosides (sphingolipids containing sugars and other molecules).

Adherence Structures: Capsules, Fimbriae, Pili, and Flagella

Some adhesins form part of an outer cell surface structure that may or may not be covalently linked to components of the cell wall. For example, some notable pathogenic bacteria form a **capsule**. In *Bacillus anthracis* (the bacterium that causes anthrax), the capsule is composed of polypeptide containing only the amino acid D-glutamate. The capsule of *B. anthracis* can be seen in cells by light microscopy, and the encapsulated cells form smooth slimy colonies when grown on agar plates (**Figure 25.3**). The electron microscope can also clearly reveal bacterial capsules (Figure 25.3*c*). The capsule surface contains specific receptors that facilitate adherence to host tissues, but the inherently sticky nature of the capsule itself also assists in the overall attachment process. Although capsules are important for adherence of some pathogens to host tissues, many important pathogens, such as *Vibrio cholerae*, the causative agent of the disease cholera (Figure 25.2*a*), lack them.

Besides adherence, capsules are important for protecting pathogenic bacteria from host defenses. For example, the only known virulence factor for *Streptococcus pneumoniae* (bacterial pneumonia) is its polysaccharide capsule (**Figure 25.4**). Encapsulated strains of *S. pneumoniae* grow voraciously in the lungs where they initiate

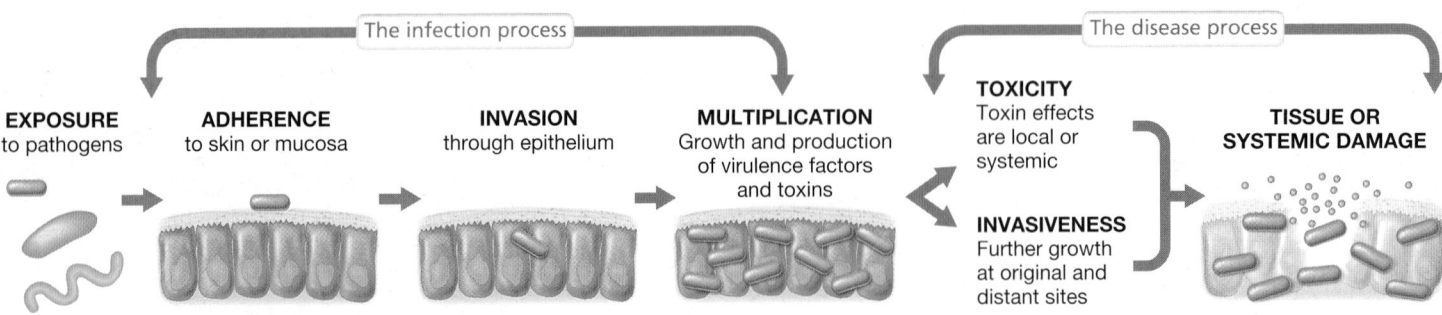

Figure 25.1 Microbial pathogenesis. Following exposure to a pathogenic microbe, a series of events leads to infection and a further series of events results in disease.

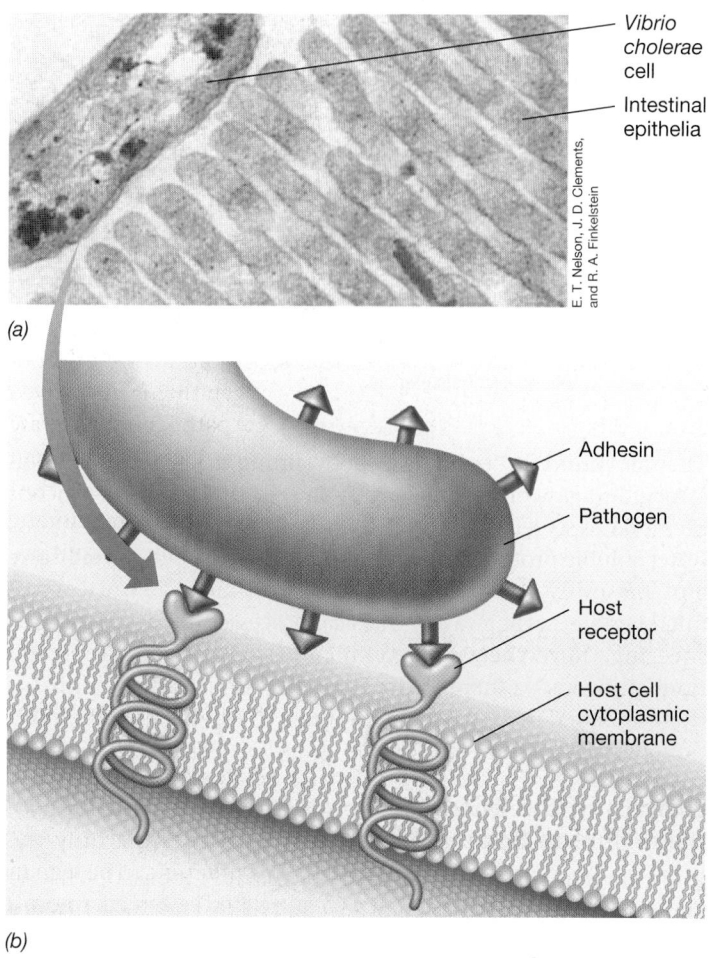

(a)

E. T. Nelson, J. D. Clements, and R. A. Finkelstein

Vibrio cholerae cell

Intestinal epithelia

Adhesin

Pathogen

Host receptor

Host cell cytoplasmic membrane

(b)

Figure 25.2 Adherence of pathogens to tissues via receptor molecules on the cell surface. *(a)* Transmission electron micrograph of a thin section of the gram-negative bacterium *Vibrio cholerae* adhering to the brush border of microvilli in the intestine. *(b)* A bacterial pathogen attaches specifically to host tissues by way of complementary receptors on the bacterial and host cell surfaces.

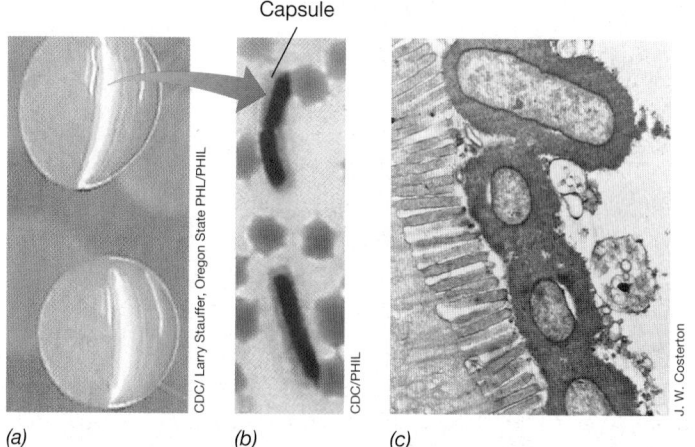

Capsule

(a) — CDC/ Larry Stauffer, Oregon State PHIL/PHIL

(b) — CDC/PHIL

(c) — J. W. Costerton

Figure 25.3 The bacterial capsule as a facilitator of pathogen attachment. *(a) Bacillus anthracis* growing on an agar plate. The mucoid colonies of encapsulated cells are about 0.5 cm in diameter. *(b)* Light micrograph of cells of *B. anthracis* growing in horse blood and stained to show the capsule (pink). Cells are about 1 μm in diameter. *(c)* Cells of enteropathogenic *Escherichia coli* attached to the brush border of intestinal microvilli by way of a distinct capsule. The *E. coli* cells are about 0.5 μm in diameter.

host responses that interfere with lung function, cause extensive host damage, and can cause death (⊘ Section 30.2). By contrast, nonencapsulated strains of *S. pneumoniae* are quickly and efficiently ingested and destroyed by phagocytes, white blood cells that ingest and kill bacteria by a process called *phagocytosis* (⊘ Sections 26.5–26.7).

Many pathogens selectively adhere to particular types of cells through cell surface structures other than adhesins, capsules, or slime layers. For example, *Neisseria gonorrhoeae*, the pathogen that causes the sexually transmitted disease gonorrhea (⊘ Section 30.13), adheres specifically to mucosal epithelial cells in the genitourinary tract, eye, rectum, and throat; by contrast, other tissues are not infected. *N. gonorrhoeae* has a cell surface protein called Opa (*o*pacity *a*ssociated *p*rotein) that binds specifically to a host protein found only on the surface of epithelial cells of these body regions, allowing adherence of the pathogen to host cells. Likewise, influenza virus targets upper respiratory tract mucosal cells and attaches specifically to these and later to lung epithelial cells by way of the protein hemagglutinin present on the virus surface (⊘ Sections 10.9 and 30.8).

Fimbriae and pili are bacterial cell surface protein structures (⊘ Section 2.7) that function in attachment (**Figure 25.5**). For instance, along with Opa, the pili of *Neisseria gonorrhoeae* play a key role in attachment to urogenital epithelia, and fimbriated strains of *Escherichia coli* are more frequent causes of urinary tract infections than strains lacking fimbriae. Among the best-characterized fimbriae are the *type I fimbriae* of enteric bacteria (*Escherichia, Klebsiella, Salmonella*, and *Shigella*), which are uniformly distributed on the surface of cells (Figure 25.5). Pili are typically longer and fewer in number than fimbriae, and in addition to attachment, some pili function in the bacterial genetic transfer process of conjugation (⊘ Section 11.8). Both pili and fimbriae function by specifically binding to host cell surface glycoproteins, thereby

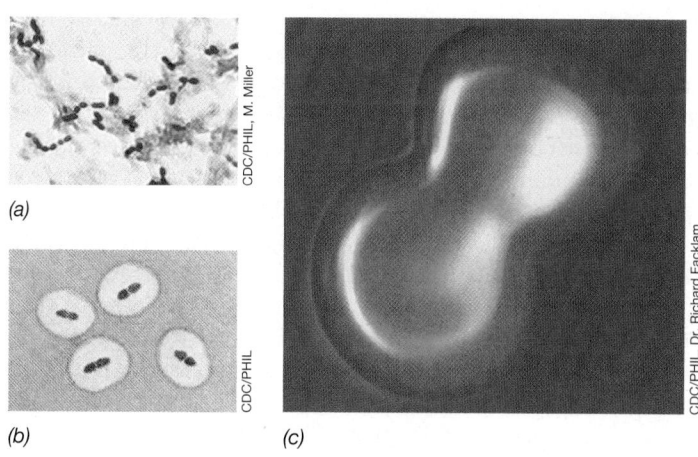

(a) — CDC/PHIL, M. Miller

(b) — CDC/PHIL

(c) — CDC/PHIL, Dr. Richard Facklam

Figure 25.4 Capsules and colonies in *Streptococcus pneumoniae*. *(a)* Gram stain of *S. pneumoniae* cells; capsules are not visible. *(b) S. pneumoniae* treated with anticapsular antibodies (Quellung reaction) that make the capsule visible. *(c)* Colonies of encapsulated *S. pneumoniae* cells grown on blood agar show a mucoid morphology with a sunken center. The colonies are about 2–3 mm in diameter and a single cell of *S. pneumoniae* is about 0.75 μm in diameter.

UNIT 6

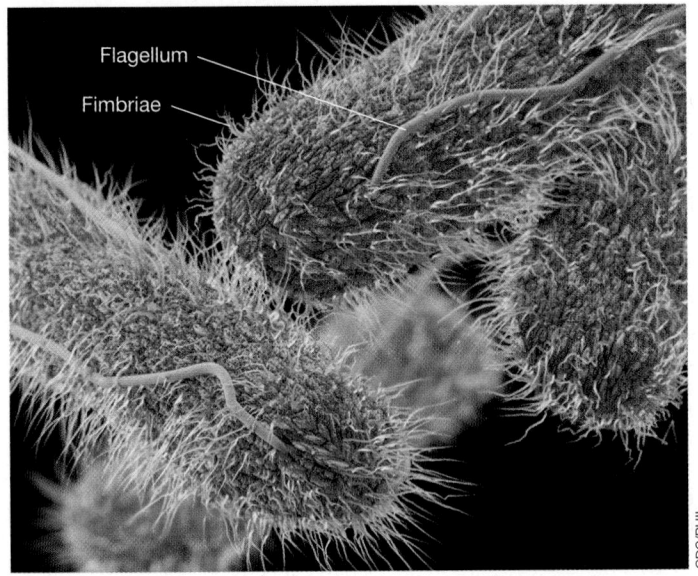

Flagellum
Fimbriae

CDC/PHIL

Figure 25.5 Fimbriae. Computer-generated image of a scanning electron micrograph of cells of *Salmonella enterica* (*typhi*) showing the numerous thin fimbriae and the much thicker peritrichously arranged flagella. A single cell is about 1 μm in diameter.

initiating adherence. Flagella may also facilitate adherence of bacterial cells to host cells (Figure 25.5) although their role is thought to be less important than that of fimbriae and pili.

MINIQUIZ

- What event is required but not sufficient to cause an infectious disease?
- Describe the molecules or structures that facilitate pathogen adherence to host tissues.

25.2 Colonization and Invasion

If a single pathogenic virus or cell attaches to its specific host tissue, it alone is insufficient to cause disease; the pathogen must establish residence there and multiply. **Colonization**, the growth of a microorganism after it has gained access to host tissues, begins at birth as a newborn is naturally exposed to a suite of harmless (and in many cases necessary) bacteria and viruses that will be the infant's initial normal microbiota (Chapter 24).

The human body is rich in organic nutrients and provides conditions of controlled pH, osmotic pressure, and temperature that are favorable for the growth of microorganisms. However, each body region such as the skin, respiratory, gastrointestinal, and genitourinary tracts differs chemically and physically from others, and thus provides a selective environment for the growth of certain microbes and not others. The result is that pathogens show rather rigid tissue specificities (⇝ Table 26.1), and this reality is often helpful in the diagnosis of microbial infections.

Colonization typically begins at sites in the **mucous membranes** (Figure 25.6). Mucous membranes consist of *epithelial cells*, tightly packed cells that interface with the external environment.

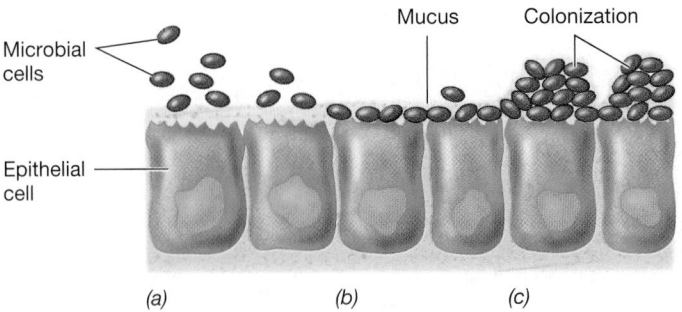

Mucus　Colonization

Microbial
cells

Epithelial
cell

(a)　(b)　(c)

Figure 25.6 Bacterial interactions with mucous membranes. (a) Loose association. (b) Adhesion. (c) Colonization.

They are found throughout the body, lining the urogenital, respiratory, and gastrointestinal tracts. The epithelial cells in mucous membranes secrete **mucus**, a thick liquid secretion that contains water-soluble proteins and glycoproteins. Mucus retains moisture and naturally inhibits microbial attachment because most microbes are swept away by physical processes like swallowing or sneezing. Nevertheless, some microbes—both pathogens and nonpathogens—adhere to the epithelial surface and colonize. If these attached microbes are pathogens, it sets the stage for infection, invasion, and disease (Figure 25.1).

Growth of the Microbial Community: An Example from Human Dental Caries

Infection requires growth of the pathogen after it has attached to and colonized a surface (Figures 25.1 and 25.6), and the actual disease process may not be the result of a single type of microbe but of a community of interacting microorganisms. An excellent example of this is found in the oral microbial disease **dental caries** (tooth decay), where attachment and infection have been well studied as models of these key events in the disease process.

Even on a freshly cleaned tooth surface, acidic glycoproteins from the saliva form a thin organic film several micrometers thick; this film provides an attachment site for bacterial cells, and oral streptococci quickly colonize it. These include in particular the two *Streptococcus* species most often implicated in tooth decay, *S. sobrinus* and *S. mutans*. Both of these organisms produce a capsule (Section 25.1). The *S. sobrinus* capsule contains adhesins (Figure 25.2a) specific for host salivary glycoproteins (**Figure 25.7a, b**), whereas *S. mutans* resides in crevices and small fissures where it relies on dextran—a strongly adhesive polysaccharide—that it produces to secure cells to the tooth and gum surface (Figure 25.7c, d). Both *S. sobrinus* and *S. mutans* are lactic acid bacteria (⇝ Section 16.6) that ferment glucose to lactic acid, the agent that destroys tooth enamel. However, the trigger for decay activities is *sucrose* (table sugar), since it is sucrose that allows these species to produce the thick capsules necessary for attachment and colonization.

Extensive bacterial growth of these oral streptococci results in a thick biofilm called **dental plaque** (Figure 25.7). Using phylogenetic probes, it has been possible to more readily explore the microbial diversity of dental plaque, and it is clear that the two *Streptococcus* species are not the entire story. Many other gram-positive and gram-negative *Bacteria* are present in plaque, including species of *Corynebacterium, Porphyromonas, Leptotrichia,*

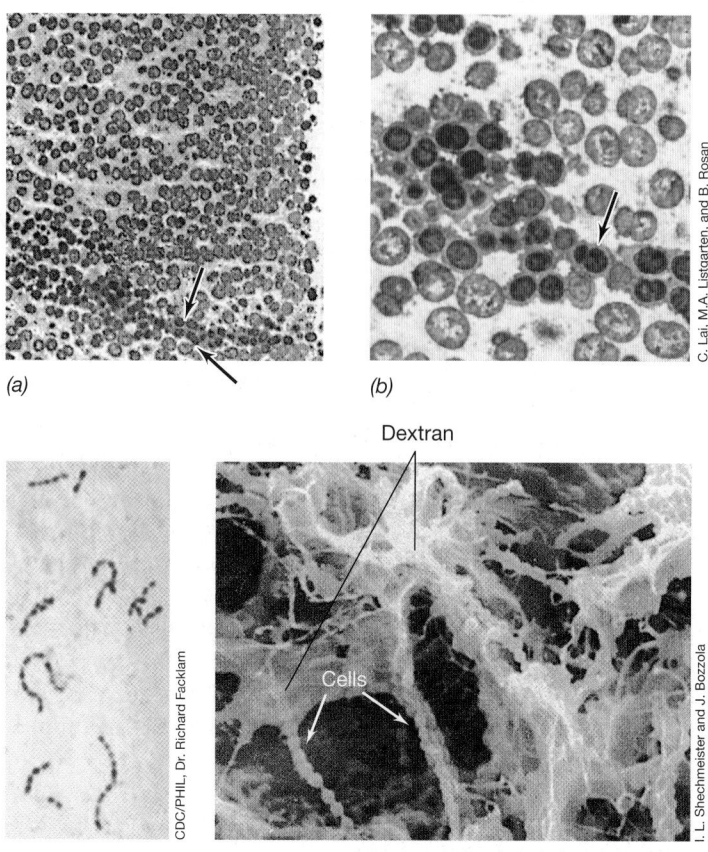

(a)

(b)

(c)

(d)

Dextran

Cells

C. Lai, M.A. Listgarten, and B. Rosan

CDC/PHIL, Dr. Richard Facklam

I. L. Shechmeister and J. Bozzola

Figure 25.7 Cariogenic *Streptococcus* spp. and dental plaque. *(a)* The low-magnification micrograph shows predominantly streptococcal cell morphology embedded in dental plaque. The species *Streptococcus sobrinus* (arrows) appears darker from a specific staining technique. *(b)* Higher-magnification micrograph showing a region in the plaque with *S. sobrinus* cells (dark, arrow). Note the extensive capsule surrounding the *S. sobrinus* cells. *(c)* Light micrograph of a *Streptococcus mutans* culture showing the characteristic cell chains of streptococci. *(d)* Scanning electron micrograph of the sticky dextran material that holds cells together in filaments. Individual cells of both *S. sobrinus* and *S. mutans* are about 1 μm in diameter.

Neisseria, the filamentous anaerobe *Fusobacterium*, and many others (**Figure 25.8**). Moreover, the different species likely play specific structural and functional roles in mature dental plaque. This can be seen in FISH-stained (⮌ Section 19.5) sections of plaque, where filamentous streamers of cells of *Corynebacterium* attached to a thin biofilm on the tooth surface anchor cells of *Streptococcus* and other bacteria a short distance away from the tooth surface (Figure 25.8). Such an arrangement probably allows *Streptococcus* cells to extend out from the tooth surface into regions of the oral cavity where saliva, sugars, and other nutrients are more abundant.

Dental plaque is thus a complex mixed-culture biofilm composed of several different genera of *Bacteria* and their accumulated products. A few *Archaea* are also present in dental plaque, primarily methanogenic species such as *Methanobrevibacter oralis*. As dental plaque accumulates, the microbiota produce locally high concentrations of lactic acid that decalcifies tooth enamel, resulting in dental caries. Tooth enamel is strongly calcified tissue, and the ability of microbes to invade this tissue plays a major role in the extent of dental caries and related more serious oral pathologies,

including periodontal conditions (disease in the tooth-supporting gum and bone tissues).

Invasion and Systemic Infection

In the case of dental caries, the bacterial infection primarily resides on the tooth and gum surfaces. By contrast, in most infectious diseases, the pathogen must invade past the tissue surface in order to promote disease. **Invasion** is the ability of a pathogen to enter into host cells or tissues, spread, and cause disease. Some pathogens remain localized after initial entry, multiplying and invading at a single focus of infection such as the boil that may arise from *Staphylococcus* skin infections (⮌ Section 30.9). However, sometimes the pathogens enter the bloodstream, from where they can travel to distant parts of the body. Depending on the pathogen and the overall health of an individual and that individual's immune system, the presence of bacteria in the blood can have mild or highly severe consequences.

The mere presence of bacteria in the blood is called **bacteremia**; this condition is typically self-limiting as the bacterial cells do not grow in the bloodstream and thus the immune system quickly removes them. The symptoms of bacteremia may be mild or none. By contrast, in **septicemia** bacteria multiply in the bloodstream and the organism spreads systemically from an initial focus and

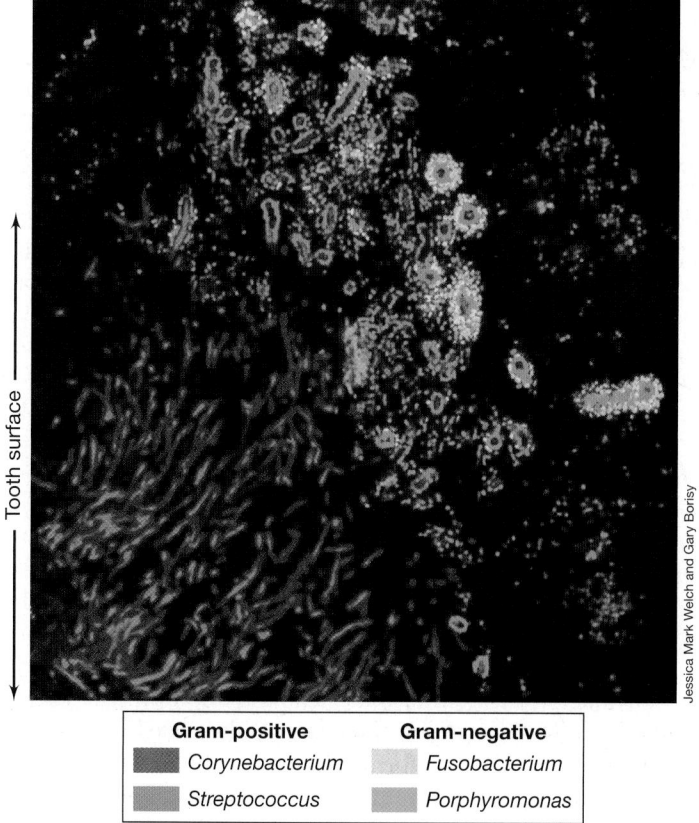

Tooth surface

Gram-positive	Gram-negative
Corynebacterium	*Fusobacterium*
Streptococcus	*Porphyromonas*

Jessica Mark Welch and Gary Borisy

Figure 25.8 Bacterial diversity of dental plaque. Confocal micrograph of a FISH-stained (⮌ Section 19.5) section through human dental plaque using a suite of phylogenetic probes containing different fluorescent tags. Color matching to specific groups is shown below the photo. The tooth surface would be on the left with the filamentous *Corynebacterium* species (red) containing attached streptococci flaring out from an attachment site on the tooth surface.

UNIT 6

produces toxins or other poisonous substances. Septicemia usually begins as an infection in a specific organ such as the intestine, kidney, or lung, and then spreads rapidly throughout the body from there. Septicemia is typically associated with major symptoms and may lead to massive inflammation, culminating in septic shock (sepsis) and death. *Viremia* is the term used to describe viruses present in the bloodstream, and measles, a highly infectious disease in those not vaccinated (⇌ Section 30.6), is a good example of a systemic viremia.

A pathogen that causes disease in a given host can trigger mild or severe outcomes depending on its inherent capacity to elicit disease. We consider the principles that govern these capacities now with a focus on the important infectious disease concept called *virulence*.

MINIQUIZ

- At what body sites do pathogens typically attach and colonize?
- Distinguish between infection and disease.
- Which is the more serious condition, bacteremia or septicemia, and why?

25.3 Pathogenicity, Virulence, and Attenuation

Unique properties of each pathogen contribute to its **pathogenicity**, the overall ability to cause disease. The measure of pathogenicity is called **virulence**, the relative ability of a pathogen to cause disease. Pathogenicity and virulence are not uniform properties of a given pathogen and can differ dramatically between different strains of the same bacterial species or virus. Highly virulent strains of a given pathogen tend to emerge every so often and when they do, they often trigger a rapid, widespread, and particularly severe course of disease (epidemic or pandemic, Chapter 29). The virulence of a given pathogen depends on a number of factors including its relative abilities to adhere, colonize, and invade (Figure 25.1), and its arsenal of virulence factors.

Virulence

Virulence is the net outcome of host–pathogen interactions, a dynamic relationship between the two organisms influenced by ever-changing conditions in the pathogen, the host, and the environment. Host damage in an infectious disease is mediated by **virulence factors**, toxic or destructive substances produced by the pathogen that directly or indirectly enhance invasiveness and host damage by facilitating and promoting infection. The second part of this chapter is devoted to major virulence factors.

Virulence is a quantifiable entity, especially if a pathogen is lethal and an experimental animal model is available. For example, the LD_{50} (LD stands for "lethal dose") is defined as the number of cells of a pathogen (or virions, for a viral pathogen) that kills 50% of the animals in a test group. Highly virulent pathogens frequently show little difference in the number of cells required to kill 100% of a test group of animals as compared with the LD_{50}. To illustrate this, recall the foundational work of the British microbiologist Frederick Griffith (⇌ Section 1.12). Griffith worked with the

gram-positive bacterium *Streptococcus pneumoniae* and discovered that strains of *S. pneumoniae* that contained a capsule ("smooth" strains because they formed smooth colonies on plates) were highly virulent for mice, whereas mutant derivatives lacking a capsule ("rough" strains) were not. The *S. pneumoniae* capsule is the primary virulence factor of this bacterium because it helps the bacterium evade immune surveillance. Griffith's key discovery was that the smooth phenotype could be transferred to rough cells by treating rough cells with an extract from smooth cells (⇌ Figure 1.34). This was the first experimental example of transformation, a bacterial genetic transfer process (⇌ Section 11.6), and the active principle in the extract was later shown (by other scientists) to be DNA.

Griffith's choice of experimental organism was fortuitous because *S. pneumoniae* is both readily transformable and highly virulent for mice. Only a few cells of an encapsulated strain of *S. pneumoniae* can establish a fatal infection and kill all mice in a test population. As a result, the LD_{50} for *S. pneumoniae* in mice is not proportional to the number of cells delivered (**Figure 25.9**). By contrast, the number of cells of a less virulent pathogen such as the gram-negative enteric bacterium *Salmonella enterica* (*typhimurium*) necessary to kill all of the mice in the test population is about 10,000-fold greater than the highly virulent *S. pneumoniae* cells, and the LD_{50} is proportionally related to the number of cells of the pathogen cells injected into the mice (Figure 25.9).

There are many examples of highly virulent human pathogens, especially among viruses. For example, some strains of the influenza virus are so highly virulent that only a few virions can initiate disease even though mortality rates are typically low. Ebola virus is

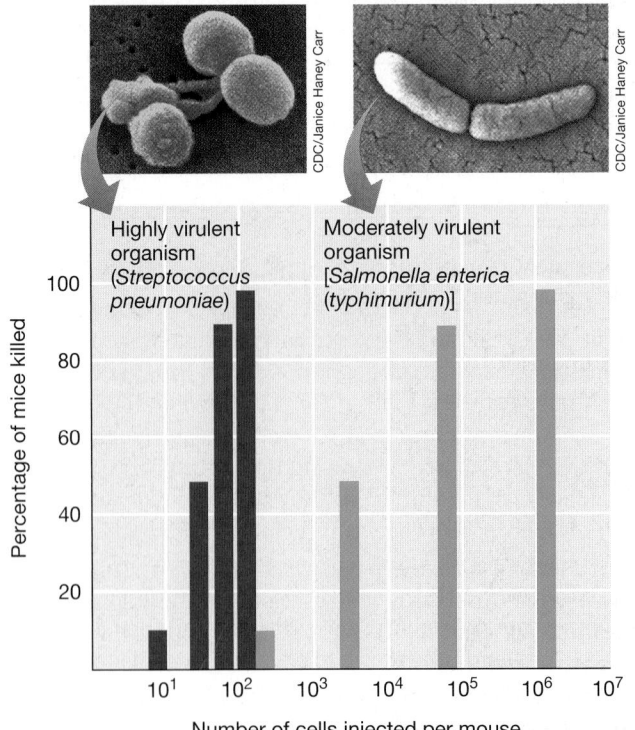

Figure 25.9 Microbial virulence. Differences in microbial virulence demonstrated by the number of cells of *Streptococcus pneumoniae* (red bars) and *Salmonella enterica* (*typhimurium*) (green bars) required to kill mice. Colorized scanning electron micrographs of each bacterium are shown above their respective graph.

also highly virulent, and the tiny inoculum necessary to initiate disease often results in a fatal infection. By contrast, the bacterial pathogen *Vibrio cholerae* (which causes cholera) is not especially virulent, as a large inoculum of this intestinal pathogen is necessary to initiate disease (⊘ Section 32.3).

Attenuation

The virulence of a pathogen can change. **Attenuation** is the decrease or loss of virulence of a pathogen. When pathogens are kept in laboratory culture rather than isolated from diseased animals, their virulence often decreases, or may be completely lost. Strains that have either a reduced virulence or are no longer virulent are said to be *attenuated*. Attenuation is thought to occur because nonvirulent or weakly virulent mutants grow faster than virulent strains in laboratory media where virulence has no selective advantage. After successive transfers in fresh media, such mutants are therefore selectively favored. However, if an attenuated culture is reinoculated into an animal, the organism may regain its original virulence, especially with continued in vivo passage as more-virulent strains are naturally selected. In some cases, though, the loss of virulence is permanent. For example, if a deletion mutation led to a major modification of a required receptor molecule (Figure 25.2*b*) or to the inability to produce a key virulence factor, such as the production of a toxin or invasive enzyme, then the mutant strain would be permanently attenuated.

Attenuated strains of various pathogens are valuable to clinical medicine because they are often used for the production of vaccines, especially viral vaccines. For example, vaccines for measles, mumps, and rubella, and rabies vaccines for animals other than humans, employ attenuated strains of each virus. Although attenuated viruses are "live" in the sense that, unlike "killed" strains, they could in principle become once again active and replicate, properly attenuated virus vaccines (those free of any unattenuated virions) typically show greater efficacy and generate an overall stronger immune response than do killed virus vaccines.

--- **MINIQUIZ** ---

- What are virulence factors? How can the LD_{50} test be used to define virulence of a pathogen?
- What circumstances can contribute to attenuation of a pathogen?

25.4 Genetics of Virulence and the Compromised Host

The virulence of a bacterial pathogen and the eventual outcome of an infectious disease are the net result of genetic and physiological features of both the pathogen and the host. In the case of the host, a pathogen may infect a healthy, well-rested young adult or an individual compromised by a physiological condition (old age, hospitalization, immune suppression); an ongoing infectious disease, for example, acquired immunodeficiency syndrome (AIDS) caused by the human immunodeficiency virus (HIV); or a genetic disease (for example, cystic fibrosis). The outcome of infection—health or disease—may be very different in these different

individuals, even if they are infected by the same strain of a viral or bacterial pathogen.

The virulence of a pathogen may be encoded by firmly entrenched chromosomal genes or by highly mobile genetic elements. For example, the gram-negative bacterium *Bordetella pertussis*, the causative agent of whooping cough (pertussis, ⊘ Section 30.3), makes several toxins including pertussis toxin, a potent AB-type exotoxin (Section 25.6); collectively, these toxins trigger the symptoms of whooping cough. Other species of *Bordetella* do not make pertussis toxin, and the chromosomal gene encoding pertussis toxin does not readily move from *B. pertussis* to other species. But in contrast to *B. pertussis*, some bacterial pathogens routinely exchange genes encoding virulence factors with different bacterial species or even genera, and thus highly related versions of their most potent weapons may appear in several different pathogens. *Salmonella* is a well-studied example of the genetic transfer of virulence factors, and we focus on this bacterium now.

Virulence in *Salmonella*: Pathogenicity Islands and Plasmids

Salmonella species infect humans, leading to various gastrointestinal illnesses (⊘ Section 32.10). *Salmonella* species encode a large number of virulence factors that are important in disease. These include type I fimbriae (Section 25.1) to facilitate attachment of cells to gastrointestinal tissues; several different classes of exotoxins (Section 25.6); antiphagocytic proteins that block engulfment of bacterial cells by host phagocytes; proteins that promote survival if the bacterium does get phagocytosed; siderophores, organic molecules that bind iron tightly and, in pathogenic bacteria, allow the bacteria to outcompete host sequestration systems for iron; and endotoxin (Section 25.8). With the exception of endotoxin, many of these virulence factors are encoded by genes present on mobile DNA rather than on the cell's chromosome (**Figure 25.10**).

Several genes that encode these virulence factors in *Salmonella* and related gram-negative pathogens such as pathogenic strains of *Escherichia coli* (⊘ Section 32.11) are found clustered together on the chromosome as *pathogenicity islands* (⊘ Section 9.7). *Salmonella* pathogenicity island 1 (SPI1) is a cluster of genes that encode over 10 distinct proteins that promote virulence and invasion. One of these is *invH*, a gene encoding a surface adhesion protein (Section 25.1). Several *inv* genes encode proteins important for trafficking of virulence factors. For example, the InvJ regulator protein controls assembly of structural proteins InvG, PrgH, PrgI, PrgJ, and PrgK that form a type III secretion system called the *injectisome*, an organelle in the bacterial envelope that allows for the direct transfer of virulence proteins into host cells through a needle-like assembly (⊘ Section 4.13 and Figures 4.42 and 4.43).

A second *Salmonella* pathogenicity island, SPI2, contains genes that are responsible for causing more systemic than localized disease and resistance to host defenses. In addition, several plasmid-encoded virulence factors such as antibiotic resistance genes encoded on R plasmids (⊘ Section 4.2) can spread between *Salmonella* species and to other genera of enteric bacteria. Pathogenicity islands and R plasmids allow for the facile and rapid transfer of virulence factors. It is thus not uncommon for genes encoding factors in one pathogen to be very similar if not

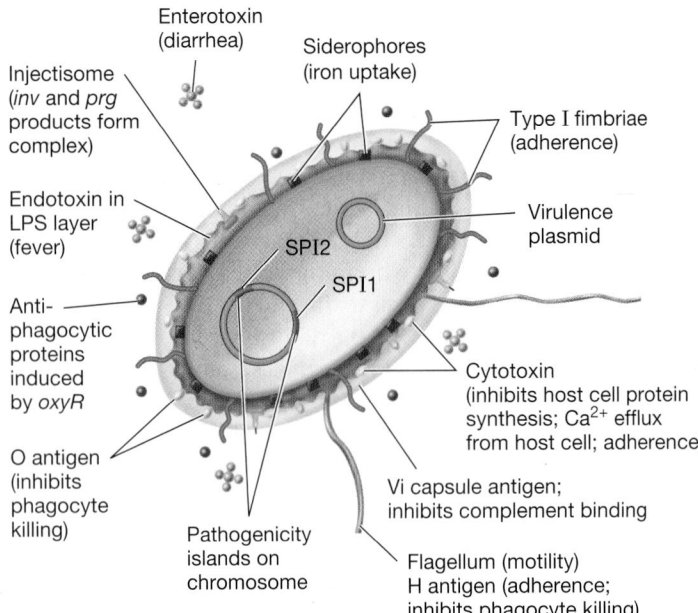

Figure 25.10 Virulence factors in *Salmonella*. Factors important for virulence and the development of pathogenesis in this gram-negative enteric pathogen are shown. Genes encoding many of the factors reside on the pathogenicity islands or plasmids.

identical to those in another because of transfer of parts or all of the islands or plasmids between species by horizontal gene exchange (Chapter 11).

In the well-known opportunistic pathogen (see next subsection) *Pseudomonas aeruginosa*, pathogenicity islands can also contain genes encoding antibiotic resistance. As for *Salmonella*, many cases of multiple antibiotic resistance in *P. aeruginosa* are linked to plasmids. However, in some strains of *P. aeruginosa*, genomic islands containing transposable elements (⮌ Section 9.6) are present. By transposition these islands have "captured" multiple antibiotic resistance genes and can then disseminate them to other organisms; these islands are present in *P. aeruginosa* in place of or in addition to resistance plasmids. However, regardless of how they are encoded, these strains have become resistant to virtually all of the clinically useful antibiotics that have traditionally been used to control *P. aeruginosa* infections. This is a particularly serious problem for the compromised host and in the hospital environment, and we consider these issues now.

The Compromised Host

Some individuals are just more susceptible to infection than others for reasons that have little to do with the virulence of the pathogen. These so-called *compromised hosts* are individuals in which one or more resistance mechanisms are inactive and in whom the probability of infection is therefore increased. Many hospital patients with noninfectious diseases (for example, cancer and heart disease) acquire microbial infections more readily because they are compromised hosts. Such *healthcare-associated infections* (also called *nosocomial infections*) affect up to 2 million individuals each year in the United States, with a nearly 5% mortality rate. Invasive medical procedures such as catheterization, hypodermic injection, spinal puncture, biopsy,

and surgery may unintentionally introduce microorganisms into the patient. The stress of surgery and the anti-inflammatory drugs given to reduce pain and swelling can also reduce host resistance (⮌ Section 28.2).

Some factors can compromise host resistance outside the hospital including lifestyle choices that affect major organs of the body, such as intravenous drug usage, tobacco, excessive alcohol, and the like, or genetic diseases that eliminate parts of the immune system. People that are physically compromised for any of a number of reasons may be more susceptible to infections, not only because they are physically weakened but also because their living conditions and lifestyle choices may put them in more continual contact with infectious agents. For example, infection with the human immunodeficiency virus (HIV) predisposes an individual to infections from **opportunistic pathogens**, microbes that cause disease only in the absence of normal host resistance. HIV causes AIDS by destroying a specific class of immune cell, the CD4 T lymphocytes (⮌ Figure 30.44), which are key to an effective immune response. The reduction in CD4 T cells reduces immunity, and an opportunistic pathogen, one that does not cause disease in a healthy, uninfected host, can then cause serious disease or even death. Individuals with immunodeficiencies from underlying genetic causes rather than infection are also more susceptible to opportunistic infections because part of their immune system is either nonfunctional or suboptimal.

The outcome of an infectious disease thus depends on several factors, both host and pathogen related. Two individuals exposed to the same pathogen in the same way may well show different outcomes. But once an infection has proceeded to the actual stage of disease, the symptoms that appear are due to products of the pathogens, and we turn our attention to these now.

MINIQUIZ

• What major virulence factors are produced by *Salmonella*?

• What is an opportunistic pathogen? What steps can a person take to help avoid opportunistic infections?

• What is a nosocomial infection?

II • Enzymes and Toxins of Pathogenesis

Bacterial pathogens damage host tissues (or the entire host) in two major ways: (1) by secreting tissue-destroying enzymes and (2) by secreting or shedding toxins that target specific host tissues or the entire host. In contrast to bacterial pathogens, most viral pathogens damage host tissues by lysing cells directly, although some viruses are nonlytic and instead introduce genes into host cells that may eventually harm the host (⮌ Section 8.8).

We turn our focus now to the enzymes and toxins of well-studied pathogenic bacteria, contrasting their efficacy and modes of action. Some of these virulence factors cause only minor disease symptoms, whereas others are some of the most poisonous substances known.

25.5 Enzymes as Virulence Factors

Following adherence, colonization, and infection by a pathogen, invasiveness requires the breakdown of host tissues and access to nutrients released from host cells. In many classical bacterial pathogens, this is accomplished through the activity of *enzymes* that attack and destroy cells in one type of tissue or another (Table 25.1).

Tissue-Destroying Enzymes

Many virulence factors are enzymes. For example, streptococci, staphylococci, and certain clostridia produce *hyaluronidase* (Table 25.1), an enzyme that promotes spreading of organisms in tissues by breaking down the polysaccharide hyaluronic acid. Among other functions, hyaluronic acid is a component of the extracellular matrix and functions as a type of "intercellular cement" in animal tissues, helping to maintain the organization of individual cells into tissues; the activity of hyaluronidase causes host cells to slough apart, allowing pathogens at an initial colonization site to spread between host cells to attack subsurface tissues (**Figure 25.11a**). Similarly, the clostridia that cause gas gangrene produce *collagenase*, an enzyme that destroys collagen (a major protein of connective tissues in muscle and other body tissues); collagenase enables these bacteria to gain access to deeper host tissues and spread through the body. Recall that clostridia are anaerobes, and colonizing deeper tissues allows them to reach less oxic conditions and provides a ready source of nutrients from destroyed tissues (gangrene clostridia are typically proteolytic species, ⊂⊃ Sections 16.8 and 31.9). Many pathogenic streptococci and staphylococci also produce proteases, nucleases, and lipases that degrade host proteins, nucleic acids, and lipids, respectively (Table 25.1).

Two virulence factors are enzymes that affect fibrin, the insoluble blood protein that triggers blood clots, but the activities of the enzymes yield opposing results (Figure 25.11b). Blood clotting is triggered by tissue injury and functions not only to stop blood loss but also, in the case of a bacterial infection, to isolate the pathogen, limiting the infection to a local region. Some pathogens counter this host protective mechanism by producing fibrinolytic enzymes, such as *streptokinase* produced by *Streptococcus pyogenes*. This bacterium is often associated with pus-forming wounds and secretes streptokinase to dissolve fibrin clots and

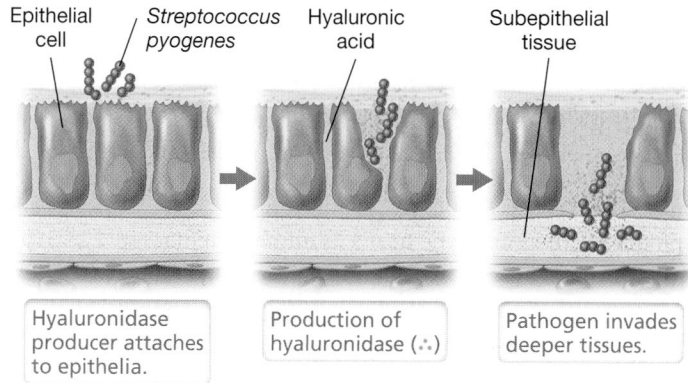

(a) **Hyaluronidase**

Hyaluronidase producer attaches to epithelia. → Production of hyaluronidase (∴) → Pathogen invades deeper tissues.

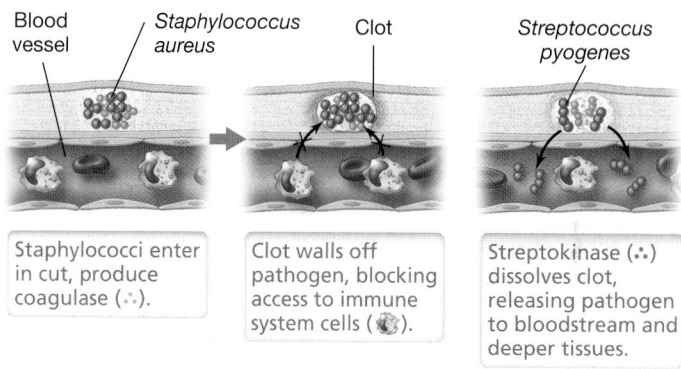

(b) **Coagulase and streptokinase**

Staphylococci enter in cut, produce coagulase (∴). → Clot walls off pathogen, blocking access to immune system cells. → Streptokinase (∴) dissolves clot, releasing pathogen to bloodstream and deeper tissues.

Figure 25.11 Activity of some enzyme virulence factors. *(a)* Hyaluronidase. *(b)* Coagulase and streptokinase. Among other bacterial pathogens, hyaluronidase and streptokinase are typical of virulent strains of *Streptococcus pyogenes* and coagulase is typical of virulent strains of *Staphylococcus aureus*.

make further invasion possible (Table 25.1, Figure 25.11b). Streptokinase specifically activates the host to produce plasmin, an enzyme that degrades fibrin blood clots. Because of this powerful activity, streptokinase also has a medically beneficial function. The protein is marketed as a pharmaceutical and administered intravenously to dissolve clots for conditions in which blood clots are blocking normal blood flow, such as from heart attacks, deep vein thromboses, or embolisms.

TABLE 25.1 Enzyme virulence factors of some well-known gram-positive bacterial pathogens

Organism	Disease	Enzyme[a]	Enzyme activity
Staphylococcus aureus	Pus-forming infections	Coagulase	Induces fibrin clotting; allows bacterial cells to remain at site of infection (prevents access to pathogens by cells of the immune response)
		Nuclease; lipase	Break down nucleic acids or lipids
Streptococcus pyogenes	Pus-forming infections; scarlet fever; strep throat	Hyaluronidase	Dissolves hyaluronic acid in connective tissues; allows bacterial cells to spread (enhances pathogen invasion)
		Streptokinase	Dissolves fibrin clots; allows bacterial cells to spread
Clostridium perfringens	Gas gangrene; food poisoning	Collagenase	Breaks down collagen (a protein), allowing the bacterium to spread to other tissues
		Protease	Breaks down proteins

[a]The activities of coagulase, hyaluronidase, and streptokinase are depicted in Figure 25.11.

In contrast to the fibrin-*destroying* activity of streptokinase, some pathogens produce enzymes that actually *promote* the formation of fibrin clots. These clots protect the pathogen from host responses. For example, *coagulase* (Table 25.1), produced by virulent *Staphylococcus aureus*, converts fibrinogen to fibrin, resulting in the clotting of blood and the formation of fibrin surrounding the *S. aureus* cells; this blanket of fibrin protects the *S. aureus* cells from attack by cells of the host's immune system (Figure 25.11*b*). The fibrin matrix produced as a result of coagulase activity may also account for the localized nature of many staphylococcal infections, as is typically seen in boils and pimples (↩ Section 30.9). Coagulase-positive *S. aureus* strains are typically more virulent than coagulase-negative strains, a likely reflection of the former's ability to evade innate immune responses such as phagocytosis (Chapter 26) and continue growth and tissue destruction for a longer period.

Enzyme Activities at the Host's Mucosal Surface

Host mucosal surfaces are bathed in immune substances including enzymes such as lysozyme, an enzyme that cleaves the peptidoglycan of bacterial cells and promotes their osmotic lysis (↩ Section 2.4). Virulence factors produced by the gram-positive bacterium *Enterococcus faecalis*—a major cause of bacteremia, surgical wound infections, and urinary tract infections—attack the protective role of lysozyme by altering the structure of the bacterium's peptidoglycan such that lysozyme can no longer recognize its substrate.

Antibodies are also present on mucosal surfaces, in particular a class of antibody called *IgA* (↩ Section 27.3). These "secretory antibodies," as they are called, help prevent pathogen adherence to host tissues (Section 25.1). However, certain pathogenic bacteria counter this protective role by producing enzymes that specifically cleave IgA (IgAases), rendering this host defense useless; *Neisseria* species such as *N. gonorrhoeae* (gonorrhea) and *N. meningitidis* (meningitis) are particularly notorious in this regard.

We thus see that pathogens can produce enzymes both as offensive weapons—to destroy host tissues—and as defensive weapons—to destroy or inactivate offensive weapons of the host. Both strategies accomplish a similar objective: The invasiveness of the pathogen is increased and this allows it to ultimately extract more resources from its host.

MINIQUIZ

- Identify host factors that limit or accelerate infection of a microorganism at selected local sites.
- How do streptokinase and coagulase promote bacterial infection and invasion?
- What is an IgAase and why would a bacterial pathogen produce one?

25.6 AB-Type Exotoxins

Toxicity is the ability of an organism to cause disease by means of a toxin that inhibits host cell function or kills host cells (or the host itself). **Exotoxins** are toxic *proteins* released from the pathogen as it grows. These toxins travel from a site of infection and cause damage at distant sites. Some exotoxins are **enterotoxins**, toxic proteins whose site of action is the small intestine, generally causing secretion of fluid into the intestinal lumen, resulting in vomiting and diarrhea. Exotoxins fall into three categories in terms of their mechanism: *AB toxins, cytolytic toxins,* and *superantigen toxins*. In this section we consider the AB toxins and in Section 25.7 we focus on cytolytic and superantigen toxins.

As the name implies, AB toxins consist of two subunits, A and B. The B component binds to a host cell surface molecule, facilitating the transfer of the A subunit across the cytoplasmic membrane, where it damages the cell. Some of the best-known exotoxins are AB toxins, including those expressed in the diseases diphtheria, tetanus, botulism, and cholera (**Table 25.2**).

Diphtheria Exotoxin: Blockage of Protein Synthesis

The diphtheria toxin produced by the aerobic gram-positive bacterium *Corynebacterium diphtheriae* is an AB toxin and an important virulence factor of the pathogen (↩ Section 30.3). Diphtheria toxin inhibits protein synthesis in eukaryotic cells. Rats and mice are relatively resistant to diphtheria toxin, whereas humans are very susceptible, with only a single molecule of toxin sufficient to

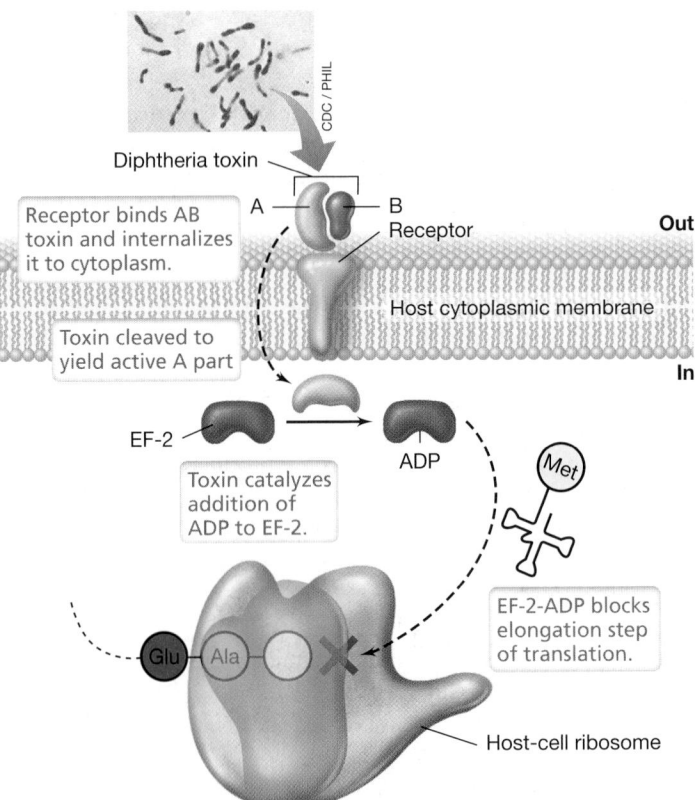

Figure 25.12 The activity of diphtheria toxin. Diphtheria toxin, an AB toxin produced by cells of *Corynebacterium diphtheriae*, binds to the host cytoplasmic membrane by way of its B subunit. Cleavage of the toxin allows the A subunit to enter the cell where it catalyzes ADP-ribosylation of elongation factor 2 (EF-2), a key factor in translation. The modified elongation factor no longer binds to the ribosome, resulting in the cessation of protein synthesis and host cell death. Inset photo: Light micrograph of Gram-stained cells of *Corynebacterium diphtheriae*. A single cell is about 0.75 μm in diameter.

TABLE 25.2 Some classic exotoxins and cytotoxins produced by human bacterial pathogens

Organism	Disease	Toxin[a]	Activity[b]
Bacillus anthracis	Anthrax	Lethal factor Edema factor Protective antigen (AB)	Combine to cause cell death
Bordetella pertussis	Whooping cough	Pertussis toxin (AB)	Blocks G protein function; kills cells
Clostridium botulinum	Botulism	Botulinum toxin (AB)	Causes flaccid paralysis
Clostridium tetani	Tetanus	Tetanospasmin (AB)	Causes rigid paralysis
Clostridium perfringens	Gas gangrene Food poisoning	α, β, γ, δ toxins (AB) Enterotoxin (CT)	Hemolysis, lecithin destruction Alters intestinal tract permeability
Corynebacterium diphtheriae	Diphtheria	Diphtheria toxin (AB)	Inhibits eukaryotic protein synthesis
Escherichia coli (enterotoxigenic strains only)	Gastroenteritis	Shiga-like (*E. coli*) (AB)	Inhibits protein synthesis, induces bloody diarrhea
Pseudomonas aeruginosa	Burn and certain wound and ear infections; cystic fibrosis lung infections	Exotoxin A (AB)	Inhibits eukaryotic protein synthesis
Salmonella sp.	Gastroenteritis	Enterotoxin (AB) Cytotoxin (CT)	Lyses cells; inhibits protein synthesis Induces fluid loss from intestine
Shigella dysenteriae	Gastroenteritis	Shiga toxin (AB)	Bloody diarrhea and hemolytic uremic syndrome
Staphylococcus aureus	Pyogenic (pus-forming) wounds; food poisoning, toxic shock	α, β, γ, δ toxins (CT) Toxic shock toxin (SA) Enterotoxins A–E (SA)	Hemolysis, leukolysis, cell death Systemic shock Vomiting, diarrhea, systemic shock
Streptococcus pyogenes	Pyogenic infections; strep throat; scarlet fever	Streptolysin O, S (CT) Erythrogenic toxin (SA)	Hemolysis Causes scarlet fever
Vibrio cholerae	Cholera	Cholera (AB)	Induces fluid loss from intestine

[a]AB, AB toxin; CT, cytotoxin; SA, superantigen.
[b]See Figures 25.11–25.16 for the mode of action of some of these toxins.

kill a cell. Diphtheria has a significant mortality rate, especially in the young, and death ensues from the destruction of tissues in vital organs such as the heart and liver from the blockage of protein synthesis by diphtheria toxin.

Cells of *C. diphtheriae* secrete diphtheria toxin as a single polypeptide. One component of the toxin, subunit B, specifically binds to a host cell receptor protein on eukaryotic cells, the heparin-binding epidermal growth factor (**Figure 25.12**). After binding, proteolytic cleavage between subunit B and the remaining portion of the protein, subunit A, allows subunit A to move across the host cytoplasmic membrane into the cytoplasm. Here subunit A disrupts protein synthesis by blocking transfer of an amino acid from tRNA to growing polypeptide chains. Diphtheria toxin specifically inactivates elongation factor 2 (EF-2), a protein that functions in growth of the polypeptide chain, by catalyzing the attachment of adenosine diphosphate (ADP) ribose from NAD⁺. Following ADP-ribosylation, the activity of the modified EF-2 decreases dramatically and protein synthesis stops (Figure 25.12).

Diphtheria toxin is not encoded by the bacterium but instead by a viral gene called *tox* present in the genome of the lysogenic bacteriophage β. Lysogenic phages are those whose genomes have become integrated into their host's chromosome

(⇄ Section 8.7). Toxigenic, pathogenic strains of *C. diphtheriae* are infected with phage β and hence produce the toxin. Nontoxigenic, nonpathogenic strains of *C. diphtheriae* can be converted to pathogenic strains by infection with phage β, a process called *phage conversion* (⇄ Section 11.7).

Exotoxin A of *Pseudomonas aeruginosa* functions similarly to diphtheria toxin, also modifying EF-2 by ADP-ribosylation. The enterotoxin produced by *Shigella dysenteriae*, called *Shiga toxin*, and the Shiga-like toxin produced by enteropathogenic *E. coli* O157:H7 (⇄ Section 32.11) are also AB toxins (Table 25.2). Shiga and Shiga-like toxins target cells of the small intestine near where the pathogen colonized, shutting down protein synthesis. This leads to cell death, bloody diarrhea, and hemolytic uremic syndrome, a kidney disease that can trigger kidney failure, especially in children.

Neurological Exotoxins: Botulinum and Tetanus Toxins

Clostridium botulinum and *Clostridium tetani* are endospore-forming bacteria commonly found in soil and which cause the serious and potentially fatal diseases botulism and tetanus, respectively; both diseases are caused by the secretion of highly poisonous AB exotoxins that function as *neurotoxins* (⇄ Sections 31.9 and 32.9). Neither

UNIT 6

C. botulinum nor *C. tetani* is highly invasive, and therefore pathogenicity is almost exclusively due to neurotoxicity. Botulinum toxin and tetanus toxin both block the release of neurotransmitters that control muscle activities, but the mode of action and disease symptoms are quite distinct (**Figures 25.13** and **25.14**).

C. botulinum sometimes grows directly in the intestine, causing infant or wound botulism. Most frequently, however, botulism results from cells of *C. botulinum* growing and producing toxin in improperly preserved foods, such as home-canned vegetables (see page 937). Thus, infection and growth of the pathogen in the body are unnecessary. Botulinum toxins, the most potent biological substances known, are composed of seven related AB toxins. One nanogram (10^{-9} g) of botulinum toxin is sufficient to kill a guinea pig. Of the seven distinct botulinum toxins known, at least two are encoded on lysogenic bacteriophages specific for *C. botulinum*. The major botulinum toxin is a protein that forms complexes with nontoxic botulinum proteins to yield a bioactive protein complex. The complex then binds tightly to presynaptic membranes on the termini of the stimulatory motor neurons at the neuromuscular junction, blocking the release of acetylcholine, a neurotransmitter. Normal transmission of a nerve impulse to a muscle cell requires acetylcholine interaction with a muscle receptor; the binding of botulinum toxin poisons the neuron, preventing it from sending the excitatory acetylcholine signal to the muscle (Figure 25.13). This prevents muscle contraction and is recognized as *flaccid paralysis* in a botulism victim. This can lead to death by suffocation if the diaphragm muscles are severely affected.

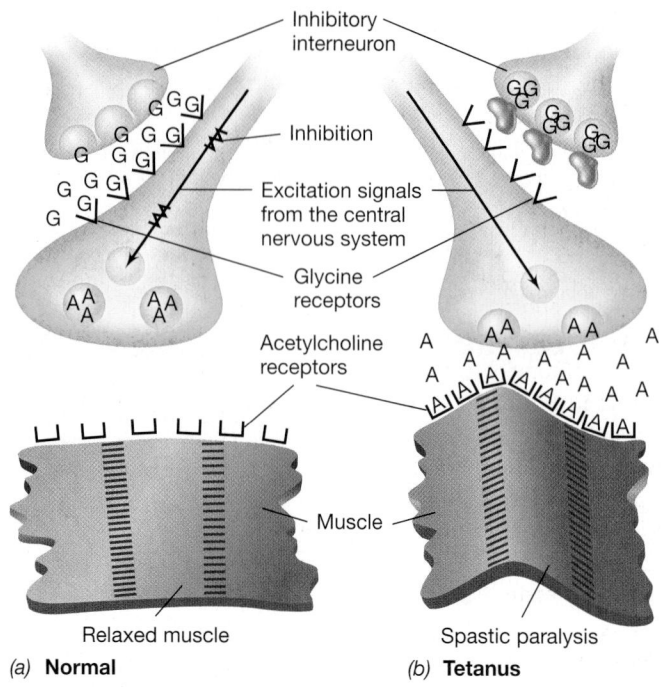

(a) **Normal**   (b) **Tetanus**

Figure 25.14 The activity of tetanus toxin. *(a)* Muscle relaxation is normally induced by glycine (G) release from inhibitory interneurons. Glycine acts on the motor neurons to block excitation and release of acetylcholine (A) at the motor end plate. *(b)* Tetanus toxin binds to the interneuron to prevent release of glycine from vesicles, resulting in a lack of inhibitory signals to the motor neurons, constant release of acetylcholine to the muscle fibers, irreversible contraction of the muscles, and spastic paralysis. For the purpose of illustration, the inhibitory interneuron is shown near the motor end plate, but it is actually in the spinal cord.

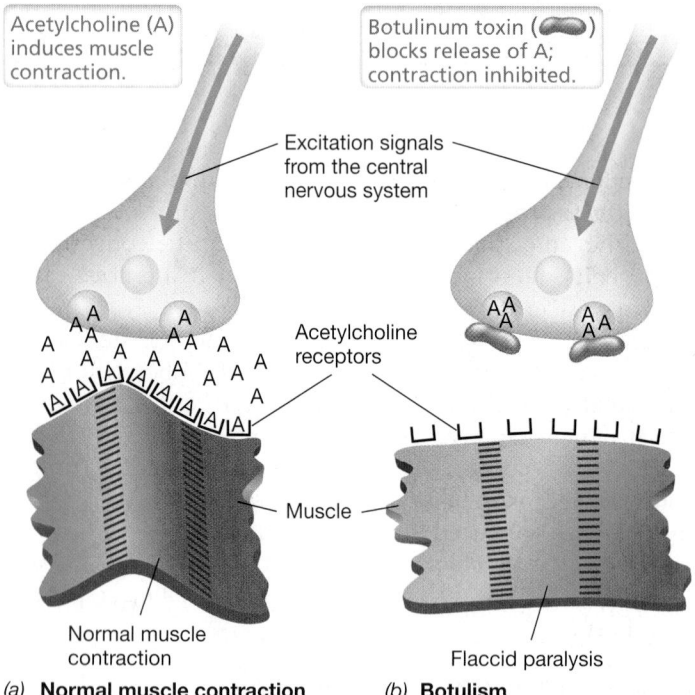

(a) **Normal muscle contraction**   (b) **Botulism**

Figure 25.13 The activity of botulinum toxin. *(a)* Upon stimulation of peripheral and cranial nerves, acetylcholine (A) is normally released from vesicles at the neural side of the motor end plate. Acetylcholine then binds to specific receptors on the muscle, inducing contraction. *(b)* Botulinum toxin acts at the motor end plate to prevent release of acetylcholine from vesicles, resulting in a lack of stimulus to the muscle fibers, irreversible relaxation of the muscles, and flaccid paralysis.

In contrast to *C. botulinum*, *C. tetani* grows in the body in deep wounds that become anoxic, such as punctures. *C. tetani* cells rarely leave the wound where they were first introduced, growing relatively slowly at the wound site. On contact with the nervous system, tetanus toxin, called *tetanospasmin*, is transported through the motor neurons to the spinal cord, where it binds specifically to ganglioside lipids at the termini of the inhibitory interneurons. The inhibitory interneurons normally work by releasing an inhibitory neurotransmitter, typically the amino acid glycine, which binds to receptors on the motor neurons. Glycine from the inhibitory interneurons then stops the release of acetylcholine by the motor neurons and inhibits muscle contraction, allowing relaxation of the muscle fibers. However, if tetanus toxin blocks glycine release, the motor neurons cannot be inhibited, resulting in the continual release of acetylcholine and uncontrolled contraction of the muscle fibers. Thus, in contrast to botulinum toxin, which prevents muscle *contraction* (Figure 25.13), tetanus toxin prevents muscle *relaxation* (Figure 25.14).

The outcome in a case of tetanus is a twitching, *spastic paralysis*—the hallmark of tetanus (⇔ Section 31.9 and Figure 31.33*b*)—as affected muscles are constantly contracting. If the muscles of the mouth are involved, the prolonged contractions restrict the mouth's movement, resulting in a condition called *lockjaw*. If the diaphragm is affected, its prolonged contraction may result in death due to asphyxiation.

Cholera Enterotoxin: Intestinal Distress

Cholera enterotoxin is an AB-type exotoxin produced by *Vibrio cholerae*, the causative agent of the waterborne disease cholera (⊄ Section 32.3). Cholera is characterized by massive fluid loss from the intestines, resulting in severe diarrhea, life-threatening dehydration, and electrolyte depletion (**Figure 25.15**). Cholera starts by ingestion of *V. cholerae* cells from food or water contaminated with human feces. The organism travels to the small intestine, where it colonizes and secretes cholera toxin. In the small intestine, the B subunit of cholera toxin, consisting of five identical monomers, binds specifically to GM1 ganglioside, a complex glycolipid found in the cytoplasmic membrane of intestinal epithelial cells (Figure 25.15).

The B subunit targets cholera toxin specifically to receptors in the intestinal epithelium but has no toxicity itself; toxicity is a function of the A subunit, which crosses the cytoplasmic membrane and activates adenylate cyclase, the enzyme that converts ATP to cyclic adenosine monophosphate (cAMP). This molecule is a cyclic nucleotide (⊄ Figure 6.13) that mediates several regulatory systems in cells, including ionic balance. The increased cAMP induced by cholera enterotoxin blocks Na^+ uptake by small intestine epithelial cells and induces secretion of chloride and bicarbonate (HCO_3^-) into the intestinal lumen. This change in ion concentrations leads to the secretion of large amounts of water; the rate of water loss into the small intestine is greater than the possible reabsorption of water by the large intestine, resulting in a large net fluid loss and watery diarrhea. If untreated, cholera victims can die within hours of the major onset. However, if lost fluids are replaced with an oral rehydration solution (the main treatment for cholera), the effects of cholera toxin can be neutralized and a cholera victim can return to normal in just a few days.

A few other enterotoxins, most notably the Shiga-like toxin of enterotoxigenic strains of *Escherichia coli* (⊄ Section 32.11) and *Shigella* and *Salmonella* enterotoxins are also of the AB type (Table 25.2), and all of these function by inhibiting protein synthesis. This leads to major bouts of diarrhea, typically bloody and foul smelling, and severe dehydration. In addition, some cytolytic enterotoxins are known and the powerful enterotoxins of *Staphylococcus aureus* (Table 25.2) are of the superantigen type.

— **MINIQUIZ** —

- What key features are shared by all AB exotoxins?
- Are bacterial growth and infection in the host necessary for the production of toxins? Explain and cite examples for your answer.
- Why do botulism and tetanus show such opposing symptoms?

25.7 Cytolytic and Superantigen Exotoxins

The pathogenesis of exotoxins such as the cytolytic and superantigen toxins differs from those of the classical AB toxins. Cytolytic and superantigen exotoxins function by destroying host cells, such

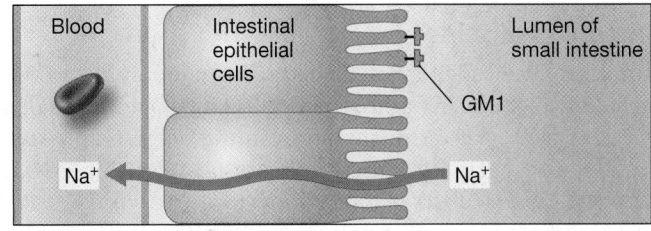

1. Normal ion movement, Na^+ from lumen to blood, no net Cl^- movement

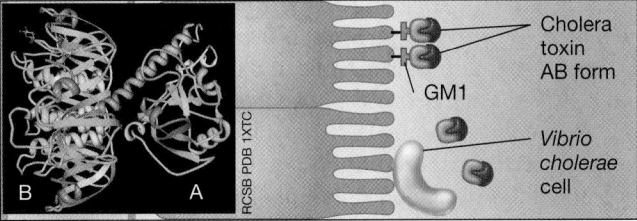

2. Infection and toxin production by *V. cholerae*

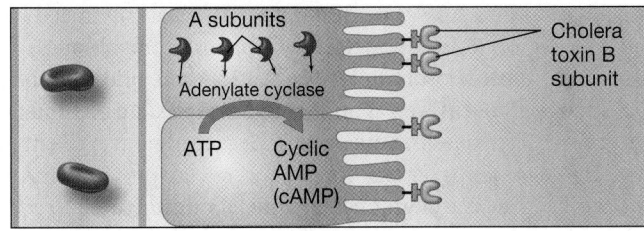

3. Activation of epithelial adenylate cyclase by cholera toxin

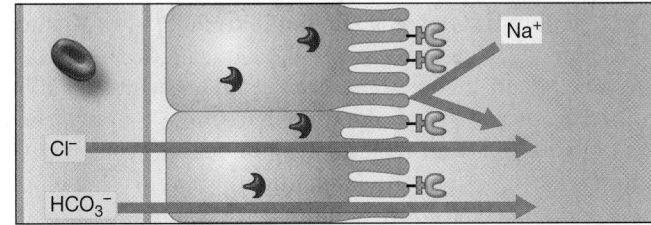

4. Elevated cAMP blocks Na^+; net anion movement to intestinal lumen

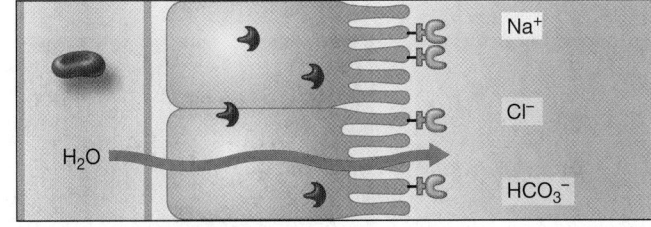

5. Massive water movement to the lumen and ion loss trigger cholera symptoms.

Figure 25.15 The activity of cholera enterotoxin. Cholera toxin is an AB enterotoxin that activates a second-messenger pathway, disrupting normal ion flow in the intestine, resulting in potentially life-threatening diarrhea. The thumbnail photo of the three-dimensional structure shows a side view of the toxin, with the separate cell-binding B subunit and the enzymatically active A subunit.

as from the activity of hemolysins, or by triggering a massive immune response, such as in the case of toxic shock exotoxin, the cause of toxic shock syndrome. As for other exotoxins, however,

UNIT 6

these proteins are produced by the pathogen to increase its invasiveness and release host cell resources that can benefit the pathogen.

Cytolytic Exotoxins

Cytotoxins (also called cytolytic exotoxins) are soluble proteins secreted by a variety of pathogens (Table 25.2). Cytotoxins damage the host cytoplasmic membrane, causing cell lysis and death. Because the lytic activity of these toxins is most easily observed in assays that use red blood cells (erythrocytes), the toxins are often called *hemolysins* (Table 25.2). However, hemolysins also lyse cells other than erythrocytes. The production of hemolysins can be demonstrated by streaking the pathogen on a blood agar plate (a rich medium containing 5% sterile blood). During growth of the colonies, hemolysin is released and lyses the surrounding red blood cells, releasing hemoglobin and creating a clear area, called a zone of *hemolysis*, around the growing colonies (**Figure 25.16a**).

Some hemolysins attack the phospholipid of the host cytoplasmic membrane. Because the phospholipid lecithin (phosphatidylcholine) is often used as a substrate, these enzymes are called *lecithinases* or *phospholipases*. An example is the α-toxin of *Clostridium perfringens*, a lecithinase that dissolves membrane lipids, resulting in cell lysis (Table 25.2, Figure 23.16b). Because the cytoplasmic membranes of all organisms contain phospholipids (⮎ Section 2.3), phospholipases can destroy bacterial as well as animal cell cytoplasmic membranes. In the case of *C. perfringens*, a major cause of gas gangrene, the activity of its lecithinase helps destroy tissues and release proteins that are fermented by the bacterium in energy metabolism. Since *C. perfringens* lecithinase is secreted immediately after synthesis and is unable to reenter the cell, the enzyme does not affect the phospholipids in *C. perfringens* cells.

Some hemolysins, however, are not phospholipases. Streptolysin O, a hemolysin produced by streptococci, affects the sterols of the host cytoplasmic membrane. *Leukocidins* lyse white blood cells and thereby decrease the host immune response. Staphylococcal α-toxin (**Figure 25.17** and Table 25.2) kills nucleated cells and lyses erythrocytes. To do this, the seven α-toxin subunits first bind to the phospholipid bilayer. The subunits then combine to form into nonlytic heptamers, now associated with the membrane. Each heptamer then undergoes conformational changes to produce a

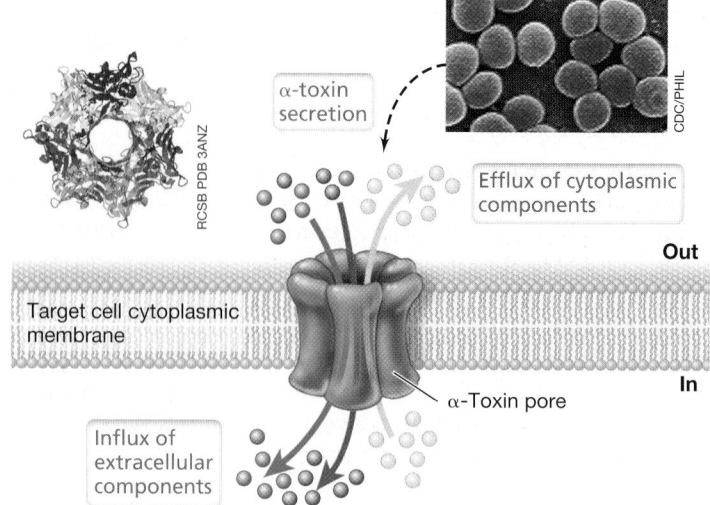

Figure 25.17 Staphylococcal α-toxin. Staphylococcal α-toxin is a pore-forming cytotoxin that is produced by growing *Staphylococcus aureus* cells. Released as monomers, seven identical protein subunits oligomerize in the cytoplasmic membrane of target cells. The oligomer forms a pore, releasing the contents of the cell. In red blood cells, hemolysis occurs, visually indicating cell lysis. The inset photo on the top left shows the structure of α-toxin looking down through the pore. Each of the seven identical subunits is shown in a different color. The inset photo on the top right is a scanning electron micrograph of *S. aureus* cells.

membrane-spanning pore that releases cytoplasmic contents and allows the influx of extracellular materials, thus killing the cell (Figure 25.17).

Superantigen Exotoxins

The gram-positive bacteria *Staphylococcus aureus* and *Streptococcus pyogenes* are major producers of exotoxin superantigens. We consider the mode of action of superantigens in Section 27.10 in the context of adaptive immunity and focus here only on some basic disease symptoms.

Superantigen poisoning can be triggered by certain types of food poisoning, in particular that caused by the enterotoxins of *S. aureus*, and also by pyrogenic fever (fever induced from an internal source, typically a small immune system protein called a cytokine, or from an external substance such as endotoxin; see Section 25.8) and toxic shock syndrome. Reactions to superantigen poisoning are severe and can even be fatal in some individuals, particularly those whose immune systems and overall health are weakened from cancer, drug treatments, HIV infection, or old age. Toxic shock syndrome (TSS) is the classic example of the systemic effects of a toxic superantigen and results from exposure to any of a series of exotoxins secreted during infection by certain strains of *S. aureus* or *S. pyogenes* (⮎ Sections 30.9 and 30.2, respectively).

S. aureus TSS commonly originates as a result of a localized rather than a generalized infection. By contrast, *S. pyogenes* TSS is typically the result of a systemic infection where bacteremia or septicemia (Section 25.2) is present and tissue damage including extensive tissue necrosis occurs (⮎ Figure 30.10); as a result, mortality rates from *S. pyogenes* TSS are considerably higher than from *S. aureus* TSS. In both cases, however, the symptoms of TSS are triggered when the immune system recognizes the superantigen

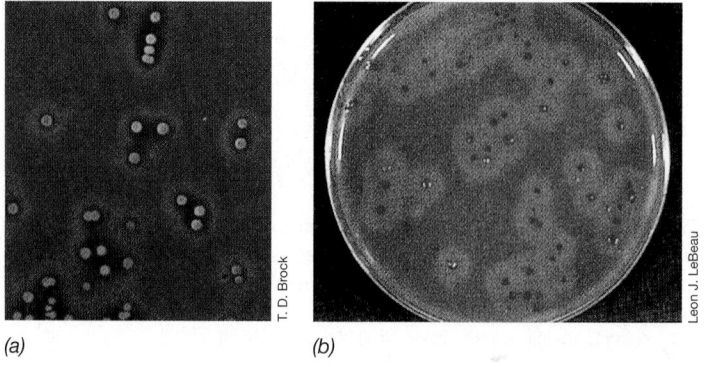

(a)   (b)

Figure 25.16 Hemolysis. *(a)* Zones of hemolysis around colonies of *Streptococcus pyogenes* growing on a blood agar plate. *(b)* Activity of lecithinase, a phospholipase, around colonies of *Clostridium perfringens* growing on an agar medium containing egg yolk, a source of lecithin. Lecithinase dissolves the cytoplasmic membranes of red blood cells, producing cloudy zones of hemolysis around each colony.

toxin but then, rather than just activating a small subset of T lymphocytes (key cells in the adaptive immune response, Chapter 27) as usually occurs in the adaptive immune response, activates a large proportion of the entire T lymphocyte pool. It is the structure of the superantigen itself that leads to this overblown immune response, a result of which is widespread inflammation that leads to hypotension (low blood pressure), intestinal disruption, organ failure, and eventually systemic shock.

In a clinical diagnosis, either staphylococcal or streptococcal TSS is suspected when an individual presents with a fever of 39°C or greater, systolic blood pressure of <90 mm Hg, and functional disruption of three or more organ systems, most often gastrointestinal, kidney, and liver. In severe cases of TSS, intensive care hospitalization may be needed with intravenous administration of antibiotics.

--- MINIQUIZ ---
- Give an example of a cytolytic exotoxin and a superantigen exotoxin, as well as the bacteria that produce each.
- How can activity of a hemolytic exotoxin be detected?

25.8 Endotoxins

Endotoxins are the toxic lipopolysaccharides found in the cell walls of most gram-negative *Bacteria*. Endotoxins are not proteins but are structural components of the gram-negative outer membrane (⊃ Section 2.5). In contrast to exotoxins, which are the secreted products of living cells, endotoxins are cell bound and released in toxic amounts only when the cells lyse. The basic properties of exotoxins and endotoxins are compared in **Table 25.3**.

Endotoxin Structure and Biology

A major component of the gram-negative cell outer membrane is lipopolysaccharide (LPS) (⊃ Figures 2.13 and 2.14). LPS consists of three covalently linked subunits: the membrane-distal O-specific polysaccharide, a membrane-proximal core polysaccharide, and lipid A—a phosphoglycolipid and the membrane-anchoring portion of LPS (**Figure 25.18**). The lipid A portion of LPS is responsible for toxicity, whereas the polysaccharide fraction by itself is nontoxic. The polysaccharide functions to make the entire LPS complex soluble and immunogenic, and thus both the lipid and polysaccharide fractions must be delivered as a unit for toxicity to occur.

Endotoxins have been well studied in the bacteria *Escherichia*, *Shigella*, and especially in *Salmonella*, where they are another of the many virulence factors that contribute to pathogenesis (Figure 25.10). The biosynthesis of the toxic component of endotoxin, lipid A, is known and is a highly conserved process among gram-negative bacteria. Nevertheless, not all lipid A is structurally the same, as the molecule can be modified using enzymes that catalyze postsynthesis modifications that control the presence, absence, or number of phosphate groups, and the chemistry and number of fatty acid side chains (Figure 25.18). These subtle but important alterations affect the properties of the LPS molecule and are a virulence strategy that certain pathogens have evolved to either evade recognition by the host immune system or increase the toxicity of the molecule. However, some lipid A alterations affect toxicity in a negative way. For example, the phosphate groups (Figure 25.18) are essential for binding lipid A to animal cell receptors, and although phosphate-free lipid A can evade immune surveillance, its toxicity is greatly reduced.

Endotoxins cause a variety of physiological effects. Fever is an almost universal result of endotoxin exposure because endotoxin stimulates host cells to release cytokines, soluble proteins secreted by certain cells of the immune system that function as *endogenous pyrogens*, proteins that affect the temperature-controlling center of the brain, causing fever. Cytokines released as a result of endotoxin exposure can also cause diarrhea, increased heart rate, a rapid decrease in the numbers of lymphocytes and platelets, and

TABLE 25.3 Properties of exotoxins and endotoxins

Property	Exotoxins	Endotoxins
Chemistry	Proteins, secreted by certain gram-positive or gram-negative *Bacteria;* generally heat-labile	Lipopolysaccharide–lipoprotein complexes, released on cell lysis as part of the outer membrane of gram-negative *Bacteria;* extremely heat-stable
Mode of action; symptoms	Specific; usually binds to specific cell receptors or structures; either cytotoxin, enterotoxin, or neurotoxin with defined, specific action on cells or tissues	General; fever, diarrhea, vomiting
Toxicity	Often highly toxic in picogram to microgram quantities, sometimes fatal	Moderately toxic in tens to hundreds of microgram amounts, rarely fatal
Immune response	Highly immunogenic; stimulate the production of neutralizing antibody (antitoxin)	Relatively poor immunogen; immune response not sufficient to neutralize toxin
Toxoid potential[a]	Heat or chemical treatment may destroy toxicity, but treated toxin (toxoid) remains immunogenic	None
Fever potential	Nonpyrogenic; does not produce fever in the host	Pyrogenic; often induces fever in the host
Genetic origin	Often encoded on extrachromosomal elements or lysogenic bacteriophages	Encoded by chromosomal genes

[a]A toxoid is a modified toxin that is no longer toxic but can still elicit an immune response against the toxin (⊃ Section 28.9).

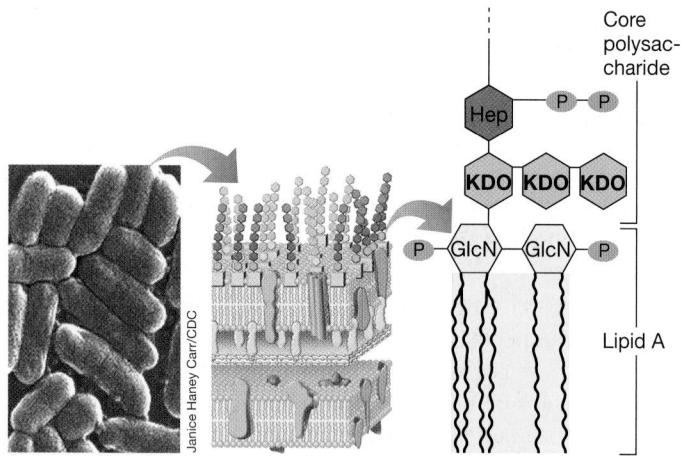

Figure 25.18 Endotoxin. Left to right: Scanning electron micrograph of cells of the gram-negative bacterium *Escherichia coli*; structure of the gram-negative cell wall including the lipopolysaccharide (LPS) outer membrane; detailed structure of lipid A, the toxic portion of LPS, along with part of the core polysaccharide of LPS.

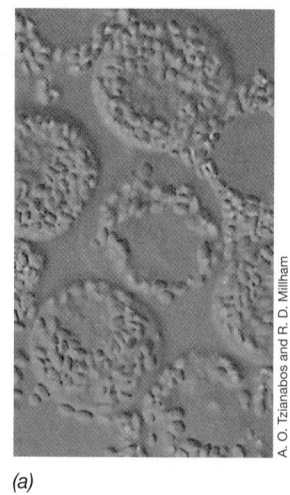
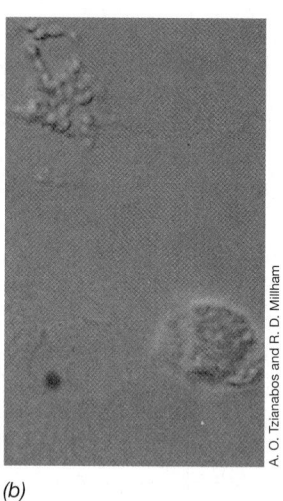

(a)   *(b)*

Figure 25.19 *Limulus* amoebocyte assay for endotoxin. *(a)* Normal amoebocytes from the horseshoe crab *Limulus polyphemus*. *(b)* Amoebocytes following exposure to bacterial lipopolysaccharide (LPS). LPS induces degranulation and lysis of the cells.

generalized inflammation (⇌ Section 26.8). Other physiological consequences of endotoxin exposure include activation of the complement cascade (complement is a series of immune system proteins, ⇌ Section 26.9), which also triggers inflammation, and activation of the blood coagulation cascade, which can lead to blood clots and reduced blood flow. Large doses of endotoxin can cause death from hemorrhagic shock and kidney failure.

Although significant virulence factors, endotoxins are generally less toxic than most exotoxins and rarely cause symptoms that can lead to death in an otherwise healthy individual if exposure is from a gastrointestinal source (Table 25.3). For instance, in mice the LD_{50} (see Figure 25.9) for endotoxin is 200–400 *micro*grams per animal, whereas the LD_{50} for botulinum exotoxin is about 25 *pico*grams, about 10 million times less. By contrast, intravenous administration of endotoxin, for example from a heavily contaminated intravenous solution, could have fatal consequences. We see how such faulty solutions can be identified now.

Limulus Amoebocyte Lysate Assay for Endotoxin

Because endotoxins induce fever and can trigger other, more serious symptoms, pharmaceuticals such as injectable antibiotics and intravenous solutions must be free of endotoxin. An endotoxin assay of very high sensitivity and specificity in widespread use employs lysates of amoebocytes from the horseshoe crab *Limulus polyphemus* (amoebocytes are mobile cells in the blood and fluids of invertebrates that are analogous to the white blood cells of vertebrates). Endotoxin specifically causes lysis of the amoebocytes

(Figure 25.19). In the *Limulus* amoebocyte lysate (LAL) assay, *Limulus* amoebocyte extracts are mixed with the solution to be tested. If endotoxin is present, the amoebocyte extract forms a gel and precipitates, causing a change in turbidity. This reaction is measured quantitatively with a spectrophotometer and can detect as little as 10 picograms of LPS in a 1-ml sample.

The LAL assay is used to detect endotoxin in clinical samples such as serum or cerebrospinal fluid. A positive test is presumptive evidence for infection by gram-negative bacteria. Drinking water, water used for formulation of injectable drugs, and injectable aqueous solutions are routinely tested using the LAL assay to identify and eliminate endotoxin contamination from gram-negative organisms. A commercially available assay uses horseshoe crab factor C made by recombinant DNA techniques (factor C is the key protein activated by endotoxin in the LAL assay). Rather than relying on amoebocytes collected from harvested horseshoe crabs, the recombinant protein is just as sensitive but less expensive, allows for a more standardized assay protocol, and is totally free of animal products.

── MINIQUIZ ──

- What part of the *Escherichia coli* cell contains endotoxin? Why do gram-positive bacteria not produce endotoxins?

- Why is it necessary to test for endotoxin in water used for injectable drug preparations?

MasteringMicrobiology® **Visualize, explore,** and **think critically** with Interactive Microbiology, MicroLab Tutors, MicroCareers case studies, and more. MasteringMicrobiology offers practice quizzes, helpful animations, and other study tools for lecture and lab to help you master microbiology.

UNIT 6

Chapter Review

I • Human–Microbial Interactions

25.1 If a pathogen gains access to the specific tissues it infects, disease will only occur if it first adheres to those tissues. Although adherence is required to initiate disease, it is not sufficient to initiate disease; colonization, invasion, and production of toxic substances are also required.

Q How is microbial adherence to host tissues facilitated by structures such as capsules and fimbriae?

25.2 Each body region differs chemically and physically from others, and thus provides a variety of selective environments for the growth of certain microbes but not others. Colonization of tissues by a pathogenic microbe followed by growth to form populations sufficient to trigger a biological effect is necessary before disease symptoms appear.

Q What common disease is a model for the study of bacterial attachment and colonization, and which species of bacteria cause the bulk of the damage in this disease?

25.3 The pathogenicity of a pathogen is a function of its virulence, its relative ability to cause disease. Virulence is a quantitative measure that can be assayed in terms of the number of cells (or virions, if a viral pathogen) required to infect or kill 50% of a given population—the LD_{50}. Attenuated pathogens are strains with diminished virulence and are quite useful for preparing vaccines.

Q What virulence factor, present in *Streptococcus pneumoniae* but absent from *Salmonella enterica*, makes *S. pneumoniae* so highly virulent for mice?

25.4 The genetics and physiology of both the pathogen and the host affect the outcome of an infectious disease. In *Salmonella* species, chromosomal islands or conjugative plasmids are present that encode several virulence factors; these mobile genetic elements can quickly spread a suite of virulence factors to other gram-negative bacteria. Susceptibility to an infectious disease is increased in a compromised host.

Q What factors might diminish the ability of a host to fight off an infectious disease?

II • Enzymes and Toxins of Pathogenesis

25.5 The virulence of a pathogen depends on the number and kinds of virulence factors it produces. Some pathogenic bacteria produce enzymes that function to either destroy host tissues or disarm host defenses. The activity of these enzymes releases nutrients to support growth of the pathogen and facilitate further invasiveness.

Q Give two reasons why a pathogen might benefit from the secretion of an enzyme that destroys host tissue integrity.

25.6 Exotoxins are toxic proteins and major virulence factors. Each exotoxin affects a specific host cell function. Enterotoxins are exotoxins that affect the small intestine. The clostridial botulinum and tetanus exotoxins are among the most poisonous substances known.

Q Assuming a person was poisoned with either botulinum toxin or tetanus toxin, how could the person's physical state signal the type of poisoning that had occurred?

25.7 Cytotoxins and superantigens are toxic proteins that lyse host cells and trigger a massive immune response against host tissues, respectively. Hemolysis on a blood agar plate is a classic cytotoxic effect, whereas toxic shock syndrome, a potentially fatal condition, is one result of superantigen activity.

Q Distinguish between the mechanism of cytotoxins and AB toxins, and provide one example of each.

25.8 Endotoxins are lipopolysaccharides derived from the outer membrane of gram-negative bacteria. Both the lipid and the polysaccharide components of endotoxin are necessary for toxicity. Symptoms of endotoxin poisoning include fever and intestinal distress.

Q Identify the structural features, origins, and major effects of endotoxins.

Application Questions

1. Coagulase is a virulence factor for *Staphylococcus aureus* that acts by causing clot formation at the site of *S. aureus* growth. Streptokinase is a virulence factor for *Streptococcus pyogenes* that acts by dissolving clots at the site of *S. pyogenes* growth. Reconcile these opposing strategies for enhancing pathogenicity.

2. Although mutants incapable of producing exotoxins are relatively easy to isolate, mutants incapable of producing endotoxins are much harder to isolate. From what you know of the structure and function of these types of toxins, explain the differences in mutant recovery.

Chapter Glossary

Adherence the enhanced ability of a microorganism to attach to a cell or surface

Adhesins glycoproteins or lipoproteins covalently bound to the outer layer of the pathogen that function in attachment to host tissues

Attenuation a decrease or loss of virulence

Bacteremia the presence of bacteria in the blood

Capsule a dense, well-defined polysaccharide or protein layer closely surrounding a cell

Colonization the growth of a microorganism after it has gained access to host tissues

Dental caries tooth decay resulting from bacterial infection

Dental plaque a bacterial biofilm consisting of a matrix of extracellular polymers and salivary products, found on the teeth

Disease an injury to a host organism, caused by a pathogen or other factor, that affects host function

Endotoxin the lipopolysaccharide portion of the cell envelope of most gram-negative *Bacteria*, which is a toxin when solubilized

Enterotoxin a protein released extracellularly by a microorganism as it grows that produces immediate damage to the small intestine of the host

Exotoxin a protein released extracellularly by a microorganism as it grows that produces immediate host cell damage

Infection an event during which a microorganism not a member of the local microbiota is established and grows in a host, regardless of whether the host is harmed

Invasion the ability of a pathogen to enter into host cells or tissues, spread, and cause disease

Mucous membrane layer of mucus-covered epithelial cells that interacts with the external environment

Mucus a liquid secretion that contains water-soluble glycoproteins and proteins that retain moisture and aid in resistance to microbial invasion on mucosal surfaces

Opportunistic pathogen an organism that causes disease only in the absence of normal host resistance

Pathogen an organism, usually a microorganism, that grows in or on a host and causes disease

Pathogenicity the ability of a pathogen to cause disease

Septicemia a bloodborne systemic infection

Toxicity the ability of an organism to cause disease by means of a preformed toxin that inhibits host cell function or kills host cells

Virulence the relative ability of a pathogen to cause disease

Virulence factors substances or strategies of a pathogen that indirectly or directly enhance invasiveness and host damage by facilitating and promoting infection

Innate Immunity: Broadly Specific Host Defenses

26

Rehabilitating a Much-Maligned Peptide: Amyloid-β

One of the hallmarks of Alzheimer's disease is the presence of amyloid plaques in the brain tissue of patients. The plaques, which develop from the aggregation of amyloid-β protein (Aβ) into insoluble fibril-like complexes, interfere with cognitive brain function and have traditionally been closely linked to Alzheimer's disease pathology. However, recent work casts Aβ in a more positive light by showing that it functions as a natural antimicrobial that protects the brain from infection and is potentially an important part of the innate immune system, the body's first-line defense against invading pathogens.

Deciphering the natural role of Aβ in the body has required a shift of prevailing thought. Despite the fact that the amino acid sequence of Aβ is highly conserved among nearly all vertebrates, it has traditionally been viewed as a functionless and even detrimental peptide that can hasten neurodegenerative disease. However, it is now evident that Aβ most certainly does have a function, and its normal role is actually protective rather than deleterious. A major clue to the true function of this protein was uncovered when scientists discovered similarities between Aβ and various ancient antimicrobial peptides, including their tendency to bind and agglutinate (clump) pathogens by aggregating into a network of fibrils (top photo). These activities suggested that Aβ might be a previously unrecognized but key component of the innate immune response, especially in the central nervous system where Aβ is abundant and efficacy of the adaptive (antibody- and cell-mediated) immune response is limited.

To examine the role of Aβ in living systems, scientists used mouse, nematode (*Caenorhabditis elegans*), and cell culture models to show that host survival following introduction of a pathogen (*Salmonella* bacteria; bottom photo, green) was enhanced by the activity of Aβ. The bacteria became entrapped in the growing mesh of Aβ fibrils (bottom photo, red), which prevented microbial adhesion to host cells and enabled more efficient pathogen destruction by immune mechanisms. Thus, the antimicrobial activity of Aβ is dependent upon its deposition and propagation of a dense fibril network. This raises the intriguing possibility that the onset of Alzheimer's-associated amyloid plaques may be precipitated by chronic microbial infection of the brain, resulting in cerebral inflammation and an overabundance of Aβ.

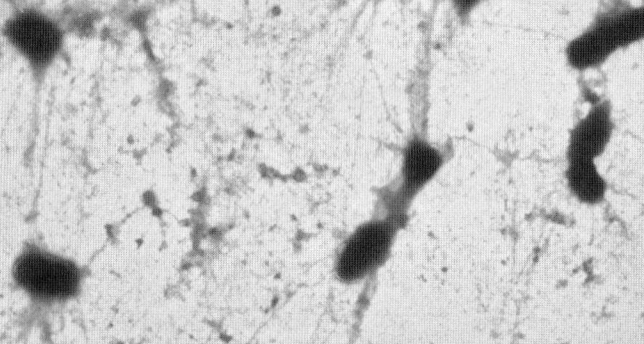
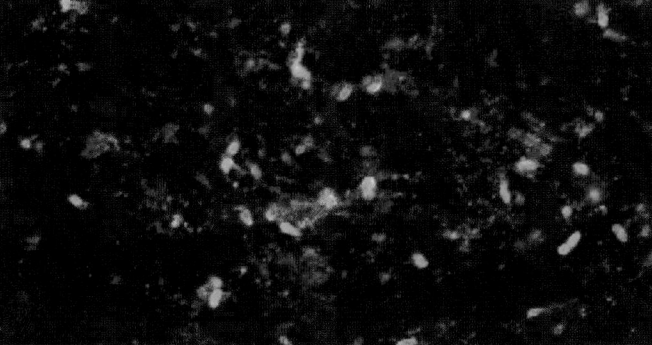

Source: Kumar, D.K.V., et al. 2016. Amyloid-β peptide protects against microbial infection in mouse and worm models of Alzheimer's disease. *Sci. Transl. Med. 8:* doi: 10.1126/scitranslmed.aaf1059.

We began this unit by considering the associations of microbes with humans, and in the last chapter we discussed pathogenicity, virulence, and risk factors for infection. In this and the next chapter, we focus on mechanisms used by vertebrates to resist pathogens and the diseases they cause, the science of *immunology*. In this chapter, we present concepts of **innate immunity**, inborn host defenses against a *broad range* of pathogens. In Chapter 27 we discuss **adaptive immunity**, the essential second tier of the immune system that targets *specific* pathogens to minimize their harmful effects.

I • Fundamentals of Host Defense

We begin with an overview of innate and adaptive immunity and follow this with a consideration of the human body's significant natural barriers to invasion by pathogens.

26.1 Basic Properties of the Immune System

Immunity is the ability of an organism to resist infection. The human immune system employs a two-pronged defense against invading pathogens. *Innate immunity*, the first of these interconnected defensive mechanisms, is the built-in capacity of the immune system of multicellular organisms to target common pathogens regardless of their identity. By contrast, *adaptive immunity* is triggered by exposure to specific pathogens that cannot be eliminated from the body by innate mechanisms alone. Each adaptive immune response is specifically targeted to a particular type of invading pathogen, such as a single strain of virus. **Figure 26.1** compares and contrasts the fundamental elements of each of these indispensable branches of the immune system.

Principles of Innate Immunity

Innate immunity is a noninducible, preexisting ability of the body to recognize and destroy a broad range of pathogens or their

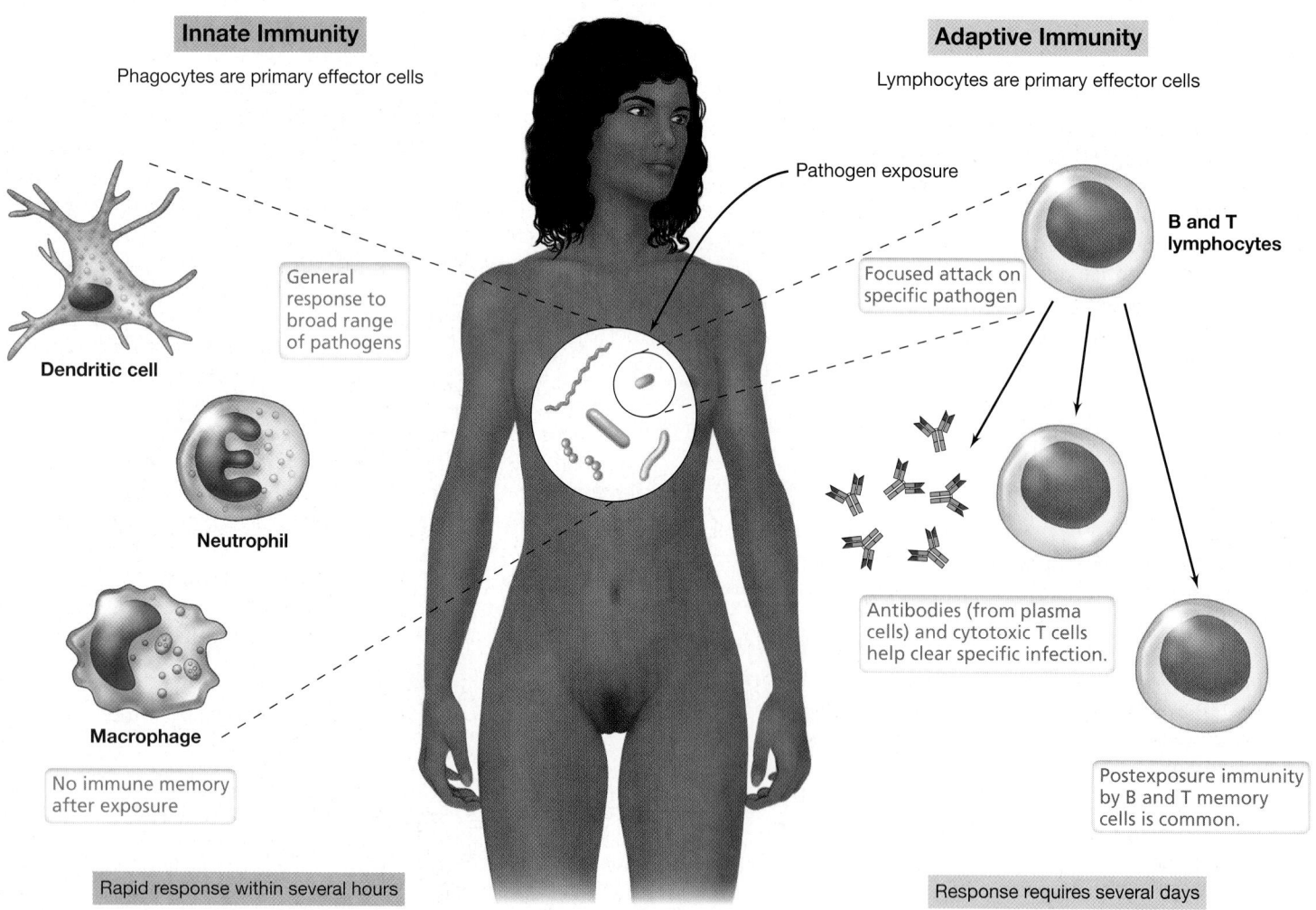

Figure 26.1 Overview of the two-pronged immune response. Principal characteristics of innate and adaptive immunity are depicted. Note the cell types that participate in each form of immunity and compare with their functional descriptions shown in Figure 26.4.

products. Innate immune mechanisms do not require previous exposure to a pathogen or its products for activation. Eukaryotes have functionally similar pathogen recognition mechanisms that lead to rapid and effective host defense. For example, pathogen recognition in the insect *Drosophila* (fruit fly) is carried out in much the same way as it is in humans, using immune cells that clearly show structural and evolutionary homology.

In addition to a variety of physical and chemical barriers to infection that we will describe in Section 26.2, innate immunity is largely dependent on the activity of **phagocytes**, cells of several types that can ingest, kill, and digest microbial pathogens. Innate immune responses develop quickly, within a few hours of exposure to a pathogen. Structural features common to a variety of bacterial pathogens, such as flagella and certain constituents of the cell wall, interact with universal receptors on phagocytes. This interaction activates genes in the phagocyte whose gene products lead to pathogen destruction.

Principles of Adaptive Immunity

Some pathogens are so virulent that innate responses are not completely effective. When this is the case, the innate response activates the adaptive immune response to deal with these infections. Adaptive immunity is the *acquired* ability to recognize and destroy a specific pathogen or its products. Adaptive responses show **specificity** because they are directed at unique pathogen surface molecules called **antigens**, which often define a particular strain of pathogen or type of foreign material. Phagocytes ingest and process antigen molecules and present them to immune cells called **lymphocytes**, essential components of the adaptive response. The presented antigens bind specific receptors on the surface of the lymphocyte, triggering genes that promote lymphocyte multiplication and production of antigen-specific proteins called **antibodies (immunoglobulins)** that interact with the pathogen and mark it for destruction.

Unlike the comparatively rapid innate response, a protective adaptive response usually takes several days to develop, and the strength of the adaptive response increases as the numbers of antigen-reactive lymphocytes increase. Although adaptive immune responses are typically slower than innate response mechanisms, they are highly specific and often result in **immune memory**, the ability to quickly produce specific immune cells or antibodies after subsequent exposure to a previously encountered antigen.

With a broad overview of inborn and acquired immune responses in place, we now consider the many natural barriers to infection that exist in the healthy human body. It is only after one or more of these barriers has been breached that pathogen invasion can occur.

MINIQUIZ

- What major class of immune cells mediates an innate immune response? What additional type of immune cells is required for an adaptive immune response?

- What term is used to describe the unique molecules found on the surface of different pathogens?

26.2 Barriers to Pathogen Invasion

The best defense against infectious diseases for the human body is to prevent pathogens from colonizing in the first place. Humans have a host of innate resistance factors that inhibit pathogen colonization and thereby prevent the onset of infectious diseases. These include a variety of physical and chemical barriers to microbial infection. In addition, the condition of the host and the composition of microorganisms that normally inhabit the host are factors that are often the tipping point between health and disease.

Natural Host Resistance

Several resistance factors common to vertebrate hosts inhibit infection by most pathogens in a nonspecific way (**Figure 26.2**). For example, the normal microbiota associated with the human body are critically important for resisting pathogen infection, especially on the skin and in the gut (Chapter 24). Pathogens do not easily infect tissues on which normal microbiota are well established because the harmless microbes limit available nutrients and sites for infection by the pathogens. Although the *competitive exclusion* of invading pathogens by resident microbes is technically not a component of the host's immune system, the nonpathogenic normal microbiota found in or on the body play a major role in preventing disease through this principle. It is not uncommon for disruptions to the composition of the normal microbiota, such as that which may occur when antibiotic drug treatments nonspecifically kill harmless and even beneficial microbes in the body, to trigger the onset of disease by opportunistic pathogens.

The ability of a particular pathogen to cause disease in an animal is highly variable. For instance, certain animal species, including raccoons and skunks, are much more susceptible to the rabies virus than others, such as the opossum, which only rarely develops the disease. Anthrax infects many species of animals, causing disease symptoms varying from fatal blood poisoning in cattle to the mild pustules of human cutaneous anthrax. Introduction of the same pathogen by other routes, however, may challenge the resistance of the host. For example, anthrax causes a localized infection when acquired through the skin but a lethal, systemic infection when acquired through the mucous membranes of the lungs (🔗 Sections 29.9 and 31.8). Another barrier to infection is the fact that diseases of endothermic ("warm-blooded") animals, such as birds and mammals, are rarely transmitted to ectothermic ("cold-blooded") animals, such as amphibians and reptiles, and vice versa. Presumably, the anatomical and metabolic features of one group are not compatible with pathogens that infect the other group.

In addition to these factors, the general condition of the host plays a role in the balance between health and disease. Infectious diseases are typically more common in the very young and the very old, as well as those suffering from nutritional or stress-related problems. For example, the intestinal microbiota matures over time and thus infants under one year of age often contract diarrhea caused by bacterial or viral pathogens more readily than do adults. Conversely, the immune system of older adults may be

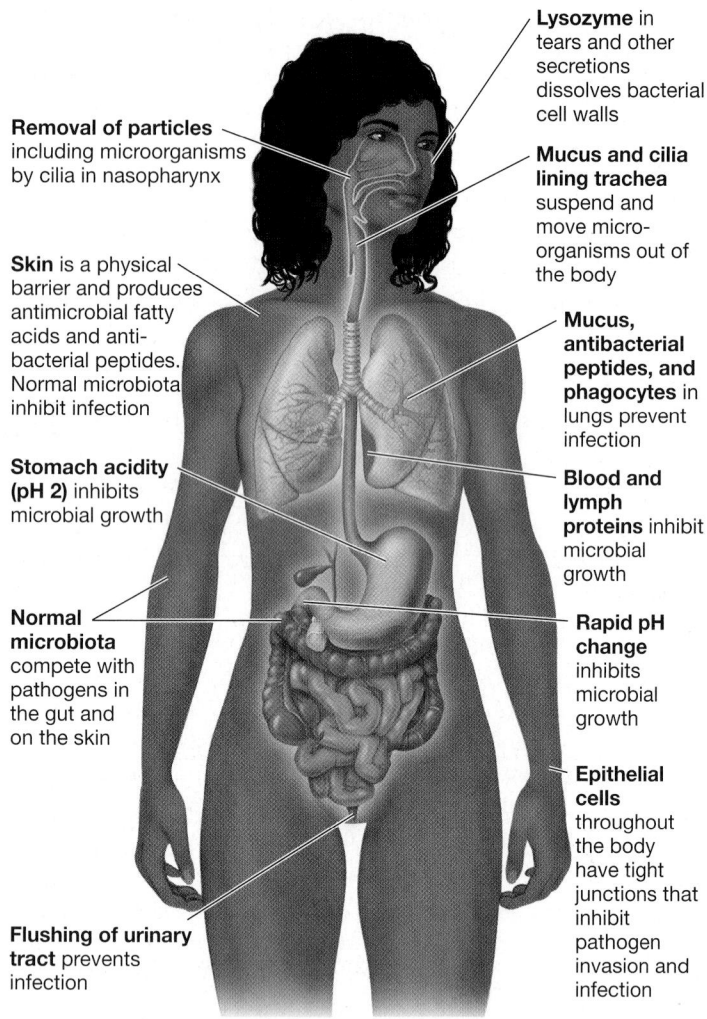

Lysozyme in tears and other secretions dissolves bacterial cell walls

Removal of particles including microorganisms by cilia in nasopharynx

Mucus and cilia lining trachea suspend and move microorganisms out of the body

Skin is a physical barrier and produces antimicrobial fatty acids and anti-bacterial peptides. Normal microbiota inhibit infection

Mucus, antibacterial peptides, and phagocytes in lungs prevent infection

Stomach acidity (pH 2) inhibits microbial growth

Blood and lymph proteins inhibit microbial growth

Normal microbiota compete with pathogens in the gut and on the skin

Rapid pH change inhibits microbial growth

Epithelial cells throughout the body have tight junctions that inhibit pathogen invasion and infection

Flushing of urinary tract prevents infection

Figure 26.2 Barriers to infection in the human body. These barriers provide natural resistance to colonization and infection by pathogens.

weakened or compromised from medical procedures or drug therapy, leading to a higher susceptibility to infectious disease. Other issues that affect both the young and the old—smoking, poor diet, intravenous drug use, excessive alcohol consumption, chronic lack of sleep—can all play a role in whether a pathogen can infect a host and cause disease.

Infection Site and Tissue Specificity

Most pathogens must adhere and infect at the site of exposure to initiate infection. Even if pathogens adhere to an exposure site, the organisms cannot colonize the host if the site is not compatible with the pathogen's nutritional and metabolic needs. For example, if cells of *Clostridium tetani* (cause of the disease tetanus) were ingested, tetanus would not normally result because the pathogen would be either killed by the acidity of the stomach or unable to compete with the well-developed normal intestinal microbiota. If, on the other hand, *C. tetani* cells or endospores were introduced into a puncture wound, the organism would grow and produce tetanus toxin in the anoxic zones created by local tissue death. Conversely, enteric bacteria, such

as *Salmonella* and *Shigella*, do not normally cause wound infections but can infect the intestinal tract and cause gastrointestinal illness.

In some cases, pathogens interact exclusively with members of a few closely related host species because the hosts share tissue-specific receptors, important in establishing infection (Section 25.2). For example, human immunodeficiency virus (HIV, the causative agent of AIDS) infects only humans and their closest primate relatives, including the great apes. This is because the HIV-binding cell surface proteins CXCR4, present on human T lymphocytes, and CCR5, present on human macrophages (T cells and macrophages are cells of the immune system), are also expressed in great apes. Other animals, even most other primates, lack CXCR4 and CCR5 and are therefore not susceptible to HIV infection. Many other pathogens exhibit tissue specificity to such a degree that they infect only a single species. For example, a rhinovirus that infects nasal epithelial cells in humans, causing symptoms of the common cold, would generally not infect and cause similar symptoms in a nonhuman host. Table 26.1 presents some other examples of tissue specificity in pathogens.

Physical and Chemical Barriers to Infection

The structural integrity of tissue surfaces poses a barrier to penetration by microorganisms (Section 25.2). The tight junctions between epithelial cells in all body tissues inhibit invasion and infection (Figure 26.2). In the skin and mucosal tissues, potential pathogens must first adhere to tissue surfaces and then grow at these sites before traveling elsewhere in the body. Mucosal surfaces are coated with a layer of protective mucus that traps microorganisms, pollen, and other foreign agents. Epithelial

TABLE 26.1 Tissue specificity in infectious disease

Disease	Tissue infected	Pathogen
Acquired immunodeficiency syndrome (AIDS)	T-helper lymphocytes	Human immunodeficiency virus (HIV)
Botulism	Motor end plate	*Clostridium botulinum*
Cholera	Small intestine epithelium	*Vibrio cholerae*
Dental caries	Oral epithelium	*Streptococcus mutans, S. sobrinus, S. mitis*
Diphtheria	Throat epithelium	*Corynebacterium diphtheriae*
Gonorrhea	Mucosal epithelium	*Neisseria gonorrhoeae*
Influenza	Respiratory epithelium	Influenza A and influenza B virus
Malaria	Blood (erythrocytes)	*Plasmodium* spp.
Pyelonephritis	Kidney medulla	*Proteus* spp.
Spontaneous abortion (cattle)	Placenta	*Brucella abortus*
Tetanus	Inhibitory interneuron	*Clostridium tetani*

cells underlying the mucus layer have cilia on their surfaces that carry out coordinated movements to expel suspended pathogens and keep them from adhering to tissues. This combination of mucus and ciliary motion plays a key role in removing inhaled microorganisms from the mucosal surfaces of the trachea and bronchial tubes, thus preventing microbial colonization of the lungs.

Sebaceous glands in the skin secrete fatty acids and lactic acid, lowering the acidity of the skin to pH 5 and inhibiting colonization by many pathogenic bacteria (blood and most internal organs are about pH 7.4). Potential pathogens ingested in food or water must survive the strong acidity (pH 2) and digestive enzymes, such as pepsin, in the stomach (Figure 26.2). Although the pH returns to neutral in the lower intestinal tract, pathogens that survive passage through the stomach must then compete with the abundant resident microbiota present in the small and large intestines (ᗒᗕ Figure 24.3). In addition to the established normal microbiota that exist at many potential infection sites, resistance to infection and invasion is enhanced by antibacterial substances called *defensins*, produced in the skin, lungs, and gut. Finally, the lumen of the kidney, the eye, the respiratory system, and the cervical mucosa are constantly bathed with tears, mucus, or other secretions that contain *lysozyme*, an enzyme that can kill bacteria by digesting the cell wall.

We have seen how the human body is naturally protected by a host of physical barriers, chemicals, and secretions, all of which combine to naturally suppress pathogen invasion and infection. Beyond this first line of defense, however, the immune system becomes activated, beginning with innate responses. We consider the tools of the innate immune response—the cells and organs of the immune system—now.

MINIQUIZ

- Describe host tissue specificity for pathogens.
- Identify physical and chemical barriers to pathogens. How might these barriers be compromised?
- What other factors may control the outcome of an infectious disease?

II • Cells and Organs of the Immune System

The immune response results from the activities of a wide variety of specialized cells that circulate throughout the body, primarily through the blood and *lymph*, a fluid similar to blood but which lacks red blood cells. Blood and lymph interact directly or indirectly with every major organ system.

26.3 The Blood and Lymphatic Systems

Circulatory systems in the human body include separate vasculature for blood and lymph. It is through circulation of these fluids that cells and proteins of the immune system are transported to the various tissues and organs of the body. All cells found in the blood and lymph are derived from **stem cells** in the bone marrow. These cells are continuously dividing and differentiating to supply the body with both erythrocytes (red blood cells) and **leukocytes** (white blood cells).

Blood and Lymph Circulation

Blood is pumped by the heart through arteries and capillaries throughout the body and is returned through the veins (**Figure 26.3a, b**). In the capillary beds, leukocytes and solutes pass to and from the blood into the lymphatic system. Lymph drains from extravascular tissues into lymphatic capillaries (lymph ducts) and then into **lymph nodes** throughout the lymphatic system (Figure 26.3c, d). Lymph nodes contain lymphocytes and phagocytes arranged to encounter microorganisms and antigens as they travel through the lymphatic circulation. **Mucosa-associated lymphoid tissue (MALT)** is another part of the lymphatic system that interacts with antigens and microorganisms that enter the body through mucous membranes, including those of the gut, the genitourinary tract, and bronchial tissues. MALT also contains phagocytes and lymphocytes. Lymph fluid with antibodies and immune cells empties into the blood circulatory system via the thoracic lymph duct (Figure 26.3a).

The *spleen* is an important organ in immunity and consists of *red pulp*—rich in red blood cells—and *white pulp*, which consists of organized lymphocytes and phagocytes arranged to filter blood in a manner similar to lymph nodes and MALT in the lymphatic system. Collectively, the lymph nodes, MALT, and spleen are called **secondary lymphoid organs** (Figure 26.3). The secondary lymphoid organs are the sites where antigens interact with antigen-presenting phagocytes and lymphocytes to trigger the adaptive immune response.

Hematopoietic Stem Cells, Blood, and Lymph

Hematopoietic stem cells are precursor cells found in bone marrow that can differentiate into any blood cell (see Figure 26.4). Stem cells grow in the bone marrow where they differentiate into a variety of cell types under the influence of soluble cytokines and chemokines, proteins that influence many aspects of immune cell differentiation (Section 26.5). The differentiated cells then travel through the blood and lymph to other parts of the body.

Blood consists of cellular and noncellular components, including many cells and molecules active in the immune response (Table 26.2). The most numerous cells in human blood are *erythrocytes*, small, nonnucleated cells that carry oxygen from the lungs to the tissues. About 0.1% of the cells in blood, however, are white blood cells—leukocytes—of which there are many different types (Table 26.2). All phagocytes of the innate immune system are leukocytes, as are lymphocytes, the cells active in the adaptive response.

Blood is composed of suspended cells and **plasma**, a liquid that contains proteins and other solutes. Outside the body, blood or plasma quickly forms an insoluble fibrin clot, remaining liquid only when an anticoagulant such as potassium citrate or heparin is added. When blood clots, the insoluble proteins trap the cells in a large, insoluble mass. The remaining fluid, called **serum**, lacks both cells and clotting proteins. Serum does, however, contain a

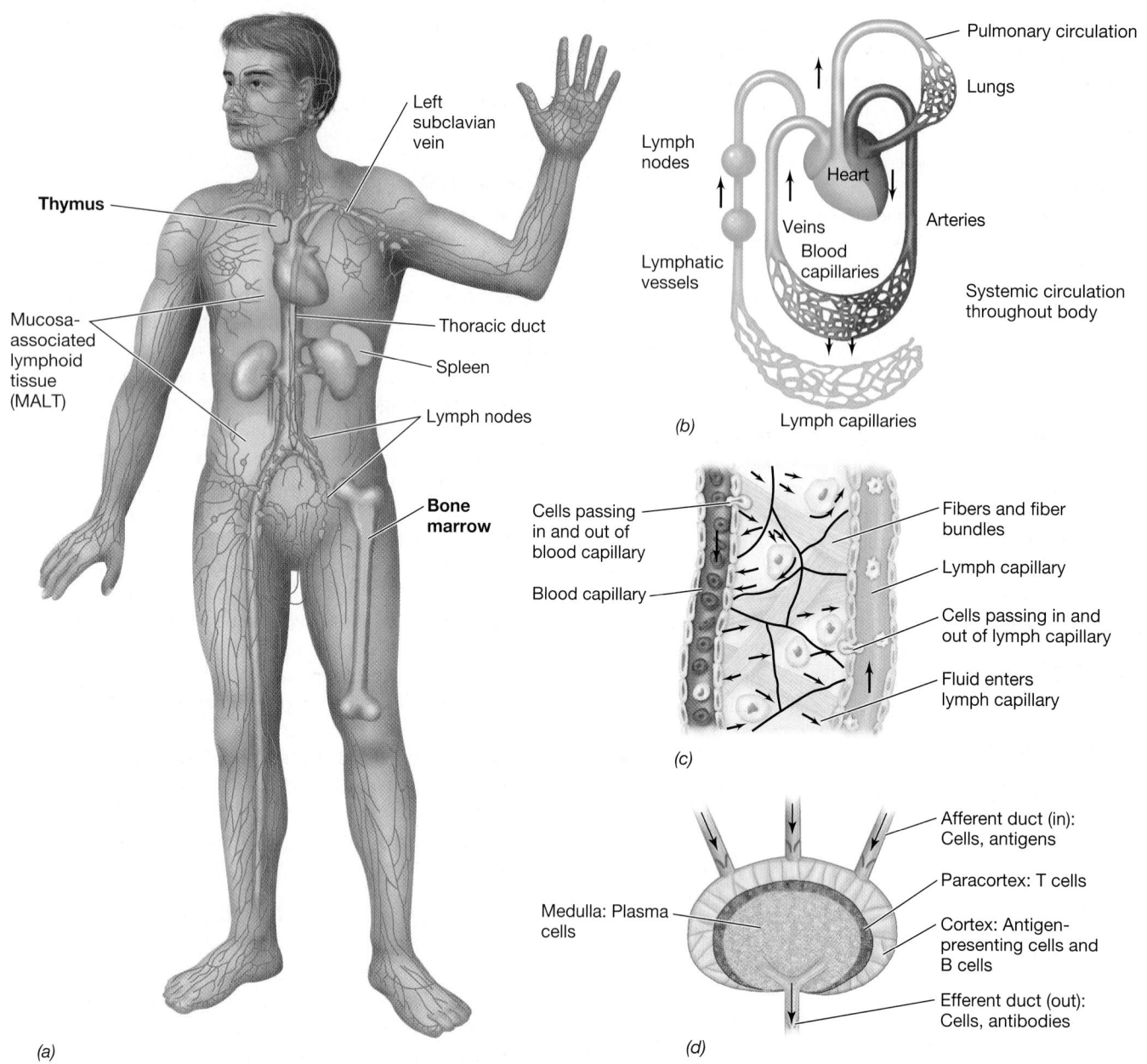

Figure 26.3 The blood and lymphatic systems. (a) Blood and lymph circulation in the body. Major blood vessels and associated organs are shown in red. Major lymphatic organs and vessels are shown in green. The primary lymphoid organs are the bone marrow and thymus. The secondary lymphoid organs are the lymph nodes, spleen, and MALT. (b) Connections between the lymphatic and blood systems. Blood flows from the veins to the heart, to the lungs, and then through the arteries to capillaries in tissues. Exchange of solutes and cells occurs between blood and lymphatic capillaries. Lymph drains from the thoracic duct into the left subclavian vein of the blood circulatory system. (c) The exchange of cells between the blood and lymphatic systems is shown in detail. Both blood and lymphatic capillaries are closed vessels, but cells pass from blood capillaries to lymphatic capillaries and back by the process of extravasation. (d) A secondary lymphoid organ, the lymph node, showing the major anatomical areas and the immune cells concentrated in each area. The anatomy of the MALT and the spleen is analogous to that of the lymph nodes.

high concentration of other proteins, in particular antibodies, key players in the adaptive immune response.

MINIQUIZ

- Describe the circulation of a leukocyte from the blood to the lymph and back to the blood.
- What soluble molecules determine whether a particular stem cell will become a phagocyte, lymphocyte, or erythrocyte?

26.4 Leukocyte Production and Diversity

Several distinct leukocytes participate in innate and adaptive immunity (Table 26.2). From its hematopoietic stem cell precursor, a leukocyte will differentiate and mature through either the *myeloid* lineage or the *lymphoid* lineage (**Figure 26.4**). Leukocytes move throughout the body and, through a process called *extravasation* (also called *diapedesis*), pass from blood to interstitial spaces. They are then collected with lymph into lymphatic

TABLE 26.2 Major cell types found in normal human blood

Cell type	Cells per milliliter
Erythrocytes	$4.2–6.2 \times 10^9$
Leukocytes[a]	$4.5–11 \times 10^6$
Lymphocytes	$1.0–4.8 \times 10^6$
Granulocytes and monocytes	Up to 7.0×10^6

(a) **Red blood cells (erythrocytes)**

(b) **Lymphocyte**

(c) **Neutrophil (a granulocyte)**

(d) **Monocyte**

[a]Leukocytes include all nucleated blood cells. They include lymphocytes and cells derived from myeloid stem cells, the monocytes and granulocytes, such as neutrophils.

vessels, where they are eventually returned to the blood circulatory system (Figure 26.3c).

Myeloid Cells—Monocytes and Granulocytes

Myeloid cells, active in innate immunity, are derived from myeloid precursor cells. Mature myeloid cells develop from one of two lineages: **monocytes** or **granulocytes** (Figure 26.4 and Table 26.2). Immature monocyte precursor cells circulate in the blood for several days before moving into other tissues and differentiating into specialized phagocytic cells called **antigen-presenting cells (APCs)**. These cells engulf, process, and present antigens to lymphocytes to initiate an adaptive immune response (Chapter 27). Phagocytic APCs derived from monocytes include **macrophages** and **dendritic cells** (Figure 26.4).

Macrophages are often the first defensive cells that interact with a pathogen and are abundant in many tissues and organs, especially the spleen, lymph nodes, and MALT, where they constitute up to 15% of total cells. Because they ingest and destroy most pathogens and foreign molecules that invade the body, macrophages are essential to the innate response. Dendritic cells are found throughout the body tissues and are especially abundant along epithelial linings, including the skin and mucous membranes. When dendritic cells ingest antigen, they migrate to nearby lymph nodes, where they are extremely efficient at presenting antigen to T lymphocytes. Thus, dendritic cells are an important

link between innate and adaptive immunity. The specialized antigen-presenting properties of macrophages and dendritic cells will be examined in Section 27.5.

Granulocytes are the second lineage of cells derived from myeloid precursors. Granulocytes contain cytoplasmic inclusions, or granules, that are visible in stained microscopic preparations. These granules contain toxins or enzymes that are released to destroy target cells. One granulocyte, the **neutrophil**, also called a *polymorphonuclear leukocyte (PMN)*, is an abundant, highly motile phagocyte that responds rapidly to pathogen challenge, an activity that is central to innate immunity. Neutrophils are found predominantly in the bloodstream and bone marrow, from where they migrate to sites of active infection. A second type of granulocyte, called a **mast cell**, functions to initiate an inflammatory response by releasing its granule contents, a process called *degranulation*. Mast cell degranulation is responsible for certain types of allergic reactions (Section 27.9).

Lymphocytes

Lymphocytes are specialized leukocytes derived from lymphoid precursor cells (Figure 26.4 and Table 26.2). There are three types of lymphocytes: *B cells* (B lymphocytes), *T cells* (T lymphocytes), and *natural killer cells* (Figure 26.4). Mature B and T cells circulate through the blood and lymph system but are concentrated in the lymph nodes and spleen where they interact with antigens.

B cells originate and mature in the *bone marrow*. Like dendritic cells and macrophages, B cells are specialized APCs, but in addition to this, they are the precursors of antibody-producing **plasma cells**. Antibodies (immunoglobulins) are soluble proteins that interact with specific antigens and are produced by B cells and plasma cells. **T cells**, which interact with antigens presented by APCs, begin their development in the bone marrow but migrate to the *thymus* to mature (Figure 26.3a and see Figure 27.2). In mammals, the bone marrow and thymus are called **primary lymphoid organs** because they are the sites where lymphoid stem cells mature into functional, antigen-reactive lymphocytes (Figure 26.3a). We detail the role of B and T lymphocytes in the adaptive immune response in Chapter 27.

Like B and T cells, natural killer (NK) cells (Figure 26.4) are derived from lymphoid precursors. However, unlike B and T lymphocytes, NK cells function primarily in innate immunity. NK cells rid the body of virus-infected and tumor cells using a mechanism that distinguishes healthy versus compromised cells based on the presence (or absence) of specific surface molecules (Section 26.10).

UNIT 6

--- **MINIQUIZ** ---

- How does the development of B, T, and NK cells differ from that of dendritic cells and macrophages?
- Distinguish between the primary lymphoid organs and the secondary lymphoid organs. What is the relationship between the components of these two groups?

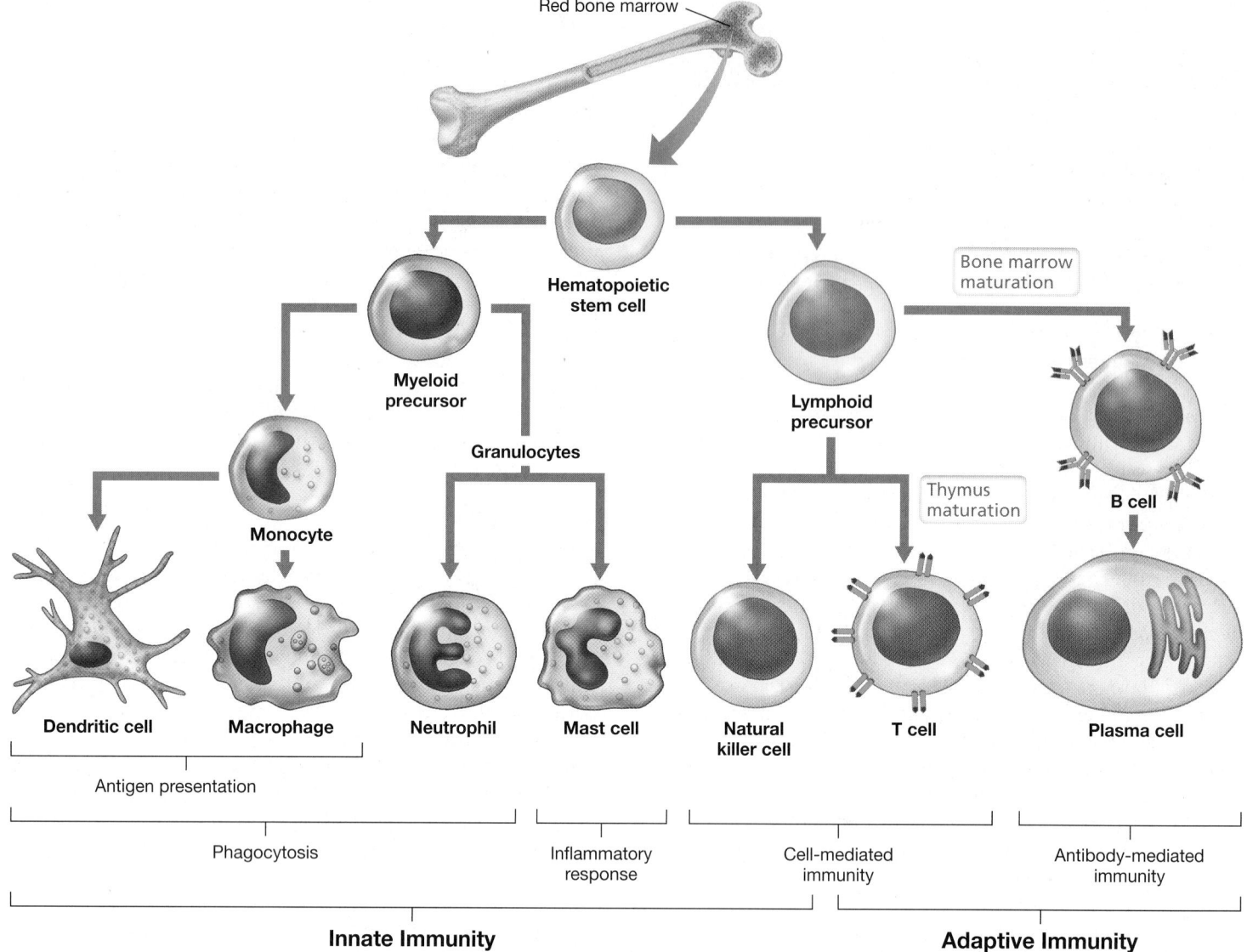

Figure 26.4 Lineage and diversity of immune response cells. Immune cells develop from hematopoietic stem cells in the bone marrow into either myeloid precursors or lymphoid precursors. These precursors, in turn, differentiate into mature cells that have various immune functions. (Erythrocytes also develop from myeloid precursors.)

III • Phagocyte Response Mechanisms

Innate immunity is primarily driven by the activities of phagocytes (Section 26.1). These cells recognize common structural features found on and in pathogens. Interactions with pathogens activate genes in the phagocytes that control the transcription, translation, and expression of proteins that destroy the pathogens.

Innate immunity develops immediately when a phagocyte contacts a pathogen, but it is not always effective enough to prevent dangerous infections. To counter these, certain phagocytes also activate adaptive immunity by processing and presenting antigens to receptors on lymphocytes (Chapter 27). However, because an adaptive response to a specific pathogen takes several days to develop, the ability of phagocytes to quickly respond to microbial invasion is indispensible for controlling infection.

26.5 Pathogen Challenge and Phagocyte Recruitment

Successful pathogens can breach physical and chemical host barriers (Figure 26.2), leading to host infection. When this happens, the immune system is mobilized to protect the host from further damage. Innate immunity is the first line of defense and is critical for host protection for about four days after an infection begins. Phagocytes engulf and destroy pathogens, often initiating complex host-mediated inflammatory reactions when they do (Section 26.8).

Microbial Invasion

The initial inoculum of a pathogen is usually insufficient to cause host damage, even if a pathogen gains access to tissues. For the pathogen to be successful, it must attach, multiply, and colonize the tissue (Figure 25.1), and these events require that the pathogen find appropriate nutrients and environmental conditions for growth.

Following colonization, a pathogen must usually invade tissues to initiate disease. **Invasion** is the ability of a pathogen to enter into host cells or tissues, multiply, spread, and cause disease. In most cases, microbial infections begin at breaks or wounds in the skin or on the mucous membranes of the respiratory, digestive, or genitourinary tract, surfaces that are normally microbial barriers (Figure 26.2). In some cases, growth may also begin on intact mucosal surfaces, especially if the composition of the normal microbiota has been altered or eliminated, for example by antibiotic therapy. Tissue damage caused by invasive pathogens or injury triggers the recruitment of large numbers of phagocytes and other immune cells to the site of infection.

Tissue Damage and Chemokine Release

When microbial invasion causes trauma to host tissues, resident phagocytes and damaged host cells release **cytokines**, a family of soluble proteins that function as chemical mediators to allow communication between different cells of the body (**Figure 26.5a**). Cytokines bind specific receptors on cells and induce a particular response from them, usually by activating a signaling pathway that controls transcription and protein synthesis. **Chemokines** are an important subclass of cytokines that have the specific role of recruiting immune cells to sites of injury.

Resident macrophages, which are found in all of the body's organ systems, are stimulated by the presence of invading pathogens to secrete chemokines that establish a gradient of *chemoattractants* (Figure 26.5b). These molecules bind receptors on other immune cells, especially neutrophils and T cells, and recruit them to the site of infection. This triggers both a neutrophil-mediated inflammatory response and, later, an adaptive immune response facilitated by lymphocytes against the specific invading agent or its toxic products.

Localized inflammation allows neutrophils, the most numerous circulating phagocytes, to migrate quickly along the chemotactic gradient to the site of infection (Figure 26.5b). Once there, chemokine-mediated binding reactions promote neutrophil adhesion to the inner wall of the blood capillary, a process called *margination*. The halted neutrophils then undergo extravasation, the movement of leukocytes from the bloodstream to surrounding infected tissues by squeezing between endothelial cells lining the capillaries (Figure 26.5b). Higher than normal numbers of neutrophils in the blood or at a site of inflammation indicate an active response to a current infection.

Neutrophils and other phagocytes that encounter pathogens in damaged areas must be able to recognize, capture, and destroy pathogens to clear infections and restore body tissues to a healthy state (Figure 26.5c). We move on to explore the molecular mechanisms that facilitate these cellular interactions in the next section.

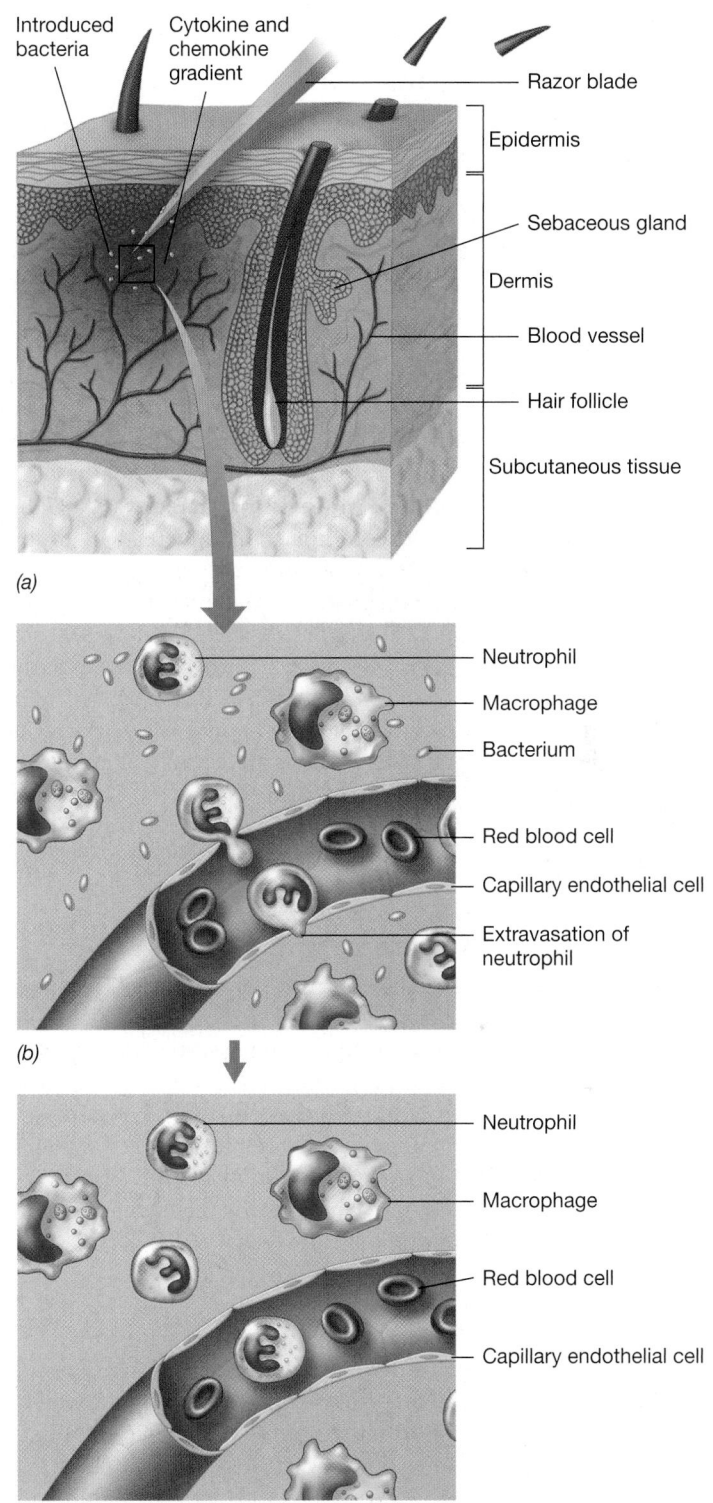

(a)

(b)

(c)

Figure 26.5 Microbial invasion and the innate immune response. (a) Tissue damage, such as that caused by a cut from a razor, can lead to invasion by microorganisms and the release of cytokines and chemokines from damaged cells. (b) Phagocytes are recruited to the site of infection by the chemokine gradient, squeezing out of dilated capillaries via extravasation (diapedesis). (c) Invading microorganisms are cleared by phagocytosis, and the tissue is restored to health. See Figure 26.4 for functional descriptions of some of the cells and components shown here.

--- **MINIQUIZ** ---

- Although technically not part of the immune system, nonpathogenic normal microbiota play a major role in preventing disease. Describe this role.

- Describe the mechanisms by which circulating phagocytic cells are recruited to a site of infection.

UNIT 6

26.6 Pathogen Recognition and Phagocyte Signal Transduction

The first type of immune cell to be activated in the innate response is typically a phagocyte, whose primary function is to engulf and destroy pathogens. Phagocytes include neutrophils and monocytes, which are found primarily in the blood, and macrophages and dendritic cells, which occur primarily in the body tissues (Figure 26.4). But how do these cells distinguish pathogens and other foreign agents from the myriad other cells in the body? Moreover, once invaders are recognized, how do phagocytes capture and destroy them? We address these topics now.

Pathogen-Associated Molecular Patterns

The macromolecules inside and on the surface of pathogens display **pathogen-associated molecular patterns (PAMPs)**, repeating structural subunits common to broadly related groups of infectious agents. PAMPs may include polysaccharides, proteins, nucleic acids, or even lipids. For example, a common PAMP is the lipopolysaccharide (LPS) common to all gram-negative bacterial outer membranes (Section 2.5). Other PAMPs include bacterial flagellin, the double-stranded RNA (dsRNA) of certain viruses, and the lipoteichoic acids of gram-positive bacteria (Section 2.4). Like all PAMPs, these molecules are found on various pathogens but are absent from host cells. Therefore, PAMPs serve as markers by which phagocytes can identify pathogenic microbes, even if the pathogens have not been encountered previously.

Pattern Recognition Receptors

Phagocytes have a pathogen-recognition system that triggers a timely and appropriate response, leading to recognition, containment, and destruction of pathogens. Phagocytes can interact quickly and effectively with pathogens because phagocyte surfaces contain numerous **pattern recognition receptors (PRRs)**, soluble or membrane-bound proteins that recognize and bind PAMPs (**Figure 26.6a**). The interaction of a PAMP with a PRR activates the phagocyte to ingest and destroy the targeted pathogen by phagocytosis (Figure 26.6b and Section 26.7).

PRRs were first observed in phagocytes of *Drosophila*, the fruit fly, where they are called *Toll receptors* (see Explore the Microbial World, "*Drosophila* Toll Receptors—An Ancient Response to Infections"). Structural, functional, and evolutionary homologs of the Toll receptors, called **Toll-like receptors (TLRs)**, are widely expressed on mammalian innate immune cells. TLRs are found associated with membranes on the surface of or in intracellular vesicles of all types of phagocytes. At least nine TLRs in humans interact with a variety of cell surface and soluble PAMPs from viruses, bacteria, and fungi (**Table 26.3**).

Each TLR on a human phagocyte recognizes and interacts with a specific PAMP. For example, TLR-2, a PRR on human phagocytes, interacts with peptidoglycan, a PAMP present in the cell wall of nearly all bacteria (**Figure 26.7**). This PAMP–PRR interaction activates the phagocytes, which then target gram-positive pathogens with exposed peptidoglycan. Access to the peptidoglycan of gram-negative bacterial cell walls is blocked by the surface lipopolysaccharides. However, another PRR found on phagocytes, TLR-4,

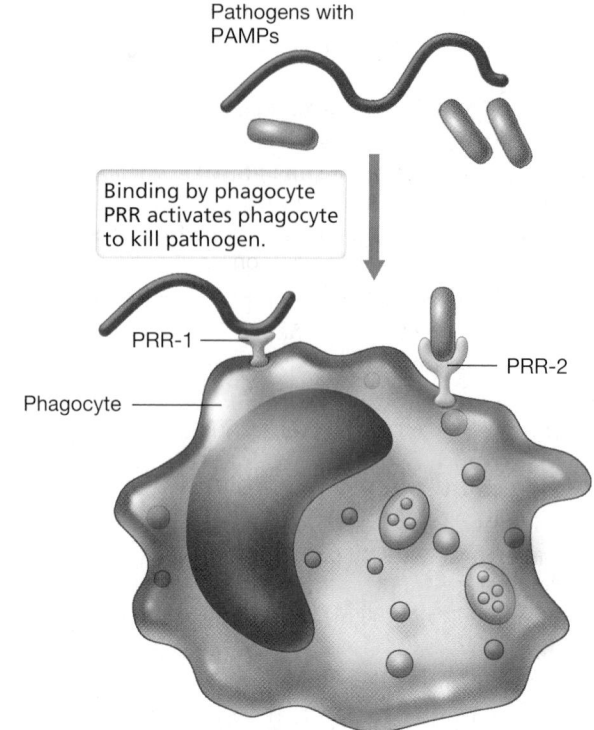

(a)

(b)

CDC/PHIL/NIAID

Figure 26.6 Pathogen recognition by phagocytes. (a) Phagocytes interact with pathogens by way of preformed pattern recognition receptors (PRRs) that bind to pathogen-associated molecular patterns (PAMPs). Binding of a PAMP by a phagocyte PRR stimulates the phagocyte to destroy the pathogen and activate other phagocytes. (b) Colorized scanning electron micrograph of a neutrophil (blue) phagocytosing several cells of methicillin-resistant *Staphylococcus aureus* (MRSA; pink). For more on MRSA, see Sections 28.12 and 30.9.

interacts with the endotoxic LPS of gram-negative bacteria, including that from all pathogenic strains of *Salmonella* spp., *Escherichia coli*, and *Shigella* spp. (Table 26.3 and see Figure 26.8).

Several soluble host molecules function similarly to these phagocyte-associated PRRs. The *NOD-like receptors* (*NLRs*) are a family of soluble PRRs found in the cytoplasm that have a nucleotide-binding domain (NOD). NOD1 and NOD2 interact

DROSOPHILA TOLL RECEPTORS—AN ANCIENT RESPONSE TO INFECTIONS

Invertebrates and plants lack adaptive immunity but have a well-developed innate immune response to a wide variety of pathogens. Virtually all multicellular organisms respond by recognizing molecules found on the pathogenic cell or virus. These molecules contain conserved, repetitive structures called pathogen-associated molecular patterns (PAMPs) that include such things as the lipopolysaccharide (LPS) and flagellin of gram-negative bacteria, the peptidoglycan of gram-positive bacteria, and the double-stranded RNA unique to certain viruses, among others. By recognizing features found in many pathogens, the innate immune mechanism provides protection against a broad range of common pathogenic agents.

Responses to pathogens by the fruit fly, *Drosophila melanogaster* (**Figure 1**), have provided insight into innate immune mechanisms in many other groups of organisms. Several proteins required for fruit fly development are also important receptors for recognizing invading bacteria. These proteins function in their immune response role as pattern recognition receptors (PRRs) that interact with PAMPs on the macromolecules produced by the pathogen. The best example of a PRR is *Drosophila* Toll, a transmembrane protein essential for dorsoventral axis formation, as well as the innate immune response of the fly.

Toll immune signaling is initiated by the interaction of a pathogen or its components with the Toll protein displayed on the surface of phagocytes. *Drosophila* Toll, however, does not interact directly with the pathogen. Signal transduction events start with the binding of a PAMP, such as the LPS of gram-negative bacteria, by one or more accessory proteins (Figure 26.8 shows the analogous TLR-4 system in humans). The LPS–accessory protein complex then binds to Toll. The membrane-integrated Toll protein initiates a signal transduction cascade, activating a nuclear transcription factor and inducing transcription of several genes that encode antimicrobial peptides. Toll-associated transcription factors induce expression of antimicrobial peptides including drosomycin, active against fungi; diptericin, active against gram-negative bacteria; and defensin, active against gram-positive bacteria. These peptides, produced in the liver-like fat body of *Drosophila*, are released into the fly's circulatory system where they interact with the target organism and cause cell lysis.

Structurally, the Toll proteins are related to lectins, a group of proteins found in all multicellular organisms, including invertebrates and plants. Lectins interact specifically with certain oligosaccharide monomers. In humans, Toll-like receptors (TLRs) react with a wide variety of PAMPs. As with *Drosophila* Toll, human TLR-4 provides innate immunity against gram-negative bacteria through indirect interactions with LPS, initiating a kinase signal cascade and activating nuclear transcription factor NFκB that activates transcription of cytokines and other phagocyte proteins key to the host's innate immune response (Figure 26.8).

Jarmo Holopainen

Figure 1 ***Drosophila melanogaster*, the common fruit fly.** The Toll protein, a homolog of the Toll-like receptors of higher vertebrates, was first discovered in the fruit fly.

Drosophila Toll is an evolutionary, structural, and functional relative of the Toll-like receptors present in vertebrates, including humans. Toll and its homologs are evolutionarily ancient, highly conserved components of the innate immune system in animals and have even been found in plants. The absence of a highly specific adaptive immune response in invertebrates points to this more specific system as appearing later in the course of evolution, possibly as a mechanism to counter disease threats that could not be controlled by the innate response alone.

---

with peptidoglycan components of gram-negative and gram-positive bacterial cell walls, respectively, stimulating production of antimicrobial peptides and inflammatory cytokines (Table 26.3). NOD-like receptor pyrin 3 (NLRP3) interacts with other proteins to form a structure called an *inflammasome*. The cytoplasmic inflammasome senses cellular stress indicators, such as the loss of potassium ions (K^+) from damaged cells, and triggers the production of proinflammatory cytokines, initiating inflammation. Later in this chapter we discuss the soluble PRRs in the context of their ability to activate proteins that enhance phagocytosis and destruction of pathogens (Section 26.9).

Signal Transduction in Phagocytes

Interaction of a PAMP with a PRR triggers transmembrane *signal transduction*, initiating gene transcription and translation of host-response proteins in a fashion similar to the two-component regulatory systems previously discussed in *Bacteria* and *Archaea* (⟳ Section 6.6). Signal transduction initiated by PAMP–PRR interaction results in enhanced phagocytosis, killing of pathogens, inflammation, and tissue healing.

For example, the binding of LPS (a PAMP derived from the cell wall of degraded gram-negative bacteria) to TLR-4 (a PRR on the surface of phagocytes) typically activates a signal transduction pathway (**Figure 26.8**). TLR-4 then binds proteins in the cytosol of the phagocyte, starting a cascade of reactions that activates transcription factors such as NFκB (nuclear factor kappa B), a protein that binds to specific regulatory sites on DNA, initiating transcription of downstream genes. Many of the NFκB-regulated genes encode host-response proteins, such as the cytokines that activate cells and initiate inflammation (Section 26.5).

TABLE 26.3 Receptors and targets in the innate immune response

Pattern recognition receptors (PRRs)	Pathogen-associated molecular patterns (PAMPs) and targets	Result of interaction
Soluble extracellular PRRs[a]		
Mannose-binding lectin (soluble)	Mannose-containing components of microbial cell surface, as in gram-negative bacteria	Complement activation
C-reactive protein (soluble)	Components of gram-positive cell walls	
Plasma membrane–associated PRRs		
TLR-1 (Toll-like receptor 1)	Lipoproteins in mycobacteria	Signal transduction, phagocyte activation, and inflammation[c]
TLR-2	Peptidoglycan on gram-positive bacteria; zymosan in fungi	
TLR-4	LPS (lipopolysaccharide) in gram-negative bacteria	
TLR-5	Flagellin in bacteria	
TLR-6	Lipoproteins in mycobacteria; zymosan in fungi	
Endosomal membrane–associated PRRs[b]		
TLR-3	Double-stranded viral RNA	Signal transduction, phagocyte activation, and inflammation
TLR-7, TLR-8	Single-stranded viral RNA	
TLR-9	Unmethylated CpG oligonucleotides in bacteria	
Cytoplasmic PRRs: NLRs (NOD-like receptors)		
NOD1	Peptidoglycan on gram-negative bacteria	Stimulate production of antimicrobial peptides and proinflammatory cytokines
NOD2	Peptidoglycan on gram-positive bacteria	
NLRP3	Inflammasome component	Triggers release of proinflammatory cytokines, increasing inflammation

[a]The extracellular soluble PRRs are produced by liver cells in response to inflammatory cytokines.
[b]TLR-3, -7, -8, and -9 are found in intracellular organelle membranes such as in lysosomes. A 10th Toll-like receptor, TLR-10, has unknown ligand specificity and function.
[c]Toll-like receptors initiate phagocyte activation via signal transduction.

TLR-4 is an integral protein having external, transmembrane, and cytoplasmic domains. The external domain of TLR-4 binds LPS complexed with a cell surface protein called CD14 (Figure 26.8). Binding of the CD14–LPS complex by the TLR-4 external domain causes a conformational change in TLR-4 that allows the cytoplasmic domain to interact with an adaptor protein called MyD88, which then binds a protein tyrosine kinase (PTK) called IRAK4. PTKs transfer energy-rich phosphates from ATP to target-protein tyrosines that are exposed when binding alters conformation. Binding of MyD88 by IRAK4 initiates a *kinase cascade* that triggers successive ATP-mediated phosphorylation of TRAF6, IκK (inhibitor of kappa kinase), and IκB (inhibitor of kappa B) proteins. Phosphorylation of IκB causes it to dissociate from, and thereby activate, NFκB.

Activated NFκB can then diffuse across the nuclear membrane, bind to NFκB-binding motifs on DNA, and initiate transcription of downstream genes.

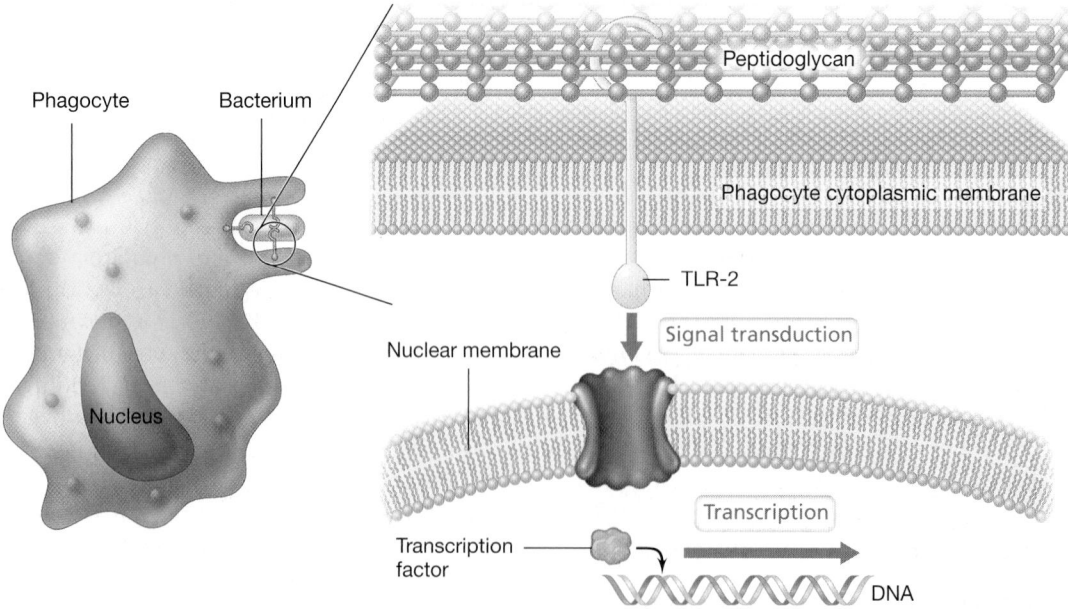

Figure 26.7 A Toll-like receptor. Membrane-spanning TLR-2 interacts with peptidoglycan from gram-positive pathogens. This interaction stimulates signal transduction, activating transcription factors in the nucleus. The result is transcription of genes encoding proteins that induce inflammation and other phagocyte activities. All Toll-like receptors have analogous mechanisms for activating innate immunity.

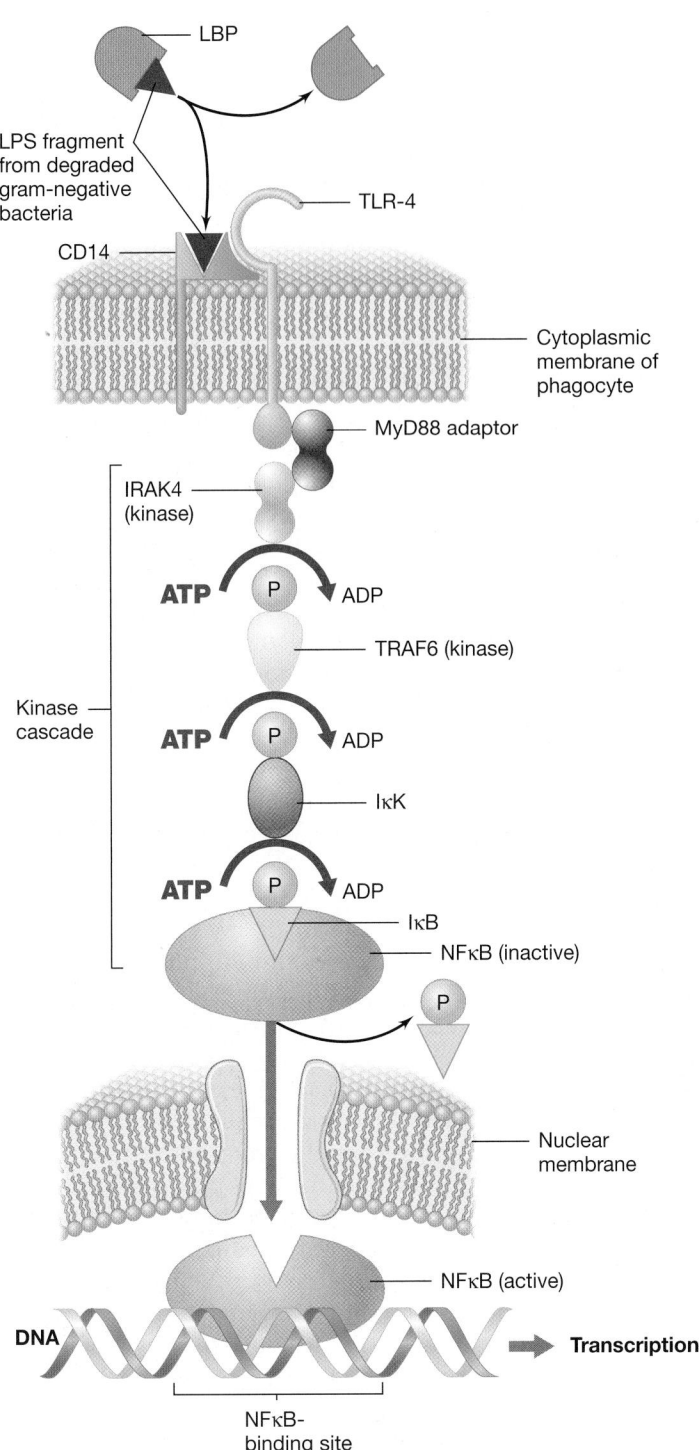

Figure 26.8 Signal transduction in innate immunity. Signal transduction is the transmission of a molecular signal across a cytoplasmic membrane by way of chemical modifications—typically involving a series of proteins in a signaling cascade—that results in a response, such as differential gene expression. In innate immunity, signal transduction is initiated when LPS, a PAMP, is bound by LBP (lipopolysaccharide-binding protein), which then transfers LPS to CD14 on the surface of a phagocyte. The LPS–CD14 complex then binds to the transmembrane TLR-4 receptor. The binding of TLR-4 initiates a series of reactions involving adaptor proteins and kinases, resulting in activation of the transcription factor NFκB. NFκB then diffuses across the nuclear membrane, binds to DNA, and initiates transcription of genes encoding proteins essential for innate immunity.

As Figure 26.8 shows, signal transduction pathways initiate activation of transcription through ligand–receptor binding on the surface of the phagocytic cell. The ligand–receptor interaction outside the cell induces the binding, recruitment, and concentration of the adaptor proteins and kinase enzymes inside the cell. A single kinase can phosphorylate many signal cascade proteins, thus amplifying the effect of a single ligand–receptor interaction. Signal transduction leading to activation of shared transcription factors and protein synthesis is also the mechanism by which lymphocytes are activated in adaptive immunity, as we discuss in Chapter 27.

> ### MINIQUIZ
>
> - Identify a PAMP shared by a group of microorganisms. Then, identify the cell types that use PRRs to provide innate immunity to pathogens.
> - Outline the general features of a signal transduction pathway starting with binding of a PAMP by a membrane-associated PRR.

26.7 Phagocytosis and Phagocyte Inhibition

When phagocytes encounter infectious agents or their harmful products, the activation of signal transduction pathways (Figure 26.8) triggers genetic responses in the phagocyte that direct the containment and removal of the threat—**phagocytosis** (**Figure 26.9**). While these mechanisms effectively protect the body from most infectious exposures, they are not foolproof; many pathogens deploy effective defenses against phagocytes in attempts to thwart the innate immune response.

Phagocytosis and the Phagolysosome

Most phagocytes contain multiple membrane-bound inclusions called *lysosomes*, cytoplasmic vacuoles containing bactericidal substances, such as toxic oxygen compounds, lysozyme, proteases, phosphatases, nucleases, and lipases. Through the molecular mechanisms we have just discussed, phagocytes identify and engage pathogens on surfaces, such as blood vessel walls or fibrin clots, before initiating phagocytosis (Figure 26.9). Activation of the phagocyte through signal transduction causes the phagocyte membrane to envelop and engulf pathogens, eventually pinching off inwardly to form a *phagosome*. The **phagosome**, a vacuole containing the engulfed pathogen, moves into the cytoplasm and fuses with a lysosome to form a *phagolysosome* (**Figure 26.10**). The toxic chemicals and enzymes within the phagolysosome combine to kill and digest the engulfed microbial cell.

Genes that control the production of oxygen compounds toxic to pathogens are highly transcribed in activated phagocytes. These toxic compounds include hydrogen peroxide (H_2O_2), superoxide anions (O_2^-), hydroxyl radicals ($OH\cdot$), singlet oxygen ($^1O_2$), hypochlorous acid ($HOCl$), and nitric oxide (NO) (Figure 26.10) ($\hookleftarrow$ Section 5.14). Phagocytic cells use these toxic oxygen compounds to kill ingested bacterial cells by

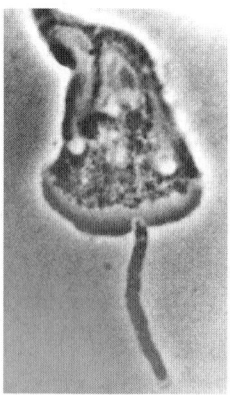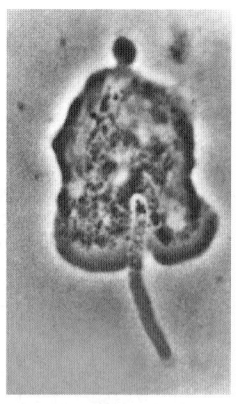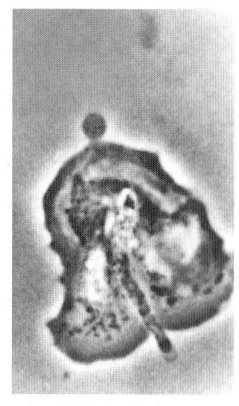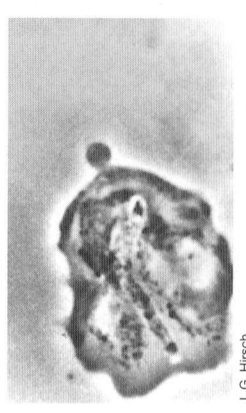

J. G. Hirsch

Figure 26.9 Phagocytosis. Time-lapse phase-contrast micrographs of the phagocytosis and digestion of a chain of *Bacillus megaterium* cells by a human macrophage. The bacterial chain is about 20 μm long.

oxidizing key cellular constituents. The lethal oxidative reactions are contained within the phagolysosome, and this prevents damage to the phagocyte itself.

Inhibiting Phagocytes

Some pathogens have mechanisms for neutralizing toxic phagocyte products, for killing the phagocytes, or for avoiding phagocytosis. For example, several species of *Mycobacterium* produce pigmented compounds called carotenoids that neutralize singlet oxygen and prevent pathogen killing. In addition, *Mycobacterium tuberculosis*, the bacterium that causes

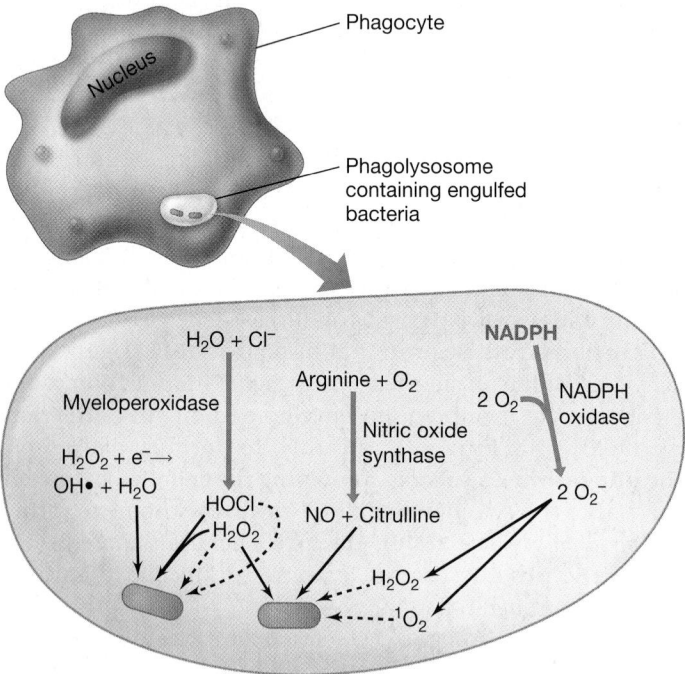

Figure 26.10 Activity of phagocyte enzymes in generating toxic oxygen compounds. These compounds include hydrogen peroxide (H_2O_2), the hydroxyl radical (OH·), hypochlorous acid (HOCl), the superoxide anion (O_2^-), singlet oxygen ($^1O_2$), and nitric oxide (NO). Formation of these toxic compounds requires a substantial increase in the uptake and utilization of molecular oxygen, O_2. This increase in oxygen uptake and consumption by activated phagocytes is called the *respiratory burst*.

tuberculosis, grows and persists within phagocytic cells (⇨ Section 30.4). Cells of *M. tuberculosis* use their cell wall glycolipids (⇨ Section 16.11) to absorb hydroxyl radicals and superoxide anions, the most lethal toxic oxygen species produced by phagocytes (Figure 26.10).

Some intracellular pathogens produce phagocyte-killing proteins called *leukocidins*. In such cases, the pathogen is ingested as usual, but the leukocidin kills the phagocyte, releasing the pathogen unharmed. Dead phagocytes make up much of the material of *pus*, and organisms such as *Streptococcus pyogenes* (scarlet and rheumatic fevers) and *Staphylococcus aureus* (skin infections) are major leukocidin producers and *pyogenic* (pus-forming) pathogens. Localized infections by pyogenic bacteria thus form boils or abscesses.

Another important pathogen defense against phagocytosis is the bacterial capsule (⇨ Section 2.7). Because the capsule prevents necessary molecular interactions between the surface of the phagocyte and that of the bacterial cell, encapsulated bacteria are often highly resistant to phagocytosis. For example, fewer than ten cells of an encapsulated strain of *Streptococcus pneumoniae* can kill a mouse within a few days (⇨ Figure 25.9), whereas strains lacking a capsule are harmless. Surface components other than capsules can also inhibit phagocytosis. For instance, pathogenic *S. pyogenes* produces M protein, a substance that alters the surface of the pathogen and inhibits phagocytosis.

Soluble PRRs and other host molecules such as antibodies (Chapter 27) can interact with capsules and other pathogen surface molecules, thereby reversing the protective effect of bacterial defense mechanisms and enhancing phagocytosis. As an example, the effective vaccine directed against *Streptococcus pneumoniae*, the agent of bacterial pneumonia, uses capsule polysaccharides to induce protective antibodies (⇨ Section 28.9). Thus, the battle between pathogen and the host innate immune system is a dynamic one, where both sides deploy multiple weapons in attempts to thwart the success of the other.

— **MINIQUIZ** —

- Identify the mechanism used by phagocytes to induce pathogen killing.

- Describe several reasons why phagocytes are not always effective at removing pathogens from the body.

IV • Other Innate Host Defenses

In addition to the physical and chemical barriers to pathogen invasion and the direct destruction of pathogens by activated phagocytes, mammalian immune systems have other inborn mechanisms that help counter infection by pathogens. Although unpleasant for the host, inflammation and fever can be effective host defense strategies for controlling microbial growth in the body and eliminating pathogens. These mechanisms, along with a consideration of the complement system and the activities of special lymphocytes called *natural killer cells*, round out our discussion of the innate immune response.

26.8 Inflammation and Fever

Inflammation is a nonspecific reaction to noxious stimuli, such as toxins and pathogens. Inflammation is characterized by redness (erythema), swelling (edema), pain, and heat, usually localized at the site of infection (**Figure 26.11** and ⟳ Figure 27.27). The mediators of inflammation are a group of cell activator and chemoattractant molecules, including cytokines and chemokines. Various cells, including those damaged by injury, produce these activators. The most important chemokines and cytokines are called *proinflammatory* because of their inflammation-inducing capacity, and they are produced in high concentrations by phagocytes and lymphocytes during pathogen challenge.

Both innate and adaptive immune responses to infection can cause inflammation, and both responses trigger the release of molecules that recruit and activate effector cells, such as neutrophils. Although it is a generalized immune response, inflammation plays a crucial role to isolate and limit tissue damage by initiating the destruction of pathogen invaders and the removal of damaged cells. As we will soon discuss, however, the inflammatory response may also inadvertently result in considerable damage to healthy host tissue.

Inflammatory Cells and Local Inflammation

Immune-mediated inflammation is an acute condition that begins at the site of pathogen entry into the body. The innate PRRs on macrophages and other tissue cells at the site of infection engage the pathogen PAMPs (Figure 26.6). This activates local cells to produce and release mediators including cytokines and chemokines that interact with receptors on other cells, such as neutrophils (Figure 26.4). For example, local tissue macrophages that are activated by PAMP–PRR interaction secrete a chemokine called CXCL8. This molecule activates neutrophils to migrate along the chemokine gradient toward the source of the CXCL8, where they begin to ingest and kill the pathogen. The neutrophils, in turn, secrete even more CXCL8, attracting more neutrophils and amplifying the response, eventually destroying the pathogens (Figure 26.11*b*).

The chemokine and cytokine mediators released by injured cells and phagocytes contribute to inflammation. For example, macrophages and other cells at the site of infection produce proinflammatory cytokines including interleukin-1 (IL-1), IL-6, and tumor necrosis factor α (TNF-α). These cytokines increase vascular permeability, causing the swelling (edema), reddening (erythema), and local heating associated with inflammation (**Figure 26.12a**). Although edema stimulates local neurons, causing pain, the pressure associated with swelling also serves to force fluids away from blood vessels and into the lymphatic system, simultaneously helping to strengthen the immune response and prevent the spread of pathogens to the bloodstream. This condition, called *bacteremia*, could trigger the much more serious *septicemia* (⟳ Section 25.2).

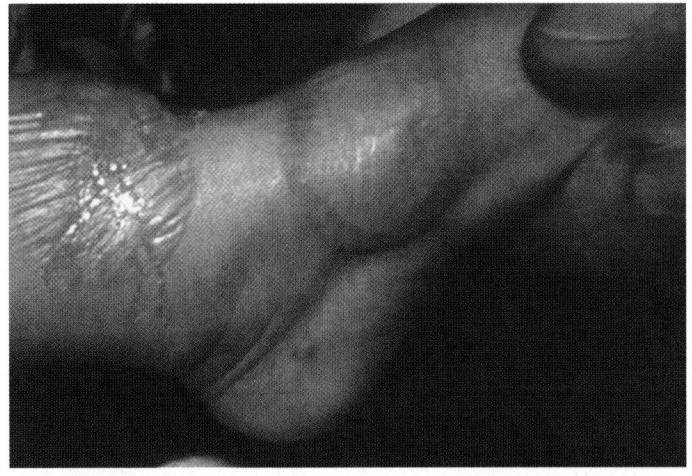

(a)

Actinomyces infection   Neutrophils

(b)

Figure 26.11 Inflammation. *(a)* Photograph of a child's foot showing swelling due to infection with vaccinia virus; fluid accumulation results from the inflammatory response. *(b)* Photomicrograph showing infection by *Actinomyces*, a filamentous bacterium. The stained cells surrounding the dark mass of bacteria in the center are neutrophils (Figure 26.4), indicating acute inflammation.

The usual outcome of the inflammatory response is a rapid localization and destruction of the pathogen by macrophages and recruited neutrophils. As the pathogens are destroyed, inflammatory cells are no longer stimulated, and as a result, their numbers at the infection site are diminished. As cytokine production decreases, the attraction of phagocytes to affected tissues lessens, and inflammation subsides.

Fever

Certain cytokines released during an inflammatory response induce **fever**, a condition of elevated body temperature. For example, the proinflammatory cytokines IL-1, IL-6, and TNF-α are *endogenous pyrogens*. These substances stimulate the hypothalamus, the temperature-controlling center of the brain, to produce *prostaglandins*, chemical signals that raise body temperature and cause fever. These same cytokines released in small amounts at local sites of infection induce localized heating, which increases blood flow and promotes healing. The release of endogenous

UNIT 6

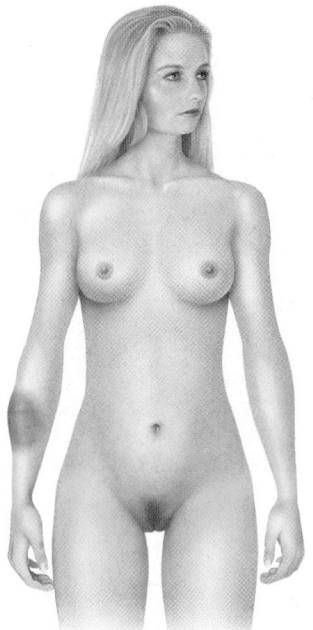
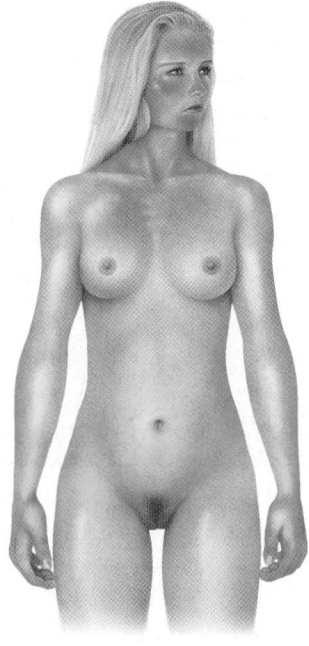

(a) Local infection leads to inflammation in a small part of the body, followed by healing.

(b) Systemic infection leads to inflammation and disease throughout the body.

Figure 26.12 Local and systemic inflammation. *(a)* Local infection, mediated by proinflammatory cytokines from local macrophages, results in inflammation that subsides as the infection is cleared. *(b)* Systemic infection causes systemic release of proinflammatory cytokines, resulting in widespread systemic inflammatory symptoms including severe edema, fever, and septic shock, even if the infection is controlled.

pyrogens is a physiological response to the presence of *exogenous pyrogens*, components of pathogens that induce fever, such as the lipid A endotoxin of gram-negative bacteria LPS (↩ Section 2.5).

Although uncomfortable for the host, fever is an important component of the innate immune response to some infections. The rise from normal human body temperature, about 37°C (98.6°F), to fever temperature, usually 38–40°C (100.4–104°F), is a beneficial response to infection because higher body temperatures inhibit growth of most pathogens. Slower growth limits multiplication of the invading pathogen, thereby minimizing tissue damage and easing the workload on other immune cells, especially phagocytes. Elevated body temperatures also increase the production of *transferrins*, molecules that bind and sequester iron in blood and lymph, thus depriving pathogens of this important nutrient (↩ Sections 3.1 and 25.4).

While low to moderate fever is an important component of the innate immune response, a continuous or uncontrolled rise in body temperature (>40°C) is a rare but life-threatening condition that may accompany certain disease conditions and requires immediate medical intervention. This usually includes the administration of antipyretic (fever-reducing) medications, such as acetaminophen or ibuprofen, which counteract the effect of endogenous pyrogens on the hypothalamus.

Systemic Inflammation and Septic Shock

In some cases, the inflammatory response fails to localize the pathogens, and the reaction spreads throughout the body. Uncontrolled *systemic inflammation* can be more dangerous than the original infection, with inflammatory cells and mediators contributing to body-wide inflammation. An inflammatory response that spreads inflammatory cells and mediators through the entire circulatory and lymphatic systems can lead to *septic shock*, a life-threatening condition.

Although there are potentially many causes of pathogen-induced septic shock, one example is systemic infection by enteric bacteria, such as *Salmonella* species or *Escherichia coli*, which can be introduced into the peritoneal cavity or the bloodstream by a ruptured or leaking bowel. The primary infection is often cleared by the activity of phagocytes and antibiotic treatment. However, the endotoxic outer membrane LPS from these gram-negative bacteria interacts with a PRR on phagocytes, stimulating production of proinflammatory cytokines that are released into the circulation. These cytokines induce systemic responses that parallel the localized inflammatory response but on a much larger scale that affects many organ systems, ultimately leading to a body-wide inflammatory event (Figure 26.12*b*). The result is a massive efflux of fluids from the central vascular tissue causing a loss of systemic blood pressure and the influx of fluids from vascular tissues into extravascular spaces. Septic shock causes death in up to 30% of affected individuals.

MINIQUIZ

- Identify the molecular mediators of inflammation and fever and define their individual roles.
- Identify the major symptoms of localized inflammation and of septic shock.

26.9 The Complement System

The **complement system**, or simply *complement*, is a group of sequentially interacting proteins, many with enzymatic activity, that functions to boost the efficiency of both innate and adaptive immune responses for the destruction of pathogens. Complement proteins are produced in the liver and found throughout the body, and their activities can be triggered by innate or adaptive mechanisms. The major outcomes of activating the complement system are enhanced phagocytosis, inflammation, and lysis of invading cells.

The individual proteins of complement are designated C1, C2, C3, and so on. Complement proteins exist in an inactive conformation until they are enzymatically split to assume their active forms. The splitting of C3 to its products, C3a and C3b, is the key event in activating complement. At least three different mechanisms lead to this outcome, and we consider each of these now.

Classical Complement Activation

The *classical pathway* of complement activation is initiated when complement proteins, attracted by bound antibodies, attach to pathogen surfaces. The antibodies are said to *fix* (bind) the ever-present complement proteins, and thus, classical complement activation depends upon the adaptive immune response. The complement proteins react in a defined sequence, or *cascade*, with activation of one complement protein leading to activation of the next, and so on.

The key steps of classical activation of complement, shown in **Figure 26.13*a***, start with binding of antibody to antigen (initiation), followed by binding of C1 components (C1q, C1r, and C1s) to the antibody–antigen complex. This complex recruits and splits C2 into its fragments, C2a and C2b, and C4 into its fragments, C4a and C4b. C2a and C4b interact and are deposited at an adjacent membrane site. The resulting C2a-C4b complex functions as a *C3 convertase*, an enzyme that cleaves C3 to C3a and C3b. C3b then binds to the convertase, forming a complex that cleaves C5 into C5a and C5b. The liberated C5b then binds C6 and C7 and inserts into the membrane of the target cell. The C5b–7 complex recruits C8 and C9, forming a large C5b–9 unit called the *membrane attack complex* (*MAC*). The MAC forms a pore through the cytoplasmic membrane of the pathogenic cell, allowing extracellular fluids to rush in and lyse the cell (Figure 26.13*a*). Dozens to hundreds of MACs may perforate a single bacterial cell at the point of lysis.

When activated by specific antibody, MAC formation lyses many gram-negative bacteria. However, gram-positive bacteria, such as *Streptococcus* species, are not usually lysed by complement because their thick cell walls make the cytoplasmic membrane less accessible to MAC proteins. Gram-positive bacteria can, however, be destroyed through *opsonization*.

Opsonization is the coating of pathogens with antimicrobial host proteins, such as antibodies or C3b, resulting in enhanced phagocytosis of target cells (**Figure 26.14***a*). Opsonization neutralizes pathogens and makes them much more likely to be identified, engulfed, and destroyed by phagocytes. This is because most phagocytes, including neutrophils and macrophages, have antibody receptors (FcR) and C3b receptors (C3R) on their surfaces, which bind antibody and C3b complement protein, respectively. Normal phagocytic processes are enhanced about 10-fold by antibody–FcR interactions and amplified another 10-fold by C3b–C3R interactions.

By-products of complement activation include chemoattractants called *anaphylatoxins*; these molecules cause inflammatory reactions at the site of complement deposition. For example, when C3 is cleaved to C3a and C3b, C3b opsonizes the target cell as described above; meanwhile, release of soluble C3a attracts and activates phagocytes, increasing phagocytosis. In addition, both C3a and the C5a cleavage product are able to bind receptors on the surface of mast cells, causing them to degranulate and release large amounts of proinflammatory histamine (Figure 26.14*c*).

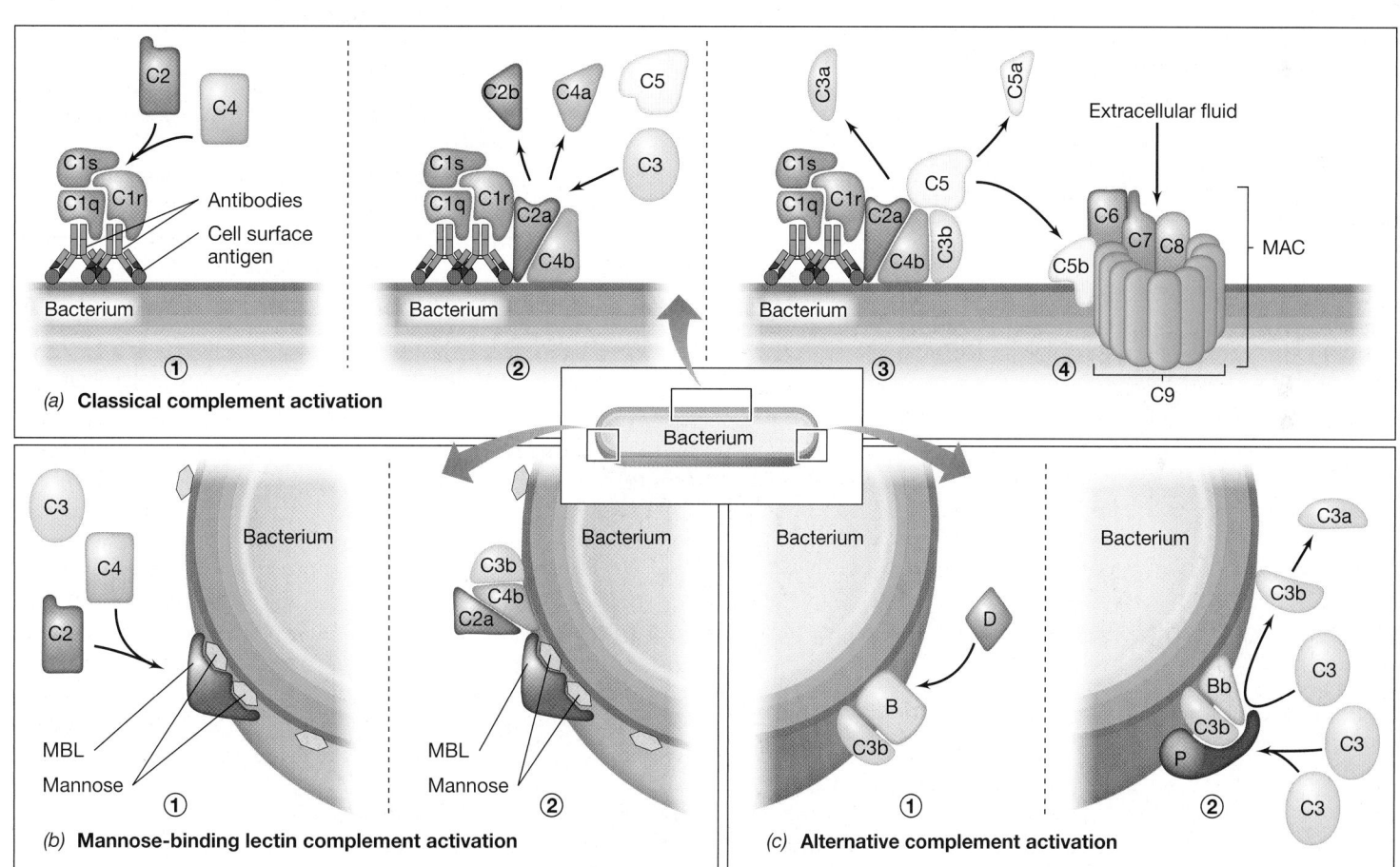

Figure 26.13 Complement proteins and complement activation in the immune response. (*a*) The sequence, orientation, and activity of the components of the classical complement pathway as they interact to lyse a cell. ① Binding of the antibody and the C1 protein complex (C1q, C1r, and C1s). ② The C2a-C4b complex is a C3 convertase that splits C3. ③ The C2a-C4b-C3b complex cleaves C5, and C5b then binds an adjacent membrane site. ④ Sequential binding of C6, C7, C8, and C9 to C5b produces a pore, the membrane attack complex (MAC), in the membrane. (*b*) The mannose-binding lectin (MBL) pathway. ① MBL binds to mannose on the bacterial membrane and recruits C2 and C4. ② MBL anchors formation of C2a-C4b-C3b. This complex activates C5, as in step 3 of part *a*, and initiates formation of the MAC (step 4 in *a*). (*c*) The alternative pathway. ① C3b bound to the cell binds protein B, which is cleaved by protein D. ② The resulting C3b-Bb complex is stabilized on the membrane by factor P (properdin) and then acts on C3 in the blood, causing more C3b to bind to the membrane. Bound C3b-Bb-P then activates C5, as in step 3 of the classical activation pathway above, and initiates formation of the MAC (step 4 in *a*).

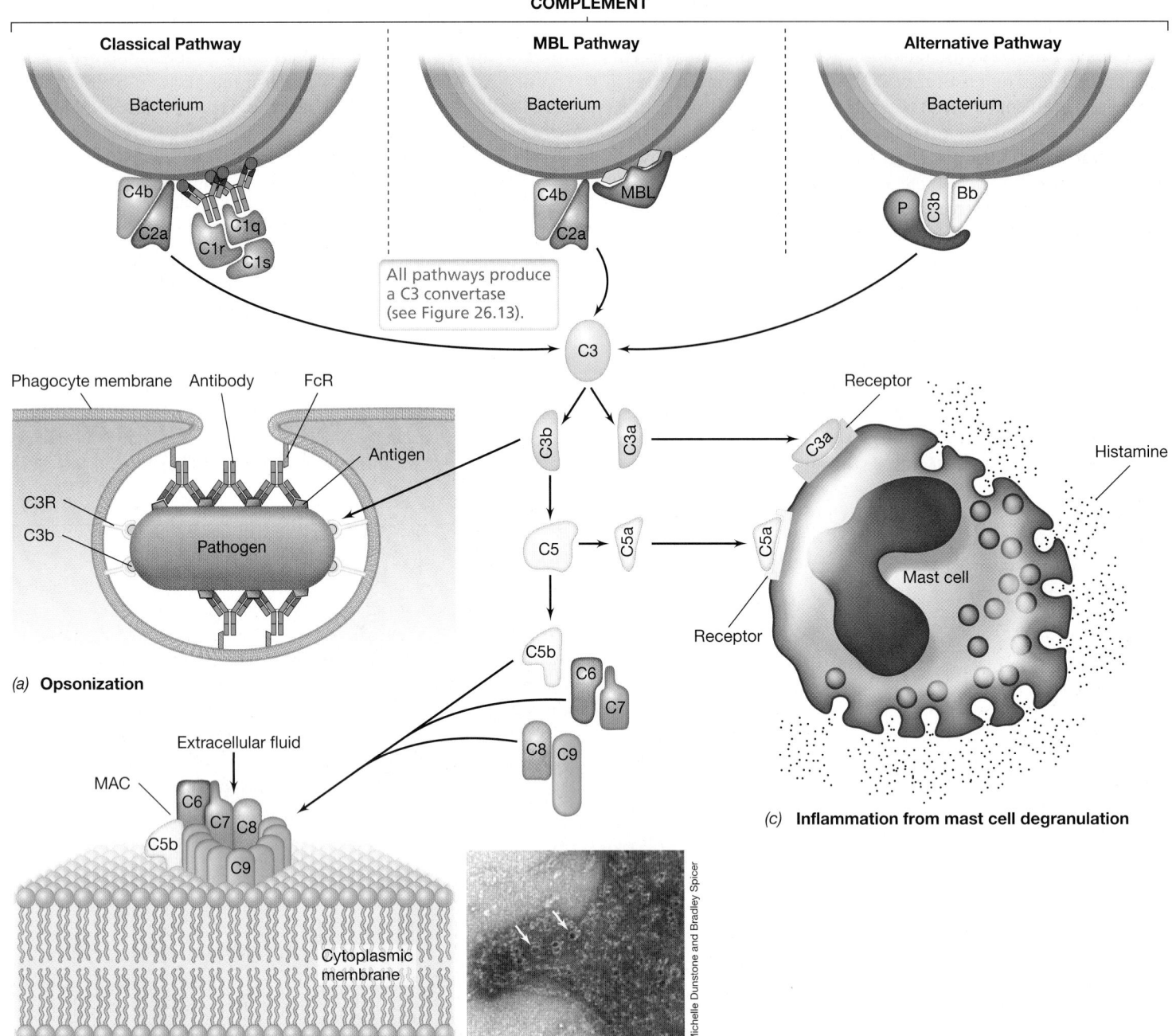

Figure 26.14 Complement proteins and the outcomes of complement activation. Each complement activation pathway leads to the production of a C3 convertase that cleaves C3 to its active products, C3a and C3b. These initiate three possible outcomes: opsonization, lysis by membrane attack complex (MAC) formation, and inflammation. *(a)* Pathogens coated with C3b (green) or with specific antibodies become opsonized, allowing for enhanced recognition by phagocytes through C3R and FcR proteins. *(b)* Transmembrane pores (arrows) formed by MAC components C5b through C9 cause lysis of target cells. The transmission electron micrograph is a negative stain image showing human MACs attacking a foreign (rabbit) erythrocyte. *(c)* During complement-activated inflammation, liberated C3a and C5a bind receptor proteins on mast cells, causing the mast cells to degranulate and release proinflammatory histamine.

Mannose-Binding Lectin and Alternative Pathways

In addition to the classical activation pathway, the *mannose-binding lectin pathway* and the *alternative pathway* can activate complement (Figure 26.13*b* and *c*). These pathways depend on recognition of shared pathogen components and are an important part of the innate immune response, especially in the initiation of inflammation.

The *mannose-binding lectin (MBL) pathway* depends on the activity of a serum MBL protein. MBL is a soluble PRR (Section 26.6) that binds to mannose-containing polysaccharides found only on bacterial cell surfaces (Figure 26.13*b*). The MBL–polysaccharide complex is similar to the C1 complex of the classical pathway in that it fixes C2a and C4b, again producing C3 convertase and binding C3b to C2a-C4b. As before, this complex catalyzes formation of the C5–9 MAC and leads to lysis or opsonization of the bacterial cell.

The *alternative pathway* is a nonspecific complement activation mechanism that uses many of the classical complement pathway components, as well as several unique serum proteins. Together they induce opsonization and activate the C5–9 MAC. The first step in alternative pathway activation is the binding of C3b to LPS on the bacterial cell surface (C3b is produced by spontaneous cleavage of C3, which occurs at low levels in the blood and tissues) (Figure 26.13c). C3b on the membrane can then bind the alternative pathway serum protein factor B, which is cleaved by factor D to yield soluble Ba and Bb. The C3b-Bb complex may then be bound by factor P (properdin) to form C3b-Bb-P, another C3 convertase. C3b-Bb-P then attracts and cleaves additional C3, depositing more C3b on the membrane and initiating the remaining steps of the complement cascade (Figure 26.14).

Both the alternative pathway and the MBL pathway nonspecifically target bacterial invaders and lead to activation of MACs and opsonization via formation of stable C3 convertases. MBL, factors B, D, and P, and classical complement proteins are part of the innate immune response, but unlike the classical pathway, neither the alternative pathway nor the MBL pathway requires prior antigen exposure or the presence of antibodies for activation.

MINIQUIZ

- In what ways does the classical pathway of complement activation differ from the mannose-binding lectin and alternative pathways?
- What is opsonization, and how does opsonization help fight bacterial infection?
- Why are the mannose-binding lectin and alternative pathways considered part of the innate immune system?

26.10 Innate Defenses against Viruses

In addition to opsonization and the activity of phagocytes, the immune system has other innate defenses that are especially important for controlling and eliminating viral infections. These include *natural killer cells* (Figure 26.4) and *interferons*. Natural killer cells are lymphocyte-like cells that recognize and kill compromised (unhealthy) cells, such as those infected with intracellular pathogens, in particular viruses. Whereas natural killer cells help rid the body of already infected cells, interferons, small proteins of the cytokine family, help healthy cells ward off viral infection. We consider both of these innate defenses here.

Natural Killer Cells

Natural killer cells (NK cells) are cytotoxic lymphocytes that are distinct from T cells and B cells (Figure 26.4). The role of NK cells is to seek out and destroy compromised cells, such as cells infected with intracellular pathogens (such as viruses) or cancer cells. When an NK cell engages an infected or otherwise diseased cell, granules in the cytoplasm of the NK cell migrate to the contact site and release their contents. These granules contain *perforin* and proteases called *granzymes*. Perforin binds the membrane of the target cell and forms a pore through which granzymes enter the target cell (**Figure 26.15**). Granzymes are cytotoxins that cause *apoptosis*, or programmed cell death, characterized by death and degradation of the target cell from within. During the killing process, NK cells remain unaffected and their membranes are not damaged by perforin. Also during this process, NK cell numbers do not increase nor do NK cells exhibit immune memory after interaction with target cells. With this in mind, how do NK cells recognize compromised host cells that should be destroyed?

Most cells in the body contain a suite of surface proteins called the **major histocompatibility complex (MHC)** (Chapter 27).

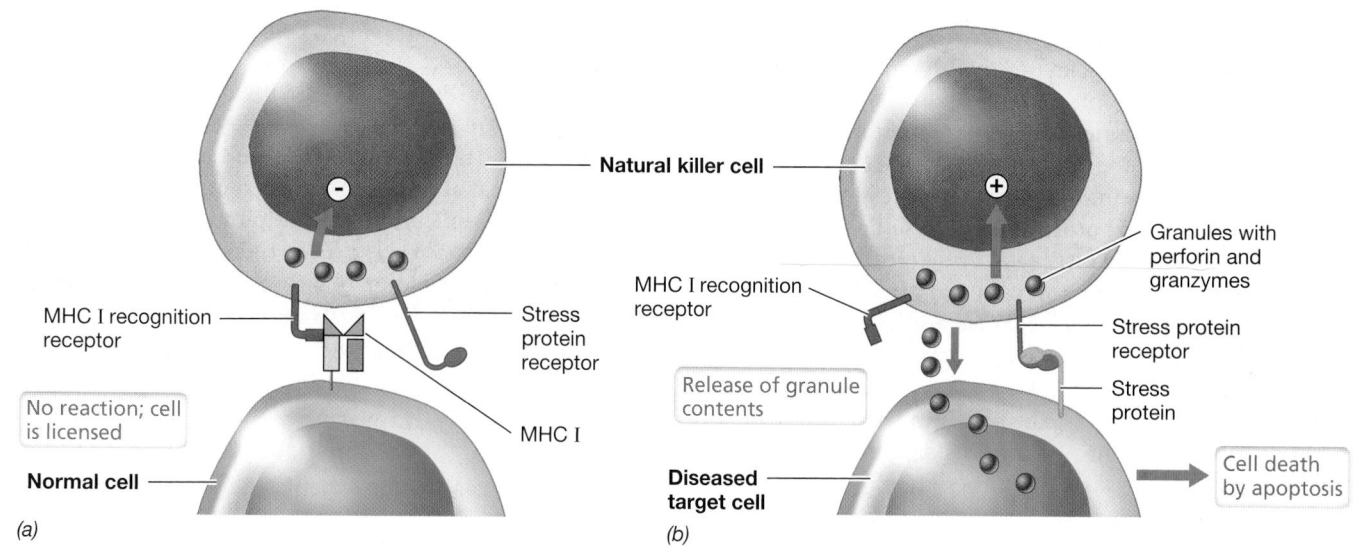

Figure 26.15 Natural killer cells. Natural killer (NK) cells have two receptors: One interacts with MHC I on healthy cells; the second one interacts with cell stress proteins, found only on tumor cells or pathogen-infected cells. *(a)* MHC I recognition licenses the healthy cells, preventing the NK cell from releasing its contents. *(b)* Pathogen-infected cells or tumor cells express stress proteins and often reduce MHC I expression. Especially in the absence of MHC I recognition, the NK cell interacts with the stress protein and releases perforin and granzymes, inducing apoptosis and killing the diseased cell.

UNIT 6

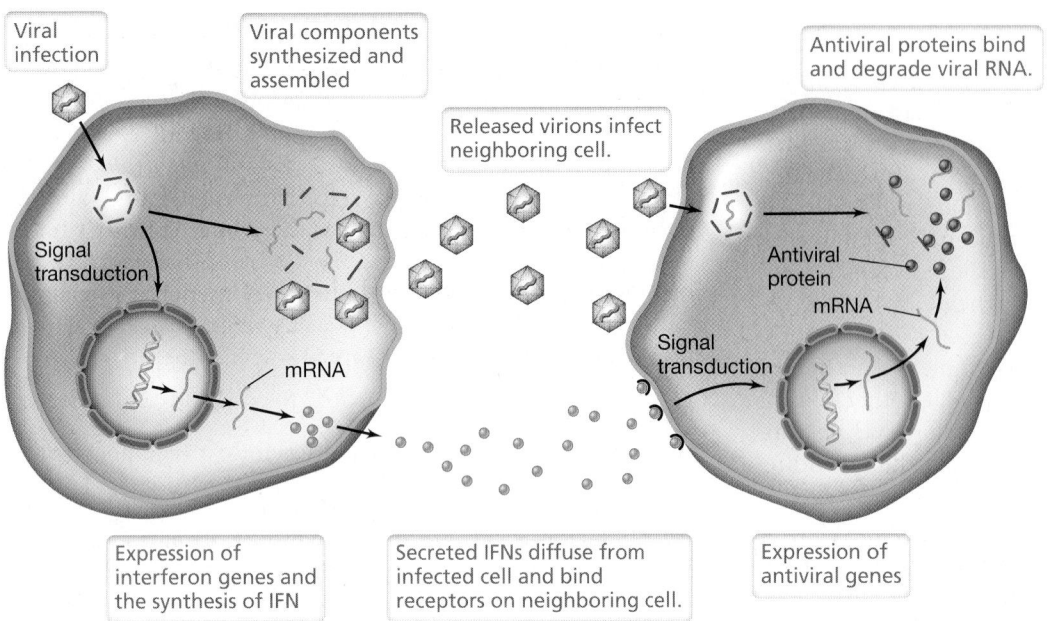

Figure 26.16 Antiviral activity of interferons. Host cells secrete interferons, a type of cytokine, in response to viral infection. The interferons bind to uninfected cells, triggering a signal transduction pathway that leads to the synthesis of proteins that bind viral nucleic acids and interfere with viral replication.

There are two classes of MHC proteins—*class I* and *class II*—and their primary function is to present antigen to various immune cells to trigger an immune response. MHC II proteins are expressed on antigen-presenting cells (APCs) only, which include B cells, macrophages, and dendritic cells (Figure 26.4). By contrast, MHC I proteins are found on the surfaces of all nucleated cells. In uninfected cells, which contain no pathogens or foreign antigens, MHC I proteins bind and present *self peptides*, protein fragments derived from the normal degradation of self proteins during growth, to the external environment. In cells that have been infected by viruses or other intracellular pathogens, MHC I proteins display peptides derived from the infectious agent. This serves as a signal to a special type of T lymphocyte called a *T-cytotoxic cell* to destroy the infected cell.

Because MHC I proteins function to interrupt multiplication of the pathogen by signaling the infected state of the host cell, many viral genomes encode proteins that suppress expression of MHC I in host cells. Without MHC I proteins on their surfaces—a condition that is also common in cancer cells—T-cytotoxic cells cannot identify and eliminate diseased host cells. This, then, is where NK cells play a major role; NK cells recognize and destroy compromised host cells that have reduced expression of MHC I on their surfaces.

NK cells recognize and destroy pathogen-infected or tumor cells by using a two-receptor system. The molecular targets of NK cells are MHC I proteins on the surface of other cells (Figure 26.15*a*). As NK cells circulate and interact with other cells in the body, they use special MHC I receptors on their surfaces to recognize MHC I proteins on normal, healthy cells. Binding of the MHC I recognition receptors on NK cells to MHC I on other cells deactivates the NK cell, turning off the perforin and granzyme

killing mechanisms. In addition to a deficiency of MHC I proteins, pathogen-infected or tumor cells frequently express *stress proteins* on their surfaces; NK cells have complementary receptors for many of these stress proteins. Especially in the absence of the MHC I interactions, the stress receptors on NK cells engage stress proteins on target cells. This interaction triggers the release of perforin and granzymes from the NK cell (Figure 26.15*b*). In this way, pathogen-infected or tumor cells that exhibit stress proteins and no longer express the MHC I proteins of healthy cells are removed from the body.

Interferons

Interferons, in particular IFN-α and IFN-β, are small proteins in the cytokine family that prevent viral replication by stimulating the production of antiviral proteins in uninfected cells (**Figure 26.16**). Host cells produce and secrete interferons in response to certain viral infections or exposure to inactivated viruses or viral nucleic acids. Interferons are produced in large amounts by cells infected with viruses of low virulence, but highly virulent viruses inhibit host protein synthesis before interferon can be produced, significantly reducing interferon production. The presence of double-stranded RNA (dsRNA) also induces interferon synthesis. In nature, dsRNA exists only in certain RNA viruses, such as rhinoviruses (one of many common cold viruses) (♂ Section 30.7); the viral dsRNA stimulates the animal cell to synthesize and release interferon.

Interferon activities are *host*-specific rather than *virus*-specific. That is, interferon produced by a species activates receptors only on cells from the same species. As a result, interferon produced by cells of an animal in response to, for example, a rhinovirus, could also inhibit multiplication of, for example, influenza viruses in cells of the same species. However, the interferon would have no effect on the multiplication of viruses, including the original rhinovirus, in other species. We examine the potential use of interferons as chemotherapeutic or prophylactic treatments in Chapter 28.

MINIQUIZ

- Identify and compare the targets and the recognition mechanisms used by T-cytotoxic cells and NK cells.

- Under what conditions are interferons produced, and how do they limit the transmission of viruses from one host cell to another?

Chapter Review

I • Fundamentals of Host Defense

26.1 Innate immunity is an inborn protective response to infection characterized in part by recognition and elimination of common pathogens, primarily through the activity of phagocytes. Adaptive immunity is the acquired ability of the immune system to eliminate specific pathogens from the body via lymphocyte-mediated responses, including the production of antibodies that bind foreign antigens on pathogens or their products.

> **Q** Compare and contrast the major features of innate immunity with those of adaptive immunity.

26.2 The human body possesses numerous protective defenses against infectious agents. Natural host resistance to infection includes physical barriers to infection posed by the skin and mucosa, as well as chemical barriers to infection including acidic secretions, defensins, and lysozyme. The specificity of pathogens for particular tissues limits which hosts and tissues might be susceptible to infection.

> **Q** Identify at least four mechanisms by which a healthy host resists infection.

II • Cells and Organs of the Immune System

26.3 Cells involved in innate and adaptive immunity originate from hematopoietic stem cells in bone marrow. The blood and lymph systems circulate cells and proteins that are important components of the immune response. Diverse leukocytes participate in immune responses in all parts of the body.

> **Q** Describe the significance of bone marrow, blood, and lymph to cells and proteins associated with the immune system.

26.4 Leukocytes are differentiated white blood cells derived from either myeloid or lymphoid precursor cells. Cells of myeloid lineage include monocytes and granulocytes. Monocytes include macrophages and dendritic cells, specialized phagocytes that function as antigen-presenting cells (APCs) to initiate an adaptive immune response. Granulocytes include neutrophils, which are also phagocytes but not APCs, and mast cells, which are important inducers of the inflammatory response but may also cause allergic reactions. Lymphocytes include B and T cells, which facilitate adaptive immunity, and natural killer cells, which play a key role in destroying virus-infected and cancerous host cells.

> **Q** What is the origin of the phagocytes and lymphocytes active in the immune response? Track the maturation of B cells and T cells.

III • Phagocyte Response Mechanisms

26.5 Pathogens may colonize host tissues when appropriate nutrients and growth conditions are present, such as on mucosal surfaces, especially where the composition of the normal microbiota has been altered. Innate responses to microbial invasion and tissue damage are initiated by the release of chemokines, which recruit phagocytes and other immune cells to sites of infection.

> **Q** Describe a scenario in which microorganisms invade body tissues. What factors allow for the migration of phagocytes to sites of infection?

26.6 Innate recognition of common pathogens occurs through pathogen-associated molecular patterns (PAMPs). Phagocytes recognize PAMPs through preformed pattern recognition receptors (PRRs). The recognition and interaction process stimulates phagocytes to destroy the pathogens through a signal transduction mechanism that induces phagocytosis of the infectious agent.

> **Q** Identify some PAMPs that are recognized by PRRs. Which cells express PRRs? How do PRRs associate with PAMPs to promote innate immunity?

26.7 Phagocytosis is the engulfing of infectious particles by phagocytes. Engulfed pathogens are bathed in toxic oxygen compounds inside the phagolysosome, killing and degrading them. However, some pathogens have developed various defense mechanisms to avoid or inhibit phagocytes, including secretion of leukocidins, the presence of a capsule, and biosynthesis of carotenoid pigments, which combat oxidative stress.

> **Q** Explain how phagocytes kill microorganisms, with particular attention to oxygen-dependent mechanisms. Then identify at least three properties of pathogens that inhibit the effectiveness of phagocytes.

IV • Other Innate Host Defenses

26.8 Fever and inflammation, characterized by pain, swelling (edema), redness (erythema), and heat, are normal and generally beneficial outcomes that result from activation of nonspecific immune response effectors. However, uncontrolled systemic inflammation, called septic shock, can lead to serious illness or death.

Q Identify the cells that initiate inflammation and the cells that are activated by inflammatory signals.

26.9 The complement system is composed of soluble proteins that catalyze bacterial opsonization and cell lysis. Complement is triggered by antibody interactions or by interactions with nonspecific activators, such as mannose-binding lectin. Complement activation may be a product of either innate or adaptive immunity. Complement may enhance phagocytosis, cause target cell lysis, or induce an inflammatory response.

Q Describe the complement system. Is the order of protein interactions important? Why or why not? Identify the components of the mannose-binding lectin pathway for complement activation. Identify the

components of the alternative pathway for complement activation. How do these complement activation pathways differ from the classical pathway?

26.10 Through an antigen-independent mechanism, natural killer cells respond to both the presence of stress proteins and the absence of MHC I on the surface of virus-infected cells or tumor cells, using perforin and granzymes to kill the compromised target cells. Interferons are cytokines produced by virus-infected cells that limit the spread of infection by stimulating the expression of antiviral proteins in uninfected cells.

Q What is the activation signal for NK cells? How does this differ from the activity of T-cytotoxic lymphocytes?

Application Questions

1. Describe the relative importance of innate immunity compared to adaptive immunity. Is one more important than the other? Can we survive in a normal environment without immunity?

2. Describe the potential problems that would arise if a person had an acquired inability to phagocytose pathogens. Could the person survive in a normal environment such as a college campus? What defects in the phagocyte might cause lack of phagocytosis? Explain.

3. Inflammation is the hallmark of an active immune response. Explain how inflammation is triggered by innate immune mechanisms. Why does inflammation subside as an infection is controlled?

4. Do you agree with the following statement? Complement is a critical component of antibody-mediated defense. Explain your answer. What might happen to persons who lack complement component C3? C5? Factor B (alternative pathway)? Mannose-binding lectin (MBL)?

Chapter Glossary

Adaptive immunity the acquired ability to recognize and destroy a particular pathogen or its products, dependent on previous exposure to the pathogen or its products

Antibody a soluble protein produced by B cells and plasma cells that interacts with antigen; also called immunoglobulin

Antigen a molecule that interacts with specific components of the immune system

Antigen-presenting cell (APC) a macrophage, dendritic cell, or B cell that takes up and processes antigen and presents it to T-helper cells

B cell a lymphocyte that has immunoglobulin surface receptors, produces immunoglobulin, and may present antigens to T cells

Chemokine a soluble protein that recruits immune cells to an injury site; a type of cytokine

Complement system a series of proteins that react sequentially with antibody–antigen complexes, mannose-binding lectin, or alternative activation pathway proteins to amplify or potentiate target cell destruction

Cytokine a soluble protein produced by a leukocyte or damaged body cell; modulates an immune response

Dendritic cell a phagocytic antigen-presenting cell found in various body tissues; transports antigen to secondary lymphoid organs

Fever an increase in body temperature resulting from infection or the presence of toxins in the body

Granulocyte a leukocyte derived from a myeloid precursor that contains cytoplasmic granules consisting of toxins or enzymes that are released to destroy target cells.

Immunity the ability of an organism to resist infection

Immunoglobulin a soluble protein produced by B cells and plasma cells that interacts with antigen; also called antibody

Inflammation a nonspecific reaction to noxious stimuli such as toxins and pathogens, characterized by redness (erythema), swelling (edema), pain, and heat (fever), usually localized at the site of infection

Innate immunity the noninducible ability to recognize and destroy an individual pathogen or its products that does not rely on previous exposure to a pathogen or its products

Interferons cytokine proteins produced by virus-infected cells that induce signal transduction in nearby cells, resulting in transcription of antiviral genes and expression of antiviral proteins

Invasion the ability of a pathogen to enter into host cells or tissues, spread, and cause disease

Leukocyte a nucleated cell in blood; also called a white blood cell

Lymph nodes organs that contain lymphocytes and phagocytes arranged to encounter microorganisms and antigens as they travel through the lymphatic circulation

Lymphocytes a subset of nucleated cells in blood involved in the adaptive immune response

Macrophage a large leukocyte found in tissues that has phagocytic and antigen-presenting capabilities

Major histocompatibility complex (MHC) a genetic region that encodes several proteins important for antigen processing and presentation. MHC I proteins are expressed on all cells. MHC II proteins are expressed only on antigen-presenting cells

Mast cell tissue cells adjoining blood vessels throughout the body that contain granules with inflammatory mediators

Memory (immune memory) the ability to rapidly produce large quantities of specific immune cells or antibodies after subsequent exposure to a previously encountered antigen

Monocyte Circulating phagocyte that contains many lysosomes and can differentiate into a macrophage or dendritic cell

Mucosa-associated lymphoid tissue (MALT) a part of the lymphatic system that interacts with antigens and microorganisms that enter the body through mucous membranes, including those of the gut, the genitourinary tract, and bronchial tissues

Natural killer (NK) cell a specialized lymphocyte that recognizes and destroys infected host cells or cancer cells in a nonspecific manner

Neutrophil a leukocyte exhibiting phagocytic properties, a granular cytoplasm (granulocyte), and a multilobed nucleus; also called polymorphonuclear leukocyte or PMN

Opsonization the deposition of antibody or complement protein on the surface of a pathogen or other antigen that results in enhanced phagocytosis

Pathogen-associated molecular pattern (PAMP) a repeating structural component of a microorganism or virus recognized by a pattern recognition receptor (PRR)

Pattern recognition receptor (PRR) a protein in a phagocyte membrane that recognizes a pathogen-associated molecular pattern (PAMP)

Phagocyte a cell that engulfs foreign particles, and can ingest, kill, and digest most pathogens

Phagocytosis the process of engulfing and killing foreign particles and cells

Phagosome an intracytoplasmic vacuole containing engulfed materials, especially pathogens or foreign particles

Plasma the liquid portion of the blood containing proteins and other solutes

Plasma cell a differentiated B cell that produces soluble antibodies

Primary lymphoid organ an organ in which antigen-reactive lymphocytes develop and become functional; the bone marrow is the primary lymphoid organ for B cells; the thymus is the primary lymphoid organ for T cells

Secondary lymphoid organ an organ at which antigens interact with antigen-presenting phagocytes and lymphocytes to generate an adaptive immune response; these include lymph nodes, spleen, and mucosa-associated lymphoid tissue

Serum the liquid portion of the blood with clotting proteins removed

Specificity the ability of the immune response to interact with particular antigens

Stem cell a progenitor cell that can develop into other cell types

T cell a lymphocyte that interacts with antigens through a T cell receptor for antigen; T cells are divided into functional subsets including Tc (T-cytotoxic) cells and Th (T-helper) cells. Th cells are further subdivided into Th1 (inflammatory) cells and Th2 cells, which aid B cells in antibody formation

Toll-like receptor (TLR) one of a family of pattern recognition receptors (PRRs) found on phagocytes, structurally and functionally related to Toll receptors in *Drosophila*, that recognize a pathogen-associated molecular pattern (PAMP)

27

Adaptive Immunity: Highly Specific Host Defenses

microbiologynow

Got (Raw) Milk? The Role of Unprocessed Cow's Milk in Protecting against Allergy and Asthma

In recent decades, the onset of allergies and asthma during childhood has become an increasingly common complication that, as we will see in this chapter, stems from hypersensitivity reactions in the adaptive immune response. However, for nearly 20 years now, researchers have been aware of one demographic of young people that has consistently shown resistance to asthma and allergies: kids that grow up on farms. Several hypotheses have been proposed to explain this phenomenon, including early childhood exposure to livestock and their feed (see photo), but while this may contribute to a protective "farm effect," it is the consumption of unprocessed cow's milk that is emerging as the major underlying explanation for the trend.

A recent report documented the correlation between the consumption of raw, unpasteurized cow's milk and the incidence of asthma in over 1100 children from rural regions of five European countries. Researchers found that compared to shop milk purchased at a supermarket, which is pasteurized (heated to 72°C for at least 15 seconds), centrifuged, and homogenized to achieve a uniform product, regular ingestion of unprocessed farm milk was inversely related to the onset of asthma in children, with the asthma-protective effect increasing over time.

To elucidate this farm milk effect, the scientists compared the fatty acid composition of unprocessed farm milk samples to that of shop milk samples. They found that the raw milk contained substantially higher levels of omega-3 (ω-3) fatty acids than shop milk, and this was attributed to both the higher overall fat content of raw milk and its lack of pasteurization, which breaks down heat-labile components of the milk. This finding was important because ω-3 fatty acids are precursors of anti-inflammatory immune mediators that suppress hypersensitivity reactions, including those that trigger allergies and asthma.

Although the implications of this research in asthma prevention are potentially significant, public health officials strongly discourage ingestion of unpasteurized milk, mainly because of the risk of foodborne illnesses, including salmonellosis, listeriosis, Q fever, staphylococcal food poisoning, and gastroenteritis. Interestingly, changing this stance may be neither warranted nor necessary since it may be possible to restore the asthma-protective effect by supplementing industrially processed milk with ω-3 fatty acids. Only time—and more research—will tell!

Source: Brick, T., et al. 2016. ω-3 fatty acids contribute to the asthma-protective effect of unprocessed cow's milk. *J. Allergy Clin. Immunol. 137(6):* 1699–1706.e13 doi:10.1016/j.jaci.2015.10.042.

In the previous chapter we discussed the key features of innate immunity and how this system protects against infection and disease from a broad range of pathogens. Here, we build on this foundation with a focus on the powerful and highly specific immune mechanisms of adaptive immunity, mechanisms that depend on both cellular and molecular components and that complement the innate response.

I • Principles of Adaptive Immunity

Innate and adaptive immunity can be viewed as different sides of the same coin that work together to protect the host from attack by foreign substances. Whereas innate immunity is characterized by broadly specific responses triggered by *common* structural features found on and in pathogens, adaptive immunity is directed toward *specific* molecular components of the pathogen (their antigens). In adaptive immunity, pathogen-specific immune receptors are produced in large numbers only after exposure to the pathogen or its products. In this way, the individual antigenic properties of different pathogens orchestrate the adaptive immune response.

We begin this part of the chapter by considering the major characteristics of the adaptive immune response and then explore the structure of the substances that trigger this response and the different forms of adaptive immunity that we experience in our everyday lives.

27.1 Specificity, Memory, Selection Processes, and Tolerance

Adaptive immunity is primarily a function of a special class of antigen-reactive leukocytes called *lymphocytes*. B lymphocytes (B cells) specialize in the production of antibodies that interact with and protect against extracellular antigens, thus conferring *antibody-mediated* (*humoral*) *immunity* to the host. T lymphocytes (T cells) express antigen-specific receptor proteins on their surfaces that defend against intracellular pathogens, such as viruses and certain bacteria, thus conferring *cell-mediated* (*cellular*) *immunity* to the host. The combination of antibody-mediated and cell-mediated defenses comprises adaptive immunity, a system that is characterized by three major features: *specificity*, *memory*, and *tolerance*. None of these features is characteristic of the innate immune response (Chapter 26).

Immune Specificity and Memory

Overall, the immune response is highly specific, but the innate and the adaptive systems differ in their degree of specificity. Innate immunity is directed against features common to a broad diversity of pathogens, such as the peptidoglycan of all gram-positive bacteria or the lipopolysaccharide (LPS) of all gram-negative bacteria. By contrast, adaptive immune mechanisms are directed against pathogen-specific macromolecules, such as a specific protein associated with a single strain of a particular pathogen.

Each B cell or T cell produces a unique protein that interacts with a single type of antigen. These proteins therefore have **specificity** for that antigen. The antigen-binding proteins of B cells are membrane-bound antibodies called **B cell receptors (BCRs)**.

T cells have antigen-binding **T cell receptors (TCRs)** on their surfaces. The specificity of the antigen–antibody or antigen–TCR interaction is dependent on the capacity of the lymphocyte cell receptor to interact with a particular antigen but not with other antigens (**Figure 27.1a**).

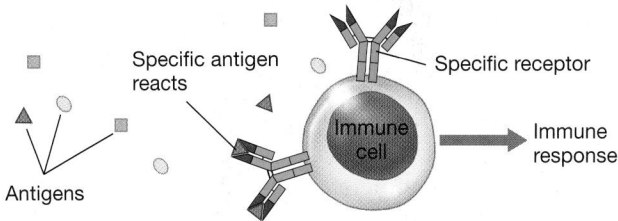

Specificity: Immune cells have surface receptors that interact with individual antigens.

(a)

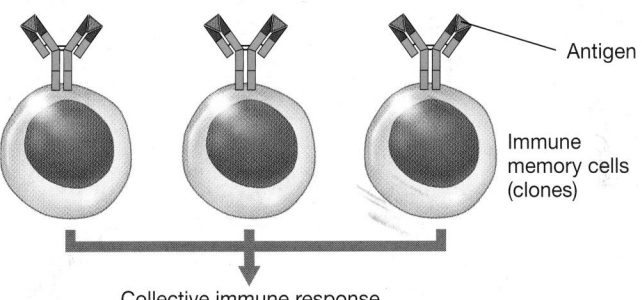

Memory: The first antigen exposure induces multiplication of antigen-reactive cells, resulting in multiple copies, or *clones*. After a subsequent exposure to the same antigen, the immune response is faster and stronger due to the large number of responding cells.

(b)

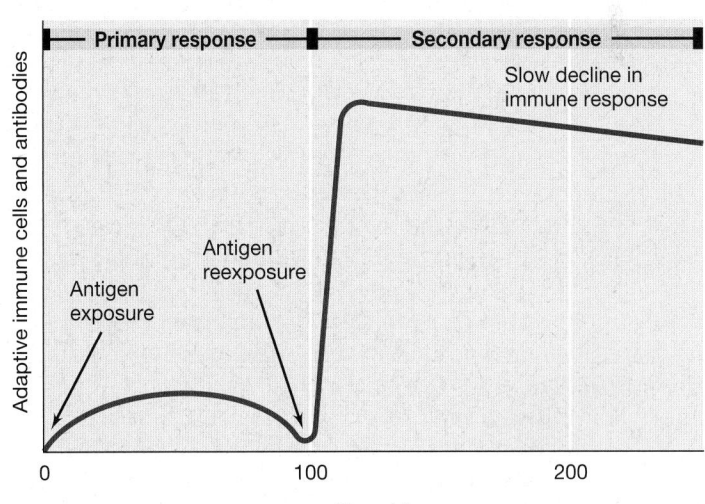

Immune responses: As a result of immune memory, antigen reexposure triggers a much stronger secondary response.

(c)

Figure 27.1 Specificity and memory in the adaptive immune response. Key features of antibody-mediated and cell-mediated immunity are *(a)* specificity and *(b)* memory. *(c)* The primary response induces both immune cells and antibodies. The antigens given at day 0 and day 100 must be identical to induce a secondary response. The secondary response may generate more than 10-fold increases in immune cells and antibody concentrations.

The adaptive immune response is induced only when triggered by a unique antigen on a pathogen. For example, polysaccharide antigen from the LPS of a particular gram-negative bacterium is unique for a genus and often even for a species within the genus. Therefore, an individual lymphocyte that interacts with an LPS polysaccharide unique to the bacterium *Salmonella* will not interact with the LPS of other bacteria, even other closely related gram-negative enteric bacteria, such as *Escherichia* or *Shigella*.

In addition to specificity, the adaptive immune system exhibits **memory** (Figure 27.1*b*). Lymphocytes must encounter antigen to stimulate production of detectable and effective antigen-activated antibodies or TCRs. The first exposure to an antigen generates a **primary immune response** in which antigen contact stimulates growth and multiplication of antigen-reactive B and T cells, thereby creating large numbers of antigen-specific **clones**. Some of these clones, called **memory cells**, persist in the body for years and confer long-term specific immunity. Subsequent exposure to the same antigen activates the clones and generates a faster and stronger **secondary immune response** that peaks within a few days (Figure 27.1*c*). This capacity to respond more quickly and vigorously to subsequent exposures to the eliciting antigen provides the host with immediate resistance to previously encountered pathogens, a topic we develop in more detail in Section 27.3. Immune memory is an important facet of clinical medicine and public health, as vaccinating humans or other animals with killed or weakened pathogens (or their products) is a major means of conferring immunity to dangerous pathogens (see Section 27.2).

T Cell Selection and Tolerance

The adaptive immune system must be able to discriminate between harmless *host* antigens ("self") and potentially dangerous *foreign* antigens ("nonself") and function to destroy only the latter. T cells undergo immune selection *for* potential antigen-reactive cells and selection *against* those cells that react strongly with self antigens. Selection against self-reactive cells results in the development of another key characteristic of the adaptive immune response: **tolerance**. Tolerance is a key component of the immune response and ensures that adaptive immunity is directed against agents that pose genuine threats to the host and not against the host's own proteins. The failure to develop tolerance may result in dangerous reactions to self antigens, a condition called *autoimmunity* (Section 27.9).

T lymphocyte precursors leave the bone marrow via the bloodstream and enter the thymus, a primary lymphoid organ (**Figure 27.2**). During the process of T cell maturation in the thymus, immature T cells undergo a two-step selection process to (1) select potential antigen-reactive cells (positive selection) and (2) eliminate cells that react strongly with self antigens (negative selection). **Positive selection** requires the interaction of immature T cells in the thymus with peptide antigens that are actually of self origin. Using their TCRs, some T cells bind to *major histocompatibility complexes* (*MHCs*) (Section 27.5) that present self peptides on the thymic epithelial tissue. The T cells that do not bind MHC–peptide complexes will be of no use in the immune response and are therefore permanently eliminated via *apoptosis* (programmed cell death). By contrast, those T cells that bind thymic MHC proteins receive survival signals and therefore remain viable. Positive selection retains T cells that recognize MHC–peptide and deletes T cells that do not recognize MHC–peptide and would therefore be unable to recognize MHC–peptide outside the thymus.

The second stage of T cell maturation is **negative selection**. In this process, the positively selected T cells continue to interact with thymic MHC–peptide. Precursor T cells that react with thymic self antigens are potentially dangerous if they react very strongly with these antigens (autoimmunity). Very strongly self-reactive T cells bind tightly to thymus epithelial cells. This binding prevents the strongly self-reactive T cells from dividing and eventually causes them to die. This process, called **clonal deletion**, prevents the propagation of precursor T cells that could potentially cause autoimmune complications. Precursor T cells having TCRs that react less strongly with self MHC bound to peptide survive this selection and live (Figure 27.2).

This two-stage thymic selection process ensures the generation of self-tolerant T cells capable of reacting strongly to foreign antigens only. Precursors of T cell clones that are either useless (do not bind) or harmful (bind too tightly) are deleted in the thymus. About 95% of all immature T cells do not survive the thymic selection process. The remaining selected T cells are destined to interact very strongly with nonself antigens. They are not destroyed in the thymus because their weak binding interactions with thymic self antigens signal them to proliferate. The selected and growing T cells leave the thymus and migrate to the spleen, mucosa-associated lymphoid tissue (MALT), and lymph nodes, where they can contact foreign antigens presented by B cells and other antigen-presenting cells (APCs).

B Cell Selection and Tolerance

To respond to the nearly infinite variety of environmental antigens, the immune system must have the capacity to generate an essentially limitless variety of antigen-specific lymphocytes. To this end, and as we shall see in Section 27.4, the body produces an enormous diversity of antigen-reactive B cells, each having BCRs specific for a single antigen on its surface. B cell selection occurs when the BCRs of a particular B cell clone interact with their corresponding antigen. The antigen-stimulated B cell can then proliferate and differentiate, a process called *clonal expansion* (**Figure 27.3**). This produces a pool of cells that expresses the same antigen-specific receptors; B cells that have not interacted with antigen do not proliferate. The newly generated pool of antigen-specific B cell clones is composed of many antibody-producing *plasma cells* and comparatively fewer long-lived, antigen-specific memory cells. We explore these concepts further in Section 27.3.

As with T cells, the development of tolerance in B cells is necessary because antibodies produced by self-reactive B cells (autoantibodies) may cause autoimmunity and damage to host tissues (Section 27.9). Therefore, B cells must undergo a similar clonal deletion process. However, unlike the thymic selection process of T cells, many self-reactive B cells are eliminated during development in the bone marrow, the primary lymphoid organ responsible for B cell maturation in mammals.

In addition to clonal deletion, **clonal anergy** (clonal unresponsiveness) also plays a role in final selection of the B cell pool. Some immature B cells are reactive to self antigens but are not activated by them. This is because B cell activation requires a

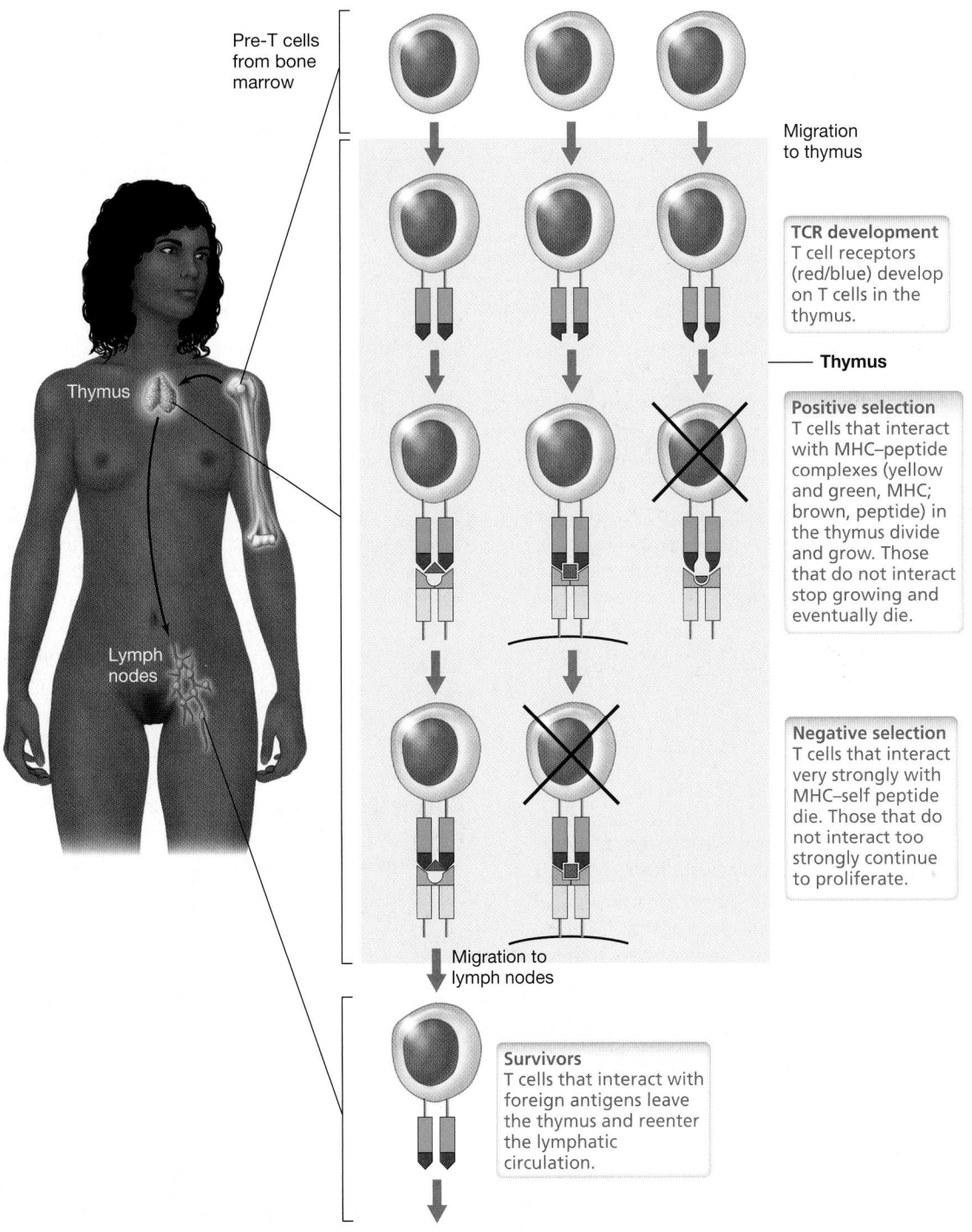

Pre-T cells from bone marrow

Migration to thymus

TCR development
T cell receptors (red/blue) develop on T cells in the thymus.

— **Thymus**

Positive selection
T cells that interact with MHC–peptide complexes (yellow and green, MHC; brown, peptide) in the thymus divide and grow. Those that do not interact stop growing and eventually die.

Negative selection
T cells that interact very strongly with MHC–self peptide die. Those that do not interact too strongly continue to proliferate.

Thymus

Lymph nodes

Migration to lymph nodes

Survivors
T cells that interact with foreign antigens leave the thymus and reenter the lymphatic circulation.

Figure 27.2 T cell selection and clonal deletion. T cells undergo selection for functionality and recognition of dangerous nonself antigens in the thymus.

second signal from a subset of T cells called *T-helper (Th) cells* (Section 27.8). The nature of this second signal is a positive interaction between proteins on the surface of B cells and Th cells that triggers the release of cytokines, especially IL-4 (*inter*leukin-4), from the Th cell that activates the B cell. If no second signal is generated because the available Th cells have been rendered tolerant to the antigen in the thymus, the B cell remains unresponsive. Using a similar mechanism, a second signal is also

required to activate T cells that are interacting with antigens presented by APCs. The requirement for a second activation signal is the key to establishing and maintaining clonal anergy in potentially dangerous self-reactive B and T lymphocytes.

With their key properties of specificity, memory, and tolerance, lymphocytes stand ready to deploy the adaptive immune response. The selection and control processes just described ensure that this response will be directed only against foreign antigens. We now

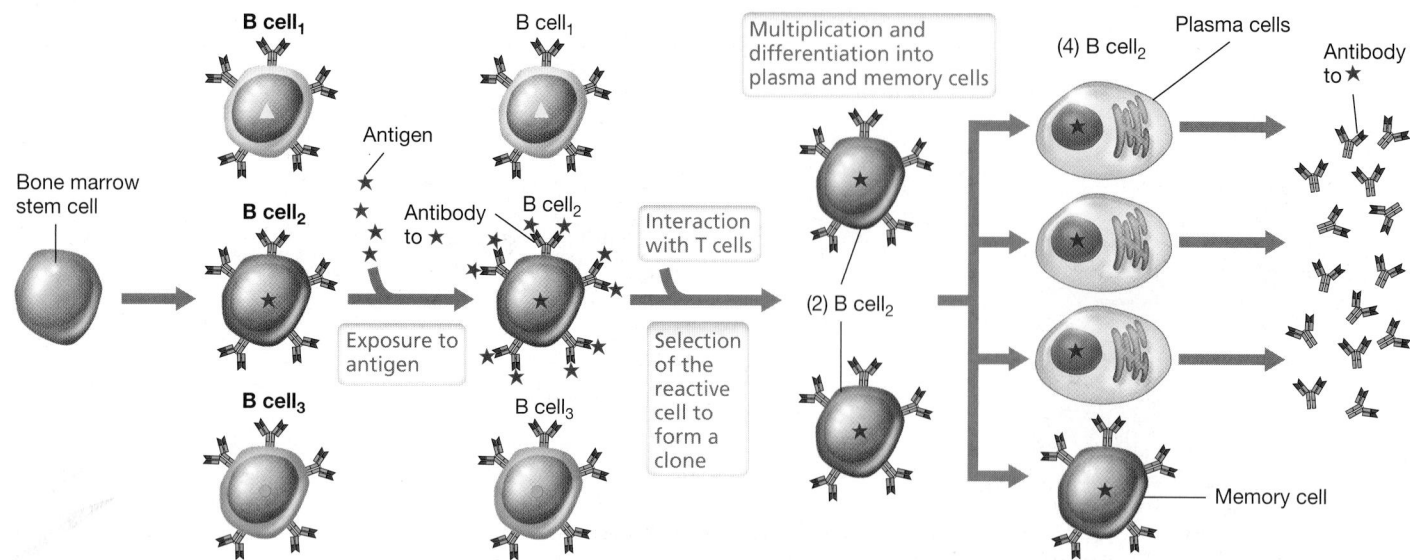

Figure 27.3 B cell clonal selection and expansion. Individual B cells, specific for a single antigen, proliferate and expand to form a clone after interaction with the specific antigen. The antigen drives selection and then proliferation of the individual antigen-specific B cell. Clonal copies of the original antigen-reactive cell have the same antigen-specific surface antibody. Continued exposure to antigen results in continued expansion of the clone.

examine the nature of antigens in more detail and consider the basic classes of adaptive immunity.

MINIQUIZ

- Distinguish between immune specificity, memory, and tolerance.
- Distinguish between positive and negative T cell selection. How do positive selection and negative selection contribute to the development of tolerance in T cells?
- Distinguish between clonal deletion and clonal anergy in B cells.

27.2 Immunogens and Classes of Immunity

The adaptive immune response recognizes a vast range of foreign macromolecules. The macromolecules are degraded and processed in host cells to produce antigens that are in turn presented to T cells. As we have discussed, **antigens** are substances that react with antibodies or TCRs. Most, but not all, antigens are **immunogens**, substances that induce an immune response. Here we examine the features of effective immunogens, define the features of antigens that promote interactions with antibodies and TCRs, and discuss the classes of immunity.

Immunogens and Antigen Binding

Immunogens share several intrinsic properties that enable them to induce an adaptive immune response. First, *molecular size* is an important property of immunogenicity; for a molecule to be immunogenic, it must be sufficiently large. Certain low-molecular-weight compounds called *haptens* do not induce immune responses themselves. However, antibodies can still bind them and they may induce an immune response if attached to a larger *carrier* molecule. Because antibodies can bind them, haptens—such as the disaccharide

lactose—are considered antigens, but they are not themselves immunogens. Proteins and complex carbohydrates are effective immunogens, whereas nucleic acids, simple polysaccharides with repeating subunits, and lipids are typically not. Thus, *sufficient molecular complexity* is another key property of immunogenicity. Large, insoluble macromolecules are usually excellent immunogens because phagocytes readily engulf and process them but not soluble molecules. Thus, *appropriate physical form* is another intrinsic property of immunogenicity.

Extrinsic factors, such as the immunogen dose and the route of administration, also influence immunogenicity. The administered dose of an immunogen can be important for an effective immune response, and in general, doses of 10 μg to 1 g are effective in most mammals. Immunogens administered parenterally

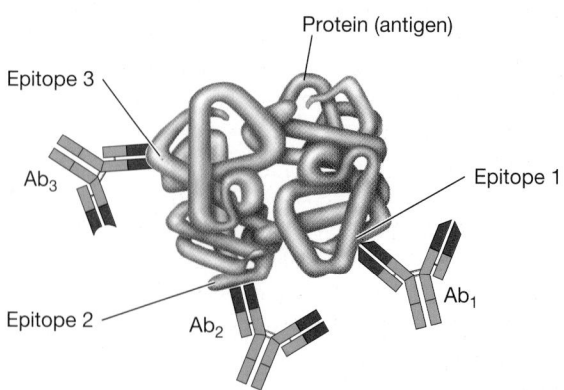

Figure 27.4 Antigens and epitopes for antibodies. Antigens may contain several different epitopes, each capable of reacting with a different antibody (Ab). The epitope 1 recognized by Ab_1 is a conformational epitope. Epitope 1 consists of two nonlinear parts of the folded polypeptide; the folding brings two distant portions of the polypeptide together to make a single epitope.

Natural Immunity

1

Active

The immune system responds to an active infection such as measles.

Antibodies are transferred from mother to infant in utero and in breast milk.

CDC/PHIL

2

Passive

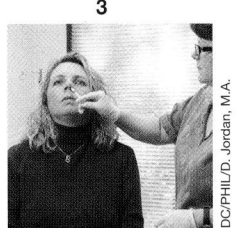

NURSE THE BABY
YOUR PROTECTION AGAINST TROUBLE

US WPA/Library of Congress

Artificial Immunity

3

Active

The immune system responds to antigens present in this vaccine.

CDC/PHIL/D. Jordan, M.A.

4

Passive

Venom from this rattlesnake can be neutralized by antibodies in rattlesnake antivenom.

CDC/PHIL/Edward J. Wozniak, D.V.M., Ph.D

Figure 27.5 Natural and artificial immunity.
Photos, left to right: (1) Childhood measles showing typical systemic measles rash. Natural active immunity requires infection with a pathogen to activate the adaptive immune response. (2) A 1934 United States government poster promoting breast-feeding. Natural passive immunity occurs when immunity is transferred from one individual to another by natural means, such as the transfer of maternal antibodies in breast milk. (3) Vaccination by nasal inhalation of antigen. Artificial active immunity occurs from exposure to particular antigens in a vaccine. (4) Timber rattlesnakes produce highly toxic venom. An antivenom consisting of purified antibodies to timber rattlesnake venom can be prepared in horses and artificial passive immunity conferred on a snakebite victim by injecting the victim with the antivenom.

(outside of the gastrointestinal tract), usually by injection, are normally more effective than those given topically or orally. When administered by oral or topical routes, antigens may be significantly degraded before contacting a phagocyte, and because of this, key immunogenic properties of the original antigen may be diminished or even lost.

Antibodies and TCRs do not interact with the immunogen as a whole but only with a distinct portion of the molecule called an **epitope** (also called an *antigenic determinant*) (**Figure 27.4**). Epitopes may include sugars, short peptides, and other organic molecules that are components of the larger immunogen. Antibodies interact with a sequence of four to six amino acids—the optimal size for an epitope—and can recognize conformational epitopes on proteins or polysaccharides expressed in their native conformations. By contrast, TCRs recognize epitopes only after the immunogens have been partially degraded, or *processed*, for example, by antigen-presenting cells. Antigen processing destroys the conformational structure of a macromolecule, breaking proteins into peptides shorter than about 20 amino acids long.

Antibodies and TCRs can discriminate between closely related epitopes. However, specificity is not absolute, and an individual antibody or TCR may react to some extent with several different but structurally similar epitopes. The antigen that induced the antibody or TCR is called the *homologous* antigen, and any noninducing antigens that react with the antibody are called *heterologous* antigens. An interaction between an antibody or TCR and a heterologous antigen is called a *cross-reaction*.

The engagement of antibodies and TCRs with naturally encountered immunogens induces an adaptive immune response that often leads to long-term immunity. But as we now discuss, immunity may also be conferred either by purposely introducing an eliciting antigen or through passive mechanisms, either in the absence of or in response to antigen exposure.

Classes of Adaptive Immunity: Active and Passive

Adaptive immunity may develop following a natural exposure to an infectious agent, or it may result from an intentional exposure to immunogens. In addition, an immune response can be either *active*—generated by actual exposure to the antigen—or *passive*, such as from the transfer of antibodies or immune cells from an immune individual to a nonimmune one. Active and passive immunity are illustrated in **Figure 27.5** and contrasted in **Table 27.1**.

Natural active immunity is the outcome of exposure to antigens through infection and typically generates protective immunity from both antibodies and T cells. By contrast, *natural passive immunity* occurs when a nonimmune person acquires immune cells or antibodies through natural transfer of cells or antibodies from an immune person, such as from mothers to the fetus before birth or from mothers to newborns in breast milk (Figure 27.5).

Artificial active immunity is conferred by **vaccination (immunization)** and is a major weapon for the prevention and treatment of many infectious diseases. The introduction of antigen into the host triggers antibody production in a primary immune response but, more importantly, produces a population of immune memory cells. A second dose, often called a "booster," of the same antigen results in a secondary immune response in

TABLE 27.1 A comparison of active and passive immunity

Active immunity	Passive immunity
Exposure to antigen; immunity achieved by purposely administering antigen or through infection	No exposure to antigen; immunity achieved by injecting antibodies or antigen-reactive T cells
Specific immune response made by individual achieving immunity	Specific immune response made by the donor of antibodies or T cells
Immunity activated by antigen; immune memory in effect	No immune system activation; no immune memory
Immunity can be maintained via stimulation of memory cells (i.e., booster immunization)	Immunity cannot be maintained and decays rapidly
Immunity develops over a period of weeks	Immunity develops immediately

which existing memory cells are quickly activated, producing much higher levels of antibodies and a larger population of immune memory cells (Figure 27.1c). Active immunity may persist throughout life as a result of immune memory.

Artificial passive immunity is conferred when an individual receives antibodies from an immune individual through injection of an *antiserum*. These antibodies gradually disappear from the body, no immune memory is conferred, and a later exposure to the antigen does not elicit a secondary response. Artificial passive immunity is often used as postexposure therapy for acute infectious diseases, such as tetanus or rabies (for individuals not already protected against such diseases due to previous vaccination), and for the treatment of bites from venomous animals (antivenom, Figure 27.5). Antisera are obtained from immunized animals, such as horses, or from humans with high levels of antibody from a natural or artificial active immune response against the antigen.

--- **MINIQUIZ** ---

- Identify the intrinsic and extrinsic properties of an immunogen.
- Describe an epitope recognized by an antibody, and compare it to an epitope recognized by a TCR.
- Give an example for each: natural and artificial active immunity and natural and artificial passive immunity.

II • Antibodies

Antibodies provide antigen-specific immunity that protects against pathogens and dangerous soluble proteins, such as toxins. Here we look at the production, structural diversity, and antigen-binding function of antibodies. We also consider the organization and recombination of genes that encode antibodies, which underlie the nearly unlimited potential for the adaptive immune response to react to foreign molecules.

27.3 Antibody Production and Structural Diversity

An antibody, or *immunoglobulin (Ig)*, is a soluble protein made by a B lymphocyte or a plasma cell (⇨ Figure 26.4) in response to antigen exposure. Each antibody binds to a specific antigen. Antibody-mediated immunity controls the spread of infection by recognizing and binding antigens from pathogens and their products in extracellular environments, such as blood and mucus secretions, and in so doing triggers the removal of these foreign substances from the body.

B Cells, Antibodies and Their Activities, and Memory

B cells are lymphocytes that specialize in antibody production and contain antibodies (B cell receptors, BCRs) on their surfaces; each B cell has an estimated 100,000 BCRs of identical antigen specificity. To make antibodies, a B cell must first bind antigen by way of its BCR (**Figure 27.6**). The surface antibody–antigen interaction induces the B cell to ingest the bound antigen, which is often part of an entire pathogen, by endocytosis. The B cell then digests

the material and generates from it a suite of pathogen-derived antigenic peptides that are affixed to proteins of the *class II major histocompatibility complex (MHC II)* (Section 27.5) and displayed (*presented*) on the surface of the B cell. The B cell now functions as an APC to initiate an antibody-mediated immune response through interaction with an antigen-specific class of T lymphocyte called a *T helper (Th) cell*, more specifically, a Th2 cell. We discuss MHCs and the various T cell subsets in Section 27.8.

Th cells do not interact directly with pathogens but instead stimulate, or "help," other cells to become activated to carry out an immune response. In this case, the cell being "helped" is the antigen-presenting B cell on which the Th2 cell recognized the MHC–peptide antigen. Th2 cells produce cytokines that stimulate antigen-reactive B cells to grow and divide, which establishes a clone of the original antigen-reactive B cell. Each activated B cell clone can then proliferate and differentiate into several thousand plasma cell clones, each with the ability to produce and secrete large amounts of antibodies of identical antigenic specificity (Figure 27.6). This primary antibody response (Figure 27.1c) is detectable within about five days after antigen exposure, and serum levels of antibodies reach their peak within several weeks.

Some of the activated B cell clones remain in circulation in the immune system as long-lived *memory B cells* (Figure 27.6). Subsequent exposure to the same antigen, for example by reinfection with the same pathogen, stimulates the antigen-reactive memory B cells, producing a secondary antibody response characterized by a faster development of higher quantities of antibodies (Figures 27.1c and 27.6). Recall that the secondary response, conferred by immune memory, is the basis for vaccination (Section 27.2).

Antibodies released from plasma cells interact with antigen, which is often located directly on the pathogen. Antibody binding may have multiple effects on a pathogen, but most antibody interactions do not directly kill pathogens but instead block interactions between pathogens (or their products) and host cells. For example, antibodies present on mucous membranes may specifically interact with influenza virus antigens able to bind to host cells, thereby blocking attachment of the influenza virus to host cells on the mucosal surface. In addition, circulating antibodies in blood and lymph serum can *neutralize* toxins by binding them and preventing their attachment to host cell receptors (**Figure 27.7**). In other cases, antibodies coat pathogens by binding to their surface antigens, a process called *opsonization*, thereby marking them for destruction by phagocytosis (⇨ Section 26.9). Phagocytes have antibody receptors called *Fc receptors (FcR)* that bind to any antibody attached to an antigen, resulting in enhanced phagocytosis of the antibody-coated cells or viruses (⇨ Figure 26.14a).

Immunoglobulin G Structure and Function

There are five classes of immunoglobulins—*IgG, IgM, IgA, IgD,* and *IgE*—distinguished from one another by their different physical, chemical, and immunological properties. Based on these distinctions, each antibody class has a defined distribution and general function (**Table 27.2**). IgG is the most common circulating antibody, comprising up to 80% of serum immunoglobulins. IgG is composed of four polypeptide chains that are interconnected by disulfide (S—S) bonds (**Figure 27.8**). In each IgG molecule, two identical short chains (called *light chains*) are paired with two

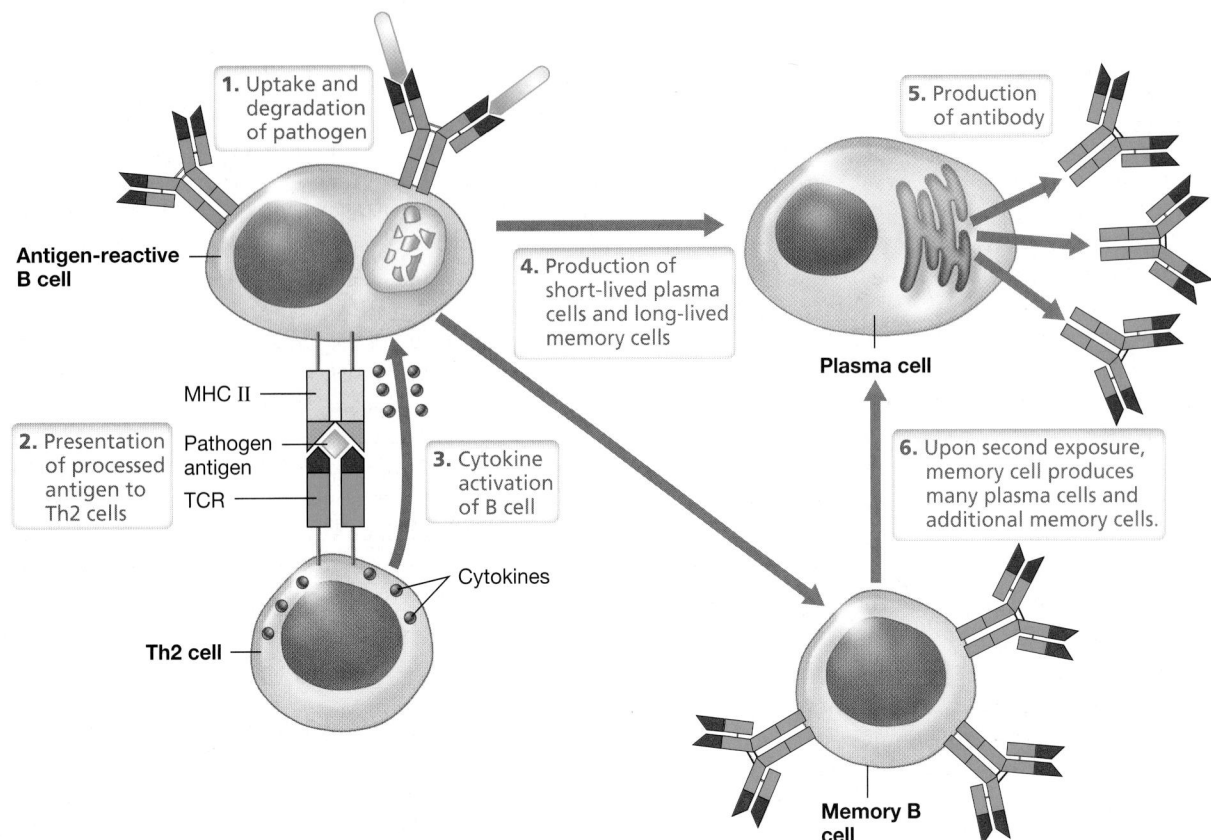

Figure 27.6 B cell–T cell interaction and antibody-mediated immunity. B cells interact with antigen and Th2 cells to produce antibodies. B cells initially function as antigen-presenting cells. First, their antigen-specific Ig receptor traps antigen. Following endocytosis, antigens are processed into peptide fragments, which are bound by MHC II and transported to the B cell surface. The MHC II–peptide complex is bound by the TCR on the Th2 cell, causing the Th2 cell to secrete IL-4 and IL-5. These cytokines activate the B cell to produce clones that differentiate into many antibody-producing plasma cells and a smaller number of memory cells. The latter are long-lived and can differentiate into antibody-producing plasma cells upon secondary exposure to the same antigen.

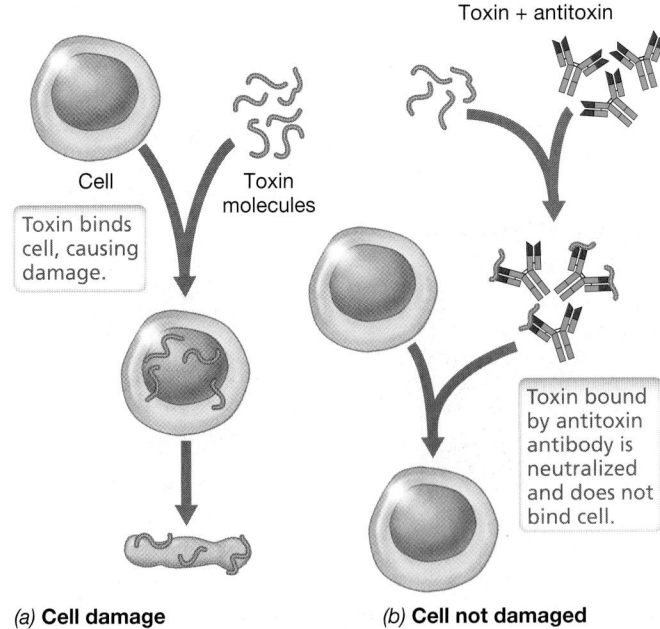

(a) Cell damage   **(b) Cell not damaged**

Figure 27.7 Neutralization of an exotoxin by an antitoxin antibody. (a) Toxin results in cell destruction. (b) Antitoxin prevents cell destruction. Examples of exotoxins include botulism and tetanus toxins and are discussed in Section 25.6.

identical longer chains (called *heavy chains*), yielding a Y-shaped, symmetrical molecule. Each light chain has about 220 amino acids, and each heavy chain has about 440 amino acids. Each heavy chain interacts with a light chain to form a functional *antigen-binding site*. Therefore, an IgG antibody molecule is *bivalent* because it has two identical sites that bind two identical antigenic determinants.

Heavy and light chains consist of a series of distinct protein domains, each composed of about 110 amino acids. A heavy-chain *variable domain* is connected to three heavy-chain *constant domains* (Figure 27.8a). The variable domain and the first constant domain (C_H1) compose part of the *Fab* (*f*ragment, *a*ntigen-*b*inding) portion of the immunoglobulin, so named because it contains the antigen-binding site. The two constant domains located most distal to the variable domain (C_H2 and C_H3) compose the *Fc* (*f*ragment, *c*rystalline) region of the antibody, named for its tendency to crystallize in solution. As previously mentioned, it is the Fc region of the antibody that binds FcR molecules on the surface of phagocytes to facilitate phagocytosis. The light chains of the antibody consist of one variable and one constant domain each and contribute to the two Fab regions only (Figure 27.8a).

Because the variable domains of a given antibody bind a specific antigenic determinant, their amino acid sequences differ

TABLE 27.2 Properties of human immunoglobulins

Class/H chain[a]	Structural conformation/ Antigen-binding sites	Serum concentration (mg/ml)/Percent of total circulating antibody	Properties	Distribution
IgG/γ	Monomer/2	13.5/70–80%	Major circulating antibody; four subclasses: IgG1, IgG2, IgG3, IgG4; IgG1 and IgG3 activate complement	Extracellular fluid; blood and lymph; intestine; crosses placenta
IgM/μ	Pentamer/10 Monomer/2	1.5/6–10% 0	First antibody to appear in primary response to extracellular pathogens or after immunization; pentamer especially effective in agglutinating antigens; strong complement activator	Blood and lymph; monomer is B cell surface receptor (BCR)
IgA/α	Monomer/2 Dimer/4	3.5/10–20% 0.05/0.2–0.3%	Important circulating antibody Major secretory antibody	Blood and lymph (monomer) and secretions, such as mucus, saliva, and colostrum (dimer)
IgD/δ	Monomer/2	0.03/0.2–0.3%	Minor circulating antibody; mostly associated with mature B cells	B cell surface receptor (BCR); blood and lymph (trace)
IgE/ε	Monomer/2	0.00005/0.0003%	Facilitates parasite immunity but also triggers allergic reactions	Blood and lymph; binds to mast cells and eosinophils

[a]All immunoglobulins may have either λ or κ light-chain types, but not both.

in each different antibody. The variable domain of a light chain (V_L) interacts with the variable domain of a heavy chain (V_H) to bind antigen. By contrast, the constant domains of each heavy chain are identical in amino acid sequence for all Ig molecules of a given class. Similarly, in light chains of the same type, the amino acid sequence of the constant domain is the same.

Other Immunoglobulin Classes and Their Functions

Immunoglobulins of the other classes differ from IgG in structure and function. Because the amino acid sequence of heavy-chain constant domains determines antibody class, the heavy chain called *gamma* (γ) defines the IgG class. Likewise, *alpha* (α) defines IgA; *mu* (μ) defines IgM; *delta* (δ) defines IgD; and *epsilon* (ε) defines IgE (Table 27.2). The constant-domain sequences constitute three-fourths of the heavy chains of IgG, IgA, and IgD and four-fifths of the heavy chains of IgM and IgE (**Figure 27.9a, b**).

The structure of IgM is shown in Figure 27.9c. Circulating IgM forms an aggregate of five immunoglobulin molecules attached by at least one J (joining) peptide. IgM is the first class of antibody produced in a typical immune response to a bacterial infection, but IgM antibodies generally have low *affinity* (binding strength) for antigen. However, overall antigen-binding strength is enhanced because pentameric IgM molecules have ten binding sites available for interaction with antigen (Table 27.2 and Figure 27.9c). The ten antigen-binding sites make IgM especially effective at agglutinating (clumping) infectious particles, thereby increasing phagocytic efficiency. IgM is also a strong activator of complement through the classical pathway (⟳ Section 26.9). Whereas up to 10% of serum antibodies are pentameric IgM, IgM monomers do not circulate in the blood but rather are one of the classes of antibodies that function as B cell receptors.

Dimers of IgA are present in high numbers in secreted body fluids, including saliva, tears, breast milk colostrum, and mucosal secretions from the gastrointestinal, respiratory, and genitourinary tracts. The mucosal surfaces in an average adult, which total about 400 m^2 (skin has about 6 m^2), contain MALT and produce large amounts (about 10 g) of secretory IgA every day. By comparison, an individual produces only about 5 g of serum IgG per day. Whereas *serum* IgA typically occurs in monomeric form (Table 27.2), *secretory* IgA consists of two IgA molecules covalently linked by a J chain peptide and complexed with a protein called the *secretory component* that aids in transport of IgA across

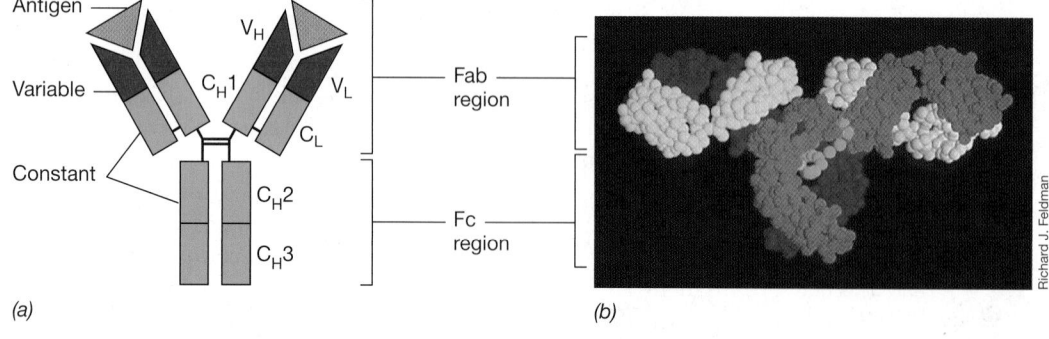

(a)

(b)

Richard J. Feldman

Figure 27.8 Immunoglobulin G structure. (a) IgG consists of two heavy chains (50,000 molecular weight) and two light chains (25,000 molecular weight), with a total molecular weight of 150,000. One heavy and one light chain interact to form an antigen-binding unit. The variable domains of the heavy and light chains (V_H and V_L) bind antigen. The constant domains (C_H1, C_H2, C_H3, C_L) are identical in all IgG proteins. The chains are covalently joined with disulfide bonds. The Fab (*fragment, antigen-binding*) region contains the antigen-binding site. The Fc (*fragment, crystalline*) stem of the antibody binds receptor molecules on phagocytes to facilitate phagocytosis of opsonized pathogens. (b) Space-filling model of an IgG molecule. The heavy chains are red and dark blue. The light chains are green and light blue.

membranes (Figure 27.9*d*). The high concentration of secretory IgA in breast milk colostrum likely plays a key role in preventing gastrointestinal disease in newborns.

IgE is found in very small amounts in serum; about 1 of every 3000 serum Ig molecules is IgE. Most IgE is bound to cells. For

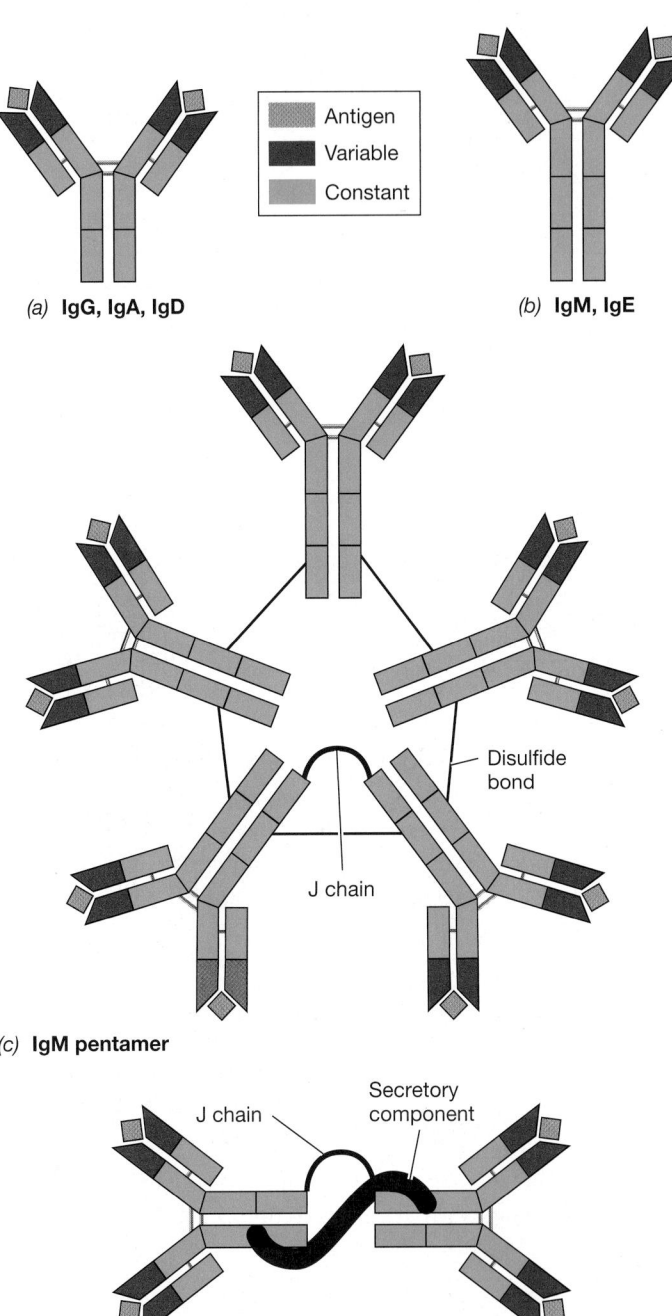

(a) **IgG, IgA, IgD**

(b) **IgM, IgE**

(c) **IgM pentamer**

(d) **IgA dimer**

Figure 27.9 Immunoglobulin classes. All classes of Igs have V_H and V_L that bind antigen. *(a)* IgG, IgA, and IgD have three constant domains. *(b)* The heavy chains of IgM and IgE have a fourth constant domain. *(c)* IgM is found in serum as a pentameric protein consisting of five IgM proteins covalently linked to one another via disulfide bonds and a joining (J) chain protein. Because it is a pentamer, IgM can bind up to ten antigens, as shown. *(d)* Secretory IgA is often found in body secretions as a dimer consisting of two IgA proteins covalently linked to one another by a J chain protein. The secretory component aids in transport of IgA across mucosal membranes.

example, through its Fc region IgE binds eosinophils, arming these granulocytes to target parasitic pathogens such as schistosomes and other helminths (⮌ Section 33.7). IgE also binds to tissue mast cells, where subsequent binding of antigen to antigen-binding sites of IgE causes degranulation of the mast cells. The ensuing release of chemical mediators, such as histamine and serotonin, triggers immediate-type hypersensitivities (a type of allergic response, Section 27.9). Like IgM, IgE has a fourth heavy-chain constant domain (Figure 27.9*b*), and this allows for binding of IgE to eosinophils and mast cells, the key step for activating the protective and allergic reactions, respectively, associated with these cell types.

IgD (Figure 27.9*a*), present in serum in low concentrations, has no known unique immune protective function. However, IgD, like IgM, is abundant on the surfaces of mature B cells, especially memory B cells, where it functions as a B cell receptor.

Antigen Exposure, Immune Memory, and the Primary and Secondary Responses

We have seen that immune memory is a major characteristic of the adaptive immune response (Figure 27.1*c*). Starting with a B cell, an antibody-mediated immune response begins with antigen exposure and culminates with the production and secretion of antigen-specific antibodies, and the route of antigen exposure influences the class of the antibodies produced. Blood and lymph, as well as the spleen and MALT (Chapter 26), are key sites for the introduction of antigens. Antigen introduced into the bloodstream by injection or natural infection travels to the spleen, where IgM, IgG, and serum IgA antibodies are formed. Antigen introduced subcutaneously, intradermally, topically, or intraperitoneally is carried through the lymphatic system to the nearest lymph nodes, again stimulating production of IgM, IgG, and serum IgA. Antigen introduced to mucosal surfaces is delivered to the nearest MALT. For example, antigen delivered orally is delivered to the MALT in the intestinal tract, preferentially stimulating production of secretory IgA in the gut.

Following initial antigen contact, each antigen-stimulated B cell multiplies and differentiates to form antibody-secreting plasma cells and memory B cells (Figure 27.6). Plasma cells in this *primary antibody response* are relatively short-lived (less than one week) and secrete large amounts of mostly IgM antibody (**Figure 27.10**). After a latent period of several days, antibody appears in the blood and a gradual increase in *antibody titer* (antibody quantity) occurs. As antigen disappears, the primary antibody response slowly diminishes.

Memory B cells generated by the initial exposure to antigen may circulate in the host for years. A second exposure to the same antigen stimulates memory B cells to rapidly differentiate into plasma cells and produce antibody; memory B cells need no T cell help. The second and each subsequent exposure to antigen causes the antibody titer to rise rapidly to a level often 10–100 times greater than the titer following the first exposure (Figure 27.10). This rise in antibody titer is called the *secondary antibody response*. The secondary response is a function of immune memory and typically results in a switch from the production of IgM to another antibody class, a transition called *class switching* (Figure 27.10). This switch is induced by a cytokine signal produced in response to the specific type of

UNIT 6

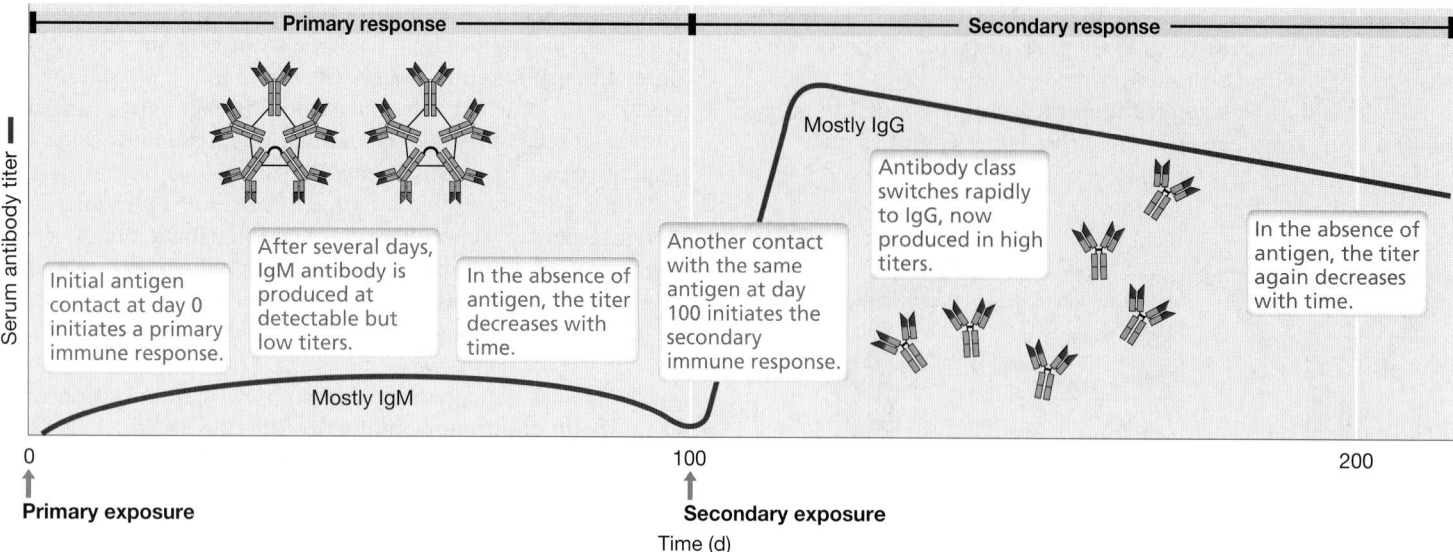

Figure 27.10 Primary and secondary antibody responses in serum. The antigen exposures at day 0 and day 100 must be to identical antigen to induce a secondary response. The secondary response, also called a booster response, may be more than 10-fold greater than the primary response. Note the class switch from IgM production in the primary response to IgG production in the secondary response.

pathogen invasion. The signal prompts a recombination event that switches the heavy-chain constant region, resulting in the production of the antibody class that can most efficiently address the threat. For example, in serum, the most common antibody class switch is from IgM to IgG, whereas in mucosal tissues, switches to IgA predominate. It is important to note that class switching does not cause a change in antigen specificity; the new class of antibody generated remains specific to the original antigen.

Serum antibody titer slowly decreases over time, but subsequent exposure(s) to the same antigen can trigger another secondary response. This rapid and strong memory response is the basis for the immunization procedure known as a "booster shot" (for example, the rabies shot given every 1–3 years to domestic animals). Periodic reimmunization maintains high levels of memory B cells and circulating antibody specific for a particular antigen, providing long-term artificial active immunity against individual infectious diseases.

— MINIQUIZ —

- Summarize antibody production starting with pathogen interaction with a B cell.
- Differentiate among antibody classes using structural characteristics, distribution patterns, and functional roles.
- Explain the rationale for periodic revaccination in children and adults.

27.4 Antigen Binding and the Genetics of Antibody Diversity

We have seen that antibodies consist of four polypeptides, two heavy (H) chains and two light (L) chains (Figure 27.8), and each chain is further divided into constant (C) and variable (V)

domains. The V domains of one H and one L chain interact to form the antigen-binding site. Here we examine the structural features of the V domains that define the antigen-binding site and then explore the genetics behind the enormous diversity of antibodies that are possible.

Variable Domains and Antigen–Antibody Interaction

The V domains of an H and an L chain interact to form a receptor that binds antigen strongly but not covalently; the measurable strength of binding is the antibody's *binding affinity*. The vertebrate immune system can recognize, or bind, countless antigens, and each antigen binds to an antibody at a unique antigen-binding site (**Figure 27.11**). To accommodate all possible antigens that a host might encounter, the synthesis of billions of different antibodies, each with its own unique antigen-binding site, is necessary. The V domains define these unique antigen-binding sites.

Amino acid sequences differ in the V domains of Igs that bind different antigens. Amino acid variability is especially apparent in several **complementarity-determining regions (CDRs)**. The three CDRs in each of the V domains provide most of the molecular contacts with antigen (Figure 27.11). The amino acid sequences of CDR1 and CDR2 differ in minor ways between different Igs, while CDR3s differ in sequence dramatically from one another. Three distinct gene segments encode CDR3 of the H chain: the carboxy-terminal portion of the V domain, followed by a short "diversity" (D) segment of about three amino acids, and a longer "joining" (J) segment about 13–15 amino acids long. The light-chain CDR3 is similar except that it lacks the D segment. The heavy- and light-chain CDRs, six in total, confer highly specific antigen binding on the antibody molecule (Figure 27.11).

The Ig three-dimensional structure was shown in Figure 27.8*b*. Each antigen–antibody interaction requires the specific binding of an antigenic *epitope* with the CDR domains of the H and

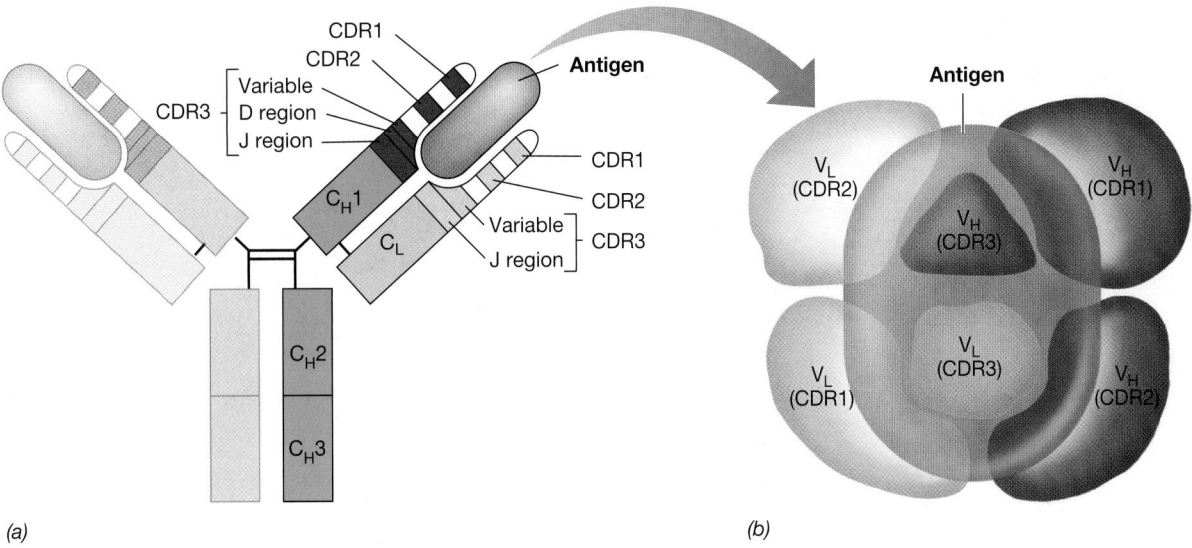

Figure 27.11 Antigen binding by immunoglobulin light and heavy chains. *(a)* An Ig is shown schematically, with bound antigen (light brown). The V domains on the H and L chains are shown in red (heavy chain) and pink (light chain), with the antigen-binding CDR1, CDR2, and CDR3. C_H1, C_H2, and C_H3

(dark blue) are constant domains in the H chain, and C_L (light blue) is the constant domain in the L chain. *(b)* Complementarity-determining regions (CDRs) from both H (red) and L (pink) chains, shown from above, are conformed to make a single antigen-binding site on the Ig. The highly variable CDR3s from both H and L chains

cooperate at the center of the site. Antigen (light brown) may contact all CDRs. The shape of the site may be a shallow groove or a deep pocket, depending on the antibody–antigen pair involved.

L chains. A single antigen-binding site of an antibody molecule measures about 2×3 nm, creating a space large enough to accommodate an epitope of 10 to 15 amino acids. Antigen binding is ultimately a function of the Ig folding pattern of the H and L polypeptide chains. The Ig folds of the V region bring all six CDRs (CDR1, 2, and 3 from both heavy and light chains) together, resulting in the formation of a unique and specific antigen-binding site (Figure 27.11).

Because an individual can recognize and bind an immense variety of antigens, the immune system must be capable of virtually unlimited antibody variation. In order to accomplish this from the relatively small genomic investment evolution has made in immunoglobulin genes, several unusual genetic mechanisms come into play. Table 27.3 summarizes these diversity-generating mechanisms, and we explore each below.

Genetic Organization of the Immunoglobulin Molecule

The gene encoding each immunoglobulin H or L chain is constructed from several gene segments. In each B cell, Ig gene segments undergo a series of random, somatic rearrangements characterized by genetic recombination followed by deletion of intervening sequences. These events produce a single, functional antibody gene derived from a relatively small pool of Ig gene segments. As a B cell matures, V, D, and J gene segments are enzymatically recombined to form a single Ig heavy-chain gene (**Figure 27.12a**). The single V gene segment encodes CDR1 and CDR2, whereas CDR3 is encoded by a mosaic of the 3′ end of the V region, followed by the D and J gene segments.

In each B cell, only one protein-producing rearrangement occurs in the heavy- and light-chain genes. Called *allelic exclusion*, this mechanism causes each B cell to express Ig genes from only one of the two inherited parental alleles, thus ensuring that all Igs

from a given clone of B cells have identical antigen specificity. Finally, separate C gene segments encode the class-defining constant domains of Igs. Therefore, four different gene segments—V, D, J, and C—recombine to form one functional heavy-chain gene (Figure 27.12a). Light chains, which lack D segments, are encoded in a similar way by recombination of light-chain V, J, and C segments (Figure 27.12b). The gene segments required for all Igs exist in all nucleated cells of the body but undergo recombination only in developing B lymphocytes.

V, D, J, and C gene segments are separated by noncoding sequences (introns) typical of gene arrangements in eukaryotes. Genetic recombination occurs in each B cell during its development. One each of the V, D, and J segments is recombined at random to form a functional heavy-chain gene, while randomly recombined V and J segments form a complete light-chain gene.

TABLE 27.3 Generation of antigen-binding receptor diversity in B cells and T cells

Diversity-generating mechanism	B cell Ig receptors, heavy and light chains	T cell receptors, α and β chains
Somatic recombination of tandem genes	Yes	Yes
Random reassortment	Yes	Yes
Imprecise V-D-J or V-J joining	Yes	Yes
Nucleotide additions at V-D-J or V-J junctions	Yes	Yes
D gene segments read in all 3 frames	No	Yes
Somatic hypermutation	Yes	No

UNIT 6

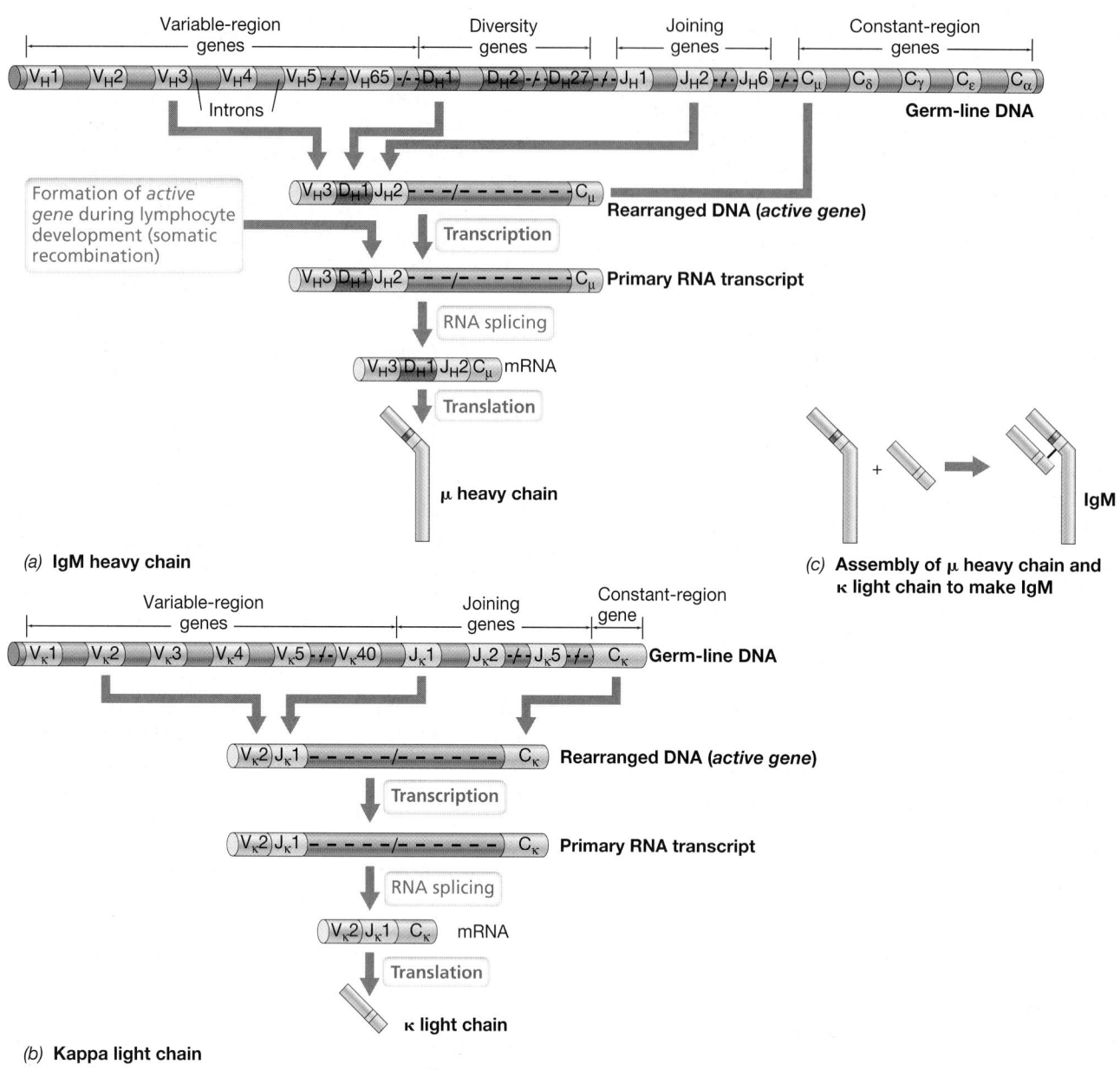

Figure 27.12 Immunoglobulin gene rearrangement in human B cells. Ig genes are arranged in tandem on three different chromosomes. *(a)* The H chain gene complex on human chromosome 14. The filled boxes represent Ig coding genes. The broken lines indicate intervening sequences and are not shown to scale. *(b)* The κ light-chain complex on human chromosome 2. The λ light-chain genes are in a similar complex on human chromosome 22. *(c)* Assembly of one-half of an antibody molecule.

The active gene, still containing an intervening sequence between the VDJ or VJ gene segments and the C gene segments, is transcribed, and the resulting primary RNA transcript is spliced to yield the final messenger RNA (mRNA). The mRNA is then translated to make the heavy and light chains of the Ig molecule.

Reassortment, VDJ Joining, and Hypermutation

If one considers all of the random reassortments possible in genes encoding immunoglobulin heavy and light chains and from these calculates the number of unique molecules that could be encoded, the total diversity is astonishing. In humans, for example, based on the numbers of kappa (κ) light-chain gene segments, there are about 40 V × 5 J possible rearrangements, or 200 possible κ light chains. For the alternative lambda (λ) light chain, the type produced instead of κ by some B cells in every individual, there are about 30 V × 4 J = 120 possible combinations. About 10,500 possible heavy chains can be formed by the rearrangement of about 65 V × 27 D × 6 J genes. Assuming that each heavy chain and light chain has an equal chance to be expressed, there are 10,500 × 200 = 2,100,000 possible immunoglobulins with κ light chains and 10,500 × 120 = 1,260,000 possible immunoglobulins with λ chains. Therefore, at least 3,360,000 possible antibodies can be expressed based on random reassortment alone!

On top of this impressive diversity, additional diversity is generated in the CDR3 regions of both heavy and light chains by several unique mechanisms. First, the DNA-joining mechanism that constructs V-D or D-J gene segments in the heavy chain or the V-J gene segments in the light chain is an imprecise process. The final nucleotide sequence at these regions occasionally varies by a few nucleotides from the original genomic sequence, and when this occurs, it will alter the amino acid sequence in this region. Even more diversity is generated at V-D and D-J coding joints in the heavy-chain genes and at V-J coding joints in light-chain genes by either random (N) or template-specific (P) nucleotide additions. Because these coding joints are contained within the sequences that encode CDR3 on both heavy and light chains, N and P diversity at V-domain coding joints changes or adds amino acids in the CDR3 of both heavy and light chains.

Immunoglobulin diversity is further expanded by the process of **somatic hypermutation**, the mutation of B cell Ig genes at much higher rates than those observed in other genes. Somatic hypermutation of Ig genes is typically evident after a second exposure to an immunizing antigen and occurs only in the V regions of rearranged heavy- and light-chain genes, creating B cells with slightly altered Ig cell surface receptors. These mutated B cells compete for available antigen, and B cells whose receptors have a higher affinity for antigen than the original B cell receptor are selected. This *affinity maturation* process is one of the factors responsible for a dramatically stronger secondary immune response (Figures 27.1*c* and 27.10). Affinity maturation also adds more diversity to the pool of antibody specificity in the adaptive immune response, thus making the potential antibody repertoire essentially infinite.

— **MINIQUIZ** —

- Draw a complete Ig molecule and identify antigen-binding sites on the antibody.
- Describe antigen binding to the CDR1, 2, and 3 regions of the heavy-chain and light-chain variable domains.
- Describe the recombination events that produce a mature heavy-chain gene and other somatic events that further enhance antibody diversity.

III • The Major Histocompatibility Complex (MHC)

The *major histocompatibility complex* (*MHC*) is a series of genes found in all vertebrates that encodes a group of proteins important in antigen presentation. The MHC proteins in humans are called **human leukocyte antigens (HLAs)** and were first identified as the major antigens responsible for immune-mediated tissue transplant rejection. We now know that MHC proteins function primarily as antigen-presenting molecules, binding pathogen-derived peptides and displaying these peptides for interaction with T cell receptors.

27.5 MHC Proteins and Their Functions

The MHC proteins consist of two protein classes encoded by about 4 megabase pairs (Mbp) of DNA (**Figure 27.13a**). **MHC class I proteins** are found on the surfaces of all nucleated cells and function to present peptide antigens to *T-cytotoxic* (*Tc*) *cells*. If a peptide antigen presented by MHC class I is recognized by the TCR of a Tc cell, the antigen-containing cell is quickly destroyed (Section 27.8). **MHC class II proteins** are found only on the surface of B lymphocytes, macrophages, and dendritic cells—the antigen-presenting cells (APCs) (ᶜᵖ Section 26.4). Through their class II proteins, APCs present antigens to *T-helper* (*Th*) *cells*, stimulating cytokine production that leads to antibody-mediated immunity or inflammatory responses (Section 27.8).

Class I and Class II MHC Proteins

Class I MHC proteins consist of two polypeptides (Figure 27.13*b, d,* and *e*). The first of these is the membrane-embedded *alpha* (α) *chain*, encoded by a gene located in the MHC gene region. The second is *beta-2 microglobulin* (β$_2$m), a smaller, noncovalently associated protein encoded by a gene not included in the MHC gene cluster. The class I α chain folds to form a peptide-binding groove that accommodates peptides of 8 to 11 amino acids. The peptides in infected cells are derived from *endogenous* (intracellular) foreign antigens (Figure 27.13*d*), for example, from viral proteins produced inside the cell. Following translocation to the cell surface, the MHC–virus-peptide complexes, which to the immune system look much like the variant MHC proteins associated with a tissue transplant, are recognized as foreign by TCRs on Tc cells and are destroyed.

Class II MHC proteins consist of two membrane-integrated polypeptides, α and β, found only on APCs. One α and one β polypeptide, expressed together, form a functional heterodimer (Figure 27.13*c, f*). The α1 and β1 domains of the class II protein interact to form a binding site for TCR–peptide similar to the class I binding site for TCR–peptide. However, the peptide-binding groove in the class II MHC is able to bind and display peptides that may be significantly longer than 8–11 amino acids. Peptides presented by class II MHCs are typically 10 to 20 amino acids long and are proteolytic fragments derived from *exogenous* (extracellular) pathogens that have been internalized and processed by APCs. The APCs use the class II–peptide complex to interact with TCRs on Th cells, leading to Th activation and an adaptive immune response (Section 27.8).

Antigen Processing and Presentation to T Cells

MHC proteins are expressed on cell surfaces only when they are complexed with a peptide, either a self peptide or a foreign peptide. In essence, then, *MHC–peptide complexes reveal to the immune system the protein composition of the cell*, and in this capacity, function to alert the immune system when a cell contains foreign antigens. T cells, through their TCRs, conduct surveillance of cell surfaces to identify any cells expressing foreign antigens. When the latter are encountered, the TCR interacts with the foreign antigen presented on the MHC protein, and this interaction targets the cell for destruction. T cells do not react with MHC complexes containing only self peptides because self-reactive T cells are eliminated from the host during the development of immune tolerance (Section 27.1).

UNIT 6

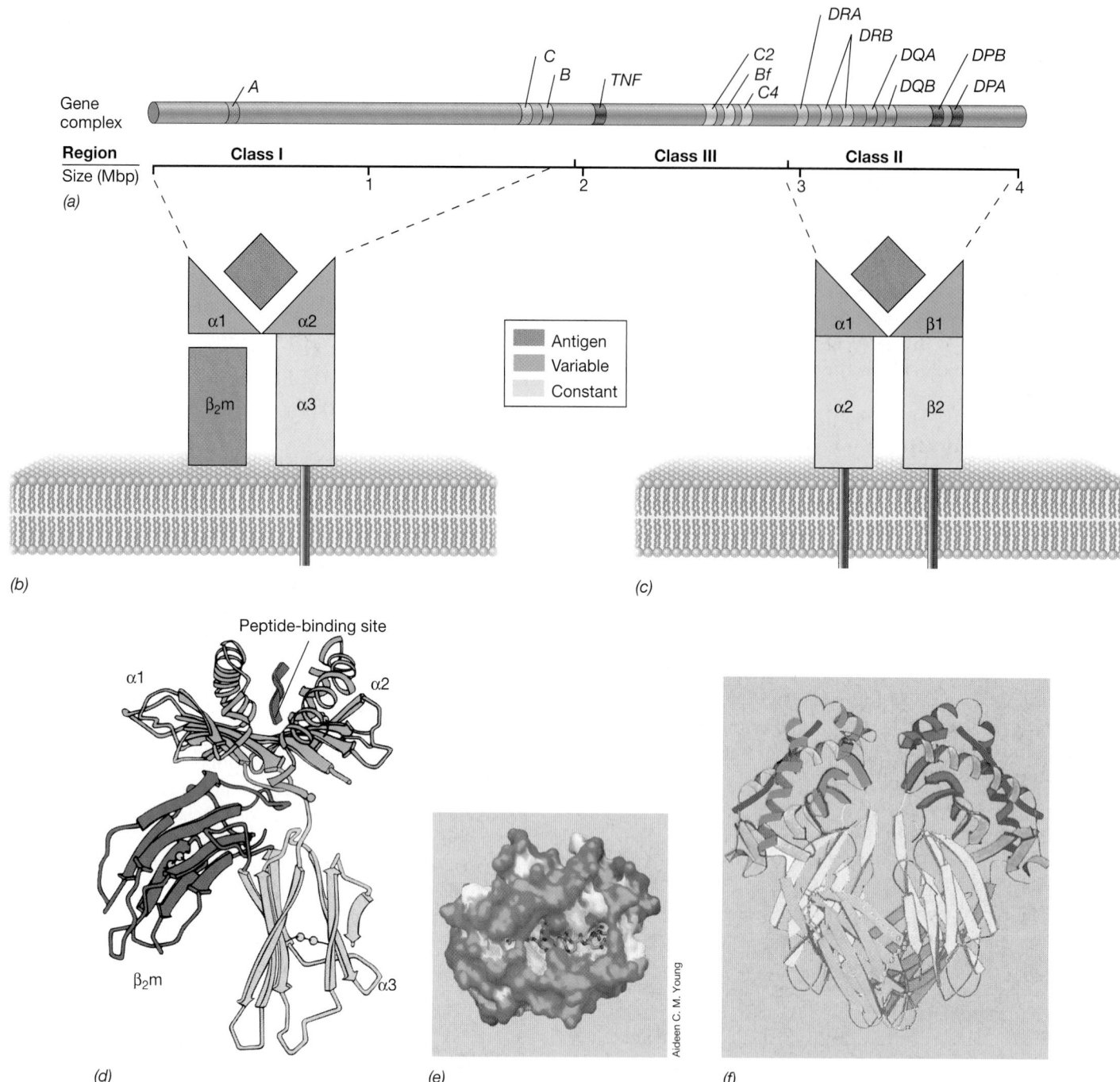

Figure 27.13 Human leukocyte antigen (HLA) genes and MHC proteins. *(a)* The HLA complex, located on human chromosome 6, contains more than 4 megabase pairs (Mbp). Class II genes *DPA* and *DPB* encode class II proteins DPα and DPβ; *DQA* and *DQB* encode DQα and DQβ; *DRA* and two *DRB* loci encode DRα and DRβ proteins. The class I MHC proteins HLA-A, HLA-C, and HLA-B are encoded by genes *A*, *C*, and *B*. The class II and class I loci are highly polymorphic and encode peptide-binding proteins. Class III MHC genes encode proteins associated with immune-related functions, such as complement proteins C4 and C2 and the cytokine TNF (tumor necrosis factor). *(b)* Schematic of MHC class I protein. The α1 and α2 domains interact to form the peptide antigen–binding site. *(c)* Schematic of MHC class II protein. The α1 and β1 domains combine to form the peptide antigen–binding site. *(d)* MHC class I protein structure. Beta-2 microglobulin (β_2m) binds noncovalently to the α chain. The antigen peptide (brown) is bound cooperatively by the α1 and α2 domains. *(e)* An MHC I protein with a bound peptide, as seen from above. A peptide of nine amino acids is shown as a carbon backbone structure, embedded in a space-filling model of a mouse MHC I protein. *(f)* Structure of MHC class II protein dimer. The peptides (brown) are shown in their positions in the binding sites of the MHC II proteins.

Two distinct antigen-processing schemes are at work to initiate adaptive immune responses: one for MHC I antigen presentation and another for MHC II antigen presentation. MHC I proteins present peptide from viruses or other intracellular pathogens; such infected cells are called *target cells* (**Figure 27.14a**). Proteins derived from infecting viruses, for example, are taken up and digested in the cytoplasm in a structure called the *proteasome*. Peptides about ten amino acids long are transported into the endoplasmic reticulum (ER)

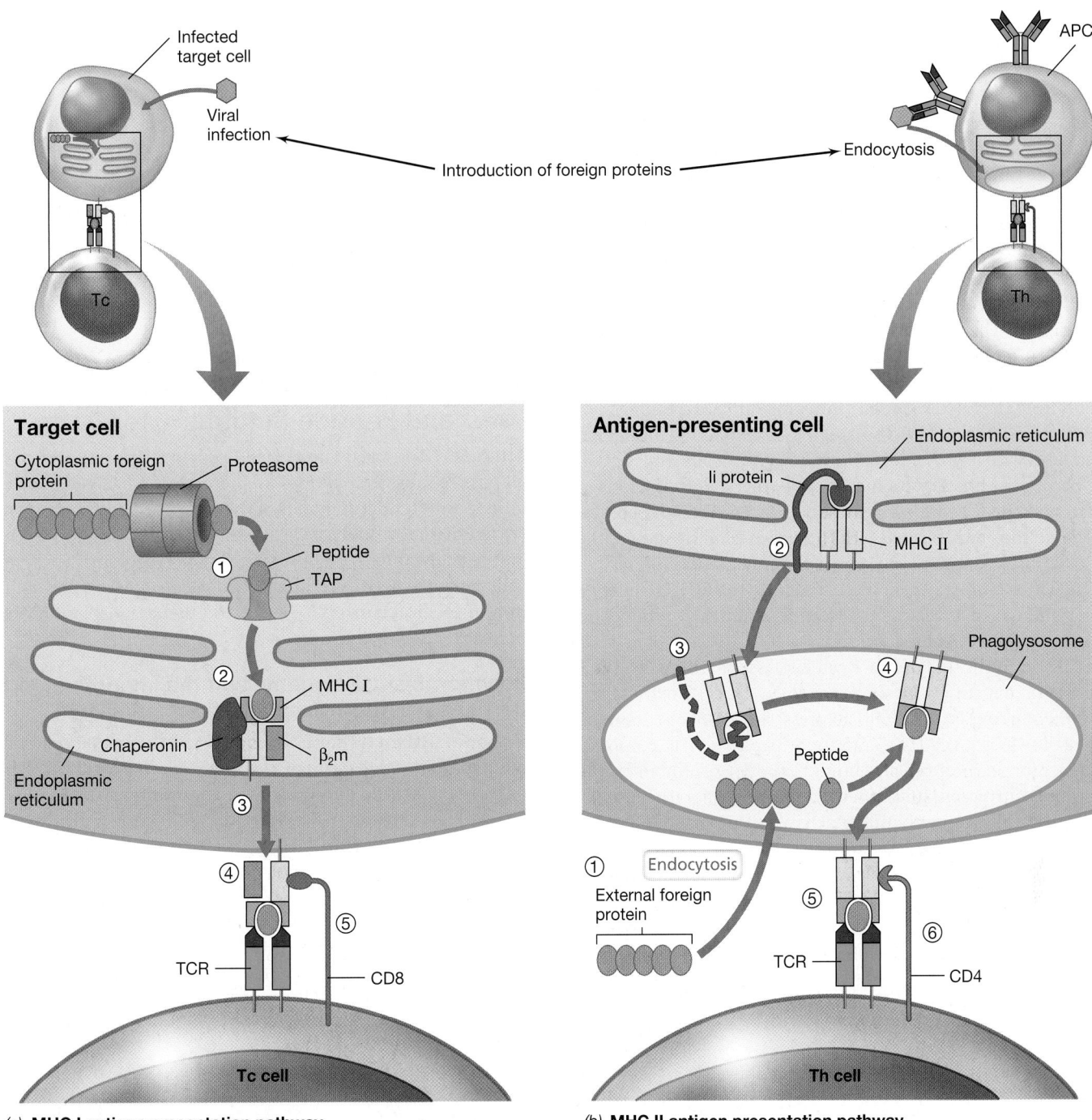

(a) **MHC I antigen presentation pathway**

(b) **MHC II antigen presentation pathway**

Figure 27.14 Antigen presentation by MHC I and MHC II proteins. *(a)* ① Protein antigens, such as virus components manufactured within the cell, are degraded by the proteasome in the cytoplasm. The peptide fragments are transported into the endoplasmic reticulum (ER) through a pore formed by the TAP proteins. ② MHC I proteins in the ER are stabilized by chaperonins until peptide fragments are bound. ③ When peptide fragments are bound by MHC I, the complex is transported to the cell surface. ④ The MHC I–peptide complex interacts with T cell receptors (TCRs) on the surface of Tc cells. ⑤ The CD8 coreceptor on the Tc cell engages MHC I, resulting in a stronger complex. The Tc cell is activated by the binding events, causing it to release cytokines and cytolytic toxins and kill the target cell. *(b)* ① External foreign proteins are imported into the cell and digested into peptide fragments in phagolysosomes. ② MHC II proteins in the ER are assembled with Ii, a blocking protein that prevents MHC II from binding with peptides in the ER. ③ The MHC II–Ii assembly is transported to the lysosome, where it remains until the lysosome fuses with the phagosome, forming a phagolysosome where Ii is degraded, ④ freeing the MHC II protein to bind the foreign peptide fragments. ⑤ The MHC II–peptide complex is transported to the cell surface, where it interacts with TCRs and ⑥ the CD4 coreceptor on Th cells. The Th cells then release cytokines that interact with other cells to promote an immune response. Note that the APC in part (b) may be either a B cell, which ingests antigen by endocytosis (shown), or a macrophage or dendritic cell, which engulf antigens through phagocytosis.

through a pore called the *transporter associated with antigen processing* (*TAP*). Once the foreign peptides have entered the ER they are bound by MHC I, and the MHC I–peptide complex moves to the cell surface where it integrates into the cell membrane. When the TCR on the surface of a Tc cell interacts with both the foreign peptide (recognized as "nonself") and the MHC I protein (recognized as "self") on the surface of the target cell, the Tc cell releases *perforin* and *granzymes*, cytotoxic proteins that kill the virus-infected target cell.

MHC II proteins are the antigen-presenting proteins in a second pathway (Figure 27.14*b*). MHC II proteins are expressed exclusively on APCs, where they function to present peptide antigens from engulfed pathogens. Similar to MHC I proteins, MHC class II proteins are assembled in the endoplasmic reticulum. However, unlike events with MHC I proteins (Figure 27.14*a*), a protein called Ii binds MHC II, and this blocks peptide loading and labels the entire complex for transport from the ER to lysosomes. After ingestion of a pathogen or pathogen product by an APC, the phagosome containing the foreign antigen fuses with a lysosome to form a phagolysosome (⊘ Section 26.7). Within the phagolysosome, enzymes digest both the foreign antigens and the Ii peptide but not MHC II. The foreign peptides then bind to the newly opened MHC II antigen-binding site and the complex is inserted on the cell surface for presentation to T-helper cells. The latter, through their TCRs, recognize the MHC II–peptide complex and secrete cytokines that stimulate antibody production by B cells or induce inflammation.

In addition to the TCR, each T cell expresses a unique cell surface protein that functions as a *coreceptor*. Th cells express the **CD4 coreceptor**, and Tc cells express the **CD8 coreceptor** (Figure 27.14). When the TCR binds to the peptide–MHC complex, the coreceptor on the T cell also binds to the MHC protein on the antigen-presenting cell, strengthening the molecular interactions between the cells and enhancing activation of the T cell. CD4 binds only to MHC II, strengthening Th cell interaction with APCs that express MHC II protein. Likewise, CD8 binds only to MHC I, enhancing the binding of Tc cells to MHC I–bearing target cells. In clinical medicine, the CD4 and CD8 proteins are used as T cell markers to differentiate Th (CD4) cells from Tc (CD8) cells in

diagnostic tests, for example, in assessing the course of disease in an AIDS patient (⊘ Section 30.15 and Figure 30.44).

27.6 MHC Polymorphism, Polygeny, and Peptide Binding

Although MHC class I and class II proteins theoretically can bind all possible antigen peptides for presentation to T cells, MHC proteins in different individuals of the same species are not identical. Different individuals typically have subtle differences in the amino acid sequence of homologous MHC proteins. These genetically encoded MHC variants, called *polymorphisms*, are the major immunological barriers for successful tissue transplantation from one individual to another.

Polymorphism, Polygeny, and the Immune Barrier to Tissue Transplantation

Polymorphism is the occurrence, within a population, of multiple alleles (alternate forms of a gene) at a specific locus (the location of the gene on the chromosome). For example, the MHC class I locus *HLA-A* (Figure 27.13*a*) has over 2000 known alleles, each of which encodes a distinct HLA-A protein that occurs within the human population. The genome of each person, however, contains only two of the *HLA-A* alleles; one allele is of paternal origin and one is of maternal origin. The two allelic protein products are expressed equally (**Figure 27.15*a***).

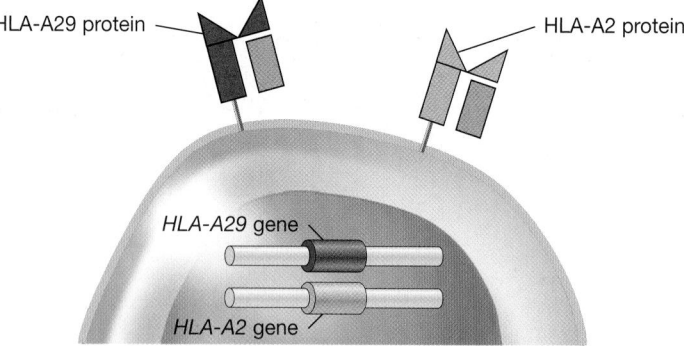

(a) **Polymorphism**

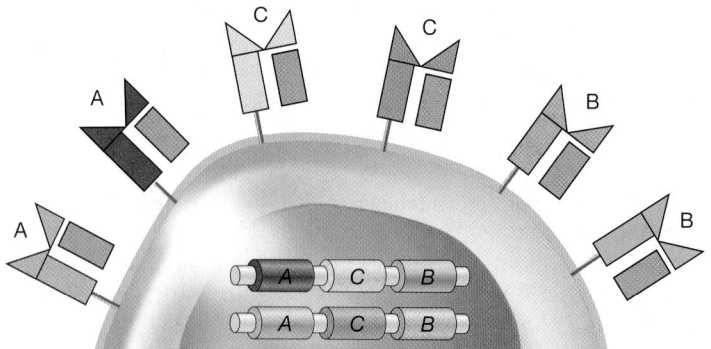

(b) **Polygeny**

Figure 27.15 Polymorphism and polygeny in MHC genes and proteins. *(a)* Polymorphism in *HLA-A* loci results in equal expression of proteins encoded by both alleles. There are over 2000 HLA-A alleles in the human population, but only two (one at each locus) are found in each individual. *HLA-B* and *HLA-C* exhibit similar levels of polymorphism. *(b)* Polygeny in MHC results in duplicated polymorphic *HLA-A*, *HLA-B*, and *HLA-C* genes that potentially encode three pairs of different MHC proteins. The colors represent alternate alleles of each gene and their respective protein.

Additional MHC class I protein diversity is a result of **polygeny**, in which evolutionary gene duplication events have led to the occurrence of two additional genetically, structurally, and functionally related loci, *HLA-B* and *HLA-C*. These gene loci are also polymorphic, having more than 2600 and 1500 allelic variants, respectively. Thus, an individual typically displays six structurally distinct proteins derived from the three polymorphic class I loci (three products derived from maternal origin and three products derived from paternal origin) (Figure 27.15*b*). Likewise, highly polymorphic and equally expressed alleles encode MHC class II proteins at the *HLA-DR, HLA-DP,* and *HLA-DQ* alpha- and beta-chain loci (Figure 27.13*a*).

As a result of polymorphism and polygeny, most individuals have unique MHC profiles. Only closely related family members are likely to have all of the same MHC genes and proteins, and this property can be used to prove (or disprove) paternity or to trace ancestral lineages (**Figure 27.16**). These highly polymorphic variations in MHC proteins are major barriers to successful tissue transplants because the MHC proteins on the donor tissue (graft) are recognized as foreign by the recipient's immune system. An immune response directed against the graft MHC proteins thus causes rejection and death of the graft tissue. However, matching MHC alleles between donors and recipients minimizes tissue graft rejection. Control of tissue rejection may also be accomplished by administering drugs that suppress the immune system.

Peptide Antigen Binding

Most of the allelic variations in MHC proteins occur as amino acid changes concentrated in the antigen-binding groove (Figure 27.13), and each polymorphic variation of the MHC protein binds a different set of peptide antigens. The peptides bound by a single MHC protein share a common structural pattern—a peptide **motif**—and each different MHC protein binds a different motif. For example, for a certain class I protein, the bound peptides contain eight amino acids with a phenylalanine (F) at position 5 and a leucine (L) at position 8 (Figure 4.28 for the structures of amino acids). All other positions in the peptide can be occupied by any amino acid (X). Thus, all peptides sharing the sequence X-X-X-X-**F**-X-X-**L** would bind that MHC protein. Another MHC class I protein encoded by a different MHC allele binds a peptide motif of nine amino acids with a tyrosine (Y) at position 2 and an isoleucine (I) at position 9 (X-**Y**-X-X-X-X-X-X-**I**).

The invariant amino acids in each motif are called *anchor residues*—they bind directly and specifically within an individual MHC–peptide binding groove. Thus, an individual MHC protein can bind and present many different peptide antigens if the peptides contain the same anchor residues. Because each MHC protein binds a different motif with different anchor residues, the six possible MHC I proteins encoded in an individual's genome bind six different motifs. In this way, each individual can present a large number of different peptide antigens using the limited number of MHC I molecules available. MHC II proteins bind peptides in a similar manner.

Because of polymorphisms and polygeny within the human species, at least a few peptide antigens from virtually any pathogen will display a motif that can be bound and presented by the MHC proteins. This system of generating antigen-binding diversity is therefore quite different from the genetic mechanisms used

Figure 27.16 Polymorphic *HLA* genes and tracing ancestral lineages. By analyzing the distribution of specific *HLA* alleles within different populations, researchers are increasingly able to reconstruct migration and interbreeding patterns of ancient humans and other hominids, including the Neanderthals. The intermixing of *HLA* genes between disparate ancient populations increased genetic diversity and likely enhanced survival by helping to strengthen and shape the modern immune response.

to synthesize Igs (Section 27.4) and TCRs (Section 27.7), in which each receptor interacts specifically with only a *single* antigen.

MINIQUIZ

- Define polymorphism and polygeny as they apply to MHC genes.
- How does a single MHC protein present many different peptides to T cells?

IV • T Cells and Their Receptors

Adaptive immunity is ultimately initiated by interactions of T lymphocytes with peptide antigens on infected cells. The infected cells that are first recognized by T cells may include the same phagocytes that participate in the innate immune response (Chapter 26). Antigen presentation activates precursor T lymphocytes to differentiate into T cells that carry out antigen-specific, cell-mediated immunity. In the absence of antigen-activated T cells, there is little antigen-specific immunity and no immune memory.

27.7 T Cell Receptors: Proteins, Genes, and Diversity

Antigen-presenting cells ingest bacteria, viruses, and other antigenic material by phagocytosis (in macrophages and dendritic cells) or through internalization of a molecular antigen bound to a BCR. Ingested antigens are then digested, complexed with MHC proteins, and moved to the cell surface for antigen presentation to T cells. The TCRs of T cells can recognize (bind) antigens only when the peptides are complexed with MHC proteins on host cell surfaces (Section 27.6). For example, a phagocyte infected with a virus will display MHC I and MHC II proteins embedded with viral peptides (**Figure 27.17**). These viral peptide–MHC complexes are the targets for T cells.

Each T cell expresses a TCR that is specific for a single peptide–MHC complex. Antigen-specific T cells are found closely associated with APCs in the spleen, lymph nodes, and MALT. T cells constantly sample surrounding APCs for peptide–MHC complexes. Peptide–MHC complexes that interact with the TCR signal the T cell to grow and divide, producing antigen-reactive clones that coordinate cell-mediated killing, induce inflammation, and activate antibody-producing B cells. Here we examine how the TCR interacts with antigens presented on an APC or an infected target cell.

TCR Structure and Diversity

The TCR is a membrane-spanning protein that extends from the T cell surface into the extracellular environment. Similar to B cells with their BCRs, each T cell has thousands of copies of its specific TCR on its surface. A functional TCR consists of two polypeptides, an α chain and a β chain. Similar to Igs, each TCR chain has a variable (V) domain and a constant (C) domain (**Figure 27.18**), and the V_α and V_β domains interact cooperatively to form an antigen-binding site that contains CDR1, CDR2, and CDR3 segments. However, unlike Igs, which can bind antigens of any composition, TCRs recognize only MHC–*peptide*; other antigens, such as complex polysaccharides, are not recognized by TCRs.

The three-dimensional structure of the TCR bound to MHC I–peptide is shown in Figure 27.18*a*. Both TCR and MHC proteins bind directly to peptide antigen. The MHC protein binds one face of the peptide—the MHC motif—and the TCR binds the other peptide face—the T cell epitope (Figure 27.18*b*). The CDR regions of the TCR bind directly to the MHC–peptide complex, and each CDR has a specific binding function. The CDR3 regions of the TCR α chain and β chain bind the antigen epitope; the CDR1 and CDR2 regions of the TCR α and β chains bind mainly to the MHC proteins.

The adaptive immune response can generate TCR diversity sufficient to bind nearly every possible peptide antigen. T cells generate receptor diversity in ways similar to the generation of Ig diversity in B cells, and Table 27.3 compares the receptor diversity–generating mechanisms for each cell type. Analogous to the H and L chains of Igs, the TCR α and β chains are encoded by distinct constant- and variable-domain gene segments. TCR V-region genes are arranged as a series of tandem segments (**Figure 27.19**). The α chain has about 80 V and 61 J gene segments, whereas the β chain has 52 V, 2 D, and 13 J gene segments. As we have seen for Igs (Section 27.4), antigen-binding diversity in TCRs is generated by somatic recombination, imprecise V-D-J (β-chain) or V-J (α-chain) joining, and random reassortment. Additional TCR diversity is generated because the D region of the β chain can be transcribed in all three reading frames (⇔ Section 4.9 and Figure 4.35), leading to production of three separate transcripts from each D-region gene.

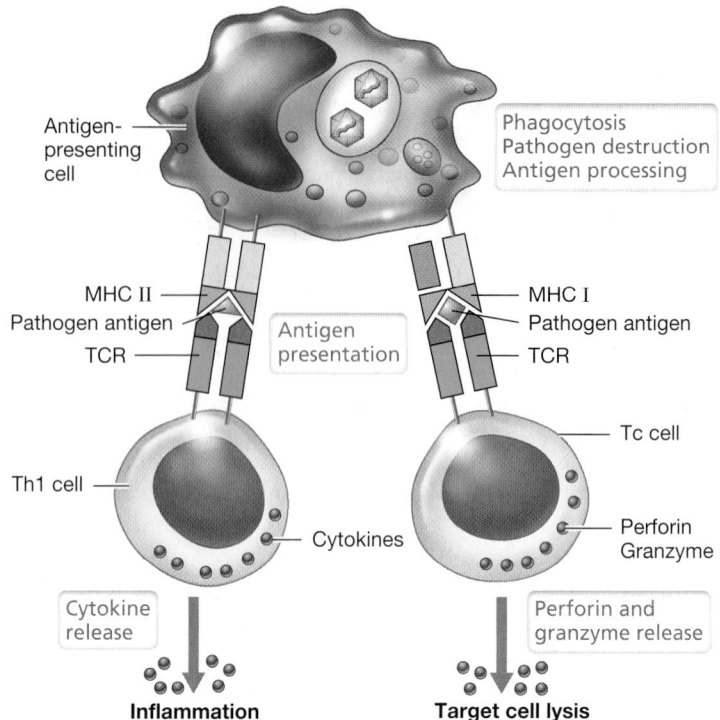

Figure 27.17 T cell immunity. Antigen-presenting cells, such as the phagocytes in innate immunity, ingest, degrade, and process antigens. They then present antigens to T cells that secrete protein cytokines that activate the adaptive immune response. T-helper 1 (Th1) cells produce cytokines that activate other cells and induce inflammation. T-cytotoxic (Tc) cells produce perforin and granzymes, proteins that destroy nearby target cells.

Similar to the assembly of Ig H and L chains, individual α and β chains are produced by each T cell at random and assembled to form a complete α:β heterodimer. The somatic hypermutation mechanisms responsible for increased receptor diversity in Ig genes do not operate in T cells, and thus additional TCR diversity from these events is not possible. However, potential TCR diversity is still enormous, as an estimated 10^{15} different TCRs can be generated.

Structural Similarities of Antigen-Binding Proteins

As we have seen, antigen-binding proteins of the adaptive immune response have common structural features. Ig, TCR, and MHC protein complexes all consist of two nonidentical polypeptides: MHC and TCR proteins are composed of α and β polypeptide chains, and Igs have a separate heavy and light chain (Section 27.4). Therefore, these protein complexes are believed to have arisen from duplication and selection of genes encoding primordial antigen receptors. Because of their shared structural, evolutionary, and functional features, genes encoding Ig, TCR, and MHC proteins are part of an extended gene family called the **immunoglobulin gene superfamily**. A comparison of Ig superfamily proteins is shown in **Figure 27.20**, with several discrete homologous domains highlighted.

The constant (C) domain of each protein in the superfamily has a highly conserved amino acid sequence consisting of about 100 amino acids with an intrachain disulfide bond spanning

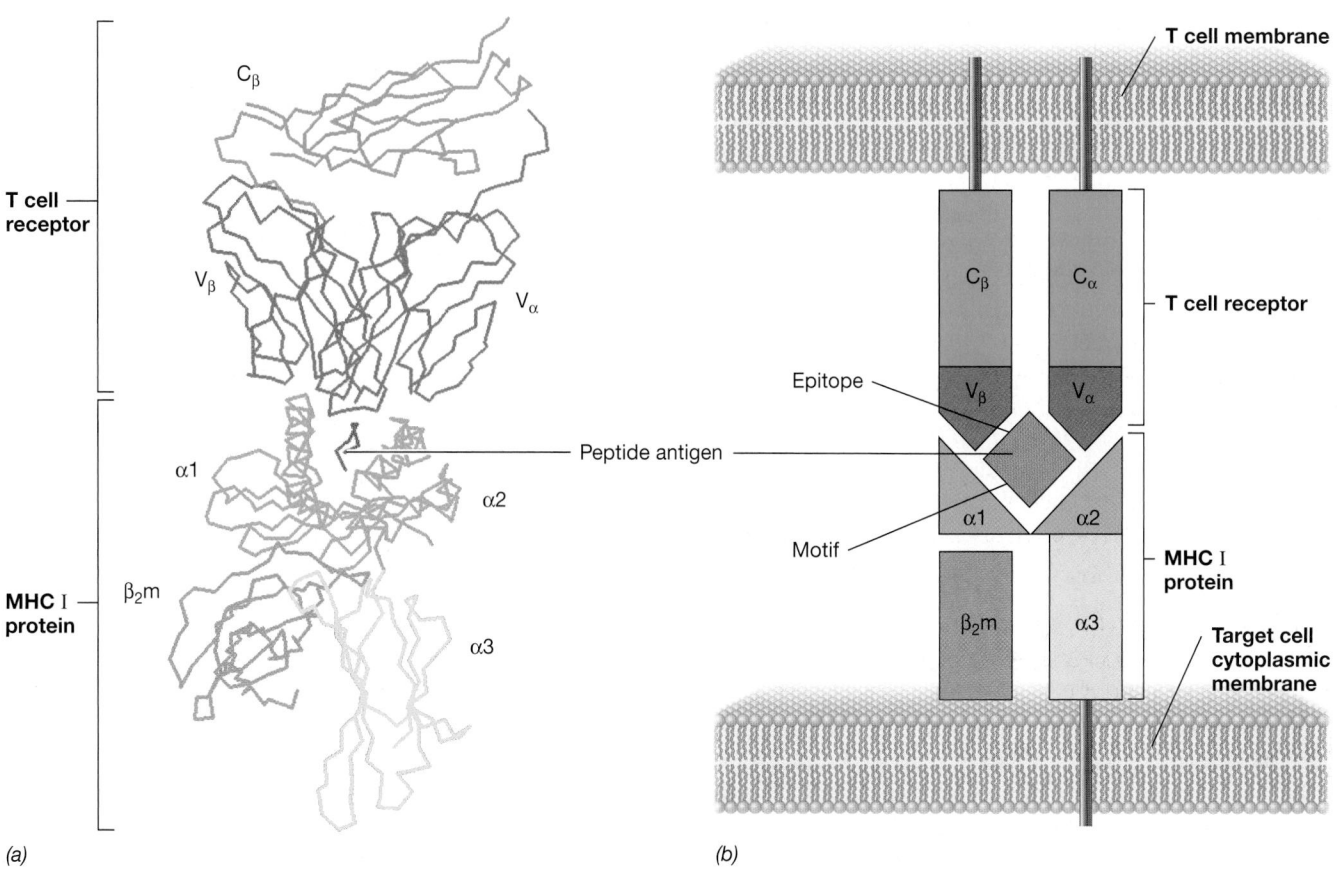

(a)   *(b)*

Figure 27.18 The TCR:MHC I–peptide complex. *(a)* A three-dimensional structure showing the orientation of TCR, peptide (brown), and MHC. This structure was derived from data deposited in the Protein Data Bank. *(b)* A diagram of the TCR:MHC–peptide structure. Note that the peptide is bound by both MHC and TCR proteins and has a distinct surface structure that interacts with each (see page 816 for discussion of this).

50–70 amino acids. C domains provide structural integrity for the antigen-binding molecules, anchor the antigen-binding V domains to the cytoplasmic membrane, and give each protein its characteristic shape. C domains can also provide recognition

sites for accessory molecules. For example, C domains of most IgG and all IgM proteins are bound by the C1q component of complement, a critical first step in initiating the classical complement activation sequence (Section 26.9). Likewise, MHC I C

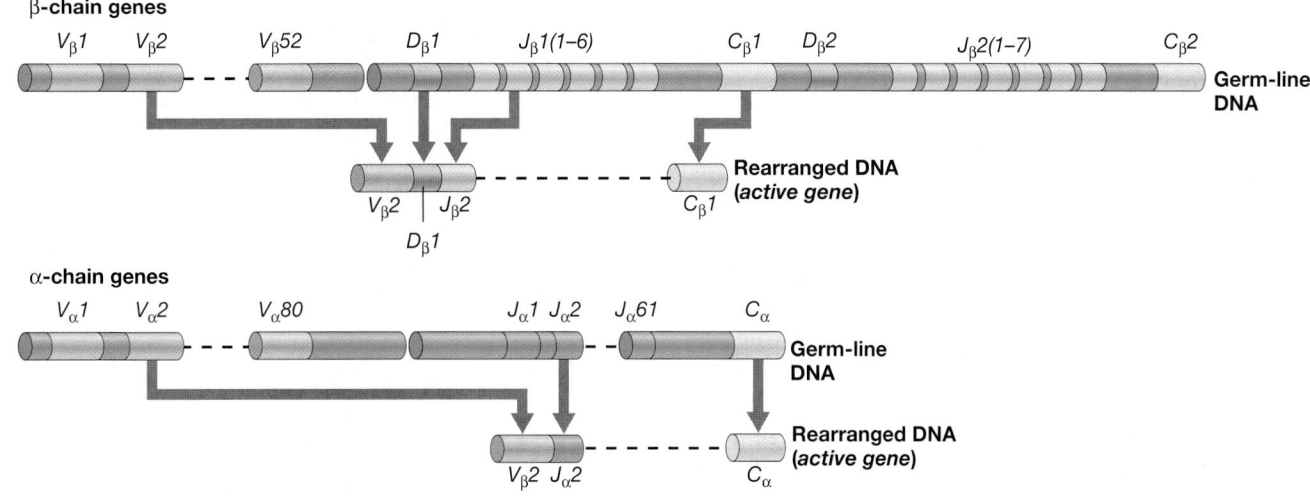

Figure 27.19 Organization of the human TCR α- and β-chain genes. The α-chain genes are located on chromosome 14 and the β-chain genes are on chromosome 6. Compare this figure with Figure 27.12.

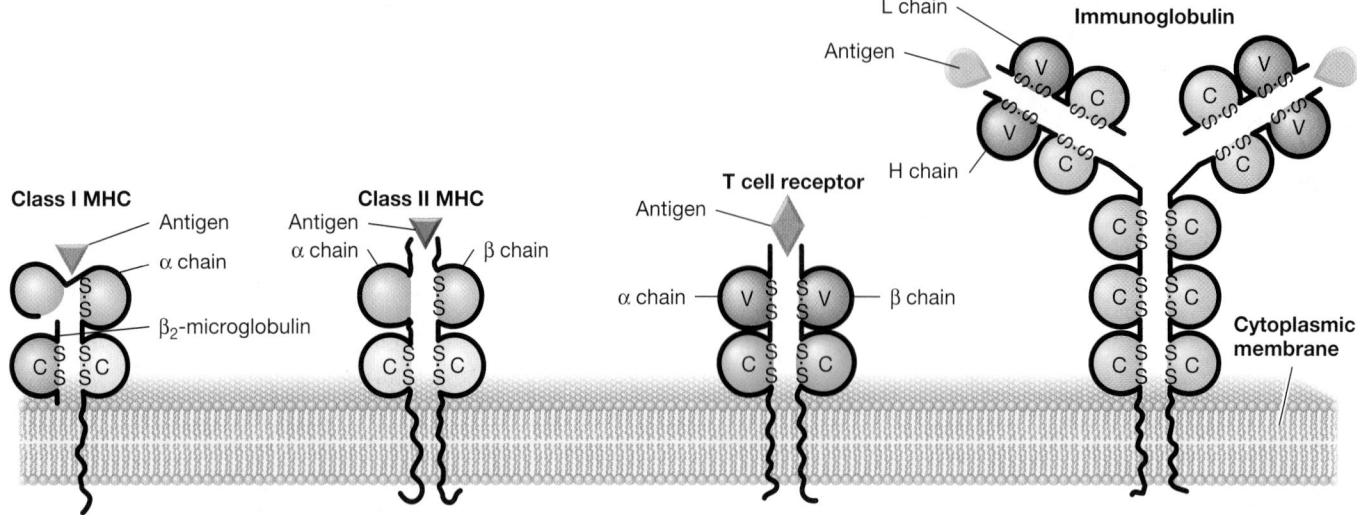

Figure 27.20 Immunoglobulin gene superfamily proteins. Constant domains have homologous amino acid sequences and higher-order structures. The Ig-like C domains in each protein chain indicate evolutionary relationships that identify the proteins as members of the Ig gene superfamily. The V domains of Igs and TCRs are also Ig domains, but the peptide-binding domains of MHC class I and class II proteins are not because their structures vary considerably from the basic features of the Ig domain.

domains bind to the accessory CD8 protein on Tc cells, and homologous MHC II C domains bind CD4 on Th cells. As we have discussed, such interactions are critical steps for T cell activation and initiation of adaptive immunity (Section 27.5).

The variable (V) domains of TCR and Ig molecules are about the same length as the C domains, but the structures of V domains can vary considerably from one another and from C domains. Ig and TCR V domains interact specifically with a nearly limitless variety of antigens. By contrast, the V domains of MHC proteins have evolved independently of Ig and TCR V domains. MHC V domains interact with foreign peptides that share a common motif (Section 27.6), resulting in the MHC–peptide complex recognized by the TCR.

MINIQUIZ

- Distinguish among the functions of the CDR1, CDR2, and CDR3 segments of the T cell receptor.
- Identify diversity-generating mechanisms unique to TCRs as compared to diversity-generating mechanisms in Igs.
- Describe and compare the structural features of Ig gene superfamily constant and variable domains.

27.8 T Cell Diversity

Antigen-reactive T cells consist of multiple T cell subsets having different functional properties. **T-cytotoxic (Tc) cells**, also called *cytotoxic T lymphocytes (CTLs)*, are CD8 T cells (Section 27.5) that recognize the peptide–MHC I complex on an infected cell. By contrast, **T-helper (Th) cells**, which can differentiate into several more specialized Th cell subsets, are CD4 T cells that interact with peptide–MHC II complexes on the surface of APCs and function to activate macrophages and stimulate antibody-mediated immunity.

T-Cytotoxic Cells

When a Tc cell interacts with a foreign peptide on an infected cell, it kills the peptide-bearing target cell through an antigen-specific mechanism. For example, a viral peptide embedded in MHC I, displayed on a virus-infected cell, marks the cell for interaction and killing by a Tc cell whose TCRs recognize the viral antigen.

Contact between a Tc cell and the target cell is required to initiate killing of the infected cell (**Figure 27.21**). The point of initial contact is between the TCR and the peptide–MHC I complex. The CD8 protein on the Tc cell then binds the MHC I protein, strengthening the interaction. On contact with the target cell, granules in the Tc cell migrate to the contact site, where the contents of the granules are released (degranulation). The

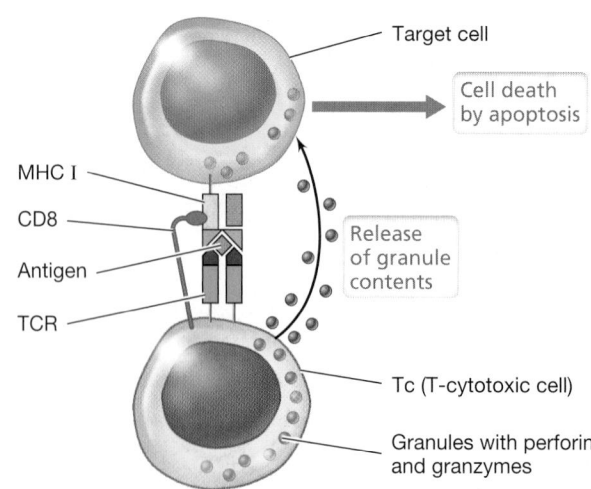

Figure 27.21 T-cytotoxic cells. When the TCR on a Tc cell binds MHC I–peptide complexes on any cell, the Tc cell releases granules that contain perforin and granzymes, cytotoxins that perforate the target cell and cause apoptosis, respectively.

TABLE 27.4 **T-helper cell subsets**

Characteristic	Th1	Th2	Th17	Treg
Antigen-presenting cell	Macrophage	B cell	Activated dendritic cell	Nonactivated dendritic cell
Major cytokines produced	IL-2, IFN-γ, TNF-α	IL-4, IL-5	IL-17, IL-6	IL-10, TGF-β
Cellular effects	Activation of T cells (IL-2) and macrophages	Activation of B cells	Activation and recruitment of neutrophils	Suppression of adaptive immune cells
Systemic effects	Cell-mediated immunity	Antibody-mediated immunity	Amplification of innate immunity	Control of Th immunity

granules contain perforin and granzymes (Section 27.5). Perforin enters the membrane of the target cell and combines to form a transmembrane pore through which granzymes then enter the target cell. Granzymes are cytotoxins that induce *apoptosis* (programmed cell death), characterized by organized killing and degradation of the target cell from within. Tc cells kill only those cells displaying the foreign antigen because the granules are released only at the contact surface between the Tc cell and the target cell bearing peptide–MHC I. Cells lacking the peptide recognized by the Tc cells do not make contact and are not killed.

Different Classes of T-Helper Cells

Interactions with APCs drive CD4 Th cells to differentiate into several subsets, each producing unique combinations of cytokines that recruit effector cells (**Table 27.4**). Macrophages (⇌ Section 26.4) play a central role as APCs in cell-mediated immunity. As illustrated in **Figure 27.22a**, macrophages engulf, process, and present antigen to **Th1 cells**. Th1 cells produce IL-2, a cytokine that promotes growth and activation of other T cells, and activate macrophages through the cytokines interferon gamma (IFN-γ), tumor necrosis factor alpha (TNF-α), and granulocyte–monocyte colony-stimulating factor (GM-CSF) (Figure 27.22a). Th1-activated macrophages take up and kill foreign cells more efficiently than nonactivated macrophages (Figure 27.22b) and also kill tumor cells because the macrophages recognize tumor-specific antigens expressed by cancer cells as nonself.

Activated macrophages also produce several cytokines and chemokines that function as proinflammatory mediators and

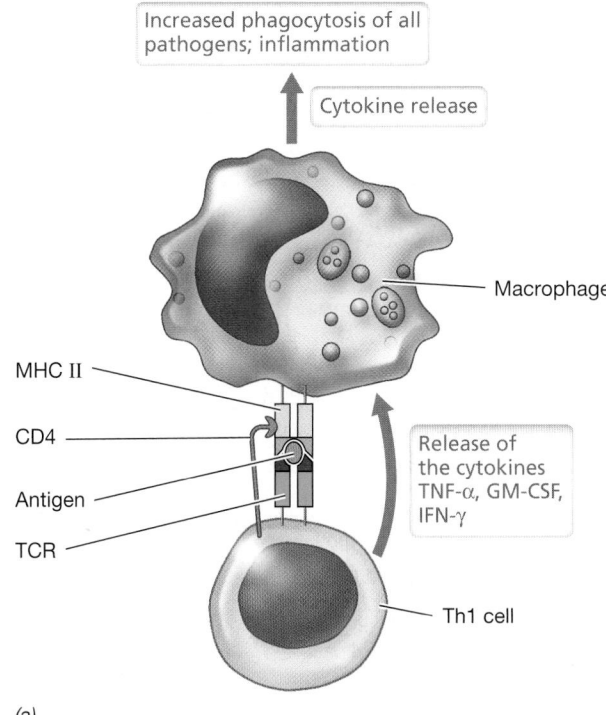

Increased phagocytosis of all pathogens; inflammation

Cytokine release

Macrophage

MHC II

CD4

Antigen

TCR

Release of the cytokines TNF-α, GM-CSF, IFN-γ

Th1 cell

(a)

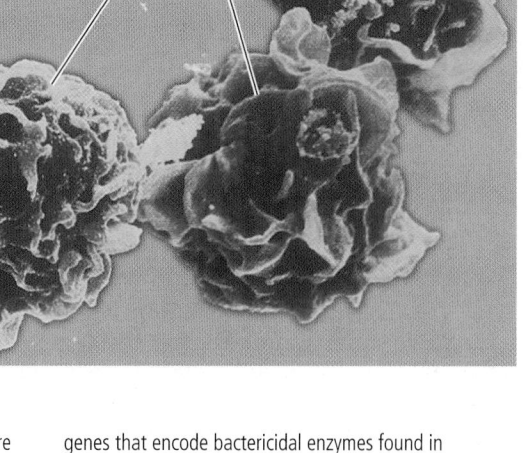

Activated macrophages

Macrophage

(b)

Figure 27.22 Th1 cells and macrophage activation. *(a)* Th1 cells (T-inflammatory cells) are activated by antigens presented on macrophages in the context of MHC II protein. Activated Th1 cells produce cytokines that activate macrophages, leading to increased phagocyte activity and inflammation. *(b)* Activated macrophages are generally larger than resting macrophages and have a ruffled surface, often with cytoplasmic extensions that "feel" for pathogens. In addition to carrying out more aggressive phagocytosis, activated macrophages express genes that encode bactericidal enzymes found in lysosomes, such as proteases and enzymes that produce reactive oxygen species, all of which are designed to quickly kill ingested microorganisms inside the phagolysosome (⇌ Figure 26.10).

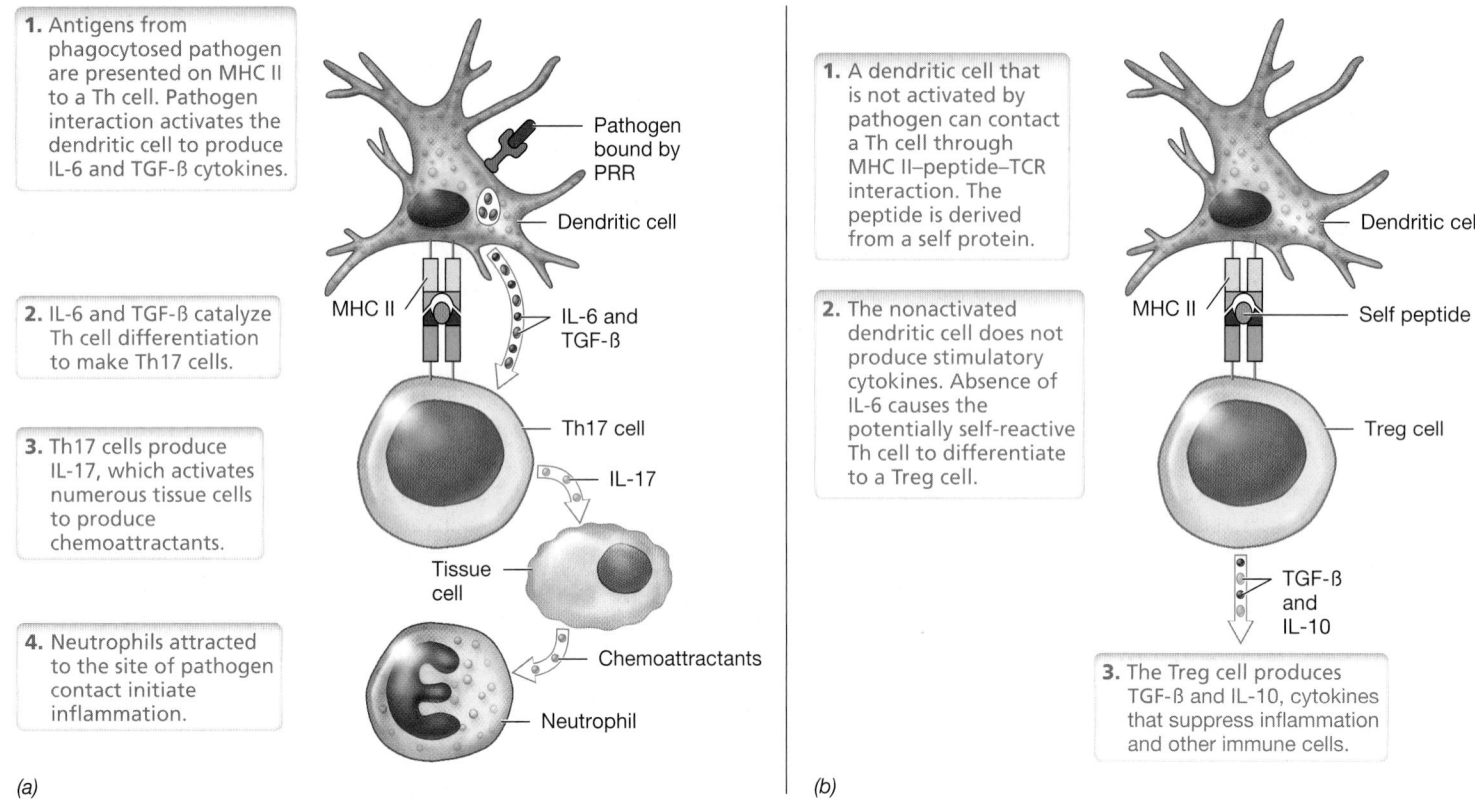

1. Antigens from phagocytosed pathogen are presented on MHC II to a Th cell. Pathogen interaction activates the dendritic cell to produce IL-6 and TGF-ß cytokines.

2. IL-6 and TGF-ß catalyze Th cell differentiation to make Th17 cells.

3. Th17 cells produce IL-17, which activates numerous tissue cells to produce chemoattractants.

4. Neutrophils attracted to the site of pathogen contact initiate inflammation.

Pathogen bound by PRR

Dendritic cell

MHC II

IL-6 and TGF-ß

Th17 cell

IL-17

Tissue cell

Chemoattractants

Neutrophil

(a)

1. A dendritic cell that is not activated by pathogen can contact a Th cell through MHC II–peptide–TCR interaction. The peptide is derived from a self protein.

2. The nonactivated dendritic cell does not produce stimulatory cytokines. Absence of IL-6 causes the potentially self-reactive Th cell to differentiate to a Treg cell.

Dendritic cell

MHC II

Self peptide

Treg cell

TGF-ß and IL-10

3. The Treg cell produces TGF-ß and IL-10, cytokines that suppress inflammation and other immune cells.

(b)

Figure 27.23 Th17 and Treg cells. *(a)* Th17 cells interact with pathogen-stimulated dendritic cells to draw neutrophils to the site of pathogen invasion, leading to inflammation and pathogen control. *(b)* Treg cells interact with nonactivated dendritic cells and respond by making immunosuppressive cytokines that control reactions to self antigens.

chemoattractants, respectively. **Table 27.5** summarizes the source and activities of these and other important immune cytokines and chemokines. In the case of organ or tissue transplants, Th1-activated macrophages can actually harm the host if the host's

Th1 cells recognize nonself MHC proteins on the transplant and trigger macrophage activation and transplant destruction (Section 27.6).

TABLE 27.5 Major immune cytokines and chemokines

Cytokine (chemokine)	Major producer cells	Major target cells	Major effect
IL-4[a]	Th2	B cells	Activation, proliferation, differentiation, IgG1 and IgE synthesis
IL-5	Th2	B cells	Activation, proliferation, differentiation, IgA synthesis
IL-2	Naive T cells, Th1, and Tc	T cells	Proliferation (often autocrine)
IFN-γ[b]	Th1	Macrophages	Activation
GM-CSF[c]	Th1	Macrophages	Growth and differentiation
TNF-α[d]	Th1	Macrophages	Activation, production of proinflammatory cytokines
	Macrophages	Vascular epithelium	Activation, inflammation
IL-1β	Macrophages	Vascular epithelium, lymphocytes	Activation, inflammation
IL-6	Macrophages, dendritic cells	Lymphocytes	Activation
IL-12	Macrophages, endothelial cells	NK cells, naive T cells	Activation, enhances differentiation to Th1
IL-17	Th17	Neutrophils	Activation
CXCL8 (chemokine)	Macrophages	Neutrophils, basophils, T cells	Chemotactic factor
CCL2 (MCP-1[e]) (chemokine)	Macrophages	Macrophages, T cells	Chemotactic factor, activator

[a]IL, interleukin; [b]IFN, interferon; [c]GM-CSF, granulocyte–monocyte colony-stimulating factor; [d]TNF, tumor necrosis factor; [e]MCP, macrophage chemoattractant protein.

Th2 cells play a pivotal role in B cell activation and antibody production. As we have discussed (Section 27.3), B cells make antibodies, and differentiated B cells are coated with antibodies (BCRs) that are antigen receptors. When antigen binds the BCR (Figure 27.6), the antibody-bound antigen is taken into the B cell by endocytosis and degraded. Peptides from the degraded antigen are then loaded into the B cell's MHC II protein for presentation to a Th2 cell. The Th2 cell responds by producing IL-4 and IL-5, cytokines that activate the B cell (Table 27.5) and cause it to proliferate and differentiate into plasma cells that produce and secrete antibodies specific to the presented antigen (Section 27.3).

Antigen presentation by dendritic cells (🔗 Section 26.4) plays a critical role in the development of other Th cell subsets, including Th17 and Treg cells. **Th17 cells** are important in the first stages of the adaptive immune response. Undifferentiated, or *naive*, Th cells differentiate into Th17 cells through the activity of dendritic cells. When dendritic cells encounter pathogens, they present antigen and secrete IL-6 and transforming growth factor-β (TGF-β), cytokines that catalyze differentiation of naive Th cells to Th17 cells (**Figure 27.23a**). Th17 cells then produce IL-17, a cytokine that activates other tissue cells to produce cytokines and chemokines that attract proinflammatory neutrophils to the site of infection (Table 27.5). Thus, the function of Th17 cells is to produce IL-17, starting a cascade that draws neutrophils to infection sites. By recruiting neutrophils, Th17 cells amplify innate immunity triggered by interaction of a pathogen with the dendritic cell.

Treg cells are important in control of immunity. Undifferentiated Th cells remain so unless they are stimulated to mature by certain cytokines, as is the case for IL-6 stimulation to produce Th17 cells. However, in the absence of a pathogen, Th cells can still interact with dendritic cells through MHC II–peptide–TCR (Figure 27.23b). In this case, the peptide is usually a self peptide, and an immune response to it could cause an autoimmune disease. However, because the dendritic cells did not interact with a pathogen, they cannot produce IL-6 to promote Th17 differentiation. Instead, the absence of IL-6 pushes differentiation to Treg cells that make IL-10 and TGF-β, two cytokines that suppress immunity and inflammation. In the presence of self antigens and in the absence of IL-6, Treg cells shut down the immune response and inhibit inflammation. This is important for controlling immune responses to self antigen and preventing autoimmunity.

MINIQUIZ

- Describe the mechanism used by Tc cells to recognize infected host cells.
- Describe the effector system (the cell-killing mechanism) used by Tc cells.
- Compare and contrast the roles and activities of the different Th cells. What cytokines do they produce, and which effector cells do these cytokines act upon?

V • Immune Disorders and Deficiencies

In some cases, reactions of the adaptive immune response can damage the host. For example, **hypersensitivity** is an immune response that results in host damage and, in some cases, even host death. Hypersensitivities are grouped according to the antigens and the mechanisms that produce disease. Likewise, exposure to *superantigens*, proteins produced by certain bacteria and viruses that initiate massive inflammatory responses, also causes severe immune reactions that result in host damage. To conclude this chapter we consider both of these along with immunodeficiency, a condition in which a host's immune response is either absent or insufficient to effectively fight infections.

27.9 Allergy, Hypersensitivity, and Autoimmunity

Antibody-mediated **immediate hypersensitivity** is more commonly called *allergy*. Cell-mediated hypersensitivities also cause allergy-like diseases, but because of the delayed onset of symptoms, cell-mediated reactions are termed **delayed-type hypersensitivity (DTH)**. **Autoimmunity** is a harmful immune reaction directed against self antigens. These hypersensitivities are categorized as type I, II, III, or IV based on immune effectors, antigens, and symptoms (Table 27.6).

Immediate Hypersensitivity

Immediate (type I) hypersensitivity is caused by the release of substances that either increase or decrease blood pressure or heart rate (vasoactive products) from mast cells coated with IgE

TABLE 27.6 **Hypersensitivity**

Classification	Description	Immune mechanism	Time of latency	Examples
Type I	Immediate	IgE sensitization of mast cells	Minutes	Reaction to bee venom (sting) Hay fever
Type II	Cytotoxic[a]	IgG interaction with cell surface antigen	Hours	Drug reactions (penicillin)
Type III	Immune complex	IgG interaction with soluble or circulating antigen	Hours	Systemic lupus erythematosus (SLE)
Type IV	Delayed type	Th1 inflammatory cell activation of macrophages	Days (24–48 h)	Poison ivy Tuberculin test

[a]Autoimmune diseases may be caused by type II, type III, or type IV reactions.

(**Figure 27.24**). Immediate hypersensitivity reactions occur within minutes after exposure to an *allergen*, the antigen that caused the type I hypersensitivity. Depending on the individual and the allergen, immediate hypersensitivity reactions can be mild allergic reactions or can cause a life-threatening reaction called *anaphylaxis*.

About 20% of the population suffers from immediate hypersensitivity allergies to pollens, molds, animal dander, certain foods (strawberries, nuts, and shellfish), insect venoms, dust mites, and other agents. Most allergens enter the body at the surface of mucous membranes, such as the lungs or the gut. Initial exposure to allergens stimulates Th2 cells to produce cytokines that induce B cells to make IgE antibodies. The allergen-specific IgE antibodies bind to IgE receptors on mast cells (Figure 27.24). Mast cells are nonmotile granulocytes (⇔ Section 26.4) associated with the connective tissue adjacent to capillaries throughout the body. With any subsequent exposure to the immunizing allergen, the mast cell–bound IgE molecules bind the antigen. Cross-linking of IgEs by an antigen triggers *degranulation*—the release of soluble allergic mediators from the mast cells. These mediators cause allergic symptoms within minutes of antigen exposure. After initial sensitization by an allergen, the allergic individual responds to each subsequent reexposure to the allergen.

The principal chemical mediators released from mast cells are histamine and serotonin, modified amino acids that cause rapid dilation of blood vessels and contraction of smooth muscle, initiating symptoms ranging from mild local discomfort to systemic *anaphylactic shock*. Local symptoms typically include mucus production, rash, sneezing, itchiness, watery eyes, and hives (**Figure 27.25**). Symptoms of anaphylactic shock may include vasodilation (causing a sharp drop in blood pressure) and asthma due to smooth muscle constriction in the lungs. Severe anaphylaxis is treated immediately with the hormone epinephrine to counter smooth muscle contraction, increase blood pressure, and promote breathing. Less serious allergic symptoms may be treated with *antihistamines*, which are drugs that neutralize histamine, or with anti-inflammatory steroids. Finally, immunization with increasing doses of the allergen may shift antibody production from IgE to IgG and IgA. The IgG and IgA interact with the allergens, thereby blocking antigen binding to IgE on sensitized mast cells. This process, called *desensitization*, inhibits IgE production and stops allergic symptoms.

Delayed-Type Hypersensitivity

Delayed-type (type IV) hypersensitivity (DTH) is cell-mediated hypersensitivity characterized by tissue damage due to inflammation initiated by Th1 cells (Table 27.6). DTH symptoms appear several hours after secondary exposure to the eliciting antigen, with a maximal response usually occurring in 24 to 48 hours. Typical DTH antigens include chemicals that are not normally immunogens but become so when they covalently bind to skin proteins, creating new antigens and eliciting a DTH response. Hypersensitivity to these newly created antigens is known as *contact dermatitis* and results in, for example, skin reactions to poison ivy (**Figure 27.26**), jewelry, cosmetics, latex, and other chemicals that react with host tissues. Several hours after a second or subsequent exposure to the antigen, the skin feels itchy at the site of contact. Erythema (reddening) and edema (swelling) appear, often with localized tissue destruction in the form of blistering, and reach a maximum in several days. The delayed onset and the progress of the inflammatory response are the hallmarks of the DTH reaction. As discussed below, certain self antigens may also elicit DTH responses, resulting in autoimmune disease.

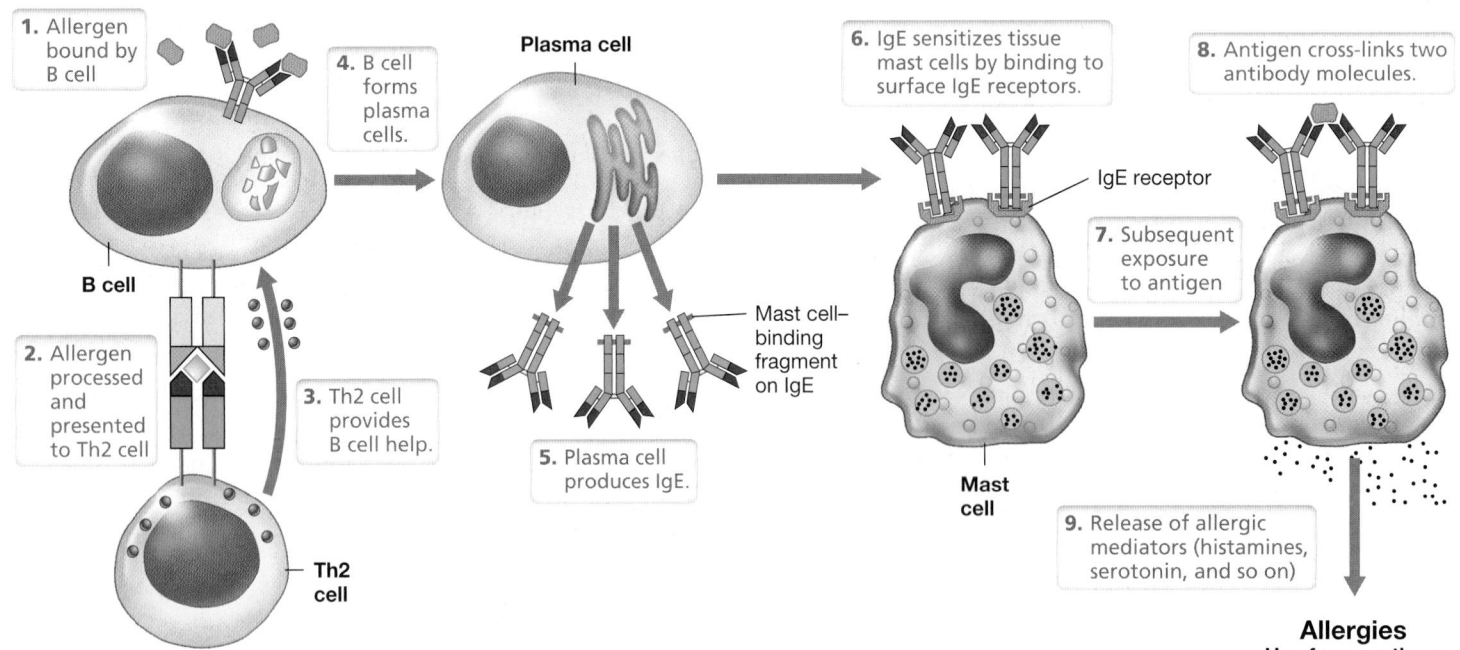

Figure 27.24 Immediate hypersensitivity. Certain antigens, such as pollens, stimulate IgE production. IgE binds to mast cells by means of a high-affinity surface receptor and arms the mast cell. Antigen cross-links surface IgE, causing release of soluble mediators, including histamine. These mediators produce symptoms ranging from mild allergic symptoms to life-threatening anaphylaxis.

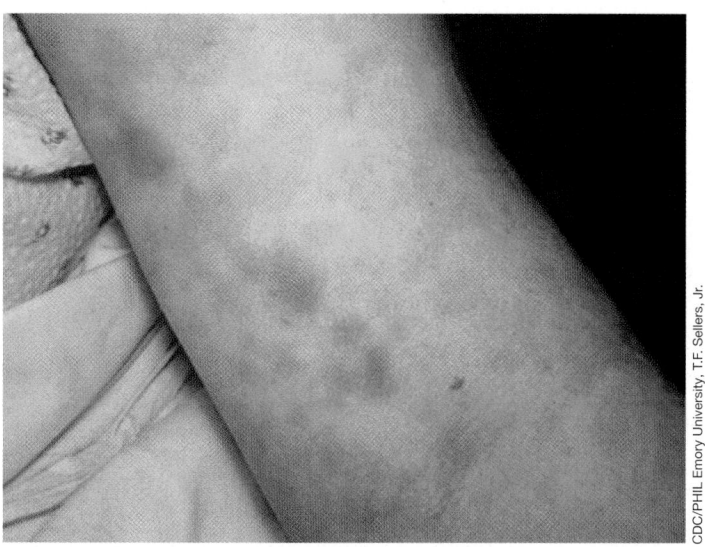

Figure 27.25 Hives due to immediate hypersensitivity. The raised, red areas are typical symptoms after contact with allergens that cause immediate hypersensitivity.

Another example of delayed-type hypersensitivity is the development of protective immunity to the bacterium that causes tuberculosis, *Mycobacterium tuberculosis* (⇨ Section 30.4). The German physician turned microbiologist Robert Koch discovered this cellular immune response during his classic studies on tuberculosis over a century ago (⇨ Section 1.10). When antigens derived from the bacterium are injected subcutaneously into a person previously infected with *M. tuberculosis*, a skin reaction called the *tuberculin reaction* develops within 24–48 hours (**Figure 27.27**). Local Th1 cells stimulated by the introduced *M. tuberculosis* antigens release cytokines that attract and activate large numbers of macrophages, which in turn produce a characteristic local inflammation, including induration (hardening), edema, erythema, pain, and heating of the skin. The activated macrophages then ingest and destroy the invading antigen. The DTH-based

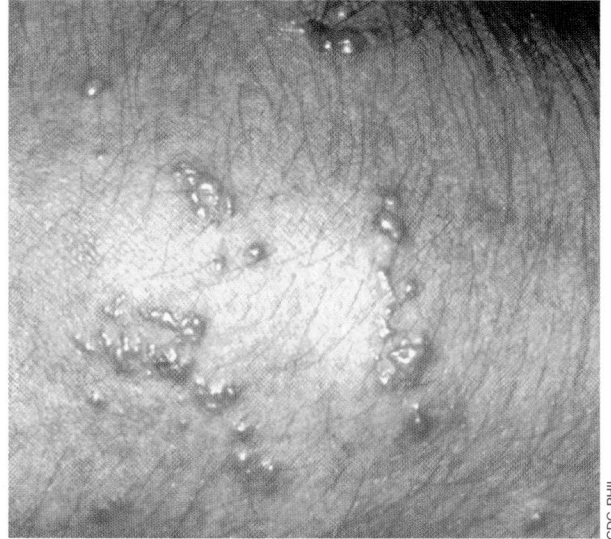

Figure 27.26 Delayed-type hypersensitivity. Poison ivy blisters on an arm. The raised rash appears 24–48 hours after exposure to plants of the genus *Rhus* as a result of macrophage activation by Th1 cells sensitized to *Rhus* antigens.

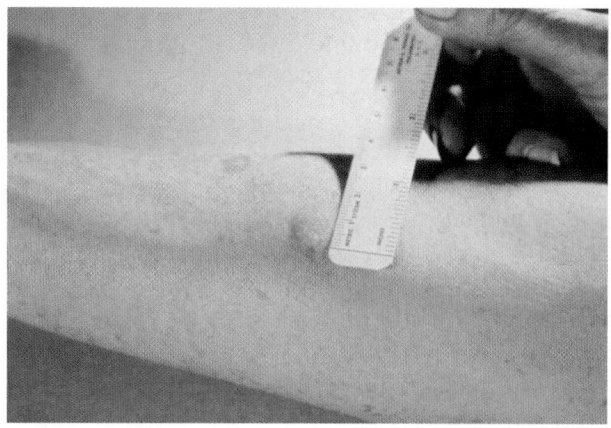

Figure 27.27 Th1 cells and macrophage activation. This tuberculin test shows a positive reaction. Macrophages activated by antigen-specific Th1 cells caused the localized, delayed-type reaction to a tuberculosis antigen, tuberculin, at the site of injection. The raised area of inflammation on the forearm is about 1.5 cm in diameter.

tuberculin skin test is used to test for a current or previous infection with *M. tuberculosis* or previous immunization with the tuberculosis vaccine (⇨ Section 28.5).

A number of other infectious diseases due to intracellular pathogens elicit DTH reactions. These include bacterial diseases such as leprosy, brucellosis, and psittacosis; viral diseases such as mumps; and fungal diseases such as coccidioidomycosis, histoplasmosis, and blastomycosis. Visible, antigen-specific skin responses similar to the tuberculin reaction occur after injection of antigens derived from the pathogens, indicating pathogen exposure and Th1-mediated immunity.

Autoimmunity

As lymphocytes develop, T and B cells that react with self antigens are normally eliminated (Section 27.1). Autoimmune diseases result when these cells are instead activated to produce immune reactions against self proteins (**Table 27.7**). For example, Th1-mediated DTH can cause autoimmune responses directed against self antigens, as in the case of the Th1-mediated response to brain-derived antigens in allergic encephalitis. In type 1 (juvenile) diabetes mellitus, Th1 cells directed to antigens on pancreatic cells cause reactions that destroy the insulin-producing beta cells in the pancreas.

Many autoimmune diseases, however, are caused by **autoantibodies**, antibodies that interact with self antigens, many of which are organ-specific. For example, in *Hashimoto's disease* (hypothyroidism), autoantibodies are made against thyroglobulin, a protein product of the thyroid gland that assists in the synthesis of thyroid hormones. In Hashimoto's disease, antibodies to thyroglobulin bind complement proteins (⇨ Section 26.9), leading to local inflammation and the destruction of host cells, the hallmarks of a type II hypersensitivity disease (Table 27.7).

Systemic lupus erythematosus (*SLE*) is an autoimmune disease caused by a type III hypersensitivity. This disease and others like it are caused by autoantibodies directed against soluble, circulating self antigens. In SLE, the antigens include nucleoproteins and DNA. Autoantibodies to these bind to soluble proteins, producing insoluble immune complexes, and disease symptoms result when these circulating antigen–antibody complexes deposit in different

TABLE 27.7 Some autoimmune diseases of humans

Disease	Organ, cell, or molecule affected	Mechanism (hypersensitivity type[a])
Type I diabetes (insulin-dependent diabetes mellitus)	Pancreas	Cell-mediated immunity and autoantibodies against surface and cytoplasmic antigens of beta cells of pancreatic islets (II and IV)
Myasthenia gravis	Skeletal muscle	Autoantibodies against acetylcholine receptors on skeletal muscle (II)
Goodpasture's syndrome	Kidney	Autoantibodies against basement membrane of kidney glomeruli (II)
Rheumatoid arthritis	Cartilage	Autoantibodies against self IgG antibodies, which form complexes deposited in joint tissue, causing inflammation and cartilage destruction (III)
Hashimoto's disease (hypothyroidism)	Thyroid	Autoantibodies to thyroid surface antigens (II)
Male infertility (some cases)	Sperm cells	Autoantibodies agglutinate host sperm cells (II)
Pernicious anemia	Intrinsic factor	Autoantibodies prevent absorption of vitamin B_{12} (III)
Systemic lupus erythematosus (SLE)	DNA, cardiolipin, nucleoprotein, blood clotting proteins	Autoantibody response to various cellular constituents results in immune complex formation (III)
Addison's disease	Adrenal glands	Autoantibodies to adrenal cell antigens (II)
Allergic encephalitis	Brain	Cell-mediated response against brain tissue (IV)
Multiple sclerosis	Brain	Cell-mediated and autoantibody response against central nervous system (II and IV)

[a]See Table 27.6.

body tissues, such as the kidney, lungs, and spleen. In the tissues, the antibodies bind complement, resulting in inflammation and local, often severe, tissue damage. Type III hypersensitivities are also called *immune complex disorders* (Table 27.6).

Organ-specific autoimmune diseases are sometimes more easily controlled clinically than diseases that affect multiple organs. For example, the product of organ function, such as thyroxine in auto-immune hypothyroidism or insulin in type I diabetes, can often be supplied in pure form from another source. SLE, rheumatoid arthritis, and other autoimmune diseases that affect multiple organs and sites can often be controlled only by general immunosuppressive therapy, such as the use of steroid drugs. Unfortunately, the immunosuppression associated with these treatments significantly increases the chance of developing opportunistic infections. Recently, however, therapies that employ monoclonal antibodies (⥁ Section 28.5) have emerged as promising alternative treatment strategies for autoimmune diseases. For example, adalimumab (Humira) is a monoclonal antibody that neutralizes tumor necrosis factor alpha (TNF-α), an inflammatory cytokine linked to several autoimmune diseases including rheumatoid arthritis, Crohn's disease, and psoriasis. Similarly, the monoclonal antibody belimumab (Benlysta) targets B cell activating factor, a cytokine that stimulates B cell maturation but is overexpressed in SLE patients, causing the persistence of B cells that produce autoantibodies.

Heredity influences the incidence, type, and severity of autoimmune diseases. Many autoimmune diseases correlate strongly with the presence of certain MHC proteins (Section 27.6). Studies of model autoimmune diseases in mice support such a genetic link, but the precise conditions necessary for developing autoimmunity may also depend on other factors such as prior infections, gender, age, and health status. Women, for example, are about ten

times more likely to develop SLE than are men. It is thus likely that the balance between a normal immune response and autoimmunity is tipped by a combination of factors in which genetic predisposition plays a central role.

— MINIQUIZ —

- Discriminate between immediate hypersensitivity and delayed-type hypersensitivity with respect to antigens and immune effectors.
- Provide examples and mechanisms for an antibody-mediated autoimmune disease directed against a specific organ and one involving circulating immune complexes.

27.10 Superantigens and Immunodeficiency

Extremes in the adaptive immune response, whether overactive or deficient, can have devastating effects on the host. Some exotoxins, called *superantigens*, damage host cells indirectly by subverting the immune system so that T cells and their cytokine products destroy host tissues through an exaggerated immune response. By contrast, some diseases—either genetic or infectious—cause *immunodeficiency*, resulting in increased susceptibility of the host to infectious diseases. We consider these opposing extremes in the adaptive immune response here.

Superantigens

Superantigens are proteins that upon exposure to the immune system activate many more T cells than normal and are therefore capable of eliciting an unusually strong immune response

(Figure 27.28). A variety of viruses and bacteria produce superantigens. For example, streptococci and staphylococci, especially certain strains of *Streptococcus pyogenes* and *Staphylococcus aureus*, produce several different extremely potent superantigens (⇔ Sections 30.2 and 30.9).

Superantigen interaction with TCRs differs from conventional antigen–TCR binding (Figure 27.18). Whereas typical foreign antigens, presented by MHC proteins, bind to a TCR at a defined antigen-binding site, superantigens bind to sites on TCR and MHC proteins that are outside the antigen-specific binding site (Figure 27.29). A superantigen binds to all TCRs with a shared common structure, and many different TCRs share the same structure outside the antigen-binding site. Whereas less than 0.01% of all available T cells interact with a conventional foreign antigen in a typical immune response, some superantigens can bind up to 25% of all T cells in the body! These interactions mimic conventional antigen presentation and stimulate large numbers of T cells to grow and divide. As in normal responses, the activated T cells produce cytokines that stimulate phagocytes and other immune cells. However, the extensive cytokine production by the large population of superantigen-activated T cells triggers a widespread cell-mediated response characterized by systemic inflammatory reactions. The resulting fever, diarrhea, vomiting, mucus production, and even systemic shock may be fatal in extreme cases. The clinical symptoms of superantigen shock are indistinguishable from those of septic shock, a condition in which a bacterial infection has spread throughout the body (⇔ Section 26.8).

One of the most common superantigen diseases is *Staphylococcus aureus* food poisoning, characterized by fever, vomiting, and diarrhea, and caused by one of several superantigen staphylococcal enterotoxins (⇔ Section 32.8). *S. aureus* also produces the superantigen responsible for *toxic shock syndrome* (Figure 27.28). *Streptococcus pyogenes* produces erythrogenic toxin, the superantigen responsible for scarlet fever (⇔ Section 30.2).

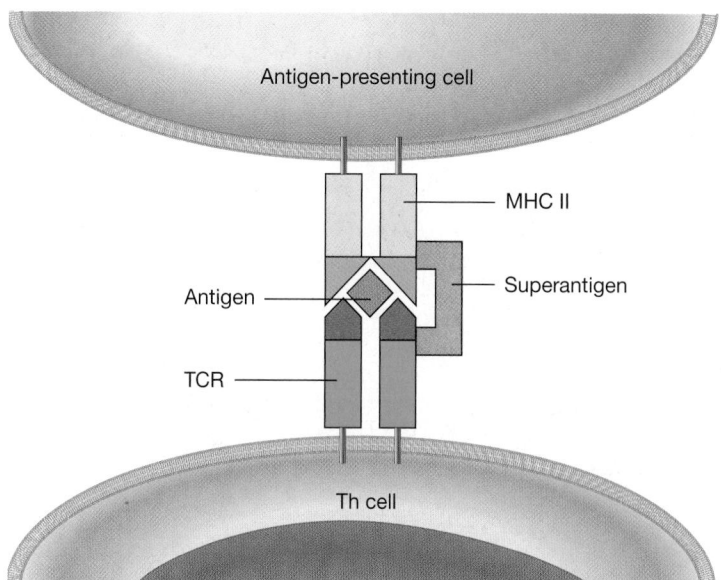

Figure 27.29 Superantigens. Superantigens bind to conserved regions of both the MHC and TCR proteins at positions outside the normal binding site. Superantigens interact with large numbers of T cells, causing large-scale T cell activation, cytokine release, and systemic inflammation.

Immunodeficiency

Active adaptive immunity is critical for infectious disease resistance. We know this because of problems caused by genetic defects and diseases that affect the adaptive immune system. For example, animals that cannot produce antibodies because of genetic defects in their B cells acquire serious infections from extracellular pathogens, especially bacteria. In addition, animals with genetic defects that prevent development of T cells suffer from recurrent infections with viruses and other intracellular pathogens.

Severe combined immune deficiency syndrome (SCID) is a genetic disorder that prevents proper formation of either B or T cells. Individuals with SCID essentially have no effective adaptive immunity. A lack of proper T cell function directly results in a deficiency of cell-mediated immunity and indirectly causes a loss of antibody-mediated immunity because B cell activation against most antigens is dependent upon the presence of functional Th cells. Unless patients receive supportive therapy, such as a bone marrow transplant and antibiotic treatments, SCID eventually causes death from multiple recurrent infections. The transplantation of compatible bone marrow tissue provides the afflicted individual with hematopoietic stem cells (⇔ Sections 26.3 and 26.4) that are free of the genetic defects that cause SCID.

Recently, gene therapy has also shown promise as a curative treatment option for some forms of SCID. In this procedure, defective genes in hematopoietic stem cells are replaced through a transduction process using viral vectors (⇔ Section 11.7). As with bone marrow transplantation, successful gene therapy restores function of the adaptive immune system. However, a major challenge in the use of this procedure is the risk of the patient developing cancer, especially leukemia. This is because the viral vectors used to introduce the functional genes are typically attenuated

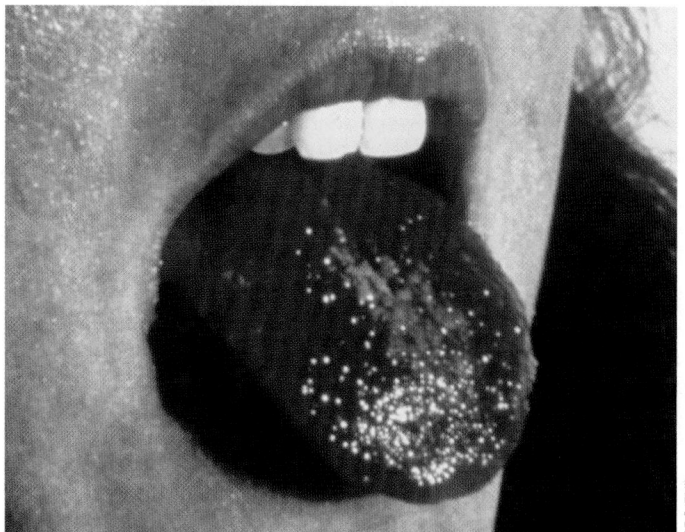

Figure 27.28 Toxic shock syndrome. This individual exhibits "strawberry tongue," a symptom of toxic shock syndrome caused by a *Staphylococcus aureus* superantigen.

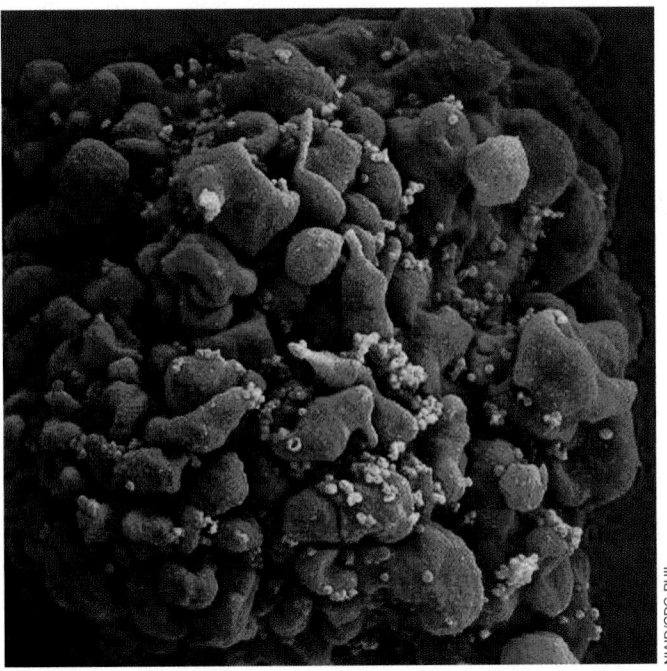

NIAID/CDC-PHIL

Figure 27.30 HIV infection of a Th cell. Colorized scanning electron micrograph of a large number of HIV virions (yellow) attacking a Th cell (red).

retroviruses or adenoviruses; such viruses readily infect host cells to facilitate the necessary transduction, but they also contain or may activate *oncogenes*, which disrupt control of cell division and induce tumorigenesis. However, promising new forms of "gene editing" that do not require viral vectors are now on the horizon (⇄ Section 12.12), and thus safe genetic therapies for SCID may be developed in the not too distant future.

In some cases, immunodeficiency is not the result of a genetic disorder but rather is caused by microbial infection. The best-studied example of this is the loss of the adaptive immune response due to *acquired immunodeficiency syndrome (AIDS)*. Human immunodeficiency virus (HIV) infects host cells that express the CD4 cell surface protein. The virus initially infects macrophages but later attacks and replicates primarily in Th cells (**Figure 27.30**). Once infected, Th cells cease growing and dividing before they eventually die. Therefore, in untreated cases, HIV infection causes gradual depletion of Th cells, resulting in a lack of effective immunity and the eventual onset of AIDS (⇄ Section 30.15 and Figure 30.44).

In most AIDS cases, the actual cause of patient death is not directly attributed to HIV but rather to any of a variety of secondary microbial infections caused by *opportunistic pathogens*. Many of these infections are caused by fungal, bacterial, and viral pathogens that only rarely cause serious disease symptoms in individuals that have a fully functional immune system. The deficiency of adaptive immunity in AIDS patients allows these opportunistic pathogens to colonize and invade body tissues, leading to additional diseases. We discuss the pathogenicity, symptoms, treatment, and other aspects of HIV/AIDS in more detail in Chapter 30.

--- MINIQUIZ ---

- Describe the binding site for superantigens on T cells and APCs. How does this relate to the activation of a much larger population of T cells than normal upon exposure to superantigens?

- Compare and contrast the immunodeficiency observed in SCID patients to that of AIDS patients. What cell types are affected by each condition?

- In the absence of treatment, what is the prognosis for individuals afflicted with SCID and AIDS?

MasteringMicrobiology® **Visualize**, **explore**, and **think critically** with Interactive Microbiology, MicroLab Tutors, MicroCareers case studies, and more. MasteringMicrobiology offers practice quizzes, helpful animations, and other study tools for lecture and lab to help you master microbiology.

Chapter Review

I • Principles of Adaptive Immunity

27.1 The adaptive immune response is characterized by *specificity* for the antigen, the ability to respond more vigorously when reexposed to the same antigen (*memory*), and the acquired inability to interact with self antigens (*tolerance*). Tolerance in lymphocytes is acquired through a selection process. Immature T cells that do not interact with MHC–peptide (positive selection) or that react strongly with self antigens (negative selection) are eliminated by clonal deletion in the thymus. T cells that survive positive and negative selection leave the thymus and can participate in the immune response. B cell reactivity to self antigens is controlled through clonal deletion and anergy.

Q Why is it necessary that all three defining characteristics of the adaptive immune response (specificity, memory, and tolerance) be in place to maintain a functional and effective adaptive immune system?

27.2 Immunogens are foreign macromolecules that induce an immune response. Immunogens initiate an immune response when introduced into a host. Antigens are molecules recognized (bound) by antibodies or TCRs. Adaptive immunity develops naturally and actively through immune responses to infections, or naturally and passively through antibody transfer from the placenta or breast milk. Artificial passive immunity occurs when antibodies or immune cells are transferred from an immune individual to a nonimmune individual. Immunization induces artificial active immunity and is widely used to prevent infectious diseases.

Q **What properties are required for a vaccine to induce an immune response? What type of immunity results from vaccination, and what role does a "booster" play in this process?**

II • Antibodies

27.3 Antibody production is initiated when an antigen contacts an antigen-specific B cell. The B cell then processes the antigen and presents it to an antigen-specific Th2 cell. The Th2 cell becomes activated, producing cytokines that signal the B cell to clonally expand and differentiate to produce soluble, antigen-specific antibodies. Each antibody (immunoglobulin) protein consists of two heavy and two light chains. The antigen-binding site is formed by the interaction of the variable regions of one heavy and one light chain. Each antibody class has different structural characteristics, expression patterns, and functional roles. Activated B and T cells can live for years as memory cells and can rapidly expand and differentiate to produce high titers of antibodies after reexposure to antigen.

Q **Describe the structural and functional differences among the five major classes of antibodies. What cellular and molecular interactions take place in the production of antibodies?**

27.4 The antigen-binding site of Ig is composed of the V (variable) domains of one heavy chain and one light chain. Each V region contains three complementarity-determining regions, or CDRs, that are folded together to form the antigen-binding site. Immunoglobulin diversity is generated by several mechanisms. Somatic recombination of gene segments allows shuffling of the various Ig gene segments. Random reassortment of the heavy- and light-chain genes, imprecise joining of VDJ and VJ gene segments, and hypermutation mechanisms contribute to nearly unlimited immunoglobulin diversity.

Q **Which Ig chains are used to construct a complete antigen-binding site? Which domains? Which CDRs? Calculate the total number of germ-line-encoded V_H and V_L domains that can be constructed from the available Ig genes.**

III • The Major Histocompatibility Complex (MHC)

27.5 T cells, with their TCRs, bind peptide antigens presented by MHC proteins on infected cells or APCs. Class I MHC proteins are expressed on all nucleated cells and present endogenous antigenic peptides to TCRs on Tc cells. Class II MHC proteins are expressed only on APCs. They function to present exogenously derived peptide antigens to TCRs on Th cells. These interactions activate T cells to

kill antigen-bearing cells or to promote inflammation or antibody production.

Q **Describe the basic structure of class I and class II major histocompatibility complex (MHC) proteins. In what functional ways do they differ?**

27.6 Class I and class II MHC genes are highly polymorphic, and the many allelic variations challenge successful tissue transplantation. Different alleles of MHC class I and class II genes encode proteins that bind and present different peptide subsets, each characterized by a specific structural motif.

Q **Polymorphism implies that each different MHC protein binds a different peptide motif. For the MHC class I polymorphisms, how many different MHC proteins are expressed in an individual? How many by the entire human population?**

IV • T Cells and Their Receptors

27.7 T cell receptors bind to peptide antigens presented by MHC proteins. The CDR3 regions of both the α chain and the β chain bind to the antigen epitope; the CDR1 and CDR2 regions bind to the MHC protein. VDJ gene segments encode the β-chain V domain of TCRs, and VJ gene segments encode the α-chain V domain. TCR diversity, generated by a variety of mechanisms, is nearly unlimited. The Ig gene superfamily encodes proteins that are evolutionarily, structurally, and functionally related to immunoglobulins. The antigen-binding Igs, TCRs, and MHC proteins are members of this family.

Q **What diversity-generating mechanisms function to produce the nearly unlimited variety of antigen-specific TCRs? What structural and functional features are common to proteins classified within the Ig gene superfamily?**

27.8 T-cytotoxic (Tc) cells recognize antigens on virus-infected host cells and tumor cells through antigen-specific TCRs. Antigen-specific recognition triggers killing via perforins and granzymes. T-helper (Th) cells differentiate into several subsets. Through the action of cytokines, Th1 inflammatory cells activate macrophage effector cells; Th2 cells activate B cells. Th17 cells are activated by pathogen-activated dendritic cells and secrete IL-17 to recruit neutrophils to the site of infection. Treg cells produce cytokines that suppress adaptive immunity.

Q **What mechanism do Tc cells use to identify and destroy infected cells in the body? How do Th cells differ from Tc cells, and how do the different subsets of Th cells differ from each other?**

V • Immune Disorders and Deficiencies

27.9 Hypersensitivity is the induction by foreign antigens of antibody-mediated or cell-mediated immune responses

UNIT 6

that damage host tissue. In autoimmunity, the immune response is directed against self antigens. Damage to host tissue is caused by the inflammation produced by immune mechanisms.

Q How do immediate and delayed-type hypersensitivites differ from one another in terms of immune effectors, target tissues, antigens, and clinical outcome?

27.10 Superantigens are components of certain bacterial and viral pathogens that bind outside the antigen-specific binding site of TCRs and, therefore, activate large numbers of T cells. Superantigen-activated T cells may produce diseases characterized by systemic inflammatory

reactions. Immunodeficiency is the inability to generate a proper immune response, resulting in recurrent, uncontrollable infections by opportunistic pathogens. Some of the most severe immunodeficiency syndromes are SCID, caused by a genetic disorder, and AIDS, caused by HIV infection.

Q How do superantigens differ from conventional antigens in terms of initial T cell activation and clinical outcome? How does immunodeficiency resulting from SCID differ from that caused by HIV infection and AIDS?

Application Questions

1. Antibodies of the IgA class are probably more prevalent than those of the IgG class. Explain this and define the benefits this may have for the host.

2. Although genetic recombination events are important for generating significant diversity in the antigen-binding site of Igs, postrecombination somatic events may be even more important in achieving overall Ig diversity. Do you agree or disagree with this statement? Explain.

3. Polymorphism implies that each different MHC protein binds a different peptide motif. However, for the MHC class I proteins, only 6 peptide motifs can be recognized in an individual, whereas over 6000 motifs can be recognized by

the entire human population. What advantage does recognition of multiple motifs have for the individual? What potential advantage does recognition of the extremely large number of motifs have for the population? Can everyone process and present the same antigens?

4. What problems would arise if a person had a hereditary deficiency that resulted in an inability to present antigens to Tc cells? What would the problems be if the person had a deficiency in presenting antigen to Th1 cells? To Th2 cells? To all T cells? What molecules might be deficient in each situation? Could a person having any one of these deficiencies survive in a normal environment? Explain for each.

Chapter Glossary

Antigen a molecule capable of interacting with specific components of the immune system and that often functions as an immunogen to elicit an adaptive immune response

Autoantibody an antibody that reacts to self antigens

Autoimmunity a harmful immune reaction directed against self antigens

B cell receptor (BCR) a cell-surface antibody that acts as an antigen receptor on a B cell

CD4 coreceptor a protein found on Th cells that interacts with MHC II on an antigen-presenting cell

CD8 coreceptor a protein found exclusively on Tc cells that interacts with MHC I on a target cell

Clonal anergy the inability to produce an immune response to specific antigens due to the neutralization of effector cells

Clonal deletion for T cell selection in the thymus, the killing of useless or self-reactive clones

Clone a copy of an antigen-reactive lymphocyte

Complementarity-determining region (CDR) a varying amino acid sequence within the variable domains of immunoglobulins or T cell receptors where contacts with antigen are made

Delayed-type hypersensitivity (DTH) an inflammatory allergic response mediated by Th1 lymphocytes

Epitope the portion of an antigen that reacts with a specific antibody or T cell receptor

Human leukocyte antigen (HLA) an antigen-presenting protein encoded by a major histocompatibility complex gene in humans

Hypersensitivity an immune response leading to damage to host tissues

Immediate hypersensitivity an allergic response mediated by vasoactive products released from IgE-sensitized mast cells

Immune memory (memory) the capacity to respond more quickly to second and subsequent exposures to an eliciting antigen

Immunogen a molecule capable of eliciting an adaptive immune response

Immunoglobulin gene superfamily a family of genes that are evolutionarily, structurally, and functionally related to immunoglobulins

Memory cell a long-lived B or T lymphocyte responsive to a specific antigen

MHC class I protein an antigen-presenting molecule found on all nucleated vertebrate cells

MHC class II protein an antigen-presenting molecule found on macrophages, B cells, and dendritic cells

Motif in antigen presentation, a specific amino acid sequence found in all peptides that bind to a given MHC protein

Negative selection in T cell selection, the deletion of T cells that interact strongly with self antigens in the thymus (see also clonal deletion)

Polygeny the occurrence of two or more genetically, structurally, and functionally related gene loci due to an evolutionary gene duplication event

Polymorphism in a population, the occurrence of multiple alleles for a gene locus at a higher frequency than can be explained by recent random mutations

Positive selection in T cell selection, the growth and development of T cells that interact with self MHC–peptide in the thymus

Primary immune response the production of antibodies or immune T cells on first exposure to antigen; the antibodies are mostly of the IgM class

Secondary immune response the enhanced production of antibodies or immune T cells on second and subsequent exposures to antigen; the antibodies are mostly of the IgG class

Somatic hypermutation the mutation of immunoglobulin genes at rates higher than those observed in other genes

Specificity the ability of the immune response to interact with particular antigens

Superantigen a pathogen product capable of eliciting an inappropriately strong inflammatory immune response by stimulating greater than normal numbers of T cells

T cell receptor (TCR) an antigen-specific receptor protein on the surface of T cells

T-cytotoxic (Tc) cell a lymphocyte that interacts with MHC I–peptide complexes through its T cell receptor and produces cytotoxins that kill the interacting target cell

T-helper (Th) cell a lymphocyte that interacts with MHC II–peptide complexes through its T cell receptor and produces cytokines that act on other cells. Th subsets include **Th1** cells that activate macrophages; **Th2** cells that activate B cells; **Th17** cells that activate neutrophils; and **Treg** cells that suppress adaptive immunity

Tolerance the acquired inability to produce an immune response to particular antigens

Vaccination (immunization) the inoculation of a host with inactive or weakened pathogens or pathogen products to stimulate protective active immunity

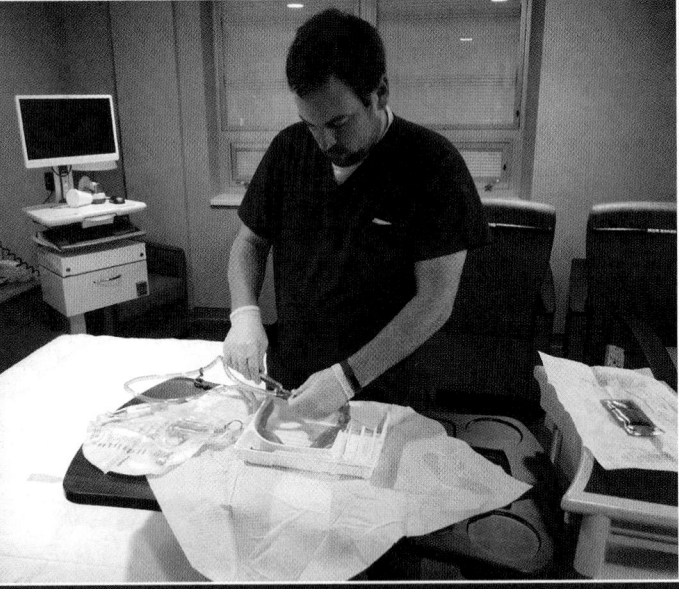

28

Clinical Microbiology and Immunology

microbiology**now**

Bacteriophages: Tiny Allies in the Fight against Antibiotic-Resistant Bacteria

For reasons that will unfold in this chapter, it is surprisingly common for patients admitted into a healthcare facility to acquire an infection during their treatment that is unrelated to their original illness or condition. In the United States, about 1 of every 10 hospitalized individuals acquires a new infection as a result of exposure to the clinical environment, and limiting these *healthcare-associated infections* (HAIs) is a constant challenge for medical personnel.

Patient treatments that require invasive medical procedures are a common cause of HAIs, and catheter-associated urinary tract infections (CAUTIs) are among the most difficult to control. Gram-negative enteric bacteria cause the vast majority of CAUTIs, with the major causative agents being *Escherichia coli* and *Proteus mirabilis*. Many strains of these bacteria produce attachment structures, such as fimbriae and pili, which allow them to colonize the surface of the catheter (see photo of technician preparing a catheter) and potentially establish a urinary tract infection. It is estimated that more than 100 million urinary catheters are fitted each year to relieve urinary retention and incontinence, creating a need for reliable and effective measures for prevention and treatment of CAUTIs.

CAUTIs are routinely treated using antibiotics, typically sulfamethoxazole–trimethoprim (SMZ-TMP), ampicillin, or a fluoroquinolone, such as ciprofloxacin, but pathogen resistance to these drugs is becoming increasingly common. Infections that were once treated successfully through a straightforward antibiotic regimen become lingering problems when the causative agents show multidrug resistance. However, a new and potentially quite effective approach to CAUTI prevention is in development and may have great practical application.

In a recent report, researchers isolated two novel, lytic bacteriophages that specifically attack strains of *Proteus* spp., especially *P. mirabilis*, the cause of more than 40% of all CAUTIs in the United States. They then combined the viruses to create a "phage cocktail" that was applied to the surface of silicone catheters. Using culture techniques, epifluorescence, and scanning electron microscopy, the researchers were able to show that colonization and biofilm formation of *P. mirabilis* was substantially reduced in phage-coated catheters. In separate studies, similar results were obtained against pathogenic strains of *Escherichia coli* using *E. coli*–specific phages, and therefore the use of phage-coated urinary catheters for the control and prevention of CAUTIs shows exciting promise.

Source: Melo, L.D.R., et al. 2016. Development of a phage cocktail to control *Proteus mirabilis* catheter-associated urinary tract infections. *Front. Microbiol.* 7: 1024 doi:10.3389/fmicb.2016.01024. Photo courtesy of Marion General Hospital, Marion Indiana.

Clinical microbiology is that subdiscipline of microbiology concerned with identifying pathogenic microbes and advising the medical provider on treatment. Clinical laboratories must identify pathogens safely, efficiently, and reliably. The clinical microbiologist examines patient samples using direct observation, culture, immunological assays, and molecular tools to identify pathogens. Identification of pathogens guides disease control by targeting antimicrobial drugs to specific pathogens.

I • The Clinical Microbiology Setting

28.1 Safety in the Microbiology Laboratory

Clinical laboratories handle dangerous materials, and thus laboratory workers must adhere to strict safety protocols to prevent the spread of infectious agents. Standard laboratory practices for handling clinical samples have been established to minimize the risk of accidental laboratory infections. One only has to be reminded of the 2014 Ebola hemorrhagic fever outbreak in West Africa (⇄ Sections 29.7 and 30.12) to appreciate how treating infected persons without paying rigorous attention to every safety detail can endanger the lives of medical personnel.

Laboratory Safety

The clinical laboratory has potential biohazards for all personnel and is especially dangerous for untrained personnel or those who do not employ the necessary precautions. All laboratories that handle human or primate tissue must have an occupational exposure control plan for handling bloodborne pathogens. This plan is specifically designed to protect workers from infection by hepatitis B virus (HBV, the cause of infectious hepatitis, ⇄ Section 30.11) and human immunodeficiency virus (HIV, the cause of acquired immunodeficiency syndrome [AIDS], ⇄ Section 30.15). The occupational exposure plan limits infection by all pathogens and typically includes the use of appropriate *personal protective equipment* (PPE), such as a lab coat, gloves, eye protection, and face mask (**Figure 28.1**).

Proper training and enforcement of established safety procedures can prevent most accidental infections, which usually do not result from identifiable exposures like culture spills but instead from routine handling of patient specimens. Infectious aerosols generated during microbiological procedures are the most common causes of laboratory infections. Clinical laboratories follow the safety rules outlined in **Table 28.1** to minimize laboratory infections. These general standards apply to all laboratories that handle potentially infectious agents and are the basis for all aspects of healthcare infection control. However, as discussed next, laboratories that handle particularly dangerous or transmissible agents adhere to additional rules and procedures to ensure a safe work environment.

Biological Containment and Biosafety Levels

The level of containment used to prevent accidental infections or accidental environmental contamination (escape) in clinical, research, and teaching laboratories must be proportional to the

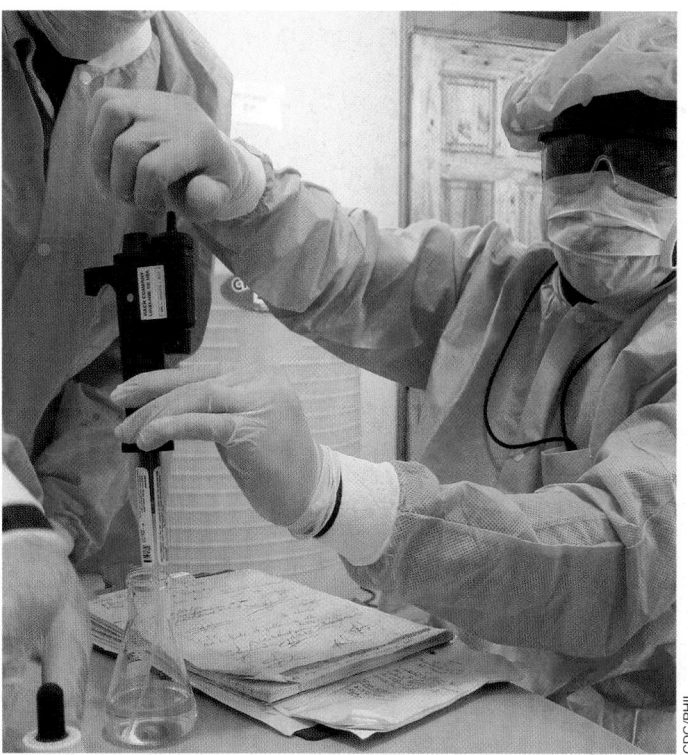

Figure 28.1 Standard apparel for clinical laboratory safety. This technician is wearing proper personal protective equipment (PPE) for a clinical laboratory, including gloves, eye protection, lab coat, and face mask.

biohazard potential of the organisms handled in the laboratory. Laboratories are classified according to their containment capabilities from least to greatest by their *biosafety level (BSL)*, designated as *BSL-1, BSL-2, BSL-3,* or *BSL-4* (**Figure 28.2**). Personnel in laboratories working at all biosafety levels must follow standard laboratory practices that ensure basic cleanliness and limit contamination (Table 28.1). The precautions, equipment, and operational costs increase with each biosafety level.

TABLE 28.1 Microbiology laboratory safety standards

Rule	Implementation
Restrict access	Only laboratory workers and trained support personnel have access.
Practice good personal hygiene	Eating, drinking, applying cosmetics, and manipulating contact lenses are forbidden in the laboratory. Hand washing prevents spread of pathogens.
Use personal protective equipment (PPE)	Lab coats, gloves, eye protection, and respirators are recommended or required depending on the pathogens being handled.
Vaccinate	Personnel must be vaccinated against agents to which they may be exposed.
Handle specimens safely	Assume all clinical specimens are infectious and handle appropriately.
Decontaminate	After use or exposure, decontaminate specimens, surfaces, and materials by disinfecting, autoclaving, or incinerating.

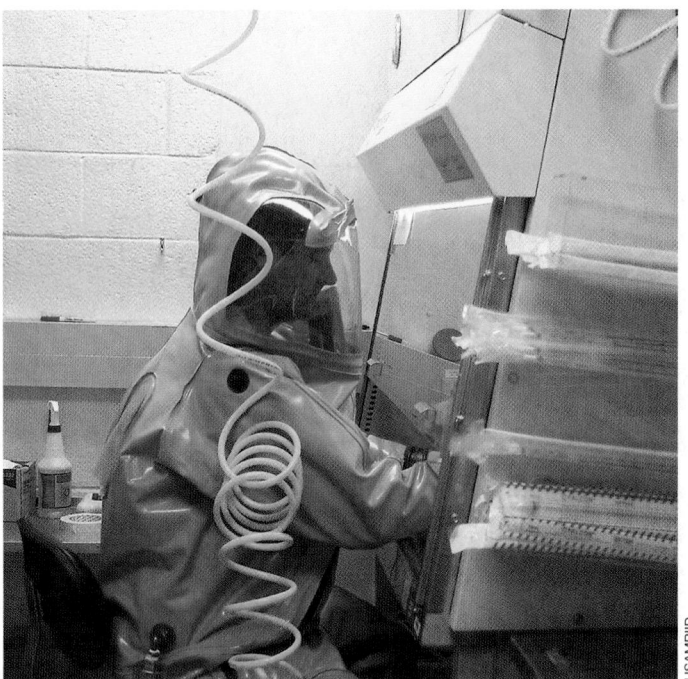

Figure 28.2 Conducting research in a BSL-4 (biosafety level 4) laboratory. BSL-4 is the highest level of biological control, affording maximum worker protection and pathogen containment. The researcher has a whole-body sealed suit with an outside air supply and ventilation system. Air locks control all access to the laboratory. All material leaving the laboratory is autoclaved or chemically decontaminated.

Most colleges and universities have BSL-1 and BSL-2 facilities for teaching and research. Standard clinical laboratories operate at BSL-2. The specialized physical requirements for BSL-3 facilities limit them to major clinical centers and research settings. Because BSL-4 facilities must ensure total isolation and physical containment of pathogens, only about fifty BSL-4 laboratories are operational worldwide. Most BSL-4 laboratories are associated with government facilities, such as the Centers for Disease Control and Prevention (CDC; Atlanta, Georgia, USA) and the U.S. Army Medical Research Institute of Infectious Diseases (USAMRIID; Fort Detrick, Maryland, USA).

— MINIQUIZ —

• The use of personal protective equipment (PPE) is required for clinical laboratory technicians. What protective apparel does PPE include?

• Identify and discuss the standard safety procedures adopted by microbiology laboratories. Under what biosafety level do most clinical laboratories operate? Where are most BSL-4 laboratories located?

28.2 Healthcare-Associated Infections

The universal safety measures described in the previous section are implemented to contain infectious agents and prevent their transmission. But despite such precautions, the accidental transfer of pathogens to individuals in healthcare facilities is rather common (see page 830 for more on this).

Mechanisms of Transfer of Healthcare-Associated Infections

A **healthcare-associated infection (HAI)**, also called a *nosocomial infection* (from the Latin *nosocomium*, meaning "hospital"), is an infection acquired by a patient during a stay at a healthcare facility (clinic, hospital, rehabilitation facility, etc.). HAIs cause significant morbidity (incidence of *disease* in a population) and mortality (incidence of *death* in a population). An estimated 10% of patients admitted to healthcare facilities in the United States acquire HAIs, and up to 2 million HAIs occur annually, leading directly or indirectly to about 75,000 deaths. Some of the common risk factors for acquiring infectious diseases in healthcare settings are summarized in **Table 28.2**.

Some HAIs are acquired from patients with communicable diseases, but others are caused by pathogens that are selected and maintained within the hospital environment, spread by cross-infection from patient to patient or from healthcare personnel. Healthcare-associated pathogens are often present as normal microbiota in either patients or healthcare staff. Therefore, healthcare facilities are high-risk environments for the spread of infections because these facilities concentrate individuals who have infectious disease or are at risk for acquiring infectious disease because of underlying health conditions. Such conditions often lead to a compromised immune system and increased susceptibility to pathogens. The frequency of HAIs at different sites of the body is shown in **Figure 28.3**.

Common Causative Agents of HAIs

Most HAIs are caused by a relatively short list of pathogens (**Table 28.3**), but many other infectious agents can cause HAIs. *Staphylococcus aureus* is one of the most important and widespread HAI pathogen (❧ Section 30.9). It is the most common

TABLE 28.2 Risk factors for hospital-acquired infections (HAIs)

Risk factor for HAI	Rationale
Patients	Patients are already ill or immunocompromised
Newborn infants and the elderly	Not fully immune competent
Infectious disease patients	Pathogen reservoirs
Patient proximity	Increases cross-infection
Healthcare personnel	Can transfer pathogens between and among patients; healthcare personnel may be asymptomatic disease carriers
Medical procedures (blood draws, etc.)	Breaching the skin barrier can introduce pathogens
Surgery	Exposes internal organs, may introduce pathogens, and causes stress, which lowers resistance to infection
Anti-inflammatory drug treatment	Lower resistance to infection
Antibiotic treatment	May select for resistant and opportunistic pathogens

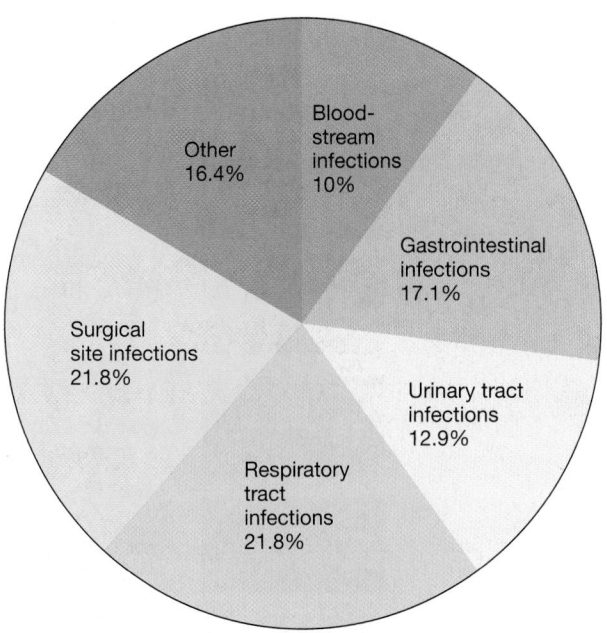

Figure 28.3 Frequency of healthcare-associated infections (HAIs) in different sites of the body. Up to 2 million HAIs occur annually in the United States. Data are from the Centers for Disease Control and Prevention.

cause of pneumonia, the third most common cause of blood infections, and is particularly problematic in nurseries. Many hospital strains of *S. aureus* are unusually virulent and are resistant to common antibiotics, making treatment especially difficult (see Explore the Microbial World: MRSA—A Formidable Clinical Challenge). The staphylococci are the most common cause of bloodborne HAIs and are also prevalent in pus-forming wound infections.

Staphylococcus and *Enterococcus*, as well as *Escherichia coli*, *Klebsiella pneumoniae*, and various other *Enterobacteriaceae*, all have the potential to cause HAIs, but they are also members of the normal microbiota of most individuals, making it essentially impossible to eliminate these potential pathogens from healthcare settings. In addition, these organisms can acquire multidrug resistance by horizontal gene flow (Chapter 11). Pathogens that are not part of the normal microbiota, such as species of *Acinetobacter* and *Mycobacterium*, can be eliminated from the healthcare environment. These pathogens are carried into the healthcare facilities by infected individuals or, in the case of some mycobacteria, as environmental contaminants that enter in dust and air.

Prevention of HAIs involves cooperation between the healthcare facility infection-control team and the rest of the facility staff, including direct healthcare workers and supporting staff, such as housekeeping. Infection control starts with management of incoming patients at the point of entry to the healthcare facility; incoming patients should be assessed for possible infections and isolated as necessary to prevent spread of infections to staff and other patients. From this point, the healthcare facility staff employs standard procedures that limit infection, applying the same general precautions outlined for laboratory technicians in Table 28.1.

II • Isolating and Characterizing Infectious Microorganisms

The growth and observation of pathogens from patient specimens are important strategies for identification of the causative agent of an infectious disease. Identification leads to antimicrobial drug susceptibility testing and development of a specific treatment plan. We begin by looking at methods for collecting, culturing, and identifying pathogens, followed by methods used to determine drug susceptibility.

28.3 Workflow in the Clinical Laboratory

Collecting specimens from infectious patients and subsequently culturing pathogens using a variety of growth media is a necessary and routine practice in clinical medicine. Combined with microscopic observation, these methods allow for direct detection and identification of causative agents of disease.

Collecting Specimens and Detecting and Culturing Pathogens

Proper clinical diagnosis of infectious diseases requires that pathogens be identified from tissue or fluid samples using a variety of microbiological, immunological, and molecular biological techniques (**Figure 28.4**). Patient specimens must be collected aseptically from the site of the infection, and the sample size must be large enough to ensure an inoculum sufficient for growth. In addition, the requirements for organism survival, such as oxic versus anoxic conditions, must be maintained at all times, and the sample should be processed quickly to avoid degradation. Sterile swabs are often used to obtain samples from infected areas, including wounds, nares, or throat (**Figure 28.5**), and the swab is then used to inoculate a suitable growth medium.

Many pathogens can be detected by direct means using one or more of several diagnostic tests. For example, the microscopic observation of gram-negative diplococci, especially inside neutrophil inclusions, in a urethral exudate sample is diagnostic for infection with *Neisseria gonorrhoeae* (**Figure 28.6a**), the causative agent of the sexually transmitted disease gonorrhea (Section 30.13). However, the reliability of any diagnostic test depends on both the *specificity* and the *sensitivity* of the test. **Specificity** is the ability of the test to recognize a single pathogen. High specificity reduces the likelihood of a *false-positive* result. **Sensitivity** defines the smallest quantity of a pathogen or a pathogen product that can be detected. High sensitivity minimizes the likelihood of a false-negative reaction. For the detection of *N. gonorrhoeae*, the specificity of Gram-stained smears of urogenital exudates is high for both men and women (≥95%), so false-positive tests for gonorrhea are rare. By contrast, the sensitivity of Gram-stained smears of

TABLE 28.3 Common healthcare-associated pathogens

Pathogen	Common infection sites and diseases	Micrographs[b]
[a]*Acinetobacter*	Wound/surgical site, bloodstream, pneumonia, urinary tract	*Acinetobacter*
Burkholderia cepacia	Pneumonia	*B. cepacia*
Clostridium difficile *C. sordellii*	Gastrointestinal tract Pneumonia, endocarditis, arthritis, peritonitis, myonecrosis	*C. difficile*
[a]*Enterobacteriaceae*, carbapenem-resistant, especially *Escherichia coli* and *Klebsiella pneumoniae*	Urinary tract, pneumonia, wound/surgical site, [c]bloodstream	*E. coli* *Klebsiella*
[a]Vancomycin-resistant *Enterococcus*	Wound/surgical site, bloodstream, urinary tract	*Enterococcus*
Hepatitis	Chronic liver infection	Hepatitis B virus
Human immunodeficiency virus (HIV)	Immunodeficiency	HIV
Influenza virus	Pneumonia	Influenza virus

[a]*Mycobacterium tuberculosis*
   M. abscessus
	Chronic lung infection
	   Skin and soft tissue infections
		M. tuberculosis

Norovirus	Gastroenteritis	Norovirus
[-a]*Staphylococcus aureus* Methicillin-resistant (MRSA) and vancomycin-intermediate and -resistant strains	Bloodstream, pneumonia, endocarditis	*S. aureus*

[a]Antibiotic-resistant organisms that exhibit multiple drug resistance. Because of the promiscuous nature by which multiple drug-resistant plasmids can be transmitted, many of the pathogens listed as well as members of the normal microbiota could be or become drug resistant in the highly selective nature of the healthcare environment where antibiotics are used routinely and extensively. In addition to the pathogens listed, other extremely pathogenic agents (such as Ebola virus) could be localized to an isolation unit of a healthcare facility, thus making that part of the facility especially dangerous for disease transmission.
[b]All inset photographs are colorized scanning or transmission electron micrographs obtained from CDC/PHIL. Additional micrograph credits (numbers run top to bottom): 1–5, 10, and 12, Janice Haney Carr; 6, Peta Wardell; 7, Erskine Palmer; 8, A. Harrison and P. Feorino; 9, Frederick Murphy; 11, Charles D. Humphrey.
[c]Many urinary tract hospital-associated infections are caused by the gram-negative enteric bacterium *Proteus mirabilis* (see page 830).

UNIT 6

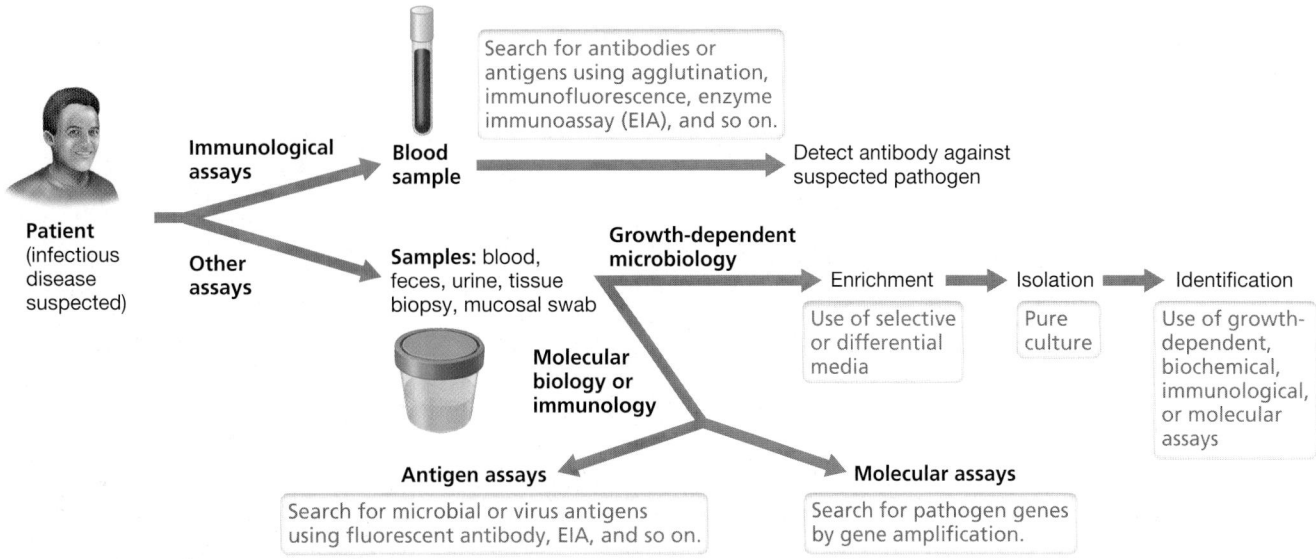

Figure 28.4 Laboratory identification of microbial pathogens. The flowchart shows alternative paths for identifying pathogens or pathogen exposure in the clinical laboratory.

urogenital exudates for the detection of *N. gonorrhoeae* is about 80% greater for men than for women. Thus, in suspected cases of gonorrhea in females, *false-negative* Gram stains are relatively common. Females must therefore be examined by more sensitive methods, including culture techniques (Figure 28.6b), to establish or confirm a diagnosis of gonorrhea.

Specific pathogens can be selectively grown, isolated, and identified from patient specimens using specialized growth media and incubation conditions to establish an *enrichment culture* (⮌ Sections 19.1 and 19.2). For primary enrichment, clinical samples are inoculated on general-purpose media, such as *blood agar* (**Figure 28.7a**) and *chocolate agar* (so called because it contains heat-lysed blood, making it brown in color; Figure 28.7b), which support the growth of a variety of microorganisms. To aid in the isolation and identification of specific pathogens, medical technologists use various *selective* and *differential* growth media. **Selective media** are specialized culture media that contain inhibitory agents for the purpose of allowing some organisms to grow but not others. **Differential media** allow identification of organisms based on the appearance of the culture after growth. For example, eosin–methylene blue (EMB) agar is a selective

medium because the methylene blue it contains inhibits the growth of gram-positive bacteria. EMB agar is also a differential medium because it distinguishes gram-negative bacteria that can ferment lactose from those that cannot. Lactose fermenters, such as *Escherichia. coli*, acidify the medium and produce dark colonies that may have a reflective metallic green sheen; non–lactose "fermenters, such as *Pseudomonas aeruginosa*, produce opaque or translucent colonies (Figure 28.7c).

Blood and Cerebrospinal Fluid Specimens

Pathogens in liquid tissue samples, such as blood and cerebrospinal fluid (CSF), are routinely detected using automated culture

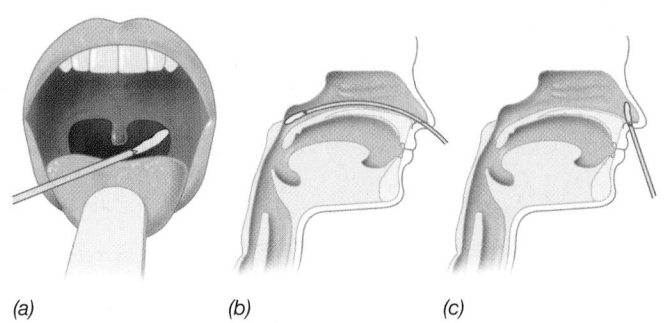

Figure 28.5 Specimens from the upper respiratory tract. *(a)* Throat swab. *(b)* Nasopharyngeal swab passed through the nose. *(c)* Swabbing the inside of the nose.

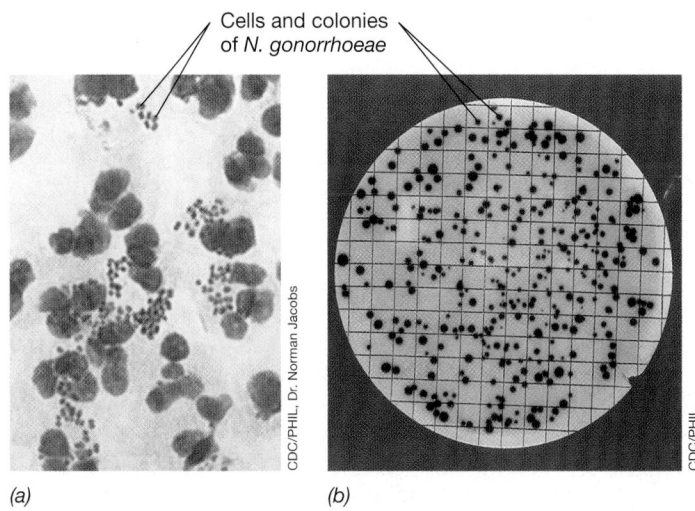

Figure 28.6 Identification of *Neisseria gonorrhoeae*, the cause of gonorrhea. *(a)* Cells of gram-negative *N. gonorrhoeae* within human polymorphonuclear leukocytes (neutrophils) from a urethral exudate. *(b)* Colonies of *N. gonorrhoeae* growing on a filter placed on Thayer–Martin agar. Oxidase reagent, which turns colonies dark purple if they contain cytochrome *c*, has been added to the filter. The dark color of the colonies shows that *N. gonorrhoeae* is oxidase-positive.

MRSA—A FORMIDABLE CLINICAL CHALLENGE

Staphylococcus aureus is an opportunistic human pathogen that colonizes an estimated one-third of the population. In such individuals, this gram-positive bacterium usually exists as an innocuous member of the normal microbiota of the skin and mucous membranes, especially in the nasal epithelium. Although most complications from staphylococci occur in the form of local skin infections, *S. aureus* may in some cases invade other tissues of the body, including the bloodstream, where a host of virulence factors it produces may transform this bacterium from a harmless commensal to a serious pathogen capable of causing life-threatening disease.

Virulence factors of *S. aureus* include coagulase and clumping factor, two proteins that precipitate blood plasma components and promote the formation of a protective matrix around the bacterial cells that inhibits the activity of phagocytes at sites of infection. In addition, protein A is a surface protein on *S. aureus* that obstructs opsonization by antibodies and complement (⟳ Section 26.9), effectively camouflaging the pathogen and further impeding phagocytosis. *S. aureus* also produces several potent exotoxins, including α-hemolysin (lyses blood cells), enterotoxin B (causes food poisoning), and toxic shock syndrome toxin-1 (causes life-threatening toxic shock).

Many *S. aureus* infections are effectively treated using β-lactam antibiotics, such as various penicillins, but the number of strains resistant to these drugs is increasing as global use of antibiotics remains extremely high (see Figure 28.28a). Infections caused by multidrug-resistant strains of *S. aureus* are often treated with *methicillin*, a powerful semisynthetic β-lactam drug that is among the last

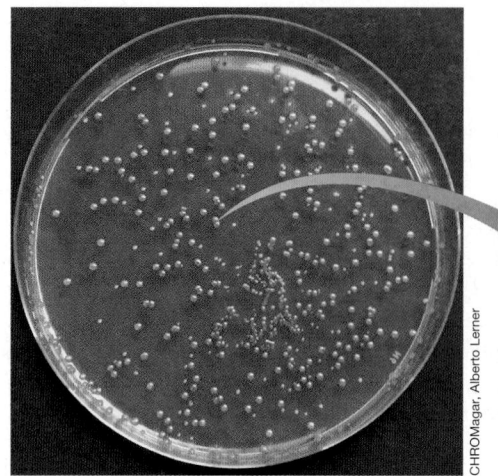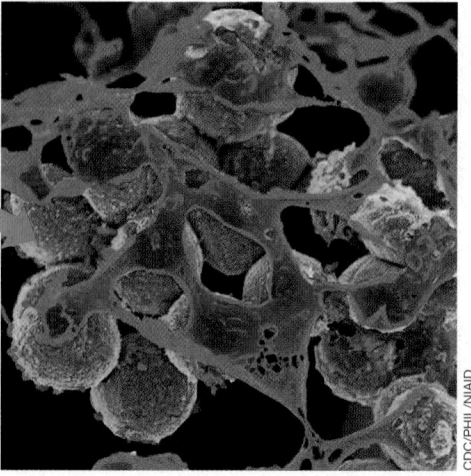

Figure 1 Methicillin-resistant *Staphylococcus aureus* (MRSA). Colonies of MRSA on this chromogenic agar medium (left) appear pink, whereas colonies of other bacteria appear blue. The colorized scanning electron micrograph on the right shows cells of MRSA (yellow) embedded in a matrix of cellular debris (orange).

treatment options available in these cases. However, an alarming percentage of *S. aureus* strains have also developed resistance to methicillin in recent years. These methicillin-resistant *S. aureus* (MRSA) strains have been especially prevalent as healthcare-associated infections.

Clinical diagnosis of MRSA infections depends upon either culture-based techniques, which often include the use of chromogenic agar media (**Figure 1**) or nucleic acid–based tests (Section 28.8). Confirmation of infection must be followed by antibiotic susceptibility testing (Section 28.4) to determine the best treatment strategy. Typically, treatments are case-specific and require a non-β-lactam antibiotic such as vancomycin, tetracycline, or sulfa drugs (Section 28.10), often administered intravenously.[1]

The global incidence of MRSA infection remains high; almost 80,000 cases are documented every year in the United States alone, and several regions of the world have reported increases in MRSA prevalence in recent years, especially in India.[2] Interestingly, the prevalence of healthcare-associated MRSA infections has steadily declined in recent years while that of community-associated MRSA infections has increased. The United States Centers for Disease Control and Prevention estimates that MRSA carriers comprise 2% of the population. Considering its prevalence and extensive multidrug resistance, MRSA may continue to prove a difficult pathogen to control.

[1]American Academy of Microbiology. 2015. FAQ: The threat of MRSA. AAM: Washington, D.C.
[2]Center for Disease Dynamics, Economics, and Policy. 2015. State of the world's antibiotics, 2015. CDDEP: Washington, D.C.

systems. For a suspected case of meningitis, a CSF specimen is obtained by a procedure called a *lumbar puncture* (spinal tap) in which 3–5 ml of fluid is collected drop-by-drop from a needle inserted between lumbar vertebrae. CSF is sterile and clear in a healthy individual, and therefore fluid turbidity and high leukocyte counts are indicators of infection. Similar to other liquid specimens (blood, urine, sputum, wound exudates, etc.), CSF is routinely examined by Gram staining and used to inoculate selective culture media.

The standard procedure for obtaining a blood sample is to aseptically draw 10–20 ml of blood from a vein and inject it into two

culture bottles containing general-purpose growth media and an anticoagulant. One bottle is incubated aerobically, while the other is incubated anaerobically (**Figure 28.8**); both are kept at 35°C for several days. Automated culture systems (Figure 28.8b) detect growth by measuring turbidity or fluorescence and by periodically monitoring the consumption of O_2 or the production of CO_2. Most clinically significant bacteria are recovered within 2 days, but growth of some pathogens, including mycobacteria and certain fungi, may take 3 to 5 days or longer. Cultures that exhibit growth are Gram stained and then inoculated onto specialized media for isolation and identification.

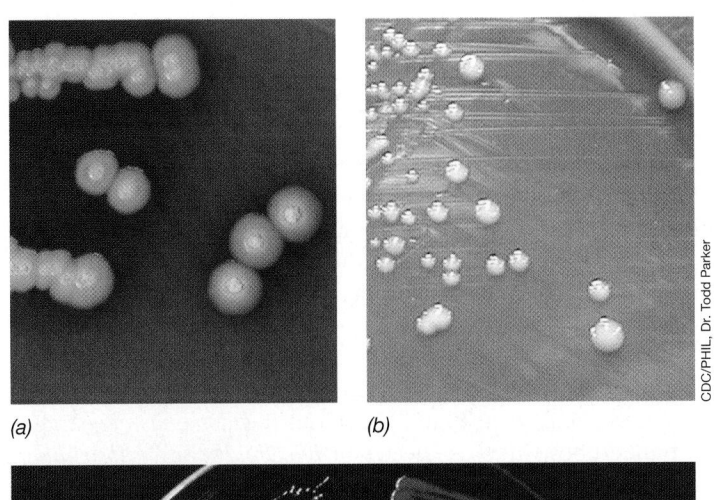

(a)   (b)

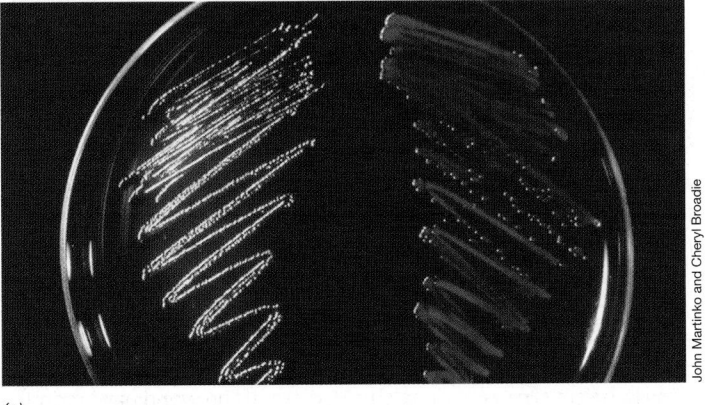

(c)

Figure 28.7 Enriched media. *(a) Burkholderia* growing on sheep blood agar (SBA); the red color is from blood suspended in the trypticase soy agar medium. *(b) Francisella tularensis* growing on chocolate agar; the brown color is due to heat-lysed blood in the trypticase soy agar medium. *(c) Escherichia coli*, a lactose fermenter (left), and *Pseudomonas aeruginosa*, a non–lactose fermenter (right) growing on eosin–methylene blue (EMB) agar. The reflective, greenish-yellow sheen of the colonies on the left identifies *E. coli* as a lactose fermenter.

The most common pathogens found in blood include species of *Staphylococcus* and *Enterococcus*, but many other bacteria may cause blood infections. Computer databases are used to unambiguously identify clinical isolates by matching their metabolic reactions in various differential media to the biochemical patterns of known pathogens. The biochemical tests incorporated into differential media evaluate the presence or absence of enzymes that catabolize a specific substrate or substrates. Although hundreds of different biochemical tests are known, just a few key tests may be sufficient to identify some pathogens.

Urinary Tract and Fecal Cultures

Urinary tract infections (UTIs) are common, especially in women. In most cases, microorganisms infect the urinary tract by ascending into the bladder from the urethra. UTIs, often introduced through the use of urinary catheters, are among the most common healthcare-associated infections. Direct microscopic examination of urine from a UTI patient usually shows the presence of abnormal numbers of bacteria in the urine. A Gram stain may be done directly on urine samples to identify the morphology of urinary tract pathogens, such as gram-negative rods (various enteric

(a)   (b)

Figure 28.8 Growth-dependent diagnostic testing for blood infections. *(a)* Cultures to assay both aerobic (left vial) and anaerobic (right vial) bacterial growth are inoculated with an aseptically drawn patient blood sample. *(b)* After inoculation, both vials in part *a* are incubated in an automated system that measures growth, for example, by turbidity, production of CO_2, or fluorescence. Photos taken courtesy of Marion General Hospital, Marion, Indiana, USA.

bacteria), gram-negative cocci (species of *Neisseria*), and gram-positive cocci (especially species of *Enterococcus*).

A significant UTI typically results in bacterial counts of 10^5 or more cells per milliliter of urine. The most common causative agents of UTIs are enteric bacteria, with *E. coli* accounting for about 90% of cases. Blood agar is often used for primary enrichment and isolation of urinary tract pathogens. Selective and differential enteric media, such as EMB or MacConkey agar, permit differentiation of gram-negative lactose fermenters from non–lactose fermenters (Figure 28.7c) and inhibit the growth of possible gram-positive contaminants, such as commensal staphylococci. Additional differential and/or selective media may be used to identify urinary tract (or other) pathogens via traditional culturing techniques (**Figure 28.9a, b**) or rapid and convenient media kits (Figure 28.9c).

Proper collection of fecal samples is important for the isolation of intestinal pathogens. Fecal specimens become more acidic during storage, so delay between sampling and processing must be minimized, especially for the isolation of acid-sensitive pathogens, such as *Shigella* and *Salmonella*. The fecal sample is placed in a sterile, sealed container for transport to the laboratory. Feces containing blood or pus, as well as feces from patients with suspected foodborne or waterborne infections, are inoculated into suitable media for isolation of potential pathogens. For example, many laboratories use selective and differential media to identify *E. coli* O157:H7 and *Campylobacter* species, important intestinal pathogens typically acquired from contaminated food or water (⌽ Sections 32.11 and 32.12). Intestinal eukaryotic pathogens, such as *Giardia intestinalis* (⌽ Section 33.4), are identified by direct microscopic observation of parasite cysts in diluted feces or through antigen-detection assays (see Figure 28.18c) rather than by culturing.

UNIT 6

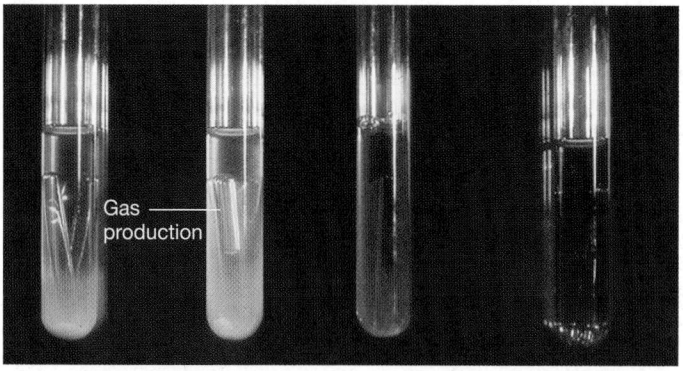

(a)

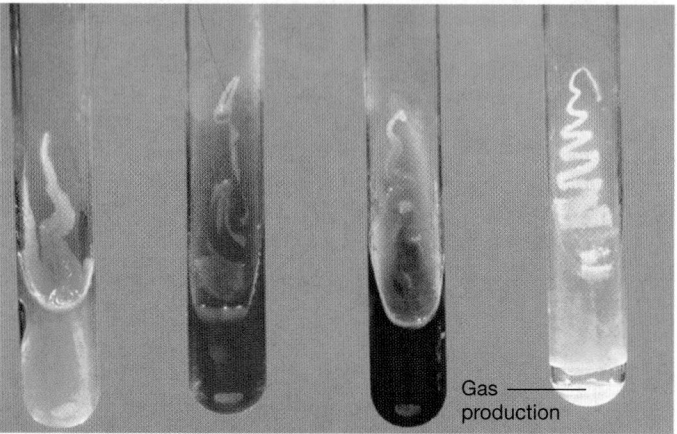

(b)

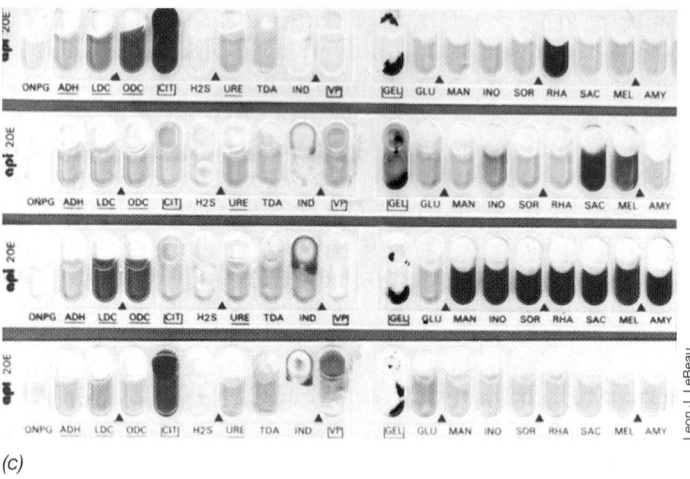

(c)

Leon J. LeBeau

Figure 28.9 Growth-dependent diagnostic tests for clinical isolates.
(a) Differential media to assess sugar fermentation. Acid production is indicated by a color change (from red to yellow) of the pH-indicating dye in the medium. The appearance of a bubble in the small, inverted inner tube indicates gas production from fermentation. *(b)* Diagnostic test for enteric bacteria using triple-sugar iron (TSI) agar. The medium contains glucose, lactose, and sucrose. Organisms able to ferment only glucose cause acid formation only in the bottom of the tube, whereas lactose- or sucrose-fermenting organisms cause acid formation throughout the slant. The breaking up of the agar in the bottom of the tube indicates gas formation. Blackening of the agar is due to the reaction of hydrogen sulfide (from either protein degradation or thiosulfate reduction) with ferrous iron in the medium. From left to right: Fermentation of glucose only (typical of *Shigella*); growth but no fermentation (typical of *Pseudomonas*); hydrogen sulfide formation (typical of *Salmonella*); fermentation of sugars with gas production (typical of *Escherichia coli*). *(c)* Miniaturized media kits allow rapid identification of clinical isolates by running many biochemical tests on specimen samples at the same time. Four separate strips, each with a different isolate, are shown.

Wounds and Abscesses

Infections associated with injuries such as animal bites, burns, or cuts are sampled to recover the relevant pathogen. The results must be interpreted carefully to differentiate between infection and contamination. Wound infections and abscesses often harbor a variety of normal microbiota, and swab samples from such lesions are frequently misleading. For abscesses and other purulent lesions, pus is aspirated with a sterile syringe and needle following disinfection of the skin surface. Internal purulent lesions are sampled by biopsy or from tissues removed in surgery. Gram stains are prepared directly from these specimens and examined microscopically.

Pathogens commonly associated with wound infections are *Staphylococcus aureus*, enteric bacteria, *Pseudomonas aeruginosa*, and anaerobes, such as species of *Bacteroides* and *Clostridium*. Because of the varied oxygen requirements of these bacteria, samples must be obtained, transported, and cultured under anoxic as well as oxic conditions. The major isolation media are blood agar, selective media for enteric bacteria, and enrichment media containing additional supplements and reducing agents for obligate anaerobes. Widely used tools for the detection of methicillin-resistant *Staphylococcus aureus* (MRSA) in skin infections are *chromogenic* agar media. These selective and differential media contain a chromogenic substrate that, when metabolized, causes MRSA to produce distinctly colored colonies (see Explore the Microbial World: MRSA—A Formidable Clinical Challenge).

Genital Specimens and Culture for Gonorrhea

Sexually transmitted infections (STIs) that cause a purulent urethral discharge, especially in males, are classified as either nongonococcal or gonococcal urethritis. Nongonococcal urethritis is usually caused by *Chlamydia trachomatis* (⇔ Section 30.14), *Ureaplasma urealyticum*, or *Trichomonas vaginalis* (⇔ Section 33.4). Gonococcal urethritis is caused by *Neisseria gonorrhoeae* (⇔ Section 30.13).

Cells of *N. gonorrhoeae* are gram-negative diplococci, a morphology not normally found in microbiota of the urogenital tract. Therefore, a Gram stain of a urethral, vaginal, or cervical smear revealing such cells, often surviving inside neutrophils (Figure 28.6*a*), is diagnostic for gonorrhea. Chocolate agar, a nonselective enriched medium, is often used for specimens suspected to contain *N. gonorrhoeae*. A selective medium used for isolation of *N. gonorrhoeae* is modified Thayer–Martin (MTM) agar (Figure 28.6*b*). This medium incorporates the antibiotics vancomycin, nystatin, trimethoprim, and colistin to suppress the growth of normal microbiota. These antibiotics have no effect on *N. gonorrhoeae* or *Neisseria meningitidis*, a cause of bacterial meningitis (⇔ Section 30.5).

Inoculated plates are incubated in a humid atmosphere containing 3–7% CO_2 for 24–48 hours and then tested for their oxidase reaction (Figure 28.6*b*). Oxidase-positive, gram-negative diplococci growing on MTM or chocolate agar are presumed to be gonococci if the inoculum was derived from the urogenital tract. Definitive identification of *N. gonorrhoeae* requires determination of carbohydrate utilization patterns and immunological or nucleic acid probe tests. Laboratory testing of urogenital samples for *N. gonorrhoeae* (and the often-associated *C. trachomatis*) is often done using DNA amplification via polymerase chain reaction (PCR) or other molecular methods.

Culture of Anaerobic Pathogens

The identification of obligately anaerobic bacteria from patient specimens requires special isolation and culture methods (⮌ Section 5.14). In general, media for anaerobes do not differ greatly from those used for aerobes, except that they (1) are usually richer in organic constituents, (2) contain reducing agents (usually cysteine or thioglycolate) to remove O_2, and (3) contain a redox indicator to show that conditions are anoxic. Collection, handling, and processing of specimens must exclude exposure to air because oxygen is toxic to obligate anaerobes. Samples collected by syringe aspiration or biopsy must be immediately transferred to a sealed tube containing O_2-free gas, usually with a dilute salt solution containing a reducing agent and a redox indicator to monitor O_2 contamination. Specimens are then used to inoculate anoxic media in an automated culture system or in an anoxic "glove box" filled with O_2-free gases, usually a mixture of N_2 and H_2.

Several habitats in the body, including portions of the oral cavity and the lower intestinal tract, are anoxic and support the growth of anaerobic normal microbiota. Other parts of the body may also become anoxic if injury or disease reduces the blood supply to certain tissues, a condition called *ischemia*. These anoxic sites can then be colonized by obligate anaerobes. Although potentially pathogenic anaerobic bacteria are part of the normal microbiota, their numbers are kept in check through competition with other members of the microbial community. Under certain conditions, however, normally benign anaerobes may become opportunistic pathogens. A key example is *Clostridium difficile* (Table 28.3); this usually harmless member of the normal microbiota of the lower intestinal tract commonly emerges as a healthcare-associated pathogen when extended antibiotic therapy destroys competing microbes (⮌ Section 24.10).

MINIQUIZ

- What are the key points necessary for proper collection of clinical specimens, and why is it important that diagnostic tests for these specimens are both highly specific and sensitive?
- Identify culture methods and conditions used for blood, wound, urine, fecal, and genital specimens. Of what importance are selective and differential growth media in pathogen detection, and what special conditions must be maintained for the isolation of anaerobic pathogens?

28.4 Choosing the Right Treatment

Pathogens isolated from clinical specimens are identified to confirm medical diagnoses and to guide antimicrobial therapy. Appropriate and effective treatment for many pathogens is based on current experience and practices. For some pathogens, however, decisions about appropriate antimicrobial therapy must be made on a case-by-case basis. Such pathogens include those for which antimicrobial drug resistance is common (for example, gram-negative enteric bacteria), those that cause life-threatening disease (for example, meningitis caused by *Neisseria meningitidis*), and those that require bactericidal rather than bacteriostatic

drugs (⮌ Section 5.17) to prevent disease progression and tissue damage. Bactericidal agents are indicated, for example, for bacterial endocarditis (infection of the inner tissues of the heart, such as the heart valves), where total and rapid killing of the pathogen is critical for patient survival.

Minimum Inhibitory Concentration

Antimicrobial susceptibility is measured by determining the smallest amount of agent needed to completely inhibit the growth of the tested organism in vitro (in laboratory culture), a value called the **minimum inhibitory concentration (MIC)**. The traditional way to determine the MIC for a given agent against a given organism is to prepare a series of culture tubes inoculated with the same number of microorganisms. Each tube contains the growth medium with an increasing concentration of the antimicrobial agent. After incubation, the tubes are checked for visible growth (turbidity), and the MIC is the lowest concentration of the agent that completely inhibits the growth of the test organism (⮌ Section 5.17 and Figures 5.37 and 5.38).

In modern practice, the MIC of a given antimicrobial agent is typically determined using microliter amounts of media and reagents. For example, a miniaturized version of the MIC test uses a microtiter method with twofold dilutions of several antibiotics in medium inoculated with a standard amount of the test organism (**Figure 28.10a**). In clinical microbiology laboratories, tests for routine MIC determinations are usually automated using instruments that also allow for species identification of pure cultures obtained from patient specimens (Figure 28.10b, c).

Measuring Antimicrobial Susceptibility

The standard assay for antimicrobial activity is the *disc diffusion test* (Figure 28.10d–g). A Petri plate containing an agar medium is inoculated by evenly spreading a suspension of a pure culture of the suspected pathogen on the agar surface. Known amounts of different antimicrobial agents infused into filter-paper discs are then placed on the surface of the agar. The agents then diffuse from the discs into the agar during incubation, establishing a gradient; the farther the chemicals diffuse away from each disc, the lower is the concentration of the agent. At some distance from each disc, the effective MIC is reached. Beyond this point the microorganism is able to grow, but closer to the disk, growth is absent. A *zone of inhibition* forms with a diameter proportional to the concentration, solubility, diffusion coefficient, and overall effectiveness of the antimicrobial agent in the disc.

In addition to the disc diffusion test, Figure 28.10g depicts antibiotic susceptibility using the *epsilometer test* (Etest). This assay uses a plastic strip infused with a predefined concentration gradient of an antimicrobial agent. When applied to the surface of an inoculated agar plate, the gradient transfers from the strip to the agar and remains stable throughout the incubation period, during which an elliptical zone of inhibition centered along the axis of the strip develops. The concentration of the antimicrobial agent (in µg/ml) is read at the point where the ellipse edge intersects the precalibrated test strip, providing a precise MIC (Figure 28.10g).

Assuming culture conditions are standardized, different antimicrobial agents can be compared to determine which is most effective

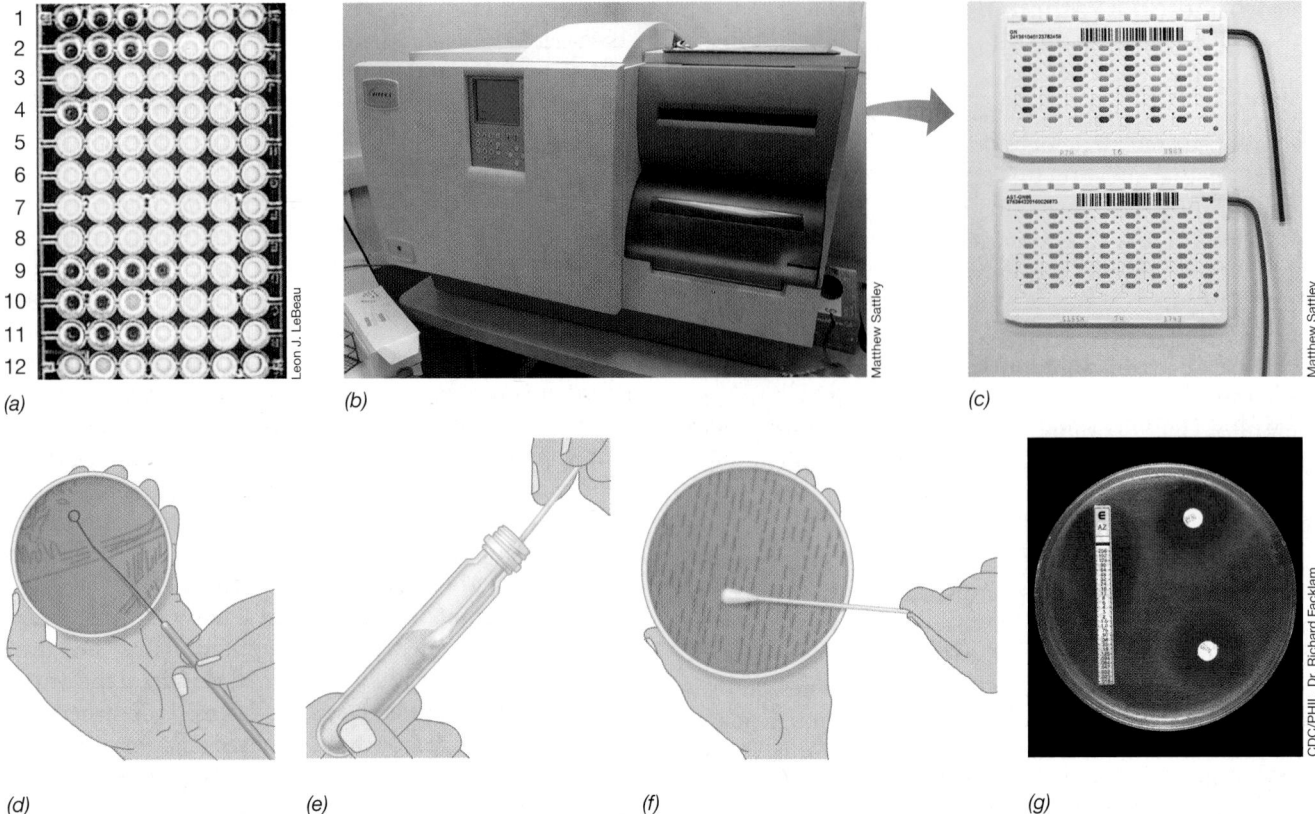

Figure 28.10 Antibiotic susceptibility testing.
(a) Antibiotic susceptibility of a pathogen as determined by the broth dilution method in a microtiter plate. The organism is *Pseudomonas aeruginosa*. Each row has a different antibiotic in a series of concentrations. The highest concentration of antibiotic is in the well at the left; serial twofold dilutions are made in the wells to the right. In rows 1 and 2, the third well has the lowest concentration of antibiotic that shows no visible growth. In row 3, the antibiotic is ineffective at the concentrations tested because there is growth in all wells. *(b, c)* An automated system for identifying clinical isolates and determining their antibiotic susceptibility. Card wells (panel *c*) are inoculated using the attached capillary tube. Following internal incubation, computer-scanned results are available in less than 24 hours. Photos taken courtesy of Marion General Hospital, Marion, Indiana (USA). *(d)* For the disc diffusion test, colonies from a pure culture of the pathogen are transferred to a liquid medium and mixed. *(e, f)* A sterile swab is dipped into the bacterial suspension and streaked evenly over the entire surface of a suitable agar medium. *(g)* Discs containing known amounts of different antibiotics are placed on the inoculated agar surface. After incubation, zones of inhibition are measured, and antibiotic susceptibility is determined using a standardized chart of zone sizes. For the epsilometer test (Etest, AB BIODISK, Solna, Sweden), a plastic strip containing an antibiotic gradient (in µg/ml) indicates the MIC at the point where the elliptical zone of inhibition meets the strip. In this example, the MIC for azithromycin (AZ) is 1.0 µg/ml.

against the isolated pathogen. The Clinical and Laboratory Standards Institute (www.clsi.org) is responsible for developing, establishing, and constantly updating consensus standards for antimicrobial testing. Hospital infection-control microbiologists produce and examine susceptibility data to generate periodic reports called *antibiograms*. These reports define the susceptibility of clinically isolated organisms to the antibiotics in current use. Antibiograms are used to monitor control of known pathogens, to track the emergence of new pathogens, and to identify the emergence of antibiotic resistance at the local level.

MINIQUIZ

- Describe the disc diffusion test and the Etest for antimicrobial susceptibility. For an individual organism and an antimicrobial agent, what do the results signify?
- What is the value of antimicrobial drug susceptibility testing for the microbiologist, the physician, and the patient?

III • Immunological and Molecular Tools for Disease Diagnosis

Culture methods for some pathogens, including many viruses and some pathogenic bacteria, are not routinely available, are unreliable, or are prohibitively difficult or expensive to perform. In such cases, *growth-independent* diagnostic methods are used in clinical, reference, and research laboratories to detect specific pathogens or their products. These include a variety of immunological and molecular assays that can yield a relatively quick and reliable means of identifying individual pathogens or host exposure to pathogens in the absence of cultured organisms.

28.5 Immunoassays and Disease

Many immunoassays use antibodies specific for pathogens or their products for in vitro tests designed to detect specific infectious agents. Patient immune responses, discussed in Chapters 26 and 27, can also be monitored to obtain evidence of pathogen exposure and infection.

Serology and Antibody Titers

The study of antigen–antibody reactions in vitro is called **serology**. Serological assays detect pathogen-induced antibodies in patient serum and are the basis for a number of diagnostic tests. The *specificity* of the antibody–antigen reactions associated with serological tests allows one to pinpoint an exposure to a single pathogen, assuming the antigen used to detect antibodies is unique to the pathogen in question. Moreover, serological tests vary considerably in their *sensitivity*, that is, in the amount of antibody necessary to detect antigen. For example, passive *agglutination* reactions (see Section 28.6) are fast and easy to perform but require antibody concentrations of up to 6 nanograms (ng, 10^{-9}g) per ml. By contrast, the very sensitive but more technically demanding *enzyme immunoassay* (*EIA*) tests require as little as 0.1 ng of antibody per ml and can detect as little as 0.1 ng of antigen (Section 28.7).

If an individual is infected with a suspected pathogen, the immune response—antibodies to that pathogen—should become elevated. Strong evidence for infection can therefore be obtained by determining the *antibody titer* directed against antigens produced by the suspected pathogen. The **titer** is a quantitative measure of antibody level and is defined as the highest dilution (lowest concentration) of serum at which an antigen–antibody reaction is observed (**Figure 28.11**).

A positive antibody titer indicates previous infection or exposure to a pathogen. For pathogens rarely found in a population, such as the life-threatening hantaviruses (⮌ Section 31.2), a single positive test for a pathogen-specific antibody may indicate active infection. In most cases, however, the mere presence of antibody does not indicate active infection. Antibody titers typically remain detectable for long periods after a previous infection has

been resolved. To link an acute illness to a particular pathogen, it is essential to show a *rise* in antibody titer in serum samples taken from a patient during the acute disease and later during the convalescent phase of the disease. Frequently, the antibody titer is low during the acute stage of the infection and rises during convalescence (Figure 28.11). A rise in antibody titer is strong circumstantial evidence that the illness is due to the suspected pathogen.

Skin Tests

A number of pathogens induce a delayed-type hypersensitivity (DTH) response mediated by Th1 cells (⮌ Section 27.9). For these pathogens, skin testing may be useful to determine exposure. As an example, a commonly used skin test is the *tuberculin test*, which consists of an intradermal injection of a soluble extract from cells of *Mycobacterium tuberculosis*. A positive inflammatory reaction at the site of injection within 48 hours indicates current infection or previous exposure to (or vaccination against) *M. tuberculosis*. This test identifies responses caused by pathogen-specific inflammatory Th1 cells (⮌ Figure 27.27). Skin tests are routinely used to aid in diagnosis of tuberculosis, Hansen's disease (leprosy), and some fungal diseases because the antibody responses for intracellular and fungal infections are often weak or undetectable.

If a pathogen is extremely localized, there may be little induction of a systemic immune response and no rise in antibody titer or skin test reactivity, even if the pathogen is proliferating profusely at the site of infection. A good example is the infection of urogenital mucosal surfaces with the bacterium *Neisseria gonorrhoeae*. Gonorrhea does not elicit a systemic or protective immune response, there is no serum antibody titer or skin test reactivity, and reinfection of individuals is common (⮌ Section 30.13).

Monoclonal Antibodies

An expanding area of research with broad application in the diagnosis and treatment of disease is the development and use of **monoclonal antibodies (mAbs)**. In contrast to *polyclonal antibodies*, which occur as a mixture of immunoglobulins produced by many individual B cells and directed at numerous antigenic determinants on a pathogen, mAbs are derived from a B cell clone sensitized to a single antigenic determinant. Thus, an in vitro B cell clone culture produces monospecific mAbs that can be collected for diagnostic or therapeutic purposes. However, antibody-producing B cells are relatively short-lived and normally die within several weeks in cell culture. To produce long-lived B cell clones for commercial mAb production, antibody-producing B cells are fused with *myelomas*, tumorigenic B cells that divide and grow indefinitely (**Figure 28.12**). The "immortal" cell lines that result from this fusion are hybrid cells, appropriately called *hybridomas*. The hybrid cell lines share the biological properties of both fusion partners; they grow indefinitely in vitro and produce antibodies.

To produce a particular mAb, a mouse is immunized with the antigen of interest. Antigen-specific B cells then proliferate over several weeks and begin to produce antibodies in the mouse. B cells are then removed from the mouse and mixed with myeloma cells (Figure 28.12), but only a small number of these fuse into antibody-producing hybridomas. The hybridomas are isolated from unfused cells using a selective *HAT* medium, so called because it contains the metabolites hypoxanthine (H) and thymidine (T),

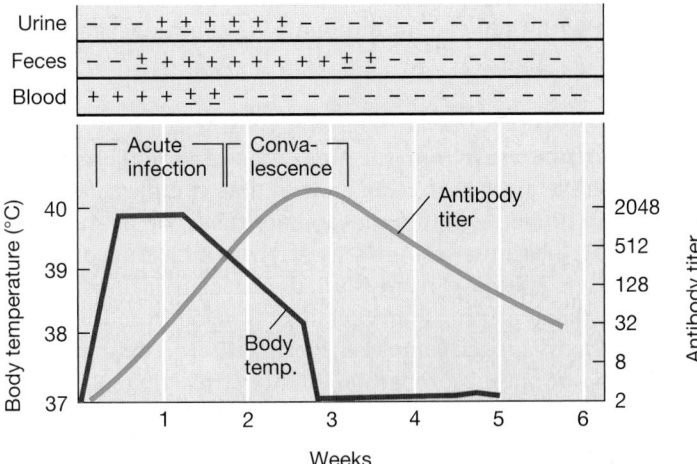

Figure 28.11 Pattern of infection and immunity in untreated typhoid fever patients. Body temperature indicates acute disease progression over time. Antibody titer is shown as the reciprocal of the highest serial dilution causing agglutination of *Salmonella enterica* (*typhi*) (⮌ Section 32.5). The presence of bacteria in blood, feces, and urine was determined from cultures (−, no bacteria; ±, low numbers of bacteria; +, high numbers of bacteria). Bacteria clear from the blood as the antibody titer rises, whereas clearance from feces and urine requires more time. Body temperature drops to normal as the antibody titer rises.

UNIT 6

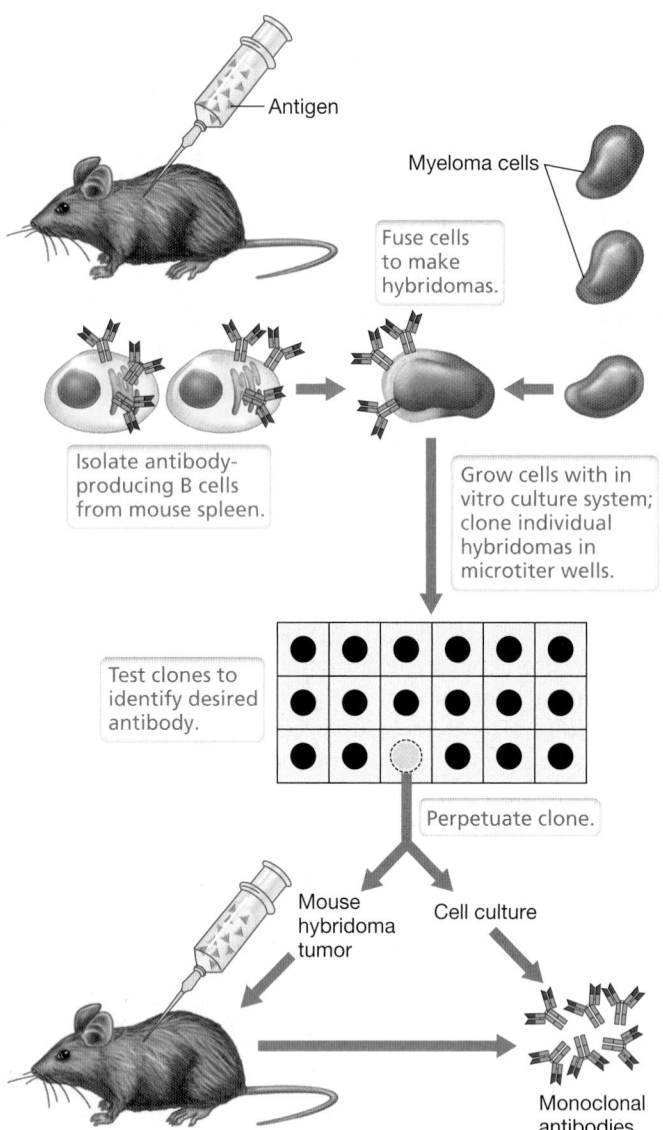

Figure 28.12 Production of monoclonal antibodies (mAbs). The hybridoma can be indefinitely cultured or passed through animals as a tumor. The hybridoma cells are stored as frozen tumor cells that can be thawed and grown in tissue culture or in a suitable animal host.

as well as the cell poison aminopterin (A). The myelomas are unable to grow in this medium because they lack an enzyme that allows them to use the H and T metabolites to circumvent the poisonous effect of aminopterin. Unfused B cells can produce the necessary enzyme, but they die after several days because they cannot divide in culture. Only the fused hybridomas, which combine the properties of both cell types, are able to both produce the necessary enzyme and grow indefinitely.

An enzyme immunoassay (Section 28.7) can be used to identify hybridomas that produce the desired mAb. From a typical fusion, several distinct clones are isolated, each making a different mAb. Once the clones of interest are identified, they can be grown either in the mouse as an antibody-producing tumor or in continuous cell culture. Antibodies can then be harvested from either source.

Commercial production of mAbs has replaced polyclonal antibodies for many immunodiagnostic applications because mAbs are highly specific bioreagents that can be generated with a high degree of reproducibility. Clinical diagnostic tests that use mAbs include immunological typing of bacterial pathogens, identification of cells containing foreign surface antigens (for example, a virus-infected cell), and highly specific blood and tissue typing. Because of their remarkable specificity, mAbs are also used to detect and treat human cancers. Malignant cells often contain tumor-specific antigens on their surfaces, and therefore mAbs prepared against these antigens specifically target the cancer cells and can be used to deliver toxins directly to them. The use of anticancer mAbs has great potential as an alternative to nonspecific chemical and radiation treatments that damage healthy cells as well as cancer cells.

MINIQUIZ

- Explain the reasons for changes in antibody titer for a single infectious agent, from the acute phase through the convalescent phase of the infection.
- Describe the method, time frame, and rationale for the tuberculin skin test. What component of the immune response does this test detect?
- What advantages do monoclonal antibodies have compared to polyclonal antibodies? How are mAbs produced?

28.6 Precipitation, Agglutination, and Immunofluorescence

Many clinically useful immunological reactions, including precipitation reactions and agglutination reactions, yield a product visible to the naked eye. Some other reactions are visualized microscopically when fluorescent dyes attached to specific antigens react with their specific antibody. We consider examples of these now.

Precipitation

Precipitation results from the interaction of a soluble antibody with a soluble antigen to form an insoluble complex. Tests can be done in liquid test tubes (or capillary tubes) or, as shown in **Figure 28.13**, in agarose gel. Antigens that have more than one antibody-binding epitope can cross-link the bivalent antibodies that recognize them, causing a precipitate to develop from the aggregated antibody–antigen complexes. Precipitation occurs maximally when there are optimal proportions of the two reacting substances. The presence of either excess antigen or excess antibody results in the formation of soluble immune complexes (Figure 28.13).

Precipitation reactions carried out in agarose gels, called *immunodiffusion tests* (Figure 28.13 inset), are especially useful for diagnosing fungal infections, including coccidioidomycosis, histoplasmosis, blastomycosis, and paracoccidioidomycosis (⊂ Section 33.2). For these tests, prepared antigen and patient antisera containing antibodies are loaded into separate wells cut into the agarose gel. The reagents diffuse outward from the wells and form precipitation bands where antibody interacts with

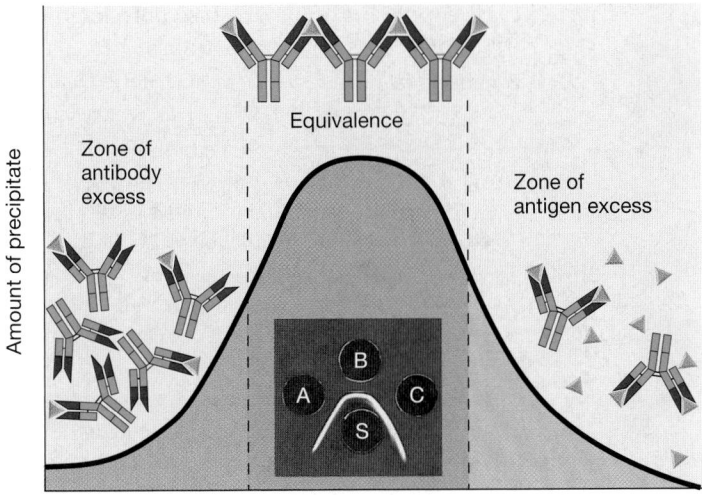

Figure 28.13 Precipitation reactions between soluble antigen and antibody. The extent of precipitation is a function of antigen and antibody concentration. Inset photo: Precipitation in agarose gel (immunodiffusion). Well *S* contains antibodies to cells of *Proteus mirabilis*. Wells *A, B,* and *C* contain soluble extracts of *P. mirabilis*. An insoluble precipitation band forms where antibody and antigen concentrations are equivalent. Immunoprecipitation photo courtesy of C. Weibull, W.D. Bickel, W.T. Hashius, K.C. Milner, and E. Ribi.

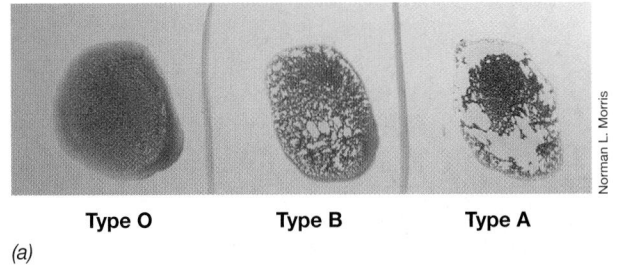

(a)

Blood type	Percentage of U.S. population	Serum	
		Anti A	Anti B
O	48	No aggl.	No aggl.
A	32	Aggl.	No aggl.
B	16	No aggl.	Aggl.
AB	4	Aggl.	Aggl.

(b)

Figure 28.14 Direct agglutination of human red blood cells: ABO blood typing. *(a)* A drop of whole blood was mixed with antigen-specific antisera for each reaction. The reaction on the left shows no agglutination with antibody, typical of blood type O. The reaction in the center shows the diffuse agglutination pattern indicative of blood type B. The reaction on the right shows the strong agglutination pattern with large, clumped aggregates typical of blood type A. *(b)* Table of expected blood typing results for people in the United States.

antigen in optimal proportions (Figure 28.13). Unfortunately, precipitation reactions are not very sensitive; visible precipitation requires microgram quantities of specific antibody rather than the nanogram quantities of more sensitive diagnostic tests. Consequently, with the exception of clinical diagnostic testing for fungal infections, precipitation assays are typically used only in research and reference laboratories.

Agglutination

Agglutination is a reaction between antibody and particle-bound antigen resulting in visible clumping of the particles. Agglutination tests can be done in test tubes, in microtiter plates, or by mixing reagents on glass or coated paper slides. Agglutination tests are quick to perform, inexpensive, highly specific, and reasonably sensitive, making them suitable for large-scale use in clinical applications. Standardized agglutination tests are used for the identification of blood group (red blood cell) antigens (**Figure 28.14a**), as well as pathogens and pathogen products. To determine blood groups, blood samples are mixed with either anti-A antisera or anti-B antisera and the agglutination of red blood cells, called *hemagglutination*, is assessed (Figure 28.14).

Agglutination is often assessed using rapid assays that employ small (0.8-μm diameter) latex beads coated with a specific antigen. The beads are mixed with patient serum on a slide and incubated for a short period. If patient antibody binds the antigen on the bead surface, the milky white latex suspension will become visibly clumped, indicating a positive agglutination reaction and exposure to the pathogen.

Latex agglutination is also used to detect bacterial surface antigens by mixing cells from a bacterial colony with antibody-coated latex beads. For example, a commercially available suspension of latex beads coated with antibodies to protein A and clumping

factor, two proteins found exclusively on the surface of *Staphylococcus aureus* cells, is specific for identification of clinical isolates of *S. aureus* (**Figure 28.15**). Latex bead assays take less than a minute and can be used directly on clinical samples, such as the exudate from a purulent infection. Latex bead agglutination assays have also been developed to identify other common pathogens, such as *Streptococcus pyogenes, Neisseria gonorrhoeae, Escherichia coli* O157:H7, and the fungus *Candida albicans*.

Immunofluorescence

Antibodies containing conjugated fluorescent dyes can be used to detect antigens on intact cells. Such **fluorescent antibodies** are

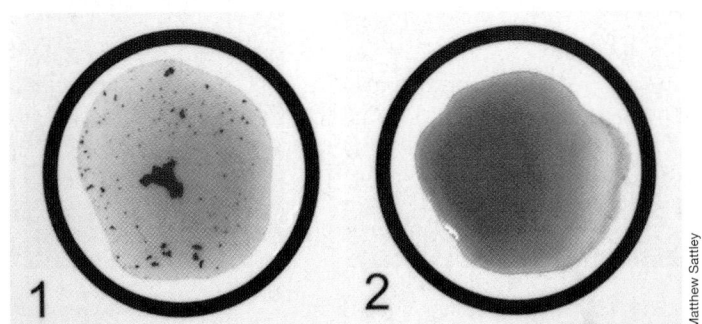

Figure 28.15 Latex bead agglutination test for *Staphylococcus aureus*. In circle 1, a loopful of material from a bacterial colony was mixed in a suspension of microscopic, red latex beads coated with antibodies to two antigens found exclusively on the surface of *Staphylococcus aureus* cells. The bright red clumps indicate positive agglutination, confirming the colony as *S. aureus*. Circle 2 is a negative control showing the uniform red color expected in the absence of agglutination.

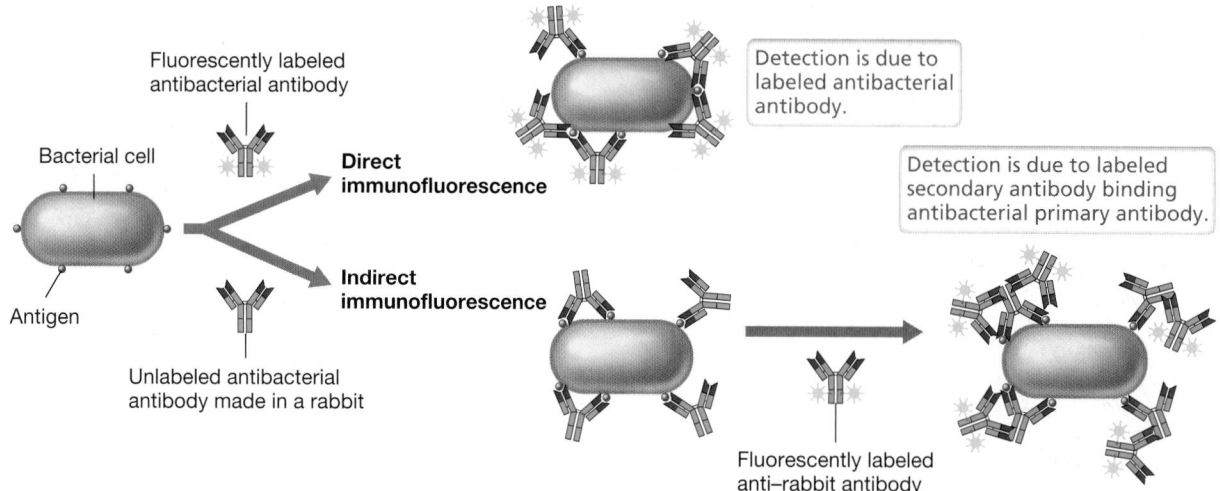

Figure 28.16 Fluorescent antibody methods for detection of microbial surface antigens. Note how indirect immunofluorescence requires a labeled secondary antibody that binds to the primary antibody.

widely used for diagnostic and research applications. Fluorescent antibody staining methods can be either direct or indirect (**Figure 28.16**). In the *direct method*, the antibody that interacts with the surface antigen is itself covalently linked to the fluorescent dye. In the *indirect method*, the presence of a nonfluorescent primary antibody on the surface of a cell is detected by the use of a fluorescent secondary antibody directed against the nonfluorescent antibody.

Fluorescent dyes typically conjugated to antibodies include rhodamine B, which fluoresces red-orange, and fluorescein isothiocyanate, which fluoresces yellow-green (**Figure 28.17**). Once the fluorescent antibodies have bound to cell surface antigens, the complex can be visualized using a fluorescence microscope (⟷ Figure 1.20). The cell-bound fluorescent antibodies emit their characteristic fluorescent color when excited with light of particular wavelengths. Fluorescent antibodies can be used for tasks as varied as identifying a microorganism directly in a patient specimen (Figure 28.17) to enumerating T cells in the blood of patients with human immunodeficiency virus/acquired immunodeficiency syndrome (HIV/AIDS).

Fluorescent antibodies applied directly to infected host tissues may permit disease diagnosis long before culture methods yield a suspected pathogen (**Figure 28.18**). For example, a presumptive diagnosis of legionellosis (Legionnaires' disease), a form of infectious pneumonia (⟷ Section 32.4), can be confirmed by staining biopsied lung tissue directly with fluorescent antibodies specific for cell wall antigens of *Legionella pneumophila* (Figure 28.17*b*), the causative agent of the disease. Immunofluorescence assays are also used to help diagnose infections from viral pathogens, such as Epstein–Barr virus (EBV) (Figure 28.18*a*); fungal pathogens, such as *Aspergillus* (Figure 28.18*b*); and gastrointestinal parasitic protozoa, such as *Giardia intestinalis* (Figure 28.18*c*).

MINIQUIZ

- How is the bivalence of antibodies significant for a precipitation reaction, and under what conditions does precipitation occur maximally?

- What are the advantages and disadvantages of agglutination tests versus fluorescent antibody assays? How are the latter used to identify specific cells in complex mixtures, such as blood?

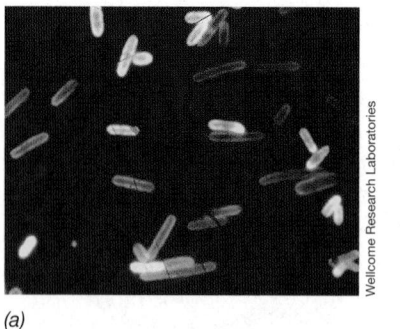

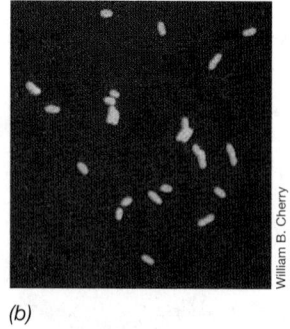

(a) *(b)*

Figure 28.17 Fluorescent antibody identification of bacteria. *(a)* Cells of *Clostridium septicum* were stained with antibody conjugated with fluorescein isothiocyanate, which fluoresces yellow-green. Cells of *Clostridium chauvoei* were stained with antibody conjugated with rhodamine B, which fluoresces red-orange. *(b)* Immunofluorescently stained cells of *Legionella pneumophila* (legionellosis) from biopsied lung tissue. Individual cells are 2–5 μm in length and were stained green with antibodies coupled to fluorescein isothiocyanate.

28.7 Enzyme Immunoassays, Rapid Tests, and Immunoblots

Enzyme immunoassays (EIAs), which include *enzyme-linked immunosorbent assays (ELISAs)*, are immunodiagnostic tools widely used in clinical microbiology and research. EIAs are particularly useful because they are inexpensive, produce no hazardous waste, and are highly specific and sensitive (they can detect as little as 0.1 nanograms of antigen or antibody). *Rapid tests* are similar to EIAs except that results can often be reported within minutes instead of hours. Many rapid tests are given as "point-of-care" tests but are generally not as specific or sensitive as EIAs.

The comparatively complex and time-consuming **immunoblot (Western blot)** uses immobilized pathogen proteins as antigens

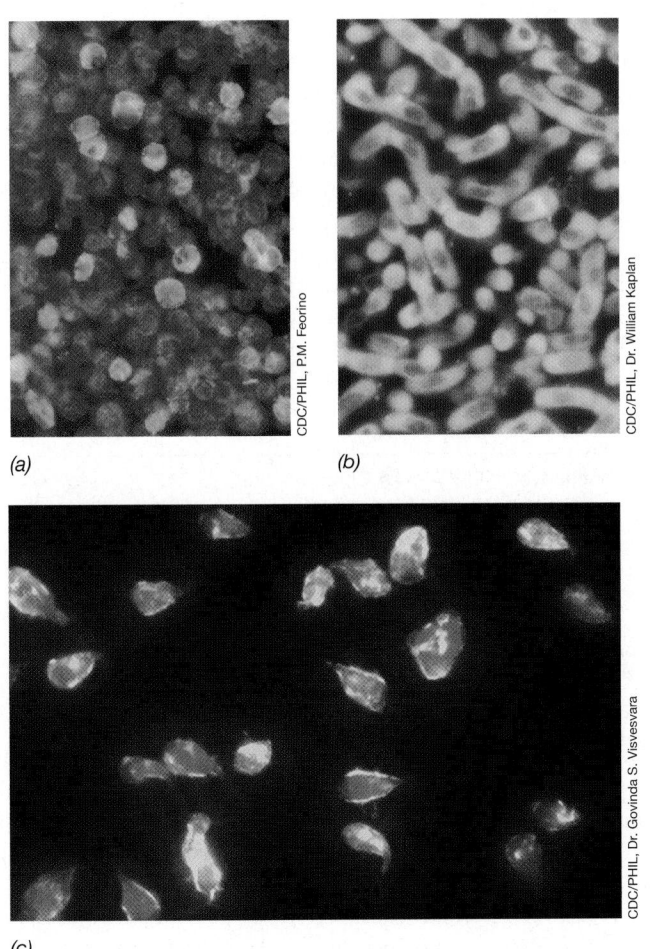

(a)

(b)

(c)

Figure 28.18 Fluorescent antibody identification of pathogens. *(a)* Detection of cells infected with Epstein–Barr virus (EBV) using indirect immunofluorescence. The green-stained cells are infected with EBV, which causes mononucleosis and lymphoma. *(b)* Detection of *Aspergillus* mold in a case of aspergillosis (⮂ Section 33.1) using fluorescein-conjugated antibodies. *(c)* Detection of the waterborne intestinal parasite *Giardia intestinalis* (⮂ Section 33.4) using indirect immunofluorescence.

to bind antibodies from patient specimens, providing highly specific evidence for pathogen exposure. Immunoblots are often used to confirm results obtained from other serological tests, including rapid tests and EIAs.

EIA

In EIA tests, an enzyme is covalently attached to an antigen or antibody molecule, creating an immunological tool with high specificity and sensitivity. Enzymes typically bound to antigen or antibody include peroxidase, alkaline phosphatase, and β-galactosidase, all of which interact with specific substrates to form colored reaction products that can be detected in low amounts. Four EIA formats are commonly used for evaluation of specimens for infectious disease: *direct EIA* (detects antigen), *indirect EIA* (detects antibodies), *antigen sandwich EIA* (detects antibodies using a sandwich technique), and *combination EIA* (detects both antigen and antibodies). The principal features of each platform are illustrated in **Figure 28.19**.

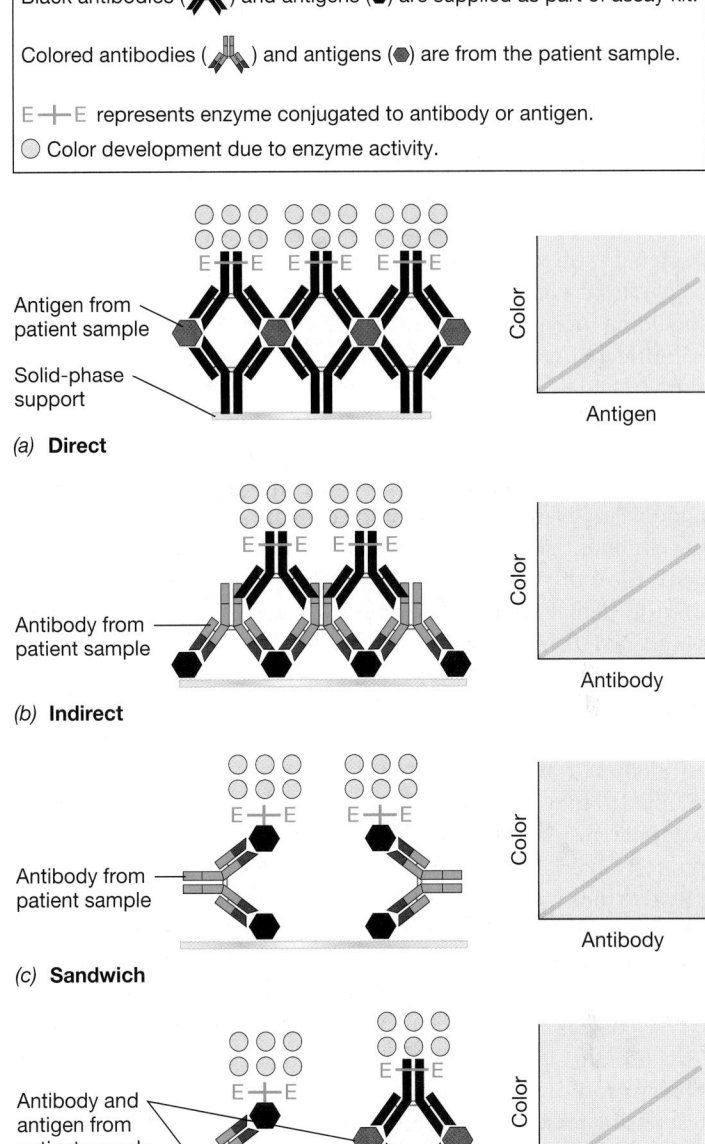

Black antibodies ( ) and antigens ( ) are supplied as part of assay kit.

Colored antibodies ( ) and antigens ( ) are from the patient sample.

E──E represents enzyme conjugated to antibody or antigen.

◯ Color development due to enzyme activity.

Antigen from patient sample

Solid-phase support

(a) **Direct**

Color / Antigen

Antibody from patient sample

(b) **Indirect**

Color / Antibody

Antibody from patient sample

(c) **Sandwich**

Color / Antibody

Antibody and antigen from patient sample

(d) **Combination**

Color / Antigen and antibody

Figure 28.19 Enzyme immunoassays (EIAs). Patient samples are in color. Assay reagents are shown in black. All assays are fixed to a solid-phase support (light blue). Enzymes bound to antigen or antibody convert substrate to a colored product, shown in yellow. In each assay, the amount of colored product is proportional to the amount of pathogen-specific antibody or antigen derived from a patient sample. *(a)* The *direct EIA* uses immobilized pathogen-specific antibody and enzyme-labeled pathogen-specific antibody to detect pathogen antigen in patient samples such as blood. *(b)* The *indirect EIA* uses immobilized pathogen antigen and enzyme-labeled antibody directed to immunoglobulin to detect pathogen-specific antibodies in patient samples such as blood. *(c)* The *sandwich EIA* uses immobilized pathogen antigen and enzyme-labeled pathogen antigen to detect pathogen-specific antibodies in patient samples such as blood. The sandwich EIA is more sensitive than the direct or indirect EIA methods. *(d)* The *combination EIA* uses both the sandwich and direct assays in one platform to identify antibody and antigen in patient samples, maximizing sensitivity. In all cases, it is the enzymatic conversion of a colorless substrate to a colored product, measured spectrophotometrically, that enables quantification of specific antigen and/or antibody.

UNIT 6

Direct EIAs are designed to detect antigens, such as virus particles in a blood or fecal sample (Figure 28.19a). Antibodies to a pathogen antigen are coated onto a support, such as a plastic microtiter plate, and the patient sample is then added. After antigen in the sample binds to the antibody, a second antibody, specific to the same antigen and coupled to an enzyme, is added. Finally, enzyme substrate is added, and the enzyme converts the substrate to its colored product in proportion to the amount of patient antigen bound by the enzyme–antibody complex. Direct EIAs are useful for detecting bacterial exotoxins, such as those produced by *Vibrio cholerae* and *Staphylococcus aureus* (↩ Section 25.6), as well as a variety of viruses, including those that cause influenza and hepatitis.

Indirect EIAs are used to detect antibodies to pathogens in body fluids (Figure 28.19b). The indirect test starts with pathogen antigen immobilized on a supportive matrix. Patient serum is added, and antibodies (if present) bind to the antigen. Next, an antibody–enzyme complex specific for the patient antibodies is added. Finally, the addition of the enzyme substrate results in the development of a colored product that is proportional to the concentration of patient antibody in the sample. Indirect EIAs are used to detect serum antibodies to a wide variety of bacterial, viral, and eukaryotic pathogens.

The antigen sandwich EIA also detects antibodies to pathogens in body fluids (Figure 28.19c). The sandwich test starts with pathogen antigen immobilized on a support. Patient serum is added, and antibodies (if present) bind to the antigen. Next, the same antigen coupled to enzyme is added, followed by the addition of the enzyme substrate, resulting in the development of colored product that is proportional to the concentration of patient antibody in the sample. This method is especially sensitive because it detects pathogen-specific antibody irrespective of antibody class. This method is often used for HIV screening (the third-generation HIV test) because it can detect IgM produced during the primary immune response to HIV as soon as four weeks after infection. By contrast, most indirect EIAs use anti-IgG as the enzyme-conjugated antibody, delaying observation of antibodies to HIV until the secondary antibody response at least five weeks after infection (↩ Figure 27.10).

The combination EIA, shown in Figure 28.19d, makes use of a direct EIA to detect pathogen antigen and the sandwich method to detect pathogen-specific antibodies, both on a single matrix. This method, used for a fourth-generation HIV test, is more sensitive than the third-generation sandwich test; antigen can be detected as little as 2.5 weeks after HIV infection, reducing the time to treatment.

Rapid Tests

Rapid immunoassay procedures use reagents adsorbed to a fixed support material, such as paper strips or plastic membranes. These point-of-care tests cause a color change on the strip within minutes and serve as rapid diagnostic aids for a variety of infectious diseases, including HIV/AIDS.

For most rapid tests, a body fluid (generally urine, blood, saliva, or sputum) is applied to a sample well in a reagent–support matrix. To detect antigen in patient samples, for example to determine infection by *Streptococcus pyogenes* (strep throat), the matrix contains soluble antibodies that are specific to the antigen in question and conjugated to a colored molecule called a *chromophore*

(Figure 28.20). As the liquid sample diffuses through the matrix, patient antigens (if present) bind the chromophore-labeled antibodies. Capillary action pulls each labeled antigen–antibody complex through the matrix, where it contacts a single line of fixed antibodies. The labeled antigen–antibody complex binds a fixed antibody and becomes immobilized. As the concentration of labeled complex increases, the chromophore becomes visible as a line of color along the fixed antibody, indicating a positive test for the antigen. Labeled antibodies not bound by antigen concentrate at a second line of fixed antibodies that are specific to the labeled antibody rather than the antigen, thus forming a second colored line that serves as the control.

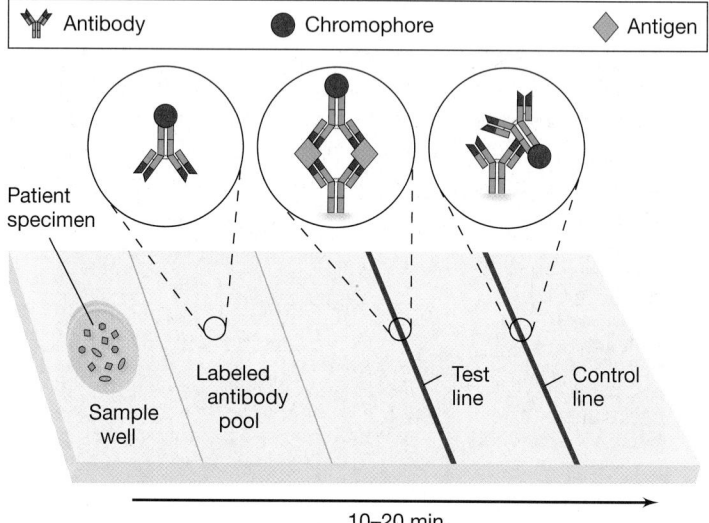

(a)

(b)

Figure 28.20 Rapid tests. *(a)* A patient specimen containing a mixture of antigens is applied to the sample well of a support matrix. Capillary action pulls the liquid sample through the matrix, and specific antigen (if present) binds soluble, chromophore-labeled antibodies. The labeled antigen–antibody complexes diffuse through the matrix and bind a line of fixed antibodies. A colored line becomes visible as the concentration of labeled complex builds, indicating a positive test for the antigen. Unbound labeled antibodies bind a second line of fixed antibody as a control. *(b)* From left to right, rapid tests for respiratory syncytial virus (RSV), group A streptococci (GAS), and influenza A/B. The RSV test and the left GAS test show a test line, indicating positive reactions. The right GAS test and the influenza A/B test show only a control line, indicating negative reactions. Photos taken courtesy of Marion General Hospital, Marion, Indiana (USA).

These tests are valuable for point-of-care analysis and provide rapid diagnostic results that can be reported almost immediately, avoiding the need for delays in patient care or for follow-up visits to obtain test results. The drawback to rapid tests, however, is that they are often less specific or less sensitive than more elaborate assays for the same pathogens. As a result, rapid tests often need to be confirmed by EIA or other tests, such as the immunoblot, discussed next.

Immunoblots

Immunoblot methodology requires the *separation* of proteins on a polyacrylamide gel, the *transfer* (blotting) of the proteins from the gel to a nitrocellulose or nylon membrane, and finally, the *identification* of the proteins using specific antibodies (**Figure 28.21a**). The HIV immunoblot (Figure 28.21b) can be used to accurately diagnose HIV infection, but although it is highly specific, it is generally not used as a screening tool because it is less sensitive, more time consuming, and more expensive than the HIV EIA. However, because the HIV EIA occasionally yields false-positive results, the immunoblot method is often used to confirm positive EIA tests.

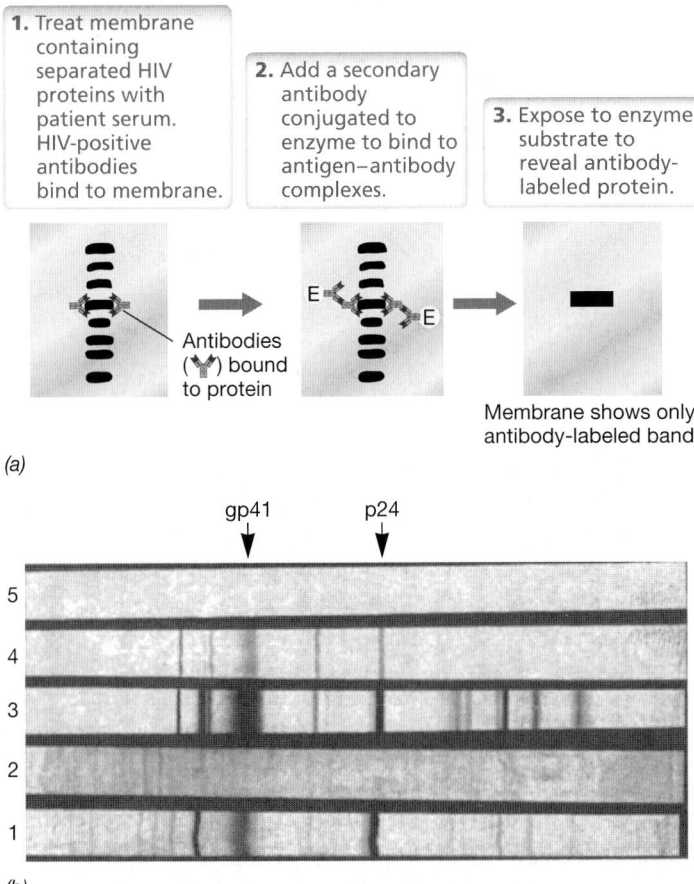

1. Treat membrane containing separated HIV proteins with patient serum. HIV-positive antibodies bind to membrane.

2. Add a secondary antibody conjugated to enzyme to bind to antigen–antibody complexes.

3. Expose to enzyme substrate to reveal antibody-labeled protein.

Antibodies (Y) bound to protein

Membrane shows only antibody-labeled bands

(a)

gp41   p24

5
4
3
2
1

(b)

Victor Tsang

Figure 28.21 The immunoblot (Western blot) and its use in the diagnosis of human immunodeficiency virus (HIV) infection. *(a)* Protocol for an immunoblot. *(b)* The molecules p24 (capsid protein) and gp41 (envelope glycoprotein) are diagnostic for HIV. Lane 1, positive control serum (from known AIDS patients); lane 2, negative control serum (from healthy volunteer); lane 3, strong positive from patient sample; lane 4, weak positive from patient sample; lane 5, reagent blank to check for background binding.

The HIV immunoblot procedure is similar to immunoblot methods used to diagnose infection by other pathogens. In general, immunoblots are used to detect pathogen-specific antibody in patient samples.

To perform the HIV immunoblot, membrane strips containing fixed HIV proteins are incubated with the patient serum sample. If the sample is HIV-positive, patient antibodies will bind to the HIV proteins on the membrane. To detect whether antibodies from the serum sample have bound to HIV antigens, a detecting antibody, anti–human IgG conjugated to an enzyme, is added to the strips. If the detecting antibody binds, the activity of the conjugated enzyme, after addition of substrate, will form a colored band on the strip at the site of antibody binding. The patient is HIV-positive if the positions of the bands in the patient sample match those of a positive control; a negative control serum is also analyzed and must show no bands (Figure 28.21b). As the test is mostly used to confirm positive EIA results for HIV (or correct false positives), variations in band intensity do not affect interpretation of the results.

MINIQUIZ

- Compare direct, indirect, sandwich, and combination EIAs with respect to their ability to identify infection with a particular pathogen.

- Compare the advantages and disadvantages of EIA, rapid tests, and immunoblots with respect to speed, sensitivity, and specificity.

28.8 Nucleic Acid–Based Clinical Assays

In Section 12.1 we discussed how the polymerase chain reaction (PCR) amplifies nucleic acids, forming multiple copies of target sequences. PCR techniques can employ primers for a pathogen-specific gene to examine DNA derived from suspected infected tissue, even in the absence of an observable or culturable pathogen. As a result, PCR-based tests are widely used in the clinical lab for pathogen identification, and they are particularly useful for identifying viruses and other intracellular pathogens that are difficult or impossible to culture using current techniques. PCR methods are extremely sensitive and do not depend on pathogen isolation or growth, and no detection of an immune response to the pathogen is required. Instead, microbe-specific nucleic acid sequences are detected in the assays.

Nucleic Acid Hybridization and Amplification

Nucleic acid hybridization (⇔ Section 12.1) is the central theme of nucleic acid–based molecular methods. In clinical medicine, hybridization methods are employed to identify specific pathogens in patient samples by using unique nucleic acid *probes* to detect the presence of specific DNA sequences. Nucleic acid probes are single-stranded DNA molecules having a sequence complementary to that of a gene of interest. A DNA probe oligonucleotide may be less than 100 base pairs or up to several kilobases in length. If a microbe from a clinical specimen contains DNA or RNA sequences complementary to the probe, the probe will hybridize (following

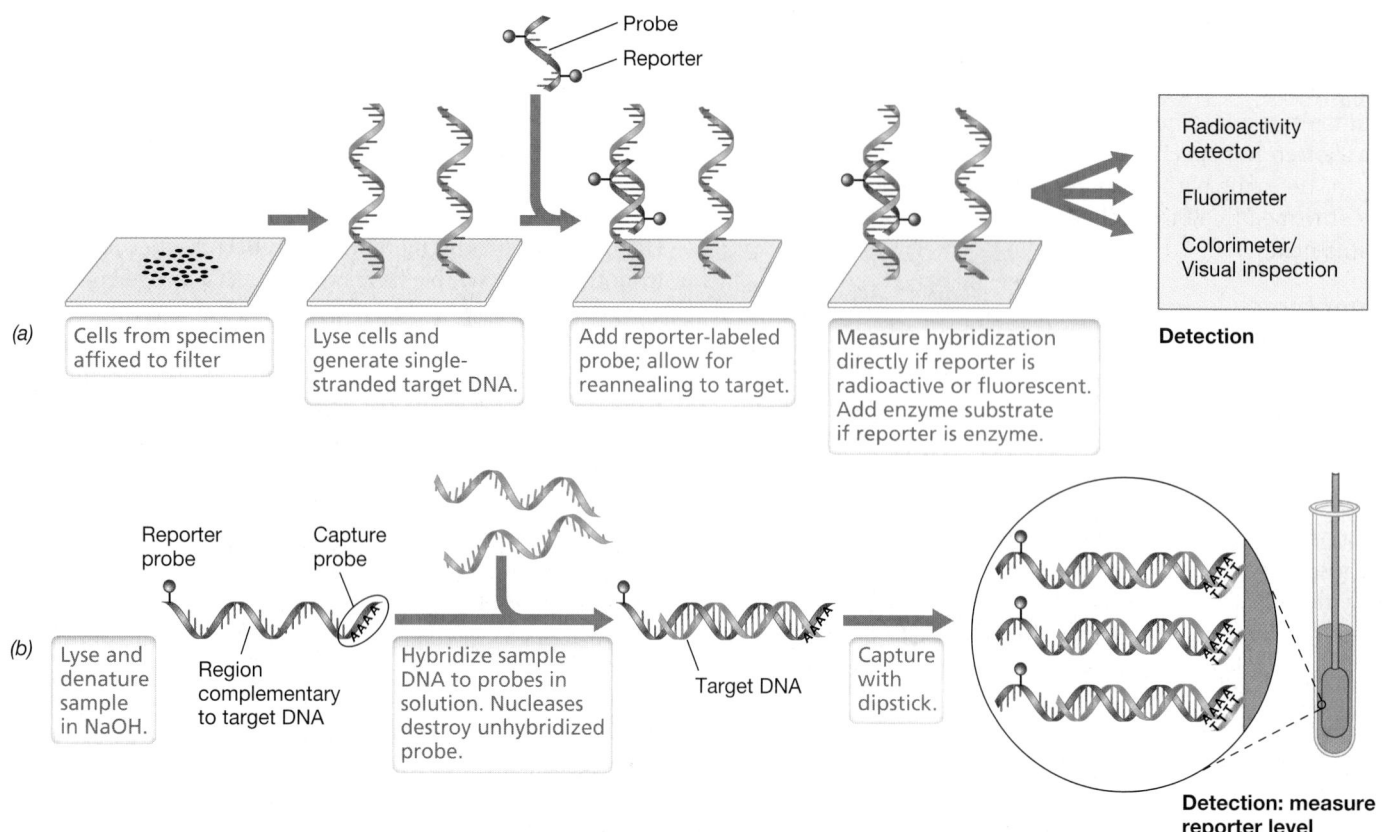

Figure 28.22 Nucleic acid probe methodology in clinical diagnostics. *(a)* Membrane filter assay. The reporter can be a radioisotope, a fluorescent dye, or an enzyme. *(b)* Dipstick assay. Dual reporter and capture probes are used. The capture probe contains a poly(A) tail that hybridizes to a poly(T) oligonucleotide affixed to the dipstick. Binding of the target DNA–reporter complex is usually detected as a visible color change.

appropriate sample preparation to yield single-stranded target molecules), forming a double-stranded molecule (**Figure 28.22**). To detect the hybridization reaction, the probe is labeled with a reporter molecule, which is usually a fluorescent compound, but radioisotopes or enzymes may also be used.

To carry out a probe assay, samples are treated with strong alkali, usually sodium hydroxide (NaOH), to lyse the cells and partially denature the pathogen DNA, forming single-stranded DNA molecules (Figure 28.22). Incubation at an appropriate temperature facilitates formation of a stable duplex between target DNA and probe DNA. The extent of hybridization is measured using the reporter molecule attached to the probe. Some assays use two-component probes that function as both a *reporter* probe and a *capture* probe; the addition of a sequence tag allows the hybridized molecule to be affixed to a matrix, usually a dipstick, for detection purposes (Figure 28.22b).

Nucleic acid hybridization also plays a critical role in the various PCR-based techniques used to amplify target DNA or RNA molecules. PCR analysis begins with the extraction of DNA or RNA from the sample to be tested. Next, the nucleic acid must be amplified using appropriate gene-specific nucleic acid *primers*. These short oligonucleotides (typically 15–27 base pairs in length) are not used as probes but instead function to jump-start DNA polymerase during PCR amplification of pathogen-specific genes. Lastly, the amplified nucleic acid product (the *amplicon*) is visualized, a procedure that may involve gel electrophoresis or, more

often in clinical laboratories, fluorescence. The presence of the appropriate amplified gene segment confirms the presence of the pathogen.

Quantitative and Reverse Transcription PCR

Many clinical PCR tests employ *quantitative real-time PCR (qPCR)*. This process uses fluorescent probes to label PCR amplicons, thereby allowing the accumulation of target DNA to be visualized. Because probe fluorescence increases upon binding to DNA, the level of fluorescence increases proportionally as the target DNA is amplified. The fluorescent probes may be either nonspecific or specific for the target DNA. For example, the dye SYBR Green binds *nonspecifically* to double-stranded DNA and fluoresces only when bound. When added to the PCR mixture, SYBR Green fluorescence indicates the presence of double-stranded DNA produced by the amplification process (**Figure 28.23a**). By contrast, *gene-specific* fluorescent probes, made by attaching a fluorescent dye to a short DNA probe specific to a target sequence, fluoresce only when double-stranded DNA of the correct sequence accumulates.

Because qPCR amplification can be monitored continuously via fluorescence, visualization by gel electrophoresis is not necessary to confirm amplification. Using modern instrumentation, such as that shown in Figure 28.23b, detection of a gene diagnostic for a particular pathogen in a clinical sample may be performed in about 2 hours. Moreover, by monitoring the *rate* of fluorescence increase in the PCR reaction, it is possible to accurately determine

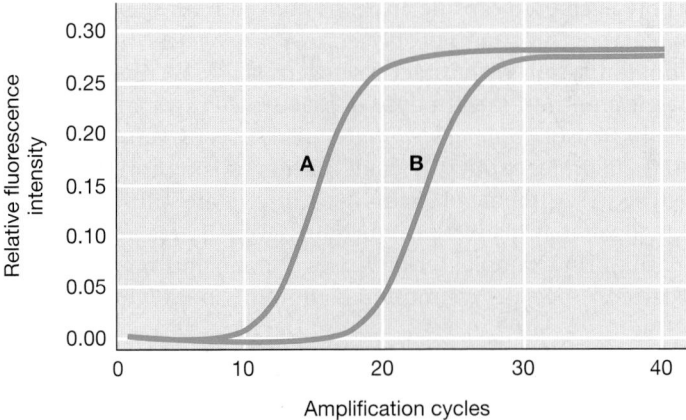

(a)

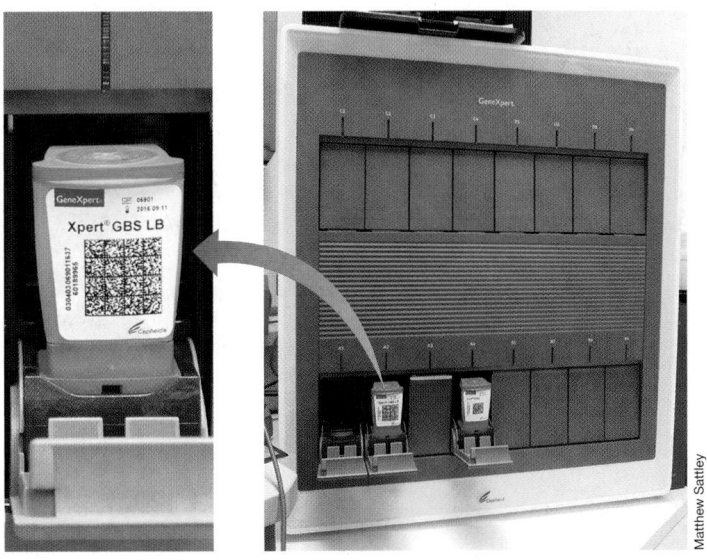

(b)

Matthew Sattley

Figure 28.23 Quantitative real-time polymerase chain reaction (qPCR) for clinical diagnostics. *(a)* DNA extracted from a gram-negative bacterial culture was monitored for expression of 16S rRNA (curve A) and *npt* (curve B), a kanamycin resistance marker, using gene-specific primers. The fluorescent dye SYBR Green was added to the PCR mixture and used to visualize amplified DNA as it formed. The curve on the left (A) had 0.15 fluorescence units after 15 cycles, while the curve on the right (B) had 0.15 fluorescence units after 22 cycles, indicating that the 16S rRNA had a higher abundance of template DNA than *npt* in this strain. *(b)* qPCR instrumentation in a clinical laboratory. The single-use cartridges contain all necessary components for the qPCR reaction, including group-specific primers and fluorescent dyes, and pathogens can be identified in less than 2 hours. The cartridge on the left is specific for detection of group B streptococci (*Streptococcus agalactiae*). Photos taken courtesy of Marion General Hospital, Marion, Indiana (USA).

the amount of target DNA present in the original sample (Figure 28.23*a*). Thus, qPCR can be used to assess the abundance of a pathogen in a sample by quantifying a gene characteristic for that particular organism.

Another variation of basic PCR is *reverse transcription PCR (RT-PCR)*, which uses pathogen-specific RNA to produce complementary DNA (cDNA) directly from patient samples (⟷ Section 12.1). This technology is especially useful for the detection of RNA viruses, including retroviruses such as HIV. The first step in RT-PCR is to use the enzyme reverse transcriptase to make a cDNA

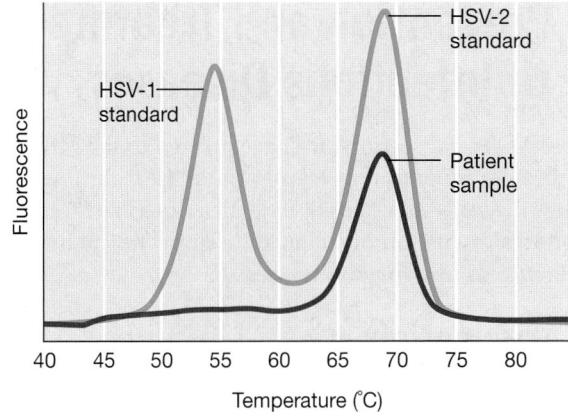

Figure 28.24 Qualitative PCR for the *pol* gene of HSV-1 (herpes simplex 1 virus) and HSV-2. DNA from a patient sample was assayed for the *pol* gene of both HSV-1 and HSV-2 following quantitative PCR (qPCR). Two fluorescent-labeled probes hybridize with an internal sequence of the amplified fragment of each viral genome during the PCR cycle. After hybridization to the template DNA, the probes are excited by a light source, and their fluorescence is measured. After the PCR cycle, each virus shows a distinct DNA melting curve. The melting profile in the patient sample (red) corresponds to the HSV-2 standard (green), indicating infection with HSV-2.

copy of an RNA sample. PCR is then used to amplify the cDNA. By isolating RNA in a sample and making cDNA copies of the corresponding gene(s), one can employ qPCR to monitor the expression of a particular gene from a pathogen. The amplified DNA can then be sequenced or probed for identification.

Qualitative PCR

Some diagnostic tests based on the qPCR format use a slightly different amplification protocol and an additional step to identify pathogen-associated genes. This method, called *qualitative PCR*, uses labeled hybridization primers that are incorporated into an amplicon product of a qPCR reaction.

In the example shown in **Figure 28.24**, the hybridization probes are targeted to the DNA *pol* gene of herpes simplex 1 and 2 viruses (HSV-1 and HSV-2) (⟷ Section 10.7). The amplicon is detected using two distinct hybridization probes labeled with fluorescent dyes. The probes hybridize to an internal sequence of the amplified fragment during the annealing phase of the PCR cycle. After hybridization to the template DNA, the probes are excited by a light source in the PCR instrument. The emitted fluorescence is then measured and, after the PCR cycle, a melting curve analysis is performed to differentiate between samples positive for HSV-1 and HSV-2. Because of nucleotide polymorphisms between the DNA *pol* genes of the two virus subtypes, the melting curve for HSV-1 is distinct from that of HSV-2 (Figure 28.24). The results are compared to internal assay control reactions, and an unambiguous diagnosis of viral infection can be obtained within hours.

--- **MINIQUIZ** ---

- What advantage(s) does nucleic acid amplification have over standard culture methods for identification of microorganisms? What are the disadvantages?

- How do quantitative PCR (qPCR) and qualitative PCR differ?

UNIT 6

IV • Prevention and Treatment of Infectious Diseases

In the past several sections we have taken a tour of the clinical microbiology laboratory to learn some of the techniques used to accurately *diagnose* infectious diseases in a timely fashion. We now move on to a discussion of strategies used in clinical medicine to both *prevent* and *treat* infectious diseases.

28.9 Vaccination

A major means of disease prevention is **vaccination (immunization)**, in which deliberate exposure to an antigen triggers an adaptive immune response intended to protect an individual against future attack by a pathogen. The immunogen used to induce this artificial active immunity is called a **vaccine**. A summary of major diseases for which vaccines are available for human use is given in **Table 28.4**.

The Nature of Vaccines

Infants are immune to many common infectious diseases during the first 6 months of life because they acquire natural passive immunity from maternal antibodies transferred across the placenta or in breast milk (⮫ Section 24.7). However, it is recommended that infants be immunized against key infectious diseases as soon as possible so that their own active immunity can replace the temporary maternal passive immunity (**Figure 28.25**). As discussed in Section 27.3, a single exposure to antigen does not lead to a high antibody titer. After an initial immunization, a series of "booster" immunizations are given to produce a secondary response and a high antibody titer.

It is well established that the introduction of effective vaccines into a population has reduced the incidence of formerly epidemic childhood diseases, such as measles and mumps (⮫ Section 30.6), and has eliminated smallpox altogether (⮫ Section 29.5). However, lifelong immunity is rarely achieved by vaccination because the immune cells and antibodies induced by immunization gradually disappear from the body. On the other hand, natural infections may stimulate immune memory. In the complete absence of antigenic stimulation, the length of effective immunity varies considerably with different vaccines. For example, tetanus toxoid vaccine provides effective immunity for 10 years, but immunity induced by inactivated influenza virus vaccine generally disappears within a year or two.

To minimize the risk of infection or other adverse reactions, pathogens or pathogen products used in vaccines must be made harmless. To achieve this, many vaccines consist of pathogens killed either by heat treatment or by chemical agents. For example, formaldehyde is used to inactivate polio virions for preparation of the Salk polio vaccine. In addition, many exotoxins can be chemically modified to create a *toxoid*, a molecule that retains its antigenicity but is no longer toxic. Toxoid vaccines, such as the vaccine for *Clostridium tetani*, safely induce long-term protective immunity against the exotoxin. In other cases, antigens extracted from pathogens are purified and injected as a vaccine. Such is the case for some pneumococcal vaccines, which consist of a mixture of polysaccharide capsule antigens derived from the most common pathogenic strains of *Streptococcus pneumoniae*.

Immunization with intact pathogens is usually more effective than immunization with dead or inactivated material. It is often possible to isolate *attenuated strains* of pathogens, those that have lost their virulence but still retain the immunizing antigens. However, because attenuated strains of pathogens are still viable, some individuals may inadvertently acquire active disease from the vaccination. For example, serious cases of polio and smallpox have occurred, especially in immunocompromised individuals, from

TABLE 28.4 Vaccines for infectious diseases in humans

Bacterial diseases	Type of vaccine used
Anthrax	Toxoid
Cholera	Killed cells or cell extract (*Vibrio cholerae*)
Diphtheria	Toxoid
Haemophilus influenzae type b meningitis	Conjugated vaccine (polysaccharide of *Haemophilus influenzae* type b conjugated to protein)
Meningitis	Purified polysaccharide from *Neisseria meningitidis*
Paratyphoid fever	Killed bacteria (*Salmonella enterica* [*paratyphi*])
Pertussis	Killed bacteria (*Bordetella pertussis*) or acellular proteins
Plague	Killed cells or cell extract (*Yersinia pestis*)
Pneumonia (bacterial)	Purified polysaccharide from *Streptococcus pneumoniae* or polysaccharide–toxoid conjugate
Tetanus	Toxoid
Tuberculosis	Attenuated strain of *Mycobacterium tuberculosis*
Typhoid fever	Killed bacteria (*Salmonella enterica* [*typhi*])
Typhus	Killed bacteria (*Rickettsia prowazekii*)
Viral diseases	Type of vaccine used
Hepatitis A	Recombinant DNA vaccine
Hepatitis B	Recombinant DNA vaccine or inactivated virus
Human papillomavirus (HPV)	Recombinant DNA vaccine
Influenza (seasonal or H1N1)	Inactivated or attenuated virus
Japanese encephalitis	Inactivated virus
Measles and mumps	Attenuated virus
Polio	Attenuated virus (Sabin) or inactivated virus (Salk)
Rabies	Inactivated virus (human) or attenuated virus (animal)
Rotavirus	Attenuated virus
Rubella	Attenuated virus
Smallpox and monkeypox	Cross-reacting virus (vaccinia)
Varicella (chicken pox/ shingles)	Attenuated virus
Yellow fever	Attenuated virus

UNIT 6

peptides. To make a vaccine, a genetic engineer can synthesize a peptide that corresponds to an antigen from an infectious agent. For example, the toxin from the foot-and-mouth disease virus, an important animal pathogen, must be modified from its native form to render it harmless for use as a vaccine. The toxin contains a peptide of 20 amino acids that is an important antigenic determinant in the protein, but the peptide is too small to be an effective vaccine by itself. Genetic engineers attached the small peptide to a larger, innocuous protein that acts as a carrier molecule, creating a *conjugate vaccine* against foot-and-mouth disease virus. This strategy has great promise for creating vaccines to a number of pathogens, especially because the complete genomic sequences of many pathogens are now known, providing the information necessary to identify the most likely antigenic parts of each.

Two widely available conjugate vaccines couple extracted bacterial polysaccharide to a protein toxoid, provoking a more robust immune response with better immune memory than injection of the polysaccharide antigen alone. One pneumococcal vaccine uses pneumococcal polysaccharide coupled to diphtheria toxoid (**Figure 28.26**). Likewise, the vaccine for *Haemophilus influenzae* type b (Hib) uses Hib polysaccharide coupled to tetanus toxoid. Although polysaccharide antigens typically provide only a primary response with little immune memory, the conjugated protein toxoids efficiently activate Th2 cells, resulting in a primary response followed by a strong secondary response and immune memory.

Genomic information is particularly useful for making viral vaccines. Genes that encode antigens from virtually any virus can be cloned into the vaccinia virus genome and expressed (∂ Section 12.8). Inoculation with the antigen-producing vaccinia virus can induce immunity to the product of the cloned gene. Such a preparation is called a *recombinant-vector vaccine*, an example of which is the recombinant vaccinia–rabies vaccine used in animals. A second immunization strategy uses proteins made from cloned DNA as immunogens. After a pathogen gene is cloned and expressed in a suitable microbial host, the pathogen protein is harvested and used to produce a *recombinant-antigen vaccine*. For example, the current hepatitis B virus vaccine is a major hepatitis surface protein antigen (HbsAg) expressed by genetically modified yeast cells. Similarly, a vaccine effective against human papillomavirus (HPV) is also a recombinant-antigen vaccine made in yeast cells.

DNA Vaccines

A novel method of immunization is based on the expression of cloned genes in host cells. *DNA vaccines* are bacterial plasmids that contain cloned DNA that encodes the antigen of interest. Typically, the vaccine is injected intramuscularly into a host animal. Once host cells take up the plasmid, the DNA is transcribed and translated to produce immunogenic proteins, triggering a conventional immune response of activated T cells and antibodies (Chapter 27) directed to the protein encoded by the cloned DNA.

DNA vaccine strategies provide considerable advantages over conventional immunization methods. For instance, because only a single pathogen gene is cloned into the plasmid and injected, there is no chance of an infection, as there might be with an attenuated

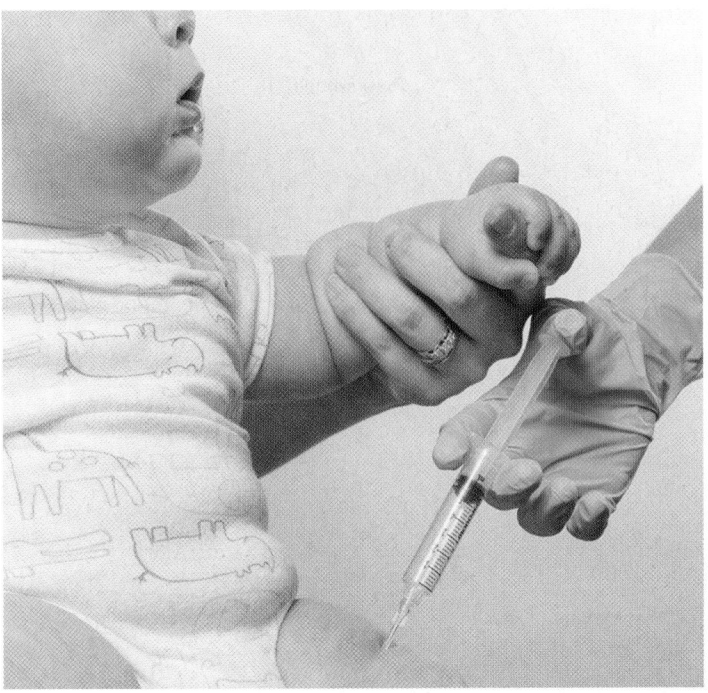

Immunizations against bacteria	Immunizations against viruses
Haemophilus Influenzae Type B (Hib)	Hepatitis A virus
	Hepatitis B virus
Meningococcal (*Neisseria meningitidis*)	Human papillomavirus (HPV)
	Influenza virus
Pneumococcal (*Streptococcus pneumoniae*)	Inactivated poliovirus (IPV)
	Measles, mumps, rubella (MMR)
Tetanus, diphtheria, pertussis (DTaP, Tdap)	Rotavirus
	Varicella virus (chicken pox)

Figure 28.25 Immunization recommendations for infants and children in the United States. The U.S. Centers for Disease Control and Prevention website (http://www.cdc.gov) has recommendations for timing and dose of immunizations for all age groups and for special populations, such as international travelers, women of childbearing age, and those with immunodeficiency or chronic disease.

attenuated vaccines. Nevertheless, although they are difficult to standardize and have a limited shelf life, attenuated vaccines tend to provide long-lasting immunity and a strong secondary booster response. By contrast, killed virus vaccines tend to provide short-lived immune responses with less long-term memory, but they are much more stable and easier to store.

Most bacterial vaccines are provided as antigens in an inactivated form, such as the toxoids that protect against tetanus and diphtheria. Inactivated bacterial vaccines induce antibody-mediated protection without exposing recipients to the risk of infection. However, variability in primary and secondary responses with each vaccine and individual make periodic reimmunization (boosters) necessary to establish and maintain immunity.

Synthetic and Genetically Engineered Vaccines

An alternative approach to vaccine development is to make use of genetic engineering tools (Chapter 12) to produce *synthetic*

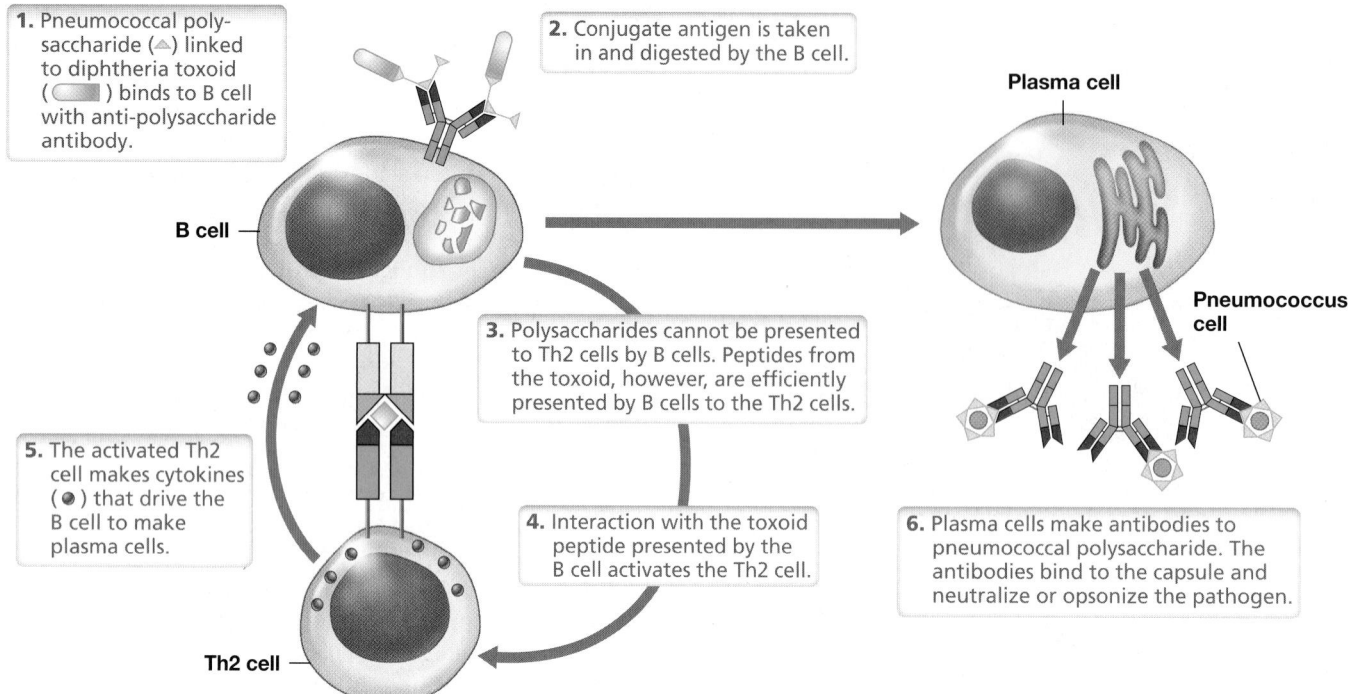

1. Pneumococcal poly-saccharide (△) linked to diphtheria toxoid (▭) binds to B cell with anti-polysaccharide antibody.

2. Conjugate antigen is taken in and digested by the B cell.

Plasma cell

B cell

3. Polysaccharides cannot be presented to Th2 cells by B cells. Peptides from the toxoid, however, are efficiently presented by B cells to the Th2 cells.

Pneumococcus cell

5. The activated Th2 cell makes cytokines (●) that drive the B cell to make plasma cells.

4. Interaction with the toxoid peptide presented by the B cell activates the Th2 cell.

6. Plasma cells make antibodies to pneumococcal polysaccharide. The antibodies bind to the capsule and neutralize or opsonize the pathogen.

Th2 cell

Figure 28.26 Conjugate vaccines. Conjugate vaccines, such as *Streptococcus pneumoniae* (pneumococcus) polysaccharide covalently linked to diphtheria toxoid (shown) or *Haemophilus influenzae* type B (Hib) polysaccharide coupled to tetanus toxoid provide effective immunity to polysaccharide antigens, which are poor immunogens in the absence of protein carriers.

vaccine. Genes for individual antigens, such as a tumor-specific antigen, can be cloned, targeting the immune response to a particular cell component. A single bioengineered plasmid encoding an antigen can be used to infect host cells and elicit a complete immune response, inducing both immune T cells and antibodies. In at least one case, an experimental DNA vaccine consisting of an engineered peptide–MHC I complex protected mice from infection with a cancer-producing papillomavirus.

— **MINIQUIZ** —

- Compare and contrast live attenuated vaccines, inactivated vaccines, and toxoids. Which of these has the greatest potential to cause active disease in the recipient? Which typically provides the longest-lasting immunity?

- Identify the advantages of alternative immunization strategies as compared to traditional immunization procedures.

28.10 Antibacterial Drugs

In cases where immunization is either not an option or fails to prevent disease, antimicrobial drug therapy is the primary weapon against infections. Antimicrobial drugs either kill or inhibit the growth of microorganisms in the host (in vivo) and are classified based on their molecular structure, mechanism of action (**Figure 28.27**), and spectrum of antimicrobial activity. **Antibiotics** are antimicrobial agents that are produced naturally by certain microorganisms, typically species of *Bacteria* or fungi.

In addition to natural antibiotics, many other drugs are produced synthetically, but regardless of their origin, it is desirable that antimicrobial drugs exhibit **selective toxicity**—they inhibit or kill pathogens without adversely affecting the host. Although thousands of antibiotics are known, fewer than 1% are clinically useful, often because of problems with host toxicity or lack of uptake by host cells.

The susceptibility of individual microbes to different antimicrobial agents varies significantly. For example, gram-positive *Bacteria* are often sensitive to natural penicillin, whereas gram-negative *Bacteria* are generally resistant; thus, natural penicillin exhibits a relatively *narrow spectrum* of activity. By contrast, antibiotics that exhibit *broad-spectrum* activity, such as tetracycline, are generally effective against both groups. Although broad-spectrum antibiotics often find wider medical use than narrow-spectrum antibiotics, antibiotics with a limited spectrum of activity may be quite useful against certain pathogens, especially those that fail to respond to other antibiotics. A good example is vancomycin, a narrow-spectrum antibiotic that is an effective bactericidal agent for penicillin-resistant, gram-positive *Bacteria*, including certain enterococci, staphylococci, and clostridia.

Important targets of antibiotics in *Bacteria* include the cell wall, ribosomes, enzymes that facilitate nucleic acid synthesis or catalyze metabolic processes, and the cytoplasmic membrane (Figure 28.27). We provided an overview of antibiotic targets in Section 7.10 and now consider the mechanisms of drug activity in further detail.

UNIT 6

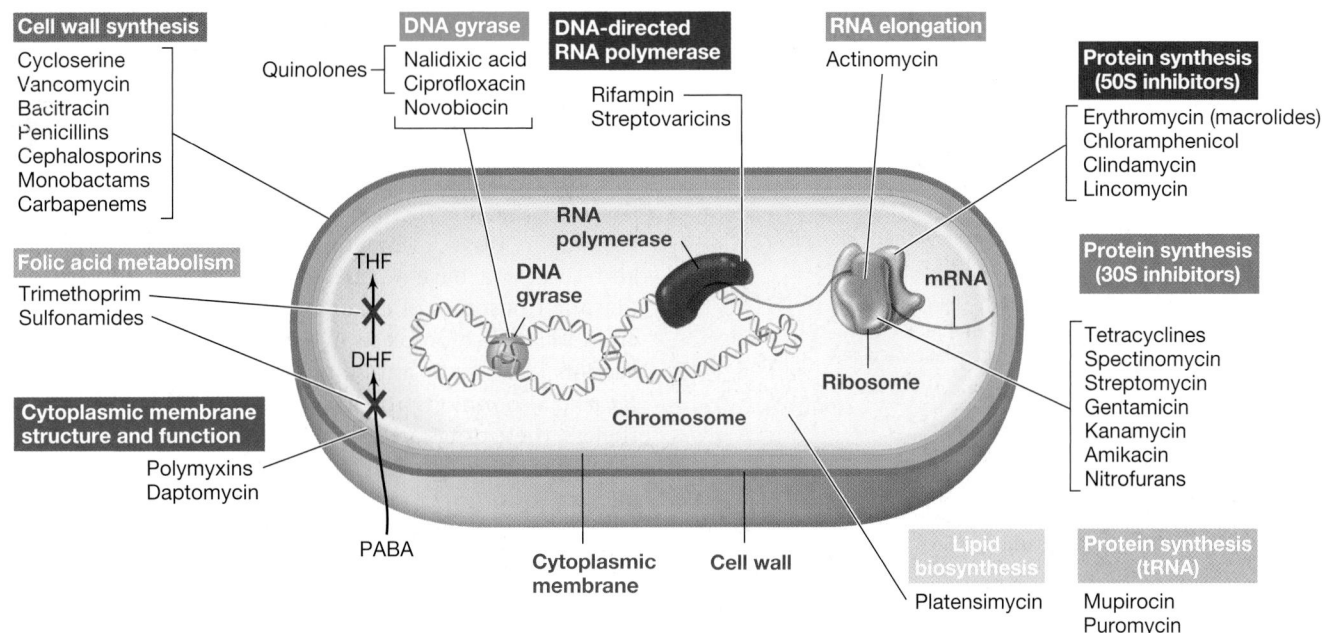

Figure 28.27 Mechanisms of action of major antibacterial agents. Agents are classified according to their target structures in the bacterial cell. THF, tetrahydrofolate; DHF, dihydrofolate; PABA, *p*-aminobenzoic acid.

The Cell Wall as a Drug Target

Worldwide, an estimated 100,000 metric tons of antibacterial drugs are manufactured and used annually (**Figure 28.28a**), and the vast majority of these target the bacterial cell wall. **β-lactam antibiotics**, which include penicillins and cephalosporins, inhibit cell wall synthesis and account for nearly two-thirds of all antibiotics produced and used worldwide (Figure 28.28b). These antibiotics share a characteristic structural component,

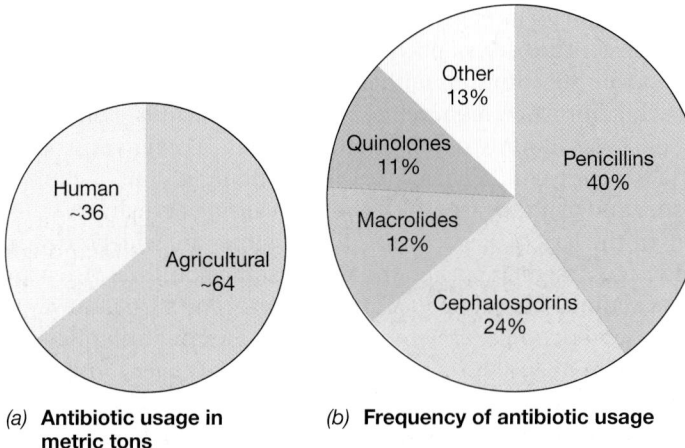

(a) **Antibiotic usage in metric tons**

(b) **Frequency of antibiotic usage**

Figure 28.28 Annual worldwide production and use of antibiotics. (a) An estimated 100,000 metric tons of antimicrobial agents are manufactured worldwide per year, and nearly two-thirds of this amount is used in an agricultural context. (b) By far, the β-lactam antibiotics (penicillins and cephalosporins) constitute the most important and widely used class of antibiotic. "Other" includes trimethoprim combinations, tetracyclines, chloramphenicol, aminoglycosides, and all other antimicrobial drug classes. Data are from the Center for Disease Dynamics, Economics, and Policy, Washington D.C.

the *β-lactam ring* (**Figure 28.29**), and historically, they comprise some of our most effective weapons against many different types of bacterial infections.

The first β-lactam antibiotic ever characterized was penicillin G, isolated from the fungus *Penicillium chrysogenum* in 1929 by Alexander Fleming. This new drug was dramatically effective for controlling staphylococcal and pneumococcal infections and was more effective for treating streptococcal infections than the synthetic sulfa drugs, discussed later in this section. Penicillin and other β-lactam antibiotics interfere with an important feature of bacterial cell wall synthesis called *transpeptidation*, the reaction that results in the cross-linking of two glycan-linked peptide chains (🔗 Section 2.4). Because peptidoglycan synthesis mechanisms are unique to *Bacteria*, the β-lactam antibiotics are highly selective and nontoxic to host cells.

Penicillin G is active primarily against gram-positive *Bacteria* because gram-negative *Bacteria* are impermeable to the antibiotic. However, chemical modification of the *N*-acyl group of penicillin G produces *semisynthetic* penicillins, such as ampicillin and carbenicillin, which have broader activity and are effective against certain gram-negative *Bacteria* (Figure 28.29). The structural differences in these semisynthetic penicillins allow them to be transported inside the gram-negative outer membrane (🔗 Section 2.5), where they inhibit cell wall synthesis. Penicillin G is also sensitive to *β-lactamases*, enzymes that destroy β-lactam antibiotics and are produced by many penicillin-resistant *Bacteria* (see Section 28.12). Oxacillin and methicillin are widely used β-lactamase-resistant semisynthetic penicillins (Figure 28.29).

Cephalosporins, produced by species of the fungus *Cephalosporium*, differ structurally from the penicillins. They retain the β-lactam ring but have a six-member dihydrothiazine ring joined to it instead of

UNIT 6

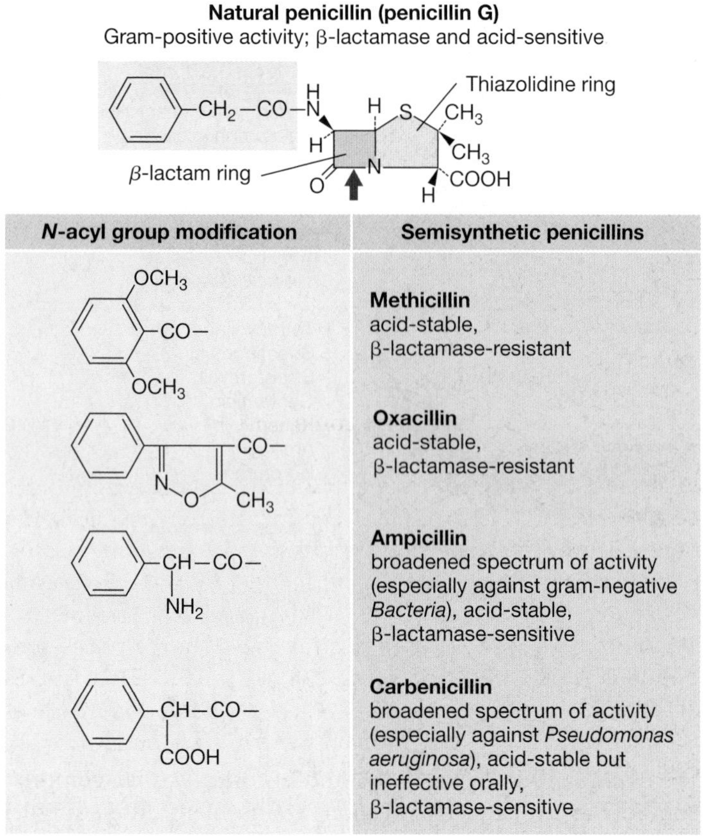

Natural penicillin (penicillin G)
Gram-positive activity; β-lactamase and acid-sensitive

N-acyl group modification	Semisynthetic penicillins
OCH₃ / OCH₃ —CO—	**Methicillin** acid-stable, β-lactamase-resistant
N, O CH₃ —CO—	**Oxacillin** acid-stable, β-lactamase-resistant
—CH—CO— NH₂	**Ampicillin** broadened spectrum of activity (especially against gram-negative *Bacteria*), acid-stable, β-lactamase-sensitive
—CH—CO— COOH	**Carbenicillin** broadened spectrum of activity (especially against *Pseudomonas aeruginosa*), acid-stable but ineffective orally, β-lactamase-sensitive

Figure 28.29 Structure of selected penicillins. The red arrow (top panel) indicates the site within the β-lactam ring cleaved by most β-lactamases, enzymes that destroy penicillin and other β-lactam antibiotics. Although the *N-acyl* group (blue shading) varies among different penicillin drugs, all penicillins have a common nucleus consisting of a β-lactam ring and a thiazolidine ring. Whereas penicillin G must be injected, most acid-stable penicillins can be administered orally (carbenicillin is an exception).

the five-member thiazolidine ring of penicillins (Table 28.5 and Figure 28.29). The cephalosporins have the same mode of action as the penicillins; they bind irreversibly to transpeptidases and prevent the cross-linking of peptidoglycan. Clinically important cephalosporins are semisynthetic antibiotics with a broader spectrum of activity than the penicillins. In addition, cephalosporins are typically more resistant to β-lactamases. For example, ceftriaxone (Table 28.5) is highly resistant to β-lactamases and has replaced penicillin for treatment of gonorrhea because many *Neisseria gonorrhoeae* strains have become resistant to penicillin (see Section 28.12).

Some antimicrobial drugs are *growth factor analogs*. Growth factors are organic nutrients obtained from an organism's environment that are required by the organism for growth and survival (⇄ Section 3.1). A **growth factor analog** is a synthetic compound that is structurally similar to a growth factor, but subtle structural differences between the analog and the growth factor prevent the analog from functioning in the cell, thereby disrupting cell metabolism. *Isoniazid* (Table 28.5) is an important growth factor analog with a very narrow spectrum of activity. Effective against mycobacteria only, isoniazid is an analog of nicotinamide, a vitamin required for mycolic acid synthesis, which is, in turn, required to construct the mycobacterial cell wall. Isoniazid is the most effective

drug for treatment of tuberculosis (⇄ Section 30.4), but isoniazid resistance in strains of *Mycobacterium tuberculosis* is increasing.

Protein Synthesis as a Drug Target

Some antibiotics inhibit bacterial pathogens by disrupting protein synthesis (translation), often through interactions with the ribosome that may include binding to ribosomal RNA (rRNA) (Figure 28.27). Most of these drugs target only bacterial ribosomes and, therefore, have no effect on the structurally distinct, cytoplasmic ribosomes of eukaryotic cells. However, because mitochondria and chloroplasts in *Eukarya* contain 70S ribosomes (⇄ Section 13.4), many antibiotics that inhibit protein synthesis in *Bacteria* also inhibit protein synthesis in these organelles. Nevertheless, these drugs are still medically useful because the eukaryotic 70S ribosomes are affected only at higher concentrations than are used for antimicrobial therapy.

The **aminoglycosides** are antibiotics that inhibit translation by targeting the 30S (small) subunit of the ribosome. Aminoglycosides contain amino sugars linked by glycosidic bonds, and clinically useful examples include streptomycin (produced by the bacterium *Streptomyces griseus*), kanamycin (Table 28.5), neomycin, and gentamicin. These broad-spectrum antibiotics are useful for treating infections caused by gram-negative *Bacteria*, but because of host toxicity effects that include kidney and hearing damage, aminoglycoside use has decreased since the development of the semisynthetic penicillins and tetracyclines (discussed next). With the exception of gentamicin, which is routinely used to combat *Pseudomonas* infections, and neomycin, which is commonly used in topical ointments, aminoglycosides are typically used only when other antibiotics fail.

Like the aminoglycoside antibiotics, **tetracyclines** interfere with the function of the 30S subunit of the ribosome. Tetracyclines are broad-spectrum antibiotics produced by several species of *Streptomyces*, and they inhibit most clinically relevant gram-positive and gram-negative *Bacteria*. The basic structure of tetracycline consists of a naphthacene ring system (Table 28.5), and side-chain substitutions (either natural or synthetic) to the various rings form new tetracycline analogs. Physicians are cautioned against the administration of tetracyclines to pregnant women and to young children because these antibiotics bind calcium in bones and teeth, weakening them and causing permanent staining of the latter (**Figure 28.30**). Tetracyclines are widely used in veterinary medicine, though, and in some countries, they are used as nutritional supplements for poultry and swine. However, because nonclinical use of medically important antibiotics can contribute to antibiotic resistance among pathogens, this practice is discouraged.

Macrolide antibiotics inhibit translation by targeting the 50S (large) subunit of the bacterial ribosome. Their basic structure contains a lactone ring bonded to sugars (as in erythromycin in Table 28.5), and variations in these constituents result in a diversity of macrolide antibiotics. Macrolides account for about 12% of global antibiotic production (Figure 28.28) and include, for example, erythromycin (produced by *Streptomyces erythreus*), clarithromycin, and azithromycin. The partial inhibition of protein synthesis by erythromycin, in particular, leads to preferential translation of some proteins and restricts translation of

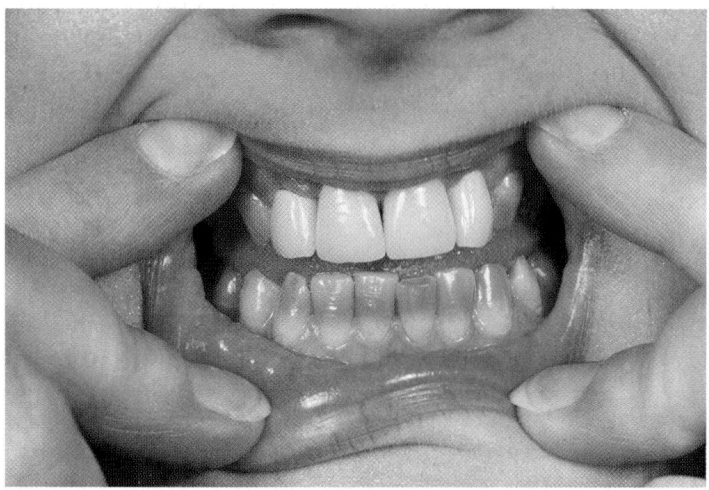

Figure 28.30 Staining of teeth from the use of tetracycline. Tetracycline binds calcium in developing bones and teeth, weakening them and causing permanent staining of tooth enamel. Therefore, tetracycline should not be administered to pregnant women or children unless absolutely necessary.

others, resulting in an imbalance in the proteome and potentially disrupting metabolic functions at all levels. Often used clinically in patients allergic to penicillin or other β-lactam antibiotics, erythromycin is particularly useful for treating legionellosis (↩ Section 32.4).

Nucleic Acid Synthesis as a Drug Target

The **quinolones** are synthetic antibacterial compounds that disrupt bacterial metabolism by interfering with DNA gyrase, thus preventing the supercoiling and packaging of DNA in the bacterial cell (↩ Section 4.1). Because DNA gyrase is found in all *Bacteria*, quinolones are effective for treating both gram-positive and gram-negative bacterial infections. *Fluoroquinolones*, such as ciprofloxacin (Table 28.5), are fluorinated derivatives of quinolones that are routinely used to treat urinary tract infections and have been widely used in the beef and poultry industries for prevention and treatment of respiratory diseases in animals. Ciprofloxacin is also the drug of choice for treating anthrax because some strains of *Bacillus anthracis*, the causative agent of anthrax (↩ Section 31.8), are resistant to penicillin. The fluoroquinolone moxifloxacin is one of only a few drugs proven effective for treatment of tuberculosis. In combination with other anti-tuberculosis drugs, moxifloxacin may significantly reduce treatment time, a major problem with isoniazid-based treatments.

Some antibiotics disrupt transcription by inhibiting RNA synthesis (Figure 28.27). For example, rifamycins, such as *rifampin* (Table 28.5), inhibit RNA synthesis by binding to the β-subunit of RNA polymerase in *Bacteria*. Rifampin, often used in concert with isoniazid for treatment of tuberculosis, has the odd side effect of causing body secretions, including tears, urine, and sweat, to turn reddish-orange (**Figure 28.31**). *Actinomycin* inhibits RNA synthesis by combining with DNA and blocking RNA elongation. This agent binds most strongly to DNA at guanine–cytosine base pairs, fitting into the major groove in the double strand where RNA is synthesized.

Other Antibacterial Drug Targets

Like isoniazids, **sulfa drugs** (also called *sulfonamides*) are synthetic growth factor analogs. However, instead of affecting the cell wall, sulfa drugs block a key biosynthetic pathway in *Bacteria*. Sulfanilamide (Table 28.5), the simplest sulfa drug, is an analog of *p*-aminobenzoic acid (PABA), which is a component of the vitamin folic acid, a nucleic acid precursor. By mimicking PABA, sulfanilamide blocks the synthesis of folic acid, thereby inhibiting nucleic acid synthesis. Sulfanilamide is selectively toxic because bacteria synthesize their own folic acid, unlike humans and most animals, which obtain folic acid from their diet. Antimicrobial therapy with sulfamethoxazole (a sulfa drug) plus trimethoprim, a related folic acid synthesis competitor, is highly effective because the drug combination blocks two sequential steps in the folic acid synthesis pathway (Figure 28.27); resistance to this drug combination therefore requires two mutations in genes of the same pathway, a relatively rare event. However, resistance to sulfonamides is increasing as many formerly susceptible pathogens develop the ability to import folic acid from their environment.

Some antibiotics have atypical structures or targets. For example, the antibiotic *daptomycin* (Table 28.5) is produced by a species of *Streptomyces* and is a cyclic lipopeptide with a unique mode of action. Used mainly to treat infections by gram-positive *Bacteria*, including pathogenic streptococci and staphylococci, daptomycin binds specifically to bacterial cytoplasmic membranes, forms a

Figure 28.31 Reddish-orange urine from rifampin usage. The antibiotic rifampin is often administered to tuberculosis patients. However, a side effect of rifampin usage is that it turns urine and other body fluids reddish-orange.

TABLE 28.5 **Selected antibacterial compounds**

Mode of action	Antibiotic class	Example(s)	Representative structures
Inhibit cell wall synthesis	β-lactams	Penicillins, cephalosporins	Ceftriaxone
	Isoniazids	Isoniazid	Isoniazid
	Polypeptide antibiotics	Vancomycin, bacitracin	Vancomycin (see Figure 28.34)
Inhibit protein synthesis	Aminoglycosides	Streptomycin, kanamycin, gentamicin	Kanamycin
	Tetracyclines	Tetracycline, doxycycline	Tetracycline
	Macrolides	Erythromycin, azithromycin	Erythromycin
	Chloramphenicol	Chloramphenicol	Chloramphenicol

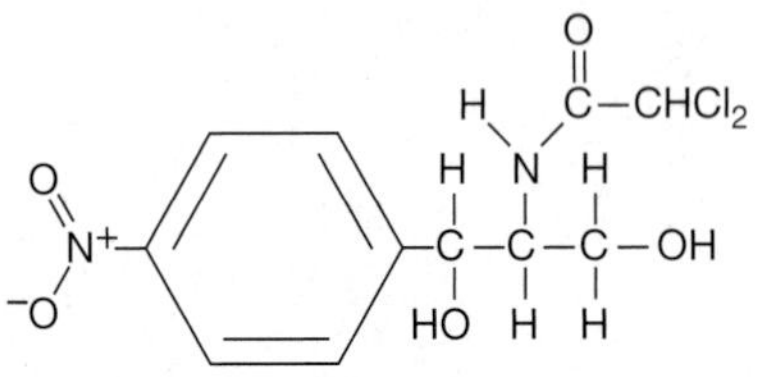

Chloramphenicol

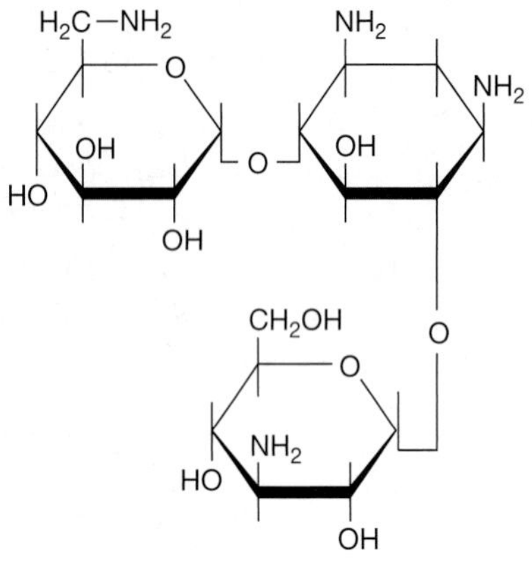

Kanamycin

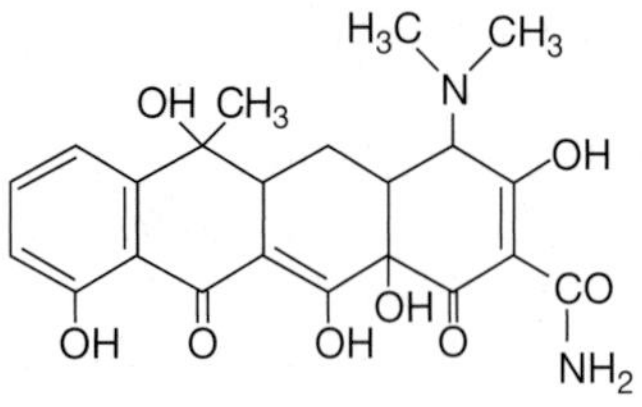

Tetracycline

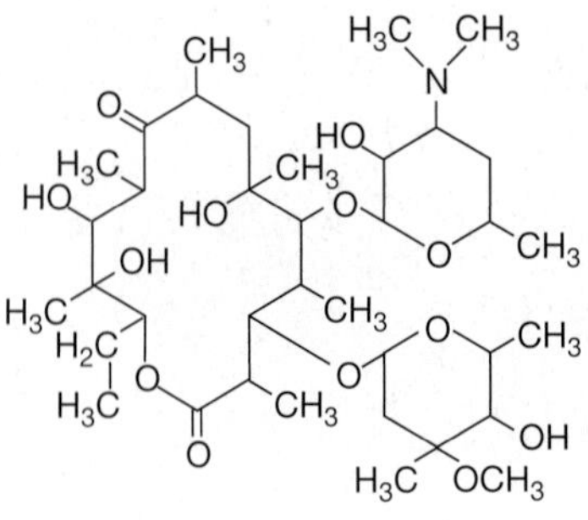

Erythromycin

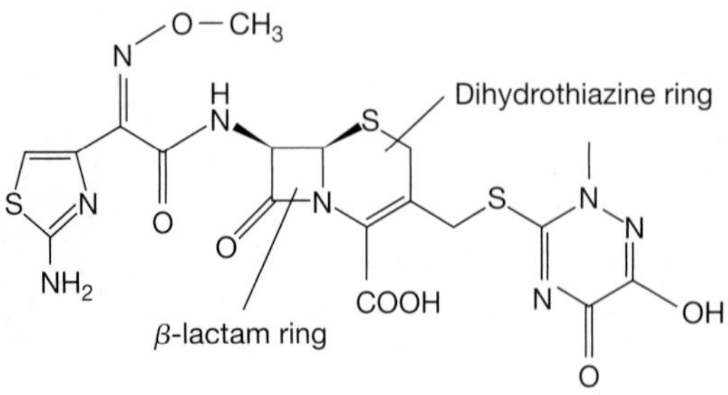

Ceftriaxone

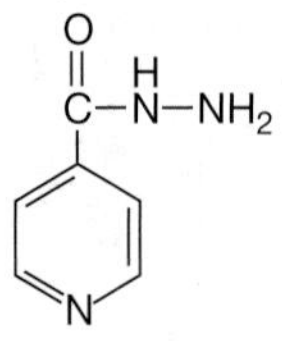

Isoniazid

TABLE 28.5 *(continued)*

Mode of action	Antibiotic class	Example(s)	Representative structures
Inhibit nucleic acid synthesis	Quinolones and fluoroquinolones	Nalidixic acid, ciprofloxacin, moxifloxacin	Ciprofloxacin
	Rifamycins	Rifampin	Rifampin
Inhibit metabolite synthesis	Trimethoprim	Trimethoprim	Trimethoprim
	Sulfa drugs	Sulfanilamide, sulfamethoxazole	Sulfanilamide
Damage to the cell membrane	Lipid biosynthesis disruptor	Platensimycin	Platensimycin
	Membrane structure disruptor	Daptomycin	Daptomycin

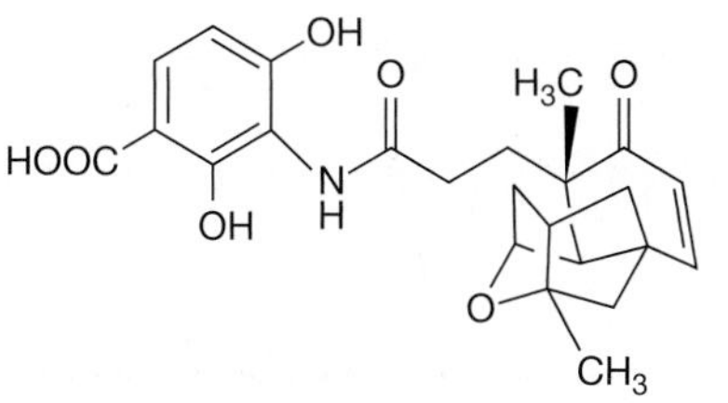

Platensimycin

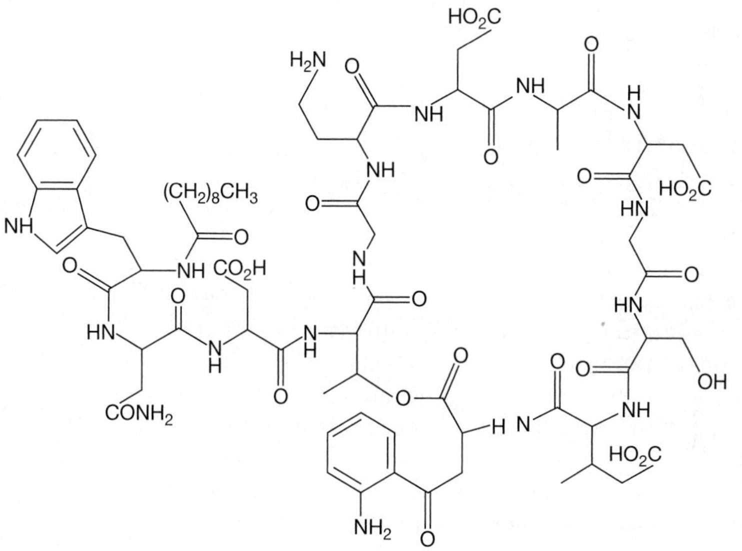

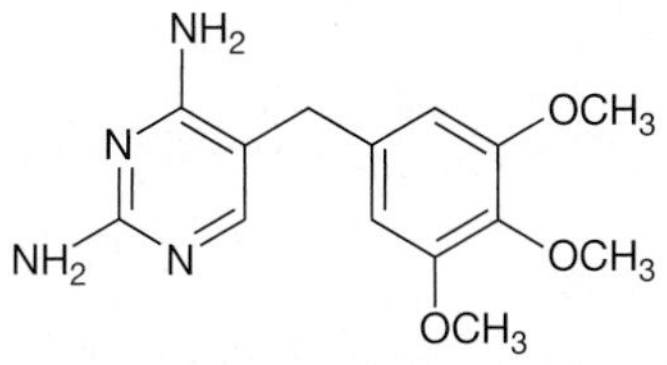

Trimethoprim

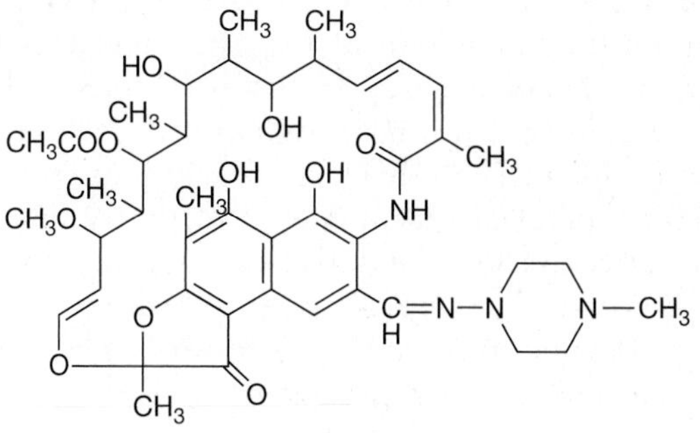

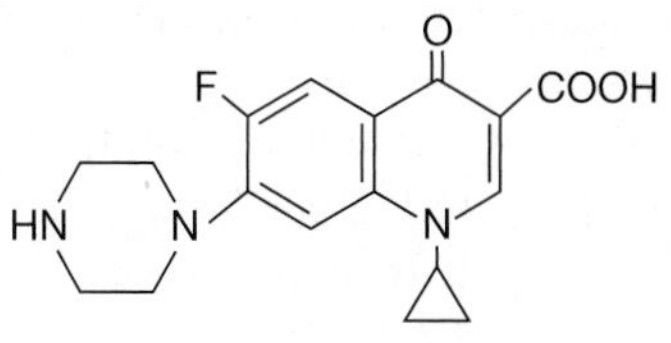

Ciprofloxacin

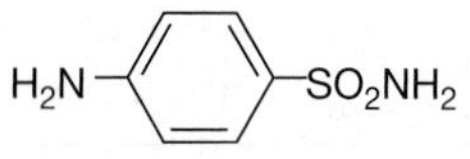

Sulfanilamide

Daptomycin

Rifampin

UNIT 6

pore, and induces rapid depolarization of the membrane. The depolarized cell quickly loses its ability to synthesize macromolecules, such as nucleic acids and proteins, resulting in cell death. However, alterations in cytoplasmic membrane structure may lead to resistance.

Platensimycin (Table 28.5), produced by *Streptomyces platensis*, is an unusual antibiotic that inhibits fatty acid and lipid biosynthesis. Platensimycin is effective against a broad range of gram-positive *Bacteria*, including nearly untreatable infections caused by MRSA and vancomycin-resistant enterococci. Platensimycin has a unique mode of action, shows no host toxicity, and there is no known potential for development of resistance by pathogens. We discuss the discovery of platensimycin in Section 28.12.

--- **MINIQUIZ** ---

- Explain the concept of selective toxicity in terms of antimicrobial therapy.

- How does the activity of each antibiotic class lead to control of the affected pathogens?

- What are the sources of aminoglycosides, tetracyclines, macrolides, daptomycin, and platensimycin?

28.11 Antimicrobial Drugs That Target Nonbacterial Pathogens

Antiviral drugs often adversely affect the host as well as the pathogen because viruses depend on host cell biosynthetic machinery for their replication. As a result, selective toxicity for viruses is achieved only with agents that preferentially affect unique viral replication pathways or the assembly of viral components. In spite of these limitations, a number of drugs are more toxic for viruses than for the host, and a few agents specifically target individual viruses. Fungi, protozoa, and helminths (worms) pose special problems for the development of effective drugs because these pathogens are eukaryotic, and therefore their cellular biology is similar to that of humans.

Antiviral Drugs

The most successful and commonly used agents for antiviral chemotherapy are the *nucleoside analogs*. These drugs are **nucleoside reverse transcriptase inhibitors (NRTIs)**, which work by inhibiting elongation of the viral nucleic acid chain by a nucleic acid polymerase. The first NRTI to be widely used was *zidovudine*, also called *azidothymidine* (*AZT*) (Table 28.6) (🔗 Figure 30.48a), which effectively blocks reverse transcription and production of complementary DNA (cDNA) in HIV and other retroviruses (🔗 Section 30.15). It is structurally similar to thymidine but is a dideoxy derivative, lacking the 3′-hydroxyl group. Another widely used nucleoside analog, *acyclovir* (Table 28.6), resembles guanosine and has been successfully used to control the symptoms of genital herpes (🔗 Section 30.14).

Some antiviral agents target the key enzyme of retroviruses, reverse transcriptase. *Nevirapine* (🔗 Figure 30.48b), a **nonnucleoside reverse transcriptase inhibitor (NNRTI)**, binds directly to

TABLE 28.6 Antiviral compounds

Examples	Mechanism of action	Virus affected
Enfuvirtide	Blocks fusion of HIV with T lymphocyte membrane	HIV (human immunodeficiency virus)
α, β, γ-Interferon	Induces proteins that inhibit viral replication	Broad spectrum (host-specific)
Oseltamivir (Tamiflu®) and zanamivir (Relenza®)	Block active site of influenza neuraminidase	Influenza A and B
Nevirapine	Reverse transcriptase inhibitor	HIV
Acyclovir (🔗 Figure 30.42)	Viral polymerase inhibitor	Herpes viruses, *Varicella zoster*
Zidovudine (AZT) (🔗 Figure 30.48a)	Reverse transcriptase inhibitor	HIV
Ribavirin	Blocks capping of viral RNA	Respiratory syncytial virus, influenza A and B, Lassa fever
Cidofovir	Viral polymerase inhibitor	Cytomegalovirus, herpesviruses
Tenofovir (TDF)	Reverse transcriptase inhibitor	HIV
Indinavir, saquinavir (Figure 28.35)	Viral protease inhibitors	HIV

reverse transcriptase and inhibits reverse transcription (Table 28.6). Phosphonoformic acid, an analog of inorganic pyrophosphate, inhibits normal internucleotide linkages, preventing synthesis of viral nucleic acids. Because their action affects normal host cell nucleic acid synthesis, both NRTIs and NNRTIs usually induce some host toxicity.

Other anti-retroviral drugs include **protease inhibitors** and *enfuvirtide*, a **fusion inhibitor** (Table 28.6). Protease inhibitors disrupt viral replication by binding the active site of HIV protease (see Figure 28.35), preventing this enzyme from processing large viral polyproteins into individual viral components (🔗 Section 10.11). Enfuvirtide is a 36-amino-acid synthetic peptide that binds to the gp41 membrane protein of HIV (🔗 Section 30.15); this stops the conformational changes necessary for the fusion of HIV with T lymphocyte membranes, thus preventing infection of host immune cells by HIV (🔗 Figure 30.43).

A single category of drugs effectively limits influenza infection. The *neuraminidase inhibitors* oseltamivir (Tamiflu) and zanamivir (Relenza) block the active site of neuraminidase in influenza A and B viruses, inhibiting virus release from infected cells. Zanamivir is used only for treatment of influenza, whereas oseltamivir is used for both treatment and prophylaxis (Table 28.6).

Recall from Chapter 26 that virus-infected cells release small cytokine proteins called *interferons* that trigger a defensive response in neighboring host cells (🔗 Section 26.10). If produced and administered properly, interferons may have potential use as prescribed antimicrobial agents. The clinical utility of interferons depends on whether they can be delivered to specific areas in the host to stimulate the production of antiviral proteins in uninfected host cells. Alternatively, appropriate interferon stimulators, such as viral nucleotides, nonvirulent viruses, or even synthetic nucleotides, if given to host cells prior to infection, may induce natural production of interferon.

Drugs That Target Eukaryotic Pathogens

Fungi cause a number of serious diseases (⇄ Sections 33.1 and 33.2), and their treatment is complicated by the fact that many antifungal agents act on metabolic pathways that are shared by fungi and their host cells, thus making the drugs toxic. As a result, many antifungal drugs can be used only for topical (surface) applications (**Table 28.7**).

A major group of antifungal compounds are *ergosterol inhibitors*, which work either by interacting directly with ergosterol or inhibiting its synthesis (Table 28.7). Ergosterol is present in fungal cytoplasmic membranes in place of the cholesterol found in animal cell cytoplasmic membranes. Important ergosterol inhibitors include the *polyene* antibiotics, which are produced by species of *Streptomyces* bacteria. Polyenes bind specifically to ergosterol, causing membrane permeability and cell death (**Figure 28.32**). By contrast, *azoles* and *allylamines* are broad-spectrum antifungal drugs that work by selectively inhibiting ergosterol biosynthesis. Treatment with these drugs causes membrane damage and alteration of critical membrane transport activities.

Echinocandins are cell wall inhibitors that block the activity of 1,3-β-D-glucan synthase, the enzyme that forms β-glucan polymers in the fungal cell wall (Figure 28.32 and Table 28.7). Because mammalian cells do not have 1,3-β-D-glucan synthase (or cell walls), the activity of these agents selectively kills fungal cells. Echinocandins are often used to treat *Candida* yeast infections, as well as some fungi that are resistant to other agents (⇄ Section 33.1).

Fungal cell walls also contain chitin, a polymer of *N*-acetylglucosamine found only in fungi and insects. Several *polyoxins* inhibit cell wall synthesis by interfering with chitin biosynthesis (Figure 28.32). Although not used clinically, polyoxins are widely used as agricultural fungicides. Other antifungal drugs inhibit folate biosynthesis, interfere with DNA topology during replication, or, in

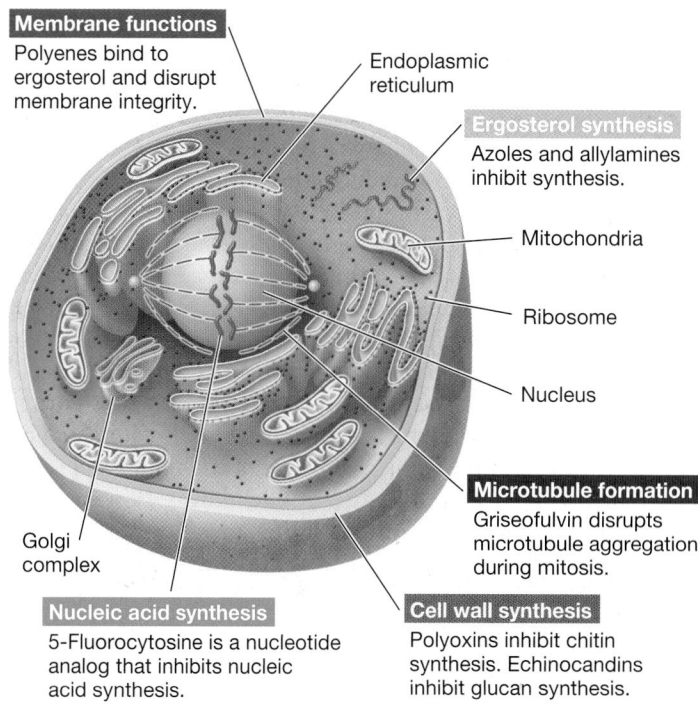

Membrane functions
Polyenes bind to ergosterol and disrupt membrane integrity.

Endoplasmic reticulum

Ergosterol synthesis
Azoles and allylamines inhibit synthesis.

Mitochondria

Ribosome

Nucleus

Golgi complex

Microtubule formation
Griseofulvin disrupts microtubule aggregation during mitosis.

Nucleic acid synthesis
5-Fluorocytosine is a nucleotide analog that inhibits nucleic acid synthesis.

Cell wall synthesis
Polyoxins inhibit chitin synthesis. Echinocandins inhibit glucan synthesis.

Figure 28.32 Targets of some antifungal agents. Traditional antibacterial agents are generally ineffective because fungi are eukaryotic cells. The cell wall and cell membrane ergosterol targets shown here are unique structures not present in vertebrate host cells.

the case of drugs such as griseofulvin, disrupt microtubule aggregation during mitosis (Figure 28.32). Moreover, the nucleic acid analog 5-fluorocytosine (flucytosine) is an effective nucleic acid synthesis inhibitor in fungi.

Historically, diseases caused by protozoans, especially malaria, have been treated with *quinine* or, more recently, quinine derivatives, such as chloroquine and mefloquine. However, extensive resistance to these drugs among the *Plasmodium* spp. that cause malaria has led to the development of artemisinin-based alternatives. Artemisinin is produced in low amounts by the Chinese

TABLE 28.7 Antifungal agents

Category	Target	Examples	Use
Allylamines	Ergosterol synthesis	Terbinafine	Oral, topical
Aromatic antibiotic	Mitosis inhibitor	Griseofulvin	Oral
Azoles	Ergosterol synthesis	Clotrimazole	Topical
		Fluconazole	Oral
		Miconazole	Topical
Chitin synthesis inhibitor	Chitin synthesis	Nikkomycin Z	Experimental
Echinocandins	Cell wall synthesis	Caspofungin	Intravenous
Nucleic acid analogs	DNA synthesis	5-Fluorocytosine	Oral
Polyenes	Ergosterol synthesis	Amphotericin B	Oral, intravenous
		Nystatin	Oral, topical
Polyoxins	Chitin synthesis	Polyoxin A and B	Agricultural

wormwood plant, which has been used traditionally to control the cyclic fever associated with malaria. A more recent discovery has shown artemisinin to also be an effective treatment for certain helminth diseases, such as schistosomiasis (⮯ Section 33.7). Using the techniques of synthetic biology, a genetically engineered yeast strain is now being used to produce artemisinin, making the drug more widely available (⮯ Section 12.11 and Figure 12.33).

The drug of choice for many parasitic protozoans is *metronidazole* (or the related drug *tinidazole*), which targets anaerobic pathogens and is therefore also useful against infections caused by clostridia. Infections with *Giardia intestinalis* (giardiasis), *Trichomonas vaginalis* (trichomoniasis), and especially *Entamoeba histolytica* (amebic dysentery) have all been treated successfully with these drugs (⮯ Sections 33.3 and 33.4). The related protist *Cryptosporidium parvum* (cryptosporidiosis) is often treated with *nitazoxanide*, a drug that has also been effective against helminths. However, perhaps the most widely used antihelminthic drugs are *praziquantel*, for treating schistosomiasis and tapeworm infections, and *mebendazole* (and similar drugs), for treating a variety of helminth infections, including tapeworms, pinworms, hookworms, trichinosis, and ascariasis (⮯ Section 33.7).

MINIQUIZ

- What steps in the viral maturation process are inhibited by nucleoside analogs? By protease inhibitors? By interferons?
- Why are there fewer clinically effective antifungal and antiparasitic agents than antibacterial agents?

28.12 Antimicrobial Drug Resistance and New Treatment Strategies

Antimicrobial drug resistance is the acquired ability of a microorganism to resist the effects of an antimicrobial agent to which it was formerly susceptible. As we have discussed, many microorganisms produce antibiotics, and genes encoding antibiotic resistance are present in virtually all of them. Horizontal gene transfer between and among microorganisms disseminates antimicrobial drug resistance.

Antimicrobial Drug Resistance

Common mechanisms of bacterial resistance to antibiotics were discussed in the context of microbial growth in Section 7.10 and depicted in Figure 7.21*b*, and some examples are listed in **Table 28.8**. Antibiotic resistance can be genetically encoded on the bacterial chromosome, but more often, antibiotic-resistant bacteria isolated from patients contain drug-resistance genes located on horizontally transmitted *R (resistance) plasmids* (⮯ Section 4.2). Enzymes encoded by genes on R plasmids confer resistance by any of three classes of mechanisms: modifying and inactivating the drug, preventing uptake of the drug, or actively pumping the antibiotic out of the cell, a process called *efflux* (Table 28.8).

The widespread use of antibiotics (Figure 28.28*a*) provides favorable conditions for the spread of R plasmids because they carry genes that confer an immediate selective advantage. The ubiquity of resistance genes limits the long-term use of any single antibiotic as an effective antimicrobial agent. A classic example is the development of multidrug resistance in the bacterium *Neisseria gonorrhoeae*. Penicillin, widely used to treat gonorrhea into the 1980s, became ineffective and was replaced by ciprofloxacin, which, in turn, lost much of its efficacy after just 10 years of use (**Figure 28.33**). This prompted a switch in treatment to the β-lactamase-resistant ceftriaxone, and more recently, a combination of ceftriaxone and azithromycin. Combining two unrelated antimicrobial agents such as this often reduces resistance because it is less likely that a mutant strain resistant to one antibiotic will also be resistant to the second antibiotic. However, certain R plasmids confer multiple drug resistance and can thwart multiple antibiotic therapy as a clinical strategy.

TABLE 28.8 Bacterial resistance to antibiotics

Resistance mechanism	Antibiotic example	Genetic basis of resistance	Mechanism present in
Reduced permeability	Penicillins	Chromosomal	Gram-negative bacteria
Inactivation of antibiotic Examples: β-lactamases; modifying enzymes, such as methylases, acetylases, phosphorylases, and others	Penicillins	Plasmid and chromosomal	*Staphylococcus aureus* Enteric bacteria
	Chloramphenicol	Plasmid and chromosomal	*Neisseria gonorrhoeae Staphylococcus aureus*
	Aminoglycosides	Plasmid	Enteric bacteria
Alteration of target Examples: RNA polymerase, rifamycin; ribosome, erythromycin and streptomycin; DNA gyrase, quinolones	Erythromycin Rifamycin Streptomycin Norfloxacin	Chromosomal	*Staphylococcus aureus* Enteric bacteria Enteric bacteria Enteric bacteria *Staphylococcus aureus*
Development of resistant biochemical pathway	Sulfonamides	Chromosomal	Enteric bacteria *Staphylococcus aureus*
Efflux (pumping out of cell)	Tetracyclines Chloramphenicol	Plasmid Chromosomal	Enteric bacteria *Staphylococcus aureus Bacillus subtilis*
	Erythromycin	Chromosomal	*Staphylococcus*

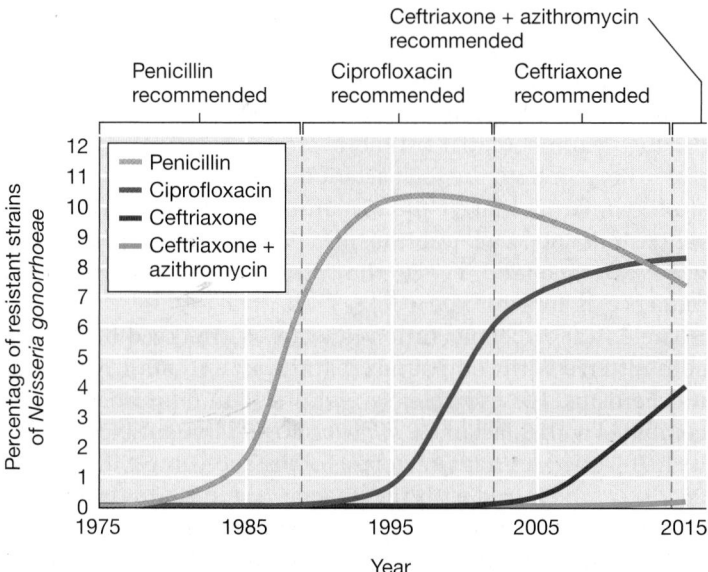

Figure 28.33 The emergence of multidrug resistance in *Neisseria gonorrhoeae*. Resistance to penicillin developed in the 1980s to the point that it could no longer be recommended for treatment of gonorrhea, at which point ciprofloxacin, a quinolone, became the drug of choice. By the early 2000s, resistance to ciprofloxacin prompted a change to ceftriaxone. The current recommendation is a combination of ceftriaxone plus azithromycin.

Overuse of antibiotics also accelerates resistance. In addition to their traditional use as a treatment for infections, antibiotics are widely used as supplements to farm animal feeds both as growth-promoting substances and as prophylactic additives to prevent the occurrence of disease. For example, broad-spectrum antibiotics widely used clinically, such as the fluoroquinolone ciprofloxacin, are used extensively as a feed additive, especially by the poultry industry. Although the trend is improving, antibiotics are also used in clinical practice far more often than necessary, and this problem is compounded by patient noncompliance—many patients stop taking antibiotics as soon as they feel better. Antibiotic resistance can be minimized if drugs are used only for treatment of susceptible pathogens and are given in sufficiently high doses and for sufficient lengths of time to eradicate the pathogen before resistant mutants can propagate.

To prevent the further emergence of multidrug-resistant pathogens, the Centers for Disease Control and Prevention (CDC) publishes guidelines stressing the importance of vaccination, rapidly and accurately diagnosing and treating infections, using antimicrobial agents prudently, and preventing pathogen transmission.

New Drugs and New Treatment Strategies

Conservative, appropriate administration of antibiotics can prolong or even resurrect the clinical usefulness of available drugs, but the long-term solution to antimicrobial drug resistance requires continuous development of new drugs through design or discovery. Developing new analogs of existing antimicrobial compounds is usually more cost effective than discovering new drugs, and analogs may actually have greater antimicrobial

activity than the parent compound. For example, using natural penicillin as the starting compound, systematic chemical substitution of the *N*-acyl group can generate hundreds of penicillin derivatives, many with broad-spectrum activity (Figure 28.29). Using this basic strategy, semisynthetic analogs of β-lactam antibiotics, tetracycline, and vancomycin (**Figure 28.34**) have been synthesized.

Novel antimicrobial compounds are more difficult to identify than analogs of existing drugs because new antimicrobial compounds must work at unique sites in metabolism or be structurally dissimilar to existing compounds to avoid resistance mechanisms. Computer-based methods accelerate novel compound design by maximizing molecular binding in a virtual environment. A dramatic success in computer-directed drug design was the development of *saquinavir*, a protease inhibitor that impedes the multiplication of HIV in infected individuals by binding the active site of the HIV protease enzyme (**Figure 28.35**). As an analog of the HIV precursor protein, saquinavir displaces the authentic protease substrate and inhibits virus maturation. Several other computer-designed protease inhibitors, including *indinavir* (Figure 28.35*b*), are currently in use for the treatment of HIV/AIDS (⊘ Section 30.15).

Another strategy to overcome antimicrobial resistance is to select for antibiotics that interact with unexploited targets. Platensimycin is the first antimicrobial drug targeted to disrupt bacterial lipid biosynthesis (Table 28.5 and Figure 28.27). Platensimycin is especially active against gram-positive pathogens, including drug-resistant staphylococci and enterococci. To select an agent for a defined target, in this case an enzyme in the bacterial lipid synthesis pathway, scientists used antisense RNA (⊘ Section 6.11) to limit the amount of accessible mRNA required

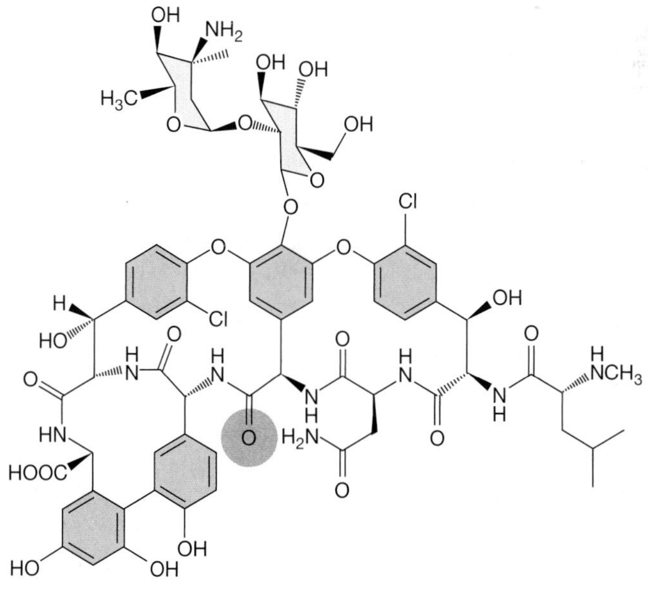

Vancomycin

Figure 28.34 Vancomycin. Many pathogenic strains now show intermediate drug resistance to the parent structure of vancomycin, but chemically substituting the carbonyl oxygen at the position shown in red with a methylene (=CH₂) group restores much of the lost activity. Like penicillin, vancomycin prevents cross-linking of peptidoglycan and is most effective against gram-positive pathogens.

UNIT 6

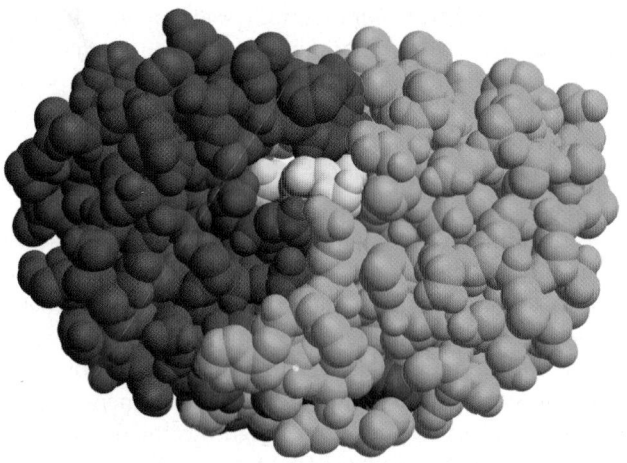

(a) **HIV protease**

Saquinavir

Indinavir

(b)

Figure 28.35 Computer-generated anti-HIV drugs. *(a)* The HIV protease homodimer. Individual polypeptide chains are shown in green and blue. A peptide (yellow) is bound in the active site. HIV protease cleaves an HIV precursor protein, a necessary step in virus maturation. Blocking of the protease site by the bound peptide inhibits precursor processing and HIV maturation. This structure is derived from information in the Protein Data Bank. *(b)* These anti-HIV drugs are peptide analogs called protease inhibitors that were designed by computer to block the active site of HIV protease. The areas highlighted in orange show the regions analogous to peptide bonds in proteins.

to produce a key lipid biosynthesis enzyme in *Staphylococcus aureus*. This reduced fatty acid synthesis in the cells and increased the sensitivity of the crippled *S. aureus* strain to antibiotics that inhibit fatty acid synthesis. After screening thousands of natural products from potential antibiotic producers, scientists isolated platensimycin from the soil bacterium *Streptomyces platensis*. This strategy identifies target-specific antibiotics present in low concentrations and is applicable to virtually any target for which the gene sequence (and, hence, the corresponding antisense RNA sequence) is known.

The efficacy of some antibiotics can be preserved if they are administered with compounds that thwart antibiotic resistance mechanisms. For example, several β-lactam *antibiotics* can be combined with β-lactamase *inhibitors* to preserve antibiotic activity in β-lactam-resistant microbes. For instance, ampicillin (Figure 28.29) can be mixed with the β-lactamase inhibitor clavulanic acid to produce the combination drug Augmentin. The inhibitor binds β-lactamase (produced by the resistant pathogen) irreversibly. This prevents ampicillin from being degraded and maintains the efficacy of the antibiotic.

Drug combination therapy has also revolutionized treatment of HIV infections. Currently, a combination therapy consisting of nucleoside analogs and a protease inhibitor is recommended. This drug treatment protocol is termed HAART, for *highly active anti-retroviral therapy*. As with antibacterial combination regimens, HAART is designed to target two independent viral functions: nucleoside analogs target virus replication, and protease inhibitors target virus maturation. Because the probability of a virus developing resistance to multiple drugs is much less than the probability of it becoming resistant to a single drug, HAART therapy has been a successful strategy in controlling HIV infections (⇄ Section 30.15).

MINIQUIZ

- Identify the basic mechanisms of antibiotic resistance in bacteria and describe what practices encourage their dissemination.
- What does vancomycin have in common with penicillin? How can native vancomycin be improved?
- Explain the advantages and disadvantages of developing new drugs based on existing drug analogs. What other methods exist for developing new drugs?

Chapter Review

I • The Clinical Microbiology Setting

28.1 Clinical laboratory safety requires training and planning to prevent contamination and possible infection of laboratory workers. Specific precautions and procedures proportional to the risk of infection by a given agent, designated by biosafety levels (BSL), must be in place to handle contaminated materials and patient specimens.

> **Q** How are most laboratory-associated infections contracted? What action can be taken to prevent laboratory infections?

28.2 Patients in healthcare facilities are unusually susceptible to infectious disease because of their compromised health and potential exposure to various pathogens in the facilities. Many healthcare-associated infections are caused by drug-resistant pathogens.

> **Q** Healthcare environments are conducive to the spread of infectious diseases. What are some of the reasons infections are spread so easily in the clinical setting?

II • Isolating and Characterizing Infectious Microorganisms

28.3 Appropriate sampling, observation, and culture techniques are necessary to isolate and identify potential pathogens. The selection of techniques requires knowledge of the ecology, physiology, and metabolism of suspected pathogens. Most pathogens exhibit unique metabolic patterns when grown on specialized selective and differential media. Growth-dependent patterns provide information helpful for unambiguous pathogen identification.

> **Q** Why is it important to process clinical specimens as rapidly as possible? What special procedures and precautions are necessary for the isolation and culture of anaerobes?

28.4 Pathogens isolated from clinical samples are often tested for antibiotic susceptibility to ensure appropriate antibiotic therapy. Testing is based on the minimum inhibitory concentration of an agent necessary to completely inhibit growth of a pathogen.

> **Q** Describe the disc diffusion test for antibiotic susceptibility. Why should potential pathogens isolated from patient samples be tested for antibiotic susceptibility?

III • Immunological and Molecular Tools for Disease Diagnosis

28.5 An immune response is a natural outcome of infection. Specific immune responses characterized by a rise in antibody titers and positive T cell–mediated skin tests can be used to provide evidence of infections and to monitor convalescence. Monoclonal antibodies reproducibly provide specificity for a wide range of diagnostic and therapeutic purposes.

> **Q** Why does antibody titer rise after infection? Is a high antibody titer indicative of an ongoing infection? Explain. Why is it necessary to obtain an acute and a convalescent blood sample to monitor infections?

28.6 Precipitation and agglutination reactions produce visible results involving antigen–antibody interactions. Fluorescent antibodies are used for quick, accurate identification of pathogens and other antigenic substances in tissue samples, blood, and other complex mixtures. Fluorescent antibody–based methods can be used for identification of a variety of microbial cell types.

> **Q** Why are agglutination tests so widely used in clinical diagnostics? How are fluorescent antibodies used to diagnose diseases? What advantages do immunofluorescent techniques have over traditional culturing?

28.7 Enzyme immunoassays (EIAs), rapid tests, and immunoblots are sensitive and specific immunological assays. These tests can be engineered to detect either antibody or antigen for diagnosis of infections by any of a large number of pathogens.

> **Q** EIAs are extremely sensitive diagnostic tools. Why, then, is the immunoblot (Western blot) procedure used to confirm screening tests that are positive for human immunodeficiency virus (HIV)?

28.8 Nucleic acid amplification (PCR) methods are applied as extremely specific diagnostic tools used for a large number of pathogens. Quantitative PCR (qPCR) and qualitative PCR techniques provide quantification and identification of pathogens, and reverse transcription PCR (RT-PCR) is especially useful for detecting RNA viruses.

> **Q** Distinguish between quantitative and qualitative PCR. How does qualitative PCR expand upon the qPCR technique?

IV • Prevention and Treatment of Infectious Diseases

28.9 Immunization induces artificial active immunity and is widely used to prevent infectious diseases. Vaccines are either attenuated or inactivated pathogens or pathogen products or are genetically engineered antigens. The latter eliminate exposure to pathogens and, in some cases, even to protein antigen. Application of this strategy is providing safer vaccines targeted to individual pathogen antigens.

> **Q** List the immunizations recommended for children in the United States. For which of these have you been immunized? List the diseases for which you may have acquired immunity naturally.

28.10 Antibiotics are chemically diverse antimicrobial compounds produced by microorganisms. Each antibiotic works by inhibiting a specific cellular process in the target microorganisms. The β-lactam antibiotics, including penicillins and cephalosporins, target bacterial cell wall synthesis and are the most important class of clinical antibiotics. The aminoglycosides, macrolides, and tetracycline antibiotics selectively interfere with protein synthesis in *Bacteria*. The quinolones are an important class of synthetic antibacterial drugs that inhibit DNA synthesis. Daptomycin and platensimycin are structurally novel antibiotics that target cytoplasmic membrane functions and lipid biosynthesis, respectively.

> **Q** What are the common sources for natural antibiotics? How do these antimicrobial drugs differ from growth factor analogs, such as the sulfa drugs? Why are β-lactam antibiotics generally more effective against gram-positive bacteria than against gram-negative bacteria?

28.11 Antiviral agents selectively target virus-specific enzymes and processes. Useful agents include analogs and compounds that inhibit nucleic acid polymerases and viral genome replication. Protease inhibitors interfere with viral maturation steps. Because fungi are *Eukarya*, antifungal agents exhibiting selective toxicity are difficult to find. Nevertheless, some effective antifungal agents are available, and they primarily target fungi-specific structures and biosynthetic processes.

> **Q** Why is host toxicity a common problem with antiviral and antifungal drugs? Identify the targets that allow for selective toxicity of antifungal agents.

28.12 The use of antimicrobial drugs inevitably leads to resistance in the targeted microorganisms. The development of resistance can be accelerated by the indiscriminate use of antimicrobial drugs, and many pathogens have developed resistance. New antimicrobial compounds must continually be discovered and developed to deal with drug-resistant pathogens and to enhance our ability to treat infectious diseases. Computer-based modeling and other novel strategies are helping to address this challenge.

> **Q** What practices contribute to the spread of antibiotic resistance? Explain how antisense RNA strategies can extend traditional methods of natural product selection for antibiotic discovery.

Application Questions

1. Define the procedures you would use to isolate and identify a new pathogen. Keep in mind Koch's postulates (⇨ Figure 1.29) as you form your answer. Be sure to include growth-dependent assays, immunoassays, and molecular assays. Which of your assays could be adapted to be used as a routine, high-throughput test for rapid clinical diagnosis?

2. Viruses and fungi present special problems for drug therapy. Explain the issues inherent in drug treatment of both groups, and explain whether you agree with the preceding statement. Give specific examples, and suggest at least one group of agents that might target both types of infectious agents.

3. Describe three important reasons why semisynthetic penicillins were first developed. Which clinical challenges does each of these reasons address? What key part of the penicillin molecule must be retained for any semisynthetic penicillin to be active?

4. Imagine yourself as a clinical microbiologist with all of the diagnostic tools described in this chapter available for your analyses. Which tool(s) would you use (and why) if (1) A patient had a life-threatening infection caused by a difficult-to-culture bacterium and where treatment of the infection was absolutely dependent on an extremely rapid identification of the pathogen? (2) A patient had a less severe bacterial infection where the pathogen was easily culturable and treatable?

Chapter Glossary

Agglutination a reaction between antibody and particle-bound antigen resulting in visible clumping of the particles

Aminoglycoside an antibiotic, such as streptomycin, containing amino sugars linked by glycosidic bonds

Antibiotic a chemical substance produced by a microorganism that kills or inhibits the growth of another microorganism

Antimicrobial drug resistance the acquired ability of a microorganism to resist the effects of an antimicrobial agent to which it was formerly susceptible

β-lactam antibiotic penicillin, cephalosporin, or a related antibiotic that contains the four-membered heterocyclic β-lactam ring

Differential media growth media that allow identification of microorganisms based on phenotypic properties

Enzyme immunoassay (EIA) a test that uses antibodies or antigens linked to enzymes to detect antigens or antibodies in body fluids

Fluorescent antibody an antibody molecule covalently modified with a fluorescent dye that makes the antibody visible under fluorescent light

Fusion inhibitor a peptide that blocks the fusion of viral and target cell cytoplasmic membranes

Growth factor analog a chemical agent that has a similar structure to and blocks the uptake or utilization of a growth factor

Healthcare-associated infection (HAI) an infection acquired by a patient in a healthcare facility, particularly during a stay in the facility. Also called a *nosocomial infection.*

Immunoblot (Western blot) the use of labeled antibodies to detect specific proteins after separation by electrophoresis and transfer to a membrane

Macrolide erythromycin or a related antibiotic that contains a lactone ring bonded to sugars

Minimum inhibitory concentration (MIC) the smallest amount of an agent needed to completely inhibit the growth of an organism in vitro

Monoclonal antibody (mAb) a single type of antibody made by a single B cell hybridoma clone

Nonnucleoside reverse transcriptase inhibitor (NNRTI) a nonnucleoside analog used to inhibit viral reverse transcriptase

Nucleoside reverse transcriptase inhibitor (NRTI) a nucleoside analog used to inhibit viral reverse transcriptase

Precipitation a reaction between antibody and a soluble antigen resulting in a visible, insoluble complex

Protease inhibitor a class of drug designed to inhibit viral protease

Quinolone a synthetic antibacterial compound that inhibits DNA gyrase and prevents supercoiling of bacterial DNA

Selective media culture media that allow the growth of certain organisms while inhibiting the growth of others through one or more added media components

Selective toxicity the ability of a compound to inhibit or kill a pathogen without adversely affecting the host

Sensitivity the lowest amount of antigen that can be detected by a diagnostic test

Serology the study of antigen–antibody reactions in vitro

Specificity the ability of an antibody or a lymphocyte to recognize a single antigen, or of a diagnostic test to identify a specific pathogen

Sulfa drugs synthetic growth factor analogs that inhibit folic acid biosynthesis in *Bacteria*

Tetracycline an antibiotic characterized by the four-member naphthacene ring structure

Titer the quantity of a substance, such as antibody, in a solution

Vaccination (immunization) the inoculation of a host with inactive or weakened pathogens or pathogen products to stimulate protective active immunity

Vaccine an inactivated or attenuated pathogen, or a harmless pathogen product, used to induce artificial active immunity

29

Epidemiology

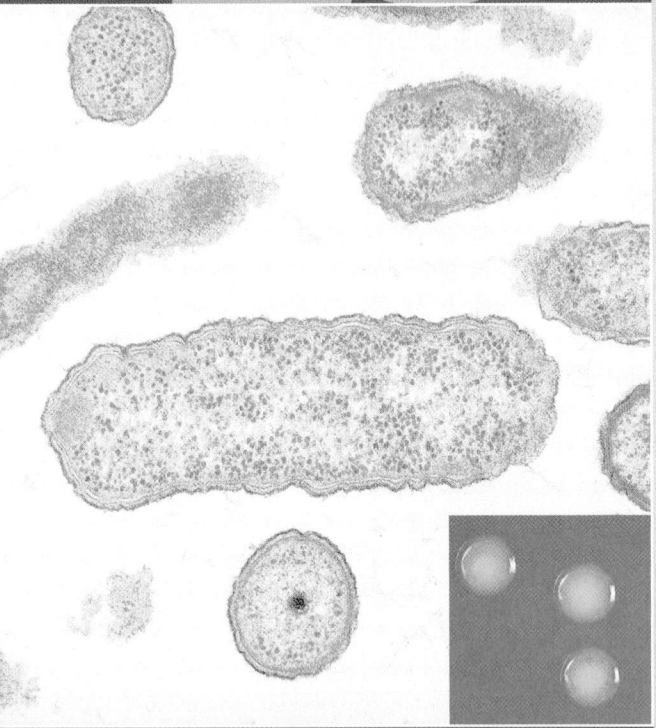

microbiology**now**

A Mysterious New Disease Outbreak

Although epidemiologists pay particular attention to large-scale disease outbreaks—cholera and the like—a disease tracker's job often entails investigating a very restricted outbreak. Such was the case with the recent *Elizabethkingia* outbreak in southern Wisconsin (USA).

Elizabethkingia anophelis is a gram-negative bacterium (main photo) of the phylum *Bacteroidetes* that forms translucent colonies on blood agar media (inset photo). The genus name *Elizabethkingia* honors American medical bacteriologist Elizabeth King, and the species epithet *anophelis* reflects the fact that this organism is the dominant bacterium in the gut of *Anopheles gambiae*, the mosquito that carries the malarial parasite. *Elizabethkingia* inhabits soil and water and is rarely linked to disease. However, in the spring of 2016, the Wisconsin Department of Public Health received reports of at least 63 confirmed cases of *E. anophelis* infection, and two additional cases, one each in Illinois and Michigan, were also reported. Of these 65 cases, 20 deaths (~30% mortality) occurred, showing that *E. anophelis* infections were indeed a serious threat.

E. anophelis is an opportunistic pathogen that can cause meningitis or bloodstream infections and can also colonize the respiratory tract. In the 2016 outbreak, the majority of infections and deaths were in patients over the age of 65 who had serious underlying health issues such as cancer, diabetes, recent surgery, or the like. This epidemiological observation suggested the possibility of a common source of infection. However, because *E. anophelis* is so widespread in the environment, many potential sources had to be checked. Common disease vehicles such as contaminated food or water were almost completely ruled out, as was person-to-person transmission. Contaminated medical equipment and a few other sources were suggested as possible explanations for the infections. However, pinpointing the primary source of the Wisconsin *E. anophelis* outbreak has thus far proven elusive.

Nevertheless, epidemiologists remain busy using their well-honed analytical skills to systematically eliminate some explanations for the Wisconsin *E. anophelis* outbreak while grouping together the most likely possibilities for further analysis. Using epidemiological methods, the source and the mode(s) of transmission of this pathogen will eventually be revealed and this should reduce *Elizabethkingia* infections to the usual number (5–10) typically reported from the entire United States each year.

 Source: Multistate outbreak of infections caused by *Elizabethkingia anophelis*. 2016. Centers for Disease Control and Prevention, Atlanta, Georgia (USA). June 16, 2016.

We begin a new unit here with a focus on infectious diseases. Everything we have learned up to this point—cell structure, metabolism, growth, genetics, and genomics; microbial evolution, diversity, and ecology; and host–microbe relationships and the immune response—will help us better understand the disease strategies and exploitable weaknesses of infectious microbial agents.

As a prelude to our coverage of the clinical aspects of infectious diseases in the following four chapters, we explore the "big picture" of how infectious diseases flow through populations. **Epidemiology** is the study of the occurrence, distribution, and determinants of health and disease in populations and also deals with **public health**, the health of the population as a whole. Although in developed countries infectious diseases are not leading causes of death, in developing countries infectious diseases can account for nearly half of all deaths. Hence, identifying and solving problems associated with infectious disease transmission is a major goal of the epidemiologist.

I • Principles of Epidemiology

Here we consider the principles of epidemiology and define key terms in the lexicon of the epidemiologist.

29.1 The Language of Epidemiology

The epidemiologist traces the spread of a disease to identify its origin and mode of transmission in a population. The population might be all people in a certain city, country, or region, or it could be the entire human population. Alternatively, the population under study could be a particular cohort of a larger population, such as only males or only those of a specific race or age group. Raw data are gathered from disease-reporting networks such as city, county, state, and national public health departments, clinical records, and patient interviews.

A major job of the epidemiologist is to carry out **disease surveillance**—the observation, recognition, and reporting of diseases as they occur—and then analyze the data provided by local and national health authorities to reveal trends and signals of disease outbreaks. The epidemiologist thus stands in contrast to the clinical health provider—the one who actually treats the infected patient. However, in order to both track a disease and predict its spread in a population, the epidemiologist must integrate clinical and surveillance results to formulate effective public health measures for disease control.

Disease Incidence and Prevalence

The epidemiologist often uses the words *incidence* and *prevalence* when discussing infectious diseases. The **incidence** of a particular disease is the *number of new cases* in a population in a given time period (**Figure 29.1**). For example, in 2013 there were 47,352 new cases of HIV infection in the United States, for an incidence of 15 new cases per 100,000 people per year. The **prevalence** of a given disease is the *total number of new and existing disease cases* in a population in a given time period (Figure 29.1). For example, within the United States there were 1,194,039 persons living with

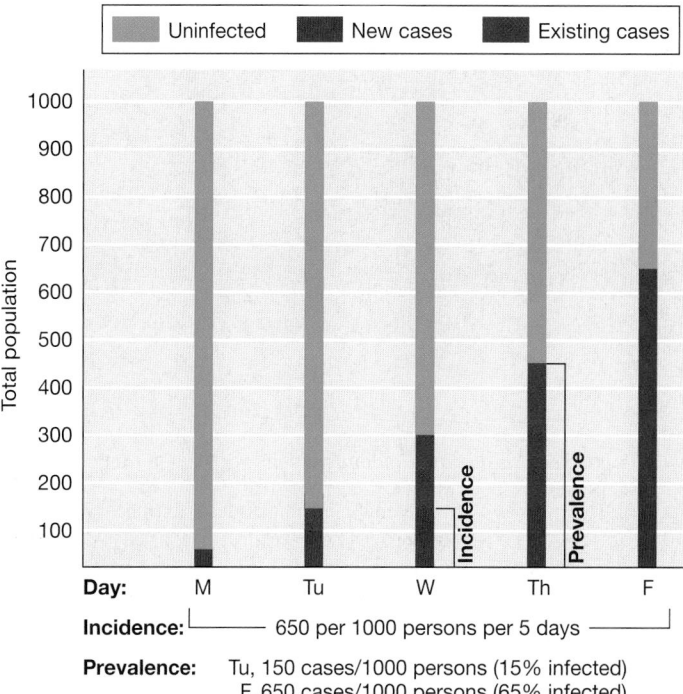

Figure 29.1 The concepts of disease incidence and disease prevalence. Disease incidence is a rate function and is defined as the number of new cases over a given time period (day[s], week, month, etc.); incidence is an indicator of infection risk. Disease prevalence is the total number of diseased individuals at some time point and is a snapshot of the extent of a disease in a population at any given time.

HIV/AIDS at the end of 2013. Expressed another way, the prevalence of HIV/AIDS in the United States was about 375 cases per 100,000 persons in 2013.

Essentially a *rate* measurement, disease incidence can be used to predict the *risk* of disease for an individual in a defined population within a specific time period. By contrast, prevalence measures the total disease burden in a population and can be thought of as a "snapshot" of the disease at a specific instant (Figure 29.1). The incidence and prevalence of disease are also major indicators of the public health of a population.

The Scope of Disease

Other common epidemiological terms speak to the scope of a disease. A disease is an **epidemic** when it simultaneously infects an unusually high number of individuals in a population; a **pandemic** is a widespread, usually global epidemic. By contrast, an **endemic disease** is one that is constantly present—typically in low numbers—in a population (**Figure 29.2**). An endemic disease implies that the pathogen may not be highly virulent or that the majority of individuals in the population may be immune, resulting in low but persistent numbers of cases. Individuals infected with a pathogen that causes an endemic disease are **reservoirs** of infection, a source of infectious agents from which susceptible individuals may be infected.

Sporadic cases of a disease occur one at a time in geographically separated areas, suggesting that the cases are not related. A disease **outbreak**, on the other hand, is the appearance of a large number of cases in a short time in an area previously experiencing only

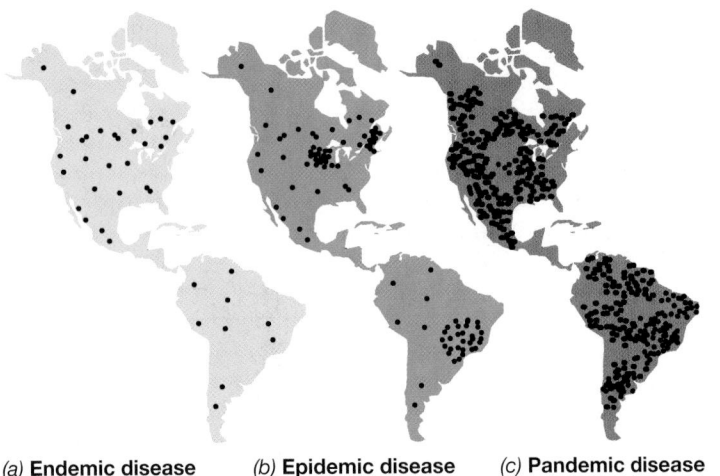

(a) **Endemic disease**   (b) **Epidemic disease**   (c) **Pandemic disease**

Figure 29.2 Endemic, epidemic, and pandemic disease. Each dot represents a disease case or outbreak. *(a)* Endemic diseases are present in the population in specific geographical areas. *(b)* Epidemic diseases show high incidence in a wider area, usually developing from an endemic focus. *(c)* Pandemic diseases are distributed worldwide.

sporadic or endemic disease. Diseased individuals that show no symptoms or only mild symptoms are said to have *subclinical infections*. Subclinically infected individuals are frequently **carriers** of the particular pathogen, with the pathogen reproducing within them and being shed into the environment where it can infect others. Finally, the term **virulence**, often used in epidemiological parlance, is a measure of the *relative ability* of a pathogen to cause disease. Some pathogens are highly virulent while others are only weakly so (Section 25.3).

Stages of Disease

A well-adapted pathogen lives in balance with its host, taking what it needs for existence and causing only minimal harm. Such pathogens may cause **chronic infections** (long-term infections) in the host. When there is a balance between host and pathogen, both host and pathogen survive. Tuberculosis (Section 30.4) is a good example of a chronic infection. On the other hand, a host whose resistance is compromised because of factors such as poor diet, age, and other stressors can be harmed or even killed; for example, a chronic tuberculosis infection can eventually kill the host.

New pathogens occasionally emerge to which specific populations or even an entire species has not developed resistance. Such emerging pathogens often cause **acute infections**, characterized by rapid and dramatic disease onset and a relatively quick return to health. Influenza caused by a new strain of influenza virus (Section 30.8) would be an example of an acute infection, as would many other infectious diseases that show a rapid onset and recovery, such as various food infections and food poisonings (Chapter 32), or even the common cold (Chapter 30). The progression of clinical symptoms for an acute infectious disease can be divided into stages, and the terms used to describe these stages are also part of the epidemiologist's lexicon:

1. *Infection:* The organism invades, colonizes, and grows in the host.
2. *Incubation period:* Some time always passes between infection and the appearance of disease signs and symptoms. Some diseases, like influenza, have very short incubation periods, measured in days; others, like AIDS, have longer ones, sometimes extending for years. The incubation period for a given disease is determined by the inoculum size, the virulence and life cycle of the pathogen, and the resistance of the host. At the end of the incubation period, the first signs and symptoms, for example, a mild cough and a feeling of general fatigue in the case of an ensuing cold, usually appear.

3. *Acute period:* The disease is at its height, with overt symptoms and signs such as fever and chills.

4. *Decline period:* Disease signs and symptoms subside. As fever subsides, usually following a period of intense sweating, a feeling of well-being develops. The decline period may be rapid (within one day), in which case decline occurs by *crisis*, or it may be slower, extending over several days, in which case decline occurs by *lysis*.

5. *Convalescent period:* The patient regains strength and returns to the normal healthy state.

After the acute period, the immune mechanisms of the host (Chapters 26 and 27) become increasingly important for complete recovery from the disease.

Mortality, Morbidity, and DALY

The terms *morbidity* and *mortality* are commonly used in epidemiology. **Mortality** is the incidence of *death* in a population. Infectious diseases were the major causes of death worldwide in 1900, but they are now less prevalent in developed countries. Noninfectious "lifestyle" diseases such as heart disease and cancer are now much more prevalent in developed regions and cause higher mortality than do infectious diseases (Figure 1.8). However, this could change rapidly if public health measures were to break down. Worldwide, and especially in developing countries, infectious diseases are still major causes of mortality (**Table 29.1** and see Figure 29.9).

Morbidity is the incidence of *disease* in a population and includes both fatal and nonfatal diseases. Morbidity statistics indicate the public health of a population more precisely than mortality statistics because many diseases have relatively low mortality. Put another way, the major causes of *illness* are quite different from the major causes of *death*. For example, high-morbidity infectious diseases include acute respiratory diseases such as the common cold and acute digestive disorders. However, seldom do these diseases cause death in populations living in developed countries. Thus, both of these diseases have high morbidity, but low mortality. On the other hand, Ebola virus infects relatively few people worldwide every year, but the mortality in some outbreaks approaches 70% and averaged 40% in the West African Ebola outbreak of 2013–2015. Thus, Ebola has low morbidity, but high mortality.

Epidemiologists tend to focus on morbidity and mortality statistics as a means of ranking the severity of pathogens and tracking disease trends. However, illness and death are not the only outcomes of an infectious disease. Lost among these statistics is the reduction in life quality and productivity due to a disease. The **disability-adjusted life year (DALY)** is a quantitative measure of disease burden and is defined as the cumulative number of

TABLE 29.1 Worldwide deaths due to infectious diseases[a]

Disease	Deaths (% of deaths from all infectious diseases)	Causative agent(s)
Respiratory infections[b]	31	Bacteria, viruses, fungi
Diarrheal diseases	15	Bacteria, viruses
Acquired immunodeficiency syndrome (AIDS)	13	Virus
Tuberculosis[c]	15	Bacterium
Malaria	6	Protist
Measles[c]	3	Virus
Meningitis, bacterial[c]	2	Bacterium
Pertussis (whooping cough)[c]	2	Bacterium
Tetanus[c]	1	Bacterium
Hepatitis (all types)[d]	1	Viruses
Other communicable diseases	11	Various agents

[a]Data show the ten leading causes of death due to infectious diseases and are representative of recent years. Worldwide in 2012 there were 56 million total deaths and 32% of these were from infectious diseases, nearly all in developing countries. In the United States in 2012, deaths from infectious diseases were about 4% of total deaths (influenza, pneumonia and septicemia were leading causes). Data adapted from data published by the World Health Organization (WHO) and the Centers for Disease Control and Prevention (CDC), Atlanta, Georgia (USA).
[b]For some acute respiratory agents such as influenza and *Streptococcus pneumoniae* there are effective vaccines; for others, such as colds, there are no vaccines.
[c]Diseases for which effective vaccines are available.
[d]Vaccines are available for hepatitis A virus and hepatitis B virus. There are no vaccines for other hepatitis agents.

years lost due to an illness itself, a disability due to an illness (whether an infectious disease or not), or premature death.

The leading causes of death are not the leading causes of disability; about one-third of all disability years lost are due to psychiatric and neurological conditions. But many infectious diseases cause chronic disability and thus such data are important measures of the overall burden of disease. This is especially true of a series of *neglected tropical diseases*, a group of infectious diseases found mainly in tropical countries that are major disablers rather than killers. These include in particular parasitic worm infections such as hookworm, filariases, and schistosomiasis (⟿ Section 33.7). Hundreds of millions of people suffer from these infections worldwide, and although some die, most do not. However, life quality and longevity of survivors is oftentimes greatly diminished, and DALY numbers attempt to quantify this often overlooked but nevertheless important aspect of epidemiological statistics.

With some of the epidemiologist's common lingo in mind, we are now able to move on to consider how infectious diseases spread (or do not spread) in susceptible populations.

--- MINIQUIZ ---

- Why do epidemiologists acquire population-based data about infectious diseases?

- Distinguish between an endemic disease, an epidemic disease, and a pandemic disease.

- Which is more severe, a disease with a high mortality or one with a high morbidity? What is a DALY?

29.2 The Host Community

The colonization of a susceptible host population by a pathogen may lead to explosive infections, transmission to uninfected hosts, and an epidemic. As the host population develops resistance, however, the spread of the pathogen is checked, and eventually a balance is reached in which host and pathogen populations reach a state of equilibrium. In an extreme case, failure to reach equilibrium could result in death and eventual extinction of the host species. If the pathogen has no other host, then the extinction of the host also results in extinction of the pathogen. The evolutionary success of a pathogen thus depends on its ability to establish an equilibrium with its host rather than destroy the host population altogether. In most cases, the evolution of the host and the pathogen affect one another; that is, the host and pathogen *coevolve*.

Coevolution of a Host and a Pathogen

A classic example of host and pathogen coevolution is a case where myxoma virus was intentionally introduced to control an exploding wild rabbit population in Australia. The virus, spread by the bite of mosquitoes and also from animal to animal by direct contact, is extremely virulent for rabbits and causes fatal infections in susceptible animals. Within several months, the infection had spread over a large area, rising to peak incidence in the summer when the mosquito vectors were present, and then declining in the winter as mosquitoes disappeared. In this experiment, over 95% of the infected rabbits died during the first year, but within six years, wild rabbit mortality dropped to about 30%, indicating that the resistance of the wild rabbit population had increased dramatically (**Figure 29.3**). When virus isolated from these wild rabbits was used to infect laboratory rabbits that had not previously been

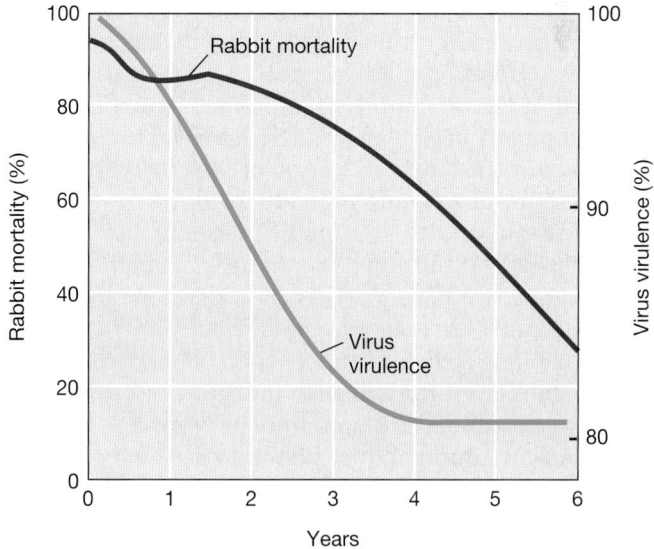

Figure 29.3 Myxoma virus and host coevolution. Myxoma virus was introduced into Australia to control the wild rabbit population. Virus virulence was measured as the average mortality in laboratory rabbits for infection with myxoma virus recovered from the field each year. Rabbit mortality was determined by removing young wild rabbits from dens and infecting them with a viral strain that killed 90–95% of control laboratory rabbits.

exposed to the virus, the virus could be seen to have lost virulence over the six-year period. This was further confirmed by the resistance observed in newborn wild rabbits exposed to the virus. Within three years, mortality of wild rabbits decreased by over 80% and maintained this resistance at a constant level (Figure 29.3). Thus, within just a few years, the rabbit population had evolved to reach an equilibrium with the pathogen.

For pathogens that do not exhibit host-to-host transmission, there is no selection for decreased virulence to support mutual coexistence, as was seen in the rabbit myxoma virus experiment. An example of this is *Clostridium tetani*, a common soil bacterium that causes tetanus when accidently introduced into flesh through a penetrating wound (Sections 25.6 and 31.9). Vector-borne pathogens transmitted solely by the bite of ticks or other arthropods, such as in spotted fever rickettsiosis (Rocky Mountain spotted fever, Section 31.3), are also under no evolutionary pressure to spare the human host. As long as the vector is only a *carrier* of the pathogen and does not contract the disease itself, there is no selection for weakened strains of the pathogen and thus the pathogen can maintain a high level of virulence.

Herd Immunity

Spread of an infectious disease through a highly susceptible population is typically much different than through a population where many, or even just some, potential hosts are immune, either from a previous natural infection with the same pathogen or by artificial means through vaccination. If a high enough proportion of the individuals in a population are immune to a pathogen, then the whole population can be protected, resulting in a collective level of resistance to infection called **herd immunity** (**Figure 29.4**).

The concept of herd immunity is easy to understand. In essence, what herd immunity amounts to is a breakage in the chain of pathogen transmission from one susceptible host to another because most hosts in the population are immune (Figure 29.4). Herd immunity is not a fixed number, and the assessment of herd immunity is important for understanding the development of epidemics. The more highly infectious a pathogen, or the longer its period of infectivity, the greater the proportion of immune individuals necessary to prevent epidemic disease spread. For a highly infectious disease such as measles, 90–95% of the population must be immune to confer herd immunity (see Table 29.3). By contrast, a lower proportion of immune individuals can prevent an epidemic of a less infectious agent or one with only a brief period of infectivity. Mumps virus, which is less infectious than measles virus, exhibits this pattern. In the absence of immunity, even poorly infectious agents can be transmitted from person to person if susceptible hosts have repeated or constant contact with an infected individual. This is the case for the transmission of H5N1 avian influenza among humans (Section 29.8).

MINIQUIZ

- Explain coevolution of host and pathogen. Cite a specific example.
- How does herd immunity prevent a nonimmune individual from acquiring a disease? Give an example.

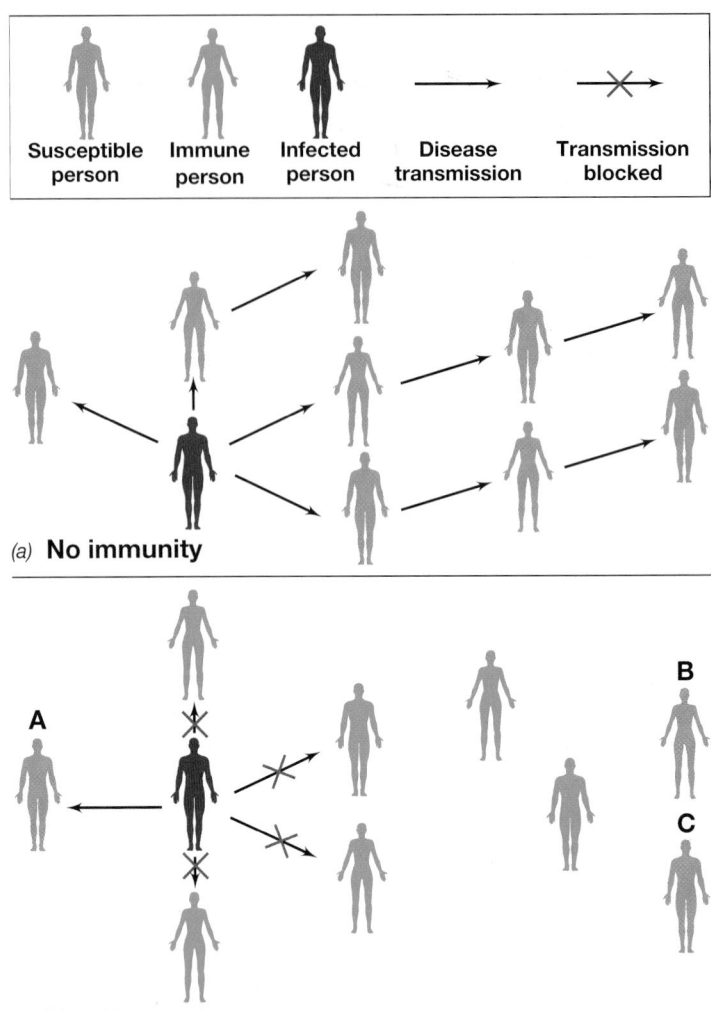

Figure 29.4 Herd immunity and transmission of infection. Immunity in some individuals protects individuals without immunity from infection. *(a)* In a population with no immunity, transfer of a pathogen from one infected individual can ultimately infect (arrows) all the individuals as newly infected individuals in turn transfer the pathogen to other individuals. *(b)* In a population that is only moderately dense and that has some immunity against a moderately transmissible pathogen such as influenza, an infected individual cannot transfer the pathogen to all susceptible individuals because resistant individuals, immune from previous exposure or immunization, break the cycle of pathogen transmission: Susceptible individual A becomes infected, but susceptible individuals B and C are protected. The proportion of a population that must be immune for herd immunity to be effective also varies with the disease; highly infectious diseases require a higher proportion of immune individuals for herd immunity to prevent transmission (see Table 29.3).

29.3 Infectious Disease Transmission and Reservoirs

Epidemiologists follow the transmission of a disease by correlating geographic, climatic, social, and demographic data with disease incidence. These correlations are then used to identify possible modes of transmission and disease patterns. Epidemiologists group infectious diseases by their *mode of transmission*. This approach reflects the ecology of the organism and is the pattern we will use in Chapters 30–33.

TABLE 29.2 Major means of human infectious disease transmission

Mode of transmission	Examples
Person to person	
Direct contact: Sexual intercourse; handshakes	Gonorrhea; *Staphylococcus aureus* infection
Indirect contact: Water glasses and other fomites	Influenza; common cold
Airborne droplets: Sneezes, coughing	Influenza; measles; tuberculosis
Vehicle	
Waterborne: Sewage-contaminated water	Cholera; giardiasis
Foodborne: Contaminated foods	Staphylococcal food poisoning; salmonellosis
Airborne: Fungal spores	Histoplasmosis; coccidioidomycosis
Soilborne: Puncture wound contaminated with soil	Tetanus
Soil aerosol or infected animal	Anthrax
Vector	
Arthropods/insects: Mites, ticks, mosquitoes	Typhus; Lyme disease; malaria

Modes of Disease Transmission

Three major modes of infectious disease transmission are known and are summarized in **Table 29.2**. These include diseases transmitted from *person to person*; diseases transmitted by some inanimate object or substance, called a *vehicle*; and diseases transmitted by *vectors*, that is, other organisms, especially those that access the bloodstream, such as ticks and biting insects. Each mechanism has three stages in common: (1) escape from the host or reservoir, (2) travel, and (3) entry into a new host.

Person-to-person disease transmission occurs when an infected host transmits a disease directly to a susceptible host without the assistance of an intermediate host or inanimate object. Upper respiratory infections such as the common cold and influenza are most often transmitted person to person by droplets resulting from sneezing or coughing. Many of these droplets, however, do not remain airborne for long, and so transmission requires close, although not necessarily intimate, person-to-person contact. Some pathogens are extremely sensitive to environmental factors such as drying and heat and are unable to survive for significant periods of time away from the host. These pathogens, transmitted only by intimate person-to-person contact such as exchange of body fluids in sexual intercourse, include those responsible for sexually transmitted diseases including syphilis (*Treponema pallidum*), gonorrhea (*Neisseria gonorrhoeae*), and HIV/AIDS (HIV, human immunodeficiency virus; AIDS, acquired immunodeficiency syndrome). Direct person-to-person contact is also how pathogens such as staphylococci (boils and pimples) and fungi (ringworm) are transmitted. Some of these pathogens (*Staphylococcus aureus* is a good example) can spread by vehicle transmission as well because when inoculated into a vehicle such as food, they grow rapidly and produce poisonous toxins.

A marked seasonality or periodicity of a disease often signals a particular mode of transmission. For example, human influenza occurs in an annual cyclic pattern, causing epidemics propagated among schoolchildren and other populations of susceptible individuals. Cases of influenza are often high in schools or crowded offices because the virus is transmitted person to person by the respiratory route; peak incidence occurs in midwinter and early spring when schools are in session and people are indoors much of the day. However, seasonality can also result from environmental factors such as weather patterns that influence the survival of the pathogen or its vector. For example, California encephalitis—a viral disease transmitted by mosquitoes—shows a pattern opposite that of influenza; the disease peaks during the summer and fall months but disappears in the winter, coinciding with the activity of its mosquito vector (**Figure 29.5**).

Diseases are often transmitted to humans by other organisms and by inanimate objects. Living disease carriers are called **vectors**, and arthropods (mites, ticks, or fleas) and vertebrates (dogs, cats, or rodents) are common disease vectors. Vectors are often not definitive hosts for the pathogen but simply carry the pathogen from one host to another. For instance, many arthropods obtain their nourishment by biting and sucking blood, and if the pathogen is present in the blood, the arthropod will ingest the pathogen and transmit it when biting another individual. In some cases viral pathogens multiply in the arthropod vector, which is then considered an *alternate host*. Such is the case for West Nile virus (in the *Culex* mosquito) and the bacterium *Yersinia pestis* (in the rat flea), the causative agent of plague (⟲ Sections 31.6 and 31.7). Such

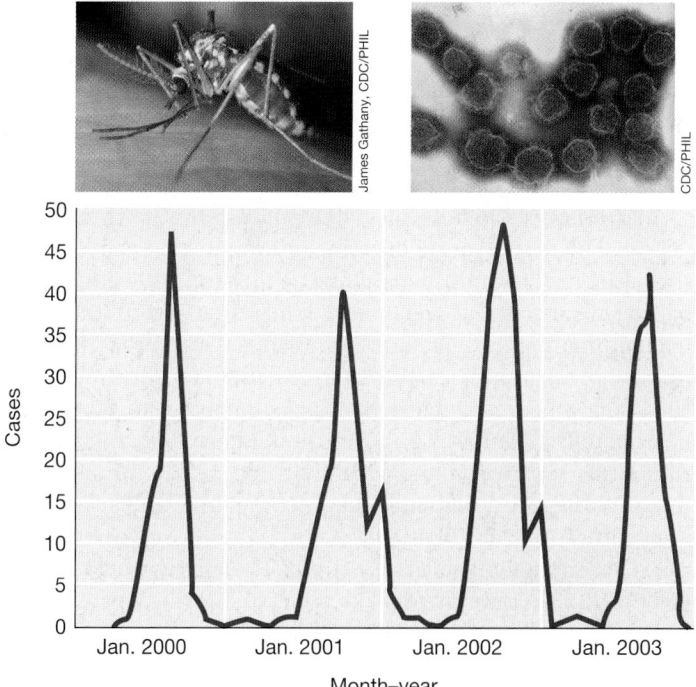

Figure 29.5 Cyclical nature of California encephalitis in the United States. California encephalitis is transmitted by the mosquito *Ochlerotatus triseriatus* (left photo) and is caused by the La Crosse encephalitis virus, a negative-strand and enveloped RNA virus (right photo). Because it depends on a seasonally available vector, disease incidence shows a sharp rise in late summer, followed by a complete decline in winter. Data are from the CDC, Atlanta, Georgia, USA.

UNIT 7

replication leads to greater pathogen abundance in the vector, and this increases the probability that a subsequent bite will lead to infection.

Inanimate agents such as bedding, toys, books, and surgical instruments can also transmit disease. Inanimate objects that, when contaminated with a viable pathogen, can transfer the pathogen to a host are called **fomites**. The term **vehicle** is used to describe nonliving sources of pathogens that, upon entering the body, may transmit disease to large numbers of individuals; common disease vehicles are contaminated food or water (Table 29.2). A key distinction here is that fomites are nonliving objects that are touched or handled by a limited number of individuals, whereas vehicle-source epidemics are typically traced to contaminated food or water—shared commodities consumed in large amounts by local or regional populations.

Disease Carriers and Disease Reservoirs and Control

As described earlier, a disease *carrier* is a pathogen-infected individual who has a subclinical infection and shows either no symptoms or only mild symptoms of the disease; carriers are thus potential sources of infection for others. Carriers may be in the incubation period of the disease, in which case the carrier state precedes the development of actual symptoms (Section 29.1). Respiratory infections such as colds and influenza, for example, are often spread via carriers who are unaware of their infection and so are not taking any precautions against infecting others. The carrier state lasts only a short time for carriers who develop acute disease. However, *chronic carriers* usually appear healthy and may spread disease for extended periods of time. Some examples here include carriers of hepatitis B, typhoid fever, HIV/AIDS, tuberculosis, and upper respiratory *Staphylococcus aureus* infections.

Disease reservoirs are sites at which infectious agents remain viable and from which individuals may become infected. Reservoirs may be either animate or inanimate. Some pathogens whose reservoirs are not in animals only incidentally infect humans and cause disease. For example, some species of *Clostridium*, common soil bacteria, occasionally infect humans, causing life-threatening diseases such as tetanus, botulism, and gangrene. In these cases, the pathogen is not dependent on the host for survival, so host–pathogen balance is not required. For many pathogens (including many human pathogens) however, living organisms are the only reservoirs. In these cases, the host is essential for the life cycle of the infectious agent; maintenance of human pathogens of this kind requires host-to-host transmission. Many viral and bacterial respiratory pathogens and sexually transmitted pathogens fall into this category. When humans are the main or only disease reservoir, infection control may be easy or not so easy. With diphtheria, for example, confirmed cases must be isolated and quarantined (Section 29.5). However, for a disease like gonorrhea, where inapparent symptoms are common in females, tracking down and treating disease carriers can be difficult if not impossible.

Some infectious diseases are caused by pathogens that reproduce in both humans and animals. A disease that primarily infects animals and is only occasionally transmitted to humans is called a **zoonosis**; rabies is a good example. The reservoir for rabies is wild mammals, primarily skunks, raccoons, foxes, and certain bats. Although person-to-person transmission of zoonoses is rare, control of zoonoses in humans is nearly impossible because of the frequent contact some humans have with wild animals and the fact that the animal reservoir can probably never be effectively controlled. Certain other infectious diseases are caused by organisms such as protists and helminths (parasitic worms) that undergo complex life cycles including an obligate transfer from a nonhuman host to a human host back to the nonhuman host; the diseases malaria and schistosomiasis (Chapter 33) are good examples here. In the case of malaria, the major reservoir other than humans is the mosquito *Anopheles gambiae*, and some control of the disease can be achieved by chemical or physical controls on the insect reservoir. In schistosomiasis, by contrast, the reservoir is an aquatic snail and therefore although treatments for the disease are possible, eliminating the reservoir is not an option.

MINIQUIZ

- What is a zoonotic disease? A disease reservoir?
- What is the difference between a disease vehicle and a disease vector?
- Why will a disease such as human rabies likely never be eliminated?

29.4 Characteristics of Disease Epidemics

Endemic infectious diseases are constantly present over long periods of time but typically occur at only low incidence in the

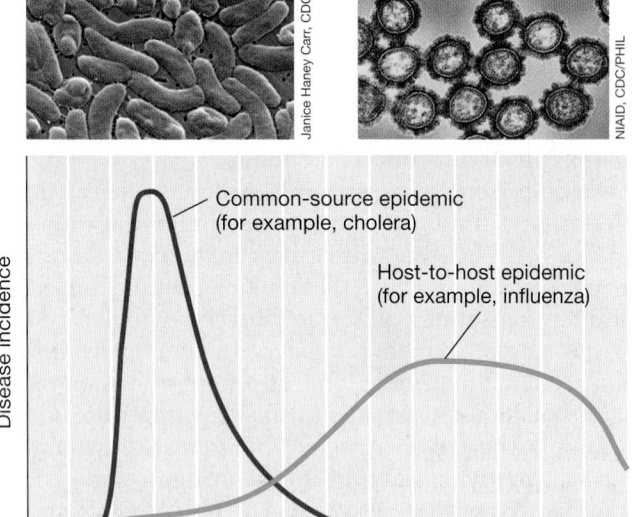

Figure 29.6 Types of epidemics. The shape of the curve that plots incidence of an epidemic disease against time identifies the likely type of the epidemic. For a common-source epidemic, such as cholera resulting from contaminated water shared by many people, the curve rises sharply to a peak and then declines rapidly. Host-to-host infectious disease incidence rises relatively slowly as new cases accumulate. Inset photos: left, scanning electron micrograph of a *Vibrio* sp. closely related to *Vibrio cholerae*, the cholera agent; right, transmission electron micrograph of virions of H1N1 influenza virus.

TEXTBOOK EPIDEMIOLOGY: THE SARS EPIDEMIC

Handling of the *severe acute respiratory syndrome* (SARS) epidemic early in this century is an excellent example of the successful application of the principles of epidemiology. Like many other rapidly emerging diseases, SARS is viral and originated in animals. Such characteristics have the potential to trigger explosive disease in humans when the infectious agents cross host species barriers.

The SARS epidemic began in late 2002 in China, and by early 2003, the virus had spread to 28 countries, primarily via international travelers. The cause of SARS was quickly traced to a coronavirus coined SARS-CoV (**Figure 1**) that most probably originated in bats. From bats, the virus infected civets (a small, nocturnal, catlike animal used as a food source in China) and from there jumped to humans.

Much like common cold viruses, SARS-CoV is a relatively hardy, very infectious RNA virus that is difficult to contain (R_0 of 3.6). Once in humans, SARS-CoV quickly spreads from person to person by sneezing and coughing or by contact with contaminated fomites or feces. Ordinarily, a new coldlike virus would be of little concern, but SARS-CoV caused infections with significant mortality. From the 8500 known SARS-CoV infections, there were over 800 deaths, for an overall mortality rate approaching 10%. In persons over 65 years of age, the mortality rate approached 50%, attesting to the virulence of SARS-CoV as a human pathogen.

About 20% of all SARS cases were in healthcare workers, demonstrating the high infectivity of the virus. Standard containment and infection-control methods practiced by healthcare personnel were not effective in controlling spread of the disease. As a result, SARS patients were placed in strict isolation in negative-pressure rooms, and healthcare staff caring for SARS patients wore respirators when working with patients or when handling fomites (bed linens, eating utensils, and so on) to prevent infection.

The recognition and containment of the SARS outbreak was the start of an international response that included clinicians, scientists, and public officials. Almost immediately, travel to and from the endemic area was restricted, limiting further outbreaks. SARS-CoV was quickly isolated and cultured, and its genome was rapidly sequenced; this allowed simple PCR tests to be developed to detect the virus in samples. As laboratory work progressed, epidemiologists traced the virus back to the civet food source in China and stopped further transmission to humans by restricting the sale of civets and other foods from wild sources. These actions collectively stopped the outbreak.

As international travel and trade expand, the chances for propagation and rapid dissemination of new exotic diseases will increase. SARS is an example of a serious infection that emerged very rapidly from a unique source. However, rapid identification and characterization of the SARS pathogen, nearly instant development of worldwide notification procedures and diagnostic tests, and a concerted effort to understand the biology and genetics of this novel pathogen controlled the disease; there have been no cases of

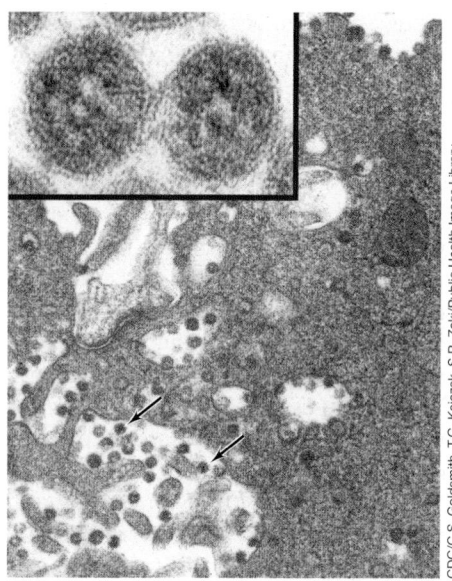

Figure 1 Severe acute respiratory syndrome coronavirus (SARS-CoV). The upper left panel shows isolated SARS-CoV virions. An individual virion is 128 nm in diameter. The large panel shows coronaviruses (arrows) within cytoplasmic membrane–bound vacuoles and in the rough endoplasmic reticulum of host cells. The virus replicates in the cytoplasm and exits the cell by way of the cytoplasmic vacuoles.

CDC/C.S. Goldsmith, T.G. Ksiazek, S.R. Zaki/Public Health Image Library

SARS since early 2004. The rapid emergence of SARS, and the equally rapid and successful international effort to identify and control the outbreak, provide a textbook example of the explosive nature of some emerging pathogens and how control of emerging epidemics is possible if strict disease surveillance and infection control are practiced.

---

population. In tropical Africa, for example, malaria is endemic; both morbidity and mortality from malaria have remained relatively constant on a long-term basis. The 2014 outbreak of Ebola hemorrhagic fever in West Africa, by contrast, ran to epidemic proportions in major districts of Sierra Leone, Liberia, and Guinea.

Disease epidemics show characteristic features (**Figure 29.6**) and require that rapid epidemiological conclusions be reached and clinical treatment instituted if the epidemic is to be contained. The characteristic features of epidemics include distinct patterns in the disease cycle and inherent properties of the pathogen that affect its virulence and herd immunity. A good example is the 2003 SARS epidemic that began in Asia (see Explore the Microbial World, "Textbook Epidemiology: The SARS Epidemic"). The SARS *epidemic* could potentially have grown to a SARS *pandemic* but was

rapidly suppressed by a combination of quick and effective epidemiological and clinical action.

Epidemics

Major epidemics are usually classified as either *common-source epidemics* or *host-to-host epidemics*. The patterns of disease incidence observed in these two types of epidemics are contrasted in Figure 29.6.

A **common-source epidemic** results from an infection (or intoxication) of a large number of people from a contaminated source such as food or water that all infected individuals have ingested. Such epidemics are often caused by a breakdown in the sanitation of a central food or water distribution system, but they can also be more local, such as contaminated food in a particular restaurant. Foodborne and waterborne common-source epidemics are primarily intestinal diseases; the pathogen leaves the body in fecal material, contaminates food or water supplies as a result of improper sanitation, and then enters the intestinal tract of the recipient during ingestion of the food or water (Chapter 32).

Common-source disease outbreaks are characterized by a rapid rise to a peak incidence because a large number of individuals become ill within a relatively brief period of time (Figure 29.6). Moreover, assuming that epidemiological surveillance quickly identifies the disease vehicle, cases of a common-source disease declines fairly rapidly, as well. Cholera is the classic example of a common-source epidemic as the disease is almost exclusively waterborne; if a sanitation breakdown occurs (or if sanitation is totally lacking, as is often the case in developing countries), the cholera bacterium can be shed from a carrier or an active infection into a water source used by many other people and quickly trigger an epidemic (Figure 29.6).

In contrast to the common-source disease pattern, in a **host-to-host epidemic** the disease incidence shows a relatively slow, progressive rise (Figure 29.6) and a gradual decline. Cases continue to be reported over a period of time equivalent to several incubation periods of the disease. A host-to-host epidemic can be initiated by the introduction of a single infected individual into a susceptible population, with this individual infecting one or more people depending on the extent of herd immunity (Figure 29.4) in that population. In a host-to-host epidemic, the pathogen replicates in susceptible individuals, reaches a communicable stage, is transferred to other susceptible individuals, and again replicates and becomes communicable; such epidemics are often controlled by effective herd immunity due to previous infection or vaccination. Influenza and chicken pox (Chapter 30) are examples of diseases that can spread in host-to-host epidemics.

Basic Reproduction Number (R_0)

The infectivity of a pathogen can be predicted using mathematical models that estimate the **basic reproduction number (R_0)** that the pathogen may trigger. The R_0 is defined as the number of expected secondary transmissions from each single case of a disease in an entirely susceptible population, and Table 29.3 lists the R_0 of selected infectious diseases. R_0 directly correlates with the herd immunity necessary to prevent spread of infection; the higher the R_0 value, the greater the herd immunity required to stop infection (Table 29.3). Unfortunately, conditions are not always ideal and the mathematical models that predict R_0 may not take into account such factors as numbers of recovered individuals, population density (close contact), length of contact time, populations of high-risk individuals, and other variables that may affect disease spread. As a result, R_0 is a theoretical construct and can only estimate infectivity. Nevertheless, R_0 is still useful as a gauge of the relative infectivity of a pathogen and helps to establish targets for immunization coverage to prevent spread of a particular infectious disease.

The *observed reproduction number, R*, calculated from studies of actual disease spread is a more empirical term because it takes into account observed transmissions from infected to susceptible individuals. Gathering the epidemiological data necessary to calculate an accurate R is often problematic, but for some diseases an empirical reproduction number has been obtained. For example,

TABLE 29.3 Basic reproduction number (R_0) and herd immunity necessary for community protection from selected infectious diseases

Disease	[a]R_0	Herd immunity[a]
Diphtheria	7	85%
Ebola	1.8	—
Influenza[b]	1.6	29%
Measles	18	94%
Mumps	7	86%
Pertussis	17	94%
Polio	7	86%
Rubella	7	85%
SARS-CoV	3.6	—
Smallpox	7	85%

[a]R_0 and herd immunity values are the highest estimates for each disease. Herd immunity values are shown only for those diseases for which vaccines are available.
[b]Values shown are for the pandemic (H1N1) 2009 influenza. Each influenza epidemic has a different R_0 and herd immunity value. Herd immunity values assume a 100% effective vaccine. Vaccine efficacy for influenza is about 60% and observed herd immunity values are 40% or greater depending on the susceptible host populations.

for the SARS epidemic of 2003 (see Explore the Microbial World, "Textbook Epidemiology: The SARS Epidemic"), the observed R was 3.6, matching its R_0 value (Table 29.3). Public health officials, recognizing the potential for a serious epidemic, instituted major infection controls including isolation of infected individuals and strict barrier protection for healthcare personnel. These measures reduced the SARS R value to 0.7 and ended the threat of further disease spread.

MINIQUIZ

- Distinguish between direct and indirect transmission of disease. Cite at least one example of each.
- By using epidemiological surveillance data, how can a common-source epidemic be recognized?
- Define the basic reproduction number for a pathogen.

II • Epidemiology and Public Health

In Part II, we focus on public health issues including some of the methods and tools used to identify, track, contain, and eradicate infectious diseases within populations. We also draw a stark contrast between the causes of mortality in developed versus developing countries. Although noninfectious diseases are the major killer in developed countries, infectious disease remains the leading cause of mortality in other countries.

29.5 Public Health and Infectious Disease

Public health refers to the health of the general population and to the activities of public health authorities in the control of disease. The incidence and prevalence of many infectious diseases dropped dramatically during the twentieth century, especially in developed countries, because of universal improvements in public health from advances in basic living conditions. Access to safe water and food, improved public sewage treatment, less crowded living conditions, and lighter workloads have all contributed immeasurably to disease control. Several historically important diseases, including smallpox, typhoid fever, diphtheria, brucellosis, and poliomyelitis, have been controlled (and in the case of smallpox, even eliminated) by active, disease-specific public health measures, and we review these here.

Controls Directed against Common Vehicles and Major Reservoirs

Common vehicles for pathogen dispersal include food, water, and air. The control of foodborne and waterborne pathogens (Chapter 32) has seen the greatest successes through improved methods of preventing microbial contamination of food and water. For example, water purification methods have dramatically reduced the incidence of typhoid fever (**Figure 29.7**), and laws controlling food purity, preparation, and storage coupled with strict monitoring of the food and water distribution network have greatly decreased the incidence of common-source disease. However, in contrast to food and water, controlling transmission of respiratory (airborne) pathogens is much more difficult. Other than wearing personal protection such as face masks and avoiding individuals you know are infected, few effective measures of airborne infection control are possible except in specialized environments such as hospital operating rooms where chemical and physical agents can treat the rather small amount of circulating air.

When the disease reservoir is primarily *domestic* animals, infection of humans can be prevented if the disease is eliminated from the infected animal population by vaccinating herds and removing diseased individuals. However, as we have seen (Section 29.3), when the disease reservoir is a *wild* animal, eradication is much more difficult. Eradication of rabies, for example, would require

the immunization or destruction of all wild animal reservoirs, a virtually impossible task. When insect vectors are involved, effective control can often be accomplished with insecticides. However, the use of chemicals must be balanced with health and environmental concerns because in some cases, the elimination of one public health problem (the disease vector) simply creates another (toxic chemical exposure).

When humans are the disease reservoir—as, for example, in HIV/AIDS—control and eradication can be difficult, especially, as mentioned previously in reference to gonorrhea, if there are asymptomatic carriers. By contrast, certain diseases that are limited to humans and have no asymptomatic phase can be prevented through immunization or treatment with antimicrobial or other drugs. However, the disease can be eradicated only if those who have contracted the disease and all possible contacts are immunized, treated, or if necessary, quarantined. Such a strategy was successfully employed by the World Health Organization to eradicate smallpox worldwide (see later) and is currently being used to eradicate polio.

Immunization

Smallpox, diphtheria, tetanus, pertussis (whooping cough), measles, mumps, rubella, and poliomyelitis have been controlled primarily by immunization. Diphtheria, for example, is no longer considered even endemic in the United States. Vaccines are routinely administered in childhood for a number of other infectious diseases (Figure 28.25). As we discussed in Section 29.4, 100% immunization is not necessary for effective disease control in a population because of herd immunity, although the percentage needed to ensure disease control varies with the infectivity and virulence of the pathogen (Table 29.3).

Measles epidemics offer an example of the power of herd immunity. The occasional resurgence of the highly contagious measles virus ($R_0 = 18$, Table 29.3) emphasizes the importance of maintaining appropriate immunization levels for a given pathogen. Until 1963, the year an effective measles vaccine was licensed, nearly every child in the United States acquired measles through natural infections, resulting in over 300,000 annual cases. However, after introduction of the vaccine, the number of annual measles infections decreased rapidly (**Figure 29.8**). Case numbers reached a low of 1497 by 1983. However, by 1990, the percentage of children immunized against measles fell to 70%, and the number of new cases rose to 27,786. A concerted effort to increase measles immunization levels to above 90% (about that needed for effective herd immunity, Table 29.3) virtually eliminated measles in the United States.

Isolation, Quarantine, and Surveillance

Isolation and quarantine are effective public health measures. **Isolation** is the separation of persons who have an infectious disease from those who are healthy. **Quarantine** is the separation and restriction of well persons who may have been exposed to an infectious disease to see if they develop the disease. The length of isolation or quarantine for a given disease varies and is typically the longest period of communicability for that disease. To be effective, these measures must prevent infected or potentially infected individuals from contacting uninfected susceptible individuals. By international agreement, six infectious diseases

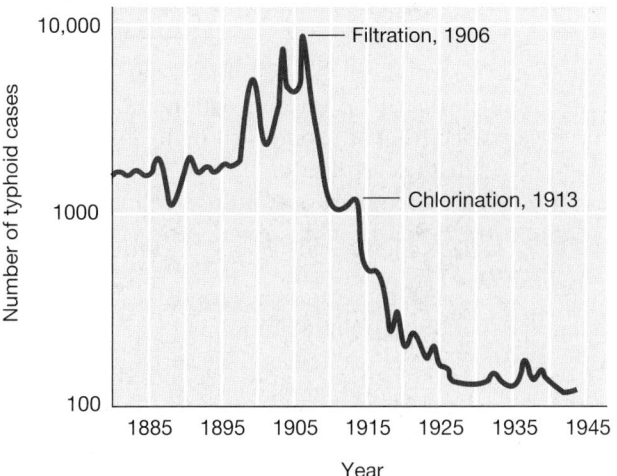

Figure 29.7 Historical progression of typhoid fever in Philadelphia. The introduction of filtration and chlorination eliminated typhoid fever in Philadelphia and other cities with well-regulated water supplies. The risk of typhoid in the United States today is very low but occasional cases are reported.

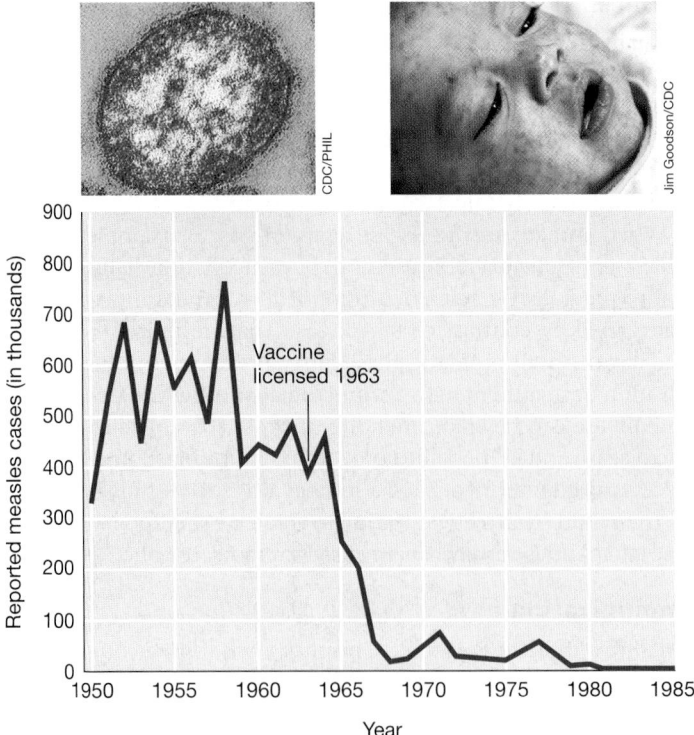

Figure 29.8 Measles immunization in the United States. The introduction of a measles vaccine eliminated measles as a common childhood infection within 20 years. Inset photos: left, transmission electron micrograph of a measles virion (a negative-strand enveloped RNA virus); right, photo of an infant showing the spotted rash characteristic of measles. With an extremely high R_0 (Table 29.3), measles outbreaks can spike quickly in unvaccinated populations.

TABLE 29.4 Reportable infectious agents and diseases in the United States, 2016

Diseases caused by bacteria

Anthrax	Q fever
Botulism	Salmonellosis
Brucellosis	Shiga toxin–producing
Chancroid	*Escherichia coli* (STEC)
Chlamydia trachomatis infection	Shigellosis
Cholera	Spotted fever rickettsiosis
Diphtheria	Streptococcal toxic shock syndrome
Ehrlichiosis/Anaplasmosis	*Streptococcus pneumoniae*, invasive disease
Gonorrhea	Syphilis, all stages
Haemophilus influenzae, invasive disease	Tetanus
Hansen's disease (leprosy)	Toxic shock syndrome (staphylococcal)
Hemolytic uremic syndrome	Tuberculosis
Legionellosis	Tularemia
Listeriosis	Typhoid fever
Lyme disease	Vancomycin-intermediate *Staphylococcus aureus* (VISA)
Meningococcal disease (*Neisseria meningitidis*)	Vancomycin-resistant *Staphylococcus aureus* (VRSA)
Pertussis	Vibriosis (non-cholera *Vibrio* infections)
Plague	
Psittacosis	

Diseases caused by viruses

Arboviruses (encephalitis, non-neuroinvasive disease, and Zika)	Rabies
	Rubella
Dengue	Severe acute respiratory syndrome (SARS-CoV)
Hantavirus pulmonary syndrome	Smallpox
Hepatitis A, B, C	Varicella (chicken pox)
HIV infection/AIDS	Viral hemorrhagic fevers
Novel influenza A	West Nile virus
Measles	Yellow fever
Mumps	
Polio	

Diseases caused by protists

Babesiosis	**Disease caused by a helminth**
Cryptosporidiosis	Trichinellosis (trichinosis)
Cyclosporiasis	**Disease caused by a fungus**
Malaria	Coccidioidomycosis/Valley fever
Giardiasis	

require isolation and quarantine: *smallpox, cholera, plague, yellow fever, typhoid fever,* and *relapsing fever.* Each is a very serious, particularly communicable disease. Spread of certain other highly contagious diseases such as Ebola hemorrhagic fever, SARS, H5N1 influenza, and meningitis may also be subject to quarantine or isolation as outbreaks emerge in particular regions.

As mentioned earlier (Section 29.1), disease surveillance is a major job of the epidemiologist. **Table 29.4** lists the infectious diseases currently under surveillance (referred to as *reportable diseases*) in the United States. The **Centers for Disease Control and Prevention (CDC)** is the agency of the United States Public Health Service that tracks disease trends reported by physicians and other health professionals, provides the latest disease information, and forms public policy regarding disease prevention. The CDC operates a number of infectious disease surveillance programs and also carries out surveillance of major noninfectious diseases, such as cancers, heart disease, and stroke. The overall practical goal of disease surveillance is to formulate and implement plans for diagnosis and treatment of infections.

Pathogen Eradication

Concerted disease eradication programs can sometimes completely eradicate an infectious disease and such was the case with naturally occurring smallpox, eradicated worldwide in 1980. Smallpox was a viral disease with a virus reservoir consisting solely of the individuals with acute smallpox infections, and transmission was

exclusively person-to-person through direct contact. Although smallpox cannot be treated once acquired, immunization practices have been very effective. The World Health Organization (WHO) implemented a smallpox eradication plan in 1967. Because of the success of previous vaccination programs, smallpox had already been confined to endemic status in parts of Africa, the Middle East, and the Indian subcontinent. WHO field health workers proceeded to vaccinate everyone in these areas they could locate with the goal of providing either direct or herd immunity (Section 29.2) to the entire population. Each subsequent outbreak or suspected outbreak was targeted by WHO teams that quickly traveled to the outbreak site, quarantined individuals with active disease, and vaccinated all contacts. To break the chain of possible infection, they then immunized everyone who had contact with the contacts, and this aggressive vaccination policy eventually eliminated smallpox.

Several other communicable diseases are candidates for global eradication. Poliomyelitis, like smallpox a viral disease with a human-only reservoir, is on its way to elimination using the same vaccination strategy used against smallpox; in 2014, a total of only 359 cases of polio were reported worldwide. Diseases caused by parasites have also been targeted, including Chagas disease (by treating active cases and destroying the insect vector) and dracunculiasis (by treating drinking water to prevent transmission of *Dracunculus medinensis*, the Guinea helminth parasite). Eradication of certain bacterial diseases is also on the horizon. For example, syphilis is a candidate because the disease is found only in humans and is readily treatable with antibiotics. Diphtheria, caused by the bacterium *Corynebacterium diphtheriae*, could also be eradicated worldwide by application of the strict immunization protocols that have virtually eliminated diphtheria from North America.

MINIQUIZ

- Compare public measures for controlling infectious disease caused by insect vectors and human carriers.
- Outline the steps taken to eradicate smallpox.
- Describe some of the public health activities of the U.S. Centers for Disease Control and Prevention.

29.6 Global Health Comparisons

The World Health Organization (WHO) has divided the world into six geographic regions for the purpose of collecting and reporting health information such as causes of morbidity and mortality. These geographic regions are Africa, the Americas (North America, the Caribbean, Central America, and South America), the eastern Mediterranean, Europe, Southeast Asia, and the western Pacific. Here we compare mortality data from a relatively developed region, the Americas, to those from a developing region, Africa, to emphasize the fact that infectious diseases are still major causes of morbidity and mortality in many regions of the world.

Infectious Disease in the Americas and Africa

Mortality statistics in developed and developing countries are significantly different, as illustrated by a comparison of data from the Americas and from Africa in 2008 when the worldwide population was nearly 6.8 billion. Worldwide, 60.8 million individuals died, giving a mortality rate of 8.8 deaths per 1000 inhabitants per year, and 15.8 million (26%) of these deaths were attributable to infectious diseases. There were 924 million people in the Americas in 2008 and there were 5.6 million deaths, or 6.1 deaths per 1000 persons per year. In Africa, there were 837 million people in 2008 and 14.1 million deaths, or 16.8 deaths per 1000 persons per year. These statistics clearly show differences in overall mortality between developed and developing countries, but a comparative examination of the *causes* of mortality is even more instructive.

Figure 29.9 indicates that infectious diseases caused the most deaths in Africa, whereas in the Americas, noninfectious diseases such as cancer and cardiovascular disease were the leading causes of mortality. In Africa, there were about 6.6 million deaths due to infectious diseases and the life expectancy was 54 years of age. The African death toll due to infectious diseases was 10% of the total deaths in the world. In stark contrast, only 672,000 died of infectious disease in the Americas and the life expectancy was 76 years

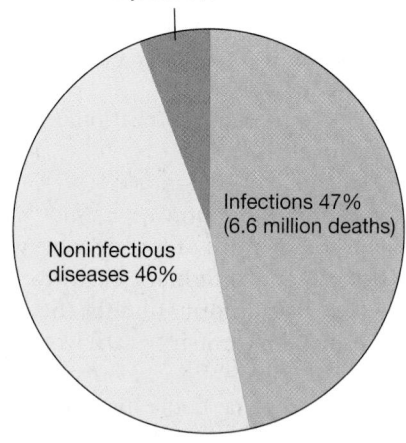

Africa: 14.1 million deaths (16.8/1000 people)

Total population of Africa in 2008: 837 million

Injuries 6%
Infections 47% (6.6 million deaths)
Noninfectious diseases 46%

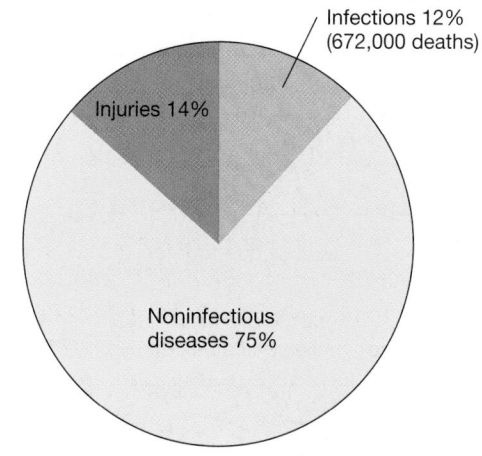

Americas: 5.6 million deaths (6.1/1000 people)

Total population of North, Central, and South America in 2008: 924 million

Infections 12% (672,000 deaths)
Injuries 14%
Noninfectious diseases 75%

Figure 29.9 Causes of death in Africa and the Americas, 2008. Noninfectious diseases include cancer, cardiovascular diseases, and diabetes. Injuries include accidents, murder, suicide, and war. Data are from the World Health Organization, Geneva.

of age. In developed countries, the increased life expectancy is a direct consequence of the reduction in death rates from infection over the last century, and most of these gains are due to the advances in public health. By contrast, lack of resources in developing countries limits access to adequate sanitation, safe food and water, immunizations, healthcare, and medicines, leading to increases in infectious diseases and, as a consequence, to dramatically shorter life expectancy.

Data for 2012 (the most recent year for which complete morbidity and mortality statistics have been compiled by the WHO) show little change in these trends. That is, the majority of deaths in sub-Saharan Africa continue to be due to infectious diseases or to perinatal, maternal, and nutritional causes, whereas in developed countries, lifestyle diseases continue to lead the way, with obesity-related health issues such as type 2 diabetes and lack of mobility rapidly emerging.

Travel to Endemic Areas

The high incidence of disease in many parts of the world is a concern for people traveling to such areas. However, travelers can be immunized against many of the diseases that are endemic in foreign countries. Specific recommendations for immunization for those traveling abroad are updated biannually and published by the U.S. Centers for Disease Control and Prevention (CDC) (http://www.cdc.gov/).

For many countries, immunization certificates for yellow fever are required for entry from areas with endemic yellow fever. These areas include much of equatorial South America and Africa. Most other nonstandard immunizations such as those for rabies and plague are recommended only for people who are expected to be at high risk, such as veterinary healthcare providers. The CDC summarizes current information for the potential for infectious disease transmission throughout the world, including diseases for which currently there are no effective immunizations (for example, HIV/AIDS, malaria, Ebola hemorrhagic fever, dengue fever, amebiasis, encephalitis, and typhus). Travelers should take precautions such as avoiding insect and animal bites, drinking only water that has been properly treated to kill all microorganisms, eating properly stored and prepared food (and avoiding fresh uncooked foods), and undergoing antibiotic and chemotherapeutic programs for prophylaxis or for suspected exposures. Although these precautions do not guarantee that one will remain disease-free, adhering to them greatly reduces the risk of infection.

MINIQUIZ

- Contrast mortality due to infectious diseases in Africa and the Americas.
- List infectious diseases for which you have not been immunized and with which you could come into contact next year.

III • Emerging Infectious Diseases, Pandemics, and Other Threats

In recent years, new infectious diseases have emerged and established diseases have reemerged with alarming frequency. In Part III of this chapter, we discuss some of these diseases and the reasons for their sudden emergence or reemergence. We also investigate the potential for the purposeful use of infectious microbes as agents of war or civilian terror.

29.7 Emerging and Reemerging Infectious Diseases

Infectious diseases are global, dynamic health problems. In this section we examine some recent patterns of infectious disease, some reasons for the changing patterns, and the methods used by epidemiologists to identify and deal with new threats to public health.

Emerging and Reemerging Diseases

The worldwide distribution of diseases can change dramatically and rapidly. Alterations in the pathogen, the environment, or the host population contribute to the spread of new diseases, with potential for high morbidity and mortality. Diseases that suddenly become prevalent are called **emerging diseases** and are not limited to "new" diseases; they also include **reemerging diseases**, diseases that were previously under control but suddenly appear as a new epidemic. Examples of global emerging and reemerging disease are shown in **Figure 29.10**, and select diseases with high potential for emergence or reemergence are described in

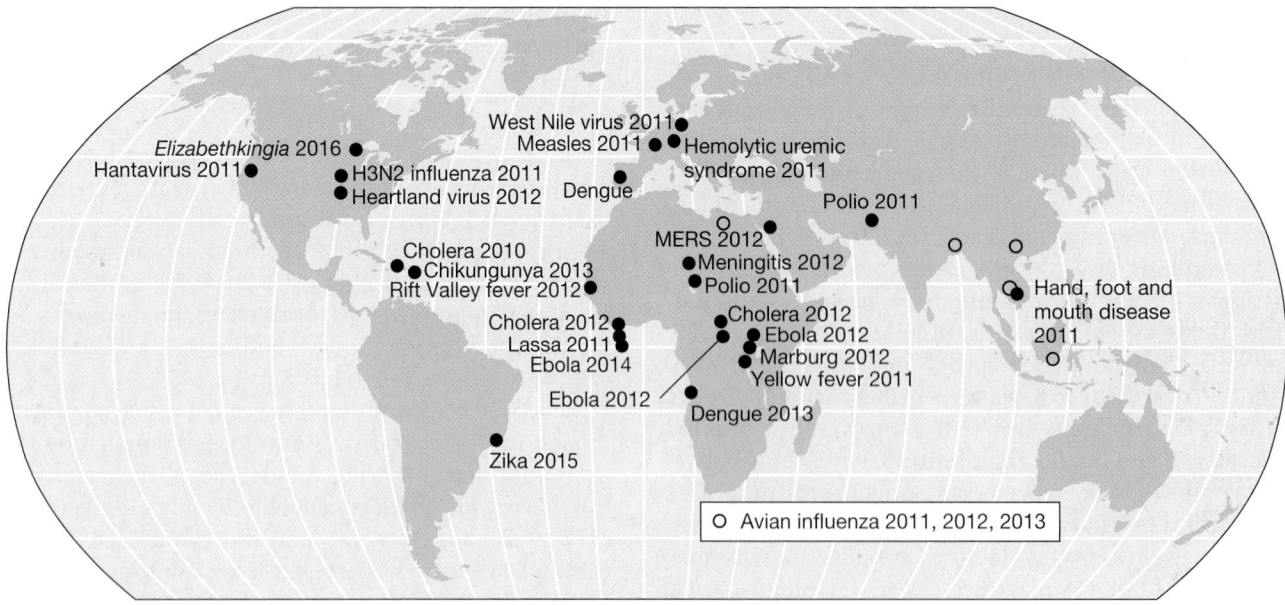

Figure 29.10 Recent outbreaks of emerging and reemerging infectious diseases. The diseases shown are local outbreaks capable of producing widespread epidemics and pandemics. Not shown are established pandemic diseases such as HIV/AIDS and cholera, and predictable annual epidemic diseases such as seasonal epidemic human influenza. MERS, Middle East respiratory syndrome. Avian influenza is caused by influenza A H5N1 (Section 29.8). For more on the *Elizabethkingia* oubreak, see page 866.

TABLE 29.5 Emerging and reemerging epidemic infectious diseases

Agent	Disease and symptoms	Mode of transmission	Cause of emergence
Borrelia burgdorferi	Lyme disease: rash, fever, neurological and cardiac abnormalities, arthritis	Bite of infective *Ixodes* tick	Increase in deer and human populations in wooded areas
Mycobacterium tuberculosis	Tuberculosis: cough, weight loss, lung lesions	Sputum droplets (exhaled through a cough or sneeze) from a person with active disease	Antimicrobial drug resistance as multidrug-resistant and extensively drug-resistant tuberculosis
Vibrio cholerae	Cholera: severe diarrhea, rapid dehydration	Water contaminated with the feces of infected persons; food exposed to contaminated water	Poor sanitation and hygiene; carried to non-endemic areas via infected travelers and commerce
Viruses			
Dengue	Hemorrhagic fever	Bite of an infected mosquito (primarily *Aedes aegypti*)	Poor mosquito control; increased urbanization in tropics; increased travel and shipping
Filoviruses (Marburg, Ebola)	Fulminant, high mortality, hemorrhagic fevers	Direct contact with infected blood, organs, secretions, and semen	Contact with vertebrate reservoirs
Influenza H5N1 (avian influenza)	Fever, headache, cough, pneumonia, high mortality	Direct contact with infected animals or humans, not easily spread via respiratory aerosols	Danger of animal–human virus reassortment; antigenic shift
Zika	Asymptomatic, or fever, rash, muscle and joint pain, headache	Bite of an infected mosquito (primarily *Aedes aegypti*), mother to fetus, sexual contact, blood transfusion	Poor mosquito control; increased urbanization in tropics
Fungi			
Candida	Candidiasis: fungal infections of the gastrointestinal tract, vagina, and oral cavity	Member of endogenous microbiota becomes an opportunistic pathogen; contact with secretions or excretions from infected persons	Immunosuppression; medical devices (catheters); antibiotic use

Table 29.5. Occasionally, new diseases emerge very unexpectedly and for totally unknown reasons; for example, an emerging infection due to the unusual bacterium *Elizabethkingia* in Wisconsin (USA) in 2016 posed a real medical mystery (see page 866).

Emerging epidemic diseases are not a new phenomenon. Among the diseases that rapidly and sometimes catastrophically emerged in the past are plague (caused by the bacterium *Yersinia pestis*) and influenza. For example, in the Middle Ages, up to one-third of all humans were killed by the periodic plague epidemics that swept Europe, Asia, and Africa. Influenza caused a devastating worldwide pandemic in 1918–1919, claiming up to 100 million lives, and the pandemic H1N1 influenza virus that emerged in 2009 killed up to a half million people in its first year. In the 1980s, HIV/AIDS and Lyme disease emerged as new diseases, and health officials worldwide are paying particular attention to the potential for rapid emergence of pandemic influenza developing from H5N1 avian influenza. More recently, isolation of patients and extra protections for their caregivers was practiced during the West African Ebola hemorrhagic fever epidemic (⮌ Section 30.12 and Figure 30.34*b*) to prevent spread of this extremely dangerous viral disease.

Emergence Factors

Many factors play into the emergence of new pathogens including human demographics and behavior, economic development, transportation, public health breakdowns, and other factors. The trend for human populations to reside in urban rather than rural areas facilitates disease transmission. For example, the high density of human hosts in cities has facilitated transmission of dengue fever, a serious viral disease spread by mosquitoes. Dengue infects nearly 400,000 people yearly, primarily in urban regions of tropical and subtropical countries including far southern reaches of the United States (**Figure 29.11**). Human behavior in large population centers also contributes to disease spread. For example, sexually promiscuous practices in population centers contribute to the spread of hepatitis and HIV/AIDS. Economic development and changes in land use also promote disease spread. For example, Lyme disease, the most common vectorborne disease in the

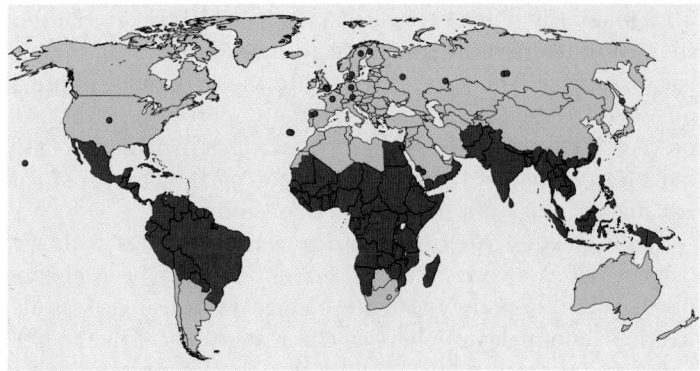

Figure 29.11 Dengue virus 2013. Dengue virus is now found in all tropical and subtropical countries as a result of the spread of its *Aedes aegypti* mosquito vector. The red areas are now endemic for the virus and mosquito vector. The red dots indicate outbreaks outside the known endemic areas. Prior to 1981, dengue virus was unknown in the Americas. Data are from the CDC, Atlanta, Georgia, USA.

United States, is on the rise largely due to residential reforestation and related changes in land use patterns. These activities increase deer habitat and thus contact between Lyme-infected deer ticks and humans, consequently increasing disease incidence.

Transportation, bulk processing, and central distribution methods have become increasingly important for quality assurance and economy in the food industry. However, these same factors have increased the potential for common-source foodborne disease epidemics when sanitation measures fail. For example, a single U.S. meat-processing plant spread *Escherichia coli* O157:H7 to people in eight states in 2009. The contaminated food source, ground beef, was recalled and the epidemic was eventually stopped, but not before several people died. International travel and commerce also affect the spread of pathogens. For example, a single person showing symptoms from an Ebola infection on an international flight could infect many other passengers because of the ease with which the Ebola virus spreads (⇔ Section 30.12). If such a situation were not immediately recognized, the disease could rapidly spread to major population centers when healthy passengers who had contact with the diseased passenger and were now carrying the virus disembarked and continued their travels.

Pathogen adaptation and change can contribute to disease emergence. For example, most RNA viruses, including influenza, HIV, and the hemorrhagic fever viruses, mutate rapidly. These mutant RNA viruses present major epidemiological problems because their altered genomes often affect their antigens, making immunity to old viral antigens ineffective for neutralizing the mutant viruses. Bacterial genetic mechanisms are also capable of enhancing virulence and promoting emergence of new epidemics. Virulence-enhancing factors are often carried by mobile genetic elements (Chapter 11) that can be transferred between and among members of the same species, and sometimes to other species and genera. Such transfers can quickly generate emerging pathogens, and multidrug-resistant strains of *Staphylococcus aureus* and *Pseudomonas aeruginosa* are good examples of this.

A breakdown of public health measures is sometimes responsible for the emergence or reemergence of diseases. For instance, cholera (caused by *Vibrio cholerae*) can be adequately controlled, even in endemic areas, by providing proper sewage disposal and water treatment. In 2010 contaminated water supplies following a major earthquake triggered a cholera outbreak in Haiti for the first time in over 100 years (Section 29.8). Inadequate public vaccination programs can also lead to the resurgence of previously controlled diseases. For example, pertussis, a vaccine-preventable childhood respiratory disease, has increased recently in Eastern Europe and in the United States partly because of inadequate immunization among adults and children.

Finally, weather patterns can also upset the usual host–pathogen balance. Disease vectors such as mosquitoes have been moving northward in response to climate change. Even a single seasonal weather abnormality can have an effect, as evidenced by the 1993 hantavirus hemorrhagic fever outbreak in the American Southwest (⇔ Section 31.2). A very mild winter coupled with record rainfall led to an explosive increase in rodents that can host hantavirus. This increased exposures for susceptible human hosts and led to the spread of this zoonotic infection.

Addressing Emerging Diseases

The keys for addressing emerging diseases are recognition of the disease and intervention to prevent pathogen transmission. Emerging diseases have, at least at first, low incidence and are usually absent from the official notifiable disease list for the United States prepared by the Centers for Disease Control and Prevention (Table 29.4). Emerging diseases are first recognized from their unique epidemic incidence, clusterings and other epidemiological patterns, and clinical symptoms unrelated to known pathogens (see page 866). Such disease patterns trigger intensive public health surveillance followed by specific interventions designed to control further outbreaks. Methods such as isolation, quarantine, immunization, and drug treatment can be applied to contain outbreaks. For vectorborne and zoonotic diseases, the nonhuman host or vector must be identified to intervene in the life cycle of the pathogen and stop human infection.

International public health surveillance and intervention programs were instrumental in controlling the emergence of severe acute respiratory syndrome (SARS), a disease that emerged rapidly, explosively, and unpredictably from a zoonotic source (Section 29.4). On the other hand, even a rapid and focused response was unsuccessful in containing the spread of pandemic (H1N1) 2009 influenza, as we will see in the next section.

MINIQUIZ

- What is the difference between an emerging and a reemerging infectious disease?
- What factors are important in the emergence or reemergence of potential pathogens?
- Indicate general and specific methods that would be useful for identifying emerging infectious diseases.

29.8 Examples of Pandemics: HIV/AIDS, Cholera, and Influenza

Through the centuries, several diseases have reached pandemic proportions. Here we consider three—HIV/AIDS, cholera, and influenza—for which epidemiological studies have been extensive.

HIV/AIDS

HIV/AIDS is a continuum of disease, starting with the infection of an individual with the human immunodeficiency virus (HIV). Eventually, infection results in acquired immunodeficiency syndrome (AIDS), a disease which if not treated cripples the immune system, leading to opportunistic infections that can be fatal (⇔ Section 30.15). The first reported cases of AIDS were diagnosed in the United States in 1981. Since then, more than 1.2 million cases have been reported in the United States with over 635,000 deaths (**Figure 29.12**); worldwide, over 25 million AIDS deaths have occurred.

Epidemiological studies in the United States in the 1980s suggested a high AIDS prevalence among men who have sex with men and among intravenous drug abusers. Individuals receiving blood or blood products were also at high risk. Collectively, these epidemiological data indicated a transmissible agent, presumably transferred

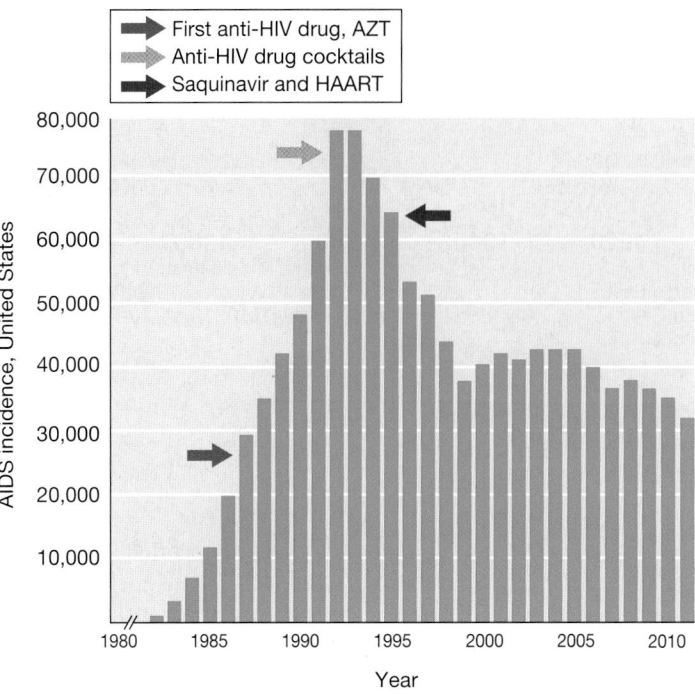

Figure 29.12 Annual new cases of human immunodeficiency virus/acquired immunodeficiency syndrome (HIV/AIDS) in the United States. Cumulatively, there were about 1.1 million cases of HIV/AIDS through 2011. In 2009, the HIV/AIDS case definition changed to include all new HIV infections and AIDS diagnoses (⇌ Section 30.15). Colored arrows indicate the introduction of different anti-HIV drugs. See Sections 28.11 and 30.15 for a discussion of anti-HIV therapies. Data are from the HIV/AIDS Surveillance Report and Division of HIV/AIDS Prevention—Surveillance and Epidemiology, CDC, Atlanta, Georgia, USA.

during sexual activity or by contaminated blood. Soon after the discovery of HIV in 1983, laboratory tests were developed to detect antibodies to the virus in blood. With this tool in hand, surveys of HIV incidence and prevalence defined the spread of HIV and showed conclusively that body fluids, primarily blood and semen, were the vehicles for transmission of the virus (**Figure 29.13**).

The HIV/AIDS data showed that in the United States, the number of AIDS cases was disproportionately high in men who have sex with men, but among women, heterosexuals were the largest risk group (Figure 29.13). Further analyses of the epidemiological data showed that the new infection rate for African American men was seven times that of Caucasian males, indicating that social and economic factors may also influence infection risk. However, regardless of gender or racial specifics, AIDS epidemiology provided a clear picture of HIV transmission: Virtually all who acquired HIV engaged in sex or intravenous drug use in which body fluids—semen or blood—were transferred and commonly had sex or exchanged syringe needles with multiple partners. We discuss the pathology and therapy of HIV/AIDS in Section 30.15.

Cholera

Cholera is primarily a waterborne infection (⇌ Section 32.3) that is normally kept in check by appropriate public health measures for water treatment (Chapter 22). Cholera is caused by ingestion of contaminated water containing *Vibrio cholerae*, a gram-negative, curved rod–shaped species of *Proteobacteria* that produces a powerful

enterotoxin that triggers severe diarrhea (⇌ Section 25.6). Cholera is endemic in Africa, Southeast Asia, the Indian subcontinent, and Central and South America. Epidemic cholera occurs frequently in areas where sewage treatment is either inadequate, altogether absent, or suffers a major breakdown, for example, from a flood or an earthquake. In 2014, the World Health Organization (WHO) reported over 190,000 cases of cholera that led to 2231 deaths. However, the WHO estimates that only 5–10% of cholera cases are actually reported because diarrheal diseases from various pathogens are so common (Table 29.1); thus total worldwide incidence of cholera likely exceeds 1 million cases per year.

Epidemic cholera may develop into pandemics when travelers from endemic areas carry the pathogen to new locations with susceptible populations and poor sanitation. Since 1817, cholera has swept the world in seven major, and nearly consecutive, pandemics (**Figure 29.14**). All but one of these originated on the Indian subcontinent, where cholera is endemic. Two distinct pandemic strains of *V. cholerae* are recognized, known as the *classic* and the *El Tor* biotypes. The *V. cholerae* O1 El Tor biotype started the seventh pandemic in Indonesia in 1961, and its spread continues to the present day. This pandemic has caused over 5 million cases of cholera and at least 250,000 deaths and continues to be a major cause of morbidity and mortality, especially in developing countries (Figure 29.14).

In October 2010 Haiti experienced its first cholera in over 100 years, and in just two years experienced nearly 600,000 cases and 8000 deaths. The outbreak began in the aftermath of the catastrophic 2010 earthquake. There were likely two triggers of this cholera outbreak, the first being a classic scenario of poor sanitation following a disaster and the second an accidental importation from an outside source. *Vibrio cholerae* is present in marine waters and as a result of the earthquake, it may have been washed into coastal freshwaters where it contaminated drinking water sources. But in addition, United Nations aid workers that arrived

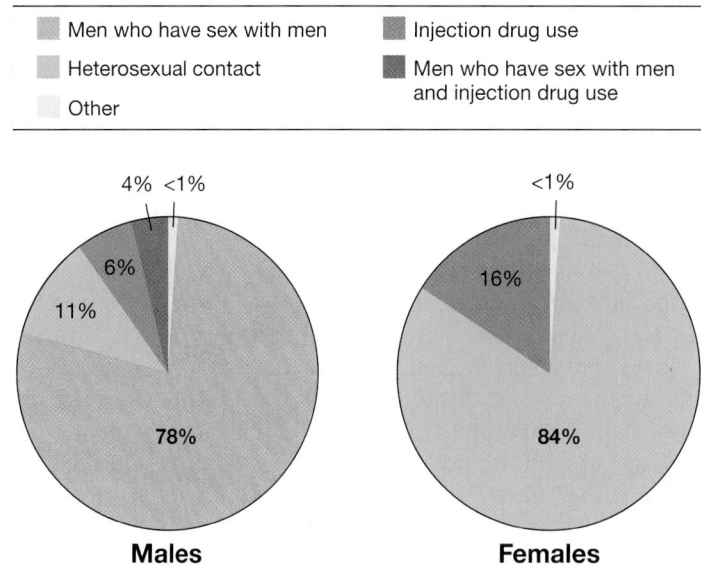

Figure 29.13 Distribution of AIDS cases by risk group and gender in the United States, 2010. Data from the CDC, Atlanta, Georgia, USA, were gathered from 38,000 males and 9,500 females diagnosed with HIV/AIDS in 2010.

UNIT 7

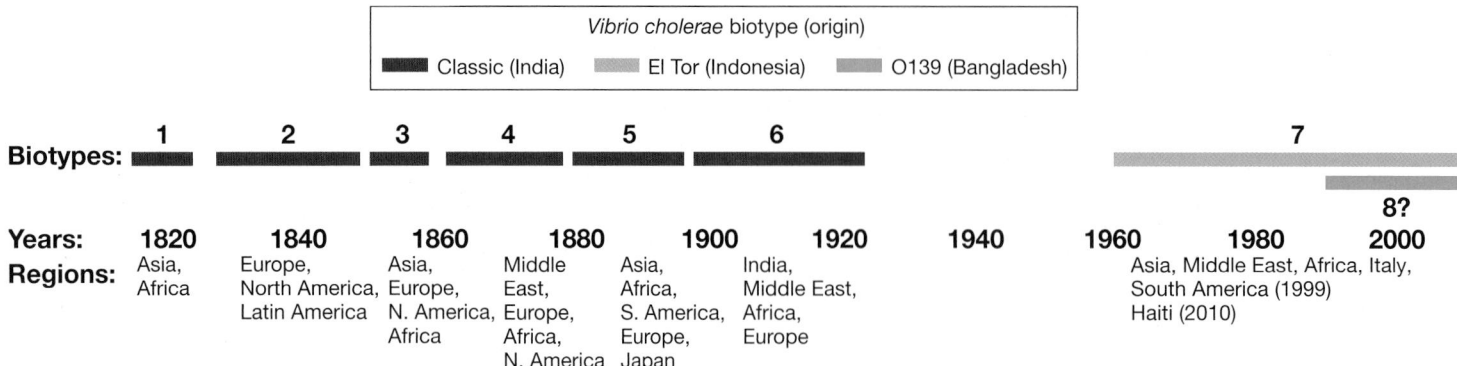

Figure 29.14 Cholera pandemic timeline. Seven cholera pandemics have been nearly consecutive for over 200 years. The seventh pandemic started in 1961 and is ongoing. The O139 strain that appeared in 1991 is endemic to Bangladesh and the Bay of Bengal and is causing epidemics that may be the prelude to an eighth pandemic.

from Nepal, where a recent cholera outbreak had occurred, are thought to have shed *V. cholerae* into sanitation streams that found their way into Haitian drinking water sources; if true, this would have contributed to the Haitian epidemic. Cholera has since spread from Haiti to the Dominican Republic and other areas of the Caribbean, and to Mexico.

Pandemic (H1N1) 2009 and Future Influenza Pandemics

Human influenza pandemics occur every 10 to 40 years as a result of major genetic changes in the influenza A virus genome that affect the virus's immune status (*antigenic drift* and *antigenic shift*, ↩ Section 30.8 and Figure 30.26). The most devastating influenza pandemic of all time occurred in 1918; this flu infected over half a *billion* people worldwide and killed approximately 50 *million* people before it ran its course. The 1918 pandemic was caused by a strain of influenza termed H1N1.

A more recent influenza pandemic began in March 2009 with the outbreak of epidemics in Mexico. The culprit virus, a strain designated (H1N1) 2009, was a hybrid of the 1918 strain and a later strain that caused a pandemic in 1957; (H1N1) 2009 contained genes from bird, swine, and human influenza viruses. Such *reassortant viruses*, as they are called (↩ Section 30.8), can be highly virulent, as they tend to produce antigens to which humans have no prior exposure and thus no immunity; the only way to obtain immunity to such a virus is to become infected or artificially immunized.

Without an effective vaccine at the ready and with the rapidity with which influenza is spread, the stage was set for the reassortant (H1N1) 2009 flu to reach pandemic proportions. Within six months of its emergence, (H1N1) 2009 had spread to almost every country in the world, qualifying it as a true pandemic (**Figure 29.15**). Although official numbers range widely, it is estimated that more than a quarter of the world's population was infected in the pandemic. In the United States, about 60 million persons were infected, with mortality confirmed as due to (H1N1) 2009 numbering about 3400 persons. By late 2010 the (H1N1) 2009 pandemic was fading, and today few cases are observed because antigens from this strain of virus are typically included in seasonal influenza vaccines (↩ Table 28.4).

Could new influenza pandemics sweep the world? Perhaps the greatest threat to global stability would be another influenza pandemic that has the virulence and infectivity of the 1918 pandemic. Because epidemiological surveillance is currently so extensive, this possibility is unlikely, but the risk can never be zero. In recent years public health officials worldwide have been following the emergence and reemergence of a potentially devastating strain of influenza virus designated influenza A H5N1, originally found in birds. This virus first appeared in Hong Kong in 1997, jumping directly from chickens and ducks to humans. Since then H5N1 has reemerged several times in small outbreaks, with the most recent occurring in Egypt, Indonesia, Cambodia, Bangladesh, and China (Figure 29.10). Through 2014, 638 cases of human H5N1 infection have been confirmed, resulting in 379 deaths, for a mortality rate of almost 60%. This high mortality rate underscores the dangerous aspects of this virus.

Besides poultry and humans, H5N1 has also infected swine. If a reassortant strain were to emerge from pigs that had the capacity

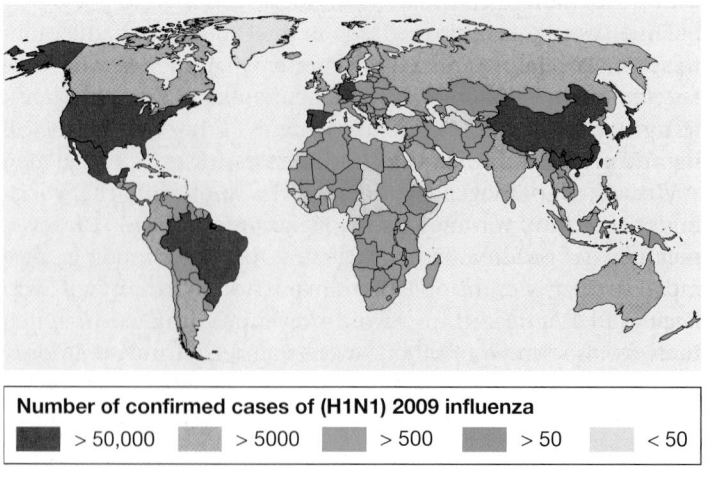

Number of confirmed cases of (H1N1) 2009 influenza

| > 50,000 | > 5000 | > 500 | > 50 | < 50 |

Figure 29.15 Pandemic (H1N1) 2009 influenza incidence. Data show minimal estimates of cases worldwide by country. It is estimated that approximately 1.7 billion people were infected by the (H1N1) 2009 pandemic flu (24% of the total population) and that deaths were between 150,000 and 575,000 (the large range for mortality estimates is because many deaths that were likely due to (H1N1) 2009 were not confirmed as such).

to spread from person to person, such a virus could trigger an influenza pandemic of unprecedented mortality. Because of this, plans are in place nationally and internationally to provide appropriate vaccines and support for potential pandemics initiated by this and other emergent influenza strains. We discuss the disease influenza in detail in Section 30.8.

MINIQUIZ

- Describe the major risk factors for acquiring HIV infection in the United States. How do these differ in males and females?
- Identify the most likely means of acquiring cholera. Why do cholera epidemics keep occurring?
- What is a reassortant influenza virus and why can such viruses be so dangerous?

29.9 Public Health Threats from Microbial Weapons

As if avoiding the wrath of pathogenic microbes that can infect us naturally is not enough, humans have researched the use of certain pathogens as weapons to be intentionally deployed on others. **Biological (microbial) warfare** is the use of microbial agents to incapacitate or kill a military or civilian population in an act of war or terrorism. Although the use and development of microbial weapons are forbidden by international law, microbial weapons have already seen use, and facilities for their production likely exist in rogue countries and perhaps also in avowed terrorist groups. Because of this, microbial weapons research continues in many peaceful nations so as to best understand the most serious threats and learn how to counter them.

Characteristics of Microbial Weapons

Effective microbial weapons are pathogens, or in a few cases toxins, that are (1) relatively easy to produce and deliver, (2) safe for use by the offensive forces, and (3) able to incapacitate or kill people in a systematic and consistent manner. Although microbial weapons are potentially useful in the hands of conventional military forces, the greatest likelihood of microbial weapons use is by terrorists because of the ready availability and low cost of producing and propagating many of the organisms.

Virtually all pathogenic bacteria or viruses are potentially useful for biological warfare, and *select agents* that have significant potential for use as microbial weapons are listed in **Table 29.6**. The most frequently mentioned candidates are smallpox virus and *Bacillus anthracis*, the bacterium that causes anthrax. Both of these microbes can be easily disseminated, are transmissible from person to person, and typically cause high mortality. Other agents have their advantages and disadvantages as microbial weapons and are categorized as to their potential risk from Category A to Category C in Table 29.6.

The United States government, through the Centers for Disease Control and Prevention, has developed the Select Agent Program surveillance system to monitor possession and use of potential bioterrorism agents. In addition, the CDC Laboratory Response

TABLE 29.6 Select agents and diseases by bioweapons threat category[a]

Category A

Highest-priority agents that pose a risk to national security. These agents are easily disseminated or transmitted and result in high mortality rates. They require special action for public health preparedness.

Disease/Pathogen

Anthrax (*Bacillus anthracis*)
Botulism (*Clostridium botulinum* toxin)
Plague (*Yersinia pestis*)
Smallpox (*Variola major*)
Tularemia (*Francisella tularensis*)
Viral hemorrhagic fevers (filoviruses [e.g., Ebola, Marburg] and arenaviruses [e.g., Lassa, Machupo])

Category B

Second-highest-priority agents. These agents are moderately easy to disseminate, result in moderate morbidity and low mortality, and require specific enhancements of public health diagnostic capacity and disease surveillance.

Disease/Pathogen

Brucellosis (*Brucella* species)
Epsilon toxin of *Clostridium perfringens*
Food safety threats (e.g., *Salmonella* spp., *Escherichia coli* O157:H7, *Shigella*)
Glanders (*Burkholderia mallei*)
Melioidosis (*Burkholderia pseudomallei*)
Psittacosis (*Chlamydophila psittaci*)
Q fever (*Coxiella burnetii*)
Ricin toxin from *Ricinus communis* (castor beans)
Staphylococcal enterotoxin B (*Staphylococcus aureus*)
Typhus fever (*Rickettsia prowazekii*)
Viral encephalitis (alphaviruses such as Venezuelan equine encephalitis, eastern equine encephalitis, western equine encephalitis)
Water safety threats (*Vibrio cholerae*, *Cryptosporidium parvum*, and others)

Category C

Third-highest-priority agents are emerging pathogens that are available, easily produced and disseminated, with high potential for high morbidity and mortality.

Pathogens

Emerging infectious diseases such as hantavirus

[a]*Source:* Centers for Disease Control and Prevention, Atlanta, Georgia, USA.

Network and the Health Alert Network have been upgraded to enhance their diagnostic capabilities and increase the reporting abilities of local and regional healthcare centers to rapidly identify bioterrorism events as well as emerging diseases.

Smallpox and Anthrax

Smallpox virus (**Figure 29.16a**) has intimidating potential as a microbial weapon because it can be spread easily by direct contact or by aerosol spray, is highly debilitating, causes a high fever, severe fatigue, and the eventual formation of pus-filled skin blisters (Figure 29.16b), and has a mortality rate of 30% or higher. Although an extremely effective smallpox vaccine is available, it has not been in use in the general population since smallpox was eradicated worldwide in 1980. Moreover, the potential of smallpox virus being deployed as a military weapon is considered low because military personnel are routinely vaccinated. Nevertheless, preparations for

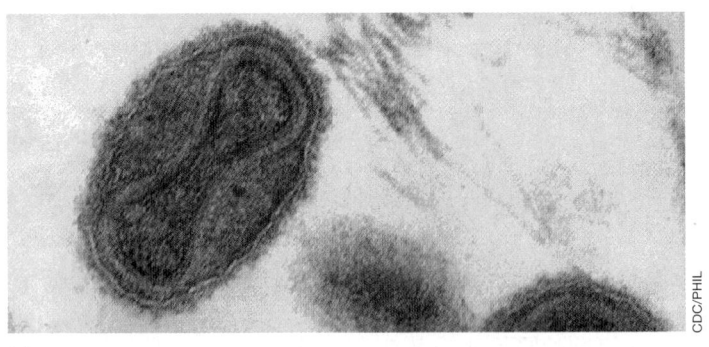

(a)

(b)

CDC/PHIL

Figure 29.16 Smallpox and its use as a microbial weapon. *(a)* Smallpox virus, a double-stranded DNA virus (⇌ Section 10.6). *(b)* Characteristic papular smallpox rash and blisters on the arm. Smallpox was officially declared eradicated worldwide in 1980.

a potential smallpox attack on civilians in the United States have been made and would include immunization of key individuals such as those evaluating, caring for, or transporting smallpox patients; laboratory personnel handling clinical specimens from smallpox patients; and other persons as necessary who might come into contact with infectious materials from smallpox patients.

Bacillus anthracis is the causative agent of anthrax, and its unique properties make it particularly attractive as a bioweapon. Chief among these is that it is easily grown aerobically, producing distinctive colonies on enriched culture media, and it differentiates into highly resistant endospores (**Figure 29.17a, b**; ⇌ Section 2.10). Once prepared, endospores can be dried and stored indefinitely and then disseminated as a weapon by aerosol or in powdered suspension. There are three clinical forms of anthrax (⇌ Section 31.8). *Cutaneous anthrax* is contracted when abraded skin is contaminated by *B. anthracis* endospores; the organism grows and kills the skin, forming a necrotic tissue lesion called an *eschar* (Figure 29.17c). The most rare form—*gastrointestinal anthrax*—is contracted from consumption of endospore-contaminated plants or meat from animals infected with anthrax. *Inhalation anthrax* (also called *pulmonary anthrax*) is the deadliest form and is contracted when *B. anthracis* endospores are inhaled. The symptoms

of inhalation anthrax include pulmonary and cerebral hemorrhage (Figure 29.17d), which makes this form of anthrax so dangerous. All forms of anthrax have the potential to become systemic infections; however, cutaneous anthrax is easily treatable with antibiotics and is fatal in only about 20% of untreated cases. Without treatment, gastrointestinal anthrax is fatal in about half those infected, whereas inhalation anthrax mortality approaches 100%.

Inhalation anthrax is the form of the disease that makers of microbial weapons would aim for, as has already transpired in the United States. At least 22 cases of anthrax leading to 5 deaths occurred in a 2001 bioterrorism attack where dense *B. anthracis* endospore preparations were mailed in envelopes to certain news outlets and government officials. Of the 22 anthrax cases, 11 were inhalational and 11 cutaneous. These *weaponized* anthrax strains were endospore preparations mixed within a fine particulate material that allowed the endospores to be spread by air currents. Thus, opening an envelope containing endospores or releasing the powder–endospore mixture into a ventilation system or other air exchange could contaminate surrounding areas and personnel.

Vaccination for anthrax is possible and is restricted to individuals who are considered at risk. This includes agricultural animal

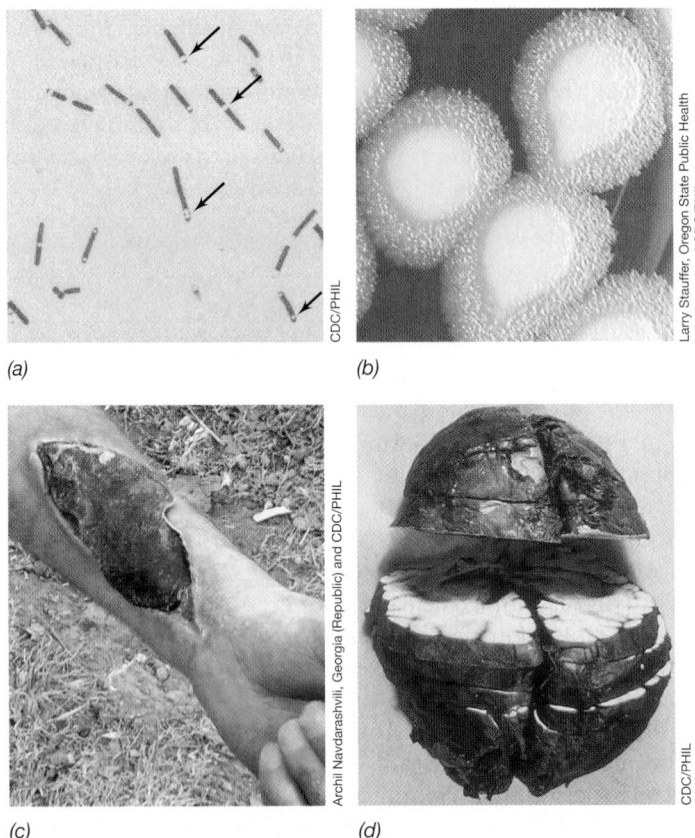

(a)

(b)

(c)

(d)

Larry Stauffer, Oregon State Public Health Laboratory, and CDC/PHIL
CDC/PHIL
Archil Navdarashvili, Georgia (Republic) and CDC/PHIL
CDC/PHIL

Figure 29.17 *Bacillus anthracis* and anthrax. *(a) Bacillus anthracis* is a gram-positive, endospore-forming rod approximately 1 μm in diameter and 3–4 μm in length. Note the developing endospores (arrows). *(b)* Characteristic "ground glass" appearance of colonies of *B. anthracis* on blood agar plates. *(c)* Cutaneous anthrax. The blackened lesion seen on this forearm is called an eschar, and results from tissue necrosis. *(d)* Inhalation anthrax can cause cerebral hemorrhage, as shown by the dark coloration in this fixed and sectioned human brain removed at autopsy.

UNIT 7

workers (livestock tenders and those working with animal products), laboratory personnel working with anthrax, veterinarians, and military personnel. As was discussed with smallpox, anthrax is an unlikely military weapon but could be a very effective means of terrorizing a civilian population because the vast majority of the population is unvaccinated.

MINIQUIZ

- What characteristics make a pathogen or its products particularly useful as a biological weapon?
- Indicate the steps you would take to identify and treat infections from smallpox virus or anthrax in a bioterror attack.

MasteringMicrobiology®

Visualize, **explore**, and **think critically** with Interactive Microbiology, MicroLab Tutors, MicroCareers case studies, and more. MasteringMicrobiology offers practice quizzes, helpful animations, and other study tools for lecture and lab to help you master microbiology.

Chapter Review

I • Principles of Epidemiology

29.1 Epidemiology is the study of the occurrence, distribution, and determinants of health and disease in populations. Epidemiologists employ surveillance measures to record the incidence and prevalence of infectious diseases and track disease outbreaks that may lead to epidemics or pandemics. The morbidity and mortality of any given disease is a function of the transmissibility of the pathogen and the severity of the disease symptoms. Disease carriers may be symptom-free or they may have chronic infections that are never resolved.

> **Q** Distinguish between disease incidence and disease prevalence. Which value is a predictor of risk of disease?

29.2 Effects on populations as well as individuals must be studied to understand infectious disease. The interactions of pathogens with hosts can be dynamic, affecting the long-term evolution and survival of all species involved. Herd immunity provides disease protection for uninfected or unimmunized hosts.

> **Q** Herd immunity does not need to be 100% to effectively stop disease transmission; why is this?

29.3 Infectious diseases can be transmitted directly from person to person, indirectly from living vectors or inanimate objects (fomites), or from common-source vehicles such as food and water. Disease reservoirs can include soil, insects, seemingly healthy people, chronic carriers, and many other sources. An understanding of disease reservoirs, carriers, and pathogen life cycles is critical for controlling disease epidemics.

> **Q** Compared with a fomite—such as a contaminated cup—why is a disease vehicle such as food a much more powerful disease transmitter?

29.4 Epidemics may be of host-to-host origin or originate from a common source. The basic reproduction number (R_0) of a pathogen gives a relative picture of its seriousness and an indication of how effective herd immunity must be to prevent its spread.

> **Q** If an emerging pathogen was found to have an R_0 of 30, would herd immunity need to be higher or lower than for an emerging pathogen with an R_0 of 3?

II • Epidemiology and Public Health

29.5 Food and water purity regulations, vector control, immunization, quarantine, isolation, and disease surveillance are all public health measures that reduce the incidence of communicable diseases. In the case of some diseases, such as smallpox, total eradication worldwide has been possible.

> **Q** If an outbreak of yellow fever occurred, who would be isolated and who would be quarantined?

29.6 Infectious diseases account for about 30% of all mortality worldwide. Most cases of infectious diseases occur in developing countries. Control of many infectious diseases can be accomplished by public health measures that include adequate sanitation, food and water protections, and widespread vaccination programs. A high infectious disease load can significantly reduce the average life expectancy of the population in a country.

> **Q** Contrast the leading causes of death in Africa and in the United States.

III • Emerging Infectious Diseases, Pandemics, and Other Threats

29.7 Changes in host, vector, or pathogen conditions, whether natural or artificial, can encourage the explosive emergence or reemergence of infectious diseases. Global surveillance and intervention programs by organizations such as the World Health Organization and the U.S. Centers for Disease Control and Prevention are especially attuned to emerging and reemerging pathogens to prevent local epidemics from spreading.

> **Q** How can bacterial genetic exchange fuel the emergence of new pathogens?

29.8 Several infectious diseases with significant mortality show pandemic characteristics. HIV/AIDS affects those who exchange bodily fluids, most often by either promiscuous unprotected sex or intravenous drug use. Cholera is primarily a waterborne infection and control can be achieved by maintaining adequate clean water and waste sanitation measures. New pandemic influenza strains resulting from bird–swine–human influenza reassortments present the biggest infectious disease threat worldwide.

Q Why is H5N1 avian influenza considered a major threat to public health?

29.9 Bioterrorism from smallpox or anthrax is a threat in a world of rapid international travel and easily accessible

technical information. Aerosols or disease vehicles are the most likely modes of delivery of microbial weapons. Prevention and containment measures rely on a well-prepared public health infrastructure.

Q Why are smallpox and anthrax more likely to be bioterrorism threats to civilians than to military personnel?

Application Questions

1. Smallpox, a disease that was limited to humans, was eradicated, whereas plague, a zoonotic disease with a reservoir in rats and related rodents, will likely never be eradicated worldwide. Explain this statement. Devise a plan to eradicate plague in a limited area such as a town or city. Be sure to consider methods that involve the reservoir, the pathogen, and the host.

2. Like smallpox, HIV/AIDS is considered to be a disease that could be eliminated worldwide because it is propagated by known means and there are no animal reservoirs. Although eliminating HIV infection is possible, why would it be much

more difficult than eliminating smallpox? What would be involved in an HIV infection eradication program?

3. H5N1 avian influenza has high potential for causing a pandemic under certain circumstances. If such a highly transmittable human–avian strain were to evolve in Asia, from the perspective of a national public health official, what measures would you employ to stop the spread of the new influenza to the United States? If you failed to contain the new virus, where would you expect to see the first cases of such pandemic influenza, in metropolitan areas or in the rural Midwest?

Chapter Glossary

Acute infection a short-term infection, usually characterized by dramatic onset

Basic reproduction number (R_0) the number of expected secondary transmissions from each single case of a disease in an entirely susceptible population

Biological (microbial) warfare the use of biological agents to incapacitate or kill a military or civilian population in an act of war or terrorism

Carrier a subclinically infected individual who may spread a disease

Centers for Disease Control and Prevention (CDC) the agency of the United States Public Health Service that tracks disease trends, provides disease information to the public and to healthcare professionals, and forms public policy regarding disease prevention and intervention

Chronic infection a long-term infection

Common-source epidemic an infection (or intoxication) of a large number of people from a contaminated common source such as food or water

Disability-adjusted life year (DALY) a quantitative measure of disease burden defined as the cumulative number of years lost due to an illness itself, a disability due to an illness, or premature death

Disease surveillance the observation, recognition, and reporting of diseases as they occur

Emerging disease an infectious disease whose incidence recently increased or whose incidence threatens to increase in the near future

Endemic disease a disease that is constantly present, usually in low numbers, in a population

Epidemic the occurrence of a disease in unusually high numbers in a localized population

Epidemiology the study of the occurrence, distribution, and determinants of health and disease in a population

Fomite an inanimate object that when contaminated with a viable pathogen can transfer the pathogen to a host

Herd immunity the resistance of a population to a pathogen as a result of the immunity of a large portion of the population

Host-to-host epidemic an epidemic resulting from person-to-person contact, characterized by a gradual rise and fall in number of new cases

Incidence the number of new disease cases reported in a population in a given time period

Isolation in the context of infectious disease, the separation of persons who have an infectious disease from those who are healthy

Morbidity the incidence of disease in a population

Mortality the incidence of death in a population

Outbreak the occurrence of a large number of cases of a disease in a short period of time

Pandemic a worldwide epidemic

Prevalence the total number of new and existing disease cases reported in a population in a given time period

Public health the health of the population as a whole

Quarantine the separation and restriction of well persons who may have been exposed to an infectious disease to see if they develop the disease

Reemerging disease an infectious disease previously under control but that produces a new epidemic

Reservoir a source of infectious agents from which susceptible individuals may be infected

Vector a living agent that transfers a pathogen (differs from genetic vector, discussed in Chapter 12)

Vehicle a nonliving source of pathogens that transmits the pathogens to large numbers of individuals; common vehicles are food and water

Virulence the relative ability of a pathogen to cause disease

Zoonosis any disease that occurs primarily in animals but can be transmitted to humans

Person-to-Person Bacterial and Viral Diseases

microbiology**now**

A New Weapon against AIDS?

Decades after it was first recognized as a communicable disease in 1981, acquired immunodeficiency syndrome (AIDS) is still a major human disease. Worldwide, nearly 80 million people are infected with the causative agent of AIDS—human immunodeficiency virus (HIV)—and AIDS-related illnesses claim over 1.1 million lives each year.

The list of viral diseases to which effective vaccines have been developed is lengthy and includes influenza, smallpox, measles, yellow fever, and rabies. However, thus far, no effective human HIV vaccine has emerged. One reason is that HIV attacks important immune cells called CD4 T lymphocytes, turning them into viral production factories (in this colorized scanning electron micrograph, green HIV virions are budding out of a pink CD4 lymphocyte) while simultaneously shutting down CD4 T cell division. This gradual dilution of CD4 T cells renders HIV-infected persons helpless to attack by opportunistic pathogens that ultimately cause death.

Some HIV-infected persons survive in good health for long periods—even decades. These individuals remain HIV-positive yet maintain adequate levels of CD4 T cells. A recent study of these patients has shown that they have an unusual immunological profile; they produce a special class of anti-HIV antibodies called broadly neutralizing antibodies (bnAbs). These antibodies keep HIV levels low and in this way prevent the progression to AIDS. However, a high proportion of bnAb-producers also make autoantibodies (antibodies that attack the body's own antigens) and show reduced levels of regulatory T (Treg) cells (T cells that function to modulate the immune response and help maintain tolerance to self antigens). These findings have suggested a new strategy to fight AIDS.

B lymphocytes are the immune cells that produce antibodies, and the activities of B cells are under the control of Treg cells. Because of their lower Treg cell levels, bnAb-producing HIV-infected individuals are thought to contain a "less regulated" pool of B cells, and this condition allows for a broad range of antibodies to be produced. These include the beneficial anti-HIV bnAbs but also undesirable antibodies, such as autoantibodies. However, if a vaccine could be developed that would temper Treg cell activities in HIV-infected persons that lack anti-HIV bnAbs, it might stimulate them to produce these helpful antibodies and be a new weapon in the fight against AIDS.

Source: Moody, M.A., et al. 2016. Immune perturbations in HIV-1-infected individuals who make broadly neutralizing antibodies. *Science Immunology 1*, aag0851.

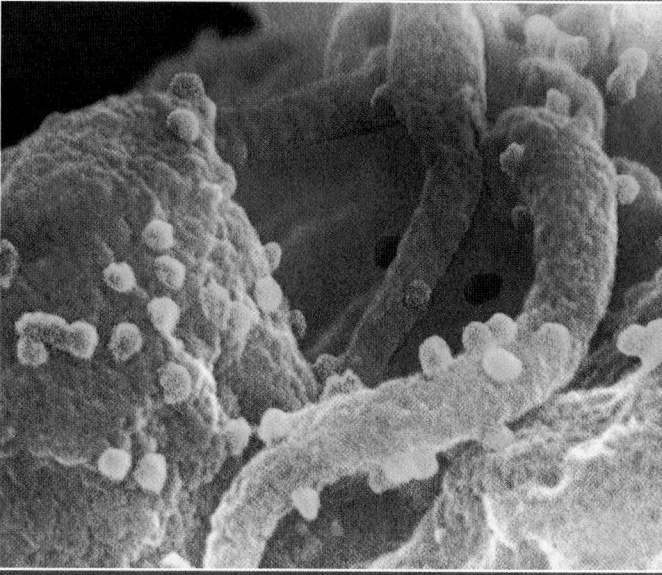

Several million species of microorganisms likely exist in nature, but only a few hundred cause disease. In this and the next three chapters we focus on this vitally important subset of the microbial world. We investigate the biology of the pathogens as well as the diseases they cause, including disease diagnosis, treatment, and prevention.

Our infectious disease coverage is ecological, being organized around each pathogen's *mode of transmission*. Although the biology of the different causative agents will certainly be highlighted, an ecological approach to infectious disease coverage emphasizes the single most important feature that links diseases caused by different microbes. Hence, tuberculosis and influenza are caused by a bacterium and a virus, respectively, but both diseases are transmitted from person to person by airborne emissions.

In this chapter we explore diseases transmitted from person to person, whether through the air, by direct contact, or through sexual contact. In Chapters 31 and 32 we focus on diseases transmitted by animal and arthropod vectors or from soil, and diseases transmitted from common sources such as water or food, respectively. We end our tour of the microbial world in Chapter 33 where we examine fungal and parasitic infections.

I • Airborne Bacterial Diseases

Worldwide, acute respiratory infections kill more than 4 million people a year, mainly in developing countries. Children and the elderly make up most of the fatalities, but in general, respiratory infections are the most common of all human diseases. Aerosols, such as those generated by a sneeze (**Figure 30.1**), as well as by coughing, talking, or breathing, are major vehicles for person-to-person transmission of respiratory diseases. Besides directly infecting a new host, infectious mucus from an aerosol can also contaminate objects, such as a door handle, and transmit infection well after the aerosol event. In these ways, respiratory diseases spread quickly, especially in congested areas, as airborne pathogens exploit a simple yet highly effective means of infecting new hosts.

30.1 Airborne Pathogens

Microorganisms found in air are derived from soil, water, plants, animals, people, surfaces, and other sources. Most microorganisms survive poorly in air. As a result, airborne pathogens are effectively transmitted between people only over short distances. Certain pathogens, however, survive drying well and can remain alive in dust or on fomites for long periods of time. For example, because of their thick, rigid cell walls, gram-positive bacteria (*Staphylococcus, Streptococcus*) are generally more resistant to drying than are gram-negative bacteria. Likewise, the waxy layer of *Mycobacterium* cell walls resists drying and promotes survival of pathogens such as *Mycobacterium tuberculosis*.

Large numbers of droplets can be expelled during a sneeze (Figure 30.1). Infectious droplets are about 10 μm in diameter and each droplet can contain one or more microbial cells or virus virions. The initial speed of the droplet movement is about 100 m/s

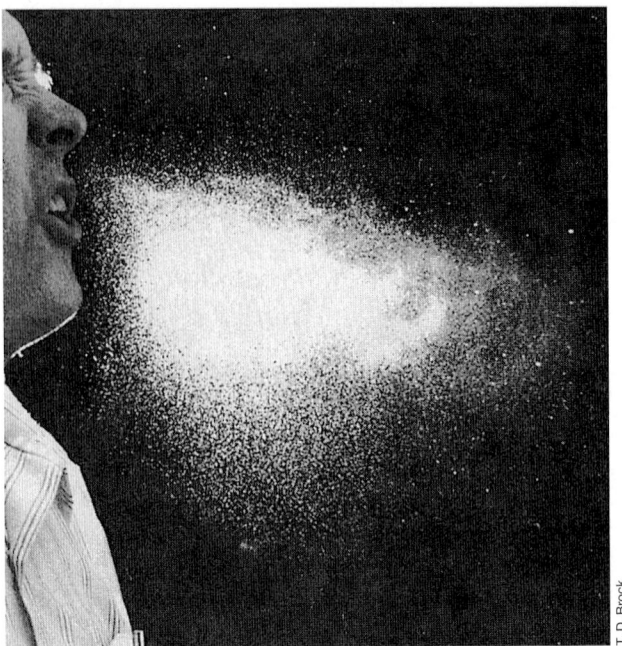

Figure 30.1 High-speed photograph of an unstifled sneeze. Effluent is emerging at over 100 m/s (200 miles/h).

(200 miles/h) in a violent sneeze and ranges from 15 to 50 m/s during coughing or shouting. The number of bacteria in a single sneeze varies from 10^4 to 10^6, and viral numbers can be much higher than this. Because of their small size, the droplets evaporate quickly in the air, leaving behind dried mucus in which the airborne pathogens remain embedded.

The human respiratory tract is divided into upper and lower regions, and specific airborne pathogens tend to exploit one region or the other, or sometimes both (**Figure 30.2**). The speed at which air moves through the human respiratory tract varies, and in the lower respiratory tract the rate is quite slow. As air slows down, particles in it stop moving and settle. Large particles settle first and the smaller ones later; only particles smaller than about 3 μm travel as far as the bronchioles in the lower respiratory tract (Figure 30.2).

Upper respiratory infections such as the common cold are typically acute and non-life-threatening. By contrast, *lower* respiratory infections, such as bacterial or viral pneumonia, are often chronic and can be quite serious, especially in the elderly or an immune-compromised person. Also, although most common respiratory infections are not serious in an otherwise healthy host, they can set the stage for *secondary infections* that can be life-threatening. For example, death of an elderly person from pneumonia following a severe case of influenza is not an uncommon event.

Most human respiratory pathogens are transmitted from person to person because humans are the only reservoir for the pathogens. However, many airborne pathogens, such as *Streptococcus* spp., cold viruses, and influenza, can also be transmitted by direct contact (for example, by a handshake) or on fomites. Accurate and rapid diagnosis and treatment of respiratory infections are well

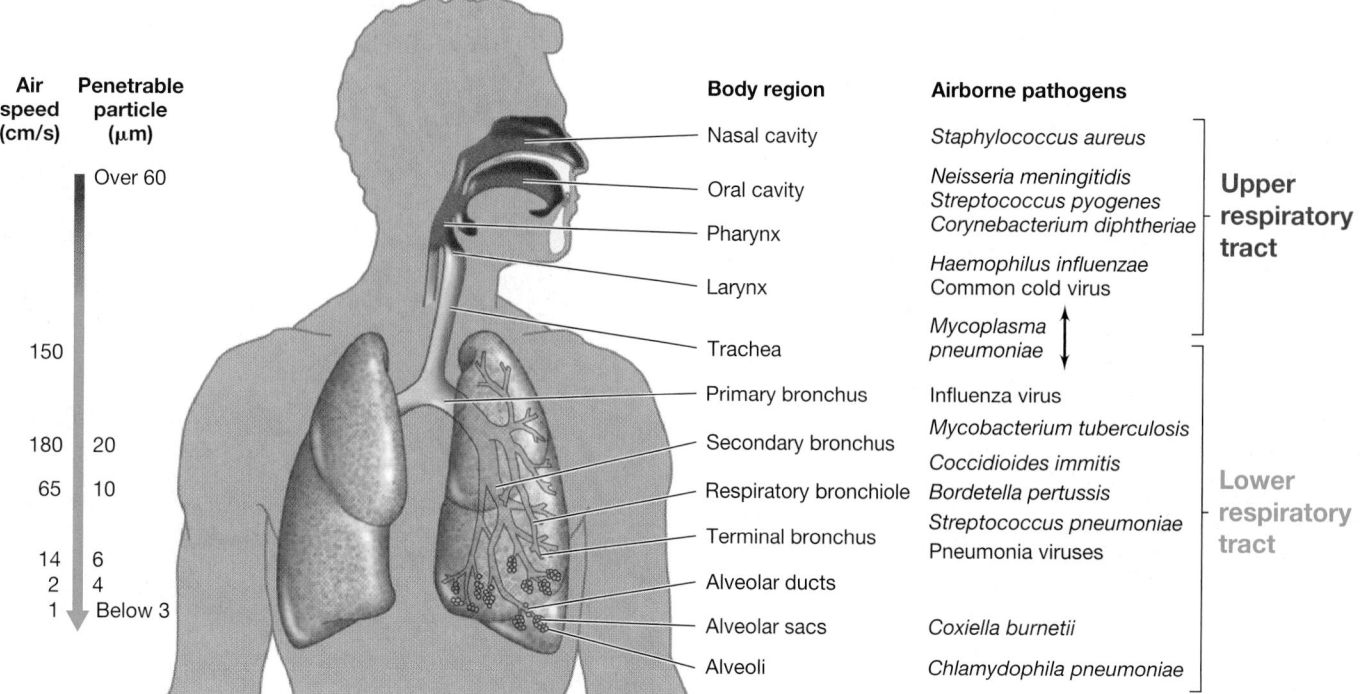

Figure 30.2 The human respiratory system. The microorganisms listed typically initiate infections at the locations indicated. The scale on the left combined with the respiratory tract diagram shows approximately where particles come to rest, determined by their size and air speed.

developed in the clinical setting and, if practiced effectively, can limit host damage. Many bacterial and viral pathogens transmitted by an airborne route can be controlled by immunization, and most respiratory bacterial pathogens respond readily to antibiotic therapy. Antiviral therapies, on the other hand, are rather limited, and recovery from viral infections is often due solely to the immune response.

MINIQUIZ

- Why can it be said that respiratory pathogens have exploited an effective means of transmission?
- Identify pathogens more commonly found in the upper respiratory tract. Identify pathogens more commonly found in the lower respiratory tract.

30.2 Streptococcal Syndromes

Streptococcal diseases are transmitted by airborne droplets or by direct contact, and the species *Streptococcus pyogenes* (**Figure 30.3**) and *Streptococcus pneumoniae* are the most important human respiratory pathogens. Streptococci are nonsporulating, homo-fermentative but aerotolerant gram-positive cocci (⊄ Section 16.6). Cells of *S. pyogenes* typically grow in elongated chains (Figure 30.3), as do many other species of the genus. Pathogenic strains of *S. pneumoniae* grow in pairs or short chains, and virulent strains produce an extensive polysaccharide capsule (see Figure 30.11). Virulent strains of *Streptococcus* can form vicious pus-forming wounds in humans and other warm-blooded (endothermic)

animals (**Figure 30.4** and see Figure 30.10). Many other serious conditions whose symptoms are less dramatic than these are also associated with streptococcal infections.

Streptococcus pyogenes

Streptococcus pyogenes (Figure 30.3), the major species in the *group A streptococci*, is frequently isolated from the upper respiratory tract of healthy adults. Although numbers are typically low, if host defenses are weakened or a new, highly virulent strain is encountered, serious infections are possible.

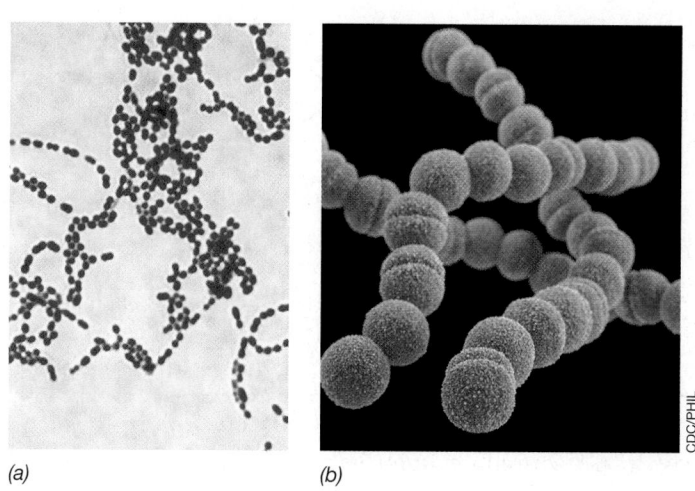

(a)   (b)

Figure 30.3 *Streptococcus pyogenes.* (a) Gram stain of cells of *Streptococcus pyogenes*. Cells grow in chains and range in size from 0.6 to 1 μm in diameter. (b) Computer-generated image of a scanning electron micrograph of cells of *S. pyogenes*.

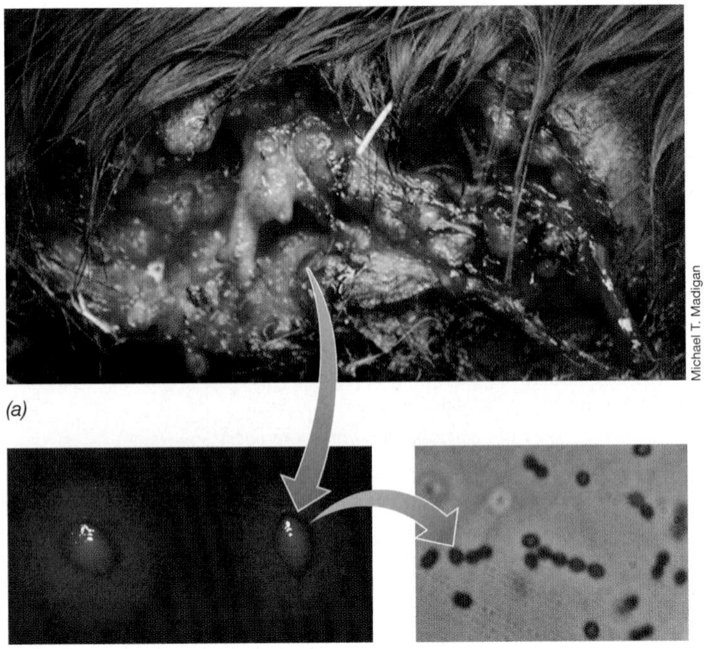

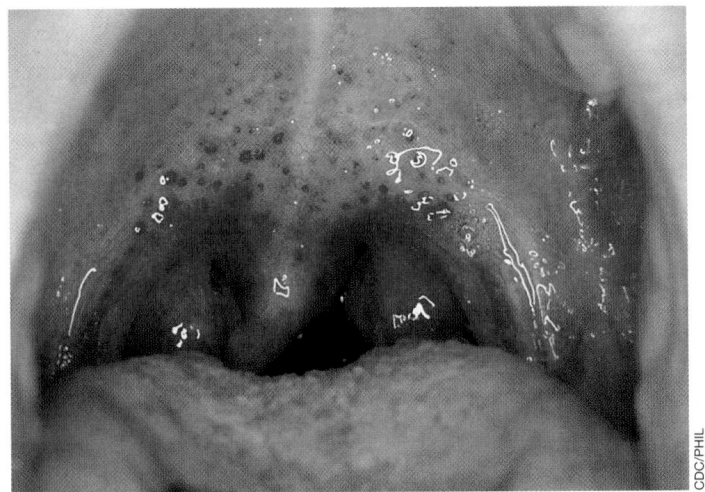

Michael T. Madigan

(a)

(b)   *(c)*

Figure 30.4 Pus-forming wound from β-hemolytic streptococci. *(a)* Pus and coagulated blood from a *Streptococcus equi* infection of a horse's salivary glands (the salivary glands have burst open from the infection). *(b)* Colonies of *S. equi* showing β-hemolysis on blood agar (compare with Figure 30.8). *(c)* Phase-contrast photomicrograph of cells of *S. equi.* Cells are 1 μm in diameter.

S. pyogenes is the cause of *streptococcal pharyngitis*, better known as *strep throat* (**Figure 30.5**). Most clinical isolates of *S. pyogenes* produce an exotoxin (⇌ Sections 25.6 and 25.7) that lyses red blood cells in culture media, a condition called *β-hemolysis* (Figure 30.4*b* and see Figure 30.8). Streptococcal pharyngitis is characterized by a severe sore throat with enlarged tonsils and red spots on the soft palate (Figure 30.5); tender cervical lymph nodes; and a mild fever and feeling of general malaise. *S. pyogenes* can also cause

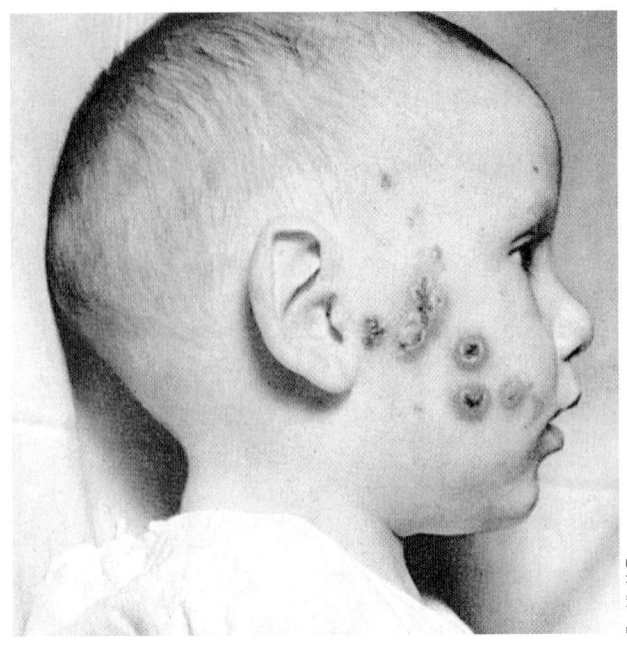

Franklin H. Top

Figure 30.6 Typical lesions of impetigo. Impetigo is commonly caused by *Streptococcus pyogenes* or *Staphylococcus aureus.*

infections of the middle ear (*otitis media*) and of the mammary glands (*mastitis*); infections of the superficial layers of the skin called *impetigo* (**Figure 30.6**); *erysipelas*, an acute streptococcal skin infection (**Figure 30.7**); and other conditions linked to the after effects of streptococcal infections.

About half of the clinical cases of severe sore throat are due to *S. pyogenes*; most others are due to viral infections. Because of this, an accurate and rapid diagnosis of a severe sore throat is important. If the sore throat is due to *S. pyogenes*, immediate treatment is important because untreated group A streptococcal infections can

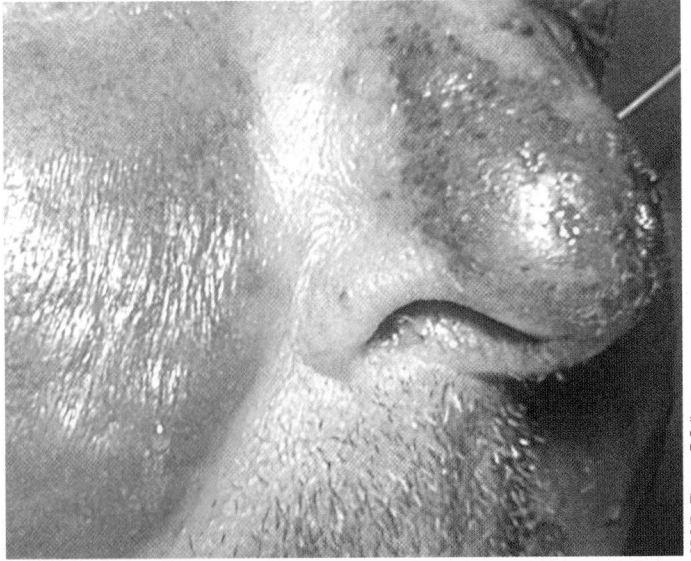

CDC/PHIL

Figure 30.5 A case of strep throat caused by *Streptococcus pyogenes*. The back of the throat is inflamed and shows small red spots, typical of streptococcal pharyngitis.

CDC/Dr. Thomas F. Sellers

Figure 30.7 Erysipelas. Erysipelas is a *Streptococcus pyogenes* infection of the skin, shown here on the nose and cheeks, characterized by redness and distinct margins of infection. Other commonly infected body sites include the ears and the legs.

UNIT 7

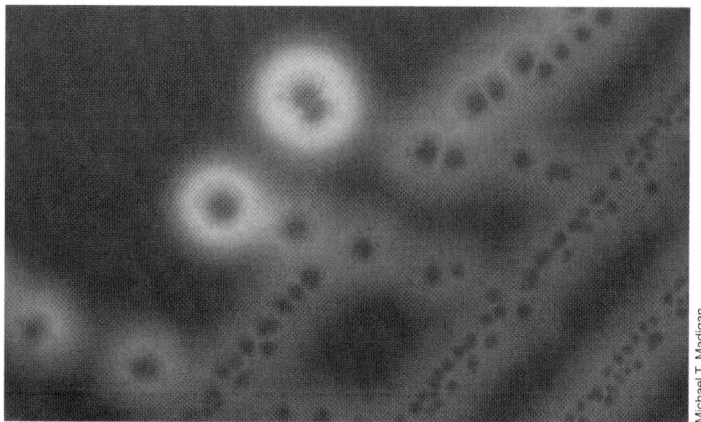

Figure 30.8 β-Hemolysis. The ability of a bacterium to lyse red blood cells and form a clear zone around a colony on a blood agar plate indicates secretion of the protein β-hemolysin. See also Figures 25.16a and 30.4b.

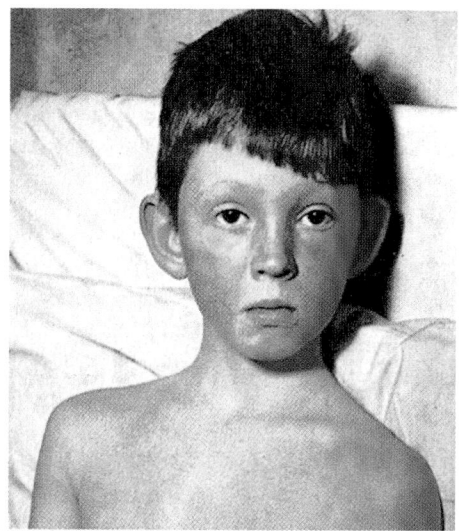

Figure 30.9 Scarlet fever. The typical rash of scarlet fever results from the activity of the pyrogenic exotoxins produced by *Streptococcus pyogenes*.

lead to serious secondary diseases such as scarlet fever, rheumatic fever, acute glomerulonephritis, and streptococcal toxic shock syndrome. On the other hand, if the sore throat is due to a virus, treatment with antibiotics will be useless and will only promote drug resistance on the part of the normal microbiota.

Clinical tools for quickly diagnosing strep throat are widely available and in routine use in primary care clinics. These tools include, in particular, rapid antigen detection systems that contain antibodies specific for cell surface proteins of *S. pyogenes* (⮑ Section 28.7). A more sensitive and accurate confirmation is possible by obtaining an actual culture of *S. pyogenes* from the throat or other suspected lesion on a blood agar plate (**Figure 30.8**). In contrast to rapid tests, however, results of a throat culture may take up to 48 h to process and such a delay in treatment can have adverse effects, as we consider now.

Scarlet and Rheumatic Fevers and Other Group A Strep Syndromes

Certain strains of group A streptococci carry a lysogenic bacteriophage that encodes streptococcal pyrogenic exotoxin A (SpeA), SpeB, SpeC, and SpeF. These exotoxins are responsible for most of the symptoms of *streptococcal toxic shock syndrome* and **scarlet fever** (**Figure 30.9**). Streptococcal pyrogenic exotoxins are superantigens that recruit large numbers of T cells to the infected tissues (⮑ Sections 25.6 and 27.10). Toxic shock results when the activated T cells secrete cytokines, which in turn activate large numbers of macrophages and neutrophils, causing severe inflammation and tissue destruction.

Scarlet fever, signaled by a severe sore throat, fever, and characteristic rash (Figure 30.9), is readily treatable with antibiotics or may be self-limiting. But treatment is always advisable because several undesirable conditions can emerge from a case of scarlet fever. Occasionally group A streptococcal infections cause fulminant (sudden and severe) invasive systemic infections, such as cellulitis, a skin infection in subcutaneous layers; or *necrotizing fasciitis*, a rapid and progressive disease resulting in extensive destruction of subcutaneous tissue, muscle, and fat (**Figure 30.10**). Necrotizing fasciitis is a clinical term for the condition caused by

"flesh-eating bacteria." In these cases, SpeA, SpeB, SpeC, and SpeF exotoxins and the bacterial cell surface M protein function as superantigens; the associated host inflammation results in extensive tissue destruction and can be fatal (Figure 30.10).

Untreated or insufficiently treated *S. pyogenes* infections may lead to other severe conditions 1 to 4 weeks after the onset of infection. For example, the immune response to the invading pathogen can produce antibodies that cross-react with host tissue antigens of the heart, joints, and kidneys, resulting in damage to these tissues. The most serious of these syndromes is **rheumatic fever** caused by rheumatogenic strains of *S. pyogenes*. These strains contain cell surface antigens that are similar in structure to heart valve and joint proteins. Thus rheumatic fever is, in effect, an *autoimmune disease* (⮑ Section 27.9) because antibodies directed against streptococcal antigens cross-react with heart valve and joint antigens, causing inflammation and tissue

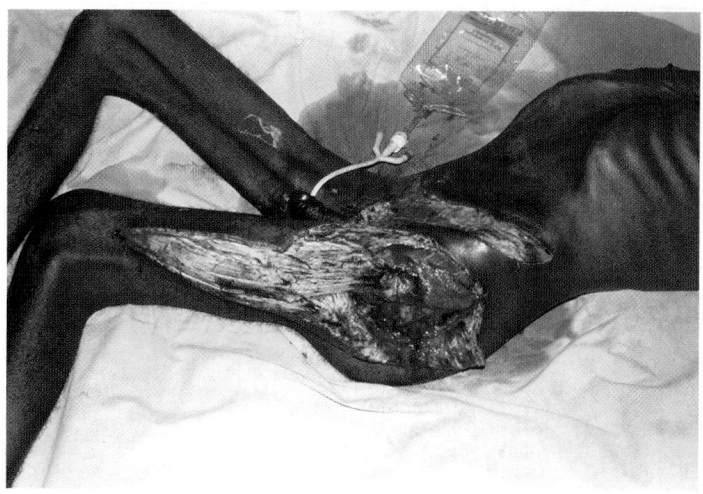

Figure 30.10 Necrotizing fasciitis (flesh-eating bacteria). Soft tissue infection of human hip and thigh by group A *Streptococcus pyogenes*. The flesh has split open to reveal muscle tissues.

UNIT 7

destruction. Damage to host tissues may be permanent, and is often exacerbated by subsequent streptococcal infections that lead to recurring bouts of rheumatic fever. Another streptococcal syndrome is *acute poststreptococcal glomerulonephritis*, a painful kidney disease. This "immune complex" disease develops transiently when streptococcal antigen–antibody complexes in the blood lodge in the glomeruli (filtration membranes of the kidney) and cause inflammation, a condition called *nephritis*.

Streptococcus pneumoniae

A second major human streptococcal pathogen is *Streptococcus pneumoniae* (**Figure 30.11**), a species that can cause invasive lung infections, typically as secondary infections to other respiratory disorders. Encapsulated strains of *S. pneumoniae* (Figure 30.11; ⮌ Figure 25.4) are particularly pathogenic because they are very invasive. Cells invade the lower respiratory tract where the capsule enables the cells to resist phagocytosis yet generate a strong host inflammatory response. Reduced lung function, called *pneumonia*, results from the accumulation of recruited phagocytic cells and fluid. Cells of *S. pneumoniae* can then spread from the focus of infection as a bacteremia, sometimes infecting the bones, middle ear, and heart valves (endocarditis). *S. pneumoniae* infection is often the cause of death in elderly persons whose death is reported to be from "respiratory failure."

Unlike the case with *S. pyogenes*, effective vaccines are available for preventing infection by the most common strains of *S. pneumoniae*. An older vaccine widely used in adults consisted of a mixture of 23 capsular polysaccharides (Figure 30.11) from the most prevalent pathogenic strains. The vaccine is recommended for those over age 60, healthcare providers, individuals with compromised immunity, and any other high-risk population. A newer vaccine, PREVNAR 13®, is an update of the traditional vaccine and is effective against the 13 *S. pneumoniae* strains most commonly seen today and is recommended for adults age 50 or older.

S. pneumoniae infections typically respond quickly to penicillin therapy, but up to 30% of pathogenic isolates now exhibit resistance to this drug. Resistance to the antibiotics erythromycin and cefotaxime is also found in some strains but thus far, all strains have been found sensitive to vancomycin, an antibiotic held in reserve for treating pneumonia and several other bacterial diseases where antibiotic resistance is widespread (⮌ Section 28.12).

─── **MINIQUIZ** ───

- How does *Streptococcus pyogenes* infection cause rheumatic fever?
- What is the primary virulence factor for *Streptococcus pneumoniae*?

30.3 Diphtheria and Pertussis

Diphtheria is a severe respiratory disease that typically infects young children. Diphtheria is caused by *Corynebacterium diphtheriae*, a gram-positive, nonmotile, and aerobic club-shaped bacterium that forms small, smooth colonies on blood agar plates (**Figure 30.12**). **Pertussis**, also known as **whooping cough**, is a serious respiratory disease caused by infection with *Bordetella pertussis*, a small, gram-negative, aerobic coccobacillus (see Figure 30.14). Pertussis mostly affects children but can cause serious respiratory disease in adults as well. Both diphtheria and pertussis can be prevented by vaccination and cured with antibiotics.

Diphtheria

Cells of *C. diphtheriae* (Figure 30.12*a*) enter the host from airborne droplets, infecting the tissues of the throat and tonsils. A child with diphtheria often displays a swollen neck (**Figure 30.13***a*), and throat tissues respond to *C. diphtheriae* infection by forming a characteristic lesion called a *pseudomembrane* consisting of damaged host cells and cells of *C. diphtheriae* (Figure 30.13*b*). Not all strains of *C. diphtheriae* cause diphtheria. *Pathogenic* strains of *C. diphtheriae* carry a lysogenic bacteriophage whose genome encodes a powerful exotoxin called *diphtheria toxin*. This toxin inhibits protein synthesis in the host, leading to cell death (⮌ Figure 25.12). Death from diphtheria is due to a combination of partial suffocation by the pseudomembrane and tissue destruction by diphtheria exotoxin. *C. diphtheriae* isolated from

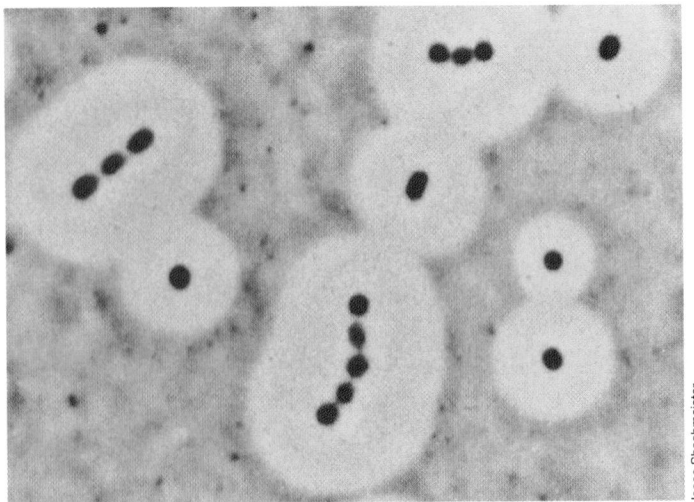

Figure 30.11 *Streptococcus pneumoniae*. India ink negative stain of cells of *Streptococcus pneumoniae*. An extensive capsule surrounds the cells, which are 1.0–1.2 μm in diameter.

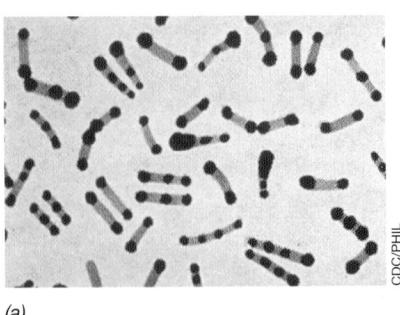
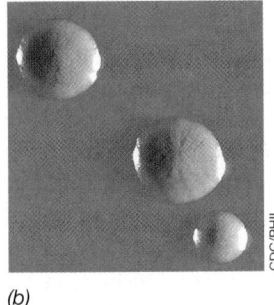

(a)   (b)

Figure 30.12 *Corynebacterium* **and diphtheria.** *(a)* Cells of *Corynebacterium diphtheriae* showing typical club-shaped appearance. The gram-positive cells are 0.5–1 μm in diameter and may be several micrometers in length. *(b)* Colonies of *C. diphtheriae* grown on a selective medium of blood agar plus tellurite.

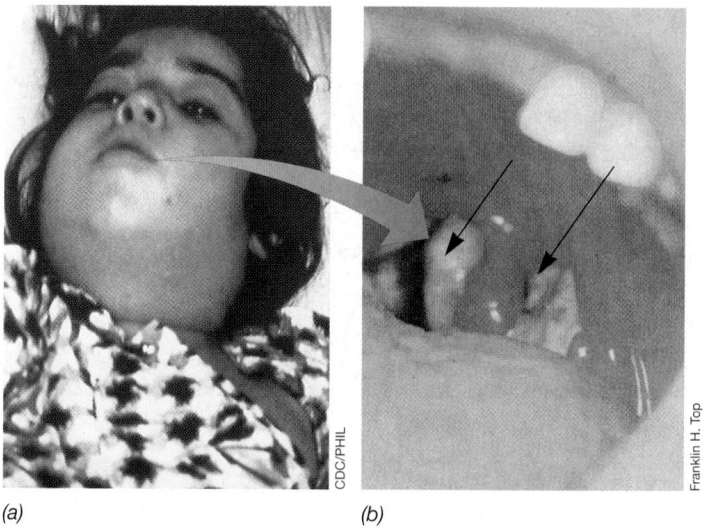

(a)   *(b)*

Figure 30.13 Diphtheria. *(a)* A swollen neck is a common symptom of diphtheria. *(b)* The pseudomembrane (arrows) in an active case of diphtheria restricts airflow and swallowing and is associated with a severe sore throat.

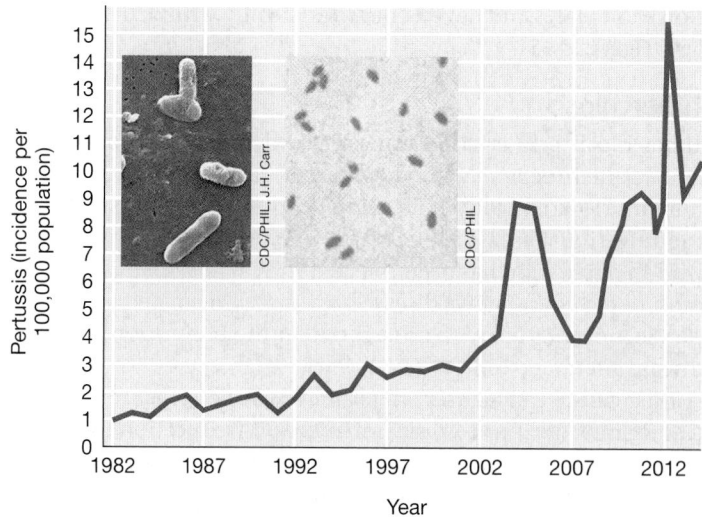

Figure 30.14 *Bordetella* **and pertussis.** Cells of *B. pertussis* are typically coccobacilli 0.2–0.5 μm in diameter and about 1 μm long. Pertussis incidence has been on the rise since 2007. Data are from the CDC. Inset photos: left, scanning electron micrograph of cells of *Bordetella*; right, Gram-stained cells of *B. pertussis*.

the throat is diagnostic for diphtheria. Nasal or throat swabs are used to inoculate blood agar containing tellurite (Figure 30.12*b*) or Loeffler's medium, a selective medium that inhibits the growth of most other respiratory pathogens.

Prevention of diphtheria is accomplished with a highly effective toxoid vaccine, part of the *DTaP* (*d*iphtheria toxoid, *t*etanus toxoid, and *a*cellular *p*ertussis) *vaccine* (⮫ Section 28.9). Diphtheria is all but absent from developed countries where this vaccine is widely used. Penicillin, erythromycin, and gentamicin are generally effective treatments for diphtheria, but in life-threatening cases, diphtheria antitoxin (an antiserum to diphtheria toxoid produced in horses) may be administered in addition to antibiotic therapy.

Pertussis

Pertussis (whooping cough) is an acute, highly infectious respiratory disease. Infants less than 6 months old, who are too young to be effectively vaccinated, have the highest incidence of disease and also have the most severe symptoms. Cells of *B. pertussis* (**Figure 30.14**) attach to host cells of the upper respiratory tract and excrete *pertussis exotoxin*. This potent toxin induces synthesis of cyclic adenosine monophosphate (cyclic AMP, ⮫ Figure 6.13), which is at least partially responsible for the events that lead to host tissue damage. *B. pertussis* also produces an endotoxin (⮫ Section 25.8), which may induce some of the symptoms of whooping cough. Clinically, whooping cough is characterized by a recurrent, violent cough that can last up to 6 weeks. The spasmodic coughing gives the disease its name; a whooping sound results from the patient inhaling deep breaths to obtain sufficient air.

Worldwide, up to 50 million cases and over 250,000 deaths occur each year from pertussis, most in developing countries. *B. pertussis* is endemic worldwide and pertussis remains a problem, even in developed countries, usually as a result of inadequate immunization. In the United States there has been a gradual upward trend of pertussis since the 1980s, with spikes in reported

cases in 2005, 2010, and 2012 (Figure 30.14); many of these have been in young adults under age 20. In 2014 there were nearly 33,000 cases of pertussis in the United States and 13 deaths, all but two in children under age 4. Pertussis is a classic endemic disease; incidence rises cyclically as populations become susceptible and are exposed to the pathogen. A combination of lax vaccination protocols and the fact that pertussis is a much more common disease than diphtheria have probably fueled the overall higher incidence of pertussis in recent years.

Whooping cough can be treated with ampicillin, tetracycline, or erythromycin, although antibiotics alone do not seem to effect a complete cure, as patients continue to show symptoms and remain infectious for up to 2 weeks after beginning antibiotic therapy. This indicates that the immune response may be as important as antibiotics in ridding the pathogen from the body.

MINIQUIZ

- Contrast the disease symptoms of diphtheria and pertussis.
- What measures can be taken to decrease the current incidence of pertussis in a population?

30.4 Tuberculosis and Leprosy

The famous pioneering microbiologist Robert Koch, the founder of the field of medical microbiology, isolated and described the causative agent of tuberculosis, *Mycobacterium tuberculosis*, in 1882 (⮫ Section 1.10). A related species, *Mycobacterium leprae*, causes leprosy (Hansen's disease). Mycobacteria are gram-positive bacteria and share the property of being *acid-fast* because of the waxy mycolic acid constituent of their cell walls (⮫ Section 16.11). Mycolic acid allows these organisms to retain the red dye carbolfuchsin after a mycobacterial smear on a slide is washed in 3% hydrochloric acid in alcohol. Colonies of *M. tuberculosis* grow slowly on plates and have a characteristically wrinkled morphology (**Figure 30.15**).

Tuberculosis

Tuberculosis (TB) is easily transmitted by the respiratory route, and at one time it was the most important infectious disease of humans. TB kills nearly 1.5 million people per year, making it the top infectious disease killer worldwide. About one-third of the world's population has been infected with *M. tuberculosis*, though most do not show active disease because cell-mediated immunity (⟲ Section 27.9) plays a critical role in the prevention of active disease after infection.

Tuberculosis can take several forms. TB can be a *primary* infection (initial infection) or *postprimary* infection (reinfection). Primary infection typically results from inhalation of droplets containing *M. tuberculosis*, after which the bacteria settle in the lungs and grow. The host mounts an immune response to *M. tuberculosis*, resulting in the formation of aggregates of activated macrophages, called *tubercles*. Bacteria are found in the sputum of individuals with active disease, and areas of destroyed tissue can be seen in chest X-rays (**Figure 30.16**). Mycobacteria survive and grow within macrophages in the tubercles, forming granulomas, and if the disease is not controlled, extensive destruction of lung tissue can occur. If the disease reaches this stage, the pulmonary infection is often fatal.

In most individuals infected with *M. tuberculosis*, however, acute disease does not occur; instead, the infection is inapparent. Nevertheless, the infection hypersensitizes the individual to *M. tuberculosis* or its products and typically protects the individual against postprimary infections. A diagnostic skin test, called the **tuberculin test**, can detect this hypersensitivity (⟲ Figure 27.27), and many healthy adults are *tuberculin-positive* as a result of previous or current inapparent infections. In most cases, the cell-mediated immune response to *M. tuberculosis* is protective and lifelong. However, some tuberculin-positive patients develop postprimary tuberculosis through reinfection from bacteria that have remained dormant in lung macrophages for years. Because

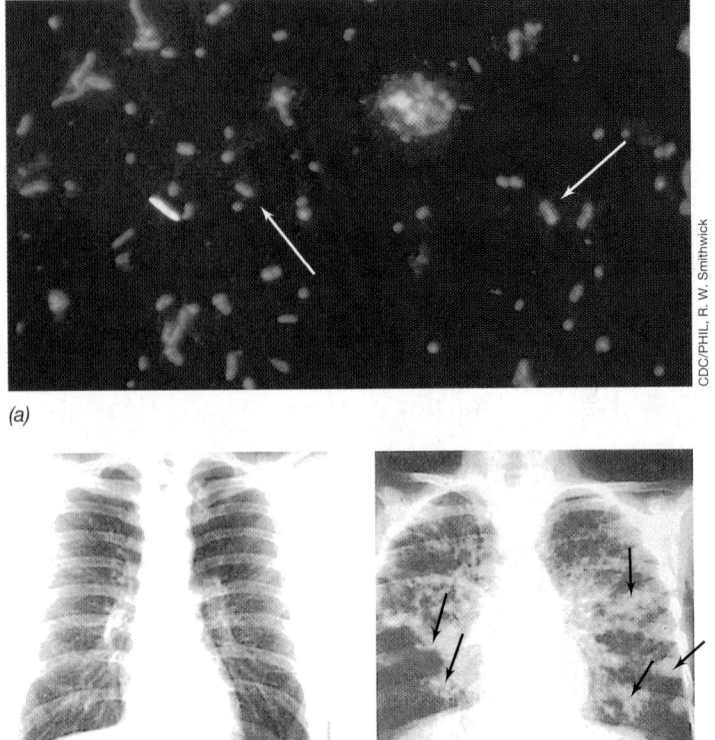

(a)

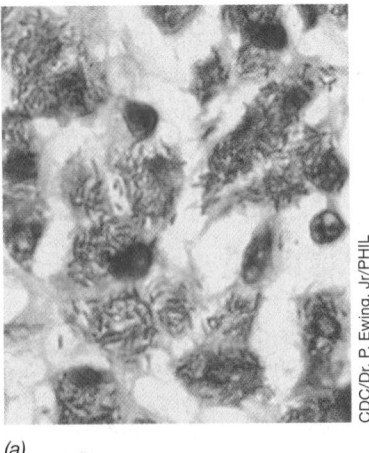

(a)

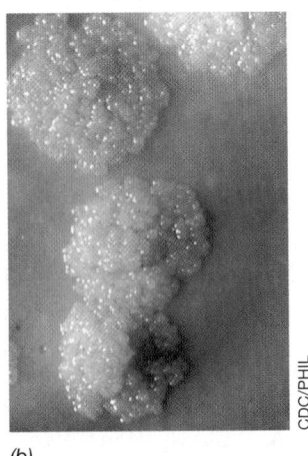

(b)

(b)   *(c)*

Figure 30.16 Tuberculosis symptoms. *(a)* Sputum sample from a patient with tuberculosis stained by the Smithwick acridine orange method. Cells of *Mycobacterium tuberculosis* are the yellow-orange rod-shaped structures (arrows). *(b)* Normal chest X-ray. The faint white lines are arteries and other blood vessels. *(c)* Chest X-ray showing an advanced case of pulmonary tuberculosis; white patches (arrows) indicate areas of tubercles that contain viable cells of *M. tuberculosis*.

Figure 30.15 Mycobacteria. *(a)* Acid-fast stained lymph node biopsy from a patient with HIV/AIDS shows cells of *Mycobacterium avium*, a relative of *M. tuberculosis*. Multiple bacilli, stained red with carbol-fuchsin and treated with 3% hydrochloric acid, are evident inside each human cell. The individual rods are about 0.4 μm in diameter and up to 4 μm in length. *(b)* Colonies of *M. tuberculosis*. The rough, wrinkled surface is typical of mycobacterial colonies.

of this, individuals who have a positive tuberculin test are typically treated with anti-tuberculosis drugs for extended periods to ensure that all mycobacteria have been killed.

Antimicrobial therapy of TB has been a major means of controlling the disease. Streptomycin was the first effective anti-tuberculosis antibiotic, but the real revolution in tuberculosis treatment came with the discovery of isonicotinic acid hydrazide, called *isoniazid* (INH). This drug is highly effective and readily absorbed when given orally. Isoniazid is a growth factor analog (⟲ Section 28.10 and Table 28.5) of the structurally related molecule nicotinamide; in mycobacteria the drug inhibits mycolic acid synthesis and this compromises cell wall integrity. Following treatment with isoniazid, mycobacteria lose their acid-fast properties, in keeping with the role of mycolic acid in this staining property.

Treatment of tuberculin-positive individuals is typically achieved with daily doses of isoniazid and the antibiotic rifampin for 2 months, followed by biweekly doses for a total of 9 months. This treatment eradicates pockets of *M. tuberculosis* cells and prevents emergence of antibiotic-resistant derivatives. Multiple drug therapy reduces the possibility that strains having resistance to more than one drug will emerge. Resistance of *M. tuberculosis* to isoniazid and other drugs, however, is increasing, especially in

HIV/AIDS patients, in whom TB is a common infection (see Figure 30.45*g*). Treatment of these strains, called *multidrug-resistant tuberculosis strains*, requires the use of second-line tuberculosis drugs that are generally more toxic, less effective, and more costly than rifampin and isoniazid.

Leprosy

Mycobacterium leprae, a relative of *M. tuberculosis*, causes the disease *leprosy*, more formally known as *Hansen's disease*. The most serious form of Hansen's disease is *lepromatous* leprosy, characterized by folded, bulblike lesions on the body, especially on cooler parts of the body such as the face and extremities (**Figure 30.17***a*). The lesions are due to the growth of *M. leprae* cells in skin Schwann cells that insulate the nerves, and the lesions contain large numbers of bacterial cells. Like cells of other mycobacteria (Figure 30.15*a*), cells of *M. leprae* from the lesions stain deep red with carbol-fuchsin in the acid-fast staining procedure, providing a definitive demonstration of active infection.

In severe untreated cases of leprosy, the disfiguring lesions lead to destruction of peripheral nerves; muscles then atrophy and motor function is impaired. The loss of sensation in the extremities leads to inapparent injuries, such as burns and cuts. Loss of bone calcium leads to a slow shrinking of the digits and their transition to clawlike forms in late-stage leprosy (Figure 30.17*b*). Pathogenicity in the disease is due to a combination of delayed-type hypersensitivity (⇄ Section 27.9) and the highly invasive activities of *M. leprae*, which can grow within macrophages and lead to the characteristic lesions (Figure 30.17*a*). Leprosy is transmitted by direct contact as well as by an airborne route, but is not as highly contagious as TB. Historically, leprosy has been associated with poverty, malnutrition, and poor sanitation and hygiene. Among other things, these factors undoubtedly affect an individual's ability to resist infection.

Many Hansen's disease patients exhibit less-pronounced lesions from which *M. leprae* cells cannot be obtained; these individuals have the *tuberculoid* form of the disease. Tuberculoid leprosy is characterized by a vigorous immune response and a good prognosis for spontaneous recovery. Hansen's disease of either form, and the continuum of intermediate forms, is treated using a multiple drug therapy protocol, which includes some combination of extended therapy of up to 1 year with *dapsone* (4,4′-sulfonylbis-benzeneamine, an inhibitor of folic acid synthesis), *rifampin*, a bacterial RNA polymerase inhibitor, and *clofazimine*, a drug that targets bacterial respiration and ion transport.

Nearly 214,000 new cases of leprosy were reported in 2014, with most cases occurring in Africa, the Indian subcontinent, and Brazil. In the United States only about 200 cases of leprosy are diagnosed per year, mainly in immigrants from countries with endemic disease. Until recently, a leprosy diagnosis relied on the identification of *M. leprae* cells from lesions. However, a quick, inexpensive, and specific blood test is now available that should greatly assist in identifying early-stage leprosy, the most treatable form.

In addition to *M. tuberculosis* and *M. leprae*, several other mycobacteria are human pathogens. These include in particular *M. bovis*, a close relative of *M. tuberculosis* and a common pathogen of dairy cattle. *M. bovis* can initiate classic symptoms of TB in humans; however, a combination of the pasteurization of milk and the culling of infected cattle has greatly reduced the incidence of bovine-to-human transmission of this form of TB.

MINIQUIZ

- Why is *Mycobacterium tuberculosis* a widespread respiratory pathogen?
- Describe three common characteristics of pathogenic mycobacteria.

30.5 Meningitis and Meningococcemia

Meningitis is an inflammation of the meninges, the membranes that are the protective covering of the central nervous system, that is, the spinal cord and brain. Several different microorganisms, including certain viruses, bacteria, fungi, and protists, can cause meningitis. Here we focus on the severe bacterial form of the disease called *infectious meningitis*, caused by the bacterium *Neisseria meningitidis*.

Pathogen and Disease Syndromes

Neisseria meningitidis, often called the *meningococcus*, is a gram-negative and obligately aerobic coccus about 0.6–1 μm in diameter (**Figure 30.18***a*); it is a relative of the bacterium that causes gonorrhea, *Neisseria gonorrhoeae*. The bacterium is transmitted to a new host, usually via the airborne route from an infected individual, and attaches to the cells of the nasopharynx. Once there, the organism quickly gains access to the bloodstream, causing widespread dissemination (bacteremia) and upper respiratory tract symptoms. Meningitis is characterized by the sudden onset of a headache accompanied by vomiting and a stiff neck, and can progress to coma and death in less than a day. Instead of or in addition to full-blown meningitis, *N. meningitidis* bacteremia sometimes leads to fulminant **meningococcemia**, a condition characterized by intravascular coagulation and tissue destruction (gangrene, Figure 30.18*b*), shock, and death in over 10% of cases.

Meningococcal meningitis often occurs in epidemics, usually in populations living in close proximity such as in military barracks or college dormitories. Anyone can get meningococcal disease, but the incidence is much higher in infants, school-age children, and young adults. Up to 30% of people carry *N. meningitidis* in their nasopharynx

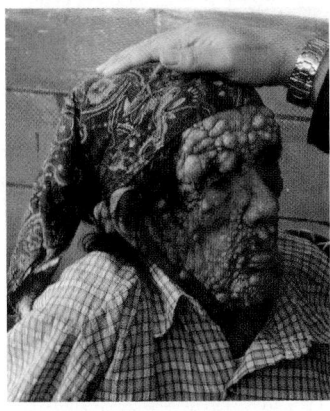
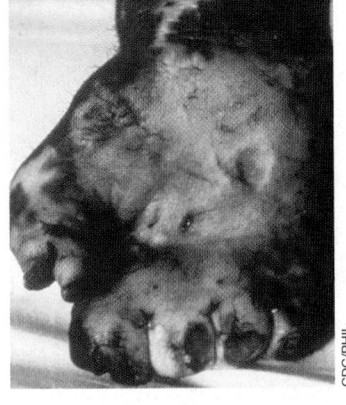

(a)   *(b)*

Figure 30.17 Lepromatous leprosy lesions on the skin. *(a)* Lepromatous leprosy is caused by infection with *Mycobacterium leprae*. The lesions can contain up to 10⁹ bacterial cells per gram of tissue, indicating an active uncontrolled infection with a poor prognosis. *(b)* The palm of the right hand of a leprosy patient showing the clawlike form and digit deformation characteristic of late-stage leprosy.

UNIT 7

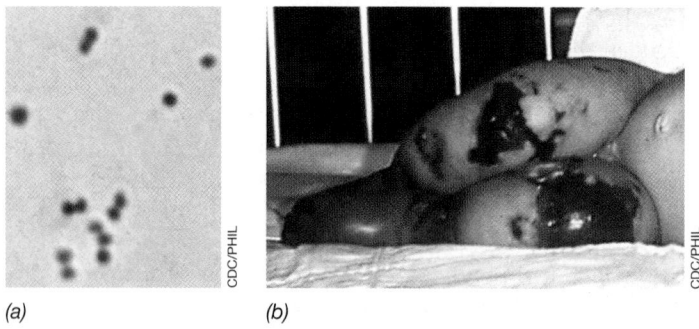

(a)   (b)

Figure 30.18 *Neisseria meningitidis.* The organism causes meningitis and meningococcemia. *(a)* Gram stain of cells of *N. meningitidis*; cocci are about 0.6–1 µm in diameter. *(b)* Four-month-old infant with gangrene on legs from meningococcemia.

with no apparent harmful effects, and the trigger for conversion from the asymptomatic carrier state to the disease state is unknown.

Diagnosis, Treatment, and Vaccines

Meningococcal meningitis is definitively diagnosed from cultures of *N. meningitidis* isolated from nasopharyngeal swabs, blood, or cerebrospinal fluid. Thayer–Martin medium, a selective medium for the growth of pathogenic *Neisseria*, including both *N. meningitidis* and *N. gonorrhoeae*, is used to isolate *N. meningitidis*, and colonies containing gram-negative diplococci (Figure 30.18*a*) are further tested. However, due to the rapid onset of life-threatening symptoms in infectious meningitis, preliminary diagnosis is often based on clinical symptoms and treatment is started before culture tests confirm infection with *N. meningitidis*. Treatment is typically with penicillin, and intravenous application is often needed to speed antibiotic infusion.

Naturally occurring antibodies acquired by subclinical infections with *N. meningitidis* are effective for preventing infectious meningitis in most adults. Vaccines consisting of purified polysaccharides or polysaccharides from the most prevalent pathogenic strains are available to immunize certain susceptible populations such as military recruits and students living in dormitories, especially if an outbreak has already occurred. In addition, the antibiotic rifampin is often used to eradicate the carrier state and prevent meningococcal disease in close contacts of infected individuals.

MINIQUIZ

- Identify the symptoms and causes of meningitis.
- Describe the infection by *Neisseria meningitidis* and the resulting development of meningococcemia.

II • Airborne Viral Diseases

30.6 MMR and Varicella-Zoster Infections

The most prevalent and difficult to treat of all human infectious diseases are those caused by viruses. This is because viruses can often remain infectious for long periods in dried mucus (Figure 30.1) or on fomites, and because viruses require host cells for replication. Hence, killing the virus often means killing the cell as well.

Most viral diseases are acute, self-limiting infections, but some can be problematic in healthy adults. We begin with measles,

rubella, mumps, and chicken pox, all common, endemic viral diseases transmitted in infectious droplets by an airborne route.

Measles and Rubella

Measles (*rubeola* or *7-day measles*) affects susceptible children as an acute, highly infectious, often epidemic disease. The measles virus (**Figure 30.19***a*) is a *paramyxovirus*, a single-stranded, minus-sense RNA virus (⇌ Section 10.9) that enters the nose and throat by airborne transmission, quickly leading to a systemic viremia. Symptoms start with nasal discharge and redness of the eyes. As the disease progresses, fever and cough appear and rapidly intensify, followed by a characteristic rash (Figure 30.19*b, c*).

Symptoms of measles generally persist for 7–10 days, and no drugs are available that will eliminate symptoms. However, the measles virus generates a strong immune response. Circulating antibodies to measles virus are measurable within 5 days of infection; these serum antibodies along with T-cytotoxic lymphocytes combine to eliminate the virus from the host. Possible postinfection complications include middle ear infection, pneumonia, and, in rare cases, measles encephalomyelitis.

Although once a common childhood illness, measles is limited to rare, isolated outbreaks in the United States because of widespread immunization programs begun in the 1960s. Those outbreaks that have occurred have been in populations that were either not immunized or inadequately immunized. In 2015, fewer than 200 cases were reported in the United States. Worldwide, measles remains

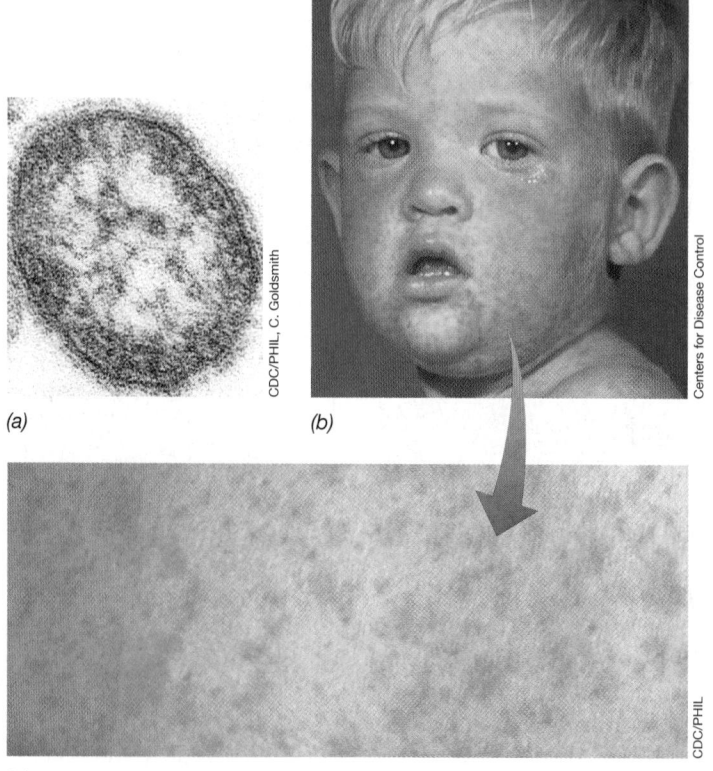

(a)   (b)

(c)

Figure 30.19 Measles in children. *(a)* Transmission electron micrograph of a measles virus virion; a virion is about 150 nm in diameter. *(b, c)* Measles rash. The light pink rash starts on the head and neck, and can spread to the chest, trunk, and limbs. Discrete papules coalesce into blotches as the rash progresses for several days.

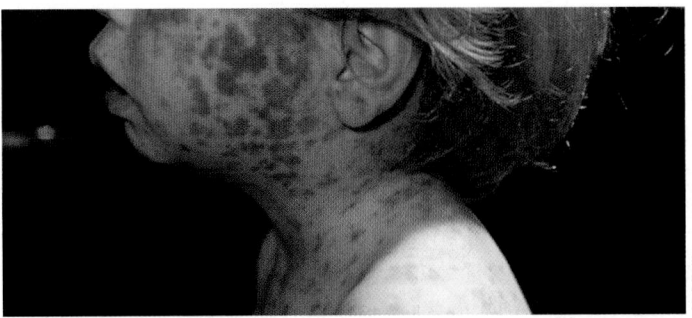

Figure 30.20 Rubella. The rash of rubella (German measles) on the face of a young child.

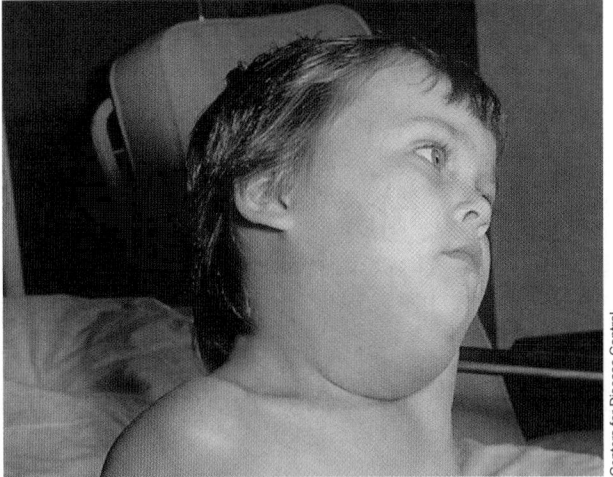

Figure 30.21 Mumps. Glandular swelling characterizes infection with the mumps virus. Mumps symptoms typically last about one week and a person is infectious both before and during the symptomatic stages.

endemic, however, and still causes over 100,000 annual deaths, mostly in children. Active immunity to measles is conferred with an attenuated virus preparation as part of the *MMR* (*m*easles, *m*umps, *r*ubella) vaccine (🔗 Section 28.9). Because the disease is highly infectious, all public school systems in the United States require proof of measles immunization before a child can enroll. A childhood case of measles generally confers lifelong immunity.

Rubella (sometimes called *German measles* or *3-day measles*) is caused by a single-stranded, positive-sense RNA virus (🔗 Section 10.8). Symptoms of rubella resemble those of measles (**Figure 30.20**) but are often restricted to just the upper torso. Rubella is less contagious than measles, and thus a significant proportion of the population has never been infected. However, during the first three months of pregnancy, rubella virus can infect the fetus by placental transmission and cause serious fetal abnormalities including stillbirth, deafness, heart and eye defects, and brain damage, events called *congenital rubella syndrome*. Thus, women should not be immunized with the rubella vaccine or contract a rubella infection during pregnancy. Also for this reason, routine childhood immunization against rubella should be practiced. An attenuated rubella virus is administered as part of the MMR vaccine. The low incidence of rubella since 2001, coupled with the high degree of protection by the vaccine and the relatively low infectivity of the virus, combine to make rubella very rare in the United States. An active rubella vaccination program worldwide is also decreasing case numbers significantly, as total reported active cases in 2009 were fewer than 125,000 and cases of congenital rubella syndrome were 165.

Mumps

Mumps, like measles, is caused by a paramyxovirus and is also highly infectious by the airborne route. Mumps is spread by airborne droplets, and the disease is characterized by inflammation of the salivary glands, typically the parotid gland, the largest of the salivary glands, leading to swelling of the jaws and neck (**Figure 30.21**). The virus spreads through the bloodstream and may infect other organs, including the testes and pancreas, and may cause encephalitis in rare severe cases. As for measles, the immune response rather than drug treatment is what cures a case of mumps. The host immune response produces antibodies to mumps virus surface proteins, and this generally leads to a quick recovery and lifelong immunity to reinfection.

An attenuated mumps vaccine (part of the MMR) is highly effective in preventing disease. Hence, the prevalence of mumps in developed countries is low, with disease generally restricted to individuals

who did not receive the vaccine. However, an outbreak of mumps in the midwestern United States in 2006 involved more than 5000 cases. The outbreak affected mainly young adults (18–34). As a result, recommendations for immunizations were revised to target school-age children, healthcare workers, and adults who had not previously had mumps. Since 2013, mumps cases in the United States have fluctuated between about 500 and 1200 annually.

Chicken Pox and Shingles

Chicken pox (*varicella*) is a common childhood disease caused by the varicella-zoster virus (VZV), a double-stranded DNA herpesvirus (🔗 Section 10.7). VZV is a mild but highly contagious disease and is transmitted by infectious droplets, especially when susceptible individuals are in close contact. In schoolchildren, for example, close confinement during the winter months leads to the spread of VZV through airborne droplets from infected classmates and through direct contact with chicken pox blisters of other children or contaminated fomites. The virus enters the respiratory tract, multiplies, and is quickly disseminated via the bloodstream, resulting in a systemic papular rash (**Figure 30.22**)

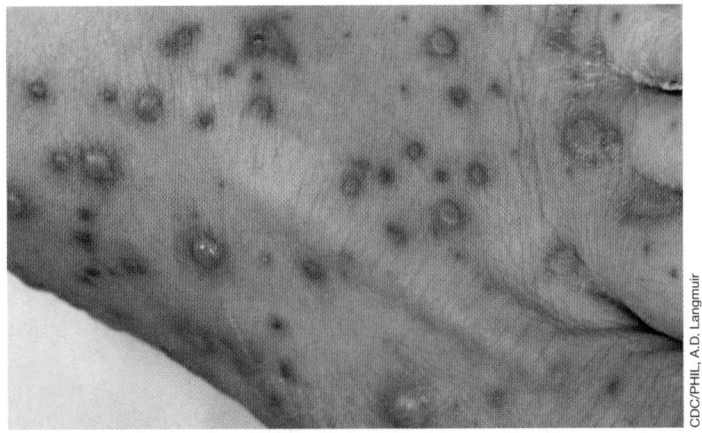

Figure 30.22 Chicken pox. Papular rash on the foot of an adult. The papules are due to infection by varicella-zoster virus, the herpesvirus that causes chicken pox.

UNIT 7

that heals quickly without scarring. An attenuated chicken pox virus vaccine (Varivax) is used in the United States but not as widely as the MMR vaccine for measles, rubella, and mumps. Consequently, the reported incidence of chicken pox in 2011 was about 15,000 cases per year, which is about 10% of the number of cases reported in 1995, the year the vaccine was first licensed. Deaths from chicken pox are extremely rare, with six deaths reported in 2011.

VZV establishes a lifelong latent (permanent) infection in nerve cells. The virus can remain dormant there indefinitely, but in some individuals the virus migrates from this reservoir to the skin surface, often years or decades later, causing a painful skin eruption called *shingles* (*zoster*). Shingles most commonly strikes immunosuppressed individuals or the elderly, causing severe blisters and a rash on the head, neck, or upper torso. A fairly effective shingles vaccine containing concentrated attenuated virus (Zostavax) is available for individuals over 50 years of age. The vaccine stimulates antibody- and cell-mediated immunity to VZV, which keeps VZV from migrating out of nerve ganglia to skin cells and triggering shingles symptoms.

MINIQUIZ

- How do the genomes of the measles virus and the German measles virus differ?
- Describe the potential serious outcomes of infection by measles, mumps, rubella, and VZV viruses.
- Identify the effects of immunization on the incidence of measles, mumps, rubella, and chicken pox.

30.7 The Common Cold

Colds are the most common of infectious diseases. Colds are typically upper respiratory tract viral infections that are transmitted via droplets spread from coughs, sneezes, and respiratory secretions. Colds are usually of short duration, lasting a week or so, and the symptoms are milder than other respiratory diseases such as influenza. **Table 30.1** contrasts the usually distinct symptoms and incidence of colds and influenza.

TABLE 30.1 Colds and influenza

Symptoms	Cold	Influenza
Fever	Rare	Common (39–40°C); sudden onset
Headache	Rare	Common
General malaise	Slight	Common; often quite severe; can last several weeks
Nasal discharge	Common and abundant	Less common; usually not abundant
Sore throat	Common	Less common
Vomiting and/or diarrhea	Rare	Common in children
Incidence[a]	340	50

[a]Cases/100 people per year in the United States for recent years. Incidence of all other infectious diseases totals about 30 cases/100 people per year.

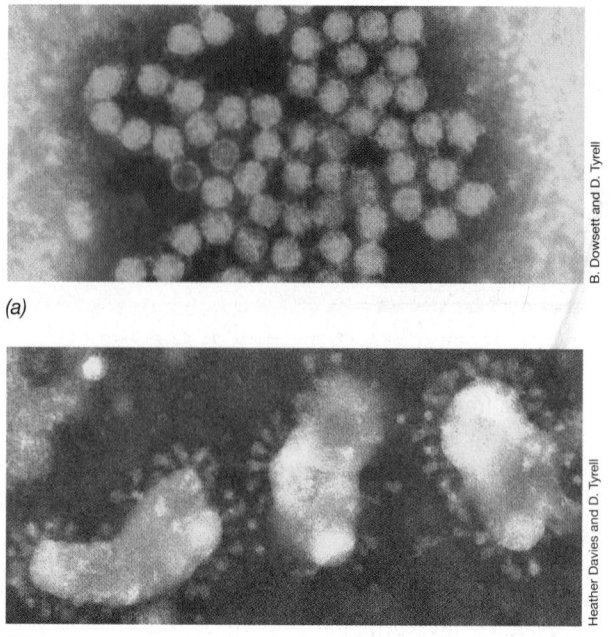

(a)

B. Dowsett and D. Tyrell

(b)

Heather Davies and D. Tyrell

Figure 30.23 Transmission electron micrographs of common cold viruses. (a) Human rhinovirus; a virion is about 30 nm in diameter. (b) Human coronavirus; a virion is about 60 nm in diameter.

Symptoms and Transmission of the Common Cold

Cold symptoms include *rhinitis* (inflammation of the nasal region, especially the mucous membranes), nasal obstruction, watery nasal discharges, muscle aches, and a general feeling of malaise, usually without fever. *Rhinoviruses* are single-stranded plus-sense RNA viruses of the picornavirus group (**Figure 30.23a**) (ꝏ Section 10.8) and are the most common causes of colds. Over 100 different rhinoviruses have been identified. About one-quarter of all colds are due to infections with other viruses. These include in particular the *coronaviruses* (Figure 30.23b). Adenoviruses, coxsackie viruses, respiratory syncytial viruses (RSV), and orthomyxoviruses are collectively responsible for only a small percentage of common colds.

Aerosol transmission is a major means of spreading colds, although experiments with volunteers suggest that direct contact and indirect contact involving fomites are also important means of transmission, perhaps even more important than aerosols. Incidence of the common cold rises when people are indoors in the winter months, although it is possible to "catch a cold" at any time of year. Most antiviral drugs are ineffective against common cold viruses, although some have shown promise for preventing the onset of symptoms following rhinovirus exposure. Moreover, new antiviral drugs are being designed based on knowledge of the three-dimensional structure of cold viruses. For example, antirhinovirus drugs that bind to the virus and change its surface properties in such a way as to prevent it from attaching to host cells have been developed. But thus far, most "cold drugs" on the market simply help to reduce the severity of symptoms—the cough, nasal discharges, headache, and the like.

Treatment

Because colds are generally self-limiting and not serious diseases, treatment is aimed at controlling symptoms, especially nasal discharges, with antihistamine and decongestant drugs. A plethora of such drugs are available without prescription, each touting its superior features. Many of these work well, but some have unwanted side effects such as drowsiness, headache, and the like. The severity of symptoms of any cold event is a function of both the virulence of the cold virus, the overall health and well-being of the person at the time of infection, and the nature of supportive factors during infection. In the long run, immunity is more important in ridding the body of a cold virus than anything drugs can achieve. Cold viruses induce an antibody-mediated immune response that targets the current cold virus. However, the number of immunologically unique strains of each type of cold virus makes any long-term immunity to the common cold impossible.

On average, a person in the United States gets about three colds per year compared with less than one case of influenza per person per year (Table 30.1). This is an indication of the ease of transmission of the common cold and the large number of different viruses that cause the same general symptoms. Thus the common cold is a recurrent event that humans must simply live with. Preventive measures such as avoiding contact with common fomites (door handles and other surfaces touched by others) and frequent hand washing are best practices for avoiding the common cold.

— MINIQUIZ —

- Define the cause and symptoms of common colds.
- Discuss the possibilities for effective treatment and prevention of colds.

30.8 Influenza

Influenza is a highly infectious airborne disease of viral origin. Influenza viruses contain a single-stranded, negative-sense, segmented RNA genome surrounded by an envelope composed of protein, a lipid bilayer, and external glycoproteins (**Figure 30.24**) (⬤ Section 10.9). There are three classes of influenza viruses: influenza A, B, and C. Here we consider only influenza A because it is the most important human pathogen.

Antigenic Drift and Antigenic Shift

Each strain of influenza A virus can be identified by a unique set of surface glycoproteins. These glycoproteins are *hemagglutinin* (HA or the "H antigen") and *neuraminidase* (NA or the "N antigen"). Each virus has one type of HA and one type of NA on its viral capsid and is named for the antigens it contains; for example, "H1N1." The HA antigen is important in *attaching* the influenza virus to host cells, while the NA antigen is instrumental in *releasing* the virus from host cells; each antigen is composed of several individual proteins (**Figure 30.25**).

Infection or immunization with influenza virus results in the production of antibodies that react with the HA and NA glycoproteins. When these antibodies bind to HA or NA, the virus is blocked from either attaching or releasing and is effectively neutralized, stopping the infection process. However, over time, the viral genes encoding the HA and NA glycoprotein antigens mutate, rendering minor changes to their amino acid sequence

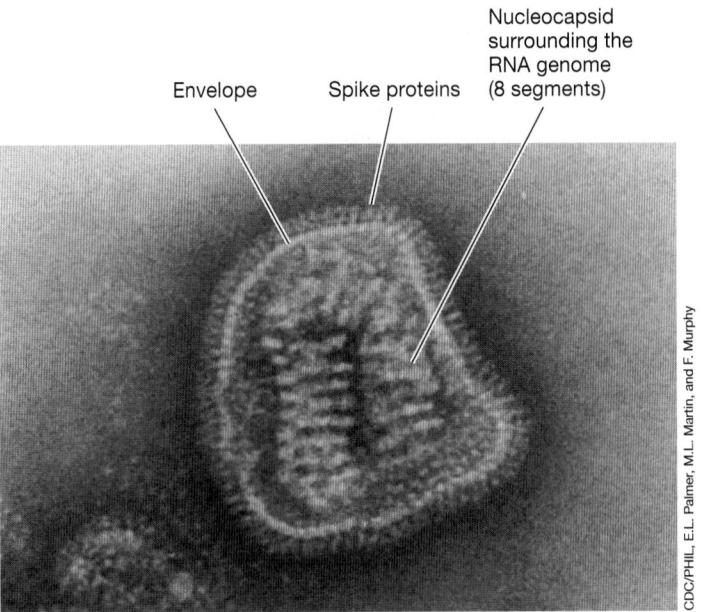

Figure 30.24 Influenza A virus. The virus contains a single-stranded, negative-sense RNA genome in eight segments; a virion is about 100 nm in diameter. Major factors in the success of influenza virus as a pathogen are antigenic drift and antigenic shift (see Figure 30.26).

and hence antigenic structure. Mutations that alter as few as one amino acid in the glycoprotein can affect how an antibody binds to these antigens. This slight variation in the structure of influenza viral surface antigens is at the heart of a phenomenon in influenza biology called **antigenic drift**. As a result of these subtle yet important changes, host immunity to a given virus strain diminishes as the strain mutates, and reinfection with the mutated strain can occur. This phenomenon is why last year's influenza vaccine may work only poorly against this year's crop of influenza viruses (**Figure 30.26a**).

In addition to antigenic drift, there is a second feature of influenza virus biology that aids virulence. The single-stranded RNA genome of influenza viruses is *segmented*, with genes found on each of eight distinct segments (⬤ Figure 10.21b). During virus maturation in the host cell, the viral RNA segments are packaged

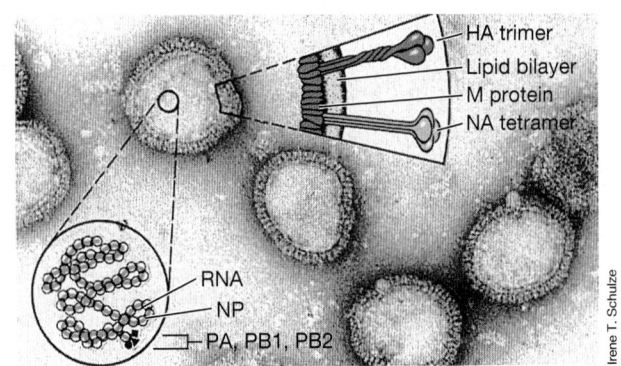

Figure 30.25 Influenza virus structure. Major viral coat proteins are: HA, hemagglutinin (three copies make up the HA coat spike); NA, neuraminidase (four copies make up the NA coat spike); M, coat protein; NP, nucleoprotein; PA, PB1, PB2, and other internal proteins, some of which have enzymatic functions.

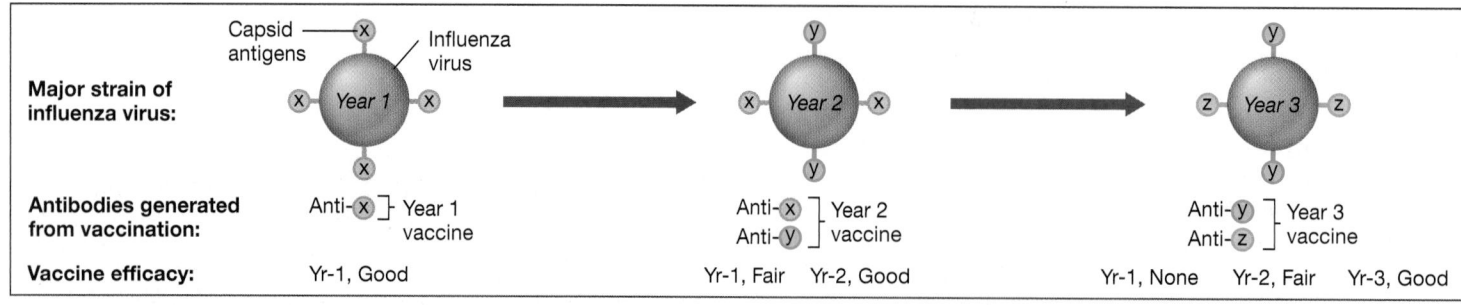

(a) **Antigenic drift**

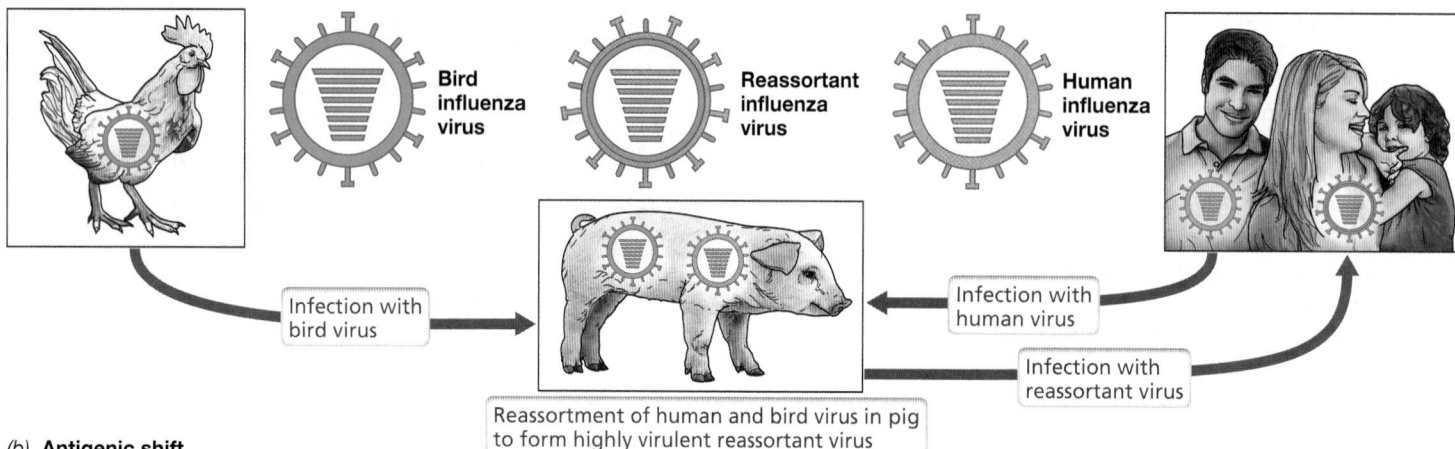

(b) **Antigenic shift**

Figure 30.26 Antigenic drift and antigenic shift in influenza virus biology. *(a)* Antigenic drift. A new vaccine is prepared each year against the major strain of influenza circulating among the population. However, vaccine efficacy wanes with time as immunologically new surface antigens appear from mutations in genes encoding viral surface proteins. *(b)* Antigenic shift. Influenza strains that originate in birds and humans can also infect swine. If a pig becomes infected with both bird and human viruses simultaneously, the viral genomes can be mixed, forming reassortant viruses. If such viruses, which now contain several unique antigens, infect humans, influenza pandemics can be triggered because of an ineffective immune response (⟲ Section 29.8).

randomly. To be infective, a virus must be packaged so it contains one copy of each of the eight gene segments. Occasionally, however, more than one strain of influenza virus infects a single animal at one time. In such cases, if the two strains infect the same cell, both viral genomes are replicated; when genome packaging occurs, the segments from the two strains may intermix. The result is a genetically unique virus that is now a *new virus strain*. This mixing of gene fragments between different strains of influenza virus is called *reassortment*.

Unique *reassortant viruses* trigger the phenomenon known as **antigenic shift** (Figure 30.26*b*), a major change in a surface antigen resulting from the total replacement of the RNA that encoded it. Antigenic shift can immediately and completely change one or both of the major HA and NA viral glycoproteins in a major way. As a result, reassortant viruses are essentially unrecognized by immune responses to previous influenza infections. Reassortant viruses also frequently display one or more unique virulence properties that help to trigger unusually strong clinical symptoms and are the usual catalysts of influenza *pandemics*, which we consider shortly.

Symptoms and Treatment of Influenza

Human influenza virus is transmitted from person to person through the air, primarily in droplets expelled during coughing and sneezing (Figure 30.1). The virus infects the mucous membranes of the upper respiratory tract and occasionally invades the lungs. Symptoms include a low-grade fever lasting up to a week, chills, fatigue, head and muscle aches, a cough and/or a sore throat, and general malaise (Table 30.1). Most of the serious consequences of seasonal influenza occur not from the disease itself but from bacterial secondary infections, especially in persons whose resistance has been lowered by the influenza infection. For example, in infants and the elderly, influenza can be followed by bacterial pneumonia (Section 30.2), sometimes in fatal form.

Most individuals develop protective immunity to the infecting strain of influenza virus, making it impossible for that strain or a very closely related strain to cause widespread infection (an epidemic) until the virus encounters another susceptible population. Immunity occurs from both antibody- and cell-mediated immune responses directed at HA and NA glycoproteins. Influenza epidemics can be controlled by immunization. Developing an effective vaccine, however, is complicated by the large number of existing influenza viral strains resulting from antigenic drift and antigenic shift (Figure 30.26). Through careful worldwide surveillance, samples of the major emerging strains of influenza virus are obtained each year before the onset of seasonal epidemics and used to prepare that year's vaccine. In most years, this approach confers adequate protective immunity.

Most human influenza viruses respond to antiviral drugs. The adamantanes—*amantadine* and *rimantadine*—are synthetic amines that inhibit viral replication, and the neuraminidase inhibitors *oseltamivir* (Tamiflu) and *zanamivir* (Relenza) (⇔ Table 28.6) block release of newly replicated human influenza virions. These drugs are often used early on to shorten the course and severity of infection, especially in immune-compromised people or the elderly.

Influenza Pandemics

Influenza pandemics—worldwide epidemics—are much less frequent than outbreaks and epidemics, occurring from 10 to 40 years apart (⇔ Section 29.8). Flu pandemics result from antigenic shift, and virtually all have been due to avian and human influenza viruses reassorting in swine (Figure 30.26*b*) because swine can propagate both avian and human influenza viruses. This results in a highly virulent influenza strain for which there is no preexisting immunity in humans.

The "Spanish flu" pandemic of 1918 was the most catastrophic in recorded history, and the extreme virulence of the 1918 H1N1 virus is thought to have triggered host production and release of unusually large amounts of inflammatory substances, resulting in systemic inflammation and more severe symptoms than those typical of yearly flu epidemics. The 1957 Asian flu was also a memorable pandemic (**Figure 30.27**), beginning in China and spreading to the United States and shortly thereafter to Europe and South America. In this case, the pandemic influenza strain was a highly virulent H2N2 virus, differing antigenically from all previous strains. Pandemic influenza A (H1N1) 2009 virus, nicknamed the "swine flu," spread much more rapidly in 2009 than even the 1957 Asian flu, starting in Mexico and spreading quickly to the United States, Europe, and Central and South America. H1N1 was a classic case of influenza virus genome reassortment in swine (Figure 30.26*b*), and from the swine reservoir, a highly virulent virus emerged to infect humans. Influenza A H5N1, nicknamed the "bird flu," emerged in Hong Kong in 1997, and is the major virus health officials are monitoring closely today. H5N1 has been detected in birds in many countries, and if this virus were to infect swine and form an easily transmissible reassortant virus that subsequently jumps to humans (Figure 30.26*b*), such a virus could initiate a very deadly influenza pandemic.

MINIQUIZ

- Distinguish between antigenic drift and antigenic shift in influenza.
- Discuss the possibilities for effective immunization programs for influenza and compare them to the possibilities for immunization for colds.

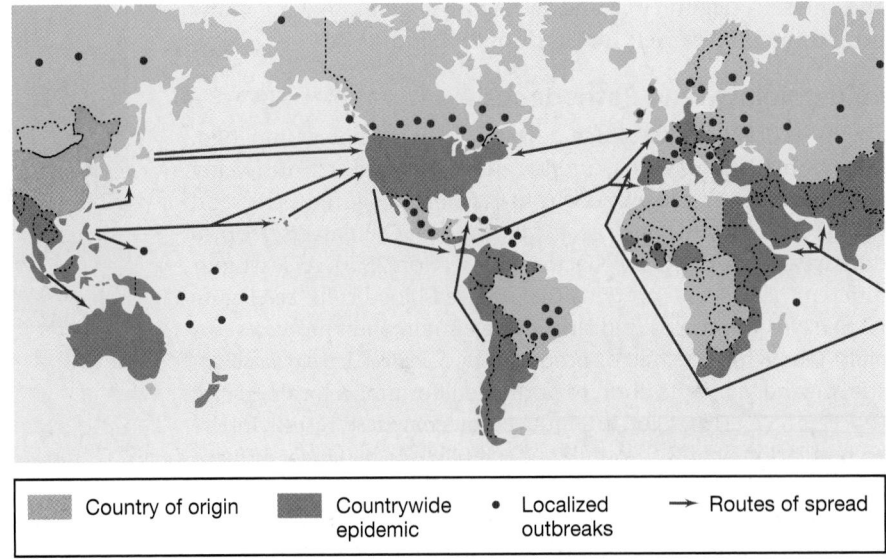

Country of origin | Countrywide epidemic | • Localized outbreaks | → Routes of spread

Figure 30.27 An influenza pandemic. Map of the Asian influenza pandemic of 1957. Lax agricultural practices with poultry and swine coupled with human interactions with these animals allowed the reassortment of influenza viral genomes from the three host species, producing a new strain for which there was no immune memory in humans. See Section 29.8 for more coverage of influenza pandemics.

III • Direct-Contact Bacterial and Viral Diseases

Some pathogens are spread primarily by direct contact with an infected person or by direct contact with blood or excretions from an infected person. Many of the respiratory diseases we have just discussed can also be spread by direct contact, but here we consider diseases spread primarily through direct contact with infected individuals rather than by an airborne route. These include staphylococcal infections, ulcers, certain types of hepatitis, and the very dangerous Ebola hemorrhagic fever.

30.9 *Staphylococcus aureus* Infections

The genus *Staphylococcus* contains pathogens of humans and other animals. Staphylococci commonly infect skin and wounds and may also cause pneumonia. Most staphylococcal infections result from the transfer of staphylococci in the normal microbiota of an infected, asymptomatic individual to a susceptible individual. Others result from toxemia following the ingestion of contaminated food ("staph food poisoning," ⇔ Section 32.8).

Staphylococci are nonsporulating, gram-positive cocci about 0.5–1.5 μm in diameter that divide in multiple planes to form irregular clumps of cells (**Figure 30.28*a*, *b***). They are resistant to drying and tolerate high concentrations of salt (up to 10% NaCl) when grown on artificial media. Staphylococci are readily dispersed in dust particles through the air and on surfaces. In humans, two species are important: *Staphylococcus epidermidis*, a nonpigmented species usually found on the skin or mucous membranes, and *Staphylococcus aureus*, a yellow-pigmented species. Both species are potential pathogens, but *S. aureus* is more commonly associated with human disease. Both species are frequently present in the normal microbiota of the upper respiratory tract and the skin (Figure 30.2 and see Figure 30.30*b*), making many people potential carriers (⇄ Sections 29.1 and 29.3).

Epidemiology and Pathogenesis

Staphylococcal diseases include acne, boils, pimples, impetigo, pneumonia, osteomyelitis, carditis, meningitis, and arthritis. Many of these diseases are *pyogenic* (pus-forming) (Figure 30.28*c* and **Figure 30.29**). Those strains of *S. aureus* that cause human disease produce a variety of virulence factors (⇄ Section 25.3). At least four different *hemolysins* (proteins that lyse red blood cells, see Figure 30.8) have been recognized, and a single strain often produces several. A key virulence factor produced by *S. aureus* is *coagulase*, an enzyme that converts fibrin to fibrinogen, forming a localized clot (⇄ Figure 25.11*b*). Clotting induced by coagulase results in the accumulation of fibrin around the bacterial cells, making it difficult for host immune cells to contact the bacteria and initiate phagocytosis. Most *S. aureus* strains also produce *leukocidin*, a protein that destroys white blood cells. Production of leukocidin in skin lesions such as boils and pimples results in host cell destruction and is one of the factors responsible for pus (Figures 30.28 and 30.29). Some strains of *S. aureus* also produce other virulence proteins including *hyaluronidase, fibrinolysin, lipase, ribonuclease,* and *deoxyribonuclease.*

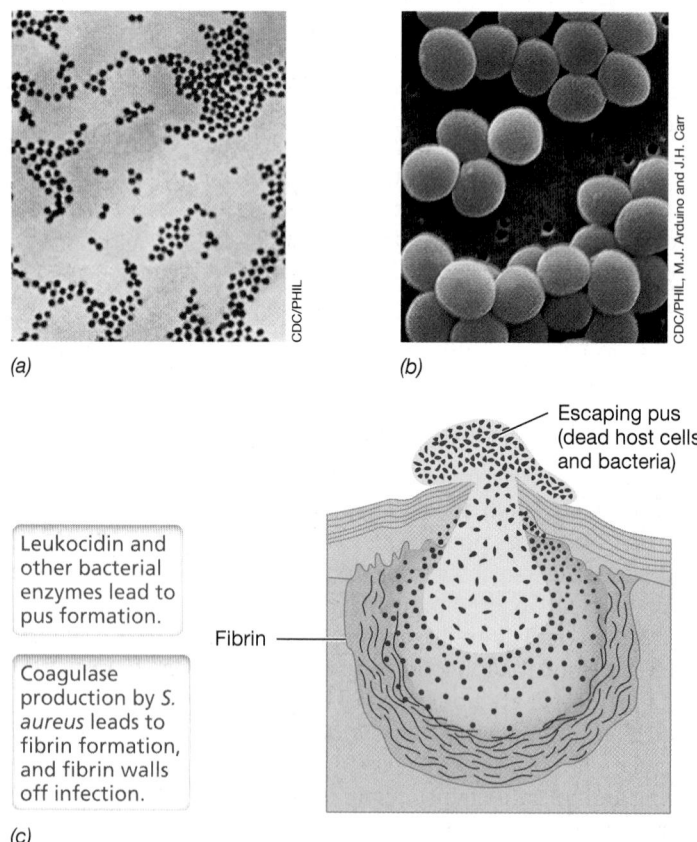

(a)

(b)

Escaping pus (dead host cells and bacteria)

Leukocidin and other bacterial enzymes lead to pus formation.

Fibrin

Coagulase production by *S. aureus* leads to fibrin formation, and fibrin walls off infection.

(c)

Figure 30.28 *Staphylococcus aureus* and *S. aureus* infections. Cells divide in several planes, giving the appearance of a cluster of grapes. *(a)* Gram stain; an individual coccus is about 1 μm in diameter. *(b)* Scanning electron micrograph of cells. *(c)* Structure of a boil. Staphylococci initiate a localized skin infection and become walled off by coagulated blood and fibrin through the activity of the enzyme coagulase, a major virulence factor. The ruptured boil releases pus, consisting of dead host cells and bacteria. See also Figure 30.29.

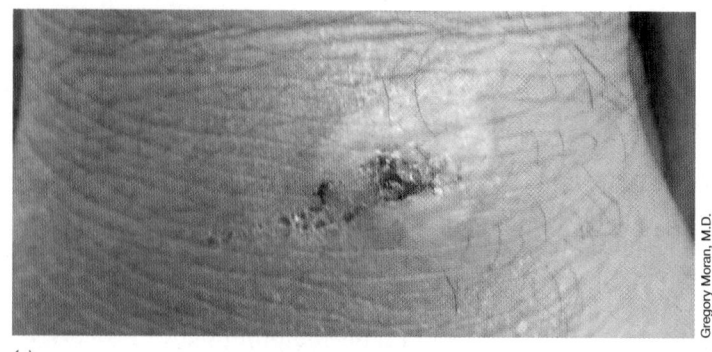

(a)

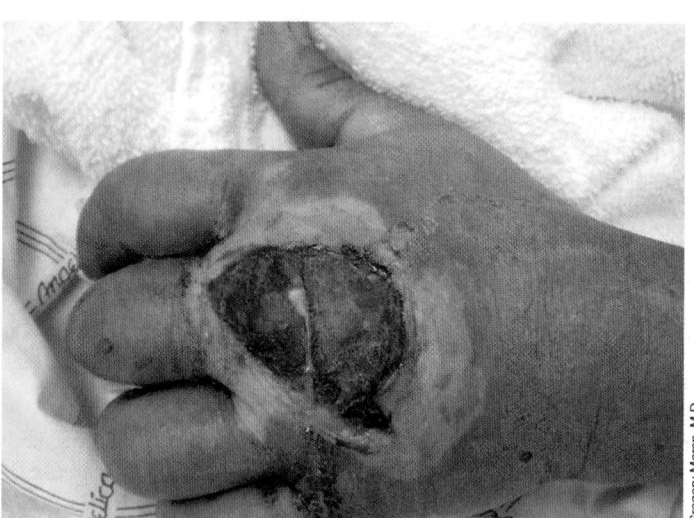

(b)

Figure 30.29 Pus-forming staphylococcal wounds. *(a)* A typical pus-forming wound on the hand. Pus lies just under the epidermal layer. *(b)* Abscess on the hand caused by a methicillin-resistant strain of *Staphylococcus aureus* (MRSA strain). If no treatment for a pus-forming wound is sought or if penicillin is administered first, MRSA infections can cause extensive tissue destruction, as shown here.

Certain strains of *S. aureus* cause **toxic shock syndrome (TSS)**, a serious outcome of staphylococcal infection, characterized by high fever, rash, vomiting, diarrhea, and death. TSS was first recognized in women and was associated with the use of highly absorbent tampons. However, TSS is now seen in both men and women and is typically initiated by staphylococcal infections following surgery. The symptoms of TSS result from an exotoxin called *toxic shock syndrome toxin-1*. This very potent toxin is a superantigen (⇄ Sections 25.7 and 27.10) that is released during cell growth and recruits large numbers of T cells to the site of infection. These then cause a major inflammatory response that is fatal in about 70% of cases. TSS can also result from superantigens from other pathogens, including *Streptococcus pyogenes* (Section 30.2).

Diagnosis and Treatment and the MRSA Epidemic

To diagnosis an *S. aureus* infection through laboratory culture, a specimen, typically from a pus-forming wound (Figure 30.29*a*), is cultured on a selective and differential medium containing 7.5% NaCl, mannitol, and phenol red, a pH indicator (mannitol–salt agar, **Figure 30.30**). The salt inhibits the growth of nonhalotolerant bacteria while allowing staphylococci to grow. In addition,

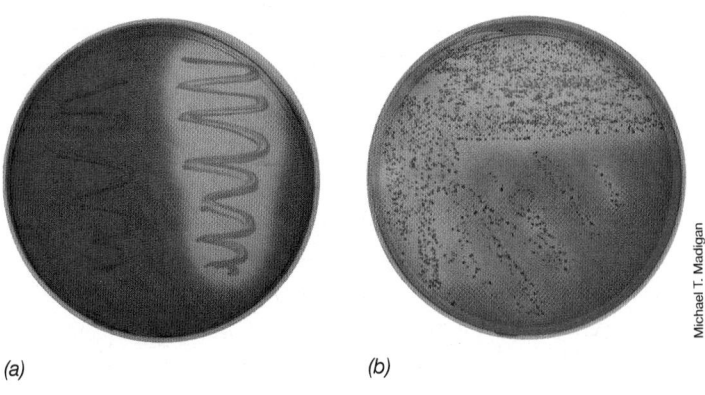

(a)   *(b)*

Michael T. Madigan

Figure 30.30 Mannitol–salt agar in the isolation of staphylococci.
(a) Mannitol–salt agar (MSA) is both selective and differential for staphylococci. The presence of 7.5% NaCl makes MSA selective and phenol red makes it differential. Left, *Staphylococcus epidermidis*; right, *Staphylococcus aureus*. *(b)* A nasal swab of the senior author of this textbook supports the observation that most humans are carriers of *S. aureus*.

because *S. aureus* ferments mannitol, it generates acidity and changes the medium from red to yellow; other staphylococci, such as *S. epidermidis*, do not (Figure 30.30).

In major clinical laboratories, the polymerase chain reaction (PCR) is used to amplify genes unique to *S. aureus* from DNA isolated from a clinical sample, and this speeds up the diagnosis (results from laboratory culture take 24 h). For specific identification of methicillin-resistant strains of *S. aureus* (MRSA), a special selective and differential medium is available as well as a PCR protocol to identify *mecA* (the gene that encodes methicillin resistance in MRSA strains) and a rapid immunological test where cells of *S. aureus* in suspension are agglutinated by antibodies to specific surface proteins (↩ Figure 28.15).

Historically, *S. aureus* infections have been treated with various penicillin and cephalosporin antibiotics. However, extensive use of these antibiotics for many years has selected for resistant strains that now predominate, especially in the clinical environment. Surgical patients, for example, may acquire staphylococci from healthcare personnel who are asymptomatic carriers of drug-resistant strains. As a result, appropriate antimicrobial drug therapy for *S. aureus* infections is a major problem in healthcare environments. The antibiotics clindamycin and various tetracycline drugs are currently used to treat MRSA infections.

MRSA infections (Figure 30.29b) are becoming more common (see Explore the Microbial World "MRSA—A Formidable Clinical Challenge" in Chapter 28). Over 80,000 cases of MRSA are reported each year in the United States, but infections are actually probably closer to ten times this number. Many of these cases are hospital-acquired (nosocomial) MRSA (↩ Section 28.2), but many others are not. Because of the potential severity of MRSA infections, it is important to rapidly identify these strains in clinical specimens so that an effective treatment is begun as soon as possible. Delayed treatment of a MRSA infection, whether due to hesitation to seek treatment or treatment with an ineffective antibiotic, can lead to extensive tissue damage (Figure 30.29b).

Prevention of staphylococcal infections is virtually impossible because many people are asymptomatic carriers of *S. aureus*, either on their skin or upper respiratory tract. However, identification

and treatment or isolation of MRSA-carrying healthcare providers who serve in surgical or nursery units has helped limit transmission of these very aggressive strains. As is true of many direct-contact diseases, MRSA transmission can also be greatly diminished by practicing good basic hygiene, avoiding contact with the personal items (including clothing and towels) of others, and keeping wounds covered.

— MINIQUIZ —

- What is the normal habitat of *Staphylococcus aureus*? How does *S. aureus* spread from person to person?
- What is MRSA, and why is it a health problem?

30.10 *Helicobacter pylori* and Gastric Diseases

Helicobacter pylori is a gram-negative, highly motile, spiral-shaped bacterium (**Figure 30.31**) associated with gastritis, ulcers, and gastric cancers. This bacterium colonizes the non-acid-secreting mucosa of the stomach and the upper intestinal tract. It is estimated that half the world's population is chronically infected with *H. pylori*. Up to 80% of gastric ulcer patients have concomitant *H. pylori* infections, and up to 50% of asymptomatic adults in developing countries are chronically infected. Although there is no known nonhuman reservoir of *H. pylori*, infection occurs in high incidence within families, suggesting person-to-person transmission. *H. pylori* infections also occasionally occur in clusters, suggesting that transmission from common sources such as food or water is also possible.

The *Helicobacter* Infection Process

H. pylori is only slightly invasive and colonizes the surfaces of the gastric mucosa, where it is protected from the effects of stomach acids by the gastric mucus layer. Cells of *H. pylori* reach these relatively

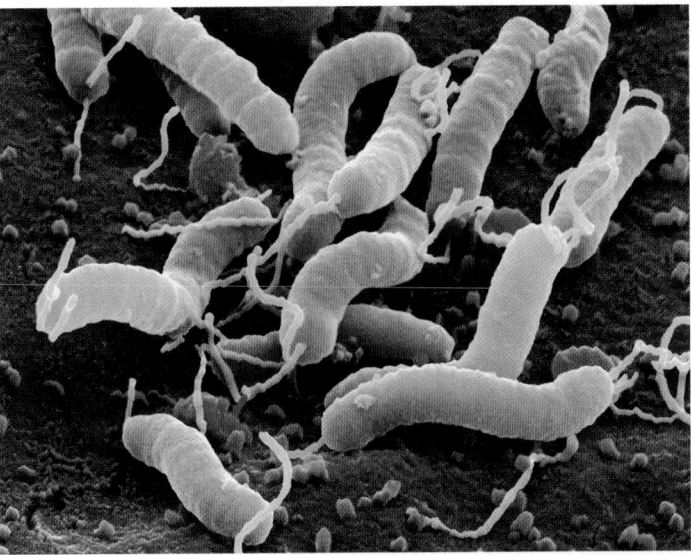

Figure 30.31 *Helicobacter pylori*. Colorized scanning electron micrograph of cells attached to the mucous lining of the stomach. Cells range in length from 3 to 5 μm and are about 0.5 μm in diameter. Note the flagella.

protected regions by employing a chemoreceptor (⟿ Sections 2.13 and 6.7) that tracks the gradient of urea produced by gut epithelia to direct flagellar rotation up the gradient. The organism is strongly ureolytic and cleaves the urea into ammonia and bicarbonate, which help buffer the region of cell colonization.

After *H. pylori* cells colonize the mucosa, a combination of virulence factors and host responses cause inflammation, tissue destruction, and ulceration. Pathogen products such as the cytotoxin VacA (an exotoxin), urease, and an autoimmune response triggered by *H. pylori* lipopolysaccharide all contribute to localized tissue destruction and ulceration. Individuals who acquire *H. pylori* tend to have chronic infections unless they are treated with antibiotics. Treatment is both simple and important, as chronic inflammation of the gastroduodenum (gastritis) due to untreated *H. pylori* infection may lead to the development of gastric cancers.

H. pylori and Clinical Disease

Clinical signs of *H. pylori* infection include belching and stomach (epigastric) pain. Definitive diagnosis requires the isolation or observation of *H. pylori* from a gastric ulcer biopsy. However, a simple diagnostic test for the *H. pylori* enzyme urease is used for a noninvasive diagnosis. In this test, a small amount of ^{13}C- or ^{14}C-labeled urea ($H_2N-CO-NH_2$) is ingested; if *H. pylori* is present, the bacterium will hydrolyze the urea, forming labeled CO_2 and ammonia. Hence, the presence of labeled CO_2 in the patient's breath is highly suggestive of *H. pylori* infection.

The best evidence for a causal association between *H. pylori* and gastric ulcers comes from antibiotic treatments for the disease. Long-term treatment with antacids helps alleviate gastric ulcer symptoms temporarily, but most patients relapse within 1 year. However, by treating the *cause* rather than the *effect* of the disease, actual cures can be obtained. *H. pylori* infection is typically treated with a combination of drugs, including the antibacterial compound metronidazole, an antibiotic such as tetracycline or amoxicillin, and a bismuth-containing antacid preparation. The combination treatment, administered for 14 days, abolishes the *H. pylori* infection and provides a true cure.

Like the link with gastric ulcers, the link between *H. pylori* infection and certain forms of gastric cancers, in particular, gastric adenocarcinoma (the most prevalent form of gastric cancer), is also strong. Gastric cancers are the second leading cause of cancer deaths worldwide. Although how *H. pylori* infection actually triggers adenocarcinomas is unclear, it is believed that long-term infection with this bacterium coupled with host and possibly environmental factors combine to predispose an individual to stomach malignancies.

For their contributions to unraveling the connection between *H. pylori* and peptic and duodenal ulcers, the Australian scientists Robin Warren and Barry Marshall were awarded the 2005 Nobel Prize in Physiology or Medicine. For an interesting story on the antiquity of *H. pylori*, see page 729.

— MINIQUIZ —

- Describe infection by *Helicobacter pylori* and the resulting development of an ulcer.
- How can gastric ulcers be diagnosed? How can they be cured?

30.11 Hepatitis

Hepatitis is a liver inflammation, commonly caused by an infectious agent. Hepatitis sometimes results in acute illness followed by destruction of functional liver anatomy and cells, a condition known as **cirrhosis**. Hepatitis due to infection can cause chronic or acute disease, and some forms lead to liver cancer.

Although many viruses and a few bacteria can cause hepatitis, a restricted group of viruses is often associated with liver disease. Hepatitis viruses A, B, C, D, and E are phylogenetically diverse viruses but share in common their ability to infect the liver (**Table 30.2**). Hepatitis A and E viruses, although occasionally transmitted person to person, are more commonly transmitted by food (hepatitis A virus) or water (hepatitis E virus). We cover hepatitis A viral disease in Chapter 32. Here our focus is on hepatitis viruses transmitted by direct contact, with the major focus on hepatitis B, the causative agent of "bloodborne hepatitis."

The incidence of hepatitis A and B, the most common forms, has decreased significantly in the past 20 years because of effective vaccines and increases in surveillance. And, by comparison to hepatitis A and B, hepatitis C infections have risen significantly in recent years (**Figure 30.32**).

Epidemiology

Infection with *hepatitis B virus* (HBV) is called bloodborne hepatitis (or serum hepatitis) because it is transmitted in blood or in body fluids in contact with blood. HBV is a hepadnavirus, a partially double-stranded DNA virus (⟿ Section 10.11). The mature virus particle containing the viral genome is called a *Dane particle* (**Figure 30.33**). HBV causes acute, often severe disease that can lead to liver failure and death. Chronic HBV infection can lead to cirrhosis and liver cancer.

TABLE 30.2 Hepatitis viruses

Disease	Virus and genome[a]	Vaccine	Clinical illness	Transmission route
Hepatitis A	*Hepatovirus* (HAV) ssRNA	Yes	Acute	Enteric (food)
Hepatitis B	*Orthohepadnavirus* (HBV) dsDNA	Yes	Acute, chronic, oncogenic	Parenteral, sexual
Hepatitis C	*Hepacivirus* (HCV) ssRNA	No	Chronic, oncogenic	Parenteral
Hepatitis D	*Deltavirus* (HDV) ssRNA	No	Fulminant, only with HBV	Parenteral
Hepatitis E	*Caliciviridae* family (HEV) ssRNA	No	Fulminant disease in pregnant women	Enteric (water)

[a]Examples and discussion of each of these genomes can be found in Chapter 10 (⟿ Figures 10.2 and 10.3).

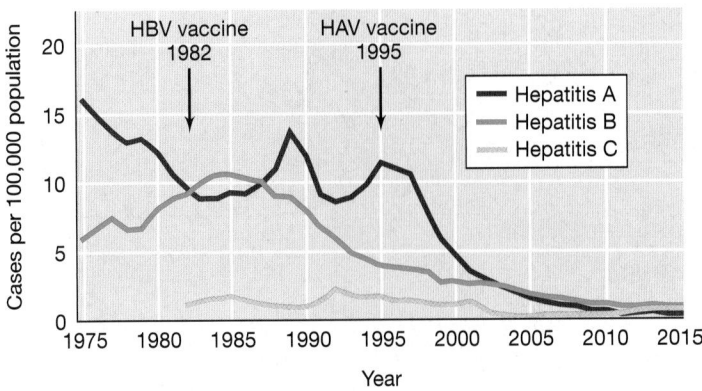

Figure 30.32 Hepatitis A, B, and C in the United States. In 2014 there were 1239 reported cases of hepatitis A, 2953 reported cases of hepatitis B, and 2194 reported cases of hepatitis C. The number of actual new cases of hepatitis A, B, or C infection is likely to be much higher than the reported new cases. Data obtained from the CDC, Atlanta, Georgia, USA.

HBV is transmitted by a *parenteral route*, which means "outside the gut." The main means of HBV transmission is from blood transfusions, contact with infected blood in a hypodermic needle, and from mother to child during childbirth. HBV may also be transmitted through exchanges of body fluids during sex. The number of new HBV infections has remained low and more or less constant since the year 2000 (Figure 30.32). Nevertheless, over 100,000 people worldwide and nearly 5000 people in the United States die yearly from liver failure or liver cancer caused by chronic HBV infection.

Hepatitis D virus (HDV) is a *defective virus* (⮌ Section 11.7) that lacks genes encoding its own capsid. HDV is also transmitted by parenteral routes, but because it is a defective virus, it cannot replicate and form an intact virion unless the cell is also infected with HBV. The HDV genome replicates independently but relies on HBV to produce capsid proteins (which are the same as those used by HBV) to form infectious virions. Thus, HDV infections are always coinfections with HBV.

Hepatitis C virus (HCV) is also transmitted parenterally. HCV generally produces a mild or even asymptomatic disease at first, but later

on up to 85% of those infected develop chronic hepatitis, with up to 20% proceeding to chronic liver disease and cirrhosis. Chronic infection with HCV leads to hepatocarcinoma (liver cancer) in 3–5% of infected individuals. The latency period for development of cancer can be several decades after the primary infection. Only a fraction of the estimated 25,000 annual new infections with HCV in the United States are recognized and formally reported (Figure 30.32). Large numbers of HCV-related deaths occur annually as a result of chronic HCV infections that develop into liver cancer. HCV-induced liver disease accounts for up to 10,000 of the 25,000 annual deaths due to liver cancer, other chronic liver diseases, and cirrhosis.

Other Aspects of Hepatitis Syndromes

Hepatitis is an acute disease of the liver, a vital organ that plays a role in several key metabolic processes, including carbohydrate, lipid, and protein syntheses, as well as detoxification and many other functions. Symptoms of hepatitis include fever, jaundice (yellowing of the skin and the whites of the eyes, Figure 30.33*b*), and liver enlargement and cirrhosis. All hepatitis viruses cause similar acute symptoms and cannot be readily distinguished based on clinical findings alone. Chronic hepatitis infections, usually caused by HBV or HCV, are often asymptomatic or produce very mild symptoms, but nonetheless cause serious liver disease, even in the absence of liver cancer.

Diagnosis of hepatitis is based on a combination of clinical symptoms and laboratory tests that assess liver function, especially key liver enzymes. Cirrhosis is diagnosed by visual examination of biopsied liver tissue. Virus-specific molecular assays are typically used to confirm a diagnosis, positively identify the infectious agent, and determine a course of treatment; isolation and culture of hepatitis viruses is usually not attempted.

Many of the immunological and molecular diagnostic tools discussed in Chapter 28 are used in hepatitis diagnoses. These include enzyme immunoassays that target viral-specific proteins or antiviral antibodies in a blood sample, immunoblots (Western blots), and immunofluorescence (microscopic) methods. Polymerase chain reaction (PCR) tests are also used to detect hepatitis viral genomes in blood or in liver tissue obtained by biopsy.

Infection with HAV or HBV can be prevented with effective vaccines. HBV vaccination is recommended and in most cases is required for school-age children in the United States. No effective vaccines are available for the other hepatitis viruses. For those unvaccinated, the practice of *universal precautions* will prevent infection. The precautions prescribe a high level of vigilance and aseptic handling and containment procedures to deal with patients, body fluids, and infected waste materials (⮌ Section 28.1). Most treatment of hepatitis is supportive, providing rest and time for the immune system to attack the infection and allow liver damage to be repaired. In some cases, in particular for HBV infections, some antiviral drugs are available that offer effective treatment.

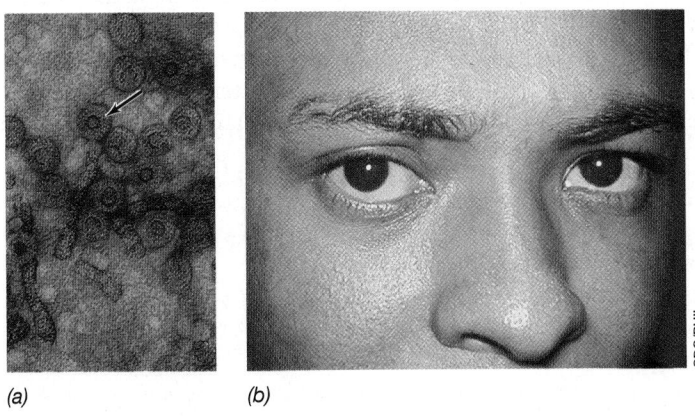

Figure 30.33 Hepatitis. *(a)* Hepatitis B virus. The arrow indicates a complete HBV virion, which is called a Dane particle. A Dane particle is about 40 nm in diameter. *(b)* Jaundice, a yellowing of the facial skin and eye conjunctiva, is a common symptom of hepatitis infections and results from the accumulation of bilirubin (a by-product of degraded red blood cells) that results from reduced liver function.

MINIQUIZ

• What host organ do hepatitis viruses attack? How are hepatitis A, B, and C viruses transmitted?

• Describe potential prevention and treatment methods for hepatitis A, B, and C viruses.

UNIT 7

30.12 Ebola: A Deadly Threat

A recent example of a highly infectious and deadly serious emerging pathogen that spreads by direct contact is Ebola virus, which ravaged parts of West Africa in 2014 and early 2015. Ebola infected about 29,000 people and killed over 11,000 of them.

Ebola emerged in 1976 in Zaire, and since then several small outbreaks have occurred in West African countries. But not until the 2014 outbreak in Guinea, Liberia, Nigeria, Senegal, and Sierra Leone did the disease kill such large numbers of people. However, nearly as fast as the disease reemerged, strong efforts to contain the spread of infection were put in place and these, along with unknown natural events that control cycles of this disease, combined to significantly reduce Ebola incidence. By the end of 2015, only a handful of cases were reported. Today, significant epidemiological surveillance for Ebola remains in place because of the ease of transmission of the virus and thus the rapidity by which a single infection can trigger an outbreak.

Ebola: The Virus and Its Transmission

Ebola hemorrhagic fever is caused by a *filovirus*, a filamentous virus that can take on many shapes (**Figure 30.34a**). The Ebola virus genome contains single-stranded and linear RNA of the negative sense, similar in this respect to influenza and rabies virus genomes. The genome contains only 19 kilobases of RNA, enough to encode just seven proteins; about a third of the genome encodes the RNA-dependent RNA polymerase (RNA replicase) needed to replicate the genome of negative-sense RNA viruses (⮎ Section 10.9).

Ebola virus is transmitted from person to person by direct contact through breaks in the skin or mucous membranes as well as by body fluids (including semen) and fomites (bedding, clothing, utensils) contaminated with the virus. The ease with which Ebola can be transmitted seems remarkable compared with other pathogens that rely on direct-contact transmission. For example, in a few documented instances of Ebola transmission to healthcare workers, significant precautions had been taken to ensure that full-body personal protection equipment (PPE, ⮎ Section 28.1) was in place to specifically prevent such transmission. Thus, the disease is not only deadly for those infected but can be an extremely dangerous risk for healthy medical providers as well. If PPE is not worn, healthcare workers who either treat patients or dispose of dead Ebola victims run a high risk of becoming infected. Thus PPE is made widely available by international health organizations to all health workers who might come in contact with an Ebola patient (Figure 30.34*b*).

The natural reservoir of Ebola virus that triggered the West African outbreak is unknown, although related filoviruses—such as hantavirus (⮎ Section 31.2)—are known to be spread from arthropods and rodents. Among the suspected reservoirs of Ebola are a variety of animals and possibly insects that inhabit tropical forests. In addition to person-to-person transmission, likely responsible for virtually all of the cases in the recent West Africa epidemic, natural Ebola infection in humans probably originates from an animal bite. In this regard, bats, and in particular fruit bats in which the virus has been documented, may be a major disease reservoir.

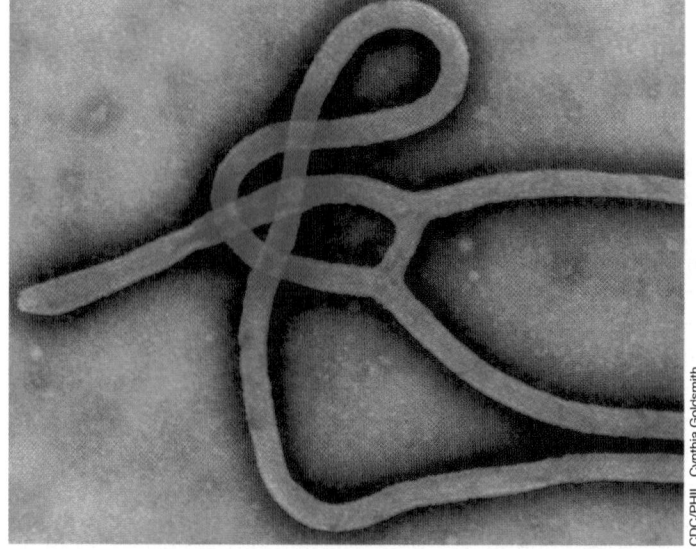

(a)

(b)

Figure 30.34 Ebola. *(a)* Colorized transmission electron micrograph of a negatively stained preparation of Ebola virus virions. A virion is about 80 nm in diameter. *(b)* Ugandan Red Cross workers donning their personal protective equipment before collecting the body of an Ebola victim.

Ebola: The Disease and Its Treatment

Ebola virus migrates from the initial site of infection to lymph nodes, from which it travels systemically to infect the liver and spleen. Once the virus has entered the body, several different types of cells can become infected. One to two weeks postinfection, an Ebola patient experiences an abrupt fever and general malaise, conditions that make Ebola difficult to distinguish from many other tropical diseases including malaria. But then more severe symptoms appear. These typically include severe fever and fatigue, diarrhea, nausea, vomiting and abdominal pain, and major loss of appetite. Bleeding through the skin and blood in vomit and feces can occur, but such bleeding is not a common symptom.

Ebola virus causes major problems in the liver, killing liver cells and disrupting normal blood clotting events. It is thought that the virus triggers host cells to release various cytokines that cause wide-spread inflammation (⊂⊃ Section 26.8) and internal bleeding; these lead to multiple organ failure, shock, and renal failure. The mortality rate in the West African Ebola outbreak averaged 35–70% depending on access to treatment, the initial state of health of those infected, viral load (abundance of the virus in the blood), and age; mortality was as high as 85% among infected people over the age of 45.

There is currently no drug treatment for Ebola, but survival rates among those that receive supportive care to help alleviate symptoms are significantly higher than in those that do not. Therapy includes the maintenance of fluids and electrolytes, oxygen supplements, and transfusions of blood to replace that lost from internal bleeding. Ebola survivors develop an antibody-mediated immune response to the virus, and some treatment success has been achieved by transfusing blood or serum from Ebola survivors into those infected. An Ebola vaccine is in development and several promising candidates have appeared. It is thus likely that putting a person's immune system to work against Ebola might be the best preventive measure against the disease. However, it is unlikely that an Ebola vaccine would help an already infected person, considering the rapidity with which the disease progresses and the major organ damage that viral infection triggers.

The 2014 Ebola outbreak is now well under control and much has been learned from it about the logistics of handling large-scale outbreaks of such a deadly disease. The ease by which Ebola is transmitted made a public education campaign about the dangers of Ebola just as important as dealing with the morbidity and mortality of the outbreak. When epidemiologists develop a better understanding of the natural reservoirs of Ebola virus, outbreaks like that in West Africa—which likely began by animal-to-person transmission—may well be preventable by reducing or eliminating the major reservoirs

and educating the populace about the dangers of encounters with known reservoirs. Also, rigorous campaigns to educate people who put themselves at risk by handling an Ebola patient or the corpse of an Ebola victim without PPE in place are also helping to reduce spread of Ebola when a case or case cluster does appear.

MINIQUIZ

- What do influenza virus and Ebola virus have in common? In what ways do their modes of transmission differ?
- Contrast mortality rates for influenza and Ebola hemorrhagic fever. Which is the more serious disease?

IV • Sexually Transmitted Infections

Sexually transmitted infections (STIs), also called *sexually transmitted diseases* (STDs), are caused by a wide variety of bacteria, viruses, protists, and even fungi (Table 30.3). Unlike respiratory pathogens that can be shed constantly in large numbers by an infected individual, sexually transmitted pathogens are typically found only in body fluids from the genitourinary tract (and blood, in the case of HIV). Because they require a protected and moist environment, sexually transmitted pathogens preferentially and sometimes exclusively colonize the genitourinary tract.

Because the transmission of STIs is limited to sexual activity, their spread can be controlled by sexual abstinence and minimized by the use of condoms that stop the exchange of body fluids during sex. With the exception of HIV/AIDS, most STIs are curable and

---

TABLE 30.3 Sexually transmitted infections and treatment guidelines

Disease	Causative organism(s)[a]	Recommended treatment[b]
Gonorrhea	*Neisseria gonorrhoeae* (B)	Cefixime or ceftriaxone, *and* azithromycin or doxycycline
Syphilis	*Treponema pallidum* (B)	Benzathine penicillin G
Chlamydia trachomatis infections	*Chlamydia trachomatis* (B)	Doxycycline or azithromycin
Nongonococcal urethritis	*C. trachomatis* (B) or *Ureaplasma urealyticum* (B) or *Mycoplasma genitalium* (B) or *Trichomonas vaginalis* (P)	Azithromycin or doxycycline Metronidazole
Lymphogranuloma venereum	*C. trachomatis* (B)	Doxycycline
Chancroid	*Haemophilus ducreyi* (B)	Azithromycin
Genital herpes	Herpes simplex 2 (V)	No known cure; symptoms can be controlled by several antiviral drugs
Genital warts	Human papillomavirus (HPV) (certain strains)	No known cure; symptomatic warts can be removed surgically, chemically, or by cryotherapy
Trichomoniasis	*Trichomonas vaginalis* (P)	Metronidazole
Acquired immunodeficiency syndrome (AIDS)	Human immunodeficiency virus (HIV)	No known cure; several drugs can stop viral replication and slow disease progression
Pelvic inflammatory disease	*N. gonorrhoeae* (B) or *C. trachomatis* (B)	Cefotetan and doxycycline
Vulvovaginal candidiasis	*Candida albicans* (F)	Butoconazole

[a]B, bacterium; V, virus; P, protist; F, fungus.
[b]Recommend.Tab Tr Td .Note Pations as of 2016 of the U.S. Department of Health and Human Services, Public Health Service.

UNIT 7

many can have only minor symptoms. These realities, combined with the fact that those infected are sometimes reluctant to seek treatment, make treatment of STIs an ongoing public health challenge. However, delaying or forgoing treatment of STIs only serves to maintain lines of transmission and can lead to long-term health problems such as infertility, cancer, heart disease, degenerative nerve disease, birth defects, stillbirth, or destruction of the immune system, any of which can result in death.

30.13 Gonorrhea and Syphilis

Gonorrhea and *syphilis* are ancient STIs, but because of major differences in their symptoms, the overall pattern of disease differs significantly between the two. In the United States, cases of gonorrhea peaked following the introduction of birth control pills in the mid-1960s, and gonorrhea is still quite prevalent today; cases of syphilis, on the other hand, have a much lower incidence (**Figure 30.35**). This is partly because syphilis exhibits very obvious symptoms in its primary stage and infected individuals usually seek immediate treatment.

Gonorrhea

Neisseria gonorrhoeae, often called the *gonococcus*, causes gonorrhea. *N. gonorrhoeae* is a gram-negative and obligately aerobic diplococcus related biochemically and phylogenetically to *Neisseria meningitidis* (Section 30.5). Cells of *N. gonorrhoeae* are killed rapidly by drying, sunlight, and ultraviolet radiation and thus normally do not survive away from the mucous membranes of the pharynx, conjunctiva, rectum, or genitourinary tract (**Figure 30.36**). Because of this, gonorrhea can only be transmitted by intimate person-to-person contact. We discussed the clinical microbiology and diagnosis of gonorrhea in Section 28.3.

The symptoms of gonorrhea are quite different in the male and female. In females, gonorrhea may be asymptomatic or cause a mild vaginitis that is difficult to distinguish from vaginal infections caused by other organisms; hence, the infection may easily go unnoticed. Complications from untreated gonorrhea in females, however, can lead to a chronic condition called *pelvic inflammatory disease* (PID), which can cause sterility. In men, *N. gonorrhoeae* causes a painful infection of the urethral canal and

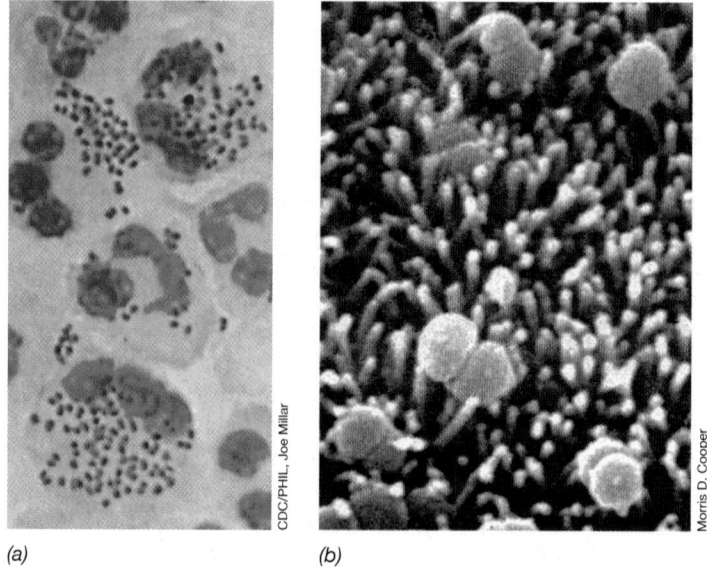

(a)   *(b)*

Figure 30.36 The causative agent of gonorrhea, *Neisseria gonorrhoeae*. *(a)* Gram stain of a urethral discharge. *(b)* Scanning electron micrograph of the microvilli of human fallopian tube mucosa with cells of *N. gonorrhoeae* attached to the surface of epithelial cells. Cells of *N. gonorrhoeae* are about 0.8 μm in diameter. *Neisseria* species are *Betaproteobacteria* (⮀ Section 16.2).

typical puslike urethral discharge. Complications from untreated gonorrhea affecting both males and females include damage to heart valves and joint tissues due to inflammatory reactions from immune complexes that deposit in these areas. In addition to disease in adults, *N. gonorrhoeae* can also cause eye infections in newborns. Infants born of infected mothers may acquire eye infections during birth. Therefore, prophylactic treatment of the eyes of all newborns with an ointment containing erythromycin is mandatory in many states in the United States to prevent gonococcal and other bacterial eye infections in infants.

Treatment of gonorrhea with penicillin was the method of choice until the 1980s when strains of *N. gonorrhoeae* resistant to penicillin emerged. The quinolones ciprofloxacin, ofloxacin, or

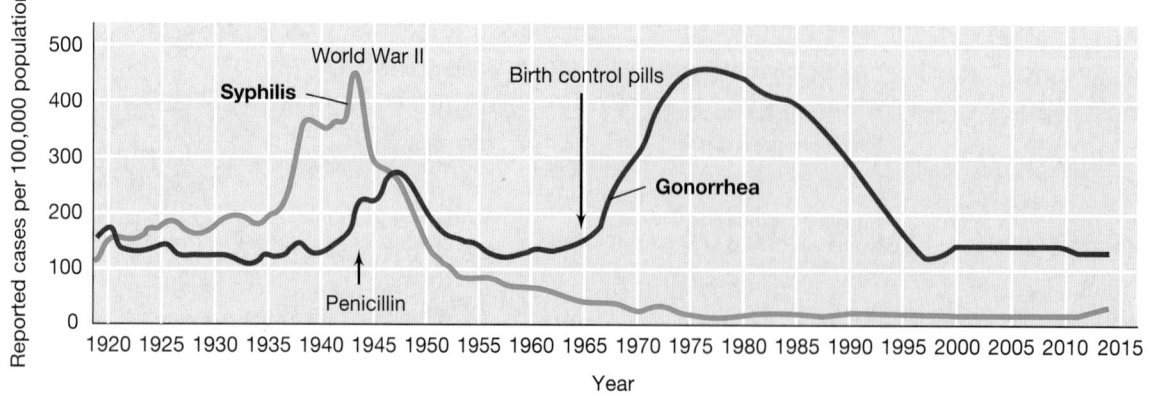

Figure 30.35 Reported cases of gonorrhea and syphilis in the United States. Note the downward trend in disease incidence after the introduction of antibiotics and the upward trend in the incidence of gonorrhea after the introduction of birth control pills. In 2014 there were 350,062 new cases of gonorrhea and 19,999 new cases of primary and secondary syphilis in the United States.

levofloxacin were also used, but by 2006, a significant fraction of *N. gonorrhoeae* strains isolated in the United States had developed resistance to these drugs as well. Strains resistant to penicillin and quinolones respond to alternative antibiotic therapy with a single dose of the β-lactam antibiotics cefixime or ceftriaxone.

Despite the fact that drugs are still effective in treating gonorrhea, incidence of gonorrhea remains relatively high (Figure 30.35) for at least three reasons. First, although anti-gonococcal antibodies are generated by an infection, they are strain-specific and provide no protection from infection by other strains of *N. gonorrhoeae*. As a consequence, gonorrhea reinfection is possible and quite common in high-risk populations (primarily sex workers and those with multiple sex partners). In addition, within a single *N. gonorrhoeae* strain, antigenic switches can thwart the immune response. For example, by mutation *N. gonorrhoeae* can alter the structure of its pilus proteins, thus creating new serotypes to challenge the immune response. Second, oral contraceptives cause a rise in vaginal pH; when this occurs, lactic acid bacteria normally found in the adult vagina fail to develop, and this reduces competition for colonization by *N. gonorrhoeae*. And finally, and most importantly, symptoms of gonorrhea in the female are often so mild that the disease may go unrecognized; a promiscuous infected female can then infect many males.

Syphilis

Syphilis is caused by a spirochete, *Treponema pallidum*, a long and extremely thin coiled cell (**Figure 30.37**). Like *N. gonorrhoeae*, *T. pallidum* is very sensitive to environmental stress and drying, and thus syphilis is only transmitted through intimate sexual contact or from mother to fetus during pregnancy. The biology of the spirochetes and the genus *Treponema* is discussed in Section 15.19.

Syphilis is often transmitted along with gonorrhea as coinfections. However, syphilis is potentially the more serious disease. For example, syphilis kills about 100,000 people per year worldwide, whereas gonorrhea kills fewer than 1000 people per year. Nevertheless, largely because of differences in the symptoms and

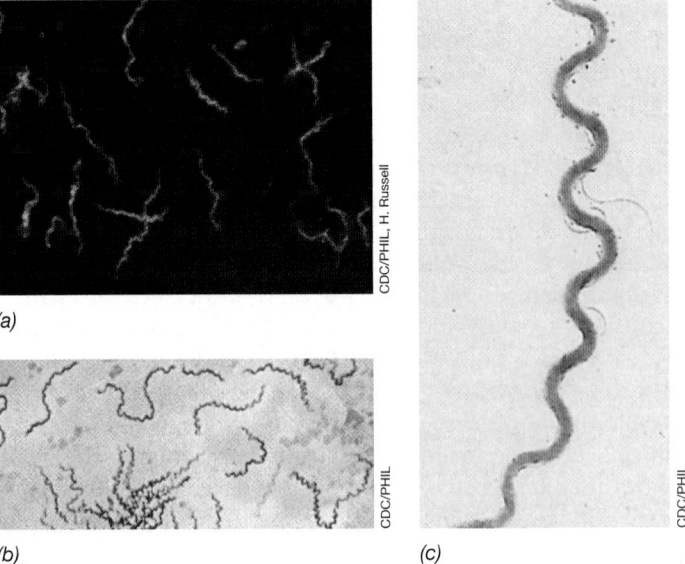

(a)

(b)

(c)

Figure 30.37 The syphilis spirochete, *Treponema pallidum*. *(a)* Cells from a chancre stained with a fluorescent antibody measure 0.15 μm wide and 10–15 μm long. *(b)* Silver-stained (Fontana method) preparation of a specimen from a syphilitic chancre. *(c)* Shadow-cast electron micrograph of a cell of *T. pallidum*. The endoflagella are typical of spirochetes (⟿ Section 15.19).

pathobiology of the two diseases, the incidence of syphilis in the United States is much lower than the incidence of gonorrhea. The incidence of syphilis in the United States, however, has increased in recent years, with over 16,000 new infections reported in 2013 from a low of around 6000 in 1997.

The syphilis spirochete (Figure 30.37) does not pass through unbroken skin, and initial infection takes place through tiny breaks in the epidermal layer. In the male, initial infection is usually on the penis; in the female it is most often in the vagina, cervix, or perineal region. In about 10% of cases, infection is extragenital, usually in the oral region (**Figure 30.38*a***). During pregnancy,

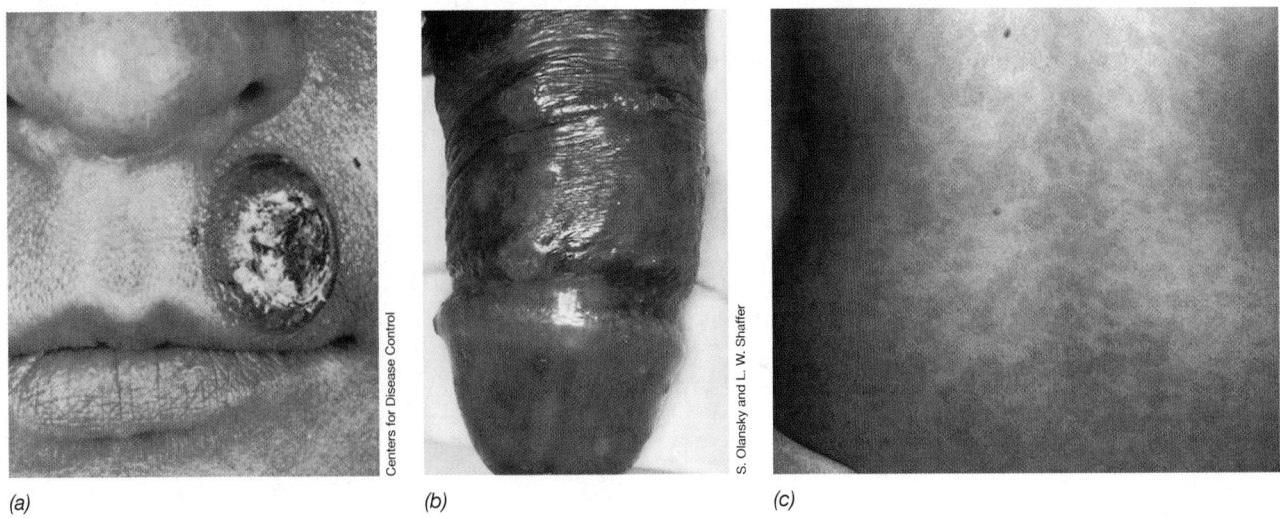

(a)

(b)

(c)

Figure 30.38 Primary and secondary syphilis. *(a)* Chancres on the lip and *(b)* the penis in cases of primary syphilis. The chancre is the characteristic lesion of primary syphilis at the site of infection by *Treponema pallidum*. *(c)* Syphilitic rash on the lower back of a patient showing secondary syphilis.

the organism can be transmitted from an infected woman to the fetus; the disease acquired by the infant is called **congenital syphilis**.

Syphilis is an extremely complex disease and can progress into increasingly serious stages. Syphilis always begins with a localized infection called *primary syphilis*. In primary syphilis, *T. pallidum* multiplies at the initial site of entry, and a characteristic lesion called a *chancre* forms within 2 weeks to 2 months (Figure 30.38*a, b*). Microscopy of a syphilitic chancre exudate reveals the actively motile spirochetes (Figure 30.37*a, b*). In most cases the chancre heals spontaneously and *T. pallidum* disappears from the site. In untreated cases, however, some cells spread from the initial site to various parts of the body, such as the mucous membranes, eyes, joints, bones, or central nervous system, where extensive multiplication occurs. A hypersensitivity reaction to the treponemes often takes place, revealed by the development of a generalized skin rash; this rash is the key symptom of *secondary syphilis* (Figure 30.38*c*).

In the absence of treatment, the subsequent course of the disease varies from case to case. About one-fourth of infected individuals undergo a spontaneous cure and are free of any further disease symptoms. Another one-fourth exhibit no further symptoms but maintain a persistent, chronic syphilitic infection. Roughly half of untreated patients develop *tertiary syphilis*, with symptoms ranging from relatively mild infections of the skin and bone to serious and even fatal infections of the cardiovascular system or central nervous system. This may occur many years after the primary infection. Involvement of the nervous system can cause paralysis or other severe neurological damage. Relatively low numbers of *T. pallidum* are present in individuals with tertiary syphilis; most of the symptoms probably result from inflammation due to delayed-type hypersensitivity reactions (⇨ Section 27.9) to the syphilis spirochetes. Tertiary syphilis can still be treated, usually with long-term intravenous antibiotic administration, but prior neurological damage from the syphilitic infection is typically irreversible.

Several laboratory tests that can be used to diagnose syphilis were discussed in Chapter 28. However, the single most important physical sign of a primary syphilis infection, the chancre (Figure 30.38*a, b*), is highly diagnostic for the disease. Infected individuals generally seek treatment for syphilis because of the chancre. Penicillin remains highly effective in syphilis therapy, and the primary and secondary stages of the disease can typically be cured by a single injection of benzathine penicillin G. Unlike the case with gonorrhea, antibiotic resistance has not seriously affected the treatment of syphilis. Resistance by *T. pallidum* to the macrolide antibiotic azithromycin has emerged, but a number of mainline antibiotics still are highly effective in treating syphilis.

— **MINIQUIZ** —

- How are gonorrhea and syphilis diagnosed?
- Explain at least one potential reason for the high incidence of gonorrhea as compared with syphilis.
- Describe the progression of untreated gonorrhea and untreated syphilis. Do treatments produce a cure for each disease?

30.14 Chlamydia, Herpes, and Human Papillomavirus

STIs caused by *Chlamydia* (a bacterium) and herpesvirus and human papillomavirus are very prevalent among sexually active adults and are often more difficult to diagnose and treat than are syphilis and gonorrhea.

Chlamydia

A number of sexually transmitted diseases can be ascribed to infection by the obligately intracellular bacterium *Chlamydia trachomatis* (**Figure 30.39**). This organism is one of a small group of parasitic bacteria that form their own phylum (the *Chlamydiae*) of *Bacteria* (⇨ Section 16.15). Because *C. trachomatis* must be grown in host cells (tissue culture), its rapid isolation and identification is not as straightforward as for *Neisseria gonorrhoeae*.

The total incidence of sexually transmitted *C. trachomatis* infections probably greatly outnumbers the incidence of gonorrhea. Over 1 million chlamydial cases are now reported annually in the United States, but because of their often inapparent nature, there may be more than 4 million new sexually transmitted chlamydial infections every year. Because of this, chlamydia is the most prevalent STI and reportable communicable disease in the United States. *C. trachomatis* also causes a serious eye infection called *trachoma*, but the strains of *C. trachomatis* responsible for STIs are distinct from those causing trachoma. Chlamydial infections may also be transmitted congenitally to the newborn in the birth canal, causing newborn conjunctivitis and pneumonia.

Nongonococcal urethritis (NGU) due to *C. trachomatis* is one of the most frequently observed sexually transmitted diseases in males and females, but the infections are often inapparent. In a small percentage of cases, chlamydial NGU leads to serious acute complications, including testicular swelling and prostate inflammation in men and cervicitis, pelvic inflammatory disease, and fallopian tube damage in women. These are due to the ability of *C. trachomatis* to

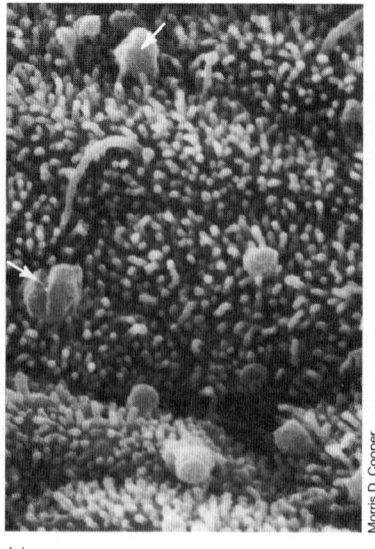
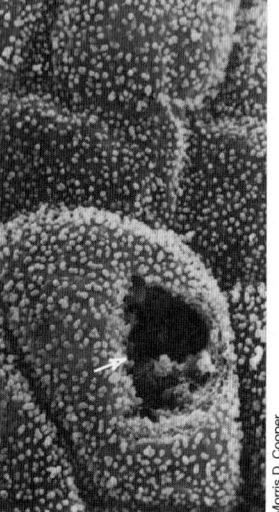

(a)   *(b)*

Figure 30.39 Cells of *Chlamydia trachomatis* (arrows) attached to human fallopian tube tissues. *(a)* Cells attached to the microvilli of a fallopian tube. *(b)* A damaged fallopian tube containing a cell of *C. trachomatis* (arrow) in the lesion.

trigger an overblown immune response and inflammation in the host. During NGU, cells of *C. trachomatis* can attach to microvilli of fallopian tube cells, enter, multiply, and eventually lyse the cells (Figure 30.39*b*). Untreated NGU in a female can thus lead to infertility, ectopic pregnancy, and chronic pelvic pain. Infections with the protist *Trichomonas vaginalis* can cause symptoms similar to those of chlamydial NGU, and we consider trichomoniasis along with other parasitic infections in Chapter 33.

Chlamydial NGU is frequently observed as a secondary infection following gonorrhea. Both *N. gonorrhoeae* and *C. trachomatis* are often transmitted to a new host simultaneously. However, treatment of gonorrhea does not eliminate the chlamydia. Although cured of gonorrhea, these patients are still infected with chlamydia and eventually experience an apparent recurrence of gonorrhea that is instead a case of chlamydial NGU. Thus, patients treated for gonorrhea with drugs such as cefixime or ceftriaxone are also given azithromycin or doxycycline to treat a potential coinfection with *C. trachomatis*. A variety of clinical techniques including nucleic acid and immunological analyses are available for making a positive diagnosis of *C. trachomatis* infection, but drug therapy in the absence of a positive diagnosis is often prescribed.

Lymphogranuloma venereum (*LGV*) is a sexually transmitted disease caused by distinct strains of *C. trachomatis* (LGV 1, 2, and 3). The disease occurs most frequently in males and is characterized by infection and swelling of the lymph nodes in and about the groin. From the infected lymph nodes, chlamydial cells may travel to the rectum and cause a painful inflammation of rectal tissues called *proctitis*. LGV has the potential to cause regional lymph node damage and the complications of proctitis. It is the only chlamydial infection that invades beyond the epithelial cell layer.

Herpes

Herpesviruses are a large group of double-stranded DNA viruses (Section 10.7), many of which are human pathogens. The herpes simplex viruses are responsible for both cold sores and genital infections.

Herpes simplex 1 virus (*HSV-1*) infects the epithelial cells around the mouth and lips, causing cold sores, also known as fever blisters (**Figure 30.40**). HSV-1 is spread via direct contact with infectious lesions or through saliva. The incubation period of HSV-1 infections is short (3–5 days), and the lesions heal without treatment in 2–3 weeks. However, latent herpes infections are common, because the virus typically persists in low numbers in nerve tissue.

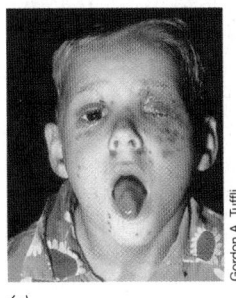
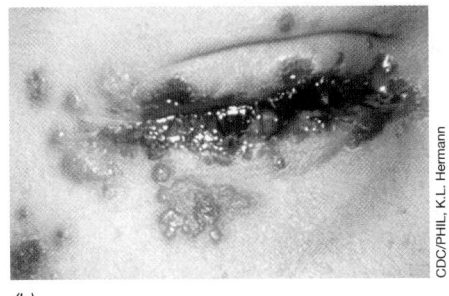

(a)   (b)

Figure 30.40 Herpes simplex 1 virus infections. *(a)* A severe case of herpes blisters on the face due to infection with herpes simplex 1 virus. *(b)* Close-up view of herpes blisters by the eye.

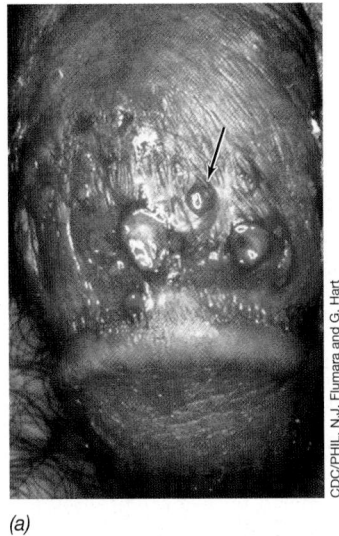
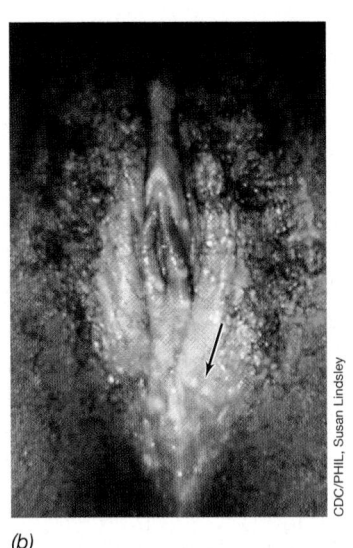

(a)   (b)

Figure 30.41 Herpes simplex 2 virus infections. Herpes simplex 2 virus blisters on the *(a)* penis and *(b)* vulva. As for herpes type 1, acute type 2 herpes infections can seemingly be cured only to reappear later from a persistent virus infection (Figure 8.20).

Recurrent acute herpes infections can then occur when the virus is triggered by coinfections with other pathogens or by bodily stress. Oral herpes caused by HSV-1 is quite common and apparently has no long-term harmful effects on the host, beyond the discomfort of the oral blisters.

Herpes simplex 2 virus (*HSV-2*) infections are associated primarily with the anogenital region, where the virus causes painful blisters on the penis of males or on the cervix, vulva, or vagina of females (**Figure 30.41**). HSV-2 infections are generally transmitted by direct sexual contact, and the disease is most easily transmitted when active blisters are present, but may also be transmitted during asymptomatic periods, even when the infection is presumably latent. HSV-2 occasionally infects other sites such as the mucous membranes of the mouth and can also be transmitted to a newborn by contact with herpetic lesions in the birth canal at birth. The disease in the newborn varies from latent infections with no apparent damage to systemic disease resulting in brain damage or even death. To avoid herpes infections in newborns, delivery by cesarean section is advised for pregnant women with genital herpes infections.

The long-term effects of genital herpes infections are not fully understood. However, studies have indicated a significant correlation between genital herpes infections and cervical cancer in females. Genital herpes infections are presently incurable, although a limited number of drugs have been successful in controlling the infectious blister stages. The guanine analog acyclovir (**Figure 30.42**), given orally and also applied topically, is particularly effective in limiting the shed of active virus from blisters and promoting the healing of blistering lesions (Figure 30.41). Acyclovir, and the related drugs valacyclovir and vidarabine, are nucleoside analogs that interfere with herpesvirus DNA polymerase, inhibiting viral DNA replication (Section 28.11).

Human Papillomavirus

As for herpesviruses, **human papillomaviruses (HPV)** comprise a family of double-stranded DNA viruses. Of more than 100

UNIT 7

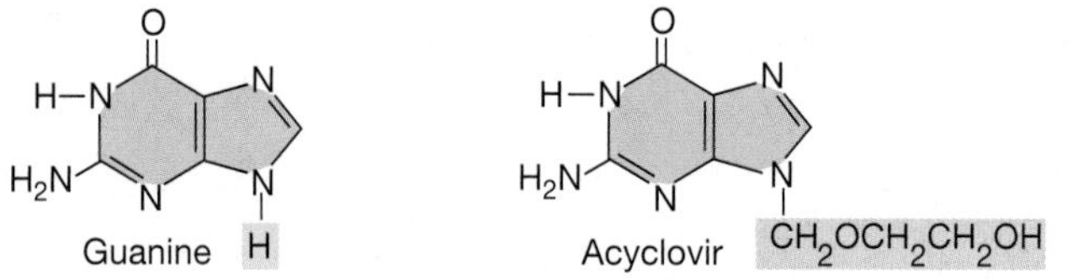

Figure 30.42 Guanine and the guanine analog acyclovir. Acyclovir has been used therapeutically to control genital herpes (HSV-2) blisters (Figure 30.41).

different strains, about 30 are transmitted sexually, and several of these cause genital warts and cervical cancer. About 20 million people in the United States are infected, and up to 80% of women over age 50 have had at least one HPV infection. Over 6 million people acquire new HPV infections annually, leading to almost 10,000 cases of cervical cancer and about 3700 deaths.

Most HPV infections are asymptomatic, with some progressing to cause genital warts. Others cause cervical neoplasia (abnormalities in cells of the cervix), and a few progress to cervical cancers. Most HPV infections resolve spontaneously but, as with many viral infections, there is no adequate treatment or cure for active infections. Because human papillomaviruses are potentially oncogenic (cancer-causing), HPV vaccines are available (a widely used one is marketed as Gardasil) and are currently recommended for use in females 11–26 years of age. The HPV vaccine is also recommended for males because immunized males no longer carry HPV and thus cannot infect females, and because HPV infection in males can lead to anal and penile cancers. In addition, the HPV vaccine should reduce incidence of certain neck and throat cancers in both males and females caused by the same strains of HPV linked to the sexually transmitted infections.

MINIQUIZ

- Describe pertinent clinical features and treatment protocols for chlamydia, herpes, and human papillomavirus.
- Why are these diseases more difficult to diagnose than gonorrhea or syphilis?

30.15 HIV/AIDS

Acquired immunodeficiency syndrome (AIDS) is caused by the human immunodeficiency virus (HIV). Worldwide, nearly 80 million people have been infected with HIV and about 34 million have died. In the United States, from a total of 5 cases diagnosed in 1981, over 1.1 million people are infected with HIV today. We covered some aspects of the epidemiology of HIV/AIDS in Section 29.8 and will pick up on that theme here.

HIV and a Definition of AIDS

HIV is of two types, *HIV-1* and *HIV-2*, but because more than 99% of global AIDS cases are due to HIV-1, we focus on HIV-1 here. HIV-1 is a retrovirus (⇔ Sections 8.8 and 10.11) that replicates in macrophages and T cells of the human immune system (Chapters 26 and 27). HIV infection eventually leads to the destruction of key immune system cells, virtually eliminating the host immune response. Death from AIDS is usually the result of a secondary

infection, typically one caused by an **opportunistic pathogen**, pathogens that in a healthy individual would be controlled by the immune system.

The current definition of a case of HIV/AIDS is a patient who tests positive for HIV in immunological and/or nucleic acid-based tests and meets at least one of the following two criteria:

1. A CD4 T cell number of less than 200/µl of whole blood (the normal count is 600–1000/µl) or a CD4 T cell/total lymphocytes percentage of less than 14%.

2. A CD4 T cell number of more than 200/µl *and* any of the following diseases: candidiasis, coccidioidomycosis, cryptococcosis, histoplasmosis, cystoisosporiasis, *Pneumocystis jirovecii* pneumonia, cryptosporidiosis, or toxoplasmosis of the brain (all fungal or protozoal diseases) (Chapter 33); pulmonary tuberculosis or other mycobacterial infections, or recurrent *Salmonella* septicemia (bacterial diseases); cytomegalovirus infection, HIV-related encephalopathy, HIV wasting syndrome, chronic ulcers, or bronchitis due to herpes simplex (viral infections); or certain malignant diseases such as invasive cervical cancer, Kaposi's sarcoma, Burkitt's lymphoma, primary lymphoma of the brain, or immunoblastic lymphoma, or recurrent pneumonia due to any agent.

Pathogenesis of HIV/AIDS

HIV infects cells that have the CD4 cell surface protein. The two cell types most commonly infected are macrophages and a class of lymphocytes called T-helper (Th) cells, both of which are important components of the immune system. Infection normally occurs first in macrophages. At the macrophage cell surface, the CD4 molecule binds to the gp120/gp41 capsid protein of HIV as the virus interacts with the macrophage receptor CCR5 (**Figure 30.43**). CCR5 is a coreceptor for HIV and, together with CD4, forms the docking site where the HIV envelope fuses with the host cytoplasmic membrane; this is required for the viral nucleocapsid to be inserted into the cell (Figure 30.43*b*). Within the macrophage, HIV replicates and makes an altered form of gp120 that recognizes a different coreceptor, CXCR4, on Th cells. HIV virions are released from macrophages and proceed to infect and replicate in Th lymphocytes; Th cells that produce HIV no longer divide and are eventually diminished by attrition.

In some HIV/AIDS patients, HIV infection does not progress immediately to killing host immune cells. HIV can exist in a dormant state as a provirus; under these conditions, the reverse-transcribed HIV genome, now in the form of DNA, is integrated into host chromosomal DNA (⇔ Figure 10.23). At this point the cell may show no outward sign of infection. Indeed, HIV DNA can remain latent for long periods, replicating only as the host cell DNA replicates. However, sooner or later, HIV begins to replicate, and progeny virus are produced and released from the cell.

Symptoms of HIV/AIDS

Ongoing HIV infection results in a progressive decline in CD4 cell numbers. In a healthy human, CD4 cells constitute about 70% of the total T cell pool. In those with HIV/AIDS, CD4 numbers steadily decrease, and by the time opportunistic infections begin

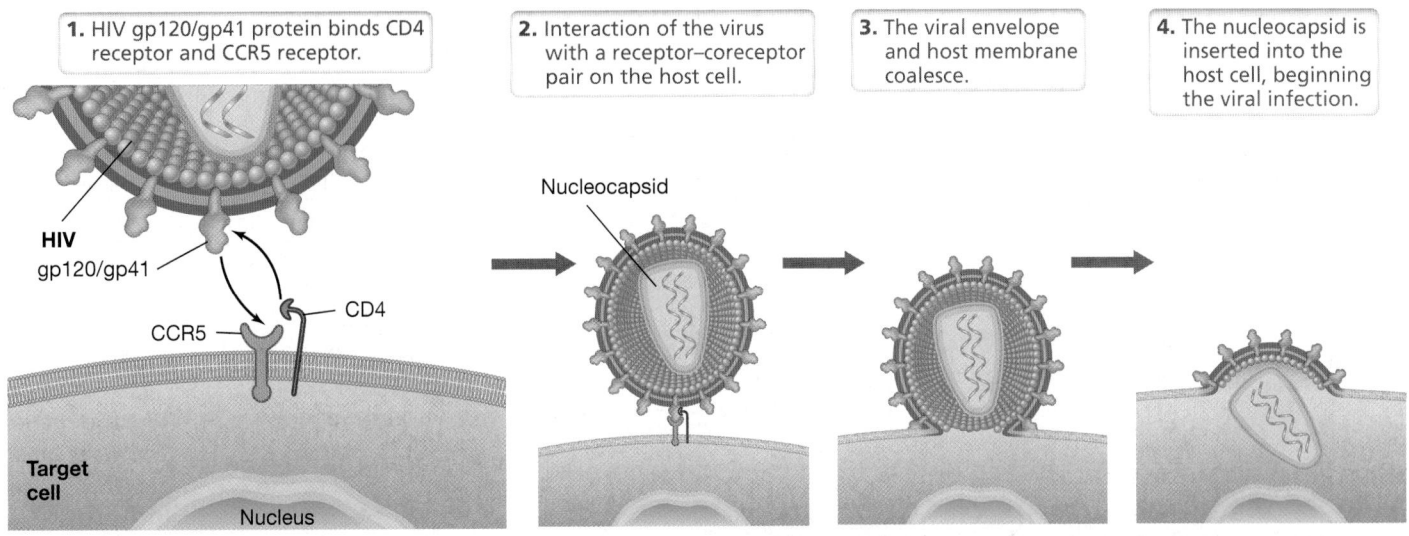

1. HIV gp120/gp41 protein binds CD4 receptor and CCR5 receptor.

2. Interaction of the virus with a receptor–coreceptor pair on the host cell.

3. The viral envelope and host membrane coalesce.

4. The nucleocapsid is inserted into the host cell, beginning the viral infection.

(a) **Interaction of HIV with a host cell**

(b) **Fusion of the HIV envelope with the host cell facilitates nucleocapsid entry**

Figure 30.43 Infection of a CD4 target cell with HIV. *(a)* Recognition and binding of HIV by CCR5 and CD4 receptors. *(b)* The viral nucleocapsid eventually enters the cell. Details of the replication of the HIV genome were shown in Figure 10.23.

to appear, CD4 cells are all but absent (**Figure 30.44**). The progression of untreated HIV infection to AIDS follows a typical pattern. First, there is an intense immune response to HIV and HIV numbers drop. But eventually, the immune response is overwhelmed and HIV levels slowly increase while CD4 T cells slowly decrease. When T cell numbers have dropped below about 200/mm³ of blood, the door is open for infections by opportunistic pathogens (Figure 30.44).

Opportunistic infections caused by normally controllable protists, fungi, bacteria, and viruses occur with high prevalence in those with HIV/AIDS and are typically the actual cause of death (**Figure 30.45**). The most common opportunistic disease in

HIV/AIDS patients is pneumonia caused by the fungus *Pneumocystis jirovecii* (Figure 30.45*d*), but infections by various molds, yeasts, protists, and bacteria are also seen (Figure 30.45). Bacterial infections are less common than those of eukaryotic pathogens, but when they occur, they are frequently of strongly antibiotic-resistant bacteria, such as multiple-drug-resistant *Mycobacterium tuberculosis*.

Eukaryotic opportunistic pathogens are difficult to treat in general because many of the drugs used to treat infections from fungi and protists have significant negative side effects on the host, which of course is also a eukaryote. A cancer frequently seen in HIV/AIDS patients is *Kaposi's sarcoma*, a cancer of the cells lining the blood vessels and characterized by purple splotches on the

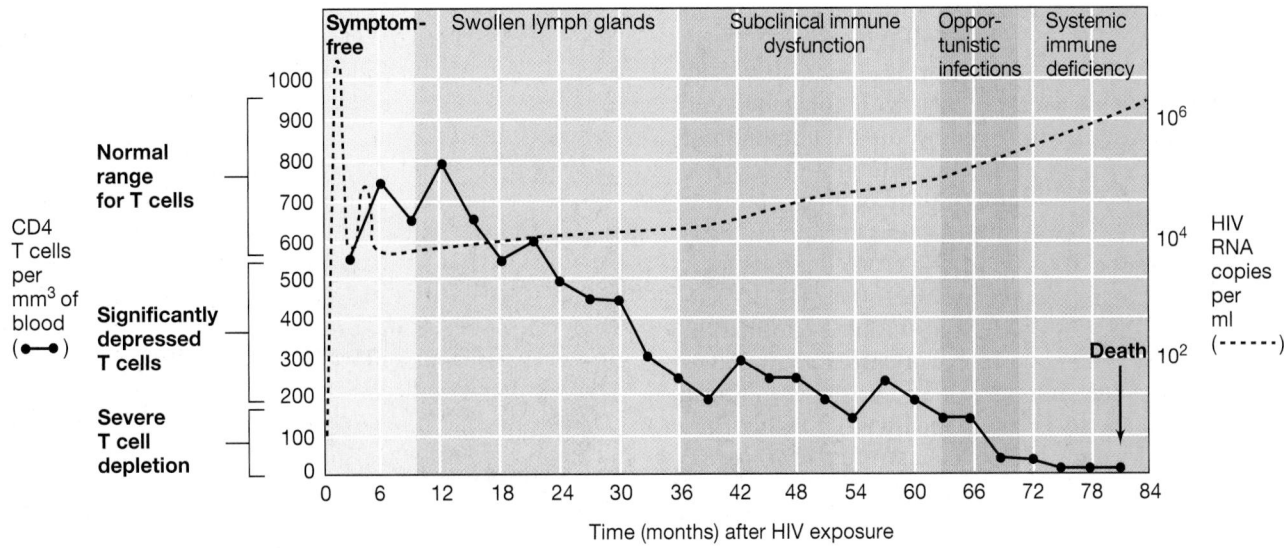

Figure 30.44 Decline of CD4 T lymphocytes and progress of HIV infection. During the typical progression of untreated AIDS, there is a gradual loss in the number and functional ability of the CD4 T cells, while the viral load, measured as HIV-specific RNA copies per milliliter of blood, gradually increases after an initial decline.

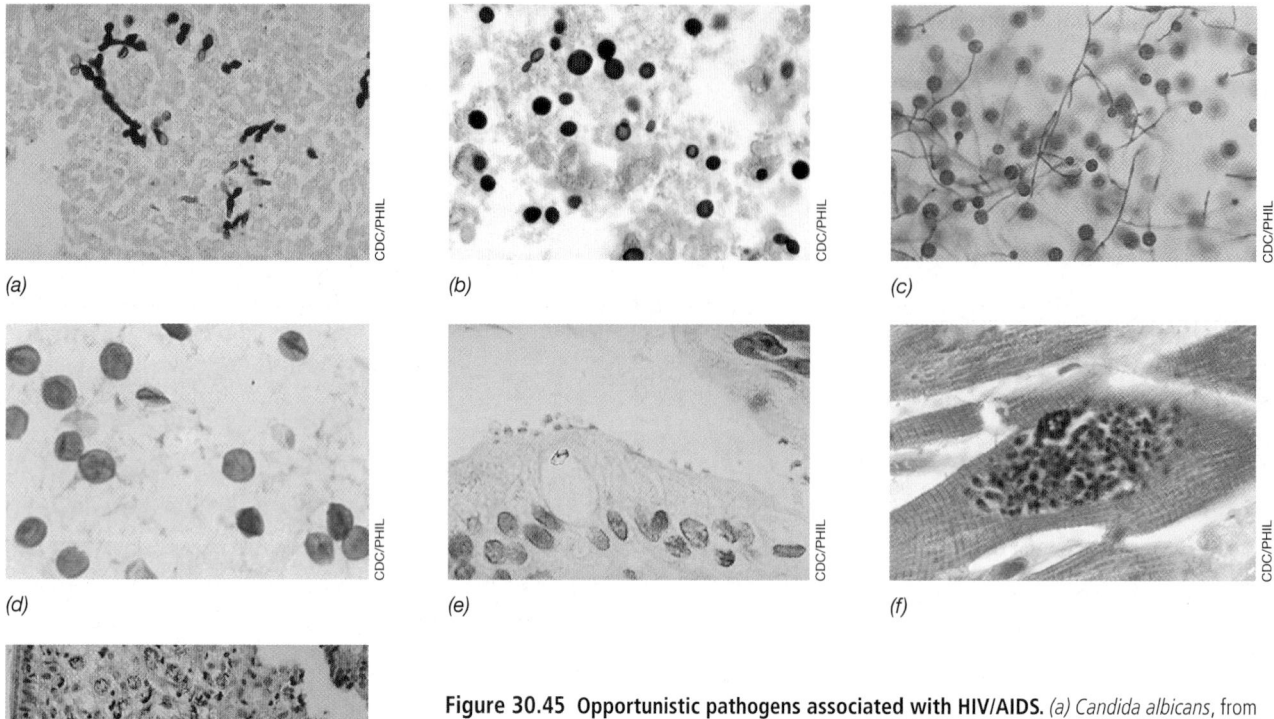

Figure 30.45 Opportunistic pathogens associated with HIV/AIDS. *(a) Candida albicans*, from heart tissue of patient with systemic *Candida* infection. *(b) Cryptococcus neoformans*, from lung tissue of an AIDS patient. *(c) Histoplasma capsulatum*, showing reproductive structures called macroconidia. *(d) Pneumocystis jirovecii*, from lungs of an immune-compromised patient. *(e) Cryptosporidium* sp. from small intestine of a patient with cryptosporidiosis. *(f) Toxoplasma gondii*, from heart tissue of patient with toxoplasmosis. *(g) Mycobacterium* spp. infection of the small bowel (acid-fast stain). *Candida, Cryptococcus, Histoplasma* and *Pneumocystis* are fungi; *Cryptosporidium* and *Toxoplasma* are protists; *Mycobacterium* is a species of *Bacteria*. Coverage of many of these fungal and parasitic diseases can be found in Chapter 33.

skin, especially in the extremities (**Figure 30.46**). Kaposi's sarcoma is caused by coinfection of HIV and human herpesvirus 8 (HHV-8) and is rarely seen outside of HIV/AIDS patients.

Diagnosing HIV/AIDS

HIV infection is typically diagnosed by identifying antibodies to the pathogen in a patient blood sample. An enzyme immunoassay (EIA, ⮂ Figure 28.19) is used for HIV screening purposes, typically for screening done on a large scale such as with donated blood. A positive HIV EIA must be confirmed by an HIV immunoblot (Western blot, ⮂ Figure 28.21) or by immunofluorescence (⮂ Section 28.6) to rule out the possibility of a false-positive screening test. Rapid and inexpensive HIV tests are also available for preliminary screening of blood in clinics. One test requires only a single drop of patient blood and detects the gp41 HIV surface antigen (Figure 30.43) by producing a visible agglutination reaction. A second uses saliva as a source of anti-HIV antibodies and yields a colored product. In general, however, these rapid tests are not as sensitive or specific as the standard HIV EIA and thus positive tests should be confirmed by more sensitive and specific tests. Unfortunately, no matter how sensitive or specific, none of the antibody tests will detect those who have recently acquired the virus and are infectious but have not yet made a detectable antibody response to HIV; this antibody response can require a period of 6 weeks or more following infection.

Diagnostic procedures also are available that directly measure the number of HIV virions in a blood sample. These tests use a virus-specific reverse transcription–polymerase chain reaction assay (RT-PCR, ⮂ Section 12.1). RT-PCR estimates the number

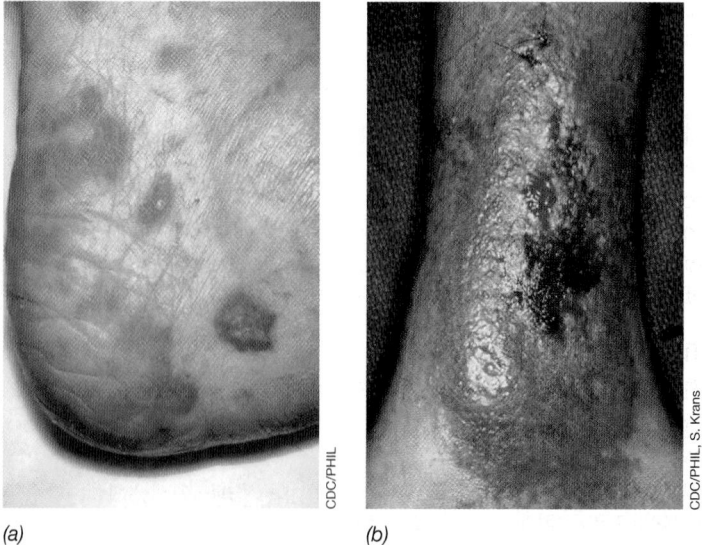

Figure 30.46 Kaposi's sarcoma. Lesions are shown as they appear on *(a)* the heel and lateral foot, and *(b)* the distal leg and ankle.

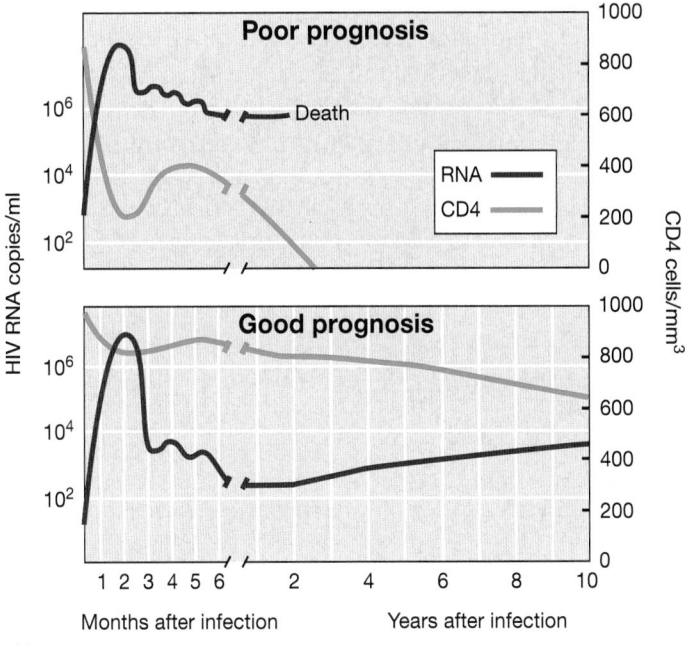

(a)

(b)

Figure 30.47 Monitoring of HIV load. *(a)* Procedure for detecting HIV by reverse transcription–polymerase chain reaction (RT-PCR) techniques. *(b)* Time course for HIV infection as monitored by HIV load and CD4 T cell counts. In the upper panel, a viral load greater than 10^4 copies/ml correlates with below normal CD4 cell numbers (normal = 600–1500/mm^3), indicating a poor prognosis and early death of the patient. In the lower panel, a viral load less than 10^4 copies/ml correlates with normal CD4 cell numbers, indicating a good prognosis and extended survival of the patient. Data are adapted from the CDC, Atlanta, Georgia, USA.

of HIV virions present in the blood, the so-called **viral load** (**Figure 30.47**). The RT-PCR test for HIV load is not routinely used to screen for HIV because it is costly and technically demanding. However, after an initial diagnosis, the test is often used to monitor progression of an HIV infection (Figure 30.47) and the effectiveness of chemotherapy.

Treatment of HIV/AIDS

The prognosis for an *untreated* HIV-infected individual is poor, as opportunistic pathogens or malignancies (Figures 30.45 and 30.46) eventually kill virtually all infected persons. Long-term studies indicate that the average person infected with HIV progresses through several stages of decreasing immune function, with CD4 cells dropping from a normal range of 600–1000/mm^3 of blood to near zero over a period of 5–7 years (Figure 30.44). Although the rate of decline varies from one HIV-infected individual to another,

it is rare for an HIV-positive individual to live for more than 10 years without anti-HIV drug therapy (see page 887 for exceptions).

Several drugs have been developed that delay the progression of HIV/AIDS and significantly prolong the life of those infected with HIV. Therapy is aimed at reducing the viral load of HIV-infected individuals to below detectable levels. The strategy to accomplish this is called *highly active anti-retroviral therapy* (HAART) and is carried out by administering at least three anti-retroviral drugs at once to inhibit the replication of HIV and prevent the development of drug-resistant strains. Multiple drug therapy, however, is not a cure for HIV infection. In individuals who have no detectable viral load after drug treatment, a significant viral load returns if therapy is interrupted or discontinued, or if multiple drug resistance develops.

Effective anti-HIV drugs fall into four categories, including two classes of *reverse transcriptase inhibitors*, various *protease inhibitors*, *fusion inhibitors*, and *integrase inhibitors*. Reverse transcriptase is the enzyme that converts the single-stranded RNA genome of HIV into cDNA and then double-stranded DNA and is essential for viral replication (⇌ Sections 8.8 and 10.11). Cells lack reverse transcriptase and thus reverse transcriptase inhibitors are viral-specific. *Azidothymidine* (AZT), also called zidovudine, closely resembles the nucleoside thymidine but lacks the correct attachment site for the next base in a replicating nucleotide chain, resulting in termination of the growing DNA chain. AZT is thus a **nucleoside reverse transcriptase inhibitor** (**Figure 30.48a**). **Nonnucleoside reverse transcriptase inhibitors**, such as *nevirapine* (Figure 30.48b), inhibit the activity of reverse transcriptase in a different way by interacting with the protein and altering the conformation of the catalytic site.

Another category of anti-HIV drugs is the **protease inhibitors**, such as *saquinavir* (⇌ Figure 28.35b). These are peptide analogs that inhibit processing of retroviral polypeptides (⇌ Figure 10.24) by binding to the active site of the processing enzyme, *HIV protease*; this effectively inhibits viral maturation. **Fusion inhibitors** include *enfuvirtide*, a synthetic peptide that functions by binding to the gp41 protein on HIV capsids (Figure 30.43); this stops fusion of the viral envelope and the CD4 cell cytoplasmic membrane. Finally, there are the **integrase inhibitors**, such as *elvitegravir* and *raltegravir*. These drugs target HIV integrase, the protein that integrates the HIV genome into host cell DNA. The interference

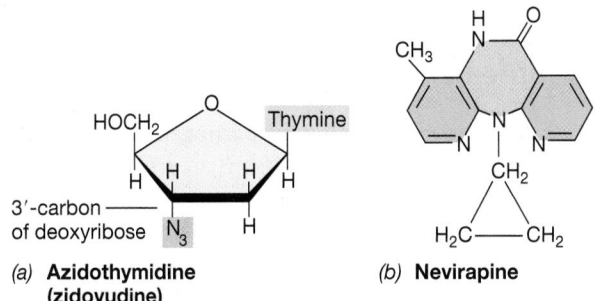

(a) **Azidothymidine (zidovudine)**

(b) **Nevirapine**

Figure 30.48 HIV/AIDS chemotherapeutic drugs. *(a)* Azidothymidine (AZT), also called zidovudine, a nucleoside reverse transcriptase inhibitor. This nucleoside analog is missing the –OH group on the 3′-carbon, causing nucleotide chain elongation to terminate when the analog is incorporated, inhibiting virus replication. *(b)* Nevirapine, a nonnucleoside reverse transcriptase inhibitor, binds directly to the catalytic site of HIV reverse transcriptase, also inhibiting elongation of the nucleotide chain.

with integration of viral DNA into the host cell genome interrupts the HIV replication cycle.

All anti-HIV drugs rapidly decrease the viral load when given to HIV-infected individuals, but drug-resistant strains of HIV arise quickly if only a single drug is administered. A typical HAART protocol for treatment of an established HIV infection includes at least one protease or nonnucleoside reverse transcriptase inhibitor plus a combination of two nucleoside reverse transcriptase inhibitors. A resistant virus would, therefore, have to develop resistance to three drugs simultaneously, and the probability of this occurring is very small. A patient receiving this combination therapy is then monitored to track changes in viral load (Figures 30.44 and 30.47). An effective HAART protocol reduces viral load to nondetectable levels within several days. Drug therapy is then continued and the patient monitored for viral load indefinitely. If the viral load again reaches detectable limits, the drug cocktail is changed because an increase in viral load indicates the emergence of drug-resistant HIV.

In addition to drug resistance, some anti-retroviral drugs are toxic to the host. In many cases, nucleoside analogs are not well tolerated by patients, presumably because they interfere with host functions such as cell division. In general, the nonnucleoside reverse transcriptase inhibitors and the protease inhibitors are better tolerated because they target virus-specific functions. However, drug resistance and host toxicity are major problems in all forms of HIV therapy. Thus, new chemotherapeutic agents and drug protocols are constantly being developed and tailored to the needs of individual patients.

HIV/AIDS Prevention

Public education about how HIV/AIDS is transmitted, sexual abstinence, and avoidance of high-risk behavior remain the major tools used to prevent HIV/AIDS. HIV spread is linked to promiscuous sexual activities and other activities that involve exchange of body fluids, which include not only men who have sex with men, but also prostitution and intravenous drug use where needles are shared. In some countries, the fastest growing mode of HIV transmission is actually between heterosexual partners. Effective prevention of HIV transmission therefore requires avoiding high-risk behaviors, regardless of sexual partners.

The United States Surgeon General has issued specific recommendations for avoiding HIV infection. These include, in addition to the avoidance of intravenous drug use where needles are shared:

1. Avoiding mouth contact with penis, vagina, or rectum.

2. Avoiding all sexual activities that could cause cuts or tears in the linings of the rectum, vagina, or penis.

3. Avoiding sexual activities with individuals from high-risk groups. These include prostitutes (both male and female); those who have multiple sex partners, particularly homosexual men and bisexuals; and intravenous drug users.

4. If a person has had sex with someone in a high-risk group, a blood test should be done to determine if infection with HIV has occurred. The blood test should be repeated at intervals for a year or longer because of the lag time in the immune response. If the test is positive, the sexual partners of the HIV-positive individual must be protected by use of a condom during sexual activities.

MINIQUIZ

- Review the definition of HIV/AIDS. Which symptoms of HIV/AIDS are shared by all HIV/AIDS patients?

- What does the enzyme reverse transcriptase do and why is it a good target for anti-HIV drugs?

- What are the current prevention guidelines for HIV/AIDS infection? Are they effective?

MasteringMicrobiology®   **Visualize, explore,** and **think critically** with Interactive Microbiology, MicroLab Tutors, MicroCareers case studies, and more. MasteringMicrobiology offers practice quizzes, helpful animations, and other study tools for lecture and lab to help you master microbiology.

Chapter Review

I • Airborne Bacterial Diseases

30.1 Bacterial and viral respiratory pathogens are transmitted in air. Most respiratory pathogens are transferred from person to person via respiratory aerosols generated by coughing, sneezing, talking, or breathing, or by direct or fomite contact. Respiratory pathogens infect either the upper or lower respiratory tracts and sometimes both.

Q **Why do gram-positive bacteria cause respiratory diseases more frequently than gram-negative bacteria?**

30.2 Streptococcal diseases include strep throat and pneumococcal pneumonia. *Streptococcus pyogenes* infections may progress into serious conditions such as scarlet and rheumatic fevers, and pneumococcal pneumonia can have high mortality. Both pathogens can be cultured and both are treatable with antimicrobial drugs including penicillin.

Q **What are the typical symptoms of a streptococcal respiratory infection? Why should streptococcal infections be treated promptly?**

30.3 Diphtheria is an acute respiratory disease caused by *Corynebacterium diphtheriae*. Early childhood immunization is effective for preventing this very serious respiratory disease. Whooping cough is an endemic disease caused by *Bordetella pertussis*. Immunization of children, adolescents, and adults can control its propagation and spread.

Q Describe the causal agents and the symptoms of diphtheria and pertussis. Why has diphtheria incidence declined in the United States, while pertussis incidence is higher than a decade ago?

30.4 Tuberculosis is one of the most prevalent and dangerous infectious diseases in the world. Its incidence is increasing in developed countries in part because of the emergence of drug-resistant strains of *Mycobacterium tuberculosis*. The pathology of tuberculosis and other mycobacterial syndromes such as Hansen's disease (leprosy) is influenced by the cell-mediated immune response.

Q Describe the process of infection by *Mycobacterium tuberculosis*. Does infection always lead to active tuberculosis? Why or why not? How is exposure to *M. tuberculosis* detected in humans?

30.5 *Neisseria meningitidis* is a common cause of meningococcemia and meningitis in young adults and occasionally occurs in epidemics in enclosed populations. Bacterial meningitis and meningococcemia can have high mortality rates, and treatment and prevention strategies including vaccines are available.

Q Describe the symptoms of meningococcemia and meningitis. How are these diseases treated? What is the prognosis for each?

II • Airborne Viral Diseases

30.6 Viral respiratory diseases are highly infectious and may cause serious health problems, although most are controllable and not life-threatening. The measles/mumps/rubella (MMR) vaccine is highly effective in controlling these diseases.

Q Compare and contrast measles, mumps, and rubella. Include a description of the pathogen, major symptoms encountered, and any potential consequences of these infections. Why is it important that women be vaccinated against rubella before puberty?

30.7 Colds are the most common infectious viral diseases. Usually caused by a rhinovirus, colds are generally mild and self-limiting diseases; "cold drugs" may help to moderate symptoms but are not a cure. Each infection induces specific, protective immunity, but the large number of cold viruses precludes complete protective immunity or vaccines.

Q Why are colds such common respiratory diseases, and why are vaccines not used to prevent colds?

30.8 Influenza is caused by an RNA virus that contains a segmented genome and is easily transmitted by the airborne route. Influenza outbreaks occur annually as a result of the plasticity of the influenza genome. Antigenic drift varies the nature of the viral envelope of influenza viruses in minor ways, causing influenza seasonal epidemics, while antigenic shift varies the virus in major ways and can trigger periodic influenza pandemics. Surveillance and immunization are used to control influenza.

Q Why is influenza such a common respiratory disease? How are influenza vaccines chosen?

III • Direct-Contact Bacterial and Viral Diseases

30.9 Staphylococci are usually benign inhabitants of the upper respiratory tract and skin, but several serious diseases can result from pyogenic infection or from the activity of staphylococcal superantigen exotoxins. Antibiotic resistance is common, even in community-acquired infections. MRSA strains of *Staphylococcus aureus* can be very difficult to treat and cause significant tissue damage.

Q Distinguish between pathogenic staphylococci and those that are part of the normal microbiota.

30.10 *Helicobacter pylori* infection is the common cause of gastric ulcers. Gastric ulcers are now treated with antibiotics as an infectious disease, promoting a permanent cure.

Q Describe the evidence linking *Helicobacter pylori* to gastric ulcers. How can these ulcers be cured?

30.11 Viral hepatitis can result in acute liver disease, which may be followed by chronic liver disease (cirrhosis). Hepatitis B and C viruses in particular are transmitted by direct contact and can cause chronic infections leading to liver cancer. Vaccines are available for hepatitis viruses A and B. Viral hepatitis is still a major public health problem because of the high infectivity of the viruses and the lack of effective treatments.

Q Describe the major hepatitis viruses. How are they related to one another? How is each spread?

30.12 Ebola hemorrhagic fever is a deadly viral disease spread by direct contact through the skin or from contaminated bodily fluids. Mortality rates from Ebola are near the highest of all diseases. Treatment is primarily supportive of symptoms, but vaccine trials have shown that effective vaccination protocols should be possible.

Q The Ebola virus cannot depend on the host to synthesize its genome; why not?

IV • Sexually Transmitted Infections

30.13 Gonorrhea and syphilis, caused by *Neisseria gonorrhoeae* and *Treponema pallidum*, respectively, are STIs with potential serious consequences if infections are not treated. In the United States, the incidence of gonorrhea has decreased in the last several years, but the incidence of syphilis has increased.

Q Why did the incidence of gonorrhea rise dramatically in the mid-1960s, while the incidence of syphilis actually decreased at the same time?

30.14 Chlamydia is the most prevalent of STIs, and if left untreated, it can cause serious complications in both males and females. Herpes simplex viruses cause incurable infections transmitted by oral or genital contact with herpes 1 or herpes 2, respectively. Human papillomaviruses cause widespread STIs that may lead to cervical and other cancers, but effective HPV vaccines are available.

Q For the sexually transmitted infections of chlamydia, herpes, and human papillomavirus, describe the organism that causes each. In each case, is treatment possible, and if so, is it an effective cure? Why or why not?

30.15 HIV is a retrovirus that destroys the immune system, leading to AIDS, and opportunistic pathogens eventually

kill the host. There is no effective cure or vaccine for HIV infection, although antiviral drugs may slow or stop the progress of AIDS. Preventing HIV infection requires education and avoidance of high-risk behaviors involving exchange of body fluids.

Q Describe how human immunodeficiency virus (HIV) effectively shuts down both antibody-mediated and cell-mediated immunity. What is HAART therapy?

Application Questions

1. Why is it that you get a cold or two each year but if you have had a case of measles, it was a one-time occurrence?

2. Your college roommate goes home for the weekend, becomes extremely ill, and is diagnosed with bacterial meningitis at a local hospital. Because he was away, university officials are not aware of his illness. What should you do to protect yourself against meningitis? Should you notify university health officials?

3. Contrast an HIV infection with an infection by any other viral pathogen considered in this chapter, regardless of mode of transmission. Why do untreated cases of HIV infection

almost always lead to death whereas untreated cases of chicken pox, influenza, or even hepatitis typically do not?

4. Discuss the molecular biology of antigenic shift in influenza viruses and comment on the immunological consequences for the host. Why has antigenic shift prevented the production of a single universally effective vaccine for influenza control? Next, compare antigenic shift to antigenic drift. Which causes the greatest antigenic change? Which creates the biggest problems for vaccine developers? Which can lead to pandemic influenza, and why?

Chapter Glossary

Antigenic drift a minor change in influenza virus antigens due to gene mutation

Antigenic shift a major change in influenza virus antigen due to gene reassortment

Cirrhosis breakdown of normal liver architecture, resulting in fibrosis

Congenital syphilis syphilis contracted by an infant from its mother during pregnancy

Fusion inhibitor a synthetic polypeptide that binds to viral glycoproteins, inhibiting fusion of viral and host cell membranes

Hepatitis liver inflammation, commonly caused by an infectious agent

Human papillomavirus (HPV) a sexually transmitted virus that causes genital warts, cervical neoplasia, and cancer

Integrase inhibitor a drug that interrupts the HIV replication cycle by interfering with integrase, the HIV protein that catalyzes the integration of viral dsDNA into host cell DNA

Meningitis inflammation of the meninges (brain tissue), sometimes caused by *Neisseria meningitidis* and characterized by sudden

onset of headache, vomiting, and stiff neck, often progressing to coma within hours

Meningococcemia a rapidly progressing severe disease caused by *Neisseria meningitidis* and characterized by septicemia, intravascular coagulation, and shock

Nonnucleoside reverse transcriptase inhibitor a nonnucleoside compound that inhibits the action of retroviral reverse transcriptase by binding directly to the catalytic site

Nucleoside reverse transcriptase inhibitor a nucleoside analog compound that inhibits the action of retroviral reverse transcriptase by competing with nucleosides

Opportunistic pathogen an organism that causes disease in the absence of normal host resistance

Pertussis (whooping cough) a disease caused by an upper respiratory tract infection with *Bordetella pertussis*, characterized by a deep, persistent cough

Protease inhibitor a compound that inhibits the action of viral protease by binding directly to the catalytic site, preventing viral protein processing

Rheumatic fever an inflammatory autoimmune disease triggered by an immune response to infection by *Streptococcus pyogenes*

Scarlet fever characteristic reddish rash resulting from an exotoxin produced by *Streptococcus pyogenes*

Sexually transmitted infection (STI) an infection that is usually transmitted by sexual contact

Toxic shock syndrome (TSS) the acute systemic shock resulting from a host response to an exotoxin produced by *Staphylococcus aureus*

Tuberculin test a skin test for previous infection with *Mycobacterium tuberculosis*

Viral load a quantitative assessment of the amount of virus in a host organism, usually in the blood

Vectorborne and Soilborne Bacterial and Viral Diseases

31

microbiology**now**

A New Look at Rabies Vaccines

Dog owners just *love* their dogs (see photo of author Michael Madigan's spouse, Nancy, with Kato), and for good reasons: Besides the loyal companionship one gets from "man's best friend," surveys have shown that dog owners lead healthier and happier lives than those who do not own dogs. Keeping your dog up to date on vaccinations is essential for the dog's health and happiness, too, and the most important vaccination is the periodic rabies booster.

Rabies can affect any mammal and is still a major human health problem. Nearly 60,000 humans die from rabies each year, mostly in developing countries in Asia and Africa where rabies vaccines are not widely administered to domestic animals. Highly effective preexposure and postexposure rabies vaccines are available for use in humans, where the incubation period is typically one to two months. In domestic animals, however, rabies symptoms appear much more quickly (in under 10 days in dogs), and once symptoms begin, death is inevitable. Hence, an effective rabies vaccine for use in animals and humans where disease symptoms have already begun would give medicine a new tool for controlling this disease.

Ongoing research is working to fill this void in the arsenal of rabies vaccines. Scientists have developed a recombinant rabies vaccine that uses parainfluenza virus (a virus that causes only mild infections in dogs or humans) as a carrier of a key rabies virus protein. The recombinant virus was genetically engineered to contain and express the gene encoding rabies virus glycoprotein, a protein that elicits a strong adaptive immune response. In experimental trials, the vaccine was found to protect half of rabies-infected mice when administered as late as 6 days postinfection, which in mice is a time when classical rabies symptoms have already begun.

Recombinant vaccines are attractive because they carry no risk of accidental infection and can be given in high doses. Although this experimental rabies vaccine has only been used in mice, it is possible that a similar vaccine strategy could work in other animals and in humans. If so, such a vaccine could help reduce the heavy toll of human rabies deaths worldwide, most of which result from seemingly minor bites or scratches from unvaccinated dogs or other domestic animals.

Source: Huang, Y., et al. 2015. Parainfluenza virus 5 expressing the G protein of rabies virus protects mice after rabies virus infection. *J. Virol.* *89:* 3427–3429. Photo courtesy of Christina Davis, Logan, Ohio.

In this chapter we focus on pathogenic bacteria and viruses transmitted to humans by animals, arthropods, or soil. Animal-transmitted pathogens have their origins in nonhuman vertebrates, and these infected animal populations can transmit infections to humans. Some arthropods are disease vectors, spreading pathogens to new hosts from a bite. Soilborne pathogens are transmitted to humans through either direct contact with soil or contact with infected animal fur or hides. A few of the diseases we will explore in this chapter produce only mild symptoms and are typically self-limiting. But most are highly dangerous with life-threatening symptoms and high mortality rates. These include such dreaded diseases as rabies, hantavirus syndromes, yellow fever, and plague.

I • Animal-Transmitted Viral Diseases

A zoonosis is an animal disease transmissible to humans, generally by direct contact, aerosols, or bites. Immunization and veterinary care control many infectious diseases in domesticated animals, reducing the transfer of zoonotic pathogens to humans. However, wild animals neither receive veterinary care nor are they immunized, making them a source of potential zoonoses. Diseases in animals may be **enzootic**, present endemically in certain populations, or **epizootic**, with incidence reaching epidemic proportions. In this part of the chapter we focus on two typically enzootic viral diseases, rabies and hantavirus syndromes, both of which can be transmitted to humans.

31.1 Rabies Virus and Rabies

Rabies occurs in wild animals, and the major enzootic reservoirs of rabies virus in the United States are raccoons, skunks, coyotes, foxes, and bats. A small number of rabies cases also occur annually in domestic animals (**Figure 31.1**).

Symptoms and Pathology of Rabies

Rabies is caused by a rhabdovirus, a single-stranded minus-sense RNA virus (⟿ Section 10.9) that infects cells of the central nervous system in most warm-blooded animals, almost invariably leading to death once symptoms have developed. The virus (**Figure 31.2a**) enters the body from virus-contaminated saliva through a wound from a bite or through contamination of mucous membranes. Rabies virus multiplies at the site of inoculation and travels to the central nervous system. The incubation period before the onset of symptoms is highly variable and depends on the host; the size, location, and depth of the inoculating wound; and the titer of rabies virions transmitted in the bite. In dogs, the incubation period for rabies is less than 2 weeks. By contrast, in humans, 9 months or more may pass before rabies symptoms become apparent.

Rabies virus proliferates in the brain, especially in the thalamus and hypothalamus. Infection leads to fever, excitation, dilation of the pupils, excessive salivation, and anxiety. A fear of swallowing (*hydrophobia*, an early name for rabies) develops from uncontrollable spasms of the throat muscles, and death eventually results from respiratory paralysis. In humans, an *untreated* rabies infection in which symptoms have begun is almost always fatal (but see page 919). Fortunately for both domestic animals and humans, a very effective rabies vaccine exists and this keeps the incidence of rabies low in domestic animals (Figure 31.1a) and a rarity in humans.

Diagnosis, Treatment, and Prevention of Rabies

Rabies is diagnosed in the laboratory by examining tissue samples for the virus. Fluorescent antibodies that bind to rabies virus in brain tissues are used to confirm a case of rabies in a postmortem examination. Infected nerve cells stained for light microscopy also

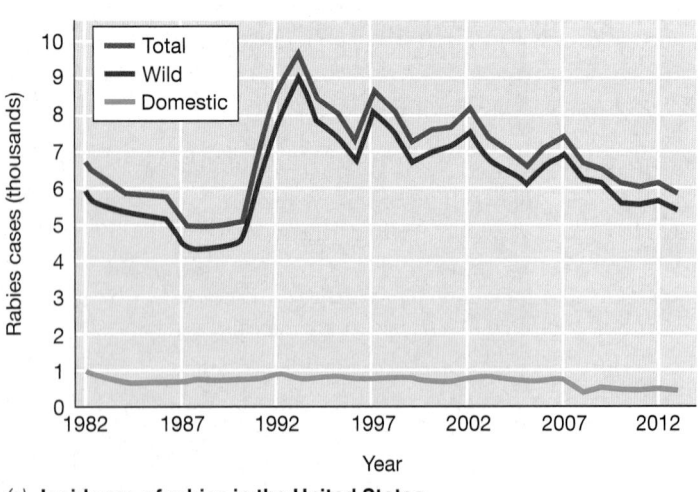

(a) **Incidence of rabies in the United States**

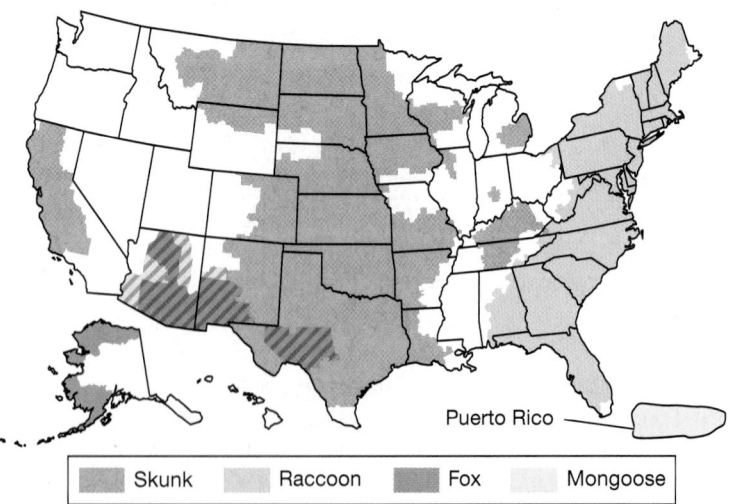

(b) **Major vectors of rabies in the United States and Puerto Rico**

Figure 31.1 Rabies cases in wild and domestic animals in the United States. *(a)* Incidence of rabies by year. Human cases are fewer than five per year. *(b)* Major vectors of rabies virus. In some areas, for example, southwest Texas, both skunks and foxes are the major vectors (shown by hatched lines). Over 90% of all reported rabies cases occur in wild animals. However, actual numbers are probably significantly higher than shown in part *a* because of undiagnosed cases and undiscovered rabid animal carcasses. Data are from the Centers for Disease Control and Prevention, Atlanta, Georgia, USA.

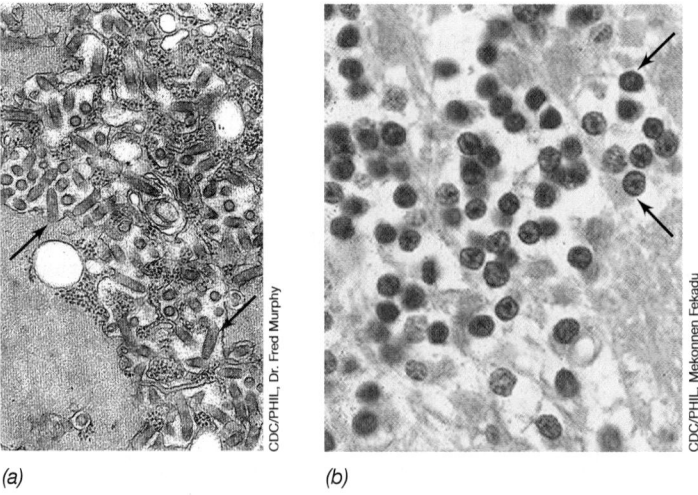

(a)   *(b)*

Figure 31.2 Rabies virus. *(a)* The bullet-shaped rabies virions (arrows) shown in this transmission electron micrograph of a tissue section from a rabid animal are about 75 × 180 nm. *(b)* Pathology of rabies in humans. In brain tissue, rabies virus causes characteristic cytoplasmic inclusions called Negri bodies (arrows), which contain rabies virus antigens. Negri bodies are about 2–10 μm in diameter.

show viral inclusions called *Negri bodies* in their cytoplasm, and these characteristic structures confirm rabies virus infection as well (Figure 31.2*b*).

Because rabies is such a serious disease, firm guidelines for treating possible human exposure to rabies have been established, and the details can be found in the rabies section of the World Health Organization website (http://www.WHO.int). In summary, the guidelines state that if a wild or stray animal is suspected of being rabid, it should be immediately examined for evidence of the rabies virus. If a domestic animal, generally a dog, cat, or ferret, bites a human, especially if the bite is unprovoked, the animal should be held in quarantine for 10 days to check for signs of rabies. If the animal exhibits rabies symptoms, or a definitive diagnosis of its illness cannot be made after 10 days, the human should be passively immunized with rabies immune globulin (purified human antibodies to rabies virus) injected at both the site of the bite and intramuscularly. The patient should also be actively immunized with a rabies virus vaccine. Because of the very slow progression of rabies in humans, this combination of passive and active immune therapy (➡ Section 27.2) is nearly 100% effective, stopping the onset of the disease.

Rabies is prevented largely through immunization. An inactivated rabies vaccine is used in the United States for both humans and domestic animals. Prophylactic rabies immunization is practiced for individuals at high risk, such as veterinarians, animal control personnel, animal researchers, and individuals who work in rabies research or rabies vaccine production laboratories. The rabies problem is primarily with wild animals (Figure 31.1), where traditional means of vaccination are impossible. However, experimental trials with an oral rabies vaccine administered in food "baits" have reduced the incidence and spread of rabies in limited geographic areas. If herd immunity (➡ Section 29.2) could be established in some of the key carriers of rabies (Figure 31.1*b*), it might be possible to reduce incidence of the disease dramatically. Some states and countries, such as Hawaii and Great Britain,

are rabies-free, and any animal imported into these areas is subject to quarantine.

Although rabies is vaccine-preventable, nearly 60,000 people per year die from rabies worldwide, primarily in developing countries in Asia and Africa where rabies is enzootic in domestic animals because of inadequate vaccination practices. Worldwide, nearly 14 million people receive prophylactic treatment for rabies after exposure annually, and in the United States, over 20,000 individuals receive such treatment. Fewer than three cases of human rabies are reported in the United States each year, nearly always the result of bites from wild animals, most frequently from bats. Because domestic animals often have exposure to wild animals, dogs and cats are routinely vaccinated against rabies beginning at 3 months of age. Large farm animals, especially horses, are often immunized against rabies as well. One other rare but possible mode of rabies transmission is from organ transplants. In 2013, a rabies death in the U.S. was linked to a transplanted kidney received from a donor who died from rabies that was misdiagnosed as severe gastroenteritis. Past cases of rabies transmission in a cornea and other transplants have also been documented where the donor had yet to show signs of clinical rabies.

— MINIQUIZ —

- What is the procedure for treating a human bitten by an animal if the animal cannot be found?
- What major advantage does an oral vaccine have over a parenteral (injected) vaccine for rabies control in wild animals?

31.2 Hantavirus and Hantavirus Syndromes

Hantaviruses cause two severe, emerging diseases, **hantavirus pulmonary syndrome (HPS)**, an acute respiratory and cardiac disease, and **hemorrhagic fever with renal syndrome (HFRS)**, an acute disease characterized by shock and kidney failure. Both diseases are caused by hantaviruses transmitted from infected rodents. Hantavirus is named for Hantaan, Korea, the site of a hemorrhagic fever outbreak where the virus was first recognized as a human pathogen.

Symptoms and Pathology of Hantavirus Syndromes

Hantaviruses are enveloped viruses with single-stranded minus-sense RNA genomes arranged in segments (**Figure 31.3***a*; ➡ Section 10.9); hantaviruses are related to other hemorrhagic fever viruses such as Lassa fever virus and Ebola virus (➡ Section 30.12). Hantaviruses infect rodents including mice, rats, lemmings, and voles, without causing disease. The virus is transmitted from these reservoirs to humans by inhalation of virus-contaminated rodent excreta. Humans are accidental hosts and are infected only when they come into contact with rodents, their waste, or their saliva.

HPS is characterized by a sudden onset of fever, muscle pain, a reduction in the number of blood platelets along with an increase in the number of circulating leukocytes, and hemorrhaging. Death (if it occurs) takes several days, and is usually a result of systemic shock and cardiac complications precipitated by leakage of fluid into the lungs, causing suffocation and heart failure. These symptoms are typical of hantaviruses, but other symptoms such

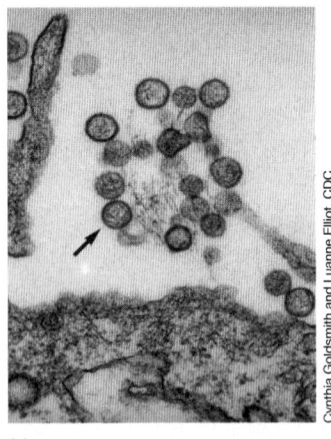
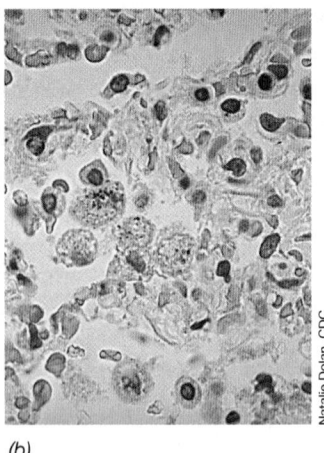

(a)   *(b)*

Figure 31.3 Hantavirus. *(a)* A transmission electron micrograph of the Sin Nombre hantavirus. The arrow indicates one of several virions that are about 100 nm in diameter. *(b)* Immunofluorescent staining of Andes hantavirus antigens in alveolar macrophages. Each granular dark blue–stained area indicates cellular infection of an individual macrophage that contains many hantavirus virions (each cell is about 15 μm in diameter).

as kidney failure are common, depending on the strain of virus causing the disease. HFRS is characterized by intense headache, back and abdominal pain, renal dysfunction, and various hemorrhagic complications. HFRS strains are more prevalent in hantavirus outbreaks in Eurasia, whereas HPS strains are more prevalent in the Americas and elsewhere in the world. HPS strains typically show a significantly higher mortality rate than HFRS strains and are found in rodents elsewhere in the world outside the Americas.

Hantaviruses can be cultured in the laboratory, but because of the danger involved, they must be handled with biosafety level 4 (BSL-4; ⊂⊃ Section 28.1) precautions. In the world of infectious diseases, hantavirus, Ebola, and other BSL-4 viral pathogens are considered "the worst of the worst" and are thus handled in the United States by the Special Pathogens Branch of the Centers for Disease Control and Prevention in Atlanta (Georgia, USA).

Epidemiology, Diagnosis, and Prevention of Hantavirus Syndromes

A significant HPS outbreak in the United States occurred near the Four Corners region of Arizona, Colorado, New Mexico, and Utah in 1993. The outbreak resulted from an enlarged population of deer mice in the spring of 1993. The previous winter was mild and was followed by abundant spring rains, triggering unusually high food levels for the mice. The HPS outbreak caused 27 deaths among 48 infected people (56% mortality), illustrating the potential danger of outbreaks due to pathogens that can be directly transmitted from animal reservoirs. In total from 1993 through 2015, there have been 659 cases of HPS in the United States, with 235 deaths (36%), mostly in western states. On a global basis, it is estimated that 200,000 infections occur annually, chiefly in China, Korea, and Russia, but mortality rates are typically very low.

Hantavirus syndromes can be diagnosed using immunological techniques that identify anti-hantavirus antibodies in a blood sample. These include immunoassays (Figure 31.3*b* and ⊂⊃ Section 28.6) that detect both exposure to the virus and the

strength of the immune response. The presence of the viral RNA genome from circulating virions can also be detected using RT-PCR (⊂⊃ Sections 12.1 and 28.8) on patient tissue or blood samples.

There is no virus-specific treatment or vaccine for hantavirus diseases. Treatment amounts to isolation, rest, rehydration, and alleviation of other symptoms. Hantavirus infection can be prevented by avoiding rodent contact and rodent habitat. Destruction of mouse habitat, restricting food supplies (for example, keeping human food in sealed containers), and aggressive rodent extermination measures are the only effective controls because areas that have experienced a hantavirus outbreak have a high proportion of mice that carry the virus, animal surveys have shown.

— MINIQUIZ —

- Why are hantaviruses considered a major public health problem in the United States?
- Describe the spread of hantaviruses to humans. What are some effective measures for preventing infection by hantaviruses?

II • Arthropod-Transmitted Bacterial and Viral Diseases

Pathogens can be spread to new hosts from the bite of an infected arthropod. In the bacterial and viral diseases we consider here—the rickettsial illnesses; yellow and dengue fevers; Lyme, Chikungunya, and Zika virus diseases; and plague—humans are only *accidental hosts* for the pathogen. The *reservoir* of the pathogen is the arthropod vector. Nevertheless, the diseases can be devastating and often fatal.

31.3 Rickettsial Diseases

The **rickettsias** are small *Bacteria* that live an obligate intracellular existence and are associated with bloodsucking arthropods such as fleas, lice, or ticks. We discussed the biology of rickettsias in Section 16.1. Of the diseases that rickettsias can cause in humans and other vertebrates, the most important are *typhus fever, spotted fever rickettsiosis (Rocky Mountain spotted fever)*, and *ehrlichiosis*. Rickettsias have not been cultured in artificial culture media but can be grown in laboratory animals, ticks and lice, mammalian tissue culture cells, and the yolk sac of chick embryos (see Figure 31.6*b*). In animals, growth takes place primarily in phagocytes, such as macrophages.

Rickettsias are divided into three groups based on the clinical diseases they cause. The groups are (1) the *typhus group*, such as *Rickettsia prowazekii*; (2) the *spotted fever group*, such as *Rickettsia rickettsii*; and (3) the *ehrlichiosis group*, characterized by *Ehrlichia chaffeensis*.

The Typhus Group: *Rickettsia prowazekii*

Typhus is transmitted from person to person by the common body or head louse (**Figure 31.4***a*), and humans are the only known mammalian host. During World War I, a typhus epidemic spread throughout Eastern Europe and caused almost 3 million deaths. Typhus has historically been a problem among troops in wartime.

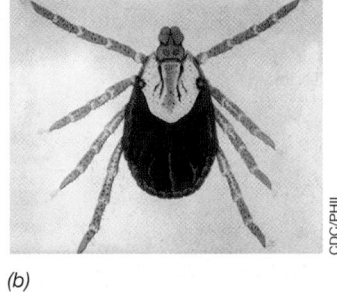

(a) CDC/PHIL, James Gathany *(b)* CDC/PHIL

Figure 31.4 Arthropod vectors of rickettsial diseases. *(a)* The female body louse, about 3 mm long, can carry *Rickettsia prowazekii*, the agent that causes typhus. In addition, the body louse can carry *Borrelia recurrentis*, the agent of relapsing fever, and *Bartonella quintana*, the agent of trench fever. *(b)* The American dog tick that carries *Rickettsia rickettsii*, the causative agent of Rocky Mountain spotted fever, is about 5 µm long, but can expand to three times this size when engorged with blood.

Because of the unsanitary, cramped conditions characteristic of wartime military operations, infected lice can spread easily among soldiers with devastating results. Up until World War II, typhus caused more military deaths than did combat.

Cells of *R. prowazekii* are introduced through the skin when a puncture caused by a louse bite becomes contaminated with louse feces that contain the rickettsial cells. During an incubation period of 1–3 weeks, the organism multiplies inside cells lining the small blood vessels. Symptoms of typhus (fever, headache, and general body weakness) then begin to appear. Several days later, a characteristic rash is observed in the armpits and generally spreads over the body, except for the face, palms of the hands, and soles of the feet. Complications from untreated typhus include damage to the central nervous system, lungs, kidneys, and heart. Epidemic typhus has a mortality rate of as much as 30%. Tetracycline and chloramphenicol are most commonly used to control infections caused by *R. prowazekii*. *Rickettsia typhi*, the organism that causes murine typhus, is another important pathogen in the typhus group and can also infect humans. A typhus vaccine is available but is typically only administered to those traveling to typhus endemic areas.

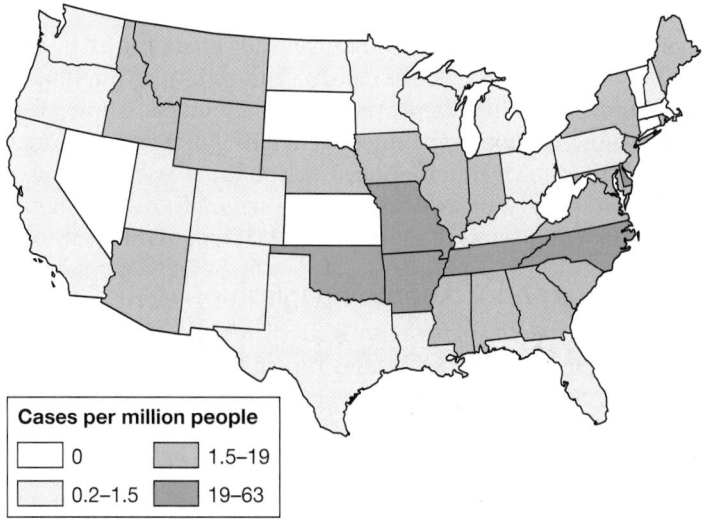

Cases per million people

☐	0	▨	1.5–19
☐	0.2–1.5	▨	19–63

Figure 31.5 Spotted fever rickettsiosis (Rocky Mountain spotted fever) in the United States, 2010. Despite the name, cases of Rocky Mountain spotted fever are currently concentrated in the eastern and mid-South states west to Oklahoma.

The Spotted Fever Group: *Rickettsia rickettsii*

Spotted fever rickettsiosis, commonly called *Rocky Mountain spotted fever* (RMSF), was first recognized in the western United States about 1900 but is more prevalent today in the central and mid-South region (**Figure 31.5**). RMSF is caused by *R. rickettsii* and is transmitted to humans by various ticks, most commonly the dog tick (Figure 31.4*b*) and wood ticks. Over 2000 people acquire RMSF yearly in the United States, nearly double the number reported in 2002, which is likely due to increased human activities in tick-infested areas. Fatalities in treated patients occur in less than 1% of those infected. Humans acquire the pathogen from the bite of an infected tick; rickettsial cells are present in the salivary glands of the tick and in the ovaries of female ticks.

Cells of *R. rickettsii*, unlike other rickettsias, grow within the nucleus of the host cell as well as in host cell cytoplasm (**Figure 31.6a, c**). Following an incubation period of 3–12 days, characteristic symptoms,

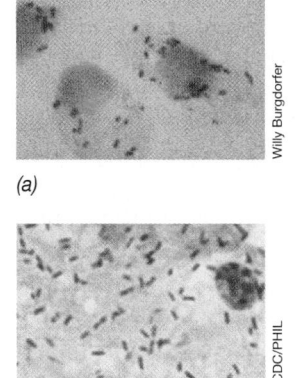
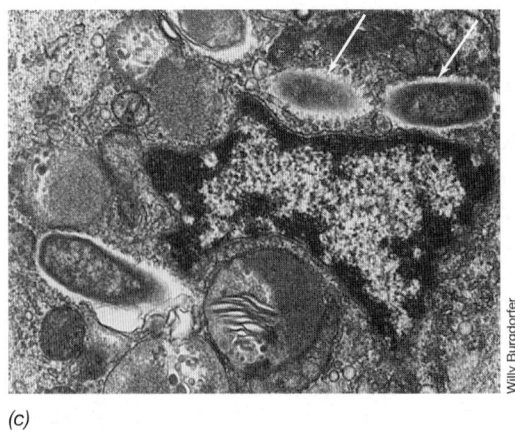
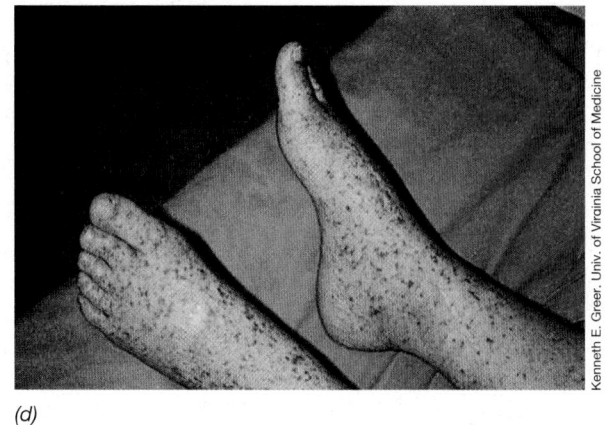

(a) Willy Burgdorfer *(b)* CDC/PHIL *(c)* Willy Burgdorfer *(d)* Kenneth E. Greer, Univ. of Virginia School of Medicine

Figure 31.6 *Rickettsia rickettsii* and spotted fever rickettsiosis. *(a)* Cells of *R. rickettsii*, growing in the cytoplasm and nucleus of tick hemocytes and *(b)* in chicken egg yolk sacs; cells are about 0.4 µm in diameter. *(c)* Transmission electron micrograph of *R. rickettsii* (arrows) in a granular hemocyte of an infected wood tick. *(d)* Rash of spotted fever rickettsiosis on the feet. The whole-body rash is indicative of spotted fever rickettsiosis and helps distinguish it from typhus, in which the rash does not cover the whole body.

UNIT 7

including fever and a severe headache, occur. A few days later, a systemic rash breaks out (Figure 31.6*d*), generally accompanied by gastrointestinal problems such as diarrhea and vomiting. The clinical symptoms of RMSF persist for over 2 weeks if the disease is untreated. Tetracycline or chloramphenicol generally promotes a prompt recovery from RMSF if administered early in the course of the infection. Mortality in untreated cases of RMSF resembles that of typhus, up to 30%. No effective vaccine against RMSF is currently available.

Ehrlichiosis and Tickborne Anaplasmosis

Ehrlichia and related genera (⊘ Section 16.1) are responsible for two emerging tickborne diseases in the United States, *human monocytic ehrlichiosis* (HME) and *human granulocytic anaplasmosis* (HGA). The pathogens that cause HME are *Ehrlichia chaffeensis* and *Rickettsia sennetsu*, and those that cause HGA are *Ehrlichia ewingii* and *Anaplasma phagocytophilum*.

The onset of these clinically indistinguishable rickettsial diseases is characterized by flulike symptoms that can include fever, headache, malaise, changes in liver function, and a reduction in white blood cell numbers. Peripheral blood leukocytes such as monocytes have visible inclusions of rickettsial cells, a diagnostic indicator for the diseases (**Figure 31.7a**). The symptoms, except for the inclusions, are similar to other rickettsial infections, and can range from subclinical to fatal. Long-term complications for progressive untreated cases may include respiratory and renal insufficiency and serious neurological involvement.

HGA and HME are spread by ticks of various species, and mammalian reservoirs of the pathogens include deer, some rodents, and humans. In the United States, HGA occurs primarily in the upper Midwest and coastal New England, while HME is concentrated in the lower Midwest and the East Coast; together, almost 2000 cases are reported each year, with cases of HGA predominating. Diagnosis of rickettsial syndromes is not straightforward because the rash observed can be mistaken for other diseases, such as scarlet fever, or even measles or syphilis. Confirmation of a rickettsial disease requires immunological tests, including fluorescent antibodies or immunoassays, or PCR-based analyses that detect pathogen DNA.

Prevention of HGA and HME is best achieved either by avoiding tick habitat or by wearing tick-proof clothing and applying insect repellents containing diethyl-*m*-toluamide (DEET). It is also good practice to examine yourself carefully for ticks after hiking in tick habitat and to remove any ticks immediately, taking care to remove all tick mouthparts if the tick has already attached. Doxycycline, a tetracycline antibiotic, is the drug of choice for the treatment of HGA and HME. Vaccines are currently unavailable for the prevention of HGA and HME.

Q Fever

Q fever (the Q stands for "query") is a pneumonia-like infection caused by the intracellular parasite *Coxiella burnetii* (Figure 31.7*b*), a bacterium related to the rickettsias (⊘ Section 16.1). Although not transmitted to humans by an insect bite, *C. burnetii* cells are transmitted to animals such as sheep, cattle, and goats by insect bites, and from these reservoirs are transmitted to humans. Domestic animals generally have inapparent infections but may shed large quantities of *C. burnetii* cells in their urine, feces, milk, and other body fluids. Infected animals or contaminated animal

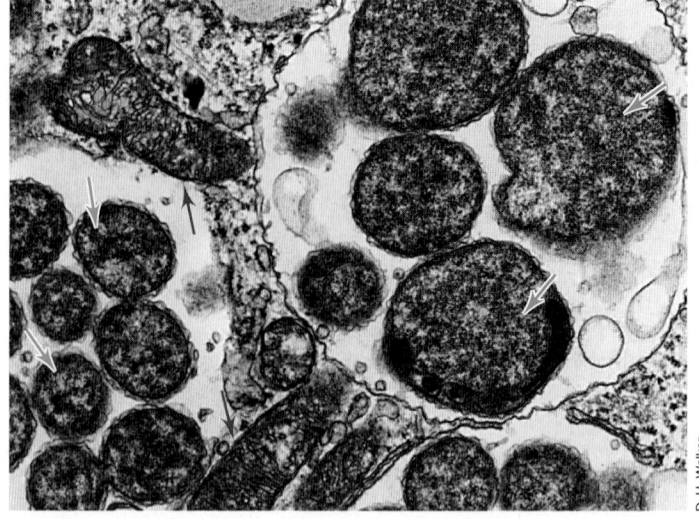

(a)

(b)

Figure 31.7 *Ehrlichia* **and** *Coxiella*. *(a) Ehrlichia chaffeensis*, the causative agent of human monocytic ehrlichiosis (HME). The electron micrograph shows inclusions in a human monocyte that contains large numbers of *E. chaffeensis* cells. The blue arrows indicate bacteria in each inclusion. The *E. chaffeensis* cells are about 0.3–0.9 μm in diameter. Mitochondria are indicated by red arrows. *(b)* Colorized scanning electron micrograph of cells of *Coxiella burnetii*, the causative agent of Q fever. The *Coxiella* cells were grown in animal cell culture and are shown inside a fractured host cell. A single *C. burnetii* cell is about 0.4 μm in diameter.

products such as wool, meat, and milk are potential sources for human infection. The resulting influenza-like illness can progress to include prolonged fever, headache, chills, chest pains, pneumonia, and endocarditis (inflammation of the inner lining of the heart). In the United States, Q fever is most prevalent in rural states with large farm or ranch animal populations, and about 100–150 cases have been reported annually in recent years.

As for rickettsial infections, laboratory diagnosis of *C. burnetii* infection is typically made by immunological tests designed to measure host antibodies to the pathogen. Q fever responds well to tetracycline, and therapy should be started quickly in any suspected case to prevent endocarditis and heart valve damage. Q fever is also a potential biological warfare agent (⊘ Section 29.9).

Unlike its relatives in the *Rickettsia* group, the Q fever pathogen *C. burnetii* can now be grown in pure culture outside a host. This was accomplished by taking into account both the resources and conditions likely to exist in the intracellular host environment along with a careful analysis of the *C. burnetii* genome sequence to reveal the metabolic capacities and limitations of this pathogen. One major discovery gleaned from genomic analysis of the *C. burnetii* complement of cytochromes was the fact that the organism was likely to be microaerophilic (⮌ Section 5.14). And indeed, one of the major secrets to its axenic culture turned out to be to incubate cultures under low oxygen tensions, which is somewhat surprising considering that host cells should be fully oxic.

MINIQUIZ

- What are the arthropod vectors and animal hosts for typhus, spotted fever rickettsiosis, ehrlichiosis, and anaplasmosis?
- What precautions can be taken to prevent rickettsial infections?

31.4 Lyme Disease and *Borrelia*

Lyme disease is a tickborne disease that affects humans and other animals. Lyme disease was named for Old Lyme, Connecticut, where cases were first recognized, and is currently the most prevalent arthropod-borne disease in the United States. Lyme disease is caused by infection with a spirochete, *Borrelia burgdorferi* (**Figure 31.8**; ⮌ Section 15.19), transmitted by a tick bite. The ticks that carry *B. burgdorferi* feed on the blood of birds, domesticated animals, various wild animals, and humans. *B. burgdorferi* is of interest in a nonmedical way as well, because it is one of only a handful of *Bacteria* that contain a linear (as opposed to a circular) chromosome (⮌ Section 4.2).

Pathology, Diagnosis, and Treatment of Lyme Disease

Cells of *B. burgdorferi* are transmitted to humans while the tick is obtaining a blood meal (**Figure 31.9a**). A systemic infection develops,

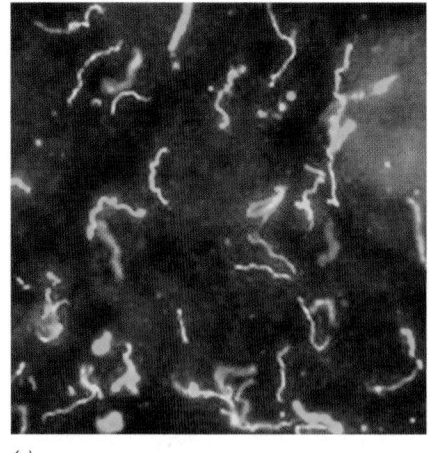 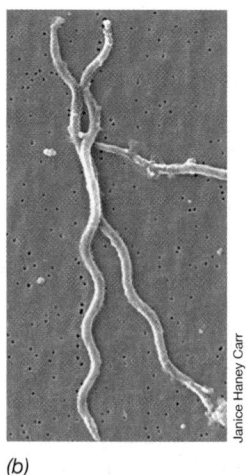

(a)   *(b)*

Figure 31.8 The Lyme spirochete, *Borrelia burgdorferi*. *(a)* Fluorescent antibody staining (⮌ Section 28.6) of cells of *B. burgdorferi* from a Lyme rash. Two different fluorescent antibodies were used, each specific for a different *B. burgdorferi* antigen and linked to either an orange or a green fluorescent tag. If both antibodies bind, cells appear yellow. *(b)* Colorized scanning electron micrograph of cells of *B. burgdorferi*. A single cell is approximately 0.4 μm in diameter and 5–20 μm long.

leading to the acute symptoms of Lyme disease: headache, backache, chills, and fatigue. In about 75% of Lyme cases, a concentric circular or "bull's-eye" rash forms within a week at the site of the tick bite (Figure 31.9b, c). During this acute stage, Lyme disease is readily treatable with tetracycline or penicillin.

Untreated cases of Lyme disease may progress to a chronic stage weeks to months after the initial tick bite, causing arthritis in about half of those infected. Neurological problems such as palsy, weakness in the limbs, and heart damage can also occur. In untreated cases, cells of *B. burgdorferi* infecting the central nervous system may lie dormant for long periods before causing additional chronic symptoms, including problems with vision and facial muscle movements, or seizures. Interestingly, the symptoms of chronic Lyme disease, especially neurological symptoms,

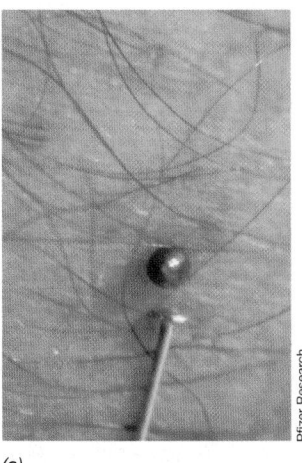 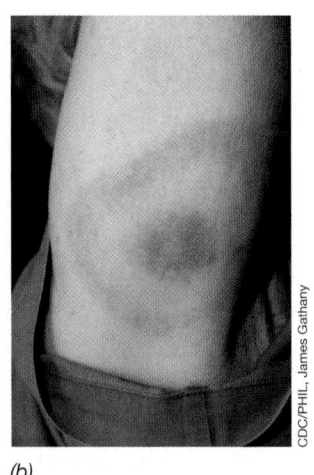 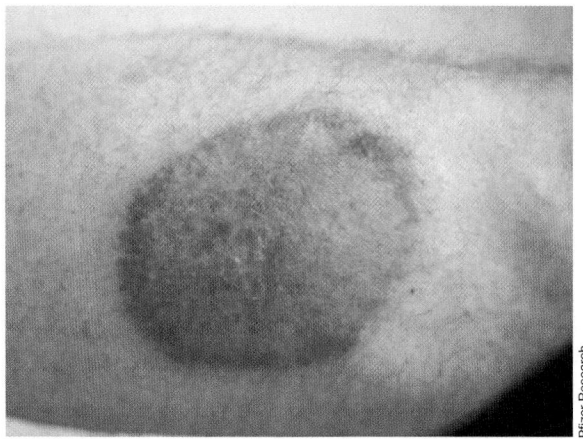

(a)   *(b)*   *(c)*

Figure 31.9 Lyme disease infection. *(a)* A blood-engorged deer tick obtaining a blood meal from a human is the route of transmission (see Figure 31.10 for photos of deer ticks). *(b, c)* Characteristic Lyme disease rashes. The rash starts at the site of a tick bite and grows in a concentric circular fashion over a period of several days. A typical rash is about 5 cm in diameter.

mimic those of chronic syphilis, caused by a different spirochete, *Treponema pallidum* (🔗 Sections 15.19 and 30.13). Unlike syphilis, however, Lyme disease is not spread person to person.

No toxins or other major virulence factors have been identified in Lyme disease pathogenesis, but the pathogen triggers a strong immune response. Antibodies to *B. burgdorferi* appear 4–6 weeks after infection and can be detected by various immunological assays. However, because antibodies to *B. burgdorferi* antigens persist for years after infection, the presence of these antibodies does not necessarily indicate a recent infection. A PCR assay (🔗 Sections 12.1 and 28.8) is also in use to detect *B. burgdorferi* DNA in body fluids and tissues. In practice, however, Lyme disease is typically diagnosed from clinical symptoms and only confirmed later by laboratory assays. If a patient has Lyme disease symptoms and other findings such as facial tics or arthritis, or has had recent tick exposure or exhibits the characteristic Lyme rash (Figure 31.9), a presumptive diagnosis of Lyme disease is made and antibiotic treatment is initiated.

Treatment of early-stage Lyme disease is usually with doxycycline or amoxicillin for 20 to 30 days. For patients having neurological or cardiac symptoms, the antibiotic ceftriaxone is administered intravenously because this drug crosses the blood–brain barrier and can thus kill spirochetes in the central nervous system.

Epidemiology and Prevention of Lyme Disease

White-footed field mice and other small rodents are the major mammalian reservoir of *B. burgdorferi* in the northeastern United States, a hotbed of Lyme infection (see Figure 31.11). These animals become infected from bites by the deer tick, *Ixodes scapularis* (**Figure 31.10**), although some other ticks can transmit the Lyme spirochete as well. Deer themselves are not *B. burgdorferi* reservoirs but are major reproductive hosts for the tick. Lyme disease has also been identified in Europe and Asia. In these countries, both the tick vector and the species of *Borrelia* differ from those in the United States, which shows that Lyme disease has a broad geographic distribution. But in all cases, Lyme disease is caused by related species of pathogenic *Borrelia* transmitted to humans by tick vectors.

Deer ticks are typically smaller than many other ticks, making them easy to overlook. Moreover, unlike the case with ticks that

Figure 31.11 Lyme disease in the United States, 2014. Each dot represents a confirmed case. Confirmed and probable cases in 2014 totaled over 33,000, with 96% of these localized to 13 states in the upper regions of the Midwest and East Coast. Lyme disease is reported through the National Notifiable Diseases Surveillance System of the Centers for Disease Control and Prevention, Atlanta, Georgia, USA.

carry other tickborne diseases (Figure 31.4*b*), a very high percentage of deer ticks carry *B. burgdorferi*. Both of these factors—small vector size and high occurrence of the pathogen—undoubtedly contribute to the fact that Lyme is the most commonly reported vector-borne disease in the United States. Most cases of Lyme disease in the United States have been reported from the Northeast and upper Midwest—areas of the country where deer are abundant—but cases have been observed in nearly every state (**Figure 31.11**). The incidence of Lyme disease in the United States is significant, with over 33,000 confirmed and probable cases reported in 2014.

As for any tickborne infection, prevention of Lyme disease begins by avoiding contact with the vector. Insect repellents containing DEET or the wearing of snug-fitting clothing is helpful, as is a thorough body exam following walks in tick-infested environments. Lyme disease vaccines are available for domestic animals, but no human Lyme disease vaccine is currently in use.

--- **MINIQUIZ** ---

- What are the primary symptoms of Lyme disease?
- In the United States, where is Lyme disease most prevalent?
- Outline methods for prevention of *Borrelia burgdorferi* infection.

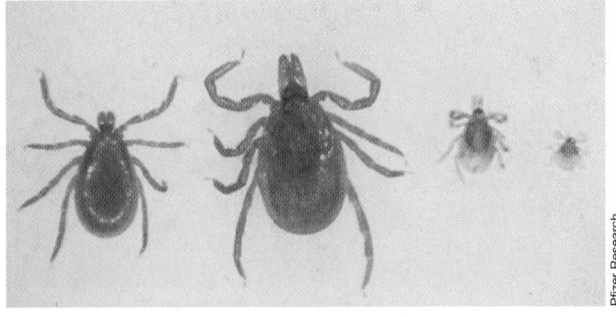

Figure 31.10 Deer ticks, the major vector of Lyme disease. Left to right, male and female adult ticks, nymph, and larva forms. The length of an adult female is about 3 mm. Although all forms feed on humans, the female nymphal and adult ticks are principally responsible for transmitting *Borrelia burgdorferi*.

31.5 Yellow Fever, Dengue Fever, Chikungunya, and Zika

Several arthropod-transmitted diseases are caused by flaviviruses. These are single-stranded plus-sense RNA viruses (🔗 Section 10.8) transmitted by the bite of an infected arthropod. Because of this characteristic mode of transmission, these viruses are also called *arboviruses* (*ar*thropod-*bo*rne viruses).

Many serious human diseases are caused by arboviruses including various types of encephalitis and hemorrhagic fevers. Here we consider two potentially fatal flavivirus diseases common in some developing countries, yellow fever and dengue. Both viruses are transmitted by the same vector, mosquitoes of the genus *Aedes* (**Figure 31.12**), and some of the disease symptoms are similar. We also consider Zika and Chikungunya, both emerging viral diseases transmitted by mosquitoes.

Yellow Fever

Yellow fever is an endemic disease of tropical and subtropical climates, especially in Latin America and Africa. Brazil, Colombia, Venezuela, and parts of Bolivia and Peru, along with most countries in sub-Saharan central Africa, experience the greatest incidence. Yellow fever is absent from the United States except in unvaccinated individuals who contract the disease through travel to an endemic area. Yellow fever virus is related to dengue virus (see later), West Nile virus (Section 31.6), and certain encephalitis viruses. Yellow fever is one of only a handful of infectious diseases for which isolation and quarantine are practiced (🔗 Section 29.5). In the case of yellow fever, although the disease is not transmitted person to person, isolation of active cases prevents local mosquitoes from taking a blood meal from the infected individual and transmitting the disease to others.

Following a bite from an infected mosquito, the yellow fever virus replicates in lymph nodes and certain immune system cells and eventually travels to the liver. Once a person is infected,

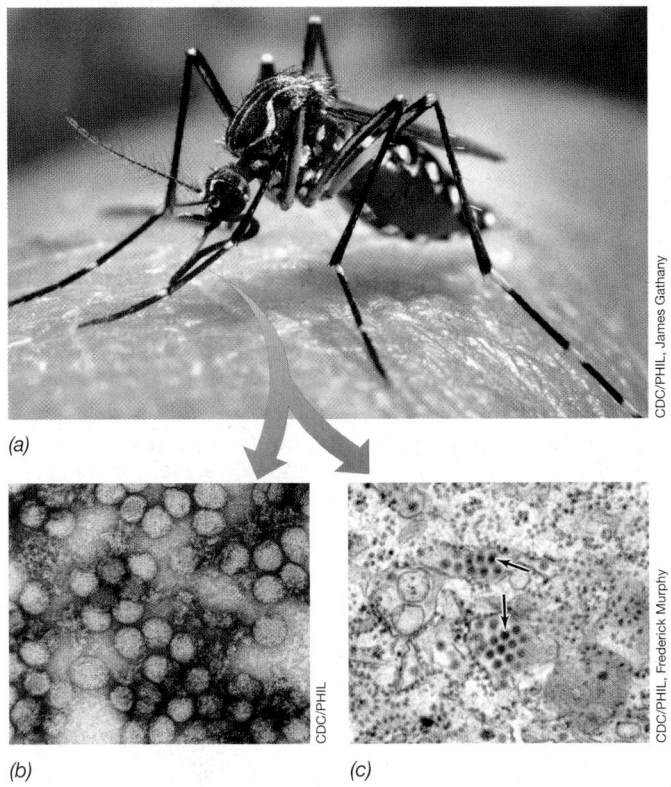

(a)

(b)   *(c)*

Figure 31.12 Yellow fever and dengue fever. *(a)* Yellow and dengue fever viruses are both transmitted by the bite of an infected *Aedes aegypti* mosquito. Transmission electron micrographs of *(b)* yellow fever virus and *(c)* dengue fever virus (arrows, in a tissue specimen). Both yellow and dengue fever viruses are about 50 nm in diameter and are plus-sense RNA viruses that replicate by way of polyprotein formation, as in poliovirus (🔗 Figure 10.18).

symptoms range anywhere from none to major organ failure and death. Most infected individuals display a mild fever with accompanying chills, a headache and back pains, nausea, and other symptoms that are not diagnostically useful. Presumably these are cases in which the immune system has the infection under control. However, in about one in five yellow fever cases, the disease enters its toxic phase, characterized by jaundice (thus the name, *yellow fever*) and by hemorrhaging from the mouth, eyes, and gastrointestinal tract. This triggers the onset of bouts of bloody vomit, and if bleeding continues, it leads to toxic shock and multiple organ failure. About 20% of cases that reach this stage are fatal. Humans and nonhuman primates are the main reservoirs for the yellow fever virus.

Yellow fever is fully preventable by an effective vaccine. A yellow fever vaccine was developed in the 1930s and widely used by military and support personnel in tropical battlefields. Historically, the disease has been controlled by a combination of vaccination and elimination of both the vector (mosquito) population by chemical agents and vector breeding grounds by draining swamps and low-lying wetlands in endemic areas.

The yellow fever vaccine is highly recommended for those traveling to endemic areas, and many countries require proof of vaccination for anyone entering their country from a foreign country where yellow fever is endemic. In addition, the World Health Organization (WHO) has initiated a mass vaccination program in Africa. Despite the availability of a vaccine, the WHO estimates that each year nearly 200,000 cases of yellow fever occur, mostly unreported, and that about 15% of all cases are fatal. No treatment for yellow fever is known. However, once the disease is diagnosed, typically by detecting anti–yellow fever virus antibodies in a blood sample, the patient is isolated and prescribed rest and drugs to control symptoms. Recovery without entering the toxemia stage is due to the immune response.

Dengue Fever

Like yellow fever, dengue (pronounced deng-gay) fever is transmitted by mosquitoes of the genus *Aedes* (Figure 31.12) and is a disease of tropical and subtropical regions. Up to 100 million cases of dengue are estimated worldwide per year with concentrations in Mexico, Latin America, India, Indonesia, and Africa (🔗 Figure 29.11).

Dengue begins with a high fever and headache or joint pains and in some patients, severe eye pain and a systemic rash. Most infected individuals show self-improvement within a week and no further symptoms, presumably because of an immune response to the dengue virus. But, as for yellow fever, a dengue infection can take a more severe course of events and proceed to *dengue hemorrhagic fever*. This condition is characterized by severe symptoms that can include bleeding from the nose and gums, bloody vomit and/or feces, intense abdominal pain, respiratory distress, and a general feeling of malaise. The blood pressure of a dengue patient can drop dramatically during the hemorrhagic fever stage, and a small percentage of these cases are fatal. Treatment for dengue is primarily to relieve symptoms, particularly dehydration from loss of blood and other fluids. Unlike yellow fever, no effective vaccine exists for dengue, and thus rest and symptom relief, even in cases of dengue hemorrhagic fever, is the only effective treatment.

CDC/PHIL, James Gathany

CDC/PHIL

CDC/PHIL, Frederick Murphy

Dengue can be controlled by eliminating either the vector or contact with the vector. Extensive chemical spraying for mosquito eradication was widely practiced in urban centers of the southern United States and kept dengue in check in the twentieth century. Now, however, spraying programs are less common and global climate change is moving tropical temperature patterns northward. In response, the *Aedes* mosquito has become entrenched in southern and central regions of the United States. Besides *Aedes aegypti* (Figure 31.12), the Asian tiger mosquito (*Aedes albopictus*), which is spreading quickly in the United States, also carries the dengue virus. Drainage of small puddles of water, such as those in discarded tires or other water traps, removes mosquito breeding grounds and greatly reduces the opportunities for a dengue outbreak. Personal protection from mosquito bites by using effective insect repellents and clothing is also a proven means of preventing infection.

The WHO estimates that nearly a half million cases of dengue hemorrhagic fever occur yearly with about 22,000 being fatal. Cases of dengue in the United States are rare, and virtually all of them are imported from endemic areas. Dengue is a prime example of an *emerging disease* (🔗 Section 29.7), as cases were geographically restricted until the mid-twentieth century when global commerce is thought to have transmitted species of the *Aedes* mosquito beyond their original range.

Zika and Chikungunya Disease

Zika and Chikungunya are both typically mild viral diseases transmitted by the same mosquito vector. Zika virus disease is a typically mild infection characterized by headache, fever, joint pain, and general malaise; a rash is occasionally seen. The disease is caused by the Zika virus (**Figure 31.13a**), a relative of the dengue virus and also transmitted by mosquitoes. Zika first emerged over 65 years ago in the Zika forest (Uganda), and occasional small outbreaks occurred periodically in west central Africa and Indonesia. However, in 2015, Zika appeared in Brazil, and by 2016, outbreaks of Zika virus disease were reported in the United States, primarily in individuals who had traveled to areas with endemic disease. However, a major outbreak in the U.S. territory of Puerto Rico has been linked to local mosquito-borne transmission, and the fear is that the disease will spread northward within the range of the *Aedes* mosquito. Zika infections were reported in Florida in late 2016 and will likely appear in other regions of the southern United States.

Besides vectorborne transmission, Zika can be transmitted through sexual contact and by contaminated blood and, of most concern, from mother to fetus. The incidence of brain abnormalities and other birth defects in infants born to Zika-infected mothers is significantly higher than in those born to uninfected mothers and so it is thought that Zika in some way affects development of the fetus. It is known that the Zika virus readily infects a type of neural cell that eventually forms the cerebral cortex, a major part of the brain that governs intellectual capacity, and this likely leads to the brain defects observed in Zika-infected newborns. This danger has led to advisories for pregnant women to take great precautions to avoid contact with mosquitoes during the entire pregnancy period. Other than in pregnant women, Zika virus disease seems to be a rather mild, self-limiting infection. However, in rare instances, Zika infection may trigger Guillain–Barré syndrome, an autoimmune disease (🔗 Section 27.9) caused by an infection with any of several different viruses and bacteria in which the immune system attacks the peripheral nervous system. For more on the Zika virus, see page 274.

Too little is known about the Zika virus and the various conditions it can or may cause to draw firm conclusions about the seriousness of this threat to public health. Deaths directly attributable to a Zika infection have been extremely rare, and thus the pathogen seems to be of only passing concern to the general public, even in the typical high-risk groups for infectious diseases such as the very young or very old. However, the birth defect connection is extremely serious. At this writing, it seems that pregnant women (or those trying to get pregnant, as Zika can also be transmitted by sexual intercourse) are the only major group known to be at risk from Zika infection.

Like Zika, Chikungunya disease is caused by single-stranded plus-sense RNA virus (Figure 31.13b), but the virus is not otherwise closely related to the Zika, dengue, and yellow fever viruses. Chikungunya virus is transmitted by species of *Aedes* mosquitoes and is currently endemic in South and Central America, Southeast Asia, central Africa, and Indonesia. Thus far, cases of Chikungunya in the United States have been limited to travelers returning from endemic areas and in 2015, a total of 679 confirmed cases were reported. As for Zika, mortality in Chikungunya disease is rare, about 0.1% of all cases. The symptoms of Chikungunya are mild and self-limiting and the immune response to the virus is strong, conferring active immunity against reinfection. Unlike with Zika, no direct connection between Chikungunya viral infection and birth defects has been observed because Chikungunya virus is not transmitted from mother to fetus.

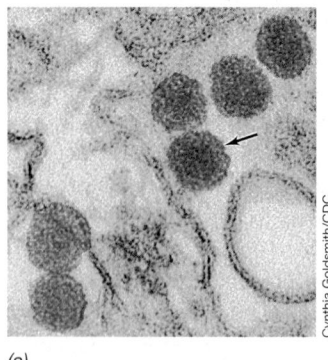
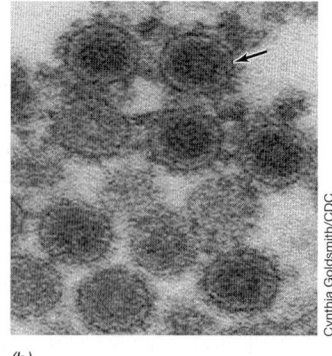

(a) (b)

Cynthia Goldsmith/CDC

Figure 31.13 Zika and Chikungunya viruses. Both viruses contain single-stranded plus-sense RNA genomes and are surrounded by an envelope (🔗 Sections 8.1, 8.2, and 10.8). *(a)* Colorized transmission electron micrograph of virions of the Zika virus embedded in a tissue sample. A single virion (shown in blue, arrow) is about 40 nm in diameter. *(b)* Transmission electron micrograph of virions of Chikungunya virus (arrow). A virion is about 50 nm in diameter.

— **MINIQUIZ** —

- Identify the vector and reservoir for yellow fever and dengue viruses.
- Contrast the procedures for preventing infection in yellow and dengue fevers.
- Why is Zika virus disease considered dangerous even though it rarely kills?

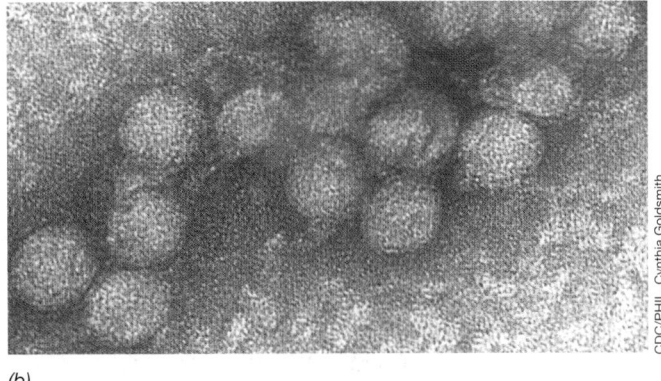

(a)   *(b)*

Figure 31.14 West Nile virus. *(a)* The mosquito *Culex quinquefasciatus*, shown here engorged with human blood, is a West Nile virus vector. *(b)* An electron micrograph of the West Nile virus. The icosahedral virion is about 40–60 nm in diameter and contains a plus-sense single-stranded RNA genome.

31.6 West Nile Fever

West Nile virus (WNV) causes **West Nile fever**, a human viral disease that is transmitted through the bite of a mosquito and thus is a seasonal disease. WNV is a flavivirus, as are the yellow fever, dengue, and Zika viruses (Section 31.5), and has an enveloped capsid (**Figure 31.14b**) containing a plus-sense single-stranded RNA genome (⟳ Section 10.8). The virus can invade the nervous system of its warm-blooded hosts, which include some species of both birds and mammals.

WNV Transmission and Pathology

WNV causes active disease in over 100 species of birds and is transferred to hosts by the bite of an infected mosquito. Over 40 species of mosquito can carry the virus including *Culex* species (Figure 31.14a), common in central and eastern states of the United States and in urban centers throughout the country. Infected birds develop a systemic viral infection (viremia) that is often fatal. Mosquitoes feeding on viremic birds are infected and can then infect other susceptible birds, renewing the cycle. In contrast to birds, humans and other animals are dead-end hosts for the virus because they do not develop the viremia necessary to infect mosquitoes.

Mortality rates for WNV infection are species-specific. For example, the human mortality rate from WNV is about 4% while that for horses is significantly higher, near 40%. Most human infections are asymptomatic or mild and are not reported. After an incubation period of 3–14 days, about 20% of infected individuals develop West Nile fever, a mild illness lasting 3–6 days. The fever may be accompanied by headache, nausea, muscle pain, rash, lymphadenopathy (swelling of lymph nodes), and malaise. Less than 1% of infected individuals develop serious neurological diseases such as *West Nile encephalitis* or *West Nile meningitis* from viral replication in neural tissues (**Figure 31.15**). These are more common in adults over age 50, and the neural effects may be permanent. About 5% of West Nile cases that progress to these forms are fatal. Diagnosis of WNV disease includes assessment of clinical symptoms followed by confirmation by immunological tests that detect WNV antibodies in blood.

Control and Epidemiology of WNV

Human WNV disease was first identified in the West Nile region of Africa in 1937 and spread from there to Egypt and Israel. In the 1990s there were WNV outbreaks in horses, birds, and humans in African and European countries. The first cases of WNV were reported in the United States in the Northeast in 1999, but in the years since, the disease has spread to every state (**Figure 31.16**). In the 1999–2014 period, the number of reported WNV cases per year has fluctuated wildly, from as few as about 20 to as many as 9800. The major U.S. foci of infection appear to be in the south central states and the Great Plains from the Texas coast to the Canadian border (Figure 31.16). West Nile disease is now enzootic in the bird population in the United States and in only low incidence in the human population, its accidental host.

Control of WNV illness is much the same as for other vectorborne diseases: Limit exposure to mosquitoes using insect repellents or wear tight-fitting clothing. Spraying for mosquitoes has limited effectiveness, but removal of mosquito breeding grounds, particularly sources of standing water, is very helpful in controlling mosquito populations. A veterinary WNV vaccine is widely used in

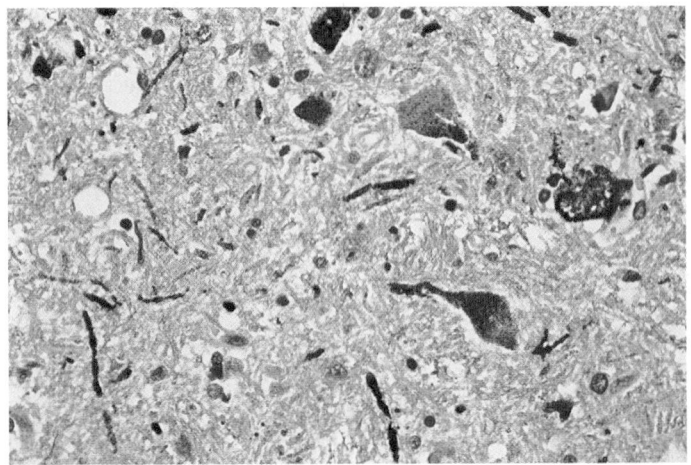

Figure 31.15 West Nile encephalitis. Brain section from a West Nile encephalitis victim. Red areas in the tissue are neurons containing West Nile virus as detected using an immunofluorescent staining technique (⟳ Section 28.6).

UNIT 7

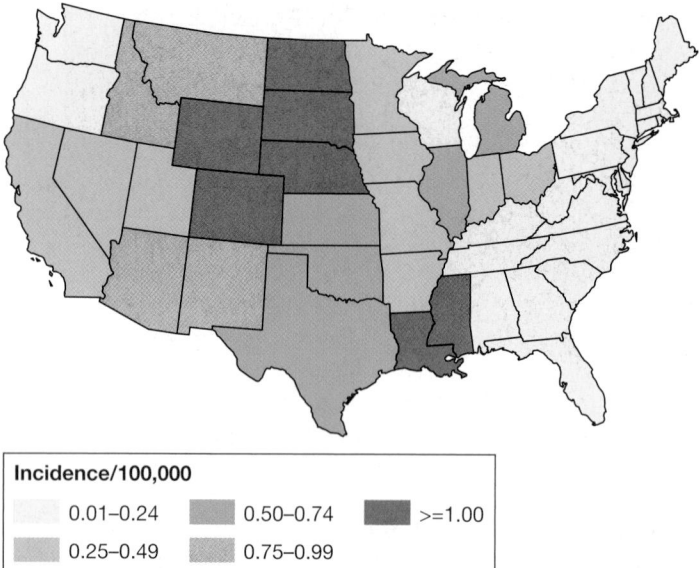

Incidence/100,000

0.01–0.24	0.50–0.74	>=1.00
0.25–0.49	0.75–0.99	

Figure 31.16 Average annual incidence of West Nile disease in the United States, 1999–2014. In 2015, 2060 cases of human West Nile disease were reported, resulting in 119 deaths (5–6% mortality). West Nile virus is now endemic in mosquitoes and birds throughout the United States. Data are from the Centers for Disease Control and Prevention, Atlanta, Georgia, USA.

horses where the mortality risk demands it, but no human WNV vaccine is currently available.

MINIQUIZ

- Identify the vector and reservoir for West Nile virus.
- Trace the progress of West Nile virus in the United States since 1999.

31.7 Plague

Plague has caused more human deaths than any other infectious disease except for malaria and tuberculosis. Plague is primarily a zoonosis of wild rodents, but humans can become accidental hosts when rodent populations experience a die-off. Plague is caused by *Yersinia pestis*, a gram-negative, facultatively aerobic, rod-shaped, and encapsulated enteric bacterium (*Gammaproteobacteria*, ⇨ Section 16.3) that is easily grown in laboratory culture (**Figure 31.17**).

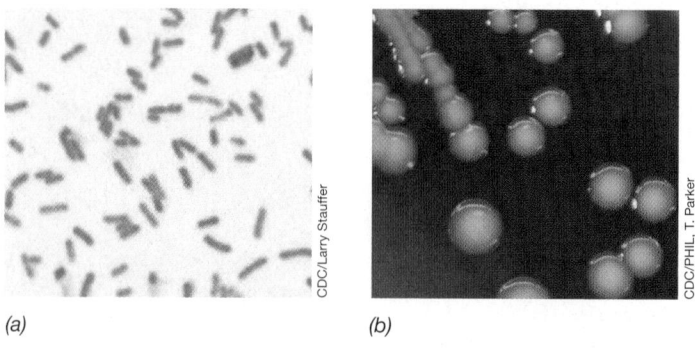

(a)   *(b)*

Figure 31.17 *Yersinia pestis.* *(a)* Gram stain of cells of *Yersinia pestis*. The cells are about 0.8 μm in diameter. *(b)* Colonies of *Y. pestis* grown on blood agar.

Pathology and Treatment of Plague

The pathogenesis of plague is not clearly understood, but cells of *Y. pestis* produce several virulence factors that contribute to the disease process. The V and W antigens in *Y. pestis* cell walls are protein–lipoprotein complexes that inhibit immune cell phagocytosis. *Murine toxin*, an exotoxin (⇨ Section 25.6) that is lethal for mice, is produced by virulent strains of *Y. pestis*. Murine toxin is a respiratory inhibitor that causes systemic shock, liver damage, and respiratory distress in mice. The toxin likely plays a role in human plague as well, because these symptoms are common in plague patients. *Y. pestis* also produces a highly immunogenic endotoxin (⇨ Section 25.8) that may play a role in the disease process.

Plague can occur in several forms (see Figures 31.19 and 31.20). *Sylvatic plague* is enzootic among wild rodents. Plague is transmitted by several species of fleas, a main one being the rat flea *Xenopsylla cheopis* (**Figure 31.18a**). Fleas ingest *Y. pestis* cells in a blood meal and the bacterium multiplies in the flea's intestine. From there, the infected flea transmits the disease to rodents or humans in the next bite. The most common form of plague in humans is *bubonic plague*. In this case, cells of *Y. pestis* travel to the lymph nodes, where they replicate and cause swelling. The regional and pronounced swollen lymph nodes are called *buboes*, and the disease gets its name from these structures (**Figure 31.19a**). The buboes become filled with *Y. pestis* cells, and the bacterium's capsule prevents phagocytosis and destruction by cells of the immune system. Secondary buboes form in peripheral lymph nodes, and cells eventually enter the bloodstream, causing septicemia. Multiple local hemorrhages produce dark splotches on the skin and eventual tissue necrosis, giving plague its historical name, the "Black Death" (Figure 31.19b). If the infection is not treated quickly, the symptoms of plague (lymph node swelling and pain, prostration, shock, gangrene, and delirium) usually progress and cause death within 3–5 days.

Pneumonic plague occurs when cells of *Y. pestis* are either inhaled directly or reach the lungs via the blood or lymphatic circulation. Significant symptoms are usually absent until the last day or two of the disease when large amounts of bloody sputum are produced. Greater than 90% of untreated cases of pneumonic plague

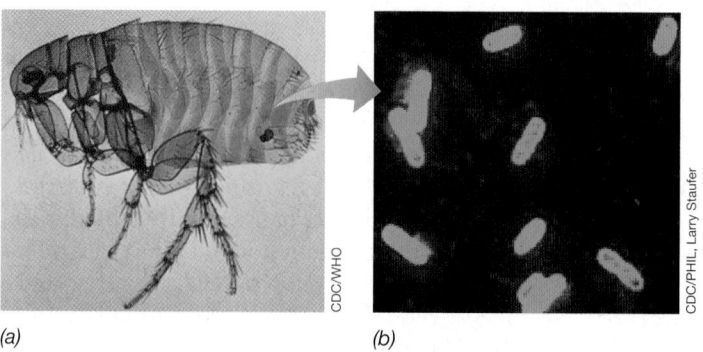

(a)   *(b)*

Figure 31.18 The rat flea, a major vector of plague. *(a)* The rat flea *Xenopsylla cheopis* carries cells of *Yersinia pestis*. The bacterium replicates in the flea gut and *(b)* cells of *Y. pestis* are transmitted to a host in a flea bite. The rat flea was the major vector for the pandemics of plague that ravaged medieval Europe in the fourteenth century. Cells in part *b* were stained with a fluorescent antibody prepared against *Y. pestis* cell surface antigens.

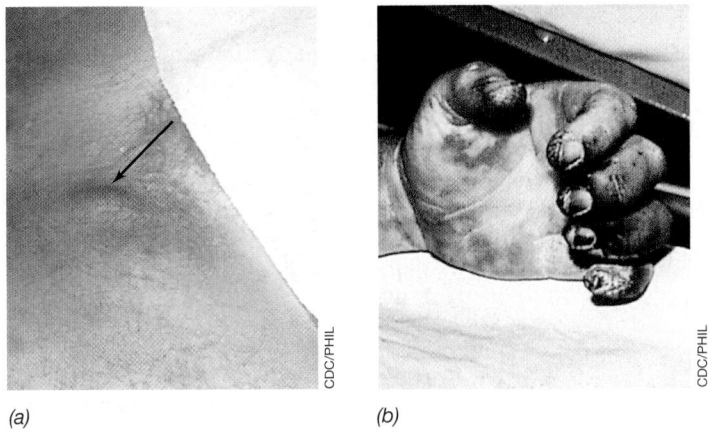

(a)   *(b)*

Figure 31.19 Plague in humans. *(a)* A bubo formed in the groin. *(b)* Gangrene and sloughing of skin in the hand of a plague victim. Human plague can manifest itself in three different forms: bubonic, pneumonic, and septicemic (see Figure 31.20).

result in death within 48 h. Moreover, pneumonic plague is highly contagious and can spread rapidly from person to person if those infected are not immediately isolated. *Septicemic plague* is the rapid spread of *Y. pestis* throughout the body via the bloodstream without the formation of buboes and is so severe that it usually causes death before a diagnosis can be made.

Bubonic plague can be successfully treated with streptomycin or gentamicin, administered by injection. Alternatively, doxycycline, ciprofloxacin, or chloramphenicol may be given intravenously. If treatment is started promptly, mortality from bubonic plague can be reduced to less than 5%. Pneumonic and septicemic plague can also be treated, but these forms progress so rapidly that antibiotic therapy, even if begun when symptoms first appear, is usually too late.

Plague Epidemiology and Control

Sylvatic plague is enzootic in a variety of rodents including ground squirrels, prairie dogs, chipmunks, and mice; rats are the primary hosts in urban communities and were typically the hosts

in episodes of sylvatic plague that triggered human pandemics during the Middle Ages. Fleas are intermediate hosts and vectors for plague (Figure 31.18), spreading the disease between rodent hosts and humans (**Figure 31.20**). Most infected rats or other rodents die soon after symptoms appear, but a small proportion of survivors develop a chronic infection, providing a persistent reservoir of *Y. pestis* cells to fuel new outbreaks.

Plague is endemic in developing countries in Africa, Asia, the Americas, and in south-central Eurasia; most cases occur in sub-Saharan Africa. Pandemic plague was historically associated with unsanitary surroundings, a major factor supporting large rat populations. In sparsely populated rural areas, this is not so great a problem as the disease runs its course when the rodent population dies off, leaving a shortage of hosts. But in urban centers where alternative hosts (humans) are plentiful, an outbreak of sylvatic plague can set the stage for a human plague epidemic. In the United States only a handful of cases of plague are diagnosed annually, mostly in southwestern states (New Mexico, Colorado, and Arizona in particular) where sylvatic plague is enzootic among wild rodents (Figure 31.20). In 2014, 14 cases of human plague were reported, with no deaths. From 2006 through mid-2015, a total of 68 cases were reported and resulted in 11 deaths.

Plague control is accomplished through good sanitation practices, surveillance and control of rodent reservoirs and vectors (fleas), isolation of active cases, and imposing quarantine on those who have had contact with diseased individuals. Improved public health practices and the control of rodent populations are the major reasons that outbreaks of plague are extremely rare in developed countries.

MINIQUIZ

- Distinguish among sylvatic, bubonic, septicemic, and pneumonic plague.
- What are the insect vector, the natural host reservoir, and the treatment for plague?

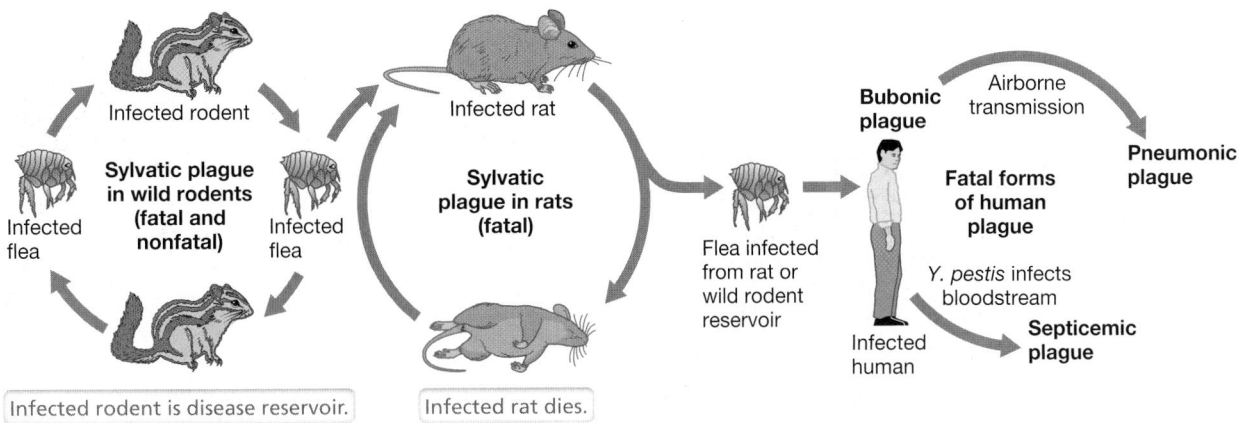

Figure 31.20 Plague epidemiology. In some wild rodents, sylvatic plague causes only a mild infection but diseased animals remain a reservoir of *Yersinia pestis*. In rodents that act as disseminating hosts, for example, rats, and in humans, plague is often fatal. When the domestic rodent reservoir dies off in an epidemic, infected fleas seek alternate hosts in humans.

III • Soilborne Bacterial Diseases

31.8 Anthrax

Some human diseases are caused by microorganisms whose major habitat is soil, and anthrax is an excellent example. We covered some aspects of the disease anthrax in Section 29.9 in the context of its use as a potential bioterrorism or biological warfare agent. Here we focus more on the biology of the organism and the disease process.

Discovery and Properties of Anthrax

The famous pioneering medical microbiologist Robert Koch (⮂ Section 1.10) first isolated the causative agent of the disease anthrax, the endospore-forming bacterium *Bacillus anthracis* (**Figure 31.21**). Using mice caught in the wild as experimental animals, Koch used the disease anthrax to develop his principles for linking cause and effect in infectious disease—Koch's postulates (⮂ Figure 1.29). Anthrax is quickly fatal in mice, but in humans, anthrax can take on several different forms, from mild to severe skin infections to respiratory failure and death.

Anthrax is an enzootic disease of worldwide occurrence. *B. anthracis* lives a saprophytic existence in soils, growing as an aerobic chemoorganotroph and forming endospores (Figure 31.21) when conditions warrant. From soil, cells or spores of *B. anthracis* can become embedded in animal hair, hides, or other animal materials, or can be ingested, and from here the disease can develop, allowing *B. anthracis* spores to be transmitted to humans. Anthrax is primarily seen in domesticated farm animals, particularly in cattle, sheep, and goats, and is transmitted from them to humans.

Forms of Human Anthrax

The disease anthrax can manifest itself in one of three forms: cutaneous (on the skin), intestinal, and respiratory (inhalation anthrax). In all forms, disease symptoms are due to two major toxins called *lethal toxin* and *edema toxin*. The different forms of anthrax show increasing severity, which is primarily a function of where in the body these toxins are excreted. Growth and toxin production by *B. anthracis* in the lymph nodes and lymphatic tissues leads to progressively worsening symptoms, beginning with a sore

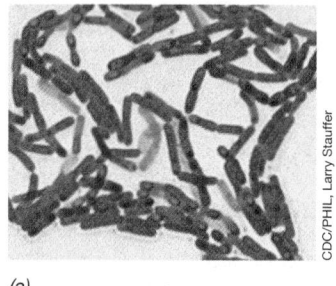

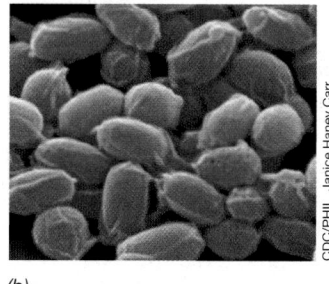

Figure 31.21 *Bacillus anthracis*. The anthrax pathogen produces endospores. (a) Light micrograph of a malachite green–stained smear of *B. anthracis* cells showing greenish-blue endospores. (b) Colorized scanning electron micrograph of *B. anthracis* endospores. Cells of *B. anthracis* are about 1.2 μm in diameter.

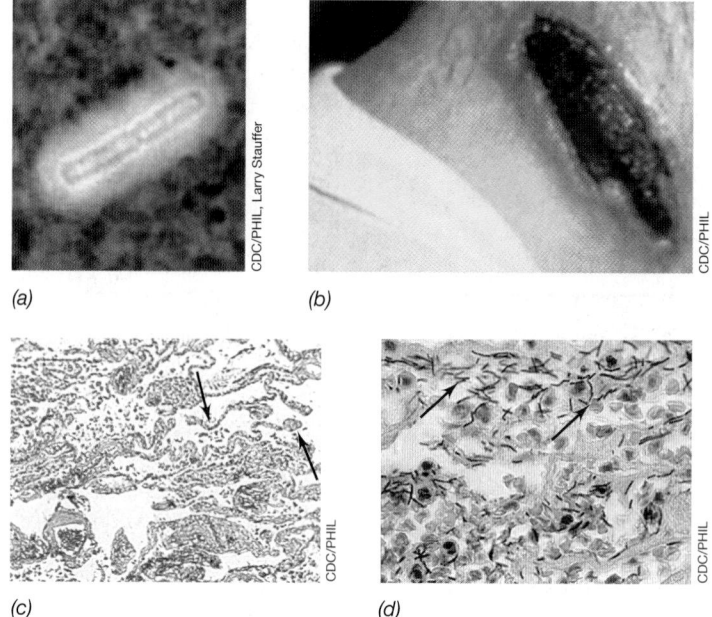

Figure 31.22 Anthrax pathology. (a) The protein capsule of *Bacillus anthracis* cells is a major virulence factor because it prevents killing by macrophages. (b) Cutaneous anthrax, with its characteristic black scabby appearance on the neck of a patient. (c, d) Inhalation anthrax. (c) The lung fills with bacterial cells (arrows) and fluids (cleared zones). (d) From the systemic infection, *B. anthracis* cells can be found almost anywhere, including the lining of the central nervous system (arrows).

throat, muscle aches, and fever, and escalating to respiratory distress and systemic shock. In addition to anthrax toxins, the unusual protein capsule that surrounds cells of *B. anthracis* (**Figure 31.22a**) is also an important virulence factor of this pathogen as it prevents destruction of the bacterium following phagocytosis by immune cells such as macrophages. Instead, cells of *B. anthracis* grow within the macrophage, eventually killing it and giving the bacterium access to the bloodstream.

Virtually all cases of human anthrax are *cutaneous anthrax*, where spores of *B. anthracis* have entered through a skin lesion, germinate, and form a painless, black and swollen pustule of dead tissue called an *eschar* (Figure 31.22b); the eschar is highly characteristic of the disease and allows for a firm diagnosis even though human anthrax is rarely seen in clinical medicine. In cutaneous anthrax the bacterium usually remains localized and the disease is readily treatable. Although cutaneous anthrax is fatal for about 20% of those untreated, most cases are treated because of the obvious symptoms, and thus fatalities are rare. *Intestinal anthrax* is very uncommon and is triggered by the ingestion of spores of *B. anthracis* (Figure 31.21b) in undercooked meat from diseased animals. Symptoms of intestinal anthrax include abdominal pain, bloody diarrhea, and ulcer-like lesions throughout the intestinal tract. The disease is still treatable at this stage but because of its rarity, diagnoses are easily missed. As a result, about half of all cases of intestinal anthrax are fatal.

Inhalation anthrax is the most severe form of the disease and is fatal in almost every case (Figure 31.22c, d). Inhalation anthrax occurs from the inhalation of endospores of *B. anthracis* and, along with cutaneous anthrax, is an occupational hazard for farm workers that process wool and hides (inhalation anthrax is also known as "woolsorter's disease"). In inhalation anthrax, the organism enters the bloodstream from inhaled dust or animal dander and multiplies to become systemic. The mounting toxemia from this runaway growth of *B. anthracis* triggers septic shock and fluid accumulation in the lungs (Figure 31.22*c*) that can kill a patient in less than a day.

Prevention and Vaccines

Complete prevention of anthrax is impossible because the reservoir of the organism is the soil. However, anthrax is avoidable by limiting close exposure to farm animals and is easily treatable with antibiotics. For the cutaneous form this is a routine treatment, but antibiotic therapy is less effective in intestinal anthrax and especially in inhalation anthrax. By the time the latter is diagnosed, the disease has progressed to the point where it is usually too late to save the patient. An anthrax vaccine is available but because the disease is so rare, it is only recommended for high-risk individuals such as scientists working with the organism, slaughterhouse or livestock workers, and military personnel (for biowarfare reasons). An effective and inexpensive anthrax vaccine is available for vaccinating livestock and is commonly used in cattle, sheep, goats, and horses.

— MINIQUIZ —

- What are the major virulence factors of *Bacillus anthracis*?
- What are the three forms of anthrax, and which is most dangerous?

31.9 Tetanus and Gas Gangrene

Tetanus is a serious, life-threatening disease. Although tetanus is completely preventable through immunization, it still causes over 150,000 deaths per year, mostly in countries in Africa and Southeast Asia. *Gas gangrene* is caused by the growth in dead tissues of bacteria related to the tetanus pathogen, leading to a gassy putrefaction and loss of an infected limb or death from systemic shock. Both diseases are caused by clostridia.

Biology and Epidemiology of Tetanus

Tetanus is caused by an exotoxin produced by *Clostridium tetani*, an obligately anaerobic, endospore-forming rod (**Figure 31.23*a***; ⇄ Section 16.8). The natural reservoir of *C. tetani* is soil, where it is a ubiquitous resident, although it is occasionally found in the gut of healthy humans, as are other *Clostridium* species.

Cells of *C. tetani* normally gain access to the body through a soil-contaminated wound, typically a deep puncture. In the wound, anoxic conditions develop around the dead tissue and allow germination of endospores, growth of the organism, and production of a potent exotoxin, the *tetanus toxin* (also called *tetanospasmin*). *C. tetani* is essentially noninvasive; its sole ability to cause disease is through toxemia, and thus tetanus is observed only as the result of untreated deep tissue injuries. The onset of tetanus symptoms may take from four days to several weeks, depending on the number of endospores inoculated at the time of the injury.

Pathogenesis of Tetanus

We have already examined the activity of tetanus toxin at the cellular and molecular level (⇄ Section 25.6). The toxin directly

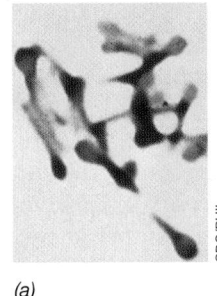
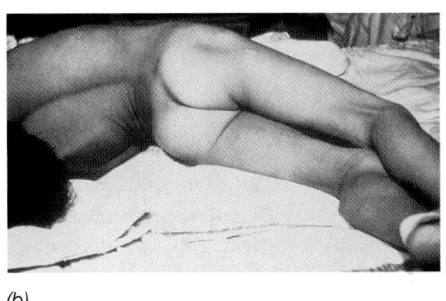
(a) *(b)*

Figure 31.23 Tetanus. *(a)* *Clostridium tetani* showing the "drumstick" appearance of sporulating cells with their terminal endospores. Cells of *C. tetani* are about 0.8 μm in diameter. *(b)* A tetanus patient showing the rigid paralysis characteristic of tetanus. Tetanus paralysis typically begins with the facial muscles ("lockjaw") and descends to lower body regions.

affects the release of inhibitory signaling molecules in the nervous system. These inhibitory signals control the "relaxation" phase of muscle contraction. The absence of inhibitory signaling molecules results in rigid paralysis of the voluntary muscles, often called *lockjaw* because it is observed first in the muscles of the jaw and face. Preceding actual lockjaw, tetanus symptoms typically include mild spasms of facial muscles and muscles of the neck and upper back. Later on, the paralysis extends to the torso and lower body (Figure 31.23*b*). When tetanus is fatal, death is usually due to respiratory failure. Mortality is relatively high, occurring in about 10% of all reported cases, and up to 50% of cases in which treatment is delayed until generalized full body tetanus has set in.

Diagnosis, Control, Prevention, and Treatment of Tetanus

Diagnosis of tetanus is based on exposure, clinical symptoms (Figure 31.23*b*), and, rarely, identification of the toxin in the blood or tissues of the patient. The natural reservoir of *C. tetani* is the soil and thus control measures must focus on disease prevention rather than pathogen removal. The tetanus toxoid vaccine is highly effective, and thus virtually all tetanus cases occur in individuals who were inadequately immunized.

A second line of tetanus protection is to administer appropriate medical care to serious cuts, lacerations, and punctures. Even though vaccination against tetanus is widely practiced, any serious wound should be thoroughly cleaned and the damaged tissue removed. If the vaccination status of the individual is unclear or the last tetanus booster was more than 10 years ago, revaccination is recommended. If a deep wound is severe or heavily contaminated by soil, treatment might also include administration of a tetanus antitoxin preparation, especially if the patient's immunization status is unknown or is out of date.

Acute symptomatic tetanus (Figure 31.23*b*) is treated with antibiotics, usually penicillin, to stop growth and toxin production by *C. tetani*, and antitoxin is injected intramuscularly (or into the sheath surrounding the spinal cord if necessary) to prevent binding of newly released toxin to cells. Supportive therapy such as sedation, administration of muscle relaxants, and mechanical respiration may be necessary to control the effects of paralysis. Treatment cannot provide a quick reversal of symptoms because toxin that is already bound to tissues cannot be neutralized. Even with antitoxin and antibiotic administration and supportive therapy, tetanus patients show significant morbidity and mortality. A complete recovery from tetanus often takes many months.

Gas Gangrene

Tissue destruction due to the growth of proteolytic and gas-producing clostridia is called **gas gangrene**. In this life-threatening condition, amino acids obtained from the breakdown of muscle proteins are fermented to the gases H_2 and CO_2 plus a variety of foul-smelling organic compounds, including short-chain fatty acids and other putrid molecules; ammonia released during amino acid fermentation (Section 14.21) adds to the stench. In addition, a variety of bacterial toxins are produced that accelerate tissue destruction.

Although *C. tetani* is a proteolytic *Clostridium* species, it does not cause gangrene but can be associated with cases of gangrene triggered by a deep tissue wound. The most common causes of gangrene are *Clostridium perfringens* (**Figure 31.24a**), which is also a common cause of foodborne illness unrelated to gangrene, *C. novyi*, and *C. septicum*. These organisms reside in soil and are also part of the normal human intestinal microbiota. When these species reach deep into tissues, typically from traumatic tissue invasion such as a war wound or other puncture wound, or occasionally from gastrointestinal tract surgery, spores and vegetative cells of proteolytic clostridia are inserted into what are now dead tissues. As the bacteria grow, they release enzymes that destroy collagen and tissue proteins and also excrete a series of toxins. *C. perfringens* (Figure 31.24a) *alpha toxin*, which is distinct from the toxins the bacterium produces in perfringens food poisoning (Section 32.9), is a major virulence factor in gangrene, as is the general ability of the pathogens to grow rapidly in the warm, moist, and protein-rich environment created by an invasive injury. Alpha toxin is a phospholipase that hydrolyzes the membrane phospholipids of host cells, leading to cell lysis and the typical accumulation of gas and fluids that accompanies gas gangrene (Figure 31.24c).

In severe cases of gas gangrene, the toxemia can become systemic and cause death. Antibiotic treatment is taken as a preventive measure in cases of gangrene in addition to the typical though dramatic treatment: amputation of the infected limb. Gangrenous tissues are dead and will not regenerate, and amputation prevents the infection from reaching healthy tissues.

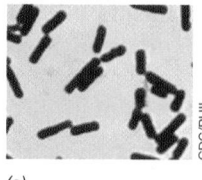
(a)

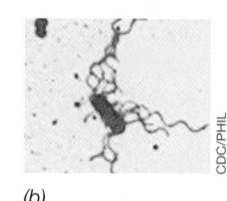
(b)

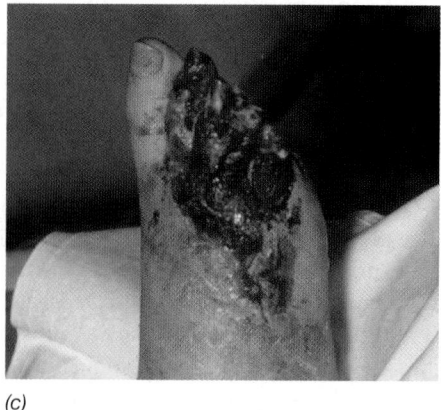
(c)

Figure 31.24 Gas gangrene. *(a)* Gram stain of cells of *Clostridium perfringens*, the most common cause of gangrene. *(b)* Flagellum stain showing a cell of *Clostridium novyi*, also an agent of gangrene. *(c)* A case of gas gangrene on the foot. Cells of both *C. perfringens* and *C. novyi* are about 1.2 μm in diameter.

Hyperbaric oxygen treatment of the infected limb is attempted in some cases to try to save it, with the high levels of O_2 inhibiting growth of the obligately anaerobic clostridia. In hyperbaric treatment, the patient sits in an enclosed chamber containing 100% O_2 at about twice atmospheric pressure. This enriches the blood in O_2 and helps still-living blood vessels seed the formation of new tissue. Several hyperbaric treatments are administered and may be accompanied by surgical removal of some of the dead tissue. If an adequate blood supply can be established in damaged tissues, a skin graft may also be done to help connect regenerating with damaged tissues.

MINIQUIZ

- Describe infection by *Clostridium tetani* and the effects of tetanus toxin. How does the mode of action of tetanus toxin differ from that of alpha toxin produced by *C. perfringens*?
- Describe the steps necessary to prevent tetanus in an individual who has sustained a puncture wound.
- How does the physiology of *C. perfringens* make it suitable for growing in puncture wounds?

MasteringMicrobiology® **Visualize**, **explore**, and **think critically** with Interactive Microbiology, MicroLab Tutors, MicroCareers case studies, and more. MasteringMicrobiology offers practice quizzes, helpful animations, and other study tools for lecture and lab to help you master microbiology.

Chapter Review

I • Animal-Transmitted Viral Diseases

31.1 Rabies occurs primarily in wild animals but can be transmitted to humans and domestic animals. In the United States, rabies is transmitted primarily from the wild animal reservoir to domestic animals or, very

rarely, to humans. Vaccination of domestic animals is key to the control of rabies. Most human deaths from rabies occur in developing countries.

Q Identify the animals most likely to carry rabies in the United States. Why is rabies so rare in humans and domesticated animals in developed countries?

31.2 Hantaviruses are present worldwide in rodent populations and cause zoonotic diseases such as hantavirus pulmonary syndrome and hemorrhagic fever with renal syndrome in humans. Hantavirus is a highly dangerous hemorrhagic fever virus related to Ebola. In the Americas, hantavirus infections have case fatality rates of over 30%.

> **Q** **Describe the conditions that may cause emergence of hantavirus pulmonary syndrome (HPS). How can HPS be prevented?**

II • Arthropod-Transmitted Bacterial and Viral Diseases

31.3 Rickettsias are obligate intracellular parasitic bacteria transmitted to hosts by arthropod vectors. The incidence of spotted fever rickettsiosis and other rickettsial syndromes is increasing as a result of several factors. Most rickettsial infections can be controlled by antibiotic therapy, but prompt recognition and diagnosis of these diseases remains difficult.

> **Q** **Identify the three major categories of organisms that cause rickettsial diseases. For typhus, spotted fever rickettsiosis, and ehrlichiosis, identify the most common reservoir and vector.**

31.4 Lyme disease is caused by the spirochete *Borrelia burgdorferi* and is the most prevalent arthropod-borne disease in the United States today. Lyme is transmitted from several mammalian host vectors to humans by ticks. Prevention and treatment of Lyme disease are straightforward, but accurate and timely diagnosis of infection is essential.

> **Q** **Identify the most common reservoir and vector for Lyme disease in the United States. How can the spread of Lyme disease be controlled? How can Lyme disease be treated?**

31.5 Yellow, dengue, and Zika fevers are caused by related flaviviruses transmitted to humans by mosquito bites; Chikungunya fever shows similar disease symptoms but is caused by a different class of virus. These diseases are widespread in tropical and subtropical countries and can show mild to very severe symptoms, including hemorrhagic fevers. An effective vaccine for yellow fever is in use, but no vaccine is currently available to prevent dengue, Zika, and Chikungunya infections.

> **Q** **Of Zika, dengue, and yellow fevers, which is usually the least serious disease?**

31.6 West Nile fever is a mosquito-borne viral disease. In the natural cycle of the pathogen, birds are infected with West Nile virus by the bite of infected mosquitoes. Humans and other vertebrates are occasional terminal hosts. Most human infections are asymptomatic and undiagnosed, but complications of some infections cause about 5% mortality.

> **Q** **Describe the spread of West Nile virus infections in the United States. Which animals are the primary hosts? Are humans productive alternate hosts?**

31.7 Plague can be transmitted to individuals who have had contact with rodent populations and their parasitic fleas, the enzootic reservoirs for the plague bacterium, *Yersinia pestis*. A systemic infection or a pneumonic infection leads to rapid death, but the bubonic form is treatable with antibiotics.

> **Q** **For a potentially serious disease like bubonic plague, vaccines are not routinely recommended for the general population; why not? Identify the public health measures used to control plague.**

III • Soilborne Bacterial Diseases

31.8 Anthrax is caused by the endospore-forming bacterium *Bacillus anthracis* and can take on three different forms: cutaneous, intestinal, or inhalation. Cutaneous anthrax is most common and along with inhalation anthrax is an occupational hazard for livestock workers, where *B. anthracis* endospores can be transmitted from animal hides to humans.

> **Q** **Which key feature of the bacterium *Bacillus anthracis* allows this organism to persist for extended periods on animal hides or other environments where growth may not occur? Which form of anthrax is the most serious?**

31.9 *Clostridium tetani* is a soil bacterium that causes tetanus, a potentially fatal disease characterized by a toxemia and rigid paralysis. Treatment for acute tetanus includes antibiotics and active and passive immunization, and the disease is preventable by toxoid immunization. Gas gangrene occurs from the growth of various proteolytic clostridia in traumatic wounds, leading to gas and toxin formation.

> **Q** **Discuss the major mechanism of pathogenesis for tetanus and define measures for prevention and treatment. Why is it possible that a traumatic puncture wound could end up causing both tetanus and gas gangrene?**

Application Questions

1. Describe the sequence of steps you would take if your child received a bite (provoked or unprovoked) from a stray dog with no collar and record of rabies immunization. Present one scenario in which you were able to capture and detain the dog and another for a dog that escaped. How would these procedures differ from a situation in which the child was bitten by a dog that had a collar and rabies tag with documented, up-to-date rabies immunizations?

2. Contrast the modes of transmission of the following diseases: rabies, Lyme disease, yellow fever, West Nile disease, anthrax, and plague. Which of these diseases could be virtually eliminated in humans by control of the disease vector and which could not, and why?

3. Devise a plan to prevent the spread of West Nile virus to humans in your community. Identify the relative costs involved in such a plan, both at the individual level and at the community level. Find out if a mosquito abatement program is active in your community. What methods, if any, are used in your area for the reduction of mosquito populations? What is a simple way to limit mosquito numbers around your residence?

Chapter Glossary

Enzootic an endemic disease present in an animal population

Epizootic an epidemic disease present in an animal population

Gas gangrene tissue destruction due to the growth of proteolytic and gas-producing clostridia

Hantavirus pulmonary syndrome (HPS) an emerging, acute disease characterized by pneumonia and caused by rodent hantavirus

Hemorrhagic fever with renal syndrome (HFRS) an emerging acute disease characterized by shock and kidney failure, caused by rodent hantavirus

Lyme disease a tick-transmitted disease caused by the spirochete *Borrelia burgdorferi*

Plague an enzootic disease in rodents caused by *Yersinia pestis* that can be transferred to humans through the bite of a flea

Rabies a usually fatal neurological disease caused by the rabies virus, which is usually transmitted by the bite or saliva of an infected animal

Rickettsias obligate intracellular bacteria of the genus *Rickettsia* and related genera responsible for diseases including typhus, spotted fever rickettsiosis, and ehrlichiosis

Spotted fever rickettsiosis a tick-transmitted disease caused by *Rickettsia rickettsii*, characterized by fever, headache, rash, and gastrointestinal symptoms; also called Rocky Mountain spotted fever

Tetanus a disease characterized by rigid paralysis of the voluntary muscles, caused by an exotoxin produced by *Clostridium tetani*

Typhus a louse-transmitted disease caused by *Rickettsia prowazekii*, characterized by fever, headache, weakness, rash, and damage to the central nervous system and internal organs

West Nile fever a neurological disease caused by West Nile virus, a virus transmitted by mosquitoes from birds to humans

Zoonosis an animal disease transmissible to humans

Waterborne and Foodborne Bacterial and Viral Diseases

32

microbiology**now**

The Classic Botulism Scenario

Botulism is a food poisoning that causes severe symptoms and occasional deaths. Botulism results from highly poisonous exotoxins produced by the endospore-forming anaerobic bacterium *Clostridium botulinum*. This soil microbe can easily travel on the surface of fresh vegetables pulled from the home garden. If the vegetables are not thoroughly washed and properly canned (if intended for later use), botulism is a possibility. Unfortunately, such circumstances unfolded at a church potluck supper in Fairfield, Ohio (USA) in April 2015 when contaminated potatoes sent 29 people to the hospital with symptoms of botulism and caused one death—the largest botulism outbreak in the United States in nearly 40 years.

Most cases of botulism not linked to restaurant foods are traced to home-canned foods. In the Ohio outbreak, potato salad similar to that shown here was the disease vehicle. The potato salad was prepared from home-canned potatoes that were heat processed in boiling water rather than a pressure cooker. Temperatures in a boiling water canner cannot exceed 100°C and this is insufficient to kill bacterial endospores, the most heat-tolerant of all biological structures. When the potatoes were later used to prepare potato salad, the stage was set for the botulism outbreak.

Quick epidemiological work and early recognition of botulism symptoms by a local clinician helped keep the Ohio botulism outbreak from claiming more than the one life it did. Once it was clear what they were facing, health officials in Fairfield were immediately sent 50 doses of botulinum antitoxin from the strategic stockpile of this life-saving drug maintain by the Centers for Disease Control and Prevention; antitoxin treatment quickly ended the Ohio botulism crisis.

The circumstances surrounding the Ohio outbreak formed the "perfect storm" for a botulism outbreak. Potato salad made from fresh potatoes and refrigerated until consumption is not a botulism threat. By contrast, home-canned potatoes not treated in a pressure cooker are a botulism threat because the sealed (and thus anoxic) container provides ideal conditions for the growth of any viable *C. botulinum* endospores that remain after heat treatment. Human botulism is a rare occurrence, but its common link to home-canned foods reminds us of the critical importance of strict adherence to proper heat-processing procedures when preparing home-canned vegetables.

Source: McCarty, C.L., et al. 2015. Large outbreak of botulism associated with a church potluck meal—Ohio, 2015. *Morbidity and Mortality Weekly Report 64:* 802–803.

In this chapter we consider microbial pathogens whose mode of transmission is either water or food. The diseases these pathogens cause are called "common-source" diseases because they occur only in those who have consumed the same contaminated water or eaten the same contaminated food. Waterborne and foodborne illnesses are common infectious diseases worldwide. While food-borne illnesses are most commonly of bacterial or viral origin, waterborne illnesses can have bacterial, viral, or parasitic causes.

I • Water as a Disease Vehicle

Water is used in enormous quantities, and its microbiological safety rests in the hands of wastewater and drinking water engineers and microbiologists. Indeed, water quality is the single most important factor for ensuring public health. In Chapter 22 we examined the microbiology of wastewater and examined how highly polluted water can be cleaned by microbial activities and reused for many purposes, including for drinking. Here we see what can happen when water intended for human use becomes a vehicle for disease.

Waterborne diseases begin as infections (or occasionally as tox-emias), and contaminated water may cause an infection even if only small numbers of the particular pathogen are present. Whether or not exposure causes disease is a function of the viru-lence of the pathogen and the ability of the host to resist infection.

32.1 Agents and Sources of Waterborne Diseases

Many different microorganisms can cause waterborne infectious diseases, and some of the major ones are summarized in Table 32.1.

TABLE 32.1 Major waterborne pathogens

Pathogen	Disease
Bacteria[a]	
Vibrio cholerae	Cholera
Legionella pneumophila	Legionellosis
Salmonella enterica (typhi)	Typhoid fever
Escherichia coli	Gastrointestinal illness
Pseudomonas aeruginosa	Nosocomial pneumonia, septicemia, and skin infections
Campylobacter jejuni	Gastrointestinal illness
Viruses	
Norovirus	Gastrointestinal illness
Hepatitis A virus	Viral hepatitis
Parasites[b]	
Cryptosporidium parvum	Cryptosporidiosis
Giardia intestinalis	Giardiasis
Schistosoma	Schistosomiasis

[a]Except for *S. enterica* (*typhi*), these bacteria have been associated with major outbreaks of waterborne illness in the United States in recent years, as have the bacteria *Shigella sonnei* and *Leptospira* sp.
[b]See Chapter 33. *C. parvum* and *G. intestinalis* are unicellular microbial parasites; *Schistosoma* is a microscopic worm 10–20 mm long.

Many different microbes can be waterborne pathogens, but here we will consider bacterial pathogens with a major focus on cholera, a waterborne disease of pandemic proportions (⮑ Section 29.8). We consider parasitic diseases in Chapter 33 and a few viral pathogens that can be either foodborne or waterborne later in this chapter.

We begin by considering the disease vehicle itself—water. Waterborne illnesses can be transmitted through untreated or improperly treated water used for drinking or food preparation or from water used for swimming and bathing (recreational water sources). The major waterborne illnesses traced to drinking water and recreational waters are typically quite different, and these different disease patterns are shown in Table 32.2.

Potable Water

Water supplies in developed countries typically meet rigid quality standards, greatly reducing the spread of waterborne diseases. Drinking water in particular undergoes extensive treatment that includes both filtration and chlorination. Although filtration removes turbidity and many microorganisms, it is *chlorination* that makes drinking water safe. Chlorine gas (Cl_2) is a strong oxi-dant and oxidizes both organic matter dissolved in the water and microbial cells themselves. Drinking water chlorination facilities add sufficient chlorine to allow a residual level to remain in the water all the way to the consumer. Water suitable for human con-sumption is called **potable** water (⮑ Sections 22.6–22.9).

Despite filtration and chlorination, waterborne disease out-breaks from potable water occasionally occur. In the United States an average of 25 outbreaks of disease associated with drinking water are recorded in a year (a waterborne outbreak is defined as two or more human illnesses specifically linked to the consumption of the same water at the same time). Nearly 80% of drinking water disease outbreaks are due to *bacterial* pathogens, most notably *Legionella*, the causative agent of legionellosis (Table 32.2 and Section 32.4).

Recreational Waters

Recreational waters include freshwater aquatic systems such as ponds, streams, and lakes, as well as public swimming and wading

TABLE 32.2 Sources of outbreaks of acute gastrointestinal illness in drinking and recreational waters, 2009–2010[a]

	Drinking water (n = 33) (%)	Recreational water (n = 81) (%)
Bacteria	76	19
Legionella pneumophila	58	5
Other	18	14
Parasites	9	35
Viruses	6	1
Chemical/toxin	3	4
Multiple causes[b]	6	1
Unknown[c]	0	40

[a]Numbers are rounded to the nearest percent and were obtained from the Centers for Disease Control Waterborne Disease and Outbreak Surveillance System.
[b]Outbreak linked to more than one cause.
[c]Suspected to be caused by one or more microbes or chemicals but not confirmed.

pools. Recreational waters can be sources of waterborne disease, and on average they cause more disease outbreaks than those due to drinking water. Moreover, in contrast to drinking water, where *bacterial* pathogens are most common, disease outbreaks from recreational waters are more frequently linked to *parasitic* pathogens. In addition, recreational waters often transmit gastrointestinal illnesses that are either of unknown microbial origin or due to chemicals or other toxic materials (Table 32.2).

In the United States, the operation of public swimming pools is regulated by state and local health departments. The United States Environmental Protection Agency (EPA) establishes limits for bacterial numbers in both potable and public recreational water sources, but local and state governments can set standards above or below these guidelines for nonpotable sources. By contrast, the water quality of *private* recreational waters, such as swimming pools, spas, and hot tubs, is totally unregulated, and these are therefore prime vehicles for waterborne disease outbreaks.

— MINIQUIZ —

- What is potable water?
- Contrast the major pathogens responsible for disease outbreaks in drinking water versus recreational waters.

32.2 Public Health and Water Quality

Water that looks perfectly transparent may still be contaminated with high numbers of microorganisms and thus pose a risk of disease. It is impractical to screen water for every pathogenic organism that may be present (Table 32.1), and so both potable and recreational waters are routinely tested for specific *indicator organisms*, the presence of which signals the potential for waterborne disease.

Coliforms and Water Quality

A widely used indicator for microbial water contamination is the **coliform** group of bacteria. Coliforms are useful because many of them inhabit the intestinal tract of humans and other animals. Thus, the presence of coliforms in water indicates likely fecal contamination. Coliforms are defined as facultatively aerobic, gram-negative, rod-shaped, nonsporulating bacteria that ferment lactose with the production of gas within 48 h at 35°C. However, this definition includes several bacteria that are not necessarily restricted to the intestine; for this reason, it is *fecal coliforms* that are important in water safety assessments. *Escherichia coli*, a coliform whose only habitat is the intestine and that survives only a relatively short time outside the intestine, is the key fecal coliform of interest. The presence of cells of *E. coli* in a water sample is taken as evidence of fecal contamination and means that the water is unsafe for human consumption. Conversely, however, the absence of *E. coli* does not ensure that a water source is potable, because other pathogenic bacteria or pathogenic viruses or protists may still be present.

Testing for Fecal Coliforms and the Importance of *Escherichia coli*

Well-developed and standardized methods are in routine use for detecting coliforms and fecal coliforms in water samples. A common method is the *membrane filter* (MF) procedure where at least 100 ml of freshly collected water is passed through a sterile membrane filter, trapping any bacteria on the filter surface. The filter is placed on a plate of eosin–methylene blue (EMB) medium, which is selective for gram-negative, lactose-utilizing bacteria. EMB medium is also differential, allowing strongly fermentative species such as *E. coli* (**Figure 32.1a**; see also Figure 32.14c) to be distinguished from weakly fermentative species such as *Proteus*.

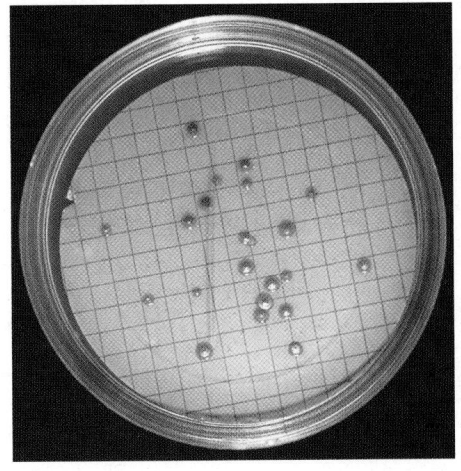

(a)

(b)

(c)

Figure 32.1 Fecal coliforms and their detection in water samples. *(a)* Colonies growing on a membrane filter. A drinking water sample was passed through the filter and the filter placed on an eosin–methylene blue (EMB) agar plate (EMB is both selective and differential for coliforms; more strongly fermentative species [such as *Escherichia coli*] form colonies with darker centers). *(b)* Total coliforms and *E. coli*. A filter exposed to a drinking water sample was incubated at 35°C for 24 hours on a medium containing special compounds that fluoresce when metabolized. The filter was then examined under ultraviolet light. The single *E. coli* colony in the sample fluoresces dark blue (arrow). Coliforms that do not metabolize the compound form colonies that fluoresce white to light blue. *(c)* The IDEXX Colilert water quality test system. When specific reagents are added to water samples and incubated for 24 h, they develop a yellow color if they contain coliforms (right). Samples containing *Escherichia coli* develop a yellow color but also fluoresce blue (left). Samples negative for coliforms remain clear (center).

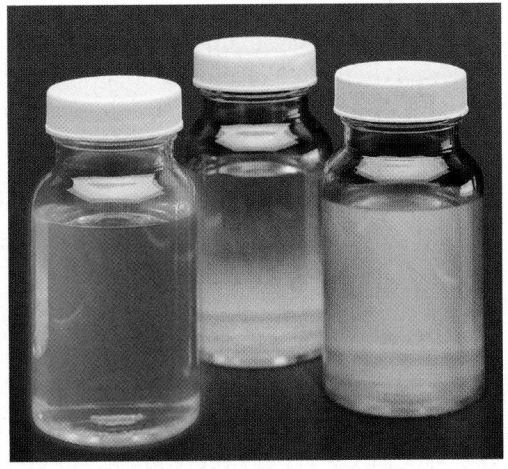

T. D. Brock

US EPA

IDEXX Laboratories

Selective media are also available that not only detect total coliforms but also specifically identify *E. coli* simultaneously. These *defined substrate tests* are typically faster and more accurate than EMB-based assays. One popular plate-based test is based on the ability of *E. coli* but not other enteric bacteria to metabolize a combination of two specific chemicals to form a fluorescent blue compound (Figure 32.1*b*). A commonly used liquid method reveals whether coliforms are present and also specifically detects *E. coli* in the water sample (Figure 32.1*c*). In addition to these colorimetric tests, dipsticks have been developed that detect ATP in a water sample. There is a strong correlation between the total number of bacteria in a sample and its ATP content, and the latter can be measured using the luciferase enzyme system in which a flash of light is emitted when ATP is hydrolyzed (⇄ Figure 9.4*b*). Using this system, the total microbial load in a water sample can be quickly assessed, and portable kits are available commercially to carry out these analyses in water purification facilities as well as remote field sites.

Reporting Water Purity Data

In properly regulated drinking water supply systems, total coliform and *E. coli* fecal coliform tests should be negative. A positive test indicates that a breakdown has occurred in either the purification or distribution system (or both). In the United States, microbiological standards for drinking water are specified in the *Safe Drinking Water Act* and are administered by the Environmental Protection Agency (EPA). Water utilities must report coliform test results to the EPA monthly, and if they do not meet the prescribed standards, the utilities must notify the public and take steps to correct the problem.

Major improvements in public health in the United States beginning in the early twentieth century were largely due to the adoption of water filtration and chlorination procedures in large-scale wastewater and drinking water treatment plants (⇄ Sections 22.7 and 22.8). Where drinking water standards have not reached this level, especially in developing countries, a variety of waterborne diseases are common. We turn our attention to these diseases now, beginning with cholera, the most widespread and devastating of all waterborne diseases.

— MINIQUIZ —

- Why is *Escherichia coli* used as an indicator organism in microbial analyses of water?
- What procedures are used to ensure the safety of potable water supplies?

II • Waterborne Diseases

32.3 *Vibrio cholerae* and Cholera

Cholera is a severe gastrointestinal diarrheal disease that is now largely restricted to countries in the developing world. Cholera is caused by ingestion of contaminated water containing cells of *Vibrio cholerae*, a gram-negative and motile curved species of

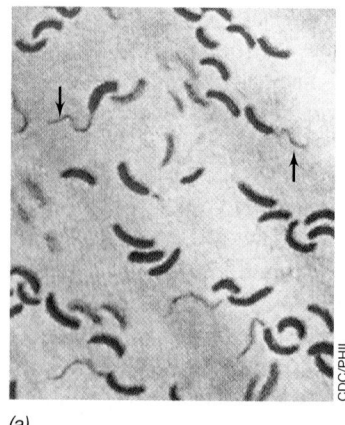
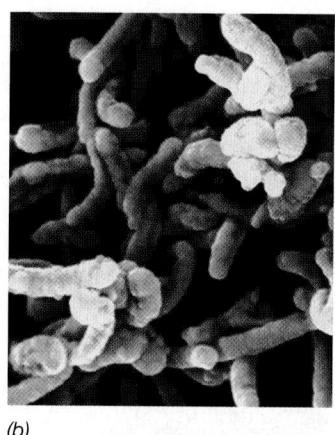

(a) CDC/PHIL   *(b)*

Figure 32.2 *Vibrio cholerae*, **the causative agent of cholera.** *(a)* Gram-stained preparation shows the typically curved (vibrio-shaped) cells of this bacterium and polar flagella (arrows). *(b)* Colorized scanning electron micrograph of cells of *V. cholerae*. A single cell is about 0.5×2 μm.

Proteobacteria (**Figure 32.2**). Cholera can also be contracted from contaminated food, especially improperly cooked shellfish.

The ingestion of a large number ($>10^8$) of *V. cholerae* cells is required to cause disease. The ingested cells attach to epithelial cells in the small intestine where they grow and release *cholera toxin*, a potent enterotoxin (⇄ Figure 25.15). Studies with human volunteers have shown that normal stomach acidity (about pH 2) is why the large inoculum of *V. cholerae* cells is needed to initiate disease. In studies with human volunteers, those given bicarbonate to neutralize gastric acidity develop cholera when given as few as 10^4 cells. Even lower cell numbers can initiate infection if *V. cholerae* is ingested with food, presumably because the food protects the vibrios from destruction by stomach acidity.

Cholera enterotoxin causes severe diarrhea that can result in dehydration and death unless the patient is given fluid and electrolyte therapy. The enterotoxin causes fluid losses of up to 20 liters (20 kg or 44 lb) per person per day, causing severe dehydration. The mortality rate from *untreated* cholera is 25–50% and can be even higher under conditions of severe crowding and malnutrition as often occurs in refugee camps or in areas that have experienced natural disasters such as floods, earthquakes, and the like. In these situations there is often a near-complete breakdown in sanitation leading to the contamination of drinking water with feces and the rapid transmission of cholera.

Diagnosis, Treatment, and Prevention of Cholera

At treatment facilities in large outbreaks of cholera, each cholera patient is placed on a "cholera cot," which is a conventional folding cot containing an opening into which feces can be voided (**Figure 32.3a**). The feces of a cholera patient are more liquid than solid, and confirmation of the disease is straightforward because the pathogen is easily cultured on selective agar media (Figure 32.3*b–d*). Cholera treatment is simple and effective. An oral (or in severe cases, intravenous) liquid and electrolyte replacement therapy is the most effective means of treatment. Oral treatment

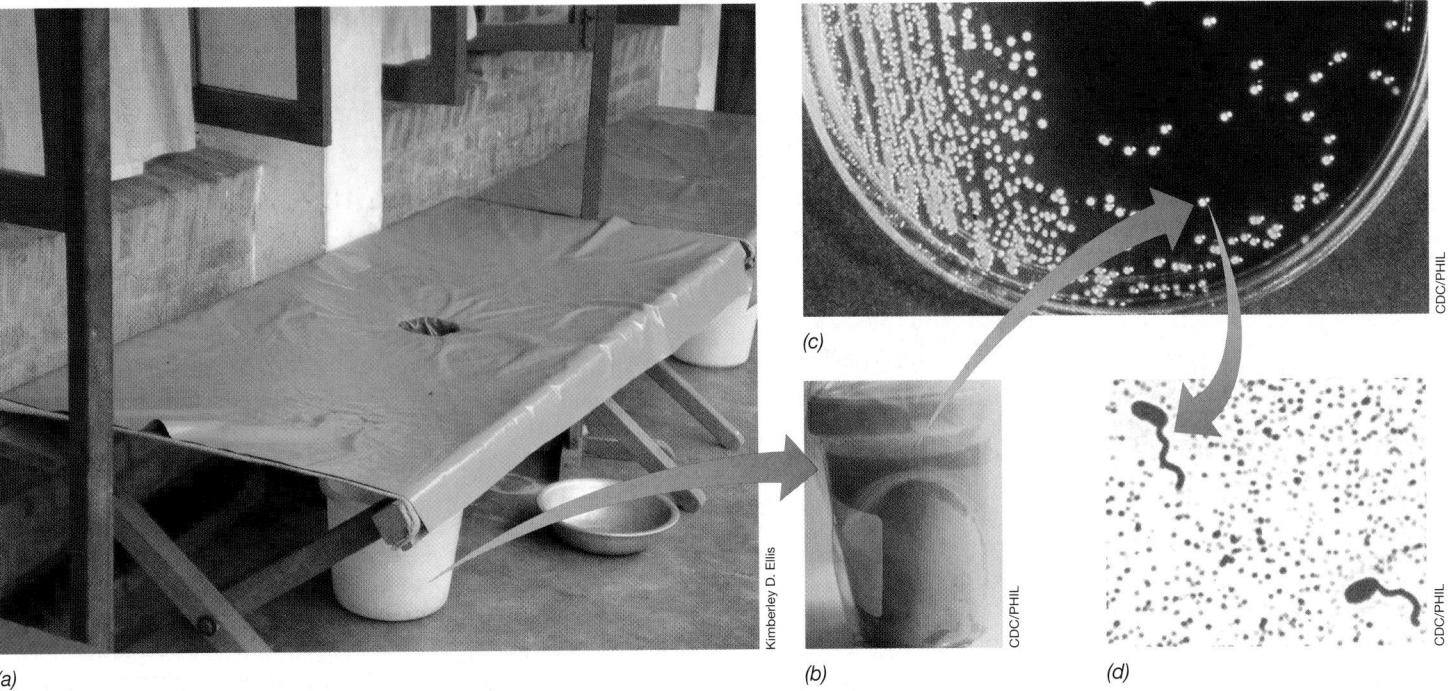

(a) *(b)* *(c)* *(d)*

Kimberley D. Ellis / CDC/PHIL / CDC/PHIL / CDC/PHIL

Figure 32.3 Cholera and its diagnosis. *(a)* A cholera cot. The cot allows a person to lie prostrate and void feces directly into a bucket. Cholera cots are used during cholera outbreaks for treating active disease cases with rehydration therapy. *(b)* Feces from a cholera patient. The "rice-water" stool is mostly liquid (the solid material in the bottom is mucus). *(c) Vibrio cholerae* is easily cultured on the medium TCBS, which is both selective and differential. TCBS contains high levels of bile salts and citrate, which inhibit enteric bacteria and gram-positive bacteria, and thiosulfate and sucrose, which cells of *V. cholerae (d)* use as a sulfur and carbon/energy source, respectively.

is preferred because no special equipment or sterile precautions are necessary. The rehydration solution is a mixture of glucose, salt (NaCl), sodium bicarbonate ($NaHCO_3$), and potassium chloride (KCl). If the solution is administered quickly during an outbreak, cholera mortality can be greatly reduced, as rehydration allows patients the time necessary to mount an immune response.

Antibiotics may shorten the course of cholera infection and the shedding of viable cells, but antibiotics are of little health benefit without simultaneous fluid and electrolyte replacement. Public health measures such as adequate sewage treatment and a reliable source of safe drinking water are the keys to preventing cholera. *V. cholerae* is eliminated from wastewater during proper sewage treatment and drinking water purification procedures (Chapter 22). For individuals traveling in cholera-endemic areas, attention to personal hygiene and avoidance of untreated water or ice, raw food, and raw or undercooked fish or shellfish that can feed on phytoplankton contaminated with *V. cholerae* (**Figure 32.4**) can prevent cholera.

Since 1817, cholera has swept the world in seven major pandemics with an eighth pandemic likely already started (↺ Section 29.8 and Figure 29.14). The World Health Organization estimates that only 5–10% of cholera cases are reported, so the total worldwide incidence of cholera probably exceeds 1 million cases per year. Only a handful of cases of cholera are reported each year in the United States, typically from imported shellfish that are eaten raw or after only minimal cooking.

— **MINIQUIZ** —

- What organism causes cholera, and what are the symptoms of the disease?
- Why does transmission of cholera usually require a large inoculum? Under what conditions can cholera be transmitted by fewer cells?
- Describe how cholera can be prevented and how it is treated.

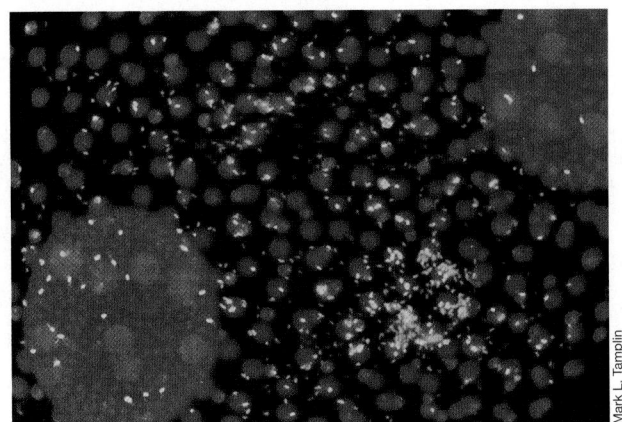

Mark L. Tamplin

Figure 32.4 Cells of *Vibrio cholerae* attached to the surface of *Volvox*, a freshwater alga. The sample was from a cholera-endemic area in Bangladesh. The *V. cholerae* cells are stained green by a fluorescently labeled antibody against a *V. cholerae* cell surface protein. The red color is from chlorophyll *a* fluorescence of the algae.

UNIT 7

32.4 Legionellosis

Legionella pneumophila, the bacterium that causes *legionellosis*, is an important waterborne pathogen whose transmission was originally linked to aerosols from evaporative cooling devices. However, *L. pneumophila* (**Figure 32.5**) is now known to be a major pathogen in residential water systems as well, where the organism persists in biofilms that form on interior surfaces of water distribution pipes and also within the cells of certain microbial parasites. In these sites, *L. pneumophila* is protected from the chlorine present in potable waters, and thus biofilms and infected parasites are reservoirs for transmitting legionellosis by a waterborne route (Section 22.9 and Figure 22.22).

Pathogenesis, Diagnosis, and Treatment

Cells of *L. pneumophila* invade the lungs and grow within macrophages and monocytes. Infections are often asymptomatic or produce only a mild cough, sore throat, mild headache, and fever; these self-limiting cases typically resolve themselves in 2–5 days. However, the elderly, whose resistance may be naturally reduced, and those with compromised immune systems often acquire more serious *Legionella* infections resulting in pneumonia. Prior to the onset of pneumonia, intestinal disorders, followed by high fever, chills, and muscle aches, are common. These symptoms precede the dry cough and chest and abdominal pains typical of legionellosis. Up to 10% of cases that reach this stage are fatal, usually as a result of respiratory failure.

Clinical detection of *L. pneumophila* infection is usually done by culture of the organism from bronchial washings, pleural fluid, or other body fluids or tissues (Figure 32.5a). Various serological tests can detect anti-*Legionella* antibodies or *Legionella* cells in

these samples and also in patient urine, and this is used to confirm a diagnosis (Figure 32.5b, d). Legionellosis can be treated with the antibiotics rifampin and erythromycin, and intravenous administration of erythromycin is the treatment of choice for life-threatening cases.

Epidemiology

L. pneumophila is a gram-negative, obligately aerobic rod-shaped species of the *Gammaproteobacteria* (Figure 32.5), and shows complex nutritional requirements including an unusually high requirement for iron. The organism can be isolated from terrestrial and aquatic habitats as well as from legionellosis patients. *Legionella pneumophila* was first recognized as the pathogen that caused an outbreak of fatal pneumonia at an American Legion convention (thus the name legionellosis) in Philadelphia (USA) in 1976. Besides legionellosis, the same bacterium can also cause a milder syndrome called *Pontiac fever*.

L. pneumophila is present in freshwaters and in soil. It is relatively resistant to heating and chlorination, so it can spread through drinking water distribution systems (Section 22.9). The pathogen is often found in large numbers in improperly sanitized cooling towers and evaporative condensers of large air conditioning systems. The pathogen grows in the water and is disseminated in humidified aerosols. Human infection is by way of airborne droplets, but the infection does not spread from person to person.

Besides its presence in evaporative coolers and domestic water systems, *L. pneumophila* has also been detected in hot water tanks and spas; in the latter, it can reach high cell numbers in warm (35–45°C), stagnant water, especially if chlorine (or other sanitizer) levels are not maintained. Many outbreaks of legionellosis have been linked to swimming pools. *L. pneumophila* can be eliminated from water supplies by hyperchlorination or by heating water to greater than 63°C. Although incidence peaks in the summer months, epidemiological studies indicate that *L. pneumophila* infections can occur at any time of year, primarily as a result of aerosols generated from heating and cooling systems and contaminated premise water (Section 22.9) used for showering or bathing. In the United States, a few thousand cases of legionellosis are typically reported each year.

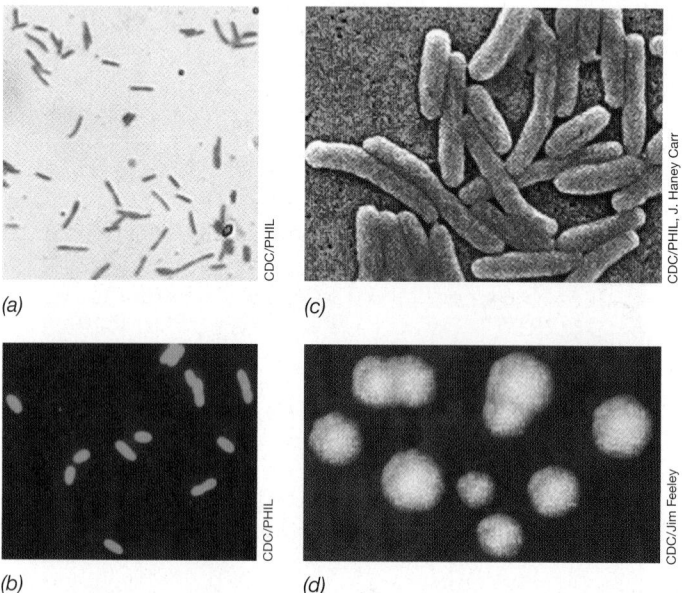

Figure 32.5 *Legionella pneumophila*. *(a)* Gram-stained cells of *L. pneumophila* from lung tissue of a legionellosis victim. *(b)* Cells of *L. pneumophila* can be positively identified using fluorescent anti–*L. pneumophila* antibodies. *(c)* Colorized scanning electron micrograph of *L. pneumophila* cells. Cells are about 0.5 × 2 μm. *(d)* Colonies of *L. pneumophila* grown on a complex enrichment medium showing their typical textured surface.

--- MINIQUIZ ---

- How is legionellosis transmitted?
- Identify specific measures for control of *Legionella pneumophila*.

32.5 Typhoid Fever and Norovirus Illness

Although cholera remains the most widespread and potentially dangerous of waterborne diseases, other waterborne pathogens also cause serious disease. We focus on two major ones here, the causative agents of typhoid (a bacterium) and norovirus gastrointestinal illness (an RNA virus).

Typhoid Fever

On a global scale, probably the most important waterborne bacterial pathogens are *Vibrio cholerae* (Section 32.3) and *Salmonella*

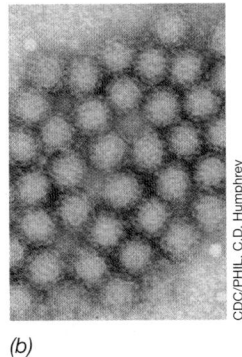

CDC/PHIL

CDC/PHIL, C.D. Humphrey

(a)   (b)

Figure 32.6 Bacterial and viral agents of severe gastrointestinal waterborne diseases. *(a)* Flagella-stained cells of *Salmonella enterica* (*typhi*) showing peritrichous flagellation (Section 2.11). A single cell measures about 1 × 2 µm. *(b)* Transmission electron micrograph of virions of norovirus. A single virion is about 30 nm in diameter.

enterica (*typhi*), the organism that causes typhoid fever. *S. enterica* (*typhi*) is a gram-negative, peritrichously flagellated bacterium related to *Escherichia coli* and other enteric bacteria (**Figure 32.6a**). The organism is transmitted in feces-contaminated water, and thus typhoid fever, like cholera, is primarily restricted to areas where sewage treatment and general sanitation are either absent or poorly maintained. Typhoid today is a well-entrenched endemic disease in sub-Saharan Africa, the Indian subcontinent, and Indonesia, but appears only sporadically in North America, Europe, northern Asia, and Australia.

Typhoid fever progresses in several stages. Cells of the pathogen (Figure 32.6a) ingested in contaminated water (or occasionally food) reach the small intestine where they grow and enter the lymphatic system and the bloodstream; from here, the pathogen can travel to many different organs. One to two weeks later, the first symptoms of typhoid appear; these include a mild fever, headache, and general malaise. During this period, the liver and spleen of the typhoid patient become heavily infected. About a week later, the fever becomes more intense (up to 40°C) and the patient typically becomes delirious; diarrhea can occur in this stage and abdominal pain can be severe. Complications can follow, including intestinal bleeding and perforation of the small intestine. The latter releases large numbers of bacterial cells into the abdomen, leading to a condition called *sepsis* (systemic infection and inflammation) and possibly also to septic shock. Both of these conditions are potentially fatal and up to 40% of sepsis cases are fatal. After about a week in this crisis stage, the symptoms of typhoid begin to wane and recovery occurs.

Treatment for typhoid consists of antibiotic therapy and fluid replacement to ward off dehydration. In some cases surgery may be necessary to repair perforated intestines. Although a variety of antibiotics can kill *S. enterica* (*typhi*), resistance to many of these has developed. Isolation of the causative strain and assessment of its antibiotic sensitivity (Section 28.4) is often necessary to ensure that antibiotics will cure the infection.

In the United States, fewer than 400 cases of typhoid occur per year, but typhoid fever used to be a major public health threat before drinking water was routinely filtered and chlorinated (Figure 29.7). However, breakdown of water treatment methods, contamination of water during floods, earthquakes, and

other disasters, or contamination of water supply pipes with leaking sewer lines can propagate epidemics of typhoid fever, even in developed countries.

In some typhoid patients, the gallbladder becomes infected with the pathogen. If these individuals also have gallstones, these can become colonized with *S. enterica* (*typhi*) cells and serve as a long-term reservoir of the pathogen from which it is continuously shed into feces and urine. Such individuals are otherwise healthy "carriers" of typhoid and can transmit the disease over long periods. The notorious cook "Typhoid Mary," who as a cook for hire spread typhoid throughout the New York City area for nearly 15 years beginning in the early twentieth century, was the classic example of a typhoid carrier.

Norovirus Illness

Viruses can be transmitted in water and cause human disease. Norovirus (Figure 32.6b) is one example and a common cause of gastrointestinal illness due to contaminated water (or food, Section 32.14). Norovirus is a single-stranded plus-sense RNA virus (Section 10.8) and is the leading cause of gastrointestinal illnesses worldwide (see Table 32.5).

Norovirus infection causes symptoms of vomiting, diarrhea, and malaise of relatively short duration. The disease is rarely fatal, although in compromised individuals (very young, elderly, or immune deficient), the significant dehydration that accompanies repeated bouts of norovirus-triggered vomiting and diarrhea can be life-threatening. A clinical diagnosis of norovirus gastrointestinal illness is made by a combination of observing symptoms and the direct detection of either viral RNA by RT-PCR (Sections 12.1 and 28.8) or viral antigens by enzyme immunoassay (Section 28.7) in samples of feces or vomit.

Norovirus disease is easily transmitted person to person or to food by the fecal–oral route. The infectious dose is very small, as exposure to as few as 10–20 norovirus virions (Figure 32.6b) is sufficient to initiate disease. The most common sources of waterborne norovirus outbreaks are well water or recreational waters that have been contaminated with sewage. Norovirus is also often the culprit when mass common-source gastrointestinal illnesses strike people on cruise ships or in long-term care facilities or other group settings. In these situations, the virus can be transmitted person to person, by contaminated food or water (usually food), or by any combination of these.

MINIQUIZ
- Contrast the causative agents of typhoid and noro gastrointestinal disease.
- What public health conditions allow for outbreaks of typhoid fever?

III • Food as a Disease Vehicle

The foods we eat, whether they are fresh, prepared, or preserved, are rarely sterile. Instead, they are almost always contaminated with spoilage microorganisms of various kinds and occasionally with pathogens. Microbial activities are key to the production of

UNIT 7

some foods, such as fermented foods, but most of the microorganisms in or on food are unwelcome and diminish either food quality or safety (or both). In this part, we explore the contrasting processes of food spoilage and food preservation, how food safety is assessed, and the transmission of pathogens in food. In the next two parts we focus on major foodborne diseases.

32.6 Food Spoilage and Food Preservation

Many foods provide an excellent medium for the growth of bacteria and fungi. Properly stored food can still undergo food spoilage but is usually not a vehicle for disease assuming that it was free of pathogens to begin with. This is because with rare exception, organisms responsible for food spoilage are not the same as those that cause foodborne illnesses.

Food Spoilage

Food spoilage is any change in the appearance, smell, or taste of a food product that makes it unacceptable to the consumer, whether or not the change is due to microbial growth. Foods are rich in organic matter, and the physical and chemical characteristics of a food determine its susceptibility to microbial activity. With respect to spoilage, a food or food product falls into one of three categories: (1) **Perishable foods** include many fresh food items such as meats and many fruits and vegetables; (2) **semiperishable foods** include foods such as potatoes, some apples, and nuts; and (3) **nonperishable foods** include items such as sugar and flour. The foods in these categories differ primarily with regard to their *moisture content*, as measured by their water activity (a_w, ⟲ Section 5.13). Nonperishable foods have low moisture levels and can generally be stored for long periods without spoilage. Perishable and semiperishable foods, by contrast, typically have higher moisture levels and hence these foods must be stored under conditions that inhibit microbial growth.

Fresh foods are typically spoiled by a wide variety of bacteria and fungi (Table 32.3). The chemical properties of foods vary widely, and each food is characterized by its moisture level and the nutrients it contains as well as other factors, such as its acidity or alkalinity. As a result, each susceptible food is typically spoiled by a specific group of microorganisms. The time required for a microbial population to reach a significant level in a given food product depends on both the size of the initial inoculum and the rate of growth during the exponential phase. Microbial numbers in a food product may initially be so low that no measurable effect can be observed, with only the last few cell doublings leading to observable spoilage. Hence, an unconsumed portion of a food product that is palatable and eaten one day can be badly spoiled the next.

The type of food spoilage and the microbial composition of the spoilage community (Table 32.3) are functions of both the food product and the storage temperature. Food spoilage microorganisms are often *psychrotolerant*, meaning that although they grow best at temperatures above 20°C, they can also grow at refrigeration temperatures (3–5°C) (⟲ Section 5.10). However, at any given storage temperature, some species grow faster than others, and thus the composition of the microbial spoilage community of the same food product stored at different temperatures can vary significantly.

TABLE 32.3 Microbial spoilage of fresh food[a]

Food product	Type of microorganism	Common spoilage organisms, by genus
Fruits and vegetables	Bacteria	*Erwinia, Pseudomonas, Corynebacterium* (mainly vegetable pathogens; rarely spoil fruit)
	Fungi	*Aspergillus, Botrytis, Geotrichum, Rhizopus, Penicillium, Cladosporium, Alternaria, Phytophthora,* various yeasts
Fresh meat, poultry, eggs, and seafood	Bacteria	*Acinetobacter, Aeromonas, Pseudomonas, Micrococcus, Achromobacter, Flavobacterium, Proteus, Salmonella, Escherichia, Campylobacter, Listeria*
	Fungi	*Cladosporium, Mucor, Rhizopus, Penicillium, Geotrichum, Sporotrichum, Candida, Torula, Rhodotorula*
Milk	Bacteria	*Streptococcus, Leuconostoc, Lactococcus, Lactobacillus, Pseudomonas, Proteus*
High-sugar foods	Bacteria	*Clostridium, Bacillus, Flavobacterium*
	Fungi	*Saccharomyces, Torula, Penicillium*

[a]The organisms listed are the most commonly observed spoilage agents of fresh, perishable foods. Many of these genera include species that are human pathogens (Chapters 30–33).

Food Preservation and Fermentation

Food storage and preservation methods are designed to slow or stop the growth of microorganisms that spoil food or that can cause foodborne disease. The major methods of food preservation include altering the temperature, acidity, or moisture level of the food, or treating it with radiation or chemicals that prevent microbial growth.

Refrigeration slows microbial growth, but a remarkable number of microorganisms, particularly bacteria, can grow at refrigeration temperatures. Storage in the household freezer reduces growth considerably, but slow growth still occurs in pockets of liquid water trapped within the frozen food. In general, a lower storage temperature results in less microbial growth and slower spoilage, but storage at temperatures below −20°C is too expensive for routine use and also can negatively affect food appearance, consistency, and taste.

Heat reduces the bacterial load and can even sterilize a food product, and is especially useful for the preservation of liquids or high-moisture foods. The limited heat treatment of **pasteurization** (⟲ Section 5.15) does not sterilize liquids but reduces microbial numbers and eliminates pathogens. *Canning*, by contrast, typically sterilizes the food but requires careful processing in a sealed container at the correct temperature for the correct length of time. If viable microorganisms remain in a can or glass jar, their growth can produce gas, resulting in bulges or even explosions (**Figure 32.7**). The environment inside a can or sealed jar is anoxic, and an important

(a)   *(b)*   *(c)*

T. D. Brock

Figure 32.7 Changes in sealed tin cans as a result of microbial spoilage.
(a) A normal can. The top of the can is pulled in a bit because of the normal slight vacuum inside. *(b)* Swelling due to gas production. *(c)* The can shown in *b* was dropped, and the gas pressure resulted in a violent explosion, tearing the lid apart.

genus of anaerobic bacteria that can grow in canned foods is the endospore-forming *Clostridium*, one species of which causes botulism (Sections 25.6 and 32.9).

Foods can be made drier by either physically removing the water or by adding solutes, such as salt or sugar. Extremely dry or solute-loaded foods help prevent bacterial growth, but spoilage can still occur and when it does, it is typically from fungi. Many foods are preserved by the addition of small amounts of antimicrobial chemicals. These chemicals, which include nitrites, sulfites, propionate, and benzoate along with a few others, find wide application in the food industry for enhancing or preserving food texture, color, freshness, or flavor. Although not widely practiced in many countries, the *irradiation* of food with ionizing radiation is also an effective means for reducing microbial contamination.

Many common foods and beverages are preserved through the metabolic activities of microorganisms; these are *fermented foods* (**Figure 32.8** and **Table 32.4**). The fermentation process (Chapters 3 and 14) yields large amounts of preservative chemicals. The major bacteria important in the fermented foods industry are organic

TABLE 32.4 Fermented foods and fermentation microorganisms

Food category/ Preservative	Primary fermenting microorganisms[a]
Dairy foods/Lactic acid, propionic acid	
Cheeses	*Lactococcus, Lactobacillus, Streptococcus thermophilus, Propionibacterium* (Swiss cheese)
Fermented milk products	
Buttermilk and sour cream	*Lactococcus*
Yogurt	*Lactobacillus, Streptococcus thermophilus*
Alcoholic beverages/Ethanol	*Zymomonas, Saccharomyces*[b]
Yeast breads/Baking	*Saccharomyces cerevisiae*[b]
Meat products/Lactic and other acids	
Dry sausages (pepperoni, salami) and semidry sausages (summer sausage, bologna)	*Pediococcus, Lactobacillus, Micrococcus, Staphylococcus*
Vegetables/Lactic acid	
Cabbage (sauerkraut)	*Leuconostoc, Lactobacillus*
Cucumbers (pickles)[c]	Lactic acid bacteria
Vinegar/Acetic acid	*Acetobacter*
Soy sauce/Lactic acid and many other substances	*Aspergillus,*[d] *Tetragenococcus halophilus,* yeasts

[a]Unless otherwise noted, these are all species of *Firmicutes* except for *Micrococcus*, which is in the *Actinobacteria*, and *Zymomonas* and *Acetobacter*, which are in the *Alphaproteobacteria*.
[b]Yeast. Various *Saccharomyces* species are used in alcohol fermentations. *S. cerevisiae* is the common baker's yeast. To make sourdough bread, species of *Lactobacillus* are used.
[c]Nonfermented pickles are cucumbers marinated in vinegar (5–8% acetic acid).
[d]A mold.

acid–producing bacteria such as the lactic acid bacteria (in fermented milks), the acetic acid bacteria (in pickling), and the propionic acid bacteria (in certain cheeses) (Table 32.4). The yeast *Saccharomyces cerevisiae* produces alcohol as the preservative in the production of alcoholic beverages. The high level of organic acids or alcohol generated from these fermentations prevents the growth of both spoilage organisms and pathogens in the fermented food product.

John M. Martinko and Cheryl Broadie

Figure 32.8 Examples of fermented foods. Bread, sausage meats, cheeses and many other dairy products, and fermented and pickled vegetables are food products that are produced or enhanced by fermentation reactions catalyzed by microorganisms (see also Table 32.4).

— **MINIQUIZ** —

- List the major food groups as categorized by their susceptibility to spoilage.
- Identify physical and chemical methods used for food preservation. How does each method limit growth of microorganisms?
- List some dairy, meat, beverage, and vegetable foods produced by microbial fermentation. What is the preservative in each case?

32.7 Foodborne Diseases and Food Epidemiology

Foodborne illnesses resemble waterborne illnesses in being *common-source* diseases. Most foodborne disease outbreaks are due to

UNIT 7

improper food handling and preparation by domestic consumers; these typically affect only a few people and are rarely reported. However, occasional disease outbreaks due to breakdowns in safe food handling and preparation at restaurants or food-processing and distribution plants can affect large numbers of people in geographically widespread regions.

Foodborne Diseases and Microbial Sampling

Foodborne diseases are of two types, *food infections* and *food poisonings*; some foodborne diseases fall into both categories. **Food poisoning**, also called **food intoxication**, results from ingestion of foods containing preformed microbial toxins. The microorganisms that produced the toxins do not have to grow in the host and may not even be alive at the time the contaminated food is consumed; ingestion and activity of the toxin is what causes the illness. We previously discussed some of these toxins, notably the exotoxin of *Clostridium botulinum* and the superantigen toxins of *Staphylococcus* and *Streptococcus* (⮂ Sections 25.6 and 27.10). In contrast to food poisoning, **food infection** occurs from the ingestion of food containing sufficient numbers of viable pathogens to cause colonization and growth of the pathogen in the host, ultimately resulting in disease.

Food infections are the most common foodborne illnesses in the United States and account for four of the top five leading foodborne illnesses. **Table 32.5** lists the major microorganisms that cause food infections and food poisonings in the United States.

Eight microorganisms account for the great majority of foodborne illness, hospitalizations, and deaths in the United States: *Salmonella* species, *Clostridium perfringens*, *Campylobacter jejuni*, *Staphylococcus aureus*, *Listeria monocytogenes*, and *Escherichia coli* (all bacteria); norovirus; and *Toxoplasma* (a protist) (Table 32.5). Four of these—norovirus, *Salmonella*, *C. perfringens*, and *Campylobacter*—account for nearly 90% of all foodborne illness, with norovirus (Sections 32.5 and 32.14) being the most common culprit (60%).

Rapid diagnostic methods that do not require culturing an organism have been developed to detect important food pathogens, and many of these were described in Chapter 28. Isolation of pathogens from foods usually requires preliminary treatment of the food to suspend microorganisms embedded or entrapped within. A standard method for this purpose employs a blender called a *stomacher* (**Figure 32.9**), a device to process food samples sealed in sterile bags. Paddles in the stomacher crush, blend, and homogenize the samples in a fashion resembling the peristaltic action of the stomach but under conditions that prevent contamination. The homogenized samples are then analyzed for specific pathogens or their products.

In addition to identifying pathogens in the food itself, disease investigators must also recover the foodborne pathogen from the diseased patient in order to establish a cause-and-effect relationship between the pathogen and the illness. In fact, identification of the *same strain* of a particular pathogen in patients and the suspected contaminated food is the "gold standard" for linking cause and effect in a foodborne disease outbreak, and a variety of microbiological, immunological, and molecular techniques are available for these purposes (Chapter 28).

Foodborne Disease Epidemiology

An outbreak of foodborne disease can occur in a home, a school cafeteria, a college dining hall, a restaurant, a military mess hall,

TABLE 32.5 Major foodborne pathogens[a]

Organism	Disease[b]	Foods
Bacteria		
Bacillus cereus	FP and FI	Rice and starchy foods, high-sugar foods, meats, gravies, pudding, dry milk
Campylobacter jejuni	FI (4)[c]	Poultry, dairy
Clostridium botulinum	FP	Improperly heat-processed nonacidic foods such as home-canned vegetables (beans, potatoes, corn, asparagus)[e]
Clostridium perfringens	FP and FI (3)[c]	Meat and vegetables held at improper storage temperature
Escherichia coli O157:H7	FI	Meat, especially ground beef, raw vegetables
Other enteropathogenic *Escherichia coli*	FI	Meat, especially ground meat, raw vegetables
Listeria monocytogenes	FI	Refrigerated "ready to eat" foods
Salmonella spp.	FI (2)[c]	Poultry, meat, dairy, eggs
Staphylococcus aureus	FP (5)[c]	Meat, desserts
Streptococcus spp.	FI	Dairy, meat
Yersinia enterocolitica	FI	Pork, milk
All other bacteria	FP and FI	
Protists[d]		
Cryptosporidium parvum	FI	Raw and undercooked meat
Cyclospora cayetanensis	FI	Fresh produce
Giardia intestinalis	FI	Contaminated or infected meat
Toxoplasma gondii	FI	Raw and undercooked meat
Viruses		
Norovirus	FI (1)[c]	Shellfish, many other foods
Hepatitis A	FI	Shellfish and some other foods eaten raw

[a]Data from the Centers for Disease Control and Prevention, Atlanta, Georgia, USA.
[b]FP, food poisoning; FI, food infection.
[c]The number in parentheses is the rank of the top five foodborne pathogens in the United States.
[d]All of these protists are discussed in Chapter 33.
[e]See page 937.

or anywhere a contaminated food is consumed by many individuals. In addition, central food-processing plants and distribution centers provide opportunities for contaminated foods to cause disease outbreaks far from where the food was originally processed. It is the job of the food epidemiologist to track disease outbreaks and determine their source, often down to the precise location in which the food was contaminated.

A good example of effective foodborne disease tracking is the outbreak caused by *Escherichia coli* O157:H7 (see Section 32.11 and Figure 32.14b) in the United States in 2006. Through culturing and molecular studies, this outbreak was linked to the consumption of contaminated packaged spinach and was quickly traced to a food-processing facility in California. The contaminated spinach was distributed nationwide from the California plant, but most disease cases were in the Midwest. In the summer of 2013, another "packaged" outbreak occurred in the Midwest but in this case was

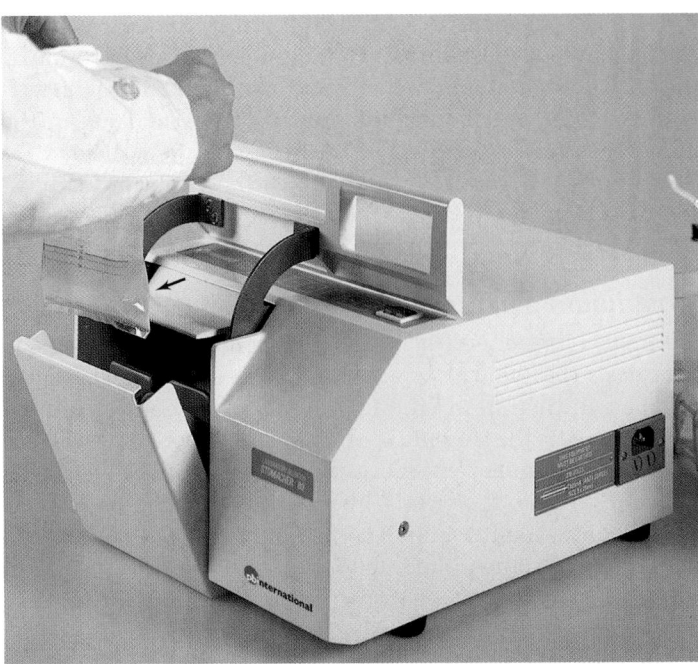

Figure 32.9 A stomacher. Paddles in this specialized blender homogenize the food sample contained in a sealed and sterile bag (arrow). The sample is first suspended in a sterile solution to form a uniform mixture.

linked to lettuce instead of spinach and to the parasite *Cyclospora cayetanensis* (⇌ Section 33.4) instead of to the bacterium *E. coli*.

To be effective, foodborne disease trackers must work quickly. For example, when the first case in the *E. coli* spinach outbreak appeared in late August, a link to the specific spinach product was made less than a month later. Because *E. coli* O157:H7 has been well studied, public health officials were able to quickly identify the strain contaminating the bagged spinach. Authorities then traced this strain back to the processing plant and eventually identified a specific agricultural field near the processing plant as the source of the pathogen. Although it remains unclear how the spinach was contaminated, domestic animal manure was the likely source. During the outbreak, two foodborne disease surveillance networks, *FoodNet* (Centers for Disease Control and Prevention) and *PulseNet* (an international molecular typing network for foodborne diseases), played important roles in exposing and ending the outbreak.

The spinach *E. coli* epidemic, although serious and even deadly for some, was discovered, contained, and stopped very quickly. However, this incident shows how centralized food-processing facilities can quickly spread disease to distant populations. Because of this, food hygiene standards and surveillance must be maintained at the highest possible level at all times in restaurants and central food-processing and distribution facilities.

MINIQUIZ

- Distinguish between food infection and food poisoning.
- Describe microbial sampling procedures for solid foods such as meat.
- Describe how a foodborne disease outbreak is tracked.

IV • Food Poisoning

Food poisoning can be caused by various bacteria and a few fungi. Here we consider the gram-positive bacteria *Staphylococcus aureus, Clostridium botulinum*, and *Clostridium perfringens*, the most common causes of bacterial food poisoning. Two of these microbes—*S. aureus* and *C. perfringens*—are part of the "top five" causes of foodborne illness (Table 32.5).

32.8 Staphylococcal Food Poisoning

A powerful form of food poisoning is caused by enterotoxins (⇌ Section 25.6) produced by the gram-positive bacterium *Staphylococcus aureus* (**Figure 32.10**; ⇌ Section 16.7). This organism is commonly associated with the skin and upper respiratory tract and is a frequent cause of pus-forming wounds (⇌ Section 30.9 and Figure 30.29). *S. aureus* can grow aerobically or anaerobically in many common foods and produces a suite of enterotoxins. When consumed, the toxins cause gastrointestinal symptoms characterized by one or more of nausea, vomiting, diarrhea, and dehydration. The onset of symptoms is rapid, within 1–6 h of ingestion depending on the amount of enterotoxin consumed, but the symptoms usually pass within 48 h.

Staphylococcal Enterotoxins

Many *S. aureus* enterotoxins are heat-stable and all are stable to stomach acidity. Most strains of *S. aureus* produce only one or two of these toxins, and some strains are nonproducers. However, any one of the staph enterotoxins can cause food poisoning. The toxins pass through the stomach to the small intestine and trigger disease symptoms from there. Besides their normal gastrointestinal activities, staph enterotoxins are also *superantigens* and can lead to potentially lethal toxic shock syndrome (⇌ Sections 25.7 and 27.10).

S. aureus enterotoxins are given acronyms beginning with "SE" (for "staphylococcus *e*nterotoxin"): SEA, SEB, SEC, and SED, which are encoded by the genes *sea, seb, sec*, and *sed*. Not all of these genes are on the *S. aureus* chromosome, but their sequences show them to be highly related. The genes *seb* and *sec* are encoded on the bacterial chromosome, *sea* on a lysogenic bacteri-ophage (⇌ Section 8.7), and *sed* on a plasmid. The phage- and plasmid-encoded genes can transfer the ability to make toxin to nontoxigenic strains of *Staphylococcus* by horizontal gene transfer (Chapter 11). SEA is the most common cause of staph food poisoning worldwide.

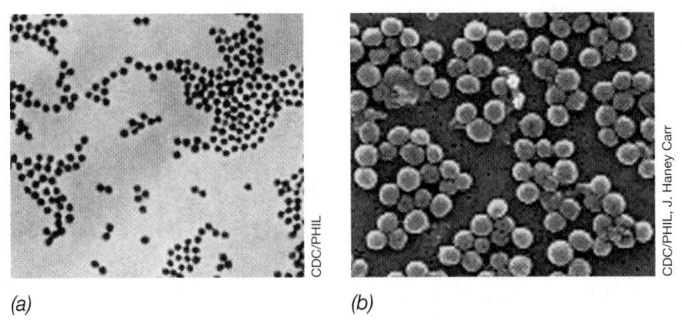

Figure 32.10 *Staphylococcus aureus*. *(a)* Gram-stained light micrograph showing the typical "cluster of grapes" morphology of staphylococci. *(b)* Colorized scanning electron micrograph of cells of *S. aureus*. A single cell is about 0.8 μm in diameter.

Disease Properties, Treatment, and Prevention

Foods may contain cells of *S. aureus* for several reasons. The organism may have been present on the food source itself, for example, on a meat product. But more commonly, cells of *S. aureus* are introduced to the food by contamination from the food preparer or by contamination of the food product with raw meat or a contaminated sauce or dressing. A common scenario for a staph food poisoning incident is when a food preparer introduces *S. aureus* from nasal secretions or from an uncovered skin wound or leaking bandage into the food during its preparation. If the contaminated food is then stored at room temperature or above, the stage is set for the rapid growth of *S. aureus* and the production of staph enterotoxins.

Each year there are an estimated nearly quarter million cases of staphylococcal food poisoning in the United States. The foods most commonly implicated are custard- and cream-filled baked goods, poultry, eggs, raw and processed meat, puddings, and creamy salad dressings. Salads prepared with mayonnaise-based dressings or those that contain shellfish, chicken, pasta, tuna, potato, egg, or meat, are also common vehicles. Salted foods such as ham can be vehicles because of the ability of *S. aureus* to grow quickly in salty environments (Section 30.9). If any of these foods are contaminated with *S. aureus* but are refrigerated immediately after preparation, they usually remain safe because the organism grows poorly at low temperatures. But if enterotoxin has already been produced, mild heating may not make the food safe, as staph enterotoxins are stable to 60°C.

Treatment of staph food poisoning with antibiotics is not useful because any ingested cells of *S. aureus* have already been killed by the acidity in the stomach and antibiotics have no effect on the enterotoxins. Rest, drinking plenty of fluids, and using antinausea drugs are the best prescription for a rapid recovery. As for any foodborne illness, staphylococcal food poisoning can be prevented by proper sanitation and hygiene in food production, preparation, and storage. In this regard, food preparers should practice thorough and frequent hand washing, prevent foods from coming into contact with nasal tissues and secretions, and routinely wear and frequently change disposable gloves when handling food products, especially if they have a bandaged hand wound.

MINIQUIZ

- Identify the symptoms and mechanism of staphylococcal food poisoning.
- Why does antibiotic treatment not affect the outcome or the severity of disease with staph food poisoning?

32.9 Clostridial Food Poisoning

The endospore-forming anaerobic bacteria *Clostridium perfringens* and *Clostridium botulinum* (Section 16.8) cause serious food poisoning. Canning and cooking procedures kill vegetative cells of these species but may not kill all endospores. If this occurs, viable endospores in the food can germinate and the resulting cells produce toxins.

There is a clear distinction in the disease process between perfringens food poisoning and botulism. In the case of botulism, the toxin is a neurotoxin and only the toxin is required for disease. Botulism does not require the growth of *C. botulinum* in the human body but growth may nevertheless occur, particularly in cases of infant botulism. By contrast, with perfringens food poisoning, a large number of cells must be ingested in order for the toxin—in this case, an enterotoxin—to be produced.

Clostridium perfringens Food Poisoning

Clostridium perfringens (**Figure 32.11a**) is commonly found in soil but can also be found in sewage, primarily because it lives in small numbers in the intestinal tract of humans and other animals. *C. perfringens* is the third most often reported cause of foodborne disease in the United States behind norovirus illnesses (Sections 32.5 and 32.14) and *Salmonella* infections (Section 32.10 and Table 32.5). In 2015, about 1 million perfringens cases were estimated to have occurred in the United States.

C. perfringens is a proteolytic bacterium; proteins are catabolized by fermentation (Section 14.21). Perfringens food poisoning requires the ingestion of a large dose ($>10^8$) of *C. perfringens* cells in contaminated cooked or uncooked foods, usually high-protein foods such as meat, poultry, and fish. *C. perfringens* can grow in meat dishes cooked in bulk where heat penetration is often insufficient. *C. perfringens* grows quickly in the food, especially if left to cool at room temperature. It is when sporulation begins that the perfringens enterotoxin is produced. The toxin alters the permeability of the intestinal epithelium, leading to nausea, diarrhea, and intestinal cramps. The onset of perfringens food poisoning typically begins 7–15 h after consumption of the contaminated food and usually resolves within 24 h; for this reason, the disease is sometimes written off as a "stomach flu" or "24-hour flu." Fatalities from perfringens food poisoning are rare, and no specific treatment is necessary other than replacing fluids lost from diarrhea or vomiting (if it occurs).

A diagnosis of perfringens food poisoning is made from isolation of *C. perfringens* from the feces or, more reliably, by an immunoassay that can detect *C. perfringens* enterotoxin in feces. Prevention of perfringens food poisoning requires that cooked foods not be contaminated with raw foods and that all foods be properly heated during cooking and home canning. The perfringens enterotoxin is

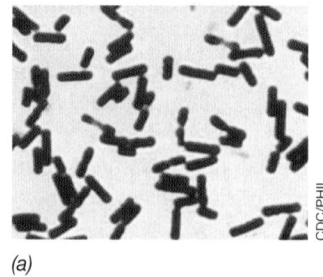
(a)

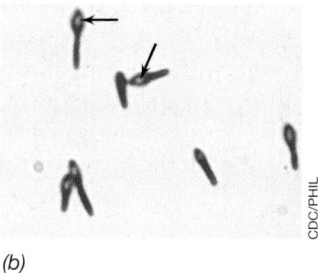
(b)

Figure 32.11 Food poisoning clostridia. *(a)* Gram stain of a growing culture of *Clostridium perfringens*, the bacterium that causes perfringens food poisoning. A cell measures about 1×3 µm. *(b)* Gram stain of a sporulating culture of *Clostridium botulinum*, the agent of botulism. A cell measures about 1×5 µm and endospores (arrows) appear red.

heat-labile and thus any toxin that may have formed in a food product is destroyed by proper heating (75°C). Cooked foods should be refrigerated as soon as possible to rapidly lower temperatures and inhibit the growth of any *C. perfringens* that may have been present.

Botulism

Botulism is a severe and potentially fatal food poisoning caused by the consumption of food containing the exotoxin produced by *C. botulinum* (Figure 32.11*b*). This bacterium normally inhabits soil or water, but its cells or endospores may contaminate raw and processed foods. If viable endospores of *C. botulinum* remain in the food, they may germinate and produce botulinum toxin; ingesting even a small amount of this highly poisonous substance can cause severe illness or death.

Botulinum toxin is a neurotoxin that affects autonomic nerves that control key body functions such as respiration and heartbeat; the typical result is a flaccid paralysis (⮂ Section 25.6 and Figure 25.13). At least seven distinct botulinum toxins are known. Because the toxins are destroyed by heat (80°C for 10 minutes), *thoroughly* cooked food, even if contaminated with toxin, is harmless. Much foodborne botulism is from improperly processed home-canned foods, especially nonacidic foods such as corn, potatoes, and beans. Any viable *C. botulinum* endospores that remain in the sealed (and now anoxic) jar may germinate during storage and produce toxin. Many of these foods are used without cooking when making cold salads, and hence any botulinum toxin present is not destroyed (see page 937 for an example). Prevention of foodborne botulism thus requires careful attention to canning and related food preservation practices.

Although infants can be poisoned by toxin-contaminated food, the majority of infant botulism cases occur from toxin produced following actual *infection* of the infant with *C. botulinum*. This occurs most commonly in newborns up to about 2 months of age because they lack a well-developed intestinal microbiota that can outcompete *C. botulinum*. Ingested *C. botulinum* endospores germinate in the infant's intestine, triggering growth and toxin production. Wound botulism can also occur from infection, presumably from endospores in contaminating material introduced via a parenteral route. Wound botulism is most commonly associated with illicit injectable drug use.

All forms of botulism are rare. In the United States about 110 cases are observed each year with about 45% being infant, 30% wound, and 25% foodborne. Botulism, however, is very serious because of the high mortality associated with untreated disease. Because most cases are diagnosed and treated, less than 5% of all botulism cases result in death. Botulism is diagnosed when either botulinum toxin or *C. botulinum* cells are detected in the patient (or in the contaminated food) coupled with clinical observations of localized paralysis (impaired vision and speech) beginning 18–24 h after ingestion of the contaminated food. Treatment for botulism is by administration of botulinum antitoxin if the diagnosis is early, and mechanical ventilation if signs of respiratory paralysis have already appeared. If the dose of toxin is not too high, infant botulism is usually self-limiting, and most infants recover with only supportive therapy, such as assisted ventilation.

— MINIQUIZ —

- Compare and contrast toxin production and toxemia in botulism and perfringens food poisoning.
- Describe differences in the transmission of botulism in adults versus infants.

V • Food Infection

Recall that food *infection* is not the same thing as food *poisoning* (Section 32.7). Food infection results from ingestion of food containing sufficient numbers of viable pathogens to allow growth of the pathogen and disease in the host. Food infections are very common, and in the United States, the sum total of food infections outnumbers cases of food poisoning by nearly 10-fold. Sections 25.1 and 25.2 reviewed the infection process, summarizing the steps by which microorganisms—friend and foe alike—attach and become established in host tissues.

32.10 · Salmonellosis

Salmonellosis is a gastrointestinal disease typically caused by ingesting food contaminated with *Salmonella* or by handling *Salmonella*-contaminated animals or animal products (**Figure 32.12**). Salmonellosis is the most common bacterial food infection in the United States and second only to norovirus in total number of cases. Symptoms of salmonellosis begin after the pathogen—a gram-negative, facultatively aerobic rod related to *Escherichia coli*

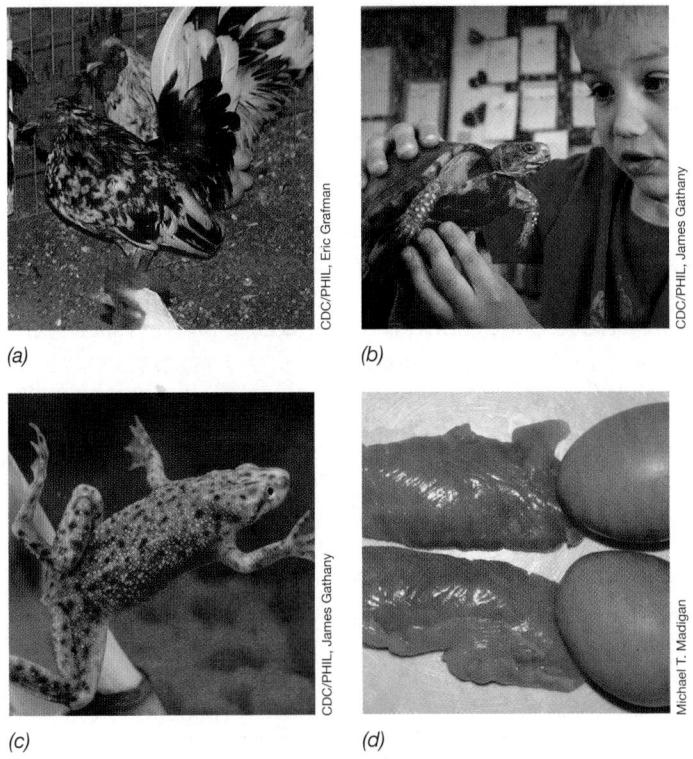

(a) CDC/PHIL, Eric Grafman   *(b)* CDC/PHIL, James Gathany   *(c)* CDC/PHIL, James Gathany   *(d)* Michael T. Madigan

Figure 32.12 Some sources of *Salmonella*. *(a)* Poultry contain *Salmonella* in their intestines and droppings. *Salmonella* can also be transferred to humans from both *(b)* reptiles and *(c)* amphibians. *(d)* Fresh chicken breasts and eggs.

(⟷ Section 16.3 and see Figure 32.13)—colonizes the intestinal epithelium. *Salmonella* species normally inhabit the intestine of warm-blooded and many cold-blooded animals (Figure 32.12) and are common in sewage. Thus, some cases of salmonellosis are waterborne rather than foodborne infections, and this is especially the case for typhoid fever (Section 32.5).

The accepted species epithet for pathogenic *Salmonella* is *enterica*, and there are seven subspecies of *S. enterica*. Most human salmonellas fall into the *S. enterica* subspecies *enterica* group. Each subspecies is also divided into *serovars* (serological variants). Thus, there are *Salmonella enterica* serovar Typhi, or *Salmonella enterica* (*typhi*) for short, and *Salmonella enterica* serovar Typhimurium, and so on. *S. enterica* serovars Typhimurium and Enteritidis are most frequently associated with foodborne salmonellosis.

Pathogenesis and Epidemiology

The most common form of salmonellosis is *enterocolitis*. Ingestion of food containing viable cells of *Salmonella* results in colonization of both the small and large intestines. From here, cells of *Salmonella* invade phagocytic cells and grow intracellularly, spreading to adjacent cells as host cells die. After invasion, pathogenic *Salmonella* deploy several virulence factors including endotoxins, enterotoxins, and cytotoxins that damage and kill host cells (Chapter 25). Symptoms of enterocolitis typically appear 8–48 h after ingestion and include a headache, chills, vomiting, and diarrhea, followed by a fever that can last for several days. The disease normally resolves without intervention in 2–5 days. After recovery, however, patients may shed *Salmonella* in their feces for several weeks and some become healthy carriers. A few serovars of *S. enterica* may also cause septicemia (a blood infection) and enteric or typhoid fever, a potentially fatal disease characterized by systemic infection and high fever lasting several weeks (Section 32.5).

The incidence of salmonellosis in the United States has been steady over the last decade, with about a million estimated cases each year. There are several routes by which *Salmonella* may enter the food supply. The bacteria may reach food through fecal contamination from food handlers. Food production animals such as chickens, pigs, and cattle harbor *Salmonella* serovars that are pathogenic to humans, and these may be carried through to fresh foods such as eggs, meat, and dairy products (Figure 32.12*d*). *Salmonella* food infections are often traced to products such as custards, cream cakes, meringues, pies, and eggnog made with uncooked eggs. Other foods commonly implicated in salmonellosis outbreaks are meats and meat products, especially poultry, cured but uncooked sausages and other meats, milk, and milk products. The simple handling of *Salmonella*-contaminated animals (Figure 32.12*b*) can also lead to salmonellosis.

Diagnosis, Treatment, and Prevention

Foodborne salmonellosis is diagnosed from a combination of clinical symptoms, a history of recent consumption of high-risk foods, and culturing of the organism from feces. Selective, differential media are used to isolate *Salmonella* and discriminate it from other gram-negative enteric bacteria (**Figure 32.13**). Tests for the presence of *Salmonella* are commonly carried out on foods of animal origin such as raw meat, poultry, eggs, and powdered milk. Tests include

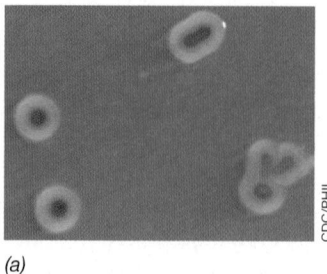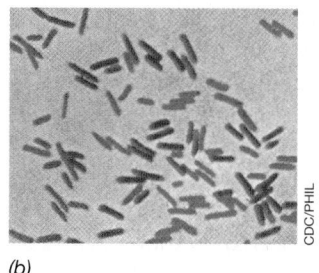

(a)   *(b)*

Figure 32.13 Isolation of *Salmonella*. *(a)* Colonies of *S. enterica* (*typhimurium*) on Hektoen agar, which contains inhibitors of gram-positive bacteria and both lactose and peptone as carbon sources. Thiosulfate in the medium is reduced to H$_2$S by *Salmonella* and complexes with iron to form black FeS. *Salmonella* thus forms white colonies with black FeS centers, a pattern unique among enteric bacteria. The blue color results from the medium turning alkaline because *Salmonella* species do not ferment lactose and instead consume the amino acids in peptone. *(b)* Gram stain of cells of *Salmonella*; an average cell is about 1 × 3 μm.

several rapid tests (Chapter 28), but even rapid tests usually rely on enrichment procedures to increase cell numbers of *Salmonella* to testable levels.

Treatment of enterocolitis is usually unnecessary, and antibiotic treatment does not shorten the course of the disease or eliminate the carrier state. Foods containing *Salmonella* but heated to at least 70°C are generally safe if consumed immediately, held at 50°C or above, or quickly refrigerated. Any foods that become contaminated by an infected food handler can support the growth of *Salmonella* if the food is held for a long enough period, especially if it is not kept very warm or refrigerated.

— MINIQUIZ —

- Describe salmonellosis food infection. How does a food infection differ from food poisoning?
- How might *Salmonella* contamination of food production animals be contained?

32.11 Pathogenic *Escherichia coli*

Most strains of *Escherichia coli* are common microbiota in the human colon and are not pathogenic. However, a few strains are potential foodborne (and occasionally waterborne) pathogens (**Figure 32.14**) and produce potent enterotoxins (⟷ Section 25.6). These pathogenic strains are grouped on the basis of the type of toxin they produce and their specific disease syndromes. We focus here on Shiga toxin–producing *E. coli* and briefly consider some other toxigenic *E. coli* strains.

Although not in the "top five" in terms of foodborne infection pathogens (Table 32.5), pathogenic *E. coli* strains cause disease symptoms so severe that they often require hospitalization. Indeed, infections with pathogenic *E. coli* may cause life-threatening diarrheal disease and urinary tract distress.

Shiga Toxin–Producing *Escherichia coli* (STEC)

Shiga toxin–producing *Escherichia coli* (STEC) strains produce *verotoxin*, an enterotoxin similar to the Shiga toxin produced by

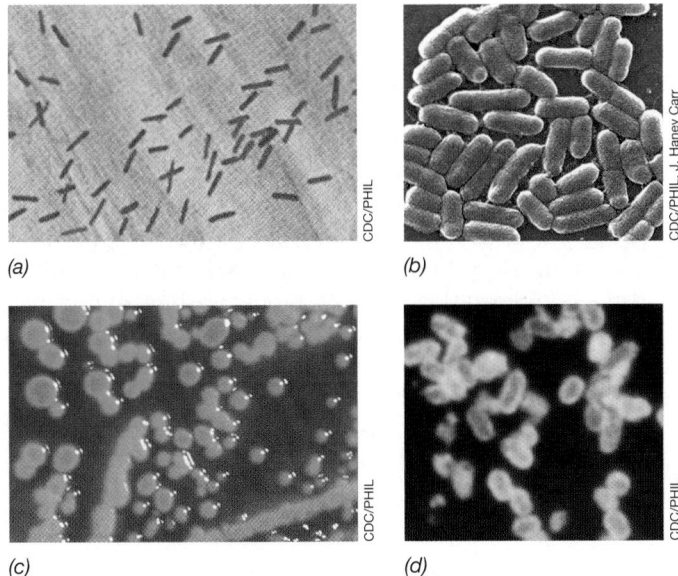

(a)   (b)   CDC/PHIL, J. Haney Carr

(c)   CDC/PHIL   (d)   CDC/PHIL

Figure 32.14 Pathogenic *Escherichia coli*. *(a)* Gram-stained cells showing the typical gram-negative, rod-shaped morphology of *E. coli*. *(b)* Colorized scanning electron micrograph of cells of *E. coli* O157:H7. Cells measure about 1 × 3 µm. *(c) E. coli* can be easily isolated on various selective and differential culture media such as Hektoen agar, where colonies of *E. coli* turn yellow because this bacterium ferments lactose and acidifies the medium (compare these with colonies of *Salmonella* on Hektoen agar in Figure 32.13a). *(d)* Enteropathogenic strains of *E. coli* can be detected, as in this fecal smear, using a specific fluorescent antibody.

Shigella dysenteriae, a close relative of *E. coli*. This toxin inhibits protein synthesis and induces a bloody diarrhea and kidney failure. STEC strains of *E. coli* are also called *enterohemorrhagic E. coli* (EHEC). The most widely distributed STEC is *E. coli* O157:H7 (Figure 32.14*b*). Following ingestion of food or water containing STEC, the bacteria infect the small intestine where they grow and produce verotoxin, which both causes a bloody diarrhea and initiates signs of kidney failure.

Nearly half of STEC infections are caused by the consumption of contaminated uncooked or undercooked meat, particularly mass-processed ground beef. *E. coli* O157:H7 is normally present in the intestines of healthy cattle and enters the human food chain if meat is contaminated with the animal's intestinal contents during slaughter and processing. STEC strains have also been implicated in food infection outbreaks caused by dairy products (especially raw milk products), fresh fruit, and raw vegetables. Contamination of the fresh foods by fecal material, typically from cattle carrying STEC strains, has been implicated in several of these cases (Section 32.7).

Other Pathogenic *Escherichia coli*

Children in developing countries often contract diarrheal disease caused by *E. coli*, and *E. coli* can also be the cause of "traveler's diarrhea," a common infection causing watery diarrhea (as opposed to the bloody diarrhea of STEC strains) in travelers to developing countries. The primary causal agents here are *enterotoxigenic E. coli* (ETEC, Figure 32.14*d*). These strains infect the small intestine and produce one of two heat-labile, diarrhea-producing enterotoxins.

In studies of United States citizens traveling in Mexico, the infection rate with ETEC is often greater than 50%. The prime vehicles are perishable foods such as fresh vegetables (for example, lettuce in salads) and public water supplies. The local population is typically resistant to the ETEC strains because of long-term contact with the organism. Other pathogenic *E. coli* strains include *enteropathogenic E. coli* (EPEC) strains that cause diarrheal diseases in infants and small children but do not cause invasive disease or produce toxins, and *enteroinvasive E. coli* (EIEC) strains, which invade the colon and cause watery and sometimes bloody diarrhea.

Diagnosis, Treatment, and Prevention

The general pattern established for the diagnosis, treatment, and prevention of STEC infection reflects current procedures used for all pathogenic *E. coli* strains. Laboratory diagnosis requires culture from the feces (Figure 32.14*c*) and identification of the O (lipopolysaccharide) and H (flagellar) antigens and toxins by immunological methods (Figure 32.14*d*). Identification and typing can also be done using various molecular analyses.

Treatment of STEC infections includes supportive care for dehydration and monitoring of renal function, blood hemoglobin, and platelets. Antibiotics may actually be harmful because they may trigger the release of large amounts of verotoxin from dying *E. coli* cells that would otherwise be voided intact in feces. For other pathogenic *E. coli* infections, treatment includes supportive therapy and, for severe cases and invasive disease, antimicrobial drugs to shorten and eliminate infection.

The most effective way to prevent infection with pathogenic *E. coli* of any type is to wash raw foods vigorously and make sure that meat, especially ground beef, is cooked thoroughly, which means that it should appear gray or brown with clear juices and have attained a temperature of greater than 70°C. In general, proper food handling, water purification, and appropriate hygiene also prevent the spread of pathogenic *E. coli*. Travelers can avoid diarrhea from pathogenic *E. coli* by drinking water only from properly sealed bottled water and avoiding any uncooked foods.

—— **MINIQUIZ** ——

- How do STEC strains of *Escherichia coli* differ from other pathogenic *E. coli*?

- Why are meats prime vehicles for pathogenic *E. coli*? How can contaminated meat be rendered safe to eat?

32.12 *Campylobacter*

Along with salmonellosis (Section 32.10) and perfringens food poisoning (Section 32.9), *Campylobacter* infections are in the top three most common bacterial foodborne diseases in the United States (Table 32.5). Cells of *Campylobacter* are gram-negative and motile spiral-shaped *Epsilonproteobacteria* (↩ Section 16.5) that grow best at reduced oxygen tension (microaerophilic). Several species of *Campylobacter* are recognized, but *C. jejuni* and *C. fetus*

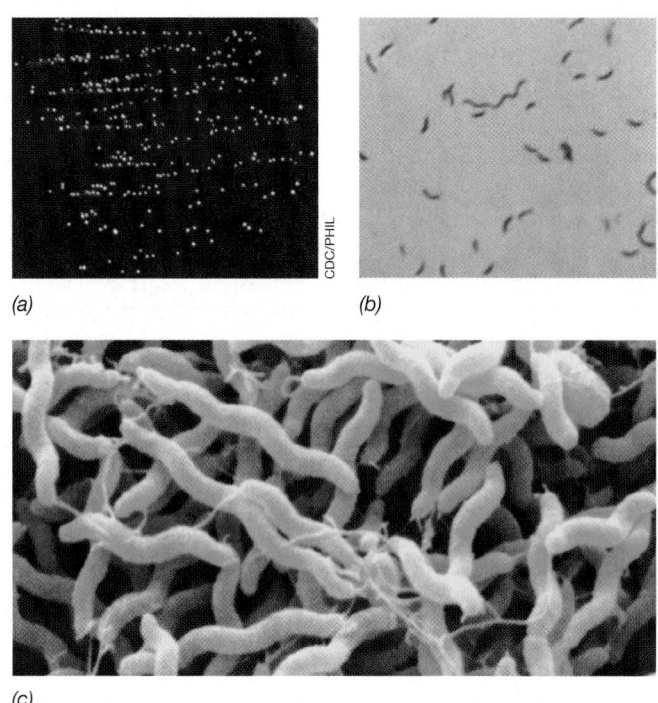

(a)

(b)

(c)

Figure 32.15 *Campylobacter.* *(a)* Colonies of *C. jejuni* grown on *Campylobacter* agar, a selective medium. The medium contains several antibiotics to which *Campylobacter* species are naturally resistant. *(b)* Gram stain and *(c)* scanning electron micrograph of cells of a *Campylobacter* species. Single cells average 0.4 × 2 μm in size.

(**Figure 32.15**) are the most commonly linked to human foodborne illnesses.

Epidemiology and Pathology

Campylobacter is transmitted to humans via contaminated food, most commonly in undercooked poultry or pork, raw shellfish, or occasionally in fecally contaminated water from surface sources. *C. jejuni* is a normal resident of the intestinal tract of poultry, and according to the United States Department of Agriculture, up to 90% of turkey and chicken carcasses are contaminated with *Campylobacter.* Pork can also carry *Campylobacter,* while beef is rarely a vehicle. *Campylobacter* species also infect domestic animals such as dogs, causing a milder form of diarrhea in the animal than that observed in humans. *Campylobacter* infections in infants in particular are often traced to infected domestic animals, especially dogs.

After cells of *Campylobacter* are ingested, the organism multiplies in the small intestine, invades the epithelium, and causes inflammation. Because *C. jejuni* is sensitive to gastric acid, cell numbers as high as 10^4 may be required to initiate infection. However, this number may be reduced to fewer than 500 cells if the pathogen is ingested in food or if the person is taking medication to reduce stomach acid production. *Campylobacter* infection causes a high fever (usually greater than 40°C), headache, malaise, nausea, abdominal cramps, and diarrhea with watery, frequently bloody emissions; symptoms subside in about a week.

Diagnosis, Treatment, and Prevention

Diagnosis of *Campylobacter* food infection requires isolation of the organism from feces and identification by growth-dependent tests, immunological assays, or genomic analyses. Culture media

containing multiple antibiotics to which campylobacters are naturally resistant have been developed for selective isolation of this organism (Figure 32.15a). Various immunological methods are also available for diagnosing a campylobacter infection.

Antibiotic treatment with the drug azithromycin is widely practiced if a confirmed diagnosis is made from culture or culture-independent evidence. In addition, severe cases of dehydration from a *Campylobacter* infection may require intravenous perfusion and hospitalization. Rigorous personal hygiene, especially by those in food preparation facilities, proper washing of uncooked poultry (and any kitchenware coming in contact with uncooked poultry), and thorough cooking of meat are the major means of preventing *Campylobacter* infections.

MINIQUIZ

- Describe the pathology of *Campylobacter* food infection. What are the major vehicles for this pathogen?
- How might *Campylobacter* contamination of food production animals be controlled?

32.13 Listeriosis

Listeria monocytogenes causes **listeriosis**, a gastrointestinal food infection that may lead to bacteremia (bacteria in the blood) and meningitis. *L. monocytogenes* is a gram-positive, nonsporulating coccobacillus (phylum *Firmicutes*) that is acid-, salt-, and cold-tolerant and facultatively aerobic (**Figure 32.16**) (Section 16.7). Although *Listeria* is a minor foodborne pathogen in terms of the number of cases observed per year, infections can be very severe and cause an estimated 20% of all deaths from foodborne illness in the United States. Listeriosis is primarily seen in the elderly, pregnant women, newborns, and adults with weakened immune systems. In 2014, 660 cases of invasive listeriosis (infection beyond the gastrointestinal tract) were reported in the United States with 107 cases being fatal (16%).

Epidemiology

L. monocytogenes is present in soil and water and although it is not common in foods, virtually no food source is safe from possible

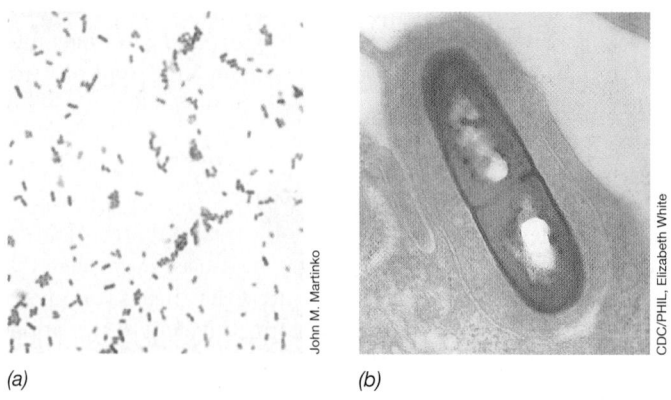

(a)

(b)

Figure 32.16 *Listeria monocytogenes.* *(a)* Gram stain and *(b)* transmission electron micrograph of cells of *L. monocytogenes,* the cause of listeriosis. The *Listeria* cell in *b* is within host tissues (see Figure 32.17).

L. monocytogenes contamination. Food can become contaminated at any stage during production or processing. Ready-to-eat meats, fresh soft cheeses, unpasteurized dairy products, and inadequately pasteurized milk are the major food vehicles for *Listeria*, even when these foods are properly stored at refrigerator temperature (4°C). Food preservation by refrigeration, which ordinarily prevents the growth of other foodborne pathogens, is ineffective in the case of *Listeria* because the organism is psychrotolerant. Cells of *L. monocytogenes* produce a series of branched-chained fatty acids that keep the cytoplasmic membrane functional at cold temperatures (⬡ Section 5.10).

Pathology

Immunity to *L. monocytogenes* is normally conferred by cell-mediated Th1 inflammatory cells (⬡ Section 27.8). However, if cells of *Listeria* evade these immune cells, as they can in hosts with compromised immune systems, the organism is taken up by intestinal phagocytic cells. Although one might think that this is good from the standpoint of host defense, it is actually not because phagocytic uptake initiates the *Listeria* infection cycle.

Listeria cells are taken up by host phagocytic cells into a vacuole called the *phagosome*. This triggers production of a major *Listeria* virulence factor, the exotoxin *listeriolysin O*, and this protein lyses the phagosome and releases *L. monocytogenes* into the cytoplasm (**Figure 32.17**). Here the bacterium multiplies and produces a second major virulence factor, *ActA*, a protein that induces host cell actin polymerization; the actin coats the bacterial cell and assists in moving the pathogen to the host cell cytoplasmic membrane. Once there, the bacterium–actin complex pushes out, forming protrusions called *filopods*, which are then taken up by surrounding phagocytic cells (Figure 32.17). Filopod formation allows cells of *L. monocytogenes* to move about host tissues without exposure to the major weapons of the immune system: antibodies, complement, and neutrophils (Chapters 26 and 27).

Cells of *Listeria* in the intestine cross the intestinal barrier and are carried by the lymph and blood to other organs, in particular the liver, and multiply there as they do in intestinal phagocytes (Figure 32.17). From here cells of *L. monocytogenes* can infect the central nervous system, where they grow in neurons and lead to inflammation of the meninges (the tissues covering the brain and spinal cord), causing meningitis. In addition to listeriolysin O, which also allows *Listeria* to establish chronic infections in many host tissues, other major virulence factors include phospholipases that can destroy host cell membranes, antioxidants that counter phagocytic cell oxidants, and an array of "stress proteins" common in many bacteria (⬡ Sections 6.9 and 6.10).

Diagnosis, Treatment, and Prevention

Listeriosis is diagnosed by culturing *L. monocytogenes* (Figure 32.16) from the blood or cerebrospinal fluid. *L. monocytogenes* can be identified in foods by direct culture or by several molecular methods. The latter methods are also used to subtype clinical isolates in order to track the source(s) of infection. Intravenous antibiotic treatment with penicillin, ampicillin, or trimethoprim plus sulfamethoxazole is used to treat invasive listeriosis.

Prevention measures include recalling contaminated food and taking steps to limit *L. monocytogenes* contamination at the

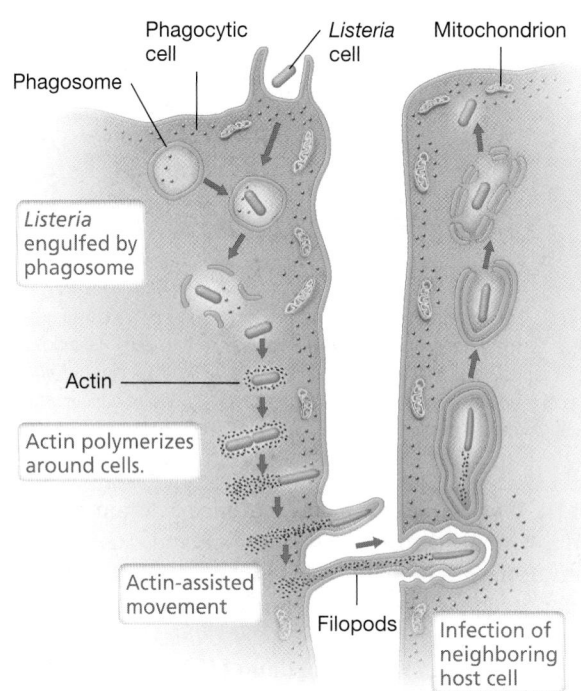

Figure 32.17 Transmission of *Listeria* during listeriosis. Cells of *Listeria* are taken up in phagosomes of phagocytic cells. These are eventually lysed by the virulence factor listeriolysin O to release *Listeria* cells. The bacterial cells then become covered with actin that assists in their movement to the cell periphery. Filopods facilitate transfer of *Listeria* cells to neighboring host cells, where the cycle repeats.

food-processing site. Because *L. monocytogenes* is susceptible to heat and radiation, raw food and food-handling equipment can be readily decontaminated. However, unless the finished food product is pasteurized (⬡ Section 5.15) or cooked, the risk of contamination cannot be eliminated because of the widespread distribution of the pathogen.

MINIQUIZ

- What is the likely outcome of *Listeria monocytogenes* exposure in normal healthy individuals?
- Which populations are most susceptible to serious disease from *L. monocytogenes* infection?

32.14 Other Foodborne Infectious Diseases

Over 200 microorganisms, viruses, and other infectious agents can cause foodborne diseases, and we have thus far summarized the major ones. Here we consider a few other bacterial pathogens that are rather uncommon compared with the "top five" (Table 32.5), and we take a second look at norovirus (previously considered as a waterborne pathogen, Section 32.5) in its more frequent context as a foodborne pathogen and overall number one cause of gastrointestinal illness in the United States.

Bacteria

Besides the major bacterial foodborne pathogens we have already considered, several other bacteria cause human gastrointestinal

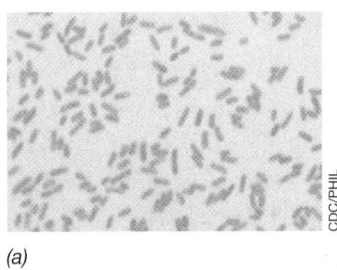
(a)

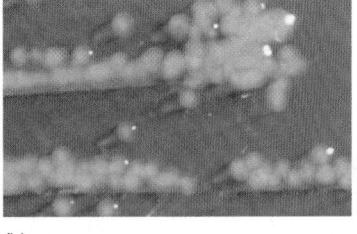
(b)

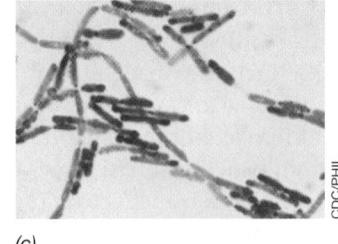
(c)

(d)

Figure 32.18 Less common foodborne bacterial pathogens: *Yersinia enterocolitica* and *Bacillus cereus*. (a) Gram-stained cells of *Y. enterocolitica*. (b) Colonies of *Y. enterocolitica* on Hektoen agar, a selective and differential medium. *Y. enterocolitica* forms white colonies because this bacterium does not ferment lactose and does not produce sulfide (compare with colonies of *Salmonella* on Hektoen agar in Figure 32.13a and colonies of *Escherichia coli* on Hektoen agar in Figure 32.14c). (c) Gram-stained cells of a sporulating culture of *B. cereus*. (d) Large crystalline-like colonies of *B. cereus* formed on blood agar. Foodborne illness due to *Y. enterocolitica* or *B. cereus* is much less common than illness due to *Salmonella*, *Campylobacter*, or *Clostridium perfringens*.

illnesses. *Yersinia enterocolitica* is an enteric bacterium commonly found in the intestines of domestic animals and causes foodborne infections from contaminated meat and dairy products. The most serious consequence of *Y. enterocolitica* infection is *enteric fever*, a severe, life-threatening infection. *Y. enterocolitica* can be isolated on the same selective, differential medium used to isolate *Salmonella* (**Figure 32.18a, b**) but is easily distinguished from this organism on plates (compare Figures 32.13a and 32.18b).

Bacillus cereus is responsible for a relatively small number of food poisoning cases. This endospore-producing bacterium (⊂⊃ Sections 2.10 and 16.8) produces two enterotoxins that cause different symptoms. In the *emetic form*, symptoms are primarily nausea and vomiting. In the *diarrheal form*, diarrhea and gastrointestinal pain are observed. *B. cereus* grows in foods such as rice, pasta, meats, or sauces that are cooked and left at room temperature to cool slowly. When endospores of this bacterium germinate, toxin is produced. Reheating may kill the *B. cereus* cells, but the toxin is heat-stable and may remain active. *B. cereus* is readily culturable and can be tentatively identified by a combination of microscopy and its typically large, grainy, and spreading colonies (Figure 32.18c, d).

The enteric bacterium *Shigella* causes the food infection *shigellosis*, and species of *Vibrio* can also cause food poisoning, primarily from consumption of contaminated shellfish. Most *Shigella* infections are the result of fecal to oral contamination, but food and water are occasional vehicles. We discussed the Shiga-like toxin produced by some pathogenic strains of *Escherichia coli* in Section 32.11.

Viruses

About 70% of annual foodborne infections in the United States are caused by norovirus (**Figure 32.19a**; Section 32.5). The virus is also known as *Norwalk virus* and is a single-stranded plus-sense RNA virus related to poliovirus (⊂⊃ Section 10.8). In general, noroviral foodborne illnesses are characterized by diarrhea, often accompanied by nausea and vomiting. Recovery from norovirus infections is typically spontaneous and rapid, usually within 24–48 h (thus the disease is often nicknamed "the 24-hour bug").

Rotavirus, astrovirus, and hepatitis A make up the bulk of the remaining foodborne *viral* infections. These viruses inhabit the gut and are often transmitted in food or water contaminated with feces. Hepatitis A virus (HAV, Figure 32.19b) is an RNA virus that, like norovirus, is related to poliovirus, but replicates in liver cells.

We considered hepatitis viruses transmitted primarily by blood in Section 30.11, but HAV is mainly a foodborne virus. HAV usually triggers mild, and in many cases subclinical, symptoms, but rare cases of severe liver disease from HAV can occur. The most significant food vehicles for HAV are shellfish, usually oysters or clams harvested from water polluted by human feces and then eaten raw. In recent years, HAV has also been seen in fresh produce served without cooking.

The general trend for incidence of both foodborne and bloodborne hepatitis has moved steadily downward and is now at record low levels, partly due to the availability of effective vaccines against both hepatitis A and hepatitis B (HBV) viruses (⊂⊃ Figure 30.32), but also because of heightened awareness of the potential danger of eating raw shellfish. Nevertheless, widespread and likely mild HAV infections continue to occur because surveys have shown that over 30% of individuals in the United States have circulating antibodies to HAV, indicating past subclinical infections. In 2014, 1239 cases of hepatitis A were reported in the United States.

Protists and Other Agents

Important foodborne protist diseases are listed in Table 32.5. The major pathogens here include *Giardia intestinalis*, *Cryptosporidium parvum*, *Cyclospora cayetanensis*, and *Toxoplasma gondii*. *G. intestinalis* and *C. parvum* are spread in foods when contaminated water is used to wash, irrigate, or spray crops. Fresh foods such as fruits are often implicated as vehicles for these protists. *Toxoplasma gondii* is a protist

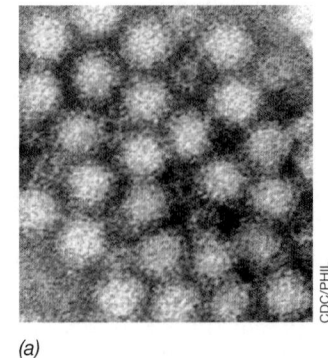
(a)

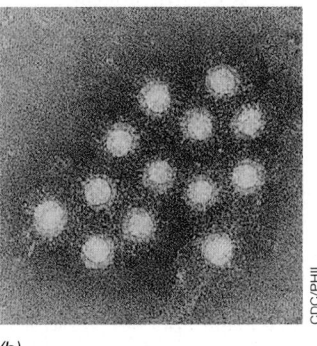
(b)

Figure 32.19 Viruses transmitted in contaminated foods. (a) Transmission electron micrograph of norovirus; an individual virion is about 30 nm in diameter. (b) Transmission electron micrograph of hepatitis A virus; a virion is 27 nm in diameter.

spread primarily through cat feces, but it can also be found in raw or undercooked meat, especially pork. The incidence of foodborne transmission of *C. cayetanensis* has remained low (fewer than 20 cases per year) in recent years, and fresh cilantro and related produce have been the major vehicles of this pathogen in the majority of outbreaks. We discuss the diseases giardiasis, cryptosporidiosis, cyclosporiasis, and toxoplasmosis in Section 33.4.

At least one type of foodborne disease agent is neither cellular nor viral; these are the prions. *Prions* are proteins that adopt novel conformations, inhibiting normal protein function and causing degeneration of host neural tissues (↩ Section 10.16). Human prion diseases are characterized by neurological symptoms including depression, loss of motor coordination, and eventual dementia. A foodborne human prion disease called *variant Creutzfeldt–Jakob*

disease (vCJD) has been linked to consumption of meat products from cattle suffering from *bovine spongiform encephalopathy* (BSE), a disease caused by a prion. Although several thousand cases of vCJD were diagnosed in Great Britain in the mid-1990s, bans on cattle feeds containing rendered cattle parts and bone meal have greatly diminished the incidence of BSE in Europe and have kept the incidence of this disease very low in the United States.

— **MINIQUIZ** —

- In what two forms can *Bacillus cereus* food poisoning manifest itself?
- Compared with all other foodborne or waterborne pathogens, what is unique about prions?

MasteringMicrobiology®   **Visualize**, **explore**, and **think critically** with Interactive Microbiology, MicroLab Tutors, MicroCareers case studies, and more. MasteringMicrobiology offers practice quizzes, helpful animations, and other study tools for lecture and lab to help you master microbiology.

Chapter Review

I • Water as a Disease Vehicle

32.1 Contaminated drinking and recreational waters are sources of waterborne pathogens. In the United States, the number of disease outbreaks due to these sources is relatively small in relation to the large exposure the population has to water. Worldwide, lack of adequate water treatment facilities and access to clean water contribute significantly to the spread of infectious diseases.

  Q What are the two main classes of water? How is water from a surface source, for example, from a lake, made safe to drink?

32.2 Drinking water quality is determined by counting coliform and fecal coliform bacteria using standardized techniques. Filtration and chlorination of water significantly decreases microbial numbers. Water purification methods in developed countries have been a major factor in improving public health, although in developing countries, waterborne illness is still a significant source of infectious disease.

  Q Define the term fecal coliform and explain the coliform test. Why is the coliform test used to assess the purity of drinking water?

II • Waterborne Diseases

32.3 The bacterium *Vibrio cholerae* causes cholera, an acute diarrheal disease associated with severe dehydration. Cholera occurs in pandemics, primarily in developing countries where sewage treatment and sanitation is lacking. Oral rehydration and electrolyte replacement can effectively treat cholera and greatly reduce disease mortality.

  Q Why are antibiotics ineffective for the treatment of cholera? What methods are useful for treating cholera victims?

32.4 *Legionella pneumophila* is a respiratory pathogen that causes Pontiac fever and legionellosis, a more serious infection that may result in pneumonia. *L. pneumophila* grows to high numbers in warm waters and is spread via cooling tower aerosols and in domestic water distribution systems where the bacterium develops in biofilms.

  Q What are the major reservoirs for the pathogen that causes legionellosis? What aspects of pathogenesis distinguish this disease from other waterborne diseases?

32.5 Typhoid fever, caused by a *Salmonella* species, and norovirus illness are important waterborne diseases. Typhoid is common in developing countries while norovirus illness is seen worldwide. Both of these diseases can be controlled by good sanitation practices and effective water treatment.

  Q Contrast the diseases typhoid and salmonellosis (Section 32.10). How are they similar and how do they differ? Which is the more serious disease?

III • Food as a Disease Vehicle

32.6 The potential for microbial food spoilage depends on the nutrients and moisture levels of the food. Growth of microorganisms in perishable foods can be controlled by refrigeration, freezing, canning, pickling, dehydration, chemicals, and irradiation. Microbial fermentations can be used to naturally preserve many foods, including dairy products, meats, fruits and vegetables, and alcoholic beverages.

Q Identify and define the three major categories of food perishability. Why is milk more perishable than sugar even though both are rich in organic matter? Identify the major methods used to preserve food and the major categories of fermented foods.

32.7 Food poisoning results from the activities of microbial toxins while food infections are due to the growth of the pathogen within the body. Identification of common characteristics of foodborne pathogens from seemingly isolated foodborne outbreaks can pinpoint the origin of foodborne contamination and track the spread of the disease. The top five foodborne pathogens in the United States in decreasing order of their appearance are: norovirus, *Salmonella* spp., *Clostridium perfringens*, *Campylobacter jejuni*, and *Staphylococcus aureus*.

Q Distinguish between food infection and food poisoning and give two examples of each.

IV • Food Poisoning

32.8 Staphylococcal food poisoning results from the ingestion of a preformed staphylococcal enterotoxin, a superantigen produced by cells of *Staphylococcus aureus* as they grow in food. Proper food preparation, handling, and storage can prevent staphylococcal food poisoning.

Q What causes the symptoms of staphylococcal food poisoning? Why are cases of staph food poisoning often linked to a food preparer?

32.9 *Clostridium* food poisoning results from ingestion of toxins produced by microbial growth in foods or from microbial growth followed by toxin production in the body. Perfringens food poisoning is quite common and is usually a self-limiting gastrointestinal disease. Botulism is a rare but serious disease, with significant mortality.

Q Identify the two major types of clostridial food poisoning. Which is most prevalent? Which is most dangerous and why?

V • Food Infection

32.10 More than a million cases of salmonellosis occur every year in the United States. Infection results from ingestion of cells of *Salmonella* introduced into food primarily from animal-derived food products or food handlers.

Q What are the possible sources of *Salmonella* spp. that cause food infections?

32.11 Toxigenic *Escherichia coli* cause many food infections, and of these, STEC strains are the most severe. Contamination of foods from animal feces spreads these pathogenic strains of *E. coli*, but good hygiene practices and specific antibacterial measures such as irradiation or thorough cooking of ground beef, a major vehicle, can control disease outbreaks.

Q How does *Escherichia coli* O157:H7 end up in ground beef? To what class of pathogenic *E. coli* does this strain belong? How does this class differ from other classes?

32.12 *Campylobacter* infection is the third most prevalent foodborne bacterial disease in the United States. Poultry is a major vehicle for *Campylobacter* illness, whereas beef and pork are not. Proper poultry preparation and cooking can prevent *Campylobacter* illness.

Q Name a food product that could transmit both *Salmonella* and *Campylobacter* simultaneously. How could this food product be rendered safe to eat?

32.13 *Listeria monocytogenes* is a ubiquitous bacterium, and in healthy individuals, it seldom causes infection. However, in immunocompromised individuals, *Listeria* can cause serious disease as it grows as an intracellular pathogen and invades the central nervous system. Listeriosis is uncommon but shows high mortality.

Q Identify the food sources of *Listeria monocytogenes* infections. How does *Listeria* evade the immune system?

32.14 Viruses, especially norovirus, cause the most foodborne illness while the bacteria *Bacillus cereus*, *Shigella* species, and *Yersinia enterocolitica* are only occasionally linked to foodborne disease outbreaks. Hepatitis A virus is also a serious foodborne pathogen. Some protists and prions also cause foodborne illness but are far less common foodborne pathogens than are bacteria and viruses.

Q What agent is the number one cause of gastrointestinal illness? What is the causative agent of vCJD? How does the structure of this agent differ from that of the agent of noro foodborne illness?

Application Questions

1. As a visitor to a country in which cholera is an endemic disease, what specific steps would you take to reduce your risk of cholera exposure? Will these precautions also prevent you from contracting other waterborne diseases? If so, which ones? Identify waterborne diseases for which your precautions may not prevent infection.

2. Argue a case for why perfringens foodborne illness can be considered both a food poisoning and a food infection.

3. Improperly prepared or handled potato salads are often the source of both staphylococcal food poisoning and salmonellosis. List some reasons why this might be the case. How do these differ from the foodborne disease incident linked to potato salad described on page 937?

Chapter Glossary

Botulism food poisoning due to ingestion of food containing botulinum toxin produced by *Clostridium botulinum*

Coliforms gram-negative, nonsporulating, facultatively aerobic rods that ferment lactose with gas formation within 48 hours at 35°C

Food infection a microbial infection resulting from the ingestion of pathogen-contaminated food followed by growth of the pathogen in the host

Food poisoning (food intoxication) a disease caused by the ingestion of food that contains preformed microbial toxins

Food spoilage a change in the appearance, smell, or taste of a food that makes it unacceptable to the consumer

Listeriosis a gastrointestinal food infection caused by *Listeria monocytogenes* that may lead to bacteremia and meningitis

Nonperishable foods foods of low water activity that have an extended shelf life and are resistant to spoilage by microorganisms

Pasteurization the use of controlled heat to reduce the microbial load, including both pathogens and spoilage organisms, in heat-sensitive liquids

Perishable foods fresh foods generally of high water activity that have a very short shelf life because of spoilage by microbial growth

Potable in water purification, drinkable; safe for human consumption

Salmonellosis enterocolitis or other gastrointestinal disease caused by any of several subspecies of the bacterium *Salmonella*

Semiperishable foods foods of intermediate water activity that have a limited shelf life because of their potential for spoilage by growth of microorganisms

33

Eukaryotic Pathogens: Fungi, Protozoa, and Helminths

microbiologynow

Environmental Change and Parasitic Diseases in the Amazon

Changes in the environment—from both human and natural causes—can have significant effects on an ecosystem. Besides the obvious changes associated with major shifts in land use, other, more subtle changes can occur, and these can be deadly if they include infectious diseases.

The Amazon is the largest drainage basin on Earth and contains a fifth of all the freshwater and a third of the tropical forests that remain on the planet. The Amazon stretches over nine countries, with the largest part located in Brazil. These are hot and moist areas of the world where major health threats exist from the many parasitic microbes that plague the region. Deforestation is a major activity in the Amazon, as agriculture, mining, ranching, road-building, logging, and oil exploration become ever more common in this ecologically sensitive area (see photo). However, lost among the environmental concern for the shrinking of the Amazon rainforests (nicknamed "the lungs of the world" for their capacity to fix CO_2 and expel O_2) is the influence of deforestation on the burden of human parasitic disease.

A recent review predicts that the incidence of malaria, leishmaniasis, soil-transmitted microscopic worm infections, Chagas disease, and schistosomiasis will increase significantly as deforestation continues in the Amazon. There are at least three reasons for this. First, deforestation leaves in its wake disturbed land loaded with water catchments that provide new habitats for the major parasitic disease vectors—mosquitoes and flies—to reproduce. Second, deforestation greatly reduces both soil and water quality. This generates more exposed soils (making transmission of soil-associated helminthic diseases more likely) and contaminated water (increasing opportunities for schistosomiasis and other waterborne infectious diseases). And third, land clearing inevitably affects demographics; land that was previously heavily forested and sparsely populated quickly transitions to densely settled communities when the trees are gone. Collectively, when numbers of both people and disease vectors increase, increases in disease incidence naturally follow.

Deforestation reaps short-term benefits to miners, ranchers, and oil explorers who descend on recently logged land. But in the long run, such major land use changes are not only ecological disasters; they set the stage for increased transmission of some of the most devastating and chronic infectious diseases known.

 Source: Confalonieri, U.E.C., C. Margonari, and A.F. Quintão. 2014. Environmental change and the dynamics of parasitic diseases in the Amazon. *Acta Tropica 129:* 33–41.

In this chapter we focus on *eukaryotic* pathogenic microorganisms. These include several fungi—both molds and yeasts—and various parasitic protists. Some small worms also cause infectious diseases and we consider the most significant of these in the final section.

A common problem in treating diseases caused by eukaryotic pathogens is the fact that their hosts are also eukaryotic. This thwarts many therapeutic strategies and often makes these diseases highly refractory and long-term chronic infections. This is especially true of systemic fungal pathogens.

I • Fungal Infections

Fungi cause a variety of human diseases. Some are mild and self-limiting, whereas others can be firmly entrenched systemic diseases. We begin by considering some of the major fungal pathogens followed by a description of some major fungal diseases, the mycoses.

33.1 Pathogenic Fungi and Classes of Infection

The fungi include the *yeasts*, which normally grow as single cells, and *molds*, which form branching filaments called *hyphae* with or without septa (cross-walls); hyphae eventually intertwine to form visible masses called *mycelia*. The diversity of the molds and yeasts was discussed in Chapter 18.

Common Fungal Pathogens

Fortunately, most fungi are harmless to humans. Most fungi grow in nature as saprophytes on dead organic material; in so doing, fungi are important catalysts in the carbon cycle, especially in oxic environments in soil. Fungi are also important in medicine both as agents of disease and in chemotherapy (antibiotic production). Only about 50 species of fungi cause human diseases, and in healthy individuals, the incidence of serious fungal infections is low, although certain superficial fungal infections (for example, athlete's foot) are fairly common. In those with compromised immune systems, however, fungal infections can be systemic, reaching even the deepest of internal tissues. Such infections can cause serious health problems and be life-threatening.

Common fungal pathogens include both yeasts and molds (**Figure 33.1**). However, many pathogenic fungi are *dimorphic*, meaning that they can exist as *either* a yeast *or* in filamentous form. In *Histoplasma*, for example, cells in laboratory culture form hyphae and mycelia and thus exist in the mold form (Figure 33.1*e*). By contrast, when *Histoplasma* causes histoplasmosis, cells grow in the host in the yeast form (see Figure 33.5*a*). In the mold form, spores are produced, either asexual spores—*conidia*—or sexual spores (↩ Section 18.9). When filamentous fungi are cultured from an infection, the morphology of these spore-bearing structures is observed and is often a major clue in reaching a diagnosis. In addition to microscopy, a variety of clinically useful molecular and immunological tools (including fluorescent antibodies, Figure 33.1*c*) are also available to diagnose fungal

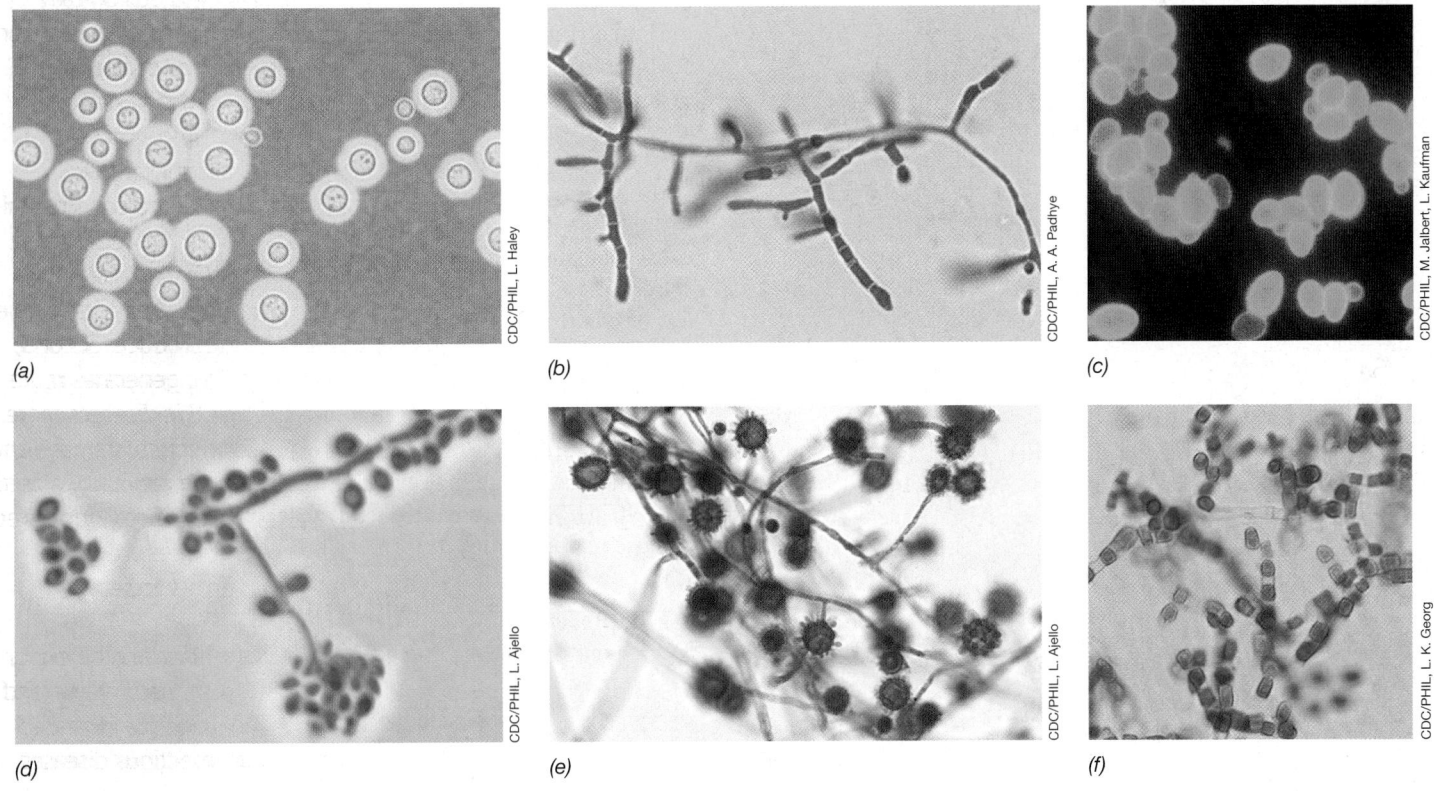

(a) CDC/PHIL, L. Haley
(b) CDC/PHIL, A. A. Padhye
(c) CDC/PHIL, M. Jalbert, L. Kaufman
(d) CDC/PHIL, L. Ajello
(e) CDC/PHIL, L. Ajello
(f) CDC/PHIL, L. K. Georg

Figure 33.1 Pathogenic fungi. These organisms range from about 4 to 20 μm in diameter. *(a) Cryptococcus neoformans* yeast cells stained to reveal the capsule. *(b) Trichophyton* spp. mycelia and conidia. *(c) Candida albicans* yeast form stained with a fluorescent antibody. *(d) Sporothrix schenckii* mycelia and conidia. *(e) Histoplasma capsulatum* mycelia and large conidia. *(f) Coccidioides immitis* conidia. See fungal disease symptoms in Figure 33.5.

infections. **Table 33.1** lists some major fungal pathogens and the types of infections they cause.

Fungal Disease Classes and Treatment

Fungi cause disease through three major mechanisms: inappropriate immune responses, toxin production, and mycoses. Some fungi trigger immune responses that result in allergic (hypersensitivity) reactions following exposure to specific fungal antigens. Reexposure to the same fungi, whether growing on the host or in the environment, may cause allergic symptoms. For example, *Aspergillus* spp. (**Figure 33.2a**), common saprophytes often found in nature as a leaf mold, produce potent allergens, triggering asthma attacks or other hypersensitivity reactions in susceptible individuals.

Fungal disease may occur from the production of *mycotoxins*, a large and diverse group of fungal exotoxins (⊂⊃ Section 25.6). The best-known examples of mycotoxins are the *aflatoxins* (Figure 33.2b) produced by *Aspergillus flavus*, a species that commonly grows on improperly stored dry foods, such as grain. Aflatoxins are highly toxic and are also carcinogenic, inducing tumors in some animals, especially in birds that feed on contaminated grain. Although aflatoxins are known to cause human liver damage including cirrhosis and even liver cancer, adults are not seriously affected by low-level aflatoxin exposure. However, chronic exposure in children can cause serious liver disease and other health effects.

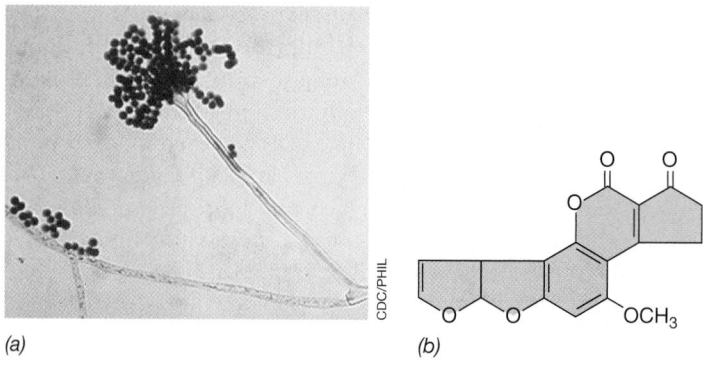

(a)   *(b)*

Figure 33.2 *Aspergillus* **and aflatoxin.** *(a)* Mycelia and conidia of an *Aspergillus* species. *(b)* Structure of aflatoxin B1. This toxin is one of a group of related compounds produced by *Aspergillus flavus*.

The final fungal disease-producing mechanism is through actual host infection. The growth of a fungus on or in the body is called a **mycosis** (plural, mycoses). Mycoses are fungal infections that range in severity from superficial to life-threatening. Mycoses fall into three classes (Table 33.1). **Superficial mycoses** are those in which the fungus infects only the surface layers of skin, hair, or nails (see Figure 33.3). **Subcutaneous mycoses** are infections of deeper layers of skin (see Figure 33.4) and are typically caused by different fungi than superficial infections (Table 33.1). The **systemic mycoses** are the most serious category of fungal infections. These are characterized by fungal growth in internal organs of the body (see Figure 33.5) and can be either primary or secondary infections. A *primary* infection occurs when an otherwise normal, healthy individual is infected with the fungal pathogen; these are rather uncommon. By contrast, a *secondary* infection occurs in a host that harbors a predisposing condition, such as antibiotic therapy or immunosuppression, that makes the individual more susceptible to infection.

Superficial and subcutaneous mycoses are for the most part easily treatable with topical drugs, including tolnaftate (applied topically), various azole drugs (applied either topically or orally), and griseofulvin, a relatively nontoxic drug that can be taken orally but passes through the bloodstream to the skin where it inhibits fungal growth. Chemotherapy against systemic fungal infections is more difficult because of issues with host toxicity (⊂⊃ Section 28.11). For example, one of the most effective antifungal agents, amphotericin B, is widely used to treat systemic fungal infections but can also reduce kidney function and have other unwanted side effects. Hence, effective treatment of the most serious of the mycoses is sometimes very difficult.

TABLE 33.1 **Major pathogenic fungal diseases**[a]

Class and disease	Causal organism	Site
Superficial mycoses		
Athlete's foot	*Epidermophyton, Trichophyton*	Between toes, skin
Jock itch	*Trichophyton, Epidermophyton*	Genital region
Ringworm	*Microsporum, Trichophyton*	Scalp, face
Subcutaneous mycoses		
Sporotrichosis	*Sporothrix schenckii*	Arms, hands
Chromoblastomycosis	*Phialophora verrucosa,* other fungi	Legs, feet, hands
Systemic mycoses		
Aspergillosis	*Aspergillus* spp.[b]	Lungs
Blastomycosis	*Blastomyces dermatitidis*	Lungs, skin
Candidiasis	*Candida albicans*[c]	Oral cavity, intestinal tract, vagina
Coccidioidomycosis	*Coccidioides immitis*[c]	Lungs
Paracoccidioidomycosis	*Paracoccidioides brasiliensis*	Skin
Cryptococcosis	*Cryptococcus neoformans*[c]	Lungs, meninges
Histoplasmosis	*Histoplasma capsulatum*[c]	Lungs
Pneumocystis pneumonia	*Pneumocystis jirovecii*[c]	Lungs

[a]Symptoms of many of these diseases are shown in Figures 33.3–33.5.
[b]*Aspergillus* can also cause allergies, toxemia, and limited infections.
[c]An opportunistic pathogen frequently implicated in the pathogenesis of HIV/AIDS.

— **MINIQUIZ** —

- Differentiate between superficial, subcutaneous, and systemic mycoses.
- What is a dimorphic fungus?
- Distinguish between a primary and a secondary fungal disease. Why do those suffering from HIV/AIDS often show secondary fungal infections of major internal organs?

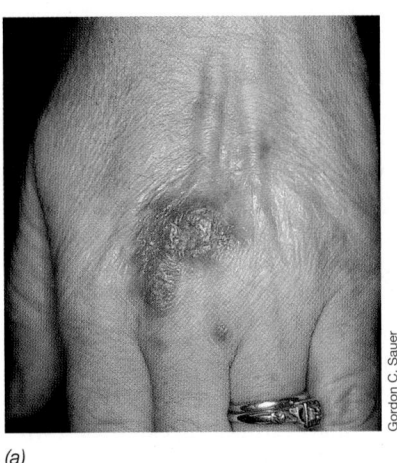

(a)   *(b)*   *(c)*

Figure 33.3 Superficial mycoses caused by *Trichophyton* spp. *(a)* Athlete's foot. *(b)* Ringworm on a child's face and *(c)* on an adult index finger. "Jock itch" (ringworm of the groin) is another common *Trichophyton* infection and can occur in females as well as males. *Tricophyton* is a filamentous fungus (see Figure 33.1*b*).

33.2 Fungal Diseases: Mycoses

The two extremes of fungal infection are the superficial mycoses and the systemic mycoses. *Superficial mycoses* are quite common, and most individuals experience at least one in their lifetime. By contrast, *systemic mycoses* are far less common, and primarily affect the elderly or otherwise immune compromised. As people age, cell-mediated immunity slowly declines as a result of surgeries, transplantations, immunosuppressive drug treatments for rheumatism and autoimmune diseases, and the onset of other conditions, such as pulmonary decline, diabetes, and cancer. Any of these can predispose the elderly to disease. Systemic mycoses also target those of any age whose immune systems have been impaired or destroyed, for example, by HIV/AIDS (⊘ Figure 30.45). Systemic mycoses are thus diseases of *opportunistic pathogens*, microbes that cause disease only in those whose immune defenses can no longer fight them off (⊘ Sections 25.4 and 30.15).

Superficial Mycoses

Table 33.1 listed some of the fungi that cause superficial mycoses; collectively, these pathogens are called *dermatophytes*. In general, superficial mycoses can be bothersome and often recurrent infections, but are not serious health concerns. Fungi such as *Trichophyton* (Figure 33.1*b*) cause infections of the feet (athlete's foot) and other moist skin surfaces, and are quite common (**Figure 33.3***a*). These infections cause flaking and itchy skin and are easily transmitted by cells or spores of the pathogen present in contaminated shower stalls, gymnasium and locker room floors, contaminated shared articles such as towels or bed linens, or from close person-to-person contact. Superficial mycoses can be treated with topical antifungal creams or liquid aerosols, although prophylactic application on a long-term basis may be necessary if constant exposure to the pathogen (for example, to *Trichophyton* on a locker room floor) is unavoidable.

Related surface mycoses include "jock itch," an itchy infection of the groin, skin folds, or anus, and *ringworm* (Table 33.1). Despite the name, ringworm is a fungal infection, typically localized to the scalp or the extremities; the infection causes hair loss and inflammation-like reactions (Figure 33.3*b, c*). These more severe superficial mycoses are usually treated topically with either miconazole nitrate or griseofulvin.

Subcutaneous Mycoses

Subcutaneous mycoses are fungal infections of deeper layers of skin than those of the superficial mycoses (Table 33.1). One disease in this class is *sporotrichosis* (**Figure 33.4***a*), an occupational hazard of agricultural workers, miners, gardeners, and others who come into close and continual contact with the soil. The causal organism, *Sporothrix schenckii* (Figure 33.1*d*), is a ubiquitous soil saprophyte whose spores can enter through a cut or abrasion and infect subcutaneous tissues (Figure 33.4*a*). *Chromoblastomycosis* is due to pathogenic fungal growth in both surface (cutaneous) *and* subcutaneous skin layers, forming crusty, wartlike lesions on the hand (Figure 33.4*b*) or leg. The disease is primarily one of tropical and subtropical countries and occurs when the fungus becomes

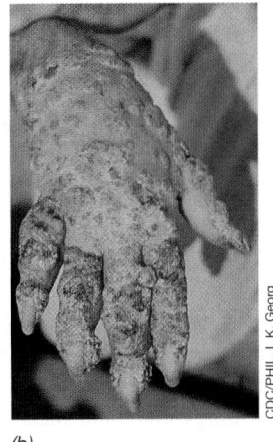

(a)   *(b)*

Figure 33.4 Subcutaneous mycoses. *(a)* Sporotrichosis, a subcutaneous infection due to *Sporothrix schenckii*. *(b)* Chromoblastomycosis on the hand caused by the fungus *Phialophora verrucosa*. Chromoblastomycosis can also be caused by species of the fungal genera *Fonsecaea* and *Cladosporium*.

UNIT 7

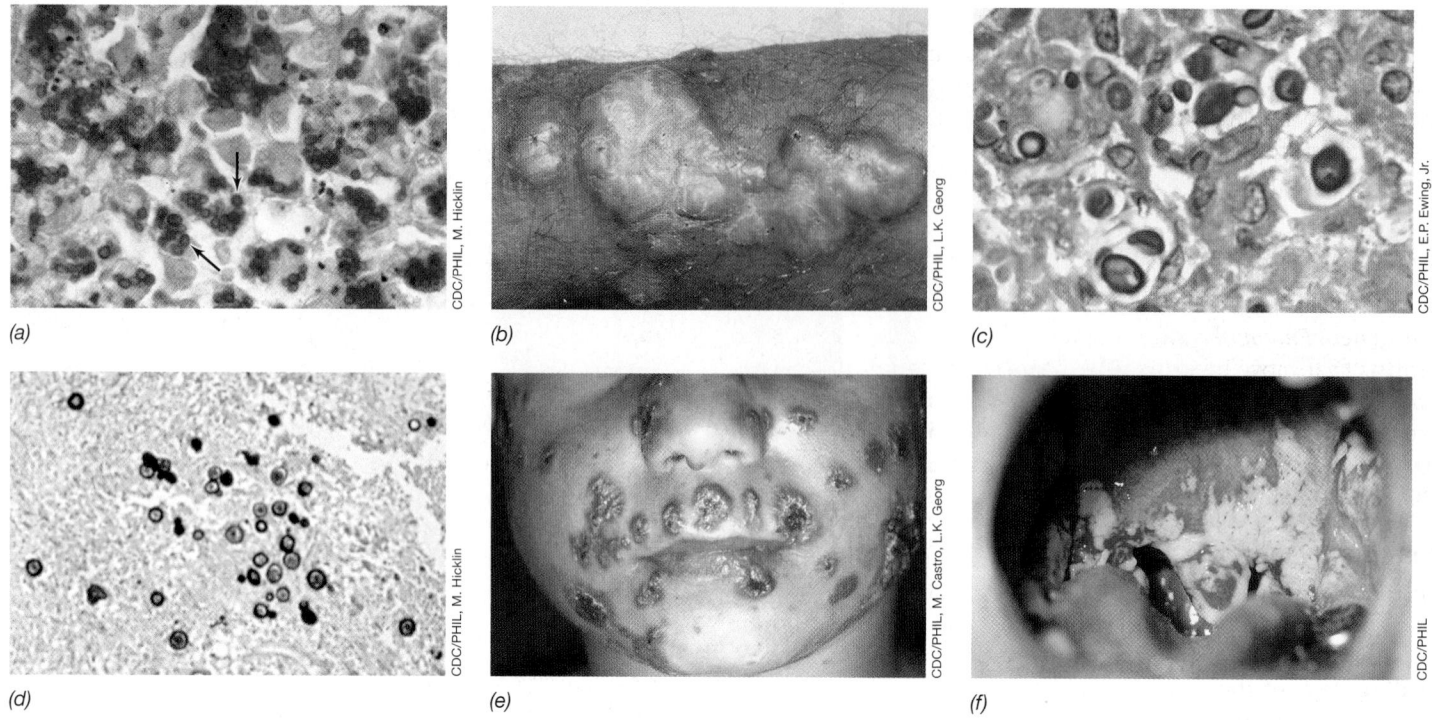

(a)   *(b)*   *(c)*

(d)   *(e)*   *(f)*

Figure 33.5 Systemic mycoses. *(a)* Histoplasmosis; yeast-form cells of *Histoplasma* (arrows) in spleen tissue. *(b)* Cutaneous blastomycosis on the arm. *(c)* Cryptococcosis; yeast-form cells (stained red) in lung tissue. *(d)* Coccidioidomycosis; yeast-form cells (stained blue-black) in lung tissue. *(e)* Paracoccidioidomycosis lesions on the face. *(f)* Oral thrush. Masses of *Candida albicans* cells (yellow) line the back of the throat. See photomicrographs of cultures of the pathogens causing most of these mycoses in Figure 33.1.

implanted under the skin from a puncture wound. Both sporotrichosis and chromoblastomycosis can be treated with oral administration of azoles.

Systemic Mycoses

Systemic fungal pathogens normally live in soil, and humans become infected by inhaling airborne spores that later germinate and grow in the lungs. From there the organism migrates throughout the body, causing deep-seated infections in the lungs and other organs and in the skin. In the United States, the three major systemic mycoses are, in order of decreasing incidence: histoplasmosis, coccidioidomycosis, and blastomycosis. Mortality from these is high, about 10%.

Histoplasmosis (**Figure 33.5a**) is caused by *Histoplasma capsulatum* (Figure 33.1*e*), and *coccidioidomycosis* (San Joaquin Valley fever, Figure 33.5*d*) is caused by *Coccidioides immitis* (Figure 33.1*f*). Histoplasmosis is primarily a disease of rural areas in midwestern states of the United States, especially in the Ohio and Mississippi River valleys, whereas coccidioidomycosis is generally restricted to the desert regions of the southwestern United States. In more tropical climates *blastomycosis,* caused by *Blastomyces dermatitidis,* is prevalent (Figure 33.5*b*). *Paracoccidioidomycosis,* caused by the fungus *Paracoccidioides brasiliensis,* is primarily a subtropical disease with lesions forming on the face (Figure 33.5*e*) or other extremities.

Cryptococcosis (Figure 33.5*c*), caused by the dimorphic yeast *Cryptococcus neoformans* (Figure 33.1*a*), can occur in virtually any organ of the body and is the major mycosis seen in HIV/AIDS patients. The dimorphic yeast *Candida albicans* (Figure 33.1*c*) is often present as a minor component of the human microbiota.

However, this fungus can cause a variety of diseases including mild vaginal infections, more serious oral infections such as thrush (Figure 33.5*f*), and systemic infection of virtually any organ in those with HIV/AIDS. Like *Histoplasma* and *Coccidioides, Candida* and *Cryptococcus* are primarily opportunistic pathogens and rarely cause life-threatening infections except in immunecompromised people.

Our discussion transitions now from fungi to pathogenic parasites. Like fungi, parasites are eukaryotic microorganisms, but the pathogenic parasites typically attack quite different body tissues and organs than do the pathogenic fungi.

MINIQUIZ

- Give an example of a superficial, a subcutaneous, and a systemic mycosis.
- Why are systemic fungal pathogens called opportunistic?

II • Visceral Parasitic Infections

Parasitism is a symbiotic relationship between two organisms, the parasite and the host (Chapters 23 and 24). The parasite derives essential nutrients from the host and may have little or no harmful effect on the host. However, in most cases, the parasite causes disease in the host. Many different phylogenetic groups of protists (Chapter 18) cause parasitic human diseases and we examine some of the key ones here.

Parasitic infections can be either visceral—inducing vomiting, diarrhea, and other intestinal symptoms—or infections of blood and internal tissues. Some of the major diseases of human history, malaria for example, are parasitic diseases. We begin here with the visceral parasites and then consider blood and tissue parasites. Table 33.2 summarizes some major parasitic human diseases.

33.3 Amoebae and Ciliates: *Entamoeba, Naegleria*, and *Balantidium*

The genera *Entamoeba* and *Naegleria* belong to a large group of protists that move by extending lobe-shaped pseudopodia, the *Amoebozoa* (⮌ Section 18.7). Both parasites can cause serious, even fatal infections, although *Naegleria* infections are very rare. *Balantidium* is a ciliated species of the alveolate group (⮌ Section 18.4) and is mainly a disease of tropical countries.

Amebic Dysentery

Entamoeba histolytica (**Figure 33.6a**) is transmitted by contaminated water or occasionally through contaminated food. *E. histolytica* is an anaerobe, and the organism's *trophozoites* (the active, motile, feeding stage of the parasite) lack mitochondria. Like another common waterborne pathogen, *Giardia* (Section 33.4), the trophozoites of *E. histolytica* produce cysts, which are the means of transmission. Ingested cysts germinate to form amoebae that grow both on and in intestinal mucosa. This leads to tissue invasion and ulceration that triggers diarrhea and severe intestinal cramps.

With further growth, the amoebae can invade the intestinal wall—a condition called *dysentery*, characterized by intestinal inflammation, fever, and the passage of intestinal blood and mucus. If the infection is not treated, *E. histolytica* can invade the liver, the lungs, and even the brain. Growth in these tissues causes abscesses that can be fatal. Nearly 100,000 people, primarily from developing countries where untreated sewage is allowed to enter surface waters, die each year from invasive amebic dysentery. *E. histolytica* amebiasis can be treated with a variety of drugs, but the host immune system plays a significant role in recovery as

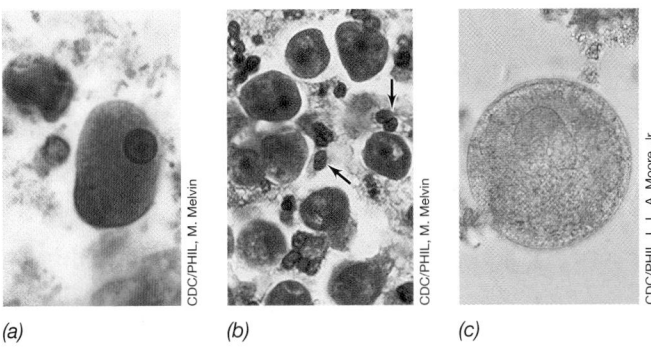

(a)   *(b)*   *(c)*

Figure 33.6 Parasitic amoebae and ciliates. *(a)* The growing stage (trophozoite) of *Entamoeba histolytica*; these can be up to 60 μm in length. *(b)* Trophozoites (arrows) of *Naegleria fowleri* in sectioned and stained brain tissue; the parasites are 10–25 μm in length. *(c) Balantidium coli* cyst (about 50 μm wide) present in a fecal sample.

well. However, protective immunity is not conferred from a primary infection, and reinfection is common.

Naegleria and *Balantidium* Infections

Naegleria fowleri can cause amebiasis, but in a very different form from that of *E. histolytica. N. fowleri* is a free-living amoeba present in soil and in runoff waters. *N. fowleri* infections result from swimming or bathing in warm, soil-contaminated waters such as warm springs or lakes and streams in summertime. *N. fowleri* enters the body through the nose and burrows directly into the brain. There the organism propagates, causing extensive hemorrhage and brain damage (Figure 33.6b), a condition called **meningoencephalitis**. Diagnosis of an *N. fowleri* infection requires observation of the amoebae in cerebrospinal fluid. If a definitive diagnosis is made quickly, which is often not the case because the incidence of *Naegleria* meningoencephalitis is so low and the symptoms resemble those of meningitis, the drug amphotericin B can save the patient; untreated infections are almost always fatal. Thus, although incidence of *Naegleria* meningoencephalitis is very low, mortality of untreated cases is very high, which makes this parasitic infection extremely dangerous. In the United States only 37 cases of *Naegleria* meningoencephalitis were reported in the 10-year period between 2006 to 2015 and virtually all were from recreational waters.

Balantidium coli is a ciliated intestinal human and swine parasite that alternates between the trophozoite and cyst (Figure 33.6c) stages; only the cysts are infective. *B. coli* is the only known ciliated parasite of humans. Cysts, typically transmitted in fecally contaminated water, germinate in the colon and infect mucosal tissues, leading to symptoms that resemble those of amebiasis, for which the disease is sometimes mistaken. An infected patient usually experiences a spontaneous recovery or may become an asymptomatic carrier, continuously shedding *B. coli* cysts in the feces. Compared with amebiasis, *B. coli* infections are uncommon, and cases are rarely fatal.

TABLE 33.2 Major parasitic human diseases

Parasitic diseases by site	Causal organism[a]
Gastrointestinal	
Amebiasis	*Entamoeba histolytica*
Giardiasis	*Giardia intestinalis*
Cryptosporidiosis	*Cryptosporidium parvum*
Toxoplasmosis	*Toxoplasma gondii*
Blood and tissue	
Malaria	*Plasmodium* spp.
Leishmaniasis	*Leishmania* spp.
Trypanosomiasis (African sleeping sickness)	*Trypanosoma brucei*
Chagas disease	*Trypanosoma cruzi*
Schistosomiasis	*Schistosoma mansoni*

[a]All are protists (Chapter 18) except for *Schistosoma*, a helminth.

— MINIQUIZ —

- Contrast an *Entamoeba* and a *Naegleria* infection in terms of tissues infected and symptoms.
- Describe a scenario for contracting a *Naegleria* infection.

33.4 Other Visceral Parasites: *Giardia, Trichomonas, Cryptosporidium, Toxoplasma,* and *Cyclospora*

The protists *Giardia intestinalis* and *Trichomonas vaginalis* are flagellated anaerobic parasites that contain either mitosomes or hydrogenosomes in place of mitochondria (⋐ Sections 2.15, 18.1, and 18.3); the parasites cause intestinal and sexually transmitted infections, respectively. The protist *Cryptosporidium* is related to *Toxoplasma*, but unlike *Toxoplasma*, which is primarily transmitted by infected food, as is the pathogenic protist *Cyclospora*, *Cryptosporidium* is transmitted primarily by contaminated water. We consider all five of these major human parasites here.

Giardiasis

Giardia intestinalis (also called *Giardia lamblia*) is typically transmitted to humans in fecally contaminated water and causes an acute gastroenteritis, *giardiasis*. The trophozoites of *Giardia* (**Figure 33.7a, c**) produce highly resistant cysts (Figure 33.7*b*) that function in transmission. Ingested cysts germinate in the small intestine to form trophozoites, and these travel to the large intestine where they attach to the intestinal wall and cause the symptoms of giardiasis: an explosive, foul-smelling, watery diarrhea, intestinal cramps, flatulence, nausea, weight loss, and malaise. The foul-smelling diarrhea and the absence of fecal blood distinguish giardiasis from diarrheas due to bacterial or viral intestinal pathogens.

G. intestinalis causes a significant number of drinking water infectious disease outbreaks in the United States (⋐ Section 32.1). The thick-walled cysts are resistant to chlorine, and most outbreaks have been associated with water systems that used only chlorination as a means of water purification. Water subjected to proper clarification and filtration followed by chlorination or other disinfection (⋐ Section 22.8) should be free of *Giardia* cysts. Most surface water sources (lakes, ponds, and streams) contain *Giardia* cysts, as beavers and muskrats are carriers of this pathogen. This is why surface waters should never be drunk untreated but instead should be filtered and disinfected with iodine or chlorine or alternatively, filtered and boiled. The drugs quinacrine, furazolidone, and metronidazole are useful for treating acute giardiasis.

Trichomoniasis

Trichomonas vaginalis (**Figure 33.8**) causes a sexually transmitted infection, *trichomoniasis*. *T. vaginalis* does not produce resting cells or cysts and as a result, *Trichomonas* transmission is typically from person to person, generally by sexual intercourse. However, unlike most sexually transmitted bacterial pathogens, cells of *T. vaginalis* can survive for several hours on moist surfaces and up to a day in urine or semen. Hence, in addition to disease transmission by intimate contact, trichomoniasis can be transmitted by contaminated toilet seats, sauna benches, and towels.

T. vaginalis infects the vagina in women, the prostate and seminal vesicles of men, and the urethra of both males and females. Trichomoniasis is often asymptomatic in males. By contrast, trichomoniasis in females is characterized by a yellowish vaginal discharge that causes a persistent vaginal itching and burning. The infection is more common in females; surveys have shown that up to 25% of sexually active women are infected with *T. vaginalis* while only about 5% of males are infected. Trichomoniasis is diagnosed by observation of the motile protists in a wet mount of fluid discharged from the patient (Figure 33.8*b*). The antiprotozoal drug metronidazole is effective for treating trichomoniasis.

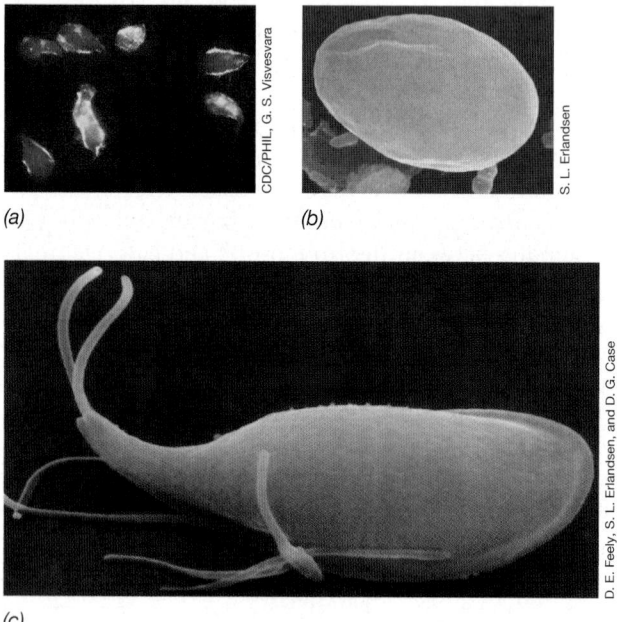

(a) (b) (c)

Figure 33.7 *Giardia*. *(a)* Fluorescently stained cells of *Giardia intestinalis*. *(b, c)* Scanning electron micrographs of *(b)* a giardial cyst and *(c)* a motile *G. intestinalis* trophozoite. The trophozoite is 15 μm long and the cyst about 11 μm wide.

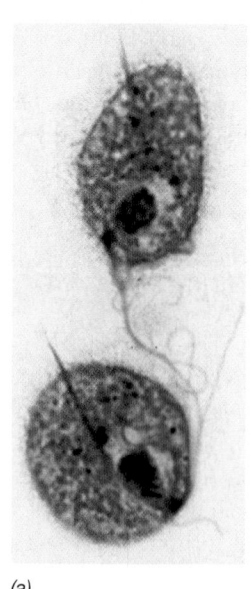

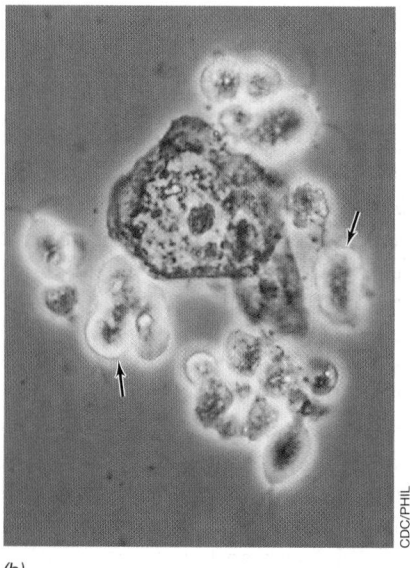

(a) (b)

Figure 33.8 *Trichomonas vaginalis*. *(a)* Light micrograph of stained cells; cells vary from 10 to 20 μm in diameter. *(b)* Vaginal discharge from a female with trichomoniasis. *T. vaginalis* cells (arrows) are present along with vaginal secretions and epithelial cells.

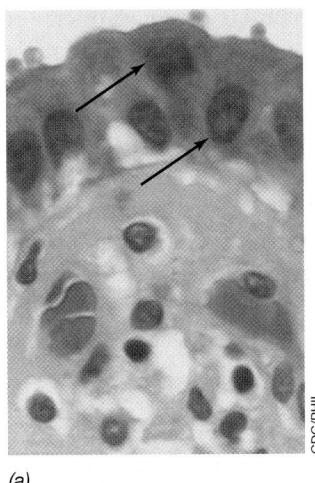
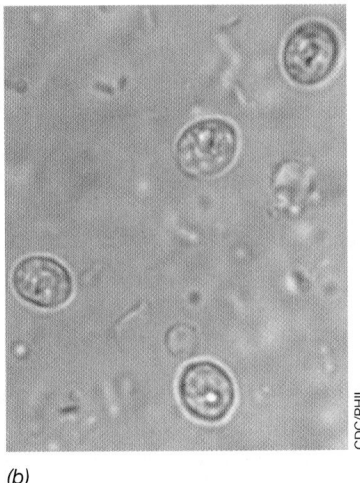

(a)   *(b)*

Figure 33.9 *Cryptosporidium parvum.* *(a)* Arrows point to intracellular trophozoites of *C. parvum* embedded in human gastrointestinal epithelium. The trophozoites are about 5 μm in diameter. *(b)* Thick-walled *C. parvum* oocysts are about 3 μm in diameter in this fecal sample.

Cryptosporidiosis, Toxoplasmosis, and Cyclosporiasis

Cryptosporidium, Toxoplasma, and *Cyclospora* are genera of parasitic coccidia (which group among the alveolates, ↩ Section 18.4). These parasites are transmitted to humans in fecally contaminated food or water and can trigger serious bouts of diarrhea, or in the case of *Toxoplasma,* serious internal organ damage.

Cryptosporidium parvum infects many warm-blooded animals, in particular cattle. The organism forms small, coccoid cells that invade and grow intracellularly in mucosal epithelial cells of the stomach and intestine (**Figure 33.9a**), resulting in the gastrointestinal illness *cryptosporidiosis. C. parvum* produces thick-walled, highly resistant cysts called *oocysts* (Figure 33.9b), which enter water from the feces of infected animals. The infection is then transmitted to other animals and humans when they consume the fecally contaminated water.

Cryptosporidium oocysts are highly resistant to chlorine, and because of this, sedimentation and filtration are the only reliable way to remove them from water supplies. In an average year, *Cryptosporidium* is responsible for the majority of recreational waterborne disease outbreaks in the United States (Chapter 32) but is only occasionally associated with drinking water outbreaks. Nevertheless, *C. parvum* was responsible for the largest single outbreak of disease associated with drinking water ever recorded in the United States. In the spring of 1993, one-quarter of the population of Milwaukee, Wisconsin (USA), developed cryptosporidiosis from consuming water from the municipal water supply. Heavy spring rains and runoff from cattle manure on farmlands had drained into Lake Michigan (the water supply for the city) and overburdened the water purification system, leading to contamination by *C. parvum.*

Cryptosporidiosis typically causes only a mild, self-limiting diarrhea, making treatment unnecessary. However, individuals with impaired immunity, such as that caused by HIV/AIDS, or the very young or old can develop serious complications from a *C. parvum* infection. The primary laboratory diagnostic method for cryptosporidiosis is the demonstration of oocysts in feces (Figure 33.9b). Immunological and molecular tools are also available for more precise identification of strains of the pathogen when such tracking is necessary.

As for *C. parvum,* the parasite *Cyclospora cayetanensis* also forms oocysts and causes a mild to occasionally severe gastroenteritis, which is called *cyclosporiasis.* However, unlike *C. parvum, C. cayetanensis* is primarily transmitted by fecally contaminated food products—typically fresh foods—rather than by contaminated water. Most cases of cyclosporiasis have been linked to contaminated fruits or vegetables. A major outbreak in the United States in the summer of 2013 was linked to packaged lettuce (↩ Section 32.7). Another major *C. cayetanensis* outbreak occurred in 2015 and affected persons in 31 states. At least some (and perhaps most) of these illnesses were caused by contaminated fresh cilantro (coriander) imported from Mexico.

Toxoplasmosis is caused by *Toxoplasma gondii* (**Figure 33.10**). This parasite infects many warm-blooded animals, and roughly half of all adults in the United States are infected but asymptomatic because their immune system keeps the organism in check. *T. gondii* is typically transmitted to humans in the form of cysts present in undercooked beef, pork, or lamb; by direct infection from cats, which are major carriers of *T. gondii*; and occasionally from contaminated water. A key step in the *T. gondii* life cycle is completed in felines and thus they are obligate hosts; humans and other animals are only incidental hosts. Most transmission to humans is thus probably from cats.

Toxoplasmosis can be associated with mild to severe symptoms. When cysts of *T. gondii* are ingested, they penetrate the wall of the small intestine. From this initial infection, symptoms can be inapparent or apparent but indistinguishable from those of a mild case of influenza (headache, muscle ache, general malaise). However, in some infected persons, *T. gondii* cysts migrate from the small intestine and circulate throughout the body. Subsequently, the parasite can penetrate nerve cells and infect tissues of the brain and eyes. Although disease symptoms in healthy adults are uncommon, in immune-compromised individuals, toxoplasmosis can damage the eyes, brain, and other internal organ systems.

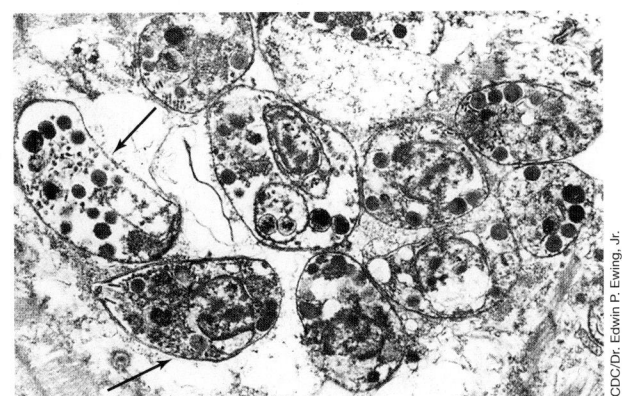

Figure 33.10 *Toxoplasma.* Tachyzoites (rapidly growing cells) of *Toxoplasma gondii,* an intracellular parasite. In this transmission electron micrograph, the tachyzoites (arrows) form a cystlike structure in a host cardiac cell. The *Toxoplasma* tachyzoites are 4–7 μm long. For a photo of *T. gondii* sporozoites (the infective phase of the parasite's life cycle), see Figure 18.11b.

UNIT 7

In addition, a first-time infection with *T. gondii* in expectant mothers can lead to birth defects in newborns; thus pregnant women who have not been in contact with cats should avoid cats until after giving birth.

III • Blood and Tissue Parasitic Infections

Several human parasites infect organs and tissues other than the gastrointestinal tract and are typically transmitted by insect vectors. We begin our consideration of these with malaria, the most devastating and widespread of parasitic diseases and one that remains a major global health problem today.

33.5 *Plasmodium* and Malaria

Malaria is caused by protists of the alveolate group (⮌ Section 18.4). Several species of the protozoal genus *Plasmodium* cause malarial diseases in warm-blooded hosts; up to 500 million people worldwide contract malaria annually and about 1 million die from the disease. Malaria is thus one of the most common causes of death worldwide from infectious disease and certainly the most prevalent of parasitic diseases.

In malaria, the complex parasite life cycle requires a mosquito vector. Four species of *Plasmodium*—*P. vivax*, *P. falciparum*, *P. ovale*, and *P. malariae*—cause most human malaria. The most widespread disease is caused by *P. vivax*, whereas the most serious disease is caused by *P. falciparum*. Humans are the only reservoirs for these four species. The protists carry out part of their life cycle in the human and part in the female *Anopheles* mosquito, the only vector that transmits *Plasmodium* spp. The vector spreads the protist from person to person.

Malarial Life Cycle

The life cycle of *Plasmodium* is complex and involves a number of stages (**Figure 33.11**). First, the human host is infected by plasmodial *sporozoites*, small, elongated cells produced in the mosquito that localize in the salivary gland of the insect. The mosquito (Figure 33.11 inset) injects saliva containing the sporozoites into the human when obtaining a blood meal. The sporozoites travel to the liver where they infect liver cells. Here they can remain quiescent for indefinite periods but eventually replicate and become enlarged in a stage called the *schizont* (see Figure 33.12*b*). The schizonts then segment into a number of small cells called *merozoites*, which exit the liver into the bloodstream. Some of the merozoites then infect red blood cells (erythrocytes).

The plasmodial life cycle in erythrocytes proceeds with repeated division, growth, and release of merozoites (**Figure 33.12**); this results in destruction of the host red blood cells. Plasmodial growth in red cells typically repeats at synchronized intervals of 48 h. During this 48-h period, the host experiences the defining clinical

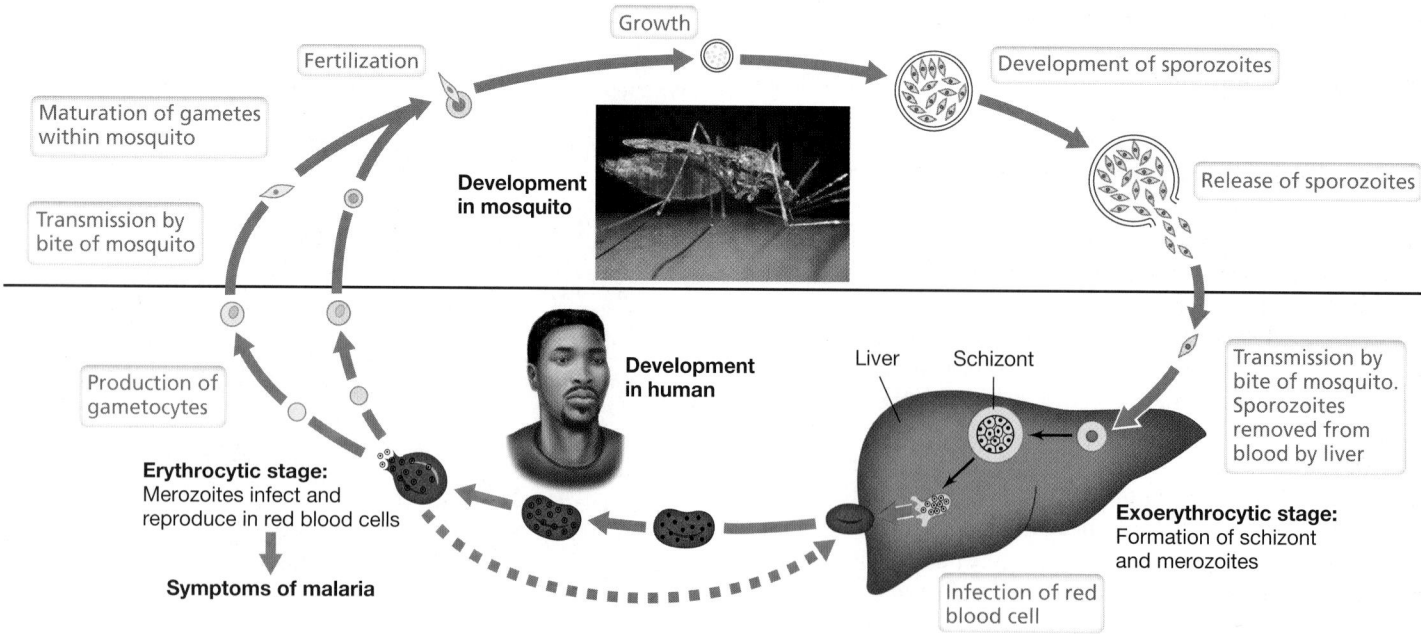

Figure 33.11 The life cycle of *Plasmodium*. The life cycle of *Plasmodium* requires both a warm-blooded host and the mosquito vector. Transmission of the protist to and from the warm-blooded (endothermic) host is done by the bite of an *Anopheles gambiae* mosquito (inset) or certain other *Anopheles* species. Mosquito photo courtesy of CDC/PHIL, J. Gathany.

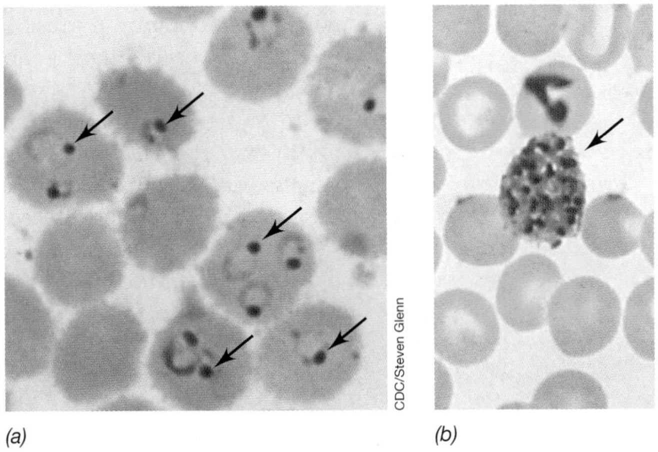

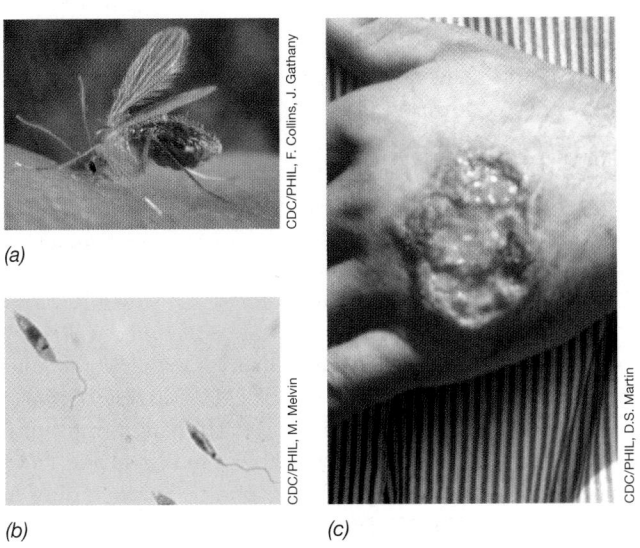

(a)   CDC/Steven Glenn

(b)   CDC/PHIL, M. Melvin

Figure 33.12 *Plasmodium* **and malaria.** *(a)* Merozoites of *Plasmodium falciparum* (arrows) growing within human red blood cells. *(b)* A schizont of *P. vivax* (arrow) along with red blood cells. When released from the schizont, the merozoites infect erythrocytes (Figure 33.11). Red blood cells are about 6 μm in diameter.

symptoms of malaria: chills followed by fever of up to 40°C (104°F). The chill–fever pattern coincides with the release of merozoites from the erythrocytes during the synchronized reproduction cycle. Vomiting and severe headache may accompany the chill–fever cycles, and over the longer term, characteristic symptomatic malaria can alternate with asymptomatic periods. Because of the destruction of red blood cells, malaria typically causes anemia and some enlargement of the spleen (splenomegaly).

Plasmodial merozoites eventually develop into *gametocytes*, cells that infect only mosquitoes. The gametocytes are ingested when an *Anopheles* mosquito takes a blood meal from an infected person, and they mature within the mosquito into *gametes*. Two gametes fuse to form a zygote, and the zygote migrates by amoeboid motility to the outer wall of the insect's intestine where it enlarges and forms several sporozoites. These are released and reach the salivary gland of the mosquito from where they can be inoculated into another human, and the cycle begins anew (Figure 33.11).

Epidemiology, Diagnosis, Treatment, and Control

Anopheles mosquitoes (Figure 33.11 inset) live predominantly in the tropics and subtropics and are the vector for malaria. Diagnosis of malaria requires the identification of *Plasmodium*-infected erythrocytes in blood smears (Figure 33.12). Fluorescent nucleic acid stains, nucleic acid probes, PCR assays, and various antigen-detection methods (Chapter 28) are also used to verify *Plasmodium* infections or to differentiate between infections with various *Plasmodium* species.

Treatment of malaria is typically accomplished with *chloroquine*. Chloroquine kills merozoites within red cells but does not kill sporozoites. The related drug *primaquine* eliminates sporozoites of *P. vivax* and *P. ovale* that may remain in liver cells. Thus treatment with both chloroquine and primaquine effectively cures most malaria. However, in some individuals, malaria recurs years after the primary infection when a few sporozoites not eliminated from the liver release a new generation of merozoites. Quinine-resistant strains of *Plasmodium* are now widespread and so *combination therapy*, where the malaria patient is treated with several antimalarial drugs at once, is now a common form of treatment.

Malaria can be controlled by either draining swamps and other breeding areas or by eliminating the mosquito with insecticides. Together, these measures have all but eliminated malaria in the United States, with most cases being imported. Several malaria vaccines are also in development, including synthetic peptide vaccines, recombinant particle vaccines, and DNA vaccines (⊘ Sections 12.8 and 28.9), but thus far no highly effective and reliable malaria vaccine has emerged for use in mass vaccination programs.

MINIQUIZ

- Which stages of the *Plasmodium* life cycle occur in humans, and which in the mosquito?
- What are the natural reservoirs and vectors for *Plasmodium* species? How can malaria be prevented or eradicated?
- What drugs are used to treat malaria?

33.6 Leishmaniasis, Trypanosomiasis, and Chagas Disease

Parasites of the genera *Leishmania* and *Trypanosoma* are transmitted by bloodsucking insect vectors. These parasites are *hemoflagellates*, organisms that reside in blood or related tissues such as the liver and spleen, and they cause major human diseases, primarily in tropical and subtropical countries.

Leishmaniasis

Leishmaniasis is a parasitic disease of various forms caused by species of the genus *Leishmania*, a flagellated protozoan related to *Trypanosoma*. The disease is transmitted to humans by a bite from the sand fly. *Cutaneous leishmaniasis*, caused by either *L. tropica* or *L. mexicana*, is the most common form of leishmaniasis. Following transmission of the parasite in a blood meal (**Figure 33.13a, b**),

(a)   CDC/PHIL, F. Collins, J. Gathany

(b)   CDC/PHIL, M. Melvin

(c)   CDC/PHIL, D.S. Martin

Figure 33.13 Leishmaniasis. *(a)* The sand fly (genus *Phlebotomus*) transmits leishmaniasis in a blood meal. *(b) Leishmania* spp. are flagellated protozoans and the cause of leishmaniasis. *(c)* Cutaneous leishmaniasis showing an open ulcer on the hand. Secondary bacterial infections of these ulcers are common. Leishmaniasis exists in over 88 tropical and subtropical countries. Occasional cases of cutaneous leishmaniasis are reported in the United States, primarily from the state of Texas.

the parasite infects and grows within human macrophage cells (⇄ Section 26.4), leading eventually (weeks or months later) to the formation of a small nodule on the skin. The nodule then becomes ulcerated and can enlarge to form a major skin lesion (Figure 33.13c) that contains active parasites. In the absence of secondary bacterial infections, which are common if the ulcerated tissue is left open, the lesions heal spontaneously over a period of several months but can leave a permanent scar.

Leishmaniasis has historically been treated with injections of pentavalent antimony (Sb^{5+}) compounds. Although the mode of action of these compounds is unknown, it is thought that Sb^{5+} in some way stimulates or activates the immune response to better attack the *Leishmania* parasites. At present, however, many *Leishmania* species are resistant to antimony compounds, but a variety of other drugs are available for treating resistant cutaneous forms of the disease. Estimates of cutaneous leishmaniasis prevalence worldwide are about 1 million.

Visceral leishmaniasis is caused by *Leishmania donovani* and is the most severe form of the disease. In visceral leishmaniasis, the parasite travels from the site of infection to internal organs, in particular the liver, spleen, and bone marrow. If left untreated, the visceral disease is almost always fatal. Common symptoms of visceral leishmaniasis include a cycling of fever and chills, a slow reduction in both red and white blood cell numbers, and significant enlargement of the spleen and liver that can lead to major distention of the abdomen. Treatment includes injections of antimony (as for the cutaneous disease), long periods of bed rest, and blood transfusions in acute cases if blood cell counts become dangerously low. Estimates of visceral leishmaniasis prevalence worldwide are about 300,000, causing about 20,000 deaths annually. In addition, because the range of the sand fly is already broad and is increasing as climate change and tropical and subtropical deforestation proceed (see page 958), it is estimated that worldwide nearly 400 million people (over 5% of the world's humans) could be at risk of some form of leishmaniasis.

Trypanosomiasis and Chagas Disease

Flagellated protozoans of the genus *Trypanosoma* (⇄ Section 18.3) cause two related forms of **trypanosomiasis**. Two subspecies of *Trypanosoma brucei* native to Africa, *T. brucei gambiense* (**Figure 33.14a**) and *T. brucei rhodesiense*, cause *African trypanosomiasis*, better known as *African sleeping sickness*. The species *T. cruzi* causes *Chagas disease*, also known as *American trypanosomiasis*. These diseases are transmitted by insect bites from either a fly or a bug.

Sleeping sickness is transmitted by the tsetse fly (genus *Glossina*), an insect similar in dimensions to a housefly and native only to tropical regions of Africa; sleeping sickness is therefore endemic only in countries of sub-Saharan Africa. The disease begins with intermittent fever, headache, and malaise. The parasite multiplies in the blood and later infects the central nervous system and grows in spinal fluid. Neurological symptoms soon begin, including sleep patterns that are no longer diel. The parasite produces the aromatic alcohol *tryptophol*, a derivative of the amino acid tryptophan, which triggers a sleep response. Without treatment, the infection gradually progresses to a coma, multiple organ failure, and eventually death after months or years depending on the case. A variety of anti-trypanosomal drugs are available for treating sleeping sickness;

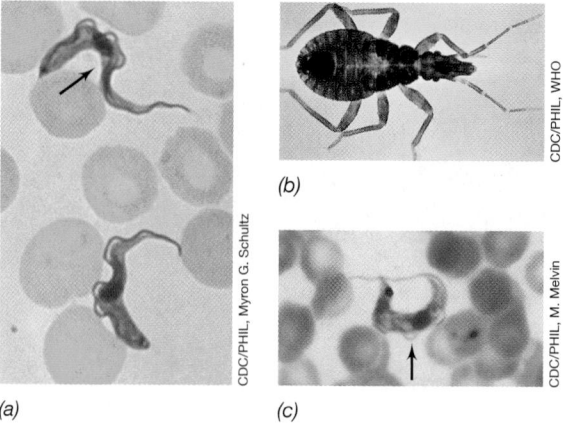

Figure 33.14 African trypanosomiasis and Chagas disease. *(a)* Two cells of *Trypanosoma brucei* (arrow), the causative agent of African sleeping sickness (African trypanosomiasis), in a blood smear. *(b)* The "kissing bug" (*Triatoma infestans*), the vector for Chagas disease (American trypanosomiasis). *(c)* A cell of *Trypanosoma cruzi* (arrow), the causative agent of Chagas disease, in a blood smear.

some are used primarily for treating the blood infection while others are used if the disease has progressed to the neurological stage. About 10,000 new cases of sleeping sickness are reported annually, but most cases are thought to go unreported.

Chagas disease, named for its discoverer, is caused by *T. cruzi*, a close relative of *T. brucei*, and is transmitted by the bite of the "kissing bug" (Figure 33.14b, c). Chagas disease mainly occurs in Latin American countries. The parasite affects several organs including the heart, gastrointestinal tract, and central nervous system, causing inflammatory reactions and tissue destruction. The acute illness is usually self-limiting, but if chronic illness develops, heart damage is significant and is the eventual cause of premature death. About 20,000 deaths due to Chagas disease occur annually in endemic Latin American countries.

Currently no vaccines are available for prevention of African or American trypanosomiases.

— MINIQUIZ —

- How are trypanosome diseases similar to malaria and how do they differ?
- How do the symptoms of cutaneous and visceral leishmaniasis differ?
- How are sleep patterns altered in cases of African trypanosomiasis?

33.7 Parasitic Helminths: Schistosomiasis and Filariases

Some parasitic diseases are caused by helminths, tiny worms that burrow into the human host and cause debilitating diseases and death. We consider the most widespread of these, schistosomiasis, along with brief coverage of other, less common helminth infections.

Schistosomiasis

Schistosomiasis, also called *snail fever*, is a chronic parasitic disease caused by species of trematodes (flatworms) of the genus

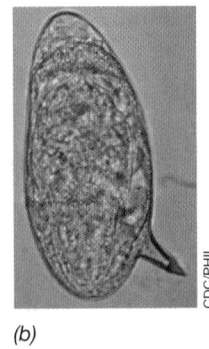
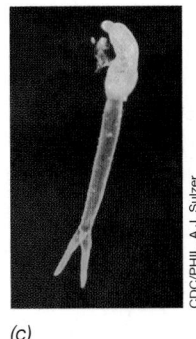
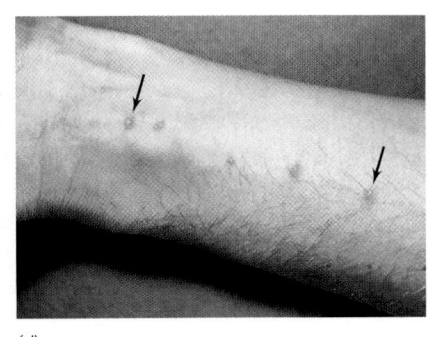

(a)   (b)   (c)   (d)

Figure 33.15 Schistosomiasis. *(a)* Adult worm of *Schistosoma mansoni*; the worm is about 1 cm in length. *(b)* An *S. mansoni* egg, about 0.15 mm long. The lateral spine is characteristic of the eggs of this species. *(c)* Fluorescently stained cercaria, the infective form of *S. mansoni*. From the head (top) to the bifurcated tail is about 1 mm. *(d)* Cercarial infection of the forearm. Five infection sites (arrows) are apparent.

Schistosoma; the major species is *S. mansoni* and adult worms can be up to a centimeter in length (**Figure 33.15a**). The life cycle of the parasite requires both snails and humans (or other mammals) as hosts. Schistosome eggs (Figure 33.15b) released into a freshwater aquatic environment hatch to generate *miracidia*, the form of the worm that infects snails. In the snail, miracidia are transformed into *cercariae* (Figure 33.15c), the motile stage of the parasite that is released and infects humans.

A cercaria burrows into the skin, leaving a small surface lesion (Figure 33.15d), and then migrates to the lungs and liver; in the process, the worm establishes a long-term infection in the blood vessels. From the liver, the parasite infects the bladder, kidneys, and urethra, and the female worm produces large numbers of eggs. The eggs are shed in the urine and also pass through the intestinal wall and are shed in the feces. Large egg masses also become trapped along with fluids in the bladder, liver, and other organs, triggering an inflammatory response and a major distention of the abdomen, a condition commonly seen in infected children (**Figure 33.16a**). Other symptoms include bloody urine, diarrhea, and abdominal pain. Eggs as well as adult worms can live in the body for years, causing chronic symptoms that can last from youth into adulthood.

Schistosomiasis is a disease of tropical countries, primarily those in Africa, but some cases also occur in subtropical countries such as Latin America and the Caribbean region; poverty, poor sanitation, and land use changes are typically associated with widespread infection (see page 958). Schistosomiasis can be effectively treated with the drug praziquantel, and the diagnosis is made relatively easily by assessing symptoms and observing parasite eggs in the urine and feces. Mortality from schistosomiasis is low, about 0.1%, but schistosomiasis is second only to malaria in terms of total parasitic infections worldwide. In 2014, over 258 million cases of the disease were treated and many others probably went untreated.

Filariases

Several other parasitic helminth infections are known, and chief among these are the *filariases*, infections by parasitic nematodes (roundworms). Unlike the schistosomiasis parasite, these worms are clearly macroscopic in the adult stage (several centimeters in length, depending on the filariasis).

Bancroft's filariasis (also called "elephantiasis") is a chronic infection of the lymphatic system by the roundworm *Wuchereria bancrofti*. The worm is transmitted to humans as tiny *microfilariae* in a mosquito bite. Once in the host, microfilariae develop into adult worms and these interrupt lymph flow, leading to major accumulation of fluids (edema). Fluid accumulation in lower regions of the body can cause massive enlargement of the legs (Figure 33.16b). Over 120 million people in the tropics suffer from *W. bancrofti* infection, but the microfilarial stage of the disease is readily treatable with antihelminthic drugs or the drug diethylcarbamazine, which kills both microfilariae and adult worms. Even simpler treatment may be available by administering antibacterial antibiotics. Although the worm itself is not sensitive to these drugs, the worm harbors an endosymbiotic bacterium, *Wolbachia* (*Alphaproteobacteria*, ⇨ Section 16.1), which is. If *Wolbachia* is eliminated by antibiotic treatment, the worms die, and so antihelminthic treatments often include antibiotics such as doxycycline (⇨ Section 28.10) to accelerate the removal of worms from the patient.

Onchocerciasis (also called *river blindness*) is due to a chronic infection by the large parasitic roundworm *Onchocerca volvulus*

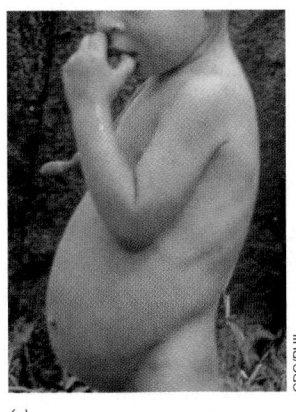
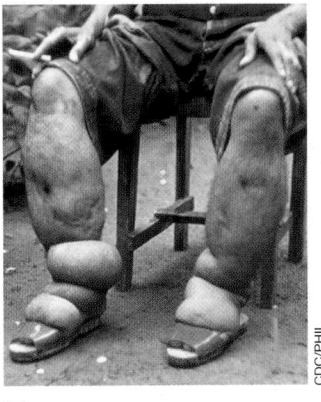

(a)   (b)

Figure 33.16 Symptoms of parasitic helminth infections. *(a)* Schistosomiasis in a small child. The swollen abdomen from the accumulation of fluids and worm eggs is characteristic of the infection. *(b)* Bancroft's filariasis (elephantiasis). The swollen legs are the result of edema from infection of lymph tissues by the roundworm *Wuchereria bancrofti*.

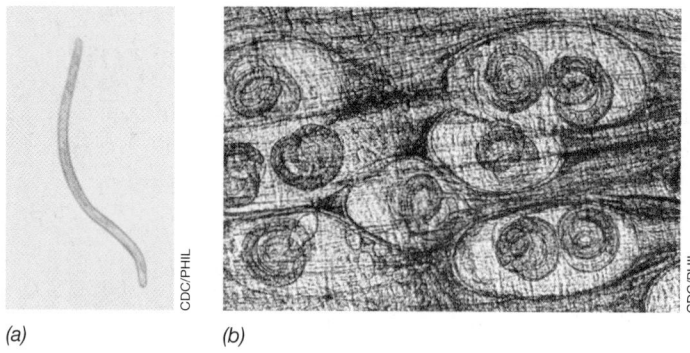

(a)   (b)

Figure 33.17 The roundworms of river blindness and trichinosis. *(a)* A worm larva of *Onchocerca volvulus*, the causative agent of river blindness. The microfilarial worm is about 0.3 millimeter long, but adult worms can be several centimeters in length. *(b)* Cysts of *Trichinella spiralis* containing worm larvae in muscle tissue. Unlike *O. volvulus*, adult worms are microscopic, just a few millimeters long.

(Figure 33.17a). Humans are the only known host for this parasite, but flies are vectors when they become infected with microfilariae in a blood meal and transmit them to uninfected humans in a bite. Blackflies of the genus *Simulium* are the major means of transmission. The microfilariae invade the cornea and from there the iris and retina, triggering an inflammatory response that causes scarring and partial to total loss of vision. *O. volvulus* infection is second only to trachoma (⇌ Section 30.14) as a cause of infectious

blindness. It is estimated that about 20 million people are infected with this parasite, primarily in equatorial Africa.

The disease *trichinosis* (also called *trichinellosis*) is caused by species of the parasitic roundworm *Trichinella* (Figure 33.17b). This worm commonly infects the muscle tissues of wild mammals and can occasionally infect domestic animals, especially swine; about 20 cases of human trichinosis are reported in the United States each year, usually from the consumption of undercooked wild game or undercooked pork. Human infection with *Trichinella* begins when worm larvae enter intestinal mucosal cells, leading to either an asymptomatic condition or mild gastroenteritis. As the larvae mature and reproduce, new larvae circulate throughout the body and lead to systemic inflammatory reactions such as malaise, facial swelling, and fever. Untreated cases of trichinosis can progress to more severe organ-specific symptoms including heart damage, encephalitis, and even death. However, if properly diagnosed, usually by immunological assays on biopsied muscle tissue, trichinosis is treatable with a variety of drugs, in particular the benzimidazole class of antihelminthics.

— **MINIQUIZ** —

- How does the pathogen causing schistosomiasis differ from all other pathogens considered in this chapter?
- From what source are most cases of human trichinosis contracted?

MasteringMicrobiology® **Visualize, explore,** and **think critically** with Interactive Microbiology, MicroLab Tutors, MicroCareers case studies, and more. MasteringMicrobiology offers practice quizzes, helpful animations, and other study tools for lecture and lab to help you master microbiology.

Chapter Review

I • Fungal Infections

33.1 Fungi include the molds and yeasts, and some fungi are dimorphic, meaning that both mycelial and yeast phases can occur. Superficial, subcutaneous, and systemic mycoses refer to fungal infections of the skin surface, skin subsurface, and internal organs, respectively. Fungal infections can be mild or serious, depending on the health and immune status of those infected.

Q Which are more common, superficial or systemic mycoses? Have you had a case of either one?

33.2 Superficial mycoses such as athlete's foot or jock itch are mild and easily treatable, whereas subcutaneous mycoses, such as sporotrichosis, or especially systemic mycoses, such as histoplasmosis, are more difficult to treat effectively. The ability of fungi that cause systemic mycoses to infect internal organs makes these pathogens particularly dangerous to the elderly or those otherwise immune compromised.

Q What are the most common systemic mycoses in the United States, and which populations are most susceptible to such infections?

II • Visceral Parasitic Infections

33.3 The genera *Entamoeba* and *Naegleria* are amoebic human parasites that cause gastrointestinal and brain infections, respectively. *Entamoeba* is transmitted in fecally contaminated waters, whereas *Naegleria* inhabits warm, soil-contaminated waters. *Balantidium* is a ciliated intestinal parasite transmitted by fecally contaminated water.

Q If you were to contract one or the other, which would you rather have, an *Entamoeba* infection or a *Naegleria* infection, and why?

33.4 The protists *Giardia intestinalis* and *Cryptosporidium parvum* are major waterborne pathogenic parasites, whereas *Toxoplasma gondii* is primarily a foodborne or cat-transmitted parasite and *Trichomonas vaginalis* is a sexually transmitted parasite. The pathogenic parasite *Cyclospora* is primarily transmitted by fresh vegetables such as lettuce and spinach contaminated with animal feces. None of these parasites cause life-threatening diseases in otherwise healthy individuals, although *T. gondii* can trigger severe and even fatal infections in immune-compromised hosts.

> Q **In contrast to disease caused by *Trichomonas*, what do giardiasis and cryptosporidiosis have in common?**

III • Blood and Tissue Parasitic Infections

33.5 Infections with *Plasmodium* spp. cause malaria, a widespread, mosquito-transmitted disease of the blood that causes significant morbidity and mortality in tropical and subtropical regions of the world. Malaria is treatable with quinines and other drugs but is not yet preventable by vaccination.

> Q **Malaria symptoms include chills followed by fever. These symptoms are related to activities of the pathogen. Describe the growth stages of *Plasmodium* spp. in the human host and relate them to the chill–fever pattern.**

33.6 Leishmaniasis is a parasitic disease caused by *Leishmania* species; the cutaneous form of the disease is most common.

Trypanosoma brucei causes African trypanosomiasis (African sleeping sickness), while the related species *Trypanosoma cruzi* causes Chagas disease. All of these diseases are transmitted by insect bites from an insect vector, either a fly or a bug.

> Q **Contrast leishmaniasis with the two types of trypanosomiasis in terms of causative agents, symptoms, and transmission vectors.**

33.7 Schistosomiasis is a major parasitic disease caused by a microscopic worm, *Schistosoma mansoni*. The life cycle of the parasite requires both snails and mammals. The worm infects the liver and kidneys and produces large egg masses that accumulate in the body, leading to systemic inflammation and abdominal distention. Other parasitic worm diseases, such as elephantiasis and river blindness, also leave readily visible signs of infection. Trichinosis is caused by a roundworm that infects the intestine and muscle tissues and is a threat from the consumption of undercooked pork or wild game.

> Q **Contrast schistosomiasis with all other parasitic infections covered in this chapter. In what major way does it differ?**

Application Questions

1. Malaria eradication has been a goal of public health programs for at least 100 years. What factors preclude our ability to eradicate malaria? If an effective vaccine was developed, could malaria be eradicated?

2. In terms of public health, what is a common problem that unites many of the visceral parasitic infections covered in this chapter? How could this problem be attacked? Why are these diseases rare in developed countries?

3. Explain why the diseases malaria, leishmaniasis, and trypanosomiasis are primarily diseases of tropical regions. How could humans be affecting the future geographical ranges of these diseases?

4. Explain why systemic fungal infections are typically seen only in certain individuals even though many people have contact with the pathogen, whereas an outbreak of giardiasis affects virtually everyone that has come in contact with the pathogen.

Chapter Glossary

Leishmaniasis a disease of the skin or viscera caused by infection with species of a parasitic flagellated protozoan, *Leishmania*

Malaria a disease characterized by recurrent episodes of fever and anemia, caused by the protist *Plasmodium* spp., usually transmitted between mammals through the bite of the *Anopheles* mosquito

Meningoencephalitis the invasion, inflammation, and destruction of brain tissue by the amoeba *Naegleria fowleri* or a variety of other pathogens

Mycosis (plural, mycoses) any infection caused by a fungus

Schistosomiasis a chronic disease caused by a parasitic worm that leads to internal organ damage and accumulation of fluids and worm egg masses

Subcutaneous mycoses fungal infections of deeper layers of skin

Superficial mycoses fungal infections of the surface layers of skin, hair, or nails

Systemic mycoses fungal growth in internal organs of the body

Trypanosomiasis any parasitic disease of the blood and internal tissues caused by species of the flagellated protozoan *Trypanosoma*; African sleeping sickness and Chagas disease are two major trypanosomiases

Photo Credits

Thomas Klose, Theodore C. Pierson, Michael G. Rossmann and Richard J. Kuhn at Purdue University; 10.5a Dr. D. Raoult, CNRS, Marseille, France; 10.10a F. Grundy and Martha Howe; 10.11a, b Mark Young; 10.11c Claire Geslin; 10.11d David Prangishvili, Institut Pasteur; 10.12 Mark Young; 10.13 Dr. Fred Murphy, Sylvia Whitfield/CDC; 10.14a Dr. G. William Gary, Jr./CDC; 10.15a Alexander Eb and Jerome Vinograd; 10.16 Dr. Fred Murphy, Sylvia Whitfield/CDC; 10.17a R.C. Valentine; 10.18a Dr. Joseph J. Esposito; F. A. Murphy/CDC; 10.18b Arthur J. Olson, Molecular Graphics Laboratory, Scripps Research Institute; 10.19a CDC; 10.20a Erskine Palmer/CDC; 10.21a C. S. Goldsmith, and T. Tumpey/CDC; 10.22a, b Timothy S. Baker and Norman H. Olson, Purdue University; 10.25a CDC; 10.26a, b, c Wei Dai, Baylor College of Medicine; 10.27 Jed Fuhrman/Matthew Sullivan; 10.29c Jeremy J. Barr; 10.30 Biao Ding & Yijun Qi; 10.33a Teresa Hammett/CDC.

Chapter 11 Chapter Opener Melanie Blokesch (Swiss Federal Institute of Technology Lausanne, EPFL, Switzerland); 11.1 Howard Shuman and Thomas Silhavy; 11.2a Thomas D. Brock; 11.2b S. R. Spilatro, Marietta College, Marietta, OH; 11.2c Shiladitya DasSarma, Priya Arora, Lone Simonsen; 11.3 Derek J. Fisher; 11.9 Patricia L. Foster and Sarita Mallik; 11.13 Melanie Blokesch; 11.16 Takehiko Kenzaka; 11.18 A.B. Westbye, P.C. Fogg, J.T. Beatty; 11.20a Charles C. Brinton, Jr., University of Pittsburgh; 11.20b Peter Graumann and Thomas Rosch; 11.20c Elisabeth Carniel; 11.21c (both) A. Babic, M. Berkmen, C. Lee, and A. D. Grossman; 11.26 Masaki Shioda and S. Takayanago; 11.27 Evelyne Marguet and Patrick Forterre; 11.32 George O'Toole, Geisel School of Medicine at Dartmouth.

Chapter 12 12.3a: Elizabeth Parker; 12.3b: Jack Parker; 12.4a: Laurie Ann Achenbach, Southern Illinois University at Carbondale; 12.4b: Megan Kempher; 12.5: Michael T. Madigan; 12.8: Daniel L. Nickrent; 12.10 left: Norbert Pfennig, University of Konstanz, German; 12.10 middle: Hans Hippe, Deutsche Sammlung von Mikroorganismen und Zellkulturen GmbH, Braunschweig, Germany; 12.10 right: Michael T. Madigan; 12.16: Jason A. Kahana and Pamela A. Silver, Harvard Medical School; 12.21: Stephen R. Padgette, Monsanto Company; 12.22a: Kevin McBride, Calgene; 12.22b: Kevin McBride, Calgene; 12.23: Aqua Bounty Technologies; 12.25: Dinesh Chandra and Claudia Gravekamp; 12.27.1: Klagyi/Shutterstock; 12.27.2: Puchan/Shutterstock; 12.27.3: Karen Lau/Shutterstock; 12.28: Cinder Biological, Concord, CA.; 12.30b: Heidi Polumbo/U.S. Department of Energy; 12.30c: Sara E. Blumer-Schuette, Jeffrey V. Zurawski, Jonathan M. Conway, and Robert M. Kelly; 12.31 top: Todd Ciche; 12.31 bottom: Marek Mis/Science Source; 12.32a, b: Rasala, B.A., Chao, S-S., Pier, M., Barrera, D.J., and Mayfield, S.P. 2014. PLoS ONE 9(4): e94028; 12.34c: Aaron Chevalier and Matt Levy; 12.36c: Hiroshi Nishimasu, Osama Nureki, and Feng Zhang; 12.37 left: Joyce Van Eck and Zachary Lippman; 12.37 right: Biao Ding & Yijun Qi; 12.38: Guo, R., Wang, H., Cui, J., Wang, G., Li, W., and Hu, J-F. 2015. *PLoS One* 10(10): e0141335.

Chapter 13 Chapter Opener: Rolf B. Pedersen; 13.2: Frances Westall, Lunar and Planetary Institute; 13.3: Michael T. Madigan; 13.6a, b: Malcolm R. Walter, Macquarie University, New South Wales, Australia; 13.6c: Daniel H. Buckley; 13.6d: Thomas D. Brock; 13.6e: Malcolm R. Walter/Macquarie University/New South Wales/Australia; 13.7a, b: J. William Schopf, University of California at Los Angeles; 13.8: John M. Hayes; 13.12.1-7: Daniel H. Buckley; 13.16.1: Norbert Pfennig, University of Konstanz, Germany; 13.16.2: Michael T. Madigan; 13.26: Carl A. Batt, Cornell University; 13.27: Jennifer Ast and Paul Dunlap; Table 13.2: Dr. Norbert Pfennig.

Chapter 14 Chapter Opener bottom: Viola Krukenberg, Katrin Knittel, and Gunter Wegener; Chapter Opener top: Viola Krukenberg, Katrin Knittel, and Gunter Wegener; 14.1.1-2: Norbert Pfennig, University of Konstanz, Germany; 14.1.3: Thomas D. Brock; 14.4b: Simon Scheuring; 14.5: Yuuji Tsukii, Protist Information Server, (protist.i.hosei.ac.jp); 14.6a, b: Michael T. Madigan; 14.7a: Niels Ulrik Frigaard; 14.10a1-3: Susan Barns and Norman R. Pace, University of Colorado; 14.10c: Kaori Ohki, Tokai University, Shimizu, Japan; 14.11a: George Feher, University of California at San

Diego; 14.11b: Marianne Schiffer, James R. Norris/Argonne National Laboratory; 14.16: Yehuda Cohen and Moshe Shilo; 14.19: Jessup M. Shively, Clemson University; 14.22a: Wael Sabra/German Research Centre for Biotechnology, Braunschweig, Germany; 14.22b: Wael Sabra, German Research Centre for Biotechnology, Braunschweig, Germany; 14.22c: Alicia M. Muro-Pastor; 14.26b: Frank Mayer, University of Gottingen, Gottingen, Germany; 14.27a, b: Thomas D. Brock; 14.29a: William Strode; 14.29b: Thomas D. Brock; 14.31a, b: Reproduced from Armin Ehrenreich and Friedrich Widdel, *Applied and Environmental Microbiology* 60:4517-4526 (1994), with permission of the American Society for Microbiology.; 14.34a: Marc Strous, University of Nijmegen, The Netherlands; 14.34b: John A. Fuerst, University of Queensland, Australia; 14.39: Dianne K. Newman and Stephen Tay, previously published in *Applied and Environmental Microbiology* 63:2022-2028 (1997).; 14.44: John A. Breznak, Michigan State University; 14.46a, b: Thomas D. Brock; 14.52a: Antje Boetius and Armin Gieseke, Max Planck Institute for Marine Microbiology, Bremen, Germany; 14.53: Laura van Niftrik and Mingliang Wu; 14.63: H.J.M. Harmsen.

Chapter 15 Chapter Opener: Holger Daims, Frank Maixner, Michael Wagner; 15.2a: Susan Barns and Norman R. Pace, University of Colorado; 15.2b-e: Daniel H. Buckley; 15.4: M.R. Edwards; 15.5: Daniel H. Buckley; 15.7a: Rachel Foster; 15.6a, b, c: Thomas D. Brock; 15.7b, c: Angel White; 15.8a-d: Daniel H. Buckley; 15.9a: Thomas D. Brock; 15.9b: Jorg Overmann, University of Munich, Germany; 15.9c: Douglas E. Caldwell, University of Saskatchewan; 15.10a-c: Norbert Pfennig, University of Konstanz, Germany; 15.10d: Johannes F. Imhoff, University of Kiel, Germany; 15.11a: Charles C. Remsen, University of Wisconsin at Milwaukee; 15.11b: Jeffrey C. Burnham and Samuel F. Conti; 15.12: Norbert Pfennig, University of Konstanz, Germany; 15.13a-e: Norbert Pfennig, University of Konstanz, Germany; 15.13f: Peter Hirsch, University of Kiel, Germany; 15.15: Daniel H. Buckley; 15.14a, b: Norbert Pfennig, University of Konstanz, Germany; 15.16ab: Daniel H. Buckley; 15.17a: Douglas E. Caldwell, University of Saskatchewan; 15.17b, c: Jorg Overmann, University of Munich, Germany; 15.17d: Douglas E. Caldwell, University of Saskatchewan; 15.18a: Daniel H. Buckley; 15.18b: Vladimir M. Gorlenko, Institute of Microbiology, Russian Academy of Sciences; 15.18c: Charles A. Abella, University of Girona, Girona, Spain; 15.18d: Deborah Jung; 15.20a: F. Rudy Turner and Howard Gest, Indiana University; 15.20b, c: Daniel H. Buckley; 15.21a: Don Bryant; 15.21b: Amaya Garcia Costas and Donald A. Bryant; 15.23a: Norbert Pfennig, University of Konstanz, Germany; 15.23b: Norbert Pfennig, University of Konstanz, Germany; 15.23c-e: Friedrich Widdel, Max Planck Institute for Marine Microbiology, Bremen, Germany; 15.23f: Norbert Pfennig, University of Konstanz, Germany; 15.23g: Daniel H. Buckley; 15.24a: Michael F. McGlannan, Florida International University; 15.24b: Andreas Teske; 15.25a: Jessup M. Shively, Clemson University; 15.25b: Hans-Dietrich Babenzien, Institute of Freshwater Ecology and Inland Fisheries, Neuglobsow, Germany; 15.26a: Verena Salman; 15.26b: Michael F. McGlannan, Florida International University; 15.27a: Michael Richard, Colorado State University; 15.27b: Markus Huttel, Max Planck Institute for Marine Microbiology, Bremen, Germany; 15.28a, b: Verena Salman; 15.29a, b: Tom Fenchel; 15.31a, b: J.-H. Becking, Wageningen Agricultural University, Wageningen, Netherlands; 15.32a, b: Harold L. Sadoff, Michigan State University; 15.34: Holger Daims; 15.33a, b: S.W. Watson; 15.35a-c: Derek R. Lovley; 15.36a: William C. Ghiorse, Cornell University; 15.36b: Reproduced with permission from W.C. Ghiorse, Biology of iron- and manganese-depositing bacteria. *Annual Review of Microbiology* 38:515-550 (1984), Fig. 1. (c) 1984 by Annual Reviews, Inc. Photo: William C. Ghiorse, Cornell University.; 15.37a, b: Douglas W. Ribbons, Technical University of Graz, Austria; 15.38a, b: Charles R. Fisher, Pennsylvania State University; 15.39a, b: Susan Koval and Ryan Chanyi; 15.40a: Susan F. Koval, University of Western Ontario; 15.41a-c: Hans Reichenbach, Gesellschaft fur Biotechnologische

Forschung mbH, Braunschweig, Germany; 15.43a, b: Herbert Voelz; 15.44a, b: Hans Reichenbach, Gesellschaft fur Biotechnologische Forschung mbH, Braunschweig, Germany; 15.45a-d: P.L. Grillone; 15.46a: Hans Reichenbach, Gesellschaft fur Biotechnologische Forschung mbH, Braunschweig, Germany; 15.46b: David White, Indiana University; 15.47a-f: Kenneth H. Nealson, University of Wisconsin; 15.48a, b: Ercole Canale-Parola, University of Massachusetts; 15.49a: Ercole Canale-Parola, University of Massachusetts; 15.50a: Noel R. Krieg, Virginia Polytechnic Institute and State University; 15.50b: Stanley L. Erlandsen, University of Minnesota Medical School; 15.50c: H.D. Raj; 15.51: A. Ryter; 15.52a: Reproduced from B.J. Paster and E. Canale-Parola, *Treponema saccharophilum* sp. nov., a large pectinolytic spirochete from the bovine rumen. *Applied and Environmental Microbiology* 50:212-219 (1985), with permission of the American Society for Microbiology.; 15.52b: Susan F. Koval & George Chaconas; 15.54a: Peter Hirsch, University of Kiel, Germany; 15.54b: Samuel F. Conti and Peter Hirsch; 15.55a: Elnar Leifson; 15.55b, c: Germaine Cohen-Bazire; 15.56a: J.L. Pate; 15.56b: James T. Staley, University of Washington; 15.56c: Heinz Schlesner, University of Kiel, Germany; 15.58: Reproduced with permission from W.C. Ghiorse, Biology of iron- and manganese-depositing bacteria. *Annual Review of Microbiology* 38:515-550 (1984), Fig. 7. (c) 1984 by Annual Reviews, Inc. Photo: William C. Ghiorse, Cornell University.; 15.59a: Thomas D. Brock; 15.59b, c: Judith F.M. Hoeniger; 15.60: Richard Blakemore, University of New Hampshire.

Chapter 16 Chapter Opener: Margarete Schüler and Harald Engelhardt; 16.4: Odile Berge; 16.6: Michael T. Madigan; 16.8: James A. Shapiro, University of Chicago; 16.11: Arthur Kelman, University of Wisconsin-Madison; 16.14: Cheryl L. Broadie and John Vercillo, Southern Illinois University at Carbondale; 16.15: Arthur Kelman, University of Wisconsin-Madison; 16.19c: Thomas D. Brock; 16.20a: Akiko Umeda, Kyushu University School of Medicine, Fukuoka, Japan; 16.21a, b: Terry J. Beveridge, University of Guelph, Guelph, Ontario; 16.22a-c: Hans Hippe, Deutsche Sammlung von Mikroorganismen und Zellkulturen GmbH, Braunschweig, Germany; 16.23: James R. Norris; 16.24: Dieter Claus, University of Gottingen, Germany; 16.25: Alan Rodwell; 16.26: Thomas D. Brock; 16.27: David L. Williamson; 16.28: Terry A. Krulwich, Mount Sinai School of Medicine; 16.29: Terry A. Krulwich, Mount Sinai School of Medicine; 16.30a-g: Hans Veldkamp; 16.31: CDC/PHIL; 16.32a: N. Rist; 16.32b: Victor Lorian; 16.32c: CDC; 16.33: CDC/PHIL; 16.34: Hubert and Mary P. Lechevalier; 16.35a: Peter Hirsch, University of Kiel, Germany; 16.35b: Hubert and Mary P. Lechevalier; 16.38a: Michael T. Madigan; 16.38b: David A. Hopwood, John Innes Centre, U.K.; 16.39a: Copyright Eli Lilly and Company. Used with permission.; 16.39b: David A. Hopwood, John Innes Centre, U.K.; 16.42a-d: Hans Reichenbach, Gesellschaft fur Biotechnologische Forschung mbH, Braunschweig, Germany; 16.44b: Morris D. Cooper, Southern Illinois University School of Medicine; 16.45: Robert R. Friis, Tiefenau Laboratory, Bern, Switzerland; 16.46: John A. Fuerst, University of Queensland, Australia; 16.47: John Bauld, Australian Geological Survey Organisation; 16.48: Heinz Schlesner, University of Kiel, Germany; 16.49a, b: Reinhard Rachel and Karl O. Stetter, University of Regensburg, Regensburg, Germany; 16.50a: Friedrich Widdel, Max Planck Institute for Marine Microbiology, Bremen, Germany; 16.51a: David Ward; 16.51b: Daniel H. Buckley; 16.51c: Reinhard Rachel and Karl O. Stetter, University of Regensburg, Regensburg, Germany; 16.52a, b: Diane Moyles and R.G.E. Murray, University of Western Ontario; 16.52c: Daniel H. Buckley.

Chapter 17 Chapter Opener: Christa Schleper; 17.1: Based on *Biochimica et Biophysica Acta* by J.H.B. Christian; Judith A. Waltho Volume 65, Issue 3. Published by Elsevier B.V, © 1965; 17.2a: Thomas D. Brock; 17.2b: NASA Headquarters; 17.2c: Daniel H. Buckley; 17.2d: Francisco Rodriguez-Valera, Universidad Miguel Hernandez, San Juan de Alicante, Spain; 17.3a, b: Mary C. Reedy, Duke University Medical Center; 17.5a-d: Alexander Zehnder, Swiss Federal Institute for Environmental Science and Technology, Dubendorf, Switzerland; 17.6a, b: J. Gregory Zeikus and V.G. Bowen; 17.7a, c: Helmut Konig and

Karl O. Stetter, University of Regensburg, Germany; 17.7b: Reinhard Rachel and Karl O. Stetter, University of Regensburg, Regensburg, Germany; 17.7d: Stephen H. Zinder, Cornell University; 17.8a: Karl O. Stetter, University of Regensburg, Regensburg, Germany; 17.9a: Thomas D. Brock; 17.9b: A. Segerer and Karl O. Stetter, University of Regensburg, Regensburg, Germany; 17.10: Thomas D. Brock; 17.12a: Helmut Konig and Karl O. Stetter, University of Regensburg, Germany; 17.12b: G. Fiala and Karl O. Stetter, University of Regensburg, Germany; 17.13a: Reinhard Rachel and Karl O. Stetter, University of Regensburg, Germany; 17.13b: Karl O. Stetter and Reinhard Rachel, University of Regensburg, Germany; 17.14a, b: Martin Könneke; 17.15a, b: Edward DeLong, Monterey Bay Aquarium Research Institute; 17.16a, b: Reinhard Rachel; 17.17a-d: Karl O. Stetter; 17.18a-d: Thomas D. Brock; 17.19a: Thomas D. Brock; 17.19b: Helmut Konig and Karl O. Stetter, University of Regensburg, Germany; 17.20a: Helmut Konig, University of Regensburg, Germany; 17.20b: Helmut Konig and Karl O. Stetter, University of Regensburg, Germany; 17.20c: Reinhard Rachel and Karl O. Stetter, University of Regensburg, Regensburg, Germany; 17.21a, b: Helmut Konig and Karl O. Stetter, University of Regensburg, Germany; 17.21c: Karl O. Stetter and Reinhard Rachel, University of Regensburg, Germany; 17.21d: Kazem Kashefi; 17.22a: Helmut Konig and Karl O. Stetter, University of Regensburg, Germany; 17.22b: Reinhard Rachel and Karl O. Stetter, University of Regensburg, Germany; 17.23: Helmut Konig and Karl O. Stetter, University of Regensburg, Regensburg, Germany; 17.24: Anna-Louise Reysenbach and Woods Hole Oceanographic Institution; 17.26: Gertraud Rieger, R. Hermann, Reinhard Rachel, and Karl O. Stetter, University of Regensburg, Germany; 17.27: Suzette L. Pereira, Ohio State University.

Chapter 18 Chapter Opener: Sergey Ivanov and Maria Harrison; 18.1: Jian-ming Li, Nancy Martin/University of Louisville School of Medicine; 18.4a: Michael Abbey/Science Source; 18.4b: Steve J. Upton/Kansas State University; 18.5: Blaine Mathison/Center for Disease Control and Prevention; 18.6a: M. I. Walker/Science Source; 18.6b: Biophoto Associates/Science Source; 18.7a: Michael T. Madigan; 18.7b: Sydney Tamm; 18.8: Steve J. Upton/Kansas State University; 18.9: Irena Kaczmarska-Ehrman/Mount Allison University; 18.10a: Rita R. Colwell/National Science Foundation; 18.10b, c: Joseph E. Kleinman/North Carolina State University Center for Applied Ecology; 18.11a: Mae Melvin/Center for Disease Control and Prevention; 18.11b: Silvia Botero Kleiven/The Swedish Institute for Infectious Disease Control; 18.12a: Jörg Piper; 18.12b-d: Irena Kaczmarska-Ehrman/Mount Allison University; 18.13a: Frank Mayer, University of Gottingen, Gottingen, Germany; 18.13b: EpicStockMedia/Fotolia; 18.13c: Michael Plewka; 18.14a: Andrew Syred/Science Source; 18.14b: Eye of Science/Science Source; 18.15: M. Haberey; 18.16: Stephen Sharnoff; 18.17a-g: Kenneth B. Raper; 18.19a: MYCOsearch, Inc.; 18.20a: Cheryl L. Broadie/Southern Illinois University at Carbondale; 18.20b: Michael T. Madigan; 18.21.1-3: Michael T. Madigan; 18.22: J. Forsdyke/SPL/Science Source; 18.24a: Hossler/Custom Medical Stock Photo; 18.24b: Forest Brem; 18.25a: Premierlight Images/Alamy Stock Photo; 18.25b: Alena Kubátová/Charles University in Prague; 18.26: Thomas D. Brock; 18.28a, b: Samuel F. Conti and Thomas D. Brock; 18.30a: Damian Palus/Shutterstock; 18.30b: U.S. Department of Agriculture; 18.30c: Ed Reschke/Photolibrary/Getty Images; 18.31: Patrick J. Lynch/Science Source; 18.32a: Christine Oesterhelt and Gerald Schonknecht; 18.32b: Richard W. Castenholz/University of Oregon; 18.33a: Arthur M. Nonomura; 18.33b: Bob Gibbons/Alamy Stock Photo; 18.33c: Thomas D. Brock; 18.33d: Ralf Wagner; 18.33e: Blickwinkel/NaturimBild/Alamy Stock Photo; 18.33f: Dr. Aurora M. Nedelcu; 18.33g: Arthur M. Nonomura; 18.34a: Guillaume Dargaud; 18.34b: Yuuji Tsukii/Hosei University, Japan.

Chapter 19 Chapter Opener: Jim Fredrickson; 19.2b: Norbert Pfennig, University of Konstanz, Germany; 19.3a: James A. Shapiro, University of Chicago; 19.3b: Marie Asao, Deborah O. Jung, and Michael T. Madigan; 19.6.1 (bottom): Steve Giovannoni; 19.6.1 (top): Excellent backgrounds/Shutterstock; 19.7: Rustem Ismagilov and Liang Ma; 19.8a, b: Marc Mussman and Michael Wagner; 19.8c: Willm Martens-Habbena; 19.9: Molecular Probes; 19.10c: Daniel Gage; 19.10b: Gage, Daniel; 19.11a, b: ASM Publications; 19.12a-c: Norman R. Pace, University of Colorado; 19.13a: David A. Stahl, Northwestern University; 19.13b: Jiri Snaidr; 19.14: Marc Mussmann and Michael Wagner; 19.16ab: Jennifer A. Fagg and Michael J. Ferris, Montana State University; 19.16c: Gerard Muyzer; 19.19: Michael Wagner; 19.20a: Jizhong (Joe) Zhou, Zhili He, and Joy Van Nostrand; 19.21: Vaughn Iverson and Ginger Armbrust; 19.27: Based on data and drawings by Niels Peter Revsbech; 19.28: Niels Peter Revsbech; 19.32.1: Colin J. Murrell; 19.33b: Jennifer Pett-Ridge, Peter K. Weber; 19.33c: Peter K. Weber; 19.33d: Jennifer Pett-Ridge, Peter K. Weber; 19.34: Eva Heinz and Matthias Horn; 19.35a-c: Michael Wagner.

Chapter 20 Chapter Opener: Andreas Teske; 20.1a-c: Hans Paerl, University of North Carolina at Chapel Hill; 20.4: Christian Jeanthon; 20.4a: Frank B. Dazzo, Michigan State University; 20.4b: Thomas D. Brock; 20.5a: C.-T. Huang, Karen Xu, Gordon McFeters, and Philip S. Stewart; 20.5b: Cindy E. Morris, INRA, Centre de Recherche d'Avignon, France. Previously published in *Applied and Environmental Microbiology* 63:1570-1576.; 20.5c: J. M. Sanchez, J. Lidel Lope and Ricardo Amils; 20.6: Tim Tolker-Nielsen and Wen-Chi Chiang; 20.7a: Jesse Dillon and David A. Stahl; 20.7b: ASM Publications; 20.8a-c: Andreas Teske and Markus Huettel; 20.9b: Michael T. Madigan; 20.11a-c: Jayne Belnap; 20.12: Winai Tepsuttinun/Fotolia; 20.13a: Esta van Heerden; 20.13b: Terry C. Hazen; 20.15: Luis R. Comolli; 20.17b: Thomas D. Brock; 20.18: Brykaylo Yuriy/Shutterstock/Pearson Asset Library; 20.18: Data assembled and analyzed by Nicolas Pinel; 20.19: Otis Brown/Robert Evans/NASA; 20.21a: U.S. Coast Guard; 20.21b: NASA; 20.22.1: Alexandra Z. Worden and Mya E. Breitbart, Scripps Institution of Oceanography, University of California at San Diego; 20.22.2: Sallie Chrisholm; 20.24a: Hans W. Paerl, University of North Carolina at Chapel Hill; 20.24b: Alexandra Z. Worden and Brian P. Palenik, Scripps Institution of Oceanography, University of California at San Diego; 20.25: Vladimir Yurkhov; 20.27: Daniela Nicastro; 20.28: Jed Fuhrman/Matt Sullivan; 20.28 (insets): Jennifer R. Brum/Matt Sullivan; 20.29: Iakov Kalinin/Shutterstock/Pearson Asset Library; 20.32: Hideto Takami, Japan Marine Science and Technology Center, Kanagawa, Japan; 20.31: Data assembled by Doug Bartlett.; 20.33a: Michael T. Madigan; 20.34a, b, inset: Andreas Teske; 20.36: Sea Research Foundation (SRF) and the Ocean Exploration Trust (OET)/NOAA; 20.38: Robert D. Ballard/Woods Hole Oceanographic Institution; 20.39: Michael T. Madigan; 20.41 (inset): Carl Wirsen/Woods Hole Oceanographic Institution.

Chapter 21 Chapter Opener: Monique Heijmans; 21.1: Data adapted from Zinder, S. 1984. Microbiology of anaerobic conversion of organic wastes to methane: Recent developments. *Am. Soc. Microbiol. News.* 50:294-298; 21.3: Evan Solomon; 21.7a: John A. Breznak, Michigan State University; 21.7b, c: Monica Lee and Stephen H. Zinder; 21.13: J. M. Sanchez, J. Lidel Lope and Ricardo Amils; 21.Explore the Microbial World-2: Lars Peter Nielsen; 21.Explore the Microbial World-1: Eye of Science/Science Source; 21.14a-g: David Emerson; 21.15a: Jörg Bollmann; 21.15b: M.L. Cros Miguel and J.M. Fortuño Alós; 21.16a: Jörg Piper; 21.19: These data are continuously collected by the Global Monitoring Division of the National Oceanic and Atmospheric Association/Earth System Research Laboratory.

Chapter 22 Chapter Opener : Mari Winkler; 22.1a-d: Thomas D. Brock; 22.3: Kodzo Gbewonyo; 22.4a: Ravin Donald; 22.4b: Thomas D. Brock; 22.5: Thomas D. Brock; 22.6: Kenneth Hurst Williams & Derek R. Lovley; 22.7a, b: U.S. Environmental Protection Agency Headquarters; 22.7c: Bassam Lahoud/Lebanese American University; 22.8: Thomas D. Brock; 22.9: Designelements/Fotolia; 22.12: Dr. Helmut Brandl/University of Zurich, Switzerland; 22.14: Michael T. Madigan; 22.15a: John M. Martinko; 22.15c: Michael T. Madigan; 22.16: Richard F. Unz/Penn State University Archives; 22.17a: Thomas D. Brock; 22.20a: Louisville Water Company; 22.22: Don Howard/Center for Disease Control and Prevention; 22.23: Norman Pace; 22.24c: Shawna Johnston and Gerrit Voordouw.

Chapter 23 Chapter Opener: Nancy Moran; 23.1a: Thomas D. Brock; 23.1b: Michael T. Madigan; 23.2: Thomas D. Brock; 23.3: Adapted from Overmann, J., and H. van Gemerden. 2000. *FEMS Microbiol. Rev.* 24:591-599; 23.4: Michael T. Madigan; 23.5a, b: Michael T. Madigan; 23.7: Joe Burton; 23.8: Ben B. Bohlool; 23.9: Joe Burton; 23.11a: M.D. Shakhawat Hossain; 23.11b, c: Howard Berg, Donald Danforth Plant Science Center; 23.16: B. Dreyfus; 23.17a: J.H. Becking; 23.17b: J.-H. Becking/Wageningen Agricultural University, Wageningen, Netherlands; 23.18a, b: J. H. Becking/Wageningen Agricultural University, Wageningen, Netherlands; 23.19ab: Heike Bucking; 23.19c: Heike Bucking; 23.22: S.A. Wilde; 23.23: Jo Handelsman/University of Wisconsin at Madison; 23.26a, b: Michael T. Madigan; 23.EMW-1a: Michael Poulsen and Cameron Currie; 23.EMW-1b: Michael Poulsen and Cameron Currie; 23.27c: Michael Pettigrew/Shutterstock; 23.29a: Chris Frazee and Margaret J. Mcfall-Ngai/University of Wisconsin; 23.29b: Margaret J. Mcfall-Ngai/University of Wisconsin; 23.30a: Dudley Foster/Woods Hole Oceanographic Institution; 23.30b: Carl Wirsen/Woods Hole Oceanographic Institution; 23.31a: American Association for the Advancement of Science; 23.31b: Macmillan Magazines Limited; 23.32a-c: Heidi Goodrich-Blair, Mengyi Cao, Matthew Stilwell, and Elle Kielar Grevstad at the Biochemistry Optical Core, Biochemistry Dept. UW-Madison.; 23.33a, c: Michael T. Madigan; 23.34a, b: Michael T. Madigan; 23.37-3: Gallas/Fotolia; 23.37.1-2: Michael T. Madigan; 23.37.4: D Photo Sudbury/Shutterstock; 23.38b: Sharisa D. Beck; 23.40: Eric Isselee/Shutterstock, Data assembled and analyzed by Nicolas Pinel.

Chapter 24 Chapter Opener : Werner Nosko/Reuters; 24.2: Adapted from Ridaura, V.K., et al. *Science* 341: DOI:10.1126/science.1241214; 24.3: Data from https://upload.wikimedia.org/wikipedia/commons/c/c5/Digestive_system_diagram_en.svg; 24.4a: Dwayne C. Savage and R.V.H. Blumershine; 24.4b: Dwayne C. Savage; 24.5: Data assembled and analyzed by Nicolas Pinel.; BOX 24.1: Hsiao, E.Y., S.W. McBride, S. Hsien, et al. 2013. Microbiota modulate behavioral and physiological abnormalities associated with neurodevelopmental disorders. Cell 155:1451-1463; BOX 24.1: Modified from Hsiao, E.Y., et al. 2013. Cell 155:1451-1463; 24.7: Data assembled and analyzed by Nicolas Pinel.; 24.9: Data assembled and analyzed by Nicolas Pinel.; 24.11b: John Durham/Science Source; 24.12: Data assembled and analyzed by Nicolas Pinel; 24.13: Data are adapted from Grice et al., 2009, Science 324:1190; 24.14: Data assembled and analyzed by Nicolas Pinel.; 24.15a-c: Amina Bouslimani & Pieter Dorrestein; 24.21a, b: Thomas J. Lie, University of Washington; 24.24: Deborah O. Jung and John Martinko.

Chapter 25 Chapter Opener: Jessica Mark Welch and Gary Borisy; 25-2a: Edward T. Nelson, J.D. Clemments, and R.A. Finkelstein; 25-3a: Larry Stauffer/Oregon State Public Health Laboratory/CDC; 25-3b: PHIL/CDC; 25-3c: J. William Costerton/Montana State University; 25-4a: M. Miller/PHIL/CDC; 25-4b: PHIL/CDC; 25-4c: Richard Facklam/PHIL/CDC; 25-5: PHIL/CDC; 25-7a, b: C. Lai, Max A. Listgarten, and B. Rosan; 25-7c: Richard Facklam/PHIL/CDC; 25-7d: Isaac L. Schechmeister and John J. Bozzola/Southern Illinois University at Carbondale; 25-8: Jessica Mark Welch and Gary Borisy; 25-9: Janice Haney Carr/CDC; 25-12: PHIL/CDC; 25-15: Zang, R.G., Scott, D.L., Westbrok, M.L., Nance, S., Spangler, B.D., Shipley, G.G., Westbrook, E.M. Journal: (1995) J.Mol. Biol. 251:563-573; 25-16a: Thomas D. Brock; 25-16b: Leon J. Le Beau/University of Illinois at Chicago; 25-17 (right): Janice Haney Carr/CDC; 25-17 (left): 2-methyl-2, 4-pentanediol induces spontaneous assembly of staphylococcal alpha-hemolysin into heptameric pore structure. Tanaka, Y. Hirano, N., Kaneko, J., Kamio, Y., Yao, M., Tanaka, I. (2011) *Protein Sci.* 20:448-456.; 25-18: Janice Haney Carr/CDC; 25-19a, b: Arthur O. Tzianabos and R.D. Millham.

Chapter 26 Chapter Opener: Dr. Deepak Kumar, Vijaya Kumar, and Dr. Robert Moir; Table 26.2a: Michael T. Madigan; Table 26.2b-d: John M. Martinko and Michael T. Madigan; 26.6b: NIAID/PHIL//CDC; 26-EMW-1: Jarmo Holopainen; 26.9: J.G. Hirsch; 26.11a: Center for Disease Control and Prevention; 26.11b: James V. Little; 26.14b: Michelle Dunstone and Bradley Spicer, Centre of Excellence for Advanced Moleular Imaging.

Chapter 27 Chapter Opener: Hubert Schriebl; 27.5.1: PHIL/CDC; 27.5.2: US WPA/Library of Congress; 27.5.3: D. Jordan, M.A./PHIL/CDC; 27.5.4: Edward J. Wozniak, D.V.M, PH.D/PHIL/CDC; 27.8b: Richard J. Feldmann, National Institutes of Health; 27.13d: Don C. Wiley, Howard Hughes Medical Institute; 27.13e: Aideen C.M. Young, Albert Einstein College of Medicine, Bronx, New York; 27.13f: Reproduced by permission from J.H. Brown et al., Three-dimensional structure of the human class II histocompatibility antigen HLA-DR1. *Nature* 364:33-39 (1993). Copyright (c) 1993 Macmillan Magazines Limited. Image by Don C. Wiley, Harvard University.; 27.16: Boensch, B./Arco Images GmbH/Alamy Stock Photo; 27.22b: Lennart Nilsson, Albert Bonniers Forlag AB; 27.25: Edward J. Wozniak/D.V.M/PH.D/CDC; 27.26: CDC; 27.27: CDC; 27.28: PHIL/CDC; 27.30: NIAID/PHIL/CDC.

Chapter 28 Chapter Opener: Matthew Sattley; 28.1: PHIL/CDC; 20.2: USAMRIID; Table 28.3.1-5: Janice Haney Carr/Center for Disease Control and Prevention (CDC); Table 28.3.6: Peta Wardell/Center for Disease Control and Prevention (CDC); Table 28.3.7: Erskine Palmer/Center for Disease Control and Prevention (CDC); Table 28.3.8: A. Harrison/Center for Disease Control and Prevention (CDC); Table 28.3.9: Frederick Murphy/Center for Disease Control and Prevention (CDC); Table 28.3.10: Janice Haney Carr/Center for Disease Control and Prevention (CDC); Table 28.3.11: Charles D. Humphrey/Center for Disease Control and Prevention (CDC); Table 28.3.12: Janice Haney Carr/Center for Disease Control and Prevention (CDC); 28.6a: Norman Jacobs//PHIL/CDC; 28.6b: PHIL/CDC; 28.7a, b: Dr. Todd Parker/PHIL/CDC; 28.7c: John M. Martinko and Cheryl L. Broadie; 28.8a, b: Matthew Sattley; 28.9a-c: Leon J. Le Beau; 28.10a: Leon J. Le Beau; 28.10b, c: Matthew Sattley; 28.10g: Richard Facklam/PHIL/CDC; 28.13: C. Weibull, W.D. Bickel, W.T. Hashius, K.C. Milner, and E. Ribi; 28.14a: Norman L. Morris; 28.15: Matthew Sattley; 28.17a: Wellcome Research Laboratories; 28.17b: Center for Disease Control and Prevention; 28.18a: P. M. Feorino/Center for Disease Control and Prevention; 28.18b: William Kaplan/PHIL/CDC; 28.18c: Govinda S. Visvesvara/PHIL/CDC; 28.20b: Matthew Sattley; 28.21b: Center for Disease Control and Prevention; 28.23b (left): Matthew Sattley; 28.23b (right): Matthew Sattley; 28.25: Amanda Mills/PHIL/CDC; 28.30: Biophoto Associates/Science Source; 28.31: Dr P. Marazzi/Science Source; 28.EMW-1a: CHROMagar, Alberto Lerner; 28.EMW-1b: NIAID/PHIL/CDC.

Chapter 29 Chapter Opener (inset): CDC/Special Bacteriology Reference Lab; Chapter Opener: Cynthia Goldsmith/CDC; 29.1: Data adapted from data published by the World Health Organization (WHO) and the Centers for Disease Control (CDC); 29.5 (right): PHIL/CDC; 29.5 (left): James Gathany/PHIL/CDC; 29.5: Based on West Nile Virus in the United States, https://stacks.cdc.gov/view/cdc/28079/cdc_28079_DS23.txt. Published by U.S. Centers for Disease Control and Prevention; 29.6 (left): Janice Haney Carr/PHIL/CDC; 29.6 (right): NIAID/PHIL/CDC; 29.6: From *Emerging Infectious Diseases*, Vol. 8, No. 2, p.226. Published by U.S. Department of Health & Human Services; 29.EMW-1: C.S. Goldsmith/T.G. Ksiazek/S.R. Zaki/CDC; 29.8 (left): CDC; 29.8 (right): Jim Goodson/CDC; 29.9: Data are from the World Health Organization, Geneva; 29.11: Data are from the CDC, Atlanta, Georgia, USA.; 29.12: Data are from the HIV/AIDS Surveillance Report and Division of HIV/AIDS Prevention—Surveillance and Epidemiology, CDC, Atlanta, Georgia, USA; 29.13: Data from the CDC, Atlanta, Georgia, USA, were gathered from 38,000 males and 9,500 females diagnosed with HIV/AIDS in 2010; 29.16a, b: PHIL/CDC; 29.17a, b: Center for Disease Control and Prevention; 29.17c: Archil Navdarashvili, Georgia (Republic) and /PHIL/CDC; 29.17d: PHIL/CDC.

Chapter 30 Chapter Opener: Cynthia Goldsmith/CDC; 30.1: Thomas D. Brock; 30.3a: Eye of Science/Science Source; 30.3b: PHIL/CDC; 30.4a-c: Michael T. Madigan; 30.5: Center for Disease Control and Prevention; 30.6: Franklin H. Top; 30.7: Center for Disease Control and Prevention; 30.8: Michael T. Madigan; 30.9: Franklin H. Top; 30.1: Biomedical Communications/Newscom; 30.11: Isaac L. Schechmeister; 30.12a, b: Center for Disease Control and Prevention; 30.13a: PHIL/CDC; 30.13b: Franklin H. Top; 30.14: Center for Disease Control and Prevention; 30.14: Data obtained from the CDC, Atlanta, Georgia, USA; 30.15a: Edwin P. Ewing/Center for Disease Control and Prevention (CDC); 30.15b: Center for Disease Control and Prevention; 30.16a: Center for Disease Control and Prevention; 30.16b, c: Aaron Friedman; 30.17a: Jorge Adorno/Reuters; 30.17b: PHIL/CDC; 30.18a, b: Center for Disease Control and Prevention; 30.19a-c: Center for Disease Control and Prevention; 30.20: Center for Disease Control and Prevention (CDC); 30.21: Centers for Disease Control and Prevention (CDC); 30.22: A. D. Langmuir/Center for Disease Control and Prevention (CDC); 30.23a: B. Dowsett; 30.23b: David A.J Tyrrell; 30.24: Centers for Disease Control and Prevention (CDC); 30.25: Irene T. Schulze; 30.28a, b: Centers for Disease Control and Prevention (CDC); 30.29a, b: Gregory Moran/Centers for Disease Control and Prevention (CDC); 30.30a, b: Michael T. Madigan; 30.31: Juergen Berger/Science Source; 30.32: Data obtained from the CDC, Atlanta, Georgia, USA; 30.33a: Eye of Science/Science Source; 30.33b: PHIL/CDC; 30.34a: Cynthia Goldsmith/Centers for Disease Control and Prevention (CDC); 30.34b: PHIL/CDC; 30.36a: Centers for Disease Control and Prevention (CDC); 30.36b: Morris D. Cooper; 30.37a-c: Centers for Disease Control and Prevention (CDC); 30.38a: Centers for Disease Control and Prevention (CDC); 30.38b: Sidney Olansky; 30.38c: Centers for Disease Control and Prevention (CDC); 30.39a, b: Morris D. Cooper; 30.40a: Gordon A. Tuffli, M.D.; 30.40b: Centers for Disease Control and Prevention (CDC); 30.41a, b: Centers for Disease Control and Prevention (CDC); 30.45a-c: Centers for Disease Control and Prevention (CDC); 30.45d: PHIL/CDC; 30.45e-g: Centers for Disease Control and Prevention (CDC); 30.46a, b: Centers for Disease Control and Prevention (CDC); 30.47: Data are adapted from the CDC, Atlanta, Georgia, USA.; 30.47a: Centers for Disease Control and Prevention (CDC).

Chapter 31 Chapter Opener: Christina Davis; 31.1: Data are from the Centers for Disease Control and Prevention, Atlanta, Georgia, USA; 31.2a, b: Centers for Disease Control and Prevention (CDC); 31.3a, b: Centers for Disease Control and Prevention (CDC); 31.4a, b: Centers for Disease Control and Prevention (CDC); 31.6a, c: Willy Burgdorfer, M.D; 31.6b: Centers for Disease Control and Prevention (CDC); 31.6d: Kenneth E. Greer/University of Virginia School of Medicine; 31.7a: Centers for Disease Control and Prevention (CDC); 31.7b: National Institute of Allergy and Infectious Diseases (NIAID); 31.8a: National Institute of Allergy and Infectious Diseases (NIAID); 31.8b: Janice Haney Carr/CDC; 31.9a: Pfizer Central Research; 31.9b: James Gathany/Centers for Disease Control and Prevention (CDC); 31.9c: Pfizer Central Research; 31.10: Pfizer Central Research; 31.12a: James Gathany/Centers for Disease Control and Prevention (CDC); 31.12b : PHIL/CDC; 31.12c: Frederick Murphy/Centers for Disease Control and Prevention (CDC); 31.13a, b: Cynthia Goldsmith/CDC; 31.14a, b: Centers for Disease Control and Prevention (CDC); 31.15: Centers for Disease Control and Prevention (CDC); 31.16: Data are from the Centers for Disease Control and Prevention, Atlanta, Georgia, USA; 31.17a: Larry Stauffer/CDC; 31.17b: T Parker/Centers for Disease Control and Prevention (CDC); 31.18a: Centers for Disease Control and Prevention (CDC); 31.18b: Larry Staufer/Centers for Disease Control and Prevention (CDC); 31.19a, b: Centers for Disease Control and Prevention (CDC); 31.21a: Larry Stauffer/Centers for Disease Control and Prevention (CDC); 31.21b: Janice Haney Carr/Centers for Disease Control and Prevention (CDC); 31.22a: Larry Stauffer/Centers for Disease Control and Prevention (CDC); 31.22b-d: Centers for Disease Control and Prevention (CDC); 31.23a, b: Centers for Disease Control and Prevention (CDC); 31.24a, b: Centers for Disease Control and Prevention (CDC); 31.24c: Biophoto Associates/Science Source.

Chapter 32 Chapter Opener: Fudio/Alamy Stock Photo; 32.1a: Thomas D. Brock; 32.1b: U.S. Environmental Protection Agency Headquarters; 32.1c: Idexx Laboratories; 32.2a: Centers for Disease Control and Prevention (CDC); 32.2b: Stem Jems/Science Source; 32.3a: Kimberley Seed, Tufts University School of Medicine; 32.3b: Centers for Disease Control and Prevention (CDC); 32.3c, d: Centers for Disease Control and Prevention (CDC); 32.4: Mark L. Tamplin; 32.5a, b: Centers for Disease Control and Prevention (CDC); 32.5c: Janice Haney Carr/Centers for Disease Control and Prevention (CDC); 32.5d: Jim Feeley/CDC; 32.6a, b: Centers for Disease Control and Prevention (CDC); 32.7a-c: Thomas D. Brock; 32.8: John M. Martinko and Cheryl Broadie; 32.5: Data from the Centers for Disease Control and Prevention, Atlanta, Georgia, USA; 32.9: International PBI S.p.A., Milano, Italy; 32.10a: Centers for Disease Control and Prevention (CDC); 32.10b: Janice Haney Carr/Centers for Disease Control and Prevention (CDC); 32.11a, b: Centers for Disease Control and Prevention (CDC); 32.12a: Eric Grafman/Centers for Disease Control and Prevention (CDC); 32.12b, c: James Gathany/Centers for Disease Control and Prevention (CDC); 32.12d: Michael T. Madigan; 32.13a, b: Centers for Disease Control and Prevention (CDC); 32.14a: Centers for Disease Control and Prevention (CDC); 32.14b: Janice Haney Carr/Center for Disease Control and Prevention (CDC); 32.14c, d: PHIL/CDC; 32.15a, b: Centers for Disease Control and Prevention (CDC); 32.15c: Mediscan/Alamy Stock Photo; 32.16a: Michael T. Madigan; 32.16b: Elizabeth White/Center for Disease Control and Prevention (CDC); 32.18a-d: Centers for Disease Control and Prevention (CDC); 32.19a, b: Centers for Disease Control and Prevention (CDC).

Chapter 33 Chapter Opener: Ricardo Funari/BrazilPhotos/Alamy Stock Photo; 33.1a, b: Centers for Disease Control and Prevention (CDC); 33.1c: M. Jalbert/L. Kaufman/Center for Disease Control and Prevention (CDC); 33.1d-f: L. Ajello/Center for Disease Control and Prevention (CDC); 33.2a: Center for Disease Control and Prevention (CDC); 33.3a: Gordon C. Sauer, M.D.; 33.3b, c: Centers for Disease Control and Prevention (CDC); 33.4a: Gordon C. Sauer, M.D.; 33.4b: L.K. Georg/Center for Disease Control and Prevention (CDC); 33.5a: M. Hicklin/Center for Disease Control and Prevention (CDC); 33.5b: L.K. Georg/Center for Disease Control and Prevention (CDC); 33.5c: E.P. Ewing, Jr./Center for Disease Control and Prevention (CDC); 33.5d: M. Hicklin/Center for Disease Control and Prevention (CDC); 33.5e: M. Castro/L.K. Georg/Center for Disease Control and Prevention (CDC); 33.5f: Centers for Disease Control and Prevention (CDC); 33.6a, b: M. Melvin/Center for Disease Control and Prevention (CDC); 33.6c: L.L.A. Moore, Jr./Center for Disease Control and Prevention (CDC); 33.7a: G.S. Visvesvara/Center for Disease Control and Prevention (CDC); 33.7b: Stanley L. Erlandsen; 33.7c: Dennis E. Feely; 33.8a: Steve J. Upton; 33.8b: Centers for Disease Control and Prevention (CDC); 33.9a, b: Centers for Disease Control and Prevention (CDC); 33.10: Edwin P. Ewing/Center for Disease Control and Prevention (CDC); 33.11: Jim Gathany/Center for Disease Control and Prevention (CDC); 33.12a: Steven Glenn/Centers for Disease Control and Prevention (CDC); 33.12b: M. Melvin/CDC/PHIL; 33.13a: F. Collins/J. Gathany/PHIL/CDC; 33.13b: M. Melvin/Center for Disease Control and Prevention (CDC); 33.13c: D.S. Martin/Center for Disease Control and Prevention (CDC); 33.14a: Myron G. Schultz/Center for Disease Control and Prevention (CDC); 33.14b: Who/Center for Disease Control and Prevention (CDC); 33.14c: M. Melvin/Center for Disease Control and Prevention (CDC); 33.15a: S.Maddison/Center for Disease Control and Prevention (CDC); 33.15b: Center for Disease Control and Prevention (CDC); 33.15c: A.J. Sulzer/Center for Disease Control and Prevention (CDC); 33.15d: Center for Disease Control and Prevention (CDC); 33.16a, b: Center for Disease Control and Prevention (CDC); 33.17a, b: PHIL/CDC.

Glossary Terms

NOTE: Page numbers indicate where the term is defined in a Chapter Glossary. A term with more than one page number is usually defined in a general way early in the book and a more detailed way later in the book, or is used in two different senses. Comprehensive glossary available at MasteringMicrobiology®.

ABC (ATP-binding cassette) transport system, 101
Acetogenesis, 450
Acid mine drainage, 694
Acid-fastness, 529
Acidophile, 172
Acridine orange, 614
Actinomycetes, 529
Activation energy, 101
Activator protein, 201
Acute infection, 886
Adaptive immunity, 796
Adenosine triphosphate (ATP), 101
Adherence, 774
Adhesins, 774
Aerobe, 172
Aerobic anoxygenic phototroph, 493
Aerotolerant anaerobe, 172
Agglutination, 864
Algae, 582
Alkaliphile, 172
Allele, 391
Allosteric protein, 201
Amino acid, 135
Aminoacyl-tRNA synthetase, 135
Aminoglycoside, 864
Anabolic reactions (anabolism), 101
Anaerobe, 172
Anaerobic respiration, 101, 450
Anaerobic treatment, 694
Anammox, 450
Anoxygenic photosynthesis, 450
Antenna pigments, 450
Antibiotic, 222, 864
Antibody, 796
Anticodon, 135
Antigen, 796, 828
Antigen-presenting cell (APC), 796
Antigenic drift, 918
Antigenic shift, 918
Antimicrobial agent, 172
Antimicrobial drug resistance, 864
Antiparallel, 135
Antiseptic (germicide), 172
Arbuscule, 728
Archaea, 391
Archaellum, 71
Aseptic technique, 33, 172
ATPase (ATP synthase), 101
Attenuation, 201, 774
Autoantibody, 828

Autoclave, 172
Autoimmunity, 828
Autoinducer, 201
Autoinduction, 493
Autotroph, 101, 450
Auxotroph, 331
Average nucleotide identity, 391
B cell, 796
B cell receptor (BCR), 828
Bacteremia, 774
Bacteria, 391
Bacterial artificial chromosome (BAC), 361
Bactericidal agent, 172
Bacteriochlorophyll, 450
Bacteriocin, 135
Bacteriocyte, 728
Bacteriome, 728
Bacteriophage, 240
Bacteriorhodopsin, 555
Bacteriostatic agent, 172
Bacteroid, 728
Banded iron formation, 391
Basal body, 71
Basic reproduction number (R_0), 886
Batch culture, 172
β-lactam antibiotic, 864
Binary fission, 172
Binomial system, 391
Biochemical oxygen demand (BOD), 650, 694
Biofilm, 172, 650
Biogeochemistry, 650
Bioinformatics, 272
Biological (microbial) warfare, 886
Bioluminescence, 493
Bioremediation, 694
Biotechnology, 361
Botulism, 957
Budding division, 172
Calvin cycle, 101, 450
Capsid, 240
Capsomere, 240
Capsule, 71, 774
Carboxysome, 450, 493
Cardinal temperatures, 172
Carotenoid, 450
Carrier, 886
Cassette mutagenesis, 361
Catabolic reactions (catabolism), 101

Catabolite repression, 201
CD4 coreceptor, 828
CD8 coreceptor, 828
Cell wall, 33
Centers for Disease Control and Prevention (CDC), 886
Chaperone, 135
Chemokine, 796
Chemolithotroph, 101, 493
Chemolithotrophy, 33
Chemoorganotroph, 101
Chemostat, 172
Chemotaxis, 71
Chitin, 582
Chloramine, 694
Chlorination, 695
Chlorophyll, 450
Chloroplast, 71
Chlorosome, 450, 493
Chromosomal island (genomic island), 272
Chromosome, 135
Chronic infection, 886
Ciliate, 582
Cirrhosis, 918
Citric acid cycle, 101
Clarifier, 695
Clonal anergy, 828
Clonal deletion, 828
Clone, 828
Coagulation, 695
Codon, 135
Codon bias, 135, 272
Coenocytic, 582
Coenzyme, 101
Coevolution, 728
Coliforms, 957
Colonial, 582
Colonization, 774
Colony, 33
Common-source epidemic, 886
Community, 650
Compatible solute, 172, 555
Complement system, 796
Complementarity-determining region (CDR), 828
Complementary, 135
Complementary DNA (cDNA), 361
Complex medium, 172
Concatemer, 240
Congenital syphilis, 918

Conidia, 582
Conjugation, 331
Consortium, 493, 728
Contrast, 33
Convergent evolution, 493
Core genome, 272, 391
Coryneform bacteria, 529
Crenarchaeota, 555
Cristae, 71
Crown corrosion, 695
Culture, 33
Culture medium, 172
Cyanobacteria, 493
Cyclic AMP, 201
Cytokine, 796
Cytoplasm, 33
Cytoplasmic membrane, 33, 71
Cytoskeleton, 71
DAPI, 614
Defined medium, 172
Delayed-type hypersensitivity (DTH), 828
Denaturation, 135
Denaturing gradient gel electrophoresis (DGGE), 614
Dendritic cell, 796
Denitrification, 450, 671
Denitrifier, 493
Dental caries, 774
Dental plaque, 774
Diazotroph, 493
Dicarboxylate/4-hydroxybutyrate cycle, 450
Differential media, 864
Differentiation, 33
Dipicolinic acid, 72
Disability-adjusted life year (DALY), 886
Disease, 774
Disinfectant, 172
Disinfection, 172
Dissimilative sulfate-reducer, 493
Dissimilative sulfur-oxidizer, 493
Dissimilative sulfur-reducer, 493
Distribution system, 695
DNA (deoxyribonucleic acid), 135
DNA cassette, 361
DNA gyrase, 135
DNA helicase, 135
DNA ligase, 135
DNA polymerase, 135

Index

NOTE: Page numbers in **boldface** indicate the location of a boldfaced term in chapter text as well as where the term is defined in a chapter glossary. A *t* following a page number indicates tabular material, an *f* indicates a figure, and a *b* indicates boxed material. Comprehensive glossary available at MasteringMicrobiology®.

Phylogeny of *Bacteria*

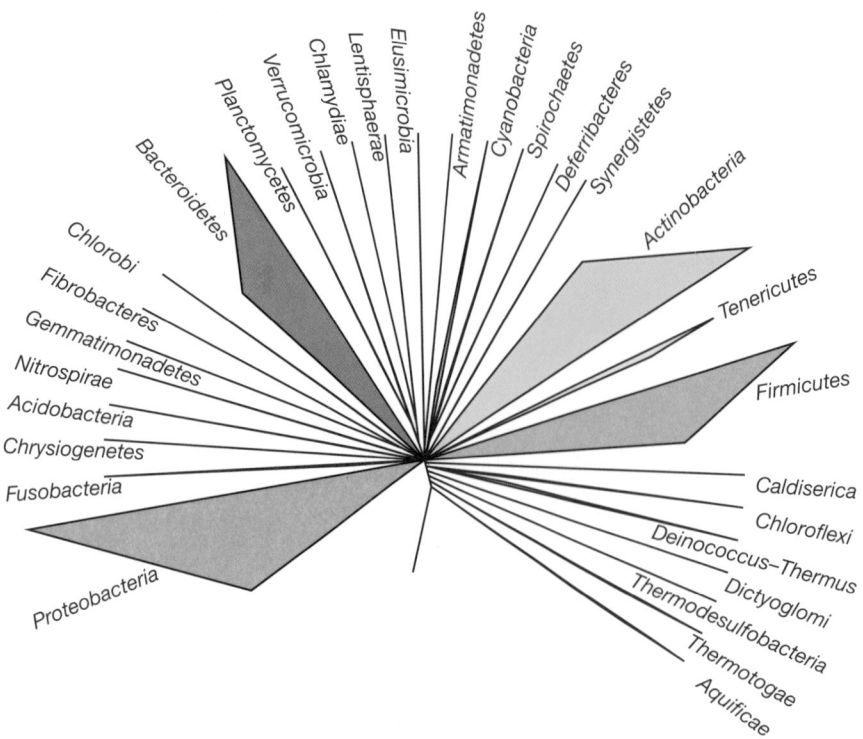